LDL	low-density lipoprotein
Leu (L)	leucine
Lys (K)	lysine
Man	mannose
MHC	major histocompatability complex
Met (M)	methionine
NAD^+	nicotinamide-adenine dinucleotide (oxidized form)
NADH	nicotinamide-adenine dinucleotide (reduced form)
$NADP^+$	nicotinamide-adenine dinucleotide phosphate (oxidized form)
NADPH	nicotinamide-adenine dinucleotide phosphate (reduced form)
NDP	nucleoside-5′-diphosphate
NAM	*N*-acetylmuramic acid
NMR	nuclear magnetic resonance
NTP	nucleoside-5′-triphosphate
Phe (F)	phenylalanine
P_i	inorganic orthophosphate
PEP	phosphoenolpyruvate
PFK	phosphofructokinase
PG	prostaglandin
2PG	2-phosphoglycerate
3PG	3-phosphoglycerate
PIP_2	phosphatidylinositol-4,5-bisphosphate
PK	pyruvate kinase
PLP	pyridoxal-5-phosphate
PP_1	inorganic pyrophosphate
Pro (P)	proline
PRPP	phosphoribosylpyrophosphate
PS	photosystem
Q	ubiquinone or plastoquinone
QH_2	ubiquinol or plastoquinol
RER	rough endoplasmic reticulum
RF	release factor or replicative form
RFLP	restriction-fragment length polymorphism
RNA	ribonucleic acid
hnRNA	heterogeneous nuclear RNA
mRNA	messenger RNA
rRNA	ribosomal RNA
snRNA	small nuclear RNA
tRNA	transfer RNA
snRNP	small ribonucleoprotein
RNase	ribonuclease
Ru1,5P	ribulose-1,5-bisphosphate
Ru5P	ribulose-5-phosphate
R5P	ribose-5′-phosphate
RSV	Rous sarcoma virus
s	Svedberg constant
SAM	*S*-adenosylmethionine
SDS	sodium dodecyl sulfate
Ser (S)	serine
S7P	sedoheptulose-7-phosphate
SRP	signal recognition particle
T	thymine
THF	tetrahydrofolate
Thr (T)	threonine
TLC	thin-layer chromatography
TMV	tobacco mosaic virus
TPP	thiamine pyrophosphate
Trp (W)	tryptophan
TTP	thymidine-5′-triphosphate
Tyr (Y)	tyrosine
U	uracil
UDP	uridine-5′-diphosphate
UDPG	UDP-glucose
UMP	uridine-5′-monophosphate
UQ	ubiquinone
Val (V)	valine
VLDL	very-low-density lipoprotein
XMP	xanthosine-5′-monophosphate
Xu5P	xylulose-5′-phosphate

LANCASTER
UNIVERSITY

LIBRARY

Biochemistry

Fourth Edition

BIOCHEMISTRY

FOURTH EDITION

GEOFFREY ZUBAY

Columbia University

Wm. C. Brown Publishers

Dubuque, IA Bogota Boston Buenos Aires Caracas Chicago
Guilford, CT London Madrid Mexico City Sydney Toronto

Project Team

Editor *Ron Worthington*
Developmental Editor *Brittany J. Rossman / Russell L. Lidberg*
Marketing Manager *Patrick E. Reidy*
Permissions Coordinator *Karen L. Storlie*
Publishing Services Coordinator *Julie Avery Kennedy*

Wm. C. Brown Publishers

President and Chief Executive Officer *Beverly Kolz*
Vice President, Director of Editorial *Kevin Kane*
Vice President, Sales and Market Expansion *Virginia S. Moffat*
Vice President, Director of Production *Colleen A. Yonda*
Director of Marketing *Craig S. Marty*
National Sales Manager *Douglas J. DiNardo*
Advertising Manager *Janelle Keefer*
Production Editorial Manager *Renée Menne*
Publishing Services Manager *Karen J. Slaght*
Royalty/Permissions Manager *Connie Allendorf*

Copyedited by *Barbara Willette*
Freelance Permissions Editor *Karen Dorman*
Production and Design *York Production Services*
Composition *York Graphic Services*

Cover:
Reprinted with permission from Lewis et al., Crystal structure of the lactose operon repressor and its complexes with DNA and inducer, *Science* 271:1247, 1996. Copyright 1996 American Association for the Advancement of Science.

The photo on the dedication page is © Roger Ressmeyer/Corbis.

The credits section for this book begins on page C-1 and is considered an extension of the copyright page.

Copyright © 1998 The McGraw-Hill Companies, Inc. All rights reserved

Library of Congress Catalog Card Number: 96-83916

ISBN 0-697-21900-3

No part of this publication may be reproduced, stored in a retrieval system, or transmitted, in any form or by any means, electronic, mechanical, photocopying, recording, or otherwise, without the prior written permission of the publisher.

Printed in the United States of America, 2460 Kerper Boulevard, Dubuque, IA 52001

10 9 8 7 6 5 4 3 2 1

98 00419

I wanted to dedicate this text to a pioneer in the field of biomolecular structure to emphasize the importance of this subject to our current understanding of biochemistry. If we look back over the twentieth century, there were many pioneers who played major roles in developing the methods and discovering the principles of biomolecular structure. Certainly, Linus Pauling stands out as one of the most colorful and possibly the greatest.

In the early part of his career, Pauling concerned himself with the structures of small molecules; this culminated in the publication of *The Nature of the Chemical Bond,* a book that was intensively used in the classroom and as a guide to researchers for more than a quarter of a century. Pauling's interest in small molecules was but a starting point. Together with R.B. Corey, Pauling studied the structures of amino acids and small peptides and applied the rules that came out of these studies to the enormously complex problems of protein structure. In the 1940s he and Corey proposed two conformations for polypeptide chains, the alpha helix and the beta sheet, which are the most common secondary structures found in most proteins.

In this picture we see Pauling juggling a fruit rich in vitamin C. He had high hopes that vitamin C in large doses might be beneficial to human health.

Brief Contents

Contents

CHAPTER 

The Building Blocks of Proteins: Amino Acids, Peptides, and Polypeptides 60

Geoffrey Zubay

CHAPTER 

The Three-Dimensional Structures of Proteins 79

Geoffrey Zubay

CHAPTER 

Functional Diversity of Proteins 115

Geoffrey Zubay

CHAPTER 

Methods for Characterization and Purification of Proteins 140

Geoffrey Zubay

PART

CATALYSIS 157

CHAPTER 

Enzyme Kinetics 159

Geoffrey Zubay

CHAPTER 

Mechanisms of Enzyme Catalysis 177

Geoffrey Zubay

CHAPTER 

Regulation of Enzyme Activities 214

Geoffrey Zubay

CHAPTER 16

Electron Transport, Proton Translocation, and Oxidative Phosphorylation 346

Geoffrey Zubay

CHAPTER 17

Photosynthesis and Other Processes Involving Light 371

Geoffrey Zubay

CHAPTER 

Structures and Metabolism of Polysaccharides and Glycoproteins 402

Pamela Stanley

PART

Metabolism of Nitrogen-Containing Compounds 561

CHAPTER 24

Amino Acid Biosynthesis and Nitrogen Fixation in Plants and Microorganisms 563

Ronald Somerville, H. Edwin Umbarger, and Geoffrey Zubay

CHAPTER 25

Amino Acid Metabolism in Vertebrates 597

Ronald Somerville, H. Edwin Umbarger, and Geoffrey Zubay

CHAPTER 26

Nucleotides 629

Raymond Blakley

CHAPTER 27

Integration of Metabolism in Vertebrates 666

Geoffrey Zubay

CHAPTER  31

DNA Replication, Repair, and Recombination 760

Geoffrey Zubay

CHAPTER  32

DNA Manipulation and Its Applications 790

Geoffrey Zubay

List of Contributors

Dr. Raymond Blakley
Chapter 26
Department of Molecular Pharmacology
332 North Lauderdale, P.O. Box 318
St. Jude Children's Research Hospital
Memphis, TN 38101

Dr. Perry A. Frey
Chapter 11
Institute for Enzyme Research
University of Wisconsin
Madison, WI 53706

Dr. Emanuel Goldman
Chapter 34
Department of Microbiology & Molecular Genetics
New Jersey Medical School-UMDNJ
185 South Orange Ave
Newark, NJ 07103

Dr. Gary R. Jacobson
Chapters 20 and 28
Department of Biology
Boston University
Boston, MA 02215

Dr. Ronald Somerville
Chapters 24 and 25
Department of Biological Sciences
Purdue University
West Lafayette, IN 47907-1153

Dr. Pam Stanley
Chapters 13 and 18
Department of Cell Biology
Albert Einstein College of Medicine
Bronx, NY 10461

Dr. H. Edwin Umbarger
Chapters 24 and 25
Department of Biological Sciences
Purdue University
West Lafayette, IN 47907

Dr. Dennis E. Vance
Chapters 19, 21, 22 and 23
Lipid and Microprotein Group
University of Alberta
Edmonton, Alberta
Canada T6G 2C2

Preface

Biochemistry is a growing discipline, closely linked to other fields of study. As biochemistry has grown, so has the need to relate biochemical phenomena to cell biology, physiology, and genetics. We must familiarize the student with the relationship of biochemistry to these other disciplines without going too far afield. The goal of this edition of *Biochemistry* is like that of the previous three editions: to provide a comprehensive, up-to-date teaching text that will enlighten today's students and equip them to deal with tomorrow's problems in biochemical research and medicine. As a discipline in college courses, graduate schools, and medical schools, biochemistry is of ever-increasing importance. The demands for a uniformly authoritative text have never been greater. We have passed the time when this subject could be properly conveyed in a one-quarter or one-semester course to a time when two or three quarters or semesters are required.

The unique concept of our text in a field that is crowded with many texts is the *team-of-experts* approach. From the start of my textbook writing, I knew I could not cope with the vast literature on the subject. That is why I involved so many people in the effort of writing a textbook. This text is written by experts whose knowledge for their chapters comes from the primary scientific literature and their scientific experiences. All of the contributors to our text are researchers who have made their mark in specific areas of biochemistry; they are also teachers of the subject. Their contributed chapters are scrutinized and sometimes edited by me so that they fit into a coordinated whole. In fact, the amount of editing has been steadily diminishing with each edition as the contributing authors have become increasingly aware of the teaching goals of this text, the role that their contributions play, and the need for good pedagogy.

The long list of contributors to previous editions has made a lasting impression on this text that has helped us to mold it into its present form. I am sure you will recognize most of the names:

Daniel Atkinson	**James P. Ferris**	**P. C. Peterson**
Wayne Becker	**Irving Geis**	**Milton H. Saier, Jr.**
James W. Bodley	**Max Gottesman**	**F. Raymond Salemme**
Frederick J. Bollum	**Lloyd Ingraham**	
Ronald Breslow	**Julius Marmur**	**Joseph Sambrook**
Ann Baker Burgess	**Richard Palmiter**	**Jack Strominger**
Richard R. Burgess	**William W. Parson**	**David A. Usher**

In addition to myself the contributors to the current edition include:

Raymond Blakley
Chapter 26
Dr. Blakley holds B.Sc. and M.Sc. degrees from Canterbury College, New Zealand; received his Ph.D. from Otago University College, University of New Zealand; and did postdoctoral work at the National Institute for Medical Research in London. His area of research is the structure and function relations in dihydrofolate reductase leading to mutants that may be useful in gene therapy. He is currently a member of the department of Molecular Pharmacology at St. Jude Children's Research Hospital.

Perry Frey
Chapter 11
Dr. Frey did his undergraduate work at Ohio State University. He received his Ph.D. at Brandeis University, and did postdoctoral work at Brandeis and at Harvard University. His primary research interests center on mechanisms of enzyme and coenzyme action. He is currently a Co-Director of the Institute for Enzyme Research at the University of Wisconsin–Madison.

Emanuel Goldman
Chapter 34
Dr. Goldman received his B.A. at Brandeis University and his Ph.D. at Massachusetts Institute of Technology. His area of specialty is regulation of gene expression in bacteria and in phage, as well as control of translational efficiency/accuracy by tRNAs. His current position is Professor, Department of Microbiology and Molecular Genetics, New Jersey Medical School.

Gary Jacobson
Chapters 20 and 28
Information of Mr. Jacobson was not available as this book went to press.

Ronald Somerville
Chapters 24 and 25
Dr. Somerville received his B.A. and M.Sc. at the University of British Columbia and his Ph.D. at the University of Michigan. He did postdoctoral work at Stanford University. His research specialty is molecular recognition, with special emphasis on protein–protein and protein–nucleic acid interactions relevant to the control of gene expression. He is currently Professor of Biochemistry at Purdue University and Adjunct Professor of Biochemistry and Molecular Biology at Indiana University School of Medicine.

Pamela Stanley
Chapters 13 and 18
Dr. Stanley received her B.Sc. and Ph.D. degrees from the University of Melbourne. She did postdoctoral work at the University of Toronto. Her research is aimed at identifying new biological functions for carbohydrates expressed at the surface of mammalian cells. She currently holds the position of Professor, Department of Cell Biology at Albert Einstein College of Medicine. She is the President-Elect of the Society of Glycobiology.

Edwin Umbarger
Chapters 24 and 25
Dr. Umbarger received B.S. and M.S. degrees from Ohio University. He received his Ph.D. from Harvard University. His research interests focus on the study of isoleucine and valine biosynthesis. Until his retirement in 1993 he was Wright Distinguished Professor of Biological Sciences at Purdue University.

Dennis Vance
Chapters 19, 21, 22, and 23
Dr. Vance received his B.S. degree at Dickinson College and his Ph.D. at the University of Pittsburgh. His research interest is on the regulation of phosphatidylcholine biosynthesis and the role of the enzymes of phosphatidylcholine biosynthesis in regulation of eukaryotic cell division. He currently works as Professor of Biochemistry at the University of Alberta.

Student Learning Aids

Chapter Opening Outline and Overview The chapter opening outline and overview were written to help students preview how the chapter is organized and the major concepts that are to be covered in the chapter.

Declarative Statement Headings Clear, informative headings help students to understand each topic.

Underlined Terms and Key Concepts Underlined important terms and key concepts are easy for students to locate.

Numbered Equations Important equations within a chapter are numbered so that students can easily locate and reference them.

Summary Tables Summary tables are designed to help students more easily understand and use important facts or characteristics presented for a specific topic.

Six Concept and Application Icons Throughout the text the student will find text sections flagged by six different graphic icons or images. These icons are intended to help the student to remember and mentally cross-reference the following concepts and applications:

Methods of Biochemical Analysis

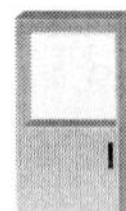

Biomedical Applications

Regulatory Aspects of Biochemistry

Plant Biochemistry

Neurochemistry

Biochemistry of Cancer Cells

Boxed Readings Thought-provoking boxed readings on relevant topics are featured in various chapters.

Molecular Graphics and Illustration Program Molecular graphics images of key molecules enable students to "see" and interpret three-dimensional structures.

End-of-Chapter Enumerated Summary This concise summary of chapter concepts is designed to serve as a guide for chapter study.

End-of-Chapter Selected Readings Each chapter concludes with carefully selected references that contain further information on the topics covered in that chapter.

End-of-Chapter Problems These problems are designed to help students assess their understanding of the chapter's basic concepts. Brief solutions for the odd-numbered problems are found in the back of the book.

Appendix A: Some Major Discoveries in Biochemistry Summarizes major research discoveries from the past and present.

Appendix B: Answers to Odd-Numbered Problems Brief solutions are provided to help the students to determine whether they are on the right track.

Glossary Over 700 important biochemical terms are defined in this end-of-book glossary.

Index An easy-to-use, comprehensive index is provided.

Endsheets with Reference Material The endsheets of the text contain useful easily accessible reference material.

Organizational Changes in the Fourth Edition

Our efforts in this edition have focused on integrating new knowledge into an already extensive body of information. Changes have been made throughout the text, with major changes being confined to areas in which the most important advances have been made. In some cases, old references have been removed from the ends of chapters, and in most cases, appropriate new references have been included.

The contents of the different parts and chapters should be self-evident from the titles and section headings. Parts 1 and 3 and the chapters therein are organized in the same way as they were in the third edition. In the remaining parts of this edition, substantial changes in organization have been made. In the third edition, part 2 covered the structures and functions of most of the major components of the cell. In the fourth edition, part 2 focuses on protein structure and function, leaving the discussion of structure and function of the remaining components to be dealt with immediately before addressing their metabolism. In this way the structure and function will be fresh in the mind as one delves into the metabolism, a clear advantage for the student.

In parts 4 and 5 of the third edition we attempted to divide the metabolism into a discussion of the catabolism and the anabolism, respectively. These parts have been supplanted by parts 4, 5, and 6 in the fourth edition, in which the intermediary metabolism is considered first for carbohydrates (part 4), then for lipids (part 5), and finally for nitrogen-containing compounds (part 6). This arrangement is considered superior because it facilitates comparison of the similar and dissimilar features of the metabolism for opposing anabolic and catabolic pathways and a discussion of their regulation, which usually focuses on strategies that prevent opposing pathways from functioning simultaneously.

The strategy of presenting structures and functions adjacent to the related metabolism sections is carried over to part 7, in which the structures and functions of nucleic acids are considered just prior to a discussion of their metabolism.

Although we believe that many instructors of biochemistry will find these changes in organization to be superior for teaching purposes, we also recognize that there are others who prefer the organizational scheme followed in the third edition. In consideration of this, the relevant materials have been flexibly packaged so that the order of teaching need not follow the order in the text. For example, if the instructor wishes to consider most of the structures first before getting into the metabolism, then after a consideration of part 2 as it appears in the fourth edition, one can turn to chapter 13 for a discussion of the structures of sugars and energy storage polysaccharides, followed by chapter 19 for a discussion of the structure and function of biological membranes, perhaps chapter 20, which deals with membrane transport, and, finally, chapter 30 for a discussion of the structures of DNA and nucleoproteins.

Physiological biochemistry, which comprised part 7 in the third edition, has been eliminated in the fourth edition, and the chapters that it contained have been relocated to the most relevant sections of the text. Vision and neurotransmission have been moved to part 6, and immunobiology and carcinogenesis, along with a new chapter on AIDS and the HIV virus, have been moved to part 7, "Storage and Utilization of Genetic Information."

Detailed Content Changes of the Fourth Edition

Part 1, "An Overview of Biochemistry and Bioenergetics," contains changes only to chapter 1, with the addition of the new section entitled "The First Living Systems Were Acellular." This section contains a brief but authoritative account of the most significant prebiotic events that are believed to have led to the origin of life.

Part 2, "Protein Structure and Function," starts with a new chapter, "The Structure and Function of Water." This chapter pulls together information that was previously scattered in several early chapters and adds new information on the general nature of buffers and the physiologically important buffers containing either phosphate or carbonate. In chapter 4 some information on the determination of amino acid composition has been added. Material in chapter 5 on the three-dimensional structures of proteins has been substantially revised in the middle and final parts. In particular, the discussion of globular protein structures has been reorganized. The structural motif is introduced as the fundamental unit of tertiary structure. Then it is shown how domains are built from single motifs or combinations of motifs and so on. Sections on nuclear magnetic resonance and optical rotatory dispersion and circular dichroism have been added to the last part of chapter 5.

In chapter 6 the discussion of the way in which various factors negatively affect oxygen binding by hemoglobin has been elaborated upon. Thanks to new structural information on myosin (see the references), it has been possible to describe the actin–myosin cycle associated with muscular contraction in greater molecular detail (see figures 6.19, 6.20, and 6.21). In box 6A a comprehensive explanation of multiple binding and the Hill plot is given. In box 6B a detailed explanation of the inheritance pattern of genetic defects is presented that will be useful for the discussion of hemoglobin in this chapter and will continue to be useful at many points throughout the text. Chapter 7 is a new chapter in which the methods of characterization and purification of proteins are presented. In the third edition this discussion was tacked onto the previous chapter, which dealt with the functional diversity of proteins. Only minor changes have been made to the content of this chapter.

Part 3 begins with chapter 8, of which substantial portions of the middle sections have been rewritten in the interest of achieving greater clarity. Some complex equations have been eliminated without any substantial loss of substance. The section entitled "The Henri-Michaelis-Menten Treatment Assumes That the Enzyme-Substrate Complex Is in Equilibrium with Free Enzyme and Substrate" is new.

The changes in chapter 9 are probably the most important changes made in this text. This is due to a new development in the field of enzymology that invites a reconsideration of many enzyme mechanisms that have been proposed. The second main section of chapter 9, which is concerned with detailed mechanisms of enzyme catalysis, has been substantially revised. In two cases, trypsin and triose phosphate isomerase, the revised mechanisms have resulted from the realization that extra-strong hydrogen bonds are probably formed in the activated complexes. Extra-strong hydrogen bonds known as low-barrier hydrogen bonds were identified in the early 1970s, but their significance in the field of enzyme catalysis has come to light only recently, partly as a result of papers published by W. W. Cleland and Kreeway and Perry Frey and his co-workers (see references and box 9B). Revision of the proposed mechanism for RNase A action has resulted from a recent paper by Ron Breslow. Several new computer-generated figures have been added to aid in visualization of the overall enzyme structure and the active site (see figures 9.6, 9.7, 9.25, and 9.26). Box 9D discusses catalytic antibodies.

Chapter 10 contains the description of how the mechanism of action of calmodulin has been significantly amplified and updated as a result of new structural information. In chapter 11 the section on coenzyme A has been expanded. The role of ascorbic acids in maintaining the enzyme that catalyzes hydroxyproline formation in collagen is explained in a new short section. Finally, a new box 11A discusses aconitase as an example of an enzyme containing an iron–sulfur complex that is involved in something other than an oxidation–reduction reaction.

In chapter 12 a new section, "A Regulated Reaction Is Effective Only if It Is Exergonic," was added to give emphasis to a principle that was made clear earlier in the text. Although chapter 13 is a new chapter, the material in it is not new; it comes from the first part of chapter 6 in the third edition, which deals with the structures and functions of simple sugars and polysaccharides.

As seen in chapter 14, it is quite remarkable how new understandings of the ancient subject of carbohydrate metabolism keep appearing as we learn more and more about the enzymes that are involved. An expanded discussion of the hexokinases found in different tissues explains why glucose is normally absorbed by most tissues but not by the liver. The hexokinases in question are assembled from different isozymes. An expanded discussion of the different isozymal forms of the bifunctional enzyme that regulates the level of fructose-2,6-bisphosphate explains the different responses of liver and muscle tissue to the hormone epinephrine. The metabolism of the two most important dietary disaccharides is presented with particular reference to the problems they can produce when there are metabolic lesions. A new box 14A describes the history of events leading to the discovery of the glycolytic pathway. A new box 14B describes factors that influence the levels of the regulatory molecule glycerate-2,3-bisphosphate in erythrocytes. This subject is taken up here because the compound in question is also an intermediate in the glycolytic pathway. Chapter 15 contains only minor changes.

The headings in chapter 16 are similar to the ones in the comparable chapter (15) of the third edition. Despite this, many sections have undergone considerable revision. The proton translocation that is powered by electron transport is described as resulting from either redox loops or proton pumps. There are still considerable mysteries concerning how these processes work. Recent crystal structure studies of the F_1 subunit of the ATP-synthase have elevated our understanding of the mechanism for ATP synthesis.

In addition to many minor changes, new information is presented in chapter 17 on the operation of the antenna systems at the photocenters and the mechanism of oxygen evolution. Chapter 18 is derived from chapter 21 of the third edition. In addition, it includes a description of the relevant structures that were presented in chapter 6 of the third edition. Except for reorganizational changes, there have been no major additions or deletions within the contents of the chapter.

Chapter 19 on lipids and membranes, the beginning of part 5, has been completely rewritten to yield a more exciting and shorter presentation. Chapter 20 on mechanisms of membrane transport has been completely rewritten, relocated, and shortened. This subject was relocated because it was considered highly desirable to present the mechanisms of transport before getting too far into the metabolism section. Although the chapter has been considerably shortened, I do not feel that this has resulted in a serious loss of substance.

In chapter 21, only a few changes have been made to accommodate recent advances. In chapter 22, many small changes have been made in the coverage of biosynthesis of membrane lipids. These changes were made to update the material, but clarity of explanation was also addressed. Chapter 23 on cholesterol metabolism includes new information on the factors that control the rate of mevalonate synthesis; the molecular basis of the inherited disease abetalipoproteinemia in which patients have no chylomicrons, VLDLs, or LDLs in the bloodstream; a description of how cholesterol suppresses the synthesis of LDL receptors; and box 3A, a description of the isoprenylation of proteins.

Chapters 24 and 25 on amino acid metabolism are closely knit, as they were in the third edition. The material has been realigned to make it more accessible and to make it easier to use either chapter independently of the other. Chapter 24 deals with amino acid biosynthesis and nitrogen fixation in plants and microorganisms. Important changes in the regulation of the enzyme glutamine synthase have been added to chapter 24. Chapter 25 deals with amino acid metabolism in vertebrates. I suspect that if time is limited, chapter 25 will be favored because of the somewhat greater interest in this subject in most college courses. An appendix that discusses detailed aspects of many catabolic pathways has been added to chapter 25.

Chapter 26 on nucleotides has been considerably updated to accommodate the rapidly changing subject of inhibitors of nucleotide synthesis and their role in chemotherapy. In chapter 27 several sections have been substantially revised to bring them up to date. These sections are "General Aspects of Cell Signaling," "The Adenylate Cyclase Pathway Is Triggered by a Membrane-Bound Receptor," "Protein Phosphorylation Is the Most Common Way in Which Regulatory Proteins Respond to Hormonal Signals," "Variability in G Proteins Adds to the Variability of the Hormone-Triggered Response," "Guanylyl Cyclase Can Be Activated by a Gas," and "Growth Factors Are Proteins That Behave Like Hormones."

In chapter 28 on neurotransmission the section entitled "The Acetylcholine Receptor Is the Best-Understood Neurotransmitter" has been revised to bring it up to date. Two new sections have been added: "Synaptic Receptors Coupled to G Proteins Produce Slow Synaptic Responses" and "Synaptic Plasticity and Learning." Minor changes have been made in chapter 29 on vision to bring the subject up to date.

In chapter 30 a new section, "Helical Structures That Use Additional Kinds of Hydrogen Bonding," deals with the conformational variants formed by telomeres. Four new sections have been added to chapter 31: "Initiation of Chromosomal Replication in Eukaryotes," "The Mismatch Repair System Is Important for Maintaining Genetic Stability," "Some Transposable Genetic Elements Encode a Reverse Transcriptase," and "Bacterial Reverse

Transcriptase Catalyzes Synthesis of a DNA-RNA Molecule." Other sections ("SV40 Is Similar to Its Host in Mode of Replication," "Several Systems Exist for DNA Repair," and "Telomerase Facilitates Replication at the Ends of Eukaryotic Chromosomes") have been significantly modified. These additions reflect important advances in our understanding of chromosome replication in eukaryotic systems and DNA repair in both prokaryotic and eukaryotic systems.

For chapter 32 there has been an enormous amount of activity in this field but surprisingly few fundamental advances. This is particularly true in the field of application to practical problems (plenty of sizzle but no steak). Two new sections have been added: "Yeast Artificial Chromosomes (YACS) Are Used for Cloning Fragments as Large as 500 kb in Length" and "Will Nucleic Acids Become Useful Therapeutic Agents?"

In chapter 33, two new sections have been added: "Comparison of *E. coli* RNA Polymerase with DNA PolI and PolIII" and "RNA Editing Involves Changing of the Primary Sequence of a Nascent Transcript." Several existing sections have been updated. The most important updated section, "Messenger RNA Transcription by Polymerase II," has resulted from genetic and biochemical studies of yeast RNA polymerase. Many heretofore unrecognized proteins have been unequivocally recognized as key components of RNA polymerase II. The popular model for stepwise assembly of the RNA polymerase II–DNA complex must now share the limelight with a model in which the holoenzyme remains largely intact in its association–dissociation cycle at the promoter.

In chapter 34, three new sections have been added: "In Addition to the P Site and the A Site for Binding tRNAs, the Ribosome May Possess a Third Site, the E Site," "Protein Folding Is Mediated by Protein Chaperones," and "ATP Plays Multiple Roles in Protein Degradation." Ten sections have been updated: "Ribosomes Are the Site of Protein Synthesis," "The Code Is Highly Degenerate," "Each Synthase Recognizes a Specific Amino Acid and Specific Regions on Its Cognate tRNA," "Aminoacyl-tRNA Synthases Can Correct Acylation Errors," "Translation Begins with the Binding of mRNA to the Ribosome," "Three Elongation Reactions Are Repeated with the Incorporation of Each Amino Acid," "Two (or Three) GTPs Are Required for Each Step in Elongation," "Targeting and Posttranslational Modification of Proteins," "Proteins Are Targeted to Their Destination by Signal Sequences," and "Ubiquitin Tags Proteins for Proteolysis."

Three new sections have been added to chapter 35: "Helix-Turn-Helix Regulatory Proteins Are Symmetrical," "DNA–Protein Cocrystals Reveal Gross Features of the Complex," and "RNA Can Function as a Repressor." The first two of these new sections reflect the progress that has been made in structural studies of DNA regulatory protein complexes. The third relates to the increasing number of examples in which DNA functions as a transcription regulator.

Regarding chapter 36, enormous progress has been made in the area of gene-regulatory proteins in eukaryotes. This has resulted in several new sections and several modified sections in the central part of this chapter. In addition to this, the section entitled "DNA Methylation Is Correlated with Inactivated Chromatin" has been updated by the finding of a long-sought-for genetic correlate. A long-overdue section on how translation controls transcription in eukaryotes has been added. The chapter ends with a new section entitled "Early Development in *Drosophila* and Vertebrates Shows Striking Similarities."

Two new sections have been added to chapter 37. The first, "T-Cell Action Is Frequently Augmented by the Secretion of Hormonelike Proteins Called Interleukins," deals with the hormonelike proteins that bias the interactions between cells of the immune system. The second, "The Immune System in Action," describes three examples of how the immune system disposes of specific invaders by choosing the most appropriate weapons from its arsenal.

Four new sections have been added to chapter 38, and another four have been substantially revised to bring them up to date. These sections comprise most of the latter half of this chapter. The changes help to delineate the distinction between protooncogenes and tumor suppressor genes. In the new chapter 39 the causes and progression of the disease known as AIDS are discussed at an introductory level. Approaches for preventing the spread of AIDS and treating AIDS patients are also discussed.

Ancillary Materials

Student Study Guide and Solutions Manual
Written by Larry Loomis-Price, Johns Hopkins University, and Gwen E. Shafer, independent scientific consultant, this manual contains study information and self-quizzes. It can help students to better understand how to solve the problems in the text and to prepare for exams.

Instructor's Manual with Test Item File and Transparency Masters
This manual contains suggestions on how to use the text in different course situations and detailed, worked-out solutions for the even-numbered problems found in the text chapters. In addition, this manual offers several objective test questions for each chapter, which can be used to generate exams, along with 300 black-and-white transparency masters from the text.

Classroom Testing Software
This software is offered free upon request to adopters of the text. It provides a database of questions for preparing exams. No programming experience is necessary to use the software. Available in IBM and Macintosh formats.

Transparency Acetates
A set of 100 full-color transparencies is available free to adopters. These acetates feature key illustrations that can be used to enhance your classroom lectures. When these acetates are coupled with the 300 masters provided with the instructor's manual, each adopter has access to 400 figures and tables from the text to use in the lecture setting.

Slides
A set of 100 full-color projection slides derived from the transparency illustrations is available free to adopters.

Biochemistry Electronic Image Bank
These computerized image files are available free to adopters upon request. These files contain images from this text along with images from other Wm. C. Brown biochemistry titles. These electronic acetates can be clearly projected on large lecture hall screens by using an LCD projection system.

Biochemical Pathways Software
Created by Bill Sofer, this is an easy-to-use tutorial review software program for Macintosh that provides quizzes and memory exercises that test students' knowledge of glycolysis and the TCA cycle. Contact your bookstore or call Wm. C. Brown Publishers at 1-800-338-5578 to place an order or request more information on this software. (ISBN 25100)

Acknowledgments

Charlotte Pratt was most helpful at editing and correcting the art. Barbara Willette was extremely helpful in finding errors and copyediting the entire manuscript and its ancillaries. She was always there and always punctual. Thank you Barbara. And last but not least I am grateful to Dennis Krebs who stuck with this project all the way despite some very hectic times. Many thanks to you Dennis.

Hassan Ahmad *University of Texas–Pan American*
Eric J. Allain *University of Illinois*
Paul Austin *Hanover College*
L. Rao Ayyagari *Lindenwood College*
Derek Baisted *Oregon State University*
Judith K. Ball *University of Western Ontario*
Tadhg Begley *Cornell University*
Erich C. Blossey *Rollins College*
John J. Brink *Clark University*
Martin L. Brock *Eastern Kentucky University*
Rebecca M. Burt *Southeast Community College–Beatrice Campus*
Kim C. Calvo *University of Akron*
Stephen W. Carper *University of Nevada, Las Vegas*
Eileen Carreiro *University of Massachusetts, Dartmouth*
Derek Cash *University of Missouri*
Scott Champney *East Tennessee State University*
David P. Chitharanjan *University of Wisconsin, Stevens Point*
Carmine J. Coscia *St. Louis University School of Medicine*
Lawrence C. Davis *Kansas State University*
S. Todd Deal *Georgia Southern University*
Jeff DeJong *University of Texas, Dallas*
Louis Delbaere *University of Saskatchewan*
Daniel V. DerVartanian *University of Georgia*
Kelsey R. Downum *Florida International University*
Lawrence K. Duffy *University of Alaska*
K. E. Ebner *University of Kansas Medical Center*
Alan D. Elbein *University of Arkansas*
William A. Elmer *Emory University*
Felicia Etzkorn *University of Virginia*
Robert Fletterick *University of California, San Francisco*
Bill Flurkey *Indiana State University*
Leslie Wo-Mei Fung *Loyola University of Chicago*
Theresa Gioannini *CUNY, Baruch College*
Darrel E. Goll *University of Arizona*
Edye E. Groseclose *Nova Southeastern University*
Lonnie J. Guralnick *Western Oregon State College*
Barbara A. Hamkalo *University of California, Irvine*
Gregory B. Hecht *Rowan College of New Jersey*
Terry L. Helser *SUNY College, Oneonta*
Linda C. Hodges *Agnes Scott College*
J. F. Honek *University of Waterloo*
Mark S. Hopkin *Life College*
H. David Husic *Lafayette College*
Larry L. Jackson *Montana State University*
George T. Javor *Loma Linda University School of Medicine*
Thomas V. Jeffries *Campbellsville University*
Warren V. Johnson *University of Wisconsin, Green Bay*
Robert M. Jonas *Texas Lutheran University*
Floyd W. Kelly *Casper College*
Ramji L. Khandelwal *University of Saskatchewan*
David Koetje *SUNY College, Fredonia*
Lawrence J. Krebaum *Missouri Valley College*
Karen Kurvink *Moravian College*
Jeffrey E. Lacy *Shippensburg University*
Daniel J. Lavoie *Saint Anselm College*
Michael A. Lea *University of Medicine & Dentistry of New Jersey*
Michael Lieberman *University of Cincinnati*
Mary Katherine K. Lockwood *University of New Hampshire*
Dennis E. Lohr *Arizona State University*
James W. Long *University of Oregon*
Larry Loomis-Price *Henry M. Jackson Foundaton and Johns Hopkins University*
Fulgentius N. Lugemwa *Murray State University*
Yinfa Ma *Northeast Missouri State University*
E. Jerome Mass *Oakton Community College*
Charles Mallery *University of Miami*
Celia Marshak *San Diego State University*
Bruce L. Martin *University of Tennessee*
Lynn M. Mason *Lubbock Christian University*
Steven Matson *University of North Carolina*
Martha McBride *Norwich University*
Joseph Mendicino *University of Georgia*
Sabeeha Merchant *University of California, Los Angeles*
Christian G. Merkel *College of Osteopathic Medicine of the Pacific*
Holly Miller *SUNY, Stony Brook*
Michael J. Minch *University of the Pacific*
Miguel O. Mitchell *Albright College*
Bruce H. Morimoto *Purdue University*
Melvyn W. Mosher *Missouri Southern State College*
Richard M. Niles *Marshall University School of Medicine*
Robert Noiva *University of South Dakota School of Medicine*
R. S. Pappas *Georgia State University*
Nancy Peterson *North Central College*
Allen T. Phillips *Pennsylvania State University*
J. Regino Perez-Polo *University of Texas Medical Branch*
Raymond Earl Poore *Jacksonville State University*
Gary L. Powell *Clemson University*
Neil P. J. Price *SUNY, College of Environmental Science & Forestry*
Michael Eugene Pugh *Bloomsburg University*

Ann Randolph *Rosemont College*
Philip Reyes *University of New Mexico*
James E. Russo *Whitman College*
Thomas Ruttledge *Earlham College*
Hildagarde K. Sanders *Villa Julie College*
R. Sarma *SUNY, Stony Brook*
Glenn R. Sauer *University of South Carolina*
Angelo M. Scanu *University of Chicago*
Edward G. Senkbeil *Salisbury State University*
Ralph Shaw *Southeastern Louisiana University*
Andrew K. Shiembe *West Virginia University*
Carl E. Shively *Alfred University*
Jessup M. Shively *Clemson University*
Bryan L. Spangelo *University of Nevada, Las Vegas*
David Speckhard *Loras College*
Larry D. Strawser *U.S. Air Force Academy*
Michael A. Sypes *Long Island University*
Lawrence A. Tabak *University of Rochester*
Jon I. Teng *University of Texas Medical Branch*
Dean R. Tolan *Boston University*
Elliott L. Uhlenhopp *Grinnell College*
George F. Uhlig *College of Eastern Utah*
William H. Voige *James Madison University*
Arthur Clover Washington *Tennessee State University*
Stephen H. Wentland *Houston Baptist University*
Steven M. Wietstock *Indiana University*
E. Brady Williams *College of St. Catherine*
Beulah M. Woodfin *University of New Mexico*
A. Edwin Woods *Middle Tennessee State University*
Kenneth H. Woodside *Nova Southeastern University*
Catherine F. Yang *Rowan College of New Jersey*
Lisa A. Zuraw *The Citadel*

PART

1

An Overview of Biochemistry and Bioenergetics

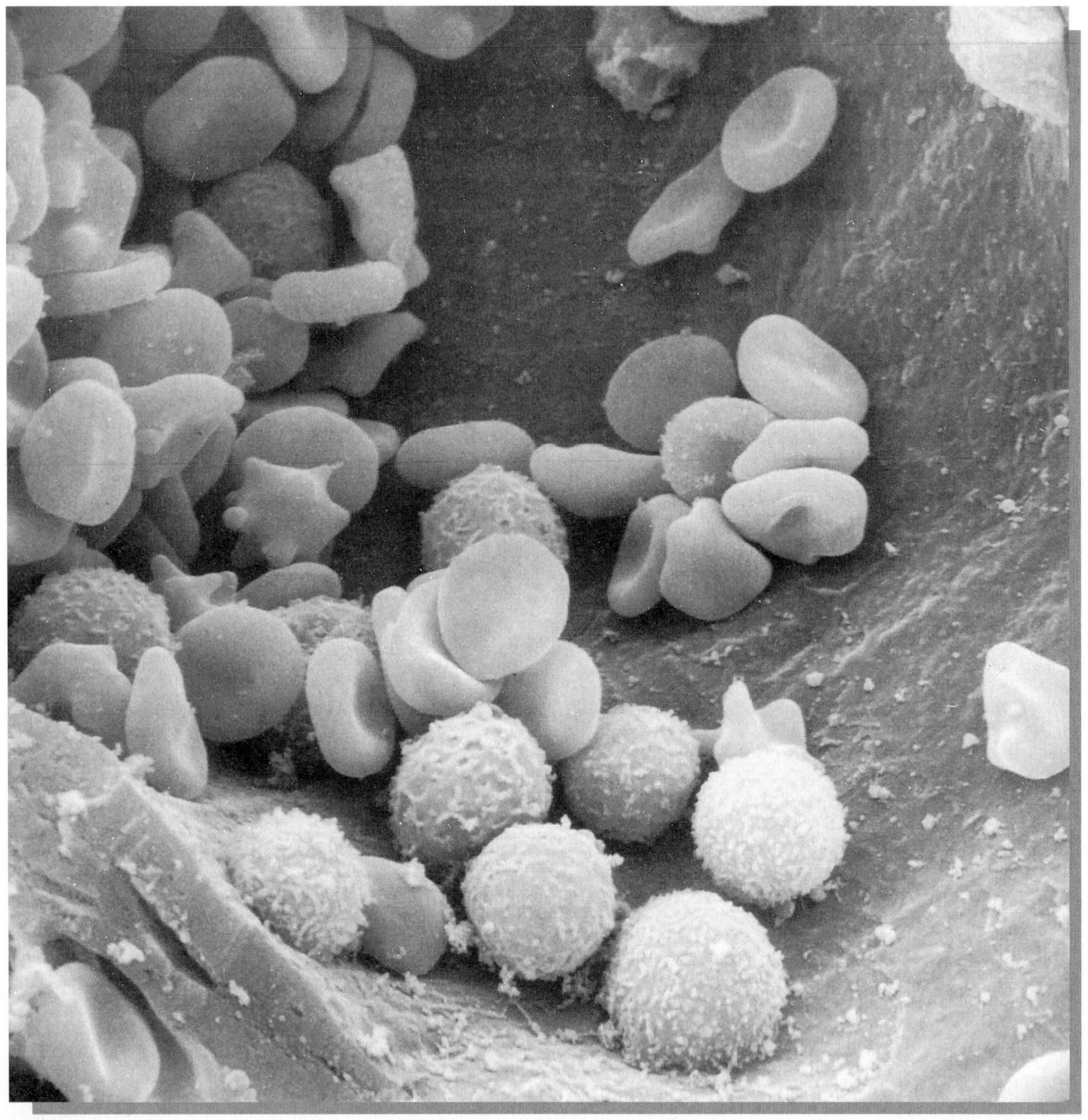

Scanning electron micrograph of erythrocytes and leukocytes in a small blood vessel. Erythrocytes are the biconcave cells with a relatively smooth outer membrane structure. Most of the protein in erythrocytes is hemoglobin, which transports oxygen from the lungs to other tissues. Leukocytes are the rounded cells that possess filamentous protrusions. They are involved in the immune response. The complex surface structures are required for interaction with other cells. Despite the dramatic difference in composition and function, these cells are both derived from the same pluripotential hematopoietic stem cells. (From R. G. Kessel and R. H. Kardon, Tissues and Organs, *W.H. Freeman, Copyright © 1979.) Photo Dr. R. G. Kessel*

C H A P T E R

Cells, Organelles, and Biomolecules

The most unique feature of the living cell is the way in which so many reactions are organized to serve a single purpose.

In biology we are introduced to the extraordinary diversity of living organisms, a diversity that is so great that taxonomists generally acknowledge that there are many more species than will ever be classified. If we limit our inspection of living things to the gross organismic level, it is easy to be overwhelmed by this complexity. Furthermore, we will not find explanations for why organisms are constructed the way they are at this level of inquiry. If we probe more deeply into organismic structure, we find that all organisms are composed of much smaller units called cells. If we probe even further, we find that cells are composed of a limited number of subcellular structures, macromolecules and small molecules. We also find that the macromolecules, which make up the bulk of the solid matter of cells, are constructed from a small number of building blocks that are closely related in structure. It is at the molecular level that we find the ultimate explanations for organismic structure and behavior. We should revel in the discovery that while living organisms are very complicated, there is a common chemistry and common rules that explain how they work.

Thanks to the investigations of thousands of biochemists over the past century, we are on the verge of a thorough understanding of life at the molecular level. This is a most satisfying and exciting time for biochemists; their investigations and accomplishments place biochemistry in the forefront of the biological sciences. The object of this text is to acquaint the student of biochemistry with the basic facts and principles of this subject.

We will begin with an overview of the structures and chemistry and strategies that are shared by all organisms.

All Organisms Are Composed of Cells

Microscopic examination of any organism will reveal that it is composed of membrane-enclosed structures called cells. The enclosing membrane is called the cell membrane or the plasma membrane. Cells vary enormously in size and shape. Figure 1.1 shows

Figure 1.1

Cells are the fundamental units in all living systems, and they vary tremendously in size and shape. All cells are functionally separated from their environment by the plasma membrane that encloses the cytoplasm. Generalized representations of the internal structures of animal and plant cells (eukaryotic cells). Plant cells have two structures not found in animal cells: a cellulose cell wall, exterior to the plasma membrane, and chloroplasts.

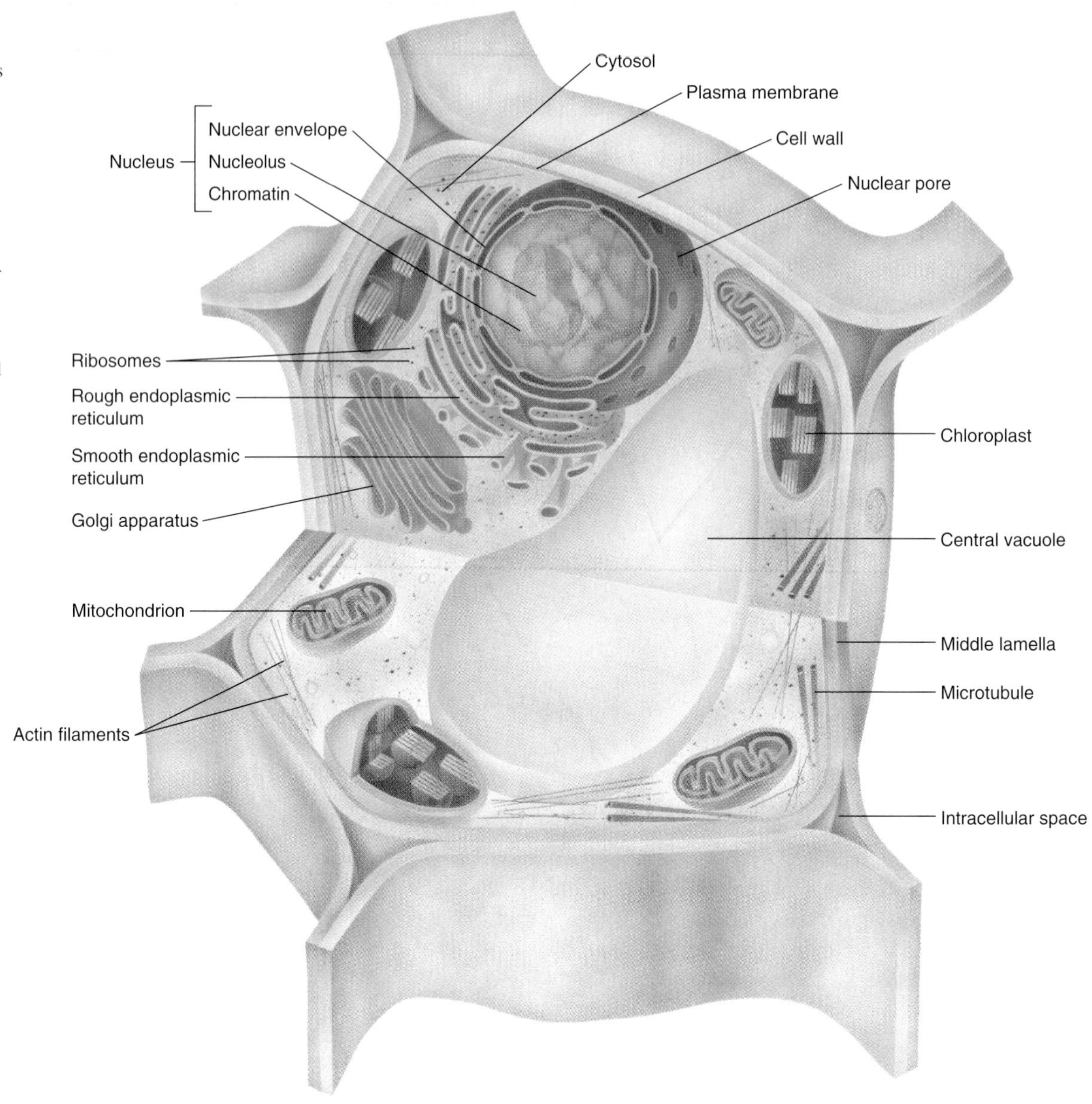

prototypical animal and plant cells. In multicellular organisms the cells associate to form specialized tissues.

The plasma membrane is a delicate, semipermeable, sheetlike covering for the entire cell. By forming an enclosure it prevents gross loss of the intracellular contents; its semipermeable character permits the selective absorption of nutrients and the selective removal of metabolic waste products. In many plant and bacterial (but not animal) cells, a cell wall encompasses the plasma membrane. The cell wall is a more porous structure than the plasma membrane, but it is mechanically stronger because it is constructed of a covalently cross-linked, three-dimensional network. The cell wall maintains a cell's three-dimensional form when it is under stress.

The contents enclosed by the plasma membrane constitute the cytoplasm. The purely liquid portion of the cytoplasm is called the cytosol. Within the cytoplasm are a number of macromolecules and larger structures, many of which can be seen by high-power light microscopy or by electron microscopy. Some of the structures are membranous and are called organelles. Organelles commonly found in plant and animal cells include the nucleus, the mitochondria, the endoplasmic reticulum, the Golgi apparatus, the lysosomes, and the peroxisomes (see fig. 1.1). Chloroplasts are an important class of organelles found in many plant cells but never in animal cells. Each type of organelle is a specialized biochemical factory in which certain biochemical products are synthesized. In addition to organelles, animal and plant cells contain a collection of filamentous structures termed the cytoskeleton, which is important in maintaining the three-dimensional integrity of the cell.

As we will see, the evolutionary tree is bisected into a lower, or prokaryotic, domain and an upper, or eukaryotic, domain. The terms prokaryote and eukaryote refer to the most basic division between cell types. The fundamental difference is that in eukaryotes the cell contains a nucleus, whereas in prokaryotes it

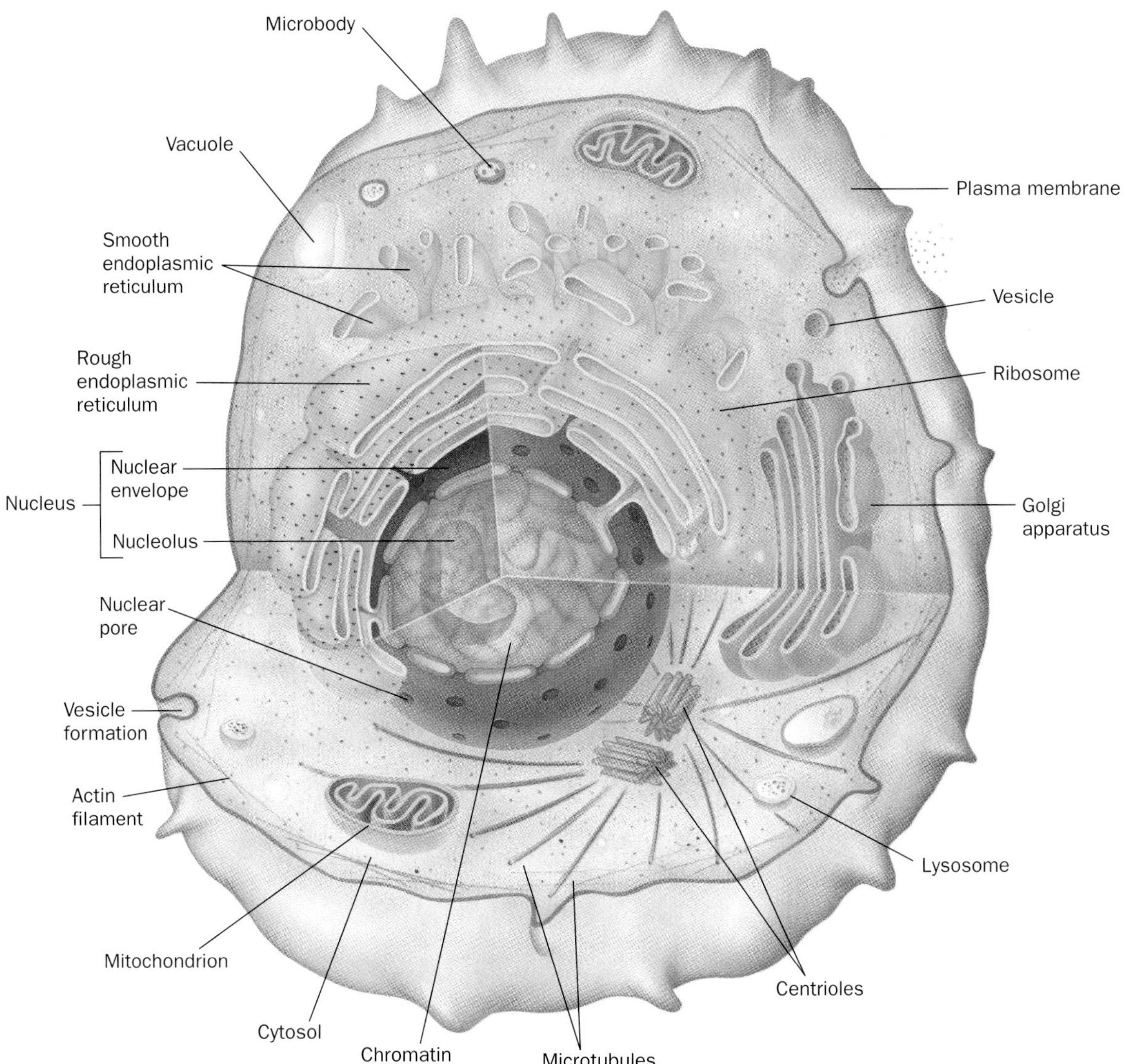

does not. The cells of prokaryotes usually lack most of the other membrane-bounded organelles as well. Plants, fungi, and animals are eukaryotes, and bacteria are prokaryotes. The biochemical functions associated with organelles are frequently present in bacteria, but they are carried out in a different way.

Cells are organized in a variety of ways in different living forms. Prokaryotes of a given type produce cells that are very similar in appearance. A bacterial cell replicates by binary fission, a process in which two identical daughter cells arise from an identical parent cell. Simple eukaryotes can also exist as single nonassociating cells. Eukaryotes of increasing complexity can contain many cells with specialized structures and functions. For example, humans contain about 10^{14} cells of more than a hundred different types. Specialized cells make up the skin, connective tissue, nervous tissue, muscle, blood, sensory functions, and reproductive organs (fig. 1.2). In such a complex organism, the capacity of different cells for replication is limited. When a skin cell or a muscle cell precursor replicates, it makes more cells of the same type. The only cells capable of reproducing an entire organism are the germ cells, that is, the sperm and the egg.

Cells Are Composed of Small Molecules, Macromolecules, and Organelles

Of the many different types of molecules in the various organelles and the cytosol that constitute the living cell, water is by far the most abundant, constituting about 70% by weight of most living matter (table 1.1). As a result, most components are essentially in an aqueous environment.

Except for water, most of the molecules found in the cell are macromolecules, which can be classified into four different categories: lipids, carbohydrates, proteins, and nucleic acids. Each type of macromolecule possesses distinct chemical properties that suit it for the functions it serves in the cell.

Lipids are primarily hydrocarbon structures (fig. 1.3). They tend to be poorly soluble in water; this is one of the reasons why they are particularly well suited to serve as a major component of the various membrane structures found in cells. Lipids also serve as a compact means of storing chemical energy to power the metabolism of the cell.

Carbohydrates, like lipids, contain a carbon backbone,

Figure 1.2

Specialized cell types found in the human. Although all cells in a multicellular organism have common constituents and functions, specialized cell types have unique chemical compositions, structures, and biochemical reactions that establish and maintain their specialized functions. Such cells arise during embryonic development by the complex processes of cell proliferation and cell differentiation. Except for the sex cells, all cell types contain the same genetic information, which is faithfully replicated and partitioned to daughter cells. Cell differentiation is the process whereby some of this genetic information is activated in some cells, resulting in the synthesis of certain proteins and not other proteins. Thus specialized cells come to have different complements of enzymes and metabolic capacities.

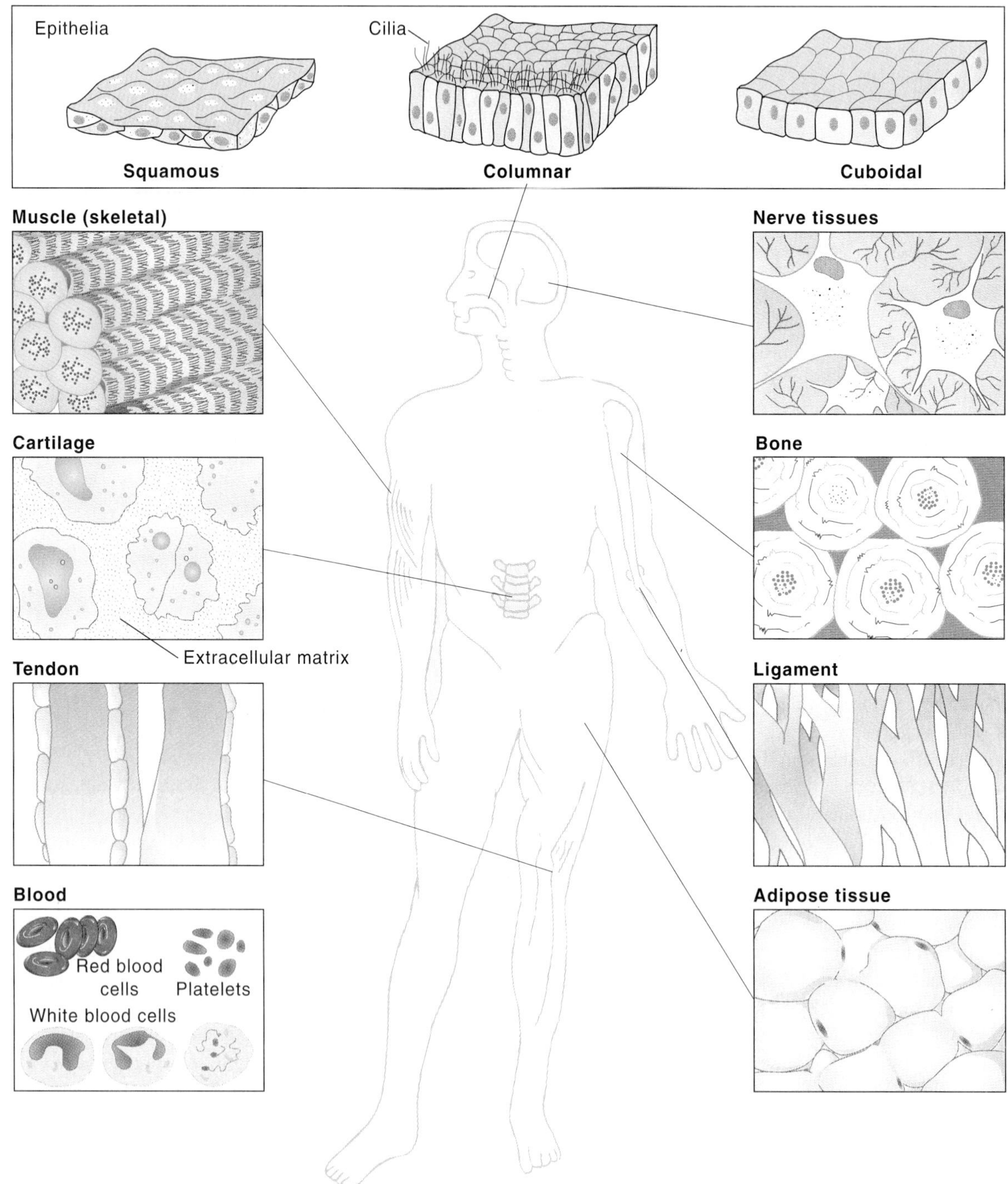

Figure 1.3

The structures of common lipids. (*a*) The structures of saturated and unsaturated fatty acids, represented here by stearic acid and oleic acid. (*b*) Three fatty acids covalently linked to glycerol by ester bonds form a triacylglycerol. (*c*) The general structure for a phospholipid consists of two fatty acids esterified to glycerol, which is linked through phosphate to a polar head group. The polar head group may be any one of several different compounds—for example, choline, serine, or ethanolamine.

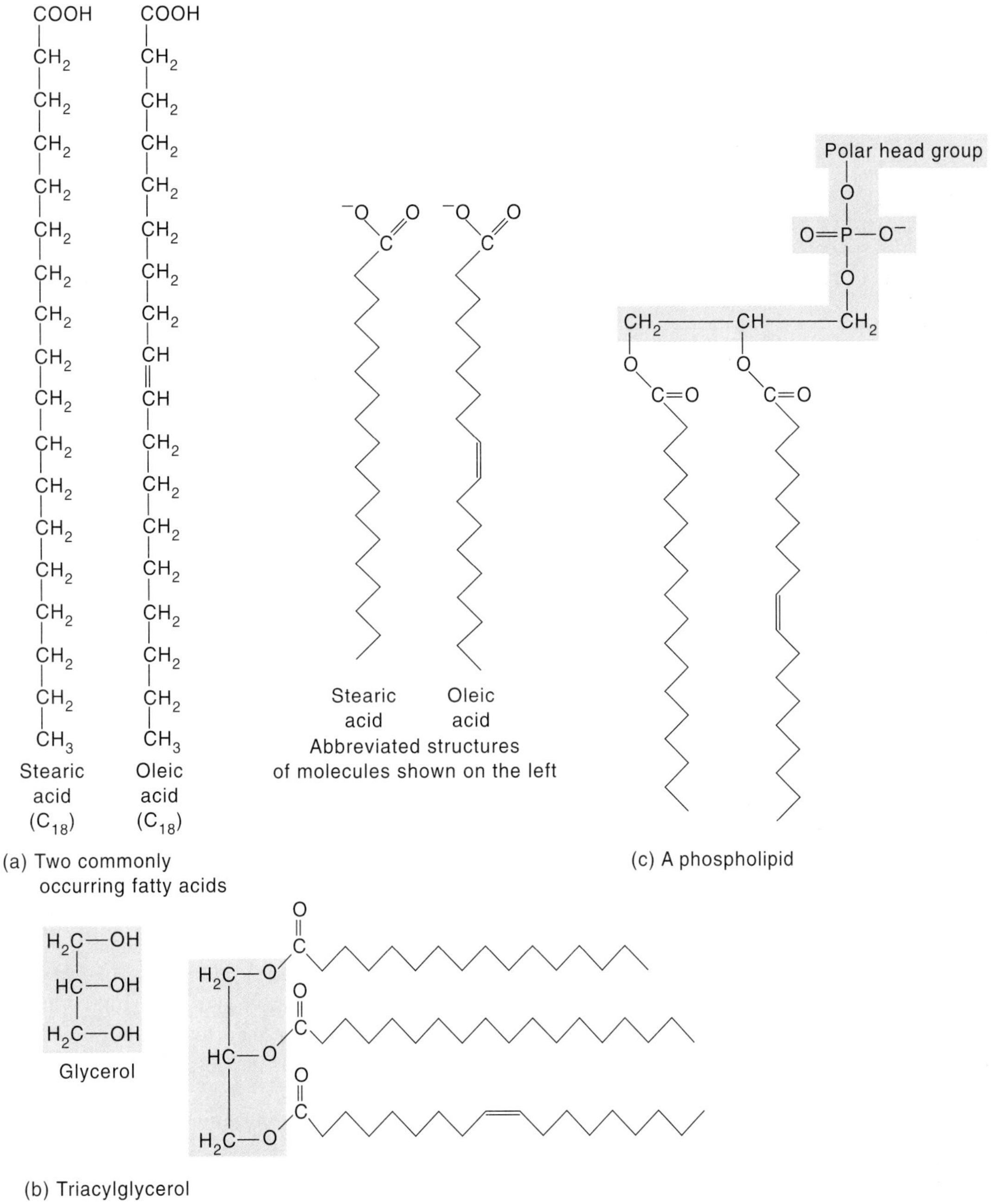

Stearic acid (C_{18})

Oleic acid (C_{18})

Stearic acid

Oleic acid

Abbreviated structures of molecules shown on the left

(a) Two commonly occurring fatty acids

Polar head group

(c) A phospholipid

Glycerol

(b) Triacylglycerol

Table 1.1
The Approximate Chemical Composition of a Bacterial Cell

	Percent of Total Cell Weight	Number of Types of Each Molecule
Water	70	1
Inorganic ions	1	20
Sugars and precursors	3	200
Amino acids and precursors	0.4	100
Nucleotides and precursors	0.4	200
Lipids and precursors	2	50
Other small molecules	0.2	~200
Macromolecules (proteins, nucleic acids, and polysaccharides)	22	~5,000

Source: S. E. Luria, S. J. Gould, and S. Singer, *A View of Life.* Copyright © 1981, Benjamin/Cummings, Menlo Park, Calif.

but they also contain many polar hydroxyl (—OH) groups making them very soluble in water. Large carbohydrate molecules called polysaccharides consist of many small, ringlike sugar molecules, the sugar monomers, attached to one another by glycosidic bonds in a linear or branched array to form the sugar polymer (fig. 1.4). In the cell, such polysaccharides often form storage granules that may be readily broken down into their component sugars. With further chemical breakdown these sugars release chemical energy and may also provide the carbon skeletons for the synthesis of a variety of other molecules. Important structural functions are also served by polysaccharides. Linear polysaccharides form a major component of plant cell walls, while bacterial cell walls are composed of linear polysaccharides that are cross-linked by polypeptide chains.

Proteins are the most complex macromolecules found in the cell. They are composed of linear polymers called polypeptides, which contain amino acids connected by peptide bonds (fig. 1.5). Each amino acid contains a central carbon atom attached to four substituents: (1) a carboxyl group, (2) an amino group, (3) a hydrogen atom, and (4) an R group. The R group gives each amino acid its unique characteristics. There are twenty commonly occurring amino acids in most proteins. Some R groups are charged, some are neutral but still polar, and some are apolar.

The linear polypeptide chains of a protein fold in a highly specific way that is determined by the sequence of amino acids in the chains. Many proteins are composed of two or more polypeptides. Certain proteins function in structural roles. Some structural proteins interact with lipids in membrane structures. Others aggregate to form part of the cytoskeleton that gives the cell its shape. Still others are the chief components of muscle or connective tissue. Enzymes constitute yet another major class of proteins; they function as catalysts that accelerate and direct biochemical reactions.

Nucleic acids are the largest macromolecules in the cell. They are very long linear polymers, called polynucleotides, composed of nucleotides. A nucleotide contains (1) a five-carbon sugar molecule, (2) one or more phosphate groups, and (3) a nitrogenous base. It is the nitrogenous base that gives each nucleotide a distinct character (fig. 1.6). There are five different types of nitrogenous bases found in the two main types of nucleic acids, deoxyribonucleic acid (DNA) and ribonucleic acid (RNA). DNA contains the genetic information that is inherited when cells divide and organisms reproduce. This genetic information is used in the cell to make ribonucleic acids and proteins.

In addition to water and the macromolecules and organelles, the cytosol contains a large variety of small molecules that differ greatly in both structure and function. These never make up more than a small fraction of the total cell mass despite their great variety (see table 1.1). One class of small molecules consists of the monomer precursors of the different types of macromolecules. These monomers are derived by chemical modification from the nutrients absorbed through the cell membrane. Rarely are the nutrients themselves the actual monomers used by the cell. As a rule each nutrient must undergo a series of enzymatically catalyzed alterations before it is suitable for incorporation into one of the biopolymers. The intermediate molecules between nutrients and monomers are also present in small concentrations in the cytosol. Another varied class of molecules found in the cytosol includes molecules formed as side products in important synthetic reactions and as breakdown products of the macromolecules. Finally, the cytosol contains small bioorganic molecules known as coenzymes, which act in concert with the enzymes in a highly specific manner to catalyze a wide variety of reactions.

Macromolecules Conceal Their Hydrophobic Parts

Some structures formed by macromolecular intraactions and interactions are shown in figures 1.7 through 1.9. Phospholipids (see fig. 1.3*c*), which have a hydrophilic polar group on one end and long hydrophobic side chains attached to it, form multimolecular aggregates in an aqueous environment (fig. 1.7). These phospholipid aggregates form monomolecular layers at the air–water interface or micelles or bilayer vesicles within the water. In all of these structures, the polar head groups of the lipid are in contact with water, whereas the apolar side chains are excluded from the solvent structure.

As another example of polarity effects on macromolecular structure, consider polypeptide chains, which usually contain a mixture of amino acids with hydrophilic and hydrophobic side chains. Enzymes fold into complex three-dimensional globular structures with hydrophobic residues located on the inside of the structure and hydrophilic residues located on the surface, where they can interact with water (fig. 1.8).

DNA forms a complementary structure of two helically oriented polynucleotide chains (fig. 1.9). The polar sugar and

Figure 1.4

Monomers and polymers of carbohydrates. (*a*) The most common carbohydrates are the simple six-carbon (hexose) and five-carbon (pentose) sugars. In aqueous solution, these sugar monomers form ring structures. (*b*) Polysaccharides are usually composed of hexose monosaccharides covalently linked together by glycosidic bonds to form long straight-chain or branched-chain structures.

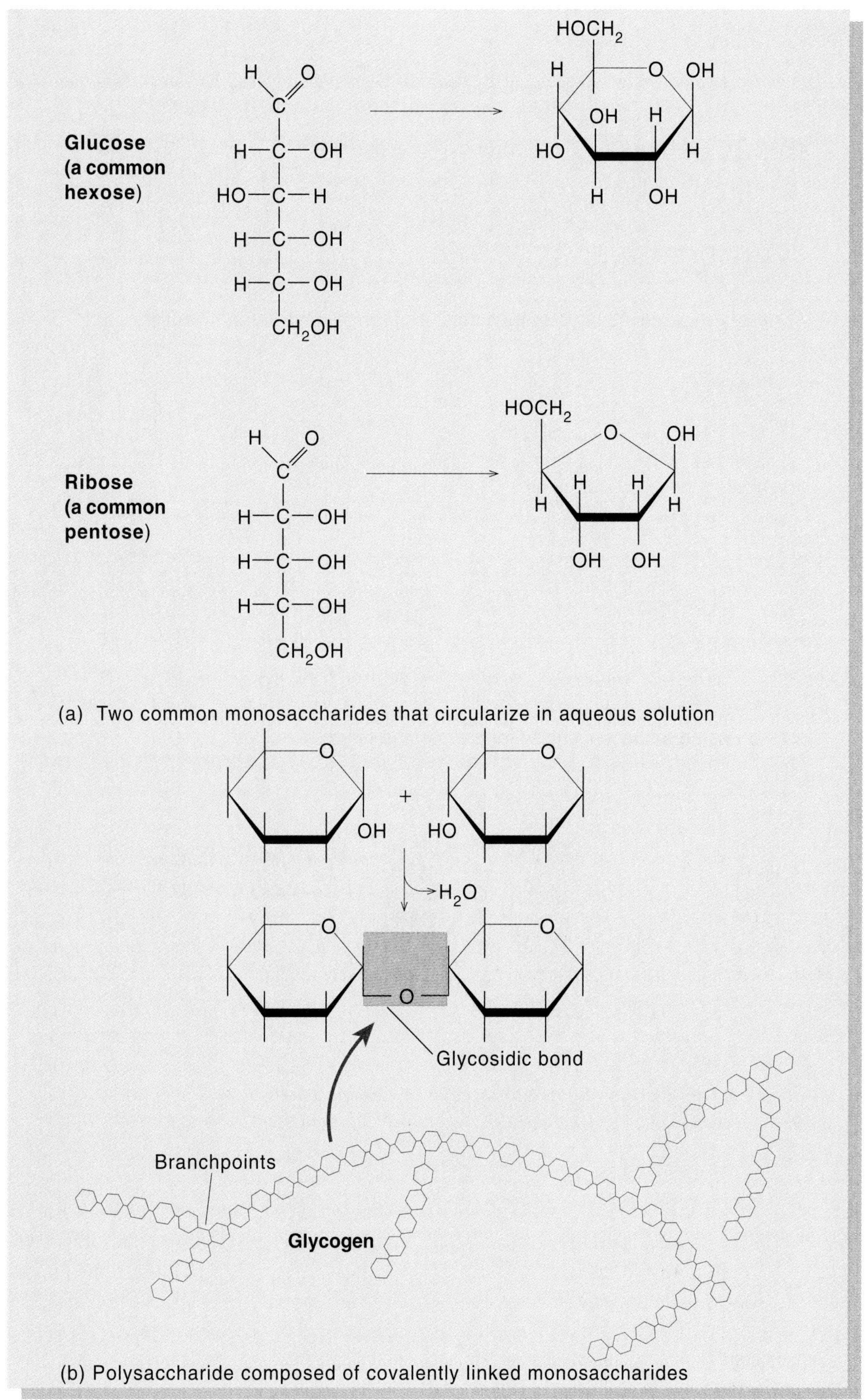

Figure 1.5

Amino acids and the structure of the polypeptide chain. Polypeptides are composed of L-amino acids covalently linked together in a sequential manner to form linear chains. (*a*) The generalized structure of the amino acid. The zwitterion form, in which the amino group and the carboxyl group are ionized, is strongly favored. (*b*) Structures of some of the R groups found for different amino acids. (*c*) Two amino acids become covalently linked by a peptide bond, and water is lost. (*d*) Repeated peptide bond formation generates a polypeptide chain, which is the major component of all proteins.

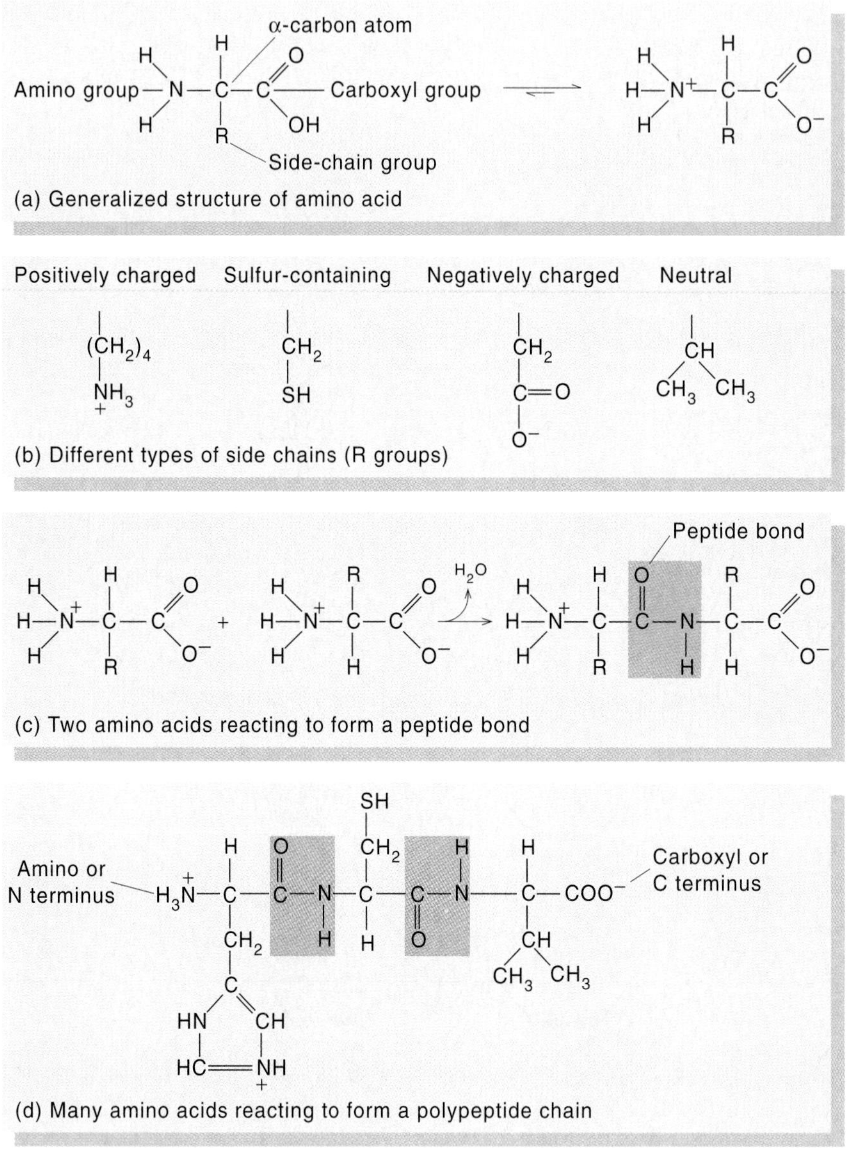

α-carbon atom

Amino group

Carboxyl group

Side-chain group

(a) Generalized structure of amino acid

Positively charged | Sulfur-containing | Negatively charged | Neutral

(b) Different types of side chains (R groups)

Peptide bond

H_2O

(c) Two amino acids reacting to form a peptide bond

Amino or N terminus

Carboxyl or C terminus

(d) Many amino acids reacting to form a polypeptide chain

Figure 1.6

The structural components of nucleic acids. Nucleic acids are long linear polymers of nucleotides, called polynucleotides. (*a*) The nucleotide consists of a five-carbon sugar (ribose in RNA or deoxyribose in DNA) covalently linked at the 5′ carbon to a phosphate and at the 1′ carbon to a nitrogenous base. (*b*) Nucleotides are distinguished by the types of bases they contain. These are either of the two-ring purine type or of the one-ring pyrimidine type. (*c*) When two nucleotides become linked, they form a dinucleotide, which contains one phosphodiester bond. Repetition of this process produces a polynucleotide.

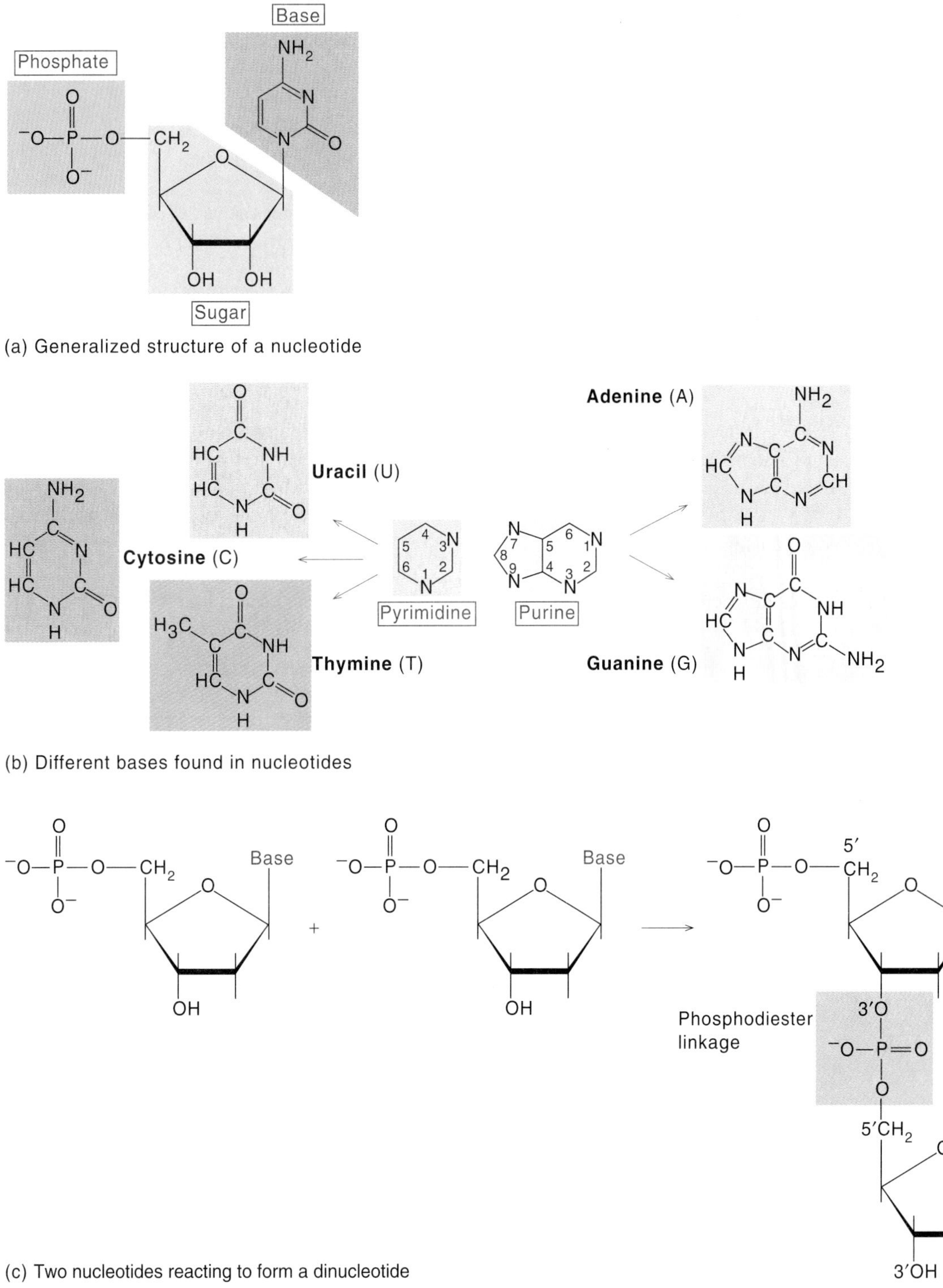

(a) Generalized structure of a nucleotide

(b) Different bases found in nucleotides

(c) Two nucleotides reacting to form a dinucleotide

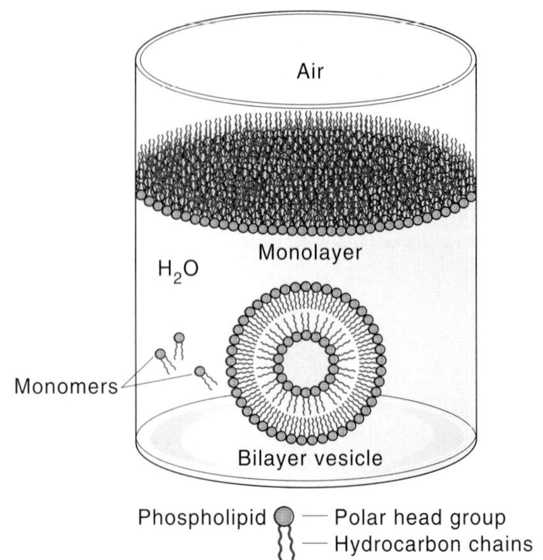

Figure 1.7

Structures formed by phospholipids in aqueous solution. Phospholipids may form a monomolecular layer at the air–water interface, or they may form spherical aggregations surrounded by water. A bilayer vesicle consists of a double molecular layer of phospholipids surrounding an internal compartment of water.

phosphate groups are on the surface, where they can interact with water; the nitrogenous bases from the two chains form intermolecular hydrogen bonds in the core of the structure.

Biochemical Reactions Form a Small Subset of Ordinary Chemical Reactions

The total number of biochemical reactions is much smaller than the potential number of reactions that occur in ordinary chemical systems. This simplification results partly from the fact that only a limited number of elements account for the vast majority of substances found in living cells. The elements of major importance in biochemical reactions, in order of decreasing numerical abundance, are hydrogen (H), carbon (C), oxygen (O), nitrogen (N), phosphorus (P), and sulfur (S). Certain metal ions are also important; these include Na^+, K^+, Mg^{2+}, Ca^{2+}, Zn^{2+}, and Fe^{2+} or Fe^{3+}. Other metals and elements that are needed in very small amounts are iodine, cobalt, molybdenum, selenium, vanadium, nickel, chromium, tin, fluorine, silicon, and arsenic. In some cases we don't know the biological roles of these "trace elements" but only that they are needed by some organisms for normal growth or development.

The types of covalent linkages most commonly found in biomolecules are also quite limited (table 1.2). Only 16 different types of linkages account for more than 95% of the linkages found in biomolecules. All the elements form single or double bonds, except for hydrogen, which can form only single bonds; all the elements exist primarily in a single valence state except for carbon and sulfur, which are commonly found in more than two valence states (table 1.3). Despite this overall simplicity, many other valence states can be found in unusual cases, and some of these are very important. For example, the biochemistry of nitrogen involves consideration of all the valence states of nitrogen from +5 to 0 to −3. A major source of nitrogen available to biosystems is gaseous nitrogen found in the atmosphere (valence state 0). Biochemical reactions convert gaseous nitrogen into other forms of nitrogen.

Table 1.2
Types of Covalent Linkages Most Commonly Found in Biomolecules

	H	C	O	N	P	S
H						
C	—C—H	—C—C— C=C				
O	—O—H	—C—O— C=O				
N	N—H	—C—N= C=N—	—			
P	—	—	P—O— P=O	—		
S	—S—H	—C—S—	S—O— S=O	—	—	—S—S—

Figure 1.8

A graphic representation of a three-dimensional model of the protein cytochrome *c*. Amino acids with nonpolar, hydrophobic side chains (color) are found in the interior of the molecule, where they interact with one another. Polar, hydrophilic amino acid side chains (gray) are on the exterior of the molecule, where they interact with the polar aqueous solvent. (Illustration copyright by Irving Geis. Reprinted by permission.)

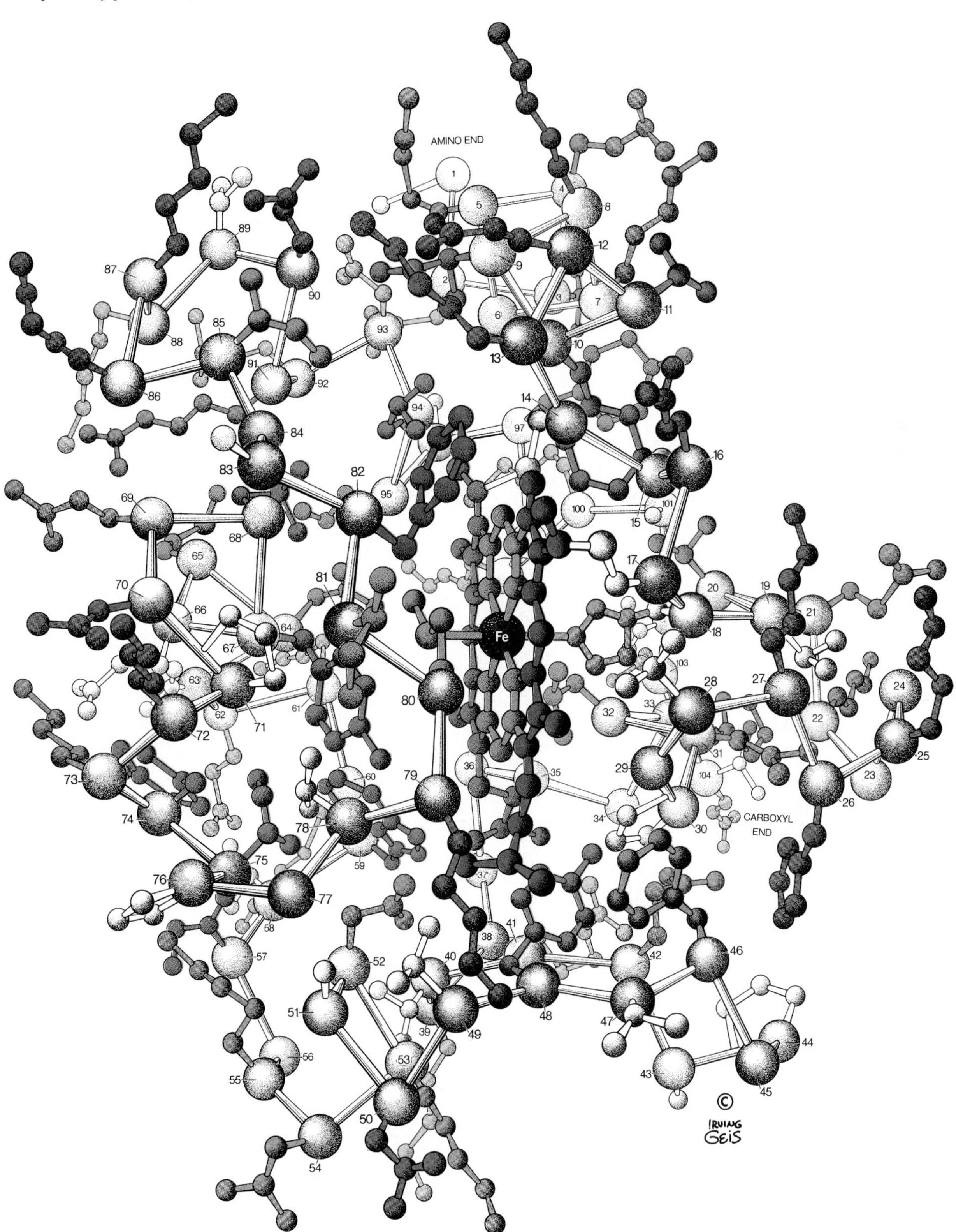

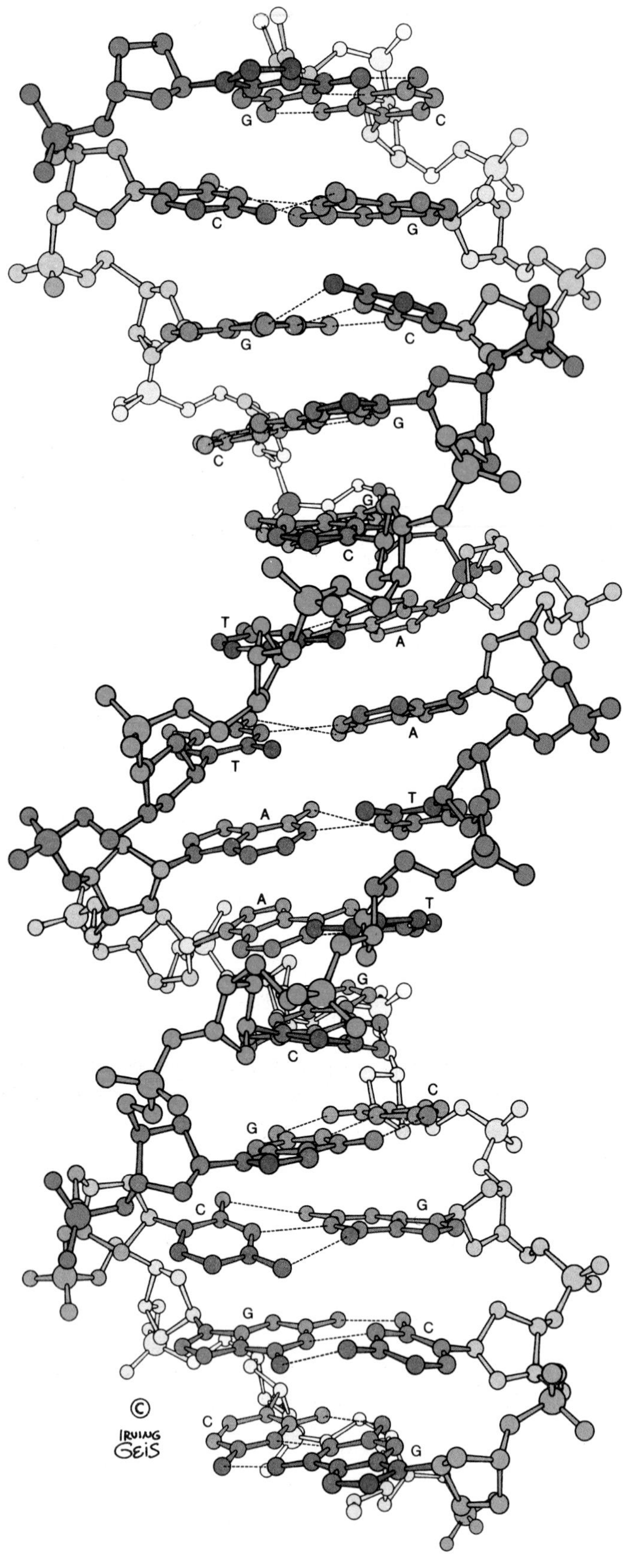

Figure 1.9

The right-handed helical structure of DNA. DNA normally exists as a two-chain structure held together by hydrogen bonds (•••) formed between the bases in the two chains. Along the chain the planar surfaces of these bases interact and, together with the hydrogen bonds, contribute to the stability of the two-chain structure. The negatively charged phosphate groups are on the outside of the structure, where they interact with water, ions, or charged molecules. (Illustration copyright by Irving Geis. Reprinted by permission.)

Table 1.3
Most Common Valences Displayed by Atoms in Biomolecules

Element	Valence
H	+1
C	−4 to +4
O	−2
P	+5
N	−3
S	+6, −2, −1

Figure 1.10

Different functional groups found in biomolecules. This figure includes the major functional groups. Other functional groups are found in minor amounts.

Ionized forms favored in aqueous solutions		Structure	Name
		>C—OH	Hydroxyl
		>C=O	Carbonyl
—C(=O)O$^-$	⇌	—C(=O)OH	Carboxyl
		>C=NH	Imino
>C—$\overset{+}{NH_3}$	⇌	>C—NH_2	Amino
		>C—SH	Thiol
—O—P(=O)(O$^-$)—O$^-$	⇌	—O—P(=O)(OH)—OH	Phosphate
—P(=O)(O$^-$)—O—P(=O)(O$^-$)—O$^-$	⇌	—P(=O)(OH)—O—P(=O)(OH)—OH	Pyrophosphate

These reactions are very complex and occur only in a select group of microorganisms that possess the necessary enzyme systems.

Biochemical reactions involving the different classes of substances use a limited number of functional groups, which are illustrated in figure 1.10. Most of the reactive groups in biomolecules contain one or more of these functional groups or closely related ones. Many cellular reactions involving these functional groups are closely related to reactions that take place outside the cell under different conditions and are studied in organic chemistry.

For example, alcohols, which contain a hydroxyl functional group, can undergo hydration reactions with either carboxylic or phosphoric acid to form esters.

$$\text{R—OH} + \text{HO—C(=O)—R}' \rightleftharpoons \text{R—O—C(=O)—R}' + H_2O$$

Alcohol **Carboxylic acid** **Ester** **Water**

$$\text{R—OH} + \text{HO—P(=O)(O}^-\text{)—O}^- \rightleftharpoons \text{R—O—P(=O)(O}^-\text{)—O}^- + H_2O$$

Alcohol **Phosphoric acid** **Phosphoric acid ester** **Water**

Thiols, containing sulfhydryl groups (—SH), can substitute for alcohols in some reactions, leading to the formation of thiol esters.

$$\text{R—SH} + \text{HO—C(=O)—R}' \rightleftharpoons \text{R—S—C(=O)—R}' + H_2O$$

Thiol **Carboxylic acid** **Thiol ester** **Water**

Two alcohols can react with one another to form an ether.

$$\text{R—OH} + \text{HO—R}' \longrightarrow \text{R—O—R}' + H_2O$$

Alcohol$_1$ **Alcohol$_2$** **Ether**

Alcohols can also undergo a dehydrogenation reaction to form a carbonyl derivative (aldehyde or ketone).

$$\text{R—CH(R)—OH} \underset{+2H}{\overset{-2H}{\rightleftharpoons}} \text{R}_2\text{C=O}$$

Alcohol **Aldehyde or ketone**

Amines undergo reactions with carboxylic acids comparable to the formation of esters from alcohols. The product is known as an amide.

$$\text{R—}\overset{+}{\text{N}}\text{H}_3 + {}^-\text{O—C(=O)—R}' \rightleftharpoons \text{R—NH—C(=O)—R}' + H_2O$$

Primary amine **Carboxylic acid** **Amide**

Amines can also undergo dehydrogenation reactions leading to the formation of imines, which are frequently unstable in water and hydrolyze to ketones or, in cases where one of the R groups is an H, to aldehydes:

$$\text{R}_2\text{CH—NH}_2 \xrightarrow{-2H} \text{R}_2\text{C=NH} \xrightarrow{+H_2O} \text{R}_2\text{C=O} + NH_3$$

Amine **Imine** **Ketone** **Ammonia**

Aldehydes and ketones both may be reduced to alcohols by hydrogenation (see the alcohol dehydrogenation reaction). Aldehydes may react with either water or alcohol to form aldehyde hydrates or hemiacetals, respectively. Reaction of an aldehyde with two molecules of alcohol leads to acetal formation.

$$\text{RCH(OH)}_2 \underset{-H_2O}{\overset{+H_2O}{\rightleftharpoons}} \text{RCH=O} \underset{-\text{ROH}}{\overset{+\text{ROH}}{\rightleftharpoons}} \text{RCH(OH)(OR)} \underset{-\text{ROH}}{\overset{+\text{ROH}}{\rightleftharpoons}} \text{RCH(OR)}_2$$

Aldehyde hydrate **Aldehyde** **Hemiacetal** **Acetal**

Dehydrogenation of an aldehyde hydrate leads to carboxylic acid formation.

$$\text{RCH(OH)}_2 \xrightarrow{-2H} \text{RC(=O)OH}$$

Aldehyde hydrate **Carboxylic acid**

Aldehydes and ketones may also isomerize to the enol form as long as the adjacent carbon atom contains at least one H atom. In the reaction a hydrogen migrates and the double bond shifts.

$$\text{R—CH(R)—C(R)=O} \rightleftharpoons \text{R}_2\text{C=C(R)—OH}$$

Keto form **Enol form**

Pyrophosphates may hydrolyze to inorganic phosphoric acid (phosphate) and an organophosphoric acid.

$$\text{R—O—P(=O)(O}^-\text{)—O—P(=O)(O}^-\text{)—O}^- \underset{-H_2O}{\overset{+H_2O}{\rightleftharpoons}} \text{R—O—P(=O)(O}^-\text{)—O}^- + \text{HO—P(=O)(O}^-\text{)—O}^- + H^+$$

Organopyrophosphate **Organophosphoric acid** **Phosphoric acid**

Hydrolysis reactions of this sort yield considerable energy, which can be utilized in biosynthesis.

All of the functional groups that we have described are electrostatically neutral in organic solvents. However, in water many of these functional groups either lose or gain protons to become charged species. Such ionization reactions are very important in biochemical systems because they frequently influence solubility and reactivity.

Carboxylic and phosphoric acids lose one or more protons in water to become negatively charged. The ionized forms are stabilized by resonance as shown:

$$R{-}C(=O){-}OH \underset{+H^+}{\overset{-H^+}{\rightleftharpoons}} \left[R{-}C(=O){-}O^- \longleftrightarrow R{-}C(-O^-){=}O \right]$$

$$(HO)_2P(=O)OH \underset{+H^+}{\overset{-H^+}{\rightleftharpoons}} \left[{}^-O{-}P(=O)(HO)(OH) \updownarrow O{=}P(O^-)(HO)(OH) \right] \underset{+H^+}{\overset{-H^+}{\rightleftharpoons}} {}^-O{-}P(=O)(O^-)(OH) \underset{+H^+}{\overset{-H^+}{\rightleftharpoons}} \left[{}^-O{-}P(=O)(O^-)(O^-) \updownarrow {}^-O{-}P(O^-)(=O)(O^-) \right]$$

Amines usually add a proton to become positively charged.

$$R_3C{-}NH_2 \underset{-H^+}{\overset{+H^+}{\rightleftharpoons}} R_3C{-}\overset{+}{N}H_3$$

Near neutrality (10^{-7} M H^+), where most biochemical systems function, the carboxyl group exists mainly in the negatively charged form, phosphoric acid exists mainly in the diionized form, and amino groups exist mainly in the positively charged form. This fact has interesting consequences for the amino acids found in proteins, since they contain one amino group and one carboxyl group. The amino acids are usually neutral overall, even though they contain two charged groups, one resulting from the deprotonation of the carboxyl group and the other resulting from the protonation of the amino group. Amino acids existing in this way as dipolar ions are called zwitterions.

$$\underset{\textbf{Uncharged}}{R{-}CH(NH_2){-}C(=O){-}OH} \rightleftharpoons \underset{\textbf{Zwitterion}}{R{-}CH(\overset{+}{N}H_3){-}C(=O){-}O^-}$$

These are some of the more important reactions involving covalent bond breakage or formation in biochemistry. By now two things should be apparent about biochemical reactions: (1) As stated at the outset, the number of reactions in biochemistry is much more limited than in ordinary chemistry; (2) as far as the reactants and products are concerned, biochemical reactions can be understood in the same terms as ordinary chemical reactions.

Biochemical Reactions Occur under Mild Conditions

Although biochemical reactions resemble ordinary chemical reactions, they differ in important ways. Chemical reactions are frequently carried out in nonaqueous solvents, using elevated temperatures and pressures, acids or bases, or other harsh reagents, conditions that would destroy the functional organization of a living cell. Biochemical reactions usually take place under very mild conditions in aqueous solution. However, many chemical reactions do not proceed at reasonable rates under such conditions. Biochemical reactions proceed at substantial rates because of the very special nature of the enzyme catalysts that accelerate them.

Enzymes are structurally complex, highly specific catalysts; each enzyme usually catalyzes only one type of reaction. The enzyme surface binds the interacting molecules, or substrates, so that they are favorably disposed to react with one another (fig. 1.11). The specificity of enzyme catalysis also has a selective effect, so that only one of several potential reactions will take place. For example, a simple amino acid can be utilized in the synthesis of any of the four major classes of macromolecules or could be simply secreted as waste product (fig. 1.12). The fate of the amino acid is determined as much by the presence of specific enzymes as by its reactive functional groups.

Many Biochemical Reactions Require Energy

Chemical energy is needed to drive many biochemical reactions, to do mechanical work, and for transport of substances across the plasma membrane. The ultimate source of energy that drives a cell's reactions is sunlight (fig. 1.13). Light energy is converted into chemical energy in the chloroplasts of plant cells or in the photosynthetic grana of certain microorganisms. The main form of chemical energy produced in the chloroplast is a nucleotide containing three phosphoric acid groups attached in sequence, adenosine triphosphate or ATP (fig. 1.14). Organisms that cannot harness the light rays of the sun themselves to make ATP are able to make ATP from the breakdown of organic nutrients originating from plants or other organisms.

Many nutrients consist of partly degraded macromolecules or various other small molecules that after absorption must be converted into a form suitable for the production of ATP. One of the simplest and yet most effective substances useful in ATP synthesis is the six-carbon sugar glucose. Degradation of a molecule of glucose can produce 38 molecules of ATP by the following overall reaction, which involves many enzymes:

$$38\ H^+ + C_6H_{12}O_6 + 6\ O_2 + 38\ ADP + 38\ P_i^* \rightarrow 6\ CO_2 + 38\ ATP + 44\ H_2O$$

Two characteristics of this equation should be noted. First, the glucose shown on the left is degraded by oxidation to CO_2 and H_2O. A substantial fraction of the energy released by this complete oxidation of glucose is used in the production of ATP. Second, ATP is being synthesized not from small molecular precursors but simply by the addition of a single phosphate (P_i) to adenosine diphos-

Figure 1.11

The structure of the complex formed between the enzyme lysozyme and its substrate. The crevice that forms the site for substrate binding (the active site) runs horizontally across the enzyme molecule. The individual hexose sugars of the hexasaccharide substrate are shown in a darker color and labeled A–F.

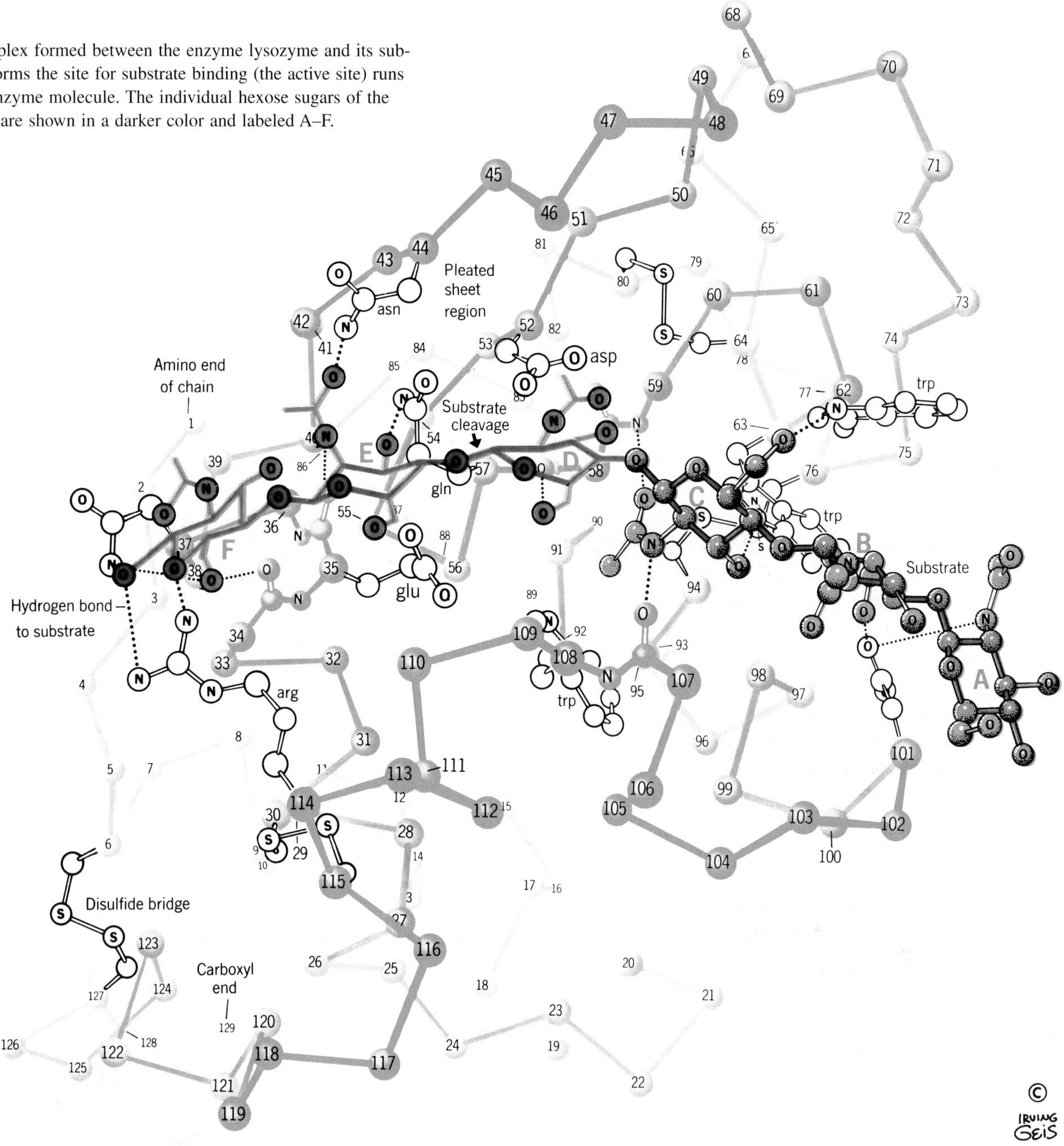

Figure 1.12

The different fates of an amino acid. Depending on which enzymes are present and active and on the needs of the organism, an amino acid can be metabolized in different ways. Each of these conversions involves one or more steps, and usually each step requires a specific enzyme.

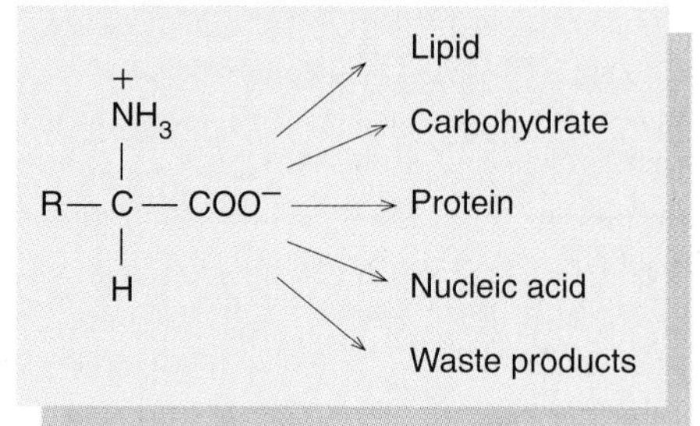

Figure 1.13

Flow of energy in the biosphere. The sun's rays are the ultimate source of energy. These rays are absorbed and converted into chemical energy (ATP) in the chloroplasts. The chemical energy is used to make carbohydrates from carbon dioxide and water. The energy stored in the carbohydrates is then used, directly or indirectly, to drive all the energy-requiring processes in the biosphere.

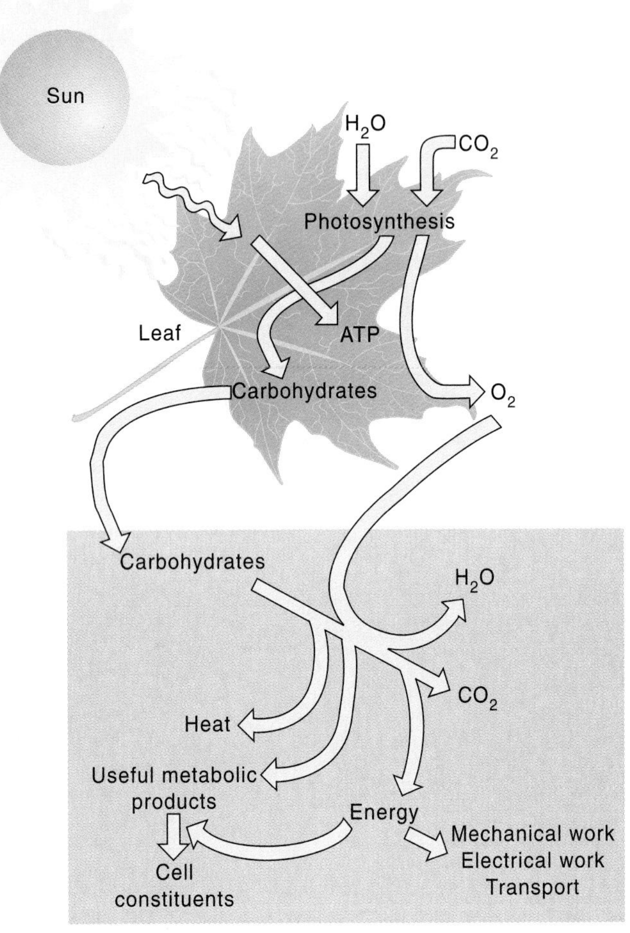

phate (ADP). Considerable energy is released when ATP is hydrolyzed to ADP and P_i, and the ADP can be reutilized many hundreds of times. These are two of the reasons that ATP is so effective as an energy source. A third reason is that ATP is quite stable in water; it does not lose its terminal phosphate readily except in enzyme-catalyzed reactions. The equation for the complete oxidation of glucose does not take into account the energy necessary to operate the shuttle systems between the cytosol and mitochondria. If this energy was accounted for, the total ATP yield would be 29.5 to 31 molecules; depending on the shuttle system (see fig. 16.28 also).

Most biochemical reactions fall into one of the two classes: degradative or synthetic. Degradative, or catabolic, reactions result in the breakdown of organic compounds to simpler substances. Synthetic, or anabolic, reactions lead to the assembly of biomolecules from simpler molecules. Most anabolic processes require energy to drive them. This energy is usually supplied by coupling the energy-requiring biosynthetic reactions to energy-releasing catabolic reactions. Most frequently, the energy-releasing reaction involves ATP hydrolysis, either directly or indirectly.

Biochemical Reactions Are Localized in the Cell

Biochemical reactions are organized so that different reactions occur in different parts of the cell. This organization is most apparent in eukaryotes, where membrane-bounded structures are visible proof for the localization of different biochemical processes. For example, the synthesis of DNA and RNA takes place in the nucleus of a eukaryotic cell. The RNA is subsequently transported across the nuclear membrane to the cytoplasm, where it takes part in protein synthesis. Proteins made in the cytoplasm are used in all parts of the cell. A limited amount of protein synthesis also occurs in chloroplasts and mitochondria. Proteins made in these organelles are used exclusively in organelle-related functions. Most

Figure 1.14

The structures of ATP and ADP and their interconversion. The two compounds differ by a single phosphate group.

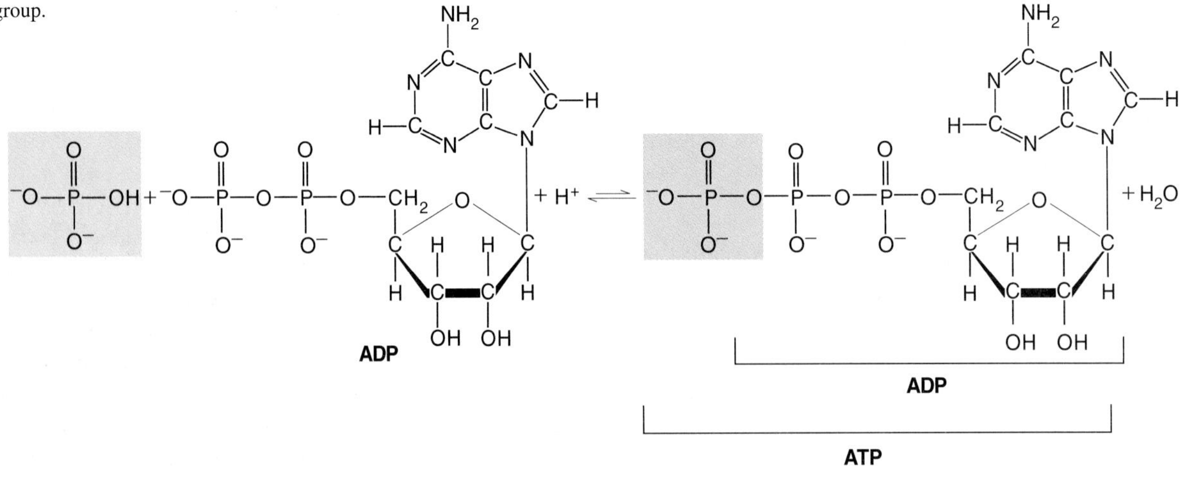

ATP synthesis occurs in chloroplasts and mitochondria. A host of reactions that transport nutrients and metabolites occur in the plasma membrane and the membranes of various organelles. The localization of functionally related reactions in different parts of the cell concentrates reactants and products at sites where they can be most efficiently utilized.

Biochemical Reactions Are Organized into Pathways

Most biochemical reactions are integrated into multistep pathways utilizing several enzymes. For example, the breakdown of glucose into CO_2 and H_2O controls a series of reactions that begins in the cytosol and continues to completion in the mitochondrion. A complex series of reactions like this is referred to as a biochemical pathway (fig. 1.15). Synthetic reactions, such as the biosynthesis of amino acids in the bacterium *Escherichia coli,* are similarly organized into pathways (fig. 1.16). Frequently, pathways have branchpoints. For example, the synthesis of the amino acids threonine and lysine starts with oxaloacetate. After three steps, a branchpoint is reached with the formation of the organic compound aspartic-β-semialdehyde. One branch of this pathway leads to the synthesis of the amino acid lysine, and another branch leads to the synthesis of the amino acids methionine, threonine, and isoleucine.

Figure 1.15

Summary diagram of the breakdown of glucose to carbon dioxide and water in a eukaryotic cell. As depicted here, the process starts with the absorption of glucose at the plasma membrane and its conversion into glucose-6-phosphate. In the cytosol, this six-carbon compound is then broken down by a sequence of enzyme-catalyzed reactions into two molecules of the three-carbon compound pyruvate. After absorption by the mitochondrion, pyruvate is completely degraded to carbon dioxide and water by a sequence of reactions that requires molecular oxygen.

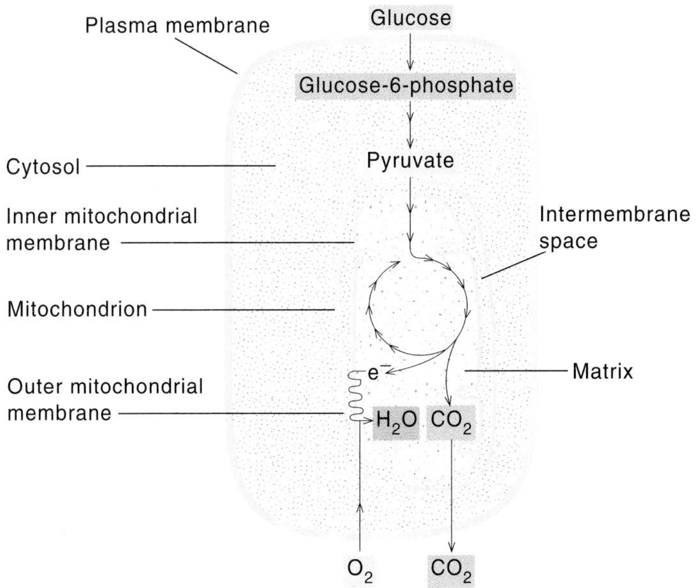

Figure 1.16

Synthesis of various amino acids from oxaloacetate. Each arrow represents a discrete biochemical step requiring a unique enzyme. Thus aspartic acid is produced in one step from oxaloacetate, whereas isoleucine is produced in five steps from threonine.

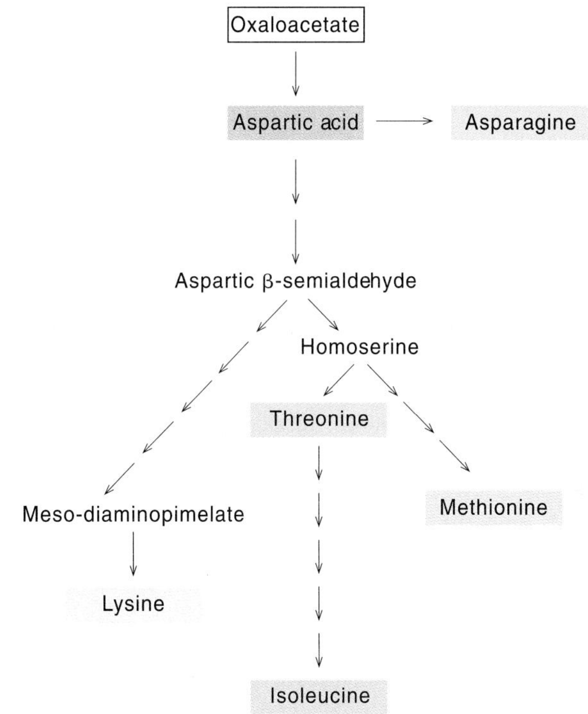

To understand the role of each biochemical reaction, we must identify its position in a pathway and also consider how that pathway interacts with others.

Biochemical Reactions Are Regulated

Hundreds of biochemical reactions take place even in the cells of relatively simple microorganisms. Living systems have evolved a sophisticated hierarchy of controls that permits them to maintain a stable intracellular environment. These controls ensure that substances required for maintenance and growth are produced in adequate amounts but without huge excesses. Control mechanisms have developed so that the cell can make adjustments in response to a changing external environment. Adjustments are needed because the temperature, ionic strength, acid concentration, and concentration of nutrients present in the external environment vary over much wider limits than could be tolerated inside the cell.

The rate of intracellular reactions is a function of the availability of substrates and enzymes. Enzyme activity is controlled at two different levels. First and foremost, the rate of a catalyzed reaction is regulated by the amount of the catalyzing enzyme that is present in the cell. Control of enzyme amounts is usually accomplished by regulating the rate of enzyme synthesis; in some cases the rate of enzyme degradation is also regulated. We can think of controls that regulate the total amount of enzyme present as coarse controls. They define the limits of possible enzyme activity as being anywhere from 0 to 100% of the full activity of the enzyme. Fine controls that act directly on enzymes are also present. Only certain special enzymes, called regulatory enzymes, are susceptible to this second type of regulation. Regulatory enzymes usually occupy key points in biochemical pathways, and their state of activity frequently is decisive in determining the utilization of the pathway.

A simple example will serve to illustrate how these two types of controls work. *E. coli* bacteria can synthesize all of the amino acids required for protein synthesis. Histidine is one of these amino acids; its synthesis starts from the sugar phosphate compound phosphoribosylpyrophosphate (PRPP) and requires ten enzymes. Each enzyme catalyzes one reaction in a ten-step pathway. The synthesis of all ten enzymes is regulated by the end product of the pathway, histidine. If there is sufficient histidine for protein synthesis, then the cell ceases to make the enzymes for this pathway. The shutdown is triggered by controls that sense the level of histidine in the cell and as a result turn off synthesis of the messenger RNA required to make the enzymes. The histidine pathway also provides an example of a regulatory enzyme. The activity of the first enzyme in the pathway is directly inhibited by histidine. Thus, when there is sufficient histidine present in the cell, the activity of the first enzyme in the pathway is inhibited, and no more material or energy is funneled into the synthesis of unneeded histidine. Finally, if there is abundant histidine available from the external environment, the extracellular histidine is absorbed into the cell, and the synthesis of the enzymes of the histidine pathway is brought to a halt. Bacterial cells grown in a histidine-rich growth medium for several generations contain only trace amounts of these enzymes.

The underlying principle in regulation is to maintain a favorable intracellular environment in the most economical manner. The cell makes products in the amounts that are needed. Each pathway is regulated in a somewhat different way, ensuring that biochemical energy and substrates are efficiently utilized.

Organisms Are Biochemically Dependent on One Another

As we stated at the beginning of this chapter, 3–4 billion years ago the first self-replicating molecules appeared on earth. These entities had to have the capacity for extracting nutrients from the chemical compounds that existed in prebiotic times. We have some general notions about what types of substances were present at that time. One of the most important substances that was not present at that time in significant amounts was molecular oxygen, O_2. Currently, this form of oxygen is required by all forms of life visible to the naked eye.

The O_2 that is used by most organisms is ultimately converted by them into CO_2. Oxygen is utilized at a rapid rate, and it would soon disappear if it were not for special classes of photosynthetic organisms that are constantly producing more O_2 by the oxidation of water.

The oxygen story is an example of the dependence of one class of organisms on another for certain chemicals. A similar situation exists with the elements carbon and nitrogen, which must be converted from gaseous forms, CO_2 and N_2, to organic forms utilizable by most organisms. Reduced carbon compounds are constantly being lost by oxidation to gaseous CO_2. The supply of organic carbon compounds required by all forms of life is replenished by photosynthetic organisms; these include most plants and certain microorganisms. Similarly, nitrogen in organic molecules is constantly being lost to the atmosphere in the form of gaseous nitrogen. The reactions required for the conversion of nitrogen to a reduced form more usable to the majority of organisms occurs in only a limited number of microorganisms; yet without these nitrogen-fixing organisms, life as we know it would soon vanish.

As we ascend the evolutionary tree, we find increasingly complex multicellular forms. Such organisms generally require more complex nutrients, which must ultimately be supplied to them by simpler living forms. Bacteria like *E. coli* can make all of their own amino acids from a reduced form of nitrogen, such as NH_3, and a reduced form of carbon, such as glucose. Humans, on the other hand, must receive most of their amino acids as nutrients. Humans and other complex organisms have gained new biochemical capacities, which permit them to synthesize the components associated with highly specialized differentiated tissues. At the same time, they have lost many of the biochemical systems required to survive on simpler nutrients.

Many biochemical reactions of great importance take place in only a limited number of organisms. This fact increases the complexity of the study of biochemistry. We must learn many reactions; we must also be aware of the biochemical potentials of different organisms. This is the only way we can understand the biochemical interdependency of organisms.

Information for the Synthesis of Proteins Is Carried by the DNA

DNA contains the genetic information transmitted to each daughter cell when cells divide. The DNA usually exists in the form of nucleoprotein (DNA-protein) complexes called chromosomes. A prokaryotic cell contains a single chromosome. Prior to cell division this chromosome duplicates and segregates so that an identical complement of DNA goes to each of two newly formed daughter cells.

Eukaryotic cells are more complex than prokaryotic cells and usually contain more DNA, which is partitioned between several chromosomes. In both prokaryotes and eukaryotes, almost all cells of the same organism contain the same number of chromosomes. In eukaryotes most of the chromosomes are localized in the nucleus. Thus the DNA is isolated from the main body of the cytoplasm—a unique feature of eukaryotes and the primary distinction between prokaryotes and eukaryotes. Some organelles, notably the mitochondria and the chloroplasts, also contain a single circular chromosome.

Eukaryotic chromosomes are detectable by light microscopy at the stage just prior to cell duplication. At this stage, called mitosis, chromosomes appear as elongated refractile structures that can be seen to segregate in equal numbers and types to each of the daughter cells before cell division (fig. 1.17). Each chromosome carries specific hereditary (genetic) information nec-

Figure 1.17

Mitosis and cell division in eukaryotes. After DNA duplication has occurred, mitosis is the process by which quantitatively and qualitatively identical DNA is delivered to daughter cells formed by cell division. Mitosis is traditionally divided into a series of stages characterized by the appearance and movement of the DNA-bearing structures, the chromosomes. (*a*) Premitosis. (*b*) through (*h*) Successive stages of mitosis. (*i*) Postmitosis.

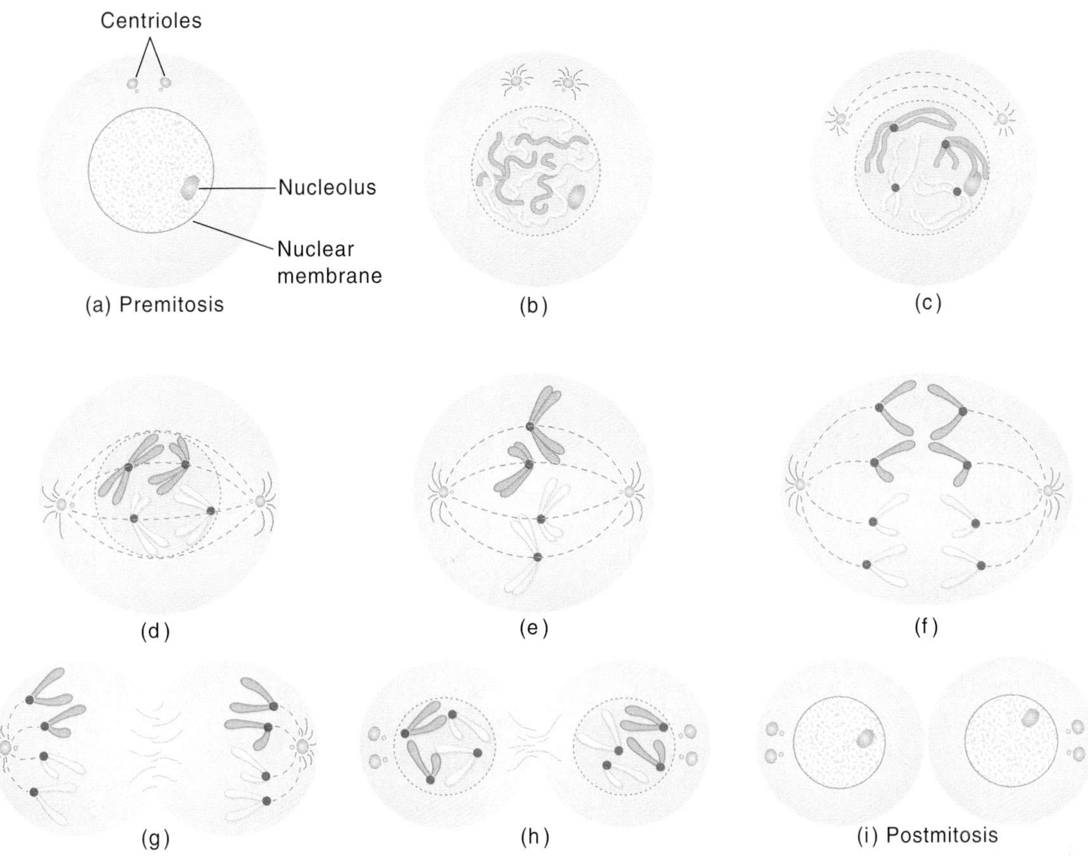

essary for synthesis of specific compounds essential for cell maintenance, growth, and replication. Each chromosome contains a single very long DNA molecule composed of 10^6 or more nucleotides in a specific sequence. The sequence of nucleotides in the chromosomal DNA determines the sequence of amino acids in the protein polypeptide chains of the organism. The relationship between base sequences and resultant amino acids is known as the genetic code. Each grouping of three bases, called a triplet, represents a specific amino acid and is called a codon. The genetic code ensures that the organism's characteristics will be reflected by the sequence of nucleotides in its DNA. When chromosomes replicate, the DNA replicates precisely, so that the same nucleotide sequence is passed along to each of the daughter cells resulting from mitosis and cell division.

The DNA does not transfer its genetic information directly to protein. Rather, this information passes through an intermediary, the messenger RNA (mRNA). The mRNA is made on a DNA template in the nucleus of a eukaryotic cell and then passes into the cytoplasm, where it serves in turn as a template for the synthesis of the polypeptide chain. The overall process of information transfer from DNA to mRNA (transcription) and from mRNA to protein (translation) is depicted in figure 1.18.

The First Living Systems Were Acellular

To appreciate why living organisms are the way they are today, we must go back to the very beginnings of the process. Whereas the records grow fainter the further back we go, there is a surprising amount of detail that we can conjecture. Isotopic records on meteorites indicate that the solar system originated 4.6 billion years ago. Initially, the earth was a molten mass that was unable to retain most of its volatiles. In the first 100 million years much of the metallic iron in this molten mass descended because of its greater density to form the core. Once the surface of the earth had cooled down to temperatures in the 300-K range, the earth was left with a largely liquid iron core and surrounding layers composed mostly of iron and magnesium oxides and silicates. The remaining elements were present in considerably smaller amounts, but many of them became overrepresented in the earth's crust (table 1.4), where they have had a major impact on the origin of

Figure 1.18

Transfer of information from DNA to protein. The nucleotide sequence in DNA specifies the sequence of amino acids in a polypeptide. DNA usually exists as a two-chain structure. The information contained in the nucleotide sequence of only one of the DNA chains is used to specify the nucleotide sequences of the messenger RNA molecule (mRNA). This sequence information is used in polypeptide synthesis. A three-nucleotide sequence in the messenger RNA molecule codes for a specific amino acid in the polypeptide chain. (Illustration copyright by Irving Geis. Reprinted by permission.)

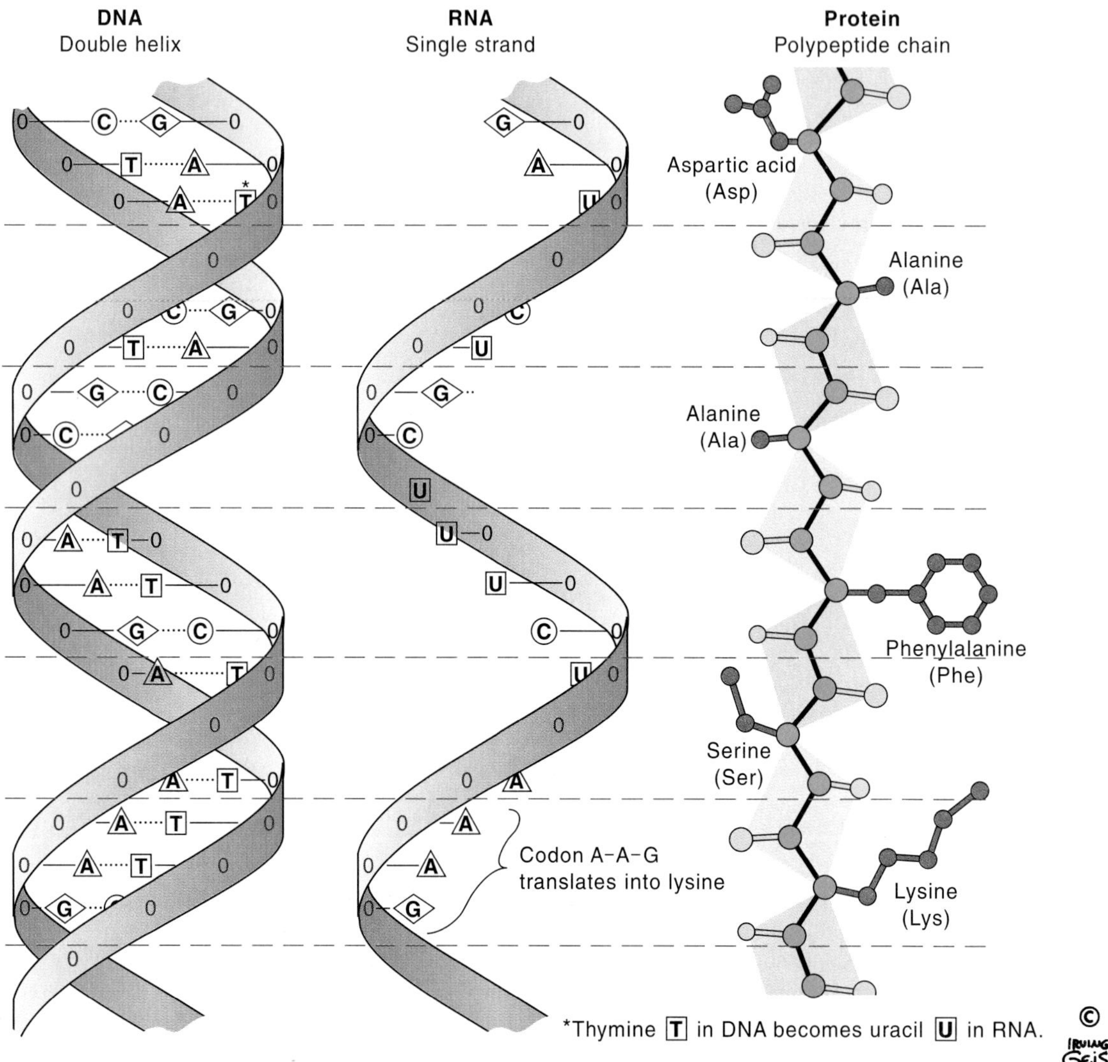

life. All gaseous substances that were originally present when the earth was formed blew away because of the intense heat. Future gases that have remained with us to a great extent were originally occluded in the solid mass and have risen to the surface primarily through volcanic emissions. These gases have always been dominated by four major elements: C, O, N, and H. The fact that these are also the four most abundant elements in living organics testifies to the importance of these gases where it is believed that the ultimate precursors to living things originated.

Two of the most important molecules to be formed in the early atmosphere from volcanic gases were hydrogen cyanide and formaldehyde. Hydrogen cyanide is a major starting point for the nucleic acid bases, while formaldehyde is a major starting point for the ribose found in nucleic acids (fig. 1.19). Together, HCN and CH_2O can give rise to amino acids, but this may be incidental to the origin of life, as it is currently considered most likely that the first living things were RNA molecules that may have existed as double-helix molecules bound to the surfaces of clay and mineral particles. RNA is considered the favored starting point for the origin of life because it can serve as a template for its own replication and because catalytic functions can also be incorporated into the RNA structure. Thus RNA in a suitable environment

Table 1.4
Distribution of the 24 Elements Used in Biological Systems*

Element	Atomic No.	Earth's Crust	Ocean	Human Body
Hydrogen (H)	1	2,882	66,200	60,562
Carbon (C)	6	56	1.4	10,680
Nitrogen (N)	7	7	<1	2,440
Oxygen (O)	8	60,425	33,100	25,670
Fluorine (F)	9	77	<1	<1
Sodium (Na)	11	2,554	290	75
Magnesium (Mg)	12	1,784	34	11
Silicon (Si)	14	20,475	<1	<1
Phosphorus (P)	15	79	<1	130
Sulfur (S)	16	33	17	130
Chlorine (Cl)	17	11	340	33
Potassium (K)	19	1,374	6	37
Calcium (Ca)	20	1,878	6	230
Vanadium (V)	23	4	<1	<1
Chromium (Cr)	24	8	<1	<1
Manganese (Mn)	25	37	<1	<1
Iron (Fe)	26	1,858	<1	<1
Cobalt (Co)	27	1	<1	<1
Nickel (Ni)	28	3	<1	<1
Copper (Cu)	29	1	<1	<1
Zinc (Zn)	30	2	<1	<1
Selenium (Se)	34	<1	<1	<1
Molybdenum (Mo)	42	<1	<1	<1
Iodine (I)	53	<1	<1	<1

*Amounts are given in atoms per 100,000.

should possess the capacity to replicate. It should also possess the capacity to evolve by mistakes in base sequence incorporation during the process of replication. As RNA systems became increasingly complex with more catalytic functions, the remainder of those components necessary for cellular life became part of the system. In time the capacity to order amino acids into proteinlike polypeptide chains developed. Because of their greater variety and flexibility most of the catalytic functions originally possessed by RNA were passed along to the proteins. And finally, as the RNA "chromosomes" became more and more complex, they transferred the job of storing useful biochemical information to the DNA. Thus, RNA, which initially managed most of the tasks of conducting living systems, passed along catalytic functions to the proteins and information storage functions to the DNA and took on an intermediary role as a messenger for transferring genetic information from the DNA for the synthesis of proteins.

Figure 1.19

A possible route leading to the origin of nucleic acids.

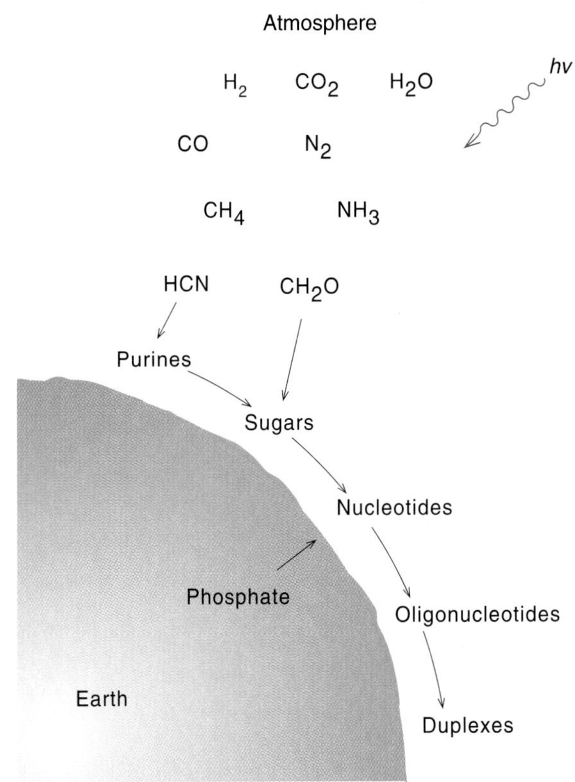

All Living Systems Are Related through a Common Evolution

The classical view of evolution, based on morphological differences among organisms, can be diagrammed as a branching tree in which all existing organisms are shown at the tips of the branches (fig. 1.20). An evolutionary tree starts from the simple ancestral prokaryotic cell, which branches off in three main directions into the archaebacteria, the eubacteria, and the eukaryotes. Each of these kingdoms has continued to branch in elaborate ways; we have shown only some of the main branch points in figure 1.20. Prokaryotes for the most part have remained as relatively undifferentiated single-celled organisms containing a single chromosome. By contrast, eukaryotes have changed dramatically. Although the organisms on many branches of the eukaryotic part of the evolutionary tree have remained as relatively undifferentiated single-cell forms, significant numbers of eukaryotic organisms have evolved into multicellular forms in which the individual cells of the total organism have differentiated to serve different functions. This process has given rise to plants, animals, and fungi.

A somewhat different view of biological evolution has arisen from a comparison of nucleic acid sequences in different organisms. The best-known sequences for such studies have come from the 16S ribosomal RNAs in different organisms. This comparative study has provided us with the evolutionary pattern shown

Figure 1.20

Classical evolutionary tree. All living forms have a common origin, believed to be the ancestral prokaryote. Through a process of evolution some of these prokaryotes changed into other organisms with different characteristics. The evolutionary tree indicates the main pathways of evolution.

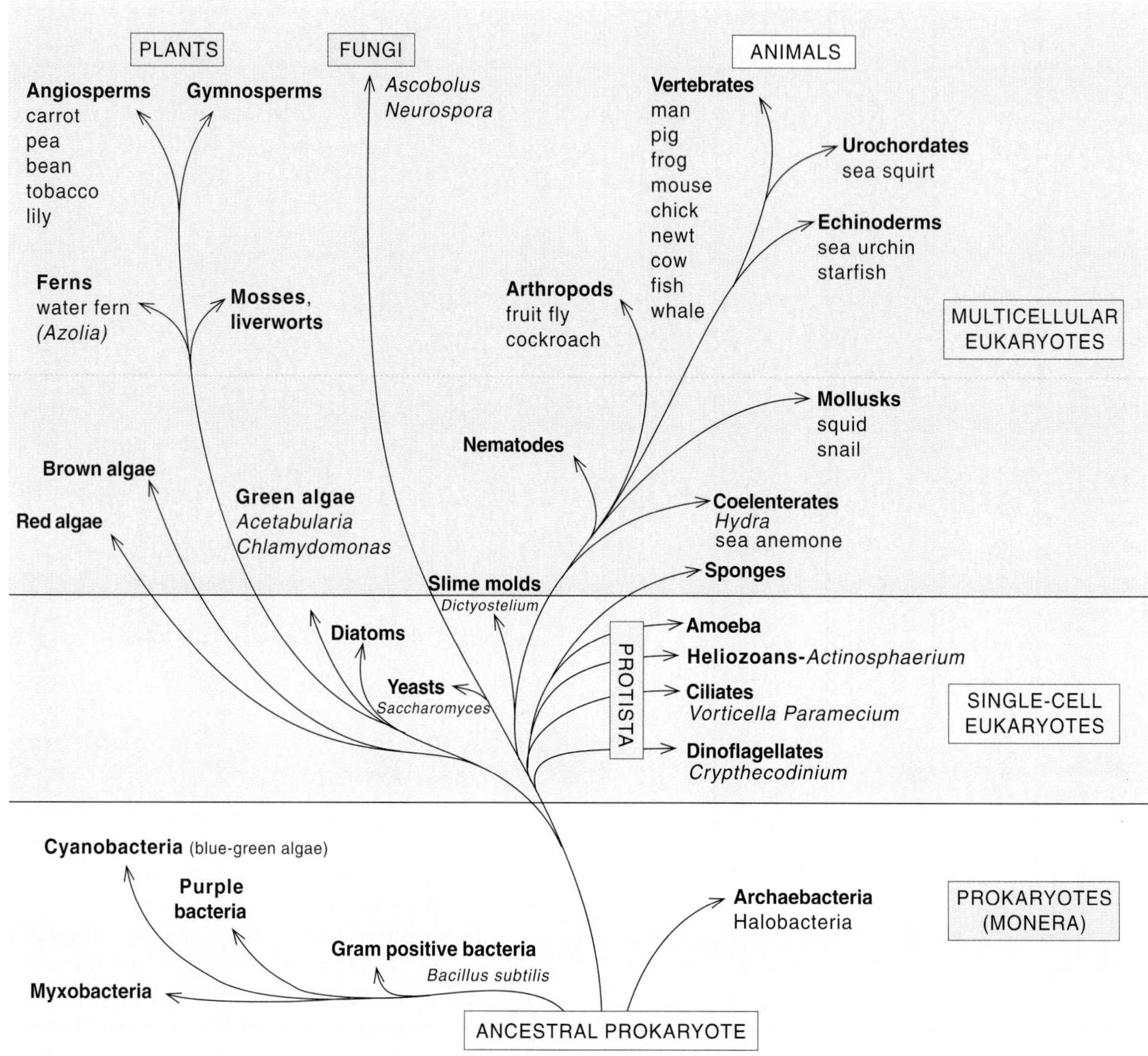

in figure 1.21, where distances are proportional to sequence differences. A most striking characteristic of this unrooted tree is the distinctness of the three primary kingdoms as evidenced by the large sequence distances that separate each of the kingdoms from the others. Although we do not know the position of the root of the tree, it seems likely that the archaebacteria are closer to the common ancestor of all three kingdoms than are the eubacteria or the prokaryotes.

The organisms most familiar to us, the multicellular plants and animals, occupy a shallow domain within the eukaryotic line of descent. True, the developmental programs of the multicellular forms have generated an incredible diversity in form and function. Nevertheless, both bacteria and unicellular eukaryotes span far greater evolutionary histories. This fact is reflected in the greater biochemical diversity that we find in the unicellular microorganisms.

Figure 1.21

An evolutionary tree can be constructed by comparing the complete sequences of 21 different 16S and 16S-like ribosomal RNAs (rRNAs). The scale bar represents the number of accumulated nucleotide differences (mutations) per sequence position in the rRNAs of the various organisms. (Source: From N. R. Pace, G. J. Olsen and C. R. Woese, *Cell* 5:325, 1986.)

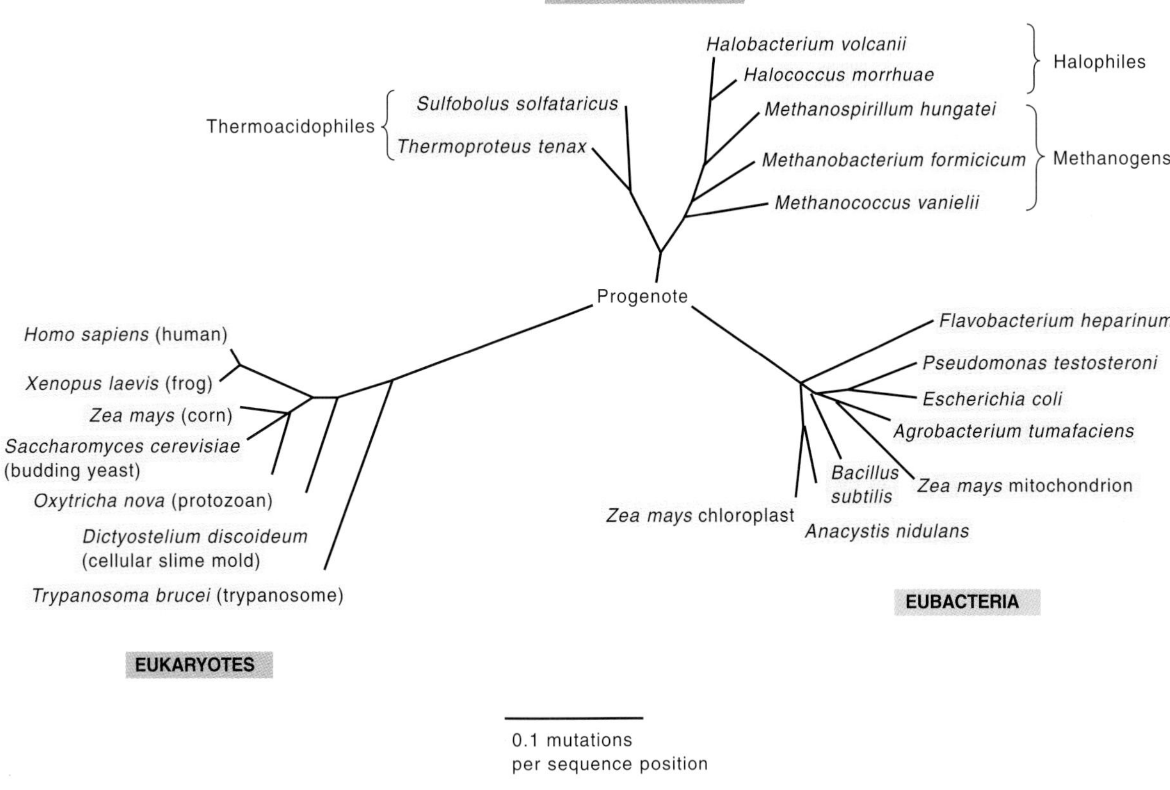

Summary

In this chapter we have discussed the ways in which biochemistry parallels ordinary chemistry and those in which it is quite different. The chief points to remember are the following.

1. The basic unit of life is the cell, which is a membrane-enclosed, microscopically visible object.
2. Cells are composed of small molecules, macromolecules, and organelles. The most prominent small molecule is water, which constitutes 70% by weight of the cell. Other small molecules are present only in quite small amounts; they are precursors or breakdown products of macromolecules or coenzymes. There are four types of macromolecules: lipids, carbohydrates, proteins, and nucleic acids.
3. In general, hydrophobic groupings are buried within the folded macromolecular structure, while hydrophilic groupings are located on the surface, where they can interact with water.
4. Biochemical reactions utilize a limited number of elements, most prominently carbon, hydrogen, oxygen, nitrogen, sulfur, and phosphorus. Many biochemical reactions are simple organic reactions.
5. Biochemical reactions are carried out under very mild conditions in aqueous solvent. The reactions can proceed under these conditions because of the highly efficient nature of protein enzyme catalysts.
6. Biochemical reactions frequently require energy. The most common source of chemical energy used is adenosine triphosphate (ATP). The splitting of a phosphate from the ATP molecule can provide the energy needed to make an otherwise unfavorable reaction go in the desired direction.
7. Biochemical reactions of different types are localized to different parts in the cell.
8. Biochemical reactions are frequently organized into multistep pathways.

9. Biochemical reactions are regulated according to need by controlling the amount and activity of enzymes in the system.
10. Most organisms depend on other organisms for their survival. Frequently, this is because a given organism cannot make all of the compounds needed for its growth and survival.
11. The specific properties of any protein are due to the specific sequence of amino acids in its polypeptide chains. This sequence is determined by the genetic information carried by the sequence of DNA nucleotides. DNA transfers the information to messenger RNA, which serves as the template for protein synthesis.
12. Shortly after the earth was formed and had cooled to a reasonable temperature, chemical processes produced compounds that would be used in the development of living cells. Nucleic acids are the most important compounds for living cells, and it is believed that they played the central role in the origin of life.
13. Evolutionary trees based on morphology or biochemical differences indicate that all living systems are related through a common evolution.

Selected Readings

Alberts, B., D. Bray, J. Lewis, M. Raff, K. Roberts, and J. D. Watson, *Molecular Biology of the Cell,* 3d ed. New York: Garland, 1994. Comprehensive, up-to-date treatment of cell biology.

de Duve, C., *Blueprint for a Cell.* Burlington, N.C.: California Biological Supply Co., 1991. A short book on the origin of life that contains an excellent reference list.

Dickerson, R. E., Chemical evolution and the origin of life. *Sci. Am.* 239(3):70–86, 1978.

Doolittle, R. F., The genealogy of some recently evolved vertebrate proteins. *Trends Biochem. Sci.* 10:233–237, 1985.

Kimura, M., The neutral theory of molecular evolution. *Sci. Am.* 241(5):98–126, 1979.

Lewis, R., *Life,* 2d ed. Dubuque, Iowa: Wm. C. Brown, 1995. Offbeat, exciting introduction to biology.

Lodish, H., D. Baltimore, A. Berk, S. L. Zipursky, P. Matsudaira, and J. Darnell, *Molecular Cell Biology,* 3d ed. New York: W. H. Freeman. This book is equal to the Alberts book and superior in some areas.

Prescott, L. M., J. P. Harley, and D. A. Klein, *Microbiology,* 2d ed. Dubuque, Iowa: Wm. C. Brown, 1993. Comprehensive, up-to-date treatment.

Schopf, J. W., The evolution of the earliest cells. *Sci. Am.* 229(3):10–138, 1978. An authoritative account from a foremost geologist.

Stillinger, F. H., Water revisited. *Science* 209:451–457, 1980. A reminder of the central importance of water to the origin of life.

Wilson, A. C., The molecular basis of evolution. *Sci. Am.* 253(4):164–173, 1985.

Zubay, G., *Origins of Life on the Earth and in the Cosmos.* Dubuque, Iowa: Wm. C. Brown, 1995. Especially focused on possible events leading to an RNA-only world.

C H A P T E R

Thermodynamics in Biochemistry

All living systems obey the laws of thermodynamics: A pathway that requires energy usually consumes ATP, while a pathway that releases energy usually produces ATP.

The primary usefulness of thermodynamics to biochemists lies in predicting whether particular chemical reactions could occur spontaneously. A simple illustration is to predict what compounds could possibly serve as energy sources for an organism. You are aware from everyday experience that oxidation of organic molecules by molecular oxygen releases energy. For example, wood or coal burns with a large output of heat. Similarly, organisms can obtain energy oxidizing carbohydrates, fats, or proteins. Some organisms oxidize hydrocarbons, some oxidize reduced forms of sulfur, and others oxidize iron. But no organisms live by oxidizing molecular nitrogen, and the explanation lies in thermodynamics. The reaction cannot occur spontaneously. This example illustrates the importance of thermodynamics in controlling all life. Because organisms live by extracting chemical energy from their surroundings, thermodynamics is not an esoteric subject. It is a matter of life or death.

We say that thermodynamics determines whether a process "could" occur, because thermodynamics tells us only whether the process is possible, not whether it actually will occur in a finite period of time. The rate at which a thermodynamically possible reaction occurs depends on the existence of a feasible pathway. For a biochemical process to occur rapidly, appropriate catalysts must be available. This distinction between spontaneity and speed is more critical for biochemists than it is for chemists, because if a chemical reaction does not proceed rapidly, a chemist can change the pressure or temperature or increase the concentration of the reactants. A living organism is under more rigid constraints: It must function within a very limited range of temperatures, pressures, and concentrations of reactants.

In this chapter we consider the principal thermodynamic quantities: energy, enthalpy, entropy, and free energy. We will then expand the discussion to show how the concept of free energy is used in predicting biochemical pathways, and we will explore the central role of ATP in providing energy for biochemical reactions.

Thermodynamic Quantities

The properties of a substance can be classified as either intensive or extensive. Intensive properties, which include density, pressure, temperature, and concentration, do not depend on the amount of the material. Extensive properties, such as volume and weight, do depend on the amount. Most of the thermodynamic properties we will be discussing are extensive properties. These include energy (E), enthalpy (H), entropy (S), and free energy (G).

Energy, enthalpy, entropy, and free energy are all properties of the state of a substance. This means that their values for a given substance do not depend on how the substance was made or how it reached a particular state. In a chemical reaction, it is the difference between the initial and final states that is important; the pathway that is taken to get from the initial state to the final state has no bearing on whether the overall reaction releases or consumes energy (fig. 2.1).

Figure 2.1

The change in energy of a system depends only on the initial and final states, not on the path by which the system gets from one state to the other. This diagram illustrates the conversion of a phosphate ester of glucose (glucose-6-phosphate) to free glucose and inorganic phosphate ion (P_i) by two different pathways. Although the two routes proceed through intermediate compounds that differ in energy (A, B, C, and D), the overall energy change (ΔE) is the same. However, the work done and the amount of heat absorbed or released generally is not the same for the two paths.

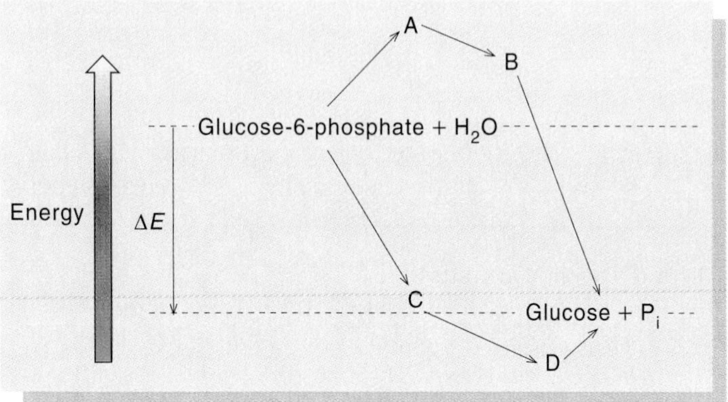

The First Law of Thermodynamics: Any Change in the Energy of a System Requires an Equal and Opposite Change in the Surroundings

Of the thermodynamic properties we have just listed, energy is probably the most familiar. Energy is the capacity to do work. The energy of a molecule includes the internal nuclear energies and the molecular electronic, translational, rotational, and vibrational energies. Electronic energies, which reflect the interactions among the electrons and atomic nuclei, usually are much larger than the translational, rotational, and vibrational energies.

In biochemistry we are concerned not so much with the absolute energies of molecules as with changes in energy that occur in the course of reactions. It is easier to evaluate a change in energy (ΔE) than to calculate the absolute energies of the reactants or products, because many of the terms that contribute to the total energy do not change much during a chemical reaction. Electronic energies usually dominate ΔE, as they do the total energies. Good estimates of the energy change resulting from a chemical reaction can usually be obtained by calculating the difference between the bond energies of the reactants and products (see table 3.2).

The first law of thermodynamics says that the total amount of energy in the universe is constant. Energy can undergo transformations from one form to another; for example, the chemical energy of a molecule can be transformed into thermal, electrical, or mechanical energy. But any change in the total energy of one part of the universe is matched by an equal and opposite change in another part, so that the overall energy of the universe remains constant.

To apply the first law to a chemical reaction, we must take into account all the energy changes that occur. This means including any changes in the surroundings, as well as in the system of interest (fig. 2.2). The system might be a reaction occurring in a test tube or a living cell; the surroundings are all the rest of the universe. In practical terms, however, we usually need to consider only the immediate surroundings, because only these are likely to be influenced by what happens in the system. According to the first law, the overall energy remains constant even though energy in some form may flow from the system to its surroundings or from the surroundings to the system.

If the amount of matter in a system is constant, there are only two means by which the system can gain or lose energy: transfer of heat or performance of work. The energy of a system increases if the system absorbs heat from its surroundings or if the surroundings do work on the system; it decreases if the system gives off heat to the surroundings or does work. A common way of stating the first law of thermodynamics is to equate the change in energy of the system (ΔE) to the difference between the heat absorbed by the system (q) and the work done by the system (w):

$$\Delta E = q - w \qquad \textbf{(1)}$$

Both q and w depend on the path of the reaction, but their difference, ΔE, is independent of the path and therefore defines a state function.

The relative energy of a compound can be measured in a bomb calorimeter, a device in which the compound is thoroughly combusted and the resulting heat is measured. Energies (and enthalpies, discussed next) are usually given in units of kilocalories per mole or kilojoules per mole. One kilojoule (kJ) is the amount of energy needed to apply 1 newton of force over 1 km; 1 kilocalorie (kcal) is the heat needed to raise the temperature of 1 kg of water from 14.5 to 15.5°C. One kcal is equivalent to 4.184 kJ.

Figure 2.2

A system and its surroundings. Heat flow into the system is designated as a positive quantity (q), and work that the system does on the surroundings is designated as a positive quantity (w). The first law of thermodynamics relates q and w to changes in the energy of the system. Any change in the energy of the system (ΔE_{sys}) is balanced by an opposite change in the energy of the surroundings (ΔE_{sur}), so that the overall energy change (ΔE_{tot}) is zero.

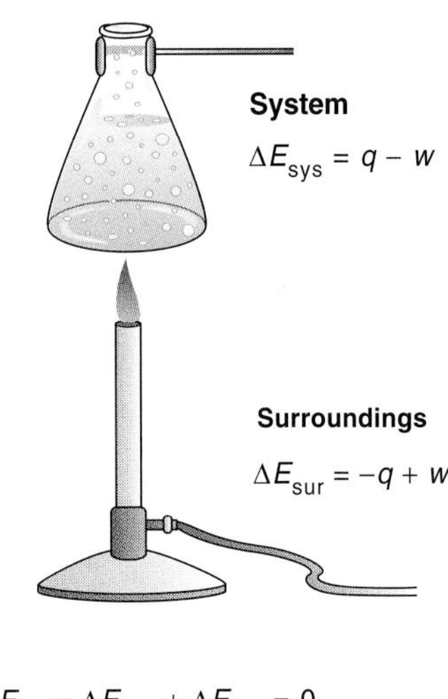

System
$\Delta E_{sys} = q - w$
Surroundings
$\Delta E_{sur} = -q + w$
$\Delta E_{tot} = \Delta E_{sys} + \Delta E_{sur} = 0$

Physicists generally prefer to use kilojoules; most chemists, including biochemists, prefer to use kilocalories.

It is customary to record energies of molecules with reference to the energies of their constituent elements. The standard energy of formation reported for a molecule, $\Delta E^{\circ}{}_{f}$, is equal to the energy change associated with the formation of the molecule from the elements in their standard states. The energy change in any other reaction can be obtained by subtracting energies of formation of the reactants from the energies of formation of the products. We will have more to say about standard states and calculations involving thermodynamic parameters later in this chapter. The complex organic molecules found in cells have relatively weak chemical bonds, and thus higher energies, than H_2O, and CO_2, and the other small molecules from which they are formed.

Enthalpy is a function of state that is closely related to energy but is usually more pertinent for describing the thermodynamics of chemical or biochemical reactions. The change in enthalpy (ΔH) is related to the change in energy (ΔE) by the expression

$$\Delta H = \Delta E + \Delta(PV) \qquad (2)$$

where $\Delta(PV)$ is the change in the product of the pressure (P) and volume (V) of the system. ΔH is the amount of heat that is absorbed from the surroundings if a reaction occurs at constant pressure and no work is done other than the work of expansion or contraction of the system. (The work done when a system expands by ΔV against a constant pressure P is $P\ \Delta V$. This type of work is generally not very useful in biochemical systems.) In most biochemical reactions, there is little change in either pressure or volume, so the difference between ΔH and ΔE can usually be ignored for all practical purposes.

In most chemical reactions that occur spontaneously, the enthalpy of the system decreases. If no work is done, the system gives off heat to the surroundings. But changes in enthalpy or energy do not provide a reliable way of determining whether a reaction can proceed spontaneously. For example, although LiCl and $(NH_4)_2SO_4$ both dissolve readily in water, the former process releases heat, whereas the latter absorbs heat. A mixture of solid $(NH_4)_2SO_4$ and water proceeds spontaneously to a state of higher enthalpy. This reaction is driven by an increase in the entropy of the system.

To assess the potential for a reaction to occur, we must consider the enthalpy as well as the entropy.

The Second Law of Thermodynamics: In Any Spontaneous Process the Total Entropy Increases

The second law of thermodynamics is that the universe inevitably proceeds from states that are more ordered to states that are less ordered. This phenomenon is measured by a thermodynamic function called entropy, which is denoted by the symbol S. A reaction in which entropy increases (ΔS is positive) will proceed in preference to one in which entropy decreases.

Entropy is an index of the number of different ways that a system could be arranged without changing its energy. If a system could be arranged in Ω different ways, all with the same energy, the absolute entropy per molecule is

$$S = k \ln \Omega \qquad (3)$$

where k is Boltzmann's constant ($k = 3.4 \times 10^{-24}$ cal/degree Kelvin). For a mole of substance,

$$S = Nk \ln \Omega = R \ln \Omega \qquad (4)$$

Here N is the number of molecules in a mole (6×10^{23}) and R is the gas constant ($R \approx 2$ cal/(degree K • mole)). Quantitative values for entropies are usually given in entropy units (1 eu = 1 cal/degree K).

The more ways a particular state could be obtained, the greater is the probability of finding a system in that state. A system that is highly disordered could be obtained in many different arrangements that are all energetically equivalent. Thus a state in which molecules are free to move about and rotate into many different orientations or conformations will be favored over a state in which motion is more restricted. The second law makes the remarkably general assertion that the total entropy change in any reaction that occurs spontaneously must be greater than zero. But note that this statement specifies the total entropy, which means that we must consider the entropy change in the surroundings as well as that in the system. The entropy of the system can decrease if the entropy of the surroundings increases by a greater amount.

The absolute entropies of small molecules can be calculated by statistical mechanical methods. Table 2.1 shows the results of such calculations for liquid propane. The largest contributions to the entropy come from the translational and rotational freedom of the molecule, and much smaller contributions come

Table 2.1 Contributions to the Entropy of Liquid Propane at 231 K	
	kcal/(degree K • mole)
Translational entropy	36.04
Rotational entropy	23.38
Vibrational entropy	1.05
Electronic entropy	0.00
Total	60.47

from vibrations; electronic terms are insignificant. Although exact calculations of this type become intractable for large biological molecules, the relative sizes of the contributions from different types of motions are similar to those in small molecules. Thus entropy is associated primarily with translation and rotation. This is very different from enthalpy, in which electronic terms are dominant and translational and rotational energies are comparatively small.

Statistical mechanical calculations show that the translational entropy of a molecule depends on $\frac{3}{2}R \ln M_r$ (plus some smaller terms), where R is the gas constant and M_r is the molecular weight. Suppose that a molecule, Y, undergoes a dimerization reaction so that its molecular weight doubles:

$$2\,Y \longrightarrow Y_2$$

Intuitively, we expect dimerization to decrease the entropy because the two monomeric units can no longer move independently. We can calculate the effect quantitatively as a function of the molecular weight as follows. If M_r is the molecular weight of the monomer, then the change in translational entropy in going from the monomeric state to the dimer is approximately

$$\begin{aligned}\Delta S &\approx \frac{3}{2}R \ln 2M_r - 2\left(\frac{3}{2}R \ln M_r\right)\\ &= \frac{3}{2}R \ln 2 - \frac{3}{2}R \ln M_r\\ &= -\frac{3}{2}R \ln (M_r/2) \qquad (5)\end{aligned}$$

The decrease of the translational entropy resulting from dimerization is a logarithmic function of the molecular weight.

Structural features that make molecules more rigid reduce rotational and vibrational contributions to entropy. As a result, the formation of a double bond or ring decreases the entropy even when the molecular weight is unchanged. The formation of comparatively rigid macromolecular structures from flexible polypeptide or polynucleotide chains also requires an entropy decrease, although this can be offset by increases in the entropy of the surrounding water molecules (see chapter 3).

The entropy of a compound depends strongly on the physical state of the material. A gas has more translational and rotational freedom than a liquid, and a liquid has more freedom than a solid. As a result, entropy increases when a solid melts or a liquid vaporizes.

It can be shown that the increase in the entropy of a system that undergoes an isothermal, reversible process is

$$\Delta S = \frac{\Delta H}{T} \qquad (6)$$

where T is the absolute temperature in degrees kelvin. An isothermal process is one that occurs at constant temperature. A reversible process is one that proceeds infinitely slowly through a series of intermediate states in which the system is always at equilibrium. For any real process occurring at a finite rate, the system is not strictly at equilibrium, and ΔS is larger than the value given by equation (6).

From equation (6), the entropy increase on vaporization or melting can be determined simply from the heat of vaporization divided by the boiling point or from the heat of fusion divided by the melting point. The entropy increase on vaporization of water is 26 eu/mole, and that on melting of ice is 5.3 eu/mole. These values are consistent with our intuition that the increase in translational and rotational freedom is much greater in going from a liquid to a gas than in going from a solid to a liquid.

The entropy of a solution is increased by mixing of solvents, and it is decreased by interactions among the solvent molecules or interactions of solutes with the solvent. The mixing of two miscible liquids is a thermodynamically favorable process because it increases the number of positions that are available to the molecules. The entropy change on going from the unmixed liquids to the mixed state can be calculated from the expression

$$\begin{aligned}\Delta S &= n_a R \ln \frac{1}{X_a} + n_b R \ln \frac{1}{X_b}\\ &= -n_a R \ln X_a - n_b R \ln X_b \qquad (7)\end{aligned}$$

where n_a and n_b are the number of moles of A and B that are mixed, and X_a and X_b are the corresponding mole fractions in the final solution. Because X_a and X_b are always less than 1, ln X_a and ln X_b will be negative. This means that the dilution of each component resulting from the mixing makes a positive contribution to the entropy. Equation (7) applies to the mixing of ideal solutions, in which there are no interactions among the molecules. Any intermolecular interactions will decrease the entropy by restricting the system's translational and rotational freedom.

Solvation, the interaction of a solute with the solvent, makes an important negative contribution to the entropy of a solution. This is because the clustering of solvent molecules around the solute restricts the movement of the solvent molecules (fig. 2.3). In general, the entropy of solvation by water becomes more negative with an increase in the charge or polarity of the solute. Small ions are solvated more strongly than large ions with the same charge, and anions are solvated more strongly than cations.

It is noteworthy that although enthalpy depends mainly on electronic interactions, whereas entropy depends mainly on translation and rotation, solvation affects both enthalpy and entropy. Enthalpies and entropies of solvation usually tend to oppose each other. For charged species, the more negative (favorable) the

Figure 2.3

The entropy decrease resulting from solvation. When a salt is dissolved in water, the entropy of the dissociated cations and anions increases because of the increased possibilities for translation and rotation. But at the same time the movement of water molecules becomes restricted in the vicinity of the ions. The net effect is frequently a decrease in the entropy of the solution. Such a decrease in entropy can occur if the solution releases heat to the surroundings, because this increases the entropy of the surroundings.

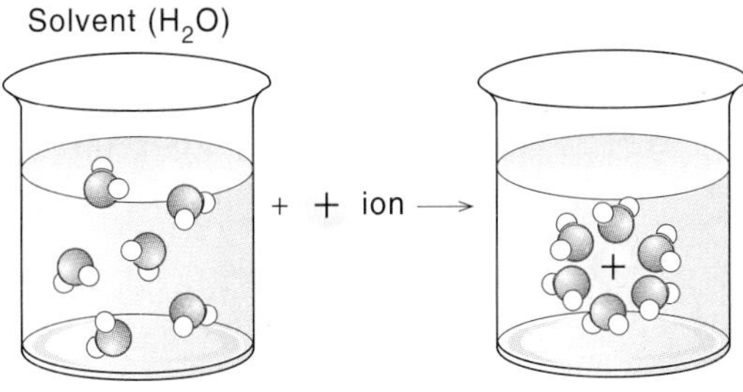

enthalpy of solvation, the more negative (unfavorable) the entropy of solvation.

From our earlier discussion, you might expect that the dissociation of a proton from a carboxylic acid, which increases the number of independent particles, would lead to an increase in entropy. However, this effect is more than counterbalanced by solvation effects. The charged anion and proton both "freeze out" many of the surrounding molecules of water (fig. 2.4). Thus the ionization of a weak acid decreases the number of mobile molecules and so leads to a decrease in entropy. The entropy of ionization of a typical carboxylic acid in water is about −22 eu/mole.

Figure 2.4

An ionization reaction often decreases the entropy of a solution, instead of increasing it as one might at first expect, because clustering of water molecules around the ions can result in a net decrease in the number of free water molecules.

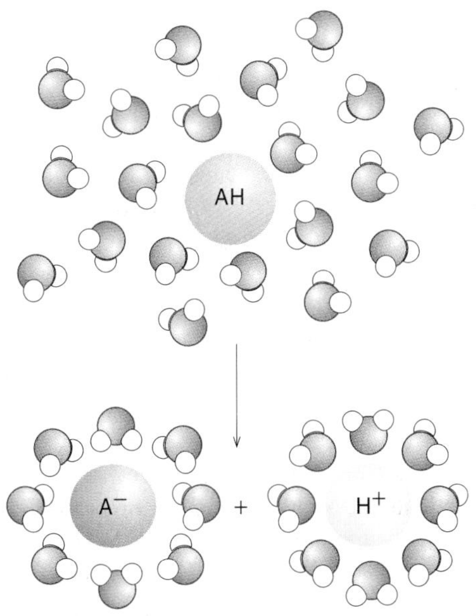

The entropy of dissociation of a proton from a quaternary ammonium group ($R—NH_3^+ \rightarrow R—NH_2 + H^+$) is usually smaller, because in this case the dissociation does not alter the number of charged species in the solution.

A different type of solvation effect occurs when an apolar molecule is added to water. The result is a decrease in entropy, but not because of favorable interactions between the molecule and the solvent. The water orients on the surface of the apolar molecule to form a relatively rigid cage held together by hydrogen bonds (see fig. 3.4). This effect plays important roles in governing the folding of proteins and determining the structure of biological membranes (see fig. 1.7).

Substantial changes in entropy can occur when a small molecule binds to a protein or other macromolecule. Of particular interest is binding of a substrate or inhibitor to an enzyme. It is instructive to compare the entropy changes here with those that accompany the binding of gas molecules to the surface of a solid catalyst. When gas molecules adsorb on a solid surface, there is a large decrease in entropy (fig. 2.5). The translational entropy of the gas disappears, and the thermodynamics of the adsorbed molecules becomes more like that of a solid. This negative change in entropy is an obstacle to the industrial use of solid catalysts for reactions of gases. To make the reaction proceed, the unfavorable entropy change must be overcome by increasing the pressure or tailoring the catalyst so that the enthalpy of interaction with the gas is strongly negative. In contrast, the binding of a substrate to an enzyme frequently has a positive ΔS. The reason is that water molecules are displaced when the substrate binds (see fig. 2.5). Binding of an ester substrate to pepsin, for example, produces an entropy increase of 20.6 eu/mole, and binding of urea to urease produces an entropy increase of 13.3 eu/mole. Because of the favorable entropy change, the binding of the substrate may occur spontaneously even if ΔH is unfavorable.

Another important entropy effect when an enzyme and substrate combine could be called the chelation effect. This effect is best understood by discussing a relatively simple case of metal

Figure 2.5

The entropy of binding of gas molecules to a solid catalyst is negative because of the restricted movements of the adsorbed molecules. By contrast, the entropy change on binding of a substrate to an enzyme is frequently positive. This effect arises because the restricted movement of the substrate is more than compensated for by the release of bound water from the enzyme and the substrate.

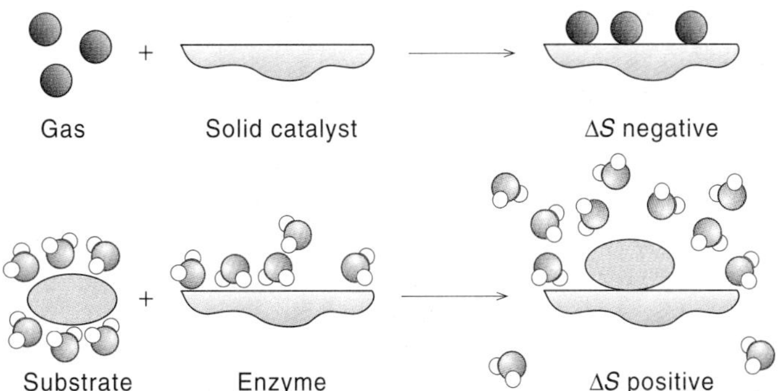

Table 2.2
Standard Enthalpy and Entropy Changes on Forming Complexes between Cadmium Ion and Methylamine or Ethylenediamine (en)

Reaction[a]	$\Delta H°$ (kcal/mole)	$\Delta S°$ (eu/mole)	$T\Delta S°$ (kcal/mole)	$\Delta G°$ (kcal/mole)
$Cd^{2+} + 4\ CH_3NH_2 \rightarrow Cd(CH_3NH_2)_4^{2+}$	−13.7	−16.0	−4.77	−8.94
$Cd^{2+} + 2\ en \rightarrow Cd(en)_2^{2+}$	−13.5	−3.3	+0.98	−14.50

[a]en = ethylenediamine; temperature = 25°C.

Source: Data from Spike and Parry, Thermodynamics of chelation I. The statistical factor in chelate ring formation, in *J. Amer. Chem. Soc.* 75:2726, 1953.

chelation. Cadmium ion tends to be tetravalent, so if there is one amino group in a ligand molecule, as in methylamine, the cadmium can bind to four molecules. If there are two amino groups, as in ethylenediamine, the cadmium will combine with two ligand molecules, as shown in table 2.2. Notice that the entropy change is much more unfavorable (negative) for the combination with four methylamines than for that with two ethylenediamines. Water molecules are released from the cadmium ion when the ligands bind, but the entropy increase from this release is about the same whether methylamine or ethylenediamine is added. The less favorable entropy change resulting from association with methylamine is due to the larger number of molecules that must attach to the cadmium in this case.

In general, the chelation effect means that a molecule with n points of attachment to another molecule will bind more strongly than n molecules with one point of attachment, even though the enthalpy change upon binding at each point is the same. The chelation effect is important in substrate binding to enzymes, where there typically are multiple points of attachment. Several weak interactions can produce an overall tight binding because of the additive contributions of the small, favorable enthalpy changes and the lack of a proportional decrease in entropy. This effect is even greater in the binding of two proteins or the binding of a protein to a nucleic acid, because there usually are many points of interaction between these large molecules.

The final column in table 2.2 indicates the changes in free energy accompanying the reactions of cadmium ion with the two amine compounds. Free energy is a function of both enthalpy and entropy; it provides the most useful criterion as to whether a reaction could proceed spontaneously, as explained in the next section.

Free Energy Provides the Most Useful Criterion for Spontaneity

We have seen that a system tends toward the lowest enthalpy and the highest entropy. The tendency for the enthalpy of a system to decrease can be explained simply by the first and second laws of thermodynamics. If no work is done, an enthalpy decrease means a transfer of heat from the system to the surroundings, and if the surroundings are at a lower temperature such a flow of heat will be driven by an increase in the entropy of the surroundings. But enthalpy changes do not afford a reliable rule for determining whether a reaction can proceed spontaneously, because the enthalpy of a system can increase if the entropy of the system also increases. The second law does provide a reliable rule, but its application is often difficult because it requires that we consider the entropy changes in the surroundings as well as the system.

A more convenient function for predicting the direction of a reaction was discovered by Josiah Gibbs. He was the first to appreciate that in reactions occurring at equilibrium and constant temperature, the change in entropy of the system is numerically equal to the change in enthalpy divided by the absolute temperature. This relationship is the one already presented in equation (6). The equation can be transposed to

$$\Delta H - T\,\Delta S = 0 \qquad \textbf{(8)}$$

In search of a criterion for spontaneity, Gibbs proposed a new function called the free energy, defined by the equation

$$\Delta G = \Delta H - T\,\Delta S \qquad \textbf{(9)}$$

Here ΔH and ΔS are the changes in enthalpy and entropy in the system alone, not including the surroundings. For a reaction occurring at equilibrium, such as the melting of ice at 0°C, the change in free energy is zero. For the same reaction occurring at a higher temperature, say 10°C, the term $T\,\Delta S$ is larger, making ΔG negative. Ice at 10°C melts spontaneously. In the reverse reaction, conversion of water to ice at 10°C, there would be a positive change in free energy. This process does not occur spontaneously. Gibbs proposed that a reaction can occur spontaneously if, and only if, ΔG is negative. If ΔG is zero, the system is in equilibrium and no net reaction will occur in either direction.

Free energy, like energy, enthalpy, and entropy, is a state function and an extensive property of a system. If the free energy change is favorable (negative), and a good pathway exists, a reaction will occur. If no pathway exists for the conversion, a catalyst may be added to provide an acceptable pathway. However, if the free energy change is unfavorable (positive), no catalyst can ever make the reaction proceed.

Applications of the Free Energy Function

The free energy function dominates most considerations of thermodynamics in biochemistry. Not only does the sign of ΔG determine the direction in which a reaction will proceed, but the magnitude of ΔG indicates just how far the reaction must proceed before the system comes to equilibrium. This is because the standard free energy change, ΔG°, has a simple relationship to the equilibrium constant. We will elaborate on these points in the following sections. Despite its usefulness, however, many people find the free energy function difficult to grasp intuitively. The reason is that ΔG is a composite of enthalpic and entropic terms, which often make opposite contributions.

Values of Free Energy Are Known for Many Compounds

The standard free energy of formation of a compound, ΔG°_f, is the difference between the free energy of the compound in its standard state and the total free energies of the elements of which the compound is composed, again when the elements are in their standard states. The standard states usually are chosen to be the states in which the elements or molecules are stable at 25°C and 1 atmosphere pressure. For oxygen and nitrogen, these are the gases O_2 and N_2; for solid elements such as carbon, they are the pure solids. For most solutes, the standard states are taken to be 1 M solutions. However, in biochemistry the standard state for hydrogen ion in solution is usually defined as a 10^{-7} M solution because this is close to the concentration in most systems of interest to biochemists.

Standard free energies of formation are known for thousands of compounds. They usually are given in units of kcal/mole or kJ/mole. The values for a few compounds of biological interest are collected in table 2.3. By subtracting the sum of the free energies of formation of the reactants from the sum of the free energies of formation of the products, it is possible to calculate the standard free energy change in any reaction for which all the free energies of formation are known.

From the values listed in table 2.3, we can calculate the standard free energy change for the reaction

$$\text{Oxaloacetate}^{2-} + H^+ \ (10^{-7}\ \text{M}) \longrightarrow CO_2(g) + \text{pyruvate}^-$$

as

$$\Delta G^\circ = -113.44 - 94.45 - (-9.87 - 190.62)$$
$$= -7.4 \text{ kcal/mole}$$

The free energy change, when all the reactants and products are in their standard states (1 M oxaloacetate dianion and pyruvate anion, 10^{-7}M hydrogen ion, and 1 atm CO_2), is −7.4 kcal/mole. The negative value of ΔG° means that the reaction would proceed spontaneously under these conditions. However, some of the concentrations are not very realistic. At pH 7, carbon dioxide would be present partly in the form of the bicarbonate anion, rather than as gaseous CO_2. To take this into account, we can add the standard free energy change for the reaction of CO_2 with water to give the bicarbonate anion plus a proton:

Table 2.3
Standard Free Energies of Formation of Some Compounds of Biological Interest

Substance	ΔG°_f (kcal/mole)	ΔG°_f (kJ/mole)
Lactate ions	−123.76	−516
Pyruvate ions	−113.44	−474
Succinate dianions	−164.97	−690
Glycerol (1 M)	−116.76	−488
Water	−56.69	−280
Acetate anions	−88.99	−369
Oxaloacetate dianions	−190.62	−797
Hydrogen ions (10^{-7} M)	−9.87[a]	−41
Carbon dioxide (gas)	−94.45	−394
Bicarbonate ions	−140.49	−587

[a]This is the value for hydrogen ions at a concentration of 10^{-7} M. The free energy of formation at unit activity (1 M) is 0.

$$CO_2(g) + H_2O \longrightarrow HCO_3^- + H^+$$

This calculation yields a correction of −140.49 − 9.87 − (−56.69 − 94.45) = 0.8 kcal/mole. The free energy change for the reaction of oxaloacetate to form pyruvate and 1M bicarbonate ions instead of CO_2 is −7.4 + 0.8 = −6.6 kcal/mole.

The preceding calculation illustrates the point that the standard free energy change for a reaction can be found by adding or subtracting the free energies of any other reactions that combine to give the desired reaction. Another example is the calculation of the standard free energy of hydrolysis of ATP at pH 7. This calculation can be done by combining the free energy change for the hydrolysis of glucose-6-phosphate with the free energy change for forming glucose-6-phosphate from glucose and ATP, as shown in table 2.4. We will return to these reactions in a later section.

The Standard Free Energy Change in a Reaction Is Related Logarithmically to the Equilibrium Constant

In biochemistry we are most concerned with reactions occurring in aqueous solution. Suppose we have a chemical reaction with the stoichiometry

$$aA + bB \rightleftharpoons cC + dD$$

where *a, b, c,* and *d* refer to the moles of A, B, C, and D, respectively. The free energy change in the reaction is

$$\Delta G = G_{\text{final state}} - G_{\text{initial state}} \tag{10}$$

If the reaction occurs at constant temperature and pressure, equa-

Table 2.4
Calculating the Standard Free Energy of ATP Hydrolysis ($\Delta G^{\circ\prime}$) by Adding the Free Energies of Two Other Reactions

Reactions[a]	$\Delta G^{\circ\prime}$ (kcal/mole)
Glucose + $ATP^{4-} \rightleftharpoons$ glucose-6-phosphate^{2-} + ADP^{3-} + H^+	−5.4
Glucose-6-phosphate^{2-} + $H_2O \rightleftharpoons$ glucose + HPO_4^{2-}	−3.0
$ATP^{4-} + H_2O \rightleftharpoons ADP^{3-} + HPO_4^{2-} + H^+$	−8.4

[a]The values of $\Delta G^{\circ\prime}$ are for reactions at pH 7 in the absence of Mg^{2+}. In the presence of 10 mM Mg^{2+}, $\Delta G^{\circ\prime}$ for ATP hydrolysis is about −7.5 kcal/mole.

tion (10) can be expressed as the difference between the standard free energies of the products and reactants, ΔG°, plus a correction for the concentrations:

$$\Delta G = \Delta G^\circ + RT \ln \frac{[C]^c[D]^d}{[A]^a[B]^b} \tag{11}$$

The last term in equation (11) is the correction for the concentration and as such is an entropic contribution to ΔG. It is derived by using equation (7) to find the entropy changes associated with diluting the reactants and products from their standard states (1 M) to the actual concentrations in the solution. (In a rigorous treatment we should use activities instead of concentrations in this formula, but for simplicity we will ignore the difference, keeping in mind that it can be substantial in some cases.) If the concentrations of the reactants exceed those of the products, so that the ratio $[C]^c[D]^d/[A]^a[B]^b$ is less than 1, the logarithm will be negative, making ΔG more negative than ΔG° and favoring the reaction in the forward direction. A concentration ratio greater than 1 will favor the reverse reaction.

When the reaction comes to equilibrium,

$$\frac{[C]^c[D]^d}{[A]^a[B]^b} = K_{eq} \tag{12}$$

where K_{eq} is the equilibrium constant for the reaction. We also know that at equilibrium $\Delta G = 0$. Therefore, from equation (11),

$$\Delta G^\circ = -RT \ln K_{eq} \tag{13}$$

Thus the standard free energy change for a reaction can be used to obtain the equilibrium constant. Conversely, if we know K_{eq}, we can find ΔG°. Because of the logarithmic relationship, K_{eq} has a very steep dependence on ΔG° (table 2.5). A reaction that proceeds to 99% completion is, for most practical purposes, a quantitative reaction. It requires an equilibrium constant of 100 but a standard free energy change of only −2.7 kcal/mole, which is little more than half the standard free energy change for the formation of a hydrogen bond.

Equations (11) and (13) are two of the most important thermodynamic relationships for biochemists to remember. If the concentrations of reactants and products are at their equilibrium values, there is no change in free energy for the reactions going in either direction. Living cells, however, maintain some compounds at concentrations far from the equilibrium values, so that their reactions are associated with large changes in free energy. We will amplify on this point in chapter 12.

Table 2.5
Relationship between ΔG° and K_{eq} (at 25°C)

ΔG° (kcal/mole)[a]	K_{eq}
−6.82	10^5
−5.46	10^4
−4.09	10^3
−2.73	10^2
−1.36	10
0	1
1.36	10^{-1}
2.73	10^{-2}
4.09	10^{-3}
5.46	10^{-4}
6.82	10^{-5}

[a]ΔG° values at 25°C are calculated from the equation

$$\begin{aligned}\Delta G^\circ &= -RT \ln K_{eq}\\ &= -1.98 \times 298 \times 2.3 \log K_{eq}\\ &= -1364 \log K_{eq}\end{aligned}$$

We have mentioned that biochemists usually define the standard state of protons as 10^{-7} M and report values of free energy and equilibrium constants for solutions at pH 7 (see equation (3) in chapter 3 for the definition of pH). These values are designated by a prime and written as $\Delta G^{\circ\prime}$, $\Delta G'$, and K'_{eq}. *Unprimed symbols are used to designate values based on a standard state of 1* M for protons (pH 0). For a reaction that releases one proton, the relationship between K'_{eq} and K_{eq} is $K'_{eq} = 10^7 K_{eq}$. In evaluating the standard free energies $\Delta G^{\circ\prime}$ and ΔG°, it is critical to use the equilibrium constants K'_{eq} and K_{eq}, respectively, because these are very different quantities in reactions where protons are consumed or produced.

Free Energy Is the Maximum Energy Available for Useful Work

The free energy change gives a quantitative measure of the maximum amount of useful work that could be obtained from a reaction that occurs at constant temperature and pressure. By "useful" work, we mean work other than the unavoidable work of expansion or contraction against the fixed pressure of the surroundings. If ΔG is zero, the system is at equilibrium, which means that we could not obtain any useful work from the process. If ΔG is less than zero, the process could yield useful work as the system proceeds spontaneously toward equilibrium. If ΔG is greater than zero, the process is headed away from equilibrium, and we would have to perform work on the system in order to drive it in this direction. The farther the reactants are from equilibrium, the larger is the value of $-\Delta G$, and the larger is the amount of work that we might obtain from the reaction. However, $-\Delta G$ gives only the maximum amount of useful work. Remember that work and heat, unlike ΔG, ΔH, and ΔS, are not functions of state. The amount of work that is actually obtained depends on the path that the process takes, and it can be zero even if $-\Delta G$ is large.

Biological Systems Perform Various Kinds of Work

To sustain and propagate life requires that cells do various types of work. This work takes three major forms, related to three broad categories of cellular activities:

1. Mechanical work: changes in location or orientation. Mechanical work is done whenever an organism, cell, or subcellular structure moves against the force of gravity or friction. As examples, consider the contracting muscles that propel a runner up a hill, the swimming of a flagellated protozoan in a pond, the migration of chromosomes toward the opposite poles of the mitotic spindle, and the movement of a ribosome along a strand of messenger RNA.
2. Concentration and electrical work: movements of molecules and ions across membranes. Concentration work, the movement of a molecule or ion across a membrane against a prevailing concentration gradient, establishes the localized concentrations of specific materials on which most essential life processes depend. Concentration work is sometimes referred to as osmotic work. Examples include the uptake of amino acids from the blood by muscle cells, pumping of sodium ions out of a marine microorganism, and movement of nitrate from the soil into the cells of a plant root. Electrical work is required to move a charged species across a membrane against an electrical potential gradient. Although the most dramatic example of this is the generation of large potential differences in the electric organ of the electric eel, electrical work is done by almost all types of cells. It underlies the mechanisms of excitation of nerve and muscle cells and the conduction of impulses along axons.
3. Synthetic work: changes in chemical bonds. Synthetic work is necessary for the formation of the complex organic molecules of which cells are composed. As we have seen, these are in general molecules of higher enthalpy and lower entropy than the simple molecules that are available to organisms from their environment, so that free energy must be expended in their synthesis. Synthetic work is most obvious during periods of growth of an organism, but it also occurs in nongrowing, mature organisms, which must continuously repair and replace existing structures. The continuous expenditure of energy to elaborate and maintain ordered structures that were created out of less-ordered raw materials is one of the most characteristic properties of living cells.

Favorable Reactions Can Drive Unfavorable Reactions

In this section we consider the value of the free energy function for understanding how energetically unfavorable reactions are coupled with energetically favorable ones to do synthetic work. In table 2.4 we made use of the principle that the free energies of all the components of a solution are additive. In general, if the free energy changes associated with two reactions $A \rightleftarrows B$ and $C \rightleftarrows D$ are ΔG_{AB} and ΔG_{CD}, the free energy change accompanying the combined process $A + C \rightleftarrows B + D$ is simply $\Delta G_{AB} + \Delta G_{CD}$. It is this principle that allows living organisms to synthesize complex molecules with high enthalpies and low entropies. Thermodynamically unfavorable reactions can be driven by coupling them to favorable processes.

From equation (13) you can see that whereas free energy changes combine additively, equilibrium constants combine multiplicatively. If the equilibrium constants for the reactions $A \rightleftarrows B$ and $C \rightleftarrows D$ are K_{AB} and K_{CD}, the equilibrium constant for $A + C \rightleftarrows B + D$ is $K_{AB}K_{CD}$.

There are numerous ways to achieve a coupling of favorable and unfavorable reactions. As an example, let's return to the formation of glucose-6-phosphate and water from glucose and inorganic phosphate ion (P_i):

$$\text{Glucose} + P_i \rightleftharpoons \text{glucose-6-phosphate} \tag{14}$$

This reaction has an unfavorable $\Delta G^{\circ\prime}$ of +3.0 kcal/mole at 298 K (K_{eq} is 0.0062); the reaction will not occur spontaneously. On the other hand, the reaction

$$\text{ATP} + H_2O \rightleftharpoons \text{ADP} + P_i + H^+ \tag{15}$$

has a highly favorable $\Delta G^{\circ\prime}$ of about -8.4 kcal/mole ($K_{eq} \approx 1.35 \times 10^6$). If reactions (14) and (15) are combined to give the reaction

$$\text{Glucose} + \text{ATP} \rightleftharpoons \text{glucose-6-phosphate} + \text{ADP} \tag{16}$$

the overall $\Delta G^{\circ\prime}$ is $3.0 - 8.4$, or -5.4 kcal/mole, and the overall equilibrium constant is $0.0062 \times 1.35 \times 10^6$ or 8.4×10^3. The combined reaction is thermodynamically favorable (fig. 2.6). A cell could use this reaction to synthesize glucose-6-phosphate, provided that it has a source of ATP.

To take advantage of such a thermodynamic combination of favorable and unfavorable processes, a cell must have a catalytic mechanism for actually linking the two reactions. The breakdown of ATP to ADP and P_i, reaction (15), would be fruitless if it occurred independently of reaction (13). In many cells the com-

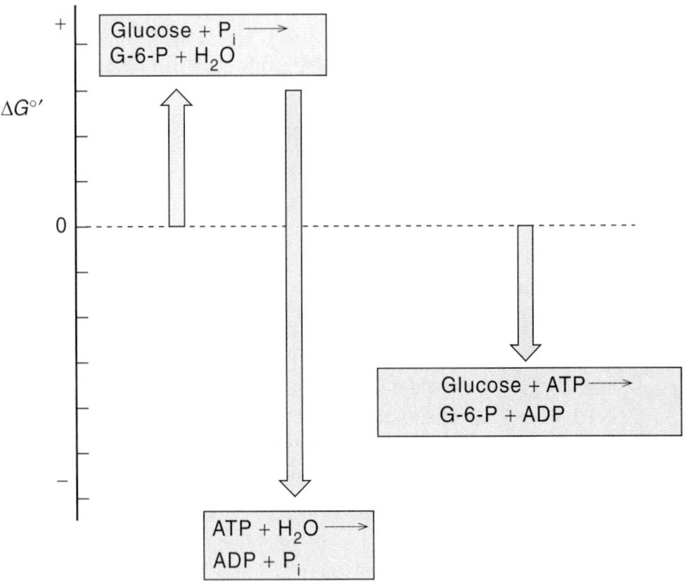

Figure 2.6

The formation of glucose-6-phosphate (G-6-P) has a positive $\Delta G^{\circ\prime}$; the hydrolysis of ATP to ADP has a negative $\Delta G^{\circ\prime}$. If the two reactions are combined, the overall $\Delta G^{\circ\prime}$ is negative.

bined reaction (16) is catalyzed by an enzyme that facilitates the transfer of phosphate from ATP directly to glucose (hexokinase). This is a common motif in biosynthetic processes. But the coupling mechanism does not have to be so direct.

Another common mechanism for coupling an unfavorable reaction to a favorable one is simply to arrange for one of the reactions to precede or follow the other. If the free energy changes for the reactions A $\rightleftarrows$ B and B $\rightleftarrows$ C are ΔG_{AB} and ΔG_{BC}, the free energy change for A $\rightleftarrows$ C is $\Delta G_{AB} + \Delta G_{BC}$. As an example, consider the following sequence of reactions:

$$\text{Acetyl-CoA} + \text{oxaloacetate} \xrightarrow{1} \text{citryl-CoA} \xrightarrow{2} \text{citrate} + \text{coenzyme A} \quad \textbf{(17)}$$

The first step has a $\Delta G^{\circ\prime}$ of -0.05 kcal/mole, which is close to zero; it will not occur to any great extent unless the concentrations of acetyl-CoA and oxaloacetate are greater than the concentration of citryl-CoA. The second step, however, has a highly favorable $\Delta G^{\circ\prime}$ of -8.4 kcal/mole. When the two steps are combined, $\Delta G^{\circ\prime}$ for the overall reaction is about -8.3 kcal/mole and the equilibrium constant lies far in the forward direction. These two reactions are catalyzed by the enzyme citrate synthase, by a mechanism that ensures that they always occur together.

In the case of reactions that occur sequentially, with one step pulling or pushing the other, it is not necessary for the two steps to be catalyzed by the same enzyme, although this can help to speed up the overall sequence. For example, in many organisms the reactions shown in (17) are preceded by a reaction in which malate is converted to oxaloacetate:

$$\text{Malate}^{2-} + \text{NAD}^{+} \longrightarrow \text{oxaloacetate}^{2-} + \text{NADH} + \text{H}^{+} \quad \textbf{(18)}$$

This step is catalyzed by a separate enzyme, malate dehydrogenase. It has a very unfavorable $\Delta G^{\circ\prime}$ of about $+7$ kcal/mole. When this reaction is followed by reaction (17), the combined $\Delta G^{\circ\prime}$ is about -1.3 kcal/mole, so the equilibrium constant for the overall sequence is favorable. You can view this as simply an illustration of the principle of mass action: (17) pulls (18) along by removing one of the products.

This last point deserves additional emphasis. It is important to keep in mind that what determines whether or not a process will occur spontaneously is ΔG, not ΔG°. As we saw in equation (11), the actual free energy change depends on the concentrations of the reactants and products. The hydrolysis of ATP (reaction 15), for example, has a $\Delta G^{\circ\prime}$ of about -8.4 kcal/mole, but the actual ΔG in the cytosol of living cells is typically more negative than this by between 5 and 6 kcal/mole because the concentration ratio, $[ADP][P_i]/[ATP]$, is usually much less than 1.

ATP as the Main Carrier of Free Energy in Biochemical Systems

Virtually all living organisms use ATP for transferring free energy between energy-producing and energy-consuming systems. Processes that proceed with large negative changes in free energy, such as the oxidative degradation of carbohydrates or fatty acids, are used to drive the formation of ATP, and the hydrolysis of ATP is used to drive biosynthetic reactions and other processes that require increases in free energy. In the human body, about 2.3 kg of ATP is formed and consumed every day in the course of these reactions.

The Hydrolysis of ATP Yields a Large Amount of Free Energy

Adenosine triphosphate (ATP) can be hydrolyzed in two different ways, as shown in figure 2.7. Hydrolysis of the linkage between the β and γ phosphate groups yields ADP and P_i. Hydrolysis between the α and β phosphates gives adenosine monophosphate (AMP) and pyrophosphate ion ($HP_2O_7^{3-}$). The standard free energy change ($\Delta G^{\circ\prime}$) is about -8.4 kcal/mole in either case. The formation of AMP and pyrophosphate, however, can be pulled forward by hydrolysis of the pyrophosphate to give two equivalents of P_i. This secondary reaction has a $\Delta G^{\circ\prime}$ of about -4.9 kcal/mole, and is catalyzed by pyrophosphatase enzymes present in most types of cells. The overall $\Delta G^{\circ\prime}$ of about -13.3 kcal/mole for breakdown of ATP to AMP and 2 P_i makes this process effectively irreversible ($K_{eq} \approx 1 \times 10^{10}$). Cells commonly use this series of reactions in biosynthetic processes in which a reversal would be intolerable, such as in the synthesis of nucleic acids. The simpler hydrolysis to produce ADP and P_i allows some reversibility but has the advantage that less free energy is needed to resynthesize ATP and ADP than from AMP.

The standard free energies for the hydrolysis of ATP, ADP, or pyrophosphate depend on the pH, on the ionic strength, and also on the concentration of Mg^{2+}, which binds to both the reactants and the products and is required as a cosubstrate by most enzymes that use ATP. At pH 7, the principal ionic forms of ATP, ADP, and P_i have net charges of -4, -3, and -2, respectively

Figure 2.7

Alternative sites of ATP hydrolysis. The charged species shown are the main ones present at physiological pH and ionic strength. The phosphate groups of ATP are referred to as α, β, and γ as indicated. Under physiological conditions, ATP and ADP also bind Mg^{2+} (not shown).

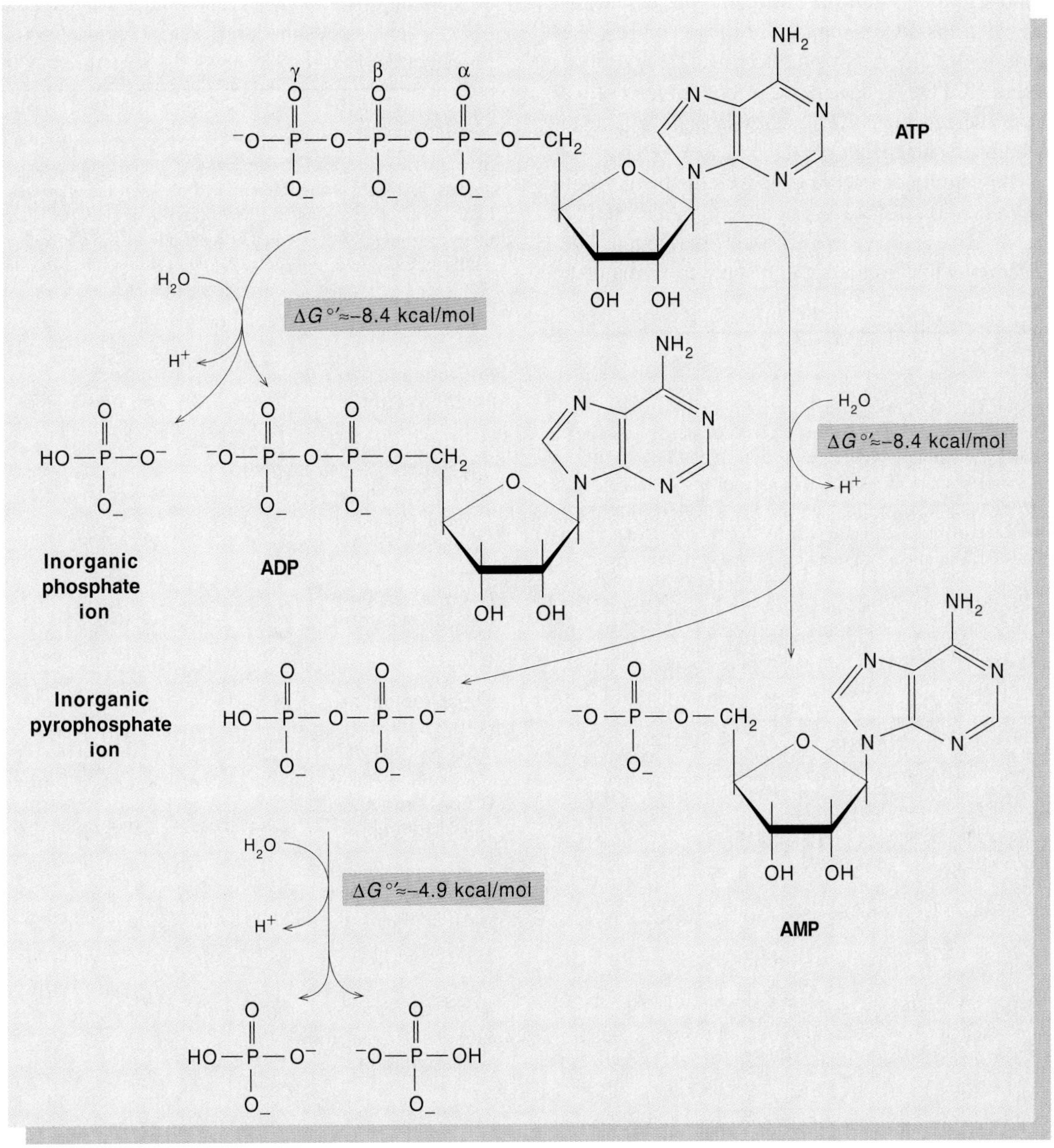

(see fig. 2.7). Because the hydrolysis of ATP^{4-} to give ADP^{3-} + $HPO_3{}^{2-}$ is accompanied by release of a proton, raising the pH makes the hydrolysis more favorable. Increasing the Mg^{2+} concentration makes the hydrolysis less favorable. $\Delta G^{\circ\prime}$ decreases in magnitude from -8.4 kcal/mole in the absence of Mg^{2+} to about -7.7 kcal/mole in the presence of 1 mM Mg^{2+} and to about -7.5 kcal/mole at 10 mM. The value -7.5 kcal/mole is probably close to the $\Delta G^{\circ\prime}$ under typical physiological conditions and will be used elsewhere in this text. But remember that the low ratio of [ADP][P_i] to [ATP] can make ΔG for hydrolysis of ATP in living cells substantially more negative than the $\Delta G^{\circ\prime}$. In resting cells the enzymatic reactions that synthesize ATP usually are more than adequate to keep up with the processes that consume it.

Why is $\Delta G^{\circ\prime}$ for the hydrolysis of ATP so negative? The answer involves several different factors. At pH 7 in the presence of 10 mM Mg^{2+}, $\Delta H^{\circ\prime}$ and $-T\,\Delta S^{\circ\prime}$ for ATP hydrolysis are both negative and the two terms make similar contributions to the overall value of $\Delta G^{\circ\prime}$. The favorable entropy change results partly from the fact that the proton released in the reaction is diluted into a solution with a very low proton concentration. In addition, solvent water molecules probably are less highly ordered around ADP and P_i than they are around ATP. The negative $\Delta H^{\circ\prime}$ results partly from the fact that the negatively charged oxygen atoms in ATP tend to repel each other. The phosphoric anhydride bond in ATP also is weakened by competition between the phosphorus atoms, which both tend to pull electrons away from the bridging oxygen. Another consideration is that the major products of ATP hydrolysis at pH 7 (ADP^{3-} and $HOPO_3^{2-}$) have a larger total number of resonance forms than the reactant ATP^{4-} does (fig. 2.8). The increased resonance lowers the energy of ADP^{3-} and $HOPO_3^{2-}$ relative to ATP^{4-}. The magnitudes of all of these effects vary with the pH, because protonation of the oxygen atoms in ATP, ADP, or P_i relieves some of the repulsive electrostatic interactions, decreases the contributions that some of the resonance forms make to the structure, and decreases the ordering of nearby water molecules. At very low pH, when ATP, ADP, and P_i are all fully protonated, ΔG° probably becomes slightly positive, favoring ATP formation rather than hydrolysis.

In addition to ATP and other nucleoside triphosphates, cells use a variety of other organic phosphate compounds in energy metabolism. These include acetyl phosphate, glycerate-1,3-bisphosphate, phosphoenolpyruvate, phosphocreatine, and phosphoarginine. Figure 2.9 shows the structures of these compounds

Figure 2.8

The phosphate groups of ATP, ADP, and P_i can be written in a variety of resonance forms. This figure shows the major resonance forms of the β phosphate group in ATP and of the same group in ADP. The hydrolysis of ATP to ADP results in an increase in the number of resonance forms available to this group. This increase contributes to the negative $\Delta H^{\circ\prime}$ of hydrolysis of ATP. At physiological pH there are also favorable contributions to $\Delta G^{\circ\prime}$ from the release of a proton and from the disordering of water around the polyphosphate chain.

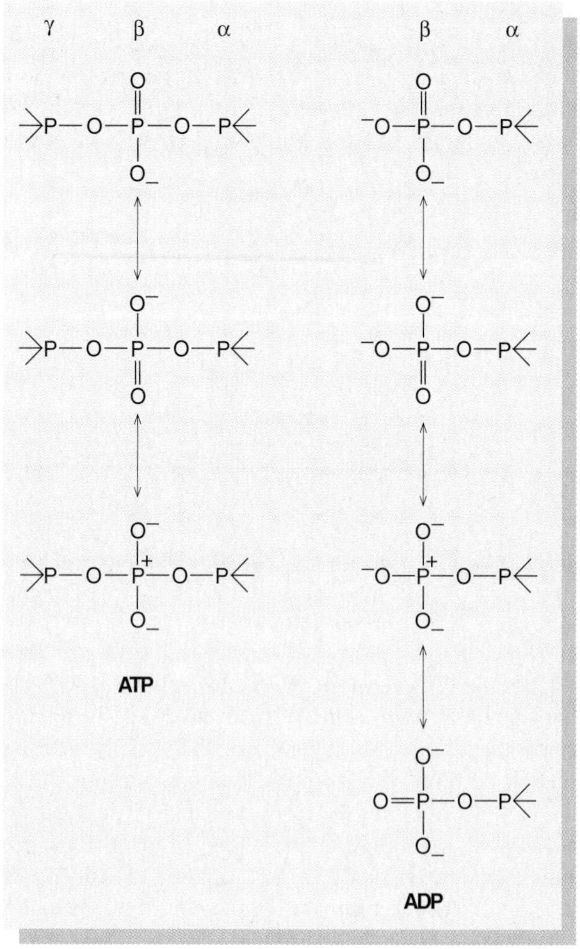

Figure 2.9

Standard free energies of hydrolysis for some common phosphorylated compounds. The $\Delta G^{\circ\prime}$ value refers to hydrolysis of the phosphate group indicated by the symbol (P). Note that ATP occupies an intermediate position among these compounds. Given equal concentrations of the reactants and products, ADP can accept a phosphate group from any of the compounds above it, and ATP can donate a phosphate to the unphosphorylated forms of the compounds below it.

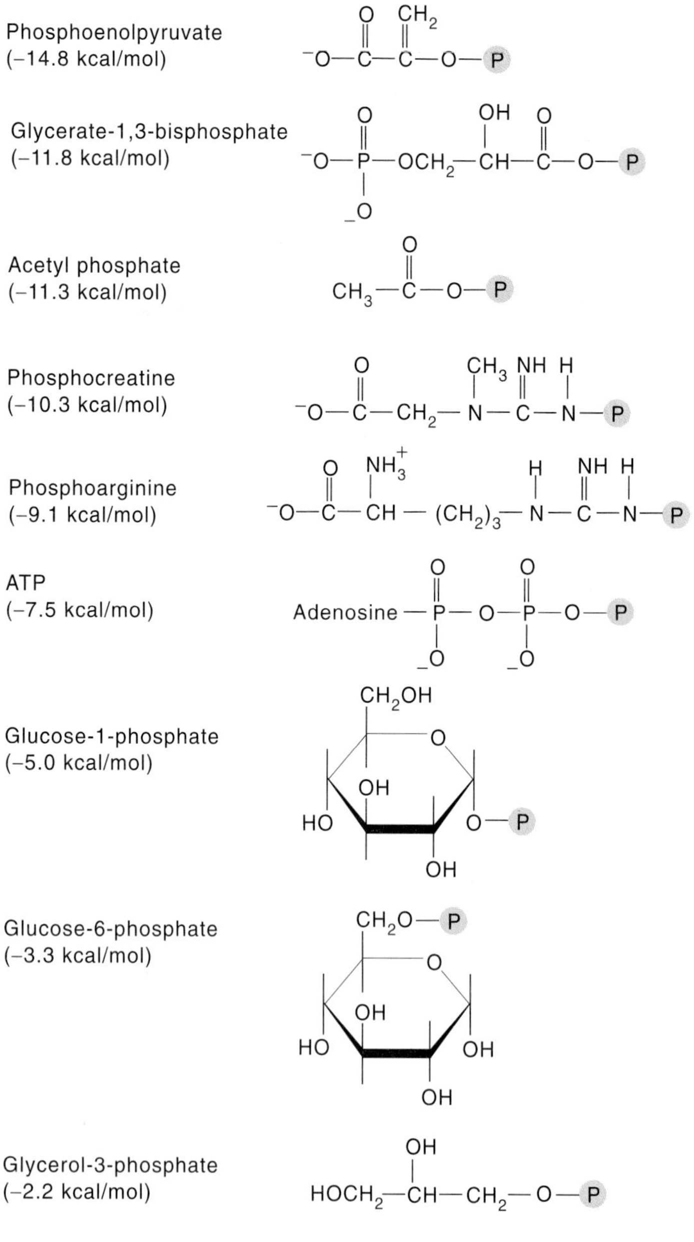

Phosphoenolpyruvate (–14.8 kcal/mol)

Glycerate-1,3-bisphosphate (–11.8 kcal/mol)

Acetyl phosphate (–11.3 kcal/mol)

Phosphocreatine (–10.3 kcal/mol)

Phosphoarginine (–9.1 kcal/mol)

ATP (–7.5 kcal/mol)

Glucose-1-phosphate (–5.0 kcal/mol)

Glucose-6-phosphate (–3.3 kcal/mol)

Glycerol-3-phosphate (–2.2 kcal/mol)

and some simpler phosphate esters. The compounds are ranked in the figure in order of their standard free energies of hydrolysis, with the materials having the most negative values of $\Delta G^{\circ\prime}$ at the top. Given equal concentrations of reactants and products, any compound in the figure could be synthesized, in principle, at the expense of any of the compounds above it. Thus ADP could be phosphorylated to ATP by phosphocreatine, glycerate-1,3-bisphosphate, or phosphoenolpyruvate, and ATP could be used to convert glucose to glucose-6-phosphate or glycerol to glycerol phosphate. The reverse processes will occur only if the ratio of reactants to products is sufficiently high, but this is not necessarily an unusual circumstance. In cells of some tissues, the reaction

$$\text{Creatine} + \text{ATP} \rightleftharpoons \text{creatine phosphate} + \text{ADP} + \text{H}^+ \quad \textbf{(19)}$$

occurs readily in either direction in response to changes in the concentrations of the reactants and products. The same is true of the reaction

$$\text{Glycerate-3-phosphate} + \text{ATP} \rightleftharpoons \text{glycerate-1,3-bisphosphate} + \text{ADP} + \text{H}^+ \quad \textbf{(20)}$$

Most of the phosphorylated materials shown in figure 2.9 participate in only a few biochemical reactions, and many of them are formed only in certain types of cells or organisms. ATP, in contrast, is formed by all living cells, and it participates in literally hundreds of different enzymatic reactions. What is it about ATP that has made this molecule such a universal currency of free energy in biology? Several considerations are relevant here. First, $\Delta G^{\circ\prime}$ for hydrolysis of ATP to ADP is large enough that this reaction releases a substantial amount of free energy, enough to drive many of the reactions that are important for biosynthetic pathways. At the same time, the $\Delta G^{\circ\prime}$ is small enough that ATP itself can be synthesized readily at the expense of available nutrients. We touched on this point above in discussing the relative merits of hydrolysis at the α-β or β-γ positions in ATP.

The second consideration returns us to the distinction between spontaneity and speed. Although the hydrolysis of ATP has a large negative $\Delta G^{\circ\prime}$, it also is critical that ATP is a relatively stable compound in aqueous solution. It does not hydrolyze rapidly under physiological conditions of pH and temperature. The hydrolysis, though far downhill thermodynamically, is slowed by a substantial activation barrier. This activation barrier is discussed in chapter 8; for the present it is sufficient to note that this barrier must be of such a nature that it can easily be overcome enzymatically. This feature allows the free energy of hydrolysis to be channeled quickly and selectively into reactions where it is needed but to be conserved when energy is not in demand.

Third, the products of the hydrolysis of ATP provide opportunities for coupling to a wide variety of chemical reactions. We have seen how the phosphate group is incorporated into glucose-6-phosphate, and later chapters will provide many illustrations of similar enzymatic processes. We also will see reactions in which the pyrophosphate group or the adenylyl group (AMP) is incorporated into the products. Such a broad array of reactions could not be driven by a molecule that decomposed to release an inert material such as N_2.

Finally, the adenine and ribosyl groups of ATP, ADP, and AMP provide additional structural features that allow these molecules to bind to enzymes and thus to participate in regulating enzymatic activities. This may be part of the reason that no known organisms base their energy-transfer reactions entirely on inorganic pyrophosphate or other polyphosphate compounds without a nucleoside moiety.

Summary

In this chapter we have discussed some principles of thermodynamics as they relate to biochemical reactions. The following points are of greatest importance.

1. Thermodynamics is useful in biochemistry for predicting whether a given reaction could occur and, if so, how much work a cell could obtain from the process.
2. The thermodynamic quantities energy, enthalpy, entropy, and free energy are properties of the state of a system. Changes in these quantities depend only on the difference between the initial and final states, not on the mechanism whereby the system goes from one state to the other.
3. Energy is the capacity to do work.
4. The first law of thermodynamics says that energy cannot be created or destroyed in a chemical reaction. If the energy of a system increases, the surroundings must lose an equivalent amount of energy, either by the transfer of heat or by the performance of work.
5. The energy of a molecule includes translational, rotational, and vibrational energy, as well as electronic and nuclear energy. Electronic terms usually account for most of the change in energy, ΔE, in a chemical reaction.
6. The change in enthalpy, ΔH, is given by the expression $\Delta H = \Delta E + \Delta(PV)$. For most biochemical reactions, ΔE and ΔH are nearly equal. The organic molecules found in cells generally have much higher enthalpies than the simpler molecules from which they are built.
7. In most reactions that proceed spontaneously, the enthalpy of the system decreases. If no work is done, the system gives off heat to the surroundings. But in some spontaneous reactions, heat is absorbed in the absence of work and the enthalpy of the system increases. Such reactions invariably show an increase in the entropy of the system.
8. Entropy is a measure of the order in a system: Systems that are highly ordered have low entropies. The entropy of a molecule depends mainly on translational and rotational freedom. Biological macromolecules generally have much lower entropies than their building blocks.
9. The second law of thermodynamics states that there must be an overall increase in the entropy of the system and its surroundings in any process that occurs spontaneously. An isolated system proceeds spontaneously to states of increasingly greater entropy (greater disorder).

10. The change in the free energy of a system is defined as $\Delta G = \Delta H - T\Delta S$, where T is the absolute temperature. A reaction at constant pressure and temperature can occur spontaneously if, and only if, ΔG is negative. The maximal amount of useful work that can be obtained from a reaction is equal to $-\Delta G$.
11. For a reaction in solution, ΔG depends on the standard free energy change ($\Delta G°$) and on the concentrations of the reactants and products. The standard free energy change is related to the equilibrium constant by the expression $\Delta G° = -RT \ln K_{eq}$. Increasing the concentration of the reactants relative to the concentration of the products makes ΔG more negative.
12. Reactions that are thermodynamically unfavorable can be coupled to favorable reactions. The coupling of the reactions may be direct or sequential, as in a biochemical pathway.
13. ATP is the main coupling agent for free energy in living cells. The free energy provided by the hydrolysis of ATP is used to drive many reactions that would not occur spontaneously by themselves.
14. Several features make ATP particularly well suited for its role. First, hydrolysis of ATP to ADP and P_i or to AMP and PP_i releases a considerable amount of free energy. Second, ATP does not hydrolyze rapidly by itself, but it can be hydrolyzed readily in enzymatically catalyzed reactions. This difference allows the free energy of hydrolysis to be channeled into reactions where it is needed, but to be conserved when energy is not in demand. Third, the products of the hydrolysis of ATP provide opportunities for coupling to a wide variety of chemical reactions. Finally, the adenine and ribosyl groups of ATP, ADP, and AMP provide additional structural features that allow these molecules to bind to a large number of enzymes and thus to participate in regulating enzymatic activities.

Selected Readings

Alberty, R. A., and F. Daniels, *Physical Chemistry,* 5th ed. New York: Wiley, 1975.

Cantor, C. R., and P. R. Schimmel, *Biophysical Chemistry.* San Francisco: Freeman, 1980.

Ingraham, L. L., and A. B. Pardee, Free energy and entropy in metabolism. In *Metabolic Pathways,* Vol. 1 (D. M. Greenberg, ed.). New York: Academic Press, 1967.

Tinoco, I., Jr., K. Sauer, and J. C. Wang, *Physical Chemistry, Principles and Applications in Biological Sciences,* 2d ed. Englewood Cliffs, N.J.: Prentice-Hall, 1985.

Van Holde, K. E., *Physical Biochemistry,* 2d ed. Englewood Cliffs, N.J.: Prentice-Hall, 1985.

Problems

1. How do intracellular conditions limit the reactions that can be utilized by nature?
2. What is meant by a state function? Why is enthalpy a state function?
3. What is the basic difference between intensive and extensive thermodynamic parameters?
4. Why can we equate energy and enthalpy for most biochemical reactions?
5. Transfer of a hydrophobic molecule (e.g., a hydrophobic amino acid side chain) from an aqueous to a nonaqueous environment is entropically favorable. Explain.
6. Indicate factors that make ATP ideally suited to transfer energy within cells.
7. As shown in chapter 15, oxaloacetate is formed by the oxidation of malate in the reaction

$$\text{L-Malate} + \text{NAD}^+ \rightleftharpoons \text{Oxaloacetate} + \text{NADH} + \text{H}^+$$

which has a $\Delta G°'$ of +7.0 kcal/mole. How can you explain the fact the reaction proceeds in the direction of oxaloacetate production in the cell?

8. You wish to measure the $\Delta G°'$ for the hydrolysis of ATP:

$$\text{ATP} \longrightarrow \text{ADP} + \text{P}_i$$

but the equilibrium for the hydrolysis lies so far toward products that analysis of the ATP concentration at equilibrium is neither practical nor accurate. However, you have the following data that will allow calculation of the value indirectly:

$$\text{Creatine phosphate} + \text{ADP} \longrightarrow \text{ATP} + \text{creatine} \quad K'_{eq} = 59.5 \qquad \textbf{(P1)}$$

$$\text{Creatine} + \text{P}_i \longrightarrow \text{creatine phosphate} \quad \Delta G°' = +10.5 \text{ kcal/mole} \qquad \textbf{(P2)}$$

Assume that $2.3RT = 1.36$ kcal/mole.
(a) Calculate the value of $\Delta G°'$ for reaction (P1).
(b) Calculate the $\Delta G°'$ for hydrolysis of ATP.

9. The hydrolysis of lactose (D-galactosyl-β-(1,4) D-glucose) to D-galactose and D-glucose occurs with a $\Delta G°'$ of −4.0 kcal/mole.
(a) Calculate K'_{eq} for the hydrolytic reaction.
(b) What are the $\Delta G°'$ and K'_{eq} for the synthesis of lactose from D-galactose and D-glucose?
(c) Lactose is synthesized in the cell from UDP-galactose plus D-glucose and is catalyzed by lactose synthase. Given that $\Delta G°'$ of hydrolysis of UDP-galactose is −7.3 kcal/mole, calculate for $\Delta G°'$ and K'_{eq} for the reaction

$$\text{UDP-Galactose} + \text{D-Glucose} \longrightarrow \text{Lactose} + \text{UDP}$$

10. For each of the following reactions, calculate $\Delta G°'$ and indicate whether the reaction is thermodynamically favorable as written.
(a) Glycerate 1,3-bisphosphate + Creatine → Phosphocreatine + 3-Phosphoglycerate
(b) Glucose-6-phosphate → Glucose-1-phosphate

(c) Phosphoenolpyruvate + ADP → Pyruvate + ATP
(d) Glycerol phosphate + ADP → Glycerol + ATP

11. Although ATP is an important phosphate donor, in biosynthetic reactions the AMP portion of the molecule is often transferred to an acceptor with the release of pyrophosphate. Such a transfer occurs as an intermediate step in the reaction

$$\text{R—COO}^- + \text{ATP} + \text{CoASH} \longrightarrow \text{R—CO—SCoA} + \text{AMP} + \text{Pyrophosphate} \quad \textbf{(P3)}$$

Inorganic pyrophosphatases hydrolyze the pyrophosphate, yielding two molecules of inorganic phosphate. Assume that the hydrolysis of R—CO—SCoA to R—COO$^-$ + CoASH proceeds with $\Delta G^{\circ\prime}$ of -10 kcal/mole and that hydrolysis of a pyrophosphate anhydride bond yields -7.5 kcal/mole. Calculate the K'_{eq} for reaction (P3) in the presence and in the absence of inorganic pyrophosphatase. What role do pyrophosphatases play in biosynthetic reactions dependent on adenylate transfer?

12. Assume that an individual needs 2,500 Calories per day (1 Calorie = 1,000 calories, or 1 kcal) to meet energy requirements. For simplicity, consider that the energy needs of this individual are met with glucose (not that unrealistic, considering that most of the world meets these needs mainly with starches). Glucose, and carbohydrates in general, contain about 4 Calories per gram. The ATP yield during catabolism is 30 moles of ATP per mole of glucose. What mass (in pounds) of K_4ATP are synthesized per day by this individual?

13. Proteins that serve as gene repressors frequently have two identical binding sites that interact with complementary sites on the DNA. If a repressor protein is cut in half without damaging either of its DNA binding sites, how is the binding to DNA affected? How is the binding affected if one of the two binding sites on the DNA is eliminated by changing the nucleotide sequence? Discuss the enthalpy, entropy, and free energy effects.

14. Notice the intermediate in the reaction involved in the formation of citrate from acetyl-CoA and oxaloacetate (equation 17). Metabolic reactions are known in which the energy content of the thioester bond in an acyl-CoA is converted into an "ATP" (e.g., see succinoyl-CoA synthetase in chapter 15). What would the consequences be if in part 2 of reaction 17, instead of a hydrolysis of the citryl-CoA the following reaction occurred:

$$\text{Citryl-CoA} + \text{ADP} + \text{P}_i \longrightarrow \text{ATP} + \text{Citrate} + \text{CoASH?}$$

PART 2

PROTEIN STRUCTURE AND FUNCTION

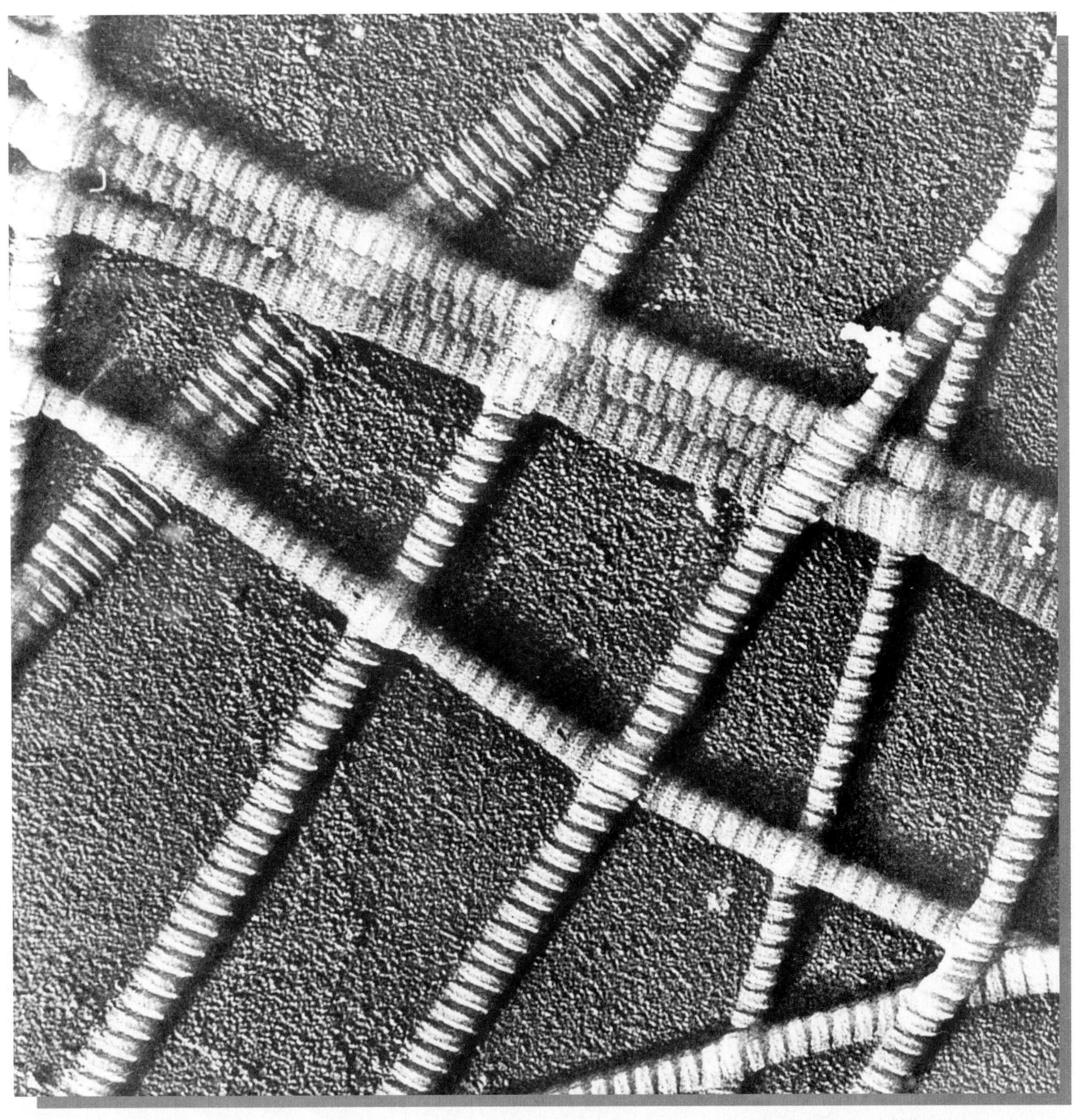

An electron micrograph of collagen fibrils from skin. The banded appearance of the fibrils arises from a staggered arrangement of collagen molecules. Each collagen molecule in turn is composed of three polypeptide chains that interact by hydrogen bonds. (Courtesy of Jerome Gross, Massachusetts General Hospital.)

C H A P T E R

3

The Structure and Function of Water

Water is the single most important factor necessary for life's origin and maintenance.

Despite the fact that the earth is only 0.5% by weight water, water has had a profound effect on the earth's climate and surface chemistry. Because of its low density, most of the water has risen to the earth's surface, where it is subject to both freezing and evaporation. As a result, over half of the earth's surface is covered by water. About 80% of this water is in the liquid state, and most of the remainder is in the solid state (ice). Liquid water has the most unusual property of expanding when it freezes, so newly formed ice floats on the surface of the liquid water (table 3.1). This provides an insulating layer so that more extensive freezing is greatly reduced. The high heats of fusion and vaporization tend to stabilize the earth's temperature between 0 and 100°C. We may think of liquid water as a "buffer" that helps to keep the earth's surface temperature in the mild range.

Although less than 1% of the water is in the gaseous phase, this water also has a significant effect on the earth's surface temperature. Rainfall, which originates from water-laden clouds, revitalizes the land areas and creates havens of fresh water, which are essential for many species of plants, animals, and microorganisms. Daytime clouds reflect a major portion of the solar radiation, which otherwise might heat up the surface unduly. Nighttime cloud cover prevents excessive heat loss at night by trapping the far-red radiation emitted by the earth.

The polar character of water encourages the solubilization of other polar molecules. Furthermore, the liquid state of water leads to mobilization of dissolved molecules; this in turn facilitates their interaction with one another. Water is such an ideal solvent for living forms that most evolutionists think that the presence of considerable surface water is the most important single factor necessary for life's origin and maintenance. We should remember that water is the dominant component of living cells (see table 1.1). In this chapter we will consider the properties of water in terms of how it affects the properties of other biomolecules that are water-soluble.

Liquid Water and Ice Have Very Similar Structures

A water molecule has a significant dipole that is due to the greater electronegativity of the oxygen atom over the hydrogen atoms (fig. 3.1). This dipole leads to strong associations between water molecules called hydrogen bonds, a noncovalent interaction between two polar molecules in which one of the interacting molecules

Table 3.1
Some Physical Properties of Water

Molecular weight	18.0 g/mole
Density of water at 4°C	1.00 g/cm^3
Density of ice	0.917 g/cm^3
Boiling point	100°C
Freezing point	0°C
Heat of fusion	80 cal/g
Heat of vaporization	540 cal/g

Figure 3.1

The structure of water and the interaction of water with other water molecules.

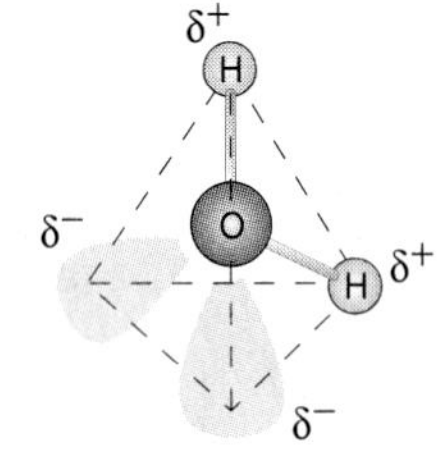

(a) Single water molecule

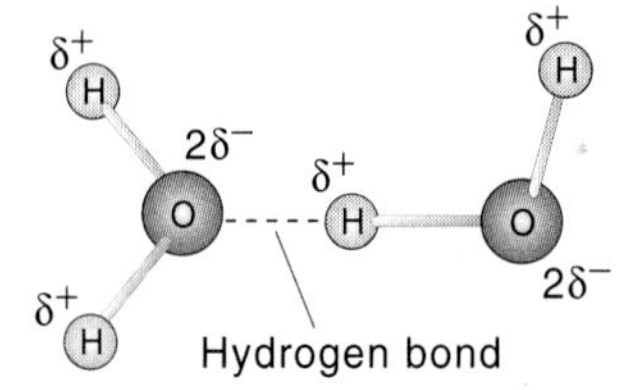

(b) Two interacting water molecules

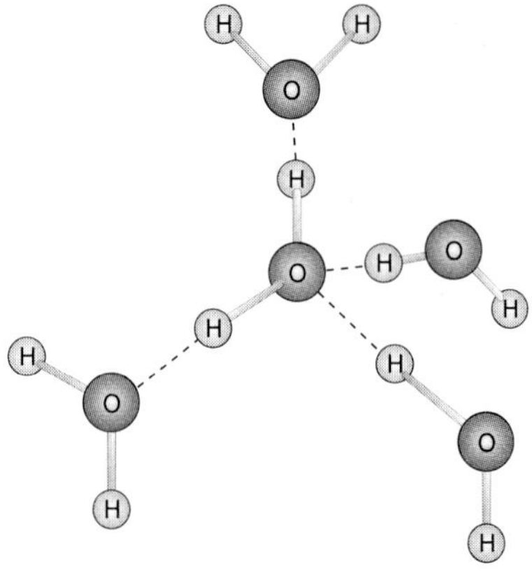

(c) Cluster of interacting water molecules

Figure 3.2

The arrangement of molecules in an ice crystal. Water molecules are oriented so that one proton along each oxygen–oxygen axis is closer to one or the other of the two oxygen atoms.

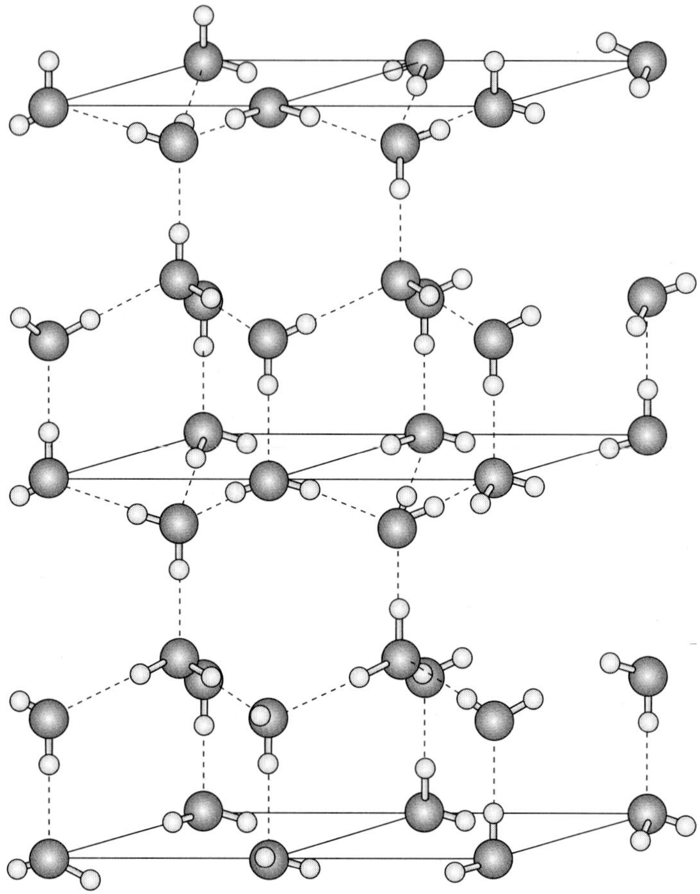

contains an unshielded proton. In ice these hydrogen bonds hold the individual molecules together in a regular three-dimensional lattice (fig. 3.2). Most of the hydrogen bonds present in ice are also present in liquid water. Hence liquid water is highly hydrogen-bonded and not too different from ice. The anomalously high electrophoretic mobility of protons in liquid water and ice are a reflection of this similar structure (box 3A). While ice and liquid water have similar structures, the less regular structure of liquid water results in a much greater mobility of molecules that are dissolved in liquid water.

The dipolar properties of water molecules also affect the interaction between water and other molecules that dissolve in water. For example, a favorable electrostatic interaction accounts for the high solubility of sodium chloride in water (fig. 3.3). The kinds of ion-dipole interactions that take place between water and simple ions such as Na^+ and Cl^- are also important in the interactions between biomolecules and water. Thus biomolecules that contain charged residues, hydrogen-bond-forming substituents, or other kinds of polar groups tend to be hydrophilic.

Nonpolar groups such as neutral hydrocarbons do not contain significant dipoles or the capacity to form hydrogen bonds.

3A BOX

The High Mobility of Protons in Water Is Due to the Hydrogen-Bonded Structure of Water

Protons migrate 4 to 5 times faster through liquid water than Na^+ or K^+ ions do. This is due to the hydrogen-bonded structure of water, which permits proton jumping from one water molecule to the next. In fact proton movement is not due to any one proton moving through the water. Instead, there is free exchange between protons and covalently linked hydrogen atoms in neutral water molecules, as suggested by the figure. Protons also have an anomalously high mobility in ice for the same reason.

$$H^+ \cdots \underset{|}{\overset{H}{}}O-H\cdots \overset{H}{O}-H\cdots \overset{H}{O}-H\cdots \overset{H}{O}-H\cdots \overset{H}{O}-H \quad H^+$$

Figure 3.3

The water molecule is composed of two hydrogen atoms covalently bonded to an oxygen atom with tetrahedral (sp^3) electron orbital hybridization. As a result, two lobes of the oxygen sp^3 orbital contain pairs of unshared electrons, giving rise to a dipole in the molecule as a whole. The presence of an electric dipole in the water molecule allows it to solvate charged ions because the water dipoles can orient to form energetically favorable electrostatic interactions with charged ions.

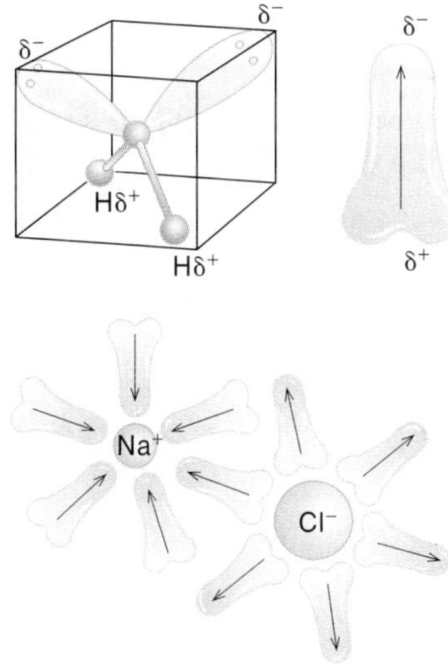

Figure 3.4

Clathrate structures are ordered cages of water molecules around hydrocarbon chains. A portion of the cage structure of $(nC_4H_9)_3S^+F^- \cdot 23\ H_2O$ is shown. The trialkyl sulfur ion nests within the hydrogen-bonded framework of water molecules. In the intact framework, each oxygen is tetrahedrally coordinated to four others. One such oxygen atom and its associated hydrogens are shown by the arrow. (Illustration copyright by Irving Geis. Reprinted by permission.)

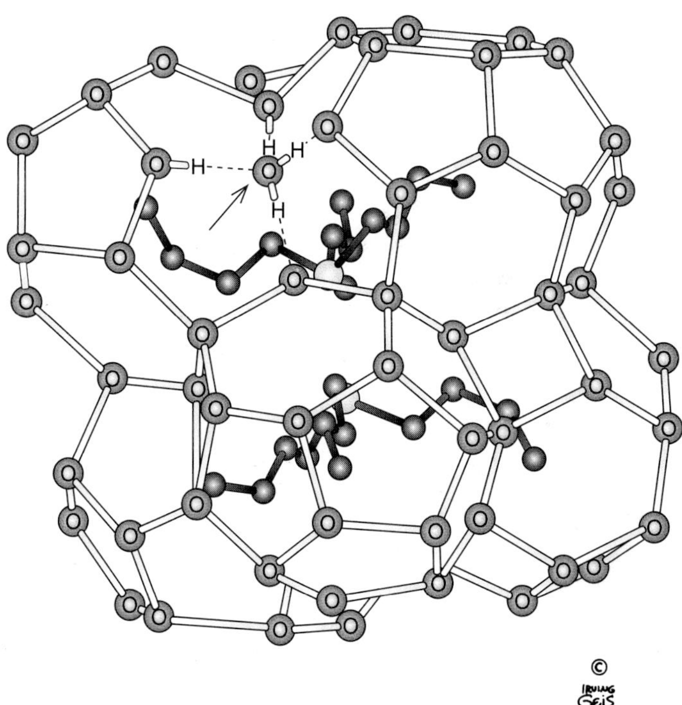

Consequently, they interact poorly with water, as evidenced by their low water solubility. When such hydrophobic molecules are present in water, the water forms a rigid clathrate (cagelike) structure around them (fig. 3.4).

This is a bird's eye view of water as a solvent. A more rigorous analysis of water as a solvent requires that we scrutinize the forces the influence water interactions in greater depth.

A Variety of Forces Affect the Interactions between Biomolecules and Water

Covalent bond energies, which account for the affinities between atoms within the same molecule, range from 30 to 230 kcal/mole between most atoms of biological interest (table 3.2). By contrast, most intermolecular bond energies between noncovalently linked atoms range in value from 0.1 to 10 kcal/mole. To appreciate the interactions that take place between water and other molecules, we must have a fuller accounting of the different intermolecular forces that prevail. The intermolecular forces that are relevant may be grouped into four categories.

Electrostatic Forces Favor Interaction between Water, Charged Molecules, and Polar Molecules

Electrostatic forces account for the high solubility of charged molecules and polar molecules in water. Each type of interaction shows a different functional dependence on distance between the interacting groups (table 3.3). For example, the energy of interaction between two charges, Q_1 and Q_2, is proportional to the product of the charges and inversely proportional to the distance R between them:

$$\text{energy of interaction} \propto \frac{Q_1Q_2}{R} \quad \textbf{(1)}$$

In solution this interaction is reduced by the dielectric constant of the medium in which the charged molecules are situated:

$$\text{energy of interaction} \propto \frac{Q_1Q_2}{\epsilon R} \quad \textbf{(2)}$$

where the dielectric constant ϵ is a dimensionless factor. The dielectric constant of a substance is greater for molecules with a high dipole moment. Not surprisingly, water has a high dielectric constant of about 80 (see table 3.1), whereas typical nonpolar organics have dielectric constants in the range 1–10. The high dielectric constant of water has the effect of screening two charge species from one another and together with the favorable ion–dipole interactions accounts for why salts like sodium chloride dissolve so readily in water.

If two charges are relatively buried within an organic molecule such as a protein, their interaction energy may be substantially increased because the dielectric constant in the regions inaccessible to water are much lower than the dielectric constant of water. If the two charges are both positive or negative, this effect can create intolerably high repulsive energies that may lead to an alteration in the structure of a molecule.

Table 3.3
Range of Some Intermolecular Interactions Expressed as the Power of the Intermolecular Separation

Range of Interaction	Type of Interaction
$1/R$	Charge–charge
$1/R^2$	Charge–dipole
$1/R^3$	Dipole–dipole
$1/R^6$	Van der Waals (dipole-induced dipole) attractive forces
$1/R^{12}$	Van der Waals repulsive forces

Van der Waals Forces Are of Two Types

The term van der Waals forces refers to two types of interactions, one attractive and one repulsive, each showing a different functional dependence on the distance between the interacting groups (see table 3.3). The attractive forces are due to favorable interactions among the induced instantaneous dipole moments that arise from fluctuations in the electron charge densities of neighboring nonbonded atoms. Such forces are small, yielding energies of 0.1 to 0.2 kcal/mole, but they add up as the number of interactions between two molecules increases. That is why a typical hy-

Table 3.2
Bond Energies between Some Atoms of Biological Interest

Energy Values for Single Bonds (kcal/mole)					
C—C	82	C—H	99	S—H	81
O—O	34	N—H	94	C—N	70
S—S	51	O—H	110	C—O	84
Energy Values for Multiple Bonds (kcal/mole)					
C=C	147	C=N	147	C=S	108
O=O	96	C=O	164	N≡N	226

Figure 3.5

Plot of the van der Waals energy of interaction as a function of the distance between two nonbonded atoms (black curve). Van der Waals attractive forces (red curve) are significant over greater distances than van der Waals repulsive forces (blue curve). At short distances the van der Waals repulsive forces become overwhelming. The distance of closest approach between two nonbonded atoms (A) defines the van der Waals radii. The most favorable interaction distance (B) is somewhat greater than this.

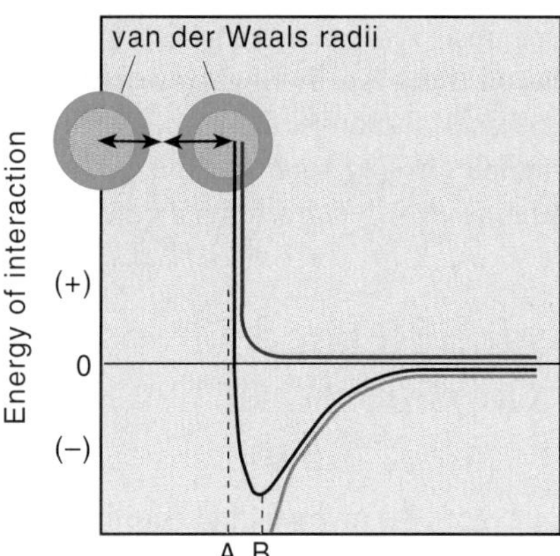

Table 3.4
Radii for Covalently Bonded and Nonbonded Atoms

Element	Single Bond	Double Bond	Triple Bond
Covalent bond radii (in Å)			
Hydrogen	0.30		
Carbon	0.77	0.67	0.60
Nitrogen	0.70		
Oxygen	0.66		
Phosphorus	1.10		
Sulfur	1.04		
Van der Waals radii (in Å)			
Hydrogen	1.2		
Carbon	2.0		
Nitrogen	1.5		
Oxygen	1.4		
Phosphorus	1.9		
Sulfur	1.8		

drophobic molecule like hexane exists primarily as a liquid rather than a gas at room temperature. This is the situation for pure hexane. When we consider hexane–hexane interactions in water, the situation changes. In aqueous solution, van der Waals attractive forces between the hydrophobic groups are far less important in determining the type of intermolecular structures formed. In fact, the interaction of hydrophobic molecules like hexane with water has a favorable enthalpy over the interaction of hexane with itself. Clearly, other factors must determine why liquid hexane and water are virtually immiscible (see the discussion of hydrophobic forces later in this part of the chapter).

Repulsive van der Waals forces arise when noncovalently bonded atoms or molecules come very close together. An electron–electron repulsion arises when the charge clouds between two molecules begin to overlap. If two molecules are held together exclusively by van der Waals forces, their average separation will be governed by a balance between the van der Waals attractive and repulsive forces (fig. 3.5). The distance is known as the van der Waals separation. Some van der Waals radii for biologically important atoms are given in table 3.4. The van der Waals separation between two nonbonded atoms is given by the sum of their respective van der Waals radii.

Hydrogen Bond Forces Involve Interactions with Unshielded Protons

We have already seen that hydrogen bonds are the main type of linkage that is formed between water molecules in liquid water and ice. Hydrogen bonds also account for a great number of the interactions that take place between biomolecules or between biomolecules and water. As already indicated, the strength of the hydrogen bond is due to the partially unshielded nature of the single proton that makes up the hydrogen nucleus. An attractive interaction exists between the lone pair of electrons on either a nitrogen atom or an oxygen atom and the hydrogen atom in either an N—H or an O—H chemical bond. The attraction is usually directed along the lone-pair orbital axis of the H-bond acceptor group. Evidence for a H bond comes from the observation of a decreased distance between the donor and acceptor groups making up the H bond. Thus from the van der Waals radii given in table 3.4 we can calculate that the distances between nonbonded H and O atoms and between nonbonded H and N atoms are 2.6 and 2.7 Å, respectively. When an H bond is present, this distance is usually reduced to 1.8 or 1.9 Å. Some of the more important H-bond donors and acceptors are shown in figure 3.6. The strength of an H bond is highly dependent on the orientation of the interacting groups. As a rule the angle between the N or O acceptor and the N—H or O—H donor is close to 180°.

Polypeptides and polynucleotides carry a large number of H-bond donor and acceptor groups. Formation of the maximum number of H bonds between either a polypeptide chain or a polynucleotide chain and water would obviously require the complete unfolding of the polypeptide or the polynucleotide chain. However, it is not obvious that such an unfolding would result in a net energy gain. One reason is that water is a highly H-bonded structure, and for every H bond that is formed between water and the biopolymer, an H bond within the water structure itself must be broken. As a rule these biopolymers form regular intermolecular or intramolecular H bonds with their own kind keeping a few H-bond-forming side chains available for interaction with water (see chapter 5 for examples of this).

Figure 3.6

Major hydrogen-bond donor and acceptor groups found in proteins. Note that the angle between the O acceptor and the N—H donor is 180° in the hydrogen-bond complex.

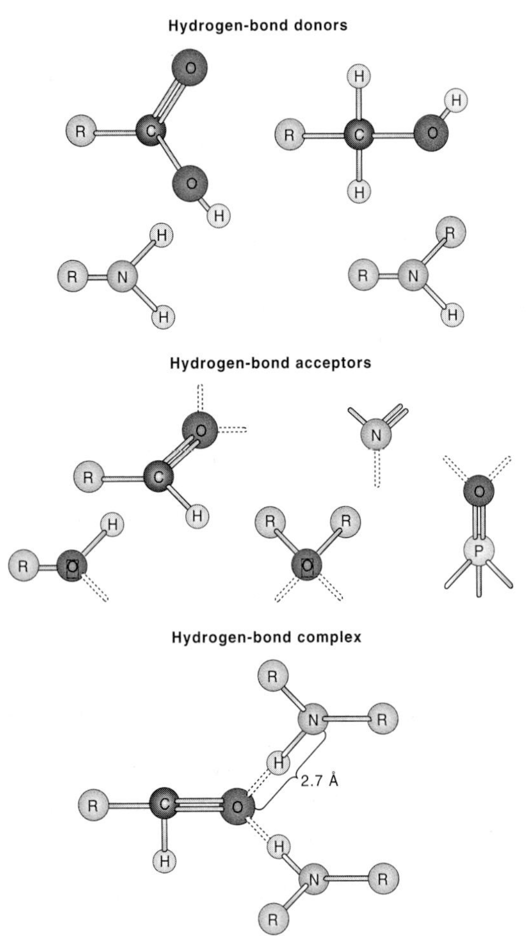

Hydrophobic Forces Are Primarily Due to Entropic Factors

Hydrophobic forces are the hardest type of interactions to appreciate. In fact the term "hydrophobic forces" is somewhat misleading. You will see why shortly. Perhaps "hydrophobic effects" would be a better term. In any event, while the previous forces discussed in this section are due primarily to enthalpic factors, hydrophobic forces relate primarily to entropic factors. Furthermore, the entropic factors mainly concern the solvent water, not the solute.

It was stated earlier that hexane has a small but favorable enthalpy for solution in water but that other factors were sufficiently unfavorable to make hexane highly insoluble in water. What are these mysterious factors? Owing to the weak enthalpic interactions between a hydrophobic molecule such as hexane and water, the water withdraws in a region surrounding the hydrophobic molecule and forms a relatively rigid H-bonded network with itself (for example, see the clathrate structure illustrated in fig. 3.4). This network effectively restricts the number of possible orientations of the water molecules forming the water–hydrophobic group interface. Now, virtually all chemical systems tend spontaneously toward a state of maximum disorder. When hydrophobic groups are exposed to water, however, some extent of ordering is introduced into the surrounding water. This energetically unfavorable entropic effect is greater than the small, favorable enthalpic effect. While one observes hydrophobic groups clustering together in aqueous solution and it is tempting to attribute this effect mainly to the van der Waals attractive forces between the hydrophobic groups, the principle driving force for this arrangement is the entropic effect relating to solvent arrangement.

Solubility and Related Phenomena Are Best Considered in Thermodynamic Terms

Solubility is an event that involves the equilibrium of a molecule between the pure state and the dispersed solution state. While equilibrium events are directly related to free energy differences between two states, the best way to get a physical picture of the factors involved for a given reaction is to consider the enthalpy and entropy effects separately. Although this approach usually does not lead to quantitative predictions of equilibrium constants or free energy changes for specific equilibrium, it does lead to a qualitative picture of the events involved. The discussion of thermodynamics in chapter 2 as well as the preceding discussion of intermolecular forces will aid us in this discussion.

In the interest of simplicity I will focus the discussion on three examples, all dealing with small molecules: a simple 1:1 salt, NaCl; an organic molecule containing polar groups, glucose; and an organic molecule that contains no polar groups, hexane.

We have already discussed some aspects of the solubility of these molecules. In the case of NaCl we saw that an ionic crystal dissolves to form hydrated ions in aqueous solution. Most 1:1 inorganic salts are quite soluble in water, so we know that the free energies of solution are often highly negative. The enthalpic factor(s) favoring the solution state result from the strong ion–dipole complexes formed between the salt ions and dipolar water (see fig. 3.3). The enthalpic factors favoring the crystalline state for NaCl result from the electrostatic interactions between the ions within the crystal. Not all 1:1 salts show a negative enthalpy (favorable) for solution. Thus in the series LiCl, NaCl, and NH_4Cl, LiCl produces a great deal of heat on dissolving. NaCl produces very little heat, and NH_4Cl produces a cooling effect. Insofar as the heat release is a measure of the enthalpy change, it would appear that enthalpy favors LiCl solubilization and opposes NH_4Cl solubilization. Very likely, that is related to the size of the cations. The smaller cation Li^+ should make the strongest ion–dipole bonds with the oxygens of water because of the shorter interatomic distances between noncovalently bonded atoms. Recall that the energy of an ion–dipole interaction is inversely proportional to the square of the distance between the interacting species (see table 3.3).

Entropic factors for salt solubilization also present a complex picture. For the salt, entropic effects favor solubilization because of the greater disorder of the salt in the solution state. For the solute this is not the case. Water molecules that form ion–dipole bonds with the salt ions have a decreased entropy because of their reduced mobility. It is noteworthy that enthalpies and entropies of solvation usually tend to oppose each other. For charged species the more negative (favorable) the enthalpy of solvation, the more negative (unfavorable) the entropy of solvation. While most inorganic salts tend to be highly soluble in water, this is by no means true of all salts. Unfortunately, it is much easier to find an explanation for the solubility properties of a particular compound than to predict them beforehand.

From an inspection of the glucose structure we see that it contains numerous polar hydroxyl groups. Since most polar molecules tend to be water soluble, we would not be surprised to find that glucose is highly soluble, which it is. However, if we are to be rigorous, the energy of glucose in the pure state should be carefully considered in a situation where we are evaluating the equilibrium between glucose in the solid state versus glucose in the solution state. In this regard it is of interest that both glycogen and cellulose are polyglucose molecules. But due to a different stereochemistry of the glycosidic linkages, cellulose forms a more stable structure in the pure state. As a result we find that glycogen is much more soluble than cellulose.

In the case of hexane we have already mentioned that the enthalpy of solution is negative (favorable). For hexane the entropy of solution must be positive (favorable) because of the dilution effect (see equation (7) in chapter 2). However, the entropy change for the solute water is negative (unfavorable). Since hexane is poorly soluble in water, it would appear that the unfavorable entropy effect dominates in this situation.

It might seem that this expanded coverage of simple molecule solubilities is far afield from considerations of biomolecule interactions with water. In fact it is highly relevant even when considering the solution behavior of biopolymers.

When considering biopolymers in aqueous solution, we are not so concerned about the equilibrium between the pure biopolymer and the dissolved biopolymer. The equilibrium we are concerned with is that involving the three-dimensional arrangement of the molecule. This is referred to as the conformation of the molecule. For a polypeptide chain we might consider the equilibrium between a completely extended chain and a highly compact folded state; for DNA we might consider the equilibrium between an extended polynucleotide chain and a double helix, and for a fatty acid chain we might consider the equilibrium between a free fatty acid chain and a highly ordered bilayer structure.

Starting with fatty acids, which are amphipathic, we see that the charged ends of the fatty acids interact favorably with water in either the free or aggregated state. The long nonpolar chains of the fatty acids tend to interact with each other to form micelles or bilayer vesicles (see fig. 19.7). By analogy with the hexane situation, it seems likely that the aggregated state is favored because of the entropic advantage of not having water in contact with these hydrophobic regions.

The situation with globular proteins shows strong parallels (see fig. 1.8). The hydrophobic groups tend to bury themselves within the protein structure mainly because of the entropic advantage, while the polar side chains of the protein are usually found on the surface because of the enthalpic advantage of direct contacts with water.

Polynucleotides have a strong tendency to form folded structures because the purine and pyrimidine bases they contain can form hydrogen bonds with each other (see figs. 1.9 and 30.8). The interactions between purines and pyrimidines are also favored by the face-to-face interactions between adjacent purine–pyrimidine base pairs. This is partly due to favorable dipole–dipole interactions and partly due to their hydrophobic character. In these folded structures the polar sugar phosphate backbones of the polynucleotide chains are exposed where they can form H bonds and electrostatic bonds with water.

The Hydrogen Ion Concentration Has a Major Impact on Biomolecular Reactions

Hydrogen ion concentration is usually expressed in terms of pH, the negative logarithm of the hydrogen ion concentration

$$\mathrm{pH} = \log\left[\frac{1}{\mathrm{H}^+}\right] = -\log[\mathrm{H}^+] \qquad \textbf{(3)}$$

The term "pH" is used for convenience, to avoid the need of writing exponentials when discussing the H^+ concentration. The pH is carefully regulated in the different body compartments. Any breakdown in this regulation leads to serious metabolic disturbances. The pH of the blood is normally kept within very narrow limits between 7.35 and 7.45. Outside of this very narrow range the organism cannot function properly. The pH of the cytosol of most cells is approximately 7.4. However, in the lysosomal organelles the pH is around 5.0. This is the pH at which the degradative enzymes of the lysosome function best. If lysosomal enzymes were to escape by accident into the cytosol, they would not wreak havoc on unsuspecting targets because these degradative enzymes are inactive at the cytosolic pH. In the gastric juices the pH is very low, below 2. The protease pepsin, which degrades ingested proteins to large peptide fragments, functions optimally at this pH. In fact pepsin is completely inactive at neutral pHs.

In the following sections we will discuss the way in which weak acids can modulate the pH and how this is used to keep the pH relatively constant in the cytosol and the plasma.

The Hydrogen and Hydroxide Ion Concentrations in Liquid Water Are Reciprocally Related

Water dissociates to a very small extent into protons and hydroxide ions, according to the following reaction:

$$\mathrm{H_2O} \rightleftharpoons \mathrm{H^+} + \mathrm{OH^-} \qquad \textbf{(4)}$$

Incidentally, protons never exist in the free state in water; they always associate with a water molecule to form a hydronium ion, H_3O^+. For convenience we will usually indicate hydronium ions by the symbol for protons.

The equilibrium expression for equation (4) is

$$K_{eq} = \frac{[H^+][OH^-]}{[H_2O]} \tag{5}$$

Because water dissociates to such a small extent, the concentration of undissociated water is high and does not vary significantly for chemical reactions in aqueous solution. Therefore, the denominator in equation (5) is effectively constant with a value of 55.5. This has led to the widespread use of the constant K_w for the dissociation of water defined by the expression

$$K_w = [H^+][OH^-] = 10^{-14}(\text{mol/l})^2 \tag{6}$$

at 25°C.

In pure water we expect equal amounts of H^+ (hydrogen ion) and OH^- (hydroxide ion). From equation (6) we can see that the concentration of H^+ or OH^- in pure water is 10^{-7}M (pH 7). A solution with a pH of 7 is referred to as a neutral solution because it contains equal numbers of hydronium and hydroxide ions. A pH lower than 7 indicates an acidic solution, and a pH greater than 7 indicates a basic solution. Values for pH and the corresponding H^+ and OH^- concentrations are given in table 3.5.

The Extent of Ionization of a Weak Acid in Water Is a Function of Its Acid Dissociation Constant, K_a

The most common equilibria that biochemists encounter are those of acids and bases. The dissociation of an acid may be written as

$$HA \rightleftharpoons H^+ + A^- \tag{7}$$

Table 3.5
The pH Scale

pH	$[H^+]$	$[OH^-]$
0	10^{0}	10^{-14}
1	10^{-1}	10^{-13}
2	10^{-2}	10^{-12}
3	10^{-3}	10^{-11}
4	10^{-4}	10^{-10}
5	10^{-5}	10^{-9}
6	10^{-6}	10^{-8}
7	10^{-7}	10^{-7}
8	10^{-8}	10^{-6}
9	10^{-9}	10^{-5}
10	10^{-10}	10^{-4}
11	10^{-11}	10^{-3}
12	10^{-12}	10^{-2}
13	10^{-13}	10^{-1}
14	10^{-14}	10^{0}

The equilibrium constant for this reaction is called the acid dissociation constant K_a and is written as

$$K_a = \frac{[H^+][A^-]}{[HA]} \tag{8}$$

Strong acids in aqueous solution dissociate completely into anions and protons. The concentration of hydrogen ion $[H^+]$ is therefore equal to the total concentration of C_{HA} of the acid HA. Thus the pH of the solution of a strong acid is simply $-\log C_{HA}$.

The pH of the solution of a weak acid, that is, an acid that does not dissociate completely, is a function of both the total acid concentration and the acid dissociation constant. The dissociation constant of a weak acid may be written in terms of the species present in the equation for acid dissociation.

First solving equation (8) for $[H^+]$ gives

$$[H^+] = \frac{K_a[HA]}{[A^-]} \tag{9}$$

Taking the logarithm of both sides and changing signs give us

$$-\log [H^+] = -\log K_a + \log \frac{[A^-]}{[HA]} \tag{10}$$

or

$$\text{pH} = \text{p}K_a + \log \frac{[\text{base}]}{[\text{acid}]} \tag{11}$$

This is known as the Henderson-Hasselbalch equation. It is useful for calculating the molar ratio of base (proton acceptor) to acid (proton donor) to achieve a given pH. It could also be used to calculate the pK_a if the pH and the ratio of bases to acid can be determined (see box 3B). From this equation it can be seen that when the concentration of anion or base is equal to the concentration of undissociated acid (i.e., when the acid is half neutralized), the pH of the solution is equal to the pK_a of the acid.

As a rule the pK_a for a weak acid such as acetic acid is determined by a titration in which the pH is measured as a function of added base as shown in figure 3.7. In this titration curve the pK_a is the point where the slope is a minimum. At this point the acid is half titrated, so the solution contains equal amounts of acetic acid and acetate. This is also an inflection point as the slope, which has been decreasing up to this point, begins increasing as the titration continues.

A simple amino acid with a nonionizable side chain gives a complex titration curve because it contains two titratable groups (fig. 3.8). At very low pH, alanine carries a single positive charge on its amino group. The first inflection point occurs at a pH of 2.3. This is the pK_a for titration of the carboxyl group. At a pH of 6.0, alanine has equal amounts of positive and negative charge. This value is referred to as the isoelectric point (pI) or the isoelectric pH. As the titration continues, a second inflection point is reached at a pH of 9.7. At this pH the amounts of uncharged and charged amino groups are equal.

More complex situations where amino acids contain charged side chains show three inflection points in their titration curves (fig. 3.9).

3B

BOX

Different Applications of the Henderson-Hasselbalch Equation

The Henderson-Hasselbalch equation relates the pH to the pK and the ratio of dissociated to undissociated acid. Accordingly, this equation can be used to calculate the pK, the pH, or the ratios of acid to base for a buffer if essential information is given.

1. If the concentration of acetic acid is 0.5, the concentration of acetate is 0.5, and the pH is 4.76, calculate the pK_a using the Henderson-Hasselbalch equation.
   Ans. Starting from the Henderson-Hasselbalch equation, which states that pH = pK_a + log $\frac{[\text{base}]}{[\text{acid}]}$, if the ratio of base/acid = 1, then the log term vanishes, and pH = pK_a. The pK_a must be 4.76 from the given pH.
2. Calculate the pH for acetic acid–acetate buffer system at the following levels of acetate and acetic acid: 0.1/0.9, 0.2/0.8, 0.3/0.7, 0.4/0.6, 0.6/0.4, 0.7/0.3, 0.8/0.2, and 0.9/0.1. Plot these values to see whether they agree with the values shown in figure 3.7.
   Ans. This problem is solved by using the pK_a determined above and the indicated ratios of base to acid. For example, when the base/acid ratio is 0.9/0.1,

   $$\text{pH} = 4.76 + \log 9 = 5.71$$

   Ans. You do the rest.
3. What ratio of acetic acid and acetate would be required to attain the physiological pH of 7.4?
   Ans. This is an interesting problem because it shows how ineffective a buffer system would be if its pK_a was almost 2.6 units away from the desired pH. Starting from the Henderson-Hasselbalch equation and using 7.40 for the pH and 4.76 for the pK_a, we get

   $$7.40 = 4.76 + \log \frac{[\text{base}]}{[\text{acid}]}$$

   Ans. Transposing, we see that the log (base/acid) = 2.64, from which we find that the base/acid ratio is 440.

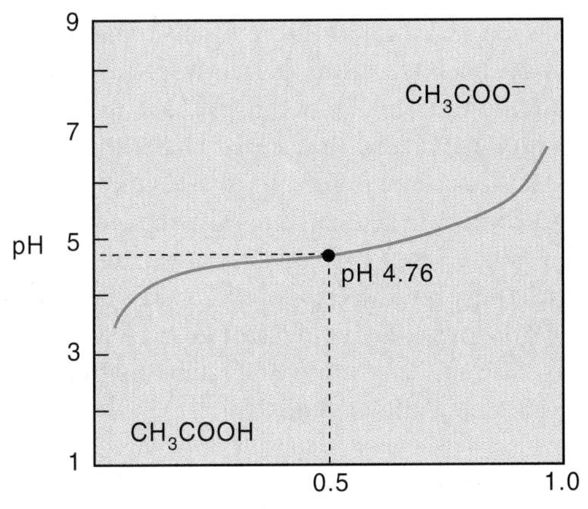

Figure 3.7

Titration curve for acetic acid. This curve has one inflection point consistent with the fact that acetic acid has one dissociable proton. At the inflection point there are equal amounts of acetic acid and acetate.

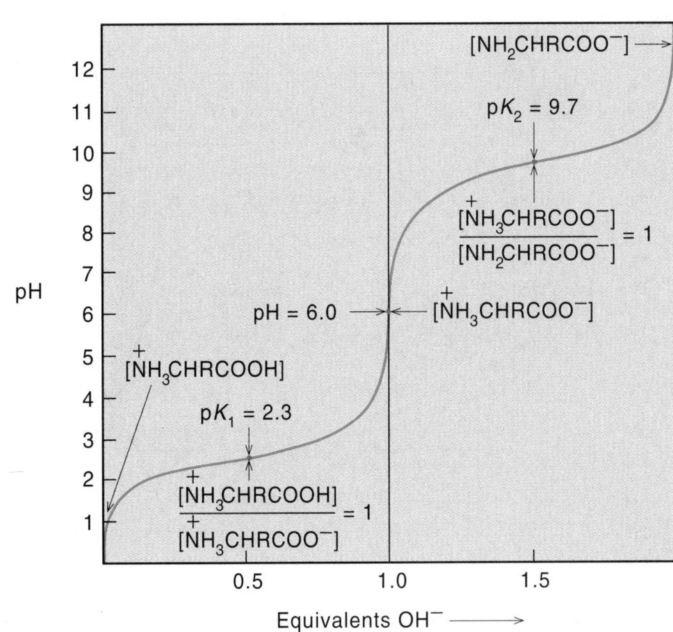

Figure 3.8

Titration curve of alanine. The predominant ionic species at each cardinal point in the titration is indicated.

Figure 3.9

Titration curves of glutamic acid, lysine, and histidine. In each case the pK of the R group is designated pK_R.

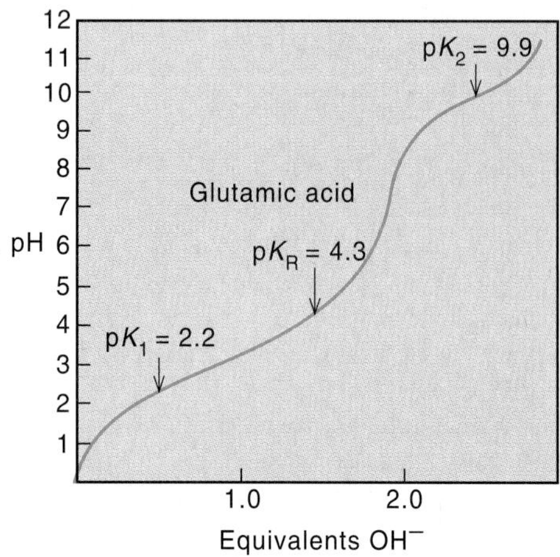

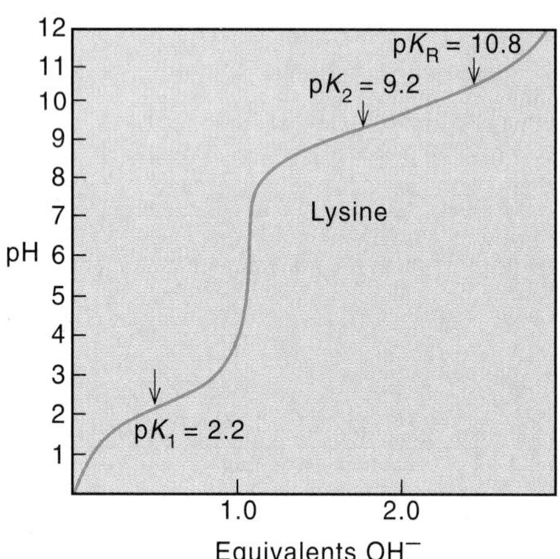

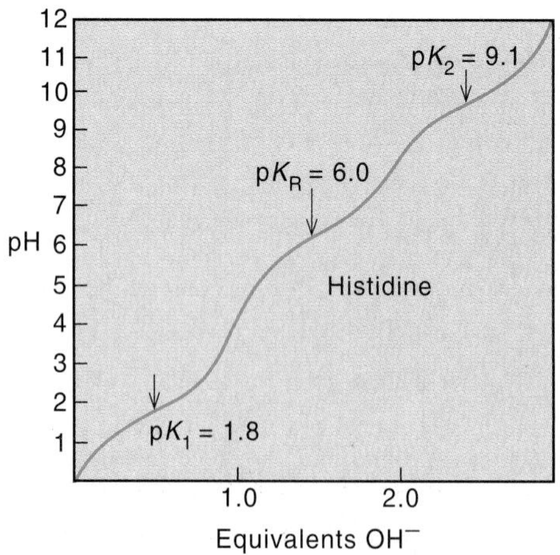

Figure 3.10

Titration curve for β-lactoglobulin. At very low pH (< 2), all ionizable groups are protonated. At pH 7.2, 51 groups (most from glutamic acid and aspartic acid residues and some from histidine residues) are deprotonated. At pH 12, most of the remaining ionizable groups (most from lysine and arginine residues and some from histidine residues) are also deprotonated. (Source: R. H. Haschenmeyer and A. E. V. Haschenmeyer, *A Guide to Study by Physical and Chemical Methods.* Copyright ©1973, John Wiley & Sons, Inc., New York, N.Y.)

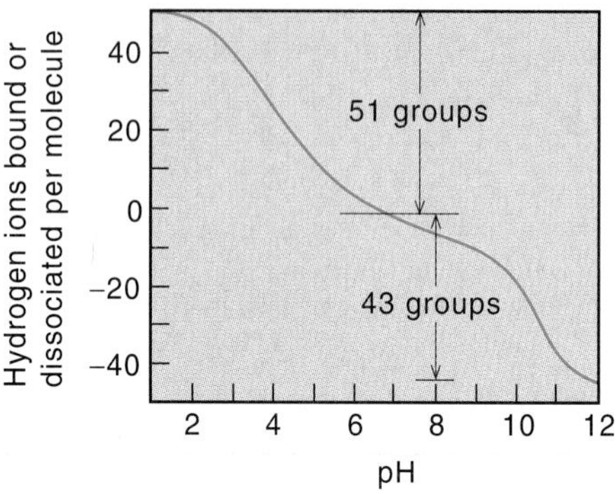

The acidic and basic groups within a protein can be titrated just like free amino acids to determine their number and their pK_a values. A titration curve for β-lactoglobulin is shown in figure 3.10. This protein contains 94 potentially ionizable groups. The protein is positively charged at low pH and negatively charged at high pH. At intermediate pH values a point is found where the sum of the positive side-chain charges exactly equals the sum of the negative charges, so that the net charge on the protein is zero. This value, as we have noted, is the isoelectric point (pI) of the protein. For β-lactoglobulin, the pI is about 5.2. The characteristic nature of the pI may be exploited to separate different proteins in a mixture (box 3C).

Buffered Solutions Are Resistant to Changes in pH

As we have seen, the titration curve for a simple weak acid shows a minimum in the slope at the pH corresponding to its pK_a (see fig. 3.7). Addition of small amounts of base or acid at this point results in a minimal change in the pH. Under these conditions the solution is said to be buffered. As we have already noted, biochemical reactions are typically highly dependent on the pH of the solution. Therefore it is frequently advantageous to study biochemical reactions in buffered solutions. The ideal buffer is one that has a pK_a numerically equivalent to the working pH. Biological systems also use this strategy to stabilize the pH in their intracellular and extracellular fluids.

Weak Acids Buffer the pH in the Fluid Compartments of the Body

Analysis of the body fluids reveals two anions of weak acids that play major roles in pH management: bicarbonate in the blood

3C

BOX

Separation of Proteins by Isoelectric Focusing Exploits the Difference in Isoelectric Points

The apparatus for this separation usually consists of a narrow tube containing a gel and a mixture of ampholytes, which are small molecules, each with both positive and negative charges. The ampholytes have a wide range of isoelectric points and are allowed to distribute in the column under the influence of an electric field. This step creates a pH gradient from one end of the gel to the other as each particular ampholyte comes to rest at a position coincident with its isoelectric point. At this stage a solution containing a mixture of proteins is introduced into the gel. Under the influence of the electric field and the pH gradient, each type of protein migrates to the point in the tube where the pH matches the pI of the protein.

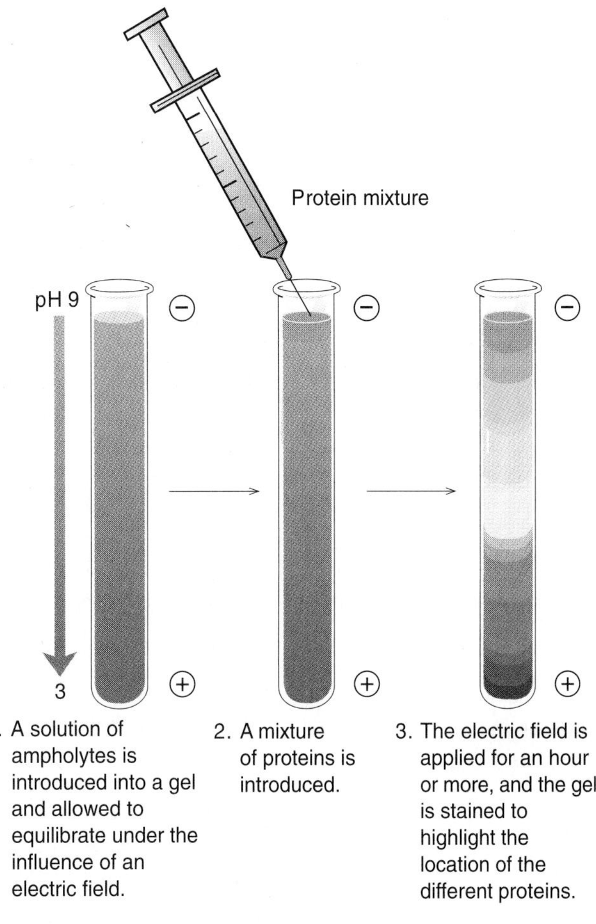
Protein mixture
pH 9
3
−
+
1. A solution of ampholytes is introduced into a gel and allowed to equilibrate under the influence of an electric field.
2. A mixture of proteins is introduced.
3. The electric field is applied for an hour or more, and the gel is stained to highlight the location of the different proteins.

plasma and interstitial fluid and monohydrogen phosphate in the intracellular fluid (fig. 3.11). Bicarbonate, in conjunction with carbonic acid and dissolved carbon dioxide, buffers the pH of the blood plasma. Monohydrogen phosphate and dihydrogen phosphate play a similar role in the intracellular fluid. We discuss both of these buffer systems in detail because of their key roles in regulating the pH of the organism.

The Intracellular pH Is Buffered by Mono and Dihydrogen Phosphates

Phosphoric acid has three inflection points in its pH titration curve arising from its three ionization protons. The p*K*s for the dissociation of the first, second, and third protons are 2.15, 7.2, and 12.4, respectively (fig. 3.12). Clearly, the only p*K* that is relevant at

Figure 3.11

Electrolyte composition of the main fluid compartments of the body. Note that HCO_3^- is abundant in the extracellular fluid while HPO_4^- is abundant in the intracellular fluid. These anions of weak acids play major roles in pH control.

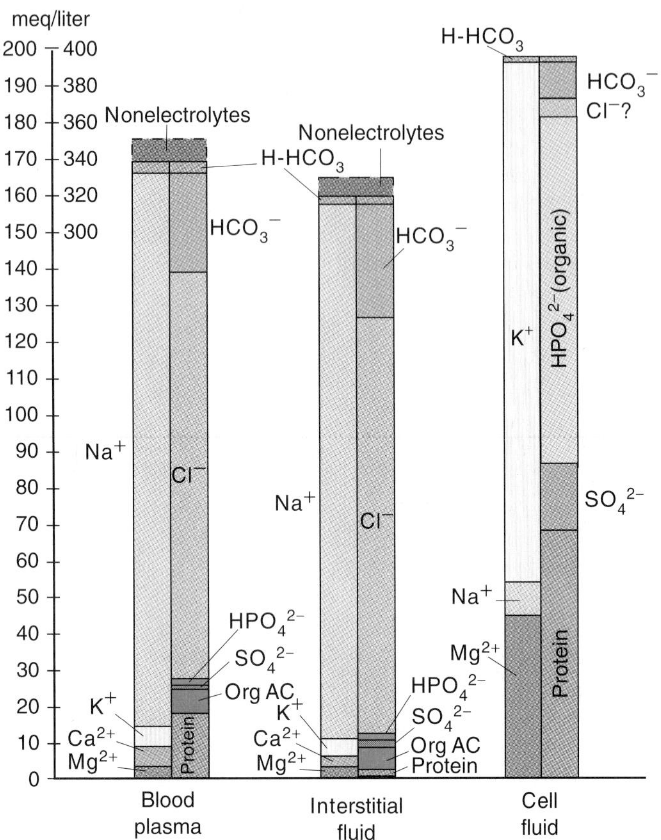

Figure 3.12

pH titration curve for inorganic phosphate.

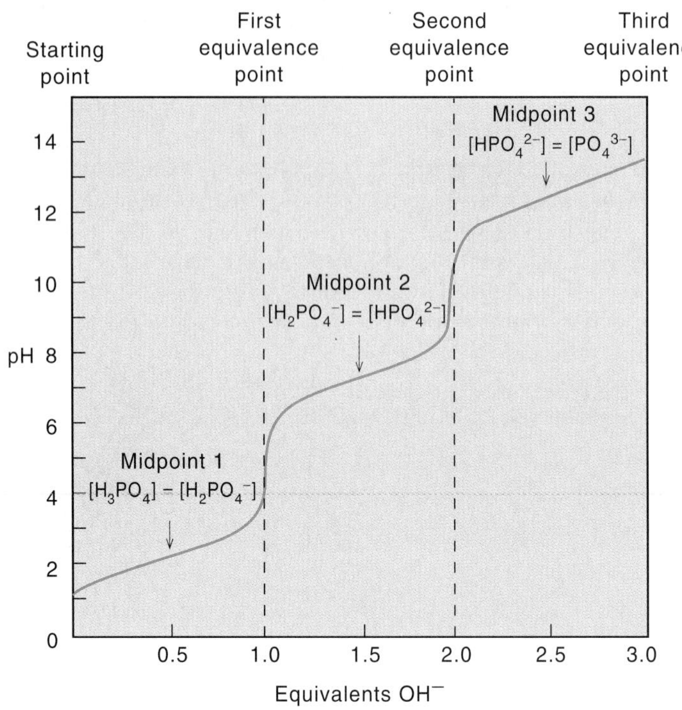

physiological pHs is the middle one (pK_2) because it is very close to the intracellular pH of most cells. Indeed, it is so close that it seems possible that biological systems evolved to operate optimally at this pH for this very reason. Using the value of 7.4 for the intracellular pH, the distribution of the major phosphate species can be calculated from the appropriate Henderson-Hasselbalch equation:

$$\text{pH} = pK_2 + \log \frac{[HPO_4^{2-}]}{[H_2PO_4^-]} \quad (12)$$

$$7.4 = 7.2 + \log \frac{[HPO_4^{2-}]}{[H_2PO_4^-]} \quad (13)$$

so that

$$\frac{[HPO_4^{2-}]}{[H_2PO_4^-]} = 1.58 \quad (14)$$

Since the total cellular phosphate is about 10 mM,

$$[HPO_4^{2-}] + [H_2PO_4^-] \cong 10 \text{ mM}$$

so that

$$[HPO_4^{2-}] = 6.13 \text{ mM} \qquad \text{and} \qquad [H_2PO_4^-] = 3.87 \text{ mM}$$

The pH of the Blood Plasma Is Stabilized by a Buffer System Involving Bicarbonate, Carbonic Acid, and Carbon Dioxide

The buffer system that regulates the pH of the blood plasma is unusual in more than one way. First of all, it involves a volatile component, dissolved carbon dioxide ($CO_2(d)$), which exchanges with gaseous CO_2 in the alveolar tissue of the lungs. Second, there is more than one equilibrium that must be considered (fig. 3.13). The equilibria relevant to the aqueous phase are

$$H_2CO_3 \xrightleftharpoons{K_a} H^+ + HCO_3^-; \qquad K_a = \frac{[H^+][HCO_3^-]}{[H_2CO_3]} \quad (15)$$

and

$$CO_2(d) + H_2O \xrightleftharpoons{K_h} H_2CO_3; \qquad K_h = \frac{[H_2CO_3]}{[CO_2(d)]} \quad (16)$$

Acid is directly produced in the first of these reactions, but it should be clear that this reaction by itself would not be conducive to good pH management because the pK_a for this reaction, 3.88, is so far removed from the pH of the cell. What makes the system effective is that carbonic acid is in equilibrium with carbon

Figure 3.13

Schematic drawing of tissues and reactions involved in carbon dioxide and acid removal. Carbon dioxide is produced in the tissues and removed in the lungs. Excess acid is eliminated by the kidneys after combining with monohydrogen phosphate or ammonia. These two elimination processes are regulated so that the plasma pH is maintained at 7.40.

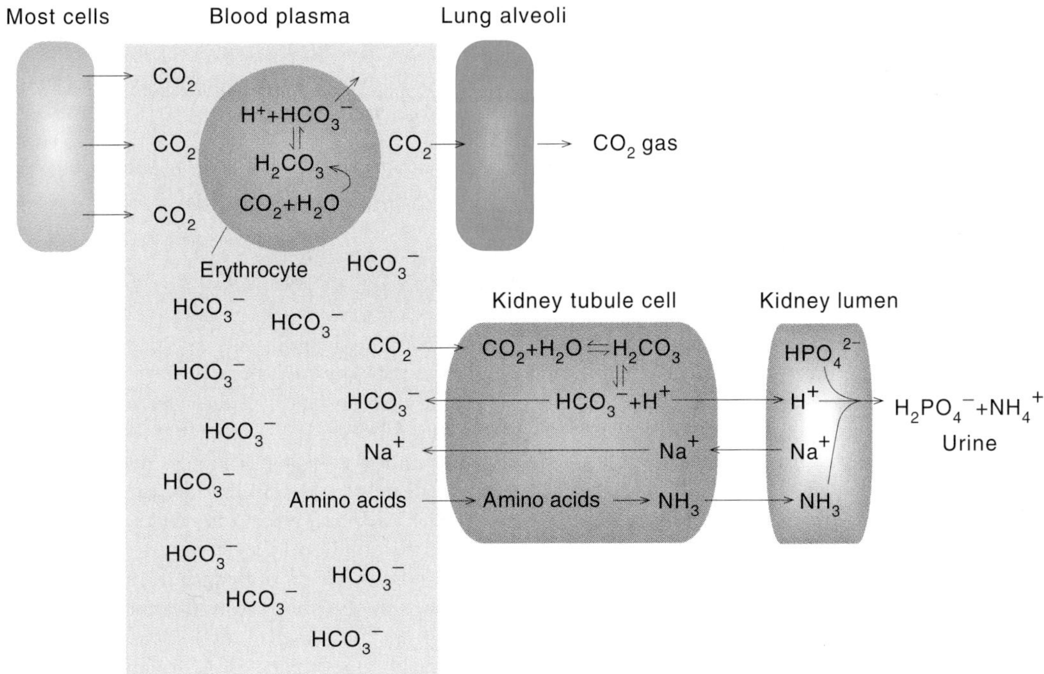

dioxide (equation 16), whose concentration is precisely controlled by physiological mechanisms. If we add equations (15) and (16), we get a better idea of the overall reaction:

$$CO_2(d) + H_2O \rightleftharpoons H^+ + HCO_3^- \quad \textbf{(17)}$$

The equilibrium constant for this reaction, designated $K_{overall}$, is computed from the product of K_h and K_a because when two reactions are added, their equilibrium constants are multiplied.

At 37°C,

$$K_{overall} = K_h \cdot K_a = 3.1 \times 10^{-3} \times 1.53 \times 10^{-4} = 4.90 \times 10^{-7} \quad \textbf{(18)}$$

so

$$pK_{overall} = 6.31 \quad \textbf{(19)}$$

When equation (17) is put in the form of the Henderson-Hasselbalch equation, we see that

$$pH = pK_{overall} + \log \frac{[HCO_3^-]}{[CO_2(d)]} \quad \textbf{(20)}$$

The pH of the plasma is 7.4 and the $pK_{overall}$ is 6.3; it should be clear from equation (20) that the ratio of bicarbonate to dissolved carbon dioxide must always be greater than 10. This ratio is controlled by the coordinated efforts of the red cells, the lungs, and the kidneys (see fig. 3.13). The red cells and the kidney cells both contain carbonic anhydrase, which catalyzes the interconversion of dissolved carbon dioxide and carbonic acid, thereby assuring the equilibrium between dissolved carbon dioxide and bicarbonate described in equation (17). Most carbon dioxide originates as an end product of catabolism. It is excreted by various tissues into the blood plasma. There in the red cells it is rapidly converted into carbonic acid, which ionizes to bicarbonate and protons. This conversion is facilitated by the buffering action of the erythrocyte hemoglobin, which has a high capacity for binding protons (see chapter 6). The dissolved carbon dioxide is ultimately eliminated by exhalation from the lungs. The breathing rate is the primary mechanism for regulating the carbon dioxide content of the blood. The kidney cells also play an important role in pH control. In the kidney tubule cells, carbon dioxide is converted into bicarbonate and protons as in the erythrocytes. The protons so formed are eliminated by excretion after first combining with monohydrogen phosphate and to a lesser extent ammonia, which is formed mostly by amino acid deamination.

Normally, the pH of the blood plasma is precisely maintained at a value of 7.4. A pH above 7.45 is associated with the condition known as alkalosis, while a pH lower than 7.35 identifies the condition known as acidosis. These conditions are quickly corrected by a change in breathing rate or a change in excretion rate. If not, the conditions that initially bring discomfort could result in serious metabolic disorders and even death.

Water Is Directly Involved in Many Biochemical Reactions

In addition to its invaluable role as a solvent, water participates directly in a multitude of biochemical reactions. We could spend a good deal of time discussing this here, but it would be more appropriate to discuss it briefly here and take it up in detail in later chapters where the focus is on the metabolism. There is a wide range of hydrolytic reactions in which water is a reactant. Reactions in which peptides are degraded to amino acids or polynucleotides are degraded to mononucleotides involve the uptake of single water molecules for each degradative reaction. Water is also produced in a number of reactions. For example, the catabolism of glucose or fatty acids results in water production. Whereas the water in such reactions is usually incidental to the coupled ATP production that is usually involved in these reactions, there are occasions where the catabolism is used to supply the animal with water. Such is the case in fatty acid breakdown in the camel or the whale, which often do not have access to fresh water for extended periods of time.

Summary

All biosystems exist in an environment that is dominated by liquid water. In this chapter we describe the properties of water and the interaction of other molecules with water.

1. Water is a highly polar molecule that forms a network of hydrogen-bonded molecules in ice and liquid water.
2. Electrostatic forces encourage bond formation between water and molecules that possess charged groups or polar groups.
3. The intermolecular distances between nonpolar molecules is determined by the van der Waals radii, which are a function of van der Waals attractive forces and van der Waals repulsive forces.
4. Hydrogen bonds form between polar molecules in which one of the interacting atoms is an unshielded proton.
5. In assessing the solubility of small molecules in liquid water and the interactions between molecules in liquid water, enthalpic and entropic factors must both be considered. Electrostatic interactions and hydrogen bonds are most important contributors to the enthalpy. Effects on solvent order usually make the major contribution to the entropy.
6. The rate of biochemical reactions is very sensitive to the pH.
7. Strong acids and weak acids make different contributions to the pH. The pH effect of a strong acid is directly related to the concentration of the acid. The pH effect of a weak acid is a function of the concentration of the acid and its acid dissociation constant. The acid dissociation constant of a weak acid may be determined by titration of the acid with base.
8. The buffering effect of a weak acid is optimal when the pH of the solution is equivalent to the pK_a of the weak acid.
9. A mixture of monohydrogen phosphate and dihydrogen phosphate is used to buffer the cell fluid. Its effectiveness is due to the presence of a substantial amount of phosphate and the fact that the pK_a for the conversion of these two forms of phosphate is very close to the pH of the cytosol for most cells.
10. The blood plasma is buffered by the carbon dioxide–carbonic acid–bicarbonate system. The effective pH regulation offered by this system is a result of physiological mechanisms that lead to the regulation of the carbon dioxide and the bicarbonate concentrations.
11. Water is a reactant or a reaction product in many biochemical reactions.

Selected Readings

Devlin, T. M., Gas transport and pH regulation. Chapter 25 in *Textbook of Biochemistry with Clinical Correlations,* 3d ed. New York: Wiley-Liss, 1993.

Eisenberg, D., and D. Crothers, *Physical Chemistry with Applications to the Life Sciences.* Menlo Park, Calif.: Benjamin/Cummings, 1979.

Pauling, L., *The Nature of the Chemical Bond,* 3d ed. Ithaca, N.Y.: Cornell University Press, 1960. A classic.

Williams, V. R., W. L. Mattice, and H. B. Williams, *Basic Physical Chemistry for the Life Sciences.* New York: Freeman, 1978.

Problems

1. Consider separately 1-butanol, butanoic acid, and 1-aminobutane as solutes in water. What types of interactions with water can you envision? How could you explain that butanoic acid and 1-aminobutane are miscible with water, while only 9 g of 1-butanol will dissolve in 100 ml of water?
2. Many amides are rather insoluble in water. This is somewhat unexpected when one considers the number of hydrogen bonds that can form between an amide and water molecules. Consider substances such as benzamide and 1,4-diketopiperazine (a cyclic dipeptide) and try to explain why these substances are not particularly soluble in water.
3. The partial molar volume of a solute is the volume that substance contributes to the total volume of a solution. The partial molar volume in water of the first four primary alcohols

are 38.38, 56.28, 70.95, and 86.08 ml/mole. When the partial molar volume of these alcohols is plotted versus the number of methylenes in $H(CH_2)_nOH$, the $n = 0$ intercept is 22.3 ml/mole. This intercept corresponds to the partial molar volume of HOH. From the density of water, calculate the partial molar volume of water and explain any discrepancy between the partial molar volume values of water and HOH.

4. What is the pH of a 0.012 M propionic acid solution if the pK_a for propionic acid is 4.87?
5. A 0.0084 M solution of an organic acid has a pH of 3.15. What is the pK_a of the acid?
6. A liter of 0.050 M pH 7.10 potassium phosphate buffer is needed for a certain experiment. The following methods were proposed for the preparation of the buffer:
   (a) A 0.050 M sample of K_2HPO_4 was dissolved in about 800 ml of water, and the pH was adjusted to 7.0 using a standardized pH meter and a 1 M HCl solution, followed by addition of water to a volume of 1.00 l.
   (b) A 0.050 M sample of K_2HPO_4 was dissolved in about 800 ml of water, and the pH was adjusted to 7.0 using a standardized pH meter and a 1 M H_3PO_4 solution, followed by addition of water to a volume of 1.00 l.
   (c) Use the Henderson-Hasselbalch equation and 0.050 mole = moles KH_2PO_4 + moles K_2HPO_4 to calculate the moles of the two salts, weigh out the salts, dissolve, and dilute to 1.00 l.
   (d) A 0.050 M sample of KH_2PO_4 was dissolved in about 800 ml of water, and the pH was adjusted to 7.0 using a standardized pH meter and a 1 M KOH solution, followed by addition of water to a volume of 1.00 l.

   Which of these methods should be used and what is wrong with the other proposals?
7. A 0.15 M pH 6.87 potassium phosphate buffer was diluted with distilled water to a 0.035 M solution. What was the pH of the resulting buffer?
8. How many millimoles of HCl can be added to 10 ml of 0.150 M pH 6.80 potassium citrate buffer before the pH drops below pH 6.50? Use a value of 6.39 for pK_{a3} of citric acid.
9. A buffer was prepared by combining 3.71 g of citric acid and 2.91 g of KOH, dissolved in water and diluted to 250 ml. What is the pH of this buffer? What is the $[H^+]$? Use 3.14, 4.77, and 6.39 for the pK_a values of citric acid.
10. A buffer was prepared by dissolving 11.7 g of KH_2PO_4 in 400 ml of water, and with the addition of a concentrated solution of KOH the pH was adjusted to 7.10. The volume was then adjusted to 500 ml. What are the $[HPO_4^{2-}]$, $[H_2PO_4^{1-}]$, and $[H_3PO_4]$? Use pK_a values of 2.12, 7.21, and 12.7 for phosphoric acid.
11. A 100-ml sample of the diluted buffer in problem 7 was combined with 15 ml of 0.076 M KOH. What was the resulting pH?
12. Phosphoric acid (2.45 g) and potassium hydroxide (2.45 g) were dissolved in water and diluted to 600 ml. Use the pK_a values listed in problem 8 to determine the pH and concentration of the buffer.
13. What are the hydrogen ion and hydroxide ion concentrations in the buffer in problem 12?

C H A P T E R

The Building Blocks of Proteins: Amino Acids, Peptides, and Polypeptides

Amino acids have common features that permit them to be linked together and uncommon features that give each polypeptide chain its unique character.

In the middle of the nineteenth century, the Dutch chemist Gerardus Mulder extracted a substance common to animal tissues and the juices of plants, which he believed to be "without doubt the most important of all substances of the organic kingdom, and without it life on our planet would probably not exist." At the suggestion of the famous Swedish chemist Berzelius, Mulder named this substance protein (from the Greek *proteios,* meaning "of first importance") and assigned to it a specific chemical formula ($C_{40}H_{62}N_{10}O_{12}$). Although he was wrong about the chemistry of proteins, he was right about their being indispensable to living organisms. The term "protein" endures.

Proteins are the most abundant of cellular components. They include enzymes, antibodies, hormones, transport molecules, and even components for the cytoskeleton of the cell itself. Proteins are also informational macromolecules, the ultimate heirs of the genetic information encoded in the sequence of nucleotide bases within the chromosomes. Structurally and functionally, they are the most diverse and dynamic of molecules and play key roles in nearly every biological process. Proteins are complex macromolecules with exquisite specificity; each is a specialized player in the orchestrated activity of the cell. Together they tear down and build up molecules, extract energy, repel invaders, act as delivery systems, and even synthesize the genetic apparatus itself.

In most of part 2 we discuss the basic structural and chemical properties of proteins. In this chapter we concentrate on the structural and chemical properties of amino acids, peptides, and polypeptides—the building blocks of proteins. From our presentation you will learn the following:

1. Certain acidic and basic properties are common to all amino acids found in proteins except for the amino acid proline.
2. Side chains give amino acids their individuality. These side chains serve a variety of structural and functional roles.
3. The alpha-carboxyl group of one amino acid can react with the alpha-amino group of another amino acid to form a dipeptide.
4. Many amino acids, reacting in a similar way, can become linked to form a linear polypeptide chain.
5. The amino acid sequence in a polypeptide can be determined by a process of partial breakdown into manageable fragments, followed by stepwise analysis proceeding from one end of the chain to the other.
6. Polypeptide chains with a prespecified sequence can be synthesized by well-established chemical methods.

Amino Acids

Every protein molecule can be viewed as a polymer of amino acids. There are 20 common amino acids. Figure 4.1*a* shows the structure of a single amino acid. At the center is a tetrahedral carbon atom called the alpha (α) carbon (C_α). It is covalently bonded on one side to an amino group (NH_2) and on the other side to a

Figure 4.1

Amino acid anatomy. (*a*) Uncharged amino acid. (*b*) Doubly charged zwitterion.

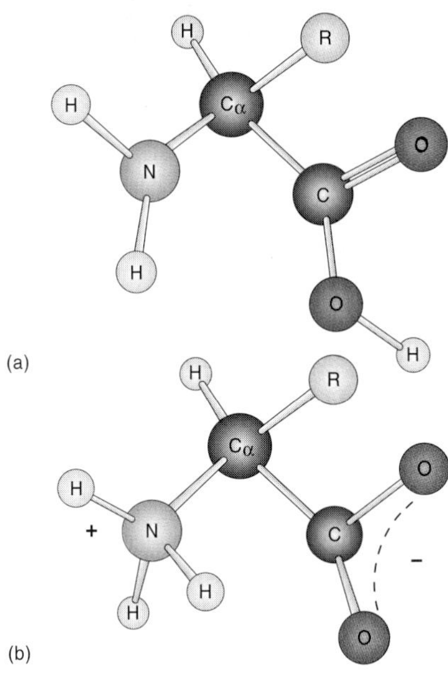

carboxyl group (COOH). A third bond is always hydrogen, and the fourth bond is to a variable side chain (R). In neutral solution (pH 7) the carboxyl group loses a proton and the amino group gains one. Thus an amino acid in solution, while neutral overall, is a double-charged species called a zwitterion (fig. 4.1*b*).

The structures of the 20 amino acids commonly found in proteins are listed in table 4.1. All of these amino acids except proline have an α ammoniun ion ($—N^+H_3$) attached to the α carbon. In proline one of the N—H linkages is replaced by an N—C linkage forming part of a cyclic structure. Various ways of classifying amino acids according to their R groups have been proposed. In table 4.1 we have divided the amino acids into three categories. The first category contains eight amino acids with relatively apolar R groups; the second category contains seven amino acids with uncharged polar R groups; and the third category contains five amino acids with R groups that normally exist in the charged state.

Amino acids are often abbreviated by three-letter symbols; when this proves to be too cumbersome (as in certain kinds of charts and figures), one-letter symbols are used. Both designations are given in table 4.1, together with the molecular weight (M_r) of each amino acid.

In addition to the 20 commonly occurring α-amino acids, a variety of other amino acids are found in minor amounts in proteins and in nonprotein compounds (fig. 4.2). The unusual amino acids found in proteins result from modification of the common amino acids. In a few cases these amino acids are incorporated directly into the polypeptide chains during synthesis (see box 34A for selenocysteine). Most frequently, the amino acid is modified after incorporation. The unusual amino acids found in nonprotein compounds are extremely varied in type and are formed by a number of different metabolic pathways.

Amino Acids Have Both Acid and Base Properties

The charge properties of amino acids are very important in determining the reactivity of certain amino acid side chains and in the properties they confer on proteins. The charge properties of amino acids in aqueous solution were considered under the general treatment of acid–base ionization theory (see chapter 3 and figures 3.8 and 3.9).

We saw that the charge properties of an amino acid are revealed by its pH titration curve. We also saw that a simple amino acid with a nonionizable R group gives a complex titration curve with two inflection points. For an example, see the titration of alanine, shown in figure 3.8.

Amino acids with an ionizable R group show even more complex titration curves, indicative of three ionizable groups (fig. 3.9). The pK for the ionizable side chain, pK_R, is usually readily

Figure 4.2

Some uncommon amino acids found in proteins. Hydroxylysine and hydroxyproline are found mainly in collagen. Thyroxine is found in thyroglobulin, a protein produced in the thyroid gland. Phosphoserine and phosphotyrosine are found in a broad range of regulatory proteins that are reversibly phosphorylated. *N*-acetyllysine is found in histones that are associated with chromosomes.

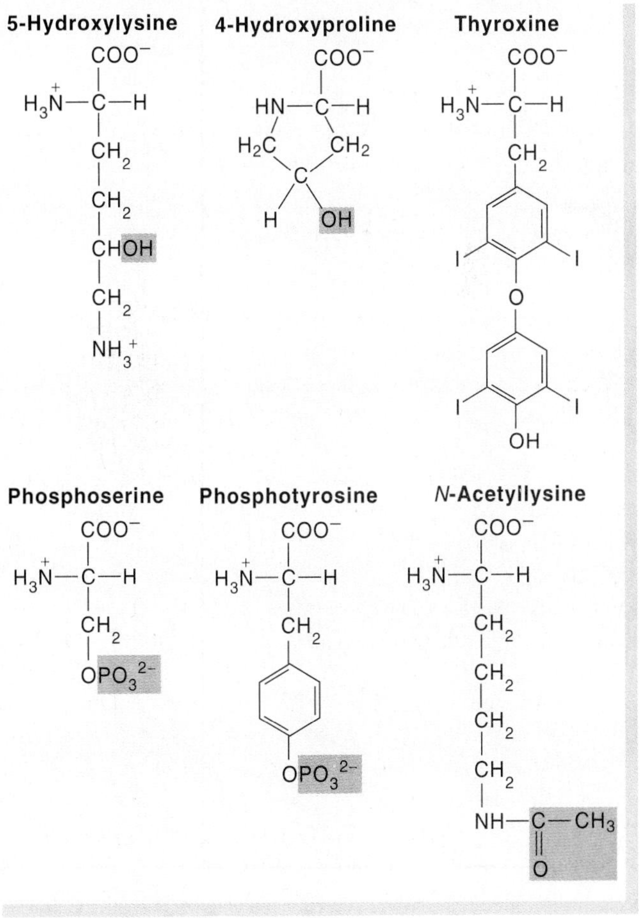

5-Hydroxylysine — COO^-, $H_3\overset{+}{N}—C—H$, CH_2, CH_2, CHOH, CH_2, NH_3^+

4-Hydroxyproline — COO^-, HN—C—H, H_2C, CH_2, C, H, OH

Thyroxine — COO^-, $H_3\overset{+}{N}—C—H$, CH_2, I, I, O, I, I, OH

Phosphoserine — COO^-, $H_3\overset{+}{N}—C—H$, CH_2, OPO_3^{2-}

Phosphotyrosine — COO^-, $H_3\overset{+}{N}—C—H$, CH_2, OPO_3^{2-}

***N*-Acetyllysine** — COO^-, $H_3\overset{+}{N}—C—H$, CH_2, CH_2, CH_2, CH_2, NH—C—CH_3, O

Table 4.1
Structure of the 20 Amino Acids Found in Proteins

Group I. Amino Acids with Apolar R Groups	**Group II. Amino Acids with Uncharged Polar R Groups**	**Group III. Amino Acids with Charged R Groups**
Alanine Ala A M_r 89[a] $CH_3-\overset{H}{\underset{\overset{+}{N}H_3}{C}}-COO^-$	Glycine Gly G M_r 75 $H-\overset{H}{\underset{\overset{+}{N}H_3}{C}}-COO^-$	Aspartic acid Asp D M_r 133 $^-O(O=)C-CH_2-CH_2-\overset{H}{\underset{\overset{+}{N}H_3}{C}}-COO^-$
Valine Val V M_r 117 $(CH_3)_2CH-\overset{H}{\underset{\overset{+}{N}H_3}{C}}-COO^-$	Serine Ser S M_r 105 $HO-CH_2-\overset{H}{\underset{\overset{+}{N}H_3}{C}}-COO^-$	Glutamic acid Glu E M_r 147 $^-O(O=)C-CH_2-CH_2-\overset{H}{\underset{\overset{+}{N}H_3}{C}}-COO^-$
Leucine Leu L M_r 131 $(CH_3)_2CH-CH_2-\overset{H}{\underset{\overset{+}{N}H_3}{C}}-COO^-$	Threonine Thr T M_r 119 $CH_3-\underset{OH}{CH}-\overset{H}{\underset{\overset{+}{N}H_3}{C}}-COO^-$	Lysine Lys K M_r 146 $H_3\overset{+}{N}-(CH_2)_4-\overset{H}{\underset{\overset{+}{N}H_3}{C}}-COO^-$
Isoleucine Ile I M_r 131 $CH_3-CH_2-\underset{CH_3}{CH}-\overset{H}{\underset{\overset{+}{N}H_3}{C}}-COO^-$	Cysteine Cys C M_r 121 $HS-CH_2-\overset{H}{\underset{\overset{+}{N}H_3}{C}}-COO^-$	Arginine Arg R M_r 174 $H_2N-\underset{\overset{+}{N}H_2}{\overset{}{C}}-NH-(CH_2)_3-\overset{H}{\underset{\overset{+}{N}H_3}{C}}-COO^-$
Proline Pro P M_r 115 $\overline{H_2C-CH_2-CH_2-\overset{+}{N}H_2}-\overset{H}{C}-COO^-$ (ring)	Tyrosine Tyr Y M_r 181 $HO-C_6H_4-CH_2-\overset{H}{\underset{\overset{+}{N}H_3}{C}}-COO^-$	Histidine (at pH 6.0) His H M_r 155 $HC{=}C(-\overset{+}{H}N{=}CH-NH-)-CH_2-\overset{H}{\underset{\overset{+}{N}H_3}{C}}-COO^-$
Phenylalanine Phe F M_r 165 $C_6H_5-CH_2-\overset{H}{\underset{\overset{+}{N}H_3}{C}}-COO^-$	Asparagine Asn N M_r 132 $NH_2(O=)C-CH_2-\overset{H}{\underset{\overset{+}{N}H_3}{C}}-COO^-$	
Tryptophan Trp W M_r 204 $\text{indole}(C{=}CH-NH)-CH_2-\overset{H}{\underset{\overset{+}{N}H_3}{C}}-COO^-$	Glutamine Gln Q M_r 146 $NH_2(O=)C-CH_2-CH_2-\overset{H}{\underset{\overset{+}{N}H_3}{C}}-COO^-$	
Methionine Met M M_r 149 $CH_3-S-CH_2-CH_2-\overset{H}{\underset{\overset{+}{N}H_3}{C}}-COO^-$		

[a]Molecular weights in this text are expressed in units of grams per mole.

distinguishable from the pK values for the ionizable α-carboxyl and α-amino groups, pK_1 and pK_2, respectively, as the latter have numerical values close to the comparable pK values of alanine (table 4.2). Note that the only ionizable R group with a pK_R in the vicinity of 7, where most biological systems function, is that for histidine. This means that although other ionizable groups are usually fully charged under biological conditions, the side chain of histidine can be fully charged, uncharged, or partially charged, depending on the precise situation. This variability has major implications for the way the histidine side chain functions in enzyme catalysis. The side chain can serve as either a proton donor or a proton acceptor (see discussion in chapter 10).

An additional point should be noted from table 4.2. Whereas the amino acid side chains (R groups) that are normally charged at physiological pH are restricted to five amino acids (aspartic acid, glutamic acid, lysine, arginine, and sometimes histidine), a number of potentially ionizable R groups serve as components of other amino acids. These include cysteine, serine, threonine, and tyrosine. The ionization reactions for all of the potentially ionizable side chains are indicated in figure 4.3.

Figure 4.3

Equilibrium between charged and uncharged forms of amino acid side chains.

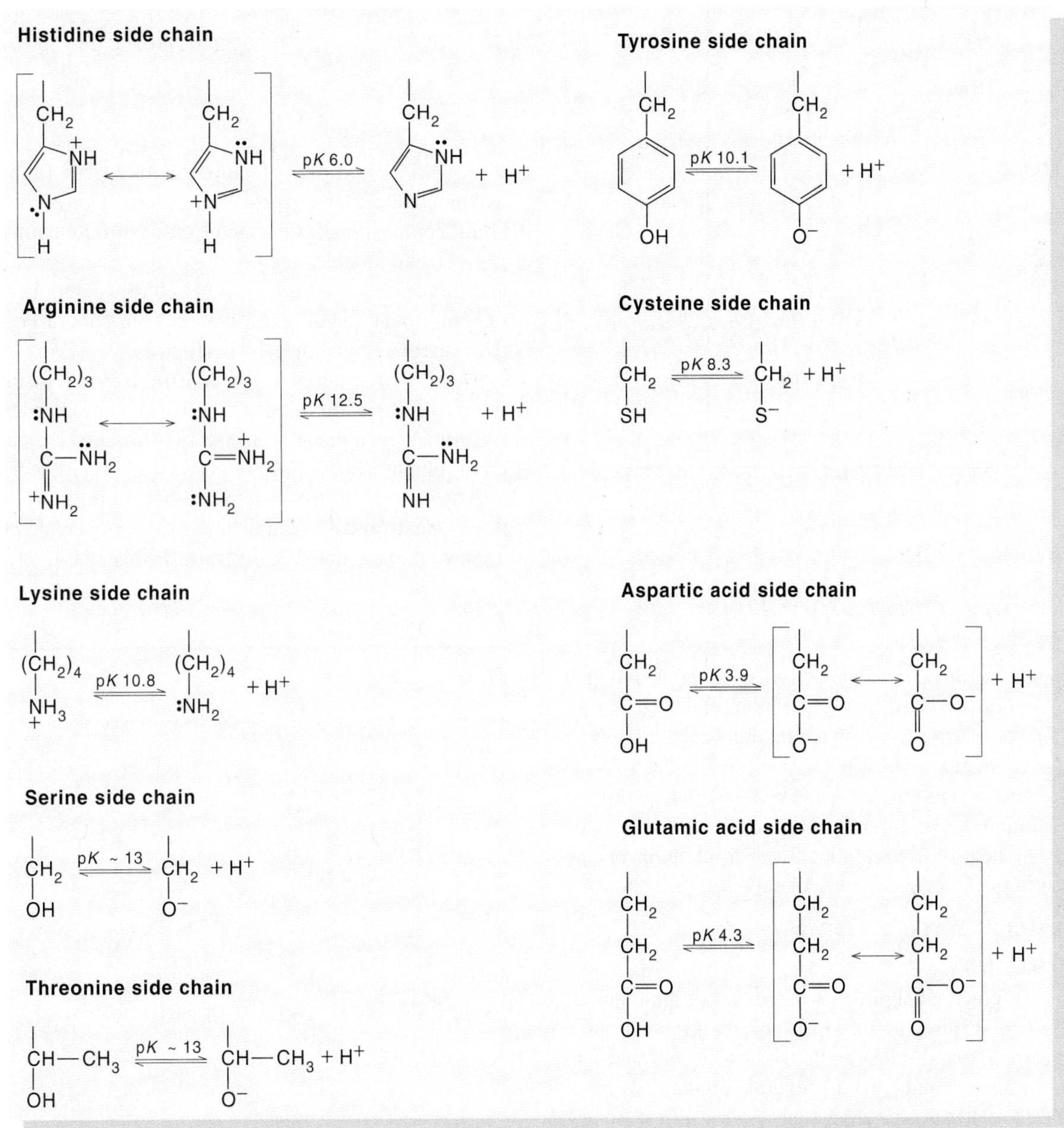

Table 4.2

Values of p*K* for the Ionizable Groups of the 20 Amino Acids Commonly Found in Proteins

Amino Acid	pK_1 (α—COOH)	pK_2 (α—NH_3^+)	pK_R (R Group)
Alanine	2.35	9.87	—
Arginine	1.82	8.99	12.48
Asparagine	2.1	8.84	—
Aspartic acid	1.99	9.90	3.90
Cysteine	1.92	10.78	8.33
Glutamic acid	2.10	9.47	4.07
Glutamine	2.17	9.13	—
Glycine	2.35	9.78	—
Histidine	1.80	9.33	6.04
Isoleucine	2.32	9.76	—
Leucine	2.33	9.74	—
Lysine	2.16	9.18	10.79
Methionine	2.13	9.28	—
Phenylalanine	2.16	9.18	—
Proline	1.95	10.65	—
Serine	2.19	9.21	~13
Threonine	2.09	9.10	~13
Tryptophan	2.43	9.44	—
Tyrosine	2.20	9.11	10.13
Valine	2.29	9.74	—

Aromatic Amino Acids Absorb Light in the Near-Ultraviolet

The aromatic amino acids phenylalanine, tyrosine, and tryptophan all possess absorption maxima in the near-ultraviolet (fig. 4.4). These absorption bands arise from the interaction of radiation with electrons in the aromatic rings. The near-ultraviolet absorption properties of proteins are determined solely by their content of these three aromatic amino acids. In solution, UV absorption can be quantified with the help of a conventional spectrophotometer and used as a measure of the concentration of proteins (see box 4A).

All Amino Acids Except Glycine Show Asymmetry

One of the most striking and significant properties of amino acids is their chirality or handedness. The word "chiral" is related to the Greek word meaning "hand." Just as the right hand is related to the left hand by a mirror image, so, in general, a naturally occurring amino acid is related to a stereoisomer by its mirror image. This observation is true of 19 out of the 20 amino acids; the one exception is glycine.

Figure 4.4

Ultraviolet absorption spectra of tryptophan (Trp), tyrosine (Tyr), and phenylalanine (Phe) at pH 6. The molar absorptivity is reflected in the extinction coefficient, with the concentration of the absorbing species expressed in moles per liter. (Source: From D. B. Wetlaufer, *Adv. Protein Chem.* 17:303–390, 1962.)

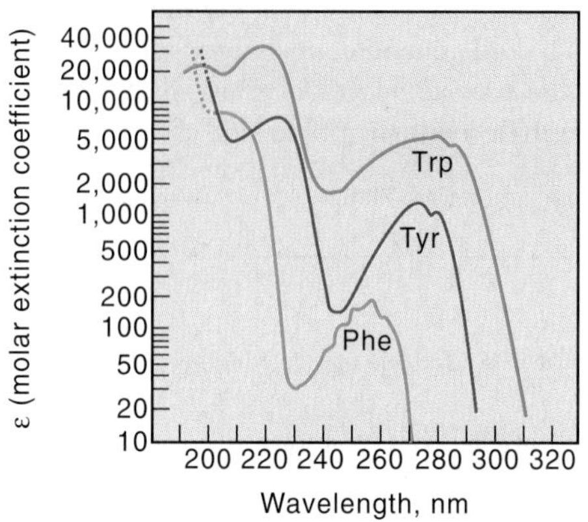

The chirality of amino acids stems from the chiral or asymmetric center, the α-carbon atom. The α-carbon atom is a chiral center if it is connected to four different substituents. Thus glycine has no chiral center. Two of the amino acids, isoleucine and threonine, possess additional chiral centers because each has one additional asymmetric carbon. You should be able to locate these carbons by simple inspection.

Two structures that constitute a stereoisomeric pair are referred to as enantiomers. The two enantiomers for alanine are illustrated in figure 4.5. These two isomers are called L-alanine and D-alanine, according to the way in which the substituents are arranged about the asymmetric carbon atom. The naming by D and L (for "dextrorotatory" and "levorotatory"; see chapter 5) refers to a convention established by Emil Fischer in 1891. According to

Figure 4.5

The covalent structure of alanine, showing the three-dimensional structure of the L and D stereoisomeric forms.

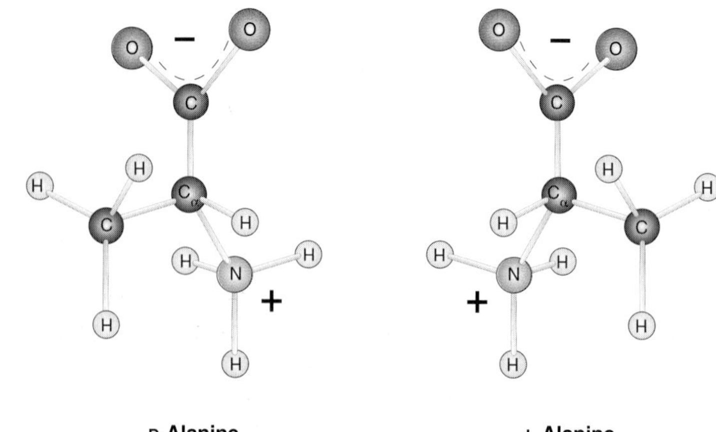

4A BOX

Measurement of Ultraviolet Absorption in Solution

The general quantitative relationship that governs all absorption processes is called the Beer-Lambert law:

$$I = I_0\ 10^{-\epsilon c d}$$

where I_0 is the intensity of the incident radiation, I is the intensity of the radiation transmitted through a cell of thickness d (in centimeters) that contains a solution of concentration c (expressed either in moles per liter or in grams per 100 ml), and ϵ is the extinction coefficient, a characteristic of the substance being investigated (see fig. 4.4).

Light absorption is measured by a spectrophotometer as shown in figure 1. The spectrophotometer is usually capable of directly recording the absorbance A, which is related to I and I_0 by the equation

$$A = \log_{10}(I_0/I)$$

Hence $A = \epsilon c d$, and A is a direct measure of concentration. We can see from figure 4.4 that the ϵ values are largest for tryptophan and smallest for phenylalanine.

Since protein absorption maxima in the near-ultraviolet (240–300 nm) are determined by the content of the aromatic amino acids and their respective values, most proteins have absorption maxima in the 280-nm region. By contrast, absorption in the far-ultraviolet (around 190 nm) is shown by all polypeptides regardless of their aromatic amino acid content. The reason is that absorption in this region is due primarily to the peptide linkage.

Figure 1

Schematic diagram of a spectrophotometer for measuring light absorption. Laboratory instruments for making measurements are much more complex than this, but they all contain the same basic components: a light source, a monochromator, a sample, and a detector. λ is the wavelength of the light; I_0 and I are the incident light intensity and the transmitted light intensity, respectively; and d is the thickness of the absorbing solution.

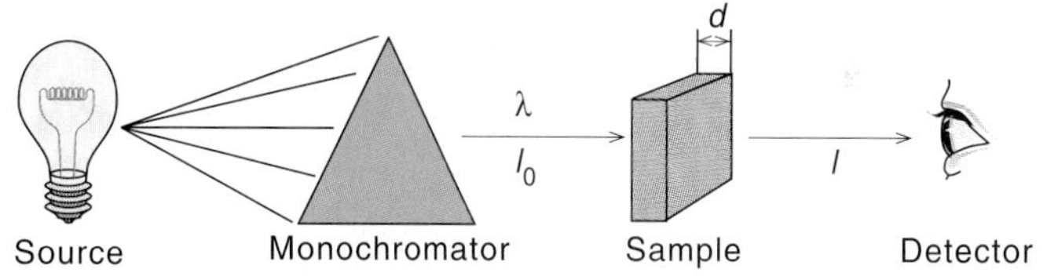

this convention all amino acids found in proteins are of the L form. Some D-amino acids are found in bacterial cell walls and certain antibiotics.

Another convention for referring to configurations is called the *R, S* convention (box 4B). The *R, S* convention is not as popular for amino acids or sugars as it is for other types of biomolecules, such as lipids.

Peptides and Polypeptides

Amino acids can link together by a covalent peptide bond between the α-carboxyl end of one amino acid and the α-amino end of another. Formally, this bond is formed by the loss of a water molecule, as shown in figure 4.6. The peptide bond has partial double-bond character owing to resonance effects; as a result, the C—N peptide linkage and all of the atoms directly connected to C and N lie in a planar configuration called the amide plane. In the following chapter we will see that this amide plane, by limiting the number of orientations available to the polypeptide chain, plays a major role in determining the three-dimensional structures of proteins.

Any number of amino acids can be joined by successive peptide linkages, forming a polypeptide chain. The polypeptide chain, like the dipeptide, has a directional sense. One end, called the N-terminal or amino-terminal end, has a free α-amino group, whereas the other end, the C-terminal or carboxyl-terminal end, has a free α-carboxyl group. The sequence of main-chain atoms from the N-terminal end to the C-terminal end is C_α—C—N—C_α, etc., and in the opposite direction it is C_α—N—C—C_α, etc. Short polypeptide chains, up to a length of about 20 amino acids, are called peptides or oligopeptides if they are fragments of whole polypeptide chains. A small protein molecule may contain a polypeptide chain of only 50 amino acids; a large protein may contain chains of 3,000 amino acids or more. One of the larger single polypeptide chains is that of the muscle protein myosin, which consists of approximately 1,750 amino acid residues. Figure 4.7 shows a section of a polypeptide chain as a linear array with α carbons and planar amides alternating as repeating units of the main chain. Different side chains are attached to each α carbon.

In addition to the covalent peptide bonds formed between adjacent amino acids within a polypeptide chain, covalent disulfide bonds can be formed within the same polypeptide chain or between different polypeptide chains (fig. 4.8). Such disulfide linkages have an important stabilizing influence on the structures formed by many proteins (see chapter 5).

The R, S Convention for Naming Stereoisomer Configurations

"Chiral" and "prochiral" are derived from the Greek word χειρ, meaning "hand," and as we have seen, they refer to "handedness" in stereoisomeric or potentially stereoisomeric centers. In the text we have used the prefixes L and D for distinguishing between stereoisomers. Another set of stereochemical symbols used for designating two stereoisomers is the pair *R* and *S*. These symbols are assigned as follows.

For a tetrahedral carbon (or other tetrahedral atom) the four different substituents are assigned relative priorities by applying rules that generally accord higher priority to groups having the larger summation of atomic weights. (You will find these rules set out in the article by G. Popják that is listed in the Selected Readings at the end of this chapter.) Once the priorities are assigned, the atom is viewed from the side opposite the lowest priority group, and the symbol *R*, for "rectus," is assigned if the remaining groups appear in clockwise order from highest to lowest priority. The symbol *S*, for "sinister," is assigned if they appear in counterclockwise order. Viewed in this way, D-alanine is equivalent to *R*-alanine.

For atoms whose substituent groups are $a < b < c < d$ in order of increasing priority, the following structures are those of the *R* and *S* isomers:

d, b—C—a, c (R) | a—C—b, d, c (S)

Atoms in which two of the groups are identical are said to be prochiral, since the elevation of one of the identical groups to a higher priority would lead to a chiral center.

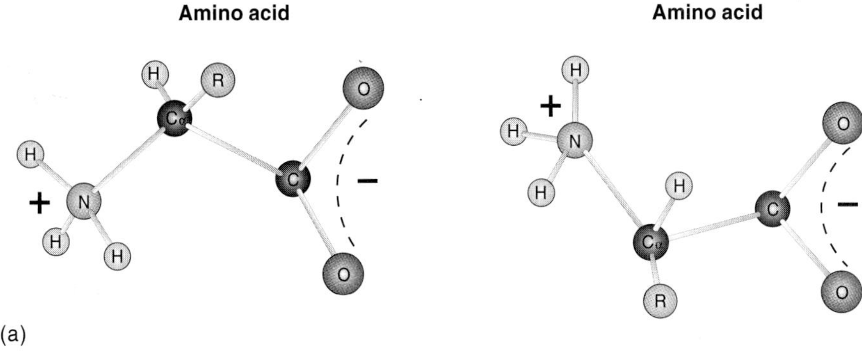

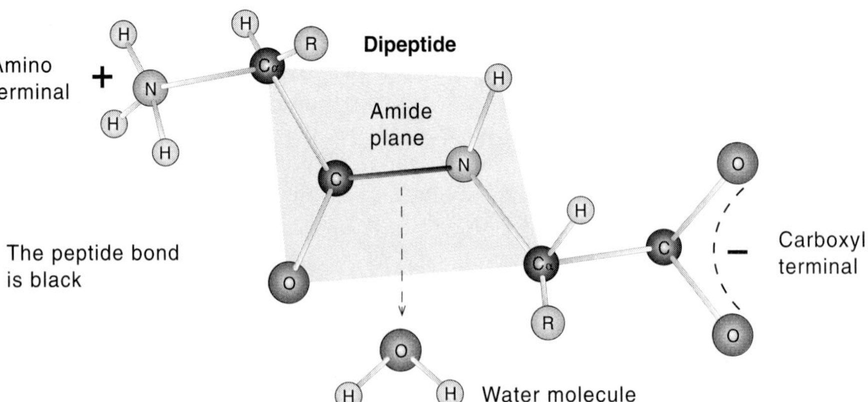

Figure 4.6

Formation of a dipeptide from two amino acids. (*a*) Two amino acids. (*b*) A peptide bond (CO—NH) links amino acids by joining the α-carboxyl group of one with the α-amino group of another. A water molecule is lost in the reaction. It is conventional to draw dipeptides and polypeptides so that their free amino terminus is to the left and their free carboxyl terminus is to the right. The amide plane refers to six atoms that lie in the same plane. (Illustration copyright by Irving Geis. Reprinted by permission.)

Figure 4.7

A polypeptide chain, with the backbone shown in color and the amino acid side chains in outline. The polypeptide chain is oriented so that the N-terminal end (not shown) is to the left. (Illustration copyright by Irving Geis. Reprinted by permission.)

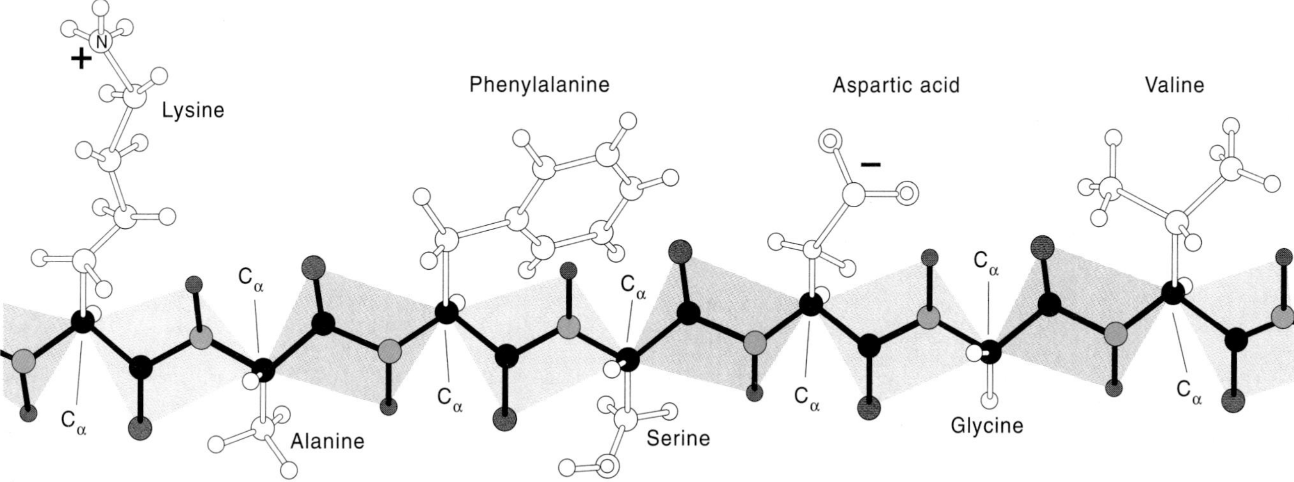

Figure 4.8

Disulfide bonds can form between two cysteines. The cysteines can exist in the cytosol as free amino acids (as shown), in which case they give rise to cystine, or they can be on polypeptide chains. In the latter instance they can be on the same polypeptide chains or different polypeptide chains. In either case the formation of covalent disulfide bonds stabilizes structural relationships.

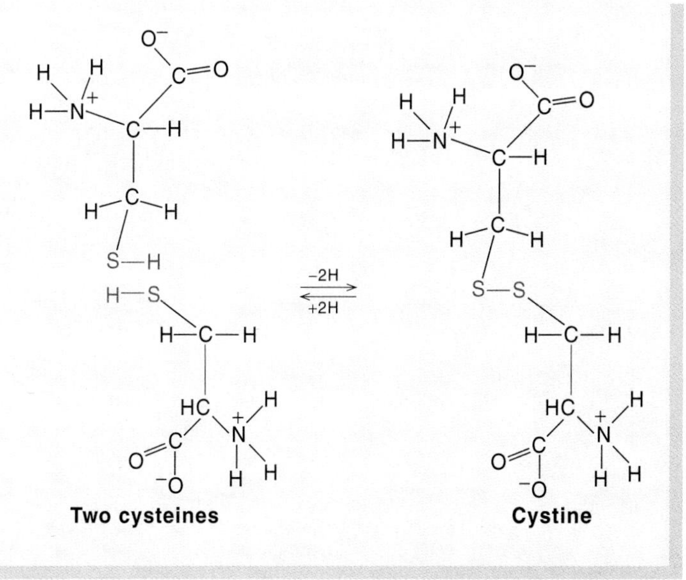

Two cysteines $\underset{+2H}{\overset{-2H}{\rightleftharpoons}}$ **Cystine**

Determination of Amino Acid Composition of Proteins

Each protein is uniquely characterized by its amino acid composition and sequence. A protein's amino acid composition is defined simply as the number of each type of amino acid composing the polypeptide chain. To determine a protein's amino acid composition, it is necessary to (1) break down the polypeptide chain into its constituent amino acids, (2) separate the resulting free amino acids according to type, and (3) measure the quantities of each amino acid.

Cleavage of the peptide bonds is usually achieved by boiling the protein in 6-*N* HCl; this treatment causes hydrolysis of the peptide bonds and the consequent release of free amino acids (fig. 4.9). Although acid hydrolysis is the most frequently used means of breaking a protein into its constituent amino acids, it results in the partial destruction of the indole ring of tryptophan. Consequently, the amount of tryptophan in the protein must be estimated by an alternative method (e.g., spectroscopic absorption) when using acid hydrolysis. In addition, acid hydrolysis results in the loss of ammonia from the side-chain amide groups of glutamine and asparagine, with the consequent production of glutamic and as-

Figure 4.9

Acid hydrolysis of a protein or polypeptide to yield amino acids.

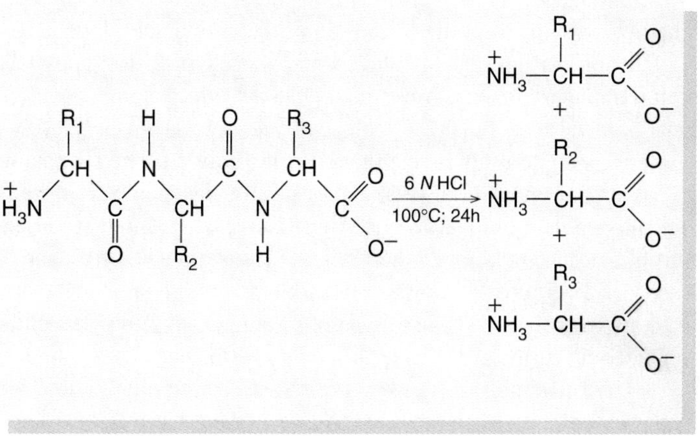

6 *N* HCl, 100°C; 24h

Figure 4.10

Acid hydrolysis of protein converts glutamine to glutamic acid. A similar reaction occurs for asparagine, which possesses an identical side-chain functional group.

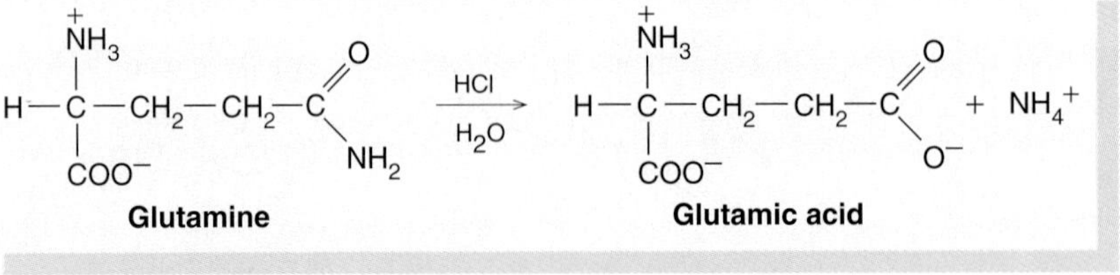

$$H-\overset{\overset{+}{NH_3}}{\underset{COO^-}{C}}-CH_2-CH_2-C(=O)NH_2 \xrightarrow[H_2O]{HCl} H-\overset{\overset{+}{NH_3}}{\underset{COO^-}{C}}-CH_2-CH_2-C(=O)O^- + NH_4^+$$

Glutamine → **Glutamic acid**

partic acids (fig. 4.10). Therefore, estimates of amino acid composition based on acid hydrolysis show glutamine and glutamic acid combined and measured as glutamic acid. Similarly, asparagine and aspartic acid are combined and measured as aspartic acid.

Quantitative analysis of an amino acid mixture is achieved by chromatography. The general efficacy of chromatographic techniques is based on differences in affinity between each compound to be separated and an immobile phase. The immobile phase usually contains some functional groups on its surface that are ideally suited to conducting a particular type of separation. In chapter 7 we will survey a range of chromatographic procedures that are in common use. Here we will describe a procedure that is ideally suited for the quantitative analysis of a mixture of amino acids present in an amino acid digest.

Amino acids are converted to derivatives that are suitable for analysis. Originally, the compound ninhydrin was used for this purpose. Reaction with ninhydrin converts most amino acids to a blue compound that can be quantitatively analyzed in a spectrophotometer. Much more sensitive techniques are now available that lead to the conversion of amino acids to fluorescent derivatives that are assayed in a fluorimeter. Once the particular reagent for amino acid conversion has been selected, a decision must be made as to whether the amino acids should be derivatized before or after chromatography. This decision depends on the particular reagent chosen for making derivatives and the chromatographic procedure to be used. A very effective procedure entails the use of reverse phase chromatography of amino acids that have been derivatized by reaction with *o*-phthalaldehyde (OPA). The chemistry involved in the formation of the amino acid derivatives is described in figure 4.11. For reverse-phase chromatography the column consists of very fine silica beads to which nonpolar alkyl chains have been linked. A polar solvent is used in the elution. First, the sample is applied at the top of the column. Then the sample is eluted under high pressure to give a reasonable flow rate. The most polar compounds elute first. The eluted sample is passed through a fluorimeter, and the fluorescence is graphed on a charge recorder. In figure 4.12, two solvent chambers are shown for occasions when it is considered useful to gradually change the eluting solvent during the course of the separation.

Conversion of the relative ratios of amino acids into an estimate of actual composition requires some additional information concerning the protein's molecular weight; for example, an analysis giving relative ratios of Ala (1.0), Gly (0.5), and Lys (2.0) could correspond to composition Ala_2-Gly-Lys_4 or any multiple thereof. The required information is usually available, and in any case an estimation of composition based on a minimum molecular weight of the protein is always possible. Results for three proteins are shown in table 4.3.

Figure 4.11

Formation of OPA-derivatized amino acids for chromatographic analysis.

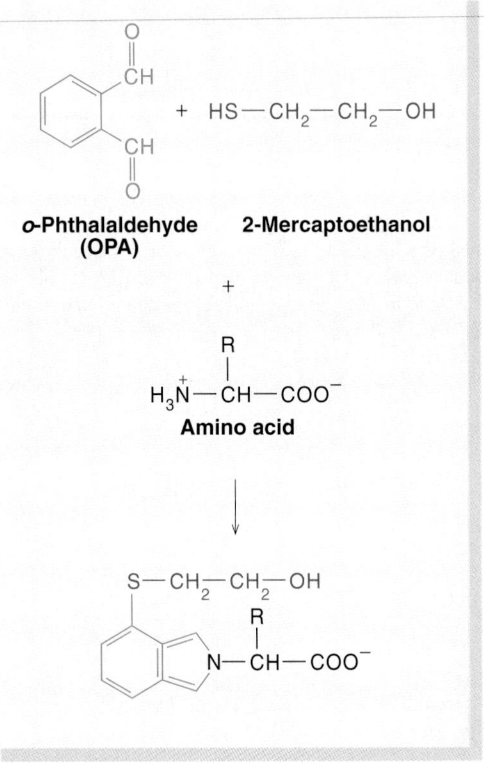

$$C_6H_4(CHO)_2 + HS-CH_2-CH_2-OH$$

o-Phthalaldehyde (OPA) **2-Mercaptoethanol**

+

$$H_3\overset{+}{N}-CH(R)-COO^-$$

Amino acid

↓

S—CH_2—CH_2—OH, R, N—CH—COO^-

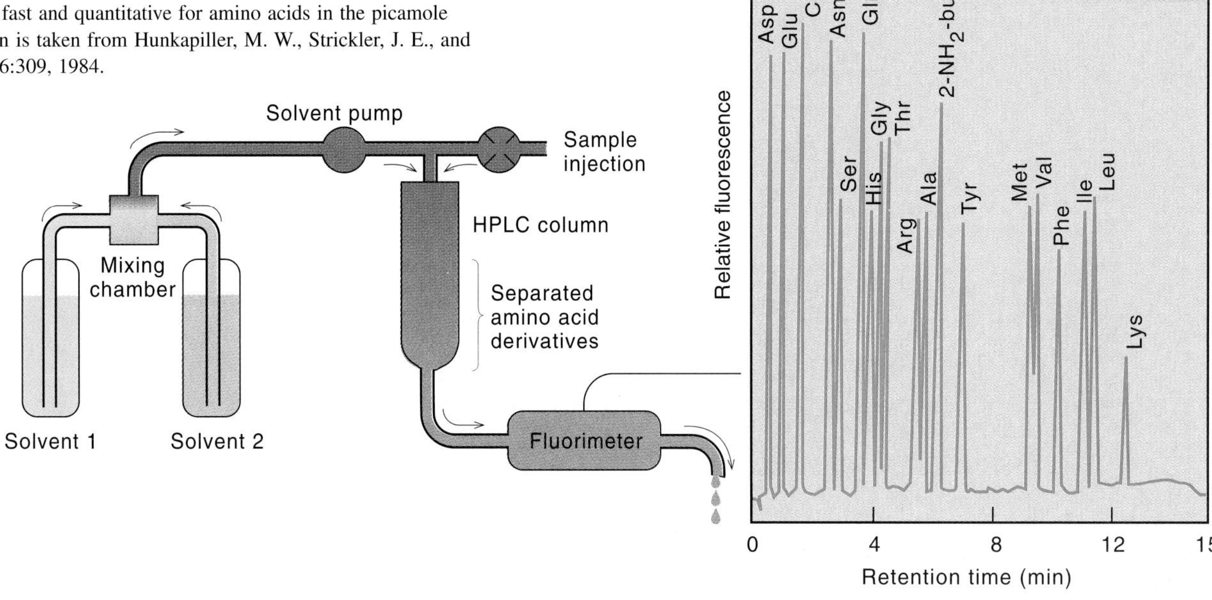

Figure 4.12

Schematic diagram of an apparatus for the chromatographic analysis of OPA-derivatized amino acids by reverse-phase high-pressure liquid chromatography (HPLC). This method is fast and quantitative for amino acids in the picamole range. The elution pattern is taken from Hunkapiller, M. W., Strickler, J. E., and Wilson, K. J. *Science* 226:309, 1984.

Table 4.3
Amino Acid Content of Proteins (in percent)

Constituent	Insulin (Bovine)	Ribonuclease (Bovine)	Cytochrome (Equine)
Alanine	4.6	7.7	3.5
Amide NH_3	1.7	2.1	1.1
Arginine	3.1	4.9	2.7
Aspartic acid	6.7	15.0	7.6
Cysteine	0	0	1.7
Cystine	12.2	7.0	0
Glutamic acid	17.9	12.4	13.0
Glycine	5.2	1.6	5.6
Histidine	5.4	4.2	3.4
Isoleucine	2.3	2.7	5.4
Leucine	13.5	2.0	5.6
Lysine	2.6	10.5	19.7
Methionine	0	4.0	2.1
Phenylalanine	8.6	3.5	4.5
Proline	2.1	3.9	3.3
Serine	5.3	11.4	0
Threonine	2.0	8.9	8.4
Tryptophan	0	0	1.5
Tyrosine	12.6	7.6	4.9
Valine	9.7	7.5	2.4

Determination of Amino Acid Sequence of Proteins

The most important properties of a protein are determined by the sequence of amino acids in the polypeptide chain. This sequence is called the primary structure of the protein. We know the sequences for thousands of peptides and proteins, largely through the use of methods developed in Fred Sanger's laboratory and first used to determine the sequence of the peptide hormone insulin in 1953. Knowledge of the amino acid sequence is extremely useful in a number of ways: (1) It permits comparisons to be made between normal and mutant proteins; (2) it permits comparisons to be made between comparable proteins in different species and thereby has been instrumental in positioning different organisms on the evolutionary tree (see fig. 1.20); (3) finally and most important, it is a vital piece of information for determining the three-dimensional structure of the protein.

Determining the order of amino acids involves the sequential removal and identification of successive amino acid residues from one or the other free terminal of the polypeptide chain. However, in practice it is extremely difficult to get the required specific cleavage reaction of the desired products to proceed with 100% yield. This obstacle becomes significant when sequencing long polypeptides, because the fraction of the total material of minimum polypeptide chain length becomes constantly smaller as the successive removal of terminal residues continues. Conversely, the amino acid released from the polypeptide chain becomes increasingly contaminated with amino acids released from previously unreacted chains.

Because of this fundamental chemical limitation, the polypeptide chain must be broken down into sequences short enough

Figure 4.13

Steps involved in the sequence determination of the B chain of insulin. Amino acids are represented here by their single-letter codes (see table 4.1).

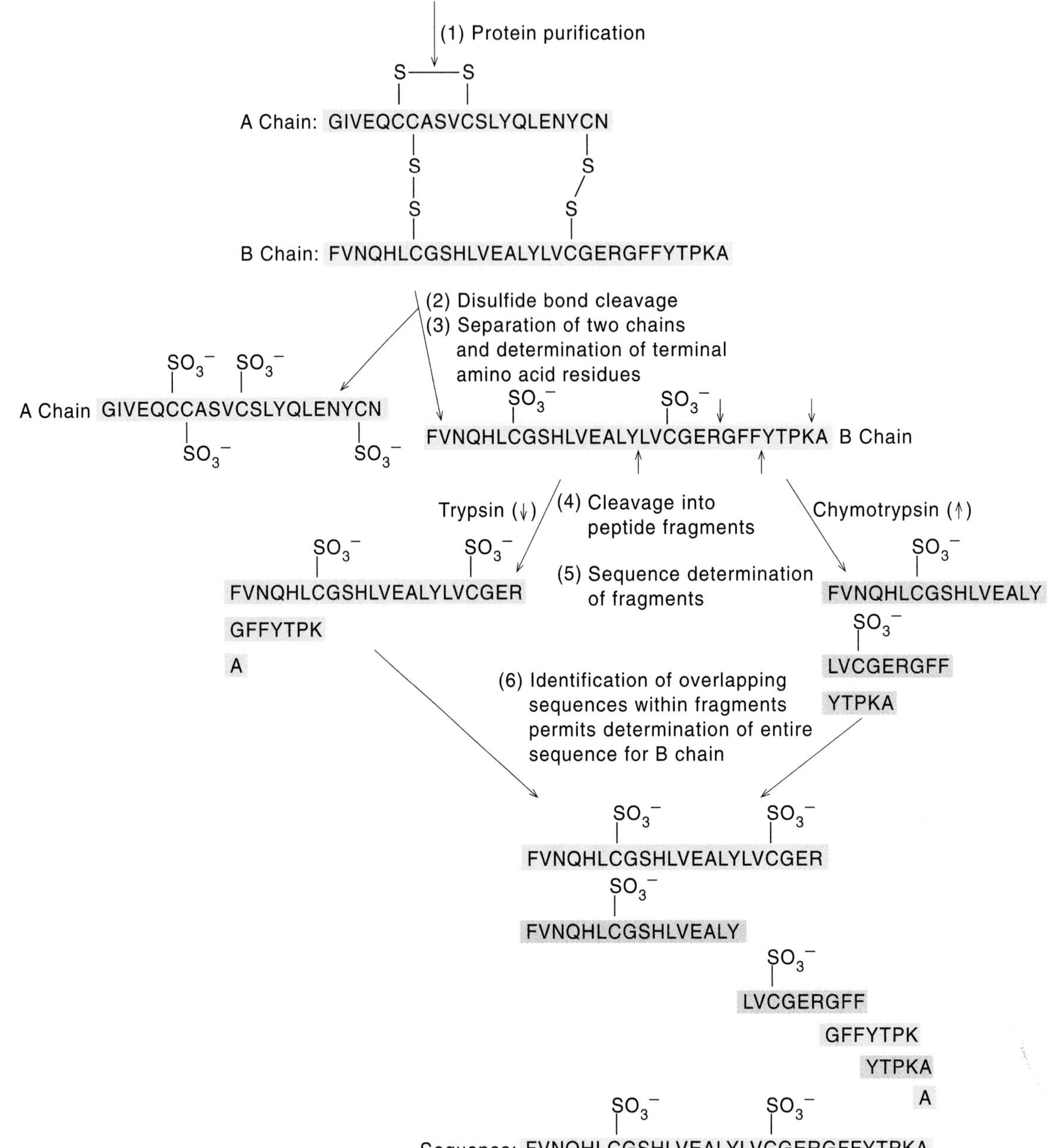

for the chemistry to produce reliable results. The short sequences are then reassembled to obtain the overall sequence. The steps actually involved in protein sequencing (fig. 4.13) are (1) purification of the protein; (2) cleavage of all disulfide bonds; (3) determination of the terminal amino acid residues; (4) specific cleavage of the polypeptide chain into small fragments in at least two different ways; (5) independent separation and sequence determination of peptides produced by the different cleavage methods; and (6) reassembly of the individual peptides with appropriate overlaps to determine the overall sequence.

The first step, protein purification, will be discussed in chapter 7. Once the protein is pure, sequence analysis can begin, with cleavage of the disulfide bonds. Cleavage is achieved by oxidizing the disulfide linkages with performic acid (fig. 4.14). Sometimes this step results in the production of two or more polypeptide chains, in which case the individual chains must be separated.

Figure 4.14

Disulfide cleavage reactions. Prior to sequence analysis inter- and intra-chain disulfide linkages are irreversibly cleaved by one of the two procedures shown.

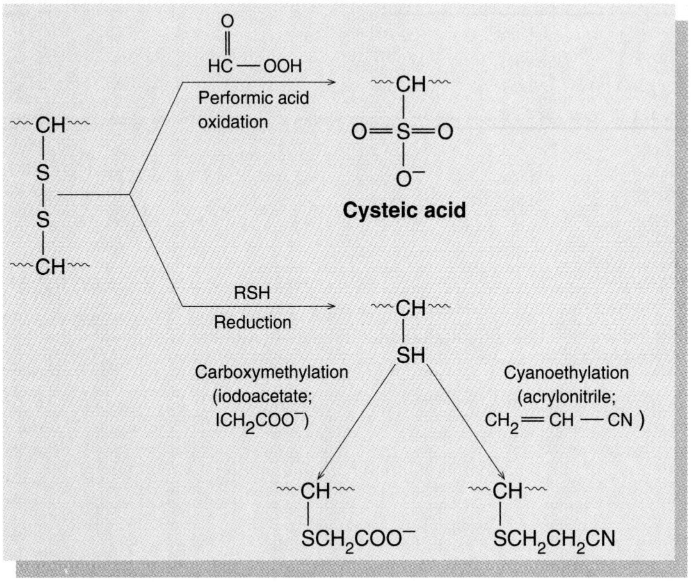

The third step is to determine the polypeptide-chain end groups. If the polypeptide chains are pure, then only one N-terminal and one C-terminal group should be detected. The amino-terminal amino acid can be identified by reaction with fluorodinitrobenzene (FDNB) (fig. 4.15). Subsequent acid hydrolysis releases a colored DNP-labeled amino-terminal amino acid, which can be identified by its characteristic migration rate on thin-layer chromatography or paper electrophoresis. A more sensitive method of end-group determination involves the use of dansyl chloride (see box 4C).

Chemical methods for carboxyl end-group determination are less satisfactory. Treatment of the peptide with anhydrous hy-

Figure 4.15

Polypeptide chain end-group analysis. (*a*) Amino-terminal group identification. A more sensitive method, the dansyl chloride method, is described in box 4C. (*b*) Carboxyl-terminal group identification. Identification of this amino acid is considerably more difficult.

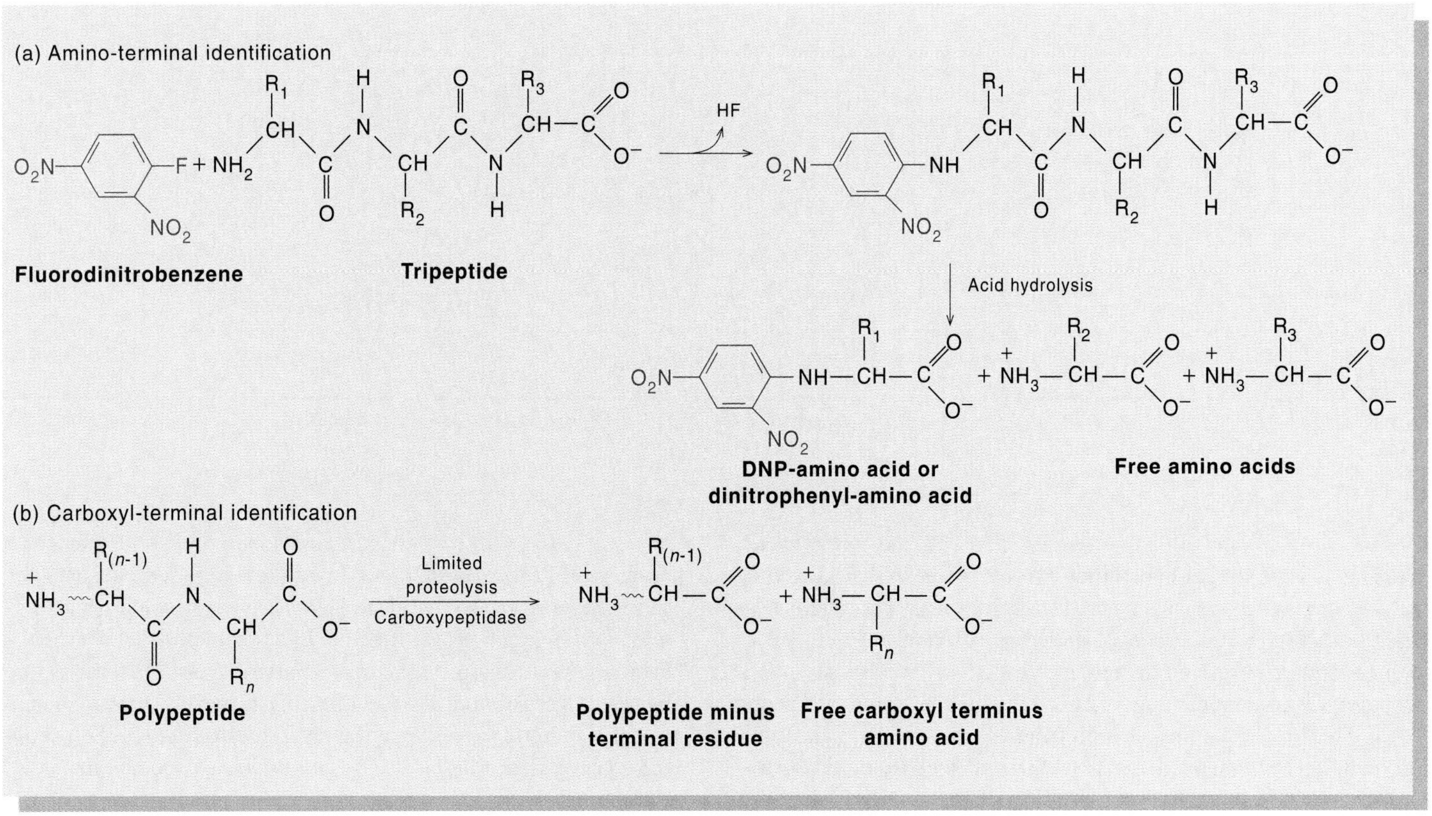

The Dansyl Chloride Method for N-Terminal Amino Acid Determination

The dansyl chloride method provides an alternative to the Sanger method for N-terminal amino acid determination. Because it is considerably more sensitive than the Sanger method, it has become the method of choice. The reaction is diagrammed in the illustration. A polypeptide is treated with dansyl chloride to give an *N*-dansyl peptide derivative. This derivative is hydrolyzed to yield a highly fluorescent *N*-dansyl-amino acid, which is detected chromatographically.

drazine at 100°C results in conversion of all the amino acid residues to amino acid hydrazides except for the carboxyl-terminal residue, which remains as the free amino acid and can be isolated and identified chromatographically. Alternatively, the polypeptide can be subjected to limited breakdown (proteolysis) with the enzyme carboxypeptidase. This results in release of the carboxyl-terminal amino acid as the major free amino acid reaction product. The amino acid type can then be identified chromatographically.

Step 4 involves breaking down the polypeptide chain into shorter, well-defined fragments for subsequent sequence analysis. Fragmentation can be achieved by the use of endopeptidases, which are enzymes that catalyze polypeptide-chain cleavage at specific peptide linkages. Figure 4.16 shows the specificity of three endopeptidases commonly used for this purpose. Another specific chemical method for polypeptide-chain cleavage involves reaction with cyanogen bromide. This reaction cleaves specifically at the methionine residues, with the accompanying conversion of free

Figure 4.16

Site of action of some endopeptidases used for polypeptide chain cleavage prior to sequence analysis. Of the three different enzymes used, trypsin is used most frequently because of its high specificity.

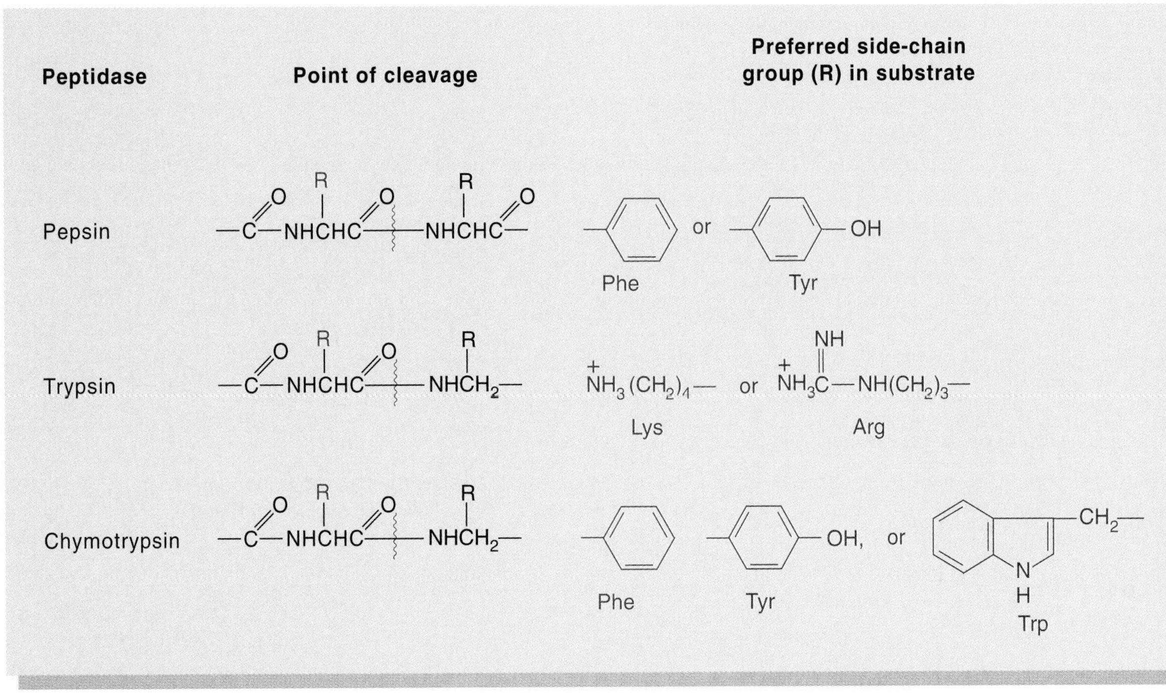

Peptidase	Point of cleavage	Preferred side-chain group (R) in substrate
Pepsin	—C(=O)—NHCH(R)C(=O)—⌇—NHCH(R)C(=O)—	Phe or Tyr (—C$_6$H$_4$—OH)
Trypsin	—C(=O)—NHCH(R)C(=O)—⌇—NHCH(R)$_2$—	$\overset{+}{N}H_3(CH_2)_4$— (Lys) or $\overset{+}{N}H_3C(=NH)$—$NH(CH_2)_3$— (Arg)
Chymotrypsin	—C(=O)—NHCH(R)C(=O)—⌇—NHCH(R)$_2$—	Phe, Tyr (—OH), or Trp (indole—CH_2—)

Figure 4.17

The cleavage of polypeptide chains at methionine residues by cyanogen bromide. The cleavage reaction is accompanied by the conversion of the newly formed free carboxyl-terminal methionine to homoserine lactone.

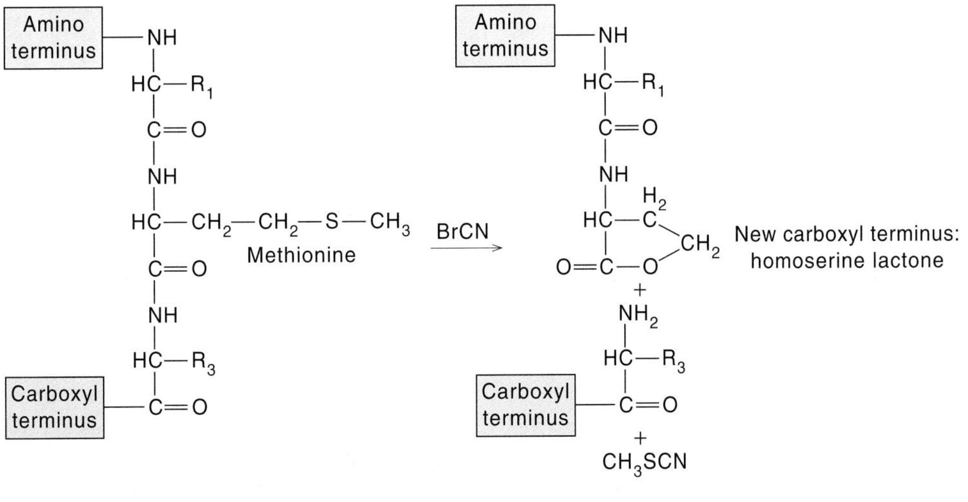

Amino terminus—NH—HC(—R_1)—C=O—NH—HC(—CH_2—CH_2—S—CH_3 Methionine)—C=O—NH—HC(—R_3)—C=O—Carboxyl terminus

$\xrightarrow{BrCN}$

Amino terminus—NH—HC(—R_1)—C=O—NH—HC—CH_2—CH_2—O—C(=O) (New carboxyl terminus: homoserine lactone)

+ $\overset{+}{N}H_2$—HC(—R_3)—C=O—Carboxyl terminus

+ CH_3SCN

carboxyl-terminal methionine to homoserine lactone (fig. 4.17). Although this methionine reaction product differs from the 20 naturally occurring amino acids, it is nevertheless readily identified by subsequent conversion to homoserine.

Peptides resulting from cleavage of the intact protein are generally separated by ion-exchange chromatographic methods (this method of chromatography is explained in chapter 7). The isolated peptides may then be analyzed (step 5) to determine both their amino acid composition and their sequence. Sequence determination involves the stepwise removal and identification of

Figure 4.18

The Edman degradation method for polypeptide sequence determination. The sequence is determined one amino acid at a time, starting from the amino-terminal end of the polypeptide. First the polypeptide is reacted with phenylisothiocyanate to form a polypeptidyl phenylthiocarbamoyl derivative. Gentle hydrolysis releases the amino-terminal amino acid as a phenylthiohydantoin (PTH), which can be separated and detected spectrophotometrically. The remaining intact polypeptide, shortened by one amino acid, is then ready for further cycles of this procedure. A more sensitive reagent, dimethylaminoazobenzene isothiocyanate, can be used in place of phenylisothiocyanate. The chemistry is the same.

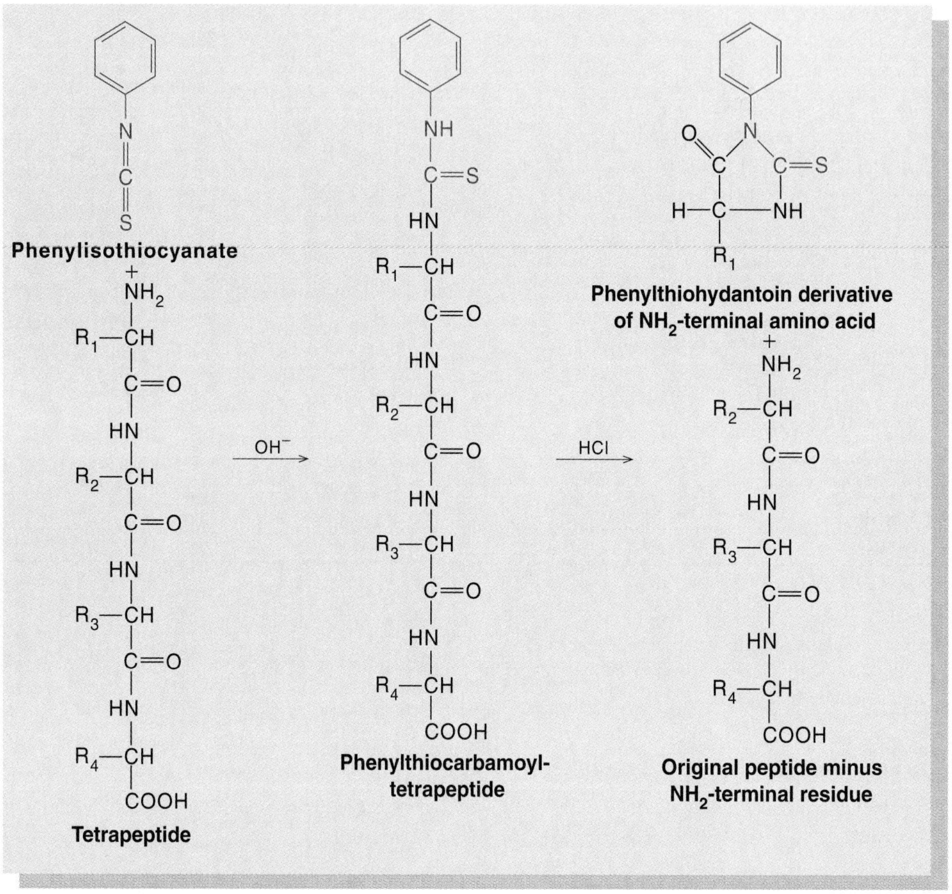

successive amino acids from the polypeptide amino terminal by means of the Edman degradation (fig. 4.18). This process is carried out by reacting the free amino-terminal group with phenylisothiocyanate to form a peptidyl phenylthiocarbamyl derivative. Gentle hydrolysis with hydrochloric acid releases the amino-terminal amino acid as a phenylthiohydantoin (PTH) derivative. The remaining intact peptide, shortened by one amino acid, is then ready for further cycles of this procedure. The PTH-amino acid can be identified by chromatography.

Devices called sequenators are available that automate the Edman degradation procedure. The success of these devices depends in large part on the technical innovation of covalently linking the peptide to be sequenced to glass beads. Attachment of the peptide through its carboxyl-terminal group to this immobile phase facilitates the complete removal of potentially contaminating reaction products during successive stages of the degradation.

Finally, having established the sequences of the individual peptides, it is necessary only to establish how they are connected together in the intact protein (step 6). It is at this stage that we see why the preceding sequence analysis was performed on peptides obtained by two different specific cleavage methods. This approach makes it possible to piece together the overall sequence, because the two sets of results produce overlapping sequences. That is, the free amino and carboxyl residues of peptides originally interconnected in the intact protein and liberated by one specific cleavage method will recur in the internal sequences of the peptides liberated by a second specific method.

Once the protein's primary sequence has been determined, the location of disulfide bonds in the intact protein can be established by repeating a specific enzymatic cleavage on another sample of the same protein in which the disulfide bonds have not previously been cleaved. Separation of the resulting peptides will show the appearance of one new peptide and the disappearance of two other peptides, when compared with the enzymatic digestion product of the material whose disulfide bonds have first been chemically cleaved. In fact, these difference techniques are generally useful in the detection of sites of mutations in protein molecules of previously known sequence, since a single substitution will generally affect the chromatographic properties of only a single peptide released during proteolytic digestion.

Great progress has been made in recent years in devising procedures for sequencing the DNA that encodes for proteins (see chapter 32). Knowing the sequence of coding triplets in DNA allows us to read off the amino acid sequence of the corresponding protein. Nevertheless, such studies have produced the remarkable observation that many eukaryotic DNA sequences coding for proteins are not continuous but instead contain untranslated intervening DNA sequences. Although these results have profound implications for protein evolution, they obviously confound the general applicability of DNA-sequencing methods for the purposes of protein primary-structure determination. In cases like this, the usual solution has been to isolate the mRNA for the protein and use this to make a DNA carrying the same sequence. This procedure circumvents the intervening-sequence problem because the mRNA carries only the coding sequences (see chapter 32).

Chemical Synthesis of Peptides and Polypeptides

Knowledge about the structure–function interrelationships in proteins and peptides has encouraged biochemists to develop techniques for synthesizing peptides and proteins with predetermined sequences. To synthesize a peptide in the laboratory, we must overcome several problems related to preventing undesired groups from reacting. The amino and carboxyl groups that are to remain unlinked must be blocked; so must all reactive side chains. Some protecting groups for carboxyl and amino groups are shown in figures 4.19 and 4.20, respectively.

After blocking those groups to be protected, we generally activate the carboxyl group; two methods for doing this are shown in figure 4.21. It is of interest that carboxyl-group activation is also employed in natural biosynthesis in the cell (see chapter 34). After peptide synthesis, the protecting groups must be removed by a mild method. The overall process—comprising protection, activation, coupling, and unblocking—is shown in figure 4.22.

An important variation of the usual methods of peptide synthesis involves attaching a protected (*t*-butyloxycarbonyl group) amino acid to a solid polystyrene resin, removal of the amino protecting group, condensation with a second protected amino acid, and so on. In the last step, the finished peptide is cleaved from the resin. This method (outlined in figure 4.23) has the advantage that cumbersome purification between steps, often resulting in serious losses, is replaced by mere washing of the insoluble resin. Since each reaction is essentially quantitative, very long peptides, and even proteins, can be synthesized by this method. Indeed, Li synthesized a 39-amino-acid protein hormone, adrenocorticotropic hormone, by this method, and Merrifield synthesized bovine pancreatic ribonuclease, which contains 129 amino acids in a single polypeptide chain. A number of variants of ribonuclease that contain one or more changes in amino acid sequence also have been made by this method. The importance of the Merrifield process was underscored by the awarding of a Nobel Prize to Merrifield in 1984.

Figure 4.19

Carboxyl protecting groups used in peptide synthesis. The symbol ▲ in the amino acid structure on the left stands for one of the protecting groups (middle), leading to the named compound indicated on the right. The protecting group prevents the carboxyl group from participating in subsequent reactions involved in peptide synthesis.

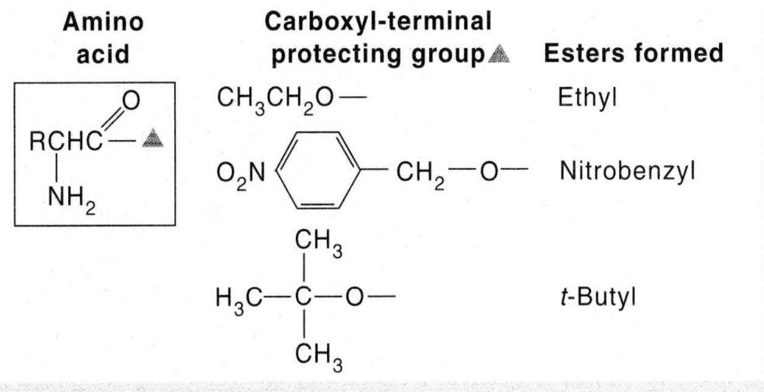

Figure 4.20

Amino protecting groups used in peptide synthesis. The symbol ■ in the amino acid structure on the left stands for one of the protecting groups (middle).

Amino acid	Amino-terminal protecting group■	Name of protecting group
RCHCOOH / NH—■	C_6H_5—$CH_2OC(=O)$—	Carbobenzoxy
	$(CH_3)_3COC(=O)$—	*t*-Butyloxycarbonyl
	CH_3—C_6H_4—$S(=O)_2$—	Tosyl
	$F_3CC(=O)$—	Trifluoroacetyl

Figure 4.21

Different ways of activating the carboxyl group for peptide synthesis. Activated amino acids will react spontaneously with most α-amino acids as illustrated in figure 4.22.

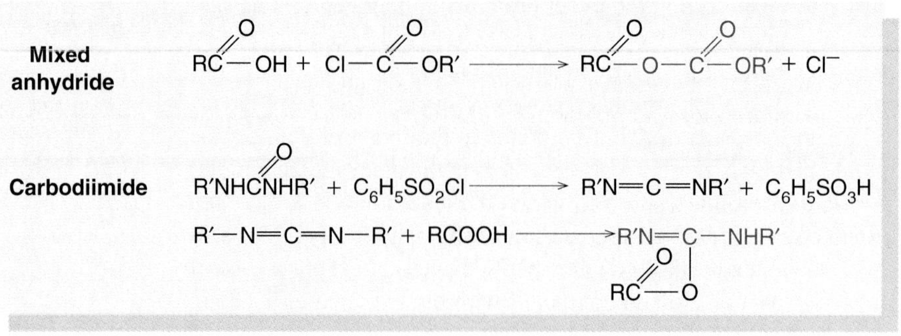

Mixed anhydride

$$RC(=O)-OH + Cl-C(=O)-OR' \longrightarrow RC(=O)-O-C(=O)-OR' + Cl^-$$

Carbodiimide

$$R'NHC(=O)NHR' + C_6H_5SO_2Cl \longrightarrow R'N{=}C{=}NR' + C_6H_5SO_3H$$

$$R'-N{=}C{=}N-R' + RCOOH \longrightarrow R'N{=}C(-O-C(=O)R)-NHR'$$

Figure 4.22

Schematic diagram illustrating the chemical method for peptide synthesis. First the amino acids to be linked are selected. The carboxyl group and the amino group that are to be excluded from peptide synthesis are protected (steps 1 and 1′). Next the amino acid containing the unprotected carboxyl group is carboxyl-activated (step 2). This amino acid is mixed and reacted with the other amino acid (step 3). Protecting groups are then removed from the product (step 4).

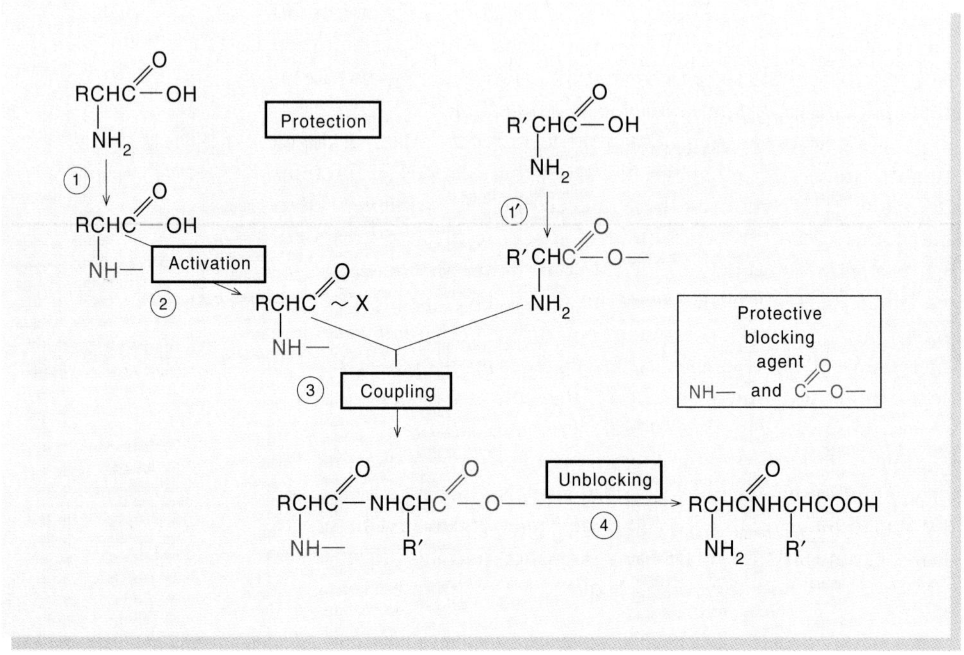

Protection

Activation

Coupling

Unblocking

Protective blocking agent NH— and C(=O)—O—

Figure 4.23

Merrifield procedure for solid-state dipeptide synthesis. (1) Polymer is activated. (2) Amino acid containing a *t*-butoxycarbonyl (BOC)-protecting group is carboxyl-linked to the polymer. This amino acid will be the carboxyl-terminal amino acid in the final peptide. (3) The BOC protecting group is removed from the polymer-linked amino acid. (4) A second amino acid, containing a BOC on its α-amino group and a dicyclohexylcarbodiimide (DCC)-activated group, is reacted with the column-bound amino acid to form a dipeptide. (5) The dipeptide is released from the polymer and the BOC-protecting group by adding hydrogen bromide (HBr) in trifluoroacetic acid.

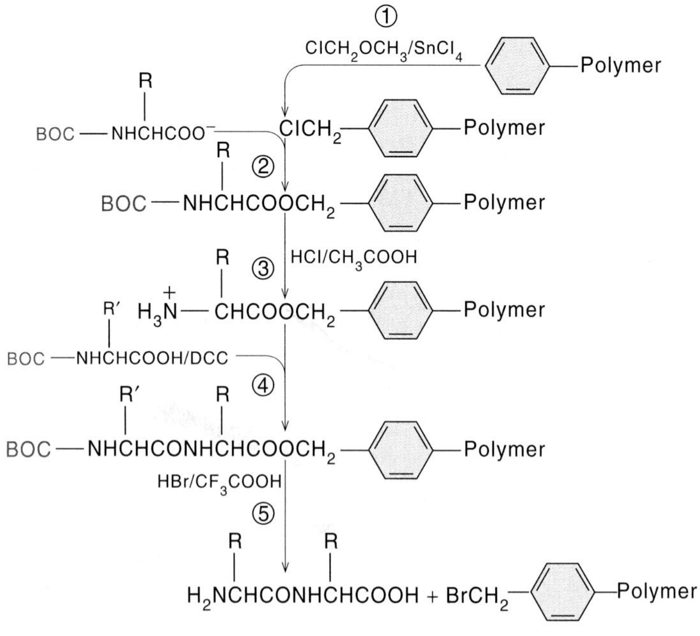

① $ClCH_2OCH_3/SnCl_4$

② BOC—NHCH(R)COO⁻

③ HCl/CH_3COOH

④ BOC—NHCH(R′)COOH/DCC

⑤ HBr/CF_3COOH

$H_2NCH(R)CONHCH(R)COOH + BrCH_2$—Polymer

BOC = *t*-Butoxycarbonyl
DCC = Dicyclohexylcarbodiimide

Summary

In this chapter we have dealt with some of the fundamental properties of amino acids and polypeptide chains. The following points are especially important.

1. Nineteen of the twenty amino acids commonly found in proteins have a carboxyl group and an amino group attached to an α-carbon atom; they differ in the side chain attached to the same α carbon.
2. All amino acids have acidic and basic properties.
3. All amino acids except glycine are asymmetric and therefore can exist in at least two different stereoisomeric forms.
4. Peptides are formed from amino acids by the reaction of the α-amino group from one amino acid with the α-carboxyl group of another amino acid.
5. Polypeptide formation involves a repetition of the process involved in peptide synthesis.
6. The amino acid composition of proteins can be discovered by first breaking down the protein into its component amino acids and then separating the amino acids in the mixture for quantitative estimation.
7. The amino acid sequences of proteins can be discovered by breaking down the protein into polypeptide chains and then partially degrading the polypeptide chains. For each polypeptide chain fragment, the sequence is determined by stepwise removal of amino acids from the amino-terminal end of the polypeptide chain. Two different methods of forming polypeptide chain fragments are used so as to produce a map of overlapping fragments, from which the sequence of undegraded polypeptide chains in the proteins can be deduced.
8. Polypeptide chains with a predetermined amino acid sequence can be synthesized by chemical methods involving carboxyl-group activation.

Selected Readings

Barrett, G. C. (ed.), *Chemistry and Biochemistry of Amino Acids*. New York: Chapman and Hall, 1985. A recent and authoritative volume on this classical subject.

Davies, J. S. (ed.), *Amino Acids and Peptides*. New York: Chapman and Hall, 1985.

Gray, W. R., End group analysis using dansyl chloride. *Methods in Enzymology* 25:121–138, 1972. This volume of *Methods in Enzymology* contains several chapters on end-group analysis.

Heiser, T., Amino acid chromatography: the "best" technique for student labs. *J. Chem. Ed.* 67:964–966, 1990.

Huheey, J. E., A novel method for assigning R, S labels to enantiomers. *J. Chem. Ed.* 63:598–600, 1986.

Hunkapiller, M. W., J. E. Strickler, and K. J. Wilson, Contemporary methodology for protein structure determination. *Science* 226:304–311, 1984.

Kent, S. B. H., Chemical synthesis of peptides and proteins. *Ann. Rev. Biochem.* 57:957–989, 1988. Comprehensive and up-to-date.

Kleinkauf, H., and H. Dohren, Nonribosomal polypeptide formation on multifunctional proteins. *Trends Biochem. Sci.* 8:281–283, 1983.

Merrifield, B., Solid phase synthesis. *Science* 232:341–347, 1986.

Mor, A., M. Amiche, and P. Nicholas, Enter a new post-transcriptional modification: D-amino acids in gene-encoded peptides. *Trends Biochem. Sci.* 17:481–485, 1992.

Popják, G., Stereospecificity of enzymic reactions. In *The Enzymes,* Vol. 2. P. D. Boyer, (ed.). New York: Academic Press, 1970, p. 115.

Sanger, F., Sequences, sequences and sequences. *Ann. Rev. Biochem.* 57:1–28, 1988.

Thompson, J., and J. A. Donkersloot, *N*-(Carboxyalkyl) amino acids: Occurrence, synthesis and functions. *Ann. Rev. Biochem.* 61:517–557, 1992.

Problems

1. The dipeptide glycylglycine has pK_a values of 3.12 and 8.17. What chemical reasons can you give for the change in pK_a values from those of glycine (2.35 and 9.78)?
2. You have 50 ml of 10 mM fully protonated histidine. How many millimoles of base must be added to bring the histidine solution to a pH that is equivalent to the pI?
3. Calculate the isoelectric point for histidine, aspartic acid, and arginine. Calculate the fractional charge for each ionizable group or aspartate at pH equal to pI. Do the results verify the isoelectric point of aspartic acid?
4. Which of the naturally occurring amino acid side chains are charged at pH 2? pH 7? pH 12? (Consider only those amino acids whose side chains have >10% charge at the pH indicated.)
5. Amino acids make excellent buffers, with only the available pK_a values and cost factors as major problems. A mixture of glutamic acid and histidine would function as a buffer at what pH range(s)?
6. Draw the structures of the major ionic forms and determine their concentrations in a 0.5 mM solution of arginine at pH 12.18. What is the average charge on arginine at this pH?
7. What is the charge on aspartyllysylcysteine at a pH of 4.4? The pK_a values of the N- and C-terminal groups of a peptide are near 9 and 3, respectively.
8. When Rose first discovered threonine in 1935, its structure was determined to be 2-amino-3-hydroxybutanoic acid. Rose (*J. Bio. Chem. 115*:721, 1936) deduced the geometry on carbon 3 in part by comparisons with threose and erythrose (see fig. 13.2). Interestingly, Rose in his 1936 paper named the amino acid "D-threonine." Can you provide an explanation for this name that does not contradict the amino acid chirality statement made in the text?

9. A mixture of valine, aspartic acid, histidine, and lysine were applied to a DEAE-cellulose ion exchange column. (The immobile phase is positively charged at pH below 8.) Determine the expected order of elution that would occur with a pH 6.2 buffer. Are all the amino acids separated from each other? If they are not separated, how might you modify the procedure to produce separation?
10. A mixture of alanine, glutamic acid, and arginine was chromatographed on a weakly basic ion-exchange column (positively charged) at pH 6.1. Predict the order of elution of the amino acids from the non-exchange column. Are the amino acids separated from each other? Explain. Suppose you have a weakly acidic ion-exchange column (negatively charged), also at pH 6.1. Predict the order of elution of the amino acids from this column. Propose a strategy to separate the amino acids using one or both columns. Explain your rationale. (Assume only ionic interactions between the amino acids and the ion exchange resin.)
11. Of the 20 amino acids found in proteins, what common "functional groups" are not represented in the side chains of the amino acid residues?
12. A heptapeptide when treated with trypsin produced two peptides: T_1 (Asp, Gly, Tyr) and T_2 (Lys, Phe, Val, Ala). When the heptapeptide was treated with chymotrypsin, three peptides were produced CT_1 (Lys, Tyr, Gly), CT_2 (Phe, Ala, Val), and CT_3 (Asp). Parentheses indicate an unknown sequence. When the peptide was treated with 2,4-dinitrofluorobenzene and hydrolyzed, DNP-Lys and DNP-Ala were recovered. What is the amino acid sequence of the heptapeptide? What is the structure of the DNP-Lys?
13. The three chymotrypsin fragments in problem 12 were subject to anion exchange chromatography at pH 7.9. (Resin is positively charged.) What elution pattern would you expect?
14. How would you detect the peptides in the eluant in problem 13? Would UV be of any value?
15. During the late 1800s, when many of the amino acids were first discovered, base ($Ba(OH)_2$) hydrolysis was commonly used. For a short time the amino acid ornithine (1,5-diaminopentanoic acid) was thought to be a protein component. In addition to ornithine, ammonia and carbonate were observed as products of base hydrolysis of proteins. Propose an explanation for these hydrolysis products.
16. An occasional double hydrogen bond has been observed in proteins between the phenolic groups of two tyrosyl residues. Propose a structure for this double hydrogen bond.
17. A heptapeptide upon HCl hydrolysis produced equimolar amounts of Asp, Cys, Glu, Lys, Phe, Tyr, Val, and ammonia. Exposure of the intact heptapeptide to FDNB followed by hydrolysis produced DNP-Val and ϵ-DNP-Lys. Treatment with trypsin produced a tripeptide, T1, and a tetrapeptide, T2. T2 had an absorbance peak near 260 nm and produced ammonia on hydrolysis. T1 had an absorbance peak near 275 nm (which was greater than that of T1) and tested positive for sulfur. Exposure of the heptapeptide to chymotrypsin produced two tripeptides (CT1 and CT2), which both had UV absorbances, and Asp. CT1 contained sulfur and had the greater absorbance of the two chymotrypsin produced tripeptides. CT2 produced ammonia upon hydrolysis. When exposed to electrophoresis at pH 6, T2 and CT1 were cations, T1 was an anion, and CT2 had essentially no charge. Deduce the sequence of the heptapeptide from the provided data.

CHAPTER 5

THE THREE-DIMENSIONAL STRUCTURES OF PROTEINS

Proteins adopt the most stable folded structures; this is a function of the way in which the individual amino acid residues interact with one another.

The enormous structural diversity of proteins begins with the amino acid sequences of polypeptide chains. Each protein consists of one or more unique polypeptide chains, and each of these polypeptide chains is folded into a complex three-dimensional structure. The final folded arrangement of the polypeptide chain in the protein is referred to as its conformation. Most proteins exist in unique conformations exquisitely suited to the highly specific function of each protein. It is the availability of a wide variety of conformations that permits proteins as a group to perform a broader range of functions than any other class of biomolecules.

Before the first x-ray diffraction results were understood, it was imagined that protein structures were relatively simple geometric arrangements of polypeptide chains, such as geometric cages, repeating zigzags, or uniform arrays of parallel rods. Indeed, the first structures determined for proteins were of this type. These structures were deduced by Pauling and Corey, using information from various sources: (1) They knew a little about the structures of peptides from small-molecule crystallography, which indicated that the peptide bond was planar and gave accurate bond lengths and angles. (2) They were already aware of the importance of hydrogen bonds in determining the orientation of amino acids, peptides, and even water in simple crystals. (3) They made shrewd guesses about the interpretation of a few spacings in the diffraction patterns of certain fibrous proteins. (4) Putting all of this information together, they experimented with molecular models until they could produce structures in reasonable agreement with all the available facts. This was a historic achievement.

It is not surprising that repeating structures with long-range order were the first protein structures to be understood. The demands on the available technology were minimal. Much more sophisticated technology was required to interpret the diffraction patterns of most proteins, which have less long-range repetition. Even Kendrew and Perutz, who led their research teams to a solution of the first such structures to be determined, myoglobin and hemoglobin, were shocked when they realized how chaotic the arrangements in such structures seemed to be. Kendrew was once introduced as the man who proved that proteins were ugly, and Perutz, addressing his initial disappointment, said, "Could the search for ultimate truth really have revealed so hideous and visceral-looking an object?" Whether Kendrew and Perutz were truly as disappointed as they appeared to be, or whether they were actually delighted that their efforts of more than a quarter century had led them to structures that were so complex that no person could have predicted them is for science historians to determine. Suffice it to say that this crowning achievement of protein structure determination by x-ray diffraction has been repeated many times since with less ado.

Enormous advances have been made in protein chemistry and in computer technology as well. These advances have systematized the necessary research and greatly reduced the amount of work and time required to determine a protein structure. Accurate structure determinations have now been made on over 500 different proteins. From this wealth of information, patterns of structure are becoming apparent that suggest, among other things, that the final folding arrangements of proteins may some day be predictable from the amino acid sequences of the polypeptide chains.

In this chapter we will see how the forces that cause proteins to fold, in concert with what are basically geometric properties of the polypeptide chain, combine to produce the highly organized structures that are typical of functionally active proteins. To start, we will look at the basic properties of the structural material of proteins, to see what sorts of local structural arrangements are possible for polypeptide chains. Then we will describe the way in which these locally organized structural units can be most efficiently assembled into progressively larger and more complicated arrangements. We will also consider the relationships among a protein's sequential, functional, and structural properties.

The Information for Folding Is Contained in the Primary Structure

The conformation of a native or highly organized protein reflects a delicate balance among a variety of interaction forces, both within the folded protein's interior and with surrounding solvent. If the protein's solvent environment is perturbed, the protein's native conformation can be disrupted, with a resulting loss of function and the production of a partially unfolded, or denatured, protein. Denaturation may be reversible or irreversible, and it may be partial or complete.

Proteins vary tremendously in their susceptibility to denaturation. Some proteins can be exposed to strong mineral acid without suffering irreversible loss of enzymatic activity, whereas most proteins would be irreversibly denatured by such conditions. The magnitude of the structural perturbation required for loss of function may also vary appreciably. There is a spectrum of intermediates between two extreme forms: the native and the so-called random-coil denatured state. The conditions giving rise to partial denaturation and consequent loss of function may be subtle, such as a small change in pH, temperature, ionic strength, or dielectric constant of the medium. Conversely, prolonged boiling or exposure to thiol-containing compounds and detergents, such as sodium dodecyl sulfate, or to hydrogen bond-breaking reagents, such as urea or guanidine hydrochloride, may be required for complete denaturation.

Figure 5.1

Schematic representation of an experiment to demonstrate that the information for folding into a biologically active conformation is contained in the protein's amino acid sequence.

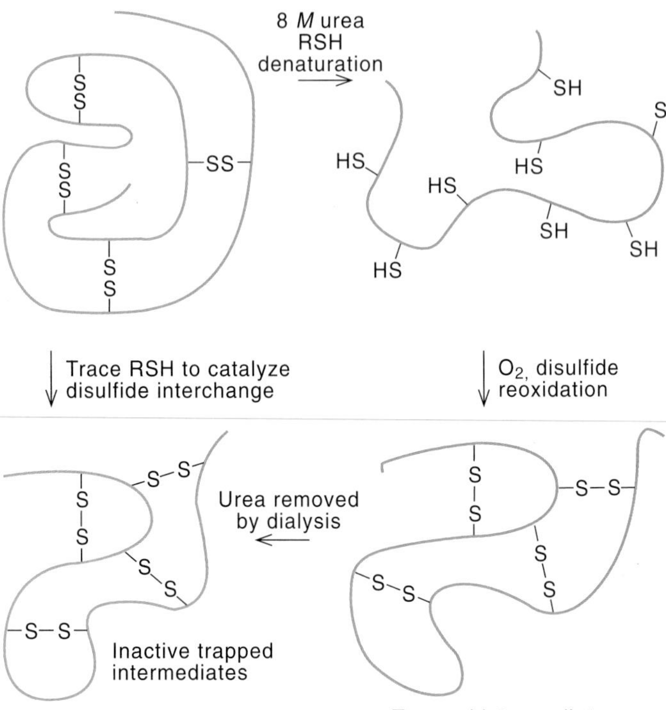

It occurred to biochemists many years ago that the native conformation of proteins might be solely determined by the amino acid sequence. This was demonstrated to be the case by a series of classic experiments performed in the early 1960s by F. White and C. Anfinsen. They chose to study bovine pancreatic ribonuclease, an enzyme containing 124 amino acid residues with four disulfide bridges. The experiment is described schematically in figure 5.1.

First, the enzyme was denatured in a solution containing 8 *M* urea and β-mercaptoethanol, a thiol reagent that reduces disulfides to sulfhydryls, thus cleaving the covalent cross-links. These conditions have been used since as a general means of denaturing proteins, by completely disrupting the conformation without giving rise to coagulation or precipitation. The reduced, denatured ribonuclease is biochemically inactive because its native structure has been destroyed.

Next, the protein was allowed to slowly air-oxidize. The result was the formation of a variety of different intermediates with randomly distributed disulfide bonds. These trapped intermediates

5A

BOX

Prion Disease: A Disease That Is Caused by Protein Misfolding

Stanley Prusiner has suggested that there are a number of diseases that can be transmitted by abnormal proteins. The list of diseases is restricted to those that affect the central nervous system and includes scrapie and bovine spongiform encephalopathy of animals and Creutzfeld-Jakob (CJD) and Gerstmann, Straussler-Scheinker (GSS) disease of humans.

The unusual nature of the scrapie agent was first remarked upon by Tikrah Alper and her colleagues at the Hammersmith Hospital in London. Experimental transmission of scrapie to mice and hamsters gave investigators a convenient laboratory model for investigating the unusual infectious agent that causes scrapie. In subcellular fractions from hamster brain enriched for scrapie infectivity, a protease-resistant protein of 27–30 kD, designated PrP 27-30, was identified. Purification of PrP 27-30 to homogeneity permitted determination of its amino acid sequence and the construction of a cDNA that was instrumental in identifying a unique chromosomal gene encoding this sequence. This gene normally produces a protein product containing 254 amino acids designated PrP^{c}. The shorter protein associated with the disease had apparently lost 67 amino acids from the amino terminus. Pulse-chase experiments with scrapie-infected cultured cells indicated that conversion of PrP^{c} to a disease-specific protein, PrP^{Sc}, is a posttranslational event. PrP^{Sc} is distinguishable from the normal PrP^{c} protein by its lower solubility and greater protease resistance. Both the normal (PrP^{c}) and abnormal versions (PrP^{Sc}) of the full-length PrP protein are synthesized in the endoplasmic reticulum and transit through the Golgi apparatus, where their asparagine-linked oligosaccharides are modified and sialic acid is added. PrP^{c} is believed to be transported in secretory vesicles to the external cell surface, where it becomes anchored by a glycosylphosphatidylinositol moiety (see fig. 18.14). In contrast, PrP^{Sc} accumulates in cells, where it is deposited in lysosomelike vesicles. PrP 27-30 is believed to arise by the protease cleavage of the PrP protein.

Several observations support the contention that PrP^{Sc} protein is the infectious agent that causes scrapie. First, the infectious properties survive stringent purification of the PrP^{Sc} protein, whereas the PrP 27-30 protein is not infectious. Second, the infectivity of extracts is greatly lowered by phenol or protease treatment but is unaffected by agents that are destructive to nucleic acids.

Turning to the question of how the PrP^{Sc} protein might cause scrapie, Prusiner has proposed that PrP^{Sc} catalyzes the conversion of PrP^{c} to PrP^{Sc} on contact. The precedent that one protein could influence the folding of another protein has already been set by the chaperons, so this is not a difficult idea to accept. The most difficult notion to accept relates to the infectivity of a protein. There is no precedent for this.

In view of this concern, convincing proof that prion diseases are caused by proteins alone may have to await further experiments in which the alleged infectious protein is synthesized de novo in a system that could not be contaminated by other potentially infectious agents.

remained inactive after removal of the denaturant urea by dialysis through a semipermeable membrane.

To recover the native structures, these inactive intermediates were exposed to a trace amount of the reducing agent mercaptoethanol. This step served to catalyze the rearrangement of the disulfide cross-links, resulting finally in the spontaneous generation of a fully active product. Analysis by several methods showed that the renatured product was indistinguishable from native ribonuclease and that all the correct disulfide pairings had been reestablished. (Renaturation can be carried out in a less cumbersome manner by removing the urea before air oxidation; in this case the native structure is a major product.)

Thus it appears that the information for folding to the native conformation is present in the amino acid sequence, for of the many possible disulfide-paired ribonuclease isomers that are possible, only the one correctly paired product was formed in major yield. Further studies have indicated that a similar result is obtained with many other proteins. However, it should be added that notwithstanding the tendency of an unfolded polypeptide to snap back to its native conformation, there is a special class of proteins known as chaperons that appear to catalyze this process (see chapter 34). There also appear to be some unusual situations where an abnormally folded protein catalyzes the folding or refolding of a homologous normal protein into the abnormal conformation (see box 5A).

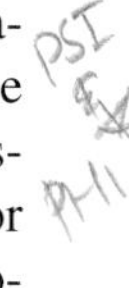

The Ramachandran Plot Predicts Sterically Permissible Structures

Figure 5.2 shows a ball-and-stick model of a short section of a polypeptide chain. Many geometric features of this structure are fixed as a result of bonded interactions between adjacent atoms. The bond lengths and bond angles are nearly constant for all proteins. Additionally, the backbone peptide bond has substantial double-bond character owing to electron delocalization over a π orbital system involving the carbonyl oxygen, carbonyl carbon, and amide nitrogen atoms of the backbone peptide bond. As a result, all the atoms of the peptide bond, together with the connected

Figure 5.2

Basic dimensions of (*a*) the peptide and (*b*) the dipeptide. Although the peptide bond is formally represented as a singly bonded interaction between N and C, it in fact has significant double-bond character owing to delocalization of oxygen electrons over a delocalized O—C—Nπ orbital system. As a result, the peptide C—N bond is shorter than a normal single C—N bond (1.32 Å instead of 1.47 Å), and the conformational degrees of freedom of a polypeptide chain are restricted to rotations about the single-bond connections between the adjacent planar transpeptide groups to C_α, i.e., the C_α—C_1 and C_α—N_1 single bonds.

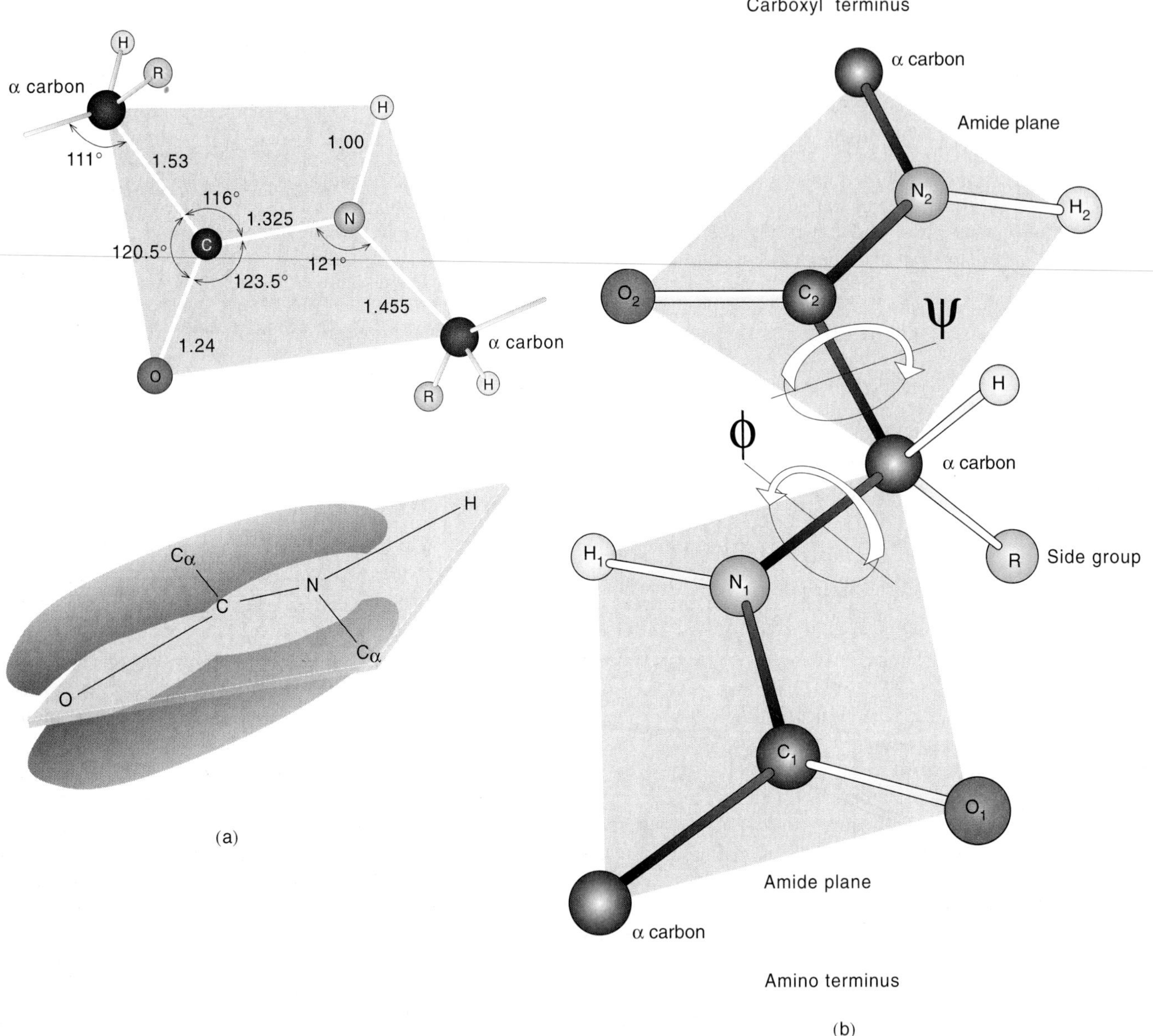

Figure 5.3

The conformation corresponding to $\phi = 0°$, $\psi = 0°$. This conformation is disallowed by the steric overlap between the H and O atoms of adjacent peptide planes. Rotation of both ϕ and ψ by 180° gives the fully extended conformation seen in figure 5.2. Curved arrows for ϕ and ψ indicate positive variations in angle.

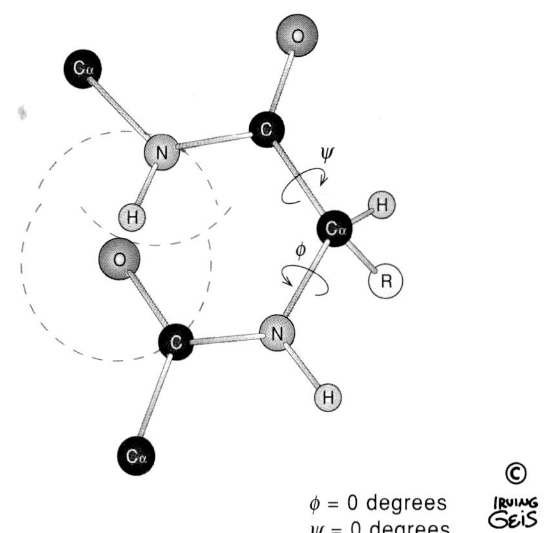

α-carbon atoms (conventionally labeled C_α), lie in a common plane with the carbonyl oxygen and amide hydrogen in the *trans* configuration. Consequently, the only adjustable geometric features of the polypeptide-chain backbone involve rotations about the single covalent bonds that connect each residue's C_α to the adjacent planar peptide groups.

Rotations about the C_α—N bond are labeled with the Greek letter ϕ (phi), and rotations about the C_α—carbonyl carbon bond are labeled ψ (psi). In principle, both ϕ and ψ can have any value between −180° and +180°, so that all possible conformations of the polypeptide chain can be described in terms of their ϕ, ψ conformational angles, a description that automatically takes account of the fixed geometric features of the polypeptide backbone. Thus any polypeptide conformation can be represented as a point on a plot of ϕ versus ψ, where ϕ and ψ have values that range from −180° to +180°. By convention the conformation corresponding to $\phi = 0°$, $\psi = 0°$ is one in which both peptide planes that are connected to a common C_α atom lie in the same plane, as shown in figure 5.3. Positive variations in ϕ correspond to clockwise rotations of the preceding peptide about the C_α—N_1 bond when viewed from C_α toward N_1 (see fig. 5.2*b*). Positive variations in ψ correspond to clockwise rotations of the succeeding peptide about the C_α—C_2 bond when viewed from C_α toward C_2 (see fig. 5.2*b*).

Experiments with models that approximate the polypeptide atoms as hard spheres, with appropriate van der Waals radii, quickly reveal that many ϕ, ψ angular combinations are impossible because of steric collisions between atoms along the backbone or between backbone atoms and the side-chain R group. For example, it is clear that the $\phi = 0°$, $\psi = 0°$ conformation shown in figure 5.3 is impossible. The reason is that this conformation results in noncovalently bonded interatomic contacts that are considerably less than the sum of the van der Waals radii of the atoms involved. In fact, of all the possible ϕ, ψ combinations, only a relatively restricted number of conformations are sterically allowed. The Ramachandran plot (fig. 5.4) shows explicitly how the accessible regions of ϕ, ψ space are limited by steric interactions among the polypeptide backbone and side-chain groups, assuming that the atomic groups behave as rigid spheres having appropriate van der Waals radii. In reality, the atoms in molecules do not behave as rigid spheres, so real proteins span a slightly greater range of values than is suggested by this plot (fig. 5.5).

Figure 5.5 shows the distribution of some observed ϕ, ψ conformational values for proteins whose three-dimensional structures are known from crystallography. The great majority of these lie within the bounds defined by allowable steric interactions. The exceptional residues are usually glycines. Glycine frequently can assume conformations that are sterically hindered in other amino acids because its R group, a hydrogen atom, is considerably smaller than the CH_2 or CH_3 groups connected to C_α in all other amino acids.

In summary, it can be seen that owing to the basic geometric properties of the polypeptide chain, its sterically allowed conformations are severely restricted by the occurrence of unfavorable steric interactions between various atomic groups.

Protein Folding Reveals a Hierarchy of Structural Organization

Although the Anfinsen experiment demonstrates that a completely unfolded protein will spontaneously refold to its native structure, it does not explain how this happens. To appreciate the magnitude of this question, consider a hypothetical example involving the spontaneous folding of a polypeptide of 100 amino acid residues. In the native folded state of such a protein, each residue is spatially fixed relative to the others to produce a unique three-dimensional structure. For the sake of simplicity, the spatial orientation of each residue in the folded structure can be described in terms of three geometric parameters. (These parameters might, for example, correspond to the values of some rotational angles about single covalent bonds in the polypeptide chain.) In order to completely describe the folded protein, it is therefore necessary to uniquely specify 300 internal geometric parameters. Even if each of the parameters could assume only two possible values, the total number of potentially possible geometric arrangements of the structure would be $2^{300} = 2 \times 10^{90}$. If folding the protein involved random sampling of all these possible arrangements at a rate corresponding to typically observed frequencies of rotational rearrangements about single covalent bonds (about 10^{-13} sec), the estimated time required to fold the protein into the correct arrangement would exceed the age of the earth.

In actuality, newly synthesized polypeptide chains typically fold in seconds. This means that protein folding must be a highly directed and cooperative process. Although much remains to be learned about the details of the process, its speed and its fa-

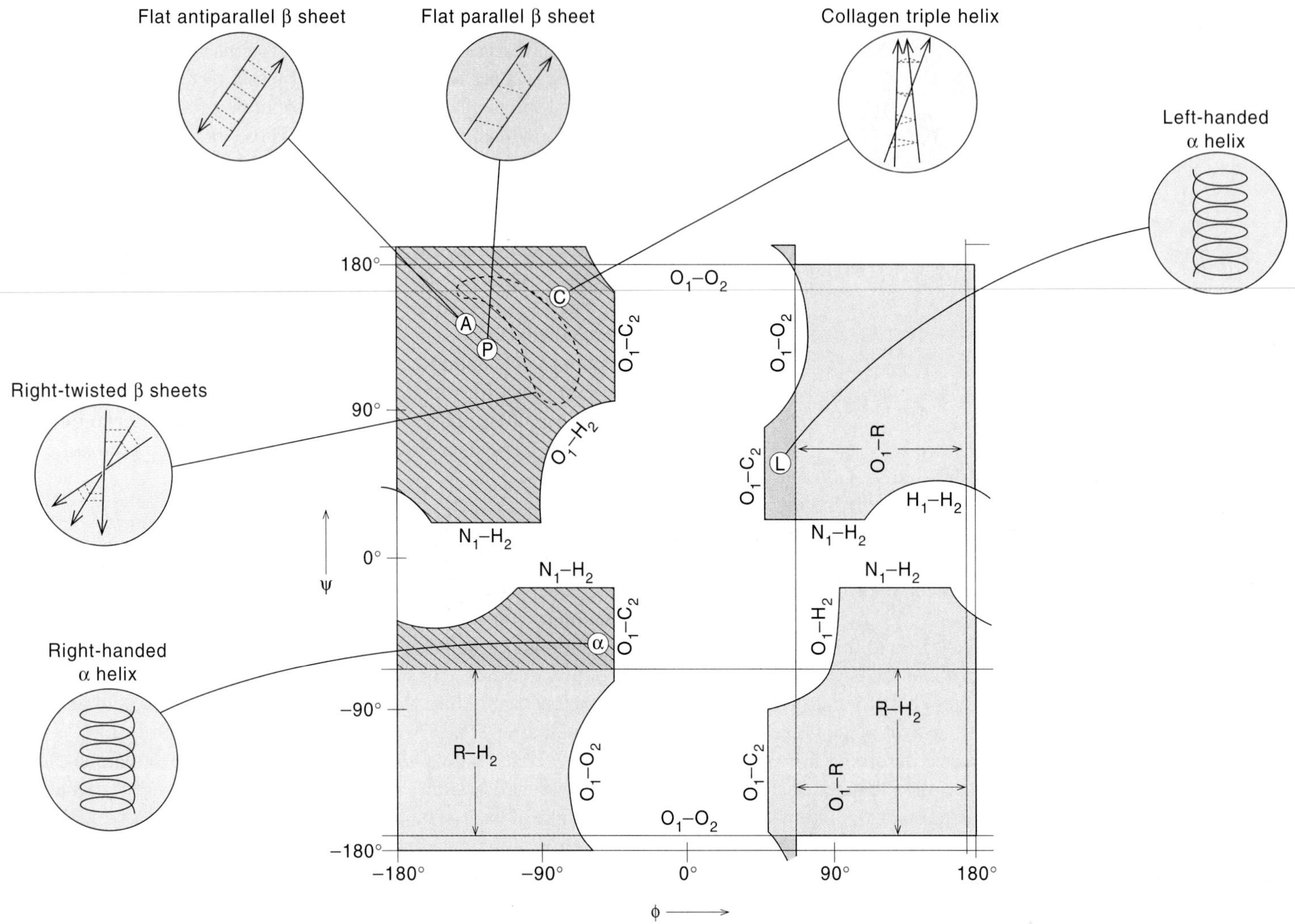

Figure 5.4

Ramachandran plot, showing which atomic collisions (using a hard-sphere approximation) produce the restrictions of the main-chain angles ϕ and ψ. The hatched regions are allowed for all residues, and each boundary of a prohibited region is labeled with the atoms that collide in that conformation. Additional shaded regions are for glycine residues only. The numbering scheme for amide atoms used in the derivation diagram is given in figure 5.2. For an explanation of the various labeled structures, see the text.

cility suggest the existence of a sequential set of folding intermediates, each being more highly organized than the one before it (for example, fig. 5.6). In what follows we will see how the forces that stabilize proteins act in concert with related energetic and geometric factors to yield successively larger and more complex protein structural arrangements (fig. 5.7).

To begin with, steric interactions restrict accessible conformations and reflect features of the protein's amino acid sequence, or primary structure. The requirements for hydrogen-bond preservation in the folded structure result in the cooperative formation of regular structural regions in proteins. This situation arises principally because of the regularly repeating geometry of the hydrogen-bonding groups of the polypeptide backbone and leads to the formation of regular hydrogen-bonded secondary structures. Association between elements of secondary structure in turn results in the formation of structural domains, whose properties are determined both by chiral properties of the polypeptide chain and packing requirements that effectively minimize the molecule's hydrophobic surface area. Further association of domains results in the formation of the protein's tertiary structure, or overall spatial arrangement of the polypeptide chain in three dimensions. Likewise, fully folded protein subunits can pack together to

Figure 5.5

Ramachandran plot of main-chain angles ϕ and ψ, experimentally determined for approximately 1,000 nonglycine residues in eight proteins whose structures have been refined at high resolution (chosen to be representative of all categories of tertiary structure).

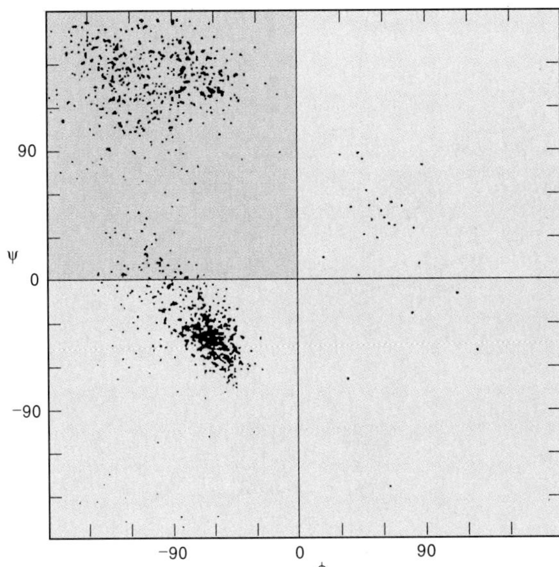

form quaternary structures, which can either serve a structural role or provide a structural basis for modification of the protein's functional properties.

Two Secondary Structures Are Found in Most Proteins

A major driving force in protein folding is the necessity to minimize the extent of exposure by the hydrophobic group to solvent. This consideration involves a sacrifice of the favorable hydrogen-bonded interactions between the unfolded polypeptide backbone and water. To preserve a favorable energy balance on folding, the backbone polypeptide groups must take part in alternative hydrogen-bonded interactions between themselves in the protein's folded state.

The α Helix

One of the most commonly observed protein secondary structures is the α helix. In an α helix the polypeptide backbone follows the path of a right-handed helical spring to form an arrangement in which each residue's carbonyl group forms a hydrogen bond with the amide NH group of the residue four amino acids farther along the polypeptide chain (fig. 5.8*a*). All residues in an α helix have nearly identical conformations, averaging $\psi = -45°$ to $-50°$ and $\phi = -60°$, so they lead to a regular structure in which each 360° of helical turn incorporates approximately 3.6 amino acid residues and rises 5.6 Å along the helix axis direction. The advance per amino acid residue along the helix axis is 1.5 Å. Although alternative helical arrangements having different hydrogen-bonding

Figure 5.6

Possible successive steps in the protein-folding process as they might apply to a typical example of each of the four major categories of structure.

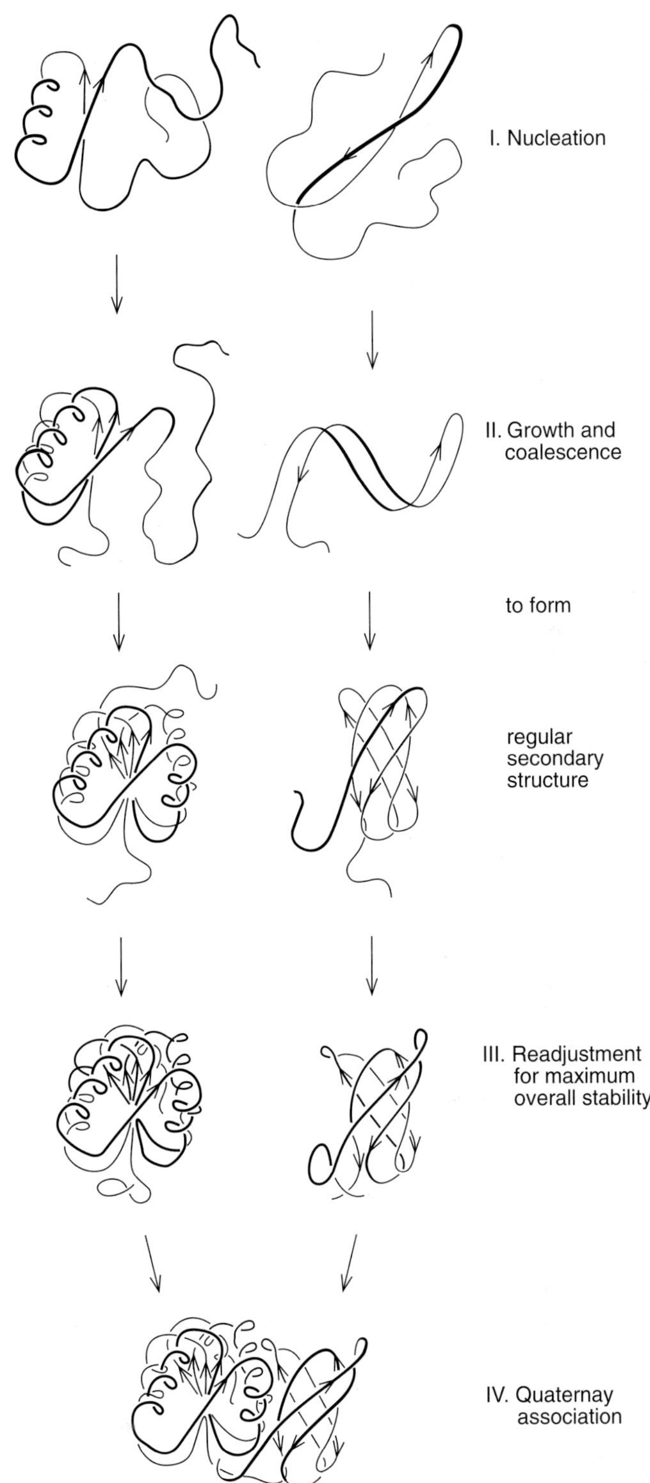

Figure 5.7

Hierarchies of protein structures.

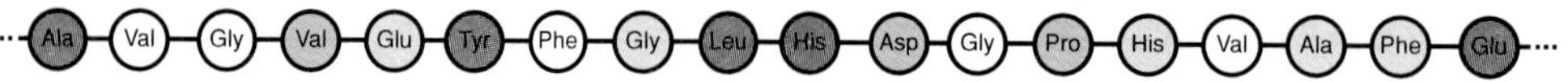

(a) Primary structure (amino acid sequence in the protein chain)

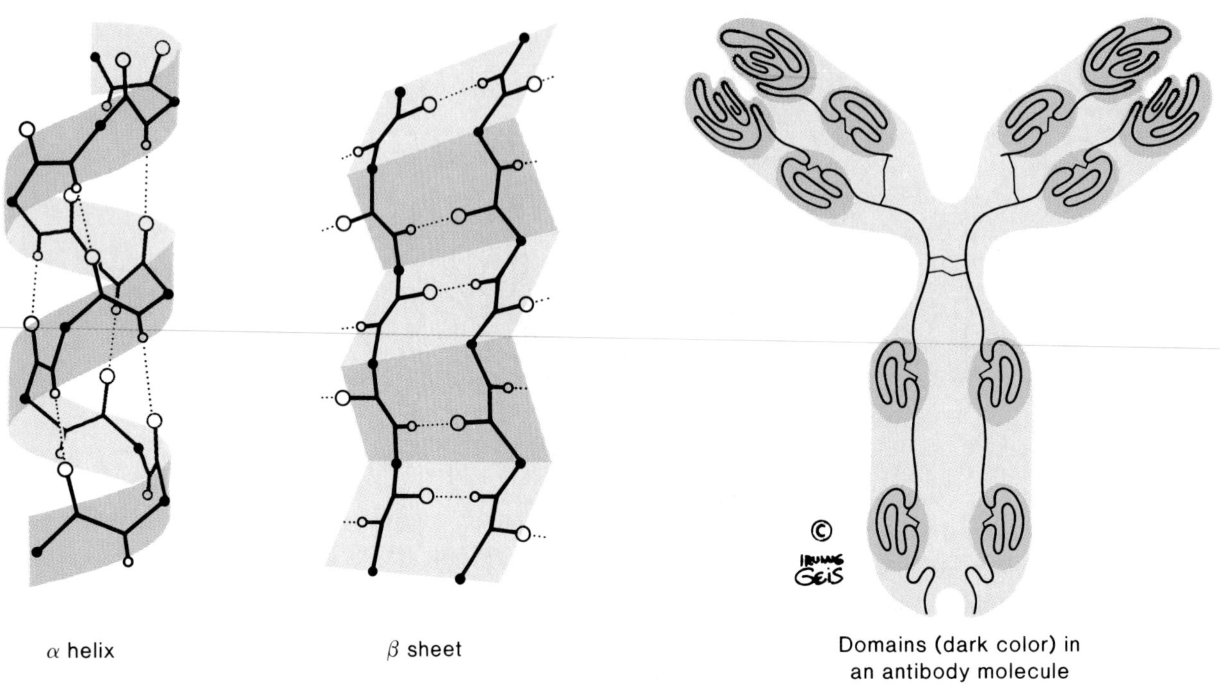

α helix

β sheet

(b) Secondary structure

Domains (dark color) in an antibody molecule

(c) Local folding

β

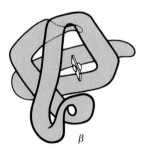

One complete protein chain (β chain of hemoglobin)

(d) Tertiary structure

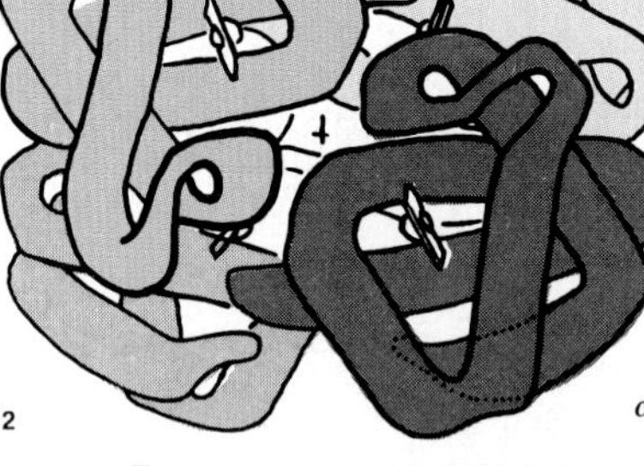

The four separate chains of hemoglobin assembled into an oligomeric protein

(e) Quaternary structure

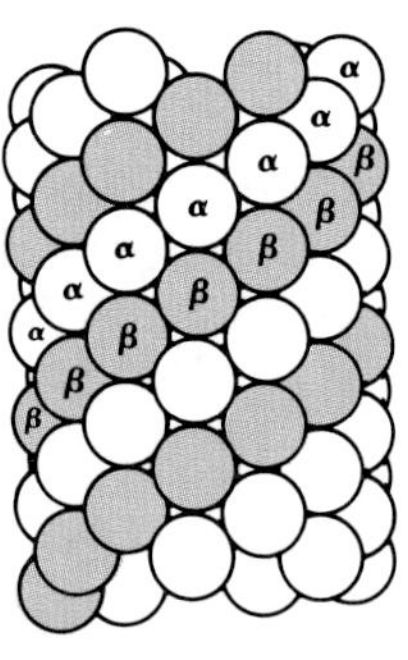

σ (white) and β (color) tubulin molecules in a microtubule

(f) Quaternary structure

Figure 5.8

Three ways of projecting the α-helix. (*a*) This simple ball-and-stick model highlights the planar peptides. The interpeptide hydrogen bonds are shown by dashed lines, and the amino acid side chains are indicated by R groups. Approximately two turnings of the helix are shown. There are about 3.6 residues per turn. (Illustration copyright by Irving Geis. Reprinted by permission.) In (*b*) and (*c*) we see identical projections of a side view using space-filling and wire models, respectively. The space-filling models use van der Waals radii for the atoms.

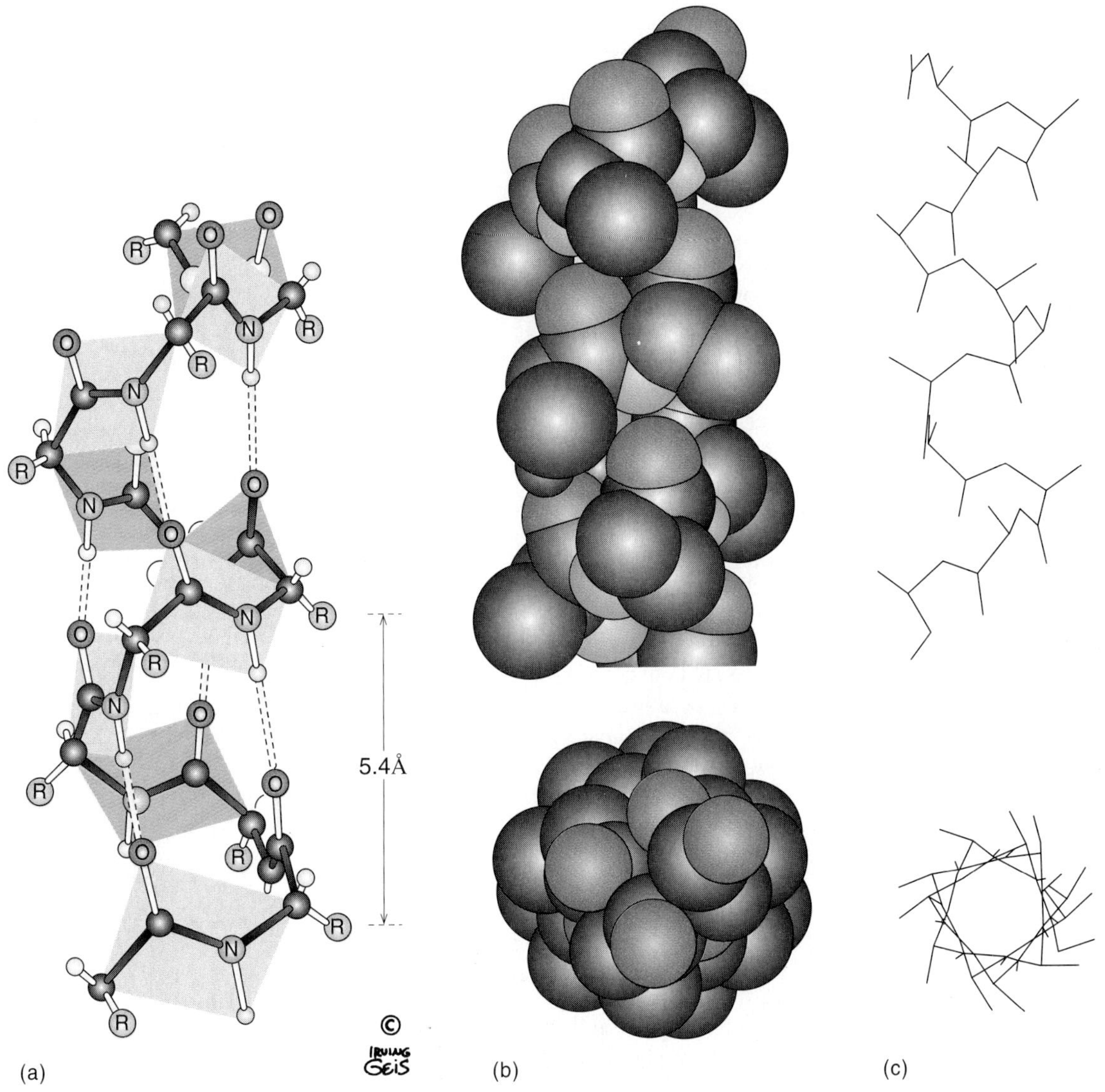

patterns and different geometries are also conformationally possible, the α helix is by far the most commonly observed helical arrangement found in proteins. The particular stability of the α helix appears to be related not only to the formation of stable hydrogen bonds between all the backbone carbonyl and NH groups, but also to the tight packing achieved in folding the chain to form the structures. This packing is most apparent when the helix is represented by space-filling van der Waals models (see fig. 5.8*b*). Note the close-packed arrangement of most of the atoms in the structure. And in the tube view (see fig. 5.8*b*), note the absence of any free space down the core of the helix. Alternative arrangements, in contrast, either have inferior hydrogen bonds or are not as tightly packed as the α helix.

Figure 5.4 shows that there is a possible but small and shallow energy minimum at the left-handed α-helical (L_α) position for nonglycine residues and that only 1 to 2% of the nonglycine residues are L_α. However, for the symmetrical glycine, whose R group is the same as its C_α H and therefore has no hand at C_α, left-handed conformations are exactly equivalent to right-handed ones, and in fact, about half the glycines have positive ϕ values. Extended L_α helices have not been observed.

An important property stemming from the conformational regularity of the α helix, which applies to other secondary structures as well, is cooperativity in folding. For example, once a single turn of α helix has been formed, addition of successive residues becomes much more likely and faster because the first turn of the helix forms a template upon which to erect successive helical residues. Owing to steric restrictions, the torsional angle ϕ is approximately correct for each additional residue. Each addition mainly involves sampling various conformations of ψ until the residue is "captured" in the correct conformation by the formation of a hydrogen bond to a group that is already fixed in the helical conformation.

The β Sheet

A second type of commonly occurring protein secondary structure is the β-pleated sheet (fig. 5.9). Sheets are formed when two or more almost fully extended polypeptide chains are brought together side by side so that regular hydrogen bonds can form between the peptide backbone amide NH and carbonyl oxygen groups of adjacent chains. Notice that since each backbone peptide group has its NH and carbonyl groups in a *trans* orientation, it is possible to extend a β sheet into a multistranded structure simply by adding successive chains to the sheet. β sheets can occur in two different arrangements. In the first of these, the chains are arranged with the same N-to-C polypeptide sense to produce a parallel β sheet. Alternatively, the chains can be aligned with opposite N-to-C sense to produce an antiparallel β sheet. As illustrated (fig. 5.10), parallel and antiparallel β sheets are both composed of polypeptide chains that have conformations pointing alternate R groups to opposite sides of the sheet but have their peptide planes nearly in the sheet plane to allow good interchain hydrogen bonding.

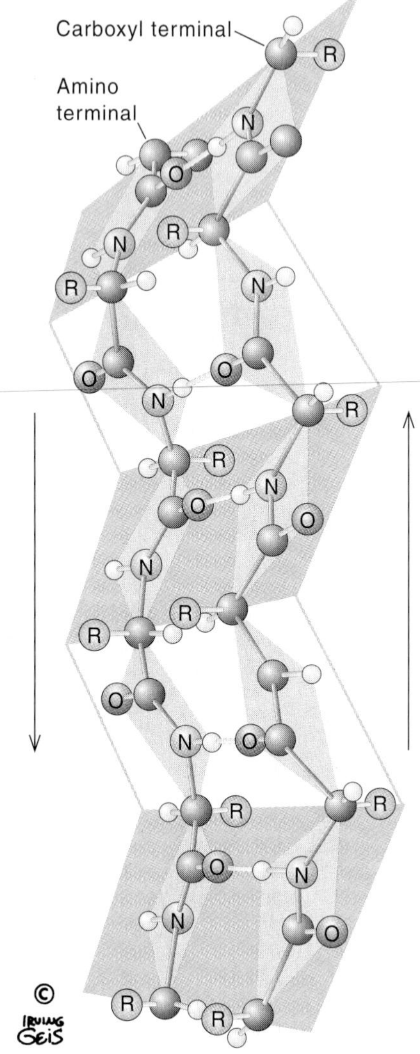

Figure 5.9

The antiparallel β sheet. This structure is composed of two or more polypeptide chains in the fully extended form, with hydrogen bonds formed between the chains. Hydrogen bonds are shown as dashed lines. (Illustration copyright by Irving Geis. Reprinted by permission.)

Pauling and Corey Provided the Foundation for Our Understanding of Fibrous Protein Structures

Linus Pauling and Robert Corey examined the structures of crystals formed by amino acids and short peptides before they ventured into the world of proteins. From their crystallographic investigations of amino acids and peptides, they formulated two rules that describe the ways in which amino acids and peptides interact with one another to form noncovalently bonded crystalline structures. These rules laid the foundations for our understanding of how amino acids in protein polypeptide chains interact with one another. Rule number one was that the peptidyl C—N linkage and the four atoms to which the C and the N atoms are directly linked always form a planar structure as we have already discussed (see fig. 5.2). From this it follows that the only flexibility in the polypeptide backbone arises from rotation about the carbon that joins adjacent peptide planar groups. Rule number two was that peptide carbonyl and amino groups always form the maximum number of hydrogen bonds. Taken together, these two rules drastically reduce the number of possible conformations available to the polypeptide chain.

Following their investigations of amino acid and peptide crystals, Pauling and Corey turned their attention to the x-ray diffraction patterns of a number of fibrous proteins. A vast number of fibrous proteins exist in nature, but the majority of them give diffraction patterns that fall into one of three types: the α pattern, the β pattern, or the collagen pattern. Fiber diffraction data give information only about the repeating units of a structure because of the lack of three-dimensional order in the fibers; but because fibrous protein polypeptide chains are arranged in simple repetitious units, much about the overall structure could be deduced. The first type of diffraction pattern, which was observed for a subgroup of the keratins, was consistent with a helical arrangement of the polypeptide chains with a monomer repeat along the helix axis of 1.5 Å.

Figure 5.10

Two forms of the β-sheet structure: (*a*) the antiparallel and (*b*) the parallel β sheet. The advance per two amino acid residues is indicated for each structure.

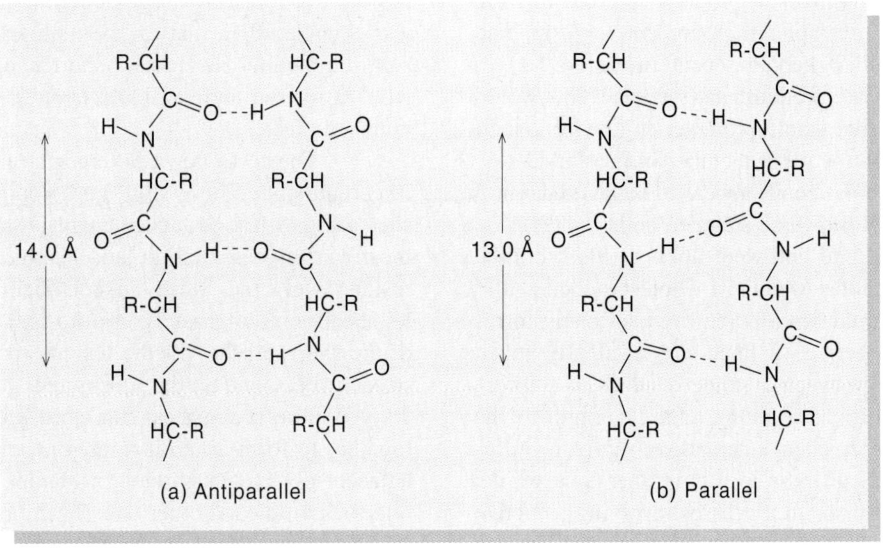

14.0 Å

(a) Antiparallel

13.0 Å

(b) Parallel

Space-filling molecular models were used in an attempt to build structures compatible with the x-ray data. These models were designed so that covalently linked atoms were accurately spaced according to known dimensions. Moreover, the individual atoms, made as hard spheres, were of a size so that nonbonded atoms could get no closer to one another than their van der Waals radii would normally allow. Pauling and Corey tried to arrange the polypeptide chains so as to maximize the number of peptide hydrogen bonds in a way that was consistent with the x-ray diffraction data. By trial and error they came to the conclusion that the right-handed α helix was a most acceptable structure (see fig. 5.8).

The x-ray diffraction pattern of α-keratin fibers supports the notion of an α-helix structure except with regard to the helix pitch. Vertically aligned helices should exhibit a 5.4-Å spacing in their diffraction pattern, but instead one finds a 5.1-Å spacing. Crick suggested that this could be explained if the individual α helices in a helix bundle were tilted about 20° with respect to the axis of the helix bundle. In fact if one examines the surface of an α helix, it can be seen that the amino acid side chains or R groups produce a knobbly surface that could interfere with side-by-side packing. This effect could be minimized by tilting two interacting helices by about 20° relative to one another. Individual helices are believed to aggregate in side-by-side fashion in pairs or triplets so as to form cables in which the individual helices are spirally twisted so that the resulting cable has an overall left-handed twist (fig. 5.11). The formation of a cable with twisted α helices results from optimization of packing among the amino acid side-chain residues. In figure 5.11 we see how the side-chain residues of an α helix are arranged in a spiral fashion so that residues falling on the same side of a helix do not lie along the line parallel to the helix axis. As a consequence the packing of helices is optimized when helices interact at an angle of about 20°. Obviously, if helices involved in such a packing interaction were straight, they would soon separate. However, their packing interaction can be preserved if the helices are slightly twisted around each other, with the resultant formation of the left-twisted cable structure characteristic of the keratins. The coiled-coil character of such fibers consequently represents a trade-off between some local deformations that coil the α helix and the optimization of extended side-chain packing interactions in the cable as a whole.

Figure 5.11

Coiling of α helices. Residues on the same side of an α helix form rows that are tilted relative to the helix axis. Packing helices together in fibers is optimized when the individual helices wrap around each other so that rows of residues pack together along the fiber axis. Helices in coiled coil (*c*) are oriented in parallel.

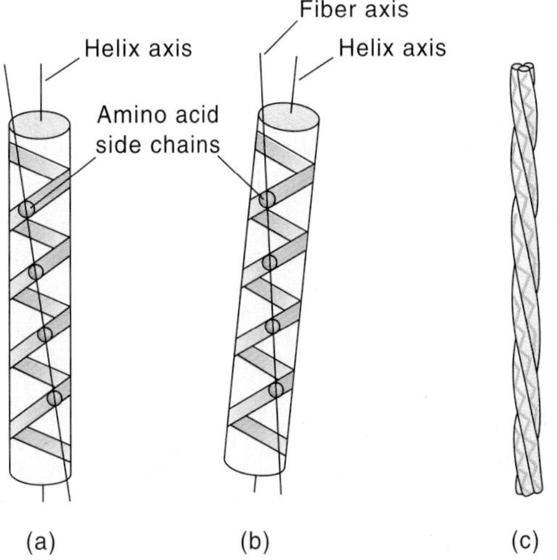

The coiled-coil conformation is believed to exist in many fibrous proteins. Such a structure would receive additional stabilization if the interacting residues between the two helices forming the coiled coil were hydrophobic on the side where they interact. In fact analysis of a number of proteins believed to form coiled-coil structures (for example, tropomyosin, myosin, and paramyosin) have a so-called heptad repeat (a-b-c-d-e-f-g) in which the a and d residues are predominantly apolar. This would create a hydrophobic strip that would promote lengthwise association with a similar strip on the other member of the coiled coil. Incidentally, two-stranded coiled coils appear to be favored, but it is clear that some proteins form three-stranded coiled coils.

The springiness of hair and wool fibers results from the tendency of the α-helical cables to untwist when stretched and to spring back when the external force is removed. In many forms of keratin the individual α helices or fibers are covalently linked by disulfide bonds formed between cysteine residues of adjacent polypeptide chains. In addition to giving added strength to the fibers, the pattern of these covalent interactions serves to influence and fix the extent of curliness in a hair fiber as a whole. Chemical reactions and mechanical processes involving reductive cleavage, reorganization, and reoxidation of these interhelical disulfide bonds form the basis of the permanent wave.

One of the lesser known facts about the discovery of the helix is why others, especially the very strong English school of crystallography, did not get there first. In fact, this was not because of lack of effort. The Cambridge crystallographers had been exploring helices as a possible solution to the protein structure problem for some time. However, they never considered structures with a nonintegral number of amino acid residues per turn of the helix. Through his work with molecular models, Pauling realized that the α helix, which contains 3.6 residues per turn, made a magnificently compact structure. This was consistent with most of the diffraction patterns, except that there was no indication of a diffraction spot representing the monomer repeat along the helix axis. In order to see this spot, it was necessary to tilt the protein fiber being x-rayed. When Pauling did this, the 1.5-Å spot showed up, confirming the prediction of the monomer repeat distance from model building.

While Pauling and Corey were working on the α helix, they recognized that there was another large class of fibrous proteins that gave a radically different diffraction pattern and behaved macroscopically as sheets rather than fibers. Diffraction patterns of these sheetlike fibrous proteins could be interpreted in terms of an extended polypeptide chain of either 13.0 and 14.0 Å, suggesting two closely related structures. Pauling and Corey interpreted these patterns as resulting from extended polypeptide chains lying side-by-side in either a parallel or an antiparallel fashion (see fig. 5.10). Although both of these structures appear to exist in nature, the antiparallel pattern is much more common in fibrous proteins. This structure is known as the antiparallel β-pleated sheet (see fig. 5.9). Regular hydrogen bonds form between the peptide backbone amide NH and carbonyl oxygen groups of adjacent chains. The β sheet can be extended into a multistranded structure simply by adding successive chains in the appropriate directions to the sheet. Parallel and antiparallel β-pleated sheets are both composed of polypeptide chains that have conformations pointing alternate R groups to opposite sides of the sheet but have their peptide planes nearly in the sheet plane to allow for good interchain hydrogen bonding. The chain confirmation that produces the best interchain hydrogen bonding in parallel sheets is slightly less extended than that for the antiparallel arrangement. As a result, the parallel sheet has a shorter repeat period of 6.5 Å (versus 7.0 for the antiparallel structure); this leads to a more pronounced pleat.

The best-known β-keratin structure in nature is that found in certain silks. These silks are composed of stacked antiparallel sheets (fig. 5.12). Sequence analysis shows them to be composed mostly of glycine, serine, and alanine, in which every alternate residue is glycine. Since the side-chain groups of a flat antiparallel sheet point alternately upward and downward from the plane of the sheet, all the glycine residues are arranged on one surface of each sheet, and all the substituted amino acids are on the other. Two or more such sheets can consequently be packed intimately together to form an arrangement of stacked sheets in which two adjacent glycine-substituted or alanine-substituted sheet surfaces interlock with each other (see fig. 5.12*b*). Owing to the extended conformations of the polypeptide chains in the sheets and the interlocking of the side chains between sheets, silk is a mechanically rigid material that resists stretching.

Collagen Forms a Unique Triple-Stranded Structure

Collagen is a particularly rigid and inextensible protein that serves as a major constituent of tendons and connective tissues. In the electron microscope it can be seen that collagen fibrils have a distinctive banded pattern with a periodicity of 680 Å (fig. 5.13). These fibrils are of varying thicknesses depending on their source and the mode of preparation. Individual fibrils are composed of collagen molecules 3,000 Å long that aggregate in a staggered side-by-side fashion (fig. 5.14).

Collagen has a most unusual amino acid composition in which glycine, proline, and hydroxyproline are the dominant amino acids. Further characterization of the polypeptide chains shows that these amino acids are arranged in a repetitious tripeptide sequence. Gly-X-Y, in which X is frequently a proline and Y is frequently a hydroxyproline. This unusual amino acid sequence and the unique diffraction pattern of collagen were strong indications of yet another totally different type of fibrous protein. The repeating proline residue excluded the possibility that the polypeptide chains in collagen could adopt either an α-helical or a β-sheet conformation. Instead, individual collagen polypeptide chains assume a left-handed helical conformation and aggregate into three-stranded cables with a right-handed twist (fig. 5.15). When viewed down the polypeptide chain axis (fig. 5.16*b*), the successive side-chain groups point toward the corners of an equilateral triangle. The glycine at every third residue is required because there is no room for any other amino acid inside the triple helix where the glycine R groups are located. The three collagen chains do not form hydrogen bonds among residues of the same chain. Instead,

Figure 5.12

(*a*) The three-dimensional architecture of silk. (*b*) The side chains of one sheet nestle quite efficiently between those of neighboring sheets. (Illustration for A and B copyright by Irving Geis. Reprinted by permission.)

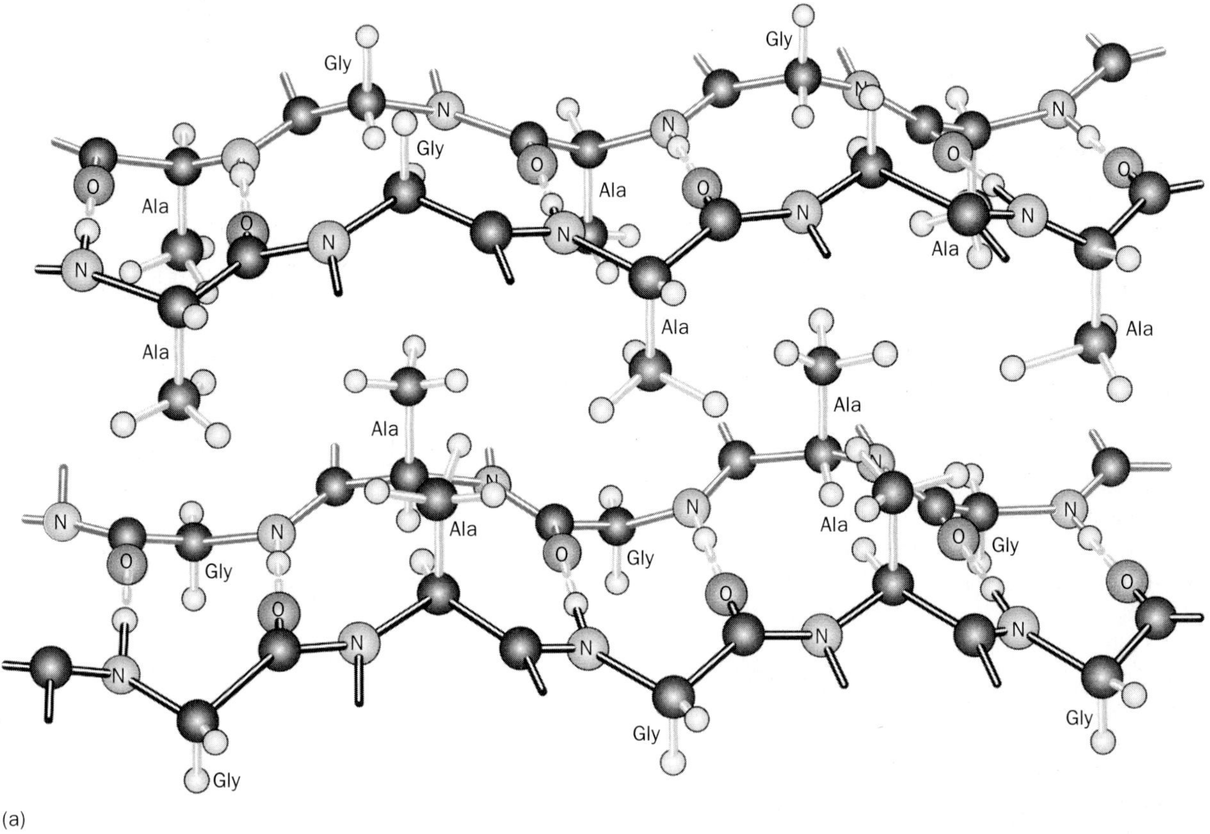

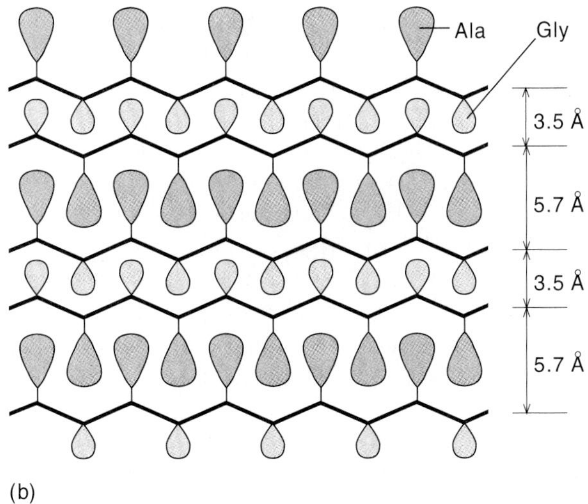

the collagen chains within each three-stranded cable form interchain hydrogen bonds. This produces a highly interlocked fibrous structure that is admirably suited to its biological role, which is to provide rigid connections between muscles and bones as well as structural reinforcement for skin and connective tissues.

Figure 5.13

An electron micrograph of collagen fibrils from skin. (Courtesy of Jerome Gross, Massachusetts General Hospital.)

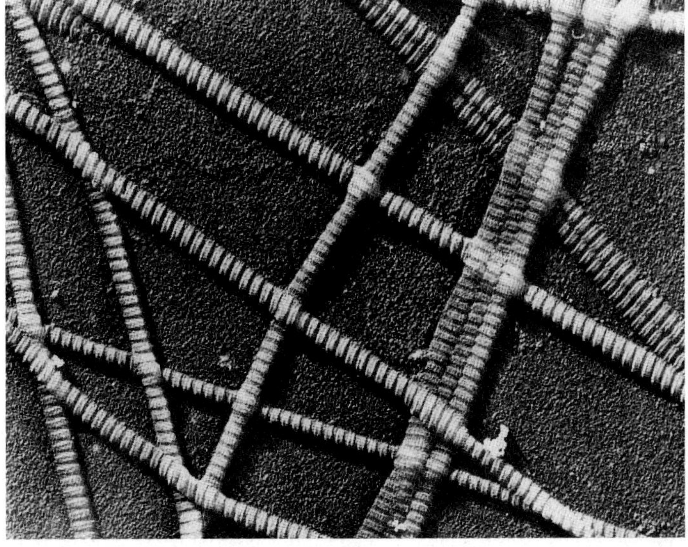

Figure 5.14

The banded appearance of collagen fibrils in the electron microscope arises from the schematically represented staggered arrangement of collagen molecules (*above*) that results in a periodically indented surface. *D*, the distance between cross striations, is ~680 Å, so the length of a 3,000-Å-long collagen molecule is 4.4*D*. (©Michael C. Webb/Visuals Unlimited.)

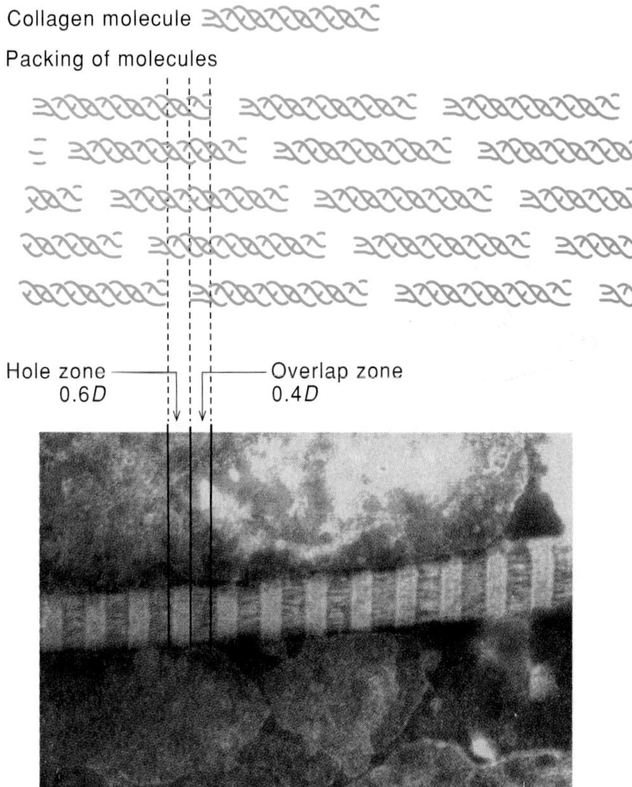

Figure 5.15

The triple helix of collagen.

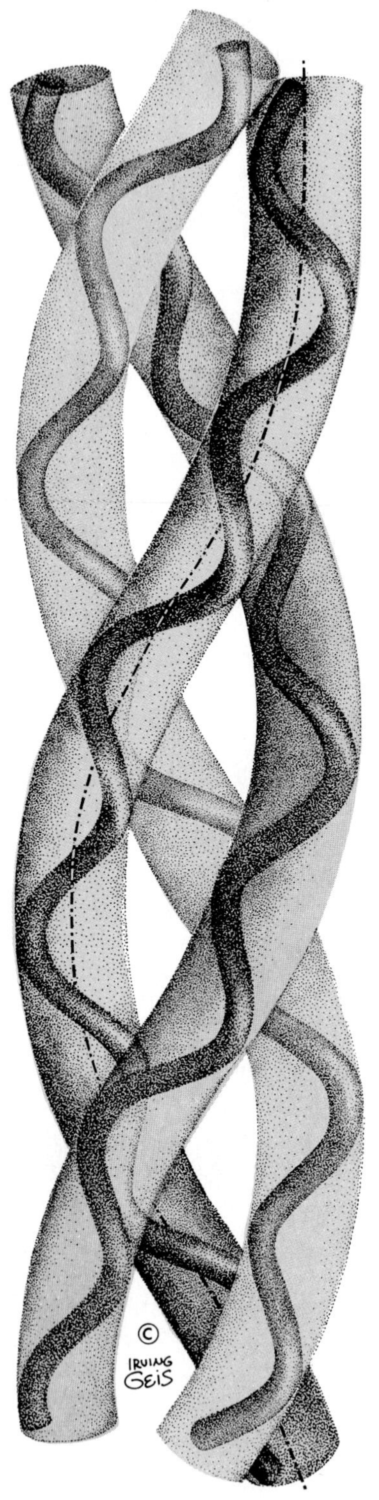

Although living organisms contain additional types of fibrous proteins, as well as polysaccharide-based structural motifs, we focused here on the three arrangements that are the most widely distributed. Two of these, the α-keratins and the β-keratins, incorporate polypeptide secondary structures that also commonly occur in globular proteins. Collagen, in contrast, is a protein that evolution has developed to play more specialized roles.

In Globular Proteins, Secondary Structure Elements Are Connected in Simple Motifs

A seemingly endless array of different folding patterns can be found for globular proteins. Despite this complexity there are rules that govern the folding process. We have already considered some basic rules that relate to the thermodynamics and inherent structural features of the polypeptide chains of fibrous proteins. It will be helpful now to consider some additional rules that relate to the folding process in globular proteins.

Globular proteins, as their name implies, differ from fibrous proteins in that they generally have shapes that approach being spherical. Nevertheless, three-dimensional structure studies show that they incorporate many of the secondary structural features that typify fibrous proteins. Figure 5.17, for example, illustrates the first enzyme whose three-dimensional structure was determined, the 120-residue protein lysozyme. This protein has local regions of ordered α-helical and antiparallel β-sheet secondary

Figure 5.16

The basic coiled-coil structure of collagen. Three left-handed single-chain helices wrap around one another with a right-handed twist. (*a*) Ball-and-stick single-collagen chain. (*b*) View from the top of the helix axis. Note that glycines are all on the inside. In this structure the C=O and N—H groups of glycine protrude approximately perpendicularly to the helix axis so as to form interchain hydrogen bonds.

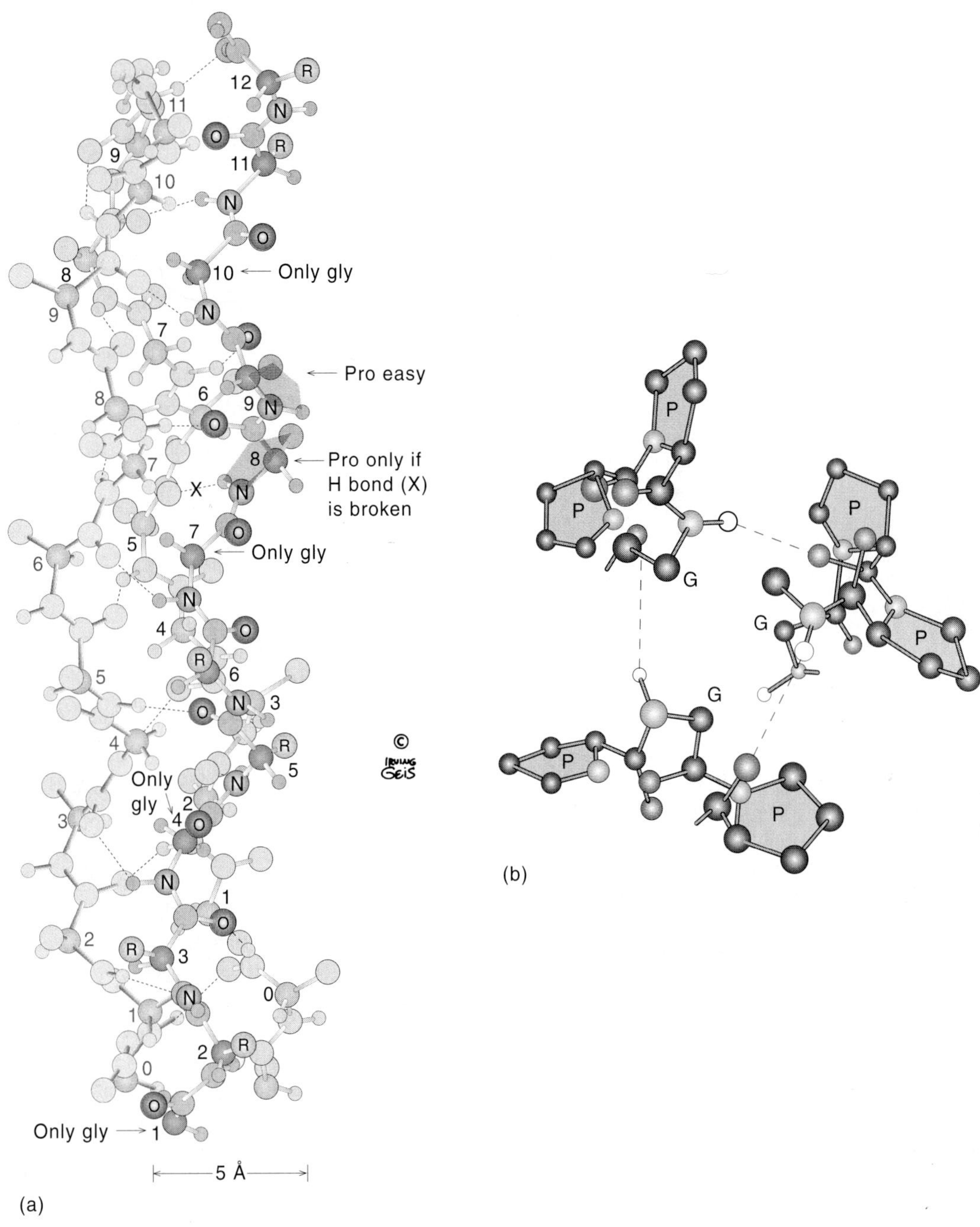

structure. In addition it has several additional regions of single-stranded loops with a less regular conformation.

At the most elementary level of structural analysis we find that simple combinations of a few secondary structure elements with specific geometric arrangements are used again and again in different protein structures. In this section we describe three of these structural motifs that are used most frequently: the helix-loop-helix, the hairpin β motif, and the β-α-β motif.

Two versions of the helix-loop-helix motif are shown in figure 5.18. The one in figure 5.18*a* is frequently associated with

Figure 5.17

Lysozyme. In this and succeeding figures the polypeptide backbone is represented as a ribbon to allow the polypeptide-chain course to be followed easily.

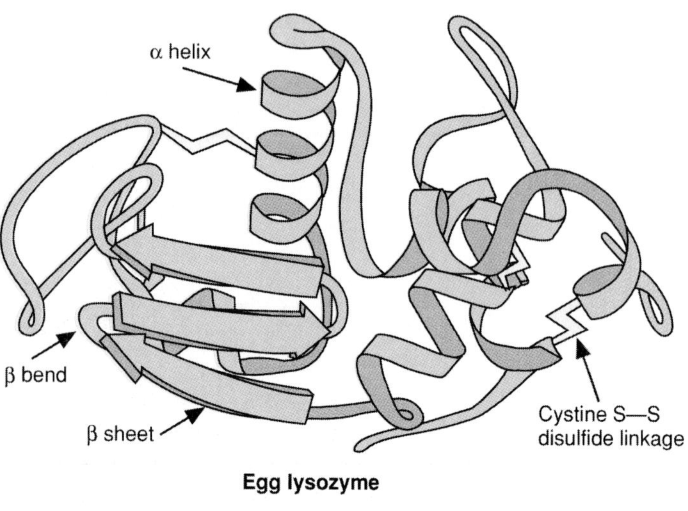

Figure 5.18

Examples of the helix-loop-helix motif. Two segments of α helix are joined by a region of single-stranded polypeptide chain of variable length. The orientation of the two segments of helix is variable.

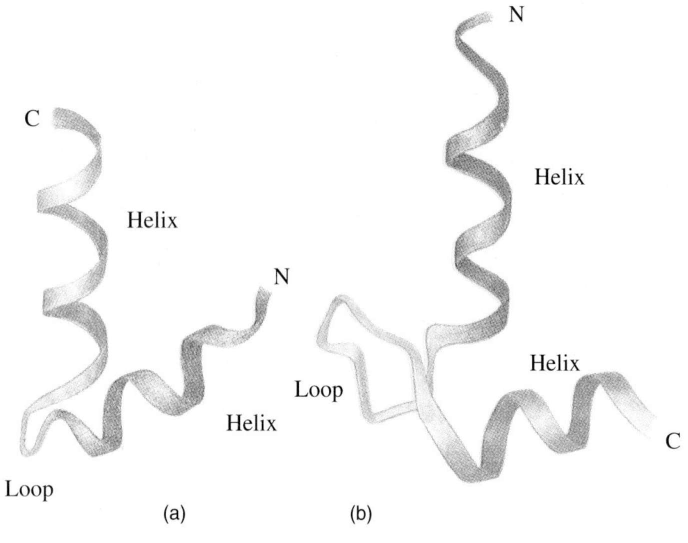

Figure 5.19

Three ways of making connections between β strands. (*a*) A hairpin same-end connection is commonly found for β strands in the antiparallel orientation. (*b*) A right-handed crossover connection is commonly found for β strands in the parallel orientation. (*c*) A left-handed crossover connection is rarely found.

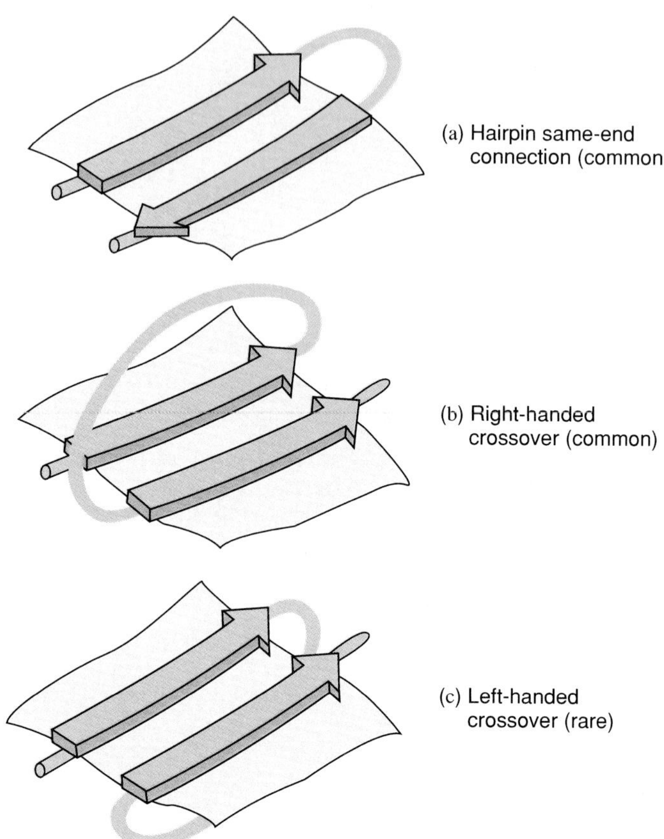

DNA-binding proteins. The part of this motif that binds to the DNA is found in one of the helix segments. The helix-loop-helix motif in figure 5.18*b* is associated with calcium-binding proteins. Usually, a single calcium divalent cation binds in the middle of the loop region.

The hairpin β motif is found between two β strands arranged in the antiparallel fashion (figure 5.19*a*). The connecting loop varies in length, the most frequent lengths containing two to five amino acid residues. The so-called β bend is a commonly observed and particularly efficient way of forming a tight loop. In the β bend a residue's carbonyl group forms a hydrogen bond with the amide NH group of the residue three positions further along the polypeptide chain (fig. 5.20).

Several conformational variations of the β bend have been observed that are a function of the amino acid sequence in the bend. In particular, it has been observed that the amino acids glycine and proline occur frequently in β bends. Because of its small size, glycine is conformationally more flexible than other amino acids. It can therefore serve as a flexible hinge between regions of polypeptide chains where steric interactions would otherwise keep them in more extended conformations. Proline, in contrast, is more conformationally restricted than other amino acids, since its cyclically bonded structure fixes its conformational degrees of freedom. In a sense, then, part of the geometry that results in bend formation is performed in proline-containing sequences. It appears probable that either situation might promote the formation of a bend during the initial stages of protein folding and so cause structures such as antiparallel β sheets to assemble cooperatively, in a manner resembling the closure of a zipper.

While we are on the subject of connections between β strands, it is appropriate to consider how connections are made

Figure 5.20

The two major types of tight turn of β bends (I and II). In type II, R_3 is generally glycine.

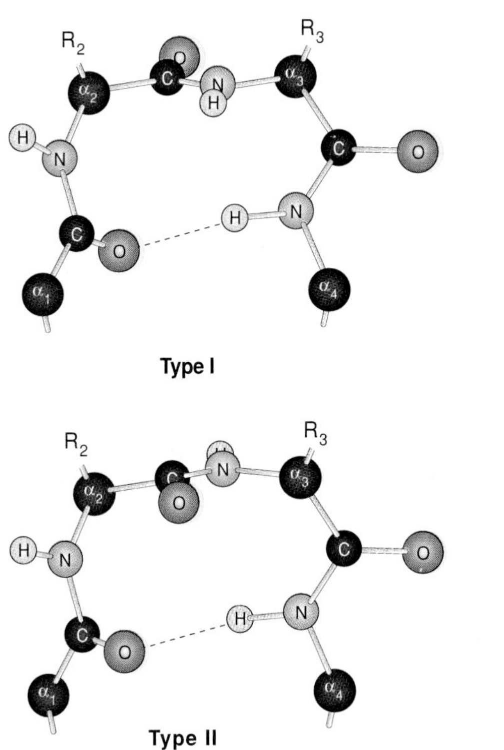

Figure 5.21

The natural tendency for the polypeptide chain to twist in the right-hand direction produces structures with an overall right-handed connectivity. The structure represents a single fully extended polypeptide chain.

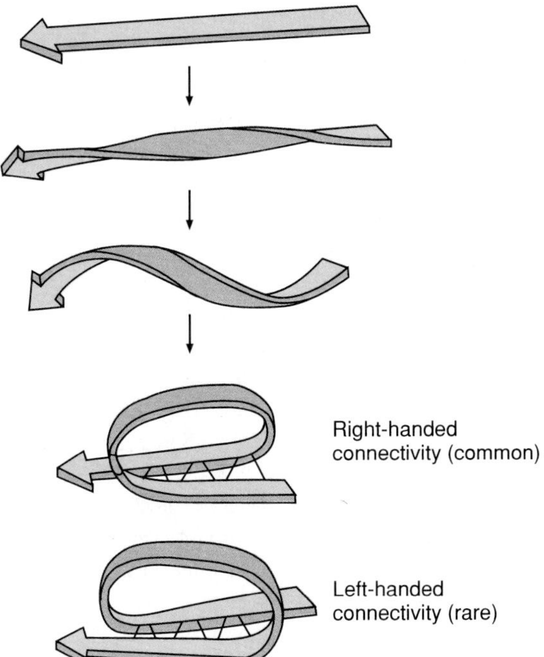

Figure 5.22

A β-α-β loop. This arrangement forms the basis of many of the more extended structural arrangements found in globular proteins.

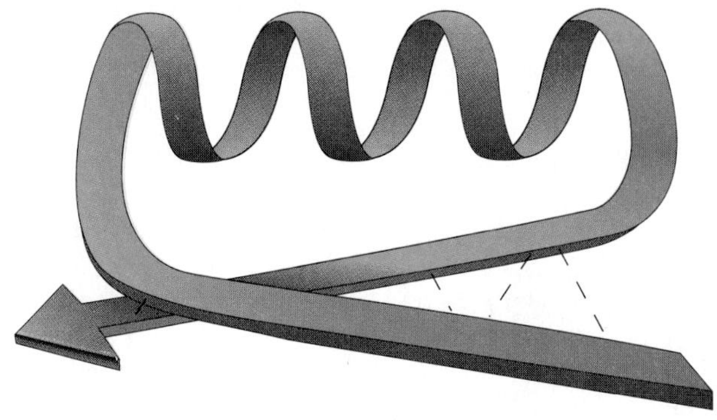

between β strands arranged in parallel. In this situation the connecting region takes on various forms from a simple extended polypeptide chain to something much more complex. In figures 5.19*b* and 5.19*c* it can be seen that two types of crossovers are possible: the right-handed crossover and the left-handed crossover. The right-handed crossover is much more common.

The strong preference for right-handed crossovers between parallel β strands is believed to be due to the natural tendency of the polypeptide chain to prefer to be slightly twisted in a right-handed sense when viewed down the polypeptide chain axis. Since the residues of straight, extended polypeptide chains alternate in position by 180°, the cumulative effect of the tendency toward right-handed twisting is to produce extended structures that are coiled in a right-handed direction (fig. 5.21).

The third motif we want to draw attention to is the β-α-β motif (fig. 5.22). This motif is a special case of the right-handed crossover between two parallel β sheets where the connecting region consists mostly of α helix.

Having presented three of the most commonly observed structural motifs found in globular proteins, we are now in a position to consider how these motifs are assembled into larger units of structure.

The Domain Is the Basic Unit of Tertiary Structure

A domain constitutes a stable unit of tertiary structure; it usually contains a combination of two or more covalently linked structural motifs. Some proteins contain a single domain. Others contain two or more domains held together by covalent linkages or noncovalent linkages. While it is clear that there is an enormous variety of domains, it is remarkable how many times we find strikingly similar domains in different proteins and how often it is possible to gain an immediate qualitative understanding of the features that account for a protein's stability and function once the structure has been determined. We will see that no protein is found as a single-layer structure. This is because it requires at least two layers to bury the hydrophobic core resulting from the hydrophobic amino acid side chains that are inevitably found in all proteins.

Figure 5.23

Schematic diagram of the four-helix bundle. (*a*) Side view showing the path of the polypeptide chain in a four-helix bundle domain. (*b*) Projection down the bundle axis. Large circles represent the main chain of the helices; small circles represent side chains. Side chains exposed on the surface of the bundle are mainly hydrophilic (red). Buried side chains tend to be hydrophobic (green).

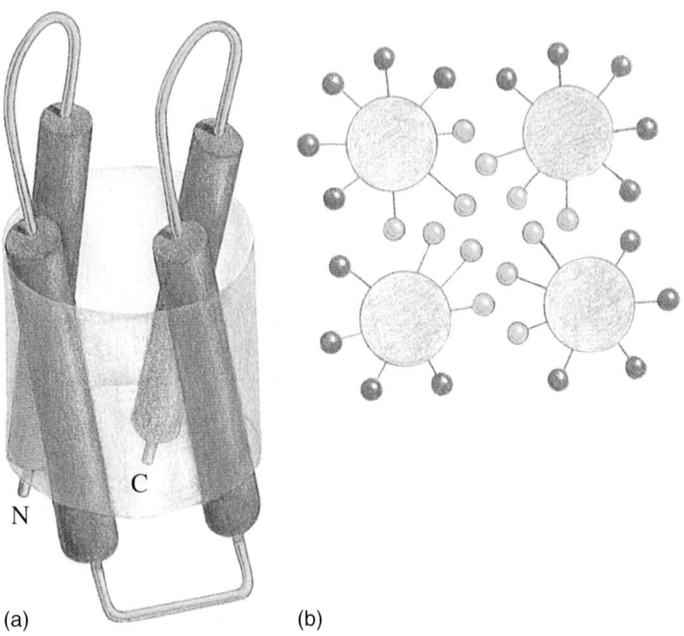

Although it is impossible to survey all major types of protein structure, we will review three of the major classes of domains. First we will deal with domains composed primarily of α helices. Then we will turn to α-β domains and next to antiparallel β domains. Finally, we will discuss a few proteins whose stability requires special features.

The Helix-Loop-Helix Motif Is the Basic Component Found in α-Domain Structures

The α-domain structures exploit the helix-loop-helix motif to create clusters of interacting α helices in which the hydrophobic side chains are buried in the connecting regions between different segments of α helix. Two commonly found domains of this type are the four-helix bundle and the globin fold.

The four-helix bundle as the name suggests contains four segments of α helix that form an intimate bundle. The segments of α helix are held together in one polypeptide chain by interconnecting loops of extended polypeptide chain. As in the case of the two- and three-stranded coiled coils found in fibrous proteins and some globular proteins, the helices that interact are oriented relative to one another so that the side chains of one helix fill the grooves of the neighboring helix. Unlike two- and three-stranded helix bundles, however, all well-characterized examples of four interacting α helices exist in an antiparallel conformation. In the four-helix bundle the interaction angle between helices is about 18° as in the helices of α-keratin. In this arrangement, four α helices, sequentially connected to their nearest neighbors, pack together to form an array with a roughly square cross section. Since each helix interacts with its neighbors at an angle of about 18°, the overall bundle has a left-handed twist (fig. 5.23).

This commonly observed folding domain clearly represents a minimum accessible surface area arrangement for four sequentially connected helices of approximately equal length. Many α-helical proteins that lack these features have more complex and irregular geometries. However, even in these cases it appears that the relative orientations of adjacently packed helices reflect geometric restrictions that accompany close packing between helices. This arrangement creates a hydrophobic core in the middle of the bundle along its length, where the side chains are so closely packed that water is excluded (see fig. 5.23). Examples of four-helix bundles are shown in figure 5.24. The active sites of the myohemerythrin and the cytochrome b_{562} are located in a hydrophobic pocket between the helices at one end of the molecule.

Figure 5.24

Examples of some proteins that share a common structural motif of four α helices. (Reprinted by permission of Jane S. Richardson.)

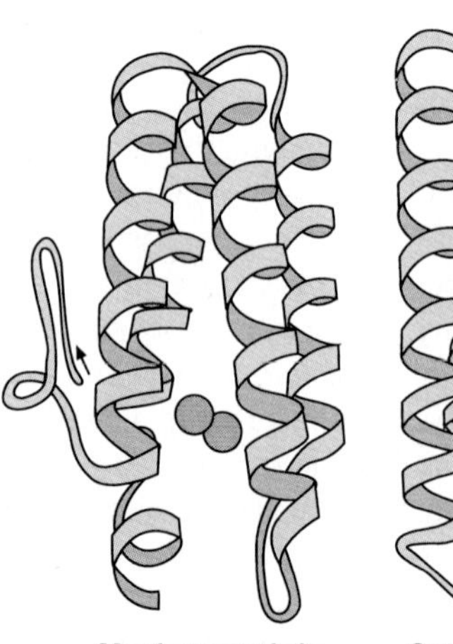

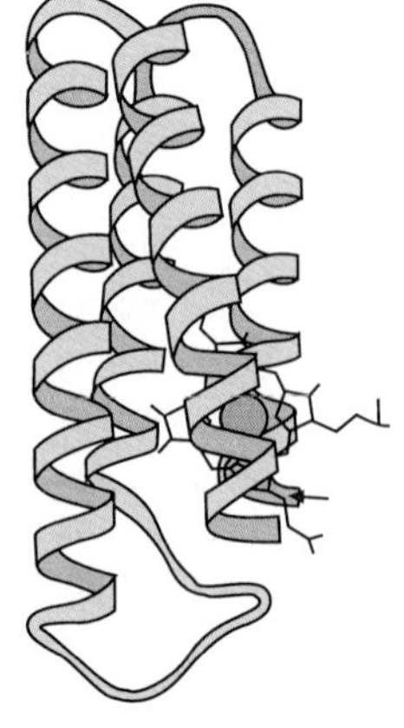

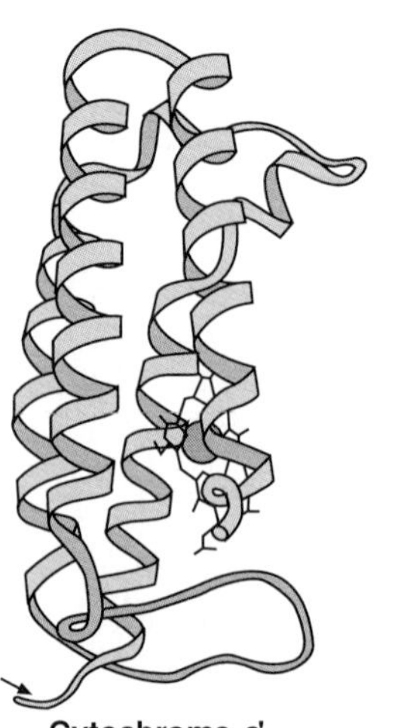

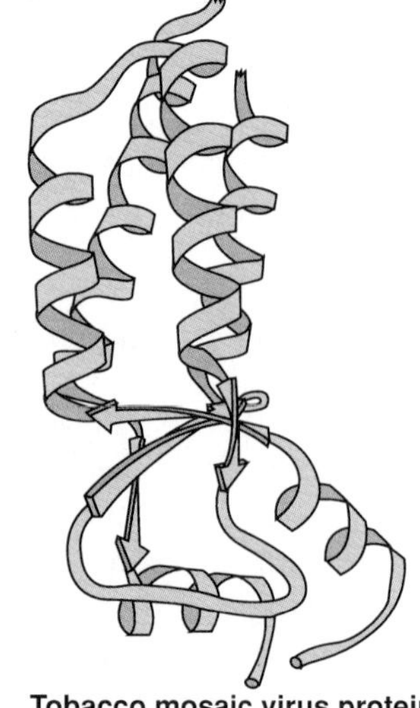

Figure 5.25

Schematic diagram of the globin fold found in myoglobin and hemoglobin. The eight helical segments are labeled A to H. The average interaction angle between helices is about 50°.

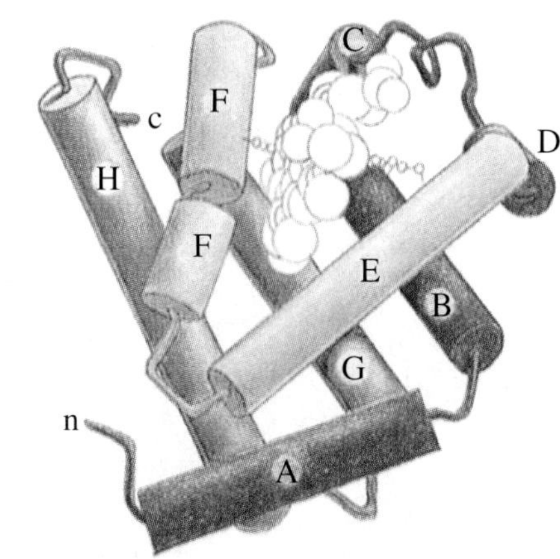

The globin fold is found in myoglobin and hemoglobin and related proteins (fig. 5.25). It comprises a bundle of eight short α-helical segments connected by short loops and arranged so that the helices form a pocket for the active site (a heme group in myoglobin and hemoglobin). In the globin fold, the interaction angle between the interacting helices is about 50°. This angle also favors efficient packing between helices so that the ridges of one helix fit into the grooves of an adjacent helix.

α/β *Domains Exploit the* β-α-β *Motif*

The most frequent and most regular of the domain structures are the α/β domains, which consist of a central parallel or mixed β sheet surrounded by α helices. Most domains of this type make extensive use of the β-α-β motif. All of the glycolytic enzymes and many other enzymes are α/β structures. In addition, many proteins that transport metabolites belong to this structural class.

Clusters containing the β-α-β motif are organized so that an array of β sheets constitutes the backbone of the tertiary structure and the loop region connecting the β polypeptides contains the helix. The clusters of β strands that arise from this motif are arranged in either barrels or saddles. The sheets are always twisted in a right-handed sense when viewed along the polypeptide chain direction. Figure 5.26 shows the polypeptide-chain folding of four proteins that incorporate twisted parallel or mixed β sheets. These include the exoprotease carboxypeptidase A, the electron transport protein flavodoxin, the glycolytic enzyme triose phosphate isomerase, and domain I of pyruvate kinase. Although these proteins all have right-twisted β sheets, it is clear that the overall geometries differ. Thus the β sheets in carboxypeptidases and flavodoxin are smoothly twisted to form saddle-shaped surfaces, while the sheets in triose phosphate isomerase and pyruvate kinase domain I take the form of a cylinder or β barrel.

Within each type of parallel β-sheet organization, the detailed hydrogen-bond pattern can be understood in terms of the forces acting within and between the polypeptide chains (fig. 5.27). In the case of the roughly rectangular sheets in carboxypeptidase and flavodoxin, the observed geometry reflects a competition between the tendency of the individual chains to twist in a right-handed direction and the tendency of the interchain hydrogen bonds to remain firm. Basically, the interchain hydrogen bonds tend to stretch when the sheet is twisted and so resist introduction of twist into the sheet. The observed saddle-shaped geometry reflects the uniform distribution of these conflicting forces through the sheet, as shown in figure 5.27.

The β sheet that forms the barrel in triose phosphate isomerase has an hourglass-shaped surface with cylindrical curvature. Twisted β strands with a staggered hydrogen-bond pattern automatically produce a cylindrical curvature; conversely, twisted strands on a cylindrical surface necessitate a staggered hydrogen-bond pattern. Again, a compromise occurs between twisting and hydrogen bonding, leading to approximately straight chains with somewhat stretched hydrogen bonds at the top and bottom, which produce the hourglass shape with somewhat stretched H bonds at the top and bottom. The differences in the geometries of rectangular and staggered plane sheets result from differences in how adjacent sheets are hydrogen-bonded together. In either case the operative forces are similar, and the final result reflects a compromise between chain twisting and preservation of good interchain hydrogen bonds.

Chiral preferences affect the connectivity as well as the sheet geometry in parallel β proteins. The right-handed crossover in β barrels cannot go down the center, which is only large enough to accommodate the hydrophobic β side chains that are found in this location. As a rule the polypeptide backbone winds in a simple right-handed spiral around the barrel, moving over by one β strand at a time and packing helices or loops around the outside. These structures are large, but their organization is very simple. Most β barrel structures contain precisely eight chains because of restricted geometric considerations that require that a barrel be a self-contained structure. The remaining residues in these domains form the loops that connect the β strands with the α helices. These loops are variable in length and conformation, and in some cases they form independent domains in the overall protein structure. In all β barrel domains the active site is situated in the bottom of a funnel-shaped pocket created by the eight loops that connect the carboxy end of the β strands with the amino end of the α helices. These structures provide very clear examples of the physical separation of the residues that contribute to the stability of the domain from those that are responsible for their specific function.

The second class of β structures, the saddle-shaped parallel sheets, such as those found in carboxypeptidase or flavodoxin (see fig. 5.26), have a layer of helices and loops on each side. To accomplish this with right-handed crossover connections, the polypeptide chain must sometimes move along the sheet in one direction and sometimes in the other direction. The saddle-shaped β sheets show much more variation in structure. From purely geometric considerations it is clear that there is scope for greater variability. Since they form an open β sheet, there are no geometric restrictions on the number of β strands, as in the case of the β barrels. In fact, the number varies from four to ten in different domains. Furthermore, the two stands that are joined by a crossover connection need not be adjacent in the β sheet. In addition, there

Figure 5.26

A comparison of parallel β-sheet structures forming the backbone structures in different enzymes (or parts of enzymes): (*a*) β-barrel arrangement, (*b*) saddle shape.

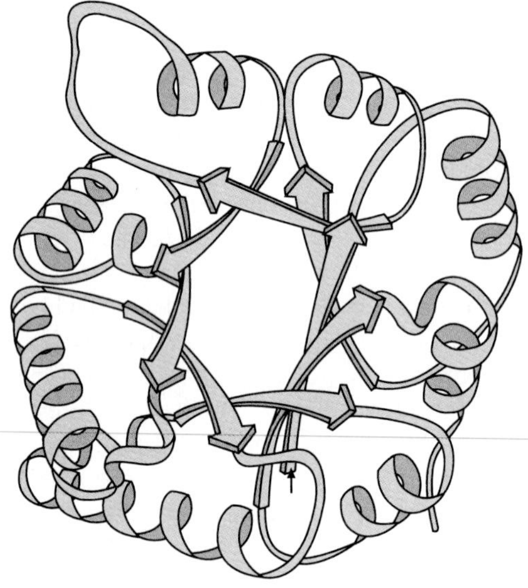

Triose phosphate isomerase

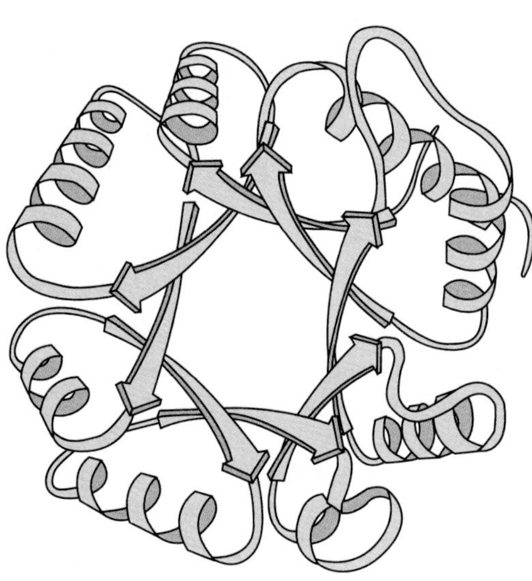

Pyruvate kinase domain 1

(a)

SADDLE SHAPE

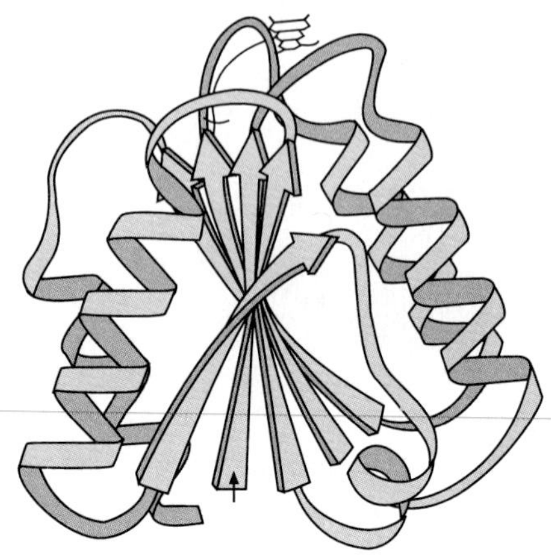

Flavodoxin

Carboxypeptidase

(b)

can be mixed sheets where hairpin connections give rise to some antiparallel β strands mixed in with the parallel β strands. All of these variations can be found in the two relevant protein structures illustrated in figure 5.26.

Antiparallel β Domains Show a Great Variety of Topologies

Antiparallel β domain structures constitute a large group of proteins with a great variety of structures and functions. Within this group we find enzymes, transport proteins, antibodies, and virus coat proteins. The cores of these domains are constructed from β strands that can vary in number from four to ten or more. The most common structural organization in antiparallel β domains has two layers of β sheets that can range from a pair of essentially separate sheets whose relative orientation is determined by complementary packing between saddle-shaped sheet surfaces to twisted barrels with continuous, staggered hydrogen bonding all the way around (fig. 5.28). Other antiparallel β proteins have a single twisted β sheet that is covered on only one side by a layer of helices and loops (fig. 5.29). In both cases the hydrophobic

Figure 5.27

Origin of β-barrel and β-sheet conformations. The observed geometry in β-sheet structures represents a competition between the tendency of the individual chains to twist in a right-handed way and the tendency of the interchain hydrogen bonds to be preserved. Rectangularly arranged sheets give rise to the saddle shape. Staggered sheets give rise to the β-barrel shape.

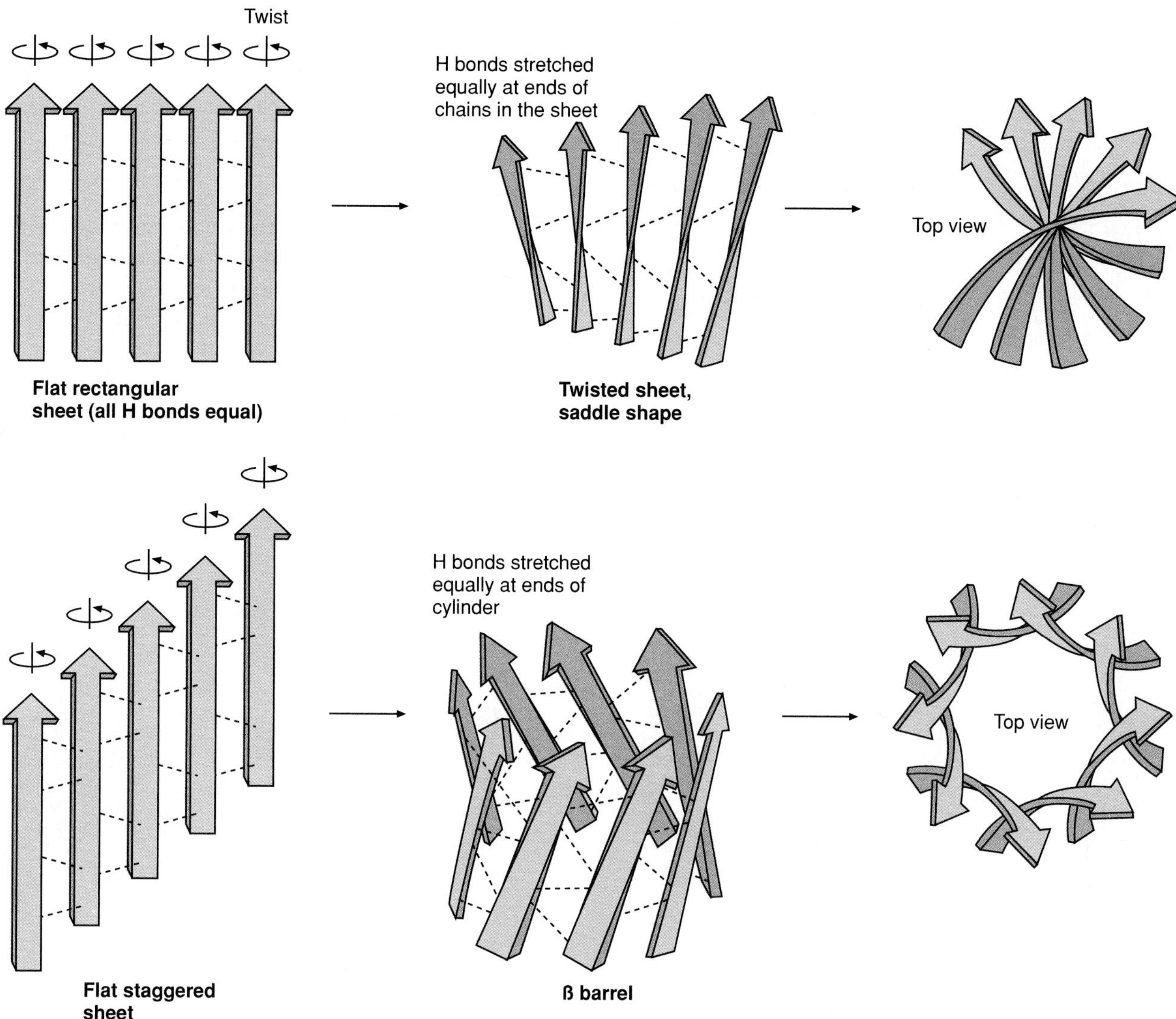

residues are buried between the layers, leaving the hydrophilic residues accessible for making solvent contacts.

Some Proteins or Domains Require Additional Features to Account for Their Stability

In addition to the packing of elements of protein secondary structure, which is a dominant feature in most proteins, there are cases, especially among the smallest structures, where the geometry and presence of disulfide bonds or nonpeptidyl groups are a dominant factor. Figure 5.30 shows examples of this sort, in which the secondary structures are short and irregular and cannot assume their native structures if the disulfides are broken (fig. 5.30*a, b*) or if the nonpeptidyl groups are missing (fig. 5.30*c*). One may wonder how some of these groups become assembled into the native conformation in the first place. In the case of insulin we have some notion of how the molecule could have formed originally and subsequently been converted into the final molecule, which requires two disulfides for stability. The insulin precursor contains 81 amino acid residues in a single polypeptide chain. During maturation the first 23 amino acids are removed. Following disulfide bond formation, additional amino acids are removed from both ends and the middle, leaving a two-chain structure with 21 and 30 amino acid residues. In this case the disulfide bonds are formed at an early stage when the tertiary structure stability is primarily a product of the noncovalent forces within the polypeptide chain.

Figure 5.28

Examples of proteins containing β-sheet domains.

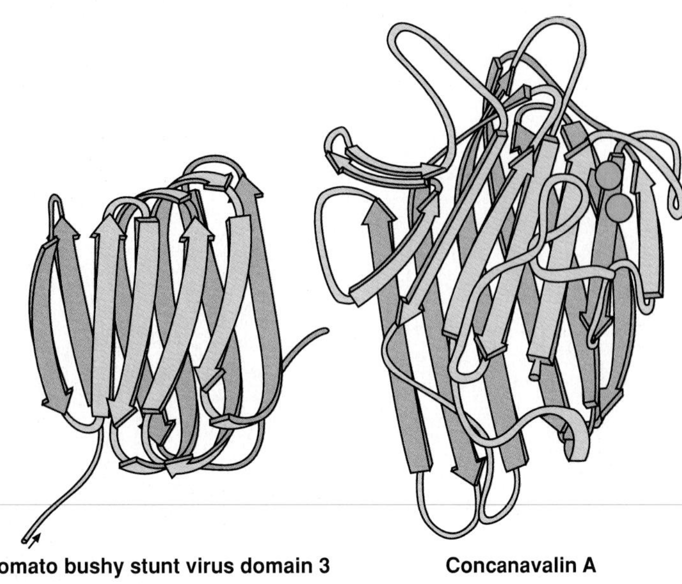

Figure 5.29

Examples of antiparallel β proteins that are covered on only one side by larger helices and loops.

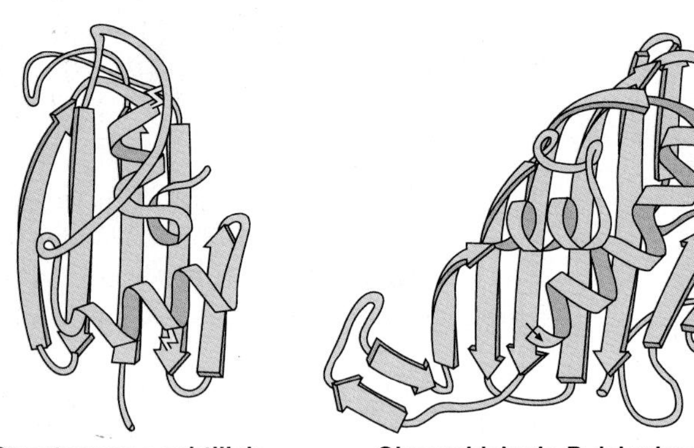

Figure 5.30

Examples of some small proteins or domains in which disulfide bonds, (*a, b*) or a porphyrin group (*c*) are a dominant factor holding the structure together. The structure of porphyrins is depicted in molecular detail in figure 11.20.

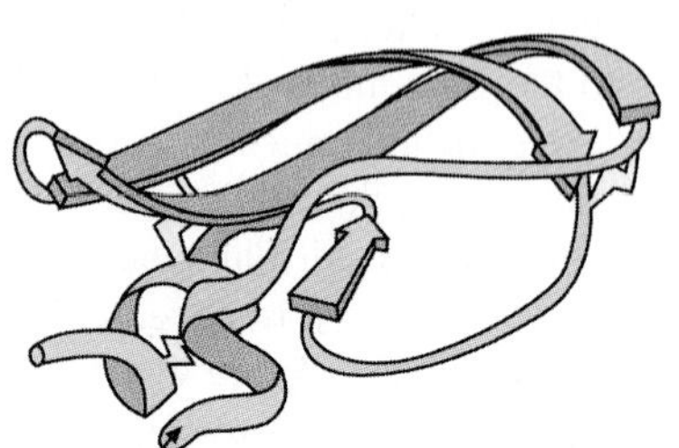

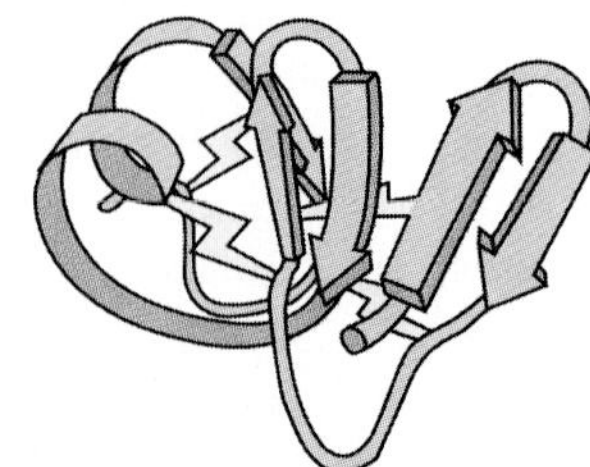

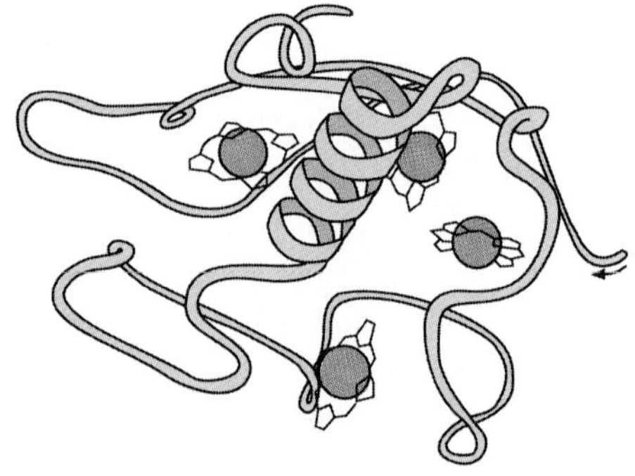

Special structural features that account for the stability of membrane-binding proteins and DNA-binding proteins will be discussed in the appropriate chapters.

Many Proteins Contain More Than One Domain

The patterns of tertiary structure just described frequently constitute an entire protein. However, within a single subunit, contiguous portions of the polypeptide chain often fold into more than one domain. Sometimes the domains within a protein are very different from one another, such as within papain (fig. 5.31), but often they resemble each other very closely, such as in rhodanase (fig. 5.32).

The separateness of two domains within a protein varies all the way from independent globular domains joined by a flexible length of polypeptide chain to domains with tight and extensive contact with a smooth globular surface (fig. 5.33). An intermediate level of domain separateness is common in known protein structures, with an elongated overall subunit shape and a definite neck or cleft between the domains such as phosphoglycerate kinase (fig. 5.34).

Domains as well as subunits can serve as modular bricks to aid in efficient assembly. Undoubtedly, the existence of sepa-

Figure 5.31

Papain, a protein in which the domains are very different from one another. (Reprinted by permission of Jane S. Richardson.)

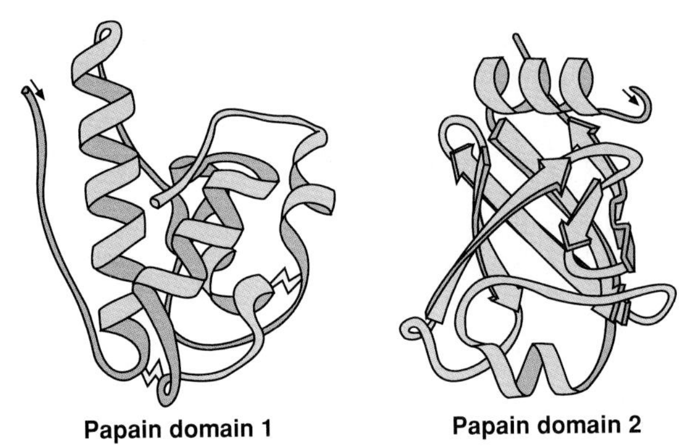

Figure 5.32

Rhodanese domains 1 and 2 as an example of a protein with two domains that resemble each other extremely closely. Rhodanese is a liver enzyme that detoxifies cyanide by catalyzing the formation of thiocyanate from thiosulfate and cyanide. (Reprinted by permission of Jane S. Richardson.)

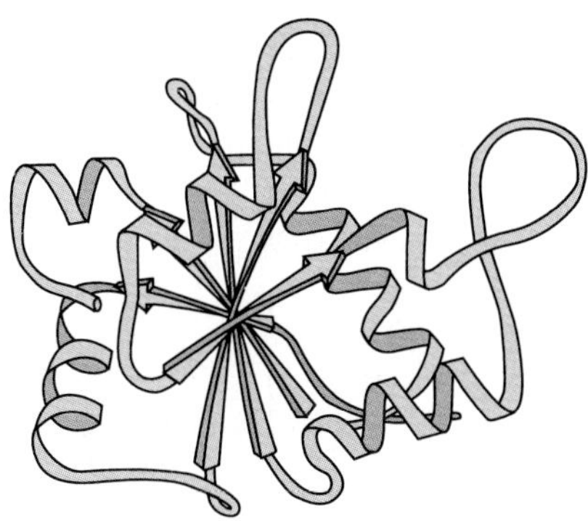

Rhodanese domain 1

Rhodanese domain 2

Figure 5.33

Schematic backbone drawing of the elastase molecule, showing the similar β-barrel structures of the two domains. (Reprinted by permission of Jane S. Richardson.)

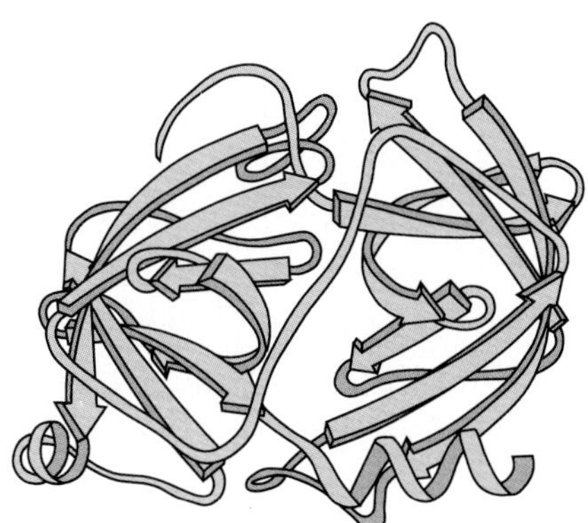

Elastase

Figure 5.34

The dumbbell domain organization of phosphoglycerate kinase, with a relatively narrow neck between two well-separated domains. Copyright 1994 by The Scripps Research Institute/Molecular Graphics Images by Michael Pique using software by Yng Chen, Michael Connolly, Michael Carson, Alex Shah and AVS, Inc. Visualization advice by Holly Miller, Wake Forest University Medical Center.

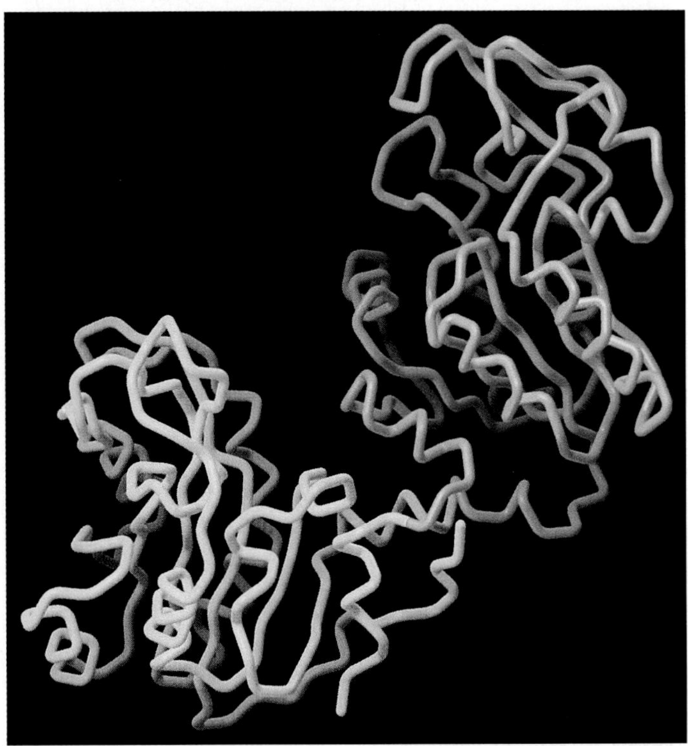

rate domains is important in simplifying the protein folding process into separable, smaller steps. The most common domain size is between 100 and 200 residues, but there is no strict upper limit on practicable folding size, since domains vary in size all the way from about 40 residues to over 400 residues.

The other important function of domains is to provide motion. Completely flexible hinges would be impossible between two polypeptide chains within a protein containing more than one polypeptide chain (subunit), because they would simply fall apart. But they can exist between covalently linked domains. Limited flexibility between domains is often crucial to substrate binding, allosteric control (see chapter 10), or assembly of large structures. In hexokinase the two domains within the individual subunits hinge toward each other upon binding of glucose, enclosing it almost completely (fig. 5.35). In this manner the substrate glucose can be bound in an environment that excludes water as a competing substrate.

Quaternary Structure Depends on the Interaction of Two or More Proteins or Protein Subunits

Although many globular proteins function as monomers, biological systems abound with examples of complex protein assemblies

Figure 5.35

Schematic representation of the change in conformation of the hexokinase enzyme on binding substrate. E and E′ are the inactive and active conformations of the enzyme, respectively. G is the sugar substrate. Regions of protein or substrate surface excluded from contact with solvent are indicated by a crinkled line. Figure 9.3 presents a more detailed view of the hexokinase molecule. (Source: From W. S. Bennett and T. A. Steitz, Glucose-induced conformational changes in yeast hexokinase, *Proc. Natl. Acad. Sci. USA,* 75:4848, 1978.)

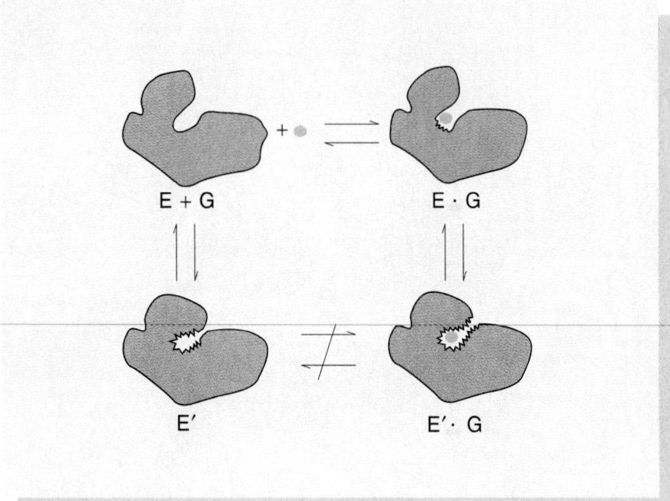

(table 5.1). This higher-order organization of globular subunits to form a functional aggregate is referred to as the quaternary structure of the protein. Protein quaternary structures can be classified into two fundamentally different types. The first involves the assembly of proteins (sometimes referred to as subunits because they constitute a part of the final structure) that are very different structures. Examples range from dimeric molecules that contain different molecular subunits to complex assemblies such as ribosomes, which contain 20 or more nonidentical protein subunits in addition to one or more RNA components. The organization of these sorts of quaternary structures depends on the specific nature of each interaction made between the different molecular subunits and their neighbors. Each intermolecular interaction generally occurs only once within a given aggregate arrangement, so the overall complex structure has a highly irregular geometry. A widely used approach for determining the state of aggregation of proteins in solution is by sedimentation and diffusion analysis (see chapter 7).

A second, commonly observed pattern of quaternary structure is typified by molecular aggregates composed of multiple copies of one or more different kinds of subunits. Owing to the recurrence of specific structural interactions between the subunits, such aggregates typically form regular geometric arrangements. Given that proteins are fundamentally asymmetrical objects (because they incorporate chiral L-amino acids), it is clear that the simplest pattern of quaternary structure involves formation of a linear aggregate. As illustrated in figure 5.36*a, b,* the formation of such an aggregate results from the repetition of one sort

Table 5.1

Molecular Weight and Subunit Composition of Selected Proteins

Protein	Molecular Weight	Number of Subunits	Function
Glucagon	3,300	1	Hormone
Insulin	11,466	2	Hormone
Cytochrome *c*	13,000	1	Electron transport
Ribonuclease A (pancreas)	13,700	1	Enzyme
Lysozyme (egg white)	13,900	1	Enzyme
Myoglobin	16,900	1	Oxygen storage
Chymotrypsin	21,600	1	Enzyme
Carbonic anhydrase	30,000	1	Enzyme
Rhodanese	33,000	1	Enzyme
Peroxidase (horseradish)	40,000	1	Enzyme
Hemoglobin	64,500	4	Oxygen transport
Concanavalin A	102,000	4	Unknown
Hexokinase (yeast)	102,000	2	Enzyme
Lactate dehydrogenase	140,000	4	Enzyme
Bacterio-chlorophyll protein	150,000	3	Enzyme
Ceruloplasmin	151,000	8	Copper transport
Glycogen phosphorylase	194,000	2	Enzyme
Pyruvate dehydrogenase (*E. coli*)	260,000	4	Enzyme
Aspartate carbamoyl-transferase	310,000	12	Enzyme
Phosphofructo-kinase (muscle)	340,000	4	Enzyme
Ferritin	440,000	24	Iron storage
Glutamine synthase (*E. coli*)	600,000	12	Enzyme
Satellite tobacco necrosis virus	1,300,000	60	Virus coat
Tobacco mosaic virus	40,000,000	2,130	Virus coat

Figure 5.36

Linear and helical quaternary aggregates of protein molecules. (*a*) A linear arrangement of hypothetical protein subunits (illustrated as simplified right shoes). The interactions in such a linear arrangement are all identical, and the structure lends itself to the formation of an indefinitely long linear structure whose subunits are related by translation in one dimension. (*b*) A helical arrangement in which equivalently interacting subunits are related by unit translations along the helix axis followed by a 180° rotation about the helix axis. (*c*) A helical arrangement in which subunits are related by unit translation plus a rotation of 360°/*n* to give an *n*-fold helix. (*d*) An *n*-fold multiple-start helix, an arrangement in which subunits form different equivalent interactions with their nearest neighbors. Red arrows indicate two types of identical contact points.

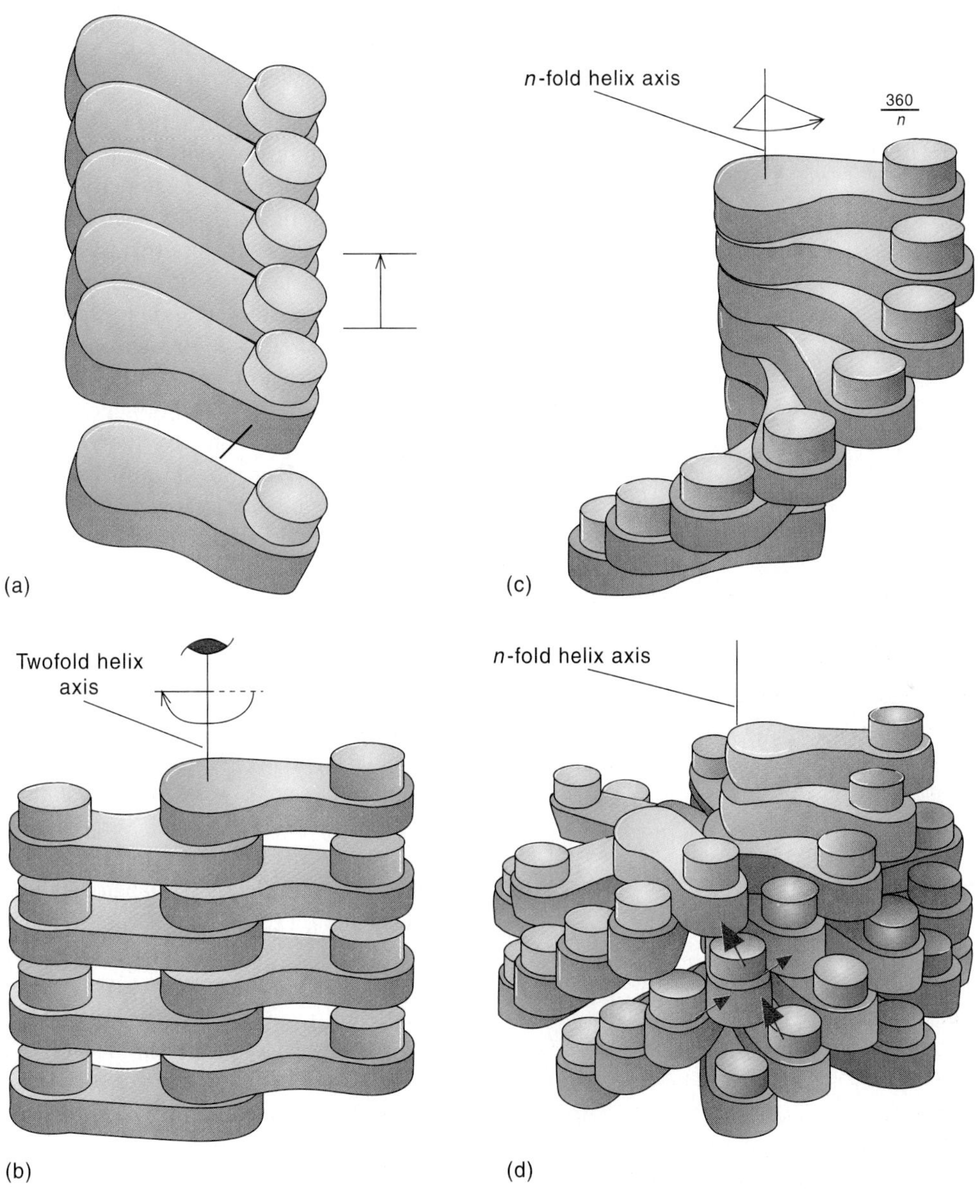

of specific structural interaction between adjacent subunits of the assembly. Structures of this type can be extended simply by the addition of successive subunits.

Somewhat more frequently observed than linear arrangements are helical arrangements of identical molecular subunits. As is the case for the amino acid residues in the α helix, in helical quaternary structures the individual subunits display different, local interactions with their nearest neighbors (see fig. 5.36*c, d*). However, the pattern of nearest-neighbor interactions is repeated for each subunit.

Figure 5.37

Tobacco mosaic virus (TMV) structure. (*a*) Diagram of TMV structure, an example of a helical virus. The nucleocapsid (protein shell) is composed of a helical assembly of 2,130 identical protein subunits (protomers) with the RNA of the virus spiraling on the inside. (*b*) An electron micrograph of the negatively stained helical capsid (400,000X). In negative staining, the virus is immersed in a pool of a heavy metal salt that is much more electron-dense than the virus. The result is that the darker portions of the figure that surround the virus appear more dense than those parts of the figure where the less dense nucleoprotein is located. (© Dennis Kunkel/Phototake.)

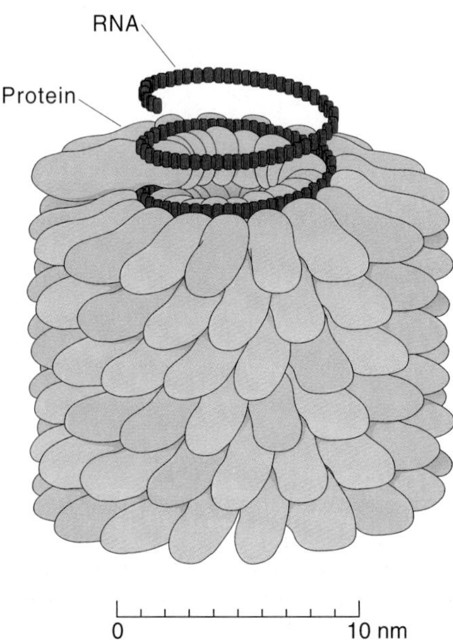

(a)

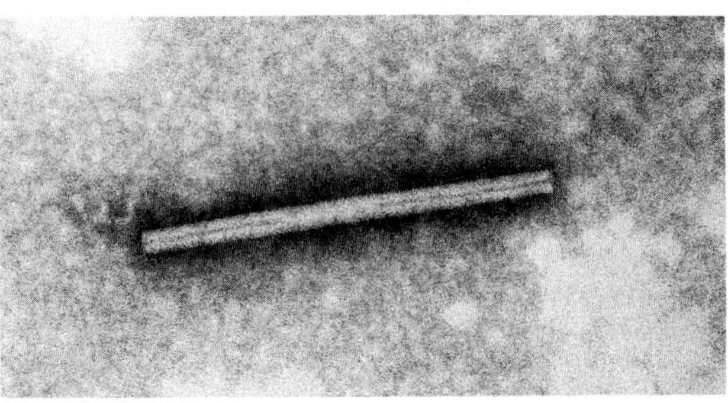

(b)

Helical molecular aggregates are frequently associated with self-assembling molecular structures. Some outstanding examples of such aggregates are the rodlike and filamentous viruses, in which the helical aggregation of the protein-coat subunits forms a cylindrical container for the virus's nucleic acid (fig. 5.37). Since the entire coat is assembled from multiple copies of the same protein, this arrangement represents a very efficient utilization of the information content of the virus nucleic acid.

Helical quaternary structures, then, are characterized by a repeating interaction that results in structures whose subunits are related both by some rise along and twist around a central axis. They are therefore similar to linear arrangements in that they are, at least potentially, indefinitely extendable. In fact, in helical viruses the length of the coat-protein structure is determined, not by the protein, but rather by the fixed length of the virus nucleic acid.

We can also imagine patterns of repeating molecular interactions that involve only twists between the subunits and so result in the formation of quaternary structures that are essentially like flat rings (fig. 5.38*a–e*). Such cyclically repeating interactions typically give rise to symmetrical molecular dimers and trimers. Larger aggregates, in contrast, most frequently do not form flat-ring structures, since flat rings would not provide enough total contact surface to stabilize an open, extended arrangement. Instead, they form arrangements that resemble geometric polyhedra. The formation of polyhedral aggregates reflects the fact that the molecular subunits can have more than one type of intermolecular interaction. Figure 5.38*f, g* illustrates some of the types of structures observed. Such arrangements can be composed of identical subunits or of different types of subunits. One property that distinguishes the polyhedral and ring quaternary structures from linear and helical types is that they incorporate fixed numbers of subunit copies.

The structures of helical and polyhedral viruses demonstrate that quaternary structures play a central role in the self-assembly of very large biological structures from individual molecular subunits. Structural stabilization occurs when all the subunits interact in geometrically similar ways, i.e., essentially like the atoms in a salt crystal. Surprisingly, however, x-ray crystallography reveals that many quaternary interactions are not symmetrical or equivalent, even when they pertain to chemically identical subunits. The simplest sort of nonequivalence occurs at some dimer contacts, where individual side chains close to the twofold axis (which in such cases is only approximate) are forced to take up different positions in order to avoid overlapping. The departures from exact symmetry are usually local, in which case they probably have no functional consequences, but sometimes the nonequivalence extends to other parts of the subunit (e.g., in insulin and in malate dehydrogenase), where it can produce such effects as different binding constants for ligands. It is even easier, of course, for contacts between nonidentical subunits to be asymmetrical. For instance, in hemoglobin the contact between the two β chains is wider than that between the two α chains and produces the binding site for several important effector molecules (see chapter 6).

Figure 5.38

Quaternary structure with rotational and polyhedral symmetry. (*a*) Arrangements of two molecules related by twofold rotational symmetry to form symmetrical dimers. (*b*) Symmetrical trimers. In these arrangements the intersubunit contacts are all identical. (*c*) The most common arrangement for tetrameric molecules (as in hemoglobin), where each subunit makes three different interactions with its neighbors: a "side-by-side" interaction, a "toe-to-toe" interaction, and a "heel-to-heel" interaction. (*d*) A common arrangement for hexameric molecules. (*e*) Octameric molecules. (*f*) A cubic quaternary structure with 24 subunits as found in some iron-storage proteins. (*g*) An icosahedral quaternary structure with 60 subunits, 3 to each triangular face. This pattern is frequently seen in viruses.

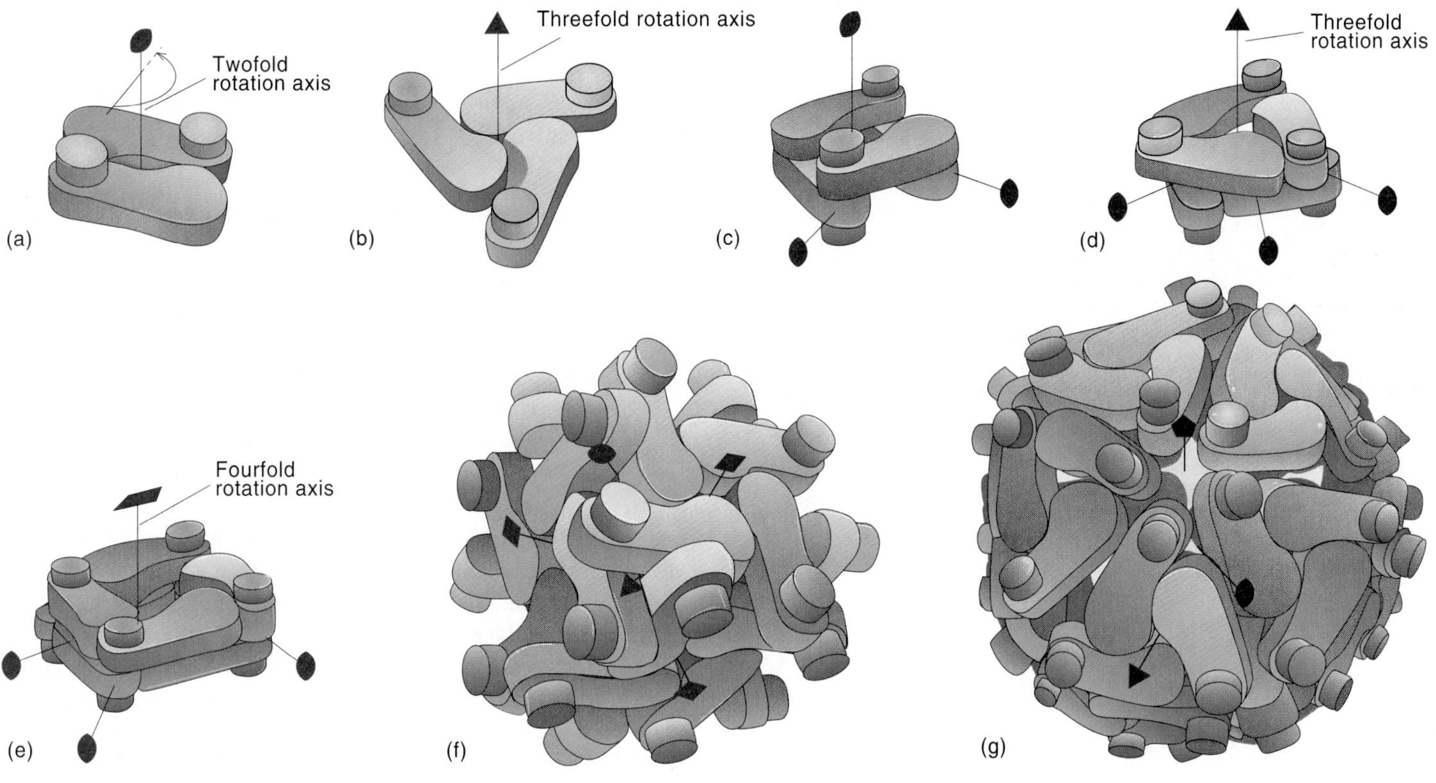

An even more extreme case of asymmetrical association occurs in the dimer of yeast hexokinase, where in place of the pure 180° rotation of a twofold axis, the subunits are related by a rotation of 156° plus a translation of 13.8 Å. Although this is basically a helical contact relationship, it cannot be extended past the dimer because if a third subunit were to bind by the same rule, it would collide with the first one, as illustrated in figure 5.39. In hexokinase the asymmetrical association creates a significant conformational difference between two initially equivalent subunits so that the two active sites in the dimer have quite different functional properties.

Yet another type of nonequivalent association occurs in icosahedral viruses, with only 60 symmetry-equivalent positions (see fig. 5.38) but with more than 60 subunits. One way of reconciling this apparent contradiction is shown by the 180 subunits of tomato bushy stunt virus, which are placed so that five subunits are in contact around each fivefold axis while six other subunits have a distinct but similar contact around each threefold axis (fig. 5.40). The versatility that permits such nonequivalent associations

Figure 5.39

A schematic drawing of a heterologous dimer interaction (blue and green structures) in which infinite polymerization is sterically prevented. Addition of further subunits (pink and beige structures) to the free binding sites on the heterologous dimer is prevented by overlap of proteins. This arrangement of subunits is observed in the hexokinase dimer. (Source: Adapted from a drawing obtained from T. A. Steitz).

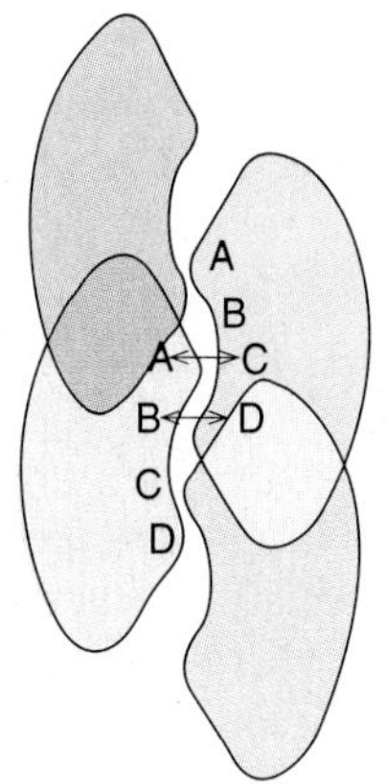

Figure 5.40

The structure of the capsid (protein shell) for an icosahedral virus such as tomato bushy stunt virus. Pentons (P) are located at the 12 vertices of the icosahedron. Hexons (H), of which there are 20, form the edges and faces of the icosahedron. Each penton is composed of five protein subunits, and each hexon is composed of six protein subunits. In all, the structure contains 180 protein subunits.

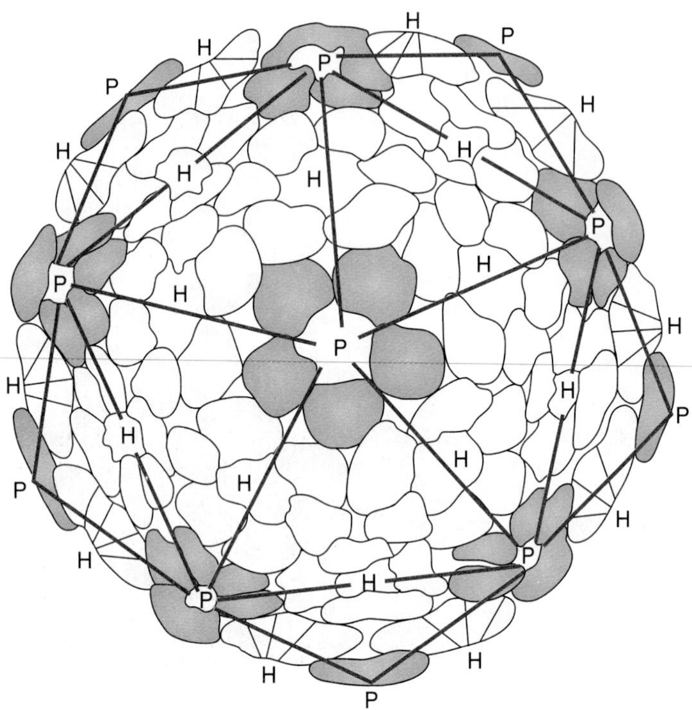

Figure 5.41

Relative probabilities that any given amino acid will occur in the α-helical, β-sheet, or β-hairpin-bend secondary structural conformations.

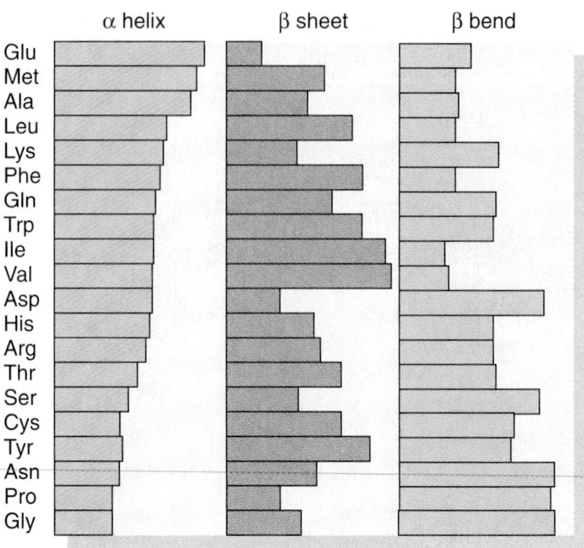

thus allows assembly of larger and more complex structures, and it may be even more common in biological structures that are too large to have been examined crystallographically.

The formation of a subunit aggregate can have extremely important functional results. In particular, as we will show later, the contact interactions provide a means of communication between the individual subunits (e.g., see chapters 6 and 10). As a result, the interaction of a ligand or substrate molecule with one subunit of an aggregate can influence the course of subsequent events in other subunits of the aggregate. Such interactions, which form the basis for cooperativity in biochemical systems, are of great importance because they provide many of the control mechanisms for regulating biochemical processes.

Predicting Protein Tertiary Stucture from Protein Primary Structure

We began the discussion of globular protein tertiary structure by pointing out that the secondary structure and the tertiary structure are determined by the primary structure and that this is probably a reflection of the fact that the native folded conformation is the most stable structure that can be formed. If this is so, then it should be possible to predict a protein's structure from its primary sequence. At this juncture, such predictions remain an elusive goal. However, most proteins are made of a limited number of domains, which tend to reappear in many different proteins. Since this is the case, it may be possible to predict the structures of many proteins in the future by using the information accumulated from x-ray diffraction studies of related proteins.

It is clear that certain amino acids tend to form particular secondary structures. As shown in figure 5.41, glutamic acid, methionine, and alanine appear to be the strongest α-helix formers, whereas valine, isoleucine, and tyrosine are the most probable β-sheet formers. Proline, glycine, asparagine, aspartic acid, and serine occur most frequently in so-called β-bend conformations—an unfolded segment that permits a sharp change in direction.

This type of information is of value in the prediction of secondary structural regions of proteins from their amino acid sequences. The observed frequencies of occurrence of an amino acid in a given conformation provide estimates of the probabilities that the same amino acid behaves similarly in a sequence the actual secondary structure of which is unknown. To predict the secondary structure from the sequence, it is consequently necessary only to plot the probabilities for the individual amino acids sequentially or, better, to plot a local average over a few adjacent residues. Plotting such an average accounts for the cooperative nature of secondary-structure formation. The sequences Gly-Pro-Ser and Ala-His-Ala-Glu-Ala, for example, give high joint probabilities for being, respectively, in β-bend and α-helical conformations. However, comparisons of predicted versus directly observed polypeptide conformations give mixed results. This situation is a consequence of two facts: that several amino acids are somewhat ambiguous in their secondary-structure-forming tendencies and that strong β-bend formers occasionally turn up in the middle of α helices.

Methods for Determining Protein Conformation

The two major techniques for determination of protein fine structure are x-ray crystallography and nuclear magnetic resonance spectroscopy (NMR). Both of these techniques can be used with protein crystals. NMR is also used on concentrated protein solutions. Two other techniques that have proved useful in making measurements on dilute solutions are optical rotatory dispersion (ORD) and circular dichroism (CD). These techniques are much simpler to use, but they do not give detailed information on protein conformation. They are most useful for determining the gross amounts of α and β secondary structure in solution. This information is valuable in two ways. First, by comparison with the results from x-ray diffraction or NMR, it is possible to see whether the structure in solution is similar to the structure in the crystalline state. For example, from x-ray crystallography we know that the majority of the polypeptide chain of hemoglobin is in the α-helix conformation. CD tells us that this is also true in solution. From this information we infer that the structure of hemoglobin does not change drastically when diluted. Since dilute solution is much closer to the environment of a protein in the organism, such information is highly relevant. Additional value of ORD and CD accrues from the rapidity and ease with which such measurements can be made. Estimates of secondary structure can be made as a function of changing solvent conditions, and the kinetics of structural transformations can be studied.

X-Ray Diffraction Analysis of Fibrous Proteins

X-ray diffraction played a major role in the discovery of the structure of fibrous proteins. In most cases the fibers under study are oriented in two dimensions by stretching. In this analysis we illustrate how the technique is used to study the α form of the synthetic polypeptide poly-L-alanine.

A stretched fiber containing many poly-L-alanine molecules is suspended vertically and exposed to a collimated monochromatic beam of CuK_α x-rays, as shown in figure 5.42*a*. Only a small percentage of the x-ray beam is diffracted; most of the beam travels through the specimen with no change in direction. A photographic film is held in back of the specimen. A hole in the center of the film allows the incident undiffracted beam to pass through.

Coherent diffraction occurs only in certain directions specified by Bragg's law: $2d \sin \theta = n\lambda$, where d is the distance between identical repeating structural elements, θ is the angle between the incident beam and the regularly spaced diffracting planes, λ is the wavelength of x-rays used, and n is the order of diffraction, which may equal any integer but is usually strongest for $n = 1$. For small θ, $\sin \theta \approx \theta$ and $d \approx 1/\theta$, so a spot far out on the photographic film is indicative of a repeating element of small dimension.

Figure 5.42*b* shows the diffraction pattern obtained when the fiber axis is normal to the beam. Note the strong off-vertical reflection at 5.4 Å (arrow). A different diffraction pattern (*c*) is obtained when the fiber axis is inclined to the beam at 31°. Note the strong reflection at 1.5 Å in the upper part of the diagram (arrow).

Figure 5.42

(*a*) Experimental arrangement for obtaining x-ray diffraction pattern shown in (*b*). (*b*) Diffraction pattern of the α form of a cluster of poly-L-alanine molecules oriented vertically. (*c*) Diffraction pattern of the same fiber bundle with the fiber axis inclined to the beam at 31°. (*b* and *c*, From L. Brown and I. F. Trotter, *Transactions of the Faraday Society,* 52:537 (1956), © The Royal Society of Chemistry, Cambridge.)

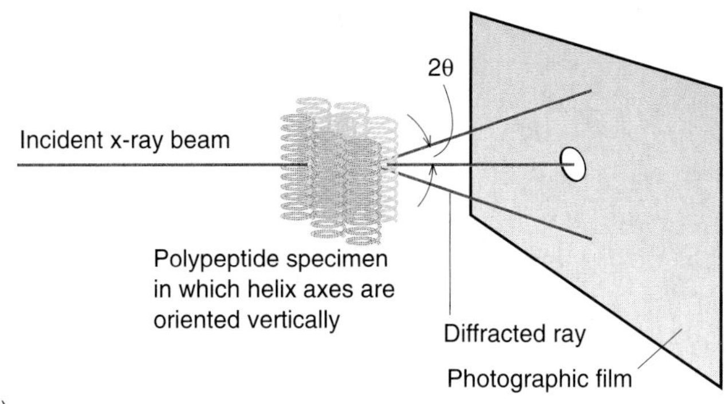

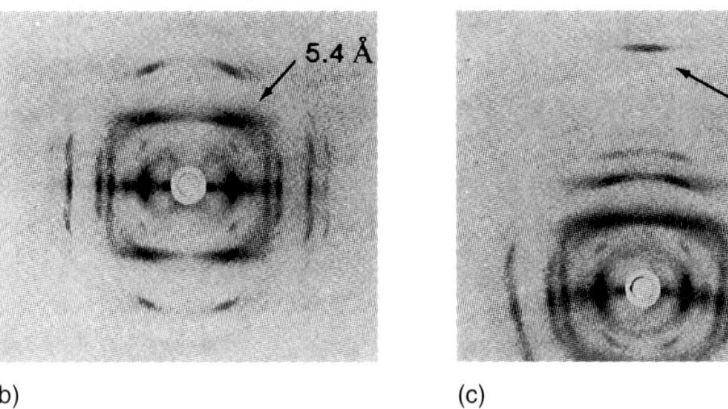

X-Ray Diffraction Analysis of Protein Crystals

The most information about a protein's structure is obtained from ordered three-dimensional protein crystals; this is the main interest of x-ray crystallographers. The goal in x-ray crystallography is to obtain a three-dimensional image of a protein molecule in its native state at a sufficient level of detail to locate its individual constituent atoms. The way this is done can most easily be appreciated by considering the more familiar problem of how we obtain a magnified image of an object in a conventional light microscope. In a light microscope, light from a point source is projected onto the object we wish to examine. When the light waves hit the object, they are scattered so that each small part of the object essentially serves as a new source of light waves. The important point is that the light waves scattered from the object contain information about its structure. The scattered waves are collected and recombined by a lens to produce a magnified image of the object (fig. 5.43).

Given this picture, we might ask what prevents us from simply putting a protein molecule in place of our object and viewing its magnified image. The basic problem here is one of resolution. The resolution, or extent of detail, that can be recovered

Figure 5.43

Schematic diagram of the procedures followed for image reconstruction in light microscopy (top) and x-ray crystallography (bottom).

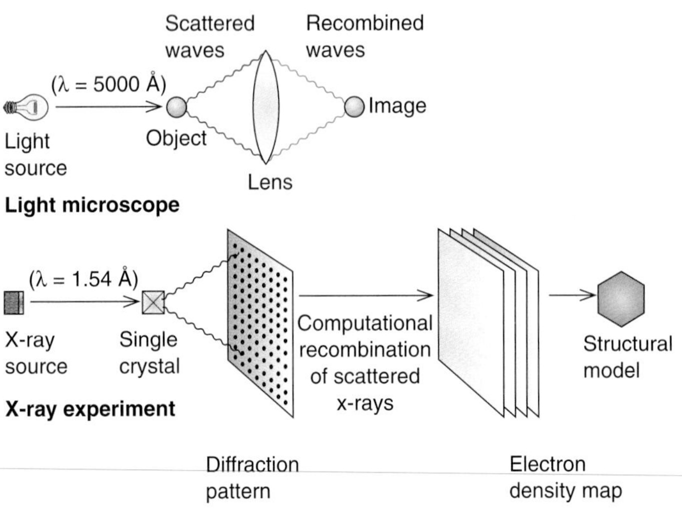

Figure 5.44

Schematic view of an x-ray diffraction pattern. The spacing of the spots is reciprocally related to the dimensions of the repeating unit cell of the crystal. The symmetry of the spots (e.g., the mirror planes in the sample shown) and the pattern of missing spots (alternating spots along the mirror axes) give information on how molecules are arranged in the unit cell. Information concerning the structure of the molecule is contained in the intensities of the spots. Spots closest to the center of the film arise from large-scale or low-resolution structural features of the molecule, whereas those farther out correspond to progressively more detailed features. Circles show 5-Å and 3-Å regions of resolution. Mirror axes are labeled *m*. Spacing of vertically oriented spots, b^*, and horizontally oriented spots, a^*, are reciprocally related to *b* and *a*, the dimensions of the unit cell.

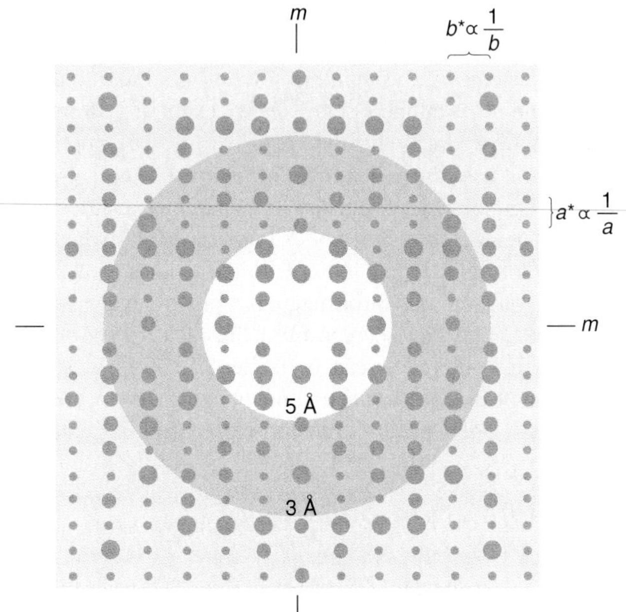

from any imaging system depends on the wavelength of light incident on the object. Specifically, the best resolution obtainable equals $\lambda/2$, or one-half the wavelength of the incident light. Because λ lies in the range of 4,000–7,000 Å for visible light, a visible-light microscope clearly does not have the resolving power to distinguish the atomic structural detail of molecules. What we need is a form of incident radiation with a wavelength comparable to interatomic distances. X-rays emitted from excited metal atoms, with wavelengths in the range of one to a few angstroms, would be most suitable.

However, simply replacing a visible-light source with an x-ray source does not solve all the problems. For example to get a three-dimensional view of a protein, some provision must be made for looking at it from all possible angles, an obvious impossibility when dealing with a single molecule. Furthermore, when x-rays interact with proteins, very few of the rays are scattered. Most x-rays pass through the protein, but a relatively large number of them interact destructively with the protein, so a single molecule would be destroyed before scattering enough x-rays to form a useful image. Both these problems are overcome by replacing a single protein molecule with an ordered three-dimensional array of many molecules that scatters x-rays essentially as if it were one molecule. The ordered array of protein molecules forms a single crystal, so the general technique is called protein x-ray crystallography.

The problems do not end here, because although the protein crystal readily scatters incident x-rays, no lens materials are available that can recombine the scattered x-rays to produce an image. Instead, the best that can be done is to directly collect the scattered x-rays in the form of a diffraction pattern. Although recording the diffraction pattern results in loss of some important information, experimental techniques have been developed for recovering the lost information. Eventually, the scattered waves can be mathematically recombined in a computational analog of a lens. By collecting the diffraction pattern of the crystal in many orientations, it is possible to construct a three-dimensional image of the protein molecule.

Crystals suitable for protein x-ray studies may be grown by a variety of techniques, which generally depend on solvent perturbation methods for rendering proteins insoluble in a structurally intact state. The trick is to induce the molecules to associate with each other in a specific fashion to produce a three-dimensionally ordered array. A typical protein crystal useful for diffraction work is about 0.5 mm on a side and contains about 10^{12} protein molecules (an array 10^4 molecules long along each crystal edge). Note especially that, because protein crystals are from 20 to 70% solvent by volume, crystalline protein is in an environment that is not substantially different from free solution.

The x-ray radiation usually employed for protein crystallographic studies is derived from the bombardment of a copper target with high-voltage (50 kV) electrons, producing characteristic copper x-rays with λ = 1.54 Å. Figure 5.44 shows, in schematic fashion, the x-ray diffraction pattern from a protein crystal. Several features about this pattern bear explanation. First, as you can see, the diffraction pattern consists of a regular lattice of spots of different intensities. The spots are due to destructive interference of waves scattered from the repeating unit of the crystal. For the crystal with the diffraction pattern shown, the repeating unit (or crystal unit cell) contains four symmetrically arranged protein molecules. Corresponding symmetrical features appear in the spot intensity pattern. Further, the lattice spacing of the diffraction spots is inversely proportional to the actual dimensions of the crystal's repeating unit or unit cell. Consequently, both the crystal's unit-cell dimensions and general molecule packing arrangement can be derived from inspection of the crystal's diffraction pattern.

Information concerning the detailed structural features of the protein is contained in the intensities of the diffraction spots. All the atoms in the protein structure make individual contributions to the intensity of each diffraction spot. Therefore, to deduce the three-dimensional structure, all the spots must be measured, either by scanning the x-ray films with a densitometer or by measuring the diffraction spots individually with a scintillation counter.

Initial studies of a protein's tertiary structure are generally carried out at low resolution, that is, using intensity data near the origin (center) of the diffraction pattern. Diffraction data near the origin reflect large-scale structural features of the molecule, whereas those nearer the edge correspond to progressively more detailed features. Figure 5.45 provides examples of electron-density maps calculated at different resolutions to show how various levels of structural detail appear at different degrees of resolution.

A powerful aspect of protein crystallography is that once the native structure is known, various cofactors or enzyme substrate analogs can be bound to the molecule in the crystal. By simply measuring the diffraction intensities, we can compute a new map that allows direct and explicit examination of the structural interactions between the native protein and its substrate or cofactor molecules. Detailed analysis of these interactions has provided much of the foundation for our current understanding of many protein catalytic and functional properties.

Figure 5.45

View of crystallographic electron-density maps, showing how the structural detail revealed depends on the resolution of the data used to compute the maps. The actual molecular structure is inserted in its true position in the electron-density maps.

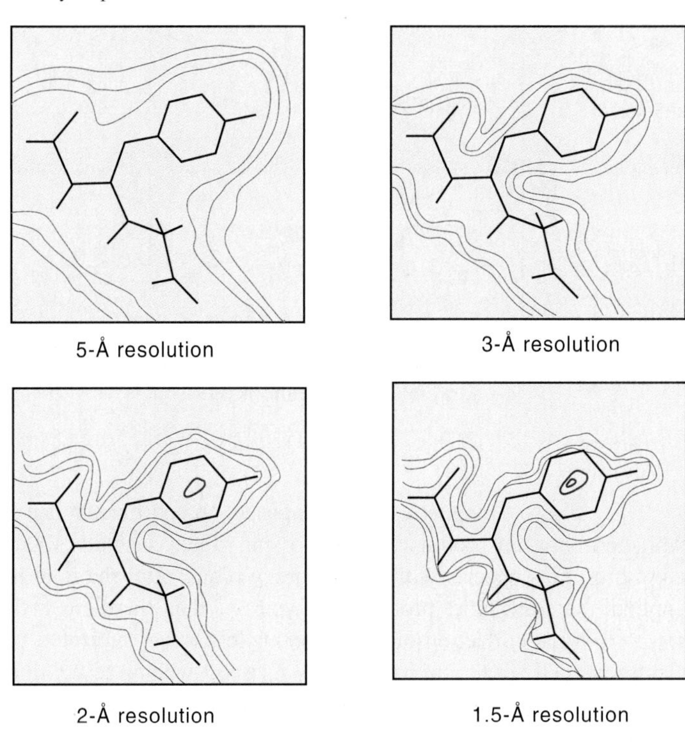

Figure 5.46

Protons align their magnetic poles (→) when an external magnetic field is applied.

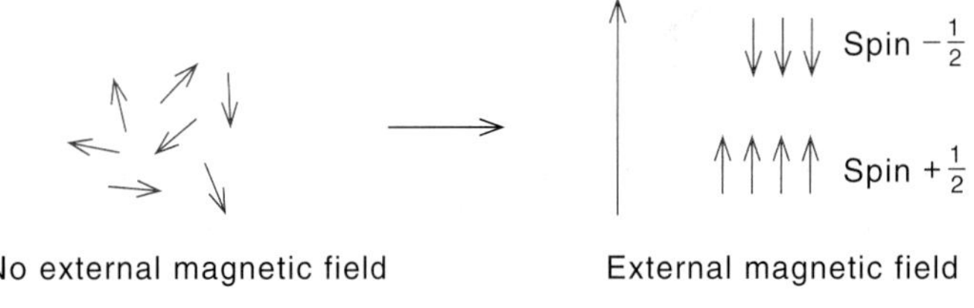

Nuclear Magnetic Resonance (NMR) Complements X-Ray Crystallography

NMR and x-ray crystallography both deal with the fine structure of a protein molecule. In many respects the two techniques complement each other. Thus while x-ray crystallography deals exclusively with the structure of proteins in the crystalline state, NMR can also be used to determine the structure in concentrated solutions and is adaptable for investigation of dynamic processes such as protein folding. Although NMR analysis possesses certain advantages, it is limited in use to the structural determination of small protein molecules. As a result, x-ray crystallography remains the only method available for determination of the structure of large protein molecules.

The NMR technique exploits the magnetic moment or spin possessed by atomic nuclei that contain an odd number of protons such as ^{1}H, ^{13}C, or ^{15}N. Most analyses of protein molecules by NMR have been confined to spin studies on ^{1}H because of the abundance of hydrogen atoms in proteins. The hydrogen nucleus ^{1}H has a nuclear spin that can assume either of two values, designated by quantum numbers of $+\frac{1}{2}$ and $-\frac{1}{2}$. Because they possess a spin, the nuclei act like magnets. We can think of a spinning proton as a little magnet with a north and south pole that can be represented by a vector. When protein molecules are placed in a strong magnetic field, the spin axis of their hydrogen atoms tends to align along the direction of the externally applied magnetic field (fig. 5.46). The extent of alignment is a function of the strength of the applied magnetic field. Protons with spins of $+\frac{1}{2}$ orient parallel to the applied field, while protons with spins of $-\frac{1}{2}$ orient antiparallel to the applied field. The presence of the external magnetic field has another effect on the nuclei with different spins. The two spin states have different energies. The $+\frac{1}{2}$ spin state has a lower energy than the $-\frac{1}{2}$ spin state. The energy difference between the two spin states, ΔE, is directly proportional to the intensity of the magnetic field H at the proton nucleus (fig. 5.47).

Figure 5.47

In the presence of a magnetic field the two spin states for the proton have different energies. The energy difference for the two spin states, ΔE, is directly proportional to the intensity of the magnetic field, H.

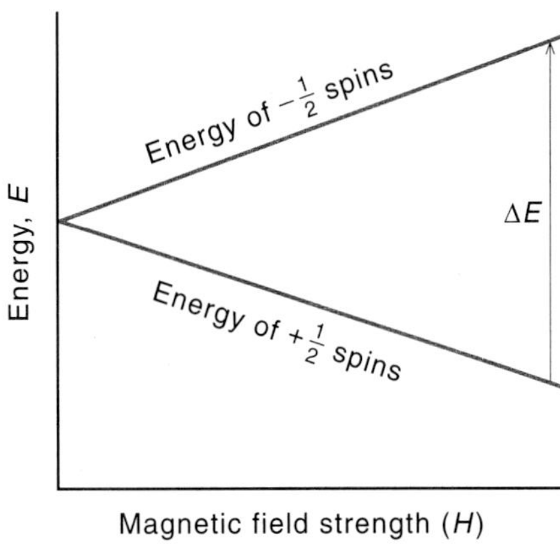

Figure 5.48

Ultraviolet absorption spectra of poly-L-lysine in different conformations: random coil, pH 6.0, 24°C; α helix, pH 10.8, 225°C; β sheet, pH 10.8, 52°C. This curve demonstrates the sensitivity of the peptide absorption band to the polypeptide chain confirmation. (Source: Adapted from K. Rosenheck and P. Doty, *Proc. Natl. Acad. Sci. U.S.A.* 47:1775, 1961.)

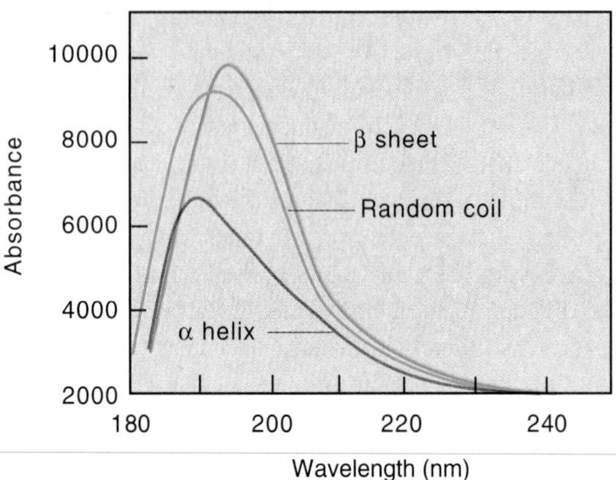

After the magnetic field is imposed, an equilibrium is established between protons in the two spin states. Usually, there are more of the species with lower energy. Thus, after the field is applied, there are more protons with spin $+\frac{1}{2}$ than there are with spin $-\frac{1}{2}$. If the collection of hydrogen nuclei is now subjected to electromagnetic radiation with energy E_0 exactly equal to ΔE, the energy is absorbed by nuclei in the $+\frac{1}{2}$ state, causing these nuclei to invert their spin to $-\frac{1}{2}$. This energy absorption by nuclei is called the nuclear magnetic resonance energy. Usually, the nuclear magnetic resonance energy is measured by emission rather than absorption. Radio frequency (RF) pulses are applied to the sample. When the excited nuclei in the $-\frac{1}{2}$ state revert to their lower energy state, they emit RF radiation precisely equal to ΔE, which can be quantitatively measured. The precise value of the observed emitted NMR radiation is a function of the molecular environment of the nucleus. By comparing the emitted RF frequencies with those of a reference compound of known structure, it is possible to determine the molecular environment in the region of each hydrogen atom. This information gives a list of distance constraints for each proton nucleus in the protein, which, in favorable cases, may be mathematically synthesized to yield a unique three-dimensional structure of the protein.

Optical Rotatory Dispersion (ORD) and Circular Dichroism (CD)

Any molecule in aqueous solution possessing one or more centers of asymmetry (chiral centers) is optically active. The extent of optical activity is quantitatively assessed in a polarimeter (see box 5B), which measures the rotation of plane-polarized light on passing through a measured amount of solution containing the optically active molecule. The molecule is described as dextrorotatory or levorotatory according to the direction (right or left, respectively) in which it rotates plane-polarized light.

A polypeptide chain is optically active because of the asymmetric nature of its individual amino acid residues. In a helical conformation a polypeptide chain has an additional asymmetric component resulting from the chiral nature of the helical conformation. Optical rotation effects resulting from the helical conformation have a maximum in the 200-nm region, which is the region of optimum absorption by the peptide linkage (fig. 5.48).

It is customary to measure optical rotation effects over a spread of wavelengths surrounding the asymmetrically oriented group that is responsible for absorption. When optical rotation is measured as a function of the wavelength of linearly polarized light, the characteristic is referred to as the optical rotatory dispersion (ORD). Circularly polarized light is also used to measure optical rotation effects. In this case the characteristic being measured is called circular dichroism (CD); it is the difference in absorption of left and right circularly polarized light. In an idealized situation, where there is a single absorption band arising from an asymmetrically oriented component, the peak in the CD curve coincides with the absorption peak of the optically active band. The ORD curve shows a value of zero at this point (fig. 5.49). ORD and CD have the same cause and are closely related. Either characteristic may be used to study conformational properties of proteins in solution, but circular dichroism has become more popular because it shows discrete spectral bands that may be either positive or negative.

The ORD and CD curves for poly-L-lysine are illustrated in figure 5.50. It can be seen that conformation has a major effect on the ORD and CD curves in the region of the peptide absorption band. Effects of this sort have led to the application of these polarimetric techniques to studies of protein conformation in solution. For example, helical contents of proteins in solution can be estimated, and denaturation temperatures can be determined.

5B
BOX

Polarized Light and Polarimetry

Light is a form of electromagnetic radiation that oscillates sinusoidally in space and time. The oscillating electric and magnetic fields of light are perpendicular to each other and are both in a plane perpendicular to the direction of the light ray. In an unpolarized beam, there are equally strong fields with all different orientations in the plane (fig. 1). A light beam is said to be polarized if the orientations of the fields in the plane are fixed. In linearly polarized light the orientation of the fields does not vary along the direction of the beam. In circularly polarized light the orientation of the fields gradually changes in either a right-handed or left-handed manner along the direction of the beam.

A polarimeter is an instrument for studying the interaction of polarized light with optically active substances (fig. 2). A cylindrical tube is filled with a solution containing the protein of interest. A monochromatic beam of polarized light is passed through the solution, and the effects on the polarized beam after passing through the solution are measured.

Figure 1

Unpolarized and polarized light.

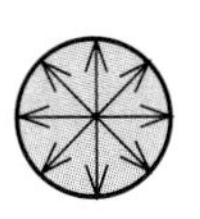

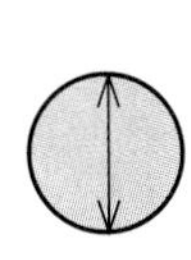

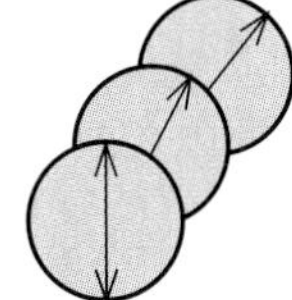

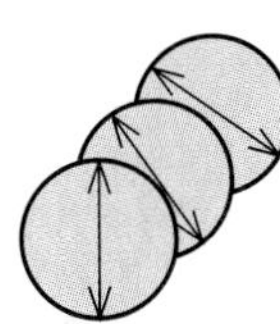

Figure 2

Simple polarimeter for measuring the rotation of linearly polarized light.

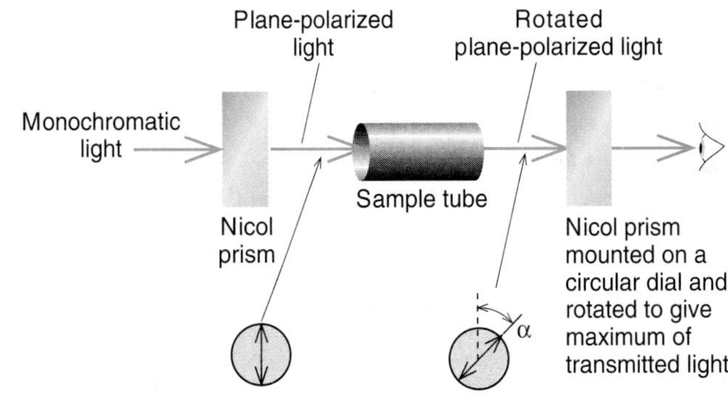

Figure 5.49

Unique spectral properties for a chiral molecule in solution as a function of wavelength, as shown by curves on the left. (*a*) The absorption band produces a symmetric curve with a maximum at the wavelength of maximum absorbance. (*b*) The optical rotatory dispersion curve shows a minimum at the λ_{max} of the optically active band. (*c*) The circular dichroism curve shows a curve similar in shape to the absorption curve. Curve (*a*) would be observed for any molecule that absorbs light, regardless of its configuration. Curves (*b*) and (*c*) are observed only if the absorbing species is optically active. For the enantiomer the curve would be the same in (*a*) but would be opposite in sign in (*b*) and (*c*), as shown on the right.

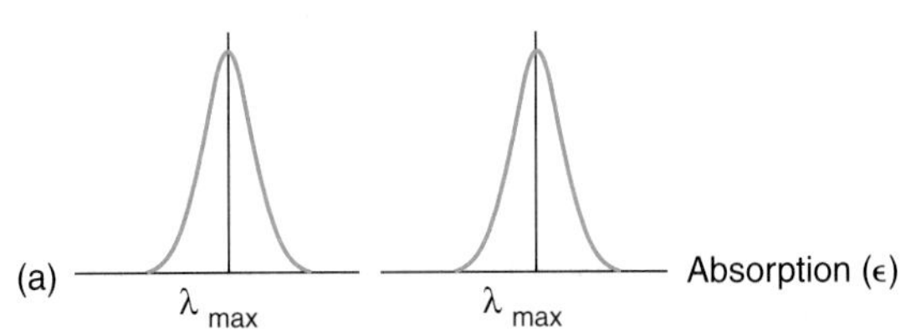

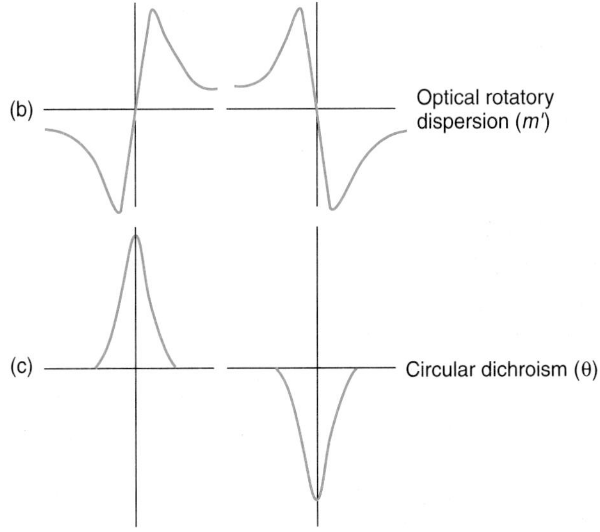

Figure 5.50

(*a*) Optical rotatory dispersion and (*b*) circular dichroism spectra for poly-L-lysine in the α helix, β sheet, and random coil (τ) conformations. These curves are more complex than those shown in figure 5.49 because more than one type of optically active component is present in the polypeptide chain. It is clear that both of these curves are a sensitive function of the polypeptide chain conformation. The precise quantitative interpretation for each curve is not known. (Adapted from A. J. Adler, W. J. Greenfield, and G. D. Fasman, 1973. *In Methods in Enzymology,* vol. 27, ed. by C. H. W. Hirs and S. N. Timasheff, New York: Academic Press.)

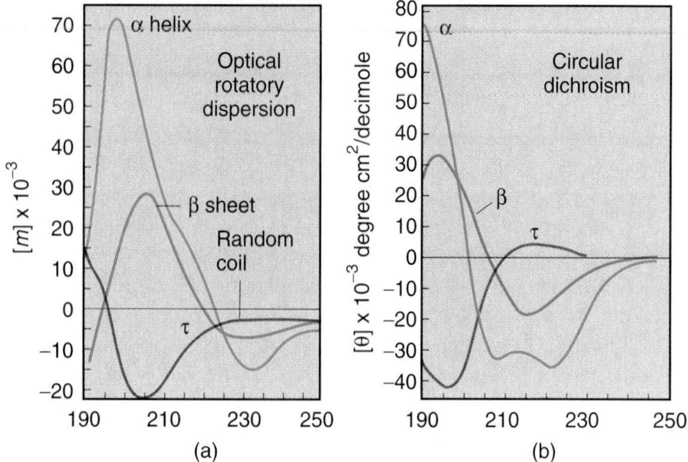

SUMMARY

In this chapter we introduce some of the basic principles that govern protein structure. The discussion of protein structures begun in this chapter is continued in many other chapters in this text in which we consider structures designed for specific purposes. In chapter 6 we examine the protein structures of two systems: the protein that transports oxygen in the blood and the proteins that constitute muscle tissue. In chapters 9 and 10 we discuss structures of specific enzymes. In chapters 19 and 20 we consider proteins that interact with membranes. In chapters 35 and 36 we study regulatory proteins that interact with specific sites on the DNA. Finally, in chapter 37 we examine the structures of immunoglobin molecules.

Our discussion in chapter 5 has focused on the following points.

1. Most proteins may be divided into two groups: fibrous and globular. Fibrous proteins usually serve structural roles. Globular proteins function as enzymes and in many other capacities.
2. The three most prominent groups of fibrous proteins are the α-keratins, the β-keratins, and collagen.
3. The α-keratins are composed of right-handed helical polypeptide chains in which all the peptide NH and carbonyl groups form intramolecular hydrogen bonds. When these helical coils interact, they form left-handed coiled coils.
4. The β-keratins consist of extended polypeptide chains in which adjacent polypeptides are oriented in either a parallel or an antiparallel fashion. Sheets formed from such extended polypeptide chains may be stacked on top of one another.
5. Collagen fibrils are composed of extended polypeptide chains that are coiled in a left-handed manner. Three of these chains interact by hydrogen bonding and coil together into a right-handed cable. Collagen fibrils are composed of a staggered array of many such cables interacting in a side-by-side manner.
6. The structures of fibrous proteins are determined by the amino acid sequence, by the principle of forming the maximum number of hydrogen bonds, and by the steric limitations of the polypeptide chain, in which the peptide grouping is in a planar conformation.
7. The forces that hold globular proteins together are the same as those that hold fibrous proteins together, but there is less emphasis on regularity and more emphasis on burying the hydrophobic regions in the interior of the protein.
8. The secondary structures found in the keratins recur in smaller patches in globular proteins. Such regions of secondary structure are folded into a seemingly endless array of tertiary structures.
9. Tertiary structures can be understood in terms of a limited number of domains.
10. Quaternary structures are formed between nonidentical subunits to give irregular macromolecular complexes or between identical subunits to give geometrically regular structures.
11. X-ray diffraction provides data from which we can deduce the dimensions of the polypeptide chains in proteins. The use of x-ray techniques is, however, limited to molecules that can be oriented to achieve two- or three-dimensional order.
12. Fibrous proteins may achieve two-dimensional order, but they usually do not achieve three-dimensional order. Therefore the diffraction pattern of fibrous proteins gives information about the regularly repeating elements along the long axis of the fibers but tells us very little about the orientation of amino acid side chains.
13. Many globular proteins can be crystallized to achieve three-dimensional order. Study of the crystals of a globular protein can lead to a complete determination of its three-dimensional structure.
14. Nuclear magnetic resonance spectroscopy (NMR) is also used for the determination of detailed three-dimensional structures. It has the advantage over x-ray diffraction that crystalline structures are not required. Only small proteins have thus far had their structures analyzed by NMR.
15. The other techniques that are useful on dilute solutions are optical rotatory dispersion and circular dichroism. These techniques are most useful for determining the gross amounts of α and β secondary structures in solution. Such measurements complement the much more detailed structural information determined on protein crystals by x-ray diffraction.

Selected Readings

Anfinsen, B. C., Principles that govern the folding of protein chains. *Science* 181:223–230, 1973. Nobel Prize recounting by the man who showed that proteins fold spontaneously into their native structures.

Baron, M., D. G. Norman, and I. D. Campbell, Protein modules. *Trends Biochem. Sci.* 16:13–17, 1991.

Branden, C., and J. Tooze, *Introduction to Protein Structure.* New York and London: Garland Publishing, 1991.

Cantor, C. R., and P. R. Schimmel, *Biophysical Chemistry,* vols. 1, 2, and 3. New York: Freeman, 1980. Includes several chapters (2, 5, 13, 17, 20, and 21) on the principles of protein folding and conformation.

Carr, C. M., and P. S. Kim, A spring-loaded mechanism for the conformational change of influenza hemagglutinin. *Cell* 73:823–832, 1993.

Chothia, C., Principles that determine the structures of proteins. *Ann. Rev. Biochem.* 53:537–572, 1984.

Chothia, C., and A. V. Finkelstein, The classification and origins of protein folding patterns. *Ann. Rev. Biochem.* 59:1007–1039, 1990.

Chothia, C., and A. Leak, Helix movements in proteins. *Trends Biochem. Sci.* 10:116–118, 1985.

Cohen, C., and D. A. D. Parry, α-Helical coiled coils—a widespread motif in protein. *Trends Biochem. Sci.* 11:245–248, 1986.

Cohen, C., and D. A. D. Parry, α-Helical coiled coils and bundles: how to design an α-helical protein. *Proteins: Structure, Function, and Genetics* 7:1–15, 1990.

Cohen, C., and D. A. D. Parry, α-Helical coiled coils: more facts and better predictions. *Science* 263:488–489, 1994.

Conway, J. F., and D. A. D. Parry, Three-stranded α-fibrous proteins: the heptad repeat and its implications for structure. *Int. J. Biol. Macromol.* 14:14–16, 1991.

Craig, E. A., Chaperones: helpers along the pathways to protein folding. *Science* 260:1902–1903, 1993.

Creighton, T. E., *Proteins, Structures and Molecular Principles.* New York: Freeman, 1984. Very readable and reasonably comprehensive.

Doolittle, R. F., and P. Bork, Evolutionarily mobile modules in proteins. *Sci. Am.* October: 50–56, 1993.

Dorit, R. L., L. Schoenbach, and W. Gilbert, How big is the universe of exons? *Science* 250:1377–1381, 1990. Predicts that there are between 1,000 and 7,000 different kinds of domains in all proteins found in nature.

Farber, G. K., and G. A. Petsko, The evolution of α/β barrel enzymes. *Trends Biochem. Sci.* 15:228–234, 1990.

Fasman, G. D., Protein conformation prediction. *Trends Biochem. Sci.* 14:295–299, 1989.

Fersht, A. R., The hydrogen bond in molecular recognition. *Trends Biochem. Sci.* 12:301–304, 1987.

Harbury, P. B., T. Zhang, P. S. Kim, and T. Alber, A switch between two-, three-, and four-stranded coiled coils in GCN4 leucine zipper mutants. *Science* 262:1401–1405, 1993.

Hogle, J. M., M. Chow, and D. J. Filman, The structure of polio virus. *Sci. Am.* 256(3):42–49, 1987.

Karplus, M., and J. A. McCannon, The dynamics of proteins. *Sci. Am.* 254(4):42–51, 1986. A reminder that proteins are not rigid inflexible structures.

Pauling, L., *The Nature of the Chemical Bond,* 3d ed. Ithaca: Cornell University Press, 1960. A classic on molecular structure.

Pauling, L., and R. B. Corey, Configurations of polypeptide chains with favored orientations around single bonds: two new pleated sheets. *Proc. Natl. Acad. Sci. USA* 37:729–740, 1953. Classic paper.

Pauling, L., R. B. Corey, and H. R. Branson, The structure of proteins: two hydrogen-bonded helical configurations of the polypeptide chain. *Proc. Natl. Acad. Sci. USA* 27:205–211, 1951. Another classic paper.

Perczel, A., B. M. Foxman, and G. D. Gasman, How reverse turns may mediate the formation of helical segments in proteins: an x-ray model. *Proc. Natl. Acad. Sci. USA* 89:8210–8214, 1992.

Prusiner, S. B., Molecular biology of prion diseases. *Science* 252:1515–1522, 1991.

Prusiner, S. B., The prion diseases. *Sci. Am.* 266(1):48–57, 1995.

Richardson, J. S., and D. C. Richardson, The *de novo* design of protein structures. *Trends Biochem. Sci.* 14:304–309, 1989.

Rose, C. D., A. R. Geselowitzt, G. J. Lesser, R. H. Lee, and M. H. Zehfus, Hydrophobicity of amino acid residues in globular proteins. *Science* 229:834–838, 1985.

Rossman, M. G., and P. Argos, Protein folding. *Ann. Rev. Biochem.* 50:497–532, 1981.

Rossman, M. G., and J. E. Johnson, Icosahedral RNA virus structure. *Ann. Rev. Biochem.* 58:533–573, 1989.

Sali, A., J. P. Overington, M. S. Johnson, and T. L. Bundell, From comparisons of protein sequences and structures to protein modelling and design. *Trends Biochem. Sci.* 15:235–240, 1990.

Seo, J., and C. Cohen, Pitch diversity in α-helical coiled coils. *Proteins, Structure, Function, and Genetics* 15:233–234, 1993.

Shulman, R. G., High resolution NMR in vivo. *TIBS* 13:37–39, 1988.

Tonegawa, S., The molecules of the immune system. *Sci. Am.* 253(4):122–130, 1985.

Valegard, K., L. Liljas, K. Fridborg, and T. Unge, The three-dimensional structure of the bacterial virus MS2. *Nature* 345:36–41, 1990.

Wuthrich, K., Protein structure determination in solution by nuclear magnetic resonance spectroscopy. *Science* 243:45–50, 1989. The most effective technique for determining protein fine structure in cases where x-ray diffraction cannot be used.

Wright, P. E., What can two-dimensional NMR tell us about proteins? *Trends Biochem. Sci.* 14:255–259, 1989.

Yang, J. T., Protein secondary structure and circular dichroism: a practical guide. *Chemtracts, Biochem. Mol. Biol.* 1:484–490, 1990.

Problems

1. Researchers interested in the "origin of life" often examine meteorites for their amino acid content. They are concerned not only with the presence of amino acids but also with the ratio of the D/L forms. The premise is that life as we visualize it would require amino acids and that these amino acids would be "all" of the L-form or "all" of the D-form but not a mixture of each form. Can you build an argument to support this premise?
2. Summarize the similarities and the differences between the α-helix and the β-pleated-sheet structures.
3. Sodium dodecyl sulfate and urea are common denaturants, have modes of action that are readily visualized, and are often reversible upon dialysis. On the other hand, heat is a very effective agent for the destruction of protein structure, but it is for practical purposes irreversible. How can you explain the "irreversible" nature of heat's impact on protein structure?
4. Examine figure 5.5 and notice how some of the nonglycine residues in proteins are found in forbidden ϕ-ψ regions in the Ramachandran plot. How can you explain this?

5. If a portion of a gene that codes for a central portion of an α helix were to mutate so that a glutamic acid residue became a glycine residue, what might happen?
6. The principal force driving the folding of some proteins is the movement of hydrophobic amino acid side chains out of an aqueous environment. Explain.
7. What is the role of loops or short segments of "random" structure in a protein whose structure is primarily α helix?
8. What are some consequences of changing a hydrophilic residue to a hydrophobic residue on the surface of a globular protein? What are the consequences of changing an interior hydrophobic to a hydrophilic residue in the protein?
9. Some proteins are anchored to membranes by insertion of a segment of the N terminal into the hydrophobic interior of the membrane. Predict (guess) the probable structure of the sequence (Met-Ala-(Leu-Phe-Ala)$_3$-(Leu-Met-Phe)$_3$-Pro-Asn-Gly-Met-Leu-Phe). Why would this sequence be likely to insert into a membrane?
10. Suppose that every other Leu residue in the peptide shown in problem 9 were changed to Asp. Would that necessarily alter the secondary structure? Explain whether insertion into the membrane would be altered.
11. Amino acid side chains coordinate to the metal cofactor in metalloproteins. Examples of these coordination ligands include Asp, Glu, His, and Cys. In most of the proteins studied, the side chains directly surrounding the ligand amino acid are highly conserved among homologous proteins isolated from different organisms, while nonconservative alterations in amino acid sequence are found at sites distant from the metal-binding site. How do these observations fit the argument that biological structure dictates function?
12. Molecular weight analysis of a protein yields the following information.

Solvent	M_r
Dilute buffer	200,000
6 M Guanidinium chloride (GuHCl)	100,000
6 M GuHCl + 100 mM 2-mercaptoethanol	75,000 and 25,000

(Guanidinium chloride is a chaotropic (denaturing) reagent, and 2-mercaptoethanol can reduce disulfide bonds.) What can you deduce about the protein's quaternary structure?

13. Using the Ramachandran diagram in the text, explain why polypeptides assume only a limited number of regular structures.
14. Many important metabolic enzymes are insoluble in water and are found attached to membranes within cells. What amino acid residues would you expect to find on the "side" of the protein that "attaches" to the membrane?
15. Often the enzymes mentioned in problem 14 are purified with the aid of detergents such as sodium dodecylsulfate [$CH_3(CH_2)_{11}OSO_3Na$]. What is the function of the detergent?

C H A P T E R

FUNCTIONAL DIVERSITY OF PROTEINS

Each protein is exquisitely suited to carry out a specific function.

Now that we have described the chief types of protein structure, let's turn to the question of how these structures relate to the function for which they were designed. We will begin by introducing the proteins that occupy the different parts of the cell or extracellular environment. The treatment will be brief, since we will be discussing many of these proteins later in the text. We will then examine two protein systems in some detail, to provide a perspective on how structure relates to function. Finally, we will consider protein design from the evolutionary viewpoint. This will show that small refinements arise from point mutations that lead to single amino acid changes, and grosser changes arise by a reshuffling of domains.

Targeting and Functional Diversity

The cell is a highly organized factory in which the constituent parts are assembled in different locations and specialized machinery exists for specific purposes. Thus single-cell organisms are compartmentalized so that specific reactions occur in unique locations. In multicellular organisms the localization of reactions is even greater. The workers in the biochemical factory of the organism are the proteins.

Proteins Are Directed to the Regions Where They Are Utilized

Our first consideration is how proteins get to their final destination, that is, the locations where they function. All proteins are made in the cytoplasm, but their final location depends on a variety of signals. We will give a brief overview of this subject here, reserving a consideration of the mechanisms for chapters 18 and 34.

All proteins are made on ribosomes. Except for a small number of ribosomes located inside the organelles themselves, the vast majority of proteins are made on ribosomes in the cytosol. Some of the ribosomes are freely floating in the cytosol, and some are attached to the endoplasmic reticulum. The ribosomes that remain free account for the proteins that are targeted to locations in the cytosol, the nucleus, the peroxisomes, the mitochondria, and the chloroplasts (fig. 6.1). Ribosomes that are bound to the endoplasmic reticulum make proteins that are deposited in the lumen of the endoplasmic reticulum. From there the newly synthesized proteins may be transferred to the Golgi apparatus while undergoing modifications of various sorts. At some point, parts of the Golgi pinch off, and the modified proteins that do not remain in the Golgi are transferred to specific locations such as the lysosomes, the plasma membrane, and the secretory granules. Those proteins targeted to the secretory granules are eventually exported.

Figure 6.1

The different routes traveled by proteins during and after synthesis. In a typical eukaryotic cell, proteins are synthesized on free polysomes or in the endoplasmic reticulum on membrane-bound polysomes. Built into the proteins are amino acid sequences that determine in which of these two locations they will be synthesized. Arrows indicate location to which proteins are transported after synthesis. Some proteins synthesized on free polysomes remain in the cytosol; others become incorporated into mitochondria, chloroplasts (not shown), peroxisomes, or the nucleus. Some proteins synthesized on membrane-bound polysomes remain in the endoplasmic reticulum; others are transported to the Golgi. Some proteins transported to the Golgi remain there; others are transported to lysosomes, secretory vesicles, or the plasma membrane. Arrows indicate the directions of protein transport.

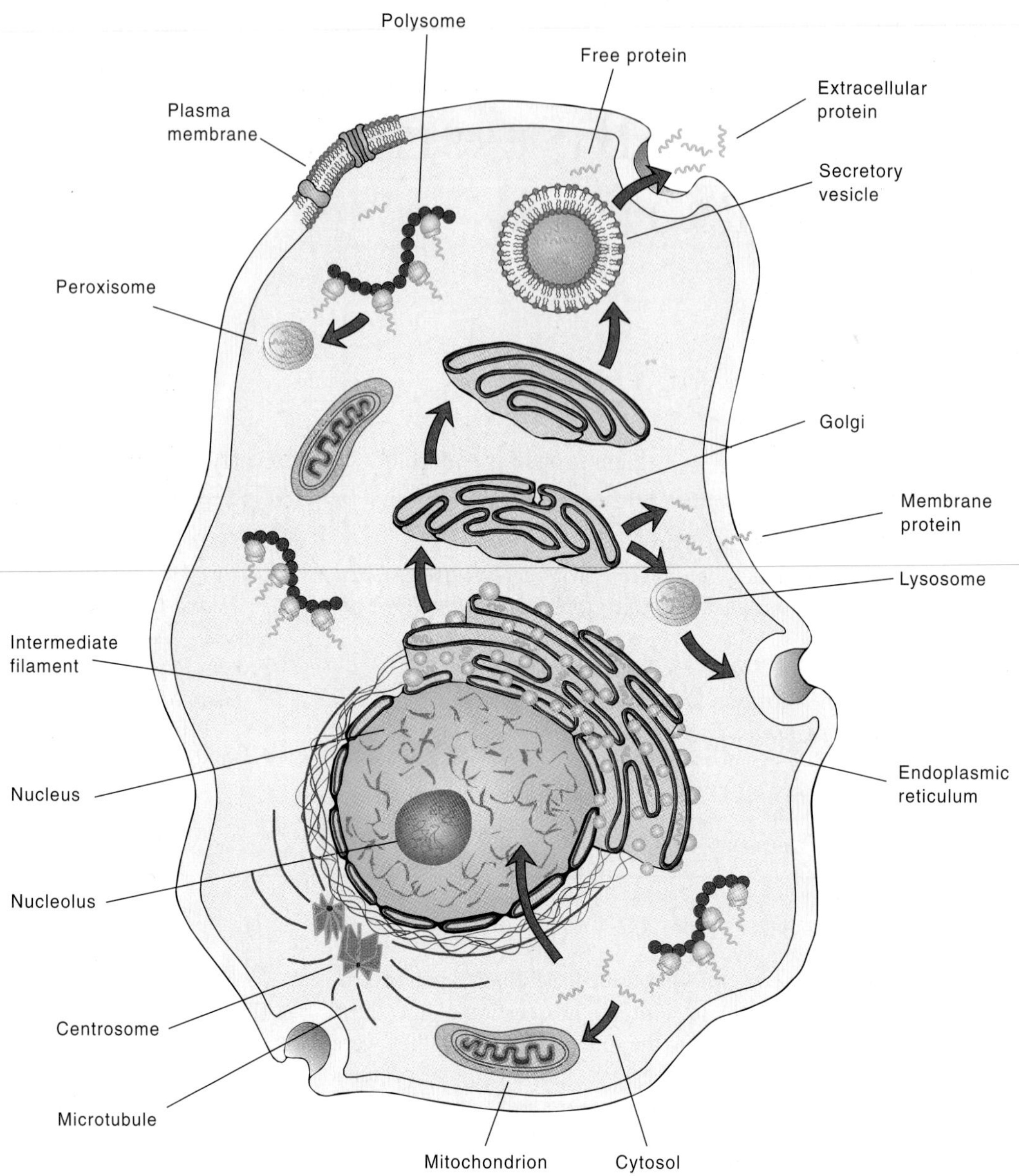

Classification of Proteins According to Location Emphasizes Functionality

Because of the great structural and functional diversity of proteins, it is difficult to capture the important features or the whole range of them within any one classification scheme. For our present purposes we will classify proteins according to the locations they occupy when they are fully functional. This is a useful classification scheme because it emphasizes functional interrelatedness—proteins that go together work together. The structures and functions of many proteins found in different locations, both inside and outside the cell, are listed in table 6.1, which also notes points in the text where the various proteins are discussed in greater detail.

Protein Structure Is Suited to Protein Function

We have seen that highly elongated fibrous proteins are well suited for compartmentalization, for giving stable form to organellar and cellular structures, and for processes involving movement of the organism. Because of their generally low mobility, fibrous proteins are rarely associated with enzyme activity or used for transport purposes. For those functions, globular proteins are more suitable. In this section we will consider two classical examples of protein assemblages that are ideally designed for the roles they play in the cell: hemoglobin and the skeletal muscle system.

Table 6.1
Some of the Main Proteins Found in Living Organisms

Protein/*Characteristics and Functions*

I. Main proteins of the cytoskeleton

Actin Bihelical filaments of aggregated globular monomers (monomer M_r = 42,000). Form cross-linked networks. In combination with myosin form actinomyosin, which functions in muscular contraction. Muscle is discussed in this chapter. Actin in the cytoskeleton is discussed in chapter 19.
Tubulin Hollow tube composed of 13 protofilaments; each protofilament contains an extensive linear aggregate of globular tubulin dimers (monomer M_r = 50,000). Tubulin exists mainly as single filaments emanating from the centrosome (see fig. 6.1) to locations throughout the cytoplasm. Involved in maintenance of cell shape. The mitotic apparatus that directs chromosomes to opposite poles of dividing cell is composed of tubulin. The cilium is a special case of tubulin in which many microtubules interact to produce a complex apparatus for cell motility (see fig. 5.7).
Intermediate filaments Interrupted α-helical proteins interacting in a side-by-side twisted manner to form ropelike structures (monomer M_r = 40,000–75,000). Intermediate filaments are more abundant in cells subject to mechanical stress; they occupy locations near membranes, where they appear to exert a protective function.
Spectrin Cytoskeletal protein particularly abundant in erythrocytes (see chapter 19).

II. Human plasma proteins

Albumin Osmotic regulation; transports acids and other substances (monomer M_r = 66,000). Most abundant serum protein.
α-Globulins A mixture of many proteins involved in transport and possibly other functions (monomer M_r = 20,000–400,000).
β-Globulins
Transferin Binds and transports iron (monomer M_r = 76,500).
β_2-Microglobulin Associated with the histocompatibility antigen (see chapter 37).
α-Globulins Antibodies (see chapter 37) (monomer M_r = 150,000).
Fibrinogen Circulating soluble protein, which, after proteolysis by thrombin, forms fibrin polymers of the blood clot (monomer M_r = 340,000) (see chapter 10).
Complement A mixture of about 11 proteins that work together to complement the immune system (see chapter 37) (monomer M_r = 80,000–200,000).

III. Some proteins of the extracellular matrix

Glycosaminoglycans Occupy large amounts of space forming hydrated gels (see chapter 18).
Proteoglycans Long glycosaminoglycans covalently linked to a core protein (see chapter 18).
Collagen (about 12 main types) The major proteins in the extracellular matrix (see chapters 5 and 34).
Types I–III Assemble into fibrils organized to meet the needs of the tissue.
Type IV Assembles into a laminar network.
Elastin Cross-linked random coil protein that gives elasticity to tissues.
Fibronectin A glycoprotein that helps to mediate cell-matrix adhesion (see chapter 19).
Integrins (several) Integral membrane protein that helps to bind cells to the extracellular matrix. Each protein usually consists of two different subunits.

IV. Digestive enzymes of the gastrointestinal tract

Amylase Degrades starch to disaccharides.
Pepsin Degrades proteins to large peptides.
Amylase As above.
Peptidases Split large peptides to small peptides.
Trypsin Degrades proteins to large peptides (see chapter 9).
Chymotrypsin Degrades proteins to large peptides (see chapter 9).
Lipase Degrades lipids into fatty acids and glycerol (see chapters 21 and 22).
Ribonuclease Degrades RNA to oligonucleotides (see chapter 9).
Peptidases Degrade peptides to amino acids.
Disaccharidases Degrade disaccharides to monosaccharides.

V. Proteins of the cytosol

Many (between 300 and 1,000) Synthesis of most small molecules required by the cell. Synthesis of proteins, carbohydrates, and lipids (see chapters 11, 22, 23, and 27).

VI. Proteins of the nucleus

Histones (5) Proteins that complex with DNA to make chromosomes (see chapter 30). There are five major histones.
Nucleic acid polymerizing enzymes (5 to 10) For DNA and RNA synthesis (see chapters 31 and 33). There are between 5 and 10 nucleic acid polymerases in different cells.

VII. Proteins of the mitochondria and the chloroplasts

Many (100 to 300) Proteins involved in energy production from metabolites or light (see chapters 14–17).

VIII. Proteins of the endoplasmic reticulum and the Golgi

Many (50 to 200) Enzymes involved in protein modification and in oligosaccharide and lipid synthesis (see chapters 18, 21, 22, and 23).

IX. Proteins of the lysosomes and the peroxisomes

Many (30 to 100) Enzymes involved in a wide variety of degradation processes for removing undesired compounds. (See chapter 21 for role of peroxisomes in fatty acid degradation. See chapter 15 for role of peroxisomes or glyoxyosomes in utilization of C-2 carbon source.)

X. Proteins of the plasma membrane

Many (100 to 500) Proteins involved in transport across membranes and for transmission of important metabolic signals across the plasma membrane (see chapters 19, 27, and 36).

Hemoglobin—An Allosteric Oxygen-Binding Protein

Hemoglobin is the best-known transport protein. Its chief function is to pick up oxygen in the lungs, where it is plentiful, and deliver it to tissues throughout the body. A central feature of hemoglobin (and myoglobin as well) is a water-free pocket for the heme, with its central iron atom located where oxygen is bound. (Note: Heme is a complex of Fe^{2+} and protoporphyrin IX; its structure is presented in figure 6.10 and in atomic detail in figure 16.3). The hydrophobic character of the heme binding cavity is dictated by the apolar side chains that line it. This is a particularly suitable environment for binding the hydrophobic porphyrin ring and where iron (Fe^{2+}) can bind oxygen reversibly without itself being oxidized to Fe^{3+}.

Hemoglobin consists of two α subunits, each with 141 amino acids, and two β subunits, each with 146 amino acids. Each subunit is capable of binding a single molecule of oxygen. In muscle cells a reserve oxygen store is provided by the myoglobin molecule, which is similar in structure to hemoglobin but exists as a monomer. While the components of myoglobin and hemoglobin are remarkably similar, their physiological responses are very different. On a weight basis, each molecule binds about the same amount of oxygen at high oxygen tensions (pressures). At low oxygen tensions, however, hemoglobin gives up its oxygen much more readily. These differences are reflected in the oxygen-binding curves of the purified proteins in aqueous solution (fig. 6.2).

The oxygen-binding curve for myoglobin (Mb) is hyperbolic in shape, as would be expected for a simple one-to-one association of myoglobin and oxygen:

$$\mathrm{Mb} + \mathrm{O_2} \rightleftharpoons \mathrm{MbO_2}$$

$$K_f = \frac{[\mathrm{MbO_2}]}{[\mathrm{Mb}][\mathrm{O_2}]} = \text{equilibrium formation constant} \quad (1)$$

If y is the fraction of myoglobin molecules saturated, and if we express the oxygen concentration in terms of the partial pressure of oxygen $[O_2]$, then

$$K_f = \frac{y}{[1-y][\mathrm{O_2}]} \quad \text{and} \quad y = \frac{K_f[\mathrm{O_2}]}{1 + K_f[\mathrm{O_2}]} \quad (2)$$

This is the equation of a hyperbola, as shown in figure 6.2. (Partial pressure is usually indicated by a lowercase p to the left. For simplicity the p has been omitted.)

Hemoglobin (Hb) behaves differently. Its sigmoidal binding curve can be fitted by an association-constant expression with a greater-than-first-power dependence on the oxygen concentration:

$$K_f = \frac{[\mathrm{HbO_2}]}{[\mathrm{Hb}][\mathrm{O_2}]^n} \quad \text{and} \quad y = \frac{K_f\mathrm{O_2}^n}{1 + K_f\mathrm{O_2}^n} \quad (3)$$

Under physiological conditions the value of n is around 2.8, indicating that the binding of oxygen molecules to the four hemes in hemoglobin is not independent and that binding to any one heme is affected by the state of the other three hemes. (A fuller discussion of the equations for multiple binding and the procedure for determining n from experimental data are presented in box 6A.) The first oxygen attaches itself with the lowest affinity, and successive oxygens are bound with a higher affinity. The exact value of n for hemoglobin is a function of the extent of oxygen binding as well as the presence of other factors discussed above. In general, a value of $n > 1$ indicates cooperative binding (or positive cooperativity) between small-molecule ligands, a value of $n < 1$ indicates anticooperative binding (or negative cooperativity), and a value of $n = 1$ indicates no cooperativity.

Figure 6.2

Equilibrium curves measure the affinity for oxygen of hemoglobin and of the simpler myoglobin molecule. Myoglobin, a protein of muscle, has just one polypeptide chain and resembles a single subunit of hemoglobin. The vertical axis gives the amount of oxygen bound to one of these proteins, expressed as a percentage of the total amount that can be bound. The horizontal axis measures the partial pressure of oxygen in a mixture of gases with which the solution is allowed to reach equilibrium. For myoglobin the equilibrium curve is hyperbolic. Myoglobin absorbs oxygen readily but becomes saturated at a low pressure. The hemoglobin curve is sigmoidal. Initially, hemoglobin is reluctant to take up oxygen, but its affinity increases with oxygen uptake. An arterial oxygen pressure, both molecules are nearly saturated, but at venous pressure, myoglobin gives up only about 10% of its oxygen, whereas hemoglobin releases roughly half. At any partial pressure, myoglobin has a higher affinity than hemoglobin, which allows oxygen to be transferred from blood to muscle.

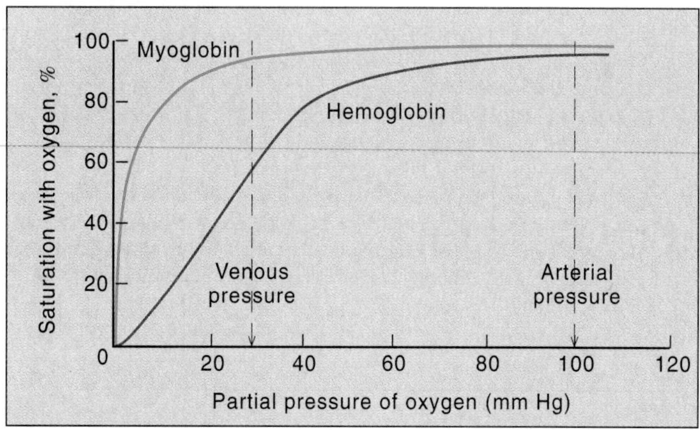

The cooperative binding of oxygen by hemoglobin is ideally suited to the conditions involved in oxygen transport. Thus in the lung, where the oxygen tension is relatively high, hemoglobin can become nearly saturated with oxygen, while in the tissues, where the oxygen tension is relatively low, hemoglobin can release about half its oxygen (see fig. 6.2). If myoglobin were used as the oxygen transporter, less than 10% of the oxygen would be released under similar conditions. The positive cooperativity associated with oxygen binding to hemoglobin in a special case of allostery in which the binding of "substrate" to one site stimulates the binding of "substrate" to another site on the same multisubunit protein. Although this is a special case of allostery, it is a type of allostery that is quite commonly observed in regulatory proteins (see chapter 10). We will consider possible mechanisms for hemoglobin allostery later in this chapter.

The Binding of Certain Factors to Hemoglobin Has a Negative Effect on Oxygen Binding

The combination of hemoglobin with oxygen depends not only on oxygen tension but also on pH, CO_2, and glycerate-2,3-bisphosphate (GBP or BPG). GBP (fig. 6.3) binds preferentially to the deoxygenated form of hemoglobin with a dissociation constant of about $10^{-5}\,M^{-1}$. Its dissociation constant with HbO is only about $10^{-3}\,M^{-1}$. Since the concentrations of GBP and hemoglobin are both about 5 mM in the erythrocyte, we expect most of the deoxy form to be complexed with GBP and most of the oxyhemoglobin to be free of GBP. The net effect of the GBP is to shift the oxygen-binding curve to higher oxygen tensions (fig. 6.4). This shift is not sufficient to lower the binding of oxygen at the high oxygen tensions in the capillaries of the lungs, but it is sufficient to cause a substantially greater release of oxygen at the lower oxygen tensions that exist in body tissues where the oxygen is utilized.

The negative influences of H^+ and CO_2 on oxygen binding are causally related and together are known as the Bohr effect. The CO_2 that diffuses into the blood plasma from many tissues is mostly in the form of dissolved CO_2, since conversion to carbonic acid (H_2CO_3) is a slow reaction with a half-time of about 10 s. Long before this conversion takes place in the plasma, the CO_2 has diffused into the red blood cells, where the conversion to carbonic acid is rapidly catalyzed by the enzyme carbonic anhydrase. This enzyme is strategically concentrated in cell types where it plays a pivotal role in carbon dioxide and pH management. The erythrocyte is one of these cells; in chapter 3 we saw that kidney cells also fall into this category. Once formed, most of the carbonic acid rapidly dissociates into H^+ and HCO_3^-:

$$CO_2 + H_2O \xrightleftharpoons[]{\text{Carbonic anhydrase}} H_2CO_3 \rightleftharpoons H^+ + HCO_3^- \quad (4)$$

Since this reaction produces protons, there must be a way of binding these protons; otherwise, this reaction would produce a serious lowering of the pH. The protons produced in the erythrocytes by this reaction are mostly taken up by the histidine groups of the hemoglobin. There are 38 histidines per hemoglobin, giving the hemoglobin, which is very abundant in the erythrocytes, a major buffering capacity. Thus the protons produced from the ionization of carbonic acid in the erythrocyte never have the opportunity to have a deleterious effect on the pH of the blood plasma, and most of the HCO_3^- produced in the erythrocytes diffuses into the plasma, where it plays a role in pH management of the plasma. The net effect of the uptake of protons by hemoglobin is to lower the affinity of hemoglobin for oxygen; this is because protons, like GBP, bind preferentially to deoxyhemoglobin. For example, at pH 7.6 and 40 mm Hg of oxygen tension, hemoglobin retains more than 80% of its oxygen; at pH 6.8 it retains only 45%. Thus the negative effect of CO_2 on oxygen binding is mainly due to the tendency of CO_2 to lower the pH (i.e., raise the H^+ concentration), and the negative effects of protons on oxygen binding are qualitatively similar to the negative effect of GBP.

Figure 6.3

The structure of glycerate-2,3-bisphosphate, a negative allosteric effector for hemoglobin oxygen release.

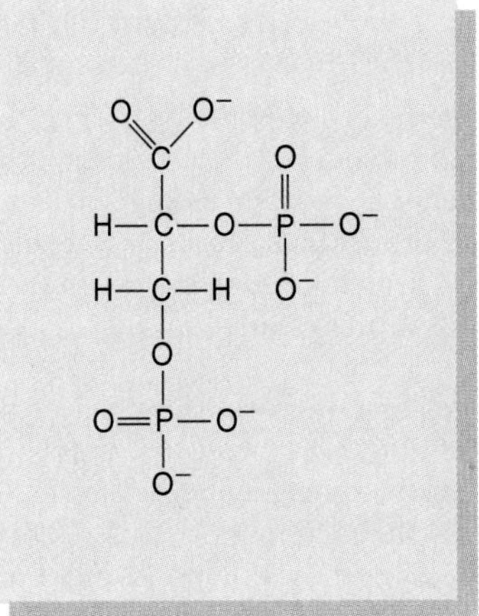

Figure 6.4

Oxygen-binding curve for hemoglobin as a function of the partial pressure of oxygen. Two curves are shown: one in the absence and one in the presence of glycerate-2,3-bisphosphate (GBP). GBP decreases the affinity between oxygen and hemoglobin, as shown by the displacement of the binding curve to high oxygen concentrations in its presence.

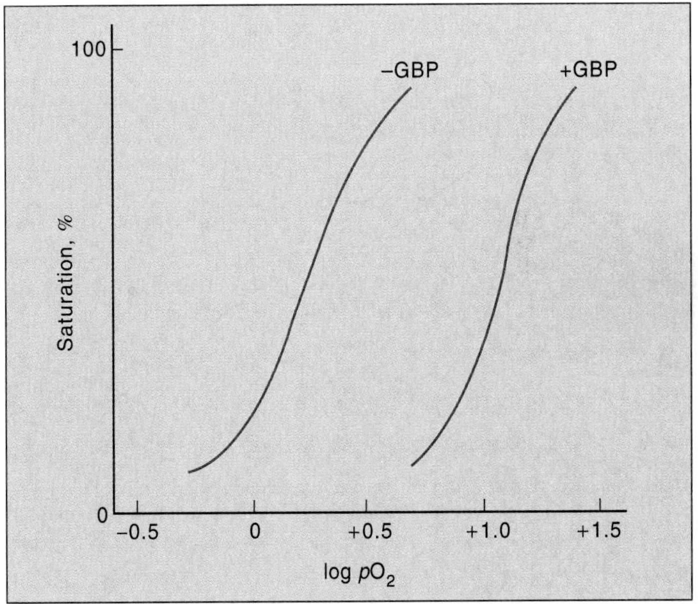

The Bohr effect also influences CO_2 disposal and regulation of the blood pH. While oxygen is being delivered to the tissues in the venous blood, the CO_2 is being absorbed from the tissues (fig. 6.5). This process would stop very quickly if it were not for the erythrocytes and the hemoglobin. As we have just seen, the CO_2 that diffuses into the plasma is rapidly converted into car-

6A

BOX

Theory of Multiple Binding and the Hill Plot

Biochemists are often faced with the problem of binding small molecules or ligands to proteins or nucleic acids. Proteins in particular commonly bind many ligands, including substrates, inhibitors, and activators. Often these ligands are bound to more than one site. Hemoglobin binds a total of four oxygen molecules, and serum albumin binds an extremely varied group of substances at a large number of sites.

Binding studies require a measurement of the free ligand, B, in the presence of the macromolecule, A. From this we may determine the average binding number, called y. The value of y is equal to the ratio of the total number of B molecules bound to the total number of binding sites. If there is only one binding site per A molecule, the expression for y is rather simple:

$$y = \frac{[AB]}{[A] + [AB]} \quad \text{(A)}$$

The value of [A] + [AB] is known from the total amount of A added to the solution. The value of [AB] is equal to the total amount of B added, minus the experimentally observed concentration of free B after A is added:

$$y = \frac{\text{total B} - \text{free B}}{\text{total A}} \quad \text{(B)}$$

Let us first consider the simple binding of one ligand B to a protein molecule A with a formation constant K_f:

$$A + B \underset{}{\overset{K_f}{\rightleftharpoons}} AB; \qquad K_f \frac{[AB]}{[A][B]} \quad \text{(C)}$$

From Equations (B) and (C) we can express y in terms of K_f, the formation constant, or K_d, the dissociation constant:

$$y = \frac{K_f[B]}{1 + K_f[B]} \quad \text{(D)}$$

or, since $K_f = 1/K_d$,

$$y = \frac{[B]}{[B] + K_d} \quad \text{(E)}$$

Taking the reciprocal of both sides, we obtain

$$\frac{1}{y} = 1 + K_d\left(\frac{1}{[B]}\right) \quad \text{(F)}$$

By plotting the experimentally determined value of $1/y$ against 1/[B], we obtain a straight line with a slope equal to the dissociation constant K_d. The intercept on the ordinate should be 1 (fig. 1).

When there is more than one site on A for binding B, the equations become more complex and in general the plots are not linear. If we consider the average binding number for n sites on a molecule, the total average binding number is the sum of the binding number for each of these sites.

Figure 1

A binding plot to determine the dissociation constant K_d for the simple situation where there is one ligand binding site, y is the average binding number, and [B] is the concentration of ligand.

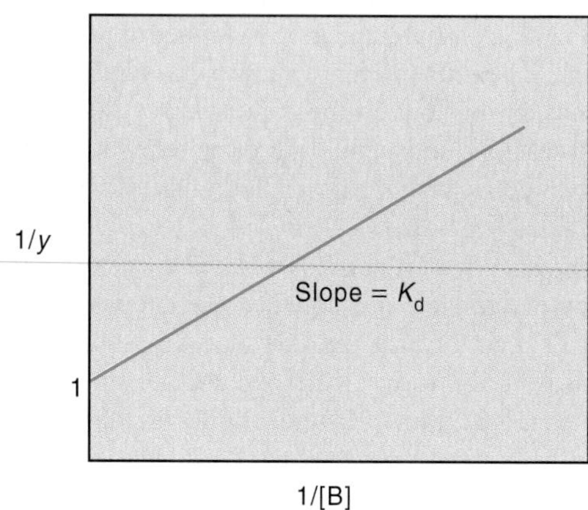

$$y = \sum_{i=1}^{n} \frac{K_{fi}[B]}{1 + K_{fi}[B]} \quad \text{(G)}$$

There are two situations in which the data for multiple binding may be treated rather simply. First, when all binding sites bind B with the same energy, all terms are equal and the solution to the sum is simply n times each term:

$$y = \frac{nK_f[B]}{1 + K_f[B]} \quad \text{(H)}$$

Again, since $K_f = 1/K_d$,

$$y = \frac{n[B]}{[B] + K_d} \quad \text{(I)}$$

and taking the reciprocal of both sides,

$$\frac{1}{y} = \frac{1}{n} + \frac{1}{n}K_d\left(\frac{1}{[B]}\right) \quad \text{(J)}$$

If $1/y$ is plotted against 1/[B], the slope is equal to K_d/n and the intercept (when 1/[B] approaches 0) is $1/n$. Alternatively, the data may be plotted as y versus y/[B] (fig. 2), in which case the slope

Figure 2

A binding plot to determine the dissociation constant K_d and the number of binding sites n for the situation where there are n ligand binding sites per macromolecule with identical binding affinities.

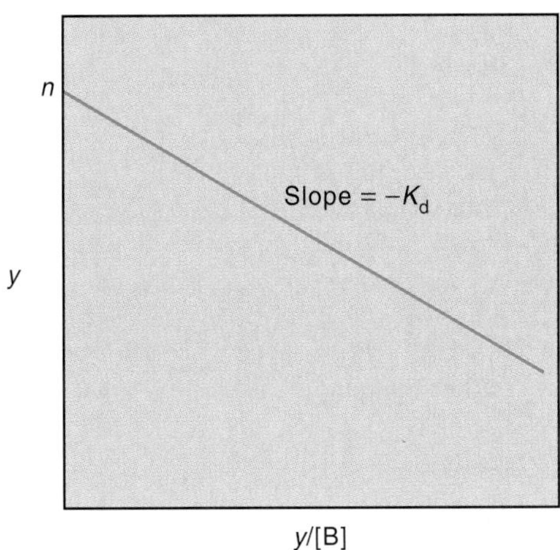

Figure 3

A Hill plot.

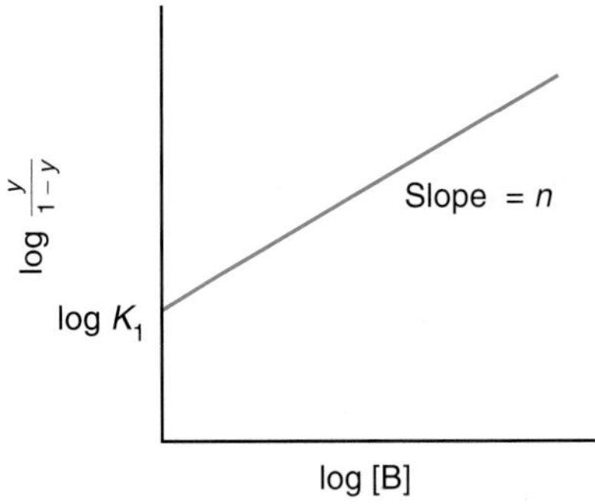

is $-K_d$ and the intercept at $y/[B] = 0$ is n, since

$$y = n - \frac{yK_d}{[B]} \tag{K}$$

Second, in one other situation, multiple binding can be treated rather simply. If the sites interact so strongly that only the fully saturated product AB_n is formed, the data may be treated in the following way. The formation of only one major product means that the binding of the ligand molecule to the macromolecule greatly enhances the further binding of additional ligands such that at equilibrium, only three species exist in significant concentrations: A, B, and AB_n:

$$A + nB \rightleftharpoons AB_n; \qquad K_f = \frac{[AB_n]}{[A][B]^n} \tag{L}$$

In this case,

$$y = \frac{[AB_n]}{[AB_n] + A} \tag{M}$$

or

$$y = \frac{K_f\,[B]^n}{1 + K_f\,[B]^n} \tag{N}$$

Taking the reciprocal of both sides,

$$\frac{1}{y} = 1 + \frac{1}{K_f[B]^n} \tag{O}$$

Under these conditions a plot of $1/y$ against $1/[B]$ is not linear. However, since

$$\frac{y}{1-y} = K_f[B]^n \tag{P}$$

and

$$\log\frac{y}{1-y} = \log K_f + n \log [B] \tag{Q}$$

a plot of $\log y/(1-y)$ versus $\log [B]$ is linear. The intercept of this plot at $\log [B] = 0$ is $\log K_f$, and the slope is n. This is called a *Hill plot,* and the experimentally determined value of n is known as the Hill coefficient (fig. 3). It usually requires a guess at the value of n. By trial and error the value of n is varied to give a linear plot. Often nonintegral values of n are reported that describe the data but lack precise meaning. For example, in the case of hemoglobin binding of oxygen, a value of 2.8 is reported for n, and this indicates some cooperativity in the binding.

If none of the preceding simplifying conditions apply, it is still possible to determine all the equilibrium constants. More advanced texts should be consulted for the use of these methods.

---

bonic acid in the erythrocytes, which in turn dissociates into H^+ and HCO_3^-. The protons produced by this dissociation would lower the pH and reverse this dissociation if it were not for the buffering action of the hemoglobin. Loss of oxygen decreases the acid dissociation constant of the hemoglobin so that it picks up the excess protons. This change serves two purposes: It helps to stabilize the pH, and it enables the system to absorb more CO_2. The additional CO_2 is ultimately disposed of when the blood reaches the lungs and the hemoglobin again becomes oxygenated. Another point to remember is that the majority of the HCO_3^- pro-

Figure 6.5

Transport of oxygen (O_2) and carbon dioxide (CO_2) in the circulatory system. In most tissues, O_2 is released and CO_2 is withdrawn by the red blood cells; in the lungs these processes are reversed.

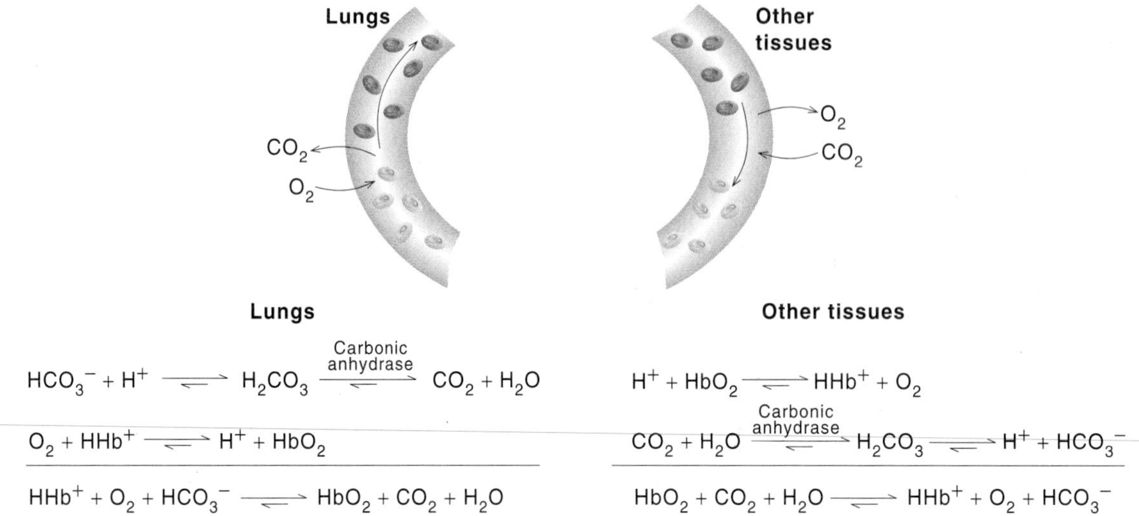

duced in erythrocytes diffuses into the venous blood. The pH of the blood system is controlled within narrow limits by the buffering action of the bicarbonate and the hemoglobin (see chapter 3), with minor assistance from other proteins in the bloodstream.

One must marvel at the way various factors work in concert so that hemoglobin can be useful in so many roles: oxygen deliverer, carbon dioxide remover, and pH stabilizer. From the explanation we have given it should be clear why the hemoglobin is confined to an intracellular location in the plasma rather than being present as a free plasma protein. Carbonic anhydrase and GBP are essential for efficient hemoglobin function, and their presence in the bloodstream at adequate concentrations would probably be unattainable or intolerable in the blood plasma. Furthermore, the protons released as a result of the carbonic anhydrase reaction are rapidly picked up by the deoxygenated hemoglobin before they have a chance to cause a potentially harmful lowering of the pH of the plasma.

X-Ray Diffraction Studies Reveal Two Conformations for Hemoglobin

X-ray diffraction studies on fully oxygenated hemoglobin and deoxygenated hemoglobin have shown that the molecule is capable of existing in two states, with significant differences in tertiary and quaternary structures (fig. 6.6). Further studies on partially oxygenated hemoglobin may some day indicate additional intermediate structures between these two extremes. Until then the two-state model serves as a useful conceptual framework for explaining the allosteric mechanism of the hemoglobin system.

The hemoglobin tetramer is composed of two identical halves (dimers), with the $\alpha_1\beta_1$ subunits in one dimer and the $\alpha_2\beta_2$ subunits in the other. The subunits within the dimers are tightly held together; however, the dimers themselves are capable of motion with respect to one another (fig. 6.7). The interface between the movable dimers contains a network of salt bridges and hydrogen bonds when hemoglobin is in the deoxy conformation (fig. 6.8). The quaternary transformation that takes place on binding of oxygen causes the breakage of these bonds.

The effects of H^+, CO_2, and glycerate-2,3-bisphosphate on oxygen binding can be understood in terms of their stabilizing effect on the deoxy conformation. The decreased oxygen binding as the pH is lowered from 7.6 to 6.8 suggests the involvement of histidine side chains, because these are the only side-chain groups in proteins that have a pK in this pH range. Certain histidines in the charged form make salt linkages that contribute to the stability of the deoxy form (see fig. 6.8). As the pH is lowered, these histidines tend to become charged, which increases the stability of the deoxy form. Such a change should inhibit a structural transition to the oxy form and thereby lower the affinity of the protein for oxygen. Similarly, glycerate-2,3-bisphosphate binds most strongly to the deoxy form (fig. 6.9) and thereby discourages the transition to the oxy form, which lowers the affinity for oxygen. Carbon dioxide binds as bicarbonate to the α-amino groups in hemoglobin; this binding also favors the deoxy conformation. Binding of CO_2 is freely reversible, being favored by the high CO_2 tensions in the tissues. As a result, hemoglobin becomes a carrier of CO_2 from tissues to the lungs, where it is discharged.

Changes in Conformation Are Initiated by Oxygen Binding

The oxygen binding at the heme group itself initiates the changes in tertiary and quaternary structure that are responsible for the cooperative effect seen on oxygen binding. The heme group contains an Fe^{2+} ion located near the center of a porphyrin ring. The Fe^{2+} makes four single bonds to the nitrogens in the heme ring and a

Figure 6.6

Three-dimensional structure of oxyhemoglobin and deoxyhemoglobin as determined by x-ray crystallography. This is a view down the twofold symmetry axis, with the β chains on top. In the oxy-to-deoxy transformation (quaternary motion), $\alpha_1\beta_1$ and $\alpha_2\beta_2$ dimers move as units relative to each other. This allows glycerate-2,3-bisphosphate to bind to the larger central cavity in the deoxy conformation. A close-up of the binding site is shown in figure 6.9. (Illustration copyright by Irving Geis. Reprinted by permission.)

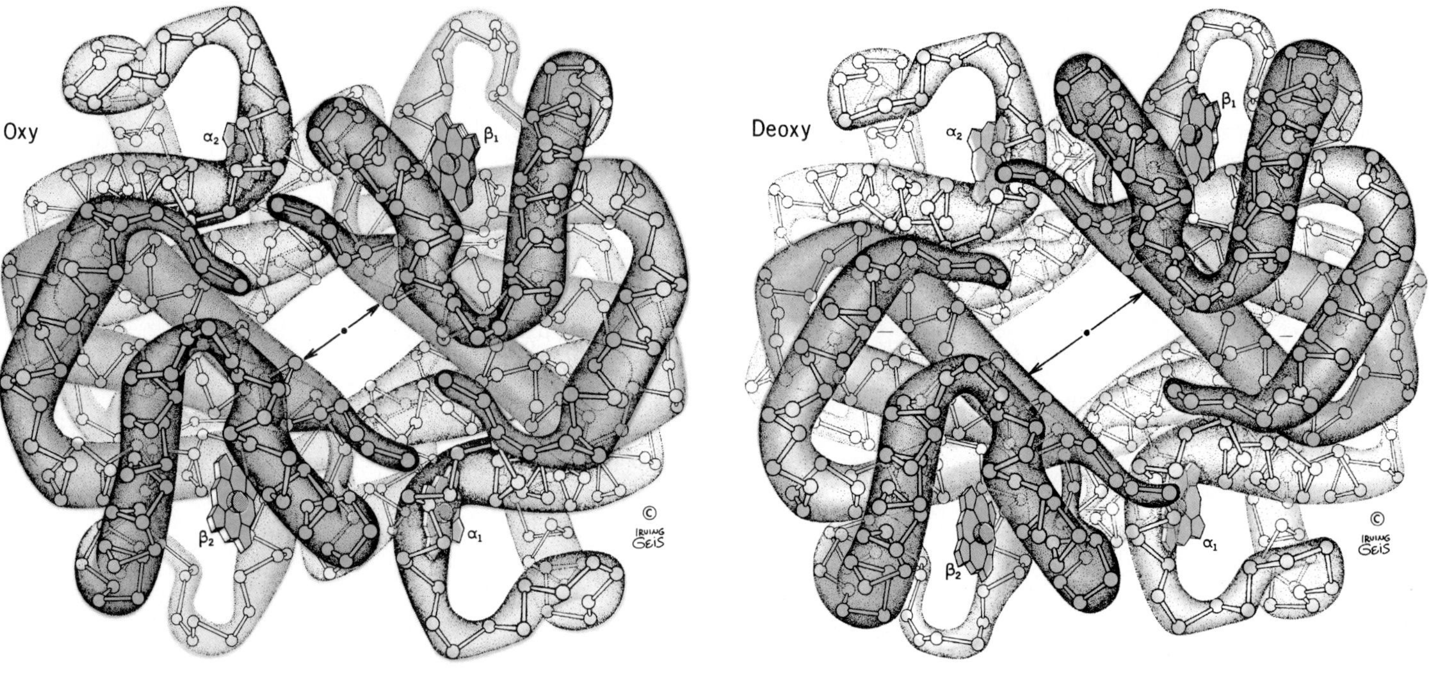

Figure 6.7

The deoxy-to-oxy shift upon binding oxygen in one hemoglobin molecule. The projection shown in this figure is approximately perpendicular to the one shown in figure 6.6. The $\alpha_1\beta_1$ dimer moves as a unit relative to the $\alpha_2\beta_2$ dimer. The interface between the two dimers is crucial to the cooperativity effect in hemoglobin. The interface is not visible in this figure (see figure 6.8). (Illustration copyright by Irving Geis. Reprinted by permission.)

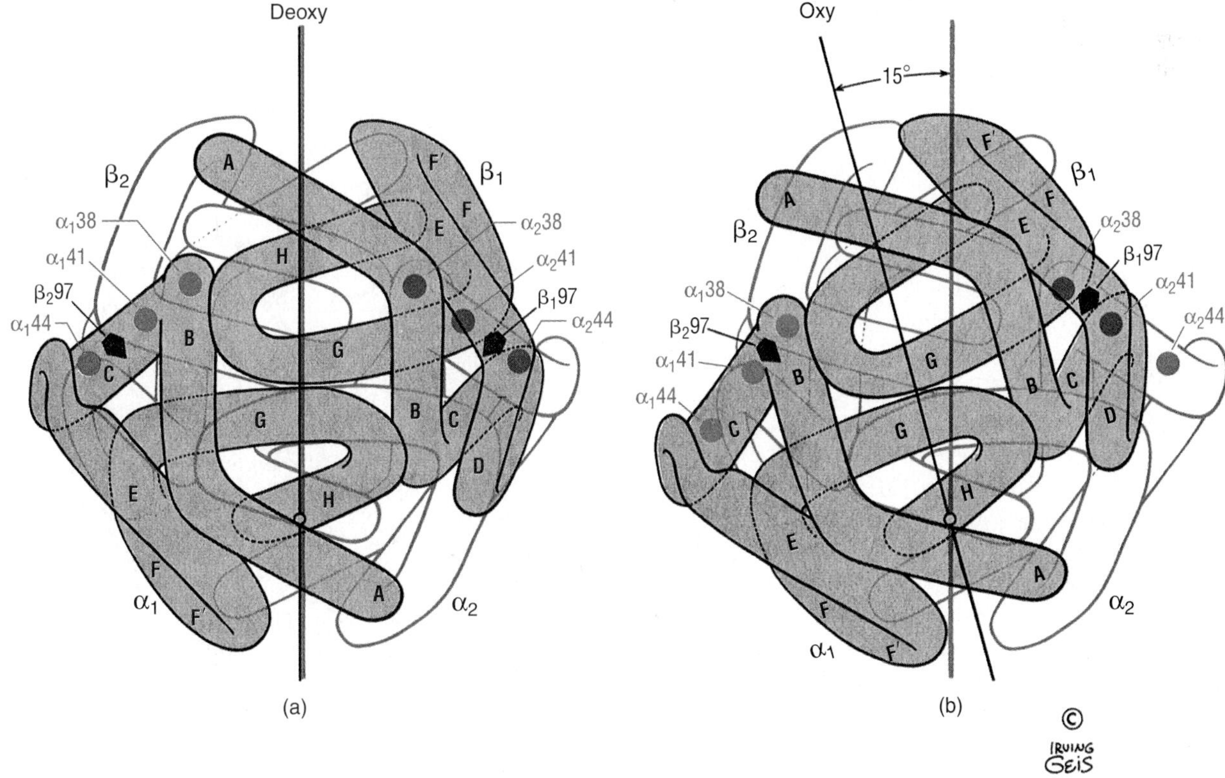

Figure 6.8

(*a*) The $\alpha_1\beta_2$ (and $\alpha_2\beta_1$) interface is shown schematically at the lower left and in detail (*b*). This is the regulatory zone of the hemoglobin molecule, which contains crucial hydrogen bonds and salt bridges. (*b*) All the hydrogen bonds and salt bridges shown here (dotted lines) exist only in the deoxy state, with the exception of $\alpha_1 41$–$\beta_2 40$ and $\alpha_1 94$–$\beta_2 102$, which exist only in the oxy state. Only the α carbons are shown in the backbone structure of the hemoglobin, except in the region of the interface. (Illustration copyright by Irving Geis. Reprinted by permission.)

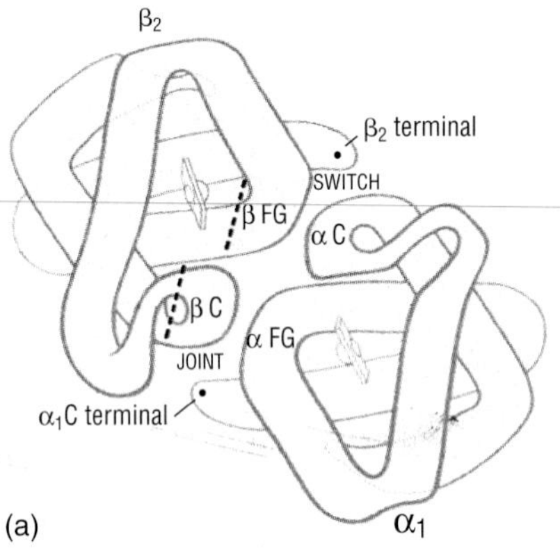

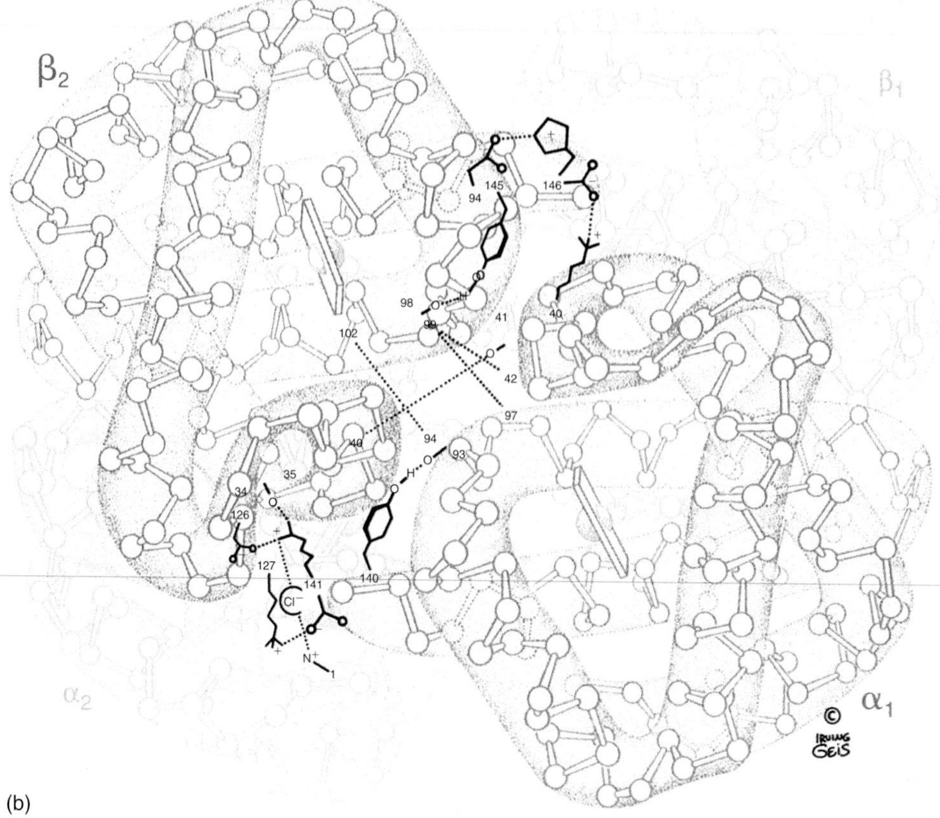

Figure 6.9

The binding of glycerate-2,3-bisphosphate in the central cavity of the deoxyhemoglobin between β chains. The surrounding positively charged residues are the amino terminal, His 2, Lys 82, and His 143. (Illustration copyright by Irving Geis. Reprinted by permission.)

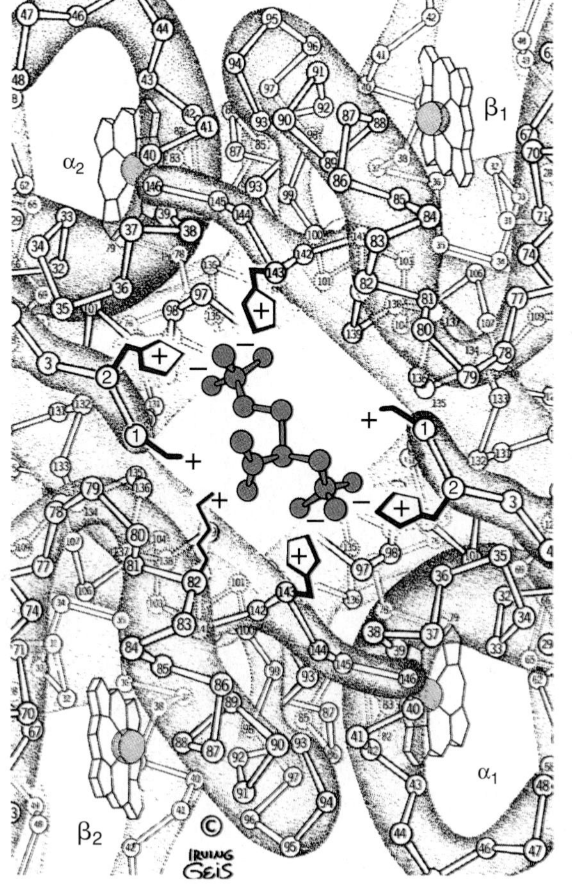

Figure 6.10

A close-up view of the iron–porphyrin complex with the F helix in deoxyhemoglobin. Note that the iron atom is displaced slightly above the plane of the porphyrin. (Illustration copyright by Irving Geis. Reprinted by permission.)

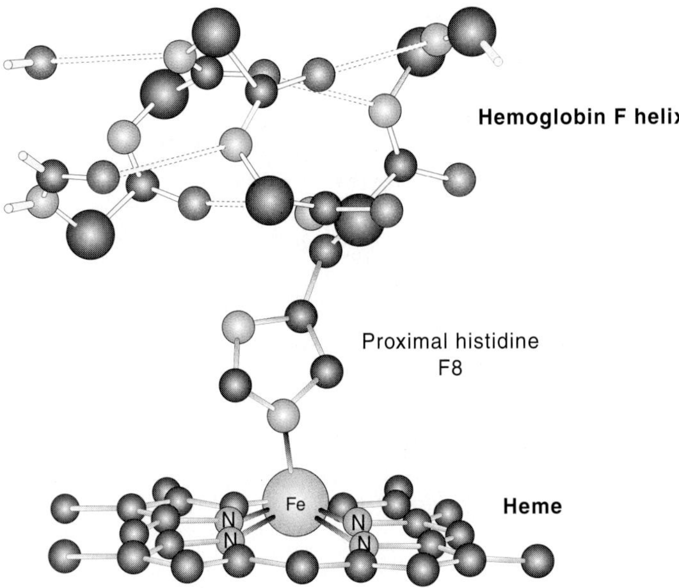

Figure 6.11

Downward movement of the iron atom and the complexed polypeptide chain on binding oxygen. The structure is shown before (black) and after (blue) binding oxygen. Movement of His F8 is transmitted to valine FG5, straining and breaking the hydrogen bond to the penultimate tyrosine. Only the α chain is shown here.

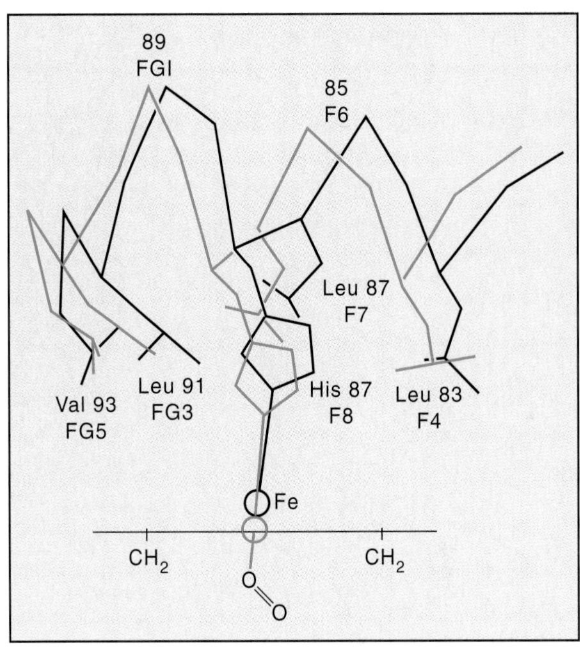

fifth bond to a histidine side chain of the F helix, F8 histidine (fig. 6.10). When oxygen is present, it binds at the sixth coordination position of the iron on the other side of the heme. Movement of the iron upon oxygen binding into the plane of the porphyrin ring pulls the F8 histidine and the F helix to which it is covalently attached (fig. 6.11). The change in location of the F helix is believed to induce a strain in the rest of the protein that facilitates the conversion of the deoxy to the oxy structure. This change favors the binding of additional oxygen at other unoccupied sites in the tetramer.

The Fe^{2+} is well suited to its job in hemoglobin, not only because it has a natural affinity for oxygen, but also because it changes its electronic structure in a highly significant way in so doing. Fe^{2+} is normally paramagnetic, having four unpaired electrons in its outer *d* electronic orbitals. In this state it is too large to sit precisely in the plane of a porphyrin, as studies with model compounds have shown. Fe^{2+} is also paramagnetic when it is pentacoordinated in deoxyhemoglobin; as expected, it is displaced from the plane of the porphyrin by a few tenths of an angstrom unit. When O_2 binds, however, the Fe^{2+} becomes hexacoordinated and diamagnetic (no unpaired electrons). This change results in a major reorganization of its outer *d* orbitals, which decreases the radius of the Fe^{2+} so that it can move to an energetically more favorable position in the center of the porphyrin (see fig. 6.11).

The structural arguments advanced here to explain oxygen binding by hemoglobin are supported by amino acid sequences of α and β chains for a large number of hemoglobins from different species. Data from 60 species of α chains and 66 species of β chains reveal 43 invariant positions in the hemoglobin molecule. These invariants are plotted on a map of the hemoglobin structure in figure 6.12, where the invariant positions are shown by colored dots. In a sense, the dots provide a diagram of the working machinery of hemoglobin, for the invariant positions line the heme pockets where oxygen is bound and the crucial $\alpha_1\beta_2$ interface, which changes its orientation when oxygen binds. Electrostatic forces and hydrogen bonds stitch the interface together in the deoxy conformation when the molecule gives up its oxygen to the tissues. If there are changes, resulting from mutation at any of these positions, then we would expect trouble to develop. This is just what happens, as can be seen in figure 6.13, which shows the positions of pathological mutations in hemoglobin. Where there are changes in the heme pockets or in the $\alpha_1\beta_2$ interface, hemoglobin abnormalities occur; many of these are associated with serious diseases. For an explanation of how genetic defects are inherited, see box 6B.

As a rule, invariant amino acids are the critical loci for hemoglobin function. One striking exception occurs at position 6 of the β chain. A hydrophobic valine residue is substituted for glutamic acid (a charged side chain) with disastrous results. The specific consequences of the β_6 alteration is to cause hemoglobin tetramers to aggregate when they are in the deoxy state. This is because the β_6 valine fits into a hydrophobic pocket in an adjoining hemoglobin molecule. The aggregates form long fibers that stiffen the normally flexible red blood cell. The resulting distortion of the red cells leads to capillary occlusion, which prevents proper delivery of oxygen to the tissues. This pathological condition is known as sickle-cell anemia.

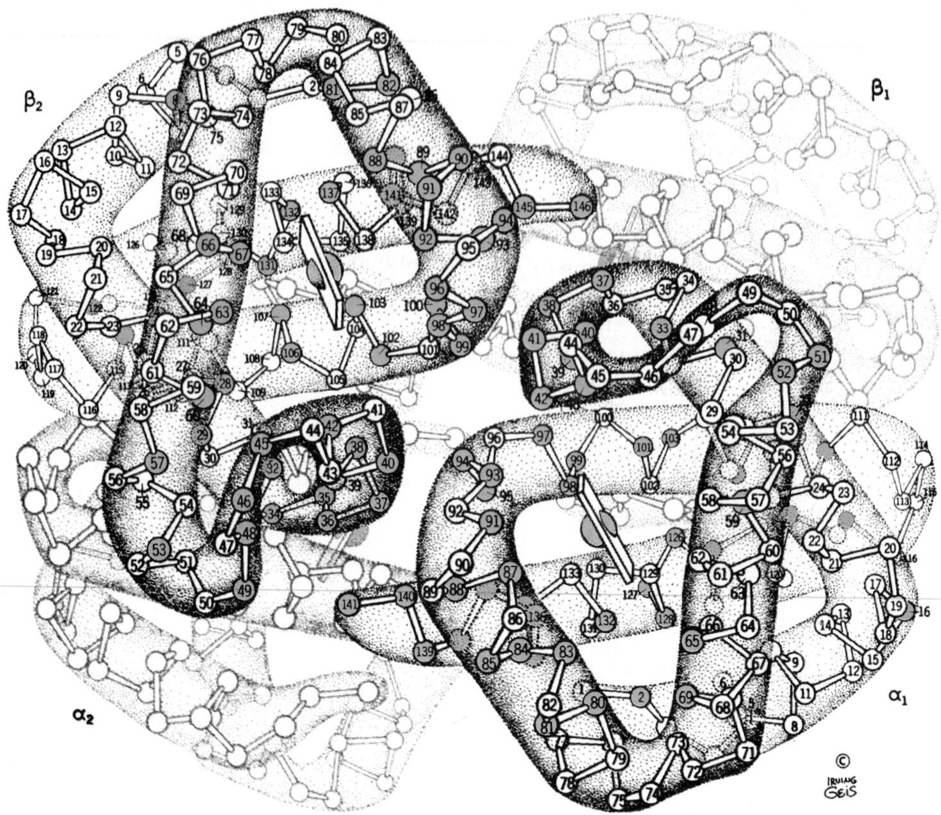

Figure 6.12

Invariant residues in the α and β chains of mammalian hemoglobin. The blue dots, indicating the positions of the invariant residues, line the heme pockets as well as the crucial $\alpha_1\beta_2$ interface. The invariant residues have been found in about 60 species. There are 43 invariant positions in the hemoglobin molecule. (Illustration copyright by Irving Geis. Reprinted by permission.)

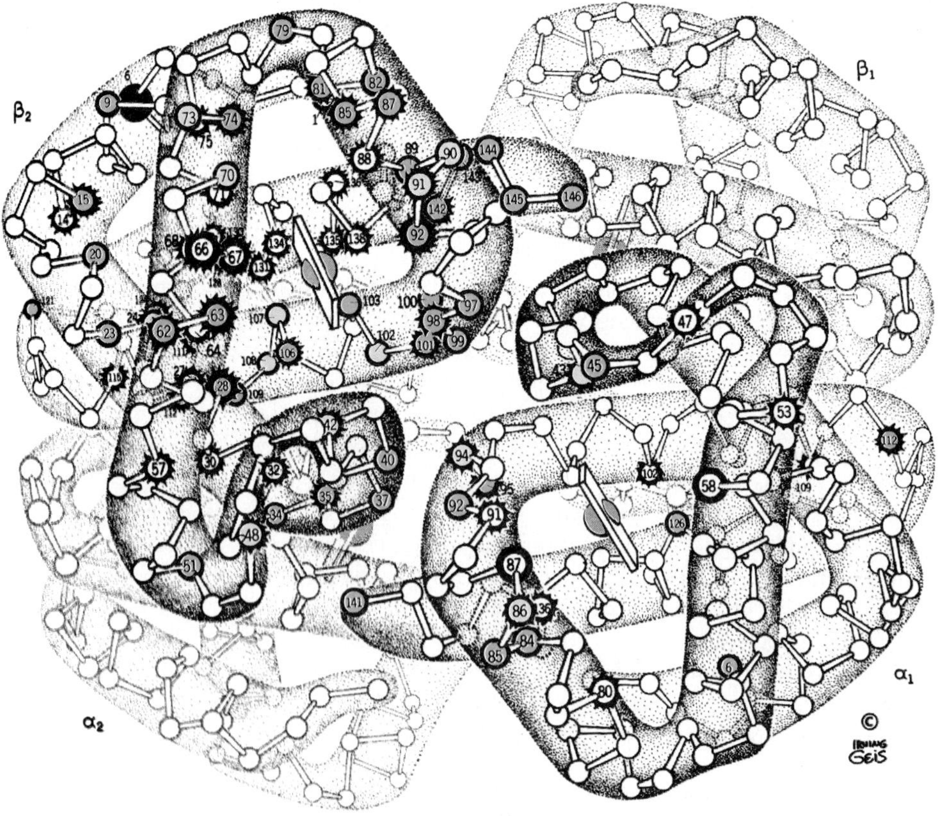

Figure 6.13

Positions of mutations in hemoglobin that produce a pathological condition. Comparison with figure 6.12 shows that these mutations, in general, show the same pattern as the distribution of the invariant positions. Dark circles around a numbered position indicate positions of abnormal residues, and a solid black dot indicates the valine β_6 mutation in sickle-cell anemia. Heavy black circles indicate a hemoglobin that (is easily) oxidized to the Fe^{3+} form (methemoglobin), and a jagged black perimeter indicates unstable hemoglobin. Dark blue indicates increased oxygen affinity; light blue indicates decreased oxygen affinity. (Illustration copyright by Irving Geis. Reprinted by permission.)

6B BOX

Inheritance Pattern for Genetic Defects

At many points throughout this text we will be considering metabolic defects that arise from inborn genetic defects. A gene is a segment of DNA that usually encodes a single polypeptide chain. Therefore a metabolic defect that arises from a faulty gene is usually caused by a defective polypeptide chain. A defective gene is usually inherited from a parent carrying the same defective gene. In prokaryotes each polypeptide gene is encoded by one and only one gene, and progeny are formed by replication of the parental genome (the total complement of genes possessed by the parent) prior to cell division. Each daughter cell produced after cell division carries the same genes as were present in the parent cell, making the inheritance pattern quite simple. Any defects in the parental genome are transferred to both daughter cells, so each of the daughter cells will produce the same defective polypeptide chain that was produced by the parent. Instead of referring to the polypeptide chain, it is customary to refer to the phenotype. The phenotype is the observable trait resulting from the genotype, and in prokaryotes there is a one-to-one correlation between genotype and phenotype.

The situation in eukaryotes is more complicated for two reasons: Most cells in eukaryotes carry two genes for the same polypeptide chain, and inheritance involves genes obtained from two parents of opposite mating type. Genes for the same trait or polypeptide chain are called alleles. If the alleles for a given organism are identical, the organism is said to be homozygous. If the alleles are nonidentical, the organism is said to be heterozygous. In the case of a homozygote in which both alleles are normal (that is, wild type), the phenotype related to the gene will be wild type. In the case of a homozygote in which both alleles bear the same defective allele, the organism will have a defective phenotype. In the case of a heterozygote in which one allele is wild type and the other is defective, the phenotype may be wild type, defective, or intermediate between wild type and defective.

If the phenotype is wild type, the wild-type allele is said to be dominant to the defective allele. If the phenotype of the heterozygote is fully defective the wild-type allele is said to be recessive to the defective allele. In the case of the heterozygote, if the phenotype is intermediate between the fully defective and the wild type, the two alleles are said to be codominant.

In the case of the diseased state known as sickle-cell anemia the organism always carries two defective alleles. In this respect the sickle-cell allele is recessive to the normal allele. If the organism carries one normal and one defective sickle-cell gene, the organism usually displays a mild case of anemia but not the full blown sickle-cell anemia; the organism is labeled as showing sickle-cell trait. From this result we can see that the normal allele is not fully dominant to the sickle-cell allele. One might say that the two alleles are codominant.

With these relationships for alleles in mind we now inquire into the inheritance patterns for sickle-cell anemia. When two eukaryotes mate, the inheritance pattern for different genes is complicated by the fact that the progeny inherit half of their genes from each parent. The mating process is not the same for all eukaryotes, but a common situation is for each parent of opposite mating type to contribute one allele for each gene to each of its progeny. The mating process starts with the formation of the mating cells or sex cells, which carry only one allele for each gene. The sex cells from the two parents fuse to reform a eukaryotic cell bearing two alleles for each gene, which goes on to divide into further cells carrying the identical genome.

A parent that is heterozygous for a given gene will usually form equal numbers of sex cells bearing the normal allele and the defective allele. If this parent is mated with a homozygous normal parent, half the progeny will be homozygous normal and half will be heterozygous with one normal and one defective gene. If both parents are heterozygous, then statistical probability leads to the prediction that one-fourth of the progeny will be homozygous normal, one-fourth will be homozygous doubly defective, and half will be heterozygous carrying one normal allele and one defective allele. Geneticists refer to this outcome as the Mendelian inheritance pattern.

For humans, where experimental matings are not conducted, the inheritance patterns are usually studied by examining the pedigree. In a pedigree chart (fig. 1) a male is indicated by a square □ and a female by a circle ○. A mating pair is indicated by a simple horizontal tie line □——○ and progeny from the mating by a vertical tie line.

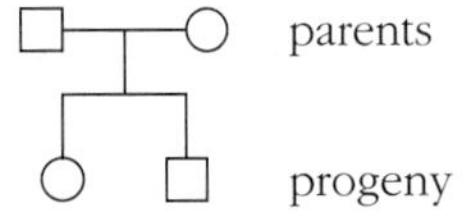

Figure 1

Pedigree charts for matings where at least one parent carries a defective gene for sickle-cell anemia. It is assumed that each mating pair has four children. From the discussion in the text we know that a completely normal parent carries two normal alleles, a parent showing sickle-cell trait must carry one normal and one defective allele, and a parent with sickle-cell anemia must carry two defective alleles. Four two-generation pedigrees are shown. Comparison of the results in (*a*) and (*b*) shows that it does not matter which parent carries the defective allele.

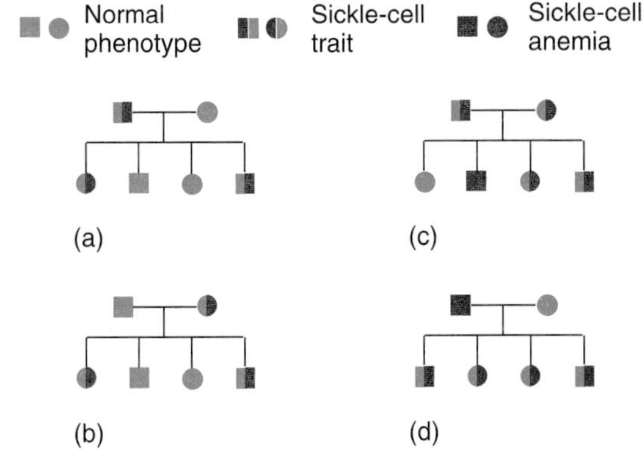

Two Models Have Been Proposed for the Way Hemoglobins and Other Allosteric Proteins Work

We have spent a good deal of time describing the function of hemoglobin because it is the best-understood regulatory protein and provides us with a model system for understanding in general terms how other regulatory proteins work. The first indication that hemoglobin was an allosteric protein came from the sigmoidal shape of its oxygen-binding curve (see fig. 6.2). Allosteric proteins are usually composed of two or more subunits. Different ligands may bind to quite different sites, or to quite similar sites as in the case of hemoglobins.

Two quite different models were proposed about 25 years ago to explain the unusual nature of the hemoglobin oxygen-binding curve. These models could also be used as a starting point for discussing other allosteric proteins (fig. 6.14). The first model, introduced by Monod, Wyman, and Changeux in 1965, is called the symmetry model. In this model, hemoglobin can exist in only two conformations, one with all four of the subunits within a given tetramer in the low-affinity form and one with all four subunits in the high-affinity form (see fig. 6.14*a*). Also, in this model, the hemoglobin molecule is always symmetrical, i.e., all the subunits are in either one state or the other, and all of the binding sites have identical affinities. The binding of oxygen to one of the subunits favors the transition to the high-affinity form. The greater the number of oxygens binding to the tetramer, the more likely it is that the transition from the low-affinity form to the high-affinity form will occur.

The second model, proposed by Koshland, Nemethy, and Filmer in 1966, is referred to as the sequential model (see fig. 6.14*b*). In this model the binding of an oxygen molecule to a given subunit causes that subunit to change its conformation to the high-affinity form. Because of its molecular contacts with its neighbors, the change increases the probability that another subunit in the same molecule will switch to the high-affinity form and bind a second oxygen more readily. The binding of the second oxygen has the same type of enhancing effect on the remaining unoccupied oxygen-binding sites.

Either of these models (or something in between) could account for the sigmoidal oxygen-binding curve of hemoglobin, and either is consistent with the fact that deoxygenated and fully oxygenated hemoglobin have different conformations. The only rigorous way to discriminate between these two models is to obtain structural information on partially oxygenated hemoglobin. This information is still lacking, so no final judgment can yet be made. Even when the situation is fully resolved for hemoglobin, there is no assurance that other allosteric proteins work in the same way.

Allostery is a most common device used by many proteins. Later in this chapter we will see that the muscle protein myosin also shows allosteric behavior, and in chapter 10 we will see that allostery is commonly found in regulatory enzymes.

Figure 6.14

Alternative models for hemoglobin allostery. (*a*) In the symmetry model, hemoglobin can exist in only two states. (*b*) In the sequential model, hemoglobin can exist in a number of different states. Only the subunit binding oxygen must be in the high-affinity form.

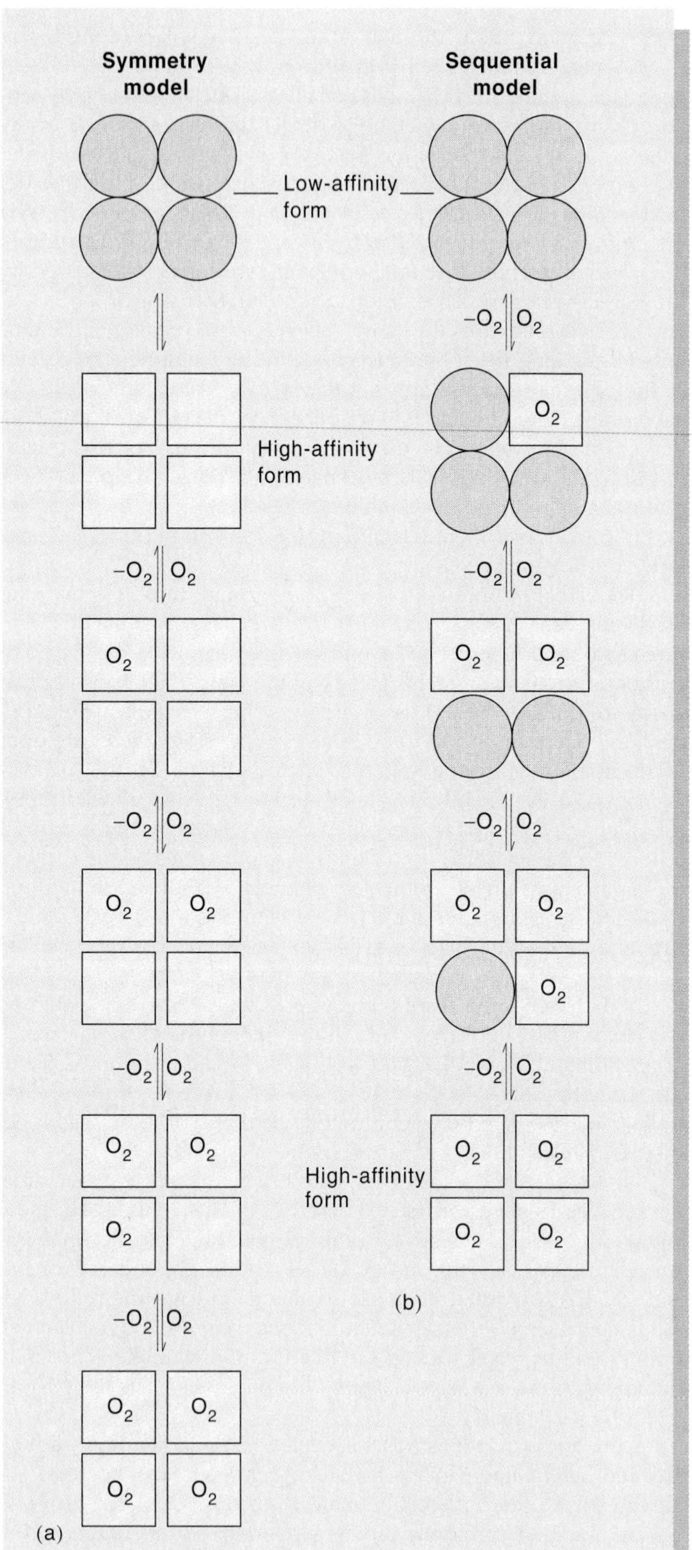

Figure 6.15

The hierarchy of muscle organization. A voluntary muscle such as the bicep is a composite of many fibers connected to tendons at both ends. Each muscle fiber is composed of several myofibrils that are surrounded by an electrically excitable membrane (sarcolemma). Myofibrils exhibit longitudinally repeating structures called sarcomeres. The fine structure of the sarcomere is described in figure 6.16.

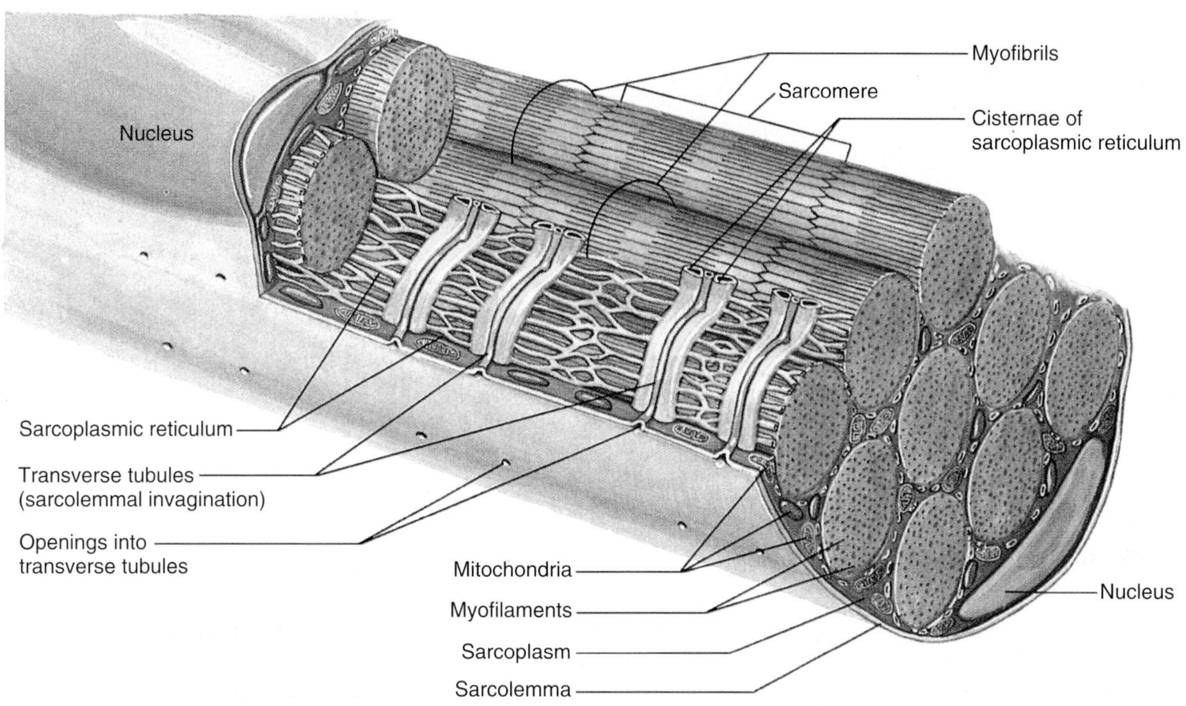

Muscle—An Aggregate of Proteins Involved in Contraction

Vertebrate skeletal muscle represents a remarkable example of a supermolecular aggregate capable of undergoing a reversible reorganization. Voluntary muscle tissue is arranged into fibers that are surrounded by an electrically excitable membrane called the sarcolemma (fig. 6.15). Each fiber is composed of many myofibrils, which when viewed in the light microscope present a striated and banded appearance. A myofibril exhibits a longitudinally repeating structure called the sarcomere (figs. 6.15 and 6.16*a*). This 23,000-Å-long repeating unit is characterized by the appearance of several distinct bands, the less optically dense band being referred to as the I band and the more dense one as the A band. Furthermore, a dense line appears in the center of the I band, called the Z line; and a dense narrow band somewhat similar in appearance also occurs in the center of the A band, called the M line. Adjacent to the M line are regions of the A band that appear less dense than the remainder; these are referred to as the H zone.

Transverse sections of the sarcomere reveal that these patterns result from the interdigitation of two sets of filaments (see fig. 6.16). For example, when a sarcomere is sectioned in the I band, a somewhat disordered arrangement of thin filaments (about 70 Å in diameter) is seen. In contrast, when sectioned in the H zone, a hexagonal array of thick filaments (about 150 Å in diameter) is apparent. The substantive observation is that a transverse section in the dense region of the A band shows a regularly packed array of interdigitating thick and thin filaments. This observation led Hugh Huxley and others to propose that the process of muscle contraction involves sliding the thick and thin filaments past each other (fig. 6.17).

Subsequent analyses have shown that the thin filaments are composed of three proteins (fig. 6.18 and table 6.2). The main filamentous structure consists of an aggregate of globular actin molecules, which takes on the form of a right-handed double helix. The individual actin molecules have a molecular weight of 42,000. Every turn of the actin helix incorporates 14 actin molecules and two molecules of the 360-Å-long filamentous protein tropomyosin (TM) that fit into the two grooves created by the actin double helix. The TM molecule is a dimer of two identical α-helical chains that wind around each other in a coiled coil. Each TM dimer spans seven actin monomers, and a succession of TM dimers extends the full length of the thin filament. Two molecules of troponin (TN) bind to the actin filament at each helical repeat. Troponin is a complex of three nonidentical subunits: TN-C, a calcium-binding subunit; TN-T, a TM-binding subunit; and TN-A, an

Figure 6.16

Electron micrograph of a striated muscle sarcomere showing the appearance of filamentous structures when cross-sectioned at the locations illustrated below. (Electron micrograph courtesy of Dr. Hugh Huxley, Brandeis University.)

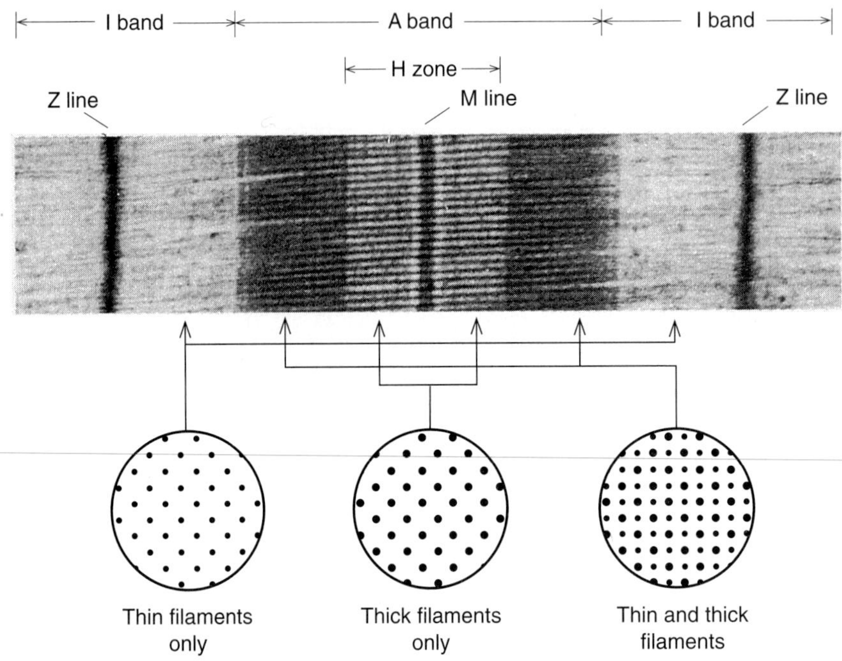

Figure 6.17

The sliding-filament model of muscle contraction. During contraction the thick and thin filaments slide past each other so that the overall length of the sarcomere becomes shorter.

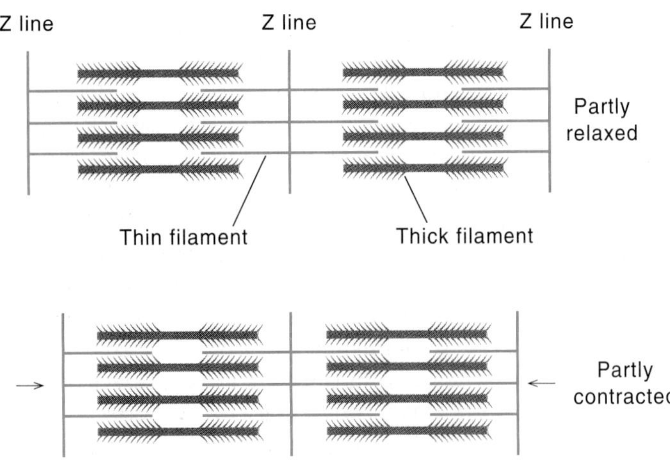

Table 6.2
Principal Proteins of Vertebrate Skeletal Muscle

Protein	M_r	Subunits	Function
Myosin	520,000	2 × 220,000 (heavy chains) 15,000–20,000 (light chains)	Major component of thick filaments
Actin	42,000	One type	Major component of thin filaments
Tropomyosin	64,000	2 × 32,000	Rodlike protein that binds along the length of actin filaments
Troponin	69,000	30,500 (TN-T) 20,800 (TN-I) 17,800 (TN-C)	Complex of three protein subunits involved in the regulation of muscle contraction

Figure 6.18

A molecular view of muscle structure. (*a*) Segment of actin-tropomyosin-troponin. (*b*) Segment of myosin. (*c*) Integration of thin filaments (actin) and thick filaments (myosin) in a muscle fiber.

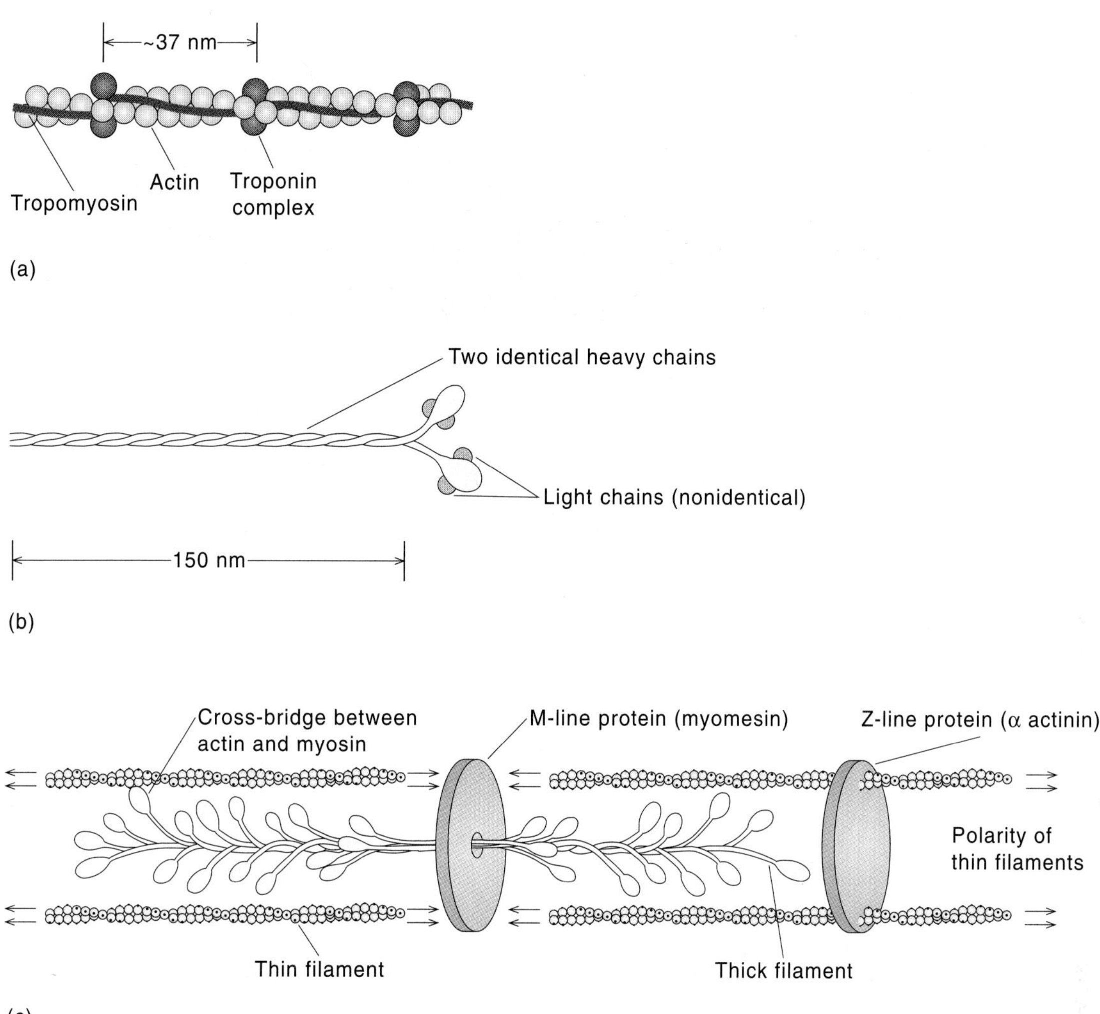

"inhibitory" subunit. The TN–TM proteins form a regulatory complex whose properties we will discuss shortly.

Thick filaments are composed of myosin, a large molecule containing two identical heavy chains (M_r = 220) and two light chains (M_r = 15 and 22). The gross structural organization of myosin is illustrated in figure 6.18*b*. The molecule has two identical globular head regions that incorporate the light chains and a significant fraction of the heavy chains. The tails of the heavy chains form very long α helices that wrap around each other to form left-handed coiled coils. The individual myosin molecules can be cleaved into fragments by partial degradation with various proteases. By separation of such fragments, it has been demonstrated that the binding sites of myosin for actin and the ATPase activity of myosin are located in the globular head regions. The α-helical coiled coils form the backbone of the thick filament, while the remainder forms an arm that can provide a flexible extension or hinge for the globular head away from the body of the thick filament. The thick filament contains many myosin molecules oriented in a staggered bipolar fashion (fig. 6.18*c*).

Granted that this is a marvelous piece of molecular architecture, how does it actually work? The answer lies in the observation that actin cyclically binds the globular myosin head group to form cross-bridges in a reaction that depends on the myosin-catalyzed hydrolysis of adenosine triphosphate (ATP). The cyclic binding of actin to myosin is driven by the energy-releasing hydrolysis of ATP, catalyzed by the myosin head group in a manner that causes rearrangement of the actin-myosin cross-bridges. When muscle is completely relaxed, there is a minimum number of cross-bridges and the muscle is fully stretched. However, when the muscle is activated and under tension, it contracts and more cross-bridges are formed as the region of overlap between actin and myosin increases. At each stage of the contraction process it

Figure 6.19

The actin–myosin cycle associated with muscle contraction. For simplicity, only a single myosin head with a shortened tail is shown. A star next to the myosin heads indicates a strong actin–myosin complex. Step 1 involves release of ADP, binding of ATP, and dissociation of the actin–myosin complex. Step 2 entails the hydrolysis of ATP to ADP and P_i. Step 3, the slowest step in the cycle, involves formation of a strong actin–myosin complex. Step 4 involves dissociation of P_i. Step 5 entails the movement of actin as a result of a gross conformational change in the myosin molecule. Less visible conformational changes are probably involved in the other four steps of the contraction cycle. (Reprinted with permission from J. A. Spudich, How molecular motors work, *Nature* 372:515–518, 1994. Copyright 1994 Macmillan Magazines Limited.)

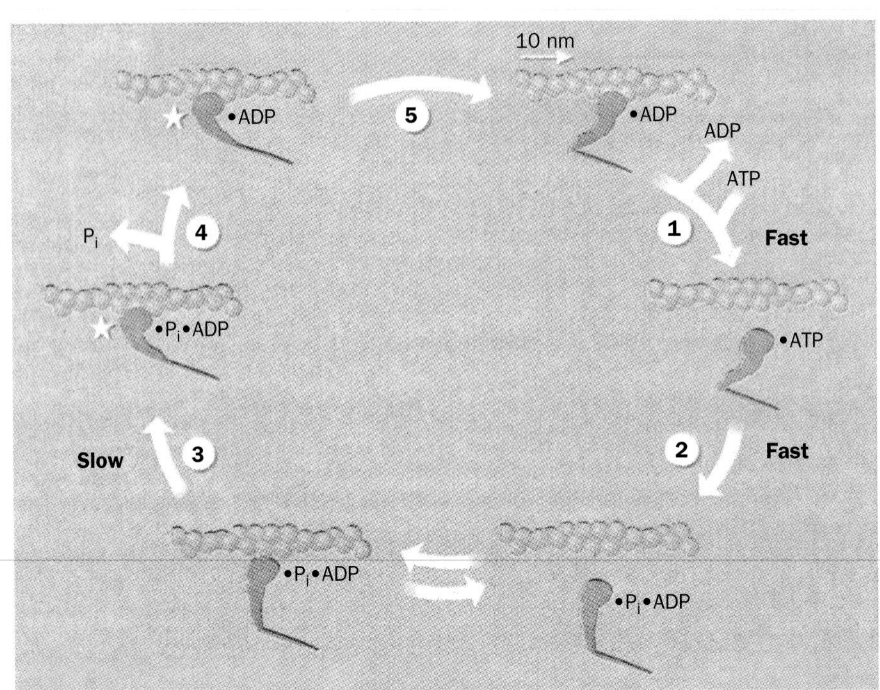

is essential to break the existing bridges with the help of ATP hydrolysis before new ones can be formed. It is important to realize that although ATP encourages more bridges to be formed, the ATP is required to break the bridges, not to form them. The breaking of bridges is required so that new and more numerous bridges can be formed. Thus cross-bridge formation is energetically favored.

A likely scenario for the contraction process is shown in figure 6.19. For clarity, only one myosin head is shown in combination with a short segment of actin. At the top right in this figure we see the actin–myosin complex just after a translation step. At this time the myosin is still complexed with actin and ADP. The conformational change that results from the translation step permits the ADP to dissociate from the complex. This makes way for the binding of an ATP, which produces a change in the conformation of myosin that results in its release from actin (step 1). The ATP is rapidly hydrolyzed to ADP and P_i, which remain firmly bound to the myosin (step 2). The myosin–ADP–P_i complex slowly forms a strong complex (indicated by a star) with actin (step 3). Formation of the strong complex with actin triggers release of the P_i (step 4), which in turn triggers a large conformational change in the myosin that leads to the translational motion of the myosin along the actin by a distance of about 10 nm (step 5).

Figure 6.20 shows a close-up of the actin–*myosin–P_i–ADP complex of figure 6.19. A portion of the S1 fragment described as the lever arm is pictured as undergoing a conformational change resulting in a translation stroke of about 10 nm.

Recent advances in our understanding of myosin structure support the scheme just described for myosin and permit us to identify key regions in the myosin molecule that are involved in the process. These advances have resulted from a structural de-

Figure 6.20

The myosin head fragment emphasizing the conformational change that occurs in the "lever" as a result of the translation step. To the right the newly formed strong actin–myosin complex is shown. (Reprinted with permission from J. A. Spudich, How molecular motors work, *Nature* 372:515–518, 1994. Copyright 1994 Macmillan Magazines Limited.)

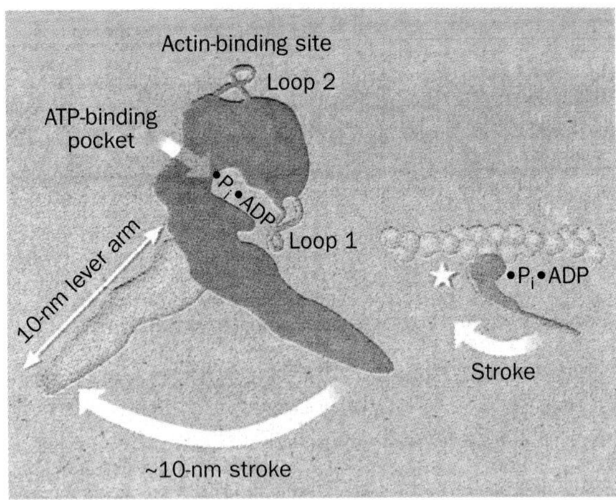

termination of the head portion of the heavy chain in a complex with the two light chains. Previous in vitro studies of Spudich and co-workers had shown that this so-called S1 fragment, obtained by limited proteolytic digestion of whole myosin, contains the binding sites for ATP and actin as well as the structural equipment for the ATP-dependent movement of actin along myosin. Thus it seems likely that the essence of the "molecular motor" involved in muscular contraction could be understood by a careful exami-

Figure 6.21

A ribbon representation of the entire model for myosin S1. In this figure, 2000 and 3000 have been added to the residue numbers of the regulatory light chains (RLC) and elective light chains (ELC), respectively, to distinguish these from the heavy chain. Heavy chain residues Asp^4 to Glu^{204}, Gly^{216} to Tyr^{626}, and Gln^{647} to Lys^{843} are colored in green, red, and blue, respectively. These segments are separated by disordered loops for which no density is evident in the current map. The A2 isozyme of the essential light chain, shown in yellow, theoretically contains 149 amino acid residues. In the model it extends from residue Asp^5 to Val^{149} and contains one ill-defined region that includes residues Leu^{50} to Ala^{60}. The regulatory light chain, which is colored in magenta, theoretically consists of 166 amino acid residues. In the current model it extends from residue Phe^{19} to Lys^{163} but is disordered between residues Pro^{142} and Asn^{147}. (Reprinted with permission from I. Rayment et al., Three-dimensional structure of myosin subfragment-1: A molecular motor. *Science* 261:50–58, 2 July 1993. Copyright 1993 American Association for the Advancement of Science.)

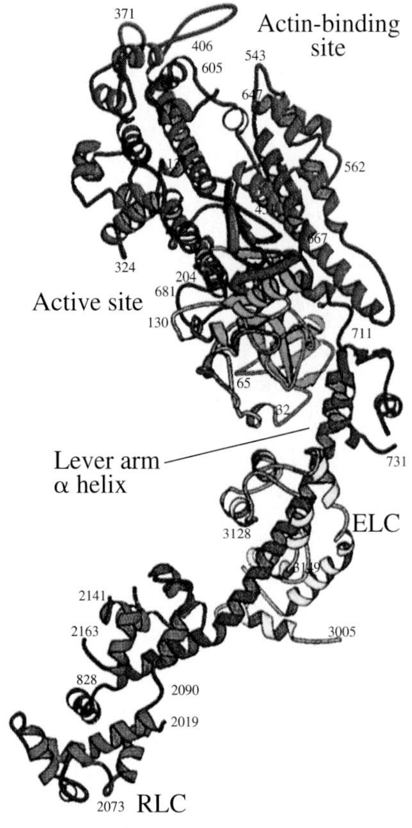

nation of the structural properties of this portion of the myosin molecule. Success in the crystallization of the S1 fragment has permitted a detailed crystallographic analysis of the myosin head structure (fig. 6.21).

A striking feature of this structure is that the nucleotide-binding site and the actin-binding site, which have to communicate with each other during contraction, are nearly 4 nm apart. Thus the region of the protein between the nucleotide-binding and actin-binding sites must act as a communication zone. Another striking feature of the S1 structure is the presence of a "lever arm" (depicted schematically in figure 6.20) comprising about 73 amino acid residues of the carboxy-terminal portion of the S1 heavy chain, in the form of a long, single α helix. The two light chains, ELC and RLC, are wrapped around this α helix. They probably provide the heavy chain α helix with sufficient rigidity to act as a lever. The lever arm is about 10 nm in length, so its swing could produce a stroke size close to that believed to result from a single step in muscular contraction. In vitro studies have shown that removal of either or both of the light chains causes a 50–90% reduction in the velocity with which myosin moves actin in vitro without significantly altering the actin-activated ATPase on the myosin. Deletion of both light-chain binding sites removes essentially all the lever arm and drastically reduces the ability of myosin to move actin in vitro, but it does not interfere significantly with the actin-activated ATPase.

Figure 6.22

The effect of calcium on muscle contraction. Binding of calcium to the TN–TM–actin complex produces a shift in the location of TM, which produces an allosteric transition in actin. The allosteric transition in actin facilitates the release of P_i from myosin, which strengthens the interaction between actin and myosin.

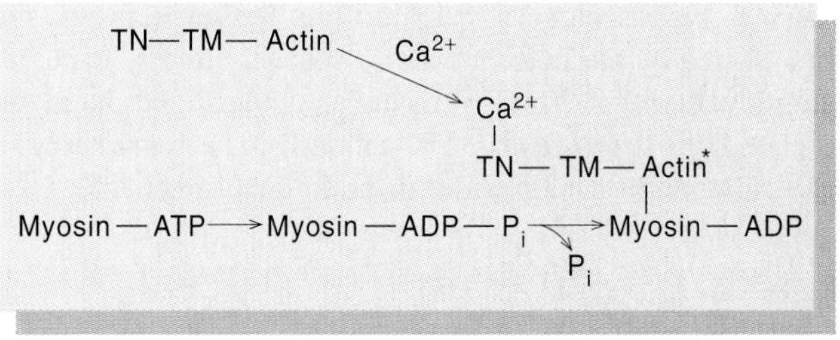

TN—TM—Actin $\xrightarrow{Ca^{2+}}$ Ca^{2+}–TN—TM—Actin*

Myosin—ATP ⟶ Myosin—ADP—P_i ⟶ Myosin—ADP + P_i

As we indicated earlier, the process of contraction is regulated or triggered by the TN–TM system. Since voluntary muscles are under the conscious control of the animal, one would expect a signal from the central nervous system to initiate the process of contraction. A nerve impulse communicated to the muscle causes a depolarization of the sarcolemma membrane that surrounds the muscle fibers. This in turn causes a release of Ca^{2+} from the endoplasmic reticulum in the cytoplasm of the muscle cell (fig. 6.22). The Ca^{2+} ions form a complex with the TN-C

Figure 6.23

Examples of structural diversification in prokaryotic cytochromes *c*. Three cytochromes are shown: cytochrome c_{550} from the denitrifying bacterium *P. denitrifican,* cytochrome c_3 from the photosynthetic bacterium *P. rubrum,* and cytochrome *c* from the mitochondria. The prokaryotic cytochromes contain more residues in their polypeptide chains (shaded) than does cytochrome *c*. Despite these variations there is a strong conservation of those amino acid residues that interact in functionally important ways with the protein's heme group.

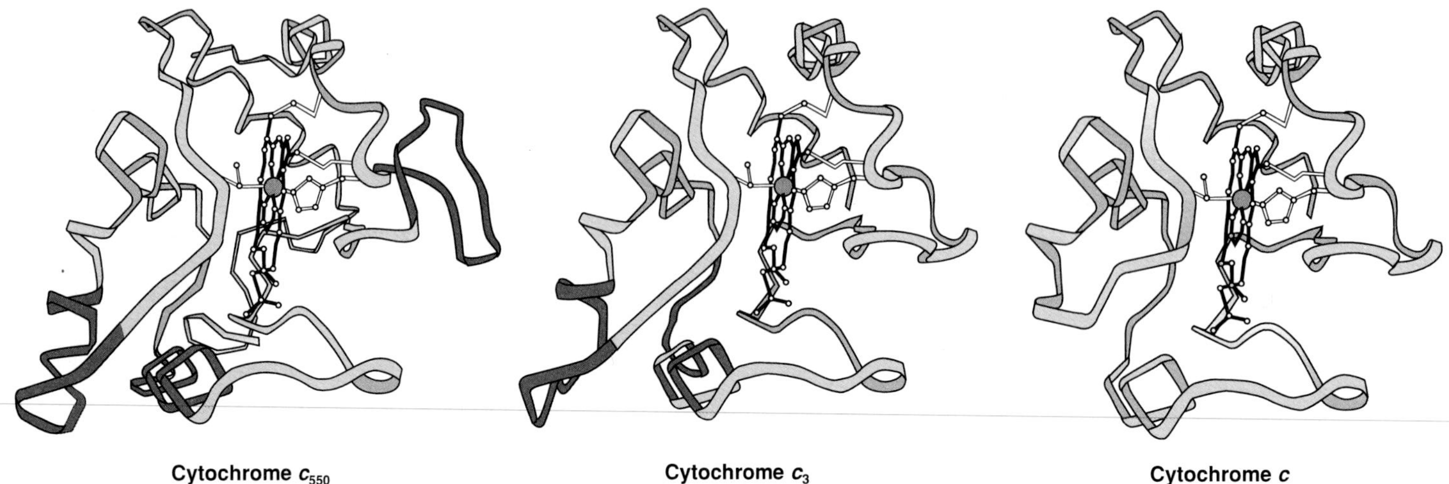

component of the troponin molecule (see table 6.2). This induces changes within the TN complex, which overcomes the inhibitory effect of the TN-I subunit. Then, through TN-T, a signal is sent to TM that triggers the contraction event. The precise nature of this signal from troponin to tropomyosin is unclear; it appears to involve a movement of the tropomyosin on the actin surface that leads to an allosteric transition of the actin that encourages more favorable contact between the complementary binding sites on actin and myosin. As a result a strong bridge is formed, and the P_i is released from the myosin. The remaining steps in muscular contraction have already been described. It is noteworthy that for some time after death, the muscle enters a state of rigor in which the muscle is fully contracted and the maximum number of bridges is formed. This is probably due to excessive neuronal firing and a considerable discharge of calcium from the sarcoplasmic reticulum. Normally, the cytosolic Ca^{2+} concentration is restored to resting levels within 30 ms of receiving a signal and the myofibrils relax. In its structure and in its action TN-C is remarkably similar to calmodulin, which is discussed in detail in chapter 10.

Protein Diversification as a Result of Evolutionary Pressures

We can gain further insight into protein function by asking the question, How do selective evolutionary pressures lead to the diversification of protein structure and function? The answer to this question is multifaceted.

Proteins that serve similar or identical biological functions in different living organisms are typically very similar in both their amino acid sequences and their tertiary structures. Such related families of molecules are generally assumed to reflect processes of divergent evolution. That is, they are thought to have evolved through gradual point-by-point modification of one ancestral molecule.

One of the most extensively studied protein families is the cytochrome *c* family. Cytochrome *c* proteins function as electron carriers in the mitochondrial electron-transport chains of all multicellular organisms (see chapter 16). Sequence comparisons of mitochondrial cytochrome *c* from organisms as diverse as humans and green plants reveal an extraordinary degree of sequence conservation. This fact suggests that the functional role of the molecule was highly refined by selective evolutionary pressures prior to the emergence of the first multicellular organisms. Some positions in the sequence are quite variable, whereas others are essentially invariant. Structural and chemical modification studies have shown that some of the invariant amino acid residues are associated with functionally important heme interactions, while others are important in governing the interactions of cytochrome *c* with its physiological oxidase and reductase.

If the sequence and structure of mitochondrial cytochrome *c* were indeed highly refined prior to the emergence of multicellular organisms, then its evolutionary precursors should still exist in prokaryotic organisms. In fact, in virtually all prokaryotic organisms that use oxidative or photosynthetic electron-transport chains to synthesize the high-energy intermediate adenosine triphosphate (ATP), we find molecules that are strikingly similar to mitochondrial cytochrome *c*. However, as might be expected, cytochrome *c* proteins from prokaryotes exhibit much more sequence diversity than the proteins typically found in higher organisms. In particular, the prokaryotic cytochrome *c* proteins often contain multiple amino acid insertions or deletions relative to mitochondrial cytochrome *c*. Nevertheless, from tertiary-structure determination of several prokaryotic proteins, we find that these molecules are all variations on a basic structural theme (fig. 6.23).

Figure 6.24

Schematic illustration of two serine proteases, elastase and subtilisin. These molecules differ totally in sequence and tertiary structure but have catalytic sites that are nearly identical. The configuration of the active site for elastase is described in chapter 9. (Reprinted by permission of Jane S. Richardson.)

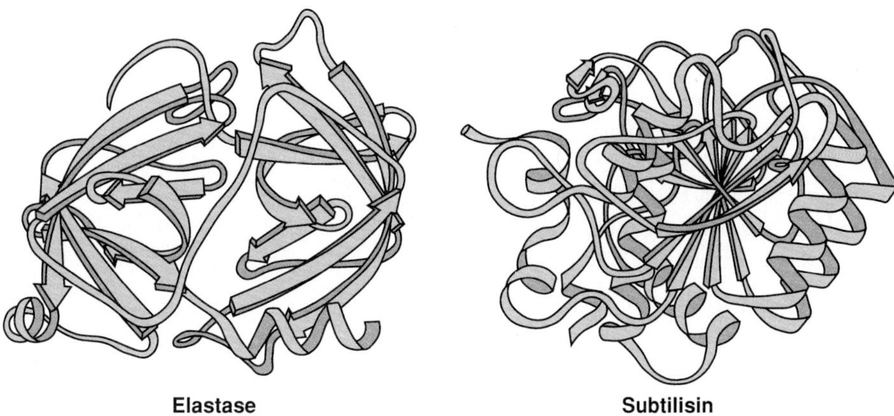

Further, the prokaryotic molecules all show a strong conservation of those amino acid residues that interact in functionally important ways with the protein's heme group. The slight variations of the basic structural theme of cytochrome *c* that are observed appear to optimize the molecule's function in different organisms.

In addition to the cytochrome *c* proteins, several other families of proteins have been found to share similarities in amino acid sequence and tertiary structure. Again, the observed differences in sequence and structure among individual members reflect evolutionary pressures that modified a basic structural arrangement in order to diversify the functional properties of the molecules. Examples of such structurally and functionally related families include the oxygen-binding globins that we discussed earlier, the serine protease enzyme families (see chapter 9), and the dehydrogenases. Generally, related members within a given enzyme family catalyze chemically similar reactions but exhibit varying specificities for structurally different substrate molecules. For example, while all serine proteases catalyze the hydrolytic cleavage of peptide bonds, different members of this molecule family cleave polypeptides at different locations, depending on the nature of the amino acid side chains adjacent to the cleavage site (see chapter 9).

Although divergent evolutionary processes usually produce gradual changes in a given protein function, some changes are less gradual, as when a mutation occurs that radically alters protein function. Such mutations frequently result in the synthesis of functionally defective molecules and so constitute one cause of inheritable disease. Alternatively, amino acid substitutions in related proteins may result in the generation of new functions. The enzyme lysozyme, which binds and subsequently cleaves polysaccharide chains, and the protein α-lactalbumin, which transports sugars, are very similar in both sequence and structure. In this case it appears that relatively slight modifications of a common ancestral precursor have resulted in selection for molecules with quite different functions.

Not all functionally related families of proteins arose by divergent evolution from a common ancestor. In some cases, proteins that have extensive functional or structural similarities appear to have arisen independently. An outstanding example of two molecules that are functionally similar but are radically different in sequence and structure occurs in the serine proteases (fig. 6.24; also see fig. 9.5). Many members of the serine protease family are closely related in sequence and structure. However, in *Bacillus subtilis* the serine protease subtilisin, while being essentially identical in its arrangement of amino acid residues at the active site to the other serine proteases, otherwise differs from them completely in sequence and tertiary structure. This situation presumably reflects convergent evolution on a particular active site arrangement required for the protein's catalytic function (also see fig. 9.6).

More frequently, proteins that differ completely in sequence and function have quite similar tertiary structures. In these cases the observed structural similarities most probably reflect selection of a particularly stable structural arrangement. Examples of such structurally related molecules include those with similarly twisted β sheets (see fig. 5.26) and proteins organized as a bundle of four closely packed α helices (see fig. 5.24).

Gene Splicing Results in a Reshuffling of Domains in Proteins

In the preceding descriptions of protein diversification we looked at evolutionary change that occurs as a consequence of the continuing selection of individual point mutations in the protein's encoding DNA. However, many proteins exhibit structural characteristics that suggest that they have resulted from processes of gene splicing. In particular, a surprisingly large fraction of known protein structures incorporate multiple copies of structurally similar domains. In many cases it appears that these molecules have arisen by the splicing together of duplicate or multiple copies of a gene coding for a given structural domain, followed by the essentially independent fixation of mutations throughout the spliced genome. The eventual result is a protein composed of sequentially different but structurally similar repeating domains.

Additional evidence for the role of gene splicing in protein evolution comes from the observation that some large proteins are composed of several different structural domains, each of which may structurally resemble parts or the entirety of other known proteins. A good example is the glycolytic enzyme pyruvate kinase (fig. 6.25). This large protein is organized as three structural domains, two of which show convincing structural similarities to, respectively, the β barrel of triose phosphate isomerase (domain 1; see fig. 5.26) and a twisted β-sheet domain common to many dehydrogenases (domain 3; see fig. 6.25). The third domain of pyruvate kinase (domain 2) also has convincing similarity to a common structural type, the antiparallel β barrel.

Evolutionary Diversification Is Directly Involved in Antibody Formation

In most cases the fixation of new mutations is a relatively infrequent event, resulting in the gradual evolution of proteins such as cytochrome *c*. By contrast, one of the important biological defense mechanisms of higher organisms, the immune response, depends on the rapid generation of structurally novel molecules that can recognize and bind foreign substances that may be harmful to the organism. The molecules responsible for the initial recognition and binding of foreign substances are the immunoglobulins. These molecules are composed of two pairs of polypeptide chains of different length that are interconnected by covalent cysteine disulfide linkages (fig. 6.26). Sequence studies of various immunoglobulins have shown that both the heavy and light polypeptide chains contain repeating homologous sequences that are about 110 residues in length. Structural studies of the immunoglobulin molecule show that the sequentially homologous regions fold individually into similar structural domains, arranged as a bilayer of antiparallel sheets. The molecule in its entirety is formed of 12 similar structural domains, of which eight are formed by the two heavy chains and four by the two light chains.

Immunoglobulins that are specific for binding to different foreign substances vary greatly in the sequences found in the amino-terminal domains of both the heavy and light chains. It is these variable regions that form the binding sites between the immunoglobulin molecule and the foreign substances that trigger the immune response. The remarkable property of this system is that it can rapidly diversify the sequence of variable regions by mutation, gene splicing, and RNA splicing. The net result is that the organism can produce an enormous variety of antibodies from a quite limited amount of informational DNA originating in the germ-line tissue.

Figure 6.25

Pyruvate kinase domains 1, 2, and 3 as an example of a protein whose domains show no structural resemblance whatsoever. (Illustration copyright by Irving Geis. Reprinted by permission.)

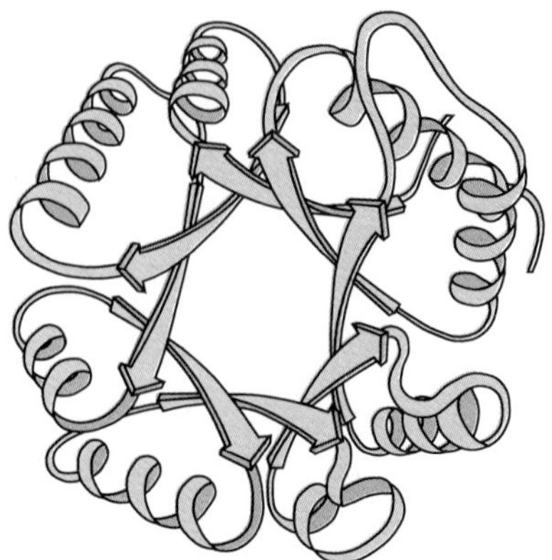

Pyruvate kinase domain 1

Pyruvate kinase domain 2

Pyruvate kinase domain 3

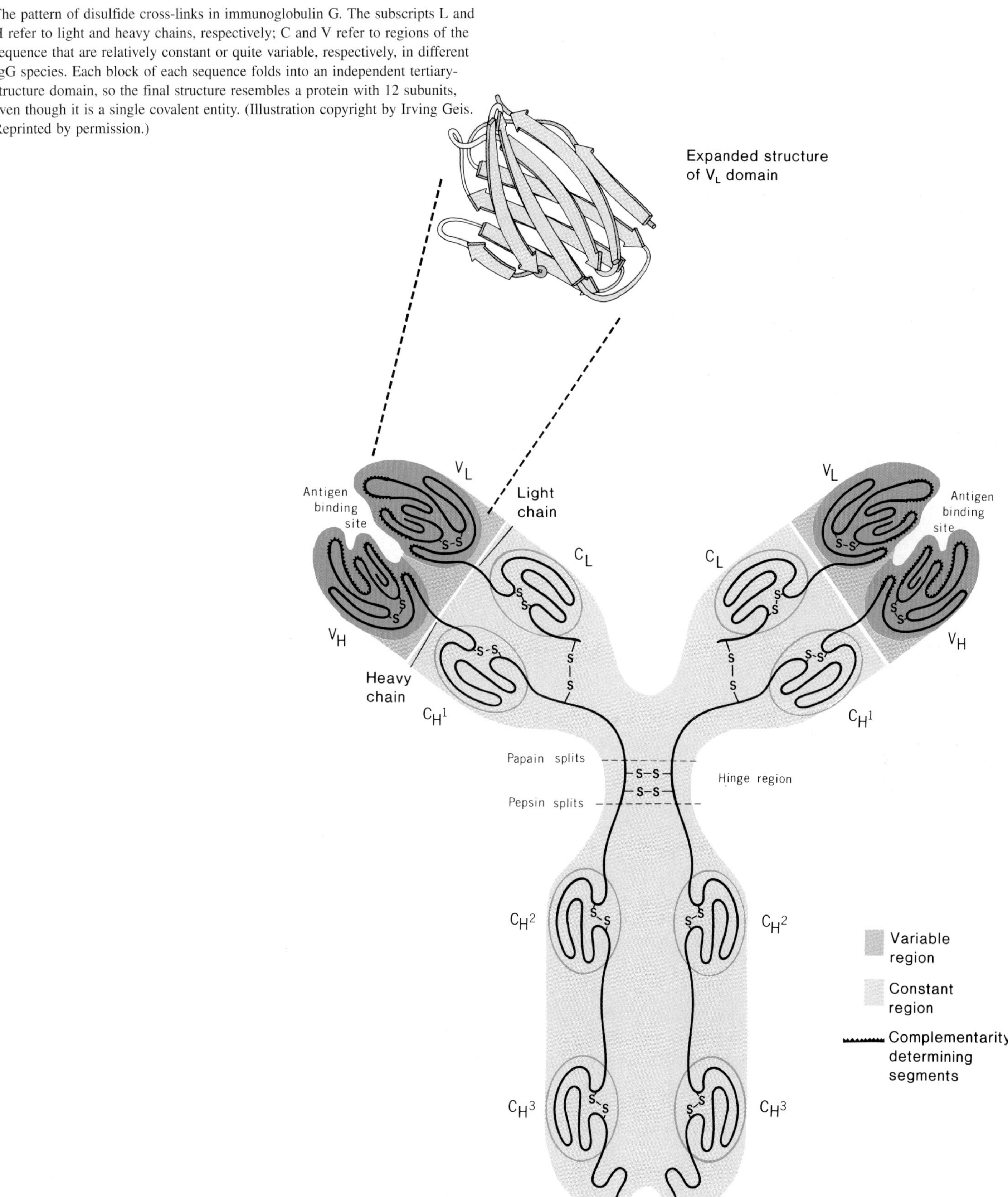

Figure 6.26

The pattern of disulfide cross-links in immunoglobulin G. The subscripts L and H refer to light and heavy chains, respectively; C and V refer to regions of the sequence that are relatively constant or quite variable, respectively, in different IgG species. Each block of each sequence folds into an independent tertiary-structure domain, so the final structure resembles a protein with 12 subunits, even though it is a single covalent entity. (Illustration copyright by Irving Geis. Reprinted by permission.)

Summary

In this chapter we have considered the relationship between structural and functional properties of proteins. The following points are the most important.

1. The polypeptide chains of proteins are synthesized on ribosomes. Ribosomes that remain free in the cytosol are associated with the synthesis of proteins that are targeted to locations in the cytosol, the nucleus, the peroxisomes, the mitochondria and the chloroplasts. Other ribosomes, which are attached to the endoplasmic reticulum, are engaged in the synthesis of proteins that are targeted to the endoplasmic reticulum, the Golgi, the lysosomes, the plasma membrane, and the secretory granules.
2. Proteins that cooperate in a particular biochemical process are usually found in the same location within the cell or extracellular milieu.
3. Each protein is carefully designed to carry out a specific function, as revealed by our discussion of hemoglobin and muscle.
4. Hemoglobin is a tetramer made of two almost identical subunits. The function of hemoglobin is threefold: to transport O_2 from the lungs to the tissues where it is consumed, to transport CO_2 from the tissues where it is produced to the lungs, where it is expelled, and to maintain the blood pH over a narrow range. The structure of hemoglobin is designed so that it will pick up the maximum amount of oxygen at high oxygen tensions in the lung tissue and will deliver the maximum amount of oxygen in the oxygen-consuming tissues.
5. Muscle is an aggregate of several different proteins involved in contraction. The main protein components of muscle are organized as overlapping filaments of two types: thin filaments, composed mainly of actin molecules, and thick filaments, composed of myosin molecules.
6. The process of muscular contraction entails a sliding of the two types of filaments past one another. In a fully contracted component of muscle tissue the actin and myosin filaments show a maximum overlap with one another. The contraction process involves the breakage and reformation of bridges between the actin and the myosin molecules in a reaction that requires the expenditure of ATP.
7. Protein diversification is the product of a long evolutionary process. It can be studied by comparing the structures of different proteins in the light of their functions. It can also be studied by comparing structurally related proteins with similar functions in different organisms or by comparing structurally related domains between proteins that have different functions.
8. Antibodies provide the unique opportunity for studying diversification within a single organism, for antibodies with unique specificities are the result of an ongoing evolutionary process that takes place within the organism.

Selected Readings

Akers, G. K., M. C. Doyle, D. Myers, and M. A. Daugherty, Molecular code for cooperativity in hemoglobin. *Science* 255:54–63, 1992. A new model for hemoglobin allostery.

Allen, R. D., The microtubule as an intracellular engine. *Sci. Am.* 238(2):42–49, 1987.

Baldwin, J., Structure and cooperativity of haemoglobin. *Trends Biochem. Sci.* 5:224–228, 1980.

Berg, H. C., How bacteria swim. *Sci. Am.* 233(2):36–44, 1975.

Cantor, C. R., and P. Schimmel, *Biophysical Chemistry*. New York: Freeman, 1980. Especially see volume 2, entitled *Techniques for Study of Biophysical Structure and Function*.

Caplan, A. I., Cartilage. *Sci. Am.* 251(4):84–94, 1984.

Dickerson, R. E., and I. Geis, *Hemoglobin*. Menlo Park, Calif.: Benjamin/Cummings, 1983. A magnificent presentation of every facet of hemoglobin biochemistry and genetics.

Doolittle, R., Proteins. *Sci. Am.* 253(4):88–96, 1985. Overview emphasizing evolutionary considerations.

Eisenberg, D., and D. Crothers, *Physical Chemistry and Its Applications to the Life Sciences*. Menlo Park, Calif.: Benjamin/Cummings, 1979.

Eyre, D. R., M. A. Pdaz, and P. M. Gallop, Cross-linking in collagen and elastin. *Ann. Rev. Biochem.* 53:717–748, 1984.

Gething, M. J., and J. Sambrook, Protein folding in the cell. *Nature* 355:33–45, 1992.

Glanz, J. Hemoglobin Reveals New Role as Blood Pressure Regulator. *Science* 271:1670, 1996.

Huxley, H. E., Sliding filaments and molecular motile systems. *J. Biological Chemistry* 265:8347–8352, 1990.

Hynes, R. O., Fibronectins, *Sci. Am.* 254(6):42–51, 1986.

Ingram, V. M., Gene mutation in human haemoglobin: the chemical difference between normal and sickle-cell haemoglobin. *Nature* 180:326–328, 1957. A classic paper.

Karplus, M., and J. A. McCammon, The dynamics of proteins. *Sci. Am.* 254(4):42–51, 1986. The atoms in a protein vibrate and rotate about average positions. This feature is essential to their function.

Lawn, R. M., and G. A. Vehar, The molecular genetics of hemoglobin. *Sci. Am.* 254(3):48–65, 1986.

Martin, G. R., R. Timpl, P. K. Muller, and K. Kuhn, The genetically distinct collagens. *Trends Biochem. Sci.* 10:285–287, 1985. A brief, authoritative account of the most abundant protein found in vertebrates.

Methods in Enzymology. New York: Academic Press. A continuing series of over 250 volumes that discuss most methods at the professional level but are still understandable for students.

Pauling, L., H. A. Itano, S. J. Singer, and I. C. Wells, Sickle-cell anemia: a molecular disease. *Science* 110:543–548, 1949. A classic paper.

Pollard, T. D., and J. A. Cooper, Actin and actin-binding proteins. *Ann. Rev. Biochem.* 55:987–1036, 1986. Discusses structure and function.

Salemme, R., Structure and function of cytochromes *c*. *Ann. Rev. Biochem.* 46:299–329, 1977.

Scopes, R., *Protein Purification: Principles and Practice,* 2d ed. New York: Springer-Verlag, 1987. A recent general treatment of this subject.

Spudich, J. A., How molecular motors work. *Nature* 372:515–518, 1994. Recent advances in myosin structure support the proposed role of myosin in muscular contraction. Excellent reference for conducting a review.

Steinert, P. M., and D. R. Roop, Molecular and cellular biology of intermediate filaments. *Ann. Rev. Biochem.* 57:593–626, 1988.

Tonegawa, S., The molecules of the immune system. *Sci. Am.* 253(4):122–130, 1985.

PROBLEMS

1. In the latter half of the 1800s one of the first suggestions that proteins were large molecules came from ashing experiments in which hemoglobin was converted to Fe_2O_3. These procedures suggested a molecular mass of hemoglobin greater than 15,900, a number unheard of at that time. Why did the researchers of the past century, who were excellent analytical chemists, deviate so significantly from the molecular mass of 64,500 now known for hemoglobin? What quantity of Fe_2O_3 results from the ashing of 1.00 g of hemoglobin?
2. Carbonic acid in the blood readily dissociates into hydrogen and bicarbonate ions. If the serum pH of 7.4 equals that inside the erythrocytes, what percentage of the carbonic acid is ionized? Use a value of 6.4 for the first pK_a of carbonic acid.
3. In addition to oxygen, hemoglobin subunits can also carry carbon dioxide. This is performed by covalent addition of CO_2 to the N termini of the hemoglobin chains to produce a carbonate structure. Propose reactions for this process utilizing (a) CO_2 and (b) HCO_3^-.
4. Figure 6.13 indicates locations of mutations that have been shown to produce pathological conditions. The majority of the types of mutations that have been discovered in human hemoglobins have been mutations in which either amino acid residues bearing charges are replaced with ones with no charge or in which uncharged amino acids are replaced with charged amino acids. Do you think this represents a basic biological principle or is it an artifact of the detection process? Explain.
5. Sickle-cell anemia becomes most apparent during a sickle-cell crisis, when the soft tissues are often acutely painful. Using information provided in the text on the molecular-cellular effects of the sickling of red blood cells, can you provide an explanation for the origin of the pain?
6. The hemoglobin present in a fetus is analogous to the $\alpha_2\beta_2$ tetramer of the adult, but the two β chains have been replaced with comparable γ chains. Considering the relevant biology, which hemoglobin type (adult or fetal) do you expect to have the greater affinity for oxygen?
7. Use the information presented in problem 6 to propose a possible "future genetic engineering solution" to sickle-cell anemia. Are there any deleterious ramifications of your proposal?
8. Carbon 2 in glycerate-2,3-bisphosphate is in the D configuration. Would you expect the L configuration of glycerate-2,3-bisphosphate to have the same effect on the biochemistry of hemoglobin? Why or why not?
9. Figures 6.10 and 6.11 show a histidyl F8 residue interacting with the heme iron on deoxy- and oxyhemoglobin. Would you expect the imidazole nitrogen of the histidyl group to be protonated or unprotonated during this interaction? Why?
10. Would you expect the blood–hemoglobin system to transport more moles of O_2 or CO_2? Why?
11. The protein tropomyosin (TM) is composed of two identical chains of α helix (see table 6.2) that are in turn twisted around each other in a helical structure. Consider the average amino acid residue weight to be 105 daltons and use typical α-helix dimensions (fig. 5.8) to calculate the length of each chain in TM. Explain any discrepancy observed between the calculated length and the observed length of 360 Å.
12. Bryan Allen made aviation history by pedaling the *Gossamer Albatross* from near Folkestone, England, to Cap Gris Nez, France, from 4:51 AM to 7:40 AM on June 12, 1979. During this flight he continually produced about a third of a horsepower (*Nat. Geographic* 156:5640, 1979). Considering that the energy available from the hydrolysis of ATP is 7.5 kcal/mole (see fig. 2.9), determine the number of moles of ATP required for this flight. For simplicity, assume that the muscles are 50% efficient in the conversion of chemical energy into mechanical energy and that one horsepower equals the energy expenditure of 178 cal/s.
13. Consider the structure of domain 1 of pyruvate kinase (fig. 6.25) and propose a simple evolutionary explanation of how the structure could have arisen.
14. Cytochrome C isolated from different organisms consists of about 100 amino acids. About 25 of the amino acids are invariant. Of those 25 amino acids, five are glycine and three are proline. If all 20 amino acids are "equal," then only one or two of the invariant amino acids would be of each amino acid type. How can you explain the high frequency of glycine and proline as invariant amino acids in the cytochrome C structure?
15. You have isolated a protein complex that sediments in the ultracentrifuge similarly to a hemoglobin marker. If the ultracentrifugation is repeated under identical conditions except that 2-M NaCl is added to the dilute buffer, the protein sediments similarly to a myoglobin marker. What conclusion can you reach about the properties of the protein complex?
16. Predict the effect on O_2 transport of a mutant form of hemoglobin with a markedly decreased affinity for glycerate-2,3-bisphosphate.

C H A P T E R

7

Methods for Characterization and Purification of Proteins

A rich variety of methods based on size, shape, and composition is available for protein purification and characterization.

In chapters 4–6 we described the structures of amino acids and proteins, and in two cases in chapter 6 we examined how these structures relate to their function. Some of the methods for structure determination have also been discussed (separation of proteins by isoelectric focusing in chapter 3; amino acid analysis, amino acid sequence analysis, and ultraviolet absorbance spectroscopy in chapter 4; and x-ray crystallography, nuclear magnetic resonance spectroscopy, optical rotatory dispersion, and circular dichroism in chapter 5). To analyze the structure of a protein, we must isolate it from the complex mixture of substances in which it exists in whole cells. The primary object of this chapter is to describe techniques and strategies used for protein purification. Because these procedures are often used for protein characterization as well, they will add to the methods already discussed for protein characterization.

In the first part of this chapter, methods for protein fractionation and characterization are discussed in isolation. Success in protein purification depends on picking a number of procedures and combining them in an effective order. The chapter concludes with a discussion of two examples in which trains of procedures are combined to purify specific proteins from crude whole-cell extracts.

Methods of Protein Characterization

First we will discuss methods used for protein characterization. This discussion supplements the methods already described in chapters 3–6.

Solubility Reflects a Balance of Protein-Solvent Interactions

The solubility of a protein reflects a delicate balance between different energetic interactions, both internally within the protein and between the protein and the surrounding solvent. Consequently, the protein's solvent or thermal environment could affect both its solubility and structure. As we have seen, extreme changes can lead to denaturation (see chapter 5). In this chapter we will, for the most part, be concerned with conditions under which the native structure is maintained. Changes in protein solubility that do not destroy the molecule's structural integrity can occur in several ways.

A Minimum in Solubility Occurs at the Isoelectric Point. Proteins typically have on their surfaces charged amino acid chains that undergo energetically favorable polar interactions with the surrounding water. The total charge on the protein is the sum of the side-chain charges. However, the actual charge on the weakly

Figure 7.1

Solubility of β-lactoglobulin as a function of pH and ionic strength. The isoelectric pH (pI) for this protein is about 5.2. This corresponds to the point of minimum solubility.

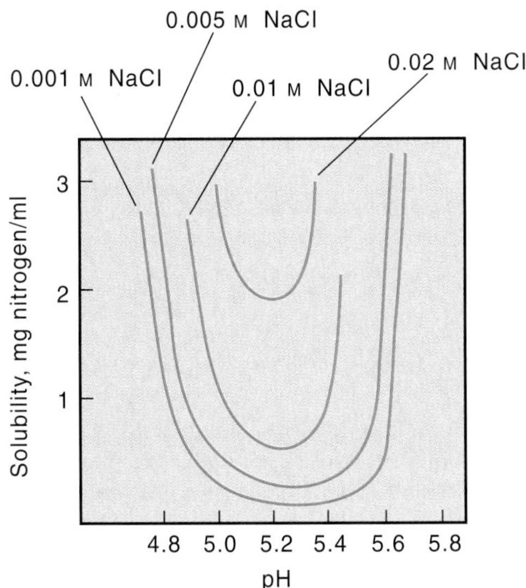

Figure 7.2

Solubility of horse carbon monoxide hemoglobin in different salt solutions. The addition of a moderate amount of salt (salting in) is required to solubilize this protein. At high concentrations, certain salts compete more favorably for solvent, decreasing the solubility of the protein and thus leading to its precipitation (salting out). (Source: E. J. Cohn and J. T. Edsall, *Proteins, Amino Acids, and Peptides as Ions and Dipolar Ions.* Copyright ©1942, Reinhold, New York, N.Y.)

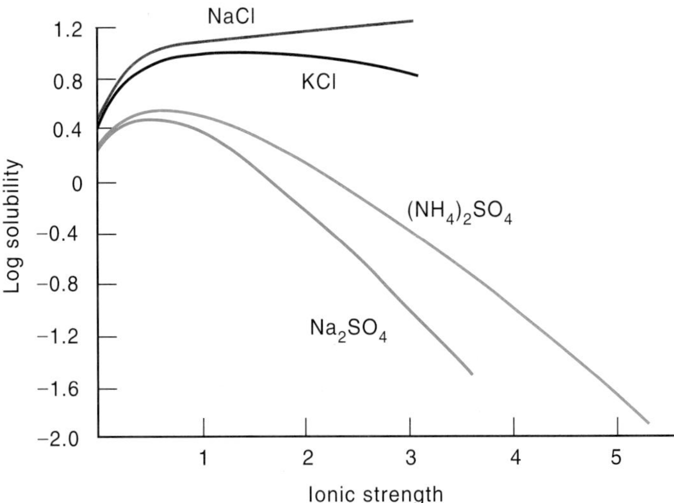

acidic and basic side-chain groups also depends on the solution pH. In fact, the acidic and basic groups within the protein can be titrated just like free amino acids (see chapter 3) to determine their number and their pK_a values. In chapter 3 we saw that the isoelectric point for β-lactoglobulin occurs at a pH of about 5.2 (see fig. 3.10). Figure 7.1 shows that β-lactoglobulin has a minimum in solubility at the same pH. Here the solubility of the protein is measured as a function of both pH and salt (NaCl) concentration. The decrease in solubility at the isoelectric pH reflects the fact that the individual protein molecules, which would all have a similar charge at pH values away from their isoelectric points, cease to repel each other. Instead, they coalesce into insoluble aggregates.

Salting In and Salting Out. Proteins also show a variation in solubility that depends on the concentration of salts in the solution. These frequently complex effects may involve specific interactions between charged side chains and solution ions or, particularly at high salt concentrations, may reflect more comprehensive changes in the solvent properties.

Figure 7.1 illustrates the effect of salt concentration on the solubility of β-lactoglobulin. Most globulins are sparingly soluble in pure water. The effect of salts such as sodium chloride on increasing the solubility of globulins is often referred to as salting in. The salting-in effect is related to the nonspecific effect the salt has on increasing the ionic strength of the solution. The higher the ionic strength, the smaller are the interactions between charged groups on the same or different proteins (see box 7A).

The effects of various salts on the solubility of hemoglobin at pH 7 are illustrated in figure 7.2. All of these salts produce the salting-in effect with this protein; two of them, sodium sulfate and ammonium sulfate, also produce a greatly decreased solubility of the protein at high salt concentrations. This result is called salting out and occurs with salts that effectively compete with the protein for available water molecules. In this case the protein molecules tend to associate with each other because at high salt concentrations, protein–protein interactions become energetically more favorable than protein–solvent interactions. Each protein has a characteristic salting-out point, and we can exploit this fact to make protein separations in crude extracts.

In the next section we will consider procedures for characterizing the size and shape of the protein in solution. For this purpose it is important to choose conditions that preserve the normal native state of the protein without undue aggregation.

Several Methods Are Available for Determination of Gross Size and Shape

Several methods are available for determining the size and shape of protein molecules in solution. Most of these give molecular weights that are accurate to a few percent. The linear dimensions of the protein are never directly given by solution measurements; rather, we obtain a parameter, the frictional coefficient, that measures the effective size. From this we may calculate the dimensions of the protein, assuming that it has a particular shape—say, that of a rod, a random coil, or a sphere. Many fibrous proteins are rodlike; globular proteins are often approximately spherical; and denatured proteins often have the structure of a random coil.

Sedimentation Rate Is a Function of Size and Shape. Information concerning the molecular weight of a protein can be obtained by observing its behavior in an intense centrifugal field. To get a qualitative understanding of how this method works, we must first

The Meaning of Ionic Strength and Its Effect on Charge-Charge Interactions in Proteins

The ionic strength μ is a measure of the effective salt strength. It is defined by the equation

$$\mu = \frac{1}{2}\sum_i M_i Z_i^2$$

where M is the molarity and Z is the charge of the ion. For a univalent salt at low concentration, μ is approximately equal to the molarity. Increasing the ionic strength decreases the "sphere of influence" of each charged site on the protein. The effective sphere of influence of a charge in solution is approximated by the Debye length $1/b$, where b is calculated from the expression

$$b^2 = \frac{4\pi e^2}{\epsilon kT}\sum_i M_i Z_i^2$$

In this equation, e is the charge of the electron, k is the Boltzmann constant, T is the absolute temperature, and ϵ is the dielectric constant of the medium (a measure of the charge-shielding effect of the medium).

The last term in the expression for b^2,

$$\sum_i M_i Z_i^2$$

is twice the ionic strength μ. Thus the Debye length $1/b$ is proportional to $1/\sqrt{\mu}$. In a 1 M aqueous solution of sodium chloride at 25°C, $1/b = 3.1$ Å. In a 0.01 M solution, $1/b = 31$ Å. As described earlier, many globular proteins precipitate at very low ionic strengths or in pure water. This happens because oppositely charged sites on different proteins are able to interact favorably, leading to an electrostatic complex. When this complex formation is extended between many protein molecules, it can lead to protein precipitation. Increasing the ionic strength tends to break up this complex because it results in a decrease of the Debye length.

recognize that protein molecules are generally slightly denser than water. However, the molecules in a protein solution seldom settle out in the earth's gravitational field (1 × g) because they are constantly being stirred up by collisions with surrounding solvent molecules. Nevertheless, protein molecules in solution can be made to settle if they are subjected to very high centrifugal force fields (~100,000 × g), such as can be attained in an ultracentrifuge (fig. 7.3).

The protein molecules slowly migrate toward the bottom of the centrifuge tube at a rate that is proportional to their molecular weight. The rate of sedimentation may be recorded by optical methods (see fig. 7.3) that do not interfere with the operation of the centrifuge. From this rate the sedimentation constant, or sedimentation coefficient, is determined. This constant equals the rate at which a molecule sediments, divided by the gravitational field (angular acceleration in a spinning rotor), and is defined by the equation

$$S = \frac{dx/dt}{\omega^2 x} \qquad (1)$$

where dx/dt is the rate at which the particle travels at distance x from the center of rotation, ω is the angular velocity of the rotor in radians per second (hence $\omega^2 x$ is the angular acceleration), and t is the time of centrifugation in seconds. The sedimentation constant is usually given in Svedberg units (S) where one S = 10^{-13} sec.

Figure 7.3

Apparatus for analytical ultracentrifugation. (*a*) The centrifuge rotor and method of making optical measurements. (*b*) The optical recordings as a function of centrifugation time. As the light-absorbing molecule sediments, the solution becomes transparent.

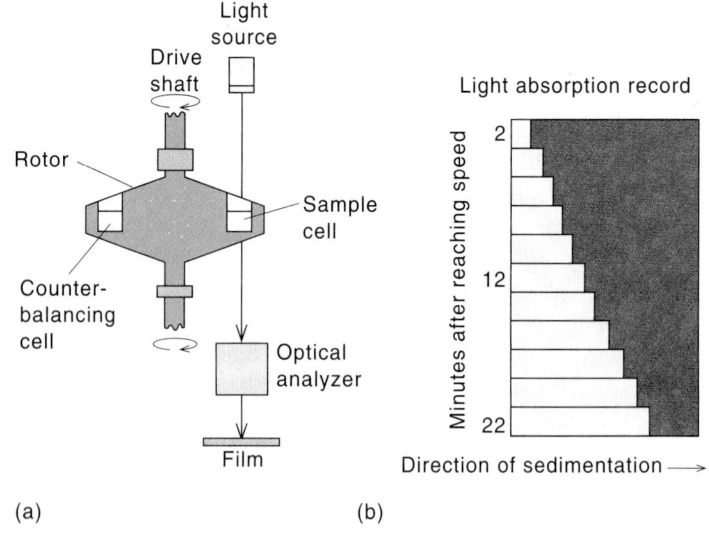

The molecular weight cannot be calculated directly from the sedimentation constant without further information; this is because the sedimentation constant is proportional to the molecular weight and inversely proportional to the frictional coefficient (f) of the protein. The coefficient f is bigger for large proteins, and, for proteins of the same molecular weight, it is larger for elongated, rodlike molecules. Another complication in calculating the molecular weight from the sedimentation constant is that the sedimentation rate is reduced by the buoyancy factor, $1 - \bar{v}_p\rho_s$, which takes into account the density difference between solvent (ρ_s) and the volume of water displaced per gram of protein ($\bar{v}_p$).* The equation that relates S, M, and f is

$$S = \frac{M(1 - \bar{v}_p\rho_s)}{Nf} \tag{2}$$

where N is Avogadro's number. Thus in order to estimate the molecular weight from the sedimentation constant we must have a means of determining $\bar{v}_p$ and f.

For most proteins, $\bar{v}_p$ is about 0.75 cc/g, so its value does not present much of a problem. If sufficient protein is available, it is an easy constant to measure. Even if the constant cannot be directly measured, not more than 15 percent error can result from using the average value.

The frictional coefficient varies over a wide range, and it must usually be determined. It can be calculated from the diffusion constant, D, with the help of the equation

$$D = \frac{RT}{Nf} \quad \text{or} \quad f = \frac{RT}{ND} \tag{3}$$

where R is the gas constant and T is the absolute temperature. Substituting this value for f in the expression for the sedimentation constant leads us to the equation

$$S = \frac{M_r \cdot D(1 - \bar{v}_p\rho_s)}{RT} \tag{4}$$

Transposing, we get

$$M_r = \frac{RTS}{D(1 - \bar{v}_p\rho_s)} \tag{5}$$

The diffusion constant needed to use equation (5) for molecular-weight calculations is usually determined with the help of Fick's first law of diffusion:

$$\frac{dn}{dt} = -DA\left(\frac{dc}{dx}\right)_t \tag{6}$$

This equation states that the amount dn of a substance crossing a given area A in time dt is proportional to the concentration gradient dc/dx across that area. The diffusion constant D, which is related to molecular weight and shape, is the proportionality constant. It can be measured by observing the spread of an initially sharp boundary between the protein solution and a solvent as the protein diffuses into the solvent layer (fig. 7.4). Once the diffusion constant is determined, the information is combined with the sedimentation data, and the molecular weight of the protein is calculated from equation (5). Some data obtained in this way for various proteins are presented in table 7.1.

*The symbol $\bar{v}_p$ is often called the partial specific volume of the protein; it is usually given in units of cc/g.

Figure 7.4

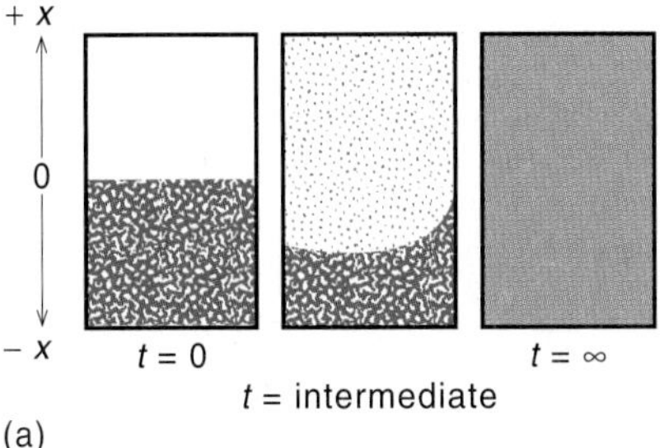

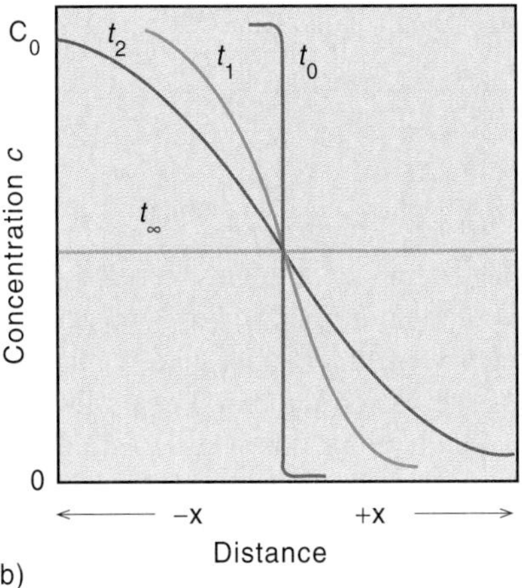

Measurement of diffusion. (*a*) A cell with a removable partition is filled on one side with pure solvent and on the other side with solvent plus solute (protein). The partition is removed, and the movement of the solute across the boundary defined by the partition is measured over time. (*b*) Graph showing concentration of solute, *c*, as a function of distance from the partition at various times. (Adapted from C. Tanford, *Physical Chemistry of Macromolecules*, New York, Wiley, 1961.)

Sometimes the technique of equilibrium sedimentation is used to measure molecular weight. In this case neither the sedimentation constant nor the diffusion constant is directly measured. Instead, the ultracentrifuge is maintained at a relatively low constant speed until the distribution of the protein molecules becomes constant. At this time the downward movement due to sedimentation is exactly counterbalanced by the upward movement due to diffusion. The equation for calculating the molecular weight from sedimentation equilibrium is

$$M_r = \frac{2\,RT \ln c}{\omega^2 x^2(1 - \bar{v}_p\rho_s)} \tag{7}$$

Equilibrium sedimentation is a most rigorous method for molecular-weight determination, precisely because it is an equilibrium method. But the technique is more difficult and time-consuming than those previously described, and also it is not effective when other macromolecules are present. Thus its use is usually

Table 7.1
Physical Constants of Some Proteins

Protein	Molecular Weight	Diffusion Constant ($D \times 10^7$)	Sedimentation Constant (S)	pI (Isoelectric)
Cytochrome *c* (bovine heart)	13,370	11.4	1.17	10.6
Myoglobin (horse heart)	16,900	11.3	2.04	7.0
Chymotrypsinogen (bovine pancreas)	23,240	9.5	2.54	9.5
β-Lactoglobulin (goat milk)	37,100	7.5	2.9	5.2
Serum albumin (human)	68,500	6.1	4.6	4.9
Hemoglobin (human)	64,500	6.9	4.5	6.9
Catalase (horse liver)	247,500	4.1	11.3	5.6
Urease (jack bean)	482,700	3.46	18.6	5.1
Fibrinogen (human)	339,700	1.98	7.6	5.5
Myosin (cod)	524,800	1.10	6.4	—
Tobacco mosaic virus	40,590,000	0.46	198	—

restricted to measurements on highly purified proteins in which a species of only one molecular weight is present.

Frequently, sedimentation analysis is carried out in linear gradients of sucrose or glycerol. First, a linear gradient is created in the centrifuge tube, with the greatest density at the bottom of the tube. A small volume of the protein solution is layered on top of the tube, and the tube is spun for the desired length of time. The protein travels down the tube at a rate roughly proportional to its sedimentation constant. This arrangement for measuring the sedimentation constant has the advantage that the gradient stabilizes the solution so that the centrifuge may be stopped at any time without disturbing the protein distribution. The solution can be carefully removed from the tube by punching a hole in the bottom and collecting fractions in separate tubes. This technique is used both for sedimentation-constant estimation and for preparative purposes, when the aim is to purify the desired protein from a mixture.

When using gradients, the sedimentation constant is measured only approximately, and the molecular weight is estimated by using a protein "marker" of known sedimentation constant in the same tube. For a large number of protein molecules the sedimentation constant is roughly proportional to the two-thirds power of the molecular weight, giving the relation

$$\frac{S(\text{of unknown})}{S(\text{of standard})} = \left[\frac{M_r(\text{of unknown})}{M_r(\text{of standard})}\right]^{2/3} \tag{8}$$

This is not a rigorous relationship, and the approximation is best for spherical molecules. Most globular proteins give a reasonable fit; fibrous proteins, which are often highly asymmetrical in shape, are likely to give a poor fit.

Gel-exclusion Chromatography Gives a Measure of Size. Gel-exclusion chromatography is used for size estimations as well as protein purifications. This popular technique exploits the availability of both natural polysaccharides and synthetic polymers (polydextrans) that can be formed into beads with varying pore sizes, depending on the extent of cross-linking between polymer chains. For example, suppose you have some polysaccharide beads with average maximum pore sizes of 30 Å and you add these to a mixed solution of proteins of molecular weights 10,000 and 50,000. The final volume of the mixture equals the sum of the volume required to hydrate and fill the beads plus the remaining excess solution volume that fills the spaces between the beads. That is, the total solution volume $v_T = v_i + v_o$, where v_i and v_o represent, respectively, the solution volumes inside and outside the beads. For fractionation purposes the critical point is that a low-molecular-weight protein is sufficiently small that it can readily penetrate into the beads; as a result it is uniformly distributed over the total volume v_T. A large protein, by contrast, cannot penetrate the beads and so is concentrated in the solution volume v_o.

By simply filtering out the beads we can achieve some separation between the molecules based on their molecular weight. By repeating this process many times, eventually the concentration of the smaller protein in v_o is vanishingly small. However, it is simpler and more efficient to construct a column of the beads and pass the protein solution through it in a continuous fashion. In this case the smaller molecule penetrates the entire column volume v_T, while the larger molecule, which is restricted to v_o, passes through the column at a more rapid rate, thus effecting a separation on the basis of molecular weight (fig. 7.5).

Although this process has been described in terms of molecules either penetrating or not penetrating the beads, by careful

Figure 7.5

Polydextran column showing separation of small and large molecules. The column material is immersed in solvent, which penetrates the gel particles. A separation is initiated by layering a small sample containing different-size proteins on the top of the column. This sample is pushed through the column by opening the stopcock at the bottom and adding additional solvent at the top to keep up with the flow. As shown, the small protein molecules can penetrate the gel particles but the big ones cannot. Therefore the big proteins move through the column much more rapidly, and a separation of the two proteins results.

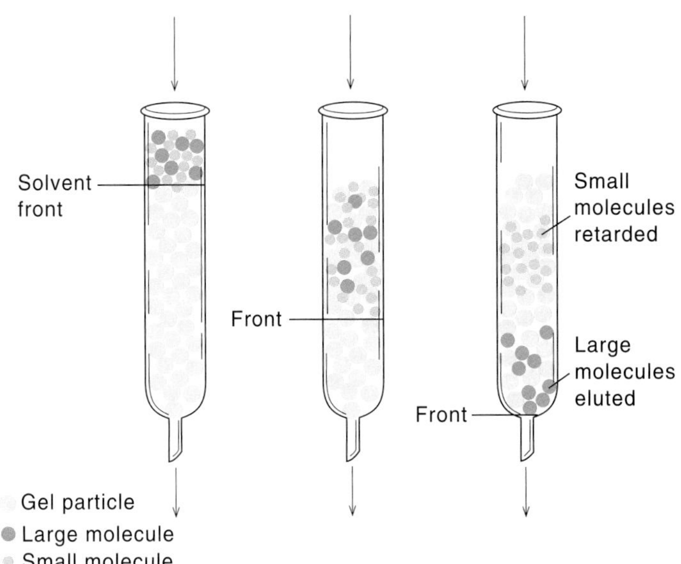

Figure 7.6

Plot of elution volume (V_e) versus the logarithm of the protein molecular weight. A cross-linked dextran (Sephadex G-2000 at pH 7.5) was used. The farther apart two proteins are on this plot, the easier it is to separate them by gel-exclusion chromatography. (Source: Adapted from P. Andrews, *Biochemical J.* 96:595, 1965.)

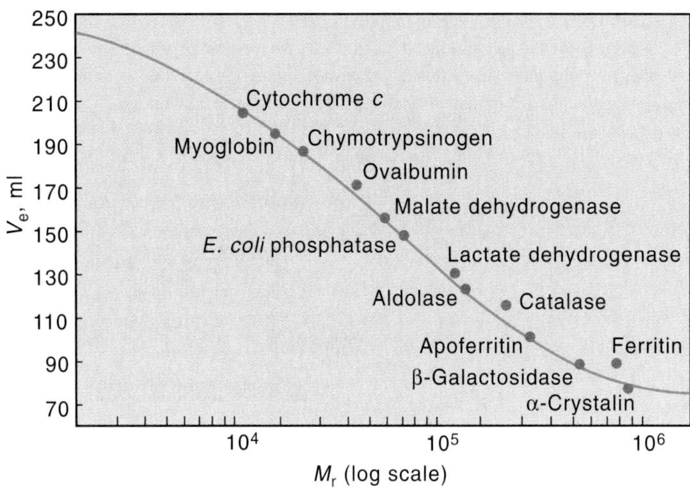

selection of the bead pore size we can create a situation in which the various molecules penetrate the beads to varying extents and, consequently, migrate through the column at varying rates. By observing the elution pattern of a mixture of proteins of known molecular weights, the behavior of the column can be standardized so that the molecular weight of an unknown protein can be estimated (fig. 7.6).

Electrophoretic Methods Are the Best Way to Analyze Mixtures

We turn now from methods in which the primary object is to determine the size and shape to electrophoretic analysis, where this goal is often secondary to determining the number of components in a mixture. The transport of particles by an electrical field is termed electrophoresis. Electrophoresis is one of the most commonly used techniques in biochemistry. The reason is, in part, that the apparatus required is quite simple and serves multiple purposes. Electrophoresis can be used to assess the isoelectric point and the molecular weight of a protein. It is also the favored approach for assessing the number of proteins in a mixture, because of the high resolving power it offers.

Electrophoresis is very much like sedimentation, since in both cases a force gradient leads to protein transport in the direction of the force. In the case of sedimentation the force is gravity, so the rate of migration depends on the effective mass of the particle, i.e., $M_r(1 - \overline{v}_p\rho_s)$. In electrophoresis the force is the electrical potential, *E*, so the rate of migration depends on the net charge on the molecule, rather than its mass.

The characteristic quantity measured in electrophoresis is called the electrophoretic mobility (μ). This is defined in terms of the ratio of the velocity of the particle (V) to the imposed electrical field (E):*

$$\mu = \frac{V}{E} = \frac{Z}{f} \tag{9}$$

That is, electrophoretic mobility is equal to the net charge on the molecule (Z) divided by its frictional coefficient (f). The frictional coefficient in this equation is closely related to the frictional coefficient that appeared in the equations for the sedimentation constant. The two quantities may, however, have quite different values in high electric fields, where the molecules tend to orient in the direction of the electric field.

Although electrophoresis can be done in pure aqueous solution, it is almost always done in an aqueous solution supported by a gel system. The gel is a loosely cross-linked network that functions to stabilize the protein boundaries between the protein and the solvent, both during and after electrophoresis, so that they may be stained or otherwise manipulated. By judicious manipulation (to be explained shortly) the gel may also be used to purify a protein in a mixture.

Electrophoretic separations are typically carried out on gels composed of the polysaccharide agarose or polyacrylamide (fig. 7.7). The percentage of gel used is gauged according to the size of the proteins being separated. Increasing the gel concentration lowers the mobility of the protein in a given electric field. For the finest separations, gradient gels are made with a continuous

*V is usually given in cm/sec, and E is usually given in volts/cm.

Figure 7.7

Gel electrophoresis for analyzing and sizing proteins. (*a*) Apparatus for slab-gel electrophoresis. Samples are layered in the little slots cut in the top of the gel slab. Buffer is carefully layered over the samples, and a voltage is applied to the gel for a period of usually 1 to 4 h. (*b*) After this time the proteins have moved into the gel at a distance proportional to their electrophoretic mobility. The pattern shown indicates that different samples were layered in each slot. (*c*) Results obtained when a mixture of proteins was layered at the top of the gel in phosphate buffer, pH 7.2, containing 0.2% SDS. After electrophoresis the gel was removed from the apparatus and stained with Coomassie blue. The protein and its molecular weight are indicated next to each of the stained bands. (*d*) The logarithm of the molecular weight against the mobility (distance traveled) shows an approximately linear relationship. (Source: Data of K. Weber and M. Osborn.)

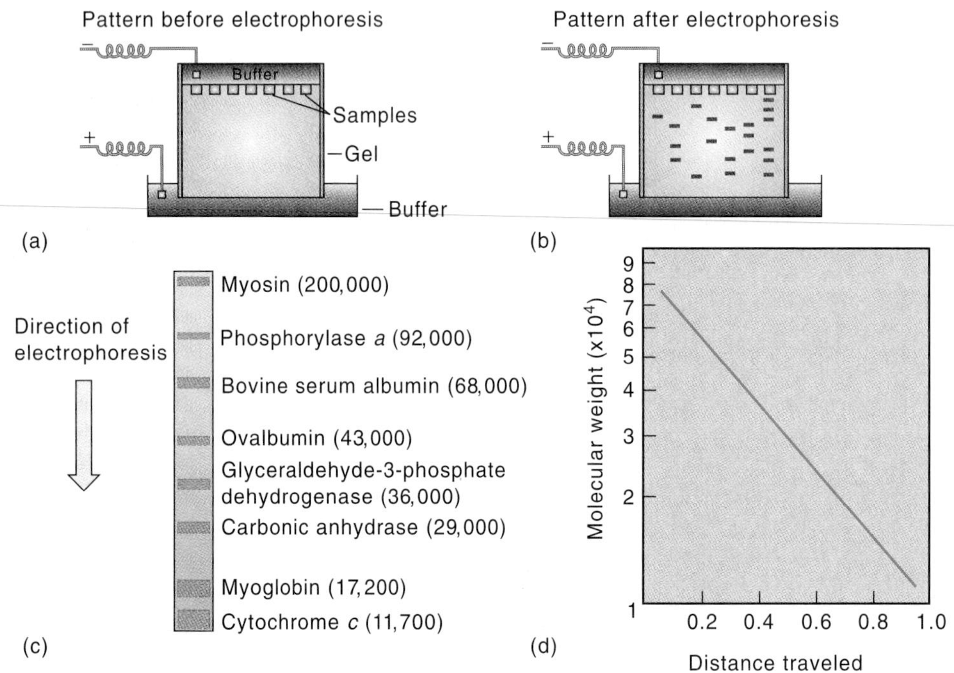

increase in gel percentage along the length of the slab. This approach leads to optimum separation of components in a mixture and the sharpest protein solvent boundaries. At the completion of a run, a dye that stains the proteins can be added to the gel. The dye establishes the locations of the protein bands.

A widely used variation of polyacrylamide gel electrophoresis involves the use of a stacking gel. The basic idea here is to increase the resolution in the gel by first compressing the protein mixture into a very narrow starting zone. This aim is achieved by forming the gel in two layers. Generally, the upper, or "stacking," gel is less extensively polymerized than the lower, or "resolving," gel, so there is a smaller pore size in the resolving gel. More important, the pH of the solution in the stacking gel is adjusted so that the mobilities of proteins to be separated are higher in the stacking gel than in the resolving gel. As a consequence of this gel arrangement a relatively dilute protein mixture moves rapidly and easily through the stacking gel but accumulates at the interface with the resolving gel because of the difference in pH and the smaller pore size.

The mixture of proteins to be characterized is first completely denatured by the addition of sodium dodecyl sulfate (a detergent) and mercaptoethanol and by a brief heating step. Denaturation is caused by the association of the apolar tails of the SDS molecules with protein hydrophobic groups. Any cystine disulfide bonds are cleaved by a disulfide interchange reaction with mercaptoethanol.

The resulting unfolded polypeptide chains have relatively large numbers of SDS molecules bound to them. The success of the technique for molecular-weight estimation depends on two facts: (1) that each bound SDS molecule contributes two negative charges to the denatured protein complex, so that the charge of the protein in its native state is effectively masked by the more numerous charged groups of the associated detergent molecules, and (2) that the total number of detergent molecules bound is proportional to the polypeptide-chain length or, equivalently, the protein's molecular weight. As a result, the SDS-denatured protein molecules acquire net negative charges that are approximately proportional to their molecular weights.

Typically, the behavior of an SDS gel electrophoresis system is calibrated by concurrently running standards of known molecular weight and then comparing the migration behavior of the unknowns with the standards. Although SDS gel electrophoresis gives only an approximate measure of protein molecular weight, with its accuracy depending on experimental conditions, the procedure offers both experimental simplicity and high resolving power. These advantages have led to its widespread use for the characterization of a variety of protein mixtures. An SDS gel electrophoresis pattern for a mixture of known proteins is illustrated in figure 7.6*c*.

A fancy extension of the electrophoretic method of separation involves combining isoelectric focusing (see box 3C) with SDS gel electrophoresis to produce a two-dimensional electrophoretogram. This technique is most valuable for the analysis of very complex mixtures. First, the sample is run in a one-dimensional pH gradient gel (isoelectric focusing). The resulting narrow strip of gel, containing the partially separated mixture of proteins, is placed alongside a square slab of SDS gel. An electrical field is imposed so that the sample moves at right angles to its motion in the first gel. Figure 7.8 shows the separation of total *E. coli* protein into more than 1000 different components.

Figure 7.8

Two-dimensional SDS isoelectric-focusing gel electrophoresis. First the sample is run in a one-dimensional pH gradient, partially separating the sample along a strip of gel. Then the strip of gel containing the sample is placed alongside an SDS gel, and the proteins are permitted to further separate by moving in the second dimension, at right angles to the first separation. Sample shown is total *E. coli* protein; individual proteins are detected by autogradiography. (Source: Photograph provided by Patrick O'Farrell. See O'Farrell, in *J. Biol. Chem.* 250:4007, 1975. American Society for Biochemistry & Molecular Biology)

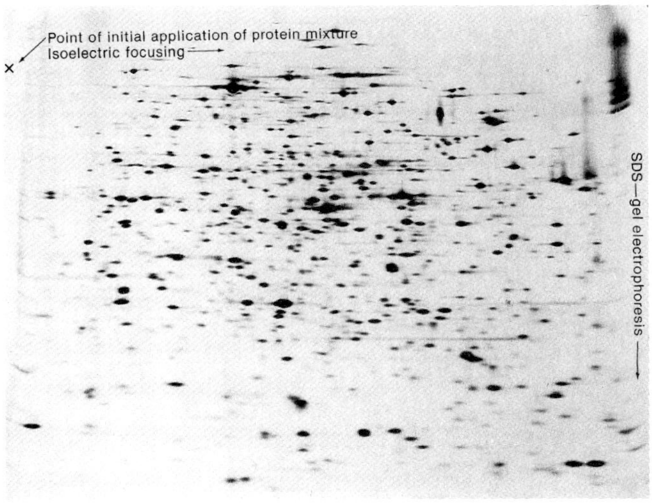

Methods of Protein Purification

Before we can fully characterize a protein, we must purify it from a natural source. Once the decision has been made to purify a particular protein, several factors must be weighed. For example, how much material is needed? What level of purity is required? The starting material should be readily available and should contain the desired protein in relative abundance. If the protein is part of a larger structure, such as the nucleus, the mitochondria, or the ribosome, then it is advisable to isolate the large structure first from a crude cell extract.

Purification must usually be performed in a series of steps, using different techniques at each step. Some purification techniques are more useful when handling large amounts of material, whereas others work best on small amounts. A purification procedure is arranged so that the techniques that are best for work with large amounts are used during early steps in the overall purification. The suitability of each purification step is evaluated in terms of the amount of purification achieved by that step and the percent recovery of the desired protein.

Combining techniques introduces new considerations and new problems. If each of two purification techniques gives a tenfold enrichment for the desired protein when executed independently on a crude extract, this does not mean that they will give 100-fold enrichment when combined. In general, they will give somewhat less. As a rule, purification techniques that combine most effectively usually are based on different properties of the protein. For example, a technique based on size fractionation is more effectively combined with a technique based on negative charge than with another technique based on size fractionation.

Throughout the purification we must have a convenient means of assaying for the desired proteins so that we can know the extent to which it is being enriched relative to the other proteins in the starting material. In addition, a major concern in protein purification is stability. Once the protein is removed from its normal habitat, it becomes susceptible to a variety of denaturation and degradation reactions. Specific inhibitors are sometimes added to minimize attack by proteases on the desired protein. During purification it is usual to carry out all operations at 5°C or below. This temperature control minimizes protease degradation problems and decreases the chances of denaturation.

In their natural habitat, proteins are usually surrounded by other proteins and organic factors. When these are removed or diluted, as during purification, the protein becomes surrounded by water on all sides. Proteins react differently to a pure aqueous environment; many are destabilized and rapidly denatured. A common remedial measure is to add 5% to 20% glycerol to the purification buffer. The organic surface of the glycerol is believed to simulate the environment of the protein in the intact cell. Two other ingredients that are most frequently added to purification buffers are mercaptoethanol and ethylenediamine tetraacetate (EDTA). The mercaptoethanol inhibits the oxidation of protein —SH groups, and the EDTA chelates divalent cations. The latter, even in trace amounts, can lead to aggregation problems or activate degradative enzymes.

In the remainder of this chapter we discuss the merits of different types of purification. We conclude by describing the purification of two different proteins.

Table 7.2
Sedimentation Conditions for Different Cellular Fractions

Fraction Sedimented	Centrifugal Force (×g)	Time (min)
Cells (eukaryotic)	1,000	5
Chloroplasts; cell membranes; nuclei	4,000	10
Mitochondria; bacteria cells	15,000	20
Lysosomes; bacterial membranes	30,000	30
Ribosomes	100,000	180

Differential Centrifugation Subdivides Crude Extracts into Two or More Fractions

A typical crude broken cell preparation contains broken cell membranes, cellular organelles, and a large number of soluble proteins, all dispersed in an aqueous buffered solution. The membranes and the organelles can usually be separated from one another and from the soluble proteins by differential centrifugation. Differential centrifugation divides a sample into two fractions: the pelleted fraction, or sediment, and the supernatant fraction, that is, the fraction that does not sediment. The two fractions may then be separated by decantation.

According to its purpose, differential centrifugation requires different speeds and different times of centrifugation (table 7.2). For example, if the protein of interest were in the mitochondrial fraction, the crude lysate should be centrifuged first at 4000 × g for 10 min to remove cell membranes, nuclei, and (in the case of plant material) chloroplasts. The supernatant from this step contains, among other elements, the mitochondria and would be decanted and recentrifuged at 15,000 × g for 20 min to obtain a sediment primarily containing mitochrondia. If ribosomes instead of mitochondria were the goal, then the crude lysate should be centrifuged at 30,000 × g for 30 min and the resulting supernatant decanted and centrifuged at 100,000 × g for 180 min to obtain a ribosomal sediment. If the soluble protein fraction were the goal, then the entire lysate would be centrifuged at 100,000 × g for 180 min and the resulting supernatant, containing the soluble protein, would be carefully decanted for further processing.

Differential Precipitation Is Based on Solubility Differences

We mentioned earlier that every protein has a characteristic salting-out point, which is reached by altering the concentration of salts such as ammonium sulfate, $(NH_4)_2SO_4$. We can use this fact to carry out a differential precipitation: Salting out is a relatively crude procedure, usually resulting in no more than a two- to threefold purification, but it is easy to do on any volume of material, and even a two- to threefold enrichment substantially decreases the bulk of an initial extract, making it more manageable in later steps.

Typically, the desired protein precipitates over a range of salt concentrations. If it precipitates in the range of 20–30% by weight, then we would first add ammonium sulfate to a concentration slightly below 20% and then centrifuge to remove by sedimentation any proteins that precipitate in the 0–20% range. To the supernatant from this centrifugation we would then add more ammonium sulfate, to 30%. Centrifugation at this point brings down the desired protein, as well as other proteins that precipitate in this range of salt concentrations. The supernatant is discarded, and the sediment is saved for further purification.

Column Procedures Are the Most Versatile Purification Methods

With two bulk steps behind us (differential centrifugation and ammonium sulfate precipitation), we are generally ready for a more sophisticated column step. Column procedures are the most important and the most varied of purification steps. Usually, a glass cylinder with a large opening at the top and a capillary opening at the bottom is used to support the column material. The cylinder is packed with the hydrated column material to be used for the fractionation; this material could be a cross-linked polydextran (see fig. 7.5), a resin, or finely divided cellulose fibers. The column material usually contains functional groups with an affinity for proteins.

Preparative Gel-exclusion Chromatography. In gel-exclusion chromatography a cross-linked dextran without any special attached functional groups is used for the column substrate. Large molecules flow more rapidly through this type of column than small ones, for the reasons explained earlier (see "Gel-exclusion Chromatography Gives a Measure of Size"). The dextrans have different degrees of cross-linking, making them effective over different size ranges. The dextran is first rinsed with a buffer. Then the protein sample, in a volume of less than 1/20 the column volume, is applied to the top of the column. Once all of the protein solution is in contact with the column, further buffer is passed through the column. The eluant appearing at the lower end of the column is collected, usually with the help of a fraction collector. The collector is equipped with an automatic device that collects the same amount of eluant in each of a series of test tubes (fig. 7.9). The various fractions are analyzed for both total protein concentration and the desired protein. Fractions containing an appreciable amount of the desired protein are combined (pooled) for further purification.

Column Chromatography with Protein Binding. In most column procedures, unlike gel-exclusion chromatography, the protein is first bound to the column material; a change in the elution buffer then leads to a differential elution of the bound protein.

Ion-exchange Chromatography. As described previously (chapter 4), amino acid derivatives can be separated in columns. Ion-exchange chromatography also can be used to separate proteins, but significant differences arise as a result of the larger size of pro-

Figure 7.9

Collecting fractions during column chromatography. Column material and elution procedure are chosen to effect optimal separation of the desired protein.

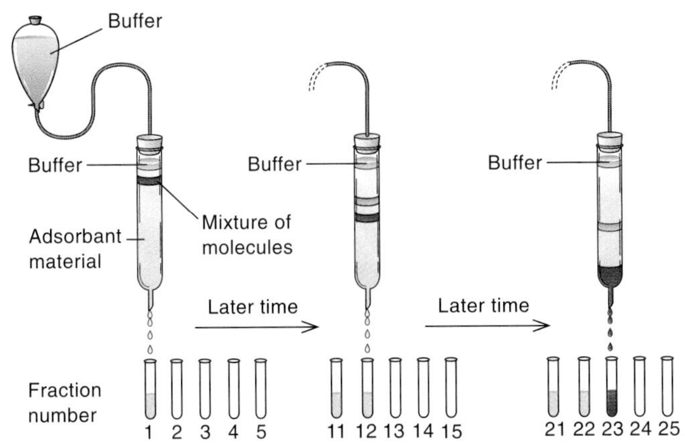

Matrix	Functional Groups on Column
Phosphocellulose (PC)	$-PO_3^-$
Carboxymethyl cellulose (CMC)	$-CH_3-COO^-$
Diethylaminoethyl cellulose (DEAE)	$-(CH_2)_2-\overset{+}{N}H(CH_2-CH_3)_2$

Table 7.3 Some Column Materials for Ion-exchange Chromatography of Proteins

teins. First, cross-linked resins are rarely used for protein separations because proteins are too large to penetrate the resin beads. Instead, finely divided celluloses containing either positively or negatively charged groups are most commonly used to make such columns (table 7.3). Second, at a given salt concentration, protein binding tends to be an all-or-nothing phenomenon, rather than an equilibrium phenomenon as it is with amino acids or resins. Consequently, the only way to achieve separations of proteins on charged cellulose columns is by changing the salt concentration. This is done in either a continuous manner (gradient elution) or a discontinuous manner (step elution).

Proteins differ enormously in their affinity for positively or negatively charged columns. This affinity is proportional to the salt concentration required to release the protein from the column material so that it will start flowing down the column. The affinity between column and protein is strongly influenced by pH. For this reason the pH of the buffers must be carefully controlled during a column step. Typically, a column is loaded with protein solution at a low ionic strength so that most of the protein binds to the column. Once the sample is loaded and rinsed, elution is initiated by gradually increasing the salt concentration of the elution buffer. Proteins are eluted in the order of increasing affinity. Those with the greatest affinity come off last. Collections, assays, and pooling procedures are similar to those used in gel-exclusion chromatography. In addition to the proteins present in various fractions, the salt concentration in each fraction is recorded.

Additional Adsorbants Used for Column Chromatography. Finely divided celluloses may also be used in the column in conjunction with attached hydrophobic groups such as octyl alcohol. In this case, proteins with exposed hydrophobic centers bind to the column with varying affinities. These proteins may be eluted in order of decreasing affinities for the column by increasing the level of free octyl alcohol in the eluting buffer.

A finely divided calcium phosphate gel known as hydroxyl apatite is used for a wide variety of separations. The eluting buffer usually consists of increasing concentrations of phosphate buffer. Many other types of adsorbants are used on occasion in column chromatography.

Affinity Chromatography. In addition to the column techniques already described for the isolation and characterization of proteins, several other methods have been developed that exploit specific binding properties of a given protein. Methods of this sort are generically referred to as affinity chromatography.

For example, many enzymes reversibly bind organic cofactor molecules, such as adenosine triphosphate or pyridine nucleotides, in order to catalyze the chemical reactions of their substrates. Often, the separation of such enzymes from other proteins can be readily achieved by preparing a chromatographic column that is first chemically reacted with a suitable cofactor derivative. As the mixture passes through the column, those protein molecules having specific binding affinities for the cofactor bind to the column, while other proteins pass through. The cofactor binding protein can subsequently be eluted with a solution containing the same soluble cofactor that is covalently attached to the column.

The power of affinity chromatography techniques lies in the great variety of specific interactions that characterize the functional properties of protein molecules. The enrichments obtainable in single steps by affinity chromatography sometimes exceed 1000-fold, a result testifying to the high selectivity of the method.

High-performance Liquid Chromatography. Thus far we have described three types of column chromatography for purification and characterization of proteins: gel-exclusion chromatography, ion-exchange chromatography, and affinity chromatography. High-performance liquid chromatography (HPLC) is not so much a new type of chromatography as a new way of looking at old chromatographic techniques. The same principles are involved, but the column support materials usually consist of more finely divided particles made of physically stronger materials, which can withstand high pressures (5000–10,000 psi) without changing their structure. The column apparatus itself also must be designed to withstand high pressures. Finely divided column materials lead to slower flow rates, but this factor can be more than compensated for by applying high hydrostatic pressures. As a rule, much better separations are achieved in a much shorter time with the proper applications of HPLC.

Table 7.4
Outline of Purification of UMP Synthase from Ehrlich Ascites Carcinoma

Fraction	Volume (ml)	Protein (mg)	OMPDase[a] Units[b]	OMPDase[a] Sp. act.[c]	OMPDase[a] Percent recovery	OPRTase[a] Units[b]	OPRTase[a] Sp. act.[c]	OPRTase[a] Percent recovery	OMPDase / OPRTase
1. Streptomycin fraction	1040	11,700	40.4	0.0034		20.5	0.0018		2.0
2. Dialyzed $(NH_4)_2SO_4$ fraction	144	311	24.3	0.0078	60	8.7	0.0028	42	2.8
3. Affinity column eluate (concentrated)	0.475	0.51	4.0	7.8[d]	10	0.35	0.69	3.3	11.4

[a]OMPDase = OMP decarboxylase; OPRTase = orotate PRTase.

[b]Units refer to total amount of enzyme activity.

[c]Specific activity refers to the units of enzyme activity divided by the total protein.

[d]This value represents a 2300-fold enrichment from fraction 1.

Electrophoretic Methods Are Used for Preparation and Analysis

Electrophoresis is used extensively to monitor protein purity during purification procedures. Typically, SDS gel electrophoresis and isoelectric focusing are used to resolve the desired protein and the contaminating proteins into discrete bands on a stained gel. Both of these procedures, which were explained in a previous section, can also be used in themselves as purification steps. Isoelectric focusing is especially useful in this regard, because it involves a principle that is unique among purification methods. SDS gel electrophoresis is used as a purification step only if there is an easy way of renaturing the denatured protein extracted from the gel. Both procedures are limited to the processing of small amounts. For this reason they are used at or near the final stages of purification, when the amounts of material being processed are relatively small.

We will now see how various purification steps are combined in actual protein purifications.

Purification of Specific Proteins Involves Combinations of Different Procedures

The following two examples of purification show how various techniques can be effectively combined to produce purified proteins with a minimum of effort and loss of activity.

The Purification of UMP Synthase from Mammalian Tumor Cells. The last two steps in the biosynthesis of the mononucleotide uridine 5′-monophosphate (UMP) are catalyzed by (1) orotate phosphoribosyltransferase (OPRTase) and (2) orotate 5′-monophosphate (OMP) decarboxylase.

Mary Ellen Jones and her colleagues set out to purify the enzyme or enzymes involved in these two reactions. Their main goal was to determine whether the two reactions are carried out by one protein or more than one. Their findings indicated that the two reactions were both catalyzed by the same enzyme, consisting of a single polypeptide chain. To demonstrate this fact, it was necessary to monitor both enzyme activities at each step in the purification and show that both activities copurified. For this purpose, Jones used specific enzyme assays for both enzyme activities. All fractions were assayed for both enzymatic activities at each stage of the purification.

The main data associated with the purification are summarized in table 7.4. This table indicates the total protein obtained in each step, the number of enzyme units* for each enzyme, and the ratio of enzyme units to total protein, called the specific activity. In the absence of enzyme inactivation the specific activity should be directly proportional to the enrichment. The percent recovery refers to the amount of enzyme activity in the indicated fraction, as compared with the amount present in fraction 1. This number is usually less than 100%. The apparent losses may reflect actual losses of enzyme during purification, or they may reflect inactivation (usually due to unknown causes) of the enzyme during purification.

The nine steps involved in the purification of UMP synthase from starting tissue are summarized in figure 7.10. All steps were carried out at 0–5°C. About 200 g of Ehrlich ascites cells, a mammalian tumor rich in the desired enzymes, was suspended in buffer and processed in a tissue homogenizer, which mechanically breaks down the tissue and the cell membranes (step 1). Then EDTA and an —SH reagent were added to this total cell lysate. Solid streptomycin sulfate was also added with stirring (step 2). Streptomycin sulfate aggregates nucleic acids so that they may be more easily removed by centrifugation. The resulting slurry was subjected to high-speed centrifugation, and the resulting super-

*Enzyme units are proportional to the amount of enzyme activity. The relationship between enzyme units and absolute amount of enzyme need not concern us here.

Figure 7.10

Outline of purification scheme for UMP synthase from Ehrlich ascites tumor cells of mice.

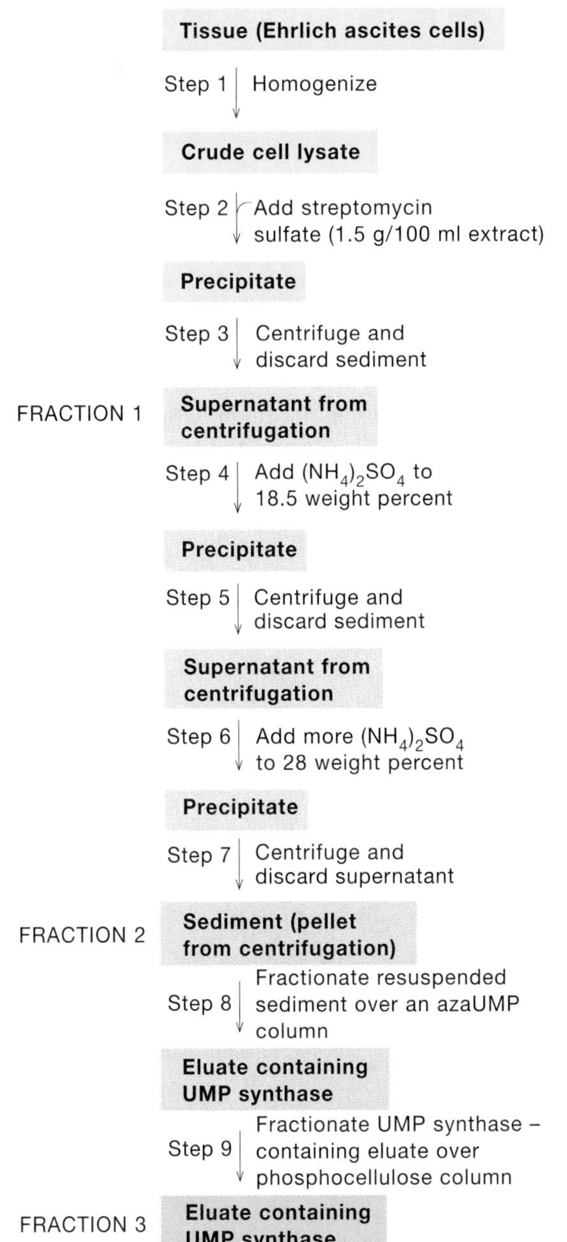

natant (table 7.4, fraction 1) was carefully decanted for further processing (step 3).

Preliminary experiments had shown that the desired enzymes were in the 18.5–28% $(NH_4)_2SO_4$ fraction. This knowledge served as the basis for the next three steps. First 239 g of solid $(NH_4)_2SO_4$ was added to the 1040 ml of supernatant (step 4). The resulting precipitate was removed by centrifugation (step 5). Then an additional 120 g of $(NH_4)_2SO_4$ was added to the supernatant (step 6). The resulting slurry was centrifuged. This time the supernatant was discarded after centrifugation, leaving the sediment containing the enzyme activity for further processing (step 7). The sediment was resuspended in a dilute buffer for column chromatography (table 7.4, fraction 2). Inspection of table 7.4 indicates only about a twofold increase in specific activity between fractions 1 and 2.

The main purification was achieved by two column steps, carried out in series. The first column was an affinity column containing an analog of UMP, 6-azauridine 5′-monophosphate (azaUMP), covalently attached to an agarose column support system. In dilute buffer, greater than 99% of the protein in fraction 2 is retained on this column. After thorough rinsing of the column with dilute buffer, 5×10^{-5} M azaUMP was added to the buffer. This addition resulted in the elution of the UMP synthase (step 8). Then the column eluant carrying the two enzyme activities associated with UMP synthase was resuspended in pure buffer and passed over a phosphocellulose column (step 9). Phosphocellulose was chosen because the negatively charged phosphate groups result in the retention of proteins by electrostatic attraction alone, but they also resemble the phosphate groups in the naturally occurring enzyme substrate, OMP. Thus the phosphocellulose column may be thought of as an ion-exchange column and an affinity column combined. Recall that in ordinary ion-exchange chromatography the protein, after column loading, is eluted by increasing the ionic strength with a simple inorganic salt. In this example, Jones used a more specific method to elute the enzyme, which involved adding 10^{-5} M azaUMP and 2×10^{-5} M OMP to the original loading buffer. The addition does not substantially increase the ionic strength of the buffer. Therefore the only phosphocellulose-bound proteins that are likely to be eluted by this treatment are those with an especially high affinity for either of these nucleotides (azaUMP and OMP). This fact should greatly favor selective elution of those enzymes that carry specific sites for binding these nucleotides. The two column steps together resulted in an enzyme preparation (table 7.4, fraction 3) that was approximately 2300-fold purified over the starting material, as measured by the increase in specific activity.

The final product (table 7.4, fraction 3) was examined by SDS gel electrophoresis and found to contain a single band with an estimated molecular weight of 51,000. The denaturing conditions of an SDS gel would be expected to dissociate a multisubunit protein. Hence the SDS gel result indicated that the enzyme contains one type of polypeptide chain, but it does not tell us whether the enzyme contains one or more of these chains. Sedimentation analysis on a sucrose density gradient in a nondenaturing buffer indicated a single band with an estimated molecular weight of about 50,000. These two results taken together demonstrate that the enzyme in its native state contains a single polypeptide chain. The purity of the enzyme was also confirmed by isoelectric focusing and two-dimensional electrophoresis with isoelectric focusing in the first direction followed by SDS gel electrophoresis in the second direction.

The conclusion drawn from these results was that both of the enzyme activities associated with UMP synthase—OMPDase and OPRTase—are contained in a single protein. The basis for the conclusion was that both enzyme activities are always present in the same fractions throughout the multistep purification. However, inspection of table 7.4 indicates a possible ob-

Table 7.5
Purification of the Lactose Carrier Protein

Fraction	Protein (mg)	Percent recovery (total protein)	Percent recovery (carrier protein)	Purification factor
1. Membrane fraction	12.5	100	100	1.0
2. Urea-extracted membrane	5.6	45	76	1.7
3. Urea/cholate-extracted membrane	2.6	21	61	2.9
4. Octylglucoside extract	0.4	3.2	38	12
5. DEAE column peak	0.056	0.4	14	35

jection to this interpretation. In the columns showing percent recovery, it can be seen that substantial amounts of enzyme activity are lost for both enzymes during purification but that considerably more activity is lost for the OPRTase. These losses could be due to actual loss of enzymes during purification or to some sort of inactivation of the enzyme sites. The preferential loss of OPRTase activity is emphasized by the last column in table 7.4, which gives the ratio of the two enzyme activities in the different fractions. Considered alone, these data could indicate that a separate catalytic unit of orotate PRTase is lost during purification. Jones thinks this is unlikely for two reasons: (1) the activity appears in no fractions other than with OMP decarboxylase during purification, and (2) the orotate PRTase activity is notably unstable. It is concluded that both enzyme activities exist at distinct sites on a single protein. The greater loss in activity of one enzyme activity over the other is attributed to a greater sensitivity of one reaction site over the other on the enzyme surface.

Purification of Lactose Carrier Protein from Escherichia coli Bacteria. The second purification procedure we will look at illustrates an unusual approach to the purification of a membrane-bound protein. The lactose carrier protein of *E. coli* is normally tightly bound to the plasma membrane. This protein is involved in the active transport of the disaccharide lactose across the cytoplasmic membrane. When lactose carrier protein is present, the intracellular concentration of lactose can achieve levels 1000-fold higher than those found in the external medium. Ron Kaback devised a simple yet elegant procedure for the purification of this protein.

Purification of the membrane-bound lactose carrier protein is a very different problem from the purification of the soluble OMP synthase. Both the approach to purification and the assays for the protein during purification are quite novel. The assay involves reconstituting a transport system with membranes that are free of lactose carrier protein, then adding the partially purified carrier protein and radioactively labeled lactose. The activity in this assay system is proportional to the transport of radioactive lactose across the membrane in the cell-free reconstituted system.

The results of the purification steps are tabulated in table 7.5, and the purification procedure is outlined in figure 7.11. In this procedure, advantage was taken of the fact that the carrier protein in its native state is firmly bound to the cytoplasmic membrane. Thus the first step consisted of isolating these membranes from the rest of the cell constituents. Starting from the membrane fraction, only 35-fold purification was required to achieve pure carrier protein. This rapid result was possible because a special strain of *E. coli,* containing about 100 times the normal carrier protein, was used as starting material for the purification. The high initial content was engineered by putting the carrier protein gene on a multicopy plasmid, which was then inserted into the cell—a procedure described in chapter 32.

Bacterial cells are much tougher than mammalian cells, requiring a more stringent procedure for cell disruption. In this case the cells were placed in a so-called French pressure cell and bled through an orifice from very high pressures (~10,000 psi) to atmospheric pressure (step 1). Under these conditions the cells literally explode, fragmenting their membranes and releasing the cytoplasmic contents. The membranes were pelleted by a brief centrifugation, leaving a supernatant containing DNA, ribosomes, and cytoplasmic protein, which was removed by decantation (step 2). The pelleted membrane fraction was next resuspended and extracted with 5 M urea and then re-extracted with 6% sodium cholate. The pellet obtained by high-speed centrifugation from these two extractions still contained most (61%) of the carrier protein in the more rapidly sedimenting membrane fraction (fraction 3), although 79% of the total membrane-bound protein was released by these treatments (steps 3 and 4).

At this point the carrier protein was released from a suspension of the membranes by addition of the hydrophobic reagent octylglucoside in the presence of *E. coli* phospholipid (step 5). It is believed that the octyl part of octylglucoside competes effectively with the membrane for binding to hydrophobic centers on the carrier protein. The *E. coli* phospholipid facilitates dissociation of the carrier protein from the membrane fraction. The solubilized octylglucoside-containing extract was fourfold enriched in carrier protein after a high-speed centrifugation to remove residual membrane and membrane-bound proteins.

Finally, the octylglucoside-containing extract was passed over a positively charged diethylaminoethyl (DEAE) sepharose column (sepharose is a form of cross-linked polydextran) in a buffer containing 10 mM potassium phosphate, 20 mM lactose, and 0.25 mg of washed *E. coli* lipid per milliliter. The carrier protein passed through this column as a symmetrical peak of protein (step 6). Most of the remaining protein in the extract adsorbed to the positively charged column. The fractions containing the bulk of the carrier protein activity were judged to be pure by SDS gel elec-

Figure 7.11

Outline of purification procedure for lactose carrier protein from *E. coli.*

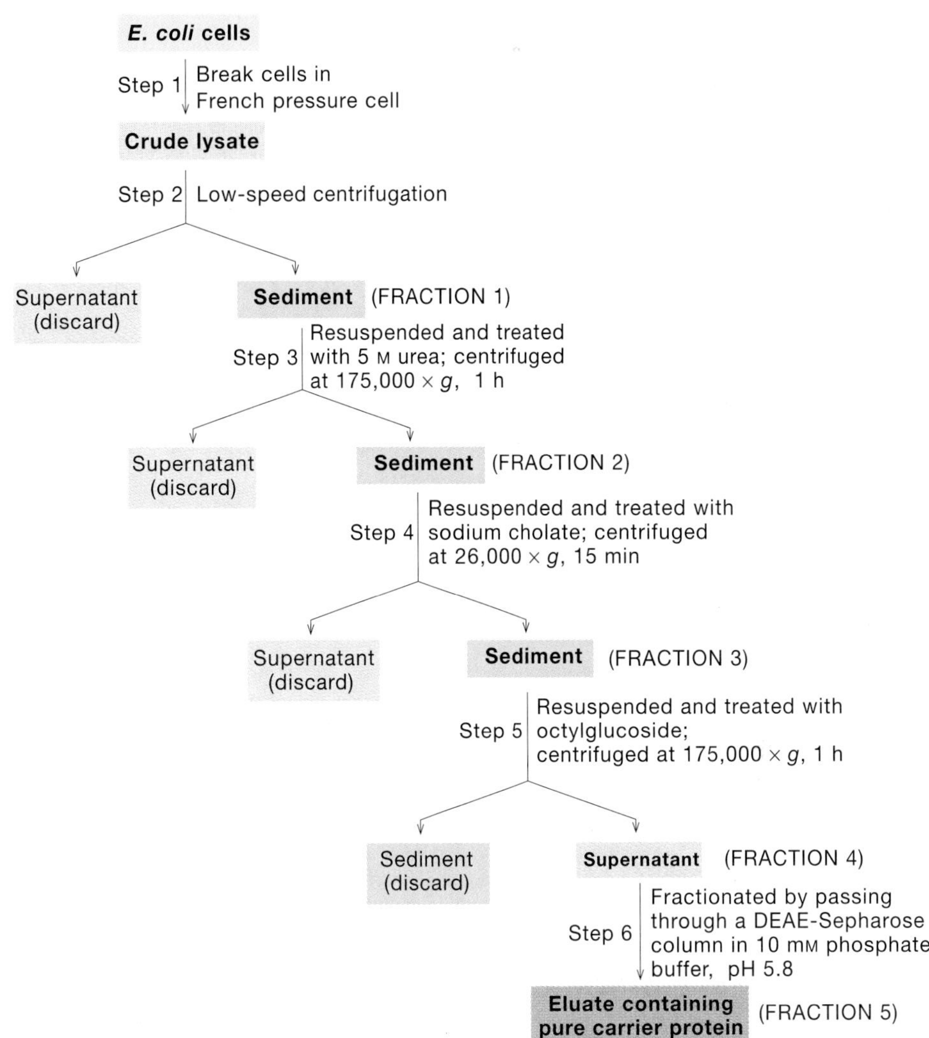

trophoretic analysis. The purified protein contained a single polypeptide chain with an estimated molecular weight of 33,000.

The two purifications described here are as different as the two proteins involved. No two purifications are exactly alike, but the principles of purification, as stated at the outset, are quite similar. Fortunately, an almost endless variety of purification techniques exist. This variety is both helpful and challenging, as a great deal of knowledge and creativity are required to exploit it. In addition to professional expertise, the two most important things required to make a purification possible are an unambiguous assay for the protein in question and a means of stabilizing the protein during purification.

Summary

This chapter discusses methods of protein characterization and protein purification. There is considerable overlap between the methods used for the two goals, but insofar as they are separable, methods of protein characterization are discussed first.

1. Protein solubility is not a fixed quantity for a given protein. Rather, it is a function of many variables. Two of these are pH and salt concentration. Proteins show a minimum solubility at their isoelectric point. Frequently, proteins require the addition of a small amount of salt to become soluble (salting in), but excessive amounts of salt lead to protein precipitation (salting out).
2. There are several methods for determination of molecular weight. These include sedimentation analysis and gel-exclusion chromatography. Sedimentation analysis may be used in two different ways: (1) by independently determining the sedimentation and diffusion rates and combining this information to calculate a molecular weight and (2) by equilibrium ultracentrifugation. Gel-exclusion chromatography uses cross-linked polydextrans and relates molecular weight to the rate of migration through a column.
3. Electrophoretic methods are used in various ways to characterize protein mixtures and purified proteins. The high resolution attainable by electrophoresis makes it ideal for determining the

number of proteins in a mixture as well as their approximate size.

4. Methods of protein purification include differential centrifugation, differential precipitation with $(NH_4)_2SO_4$, gel-exclusion chromatography, differential electrophoretic mobility, and differential affinities for column matrices containing different functional groups. Column procedures are particularly versatile because of the large number of functional groups that can be used to bind proteins in different ways and because of the variety of conditions for differential column elution.
5. The chapter concludes with a description of the purification of two proteins, UMP synthase from eukaryotic cells and lactose carrier protein from *E. coli*. Each of these purifications requires the intelligent combination of several purification procedures performed in series. Throughout purification the presence and amount of the desired proteins must be monitored by a specific protein assay. A successful purification requires careful selection of starting material and working under conditions that minimize loss of functional protein through denaturation.

SELECTED READINGS

Cantor, C. R., and P. Schimmel, *Biophysical Chemistry,* New York: W. H. Freeman, 1980. See especially volume 2, entitled *Study of Biophysical Structure and Function.*

Chaif, B. T., and S. B. H. Kent, Weighing naked proteins: Practical, high-accuracy mass measurement of peptides and proteins. *Science* 257:1885–1893, 1992. Proteins with molecular masses of as much as 100 kD or more can be analyzed as picomole sensitivities to give simple mass spectra corresponding to the intact molecule. This development has allowed unprecedented accuracy in the determination of protein molecular weights.

Eisenberg, D., and D. Crothers, *Physical Chemistry and Its Applications to the Life Sciences,* Menlo Park, Calif.: Benjamin/Cummings, 1979.

Methods in Enzymology, New York: Academic Press. A continuing series of over 250 volumes that discusses most methods at the professional level but is still understandable for students.

Scopes, R., *Protein Purification: Principles and Practice,* 2d ed. New York: Springer-Verlag, 1987. A recent general treatment of this subject.

PROBLEMS

1. A method for the purification of 6-phosphogluconate dehydrogenase from *E. coli* is summarized in the table. For each step, calculate the specific activity, percentage yield, and degree of purification (*n*-fold). Indicate which step results in the greatest purification. Assume that the protein is pure after gel exclusion (Bio-Gel A) chromatography. What percentage of the initial crude cell extract protein was 6-phosphogluconate dehydrogenase?

Step	Volume (ml)	Total Protein (mg)	Total Units	Specific Activity (U/mg)	Yield (%)	Purification (*n*-fold)
Cell extract	2,800	70,000	2,700			
$(NH_4)_2SO_4$ fractionation	3,000	25,400	2,300			
Heat treatment	3,000	16,500	1,980			
DEAE chromatography	80	390	1,680			
CM-cellulose chromatography	50	47	1,350			
Bio-Gel A chromatography	7	35	1,120			

2. Although used effectively in the 6-phosphogluconate dehydrogenase isolation procedure, heat treatment cannot be used in the isolation of all enzymes. Explain.
3. Assume that the isoelectric point (pI) of the 6-phosphogluconate dehydrogenase is 6. Explain why the buffer used in the DEAE cellulose chromatography must have a pH greater than 6 but less than 9 in order for the enzyme to bind to the DEAE.
4. Will the 6-phosphogluconate dehydrogenase bind to the CM cellulose in the same buffer pH range used with the DEAE cellulose? Explain. In what pH range might you expect the dehydrogenase to bind to CM cellulose? Explain.
5. Examine the isolation procedure shown in problem 1 and explain why gel exclusion chromatography is used as the final step rather than as the step following the heat treatment.
6. A student isolated an enzyme from an anaerobe and subjected a sample of the protein to SDS-polyacrylamide gel electrophoresis. A single band was observed upon staining the gel for protein. His advisor was excited about the result but suggested that the protein be subjected to electrophoresis under nondenaturing (native) conditions. Electrophoresis under nondenaturing conditions revealed two bands after the gel was stained for protein. Assuming that the sample had not been mishandled, offer an explanation for the observations.
7. A salt-precipitated fraction of ribonuclease contained two contaminating protein bands in addition to the ribonuclease. Further studies showed that one contaminant had a molecular weight of about 13,000 (similar to ribonuclease) but an isoelectric point 4 pH units more acidic than the pI of ribonucle-

ase. The second contaminant had an isoelectric point similar to ribonuclease but had a molecular weight of 75,000. Suggest an efficient protocol for the separation of the ribonuclease from the contaminating proteins.

8. You have a mixture of proteins with the following properties:
   Protein 1: M_r 12,000, pI = 10
   Protein 2: M_r 62,000, pI = 4
   Protein 3: M_r 28,000, pI = 8
   Protein 4: M_r 9,000, pI = 5

   Predict the order of emergence of these proteins when a mixture of the four is chromatographed in the following systems.
   (a) DEAE cellulose at pH 7, with a linear salt gradient elution.
   (b) CM cellulose at pH 7, with a linear salt gradient elution.
   (c) A gel exclusion column with a fractionation range of 1,000–30,000 M_r, at pH 7.
9. Individual proteins of known subunit molecular weight and pyrophosphatase whose native molecular weight is about 39,000 were denatured in buffer containing both SDS and 2-mercaptoethanol. A sample of each protein was electrophoretically separated by SDS-polyacrylamide gel electrophoresis. The marker dye migrated 12 cm. Staining the gel for protein revealed the following:

Protein	Molecular Weight (subunit)	Distance Migrated (cm)
Serum albumin	69,000	1.6
Catalase	60,000	2.2
Ovalbumin	43,000	3.5
Carbonic anhydrase	29,000	5.2
Myoglobin	17,000	7.4

   The pyrophosphatase migrated 6.7 cm. However, when the 2-mercaptoethanol was omitted from the sample buffer, the pyrophosphatase migrated 3.8 cm. Determine the subunit molecular weight and comment on the quaternary structure of the enzyme.
10. Why are "salting out" procedures often used as an initial purification step following the production of a crude extract by centrifugation?
11. Rarely is DEAE cellulose used above a pH of about 8.5. Can you provide a reason(s) why?
12. Given that the only structural difference between phosphorylase *a* and phosphorylase *b* is that phosphorylase *a* has a covalently bound phosphate on serine 14, do you expect phosphorylase *a* and phosphorylase *b* to elute as a single peak on DEAE-cellulose chromatography? What if gel filtration were utilized?
13. You wish to purify an ATP-binding enzyme from a crude extract that contains several contaminating proteins. To purify the enzyme rapidly and to the highest purity, you must consider some sophisticated strategies, among them affinity chromatography. Explain how affinity chromatography can be applied to this separation, and explain the physical basis of the separation.
14. What conclusions can you reach on the amino acid composition of protein 1 relative to protein 2 in problem 8?
15. Buffer used to elute proteins from DEAE- or CM-cellulose columns often contains increasing amounts of KCl (a salt gradient). What is the function of the KCl?
16. In organic chemistry you purified substances by extraction with organic solvents and fractional crystallization. Why don't biochemists use these methods for protein purification?
17. The following enzyme purification data are from one of the first papers on the partial purification of DNA polymerase (*J. Biol. Chem.* 233:163, 1958). Complete the table.

Purification Step	Volume (ml)	Protein (mg)	Total Units	Spec. Act. (U/mg)	Yield (%)	Fold Purified
Extract	1500	4500	19,500	______	______	______
DNase/ dialysis	1496	2693	18,100	______	______	______
Alumina gel	800	600	12,300	______	______	______
Concen- tration	90	441	9,900	______	______	______
Ammonium sulfate	9	75.6	6,030	______	______	______
DEAE cellulose	30	18	3,600	______	______	______

18. Thiols are often added to buffers used in protein purification to reduce disulfides formed by air-promoted oxidation of protein thiols. Mercaptoethanol is often used for this purpose, but when cost is not a factor and there are concerns about the deleterious effects of high concentrations of mercaptoethanol, the preferred reagent is DTT (dithiothreitol, or 2,3-dihydroxy-1,4-dithiolbutane). Can you propose reason(s) why DTT might be preferred?

PART

3

CATALYSIS

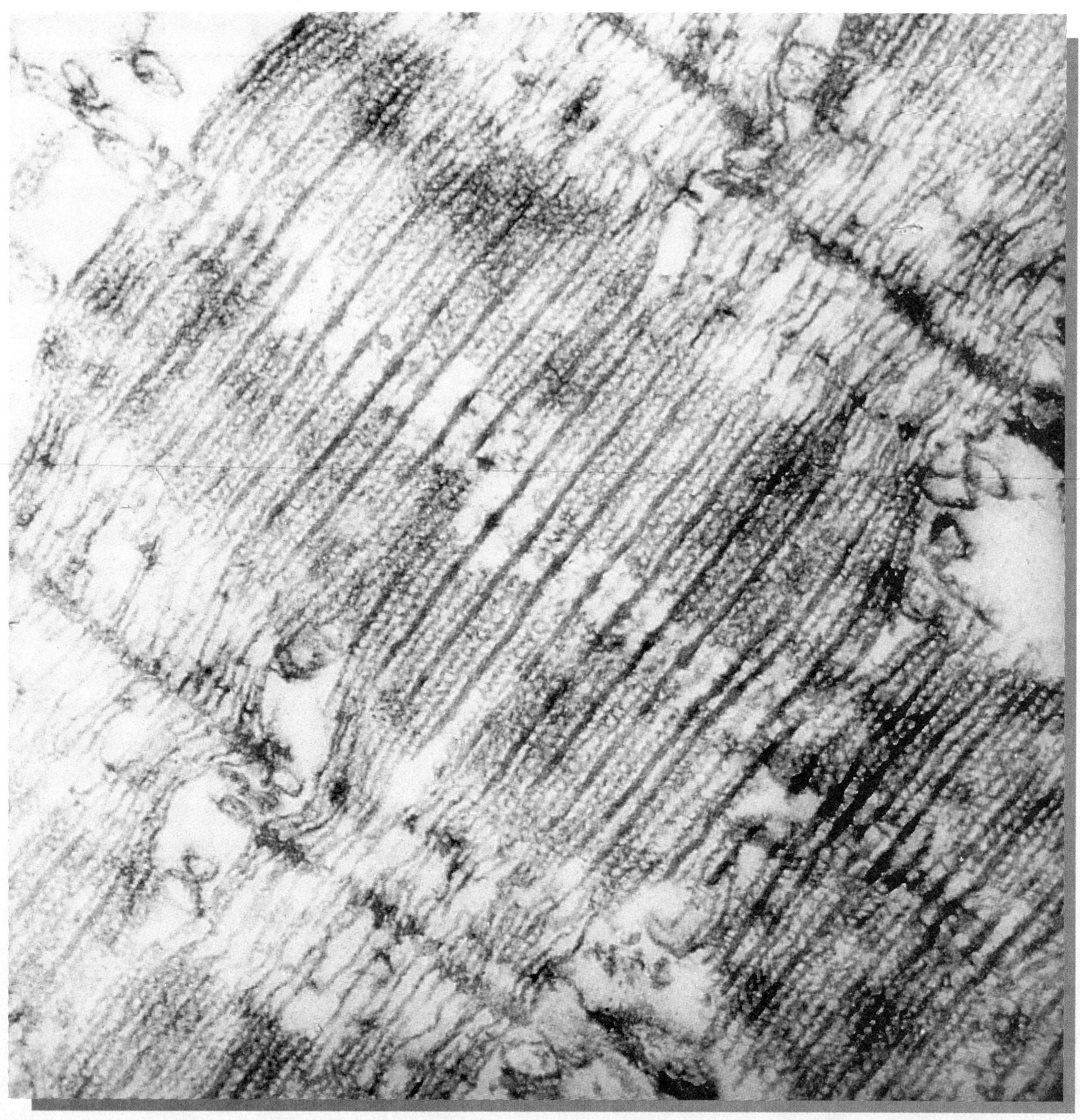

Electron micrograph of a very thin longitudinal section of a striated muscle from a rabbit. The basic repeat unit is the sarcomere, which is bounded by the Z line. The relatively dark areas comprise the A band, and the light areas bracketing the Z line make up the I band. The I band consists of thin filaments composed of actin, while the thick filaments of the A band are myosin. Bridges connect the thin actin filaments to the thicker myosin filaments. Muscle protein is an unusual example of a structural protein with enzymatic activity. The enzyme activity is localized in the myosin bridges. (Courtesy of Dr. Hugh Huxley)

C H A P T E R

8

Enzyme Kinetics

An enzyme-catalyzed reaction proceeds rapidly under mild conditions because it lowers the activation energy for a reaction; enzymes are usually highly specific for the reactions they catalyze.

Most of this book is concerned with the reactions that occur in living cells. These reactions are catalyzed by enzymes. In this and the following three chapters we focus on the ways in which enzymes function. The present chapter deals with the kinetics of enzyme-catalyzed reactions; the next two explore the mechanisms of enzymatic catalysis. Finally in chapter 11 we examine the small cofactors that work together with many enzymes.

A catalyst is a substance that accelerates a chemical reaction without itself undergoing any net change. The rate enhancements achieved by many enzymes are extraordinarily high. For example, carbonic anhydrase, an enzyme found in red blood cells, catalyzes the reaction

$$CO_2 + H_2O \longrightarrow H_2CO_3 \qquad (1)$$

In the presence of the enzyme this reaction occurs about 10^7 times as rapidly as it does in the absence of the enzyme. One molecule of carbonic anhydrase can hydrate about 10^6 molecules of CO_2 a second.

Kinetic analysis is one of the most basic topics of enzymology. Such studies reveal not only how fast an enzyme can function, but also its preferences for various reactants (or substrates, as they usually are called), the effect of substrate concentration on the reaction rate, and the sensitivity of the enzyme to specific inhibitors or activators. By studying the alterations in rate under different conditions we often gain clues to the mechanism of the reaction.

The Discovery of Enzymes

The existence of catalysts in biological materials was recognized as early as 1835 by Jöns Jakob Berzelius, the Swedish chemist who discovered several elements, introduced the way of writing chemical symbols, and coined the term "catalysis." Berzelius noted that potatoes contained something that catalyzed the breakdown of starch, and he suggested that all natural products are formed under the influence of such catalysts. But the chemical nature of biological catalysts was unknown, and it remained a mystery for many years. In the period between 1850 and 1860, Louis Pasteur demonstrated that fermentation, the anaerobic breakdown of sugar to CO_2 and ethanol, occurred in the presence of living cells and did not occur in a flask that was capped after any cells that it contained had been killed by heat. Then in 1897, Eduard Buchner discovered by accident that fermentation was catalyzed by a clear juice that he had prepared by grinding yeast with sand and filtering out the unbroken cells. Looking for a way to preserve the juice, Buchner had added sugar. It probably was a disappointment to him that the sugar was broken down rapidly and the mixture frothed with CO_2. But Buchner's discovery made it possible to explore metabolic processes such as fermentation in a greatly simplified system, without having to deal with the complexities of cell growth and multiplication, and without the barriers imposed by cell walls or membranes. Arthur Harden and William Young soon showed that yeast extracts contained two different types of molecules, both of which were necessary for fermentation to occur. Some were small, dialyzable, heat-stable molecules such as inorganic phosphate; others were much larger, nondialyzable molecules—the enzymes—that were destroyed easily by heat.

Although early investigators surmised that enzymes might be proteins, this remained in dispute until 1927, when James Sumner succeeded in purifying and crystalizing the enzyme urease from beans. In the 1930s, John Northrop isolated and characterized a series of digestive enzymes, generalizing Sumner's conclusion that enzymes are proteins. Since then, thousands of different enzymes have been purified, and the structures of many of them have been solved to atomic resolution; almost all of these molecules have proved to be proteins. Surprisingly, however, recent work has shown that some RNA molecules also have enzymatic activity.

Enzyme Terminology

Enzymes often are known by common names obtained by adding the suffix "-ase" to the name of the substrate or to the reaction that they catalyze. Thus, glucose oxidase is an enzyme that catalyzes the oxidation of glucose; glucose-6-phosphatase catalyzes the hydrolysis of phosphate from glucose-6-phosphate; and urease catalyzes the hydrolysis of urea. Common names also are used for some groups of enzymes. For example, an enzyme that transfers a phosphate group from ATP to another molecule is usually called a "kinase," instead of the more formal "phosphotransferase."

A systematic scheme for classifying enzymes was adopted in 1972 by the International Union of Biochemistry. In this scheme, each enzyme is designated by four numbers that indicate the main class, subclass, subsubclass, and serial number of the enzyme in its subsubclass. The six main classes (fig. 8.1) are (1) oxidoreductases, (2) transferases, (3) hydrolases, (4) lyases, (5) isomerases, and (6) ligases, or synthases. Oxidoreductases catalyze oxidation–reduction reactions. Transferases catalyze the transfer of a functional group from one molecule to another. Hydrolases catalyze bond cleavage by the introduction of water. Lyases catalyze the removal of a group to form a double bond or the addition of a group to a double bond. Isomerases catalyze intramolecular rearrangements, and ligases catalyze reactions that join two molecules.

Figure 8.1

The six main classes of enzymes and the reactions they catalyze.

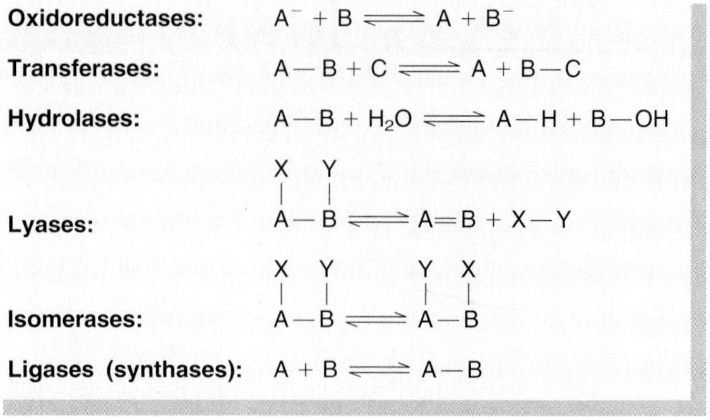

Oxidoreductases:	$A^- + B \rightleftharpoons A + B^-$
Transferases:	$A{-}B + C \rightleftharpoons A + B{-}C$
Hydrolases:	$A{-}B + H_2O \rightleftharpoons A{-}H + B{-}OH$
Lyases:	$A(X){-}B(Y) \rightleftharpoons A{=}B + X{-}Y$ (X on A, Y on B)
Isomerases:	$A(X){-}B(Y) \rightleftharpoons A(Y){-}B(X)$
Ligases (synthases):	$A + B \rightleftharpoons A{-}B$

Many enzymes require additional small molecules called cofactors for their activity. Cofactors can be simple inorganic ions such as Mg^{2+} or complex organic molecules known as coenzymes. The cofactor usually binds tightly to a special site on the enzyme. An enzyme lacking an essential cofactor is called an apoenzyme, and the intact enzyme with the bound cofactor is called the holoenzyme.

Basic Aspects of Chemical Kinetics

Before we delve into enzyme kinetics, we need to discuss some of the basic principles that apply to the kinetics of both enzymatic and nonenzymatic reactions. Let's first consider a nonenzymatic reaction that converts a single reactant (R) into a product (P):

$$R \longrightarrow P \tag{2}$$

To measure the velocity of the reaction (v), we plot the concentration of R as a function of time (fig. 8.2). The rate at any particular time (t) is

$$v = -\frac{d[R]}{dt} \tag{3}$$

where $d[R]/dt$ is the slope of the plot at that time. The minus sign is needed because v is defined by convention to be a positive number, whereas $d[R]/dt$ is always negative ([R] decreases with time). However, we might equally well choose to measure the increase in the concentration of the product as a function of time, in which case we express the rate as $v = d[P]/dt$.

For a simple reaction of this type, the rate at any given time usually is found to be proportional to the remaining concentration of the reactant:

$$v = k[R] \tag{4}$$

Figure 8.2

The kinetics of a first-order reaction in which a single reactant (R) is converted irreversibly to a product (P). The concentrations of R and P are plotted as functions of time. The rate (v) at any given time can be obtained from the slope of either curve:

$$v = -\frac{d[R]}{dt} = \frac{d[P]}{dt}$$

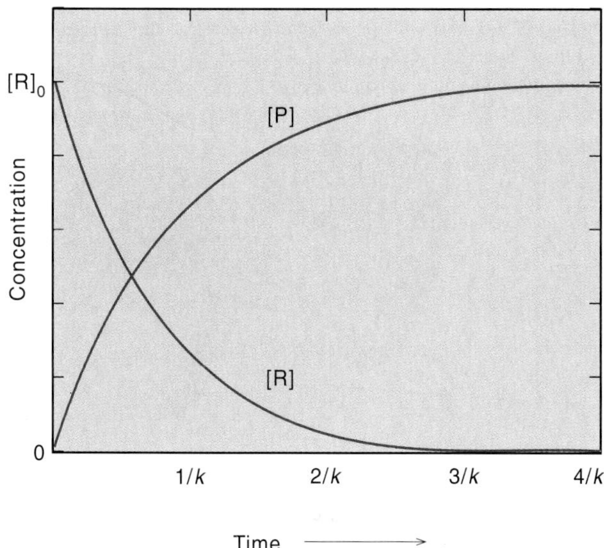

The proportionality constant k is the rate constant. The rate constant is independent of the concentration of the reactant, but it can depend on other parameters, such as temperature or pH, and, as we shall see, it may be altered by a catalyst. It has dimensions of reciprocal seconds (s^{-1}). Combining equations (3) and (4), we have

$$\frac{d[R]}{dt} = -k[R] \quad (5)$$

A reaction of this type is said to follow first-order kinetics because the rate is proportional to the concentration of a single species raised to the first power (see fig. 8.2). An example is the decay of a radioactive isotope such as ^{14}C. The rate of decay at any time (the number of radioactive disintegrations per second) is simply proportional to the amount of ^{14}C present. The rate constant for this extremely slow nuclear reaction is $8 \times 10^{-12}\, s^{-1}$. Another example is the initial electron-transfer reaction that occurs when photosynthetic organisms are excited with light (see chapter 17). In this case there actually are two reactants, the electron donor and the acceptor, but they are held close together on a protein so that they react as a unit; the excited complex simply decays spontaneously to a more stable state. This process occurs with a rate constant of $3 \times 10^{11}\, s^{-1}$, which makes it one of the fastest reactions known.

A slightly more complicated situation arises if the reaction is reversible:

$$R \underset{k_{-1}}{\overset{k_1}{\rightleftharpoons}} P \quad (6)$$

Here k_1 is the rate constant for the forward reaction, and k_{-1} is that for the back reaction. In a reversible reaction the rate equation becomes

$$\frac{d[R]}{dt} = -k_1[R] + k_{-1}[P] \quad (7)$$

The term $-k_1[R]$ is the same as in equation (5); the term $k_{-1}[P]$ describes the formation of R from P. In this case, [R] and [P] proceed from their initial values to their final, equilibrium values ($[R]_{eq}$ and $[P]_{eq}$), which generally are not zero. At equilibrium, $d[R]/dt$ must go to zero, which requires that

$$k_1[R]_{eq} = k_{-1}[P]_{eq} \quad (8)$$

Rearranging this expression gives

$$\frac{[P]_{eq}}{[R]_{eq}} = \frac{k_1}{k_{-1}} = K_{eq} \quad (9)$$

where K_{eq} is the equilibrium constant.

Now consider a reaction involving two reactants, A and B:

$$A + B \xrightarrow{k} P \quad (10)$$

The rate of such a bimolecular reaction usually depends on the concentrations of both reactants:

$$\frac{d[P]}{dt} = k[A][B] \quad (11)$$

The reaction is said to follow second-order kinetics because its rate is proportional to a product of two concentrations. The kinetics are first order in either [A] or [B] alone but second order overall. The rate constant for a second-order reaction has the dimensions of $M^{-1}s^{-1}$.

A Critical Amount of Energy Is Needed for the Reactants to Reach the Transition State

The equilibrium constant K_{eq} for a reaction is strictly a function of thermodynamic factors. The most pertinent thermodynamic quantity here is the change in free energy that occurs in the reaction ΔG. As we saw in chapter 2, ΔG and K_{eq} are related by the expression $\Delta G^\circ = -RT \ln K_{eq}$. If the free energy of the products is less than the free energy of the reactants ($\Delta G < 0$), then the equilibrium constant favors the formation of products; if the free energy of the products exceeds that of the reactants ($\Delta G > 0$), the reverse reaction is favored. We noted above that the equilibrium constant for a one-step reaction is equal to the ratio of the rate constants in the forward and reverse directions [equation (9)]. However, the value of K_{eq} does not tell us how long it takes the reaction to reach equilibrium because it says nothing about the magnitudes of the individual rate constants. These may be very large or very small. The overall ΔG for a reaction therefore indicates only whether the reaction is possible, not how rapidly the reaction occurs. To understand the kinetics, we must look into the mechanism of the reaction.

The rate at which two molecules react depends partly on how frequently the molecules collide. Collisions occur as a result of random diffusion of the reactants in the solution. The number of collisions per second is proportional to the product of the two concentrations, and the second-order rate constant k for the reaction includes a proportionality factor for this relationship. The rate constant also includes a factor that gives the fraction of the collisions that are effective. If every collision were to result in a reaction, this second factor would be 1 and the rate constant for the reaction of two small molecules in aqueous solution would be about 10^{11} $\text{M}^{-1}\text{s}^{-1}$. Because of their large masses, proteins diffuse relatively slowly, so the frequency at which a protein and a small molecule collide is lower than the collision frequency for two small molecules. As a consequence, a reaction between a protein and a small molecule has a maximum rate constant on the order of 10^8 to 10^9 $\text{M}^{-1}\text{s}^{-1}$.

But not every collision results in a reaction. What determines the fraction of the collisions that are effective? A partial answer is that the colliding species must have a certain critical energy in order to surmount a barrier that separates the reactants from the products. This is illustrated schematically in figure 8.3. The surface in figure 8.3*a* represents the energy of a system in which a proton can be bound to either of two molecules. Suppose the proton is initially on molecule A, and we are interested in how rapidly it moves to molecule B. As the proton moves from one place to the other, its electrostatic interactions with molecule A become less favorable, and the interactions with B improve. The energy of the system goes through a maximum when the proton is at an intermediate position. At this point, the system is said to be in the transition state. The probability that a collision leads to a reaction depends, in part, on the probability that the molecules collide with enough energy to reach this state.

The amount of free energy required to reach the transition state is called the activation free energy, $\Delta G^{\ddagger}$. From equation (13) of chapter 2, we can equate $\Delta G^{\ddagger}$ to $-RT \ln K^{\ddagger}$, where $K^{\ddagger}$ is an equilibrium constant for the formation of the transition state from the reactants. The fraction of the reactants that are in the transition state at any given moment is given approximately by $K^{\ddagger}$, or $e^{-\Delta G^{\ddagger}/RT}$. We therefore can write the overall rate constant for a reaction as

$$k = k_0 e^{-\Delta G^{\ddagger}/RT} \qquad \textbf{(12)}$$

Here k_0 is an intrinsic rate constant for conversion of the transition state into the products and typically depends on the nuclear vibration frequency of a bond that is being formed or broken.

Catalysts Speed up Reactions by Lowering the Free Energy of Activation

Equation (12) indicates that if $\Delta G^{\ddagger}$ is greater than zero, as is most often the case, a reaction can be sped up either by decreasing $\Delta G^{\ddagger}$ or by raising the temperature. Little can usually be done to change k_0. For living organisms it generally is impractical to raise the temperature because many organisms have little or no control over the ambient temperature and can survive only within a narrow range of temperatures. Also, changing the temperature is not a very selective way to control reaction rates because most reactions speed up when the temperature is raised. The remaining alternative is to decrease $\Delta G^{\ddagger}$.

Figure 8.3

Schematic energy diagrams for a reaction in which a proton moves from one molecule to another. In (*a*), coordinates in the plane at the bottom represent the location of the proton. The free energy of the system for a particular set of coordinates is represented by the distance of the cuplike surface above the plane. The two minima in the energy surface indicate the positions of the proton when it is bound optimally to one molecule or the other. The best route along the surface from one of these minima to the other goes through a pass, or saddle point. (*b*) A plot of the free energy as a function of distance along the optimal route over this pass. The activation free energy of the reaction ($\Delta G^{\ddagger}_a$) is the difference between the energies at the pass and at the starting point.

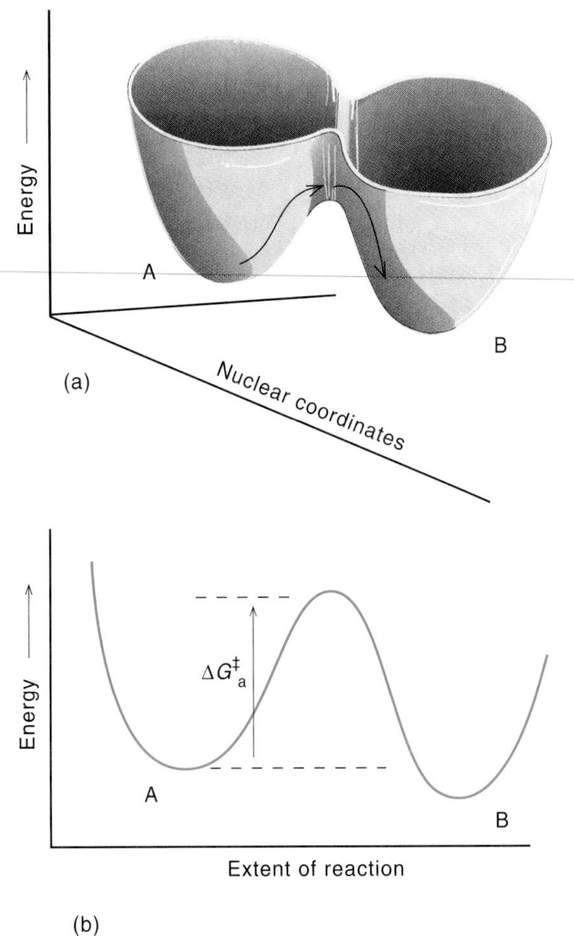

Because rate constants depend exponentially on $-\Delta G^{\ddagger}/RT$ [equation (12)], small changes in $\Delta G^{\ddagger}$ will have a large effect on the reaction rate. At physiological temperatures it takes a decrease of only 1.36 kcal/mole to speed up a reaction by a factor of 10, and a decrease by 8.16 kcal/mole increases the rate by a factor of 10^6. These are relatively modest free energy changes because forming a single H bond can release anywhere from 4 to 10 kcal/mole. In addition, because $\Delta G^{\ddagger}$ depends on the structures of the reactants and products and on the detailed nature of the reaction, it affords an opportunity to control reaction rates with a great deal of specificity. Enzymes, then, must work by decreasing the activation free energies for the specific reactions that they catalyze

Figure 8.4

An enzyme speeds up a reaction by decreasing $\Delta G^{\ddagger}$. The enzyme does not change the free energy of the substrate (S) or product (P); it lowers the free energy of the transition state. The two vertical arrows indicate the activation free energies ($\Delta G^{\ddagger}$) of the catalyzed and uncatalyzed reactions.

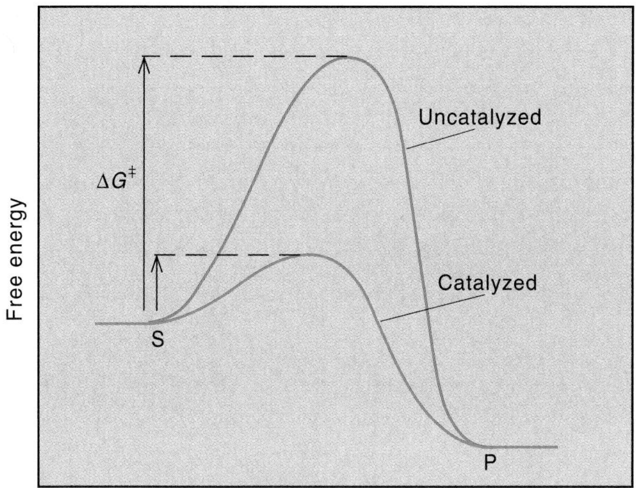

Figure 8.5

The stopped-flow apparatus for measuring enzyme-catalyzed reactions very soon after mixing enzyme and substrate.

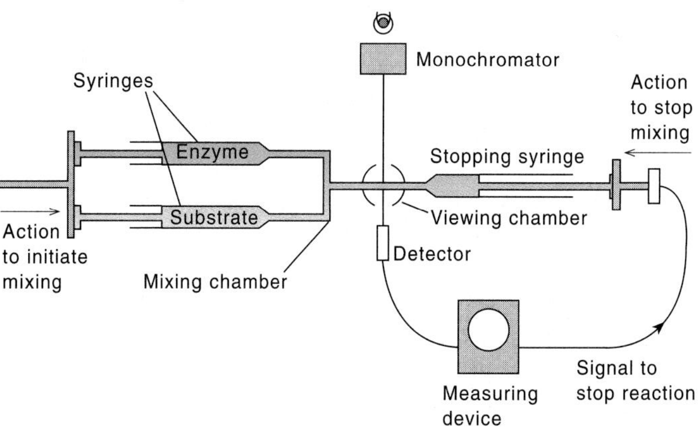

(fig. 8.4). They do this in a variety of ways that we discuss in detail in the next chapter. Our main concern here is to describe in more general terms how the kinetic properties of enzymatic reactions are measured and characterized.

It should be kept in mind that although an enzyme can increase the rate at which a reaction occurs, it (or any other catalyst) cannot alter the overall equilibrium constant. Since K_{eq} is equal to the ratio of the rate constants for the forward and reverse processes, the catalyst increases *both* of these rate constants without changing their ratio.

Kinetics of Enzyme-Catalyzed Reactions

Kinetic analysis was used to characterize enzyme-catalyzed reactions even before enzymes had been isolated in pure form. As a rule, kinetic measurements are made on purified enzymes in vitro. But the properties so determined must be referred back to the situation in vivo to ensure that they are physiologically relevant. This is important because the rate of an enzymatic reaction can depend strongly on the concentrations of the substrates and products and also on temperature, pH, and the concentrations of other molecules that activate or inhibit the enzyme. Kinetic analysis of such effects is indispensable to a comprehensive picture of an enzyme.

Kinetic Parameters Are Determined by Measuring the Initial Reaction Velocity as a Function of the Substrate Concentration

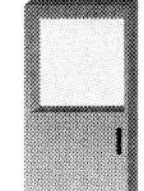

The usual procedure for measuring the rate of an enzymatic reaction is to mix enzyme with substrate and observe the formation of product or disappearance of substrate as soon as possible after mixing, when the substrate concentration is still close to its initial value and the product concentration is small. The measurements usually are repeated over a range of substrate concentrations to map out how the initial rate depends on concentration. Spectrophotometric techniques are commonly used in such experiments because in many cases they allow the concentration of a substrate or product in the mixture to be measured continuously as a function of time.

Measurements of reactions that occur in less than a few seconds require special techniques to speed up the mixing of the enzyme and substrate. One way to achieve this is to place solutions containing the enzyme and the substrate in two separate syringes. A pneumatic device then is used to inject the contents of both syringes rapidly into a common chamber that resides in a spectrophotometer for measuring the course of the reaction (fig. 8.5). Such an apparatus is referred to as a "stopped-flow" device because the flow stops abruptly when the movement of the pneumatic driver is arrested. In this type of apparatus it is possible to make kinetic measurements within about 1 ms after mixing of enzyme and substrate.

Equations (5) and (10) imply that the velocity of an uncatalyzed reaction increases indefinitely with an increase in the concentration of the reactants. With enzyme-catalyzed reactions, something very different is observed. The rate usually increases linearly with substrate concentration at low concentrations but then levels off and becomes independent of the concentration at high concentrations (fig. 8.6). The explanation for this hyperbolic dependence on substrate concentration is straightforward. For an enzyme to affect $\Delta G^{\ddagger}$, the substrate must bind to a special site on the protein, the active site (fig. 8.7). At very low concentrations of substrate the active sites of most of the enzyme molecules in the solution are unoccupied. Increasing the substrate concentration brings more enzyme molecules into play, and the reaction speeds up. At high concentrations, on the other hand, most of the enzyme molecules have their active sites occupied, and the observed rate depends only on the rate at which the bound reactants

Figure 8.6

The reaction velocity v as a function of the substrate concentration [S] for an enzyme-catalyzed reaction. At high substrate concentrations the reaction velocity reaches a limiting value, V_{max}. K_m is the substrate concentration at which the rate is half maximal.

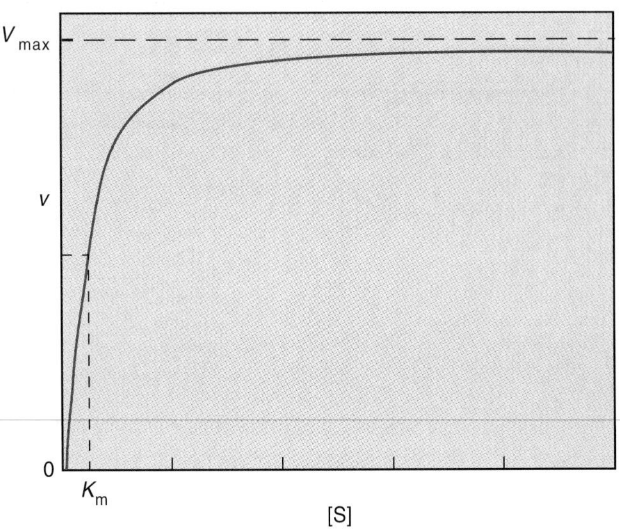

Figure 8.7

Thermolysin is an enzyme that hydrolyzes peptide bonds. Close-up view of the active site of thermolysin (beige) illustrating the enzyme–substrate interaction between a tight binding substrate analog, phosphoramidon (yellow), and the enzyme. (Based on the crystal structure described by D. E. Tronrud, A. F. Monzingo, and B. W. Matthews. Copyright 1994 by the Scripps Research Institute/Molecular Graphics Images by Michael Pique using software by Yng Chen, Michael Connolly, Michael Carson, Alex Shah, and AVS, Inc. Visualization advice by Holly Miller, Wake Forest University Medical Center.)

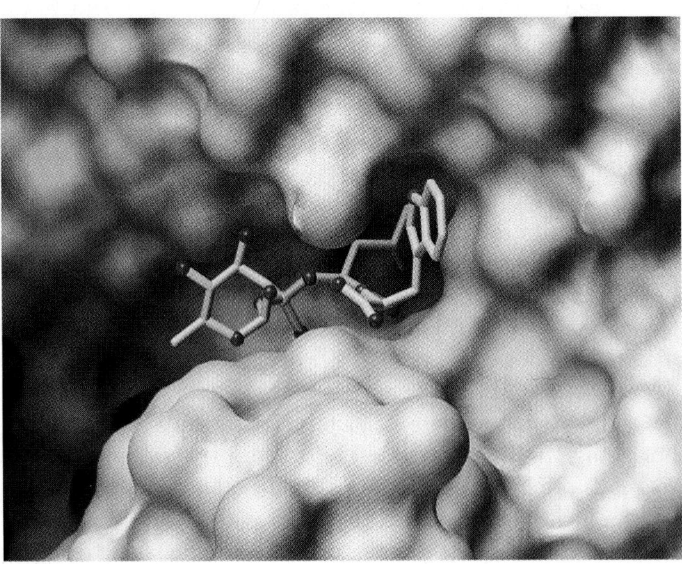

are converted into products. Further increases in the substrate concentration then have little effect.

The Henri-Michaelis-Menten Treatment Assumes That the Enzyme-Substrate Complex Is in Equilibrium with Free Enzyme and Substrate

The hyperbolic saturation curve that is commonly seen with the enzymatic reactions led Leonor Michaelis and Maude Menten in 1913 to develop a general treatment for kinetic analysis of these reactions. Following earlier work by Victor Henri, Michaelis and Menten assumed that an enzyme–substrate (ES) complex is in equilibrium with free enzyme (E) and substrate (S) and that the formation of products (P) proceeds only through the ES complex:

$$E + S \underset{k_{-1}}{\overset{k_1}{\rightleftharpoons}} ES \xrightarrow{k_2} E + P \tag{13}$$

Their object was to relate the reaction rate of observable quantities and interpretable molecular parameters.

Because the slow step in the reaction described by equation (13) is assumed to be formation of E and P from ES, the velocity of the reaction should be

$$v = \frac{d[P]}{dt} = k_2[ES] \tag{14}$$

A maximum velocity (V_{max}) is obtained when all of the enzyme is in the form of the enzyme–substrate complex. Since the total concentration of enzyme $[E_t]$ is equal to [E] + [ES],

$$V_{max} = k_2[E_t] = k_2([E] + [ES]) \tag{15}$$

Dividing equation (14) by equation (15) gives

$$\frac{v}{V_{max}} = \frac{[ES]}{[E] + [ES]} \tag{16}$$

The fraction on the right-hand side can be evaluated by making use of the dissociation constant of the ES complex in equation (13), K_s:

$$K_s = \frac{[E][S]}{[ES]} \quad \text{and} \quad [ES] = \frac{[E][S]}{K_s} \tag{17}$$

Substituting this value of [ES] into equation (16) and rearranging give

$$v = \frac{V_{max}[S]}{K_s + [S]} \tag{18}$$

Equation (18) is the Henri-Michaelis-Menten equation, which relates the reaction velocity to the maximum velocity, the substrate concentration, and the dissociation constant for the enzyme–substrate complex. Usually, substrate is present in much higher molar concentration than enzyme, and the initial period of the reaction is examined so that the free substrate concentration [S] is approximately equal to the total substrate added to the reaction mixture.

Steady-State Kinetic Analysis Assumes That the Concentration of the Enzyme-Substrate Complex Remains Nearly Constant

Rather than discussing the implications of equation (18) at this point, it is useful to develop a more general expression that avoids the assumption that the enzyme–substrate complex is in equilibrium with free enzyme and substrate. To develop this expression, we introduce the concept of the steady state, which was first proposed by G. E. Briggs and J. B. S. Haldane in 1925. The steady state constitutes the time interval when the rate of the reaction is approximately constant. The system usually reaches a steady state soon after enzyme and substrate are mixed, following a brief period when the concentration of the ES complex builds up (fig. 8.8). The ES concentration then remains almost constant for the duration of the steady state, while the concentrations of the substrate and product continue to change substantially.*

Let's now write out the reaction of equation (13) in more detail:

$$E + S \underset{k_{-1}}{\overset{k_1}{\rightleftharpoons}} ES \underset{k_{-2}}{\overset{k_2}{\rightleftharpoons}} E + P \quad \textbf{(19)}$$

In this more complete scheme, the ES complex forms from E and S with rate constant k_1. ES can either dissociate again with rate constant k_{-1} or go on to P with k_2. If we confine ourselves to measuring the initial rate of the reaction in the steady state, we can continue to neglect regeneration of ES from E and P (the step involving k_{-2}) because the concentration of P will be too small for this back-reaction to occur at a significant rate. For the rate of formation of ES, v_f, we then can write

$$v_f = k_1[E][S] \quad \textbf{(20)}$$

Similarly, the rate of disappearance of ES, v_d, is

$$v_d = k_{-1}[ES] + k_2[ES] \quad \textbf{(21)}$$

If the concentration of ES is virtually constant during the steady state, the rates of formation and disappearance of ES must be nearly equal, $v_f = v_d$. We therefore can describe the situation in the steady state by combining equations (20) and (21):

$$k_1[E][S] = (k_{-1} + k_2)[ES] \quad \textbf{(22)}$$

or

$$\frac{[E][S]}{[ES]} = \frac{k_{-1} + k_2}{k_1} = K_m \quad \textbf{(23)}$$

The constant K_m defined in equation (23) is called the Michaelis constant and is one of the key parameters in enzyme kinetics. It is a simple matter to proceed from this point to an expression comparable to the Henri-Michaelis-Menten equation (18), but with K_m in place of K_s. First, rearranging equation (23) gives

$$[ES] = \frac{[E][S]}{K_m} \quad \textbf{(24)}$$

With this expression for [ES] we can follow the same procedure that led to equation (18), this time arriving at the Briggs-Haldane equation for the reaction velocity:

$$v = \frac{V_{max}[S]}{[S] + K_m} \quad \textbf{(25)}$$

Equation (25) is identical to (18) except that the more complex constant K_m has replaced K_s. The term "Michaelis-Menten equation" is often used for either expression.

For the purpose of graphical representation of experimental data it is convenient to rearrange equation (25). Taking the

*The steady state is no stranger to the living cell. Most reactions in a living cell are in a steady state most of the time. Thus the steady state, used originally as a convenience by kineticists, is also an appropriate way to analyze enzymatic reactions in vivo.

Figure 8.8

Concentrations of free enzyme (E), substrate (S), enzyme–substrate complex (ES), and product (P) over the time course of a reaction. The shaded portion of the top graph is shown in expanded form in the bottom graph. After a brief initial period (usually less than a few seconds) the concentration of ES remains approximately constant for an extended period. The steady-state approximation is applicable during this second period. Most measurements of enzyme kinetics are made in the steady state.

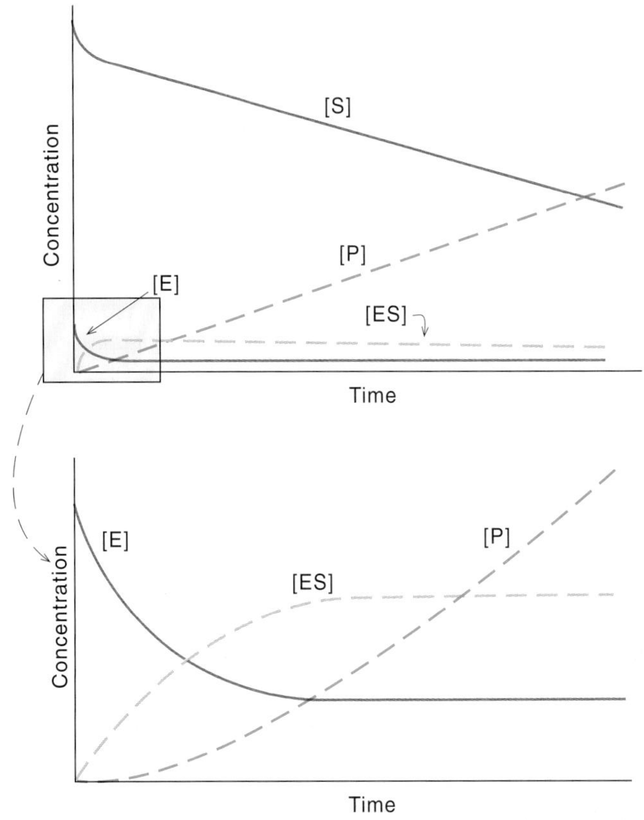

Figure 8.9

A plot of the reciprocal of the rate ($1/v$) as a function of the reciprocal of the substrate concentration ($1/[S]$) fits a straight line. Extrapolating the line to its intercept on the ordinate (infinite substrate concentration) gives $1/V_{max}$. Extrapolating to the intercept on the abscissa gives $-1/K_m$.

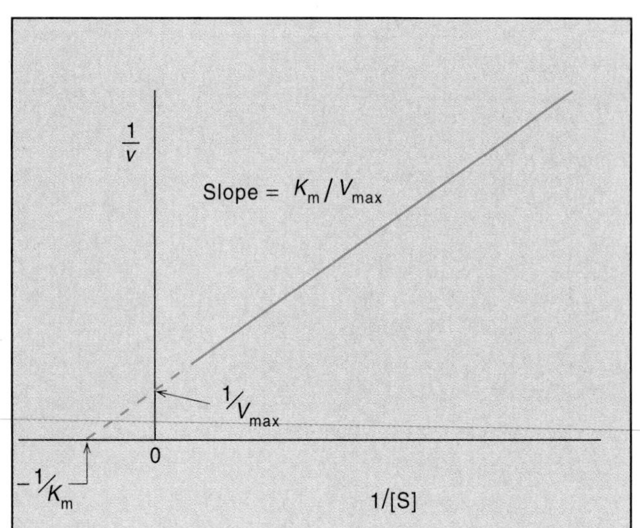

Table 8.1
The Michaelis Constants for Some Enzyme–Substrate Pairs

Enzyme and Substrate	K_m (M)
Catalase	
H_2O_2	1.1
Hexokinase	
Glucose	1.5×10^{-4}
Fructose	1.5×10^{-3}
Chymotrypsin	
N-Benzoyltyrosinamide	2.5×10^{-3}
N-Formyltyrosinamide	1.2×10^{-2}
N-Acetyltyrosinamide	3.2×10^{-2}
Glycyltyrosinamide	1.2×10^{-1}
Aspartate aminotransferase	
Aspartate	9.0×10^{-4}
α-Ketoglutarate	1.0×10^{-4}
Fumarase	
Fumarate	5.0×10^{-6}
Malate	2.5×10^{-5}

reciprocals of both sides of equation (25) gives

$$\frac{1}{v} = \frac{1}{V_{max}} + \frac{K_m}{V_{max}}\frac{1}{[S]} \quad \textbf{(26)}$$

This expression indicates that a plot of $1/v$ versus $1/[S]$ fits a straight line with a slope of K_m/V_{max} (fig. 8.9). Such a plot is known as a Lineweaver-Burk, or double-reciprocal, plot. The intercept of the line on the ordinate occurs at $1/v = 1/V_{max}$, and the intercept on the abscissa occurs at $1/[S] = -1/K_m$. V_{max} and K_m thus can be determined readily from the graph.

Significance of the Michaelis Constant, K_m. The Michaelis constant K_m has the dimensions of a concentration (molarity), because k_{-1} and k_2, the two rate constants in the numerator of equation (23), are first-order rate constants with units expressed per second (s^{-1}), whereas the denominator k_1 is a second-order rate constant with units of $M^{-1}s^{-1}$. To appreciate the meaning of K_m, suppose that $[S] = K_m$. The denominator in equation (25) then is equal to 2[S], which makes the velocity $v = V_{max}/2$. Thus the K_m is the substrate concentration at which the velocity is half maximal (see fig. 8.6).

K_m values for several enzyme–substrate pairs are given in table 8.1. The values vary over a wide range but typically lie between 10^{-6} and 10^{-1} M. With enzymes that can act on several different substrates, K_m can vary substantially from substrate to substrate. With the enzyme chymotrypsin, for example, the K_m for the substrate glycyltyrosinamide is about 50 times that for the substrate *N*-benzoyltyrosinamide.

Does the K_m indicate how tightly a particular substrate binds to the active site of an enzyme? Not necessarily. According to equation (23),

$$K_m = \frac{k_{-1} + k_2}{k_1} \quad \textbf{(23')}$$

whereas the dissociation constant of the ES complex is

$$K_s = \frac{k_{-1}}{k_1} \quad \textbf{(27)}$$

Comparing these two expressions, we see that K_m must always be larger than K_s. If $k_2 \gg k_{-1}$, then $K_m \approx k_2/k_1$, which is much greater than K_s. On the other hand, if $k_{-1} \gg k_2$, then K_m is approximately equal to K_s. In the latter circumstance a smaller K_m means a smaller dissociation constant, which implies tighter binding to the enzyme. (This is the limiting case assumed in the Henri-Michaelis-Menten treatment, for when $k_{-1} \gg k_2$, the enzyme–substrate complex is in equilibrium with the free enzyme and substrate.) But whether or not this is a valid approximation depends on the particular enzyme and the substrate. For a multistep reaction the relationship between K_m and the rate constants can be considerably more complex.

Significance of the Turnover Number, k_{cat}. The turnover number of an enzyme, k_{cat}, is the maximum number of molecules of substrate that could be converted to product each second per active site. Because the maximum rate is obtained at high substrate concentrations, when all the active sites are occupied with substrate, the turnover number is a measure of how rapidly an enzyme can operate once the active site is filled. This is given simply by

$$k_{cat} = \frac{V_{max}}{[E_t]} \quad \textbf{(28)}$$

Table 8.2
Values of k_{cat} for Some Enzymes

Enzyme	k_{cat} (s^{-1})
Catalase	40,000,000
Carbonic anhydrase	1,000,000
Acetylcholinesterase	14,000
Penicillinase	2,000
Lactate dehydrogenase	1,000
Chymotrypsin	100
DNA polymerase I	15
Lysozyme	0.5

Turnover numbers for some representative enzymes are listed in table 8.2. The enormous value of 4×10^7 molecules/s achieved by catalase is among the highest known; the low value for lysozyme is at the other end of the spectrum. As is the case with K_m, the relationship of k_{cat} to individual rate constants, such as k_2 and k_3, depends on the details of the reaction mechanism.

Significance of the Specificity Constant, k_{cat}/K_m. Under physiological conditions, enzymes usually do not operate at saturating substrate concentrations. More typically, the ratio of the substrate concentration to the K_m is in the range of 0.01–1.0. If [S] is much smaller than K_m, the denominator of the Briggs-Haldane equation [equation (25)] is approximately equal to K_m, so the velocity of the reaction becomes

$$v \approx \frac{V_{max}}{K_m}[S] \qquad (\text{when } [S] \ll K_m)$$

$$= \frac{k_{cat}}{K_m}[E_t][S] \qquad \textbf{(29)}$$

The ratio k_{cat}/K_m is referred to as the specificity constant. Equation (29) indicates that the specificity constant provides a measure of how rapidly an enzyme can work at low [S]. Table 8.3 gives the values of the specificity constants for some particularly active enzymes.

The specificity constant k_{cat}/K_m is useful for comparing the relative abilities of different compounds to serve as a substrate for the same enzyme. If the concentrations of two substrates are the same, and are small relative to the K_m values, the ratio of the rates when the two substrates are present is equal to the ratio of the specificity constants.

Another use of the specificity constant is for comparing the rate of an enzyme-catalyzed reaction with the rate at which random diffusion brings the enzyme and substrate into contact. We mentioned previously that if every collision between a protein and a small molecule results in a reaction, the maximum value of the second-order rate constant is on the order of 10^8 to 10^9 $M^{-1}s^{-1}$. Some of the values of k_{cat}/K_m in table 8.3 are in this range. The reactions these enzymes catalyze proceed at nearly the maximum possible speed, given a fixed, low concentration of substrate and given the restriction that the enzyme and substrate have to find each other by diffusion. The only practical way to go much faster is to have the substrate generated right on the enzyme or in its immediate vicinity, so that little diffusional motion is necessary.

Table 8.3
Enzymes for Which k_{cat}/K_m Is Close to the Diffusion-Controlled Association Rate

Enzyme	Substrate	k_{cat} (s^{-1})	K_m (M)	k_{cat}/K_m ($M^{-1}s^{-1}$)
Acetylcholinesterase	Acetylcholine	1.4×10^4	9×10^{-5}	1.6×10^8
Carbonic anhydrase	CO_2	1×10^6	0.012	8.3×10^7
	HCO_3^-	4×10^5	0.026	1.5×10^7
Catalase	H_2O_2	4×10^7	1.1	4×10^7
Crotonase	Crotonyl-CoA	5.7×10^3	2×10^{-5}	2.8×10^8
Fumarase	Fumarate	800	5×10^{-6}	1.6×10^8
	Malate	900	2.5×10^{-5}	3.6×10^7
Triosephosphate isomerase	Glyceraldehyde 3-phosphate	4.3×10^3	4.7×10^{-4}	2.4×10^8
β-Lactamase	Benzylpenicillin	2.0×10^3	2×10^{-5}	1×10^8

Source: From Enzyme Structure and Mechanism 2/E by Fersht. Copyright © 1985 W. H. Freeman and Company. Used with permission.

Kinetics of Enzymatic Reactions Involving Two Substrates

Enzymes that catalyze reactions with two or more substrates work in a variety of ways. In some cases the intermolecular reaction occurs when all the substrates are bound in a common enzyme–substrate complex; in others the substrates bind and react one at a time. A frequent application of kinetic measurements is to distinguish between such alternatives.

Consider a reaction in which two substrates, S_1 and S_2, are converted to products P_1 and P_2. One way for the reaction to occur is for S_1 to bind to the enzyme first, forming the binary complex ES_1. Binding of S_2 can then form the ternary complex ES_1S_2, which gives rise to the products:

$$E \xrightarrow{S_1} ES_1 \xrightarrow{S_2} ES_1S_2 \xrightarrow{P_1 + P_2} E \qquad (30)$$

This process is referred to as an ordered pathway. An alternative is a random-order pathway, in which the two substrates can bind to the enzyme in either order. Still another scheme is for S_1 to bind to the enzyme and be converted to P_1, leaving the enzyme in an altered form, E′. S_2 then binds to E′ and is converted to P_2, returning the enzyme to its original form:

$$E \xrightarrow{S_1} ES_1 \xrightarrow{P_1} E' \xrightarrow{S_2} E'S_2 \xrightarrow{P_2} E \qquad (31)$$

This process is called the Ping-Pong mechanism to emphasize the bouncing of the enzyme between two states, E and E′. Ping-Pong pathways are commonly observed with enzymes that contain bound coenzymes. Interconversion of the enzyme between the two forms usually involves modification of the coenzyme.

Kinetic equations for these and other mechanisms can be worked out just as we have done for the reactions involving only one substrate. Techniques for doing this are described in the references at the end of the chapter; here we simply illustrate a few of the results. For the Ping-Pong mechanism [equation (31)] the double-reciprocal form of the final expression is

$$\frac{1}{v} = \frac{1}{V_{max}}\left(1 + \frac{K_{m2}}{[S_2]}\right) + \frac{K_{m1}}{V_{max}}\frac{1}{[S_1]} \qquad (32)$$

where K_{m1} is the Michaelis constant for S_1, and K_{m2} is that for S_2. This expression is similar to that for a one-substrate reaction (equation 26), except that the first term on the right is multiplied by the factor $(1 + K_{m2}/[S_2])$. If we measure the rate as a function of $[S_1]$, keeping $[S_2]$ constant, a plot of $1/v$ versus $1/[S_1]$ is linear, but the intercept on the ordinate (the apparent V_{max}) depends on $[S_2]$. Increasing $[S_2]$ increases the apparent V_{max} (fig. 8.10). From a series of such plots, measured at different values of $[S_2]$, we can find the true V_{max} in addition to K_{m1} and K_{m2}.

An ordered pathway [equation (30)] results in the expression

$$\frac{1}{v} = \frac{1}{V_{max}}\left(1 + \frac{K_{m2}}{[S_2]} + \frac{K_{m1}}{[S1]} + \frac{K_{m2}}{[S_2]}\frac{K_{s1}}{[S_1]}\right) \qquad (33)$$

where K_{s1} is the dissociation constant for ES_1. A plot of $1/v$ versus $1/[S_1]$ at constant $[S_2]$ is still linear, but now both the slope and the intercept depend on $[S_2]$ (fig. 8.11). Again, all of the kinetic parameters can be obtained from a series of plots measured at different S_2 concentrations. However, additional measurements must be made to determine which substrate binds to the enzyme first. In some cases the substrate that binds first can be shown to

Figure 8.10

Double-reciprocal plots ($1/v$ versus $1/[S_1]$) for the Ping-Pong mechanism. Measurements made at different values of $[S_2]$ give a set of parallel straight lines. The intercepts on the ordinate depend on $[S_2]$. K_{m1}, K_{m2}, and V_{max} can be obtained by replotting the intercepts as a function of $1/[S_2]$.

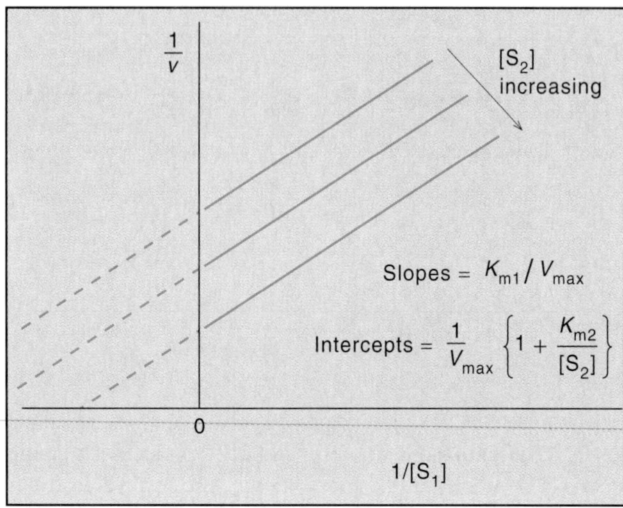

Figure 8.11

Double-reciprocal plots for an ordered pathway. Measurements made at different fixed values of $[S_2]$ give a set of lines that intersect to the left of the ordinate. The two values of K_m, as well as V_{max}, and K_{s1} can be obtained by replotting the slopes and intercepts of these lines as functions of $1/[S_2]$. A random pathway gives similar results but can be distinguished by making such measurements for the reverse reaction ($P_1 + P_2 \rightarrow S_1 + S_2$) in addition to the forward reaction.

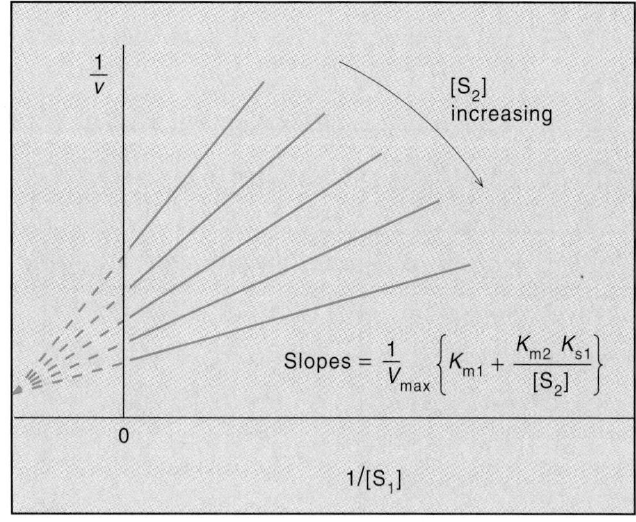

form a stable enzyme–substrate complex in the absence of the other substrate.

Effects of Temperature and pH on Enzymatic Activity

For most enzymes the turnover number increases with temperature until a temperature is reached at which the enzyme is no longer stable (fig. 8.12). Above this point there is a precipitous, and usually irreversible, drop in activity. At lower temperatures the temperature dependence of k_{cat} can be related to the activation energy of the slowest (rate-limiting) step in the catalytic pathway [see equation (12)]. With many enzymes a 10°C rise in temperature increases k_{cat} by about a factor of 2, which translates into an activation energy of about 12 kcal/mole.

Enzymes, like other proteins, are stable over only a limited range of pH. Outside this range, changes in the charges on ionizable amino acid residues result in modifications of the tertiary structure of the protein and eventually lead to denaturation. But within the range at which an enzyme is stable, both k_{cat} and K_m often depend on pH. The effects of pH can reflect the pK_a of ionizing groups on either the enzyme or the substrate. A substrate that has an amine group, for example, may bind to the enzyme best when this group is protonated. In many cases, however, the pH dependence reflects ionizable residues that constitute the active site on the enzyme or are essential for maintaining the structure of the active site, and the optimum pH is a characteristic more of the enzyme than of the substrate. Thus the maximum activity of chymotrypsin always occurs around pH 8, the activity of pepsin peaks around pH 2, and acetylcholinesterase works best at pH 7 or higher (fig. 8.13). The activity of papain, on the other hand, is essentially independent of pH between 4 and 8.

Enzyme Inhibition

Most enzymes are sensitive to inhibition by specific agents that interfere with the binding of a substrate at the active site or with conversion of the enzyme–substrate complex into products. Two of the major applications of kinetic measurements are in distinguishing between different types of inhibition and in providing quantitative information on the effectiveness of various inhibitors. Such information is essential for an understanding of how cells regulate their enzymatic activities. Comparisons of a series of inhibitors also help to map the structure of an enzyme's active site and are a key step in the design of therapeutic drugs.

Competitive Inhibitors Bind at the Active Site

In many cases an inhibitor resembles the substrate structurally and binds reversibly at the same site on the enzyme. This activity is called competitive inhibition because the inhibitor and the substrate compete for binding (fig. 8.14*a*). The inhibitor is prevented from binding if the active site is already occupied by the substrate. As an example, consider the proteolytic enzyme trypsin, which cleaves polypeptide chains at peptide linkages adjacent to basic amino acid residues. Trypsin is inhibited competitively by benza-

Figure 8.12

The activity of a typical enzyme as a function of temperature. The turnover number (k_{cat}) increases with temperature until a point is reached at which the enzyme is no longer stable. The temperature at which k_{cat} is greatest should not be interpreted as the "optimum temperature" for the enzyme. Because denaturation of the enzyme occurs continuously during the measurement, the position of the maximum depends on how quickly the experimenter is able to assay the enzyme activity.

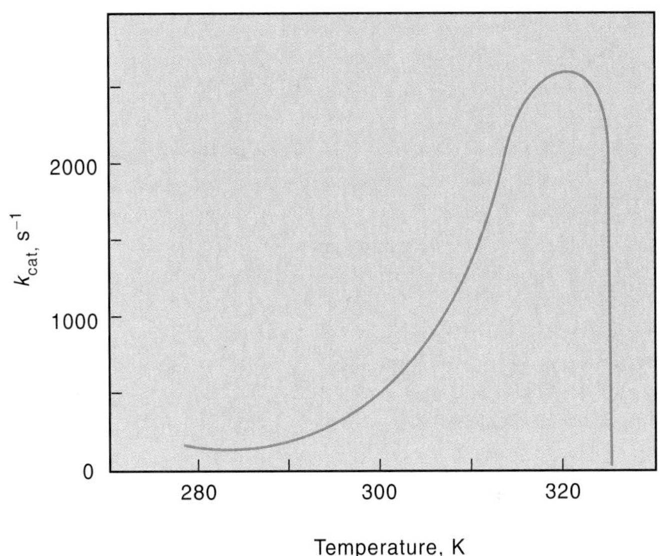

Figure 8.13

Enzyme activity (k_{cat}/K_m) as a function of pH for three different enzymes. The optimum pH usually is a characteristic of the enzyme and not the particular substrate. Often the pH sensitivity is an indication of an ionizable group at the active site.

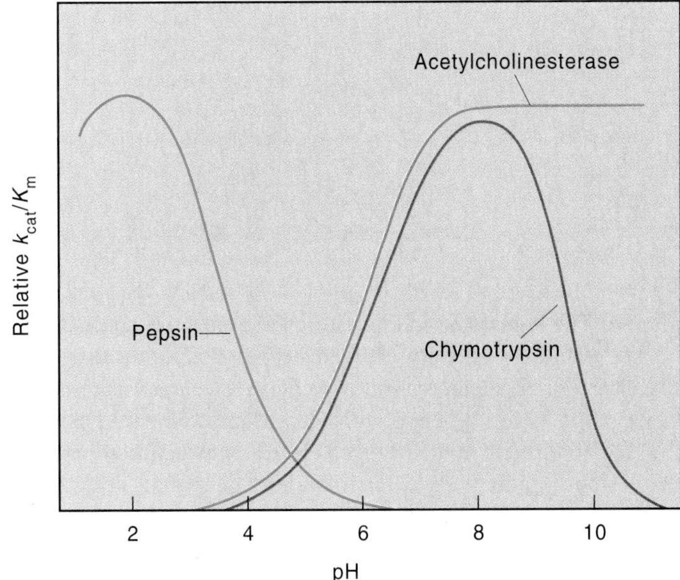

Figure 8.14

Types of enzyme inhibition. (*a*) A competitive inhibitor competes with the substrate for binding at the same site on the enzyme. (*b*) A noncompetitive inhibitor binds to a different site but blocks the conversion of the substrate to products. (*c*) An uncompetitive inhibitor binds only to the enzyme–substrate complex. (E = enzyme; S = substrate.)

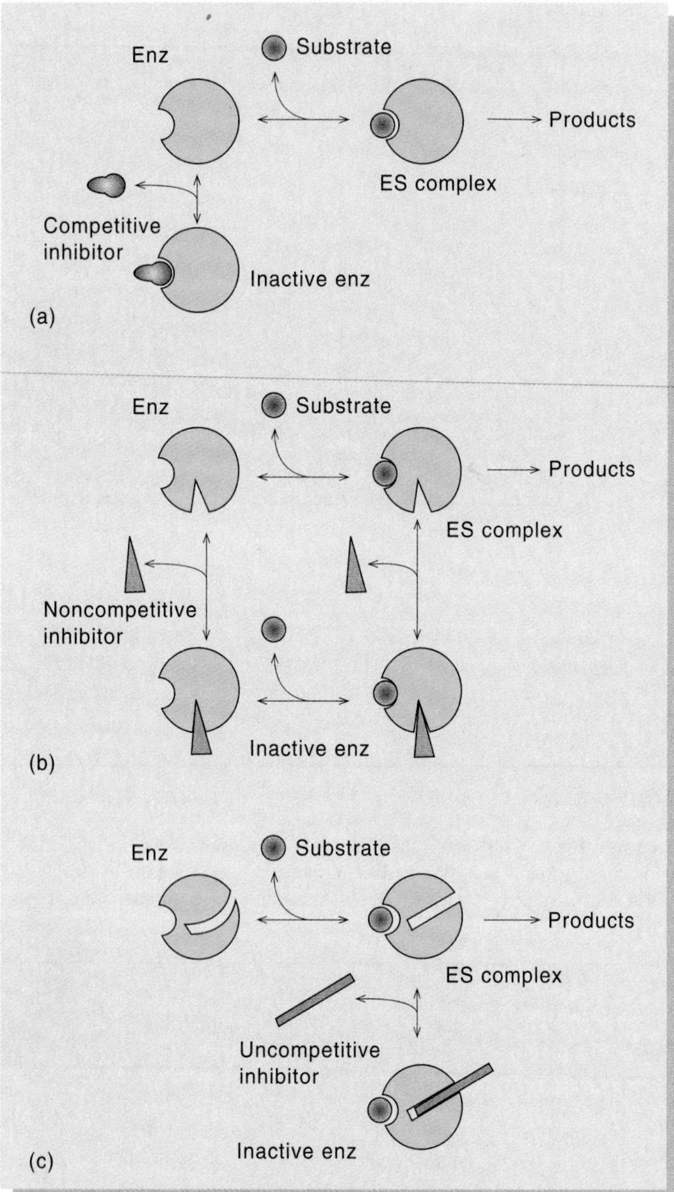

Figure 8.15

The specificity pocket of trypsin can accommodate an arginine side chain of a polypeptide substrate or a benzamidine ion, which acts as a competitive inhibitor. Asp = aspartic acid.

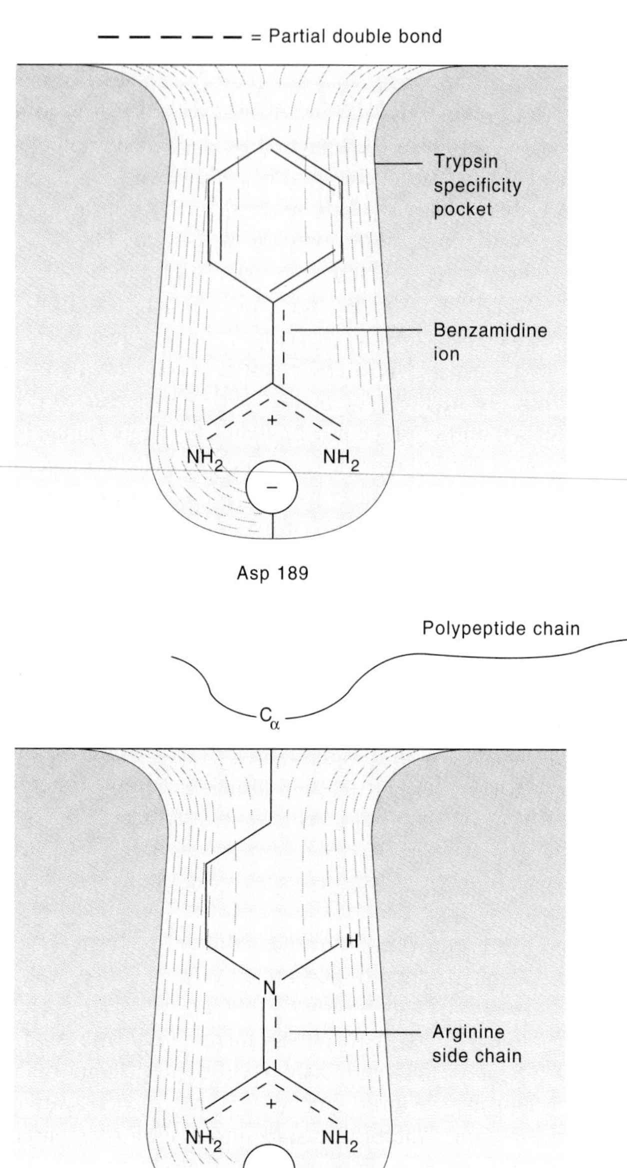

midine. The substrate-binding site on the enzyme consists of a pocket where a positively charged lysine or arginine side chain fits snugly and interacts with a negatively charged carboxylate group (fig. 8.15). When protonated, benzamidine is positively charged and has a flat, delocalized electronic structure resembling that of an arginine side chain. The binding pocket accepts and binds a benzamidine ion reversibly in place of its regular substrate.

An expression describing enzyme kinetics in the presence of a competitive inhibitor can be derived straightforwardly. Consider the reaction that we treated previously:

$$\mathrm{E + S} \underset{k_{-1}}{\overset{k_1}{\rightleftharpoons}} \mathrm{ES} \underset{k_{-2}}{\overset{k_2}{\rightleftharpoons}} \mathrm{E + P} \qquad \textbf{(19')}$$

We now have the additional feature that the enzyme also reacts reversibly with the inhibitor (I) to give an inactive complex (EI):

$$\mathrm{E + I} \rightleftharpoons \mathrm{EI} \qquad \textbf{(34)}$$

The derivation proceeds just as the derivation that led to equations (25) and (26), except that the total enzyme concentration $[E_t]$ is now [E] + [ES] + [EI], instead of just [E] + [ES]. As a result, in place of equation (26) we end up with

$$\frac{1}{v} = \frac{1}{V_{max}} + \frac{K_m}{V_{max}} \frac{1}{[S]} \left(1 + \frac{[I]}{K_i}\right) \quad \textbf{(35)}$$

where K_i is the dissociation constant of the enzyme–inhibitor complex.

$$K_i = \frac{[E][I]}{[EI]} \quad \textbf{(36)}$$

According to equation (35), a plot of $1/v$ versus 1/[S] is linear and passes through the same intercept on the ordinate as the plot obtained in the absence of the inhibitor ($1/V_{max}$). This is because the effect of the inhibitor disappears at very high substrate concentrations for a competitive inhibitor. The slope of the double-reciprocal plot, however, depends on the product

$$\frac{K_m}{V_{max}} \left(1 + \frac{[I]}{K_i}\right)$$

instead of simply K_m/V_{max} (fig. 8.16). As a result, the apparent K_m is altered to K'_m, where

$$K'_m = K_m \left(1 + \frac{[I]}{K_i}\right) \quad \textbf{(37)}$$

By measuring the slope of the plot as a function of [I] we can determine K_i.

Figure 8.16

Competitive inhibition. A series of double-reciprocal plots ($1/v$ versus 1/[S]) measured at different concentrations of the inhibitor (I) all intersect at the same point ($1/V_{max}$) on the ordinate. The slopes of the plots and the intercepts on the abscissa are simple, linear functions of $[I]/K_i$, where K_i is the dissociation constant of the inhibitor–enzyme complex.

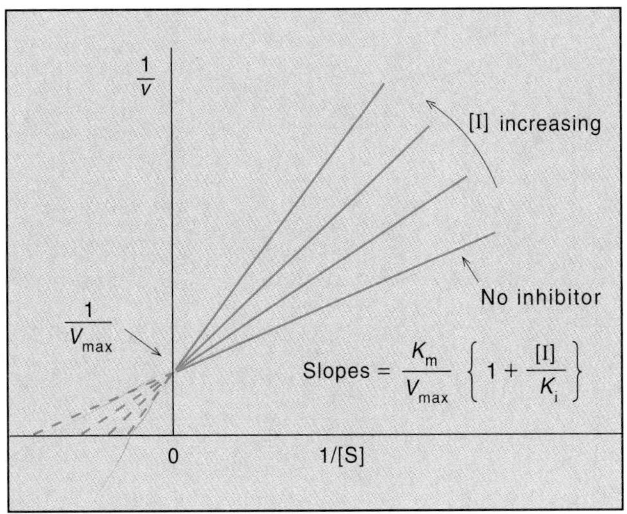

Figure 8.17

Noncompetitive inhibition. The double-reciprocal plots pass through different points on the ordinate but intersect at the same point ($-1/K_m$) on the abscissa. The slopes and the intercepts on the ordinate are linear functions of $[I]/K_i$.

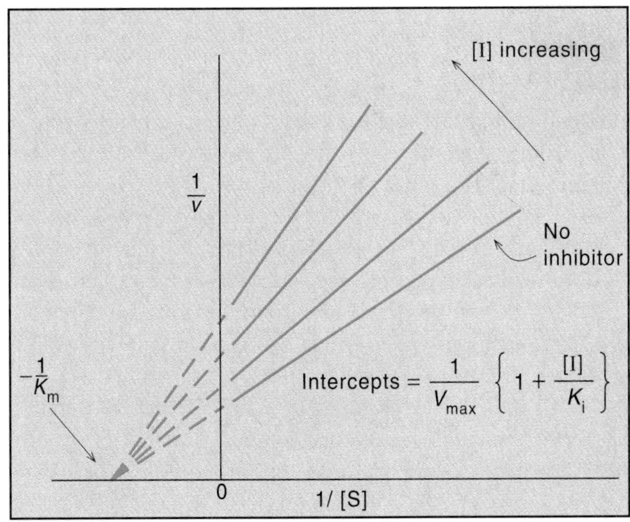

Noncompetitive and Uncompetitive Inhibitors Do Not Compete Directly with Substrate Binding

Some inhibitors bind at sites other than the enzyme's active site and do not compete directly with binding of the substrate. Instead, they act by interfering with the reaction of the enzyme–substrate complex. An inhibitor that binds to an enzyme whether or not the active site is occupied by the substrate is termed a noncompetitive inhibitor (see fig. 8.14*b*). A noncompetitive inhibitor decreases the maximum velocity of an enzymatic reaction without affecting the K_m. The inhibitor removes a certain fraction of the enzyme from operation, no matter the concentration of the substrate. Plots of $1/v$ versus 1/[S] in the presence of different concentrations of a noncompetitive inhibitor intersect at the same point on the abscissa ($-1/K_m$) but pass through the ordinate at different points (fig. 8.17).

Still another possibility is that the inhibitor binds only to the enzyme–substrate complex and not to the free enzyme (see fig. 8.14*c*). This reaction is called uncompetitive inhibition. Uncompetitive inhibition is rare in reactions that involve a single substrate but more common in reactions with multiple substrates. Plots of $1/v$ versus 1/[S] at different concentrations of an uncompetitive inhibitor give a series of parallel lines (fig. 8.18).

Irreversible Inhibitors Permanently Alter the Enzyme Structure

The various types of inhibition that we have been discussing are all reversible. If the inhibited enzyme is dialyzed to remove the inhibitor, its activity increases again. Reversibility of the binding of the inhibitor is implicit in our use of a dissociation constant,

Figure 8.18

Uncompetitive inhibition. The double-reciprocal plots are parallel.

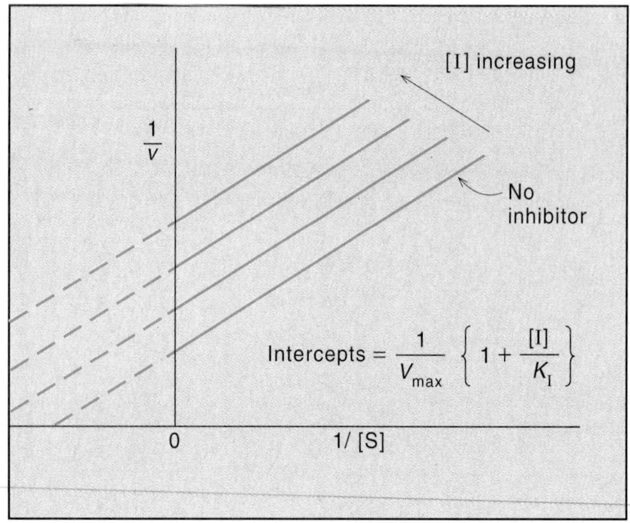

K_I. There are, however, numerous inhibitors that react essentially irreversibly with enzymes, usually by the formation of a covalent bond to the functional group of an amino acid side chain or to a bound coenzyme. Some examples of such inhibitors are given in table 8.4. The effect of an irreversible inhibitor can be to change either V_{max} or K_m, or both.

Irreversible inhibitors often provide clues to the nature of the active site on an enzyme. Enzymes that are inhibited by organic mercurial compounds or by iodoacetate, for example, frequently have a cysteine in the active site, and the cysteinyl sulfhydryl group often plays an essential role in the catalytic mechanism (fig. 8.19). An example is glyceraldehyde-3-phosphate dehydrogenase, in which the catalytic mechanism begins with a reaction of the cysteine with the aldehyde substrate to form a thiohemiacetal. The mechanism of action of this enzyme will be discussed in more detail in chapter 14. Diisopropylfluorophosphate reacts irreversibly with a critical serine residue in many proteolytic enzymes, including trypsin and chymotrypsin (see fig. 8.19). The reaction of the serine group destroys the catalytic activity. The mechanism of action of trypsin and chymotrypsin is discussed in the following chapter.

In the case of glyceraldehyde-3-phosphate dehydrogenase, further studies have confirmed that the active site does contain an essential cysteine residue, and the same is true of the essential serine in trypsin and chymotrypsin. But in exploring a new enzyme it is important to keep in mind that the chemical modification of an amino acid side chain generally causes some perturbation of the secondary or tertiary structure of a protein. A reaction involving an amino acid residue well outside the active site thus could have a long-range disruptive effect that alters the structure of the active site sufficiently to inhibit the enzyme. This possibility is less of a concern with a reversible, competitive inhibitor because in that case the competition with the substrate supports the conclusion that the inhibitor binds directly in the active site.

Table 8.4

Some Inhibitors That Form Covalent Linkages with Functional Groups on the Enzyme

Inhibitor	Enzyme Group That Combines with Inhibitor
Cyanide	Fe, Cu, Zn, other transition metals
p-Mercuribenzoate	Sulfhydryl
Diisopropylfluorophosphate	Serine hydroxyl
Iodoacetate	Sulfhydryl, imidazole, carboxyl, thioether

Figure 8.19

Iodoacetamide is an irreversible inhibitor of many enzymes that contain a cysteine residue in the active site. Diisopropylfluorophosphate is an irreversible inhibitor of trypsin, chymotrypsin, and several related enzymes. It reacts with a serine residue at the active site.

$$ICH_2C(=O)NH_2 + Enz—SH \xrightarrow{-HI} Enz—S—CH_2C(=O)NH_2$$

Iodoacetamide → **Alkylated enz**

$$F—P(=O)(—O—CH(CH_3)_2)_2 + Enz—OH \xrightarrow{-HF} Enz—O—P(=O)(—O—CH(CH_3)_2)_2$$

Diisopropyl-fluorophosphate → **Diisopropylphosphate derivative of enz**

An irreversible inhibitor often can be designed for the active site of a particular enzyme by incorporating a reactive group in a molecule that resembles a substrate. For example, 3-bromoacetol phosphate is a structural analog of dihydroxyacetone phosphate, which is a substrate for triosephosphate isomerase (fig. 8.20*a*). The inhibitor binds to the active site of the enzyme and then reacts irreversibly with the carboxyl group of a nearby glutamic acid residue. Binding to the active site greatly increases the selectivity of the inhibitor for reaction with this particular residue, in preference to glutamyl residues elsewhere in the protein. Labeling the inhibitor with a radioisotope such as ^{14}C or ^3H facilitates the identification of the derivatized amino acid residue after the protein has been split into smaller peptides, making it possi-

Figure 8.20

Affinity labels and suicide inhibitors. (*a*) 3-Bromoacetol phosphate is a structural analog of dihydroxyacetone phosphate. It binds to the active site of triosephosphate isomerase and then reacts to form a covalent bond with the carboxyl group of a nearby glutamyl residue. Bromoketone groups have been incorporated into many molecules to make similar affinity labels for other enzymes. (*b*) A photoaffinity label can be made by attaching a diazoacetyl group to a molecule (R) that resembles the substrate for a particular enzyme. After the reagent binds to the active site, it is exposed to light. This causes it to break down, forming a carbene derivative. The carbene reacts rapidly with any of several amino acid residues to form a covalent bond to the enzyme. (*c*) Vinylglycine can be used as a mechanism-based inhibitor for some enzymes that catalyze modifications of amino acids. It is not intrinsically a reactive compound but is converted into a reactive allyl-imine in the course of the reaction catalyzed by the enzyme.

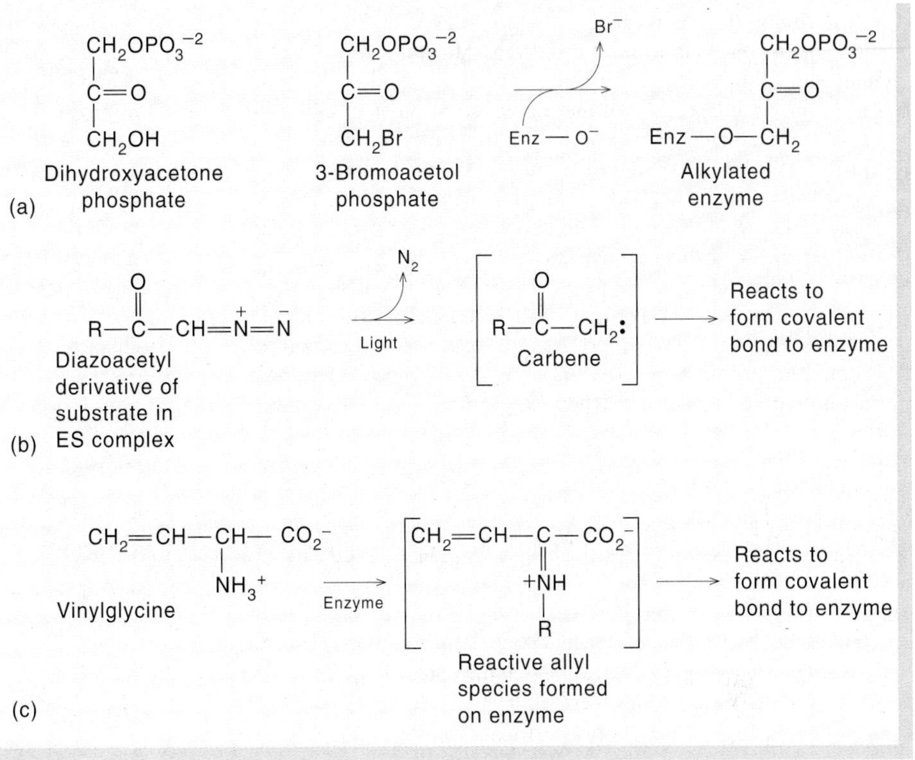

ble to locate the reactive residue in the amino acid sequence. Such a reagent is called an affinity label. Photoaffinity labels are a particularly useful group of reagents of this type, in which the covalent attachment to the protein can be triggered by light after the reagent has bound to the enzyme (see fig. 8.20*b*). Another related technique is to use a reagent that is not intrinsically reactive but becomes reactive after it has been modified chemically by the enzyme itself (see fig. 8.20*c*). Such a reagent is termed a mechanism-based inhibitor or suicide substrate, to emphasize that the enzyme brings about its own inhibition.

Metal-Ion Chelators. Enzymes that require metal ions as cofactors often are inhibited by chelators that bind to the metal. Examples of such metalloenzymes are lactate dehydrogenase from muscle and aldolase from yeast, both of which contain Zn^{2+}. Chelators inhibit aldolase by removing the required metal. The inhibition is not reversed simply by dialysis, but it can be reversed by adding Zn^{2+} to the depleted enzyme. In the case of lactate dehydrogenase the enzyme holds the Zn^{2+} more tightly, and the metal-chelator complex remains attached to the inhibited enzyme.

Summary

Enzymes are biological catalysts. Kinetic analysis is one of the most broadly used tools for characterizing enzymatic reactions.

1. The rate of a reaction depends on the frequency of collisions between the reacting species and on the fraction of the collisions that produce products. The former depends on the concentrations of the reactants; the latter depends on the temperature and activation free energy $\Delta G^\ddagger$. $\Delta G^\ddagger$ can be interpreted as the free energy needed to convert the reactants to a transition state. A catalyst increases the reaction rate by lowering $\Delta G^\ddagger$.
2. Enzyme kinetics usually are studied by mixing the enzyme and substrates and measuring the initial rate of formation of product or the disappearance of a reactant. Special techniques are necessary to measure very fast reactions. It is common to measure the rate as a function of substrate concentration, pH, and temperature.
3. Enzymes have localized catalytic sites. The substrate (S) binds at the active site to form an enzyme–substrate complex (ES). Subsequent steps transform the bound substrate into product and regenerate the free enzyme. The overall speed of the reaction depends on the concentration of ES. Shortly after the enzyme and substrate are mixed, [ES] becomes approximately constant and remains so for a period of time termed the steady state. The rate (v) of the reaction in the steady state usually has a hyperbolic dependence on the substrate concentration. It is proportional to [S] at low concentrations but approaches a maximum (V_{max}) when the enzyme is fully charged with substrate. The Michaelis constant K_m is the substrate

concentration at which the rate is half maximal. K_m and V_{max} often can be obtained from a plot of $1/v$ versus $1/[S]$. If ES is in equilibrium with the free enzyme and substrate, K_m is equal to the dissociation constant for the complex (K_s). More generally, K_m depends on at least three rate constants and is larger than K_s.

4. The turnover number k_{cat} is the maximum number of molecules of substrate converted to product per unit time per active site and is V_{max} divided by the total enzyme concentration. The specificity constant k_{cat}/K_m is a measure of how rapidly an enzyme can work at low substrate concentrations. This is usually the best index of the effectiveness of an enzyme.
5. Enzymes that catalyze reactions of two or more substrates work in a variety of ways that can be distinguished by kinetic analysis. Some enzymes bind their substrates in a fixed order; others bind in random order. In some cases, binding of one substrate gives a partial reaction before the second substrate binds.
6. Enzymes can be inhibited by agents that interfere with the binding of substrate or with conversion of the ES complex into products. Reversible inhibitors are classified as competitive, noncompetitive, or uncompetitive. A competitive inhibitor competes with substrate for binding at the active site. Consequently, a sufficiently high concentration of substrate can eliminate the effect of a competitive inhibitor. Noncompetitive inhibitors bind at a separate site and block the reaction regardless of whether the active site is occupied by substrate. An uncompetitive inhibitor binds to the ES complex but not to the free enzyme. These three forms of inhibition are distinguishable by measuring the rate as a function of the concentrations of the substrate and inhibitor. Irreversible inhibitors often provide information on the active site by forming covalently linked complexes that can be characterized.

Selected Readings

Abeles, R. H., P. A. Frey, and W. P. Jencks, *Biochemistry*. Boston: Jones and Bartlett, 1992. Chapters 3, 4, and 5 present a most detailed and authoritative account of how chymotrypsin works. But also see chapter 9 here for an update on the importance of low-barrier hydrogen bonds.

Advances in Enzymology. New York: Academic Press. An annually published volume containing monographs on selected topics.

Benkovic, S. J., Catalytic antibodies. *Ann. Rev. Biochem.* 61:29–54, 1992.

Boyer, P. D. (ed.), *The Enzymes*. New York: Academic Press. A continuing series of monographs on selected enzymes. See particularly the chapter entitled "Steady State Kinetics" by W. W. Cleland in vol. 2.

Fersht, A., *Enzyme Structure and Mechanism,* 2d ed. New York: W. H. Freeman & Co., 1985.

Frost, A. A., and R. G. Pearson, *Kinetics and Mechanism,* 2d. ed. New York: Wiley, 1961. An excellent introduction to general chemical kinetics.

Jencks, W. P., Economics of enzyme catalysis. *Cold Spring Harbor Symposium* 52:65, 1987.

Lolis, E., and G. A. Petsko, Transition-state analogues in protein crystallography: Probes of the structural source of enzyme catalysis. *Ann. Rev. Biochem.* 59:597–630, 1990.

Murphy, D. J., Revisiting ground-state and transition-state effects: The split-site model, and the "fundamentalist position" of enzyme catalysis. *Biochem.* 34:4507–4510, 1995.

Purich, D. L., *Contemporary Enzyme Kinetics and Mechanism*. New York: Academic Press, 1983. Selected chapters from *Methods in Enzymology*. Detailed information on how to analyze kinetic data and on effects of temperature, pH, and inhibitors.

Segal, I. H., *Enzyme Kinetics*. New York: Wiley, 1975.

Problems

1. Explain what is meant by the order of a reaction, using the reaction below as an example. What is the reaction order for each reactant? (Consider the forward and reverse reactions.) What is the order for the overall reaction?

$$\mathrm{A + B \rightleftharpoons 2C}$$

2. In a first-order reaction a substrate is converted to product so that 87% of the substrate is converted in 7 min. Calculate the first-order rate constant. In what time would 50% of the substrate be converted to product?
3. K_m is frequently equated with K_s, the [ES] dissociation constant. However, there is usually a disparity between those values. Why? Under what conditions are K_m and K_s equivalent?
4. Differentiate between the enzyme–substrate complex and the transition-state intermediate in an enzymatic reaction.
5. What is the steady-state approximation and under what conditions is it valid?
6. Assume that an enzyme-catalyzed reaction follows the scheme shown:

$$\mathrm{E + S} \underset{k_2}{\overset{k_1}{\rightleftharpoons}} \mathrm{ES} \underset{k_4}{\overset{k_3}{\rightleftharpoons}} \mathrm{E + P}$$

where $k_1 = 10^9\ \text{M}^{-1}\ \text{s}^{-1}$, $k_2 = 10^5\ \text{s}^{-1}$, $k_3 = 10^2\ \text{s}^{-1}$, $k_4 = 10\ \text{M}^{-1}\ \text{s}^{-1}$ and $[E_T]$ is 0.1 nM. Determine the value of each of the following.

(a) K_m
(b) V_{max}
(c) Turnover number
(d) Initial velocity when $[S]_0$ is 20 μM.

7. A colleague has measured the enzymatic activity as a function of reaction temperature and obtained the data shown in this graph. He insists on labeling point A as the "temperature optimum" for the enzyme. Try, tactfully, to point out the fallacy of that interpretation.

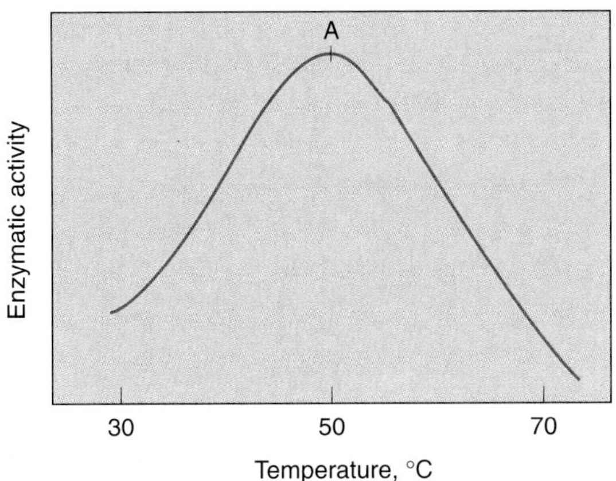

8. Prove that the K_m equals the substrate concentration at one-half maximal velocity.
9. When quantifying the activity of an enzyme, does it matter whether you measure the appearance of a product or the disappearance of a reactant?
10. You measured the initial velocity of an enzyme in the absence of inhibitor and with inhibitor A or inhibitor B. In each case the inhibitor is present at 10 μM. The data are shown in the table.

[S], (mM)	Velocity, (M s^{-1}) × 10^7 Uninhibited	Velocity, (M s^{-1}) × 10^7 Inhibitor A	Velocity, (M s^{-1}) × 10^7 Inhibitor B
0.333	1.65	1.05	0.794
0.40	1.86	1.21	0.893
0.50	2.13	1.43	1.02
0.666	2.49	1.74	1.19
1.0	2.99	2.22	1.43
2.0	3.72	3.08	1.79

(a) Determine K_m and V_{max} of the enzyme.
(b) Determine the type of inhibition imposed by inhibitor A and calculate K_I(s).
(c) Determine the type of inhibition imposed by inhibitor B and calculate K_I(s).

11. The initial velocity data shown in the table were obtained for an enzyme.

[S], (mM)	Velocity, (M s^{-1}) × 10^7
0.10	0.96
0.125	1.12
0.167	1.35
0.250	1.66
0.50	2.22
1.0	2.63

Each assay at the indicated substrate concentration was initiated by adding enzyme to a final concentration of 0.01 nM. Derive K_m, V_{max}, k_{cat}, and the specificity constant.

12. What is the primary advantage of the use of the Lineweaver-Burk plot (figure 8.9) over a velocity versus [S] plot (fig. 8.6) and the use of equation 25?
13. What is a disadvantage of the use of the Lineweaver-Burk plot (fig. 8.9)?
14. The enzyme deoxyribokinase catalyzes the following reaction:

$$\text{2-Deoxyribose} + \text{ATP} \longrightarrow \text{ADP} + \text{5-Phospho-2-deoxyribose}$$

The following data (derived from A. Ginsburg, *J. Biol. Chem.* 234:481, 1959) were obtained in a study on the impact of the concentration of 2-deoxyribose on the rate of product formation by deoxyribokinase (the reaction conditions such as pH, buffer identity, amount of enzyme, and sample volumes were held constant for all samples):

Sample No.	[2-Deoxyribose], mM	ADP Formation, Change in [ADP], μM/10 min
1	1.00	0.115
2	1.00	0.124
3	1.60	0.147
4	2.56	0.191
5	5.12	0.255
6	6.88	0.274
7	15.0	0.294
8	20.0	0.300

Use (a) *V* versus [S] and (b) 1/*V* versus 1/[S] (Lineweaver-Burk plot) to graphically determine the K_m and V_{max} for 2-deoxyribose.

15. Consider the structure of RNA and then comment on why the word "surprisingly" was used in the text statement "Surprisingly, however, recent work has shown that some RNA molecules also have enzymatic activity."
16. Why does an organic chemist typically use terms such as "collision, productive collision, velocity, or attack" when describing a reaction (mechanism), while a biochemist is more inclined to use the terms "approach, contact, touch, abut, or meet" when describing a reaction?
17. Enzymologists refer to the specificity of an enzyme, meaning the "preferred" or most likely substrate. Consider table 8.1 to determine the specificity of hexokinase. Also consider the synthetic substrates listed in the same table for chymotrypsin and determine the relative specificity for these substrates. Consider the turnover number for the enzymes to be independent of the substrate listed.
18. It is often disturbing for biochemistry students to learn that ultimately, diffusion is the factor that delivers substrates to the enzyme and removes the products. Why is the idea of giving diffusion such an important role a disturbing thought?

19. The following data were obtained on isocitrate lyase from an algae species (*Biochem. J.* 110:481, 1968). Deduce the K_m and V_{max} for the enzyme and determine the nature of the inhibition by oxaloacetate.

[S], mM	Glyoxylate Formed after 10 min, μM	Glyoxylate Formed after 10 min with 0.5 mM Oxaloacetate, μM
0.0318	0.0420	0.0040
0.0464	0.0583	0.0055
0.0593	0.0700	0.0075
0.1185	0.0955	0.0131
0.2222	0.1167	0.0233

The enzyme was supplied with a buffer, and after 10 min the enzyme was denatured and the amount of glyoxylate determined spectophotometrically.

C H A P T E R

9

Mechanisms of Enzyme Catalysis

Enzymes bring the reactants (substrates) together in a location where they are ideally oriented with respect to each other and the catalytic groups of the enzyme.

In chapter 8 we saw that enzymes can increase the rates of reactions by many orders of magnitude. We noted that enzymes work under mild conditions of temperature, pH, and pressure and that they are highly specific in the types of reactions they catalyze and in the particular substrates they accept. In this chapter we will explore the mechanisms of several enzyme-catalyzed reactions in greater detail. Our goal is to relate the activity of each of these enzymes to the structure of the active site, where the functional groups of amino acid side chains, the polypeptide backbone, or bound cofactors must interact with the substrates in such a way as to favor the formation of the transition state. We will be exploring enzyme catalytic mechanisms in many subsequent chapters as well, but usually in less detail than here.

Five Themes That Recur in Discussing Enzymatic Reactions

Several broad themes recur frequently in discussing enzymatic reaction mechanisms. Among the most important of these are (1) the proximity effect, (2) electrostatic effects, (3) general-acid and general-base catalysis, (4) nucleophilic or electrophilic catalysis by enzymatic functional groups, and (5) structural flexibility. For all known enzymes at least one of these themes is relevant, and in most cases more than one. We will start by discussing the five themes in general terms and then see how they apply to some representative enzymes.

The Proximity Effect: Enzymes Bring Reacting Species Close Together

The idea of the proximity effect is that an enzyme can accelerate a reaction between two species simply by holding the two reactants close together in an appropriate orientation. It has long been known that intramolecular reactions between groups that are tied together in a single molecule are faster than the corresponding intermolecular reactions between two independent molecules. The cyclization of succinic acid to form succinyl anhydride (equation 1), for example, is much more rapid than the formation of acetic anhydride from two molecules of acetic acid (equation 2):

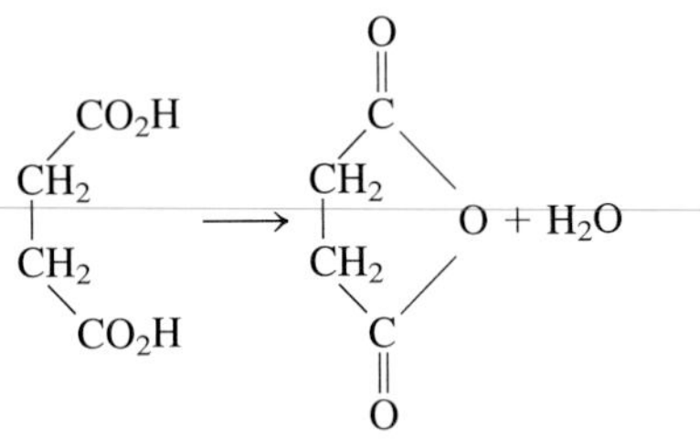

$$\begin{matrix} & CO_2H & & & & O \\ / & & & & & \| \\ CH_2 & & & CH_2 & - & C \\ | & & \longrightarrow & | & & & \diagdown \\ CH_2 & & & CH_2 & - & C & O + H_2O \\ \diagdown & & & & & \| \\ & CO_2H & & & & O \end{matrix} \qquad (1)$$

$$\begin{matrix} CH_3CO_2H & & CH_3C(=O) \\ & \longrightarrow & \quad \diagdown\; O + H_2O \\ CH_3CO_2H & & CH_3C(=O) \diagup \end{matrix} \qquad (2)$$

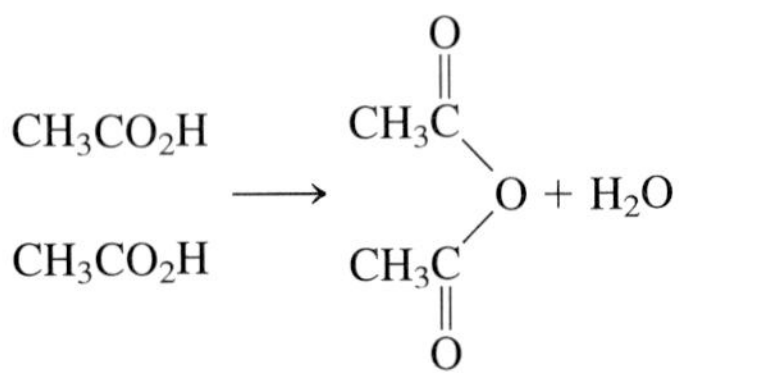

It is not possible to compare the rate constants for these two reactions directly, because they are expressed in different units. The intramolecular reaction (equation 1) is kinetically first order, while the intermolecular reaction (equation 2) is second order. But suppose that one of the two reactants in the intermolecular reaction is present in great excess over the other reactant, so that the process is effectively first order in the concentration of the second, limiting reactant. Then we can ask what the molarity of the more abundant species would have to be to make the effective first-order rate constant of the intermolecular reaction the same as the measured first-order rate constant for the intramolecular reaction. For the reactions of succinic and acetic acids (equations 1 and 2) the answer is 3×10^5 M. This is far above any concentration that could be obtained, even if the intermolecular reaction were carried out in pure acetic acid. (The concentration of CH_3COOH in glacial acetic acid is only 17.5 M.) In the related reaction the result is even more dramatic:

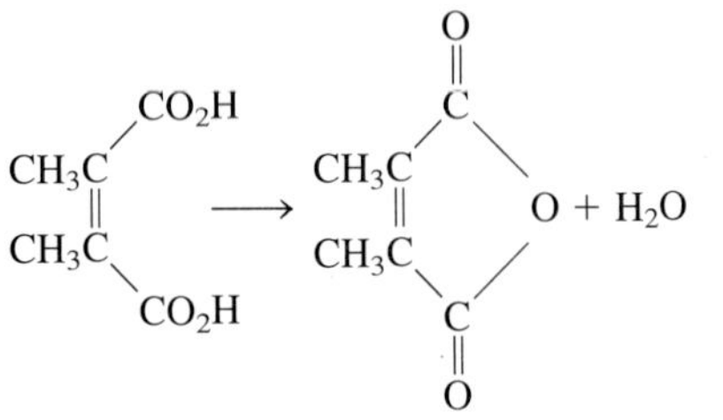

$$\begin{matrix} & CO_2H & & & & O \\ / & & & & & \| \\ CH_3C & & & CH_3C & - & C \\ \| & & \longrightarrow & \| & & & \diagdown \\ CH_3C & & & CH_3C & - & C & O + H_2O \\ \diagdown & & & & & \| \\ & CO_2H & & & & O \end{matrix} \qquad (3)$$

For the corresponding intermolecular reaction to match this intramolecular process the concentration of the fixed reactant would have to be 2×10^{12} M!

These examples from organic chemistry show that tying two reactants together in a single molecule can have an enormous effect on the rate of a reaction. This effect is, for the most part, due simply to differences between the entropy changes that accompany the inter- and intramolecular reactions. The formation of the product involves a much larger loss of translational and rotational entropy in the intermolecular molecular reaction than it does in the corresponding intramolecular reaction. A negative change in entropy increases both the overall free energy change in the reaction ($\Delta G = \Delta H - T\,\Delta S$) and the activation free energy ($\Delta G^{\ddagger} = \Delta H^{\ddagger} - T\,\Delta S^{\ddagger}$) for the formation of the transition state (see equation (9) in chapter 2 and equation (12) in chapter 8). In the intramolecular reaction much of this entropy decrease has already occurred during the preparation of the reactant.

Enzymes that catalyze intermolecular reactions take advantage of the proximity effect by binding the reactants close together in the active site, so the reactive groups are oriented appropriately for the reaction. Once the substrates are fixed in this way, the subsequent reaction behaves kinetically like an intramolecular process. The entropy decrease associated with the formation of the transition state has been moved to an earlier step, the binding of the substrates to form the enzyme–substrate complex. This step often is driven by an enthalpy decrease associated with electrostatic interactions between polar or charged groups of the substrates and the enzyme. There are, however, exceptions to this generalization, particularly in reactions involving hydrophobic substrates. As we discussed in chapter 3, the removal of a hydrophobic molecule from aqueous solution is favored by an entropy increase, and the binding of hydrophobic substrates to enzymes can be driven in this way.

General-Base and General-Acid Catalysis Provide Ways of Avoiding the Need for Extremely High or Low pH

Chemical bonds are formed by electrons, and the rearrangement or breakage of bonds requires the migration of electrons. In broad terms, reactive chemical groups can be said to function either as electrophiles or as nucleophiles. Electrophiles are electron-deficient substances that react with electron-rich substances; nucleophiles are electron-rich substances that react with electron-deficient substances. The task of a catalyst often is to make a potentially reactive group more reactive by increasing its intrinsic electrophilic or nucleophilic character. In many cases the simplest way to do this is to add or remove a proton. As an example, consider the hydrolysis of an ester (fig. 9.1). Because the electronegativity of the oxygen atom in the C=O group is greater than that of the carbon, the oxygen has a fractional negative charge, δ^-, and the carbon has a fractional positive charge, δ^+. Hydrolysis of an ester in neutral aqueous solution can occur if the oxygen atom of H_2O, acting as a nucleophile, attacks the positively charged carbon. The initial product is an intermediate in which the carbon atom has four substituents in a tetrahedral arrangement. The re-

Figure 9.1

Several ways in which the hydrolysis of an ester can occur. The formation of a chemical bond is described by a formal "flow" of an electron pair from an electron donor to an electron acceptor. This electron flow is indicated by a curved (red) arrow drawn from the electron source to the position of the newly formed bond at the electron acceptor.

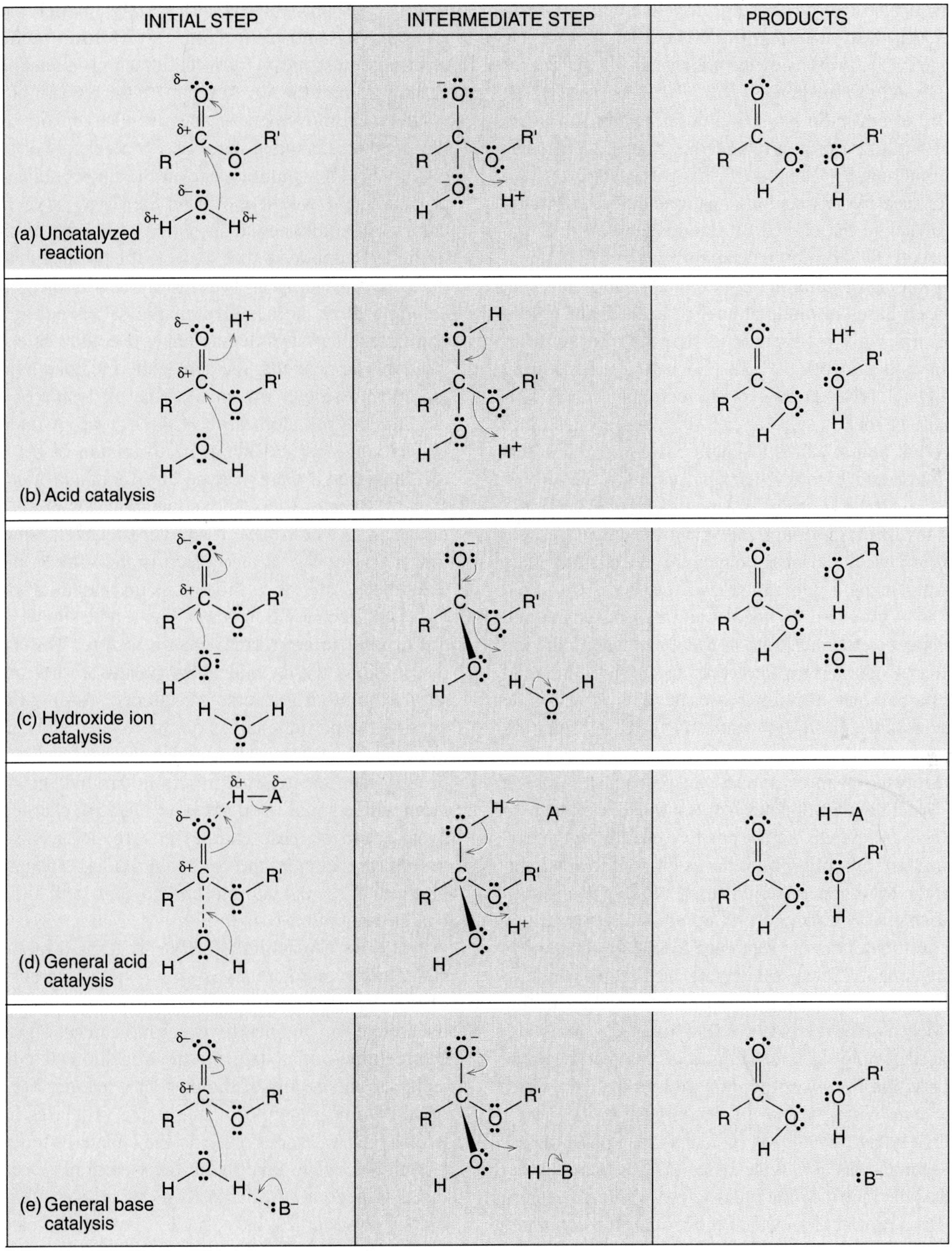

INITIAL STEP	INTERMEDIATE STEP	PRODUCTS
(a) Uncatalyzed reaction		
(b) Acid catalysis		
(c) Hydroxide ion catalysis		
(d) General acid catalysis		
(e) General base catalysis		

action is completed by the rapid breakdown of the tetrahedral intermediate to release the alcohol.

Water is intrinsically a comparatively weak nucleophile, and its reaction with esters in the absence of a catalyst is very slow. The hydrolysis of esters occurs much more rapidly at high pH, when the negatively charged hydroxide ion replaces water as the reactive nucleophile (see fig. 9.1*c*). But the nucleophilic character of water itself also can be increased by interaction with a basic group other than OH^- (see fig. 9.1*e*). The base offers a pair of electrons to one of the protons of the water and thus increases the electron density on the oxygen.

The term general base is used to describe any substance that is capable of binding a proton in aqueous solution. Enzymes use a variety of functional groups to fill this role. There are two factors that make free hydroxide ions unsuitable for enzymatic catalysis and that dictate the choice of a general base. First, the low concentration of OH^- limits its availability at physiological pH. In contrast, proteins contain numerous functional groups that can serve as general bases at moderate pH or even under mildly acidic conditions. The only requirement is that the base start out mainly in its unprotonated form, which will be the case as long as the ambient pH is above the pK_a of the conjugate acid. This condition can easily be met by selecting a basic group from among the ionizable or polar amino acid side chains, from an amino-terminal $—NH_2$ group or a carboxyl-terminal carboxylate ion, or from the oxygen or nitrogen atom of a peptide bond (see table 4.2). The pK_a of any of these groups can vary over a considerable range, depending on the local environment in the enzyme. The second advantage of using a general base instead of OH^- is that a basic group that is provided by the protein can be positioned precisely with respect to the substrate in the active site, allowing the proximity effect to come into play. Free hydroxide ions tend to be much more mobile. In exceptional cases in which a hydroxide ion acts as a nucleophile in an enzymatic reaction, it usually is tightly bound to a metal ion.

The hydrolysis of an ester also can be catalyzed by an acid (see fig. 9.1*b*). The acid donates a proton to the oxygen of the ester's C═O group, increasing the positive charge on the carbon and increasing the susceptibility of the ester to attack by a nucleophile. Again, the term general acid is used to refer to any substance that is capable of releasing a proton, and enzymes almost always use such proton donors in preference to free protons or hydronium ions, presumably because a general acid can operate at moderate pH and is easy to fix in position (see fig. 9.1*d*). In this case the requirement is that the pH be below the pK_a.

An important point to note in figure 9.1 is that the same general acid or base that catalyzes the formation of the tetrahedral intermediate also can participate in the decomposition of the intermediate. When a general acid (HA) donates a proton to the ester oxygen, it becomes the conjugate base (A^-), which can retrieve the proton as the intermediate breaks down. When a general base (B^-) removes a proton from water, it becomes the conjugate acid (BH), which can provide a proton to the alcohol. Note also that general-acid and general-base catalysis are not mutually exclusive: They could both occur in a concerted manner in the same step of a reaction.

Electrostatic Interactions Can Promote the Formation of the Transition State

The frequent use of general acids and general bases in enzymatic reaction mechanisms illustrates the underlying principle that enzymes act by stabilizing the distribution of electrical charge in transition states. In the enzymatic hydrolysis of an ester, the key transition state probably is structurally similar to the tetrahedral intermediates shown in figure 9.1. To form such an intermediate, electrons must move from the attacking nucleophile, through the carbon atom of the C═O group, to the oxygen of the C═O. There is thus a net movement of negative charge from the nucleophile to the substrate. In the absence of a general acid or base, a charge approaching +1 would appear on the nucleophile and a charge approaching −1 would appear on the C═O oxygen. A general base can stabilize this new distribution of charge by offering electrons to the nucleophile so that some of the positive charge moves to the base. By providing a proton to the C═O oxygen, a general acid can delocalize the negative charge at this end of the system. But there are other ways that an enzyme could achieve a similar stabilization. Suppose that the active site included a positively charged amino acid side chain, such as that of lysine or arginine, located near the oxygen atom of the C═O group. A fixed positive charge in this region would favor the formation of the tetrahedral intermediate, even if there were no transfer of a proton from the charged species to the oxygen. A fixed negative charge in the region of the nucleophile would have a similar effect. The interactions of such fixed charges are termed electrostatic effects. The magnitude of electrostatic effects in proteins is discussed in box 9A.

Electrostatic interactions can be significant even between groups whose net formal charge is zero. This is because charge distributions within molecular groups are not uniform but rather vary from atom to atom. We alluded to this point earlier in discussing the partial charges on the oxygen and carbon atoms of an ester (see fig. 9.1). Similar considerations apply to other functional groups: The electron distributions around the nuclei leave each atom with a small net positive or negative charge, even though the overall sum of these charges is zero. In an alcoholic $—CH_2OH$ group, for example, the oxygen atom has a negative charge of approximately −0.4 atomic charge unit, and the hydrogen has a charge of about +0.4.

As a reacting substrate is transformed into a transition state, the changing charges on its atoms interact with the charges on all of the other atoms in the surrounding protein and also with the charges on any nearby water molecules. The energy difference between the initial state and the transition state thus depends critically on the details of the protein structure. We will see illustrations of this principle in the serine proteases and the other enzymes that we discuss later in the chapter. Modern computational techniques, when taken with the wealth of structural information that has become available from x-ray crystallography and other biophysical studies, have made it possible to calculate the contributions that various components of an enzyme's active site make to the activation free energy $\Delta G^{\ddagger}$ and to predict quantitatively how $\Delta G^{\ddagger}$ might be altered by modifications of the protein. These predictions can be tested experimentally by modifying the gene that

9A
BOX

The Magnitude of Electrostatic Effects in Proteins

The energy of the electrostatic interaction between two charges Q_1 and Q_2 separated by a distance r Å is (in kcal/mole)

$$V = 332\frac{Q_1Q_2}{r} \quad \textbf{(B1)}$$

From this expression it is clear that electrostatic effects can be appreciable even at relatively large distances; the interaction energy of a set of opposite charges 10 Å apart would be −33.2 kcal/mole. However, equation (B1) refers to charges in a vacuum. In a polar solvent such as water, electrostatic interactions are weaker because they are screened by the dielectric effect of the solvent. To take this into account, equation (B1) is often replaced by an expression of the form

$$V = 332\frac{Q_1Q_2}{\epsilon r} \quad \textbf{(B2)}$$

where ϵ is the dielectric constant. The dielectric constant of pure water at 25°C is 78. Dielectric effects arise because solvent molecules near a charged species become oriented and electrically polarized, so each charged species is effectively surrounded by a cloud of opposite charges.

In the interior of a protein, molecular reorientation is relatively restricted, so the effective dielectric constant is considerably smaller than that of the surrounding solvent water. Effective values of ϵ are in the range of 2 to 10, depending on the details of the structure in the region of the charged groups. Electrostatic interactions in the interior of proteins thus can be comparatively strong. Charged groups on the surface of a protein interact less strongly because of the dielectric effect of the surrounding water; the effective dielectric constant in this region is probably about 40 in most cases.

encodes the protein, a technique termed "site-directed mutagenesis" (see chapter 32). This combination of biophysical, computational, and molecular biological techniques has opened exciting new frontiers for exploring the detailed mechanisms of enzymatic catalysis.

Enzymatic Functional Groups Provide Nucleophilic and Electrophilic Catalysts

Another strategy for catalyzing the hydrolysis of an ester or an amide is to replace water by a stronger nucleophilic group that is part of the enzyme's active site. The $HOCH_2$— group of a serine residue is often used in this way. In such cases the reaction of the serine with the substrate splits the overall reaction into a two-step process. Instead of immediately yielding the free carboxylic acid, the breakdown of the initial tetrahedral intermediate yields an intermediate ester that is covalently attached to the enzyme:

$$\text{R—C(=O)—NHR}' + \text{HOCH}_2\text{—Enzyme} \longrightarrow \text{R—C(=O)—OCH}_2\text{—Enzyme} + \text{R}'\text{NH}_2 \quad (4)$$

The acyl-enzyme ester intermediate must be hydrolyzed by a second reaction, in which water becomes the nucleophile:

$$\text{R—C(=O)—OCH}_2\text{—Enzyme} + \text{H}_2\text{O} \longrightarrow \text{R—C(=O)—OH} + \text{HOCH}_2\text{—Enzyme} \quad (5)$$

The proteolytic enzymes trypsin, chymotrypsin, and elastase, discussed in a later section, all work in this way. The two-step pathway requires that the intermediate be more susceptible to nucleophilic attack by water than the original substrate. This is likely if the original substrate is an amide, because esters are generally more reactive than amides.

Nucleophilic groups on enzymes participate in a variety of other types of reactions in addition to hydrolytic reactions. An example is acetoacetic acid decarboxylase, which catalyzes the reaction

$$\text{CH}_3\text{—C(=O)—CH}_2\text{—CO}_2\text{H} \longrightarrow \text{CH}_3\text{—C(=O)—CH}_3 + \text{CO}_2 \quad \textbf{(6)}$$

The reaction proceeds by the formation of a Schiff base intermediate, in which the substrate is covalently attached to the ϵ-amino group of a lysine residue at the enzyme's active site (fig. 9.2). This intermediate is formed by a nucleophilic attack of the amino group on the carbonyl carbon, followed by the splitting out of water. The protonated nitrogen atom of the Schiff base introduces a positive charge that pulls electrons from the nearby carbon–carbon bond, causing decarboxylation. Aldolase and transaldolase, two enzymes that catalyze steps in the breakdown of carbohydrates, use lysine residues in a similar manner.

A basic feature of both of the mechanisms outlined in equations (4)–(6) and figure 9.2, and of other instances of nucleophilic catalysis by enzymes, is the formation of an intermediate state in which the substrate is covalently attached to a nucleophilic group on the enzyme. In addition to the $—CH_2OH$ group of serine and the ϵ-amino group of lysine, the $—CH_2SH$ of cysteine is

Figure 9.2

In acetoacetic acid decarboxylase the positive charge of a protonated Schiff base intermediate pulls electrons from a nearby carbon–carbon bond thereby releasing CO_2.

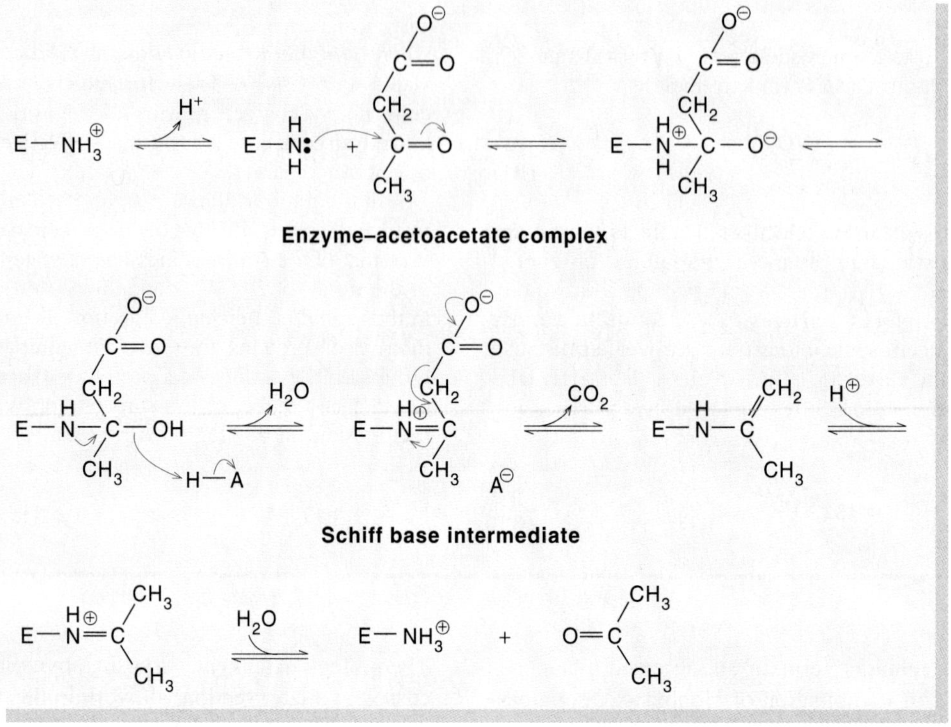

Enzyme–acetoacetate complex

Schiff base intermediate

often used as a nucleophile. The carboxylate of aspartate or glutamate participates in reactions involving the hydrolysis of ATP, and the imidazole group of histidine can play a similar role. Some enzymes take advantage of bound coenzymes such as thiamine, biotin, pyridoxamine, or tetrahydrofolate to obtain additional nucleophilic reagents (see chapter 11).

There also are numerous enzymes that use bound metal ions to form complexes with substrates. In these enzymes the metal ion generally serves as an *electrophilic,* rather than a nucleophilic, functional group. Carbonic anhydrase, for example, contains a Zn^{2+} ion that binds one of the substrates, hydroxide ion, as a ligand. The bound OH^- reacts with the other substrate, CO_2. In alcohol dehydrogenase, and in the proteolytic enzymes thermolysin and carboxypeptidase A, a Zn^{2+} ion in the active site forms a complex with the carbonyl oxygen atom of the aldehyde or peptide substrate. The withdrawal of electrons by the Zn^{2+} increases the partial positive charge on the carbonyl carbon atom and thus promotes the reaction of the carbon with a nucleophile. We discuss such enzymes in more detail in a later section (see figs. 9.14 and 9.15).

Structural Flexibility Can Increase the Specificity of Enzymes

Although precise positioning of the reactants is a fundamental aspect of enzyme catalysis, some enzymes undergo major structural rearrangements when they bind substrates or inhibitors. An example is hexokinase, which catalyzes the transfer of a phosphate group from ATP to glucose:

$$\text{ATP} + \text{Glucose} \longrightarrow \text{ADP} + \text{Glucose-6-phosphate} \qquad (7)$$

When hexokinase binds glucose, it undergoes a structural reorganization that brings together the elements of the active site (fig. 9.3). The enzyme literally closes like a set of jaws around the substrate! Such a structural change is often referred to as an induced fit.

Carboxypeptidase A, which we will discuss in more detail later, is another enzyme that undergoes a major structural change when it binds its substrate. In this case the rearrangement of the protein effectively pulls the hydrophobic part of the substrate out of the aqueous solution by surrounding it with nonpolar portions of the protein. Enfolding a substrate in this way can be beneficial in several ways. First, it can serve to maximize the favorable entropy change associated with removing a hydrophobic molecule from water. Second, it should allow the enzyme to control and intensify the electrostatic effects that promote the formation of the transition state. The substrate is forced to respond to the directed electrostatic fields from the enzyme's functional groups instead of the disordered fields from the solvent.

Structural changes also can help to explain the high specificity of some enzymatic reactions. In hexokinase, for example, the structural change induced by glucose promotes the binding of the other substrate, ATP (see fig. 9.3). ATP does not bind to the

Figure 9.3

Models of the crystallographic structure of hexokinase in the (*a*) "open" and (*b*) "closed" conformations. The enzyme (shown in blue) adopts the open conformation in the absence of substrates but switches to the closed conformation when it binds glucose (red). Hexokinase also has been crystallized with a bound analog of ATP. In the absence of glucose the enzyme with the bound ATP analog remains in the open conformation. The structural change caused by glucose results in the formation of additional contacts between the enzyme and ATP. This can explain why the binding of glucose enhances the binding of ATP. (Courtesy of Dr. Thomas A. Steitz.)

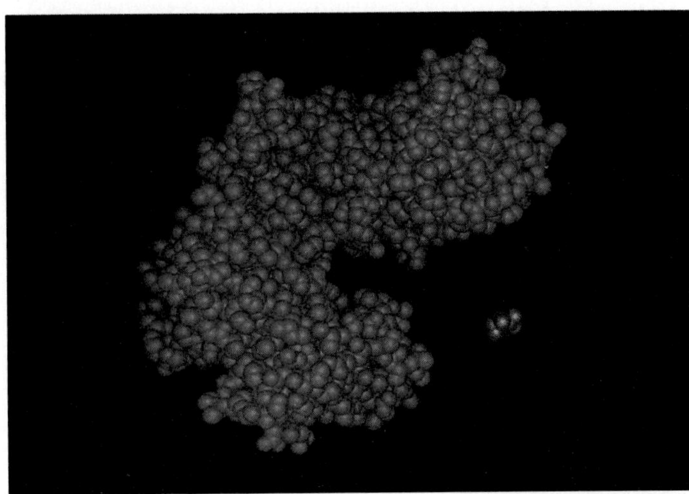

(a)

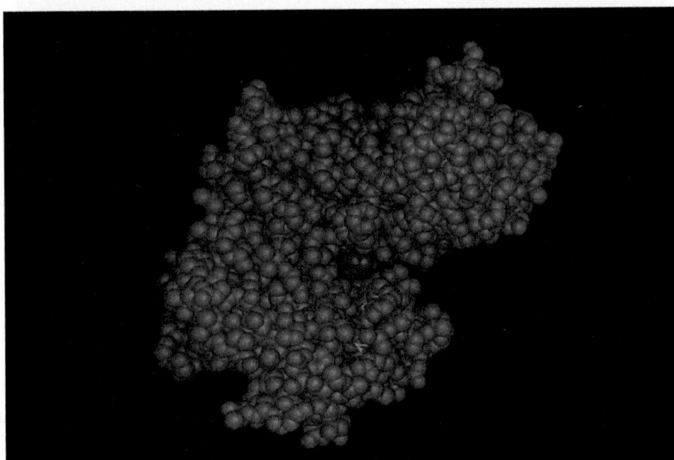

(b)

enzyme properly unless glucose is already present in the catalytic site. If ATP were to bind in the absence of glucose, the enzyme might have a tendency to catalyze the transfer of phosphate from ATP to water, resulting in a wasteful loss of ATP:

$$ATP + H_2O \longrightarrow ADP + P_i \quad \textbf{(8)}$$

Hexokinase does not catalyze this side-reaction; it waits for glucose to bind first.

As another example, the enzyme serine hydroxymethylase (see chapter 24) catalyzes the removal of formaldehyde from serine, forming glycine. In a second step, the enzyme transfers the formaldehyde to a bound coenzyme, tetrahydrofolic acid (THF). If the removal of formaldehyde from serine proceeded even in the absence of THF, the formaldehyde might be set free in the cytosol, where it could enter into other, undesirable reactions. (Free formaldehyde is highly reactive and is toxic to cells.) To prevent this potential disaster, serine hydroxymethylase does not catalyze the first reaction until after THF is bound. The binding of the coenzyme causes the enzyme to fold so that the first step can occur, even though the THF is not yet involved directly in the chemistry.

The structural changes that occur in hexokinase, carboxypeptidase A, and serine hydroxymethylase bring home the point that enzyme crystal structures give static snapshots of molecules that, in many cases, actually are highly flexible. In solution, the structure of an enzyme undergoes fluctuations that vary widely in amplitude and frequency from place to place in the protein. Vibrations and rotations involving only a few atoms occur on time scales of 10^{-13} to 10^{-11} s. Somewhat larger motions, such as the flipping of the aromatic ring of a tyrosine or tryptophan, typically occur on scales of 10^{-9} to 10^{-8} s. Major reorganizations may take 10^{-6} to 10^{-3} s. All of these types of motions can be important in catalysis.

Detailed Mechanisms of Enzyme Catalysis

In the foregoing sections we discussed five themes that recur frequently in enzyme reaction mechanisms. We will now examine a few enzymes in finer detail. We will focus on enzymes for which crystal structures have been obtained, because the most decisive advances in our understanding of enzyme reaction mechanisms have come by inspecting such structures. Crystals of many enzymes have been shown to be enzymatically active, and it appears that in most cases the three-dimensional structures of crystalline enzymes are close to the structures of the proteins in solution. It is important to keep in mind, however, that crystal structures provide pictures of enzymes in relatively stable states. To fill in the intermediates and transition states between these resting states requires a variety of other techniques, including studies of related nonenzymatic reaction mechanisms.

There are over 1,500 known enzymes, each with its own unique structure, specificity, and catalytic mechanism. However, the situation is less complicated than this number might suggest, because many enzymes can be grouped in families that share certain basic features. In some cases the enzymes that make up a family appear to have diverged from a common evolutionary ancestor. Family members that arose in this way are apt to retain similar secondary and tertiary structures, and they typically have the same amino acid residue at between 20% and 50% of the corresponding positions in their primary sequences. In other cases, enzymes with diverse ancestral origins appear to have converged on structural features that are well suited for catalyzing particular types of reactions. Such enzymes resemble each other in their active sites but may have little in common elsewhere in their structures.

Serine Proteases Are a Diverse Group of Enzymes That Use a Serine Residue for Nucleophilic Catalysis

The serine proteases are a large family of proteolytic enzymes that use the reaction mechanism for nucleophilic catalysis outlined in equations (4) and (5), with a serine residue as the reactive nucleophile. The best-known members of the family are three closely related digestive enzymes, trypsin, chymotrypsin, and elastase. These enzymes are synthesized in the mammalian pancreas as inactive precursors termed zymogens. They are secreted into the small intestine, where they are activated by proteolytic cleavage in a manner that will be described in chapter 10. Many of the enzymes that participate in blood coagulation also are serine proteases; these enzymes circulate in the blood as inactive zymogens and are activated by proteolytic cleavage when blood vessels are damaged (see chapter 10). The serine protease family also includes many enzymes from bacteria and other nonmammalian organisms.

In the digestive system, trypsin, chymotrypsin, and elastase work as a team. They are all endopeptidases, which means that they cleave protein chains at internal peptide bonds, but each preferentially hydrolyzes bonds adjacent to a particular type of amino acid residue (fig. 9.4). Trypsin cuts just past the carbonyl groups of basic residues (lysine or arginine); chymotrypsin cuts next to aromatic residues (phenylalanine, tyrosine, or tryptophan); elastase is less discriminating but prefers small, hydrophobic residues such as alanine.

About half of the amino acid residues of trypsin are identical to the corresponding residues in chymotrypsin, and about a quarter of the residues are conserved in all three of the pancreatic endopeptidases (fig. 9.5). The structural similarities of trypsin, chymotrypsin, and elastase are even more evident in the crystal structures. As is shown in figure 9.6*a* the folding of the polypeptide chain is essentially the same in all three enzymes, with the only substantial variations occurring in the external loops. These enzymes are classic illustrations of diverging evolution from a common ancestor. The structures of some of the bacterial serine proteases also are homologous to those of the mammalian enzymes. On the other hand, subtilisin, a serine protease obtained from *Bacillus subtilis,* has an amino acid sequence that seems totally unrelated to the mammalian sequences. Its three-dimensional structure also is very different from those of the mammalian enzymes (fig. 9.6*b*). Subtilisin therefore is likely to have joined the serine protease family by convergent evolution. Remarkably, there is a small set of critical amino acid residues that come together in the folded structure to form the essential elements of the active site in all of these proteases. In the chymotrypsin numbering system these are His 57, Asp 102, and Ser 195. The locations of these residues in the three-dimensional structures of trypsin, chymotrypsin, and elastase can be seen in figures 9.6*a* and 9.7.

That Ser 195 played an important role in the catalytic mechanism was known from early studies on the enzyme inhibitor diisopropylfluorophosphate (see chapter 8). This inhibitor reacts irreversibly with chymotrypsin or trypsin to form an inactive derivative in which the diisopropylphosphate group is covalently attached to the serine residue (see fig. 8.19). The derivative, a phosphate ester of the serine, is similar in structure to the acyl-enzyme

$$(\mathrm{R{-}\overset{\overset{\displaystyle O}{\|}}{C}{-}OCH_2{-}Enzyme})$$

intermediate in equations (4) and (5). A variety of other inhibitors have been found to react in a parallel manner with this particular serine residue. As a rule, these inhibitors do not react with other serines in the enzyme or with serines in enzymes that are not part of the serine protease family. Exceptions to the rule are a number of enzymes that catalyze the hydrolysis of esters; these enzymes have a similar, reactive serine residue, and their catalytic mechanism appears to be very similar to that of the serine proteases. The reactivity of Ser 195 thus is not a property of serine residues in general but depends on the special surroundings of this residue in the protein. We will see shortly that it is the juxtaposition of Ser 195 with His 57 and Asp 102 that makes this serine especially reactive.

Figure 9.4

Trypsin, chymotrypsin, and elastase, three members of the serine protease family, catalyze the hydrolysis of proteins at internal peptide bonds adjacent to different types of amino acids. Trypsin prefers lysine or arginine residues; chymotrypsin, aromatic side chains; and elastase, small, nonpolar residues. Carboxypeptidases A and B, which are not serine proteases, cut the peptide bond at the carboxyl-terminal end of the chain.

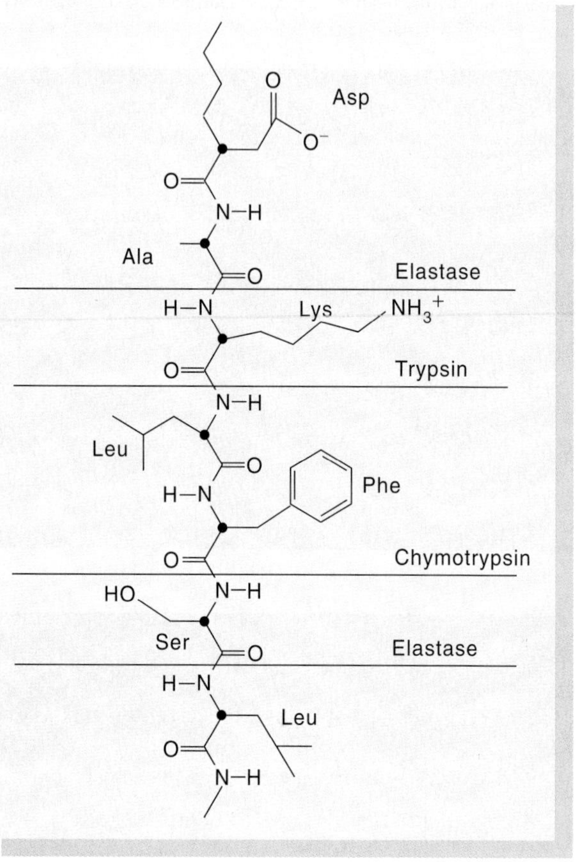

Figure 9.5

Schematic diagrams of the amino acid sequences of chymotrypsin, trypsin, and elastase. Each circle represents one amino acid. Amino acid residues that are identical in all three proteins are in solid color. The three proteins are of different lengths but have been aligned to maximize the correspondence of the amino acid sequences. All of the sequences are numbered according to the sequence in chymotrypsin. Long connections between nonadjacent residues represent disulfide bonds. Locations of the catalytically important histidine, aspartate, and serine residues are marked. The links that are cleaved to transform the inactive zymogens to the active enzymes are indicated by parenthesis marks. After chymotrypsinogen is cut between residues 15 and 16 by trypsin and is thus transformed into an active protease, it proceeds to digest itself at the additional sites that are indicated; these secondary cuts have only minor effects on the enzyme's catalytic activity. (Illustration copyright by Irving Geis. Reprinted by permission.)

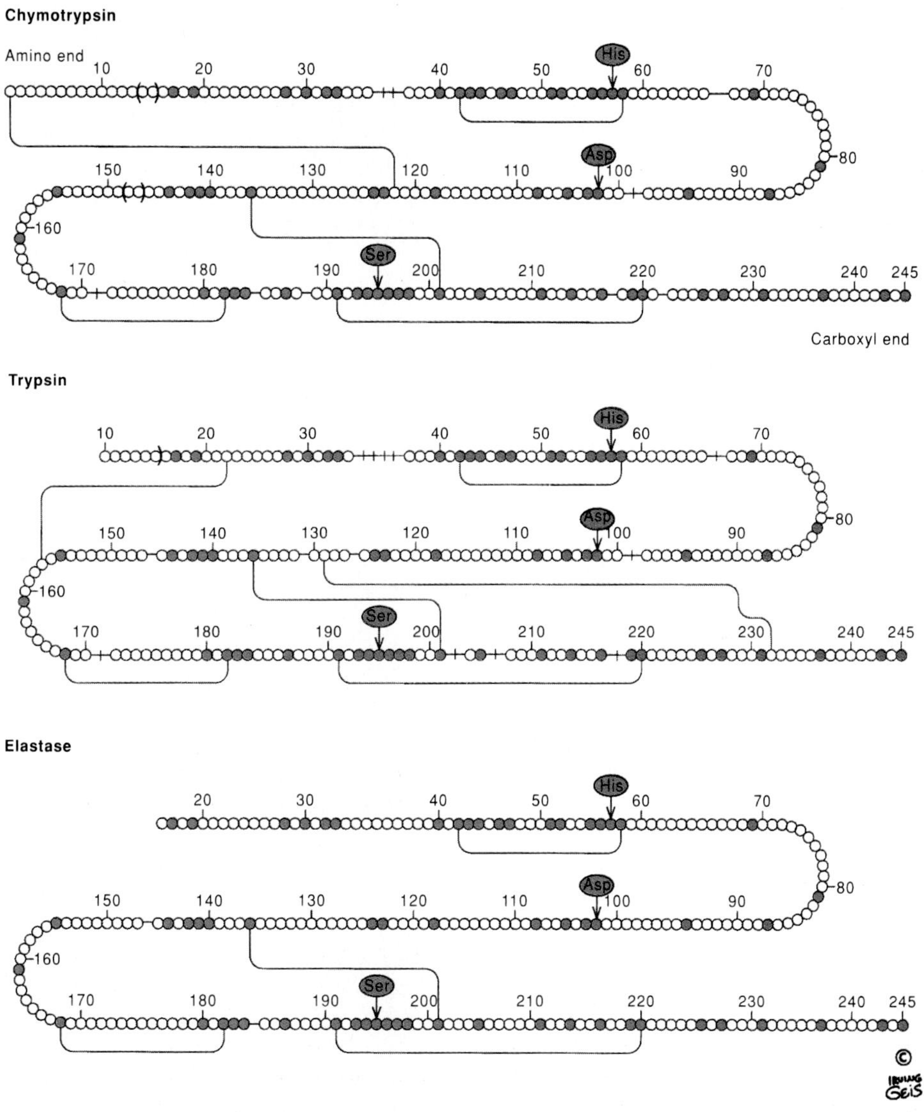

Although His 57 is far removed from Ser 195 in the primary sequence, studies with affinity labels showed that it must be near the serine residue in the active site. A derivative of phenylalanine containing a reactive chloromethyl ketone group inhibits chymotrypsin irreversibly by reacting with the histidine. The inhibition can be prevented by the presence of other aromatic molecules that bind competitively at the active site.

The initial evidence for the formation of an acyl-enzyme ester intermediate came from studies of the kinetics with which chymotrypsin hydrolyzed various analogs of its normal polypep-

Figure 9.6

(*a*) Superimposed computer-generated *C*-alpha tube structures for trypsin (green), chymotrypsin (yellow-gold), and elastase (pale pink). The side chains for the three key catalytic residues, Asp 102 (red), His 57 (light blue), and Ser 195 (orange) are shown for trypsin. The structure for trypsin also includes a bound inhibitor, benzamidine, in yellow. (*b*) Superimposed *c*-alpha tube structures for subtilisin (lavender) and trypsin (green), both with catalytic side chains showing. The three key catalytic residues, Asp 102 (red), His 57 (light blue), and Ser 195 (orange) for trypsin and Asp 32 (red), His 64 (light blue), and Ser 221 (orange) for subtilisin, are shown. (Images (*a*) and (*b*): Copyright 1994 by the Scripps Research Institute/Molecular Graphics Images by Michael Pique using software by Yng Chen, Michael Connolly, Michael Carson, Alex Shah, and AVS, Inc. Visualization advice by Holly Miller, Wake Forest University Medical Center.)

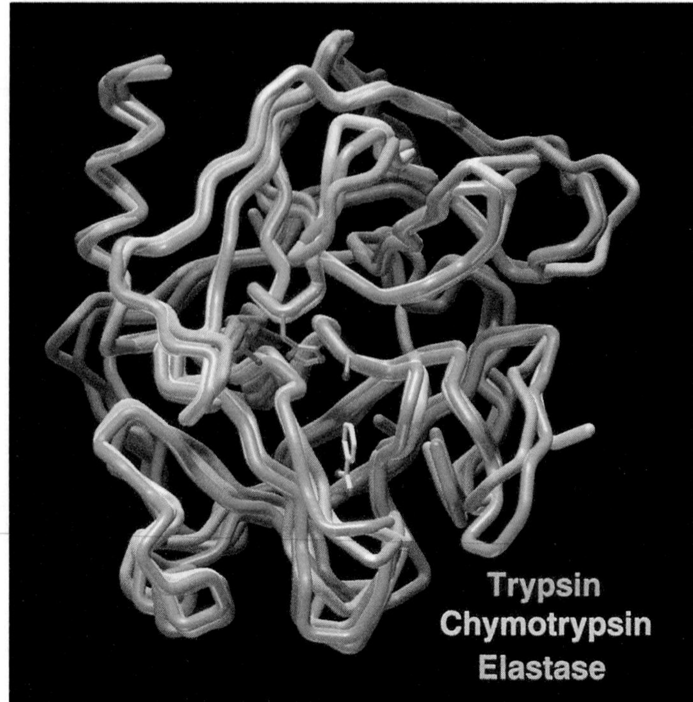

(a)

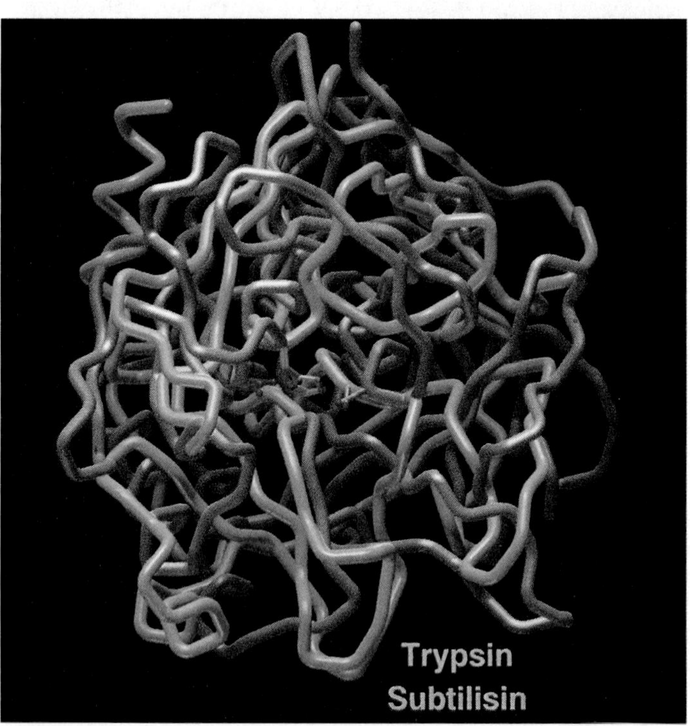

(b)

tide substrates. The enzyme turns out to hydrolyze esters, as well as peptides and simpler amides. Of particular interest was the reaction with the ester *p*-nitrophenyl acetate. This substrate is well suited for kinetic studies because one of the products of its hydrolysis, *p*-nitrophenol, has a characteristic yellow color in aqueous solution, whereas *p*-nitrophenyl acetate itself is colorless. The change in the absorption spectrum makes it easy to follow the progress of the reaction spectrophotometrically. When rapid mixing techniques were used to add the substrate to the enzyme, it was found that an initial burst of *p*-nitrophenol is released within the first few seconds, before the reaction settles down to a constant rate (fig. 9.8). The amount of *p*-nitrophenol that appeared in the burst was approximately equal to the amount of enzyme present in the solution. These observations suggested that the overall enzymatic reaction occurs in two distinct steps, as shown in figure 9.9. In the first step, *p*-nitrophenol is released and the acetyl group is transferred to the enzyme, forming an acyl–enzyme intermediate. In the second step, the intermediate is hydrolyzed, and acetate is released. Diisopropylfluorophosphate prevents the initial burst of *p*-nitrophenol, as well as the subsequent steady-state reaction, a fact suggesting that the enzymatic group that forms the ester intermediate is Ser 195.

When the crystal structures of trypsin, chymotrypsin, and elastase with bound substrate analogs were solved, the substrate analogs were indeed found to be located close to Ser 195 (see figs. 9.6 and 9.7). Histidine 57, the second of the three residues mentioned above, is located nearby, in an orientation suggesting that the OH group of Ser 195 forms a hydrogen bond to the imidazole side chain of the histidine. Aspartic acid 102 sits on the opposite edge of the imidazole ring, where its negatively charged carboxylate group could interact with the proton on the other nitrogen of the ring. The side chains of the aspartate and histidine residues thus appear to be oriented so as to facilitate removal of the proton from the serine's OH group:

$$\text{—}\overset{\overset{\displaystyle O}{\|}}{C}\text{—}O^{-}\cdots HN\underset{C}{\overset{C=C}{\diagup\quad\diagdown}}N\cdots HOCH_2\text{—} \qquad \textbf{(9)}$$

Because withdrawing the proton would increase the nucleophilic character of the oxygen, this arrangement could help to explain

Figure 9.7

Computer-generated *C*-alpha tube model of the crystal structure of trypsin, seen from the same perspective as in figure 9.6*a*. The coloring of the amino acid side chains and the inhibitor (benzamidine) is as in figure 9.6*a*. (*b*) A close-up view of the active site of trypsin with bound benzamidine. (Images (*a*) and (*b*): Copyright 1994 by the Scripps Research Institute/Molecular Graphics Images by Michael Pique using software by Yng Chen, Michael Connolly, Michael Carson, Alex Shah, and AVS, Inc. Visualization advice by Holly Miller, Wake Forest University Medical Center.)

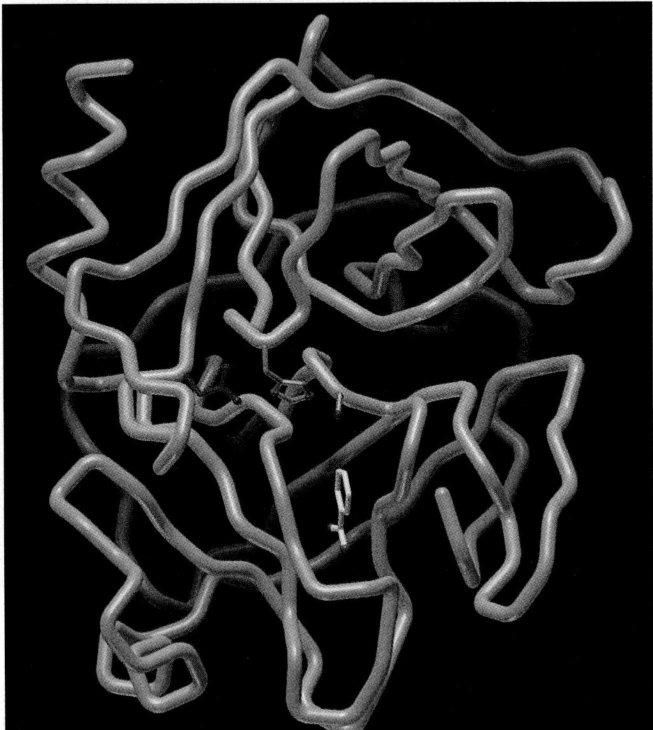

(a)

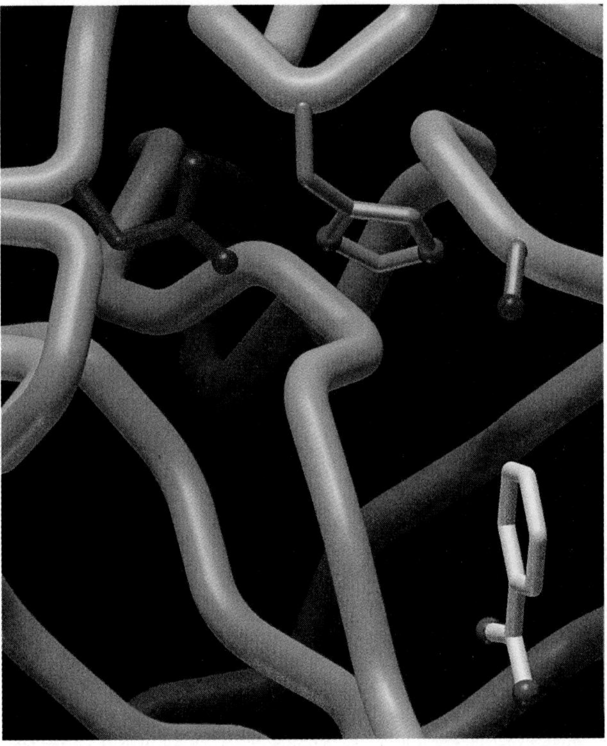

(b)

Figure 9.8

p-Nitrophenol formation as a function of time during the hydrolysis of *p*-nitrophenyl acetate by chymotrypsin. A rapid initial burst of *p*-nitrophenol is followed by a slower, steady-state reaction. The amount of *p*-nitrophenol released in the burst is approximately equal to the amount of enzyme present.

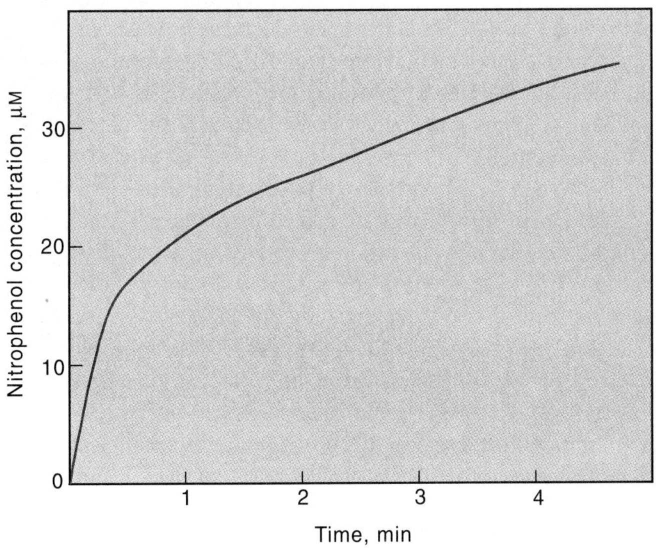

why Ser 195 is exceptionally reactive. The same arrangement of aspartate, histidine, and serine residues has been found in all of the serine proteases that have been examined. It is often referred to as a "charge relay system." The dependence of the enzyme kinetics on pH agrees with this picture. Enzymatic activity appears to depend on the presence of a basic group with a pK_a of about 6.8, which is in the range consistent with a histidine side chain. The k_{cat} decreases abruptly if this group is protonated (fig. 9.10). This is because protonating His 57 prevents the histidine from forming a hydrogen bond to Ser 195 and thus greatly decreases the nucleophilic reactivity of the serine.

The crystal structures also provided a simple explanation for the different substrate specificities of trypsin, chymotrypsin, and elastase. In both trypsin and chymotrypsin the side chain of the substrate fits snugly into a pocket. At the far end of the pocket in trypsin is the carboxylate group of an aspartic acid residue. The negative charge of the carboxylate would favor the binding of the positively charged side chain of lysine or arginine. In chymotrypsin the aspartic acid residue is replaced by serine, creating a less polar environment that suits the side chain of tyrosine, phenylalanine, or tryptophan. In elastase the binding pocket is obstructed by the bulky side chains of a valine and a threonine residue, so it can accommodate only small substrates such as alanine.

Figure 9.9

Steps in the hydrolysis of *p*-nitrophenyl acetate by chymotrypsin. In the hydrolysis of this and most other esters the breakdown of the acyl–enzyme intermediate is the rate-determining step. In the hydrolysis of peptides and amides the rate-determining step usually is the formation of the acyl–enzyme intermediate. This makes the transient formation of the intermediate more difficult to study, because the intermediate breaks down as rapidly as it forms.

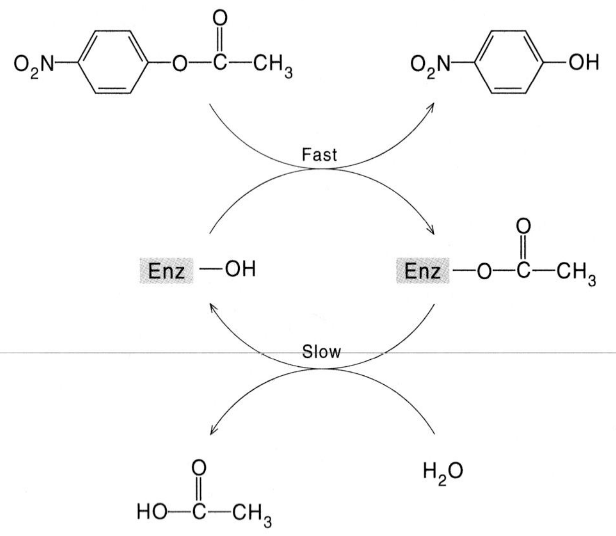

$O_2N–C_6H_4–O–C(=O)–CH_3$ → (Fast) → $O_2N–C_6H_4–OH$

Enz —OH → Enz —O—C(=O)—CH_3

Enz —O—C(=O)—CH_3 + H_2O → (Slow) → HO—C(=O)—CH_3 + Enz —OH

Figure 9.10

The turnover number (k_{cat}) and the Michaelis constant (K_m) as a function of pH for the hydrolysis of *N*-acetyl-L-tryptophanamide by chymotrypsin at 25°C. The decrease in k_{cat} as the pH is lowered from 8 to 6 probably reflects the protonation of His 57. The increase in K_m above pH 9 probably reflects the deprotonation of Ile 16, which results in the rotation of Gly 193 out of the substrate-binding site (see fig. 10.1).

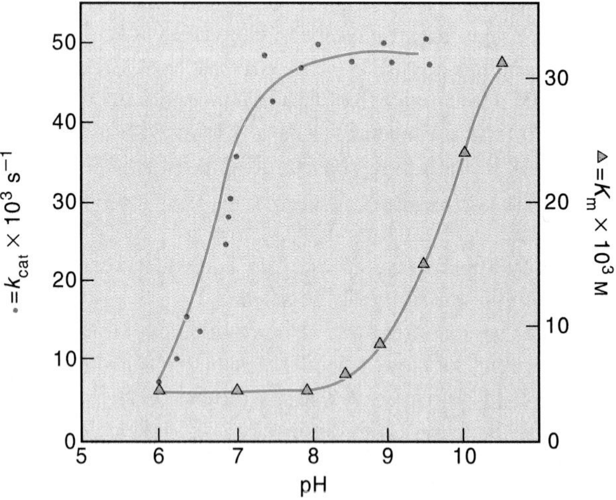

Another important feature of the substrate-binding site in serine proteases is that the carbonyl oxygen atom of the scissile peptide bond is hydrogen-bonded to one or more NH groups. In trypsin, chymotrypsin, and elastase these are the amide NH groups of Ser 195 and of Gly 193. The interaction with the two protons would favor an increase in the negative charge on the oxygen, facilitating the formation of a tetrahedral intermediate state, as we discussed earlier in connection with general-acid catalysis and electrostatic effects.

An overall reaction mechanism for chymotrypsin is sketched in figure 9.11. Part (*a*) shows the enzyme–substrate complex, with an aromatic side chain of the substrate seated in the binding pocket and the carbonyl oxygen atom hydrogen-bonded to the amide NH hydrogens of Gly 193 and Ser 195. Aspartate 102 and His 57 are aligned as described above, with Ser 195 forming a hydrogen bond with the histidine. Part (*b*) shows a tetrahedral intermediate state. Serine 195 has released its proton to His 57 and launched a nucleophilic attack on the carbonyl carbon of the substrate. The histidine residue thus acts as a general-base catalyst in the formation of the intermediate. The movement of negative charge to the carbonyl oxygen creates what is often called an "oxyanion," which is stabilized largely by electrostatic interactions with the amide protons of Ser 195 and Gly 193. As is frequently the case with high-energy, transition-state intermediates, the tetrahedral oxyanion intermediate is not stable enough to be isolated, and the evidence for its existence is inferential. In part (*c*) this fleeting intermediate has decomposed to form the more stable acyl–enzyme intermediate, in which the serine is linked as an ester to the carboxylic part of the substrate and the amine product has been released. Histidine 57 probably facilitates this decomposition by acting as a general acid.

To complete the reaction, the breakdown of the acyl–enzyme probably occurs as shown in parts (*d*), (*e*), and (*f*) of figure 9.11. The steps here are essentially a reversal of the steps through parts (*a*), (*b*), and (*c*), except that water replaces the amino part of the substrate. The reaction probably proceeds by way of a tetrahedral intermediate (*e*) that is similar to one shown in part (*b*) except for this substitution.

One way to test a scheme such as that shown in figure 9.11 is to investigate the effects of modifying the amino acid residues that play important roles. We have already mentioned the inhibitory effect of the reaction of Ser 195 with diisopropylfluorophosphate. To examine the importance of His 57, this residue also was modified by methylation, which would disrupt the interaction with Asp 102. This treatment decreased the activity of chymotrypsin by a factor of more than 10^3. The importance of Asp 102 was demonstrated by site-directed mutagenesis of trypsin and subtilisin. Replacing the aspartate by asparagine reduced k_{cat} by a factor of about 10^4.

When the importance of the catalytic triad (Asp-His-Ser) was first recognized, it was suggested that the triad functioned as a charge relay system in which a proton was actually transferred from the His to the Asp in the activated complex. However, the available spectroscopic evidence did not support this, and the idea was abandoned. For some time after this, it was believed that Asp 102 had a purely inductive electrostatic effect with no charge transfer. However, Perry Frey and his co-workers recently suggested an alternative role for Asp 102 in which an unusually strong hydrogen bond is formed between Asp 102 and His 57 in the activated complex (fig. 9.12). The formation of an extra-strong hydrogen bond in the activated complex would significantly lower

Figure 9.11

The probable mechanism of action of chymotrypsin. The six panels show (*a*) the initial enzyme–substrate complex, (*b*) the first tetrahedral (oxyanion) intermediate, (*c*) the acyl–enzyme (ester) intermediate with the amine product departing, (*d*) the same acyl–enzyme intermediate with water entering, (*e*) the second tetrahedral (oxyanion) intermediate, and (*f*) the final enzyme–product complex. In the transition states between these intermediates there probably is a more even distribution of negative charge between the different oxygen atoms attached to the substrate's central carbon atom.

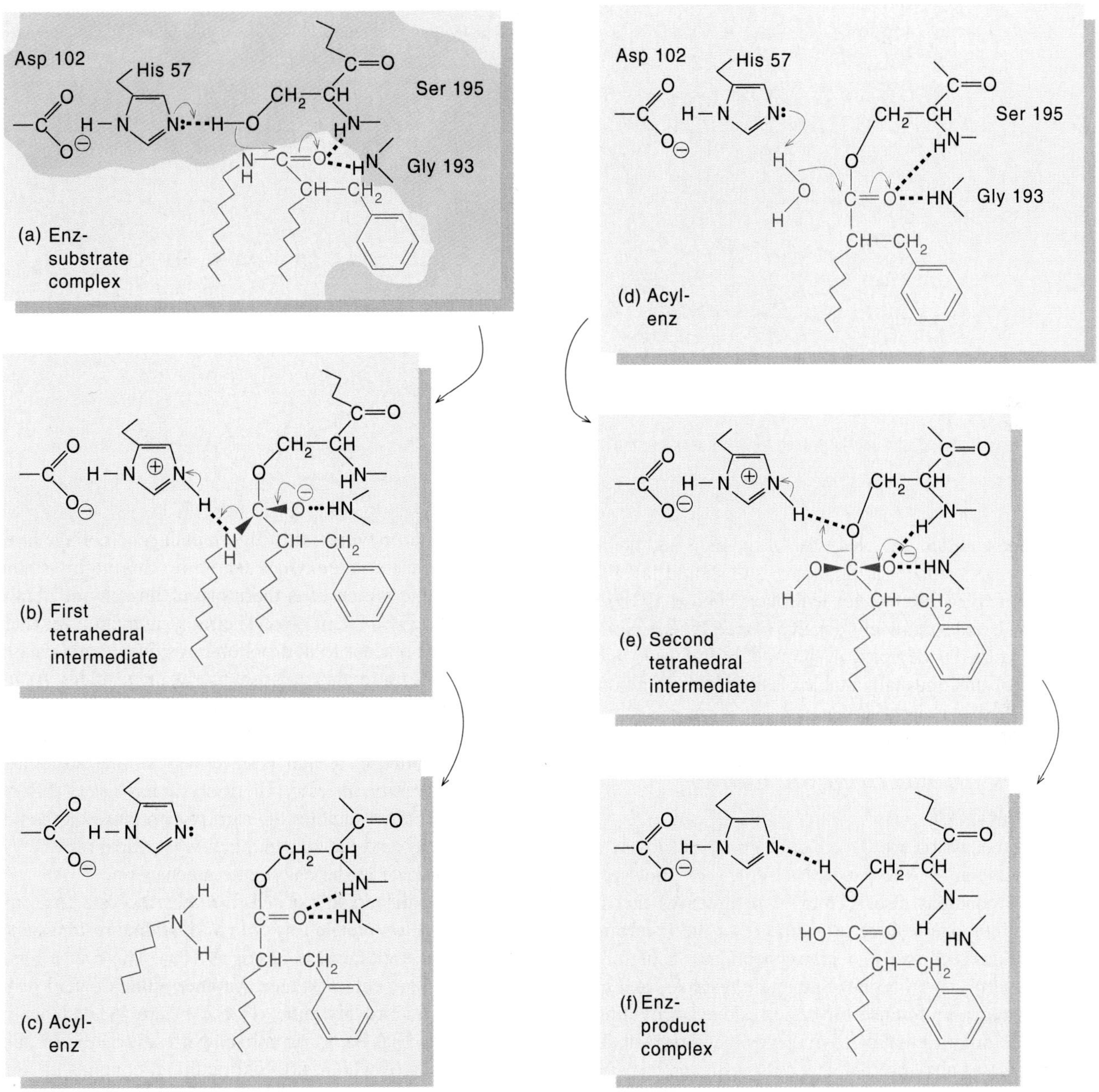

the energy for the formation of the activated complex and therefore should have a significant impact on increasing the reaction rate. The first indication that an especially strong hydrogen bond could form between Asp 102 and His 57 came from spectroscopic measurements on trypsin at low pH (about pH 3). At such a low pH, His 57 should be protonated, as it is postulated to be in the activated complex (see fig. 9.11). A protonated histidine would ordinarily have a pK_a of about 7, and a negatively charged aspartate should have a pK_a of about 5. In the vicinity of the relatively water-free active site these two pK_a values would probably be

Figure 9.12

A modified view of the first three steps in the trypsin-catalyzed peptide hydrolysis suggested by Perry Frey and his colleagues. Formation of the high-energy tetrahedral adduct shown in the middle frame is greatly favored by the formation of an unusually strong hydrogen bond between His 57 and Asp 102. Spectroscopic measurements indicate that this hydrogen bond is two to three times stronger than a conventional hydrogen bond. The proton implicated in this hydrogen bond is approximately equally shared by the N and O atoms from His 57 and Asp 102, respectively. The charges on the histidine and the aspartate of +1 and −1, respectively, are altered when the extra-strong hydrogen bond is formed between them. The magnitude of this charge is designated by $y+$ and $y-$, where $1 \geq y \geq 0.5$. The value of y would be 0.5 if the proton is shared equally by the participating N and O atoms.

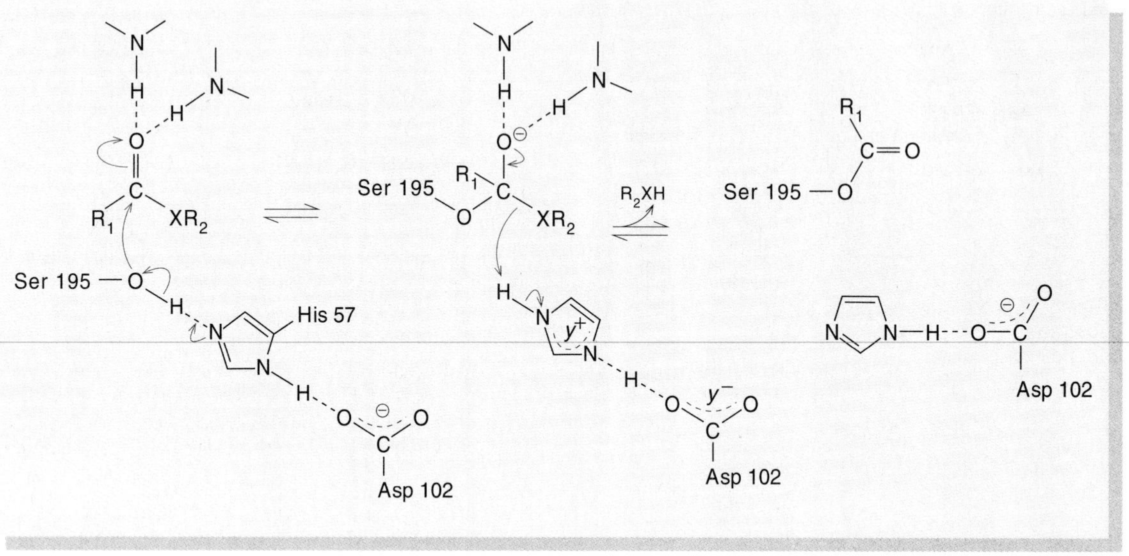

much closer to one another. The closeness of the pK_as and the relatively water-free environment at the active site should favor the formation of a so-called low-barrier hydrogen bond (LBHB), in which the proton is approximately equally shared by the participating N and O atoms as depicted in the middle frame of figure 9.12. The nature of this unusually stable class of hydrogen bonds is elaborated on further in box 9B.

Zinc Provides an Electrophilic Center in Some Proteases

In the preceding discussion we described a group of closely related proteases secreted by the pancreas: trypsin, chymotrypsin, and elastase. The pancreas also secretes two proteases that hydrolyze oligopeptides one residue at a time from the C-terminal end, carboxypeptidases A and B. Carboxypeptidase A preferentially catalyses hydrolysis adjacent to aromatic residues, and carboxypeptidase B has a preference for basic residues, providing a complementarity similar to that of chymotrypsin and trypsin. The two carboxypeptidases also resemble chymotrypsin and trypsin in being secreted as zymogens that must be processed to form the active enzymes in the intestine. Kinetically, the carboxypeptidases are considerably more effective catalysts than the serine proteases. Values of k_{cat} for carboxypeptidase A with its best peptide substrates are on the order of 100 times greater than the k_{cat} values for trypsin with *its* best peptide substrates.

Carboxypeptidases A and B are members of a family of proteolytic enzymes that contain bound zinc. The zinc proteases also include enzymes that digest collagen (collagenases) and a number of bacterial enzymes. The most thoroughly studied of the bacterial zinc proteases is thermolysin, an enzyme obtained from *Bacillus thermoproteolyticus.* Thermolysin hydrolyzes internal peptide bonds adjacent to hydrophobic residues; collagenases cut internal bonds preferably adjacent to glycine residues. As in the case of serine proteases, the zinc protease family appears to have resulted from a combination of divergent and convergent evolution. Carboxypeptidases A and B have very similar structures indicative of a common ancestry; thermolysin has a very different structure overall but resembles the carboxypeptidases in the amino acid residues that bind the zinc and in a few critical residues that probably participate in the enzymatic mechanism.

In the crystal structure of thermolysin the Zn^{2+} ion is bound to the imidazole rings of two histidine residues and the carboxylate of a glutamic acid (fig. 9.13*a*). The fourth ligand of the Zn^{2+} is a molecule of water. Another glutamic acid residue (Glu 143) and a third histidine (His 231) are located nearby. Thermolysin also has been crystallized with several different bound inhibitors, one of which, phosphoramidon, contains an amide bond between phosphoric acid and an amino acid (see figs. 8.7 and 9.13*b*). Because the oxygen and nitrogen substituents of the phosphorus atom are arranged in a tetrahedron, phosphoramidon seems likely to mimic the tetrahedral intermediate that would be formed in the course of the hydrolysis of a peptide (see box 9C located on page 193). In the crystal, one of the oxygen atoms of the phosphate is linked directly to the Zn^{2+} ion, displacing the molecule of water. Glutamic acid 143 appears to form a hydrogen bond to

9B
BOX

The Nature of Low-Barrier Hydrogen Bonds Found in Enzyme Substrate Complexes

Any factor that lowers the energy of formation of the activated complex is likely to increase the reaction rate. There are strong indications that in a number of enzyme–substrate complexes, extra-strong hydrogen bonds called low-barrier hydrogen bonds (LBHBs) are uniquely formed in the activation complex. Because of their potential importance for rates and mechanisms of enzyme-catalyzed reactions, it is essential that we consider the nature of these hydrogen bonds in this chapter.

In conventional hydrogen bonds the participating hydrogen atom is much more strongly associated with one atom, the hydrogen donor, than the other atom, the hydrogen acceptor. There is an appreciable energy barrier that prevents the hydrogen atom from shifting its location so that the roles of proton donor and proton acceptor are reversed (fig. 1*a*). For purposes of this discussion we will call these conventional hydrogen bonds high-barrier hydrogen bonds (HBLBs). HBLBs have a free energy of formation of between 2.4 and 12 kcal/mole. The importance of hydrogen bonds of this sort in determining the associations in water and the effects on polymer conformations were discussed in chapters 3 and 5. In the case of low-barrier hydrogen bonds (LBHBs) the two potential wells for the hydrogen atom are separated by a very low-energy barrier (see fig. 1*b*) so that the hydrogen atom can shift from one location to the other quite readily. The average location of the hydrogen is approximately midway between the two participating heteroatoms. Hydrogen bonds of this type have free energies of formation between 12 and 24 kcal/mole and the distances between the heteroatoms is less than 2.55 Å for O—H—O hydrogen bonds and less than 2.65 Å for O—H—N hydrogen bonds. Super-strong hydrogen bonds have a single energy potential well, so the proton is fixed midway between the participating heteroatoms (see fig. 1*c*). F—H—F hydrogen bonds typify this class of hydrogen bonds. The hydrogen bond energy in this case is 37 kcal/mole, and the distance between the fluorine atoms is only 2.26 Å.

Super-strong hydrogen bonds are not of particular interest in enzymology because they are very rare. However, HBHBs and LBHBs are of great interest because they are frequently interconvertible by subtle changes in the environment that occur during enzyme-catalyzed reactions. The formation of LBHBs from HBHB is favored by moving the interacting groups to a nonaqueous aprotic environment and by any transition that results in a near equality between the pK_a values of the participating heteroatoms.

A LBHB should be detectable by crystallographic determination of the distances between the groups forming the hydrogen bonds. However, a more versatile method of detection for these extra-strong hydrogen bonds is nuclear magnetic resonance spectroscopy, which measures the chemical shift of the proton (see chapter 5). Typical chemical shifts by protons involved in LBHBs reside in the 20-ppm range.

In this chapter we deal with two cases in which LBHBs are possibly major factors in increasing the reaction rates by lowering the activation energies. The recently appreciated importance of LBHBs in activated complexes could have a sweeping effect in the field of enzyme mechanisms, and many changes in proposed mechanism should be forthcoming in the near future. Despite the excitement that the proposal of LBHBs has generated in the field of enzymology, further documentation is needed to gain final acceptance.

Figure 1

Energy diagram for hydrogen bonds formed between two atoms as a function of the distance between them, $R_{A\text{- -}B}$. (*a*) The heteroatoms A and B are separated by a distance corresponding to a weak hydrogen bond (HBHB). The energy well separating atom A from hydrogen is deep and narrow, so the hydrogen atom is covalently bonded to atom A and hydrogen bonded to atom B. (*b*) The heteroatoms are close together, so the energy barrier for the migration of the hydrogen between them is low. The energy for binding the hydrogen atom places it at or slightly above the barrier between the heteroatoms in a strong low-barrier hydrogen bond (LBHB). (*c*) The heteroatoms are so close together that there is no energy barrier, and the hydrogen atom is equally bonded to each heteroatom in a very strong, no-barrier hydrogen bond (NBHB).

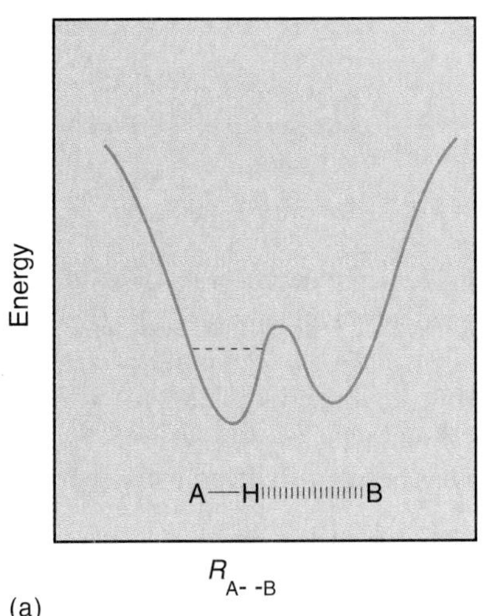

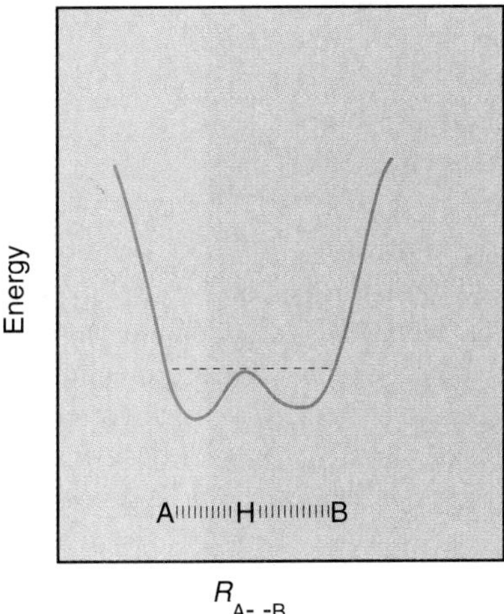

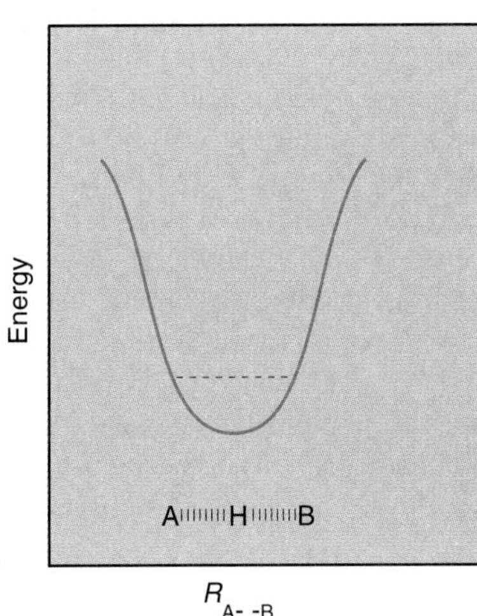

Figure 9.13

Functional groups in the active site of thermolysin. (*a*) The Zn^{2+} ion is bound to two histidines and a glutamic acid and has a molecule of water as its fourth ligand. (*b*) Phosphoramidon, an inhibitor that probably mimics the tetrahedral intermediate formed during hydrolysis of a peptide, binds to the Zn^{2+} by displacing the molecule of water. Residue Glu 143 forms a hydrogen bond with the hydroxyl group of the bound inhibitor, and His 231 is located close to the NH group.

Figure 9.14

The reaction mechanism of thermolysin probably involves polarization of the peptide carbonyl group by interaction with the Zn^{2+} ion and general-base catalysis by Glu 143 to form a tetrahedral intermediate. His 231 probably acts as a general acid to promote the decomposition of the intermediate.

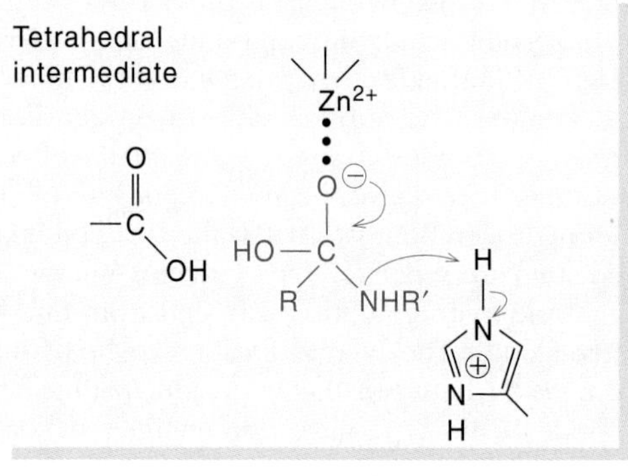

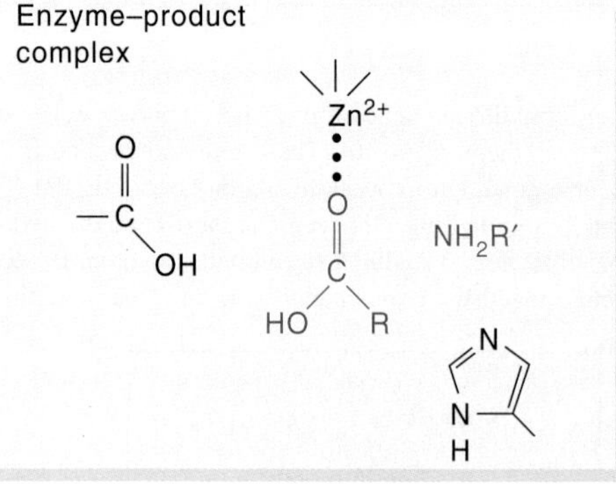

the hydroxyl oxygen, and one of the nitrogen atoms of His 231 forms a hydrogen bond with the inhibitor's NH group (see fig. 9.13).

The structure of the enzyme–phosphoramidon complex suggests that the enzymatic hydrolysis of a peptide by thermolysin proceeds as shown in figure 9.14. In this scheme, binding of the peptide carbonyl oxygen atom to the Zn^{2+} withdraws electrons from the carbonyl group, increasing its susceptibility to nucleophilic attack by water. Glutamic acid 143 acts as a general base to remove a proton from the water. Histidine 231 then could act as a general acid, providing a proton to the amine group as the tetrahedral intermediate decomposes.

Several variations on this mechanism have been proposed. One suggestion is that the water molecule that attacks the peptide carbonyl is the one that starts out as a ligand of the Zn^{2+} and that the Zn^{2+} goes through a stage in which it has five ligands—both the peptide oxygen and the water in addition to the original two histidines and glutamic acid. It also has been suggested that the general acid that facilitates breakdown of the tetrahedral intermediate is Glu 143, rather than His 231. At present, the experimental data do not distinguish clearly between these alternatives. However, the central feature of all the proposed mechanisms is the electrostatic effect of the Zn^{2+} on the peptide carbonyl group.

The Zn^{2+} atom of carboxypeptidase A or B resembles that of thermolysin in having two histidines (His 69 and His 196), a glutamic acid (Glu 72), and a molecule of water as its ligands. Again a second glutamic acid residue (Glu 270) is found nearby, in a very similar position with respect to the metal ion. Carboxypeptidase A also has been crystallized with a bound inhibitor,

Transition-State Analogs

Important clues to an enzyme mechanism often can be obtained by studying the inhibitory effects of compounds that resemble possible intermediates or transition states. Since the enzyme must stabilize the transition state of the reaction, a compound that mimics this state might be expected to bind to the enzyme particularly tightly. Such transition-state analogs frequently do prove to be strong competitive inhibitors. Thus the inhibitory effect of phosphoramidon on thermolysin supports the view that the enzymatic reaction proceeds by way of a state or intermediate in which a tetrahedral carbon atom is surrounded by a nitrogen and several oxygen atoms. A closely related example is the compound

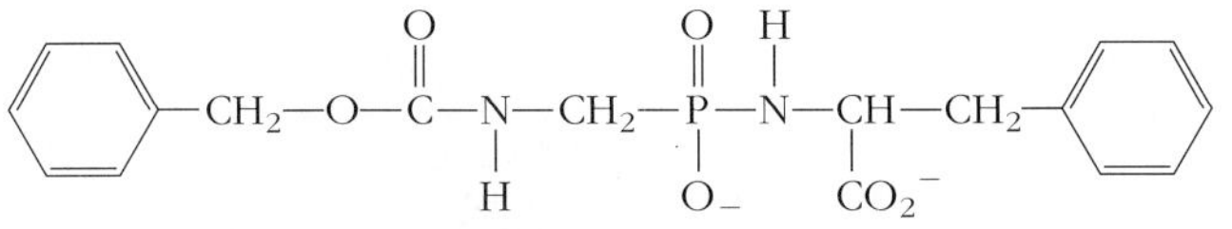

$$C_6H_5{-}CH_2{-}O{-}C(=O){-}NH{-}CH_2{-}P(=O)(O^-){-}NH{-}CH(CO_2^-){-}CH_2{-}C_6H_5$$

which is a potent competitive inhibitor of carboxypeptidase A. Here again, the tetrahedral phosphorus atom resembles the tetrahedral carbon in the intermediate that probably forms in the course of the enzymatic reaction. Similarly, we will see in this chapter that uridine vanadate, which contains a vanadium atom surrounded by five oxygen atoms, is a competitive inhibitor of ribonuclease. This behavior is consistent with the formation of an intermediate in which a phosphorus atom of the substrate takes on a pentavalent geometry (see fig. 9.23).

Figure 9.15

The functional groups surrounding the Zn^{2+} ion in carboxypeptidase A are similar to those in thermolysin. Glycyl-L-tyrosine, shown here in color, is a competitive inhibitor of carboxypeptidase A. The tyrosine side chain of glycyl-L-tyrosine fits into a hydrophobic pocket on the enzyme. The peptide oxygen binds directly to the Zn^{2+} ion, the carboxyl group forms a salt bridge with Arg 145, and the amide NH forms a hydrogen bond with Tyr 248. The salt bridge between the amino group and Glu 270 probably explains why glycyl-L-tyrosine is not a good substrate, because it would prevent the glutamic acid residue from acting as a general base.

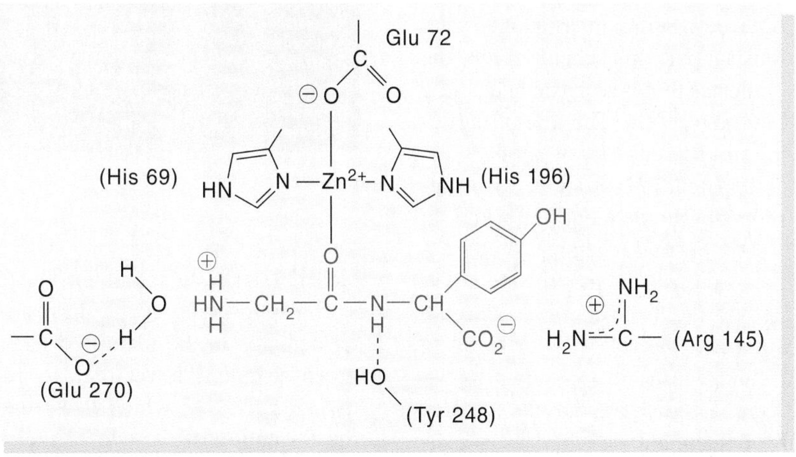

glycyl-L-tyrosine. This dipeptide is hydrolyzed only very slowly, possibly because it interacts with the enzyme in a somewhat more complicated manner than the oligopeptides that are the normal substrates. The free N-terminal amino group of the dipeptide is linked indirectly to the carboxyl group of Glu 270 by way of a hydrogen-bonded molecule of water, as shown in figure 9.15. This could prevent the glutamic acid from participating as a general base in the hydrolytic reaction. In a normal oligopeptide substrate the N-terminal amino group would be too far away to interact with the glutamic acid.

As in thermolysin, the carbonyl oxygen of the peptide binds directly to the Zn^{2+} atom of carboxypeptidase A, displacing the water (see fig. 9.15). The active site also includes an arginine residue (Arg 145), which interacts closely with the negatively charged, C-terminal carboxyl group of glycyl-L-tyrosine and presumably would do the same with a better substrate. This explains the specificity of the carboxypeptidases for pruning off carboxyl-terminal amino acid residues. Carboxypeptidase A differs from thermolysin in that there is no histidine imidazole close enough to the bound dipeptide to serve as a general acid in the hydrolysis. The phenolic OH group of Tyr 248 *is* located nearby (see fig. 9.15), but we will see shortly that the role of the tyrosine residue appears to be rather different.

A remarkable finding in the carboxypeptidase A crystal structures, which was not seen with the serine proteases, is that the binding of the substrate analog (glycyl-L-tyrosine) causes the protein to undergo a major structural rearrangement (fig. 9.16). Arginine 145 and Glu 270 both move by about 2 Å to come into position near the substrate, and Tyr 248 swings down by about 12 Å so that its phenolic group is close to the NH group of the scissile bond. At least four molecules of H_2O are expelled in the process, so the substrate finds itself surrounded largely by hydrophobic groups of the enzyme instead of by water. The binding of the substrate to the enzyme thus is favored by the entropy increase associated with removing the substrate's aromatic side chain from water, in addition to the electrostatic interactions between the terminal carboxyl group and Arg 145.

The most plausible mechanism for the operation of the carboxypeptidases is essentially the same as the scheme we just discussed for thermolysin: The Zn^{2+} ion pulls electrons from the oxygen atom of the peptide bond, increasing the positive charge on the carbonyl carbon, and Glu 270 acts as a general base, increasing the nucleophilic character of a molecule of water by removing a proton. Tyrosine 248 then might serve as a general acid to facilitate the decomposition of the tetrahedral intermediate state. Several experimental observations support this scheme. First, the pH dependence of the kinetics indicates that both basic and acidic functional groups are involved in the reaction. The rate of the reaction decreases if a group with an apparent pK_a of about 6.2 is protonated or if a group with an apparent pK_a of about 9.0 is deprotonated. The former group probably can be identified with Glu 270, and the latter could be Tyr 248. Second, chemical modifications of either Glu 270 or Tyr 248 destroy or modify the enzymatic activity. Acetylation of the tyrosine, for example, makes the enzyme essentially inactive. But when Tyr 248 was changed to phenylalanine by site-directed mutagenesis, it was found to decrease k_{cat} by only a factor of about 2, indicating that the tyrosine phenolic group is not essential for enzymatic activity. The substitution of phenylalanine for Tyr 248 did increase the K_m, suggesting that the tyrosine contributes primarily to the binding of the substrate.

Ribonuclease A: An Example of Concerted Acid-Base Catalysis

Ribonucleases are a widely distributed family of enzymes that hydrolyze RNA by cutting the P—O ester bond attached to a ribose 5′ carbon. The reaction occurs in two steps, with a 2′,3′-phosphate cyclic diester intermediate (fig. 9.17). The intermediate can be identified relatively easily, because its breakdown is much slower than its formation. Ribonucleases do not hydrolyze DNA, which lacks the 2′-hydroxyl group needed for the formation of the cyclic intermediate. The best-studied of the ribonucleases, the pancreatic enzyme ribonuclease A, is specific for a pyrimidine base (uracil or cytosine) on the 3′ side of the phosphate bond that is cleaved.

When the amino acid sequence of bovine ribonuclease A was determined in 1960 by Sanford Moore and William Stein, it was the first enzyme and only the second protein to be sequenced. Ribonuclease A also was one of the first enzymes whose three-dimensional structure was elucidated by x-ray diffraction, and it was

Figure 9.16

Crystal structures of carboxypeptidase A (*a*) with no substrate bound and (*b*) with bound glycyl-L-tyrosine. The glycl-L-tyrosine or, in (*a*), the molecule of water that binds to the zinc in place of the substrate is shown in yellow. The side chains of Glu 270 and Tyr 248 are in orange, and the histidine ligands of the zinc are in red. Note how the large change in the position of the tyrosine, together with smaller changes in other residues, folds the enzyme around the substrate. The figures are based on the crystal structures described by W. N. Lipscomb.

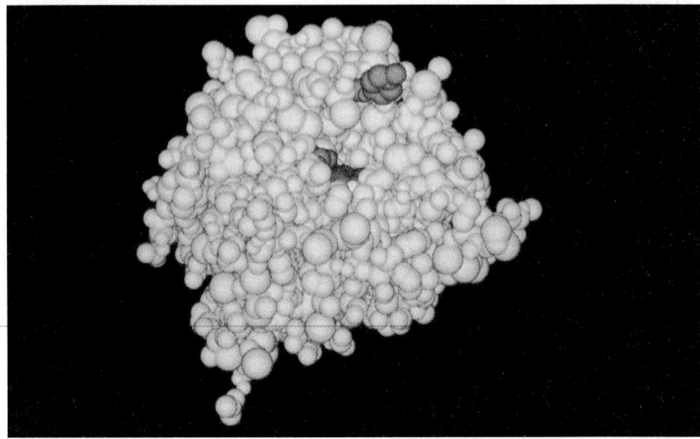

(a)

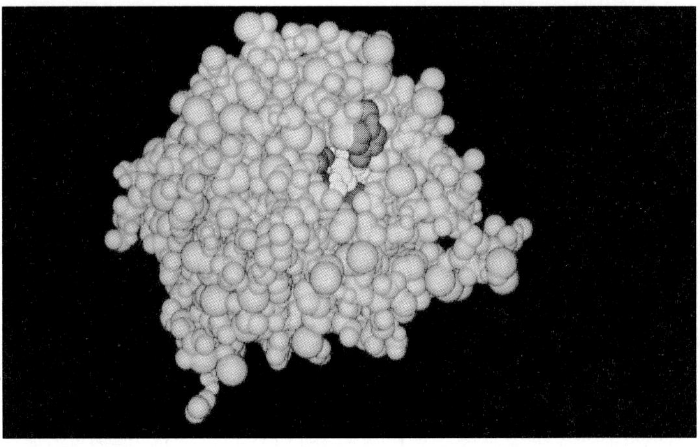

(b)

Figure 9.17

A portion of an RNA chain, indicating points of cleavage by pancreatic ribonuclease. "Pyr" refers to a pyrimidine; "Base" can be either a purine or a pyrimidine. The 2′, 3′, and 5′ carbon atoms are labeled. The enzymatic reaction proceeds in two steps, with a cyclic 2′, 3′-phosphate diester as an intermediate.

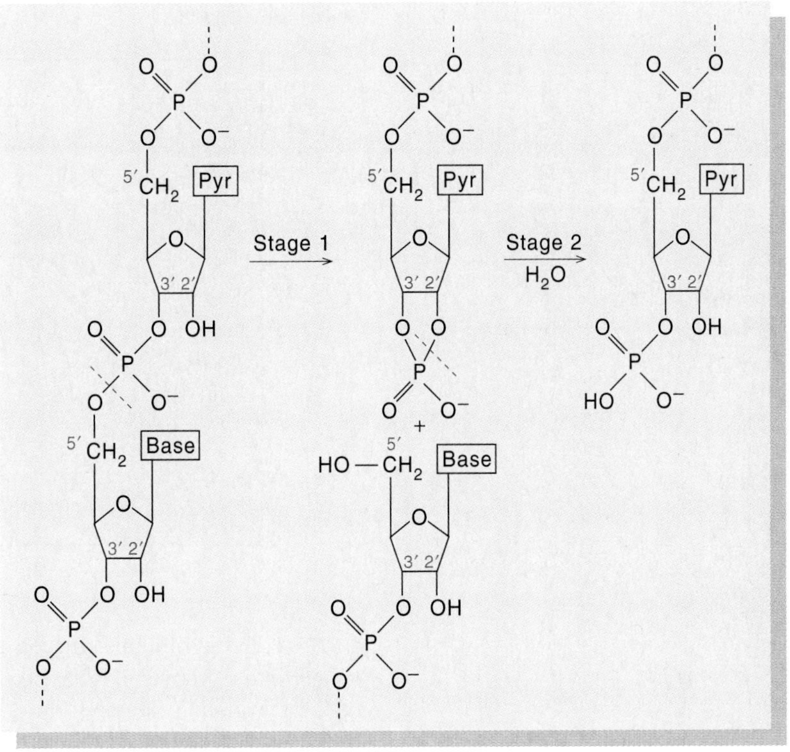

the first to be synthesized completely from its amino acids. The synthetic protein proved to be enzymatically indistinguishable from the native enzyme.

Ribonuclease A (RNase A) is a relatively small protein, with a molecular weight of 13,680 and a single polypeptide chain of 124 amino acid residues. An early discovery that turned out to be very useful was that the protein could be cleaved specifically between residues 20 and 21 by the bacterial serine protease subtilisin. The resulting two polypeptides could be separated and purified. They were found to be enzymatically inactive individually but to regain the complete activity of the native enzyme when they were recombined. This work showed that there are strong, noncovalent interactions that can hold protein chains together even when one of the peptide links is cut. It also made it possible to modify specific amino acid residues of the two polypeptide chains independently and to explore how each residue contributed to the reassembly of the protein and the recovery of enzymatic activity.

RNase A is completely inhibited if either of two histidine residues (His 12 or His 119) is modified by carboxymethylation with iodoacetate. This observation suggested that both of these histidines play important roles in the active site (fig. 9.18). In support of this conclusion the reaction of iodoacetate with His 12 or His 119 is inhibited by cytidine-3′-phosphate or other small molecules that bind at the active site. Lysine 41 has been implicated similarly in the active site by the observation that the reaction of fluorodinitrobenzene with the ϵ-amino group of this residue destroys enzymatic activity (see fig. 9.18). A second lysine (Lys 7) also appears to be important for substrate binding, although the enzyme retains some activity when this residue is modified.

The enzymatic activity of RNase A shows a bell-shaped dependence on pH, with an optimum near pH 7 (fig. 9.19). The operation of the enzyme appears to require that a dissociable group with a pK_a of about 6.3 be in the protonated form and that a group with a pK_a of about 8 be unprotonated. These groups have been identified as His 12 and His 119 by measuring the nuclear magnetic resonance (NMR) spectrum of the enzyme as a function of pH. The spectral peaks due to histidine residues undergo characteristic shifts as the imidazole ring is protonated. Analysis of the NMR spectra, though not straightforward, was aided by the fact that bovine RNase A contains only four histidine residues, one of which (His 12) can be separated from the others by cutting the protein with subtilisin. Histidines 12 and 119 have pK_a values of 5.8 and 6.2 in the free enzyme, but both pK_a values shift to a more alkaline range when the enzyme binds its substrate.

The pH dependence of the kinetics, taken with the information that histidines 12 and 119 are essential for enzymatic activity, suggests that the two histidines participate in the reaction by concerted general-base and general-acid catalysis. This suppo-

Figure 9.18

When RNase A is treated with iodoacetate ($ICH_2CO_2^-$), the two major products obtained are carboxymethylated derivatives of His 12 and His 119. Both of these enzymes are severely inhibited. This result suggests that both His 12 and His 119 are important in the active site. The enzyme also is completely inhibited by the reaction of Lys 41 with fluorodinitrobenzene.

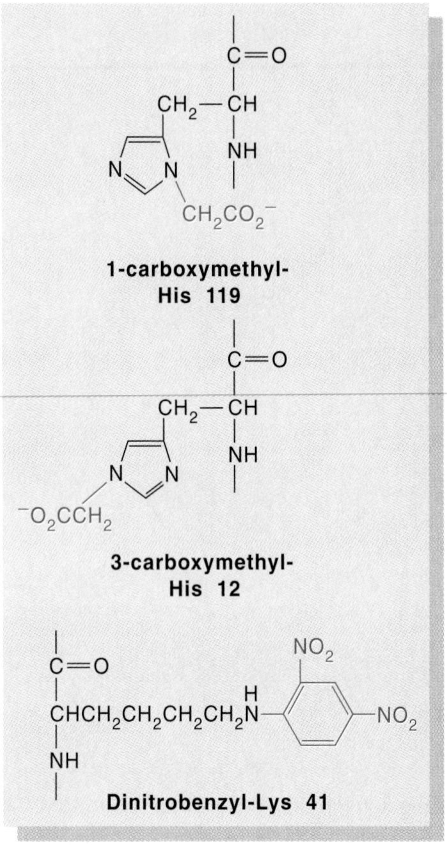

1-carboxymethyl-His 119

3-carboxymethyl-His 12

Dinitrobenzyl-Lys 41

Figure 9.19

The dependence of k_{cat} of RNase A on pH. The bell-shaped curve suggests that one histidine residue must be in the protonated state and another must be unprotonated. Similar pH dependences are found for the hydrolysis of either RNA or pyrimidine nucleoside-2′,3′-phosphate cyclic diesters, a fact indicating that the two steps of the overall reaction both require concerted general-acid and general-base catalysis.

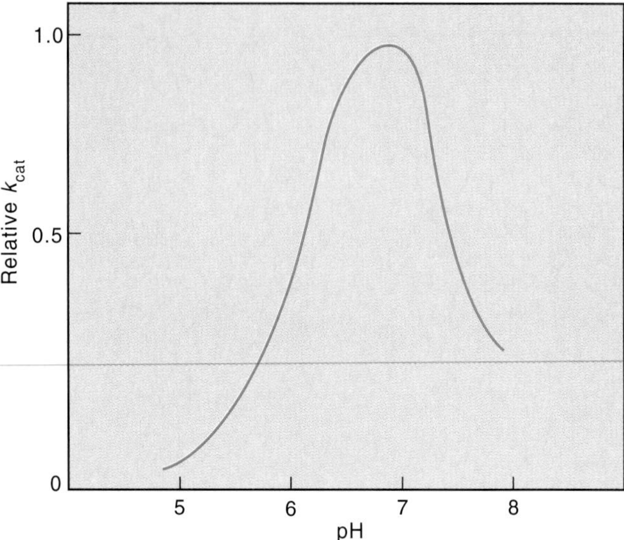

sition is supported by the crystal structure of the enzyme. Figure 9.20 shows a model of the crystal structure of RNase S (the active enzyme obtained by cutting ribonuclease with subtilisin and recombining the two polypeptides) with a bound substrate analog, "$UpCH_2A$." The analog is a competitive inhibitor of the enzyme. It resembles the dinucleotide UpA but cannot be hydrolyzed because it has a methylene carbon in place of the oxygen between the P and the 5′-CH_2 of the ribose of adenosine (fig. 9.21). Similar crystal structures have been obtained with a variety of other substrate analogs. In all of the structures, His 12 and His 119 are found on opposite sides of the substrate's phosphate group, with His 12 positioned close to the 2′ OH group. Lysine residues 7, 14, and 66 also are located nearby, where their positively charged amino groups would have favorable electrostatic interactions with the negatively charged phosphate. Near His 119 is the carboxyl group of an acidic residue, Asp 121. The electrostatic effect of Asp 121 thus would favor the transfer of a proton from a molecule of water to His 119, much as Asp 102 in chymotrypsin induces a proton to move from Ser 195 to His 57 (see figs. 9.11 and 9.12).

The pyrimidine ring on the 3′ side of the substrate fits snugly into a groove on RNase A. Valine 43 is located on one side of the pyrimidine ring and Phe 120 on the other. The specificity of the enzyme for pyrimidine nucleotides appears to result largely from hydrogen bonding to Thr 45. The threonine side chain can form a pair of complementary hydrogen bonds with either uracil or cytosine (fig. 9.22). Crystallographic studies have shown that if the pyrimidine is replaced by a purine, the compound can still bind to the enzyme, but the distance between His 12 and the 2′ OH of the ribose increases by about 1.5 Å. This increase evidently is enough to prevent the catalytic reaction from occurring.

Ronald Breslow has proposed a detailed mechanism for the hydrolysis of RNA by RNase A that is consistent with the formation of a cyclic phosphate intermediate as shown in figure 9.17 (fig. 9.23). Although several of the details of this mechanism have not been proven, the mechanism is supported by all that is known about the reaction. First His 119 and His 12 function as proton donor (acid) and proton acceptor (base), respectively, in a two-proton shift. The loss of the proton of the 2′-hydroxyl group to His 12 increases the nucleophilic character of the oxygen atom, precipitating a nucleophilic attack on the phosphate, which leads to a transient intermediate state in which the phosphorus atom is pentavalent. The formation of the pentavalent intermediate is promoted by His 119, which protonates one of the phosphate oxygens. The pentavalent phosphorus oxyanion is stabilized by electrostatic interaction with Lys 41. In step 2, His 12 donates its newly acquired proton to the remaining negatively charged phosphate oxygen. This facilitates step 3, in which His 119 acting as a base removes a proton from a phosphate hydroxyl group on the other side of the phosphate. The two histidines are thereby returned to

Figure 9.20

A model of the binding of the substrate analog $UpCH_2A$ (blue) to RNase S. (The structure of $UpCH_2A$ is shown in figure 9.21.) RNase S is an active enzyme, although it differs from the native enzyme because of a peptide bond cleavage between residues 20 and 21 (light dashed line). The two histidines and the lysine that are crucial to the active site (His 12, His 119, and Lys 41) are indicated in gray. The dotted lines indicate some of the hydrogen bonds that maintain the protein structure. (Illustration copyright by Irving Geis. Reprinted by permission.)

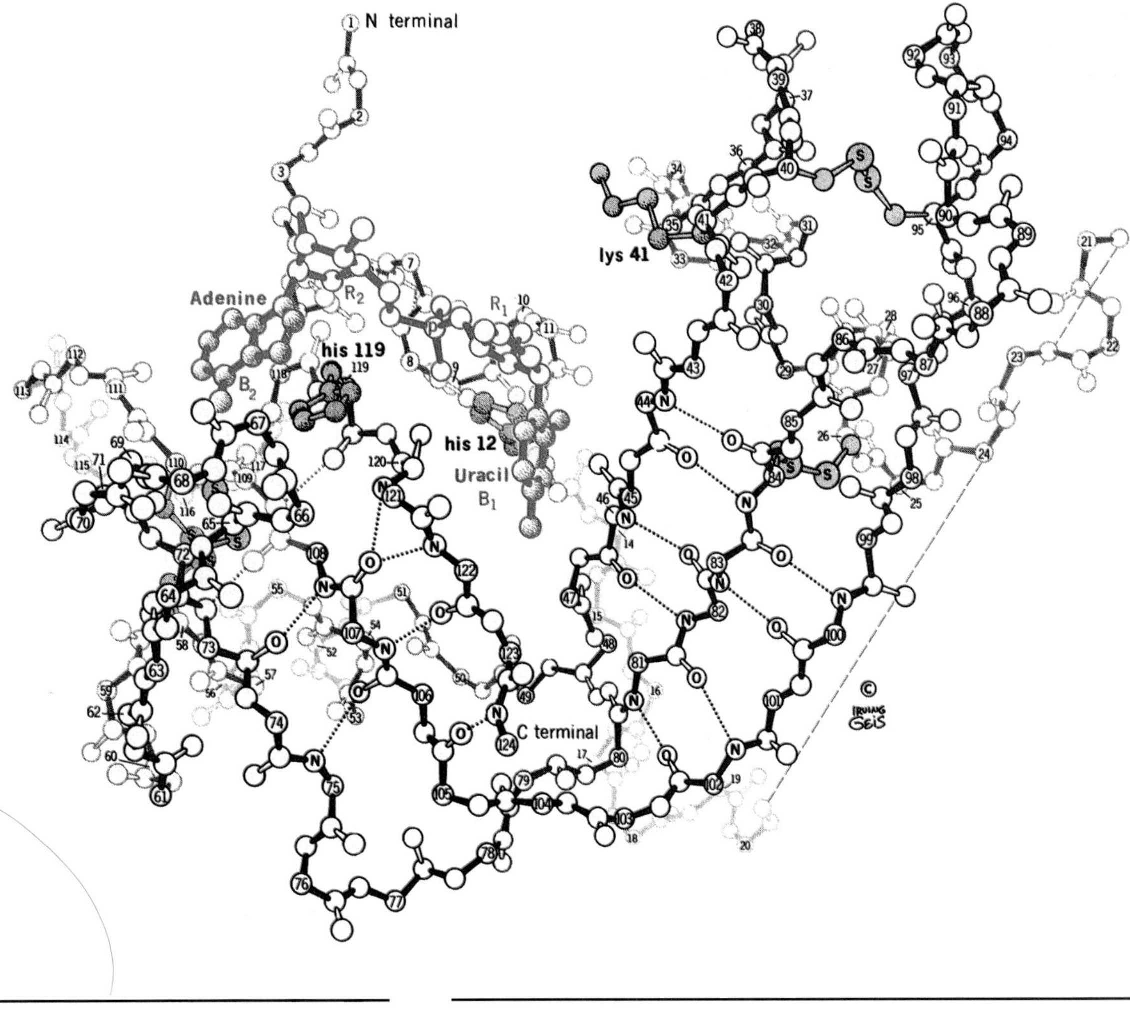

Figure 9.21

Structure of $UpCH_2A$, a competitive inhibitor of RNase A. $UpCH_2A$ differs from the dinucleotide substrate UpA in having a methylene group instead of oxygen between the phosphorus and C-5′ of the adenosine.

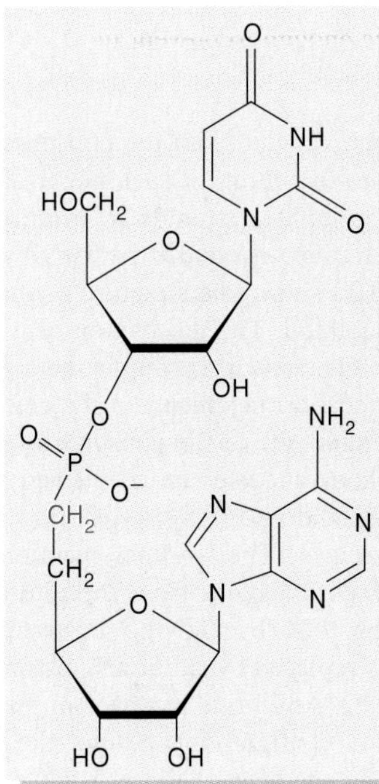

Figure 9.22

The side chain and amide NH group of Thr 45 in RNase A can form a pair of complementary hydrogen bonds with either uracil or cytosine. These probably account for the specificity of RNase A for pyrimidine nucleotides.

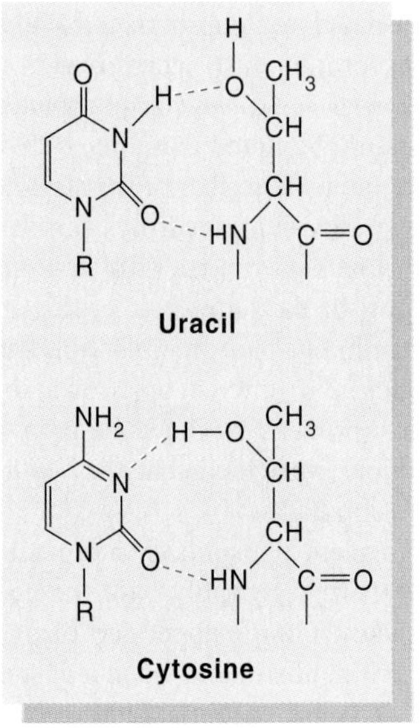

Uracil

Cytosine

Figure 9.23

Proposed mechanism for the hydrolysis of RNA by ribonuclease A. Steps leading to the formation of the 2′,3′-cyclic phosphate intermediate are shown. For the hydrolysis of the cyclic phosphate the same mechanism is run backwards, with H_2O substituting for ROH. Histidines 12 and 119 are directly involved in the reaction. Lysine 41 most likely stabilizes the intermediate phosphorane anions consistent with x-ray data. Note that after cyclization the enzyme ends up in a protonation state reversed from that of the starting enzyme but would be restored to its original state after hydrolysis of the cyclic phosphate.

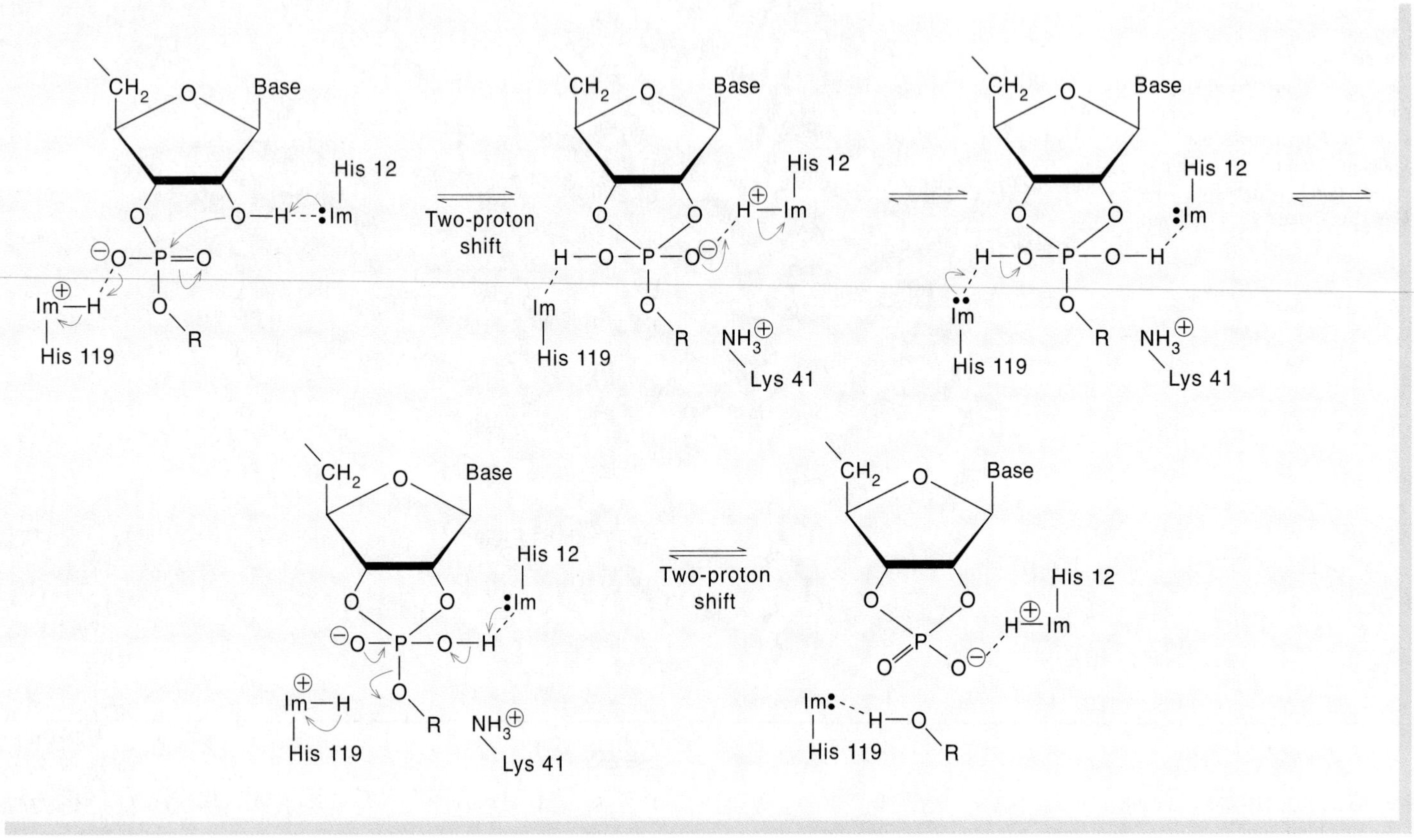

their original protonated and unprotonated states, and the stage is set for another two-proton shift, which results in cleavage of the RNA and formation of a 2′,3′-cyclic phosphate (step 4). We are now halfway. It should be noted that after step 4 the enzyme is in a protonation state opposite to that of the starting enzyme. The final steps (not shown) involving hydrolysis of the cyclic phosphate go by the same mechanism run backwards with H_2O substituting for ROH. In support of the latter half of this mechanism, which entails running essentially the same reaction backwards, Breslow alludes to the principle of microscopic reversibility, which states that a reaction must proceed by the same pathway in the forward and reverse directions, provided that they both occur under essentially the same conditions.

The pentavalent phosphoryl intermediates that figure in the reactions of RNase A probably have the geometry of a trigonal bipyramid, as shown in the upper part of figure 9.24. Three of the phosphorus atom's substituents lie in a plane. The entering and leaving oxygen atoms are on either side of the plane, forming a straight line with the phosphorus. One observation that supports the formation of such pentavalent intermediates is that RNase A is inhibited strongly by uridine vanadate, in which the vanadium atom is surrounded by five oxygens in a similar geometry. Figure 9.25 shows the structure of the complex of the enzyme with the inhibitor. The disposition of histidines 12 and 119 on either side of the vanadate group in the crystal structure is consistent with the linear arrangements of the entering and leaving oxygen atoms on either side of the phosphorus atom in figures 9.23 and 9.24. Uridine vanadate can be viewed as a transition-state analog (boxes 9C and 9D).

The "in-line" reaction mechanism can be distinguished experimentally from the alternative "adjacent" mechanism (see fig. 9.24) by studying substrates in which one of the oxygen atoms is replaced by sulfur and another is labeled stereospecifically with ^{18}O. An in-line mechanism results in stereochemical inversion of the configuration around the phosphorus, whereas an adjacent arrangement of the entering and departing groups results in re-

Figure 9.24

Two routes leading to phosphoryl group transfer. Both routes pass through a pentavalent intermediate, in which three of the oxygen atoms lie in a plane with the phosphorus atom. In the "in-line" mechanism, the entering and leaving groups ($^{-}$OR and RO^{-}) are on opposite sides of this plane. In the "adjacent" mechanism the entering and leaving groups are in neighboring positions. The in-line mechanism results in stereochemical inversion; the adjacent mechanism retains the stereochemistry. Three of the oxygen atoms have been labeled with numbers to indicate their stereochemical relationships in the figure. Experimentally, the stereochemical outcome of the RNase reaction can be tested by using a substrate that is labeled stereospecifically with a combination of ^{17}O and ^{18}O. Substrates in which one of the oxygens is replaced by sulfur also can be used. (The adjacent mechanism requires one additional step not shown here: Before it can depart, the leaving group has to move to an axial orientation perpendicular to the plane. This rearrangement is called *pseudorotation.* Several of the substituents move in a concerted manner in the pseudorotation step, so that their stereochemical relationships are maintained.)

In-line mechanism

R—O—P(O^1)(O^2)(OH^3) + $^{-}$OR′ ⇌ R—O—P(O^1)(O^2)(OH^3)—OR′ ⇌ RO^{-} + P(O^1)(O^2)(HO^3)—OR′

Adjacent mechanism

R—O—P(O^1)(O^2)(OH^3) + $^{-}$OR′ ⇌ R—O—P(O^2)(O^1)(OH^3)(OR′) ⇌ RO^{-} + P(O^2)(O^1)(OH^3)—OR′

tention of the configuration (see fig. 9.24). The reactions catalyzed by RNase A and a number of other ribonucleases have been shown to proceed with inversion. Many of the other enzymes that transfer phosphoryl groups from one substrate to another exhibit retention.

Triosephosphate Isomerase Has Approached Catalytic Perfection

Triosephosphate isomerase catalyzes the interconversion of dihydroxyacetone phosphate and glyceraldehyde-3-phosphate:

$$\underset{\text{Dihydroxyacetone phosphate}}{\begin{array}{c} CH_2OH \\ | \\ C{=}O \\ | \\ CH_2OPO_3^{\,2-} \end{array}} \rightleftharpoons \underset{\text{Glyceraldehyde-3-phosphate}}{\begin{array}{c} H{-}C{=}O \\ | \\ H{-}C{-}OH \\ | \\ CH_2OPO_3^{\,2-} \end{array}} \qquad (10)$$

The reaction requires moving a proton from carbon 1 of dihydroxyacetone phosphate to carbon 2, removing another proton

Figure 9.25

(*a*) Ribbon diagram representation of the crystal structure of the complex between RNase A and uridine vanadate. The catalytic histidine side chains, histidines 12 and 119, are shown in light blue. Uridine vanadate, which contains a vanadium atom (pink) surrounded by five oxygen atoms (red), is thought to mimic the pentavalent transition state formed by a phosphoryl substrate. Uridine vanadate binds to RNase A about 10^4 times more tightly than the corresponding uridine 2′,3′-cyclic phosphate ester. (*b*) Close-up view of the RNase A and uridine vanadate complex showing the proximity and position of histidines 12 and 119. (Copyright 1994 by the Scripps Research Institute/Molecular Graphics Images by Michael Pique using software by Yng Chen, Michael Connolly, Michael Carson, Alex Shah, and AVS, Inc. Visualization advice by Holly Miller, Wake Forest University Medical Center.)

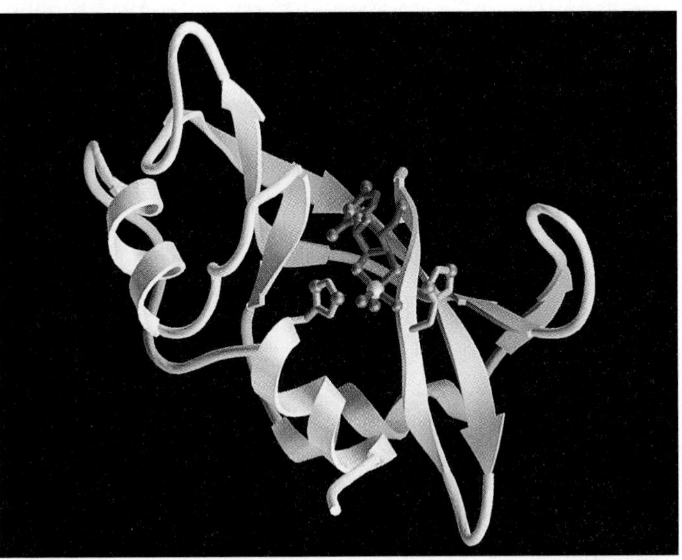

(a)

(b)

from the oxygen atom at C-1, adding another proton to the oxygen at C-2, and moving a pair of electrons in the same direction. The interconversion of the two triosephosphates is an essential step in the catabolism of carbohydrates. Triosephosphate isomerase is among the enzymes that have approached evolutionary perfection

Catalytic Antibodies

According to kinetic theory, formation of the transition state is the rate-limiting step in a reaction. Whereas enzyme binds substrate, it is believed to bind most strongly to the transition state intermediate, thereby favoring the conversion of the substrate to the transition state. In 1969, William Jencks conceived of a way to test this theory by making antibodies that bind specifically to the transition state. To do this, it is necessary to first synthesize a stable transition-state analog that resembles the hypothesized transition-state intermediate. This approach has been used in a number of cases, and it has been found that antibodies that function as catalysts can be made in this way that accelerate reaction rates by factors of 10^4 to 10^6. Although this is far less than the rate effects often seen with bonafide enzymes, it lends support to the notion that a major factor in enzyme catalysis is the ability of the enzyme to favor the formation of the transition state from the substrate.

A case in point is the conversion of chorismate to prephenate, a step in the synthesis of phenylalanine and tyrosine (see fig. 1 and chapter 24). This reaction is believed to proceed by the so-called Claisen rearrangement, a reaction that involves the cyclic flow of six electrons and a six-membered ring intermediate. A stable compound resembling the transition-state intermediate (fig. 2) was synthesized and used to stimulate antibody synthesis. The resulting antibody was found to stimulate the rate of conversion of chorismate to prephenate by a factor of 10^4 over the uncatalyzed reaction.

Figure 1

Hypothesized route for the enzyme-catalyzed conversion of chorismate to prephenate.

Chorismate → Transition state → Prephenate

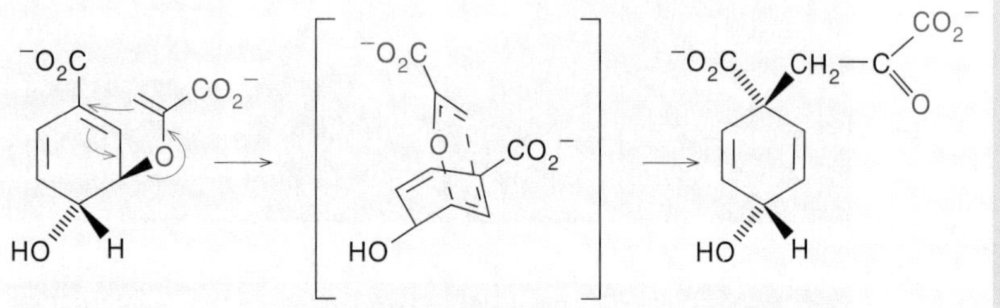

Figure 2

Transition-state analog used to stimulate the synthesis of antibody.

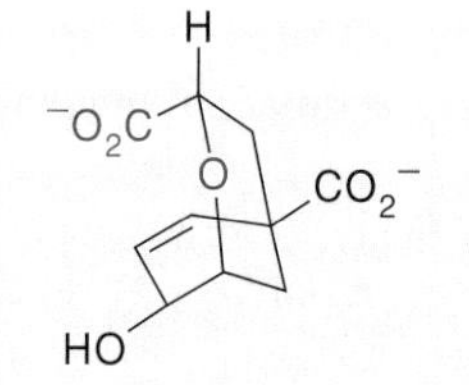

in the sense that the specificity constant k_{cat}/K_m is greater than 10^8 $M^{-1}s^{-1}$, which is near the limit set by the rates of diffusion of the substrate and the protein (see table 8.4). The enzyme achieves this high value for the specificity constant by having both a high k_{cat} and a low K_m.

Crystal structures have been obtained for triosephosphate isomerase purified both from yeast and from chicken muscle. Although the proteins from the two organisms differ at about half of the positions in their amino acid sequences, their three-dimensional structures are nearly identical, and the same sets of amino acids have been implicated in their catalytic mechanisms. The overall structure is a β barrel, with eight parallel β sheets linked by eight α helices (fig. 9.26). In both enzymes a glutamic acid residue (Glu 165) reacts with affinity labels that bind at the active site, such as chloroacetone phosphate. (Chloroacetone phosphate is the same as dihydroxyacetone phosphate except that a chlorine atom replaces the hydroxyl group on C-1.) Crystal structures of the complexes of the enzyme with bound dihydroxyacetone phosphate or a competitive inhibitor show that the carboxyl group of Glu 165 is located close to C-1 of the substrate. When Glu 165 is changed to aspartate by site-directed mutagenesis, the rate of the isomerization reaction catalyzed by the enzyme is decreased approximately 1,000-fold. This is a remarkable effect, considering that the substitution of Asp for Glu probably changes the position of the carboxyl group by less than 1 Å.

The rate of the isomerase reaction decreases if a basic group is protonated by bringing the pH below 6.5. It seems clear that this is the carboxyl group of Glu 165, because the reaction of

Figure 9.26

(*a*) Ribbon diagram of the crystal structure of triosephosphate isomerase from chicken muscle. This figure shows one of the two subunits of the enzyme. The active site amino acid side chains Glu 165 (red), His 95 (light blue), and Lys 13 (dark blue) are shown. (*b*) Close-up view of the active site of triosephosphate isomerase. (Copyright 1994 by the Scripps Research Institute/Molecular Graphics Images by Michael Pique using software by Yng Chen, Michael Connolly, Michael Carson, Alex Shah, and AVS, Inc. Visualization advice by Holly Miller, Wake Forest University Medical Center.)

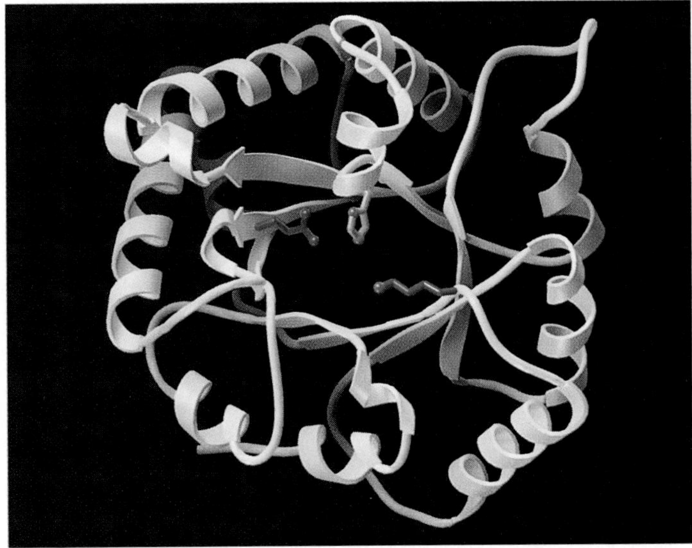

(a)

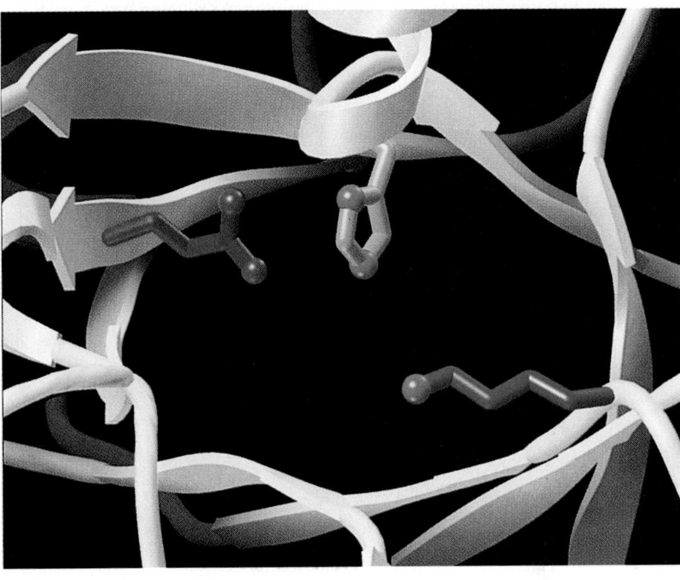

(b)

Glu 165 with affinity labels drops out in parallel with the enzymatic activity. This suggests that, in the enzymatic reaction, Glu 165 acts as a general base to remove the proton from C-1 of the dihydroxyacetone phosphate, facilitating conversion of the substrate into an enediolate intermediate as shown in figure 9.27. The reaction involving the protonated glutamic acid residue could be completed by returning the proton to C-2 of the enediolate, instead of to C-1.

Independent evidence in support of this scheme has come from experiments in which the enzymatic reaction is run in D_2O. Deuterium is incorporated stereospecifically onto carbon 2 of the product:

$$\begin{matrix} CH_2OH \\ | \\ C{=}O \\ | \\ CH_2OPO_3^{-2} \end{matrix} \xrightarrow{D_2O} \begin{matrix} H{-}C{=}O \\ | \\ D{-}C{-}OH \\ | \\ CH_2OPO_3^{-2} \end{matrix} \qquad \textbf{(11)}$$

In the absence of the enzyme this carbon-bound hydrogen atom on glyceraldehyde-3-phosphate does not exchange with deuterium atoms of the solvent at any significant rate. The incorporation of the deuterium isotope at C-2 can be explained as follows: After a proton is transferred from C-1 of the substrate to the carboxyl group of Glu 165, it has an opportunity to escape into the solution and to be replaced by a deuteron. A proton attached to a carboxylic acid oxygen usually exchanges very rapidly with the solvent. The deuteron on Glu 165 then can be transferred to C-2 of the product.

Additional studies of the kinetics of such exchange processes indicate that the proton probably moves from Glu 165 directly back to the enediolate in one step, rather than hopping first to another acid–base group and proceeding from there to the enediolate. This is consistent with the crystal structure, which shows that the carboxylate group of Glu 165 is in a good position to interact with a proton on either C-1 or C-2 of the substrate.

Now let us consider the fate of the other proton that must be removed from the oxygen of C-1 and the origin of the proton that finds its way to the oxygen of C-2. In the crystal structure there are two nearby acid–base groups that could participate in this process; these are His 95 and Lys 13 (see fig. 9.26). His 95 appears to play an important role in the enzymatic reaction, because replacing this residue with glutamine by site-directed mutagenesis decreases k_{cat} by a factor of about 400. Considering the position of His 95 in the crystal structure a conceivable scheme for the overall reaction would be for this residue to act together with Glu 165 to provide concerted acid–base catalysis in both steps of the reaction. In the formation of the intermediate state, Glu 165 would remove a proton from C-1 of dihydroxyacetone phosphate, and a protonated His 95 would add a proton to the carbonyl oxygen. This means that the intermediate would be a neutral enediol, rather than an enediolate ion as shown in figure 9.27. In the breakdown of the enediol, Glu 165 would return a proton to C-2, and a neutral His 95 would extract a proton from the oxygen atom on C-1.

Figure 9.27

The probable reaction mechanism of triosephosphate isomerase. The γ-carboxylate group of Glu 165 acts as a general base to remove a proton from C-1 of the substrate, dihydroxyacetone phosphate (DHAP). This generates a planar enediolate intermediate and the transformation of the hydrogen bond formed between the substrate and His 95 to an extra-strong low barrier hydrogen bond (LBHB). The proton participating in this hydrogen bond becomes transferred to C-2 of the substrate, and simultaneously, His 95 forms a comparable LBHB with C-1, which results in the loss of the participating proton from C-1 to His 95 and the return of the proton originally captured by Glu 165 to the C-2 position. The positively charged side chain of Lys 13 has a stabilizing influence on the binding of the negatively charged substrate and its intermediates.

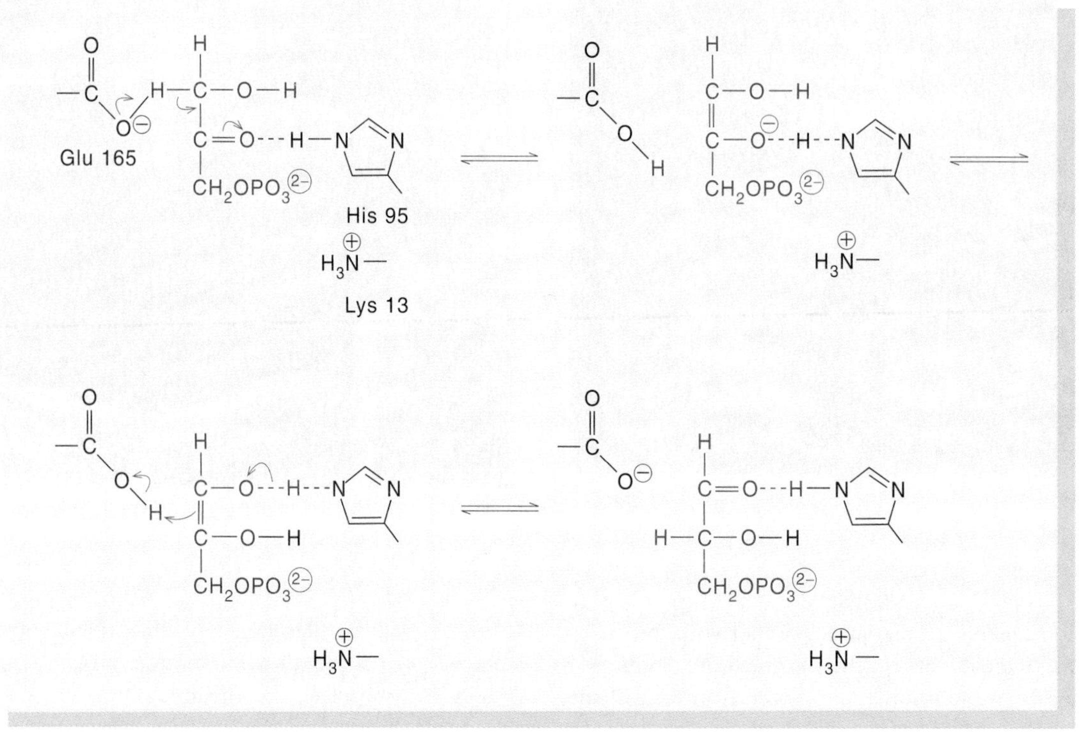

There are serious difficulties with the idea that His 95 works as a general acid in this way. Thus NMR measurements indicate that His 95 is not protonated between pH 5 and 9.9. Since the enzyme does not function outside of this pH range, it appears that if His 95 is involved in the active site, then it must be involved in the neutral form. This could be explained by the formation of an unusually strong low-barrier hydrogen bond (LBHB) between the histidine and the enolate intermediate(s). Since the enzyme–substrate complex initially probably contains a standard weak hydrogen bond, the formation of an LBHB uniquely in the transition-state intermediate would substantially lower the activation energy for the reaction. It should be remembered that one of the requirements for the formation of an LBHB is that the pK_a values of the participating groups match (see box 9B). The pK_a of the oxygen on C-2 of the enediol should be about 14, which would nicely match the pK_a of neutral imidazole but not the pK_a of protonated imidazole (about 6); actually, the pK_a values for both the enediol and the neutral histidine are decreased somewhat in the local environment of the enzyme. In this reaction there are two enediolate intermediates, one with the LBHB to the oxygen at C-2 and the other with an LBHB to the oxygen at C-1. The LBHBs that are found uniquely in these high-energy intermediates should substantially lower the k_{cat} for the reaction if they are involved in the rate-limiting step(s) for the reaction.

Assuming that the scheme shown in figure 9.27 is correct, how could we determine whether the formation or the breakdown of the enediolate intermediate is the rate-limiting step in the overall reaction? A reasonable approach to determining this would be to study the kinetics using substrates that are specifically labeled with deuterium. With [1(R)-^{2}H]-dihydroxyacetone phosphate (fig. 9.28), k_{cat} is about three times smaller than the k_{cat} obtained with the [^{1}H] analog. An isotope effect of this magnitude is consistent with the view that the bond holding the ^{1}H or ^{2}H atom to C-1 must be broken in the rate-limiting step. This result demonstrates that formation of the enediolate is indeed rate-limiting.

Lysozyme Hydrolyzes Complex Polysaccharides Containing Five or More Residues

Lysozyme catalyzes the hydrolysis of polysaccharide chains that form structural elements of bacterial cell walls. It was discovered

Figure 9.28

The substrate of triosephosphate isomerase can be labeled specifically with deuterium. A kinetic isotope effect of about 3 is obtained with [1(R)-^{2}H]-dihydroxyacetone phosphate. This indicates that the formation of the enediolate intermediate is the rate-limiting step in the conversion of dihydroxyacetone phosphate to glyceraldehyde-3-phosphate. Note that although the two hydrogen atoms on C-1 of dihydroxyacetone phosphate are chemically identical, they are stereochemically distinguishable. The enzyme always removes the one that has been replaced by deuterium in the figure.

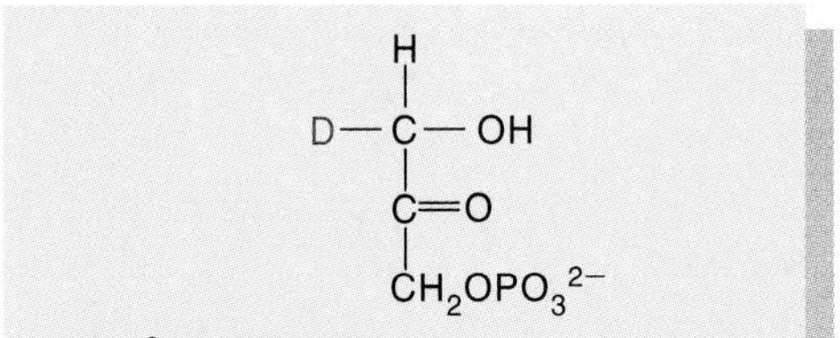

first in the whites of chicken eggs and subsequently was found to occur widely in biological tissues and secretions such as tears. Related glycosidases have been obtained from a variety of organisms including fungi, bacteria, and the bacteriophage T4. The enzyme from hen eggwhite, a small, readily purified protein with a molecular weight of 14,500, was the first enzyme to have its crystal structure solved by x-ray diffraction. Although essentially nothing was known at the time about how lysozyme worked, the structure quickly suggested a likely mechanism.

Lysozyme's principal substrate, the bacterial cell-wall polysaccharide, is a polymer of alternating *N*-acetylglucosamine (GlcNAc) and *N*-acetylmuramic acid (MurNAc) residues connected by $\beta(1,4)$ glycosidic linkages (fig. 9.29). In the cell walls the MurNAc residues are cross-linked by short polypeptides to form a two-dimensional network termed a peptidoglycan (see chapter 18). Lysozyme cuts the polysaccharide chain at C-1 of a MurNAc residue. It also will hydrolyze some shorter oligosaccharides such as (GlcNAc-MurNAc)$_3$ or (GlcNAc)$_6$, but it does not accept (MurNAc)$_6$ or other homopolymers of MurNAc. Thus the active site appears to be specific for a GlcNAc residue next to the bond that is cleaved. Very short oligomers of GlcNAc, such as (GlcNAc)$_3$, bind to the enzyme but are poor substrates (table 9.1).

The crystal structure of eggwhite lysozyme was solved both with and without bound (GlcNAc)$_3$. Although longer oligosaccharides were hydrolyzed too rapidly to afford crystal structures, it was possible to build a model for the enzyme with bound (GlcNAc)$_6$ by starting with the structure of the complex with (GlcNAc)$_3$ (see fig. 1.11). The oligosaccharide sits in a shallow crevice that contains recognizable binding sites for six hexose units. These are labeled A through F in the figure. All of the sites appear to be tailored to bind the acetamide

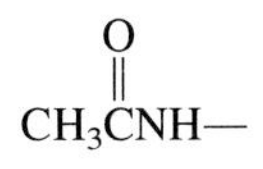

$$CH_3\overset{\overset{\displaystyle O}{\|}}{C}NH—$$

side chains, but sites A, C, and E would not be spacious enough to accommodate the lactyl

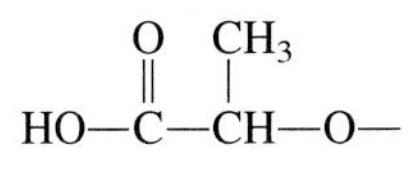

$$HO—\overset{\overset{\displaystyle O}{\|}}{C}—\overset{\overset{\displaystyle CH_3}{|}}{C}H—O—$$

Figure 9.29

Cell-wall polysaccharide, the substrate of lysozyme. (*a*) Conventional drawing of the hexose rings. (*b*) Drawing showing the conformations of the rings. In (*b*), alternating hexose units are flipped over by 180° relative to their neighbors; this is the preferred conformation for polysaccharides with $\beta(1,4)$ linkages (see chapter 13). The bond that is cleaved by lysozyme is indicated.

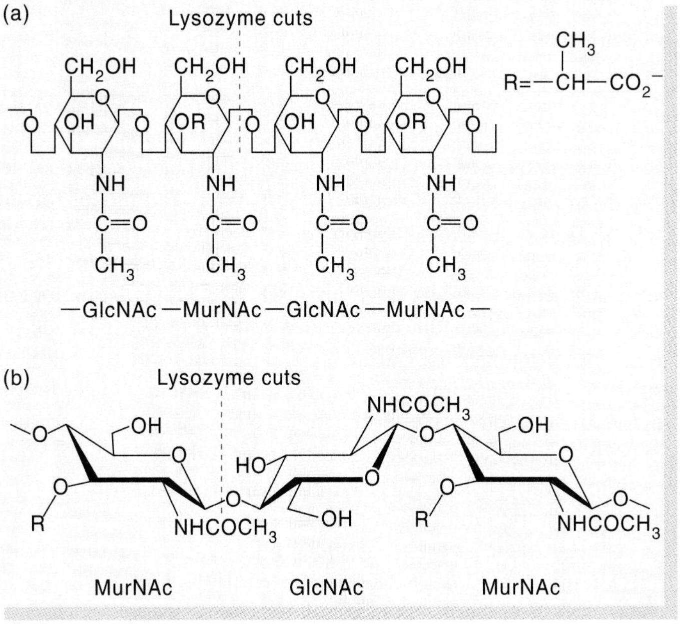

Table 9.1
Rates of Reaction and Cleavage Patterns Shown by Different Substrates of Lysozyme

Compound	Turnover Number k_{cat} (s^{-1})	Cleavage Pattern
(GlcNAc)$_3$	8.3×10^{-6}	$X_1 \overset{\downarrow}{—} X_2 \overset{\downarrow}{—} X_3$
(GlcNAc)$_4$	6.6×10^{-5}	$X_1 \overset{\downarrow}{—} X_2 \overset{\downarrow}{—} X_3 \overset{\downarrow}{—} X_4$
(GlcNAc)$_5$	0.033	$X_1 \overset{\downarrow}{—} X_2 \overset{\downarrow}{—} X_3 \overset{\downarrow}{—} X_4 \overset{\downarrow}{—} X_5$
(GlcNAc)$_6$	0.25	$X_1 — X_2 — X_3 — X_4 \overset{\downarrow}{—} X_5 — X_6$
(GlcNAc-MurNAc)$_3$	0.50	$X_1 — X_2 — X_3 — X_4 \overset{\downarrow}{—} X_5 — X_6$

group of MurNAc. The restrictions are particularly severe in site C because of the bulky side chain of Ile 98. This suggests that the normal substrate of repeating (GlcNAc-MurNAc) units would bind with GlcNAc units occupying sites A, C, and E and with MurNAc units in sites B, D, and F, as shown schematically in figure 9.30. When taken with the observation that lysozyme hydrolyzes $(GlcNAc)_6$ but not $(GlcNAc)_3$ and the fact that $(GlcNAc\text{-}MurNAc)_3$ is hydrolyzed at C-1 of a MurNAc residue (see fig. 9.29 and table 9.1), the model indicates that the glycosidic linkage that is cleaved must fall between sites D and E.

There was another reason for focusing attention on the region of the enzyme between binding sites D and E. The amino acid side chains here included two potentially reactive groups, the carboxyl groups of Asp 52 and Glu 35. When the model for the $(GlcNAc)_6$ substrate was positioned so that sites A–E were all occupied, the glycosidic bond that would be cleaved was located in between the two carboxyl groups (see fig. 1.11). Homologous acidic amino acid residues subsequently were found in the active site of bacteriophage T4 lysozyme, and mutations that perturb the activity of the bacteriophage enzyme were found to cluster in this region. In eggwhite lysozyme, chemical conversion of the carboxyl group of Asp 52 to $—CH_2OH$ destroys enzymatic activity. Studies with affinity labels also have implicated Asp or Glu residues in other bacterial and fungal glucosidases.

Lysozyme works best under mildly acidic conditions with a pH optimum of 5. The pH dependence of the kinetics indicates that the reaction depends on an acidic group with a pK_a of about 6 and on a basic group with a pK_a of about 4.5. The latter group was identified as Asp 52 by studying how specific chemical modification of this residue affects the pH titration curve of the protein. The group with a pK_a of about 6 is probably Glu 35.

Given the information that the enzymatic reaction requires the carboxyl group of Glu 35 to be protonated and that of Asp 52 to be unprotonated, the first reaction mechanism that comes to mind might be a direct nucleophilic attack by H_2O, aided by concerted general-acid and general-base catalysis (fig. 9.31). In such a scheme, Asp 52 could remove a proton from the water molecule, and Glu 35 could provide a proton to the departing alcoholic group. But this mechanism is at odds with the stereochemistry of the reaction. If H_2O or HO^- attacked C-1 of the MurNAc residue from one side of the tetrahedral carbon atom, and the alcoholic group of GlcNAc departed from the other side, the result would be an inversion of configuration around the C-1 carbon. The β-glycosidic bond would be replaced by a hydroxyl group in the α configuration. Contrary to this expectation, the product is found to have the hydroxyl group in the β configuration. This indicates that the reaction probably proceeds in two distinct steps. The first step evidently removes the alcohol from the β side of the MurNAc residue but leaves an intermediate derivative of the MurNAc residue associated with the enzyme. Water or HO^- then must enter the active site on the same side of the MurNAc residue, replacing the alcohol.

Figure 9.32 shows a mechanism that meets these requirements. In this scheme, the intermediate is a carboxonium ion, in which a positive charge is distributed between C-1 of the MurNAc and the attached oxygen atom. The formation of this

Figure 9.30

A schematic diagram showing the specificity of the hen eggwhite lysozyme for its cell-wall polysaccharide substrate. Six subsites (A–F) on the enzyme bind the hexose units. Alternate sites (A, C, and E) interact with the acetamide side chains (*a*). These sites are unable to accommodate MurNAc residues with their lactyl side chains (Lac). The glycosidic linkage that is cleaved is between sites D and E.

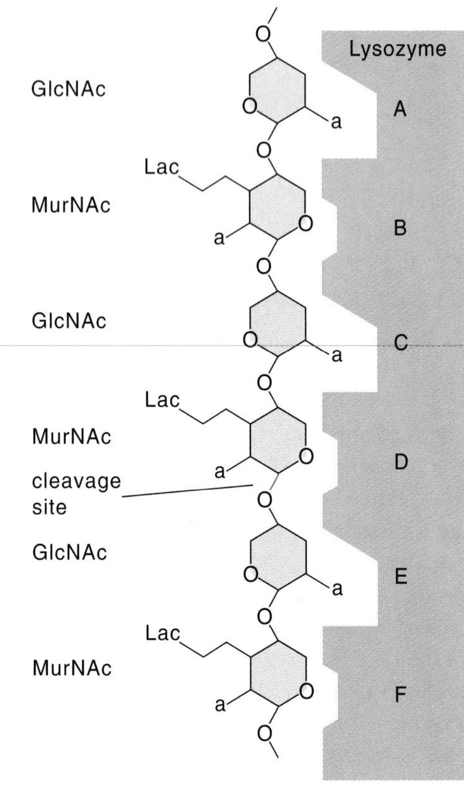

Figure 9.31

A direct nucleophilic (S_N2) attack by a hydroxyl ion on the β-glycosidic bond in the cell-wall polysaccharide would leave the MurNAc product with an α-hydroxyl group. The observed product has a β-hydroxyl group. This suggests that the reaction involves an intermediate step in which a derivative of the MurNAc residue remains associated with the enzyme while the alcohol product is replaced by water, as shown in figure 9.32.

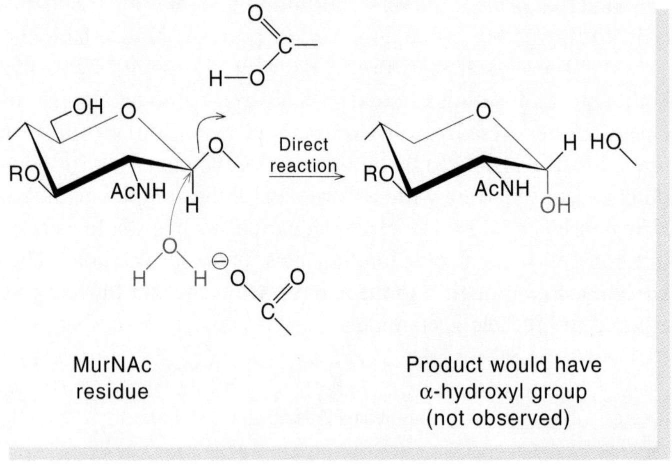

Direct reaction

MurNAc residue

Product would have α-hydroxyl group (not observed)

Figure 9.32

Proposed mechanism for the reaction catalyzed by lysozyme. In the carboxonium ion intermediate, carbons 1 and 2 and the internal oxygen of the pyranose ring all lie in a plane. The formation of this intermediate is favored by electrostatic interactions with Asp 52, along with general-acid catalysis by Glu 35.

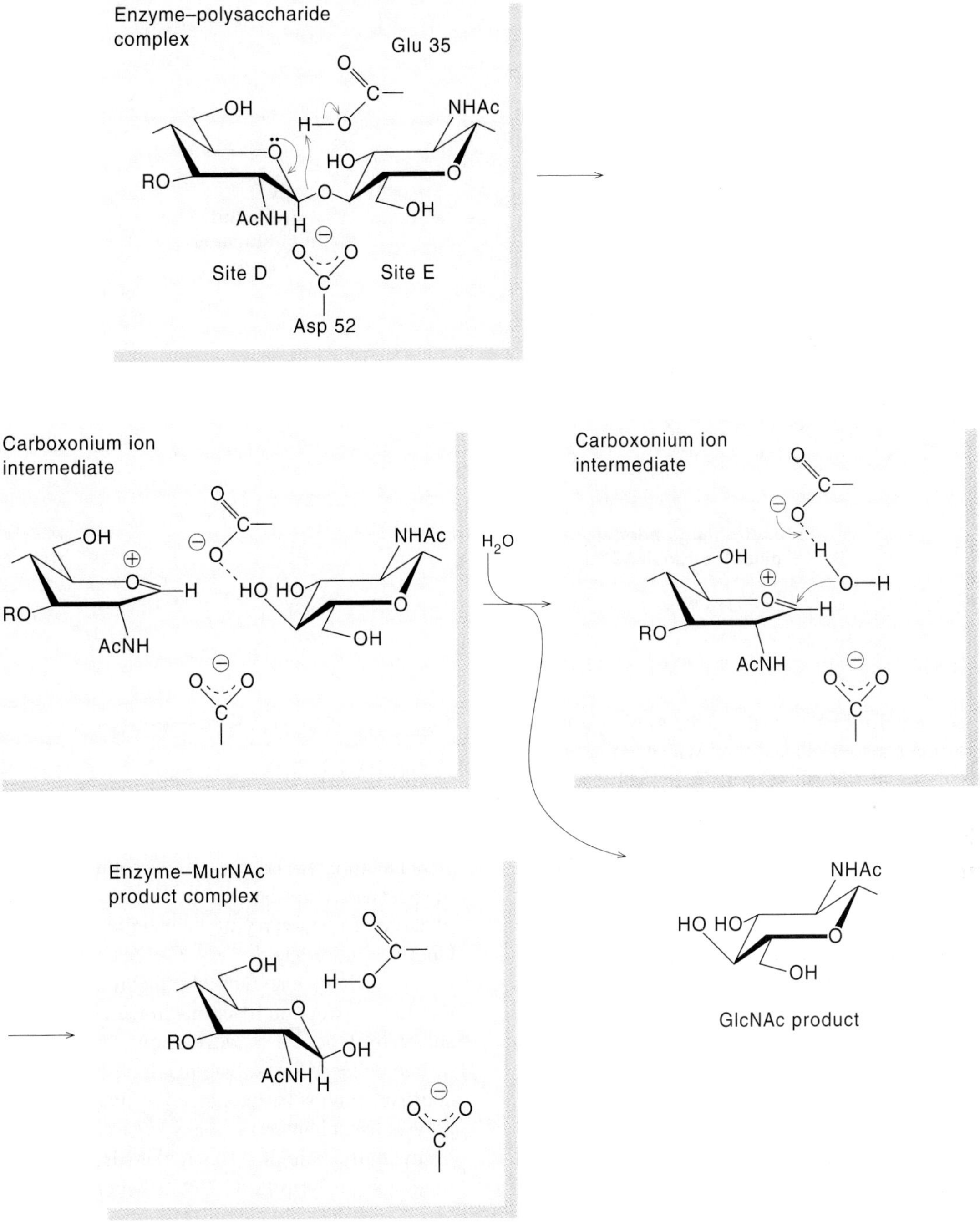

species would involve general-acid catalysis by Glu 35, supported by a favorable electrostatic interaction between the carboxonium ion and the negatively charged carboxylate of Asp 52. In the breakdown of the carboxonium intermediate, Glu 35 could act as a general base to remove a proton from water.

The original formulation of the carboxonium-intermediate mechanism (see fig. 9.32) included an additional feature that was suggested by the crystallographic structure. In the model of $(GlcNAc)_6$ bound to the enzyme, it seemed difficult to squeeze even a GlcNAc residue into site D without some distortion of the

Figure 9.33

Structures of NAD^+ and NADH.

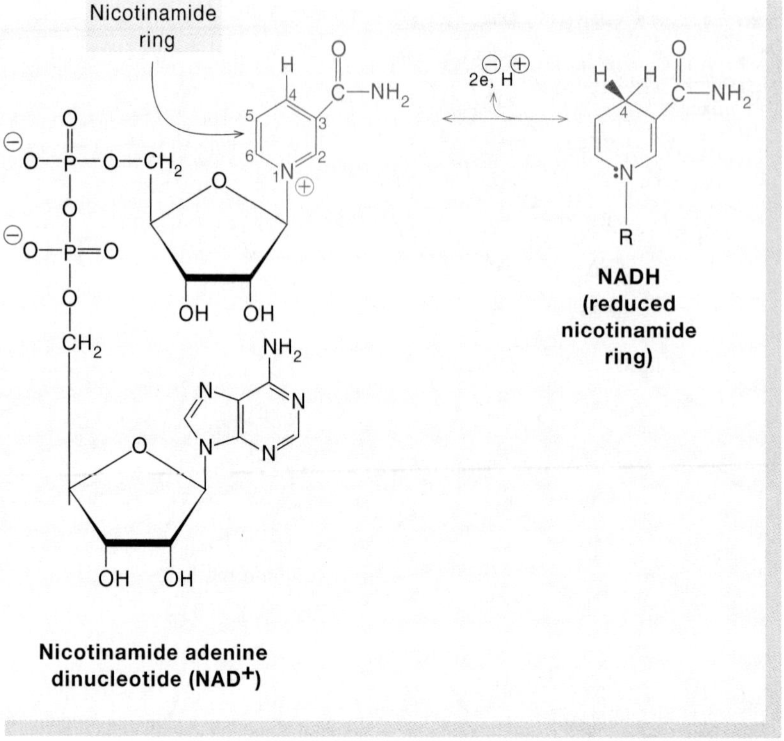

hexose ring. Experimental measurements of the binding energies for a series of substrates also indicated that, whereas the binding of a GlcNAc residue to any of sites A, B, C, E, or F was energetically favorable, the binding to site D was very weak or even energetically unfavorable. These observations suggested that the tight binding of hexose units to sites A, B, C, E, and F forced the hexose unit at site D to sit uncomfortably in a distorted geometry. The geometric distortion appeared to push the hexose ring from its normal chain configuration into a sofa-like configuration, in which carbons 1 and 2 and the internal oxygen atom of the pyranose ring all lie in a plane with carbons 3 and 5. Because these atoms would have a similar, planar configuration in the carboxonium intermediate (see fig. 9.32), it was suggested that geometric "strain" in the bound substrate contributes significantly to pushing the substrate in the direction of the transition state. The enzyme thus would decrease the activation free energy for the reaction ($\Delta G^{\ddagger}$) partly by raising the energy of the bound substrate.

The possibility that substrate binding leads to distortion of the reactive sugar is strongly supported by high-resolution x-ray diffraction observations on a complex of the enzyme and the polysaccharide NAM-NAG-NAM. This trisaccharide binds to subsites B, C, and D, and the NAM sugar ring at the D site shows the distortion that would favor formation of the proposed carboxonium intermediate.

Lactate Dehydrogenase: A Bisubstrate Enzyme

Lactate dehydrogenase (LDH) is a tetramer of M_r 140,000. The enzyme is widely distributed in tissues and plays a key role in energy metabolism. Lactate dehydrogenase is a bisubstrate enzyme; the substrates consist of either lactate or pyruvate and a derivative of the hydrogen-carrying coenzyme, nicotinamide adenine dinucleotide (fig. 9.33).

Many enzymes use nicotinamide adenine dinucleotide (NAD) in oxidation reactions in which NAD^+ reacts with another substrate to accept a hydride ion (two electrons and one proton) in one or more steps. The resulting NADH (reduced form of the coenzyme) must be reoxidized so that it can be used again. In glycolysis (see chapter 14) many organisms accomplish the reoxidation with the help of L-lactate dehydrogenase, which catalyzes the reduction of pyruvate to lactate according to the reaction

$$\underset{\text{Pyruvate}}{CH_3-C(=O)-COOH} + NADH + H^+ \rightleftharpoons \underset{\text{L-Lactate}}{CH_3-HCOH-COOH} + NAD^+ \qquad (12)$$

The equilibrium for this reaction is strongly favored to the right at neutral pH. In spite of this, the initial rate of the reaction can be studied in the leftward direction by addition of the appropriate substrates at elevated values of pH. Physiologically, it also appears, as we will see, that LDH can play useful roles in both directions, although the conversion of pyruvate to lactate under anaerobic conditions is certainly the best-understood function. Lactate produced in tissues functioning anaerobically is secreted and absorbed by other tissues functioning aerobically. LDH can then convert the lactate back to pyruvate for further utilization in the Krebs cycle (see chapter 15).

Binding of Coenzyme Occurs before Binding of Sugar

The strategy of all NAD-dependent dehydrogenases is to orient the coenzyme and the substrate on the enzyme surface so that the C-4 atom on the nicotinamide is pointed toward the reactive carbon of the substrate. Different dehydrogenases have remarkably similar protein domains for binding NAD^+ but dissimilar sites for binding the cosubstrate, the latter being dependent on the structure of the substrate. The coenzyme-binding domain for lactate dehydrogenase is shown in figure 9.34.

In LDH, binding of NADH is about 400 times stronger than the binding of NAD^+ and about 50 times stronger than that of NADPH (the same molecule with an additional phosphate: see figure 9.33). Some enzymes can use either NAD^+ or NADP as coenzymes, but for LDH the binding data show that there is a strong preference for the former. A conformational change takes place in the enzyme when it interacts with the coenzyme NAD^+ (or NADH) and a suitable substrate, usually lactate (or pyruvate). A minor movement of Asp 53 and its associated polypeptide chain occurs on binding the adenosine part of the coenzyme. There is a larger movement of the loop connecting D and E, including the helix αD (residues 98–120; see figure 9.34), involving the main-chain displacement of up to 11 Å. In ternary complexes this flexible loop drops down and encloses the coenzyme and substrate. In the free enzyme it extends into the solvent. Arg 101 forms an ion pair with the pyrophosphate group of the coenzyme in the ternary complex (fig. 9.35). The formation of this ion pair may be the driving force for the conformational change of the loop and the subsequent rearrangements in the subunit. The guanidinium group of Arg 109 moves 14 Å and changes from being completely exposed to the solvent to having a close interaction with groups around the substrate site. The two connected helixes αD and αE are more angled to each other in the ternary complex.

The movement of the N-terminal part of helix αD is associated with a movement of the neighboring C-terminal part of helix αH (located in the substrate-binding domain, not shown). The overall effect of these conformational changes in the protein involves movement toward the substrate and a contraction of the subunit on binding coenzyme and substrate. The enzyme folds up around the substrates, so the fit of enzyme and substrates has been induced by substrate binding.

In the native state, lactic dehydrogenase exists as a tetramer containing four identical subunits. With all the conformational changes taking place in going from a free enzyme to a ternary complex, it is natural to suspect that there might be some effects between different binding sites. However, careful equilibrium binding studies of the interaction between enzyme and NAD^+ or NADH have led to the conclusion that the intact enzyme contains four independent noninteracting binding sites. Existence of the enzyme as a tetramer must have other advantages. Sometimes dimers or tetramers are preferable simply because they are more stable and therefore less susceptible to denaturation or degradation.

Figure 9.34

Nucleotide-binding domain and related tertiary structure of lactate dehydrogenase.

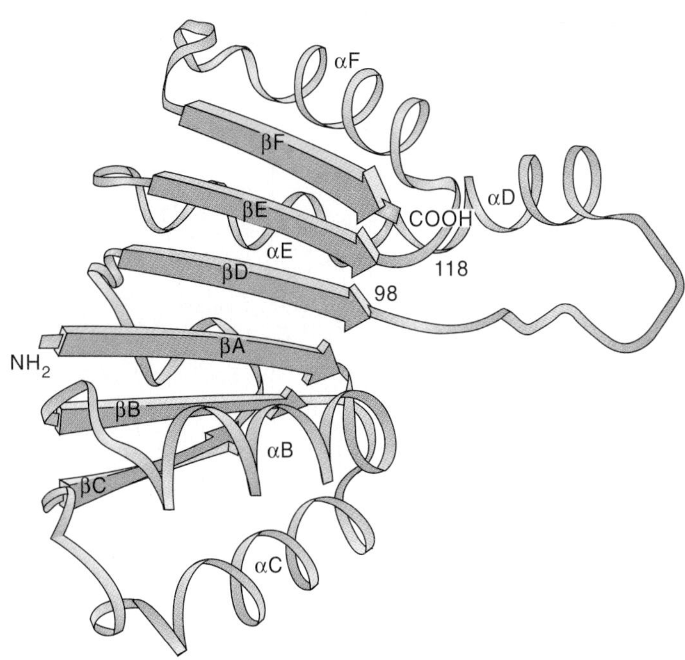

Kinetic Studies Reveal Intermediates and Slow Step in the Reaction

The forward reaction for the oxidation of lactate by LDH and NAD^+ fits the general form of the steady-state kinetic expression for the ordered pathway described in chapter 8. Equilibrium binding experiments have been used to demonstrate the order of binding. The coenzyme NAD^+ will bind to the enzyme even without lactate, so NAD^+ binds first; this by itself is believed to produce an alteration in the enzyme structure that creates a binding site for the lactate. Lactate won't bind to the enzyme unless NAD^+ is present and bound. The same is true for the reverse reaction; i.e., the NADH binds before pyruvate. In the release of products the sugar is always released before the coenzyme.

Optical properties of the protein and the coenzyme have been instrumental in making pre-steady-state kinetic measurements, which have been most important in determining the steps that take place after binding. NADH fluoresces weakly at 470 nm

Figure 9.35

Diagrammatic representation of binding of a covalent adduct formed between pyruvate and NAD$^+$ to lactate dehydrogenase. This compound (shown in red) is a competitive inhibitor and therefore it is presumed to bind similarly to the substrates. The region colored in blue corresponds to the pyruvate substrate/product.

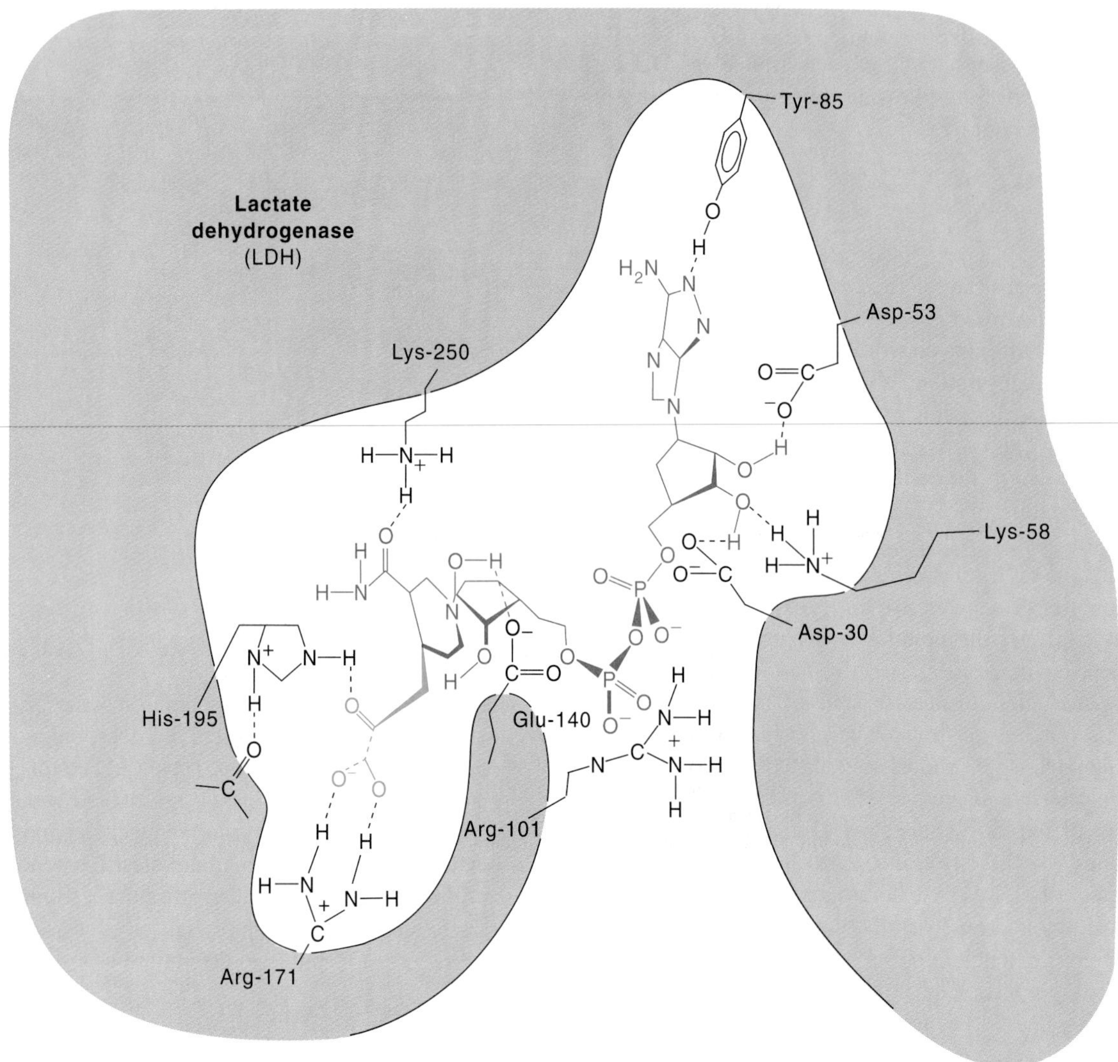

when its absorption band at 340 nm is excited. Its fluorescence is greatly increased on combination with the enzyme, so fluorescence may be used to measure coenzyme binding. The enzyme fluoresces at about 350 nm when excited by radiation at 270–305 nm. This fluorescence is due to the tryptophan residues in the enzyme. In LDH there is a substantial drop in intrinsic fluorescence on binding NADH, which is probably due to the transfer of some of the tryptophan residues to a more hydrophilic environment. Despite the fact that the exact cause of the drop in fluorescence is unclear, the change may be used to determine whether or not the protein is bound to coenzyme and substrate.

In addition, the release of H$^+$, a product of the reaction, can be measured optically by combination of H$^+$ with a dye molecule. All these optical changes have been measured as a function of time immediately after mixing the enzyme with saturating amounts of NAD$^+$ and lactate. Such measurements suggest at least three phases in the approach to the steady state. In the first phase, which takes less than 1 ms, a small amount of NADH can be detected with no release of proton. In the second phase, which follows a first-order rate with respect to lactate concentration, NADH and H$^+$ are produced in equimolar amounts, and only the proton is liberated. The third phase, which follows zero-order kinetics, begins after about 40 ms. During this steady-state phase of the reaction, NADH and H$^+$ are liberated at the same rate. The proposed phases in the reaction after binding substrate and coenzyme are indicated in figure 9.36. One of the important conclusions from this experiment is that the rate-limiting step after reaching the steady state is the release of the NADH product from the enzyme.

Figure 9.36

Proposed phases in the reaction of NAD^+ and lactate after enzyme binding. Here >CHOH represents lactate and >C=O represents pyruvate.

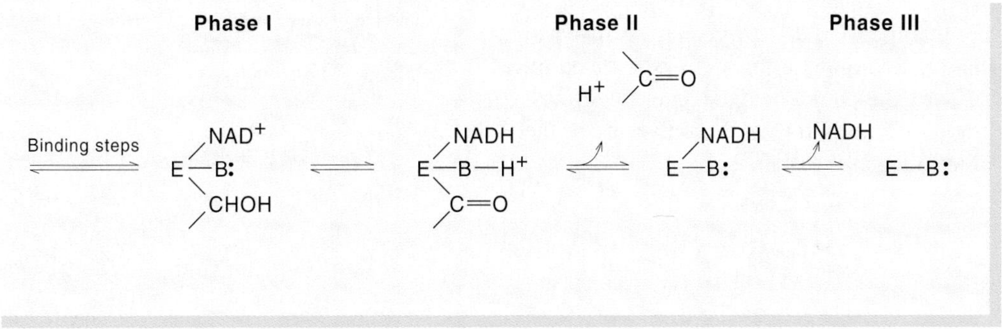

Figure 9.37

The catalytic mechanism of lactate dehydrogenase. His 195 probably acts as a general base in the conversion of lactate to pyruvate and as a general acid in the conversion of pyruvate to lactate. Arg 171 probably serves to orient the substrate.

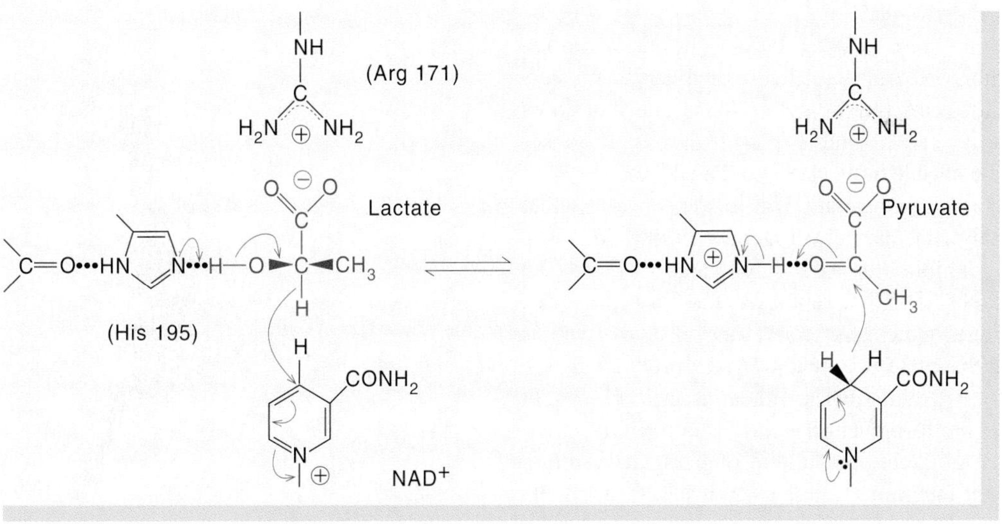

Reaction Results from Concerted Catalysis

An expanded view of the active site in the ternary complex, as inferred from the x-ray-determined structure, is shown in figure 9.37. The substrates, lactate and pyruvate, are oriented so that they are favorably disposed for a concerted reaction with the nicotinamide ring of the coenzyme and the imidazole ring of His 195. In both forward and reverse reactions, Arg 171 helps to anchor the substrate through H bonds and electrostatic bonds formed with the nonreacting substrate carboxylate group. In both reactions a substrate-orienting H bond is also formed with the imidazole group of His 195. His 195 acts as an acid–base catalyst, removing the proton from lactate during oxidation (rightward reaction in figure 9.37) or donating it to pyruvate in the reverse reduction reaction. The nicotinamide accepts a hydride from lactate in the oxidation reaction and donates a hydride to the pyruvate in the reduction reaction.

Isoenzymes of Lactate Dehydrogenase Serve Different Functions

Most vertebrates possess at least two genes for lactate dehydrogenase, which make similar but nonidentical polypeptides called M and H. In embryonic tissue both genes are equally active, resulting in equimolar amounts of the two gene products and a statistical array of tetramers (M_4, M_3H_1, M_2H_2, H_3M_1, and H_4 in the ratios of 1:4:6:4:1). These forms are called isoenzymes, or isozymes. They can usually be detected by differing electrophoretic mobilities. As embryonic tissue multiplies and differentiates, the relative amounts of the M and H forms change. In pure heart tissue, which is considered aerobic, the H_4 tetramer predominates. In skeletal muscle, which functions anaerobically under stress, the M_4 isozyme predominates.

It seems likely that the M and H forms have evolved to serve different functions. A clue to these functions is revealed by the inhibiting effect of pyruvate on the dehydrogenase. Pyruvate can form a covalent complex with NAD^+ at the active site of the enzyme according to the following reaction:

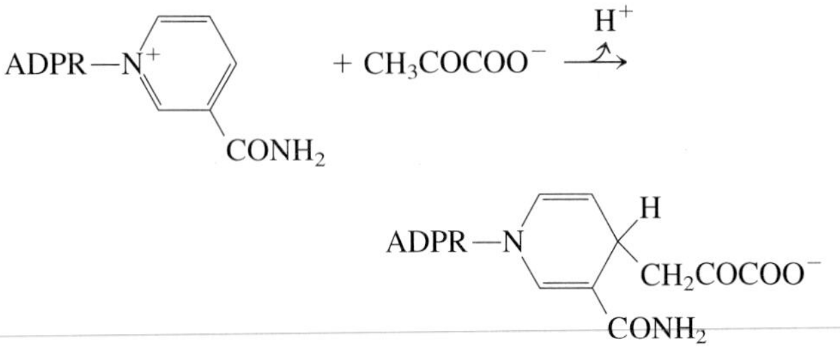

$$\text{ADPR—N}^+\text{(pyridine ring, CONH}_2\text{)} + CH_3COCOO^- \xrightarrow{H^+} \text{ADPR—N(dihydropyridine ring, H, } CH_2COCOO^-\text{, CONH}_2\text{)} \qquad (13)$$

Indeed, this is the compound whose binding was the basis for figure 9.35. The H_4 tetramer shows a much greater inhibition by this compound than the M_4 tetramer.

Active muscle tissue is anaerobic and produces a good deal of pyruvate. Inhibition of lactate dehydrogenases under anaerobic conditions would shrink the supply of NAD^+ and shut down the glycolytic pathway (see chapter 14), with disastrous consequences. In fact, active muscle tissue has augmented levels of pyruvate but converts this readily to lactate. The reason is possibly that the predominant form of lactate dehydrogenase in muscle is M_4, which is only poorly inhibited by excess pyruvate.

The function of LDH in aerobic tissue is less clear. Heart muscle is aerobic tissue, and consequently, most of its pyruvate is funneled into the Krebs cycle for greater energy production (see chapter 15). The lactate dehydrogenase of heart muscle, H_4, might be inhibited by pyruvate to prevent waste of this potential high-energy carbon source or excessive buildup of pyruvate resulting from the conversion of incoming lactate to pyruvate. Possibly the H_4 enzyme is used in such tissues to convert absorbed lactate into pyruvate.

The control of enzyme activity by isozyme type is very beneficial in certain cases. A more widespread form of control of enzyme activity involves modulation in their activity through binding by regulatory small molecules. This topic will be the main subject of chapter 10.

Alcohol Dehydrogenase Uses Zinc as an Electrophilic Catalyst

Alcohol dehydrogenase also follows an ordered kinetic pathway, in which the nucleotide binds to the enzyme first and dissociates last. Again, the binding of the nucleotide causes substantial structural changes that enhance the enzyme's ability to bind ethanol or acetaldehyde. However, alcohol dehydrogenase differs from lactate dehydrogenase in having a Zn^{2+} ion at its active site, and its enzymatic mechanism appears to be based on electrostatic effects of the metal ion.

Figure 9.38

Functional groups at the active site of alcohol dehydrogenase from horse liver. (*a*) In the absence of substrate the ligands of the Zn^{2+} ion are two cysteines, a histidine, and a molecule of water that is hydrogen-bonded to Ser 48. (*b*) The substrate ethanol probably binds to the Zn^{2+} as the alcoholate anion, displacing the molecule of water and forming a hydrogen bond to Ser 48. A cascade of electron migrations and a hydride ion shift result in the formation of acetaldehyde and the reduced form of the coenzyme shown in (*c*).

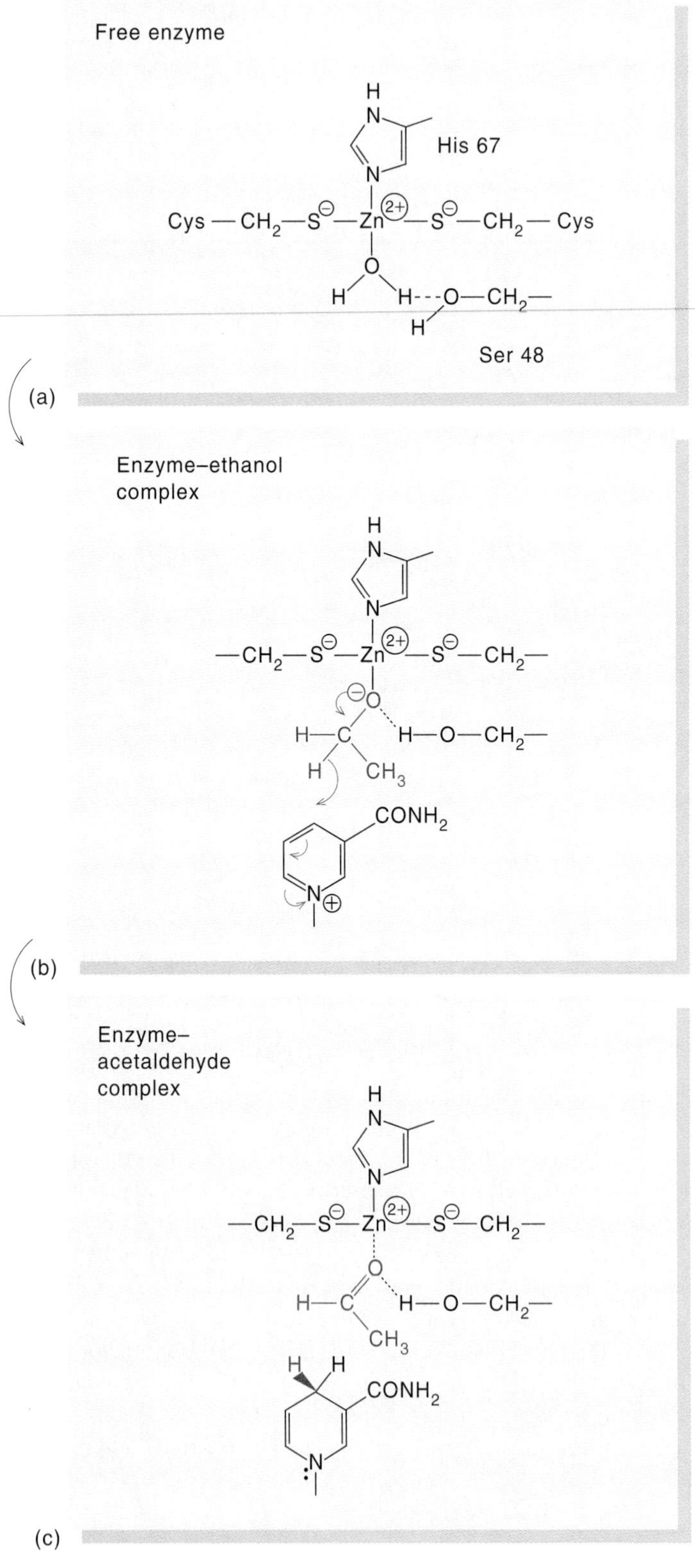

Alcohol dehydrogenases have been crystallized from a variety of sources, including horse liver and yeast. The crystal structures show that the Zn^{2+} is attached to a histidine residue and the thiol groups of two cysteines (fig. 9.38). In the absence of the substrate the fourth ligand of the Zn^{2+} is a molecule of water that forms a hydrogen bond to a nearby serine residue. When the substrate adds to the enzyme, the alcoholic or carbonyl oxygen atom is believed to coordinate directly to the Zn^{2+}, displacing the molecule of water. The alcohol probably binds in the form of the alcoholate anion, as shown in figure 9.38. The negative charge on the oxygen atom makes the alcoholate a better hydride ion donor for the reduction of NAD^+, compared with the undissociated alcohol. The NAD^+ and the substrate are oriented for direct hydride ion transfer when they are bound to the enzyme except that they are slightly too far apart. It has been suggested that the hydride ion shift could be facilitated by dynamic fluctuations of the protein structure bringing the reactants closer together than they are in the average structure. An effect of this nature is referred to as tunneling.

Summary

In this chapter we have looked at mechanisms of enzyme catalysis, first from a general standpoint and then in some detail. The following points are the most important.

1. Like ordinary chemical catalysts, enzymes associate directly with the reacting species and interact with them in a manner that lowers the free energy of the transition state. Enzyme–substrate interactions usually are highly specific and can depend on critically positioned amino acid side chains, the atoms of the polypeptide backbone, or bound cofactors such as metal ions.
2. All known enzymatic reaction mechanisms depend on one or more of the following themes: proximity effects (enzymes hold the reactants close together in an appropriate orientation); electrostatic effects (charged, polar, or polarizable groups of the enzyme are positioned to favor the redistribution of electrical charges that occur as the substrate evolves into the transition state); general-acid or general-base catalysis (acidic or basic groups of the enzyme donate or remove protons and often do first one and then the other); nucleophilic or electrophilic catalysis (nucleophilic or electrophilic functional groups of the enzyme or a cofactor react with complementary groups of the substrate to form covalently linked intermediates); and structural flexibility (changes in the protein structure can increase the specificity of enzymatic reactions by ensuring that substrates bind or react in an obligatory order and by sequestering bound substrates in pockets that are protected from the solvent).
3. Trypsin, chymotrypsin, and elastase are very similar in structure but have substrate-binding pockets that are tailored for different types of amino acid side chains. The active site of each enzyme contains three critical residues: serine, histidine, and aspartate. The side chains of these residues are oriented so that the serine hydroxyl group becomes a strong nucleophilic reagent for attacking the substrate's peptide carbonyl group adjacent to the preferred amino acid. This reaction displaces the amine member of the peptide bond, and generates an acyl-ester intermediate, in which the carboxyl member is linked covalently to the serine group of the enzyme. Formation and hydrolysis of the acyl-ester intermediate probably are promoted both by electrostatic interactions that stabilize tetrahedral transition states and by general-acid and general-base catalysis.
4. Thermolysin and carboxypeptidase A contain a bound Zn^{2+} ion, which interacts with the carbonyl oxygen of the peptide bond that is to be cleaved. The electrostatic effect of the Zn^{2+} probably facilitates formation of a tetrahedral transition state in the direct nucleophilic attack of H_2O or OH^- at the carbonyl carbon. Acidic and basic amino acid side chains again may serve as general-acid and general-base catalysts in the reaction. Carboxypeptidase A undergoes a substantial change in structure when it binds its substrate.
5. Ribonuclease A hydrolyzes RNA adjacent to pyrimidine bases. The reaction proceeds through a 2′,3′-phosphate cyclic diester intermediate. The formation and breakdown of the cyclic diester appear to be promoted by concerted general-base and general-acid catalysis by two critical histidine residues, and by electrostatic interactions with two lysines. These reactions proceed through pentavalent phosphoryl intermediates. The geometry of the oxygens surrounding the phosphorus atom in these intermediates resembles the geometry of vanadate compounds that act as inhibitors of the enzyme.
6. Triosephosphate isomerase interconverts dihydroxyacetone phosphate and glyceraldehyde-3-phosphate. A glutamic acid residue probably acts as a general base to remove a proton from the substrate, forming an enediolate intermediate. This generates a planar enediolate intermediate and the transformation of the hydrogen bond formed between the substrate and a nearby histidine to an extra-strong low-barrier hydrogen bond. The proton participating in this hydrogen bond becomes transferred to C-2 of the substrate, and simultaneously the nearby His forms a comparable LBHB with C-9, which results in the loss of the participating protons from C-1 to the His and the return of the proton originally captured by a Glu to the C-2 position. Triosephosphate isomerase appears to have reached catalytic perfection, in the sense that it catalyzes its reaction at the maximum possible rate, given the concentration of the substrate in the cell. The rate-limiting step is the collision of the enzyme and substrate by diffusion through the solution.
7. Lysozyme hydrolyzes complex polysaccharides containing five or more hexose residues. Its substrate-binding site includes a crevice that can accommodate six such residues. The enzyme probably first releases the truncated polysaccharide from one side of the bond that is cleaved, leaving the polysaccharide on the other side bound noncovalently to the enzyme in the form of a positively charged carboxonium intermediate. The formation of this intermediate is promoted by general-acid catalysis by a glutamic acid residue, and by the electrostatic effect of a nearby aspartate.

8. Lactate dehydrogenase and alcohol dehydrogenase catalyze the reversible transfer of a hydride ion (H^-) to NAD^+ from, respectively, lactate and ethanol. Both enzymes bind their substrates in an obligatory order. NAD^+ binds first, causing a structural change that sets up the binding site for the alcohol. The alcohol substrate then binds with the hydrogen that will be transferred close to C-4 of the nicotinamide ring. In lactate dehydrogenase a histidine residue probably acts as a general base to remove a proton from the —OH group. This would facilitate the release of the negatively charged hydride ion. In alcohol dehydrogenase a Zn^{2+} ion attached to the enzyme probably encourages the ethanol to bind in the form of the alcoholate anion, which again is a good donor of a hydride ion.

Selected Readings

Albery, W. J., and J. R. Knowles, Free energy profile of the reaction catalyzed by triosephosphate isomerase. *Biochem.* 15:5588, 5627, 1976.

Blackburn, P., and S. Moore, Pancreatic ribonuclease. *The Enzymes* 15:317, 1982.

Blow, D., Structure and mechanism of chymotrypsin. *Acc. Chem. Res.* 9:145, 1976.

Branden, C.-I., H. Jornvall, H. Eklund, and B. Furugren, Alcohol dehydrogenases. *The Enzymes* 11:104, 1975.

Breslow, R., How do imidazole groups catalyze the cleavage of RNA in enzyme models and enzymes? Evidence from "negative catalysis." *Acc. Chem. Res.* 24:317, 1991.

Chen, Y.-Q., J. Kraut, R. Blakeley, and B. Callendar, Determination by Raman spectroscopy of the pK_a of N5 of dihydrofolate reductase: Mechanistic implications. *Biochem.* 33:7021–7026, 1994.

Cleland, W. W., and M. M. Kreevay, Low-barrier hydrogen bonds and enzymic catalysis. *Science* 264:1887–1890, 1994.

Eklund, H., B. V. Plapp, J.-P. Samama, and C.-I. Branden, Binding of substrate in a ternary complex of horse liver alcohol dehydrogenase. *J. Biol. Chem.* 257:14349, 1982.

Fersht, A., *Enzyme Structure and Mechanism,* 2d ed. New York: Freeman, 1985.

Fersht, A. R., D. M. Blow, and J. Fastrez, Leaving group specificity in chymotrypsin-catalyzed hydrolysis of peptides: A stereochemical interpretation. *Biochem.* 12:2035, 1973.

Findlay, D., D. G. Herries, A. P. Mathias, B. R. Rabin, and C. A. Ross, The active site and mechanism of pancreatic ribonuclease. *Nature* 190:781, 1961.

Frey, P. A., S. A. Whitt, and J. B. Tobin, A low-barrier hydrogen bond in the catalytic triad of serine proteases. *Science* 264:1927–1930, 1994.

Hardy, L. W., and K. Poteete, Reexamination of the role of Asp 20 in catalysis by bacteriophage T4 lysozyme. *Biochem.* 30:9457, 1991.

Holbrook, J. J., A. Liljas, S. J. Steindel, and M. G. Rossmann, Lactate dehydrogenase. *The Enzymes* 11:191, 1975.

Holmes, M. A., D. E. Tronrud, and B. W. Matthews, Structural analysis of the inhibition of thermolysin by an active-site-directed irreversible inhibitor. *Biochem.* 22:236, 1983.

Imoto, T., L. N. Johnson, A. C. T. North, D. C. Phillips, and J. A. Rupley, Vertebrate lysozymes. *The Enzymes* 7:665, 1972.

Kelly, J. A., A. R. Sielecki, B. D. Sykes, M. N. G. James, and D. C. Phillips, X-ray crystallography of the binding of the bacterial cell wall trisaccharide NAM-NAG-NAM to lysozyme. *Nature* 282:875, 1979.

Kraut, J., How do enzymes work? *Science* 242:533, 1988.

Lolis, E., T. Alber, R. C. Davenport, D. Rose, F. C. Hartman, and G. A. Petsko, Structure of yeast triosephosphate isomerase at 1.9-Å resolution. *Biochem.* 29:6609, 1990.

Markley, J. L., Correlation proton magnetic resonance studies at 250 MHz of bovine pancreatic ribonuclease. I. Reinvestigation of histidine peak assignments. *Biochem.* 14:3546, 1975.

Page, M. I. (ed.), *Enzyme Mechanisms.* London: Royal Society of Chemistry, 1987. See particularly the chapters by M. I. Page (Theories of Enzyme Catalysis), A. L. Fink (Acyl Group Transfer—The Serine Proteinases), D. S. Auld (Acyl Group Transfer—Metalloproteinases), P. M. Cullis (Acyl Group Transfer—Phosphoryl Transfer), M. L. Sinnott (Glycosyl Group Transfer), and J. P. Richard (Isomerization Mechanisms through Hydrogen and Carbon Transfer).

Ramaswany, S., H. Eklund, and B. V. Plapp, Structures of horse liver alcohol dehydrogenase complexed with NAD^+ and substituted benzyl alcohols. *Biochem.* 33:5230–5237, 1994.

Reeke, G. N., J. A. Hartsuck, M. L. Ludwig, F. A. Quiocho, T. A. Steitz, and W. N. Lipscomb, The structure of carboxypeptidase A: Some results at 2.0-Å resolution, and the complex with glycyl tyrosine at 2.8-Å resolution. *Proc. Natl. Acad. Sci. USA* 58:2220, 1967.

Rose, I. A., Mechanism of the aldose-ketose isomerase reactions. *Adv. Enzymol.* 43:491, 1975.

Schultz, P. G., and R. A. Lerner, From molecular diversity to catalysis: Lessons from the immune system. *Science* 269:1835–1842, 1995.

Shoham, M., and T. Steitz, Crystallographic studies and model building of ATP at the active site of hexokinase. *J. Mol. Biol.* 140:1, 1980.

Strynadka, N. C. J., and M. N. G. James, Lysozyme revisited: Crystallographic evidence for distortion of an N-acetyl-muramic acid residue bound in Site D. *J. Mol. Biol.* 220:401, 1991.

Warshel, A., G. Naray-Szabo, F. Sussman, and J.-K. Hwang. How do serine proteases really work? *Biochem.* 28:3629, 1989.

Problems

1. Detailed mechanisms are shown in this chapter for several specific reactions. The involvement of specific amino acyl groups on the surface of several enzymes is indicated. Do each of the 20 amino acids participate equally in this process?
2. Indicate why the flexibility of enzymes is often important for enzymes that have two substrates.
3. Trypsin contains several seryl residues besides Ser 195. Why does diisopropylfluorophosphate form only the diisopropylphosphate derivative of Ser 195?
4. What geometric constraints are the amide N—H groups of Ser 195 and Gly 193 subject to in trypsin, chymotrypsin, and elastase?
5. Assume that the space allotted for the water molecule that donates a hydrogen to Glu 143 in thermolysin (fig. 9.14) is sufficient for the rotation of the water. A spherical space 0.3 nm in diameter is more than adequate. The reaction mechanism can be drawn with a hydroxide intermediate. Calculate the hydroxide ion concentration and the pH in the spherical space indicated above.

6. (a) In what ways are the mechanistic features of chymotrypsin, trypsin, and elastase similar?
   (b) If the mechanisms of these enzymes are similar, what features of the enzyme active site dictate substrate specificity?
7. (a) If you monitor the lactate dehydrogenase reaction by the formation of NADH (increase in absorbance at 340 nm), should increasing pH make it easier to measure the dehydrogenation of lactate to pyruvate?
   (b) Would a chemical trapping agent for pyruvate serve the same purpose at lower pH values? Why?
   (c) Dehydrogenase activity can be measured by reoxidizing the NADH and reducing a tetrazolium dye. The reduced dye is intensely colored. What are the advantages of measuring the LDH reaction by means of the tetrazolium dye system?
8. Carboxypeptidase A preferentially cleaves C-terminal aromatic residues from proteins. When the aromatic substrate side chain is bound, water is expelled from the active site. How does the release of water stabilize binding of substrate in the active site?
9. You have isolated a metalloenzyme that preferentially cleaves basic amino acids from the carboxyl terminal of proteins. Would you expect the enzyme to retain an arginine in the active site as does carboxypeptidase A? Why or why not? What other residues would you predict to be in the substrate binding site for the new enzyme? How would these residues dictate cleavage specificity?
10. RNase can be completely denatured by boiling or by treatment with chaotropic agents (e.g., urea), yet can refold to its fully active form upon cooling or removal of the denaturant. By contrast, when enzymes of the trypsin family and carboxypeptidase A are denatured, they do not regain full activity upon renaturation. What aspects of trypsin and carboxypeptidase A structure preclude their renaturation to the fully active form?
11. (a) The amino acids in the active site of the protease papain are shown. Predict a feasible reaction mechanism for papain.

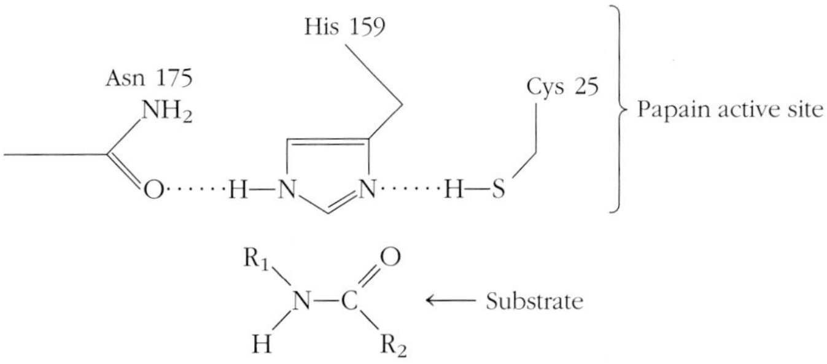

His 159
Asn 175
NH_2
Cys 25
Papain active site
O·····H—N ... N·····H—S
R_1 N—C(=O) R_2, H ← Substrate

   (b) *N*-ethylmaleimide (NEM) reacts rapidly with cysteine thiolate anion via a Michaelis addition. What is the product of the reaction between NEM and cysteine? Would you expect the rate of R—SH reaction with *N*-ethylmaleimide to be more rapid at pH 5 or pH 7.5? Why?
   (c) Would you expect cysteine 25 (see part a) to be more reactive with *N*-ethylmaleimide than any of the other cysteine residues in the protein? Explain.
12. Why do structural analogs of the transition-state intermediate of an enzyme inhibit the enzyme competitively and with low K_i values?
13. Transition-state analogs of a specific chemical reaction have been used to elicit antibodies with catalytic activity. These catalytic antibodies have great promise as experimental tools as well as having commercial value. Why is it reasonable to assume that the binding site for the transition-state analog on the antibody would mimic the enzyme active site? What difficulties might be encountered if a catalytic antibody were sought for a reaction requiring a cofactor (coenzyme)?

C H A P T E R

Regulation of Enzyme Activities

Strategically located enzymes are regulated by one of two strategies: modification of the covalent structure of the enzyme or modification of their structure by the reversible binding of effector molecules.

The living cell resembles a complex factory with many different assembly lines of worker enzymes performing specific tasks. The activities of these workers have to be regulated precisely so that resources are used efficiently and a steady flow of parts keeps all the assembly lines moving smoothly. Cells use two basic strategies for regulating their enzyme activities. The first strategy is to adjust the amount and location of key enzymes. This requires mechanisms for the control of synthesis, degradation, and transport of proteins, which we discuss in chapters 34, 35, and 36. The second strategy is to regulate the activities of the enzymes that are on hand. We deal exclusively with the second strategy in this chapter.

Enzymes subject to regulation are a select few of the total enzymes in a cell and are so subject for two reasons. First, it is not necessary to regulate the activity of every enzyme to achieve the desired level of control. In many cases an entire metabolic pathway can be controlled by regulating only the enzyme that catalyzes the first step in the pathway (see chapter 12). Second, elaborate structural properties are required to create an enzyme that can be regulated.

In the first part of this chapter we survey the different methods that cells use to regulate enzyme activities. Then we consider in detail how these methods apply to four of the most thoroughly characterized regulatory enzymes.

In principle, the activities of many enzymes can be altered by changes in pH. Cells do take advantage of this possibility in a few cases. Lysozyme, for example, is most active in the pH 5 region, which is characteristic of some extracellular secretions and is much less active in the intracellular pH region near 7. The activity of lysozyme thus remains low until the enzyme is secreted. But this is not a very practical solution to the problem of regulating the activities of intracellular enzymes, because most cells must hold their pH within narrow limits. Two strategies are much more widely applicable. The first is to modify the covalent structure of the enzyme in such a way as to alter either K_m or k_{cat}. The second is to use an inhibitor or activator, an *effector,* that binds reversibly to the enzyme and, again, alters either the K_m or k_{cat}. Such an effector may bind either at the active site itself or at some more distant site on the enzyme. In the latter case it is termed an allosteric effector (from the Greek words *allos,* meaning "other," and *stereos,* meaning "space,") and enzymes that are regulated by such effectors are called allosteric enzymes. Although allosteric effectors usually are small molecules such as ATP, some proteins are inhibited or activated when they bind to other proteins.

Partial Proteolysis Results in Irreversible Covalent Modifications

In chapter 9 we mentioned that the pancreas secretes trypsin, chymotrypsin, and elastase as inactive zymogens, which are activated by extracellular proteases. Trypsin is activated when the intestinal enzyme enteropeptidase cuts off an N-terminal hexapeptide. Trypsin in turn activates chymotrypsin by cutting it at the N-terminal end between Arg 15 and Ile 16. This type of change in the covalent structure of an enzyme is termed partial proteolysis. Delaying the activation prevents the digestive enzymes from destroying the pancreatic cells in which they are synthesized.

The crystal structures of chymotrypsin and its zymogen precursor *chymotrypsinogen* suggest an explanation for the increase in catalytic activity that results from trimming off the N-terminal end of the zymogen. Although the catalytic triad of Asp 102, His 57, and Ser 195 has a similar structure in chymotrypsinogen and chymotrypsin, the substrate-binding pocket is not properly formed in the zymogen (fig. 10.1*a*). The NH group of Gly 193 is not in position to form a hydrogen bond with the carbonyl oxygen of the substrate. The importance of this bond for the activity of the enzyme was discussed in chapter 9 (see fig. 9.11). The major constraint preventing the completion of the binding pocket appears to be a hydrogen bond between the carboxylate group of the neighboring residue, Asp 194, and His 40. When the zymogen is cleaved, the new N-terminal —NH_3^+ group of Ile 16 forms a salt bridge with Asp 194, thus allowing Gly 193 to rotate into the correct orientation (fig. 10.1*b*). A similar bridge between the N-terminal group and Asp 194 occurs in trypsin.

The enzymes that participate in blood clotting also are activated by partial proteolysis, which again serves to keep them in check until they are needed. The blood coagulation system involves a cascade of at least seven serine proteases, each of which activates the subsequent enzyme in the series (fig. 10.2). Because each molecule of activated enzyme can, in turn, activate many molecules of the next enzyme, initiation of the process by factors that are exposed in damaged tissue leads explosively to the conversion of prothrombin to thrombin, the final serine protease in the series. Thrombin then cuts another protein, fibrin, into peptides that stick together to form a clot.

Table 10.1 lists some of the other enzymes activated by partial proteolysis. A common pattern is for proteolysis to occur in a loop that connects two different domains of the protein, relieving a constraint that interferes with the formation of the active site.

The activation of an enzyme by partial proteolysis is an irreversible process. Once the enzyme is cut, it remains active until it is degraded or inhibited by some other means. This is fine for the digestive enzymes, but it clearly raises a problem in blood coagulation: A mechanism must exist to prevent the clot from spreading away from the site of the injury and taking over the entire blood supply. Several mechanisms work to this end. The activated enzymes are diluted by the flow of the blood, degraded in the liver, and inhibited by other blood proteins that bind tightly to the active enzymes and occlude their active sites. Blood coagulation thus depends on a delicate balance between a rapid, irreversible mechanism of enzyme activation and rapid, irreversible mechanisms for disposing of the active enzymes.

Figure 10.1

A schematic drawing of the structural changes that occur when chymotrypsinogen (*a*) is converted to chymotrypsin (*b*). In chymotrypsinogen the carboxylate group of Asp 194 forms a salt bridge to His 40; in chymotrypsin the bridge goes to the new N terminal, the —NH_3^+ group of Ile 16. This change evidently allows Gly 193 to swing around so that its amide NH comes closer to the NH of Ser 195. The two amide groups form essential hydrogen bonds to the substrate in the enzyme–substrate complex (see fig. 9.11).

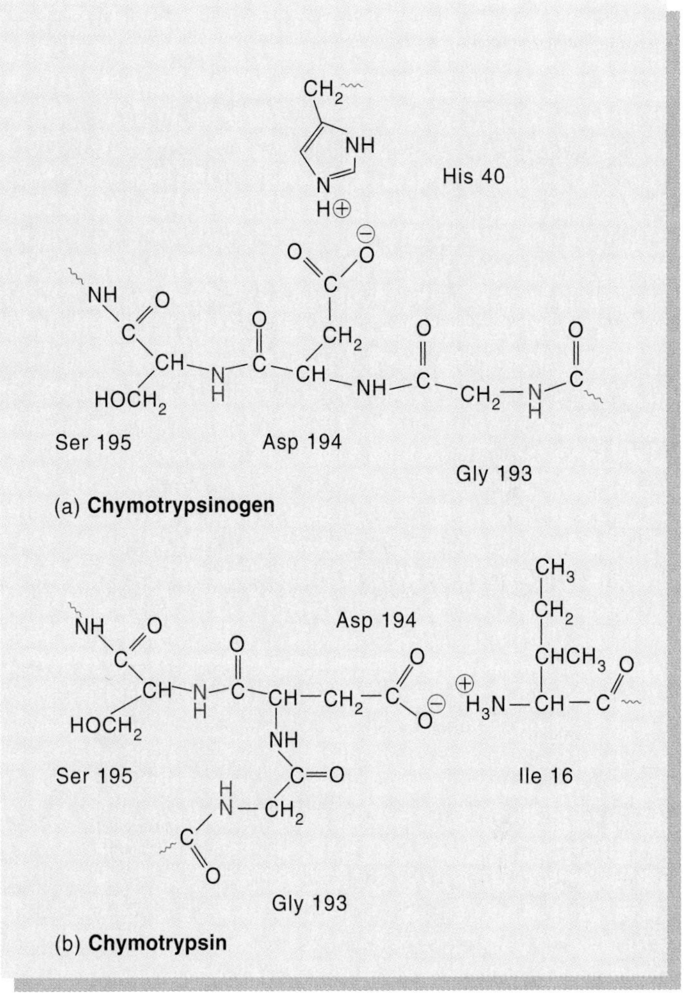

Figure 10.2

The blood coagulation cascade. Each of the curved red arrows represents a proteolytic reaction, in which a protein is hydrolyzed at one or more specific sites. With the exception of fibrinogen the substrate in each reaction is an inactive zymogen; except for fibrin each product is an active protease that proceeds to cleave another member in the series. Many of the steps also depend on interactions of the proteins with Ca^{2+} ions and phospholipids. The cascade starts when factor XII and prekallikrein come into contact with materials that are released or exposed in injured tissue. (The exact nature of these materials is still not fully clear.) When thrombin cleaves fibrinogen at several points, the trimmed protein (fibrin) polymerizes to form a clot.

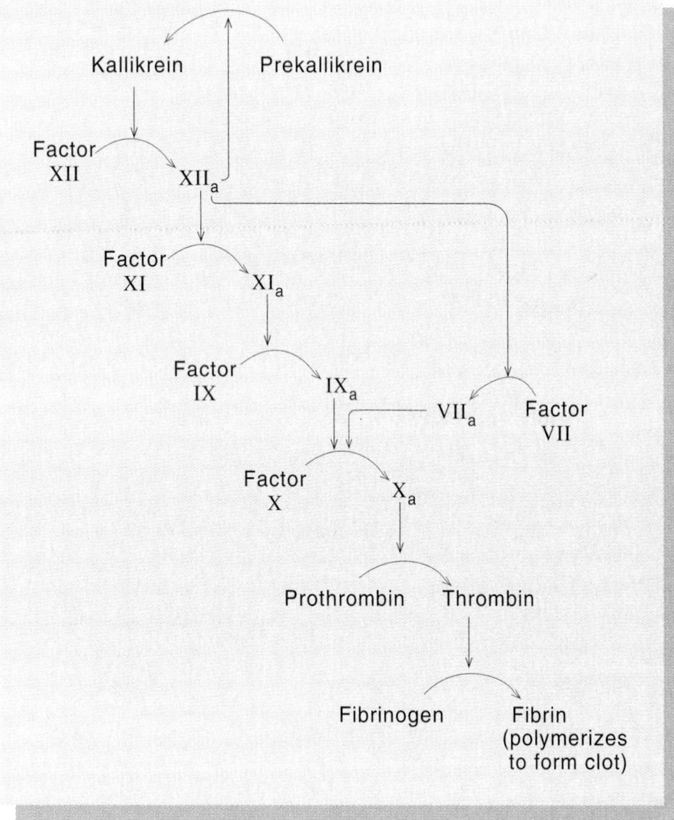

Table 10.1

Some Enzymes Regulated by Partial Proteolysis

Digestive Enzymes
Trypsin, chymotrypsin, carboxypeptidase A and B, elastase, pepsin, phospholipase

Blood Coagulation Enzymes
Factors VII, IX, X, XI, and XIII, kallikrein, thrombin

Enzymes Involved in Dissolving Blood Clots
Plasminogen, plasminogen activator

Enzymes Involved in Programmed Development
Chitin synthetase, cocoonase, collagenase

Phosphorylation, Adenylylation, and Disulfide Reduction Lead to Reversible Covalent Modifications

A type of covalent modification used more widely than partial proteolysis is phosphorylation of the side chains of serine, threonine, or tyrosine residues. Phosphorylation differs from partial proteolysis in being reversible. The introduction and removal of the phosphate group are catalyzed by separate enzymes (phosphorylation by a protein kinase and dephosphorylation by a phosphatase), which are themselves usually under metabolic regulation (fig. 10.3).

In eukaryotic organisms, phosphorylation is used to control the activities of literally hundreds of enzymes, a few of which are listed in table 10.2. These enzymes are phosphorylated or de-

Figure 10.3

Phosphorylations of serine side chains in enzymes are catalyzed by kinases, and dephosphorylation is catalyzed by phosphatases. Threonine and tyrosine side chains undergo similar reactions, but these are less common than the phosphorylation of serine.

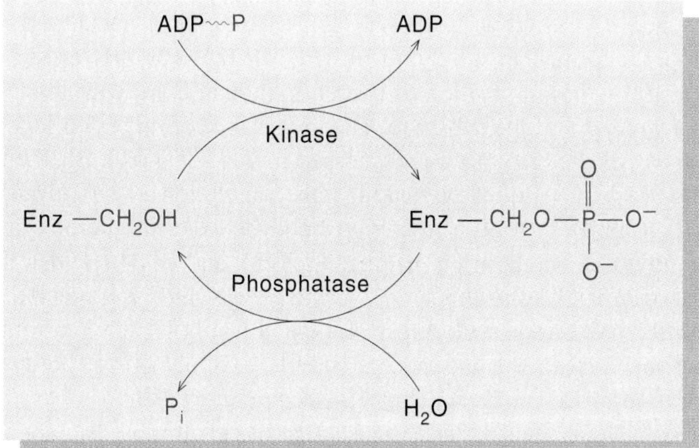

Table 10.2

Some Enzymes Regulated by Phosphorylation

Enzymes of Carbohydrate Metabolism
Glycogen phosphorylase
Phosphorylase kinase
Glycogen synthase
Phosphofructokinase-2
Pyruvate kinase
Pyruvate dehydrogenase

Enzymes of Lipid Metabolism
Hydroxymethylglutaryl-CoA reductase
Acetyl-CoA carboxylase
Triacylglycerol lipase

Enzymes of Amino Acid Metabolism
Branched-chain ketoacid dehydrogenase
Phenylalanine hydroxylase
Tyrosine hydroxylase

Figure 10.4

The transfer of the adenylyl group from ATP to a tyrosine residue is used to regulate some enzymes. The other product of the adenylylation reaction (*top*) is inorganic pyrophosphate. Hydrolysis of the adenylyl-tyrosine ester bond releases AMP (*bottom*).

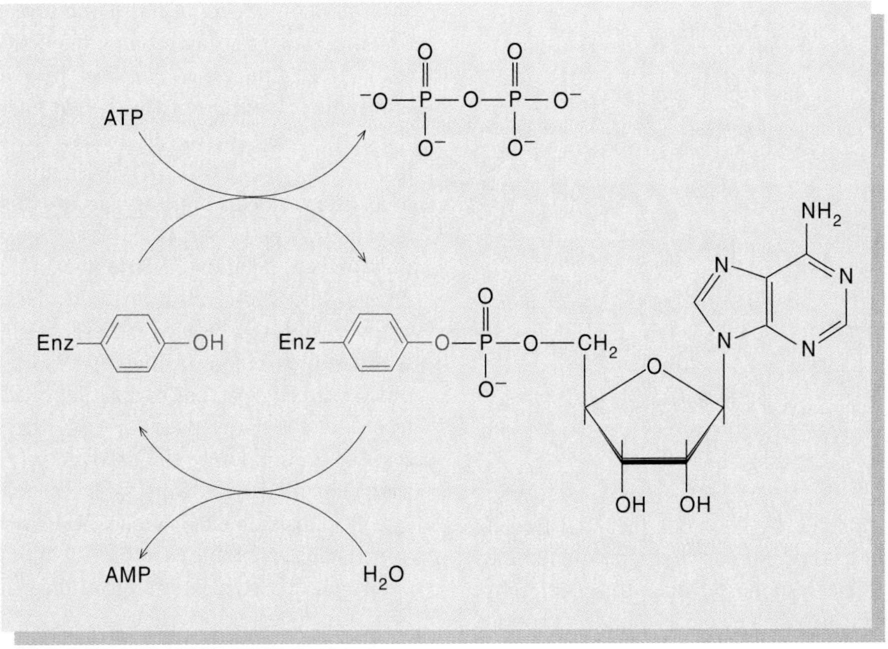

phosphorylated in response to extracellular signals, such as hormones or growth factors. The adrenal hormone epinephrine, for example, is transmitted through the blood to muscle and adipose tissue when there is a need for muscular exertion. On reaching the target tissue, epinephrine initiates a chain of events that leads to the activation of a protein kinase. The kinase [cyclic adenosine monophosphate (cAMP)-dependent protein kinase] catalyzes the phosphorylation of one or more specific enzymes, depending on the tissue. Some enzymes are activated when they are phosphorylated and inactivated when the phosphate is removed; others are inactivated by phosphorylation. In adipose tissue, phosphorylation activates triacylglycerol lipase, an enzyme that breaks down esters of fatty acids. In muscle, phosphorylation activates glycogen phosphorylase, an enzyme that breaks down glycogen, and it stops the synthesis of glycogen by switching off glycogen synthase.

Phosphorylation also can modify an enzyme's sensitivity to allosteric effectors. Phosphorylation of glycogen phosphorylase reduces its sensitivity to the allosteric activator adenosine monophosphate (AMP). Thus a covalent modification triggered by an extracellular signal can override the influence of intracellular allosteric regulators. In other cases, variations in the concentrations of intracellular effectors can modify the response to the covalent modification, depending on the metabolic state of affairs in the cell.

The covalent addition of an adenylyl (AMP) group to a tyrosine residue is another form of reversible, covalent modification (fig. 10.4). In *Escherichia coli,* adenylylation is used to regulate glutamine synthase, a key enzyme in nitrogen metabolism (see chapter 24). The tyrosine residue that accepts the adenylyl group is located close to the active site. The addition of the bulky and negatively charged adenylyl group inhibits the enzyme, perhaps simply by occluding the active site.

Other groups that can be attached covalently to enzymes include fatty acids, isoprenoid alcohols such as farnesol, and carbohydrates. Although such modifications are widespread, our understanding of how cells use them to regulate enzymatic activities is still fragmentary.

A reversible covalent modification that plants use extensively is the reduction of cystine disulfide bridges to sulfhydryls. Many of the enzymes of photosynthetic carbohydrate synthesis are activated in this way (table 10.3). Some of the enzymes of carbo-

Table 10.3
Some Enzymes Regulated by Disulfide Reduction in Plants

Activated by Reduction
Fructose-1,6-bisphosphatase
Sedoheptulose-1,7-bisphosphatase
Glyceraldehyde-3-phosphate dehydrogenase
NADP-malate dehydrogenase
Phosphoribulokinase
Thylakoid ATP-synthase
Inhibited by Reduction
Phosphofructokinase

Figure 10.5

Reduction of the disulfide bond of cystine is used to activate enzymes of photosynthetic carbohydrate biosynthesis in plants. The reductant is a small protein called thioredoxin. Thioredoxin also serves as a reductant for the biosynthesis of deoxynucleotides in animals and microorganisms as well as in plants.

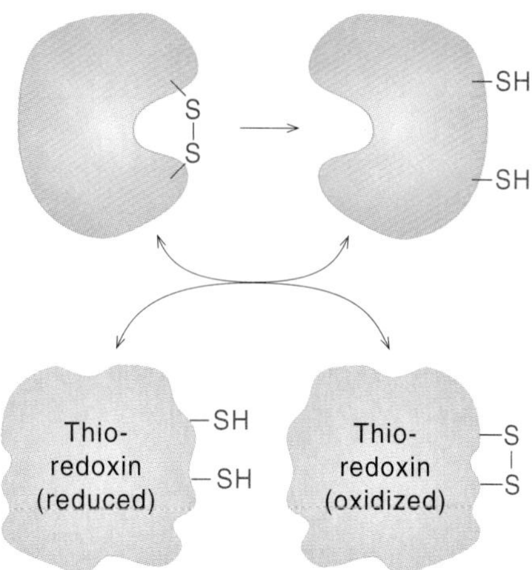

hydrate breakdown are inactivated by the same mechanism. The reductant is a small protein called thioredoxin, which undergoes a complementary oxidation of cysteine residues to cystine (fig. 10.5). Thioredoxin itself is reduced by electron-transfer reactions driven by sunlight, which serves as a signal to switch carbohydrate metabolism from carbohydrate breakdown to synthesis. In one of the regulated enzymes, phosphoribulokinase, one of the freed cysteines probably forms part of the catalytic active site. In nicotinamide-adenine dinucleotide phosphate (NADP)-malate dehydrogenase and fructose-1,6-bisphosphatase, the cysteines both appear to be some distance from the catalytic site. How the structural changes that result from the reduction are transmitted to the active sites of these enzymes is not yet known.

Covalent modifications of enzymes allow a cell to regulate its metabolic activities more rapidly and in much more intricate ways than is possible by changing the absolute concentrations of the same enzymes. They still do not provide truly instantaneous responses to changes in conditions, however, because each modification requires the action of another enzyme, which must itself be regulated. A lag also occurs in responding to the removal or inhibition of the enzyme that causes the modification, because reversing the modification requires still another enzyme.

Allosteric Regulation Allows an Enzyme to Be Controlled Rapidly by Materials That Are Structurally Unrelated to the Substrate

Whereas eukaryotic organisms use phosphorylation to handle responses to extracellular signals, both prokaryotic and eukaryotic organisms commonly use allosteric regulation in responding to changes in conditions within a cell. A typical circumstance that might demand such a response is a surplus or deficit of ATP or some other metabolic intermediate. Allosteric regulation enables a cell to adjust an enzymatic activity almost instantaneously in response to changes in the concentration of a metabolite because, unlike covalent modification, it does not require an intermediate enzyme. If the metabolite acts as an allosteric effector, the activity of the enzyme can increase (or decrease) as soon as the concentration of the effector rises and can decrease (or increase) again as soon as the concentration falls.

Regulation of enzymes by allosteric effectors is considerably more common than regulation by compounds that bind at the active site for two reasons: First, whereas an agent that binds at the active site usually acts as an inhibitor, a compound that binds at an allosteric site can serve as either an inhibitor or an activator, depending on the structure of the enzyme. Second, a substance that binds to an allosteric site does not need to have any structural relationship to the substrate. Consider, for example, the metabolic pathway of histidine biosynthesis in plants and bacteria. The pathway requires nine enzymes that work one after another. If histidine is already present in abundance, it is advantageous for a cell to cut off the entire pathway at the first step to avoid wasting energy or accumulating the products of the intermediate steps. The first step in the pathway is the reaction between phosphoribosylpyrophosphate and adenosine triphosphate (ATP), neither of which even vaguely resembles histidine, and yet the enzyme that catalyzes this step is strongly inhibited by histidine. Binding of histidine to an allosteric site causes a structural change that is transmitted to the active site.

Inhibition of the initial step of a biosynthetic pathway by an end product of the pathway is a recurrent theme in metabolic regulation. In addition, many key enzymes are regulated by ATP, adenosine diphosphate (ADP), AMP, or inorganic phosphate ion (P_i). The concentrations of these materials provide a cell with an index of whether energy is abundant or in short supply. Because ATP, ADP, AMP, or P_i often are chemically unrelated to the substrate of the enzyme that must be regulated, they usually bind to an allosteric site rather than to the active site.

Allosteric Enzymes Typically Exhibit a Sigmoidal Dependence on Substrate Concentration

In chapter 6 we saw that the binding of O_2 to any one of the four subunits of hemoglobin increases the affinity of the other subunits for O_2. This effect reflects cooperative changes in the tertiary and quaternary structure of the protein. Binding of O_2 alters the interactions among the subunits in such a way that the entire protein tends to flip into a state with increased O_2 affinity; binding of glycerate-2,3-bisphosphate favors a transition in the opposite direction. When Jacques Monod, Jeffreys Wyman, and Jean-Pierre Changeux first advanced the idea of allosteric enzymes in 1963, they suggested that these enzymes might contain multiple subunits and that the changes in catalytic activity caused by allosteric effectors can reflect alterations in quaternary structure. This suggestion turned out to be remarkably accurate. Although there is no reason why an enzyme consisting of a single subunit cannot be sensitive to allosteric effectors, most of the enzymes regulated in this way do have multiple subunits, and the changes in activity often can be related to interactions among the subunits.

One indication of the importance of intersubunit interactions in allosteric enzymes is that many such enzymes do not obey the classical Michaelis-Menten kinetic equation. A plot of the rate of reaction as a function of substrate concentration is not hyperbolic, as described by the Michaelis-Menten equation, but rather sigmoidal, resembling the curve for the binding of O_2 to hemoglobin (see fig. 6.2). Furthermore, allosteric effectors often cause the kinetics to change from one of these forms to the other, much as glycerate-2,3-bisphosphate affects the degree of cooperativity in the binding of O_2 to hemoglobin. Figure 10.6 shows an illustration of these effects for phosphofructokinase, which catalyzes the formation of fructose-1,6-bisphosphate from fructose-6-phosphate and ATP.

$$\text{Fructose-6-phosphate} + \text{ATP} \xrightarrow{\text{phosphofructokinase}} \text{Fructose-1,6-bisphosphate} + \text{ADP} \quad (1)$$

In the presence of 1.5 mM ATP the kinetics have a sigmoidal dependence on the concentration of fructose-6-phosphate; at very low concentrations of ATP, or in the presence of AMP the kinetics become hyperbolic.

Phosphofructokinase has four identical subunits. To explore how sigmoidal kinetics can arise in such an enzyme, consider an enzyme that has just two such subunits, each with its own catalytically active site. We can schematize the binding of substrate to the enzyme as follows:

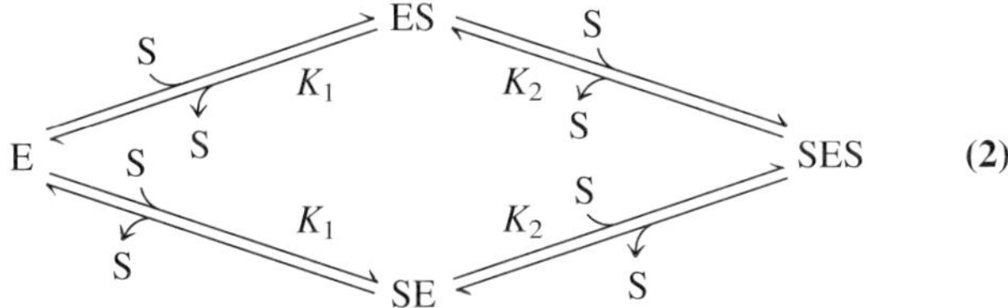

(2)

Here ES and SE represent complexes of the enzyme (E) with substrate (S) on the two different subunits, and SES represents the complex in which both binding sites are occupied. The two dissociation constants K_1 and K_2 are not necessarily identical because the binding of substrate to one subunit can affect the dissociation constant for the other subunit. In fact, this is just the point we want to explore.

Let's assume that the rate constant k_{cat} for the formation of products on either subunit is the same, whether only that site or both catalytic sites are occupied. Suppose also that ES, SE, and SES are in equilibrium with the free enzyme and substrate. By following the same procedure that led to the Henri-Michaelis-Menten equation in chapter 8, we can derive an expression for the rate of the enzymatic reaction in terms of [S], K_1, and K_2. Here we just give the result:

$$v = \frac{V_{max}\dfrac{[S]}{K_1}\left(1 + \dfrac{[S]}{K_2}\right)}{\left(1 + \dfrac{[S]}{K_1}\right) + \dfrac{[S]}{K_1}\left(1 + \dfrac{[S]}{K_2}\right)} \quad (3)$$

Figure 10.6

Kinetics of the reaction catalyzed by phosphofructokinase. In the presence of 1.5 mM ATP the rate has a sigmoidal dependence on the concentration of the substrate fructose-6-phosphate. Although ATP also is a substrate for the reaction, the sigmoidal kinetics seen under these conditions are associated with the binding of ATP to an inhibitory allosteric site. The kinetics become hyperbolic if a low concentration of AMP is added.

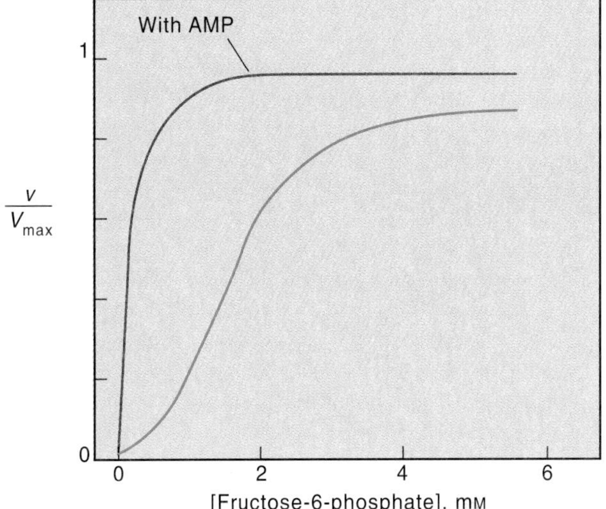

If the binding of S to one subunit does *not* affect binding to the other, then $K_2 = K_1$, and equation (3) reduces to

$$v = \frac{V_{max}[S]}{K_s + [S]} \quad (4)$$

where $K_s = K_1 = K_2$. This is simply the Henri-Michaelis-Menten equation [equation (18) in chapter 8]. A plot of v versus [S] according to equation (4) is hyperbolic. Such a plot is shown as the solid curve in figure 10.7. The dashed curve in figure 10.7 is a plot of equation (3) when $K_2 = K_1/25$. To facilitate comparison with the solid curve, K_1 is taken to be $5K_d$ and K_2 to be $K_d/5$, where K_d is the same for the two curves. This means that the overall $\Delta G°$ for the formation of SES from E + 2S is the same. (The overall equilibrium constant is $(1/K_1)(1/K_2)$, or $1/K_d^2$.) Note that the dashed curve is sigmoidal in shape. It starts out rising more slowly than the solid curve, rises steeply in the region of $[S] \approx K_d$, where $v/V_{max} \approx 0.5$, and then continues rising more slowly.

A ratio of 25 between K_1 and K_2 means that the $\Delta G°$ for binding the second molecule of substrate is only 1.9 kcal/mole more favorable than the $\Delta G°$ for binding the first molecule. This is the order of magnitude of the $\Delta G°$ for forming a single hydrogen bond. Evidently, sigmoidal kinetics might be obtained if binding the substrate causes even a relatively minor change in the conformation of the enzyme. In chapter 9 we noted that binding of glucose to hexokinase results in a pronounced conformational change that can be seen in the crystal structure. But to yield sigmoidal kinetics, it is essential that events occurring at two different binding sites be linked: Binding at one site must decrease the dissociation constant at the other site. If no such coupling occurs, the kinetics follow equation (4) and are hyperbolic even if the en-

Figure 10.7

Kinetics of an enzyme with two identical subunits. The curves were calculated with equation (3). The abscissa is the ratio of the substrate concentration [S] to K_d, where $K_d = \sqrt{K_1 K_2}$ and K_1 and K_2 are the dissociation constants for the first and second molecule of substrate. K_d is taken to be the same for all three curves. For the *solid curve*, K_1 and K_2 are assumed to be identical ($K_1 = K_2 = K_d$); equation (3) then reduces to equation (4). For the *dashed curve*, $K_2 = K_1/25$ ($K_1 = 5K_d$ and $K_2 = K_d/5$). For the *dotted curve*, $K_2 = 10^{-4}K_1$ ($K_1 = 100K_d$ and $K_2 = K_d/100$); equation (3) then reduces to equation (5).

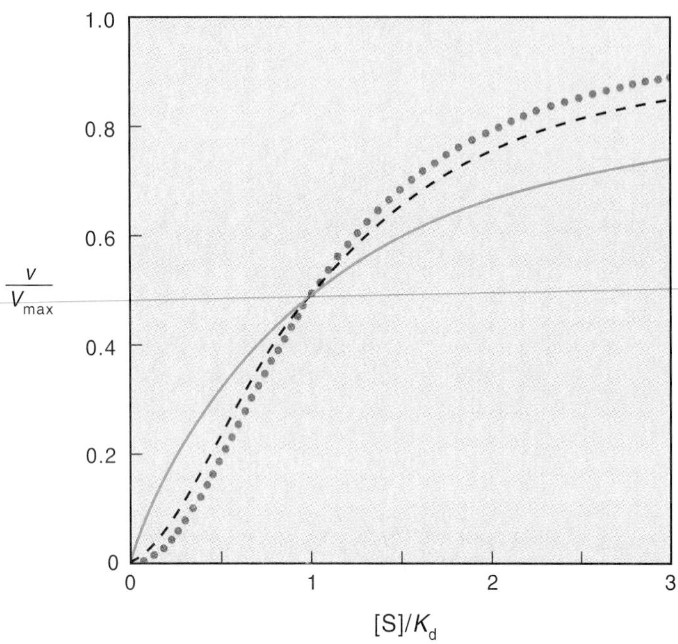

zyme has multiple subunits. This situation is observed with many enzymes, including one we discussed in chapter 9. Triosephosphate isomerase exists as a dimer, but its two subunits work independently, and its kinetics are hyperbolic.

The dotted curve in figure 10.7 shows a plot of equation (3) with $K_2 = 10^{-4} K_1$, which is equivalent to a difference of 5.5 kcal/mole between the $\Delta G°$ values for binding the first and second molecules of substrate. When the ratio of the dissociation constants is this large, the kinetics can be described equally well by the simpler expression

$$v = \frac{V_{max}[S]^2}{K_1 K_2 + [S]^2} \tag{5}$$

This is the limiting form of equation (3) when $K_2 \ll [S] \ll K_1$.

It is not uncommon for enzymes that are regulated allosterically to have four or even more subunits. The active form of acetyl-coenzyme A (acetyl-CoA) carboxylase consists of linear strings of 10 or more identical subunits (see fig. 21.15). In general, the larger the number of subunits that interact such that the binding of substrate to one subunit promotes binding to others, the more steeply the enzyme's kinetics depend on [S] in the region where $v/V_{max} \approx 0.5$. To derive a more general form of equation (5), consider a protein E with n identical subunits, each of which has a binding site for a substrate S. Suppose that binding of the first molecule of S strongly favors binding to all n subunits, giving the overall reaction

$$E + nS \rightleftharpoons ES_n \tag{6}$$

The concentration of ES_n then is

$$[ES_n] \approx \frac{[E][S]^n}{K_h} \tag{7}$$

where K_h is the product of the individual dissociation constants for all n steps leading to ES_n. The fraction of the protein that has taken up the substrate is

$$y \approx \frac{[ES_n]}{[E] + [ES_n]} = \frac{[S]^n}{K_h + [S]^n} \tag{8}$$

This is called the Hill equation, and the exponent n is the Hill coefficient.

Strictly speaking, equation (8) applies only to situations where the enzyme is substrate free or has all substrate-binding sites occupied. Despite this, the Hill coefficient provides a useful measure of cooperativity. The binding of O_2 to hemoglobin is described well by the Hill equation with $n \approx 2.8$. In the case of phosphofructokinase, which has four subunits, the dependence of the rate on the fructose-6-phosphate concentration at a fixed, relatively high concentration of ATP is described well with $n \approx 3.8$.

The Symmetry Model Provides a Useful Framework for Relating Conformational Transitions to Allosteric Activation or Inhibition

Several theoretical models have been developed for relating changes in dissociation constants to conformational changes in oligomeric proteins. We discussed two such models in connection with the oxygenation of hemoglobin (see fig. 6.14), and one of them is shown again in a slightly different form in figure 10.8. The basic idea is that the protein's subunits can exist in either of two conformations: R ("relaxed") and T ("tight" or "tense"). The substrate is assumed to bind more tightly to the R form than to the T form, which implies that binding of the substrate favors the transition from T to R. Conformational transitions of the individual subunits are assumed to be tightly linked, so that if one subunit flips from T to R, the others must do the same. Binding of the first molecule of substrate thus promotes the binding of a second. Because the concerted transition of all of the subunits from T to R or back preserves the overall symmetry of the protein, this model is called the "symmetry" model. The symmetry model can account for the behavior of allosteric activators if these compounds also bind preferentially to the R state and thus stabilize the conformation that is more effective at binding the substrate. Allosteric inhibitors act by stabilizing the T state.

Using the symmetry model, the fraction of the binding sites occupied at any given substrate concentration can be described with an expression that includes the substrate dissociation constants for the two conformations (K_R and K_T) and the equilibrium constant between the T and R conformations in the absence of substrate, $L = [T]/[R]$. Thus the symmetry model attempts to

Figure 10.8

Symmetry model for allosteric transitions of a dimeric enzyme. The model assumes that the enzyme can exist in either of two different conformations (T and R), which have different dissociation constants for the substrate (K_T and K_R). Structural transitions of the two subunits are assumed to be tightly coupled, so both subunits must be in the same state. L is the equilibrium constant (T)/(R) in the absence of substrate. If the substrate binds much more tightly to R than to T ($K_R \ll K_T$), the binding of a molecule of substrate to either subunit pulls the equilibrium between T and R in the direction of R. Because both subunits must go to the R state, the binding of the second molecule of substrate is promoted.

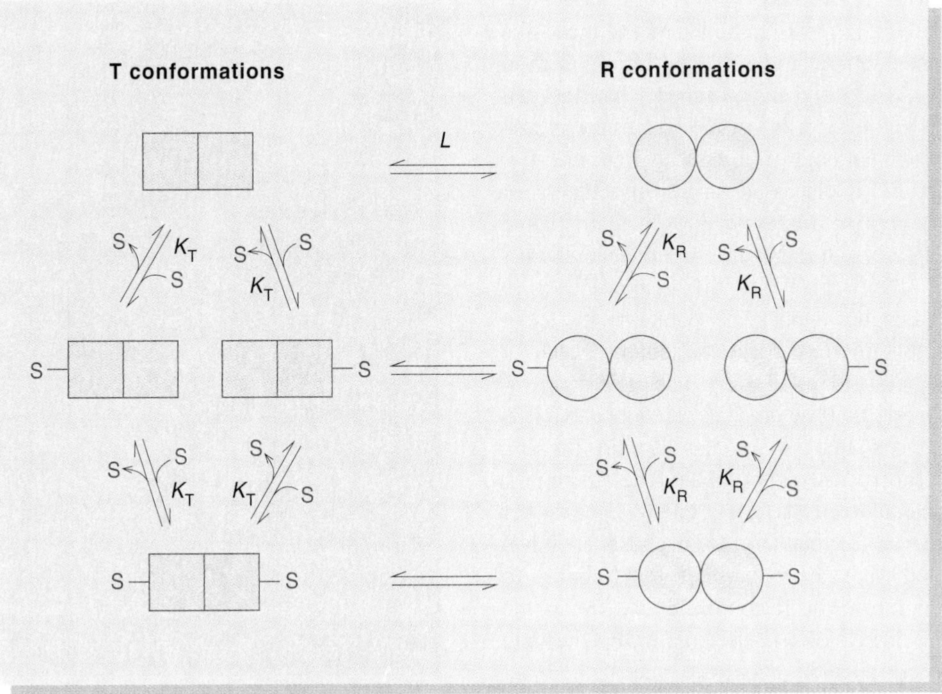

explain the difference between K_1 and K_2 in equation (3) by introducing a third independent parameter. Considering that equation (3) can fit the experimental data for a dimeric enzyme with only two parameters, what do we gain by using a more complicated equation? The usefulness of the symmetry model is that it can be generalized to enzymes that have a larger number of subunits. The general expression still contains only four independent parameters: K_R, K_T, L, and the total number of subunits (n). But this simplicity comes with a cost: We have to assume that the entire oligomeric protein has just two different conformational states. Is it not more realistic to assume that the protein also can adopt a variety of intermediate conformations? The answer depends in part on the particular protein and in part on our goals in using any type of model. Models that allow intermediate conformations have been developed, but they of course require more independent parameters. The "sequential" model shown in figure 6.14 is one of these more elaborate models. Although the sequential model is more general than the symmetry model and may be more realistic, the available experimental data in most cases do not justify a distinction between the two.

The symmetry model is useful even if it does oversimplify the situation, because it provides a conceptual framework for discussing the relationships between conformational transitions and the effects of allosteric activators and inhibitors. In the following sections we consider three oligomeric enzymes that are under metabolic control and see that substrates and allosteric effectors do tend to stabilize each of these enzymes in one or the other of two distinctly different conformations.

Allosteric Control of Phosphofructokinase Is Consistent with the Symmetry Model

Phosphofructokinase catalyzes the transfer of a phosphate group from ATP to fructose-6-phosphate (equation 1). This is a major site of regulation of glycolysis, the metabolic pathway by which glucose breaks down to pyruvate (see chapter 14). As we saw in figure 10.6, the kinetics of phosphofructokinase are strongly cooperative with respect to fructose-6-phosphate. The kinetics are noncooperative with respect to the other substrate, ATP, at low concentrations, but at concentrations above 0.5 mM, ATP acts as an inhibitor. The inhibitory effect results from binding to an allosteric site that is distinct from the substrate-binding site for ATP. Phosphofructokinase also is inhibited by phosphoenolpyruvate and by citrate, an intermediate in two other metabolic pathways that embark from pyruvate. On the other hand, it is stimulated by ADP, AMP, guanosine diphosphate (GDP), cAMP, fructose-2,6-bisphosphate, and a variety of other compounds. Fructose-2,6-bisphosphate is not present in bacteria, but it is probably the main activator of phosphofructokinase in yeast and animals. Most if not all of the activators bind to the same allosteric site as ATP and may work simply by preventing the inhibitory effect of ATP. The effects of ATP, ADP, and AMP are such that phosphofructokinase is restrained when the cell's needs for energy have been satisfied, as reflected in a high ratio of [ATP] to [ADP] and [AMP], and is unleashed when the cell needs additional energy. The effects of cAMP and fructose-2,6-bisphosphate relate to hormonal control of carbohydrate metabolism in higher organisms and is discussed further in chapters 14 and 27. Our focus here is on how fructose-6-phosphate and some of the major allosteric effectors alter the structure of the enzyme.

Phosphofructokinase was one of the first enzymes to which Monod and his colleagues applied the symmetry model of allosteric transitions. It contains four identical subunits, each of which has both an active site and an allosteric site. The cooperativity of the kinetics suggests that the enzyme can adopt two different conformations (T and R) that have similar affinities for ATP but differ in their affinity for fructose-6-phosphate. The binding for fructose-6-phosphate is calculated to be about 2,000 times tighter in the R conformation than in T. When fructose-6-phosphate binds to any one of the subunits, it appears to cause all four subunits to flip from the T conformation to the R conformation, just as the symmetry model specifies. The allosteric effectors ADP, GDP, and phosphoenolpyruvate do not alter the maximum rate of the reaction but change the dependence of the rate on the fructose-6-phosphate concentration in a manner suggesting that they change the equilibrium constant (L) between the T and R conformations.

Philip Evans and his co-workers have determined the crystal structures of phosphofructokinase from two species of bacteria, *E. coli* and *Bacillus stearothermophilus*. By crystallizing the enzyme in the presence and absence of the substrate and several allosteric effectors, they obtained detailed views of both the T and R conformations. This work has led to an explanation for why phosphofructokinase appears to be constrained largely to all-or-nothing transitions between these two states, rather than adopting a series of intermediate conformations.

Figure 10.9 shows the crystal structure of two of the subunits of phosphofructokinase from *B. stearothermophilus*. In the complete enzyme the subunits are disposed symmetrically about three mutually perpendicular axes. Each of these axes is a twofold symmetry axis, which means that rotating the entire structure by half of a full circle (180°) around the symmetry axis results in an identical structure. This rotation is shown diagrammatically in figure 10.10. Because ADP is a product of the enzymatic reaction as well as an allosteric activator, it binds at both the catalytic and allosteric sites (see figs. 10.9 and 10.10). The catalytic site for fructose-6-phosphate in each subunit is at the interface of the subunit with one of its neighbors, and the allosteric site is at the interface with a different neighbor.

Figure 10.9

Computer-generated structure of two of the four subunits of phosphofructokinase from *Bacillus stearothermophilus*. The enzyme, shown as yellow and light blue tubes, was crystallized in the R conformation in the presence of the substrate fructose-6-phosphate (dark blue) and the allosteric activator ADP (pink). The magnesium ions (white/silver spheres), Mg^{2+}, bound to the ADP molecules are also shown. (Copyright 1994 by the Scripps Research Institute/Molecular Graphics Images by Michael Pique using software by Yng Chen, Michael Connolly, Michael Carson, Alex Shah, and AVS, Inc. Visualization advice by Holly Miller, Wake Forest University Medical Center.)

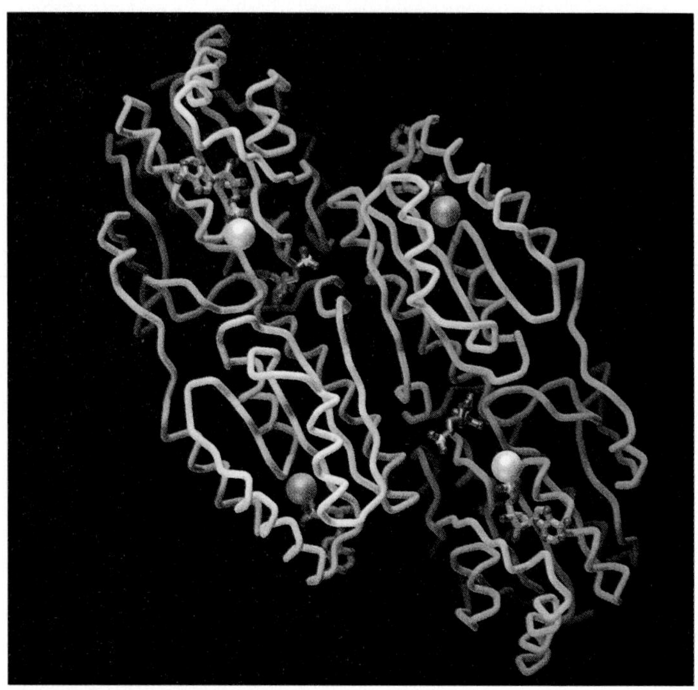

In the transition between the T and R conformations the four subunits rotate by about 7° with respect to each other (see fig. 10.10). This rotation is associated with coupled rearrangements of the structures at the interfaces between adjacent subunits. Figure 10.11 shows how these rearrangements affect the binding site for fructose-6-phosphate. The most significant structural change in this region is an inversion of the orientation of the side chains of Glu 161 and Arg 162. In the R conformation, Arg 162 forms a hydrogen bond to the phosphate group of fructose-6-phosphate, whereas Glu 161 points in the opposite direction. In the T conformation, Arg 162 points away from the binding site, and Glu 161 inserts a negative charge into the site, where it forms a hydrogen bond with Arg 243. The change in the orientations of the negatively charged Glu 161 and the positively charged Arg 162 probably accounts for most of the difference between the dissociation constants for fructose-6-phosphate in the two states. Note that although figure 10.11*b* shows the molecule of fructose-6-phosphate bound on subunit A, Glu 161 and Arg 162 are residues of subunit D. Arginines 252 and 243 also contribute to the binding site from opposite sides of the boundary. Structural changes

Figure 10.10

Outlines of phosphofructokinase in the T (solid lines) and R (dashed lines) conformations. The enzyme contains four identical subunits (A, B, C, and D). The locations of the catalytic and allosteric sites are indicated in the two subunits closest to the viewer (A and D). The binding sites for fructose-6-phosphate (F6P) are at the interface of these subunits; the allosteric sites are at the interfaces of A with B and of D with C. Two of the three perpendicular symmetry axes are labeled *p* and *q*. A 180° rotation about axis *q* interchanges the positions of subunits A and C and also interchanges B and D. A similar rotation about *p* interchanges A with B and C with D. (Source: From T. Schirmer and P. R. Evans, Structural basis of the allosteric behaviour of phosphofructokinase, *Nature* 343:140, 1990.)

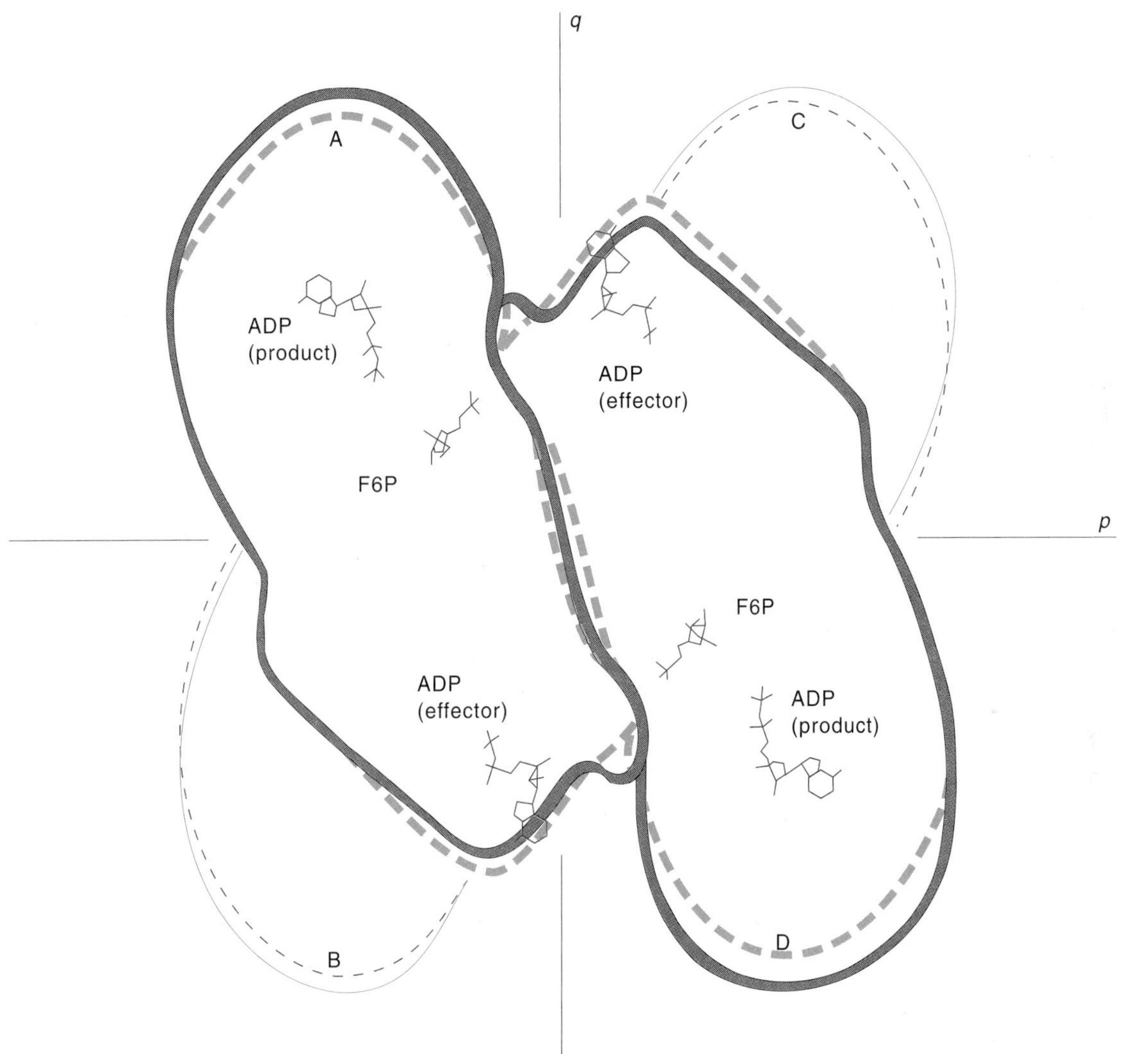

that occur on one of the subunits thus are intricately linked to changes on the other. The substrate-binding site for ATP, on the other hand, is made up of residues from only one subunit (subunit A in figure 10.11). This probably explains why the binding of fructose-6-phosphate to the enzyme is strongly cooperative, whereas the binding of ATP as a substrate is not cooperative.

In the T conformation, Glu 161 and Arg 162 are located at the end of a stretch of polypeptide that winds up into a helical turn in the transition to the R structure (see fig. 10.11). This coiling is linked to a major structural change in an adjacent region of the interface between subunits A and D. The interface here includes a pair of antiparallel β strands, each of which is hydrogen-bonded to a parallel strand in its own subunit, as shown in figure 10.12. In the T conformation (fig. 10.12*a*) the β strands from the different subunits are hydrogen-bonded together directly across the interface. In the R conformation (fig. 10.12*b*) the strands have

Figure 10.11

Interface between subunits A and D of phosphofructokinase near the catalytic site in (*a*) the T and (*b*) the R structures. Crystals of the enzyme in the R state were obtained in the presence of fructose-6-phosphate and ADP (see fig. 10.9); crystals in the T state were obtained in the presence of a nonphysiological allosteric inhibitor, 2-phosphoglycolate. The wavy green line represents part of the boundary between subunits A and D. The heavy green line indicates the polypeptide backbone. The side chains of Glu 161 and Arg 162 are shown in red. Note the inversion of the positions of these side chains in the two structures. (Source: From T. Schirmer and P. R. Evans, Structural basis of the allosteric behaviour of phosphofructokinase, *Nature* 343:140, 1990.)

(a) T state

(b) R state

Figure 10.12

Hydrogen bonds of the peptide backbone and the side chain of threonine 245 at the interface between subunits A and D of phosphofructokinase, in (*a*) the T and (*b*) the R structures. Note the additional molecules of water (red) between the two subunits in the R structure. (Source: From T. Schirmer and P. R. Evans, Structural basis of the allosteric behaviour of phosphofructokinase, *Nature* 343:140, 1990.)

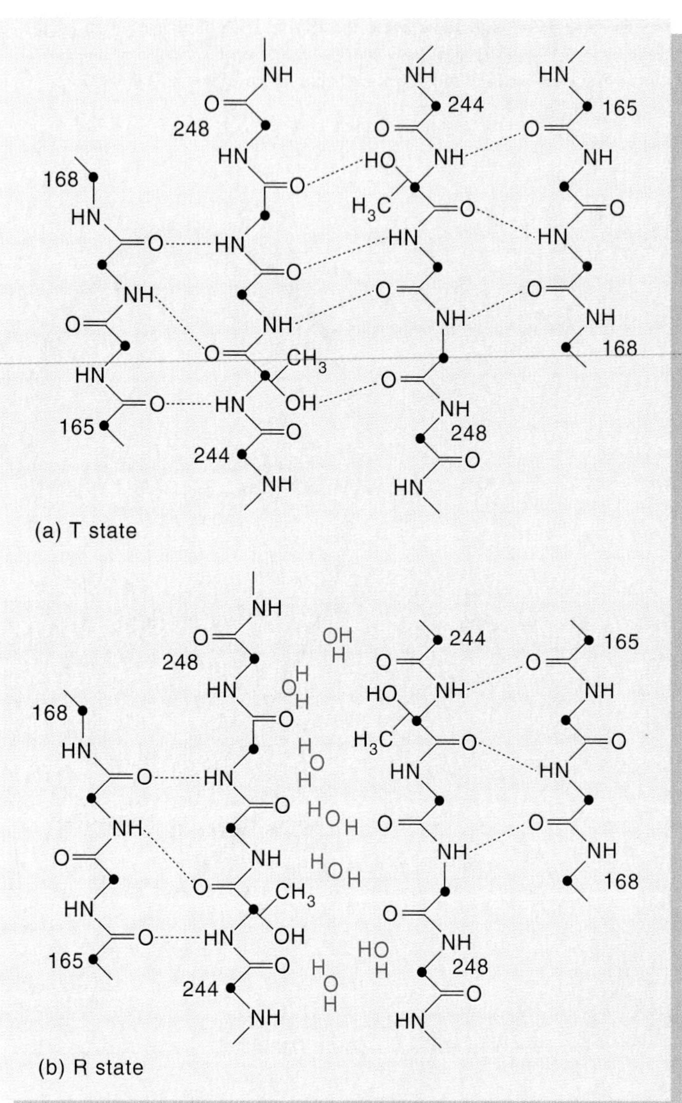

moved apart, and the region between them is filled by a row of hydrogen-bonded water molecules.

The insertion of water at the interface between subunits A and D is an essential component of the rotation of the subunits with respect to each other, and it appears to be an all-or-nothing effect. Intermediate conformations in which only some of the water molecules are present would have a less extensive network of hydrogen bonds and thus are probably less stable than either the R or the T conformation. The same might be said of the winding of the helical turn between residues 155 and 161; intermediates in which the helical turn is partially unwound probably would be destabilized by steric crowding or force the structure to expand in a way that leaves empty spaces in other regions. These considerations, taken with the close coupling between the individual subunits, seem to explain why all of the subunits undergo concerted transitions from the R to the T state or back, without giving appreciable concentrations of intermediate states.

Figure 10.13

The reaction catalyzed by aspartate carbamoyl transferase and the feedback inhibition of this enzyme in *E. coli* by the end product of the pathway, CTP. Additional reaction steps in the pathway from carbamoyl aspartate to CTP are discussed in chapter 26. The upward arrow with the negative sign indicates the feedback inhibition.

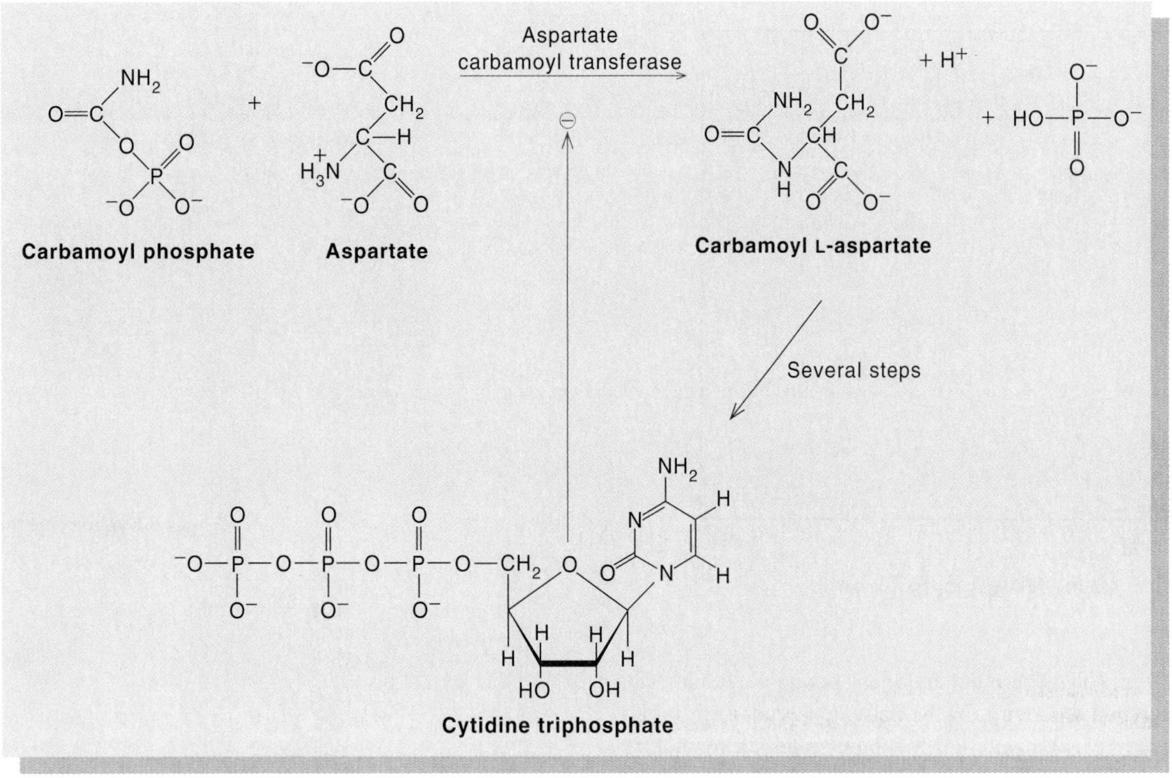

Although the crystal structures offer a very plausible explanation for the cooperative binding of fructose-6-phosphate, it is still not clear why various allosteric effectors stabilize the protein in different conformational states. However, it is significant that the allosteric binding site lies at the interface of different subunits. The equilibrium between the R and T conformations appears to be sensitive to subtle structural changes in this region.

In Aspartate Carbamoyl Transferase the Catalytic and Regulatory Sites Are Located on Different Subunits

Aspartate carbamoyl transferase, or aspartate transcarbamylase, catalyzes the transfer of a carbamoyl group

$$(H_2N{-}\overset{\displaystyle O}{\overset{\|}{C}}{-})$$

from carbamoyl phosphate to aspartic acid to form carbamoyl aspartate (fig. 10.13). This step commits aspartate to the biosynthetic pathway for pyrimidines (see figure 26.18). Aspartate carbamoyl transferase is inhibited by cytidine triphosphate (CTP) and uridine triphosphate (UTP), the end products of the pathway, and it is stimulated by a purine, ATP. The opposing effects of CTP, UTP, and ATP serve to keep the biosynthesis of pyrimidines in balance with that of purines. This balance is important because cells need purines and pyrimidines in approximately equal amounts for the synthesis of nucleic acids.

The kinetics and physical properties of aspartate carbamoyl transferase from *E. coli* were studied in considerable detail by Howard Schachman and his colleagues. The kinetics have a sigmoidal dependence on the concentration of aspartate, as shown in figure 10.14. CTP shifts the kinetic curve to the right and thus inhibits the enzyme strongly at low concentrations of aspartate but not at high concentrations. ATP reverses the effect of CTP or, in the absence of CTP, eliminates the cooperativity altogether, making the kinetics hyperbolic instead of sigmoidal.

The behavior of aspartate carbamoyl transferase changed dramatically when the enzyme was treated with the organic mercurial compound *p*-hydroxymercuribenzoate (see fig. 10.14). The binding of aspartate no longer showed positive cooperativity, and ATP or CTP was without effect. Exposure to mercurials was found to cause the enzyme to dissociate into two types of fragments, one of which retained the enzymatic activity but was no longer af-

Figure 10.14

Effects of CTP, ATP, and mercurials on the rate of the reaction catalyzed by aspartate carbamoyl transferase. In the absence of CTP and ATP the sigmoidal kinetics show positive cooperativity with respect to aspartate. CTP augments the positive cooperativity; ATP reverses the effect of CTP. An organic mercurial, or ATP in the absence of CTP, eliminates the cooperativity, converting the curve from sigmoidal to hyperbolic.

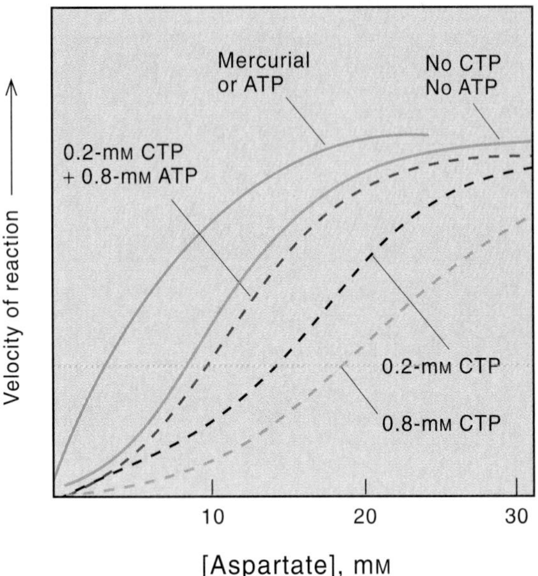

Figure 10.15

Subunit structure of aspartate carbamoyl transferase and the fragments produced by treating the enzyme with mercurials. In the complete enzyme (*top*) the three sets of regulator dimers are sandwiched between two trimers of catalytic subunits (see fig. 10.17). The approximate location of the active site in each c subunit of the trimer facing the viewer is indicated with a c.

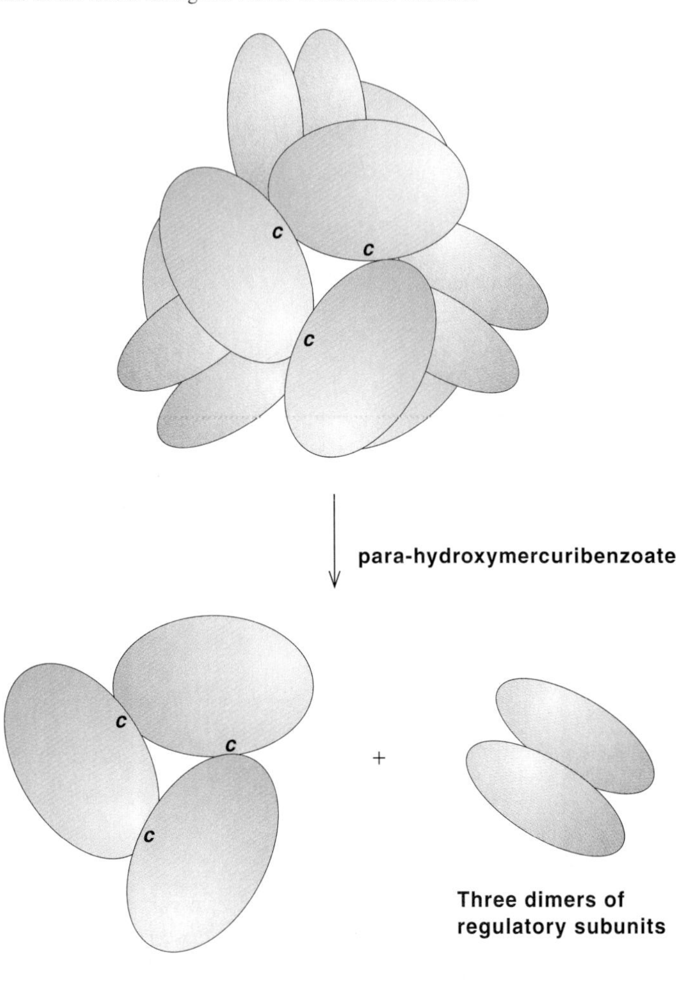

fected by CTP or ATP. The other fragment bound CTP and ATP but had no enzymatic activity. In the native enzyme, each molecule is a complex of six catalytic (c) subunits and six regulatory (r) subunits (fig. 10.15). The fragments resulting from treatment with mercurials consist of trimers of the c subunit and dimers of r. When the c_6r_6 complex was reconstituted from the fragments, the enzyme regained its sigmoidal kinetics and its sensitivity to CTP and ATP. These observations provided a convincing demonstration that the regulatory agents bind at an allosteric site and not at the active site. The two sites are on totally different subunits!

A substrate analog that proved particularly useful for studying aspartate carbamoyl transferase is N-phosphonacetyl-L-aspartate (PALA). PALA is structurally similar to a covalently linked adduct of carbamoyl phosphate and aspartate and thus resembles a likely intermediate in the enzymatic reaction (fig. 10.16). It binds to the enzyme in a highly cooperative manner, and its binding is promoted by ATP and opposed by CTP. The binding of PALA decreases the sedimentation and diffusion coefficients of the enzyme, indicating that the protein expands or changes shape. It also decreases the chemical reactivity of a cysteine residue in each c subunit and increases the reactivity of several cysteines in each r subunit. CTP opposes these effects.

The changes in sedimentation coefficient and chemical reactivity caused by PALA fit the model that the enzyme exists in two distinct conformations (T and R) and that the binding of PALA to only one or two of the c subunits causes the entire c_6r_6 complex to flip from the T to the R state. The dissociation constant for PALA is higher in the T state than in the R state. The equilibrium constant $L = [T]/[R]$ has been calculated to be 250 in the absence of substrates and allosteric effectors, 70 in the presence of ATP alone, and 1,250 in the presence of CTP alone. ATP thus shifts the equilibrium toward the conformational state that favors binding of the substrate, and CTP shifts the equilibrium in the direction of weaker binding.

Crystallization of aspartate carbamoyl transferase with and without bound PALA or CTP, by William Lipscomb and his colleagues, led to detailed pictures of the R and T conformations. Figures 10.17 and 10.18 show the crystal structures from two different perspectives. In both the R and the T conformations the six c subunits are arranged in two equilateral trimers, one of which is inverted and stacked on top of the other (see figs. 10.15 and 10.17). The three c units in each trimer are related to each other by a threefold axis of symmetry. (Rotating the structure by one-third of a circle about this axis results in an identical structure.) The sub-

Figure 10.16

(*a*) The reaction catalyzed by aspartate transcarbamylase and the presumed intermediate that is formed. (*b*) The chemical structure of the inhibitor N-phosphonacetyl-L-aspartate (PALA).

Aspartate **Carbamoyl phosphate** → **Intermediate** → ***N*-carbamoyl-L-aspartate** + P_i

(a)

***N*-phosphonacetyl-L-aspartate (PALA)**

(b)

Figure 10.17

Structures of aspartate carbamoyl transferase in (*a*) the T conformation and (*b*) the R conformation viewed along the threefold symmetry axis. The enzyme contains two c_3 clusters and three r_2 clusters. The α-carbon chains of one of the c_3 groups are shown in aqua; those of the other c_3 group are in blue. One of the *r* subunits in each of the r_2 groups is shown in orange; the other in red. The enzyme was crystallized in the T form in the absence of substrate or allosteric effectors; the R structure was obtained with bound PALA. The PALA molecules are seen in yellow in (*b*). Zinc ions bound to the *r* subunits are shown in white.

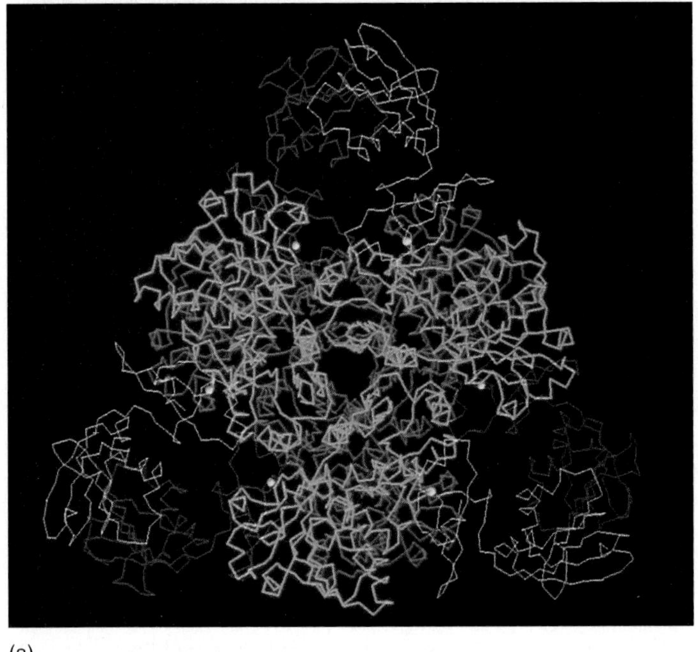

(a)

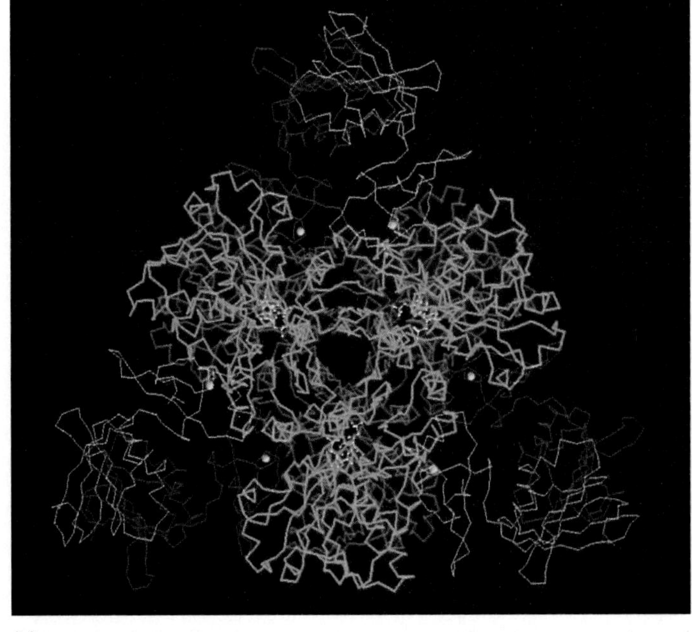

(b)

Figure 10.18

Structures of aspartate carbamoyl transferase in (*a*) the T conformation and (*b*) the R conformation viewed along an axis perpendicular to the threefold symmetry axis. The structures and the color coding are the same as in figure 10.17. Note the expansion of the cavity between the upper and lower c_3 groups in the R structure. (After W. N. Lipscomb)

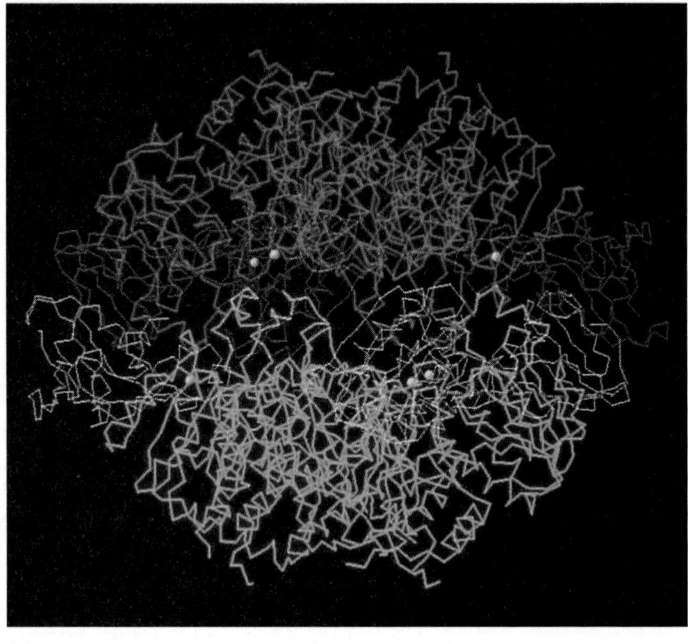

(a)

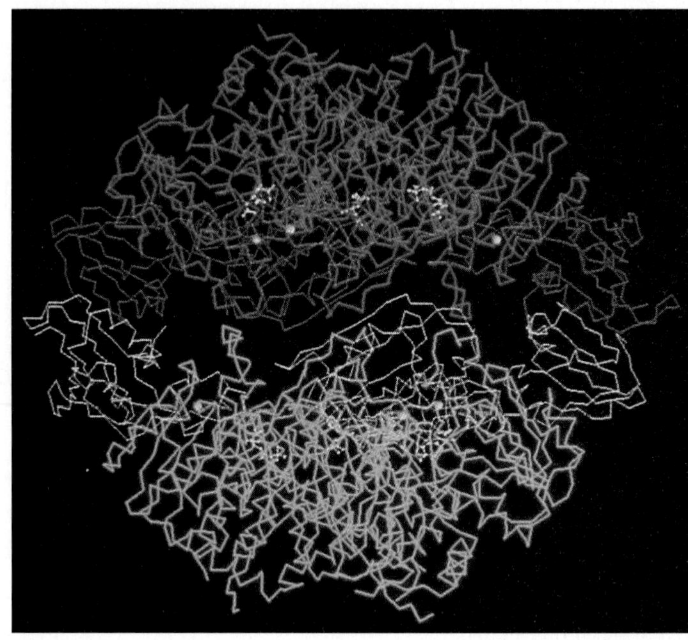

(b)

strate-binding site on each *c* subunit is located in a pocket between two domains of the polypeptide. At the end of one of these domains the *c* subunit interacts with another *c* subunit in the same trimer, close to *its* substrate-binding site. In the other domain it interacts more extensively with a *c* subunit in the other c_3 trimer. The six *r* subunits are arranged in three sets of dimers that form another equilateral triangle about the same axis of symmetry. Like the *c* subunits, each *r* subunit is folded into two domains: a peripheral domain, where it interacts with its companion *r* subunit in the dimer, and a smaller domain that interacts with two adjacent *c* subunits. In the latter region each *r* subunit binds an atom of Zn^{2+} that evidently plays a purely structural role. The binding site for the allosteric effectors CTP and ATP is located in the peripheral domain of the *r* subunit, at a considerable distance from the active sites.

In the transition from the T to the R conformation the two *c* trimers rotate slightly with respect to each other about the threefold symmetry axis, so that they come into a more eclipsed alignment (see fig. 10.17). The *r* dimers rotate with respect to each other about a perpendicular axis and appear to act as a lever that moves the two *c* trimers apart by about 12 Å and opens up a cavity at the center of the entire structure (see fig. 10.18b).

Figure 10.19 shows some of the details of the substrate-binding site in the R structure. As we mentioned above, the binding site consists of a pocket between two domains of a *c* subunit. Arginines 167 and 229 from one domain interact with the two car-

Figure 10.19

The binding of PALA to the R conformation of aspartate carbamoyl transferase. Arg 105 and His 134 are provided by one domain of a *c* subunit, and Arg 167 and Arg 229 by the other domain. Ser 80 and Lys 84 are part of a loop of protein from a different *c* subunit. The PALA is indicated by red. The wavy green lines indicate polypeptide backbone structure.

Arg 167
His 134
Arg 105
Arg 229
Ser 80
Lys 84
Adjacent *c* subunit

boxylate groups of PALA. Arginine 105 of the other domain interacts with the phosphonate group and presumably does the same with the phosphate of carbamoyl phosphate. Histidine 134 appears to provide a hydrogen bond to the peptide oxygen atom of PALA. If it does the same to the corresponding oxygen of carbamoyl phosphate, it can serve as a general acid in the catalytic mechanism. In addition to these residues the active site also includes two residues from a different *c* subunit, Ser 80 and Lys 84 (see fig. 10.19). As with phosphofructokinase, the location of the active site at the interface between two subunits provides a clue to how binding of substrate to one subunit can affect the binding at another.

The structural transition to the T state disrupts the active site in two major ways. First, the domain of the *c* subunit that includes Arg 105 and His 134, which interact with carbamoyl phosphate, is pulled away from the domain that interacts with aspartate, because some of the residues in both domains are tied up in an alternative set of hydrogen bonds. Arginine 105 is hydrogen-bonded to Glu 50 in the same domain, instead of to the substrate; His 134 interacts with a residue in the other *c* trimer. In addition, the loop of the *c* subunit that contains Ser 80 and Lys 84 is pulled out of the active site by hydrogen bonds to still another *c* subunit. A complex net of interrelationships thus links the catalytic sites of all the *c* subunits in the complex.

The transition between the R and T states also involves large changes in the conformation of the *r* subunits. These changes include both the peripheral domain where CTP or ATP binds and the domain that interfaces with the *c* subunits (see figs. 10.17 and 10.18). However, it still is not clear how the binding of CTP to the peripheral domain tips the conformational equilibrium in favor of T, whereas ATP, which binds to the same site as CTP, favors the formation of R. The crystal structures of the enzyme with bound ATP or CTP are very similar, both at the catalytic site and at the allosteric site.

The Advantages of Positive Cooperativity

In hemoglobin and aspartate carbamoyltransferase we see two classic examples of allostery that involve positive cooperativity in substrate binding (here we are speaking loosely of oxygen as a substrate). In both cases the binding of one oxygen or one aspartate molecule encourages the binding of a second and so on. If the communication between subunits is interrupted, as by dissociation of the subunits, the binding at low levels of substrate is increased and the binding changes from a sigmoidal to a hyperbolic dependence on substrate concentration. From this result it seems unlikely that the function of the allostery in these examples of positive cooperativity is merely to increase the binding. The unique advantage of the positive cooperativity resulting from substrate binding is that it encourages the binding to occur over a narrow range of concentrations of small molecule. This makes the system extremely sensitive to small changes in concentration of small molecule. In the case of hemoglobin we have seen that this sensitivity improves the net transport of oxygen from the lungs (where the oxygen tension is comparatively high) to other tissues (where the oxygen tension is comparatively low). In the case of aspartate carbamoyl transferase the enzyme will tend towards low activity unless the aspartate concentration is quite high; then it will rise rapidly. The advantage of sensitivity to substrate concentration is that it permits control of metabolic processes over a narrow range of concentrations. We will elaborate on the merits of this feature in Chapter 12.

Negative Cooperativity

We have seen that aspartate carbamoyl transferase shows positive cooperativity for the binding of aspartate, just as hemoglobin shows positive cooperativity for the binding of oxygen. The ACTase enzyme also shows negative cooperativity in the binding of CTP to the R subunits. The advantage of this negative cooperativity is that it limits the production of CTP according to need by inhibiting the first reaction in the pathway when CTP is present in excess. Quite remarkably, ACTase also shows another type of negative cooperativity in the binding of carbamoyl phosphate to the C subunits. Careful pre-steady-state measurements on purified enzyme show that only three of the six subunits participate readily in the binding of either of these molecules. Apparently, the binding of the small molecule to one subunit causes a structural transition so that a nearby subunit cannot participate similarly in the binding process.

A number of other enzymes show the same type of negative cooperativity, which is known as half-of-the-sites reactivity. They include acetoacetate decarboxylase, aldolase, alkaline phosphatase, some aminoacyl-tRNA synthases, and many other enzymes. There is not yet a convincing argument for the physiological usefulness of this widespread phenomenon, but Paul Boyer has made an interesting proposal. Ideally, after an enzyme has catalyzed a reaction, it should open up to release the products rapidly. If two subunits are coupled so that closing one will open the other one, like a see-saw, then the two can alternate as catalysts. One binds the substrate and closes, while the other opens up to release products, then the roles reverse. It has also been suggested that half-of-the-sites reactivity might assist the catalytic steps, not just the product release steps.

Glycogen Phosphorylase Activity Is Regulated by Allosteric Effectors and by Phosphorylation

Glycogen phosphorylase is an essential enzyme in mammals that provides the fuel needed to sustain life between feedings. The enzyme senses the metabolic state of the cell or organism and responds by liberating glucose from stored glycogen as needed. Phosphorylase is studied both for its central role in a vital biological process and also as a model for allosteric behavior in proteins.

Primary sequence data available for glycogen phosphorylases from mammals, potatoes, bacteria, and yeast indicate that these enzymes all had a common origin. Despite this and the similarity in their catalytic activities the enzymes differ markedly in their regulatory properties. This is a common occurrence for regulatory enzymes and results from the variable needs of different organisms. Plant and bacterial phosphorylases are completely unregulated. The yeast enzyme is regulated by phosphorylation alone. The mammalian enzymes are the most highly regulated, with a phosphorylation site and multiple allosteric sites. Most of the discussion here is related to the mammalian enzymes.

Glycogen phosphorylase catalyzes the removal of a terminal glucose residue from glycogen. (The structure of glycogen is shown in fig. 13.14.) The glycosidic bond is cleaved by a reaction with inorganic phosphate, so the product is glucose-1-phosphate instead of free glucose:

$$(\text{Glucose})_n + \text{P}_\text{i} \longrightarrow (\text{Glucose})_{n-1} + \text{Glucose-1-phosphate} \quad \textbf{(9)}$$

This is the first step in the metabolic breakdown of glycogen to pyruvate.

In the early 1940s, Carl Cori and Gerty Radnitz Cori discovered that phosphorylase exists in two forms, *a* and *b*, which differ greatly in their catalytic activities. As shown in figure 10.20, phosphorylase *b* has virtually no activity in the absence of AMP. It is activated by low concentrations of AMP, but the activation is inhibited competitively by ATP. At the concentrations of AMP and ATP that prevail in resting muscle tissue, phosphorylase *b* is essentially inactive. Phosphorylase *a*, on the other hand, has about 80% of its maximal activity in the absence of AMP and becomes fully active at very low concentrations of AMP. It also is relatively insensitive to inhibition by ATP (see fig. 10.20).

The Coris found that the interconversion of phosphorylases *a* and *b* is catalyzed by another enzyme, and subsequent work by Earl Sutherland showed that this process is under hormonal control. In muscle, conversion of phosphorylase *b* to *a* is stimulated by epinephrine; in liver it is stimulated by both epinephrine and the pancreatic hormone glucagon. The structural basis for the difference between the two forms of phosphorylase remained unknown until the late 1950s, when Edwin Krebs and Edmund Fischer showed that phosphorylase *a* has a phosphate on serine 14. This phosphate is absent in the *b* form of the enzyme. Krebs and Fischer also showed that the kinase that catalyzes the addition of the phosphate is itself regulated by a phosphorylation catalyzed by another enzyme, the cAMP-dependent protein kinase!

The main effect of AMP on either phosphorylase *b* or phosphorylase *a* is to decrease the K_D for P_i. This change can be interpreted as we have interpreted the actions of allosteric effectors on phosphofructokinase and aspartate carbamoyl transferase, on the model that the enzyme can exist in two conformational states (R and T) with different affinities for the substrate. However, phosphorylase presents the additional complexity that the equilibrium constant (*L*) between the two conformational states can be altered by a covalent modification of the enzyme. In the absence of substrates, [T]/[R] appears to be greater than 3,000 in phosphorylase *b* but to decrease to about 10 in phosphorylase *a*.

Both forms of phosphorylase are inhibited by glucose or glucose-6-phosphate. Glucose inhibits by binding at the catalytic site while glucose-6-phosphate binds at the same allosteric site as AMP and ATP. A separate inhibitory allosteric site binds adenine, adenosine, or (much more weakly) AMP.

Louise Johnson and her co-workers have determined the crystal structures of T and the R forms of muscle phosphorylase *b* and the R form of phosphorylase *a*. In parallel with this work, Robert Fletterick and co-workers determined the structure of the T form of muscle phosphorylase *a*. The crystal structures provide an incisive look at the structural changes that accompany the transitions from the T to the R conformation and from the nonphosphorylated form of the enzyme to the phosphorylated form.

Figure 10.20

The rate of the reaction catalyzed by glycogen phosphorylase, as a function of the concentration of its main allosteric activator, AMP. The curves shown in red were obtained in the presence of 9 mM of ATP. Phosphorylase *b (lower two curves)* is almost completely inactive in the absence of AMP. Its activity is half maximal at an AMP concentration of about 40 μM. ATP greatly increases the concentration of AMP required for activity. Phosphorylase *a (upper two curves)* has about 80% of its maximal activity in the absence of AMP and reaches full activity at very low AMP concentrations; it also is relatively insensitive to inhibition by ATP.

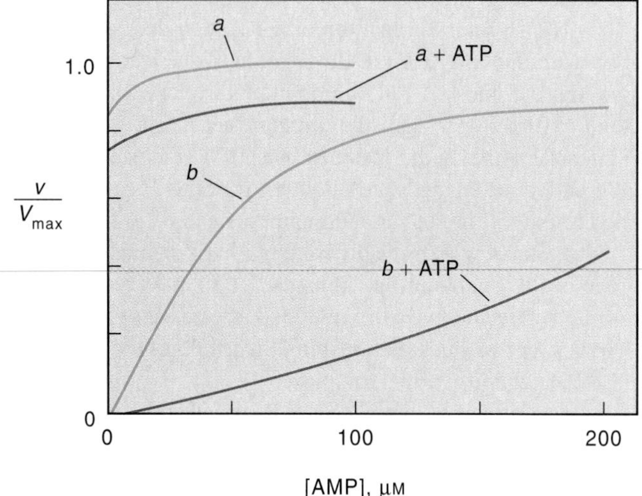

In keeping with the complexity of its allosteric and covalent regulation, phosphorylase is a large enzyme. It consists of a dimer of two identical subunits, each with a molecular weight of about 97,400 (fig. 10.21). The catalytic site is buried near the center of each subunit, at the end of a tunnel that opens to a concave surface. A binding site for glycogen on the surface can be recognized by the location of a small oligosaccharide in the crystal structure. Because the glycogen attachment site is about 30 Å from the catalytic site, the enzyme evidently clings to one branch of a glycogen particle while it chews on another branch. Near the catalytic site is a covalently bound molecule of the coenzyme pyridoxal phosphate (fig. 10.22), which probably participates as a general acid in the catalytic mechanism. The binding site for the allosteric effectors AMP, ATP, and glucose-6-phosphate is about 30 Å from the catalytic site, at one of the interfaces of the two subunits (see fig. 10.21). Serine 14, the locus of the covalent modification that converts phosphorylase *b* to *a*, is in the same region.

When phosphorylase *b* undergoes the transition from the T to the R conformation, major structural changes occur in a loop between residues 282 to 286, which connects two α-helical chains (fig. 10.23). In the T form, this loop obstructs the substrate-binding site for P_i; in the R form, the loop is pulled out of the P_i site. Some of the structural changes are shown in figure 10.24. In the

Figure 10.21

(*a*) Ribbon diagram of the crystal structure of phosphorylase *a* in the R state. The view is along the twofold rotational symmetry axis of the dimer, with the allosteric sites and the phosphoserine (Ser-P) on each subunit facing forward. Regions where the positions of the C_α carbons differ by more than 1 Å between the R and T states are shown in orange for subunit 1 (*bottom*) and in pink for subunit 2 (*top*). Less mobile regions of the polypeptide chain are in yellow for subunit 1 and in blue for subunit 2. The N-terminal residues (10–23) and the C-terminal residues (837–842) are in white. (Residues 1–9 are disordered and cannot be seen in the crystal structure.) The bound pyridoxal phosphate (PLP) indicates the location of the catalytic site. The allosteric effector site is occupied by AMP. The first two helices at the N-terminal end (labeled α_1 and α_2 in subunit 1) are connected by a loop (Cap) that forms one of the interfaces between the two subunits. (*b*) Ribbon diagram of phosphorylase *b* in the T state. The orientation of the dimer is the same as in (*a*). Regions where the C_α positions differ by more than 1 Å between the R and T states are represented in red for subunit 1 (*bottom*) and in yellow for subunit 2 (*top*); more fixed regions are in cyan and purple. The N-terminal residues and the C-terminal residues are in white. PLP is at the catalytic site and AMP at the allosteric effector site, as in (*a*). Maltopentaose is bound at the glycogen storage site. (From D. Barford, S.-H. Hu, and L. N. Johnson, *J. Mol. Biol.* 218:233, 1991. © 1991 Academic Press LTD., London, England.)

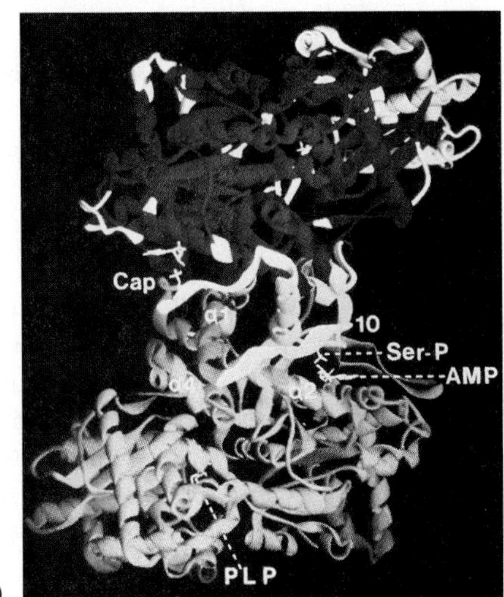

(a)

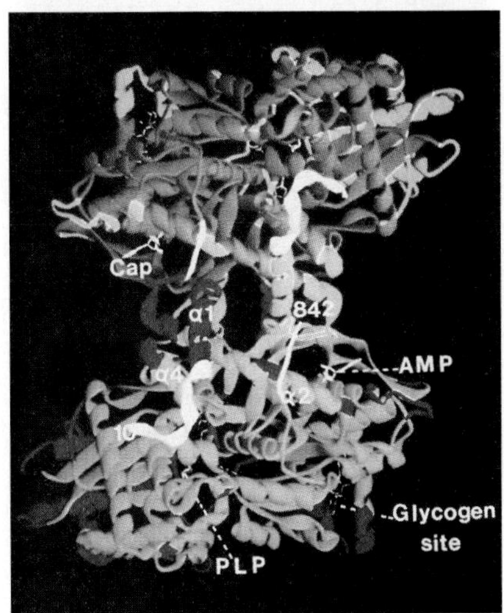

(b)

Figure 10.22

Pyridoxal phosphate, a prosthetic group in glycogen phosphorylase, is covalently attached to a lysine side chain of the enzyme. The phosphate group of the pyridoxal phosphate probably acts as a general acid to transfer a proton to inorganic phosphate in the enzymatic mechanism.

Figure 10.23

(*a*) In the T state of phosphorylase *b*, a loop of the polypeptide chain between residues 282 and 286 obstructs the catalytic site. Asp 283, near the middle of this loop, inserts its negatively charged side chain into the binding site for phosphate. The locations of the aspartate and the pyridoxal phosphate prosthetic group (PLP) are indicated in red. (*b*) In the R state, the subunits rotate with respect to each other. The loop from residues 282 to 286 is disordered, and Asp 283 leaves the phosphate-binding site. (Source: From D. Barford and L. N. Johnson, The allosteric transition of glycogen phosphorylase *Nature,* 340:609, 1989.)

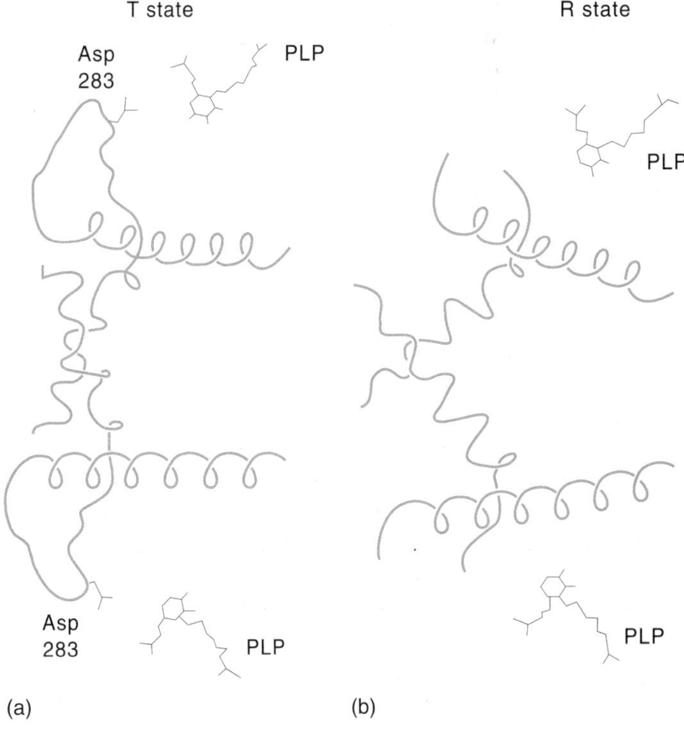

Figure 10.24

Residues in the region of the substrate-binding site for phosphate in phosphorylase *b* in (*a*) the T state and (*b*) the R state. The side chain of Asp 283 leaves the binding site in the R structure, and the side chain of Arg 569 becomes available to interact with the phosphate. Key side chains and bound phosphate anion are shown in red. Portions of the polypeptide backbone are drawn in heavy green. (Source: From D. Barford and L. N. Johnson, The allosteric transition of glycogen phosphorylase, *Nature* 340:609, 1989.)

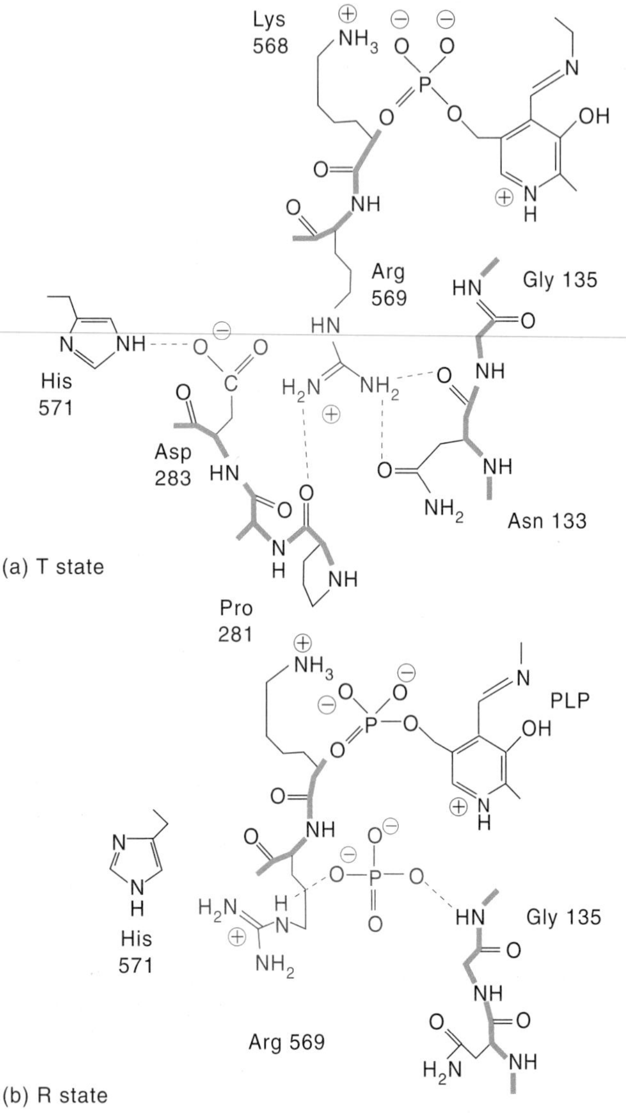

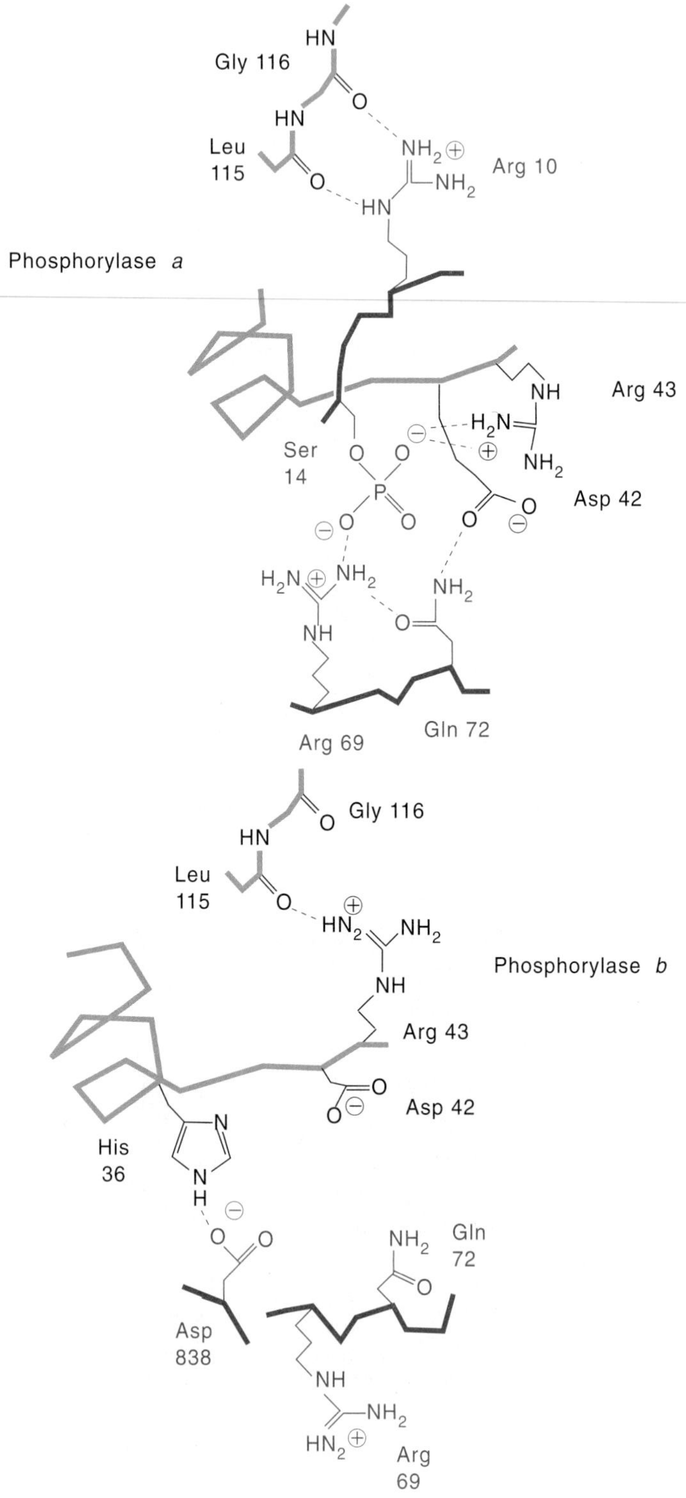

R form, the substrate P_i is bound to Arg 569, Gly 135, Lys 574 (not shown in the figure), and the pyridoxal phosphate. In the T form, the side chain of Asp 283 sits in this region, and Arg 569 is pulled away by hydrogen bonding to Pro 281. The replacement of the positively charged arginine side chain by the negatively charged aspartate explains why the affinity for P_i is much lower in the T conformation.

The transformation from phosphorylase *b* to phosphorylase *a* includes structural changes that tighten the interactions between the two subunits. Figures 10.21 and 10.25 show some of these changes. In phosphorylase *a* the phosphate group attached

Figure 10.25

Structural changes that accompany the conversion of phosphorylase *b* to phosphorylase *a*. This figure shows the interface between the two subunits in the region of Ser 14, the residue that gains a phosphate group in phosphorylase *a*. Portions of the polypeptide backbone and the side chains of some residues from one of the subunits are drawn in red; the other subunit is drawn in green. A larger number of hydrogen bonds link the two subunits in phosphorylase *a* (*top*) than in phosphorylase *b* (*bottom*). (Source: From S. R. Sprang et al., Structural changes in glycogen phosphorylase induced by phosphorylation, *Nature* 336:215, 1988.)

Figure 10.26

The structure of CaM in isolation and in a complex with the myosin light-chain kinase binding site. The carboxy terminus of CaM is shown in green, the amino terminal domain is shown in yellow, and the target is shown in red. From "Calmodulin-binding domains: just two faced or multi-faceted?" by Peter James, Thomas Vorherr and Ernesto Carafoli, *Trends in Biochemical Sciences,* © 1995, Elsevier Science Ltd., Oxford

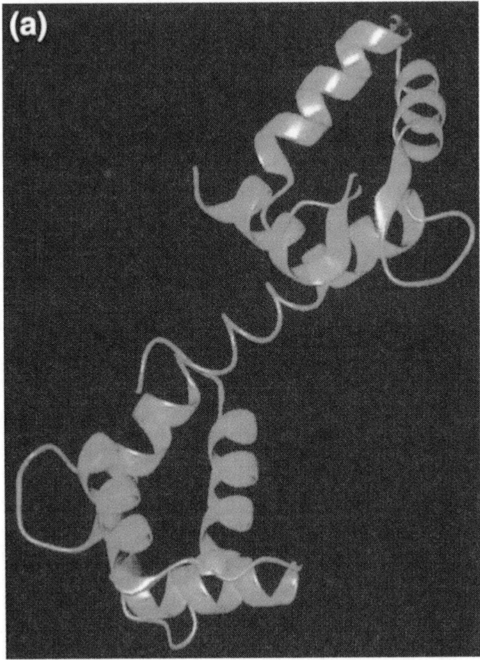

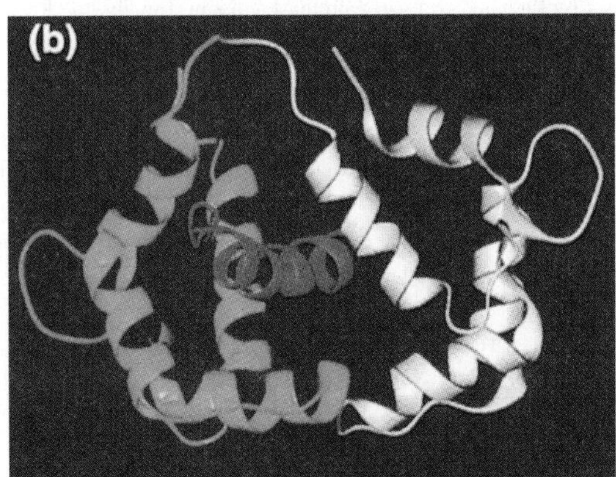

to each serine 14 is hydrogen-bonded to Arg 69 of its own subunit but also to Arg 43 of the other subunit. Hydrogen bonds also exist between Arg 10 and Leu 115 of different subunits and between Gln 72 and Asp 42. All of these intersubunit hydrogen bonds are missing in phosphorylase *b*. Instead, a bond exists between Arg 43 and a leucine in the same subunit, and one intersubunit hydrogen bond exists between His 36 and Asp 838. The N-terminal portion of the chain is ejected from this region in phosphorylase *b* and is replaced by residues from the C-terminal end, including Asp 838. At the nearby allosteric effector site the pulling together of the two subunits in phosphorylase *a* enhances the binding of AMP but disfavors the binding of the inhibitor glucose-6-phosphate.

Calmodulin Regulates Other Regulatory Proteins by Protein–Protein Interaction

Many regulatory proteins are activated by interaction with calmodulin (CaM). This list includes enzymes that stimulate cyclic nucleotide metabolism, phosphorylation, dephosphorylation, and Ca^{2+} transport itself.

Calcium levels in the cytosol are much lower than in the extracellular fluid or the lumen of the endoplasmic reticulum where it is concentrated. Specific pumps situated in the plasma membrane and the membrane of the endoplasmic reticulum can raise the Ca^{2+} concentration in the cytosol where CaM is found. The rise in Ca^{2+} concentration triggers CaM. In the relaxed state, CaM assumes a dumbbell shape with two globular domains arranged in a *trans* configuration (fig. 10.26). These domains are connected by a long extended central α-helix. Each of the globular domains has two binding sites for Ca^{2+}. The carboxy-terminal domain binds Ca^{2+} with a higher affinity ($K_d \cong 10^{-7}$ M) than the amino-terminal domain ($K_d \cong 10^{-5}$ M). When the free Ca^{2+} concentration is around 10^{-7} M only the carboxy domain is complexed with Ca^{2+} and the protein retains its dumbbell shape. If the Ca^{2+} concentration rises sufficiently, the amino-terminal domain also becomes complexed with Ca^{2+} and a gross conformational change takes place that exposes hydrophobic segments in each of the globular domains. The exposed hydrophobic segments in CaM have a high affinity for specific regions in Ca^{2+} target proteins. The structure of a complex of CaM bound to the synthetic CaM-binding domain of skeletal muscle light chain is believed to typify the way in which CaM binds to a target protein (see fig. 10.26*b*). It can be seen that the extended structure of CaM has collapsed around the target.

Binding of a target protein by CaM usually leads to its activation because the domain to which CaM binds functions as an autoinhibitor of the target protein. Many target proteins can also be activated by limited proteolysis or phosphorylation. The plasma membrane Ca^{2+} pump exemplifies all three modes of activation (fig. 10.27).

In vivo phosphorylation of rat liver CaM lowers the affinity for a number of target proteins. It is not yet known whether this phosphorylation of CaM is physiologically significant.

Figure 10.27

Diagram of the plasma membrane Ca^{2+} pump indicating three pathways for activation. The first route to activation involves removal of the autoinhibitory domain from the active site by Calmodulin (CaM) binding. The second route entails limited proteolysis by the Ca^{2+}-activated protease calpain. Finally, the third route to activation involves phosphorylation of the plasma membrane pump by Protein Kinase C (PKC).

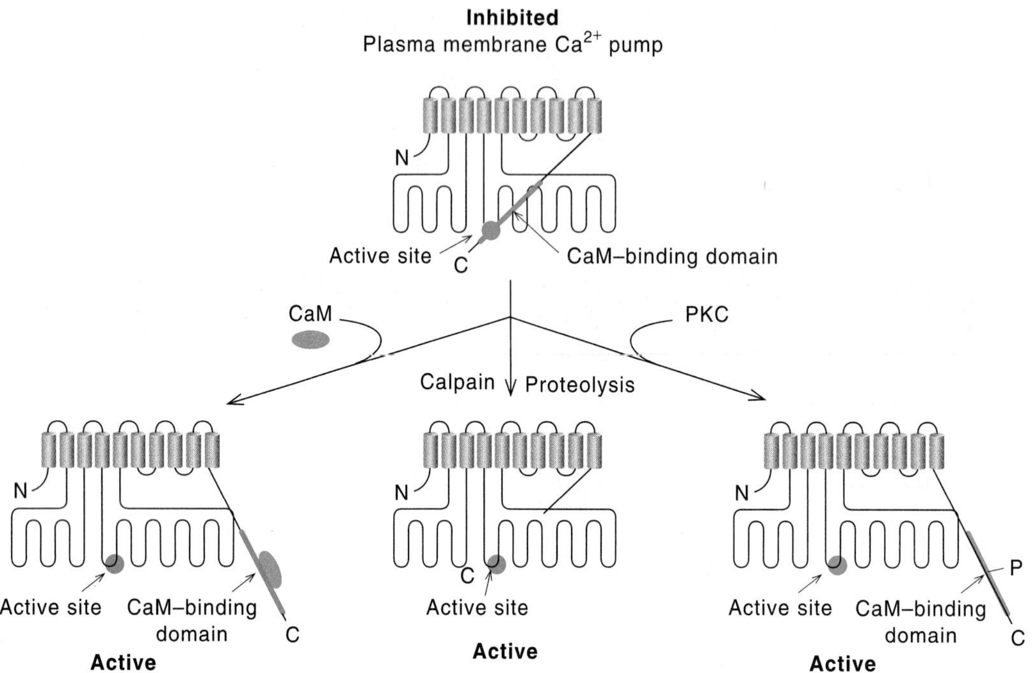

Summary

Cells regulate their metabolic activities by controlling rates of enzyme synthesis and degradation and by adjusting the activities of specific enzymes. Enzyme activities vary in response to changes in pH, temperature, and the concentrations of substrates or products but also can be controlled by covalent modifications of the protein or by interactions with activators or inhibitors.

1. Partial proteolysis, an irreversible process, is used to activate proteases and other digestive enzymes after their secretion and to switch on enzymes that cause blood coagulation. Common types of reversible covalent modification include phosphorylation, adenylylation, and disulfide reduction.
2. Allosteric effectors are inhibitors or activators that bind to enzymes at sites distinct from the active sites. Allosteric regulation allows cells to adjust enzyme activities rapidly and reversibly in response to changes in the concentrations of substances that are structurally unrelated to the substrates or products. The initial steps in a biosynthetic pathway commonly are inhibited by the end products of the pathway, and numerous enzymes are regulated by ATP, ADP, or AMP.
3. The kinetics of allosteric enzymes typically show a sigmoidal dependence on substrate concentration rather than the more common hyperbolic saturation curves. These enzymes usually have multiple subunits, and the sigmoidal kinetics can be ascribed to cooperative interactions of the subunits. Binding of substrate to one subunit changes the dissociation constant for substrate on another subunit. The extent of the cooperativity can be described by the Hill equation or by the "symmetry" model or the more general "sequential" model. The symmetry model postulates that the enzyme can exist in two conformations, T and R. It is assumed that the substrate binds more tightly to the R conformation than to the T conformation, that binding of the substrate or an allosteric effector changes equilibrium between these conformations, and that cooperative interactions make all of the subunits switch from one conformation to the other in a concerted manner. The symmetry model provides a useful conceptual framework that is consistent with the behavior of many allosteric enzymes.
4. Phosphofructokinase, the key regulatory enzyme of glycolysis, has four identical subunits. It exhibits sigmoidal kinetics with respect to fructose-6-phosphate and is inhibited by ATP and stimulated by ADP. In the transition between the R and T conformations the subunits rotate with respect to each other and there is a rearrangement of the binding site for fructose-6-phosphate, which is located at an interface between subunits. At another interface, antiparallel β strands of two subunits are hydrogen-bonded together in the T structure but are separated by a row of water molecules in the R structure. This structural

feature explains the cooperative nature of the conformational transition in all of the subunits: Intermediate structures probably would be less stable than either the R or the T structure.

5. Aspartate carbamoyl transferase, the first enzyme in the biosynthesis of pyrimidines, is inhibited allosterically by CTP, an end product of the pathway, and is stimulated by ATP. It has six identical catalytic (*c*) subunits and six regulatory (*r*) subunits. The *c* and *r* subunits can be separated by treating the enzyme with mercurials. Binding of a substrate analog to one of the *c* subunits causes the entire c_6r_6 complex to flip to the R conformation; binding of CTP to an *r* subunit favors the T conformation. Again, the substrate-binding sites are located at interfaces between subunits, and the T to R transition results in a rotation of the subunits and brings together components of the active site.
6. Glycogen phosphorylase breaks down glycogen to glucose-1-phosphate. It exists in two forms that differ by a covalent modification. Phosphorylase *a* is phosphorylated on a serine residue and is the more active form under cellular conditions. Phosphorylase *b*, which lacks the phosphate, is stimulated allosterically by AMP but inhibited by ATP. The enzyme has two identical subunits, with the binding site for allosteric effectors at the interface. In the T to R transition there is a repositioning of arginine and aspartate residues at the binding site for P_i. Conversion of phosphorylase *b* to phosphorylase *a* tightens the interactions between the subunits and favors the transition to the R form.
7. Calmodulin undergoes a major conformational change at elevated levels of Ca^{2+}. This change exposes hydrophobic regions that bind specifically to a number of different target proteins. As a rule the binding of CaM activates the target protein.

Selected Readings

Babu, Y. S., J. S. Sack, T. J. Greenough, C. E. Bugg, A. R. Means, and W. J. Cook, Three-dimensional structure of calmodulin. *Nature* 315:37, 1985.

Barford, D., S.-H. Hu, and L. N. Johnson, Structural mechanism for glycogen phosphorylase control by phosphorylation and AMP. *J. Mol. Biol.* 218:233, 1991.

Barford, D., and L. N. Johnson, The allosteric transition of glycogen phosphorylase. *Nature* 340:609, 1989.

Browner, M. F., and R. J. Fletterick, Phosphorylase: A biological transducer. *TIBS* 17:66–71, 1992.

da Silva, A. C. R., and F. C. Reinach, Calcium binding induces conformational changes in muscle regulatory proteins. *TIBS* 16:53–57, 1991.

Furie, B., and B. C. Furie, The molecular basis of blood coagulation. *Cell* 53:505, 1988.

James, P., T. Vorherr, and E. Carafoli, Calmodulin-binding domains: Just two faced or multi-faceted? *TIBS* 20:38–42, 1955. A short, excellent review on how calmodulin works.

Laskowski, M., Jr., and I. Kato, Protein inhibitors of proteinases. *Ann. Rev. Biochem.* 49:593, 1980.

Lipscomb, W. N., Structure and function of allosteric enzymes. *Chemtracts-Biochem. Mol. Biol.* 2:1, 1991.

Lipscomb, W. N., Aspartate transcarbamylase from *Escherichia coli*: Activity and regulation. *Adv. Enzymol.* 68:67–198, 1994.

Perutz, M. F., Mechanisms of cooperativity and allosteric regulation in proteins. *Quart. Revs. Biophys.* 22:139–51, 1989.

Schachman, H. R., Can a simple model account for the allosteric transition of aspartate transcarbamoylase? *J. Biol. Chem.* 263:18583, 1988.

Schirmer, T., and P. R. Evans, Structural basis of the allosteric behaviour of phosphofructokinase. *Nature* 343:140, 1990.

Sprang, S. R., K. R. Acharya, E. J. Goldsmith, D. I. Stuart, K. Varvill, R. J. Fletterick, N. B. Madsen, and L. N. Johnson, Structural changes in glycogen phosphorylase induced by phosphorylation. *Nature* 336:215, 1988.

Problems

1. Many hemophiliacs are sustained by regular injections of one of the cascade components, e.g., Factor VIII (fig. 10.2). Why doesn't the addition of Factor VIII cause an uncontrolled clotting of the blood in these individuals?
2. While reading this chapter, you should notice that protein regulation mechanisms fall into two major categories: covalent modifications and allosteric regulation. Does one approach seem to be inherently less complicated? Why?
3. Is the interaction between an allosteric effector and an allosteric enzyme always an equilibrium?
4. Frequently, the first enzyme in a metabolic pathway is regulated allosterically by the final product of the pathway. What is meant by the "first enzyme in the pathway"? Why is the first enzyme of a pathway often the enzyme that is regulated?
5. How could you determine whether an activation/inactivation involving thiols and disulfides utilized inter- or intra-protein disulfides?
6. Why do you think the researchers who initially prepared PALA (fig. 10.16) utilized a methylene between the carbonyl and phosphoryl group? Why didn't they "just" use a "regular" phosphate group?
7. Notice the salt bond in figure 10.1 between the β-carboxyl group of Asp 194 and the N-terminus (Ile 16) in the structure of chymotrypsin. Why is the N-terminus amino acid number 16?
8. What similarities or differences can you find in the covalent modification of proteins shown in figures 10.3 and 10.4?
9. Show by writing the reactions how the combination of protein kinase and the corresponding phosphoprotein phosphatase, if unregulated, theoretically forms a futile cycle.
10. The cAMP-dependent protein kinases phosphorylate specific Ser (Thr) residues on target proteins. Given the availability of serine and threonine residues on the surface of globular proteins, how might a protein kinase select the "correct" residues to phosphorylate?
11. Assume that the flow diagram shown represents an amino acid biosynthetic pathway where G, J, and H are amino acids and A is a common precursor. Products G, H, and J are required by the cell. Enzymes catalyzing the steps are numbered. Sug-

gest a plausible scheme for the regulation of specific enzymes by their products.

$$A \xrightarrow[1]{} B \xrightarrow[2]{} C \xrightarrow[3]{} D \xrightarrow[4]{} E \xrightarrow[5]{} F \xrightarrow{6} G$$

$$D \xrightarrow{7} I \xrightarrow{8} J$$

$$F \xrightarrow{9} H$$

12. Aspartate carbamoyltransferase is an allosteric enzyme in which the active sites and the allosteric effector binding sites are on different subunits. Explain how it might be possible for an allosteric enzyme to have both kinds of sites on the same subunit.
13. ATP is both a substrate and an inhibitor of the enzyme phosphofructokinase (PFK). Although the substrate fructose-6-phosphate binds cooperatively to the active site, ATP does not bind cooperatively. Explain how ATP may be both a substrate and an inhibitor of PFK.
14. Light-dependent activation of key enzymes in photosynthetic CO_2 fixation involves activation by thioredoxin-mediated reduction of critical disulfides on the enzymes. Write a reaction linking the reductant ferredoxin (a single-electron donor) to the reduction of protein disulfides using thioredoxin as an intermediate.
15. If you separated hemoglobin into dimers of $(\alpha + \beta)$ subunits, would you expect the dimers to bind more or less O_2 at low O_2 tensions? Explain. What effect would 2,3-bisphosphoglycerate have on oxygen binding? (Review O_2 binding to hemoglobin in chapter 6.)

C H A P T E R

11

Vitamins and Coenzymes

Coenzymes are small organic molecules that act in concert with the enzyme to catalyze biochemical reactions.

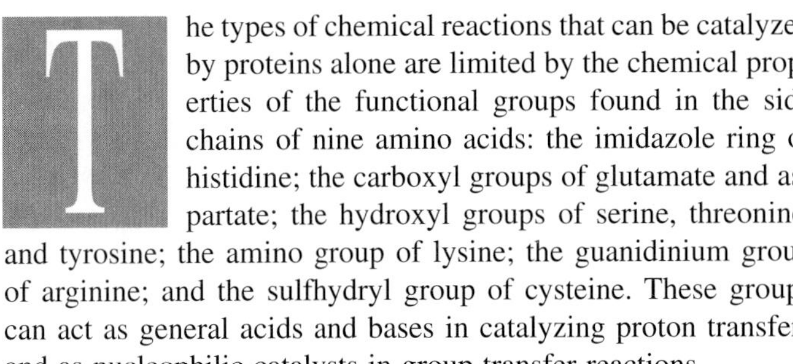

The types of chemical reactions that can be catalyzed by proteins alone are limited by the chemical properties of the functional groups found in the side chains of nine amino acids: the imidazole ring of histidine; the carboxyl groups of glutamate and aspartate; the hydroxyl groups of serine, threonine, and tyrosine; the amino group of lysine; the guanidinium group of arginine; and the sulfhydryl group of cysteine. These groups can act as general acids and bases in catalyzing proton transfers and as nucleophilic catalysts in group transfer reactions.

Many metabolic reactions involve chemical changes that cannot be brought about by the structures of the amino acid side chain functional groups in enzymes acting by themselves. In catalyzing these reactions, enzymes act in cooperation with other smaller organic molecules or metallic cations, which possess special chemical reactivities or structural properties that are useful for catalyzing reactions. In this chapter we introduce these small molecules and survey the range of reactions they catalyze.

It has been known since the middle of the nineteenth century that small amounts of certain substances are very important to healthy nutrition. Vitamins are organic molecules that are essential in small quantities for healthy nutrition in rats or humans. The list of such molecules grew as they were purified from foodstuffs and shown to cure various disorders in animals maintained on deficient diets. The name "vitamine" was given in 1911 to the first vitamin to be isolated, thiamine. When it became clear that a number of essential organic micronutrients were not amines, the *-e* was dropped.

Vitamins are divided into water-soluble and lipid-soluble groups. In addition to vitamins, vitaminlike nutrients are required in small amounts by the organism and frequently function in similar capacities to vitamins. These vitaminlike compounds are not classified as vitamins because rats and humans have a limited capacity to synthesize them, provided that the diet contains the essential precursors. Table 11.1 lists both vitamins and vitaminlike nutrients.

In this chapter we are primarily concerned with coenzymes, which are frequently modified forms of vitamins. The modifications take place in the organism after ingestion of the vitamins. Coenzymes act in concert with enzymes to catalyze

Table 11.1
Vitamins and Vitaminlike Nutrients

Vitamin	Function
Water-Soluble Vitamins	
Thiamine (B_1)	Precursor of the coenzyme thiamine pyrophosphate. Deficiency can cause beriberi.
Riboflavin (B_2)	Precursor of the coenzymes flavin mononucleotide and flavin adenine dinucleotide. Deficiency leads to growth retardation.
Pyridoxine (B_6)	Precursor of the coenzyme pyridoxal phosphate. Deficiency causes dermatitis in rats.
Nicotinic acid (niacin)	Precursor of the coenzymes nicotinamide adenine dinucleotide and nicotinamide adenine dinucleotide phosphate. Deficiency leads to pellagra.
Pantothenic acid	Precursor of coenzyme A (CoA). Deficiency leads to dermatitis in chickens.
Biotin	Precursor of the coenzyme biocytin. Deficiency leads to dermatitis in humans.
Folic acid	Precursor of the coenzyme tetrahydrofolic acid. Deficiency causes anemias.
Vitamin B_{12}	Precursor of the coenzyme deoxyadenosyl cobalamin. Deficiency leads to pernicious anemia.
Vitamin C	Cosubstrate in the hydroxylation of proline in collagen. Deficiency leads to scurvy.
Lipid-Soluble Vitamins	
Vitamin A	Vision, growth, and reproduction (chapter 29).
Vitamin D	Regulation of calcium and phosphate metabolism (see chapter 27).
Vitamin E	Antisterility factor in rats.
Vitamin K	Important for blood coagulation.
Vitaminlike Nutrients	
Inositol	Mediator of hormone action (see chapters 22 and 27).
Choline	Important for integrity of cell membranes and lipid transport (see chapter 22).
Carnitine	Essential for transfer of fatty acids to mitochondria (see chapter 21).
α-Lipoic acid	Coenzyme in the oxidative decarboxylation of keto acids (this chapter).
p-Aminobenzoate (PABA)	Component of folic acid (this chapter).
Coenzyme Q (ubiquinones)	Important for electron transport in mitochondria (see chapter 16).

biochemical reactions. Tightly bound coenzymes are often referred to as prosthetic groups. A coenzyme usually functions as a major component of the active site on the enzyme, which means that understanding the mechanism of coenzyme action usually requires an understanding of the catalytic process.

Because of the large amount of information in this chapter, the student may wish to read this chapter in snatches. To facilitate this, cross-references are given in later chapters.

Water-Soluble Vitamins and Their Coenzymes

As we have noted, some vitamins are soluble in water, and others are soluble in lipids. In the following sections we survey the range of biochemical reactions in which water-soluble coenzymes participate.

Thiamine Pyrophosphate Is Involved in C—C and C—X Bond Cleavage

The structure of thiamine pyrophosphate (TPP) is given in figure 11.1*a*. The vitamin thiamine, or vitamin B_1, lacks the pyrophos-

Figure 11.1

(*a*) Structure of thiamine pyrophosphate and (*b, c*) the bonds it cleaves or forms. Reactive part of the coenzyme and the bonds subject to cleavage in (*b*) and (*c*) are indicated in red.

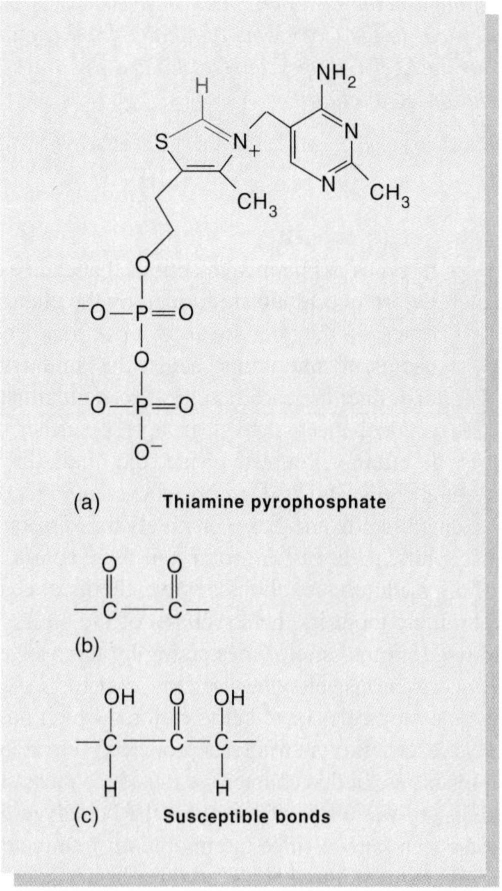

Figure 11.2

Mechanism of thiamine pyrophosphate action. Intermediate (*a*) is represented as a resonance-stabilized species. It arises from the decarboxylation of the pyruvate-thiamine pyrophosphate addition compound shown at left of (*a*) and in equation (2). It can react as a carbanion with acetaldehyde, pyruvate, or H^+ to form (*b*), (*c*), or (*d*), depending on the specificity of the enzyme. It can also be oxidized to acetyl-thiamine pyrophosphate (*e*) by other enzymes, such as pyruvate oxidase. The intermediates (*b*) through (*e*) are further transformed to the products shown by the actions of specific enzymes.

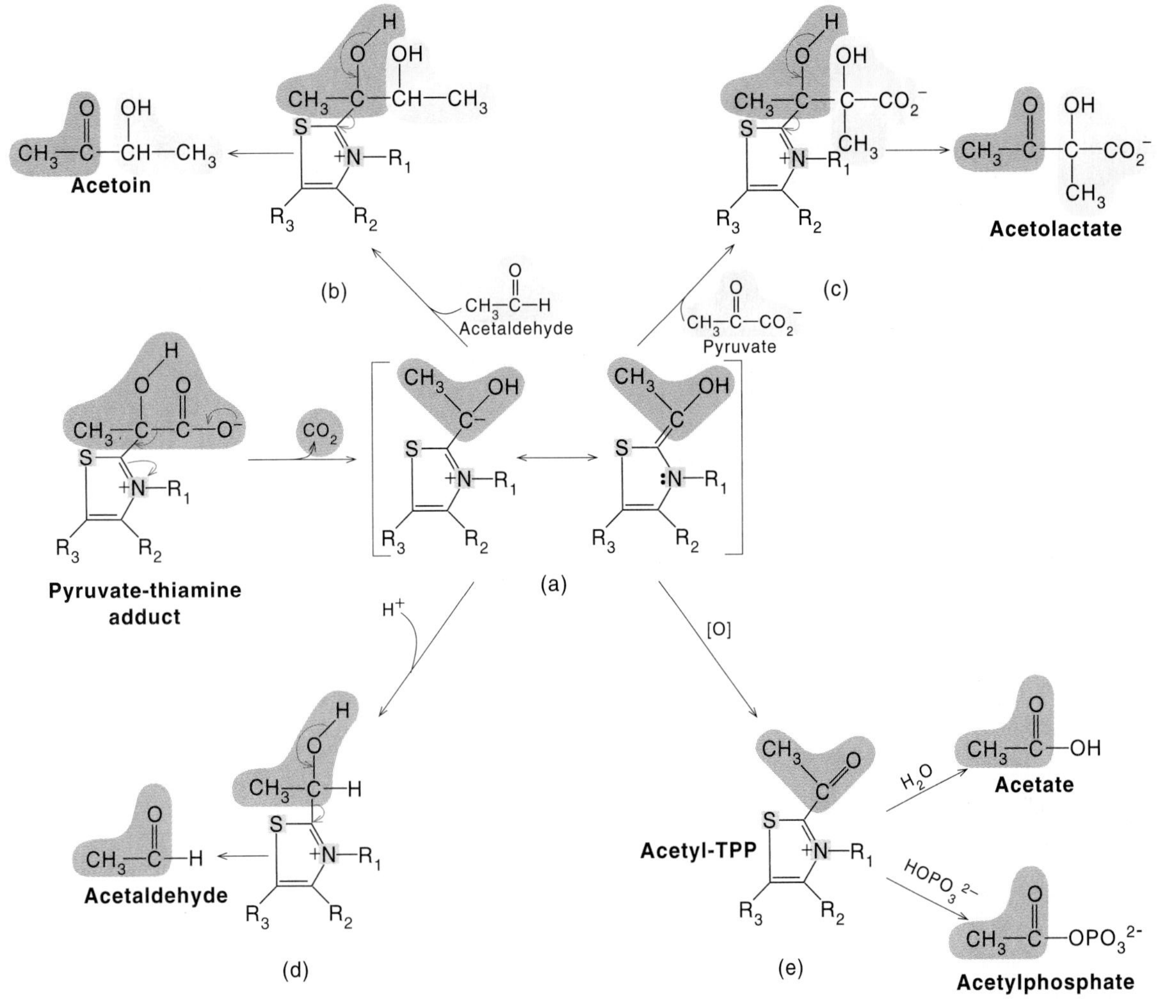

phoryl group. Thiamine pyrophosphate is the essential coenzyme involved in the actions of enzymes that catalyze cleavage of the bonds indicated in red in figure 11.1. The bond scission in figure 11.1*b* is representative of those in many α-keto acid decarboxylations, nearly all of which require the action of TPP. The phosphoketolase reaction involves both cleavages shown in figure 11.1*c*, whereas the transketolase reaction (see fig. 14.24) involves the cleavage of the carbon–carbon bond but not the elimination of —OH. Acetolactate and acetoin arise by the formation of the carbon–carbon bond in figure 11.1*c*. (The structures of these two compounds are shown in figure 11.2.)

The mechanism for the bond cleavages indicated in figure 11.1*b* was elucidated by Ronald Breslow. In one of the earliest applications of nuclear magnetic resonance to biochemical mechanisms, he demonstrated that the proton bonded to C-2 in the thiazolium ring is readily exchangeable with the protons of H_2O and deuterons of D_2O in a base-catalyzed reaction:

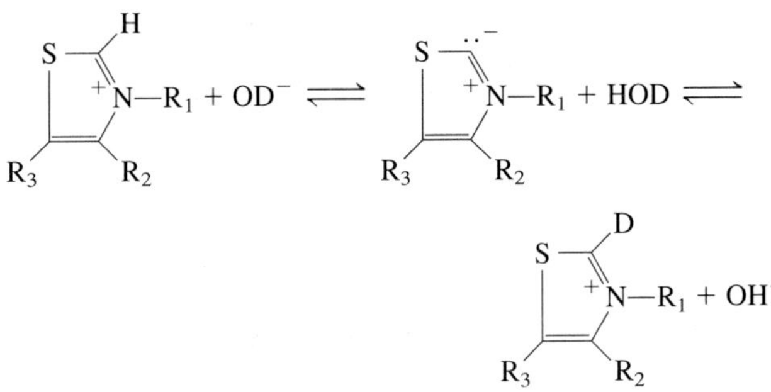

$$\text{(C-2–H thiazolium)}\ {}^{+}\text{N—R}_1 + \text{OD}^- \rightleftharpoons \text{(C-2 carbanion)}\ {}^{+}\text{N—R}_1 + \text{HOD} \rightleftharpoons \text{(C-2–D thiazolium)}\ {}^{+}\text{N—R}_1 + \text{OH}^- \quad (1)$$

Figure 11.3

Structures of vitamin B_6 derivatives and the bonds cleaved or formed by the action of pyridoxal phosphate (*a*). The reactive part of the coenzyme is shown in red in (*a*). The bonds shown in red in (*d*) are the types of bonds in substrates that are subject to cleavage.

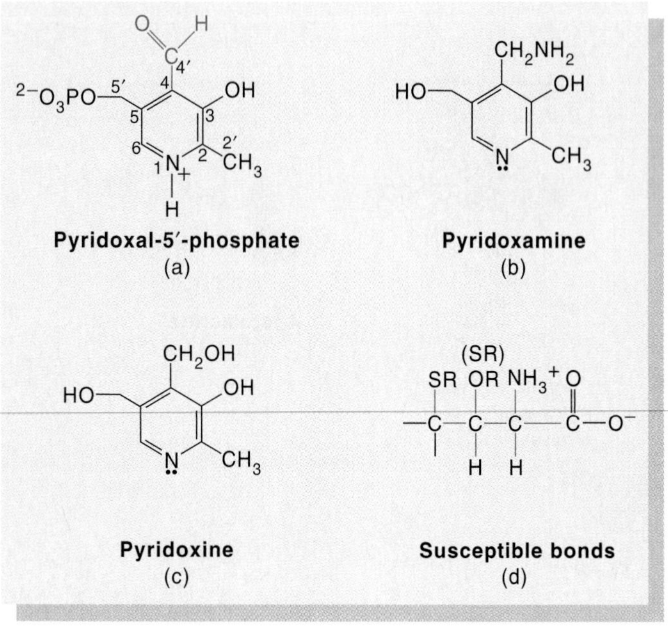

Pyridoxal-5′-phosphate
(a)

Pyridoxamine
(b)

Pyridoxine
(c)

Susceptible bonds
(d)

The active intermediate shown in equation (1) undergoes nucleophilic addition to the bond of polar carbonyl groups in substrates to produce intermediates, such as

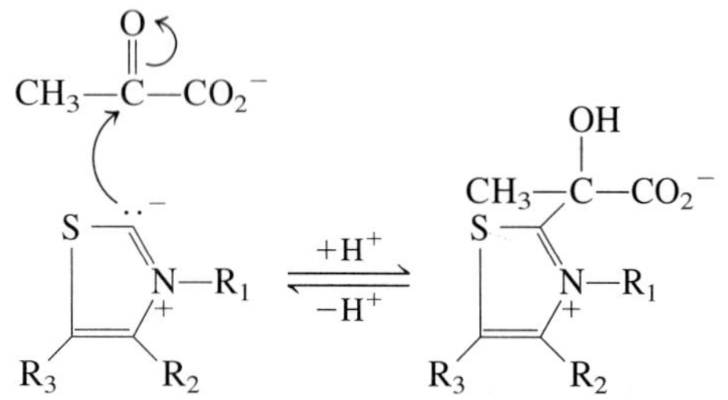

$$\mathrm{CH_3{-}CO{-}CO_2^-} + \text{thiazolium ylide } (\mathrm{S, N{-}R_1, R_2, R_3}) \underset{-\mathrm{H^+}}{\overset{+\mathrm{H^+}}{\rightleftharpoons}} \mathrm{CH_3{-}C(OH)(CO_2^-){-}}\text{thiazolium } (\mathrm{S, N^+{-}R_1, R_2, R_3}) \qquad (2)$$

Intermediates of this type have the necessary chemical reactivity for cleaving the bonds indicated in figure 11.1*b* and *c*. The decarboxylated product of the pyruvate adduct shown in equation (2) is resonance-stabilized by the thiazolium ring (fig. 11.2*a*). This intermediate may be protonated to α-hydroxyethyl thiamine pyrophosphate (fig. 11.2*d*); alternatively, it may react with other electrophiles, such as the carbonyl groups of acetaldehyde or pyruvate, to form the species in figure 11.2*b* and *c*; or it may be oxidized to acetyl-thiamine pyrophosphate (fig. 11.2*e*). The fate of the intermediate depends on the reaction specificity of the enzyme with which the coenzyme is associated.

Pyridoxal-5′-Phosphate Is Required for a Variety of Reactions with α-Amino Acids

Pyridoxal-5′-phosphate is the coenzyme form of vitamin B_6 and has the structure shown in figure 11.3. The name vitamin B_6 is applied to any of a group of related compounds lacking the phosphoryl group, including pyridoxal, pyridoxamine, and pyridoxine.

Pyridoxal-5′-phosphate participates in many reactions with α-amino acids, including transaminations, α decarboxylations, racemizations, α,β eliminations, β,γ eliminations, aldolizations, and the β decarboxylation of aspartic acid. The following equations illustrate several reactions in which pyridoxal-5′-phosphate acts as a coenzyme:

$$\mathrm{R_1{-}CH(NH_3^+){-}CO_2^-} + \mathrm{R_2{-}C(=O){-}CO_2^-} \xrightleftharpoons{\text{Transaminase}} \mathrm{R_1{-}C(=O){-}CO_2^-} + \mathrm{R_2{-}CH(NH_3^+){-}CO_2^-} \qquad (3)$$

$$\mathrm{R{-}CH(NH_3^+){-}CO_2^-} + \mathrm{H^+} \xrightleftharpoons{\text{Decarboxylase}} \mathrm{CO_2} + \mathrm{R{-}CH_2{-}NH_3^+} \qquad (4)$$

$$\mathrm{R{-}C(H)(NH_3^+){-}CO_2^-} \xrightleftharpoons{\text{Racemase}} \mathrm{R{-}C(NH_3^+)(H){-}CO_2^-} \qquad (5)$$

$$\mathrm{R{-}CH(OH){-}CH(NH_3^+){-}CO_2^-} \xrightleftharpoons{\text{Aldolase}} \mathrm{R{-}CHO} + \mathrm{CH_2(NH_3^+){-}CO_2^-} \qquad (6)$$

$$\mathrm{{}^-O_2C{-}CH_2{-}CH(NH_3^+){-}CO_2^-} + \mathrm{H^+} \xrightleftharpoons{\text{Aspartate-}\beta\text{-decarboxylase}} \mathrm{CO_2} + \mathrm{CH_3{-}CH(NH_3^+){-}CO_2^-} \qquad (7)$$

These equations involve bond cleavages of the type shown in color in figure 11.3*d*. Pyridoxal-5′-phosphate promotes these heterolytic bond cleavages by stabilizing the resulting electron pairs at the α- or β-carbon atoms of α-amino acids. To do this, the aldehyde group of the coenzyme first reacts with the α-amino group of an amino acid to produce an aldimine (fig. 11.4*a*) or Schiff's base, which is internally stabilized by H bonding. Loss of the α hydrogen as H^+ produces a resonance-stabilized species (fig. 11.4*b*) in which the electron pair is delocalized into the pyridinium system. This active intermediate may undergo further reactions at the carbon to form products determined by the reaction specificity of the enzyme. If, for example, the enzyme is a racemase, the species resulting from the loss of the proton from the α carbon may accept a proton from the opposite side to produce, ultimately, the enantiomer of the amino acid.

When the substrate is substituted at the β carbon with a potential leaving group, such as —OH, —SH, or $\mathrm{—OPO_3^{3-}}$ (see fig. 11.3*d*), the corresponding α-carbanion intermediate (see fig. 11.4*b*) can eliminate the group. This is an essential step in α,β

Figure 11.4

Structures of catalytic intermediates in pyridoxal-phosphate–dependent reactions. The initial aldimine intermediate resulting from Schiff's base formation between the coenzyme and the α-amino group of an amino acid (*a*). This aldimine is converted to the resonance-stabilized intermediate (*b*) by loss of a proton at the α carbon. Further enzyme-catalyzed proton transfers to intermediates (*c*) and (*d*) may occur, depending on the specificity of a given enzyme. The enzymes use their general acids and bases to catalyze these proton transfers. Reactive protons are colored to facilitate tracking.

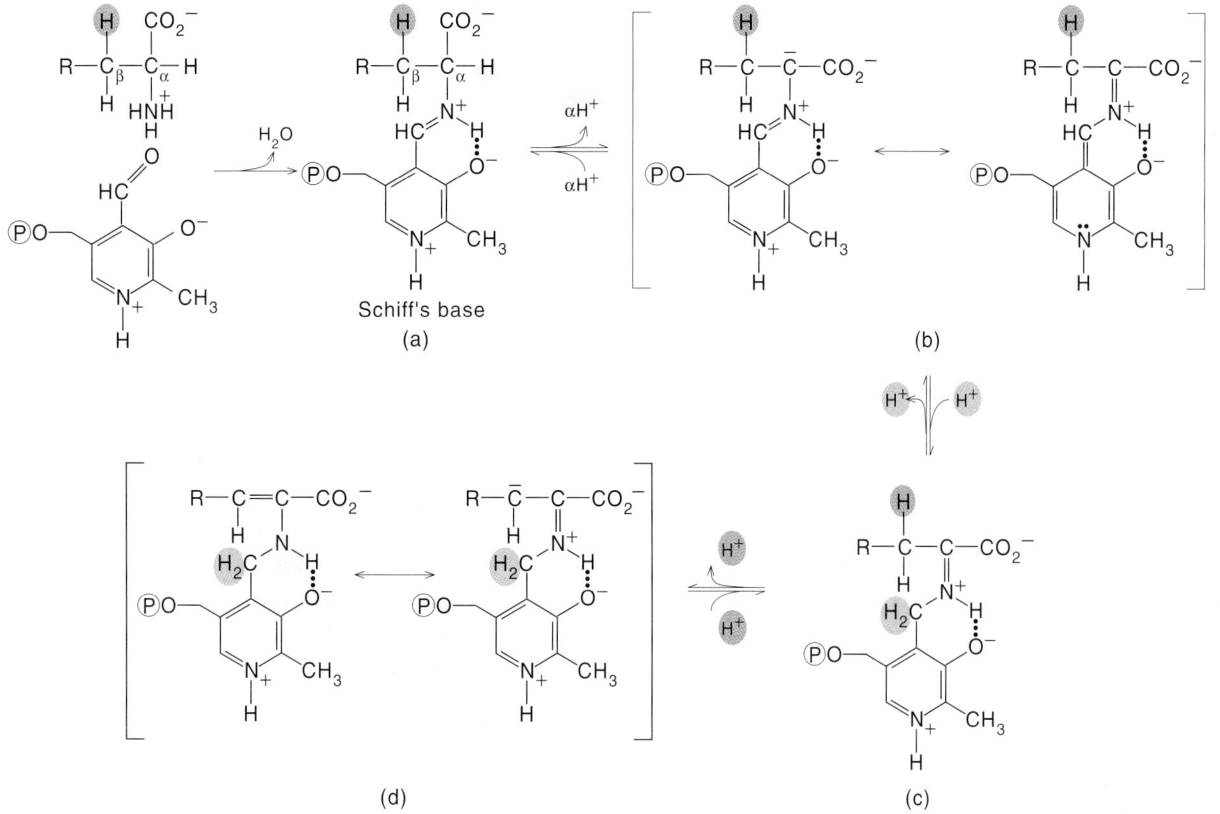

eliminations. Upon hydrolysis the elimination intermediate produces pyridoxal-5′-phosphate and the substrate-derived enamine, which spontaneously hydrolyzes to ammonia and an α-keto acid.

The full series of intermediates in a transamination is shown in figure 11.5*a*. After protonation at the aldimine carbon of pyridoxal-5′-phosphate (step 3), hydrolysis (step 4) forms an α-keto acid and pyridoxamine-5′-phosphate. The reverse of this sequence with a second α-keto acid (steps 5 through 8) completes the transamination reaction.

An intermediate analogous to that in figure 11.4*b* but generated from glycine and so lacking the β and γ carbons, can react as a carbanion with an aldehyde to produce a β-hydroxy-α-amino acid. These reactions are catalyzed by aldolases, such as threonine aldolase or serine hydroxymethyl transferase (see box 25A, fig. 1).

β-Decarboxylases (fig. 11.5*b*) generate β-carbanionic intermediates analogous to those in figure 11.4*d* by catalyzing the elimination of CO_2 instead of H^+ from the intermediate in figure 11.4*a* (step 4, fig. 11.5*b*). Protonation of the β-carbanionic intermediates by protons from H_2O, followed by hydrolysis of the resulting imines, produces the amines corresponding to the replacement of the carboxylate group in the substrate by a proton (steps 5 through 8, fig. 11.5*b*).

The stability of the resonance hybrid (see fig. 11.4*b*) accounts for the catalytic effectiveness of pyridoxal-5′-phosphate in the reactions shown in equations (3) through (6).

The β decarboxylation of aspartate (equation 7) proceeds by elimination of a β-carbanionic intermediate like that in figure 11.4*d* from the ketimine, analogous to the intermediate produced by loss of the α proton from the aldimine of aspartate with pyridoxal-5′-phosphate.

The fundamental biochemical function of pyridoxal-5′-phosphate is the formation of aldimines with α-amino acids that stabilize the development of carbanionic character at the α and β

Figure 11.5

Mechanisms of action of pyridoxal phosphate: (*a*) in glutamate-oxaloacetate transaminase and (*b*) in aspartate β-decarboxylase.

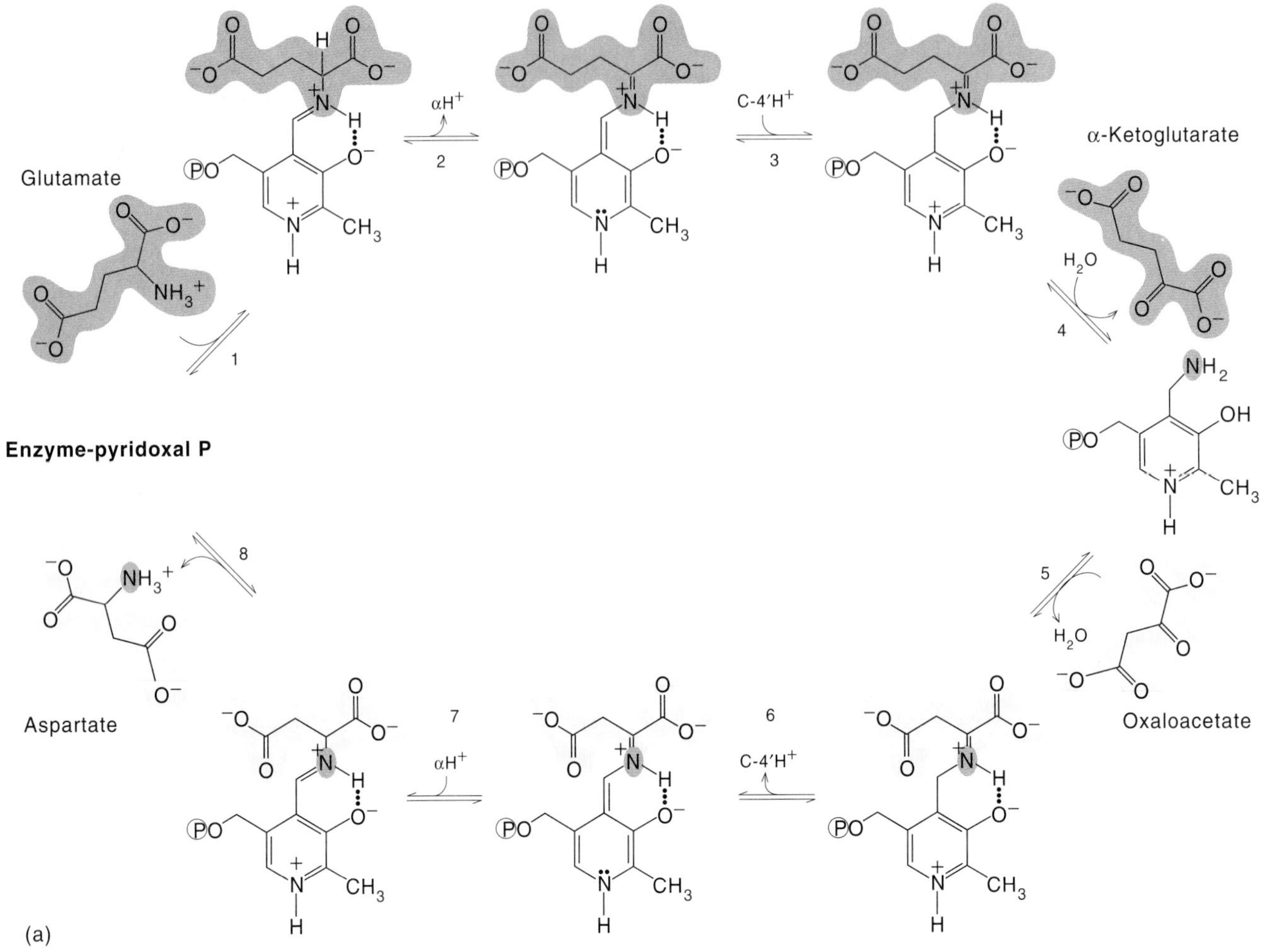

carbons of α-amino acids in intermediates, such as those in figures 11.4*b* and *c*. Enzymes acting alone cannot stabilize these carbanions and so cannot, by themselves, catalyze reactions requiring their formation as intermediates.

Nicotinamide Coenzymes Are Used in Reactions Involving Hydride Transfers

Nicotinamide adenine dinucleotide (NAD^+) is one of the two coenzymatic forms of nicotinamide (fig. 11.6). The other is nicotinamide adenine dinucleotide phosphate ($NADP^+$), which differs from NAD^+ by the presence of a phosphate group at C-2′ of the adenosyl moiety.

The nicotinamide coenzymes are biological carriers of reducing equivalents (electrons). The most common function of NAD^+ is to accept two electrons and a proton (H^- equivalent) from a substrate undergoing metabolic oxidation to produce NADH, the reduced form of the coenzyme. This then diffuses or is transported to the terminal-electron transfer sites of the cell and reoxidized by terminal-electron acceptors, O_2 in aerobic organisms, with the concomitant formation of ATP (chapter 15). Equations (8), (9), and (10) are typical reactions in which NAD^+ acts as such an acceptor.

$$NAD^+ + CH_3CH_2OH \xrightleftharpoons{\text{Alcohol dehydrogenase}} CH_3{-}\overset{\overset{\large O}{\|}}{C}H + NADH + H^+ \quad (8)$$

$$^-O_2C(CH_2)_2\overset{\overset{NH_3^+}{|}}{C}HCO_2^- + NAD^+ + H_2O \xrightleftharpoons{\text{Glutamate dehydrogenase}} {}^-O_2C(CH_2)_2\overset{\overset{\large O}{\|}}{C}CO_2^- + NADH + NH_4^+ + H^+ \quad (9)$$

$$HPO_4^{2-} + {}^{2-}O_3POCH_2\overset{\overset{OH}{|}}{C}H-\overset{\overset{\large O}{\|}}{C}H + NAD^+ \xrightleftharpoons{\text{Glyceraldehyde-3P dehydrogenase}} {}^{2-}O_3POCH_2\overset{\overset{OH}{|}}{C}H\overset{\overset{\large O}{\|}}{C}OPO_3^{2-} + NADH + H^+ \quad (10)$$

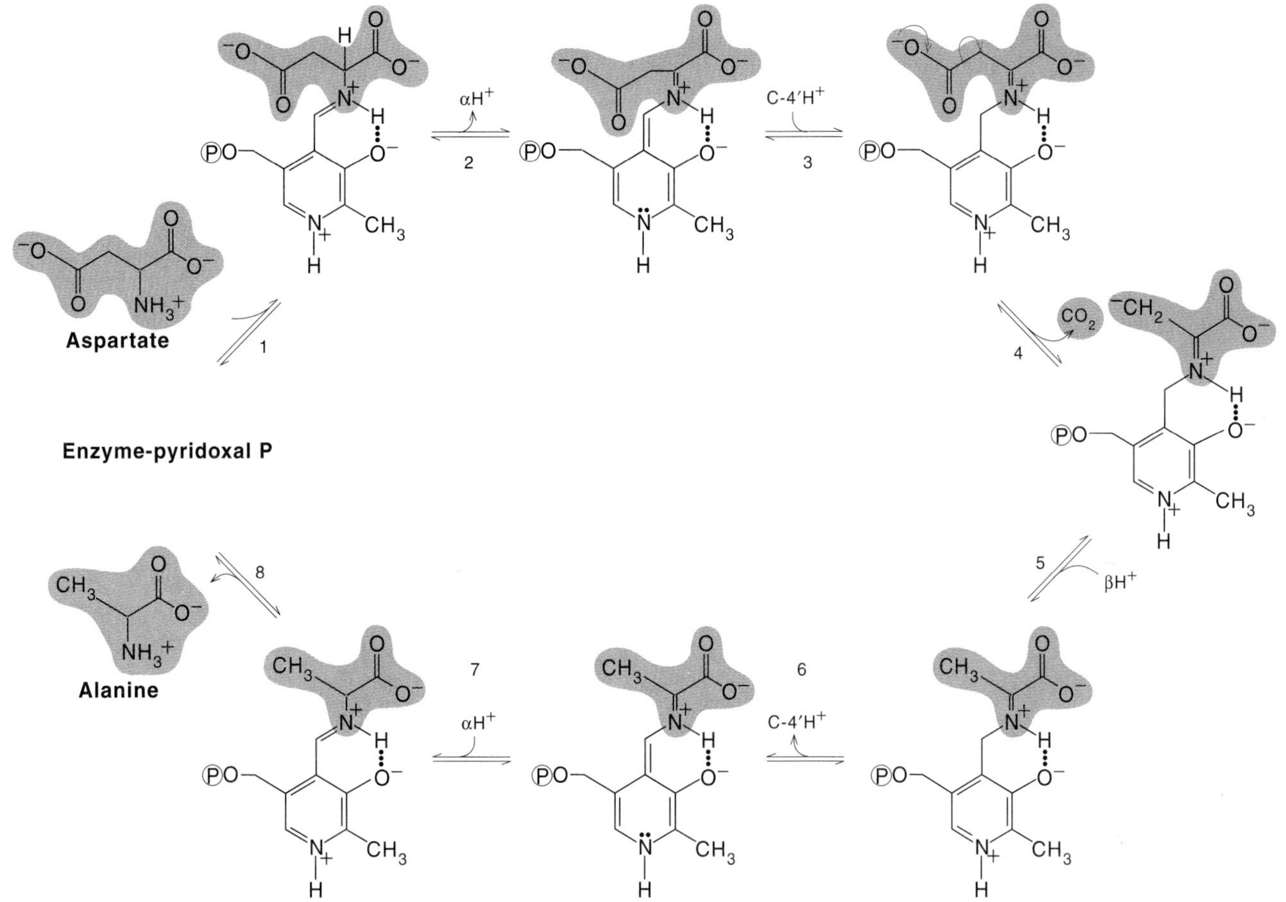

(b)

Figure 11.6

Structures of nicotinamide and nicotinamide coenzymes. The reactive sites of the coenzymes are shown in red.

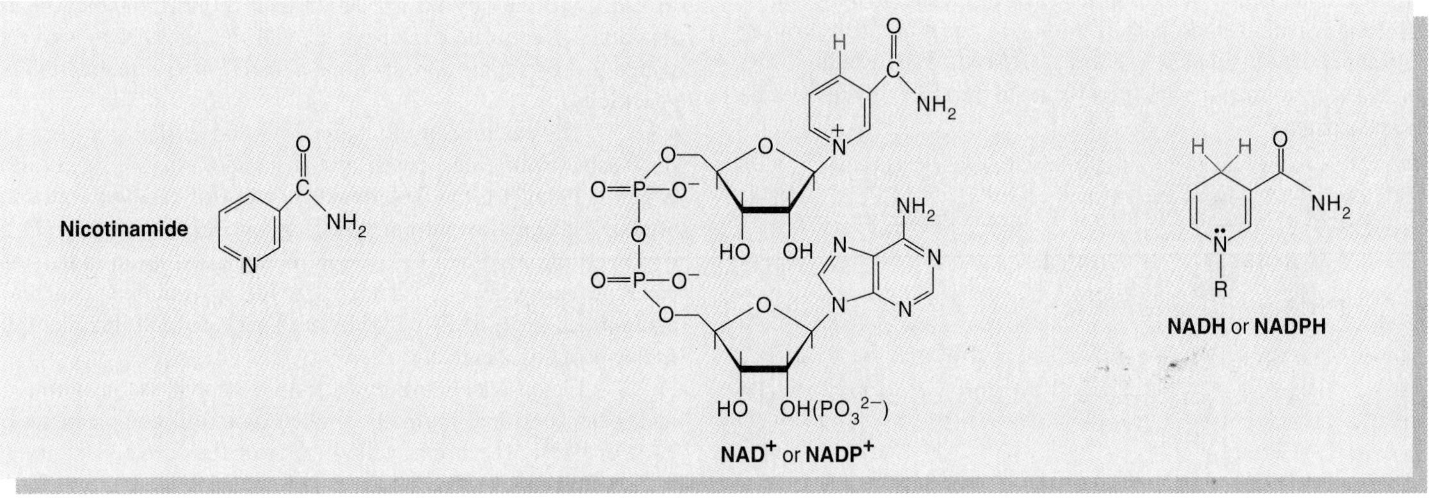

Figure 11.7

Mechanism of NAD^+ action in UDP-galactose-4-epimerase. No net oxidation or reduction occurs. Only the intermediate is oxidized.

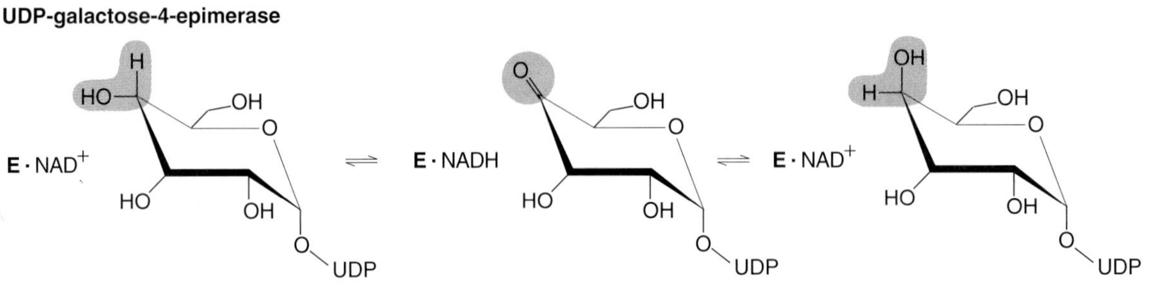

UDP-galactose-4-epimerase

$$\text{E}\cdot\text{NAD}^+ \ (\text{UDP-galactose}) \rightleftharpoons \text{E}\cdot\text{NADH}\ (\text{4-keto intermediate}) \rightleftharpoons \text{E}\cdot\text{NAD}^+\ (\text{UDP-glucose})$$

The chemical mechanisms by which NAD^+ is reduced to NADH in equations (8) through (10) are probably similar, as represented in generalized forms in equation (11):

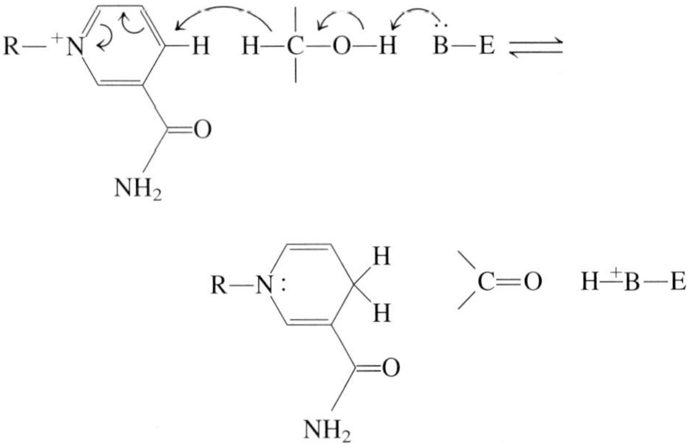

$$\text{R}{-}^{+}\text{N}(\text{pyridinium, } C{=}O\text{-}NH_2)\text{—H} \quad \text{H—}\overset{|}{\underset{|}{\text{C}}}\text{—O—H} \quad \text{B—E} \rightleftharpoons \text{R—N}{:}(\text{dihydropyridine, } \text{H}, \text{H}, C{=}O\text{-}NH_2) \quad {>}\text{C=O} \quad \text{H—}^{+}\text{B—E} \qquad \textbf{(11)}$$

According to this formulation, the immediate oxidation product in equation (9), where $—NH_2$ replaces —OH in equation (11), is the imine of α-ketoglutarate, which quickly undergoes hydrolysis to α-ketoglutarate and ammonia in aqueous solution. The oxidation of an aldehyde group catalyzed by glyceraldehyde-3-phosphate dehydrogenase [equation (10)] also can be understood on the basis of this formulation once it is appreciated that an essential —SH group at the active site is transiently acylated during the course of the reaction. The —SH group reacts with the aldehyde group of glyceraldehyde-3-phosphate according to equation (12), forming a thiohemiacetal, which becomes oxidized. The resulting acyl enzyme then reacts with phosphate to produce glycerate-1,3-bisphosphate:

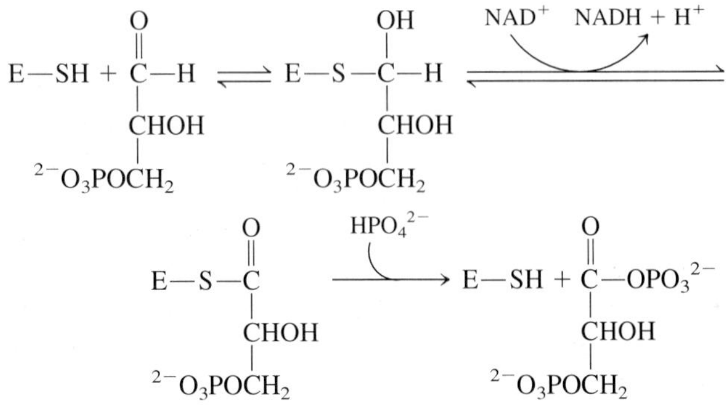

$$\text{E—SH} + \underset{\underset{^{2-}O_3POCH_2}{|}}{\underset{\text{CHOH}}{|}}{\overset{O}{\overset{\|}{\text{C}}}}\text{—H} \rightleftharpoons \text{E—S—}\underset{\underset{^{2-}O_3POCH_2}{|}}{\underset{\text{CHOH}}{|}}{\overset{OH}{\overset{|}{\text{C}}}}\text{—H} \xrightarrow{NAD^+ \;\to\; NADH + H^+}$$

$$\text{E—S—}\underset{\underset{^{2-}O_3POCH_2}{|}}{\underset{\text{CHOH}}{|}}{\overset{O}{\overset{\|}{\text{C}}}} \xrightarrow{HPO_4^{2-}} \text{E—SH} + \underset{\underset{^{2-}O_3POCH_2}{|}}{\underset{\text{CHOH}}{|}}{\overset{O}{\overset{\|}{\text{C}}}}\text{—}OPO_3^{2-} \qquad \textbf{(12)}$$

In addition to acting as a cellular electron carrier, NAD^+ also acts as a true coenzyme with certain enzymes. Enzymes are sometimes confronted with the problem of catalyzing such reactions as epimerizations, aldolizations, and eliminations on substrates lacking the intrinsic chemical reactivities required for these reactions to occur at significant rates. Sometimes such reactivities can be introduced into the substrate by oxidizing an appropriate alcohol group to a carbonyl group, and the enzyme is then found to contain NAD^+ as a tightly bound coenzyme. NAD^+ functions coenzymatically by transiently oxidizing the key alcohol group to the carbonyl level, producing an oxidatively activated intermediate, the further transformation of which is catalyzed by the enzyme. In the last step, the carbonyl group is reduced back to the hydroxyl group by the transiently formed NADH. A reaction of this type is illustrated in figure 11.7 for the enzyme UDP-galactose-4-epimerase, which contains tightly bound NAD^+.

Flavins Are Used in Reactions Involving One or Two Electron Transfers

Flavin adenine dinucleotide (FAD) (fig. 11.8) and flavin mononucleotide (FMN) are the coenzymatically active forms of vitamin B_2, riboflavin. Riboflavin is the N^{10}-ribityl isoalloxazine portion of FAD, which is enzymatically converted into its coenzymatic forms first by phosphorylation of the ribityl C-5′ hydroxy group to FMN and then by adenylylation to FAD. FMN and FAD are functionally equivalent coenzymes, and the one that is involved with a given enzyme appears to be a matter of enzymatic binding specificity.

The catalytically functional portion of the coenzymes is the isoalloxazine ring, specifically N-5 and C-4a (see fig. 11.8*b*), which is thought to be the immediate locus of catalytic function, although the entire chromophoric system extending over N-5, C-4a, C-10a, N-1, and C-2 should be regarded as an indivisible catalytic entity, as are the nicotinamide, pyridinium, and thioazolium rings of NAD^+, pyridoxal phosphate, and thiamine pyrophosphate, respectively.

Flavin-containing enzymes are known as flavoproteins and, when purified, normally contain their full complements of FAD or FMN. The bright yellow color of flavoproteins is due to the isoalloxazine chromophore in its oxidized form. In a few flavo-

Figure 11.8

Structures of the vitamin riboflavin (*a*) and the derived flavin coenzymes (*b*). Like NAD^+ and $NADP^+$, the coenzyme pair FMN and FAD are functionally equivalent coenzymes, and the coenzyme involved with a given enzyme appears to be a matter of enzymatic binding specificity. The catalytically functional portion of the coenzymes is shown in red.

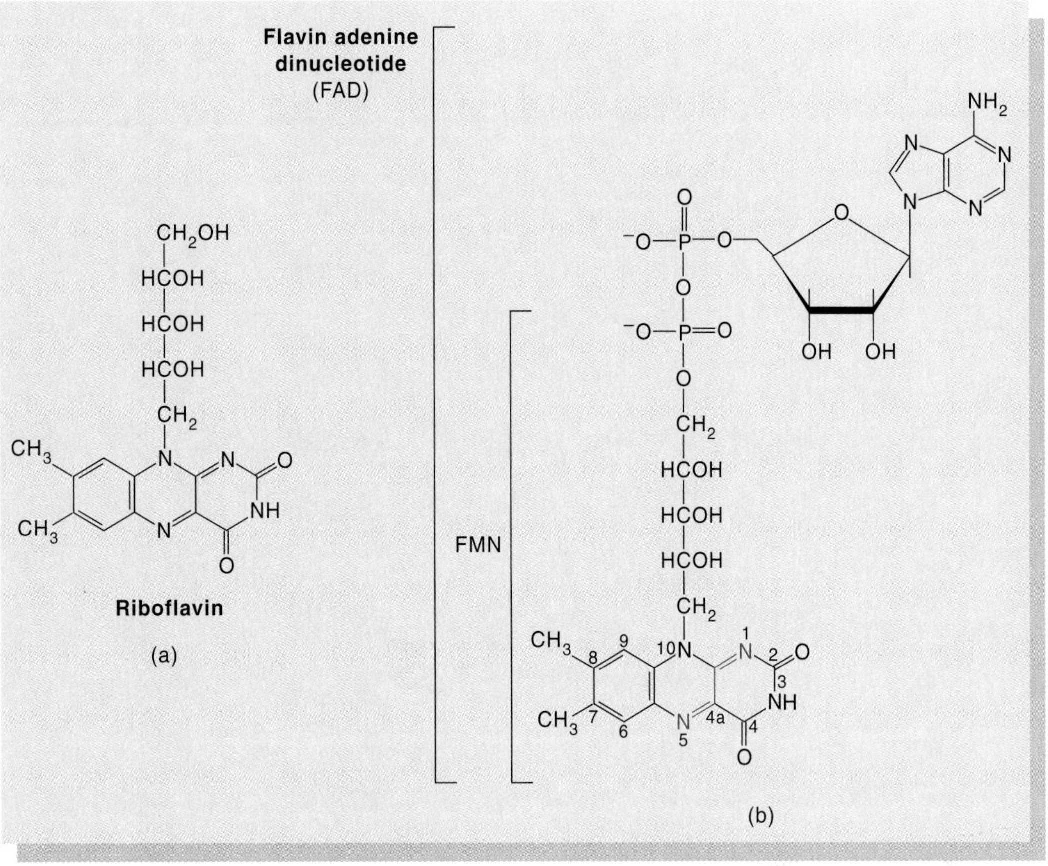

proteins the coenzyme is known to be covalently bonded to the protein by means of a sulflhydryl or imidazole group at the C-8 methyl group and in at least one case at C-6. In most flavoproteins the coenzymes are tightly but noncovalently bound, and many can be resolved into apoenzymes that can be reconstituted to holoenzymes by readdition of FAD or FMN.

Flavin coenzymes exist in four spectrally distinguishable oxidation states that account in part for their catalytic functions: the yellow oxidized form, the red or blue one-electron reduced form, and the colorless two-electron reduced form. Their structures are depicted in figure 11.9. These and other less well defined forms often have been detected spectrally as intermediates in flavoprotein catalysis.

Flavins are very versatile redox coenzymes. Flavoproteins are dehydrogenases, oxidases, and oxygenases that catalyze a variety of reactions on an equal variety of substrate types. Since these classes of enzymes do not consist exclusively of flavoproteins, it is difficult to define catalytic specificity for flavins. Biological electron acceptors and donors in flavin-mediated reactions can be two-electron acceptors, such as NAD^+ or $NADP^+$, or a variety of one-electron acceptor systems, such as cytochromes (Fe^{2+}/Fe^{3+}) and quinones, and molecular oxygen is an electron acceptor for flavoprotein oxidases as well as the source of oxygen for oxygenases. The only obviously common aspect of flavin-dependent reactions is that all are redox reactions.

Some of the best-known reactions catalyzed by flavoproteins are listed in table 11.2, which groups flavoproteins into those that do not utilize molecular oxygen as a substrate and those that do. You can best appreciate the significance of this difference when you realize that the reduced form of FAD ($FADH_2$), a likely intermediate in many flavoprotein reactions, spontaneously reacts with O_2 to produce H_2O_2. In the case of the dehydrogenases, therefore, $FADH_2$ either is not an intermediate or is somehow prevented from reacting with O_2. Among the dehydrogenases are two that utilize the two-electron acceptor substrates NAD^+ or $NADP^+$, and it is reasonable to suppose that the two-electron reduction of NAD^+ by an intermediate $E \cdot FADH_2$ is involved. Also listed in table 11.2 are other dehydrogenases for which the electron acceptors from $E \cdot FADH_2$ are not given. These enzymes are membrane-bound and transfer electrons directly to membrane-bound acceptors, mainly

Figure 11.9

Oxidation states of flavin coenzymes. The flavin coenzymes exist in three spectrally distinguishable oxidation states that account in part for their catalytic functions. They are the yellow oxidized form, the red or blue one-electron reduced form, and the colorless two-electron reduced form. Groups in red are those which are centrally involved in oxidation–reduction reactions.

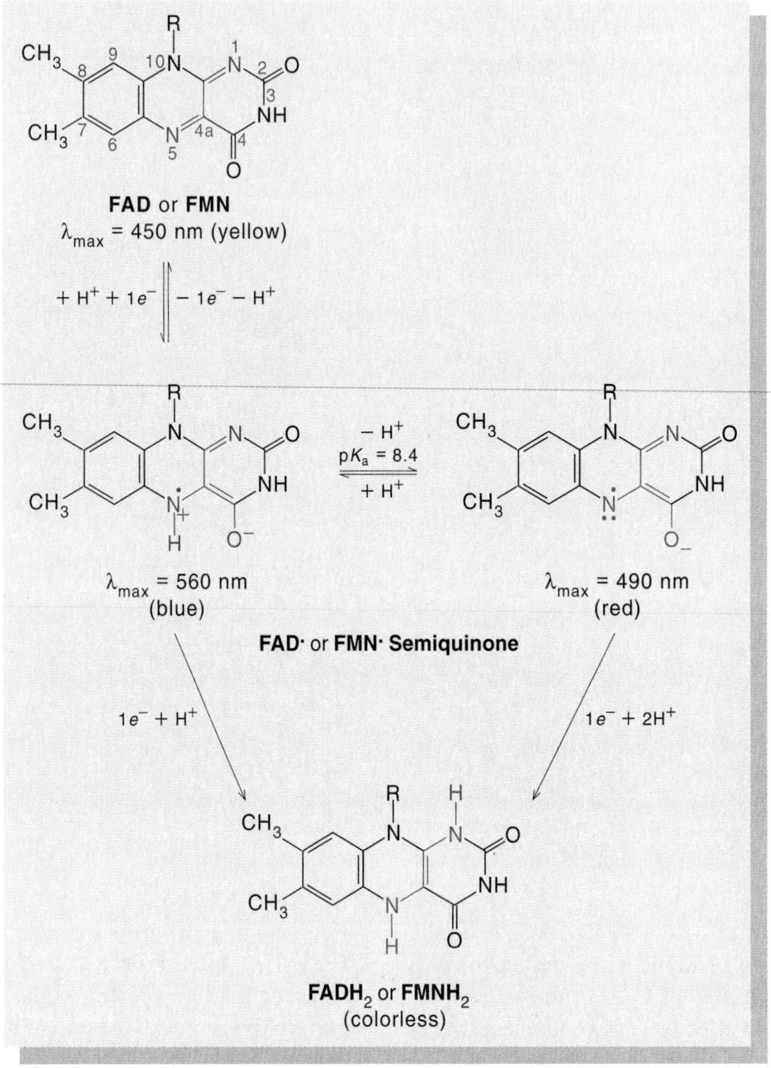

FAD or **FMN**
λ_{max} = 450 nm (yellow)

$+ H^+ + 1e^-$ / $- 1e^- - H^+$

$- H^+$, $pK_a = 8.4$, $+ H^+$

λ_{max} = 560 nm (blue)

λ_{max} = 490 nm (red)

FAD· or **FMN· Semiquinone**

$1e^- + H^+$

$1e^- + 2H^+$

$FADH_2$ or **$FMNH_2$**
(colorless)

one-electron acceptors, such as quinones and cytochromes (Fe^{2+}/Fe^{3+}). The stability of the flavin semiquinone, FAD • and FMN • in figure 11.9, gives flavins the capability of interacting with one-electron acceptors in electron-transport systems.

The other classes of flavoproteins in table 11.2 interact with molecular oxygen either as the electron-acceptor substrates in redox reactions catalyzed by oxidases or as the substrate sources of oxygen atoms for oxygenases. Molecular oxygen also serves as an electron acceptor and source of oxygen for metalloflavoproteins and dioxygenases, which are not listed in the table. These enzymes catalyze more complex reactions, involving catalytic redox components, such as metal ions and metal–sulfur clusters in addition to flavin coenzymes.

A recurrent theme in many flavoprotein reactions is the probable involvement of $FADH_2$ or the reduced form of FMN ($FMNH_2$) as transient intermediates. Figure 11.10 illustrates a reasonable catalytic pathway for the first enzyme listed in table 11.2; this reaction shows the likely involvement of E · $FADH_2$ in each case. The mechanisms by which E · FAD is reduced to E · $FADH_2$ by NADPH in the forward direction and by glutathione in the reverse direction are undoubtedly different.

The biochemical importance of flavin coenzymes resides in their versatility in mediating a variety of redox processes, including electron transfer and the activation of molecular oxygen for oxygenation reactions. An especially important manifestation of their redox versatility is their ability to serve as the switch point

Table 11.2
Reactions Catalyzed by Flavoproteins

Flavoprotein	Reaction
Dehydrogenases	
Glutathione reductase	$H^+ + GSSG + NADPH \rightleftharpoons 2\ GSH + NADP^+$
Acyl-CoA dehydrogenases	$RCH_2CH_2COSCoA + NAD^+ \rightleftharpoons RCH\text{=}CHCOSCoA + NADH + H^+$
Succinate dehydrogenase	$^-O_2CCH_2CH_2CO_2^- + E \cdot FAD \rightleftharpoons {}^-O_2CCH\text{=}CHCO_2^- + E \cdot FADH_2$
D-Lactate dehydrogenase	$CH_3\text{—}CHOH\text{—}CO_2^- + E \cdot FAD \rightleftharpoons CH_3\text{—}CO\text{—}CO_2^- + E \cdot FADH_2$
Oxidases	
Amino acid oxidases	$R\text{—}CH(NH_3^+)\text{—}CO_2^- + O_2 + H_2O \longrightarrow R\text{—}CO\text{—}CO_2^- + H_2O_2 + NH_4^+$
Monoamine oxidase	$R\text{—}CH_2\overset{+}{N}H_3 + O_2 + H_2O \longrightarrow R\text{—}CHO + H_2O_2 + \overset{+}{N}H_4$
Monooxygenases	
Lactate oxidase	$CH_3\text{—}CHOH\text{—}CO_2^- + O_2 \longrightarrow CH_3\text{—}CO_2^- + CO_2 + H_2O$
Salicylate hydroxylase	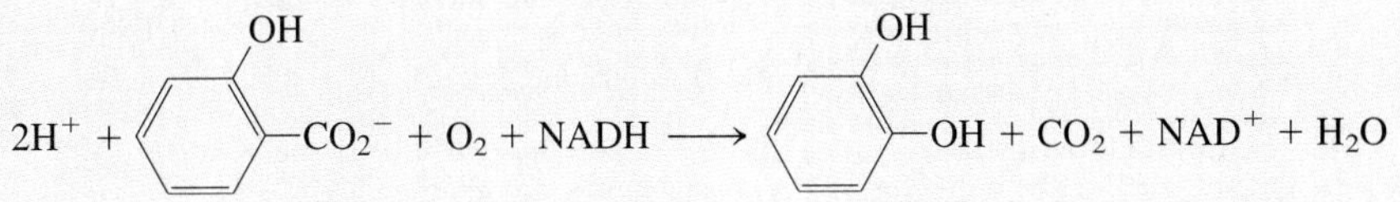 $2H^+ + C_6H_4(OH)\text{—}CO_2^- + O_2 + NADH \longrightarrow C_6H_4(OH)\text{—}OH + CO_2 + NAD^+ + H_2O$

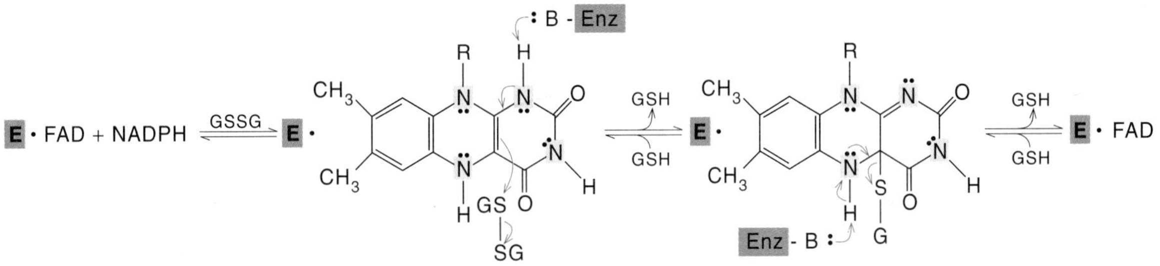

Figure 11.10

Mechanism of the flavin-dependent glutathione reductase reaction. The first steps, not shown, involve the reduction of FAD to $FADH_2$ by NADPH and the binding of glutathione (glutathione is a sulfhydryl compound, see figure 25.15). The mechanism by which oxidized glutathione (GSSG) is reduced by the $E \cdot FADH_2$ is shown.

E · FAD + NADPH ⇌(GSSG) E · [reduced flavin] ⇌(GSH) E · [flavin–SG adduct] ⇌(GSH) E · FAD

from the two-electron processes, which predominate in cytosolic carbon metabolism, to the one-electron transfer processes, which predominate in membrane-associated terminal electron-transfer pathways. In mammalian cells, for example, the end products of the aerobic metabolism of glucose are CO_2 and NADH (see chapter 14). The terminal electron-transfer pathway is a membrane-bound system of cytochromes, nonheme iron proteins, and copper-heme proteins — all one-electron acceptors that transfer electrons ultimately to O_2 to produce H_2O and NAD^+ with the concomitant production of ATP from ADP and P_i. The interaction of NADH with this pathway is mediated by NADH dehydrogenase, a flavoprotein that couples the two-electron oxidation of NADH with the one-electron reductive processes of the membrane.

Reactions Requiring Acyl Activation Frequently Use Phosphopantetheine Coenzymes

4′-Phosphopantetheine coenzymes are the biochemically active forms of the vitamin pantothenic acid. In figure 11.11, 4′-phosphopantetheine is shown as covalently linked to an adenylyl group

Figure 11.11

Structures of the vitamin pantothenic acid (in red) and coenzyme A. The terminal —SH (in blue) is the reactive group in coenzyme A (CoASH).

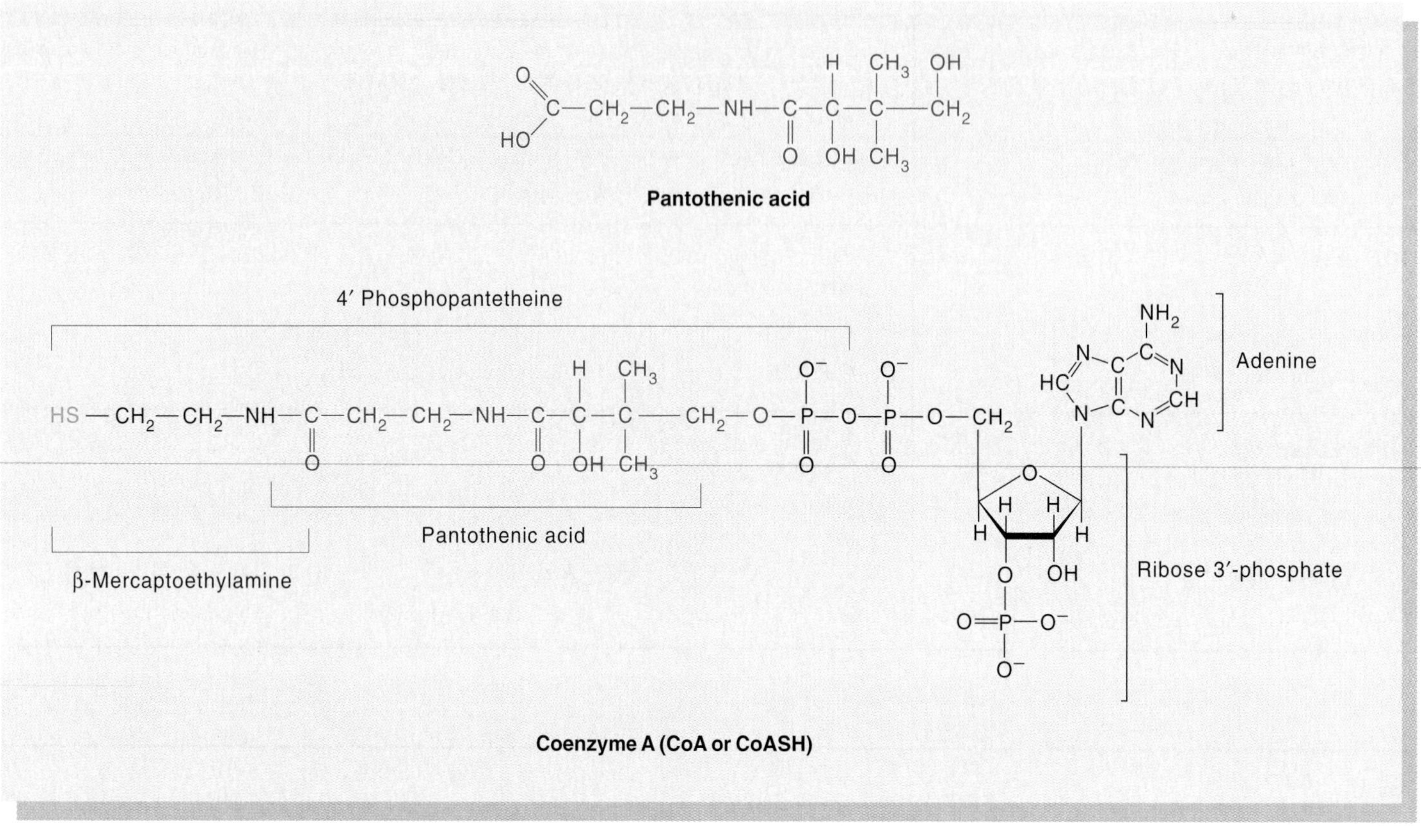

Pantothenic acid

4′ Phosphopantetheine

Adenine

Pantothenic acid

β-Mercaptoethylamine

Ribose 3′-phosphate

Coenzyme A (CoA or CoASH)

in coenzyme A; or it can also be linked to a protein such as a serine hydroxyl group in acyl carrier protein (ACP). It is also found bonded to proteins that catalyze the activation and polymerization of amino acids to polypeptide antibiotics. Coenzyme A was discovered, purified, and structurally characterized by Fritz Lipmann and colleagues in work for which Lipmann was awarded the Nobel Prize in 1953.

The sulfhydryl group of the β-mercaptoethylamine (or cysteamine) moiety of phosphopantetheine coenzymes is the functional group directly involved in the enzymatic reactions for which they serve as coenzymes. From the standpoint of the chemical mechanism of catalysis, it is the essential functional group, although it is now recognized that phosphopantetheine coenzymes have other functions as well. Many reactions in metabolism involve acyl-group transfer or enolization of carboxylic acids that exist as unactivated carboxylate anions at physiological pH. The predominant means by which these acids are activated for acyl transfer and enolization is esterification with the sulfhydryl group of pantetheine coenzymes.

The mechanistic importance of activation is exemplified by the condensation of two molecules of acetyl-coenzyme A to acetoacetyl-coenzyme A catalyzed by β-ketothiolase:

$$CH_3-\overset{O}{\overset{\|}{C}}-SCoA + CH_3-\overset{O}{\overset{\|}{C}}-SCoA \rightleftharpoons CH_3-\overset{O}{\overset{\|}{C}}-CH_2-\overset{O}{\overset{\|}{C}}-SCoA + CoASH \quad \textbf{(13)}$$

The two important steps in this reaction depend on both acetyl groups being activated, one for enolization and the other for acyl-group transfer. In the first step, one of the molecules must be enolized by intervention of a base to remove an α proton, forming an enolate:

$$B\!: H-CH_2-\overset{O}{\overset{\|}{C}}-SCoA \rightleftharpoons \overset{+}{B}-H + CH_2\!\!=\!\!\overset{O^-}{\overset{|}{C}}-SCoA \quad \textbf{(14)}$$

The enolate is stabilized by delocalization of its negative charge between the α carbon and the acyl oxygen atom and by interactions with enzymatic groups, making it thermodynamically accessible as an intermediate. Moreover, this developing charge is also stabilized in the transition state preceding the enolate, so it is also kinetically accessible; that is, it is rapidly formed. If, by contrast, the same enolization reaction were carried out by the acetate anion, it would result in the generation of a second negative charge in the enolate, an energetically and kinetically unfavorable process.

The second stage of the condensation is the reaction of the enolate anion with the acyl group of a second molecule of acetyl-CoA:

$$\mathrm{CH_3C(=O)\text{—}S\text{-}CoA} + \mathrm{CH_2{=}C(O^-)\text{—}S\text{-}CoA} \rightleftharpoons \left[\mathrm{CH_3C(O^-)(S\text{-}CoA)\text{—}CH_2\text{—}C(=O)\text{—}S\text{-}CoA}\right] \xrightleftharpoons{H^+} \mathrm{CH_3C(=O)\text{—}CH_2\text{—}C(=O)\text{—}S\text{-}CoA} + \mathrm{CoASH} \quad (15)$$

Nucleophilic addition to the neutral activated acyl group is a favored process, and coenzyme A is a good leaving group from the tetrahedral intermediate. The occurrence of this process with the acetate anion, that is, acetate reacting with an enolate anion, again provides a sharp contrast with the process of equation (15), for it would entail the nucleophilic addition of an anion to an anionic center, generating a dianionic transition state—an unfavorable process from both thermodynamic and kinetic standpoints. Moreover, the resulting intermediate would not have a very good leaving group other than the enolate anion itself, so the transition-state energy for acetoacetate formation would be high. Finally, the K_{eq} for the condensation of 2 moles of acetate to 1 mole of acetoacetate is not favorable in aqueous media, whereas the condensation of 2 moles of acetyl-CoA to produce acetoacetyl-CoA and coenzyme A is thermodynamically spontaneous. The maintenance of metabolic carboxylic acids involved in enolization and acyl-group transfer reactions as coenzyme A esters provides the ideal lift over the kinetic and thermodynamic barriers of these reactions.

The foregoing discussion, in emphasizing the purely electrostatic energy barriers, does not address the question of whether there is an activation advantage in thiol esters relative to oxygen esters. Why thiol esters in preference to oxygen esters? Thiol esters are more readily enolized than oxygen esters. They are more "ketonelike" because of their electronic structures, in which the degree of resonance-electron delocalization from the sulfur atom to the acyl group is less than that of oxygen esters. As a result the charge-separated resonance form is a smaller contributor to the electronic structure in thiol esters than in oxygen esters:

$$\left[\mathrm{R_1\text{—}C(=O)\text{—}\ddot{S}\text{—}R_2} \longleftrightarrow \mathrm{R_1\text{—}C(O^-){=}\overset{+}{S}\text{—}R_2}\right]$$
$$\left[\mathrm{R_1\text{—}C(=O)\text{—}\ddot{O}\text{—}R_2} \longleftrightarrow \mathrm{R_1\text{—}C(O^-){=}\overset{+}{O}\text{—}R_2}\right] \quad (16)$$

CoA thioesters allow the basic chemical process of carbon–carbon bond formation and cleavage as depicted in equation (15) and analogous reactions to take place. While such processes are chemically feasible for esters and thioesters, alcohols and alkanes or alkenes could not react in analogous ways. However, the foregoing discussion of the reasons for the reactivities of thioesters does not consider the role of the enzyme in catalyzing the reaction. There are very fundamental chemical barriers to reactions like the ones shown in equation (15), even in the reaction of a CoA-thioester. The pK_a for a thioester such as acetyl-CoA in the formation of an enolate according to equation (17) is about 23:

$$\mathrm{HO^-} + \mathrm{H_3C\text{—}C(=O)\text{—}S\text{-}CoA} \xrightleftharpoons{K_a = 10^{-23}} \mathrm{H_2C{=}C(O^-)\text{—}S\text{-}CoA} + \mathrm{H_2O} \quad (17)$$

Although this is much lower than that of an alcohol or alkane (40–50), it is still too high for the reaction to take place at a significant rate at pH 7.0. A simple calculation using $K_a = 10^{-23}$ shows that the ratio of enolate to un-ionized acetyl-CoA would be 10^{-16} at pH 7.0. The rate at which such a tiny fraction of enolate could react would be much too slow, even if the rate constant for its reaction were as large as bond vibrational frequencies. It must be concluded that the binding interactions between the substrate and the enzyme must decrease the pK_a values significantly in order for the reaction to take place. In addition to this thermodynamic problem, enolization rates are typically orders of magnitude slower than ionizations for normal acids with the same pK_a values. The kinetic problem must also be overcome by binding interactions at the active site of the enzyme. The mechanism of enolization at active sites is an important current issue on the mechanism of enzyme action.

While the pantetheine sulfhydryl group has the appropriate chemical properties for activating acyl groups, this characteristic is not unique to pantetheine coenzymes in the biosphere. Both glutathione and cysteine, as well as cysteamine, would serve, so the chemistry does not itself explain the importance of these coenzymes. Coenzyme A has many binding determinants in its large structure, especially in the nucleotide moiety, so it may serve a specificity function in the binding of coenzyme A esters by enzymes. It also may serve as a binding "handle" in cases in which the acyl group must have some mobility in the catalytic site, that is, if it must enolize at one site and then diffuse a short distance to undergo an addition reaction to a ketonic group of a second substrate.

One system in which pantetheine almost certainly performs such a carrier role is the fatty acid synthase from *E. coli*, in which 4′-phosphopantetheine is a component of the acyl carrier protein (see chapter 21).

α-Lipoic Acid Is the Coenzyme of Choice for Reactions Requiring Acyl-Group Transfers Linked to Oxidation-Reduction

α-Lipoic acid is the internal disulfide of 6,8-dithiooctanoic acid, the structural formula of which is given in figure 11.12. It couples

Figure 11.12

(a) Lipoic acid. (b) Reduced lipoid acid. (c) Lipoic acid bound to the ϵ-amino group of a lysine residue. Structures shown in red are those centrally involved in coenzyme reactions.

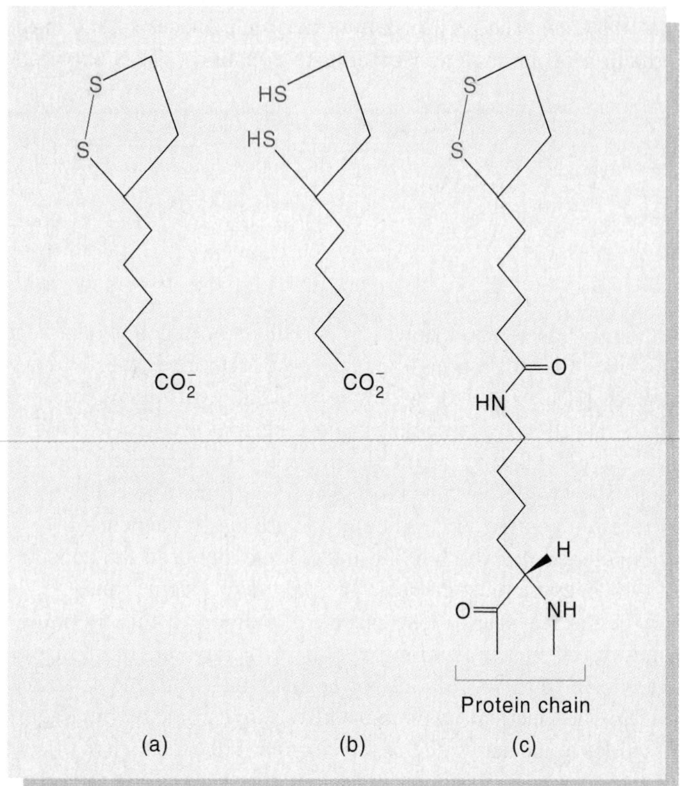

electron and group transfers catalyzed by α-keto acid dehydrogenase multienzyme complexes. The pyruvate and α-ketoglutarate dehydrogenase complexes are centrally involved in the metabolism of carbohydrates by the glycolytic pathway (chapter 14) and the tricarboxylic acid cycle (chapter 15). They catalyze two of the three decarboxylation steps in the complete oxidation of glucose, and they produce NADH and activated acyl compounds from the oxidation of the resulting ketoacids:

$$R{-}\overset{\overset{\displaystyle O}{\|}}{C}{-}CO_2^- + NAD^+ + CoASH \longrightarrow$$

$$CO_2 + NADH + R{-}\overset{\overset{\displaystyle O}{\|}}{C}{-}S\text{-}CoA \quad \textbf{(18)}$$

The chemical role of α-lipoic acid is to mediate the transfer of electrons and activated acyl groups resulting from the decarboxylation and oxidation of α-keto acids within the complexes. In this process, lipoic acid itself is transiently reduced to dihydrolipoic acid (see fig. 11.12); this reduced form is the acceptor for activated acyl groups. Its dual role of electron and acyl-group acceptor enables lipoic acid to couple the two processes.

The coenzymatic capabilities of α-lipoyl groups result from a fusion of its chemical and physical properties, the ability to act simultaneously as both electron and acyl-group acceptor, the ability to span long distances to interact with sites separated by up to 2.8 nm, and the ability to act cooperatively with other α-lipoyl groups by disulfide interchange to relay electrons and acyl groups through distances that exceed its reach. A reaction of this type is discussed in chapter 15.

Biotin Mediates Carboxylations

The biotin structure shown in figure 11.13 is an imidazolone ring *cis*-fused to a tetrahydrothiophene ring substituted at position 2 by valeric acid. In carboxylase enzymes, biotin is covalently bonded to the proteins by an amide linkage between its carboxyl group and a lysyl-ϵ-NH_2 group in the polypeptide chain. This arrangement places the imidazolone ring at the end of a flexible chain of atoms extending a maximum of about 1.4 nm from the α carbon of lysine.

Biotin is the essential coenzyme for carboxylation reactions involving bicarbonate as the carboxylating agent. Several reactions have been described in which ATP-dependent carboxylation occurs at carbon atoms activated for enolization by ketonic

Figure 11.13

Structures of biotinyl enzyme and $N^{1'}$-carboxybiotin. The reactive portions of the coenzyme and the active intermediate are shown in red. In carboxylase enzymes, biotin is covalently bonded to the proteins by an amide linkage between its carboxyl group and a lysyl-ϵ-NH_2 group in the polypeptide chain.

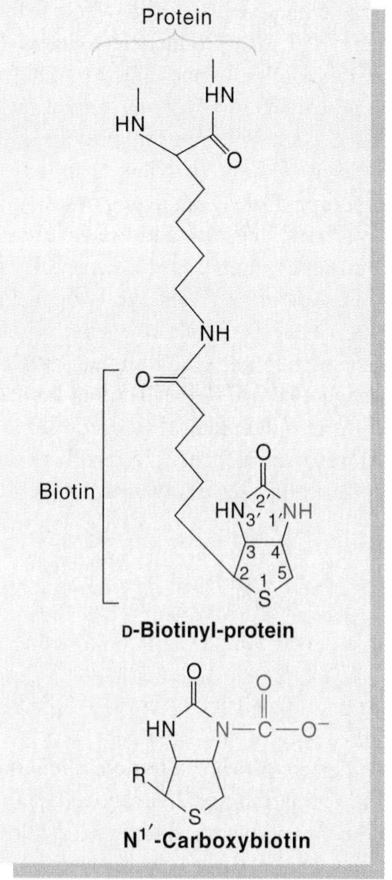

or activated acyl groups. One reaction is known in which a nitrogen atom of urea is carboxylated.

A general formulation of the ATP-dependent carboxylation of an α carbon by ^{18}O-enriched bicarbonate is

$$RCH_2{-}\overset{\overset{\Large O}{\|}}{C}{-}SCoA + ATP + HC^{18}O_3^{-} \xrightarrow{\text{Biotinyl carboxylase}}$$

$$H^+ + R{-}\underset{\underset{\Large C^{18}O_2^{-}}{|}}{CH}{-}\overset{\overset{\Large O}{\|}}{C}{-}SCoA + ADP + H^{18}OPO_3^{2-} \qquad \textbf{(19)}$$

The appearance of ^{18}O in inorganic phosphate verifies that the function of ATP in the reaction is essentially to dehydrate the bicarbonate.

The ATP-dependent carboxylation of biotin by bicarbonate is believed to control the transient formation of carbonic-phosphoric anhydride, or "carboxyphosphate," as an active carboxylation intermediate:

$$HO{-}\overset{\overset{\Large O}{\|}}{C}{-}O^- \xrightarrow[\text{ADP}]{\text{ATP}} {}^-O{-}\underset{\underset{\Large OH}{|}}{\overset{\overset{\Large O}{\|}}{P}}{-}O{-}\overset{\overset{\Large O}{\|}}{C}{-}O^- \xrightarrow[\text{P}_\text{i}]{\text{Biotinyl-E}} N^1\text{-carboxybiotinyl-E} \qquad \textbf{(20)}$$

The coenzymatic function of biotin appears to be to mediate the carboxylation of substrates by accepting the ATP-activated carboxyl group and transferring it to the carboxyl acceptor substrate. There is good reason to believe that the enzymatic sites for the ATP-dependent carboxylation of biotin are physically separated from the sites at which N^1-carboxybiotin transfers the carboxyl group to acceptor substrates, that is, the transcarboxylase sites. In fact, in the case of the acetyl-CoA carboxylase from *E. coli* (see chapter 21), these two sites reside on different subunits, while the biotinyl group is bonded to a third, a small subunit designated biotin carboxyl carrier protein.

Biotin appears to have just the right chemical and structural properties to mediate carboxylation. It readily accepts activated carboxyl groups at N^1 and maintains them in an acceptably stable yet reactive form for transfer to acceptor substrates. Since biotin is bonded to a lysyl group, the N^1-carboxyl group is at the end of a 1.6-nm chain with bond rotational freedom about nine single bonds, giving it the capability to transport activated carboxyl groups through space from the carboxyl activation sites to the carboxylation sites.

Folate Coenzymes Are Used in Reactions for One-Carbon Transfers

Tetrahydrofolate and its derivatives N^5,N^{10}-methylenetetrahydrofolate, N^5,N^{10}-methenyltetrahydrofolate, N^{10}-formyltetrahydrofolate, and N^5-methyltetrahydrofolate are the biologically active forms of folic acid, a four-electron oxidized form of tetrahydrofolate. The structural formulas are given in figure 11.14, which also shows how they arise from tetrahydrofolate. The structures are shown glutamylated on the carboxyl group of the *p*-aminobenzoyl group; the most active forms contain oligo- or polyglutamyl groups, linked through the γ-carboxyl groups.

The tetrahydrofolates do not function as tightly enzyme-bound coenzymes. Rather, they function as cosubstrates for a variety of enzymes associated with one-carbon metabolism. N^{10}-formyltetrahydrofolate is produced enzymatically from tetrahydrofolate and formate in an ATP-linked process in which formate is activated by phosphorylation to formyl phosphate; the formyl group of formyl phosphate is then transferred to N^{10} of tetrahydrofolate. N^{10}-Formyltetrahydrofolate is a formyl donor substrate for some enzymes and is interconvertible with N^5,N^{10}-methenyltetrahydrofolate by the action of cyclohydrolase.

N^5,N^{10}-Methylenetetrahydrofolate is a hydroxymethyl-group donor substrate for several enzymes and a methyl-group donor substrate for thymidylate synthase (fig. 11.15). It arises in living cells from the reduction of N^5,N^{10}-methenyltetrahydrofolate by NADPH and also by the serine hydroxymethyltransferase-catalyzed reaction of serine with tetrahydrofolate.

N^5-Methyltetrahydrofolate is the methyl-group donor substrate for methionine synthase, which catalyzes the transfer of the five-methyl group to the sulfhydryl group of homocysteine. This and selected reactions of the other folate derivatives are outlined in figure 11.15, which emphasizes the important role that tetrahydrofolate plays in nucleic acid metabolism by serving as the immediate source of one-carbon units in purine and pyrimidine biosynthesis.

Formaldehyde is a toxic substance that reacts spontaneously with amino groups of proteins and nucleic acids, hydroxymethylating them and forming methylene-bridge crosslinks between them. Free formaldehyde therefore wreaks havoc on living cells and could not serve as a useful hydroxymethylating agent. In the form of N^5,N^{10}-methylenetetrahydrofolate, however, its chemical reactivity is attenuated but retained in a potentially available form when needed. Formate, however, is quite unreactive under physiological conditions and must be activated to serve as an efficient formylating agent. As N^{10}-formyltetrahydrofolate it is in a reactive state suitable for transfer to appropriate substrates. The fundamental biochemical importance of tetrahydrofolate is to maintain formaldehyde and formate in chemically poised states, not so reactive as to pose toxic threats to the cell but available for essential processes by specific enzymatic action.

Ascorbic Acid Is Required to Maintain the Enzyme That Forms Hydroxyproline Residues in Collagen

Ascorbic acid (vitamin C; fig. 11.16) is the reducing agent required to maintain the activity of a number of enzymes, most notably proline hydroxylase, which forms 4-hydroxyproline residues in collagen. Hydroxyproline (see fig. 11.16*c*) is not synthesized biologically as a free amino acid but rather is created by modification of proline residues already incorporated into the precollagen polypeptide chain. The hydroxylation reaction occurs as the protein is synthesized in the endoplasmic reticulum. At least a third of the numerous proline residues in collagen are modified in this way, substantially increasing the resistance of the protein to thermal denaturation.

Proline hydroxylase is a diooxygenase that requires ferrous iron as a cofactor; it uses α-ketoglutarate as its second sub-

Figure 11.14

Structures and enzymatic interconversions of folate coenzymes. The reactive centers of the coenzymes are shown in red. The most active forms of the coenzyme contain oligo- or polyglutamyl groups.

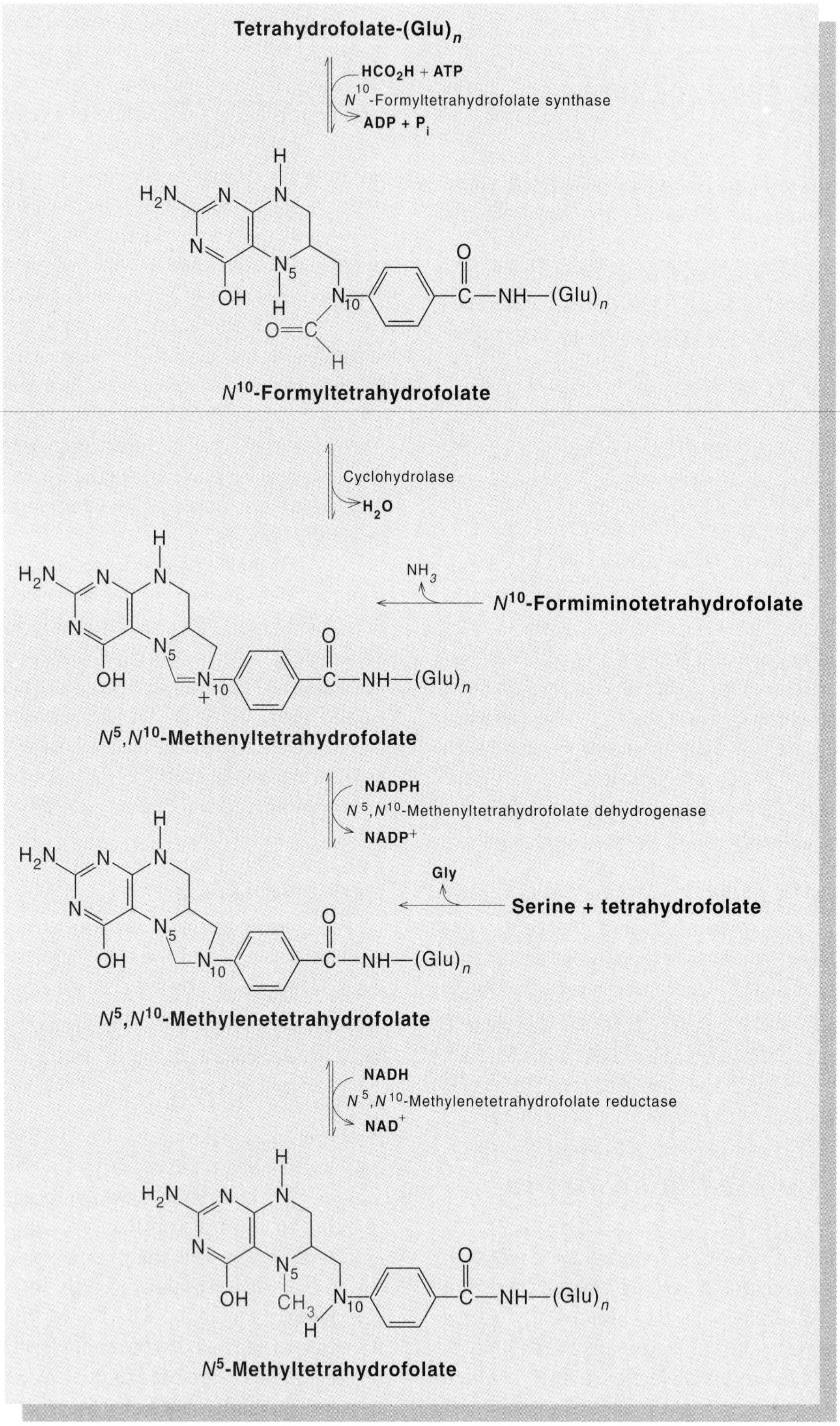

Figure 11.15

Involvement of folate coenzymes in one-carbon metabolism. Shown in red are the one-carbon units of the end products that originate with the reactive one-carbon units of the folate coenzymes.

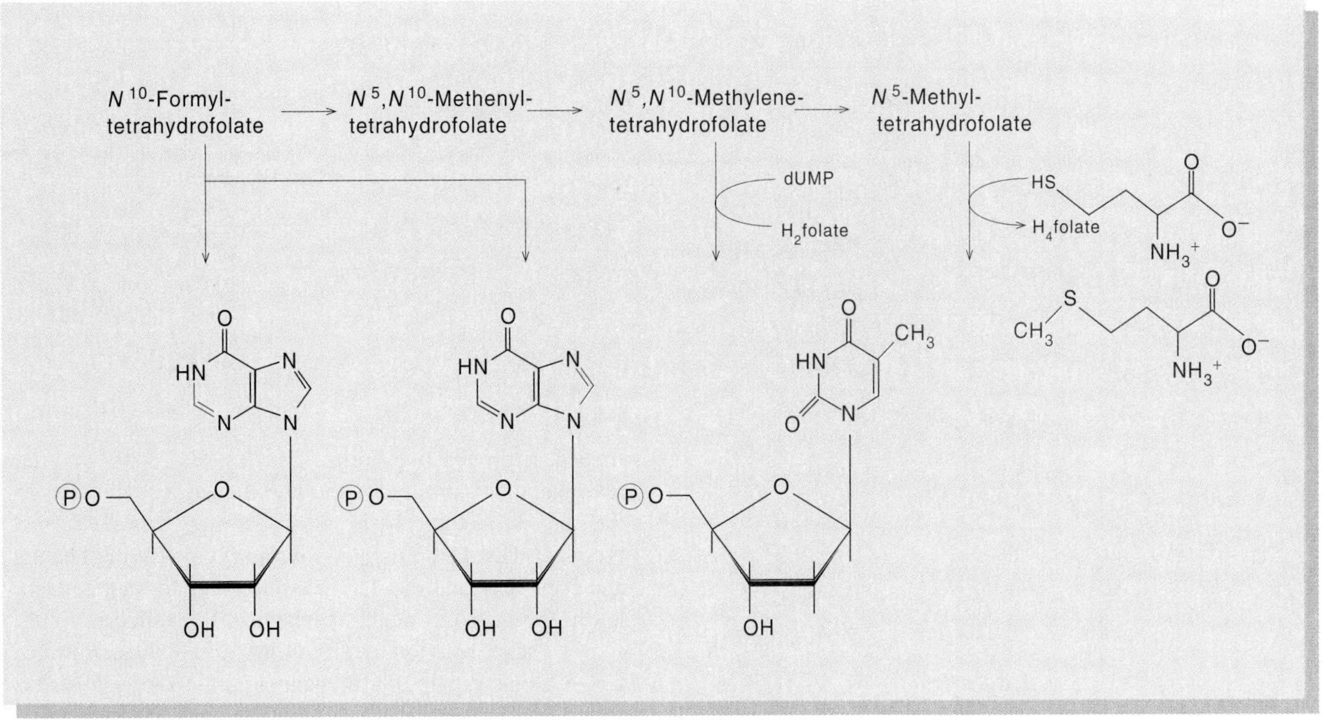

N^{10}-Formyl-tetrahydrofolate → N^5,N^{10}-Methenyl-tetrahydrofolate → N^5,N^{10}-Methylene-tetrahydrofolate → N^5-Methyl-tetrahydrofolate

dUMP → H_2folate

HS ... → H_4folate

Figure 11.16

(*a*) Ascorbic acid is a reductant that appears to be required to keep some enzymes in their active states. Oxidation converts ascorbic acid to (*b*) dehydroascorbic acid, which decomposes irreversibly by hydrolysis of the lactone ring. One enzyme that requires ascorbic acid is proline hydroxylase, which converts proline residues in collagen to (*c*) 4-hydroxyproline (hydroxy group shown in red). Hydroxyproline residues help to stabilize the structure of collagen.

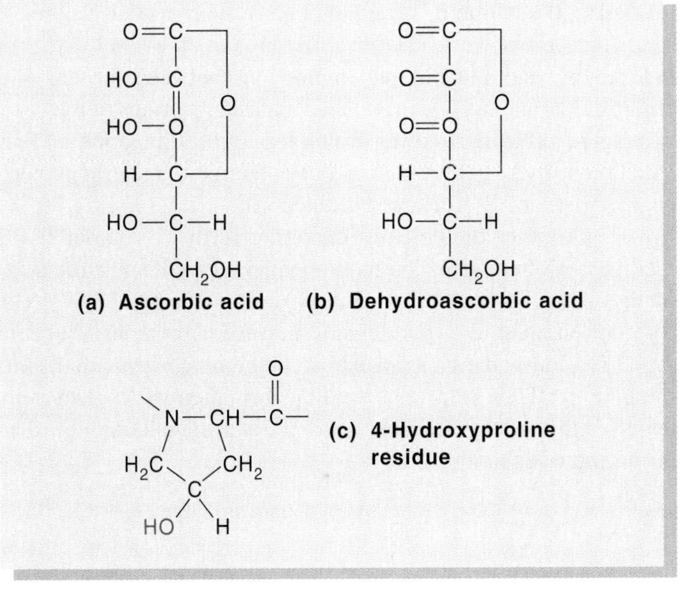

(a) Ascorbic acid (b) Dehydroascorbic acid

(c) 4-Hydroxyproline residue

strate. One oxygen atom from O_2 is incorporated into hydroxyproline, while the other goes to the α-ketoglutarate, which decomposes to succinate and CO_2:

$$\text{prolyl-peptide} + {}^{-}O_2CCH_2CH_2\overset{\overset{O}{\|}}{C}{-}CO_2^{-} + O_2 \longrightarrow \text{hydroxyprolyl-peptide} + {}^{-}O_2CCH_2CH_2CO_2^{-} + CO_2 \qquad \textbf{(21)}$$

Ascorbic acid is synthesized by plants and many animals but not by primates or guinea pigs. In scurvy, the disease associated with a severe deficiency of ascorbic acid, connective tissues throughout the body deteriorate. Weakening of the capillary walls results in hemorrhages, wounds heal poorly, and lesions occur in the bones.

Vitamin B_{12} Coenzymes Are Associated with Rearrangements on Adjacent Carbon Atoms

The principal coenzymatic form of vitamin B_{12} is 5′-deoxyadenosylcobalamin, the structural formula of which is given in figure 11.17. The structure includes a cobalt–carbon bond between the 5′ carbon of the 5′-deoxyadenosyl moiety and the cobalt (III) ion of cobalamin. [Note: The metal oxidation state may be denoted either as Co^{3+} or as Co(III). The former stresses the free-ion character, while the latter stresses the bound character.] Vita-

Figure 11.17

Structure of 5′-deoxyadenosylcobalamin coenzyme (vitamin B_{12}). The reactive groups are shown in red.

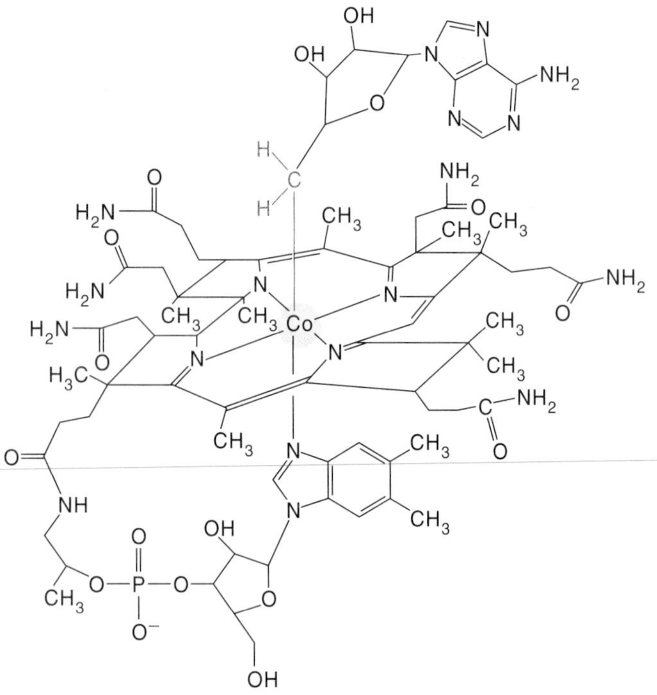

min B_{12} itself is cyanocobalamin, in which the cyano group is bonded to cobalt in place of the 5′-deoxyadenosyl moiety. Other forms of the vitamin have water (aquocobalamin) or the hydroxyl group (hydroxycobalamin) bonded to cobalt.

The vitamin was discovered in liver as the antipernicious anemia factor in 1926, but discovery of its complete structure had to await its purification, chemical characterization, and crystallization, which required more than 20 years. Even then the determination of such a complex structure proved to be an elusive goal by conventional approaches of that day and had to await the elegant x-ray crystallographic study of Lenhert and Hodgkin in 1961, for which Dorothy Crowfoot Hodgkin was awarded the Nobel Prize in 1964.

Most 5′-deoxyadenosylcobalamin-dependent enzymatic reactions are rearrangements that follow the pattern of equation (22), in which a hydrogen atom and another group (designated X) bonded to an adjacent carbon atom exchange positions, with the group X migrating from C_α to C_β:

$$a\text{—}\underset{\text{X}}{\overset{b}{C_\alpha}}\text{—}\underset{\text{H}}{\overset{c}{C_\beta}}\text{—}d \rightleftharpoons a\text{—}\underset{\text{H}}{\overset{b}{C_\alpha}}\text{—}\underset{\text{X}}{\overset{c}{C_\beta}}\text{—}d \qquad \textbf{(22)}$$

Three specific examples of rearrangement reactions are given in equations (23) through (25). It is interesting and significant that the migrating groups —OH, —COSCoA, and —CH $(NH_3^+)CO_2^-$ have little in common and that the hydrogen atoms migrating in the opposite direction are often chemically unreactive.

$$CH_3CH(OH)CH_2OH \xrightarrow{\text{Dioldehydrase}} CH_3CH_2CH(OH)\text{—}OH \xrightarrow{\text{Dioldehydrase}} CH_3CH_2CHO + H_2O \qquad \textbf{(23)}$$

$$^-O_2C\text{—}CH_2CH_2\text{—}\overset{O}{\overset{\|}{C}}\text{—}SCoA \underset{\text{mutase}}{\overset{\text{Methylmalonyl-CoA}}{\rightleftharpoons}} {}^-O_2C\text{—}\overset{CH_3}{\overset{|}{C}H}\text{—}\overset{O}{\overset{\|}{C}}\text{—}SCoA \qquad \textbf{(24)}$$

$$^-O_2C\text{—}CH_2CH_2\text{—}\underset{NH_3^+}{\underset{|}{C}H}\text{—}CO_2^- \underset{\text{mutase}}{\overset{\text{Glutamate}}{\rightleftharpoons}} {}^-O_2C\text{—}\overset{CH_3}{\overset{|}{C}H}\text{—}\underset{NH_3^+}{\underset{|}{C}H}\text{—}CO_2^- \qquad \textbf{(25)}$$

Indeed, hydrogen migrations in all the B_{12} coenzyme-dependent rearrangements proceed without exchange with the protons of water; that is, isotopic hydrogen in substrates is conserved in the products. This fact plus spectroscopic evidence implicating Co(II) and organic radicals as catalytic intermediates in the reaction have led to the proposal of the mechanism illustrated in figure 11.18.

The reaction begins with homolytic cleavage of the Co—C bond (see fig. 11.18), generating Co(II) and 5′-deoxyadenosyl free radical (step 1). The radical abstracts a hydrogen atom from the substrate, the migrating hydrogen in equation (22), generating 5′-deoxyadenosine and a substrate-derived free radical as intermediates (step 2). The substrate-radical undergoes rearrangement to a product-derived free radical, which abstracts a hydrogen atom to form the final product and regenerate the coenzyme (steps 3–5).

The most fundamental property of 5′-deoxyadenosylcobalamin leading to its unique action as a coenzyme is the weakness of the Co—C bond. This bond has a low dissociation energy, less than 30 kcal/mole, strong enough to be essentially stable in free solution but weak enough to be broken as a result of strain induced by multiple binding interactions between the enzyme-binding sites and the adenosyl and cobalamin portions of the coenzyme. The radicals resulting from cleavage of this bond and abstraction of hydrogen from substrates undergo the rearrangements characteristic of B_{12}-dependent reactions.

Another biologically important form of vitamin B_{12} is methylcobalamin, in which the upper axial cobalt substituent is a methyl group instead of an adenosyl group (see fig. 11.17). This form of vitamin B_{12} participates in several complex bacterial processes and in the biosynthesis of methionine in mammals. Methionine synthase catalyzes the transformation of homocysteine and 5-methyltetrahydrofolate into methionine and tetrahydrofolate according to equation (26):

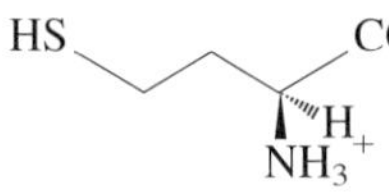

$$HS\text{—}CH_2CH_2\text{—}\underset{NH_3^+}{\underset{|}{C}H}\text{—}CO_2^- + N^5\text{-Methyltetrahydrofolate} \longrightarrow$$

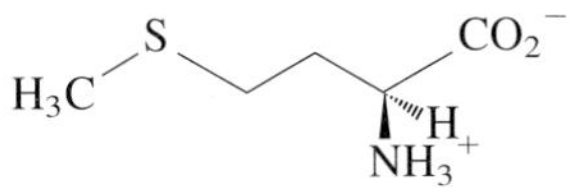

$$H_3C-S-CH_2-CH_2-CH(NH_3^+)-CO_2^- + \text{Tetrahydrofolate} \quad (26)$$

The mammalian enzyme contains cobalamin, which mediates the transfer of the methyl group from 5-methyltetrahydrofolate to methionine as illustrated in figure 11.19. In this process, the cobalamin exists normally as methylcobalamin, which acts as an alkylating agent to transfer its methyl group to homocysteine. This generates a reduced form of cobalamin, in which cobalt is in its +1 oxidation state, designated Co(I). In this form, cobalt is a highly reactive nucleophile. It is methylated by 5-methyltetrahydrofolate to regenerate methylcobalamin for another reaction cycle. The Co(I) form of cobalamin is very sensitive to oxygen (O_2) and has a fleeting existence as an intermediate. It is sufficiently protected from oxygen at the active site to prevent its oxidation to Co(II).

Methionine synthase is required for methionine biosynthesis in mammals. Its dependence on both tetrahydrofolate and cobalamin accounts for the nutritional relationship between vitamin B_{12} and folate. Individuals who are deficient in vitamin B_{12} also often exhibit the symptoms of a folate deficiency. These individuals have normal amounts of total folates in their tissues, but most of the folates are in the form of N^5-methyltetrahydrofolate, which accumulates because of the inability of cells to convert it into tetrahydrofolate through the action of methionine synthase. 5-Methyltetrahydrofolate is not easily transformed into other folate derivatives except by the methionine synthase reaction, so in the absence of vitamin B_{12} the cells are starved for N^{10}-formyltetrahydrofolate and N^5,N^{10}-methylenetetrahydrofolate.

Figure 11.18

Partial mechanism of vitamin B_{12}-dependent rearrangements. The designations Co(III) and Co(II) refer to species that are spectrally and magnetically similar to Co^{3+} and Co^{2+}, respectively. Co(III) is diamagnetic and red, and Co(II) is paramagnetic (unpaired electron) and yellow. The metal does not undergo a change in electrostatic charge when the cobalt–carbon bond breaks homolytically (i.e., without charge separation), because one electron remains with the metal and the other with 5′-deoxyadenosine.

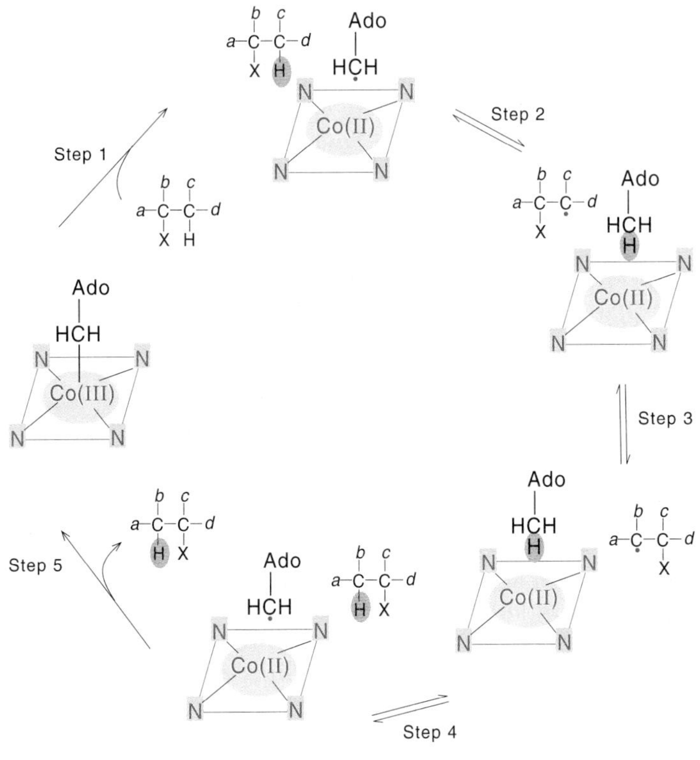

Figure 11.19

The chemical transformations that take place in the active site of mammalian methionine synthase are illustrated here. Most of the details of the participation of the enzyme are not known. Methylcobalamin is normally present in the active site. Upon binding homocysteine, the enzyme catalyzes its methylation by methylcobalamin to form methionine and cobalamin(I), in which cobalt is in the +1 oxidation state. This form of cobalamin is a powerful nucleophile that accepts a methyl group from N^5-methyltetrahydrofolate to regenerate methylcobalamin and produce tetrahydrofolate.

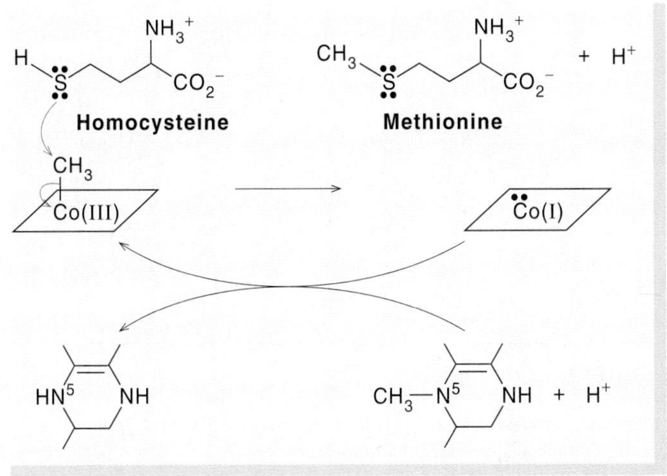

Iron-Containing Coenzymes Are Frequently Involved in Redox Reactions

Iron as a cofactor in catalysis is receiving increasing attention. The most common oxidation states of iron are Fe^{2+} and Fe^{3+}. Iron complexes are nearly all octahedral, and practically all are paramagnetic (as a result of unpaired electrons in the 3*d* orbital). The most common form of iron in biological systems is heme. Heme groups (Fe^{2+}) and hematin (Fe^{3+}) most frequently involve a complex with protoporphyrin IX (fig. 11.20). They are the coenzymes (prosthetic groups) for a number of redox enzymes, including catalase, which catalyzes the dismutation of hydrogen peroxide (equation (27)), and peroxidases, which catalyze the reduction of alkyl hydroperoxides by reducing agents such as phenols, hydroquinones, and dihydroascorbate (represented as AH_2 in equation (28)):

$$2\,H_2O_2 \rightleftharpoons 2\,H_2O + O_2 \quad (27)$$

$$R{-}O{-}O{-}H + AH_2 \longrightarrow A + R{-}O{-}H + H_2O \quad (28)$$

Heme proteins have characteristic visible absorption spectra as a result of protoporphyrin IX; their spectra differ depending on the

Figure 11.20

Structure of iron protoporphyrin IX. This coenzyme acts in conjunction with a number of different enzymes involved in oxidation and reduction reactions.

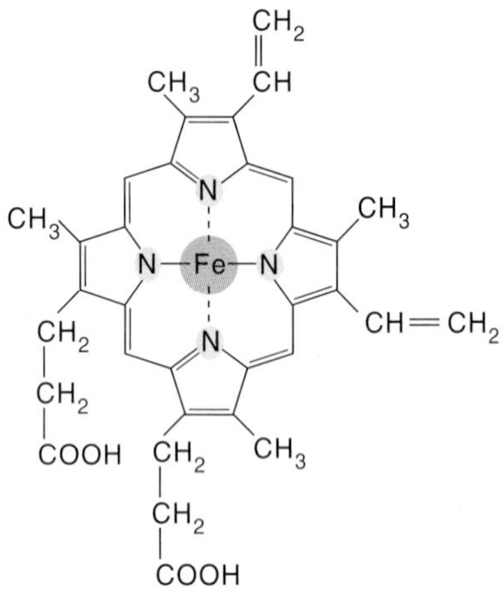

Figure 11.21

Structures of iron–sulfur clusters. Many redox enzymes contain iron–sulfur clusters that mediate one-electron transfer reactions.

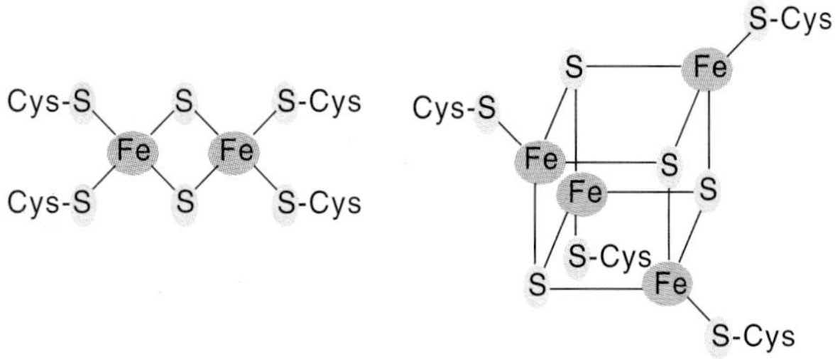

identities of the lower axial ligand donated by the protein and the oxidation state of the iron as well as the identities of the upper axial ligands donated by the substrates. Spectral data show clearly that the heme coenzymes participate directly in catalysis; however, the mechanisms of action for most heme coenzymes are not as well understood as those for other coenzymes.

Many redox enzymes contain iron–sulfur clusters that mediate one-electron transfer reactions. These clusters consist of two or four irons and an equal number of inorganic sulfide ions clustered together with the iron, which is also liganded to cysteinyl-sulfhydryl groups of the protein (fig. 11.21). The enzyme nitrogenase, which catalyzes the reduction of N_2 to 2 NH_3 contains such clusters in which some of the iron has been replaced by molybdenum (see chapter 24). Electron-transferring proteins involved in one-electron transfer processes often contain iron–sulfur clusters. These proteins include the mitochondrial membrane enzymes NADH dehydrogenase and succinate dehydrogenase (chapter 14), which are flavoproteins, and the small-molecular-weight proteins ferredoxin, rubredoxin, adrenodoxin, and putidaredoxin (chapters 16, 17, and 26). Despite their common use in electron transfer reactions, iron–sulfur clusters are involved in numerous reactions that do not involve electron transfer (e.g., see box 11A).

Heme coenzymes, iron–sulfur clusters, flavin coenzymes, and nicotinamide coenzymes cooperate in multienzyme systems to catalyze the chemically remarkable hydroxylations of hydrocarbons such as steroids (chapter 23). In these hydroxylation systems the heme proteins constitute a family of proteins known as cytochrome P450, named for the wavelength corresponding to the most intense absorption band of the carbon monoxide–liganded heme, an inhibited form. The reactions catalyzed by these systems are represented in generalized form by equation (29), which also shows the fate of the two oxygens from $^{18}O_2$:

$$H^+ + NADPH + {}^{18}O_2 + R{-}CH_2{-}R \longrightarrow R{-}CH(^{18}OH){-}R + NADP^+ + H_2{}^{18}O \quad \textbf{(29)}$$

One oxygen atom of O_2 is incorporated into the hydroxyl group of the product, whereas the other is incorporated into water. The enzymes usually include a cytochrome P450 and a flavoprotein reductase. An iron–sulfur cluster containing protein is often also involved in electron transfer.

In the mechanism of oxygenation by these enzymes, the flavoproteins and iron–sulfur proteins supply reducing equivalents in one-electron units from NADPH to cytochrome P450, and reduced cytochrome P450 reacts directly with O_2. These reactions generate the Fe(III)–peroxide complex, shown in figure 11.22, which undergoes a further dehydration to an oxygenating species of cytochrome P450. The oxygenating species is thought to be an oxo-complex of Fe(IV), in which the porphyrin ring has been oxidized to a radical-cation by loss of an electron. The positive charge and unpaired electron in figure 11.22 are stabilized by delocalization through the conjugated π-electron system of the porphyrin ring (see fig. 11.20).

The most widely accepted mechanism for cytochrome P450–catalyzed oxygenation of substrates is illustrated for a hydrocarbon substrate in figure 11.23. The oxygenating species, the oxo-Fe(IV) radical-cation, abstracts a hydrogen atom from the hydrocarbon to form a hydrocarbon radical and a hydroxy–Fe complex. Hydrogen abstraction proceeds by the pairing of one electron from the oxo-Fe(IV) species and one electron from the carbon–hydrogen bond of the hydrocarbon. In the second step of oxygenation the hydrocarbon radical abstracts the hydroxy group from the hydroxy–Fe(III) complex by a pairing of the unpaired radical electron with one electron of the hydroxyl group, leaving cytochrome P450 in the +3 oxidation state, where it began in figure 11.23. The mechanism shown in figure 11.23 is known as the "rebound" mechanism of oxygenation. The second step of the mechanism is the rebound step, and the rate constant for this step in a microsomal cytochrome P450 has been estimated to be greater

Aconitase and Iron-Sulfur Clusters

Iron–sulfur clusters most often participate in electron transfer processes by serving as one-electron donors and acceptors. They play essential roles in the terminal electron transport pathways in the membranes of cells (see chapter 16). In recent years, iron–sulfur clusters have been found in enzymes that catalyze reactions in which electron transfer does not take place. In these enzymes the iron–sulfur complexes have other functions. Aconitase is an example of an enzyme in which an iron–sulfur cluster participates in catalysis but not in electron transfer.

Aconitase is one of the important enzymes in the TCA cycle of metabolism (see chapter 15). The reaction it catalyzes is

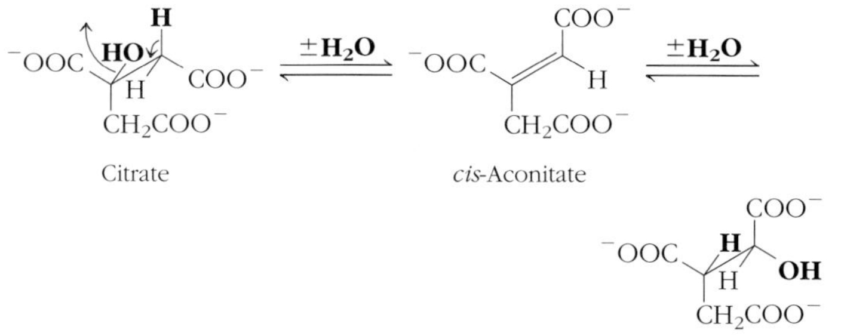

Citrate $\underset{}{\overset{\pm H_2O}{\rightleftharpoons}}$ *cis*-Aconitate $\underset{}{\overset{\pm H_2O}{\rightleftharpoons}}$ Isocitrate

This is a dehydration/hydration in which the hydroxyl group leaves with its bonding electron pair, that is, as if it were OH^-. Hydroxide is a very poor leaving group at pH 7 and cannot leave by itself; it must be derivatized in some way and leave in the form of some other entity. Dehydration reactions are catalyzed by strong acids, which protonate a hydroxyl group to an oxonium ion $R{-}OH_2^+$ and allow it to depart as a molecule of water. Strong acids are incompatible with cellular conditions, so alternative means for stabilizing the departing hydroxyl group have evolved at enzymatic sites. In the case of aconitase the hydroxyl group is coordinated to a metal ion, one of the irons in an iron–sulfur cluster. Purified aconitase contains an $[Fe_3S_4]$ cluster, in which one of the four iron positions is vacant. The enzyme lacks a cysteine residue that would normally coordinate to the fourth iron. The enzyme is activated by Fe^{2+}, which binds to the vacant iron site. This fourth iron participates in binding the substrate, which provides two ligands to it in the cluster. One of the ligands is the hydroxyl group that will depart.

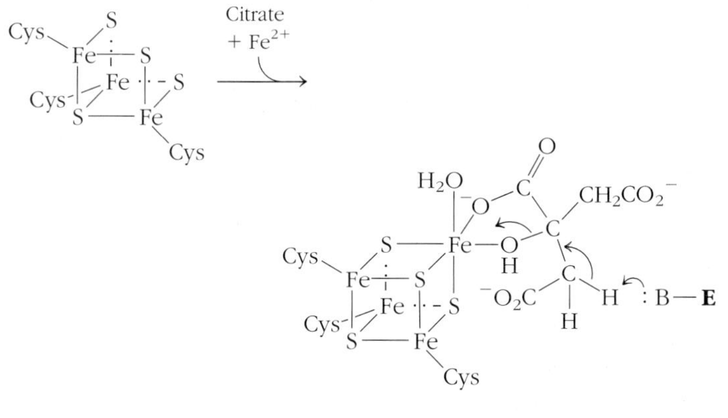

Coordination of the departing hydroxyl group to Fe^{2+} in the iron–sulfur cluster makes it a good leaving group. Metal coordination is analogous to protonation in this case. The metal ion makes the hydroxyl group into a good leaving group. In the second step, the hydroxyl group adds to the other carbon of *cis*-aconitate to form isocitrate:

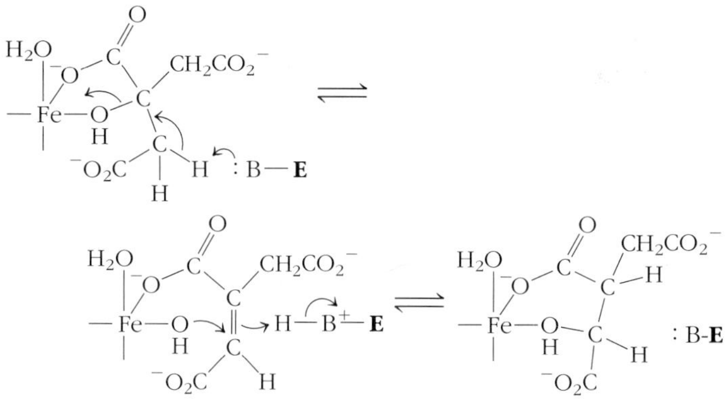

As product and substrate, respectively, isocitrate and citrate rapidly dissociate from their complexes with the iron–sulfur complex. However, *cis*-aconitate is transformed into either isocitrate or citrate faster than it dissociates from the active site, so its dissociation is a rare event.

Metal ions acting in this capacity are said to act as *Lewis acids*. The Lewis definition of acids and bases extends the Bronsted acid–base concept (H^+) to species other than protons by defining an acid as a molecule or ion that will form a bond with a nonbonding electron pair. A base is defined as a molecule or ion that will donate an electron pair to form a bond with a Lewis acid. The Lewis definition includes protons as acids as well as other species such as metal ions, BCl_3, $FeCl_3$, etc. You will recall from organic chemistry that Lewis acids such as $AlCl_3$ are used as catalysts in electrophilic aromatic substitution reactions, for example in the acylation of benzene.

Figure 11.22

Steps in the formation of the oxygenating species of cytochrome P450. The oxygenating species of cytochrome P450 is generated by the transfer of two electrons in discrete steps from NADPH via the flavoprotein reductase and iron–sulfur clusters to the iron–porphyrin complex, together with the reaction with oxygen. The Fe(III)–peroxide complex then undergoes a further expulsion of water to form the oxygenating species shown in brackets. The oxygenating species is not directly observed, since it reacts quickly, but it is thought to contain Fe(IV) and a delocalized radical-cation in the porphyrin ring.

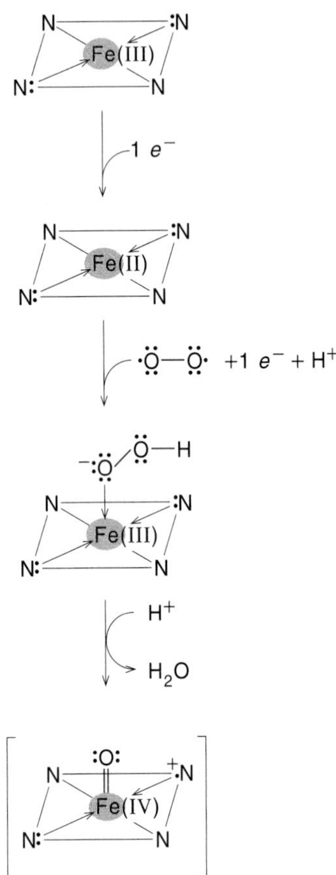

Figure 11.23

The rebound mechanism for oxygenation of hydrocarbons by cytochrome P450. The oxygenating species of cytochrome P450 in figure 11.22 is a highly reactive paramagnetic species that can abstract an unactivated and unreactive hydrogen from a hydrocarbon in the first step to form a hydrocarbon radical. In the rebound step the hydrocarbon radical abstracts a hydroxyl radical from the hydroxyporphyrin intermediate.

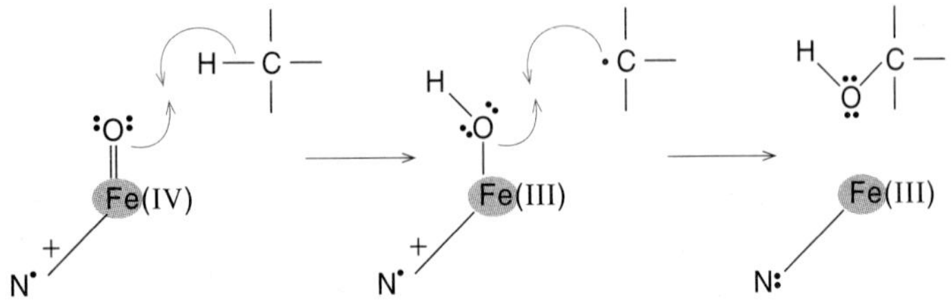

than $10^{10}\ s^{-1}$. The magnitude of this rate constant explains why the radical and hydroxy-Fe intermediates have not been detected by presently available spectroscopic methods.

Metal Cofactors

In addition to cobalt and iron (discussed above), other metals frequently function as cofactors in enzyme-catalyzed reactions. Like coenzymes, they are useful because they offer something not available in amino acid side chains. The most important of such features of metals are their high concentration of positive charge, their directed valences for interacting with two or more ligands, and their ability to exist in two or more valence states.

The alkali metal and alkaline-earth metal ions are spherically symmetrical with respect to charge distribution, so they do not usually show directed valences. There are notable exceptions, such as the case of Mg^{2+} in chlorophyll (see chapter 17), which adopts an approximately square planar distribution of orbitals. Alkali metals, Na^+ and K^+, with their single positive charges and lack of *d*-electronic orbitals for sharing, almost never make tight complexes with proteins. On rare occasions the alkaline-earth divalent cations, Mg^{2+} and Ca^{2+}, can make strong complexes. The alkali metal and alkaline-earth cations are asymmetrically distributed in the organism, with K^+ and Mg^{2+} being concentrated in the cytosol and Na^+ and Ca^{2+} being concentrated in the endoplasmic reticulum and the blood.

Most of the remaining metals found in biological systems are from the first transition series, in which the 3*d* orbitals are only partially filled. These metals all prefer structures with multiple coordination numbers, and, except for Zn^{2+}, which has a filled 3*d* orbital, they can exist in more than one oxidation state, a feature making them potentially useful in oxidation–reduction reactions. The stability of zinc's electronic state combined with its preference for a coordination number of 4–6 makes zinc singularly useful in enzyme reactions, in which it acts as a Lewis acid (e.g., fig. 11.24), and in structural situations, in which it chelates nitrogen

Figure 11.24

Proposed role of Zn^{2+} in carbonic anhydrase. Carbonic anhydrase catalyzes the reaction $H_2O + CO_2 \rightleftharpoons HCO_3^- + H^+$. In the enzyme, Zn^{2+} forms a complex with H_2O. Proton displacement generates an OH^- ion still bound to the Zn^{2+}. Nucleophilic attack of CO_2 by the OH^- ion generates bicarbonate.

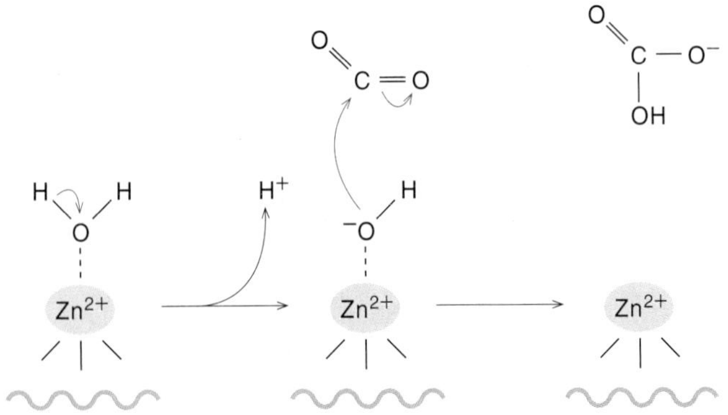

(usually in histidine side chains) and/or sulfur-containing amino acid side chains (usually cysteines) to rigidify the structural domain of a protein (see fig. 36.18).

The involvement of transition metals in biochemical reactions is discussed in later chapters, as indicated in table 11.3, which also gives the coordination numbers and stereochemistries of the most significant oxidation states.

Lipid-Soluble Vitamins

We have seen that most water-soluble vitamins are converted by single or multiple steps to coenzymes. Our understanding of the lipid-soluble vitamins (see table 11.1) and of how they are utilized by the organism is much less extensive. We briefly discuss the structure and functions of some lipid-soluble vitamins.

Vitamin D_3 (cholecalciferol) can be made in the skin from 7-dehydrocholesterol in the presence of ultraviolet light (see fig. 27.13). Vitamin D_3 is formed by the cleavage of ring 3 of 7-dehydrocholesterol. Vitamin D_3 made in skin or absorbed from the small intestine is transported to the liver and hydroxylated at C-25 by a microsomal mixed-function oxidase. 25-Hydroxyvitamin D_3 appears to be biologically inactive until it is hydroxylated at C-1 by a mixed-function oxidase in kidney mitochondria. The 1,25-dihydroxyvitamin D_3 is delivered to target tissues for the regulation of calcium and phosphate metabolism. The structure and mode of action of 1,25-dihydroxyvitamin D_3 are analogous to those of the steroid hormones (see chapter 27).

Vitamin K was discovered by Henrik Dam in Denmark in the 1920s as a fat-soluble factor important in blood coagulation (K is for "koagulation"). The structures of vitamins K_1 and K_2 (fig. 11.25) were elucidated by Edward Doisy. Vitamin K_1 is found in plants, vitamin K_2 in animals and bacteria. How this vitamin functions in blood coagulation eluded scientists until 1974, when vitamin K was shown to be needed for the formation of γ-carboxyglutamic acid (fig. 11.26) in certain proteins. γ-Carboxyglutamic acid specifically binds calcium, which is important for blood coagulation. Such modified glutamic acid residues appear to be important in many other processes involving calcium transport and calcium-regulated metabolic sequences.

Vitamin E (α-tocopherol) (fig. 11.27) was recognized in 1926 as an organic-soluble compound that prevented sterility in rats. The function of this vitamin still has not been clearly established. A favorite theory is that it is an antioxidant that prevents peroxidation of polyunsaturated fatty acids. Tocopherol certainly prevents peroxidation in vitro, and it can be replaced by other antioxidants. However, other antioxidants do not relieve all the symptoms of vitamin E deficiency.

Vitamin A (*trans*-retinol) is called an isoprenoid alcohol because it consists, in part, of units of a single five-carbon compound called isoprene:

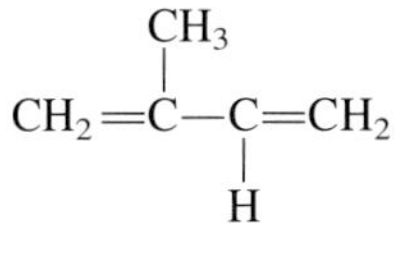

$$CH_2{=}C(CH_3){-}CH{=}CH_2$$

Figure 11.25

Structures of vitamins K_1 and K_2. K_1 is found in plants, K_2 in animals and bacteria.

Vitamin K_1
(phylloquinone)

Vitamin K_2
(menaquinone)

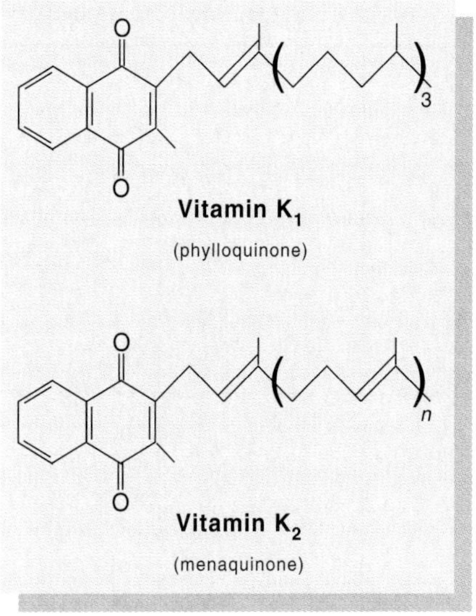

Figure 11.26

Vitamin K–dependent carboxylation of a glutamic acid residue in a protein. This reaction is essential to blood clotting.

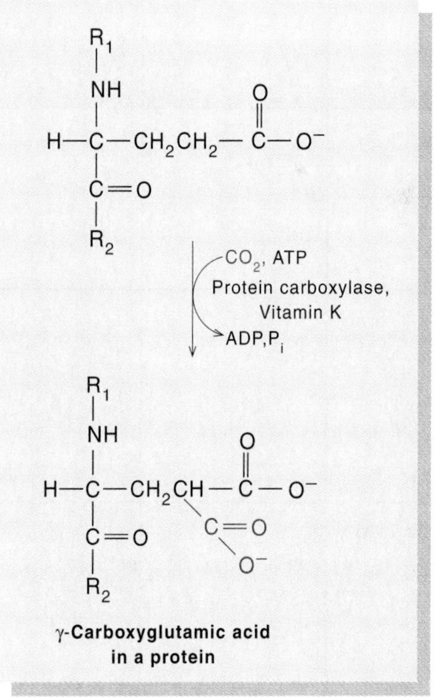

CO_2, ATP
Protein carboxylase, Vitamin K
ADP, P_i

γ-Carboxyglutamic acid in a protein

Isoprene is also a precursor of steroids and terpenes. This relationship becomes clear when we examine the biosynthesis of these compounds (see chapter 23). Vitamin A is either biosynthesized

Table 11.3
Complexing Properties of Transition Metals of Biologic Importance

Metal and Valence State	Electron Configuration	Coordination Number	Stereochemistry	Examples
Manganese (Mn)				
Mn(II)	$3d^5$	4	Tetrahedral 	Pyruvate decarboxylase (see chapters 11, 15, and 16)
		4	Square planar 	
Mn(III)	$3d^4$	6	Octahedral	
Iron (Fe)	$3d^6$	4	Tetrahedral	
Fe(II)		5	Trigonal bipyramidal 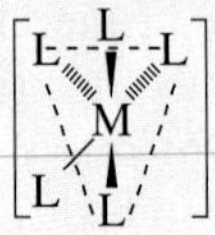	
		6	Octahedral	Hemoglobin (chapter 6)
Fe(III)	$3d^5$	4	Tetrahedral	Iron–sulfur coenzymes and cytochromes (see chapters 11 and 16)
		6	Octahedral 	
Cobalt (Co)				
Co(I)	$3d^8$	4 5	Tetrahedral Trigonal bipyramidal	
Co(II)	$3d^7$	4 5 6	Tetrahedral; square planar Trigonal bipyramidal Octahedral	
Co(III)	$3d^6$	6	Octahedral	Vitamin B_{12} (this chapter)
Copper (Cu)				
Cu(I)	$3d^{10}$	2 4	Linear [L—M—L] Tetrahedral	
Cu(II)	$3d^9$	4 6	Tetrahedral; square pyramidal	Cytochrome (see chapter 16)
Zinc(Zn)				
Zn(II)	$3d^{10}$	4	Tetrahedral	Thermolysin (chapter 9) Alcohol dehydrogenase (chapter 9)
		5 6	Trigonal bipyramidal	Nucleic acid polymerase (chapter 31)
			Octahedral	Regulatory proteins (chapters 35 and 36)
Molybdenum (Mo)				
Mo(III)	$4d^3$	6 8	Octahedral Dodecahedral	
Mo(V)	$4d^1$	5 6 8	Trigonal bipyramidal Octahedral Dodecahedral	Nitrogenase (chapter 24)
Mo(VI)	$4d^0$	4 6	Tetrahedral Octahedral	

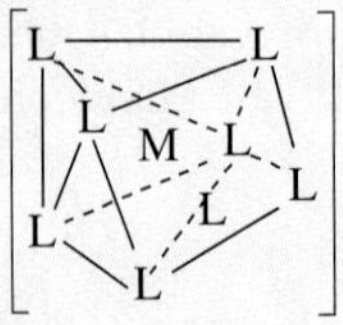

Figure 11.27

Structure of vitamin E (α-tocopherol).

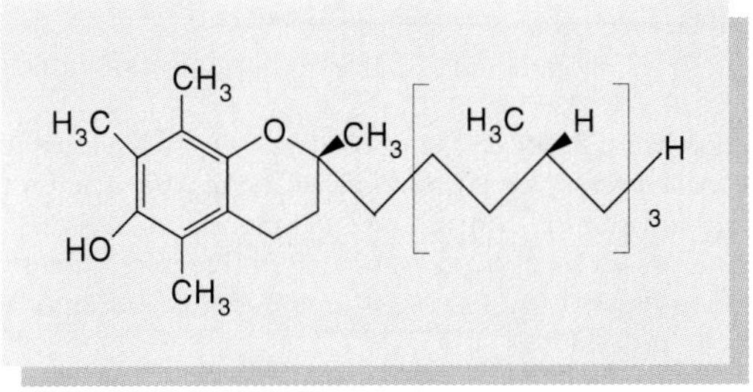

from β-carotene (see fig. 29.4) or absorbed in the diet. Vitamin A is stored in the liver predominantly as an ester of palmitic acid. For many decades it has been known to be important for vision and for animal growth and reproduction. The form of vitamin A active in the visual process is 11-*cis*-retinal, which combines with the protein opsin to form rhodopsin. Rhodopsin is the primary light-gathering pigment in the vertebrate retina (see chapter 28).

Summary

Coenzymes are molecules that act in cooperation with enzymes to catalyze biochemical processes, performing functions that enzymes are otherwise chemically not equipped to carry out. Most coenzymes are derivatives of the water-soluble vitamins, but a few, such as hemes, lipoic acid, and iron–sulfur clusters, are biosynthesized in the body. Each coenzyme plays a unique chemical role in the enzymatic processes of living cells.

1. Thiamine pyrophosphate promotes the decarboxylation of α-keto acids and the cleavage of α-hydroxy ketones.
2. Pyridoxal-5′-phosphate promotes decarboxylations, racemizations, transaminations, aldol cleavages, and elimination reactions of amino acid substrates.
3. Nicotinamide coenzymes act as intracellular electron carriers to transport reducing equivalents between metabolic intermediates. They are cosubstrates in most of the biological redox reactions of alcohols and carbonyl compounds and also act as cocatalysts with some enzymes.
4. Flavin coenzymes act as cocatalysts with enzymes in a large number of redox reactions, many of which involve O_2.
5. Phosphopantetheine coenzymes form thioester linkages with acyl groups, which they activate for group transfer reactions.
6. Lipoic acid mediates electron transfer and acyl-group transfer in α-keto acid dehydrogenase complexes.
7. Biotin mediates carboxylation of activated methyl groups.
8. Phosphopantetheine, lipoic acid, and biotin, by virtue of their long, flexible structures, facilitate the physical translocation of chemically reactive species among separate catalytic sites.
9. Tetrahydrofolates are cosubstrates for a variety of one-carbon transfer reactions. Tetrahydrofolates maintain formaldehyde and formate in chemically poised states, making them available for essential processes by specific enzymatic action.
10. Heme coenzymes participate in a variety of electron-transfer reactions, including reactions of peroxides and O_2. Iron–sulfur clusters, composed of Fe and S in equal numbers with cysteinyl side chains of proteins, mediate other electron-transfer processes, including the reduction of N_2 to 2 NH_3. Nicotinamide, flavin, and heme coenzymes act cooperatively with iron–sulfur proteins in multienzyme systems that catalyze hydroxylations of hydrocarbons and also in the transport of electrons from foodstuffs to O_2.
11. 5′-Deoxyadenosylcobalamin (a vitamin B_{12} coenzyme) is involved as the essential cofactor in intramolecular rearrangements in which an unreactive hydrogen exchanges positions with a group bonded to an adjacent carbon. These are radical rearrangements in which the coenzyme initiates the formation of substrate radicals by virtue of the weakness of the cobalt–carbon bond in the coenzyme. Homolytic scission of this bond generates a 5′-deoxyadenosyl-5′ radical that initiates radical formation by abstraction of a hydrogen atom.

    Another biologically important form of vitamin B_{12} is methylcobalamin, in which the upper axial cobalt substituent is a methyl group instead of an adenosyl group (see fig. 11.17). This form of vitamin B_{12} participates in many reactions in bacteria and in the biosynthesis of methionine in mammals.
12. Metals often perform the role of enzyme cofactors. The most important features of metals as cofactors are their high concentration of positive charge, their directed valences for interacting with two or more ligands, and their ability to exist in two or more valence states.
13. Ascorbic acid is needed as a reductant to maintain some enzymes in their active forms.
14. In general, less is known about the mechanisms of action of the lipid-soluble vitamins than about the coenzymes derived from water-soluble vitamins. The structures and functions of vitamins D, K, E, and A are discussed briefly.

Selected Readings

Bruice, T. C., and S. J. Benkovic, *Bioorganic Chemistry,* vols. 1 and 2. Menlo Park, Calif.: Benjamin, 1966. A detailed discussion of the mechanisms of bioorganic reactions, including those involving coenzymes.

DiMarco, A. A., T. A. Bobik, and R. S. Wolfe, Unusual coenzymes of methanogenesis. *Ann. Rev. Biochem.* 59:355 (1990). A review of the recently characterized coenzymes required for the biosynthesis of methane in methanogenic bacteria.

Dolphin, D. (ed.), *B₁₂*, vols. 1 and 2. New York: Wiley-Interscience, 1982. Chemistry and mechanism of action of vitamin B_{12}.

Dolphin, D., R. Poulson, and O. Avamovic (eds.), *Pyridine Nucleotide Coenzymes,* part A & part B. New York: Wiley-Interscience, 1987. Chemical, biochemical, and medical aspects of pyridine nucleotide coenzymes.

Dolphin, D., R. Poulson, and O. Avamovic (eds.), *Pyridoxal Phosphate,* part A & part B. New York: Wiley-Interscience, 1987. Chemistry and mechanism of action of pyridoxal phosphate.

Drennan, C. L., S. Huang, J. T. Drummond, R. G. Matthews, and M. L. Ludwig, How a protein binds B_{12}: A 3.0 Å x-ray structure of B_{12}-binding domains of methionine synthase. *Science* 266:1669–1674, 1994. Structure of methylcobalamin provides unanticipated mechanistic insight into the normal methylation cycle.

Frausto de Silva, J. J. K., and R. J. P. Williams, *The Biological Chemistry of the Elements.* Oxford: Clarendon Press, 1991.

Frey, P. A., The importance of organic radicals in enzymatic cleavage of unactivated C—H bonds. *Chem. Rev.* 90:1343, 1990. A brief review of coenzymes required to cleave unreactive C—H bonds.

Jencks, W. P., *Catalysis in Chemistry and Enzymology.* New York: McGraw-Hill, 1969. A detailed analysis of mechanisms of enzymatic and nonenzymatic reactions, including those involving coenzymes.

Knowles, J. R., The mechanism of biotin-dependent enzymes. *Ann. Rev. Biochem.* 58:195, 1989. Review of the chemical mechanism of biotin-dependent carboxylation reactions.

Kovacs, J. A., S. C. Shoner, and J. J. Ellison, Metal-carbon bonds in nature. *Science* 270:587, 1995.

Ortiz de Montellano, P. R. (ed.), *Cytochrome P-450: Structure, Mechanism and Biochemistry.* New York: Plenum, 1986. A treatise on the structure and function of cytochrome P450 monooxygenases.

Phipps, D. A., *Metals and Metabolism.* Oxford Chemistry Series. Oxford: Clarendon Press, 1976. Examines the importance of metal ions in metabolic processes.

Popják, G., Stereospecificity of enzymic reactions. In P. D. Boyer (ed.), *The Enzymes,* vol. 2. New York: Academic Press, 1970. p. 115.

Walsh, C. T., *Enzymatic Reaction Mechanisms.* San Francisco: Freeman, 1977. Provides discussion of the mechanisms of enzymatic reactions. An in-depth treatment of coenzymes.

PROBLEMS

1. When amino acid 1 labeled with deuterium in the $\alpha = -1$ hydrogen position is supplied as a substrate for a transaminase, the product, amino acid 2, has about 40% of the molecules labeled with deuterium in the α position. How can you explain this transfer of the deuterium?
2. If lysine deuterated in the α position is exposed to lysine decarboxylase, would you expect the resulting cadaverine (1,5-diaminopentane) to contain the deuterium?
3. Why does the frequent use of the abbreviation $NADH_2$ raise the blood pressure of a biochemist?
4. What structural features of biotin and lipoic acid allow these cofactors to be covalently bound to a specific protein in a multienzyme complex yet participate in reactions at active sites on other enzymes of the complex?
5. The following reactions are catalyzed by pyridoxal-5′-phosphate-dependent enzymes. Write a reaction mechanism for each, showing how pyridoxal-5′-phosphate is involved in catalysis.

(a) $$CH_3-\overset{\overset{O}{\|}}{C}-COO^- + R-\underset{\underset{NH_3^+}{|}}{CH}-COO^- \longrightarrow CH_3-\underset{\underset{NH_3^+}{|}}{CH}-COO^- + R-\overset{\overset{O}{\|}}{C}-COO^-$$

(b) $$H_2N-(CH_2)_4-\underset{\underset{NH_2}{|}}{CH}-CO_2^- \longrightarrow CO_2 + H_2N-(CH_2)_5-NH_2$$

(c) $$^-O_2C-CH_2-\underset{\underset{NH_2}{|}}{CH}-CO_2^- \longrightarrow CO_2 + CH_3-\underset{\underset{NH_2}{|}}{CH}-CO_2^-$$

6. $NADP^+$ differs from NAD^+ only by phosphorylation of the C-2′ OH group on the adenosyl moiety. The redox potentials differ only by about 5 mV. Why do you suppose it is necessary for the cell to employ two such similar redox cofactors?
7. Thiamine pyrophosphate–dependent enzymes catalyze the reactions shown below. Write a chemical mechanism that shows the catalytic role of the coenzyme.

(a) $$CH_3-\overset{\overset{O}{\|}}{C}-CO_2H \longrightarrow CO_2 + CH_3-\overset{\overset{O}{\|}}{C}-H$$

(b) $$ⓅOCH_2-(CHOH)_2-\overset{\overset{OH}{|}}{CH}-\overset{\overset{O}{\|}}{C}-CH_2OH + HOPO_3^{2-} \longrightarrow ⓅO-CH_2-(CHOH)_2-CHO + CH_3-\overset{\overset{O}{\|}}{C}-OPO_3^{2-} + H_2O$$

8. Malate synthase catalyzes the condensation of acetyl-CoA with glyoxalate to form L-malate.
   (a) Write a chemical reaction illustrating the reaction catalyzed by malate synthase.
   (b) Explain the chemical basis for using acetyl-CoA rather than acetate in the condensation reaction.
   (c) The product released from the enzyme is L-malate rather than malyl-CoA. Explain the role of thioester hydrolysis in the condensation reaction.
9. Write the mechanism showing how $NAD(P)^+$ is involved in the following reactions.
   (a) Malate dehydrogenase:

$$\begin{array}{l}COOH\\ |\\ CHOH\\ |\\ CH_2\\ |\\ COOH\end{array} + NAD^+ \longrightarrow \begin{array}{l}COOH\\ |\\ C{=}O\\ |\\ CH_2\\ |\\ COOH\end{array} + NADH + H^+$$

   (b) Malate dehydrogenase (decarboxylating; malic enzyme)

$$\begin{array}{l}COOH\\ |\\ CHOH\\ |\\ CH_2\\ |\\ COOH\end{array} + NADP^+ \longrightarrow \begin{array}{l}COOH\\ |\\ C{=}O\\ |\\ CH_3\end{array} + NADPH + CO_2 + H^+$$

10. Write the mechanisms that show the involvement of biotin in the following reactions:

(a) $$CH_3-\overset{\overset{O}{\|}}{C}-SCoA + HCO_3^- + ATP \longrightarrow$$

$$HO_2C—CH_2—\overset{O}{\overset{\|}{C}}—SCoA + ADP + HOPO_3{}^{2-}$$

(b) 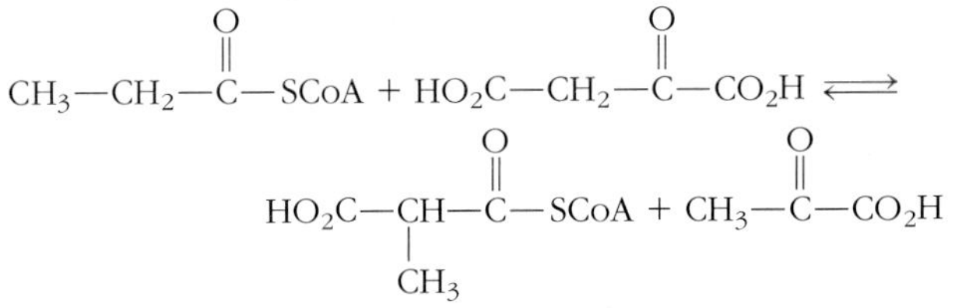 $CH_3—CH_2—\overset{O}{\overset{\|}{C}}—SCoA + HO_2C—CH_2—\overset{O}{\overset{\|}{C}}—CO_2H \rightleftharpoons$

$$HO_2C—\underset{CH_3}{\underset{|}{CH}}—\overset{O}{\overset{\|}{C}}—SCoA + CH_3—\overset{O}{\overset{\|}{C}}—CO_2H$$

11. (a) What metabolic advantage is gained by having flavin cofactors covalently or tightly bound to the enzyme?
    (b) Would covalently bound NAD^+ ($NADP^+$) be a metabolic advantage or disadvantage?
12. Given the amino acid of the general structure

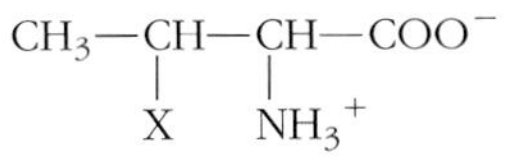

$$CH_3—\underset{X}{\underset{|}{CH}}—\underset{NH_3{}^+}{\underset{|}{CH}}—COO^-$$

we could use pyridoxal-5′-phosphate to eliminate X, decarboxylate the amino acid, or oxidize the α carbon to a carbonyl with formation of pyridoxamine-5′-phosphate. The metabolic diversity afforded by PLP, unchanneled, could wreak havoc in the cell. What other components are required to channel the PLP-dependent reaction along specific reaction pathways?
13. Amino acid oxidases catalyze the flavin-dependent reaction

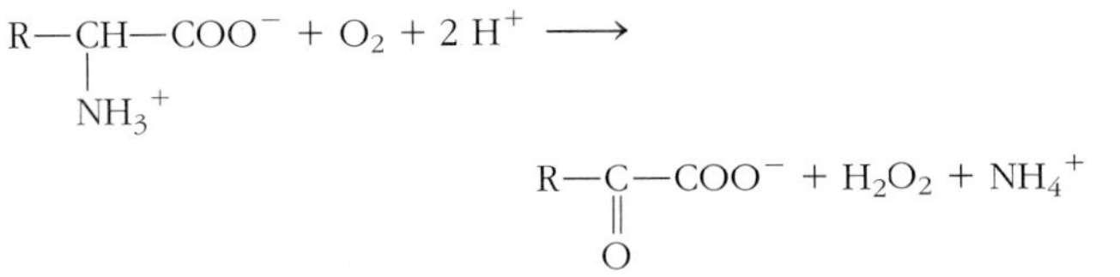

$$R—\underset{NH_3{}^+}{\underset{|}{CH}}—COO^- + O_2 + 2\,H^+ \longrightarrow$$

$$R—\underset{O}{\underset{\|}{C}}—COO^- + H_2O_2 + NH_4{}^+$$

What advantages are gained by the cell in using the PLP-dependent transamination reaction rather than the FAD-dependent deamination to convert α-amino acids to the corresponding α-keto acids?
14. For each of the following enzymatic reactions, identify the coenzyme involved.

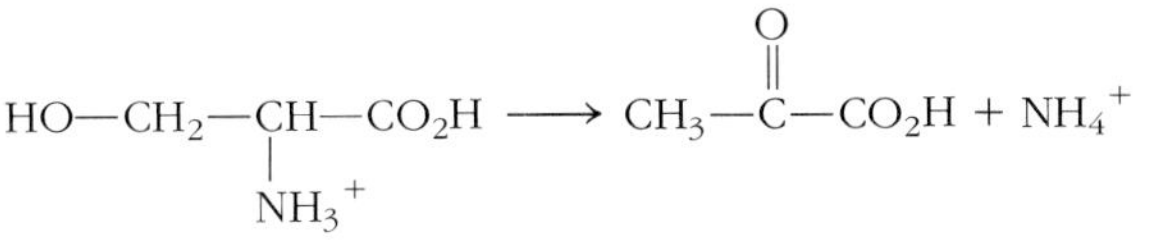

(a) 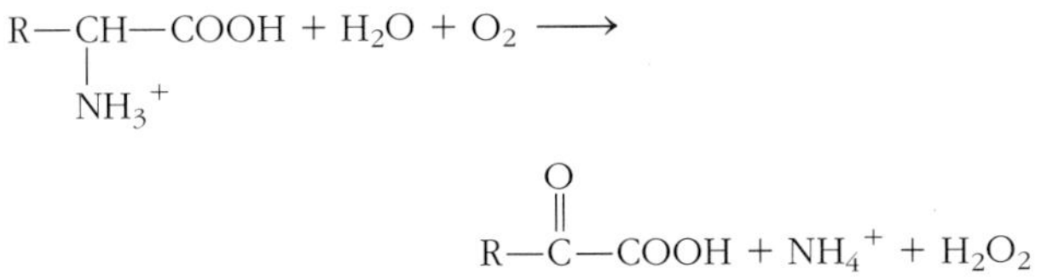 $R—\underset{NH_3{}^+}{\underset{|}{CH}}—COOH + H_2O + O_2 \longrightarrow$

$$R—\overset{O}{\overset{\|}{C}}—COOH + NH_4{}^+ + H_2O_2$$

(b) $HO—CH_2—\underset{NH_3{}^+}{\underset{|}{CH}}—CO_2H \longrightarrow CH_3—\overset{O}{\overset{\|}{C}}—CO_2H + NH_4{}^+$

(c) $CH_3—CH_2—\overset{O}{\overset{\|}{C}}—SCoA + HCO_3{}^- + ATP \longrightarrow$

$$HO_2C—\underset{CH_3}{\underset{|}{CH}}—\overset{O}{\overset{\|}{C}}—SCoA + ADP + P_i$$

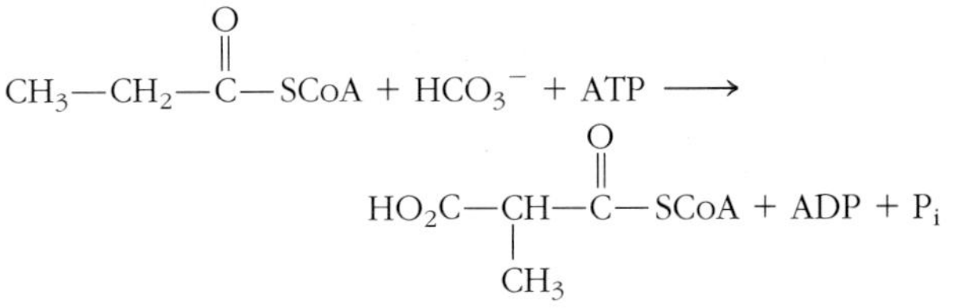

(d) 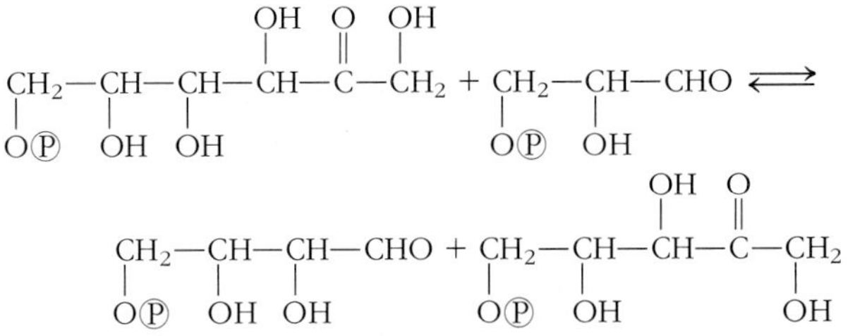 $\underset{OⓅ}{\underset{|}{CH_2}}—\underset{OH}{\underset{|}{CH}}—\underset{OH}{\underset{|}{CH}}—\overset{OH}{\overset{|}{CH}}—\overset{O}{\overset{\|}{C}}—\overset{OH}{\overset{|}{CH_2}} + \underset{OⓅ}{\underset{|}{CH_2}}—\underset{OH}{\underset{|}{CH}}—CHO \rightleftharpoons$

$$\underset{OⓅ}{\underset{|}{CH_2}}—\underset{OH}{\underset{|}{CH}}—\underset{OH}{\underset{|}{CH}}—CHO + \underset{OⓅ}{\underset{|}{CH_2}}—\underset{OH}{\underset{|}{CH}}—\overset{OH}{\overset{|}{CH}}—\overset{O}{\overset{\|}{C}}—\underset{OH}{\underset{|}{CH_2}}$$

15. Lipid-soluble polycyclic aromatic hydrocarbons are in part excreted from the body conjugated to one of several hydrophilic substituents, principally glucuronic acid. What role does the liver microsomal cytochrome P450 (polysubstrate monooxygenase) system play in converting a polycyclic aromatic hydrocarbon to forms that could be conjugated to glucuronic acid?
16. Some bacterial toxins use NAD^+ as a true substrate rather than as a coenzyme. The toxins catalyze the transfer of ADP-ribose to an acceptor protein. Examine the structure of NAD^+ and indicate which portion of the molecule is transferred to the protein. What is the other product of the reaction?
17. Which of the coenzymes listed in table 11.1 can be considered to be derivatives of AMP?
18. The structure of ascorbic acid is shown in figure 11.16. Which hydrogen is the most acidic? Why?
19. Consider the structure of riboflavin and flavin mononucleotide (fig. 11.8). Are these substances actually a nucleoside and a nucleotide, respectively?
20. Both lipoic acid and biotin are coenzymes that are covalently bound to apoenzymes by amide linkages to lysyl side chains (figs. 11.11 and 11.13). Can you provide an explanation of why these similarities might occur in nature?
21. α-Keto acid dehydrogenases have been, on occasion, called "multivitamin pills." Can you explain this statement?
22. What coenzymes covered in this chapter are (a) biological redox agents, (b) acyl carriers, (c) both redox agents and acyl carriers?
23. Iron–sulfur proteins (fig. 11.21) are considered by many biochemists to be proteins left over from earlier forms of life. They are among the few proteins to contain inorganic sulfur. This category of proteins has been known to biochemists since the early days of protein chemistry. What feature would you surmise led to their discovery and classification as a distinct protein type?

PART

4

Metabolism of Carbohydrates

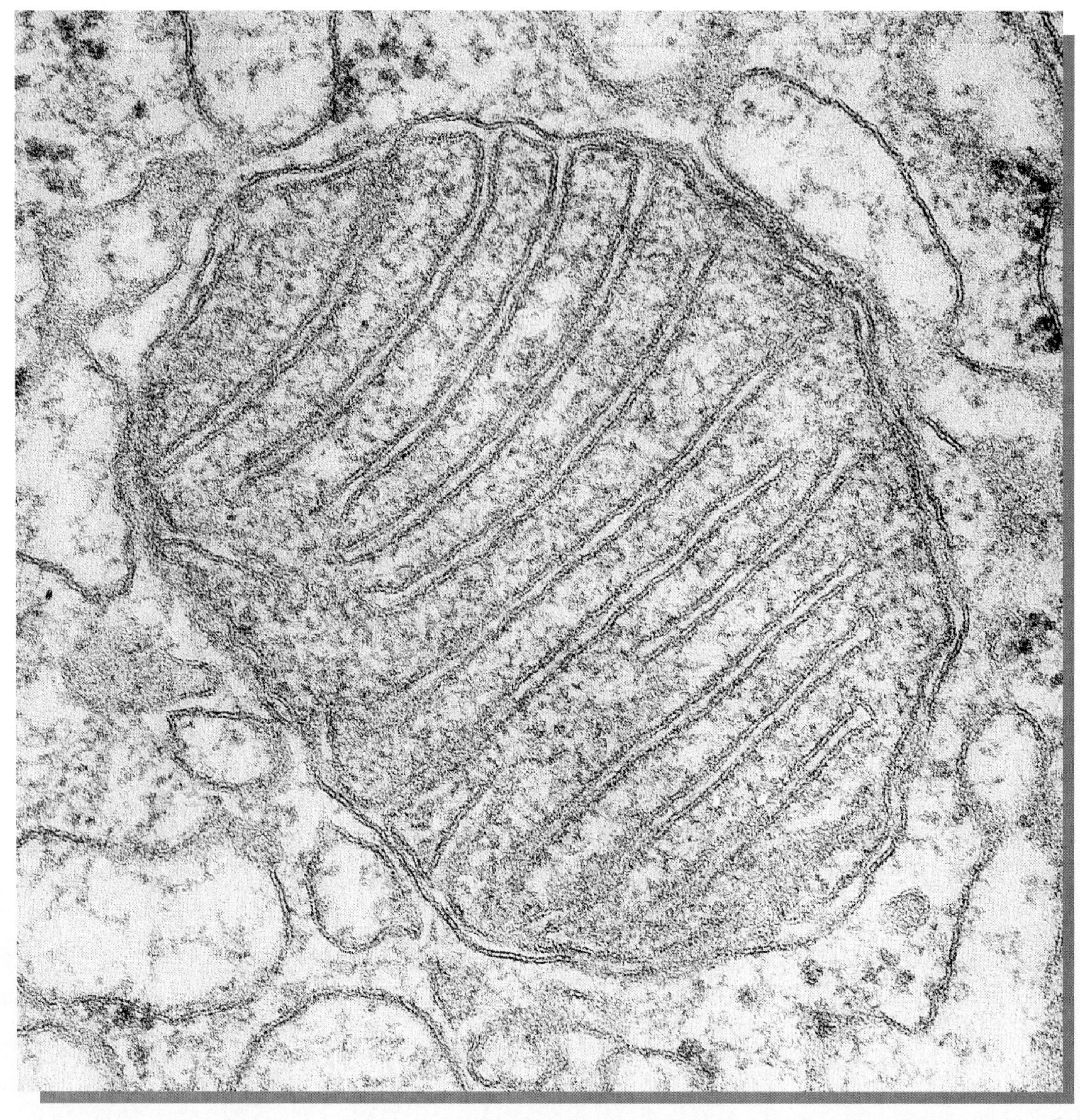

Electron micrograph of a thin section of a liver mitochondrion. Courtesy of Dr. Daniel S. Friend

C H A P T E R

METABOLIC STRATEGIES

12

Biochemical reactions are organized into catabolic pathways that produce energy and reducing power and anabolic pathways that consume these products in the process of biosynthesis.

In this and most of the next 16 chapters we consider the synthesis and degradation of small molecules in the living cell. These aspects of biochemistry are collectively referred to as intermediary metabolism because they focus on the small-molecule intermediates in metabolic pathways. In this chapter the principles governing intermediary metabolism are discussed.

Living Cells Require a Steady Supply of Starting Materials and Energy

Intermediary metabolism, the synthesis and degradation of small molecules, serves two functions: It supplies the energy needed for the synthesis of macromolecules and other energy-requiring processes, and it furnishes these processes with the necessary starting materials—amino acids for protein synthesis, fatty acids for lipid synthesis, nucleoside triphosphates for nucleic acid synthesis, and sugars for polysaccharide synthesis.

The demand for energy and starting materials varies widely in different biological processes. To meet these fluctuating needs, the rates of the reaction sequences in intermediary metabolism must be adjustable over broad ranges. Living organisms adjust these rates so that the concentrations of key metabolites are remarkably stable. This constancy is achieved by maintaining a rate of synthesis for each intermediate that balances its rate of utilization. The term steady state is used to describe this nonequilibrium situation. Unfortunately, the term "steady state" is commonly used to describe a somewhat different situation in enzyme kinetics (see chapter 8).

Organisms Differ in Sources of Energy, Reducing Power, and Starting Materials for Biosynthesis

To maintain a steady state and to permit growth and reproduction as well, all living cells require energy (ATP) and starting materials for biosynthesis. Reducing power (NADPH) is also required, since most biosynthesis involves converting compounds to a more reduced state. Whereas the needs of different organisms are similar, the ways in which organisms satisfy their needs can be quite

Table 12.1
Metabolic Classification of Organisms

Type of Organism	Source of ATP[a]	Source of NADPH[b]	Source of Carbon	Examples
Chemoautotroph	Oxidation of inorganic compounds	Oxidation of inorganic compounds	CO_2	Hydrogen, sulfur, iron, and denitrifying bacteria
Photoautotroph	Sunlight	H_2O	CO_2	Cells of higher plants, blue-green algae, photosynthetic bacteria
Photoheterotroph	Sunlight	Oxidation of organic compounds	Organic compounds	Nonsulfur purple bacteria
Heterotroph	Oxidation of organic compounds	Oxidation of organic compounds	Organic compounds	All higher animals, most microorganisms, nonphotosynthetic plant cells

[a]ATP = adenosine triphosphate.

[b]NADPH = The reduced form of nicotinamide adenine dinucleotide phosphate is the main source of reducing power.

different. Indeed, the most fundamental metabolic distinction between organisms relates to the ways in which they satisfy their basic metabolic needs (table 12.1).

Autotrophs exploit the inorganic environment without recourse to compounds produced by other organisms. Thus all of their carbon compounds must be synthesized from inorganic compounds of carbon, carbonates, or carbon dioxide. Chemoautotrophs obtain reducing power by oxidation of inorganic materials, such as hydrogen, or reduced compounds of sulfur or nitrogen. Each species is usually specific in the electron donors it can utilize. The same reduced compounds supply electrons for regeneration of ATP by electron-transfer phosphorylation. Photoautotrophs obtain ATP by electron-transfer phosphorylation during the cycling of photochemically excited electrons. Some photoautotrophs, including plants, are also able to use the energy of sunlight to extract electrons from water; thereby photoautotrophs have achieved independence of all sources of energy except for the sun, and of all sources of electrons and carbon except for water and carbon dioxide (CO_2). For this reason, photoautotrophic fixation of carbon dioxide is the predominant base of the food chain in the biosphere.

Common heterotrophs depend on preformed organic compounds for all three primary needs. Although some carbon dioxide is fixed in heterotrophic metabolism, the heterotrophic cell thrives at the expense of compounds formed by other cells and is not capable of the net conversion (fixation) of carbon dioxide into organic compounds. Some bacteria referred to as photoheterotrophs are able to regenerate ATP photochemically but cannot use photochemical reactions to supply electrons to $NADP^+$. Such organisms are like other heterotrophs in their dependence on preformed organic compounds. But because of their photochemical apparatus, the photoheterotrophs are able to use available organic food more efficiently than common heterotrophs.

Although all heterotrophs depend on preformed organic compounds, they differ markedly in the numbers and types of compounds they require. Some species can make all required compounds when supplied with a single carbon source; others have lost some or many biosynthetic capabilities. Mammals must obtain about half of their amino acids from external sources. They also are unable to make several metabolic cofactors, as evidenced by their well-known nutritional requirements for vitamins. Some bacteria and some parasites have even more extensive nutritional requirements than mammals.

Reactions Are Organized into Sequences or Pathways

To appreciate how biochemical reactions serve common functions, we must examine the organization of biochemical reactions in some detail. Most of the enzyme-catalyzed reactions in living cells are organized into sequences or pathways. In a pathway it is the overall sequence that serves a function, not the individual reactions. For example, the conversion of the organic compound chorismate to tryptophan occurs in five discrete steps, each requiring a specific enzymatic activity (fig. 12.1). The intermediates between chorismate and tryptophan serve no function except as precursors of tryptophan. The end product, tryptophan, has many uses, including its role as a building block in protein synthesis.

A central role of breakdown of the six-carbon sugar, glucose, to the three-carbon compound, pyruvate, is to supply ATP to the cell. But as in the case of the tryptophan pathway, the function of each reaction in glucose catabolism can be appreciated only by considering the overall sequence. This pathway involves ten enzymatically catalyzed steps in going from glucose to pyruvate (fig. 12.2). In the first and third steps, ATP is actually consumed. It is only in the seventh and tenth steps that the starting amount of ATP is recovered and further ATP is synthesized. Hence the function of this sequence is not fully served until each substrate has passed through the entire sequence.

Figure 12.1

The biosynthesis of tryptophan. Tryptophan is synthesized in five steps from chorismate. Each step requires a specific enzyme activity. The four intermediates between chorismate and tryptophan serve no function other than as precursors of tryptophan. The detailed steps of each reaction in this sequence are described in chapter 24.

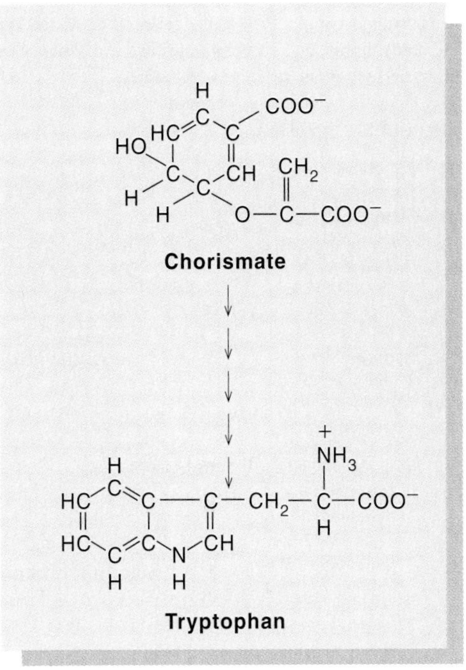

Figure 12.2

The breakdown of glucose to pyruvate. The conversion of glucose to pyruvate requires ten steps and one enzyme for each step. In steps 1 and 3, ATP is consumed. In steps 7 and 10, ATP is produced. The net production of ATP is 2 moles for each mole of glucose consumed. Unlike the intermediates in the tryptophan pathway, many of the intermediates between glucose and pyruvate serve as useful substances for other purposes. A full description of all the steps in this sequence is given in chapter 14.

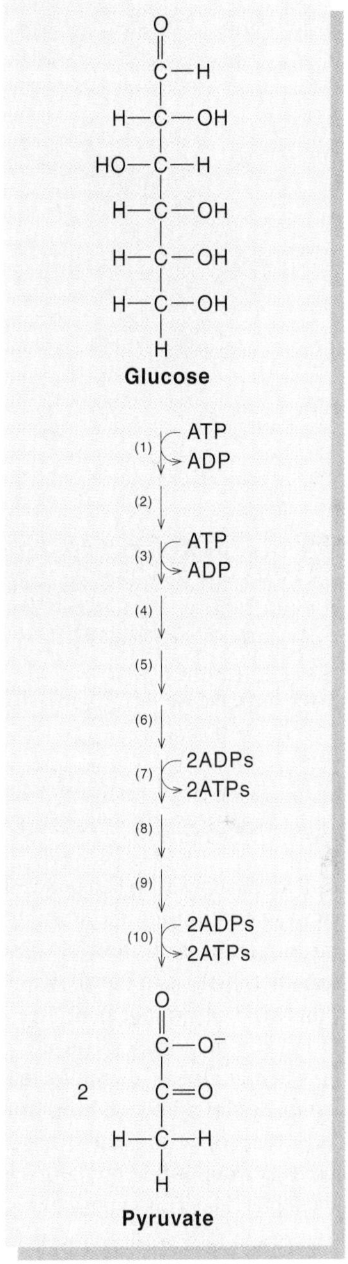

Glucose breakdown and tryptophan biosynthesis illustrate pathways consisting of a linear series of reactions. In chapter 14 we will see that the glycolytic pathway also contains branches so that common intermediates can proceed along diverging routes after the branchpoint. We will also see that the glycolytic pathway is organized so that some of the reactions within the pathway can go in either the forward or the reverse direction.

Sequentially Related Enzymes Are Frequently Clustered

A fundamental aspect of biochemical organizations is that the enzymes that catalyze a sequence are often clustered together in the cell. As a result, an intermediate produced in the first reaction in a pathway is passed directly to the second enzyme in the pathway, and so on. Such an arrangement might be expected to accelerate synthesis of the end product, minimize the loss of intermediates, and facilitate regulation of the pathway.

Three types of enzyme clustering are found (fig. 12.3). In the simplest situation, all of the catalytic activities for a particular pathway are found in proteins that exist as independent soluble proteins in the same cellular compartment. In such cases the intermediates must get from one enzyme to the next in the sequence by free diffusion through the cytosol or by transfer after contact between two sequentially related enzymes—one carrying the reactive intermediate and the other ready to receive it. Such is the situation for the enzymes involved in the breakdown of glucose to pyruvate.

Enzymes with related functions often are compartmentalized into specific organelles, a process that enhances metabolic efficiency in various ways. First, the higher concentrations of enzymes and intermediates resulting from localization lead to faster

Figure 12.3

The organization of functionally related enzymes. Functionally related enzymes are organized in three ways: (*a*) as unlinked proteins soluble in the aqueous milieu of the same cellular compartment, (*b*) as components in a multiprotein complex, or (*c*) as components on a membrane.

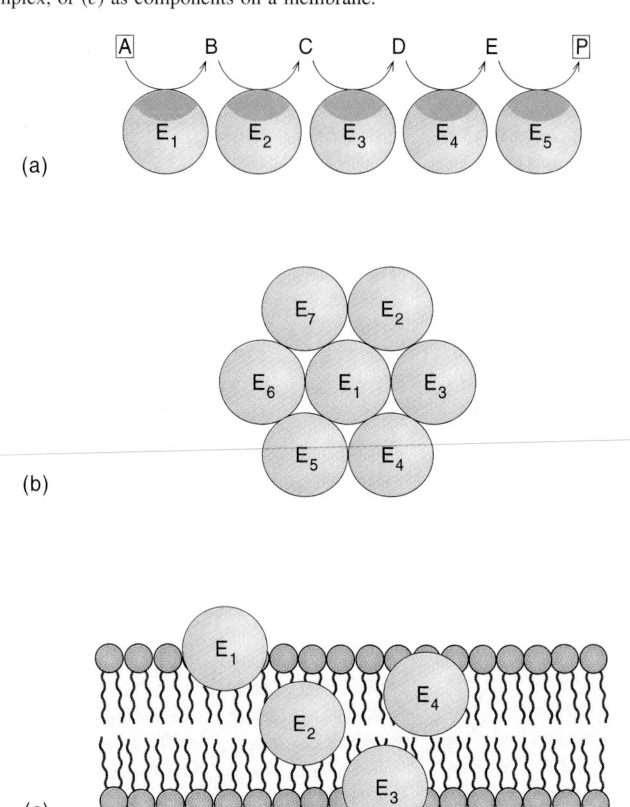

Figure 12.4

A schematic block diagram of the metabolism of a typical aerobic heterotroph. The block labeled "Catabolism" represents pathways by which nutrients are converted to small-molecule starting materials for biosynthetic processes. Catabolism also supplies the energy (ATP) and reducing power (NADPH) needed for activities that occur in the second block; these compounds shuttle between the two boxes. The block labeled "Biosynthesis" represents the synthesis of low- to medium-molecular-weight components of the cell as well as the synthesis of proteins, nucleic acids, lipids, and carbohydrates and the assembly of membranes, organelles, and the other structures of the cell.

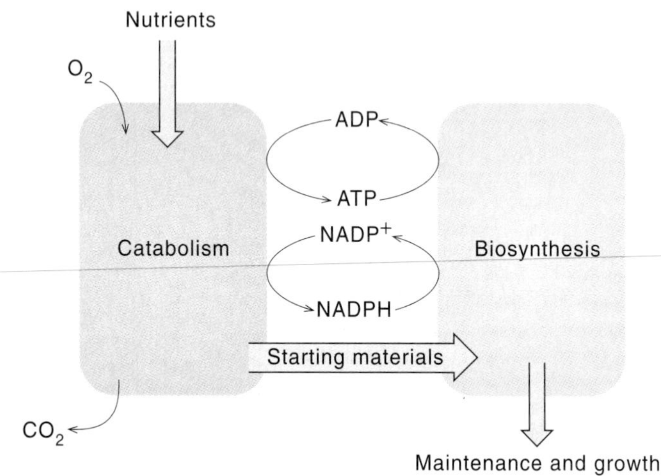

rates of reactions. Second, these higher concentrations facilitate regulation of a pathway by limiting the starting materials or key intermediates to an organelle and allowing direct interaction of one enzyme with another. Fatty acid metabolism in eukaryotes superbly illustrates this type of arrangement (see chapter 21). The enzymes involved in fatty acid catabolism are all located inside the mitochondrion, whereas all enzymes involved in fatty acid synthesis are located outside the mitochondrion in the cytosol.

The second type of arrangement observed for sequentially related enzymes is exemplified by the *Escherichia coli* fatty acid synthase (see chapter 21). This synthase is a complex of most of the enzymes involved in fatty acid synthesis. The intermediates in this case are bound to the enzyme complex until synthesis is complete.

In the third type of arrangement, functionally related enzymes are membrane-bound. This is the case for the enzymes of electron transport and oxidative phosphorylation (see chapter 16). These enzymes are all bound to the inner mitochondrial membrane.

Pathways Show Functional Coupling

Large metabolic charts have been designed to display all the major biochemical pathways. Such charts present a bewildering array of interconnected pathways, making it difficult to appreciate relationships between different pathways. The overall operational aspects of metabolism may be clarified by simpler block diagrams that omit details and focus on functional relationships. Such a functional block diagram for a typical heterotrophic aerobic cell is shown in figure 12.4. The metabolism of such a system is symbolized by two functional blocks:

1. Catabolism or degradative metabolism. Foods are oxidized to carbon dioxide. Most of the electrons liberated in this oxidation are transferred to oxygen, with concomitant production of ATP (electron-transfer phosphorylation). Other electrons are used in the regeneration of NADPH, the most frequently used reducing agent for biosynthesis. The major pathways in this block are the glycolytic sequence and the tricarboxylic acid (TCA) cycle. These same sequences, with the pentose phosphate pathway, also supply carbon skeletons for the cell's biosynthetic processes.
2. Biosynthesis, or anabolism. This block includes much greater chemical complexity than the first. The starting materials produced by glycolysis and the TCA cycle are converted into hundreds of cell components, with NADPH serving as a reducing agent when necessary, and with ATP serving as the universal coupling agent or energy-transduc-

ing compound. Synthesis of starting materials is followed by the synthesis of macromolecules and growth.

The starting materials consumed in biosynthetic sequences must be continually replaced by catabolic processes. The energy sources and reducing sources, ATP and NADPH, are used in very different ways. When they contribute to biosynthesis, ATP is converted to ADP or AMP, and NADPH is converted to $NADP^+$. They (ADP, AMP, or $NADP^+$) must be regenerated, reduced, or rephosphorylated at the expense of substrate oxidation in the catabolic block of reactions. As figure 12.4 indicates, ATP and NADPH are the only compounds that have primary roles in coupling major functional blocks. Other coupling agents are essential but narrower in scope. For example, NADH and $FADH_2$ participate within the catabolic block in the transfer of electrons from substrate to oxygen, but they are not significantly involved in coupling between major functional blocks.

The block diagram of figure 12.4 represents metabolism in aerobic heterotrophs. The photochemical production of ATP in photoheterotrophs requires the addition of a photochemical block, with ATP production in the catabolic block being deleted. For a photoautotroph, two blocks are needed in addition to those shown in figure 12.4: (1) a photochemical block that regenerates ATP from ADP and reduces $NADP^+$ to NADPH while consuming water and producing oxygen, and (2) a block that uses the ATP and NADPH for the reduction of carbon dioxide to various products. These photosynthetic products provide the input to the catabolic block, which in this case serves only one function—the production of the starting materials for synthesis.

Another way of looking at metabolism, which emphasizes the precursor–product relationship, is depicted in figure 12.5. Here metabolism is pictured as three concentric boxes. The inner box shows the central metabolic pathways and the interconversions of various small molecules. These substances serve as the starting materials for all other metabolism. Some of the compounds formed from these intermediates are shown in the middle concentric box. These compounds serve as building blocks for the final products indicated in the outer box. Each of the next 16 chapters deals with an expanded segment of the metabolism depicted in this figure.

The ATP–ADP System Mediates Conversions in Both Directions

Figure 12.4 emphasizes the coupling of catabolic sequences and biosynthetic (anabolic) sequences to the ATP–ADP cycles. In fact, the involvement of the ATP–ADP system is much greater than indicated. Most metabolic sequences within the two major blocks are tightly linked by the ATP–ADP system. To appreciate the way in which this is done, we must look at individual sequences in greater detail.

In analyzing metabolism we must distinguish between a sequence and a conversion. A conversion might be the transformation of starting material A to end product Z; a sequence is the specific set of reactions by which such a conversion is carried out. The conversion of glucose to pyruvate involves a 10-step sequence (see figs. 12.2 and 14.1). As a rule, a conversion in one direction is catabolic and produces energy in the form of ATP, whereas a conversion in the reverse direction is anabolic and requires the input of energy. Any conversion can be carried out by many different hypothetical sequences of reactions, which might be coupled to nearly any number of ATP-to-ADP or ADP-to-ATP interconversions, depending on the specificities of the enzymes that catalyze the component reactions. The value of the overall free energy change and the overall equilibrium constant depend strongly on the stoichiometry of this coupling, that is, the number of molecules of ATP converted to ADP (or of ADP converted to ATP) per mole of Z that is produced.

In fact any conversion can be made thermodynamically favorable by coupling it to a sufficient number of ATP-to-ADP conversions. This strategy is exploited in the design of metabolically paired sequences. A simple example is the interconversion of fructose-1,6-bisphosphate (F-1,6-bisP) into fructose-6-phosphate (F-6-P). This is the simplest possible type of interconversion, since it involves only one reaction in either direction (fig. 12.6):

$$\text{F-6-P} + \text{ATP} \longrightarrow \text{F-1,6-bisP} + \text{ADP}; \quad K_{eq} \approx 10^4 \qquad \textbf{(1)}$$

$$\text{F-1,6-bisP} + \text{H}_2\text{O} \longrightarrow \text{F-6-P} + \text{P}_i; \quad K_{eq} \approx 10^4 \qquad \textbf{(2)}$$

$$\overline{\text{ATP} + \text{H}_2\text{O} \longrightarrow \text{ADP} + \text{P}_i}$$

The conversion of fructose-1,6-bisphosphate to fructose-6-phosphate is an energetically favorable reaction, so it does not require any ATP-to-ADP conversions. The oppositely directed conversion of fructose-6-phosphate to fructose-1,6-bisphosphate requires the conversion of a single ATP to ADP to make it thermodynamically favorable. Since the interconversions have dissimilar starting materials and products, it is not surprising that the two conversions use different enzymes.

From this simple example we can see that paired oppositely directed conversions can be made thermodynamically feasible in either direction by coupling to an appropriate number of ATP-to-ADP conversions. Metabolic sequences commonly occur in pairs. The two sequences of a pair connect the same compounds, say A and Z, but run in opposite directions (fig. 12.7). Invariably the reactions of two oppositely directed sequences are coupled to different numbers of ATP-to-ADP conversions, as in the previous examples, so they are thermodynamically favorable in either direction. Usually $x > y$, where x is the number of ATPs consumed in the anabolic direction and y is the number of ATPs produced in the catabolic direction.

Conversions Are Kinetically Regulated

Pairs of oppositely directed sequences appear to be cycles (see figs. 12.6 and 12.7). Indeed, if both sequences were simultaneously active, they would form a true cycle. Because the end products of one sequence are the starting materials for the other, such a cycle has no metabolic consequences except that ATP is expended. This is because more ATP is used in one

Figure 12.5

The overall metabolism of an aerobic heterotroph represented as three concentric boxes. The central metabolic pathways are represented in the innermost box. It is here that carbohydrate compounds are degraded to produce energy, reducing power and starting materials for biosynthesis. These pathways will be discussed in chapters 13 through 16. Only some key intermediates are shown. In general, in moving out from the central box to the outer two boxes, reactions require energy and reducing power, while moving in the opposite direction leads to the production of energy and reducing power. The middle box shows some of the building blocks for the larger molecules found in the outer box. The reactions involved in fatty acid metabolism, mevalonate synthesis, amino acid metabolism, and nucleotide synthesis are discussed in chapters 21, 23, 24 and 25, and 26, respectively. The reactions involved in lipid synthesis, cholesterol synthesis, nucleic acid synthesis, and protein synthesis are discussed in chapters 22 and 23, 31, 33, and 34, respectively.

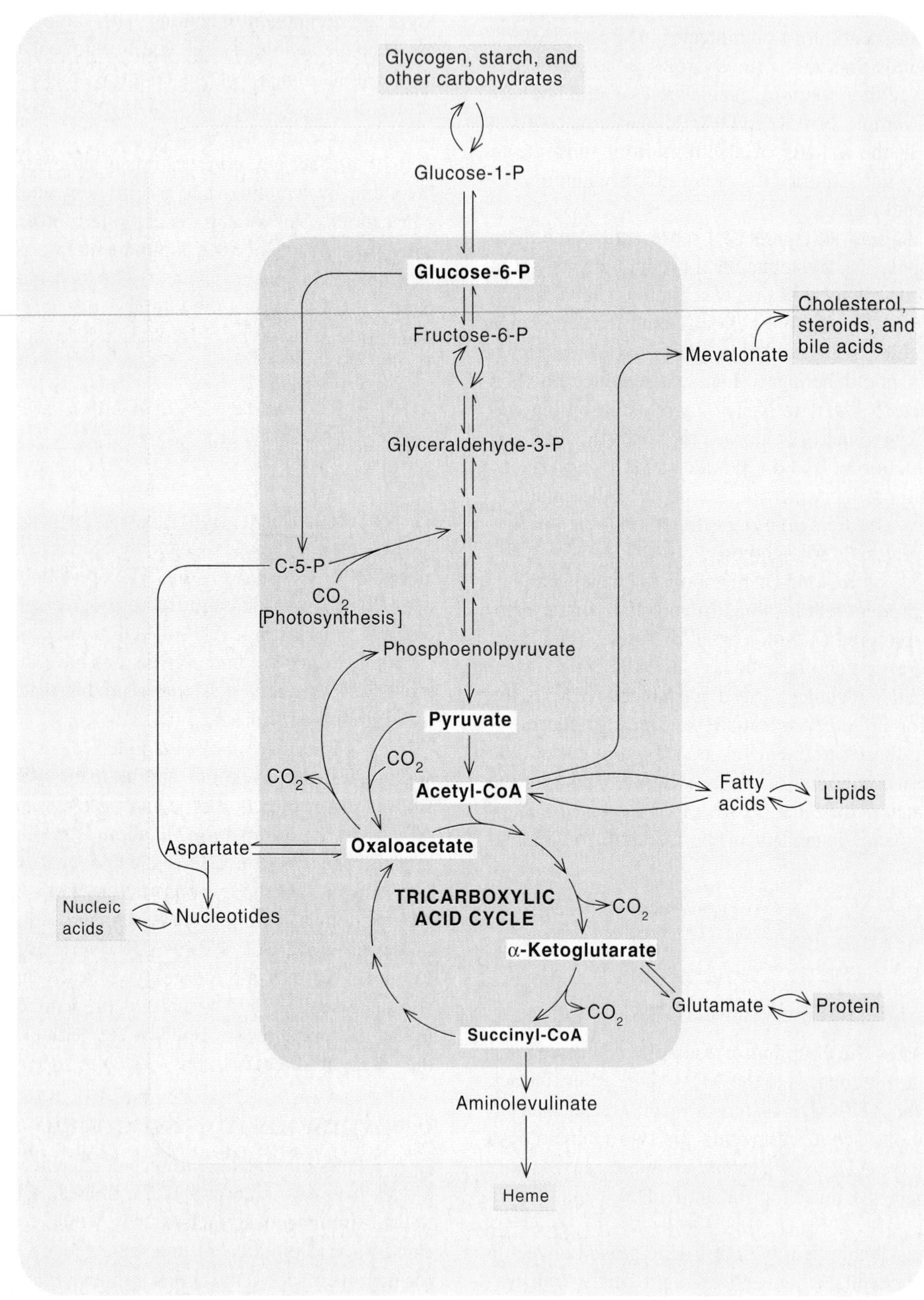

Figure 12.6

The interconversion of fructose-6-phosphate and fructose-1,6-bisphosphate. The enzymes that catalyze these conversions are usually regulated so that at any given time, conversions take place in only one direction.

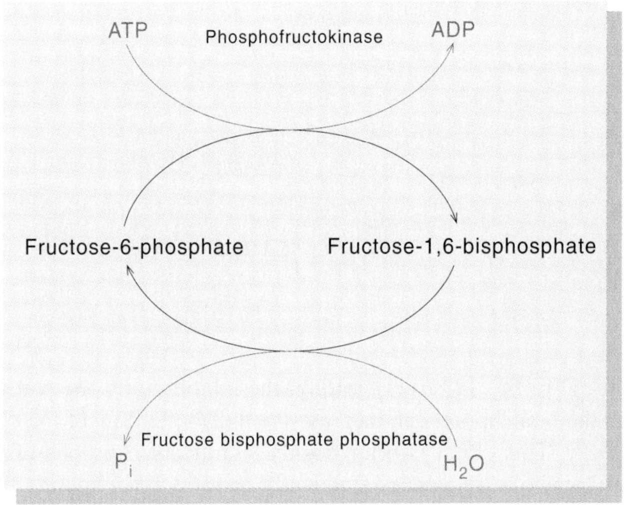

Figure 12.7

Schematic illustration of the organization of metabolic sequences into oppositely directed pairs. The two sequences result in opposite conversions. The values for the overall equilibrium constant are a function of the conversion and the number of ATP-to-ADP conversions to which each sequence is coupled. Note the similarity to figure 12.6.

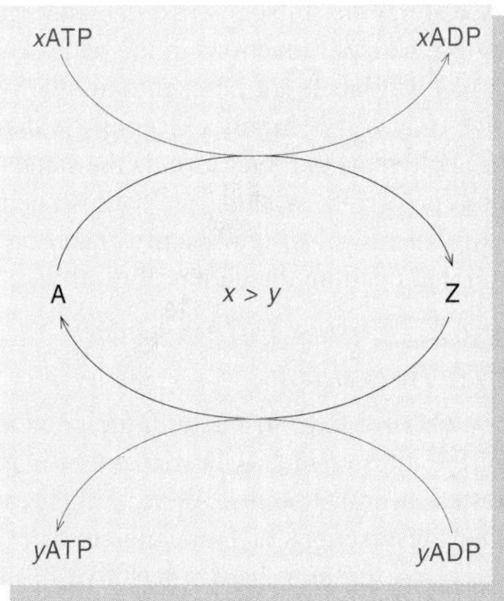

sequence than is regenerated in the other. Because a cyclic operation can waste energy, such oppositely directed paired sequences are often called futile cycles. Fortunately for the metabolism, kinetic constraints usually prevent futile cycles from operating. For this reason, futile cycles are commonly referred to as pseudocycles. These constraints result from the regulation of one or more enzymes for each sequence of a pair so that conversions occur only in one direction at any given time.

The interconversion of fructose-6-phosphate and fructose-1,6-bisphosphate provides an instructive example of the regulation of an interconversion (see fig. 12.6). Each "sequence" of this pseudocycle consists of only one reaction, and the "sequences" differ by one ATP-to-ADP conversion. Clearly, if both reactions in this conversion were allowed to function simultaneously, a metabolic mishap would occur, leading to nothing except the loss of ATP. This is prevented from occurring by the design of the enzymes that carry out these two conversions. Both of these enzymes are regulatory enzymes whose activities are modulated by small molecules that serve as signals of the metabolic state of the cell. The system is designed so that only one of the enzymes is active at any given time. Which enzyme is active depends on the metabolic state of the cell. The precise mechanisms for the regulation of these particular enzymes are discussed in chapters 10 and 14.

Pathways Are Regulated by Controlling Amounts and Activities of Enzymes

The two most important principles governing the regulation of biochemical pathways are economy and flexibility. Organisms regulate their metabolic activities to avoid deficiencies or excesses of metabolic products. If a significant change occurs in the environment, such as a shift in the concentration or kinds of nutrients, the organism must quickly adjust its metabolism if it is to survive.

Pathways are regulated at two levels: (1) They are regulated by controlling the amounts of enzyme (see chapters 35 and 36), and (2) they are regulated by controlling the activity of existing enzymes. In this chapter we focus on the latter. Important aspects of this subject were already dealt with in chapter 10.

Enzyme Activity Is Regulated by Interaction with Regulatory Factors

The most common way of regulating metabolic activity is by direct control of enzyme activity. Enzyme activities are usually regulated by noncovalent interaction with small-molecule regulatory factors (see chapter 10) or by a reversible covalent modification, such as phosphorylation or adenylation (also see chapters 10, 14, and 27) of an amino acid side chain. The effect of the regulatory factor in specific instances may be to increase or decrease the activity of the enzyme. What sets these regulatory factors in motion is a separate question. Sometimes the metabolism inside the cell is the major player in determining which regulatory factors are produced; at other times, external sources control the action as in the case of hormones that are made elsewhere and influence special target cells. One instance of the effectiveness of hormonal signals is discussed in the next chapter, and the subject is discussed in general terms in chapter 27.

Regulatory Enzymes Occupy Key Positions in Pathways

Enzymes that are susceptible to direct regulation occupy key positions in metabolic pathways. In a multistep pathway the first enzyme in the pathway is usually regulated, and the others are not (fig. 12.8*a*). In the case of CTP synthesis (see chapter 10) we saw

Figure 12.8

Three patterns for end-product inhibition. In end-product inhibition the first reaction in the pathway (*a*) or the first reactions after a branchpoint (*b*) are inhibited by the specific end products (*c*). Sometimes the first step in the overall pathway is inhibited by the end products from both branchpoints.

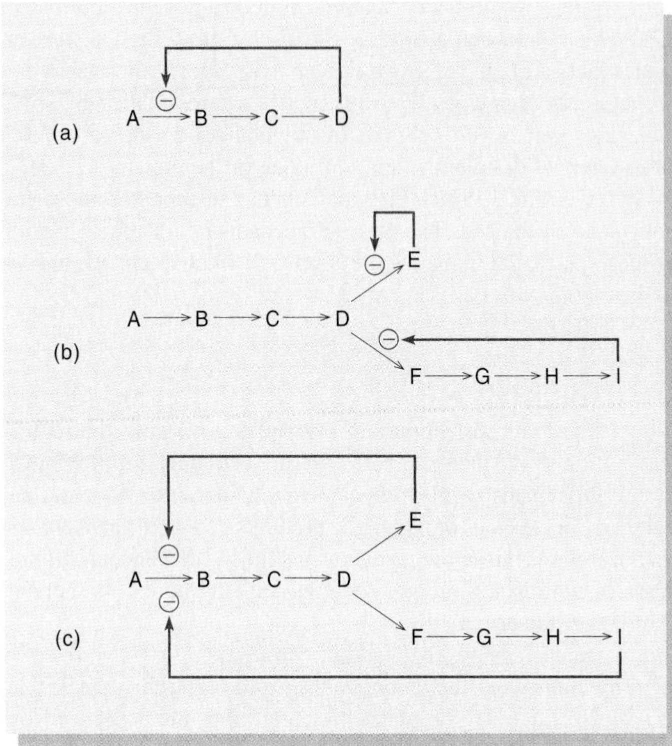

Figure 12.9

Branchpoints in metabolic pathways sometimes occur at locations where an intermediary, S, can follow a catabolic sequence or an anabolic sequence. The first reactions after the branchpoints are catalyzed by enzymes B and A. Once the first step after the branchpoint has been taken, the metabolic intermediate is irreversibly committed to follow that pathway.

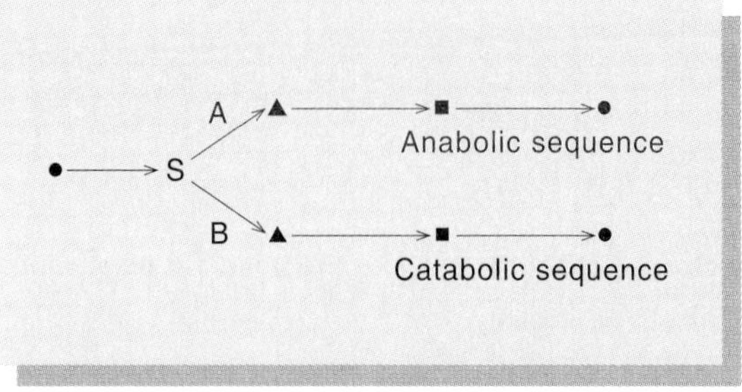

that the enzyme aspartate carbamoyl transferase is negatively regulated by CTP. That is, excess CTP binds to the regulatory sites on this multiprotein enzyme, thereby inhibiting its activity. Aspartate carbamoyl transferase catalyzes the first step in a multistep pathway leading to CTP. Because CTP is the end product of the pathway, this type of negative feedback regulation is often referred to as end-product inhibition. End-product inhibition is very common in anabolic pathways.

The mechanism is clearly flexible and economical. If there is sufficient end product, there is no point in processing substrates down the pathway. To do so is a waste of both energy and materials. Furthermore, it is most effective to block the first enzyme in the pathway; this by itself rapidly reduces activity of the entire pathway. It is redundant to block any other enzymes in the pathway, because no intermediates are available to these enzymes if the first step is blocked. Blocking a middle enzyme in the pathway, instead of the first enzyme, also is quite wasteful. The result is accumulation of an intermediate that serves no useful function. Indeed, such intermediates in excess frequently have harmful effects.

In a branched-chain anabolic pathway, end-product inhibition usually regulates the first enzyme after the branchpoint (fig. 12.8*b*). This arrangement leads to control of either pathway after the branchpoint (fig. 12.8*b*). If the supply of the branchpoint substrate (D in fig. 12.8*b*) is limiting, the inhibition of one pathway after the branchpoint can increase the metabolic flow of the other pathway. This effect is often highly significant. Apropos of this point, anabolic sequences frequently arise as branchpoints from catabolic pathways (fig. 12.9). In such instances the control factors affect whether a catabolic intermediate is further degraded or serves as a substrate for biosynthesis.

A Regulated Reaction Is Effective Only If It Is Exergonic

Whereas functional pathways are always designed to be highly exergonic, not all steps in a pathway need to be exergonic. Indeed, we will study cases in which a pathway includes individual steps that are highly endergonic. In such cases the endergonic reaction is pulled along by the other reactions in the pathway. Whereas all reactions in a pathway need not be exergonic, the regulated step will be effective only if it is highly exergonic. A dam–river analogy is appropriate here. If the river level is the same on both sides of a dam, it does not matter whether the dam is open or closed. It is only when the water level is significantly higher on the upstream side of the dam that opening the dam will result in a greater flow.

Regulatory Enzymes Often Show Cooperative Behavior

It is clear from what we have just said in the previous subsection that control of the rates of metabolic sequences is most effective when it is exerted at branchpoints. As you might expect, the enzymes that catalyze reactions at branchpoints have evolved features to make the partitioning highly sensitive to metabolic signals. A striking feature of a typical branchpoint enzyme (the first enzyme in a biosynthetic sequence, for example) is that the reaction that it catalyzes is of high kinetic order with respect to its substrate concentration (see chapter 10). Whereas other enzymes along the pathway typically exhibit normal Michaelis kinetic responses to the concentrations of their substrates (first order or hyperbolic kinetics), branchpoint enzymes usually catalyze reactions of higher order (cooperative or sigmoidal kinetics). A fourth-order response is quite common. The two types of behavior are shown in figure 12.10. Any enzyme-catalyzed reaction must level

Figure 12.10

Reaction velocity as a function of substrate concentration (*a*) for a first-order enzymic reaction and (*b*) for a cooperative enzyme with fourth-order kinetics. Substrate concentration is expressed as $[S]/[S_{0.5}]$ and velocity as a fraction of maximum velocity. At a $[S]/[S_{0.5}]$ value of 1 the substrate concentration is equal to $[S_{0.5}]$ and the reaction velocity is half of the maximum velocity. Note the much greater rate of increase over the same interval for the cooperative enzyme.

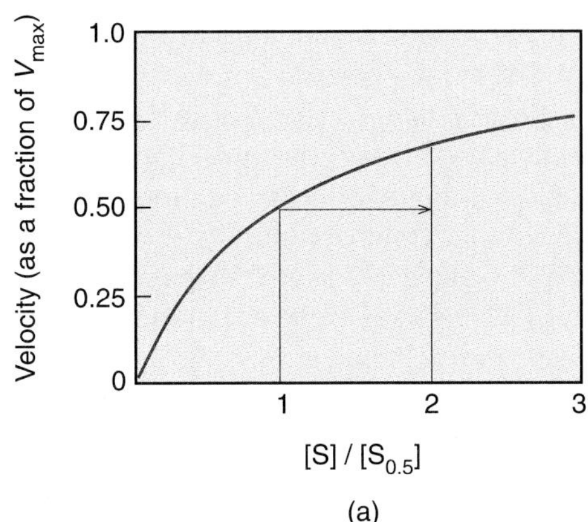

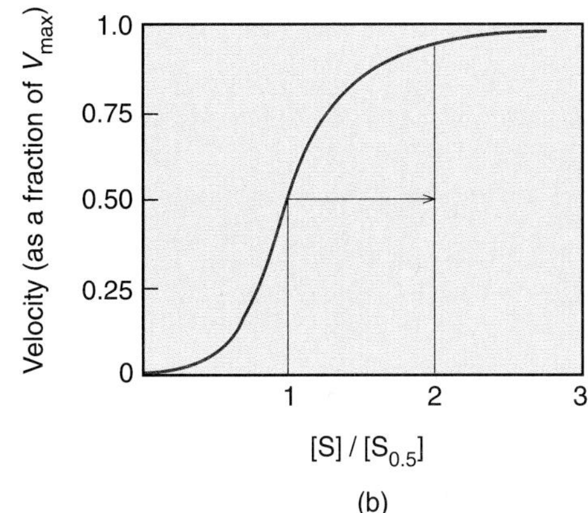

off (to a kinetic order of 0) at high concentration because of saturation of the catalytic sites, but for the "normal Michaelis" enzyme the kinetic order at very low concentrations of substrate is 1, whereas for the cooperative enzyme it has a higher value, frequently 4.

A major advantage of this higher order is that it makes the system very sensitive to changes in substrate concentration. If the first reaction in the sequence responds to the fourth power of the concentration of its substrate (the starting material for the sequence), the flow of material into the sequence is much more sensitive to small changes in the concentration of the starting material than if the reaction were first order. This will help to maintain the steady state of the system.

Perhaps an even more important advantage of high kinetic order is that the regulatory responses of the enzyme are greatly strengthened. We discussed kinetic regulation of enzymes at metabolic branchpoints by negative feedback, in which the end product of the sequence binds to a regulatory site and decreases the affinity of the catalytic site for the substrate (the branchpoint metabolite). When the enzyme is constructed so as to have cooperative interactions between sites, the sensitivity to negative feedback control is greatly sharpened.

Both Anabolic and Catabolic Pathways Are Regulated by the Energy Status of the Cell

It is not always an advantage for the concentration of an amino acid or other end product to be the only factor that determines the rate at which that end product is synthesized. If a cell is low on energy or catabolic intermediates, it may not be able to afford to synthesize amino acids, even if their concentrations are quite low. Continued biosynthesis under such circumstances might deplete the already scarce supply of energy to the point where essential functions, such as maintenance of concentration gradients across membranes, is impaired. It is clearly advantageous for the rate of biosynthesis to be regulated by the general energy status of the cell as well as the need for a specific end product.

Because the adenine nucleotide system ATP–ADP (and occasionally ATP–AMP) couples energy into biosynthetic sequences, we might expect that this system also supplies the necessary signal indicating the energy status of the cell, to be sensed by the kinetic control mechanisms of biosynthesis. It is helpful to have a term that quantitatively expresses the energy status of the cell. The term most often used is the energy charge, which is defined as the effective mole fraction of ATP in the ATP–ADP–AMP pool:

$$\text{Energy charge} = \frac{(\text{ATP} + 0.5\ \text{ADP})}{(\text{ATP} + \text{ADP} + \text{AMP})} \tag{3}$$

In this equation the 0.5 in the numerator takes into account the fact that ADP is about half as effective as ATP at carrying chemical energy. Thus two ADPs can be converted into one ATP and one AMP by a reaction with an equilibrium constant of about 1. The highly active enzyme that catalyzes this reaction (adenylate kinase) is universally distributed:

$$\text{ATP} + \text{AMP} \rightleftharpoons 2\ \text{ADP}, \qquad K_{eq} \approx 1 \tag{4}$$

The values for the energy charge could conceivably vary from 0 to 1, as illustrated in figure 12.11. In fact, however, the values for real cells are usually held within very narrow limits. The result is a general stabilizing effect on the cellular metabolism. In much the same way, voltage regulation is necessary for the effective operation of many electronic devices. Now let us see how the limits are maintained.

Several reactions in anabolic and catabolic pathways have been found to respond to variation in the value of the energy charge. As might be expected, the enzymes in catabolic pathways respond in a direction opposite to that of enzymes in anabolic path-

Figure 12.11

Relative concentrations of ATP, ADP, and AMP as a function of the adenylate energy charge. The adenylate kinase reaction was assumed to be at equilibrium, and a value of 1.2 was used for its effective equilibrium constant in the direction shown in the equation in the text.

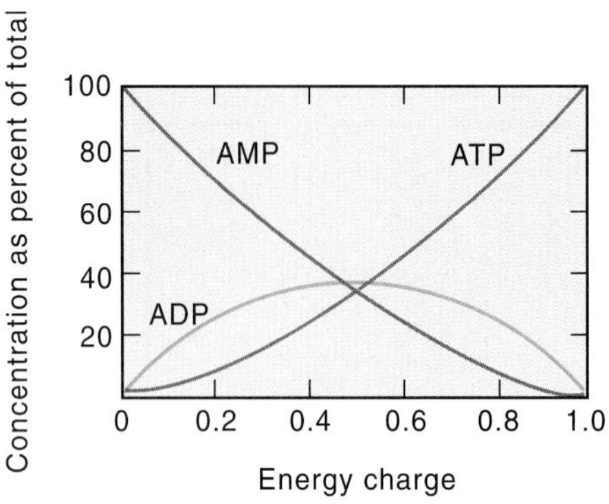

Figure 12.12

Variation in reaction rates as a function of the energy charge. As the energy charge increases, the rate of catabolic reactions decreases. Meanwhile, the rate of anabolic reactions increases. The combined effect is to stabilize the energy charge at a value around 0.9.

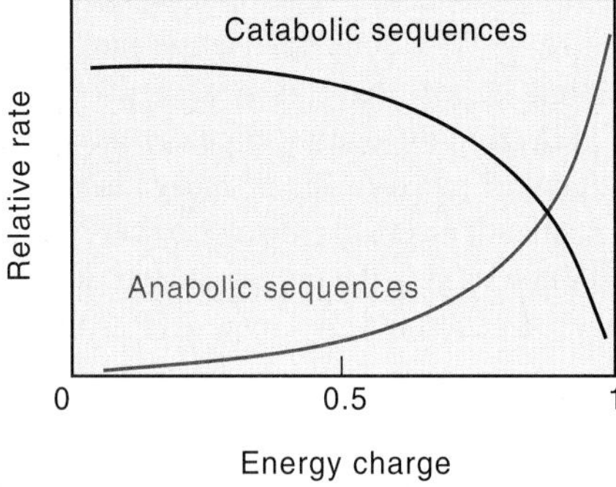

ways. The two responses are compared in figure 12.12. It is clear that if catabolic sequences, which lead to the generation of ATP, respond as shown by the upper curve, and biosynthetic sequences, which use ATP, respond as shown by the lower curve, then the charge is strongly stabilized at a value at which the two curves intersect. A tendency for the charge to fall is resisted by the resulting increases in the rates of catabolic sequences and the decreases in the rates of biosynthetic sequences. A tendency for the charge to rise is resisted by opposing effects.

Regulation of Pathways Involves the Interplay of Kinetic and Thermodynamic Factors

The general properties of paired unidirectional sequences and their advantages to the organism are in one sense very simple. But on further consideration we see that they depend on a rather complex interplay between kinetic and thermodynamic factors and effects.

In any kind of negative feedback system, cause-and-effect relationships are circular. We see the circularity of negative feedback at a simple level in the response of a biosynthetic sequence to the concentration of its end product. The concentration of the end product is a major factor controlling the rate of synthesis, and the rate of synthesis is a major factor controlling the concentration of the end product.

For pairs of oppositely directed metabolic sequences, greater sophistication is needed than for simple feedback control of a synthetic pathway. In our discussion of oppositely directed sequences thus far, we have assumed that the ATP/ADP ratio is held at a value very far from equilibrium. But of course, in the absence of highly effective controls, the ATP/ADP ratio would not remain far from equilibrium; it would rapidly approach its equilibrium value. What are the controls that keep this from happening? The answer is the enzymes that catalyze first steps in oppositely directed sequences—the same controls that we have already discussed.

The rates of reactions in which there is net conversion of ATP to ADP are high when the energy charge (or the ATP/ADP ratio) is high, and they decrease sharply with a decrease in those parameters. Rates of regulatory enzymes in sequences in which ATP is regenerated are high when the energy charge or ATP/ADP ratio is low and decrease sharply as the energy charge increases. The curves intersect at an energy charge value of about 0.9 (see fig. 12.12). These kinetic effects play a major role in stabilizing the energy charge and the ATP–ADP system in vivo. So kinetic control of rates of conversion depends on the value of the ATP/ADP ratio being far from equilibrium (that is, on thermodynamic factors); but at the same time the ATP/ADP ratio itself depends on reciprocal regulation of oppositely directed sequences (that is, on kinetic factors).

As we have seen, the characteristic and essential feature of paired, oppositely directed sequences is that they differ in their ATP stoichiometries. But that difference is meaningful only because the ATP/ADP ratio is far from equilibrium in the cell. If ATP, ADP, and P_i were at equilibrium, it would not matter how many ATP-to-ADP conversions were coupled to any metabolic sequence. For any system at equilibrium, ΔG is zero, and coupling to another reaction or sequence would have no effect on the free energy change or the position of equilibrium of the reaction or sequence. So the differential ATP stoichiometry would mean nothing if it were not for the kinetic controls that hold the ATP/ADP ratio at a steady-state value far from equilibrium. But kinetic regulation could not control directions of conversions and thus regulate the balance between ATP utilization and regeneration if it were not for the thermodynamic difference between the sequences of each pair, which depends on the ATP/ADP ratio. Neither the thermodynamic nor the kinetic features can be said to be more fundamental. Metabolic correlation and control, and hence life, depend on an intricate interplay between thermodynamic and kinetic factors.

Strategies for Pathway Analysis

Biochemical precursor–product relationships vary in complexity. Some pathways, such as the conversion of phenyl-

Figure 12.13

The tryptophan biosynthetic pathway in *E. coli*. There are five enzymatically catalyzed reactions involved in tryptophan biosynthesis and five different polypeptides associated with these reactions. Polypeptides E and D normally make a tetrameric complex, which catalyzes the first two reactions in the pathway. The next two reactions are catalyzed by a single polypeptide, C. The final reaction is catalyzed by a tetrameric complex composed of the B and A polypeptides in equal numbers.

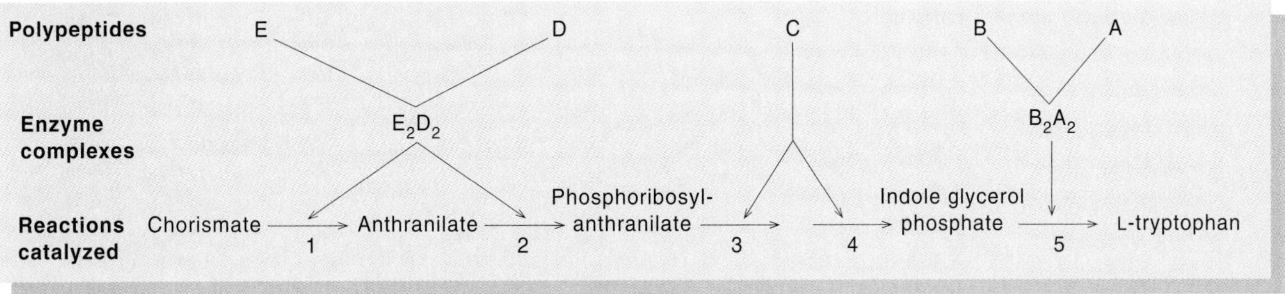

alanine to tyrosine, involve only one enzyme-catalyzed reaction: one precursor and one product. Other pathways, such as the utilization of CO_2 in autotrophic organisms, illustrate the most complex type of relationship between precursor (CO_2) and products (carbon-containing compounds of the cell), a multistep pathway with many branchpoints; the complete description of all of the reactions involved in CO_2 utilization would include most of the synthetic reactions of the organism. Different procedures are needed to analyze simple and complex pathways.

Analysis of Single-Step Pathways

The investigation of a single-step pathway usually begins with a crude cell-free extract from a source abundant in the enzyme that catalyzes the conversion. Two of the most popular sources of cells for many biochemical studies are rat liver and *E. coli*. To investigate a particular reaction, the precursor (substrate) is added to the extract, and the amount of product formed as a function of time is determined. Once an assay for disappearance of precursor and appearance of product has been developed, the crude extract can be processed into fractions that can be tested to see which are active in the conversion. Through further fractionation and assays it should ultimately be possible to purify the enzyme of interest. Some procedures followed in enzyme purification were discussed in chapter 7, and many procedures used to determine the mechanisms of action of the purified enzymes were considered in chapters 9 and 10.

The process of assaying for a particular reaction during purification may be complicated in cases requiring cosubstrates, coenzymes, or cofactors. Usually, these additional requirements are discovered by the trial-and-error procedure of adding test substances to the reaction mixture and observing whether they accelerate the reaction or lead to the formation of more product(s). If more than one step and several enzymes are involved, then intermediates between precursor and product must also be considered, and the whole process of analysis increases in complexity.

Analysis of Multistep Pathways

Most pathways comprise a series of enzyme-catalyzed steps exemplified by the five-step pathway leading from chorismate to tryptophan (fig. 12.13). Because tryptophan synthesis does not occur in vertebrates, this reaction cannot be studied in rat liver extracts, but it can be studied in extracts made from *E. coli* bacteria. Bacteria and many other microorganisms have the additional advantage that powerful genetic methods are applicable. In general, genetic methods are most useful at the beginning of a study, when the goal is to determine the number of genes and related enzyme-catalyzed reactions of a pathway. Genetic methods are also useful to test conclusions drawn from in vitro studies by making parallel measurements in vivo. In the case of the tryptophan biosynthetic pathway, genetic methods were used to determine the number of gene-encoded proteins required to synthesize tryptophan. This was done by a procedure known as complementation analysis. First, a large number of bacterial mutants were isolated, which could not grow unless tryptophan was added to the growth medium. Such mutants usually have defects in one or more of the genes required for tryptophan biosynthesis. Pairwise matings are done between mutants to create partial diploid cells that carry complete sets of the tryptophan pathway genes from two different mutants. Such diploid cells no longer require added tryptophan in cases in which the defects in the two genomes are in different genes. With some mated pairs, the need for tryptophan is not eliminated. This occurs in cases where the paired genomes carry defects in the same gene (fig. 12.14). Such mutants are said to belong to the same complementation group. An analysis of this sort led to a classification of most of the tryptophan-requiring mutants into a limited number of complementation groups. In an exhaustive analysis the total number of complementation groups should correspond to the number of genes required to make a functional tryptophan pathway. Since most genes encode a single polypeptide chain, the number of groups should reflect the number of polypeptide chains involved in the biosynthesis of tryptophan.

Figure 12.14

Complementation occurs only between genomes that are defective in different genes. On the left are shown two different types of gene pairs that might arise in a mating to test for complementation. The genes *E, D, C, B,* and *A* are intended to represent the five genes of the tryptophan pathway, which are clustered in *E. coli.* A minus sign indicates a defective gene. On the right are shown the results of plating 10 cells on a minimal medium with and without tryptophan added. Growth in the absence of added tryptophan indicates complementation. Growth is observed after overnight incubation. Under favorable growth conditions each cell divides many times, forming a clone of identical cells.

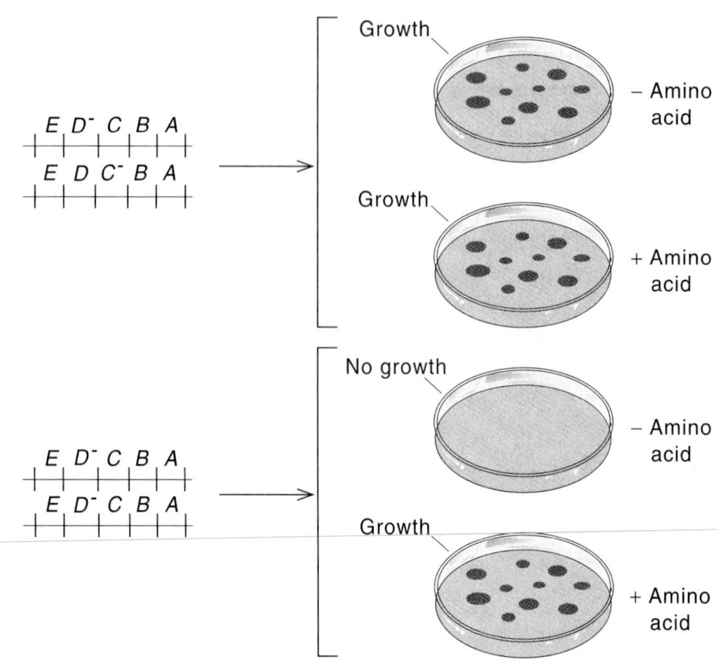

Figure 12.15

Comparison of the relative amounts of pathway intermediates and final products in normal and mutant organisms containing a defective enzyme in the pathway. The precursor, intermediates, and final product of the pathway are labeled A, B–D, and E, respectively. The enzymes 1, 2, 3, and 4 are indicated by the numbers over the reaction arrows. A cross through the reaction arrow indicates a defective enzyme. In the wild-type organism with no defective enzymes (*top*), intermediates are present at low concentrations compared with final product. In the mutant organism with a defective enzyme in the pathway (*bottom*), the intermediate just before the defective enzyme accumulates to an abnormally high concentration. Intermediates and products after the block are practically nonexistent. If the final product of the pathway is required for viability, it has to be supplied directly from external sources.

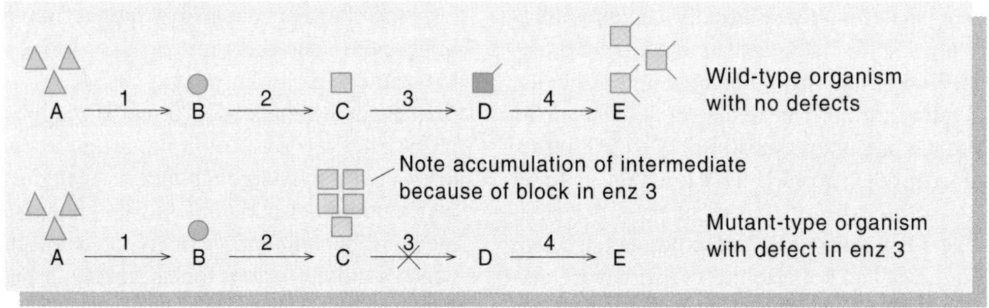

The next phase in pathway analysis involves biochemical procedures. As many metabolic intermediates as possible are isolated, their structures are determined, the order of reactions is determined, and the enzymes that catalyze the different reaction steps are isolated and characterized. The mutants isolated for the complementation analysis are also valuable aids for these biochemical experiments. The usefulness of the mutants at this stage in the investigation stems from the fact that each mutation introduces a specific block at some point in the pathway. A bacterium with one nonfunctioning enzyme in a pathway accumulates the intermediate just before the defect and makes little or no intermediates (or final product) for the remaining steps in the pathway (fig. 12.15). Intermediates often are much easier to isolate from mutants than from normal cells because of their greater abundance and stability. Ideally, the structures of all intermediates isolated from different mutant cells are determined. The information resulting from such observations should give a step-by-step picture of the chemical reactions that occur in the pathway.

From this point on there are numerous ways of proceeding with the analysis. Crude extracts can be made from different mutant cells. The extract from one mutant can be used to complement the extract from another mutant in tryptophan synthesis, which can lead to an assay for a particular enzyme carried by one mutant that is missing in the other mutant. The goal at this juncture is to purify each enzyme of the pathway so that the properties of the enzymes and the reactions they catalyze can be individually scrutinized. All of the enzymes belonging to the tryptophan biosynthetic pathway have been isolated from *E. coli*. Studies of other systems indicate a remarkable similarity for the operation of this pathway in different microorganisms and plants.

Radiolabeled Compounds Facilitate Pathway Analysis

Before radioactively labeled substrates became available, detection of intermediates and products was often quite difficult. Radiolabeled compounds greatly increase the sensitivity of detection of products, especially the intermediates in a pathway. Moreover, they also facilitate determination of the order of reactions in a pathway. Thus a radiolabeled precursor administered to a reaction mixture for a short time is expected to label the early intermediates preferentially. The same compound administered for a long time, however, is expected to accumulate in the final product of the pathway. This is the situation for an extract prepared from a wild-type cell. If the extract were prepared from a mutant cell, we would expect a difference in the labeling pattern, as suggested in figure 12.15.

Specific inhibitors can serve the same role as genetic blocks in the analysis of a pathway. Genetic methods were rarely used for pathway analysis until the 1940s. Radiolabeling was not used until the 1950s. In spite of this, many important pathways were elucidated before this. For example, Hans Krebs discovered the complex multistep pathway known as the tricarboxylic acid cycle in the 1930s. Krebs used cell-free extracts from pigeon flight muscle, which are especially rich in the enzymes of the cycle (see chapter 15). Although mutants were not available for imposing specific blocks on the passage of intermediates in this pathway, a strong inhibitor of one of the reactions was discovered. Frequently, even today, inhibitors are useful for pathway analysis, especially in cases in which mutants are not available. Inhibitors sometimes have an advantage over the use of mutant extracts for in vitro analysis because they can be added at various times after starting the reaction.

Pathways Are Usually Studied Both in Vitro and in Vivo

Although biochemists spend most of their time studying reactions in vitro, the ultimate proof of the significance of a reaction or a series of reactions is that it is used in vivo. For the purpose of making parallel measurements, in vivo and in vitro, isotopes and genetic blocks are invaluable. Isotopes permit intermediates and products to be detected in vivo as well as in vitro. Genetic mutants permit parallel observations to be made of the effect of a specific enzyme deficiency on the specified pathway. In cases in which in vitro and in vivo analyses lead to different conclusions, the significance of the in vitro analysis may not have been correctly assessed, and further studies are necessary. A classic example of this occurred in early investigations of *E. coli* DNA polymerase (chapter 31). An enzyme was isolated that catalyzes the replication of DNA in vitro. However, subsequent mutant studies showed that this activity was not required for replication of DNA in vivo.

The brief description given here has emphasized the power of genetic approaches for elucidating metabolic pathways. In the last 15 years the fields of genetics and biochemistry have been revolutionized by the techniques of DNA recombinant technology. DNA recombinant technology has made it possible to isolate virtually any gene from any cell type, even those for which genetic methods have not been accessible in the past. This means that in the future, biochemists will place even greater emphasis on the genetic approach in their work. DNA recombinant technology methods are discussed in chapter 32.

Summary

In this chapter the main principles governing the intermediary metabolism are discussed.

1. All cells need energy and starting materials for synthesis. Ultimately, these are supplied by autotrophic organisms, especially green plants; in plants the starting materials are made from CO_2, and the supply of chemical energy and reducing power is dependent on the absorption of light energy. In a heterotrophic organism the role of catabolism is to supply those basic needs from conversions (usually oxidation) of foodstuffs.
2. Metabolic chemistry is characterized by functionality. Each reaction is important because of its participation in a sequence of reactions, and each sequence interacts functionally with other sequences.
3. Sequences may be broadly classified into two main types: biosynthetic (anabolic) and degradative (catabolic). Anabolic sequences usually require energy, and catabolic sequences usually produce energy.
4. Metabolic regulatory mechanisms have evolved so as to stabilize concentrations of key metabolites under a broad range of conditions.
5. Energy is coupled from catabolic sequences of energy-requiring activities of a cell by the ATP–ADP system. In a similar manner, reducing power is coupled by the NADPH–$NADP^+$ system.
6. The stoichiometry of coupling to ATP-to-ADP conversions contributes to the overall equilibrium constant of a sequence and therefore can determine the direction of conversion that is thermodynamically favorable. Any conversion can be made favorable by coupling to an appropriate number of ATP-to-ADP conversions.
7. Metabolic sequences occur in oppositely directed pairs, which are controlled by regulatory enzymes.
8. Regulatory enzymes respond to signals in such a way that rates of biosynthesis are controlled by the need for product.
9. Metabolism is regulated primarily by adjustment of the ratios by which intermediates are partitioned at metabolic branchpoints.
10. Different procedures are used for analysis of simple and com-

plex pathways. Analysis of single-step pathways often begins with the isolation and characterization of the enzyme involved. During enzyme isolation each purification step is monitored by a specific assay that measures the conversion of substrate to product. Multistep pathway analysis ideally begins with complementation analysis, a genetic technique that entails the isolation of mutants with genetic blocks in each step of the pathways. Once the numbers of enzymes and intermediates are established, each enzyme can be isolated with the help of a specific assay.

11. Radiolabeled compounds are most useful for pathway analysis in two respects: First, they permit detection of pathway intermediates with great sensitivity, and second, they can be used as tracers for determining the order of intermediates in a pathway.
12. Inhibitors that block specific steps in a pathway play the same role as genetic blocks in the in vitro analysis of a pathway.
13. A complete understanding of a pathway requires parallel investigations in vitro and in vivo. In vitro analyses permit a detailed study of isolated components. In vivo observations substantiate the biologic significance of in vitro observations.

Selected Readings

Atkinson, D. E., *Cellular Energy Metabolism and Its Regulation*. New York: Academic Press, 1977. General discussion covers some topics in this and the two following chapters in somewhat greater depth than treated in our book.

Cohen, P., *Control of Enzyme Activity*, 2d ed. London and New York: Chapman and Hall, 1983. Brief discussion of some types of regulation of activity of metabolic enzymes, emphasizing regulation by covalent modification of the enzymes.

Herman, R. H., R. M. Cohn, and P. D. McNamara, *Principles of Metabolic Control in Mammalian Systems*. New York and London: Plenum Press, 1980. Discusses various aspects of metabolic control in mammals, mainly at the intracellular level. Many references.

Hochachka, P. W., and G. N. Somera, *Biochemical Adaptation*. Princeton, N.J.: Princeton University Press, 1984. An excellent and extensive discussion of how biochemical processes, including many discussed in this book, are adapted by various types of organisms in fitting themselves for survival under specific and often difficult conditions.

Jencks, W. P., How does ATP make work? *Chemtracts-Biochem. Mol. Biol* 1:1–13, 1990.

Westheimer, F. H., Why nature chose phosphates. *Science* 235:1173–1178, 1987.

Problems

1. Use the data presented in figure 2.7 to calculate K_{eq} for equation (4). How does your value compare to that given in figure 12.11 and the text associated with equation (4)? As indicated in figure 12.12, the energy charge is stabilized near 0.9. Calculate the molar ratio of ATP:ADP:AMP relative to ATP in a cell in which adenylate kinase is functioning.
2. How can nature use organelles to prevent futile cycles?
3. What isotopes are useful for the analysis of metabolic pathways? What difficulties are encountered in terms of the isotopes provided by nature? Consult a chemical handbook for primary data to answer this question.
4. Examine the mutant-type organism (fig. 12.15) with a defect in enzyme 3. Will compound C always be the substance that accumulates?
5. If in a certain anabolic pathway an enzymatic step involves the phosphorylation of the substrate, the process is typically coupled with the conversion of an ATP to an ADP (i.e., a kinase). However, the catabolic version of the pathway often contains a hydrolysis reaction at this point. Why doesn't the organism just reverse the pattern involved in the anabolic pathway and form an ATP?
6. Consider the following relationships among the four major classes of biological molecules. What similarities and differences can you see in the chemical relationships between the right and left columns?

Small Molecule	Large Molecule
Nucleotide	Nucleic acid
Amino acid	Protein
Monosaccharide	Polysaccharide
Fatty acid	Lipid

7. What is a metabolic pathway?
8. What is the relationship between catabolism and anabolism?
9. Besides a difference in the number of phosphate moieties, what are the major biochemical differences between NAD^+ and $NADP^+$?
10. What is the primary advantage of subcellular compartments to intermediary metabolism?
11. Using only figure 12.12, determine whether the conversion of glucose to pyruvic acid is an oxidation or a reduction or neither.
12. Explain why each of the following statements is false in terms of efficient metabolic regulation:
    (a) Most enzymes operate in vivo near V_{max}.
    (b) End-product inhibition usually occurs at the last or next-to-last enzyme in a metabolic pathway.

(c) Catabolic pathways tend to diverge from a single metabolite.
(d) The enzymes regulated in a metabolic pathway usually exhibit simple Michaelis-Menten kinetics.
(e) Energy charge is unimportant in the regulation of anabolic sequences but is of primary importance in the regulation of catabolic sequences.
(f) Enzymes that catalyze a sequence of reactions are rarely grouped in multienzyme complexes.
(g) Compartmentalization of metabolic pathways is seldom a regulatory strategy the cell uses.

13. How is it possible that both the glycolytic degradation of glucose to lactate and the reverse process, formation of glucose from lactate (gluconeogenesis), are energetically favorable?
14. What is the metabolic advantage of having the "committed step" of a pathway under strict regulation?
15. Theoretically, the following reactions constitute a futile cycle. Explain.

$$\text{Glucose} + \text{ATP} \xrightarrow{\text{Glucokinase}} \text{Glc-6-P} + \text{ADP}$$

$$\text{Glc-6-P} + \text{HOH} \xrightarrow[\text{}]{\text{Glucose-6-phosphatase}} \text{Glucose} + \text{Phosphate}$$

In the liver cell the enzymes are spatially separated: glucokinase in the cytosol and glucose-6-phosphatase in the endoplasmic reticulum. Does this separation influence the futile cycle?

C H A P T E R

13

Structures of Sugars and Energy-Storage Polysaccharides

Most sugars contain five or six carbons, which form ring structures that become linked into linear or branched structures in polysaccharides.

Carbohydrates are made from carbon skeletons richly laced with hydroxyl groups. The hydrophilic hydroxyl groups give carbohydrates the potential for strong interaction with water; they also give them functional groups to which various substituents can add. Finally, the hydroxyl groups provide the possibility of strong intra- or interchain interaction via hydrogen bonds. The simplest carbohydrates contain only carbon, hydrogen, and oxygen. Derivatives of carbohydrates contain nitrogen, phosphorus, and even sulfur compounds. Carbohydrates are often attached to other macromolecules to form glycoproteins or glycolipids, and they are also an important component of nucleic acids.

Small carbohydrates such as glucose, fructose, and pyruvate occupy key roles in energy metabolism and supply carbon skeletons for the synthesis of other compounds. Polymeric carbohydrates are important as short-term energy-storage compounds and also as major structural compounds in plant and bacterial cell walls and in the extracellular matrix. Branched-chain polymeric carbohydrates covalently linked to proteins are used to give proteins specific surface features that may be exploited to signal protein targeting and in a variety of cell–cell recognition processes.

Carbohydrates play so many roles that it is almost easier to discuss the things they can't do than the things they can do. There are two areas where they seem to be lacking in potential, at least by themselves. They cannot store genetic information, and they cannot function as enzymes (so far as is known). In this chapter we will describe the structural and functional properties of carbohydrates involved in energy metabolism. In the following chapters we will deal with the synthesis and energy metabolism of these carbohydrates.

Monosaccharides and Related Compounds

The simplest carbohydrates, sometimes referred to as monosaccharides or sugars, are either polyhydroxyaldehydes (aldoses) or polyhydroxyketones (ketoses). They can be derived from polyalcohols (polyols) by oxidation of one carbinol group to a carbonyl group. For example, the simple three-carbon triol glycerol can be converted either to the aldotriose, glyceraldehyde, or to the ketotriose, dihydroxyacetone, by loss of two hydrogens (fig. 13.1).

Since the middle carbon of glyceraldehyde is connected to four different substituents, it is a chiral center leading to two possible forms of glyceraldehyde. D-Glyceraldehyde is illustrated in figure 13.1 in the Fischer projection formula, in which the —OH group attached to the central carbon atom points to the right. If the central carbon were in the plane of the paper with tetrahedrally arranged substituents, the H and OH connected to it would project above the plane of the paper and the other two substituents

Figure 13.1

Loss of two hydrogens by glycerol leads to the formation of glyceraldehyde or dihydroxyacetone, depending on whether the two hydrogens are lost from the end or middle position, respectively.

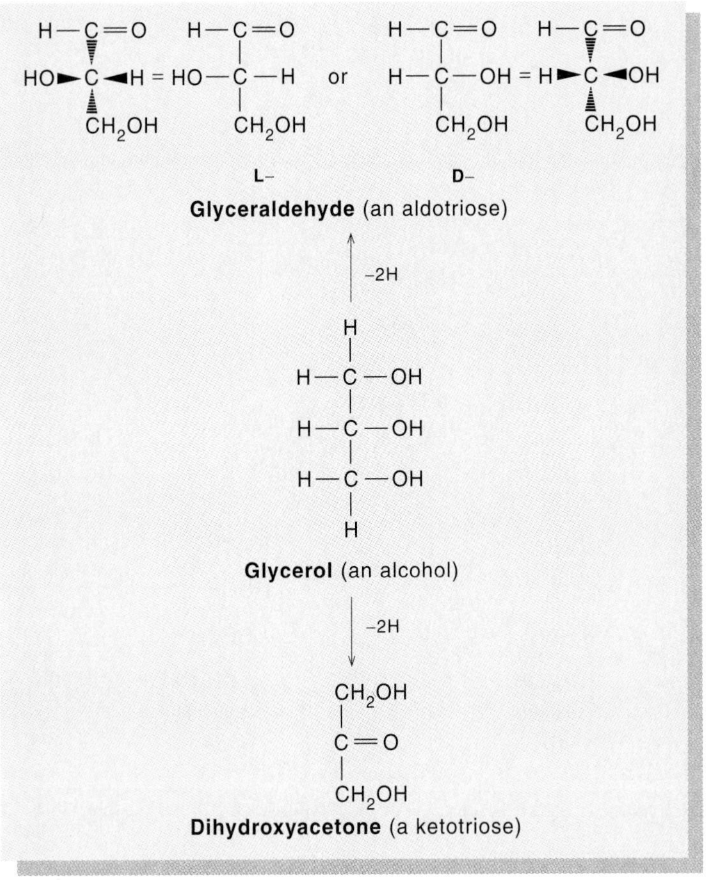

L- D-

Glyceraldehyde (an aldotriose)

−2H

Glycerol (an alcohol)

−2H

Dihydroxyacetone (a ketotriose)

would project below the plane of the paper. For larger sugars, Fischer projections are written with the most highly oxidized carbon, C-1, at the top.

A molecule such as glyceraldehyde, having one center of asymmetry (chiral center), is optically active, and the two forms of the molecule can be described as the dextrorotatory and levorotatory forms, according to the way in which they rotate plane-polarized light. The symbol *d* or (+) refers to dextrorotatory rotation, and the symbol *l* or (−) refers to levorotatory rotation. Rotation can be measured by making a solution of an optically active compound and measuring the rotation of plane-polarized light that passes through it (see chapter 5). A mixture containing equal amounts of the two forms is optically inactive and is referred to as a racemic mixture.

The stereochemistry of sugars is based on configurational properties. The actual sign of rotation may still be indicated by the italic letters *d* and *l*, but the absolute configuration of the four different substituents around the asymmetric carbon atom is designated by the prefix symbols D and L. For the common sugars the prefixes D and L refer to that center of asymmetry most remote from the aldehyde or ketone end of the molecule. By convention, all optically active centers are related to the asymmetric carbon of glyceraldehyde. Isomers stereochemically related to D-glyceraldehyde are designated D, and those related to L-glyceraldehyde are designated L. We may visualize the four-, five-, and six-carbon sugars as arising from the trioses through the stepwise condensation of formaldehyde to either glyceraldehyde or dihydroxyacetone.

Families of Monosaccharides Are Structurally Related

A phosphorylated derivative of D-glyceraldehyde is an intermediate in the degradation of carbohydrates. This aldose is reversibly converted to its ketose isomer, dihydroxyacetone (see fig. 13.1), by an enzyme. The tetroses, pentoses, and hexoses related to D-glyceraldehyde are shown in figure 13.2. The ketoses (e.g., fructose) are similarly related to dihydroxyacetone (fig. 13.3).

Monosaccharides Cyclize to Form Hemiacetals

Aldehydes can add hydroxyl compounds to the carbonyl group. If a molecule of water is added, the product is an aldehyde hydrate, as shown in figure 13.4. If a molecule of alcohol is added, the product is a hemiacetal; the addition of a second alcohol results in an acetal. Sugars form intramolecular hemiacetals in cases where the resulting compound has a five- or six-membered ring.

Hemiacetal formation first became apparent from optical studies of D-glucose. The optical rotation of a freshly dissolved sample of D-glucose changes with time because there are two different hemiacetals that are convertible in solution (fig. 13.5). They are referred to as α-D-glucose, where $[\alpha]_D = 112°$, and β-D-glucose, where $[\alpha]_D = 19°$. A freshly prepared solution of either of these compounds will eventually approach an intermediate value that depends on the equilibrium between the two forms.

The convention for numbering hexoses is shown in the central structure of figure 13.5. The α designation for the D series indicates that the aldehyde or C-1 hydroxyl group is on the same side of the structure as the ring oxygen in the Fischer projection, and the β designation indicates that it is on the opposite side. For the L series the reverse is the case. When the sugar is dissolved in water, the hemiacetal is in equilibrium with the straight-chain hydrated form. The straight-chain form can produce either hemiacetal, α or β. Conversion of one isomer to another in solution is referred to as mutarotation. The different forms are referred to as anomers, and the anomeric carbon is the carbon that contains the reactive carbonyl. In the case of glucose this is the C-1 carbon. Equilibrium is reached without added catalyst in a few hours at room temperature. The open-chain form usually represents only a small fraction of the total (see figure 13.5 for actual percentages).

Hemiacetals with five-membered rings are called furanoses, and hemiacetals with six-membered rings are called pyranoses. In cases where either five- or six-membered rings are possible, the six-membered ring usually predominates. For example, for glucose less than 0.5% of the furanose forms exist at equilibrium (see fig. 13.5, bottom). The reason for the general prepon-

Figure 13.2

Configurational relationships of the D-aldoses. The most important sugars are starred. Note that in the D series the configuration about the chiral center farthest from the carbonyl is the same.

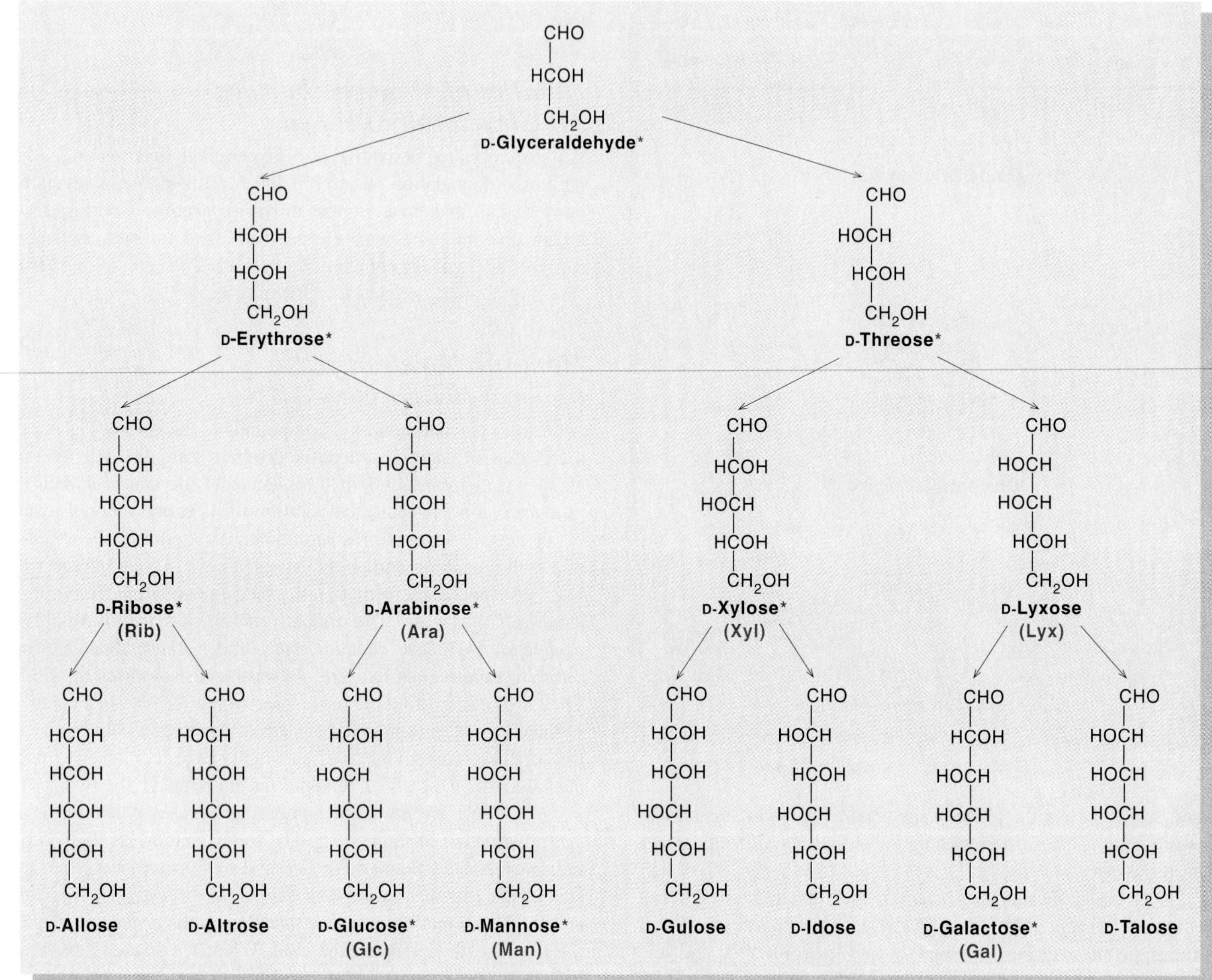

derance of the pyranose is not known. Both furanoses and pyranoses are more realistically represented by pentagons or hexagons in the Haworth convention, as shown for glucose in figure 13.6.

Haworth structures are unambiguous in depicting configurations, but even they do not show correctly the spatial relationship of groups attached to rings. The normal valence angle of saturated carbon (109°) prevents a stable planar arrangement for cyclohexane or the related pyranose molecule. The two most likely conformations* are the so-called chair and boat forms (fig. 13.7). Usually, the chair form is considerably more stable than the boat form. The 12 substituent atomic groups of the ring carbons fall into two classes: those that are approximately perpendicular to the plane of the ring, i.e., axial, and those that are parallel to the plane of the ring, i.e., equatorial. As a rule, a substituent is at a lower energy state in the equatorial position because there is less chance of steric hindrance with other substituents. This fact becomes more important with larger substituents. The two anomers for the favored chair form of D-glucose are shown in figure 13.8. In sugar chemistry, formulating the conformations of aldohexoses is important in interpreting reactivity of hydroxyl groups and other siz-

*When discussing sugars, we make frequent use of the terms "configuration" and "conformation." These terms have different meanings. Two conformations of the same molecule are interconvertible without breakage of any chemical bonds; two configurations of the same molecule are not. For example, the chair and boat forms of α-D-glucose represent two different conformations of the same molecule. But α-D-glucose and β-D-glucose represent two different configurations of D-glucose.

Figure 13.3

Configurational relationships of the D-ketoses. The most important sugars are starred.

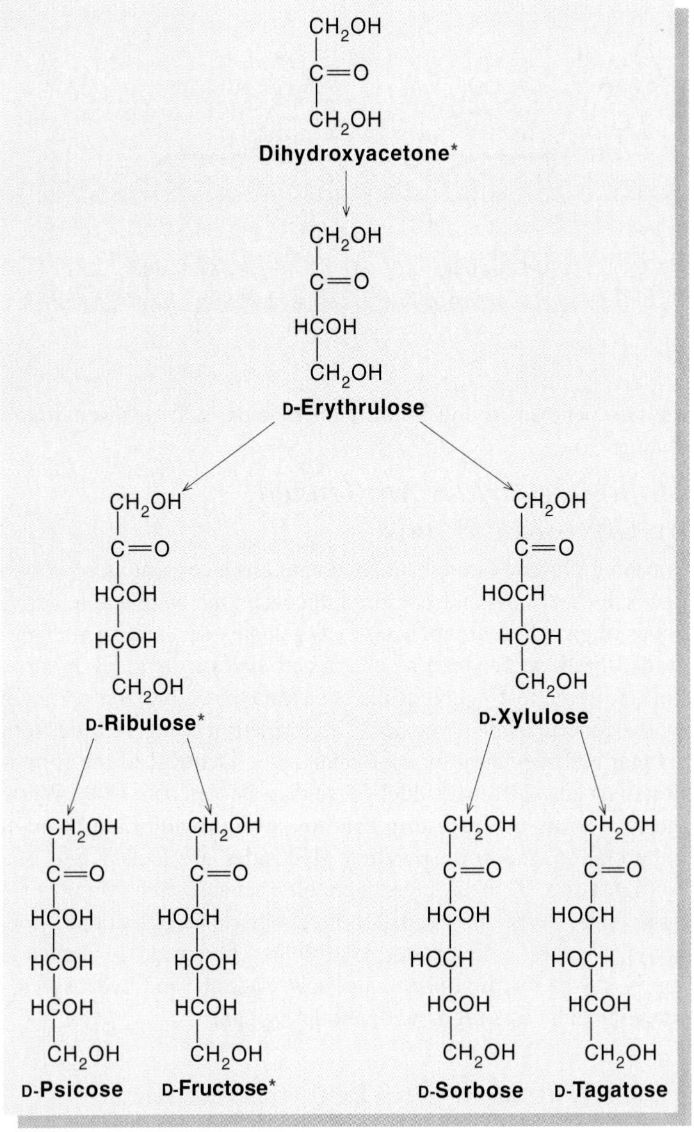

Figure 13.5

Different forms of glucose that result from dissolving glucose in water. At 25°C in water, glucose reaches an equilibrium containing about 0.02% free aldehyde, 38% α-pyranose form ($[\alpha]_D = 112°$), 62% β-pyranose form ($[\alpha]_D = 19$), and less than 0.5% of the furanose forms. The anomeric carbon is shown in color.

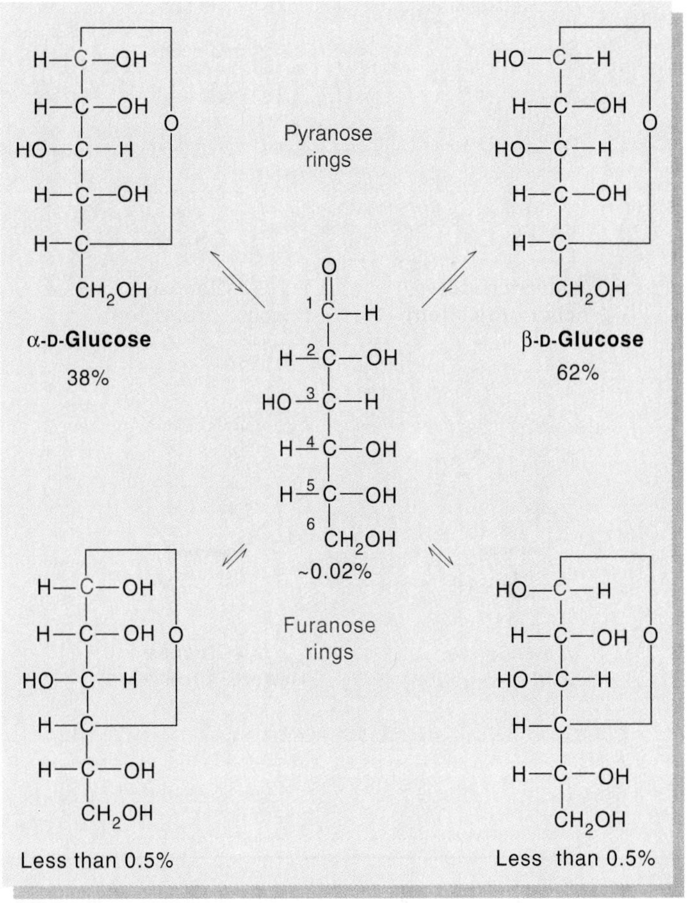

able substituents. The most stable conformation for a particular aldohexose is the chair form, which places the maximum number of substituents larger than hydrogen in the equatorial position.

The furanose ring is nonplanar also and can exist in more than one conformation. The conformations for D-ribose (β-D-ribofuranose) and D-2-deoxyribose (β-D-2-deoxyribofuranose),

Figure 13.4

Aldehydes can add H_2O to form hydrates or can add alcohols to form hemiacetals and acetals.

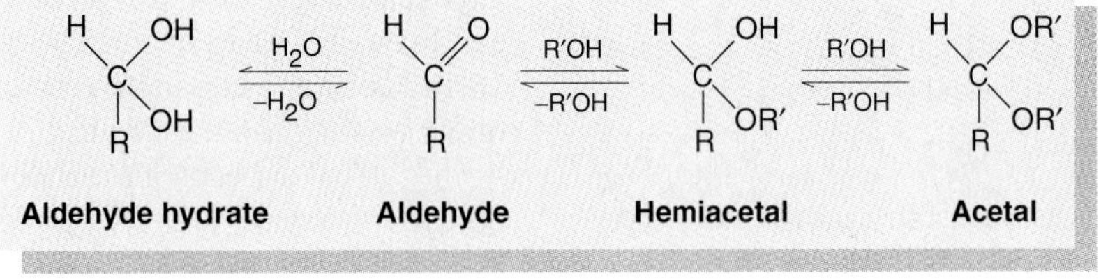

Figure 13.6

Comparison of the Fischer and Haworth projections for α- and β-D-glucose. The Haworth projection is a step closer to reality.

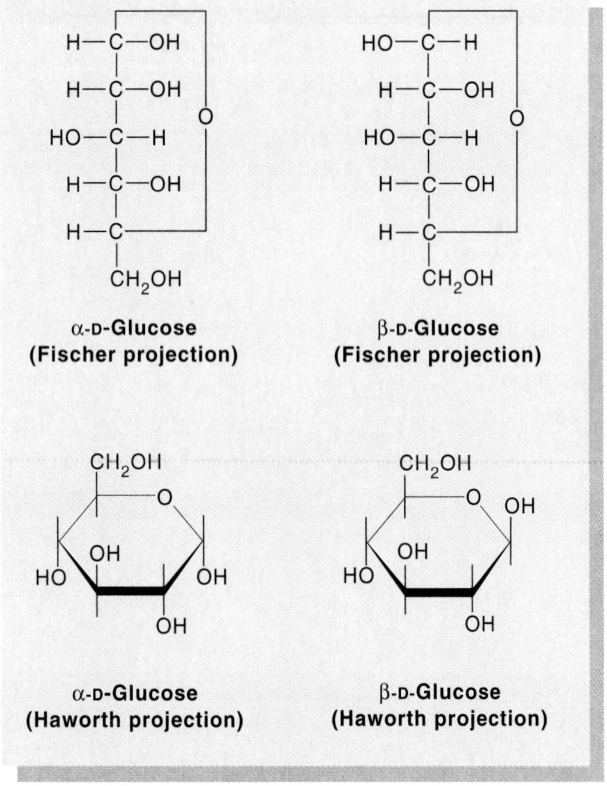

α-D-Glucose (Fischer projection)

β-D-Glucose (Fischer projection)

α-D-Glucose (Haworth projection)

β-D-Glucose (Haworth projection)

Figure 13.7

Chair and boat forms for a generalized pyranose ring structure. Structures of this type are more realistic than the Haworth structure, as the carbon–carbon bond angle is correct. The chair form is usually favored over the boat form. The substituents are labeled "a" (axial) and "e" (equatorial). The axis of symmetry is labeled for both forms. Axial bonds are parallel to this axis of symmetry. Equatorial bonds are parallel to the nonadjacent sides of the rings. A large substituent generally prefers to be in an equatorial location.

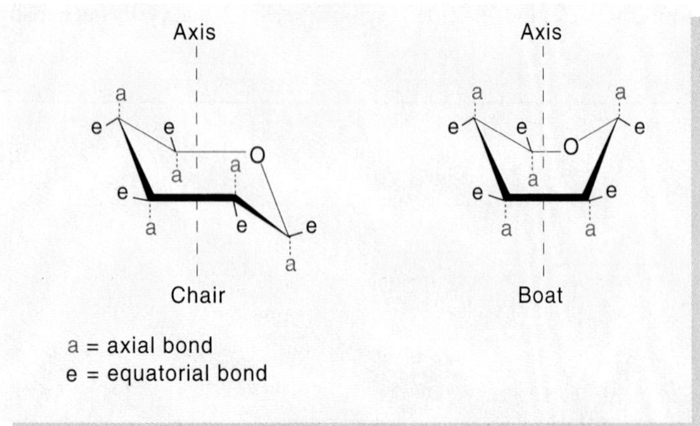

Figure 13.8

Chair configurations for the two anomers of D-glucose. Note that the largest substituent, $—CH_2OH$, is in an equatorial location in both structures. The differences between the two anomers are shown in red.

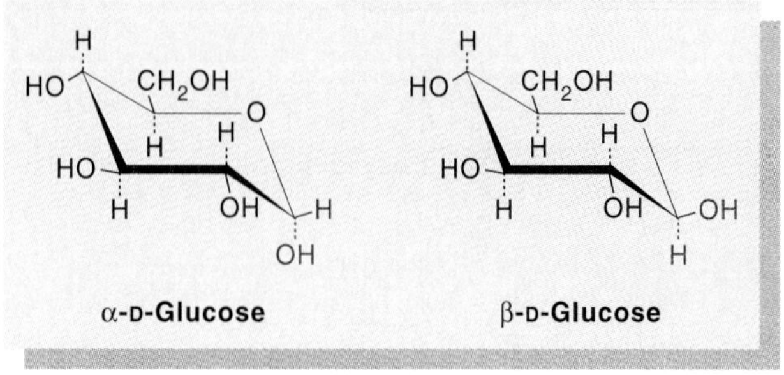

α-D-Glucose

β-D-Glucose

the two pentoses found in all nucleic acids, will be discussed in chapter 30.

Monosaccharides Are Linked by Glycosidic Bonds

Warming glucose in methanol and acid produces a mixture of two new substances, α- and β-methylglucoside; the comparable derivatives of galactose are referred to as galactosides, and so on. Generally, the bond between a sugar's carbonyl oxygen and an alcohol is referred to as a glycosidic bond, and the compound is known by the generic name glycoside. The formation of a glycoside from a sugar and methanol by acid catalysis is identical to the formation of an acetal from an aldehyde and an alcohol (fig. 13.9). While the two forms of glucose in solution are in equilibrium through mutarotation, the corresponding glycosides are locked into one configuration. This is understandable, because mutarotation requires that, in the intermediate, the anomeric carbon adopt a carbonyl structure, which is not possible in a glycoside. A glycoside can be formed with aliphatic alcohols, phenols, and hydroxy carboxylic acids, as well as with another sugar.

Disaccharides and Polysaccharides

The most important glycosides are those formed with other sugars. Monosaccharides are glycosidically linked to form disaccharides (fig. 13.10). For instance, the disaccharide maltose contains a glycosidic bond between the C-1 of one glucose molecule and the C-4 of another glucose molecule. The compound is said to have an $\alpha(1,4)$ glycosidic linkage because the anomeric C-1 carbon of one sugar is connected to the C-4 of another sugar and the configuration about the anomeric carbon is α. Maltose possesses one potentially free aldehyde group and is therefore referred to as a reducing sugar (box 13A). The configuration about the hemiacetal hydroxyl group of maltose has not been specified in figure 13.10 because it can undergo mutarotation. Maltose is most familiar as a degradation product of starch.

The disaccharide cellobiose is identical with maltose except for having a $\beta(1,4)$ glycosidic linkage. Cellobiose is a

Figure 13.9

Formation of methyl glycosides. Glycosides (in this case glucosides) are quite stable in alkali, but they hydrolyze readily in dilute acid.

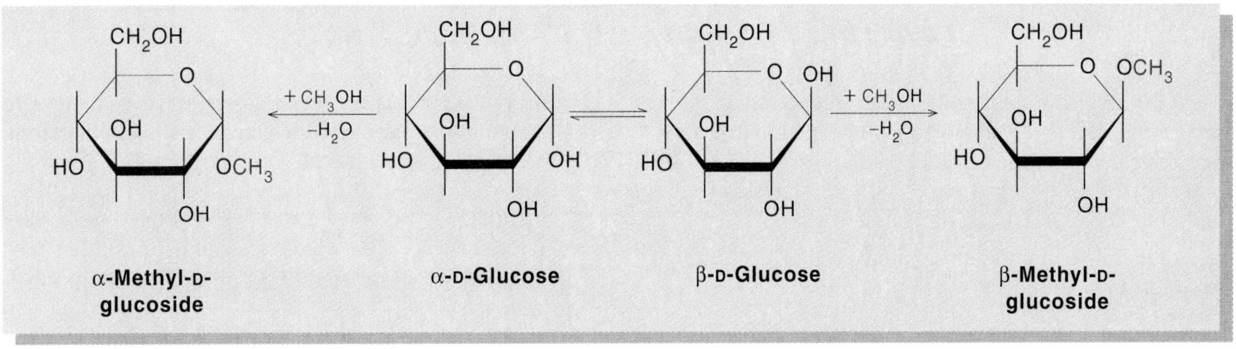

Figure 13.10

Four commonly occurring disaccharides. The configuration about the hemiacetal group has not been specified for lactose, maltose, or cellobiose because both anomers exist in equilibrium.

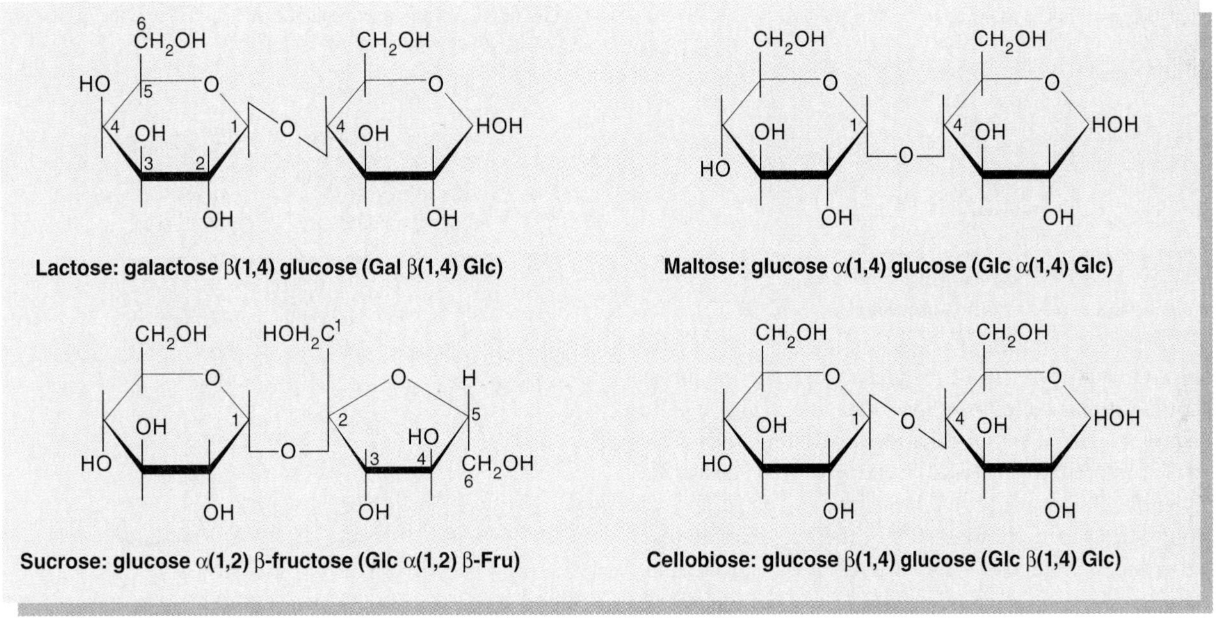

degradation product of cellulose. Lactose is a disaccharide found exclusively in the milk of mammals. Lactose contains a $\beta(1,4)$ glycosidic linkage between galactose and glucose. Sucrose is found in abundance in sugar beets and sugar cane. On acid hydrolysis it yields equivalent amounts of D-glucose and D-fructose. Sucrose contains an $\alpha(1,2)\beta$ glycosidic linkage.

Most carbohydrates in nature exist as high-molecular-weight polymers called polysaccharides. Polysaccharides are composed of simple or derived sugars connected by glycosidic bonds. The most common building block used in polysaccharides is D-glucose. Polymers composed of a single type of building block are called homopolymers; those composed of more than one type are called heteropolymers.

Polysaccharides function in two quite distinct roles; some serve as a means for storage of chemical energy and others serve a structural function.

Cellulose Is a Major Homopolymer Found in Cell Walls

Cellulose is a structural polysaccharide found as the major component of cell walls in plants. It is the most abundant of organic compounds, constituting approximately 50% of all the carbon

Detection of Sugars with Aldehyde Groups

The test for an aldose is based on the unique reducing capacity of the aldehyde among organic functional groups. Hence a sugar containing an aldehydic group is called a reducing sugar.

Two different chemical tests are in common use:

1. Reduction of the silver ammonium complex, $Ag(NH_3)_2^+$, known as Tollen's reagent. The reaction is

$$RCHO + 2\ Ag(NH_3)_2^+ + 2\ OH^- \longrightarrow RCO_2^- + 2\ Ag + 3\ NH_3 + NH_4^+ + H_2O$$

Deposition of silver metal is a positive qualitative test.

2. Fehling's solution. Cupric sulfate and sodium carbonate with a tartrate buffer are used. The reaction is

$$RCHO + 2\ Cu^{2+} + 5\ OH^- \longrightarrow RCO_2^- + Cu_2O + 3\ H_2O$$

Deposition of brick-red Cu_2O precipitate is a positive qualitative test.

Figure 13.11

Structure of 2,3,6-tri-*O*-methylglucose. This is the main product resulting from exhaustive methylation of cellulose followed by acid hydrolysis.

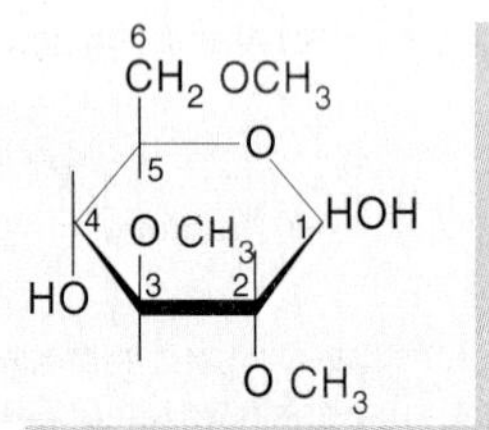

2,3,6-tri-*O*-methylglucose

Figure 13.12

Structure of the repeating unit of cellulose (top) and methylation linkage analysis by reaction with dimethylsulfate. As can be seen, the residues in the native structure are connected by $\beta(1,4)$ linkages.

found in plants. On acid hydrolysis, cellulose yields the monomer glucose and some dimer cellobiose, the latter due to incomplete hydrolysis. The reaction of polysaccharides with dimethylsulfate is often useful in structural analysis. Treatment of a saccharide with dimethylsulfate in alkali results in the conversion of all free hydroxyl groups to *O*-methyl ethers. Fully methylated cellulose gives 2,3,6-tri-*O*-methylglucose on acid hydrolysis (fig. 13.11). The absence of a methyl on the anomeric C-1 says nothing about the structure of cellulose, since such methyl derivatives are susceptible to mild acid hydrolysis. However, the absence of a methyl group at the C-4 position indicates that this position is inaccessible in the cellulose structure. This fact proves that in cellulose the glycosidic linkages are of the 1,4 type. Other measurements show that the 1,4 linkages are about the anomeric carbon. The repeating unit of cellulose is indicated in figure 13.12.

Cellulose is insoluble in water because of the high affinity of the polymer chains for one another. Individual polymeric chains have a molecular weight of 50,000 or greater. The molecular chains of cellulose interact in parallel bundles of about 2,000 chains, a bundle having a diameter of 100–250 Å. Each bundle of 2,000 comprises a single microfibril. Many microfibrils arranged in parallel comprise a macrofibril, which can be seen under the light microscope. Figure 13.13 shows the inner secondary walls of the plant *Valonia;* the fibrils in the secondary wall are almost pure cellulose.

Figure 13.13

Fibril arrangements in the cell wall of *Valonia* (12,000×). (Electron micrograph from *Biophoto Associates/Photo Researchers.*)

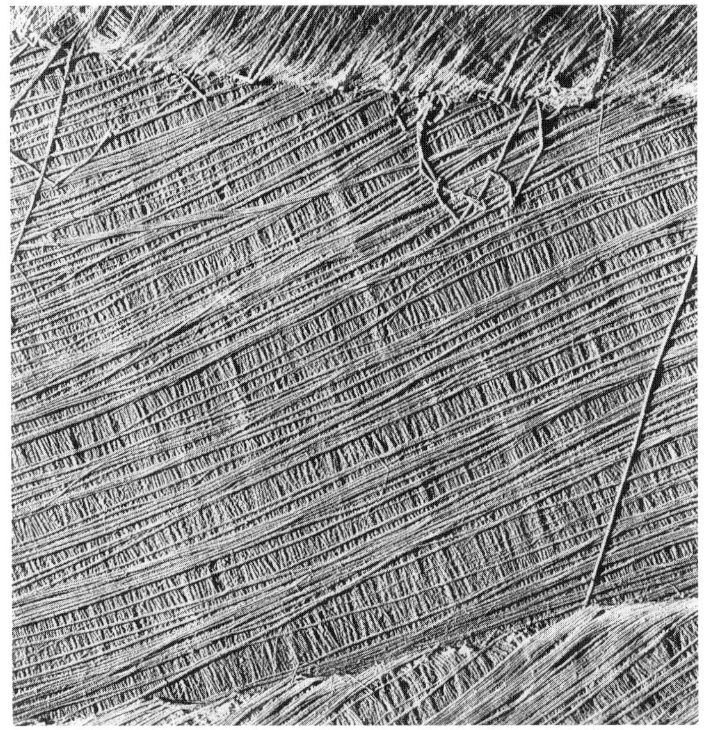

Starch and Glycogen Are Major Energy-Storage Polysaccharides

Although glucose is the most important sugar involved in energy metabolism in most cells and tissues, it is not present in the cell to any large extent as the free monosaccharide. Cells store glucose for future use in the form of simple homopolymers and thereby reduce the osmotic pressure of the stored sugar. A polysaccharide consisting of 1,000 glucose units exerts an osmotic pressure that is only 1/1,000 of the pressure that would result if the glucose units were all present as separate molecules. In the polymeric form, glucose can be stored compactly until needed.

The two major polysaccharides used for energy storage are starch in plant cells and glycogen in animal cells. Both are $\alpha(1,4)$ homopolymers with occasional $\alpha(1,6)$ linkages to make branchpoints (fig. 13.14). The two polysaccharides, starch and glycogen, differ primarily in their chain lengths and branching patterns. Glycogen is highly branched, with an $\alpha(1,6)$ linkage occurring every 8 to 10 glucose units along the backbone, giving rise in each case to short side chains of about 8 to 12 glucose units each. Starch occurs both as unbranched amylose and as branched amylopectin. Like glycogen, amylopectin has $\alpha(1,6)$ branches, but these occur less frequently along the molecule (once every 12 to 25 glucose residues) and give rise to longer side chains (lengths of 20 to 25 glucose units are common). Starch deposits are usually about 10–30% amylose and 70–90% amylopectin.

Because all of the branches in both starch and glycogen terminate with a nonreducing sugar, these polymers have only one reducing end originating from the original unbranched homopolymer and as many nonreducing ends as they have branches.

Figure 13.14

Structure of the storage polysaccharides glycogen and starch. The main chain is $\alpha(1,4)$-linked. Side chains are connected to the main chain by $\alpha(1,6)$ linkages.

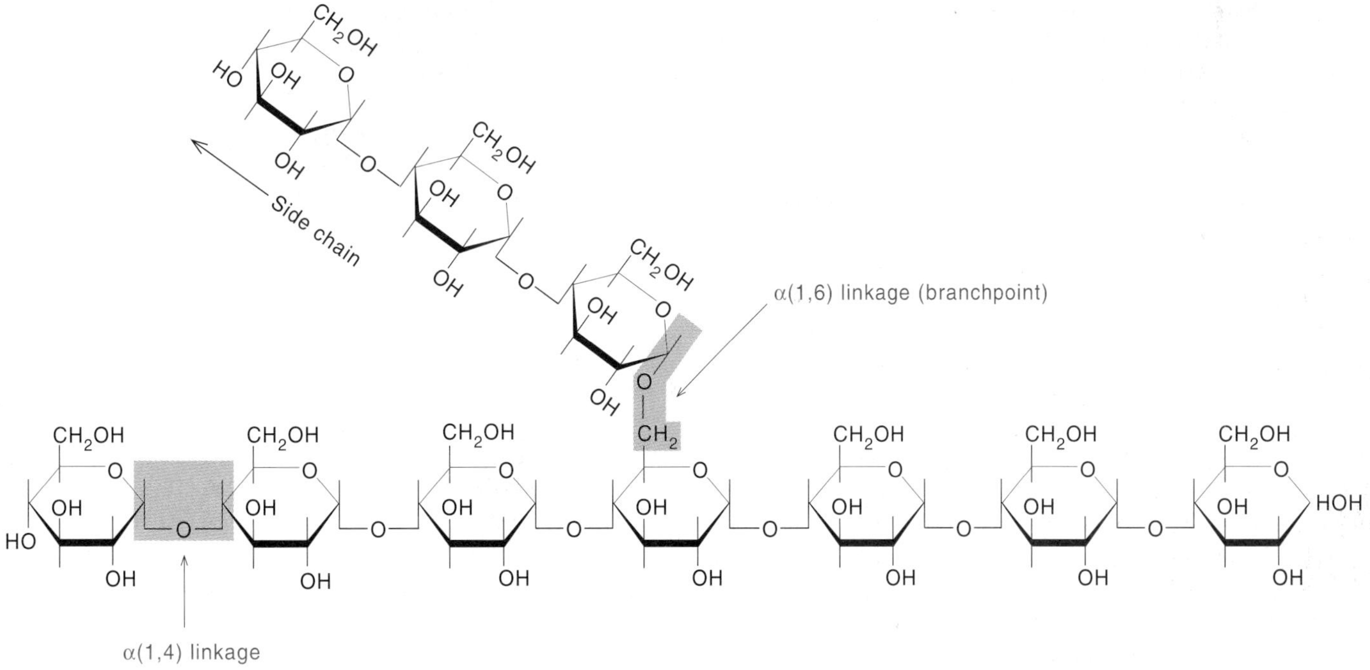

Figure 13.15

Energetically favored conformations of β(1,4)-linked D-glucose (*a*) found in cellulose and α(1,4)-linked D-glucose (*b*) found in starch and glycogen. Note that in the β(1,4) configuration in (*a*), alternating residues are flipped 180° relative to one another so that long straight chains result. Stabilizing H bonds are indicated by dashed lines. In the α(1,4) configuration (*b*) the chain has a natural curvature. (Illustration copyright by Irving Geis. Reprinted by permission.)

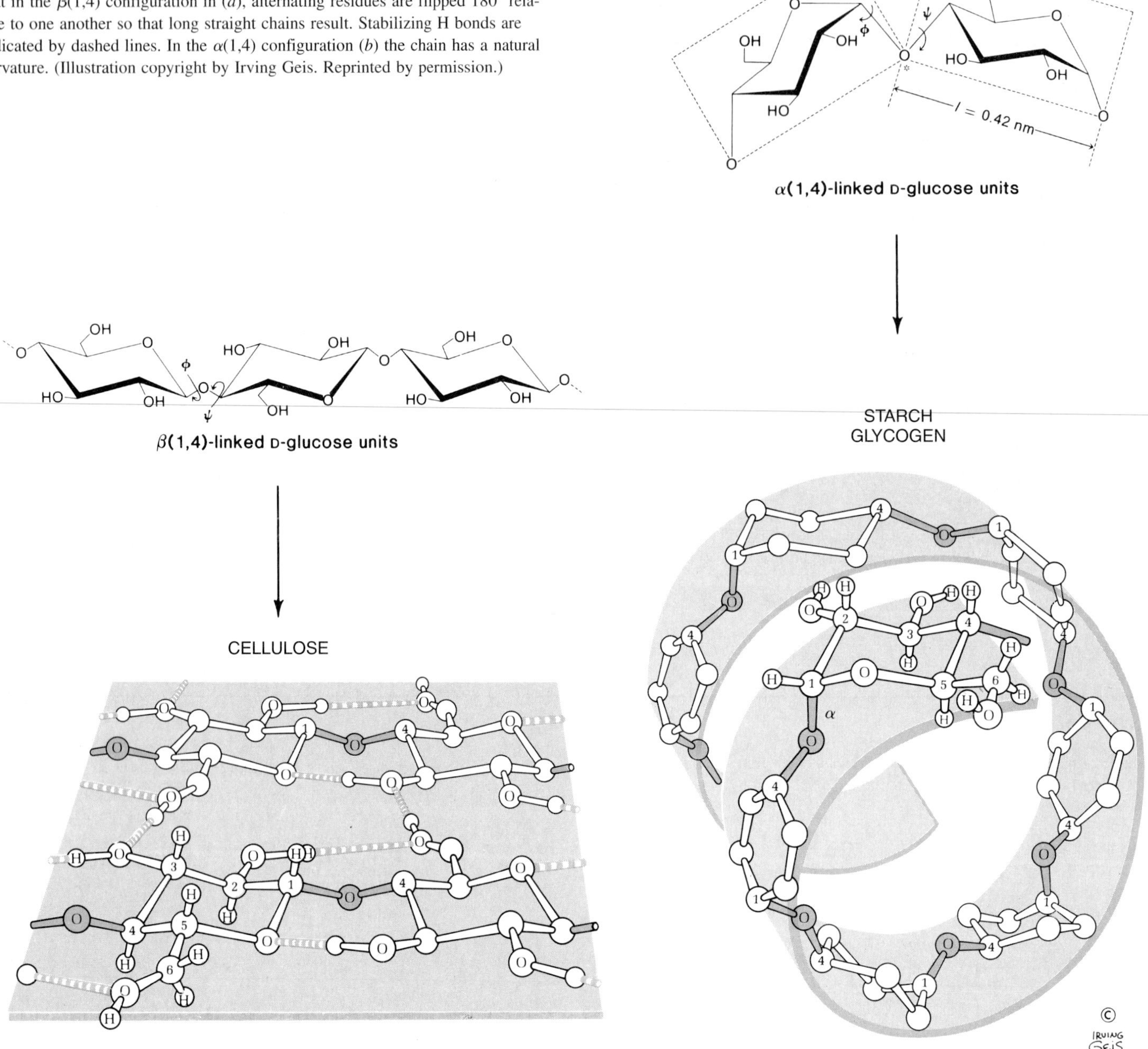

The Configurations of Glycogen and Cellulose Dictate Their Roles

It is a remarkable fact that the main energy-storage polysaccharides and the main structural polysaccharides found in nature both have a primary structure of (1,4)-linked polyglucose. Why should two such closely related compounds be used in totally different roles? A closer look at the stereochemistry of the α and β glycosidic linkage for polyglucose indicates why this is so.

Recall that D-glucose exists in the chair form of a pyranose ring (see figs. 13.7 and 13.8). The ring has a rigid character to it. We can think of it as a structural building block in a polysaccharide chain, just as we think of the rigid planar peptide grouping as a structural building block in a polypeptide chain. It is also possible to specify two torsional angles ϕ and ψ for rotation about the glycosidic C—O linkage (fig. 13.15). These angles are used extensively to discuss polypeptide configuration in proteins (see chapter 5). They have limited value in discussing polysaccharide structures, because much less information is available about such structures. Nevertheless, it is clear that only the $\beta(1,4)$-linked polyglucose has the capacity to form straight chains (see fig. 13.15*a*). A straight chain can be created by flipping each glucose unit by 180° relative to the previous one. This process should result in an almost fully extended polysaccharide chain, which is known to be characteristic of the structure of cellulose. Evidently, this conformation is energetically favored. By contrast, $\alpha(1,4)$-linked units in a polyglucose cause a natural turning of the chain (see fig. 13.15*b*). Consistent with this fact is the observation that amylose adopts a coiled helical configuration. Indeed, one of the first helical structures to be discovered (1943) was the left-handed helix of amylose wound around molecules of iodine (fig. 13.16). This structure is responsible for the characteristic blue color of the amylose–iodine complex.

The extended-chain form of polyglucose has been exploited in nature for structural purposes, leaving by default the coiled form for use as an energy-storage macromolecule. Correlated with this functional difference is the omnipresence of degrading enzymes for glycogen and starch and the very limited phylogenetic distribution of comparable enzymes for cellulose. Cellulose is degraded in the gastrointestinal tract of herbivores, such as the cow, or in insects, such as termites, by a protozoan that synthesizes the enzyme cellulase. Humans do not possess this enzyme and hence cannot degrade cellulose.

Figure 13.16

Structure of the helical complex of amylose with iodine (I_2). The amylose forms a left-handed helix with six glucosyl residues per turn and a pitch of 0.8 nm. The iodine molecules (I_2) fit inside the helix parallel to its long axis.

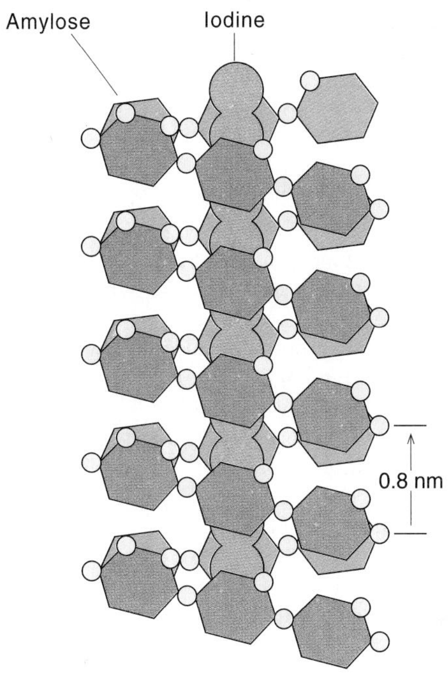

SUMMARY

Carbohydrates are the most abundant organic substances. They are an important source of carbon compounds for all biomolecules, and they also serve important energy and structural needs. In this chapter we have focused on the following points.

1. Carbohydrates may be divided into monosaccharides, oligosaccharides, and polysaccharides.
2. The monosaccharides are either polyhydroxyaldehydes or polyhydroxyketones. All monosaccharides are optically active because of the presence of one or more asymmetrical carbon atoms (chiral centers).
3. Straight-chain sugars in aqueous solution tend to form ring structures known as intramolecular hemiacetals, especially when the resultant is a five-membered (furanose) ring or a six-membered (pyranose) ring. Depending on which way the ring forms about the anomeric carbon, the structure is called an α or β hemiacetal.
4. For a given hemiacetal several conformations are possible. Usually, the chair form is favored over the boat form because of the lower steric repulsion produced by side chains in the chair form. Polymer synthesis locks in a particular hemiacetal configuration.
5. Monosaccharides are linked by glycosidic bonds to form oligosaccharides and polysaccharides.
6. Polysaccharides function in two quite distinct roles: Some serve as a means for storage of chemical energy, and others serve a structural function.
7. Polymers that use one type of building block (monomer) are called homopolymers, and those that use more than one are called heteropolymers.
8. Cellulose is the best known and most abundant structural polysaccharide. It is a homopolymer of glucose with $\beta(1,4)$ linkages between adjacent monomeric residues.
9. Starch and glycogen are energy-storage polysaccharides. They also are homopolymers of glucose but with $\alpha(1,4)$ linkages between adjacent residues. In addition, they contain branches with $\alpha(1,6)$ linkages.

SUGGESTED READINGS

See chapters 14 and 18 for related readings.

PROBLEMS

1. What is the significance of the chirality of the penultimate carbon of a monosaccharide?
2. Consider the aldohexoses presented in figure 13.2. Can you propose a reason why, in addition to glucose, the most significant (common) sugars in nature are mannose and galactose?
3. What is the simplest explanation you can give to an organic chemist that would explain why there are more D-aldohexoses than there are D-ketohexoses?
4. Draw the structure of the ketoheptopyranose that has the most bulky groups in the equatorial positions. Is the sugar an α or a β?
5. The legend of figure 13.6 contains the sentence "The Haworth projection is a step closer to reality." What is the meaning of this statement?
6. Is it possible that carbohydrates could have produced the diversity required to catalyze the myriad cellular reactions now relegated primarily to proteins?
7. In solution, D-glucose has a specific rotation of $[\alpha]_D^{20} = +52.7°$. The specific rotation of pure β-D-glucose is +18.7°, and that of pure α-D-glucose is +112.2°. Calculate the fraction of the α and β anomers in solution.
8. If either pure crystalline α- or β-D-glucose is dissolved in water, the final solution will contain the same fraction of α and β anomers as determined in problem 7. Explain chemically how this mutarotation occurs.
9. You are given two containers of polysaccharide by a colleague who has labeled one container as cellulose and one as glycogen. In his haste he may have mislabeled them and has asked you to verify which is which by methylation analysis. Indicate what product(s) you would expect from exhaustive methylation and mild acid hydrolysis of each polysaccharide and how the products could be used to differentiate the samples.
10. Draw the structures of α-sophorose, α-melibiose, and β-lactulose. Sophorose is two glucoses linked $\beta(1,2)$. Melibiose is a galactose linked to a glucose by an $\alpha(1,6)$ bond. Lactulose is galactose linked $\beta(1,4)$ to fructose.
11. A tool that has often been used to deduce ring sizes and linkages between sugars is methylation with methyl iodide. The methylated sugar is then hydrolyzed and the location of the methyl groups determined. A disaccharide called trehalose has been obtained from a number of vegetable sources. Trehalose is composed only of glucose, and when it is methylated and hydrolyzed, only 2,3,4,6-tetramethylglucose results. By using the same procedure, maltose (fig. 13.10) gives 2,3,4,6-tetramethylglucose and 2,3,6-trimethylglucose in equal molar ratios. Propose a structure(s) for trehalose.
12. The substance glucuronic acid can be considered to be glucose with carbon 6 oxidized to the carboxylic acid level. Glucuronic acid β-glycosides are the metabolic fate of a number of obnoxious compounds found in plants. Draw the structure of glucuronic acid. From its name, can you guess where it was first found?
13. Aldoses and ketoses that have the potential to form open chains (i.e., have hemiacetal or hemiketal groups) are referred to as reducing sugars. This name is derived from the fact they can readily reduce Ag^+ to Ag or Cu^{2+} to Cu^{1+}. Examine the disaccharides in figure 13.10 and determine which substances are reducing sugars. Is trehalose (problem 11) a reducing sugar?
14. The polysaccharide inulin is an energy-storage substance found in the tubers of a number of plants. The major part of inulin is composed of fructoses linked $\beta(2,1)$. Draw the structure of inulin.

CHAPTER

14

GLYCOLYSIS, GLUCONEOGENESIS, AND THE PENTOSE PHOSPHATE PATHWAY

Glycolysis entails the breakdown of hexoses to three-carbon compounds while producing ATP. The reverse process, which consumes ATP, is called gluconeogenesis.

Having covered the basic structural features of carbohydrates, we can now discuss sugar metabolism. First we describe how sugars are degraded to produce energy and carbon compounds that can be used in biosynthesis. Then we consider the reverse process of gluconeogenesis, synthesis of sugars from smaller carbon compounds. Next we consider how these two oppositely directed pathways are regulated, with a particular focus on cellular and organismic needs. Finally, we consider an alternative mode of sugar breakdown that is most important in supplying reducing power for biosynthesis and ribose for nucleic acids.

Pivotal Events in the Elucidation of the Glycolytic Pathway

The study of glycolysis was a powerful stimulus for the detailed study of enzymes, guided by a strategic concept, the elucidation of an entire pathway. Methods for the isolation and crystallization of the glycolytic enzymes were established primarily in the laboratories of Warburg and the Coris to become a standard for all enzymologists, and the mechanistic and kinetic studies on them paved the way to a growing understanding of the catalytic mechanisms of enzymes.

Meaningful work on the glycolytic pathway dates back to the turn of the century when the Buchners discovered that glycolysis could proceed even after the destruction of the integrity of the cell. For many years after this the main systems used in metabolic research consisted of yeast juices or muscle pulp and its cell-free extracts. Harden and Young's insight that phosphate is required for glycolysis focused early research on the routes followed by phosphate. The identification of phosphocreatine by Fiske and Subbarow and of ATP by Lohmann combined with the introduction of thermochemistry by Meyerhof led to the concept of energy-rich bonds and to the discovery of transphosphorylation. The adenylate system was pictured as a coenzyme of glycolysis as well as a transmitter of energy. The accidental discovery of 3-phosphoglycerate by Nilsson and the work of Lohmann and Lipmann on the conversion of fructose-1,6-bisphosphate to more hydrolysis-resistant phosphate esters was the final trigger that led Embden to propose the unified scheme of glycolysis basically as we know it now. From this point it was only a few years until the remaining postulated steps were to be revealed.

Overview of Glycolysis

Glycolysis is the process by which glucose or other hexoses are converted first to pyruvate and then either to lactate or to ethanol and CO_2. The glycolytic pathway is amphibolic (from the Greek, *amphi,* meaning "both"), as it can provide the cell with energy from glucose catabolism and can also serve an anabolic function by yielding C_3 precursors for the synthesis of amino acids, fatty acids, and cholesterol.

Glycolysis is the most ubiquitous pathway in all energy metabolism, occurring in almost every living cell. It is regarded as a primitive process because it probably arose early in biological history long before there was a significant amount of molecular oxygen in the atmosphere and well before the advent of eukaryotic organelles.

The glycolytic pathway was the first major metabolic sequence to be elucidated. Much of the definitive work was done in the 1930s by the German biochemists Gustav Embden, Otto Meyerhof, and Otto Warburg. Because of their contributions the alternative name, Embden-Meyerhof pathway, is sometimes used for the glycolytic pathway (box 14A).

The glycolytic pathway appears in detail in figure 14.1. The essence of the process is suggested by the name, since glycolysis comes from the Greek roots *glykos,* meaning "sweet," and *lysis,* meaning "loosing." Literally, glycolysis is the loosing, or splitting, of something sweet, that is, the starting sugar. From figure 14.1 it is clear that the actual splitting occurs at step 4 (starting from glucose). At this point, a six-carbon sugar is cleaved to yield two three-carbon compounds, one of which, glyceraldehyde-3-phosphate, is the only oxidizable molecule in the entire pathway. After the cleavage of step 4, two successive ATP-generating steps occur: one at step 7 and the other at step 10. These steps represent the energy payoff of the process, since these are the only ATP-yielding reactions of the pathway under anaerobic conditions.

Except for glycerate-1,3-bisphosphate all of the intermediates in the pathway are pictured as belonging to one of three metabolic pools. Within each metabolic pool the intermediates are readily interconvertible and usually present in relative concentrations close to their equilibrium values. Between the pools the concentrations of the intermediates can be very different because of the lack of rapid interconversion and also because the equilibrium values are often very large or very small.

Three Hexose Phosphates Constitute the First Metabolic Pool

The three hexose phosphates—glucose-1-phosphate, glucose-6-phosphate, and fructose-6-phosphate—constitute a single metabolic pool (fig. 14.2). This pool can be replenished by generation of any of its components (also see fig. 18.1 for conversions of other hexoses to members of this pool). Glucose-1-phosphate is the first product in the utilization of storage polysaccharides; glucose-6-phosphate is the first hexose phosphate formed when free glucose is metabolized; and fructose-6-phosphate is the first hexose phosphate formed when carbohydrate is made de novo by gluconeogenesis or photosynthesis.

Phosphorylase Converts Storage Carbohydrates to Glucose Phosphate

The first step in the cell's utilization of stored starch or glycogen involves the removal of a terminal glucose residue, a reaction catalyzed by the enzyme glycogen phosphorylase. Overall, the reaction involves same-side displacement at C-1 of the terminal residue, with an incoming phosphate group replacing the remainder of the polysaccharide molecule. The product generated is therefore glu-

Figure 14.1

The glycolytic pathway from glucose to pyruvate, indicating two anaerobic options (ethanol or lactate) and one aerobic option (TCA cycle). The red arrows indicate how NADH formed in glycolysis could be reoxidized to NAD^+ when pyruvate is converted to an end product of ethanol or lactate under anaerobic conditions. The three metabolic pools are shown in color panels.

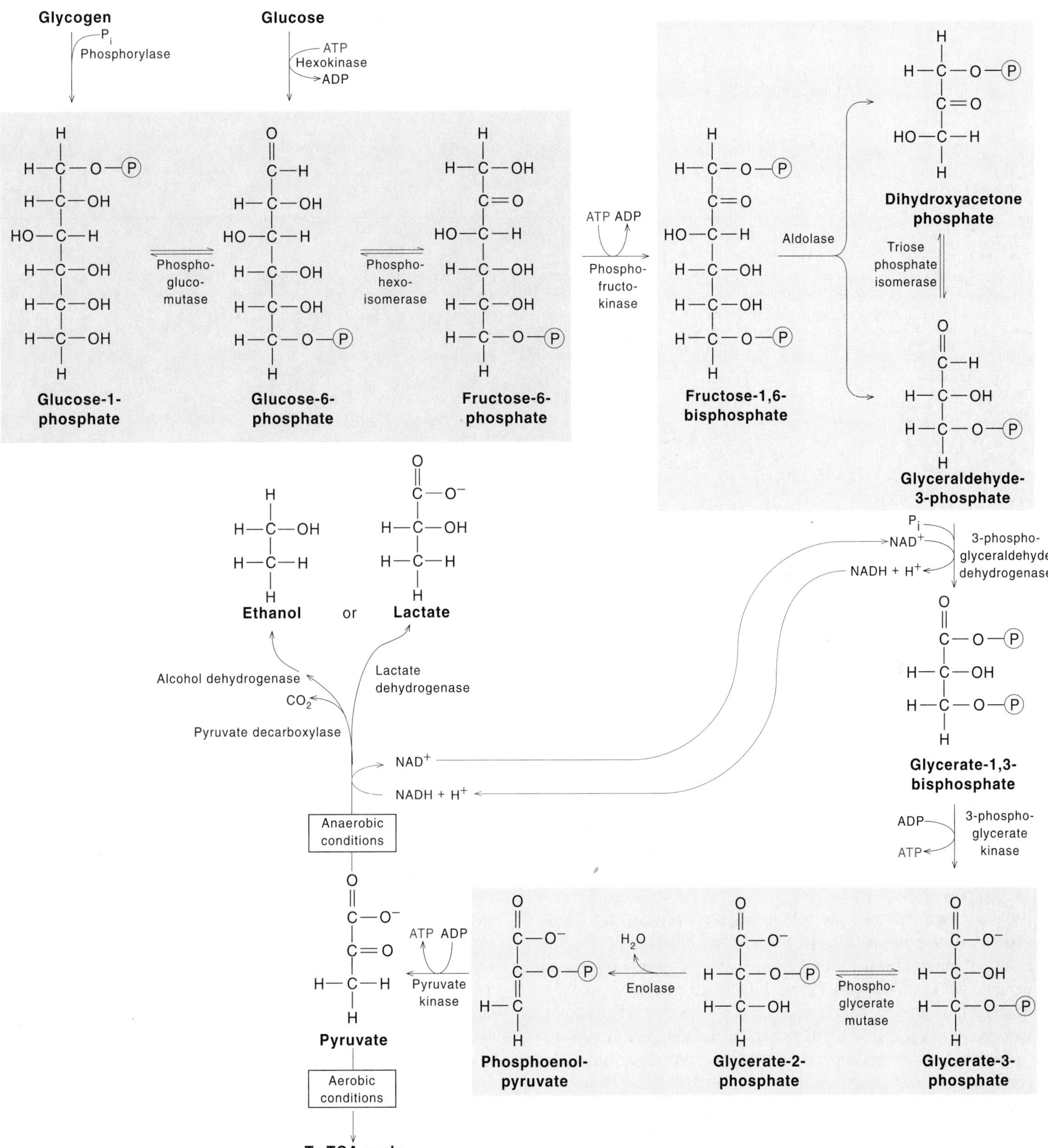

Figure 14.2

The hexose monophosphate pool. The equilibrium percentages of the three hexose monophosphates are indicated. Horizontal flow arrows indicate major routes of replenishment and utilization of the three hexose monophosphates.

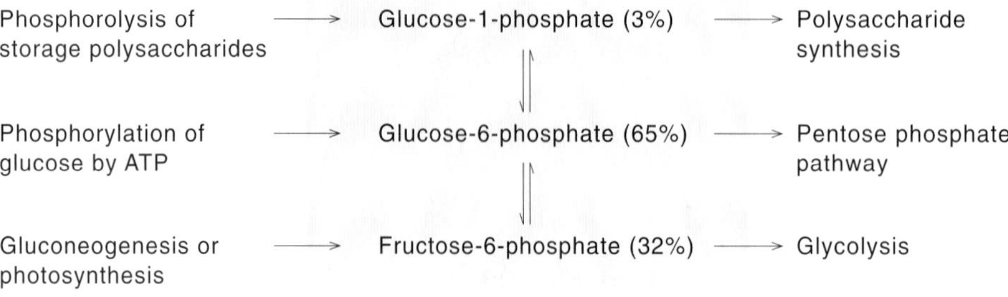

Figure 14.3

The reaction catalyzed by phosphorylase. An oxygen of a phosphate ion attacks C-1 of the terminal glucosyl unit of starch or glycogen, displacing the macromolecule and generating glucose-1-phosphate. The reaction proceeds with retention of configuration, a fact suggesting that it may involve an oxonium ion intermediate. Phosphorylase is a complex regulatory enzyme, the allosteric properties of which were discussed in chapter 10.

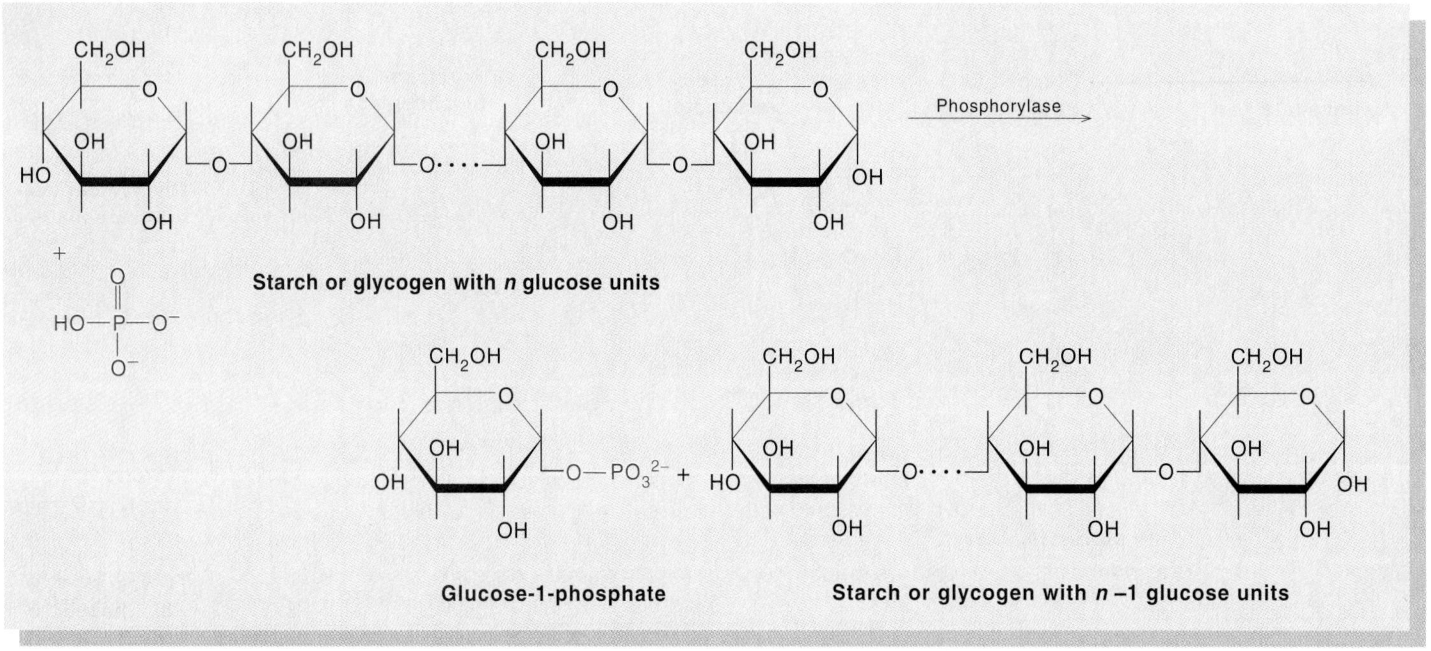

Phosphorylase

Starch or glycogen with *n* glucose units

Glucose-1-phosphate

Starch or glycogen with *n* –1 glucose units

cose-1-phosphate with retention of configuration about the C-1 residue (fig. 14.3).

Since glycogen has a branched structure (see fig. 13.1), there must be a mechanism for removing the branches prior to the action of phosphorylase. In the process of glycogen breakdown a debranching enzyme transfers $\alpha(1,4)$-linked residues from the branchpoints to the nonreducing end of another branchpoint, thereby forming a new $\alpha(1,4)$ linkage and making more glucose units available for phosphorylase-catalyzed phosphorolysis. The remaining $\alpha(1,6)$ bond linking the ultimate glycosyl residue at the branch to the main chain is hydrolyzed by the same debranching enzyme.

Digestion of dietary starch or glycogen in the intestine follows a different course. Interior bonds of the polysaccharide are hydrolyzed by enzymes secreted into the intestine by the pancreas and the intestinal mucosal cells; the macromolecule is split into progressively smaller fragments. Individual sugar residues are not removed from the polysaccharide or its major fragments. The degradation proceeds to the disaccharide maltose. The final step entails hydrolysis of maltose to free glucose.

The difference between the modes of hydrolysis inside cells and in the intestine reflects biological needs and functions. Phosphorylated sugars do not cross biological membranes readily. In the cell, phosphorylated hexoses are desirable because they

Figure 14.4

The reaction catalyzed by hexokinase. Attack on the terminal phosphorus atom of ATP is probably facilitated by proton removal by a negatively charged group in the catalytic site of the enzyme (:**B** in the figure). It is also facilitated by the fact that the terminal phosphate of ATP is an excellent leaving group.

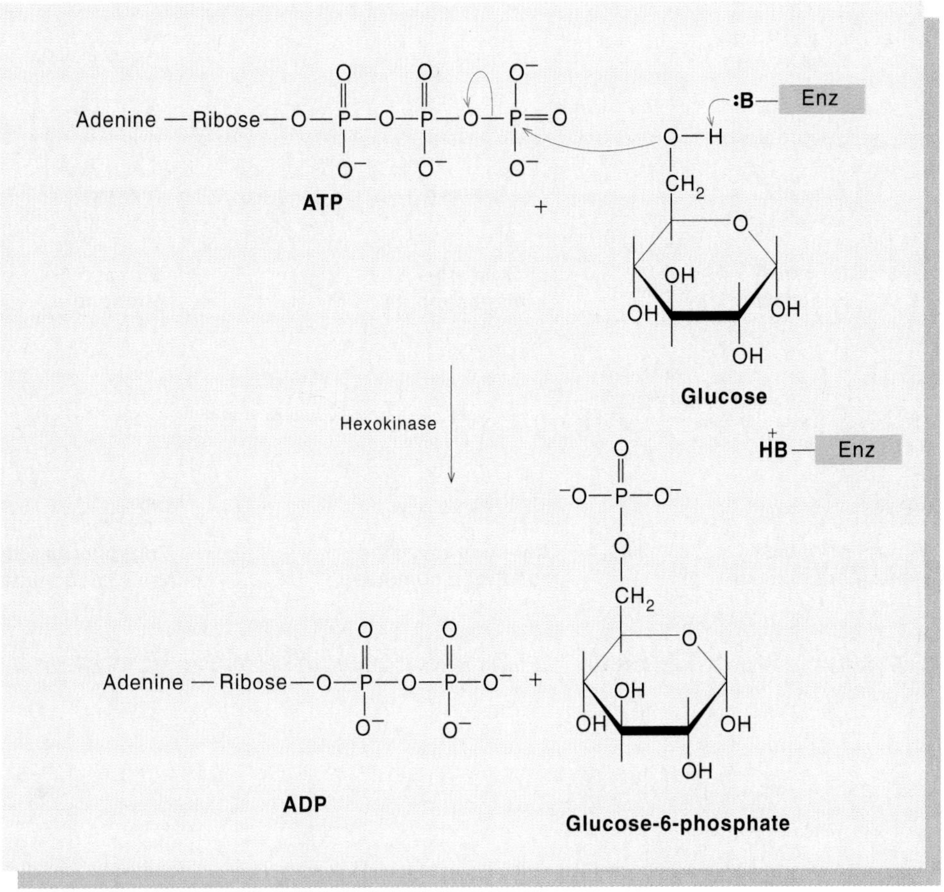

are not lost by diffusion out of the cell. In contrast, most cells are able to take up nonphosphorylated sugars from their surroundings. Because the very reason for hydrolysis of polysaccharides in the intestine is to make their component residues available for absorption into the body, unphosphorylated glucose is preferable.

Hexokinase Converts Free Sugars to Hexose Phosphates

Glucose obtained by mammals from the diet through intestinal hydrolysis of lactose, sucrose, glycogen, or starch is brought into the hexose phosphate pool through the action of hexokinase. This enzyme catalyzes phosphorylation at the oxygen attached to C-6 of glucose (fig. 14.4). As usual in metabolism, the source of the phosphate group is ATP:

$$\text{Glucose} + \text{ATP} \longrightarrow \text{Glucose-6-phosphate} + \text{ADP} + \text{H}^+ \quad \textbf{(1)}$$

The standard free energy change for this reaction is about −5 kcal/mole (20 kJ), and the equilibrium constant is about 5,700. Thus equilibrium considerations indicate a potential for this reaction to occur. The potential can be converted to reality only by an enzyme with appropriate kinetic properties. Hexokinases purified from various tissues typically have a Michaelis constant for glucose between 10 and 20 μM. Thus by the expenditure of ATP, hexokinase can convert glucose in the micromolar range to glucose-6-phosphate in the millimolar range for utilization in metabolism.

It is not uncommon in vertebrate tissues to find multiple forms of an enzyme that catalyze the same reaction but that differ from each other in their affinity for substrate, regulatory properties, or other important characteristics in a way that best meets the needs of a particular tissue. Such closely related enzymes usually have striking similarities in amino acid sequence and are referred to as isozymes because they are almost certainly evolutionarily related. We saw an example of isozymes in our discussion of lactate dehydrogenase in chapter 9. Mammals possess four different isozymes of hexokinase. Hexokinase D (or type IV), the predominant isozyme in the liver, is a low-affinity hexokinase that differs significantly from the other three isozymes for hexokinase found in mammals. Its level varies markedly with dietary and hormonal status. It requires much higher glucose concentrations (about 10 mM) for half-saturation, with a sigmoidal de-

Figure 14.5

The reaction catalyzed by phosphoglucomutase. The enzyme apparently can bind glucose phosphates in two ways, allowing it to transfer phosphoryl groups to and from either the oxygen atom at C-1 or the oxygen at C-6. The direction of reaction is driven by mass action, depending on the relative concentrations of glucose-1-phosphate and glucose-6-phosphate.

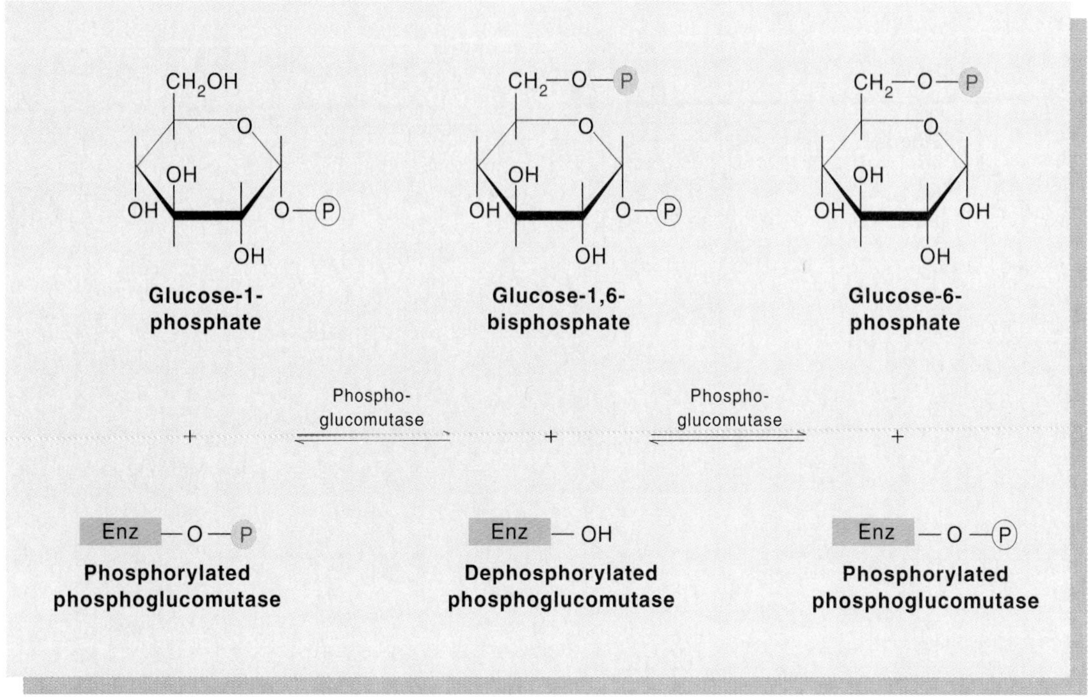

pendence, and it is insensitive to physiological concentrations of glucose-6-phosphate. These properties are well suited to one of liver's unique metabolic roles, namely, to stabilize the level of glucose in the blood. Liver hexokinase (sometimes referred to as glucokinase) is likely to take up glucose from the blood only when the level is high, as after a meal. In this situation the liver takes up glucose for the conversion to glycogen. When the liver requires glucose to satisfy its own metabolic needs, it obtains this glucose from its glycogen reserves, or, barring this, it synthesizes the glucose from smaller C_3 carbon units.

Phosphoglucomutase Interconverts Glucose-1-phosphate and Glucose-6-phosphate

The enzyme phosphoglucomutase catalyzes the interconversion of glucose-1-phosphate and glucose-6-phosphate (fig. 14.5). The interconversion occurs in two discrete steps by way of a phosphorylated enzyme and glucose-1,6-bisphosphate. The reactions are

$$\text{Glucose-1-phosphate} + \text{Enz-P} \rightleftharpoons \text{Glucose-1,6-bisphosphate} + \text{Enz} \quad (2)$$

$$\text{Glucose-1,6-bisphosphate} + \text{Enz} \rightleftharpoons \text{Glucose-6-phosphate} + \text{Enz-P} \quad (3)$$

$$\text{Sum:} \quad \text{Glucose-1-phosphate} \rightleftharpoons \text{Glucose-6-phosphate} \quad (4)$$

where Enz represents phosphoglucomutase, and Enz-P represents phosphoglucomutase phosphorylated at a specific serine residue in the active site. The enzyme cycles between the free and the phosphorylated states at each conversion of glucose-1-phosphate to glucose-6-phosphate (or of glucose-6-phosphate to glucose-1-phosphate).

Phosphohexoisomerase Interconverts Glucose-6-phosphate and Fructose-6-phosphate

Glucose-6-phosphate and fructose-6-phosphate interconvert readily in weak alkaline solution. This is because the intermediate enediol is stabilized by ionization in an alkaline medium (fig. 14.6). In the cytosol, glucose-6-phosphate and fructose-6-phosphate are interconverted by the action of the enzyme phosphohexoisomerase. The reaction resembles the nonenzymic conversion by going via the enediol intermediate.

Formation of Fructose-1,6-bisphosphate Signals a Commitment to Glycolysis

Fructose-6-phosphate is converted to fructose-1,6-bisphosphate by transfer of a phosphoryl group from ATP in a reaction catalyzed by phosphofructokinase (fig. 14.7). Since net regeneration of ATP

Figure 14.6

Mechanism of the interconversion of glucose-6-phosphate and fructose-6-phosphate. Loss of a proton from the oxygen attached to C-2 of the intermediate enediol leads to fructose-6-phosphate. **A** and **B** represent catalytic groups on the enzyme. It is not always known what specific groups are involved in a catalysis. In this case the HA group originates from a glutamate on the enzyme.

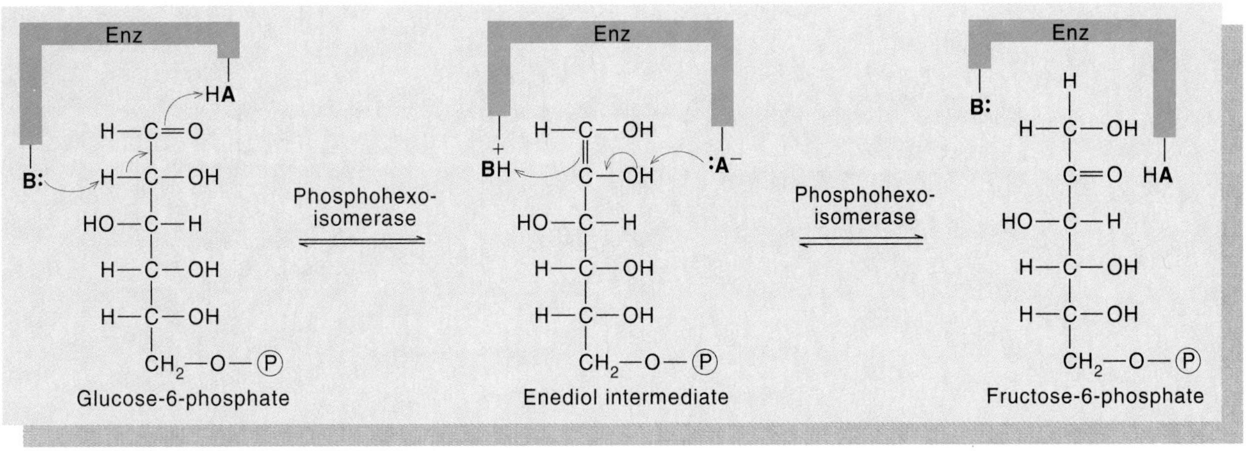

Figure 14.7

The reaction catalyzed by phosphofructokinase. The mechanism of this reaction is very similar to the hexokinase reaction shown in figure 14.4. :**B** is a proton acceptor at the active site.

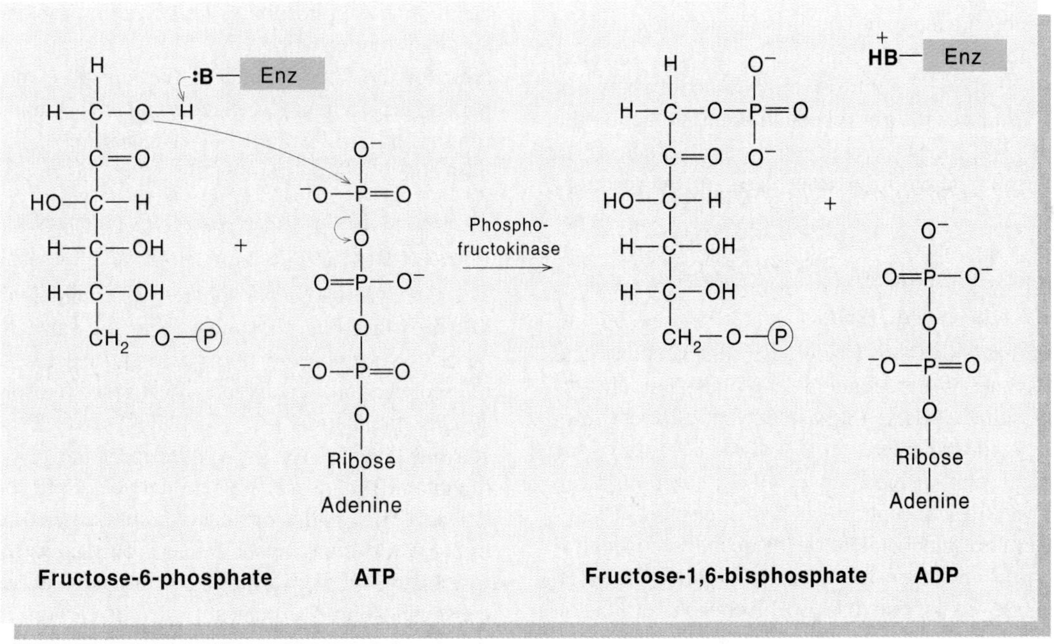

is a major function of the catabolism of carbohydrates, it may seem strange that an early step of glucose breakdown consumes ATP. However, there are good reasons for this. If fructose-6-phosphate were the substrate for aldolase, which cleaves the six-carbon sugar to two trioses, only one of the products would be phosphorylated. The unphosphorylated product might need to be phosphorylated immediately for protection against loss by diffusion. Phosphorylation before cleavage is even more effective.

Another, perhaps even more important, reason why phosphorylation occurs before cleavage has to do with regulation. The reaction by which material is removed from the hexose phosphate pool is the point at which control must be ex-

Figure 14.8

Cleavage of fructose-1,6-bisphosphate, an aldolase-catalyzed reaction. The aldolase reaction entails a reversal of the familiar aldol condensation. The first step involves abstraction of the hydrogen of the C-4 hydroxyl group, followed by elimination of an enolate anion.

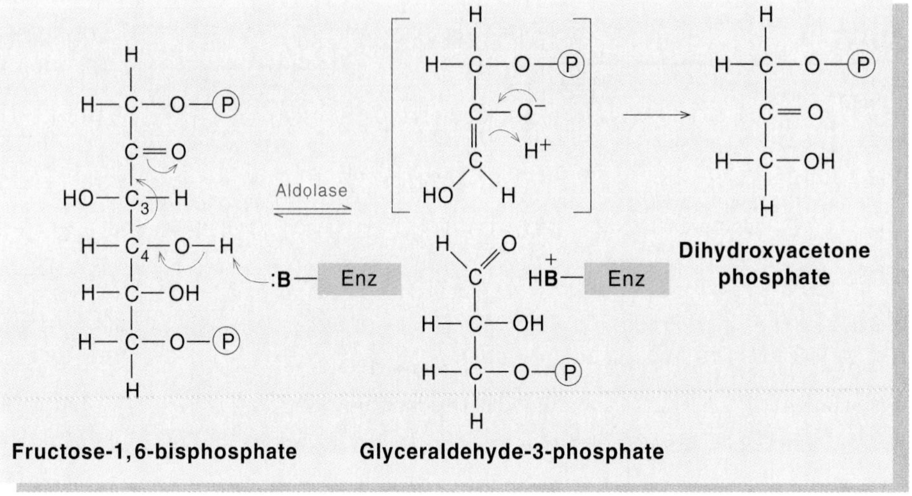

erted for maximal effectiveness. Metabolic regulation is possible only at steps for which the physiological ratio of precursor to product is far from the equilibrium ratio, so that kinetic control mechanisms can cause increases and decreases in the rates of reactions without thermodynamic constraints.

Under physiological conditions the concentration ratio of fructose-bisphosphate to fructose-6-phosphate varies considerably, depending on metabolic conditions, but is probably between 5 and 0.2. At equilibrium for the phosphofructokinase reaction the ratio would be about 10^4. Since the concentration ratio is far from equilibrium, no thermodynamic limitations exist for the reaction.

Fructose-1,6-bisphosphate and the Two Triose Phosphates Constitute the Second Metabolic Pool in Glycolysis

The metabolic pool that consists of fructose-1,6-bisphosphate and the two triose phosphates—glyceraldehyde-3-phosphate and dihydroxyacetone phosphate (DHAP)—is somewhat different from the other two pools of intermediates in glycolysis because of the nature of the chemical relationships between these compounds. In the other pools the relative concentrations of the component compounds at equilibrium are independent of the absolute concentrations. Because of the cleavage of one substrate into two products, the relative concentrations of fructose-1,6-bisphosphate and the triose phosphates are functions of the actual concentrations. For such reactions the relative concentrations of the split products must increase with dilution. (For the reaction A $\rightleftharpoons$ B + C the equilibrium constant is equal to [B][C]/[A]. If the concentration of A decreases, for example, by a factor of 4, equilibrium is restored when the concentrations of B and C decrease by a factor of only 2). Except for that difference, interconversions of the compounds in this pool are similar to those in the other pools because reactions within the pool are close to equilibrium and can go rapidly in either direction.

Aldolase Cleaves Fructose-1,6-bisphosphate

Fructose-1,6-bisphosphate is cleaved by aldolase into two molecules of triose phosphate. This reaction represents the reversal of an aldol condensation (fig. 14.8). Most aldolases are highly specific for the "upper" end of the substrate molecule, requiring a phosphate group at C-1, a carbonyl at C-2, and specific steric configurations at C-3 and C-4. The nature of the remainder of the molecule is unimportant as far as the enzyme action is concerned.

Triose Phosphate Isomerase Interconverts the Two Trioses

The isomeric triose phosphates, glyceraldehyde-3-phosphate and dihydroxyacetone phosphate, bear the same relationship to each other as do glucose-6-phosphate and fructose-6-phosphate. Their interconversion, catalyzed by triose phosphate isomerase, is equally facile (see fig. 14.1). Dihydroxyacetone phosphate is a starting material for the synthesis of the glycerol moiety of fats (chapter 22), but only glyceraldehyde-3-phosphate is used in glycolysis. Thus under ordinary circumstances, nearly all of the dihydroxyacetone phosphate that is formed in the cleavage of fructose bisphosphate is converted to glyceraldehyde-3-phosphate by triose phosphate isomerase. All six carbons of the hexoses are thereby made available for the later steps of carbohydrate catabolism. The mechanism of action of triosephosphate isomerase is described in chapter 9.

The Conversion of Triose Phosphates to Phosphoglycerates Occurs in Two Steps

The oxidation of glyceraldehyde-3-phosphate to glycerate-3-phosphate is the first energy-yielding (ATP-producing) reaction in the

Figure 14.9

The reaction catalyzed by glyceraldehyde-3-phosphate dehydrogenase (3-phosphoglyceraldehyde dehydrogenase). This interesting and complex reaction consists of several steps. The enzyme first catalyzes a reaction of the substrate with a sulfhydryl group of a cysteine residue of the enzyme itself. The substrate is then oxidized from the aldehyde level of oxidation to the carboxylic acid level while still attached covalently to the enzyme. Displacement of the enzyme by inorganic phosphate ion liberates the product, glycerate-1,3-bisphosphate. The bound NADH of the enzyme, which became reduced when the substrate was oxidized, then transfers a pair of electrons to an unbound NAD^+, and the enzyme is ready for another catalytic cycle.

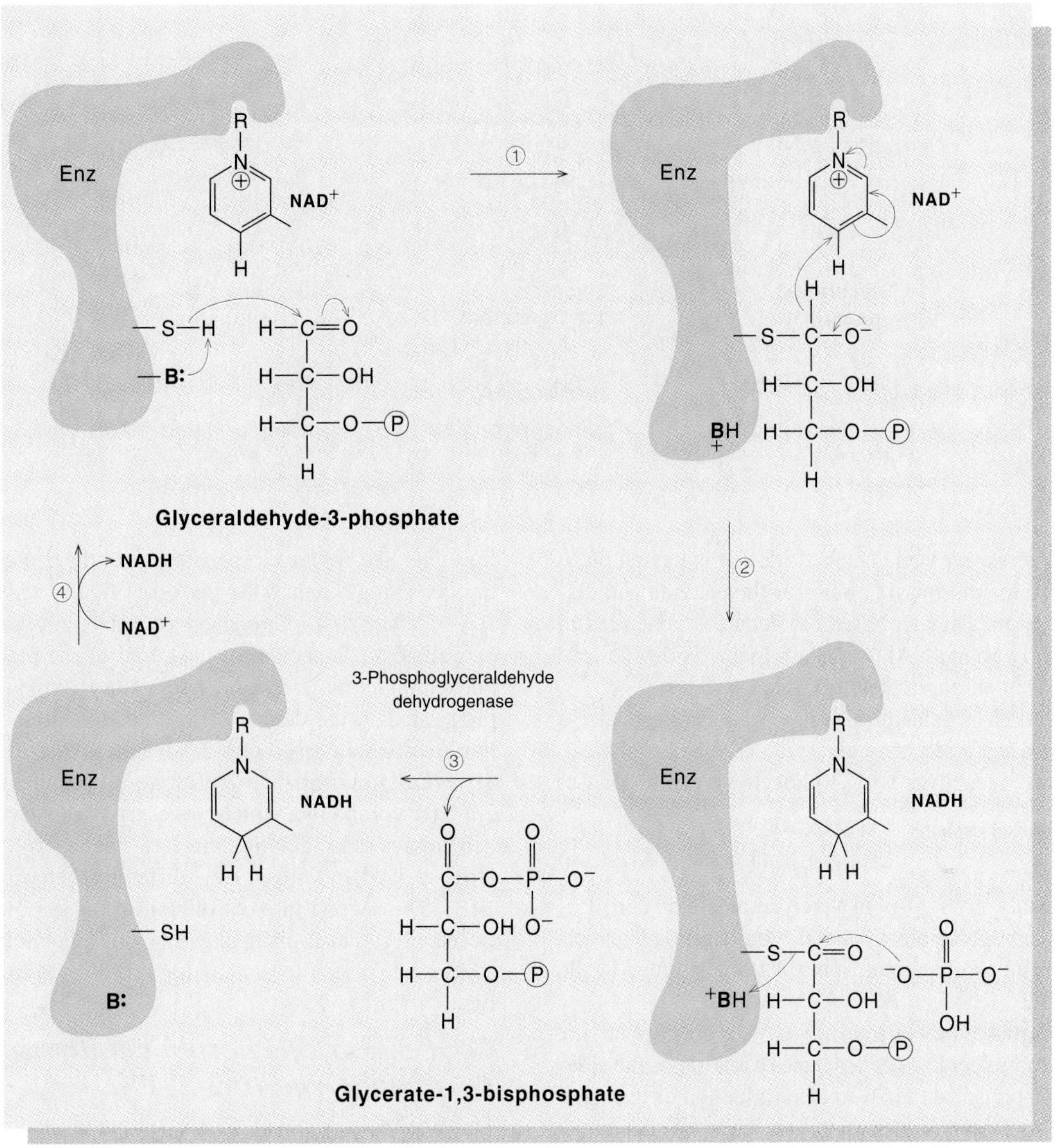

glycolytic pathway. The mechanism by which an inorganic phosphate ion is taken up and transferred to adenosine diphosphate (ADP) during the conversion of glyceraldehyde-3-phosphate to glycerate-3-phosphate requires two separate enzyme-catalyzed reactions.

The first reaction is catalyzed by the enzyme 3-phosphoglyceraldehyde dehydrogenase.

$$\text{Glyceraldehyde-3-phosphate} + NAD^+ + P_i \longrightarrow \text{Glycerate-1,3-bisphosphate} + NADH + H^+ \quad \textbf{(5)}$$

This reaction is initiated by the condensation of an —SH group of a specific cysteine residue at the catalytic site of the enzyme with the aldehyde carbonyl, forming a sulfhydryl adduct, or thiohemiacetal (fig. 14.9, step 1). A pair of electrons, along with a proton (thus in effect a hydride ion), are then donated to a NAD^+ molecule that is tightly bound nearby, converting the tetrahedral hemiacetal into a thioester (step 2). Thioesters provide considerably more free energy on hydrolysis than comparable oxygen esters, facilitating the next step (step 3), an attack by an oxygen of a phosphate ion on the carbonyl carbon of the thioester. The sulfur atom, and thus the enzyme molecule, is displaced from covalent linkage to the reaction intermediate, and the product, glycer-

Figure 14.10

Interconversion of glycerate-3-phosphate and glycerate-2-phosphate, catalyzed by phosphoglyceromutase. The reaction closely resembles that catalyzed by phosphoglucomutase (see fig. 14.5) except that the phosphate binds to a histidine side chain instead of a serine side chain. The enzyme can transfer a covalently bound phosphoryl group either to the oxygen on C-2 of glycerate-3-phosphate or to the oxygen on C-3 of glycerate-2-phosphate. The resulting glycerate-2,3-bisphosphate, in turn, can donate either of its phosphoryl groups to the enzyme. These catalytic capabilities provide for interconversion of the two monophosphoglycerates in either direction.

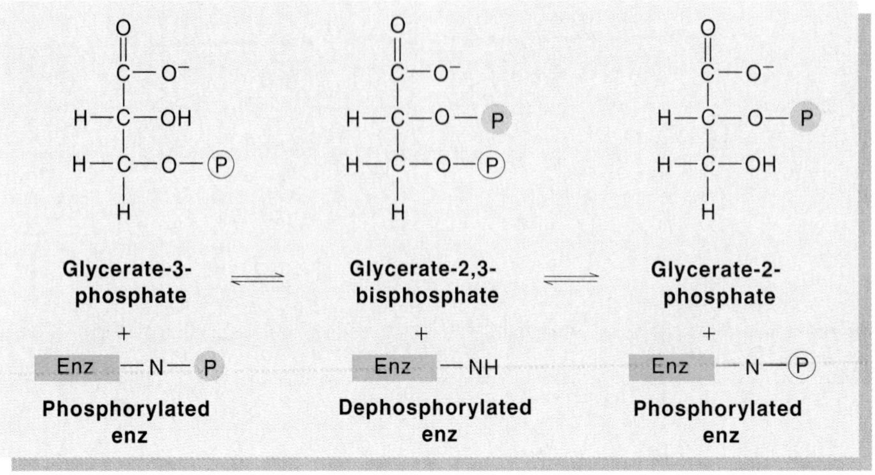

ate-1,3-bisphosphate, is released. The hydride ion is passed on to an NAD^+ molecule in solution that binds to the enzyme and dissociates after reduction. Thus an NADH molecule is generated in solution, and a tightly bound NAD^+ is again in the oxidized form, ready to participate in another catalytic cycle.

In the second step leading to glycerate-3-phosphate, a phosphate group is transferred from glycerate-1,3-bisphosphate to ADP. This reaction is catalyzed by 3-phosphoglycerate kinase:

$$H^+ + \text{Glycerate-1,3-bisphosphate} + \text{ADP} \rightleftharpoons \text{Glycerate-3-phosphate} + \text{ATP} \quad \textbf{(6)}$$

In the two steps catalyzed by 3-phosphoglyceraldehyde dehydrogenase and 3-phosphoglycerate kinase the oxidation of glyceraldehyde-3-phosphate to glycerate-3-phosphate is coupled to the regeneration of ATP.

Since two molecules of glyceraldehyde-3-phosphate are produced from each molecule of hexose, the oxidation of the aldehyde to glycerate-3-phosphate leads to the production of two molecules of ATP per molecule of glucose or other hexose consumed. At this point the energy (ATP) gained is equal to the energy invested if the starting material was a free hexose. The two phosphate groups supplied by ATP in the hexokinase and phosphofructokinase reactions are still contained in glycerate-3-phosphate. If storage glycogen or starch was the starting material, only one phosphate group is supplied by ATP, and one of the two molecules of ATP regenerated in the phosphoglycerate kinase step represents a net gain.

The Three-Carbon Phosphorylated Acids Constitute a Third Metabolic Pool

Glycerate-3-phosphate, glycerate-2-phosphate, and phosphoenolpyruvate (PEP) make up another equilibrium group of metabolites that, like the hexose phosphates or the triose phosphates, function as a single metabolic pool (see fig. 14.1).

The first interconversion between glycerate-3-phosphate and glycerate-2-phosphate is similar to the reaction catalyzed by phosphoglucomutase (fig. 14.10). The enzyme is transiently phosphorylated in the course of the reaction. It should be noted that the intermediate glycerate-2,3-bisphosphate in the conversion of glycerate-3-phosphate to glycerate-2-phosphate is identical to the 2,3 GBP compound, which serves as an allosteric effector for oxygen transport by hemoglobin (see chapter 6). Despite this similarity, 2,3 GBP is made by a different route in erythrocytes (box 14B). The second interconversion in the three-carbon pool is between glycerate-2-phosphate and phosphoenolpyruvate. This reaction, which entails a dehydration, is catalyzed by enolase (see fig. 14.1).

Conversion of Phosphoenolpyruvate to Pyruvate Generates ATP

The final energy payoff in the glycolytic pathway occurs in the hydrolysis of phosphoenolpyruvate to pyruvate and the concomitant phosphorylation of ADP to ATP. Two molecules of ATP are produced for each molecule of hexose phosphate consumed, bringing the net yield of ATP to two molecules for each molecule of glucose (two molecules of ATP are regenerated in the phosphoglycerate kinase step and two in this step, and two are consumed in the hexokinase and phosphofructokinase steps):

$$\text{Glucose} + 2\ NAD^+ + 2\ \text{ADP} + 2\ P_i \longrightarrow 2\ \text{Pyruvate} + 2\ \text{NADH} + 2\ H^+ + 2\ \text{ATP} + 2\ H_2O \quad \textbf{(7)}$$

The net yield is three ATP molecules for each glucose molecule catabolized when the substrate is storage glycogen or starch. These are the total yields of ATP from the anaerobic catabolism of sug-

14B BOX

Glycolysis Influences Oxygen Transport

In chapter 6 we saw that the glycerate-2,3-bisphosphate (2,3 GBP) in erythrocytes facilitates the delivery of oxygen from the lungs to oxygen-consuming tissues. The concentration of 2,3 GBP is unusually high in erythrocytes (~5 mM), owing to the presence of a bifunctional enzyme that catalyzes its formation and breakdown from glycolytic intermediates. Increased rates of glycolysis in the erythrocytes results in elevated levels of 2,3 GBP, which augments the delivery of oxygen to consumer tissues.

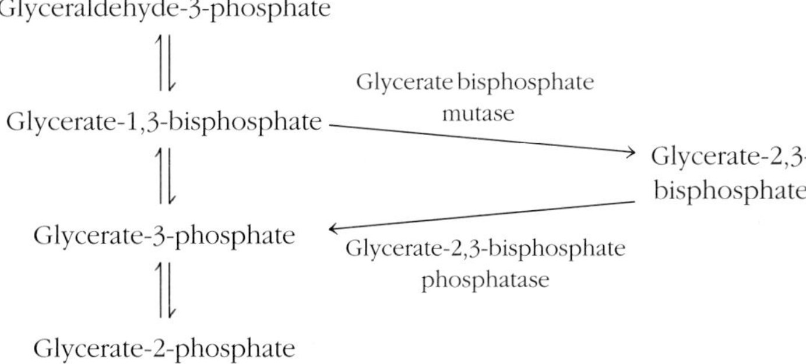

Table 14.1
Reactions, Enzymes, and Standard Free Energies for Steps in the Glycolytic Pathway

Step	Reaction	Enzyme	$\Delta G^{\circ\prime}$	$\Delta G'^{a}$
1	Glucose + ATP $\longrightarrow$ Glucose-6-phosphate + ADP + H^+	Hexokinase	−4.0[c]	−8.0
2	Glucose-6-phosphate $\rightleftharpoons$ Fructose-6-phosphate	Phosphohexose isomerase	+0.4	−0.60
3	Fructose-6-phosphate + ATP $\longrightarrow$ Fructose-1,6-bisphosphate + ADP + H^+	Phosphofructokinase	−3.4	−5.3
4	Fructose-1,6-bisphosphate $\rightleftharpoons$ Dihydroxyacetone phosphate + Glyceraldehyde-3-phosphate	Aldolase	+5.7	−0.31
5	Dihydroxyacetone phosphate $\rightleftharpoons$ Glyceraldehyde-3-phosphate	Triose phosphate isomerase	+1.8	+0.60
6	Glyceraldehyde-3-phosphate + P_i + NAD^+ $\rightleftharpoons$ Glycerate-1,3-bisphosphate + NADH + H^+	Phosphoglyceraldehyde dehydrogenase	+1.5	−0.41
7	H^+ + Glycerate-1,3-bisphosphate + ADP $\rightleftharpoons$ Glycerate-3-phosphate + ATP	3-Phosphoglycerate kinase	−4.5	+0.31
8	Glycerate-3-phosphate $\rightleftharpoons$ Glycerate-2-phosphate	Phosphoglyceromutase	+1.1	+0.19
9	Glycerate-2-phosphate $\rightleftharpoons$ Phosphoenolpyruvate + H_2O + H^+	Enolase	+0.4	−0.79
10	Phosphoenolpyruvate + ADP + H^+ $\longrightarrow$ Pyruvate + ATP	Pyruvate kinase	−7.5	−4.0
Net reaction:	$C_6H_{12}O_6 + 2\ NAD^+ + 2\ ADP + 2\ P_i \longrightarrow 2\ C_3H_4O_3 + 2\ NADH + 2\ H^+ + 2\ ATP + 2\ H_2O$		−27.8[b]	−23.0

[a]ΔG values are for the human erythrocyte using approximate values for concentrations.

[b]To obtain this sum for the standard free energy of the overall reaction per mole of glucose, the indicated free energies for reactions 6–10 must all be doubled. This is because each of these reactions occurs twice for every glucose consumed.

[c]Values are given in units of kcal/mole; to convert this to kJ, merely multiply the given values by 4.18.

ars. The 10 steps from glucose to pyruvate are reviewed in table 14.1. It can be seen that overall, glycolysis goes with a significant drop in free energy, about −25 kcal/mole (104 kJ) of glucose. In calculating the free energy drop at each step, we do not get a realistic picture from the standard free energies. This is because the standard free energies do not take into account the concentrations of the intermediates. If we take this into account, we see that a sizeable drop in free energy occurs at only three steps (see table 14.1 and fig. 14.11). Most of the other steps proceed with little or no change in free energy. Our discussion of regulation will show that the process of glycolysis is regulated at precisely these three steps.

Figure 14.11

Energy profile in going from glucose to pyruvate by the glycolytic pathway. The three large energy drops are between glucose and glucose-6-phosphate, fructose-6-phosphate and fructose-1,6-bisphosphate, and phosphoenolpyruvate (PEP) and pyruvate. The actual reactions are described in table 14.1.

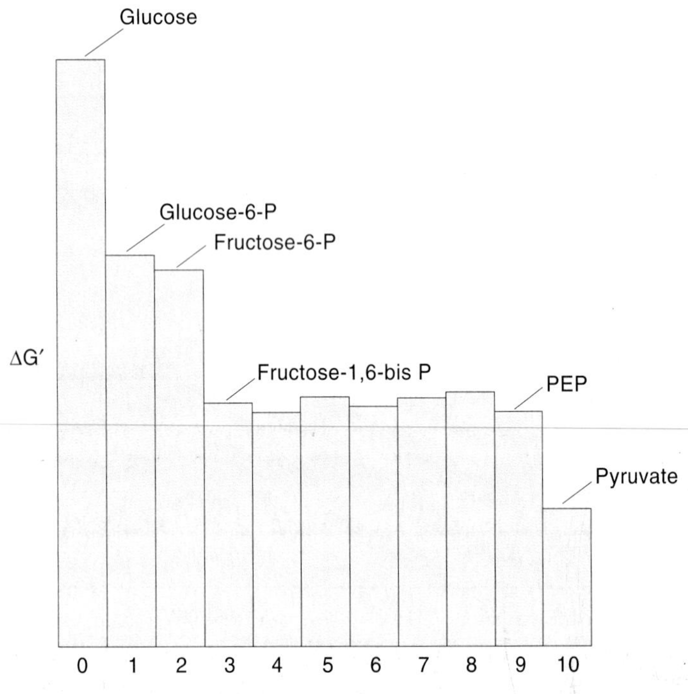

The NAD^+ Reduced in Glycolysis Must Be Regenerated

The pyruvate produced in glycolysis must be metabolized further if glycolysis is to continue. This is essential because the NAD^+ reduced for each molecule of phosphoglyceraldehyde that was oxidized needs to be regenerated (see fig. 14.1). Because equimolar amounts of NADH and pyruvate are produced in glycolysis, a simple way to reoxidize the NADH is by transfer of electrons to pyruvate, forming lactate (fig. 14.12). This is the solution employed by many kinds of organisms. Indeed, even skeletal muscle tissue uses this strategy in times of exertion when the oxygen supply cannot keep up with the energy needs of the tissue. The reduction of pyruvate is catalyzed by the enzyme lactate dehydrogenase:

$$\text{Pyruvate} + \text{NADH} + \text{H}^+ \longrightarrow \text{Lactate} + \text{NAD}^+ \quad \textbf{(8)}$$

Yeast and some other kinds of organisms use a different strategy to regenerate NAD^+. They produce ethanol and CO_2, rather than lactate, from the anaerobic catabolism of sugars. Unlike true facultative anaerobes such as *E. coli,* which can live anaerobically for an indefinite period, yeasts can live only for a few generations in the total absence of oxygen because they require molecular oxygen for synthesis of membrane components. Rather than a true alternative life style, yeasts use alcohol fermentation as part of a very effective competitive strategy. When fruits ripen and fall from the tree or bush, yeasts are likely to be among the first invaders. In the anaerobic interior of the fruit, yeasts rapidly convert sugars to ethanol, which they excrete. As ethanol accumulates, the growth of most other microorganisms is discouraged, but yeasts can continue to grow until the concentration of ethanol reaches about 12%. When the softened fruit breaks open, a thoroughgoing change in the energy metabolism of the yeast cells occurs. They begin to take up the ethanol they had previously excreted and oxidize it to CO_2, with a much larger production of ATP than in the anaerobic stage (about 14 moles of ATP per mole of ethanol, compared with 2 moles of ATP per mole of glucose in the anaerobic stage). By rapidly converting the available sugars into a compound that cannot be metabolized for energy by most other organisms and that is in fact toxic to them, yeasts gain an advantage over competing microorganisms and also gain energy in the process. They then can exploit the waste product of the first stage of growth as the main nutrient for the later aerobic stages of growth. This example illustrates that the organization of even the most central metabolic pathways can, like other properties of an organism, be shaped by competition.

Summary of Glycolysis

Glycolysis consists of a chain of 10 reactions that starts from glucose and ends with the 3-carbon compound pyruvate (see fig. 14.1 and table 14.1). All of the intermediates in this pathway are phosphorylated, and in the process of degradation, two ATP molecules are made from two ADP molecules. This is the energy payoff of the pathway. In effect, the energy of the carbon–carbon bond is expended to create the immediately useful form of chemical energy, ATP. This pathway does not require oxygen, and it is found in nearly all cells, whether they are aerobic or anaerobic. Glucose is frequently not the starting material for glycolysis, and so differences between organisms and differences in the same organism in different environments show up at the starting point and the termination point of the pathway. One oxidation step in the pathway results in the conversion of NAD^+ to NADH. This NAD^+ must be regenerated if glycolysis is to continue. Under anaerobic conditions some reaction must occur involving the reduction of pyruvate to regenerate the NAD^+.

The glycolytic pathway resembles a series of "lakes" (metabolic pools) connected by short "rivers" (the reactions between the pools). This pattern is reflected in the ways that functional metabolic relationships have evolved. Reactions involving ATP and ADP occur in the interconnecting reactions, or rivers. Clearly, this is where they are expected, because an ATP-linked reaction within a metabolic pool makes no more sense than a hydroelectric power plant in the middle of a lake.

Catabolism of Other Sugars

A variety of sugars other than glucose is available to cells, either by ingestion or upon degradation of storage carbohydrates. Most of these are either hexoses or pentoses, with the former predominating. Ordinary table sugar (sucrose), for example, consists of the hexoses glucose and fructose, and milk sugar (lactose) contains glucose and galactose. Organisms consuming these disaccharides must cope not only with the glucose, but with the fructose or the galactose as well. Other sugars with which cells must

Figure 14.12

Regeneration of NAD^+ by reduction of pyruvate to lactate. Because NAD^+ is a necessary participant in the oxidation of glyceraldehyde-3-phosphate to glycerate-1,3-bisphosphate, glycolysis is possible only if there is a way by which NADH can be reoxidized.

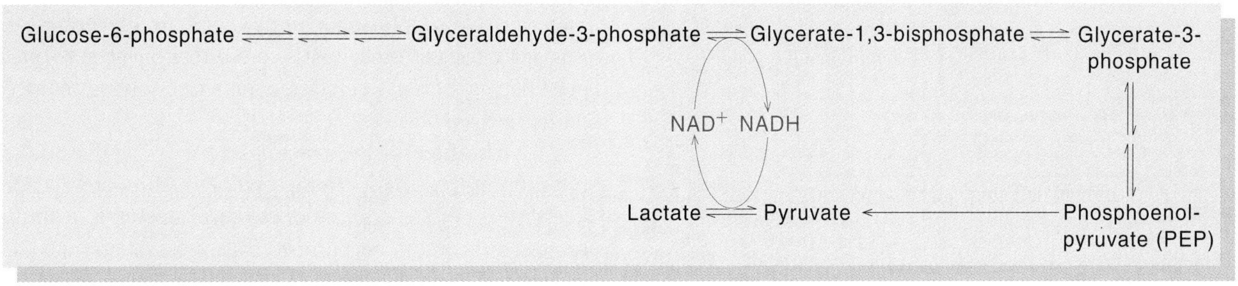

Glucose-6-phosphate ⇌ ⇌ ⇌ Glyceraldehyde-3-phosphate ⇌ Glycerate-1,3-bisphosphate ⇌ Glycerate-3-phosphate
NAD⁺ NADH
Lactate ⇌ Pyruvate ← Phosphoenolpyruvate (PEP)

frequently contend include the hexose mannose and the pentoses ribose, ribulose, and xylulose. In general, each of these has a specific reaction sequence that brings it as quickly as possible into the glycolytic sequence. We shall limit ourselves to a discussion of two of the most abundant hexoses found in the mammalian diet: fructose and galactose.

Fructose

Fructose is most commonly obtained by ingestion and hydrolysis of sucrose. Its entry into the glycolytic pathway is direct indeed, since like glucose and mannose, it also can be phosphorylated at position 6 by hexokinase.

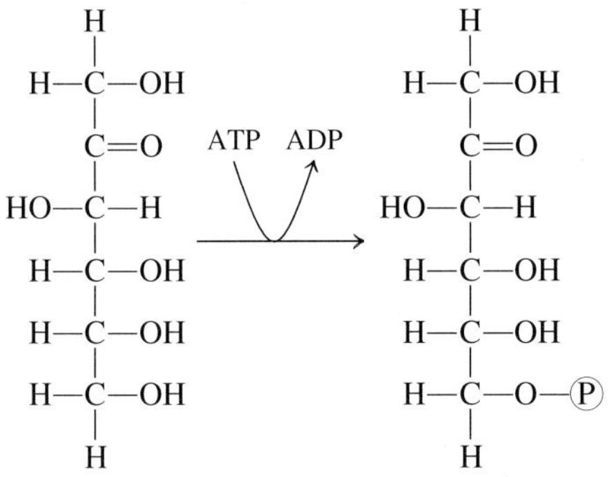

$$\text{D-Fructose (H–C(H)–OH; C=O; HO–C–H; H–C–OH; H–C–OH; H–C(H)–OH)} \xrightarrow{\text{ATP} \;\; \text{ADP}} \text{Fructose-6-phosphate (H–C(H)–OH; C=O; HO–C–H; H–C–OH; H–C–OH; H–C(H)–O–Ⓟ)} \quad (9)$$

An alternative pathway exists in mammalian liver, where phosphorylation occurs on carbon 1 instead, catalyzed by the enzyme fructokinase. The fructose-1-phosphate that results can then undergo cleavage by a specific aldolase to dihydroxyacetone phosphate and free glyceraldehyde. The latter compound is phosphorylated to glyceraldehyde-3-phosphate by glyceraldehyde kinase, so the overall process from hexose to two triose phosphates requires two ATPs, just as for glucose in the initial steps of glycolysis, but with the phosphorylation events occurring in this case in a different sequence:

$$\begin{aligned}
\text{D-Fructose} + \text{ATP} &\longrightarrow \text{Fructose-1-phosphate} + \text{ADP}\\
\text{Fructose-1-phosphate} &\rightleftharpoons \text{Glyceraldehyde} + \text{Dihydroxyacetone phosphate}\\
\text{Glyceraldehyde} + \text{ATP} &\longrightarrow \text{Glyceraldehyde-3-phosphate} + \text{ADP}\\
\hline
\text{D-Fructose} + 2\,\text{ATP} &\longrightarrow \text{Glyceraldehyde-3-phosphate} + \text{dihydroxyacetone phosphate} + 2\,\text{ADP}
\end{aligned}$$

There are legitimate concerns about diets that are high in fructose. These concerns have been heightened by the use of fructose as a substitute sweetener in many food products. Humans in general have a limited capacity to handle fructose; this is because the capacity of the normal liver to phosphorylate fructose greatly exceeds its capacity to split fructose-1-phosphate. As a result, fructose used by the liver is poorly controlled, and excessive fructose decreases the capacity of liver cells to synthesize ATP. Some individuals possess a hereditary deficiency known as fructose intolerance. These individuals are deficient in the liver aldolase that splits fructose-1-phosphate into dihydroxyacetone phosphate and glyceraldehyde. This leads to fructose-1-phosphate accumulation and a depletion of P_i and ATP in the liver that can result in severe hypoglycemia. Prolonged ingestion of fructose by young children with this condition can have serious consequences.

Galactose

The most common dietary source of galactose is the disaccharide lactose. Galactose is metabolized by phosphorylation and conversion of glucose, but the reaction sequence is somewhat complicated because the conversion to glucose (an *epimerization* reaction on carbon 4) occurs while the sugar is attached to the carrier uridine diphosphate (UDP; fig. 14.13). The reaction begins with phosphorylation of galactose on carbon 1, catalyzed by galactokinase, a liver enzyme. The galactose is then exchanged for the glucose of UDP-glucose in a reaction catalyzed by phosphogalactose uridyl transferase. This results in UDP-bound galactose, which can undergo epimerization on carbon 4 to form UDP-glucose, mediated by the enzyme UDP-glucose epimerase. The glucose-1-phosphate that results from the transferase reaction is in the meantime converted by the enzyme phosphoglucomutase into glucose-6-phosphate, at which point entry is gained into the glycolytic pathway:

$$\begin{aligned}
\text{D-Galactose} + \text{ATP} &\longrightarrow \text{Galactose-1-phosphate} + \text{ADP}\\
\text{Galactose-1-phosphate} + \text{UDP-glucose} &\rightleftharpoons \text{Glucose-1-phosphate} + \text{UDP-galactose}\\
\text{UDP-galactose} &\rightleftharpoons \text{UDP-glucose}\\
\text{Glucose-1-phosphate} &\rightleftharpoons \text{Glucose-6-phosphate}\\
\hline
&\longrightarrow \text{Glucose-6-phosphate} + \text{ADP}
\end{aligned}$$

Figure 14.13

The structure of uridine diphosphogalactose (UDP-galactose).

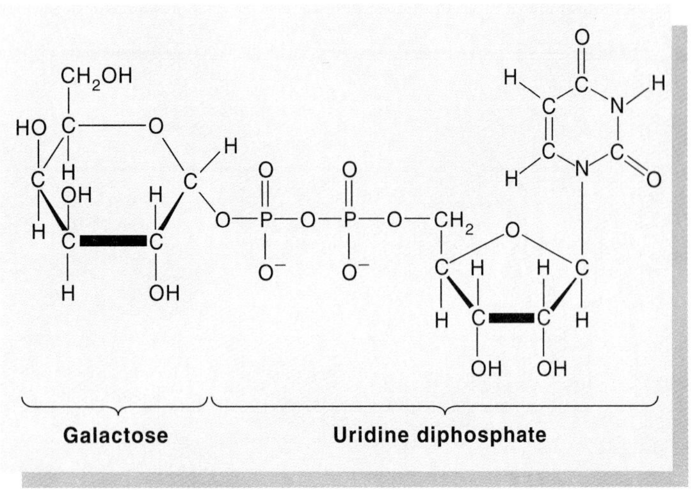

The congenital disease galactosemia is most frequently due to a genetic absence of the transferase enzyme. The result in infants ingesting significant quantities of milk is an accumulation of high levels of galactose in the blood (designated by the *-emia* ending). This gives rise to mental disorders, cataracts of the eye, and other characteristic symptoms. Treatment of galactosemic infants involves the obvious remedy of eliminating milk and other galactose sources from the diet during childhood.

Gluconeogenesis

Gluconeogenesis embraces the pathways from C-3 carbon sources, such as lactate, pyruvate, or amino acids, to hexoses or storage polysaccharides.

Glycolysis and gluconeogenesis constitute a set of oppositely directed conversions with different ATP-to-ADP stoichiometries. In most oppositely directed conversions the two sequences and the associated enzymes are entirely separate. Not so in these two sequences. Only at three points, all outside of the three pools, are the oppositely directed reactions of glycolysis and gluconeogenesis catalyzed by different enzymes (fig. 14.14).

Gluconeogenesis Consumes ATP

The organization of glycolysis and gluconeogenesis as a series of connected metabolic pools makes it possible for most of the same enzymes to function in both directions. Only the reactions connecting the metabolic pools require different enzymes and a coupling to the ATP–ADP system to make them thermodynamically feasible in the direction of gluconeogenesis.

Flux within a pool is determined by simple mass-action considerations. For example, when glycerate-3-phosphate is being produced in the phosphoglycerate kinase reaction and phosphoenolpyruvate is being removed by the action of pyruvate kinase, the flux in the 3-carbon carboxylate pool is in the glycolytic direction, from glycerate-3-phosphate to phosphoenolpyruvate. When the pattern of activation of the regulatory enzymes is different, so that phosphoenolpyruvate is being produced and glycerate-3-phosphate is being consumed, the flux in the same pool, catalyzed by the same enzymes, proceeds in the gluconeogenic direction, from phosphoenolpyruvate to glycerate-3-phosphate. Similarly for the other pools, the direction of net flux within the pool depends on which components are flowing into the pool and which are leaving it.

In three of the reactions connecting the pools we find a big drop in free energy in the glycolytic direction (see table 14.1). Clearly, if cells are to conduct these reactions in the reverse direction, ATP must be pumped into the system, and different enzymes will be required.

Conversion of Pyruvate to Phosphoenolpyruvate Requires Two High-Energy Phosphates

The first step in the gluconeogenic direction involves the formation of phosphoenolpyruvate (PEP) from pyruvate. Reversal of the pyruvate kinase reaction requires at least two ATP-to-ADP conversions. The means by which this is done is shown in figure 14.15. Both of the reactions involved in PEP synthesis from pyruvate normally occur in the mitochondrial matrix. The PEP so formed is transported to the cytosol, where further reactions take place.

The first step in the conversion of pyruvate to PEP entails the carboxylation of pyruvate to form oxaloacetate. This reaction is catalyzed by pyruvate carboxylase. As in many other enzymic carboxylation reactions the immediate CO_2 donor is a carboxylated derivative of the coenzyme biotin (see chapter 11). Equation (10) depicts the reaction for the production of oxaloacetate:

$$\text{Pyruvate}^- + \text{ATP}^{4-} + \text{HOCO}_2^- \xrightarrow{\text{Enz-biotin}} \text{Oxaloacetate}^{2-} + \text{ADP}^{3-} + \text{P}_i^{2-} + \text{H}^+ \quad \textbf{(10)}$$

The formation of oxaloacetate supplies a significant portion of the thermodynamic push for the next step in the sequence. This is because the free energy change for decarboxylation of β-keto carboxylic acids such as oxaloacetate is large and negative. The oxaloacetate formed from pyruvate by carboxylation is converted to phosphoenolpyruvate in a reaction catalyzed by phosphoenolpyruvate carboxykinase. In many species, including mammals, this reaction involves a GTP-to-GDP conversion:

$$\text{Oxaloacetate} + \text{GTP} \xrightarrow{\substack{\text{Phosphoenolpyruvate} \\ \text{carboxykinase}}} \text{Phosphoenolpyruvate} + \text{GDP} + \text{CO}_2 \quad \textbf{(11)}$$

The use of GTP, UTP, or CTP in a metabolic reaction or sequence is energetically equivalent to the use of ATP, because the nucleoside diphosphate that is produced is rephosphorylated at the expense of ATP in the reaction catalyzed by nucleoside diphosphate (NDP) kinase:

$$\text{ATP} + \text{NDP} \rightleftharpoons \text{ADP} + \text{NTP} \quad \textbf{(12)}$$

NDP may be CDP, GDP, or UDP, and NTP is the corresponding

Figure 14.14

Relationships in glycolysis and gluconeogenesis. Points at which ATP is produced or consumed are indicated. Compounds in the same metabolic pools are indicated by purple boxes. Three small pseudocycles (Ia, II, III) in the paired sequences occur between glycogen and pyruvate, or between glycogen and glucose (Ib, II, III). Only enzymes that are unique to either glycolysis or gluconeogenesis are indicated (screened in blue).

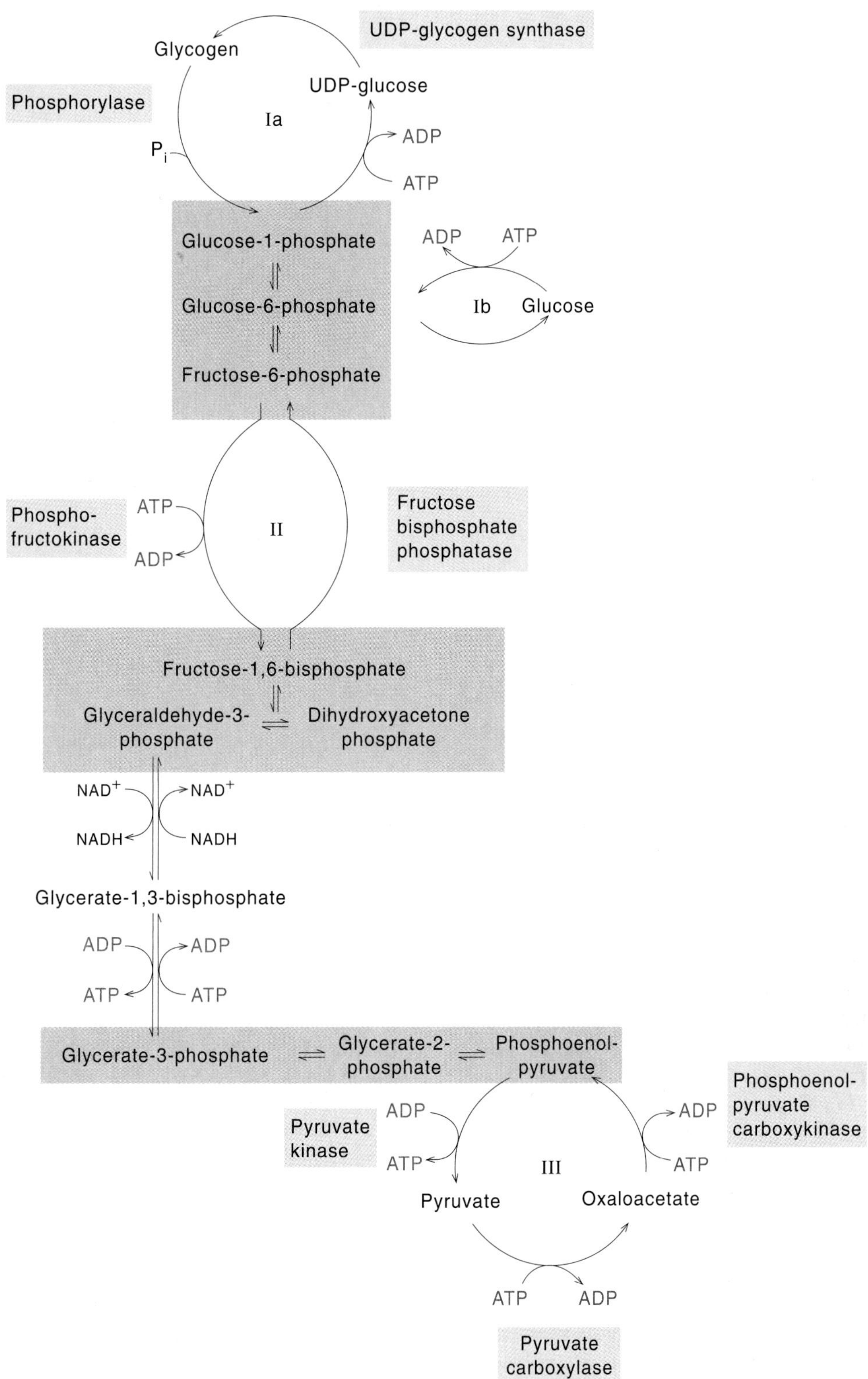

Figure 14.15

The pyruvate–phosphoenolpyruvate pseudocycle. One molecule of ATP is regenerated in the conversion of phosphoenolpyruvate to pyruvate, catalyzed by pyruvate kinase, which occurs in glycolysis. Conversion of pyruvate to phosphoenolpyruvate, which is necessary in gluconeogenesis, must be coupled to two ATP-to-ADP conversions if it is to be thermodynamically feasible. The first ATP is used indirectly, in the production of the active carboxyl transfer agent, biotin-COO^- (chapter 11). Transfer of a carboxyl group to pyruvate produces oxaloacetate. Decarboxylation of oxaloacetate aids in the attack by the carbonyl oxygen on the terminal phosphorus atom of ATP or GTP, producing phosphoenolpyruvate and ADP or GDP. This figure expands on the reactions shown in the lower part of figure 14.14 (III).

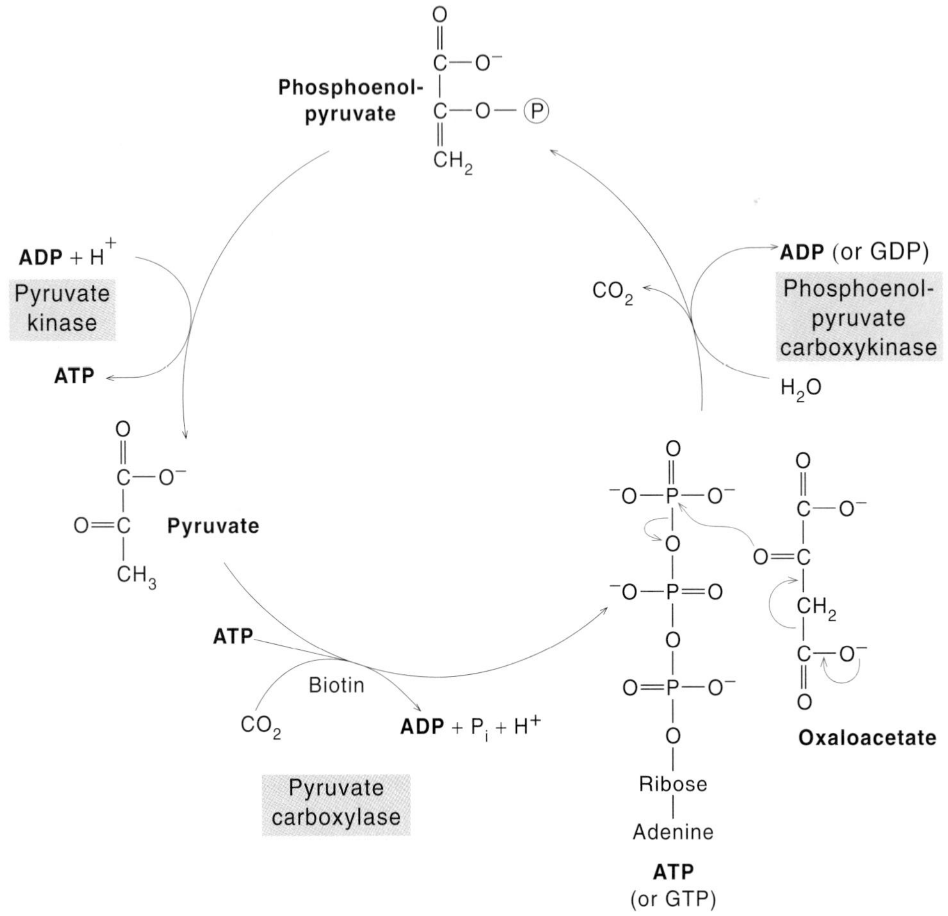

triphosphate. The enzyme is unusual in its lack of specificity for the acceptor nucleoside diphosphate.

If we add the equations for the reactions catalyzed by pyruvate carboxylase, phosphoenolpyruvate carboxykinase, and nucleoside diphosphate kinase, we obtain the overall reaction for conversion of pyruvate to phosphoenolpyruvate:

$$H_2O + \text{Pyruvate} + 2\ \text{ATP} \longrightarrow \text{Phosphoenolpyruvate} + 2\ \text{ADP} + P_i + 2\ H^+ \quad \textbf{(13)}$$

Thus, in effect, two molecules of ATP are used to reverse the conversion that yields one molecule of ATP in the glycolytic direction.

The conversion of pyruvate to PEP is complicated by the fact that the required reactions can follow either of two pathways. In the pathway just described, the reactions occur exclusively in the mitochondrion (fig. 14.16*a*). Pyruvate made in the cytosol is transported into the mitochondrion, and PEP synthesized in the mitochondrion is transported out. This route is favored when lactate is the gluconeogenic precursor as long as both enzymes, pyruvate carboxylase and PEP carboxykinase, are present in the mitochondrion. In human liver cells, pyruvate carboxylase is present only in the mitochondrion, but PEP carboxykinase is present in both the cytosol and the mitochondrion. The presence of carboxykinase in both the mitochondrion and the cytosol permits an alternative route to PEP in which the first reaction takes place in the mitochondrion and the second reaction takes place in the cytosol. However, there is a complication because the oxaloacetate cannot pass across the mitochondrial membrane. It must first be converted to malate, which can move freely across the mitochondrial membrane. Once in the cytosol, the malate can be converted back to oxaloacetate, and the oxaloacetate can subsequently be converted to PEP. In the human liver the second pathway would probably be favored when the cytosolic NADH, which is required for later steps in gluconeogenesis, is in short supply. The malate–oxaloacetate interconversion in effect results in the transfer of NADH into the cytosol (see fig. 14.16*b*).

Figure 14.16

Alternative routes from pyruvate to phosphoenolpyruvate (PEP). In (*a*) the conversion of pyruvate to PEP occurs exclusively in the mitochondrion. In (*b*) the first step in this conversion occurs in the mitochondrion, and the second step occurs in the cytosol.

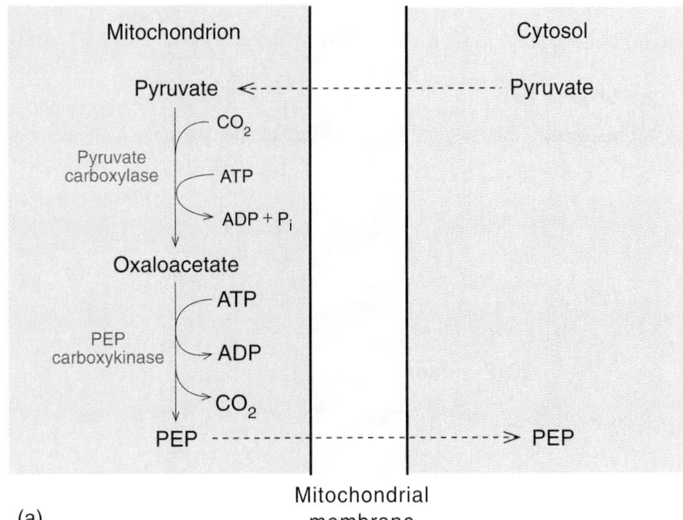

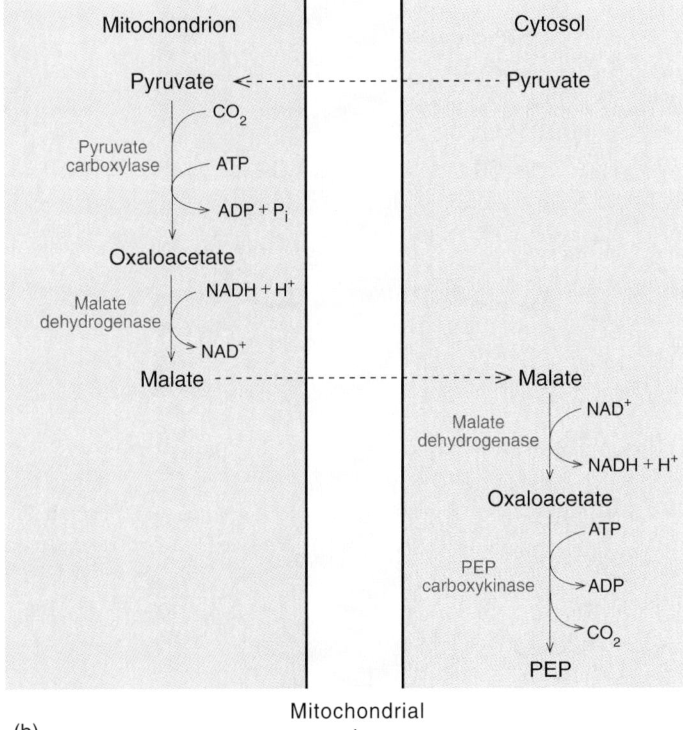

Conversion of Phosphoenolpyruvate to Fructose-1,6-bisphosphate Uses the Same Enzymes as Glycolysis

The interconversions of the phosphorylated three-carbon acids—phosphoenolpyruvate, glycerate-2-phosphate, and glycerate-3-phosphate—have already been discussed in this chapter. This pool is linked to the fructose-1,6-bisphosphate/triose phosphate pool by the two reactions that lead to the conversion of glycerate-3-phosphate to glyceraldehyde-3-phosphate. The linkage is organized differently from the reactions that connect most other metabolic pools. In this case the same enzymes function in both directions.

A glance at the energy chart (see table 14.1) shows that this is energetically feasible. Thus free energy changes very little in any of the steps between phosphoenolpyruvate and fructose-1,6-bisphosphate. Furthermore, in the glycolytic direction, one of these reactions results in the reduction of an NAD^+ to an NADH, and the other results in the conversion of an ADP to an ATP. In the gluconeogenic direction these conversions are reversed. Low ratios of ATP to ADP and NADH to NAD^+ favor glycolysis, while high ratios favor gluconeogenesis.

Fructose-bisphosphate Phosphatase Converts Fructose-1,6-bisphosphate to Fructose-6-phosphate

Fructose-1,6-bisphosphate is converted to fructose-6-phosphate by hydrolysis of the phosphoryl ester bond at C-1 in a reaction catalyzed by fructose bisphosphate phosphatase. The standard free energy change for this reaction is about −4 kcal/mole, corresponding to an equilibrium constant of about 10^3. Thus the two conversions (the phosphorylation of fructose-6-phosphate to form fructose-1,6-bisphosphate with ATP as the phosphate donor, and the hydrolysis of fructose-bisphosphate to form fructose-6-phosphate) are both thermodynamically favored under any conditions that are likely to exist in a living cell. These two reactions constitute a pseudocycle and, consistent with the principles enunciated in the previous chapter, the pathways have evolved so the number of ATP-to-ADP conversions is greater in one direction than in the other.

Hexose Phosphates Can Be Converted to Storage Polysaccharides

When fructose-6-phosphate is generated by gluconeogenesis or photosynthesis (see chapter 17), an equimolar amount of glucose-1-phosphate is usually removed from the hexose monophosphate pool by conversion to storage polysaccharide (glycogen in animals and many kinds of microorganisms; starch in green plants).

The first step in the conversion of glucose-1-phosphate to starch or glycogen is activation at the expense of a nucleoside triphosphate. In plants the activated form is ADP-glucose, formed by transfer of an AMP moiety from ATP to the phosphate of glucose-1-phosphate by the enzyme ADP-glucose synthase. An oxygen atom of the phosphoryl group of glucose-1-phosphate attacks the α phosphorus atom of ATP, displacing pyrophosphate (fig. 14.17). Since the bond broken and the bond formed are both pyrophosphate bonds, the change in standard free energy for this reaction is close to 0, and the equilibrium constant is near 1. However, if the pyrophosphate is hydrolyzed by the enzyme pyrophosphatase, as probably occurs under most conditions in living cells, the standard free energy change for the two reactions combined is large and negative, about −4.9 kcal/mole (−20.4 kJ).

In the reaction catalyzed by starch synthase, ADP-glucose decomposes to generate an oxonium ion. This species reacts with

Figure 14.17

The elongation step in starch synthesis. (*a*) The activated form of glucose in starch synthesis is ADP-glucose. This is formed from glucose and ATP as shown. (*b*) The ADP-glucose does not react directly with the elongating starch molecule. Rather, the starch synthase produces an oxonium ion intermediate, which is attacked by the terminal C-4 hydroxyl on the growing polymer.

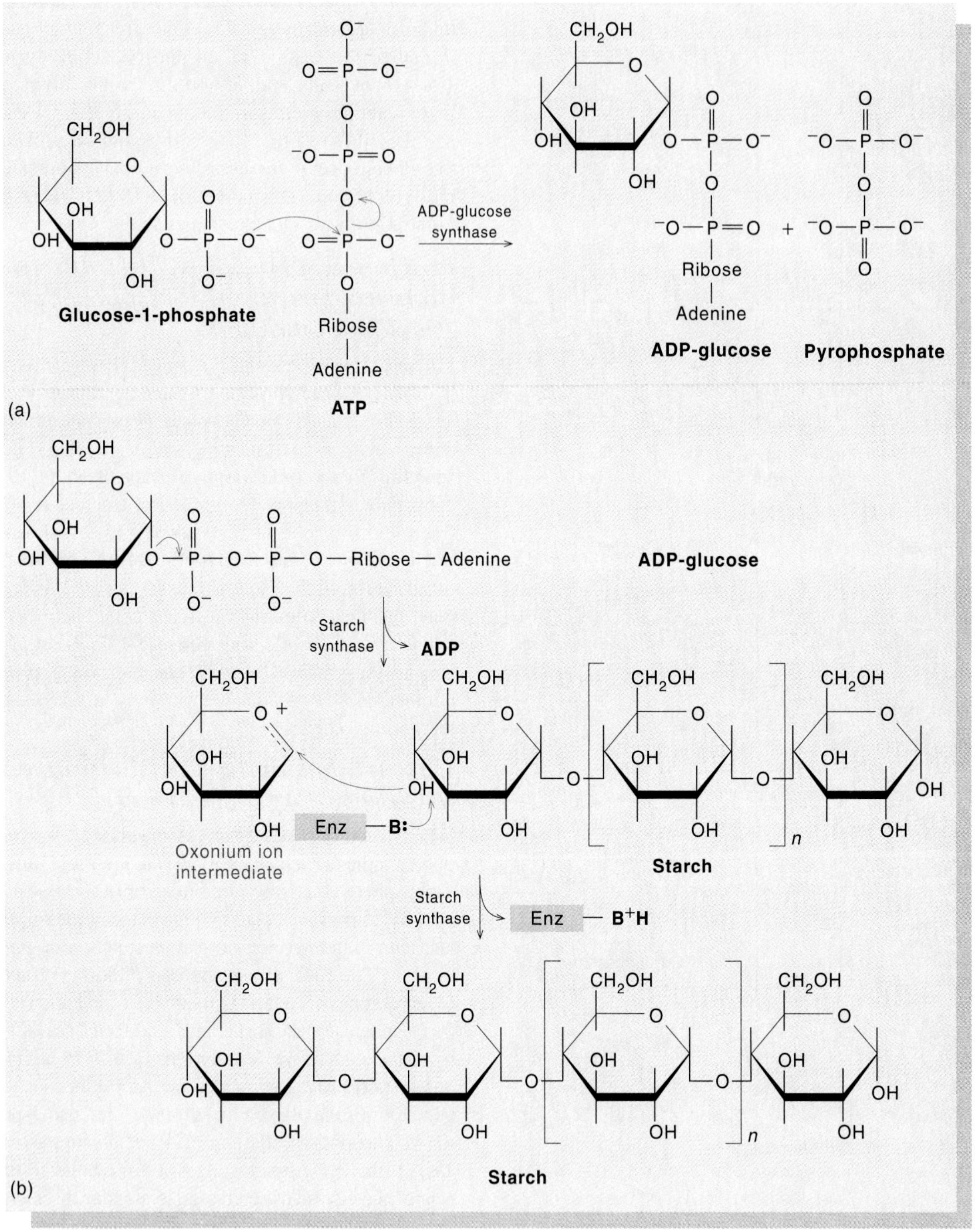

the oxygen of C-4 of the terminal glucosyl residue of a starch chain, resulting in the addition of a glucosyl unit to the polymeric starch molecule (see fig. 14.17). The standard free energy for the elongation step in starch synthesis is about −3 kcal/mole (−12 kJ).

Assuming that the pyrophosphate formed is hydrolyzed, the equation for the incorporation of the glucosyl residue of glucose-1-phosphate into starch is the reverse of the equation for the phosphorolysis of a polysaccharide, except for the conversion of

ATP to ADP and phosphate ion:

$$\text{Glucose-1-phosphate} + \text{ATP} + \text{Starch}^n \longrightarrow \text{Starch}^{n+1} + \text{ADP} + 2\ \text{P}_i + \text{H}^+ \quad (14)$$

In this equation, Starch^{n+1} represents the starch molecule after addition of a glucosyl residue. The reactions in this conversion, which include cleavage of both of the pyrophosphate bonds of ATP and the formation of a new pyrophosphate bond, are a bit more complex than in the case of a simple kinase reaction, but the thermodynamic effect is merely that of adding an ATP-to-ADP conversion in the direction of polysaccharide synthesis. Thus the pseudocycle that connects glucose-1-phosphate and starch is energetically equivalent to any other in which two oppositely directed conversions differ by one ATP-to-ADP conversion.

In most animals, many bacterial species, yeasts, and other fungi, glycogen serves the same function as starch does in plants. Glycogen resembles starch in consisting of glucose residues linked primarily by $\alpha(1,4)$-acetal bonds, but it contains a larger number of $\alpha(1,6)$ branches. Bacteria use ADP-glucose as the glucosyl donor in glycogen synthesis, while vertebrates use UDP-glucose. Energetically, this makes no difference. Since UTP is regenerated from UDP at the expense of ATP, the net reaction is the same as when ADP-glucose is used:

$$\text{H}^+ + \text{Glucose-1-phosphate} + \text{UTP} \rightleftharpoons \text{UDP-glucose} + \text{PP}_i + \text{H}_2\text{O} \quad (15)$$

$$\text{H}_2\text{O} + \text{PP}_i \longrightarrow 2\ \text{P}_i + \text{H}^+ \quad (16)$$

$$\text{UDP-glucose} + \text{Glycogen}^n \longrightarrow \text{Glycogen}^{n+1} + \text{UDP} + \text{H}^+ \quad (17)$$

$$\text{UDP} + \text{ATP} \rightleftharpoons \text{UTP} + \text{ADP} \quad (18)$$

$$\text{Sum:} \quad \text{Glucose-1-phosphate} + \text{ATP} + \text{Glycogen}^n \longrightarrow \text{Glycogen}^{n+1} + \text{ADP} + 2\ \text{P}_i + \text{H}^+ \quad (19)$$

In these equations, Glycogen^{n+1} represents the glycogen molecule after addition of a glucosyl residue.

Because of its roles in the synthesis of glycogen, in isomerization of hexose phosphates, and as a precursor for numerous biosynthetic intermediates, UDP-glucose is regarded as a central hexose derivative in mammalian metabolism. In bacteria and plants, both ADP-glucose (production of storage polysaccharide) and UDP-glucose (sugar interconversions and biosynthesis) play important roles as precursors.

Summary of Gluconeogenesis

The conversions in gluconeogenesis are the reverse of what we observed in glycolysis. Thus glucose or storage polysaccharides are produced from pyruvate. Overall, the conversion of pyruvate to storage polysaccharide "costs" 7 ATPs per six-carbon unit. Most of the reactions in gluconeogenesis simply involve a reversal of identical reactions in glycolysis. Only at three points, all outside the metabolic pools, do we find reactions in gluconeogenesis that use different enzymes: (1) the conversion of pyruvate to phosphoenolpyruvate, (2) the conversion of fructose-1,6-bisphosphate to fructose-6-phosphate, and (3) the conversion of hexose phosphate to storage polysaccharide (or hexose phosphate to glucose). These three reactions have different ATP-to-ADP stoichiometries in the two directions, which ensures that the equilibrium is favorable in either direction.

Regulation of Glycolysis and Gluconeogenesis

Several features of the glycolysis–gluconeogenesis system indicate that regulation of these conversions are of special importance to the organism. In many organisms these conversions have the highest fluxes of all metabolic sequences. Not only are the fluxes high, but they change direction frequently. In the mammalian liver, for example, massive glycogen synthesis takes place after a meal. If the meal was high in protein, much of the glucose that is incorporated into the glycogen is produced from pyruvate and oxaloacetate via gluconeogenesis. During a period of fasting, much or all of the glycogen is converted to glucose-1-phosphate and metabolized to pyruvate.

When focusing on the energetics of these pathways, it is appropriate to consider the pathway as a whole. But when considering problems of regulation, it is more useful to look at the small pseudocycles where the regulatory enzymes are located (see figs. 14.14 and 14.18). At any one of these crucial points it would be a metabolic disaster for the conversions to operate simultaneously in both directions because this would merely result in the degradation of ATP. Furthermore, these control points must respond promptly to changing metabolic needs to preserve the metabolic harmony of the steady state.

Sites of metabolic control in intact cells can be identified by examining changes in metabolite concentrations under conditions when metabolic fluxes change abruptly. Experiments of this nature were carried out by Oliver Lowry in the 1960s. Lowry showed that depriving mouse brain tissue of O_2 resulted in a dramatic increase in the rate of glycolysis. Since the glycolytic pathway is a linear series of reactions, the flux through each enzymatic step in the pathway must have increased. Measurements of the metabolites showed that the concentration of fructose-6-phosphate, the substrate of the phosphofructokinase reaction, had decreased, whereas the concentration of fructose-6-phosphate, the product of this reaction, had increased. By themselves these changes in the concentrations of its substrate and product would tend to decrease the rate of phosphofructokinase activity. Because the rate of the reaction increased, it is clear that phosphofructokinase must have been influenced by factors other than its substrate and product concentrations.

As an analogy, consider a river with a dam somewhere between its source and its terminus. If the flood gates of the dam are opened, the flow of water transiently increases everywhere in the river. At sites near the end of the river, the increase in the flow can be explained simply by the increased pressure of water coming from upstream; at sites near the source it can be attributed to a reduced pressure downstream. Only at the site of the dam is the situation reversed: The flow of water increases here in spite of the fact that the pressure downstream has risen and the pressure upstream has fallen.

Phosphofructokinase and the other enzymes that regulate pseudocycles I, II, and III of glycolysis are influenced by both intracellular and extracellular signals. We consider some of the intracellular signals first.

Figure 14.18

Some major points of regulation in the glycolytic-gluconeogenic pathways. Intracellular controls favor glycogen breakdown and glycolysis when the energy charge is low. However, when the blood glucose level is low, hormonal controls on the liver cells favor glycogen breakdown and gluconeogenesis. The target sites for control factors are indicated by red arrows, with a plus or a minus at the arrowhead to indicate activation or inhibition, respectively.

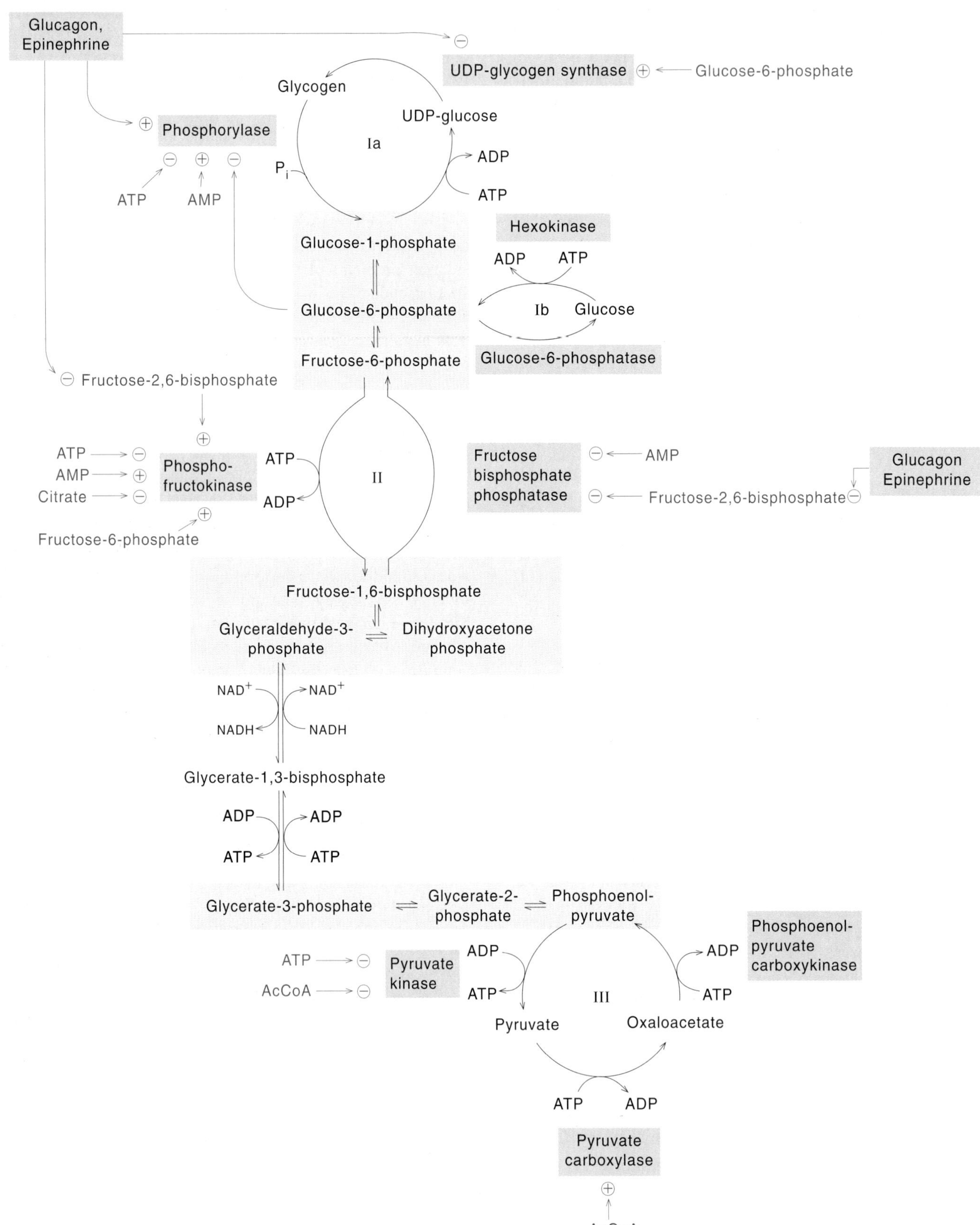

How Do Intracellular Signals Regulate Energy Metabolism?

The intracellular signals for the glycolysis–gluconeogenesis pathways consist of small-molecule allosteric effectors, the concentrations of which reflect the energy charge: ATP and citrate are two of the effectors that reflect a high energy charge (we will hear more about citrate in the next chapter); AMP is one of the effectors that indicates a low energy charge. The way in which these signals operate is exemplified by a consideration of the well-studied case of phosphofructokinase (see chapter 10). This enzyme has a site(s), separate from the catalytic site, for binding small-molecule allosteric effectors. Binding of allosteric effectors at the regulatory site changes the conformation of the enzyme. This conformational change affects the catalytic site. There are two conformations: One favors catalytic activity, the other does not. ATP and citrate, which indicate a high energy charge, favor the less active form of the enzyme, whereas AMP, an indication of a low energy charge, favors the more active form of the enzyme.

Fructose bisphosphate phosphatase, the enzyme paired with phosphofructokinase at pseudocycle II, is affected in just the opposite way by some of the same allosteric effectors. The overall effect of these allosteric effectors is to encourage glycolysis when the energy charge is low and to encourage gluconeogenesis when the energy charge is high.

The paired enzymes that regulate the flux between glycogen and glucose-1-phosphate (pseudocycle Ia) are affected in a parallel way by the same small-molecule allosteric effectors that operate at pseudocycle II. Thus AMP stimulates phosphorylase and inhibits glycogen synthase, while ATP inhibits phosphorylase and stimulates glycogen synthase. In addition, a high level of glucose-6-phosphate stimulates the synthase while it inhibits the phosphorylase.

At pseudocycle III we find that the paired enzymes that regulate the flux are influenced in opposite ways by the same small-molecule allosteric effectors. Thus pyruvate kinase is strongly inhibited by AcCoA as well as ATP, while pyruvate carboxylase is strongly stimulated by AcCoA. A high level of AcCoA indicates a high energy charge, so it is appropriate to direct the available pyruvate in the gluconeogenic direction when the level of AcCoA is high. Effects on the TCA cycle are discussed in Chapter 15.

Finally, turning to enzymes that regulate the flux between glucose and glucose-6-phosphate, we find that there are sharp differences between liver and other mammalian tissues. We have already indicated that there are several isozymes for hexokinase in mammalian tissues. The isozyme that predominates in liver has a very low affinity for glucose, reflecting the fact that liver cells take up glucose only when the blood glucose level is quite high. Liver is also unusual in being one of the few tissues that bears the opposing enzyme of pseudocycle Ib, glucose-6-phosphatase. This is consistent with the role of liver in maintaining the blood glucose level.

The differences between liver and other mammalian tissues become more apparent as we consider the hormonal effects on regulation.

Hormonal Controls Can Override Intracellular Controls

Hormonal signals coming from other cells are more difficult to understand than allosteric effectors because they sometimes seem to go against the interests of the cell they influence. This influence is possible because hormonal signals respond to systemic needs.

In liver cells the response to energy charge, or to signals that reflect the energy charge, can be quite different from the responses of other cells because liver cells are involved in the regulation of the energy needs for the entire organism. In particular, liver cells secrete glucose so as to maintain a reasonably constant level of blood glucose. A low blood glucose level indicates a systemic need for glucose or energy. This need triggers an extracellular signal (discussed later on), which elevates liver cell activities of both glycogen phosphorylase and fructose bisphosphate phosphatase and simultaneously lowers the activities of glycogen synthase and phosphofructokinase. The net effect of this stimulation is to encourage glucose production both by breakdown of glycogen and by gluconeogenesis (table 14.2).

Since the liver's metabolic functions are not dictated by local needs so much as by the general need for controlling the blood glucose level, we might ask what additional regulatory inputs are required to respond to these systemic needs. As in many other cases in which the activities of different tissues are correlated for the good of the whole organism, hormones are involved at this additional level of control. A low blood glucose level is first sensed by the pancreas, whose α cells respond by releasing the peptide hormone glucagon into the blood. Glucagon binds to specific receptors in the plasma membranes of liver cells. This binding initiates a chain of reactions that leads to the conversion of glycogen phosphorylase from its less active form (phosphorylase *b*) to its hormone-activated form (phosphorylase *a*). Phosphorylase *a* is not only more active than phosphorylase *b,* it is largely insensitive to AMP (see chapter 10). The net result is that glucagon directs the liver cells to respond to the systemic (organismwide) need for raising the blood glucose level by breaking down glycogen to hexose phosphate intermediates.

Hormonal Effects of Glucagon Are Mediated by Cyclic AMP

The hormone glucagon itself never enters the liver cells, so it must somehow trigger an event to occur on the cytosolic side of the plasma membrane. A breakthrough in our understanding of this phenomenon came in the 1950s when Earl Sutherland found that the membrane-containing fraction of broken liver cells produces a soluble factor in the presence of ATP and glucagon, which stimulates the activity of glycogen phosphorylase. Further analyses showed that the factor was 3′,5′-cAMP-synthesized by a membrane-bound enzyme (fig. 14.19). The detailed mechanism for how cAMP synthesis is triggered will be discussed in chapter 27. Here we focus on the effects of the cAMP so produced.

Cyclic AMP triggers a cascade of reactions that ultimately lead to glycogen breakdown. The immediate action of

Table 14.2
Activities of Some of the Enzymes That Control the Fluxes in the Glycolysis–Gluconeogenesis Pathways in the Liver

	Pseudocycle I		**Pseudocycle II**	
Intracellular Controls	***Glycogen Phosphorylase***	***Glycogen Synthase***	***Phosphofructokinase***	***Fructose Bisphosphate Phosphatase***
High energy charge	↓	↑	↓	↑
Low energy charge	↑	↓	↑	↓
Hormonal Controls				
Low blood glucose	↑	↓	↓	↑

↑ = increase. ↓ = decrease.

Figure 14.19

Synthesis of cyclic AMP. A catalytic site on adenylate cyclase (**:B**) removes a proton from the C-3 oxygen, which then attacks the α-phosphate and displaces the pyrophosphate group. This reaction occurs on the inner plasma membrane (see fig. 14.20).

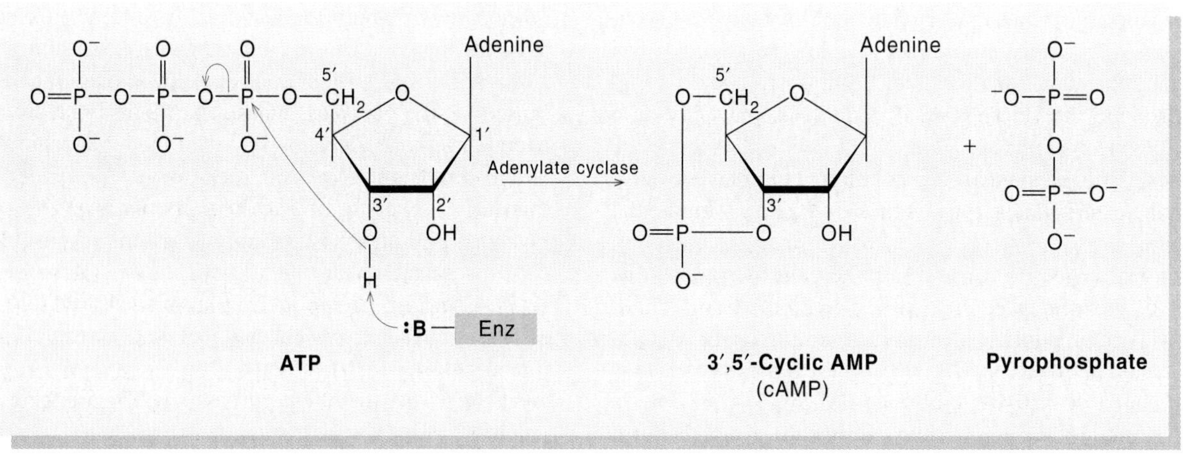

cAMP is to activate a protein kinase that phosphorylates a number of proteins, including phosphorylase kinase. Phosphorylation of phosphorylase kinase converts it from an inactive to an active form, which catalyzes the conversion of phosphorylase *b* to phosphorylase *a* (see chapter 10). The cascade of effects triggered by glucagon is shown in figure 14.20.

The same cAMP-dependent protein kinase that is responsible for phosphorylating phosphorylase kinase also catalyzes the phosphorylation of glycogen synthase. Whereas phosphorylation of glycogen phosphorylase leads to increased activity, the phosphorylation of glycogen synthase decreases its activity. As a result, when glycogen breakdown is stimulated in response to glucagon, glycogen synthesis is inhibited. In this way the simultaneous operation of both enzymes associated with pseudocycle Ia is prevented.

When the systemic need has been satisfied there should be a way of reversing the hormone triggered process. All of the effects we have seen are promptly reversed when the need for them has ended. Cyclic AMP is hydrolyzed to 5′-AMP by an enzyme called cyclic AMP phosphodiesterase, and protein phosphatases catalyze the hydrolytic removal of the phosphoryl groups that were attached to proteins by protein kinase and phosphorylase kinase.

The Hormone Epinephrine Stimulates Glycogenolysis in Both Liver Cells and Muscle Cells

The hormone epinephrine responding to somewhat different needs of the organisms affects glycogen metabolism in the liver in the same way as glucagon. Epinephrine is a small tyrosine-derived hormone that is made in the adrenal glands (see chapter 27). It is

Figure 14.20

Molecular basis for the stimulation of glycogen breakdown by the hormones epinephrine and glucagon. The epinephrine or the glucagon activates adenylate cyclase, which activates protein kinase, which activates phosphorylase kinase. Phosphorylase kinase in turn activates phosphorylase, which cleaves glucose from glycogen. This cascade of regulatory interactions results in a very large amplification, so a small amount of hormone can have a large effect on the rate of breakdown of glycogen.

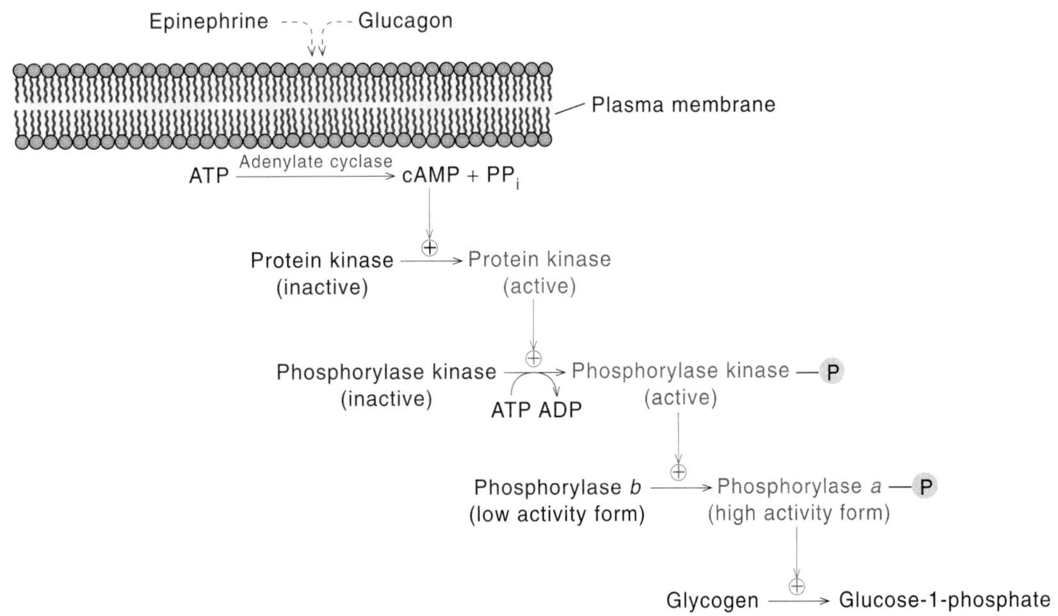

liberated into the blood in response to distress signals from the central nervous system. The two hormones, glucagon and epinephrine, have similar effects on liver cells even though they originate from different sources and bind to different plasma membrane receptors.

Despite some similarities in their effects, the physiological significance of these two hormones is quite different. Glucagon is part of a system designed to stabilize the blood glucose level and is therefore involved in maintaining the status quo. By contrast, epinephrine is an alarm signal that leads to deviations from the status quo. Its release, triggered by the central nervous system in response to fear or anger, prepares the body for a sudden high energy output. Epinephrine is often termed the "fight-or-flight" hormone. Its function in activating phosphorylase is to build up glycolytic intermediates, thereby eliminating any time lag in energy mobilization if and when the organism begins to fight or to flee.

Whereas epinephrine and glucagon have very similar effects on glycogen metabolism in liver cells, only epinephrine affects glycogen metabolism in skeletal muscle cells. This difference occurs because muscle cells carry plasma receptors for epinephrine but not for glucagon. Muscle is strictly a consumer tissue and does not contribute to the stabilization of the blood glucose level; thus it should not, and does not, break down its glycogen reserves in response to low blood glucose. By contrast it is appropriate that muscle tissue should respond to epinephrine because a buildup of glycolytic intermediates in the muscle tissue itself prepares the organism for a sudden burst of muscle activity.

Another indication of the difference between the metabolic roles of muscle, a consumer tissue, and liver, a contributer tissue, is the enzyme glucose-6-phosphatase. This enzyme is required for the production of neutral glucose. It is present in liver and kidney but not in strict consumer tissues, such as muscle and brain. Without glucose-6-phosphatase the glucose-6-phosphate cannot be converted to dephosphorylated glucose, which is necessary for secretion into the blood stream. This enzyme and hexokinase constitute a pseudocycle for the interconversion of glucose and glucose-6-phosphate (see fig. 14.18). It is clear that glucose-6-phosphatase should be active only when the blood glucose level is low. A deficiency of this enzyme is associated with von Gierke's disease, a condition in which there is massive liver enlargement and severe hypoglycemia (low blood sugar) after a fast.

Hormonal Regulation of the Flux between Fructose-6-phosphate and Fructose-1,6-bisphosphate in the Liver Is Mediated by Fructose-2,6-bisphosphate

In eukaryotes the most potent regulator of the flux between fructose-6-phosphate and fructose-1,6-bisphosphate is the allosteric effector fructose-2,6-bisphosphate (fructose-2,6-bisP). We have

Figure 14.21

The bifunctional enzyme PFK-2/FBPase-2 catalyzes the synthesis of fructose-2,6-bisphosphate from fructose-6-phosphate and its breakdown to fructose-6-phosphate. In mammalian tissues this enzyme exists in different isozymal forms, which show different relative activities and different activities to hormonally stimulated phosphorylation. In liver the phosphorylated form of the enzyme is active in the breakdown of fructose-2,6-bisphosphate, while the dephosphoenzyme is active in its synthesis.

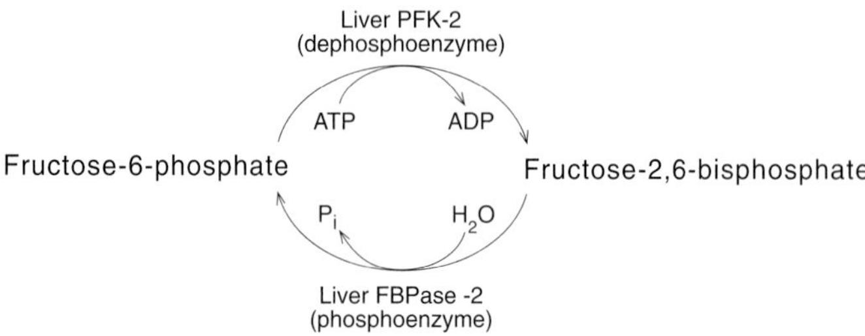

not mentioned this effector thus far because it is found only in eukaryotes and its concentration is hormonally controlled. Fructose-2,6-bisP stimulates phosphofructokinase while inhibiting the oppositely directed enzyme of pseudocycle II, fructose-bisphosphate phosphatase. As a result, the presence of fructose-2,6-bisP favors glycolysis over gluconeogenesis. The concentration of fructose-2,6-bisP is regulated by a bifunctional enzyme, PFK-2/FBPase-2. The PFK-2 activity catalyzes the synthesis of fructose-2,6-bisP from fructose-6-phosphate, while the FBPase-2 activity catalyzes its phosphorolysis back to fructose-6-phosphate (fig. 14.21). In mammalian tissues this bifunctional enzyme exists in different isozymal forms that are tissue-specific. These different isozymal forms show very different relative activities and sensitivities to hormone-induced phosphorylation. For example, in skeletal muscle the enzyme is insensitive to hormone-induced phosphorylation. By contrast, in liver the same cAMP-dependent protein kinase that phosphorylates the enzymes of pseudocycle Ia phosphorylates PFK-2/FBPase-2 at a unique serine residue. This has a major effect on both of the enzyme's catalytic activities. Thus the dephosphorylated enzyme is active in synthesis of fructose-2,6-bisP, while the phosphorylated enzyme is active in its breakdown. As a result, in liver the hormonal signals originating from glycogen and/or epinephrine block glycolysis of hexose monophosphates while encouraging their synthesis from C-3 substrates. The combined effects of these hormones in liver on pseudocycle Ia and II is to stimulate hexose monophosphate production and the subsequent formation of glucose for export to other tissues. By contrast, in skeletal muscle the effect of epinephrine is confined to pseudocycle Ia, so glycogen breakdown and the subsequent glycolysis of the hexose monophosphates is favored. It should be recalled that glucagon does not act on these latter tissues because they do not possess glucagon receptors.

Summary of the Regulation of Glycolysis and Gluconeogenesis

Control of glycolysis and gluconeogenesis is confined to three points, all located outside the pools. These control points are organized into three small pseudocycles (see fig. 14.18). The ATP stoichiometries are different for going in opposite directions in these pseudocycles. This difference ensures that the reactions are thermodynamically feasible in either direction, but it does not dictate which reactions are active and which are inactive under any particular set of metabolic conditions. The activity status is determined by small-molecule allosteric factors, which interact with the regulatory enzymes that catalyze the reactions within each of the pseudocycles. The enzymes are designed in such a way that only one arm of a pseudocycle is active at any given time.

In all unicellular organisms and in most cells of multicellular organisms, these regulatory enzymes are sensitive to inhibition and activation by small molecules that reflect the metabolic state of the cell. When an effector binds to the regulatory site of an enzyme, it encourages the enzyme to adopt one of two possible conformations. The conformation adopted may be the active or the inactive conformation, depending on the enzyme and the allosteric effector. For example, phosphofructokinase, which converts fructose-6-phosphate to fructose-1,6-bisphosphate, is active when binding to AMP. This is likely to happen when the AMP level is relatively high and energy is in short supply. The oppositely directed enzyme in the same pseudocycle, fructose bisphosphate phosphatase, adopts an inactive conformation when binding to the same allosteric effector. In this way the pseudocycle is controlled so that it is active only in the direction favoring the metabolic state of the cell.

In some tissues of multicellular organisms, cells have a dual responsibility: maintaining themselves and serving other cells in the same organism. For example, liver cells must maintain themselves, but they also have an organismwide responsibility to maintain the blood glucose level. If the blood glucose level falls below normal, a signal is sent to the liver from the pancreas in the form of the hormone glucagon. Glucagon binds to specific receptors on the outside surface of the plasma membrane of the liver cell, which stimulates the formation of cAMP by an enzyme binding to the inner side of the plasma membrane. The cAMP triggers a series of phosphorylations inside the cell, which ultimately results in the phosphorylation of both glycogen phosphorylase and glycogen synthase. While the phosphorylation activates phosphorylase, it inhibits the synthase, thereby promoting the formation of hexose phosphates. In general the metabolic signal delivered by a hormonal signal overrides any metabolic signals that may have originated from inside the cell.

Glucagon and epinephrine also regulate pseudocycle II in the liver so as to stimulate gluconeogenesis while inhibiting glycolysis. They do this through a chain of reactions that results in a lowering of the concentration of the allosteric effector fructose-2,6-bisphosphate. This effector stimulates phosphofructokinase while it inhibits fructose bisphosphate phosphatase.

The Pentose Phosphate Pathway

Many kinds of organisms and some mammalian organs, notably liver, possess an alternative pathway for the oxidation of hexoses, which results in a pentose phosphate and carbon dioxide. This pentose can be used as a precursor of the ribose found in nucleic acids or other sugars containing from three to seven carbon atoms, which

Figure 14.22

Stage 1 of the pentose phosphate pathway. Net reaction: Glucose-6-phosphate + 2 $NADP^+$ $\rightarrow$ Ribulose-5-phosphate + CO_2 + 2 NADPH + 2 H^+.

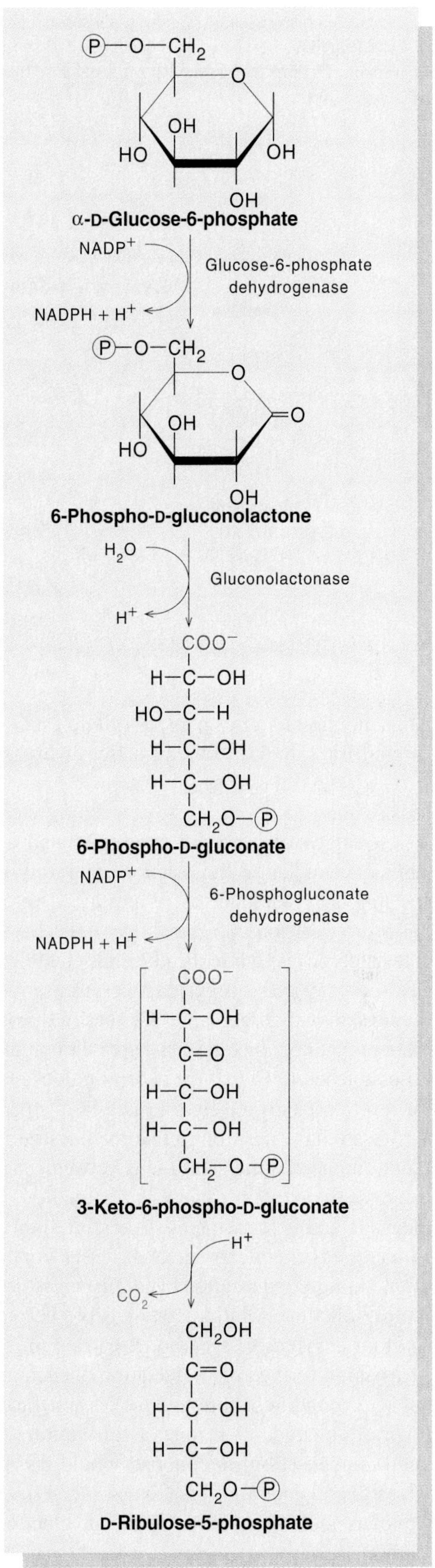

are needed in smaller amounts. The first and third reactions in the pentose phosphate pathway generate NADPH, which is a major source of reducing power in many cells.

Cells differ considerably in their use of the pentose phosphate pathway. In muscle, a tissue in which carbohydrates are utilized almost exclusively for generation of mechanical energy, the enzymes of the pentose phosphate pathway are lacking. By contrast, red blood cells are totally dependent on the pentose phosphate pathway as a source of NADPH, which they need to keep the iron of hemoglobin in its normal +2 valence state. A deficiency in glucose-6-phosphate dehydrogenase, the first enzyme in the pentose phosphate pathway, can lead to the wholesale destruction of red blood cells and a condition known as hemolytic anemia.

Two NADPH Molecules Are Generated by the Pentose Phosphate Pathway

NADPH is required for many biosynthetic sequences. It is generated in different kinds of cells by a variety of reactions, including an $NADP^+$-linked oxidation of malate to pyruvate and CO_2 and transfer of hydride ion from NADH to $NADP^+$ in a mitochondrial reaction that is driven by metabolic energy. However, in many cases, including in the mammalian liver, a major part of the NADPH requirement is met by oxidation of glucose-6-phosphate to ribulose-5-phosphate and CO_2. The four electrons that are released by the oxidation are transferred to two molecules of $NADP^+$.

The four reactions involved in this conversion are shown in figure 14.22. The first oxidation, catalyzed by glucose-6-phosphate dehydrogenase at C-1, converts the hemiacetal derivative of the aldehyde group to the lactone of the corresponding acid, 6-phosphogluconic acid. After hydrolysis of the lactone the second oxidation, at C-3, converts the secondary alcohol to a ketone. The expected product, 3-keto-6-phosphogluconic acid, is decarboxylated, yielding ribulose-5-phosphate.

At this point the metabolic function of the sequence, when it is serving to supply electrons for biosynthesis, is fulfilled. Two molecules of NADPH are generated for each molecule of glucose-6-phosphate oxidized. It is necessary only to convert ribulose-5-phosphate, the end product of the oxidative sequence, to compounds in the mainstream of metabolism.

Transaldolase and Transketolase Catalyze the Interconversion of Many Phosphorylated Sugars

The key enzymes involved in these conversions are transaldolase and transketolase. The two enzymes are similar in their substrate specificities. Both require a ketose as a donor and an aldose as an acceptor. The steric requirements at positions C-1 through C-4 are the same as the requirements of aldolase in the glycolytic path-

Figure 14.23

The transaldolase-catalyzed conversion of fructose-6-phosphate and erythrose-4-phosphate to glyceraldehyde-3-phosphate and sedoheptulose-7-phosphate. This is a two-step conversion. The first step is similar to the aldolase reaction except that the dihydroxyacetone produced is held at the catalytic site while the aldose product diffuses away and is replaced by another aldose molecule. The second step involves an aldol condensation.

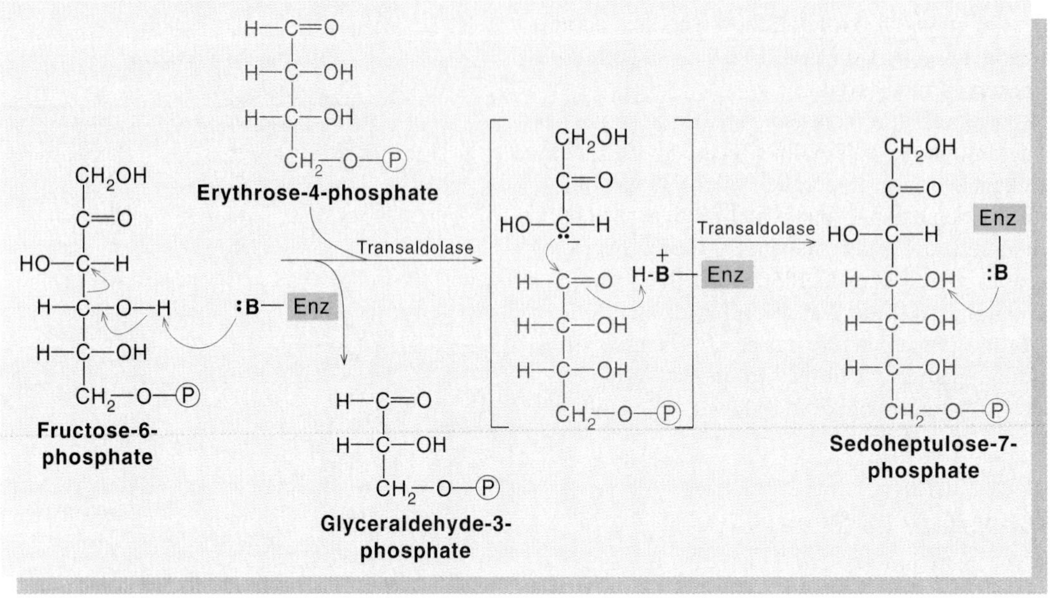

way, except that aldolase requires phosphorylation at C-1, and both transaldolase and transketolase require a free hydroxyl group at C-1.

Transaldolase catalyzes a two-step conversion. The first step, an aldol cleavage of the bond between C-3 and C-4 of a ketose, is essentially identical to the reaction catalyzed by aldolase. However, the dihydroxyacetone that is produced in the transaldolase reaction from carbons 1, 2, and 3 is not released. Rather, it is held at the catalytic site while the glyceraldehyde-3-phosphate produced diffuses away and is replaced by erythrose-4-phosphate. An aldol condensation then generates the second product of the reaction, a ketose that contains the first three carbon atoms of the original ketose attached to C-1 of the acceptor aldose (fig. 14.23).

The reaction catalyzed by transketolase superficially resembles the transaldolase reaction in that the substrate specificity is identical, but in this case the cleavage is between carbons 2 and 3 of the ketose. A two-carbon moiety is retained on the enzyme following cleavage and is subsequently transferred to an acceptor aldose. This reaction is chemically quite different from the transaldolase reaction. Aldol condensations and cleavages, which occur readily in mildly alkaline solution, are feasible because protons on a carbon atom adjacent to a carbonyl carbon are moderately acidic. The carbanion formed on dissociation of such a proton can participate in a nucleophilic addition to the carbonyl carbon of another molecule of aldehyde or ketone, as in the transaldolase reaction. A ketol condensation, in contrast, would involve the carbonyl carbon as the nucleophile. That is not energetically feasible because the polarity of the carbonyl bond precludes a negative charge on carbon.

Transketolase is one of several enzymes that catalyze reactions of intermediates with a negative charge on what was initially a carbonyl carbon atom. All such enzymes require thiamine pyrophosphate (TPP) as a cofactor (see chapter 11). The transketolase reaction is initiated by addition of the thiamine pyrophosphate anion to the carbonyl of a ketose phosphate, for example xylulose-5-phosphate (fig. 14.24). The adduct next undergoes an aldol-like cleavage. Carbons 1 and 2 are retained on the enzyme in the form of the glycolaldehyde derivative of TPP. This intermediate condenses with the carbonyl of another aldolase. If the reactants are xylulose-5-phosphate and ribose-5-phosphate, the products are glyceraldehyde-3-phosphate and the seven-carbon ketose, sedoheptulose-7-phosphate (see fig. 14.24).

Production of Ribose-5-phosphate and Xylulose-5-phosphate

Both transaldolase and transketolase require a ketose phosphate as the donor molecule and an aldose phosphate as the acceptor. Furthermore, both enzymes require the same steric configuration at carbons 3 and 4 as is found in glucose and fructose. Ribulose-5-phosphate, the first pentose phosphate to be formed in the pentose phosphate pathway, does not have the correct configuration to serve as a substrate for either transaldolase or transketolase. However, both a suitable donor ketose and an acceptor aldose can be made by isomerizations of ribulose-5-phosphate, and enzymes that catalyze those isomerizations are found in cells that possess the pentose phosphate pathway (fig. 14.25).

Because of the possible alternative pathways, which probably occur simultaneously, no single set of reactions can uniquely

Figure 14.24

The transketolase-catalyzed conversion of xylulose-5-phosphate and ribose-5-phosphate to glyceraldehyde-3-phosphate and sedoheptulose-7-phosphate. Although the aldolase and ketolase reactions superficially resemble each other, they proceed by very different mechanisms. This is because in the aldolase reaction the carbon adjacent to a carbonyl acts as a nucleophilic agent, whereas the ketolase reaction involves an intermediate with a negative charge on what was originally a carbonyl carbon. The latter type of reaction is more complex and requires thiamine pyrophosphate.

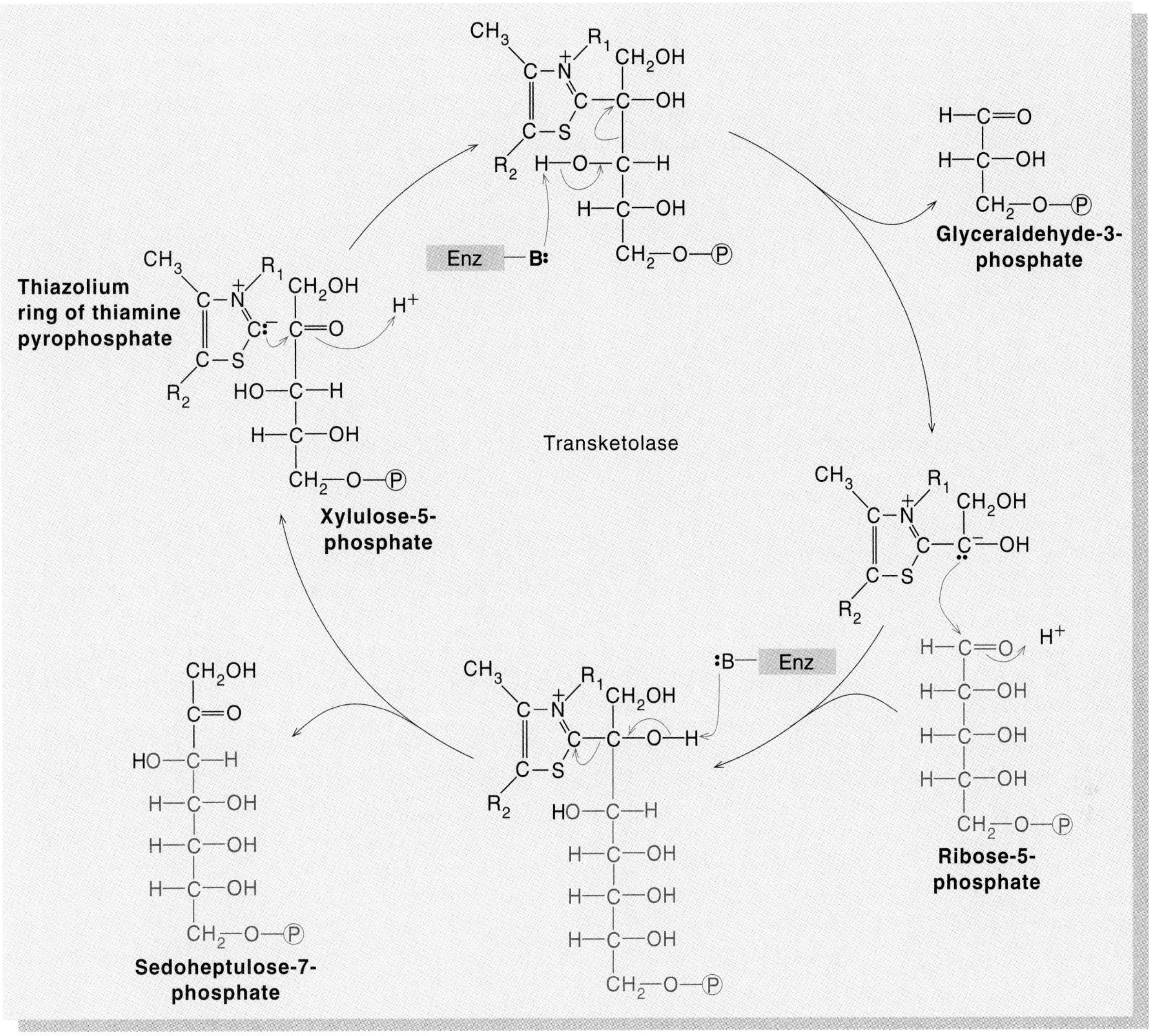

describe the pentose phosphate pathway. One possible set of pathways is shown in figure 14.25. If the triose phosphate formed is converted to hexose phosphate, the overall pathway can be seen as regenerating five molecules of hexose phosphate for each six used initially:

$$\begin{aligned} &6 \text{ Glucose-6-phosphate} + 12 \text{ NADP}^+ \longrightarrow \\ &\qquad 6\ CO_2 + 5 \text{ Fructose-6-phosphate} \\ &\qquad\qquad + 12 \text{ NADPH} + 10\ H^+ \end{aligned} \quad \textbf{(20)}$$

The pathway was first formulated in this form.

Alternatively, it is possible to write a sequence of reactions, including the action of phosphofructokinase and aldolase on seven-carbon intermediates, in which the carbon of ribulose-5-phosphate is converted mainly to glyceraldehyde-3-phosphate. Such a pathway, with the triose phosphate entering the glycolytic sequence, amounts to a bypass, or shunt, around the first reactions of glycolysis, and the name hexose monophosphate shunt is sometimes used. Any amount of ribose-5-phosphate or erythrose-4-phosphate that may be needed for biosynthetic sequences can also be obtained from this oxidative pentose phosphate pathway.

Figure 14.25

Stage 2 of the pentose phosphate pathway. The groups in red are those transferred in transketolase-catalyzed reactions. The groups in bold type are transferred in the transaldose-catalyzed reactions. All of the reaction arrows in this figure are double-headed to indicate that the reactions can go in either direction with little change in free energy.

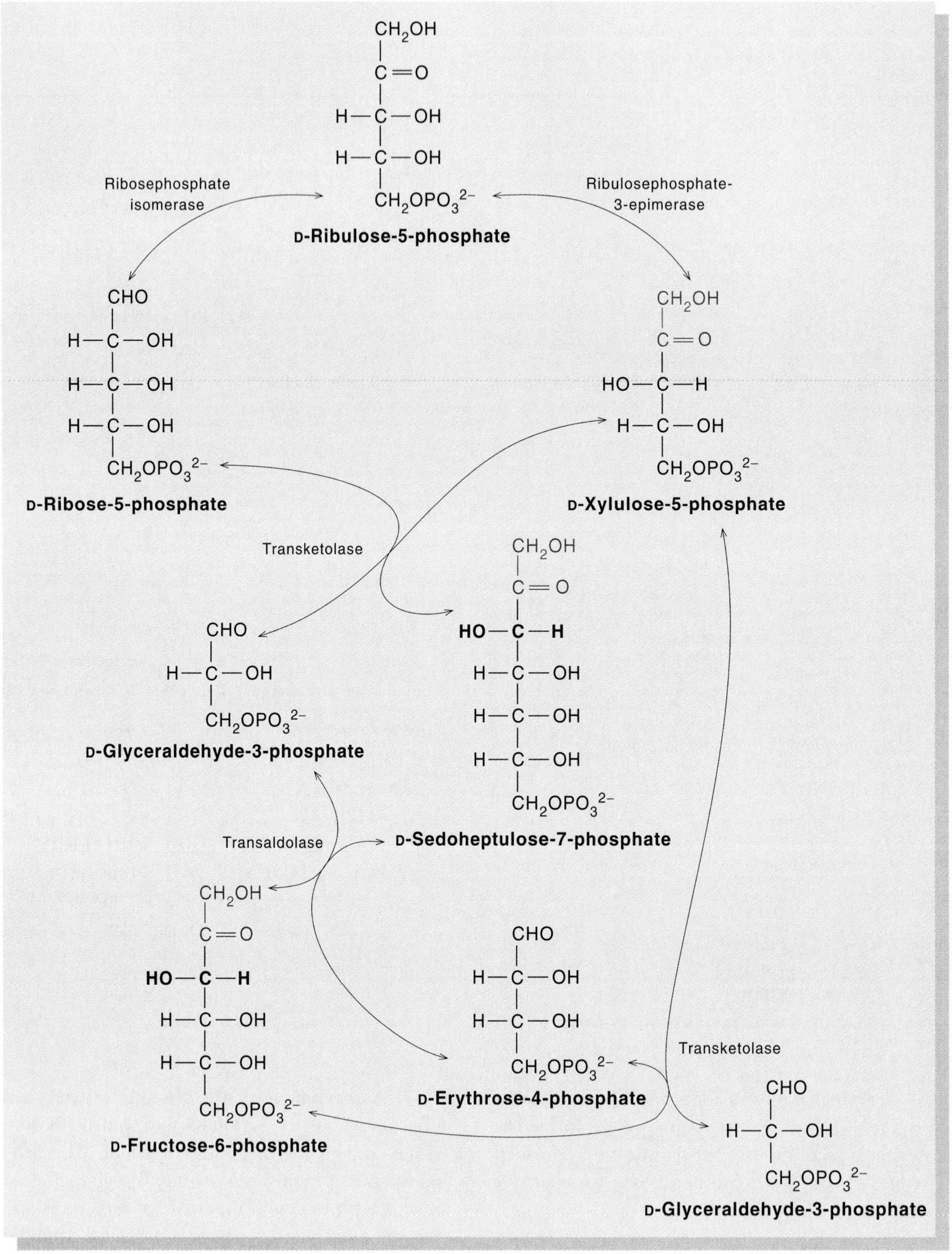

D-Ribulose-5-phosphate

Ribosephosphate isomerase

Ribulosephosphate-3-epimerase

D-Ribose-5-phosphate

D-Xylulose-5-phosphate

Transketolase

D-Glyceraldehyde-3-phosphate

D-Sedoheptulose-7-phosphate

Transaldolase

D-Erythrose-4-phosphate

Transketolase

D-Fructose-6-phosphate

D-Glyceraldehyde-3-phosphate

Because the transaldolase and transketolase reactions are symmetrical with respect to types of bonds cleaved and formed, their equilibrium constants are near 1. The pool of sugar phosphates is thus near equilibrium in cells that contain these enzymes. In a water-flow analogy, the sugar phosphates and the reactions that interconvert them resemble a large swamp, with ill-defined flows along many interconnecting channels. Water may be fed in from any direction and may leave the swamp in any direction.

Summary

In this chapter we discuss the anaerobic metabolism of the carbohydrates involved in energy metabolism. We focused on the following points.

1. Glycolysis involves the breakdown of energy-storage polysaccharides or glucose to the 3-carbon acid, pyruvate. Its function is to produce energy in the form of ATP and 3-carbon intermediates for further metabolism. Glycolysis occurs under both aerobic and anaerobic conditions.
2. Except for glycerate-1,3-bisphosphate all of the intermediates in glycolysis are pictured as belonging to one of three metabolic pools: the hexose monophosphate pool, the aldolase pool and the 3-carbon pool. Within each metabolic pool the intermediates are readily interconvertible and usually present in relative concentrations close to their equilibrium values.
3. The three sugar phosphates, glucose-1-phosphate, glucose-6-phosphate, and fructose-6-phosphate make up the hexose monophosphate pool. This pool is a major crossroad of energy metabolism. Glucose-1-phosphate is produced from storage polysaccharides. Glucose-6-phosphate is produced from glucose and is the precursor for blood glucose in mammals and the starting material for the pentose phosphate pathway. Fructose-6-phosphate is produced by gluconeogenesis, the reverse of glycolysis in which 6-carbon sugars are synthesized from pyruvate.
4. Glycolysis and the oppositely directed process gluconeogenesis share common intermediates and many enzymes. Only reactions outside of the three aforementioned pools are unique in different directions.
5. Most of the enzymes outside of the pools are regulated so that glycolysis and gluconeogenesis do not take place simultaneously. The regulated enzymes in glycolysis are activated only when there is an energy shortage. And the regulated enzymes in gluconeogenesis are activated only when there is an energy surplus. The activity of these enzymes is regulated by small molecule allosteric effectors that reflect the energy status of the cell.
6. In multicellular organisms there are extracellular hormone regulators that control carbohydrate metabolism. These hormones respond to the systemic needs of the organism. For example in response to a low blood glucose level, the pancreas secretes the hormone glucagon. This hormone binds to a specific receptor on liver cells which leads to a chain of reactions that results in glucose synthesis and secretion.
7. An alternative pathway for glycolysis is provided by the pentose phosphate pathway. In this pathway glucose is degraded to a pentose phosphate and carbon dioxide and in the process $NADP^+$ is reduced to NADPH. This pathway serves a variety of functions in different cell types: (1) the production of NADPH for biosynthesis, (2) the production of ribose required mainly for nucleic acid synthesis, and (3) the interconversion of a variety of phosphorylated sugars.

Selected Readings

Atkinson, D. E., *Cellular Energy Metabolism and Its Regulation.* New York: Academic Press, 1977. A general discussion, covering some topics in this and the preceding and following chapters in somewhat greater depth than the treatment in our book.

Beitner, R., *Regulation of Carbohydrate Metabolism.* Boca Raton, Fla.: CRC Press, 1985. Two-volume discussion, at a rather advanced level, of many aspects of the regulation of mammalian carbohydrate metabolism with frequent discussions of clinical conditions. Many references.

Browner, M. F., and R. J. Fletterick, Phosphorylase: A biological transducer. *Trends Biochem. Sci.* 17:66–71, 1992.

Cleland, W. W., and M. M. Kreevoy, Low-barrier hydrogen bonds and enzymic catalysis. *Science* 264:1887–1890, 1994.

Cornish-Bowden, A., and M. L. Cardenas, Hexokinase and glucokinase in liver metabolism. *Trends Biochem. Sci.* 16:281–282, 1991.

Dawes, E. A., *Microbial Energetics.* Glasgow and London: Blackie [New York: Chapman and Hall], 1986. Discussion of bacterial metabolism, with emphasis on the central pathways and the generation and use of ATP. A good source for learning something of the diversity of metabolic adaptations among different groups of bacteria.

Fothergill-Gilmore, L. A., The evolution of the glycolytic pathway. *Trends Biochem. Sci.* 11:47–51, 1986.

Gerlt, J. A., and P. G. Gassman, Understanding the rates of certain enzyme-catalyzed reactions: Proton abstraction from carbon acids acyl-transfer reactions, and displacement of phosphodiesters. *Biochemistry* 32:11943–11952, 1993.

Hers, H. G., and L. Hue, Gluconeogenesis and related aspects of glycolysis. *Ann. Rev. Biochem.* 52:617–653, 1983.

Hochachka, P. W., and G. N. Somera, *Biochemical Adaptation.* Princeton, N.J.: Princeton University Press, 1984. An excellent and extensive discussion of how biochemical processes, including many discussed in this book, are adapted by various types of organisms in fitting themselves for survival under specific and often difficult conditions.

Hoffman, E., Phosphofructokinase—A favorite of enzymologists and students of metabolic regulation. *Trends Biochem. Sci.* 3:145–147, 1978.

Hue, L., and M. H. Rider, Role of fructose 2,6-bisphosphate in the control of glycolysis in mammalian tissues. *Biochem. J.* 245:313–324, 1987.

Katz, J., and R. Rognstad, Futile cycling in glucose metabolism. *Trends Biochem. Sci.* 3:171–174, 1978.

Pilkis, S. J., M. R. El-Maghrabi, and T. H. Claus, Hormonal regulation of hepatic gluconeogenesis and glycolysis. *Ann. Rev. Biochem.* 57:755–784, 1988.

Roehrig, K. L., *Carbohydrate Biochemistry and Metabolism.* Westport, Ct.: Avi Publishing Co., 1984. Carbohydrate metabolism discussed at a rather elementary level. Covers several topics that are not included in this book, such as disorders of carbohydrate metabolism with brief discussions of many types of human genetic diseases in which carbohydrate metabolism is impaired.

Rousseau, G. G., and L. Hue, Mammalian 6-phosfructo-2-kinase/fructose-2,6-bisphosphatase: A bifunctional enzyme that controls glycolysis. *Progr. Nucleic Acid Res.* 45:99–127, 1993.

Srivastava, D. K., and S. A. Bernhard, Metabolite transfer via enzyme–enzyme complexes. *Science* 234:1081–1087, 1986. Unusual mode of transfer observed for intermediates in the glycolytic pathway.

Van Schaftingen, E., Fructose-2,6-bisphosphate. *Adv. Enzymol.* 59:315–395, 1987. An allosteric effector that regulates the flow between fructose-6-phosphate and fructose-1,6-bisphosphate.

Problems

1. Homemade beer is produced by combining a mixture of malted barley, hops, and yeast in a container that excludes air but allows for any pressure release via an open-ended rubber tube in a glass of water. When the bubbling from the rubber tube ceases, the upper phase of the mixture is placed in bottles, a tablespoon of syrup is added, and the bottles are capped. Why do the contents of the container need to be isolated from the environment? What are the bubbles? Why do you wait until they cease? What is the function of the syrup added prior to capping?
2. In the conversion of pyruvate to phosphoenolpyruvate in gluconeogenesis (fig. 14.16) an addition of CO_2 is followed by a decarboxylation. Why would nature add an item only to remove it in the next step? Is the carbon added the same as the one removed?
3. Glucose can be purchased with essentially any specific carbon labeled with ^{14}C. If your objective is to assess the relative importance of glycolysis versus the pentose phosphate pathway in a particular tissue, how might you use radioactive glucose to answer this question?
4. After reading about glycolysis and gluconeogenesis, do you find anything unusual about the name pyruvate kinase?
5. An intermediate in the interconversion of glycerate-3-phosphate to glycerate-2-phosphate is glycerate-2,3-bisphosphate. Where have we encountered this bisphosphate before?
6. The common yeast genus *Rhodotorula* is thought to be missing the enzyme phosphofructokinase yet seems to survive nicely. Can you propose a pathway around phosphofructokinase?
7. Why are the mechanisms of the enzymes that interconvert glucose-6-phosphate and fructose-6-phosphate and the enzyme that interconverts dihydroxyacetone phosphate and glyceraldehyde-3-phosphate virtually identical? Compare the mechanisms of phosphoglycerate mutase and phosphoglucomutase. Why are they so similar?
8. Suppose that you have isolated a facultative microorganism that you are growing anaerobically in a medium containing a carbohydrate. Explain, on the basis of your knowledge of metabolism, why each of the following statements about the fermentation is false:
   (a) The culture must be growing on glucose because bacteria ferment few other compounds.
   (b) The products of the fermentation must be more highly oxidized than the substrates; otherwise, no energy is conserved.
   (c) The culture cannot be producing any CO_2.
9. Assume that a mutant form of glyceraldehyde-3-phosphate dehydrogenase was found to hydrolyze the oxidized enzyme-bound intermediate with water rather than phosphate.
   (a) Write a chemical reaction that describes the hydrolysis, showing the structures of the products.
   (b) What would be the effect, if any, on the ATP yield from glycolysis of glucose to lactate?
   (c) What would be the effect of the mutation on an obligately aerobic microorganism?
10. Suppose that you are seeking bacterial mutants with altered triose phosphate isomerase (TPI). The organism of interest is known to use the glycolytic pathway with the production of lactate.
    (a) Explain why the absence of TPI would be lethal to an organism fermenting glucose exclusively through the glycolytic pathway.
    (b) Suppose that you have an organism that uses glycolysis and an oxidative pathway as energy sources. Mutants of that organism having only 10% of the TPI activity present in the wild-type cells grew slowly on glucose under anaerobiosis but grew faster aerobically. Explain the metabolic basis of the observation.
    (c) You have constructed a plasmid that would direct the synthesis of dihydroxyacetone phosphate phosphatase and have introduced the plasmid into the mutant organism described in part (b). Predict whether the plasmid-bearing organism with an active DHAP phosphatase could grow on glucose or glycerol either anaerobically or aerobically. What are the metabolic considerations you used to make your predictions?
11. There are two sites of ADP phosphorylation in glycolysis. These processes are called substrate-level phosphorylations. Arsenate (AsO_4^{3-}), an analog of phosphate, uncouples ATP formation resulting from glyceraldehyde-3-phosphate oxidation but not that resulting from dehydration of glycerate-2-phosphate. Explain.
12. The disaccharide sucrose can be cleaved by either of two methods:

$$\text{Sucrose} + H_2O \xrightarrow{\text{Invertase}} \text{Glucose} + \text{Fructose}$$

or

$$\text{Sucrose} + P \xrightarrow[\text{}]{\text{Sucrose phosphorylase}} \text{Glucose-1-phosphate} + \text{Fructose}$$

    (a) Given that the $\Delta G^{\circ\prime}$ value for the invertase reaction is -7.0 kcal/mole, calculate the $\Delta G^{\circ\prime}$ value for sucrose phosphorylase. Assume that the $G^{\circ\prime}$ for hydrolysis of glucose-1-phosphate is -5.0 kcal/mole. From the calculated value of $\Delta G^{\circ\prime}$, calculate the equilibrium constant for sucrose phosphorylase at 25°C.
    (b) Explain the metabolic advantage to the cell of cleaving sucrose with phosphorylase rather than with invertase.
13. 2-Phosphoglycerate and phosphoenolpyruvate differ only by dehydration between C-2 and C-3, yet the difference in $\Delta G^{\circ\prime}$ of hydrolysis is about -12 kcal/mole. How does dehydration "trap" so much chemical energy?
14. Beginning with pyruvate, show which reactions of gluconeogenesis introduce the four chiral centers into glucose.
15. In gluconeogenesis the thermodynamic barrier imposed by pyruvate kinase is overcome by coupling two separate reactions for the synthesis of PEP from pyruvate.
    (a) Write the two chemical reactions used to bypass the pyruvate kinase reaction.
    (b) Calculate the overall $\Delta G^{\circ\prime}$ of the two reactions you wrote in part (a). (Assume that GTP is the thermodynamic equivalent of ATP.) What can you now surmise about the feasibility of PEP formation from pyruvate by this route?
16. Write a chemical reaction for the $NADP^+$-dependent oxidation of 6-phosphogluconate to ribulose-5-phosphate.

17. Fructose-2,6-bisphosphate is a potent activator of the liver phosphofructokinase (PFK-1) and a potent inhibitor of liver fructose-1,6-bisphosphate phosphatase (FBPase-1). Fructose-2,6-bisphosphate is the product of a second phosphofructokinse (PFK-2) and is hydrolyzed to fructose-6-phosphate of FBPase-2. The activities of PFK-2 and FBPase-2 reside on a single, bifunctional protein in liver. The bifunctional protein is under glucagon control imposed via cAMP. (Reference: H.-G. Hers, *Arch. Biol. Med. Exp.* 18:243–251, 1985.)
    (a) Under what metabolic conditions would PFK-2 be active? FBPase-2?
    (b) Gluconeogenesis in liver is stimulated by the hormone glucagon. The activity of PFK-2/FBPase-2 bifunctional enzyme under glucagon regulation shifts from an active PFK-2 to an inactive PFK-2. Inactivation of the PFK-2 alone would still not be adequate to stimulate gluconeogenesis sufficiently for the organism. Explain.
    (c) cAMP-dependent phosphorylation of PFK-2/FBPase-2 not only inhibits PFK-2 but also stimulates FBPase-2. Under these conditions, gluconeogenesis is sufficiently rapid to meet cellular demand. Explain.
    (d) What would you predict as the relative activities of the following enzymes in the liver of a rat made diabetic through chemical means (administration of alloxan or streptozotocin)? PFK-2, FBPase-2, PFK-1, FBPase-1, pyruvate carboxylase, PEP carboxykinase.
    (e) If the diabetic animal were treated with insulin, what changes would you predict in the activities of the liver enzymes cited in part (d)? In the concentration of cAMP?

C H A P T E R

The Tricarboxylic Acid Cycle

The tricarboxylic acid (TCA) cycle catabolizes pyruvate to CO_2 and H_2O in an oxygen-consuming process.

Only a small fraction of the total free energy content of glucose is released under anaerobic conditions. This is because no net oxidation of organic substrates can occur in the absence of oxygen. Catabolism under anaerobic conditions means that every oxidative event in which electrons are removed from an organic compound must be accompanied by a reductive event in which electrons are returned to another organic compound, often closely related to the first compound. The cell operating under anaerobic conditions must content itself with the generation of only two ATP molecules per molecule of glucose fermented. Most of the energy of the glucose molecule remains untapped. Given access to oxygen, however, the cell can do much more with the oxidizable organic molecules available to it, and the energy yield increases dramatically. In this chapter the lower half of the central metabolic pathways requiring oxygen is explored (fig. 15.1). With oxygen available as the electron acceptor, the carbon atoms of glucose (or another substrate) can be oxidized fully to CO_2, and all the electrons that are removed during the multiple oxidation events are transferred ultimately to oxygen. In the process, the ATP yield per glucose is close to 15 times greater than that possible under anaerobic conditions. Therein lies the advantage of the aerobic way of life.

It is not surprising then, that aerobic processes capable of extracting further energy from pyruvate have come to play so prominent a role in energy metabolism. The overall process of aerobic respiratory metabolism and its distinguishing characteristics are (1) the use of oxygen as the ultimate electron acceptor, (2) the complete oxidation of organic substrates to CO_2 and water, and (3) the conservation of much of the free energy as ATP.

The oxygen required for aerobic metabolism actually serves as the terminal electron acceptor only, providing for the continuous reoxidation of reduced coenzyme molecules (the most prominent of which are NADH and $FADH_2$). It is these coenzyme

Figure 15.1

Outline of the main reactions of carbohydrate metabolism considered in this chapter. Most of these reactions (in black) belong to the tricarboxylic acid cycle, which provides a means for catabolizing two carbon units all the way to CO_2. This process, which requires oxygen, yields considerably more energy and reducing power than simple glycolysis. Two of these reactions are part of the glyoxylate bypass, a means of synthesizing four-carbon and six-carbon units from the two-carbon level of acetyl-CoA. The reactions of the glyoxylate bypass are found in plants, fungi, and microorganisms but not in vertebrates. Citrate formed in the mitochondrion can be transferred to the cytosol, where it leads to the formation of acetyl-CoA and NADPH (not shown) used in biosynthesis.

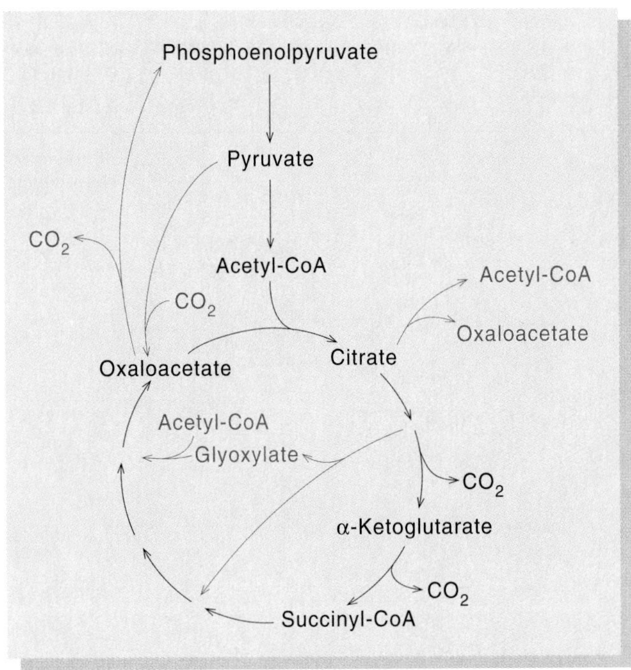

molecules (in their oxidized form) that serve as electron carriers for the stepwise oxidation of organic intermediates derived from pyruvate. Aerobic respiratory metabolism can therefore be thought of in terms of two separate but intimately linked processes: the actual oxidative metabolism, in which electrons are removed from organic substrates and transferred to coenzyme carriers, and the concomitant reoxidation of the reduced coenzymes by transfer of electrons to oxygen, accompanied indirectly by the generation of ATP.

Under aerobic conditions the glycolytic pathway becomes the initial phase of glucose catabolism (fig. 15.2). The other three components of respiratory metabolism are the tricarboxylic acid (TCA) cycle, which is responsible for further oxidation of pyruvate, the electron-transport chain, which is required for the reoxidation of coenzyme molecules at the expense of molecular oxygen, and the oxidative phosphorylation of ADP to ATP, which is driven by a proton gradient generated in the process of electron transport. Overall, this leads to the potential formation of approximately 30 molecules of ATP per molecule of glucose in the typical eukaryotic cell.

In this chapter, discussion focuses on the TCA cycle and its central role in the aerobic catabolism of carbohydrates. Chapter 16 explains how the free energy present in the reduced coenzymes that are generated by glycolysis and the TCA cycle is conserved as ATP during the companion process of electron transport and oxidative phosphorylation.

Discovery of the TCA Cycle

The discovery of the TCA cycle began with a series of biochemical experiments performed in the early 1900s on anaerobic suspensions of minced animal tissues. The experiments established that the suspensions contained enzymes that could transfer hydrogen atoms from various low-molecular-weight organic acids to other reducible compounds, such as the dye methylene blue. (Methylene blue was a convenient indicator in these experiments because it is converted from a blue to a colorless form by reduction.) Only a few organic acids were active in the reduction: succinate, fumarate, malate, and citrate. It was later observed that in the presence of oxygen the same suspensions oxidized these acids to CO_2 and water.

Albert Szent-Györgyi found that when small amounts of these organic acids were added to a tissue suspension in the presence of glucose, considerably more oxygen was consumed than was required to oxidize the added acid. He concluded that in the presence of oxygen the organic acids had a catalytic effect on the oxidation of glucose or other carbohydrates in the tissue slices. Szent-Györgyi also observed that a specific inhibitor of succinate dehydrogenase, malonate, blocked the utilization of oxygen (respiration) by muscle suspensions. This finding suggested that oxidation of succinate is an indispensable reaction in muscle oxidative metabolism. When the same response was found in many other tissues, the phenomenon seemed to have fundamental and perhaps universal significance.

Building on these results, Hans Krebs studied the interrelationships between the oxidative metabolism of different organic acids. For his experiments, slices of pigeon flight muscle, which are particularly active in oxidative metabolism, were used. He found a select group of organic acids that was oxidized very rapidly by extracts from the muscle tissue. His list included the organic acids that Szent-Györgyi had found plus oxaloacetate, α-ketoglutarate, isocitrate, and *cis*-aconitate. Like Szent-Györgyi, Krebs found that catalytic amounts of the same organic acids stimulated the oxidation of pyruvate or carbohydrate and that their catalytic effect was blocked by malonate. Because malonate specifically inhibited the conversion of succinate to fumarate by succinate dehydrogenase, he concluded that this reaction was one of a series of steps essential for the complete oxidation of pyruvate. Additional steps in the process were presumed to involve the other organic acids that had been found to catalyze the process. The organic acids with catalytic potential could be arranged in a chain related by biochemical conversions:

$$\left(\begin{array}{l}\text{Citrate} \longrightarrow \textit{cis}\text{-Aconitate} \longrightarrow \text{Isocitrate} \longrightarrow \alpha\text{-Ketoglutarate} \longrightarrow \\ \quad \text{Succinate} \longrightarrow \text{Fumarate} \longrightarrow \text{Malate} \longrightarrow \text{Oxaloacetate}\end{array}\right) \quad \textbf{(1)}$$

But if this chain of reactions were to act catalytically, then some way must exist for regenerating all of the intermediates of the chain from a single intermediate. A cyclical process would be the

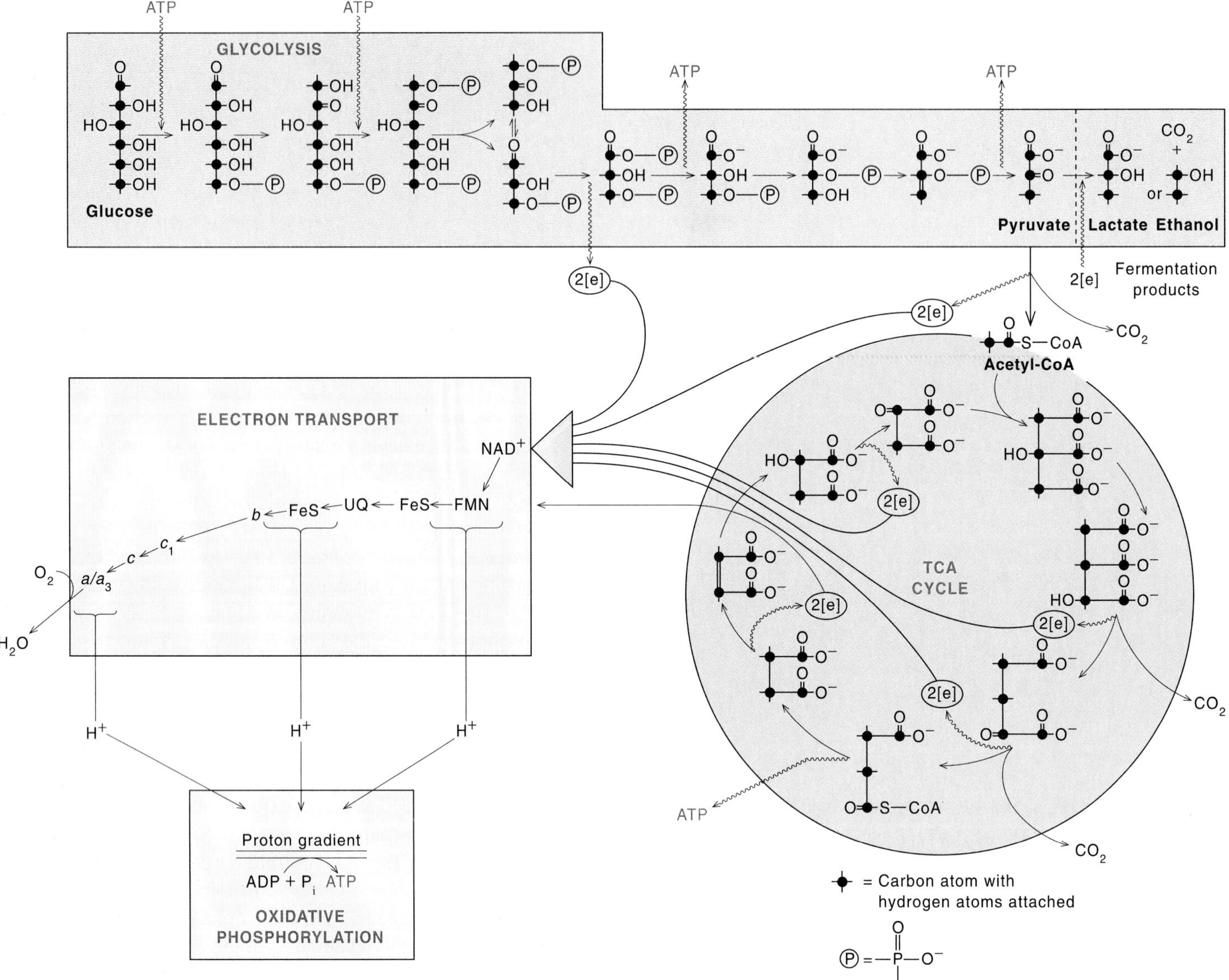

Figure 15.2

The components of respiratory metabolism include glycolysis, the tricarboxylic acid (TCA) cycle, the electron-transport chain, and the oxidative phosphorylation of ADP to ATP. Glycolysis converts glucose to pyruvate; the TCA cycle fully oxidizes the pyruvate (by means of acetyl-CoA) to CO_2 by transferring electrons stepwise to coenzymes; the electron-transport chain reoxidizes the coenzymes at the expense of molecular oxygen; and the energy of coenzyme oxidation is conserved in the form of a transmembrane proton gradient that is then used to generate ATP.

simplest way of doing this. For the chain to operate as a cycle, one or more additional reactions needed to be discovered. Krebs found that in the absence of oxygen, small amounts of citrate could be formed from added oxaloacetate and pyruvate. Could this reaction be the missing link in the cyclical process? Krebs thought so and proposed that the first step in pyruvate oxidation involved its condensation with oxaloacetate to form citrate and CO_2. The oxidation of more complex carbohydrates then could be explained by their prior conversion to pyruvate through the glycolytic pathway.

Further support for the cyclical nature of the chain came from observations of malonate-inhibited muscle slices. In such preparations the addition of any of the intermediates of the chain led to accumulation of succinate. Even the addition of fumarate, the immediate product of succinate oxidation, led to accumulation of succinate.

When the oxidation of substantial amounts of pyruvate was blocked by malonate, stoichiometric amounts of pyruvate would react if either oxaloacetate or its precursors, malate or fumarate, was added. None of the precursors to succinate in the chain

Figure 15.3

Original tricarboxylic acid (TCA) cycle proposed by Krebs. This cycle is also called the citrate cycle or the Krebs cycle. To start the cycle in operation, pyruvate loses one of its carbons and condenses with a four-carbon dicarboxylic acid, oxaloacetic acid, to form a six-carbon tricarboxylic acid, citrate. In one turning of the cycle, two carbons are lost as CO_2, thus returning the citrate to oxaloacetate. The conversion blocked by malonate is indicated by a red bar.

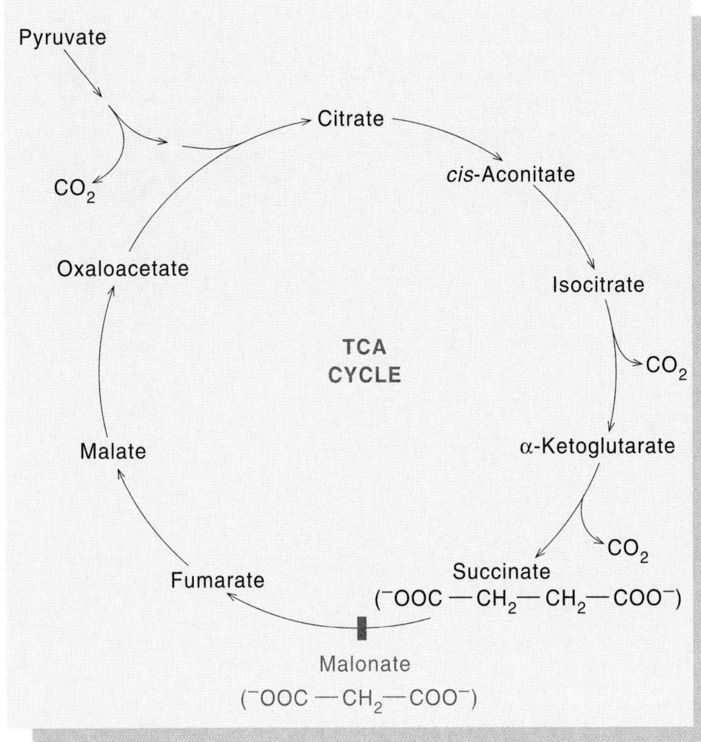

was effective in this regard. This strongly supported Krebs's hypothesis that oxidation of pyruvate involved its condensation with oxaloacetate and the subsequent series of conversions in the chain described by Krebs. The Krebs cycle in its originally proposed form is shown in figure 15.3. It is also known as the tricarboxylic acid (TCA) cycle or the citrate cycle.

Steps in the TCA Cycle

We now know that the tricarboxylic acid cycle (fig. 15.4) begins with acetyl-coenzyme A, which results from the oxidative decarboxylation of pyruvate (acetyl-CoA) or the oxidative cleavage of fatty acids (see chapter 21). Regardless of its source, acetyl-CoA transfers its acetyl group to oxaloacetate, thereby generating citrate. In a cyclic series of reactions the citrate is subjected to two successive decarboxylations and several oxidative events, leaving a four-carbon compound from which the starting oxaloacetate is eventually regenerated. Each turn of the cycle involves the entry of two carbons from acetyl-CoA and the release of two carbons as CO_2. As a result, the cycle is balanced with respect to carbon flow and functions without net consumption or buildup of oxaloacetate (or any other intermediate) unless side reactions occur that either feed carbon into the cycle or drain it off into alternative pathways.

The Oxidative Decarboxylation of Pyruvate Leads to Acetyl-CoA

Although the acetyl-CoA with which the cycle begins may be derived catabolically from fatty acids or amino acids, the major source of acetyl-CoA in most cells is the pyruvate available from the glycolytic breakdown of carbohydrate. The gap from the glycolytic pathway to the TCA cycle is bridged by the oxidative decarboxylation of pyruvate (with which glycolysis ends) to yield acetate in the form of acetyl-CoA (with which the TCA cycle commences). The decarboxylation of pyruvate is catalyzed by a cluster of three enzymes called the pyruvate dehydrogenase complex. In eukaryotic cells this enzyme complex is located in the mitochondria, as are all the other reactions of aerobic energy metabolism beyond pyruvate. Since the glycolytic pathway occurs in the cytosol, it is as pyruvate that the carbon derived from glucose (or other carbohydrate substrates) enters the mitochondria.

Decarboxylation of an α-keto acid like pyruvate is a difficult reaction for the same reason as are the ketol condensations (see fig. 14.24): Both kinds of reactions require the participation of an intermediate in which the carbonyl carbon carries a negative charge. In all such reactions that occur in metabolism, the intermediate is stabilized by prior condensation of the carbonyl group with thiamine pyrophosphate. In figure 15.5, thiamine pyrophosphate and its hydroxyethyl derivative are written in the doubly ionized ylid form rather than the neutral form because this is the form that actually participates in the reaction, even though it is present in much smaller amounts.

In the first step of the conversion catalyzed by pyruvate decarboxylase, a carbon atom from thiamine pyrophosphate adds to the carbonyl carbon of pyruvate. Decarboxylation produces the key reactive intermediate, hydroxyethyl thiamine pyrophosphate (HETPP). As shown in figure 15.5, the ionized ylid form of HETPP is resonance-stabilized by the existence of a form without charge separation. The next enzyme, dihydrolipoyltransacetylase, catalyzes the transfer of the two-carbon moiety to lipoic acid. A nucleophilic attack by HETPP on the sulfur atom attached to carbon 8 of oxidized lipoic acid displaces the electrons of the disulfide bond to the sulfur atom attached to carbon 6. The sulfur then picks up a proton from the environment as shown in figure 15.5. This simple displacement reaction is also an oxidation–reduction reaction, in which the attacking carbon atom is oxidized from the aldehyde level in HETPP to the carboxyl level in the lipoic acid derivative. The oxidized (disulfide) form of lipoic acid is converted to the reduced (mercapto) form. The fact that the two-carbon moiety has become an acyl group is shown more clearly after dissociation of thiamine pyrophosphate (TPP), which generates acetyl lipoic acid.

Further transfer of the acyl group to coenzyme A is catalyzed by the same enzyme. This displacement reaction produces reduced lipoic acid. A third enzyme, dihydrolipoyl dehydrogenase,

Figure 15.4

The TCA cycle. In each turn of the cycle, acetyl-CoA from the glycolytic pathway or from β oxidation of fatty acids enters and two fully oxidized carbon atoms leave (as CO_2). ATP is generated at one point in the cycle, and coenzyme molecules are reduced. The two CO_2 molecules lost in each cycle originate from the oxaloacetate of the previous cycle rather than from incoming acetyl from acetyl-CoA. This point is emphasized by the use of color.

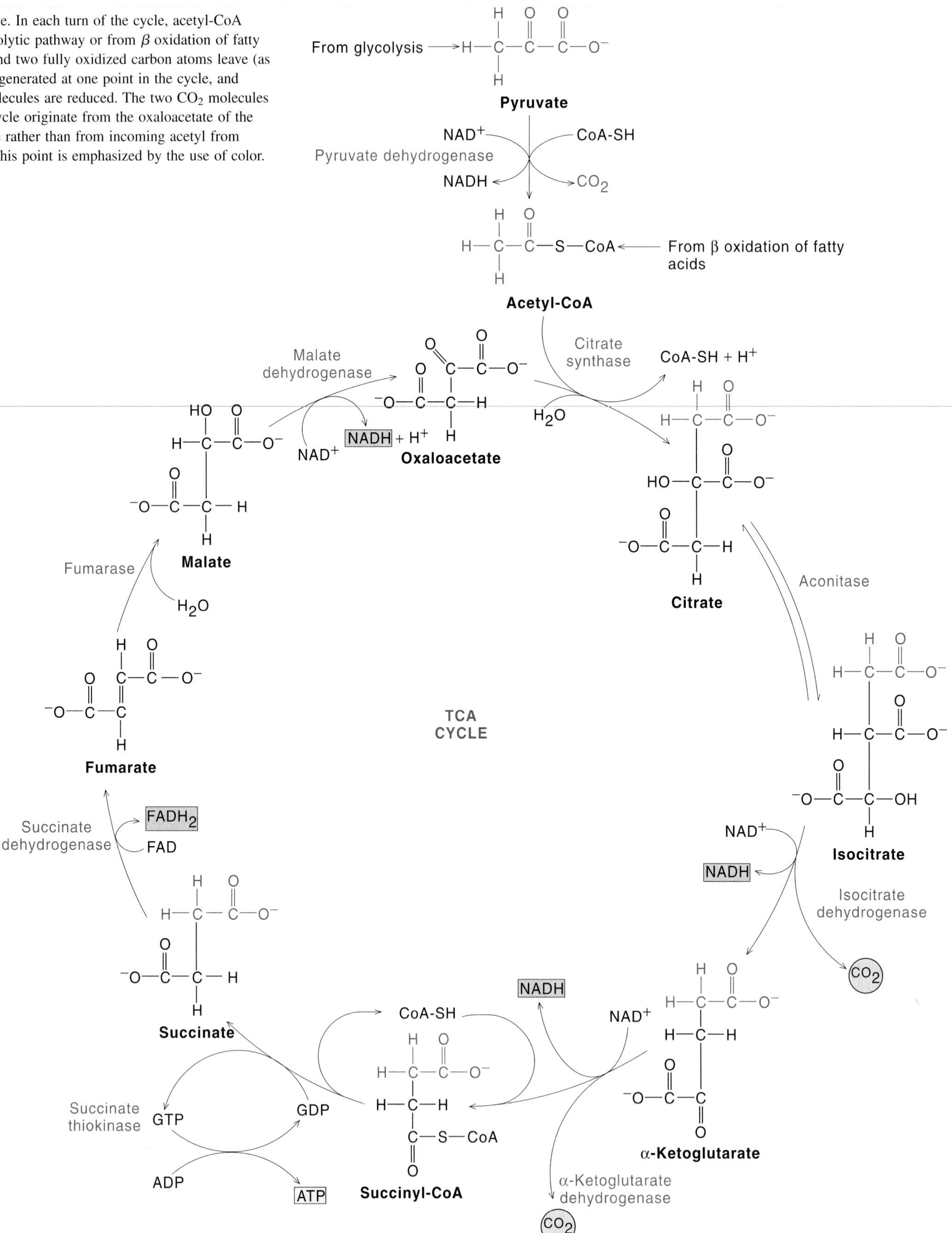

Figure 15.5

The conversion of pyruvate to acetyl-CoA. The reactions are catalyzed by the enzymes of the pyruvate dehydrogenase complex. This complex has three enzymes: pyruvate decarboxylase, dihydrolipoyl transacetylase, and dihydrolipoyl dehydrogenase. In addition, five coenzymes are required: thiamine pyrophosphate, lipoic acid, CoASH, FAD, and NAD^+. For further information of these five coenzymes, the reader should refer back to chapter 11. Lipoic acid is covalently attached to the transacetylase component of the complex by an amide bond between the carboxyl group of lipoic acid and the terminal amino group of a lysine residue of the enzyme. In fact, one or more lipoic acids are involved in each transfer reaction (see fig. 15.6). Reactants and products are shown in red.

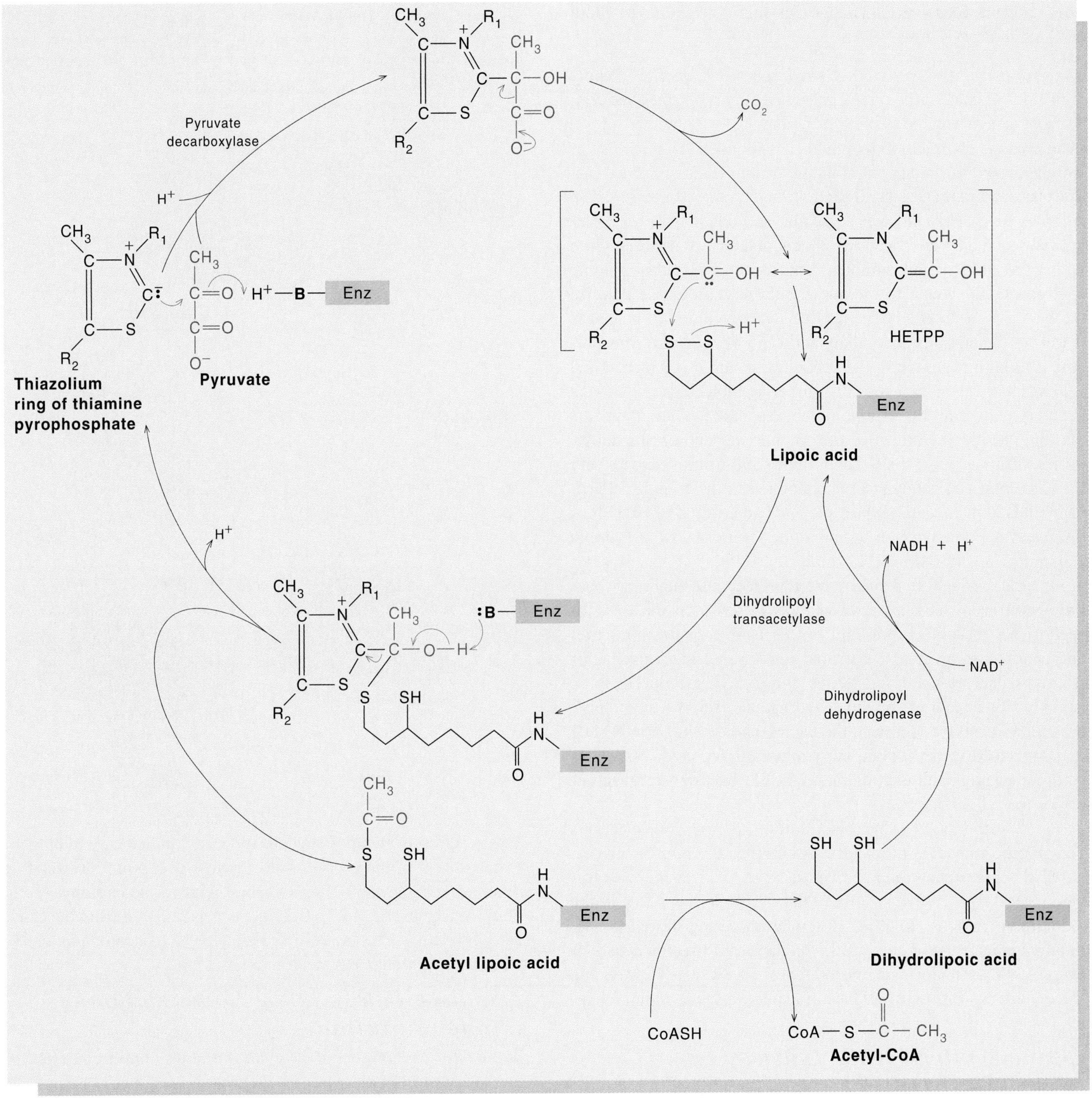

Pyruvate decarboxylase

CO_2

H^+

H^+—B— Enz

Thiazolium ring of thiamine pyrophosphate

Pyruvate

HETPP

Lipoic acid

H^+

NADH + H^+

:B— Enz

Dihydrolipoyl transacetylase

NAD^+

Dihydrolipoyl dehydrogenase

Acetyl lipoic acid

Dihydrolipoic acid

CoASH

CoA—S—C(=O)—CH_3

Acetyl-CoA

catalyzes oxidation of this product back to the disulfide form. The electrons lost in that oxidation are transferred first to an enzyme-bound flavin (not shown in the figure) and then to NAD^+.

The overall equation for the conversion catalyzed by the pyruvate dehydrogenase complex is

$$\text{Pyruvate} + NAD^+ + \text{CoA} \longrightarrow \text{Acetyl-CoA} + \text{NADH} + CO_2 \quad (2)$$

The standard free energy change for this conversion is about −8 kcal/mole.

The Nature of the Pyruvate Dehydrogenase Complex Many molecules of each of the three enzymes that participate in the conversion of pyruvate to acetyl-coenzyme A are organized into a giant enzyme complex. In mammals the complex has a molecular weight of about 9×10^6; it contains 60 molecules of the transacetylase and perhaps 20–30 each of the other two enzymes. The complex in *Escherichia coli* is smaller and contains fewer molecules of each enzyme (fig. 15.6). Each molecule of transacetylase within the multienzyme complex contains two molecules of lipoic acid covalently bound to the enzyme through an amide bond to the ϵ-amino group of a lysine residue. The disulfide of lipoic acid is thus at the end of a long chain and can sweep over a considerable area on the surface of the subunit to which it is attached and also reach the catalytic sites of neighboring molecules of pyruvate decarboxylase and dihydrolipoyl dehydrogenase. Thus it appears that the catalytic activities of the complex depend on the ability of the sulfur atoms of the tethered lipoic acid to visit successively the three types of catalytic sites contained in the complex. Pyruvate binds at the decarboxylase site and is converted to HETPP as discussed previously; then the disulfide group of oxidized lipoic acid picks up the two-carbon moiety, oxidizing it to an acetyl group, and carries it to a site on the transacetylase subunit, where the acetyl group is transferred to coenzyme A. In the form of acetyl-CoA the two-carbon unit is free from the enzymic tether and can diffuse away as the primary product of the sequence of reactions that are catalyzed by the complex. The reduced lipoic acid is oxidized at a site on the dehydrogenase and is ready to pick up another acetyl group from the decarboxylase site. The NADH that is formed in the oxidation of dihydrolipoic acid dissociates from the enzyme and is available to the electron-transfer machinery of the mitochondrion.

At this point in the oxidation of glucose, four electrons per glucose molecule have been lost in the oxidation of glyceraldehyde-3-phosphate and four more in the conversion of pyruvate to acetyl-CoA. Thus, of the total of 24 electrons lost in the oxidation of glucose to CO_2, 16 remain to be transferred to oxidizing agents in the course of the oxidation of two molecules of acetyl-CoA. A major function of the TCA cycle is to mobilize these electrons for use in electron-transfer phosphorylation (chapter 16).

Citrate Synthase Is the Gateway to the TCA Cycle

The final step leading to the TCA cycle is catalyzed by citrate synthase, in which acetyl-CoA enters the cycle and citrate is formed:

$$\text{Acetyl-CoA} + \text{Oxaloacetate} \longrightarrow \text{Citrate} + \text{CoA} \quad (3)$$

Figure 15.6

Interactions of α-lipoyl groups in the pyruvate dehydrogenase complex of *E. coli*. The cubic structure represents the 24 subunits of dihydrolipoyl transacetylase, which constitutes the core of the complex. Two of the 48 lipoyl groups in the core are shown interacting with one of the 24 pyruvate decarboxylases ($E_1 \cdot$ TPP) and one of the 12 dihydrolipoyl dehydrogenase ($E_3 \cdot$ FAD) subunits. Note the interaction of the lipoyl groups in relaying electrons over the long distance between TPP and FAD. The lipoic acid must interact at active sites over distances too long to be covered by a single fully extended coenzyme. This problem is solved by use of a shuttle system involving two α-lipoyl groups for each transfer.

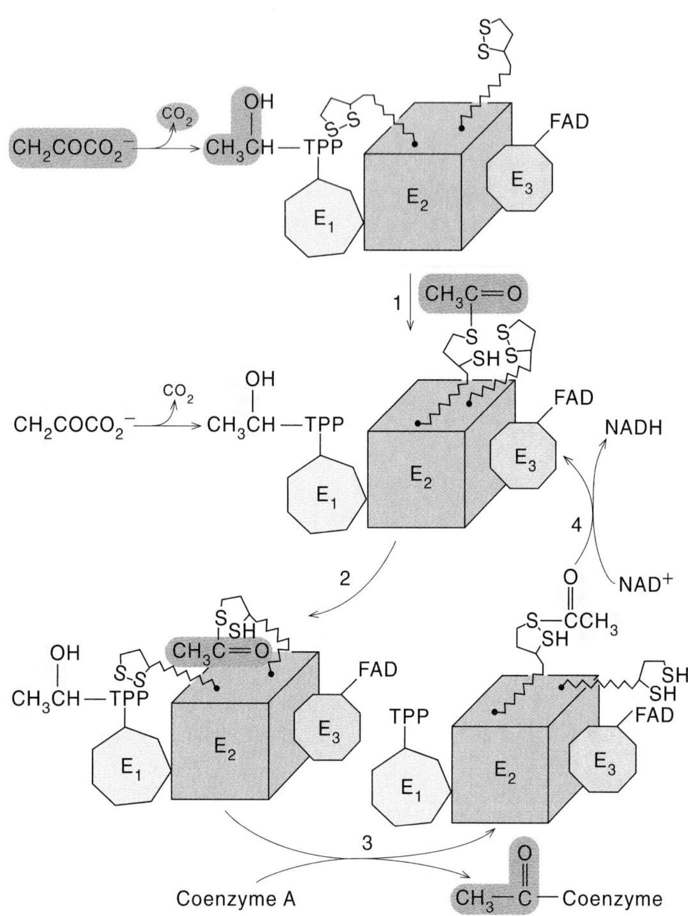

This reaction is an aldol condensation, in which a carbanion generated at C-2 of the acetyl group (by loss of a proton to water or to an acceptor group on the enzyme) adds to the carbonyl group of oxaloacetate (fig. 15.7). Coenzyme A is released from the product while it is still bound to the enzyme, so the products of the reaction are free coenzyme A and citrate.

Aconitase Catalyzes the Isomerization of Citrate to Isocitrate

In the first reaction within the cycle, citrate is converted to its isomer isocitrate (fig. 15.8):

$$\text{Citrate} \rightleftharpoons \textit{cis}\text{-Aconitate} \rightleftharpoons \text{Isocitrate} \quad (4)$$

Aconitase, the enzyme that catalyzes this isomerization, is named for the fact that the unsaturated compound formed by removing

Figure 15.7

Formation of citrate, catalyzed by citrate synthase. A carbanion of acetyl-CoA, generated by loss of a proton to water (or to an acceptor group on the enzyme), adds to the carbonyl group of oxaloacetate. The immediate product of the condensation is probably citryl-coenzyme A, but the products that dissociate from the catalytic site of the enzyme are citrate and free coenzyme A.

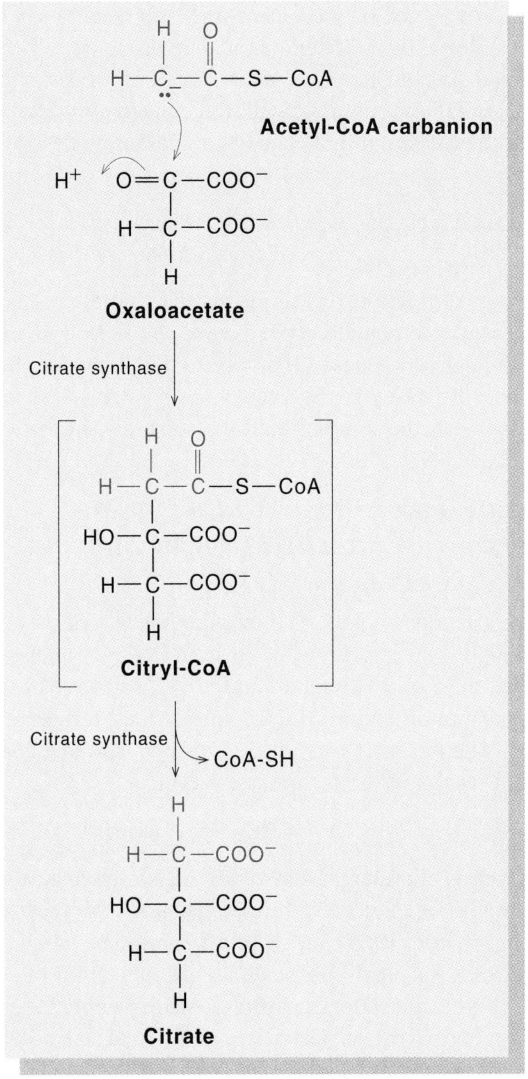

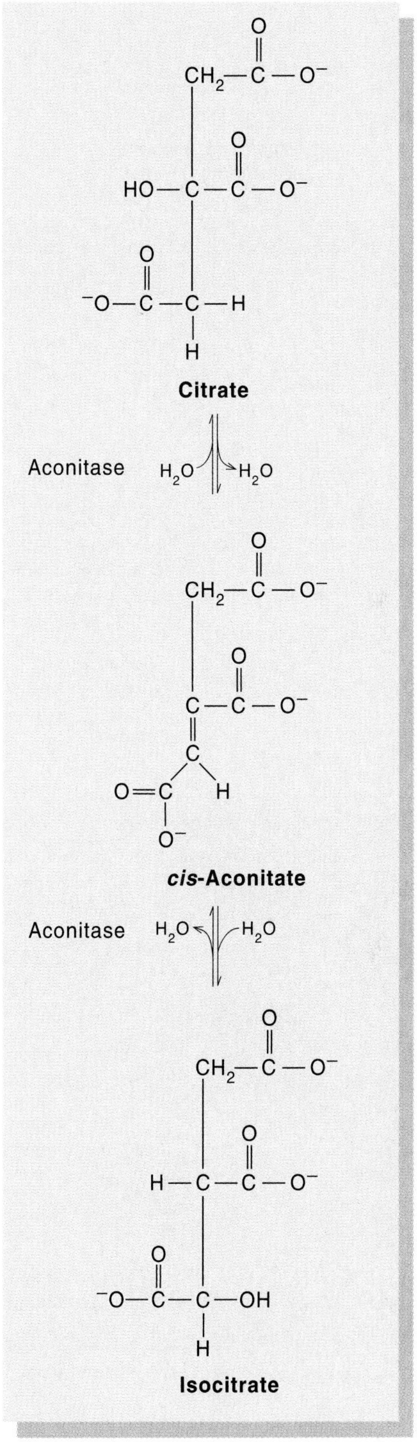

H_2O from either citrate or isocitrate, *cis*-aconitate, can also serve as substrate or product. Aconitase catalyzes the attainment of equilibrium between citrate, isocitrate, and *cis*-aconitate. The three compounds may be considered as belonging to the same metabolic pool. The mechanism for aconitase catalysis is described in detail in box 11A.

Isocitrate Dehydrogenase Catalyzes the First Oxidation in the TCA Cycle

The first oxidative conversion within the TCA cycle is catalyzed by isocitrate dehydrogenase. This conversion takes place in two steps: oxidation of the secondary alcohol isocitrate to a ketone (oxalosuccinate), followed by a β decarboxylation to produce

Figure 15.8

Interconversion of citrate, *cis*-aconitate, and isocitrate, catalyzed by aconitase. At equilibrium the relative concentrations of these compounds are about 90, 6, and 4, respectively. The mechanism of action of aconitase is described in box 11A.

α-ketoglutarate (fig. 15.9):

$$\text{Isocitrate} + \text{NAD}^+ \longrightarrow \alpha\text{-Ketoglutarate} + \text{NADH} + \text{CO}_2 \qquad (5)$$

NAD^+ is the electron acceptor for the oxidative step, and Mg^{2+}

Figure 15.9

The oxidative decarboxylation of isocitrate to α-ketoglutarate, catalyzed by mitochondrial isocitrate dehydrogenase. The intermediate, oxalosuccinate, is not released from the enzyme. :**B** represents a catalytic side chain from the enzyme.

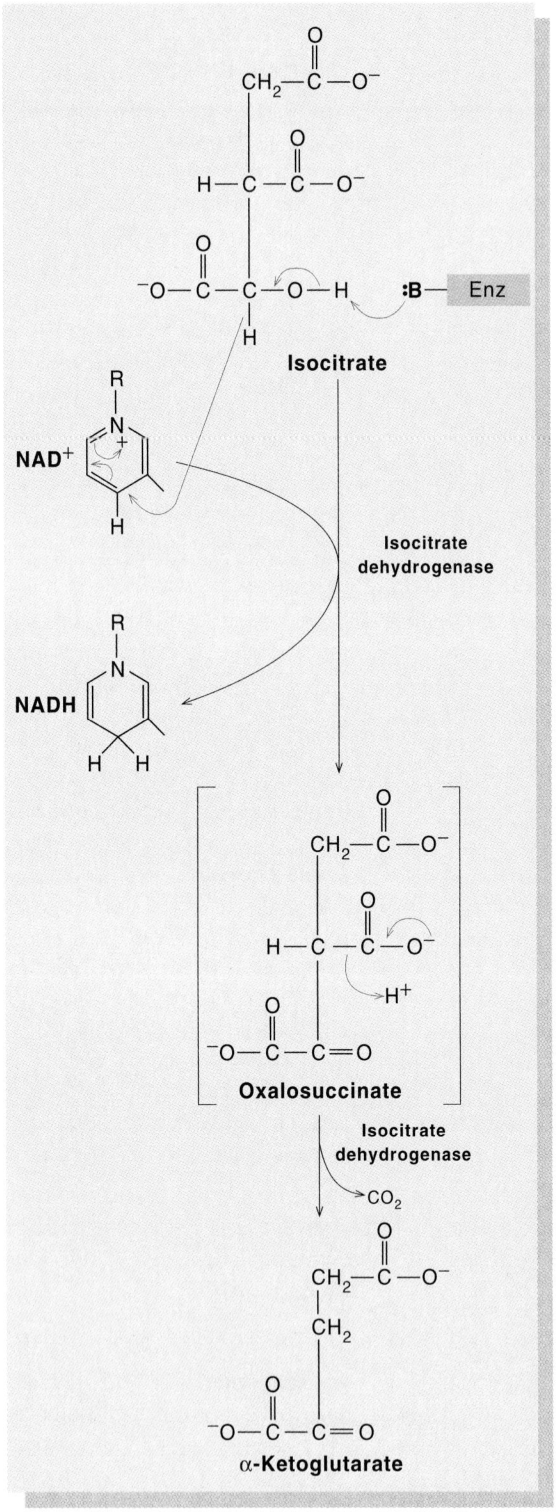

Isocitrate

NAD$^+$

NADH

Isocitrate dehydrogenase

Oxalosuccinate

Isocitrate dehydrogenase

CO_2

α-Ketoglutarate

or Mn^{2+} is required for the decarboxylation. The oxalosuccinate that is presumably an intermediate does not dissociate from the enzyme.

α-Ketoglutarate Dehydrogenase Catalyzes the Decarboxylation of α-Ketoglutarate to Succinyl-CoA

The second of two oxidative decarboxylation reactions of the TCA cycle is catalyzed by α-ketoglutarate dehydrogenase. Steps in this reaction run parallel to those catalyzed by pyruvate dehydrogenase (see fig. 15.5), complete with the conversion of some of the energy of the oxidation in a coenzyme-containing derivative, succinyl-CoA:

$$\alpha\text{-Ketoglutarate} + NAD^+ + \text{CoA} \longrightarrow \text{Succinyl-CoA} + NADH + CO_2 \quad \textbf{(6)}$$

The enzyme complex involved in this reaction also is very similar to the pyruvate dehydrogenase complex. Indeed, the same dihydrolipoyl dehydrogenase subunit is used in both complexes. The product in both cases is the coenzyme A ester of the acid containing one less carbon atom than the substrate, in this case succinyl-CoA (see fig. 15.4).

Succinate Thiokinase Couples the Conversion of Succinyl-CoA to Succinate with the Synthesis of GTP

The thioester bond makes succinyl-CoA an activated intermediate. Although some succinyl-CoA is used in the synthesis of heme in animals, most of it is retained in the TCA cycle, where it leads to the regeneration of the oxaloacetate needed to keep the cycle operating. The thioester is converted to succinate in a coupled reaction that results in the formation of GTP:

$$P_i + \text{Succinyl-CoA} + \text{GDP} \longrightarrow \text{Succinate} + \text{GTP} \quad \textbf{(7)}$$

The reaction is complex and involves an intermediate in which a phosphate is attached to a histidine residue of the succinate thiokinase enzyme. Probably, CoA is first displaced by inorganic phosphate, forming succinyl phosphate. A nitrogen atom of a specific histidine residue then attacks phosphorus, displacing succinate and forming an *N*-phosphoryl derivative. In the final step, GDP attacks the phosphorus atom of that derivative, forming GTP. The role of GTP in this reaction is played by ATP in some organisms.

Succinate Dehydrogenase Catalyzes the Oxidation of Succinate to Fumarate

In terms of carbon atoms we might say that because two carbons entered the cycle as the acetyl group and two left as CO_2, the functional part of the cycle is finished at this stage, and thus the four carbons of succinate must merely be converted to oxaloacetate to serve again as an acetyl acceptor. However, that assumption would be erroneous. When the TCA cycle serves as a means of oxidizing acetyl groups, that is, when it is not producing biosynthetic starting materials at a significant rate, its function is to supply electrons to the electron-transfer phosphorylation system. When we reach succinate, four electrons have been transferred to NAD$^+$. Thus, of the eight electrons that are removed in the oxidation of acetate to CO_2, four remain in succinate. In terms of its oxidative function the cycle is only half finished at this point.

The next step in the TCA cycle, the oxidation of succinate to fumarate, involves insertion of a double bond into a saturated hydrocarbon chain:

$$\text{Succinate} + \text{FAD} \longrightarrow \text{Fumarate} + \text{FADH}_2 \quad \textbf{(8)}$$

This is not an easy reaction in organic chemistry. It is, however, a very important type of reaction in metabolic chemistry and is an integral step in the oxidation of carbohydrates, fats, and several amino acids.

Because of the nature of the reaction, a strong oxidizing agent is required. The electron acceptor that is used in most oxidative steps of catabolism, NAD^+, is not a strong enough oxidizing agent to give a reasonable equilibrium constant. Flavoproteins are stronger oxidizing agents than NAD^+, and succinate dehydrogenase, which catalyzes the oxidation of succinate to fumarate, is a flavoprotein enzyme. The oxidation of succinate by the electron acceptor of succinate dehydrogenase is about 16 kcal (67 kJ) more favorable than it would be if the electron acceptor were NAD^+. This makes the equilibrium constant more favorable by a factor of about 10^{11}. In this example we see how important the choice of cofactors can be. In general, FAD is a better oxidizing agent than NAD^+, and NADH is a better reducing agent than $FADH_2$.

Fumarase Catalyzes the Addition of Water to Fumarate to Form Malate

Fumarate is converted to L-malate by stereospecific addition of water across the double bond:

$$\text{Fumarate} + \text{H}_2\text{O} \longrightarrow \text{L-Malate} \quad \textbf{(9)}$$

Malate Dehydrogenase Catalyzes the Oxidation of Malate to Oxaloacetate

The final oxidation step of the cycle involves the conversion of malate to oxaloacetate by malate dehydrogenase:

$$\text{L-Malate} + \text{NAD}^+ \longrightarrow \text{Oxaloacetate} + \text{NADH} + \text{H}^+ \quad \textbf{(10)}$$

This enzyme uses NAD^+ as the oxidizing agent.

Stereochemical Aspects of TCA Cycle Reactions

Some early applications of isotopic tracer techniques to the TCA cycle led to a new generalization concerning the stereochemistry of interaction between enzymes and certain types of substrates.

In the early 1940s, before the discovery of ^{14}C and when acetyl-coenzyme A was unknown, two research groups used the stable isotope ^{13}C and mass spectrometers to study carbon flow in the TCA cycle. Pyruvate and $^{13}CO_2$ were added to pigeon liver preparations to form carboxy-labeled oxaloacetate. Malonate was added to stop the TCA cycle at succinate. The expected result was that half of the ^{13}C would be found in succinate and the other half in CO_2 (fig. 15.10).

To the surprise of the biochemical community it was found that although pyruvate was consumed and succinate was produced, there was no ^{13}C in the succinate. Further experiments showed that the isotope had been incorporated but was all lost in the oxidative decarboxylation of α-ketoglutarate to succinate. Since the citrate molecule has a plane of symmetry, it was taken for granted that the two $—CH_2—COO^-$ arms of the molecule must be chemically equivalent and that the —OH group would have an equal chance of migrating into either arm in the aconitase reaction. In that case the label also would have an equal chance of being lost or retained in the decarboxylation of α-ketoglutarate. For a few years this experimental result was thought to prove that citrate could not be an intermediate of the cycle and that some derivative of citrate must be the immediate product of the condensation of oxaloacetate with an unknown active two-carbon intermediate.

Then, in 1948, Ogston suggested that citrate was not necessarily excluded by the isotopic evidence, because the two $—CH_2—COO^-$ arms might actually not be equivalent when citrate was the substrate for an enzymic reaction. He pointed out that if the substrate were attached to the enzyme at three points, its orientation would be fixed by those attachments, and it would be impossible for the two identical arms to exchange positions. Thus only one of them occupies the position that allowed it to participate in the reaction (fig. 15.11).

Ogston's concept is a valid and important generalization. Three constraints are necessary to fix an object in three-dimensional space, but they need not all be points of attachment. In principle the two *a* groups of a molecule of the type Ca_2bd could be distinguished by an enzyme if the three constraints were one point of attachment, one pocket, into which *b* could fit but *d* could not, and the position of the reactive groups of the catalytic site.

Ogston's contribution led to an interesting extension of concepts concerning stereochemistry of enzyme action. Compounds of the type Ca_2bd are termed prochiral, and it is recognized that an enzyme that either synthesizes such a compound or uses it as a substrate nearly always does so stereospecifically. In the case of citrate synthase, for example, it is inherently likely that the planar carbonyl carbon of oxaloacetate lies flat on an enzyme surface and that only one side of the atom is available for attack by acetyl-coenzyme A.

ATP Stoichiometry of the TCA Cycle

By summing the component reactions of the TCA cycle, we arrive at the following overall reaction:

$$\text{Acetyl-CoA} + 2\,\text{H}_2\text{O} + 3\,\text{NAD}^+ + \text{FAD} + \text{ADP} + \text{P}_i \longrightarrow 2\,\text{CO}_2 + 3\,\text{NADH} + 2\,\text{H}^+ + \text{FADH}_2 + \text{CoA} + \text{ATP} \quad \textbf{(11)}$$

Looking at this summary reaction, you may wonder why it doesn't reflect the substantially greater ATP yield that is supposed to be characteristic of aerobic metabolism. The answer is that energy is stored in the reduced coenzyme molecules on the right-hand side of the reaction. Reoxidation of these compounds liberates a large amount of free energy. It is only as the electrons are transferred stepwise from the coenzymes to molecular oxygen that the

Figure 15.10

Stereochemical relationships in the synthesis and metabolism of citrate. When oxaloacetate labeled with ^{13}C in the carbonyl group β to the keto group (*) was used as substrate, the researchers expected that half of the label would be found in succinate and half in CO_2. That prediction was based on the assumption that the two $—CH_2—COO^-$ arms of citrate must be equivalent in every way. In fact, all of the label was found in CO_2. Thus only the intermediates shown on the left were produced. This result shows that both the condensation of acetyl-CoA with oxaloacetate and the isomerization of citrate are stereospecific reactions. The carbon atoms supplied by acetyl-CoA are shown in red. Neither of those atoms is lost in the first turn of the cycle after their entry.

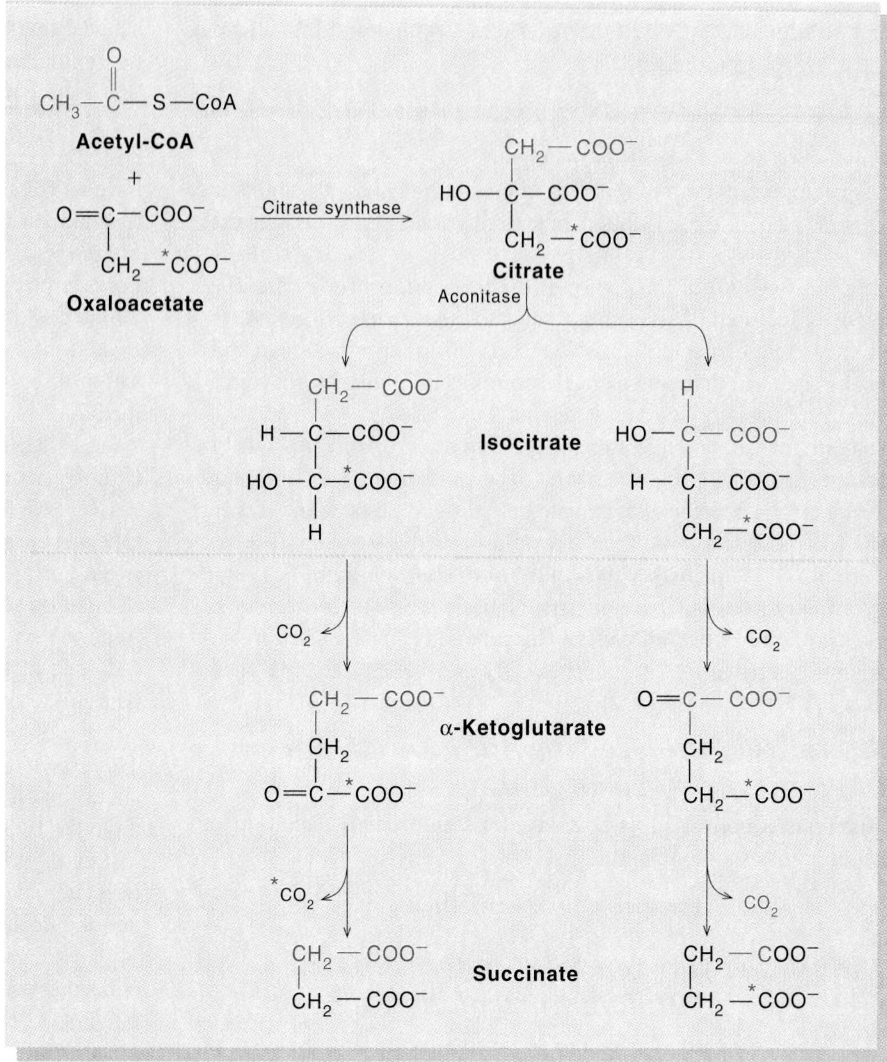

Figure 15.11

Citrate is shown with three of the substituents from the central carbon atom making contact with the enzyme surface. Binding in this way makes the two $—CH_2—COO^-$ groups nonequivalent, as they must be in the TCA cycle.

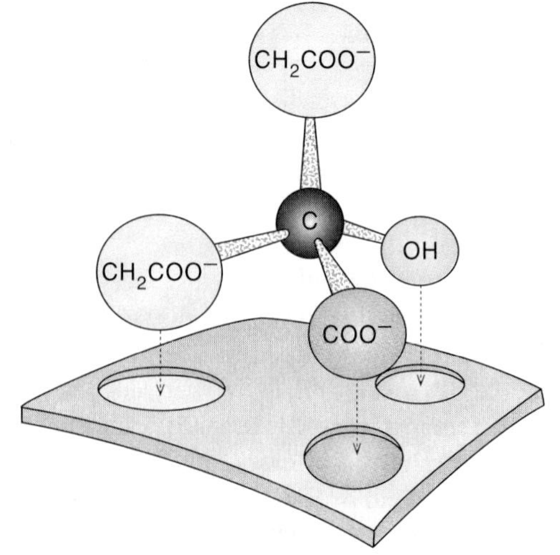

Table 15.1
Reactions of the TCA Cycle

Reaction	Enzyme	$\Delta G^{o\prime}$ (kcal/mole)
1. Acetyl-CoA + Oxaloacetate + H_2O $\longrightarrow$ Citrate + CoA	Citrate synthase	−7.7 (−32 kJ)
2. Citrate $\rightleftharpoons$ *cis*-Aconitase $\rightleftharpoons$ Isocitrate	Aconitase	+1.5 (6.3 kJ)
3. Isocitrate + NAD^+ $\longrightarrow$ α-Ketoglutarate + NADH + CO_2	Isocitrate dehydrogenase	−5.0 (−20.9 kJ)
4. α-Ketoglutarate + NAD^+ + CoA $\longrightarrow$ Succinyl-CoA + NADH + CO_2	α-Ketoglutarate dehydrogenase	−8.0 (−33 kJ)
5. Succinyl-CoA + GDP + P_i $\longrightarrow$ Succinate + GTP + CoA	Succinate thiokinase	−0.7 (−2.9 kJ)
6. Succinate + FAD $\longrightarrow$ Fumarate + $FADH_2$	Succinate dehydrogenase	0.0 (0.0 kJ)
7. Fumarate + H_2O $\longrightarrow$ L-Malate	Fumarase	−0.9 (−3.7 kJ)
8. L-malate + NAD^+ $\longrightarrow$ Oxaloacetate + NADH + H^+	Malate dehydrogenase	+7.1 (29.7 kJ)
Acetyl-CoA + 2 H_2O + 3 NAD^+ + FAD + ADP + P_i $\longrightarrow$ 2 CO_2 + 3 NADH + 2 H^+ + $FADH_2$ + CoA + ATP		−13.7 (−57.3 kJ)

coupled generation of ATP occurs. We will see that the oxidative phosphorylation system regenerates approximately 2.5 molecules of ATP for each pair of electrons from NADH and approximately 1.5 molecules of ATP for each pair of electrons from $FADH_2$ (see chapter 16). Thus in aerobic mitochondrial metabolism, oxidation of 1 mole of acetyl groups leads to regeneration of 10 moles of ATP.

If we take pyruvate rather than acetyl-CoA as the starting point, the equation for the TCA cycle is

$$\text{Pyruvate} + 4\ NAD^+ + FAD + ADP\ (\text{or GDP}) + P_i + 2\ H_2O \longrightarrow 3\ CO_2 + 4\ NADH + 2\ H^+ + FADH_2 + ATP\ (\text{or GTP}) \quad \textbf{(12)}$$

and we see that the mitochondrial part of carbohydrate metabolism yields about 12.5 moles of ATP per mole of pyruvate or 25 moles per mole of glucose.

Thermodynamics of the TCA Cycle

In the metabolic pathway for the oxidation of pyruvate, four reactions occur (those catalyzed by pyruvate dehydrogenase, citrate synthase, isocitrate dehydrogenase, and α-ketoglutarate dehydrogenase) for which the equilibrium constants are large. For four others (those catalyzed by aconitase, succinate thiokinase, succinate dehydrogenase, and fumarase) the equilibrium constant is fairly close to 1, and for one reaction (catalyzed by malate dehydrogenase) it is much less—about 10^{-5} (table 15.1). It seems strange that such an unfavorable reaction should occur in a sequence (the TCA cycle) that has one of the largest fluxes in metabolism. This is possible only because the following reaction, catalyzed by citrate synthase, is so thermodynamically favorable. The equilibrium constant for this reaction is about 5×10^5.

The highly unfavorable free energy for the conversion of malate to oxaloacetate may actually serve an important function in facultative aerobes. In facultative aerobes operating in the absence of O_2, the Krebs cycle cannot operate in the normal way. In spite of this the organism still has a need for the intermediates for biosynthesis normally supplied by the Krebs cycle. This situation is remedied by operating the first part of the cycle to α-ketoglutarate or succinyl-CoA in the forward direction while operating the last part of the cycle from oxaloacetate to succinate or succinyl-CoA in the reverse direction. In this way the organism is able to synthesize the intermediates for biosynthesis while balancing its needs for oxidation and reduction. The highly unfavorable free energy for the conversion of malate to oxaloacetate becomes a highly favorable free energy for the conversion of oxaloacetate to malate so that the reactions between oxaloacetate and succinyl-CoA have a favorable overall free energy for operating in the reverse direction (see table 15.1).

Before we leave the subject of thermodynamics, it should be pointed out that most of the arguments in this section are based on standard free energy data rather than the actual free energies, which would require knowing the concentrations of reactants and products in the mitochondria. This is a serious limitation we must live with until there is a reliable means for estimating these concentrations.

The Amphibolic Nature of the TCA Cycle

A major function of the TCA cycle is the oxidation of acetate to CO_2 with concomitant conservation of the energy of oxidation as reduced coenzymes and eventually as ATP. Strictly speaking, then, the TCA cycle has but a single substrate, acetyl-CoA. In most cells, however, there is considerable flux of four-, five-, and six-carbon intermediates into and out of the cycle, which occurs in addition to the catabolic function of the cycle. Such branchpoint reactions serve two main purposes: (1) to provide for the synthesis of compounds derived from any of several intermediates of the cycle and (2) to replenish and augment the supply of intermediates in the cycle as needed. Because the TCA cycle can function both in a catabolic mode and as a source of precursors for anabolic pathways, it is often called an amphibolic pathway. Some of the main biosynthetic pathways that begin with intermediates in the TCA cycle, as well as the ways in which the supply of intermediates in the cycle are replenished, are indicated in figure 15.12. We discuss these pathways briefly here; they are discussed in greater detail in subsequent chapters (for fats, see chapters 21, 22, and 23; for amino acids, see chapters 24 and 25).

Four intermediates in the TCA cycle serve as starting points for compounds outside of the cycle. Oxaloacetate and α-ketoglutarate are used in the synthesis of several amino acids, succinyl-CoA is used in heme synthesis; and citrate is the source of the acetyl-CoA in the cytosol, which is used for the synthesis of fats and other lipids and some amino acids. These are the major drains on the TCA cycle.

Reactions that replenish the intermediates in the TCA cycle are termed anaplerotic, from a Greek root that means "filling up." It is not necessary to replenish the intermediate that is used in a biosynthetic pathway directly, because any intermediate can be replenished by a feeding-in process from any point in the cycle.

When the carbohydrates are being metabolized, TCA cycle intermediates are replenished by production of oxaloacetate from pyruvate. In mammals this reaction is catalyzed by pyruvate carboxylase, and one ATP-to-ADP conversion is associated with the carboxylation. Other properties of this reaction are discussed later in this chapter in connection with regulation of the TCA cycle and related metabolic sequences.

In prokaryotic organisms and some eukaryotes, oxaloacetate is fed into the cycle by carboxylation of phosphoenolpyruvate. Energetically, the carboxylation of phosphoenolpyruvate, which is catalyzed by phosphoenolpyruvate carboxylase, is equivalent to the sum of the pyruvate kinase and pyruvate carboxylase reactions, which are used by mammals (see fig. 14.15):

$$\text{H}^+ + \text{Phosphoenolpyruvate} + \text{ADP} \longrightarrow \text{Pyruvate} + \text{ATP} + \text{H}_2\text{O} \quad \textbf{(13)}$$

$$\text{H}_2\text{O} + \text{Pyruvate} + \text{ATP} + \text{CO}_2 \longrightarrow \text{Oxaloacetate} + \text{ADP} + \text{P}_\text{i} + \text{H}^+ \quad \textbf{(14)}$$

$$\text{Sum:}\quad \text{Phosphoenolpyruvate} + \text{CO}_2 \longrightarrow \text{Oxaloacetate} + \text{P}_\text{i} \quad \textbf{(15)}$$

The Glyoxylate Cycle Permits Growth on a Two-Carbon Source

Usually, condensation of acetyl-coenzyme A with oxaloacetate to form citrate is a signal that the metabolic fate of the acetyl carbons is sealed; the inevitable result, by means of the TCA cycle, is their oxidation and eventual release as CO_2. However, the glyoxylate cycle, shown in figure 15.13, represents an alternative pathway that also begins with citrate formation but results in anabolism to the four-carbon level rather than catabolism to the one-carbon level. Comparison of the glyoxylate cycle with the TCA cycle (see fig. 15.13) reveals that two of the five reactions of the glyoxylate cycle are unique to this pathway, while the other three are also part of the TCA cycle. Specifically, the glyoxylate cycle effectively bypasses the two steps of the TCA cycle in which CO_2 is released. Furthermore, two molecules of acetyl-CoA are taken in per turn of the cycle rather than just one, as in the TCA cycle. The net result is the conversion of two molecules of two carbons each (i.e., the acetate of acetyl-CoA) into one four-carbon compound, succinate.

The glyoxylate cycle is an indispensable metabolic capability for those species of bacteria, protozoans, fungi, and algae that grow on a two-carbon substrate such as acetate or ethanol. It is also an essential reaction sequence for seedlings of fat-storing plant species that must effect net synthesis of sugars and other cellular components from the acetyl-CoA produced by oxidation of storage triglycerides. In such plant seedlings and many other eukaryotic organisms that possess this capability, the enzymes of the glyoxylate cycle (and those of related metabolic pathways to be discussed later) are compartmentalized together in specialized organelles called glyoxysomes.

Species capable of growth on two-carbon substrates need, in addition to the glyoxylate cycle, some preparatory sequence for converting the substrate into acetyl-CoA. If the substrate is acetate, activation requires only formation of the CoA derivative, catalyzed by acetate thiokinase, with ATP hydrolysis as the driving force:

$$\text{CH}_3\overset{\overset{\displaystyle\text{O}}{\|}}{\text{C}}\text{O}^- + \text{CoASH} + \text{ATP} \longrightarrow \text{CH}_3\overset{\overset{\displaystyle\text{O}}{\|}}{\text{C}}\text{SCoA} + \text{AMP} + \text{PP}_\text{i} \quad \textbf{(16)}$$

If the substrate is ethanol, it must first be oxidized in two steps to the level of acetate:

$$\underset{\text{Ethanol}}{\text{CH}_3\text{CH}_2\text{OH}} \xrightarrow{\text{NAD}^+ \searrow\nearrow \text{NADH} + \text{H}^+} \underset{\text{Acetaldehyde}}{\text{CH}_3\text{CHO}} \xrightarrow{\text{NAD}^+ \searrow\nearrow \text{NADH} + \text{H}^+} \underset{\text{Acetate}}{\text{CH}_3\text{COO}^-} \quad \textbf{(17)}$$

Other two-carbon substrates are also possible; each requires specific processing to convert it to the acetyl-CoA with which the glyoxylate cycle itself begins.

Both the initial formation of citrate from oxaloacetate and acetyl-CoA and its subsequent conversions by means of aconitate to isocitrate are already familiar from the TCA cycle. The only difference is that in eukaryotic cells the citrate synthase and aconitase enzymes that carry out these reactions in the glyoxysomes are organelle-specific isozymes, differing in physical and enzymatic properties from the enzymes responsible for the same reactions that occur in the mitochondria as part of the TCA cycle.

The two reactions unique to the glyoxylate cycle are those responsible for generation and subsequent utilization of glyoxylate, the two-carbon compound from which the cycle (and the organelle) derives its name. Isocitrate is split into two molecules rather than being oxidatively decarboxylated, as would occur in the TCA cycle (see fig. 15.13). The products are glyoxylate and succinate, with two and four carbons, respectively. The reaction is catalyzed by isocitrate lyase, an enzyme found only in those microbial and plant species that are able to carry out net growth (and hence synthesis of higher-carbon compounds) from the two-carbon level.

The succinate arises from the "upper" four carbon atoms of isocitrate (see fig. 15.13), while the glyoxylate corresponds to

Figure 15.12

The TCA cycle, showing some of the branchpoint pathways (in red) that either drain or replenish the TCA intermediates.

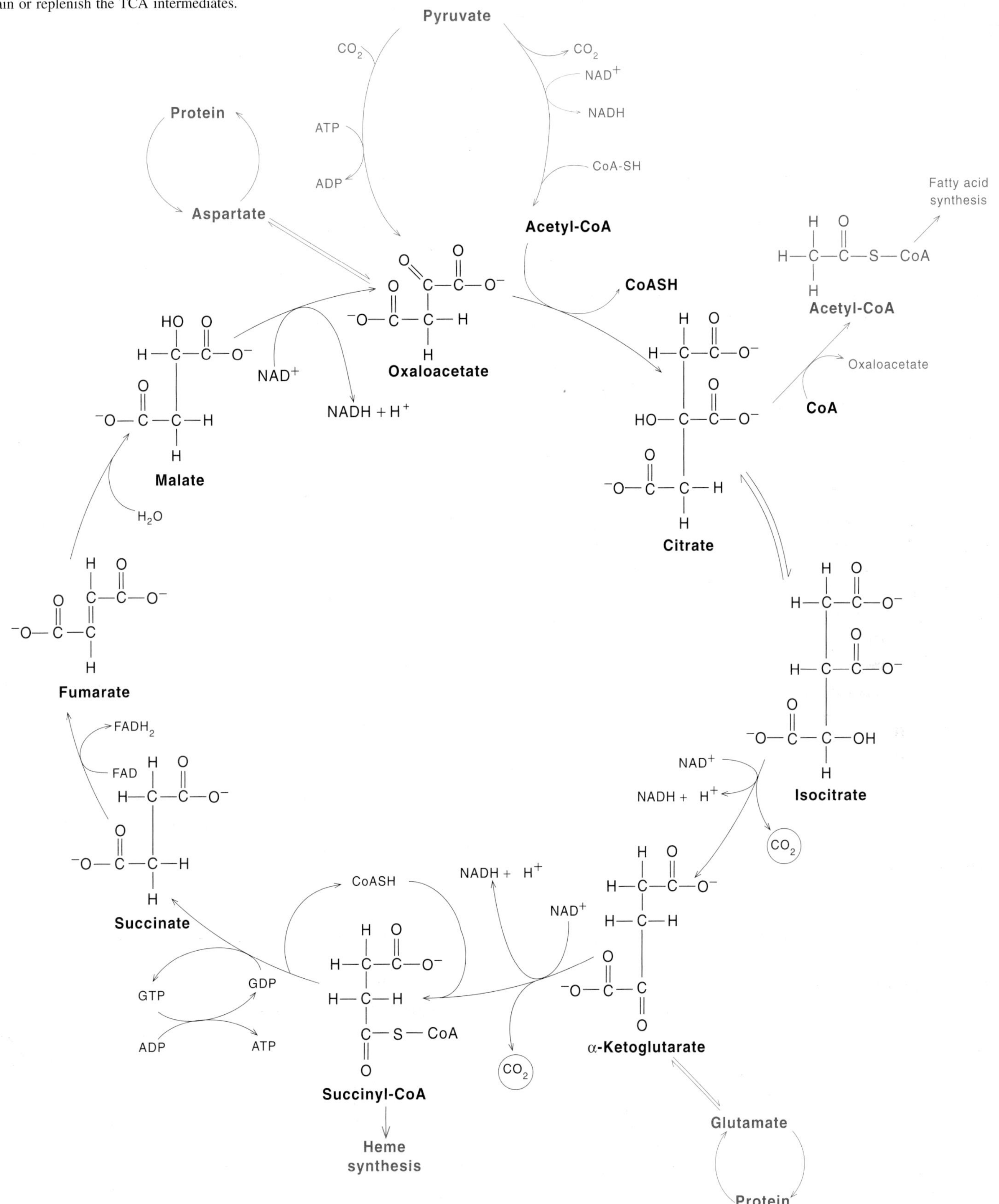

Figure 15.13

Comparison of the TCA cycle and the glyoxylate cycle. In the TCA cycle, one molecule of acetyl-CoA is oxidized to two molecules of CO_2. In the glyoxylate cycle (red), two molecules of acetyl-CoA are converted to one molecule of oxaloacetate. As indicated, the glyoxylate cycle uses some of the enzymes of the TCA cycle. Only enzymes operative in the glyoxylate cycle are shown. In plant cells, enzymes of the glyoxylate cycle are located in specialized organelles called glyoxysomes. In yeasts and other eukaryotic organisms these enzymes are located in the cytosol.

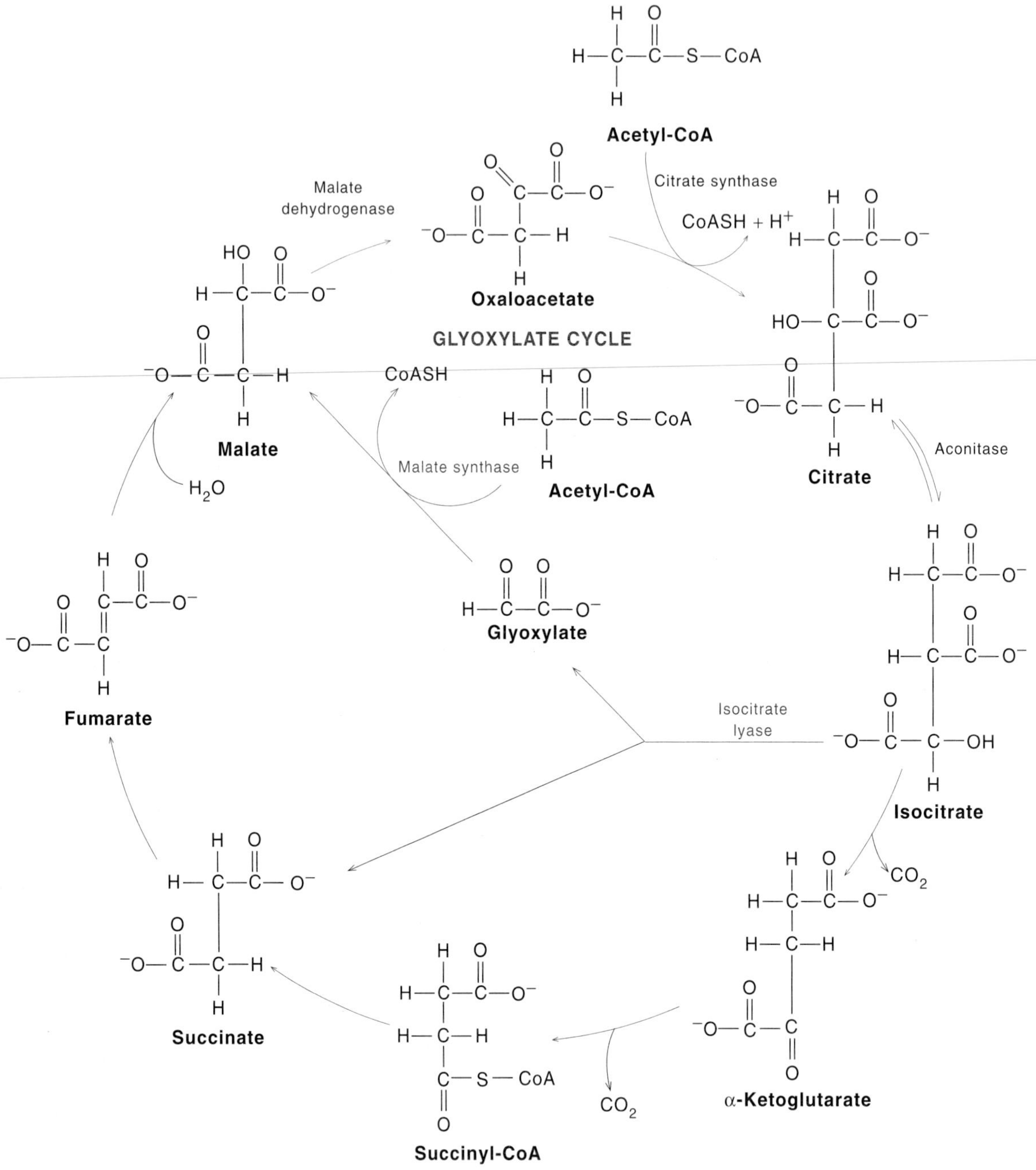

Acetyl-CoA

Citrate synthase

CoASH + H^+

Citrate

Aconitase

Isocitrate

Isocitrate lyase

CO_2

α-Ketoglutarate

CO_2

Succinyl-CoA

Succinate

Fumarate

H_2O

Malate

Malate dehydrogenase

Oxaloacetate

GLYOXYLATE CYCLE

CoASH

Acetyl-CoA

Malate synthase

Glyoxylate

Figure 15.14

Role of the glyoxylate cycle in gluconeogenesis from (*a*) two-carbon compounds or (*b*) fatty acids. The pathway involves reaction sequences that in plant cells are localized within lipid bodies, glyoxysomes, mitochondria, and the cytoplasm.

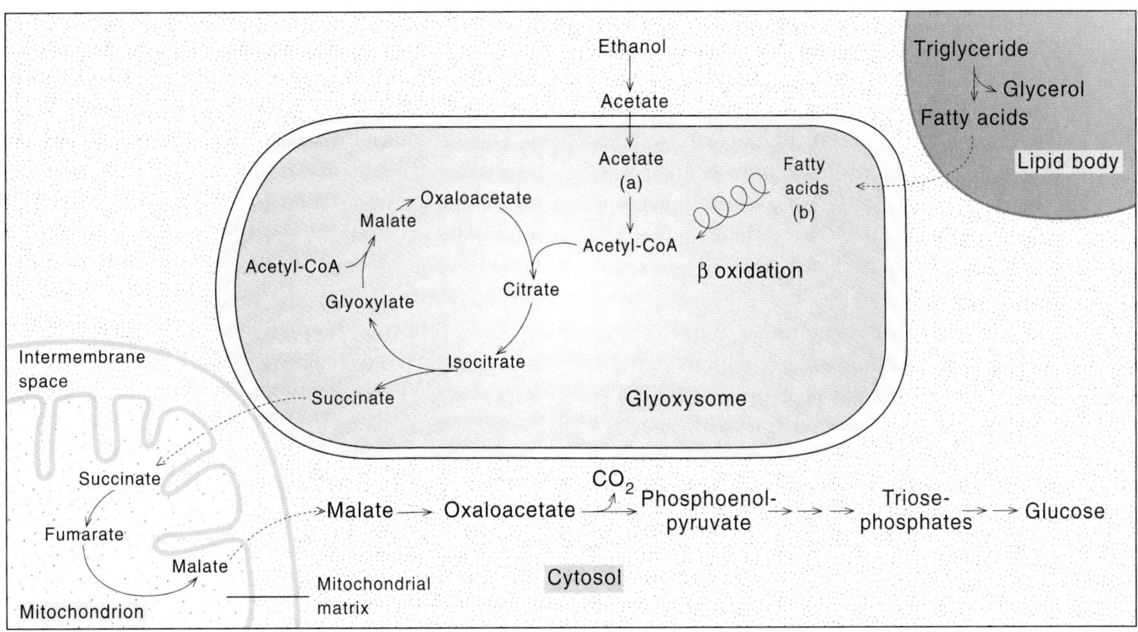

the "lower" two carbons. The succinate represents the immediate product of the glyoxylate cycle and becomes in turn the starting point for synthesis of other compounds that the cell or organism needs. The glyoxylate, however, becomes the acceptor for the acetate group from the second acetyl-CoA molecule that enters the cycle. The enzyme responsible for this reaction is malate synthase. Like its companion enzyme isocitrate lyase, malate synthase is found only in species capable of net growth from the two-carbon level.

The glyoxylate cycle is completed by oxidative conversion of malate to oxaloacetate, a reaction catalyzed here, as in the TCA cycle, by malate dehydrogenase with NAD^+ as the electron acceptor. In fat-storing plant species the isozyme of malate dehydrogenase that is involved in the glyoxylate cycle is specific for the glyoxysomes in which the process is localized.

The glyoxylate cycle itself can be summarized by the following overall reaction:

$$2 \text{ Acetyl-CoA} + NAD^+ \longrightarrow \text{Succinate} + NADH + 3\,H^+ + 2\,\text{CoASH} \quad \textbf{(18)}$$

Utilization of the Succinate Requires Passage from the Glyoxysome to the Mitochondria

Since the glyoxylate cycle is the mechanism by which biosynthesis of more complex molecules from the two-carbon level is initiated, and succinate is the immediate product of the cycle, let's examine the pathways by which other compounds can be synthesized from succinate. As you are already aware, succinate is also an intermediate in the TCA cycle, so we are in effect dealing with an aspect of the amphibolic nature of the TCA cycle.

Of greatest significance to an organism that depends on a two-carbon substrate for all its carbon needs is the gluconeogenic (sugar-synthesizing) route from succinate to the hexose level, since access to the six-carbon level essentially guarantees access to all other biosynthetic pathways in the cell. Gluconeogenesis in organisms that do not possess the enzymes specific to the glyoxylate cycle usually proceeds from molecules with at least three carbon atoms, such as pyruvate or lactate (see chapter 14).

Synthesis of hexoses from succinate proceeds by means of phosphoenolpyruvate by the pathway shown in figure 15.14. The succinate arising from glyoxylate cycle activity is converted by TCA enzymes located in the mitochondria, first to fumarate and then to malate, which can diffuse from the mitochondria into the cytosol for the further conversion steps. This transfer of succinate to the mitochondria is necessary for further processing because the glyoxysome does not possess the necessary TCA cycle enzymes. The steps from oxaloacetate to glucose were discussed in chapter 14 (see fig. 14.14).

Finally, note that the glyoxylate bypass permits the synthesis of sugars from fatty acid degradation, since acetyl-CoA is the ultimate product of this pathway (see fig. 15.4). The subject of fatty acid degradation is treated in chapter 21.

Oxidation of Other Substrates by the TCA Cycle

The TCA cycle, strictly speaking, has only one input fuel: acetyl-CoA. Catabolism of carbohydrates and fats leads to the production of acetyl-CoA, so the TCA cycle is ideally suited to serve as the major oxidative sequence in the catabolism of those types of compounds. However, degradation of the amino acids that result from the hydrolysis of protein produces a number of intermediates, among which are α-ketoglutarate, succinyl-CoA, and oxaloacetate (chapter 25). α-Ketoglutarate and succinyl-CoA can be oxidized to oxaloacetate, but the cycle as such cannot oxidize oxaloacetate further. In the presence of acetyl-CoA, each molecule of oxaloacetate used in the synthesis of citrate is regenerated in the cycle; thus no net oxidation of oxaloacetate occurs in that case either.

The problem of how to oxidize oxaloacetate is solved by the action of phosphoenolpyruvate carboxykinase, which we discussed in connection with gluconeogenesis (see equation (11) in chapter 14). This enzyme catalyzes the conversion of oxaloacetate to phosphoenolpyruvate with the help of ATP or GTP and thus permits the total oxidation of oxaloacetate to CO_2 by the enzymes of the TCA cycle.

In addition to the amino acids that are converted to intermediates of the TCA cycle, others are converted to pyruvate or acetyl-CoA and thus enter the cycle in the usual way. In fact, all 20 of the protein amino acids are metabolized by way of the TCA cycle (see fig. 25.11). Thus although it is often thought of as part of the pathway for carbohydrate metabolism, the TCA cycle is actually the central oxidative sequence for all three of the major types of carbon and energy sources: carbohydrates, fats, and proteins.

The TCA Cycle Activity Is Regulated at Metabolic Branchpoints

The TCA cycle is carefully regulated to ensure that its level of activity relates to cellular needs. The cycle serves two functions: (1) furnishing reducing equivalents (as NADH and to a lesser extent as $FADH_2$) to the electron-transport chain and (2) by means of side reactions, providing substrates for biosynthesis reactions. Both of these functions are reflected in the regulation of the cycle.

In its primary role as a means of oxidizing acetyl groups to carbon dioxide and water, the TCA cycle is sensitive both to the availability of its substrate, acetyl-CoA, and to the accumulated levels of its principal end products, NADH and ATP. Actually, the ratio of NADH to NAD^+ and the energy charge or the ATP-to-ADP ratio are more important than the individual concentrations. Other regulatory parameters to which the TCA cycle is sensitive include the ratios of acetyl-CoA to free CoA, acetyl-CoA to succinyl-CoA, and citrate to oxaloacetate. The major known sites for regulation are shown in figure 15.15. These include two enzymes outside the TCA cycle (pyruvate dehydrogenase and pyruvate carboxylase) and three enzymes inside the TCA cycle (citrate synthase, isocitrate dehydrogenase, and α-ketoglutarate dehydrogenase). As might be suspected, each of these sites of regulation represents an important metabolic branchpoint.

The Pyruvate Branchpoint Partitions Pyruvate between Acetyl-CoA and Oxaloacetate

Acetyl-CoA is the only compound that can enter the TCA cycle when the cycle is operating purely oxidatively, but one molecule of oxaloacetate must enter for each molecule of citrate, α-ketoglutarate, or succinyl-CoA that is removed for use in biosynthesis. It follows that pyruvate is a major metabolic branchpoint in a cell that is living on carbohydrate. The partitioning of pyruvate between decarboxylation to acetyl-CoA and carboxylation to oxaloacetate is, in effect, partitioning between the two major metabolic uses of pyruvate: oxidation of carbon for regeneration of ATP and conversion to starting materials for biosynthesis.

Some of the regulatory interactions that affect partitioning at pyruvate are shown in figure 15.15. We focus first on the effects of acetyl-CoA, which is a negative modifier for pyruvate dehydrogenase and a very strong positive modifier for pyruvate carboxylase. To illustrate the regulatory roles of those effects, consider a mitochondrion in which the TCA cycle is functioning only oxidatively, so no input of oxaloacetate is required. If conditions change and α-ketoglutarate begins to be removed for biosynthesis, the rate at which oxaloacetate is regenerated by the cycle is reduced by the rate of α-ketoglutarate removal. The citrate synthase reaction can proceed no more rapidly than the rate at which oxaloacetate, one of its substrates, is supplied, so the rate of citrate synthesis, too, decreases. Consequently, acetyl-CoA is used more slowly, causing its concentration to increase slightly. The increase in acetyl-CoA concentration leads to a decrease in the activity of the pyruvate dehydrogenase complex and an increase in the activity of pyruvate carboxylase. As a result of those effects, the partitioning of pyruvate is changed so as to produce more oxaloacetate, thus replenishing the cycle and allowing the cycle to continue to function in spite of the removal of α-ketoglutarate.

Since the system responds to the concentration of acetyl-CoA and thus indirectly to the concentration of oxaloacetate, it adjusts automatically to maintain a functional concentration of oxaloacetate regardless of whether the intermediate removed from the cycle is citrate, α-ketoglutarate, succinyl-CoA, oxaloacetate itself, or any combination of these. The same effects cause the rate of conversion of pyruvate to oxaloacetate to decrease when the rate of removal of biosynthetic precursors decreases.

The negative effects of acetyl-CoA on pyruvate dehydrogenase activity are supplemented by ATP and NADH. These effects are in the correct direction to cause the rate of acetyl-CoA synthesis to vary with the need for electrons and for regeneration of NADH and ATP.

In addition to regulation by these small-molecule modifiers, the pyruvate dehydrogenase complex, at least in mammals, is subject to regulation by covalent modification. Each of the giant complexes contains several molecules of a protein kinase and a protein phosphorylase. The kinase catalyzes phosphorylation of specific serine hydroxyl groups of the pyruvate decarboxylase por-

Figure 15.15

Major regulatory sites of the TCA cycle, with activators (⊕) and inhibitors (⊖) of specific reactions shown in red.

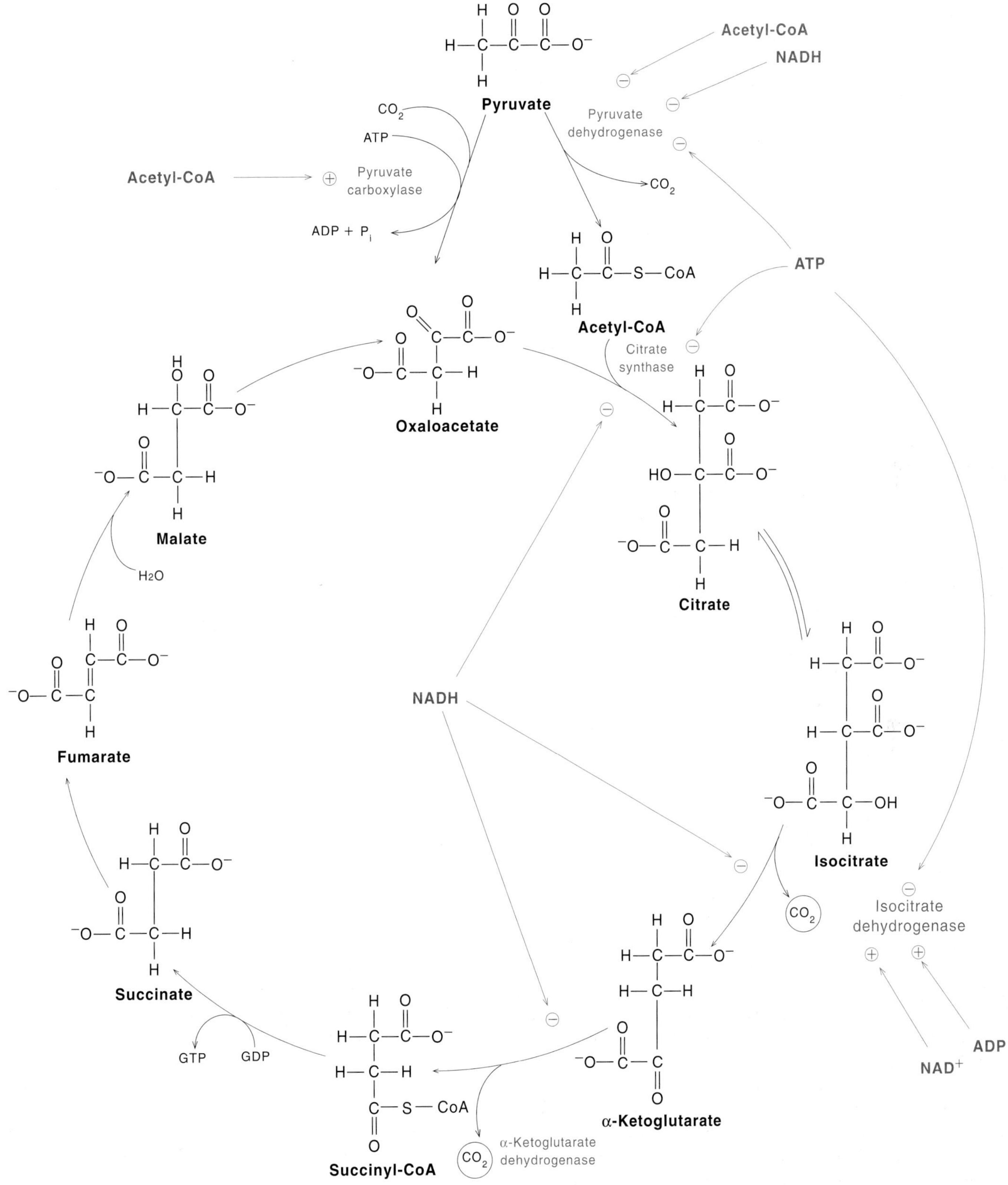

Figure 15.16

Cycling between mitochondria and cytosol is the supply of acetyl-CoA for use in biosynthetic sequences in the cytosol. Citrate moves from the interior of the mitochondria to the cytosol. In the cytosol it is cleaved to acetyl-CoA and oxaloacetate by citrate lyase. The equilibrium constant for this reaction is favorable because an ATP-to-ADP conversion is involved. Most of the oxaloacetate is reduced to malate. The malate may be taken up by mitochondria or oxidized to pyruvate and CO_2, generating NADPH for use in biosynthetic sequences in the cytosol. The pyruvate enters the mitochondria, where it may be converted to oxaloacetate or acetyl-CoA by the usual routes (see fig. 15.15).

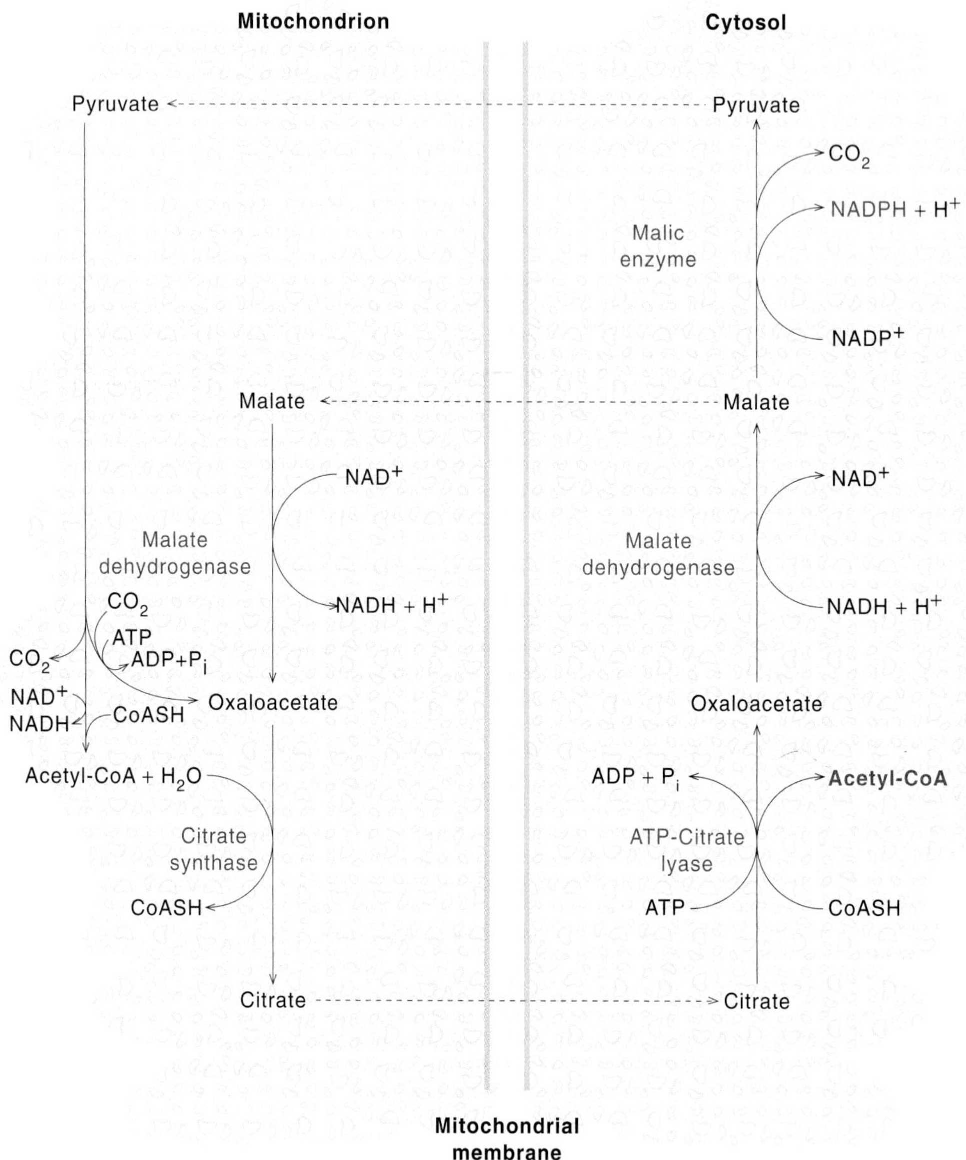

tion of the complex, and the phosphorylase catalyzes hydrolytic removal of these phosphoryl groups. The phosphorylated enzyme is relatively inactive. Thus the phosphorylation–dephosphorylation system also contributes to regulation of the rate of conversion of pyruvate to acetyl-CoA. The action of the kinase, and the resultant decrease in the activity of the pyruvate dehydrogenase complex, is favored by high ATP-to-ADP and NADH-to-NAD^+ ratios and by high acetyl-CoA. These effects act in the same direction as the direct effects of the modifier metabolites on the enzymes of the complex, so the two types of regulation reinforce one another.

Citrate Synthase Is Negatively Regulated by NADH and the Energy Charge

Thus far we have discussed the two most important enzymes regulating the supplies of acetyl-CoA and oxaloacetate for the TCA cycle. This leaves us with the three enzymes, all within the cycle, that regulate the activity of the cycle. The first of these, citrate synthase, catalyzes the formation of citrate from acetyl-CoA and oxaloacetate. Regulation, by energy charge and other parameters, of the rate of glycolysis and of the pyruvate dehydrogenase reaction plays important roles in controlling the rate of citrate synthesis. For example, yeast citrate synthase has been shown to respond sensitively to variation in the value of the energy charge. Synthesis of citrate is also favored by a low NADH-to-NAD^+ ratio. This is clearly metabolically desirable, since the operation of the cycle leads to an increase in this ratio.

Isocitrate Dehydrogenase Is Regulated by the NADH-to-NAD^+ Ratio and the Energy Charge

The equilibrium constant for the conversion of citrate to isocitrate is small, and the interconversion is rapid, so these two intermediates make up a metabolic pool. Since regulation is not expected within a metabolic pool for reasons discussed in chapters 12 and 14, the next possible regulatory site as we go around the TCA cycle is the conversion of isocitrate to α-ketoglutarate. This reaction is catalyzed by isocitrate dehydrogenase, which is a highly regulated enzyme. In yeast the activity of isocitrate dehydrogenase is strongly cooperative (sigmoidal) with respect to the isocitrate concentration. The enzyme is stimulated by NAD^+ and AMP and inhibited by NADH, making it very sensitive to the NADH-to-NAD^+ ratio. The properties of the mammalian enzyme are similar to those of the yeast enzyme, with the exception that ADP replaces AMP as a positive modifier.

To understand why isocitrate dehydrogenase is so intensely regulated, we must consider reactions beyond the TCA cycle, and indeed beyond the mitochondrion (fig. 15.16). Of the two compounds citrate and isocitrate, only citrate is transported across the barrier imposed by the mitochondrial membrane. Citrate that passes from the mitochondrion to the cytosol plays a major role in biosynthesis, both because of its immediate regulatory properties and because of the chain of covalent reactions it initiates. In the cytosol, citrate undergoes a cleavage reaction in which acetyl-CoA is produced. The other cleavage product, oxaloacetate, can be utilized directly in various biosynthetic reactions, or it can be converted to malate. The malate so formed can be returned to the mitochondrion, or it can be converted in the cytosol to pyruvate, which also results in the reduction of $NADP^+$ to NADPH. The pyruvate is either utilized directly in biosynthetic processes or, like malate, can return to the mitochondrion.

The acetyl-CoA produced in the cytosol from citrate breakdown is used in biosynthetic reactions, including the synthesis of lipids, some amino acids, cofactors, and pigments. This is the only source of acetyl-CoA in the cytosol because the acetyl-CoA produced in the mitochondrion cannot diffuse across the mitochondrial membrane. The NADPH produced indirectly in the cytosol from the citrate is also of great importance in biosynthetic reactions.

In addition to its importance in providing cytosolic acetyl-CoA and NADPH, citrate also serves as a major regulator of the rate of fatty acid synthesis. As we shall see (chapter 21), citrate is a strong positive modifier of the first reaction in fatty acid synthesis. It should be remembered (see chapter 14) that citrate also is a negative modifier of phosphofructokinase and thereby exerts a negative effect on glycolysis, which also occurs in the cytosol.

The effect of small-molecule modifiers on isocitrate dehydrogenase is appropriate in that a high-energy charge favors inhibition of isocitrate dehydrogenase and thus favors an accumulation of mitochondrial citrate. This leads to an increased flow of citrate from the mitochondrion to the cytosol, where the citrate can exert its multiple positive effects on biosynthesis and its negative effects on glycolysis.

α-Ketoglutarate Dehydrogenase Is Negatively Regulated by NADH

α-Ketoglutarate is also a branchpoint metabolite, since it can be transaminated to form glutamate (see fig. 15.12 and chapter 24). Glutamate is needed in protein synthesis directly and also is a precursor of a number of other amino acids and peptides. Thus it is consumed in the cytosol in large amounts in a cell that is synthesizing protein rapidly. We expect that the reaction catalyzed by α-ketoglutarate dehydrogenase is regulated so as to retain carbon in the TCA cycle when energy is in short supply. Conversely, when the energy supply is high, it is advantageous to allow the concentration of α-ketoglutarate to rise, facilitating the formation of glutamate and exit from the cycle. The balance is achieved by negative regulation of α-ketoglutarate dehydrogenase in response to NADH. When NADH is low, the enzyme competes more strongly for α-ketoglutarate and thus maximizes the regeneration of ATP while tending to decrease biosynthetic activities that use ATP.

Summary

This chapter is mainly concerned with the contribution of the tricarboxylic acid cycle to carbohydrate metabolism. The TCA cycle is the main source of electrons for oxidative phosphorylation and thereby the major energetic sequence in the metabolism of aerobic cells or organisms. It serves as the main distribution center of metabolism, receiving carbon from the degradation of carbohydrates, fats, and proteins and, when it is appropriate, supplying carbon compounds for the synthesis of carbohydrates, fats, or proteins. Every aspect of the metabolism of an aerobic organism is directly dependent on the TCA cycle.

1. The TCA cycle begins with acetyl-CoA, which is obtained either by oxidative decarboxylation of pyruvate available from glycolysis or by oxidative cleavage of fatty acids.
2. The acetyl-CoA transfers its acetyl group to oxaloacetate, thereby generating citrate. In a cyclic series of reactions the citrate is subjected to two successive decarboxylations and four oxidative events, leaving a four-carbon compound malate from which the starting oxaloacetate is regenerated.
3. Only a single ATP is directly generated by a turn of the TCA cycle. Most of the energy produced by the cycle is stored in the form of reduced coenzyme molecules, NADH and $FADH_2$. Reoxidation of these compounds (see chapter 16) liberates a large amount of free energy, which is captured in the form of ATP.
4. Some of the main biosynthetic pathways begin with intermediates in the TCA cycle. When intermediates in the cycle are used as starting materials for biosynthesis, they must be re-

plenished to keep the cycle operating. When carbohydrates are being metabolized, TCA cycle intermediates are replenished by production of oxaloacetate from pyruvate.

5. The glyoxylate cycle permits growth on a two-carbon source. The glyoxylate cycle bypasses the two steps of the TCA cycle in which CO_2 is released. Furthermore, two molecules of acetyl-CoA are taken in per turn of the cycle rather than just one, as in the TCA cycle. The net result is the conversion of two molecules of two carbons each into one four-carbon compound, succinate. Two additional enzymes are needed for operation of the glyoxylate cycle: isocitrate lyase and malate synthase.
6. Degradation of amino acids produces a number of intermediates, among which are α-ketoglutarate, succinyl-CoA, and oxaloacetate. α-Ketoglutarate and succinyl-CoA can be oxidized to oxaloacetate, but the cycle as such cannot oxidize oxaloacetate further. Oxaloacetate is oxidized further by first converting it to phosphoenolpyruvate. This permits the total oxidation of oxaloacetate to CO_2 by the enzymes of the TCA cycle.
7. The TCA cycle is regulated to ensure that its level of activity corresponds closely to cellular needs. In its primary role as a means of oxidizing acetyl groups to CO_2 and water, the TCA cycle is sensitive both to the availability of its substrate, acetyl-CoA, and to the accumulated levels of its principal end products, NADH and ATP. Other regulatory parameters to which the TCA cycle is sensitive include NAD^+, ADP, acetyl-CoA, succinyl-CoA, and citrate. The major known sites for regulation of the cycle include two enzymes outside the cycle (pyruvate dehydrogenase and pyruvate carboxylase) and three enzymes inside the cycle (citrate synthase, isocitrate dehydrogenase, and α-ketoglutarate dehydrogenase). All of these sites of regulation represent important metabolic branchpoints.

Selected Readings

Atkinson, D. E., *Cellular Energy Metabolism and Its Regulation*. New York: Academic Press, 1977. A solid, detailed discussion of energy metabolism by an author who is prominent in this area.

Broda, E., *The Evolution of Bioenergetic Processes*. New York: Pergamon Press, 1975. An excellent discussion of cellular energetics from an evolutionary perspective.

de Kok, A. H., and W. G. J. Hol, Atomic structure of the cubic core of the pyruvate dehydrogenase multienzyme complex. *Science* 255:1544–1550, 1992.

Gest, H., Evolutionary roots of the citric acid cycle in prokaryotes. *Biochem. Soc. Symp.* 54:3–16, 1987. A fascinating recount of how the evolution of the cycle is traced by studying the way in which enzymes of the cycle are used in present day microorganisms.

Hers, H. G., and Hue, L., Gluconeogenesis and related aspects of glycolysis. *Ann. Rev. Biochem.* 52:617–653, 1983.

Krebs, H. A., The history of the tricarboxylic acid cycle. *Perspect. Biol. Med.* 14:154, 1970. An engaging historical account by the man who masterminded much of it.

Lauble, H., M. C. Kennedy, H. Beinert, and D. C. Stout, Crystal structures of aconitase with isocitrate and nitroisocitrate bound. *Biochemistry* 31:2735–2748, 1992.

Mehlman, M., and Hanson, R. W., *Energy Metabolism and the Regulation of Metabolic Processes in Mitochondria*. New York: Academic Press, 1972. The title speaks for itself—A good reference.

Williamson, J. R., and Cooper, R. V., Regulation of the citric acid cycle in mammalian systems. *FEBS Lett.* 117(Suppl.):K73, 1980. A well-rounded review of TCA cycle regulation as a contemporary research theme from a symposium dedicated to Hans Krebs.

Problems

1. As you will see later, the sequence of reaction types used in the citric acid cycle are utilized in many other pathways. Many authors interested in the possibility of life evolving under different conditions than those on the earth have proposed a myriad of "citric acid cycle look-alikes." Draw upon the reactions used in the citric acid cycle to produce a "3-hydroxybutyric acid cycle" in which acetaldehyde initially condenses with acetyl-CoA. Does your scheme produce the same amount of NADH, ATP, and FAD per acetyl-CoA as the "real cycle"?
2. The catabolism of aspartate involves conversion to the corresponding α-ketoacid, which enters the citric acid cycle. Which carbons of aspartate are the first ones lost as CO_2?
3. Examine the reaction sequence from succinate thiokinase to and including malate dehydrogenase. What are the simplest terms you can use to describe the biological function of this four-enzyme sequence?
4. In the conversion of isocitrate to α-ketoglutarate, oxidation and decarboxylation steps occur. Figure 15.9 shows the oxidation step first. Can you suggest a reason why the oxidation step is first?
5. In the late 1930s the tricarboxylic acid cycle (compare figs. 15.3 and 15.4) contained *cis*-aconitate, but more modern versions of the cycle (since the late 1950s) do not. The involvement of *cis*-aconitate is still shown in many nonbiochemistry texts. Can you explain why *cis*-aconitate is deleted from modern versions of the tricarboxylic acid cycle?
6. Notice that in the initial steps of the tricarboxylic acid cycle, citrate is converted to isocitrate, which is then oxidized. Why didn't nature "just" oxidize citrate and save an enzymatic step?
7. Ethylene glycol (ethane-1,2-diol), a major component of antifreeze, is readily metabolized by many organisms that have the glyoxylate cycle. Can you propose a sequence of catabolic steps that can explain the catabolism of ethylene glycol?
8. Many organisms can live on glutamate as their sole carbon and nitrogen source. Assuming that glutamate is converted into α-ketoglutarate, produce a scheme that completely oxidizes glutamate.
9. Notice the intermediate in the reaction of citrate synthase (fig. 15.7). Do you think that at some time in the future, evolution will produce a variety of citrate synthase that recovers the energy in the thioester, analogous to the production of GTP (ATP) by succinate thiokinase? Would this energy recovery have any effect on the thermodynamics of the tricarboxylic acid cycle?

10. Summarize in the simplest words the portion of the tricarboxylic acid cycle that is bypassed by the glyoxylate cycle.
11. The reactions catalyzed by isocitrate dehydrogenase and α-ketoglutarate dehydrogenase are both oxidative decarboxylation reactions. How similar are the reactions?
12. What effect would the following have on the control of the citric acid cycle:
    (a) A sudden influx of acetyl-CoA from the degradation of fatty acids.
    (b) A sudden need for heme biosynthesis?
13. Do all the metabolites associated with the tricarboxylic acid cycle have free access across the mitochondrial membrane?
14. Using your knowledge of metabolism, determine whether each of the following statements is true or false and explain the reasoning behind your decision.
    (a) Dihydrolipoamide dehydrogenase catalyzes the only oxidation–reduction reaction in the pyruvate dehydrogenase complex.
    (b) Hydrolysis of the thioester bond of acetyl-CoA yields insufficient energy to drive phosphorylation of ADP.
    (c) The methyl group of each acetyl-CoA molecule entering the TCA cycle is derived from the methyl group of pyruvate.
    (d) Even if aconitase were unable to discriminate between the two ends of the citrate molecule, the CO_2 released would still come from the oxaloacetate rather than the acetyl-CoA substrate of the citrate synthase reaction.
    (e) Malate cannot be converted to fumarate because the TCA cycle is unidirectional.
15. Assume that you have a buffered solution containing pyruvate dehydrogenase and all the enzymes of the TCA cycle but none of the cycle intermediates.
    (a) If you add 3 μmoles each of pyruvate, CoASH, NAD^+, GDP, and P_i, how much CO_2 evolves? What other products form?
    (b) In addition to the reagents in part (a), you add 3 μmoles each of the TCA cycle intermediates. How much CO_2 evolves? Explain.
    (c) If you were to add an electron acceptor that reoxidized NADH to the system described in part (a), would there be increased CO_2 evolution? Why or why not?
    (d) Explain the effect on CO_2 evolution of adding the NADH-reoxidizing system to the system described in (b), assuming that you also added excess GDP and P_i.
16. What would you expect to be the metabolic consequences of the following mutations in yeast?
    (a) Inability to synthesize malate synthase.
    (b) Pyruvate carboxylase that is not activated by acetyl-CoA.
    (c) Pyruvate dehydrogenase that is inhibited by acetyl-CoA more strongly than is the wild-type enzyme.
17. The substrate hydroxypyruvate

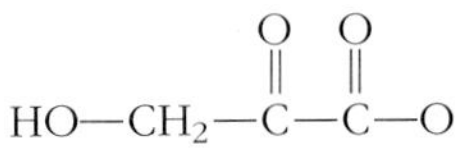

$$HO-CH_2-\overset{O}{\overset{\|}{C}}-\overset{O}{\overset{\|}{C}}-O$$

Hydroxypyruvate

is metabolized to pyruvate in a five-step process requiring the four intermediates whose structures are shown here:

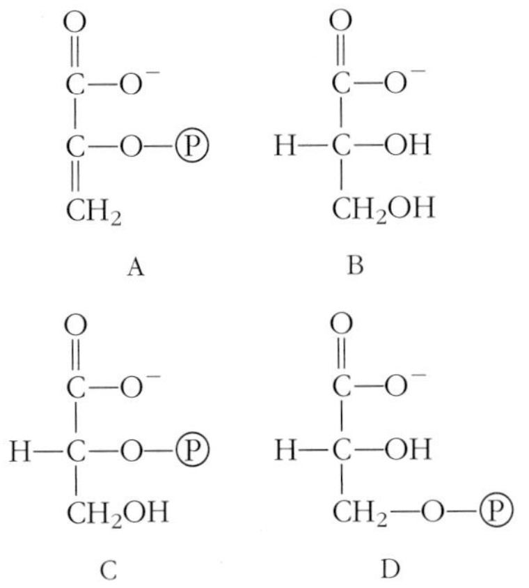

A: $^-O-\overset{O}{\overset{\|}{C}}-C(-O-Ⓟ)=CH_2$

B: $^-O-\overset{O}{\overset{\|}{C}}-CH(OH)-CH_2OH$

C: $^-O-\overset{O}{\overset{\|}{C}}-CH(-O-Ⓟ)-CH_2OH$

D: $^-O-\overset{O}{\overset{\|}{C}}-CH(OH)-CH_2-O-Ⓟ$

The letter designating the intermediate does not necessarily reflect the order in which it is used. The metabolic conversion requires NADH and catalytic quantities of both ATP and ADP. Assume that the pathway begins with an NADH-mediated reaction.
    (a) Designate the order in which the intermediates are used in the metabolism of hydroxypyruvate to pyruvate.
    (b) Write an overall equation for the metabolism of hydroxypyruvate to pyruvate.
    (c) Name each of the intermediates (A–D) and indicate which are intermediates in the glycolytic pathway.
    (d) Explain why only catalytic rather than stoichiometric amounts of ADP and ATP are required in the pathway.
18. Under anaerobic conditions, *E. coli* synthesizes an NADH-dependent fumarate reductase rather than succinate dehydrogenase, the flavoprotein that oxidizes succinate to fumarate.
    (a) Write an equation for the reaction catalyzed by fumarate reductase.
    (b) NADH produced by the glyceraldehyde-3-phosphate dehydrogenase reaction is reoxidized by reducing an organic intermediate. Rather than reducing pyruvate to lactate, anaerobic *E. coli* utilizes the fumarate reductase. However, under anaerobiosis the activity of α-ketoglutarate dehydrogenase is virtually nonexistent. Show how fumarate is formed, using reactions beginning with PEP and including the necessary TCA cycle enzymes. (Spiro, S., and J. R. Guest, *TIBS* 16:310–314, 1991.)
    (c) What is the metabolic advantage of anaerobic *E. coli* in using the fumarate reductase pathway rather than lactate dehydrogenase to reoxidize NADH?
19. Consider the glyceraldehyde-3-phosphate dehydrogenase-phosphoglycerokinase enzymes of glycolysis and the succinate thiokinase of the TCA cycle. Compare the mechanisms of incorporation of inorganic phosphate into the respective nucleoside diphosphates.
20. The pyruvate dehydrogenase complex may have been regulated by phosphorylation of any one of the three different enzymes in the complex, yet regulation occurs on the first enzyme of the complex. How is regulation of the complex consistent with the regulation observed in metabolic pathways whose enzymes are not physically associated?
21. Although there is no net synthesis of glucose from acetyl-CoA in mammals, acetyl-CoA has two major functions in gluconeogenesis. Explain the functions of acetyl-CoA in the synthesis of glucose from lactate in mammalian liver.

CHAPTER 16

Electron Transport, Proton Translocation, and Oxidative Phosphorylation

Potential energy stored in the reduced coenzymes NADH and $FADH_2$, produced in the reactions of the TCA cycle, is used to drive the synthesis of large amounts of ATP.

In the two preceding chapters we described the catabolism of glucose to pyruvate by glycolysis and the further breakdown of pyruvate to CO_2 and H_2O in the TCA cycle. These are oxidative processes in which the electrons obtained from carbohydrate oxidation are transferred to the coenzymes NAD^+ and FAD. Glucose catabolism involves six such oxidative reactions: one in glycolysis, one in the conversion of pyruvate to acetyl-CoA, and the remaining four in the TCA cycle. In five of these reactions, NAD^+ is reduced to NADH; in the sixth, FAD bound to succinate dehydrogenase is reduced to $FADH_2$. In later chapters we see that fatty acids and amino acids also are catabolized by oxidative pathways that use these same coenzymes as electron acceptors.

The continued availability of NAD^+ and FAD for use as electron acceptors depends on reoxidation of the reduced coenzymes. It is in this process of coenzyme reoxidation that molecular oxygen finally enters the picture, because the terminal electron acceptor for NADH and $FADH_2$ oxidation is O_2.

Reoxidation of NADH and $FADH_2$ at the expense of molecular oxygen can be summarized by the following overall reactions:

$$NADH + H^+ + \frac{1}{2}O_2 \longrightarrow NAD^+ + H_2O \qquad \Delta G^{\circ\prime} = -52.6 \text{ kcal/mole} \quad (1)$$

$$FADH_2 + \frac{1}{2}O_2 \longrightarrow FAD + H_2O \qquad \Delta G^{\circ\prime} = -43.4 \text{ kcal/mole} \quad (2)$$

The most important aspect of these reactions is the large amount of free energy they release. In fact, most of the free energy available from the oxidation of glucose is still present in the reduced coenzymes generated during glycolysis and the TCA cycle. Complete oxidation of 1 mole of glucose to CO_2 and H_2O results in the formation of 10 moles of NADH and 2 moles of $FADH_2$, which release $10 \times 52.6 + 2 \times 43.4$, or a total of 613 kcal when they are reoxidized. Since the total $\Delta G^{\circ\prime}$ for glucose oxidation is about -686 kcal/mole, it is clear that about 90% of this free energy is tapped by the cell only when the reduced coenzymes are reoxidized. Cells that live aerobically use the energy released in the transfer of electrons from NADH or $FADH_2$ to O_2 to drive the formation of ATP, and this process accounts for most of the ATP yield in aerobic metabolism.

Electron transport does not occur by a direct reaction of the coenzymes with O_2 but rather by a stepwise flow of electrons through a chain of intermediate electron carriers in the mitochondrial inner membrane (fig. 16.1). In the course of these reactions, protons are translocated from the solution on one side of the membrane and released on the opposite side. ATP synthesis is driven by the flow of protons back across the membrane. ATP synthesis that is linked to electron-transfer reactions in this way is called oxidative phosphorylation.

Electron Transport Is a Membrane-Localized Process

In eukaryotic cells, electron transport and oxidative phosphorylation occur in mitochondria. Mitochondria have both an outer membrane and an inner membrane with extensive infoldings called cristae (fig. 16.2). The inner membrane separates the internal matrix space from the intermembrane space between the inner and outer membranes. The outer membrane has only a few known enzymatic activities and is permeable to molecules with molecular weights up to about 5,000. By contrast the inner membrane is impermeable to most ions and polar molecules, and its proteins include the enzymes that catalyze oxygen consumption and formation of ATP. The role of mitochondria in O_2 uptake, or respiration, was demonstrated in 1913 by Otto Warburg; in 1948, Eugene Kennedy and Albert Lehninger showed that mitochondria carry out the reactions of the TCA cycle, the transport of electrons to O_2, and the formation of ATP.

Figure 16.1

Overview of electron transport, proton translocation, and oxidative phosphorylation. Multiprotein electron-transfer complexes in the mitochondrial inner membrane reoxidize the NADH and $FADH_2$ that are generated by the TCA cycle and other mitochondrial reactions. These complexes pass electrons stepwise to molecular oxygen, and this flow is coupled indirectly but tightly to the synthesis of ATP. As electrons move through some of the electron-transfer complexes, protons are taken up from the solution in the mitochondrial matrix space and released on the other side of the inner membrane, setting up a gradient of pH and electric potential across the membrane. Protons flow back into the mitochondrion through an ATP-synthase enzyme, which uses the free energy liberated from the proton influx to drive ATP synthesis.

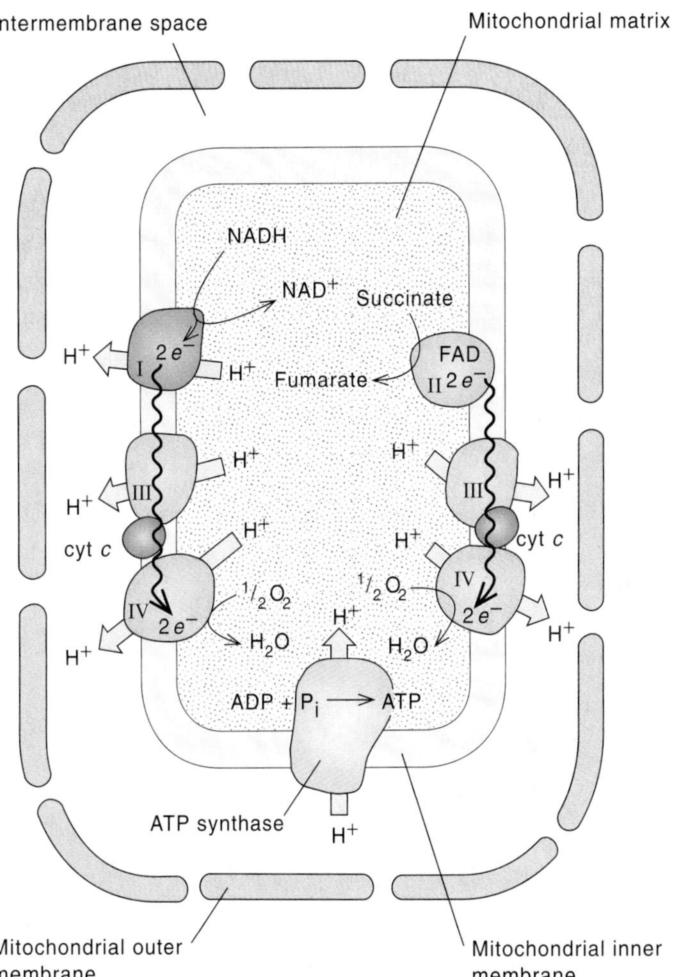

In bacteria the plasma membrane plays the same role as the inner mitochondrial membrane in electron transport and oxidative phosphorylation.

A Bucket Brigade of Molecules Carries Electrons from the TCA Cycle to O_2

The electron carriers that participate in the flow of electrons to O_2 are a structurally diverse group. Occupying a central position are a series of heme-containing proteins, the cytochromes. Hemes are porphyrins with iron at the center (see chapter 11). Three main

Figure 16.2

(*a*) Thin-section electron micrograph of a mitochondrion in a frog kidney cell. (Courtesy of Dr. John Luft.) (*b*) Schematic cross-sectional view of a mitochondrion. The cristae are extensions of the inner membrane, which separates the matrix from the intermembrane space. The intermembrane space communicates with the cytosol through pores in the outer membrane.

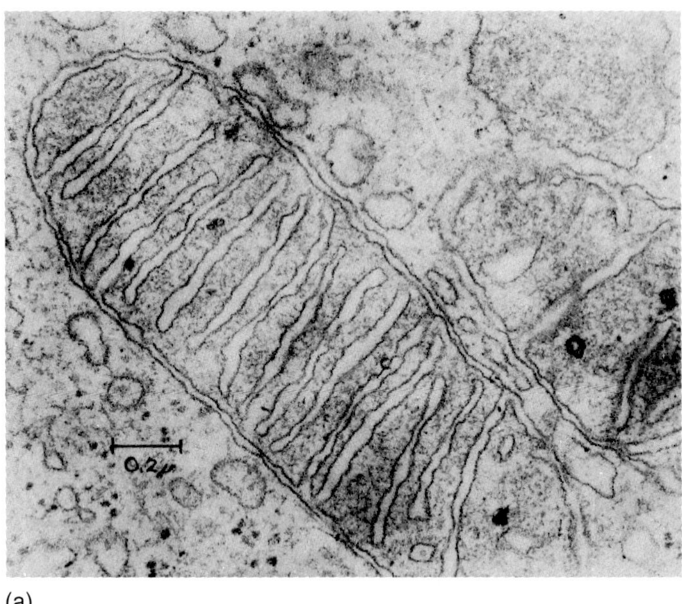

(a)

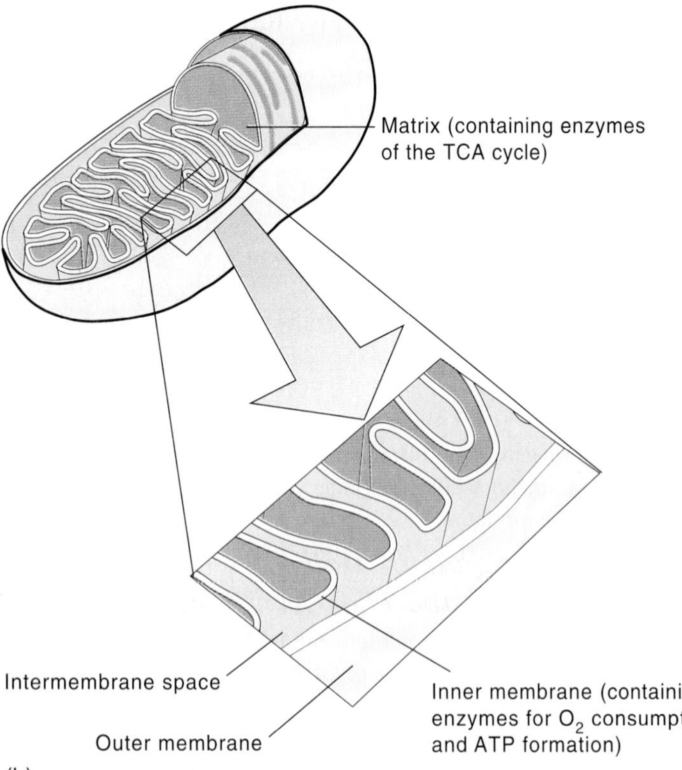

(b)

Figure 16.3

Iron protoporphyrin IX (heme) is found in the *b*-type cytochromes and in hemoglobin and myoglobin. In heme *c*, cysteine residues of the protein (R) are attached covalently by thioether links to the two vinyl (—CH=CH_2) groups of protoporphyrin IX. Heme *c* is found in the *c* cytochromes. In heme *a*, which is found in the *a* cytochromes, a 15-carbon isoprenoid side chain is attached to one of the vinyls, and a formyl group replaces one of the methyls.

Iron protoporphyrin IX

Heme *c*

Heme *a*

types of cytochromes, *a, b,* and *c,* exist and are distinguished by different substituents on the periphery of the porphyrin ring and different modes of attachment of the porphyrin to the protein (fig. 16.3). Cytochrome *c* is a small, water-soluble protein associated loosely with the inner mitochondrial membrane. Another *c* cytochrome (c_1), two *b* cytochromes (b_L and b_H), and two *a* cytochromes (*a* and a_3) are embedded in the membrane as parts of large complexes described below.

The Fe atoms of the cytochromes undergo oxidation and reduction during respiration, cycling between the ferrous (Fe^{2+}) and ferric (Fe^{3+}) oxidation states. The absorption spectra of the oxidized and reduced forms differ (fig. 16.4). In the 1930s, David Keilin used this property to measure the oxidation–reduction states of cytochromes in living cells. Under anaerobic conditions the cytochromes rapidly became reduced; in the presence of O_2 they became oxidized. Certain molecules that inhibited respiration (CO, N_3^-, or CN^-) blocked the oxidation; other inhibitors (amytal, rotenone, and malonate) blocked the reduction. Keilin found that the transfer of electrons from cytochrome *c* to O_2 required another component, which he called "cytochrome oxidase." By 1940 it was clear that cytochrome oxidase was identical to an enzyme that Warburg had discovered and that it involved cytochromes *a* and a_3. The two *a* cytochromes appeared to work in series in passing electrons from cytochrome *c* to O_2:

$$\begin{array}{cccc} \text{cytochrome } c\ (Fe^{2+}) & \text{cytochrome } a\ (Fe^{3+}) & \text{cytochrome } a_3\ (Fe^{2+}) & \frac{1}{4}O_2 + H^+ \\ \text{cytochrome } c\ (Fe^{3+}) & \text{cytochrome } a\ (Fe^{2+}) & \text{cytochrome } a_3\ (Fe^{3+}) & \frac{1}{2}H_2O \end{array} \qquad (3)$$

The *b* cytochromes and cytochrome c_1 fit into this scheme between reducing substrates and cytochrome *c*. The idea thus developed that the respiratory apparatus includes a chain of cytochromes that operate in a defined sequence. The next question was whether the cytochromes are all bound together in a giant complex or whether they diffuse independently in the membrane. Before we address this point, we need to consider three other types of electron carriers that participate in the electron-transport chain: flavoproteins, iron–sulfur proteins, and ubiquinone.

The dehydrogenases that remove electrons from succinate or NADH contain flavins as prosthetic groups. NADH dehydrogenase contains flavin mononucleotide (FMN); succinate dehydrogenase contains covalently bound flavin adenine dinucleotide (FAD). Flavins can undergo one-electron reduction to semiquinone forms or two-electron reduction to the dihydroflavins, $FMNH_2$ and $FADH_2$ (see figs. 11.8 and 11.9).

The mitochondrial inner membrane also has a flavoprotein, glycerol-3-phosphate dehydrogenase, that oxidizes glycerol-3-phosphate to dihydroxyacetone phosphate, reducing a bound FAD. In the matrix space are at least eight other flavoprotein dehydrogenases, including enzymes that participate in the oxidation of fatty acids (see chapter 21). These dehydrogenases feed electrons to still another flavoprotein in the membrane.

NADH dehydrogenase and succinate dehydrogenase also contain Fe atoms that are bound by the S atoms of cysteine residues of the protein, in association with additional, inorganic sulfide atoms. Structures of these complexes are shown in figure 11.21. Succinate dehydrogenase has three iron–sulfur centers, one with a [2Fe–2S] cluster, one with [4Fe–4S], and one with a cluster containing 3 Fe atoms and 3 (or possibly 4) sulfides. Iron–sulfur centers undergo one-electron oxidation–reduction reactions.

The final electron carrier to be considered is ubiquinone (UQ), a benzoquinone with a long, hydrophobic side chain (fig. 16.5). The concentration of UQ in the mitochondrial inner membrane far exceeds that of the cytochromes. In heart mitochondria, for example, the concentration of UQ is about seven times that of cytochrome a_3. Because of its hydrophobic character, the UQ is able to move freely in the phospholipid bilayer of the membrane.

Figure 16.4

Optical absorption spectra of a 10 μM solution of cytochrome *c* in the reduced (blue) and oxidized (red) states.

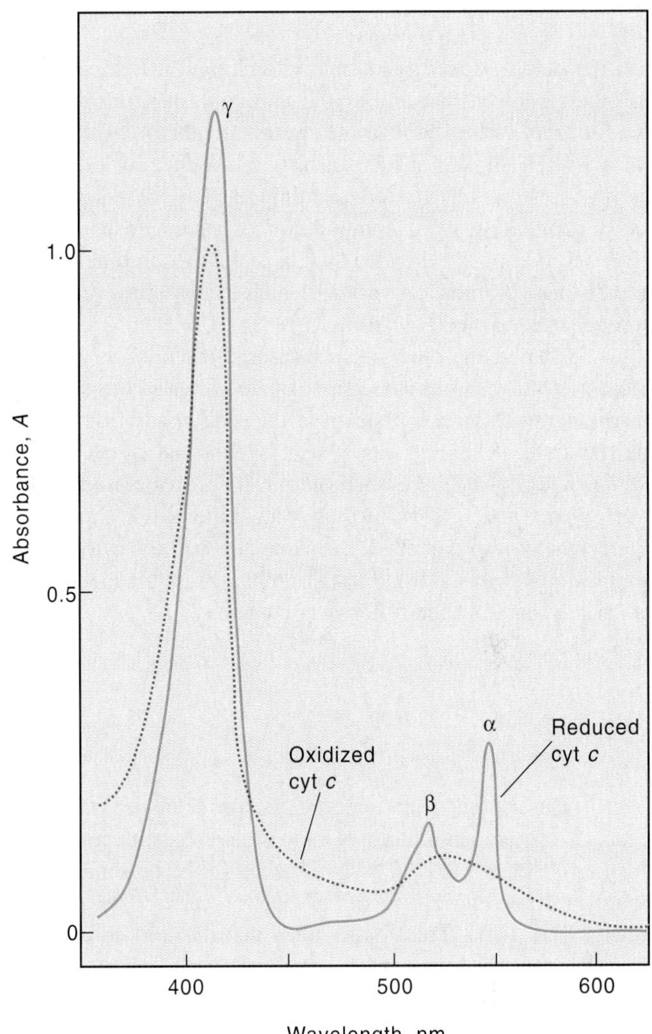

Ubiquinone undergoes a two-electron reduction to the dihydroquinone or quinol, UQH_2 (see fig. 16.5). It also can accept a single electron and stop at the semiquinone, which can be either anionic ($UQ\bar{\bullet}$) or neutral (UQH•), depending on the pH and on the nature of the binding site when the semiquinone is bound to a protein. The • in $UQ\bar{\bullet}$ or UQH• indicates that the semiquinone contains an unpaired electron.

The reduction of UQ can be measured by the disappearance of an absorption band at 275 nm. By using this technique, it was shown that adding a substrate such as succinate caused a rapid reduction of essentially all the UQ present in the inner membrane; the resulting UQH_2 could be reoxidized by the cytochrome system in the presence of O_2. To determine whether UQ is a *necessary* participant in electron transport from succinate to O_2, the quinone was removed from mitochondria by selective extraction with an organic solvent. The depleted mitochondria were incapable of respiration but recovered this activity when UQ was added back.

The Sequence of Electron Carriers Was Deduced from Kinetic Measurements

With the help of spectrophotometric techniques, Britton Chance and others showed that any of the normal respiratory substrates can reduce all of the cytochromes in the membrane. Either succinate or NADH-linked substrates also can reduce all of the UQ. The flavins, however, are reduced only partially by succinate but almost completely by a combination of succinate and malate. These observations fit the idea that all of the cytochromes and UQ are part of a common network to which various flavoprotein dehydrogenases can feed electrons.

Chance and others also measured the rates at which the different electron carriers became oxidized after addition of O_2 to anaerobic mitochondria or became reduced after addition of a suitable substrate. When O_2 was added, cytochrome a_3 became oxidized first, followed by cytochrome a, the c cytochromes, the b cytochromes, and flavins, in that order. Later work showed that UQ became oxidized at about the same rate as the b cytochromes. Observations of this sort led to the proposal that the electron carriers are arranged in the following sequence:

Succinate ⟶ FAD ↘

NADH ⟶ FMN ⟶ [UQ, cyt b] ⟶

[cyt c_1, cyt c] ⟶ cyt a ⟶ cyt a_3 ⟶ O_2 **(4)**

This scheme was supported and refined by examining the effects of specific inhibitors of individual steps in the electron-transport chain. If CO or CN^- was added in the presence of a reducing substrate and O_2, all of the electron carriers became more reduced (box 16A). This fits the idea that these inhibitors act at the end of the respiratory chain, preventing the transfer of electrons from cytochrome to O_2. If amytal (a barbiturate) or rotenone (a plant toxin long used as a fish poison) was added instead, NAD^+ and the flavin in NADH dehydrogenase were reduced, but the carriers downstream became oxidized. The antibiotic antimycin caused NAD^+, flavins, and the b cytochromes to become more reduced, but cytochromes c, c_1, a, and a_3 all became more oxidized. The situation here is analogous to the construction of a dam across a stream: When the gates are closed, the water level rises upstream from the dam and falls downstream. The observation that antimycin did not inhibit reduction of UQ showed that the quinone fits into the chain upstream of cytochromes c, c_1, a, and a_3.

Redox Potentials Give a Measure of Oxidizing and Reducing Strengths of the Different Electron Carriers

We now see that mitochondria contain a variety of molecules—cytochromes, flavins, ubiquinone, and iron–sulfur proteins—all of which can act as electron carriers. To discuss how these carriers cooperate to transport electrons from reduced substrates to O_2, we need to have a measure of each molecule's tendency to release or accept electrons. The standard redox potential, $E°$, provides such a measure. Redox potentials are thermodynamic properties that depend on the differences in free energy between the oxidized and reduced forms of a molecule. Like the electric potentials that govern electron flow from one pole of a battery to another, $E°$ values are specified in volts. Because electron-transfer reactions frequently involve protons also, an additional symbol is used to indicate that an $E°$ value applies to a particular pH; thus $E°'$ refers to an $E°$ at pH 7.

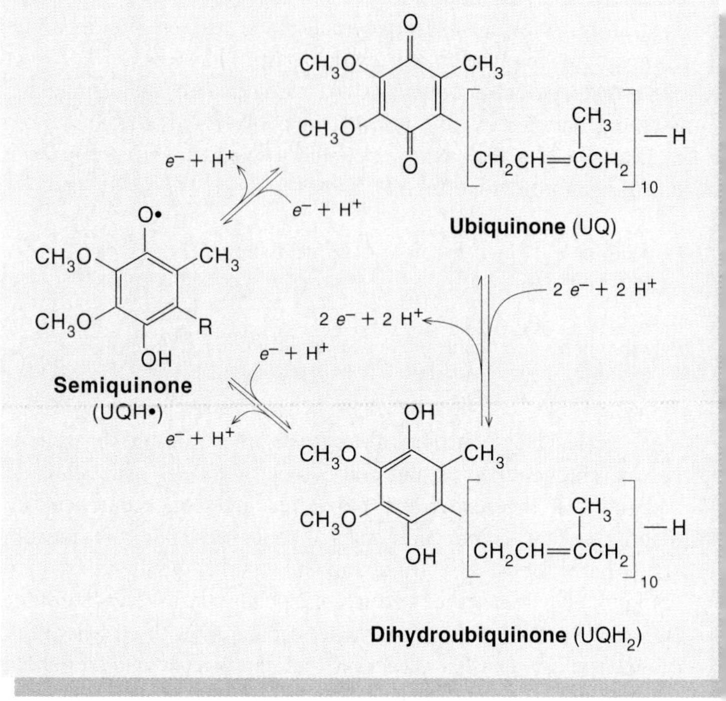

Figure 16.5

Structures of ubiquinone in its oxidized state (UQ), in its half reduced form UQH•, and in its fully reduced dihydroquinone or ubiquinol state (UQH_2). The UQ found in eukaryotes has 50 carbons (10 prenyl units) in the side chain; shorter side chains are found in some bacteria.

Consider the transfer of electrons from a reduced molecule, D_{red}, to an oxidized acceptor, A_{ox}. $\Delta G°$ for the reaction is proportional to the difference between the $E°$ values of the two molecules ($\Delta E°$):

$$D_{red} + A_{ox} \longrightarrow D_{ox} + A_{red} \qquad (5)$$

$$\Delta G° = -n\mathcal{F}\,\Delta E° = -n\mathcal{F}(E_A° - E_D°) \qquad (6)$$

Here n is the number of electrons transferred from the donor to the acceptor, and $\mathcal{F}$ is the Faraday constant (23,060 cal $volt^{-1}$ $mole^{-1}$ or 96.5 kJ $volt^{-1}$ $mole^{-1}$). Note that the overall reaction is spontaneous ($\Delta G°$ negative) if $\Delta E°$ is positive, that is, if electrons move from the molecule with the more negative $E°$ value to that with the more positive value. In other words, negative $E°$ values are associated with strong reductants, and positive values with strong oxidants. Electrons flow spontaneously in the direction of more positive potential. With $n = 2$, a $\Delta E°$ of 0.1 V corresponds to a $\Delta G°$ of −4.61 kcal/mole.

16A
BOX

Cyanide Poisoning

Cyanide may be inhaled as the gas HCN or ingested as the anion. If the person is not treated, mitochondrial respiration and energy production cease, and cell death occurs rapidly. This is not surprising, as we know that cyanide binds strongly to the central Fe^{3+} ligand in the hemes of cytochromes *a* and a_3. If cyanide poisoning is diagnosed immediately, it may be treated with thiosulfate. The thiosulfate reacts with cyanide to form thiocyanate. Rhodanase catalyzes this reaction:

$$CN^- + S_2O_3^{2-} \xrightarrow[\text{Rhodanase}]{\nearrow SO_3^{2-}} SCN^-$$

The greatest hope is that thiosulfate is administered before the cyanide has reacted extensively with the cytochromes.

Alternative treatments for cyanide poisoning include administration of high concentrations of oxygen and administration of methylene blue. Methylene blue has an oxidation potential close to that of the cytochromes, so it acts as a biochemical detour for electrons around the poisoned respiratory chain.

$\Delta E°$ values can be measured by separating the two redox couples into different solutions connected by a salt bridge and a voltmeter (fig. 16.6). If both solutions are under standard conditions, the meter senses a voltage difference equal to the $\Delta E°$. Standard conditions are chosen to be 1 M concentrations of all reactants and products except for H^+, OH^-, and H_2O. By the law of mass action a solution containing a redox couple D_{ox}/D_{red} can be made more oxidizing by increasing the concentration ratio $[D_{ox}]/[D_{red}]$ or more reducing by decreasing this ratio.

Equation (6) defines the *difference* between two $E°$ values. To set the individual values, we need to choose a particular redox couple as a reference. The reference that biochemists use most commonly is the standard hydrogen half-cell, in which protons at pH 0 are reduced to H_2 at a pressure of 1 atm ($2\ H^+ + 2\ e^- \longrightarrow H_2$). By convention this is assigned an $E°$ value of zero. Relative to the standard hydrogen half-cell, the NAD^+/NADH couple at pH 7.0 has an $E°'$ of -0.32 V, and the pyruvate/lactate couple has an $E°'$ of -0.19 V. Table 16.1 gives the $E°'$ values of additional redox couples, including components of the respiratory chain. These values are listed in order of increasing $E°'$, which means that under standard conditions a given couple reduces any of the couples below it in the table.

The sequence of carriers in the respiratory chain should be generally consistent with the relative $E°$ values of the carriers because, given equal concentrations of reactants and products, electrons flow from a carrier with a more negative $E°$ value to one with a more positive value. The sequence of flavoproteins, UQ, and cytochromes that we presented in equation (4) agrees with this expectation. This proposition is demonstrated by plotting the $E°'$ values as a function of the carriers' suggested positions in the chain (fig. 16.7). However, the $E°$ values do not dictate the detailed organization of the respiratory chain because electrons can move in the direction of more negative $E°$ if the reactants are present at sufficiently high concentrations relative to the products. Furthermore, even if electron transfer from one particular carrier to another is thermodynamically favorable, it does not occur unless the two carriers are able to make contact. The contacts between carriers in the respiratory chain depend on how the hemes, flavins, quinones, and iron–sulfur centers are positioned in the proteins that bind them.

Figure 16.6

Apparatus for measuring the difference between the $E°$ values of two redox couples. The cell on the left contains equimolar concentrations of NADH and NAD^+; that on the right, equimolar concentrations of $FMNH_2$ and FMN. If both solutions are at pH 7, the voltmeter senses the difference between the two $E°'$ values (0.10 V, negative on the left, in this example). To determine the $E°'$ of one of the redox couples, the other couple is replaced by a standard redox couple.

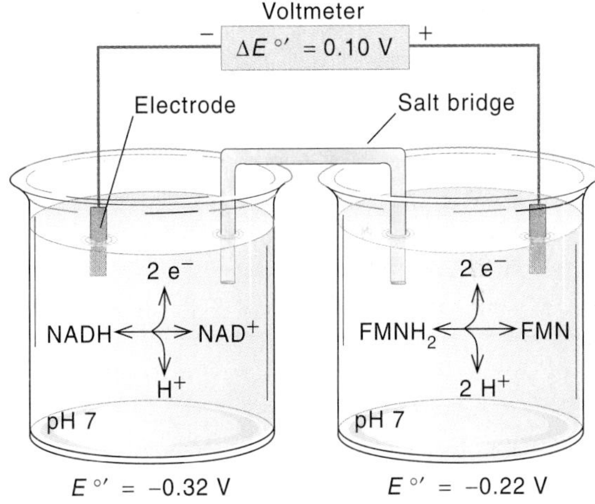

Most of the Electron Carriers Exist in Large Complexes

By disrupting the mitochondrial inner membrane with detergents, Yousef Hatefi, David Green, and others found that all of the major electron carriers except for cytochrome *c* and UQ occur in the

Table 16.1
$E^{\circ\prime}$ Values of Biochemical Redox Couples

Redox Couple	$E^{\circ\prime}$ (V)	n[a]
Succinate + CO_2 + 2 H^+ + 2 e^- $\rightleftharpoons$ α-Ketoglutarate + H_2O	−0.67	2
Glycerate-3-phosphate + 2 H^+ + 2 e^- $\rightleftharpoons$ Glyceraldehyde-3-P + H_2O	−0.55	2
α-Ketoglutarate + CO_2 + 2 H^+ + 2 e^- $\rightleftharpoons$ Isocitrate	−0.38	2
NAD^+ + H^+ + 2 e^- $\rightleftharpoons$ NADH	−0.32	2
FMN + 2 H^+ + 2 e^- $\rightleftharpoons$ $FMNH_2$	−0.22[b]	2
FAD + 2 H^+ + 2 e^- $\rightleftharpoons$ $FADH_2$	−0.22[b]	2
Pyruvate + 2 H^+ + 2 e^- $\rightleftharpoons$ Lactate	−0.19	2
Oxaloacetate + 2 H^+ + 2 e^- $\rightleftharpoons$ Malate	−0.17	2
Fumarate + 2 H^+ + 2 e^- $\rightleftharpoons$ Succinate	−0.03	2
Cytochrome b_L (Fe^{+3}) + e^- $\rightleftharpoons$ Cytochrome b_L (Fe^{+2})	−0.03	1
UQ + H^+ + 2 e^- $\rightleftharpoons$ UQH•	+0.03[c]	1
Cytochrome b_H (Fe^{+3}) + e^- $\rightleftharpoons$ Cytochrome b_H (Fe^{+2})	+0.05	1
UQ + 2 H^+ + 2 e^- $\rightleftharpoons$ UQH_2	+0.11[c]	2
UQH• + H^+ + e^- $\rightleftharpoons$ UQH_2	+0.19[c]	1
Rieske Fe–S (Fe^{+3}) + e^- $\rightleftharpoons$ Fe–S (Fe^{+2})	+0.28	1
Cytochrome c_1 (Fe^{+3}) + e^- $\rightleftharpoons$ Cytochrome c_1 (Fe^{+2})	+0.23	1
Cytochrome c (Fe^{+3}) + e^- $\rightleftharpoons$ Cytochrome c (Fe^{+2})	+0.24	1
Cytochrome a (Fe^{+3}) + e^- $\rightleftharpoons$ Cytochrome a (Fe^{+2})	+0.28	1
Cytochrome a_3 (Fe^{+3}) + e^- $\rightleftharpoons$ Cytochrome a_3 (Fe^{+2})	+0.35	1
O_2 + 4 H^+ + 4 e^- $\rightleftharpoons$ 2 H_2O	+0.82	4

[a]n is the number of electrons transferred.

[b]This value is for the free coenzyme. $E^{\circ\prime}$ values for flavoproteins range from −0.3 to 0 V.

[c]For UQ in aqueous ethanol.

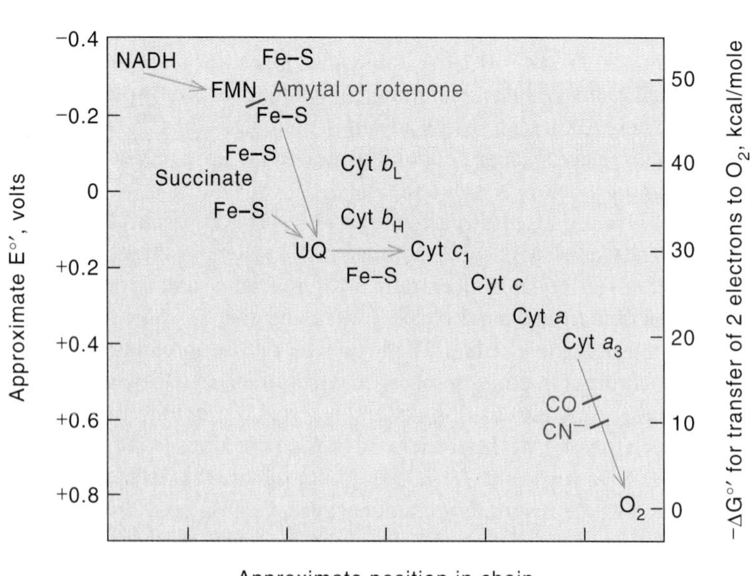

Figure 16.7

$E^{\circ\prime}$ values of some of the electron carriers in the respiratory chain, plotted as a function of the positions of the carriers in the chain. The diagram includes components of the cytochrome bc_1 complex that are discussed later in the chapter. For simplicity it shows a single $E^{\circ\prime}$ value for ubiquinone, although the UQH_2/UQH• and UQH•/UQ redox couples have different values. The $E^{\circ\prime}$ for the FMN associated with NADH dehydrogenase is not known accurately; the value shown is for free FMN. The $E^{\circ\prime}$ values and positions for the iron–sulfur centers in NADH dehydrogenase also are uncertain; only two of the six to eight iron–sulfur centers in this region of the respiratory chain are shown. The right-hand scale gives the standard free energy change ($\Delta G^{\circ\prime}$) for transferring two equivalents of electrons from a carrier with the corresponding $E^{\circ\prime}$ value to O_2. Sites of action of some inhibitors (amytal, rotenone, CO, and CN^-) are shown.

Figure 16.8

The multisubunit complexes of the respiratory chain. Complexes I (NADH dehydrogenase) and II (succinate dehydrogenase) transfer electrons from NADH and succinate to UQ. Complex III (the cytochrome bc_1 complex) transfers electrons from UQH_2 to cytochrome *c*, and complex IV (cytochrome oxidase) transfers electrons from cytochrome *c* to O_2. The arrows represent paths of electron flow. NADH and succinate provide electrons from the matrix side of the inner membrane, and O_2 removes electrons on this side. Cytochrome *c* is reduced and oxidized on the opposite side of the membrane, in the lumen of a crista or in the intermembrane space.

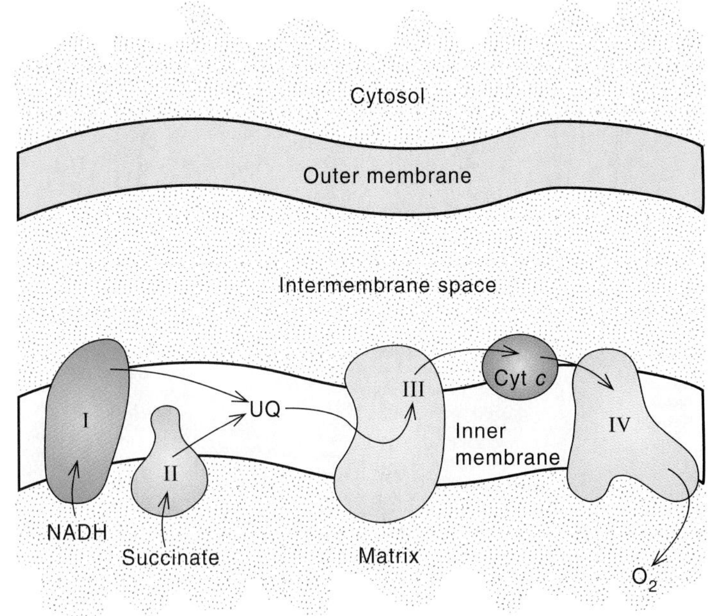

form of four large complexes (fig. 16.8). The isolated NADH dehydrogenase complex (complex I) catalyzes the reduction of UQ by NADH; the succinate dehydrogenase complex (complex II) catalyzes reduction of UQ by succinate. The cytochrome bc_1 complex (complex III) transfers electrons from reduced ubiquinone (UQH_2) to cytochrome *c*. And finally, cytochrome oxidase (complex IV) transfers electrons from cytochrome *c* to O_2. Each of these complexes contains numerous polypeptide subunits (table 16.2).

The Main Function of the Mitochondrial Electron Transport Complexes Is to Translocate Protons

The main function of the electron transport complexes residing in the inner mitochondrial membrane is to provide a means for translocating protons from the mitochondrial matrix to intermembrane space so as to create a proton gradient across the inner membrane. This energy-requiring process is fueled by the energy-releasing process of electron transport. Before we get into a detailed description of the electron transport complexes, we should indicate the general ways in which protons can be translocated, since the design of the electron transport complexes is intimately related to this.

It appears that there are two ways, one that involves a redox loop and one that involves a proton pump. The redox loop works by having an enzyme with two spatially separated, active sites with the proton-requiring reaction occurring on the proton-poor side of the membrane and the proton-releasing reaction occurring on the proton-rich side of the membrane. This is easiest to appreciate by example. The cytochrome *bd* ubiquinol oxidase of *E. coli* exemplifies the redox loop mechanism (fig. 16.9). The ubiquinol oxidation site is located on the outer surface of the plasma membrane, and the site for oxygen reduction is located on the inner, cytosolic surface of the membrane. The two-electron oxidation of ubiquinol releases two substrate-derived protons to the periplasm, and the electrons are passed along to the second active center, where the oxygen is reduced, utilizing protons from the cytoplasm. For each electron used to form water, one proton appears in the periplasm and one disappears from the cytoplasm, yielding a H^+/e^- ratio of 1. The cytochrome *bo* ubiquinol oxidase of *E. coli* appears to use both a redox loop and a proton pump to

Table 16.2
Components of the Mitochondrial Electron-Transport Complexes

Complex	$M_r \times 10^6$ (Monomer)[a]	Polypeptides	Prosthetic Groups	Ratio in Mitochondria[b]
I NADH dehydrogenase	0.7–0.9	>30	FMN, Fe–S clusters	1
II Succinate dehydrogenase	0.14	4–5	FAD, Fe–S clusters, b_{560} heme	2
III Cytochrome bc_1 complex	0.25	11	b_{562}, b_{566}, c_1 hemes, [2Fe–2S] cluster	3
IV Cytochrome oxidase	0.16–0.17	13	aa_3 hemes, Cu_a Cu_{a_3}	6–7

[a]Protein only.

[b]Based on bovine SMP content of FMN (I), covalently bound FAD (II), cytochrome c_1 (III), and cytochrome aa_3 (IV), corrected to nearest integer relative to I.

Figure 16.9

Two schemes by which respiratory cytochrome complexes generate a transmembrane proton electrochemical gradient. (*a*) In the cytochrome *bd* ubiquinol oxidase of *E. coli* the site where oxygen reacts is believed to be close to the inner surface, so it is likely to pick up protons from the cytosol. Protons that appear on the outside are derived from the ubiquinol substrate. (*b*) In the cytochrome *bo* ubiquinol oxidase from *E. coli*, half of the protons that appear on the outside are transported through a proton-conducting channel through the enzyme, and half originate from the substrate as in (*a*).

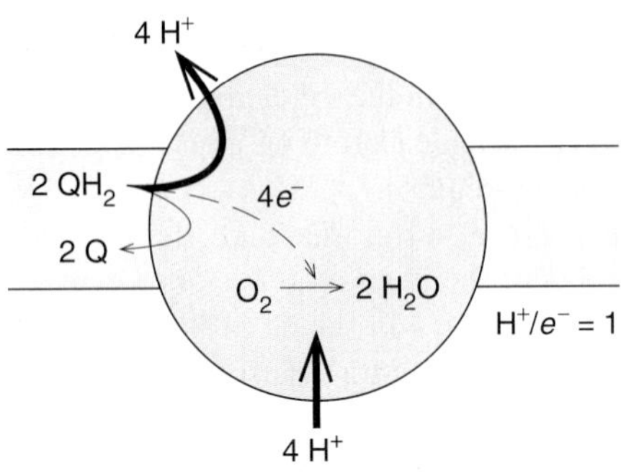

(a) **Cytochrome *bd* ubiquinol oxidase**

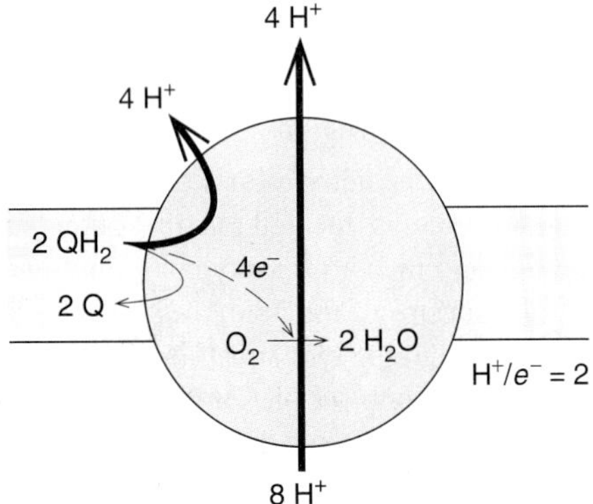

(b) **Cytochrome *bo* ubiquinol oxidase**

Figure 16.10

Electron transfer through complex I. The reduction of FMN to $FMNH_2$ by NADH requires the uptake of one proton from the matrix. $FMNH_2$ subsequently transfers electrons to a series of iron–sulfur centers and releases protons to the solution in the intermembrane space. When the iron–sulfur centers reduce UQ to UQH_2, two more protons are taken up from the matrix.

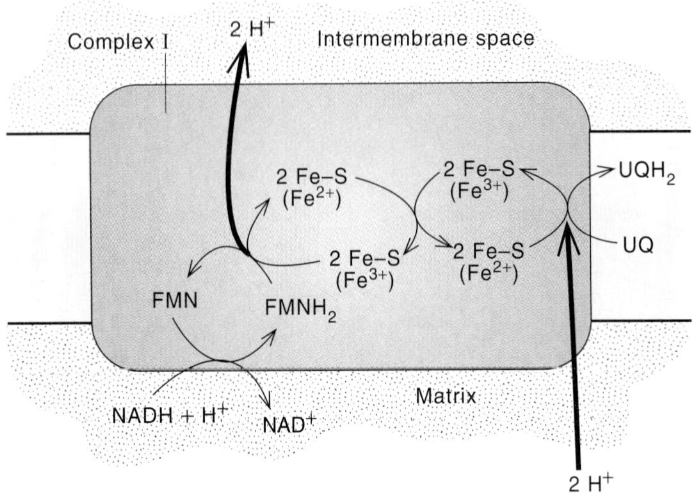

translocate protons. As does cytochrome *bd*, cytochrome *bo* deposits one proton per electron into the periplasm from oxidation of the quinol and consumes protons from the cytoplasm in forming water. Cytochrome *bo* pumps one additional proton per electron through a proton-conducting channel. These two mechanisms combined yield a H^+/e^- ratio of 2. We will have more to say about how proton pumps work later in this chapter and especially in chapter 20.

Complexes I and II Mediate the Transfer of Electrons from NADH and $FADH_2$ to Ubiquinone

In complexes I and II we find the first steps in the extended pathway of electron transfer from the TCA cycle coenzymes, NADH and $FADH_2$, to molecular oxygen. If we were focusing on the electron transfer parts of these reactions only, we might refer to these two complexes as NADH–coenzyme Q reductase and $FADH_2$–coenzyme Q reductase, since they both reduce ubiquinone (coenzyme Q) to dihydroquinone. However, complex I is also involved in the concomitant translocation of protons by a redox loop mechanism.

The first reaction in complex I involves a two-electron reduction of a tightly bound FMN to $FMNH_2$ (fig. 16.10). The reduced flavin then transfers two electrons, one at a time, to an iron–sulfur center, and in the process, two protons are released into intermembrane space. These electrons move from one iron–sulfur center to another and eventually to UQ, which picks up two electrons from the Fe–S center and two protons from the matrix side of the mitochondrion. One important point about organization of the electron transfer processes that occur in complex I is necessitated by the nature of the carriers. NADH is a two-electron carrier, whereas Fe–S centers are one-electron carriers. FMN and ubiquinone can transfer electrons in either one or two electron steps. One can see that the four electron carriers in complex I are arranged in the order NAD, FMN, Fe–S, and U, so NAD and Fe–S do not make direct contact.

The mechanism for electron transport and proton translocation suggests that one proton is translocated per electron transferred through complex I. On the basis of the high yield of ATP formed when electrons pass through complex I (see fig. 16.17), this mechanism will almost certainly have to be modified to accommodate a higher H^+/e^- ratio.

Complex II, the succinate dehydrogenase complex, is active in electron transport but not in proton translocation. Succinate dehydrogenase is the only enzyme of the TCA cycle that is embedded in the inner membrane. Within its four subunits it has three iron–sulfur centers. As in NADH dehydrogenase, the substrate-oxidation site is on the matrix side of the membrane (fig. 16.11).

Figure 16.11

Complex II: The succinate dehydrogenase complex.

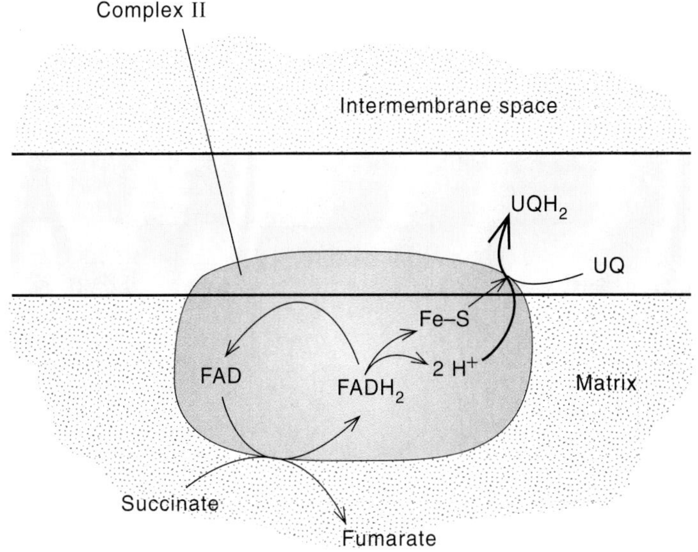

There is no translocation of protons in the reactions of complex II. The two protons obtained from the conversion of succinate to fumarate are transferred to the FAD. Next the two protons in the $FADH_2$ are released when $FADH_2$ transfers two electrons to the iron–sulfur centers; these protons are needed for the conversion of UQ to UQH_2.

The UQH_2 formed in complexes I and II diffuses in the membrane to complex III, where it becomes reoxidized.

Complex III, the Cytochrome bc$_1$ *Complex, Transfers Electrons from QH$_2$ to Cytochrome While Translocating Protons by a Redox Loop*

The cytochrome bc_1 complex contains two b-type cytochromes: b_L and b_H. Both of the b hemes are on a single 30-kD polypeptide. The complex also contains cytochrome c_1, an iron–sulfur protein, and between four and six additional subunits.

Electrons moving through complex III from UQH_2 to cytochrome c follow a circuitous path (fig. 16.12). UQH_2 must transfer one of its electrons to the iron–sulfur protein before it can transfer an electron to cytochrome b_L:

$$UQH_2 \xrightarrow[\text{Fe–S }(Fe^{+3}) \to \text{Fe–S }(Fe^{+2})]{H^+} UQH\bullet \xrightarrow[\text{cyt } b_L\,(Fe^{+3}) \to \text{cyt } b_L\,(Fe^{+2})]{H^+} UQ \qquad (7)$$

Oxidation of UQH_2 by the iron–sulfur protein generates the semiquinone UQH•, which then serves as the reductant for cytochrome b_L. The reduced iron–sulfur protein transfers an electron to cytochrome c_1 and on to cytochrome c. Meanwhile, the reduced cytochrome b_L passes an electron to cytochrome b_H, which then contributes the electron for reduction of another molecule of UQ at a second site in the complex:

$$\begin{matrix} \text{cyt } b_L\,(Fe^{+2}) & & \text{cyt } b_H\,(Fe^{+3}) & & \frac{1}{2}UQH_2 \\ & \times & & \times & \\ \text{cyt } b_L\,(Fe^{+3}) & & \text{cyt } b_H\,(Fe^{+2}) & & \frac{1}{2}UQ + H^+ \end{matrix} \qquad (8)$$

If all of these steps occur twice, two electrons pass through the b cytochromes, one molecule of UQ is reduced to UQH_2 and one molecule of UQ diffuses back to either complex I or complex II. At this point we may seem to be back to where we started. But note that for each UQH_2 that is regenerated, two molecules of UQH_2 are oxidized. Thus there is a net oxidation of one UQH_2 to UQ, and two electrons proceed to cytochrome c molecules on their way to O_2 (see fig. 16.12).

This scheme for electron transport through the cytochrome bc_1 complex is known as the Q cycle. The role of the iron–sulfur protein was confirmed by extracting this protein from the cytochrome bc_1 complex. If the iron–sulfur protein is removed, electron transfer from UQH_2 to cytochrome c_1 is prevented. This shows that the iron–sulfur protein operates upstream of cytochrome c_1, even though its $E°$ is more positive than that of the cytochrome (see fig. 16.7 and table 16.1).

As shown in figure 16.12, the sites at which UQH_2 and UQH• undergo oxidation face the intermembrane space, whereas the UQ reduction site is on the matrix side. UQ and UQH_2 evidently diffuse through the membrane from one site to the other. Antimycin blocks electron transfer from cytochrome b_H to UQ at the reduction site; this inhibitor was particularly helpful in clarifying the steps of the Q cycle.

Oxidation and reduction of UQ by the cytochrome bc_1 complex results in the uptake of protons from the solution on the matrix side of the inner membrane and the release of protons to intermembrane space. Two protons are taken up on the matrix side, and four protons are released on the intermembrane side for each pair of electrons that proceed on to cytochrome c (see fig. 16.12).

Complex IV, the Cytochrome Oxidase Complex, Transfers Electrons from Cytochrome c *to O$_2$ While Pumping Protons across the Membrane*

In the cytochrome oxidase complex, molecular oxygen is reduced to water. It takes four electrons to reduce one molecular oxygen to water. There are no substrate protons released into intermembrane space concomitant with the oxidation of cytochrome c, but four protons from the matrix are required for the formation of two water molecules, and an additional four are pumped from the matrix to the intermembrane space (fig. 16.13). Thus the transfer of one electron from cytochrome to O_2 results in translocation of one proton, $H^+/e^- = 1$.

Cytochrome oxidase contains three atoms of copper in addition to the hemes of cytochromes a and a_3. The copper atoms undergo one-electron oxidation–reduction reactions between the cuprous (Cu^+) and cupric (Cu^{2+}) states. One of the copper atoms (Cu_B) is close to the iron atom of cytochrome a_3 (fig. 16.14). The other two (Cu_A) are associated with cytochrome a but not as intimately. Cytochrome a and Cu_A both have effective $E°'$ values of

Figure 16.12

Electron transport through complex III: the Q cycle. UQH_2 is oxidized to UQ in two steps at an enzyme site on the side of the inner membrane facing the intermembrane space (top part of the figure), transferring the first electron to the Fe–S protein and the second electron to cytochrome b_L. The stoichiometries indicated are for two molecules of UQH_2 undergoing these reactions. One of the two molecules of UQ that are produced diffuses to a site on the matrix side of the membrane (bottom), where it is reduced back to UQH_2. The two electrons required for the reduction come through the *b* cytochromes. Antimycin inhibits this reaction. The UQH_2 generated at the reduction site diffuses back to the oxidation site, where it joins the pool of UQH_2 coming from complexes I and II. Electrons leaving the UQH_2 oxidation site reduce cytochrome *c* (top), which dissociates from the complex to go to cytochrome oxidase. Protons also are released to the solution in the intermembrane space. Protons are taken up from the matrix at the UQ reduction site. For clarity complex III is drawn as four separate compartments. In reality these functional units are all fused into one large multiprotein complex.

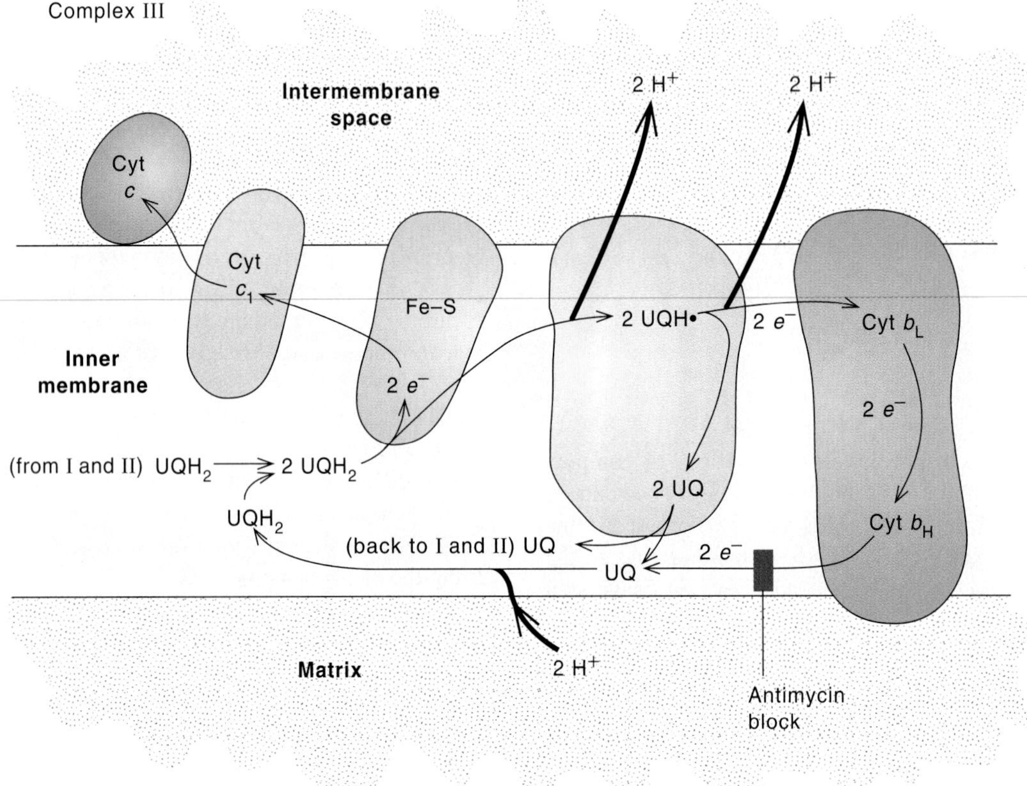

Figure 16.13

Cytochrome *c* oxidase complex IV. It takes four electrons to reduce one oxygen molecule. In the process, eight protons are picked up from the matrix side of the membrane. Four of these protons are used to provide the hydrogens for water. The remaining four are pumped to intermembrane space.

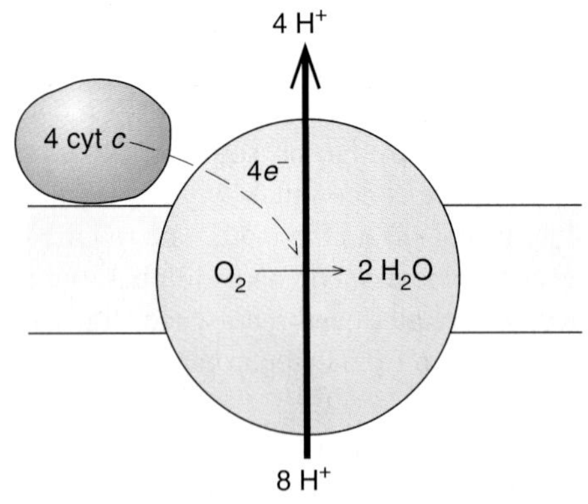

about 0.28 V, somewhat more positive than that of cytochrome *c* (0.22 V). Cytochrome a_3 and Cu_B have $E^{\circ\prime}$ values near 0.35 V in the absence of O_2, but these values become much more positive when the complex binds O_2.

Reduction of O_2 to 2 H_2O requires the addition of four electrons. Addition of one electron to O_2 to produce the superoxide radical ($O_2\bullet^-$) is thermodynamically unfavorable and has to be pulled along by a further reduction to the level of peroxide (H_2O_2) or H_2O. (Although the $E^{\circ\prime}$ for the overall reduction of O_2 to H_2O is +0.82 V, the redox couple $O_2/O_2\bullet^-$ has an $E^{\circ\prime}$ of −0.33 V.) This is one of the reasons that O_2 is unreactive in the absence of the enzyme. Cytochrome oxidase appears to get over this thermodynamic hurdle by binding O_2 to Cu_B and the Fe of cytochrome a_3 and then transferring two electrons at a time (see fig. 16.14).

Reconstitution Experiments Demonstrate the Key Roles of Ubiquinone and Cytochrome c *as Mobile Electron Carriers between the Giant Complexes*

Having described the four complexes that make up the respiratory chain, we now turn to the question of how electrons move between the complexes. To address this question, purified complexes were recombined so that they reconstituted longer stretches of the res-

Figure 16.14

A schematic drawing of heme a_3 and Cu_B in cytochrome oxidase (complex IV). The other two electron carriers (heme *a* and Cu_A) are not shown. (*a*) The oxidized enzyme, in which heme a_3 is in the ferric (Fe(III)) state, and Cu_B is cupric (Cu(II)). The heme is viewed edge-on. One of its axial ligands is a histidine; the other axial ligand probably is an hydroxide ion that is shared with Cu_B. (*b*) When heme a_3 and Cu_B are reduced to the ferrous and cuprous states by two electrons provided by heme *a* and Cu_A, the bridging ligand is replaced by a molecule of O_2. (*c*) Two electrons now can be transferred to the O_2—one from the Fe and one from the Cu—generating a peroxy intermediate ($O^-—O^-$). (*d*) The uptake of another electron and two protons causes the bridging O—O bond to split, releasing one of the O atoms as H_2O. The other O remains attached to the Fe as an O^{2-}—Fe(IV) complex. With the transfer of a fourth electron and one more proton, the second H_2O molecule is released, and the enzyme returns to its original state. The protons that are picked up in the course of these reactions come from the matrix side of the membrane. This scheme does not account for the additional protons that the enzyme pumps across the membrane.

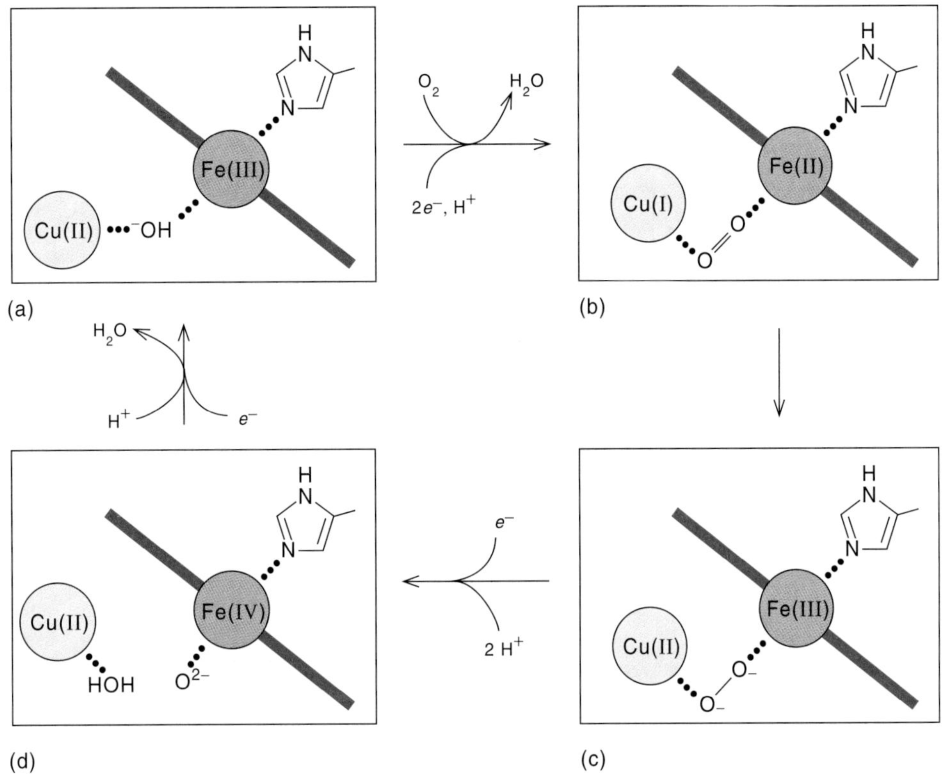

piratory chain. First, the complexes are mixed in the presence of phospholipids and a detergent. When the detergent is removed by dilution, the phospholipids assemble into vesicles and the proteins are incorporated into the phospholipid bilayer. If UQ is added, vesicles that contain complex I and complex III catalyze electron transfer from NADH to cytochrome *c*. In the presence of cytochrome *c*, vesicles containing complex III and complex IV transfer electrons from UQH_2 to O_2.

These reconstitution experiments supported the model for electron transfer shown in figure 16.8. In this model the complexes do not bind to each other directly. Instead, movement of electrons from complexes I and II to complex III is mediated by diffusion of UQH_2 from one complex to the other within the phospholipid bilayer. Similarly, electrons move from complex III to complex IV by the diffusion of reduced cytochrome *c* along the surface of the membrane. Remember that cytochrome *c* differs from the other cytochromes in being a water-soluble protein. It is attached loosely to the membrane surface by electrostatic interactions.

There is additional evidence that the electron-transfer complexes are not connected in fixed chains. Thus if most of the cytochrome oxidase complexes in the membrane are inhibited with CO, the few molecules that remain uninhibited are still able to catalyze oxidation of all the cytochrome *c* by O_2. This suggests that cytochrome *c* can diffuse from one cytochrome oxidase complex to another, rather than remaining bound to an individual complex. Also, cytochrome *c*, UQ, and the complexes themselves move about at different rates, which means that they cannot all stay stuck together.

Experiments on Mitochondrial Suspensions Demonstrate That Electron Transport Creates an Electrochemical Potential Gradient for Protons across the Inner Membrane

We have described how electron transport down the pathway from NADH or succinate to molecular oxygen results in protons being translocated across the inner membrane. Direct experimental evidence that protons are pumped out of mitochondria during respiration was first obtained by Peter Mitchell and Jennifer Moyle. They found that when O_2 was added to a suspension of mitochondria, the pH decreased in the solution surrounding the mito-

Figure 16.15

Respiring mitochondria extrude protons. A weakly buffered suspension of mitochondria is provided with an oxidizable substrate and allowed to use up all the O_2 in the solution. The traces show measurements of the pH of the suspension, with pH decreases plotted upward. When a small amount of O_2 is added (upward-pointing arrow), respiration can occur for a few seconds. The pH of the solution decreases suddenly at this point (Experiment A). The pH change persists for a period of several minutes, long after the O_2 has been used up. If an uncoupler is added, the pH returns abruptly to nearly its original level. Experiment B was done in the presence of a detergent that disrupted the inner membrane, making the membrane permeable to protons. Respiration still occurs under these conditions, but the pH change is not seen. In Experiment A the number of protons that move across the membrane can be estimated from the pH change and can be related to the number of O atoms consumed. In this experiment, about six protons appear to be translocated per O atom. However, this is an underestimate because part of the pH gradient is dissipated by the phosphate/OH^- transport system discussed later in the chapter. Larger ΔH^+ to O ratios are measured if the phosphate transporter is inhibited. For more details, see P. Mitchell and J. Moyle, *Nature,* 208:147–151, 1965; and *Biochem. J.* 105:1147–1162, 1967; also A. Alexandre et al., *J. Biol. Chem.* 255:10721–10730, 1980.

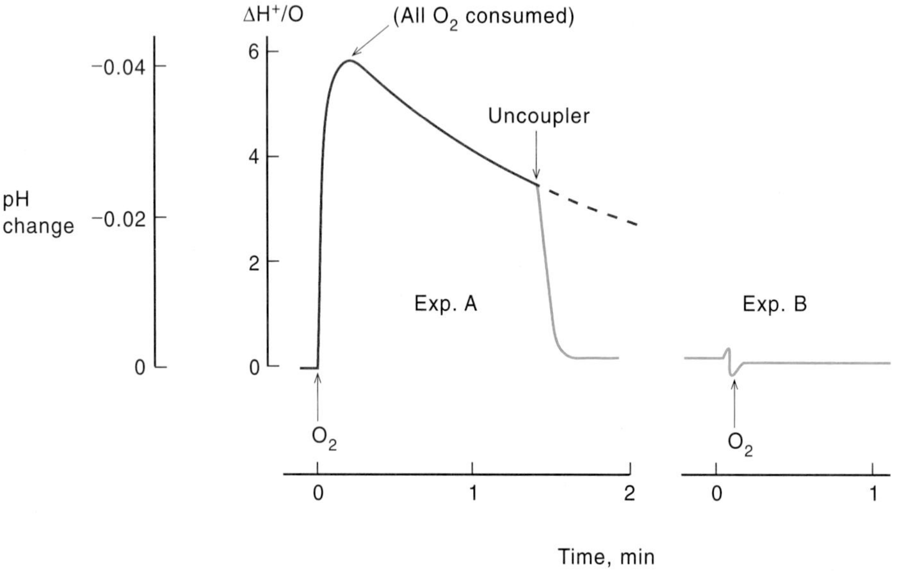

chondria (fig. 16.15). If the integrity of the inner membrane was broken by the addition of a detergent, this pH change was not seen, indicating that the protons that appeared outside the mitochondria probably emerged from the matrix space inside. A detergent makes the membrane leaky, allowing any protons that are pumped out to go right back in.

The pH gradient created by respiration collapses if an uncoupler is added (see figs. 16.15 and 16.16). An uncoupler is a molecule that inhibits mitochondrial ATP synthesis without inhibiting electron transport. The presumption is that the pH gradient is essential for phosphorylation. We will have more to say about this later. In the absence of an uncoupler the pH change caused by a brief period of respiration decays slowly, in agreement with the view that the mitochondrial inner membrane blocks the free diffusion of protons between the matrix and the intermembrane space. Mitchell and Moyle reinforced this conclusion by experiments in which they added a small amount of HCl to a suspension of mitochondria and measured the rate at which protons leaked through the membrane. The leakage occurs on a time scale of minutes, which is much slower than the milliseconds-to-seconds time scale of oxidative phosphorylation. The idea that the membrane is largely impermeable to protons fits well with what we know about the structure of biological membranes (see chapter 19). An ion such as H_3O^+ cannot pass readily through the hydrocarbon region of the phospholipid bilayer. In the absence of an uncoupler the conductance of the phospholipid bilayer to protons is about 10^6 times lower than that of the aqueous phases on either side of the membrane.

Figure 16.16

Structures of two uncouplers of oxidative phosphorylation, 2,4-dinitrophenol (DNP) and carbonylcyanide-*p*-trifluoromethoxyphenylhydrazone (FCCP). DNP has a weakly dissociable proton on the oxygen atom; FCCP has a similar proton on one of the nitrogens.

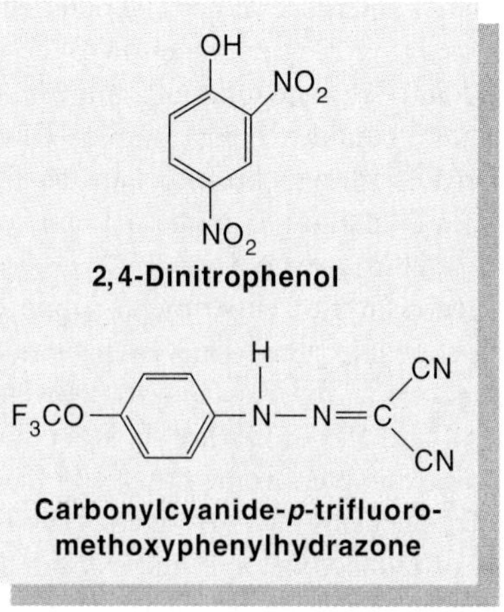

OH, NO_2, NO_2

2,4-Dinitrophenol

F_3CO—, H, —N—N=C, CN, CN

Carbonylcyanide-*p*-trifluoro-methoxyphenylhydrazone

The fact that uncouplers are lipophilic weak acids (see above) explains their ability to collapse transmembrane pH gradients. Their lipophilic character allows uncouplers to diffuse relatively freely through the phospholipid bilayer. Because they are weak acids, uncouplers can release a proton to the solution on one side of the membrane and then diffuse across the membrane to fetch another proton.

The pumping of protons across the membrane by the respiratory chain creates a transmembrane electrical potential gradient, in addition to a pH gradient. This was shown by examining the movements of lipophilic anions and cations. If the matrix space becomes negatively charged with respect to the region outside of the mitochondrion, the electrical difference tends to pull positively charged ions into the matrix and to push negatively charged ions out. Whether or not a particular ion actually moves across the membrane depends on whether the ion can pass through the phospholipid bilayer. In some lipophilic ions such as the triphenylmethylphosphonium ion $(C_6H_5)_3CH_3P^+$ the charge is sufficiently buried by apolar groups that the ion can move across the membrane relatively freely. Another lipophilic cation that passes rapidly through phospholipid bilayers is the complex of the ionophore valinomycin with K^+. Respiring mitochondria take up lipophilic cations, and they extrude lipophilic anions. These observations support the idea that the efflux of protons driven by electron transport makes the interior of the mitochondrion negatively charged relative to the solution outside.

One way to estimate the electrical potential difference across the mitochondrial membrane is to measure the concentrations of a lipophilic ion after the ion has come to equilibrium. The larger the electrical potential difference ($\Delta\psi$), the larger the ratio of the concentrations on the two sides. Such measurements indicate respiring mitochondria generate a $\Delta\psi$ on the order of -0.15 V, inside negative.

Figure 16.17

Approximately 2.5 molecules of ADP can be phosphorylated to ATP for each pair of electrons that traverse the electron-transport chain from NADH to O_2. About 1.5 molecules of ATP are formed for a pair of electrons that enter the chain via succinate dehydrogenase or other flavoproteins such as glycerol-3-phosphate dehydrogenase. Approximately 1.0 ATP is formed for a pair of electrons that enters via cytochrome *c*. These results suggest that there are coupling sites associated with complexes I, III, and IV.

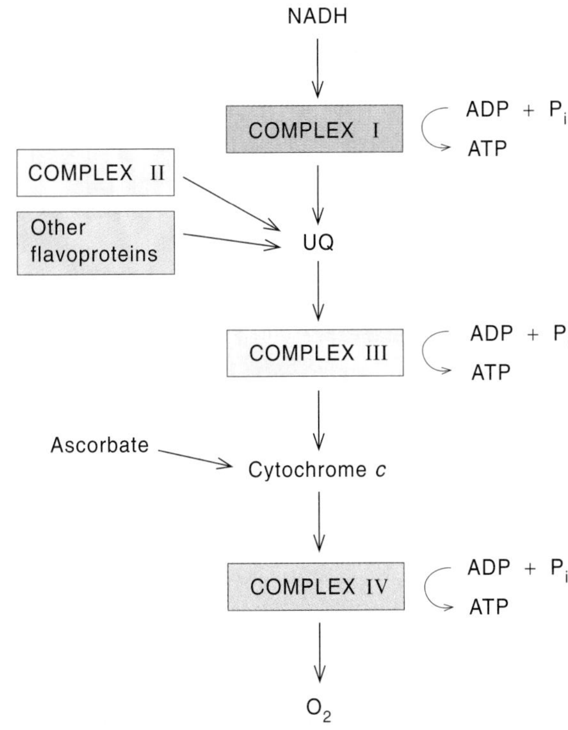

Oxidative Phosphorylation

Evidence that oxidative reactions can drive the formation of ATP was obtained about 1940 by Herman Kalckar, who showed that aerobic cells make ATP from ADP and P_i by a process that depends on respiration. At that time it was unclear that this oxidative phosphorylation occurred in mitochondria and that it involved NADH. Resolution of these questions had to await development of methods for preparing mitochondria free of other cellular constituents and for presenting NADH on the matrix side of the inner membrane. When these methods were devised in the late 1940s, Kennedy and Lehninger found that mitochondria oxidize endogenous NADH and that at least two ATPs are generated for each NADH oxidized.

The Respiratory Chain Contains Three Coupling Sites for ATP Formation

The number of molecules of ATP formed per pair of electrons transferred down the respiratory chain to O_2 is termed the P-to-O ratio. When NADH is used as the reducing substrate, the measured P-to-O ratio is about 2.5; with succinate it is about 1.5. Reductants that feed electrons directly to cytochrome *c* give a P-to-O ratio of about 1.0. This last observation indicates that complex IV, which conducts electrons from cytochrome *c* to O, has a site where the energy of the electron-transfer reactions is coupled to ATP synthesis. Because the electron-transport chains for NADH and succinate converge at UQ, the higher P-to-O ratio obtained with NADH suggests that another coupling site exists in complex I, between NADH and UQ (fig. 16.17). Finally, the intermediate P-to-O ratio obtained with succinate indicates that a coupling site occurs in complex III but probably not in complex II. Thus the respiratory chain appears to have three distinct coupling sites for ATP synthesis.

Electron Transfer Is Tightly Coupled to ATP Formation

By incorporating the purified electron-transport complexes into phospholipid vesicles along with the mitochondrial ATP-synthase enzyme that is described below, Efraim Racker and his co-workers verified the capacity of the individual complexes I, III, and IV to support the formation of ATP. In figure 16.7 you can see that the flow of two electrons through each of these complexes involves a sufficiently negative $\Delta G°'$ to drive the phosphorylation of ADP to ATP. ($\Delta G°'$ for the reaction ADP +

Figure 16.18

The rate of respiration by a suspension of mitochondria can be increased dramatically by the addition of a small amount of ADP. An oxidizable substrate (succinate) and P_i first are added at the times indicated by the vertical arrows, but the respiratory rate remains low until ADP is added. A second period of rapid respiration is obtained when more ADP is added, indicating that essentially all of the added ADP gets used up. The P-to-O ratio can be determined by dividing the amount of ADP added by the amount of oxygen taken up during the period of rapid respiration.

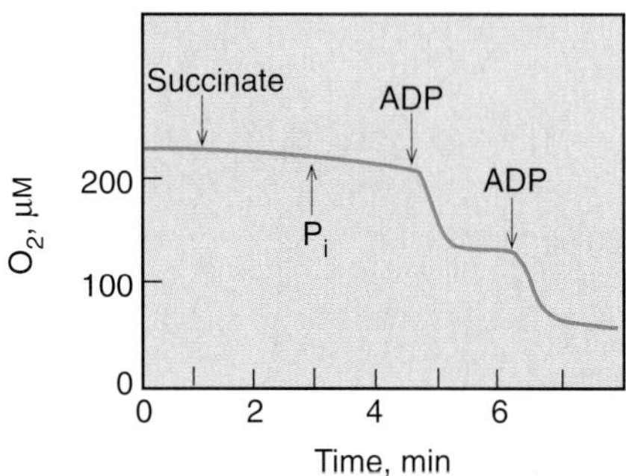

Figure 16.19

(*a*) Addition of an uncoupler to a suspension of mitochondria causes brisk O_2 consumption, which continues until all the O_2 in the solution is used up. An oxidizable substrate (succinate) is added before the uncoupler, but P_i is not required. (*b*) Oligomycin, an inhibitor of the ATP-synthase, blocks the stimulation of respiration caused by ADP. It does not block the stimulation caused by an uncoupler.

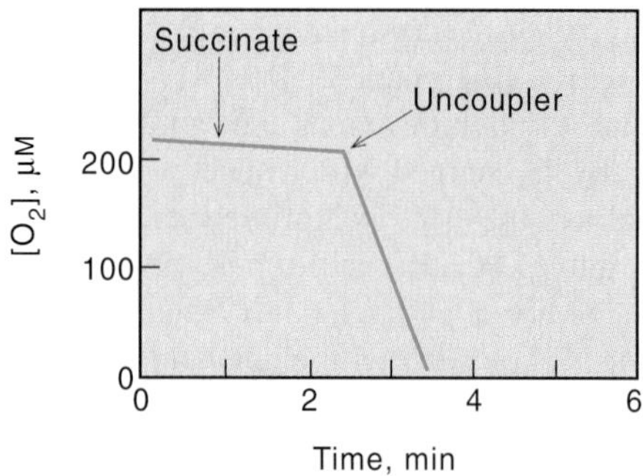

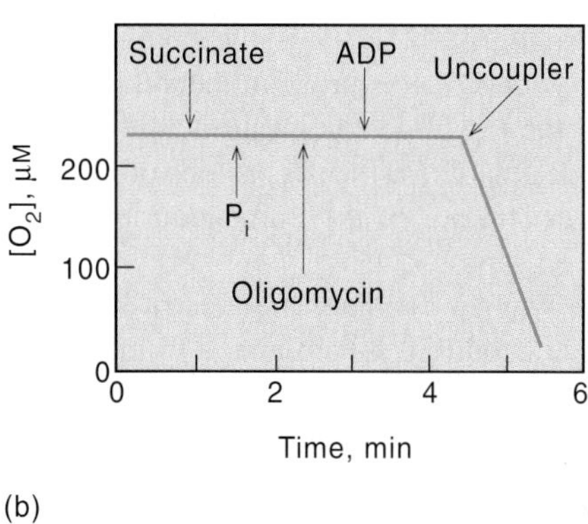

$P_i \longrightarrow ATP + H_2O$ in the presence of 10 mM Mg^{2+} is about 7.5 kcal/mole, which is equivalent to a $\Delta E^{\circ\prime}$ of about 0.16 V for n = 2 electrons/molecule.)

Respiration and phosphorylation usually are tightly coupled processes; if phosphorylation stops, so, as a rule, does respiration and vice versa. As an example, respiration slows down greatly if mitochondria run out of ADP (fig. 16.18). Addition of a small amount of ADP in the presence of an oxidizable substrate and excess P_i causes a brief period of brisk respiration that can be repeated by adding more ADP. Experiments of this type provide one means for estimating P-to-O ratios. The P-to-O ratio is proportional to the amount of ADP consumed to reduce one oxygen of O_2 to one oxygen of water.

As was already mentioned, the tight coupling of respiration and phosphorylation can be disrupted by molecules known as uncouplers. Uncouplers include a diverse group of molecules structurally, but they are all lipophilic weak acids (see fig. 16.16). If an uncoupler is added to mitochondria in the presence of an oxidizable substrate, O_2 uptake commences immediately and continues until essentially all of the O_2 in the solution is used up (fig. 16.19*a*). This happens even in the absence of added ADP or P_i. The free energy released in the electron-transfer reactions is lost as heat rather than being captured in ATP.

If an uncoupler somehow caused the breakdown of an intermediate form of an electron carrier, the electron carrier would be set free and electron transport to O_2 could continue. However, something evidently is different about oxidative phosphorylation compared with the substrate-level phosphorylation catalyzed by 3-phosphoglyceraldehyde dehydrogenase (see chapter 14), because uncouplers have no effect on the latter reaction. Nor do they affect other soluble enzymes that make or use ATP. On the other hand, a molecule that acts as an uncoupler at any one of the three coupling sites of oxidative phosphorylation invariably has a similar effect at the other two sites. This suggests that uncouplers cause the breakdown of something that is generated at all three sites.

Another class of phosphorylation inhibitors functions by blocking the ATP-synthase enzyme directly. Inhibitors of this type include the antibiotic oligomycin. If mitochondria are treated with oligomycin, ADP no longer is able to increase the rate of respiration (see fig. 16.19*b*). Oligomycin does not block the stimulation of respiration caused by an uncoupler, demonstrating that the two types of inhibitors of phosphorylation act by different mechanisms.

The Chemiosmotic Theory Proposes That Phosphorylation Is Driven by Proton Movements

In 1961, Peter Mitchell suggested a radically new theory to explain the coupling mechanism of oxidative phosphorylation. Mitchell proposed that the component generated at all three coupling sites is not a high-energy chemical species but rather an electrochemical potential gradient for protons across the mitochondrial inner membrane. Figure 16.20 illustrates the basic idea. Electron transport down the respiratory chain results in the movement of protons across the membrane, from the mitochondrial matrix to the intermembrane space. The removal of protons from the

Figure 16.20

According to the chemiosmotic theory, flow of electrons through the electron-transport complexes pumps protons across the inner membrane from the matrix to the intermembrane space. This raises the pH in the matrix and leaves the matrix negatively charged with respect to the intermembrane space and the cytosol. Protons flow passively back into the matrix through a channel in the ATP-synthase, and this flow drives the formation of ATP.

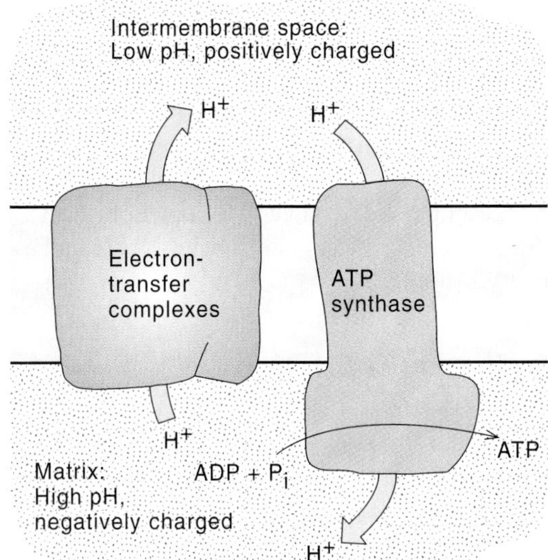

matrix causes the pH in this region to rise. Because protons are positively charged, the matrix also becomes negatively charged with respect to the intermembrane space and the cytosol. The differences in pH and electrical potential across the membrane provide a driving force that tends to pull protons back into the matrix. Part of the free energy decrease associated with the electron-transfer reactions thus is exchanged for an electrochemical potential difference between the solutions on the two sides of the inner membrane. Mitchell suggested that protons move back across the membrane through special channels in an ATP-synthase, which uses this inward flow of protons to drive the formation of ATP. This is called the chemiosmotic theory because it emphasizes that chemical reactions can drive, or be driven by, movements of molecules or ions between osmotically distinct spaces separated by membranes.

Flow of Protons Back into the Matrix Drives the Formation of ATP

The chemiosmotic theory postulates that protons moving back into the matrix via an ATP-synthase drive the formation of ATP. This theory is supported by the finding that an electrochemical potential gradient for protons can support the formation of ATP in the absence of electron-transfer reactions. A transient pH gradient that pulls protons into the matrix can be set up by first incubating mitochondria at pH 9, so that the inside becomes alkaline, and then quickly lowering the pH of the suspension medium to 7 (fig. 16.21).

Current estimates are that three protons move into the matrix through the ATP-synthase for each ATP that is synthesized.

Figure 16.21

An electrochemical potential gradient for protons can be set up across the mitochondrial inner membrane in the absence of electron transfer. In step 1, mitochondria are incubated at high pH in the presence of KCl to reduce the proton concentration inside and to load the inside with K^+ and Cl^-. In step 2 the pH and the KCl concentration outside the mitochondria are decreased. Valinomycin (Val) is added to carry K^+ ions out. Because the counterion Cl^- cannot move rapidly across the membrane, the efflux of K^+ down its concentration gradient leaves the inside of the mitochondrion with a net negative charge relative to the outside. This difference in electric potential together with the higher pH inside favors the influx of protons. Protons moving inward through the ATP-synthase can drive the formation of ATP.

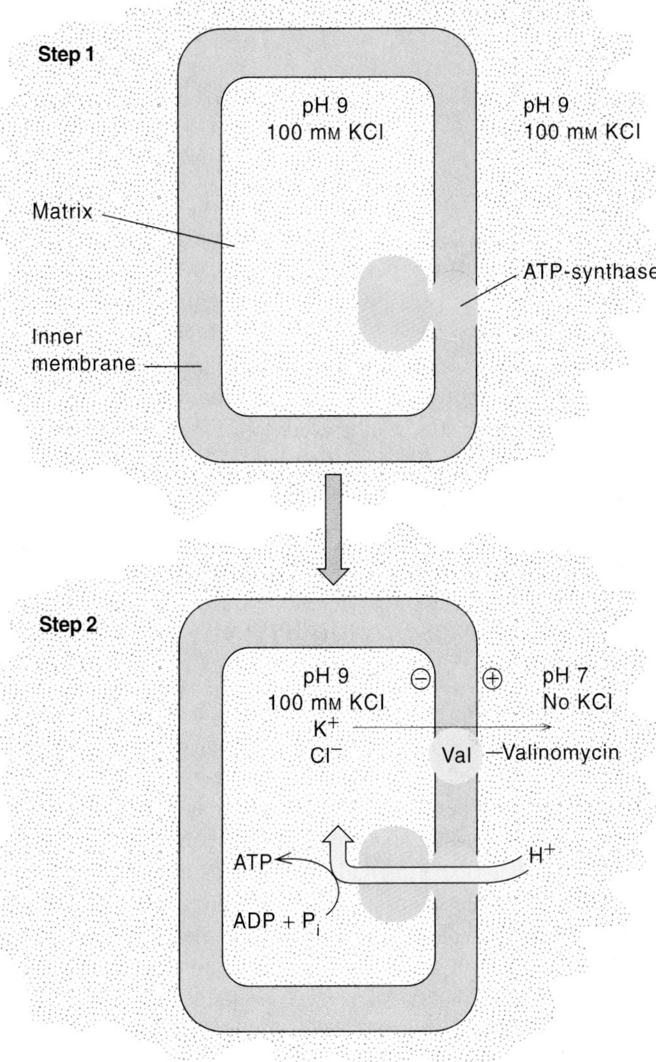

We see below that one additional proton enters the mitochondrion in connection with the uptake of ADP and P_i and export of ATP, giving a total of four protons per ATP. How does this stoichiometry relate to the P-to-O ratio? When mitochondria respire and form ATP at a constant rate, protons must return to the matrix at a rate that just balances the proton efflux driven by the electron-transport reactions. Suppose that 10 protons are translocated for each pair of electrons that traverse the respiratory chain from NADH to O_2, and 4 protons move back in for each ATP molecule that is synthesized.* Because the rates of proton efflux and influx must balance, 2.5 molecules of ATP (10/4) should be formed for each pair of electrons that go to O_2. The P-to-O ratio thus is given by the ratio of the proton stoichiometries. If oxidation of succinate extrudes six protons per pair of electrons, the P-to-O ratio for this substrate is 6/4, or 1.5. These ratios are in approximate agreement with the measured P-to-O ratios for the two substrates.

Let's now consider how much free energy is released by moving protons into the mitochondrion. Is it really enough to drive the synthesis of ATP? The free energy change depends both on the ratio of the proton concentrations on the two sides of the membrane and on the difference between the electric potentials on the two sides:

$$\Delta G_{H^+} = G_{H^+(in)} - G_{H^+(out)} = RT \ln \left\{ \frac{[H^+]_{in}}{[H^+]_{out}} \right\} + \mathcal{F}\,\Delta\psi$$
$$= -2.3RT\,\Delta pH + \mathcal{F}\,\Delta\psi \qquad (9)$$

Here $\mathcal{F}$ is the Faraday constant, $\Delta pH = pH_{in} - pH_{out}$, and $\Delta\psi$ is the electric potential difference (the electric potential inside minus that outside). If the pH in the matrix space is 0.05 pH units above that in the cytosol, the term $-2.3RT\,\Delta pH$ amounts to -0.07 kcal/mole. If $\Delta\psi$ is -0.15 V, $\mathcal{F}\,\Delta\psi$ is -3.46 kcal/mole. This gives a total ΔG_{H^+} of -3.53 kcal/mole, with the term $\mathcal{F}\,\Delta\psi$ making the dominant contribution. If four protons move across the membrane for each ATP molecule synthesized, the ΔG associated with proton translocation is $4 \times (-3.53)$, or -14.1 kcal/mole of ATP (-58.4 kJ/mole), which is about twice as large as the $\Delta G^{\circ\prime}$ of ATP hydrolysis (-7.5 kcal/mole, or -31.4 kJ/mole). Thus the free energy decrease is large enough to account for the [ATP]/[ADP][P_i] ratio that mitochondria generate under physiological conditions.

The Proton-Conducting ATP-Synthase or ATPase: F_1 and F_o

Let us now turn to the ATP-synthase. What is its structure, and how is it able to use a proton-motive force to drive the formation of ATP? Studies of the ATP-synthase began in the early 1960s, when Efraim Racker and his colleagues identified several enzymatic components that appeared to participate in oxidative phosphorylation. The experimental strategy was to disrupt mitochondria and fractionate the proteins released from the membranes. In some cases the depleted membranes lost their ability to synthesize ATP, although the electron-transfer reactions were undisturbed. Recombining the solubilized proteins, or coupling factors, with the membranes restored oxidative phosphorylation.

The first coupling factor to be purified, F_1, was an active ATPase when it was removed from the mitochondria. Unlike the membrane-bound ATPase in intact mitochondria, the solubilized F_1 ATPase was not inhibited by oligomycin. However, it recovered its sensitivity to oligomycin if it was reattached to the membranes in combination with another coupling factor, F_o. (The subscript "o" stands for oligomycin.)

*If the mechanisms proposed in figures 16.10, 16.12, and 16.13 accounted for everything, this number would be 8 instead of 10.

*Much higher ΔpH values are commonly used in this calculation but the membrane potential term still dominates.

F_o and F_1 both turned out to be parts of the proton-conducting ATP-synthase. Together they form a multiprotein complex that, like the electron-transfer complexes, is partially embedded in the mitochondrial inner membrane. F_1, which contains the catalytic sites for ATP formation or hydrolysis, consists of five different polypeptide subunits (α, β, γ, δ, and ϵ) with the stoichiometry $\alpha_3\beta_3\gamma\delta\epsilon$. Its total molecular weight is about 360,000. In electron micrographs of mitochondria the F_1 complexes can be seen as spheres with diameters of about 85 Å, protruding from the surface of the inner membrane (fig. 16.22). F_o, a complex of approximately five hydrophobic polypeptides, is an integral component of the inner membrane and acts as a base-piece and stalk that holds F_1 to the membrane (fig. 16.23). Complexes similar to the mitochondrial F_o and F_1 have been isolated from the chloroplast thylakoid membrane (see chapter 16) and from bacteria. In the enzyme from *E. coli,* F_o has three subunits (*a, b,* and *c*) with the stoichiometry $a_1b_2c_{10}$.

The Mechanism of Action of the ATP-Synthase

If the purified F_1–F_o complex is incorporated into a phospholipid vesicle and an electrochemical potential gradient for protons is set up across the membrane, the complex can synthesize ATP. To show this, Racker and Stoeckenius generated the proton gradient by incorporating a light-driven proton pump, bacteriorhodopsin from *Halobacterium halobium,* into membrane vesicles along with the F_1–F_o complex. When the vesicles were illuminated, ATP was formed (fig. 16.24). This experiment provided a dramatic demonstration that the ATP-synthase does not have to be attached directly to the electron-transport complexes of the respiratory chain, for no electron carriers were included in the liposomes. It added strong support to the idea that the coupling of phosphorylation to electron transport is mediated by a proton-motive force that can be generated at one site on the membrane and used at relatively distant sites.

F_o appears to contain the proton-conducting channel of the ATP-synthase. If F_o alone is incorporated into liposomes, it makes the membranes leaky specifically to protons. The leaks clearly are relevant to oxidative phosphorylation because they can be plugged by either oligomycin or dicyclohexylcarbodiimide (DCCD). DCCD is particularly useful here because it reacts irreversibly with a critical aspartate residue in one of the subunits. In the enzyme from *E. coli* the site of inhibition is in subunit *c*, which is a small, hydrophobic polypeptide ($M_r \approx 8{,}000$). Analysis of the amino acid sequence suggests that each of the 10 copies of the *c* subunit in the F_1–F_o complex folds into a hairpinlike structure with two transmembrane α helices. The multiple copies of the subunit may assemble to form a tubular channel across the membrane at the base of F_1 (see fig. 16.23).

The crystal structure of F_1 was determined recently by John Walker and his colleagues, making this the largest protein for which a high-resolution structure has been obtained. The α and β subunits are arranged alternately around a coil of α helices formed by the γ subunit (fig. 16.25). A catalytic site is located on

Figure 16.22

(*a*) Negatively stained electron micrograph of part of a mitochondrion, showing head-pieces of the F_1 ATP-synthase lining the inner membrane. The wormlike white area is a crista; the large dark areas are the matrix. The F_1 head-pieces show up as white spots projecting from the membrane of the crista into the matrix. (From B. Chance and D. Parsons, Cytochrome function in relation to inner membrane structure of mitochondria, *Science* 142:1176, 1963, © 1963 by the AAAS.) (*b*) Negatively stained electron micrograph of mitochondrial membrane vesicles. These vesicles are capable of oxidative phosphorylation. F_1 head-pieces line the outer surfaces of the membranes. (*c*) Electron micrograph of purified F_1. (*d*) Mitochondrial membrane vesicles that were depleted of F_1 by treatment with urea and trypsin. These membranes contain a functional electron-transfer chain but do not form or hydrolyze ATP. Note that the surfaces of the membranes appear smooth. (Micrographs, *b, c,* and *d* courtesy of Dr. E. Racker.)

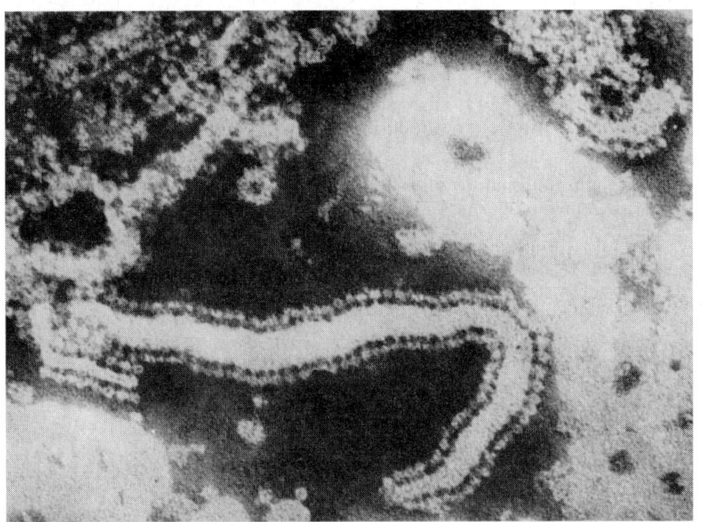

(a)

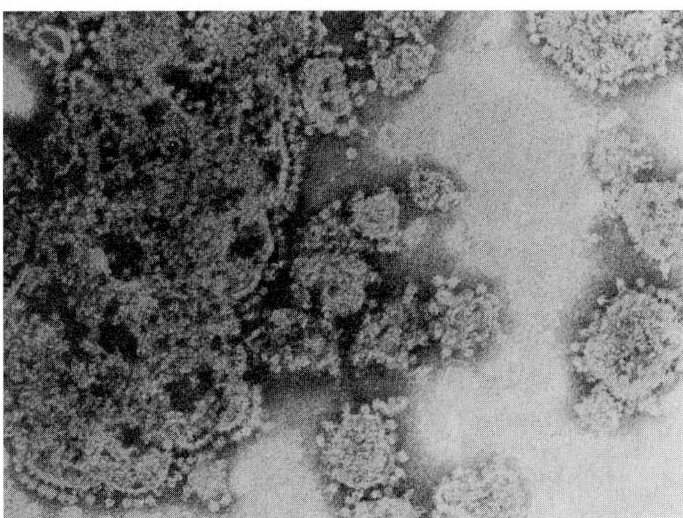

(b)

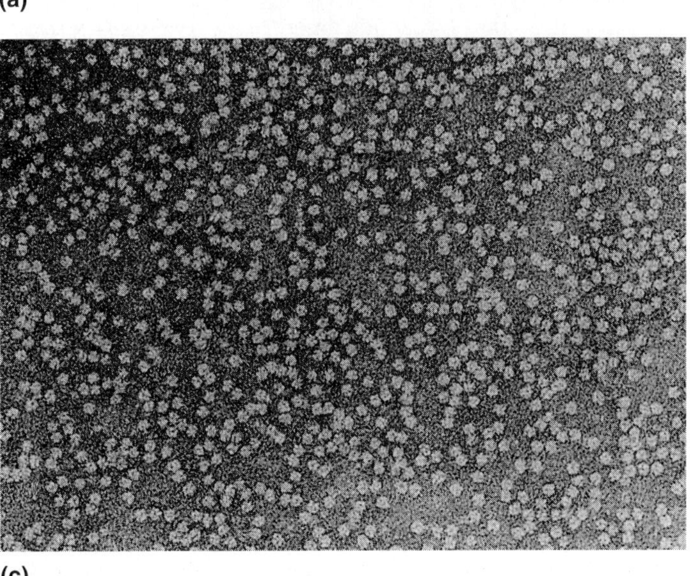

(c)

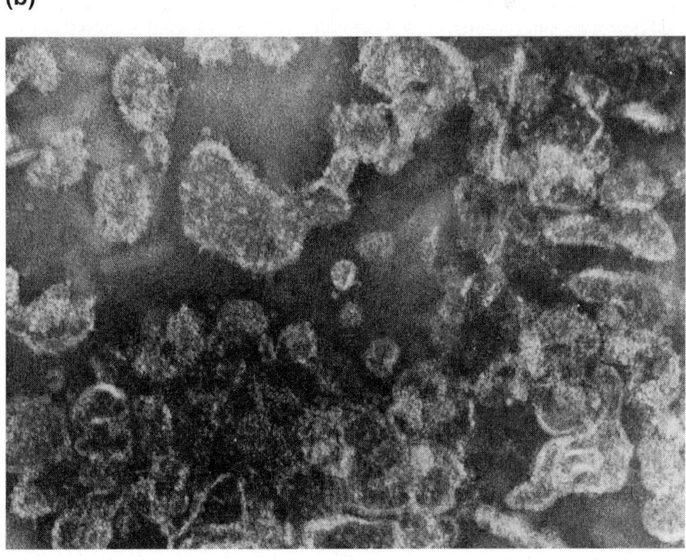

(d)

each of the β subunits near one of the interfaces with an α subunit. Studies by Paul Boyer, Harvey Penefsky, and others have shown that at any given time the catalytic site of one of the three β subunits binds ADP and ATP very tightly. Remarkably, the reaction

$$ADP + P_i \rightleftharpoons ATP + H_2O \quad \textbf{(10)}$$

occurs readily between the bound nucleotides in this site, even in the absence of a proton-motive force. The equilibrium constant for the formation of ATP from the bound nucleotides is close to 1. (Remember that the equilibrium constant for free ADP, P_i, and ATP in solution is about 10^{-5}!) Meanwhile, the catalytic site on one of the other β subunits is unoccupied, and the third site binds ADP and P_i loosely. ADP and P_i bound in this last site behave much like ADP and P_i in solution, without any special tendency to form ATP. The crystal structure suggests that the tightly binding, loosely binding, and empty sites reflect different conformational states of the β subunits ("tight," "loose," and "open") imposed by asymmetric interactions with the central γ subunit (see fig. 16.25).

Figure 16.23

Components of the proton-conducting ATP-synthase. The F_1 head-piece includes three α and three β subunits and one copy each of three other subunits (γ, δ, and ϵ). F_o includes a cluster of 9–12 copies of a small peptide, which form a transmembrane channel for protons.

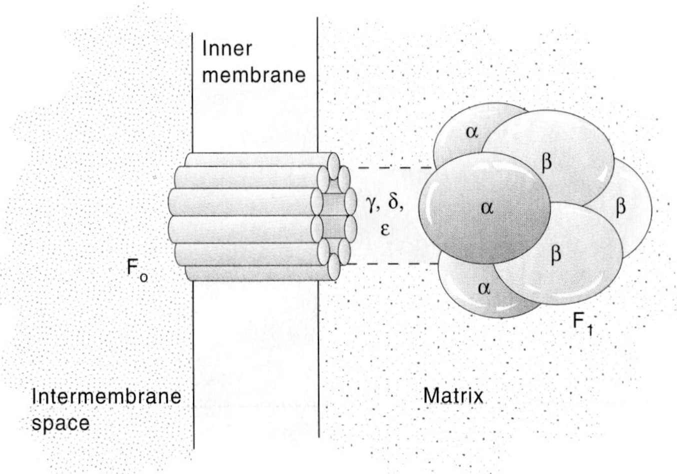

Figure 16.24

ATP synthesis can be obtained in an artificial system in which the mitochondrial ATP-synthase is incorporated into membrane vesicles together with bacteriorhodopsin, a membrane protein obtained from *Halobacterium halobium.* When bacteriorhodopsin is excited with light, it pumps protons across the membrane (see chapter 17). The solution inside the vesicle thus becomes acidic and positively charged relative to the external solution. Protons can move back out through the ATP-synthase, thereby driving the formation of ATP. The formation of ATP is blocked by uncouplers or oligomycin. Note that the F_1 head-piece of the ATP-synthase, which is on the inner surface of the mitochondrial inner membrane, is on the outside of the membrane in this artificial system. The ATP-synthase also faces outward in submitochondrial particles made by breaking the inner membrane with sonic oscillations. In all cases, ATP synthesis requires protons to move through F_o in the direction of F_1.

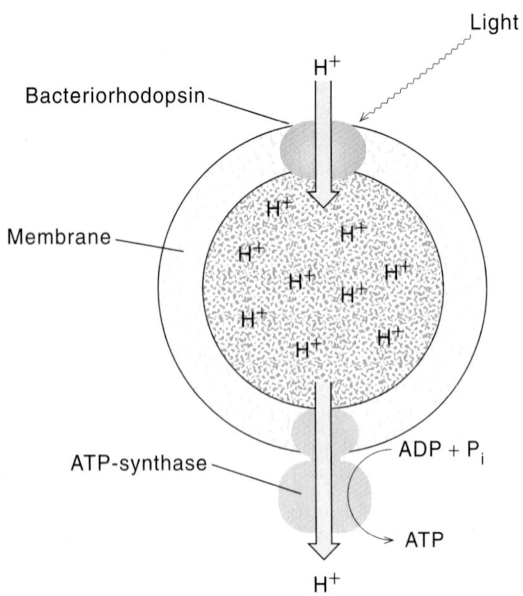

Figure 16.25

The nucleotide-binding domains of the mitochondrial ATP synthase and a model of the enzymatic reaction. This end-on view of F_1 shows the approximate outlines of the protein's α, β, and γ subunits in the region of the binding sites for ADP, P_i, and ATP. The three catalytic sites lie at interfaces of the α and β subunits. The β subunits adopt three different conformational states, probably as a result of asymmetric interactions with the central γ subunit. In this drawing, the catalytic site of the subunit in the "loose" conformation has bound ADP and P_i, while the subunit in the "tight" conformation holds tightly bound ATP, and the catalytic site of the subunit in the "open" conformation is unoccupied (dashed circle). In the regions of the catalytic sites the α and β subunits are juxtaposed closely in the tight conformation and are much farther apart in the open conformation. Protons moving through the channel in the F_o base-piece may cause the $\alpha_3\beta_3$ cluster to rotate with respect to the γ subunit, so that the conformational states of the β subunits change in tandem. ATP is released from the subunit that switches from tight to open, fresh molecules of ADP and P_i are taken up by the subunit changing from open to loose, and the bound ADP and P_i are converted to tightly bound ATP on the subunit changing from loose to tight. The drawing is based on the crystal structure described by J. P. Abrahams, A. G. W. Leslie, R. Lutter, and J. E. Walker, Structure at 2.8-Å resolution of F_1-ATPase from bovine heart mitochondria. *Nature* 370:621, 1994.

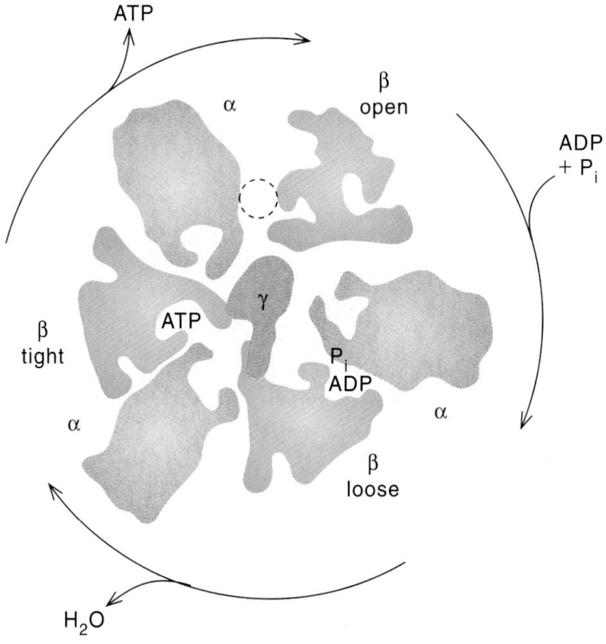

The movement of protons through the F_o base-piece does not have much effect on the equilibrium constant for the formation of ATP at the tight binding site on the F_1 headpiece. Instead, it evidently drives a cycling of the β subunits between their three conformational states. As protons move through the channel in F_o, the conformational states of the three β subunits appear to switch in tandem, so that the subunit in the tight conformation changes to open, loose changes to tight, and open becomes loose (see fig. 16.25). When this happens, ATP is released from the new open site and another molecule of ATP is formed at the new tight site. The concerted conformational changes in the subunits could be driven by rotation of the $\alpha_3\beta_3$ cluster with respect to the central γ subunit.

16B
BOX

Natural Uncouplers: Thermogenesis in Brown Fat and Skunk Cabbage

Several unusual types of cells use the respiratory chain for thermogenesis, the generation of heat, rather than the formation of ATP. Newborn mammals and the adults of some mammals that are adapted to live in cold climates do this in a specialized adipose tissue called brown fat. The brown color is due largely to the cytochromes in the tissue's abundant mitochondria. Heat produced by brown-fat mitochondria is important for maintaining body temperature in the newborn and for the arousal of hibernating animals. Brown fat generates heat at a rate of about 400 W/kg, far above the rate of about 1 W/kg that is typical of other resting mammalian tissues.

The inner membrane of brown-fat mitochondria contains a protein, thermogenin, that acts as a channel for anions. The protein allows OH^- or Cl^- ions to pass rapidly across the membrane. Because a movement of OH^- from the mitochondrial matrix to the cytosol has the same consequences as a movement of H^+ in the opposite direction, thermogenin acts as an uncoupler. The free energy released in the electron-transfer reactions is stored transiently as an electrochemical potential gradient for protons but then is degraded largely to heat.

The uncoupling activity of thermogenin is regulated by ATP, ADP, GTP, and GDP, which bind tightly to the protein and inhibit anion transport. Very low concentrations of fatty acids increase anion permeability. Fatty acids released from triacylglycerol stores in the tissue could be important in the hormonal control of thermogenesis. The amount of thermogenin in the brown-fat mitochondria also changes in response to physiological conditions. In animals that live at low temperatures it can represent as much as 15% of the protein in the inner membrane.

Skunk cabbage uses a different strategy for thermogenesis. It oxidizes UQH_2 by an alternative pathway that bypasses the cytochrome bc_1 complex and cytochrome oxidase. Electron transport by this pathway does not result in proton translocation. Heat produced by respiration may help the plant thrive early in spring, when the climate is cool.

Soluble enzymes offer several precedents for a substantial difference in the equilibrium constant of a reaction that forms ATP, depending on whether the reactants and products are bound to the enzyme or are free in solution. With both pyruvate kinase and 3-phosphoglycerate kinase the equilibrium constant for the formation of bound ATP on the enzyme is much more favorable than the equilibrium constant for the reaction of the free materials. This difference could reflect the fact that the $\Delta G^{\circ\prime}$ for ATP formation in solution includes an entropy decrease associated with the ordering of water molecules around the polyphosphate chain. Such an entropy decrease can be much smaller if the ADP, P_i, and ATP are bound tightly on an enzyme.

Although there is much evidence that movement of protons through the ATP-synthase drives the formation of ATP, changes in the [ATP]/[ADP][P_i] ratio have been measured in mitochondria under conditions when there is little change in the proton-motive force. One way to explain such observations is to postulate that some of the protons that are pumped by the electron-transport reactions are not released into the aqueous solution at the surface of the membrane, but rather are conducted more directly to the ATP-synthase. The proton-motive force between the solutions on the two sides of the membrane thus might reflect only a part of the driving force that is expressed at the ATP-synthase.

In some unusual cell types, natural uncouplers exist whose function is to channel the energy of the respiratory chain to heat production rather than ATP formation (box 16B).

Transport of Substrates, P_i, ADP, and ATP into and out of Mitochondria

One of the principles underlying the chemiosmotic theory is that the mitochondrial inner membrane is basically impermeable to charged and highly polar molecules. As we have discussed, NADH, NAD^+, and H_3O^+ cannot pass freely across the membrane. How, then, can P_i, adenine nucleotides, and substrates such as pyruvate and citrate move into and out of mitochondria? The answer is that the inner membrane has a set of transport systems that specifically catalyze the movements of these materials.

Uptake of P_i and Oxidizable Substrates Is Coupled to the Release of OH^- Ions

The transport of materials across biological membranes is discussed in detail in chapter 20, where we will develop the idea that the movement of one substance across a membrane can be coupled to movement of another substance in the same or the opposite direction. The uptake of β-galactosides by some bacteria, for example, is linked to the movement of protons into the cell. Like the proton-conducting ATP-synthase, transport systems in the cell membrane are able to harness the free energy decrease associated with moving protons down an electrochemical potential gradient. They use this free energy to move sugars and other nutrients thermodynamically uphill, against the concentration gradient of the nutrient. This concept was first advanced by Mitchell as an extension of the chemiosmotic theory, and it has turned out to apply to the movements of many materials into and out of mitochondria.

The uptake of P_i by mitochondria is coupled to an outward movement of OH^-. If the phosphate moves in the form of $H_2PO_4^-$, carrying one negative charge, an exchange of P_i for OH^- will be electrically neutral: There is no net movement of charge in either direction. The transport of P_i thus will be relatively insensitive to the electrical potential gradient $\Delta\psi$ that the respiratory chain sets up across the inner membrane. It will, however, be sensitive to the pH gradient. An outward movement of OH^- is essentially equivalent to an inward movement of H^+ and is thermodynamically downhill in respiring mitochondria because the pH is higher in the matrix than in the cytosol (fig. 16.26*a*). Mitochondria thus are capable of taking up P_i from the cytosol even when the concentration of P_i inside exceeds that outside.

Pyruvate, which is generated in the cytosol by glycolysis but consumed in the mitochondria by the TCA cycle, also is transported into the mitochondria in exchange for OH^-. Succinate and malate are taken up by a transport system that can exchange either of these dicarboxylic acids for P_i (see fig. 16.26*b*). The same transport system also can exchange one of the dicarboxylic acids for another. A separate system exchanges the tricarboxylic acids citrate and isocitrate for each other or for a dicarboxylic acid. Mitochondria thus can achieve the uptake of any of these substrates by a series of exchanges of one carboxylic acid for another, of an acid for P_i, and ultimately of P_i for OH^-.

Export of ATP Is Coupled to ADP Uptake

ATP, which is produced in the mitochondria largely for use elsewhere in the cell, is exported by a transport system that exchanges ATP for ADP. This exchange is not electrically neutral. At pH 7 the net charge on ATP is approximately -4, whereas that on ADP is about -3. The outward movement of a molecule of ATP in exchange for uptake of an ADP removes one negative charge from the matrix and is driven by the electrical gradient, $\Delta\psi$ (see fig. 16.26*c*). The exchange is said to be electrogenic. The ATP/ADP exchange protein, or adenine nucleotide translocator, is actually the most abundant protein in the mitochondrial inner membrane. It is sensitive to several specific inhibitors, atractyloside, and bongkrekic acid (fig. 16.27).

The combined effect of exchanging extramitochondrial ADP^{3-} and $H_2PO_4^-$ for mitochondrial ATP^{4-} and OH^- is to move one proton into the mitochondrial matrix for every molecule of ATP that the mitochondria release into the cytosol. This proton translocation must be considered, along with the movement of protons through the ATP synthase, in order to account for the P/O ratio of oxidative phosphorylation. If three protons pass through the ATP synthase and the adenine nucleotide and P_i transport systems move one additional proton, then in total, four protons move into the matrix for each ATP provided to the cytosol. As we explained earlier, translocation of four protons per ATP is consistent with the relationship between the proton-motive force and the measured ΔG for ATP formation under physiological conditions.

Figure 16.26

(*a*) The phosphate/hydroxide exchange protein carries a phosphate ion ($H_2PO_4^-$) in one direction across the mitochondrial inner membrane in exchange for an OH^- ion moving in the opposite direction. These movements are reversible, but a net efflux of OH^- is thermodynamically downhill because the respiratory chain pumps protons out of the matrix and raises the pH there. Because OH^- efflux is linked to $H_2PO_4^-$ uptake, phosphate is concentrated in the matrix. (*b*) The phosphate/dicarboxylic acid exchange protein exchanges succinate (or some other dicarboxylic acids) for phosphate (HPO_4^{2-}). Phosphate efflux is thermodynamically favorable because the phosphate concentration in the matrix exceeds that in the cytosol, as a result of the action of the phosphate/hydroxide exchange protein. (*c*) The ADP/ATP exchange protein exchanges ADP^{3-} for ATP^{4-}. An outward movement of ATP removes one negative charge from the matrix and is favored because proton pumping by the respiratory chain gives the matrix a negative charge relative to the cytosol.

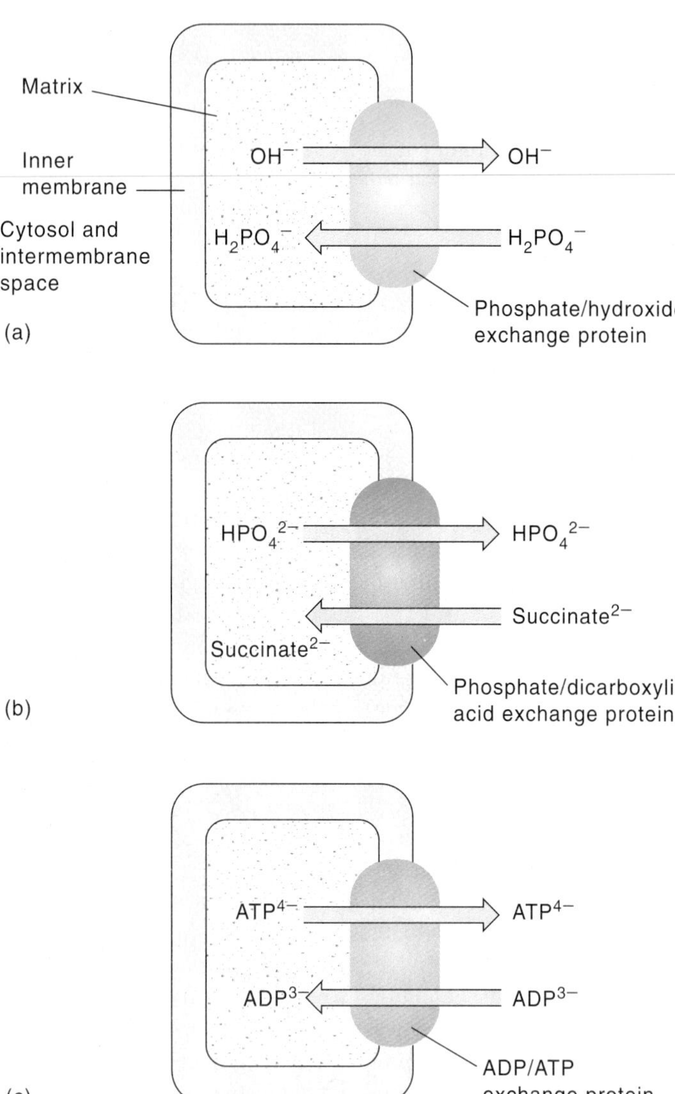

Electrons from Cytosolic NADH Are Imported by Shuttle Systems

The mitochondrial inner membrane does not contain a transport system for NAD^+ or NADH. In animal cells, most of the NADH

Figure 16.27

Structures of two inhibitors of the ATP/ADP exchange protein. Atractyloside is obtained from a species of thistle; bongkrekic acid is obtained from a fungus found in decaying coconuts (*bongkrek* in Indonesian). The ATP/ADP exchange protein has two distinct conformational states. In one of these states it binds ADP and atractyloside; in the other it binds ATP and bongkrekic acid.

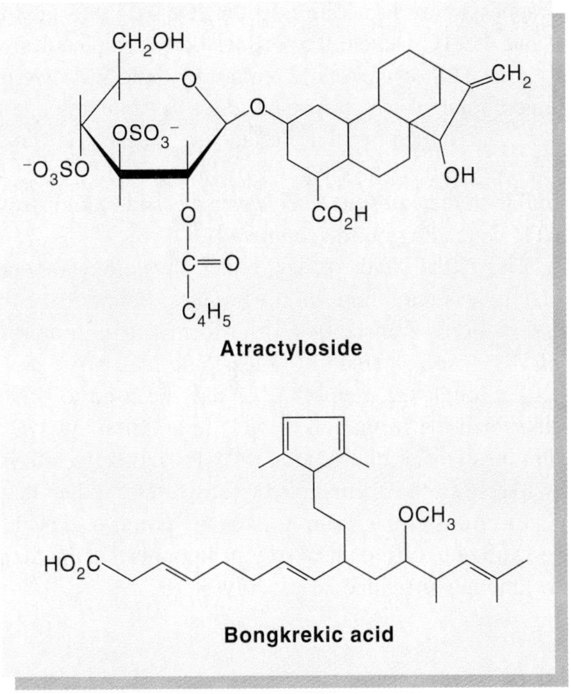

Atractyloside

Bongkrekic acid

that must be oxidized by the respiratory chain is generated in the mitochondrial matrix by the TCA cycle or by the oxidation of fatty acids. However, NADH also can be generated by glycolysis in the cytosol, and this NADH must be reoxidized to NAD^+ in some manner. If O_2 is available, it clearly is advantageous to reoxidize the NADH by the respiratory chain, rather than by the formation of lactate or ethanol. Approximately 2.5 molecules of ATP are formed for each NADH that is oxidized in the mitochondria, whereas no ATP is made when NADH is oxidized by the cytosolic lactate dehydrogenase or alcohol dehydrogenase.

Plant mitochondria have a second NADH dehydrogenase that is distinct from complex I and can oxidize cytosolic NADH, but this enzyme is not found in animals. Instead, animal cells have several shuttle systems that transfer electrons from cytosolic NADH to the respiratory chain. The simplest shuttle involves the reduction of dihydroxyacetone phosphate to glycerol-3-phosphate in the cytosol, followed by reoxidation of the glycerol-3-phosphate by the mitochondrial glycerol-3-phosphate dehydrogenase:

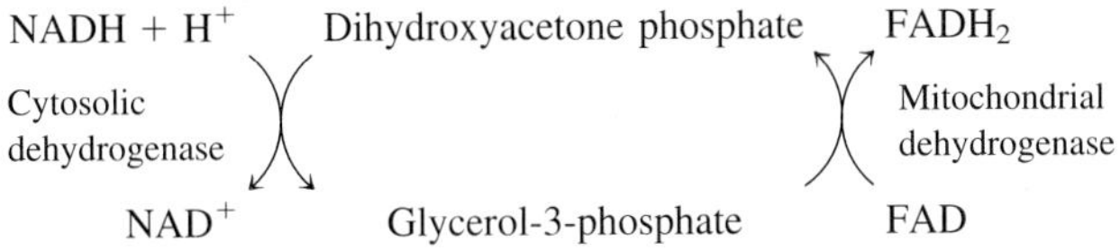

$$\text{NADH} + \text{H}^+ + \text{Dihydroxyacetone phosphate} \xrightarrow{\text{Cytosolic dehydrogenase}} \text{NAD}^+ + \text{Glycerol-3-phosphate}$$
$$\text{Glycerol-3-phosphate} + \text{FAD} \xrightarrow{\text{Mitochondrial dehydrogenase}} \text{Dihydroxyacetone phosphate} + \text{FADH}_2 \qquad (11)$$

The catalytic site of the mitochondrial glycerol-3-phosphate dehydrogenase is on the cytosolic surface of the inner membrane, so the glycerol-3-phosphate does not have to pass through this membrane in order to be reoxidized.

The dihydroxyacetone phosphate/glycerol-3-phosphate shuttle has the shortcoming that oxidation of glycerol-3-phosphate by the respiratory chain generates only 1.5 ATP, instead of the 2.5 that can be generated by the oxidation of mitochondrial NADH (see fig. 16.17). The mitochondrial glycerol-3-phosphate dehydrogenase passes electrons to UQ, below the coupling site in complex I. There is another, more complicated shuttle system that allows the electrons to pass through complex I (see fig. 13.16). In this scheme, cytosolic NADH reduces oxaloacetate to malate, which is carried across the inner membrane by a specific transporter. Inside the mitochondria, the malate is reoxidized to oxaloacetate, reducing mitochondrial NAD^+ to NADH:

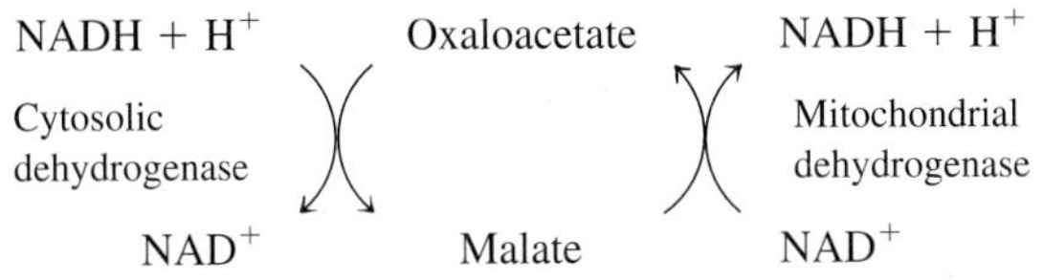

$$\text{NADH} + \text{H}^+ + \text{Oxaloacetate} \xrightarrow{\text{Cytosolic dehydrogenase}} \text{NAD}^+ + \text{Malate}$$
$$\text{Malate} + \text{NAD}^+ \xrightarrow{\text{Mitochondrial dehydrogenase}} \text{Oxaloacetate} + \text{NADH} + \text{H}^+ \qquad (12)$$

The internal NADH then can be oxidized by the respiratory chain.

To continue the oxidation of cytosolic NADH by this shuttle system, oxaloacetate must return from the mitochondria to the cytosol. Oxaloacetate itself cannot cross the mitochondrial inner membrane; there is no transporter for it. However, oxaloacetate can react with the amino acid glutamate to form aspartate and α-ketoglutarate (see chapter 24). The inner membrane has transporters that couple exchange of α-ketoglutarate for malate and exchange of glutamate for aspartate. Working together, these enzymes can complete the shuttle. Because the glutamate/aspartate transporter also moves a proton into the mitochondrion, decreasing the proton-motive force across the membrane, oxidation of cytosolic NADH by the oxaloacetate/malate shuttle still does not generate as much ATP as can be obtained from mitochondrial NADH.

Complete Oxidation of Glucose Yields about 30 Molecules of ATP

The shuttle systems that operate between the cytosol and mitochondria must be taken into account when we calculate the total yield of ATP from a molecule of glucose that is oxidized to CO_2 and H_2O. To make such a calculation, recall first that two molecules of ATP are formed from each glucose in glycolysis and two more are formed via GTP in the TCA cycle (fig. 16.28). At the same time, two molecules of NADH are produced in the cytosol by glycolysis, and eight molecules of NADH are generated in the mitochondrial matrix by the pyruvate dehydrogenase complex and the TCA cycle. The two molecules of succinate proceeding through the succinate dehydrogenase step of the TCA cycle reduce two molecules of FAD to $FADH_2$. Reoxidation of the eight mitochondrial NADH molecules and two $FADH_2$ molecules by the respiratory chain can generate approximately 23 molecules of ATP (2.5 for each NADH and 1.5 for each $FADH_2$). The two NADH molecules that are formed in the cytosol could contribute their reducing equiv-

Figure 16.28

Complete oxidation of a molecule of glucose to CO_2 and H_2O generates approximately 30 molecules of ATP. Reoxidation of the 8 molecules of NADH formed in the mitochondrion yields about 20 ATP (2.5 ATP per NADH). Reoxidation of two molecules of $FADH_2$ bound to succinate dehydrogenase yields 3 ATP (1.5 per $FADH_2$). If the dihydroxyacetone phosphate/glycerol phosphate shuttle is used as indicated here, the two molecules of cytosolic NADH formed in glycolysis reduce two additional molecules of bound FAD, which yield 3 molecules of ATP on reoxidation. Two molecules of ATP are produced directly in glycolysis, and the TCA cycle produces two more in the mitochondrion via GTP. Each of the latter two molecules of ATP is equivalent to about 3/4 molecule of ATP in the cytosol because exporting an ATP from the mitochondria requires uptake of one proton (1/4 of the proton uptake required for synthesis and export of ATP by oxidative phosphorylation).

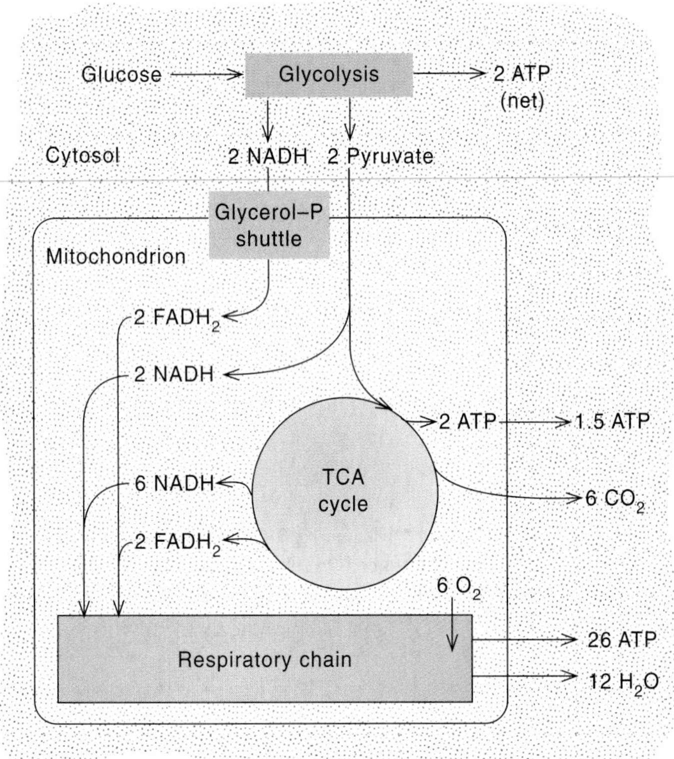

alents to the mitochondria in the form of either malate or glycerol-3-phosphate. Operation of the glycerol-3-phosphate shuttle would provide two pairs of electrons at the level of UQH_2, yielding three molecules of ATP (2×1.5). The malate shuttle allows electrons to pass through complex I also, leading to the formation of about 4.5 ATP (2×2.25). Thus, depending on which shuttle system is in operation, a total of either 26 or 27.5 ATPs can be formed by oxidative phosphorylation in addition to the four ATPs formed in glycolysis and the TCA cycle. Exporting the two molecules of ATP formed in the TCA cycle would require the uptake of two protons by the mitochondria, which would reduce the total amount of ATP by about 0.5. (Remember that the adenine nucleotide transporter and the ATP-synthase together move four protons for each ATP they provide to the cytosol.) This leaves a grand total of from 29.5 to 31 ATP, depending on the shuttle.

The actual yields of ATP under physiological conditions are probably less than these limiting values. Some of the free energy that is stored transiently in the form of the proton-motive force may be lost as a result of nonspecific leakage of protons or other ions through the membrane or may be used to drive reactions other than the formation of ATP. In addition, NADH that is formed in the cytosol may be used there for reductive biosynthetic reactions such as the formation of fatty acids, rather than contributing electrons to the respiratory chain. Even so, it is clear that respiration allows cells to make ATP in amounts that are far greater than the amounts provided by glycolysis.

Summary

In this chapter we have focused on the oxidation–reduction reactions that are coupled to the synthesis of ATP. The following points are highlights of our discussion.

1. In eukaryotes the TCA cycle and most of the reactions of aerobic energy metabolism occur in mitochondria. An inner membrane separates the mitochondrion into two distinct spaces: the internal matrix space and the intermembrane space. All but one of the enzymes of the TCA cycle are in the matrix space; succinate dehydrogenase is in the inner membrane.
2. In the TCA cycle, electrons are passed to NAD^+ and FAD. An electron-transport system in the inner membrane reoxidizes the reduced coenzymes (NADH and $FADH_2$) at the expense of molecular oxygen. Flow of electrons to O_2 releases the free energy that supports the synthesis of most of the ATP formed during aerobic metabolism.
3. Electron transfer to O_2 occurs stepwise, via a series of flavoproteins, cytochromes (heme-proteins), iron–sulfur proteins, quinones, and copper atoms. Most of the electron carriers are collected in four large complexes, which communicate via two mobile carriers, ubiquinone (UQ) and cytochrome *c*.
4. Complex I transfers electrons from NADH to UQ, and complex II transfers electrons from succinate to UQ. Both of these complexes contain flavins and numerous iron–sulfur centers.
5. Complex III, which contains three cytochromes (cytochromes b_L, b_H, and c_1) and one iron–sulfur protein, passes electrons from reduced ubiquinone (UQH_2) to cytochrome *c*. Complex IV contains two cytochromes (a and a_3) and two copper

atoms and transfers electrons from cytochrome c to O_2. The transfer of electrons through complex III involves a cyclic series of reactions (the Q cycle), in which UQH_2 and UQ undergo oxidation and reduction at two distinct sites.

6. As electrons move through complexes I, III, and IV toward O_2, protons are taken up from the solution on the matrix side of the membrane and released on the cytosolic side. This raises the pH of the matrix slightly above that of the cytosol and leaves the matrix negatively charged relative to the cytosol, creating an electrochemical potential difference or proton-motive force that tends to pull protons from the cytosol back into the matrix.
7. Proton movements through the F_o base-piece of the ATP-synthase in the inner membrane drive the formation of ATP by causing the release of bound ATP from the catalytic site on the F_1 head-piece of the enzyme.
8. Respiration is tightly coupled to the formation of ATP. Approximately 2.5 molecules of ATP are synthesized for each pair of electrons that pass down the electron-transport chain from NADH to O_2. Uncouplers, which are lipophilic weak acids, can dissipate the proton-motive force by carrying protons across the membrane. Respiration then occurs rapidly even in the absence of the phosphorylation reactions.
9. The proton-motive force also drives the uptake of P_i and ADP into the mitochondrial matrix and the export of ATP to the cytosol. By exchanging P_i for an organic acid and exchanging one organic acid for another, mitochondria concentrate pyruvate and the other substrates that provide electrons to the electron-transport chain.

SELECTED READINGS

Abrahams, J. P., A. G. W. Leslie, R. Lutter, and J. E. Walker, Structure at 2.8 Å resolution of F_1-ATPase from bovine heart mitochondria. *Nature* 370:621–628, 1994. Structure and mechanism of the ATP synthase.

Calhoun, M. W., J. W. Thomas, and R. B. Geneis, The cytochrome oxidase superfamily of redox-driven proton pumps. *Trends Biochem. Sci.* 19:325–330, 1994. A review of the structure and operation of mitochondrial complex IV and of homologous bacterial enzymes.

Chance, B., and G. R. Williams, Respiratory enzymes in oxidative phosphorylation II. Difference spectra. *J. Biol. Chem.* 217:395, 1955. One of a series of papers developing kinetic techniques for elucidating the sequence of electron carriers in the respiratory chain.

Ernster, L. (ed.), *Bioenergetics*. Elsevier: Amsterdam, 1984. A collection of reviews covering electron transport, the ATP-synthase, translocation of ions across the mitochondrial inner membrane, thermogenesis in brown fat, and other topics in bioenergetics.

Hatefi, Y., The mitochondrial electron transport and oxidative phosphorylation system. *Ann. Rev. Biochem.* 54:1015–1069, 1985.

Hinkle, P. C., A. Kumar, A. Resetar, and D. L. Harris, Mechanistic stoichiometry of mitochondrial oxidative phosphorylation. *Biochem.* 30:3576, 1991. Measurements of P-to-O ratios and a discussion of the amount of ATP synthesized during the oxidation of glucose.

Mitchell, P., and J. Moyle, Stoichiometry of proton translocation through the respiratory chain and adenosine triphosphatase systems of rat liver mitochondria. *Nature* 208:147, 1965. The initial observations that electron transport moves protons outward across the mitochondrial inner membrane and that ATP hydrolysis does the same.

Reynafarje, B., and A. L. Lehninger, The K^+/site and H^+/site stoichiometry of mitochondrial electron transport. *J. Biol. Chem.* 254:6331, 1978. Valinomycin is used to allow K^+ to move inward across the inner membrane in response to the electrical potential difference created by H^+ efflux. The number of protons pumped out is found to be larger than previously estimated.

Trumpower, B., The proton motive Q cycle. *J. Biol Chem.* 265: 11409, 1990. A review of the structure and operation of complex III and the Q cycle.

Trumpower, B. L., and R. B. Gennis, Energy transduction by cytochrome complexes in mitochondrial and bacterial respiration: The enzymology of coupling electron transfer reactions to transmembrane proton translocation. *Ann. Rev. Biochem.* 63:675–716, 1994.

PROBLEMS

1. The terms "coupled" and "uncoupled" are used throughout this chapter. What do these terms mean?
2. Figure 16.17 indicates that a pair of electrons can feed into the electron transport scheme at the level of cytochrome c. How many ATP molecules result from the oxidation of an ascorbate? Is this the normal function of ascorbate?
3. Compare iron–sulfur proteins, flavoproteins, and quinones with respect to the following:
   (a) Chemical nature of the functional group that undergoes oxidation–reduction.
   (b) Number of reducing equivalents per redox center involved in electron donor–acceptor reactions of physiological importance. If semiquinones are formed, so indicate and include them in the reduction scheme.
   (c) Stoichiometry of protons taken up per electron.
4. (a) Describe how heme is bound to the protein portion of the a/a_3-, b-, and c-type cytochromes.
   (b) Although there are three types of cytochromes in rat liver mitochondria, CO and CN^- inhibit electron transfer only at the cytochrome a/a_3 complex. Why do these inhibitors interact with cytochrome a/a_3 but not with cytochrome b or cytochrome c?
5. Calculate the standard redox potential change ($\Delta E°'$) and the standard free energy change ($\Delta G°'$) for the following reactions at pH 7.0. Write a balanced equation for each reaction.
   (a) cyt $c(Fe^{2+})$ + cyt $a_3(Fe^{3+}) \rightarrow$ cyt $c(Fe^{3+})$ + cyt $a_3(Fe^{2+})$
   (b) 4 cyt $c(Fe^{2+}) + O_2 + 4\ H^+ \rightarrow$ 4 cyt $c(Fe^{3+})$ + 2 HOH
   (c) Oxidation of succinate by succinate: cytochrome c reductase.
   (d) Reduction of extramitochondrial NAD^+ by dihydrobiquinone, via the α-glycerolphosphate shuttle.
6. (a) In biological oxidation–reduction reactions, does the stoichiometry of electron transfer (reducing equivalents/

mole) differ among the 1 Fe, 2 Fe–2 S, and 4 Fe–4 S centers?

(b) 4 Fe–4 S centers function in electron transport over a wide range of reduction potentials. There is nothing inherent in the iron–sulfur cluster to suggest this range of reduction potentials. Therefore what other component(s) must dictate reduction potential?

7. What percent of cytochrome *c* will be in the oxidized form in a solution held at +0.30 V and pH 7.0?

8. Given the standard reduction potentials for cytochrome *c* and ubiquinone at pH 7.0 (see text), calculate the corresponding values at pH 6.0 and 8.0.

9. (a) Acetylation of one or more lysines near the edge of the heme in cytochrome *c* decreases both the rate of electron transfer to the cytochrome *c* from complex III and transfer of electrons from the reduced cytochrome to cytochrome oxidase. What does this suggest concerning operation of the respiratory chain?

(b) What types of amino acid residues on complex III and complex IV would you expect to interact with cytochrome *c*?

10. We can estimate the overall ATP yield for the oxidation of specific metabolic intermediates by considering both oxidative and substrate-level phosphorylation. In principle, total molar yields of "high-energy phosphate" (ATP or equivalent) from cytosolic and mitochondrial processes divided by the molar consumption of O in the mitochondria yield theoretical P/O (or ADP/O) ratios. Using the value of P/O = 2.5 for NADH oxidation and P/O = 1.5 for succinate oxidation in the mitochondria, calculate theoretical P/O ratios for the oxidations given below. Assume that all required enzymes and cofactors are present and that extramitochondrial NADH is oxidized via the α-glycerolphosphate shuttle.

(a) Oxidation of lactate to CO_2 and HOH.

(b) Oxidation as in part (a), with 2,4-dinitrophenol present.

(c) Oxidation of dihydroxyacetone phosphate to CO_2 and HOH.

(d) Oxidation as in part (c), with 2,4-dinitrophenol present.

11. (a) Explain what is meant by "tightly coupled" mitochondria. How can we determine whether mitochondria are tightly coupled?

(b) What is the importance of "respiratory control" in oxidation of metabolites?

(c) In what metabolic circumstance is it advantageous to the organism to have mitochondria uncoupled?

C H A P T E R

17

Photosynthesis and Other Processes Involving Light

Light energy is transformed into chemical energy, and the reciprocal process also occurs in certain living systems.

In chapter 16 we saw that mitochondria convert the chemical free energy of electron-transfer reactions into a transmembrane electrochemical potential gradient for protons. The vectorial aspects of metabolism presented new ideas that biochemists initially found difficult to grasp. It was necessary to consider free energy changes that accompany movements of ions between different compartments of a cell, in addition to the free energy changes associated with the electron-transfer reactions themselves. In this chapter we come upon a transfer of energy that initially may seem even more esoteric: the transfer of energy between light and matter. We will examine this phenomenon in three different biological situations: (1) the conversion of light into chemical energy in photosynthesis; (2) the effects of light on phytochrome, a pigment that regulates circadian and seasonal cycles of growth and flowering in plants; and (3) the production of light by chemical reactions in bioluminescence. We begin with photosynthesis, which will be the major concern of the chapter.

Photosynthesis

Heterotrophic organisms, including animals, fungi, and most types of bacteria, live by degrading complex molecules provided by other organisms. Life on earth obviously could not continue indefinitely in this manner without an independent mechanism for synthesizing complex molecules from simple ones. The energy that sustains this synthesis comes almost entirely from the sun and is captured in the process of photosynthesis (fig. 17.1). Plants and other photosynthetic organisms convert, or "fix," about 10^{11} tons of carbon from CO_2 into organic compounds annually. In spite of the enormous magnitude of this conversion, the total amount of fixed carbon on earth is decreasing as a result of consumption. As

Figure 17.1

Photosynthesis harnesses the energy of sunlight for biosynthesis. In plants the reactions of photosynthesis occur in chloroplasts. Pigment-protein antenna complexes in the thylakoid membrane absorb light and pass energy to the reaction centers of two distinct photosystems, where excited chlorophyll molecules (P680 in photosystem II, P700 in photosystem I) transfer electrons to a series of electron acceptors. Photosystem I generates strong reductants that ultimately reduce $NADP^+$; photosystem II generates a strong oxidant that splits H_2O, releasing O_2. Electrons flow from photosystem II to photosystem I through the cytochrome b_6f complex. The electron-transfer reactions release protons in the thylakoid lumen and take up protons from the stroma, setting up a transmembrane pH gradient. Flow of protons back across the membrane through an ATP-synthase drives the formation of ATP. NADPH and ATP provided by the electron-transfer reactions are used to convert CO_2 to carbohydrates. CO_2 combines with ribulose-1,5-bisphosphate to form two molecules of glycerate-3-phosphate. ATP and NADPH are used to reduce the glycerate-3-phosphate to glyceraldehyde-3-phosphate and to recycle some of the glyceraldehyde-3-phosphate to ribulose-1,5-bisphosphate. The remaining glyceraldehyde-3-phosphate is exported from the chloroplast or converted to other carbohydrates. Many of the photosynthetic reactions resemble reactions seen in earlier chapters: The formation of hexoses and pentoses from trioses resembles reactions of gluconeogenesis and the pentose phosphate pathway (see chapter 14); the cytochrome b_6f complex is similar to the mitochondrial complex III, and the proton-conducting ATP-synthase is similar to the mitochondrial ATP-synthase (see chapter 16). The major difference between the operation of mitochondria and chloroplasts is the direction of electron flow. In mitochondria, electrons flow from reduced organic compounds to O_2; photosynthetic organisms use the energy of light to push electrons in the opposite direction.

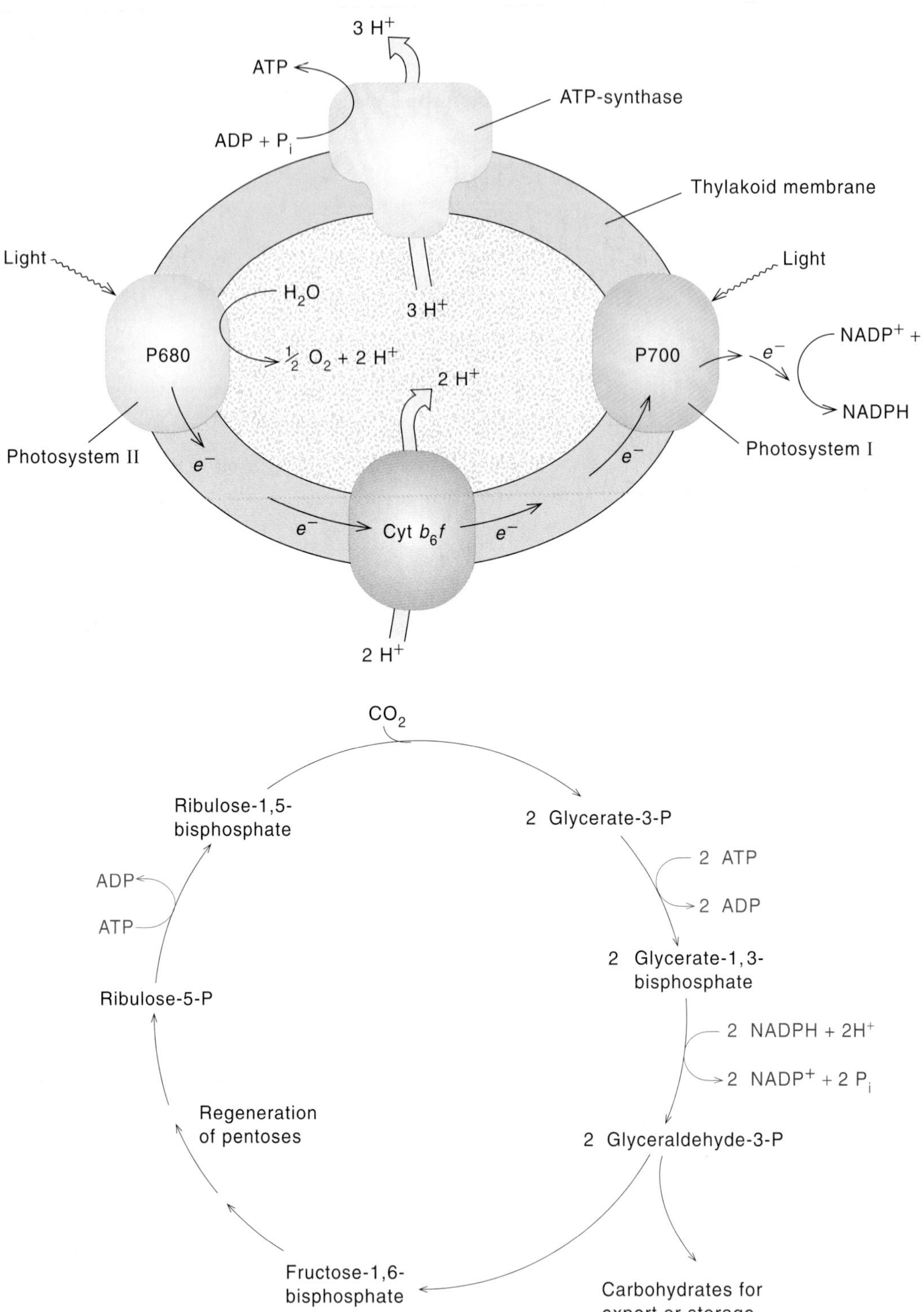

our reserves of energy diminish, it becomes increasingly important that we understand how photosynthesis works and how our activities affect it.

When most of us think of photosynthesis, we think of green trees or perhaps of fields of grain. Actually, about one-third of the carbon fixation that occurs on earth takes place in the oceans and is carried out by microorganisms. In addition to plants, several groups of bacteria are capable of photosynthesis. Bacteria have been extremely useful in the study of photosynthesis because their photosynthetic apparatus is simpler than that of plants, and it has

Figure 17.2

(a) Electron micrograph of a chloroplast in a lettuce leaf. The organelle is shaped like a flattened sausage with a width of about 10 μm and a thickness of about 3 μm. It is surrounded by a double outer membrane or envelope (E). The stroma (S) contains DNA, ribosomes, and the soluble enzymes of CO_2 fixation. Extending throughout the stroma is the thylakoid membrane, which is differentiated into stacked regions or grana (G) and unstacked stromal lamellae (SL). (Courtesy of Dr. Charles Arntzen.) (b) A schematic drawing of the chloroplast membrane systems. The highly folded thylakoid membrane separates the thylakoid lumen from the stroma.

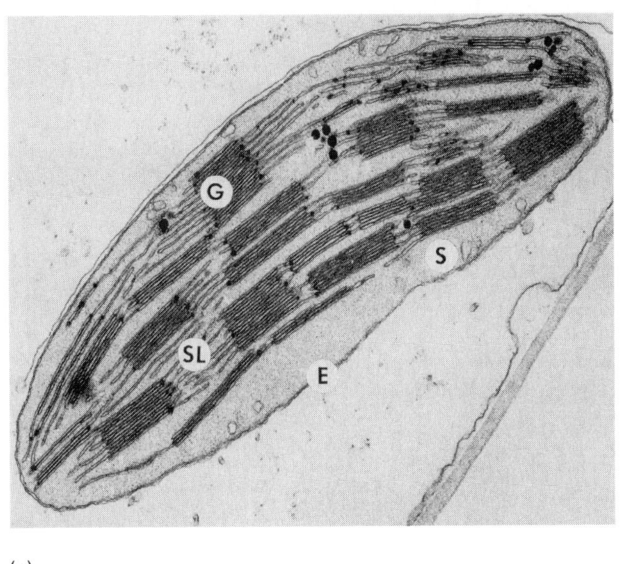

(a)

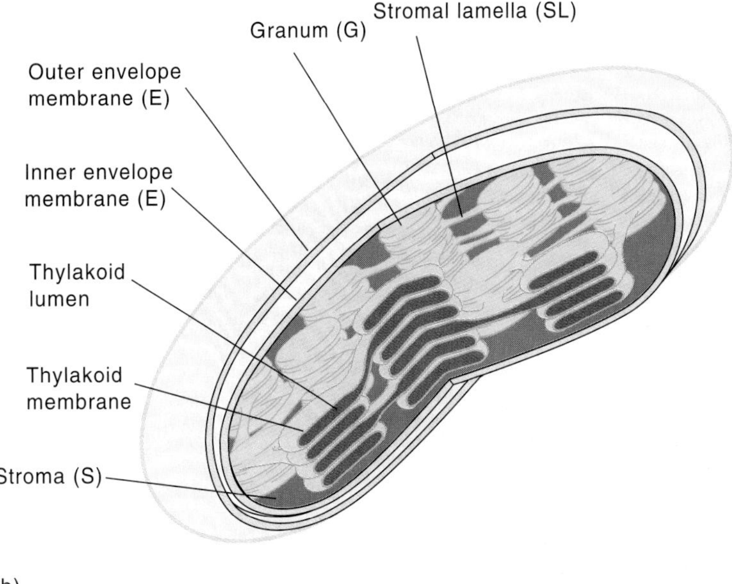

(b)

been easier to purify the pigment–protein complexes that are components of this apparatus in the bacteria. Although there are important differences between photosynthesis in bacteria and plants, the photochemical reactions that capture the energy of light are basically the same.

An overall equation for CO_2 fixation as it occurs in plants is

$$6\ CO_2 + 6\ H_2O + \text{Light} \longrightarrow C_6H_{12}O_6 + 6\ O_2 \qquad (1)$$

or, more generally,

$$CO_2 + H_2O + \text{Light} \longrightarrow (CH_2O) + O_2 \qquad (2)$$

where (CH_2O) represents part of a carbohydrate molecule. Electrons and protons are removed from H_2O, O_2 is evolved, and CO_2 is reduced to the level of a carbohydrate. One group of photosynthetic bacteria, the cyanobacteria, carry out the same process. Other types of photosynthetic bacteria carry out similar overall processes except that they do not evolve O_2 because they use materials other than H_2O as a source of electrons.

If we leave out the light, the equilibrium for the synthesis of glucose from CO_2 and H_2O lies vanishingly far to the left. The equilibrium constant at 27° C is 10^{-496}! Our goal is to explore how photosynthetic organisms use light to drive the reaction in the direction of carbohydrates, against this enormous thermodynamic gradient. How can light do chemistry?

The Photochemical Reactions of Photosynthesis Take Place in Membranes

In plants the reactions of photosynthesis take place in specialized subcellular organelles, the chloroplasts. Figure 17.2a shows an electron micrograph of a chloroplast from a lettuce leaf. Chloroplasts are bounded by an envelope of two membranes, and they have an extensive internal membrane called the thylakoid membrane. In electron micrographs of thin-sectioned chloroplasts, the thylakoid membrane gives the appearance of a large number of separate sheets or flattened vesicles. It actually is a single membrane that is highly folded and encloses a distinct compartment, the thylakoid lumen (fig. 17.2b). In places, the folded membrane is tightly stacked into disklike structures called grana. The chlorophyll found in chloroplasts is bound to proteins that are integral constituents of the thylakoid membrane, and it is here that the initial conversion of light into chemical energy occurs. The thylakoid membrane also contains a collection of electron carriers and an ATP-synthase similar to the proton-translocating ATP-synthase of mitochondria. The enzymes responsible for the actual fixation of CO_2 and the synthesis of carbohydrates are soluble proteins and reside in the stroma that surrounds the thylakoid membrane (see fig. 17.2). The stroma also contains DNA and ribosomes, which are responsible for synthesizing some of the proteins found in the chloroplast.

Figure 17.3

Longitudinal view of a cell of *Rhodospirillum rubrum*, a purple photosynthetic bacterium. A double-membrane system surrounds the cell. The inner membrane is extensively invaginated into tubules (arrows). These look circular when they are cut in cross section. (Courtesy of Dr. Gerald Peters.)

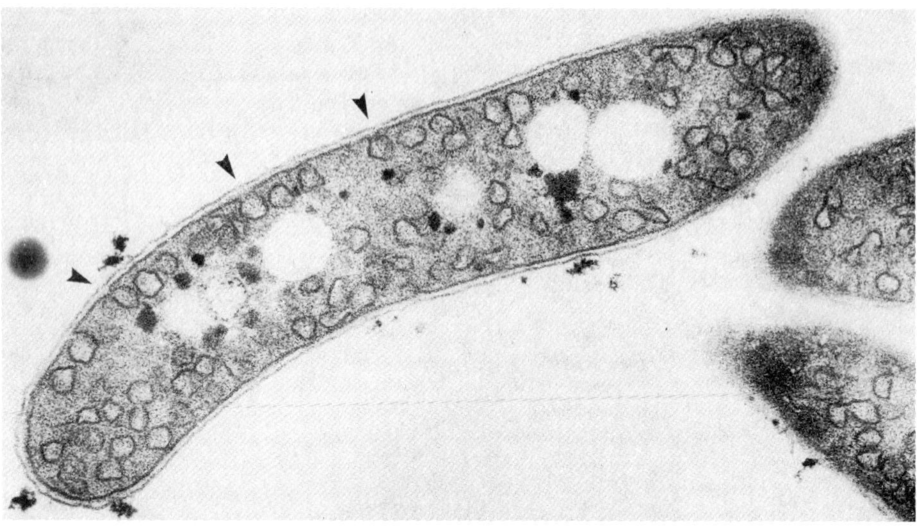

Table 17.1
Properties of Photosynthetic Organisms

Group	Evolve O_2	Contain Chloroplasts	Type of Chlorophyll	Number of Photosystems
Algae and higher plants	Yes	Yes	Chlorophyll *a* and *b*	2
Cyanobacteria	Yes	No	Chlorophyll *a* and *b*	2
Purple bacteria	No	No	Bacteriochlorophyll *a* or *b*	1

Algae are members of the plant kingdom and contain chloroplasts similar to those of higher plants. Prokaryotic photosynthetic organisms, which include the cyanobacteria and several groups of purple or green bacteria, do not have chloroplasts. In prokaryotes the photochemical reactions of photosynthesis take place in the membrane that encloses the cell. This membrane has extensive invaginations resembling the cristae of the mitochondrial inner membrane (fig. 17.3). Table 17.1 summarizes the distinctions between some of the major groups of photosynthetic organisms.

Photosynthesis Depends on the Photochemical Reactivity of Chlorophyll

With the exception of certain halophilic bacteria (see chapter 29), all known photosynthetic organisms take advantage of the photochemical reactivity of one or another type of chlorophyll. Figure 17.4 shows the structures of several of the different types of chlorophyll that occur in nature. Chlorophylls resemble hemes, the prosthetic groups of the cytochromes and hemoglobin, and they are derived biosynthetically from protoporphyrin IX. However, they differ from hemes in four major respects: (1) Chlorophylls have an additional ring (ring V) with carbonyl and carboxylic ester substituents. (2) The central metal atom is magnesium rather than iron. (3) In the case of chlorophyll *a* and chlorophyll *b*, the two major chlorophylls in plants and cyanobacteria, one of the pyrrole rings (ring IV) is reduced by the addition of two hydrogens. In bacteriochlorophyll *a* and bacteriochlorophyll *b*, which occur in the purple and green bacteria, two of the rings are reduced (rings II and IV). Finally (4), the propionyl side chain of ring IV is esterified with a long-chain isoprenoid alcohol. Chlorophylls *a* and *b* contain the alcohol phytol; bacteriochlorophylls *a* and *b* have either phytol or geranylgeraniol, depending on the species of bacteria. Photosynthetic organisms also contain small amounts of pheophytins or bacteriopheophytins, which are the same as the corresponding chlorophylls or bacteriochlorophylls except that two hydrogens replace the magnesium (see fig. 17.4). We will see that the pheophytins and bacteriopheophytins play special roles as electron carriers in photosynthesis.

Figure 17.4

Structures of protoporphyrin IX (the prosthetic group of hemoglobin, myoglobin, and the *c*-type cytochromes) and several types of chlorophyll and bacteriochlorophyll. Chlorophyll *a* and chlorophyll *b* are the main types of chlorophyll in plants and the cyanobacteria. Purple photosynthetic bacteria contain either bacteriochlorophyll *a* or bacteriochlorophyll *b*, depending on the bacterial species. (The green photosynthetic bacteria contain still another form of bacteriochlorophyll.) Chlorophyll *b* and bacteriochlorophyll *b* are the same as chlorophyll *a* and bacteriochlorophyll *a* except for the substituents on ring II. Pheophytins and bacteriopheophytins are the same as the corresponding chlorophylls or bacteriochlorophylls except that two hydrogen atoms replace Mg.

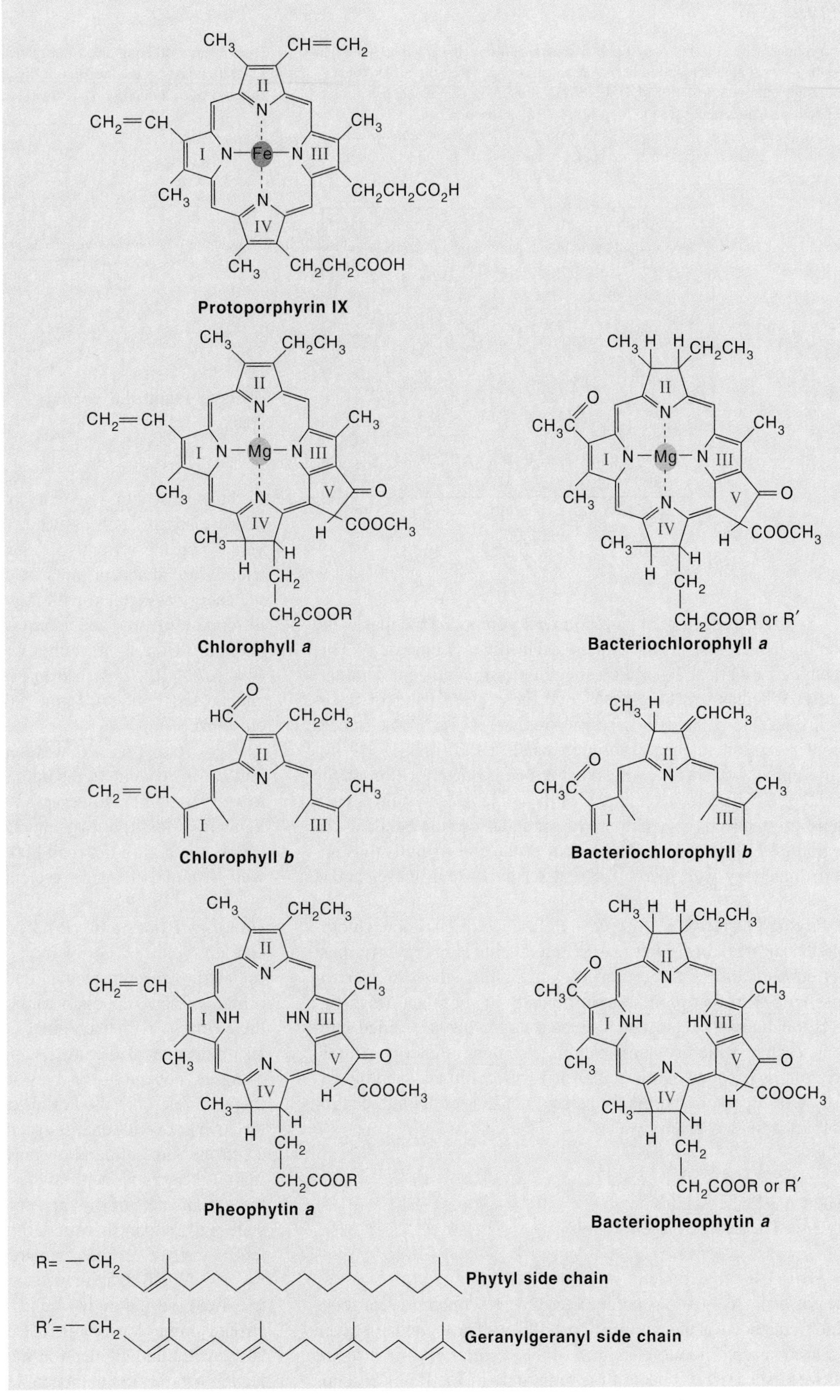

Figure 17.5

Absorption spectra of chlorophyll *a* and bacteriochlorophyll *a* in ether. Note that the long-wavelength absorption bands are much stronger than the α band of reduced cytochrome *c* (see fig. 16.4). In the chlorophyll–protein complexes found in photosynthetic organisms, the long-wavelength absorption band generally is shifted to even longer wavelengths. This probably reflects interactions between neighboring chlorophylls, which are bound to the proteins as dimers or larger groups.

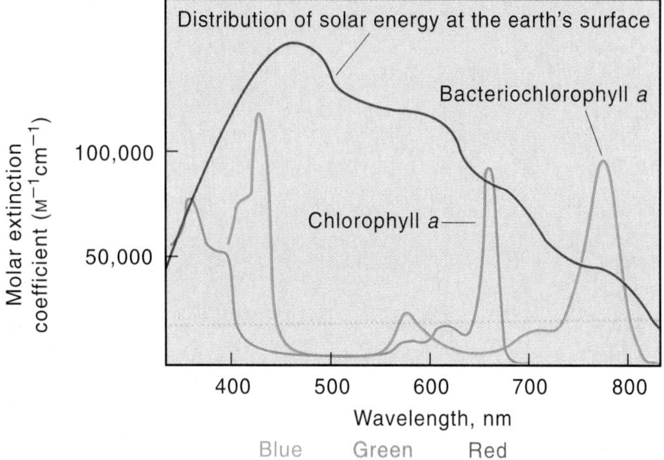

Figure 17.6

Light is an oscillating electromagnetic field. The lengths of the arrows in this diagram represent the strength of the electric field at a particular time as a function of position in a ray of light proceeding along the *y* axis.

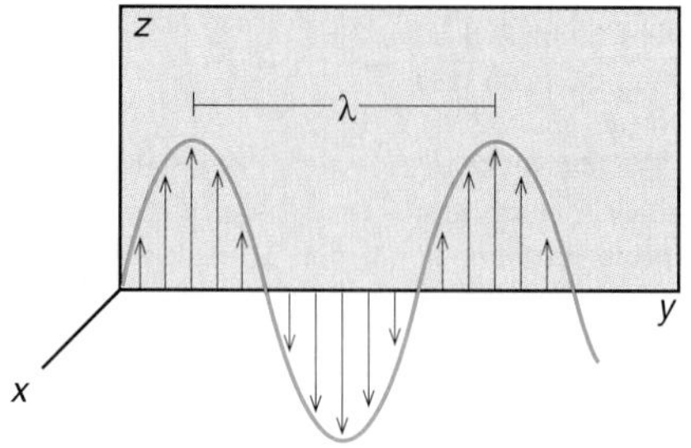

The reduction of ring IV in chlorophyll *a* or *b* makes the conjugated aromatic ring system decidedly asymmetrical. This changes the optical absorption spectrum of the molecule dramatically. Whereas the long-wavelength absorption band of a cytochrome (the α band) is relatively weak (see fig. 16.4), chlorophyll *a* has an intense absorption band at 676 nm (fig. 17.5). Chlorophyll *b* has a similar band at 642 nm. Bacteriochlorophylls *a* and *b*, in which the asymmetry of the conjugated system is even more pronounced, have extremely strong absorption bands in the region of 770 nm (see fig. 17.5). All of the chlorophylls thus absorb light very well, particularly at relatively long wavelengths.

A Physical Definition of Light. Before we discuss how chlorophylls can transform light energy into chemical energy, we must review some of the basic properties of light. Light is an electromagnetic field that oscillates sinusoidally in space and time (fig. 17.6). It interacts with matter in packets, or quanta, called photons, each of which contains a definite amount of energy. A mole of photons is called an einstein. The relationship between the energy ϵ of a photon and the frequency v of the oscillating electromagnetic field is given by

$$\epsilon = hv \tag{3}$$

where h is Planck's constant (6.63×10^{-27} erg · s or 4.12×10^{-15} eV · s). The energy per einstein is

$$E = N\epsilon = Nhv \tag{4}$$

where N is Avogadro's number (6.02×10^{23} photons/einstein). The frequency v is the number of oscillations per second at a given point in space. The wavelength λ of the oscillations (the distance between successive peaks in the amplitude of the field) depends on both v and the velocity c at which the peaks move through space:

$$\lambda = c/v \tag{5}$$

Light travels with a velocity of 3×10^{10} cm/s in a vacuum. In a dense medium such as water the velocity is less, depending inversely on the refractive index of the medium. Photon energies usually are stated in units of electron volts (eV), where 1 eV is the energy associated with moving one electron across a potential difference of 1 V and is equivalent to 23,060 cal/mole (96,480 kJ/mole). Blue light, with a wavelength in the region of 450 nm ($v = 6.7 \times 10^{14}\ s^{-1}$), has an energy of 2.75 eV, or 64 kcal/einstein, and far-red light (700 nm) has an energy of 1.77 eV (41 kcal/einstein).

Radiation with wavelengths much below 400 nm or above 750 nm is invisible to the human eye, and some authors prefer not to call it "light." However, such radiation can be important biologically. Many photosynthetic bacteria are adapted to use radiation in the region between 800 and 900 nm, and some species do well with even longer wavelengths.

How Light Interacts with Molecules. The electrons in a molecule are held in a set of molecular orbitals, each of which is associated with a particular energy. In an isolated molecule the orbital energies depend mainly on the electrostatic interactions of the electrons with each other and with the nuclei and are more or less independent of time. A molecule thus can have a variety of energies, depending on how its electrons are distributed among the available orbitals. For an organic molecule with $2n$ electrons the lowest overall energy usually is obtained when there are two electrons with antiparallel spins in each of the first n orbitals, leaving all the orbitals with higher energies empty (fig. 17.7). This is the ground state of the molecule. A molecule that is placed in this state will remain there indefinitely, as long as the electronic potential energies do not change.

When light interacts with a molecule, the oscillating electric field of the light makes the electronic potential energies strongly time-dependent. The result of this can be that a photon is absorbed and an electron moves from one of the occupied molecular orbitals to an unoccupied orbital with a higher energy (see

Figure 17.7

When a molecule absorbs light, an electron is excited to a molecular orbital with higher energy. The horizontal bars in this diagram represent molecular orbitals for electrons. Each orbital can hold two electrons with antiparallel spins (arrows pointing upward or downward). Only the top few of the filled orbitals are shown here. Absorption of light raises an electron from one of these orbitals to an orbital that is normally unoccupied. For this to occur the energy of the photon must match the difference between the energies of the two orbitals. The choice of an upward or downward arrow is arbitrary, but no change of spin occurs during the excitation. As long as the spins of the two unpaired electrons remain antiparallel, so the molecule has no net electronic spin, the molecule is said to be in an excited "singlet" state. An excited molecule can return directly to the ground state by giving off energy as fluorescence or heat or by transferring energy to another nearby molecule, or it can transfer an electron to another molecule (see fig. 17.8).

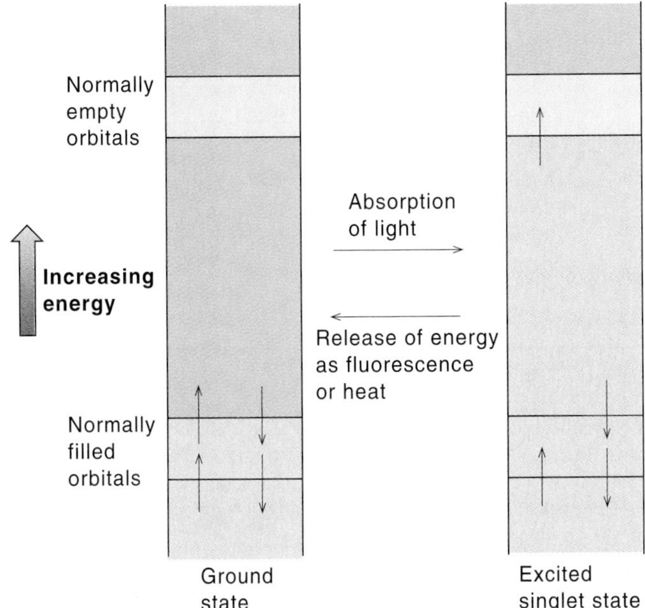

fig. 17.7). Two requirements must be met in order for this to occur. First, the difference between the energies of the two orbitals must be the same as the photon's energy, $h\nu$. This is why a given type of molecule absorbs light of some wavelengths and not of others. The second, more complex requirement has to do with the shapes of the two orbitals and the orientation (polarization) of the oscillating field relative to the disposition of the orbitals in space. The two orbitals must have different geometrical symmetries and must be oriented in an appropriate way with respect to the field. This requirement explains why some absorption bands are stronger than others.

If chlorophyll (or any other molecule) absorbs a photon, it is excited to a state that lies above the ground state in energy. This usually occurs with no change in electronic spin (see fig. 17.7). As long as the spins of the two unpaired electrons remain antiparallel, so that the molecule has no net electronic spin, the molecule is said to be in an excited *singlet* state.

A molecule in an excited singlet state can decay back to the ground state by releasing energy in several different ways. One possibility is simply to emit a photon; this is fluorescence (see fig. 17.7). The wavelength of the fluorescence generally is longer than that of the light that was originally absorbed, because readjustments of the molecular geometry decrease the energy of the excited molecule somewhat before the molecule fluoresces. The extra energy is given off to the environment as heat. If the molecule has several absorption bands, as the chlorophylls do, the wavelength of the fluorescence is usually slightly longer than that of the longest-wavelength absorption band. The molecule relaxes to the lowest, or "first," excited singlet state before the emission occurs. In some cases the molecule can decay all the way to the ground state by radiationless processes, converting the excitation energy entirely into heat. Another decay mechanism is to transfer the energy to a neighboring molecule by the process of resonance energy transfer. This phenomenon plays an important role in photosynthesis, and we will return to it later. A fourth possibility is for the excited molecule to transfer an electron to a neighboring molecule.

Light Causes an Electron-Transfer Reaction. Electron transfer can be a favorable path for the decay of an excited molecule, because an electron in the upper, normally unoccupied orbital is bound less tightly than one in a lower, normally filled orbital. If the absorption of a photon increases the energy of the molecule by ϵ electron volts, where $\epsilon \approx h\nu$, it will lower the standard redox potential for removing an electron from the excited molecule by ϵ volts, compared with the $E°$ for the molecule in the ground state. In the case of chlorophyll *a*, $E°$ for oxidation in the ground state is approximately +0.5 V, and $h\nu$ for the long-wavelength absorption band is about 1.7 eV. The $E°$ of the excited molecule is therefore about −1.2 V. This means that in the excited state, chlorophyll *a* is an extremely strong reductant. For comparison, recall that NAD^+/NADH has an $E°'$ of only −0.32 V. The basic principle underlying the photochemistry of photosynthesis is that excitation causes a molecule of chlorophyll or bacteriochlorophyll (or a complex of several such molecules) to release an electron. The chlorophyll complex is oxidized, and another molecule becomes reduced (fig. 17.8). The oxidized chlorophyll species that is formed is a relatively strong oxidant and can extract an electron from a third molecule.

The idea that light drives the formation of oxidants and reductants was first advanced by C. B. van Niel in the 1920s. It was known at the time that the purple photosynthetic bacteria thrive only if they are provided with a reduced substrate such as an organic acid. Some species grow well on a reduced inorganic material such as H_2S. Van Niel noticed that although the bacteria do not evolve O_2, the reactions they carry out have a formal resemblance to the process that occurs in plants. If we let H_2B represent a reduced substrate and let B represent an oxidized product, we can write the process that occurs in the purple bacteria as

$$CO_2 + 2\,H_2B \xrightarrow{\text{Light}} (CH_2O) + H_2O + 2\,B \qquad (6)$$

The equation we presented earlier for CO_2 fixation in plants and cyanobacteria (equation 2) can be put in a similar form by replacing H_2B and 2 B with H_2O and O_2:

$$CO_2 + 2\,H_2O \xrightarrow{\text{Light}} (CH_2O) + H_2O + O_2 \qquad (7)$$

Figure 17.8

The photochemical process that initiates photosynthesis is an electron-transfer reaction. The horizontal bars and vertical arrows represent molecular orbitals and electrons, as in figure 17.7. Absorption of light increases the free energy of a chlorophyll complex (Chl) by $h\nu$, making the transfer of an electron to an acceptor (A) thermodynamically favorable. The oxidized chlorophyll complex (Chl^+) extracts an electron from a donor (D).

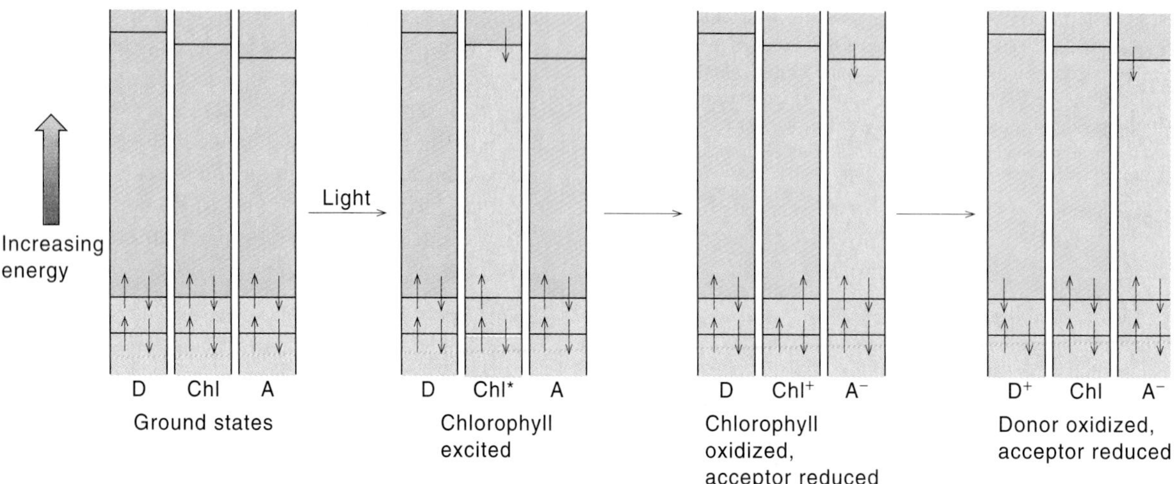

To van Niel this suggested that the essence of photosynthesis in both plants and bacteria is the photochemical separation of oxidizing and reducing power. The substance that is reduced in the photochemical reaction could be used to reduce CO_2 to carbohydrates in enzyme-catalyzed reactions that do not require light. The material that is oxidized photochemically could be discharged by the oxidation of H_2O to O_2 in plants or by the oxidation of some other material in bacteria (fig. 17.9).

Support for these ideas came from experiments done by Robin Hill in 1939. Hill discovered that isolated chloroplasts would evolve O_2 if they were illuminated in the presence of an added nonphysiological electron acceptor such as ferricyanide [$Fe(CN)_6^{3-}$]. The electron acceptor became reduced in the process. Because no fixation of CO_2 occurred under these conditions, it was clear that the photochemical reactions of photosynthesis can be separated from the reactions that involve CO_2.

Photooxidation of Chlorophyll Generates a Cationic Free Radical

When chlorophyll or bacteriochlorophyll is oxidized, the product is a positively charged free radical. The situation differs subtly from the oxidation of a cytochrome. When a cytochrome is oxidized, the electron that is removed comes from the iron, which changes from ferrous (Fe^{2+}) to ferric (Fe^{3+}). In chlorophyll the electron is removed not from the magnesium, but rather from the aromatic π-electron system of the molecule. The positive charge of the oxidized chlorophyll and the spin of the unpaired electron that remains behind are delocalized extensively over the π-electron system. This can be shown by studying the electron spin resonance (ESR) and electron nuclear double-resonance (ENDOR) spectra of the radical.

Figure 17.9

Van Niel proposed that the reductant generated in a photochemical electron-transfer reaction (A^-) is used to reduce CO_2 to carbohydrate. He suggested that the oxidant (D^+) oxidizes to H_2O to O_2 in plants and oxidizes some other material (H_2B) in the purple bacteria. Only the initial charge separation requires light. (Chl = chlorophyll.)

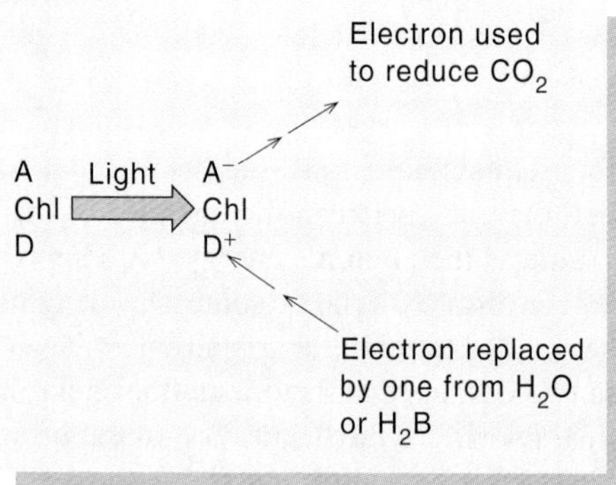

The photooxidation of chlorophyll can be detected by measuring changes in the optical absorption spectrum of the molecule. Oxidation results in the loss of the chlorophyll's characteristic absorption bands. The first measurements of this sort were made in the 1950s by Bessel Kok and Louis Duysens. Kok found that illumination of chloroplasts caused an absorbance decrease at 700 nm (fig. 17.10). He suggested that this reflected photooxidation of a reactive chlorophyll complex, which he called P700. (P stood for "pigment," and 700 for the wavelength at which the unoxidized complex had its main absorption band.) Duysens made similar observations on *Rhodospirillum rubrum,* a purple photo-

Figure 17.10

When a suspension of chloroplasts is illuminated, its optical absorbance decreases in the regions around 430 and 700 nm. The absorbance changes reflect the oxidation of a special chlorophyll complex (P700). Addition of a chemical oxidant such as potassium ferricyanide causes similar absorbance changes. (Source: From B. Kok, Partial purification and determination of oxidation reduction potential of the photosynthetic chlorophyll complex absorbing at 700 mμ, *Biochim. Biophys. Acta* 48:527, 1961.)

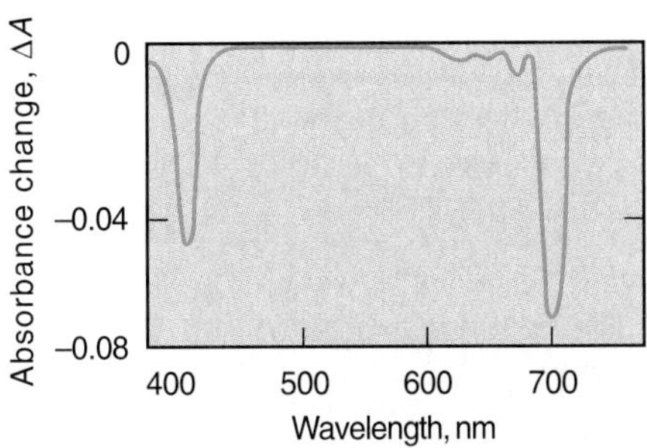

Figure 17.11

The structures of ubiquinone, menaquinone (vitamin K_2), plastoquinone, and phylloquinone (vitamin K_1). Purple photosynthetic bacteria contain ubiquinone, menaquinone, or both, depending on the bacterial species; chloroplasts contain plastoquinone and phylloquinone.

Ubiquinone

Menaquinone

Plastoquinone

Phylloquinone

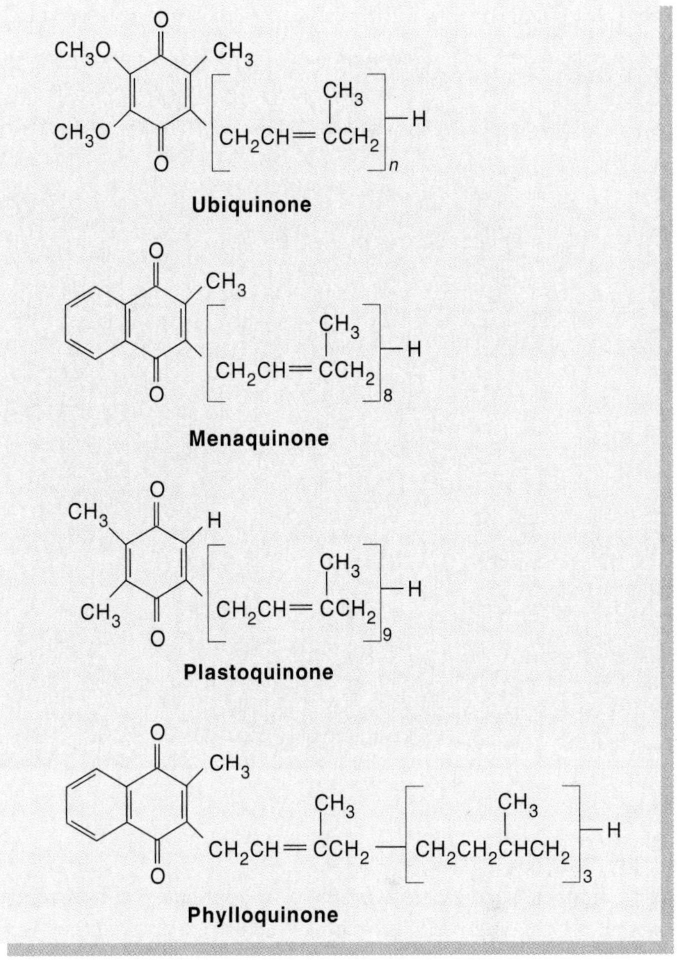

synthetic bacteria. Here the reactive bacteriochlorophyll complex absorbed at 870 nm, and Duysens called it P870. A second type of reactive complex in chloroplasts, P680, was discovered subsequently by H. Witt and his colleagues. We will see below that P700 and P680 are parts of two distinct photochemical systems, photosystem I and photosystem II. P700 and P680 also are found in the cyanobacteria.

The photooxidation of P870, P700, or P680 generates a cationic radical ($P870^+$, $P700^+$, or $P680^+$) in which the charge and the unpaired electron are delocalized over the macrocyclic ring system. In fact, the ESR and ENDOR spectra of the $P870^+$ radical indicate that the unpaired electron is delocalized over *two* bacteriochlorophyll *a* molecules, which form a closely interacting pair. The strong interaction between the two bacteriochlorophylls probably explains why the absorption band of the complex is at 870 nm, whereas the long-wavelength band of monomeric bacteriochlorophyll *a* in solution is at 770 nm (see fig. 17.5). In bacterial species that contain bacteriochlorophyll *b* instead of bacteriochlorophyll *a*, the reactive dimers absorb at 960 nm instead of 870 and are often called P960. For simplicity we will neglect the differences in wavelength among the various species of purple bacteria and will use the term P870 in a general sense. P700 and P680 probably consist of similar dimers of chlorophyll *a*, but the evidence on this point is not entirely conclusive.

The Reactive Chlorophyll Is Bound to Proteins in Complexes Called Reaction Centers

The chlorophyll or bacteriochlorophyll that undergoes photooxidation is bound to a protein in a complex called a reaction center. Reaction centers have been purified by disrupting chloroplasts or bacterial membranes with detergents. This was first achieved with purple photosynthetic bacteria, particularly by Roderick Clayton, and the structure of the bacterial complex is still much better understood than are the structures of the plant reaction centers. When they are excited with light, purified bacterial reaction centers can carry out the initial photochemical transfer of an electron from P870 to a series of electron acceptors.

Reaction centers of the purple bacteria generally contain three polypeptides with a total molecular weight of about 100,000. Bound noncovalently to the protein are four molecules of bacteriochlorophyll and two bacteriopheophytins. The reaction center also contains two quinones and one nonheme iron atom. In some bacterial species, both quinones are ubiquinone. In others, one of the quinones is menaquinone (vitamin K_2), a naphthoquinone that resembles ubiquinone in having a long isoprenoid side chain (fig. 17.11). Reaction centers from some purple bacteria, such as *Rhodopseudomonas viridis,* also have a cytochrome subunit with four *c*-type hemes.

The crystal structure of reaction centers from *R. viridis* was solved by Hartmut Michel, Johann Deisenhofer, Robert Huber, and their colleagues in 1984. This was a landmark achievement because it was the first high-resolution crystal structure to be obtained of an integral membrane protein. Reaction centers

Figure 17.12

(a) Structure of the reaction center of *Rhodopseudomonas viridis*. The α-carbon chains of the two main polypeptides are shown in yellow and gold. Each of these subunits has five α-helical regions that pass back and forth across the phospholipid bilayer of the cell membrane. A third subunit, shown in light yellow, sits on the cytosolic side of the membrane but also has one transmembrane α helix. The cytochrome subunit, which resides on the outer (periplasmic) surface of the membrane, is shown in purple at the top, with its four heme groups in red. The four bacteriochlorophylls are represented in shades of blue and green, the two bacteriopheophytins in white, the iron atom in red, the two quinones in purple, and a carotenoid molecule in orange. The phytyl side chains of the bacteriochlorophylls and bacteriopheophytins have been truncated for clarity; the isoprenoid tail of one of the quinones (Q_B) is disordered and not visible in the crystal structure. (Based on the crystal structure described by J. Deisenhofer, O. Epp, K. Miki, R. Huber, and H. Michel.) (b) An expanded view of some of the components of the *R. viridis* reaction center. The color coding is as in (a). The bacteriochlorophyll dimer that undergoes photooxidation is at the top. These two molecules are about 3 Å apart where they overlap in ring I. In *R. viridis,* one of the two quinones (Q_A, on the right in the figure) is menaquinone and the other (Q_B, left) is ubiquinone.

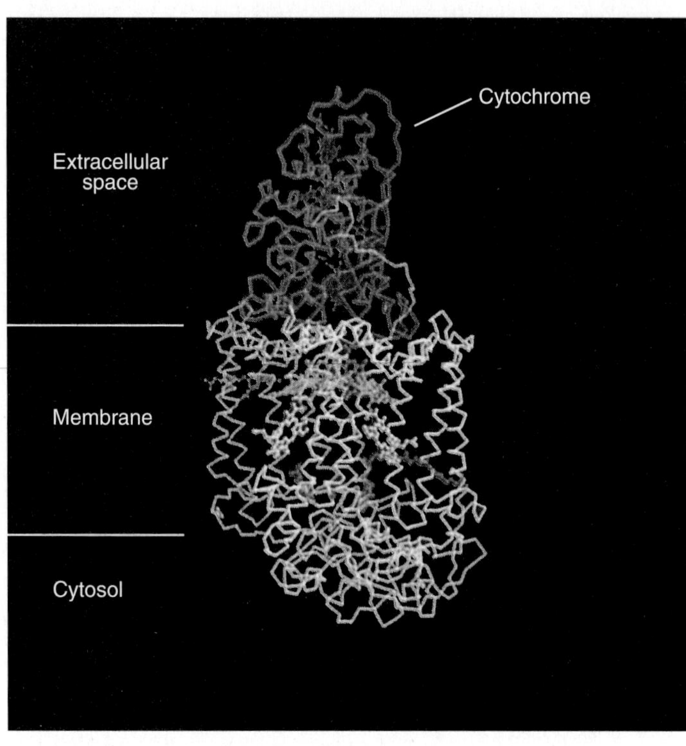

(a)

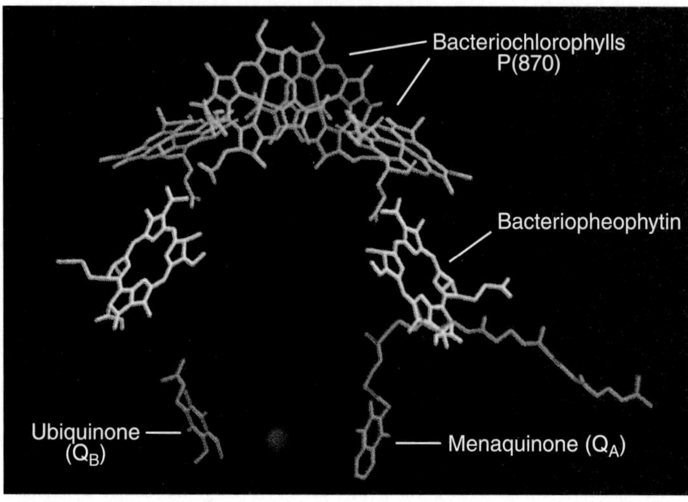

(b)

from another species, *Rhodobacter sphaeroides,* subsequently proved to have essentially the same structure except that they lack the bound cytochrome. In both species the bacteriochlorophyll and bacteriopheophytin, the nonheme iron atom and the quinones are all bound to two of the polypeptides, which are folded into a series of α helices that pass back and forth across the cell membrane (fig. 17.12a). The third polypeptide resides largely on the cytoplasmic side of the membrane, but it also has one transmembrane α helix. The cytochrome subunit in *R. viridis* sits on the external (periplasmic) surface of the membrane.

Figure 17.12b shows the arrangement of the pigments in greater detail. Two of the four bacteriochlorophylls are packed closely together. Studies of the reaction center's optical absorption spectrum indicate that this special pair of bacteriochlorophylls is P870, the reactive complex that releases an electron when the reaction center is excited with light.

Although the structures of the plant reaction centers are not yet known in detail, photosystem II reaction centers resemble the reaction centers of purple photosynthetic bacteria both functionally and structurally. The amino acid sequences of their two major polypeptides are homologous to those of the two polypeptides that hold the pigments in the bacterial reaction center. Also, the reaction centers of photosystem II contain a nonheme iron atom and two molecules of plastoquinone, a quinone that is closely related to ubiquinone (see fig. 17.11), and they contain one or more molecules of pheophytin a and several molecules of chlorophyll a in addition to the two that probably make up P680.

The reaction center of photosystem I is larger and more complex. It contains two polypeptides with molecular weights of about 80,000 and at least seven other polypeptides ranging in size from 8,500 to 21,000. The reactive chlorophyll a complex P700 resides on the two large polypeptides, along with about 60 additional molecules of chlorophyll a, two quinones, and an iron–sulfur center. Some types of green photosynthetic bacteria have reaction centers that resemble those of photosystem I.

In Purple Bacterial Reaction Centers, Electrons Move from P870 to Bacteriopheophytin and Then to Quinones

Let's now consider the acceptors that extract an electron from P870 in the purple bacterial reaction center. The first acceptor whose reduction can be well resolved kinetically is a bacteriopheophytin. Note that the reaction center contains two bacteriopheophytins and two additional bacteriochlorophylls, in addition to the special pair

Figure 17.13

The initial electron-transfer steps in reaction centers of purple photosynthetic bacteria. (P870* = the first excited singlet state of P870; BPh = bacteriopheophytin; Q_A = menaquinone or ubiquinone, depending on the bacterial species.) In this scheme, various states of the photosynthetic apparatus are positioned vertically according to their free energies, with the states that have the highest free energies at the top.

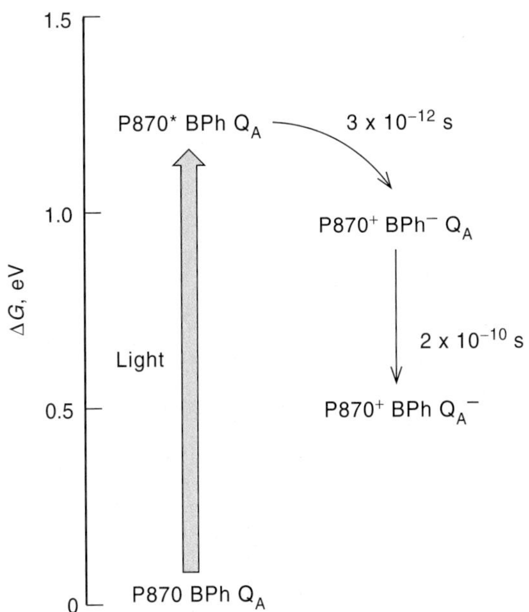

of bacteriochlorophylls that make up P870 (see fig. 17.12*b*). When isolated reaction centers are excited with a short flash of light, an excited singlet state of the reactive bacteriochlorophylls (P870*) is formed essentially instantaneously. This state decays in about 3×10^{-12} s, and as it does, a $P870^+BPh^-$ radical-pair is created. Here $P870^+$ is the cationic radical formed by removing an electron from P870, and BPh^- is the anionic radical formed by adding an electron to a bacteriopheophytin (fig. 17.13). The creation of the radical-pair state can be detected spectrophotometrically by the disappearance of absorption bands of P870 and the bacteriopheophytin and the formation of new absorption bands attributable to the two radicals. (The bacteriopheophytin that undergoes reduction is the one on the right in figures 17.12*a* and *b*. The two bacteriopheophytins are bound to the polypeptides in slightly different ways and are distinguishable by their different absorption spectra.) It is likely that the electron that moves from P870 to the BPh passes through the intervening molecule of bacteriochlorophyll, but if a $P870^+BChl^-$ intermediate is formed, this radical-pair lives for less than 1×10^{-12} s before the electron moves on to the BPh.

The $P870^+BPh^-$ radical-pair lasts for about 2×10^{-10} s, decaying by the movement of an electron from BPh^- to one of the quinones (Q_A). This leaves the reaction center in the state $P870^+Q_A^-$, where Q_A^- is the anionic semiquinone radical (see fig. 17.13).

The electron carriers that participate in these first few steps are all fixed in position close together in the reaction center and have little freedom of motion. One indication of this fact is that the electron-transfer reactions, in addition to being phenomenally fast, are almost independent of temperature. They actually increase slightly in speed with decreasing temperature. This behavior indicates that the reactants do not need to diffuse together in order for an electron to move from one molecule to another, and they do not have to acquire any additional thermal energy from their surroundings. Further, the reactions are amazingly efficient. Their efficiency can be expressed in terms of the quantum yield of $P870^+Q_A^-$, which is the number of moles of $P870^+Q_A^-$ formed per einstein of light absorbed. The measured quantum yield in purified reaction centers is 1.02 ± 0.04. Essentially, every time the reaction center is excited, an electron moves from P870 to Q_A.

A Cyclic Electron-Transport Chain Returns Electrons to P870 and Moves Protons Outward across the Membrane; Flow of Protons Back into the Cell Drives the Formation of ATP

From Q_A^-, an electron moves to the second quinone that is bound to the reaction center (Q_B in figure 17.14). This step takes about 10^{-4} s, considerably longer than the earlier steps. In the meantime a *c*-type cytochrome replaces the electron that was removed from P870, preparing the reaction center to operate again. Electron transfer between the cytochrome and $P870^+$ takes between 10^{-7} and 10^{-3} s, depending on the bacterial species. In *R. viridis* the electron donor is the bound cytochrome with four hemes shown at the top of figure 17.12*a*. In other species it often is a soluble *c*-type cytochrome with a single heme, more like mitochondrial cytochrome *c*.

When the reaction center is excited a second time, a second electron is pumped from P870 to the bacteriopheophytin and on to Q_A and Q_B. This change places Q_B in the fully reduced form, Q_B^{2-}. The uptake of two protons transforms the reduced quinone to the uncharged quinol, QH_2, which dissociates from the reaction center into the phospholipid bilayer of the cell membrane. In intact bacteria the protons that are taken up in the formation of QH_2 come from the cytosol of the cell (see fig. 17.14). Like the mitochondrial inner membrane, the bacterial membrane contains a relatively high concentration of quinone, which again can be either ubiquinone or menaquinone, depending on the species of bacteria.

QH_2 is reoxidized by a cytochrome bc_1 complex that is very similar to complex III of the mitochondrial respiratory chain. Electrons move through the cytochrome bc_1 complex to a *c*-type cytochrome, which then diffuses to the reaction center and provides an electron for the reduction of $P870^+$ (see fig. 17.14). As it does in mitochondria, movement of electrons through the cytochrome bc_1 complex results in the release of protons on the extracellular surface of the membrane. The proton extrusion can be explained well by the Q cycle that we discussed in connection with the mitochondrial complex (see fig. 16.12). The flow of electrons from QH_2 back to P870 thus drives the movement of protons outward across the cell membrane, generating a transmembrane electrochemical potential gradient for protons. The inside of the cell becomes negatively charged relative to the external

Figure 17.14

In purple photosynthetic bacteria, electrons return to $P870^+$ from the quinones Q_A and Q_B via a cyclic pathway. When Q_B is reduced with two electrons, it picks up protons from the cytosol and diffuses to the cytochrome bc_1 complex. Here it transfers one electron to an iron–sulfur protein and the other to a *b*-type cytochrome and releases protons to the extracellular medium. The electron-transfer steps catalyzed by the cytochrome bc_1 complex probably include a Q cycle similar to that catalyzed by complex III of the mitochondrial respiratory chain (see fig. 16.12). The *c*-type cytochrome that is reduced by the iron–sulfur protein in the cytochrome bc_1 complex diffuses to the reaction center, where it either reduces $P870^+$ directly or provides an electron to a bound cytochrome that reacts with $P870^+$. In the Q cycle, four protons probably are pumped out of the cell for every two electrons that return to P870. This proton translocation creates an electrochemical potential gradient across the membrane. Protons move back into the cell through an ATP-synthase, driving the formation of ATP.

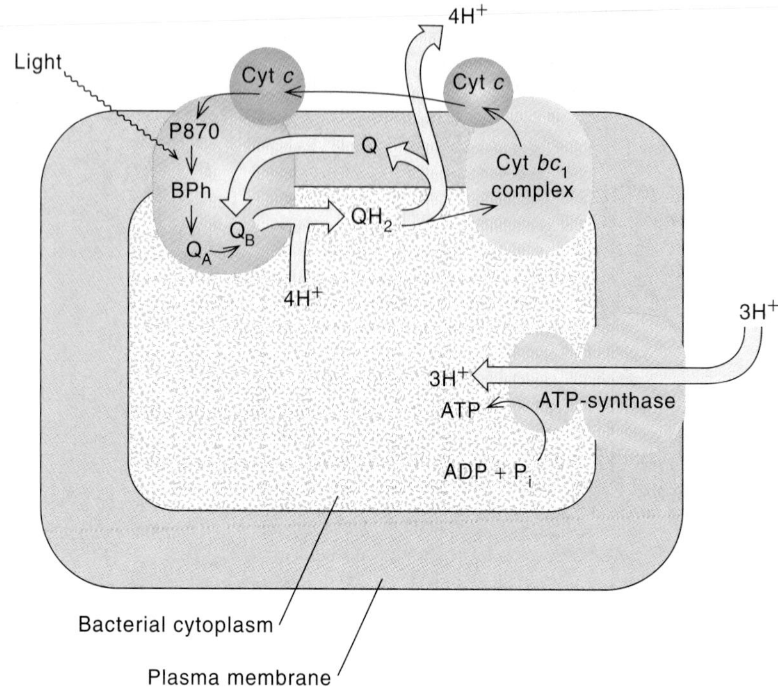

medium, and the pH of the cytosol becomes higher than the external pH.

The flow of protons back into the bacterial cell, down the electrochemical potential gradient, is mediated by an ATP-synthase resembling the proton-conducting ATP-synthase of the mitochondrial inner membrane (see chapter 16). As in mitochondria, the movement of protons through a channel in the F_o base-piece of the enzyme is linked to the formation of ATP (see fig. 17.14).

Note that the electron-transport system of purple photosynthetic bacteria is cyclic. Excitation of the reaction center with light creates a strong reductant (BPh^-) and a strong oxidant ($P870^+$), and electrons return from BPh^- to $P870^+$ via quinones, the cytochrome bc_1 complex, and *c*-type cytochromes. The cyclic flow of electrons results in the formation of ATP but no net oxidation or reduction. How, then, can we explain van Niel's observation that the bacteria carry out a net transfer of electrons from organic acids to carbohydrates (see fig. 17.9)? The biosynthesis of carbohydrates requires a reductant such as NADPH in addition to ATP. The answer to this puzzle is that the bacteria use ATP to support the transfer of electrons from succinate or other substrates to $NADP^+$. These reactions are carried out by dehydrogenase complexes in the cell membrane. As we discussed in chapter 16, a similar reduction of NAD^+ by succinate can occur in mitochondria if ATP is added.

An Antenna System Transfers Energy to the Reaction Centers

The reactive chlorophyll or bacteriochlorophyll molecules of P700, P680, or P870 account for only a small fraction of the total pigment in photosynthetic membranes. Chloroplasts contain on the order of 300 chlorophylls per P700 and P680, and the cell membranes of purple photosynthetic bacteria have from 25 to several hundred bacteriochlorophylls per P870, depending on the species. Most of the chlorophyll or bacteriochlorophyll is not photochemically active. Instead, it serves as an antenna. When one of the molecules in the antenna system is excited with light, it can transfer its energy to a neighboring molecule by resonance energy transfer. Energy absorbed anywhere in the antenna migrates rapidly from molecule to molecule until it is trapped by an electron-transfer reaction in a reaction center (fig. 17.15). Measurements of the lifetime of fluorescence from the antenna system indicate that after the antenna absorbs a photon, the energy is trapped in a reaction center within about 10^{-10} s in purple bacteria and within about 5×10^{-10} s in chloroplasts.

The distinction between the antenna system and the reaction centers grew out of experiments done by Robert Emerson and William Arnold in the 1930s. Emerson and Arnold measured the amount of O_2 evolution that occurred when they excited suspensions of green algae with a train of short flashes of light. To obtain the maximum O_2 per flash, they found that they had to allow a period of about 2×10^{-2} s of darkness between successive flashes. The amount of O_2 evolved per flash decreased if the flashes were spaced more closely together (fig. 17.16*a*). Each flash evidently generated a product that had to be consumed before the photosynthetic apparatus was ready to work again. If the next flash arrived too soon, its energy was wasted, mainly as heat and fluorescence. We know now that the chlorophyll complexes that undergo oxidation in the reaction centers (P700 and P680 in algae)

Figure 17.15

(*a*) A molecule in an excited state (Chl^*_D) can transfer its energy to another molecule (Chl_A). The donor molecule returns to its ground state (Chl_D) and the acceptor is elevated to an excited state (Chl^*_A). This requires that the difference in energy between Chl^*_D and Chl_D (the excitation energy of Chl_D) be the same as the difference between Chl^*_A and Chl_A (the excitation energy of Chl_A). When the energies match in this way, a resonance can occur between the two states {Chl^*_D Chl_A} and {Chl_D Chl^*_A}. (*b*) The energy of light absorbed by pigment–protein complexes in the antenna system hops rapidly from complex to complex by resonance energy transfer until it is trapped in an electron-transfer reaction in a reaction center.

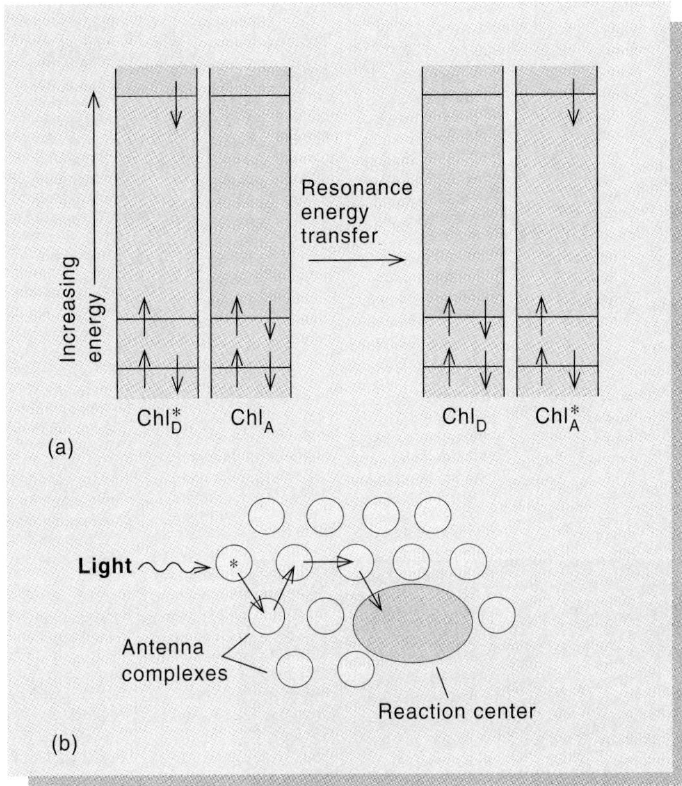

Figure 17.16

The amount of O_2 released when a suspension of algae (*Chlorella pyrenoidosa*) was excited with a train of short flashes of light, as a function of (*a*) the time between the flashes and (*b*) the intensity of the flashes. Each flash lasted about 10 μs. The total amount of O_2 released was measured after several thousand flashes and was divided by the number of flashes and by the amount of chlorophyll in the suspension. (What happens on the first few flashes is discussed later.) The measurements in (*a*) were made with flashes comparable to the strongest flashes used in (*b*). For the measurements shown in (*b*) the flashes were spaced 20 ms apart. Note that the maximum amount of O_2 released per flash was only one molecule of O_2 per several thousand molecules of chlorophyll. With saturating flashes the amount of O_2 evolution is limited by the concentration of reaction centers, whereas most of the chlorophyll in the algae is part of the antenna system.

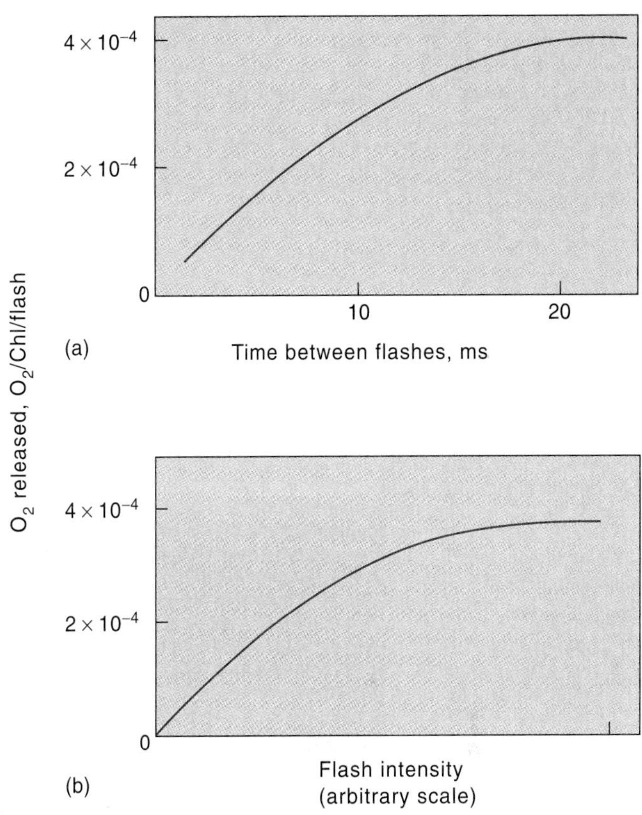

have to be returned to their reduced states before they can react again. The acceptors that remove electrons from the chlorophyll complexes also have to be reoxidized.

Emerson and Arnold found that the amount of O_2 evolved per flash increased as they raised the strength of the flashes (i.e., as they excited the algae with a larger and larger number of photons on each flash) but reached a plateau when they made the flashes sufficiently strong (see fig. 17.16*b*). Even with flashes of saturating intensity and optimal timing, the amount of O_2 was very small relative to the chlorophyll content of the algae. The cells contained about 2,500 molecules of chlorophyll for each molecule of O_2 that they evolved. The low yield of O_2 was extremely puzzling at the time because it was generally accepted that chlorophyll was intimately involved in the photochemistry of CO_2 fixation and O_2 evolution. Now it is clear that most of the chlorophyll is part of the antenna system. When the flash intensity is high, the amount of O_2 evolution that can occur on each flash is limited by the concentration of reaction centers, and this is much smaller than the total concentration of chlorophyll. The apparent discrepancy between the number of chlorophyll molecules per P700 or P680 (about 300) and the number of chlorophylls per O_2 (several thousand) will make more sense after we have discussed the fact that chloroplasts have to absorb about eight photons for each molecule of O_2 that is evolved.

In most photosynthetic organisms the chlorophyll or bacteriochlorophyll molecules that comprise the antenna are bound to integral membrane proteins. The major antenna protein in chloroplasts, the *light-harvesting chlorophyll a/b protein* or *LHC-II,* has a single polypeptide with a molecular weight of about 25,000. Its structure has been determined by electron microscopy of two-dimensional crystals. As shown in figure 17.17*a*, LHC-II has three transmembrane α helices and an additional short helix running parallel to the membrane surface. Of the 12 molecules of chlorophyll

Figure 17.17

(*a*) Structure of LHC-II, the major antenna complex of chloroplasts. Blue bands indicate the approximate boundaries of the phospholipid bilayer of the thylakoid membrane; the chloroplast stroma is at the top and the thylakoid lumen at the bottom. The 12 chlorophyll molecules are identified tentatively as either chlorophyll *a* (dark green) or chlorophyll *b* (light green). Two carotenoid (lutein) molecules are shown in yellow. α-Helical regions of the protein are represented by ribbons. (From W. Kühlbrandt, D. N. Wang, and Y. Fujiyoshi, Atomic model of plant light-harvesting complex by electron crystallograph, *Nature* 367:614, 1994.) (*b*) Structure of the B800-850 complex of the photosynthetic bacterium *Rhodopseudomonas acidophila.* The complex consists of concentric cylinders formed by two types of α-helical polypeptides (red). This figure shows a view along the axis of the cylinders from the cytoplasmic side of the membrane. The inner cylinder, which is composed of nine copies of the α polypeptide, has a diameter of about 18 Å; the outer ring has nine copies of the β polypeptide and a diameter of about 34 Å. The complex has nine relatively isolated molecules of bacteriochlorophyll (light blue) about 11 Å from the cytoplasmic face of the membrane and a ring of 18 bacteriochlorophylls (green) sandwiched between the polypeptides near the opposite face of the membrane. It also contains nine molecules of a carotenoid (rhodopin glucoside, white), which are aligned approximately perpendicular to the plane of the membrane. (Figure courtesy of Dr. Richard Cogdell, University of Glasgow.)

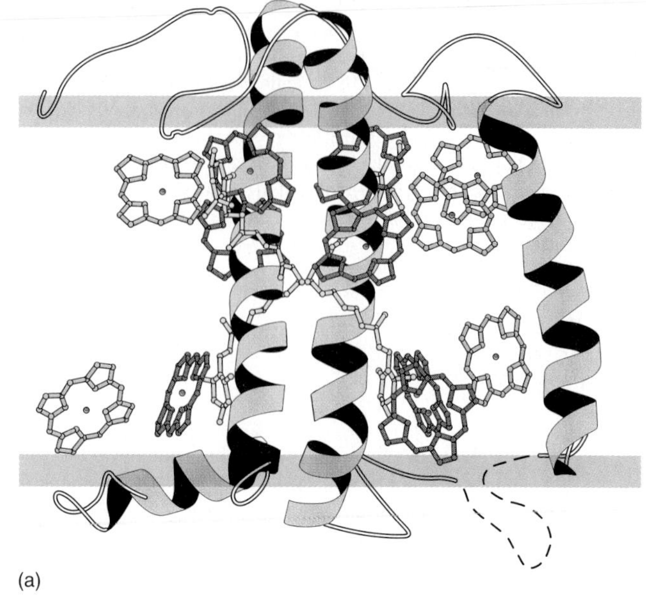

(a)

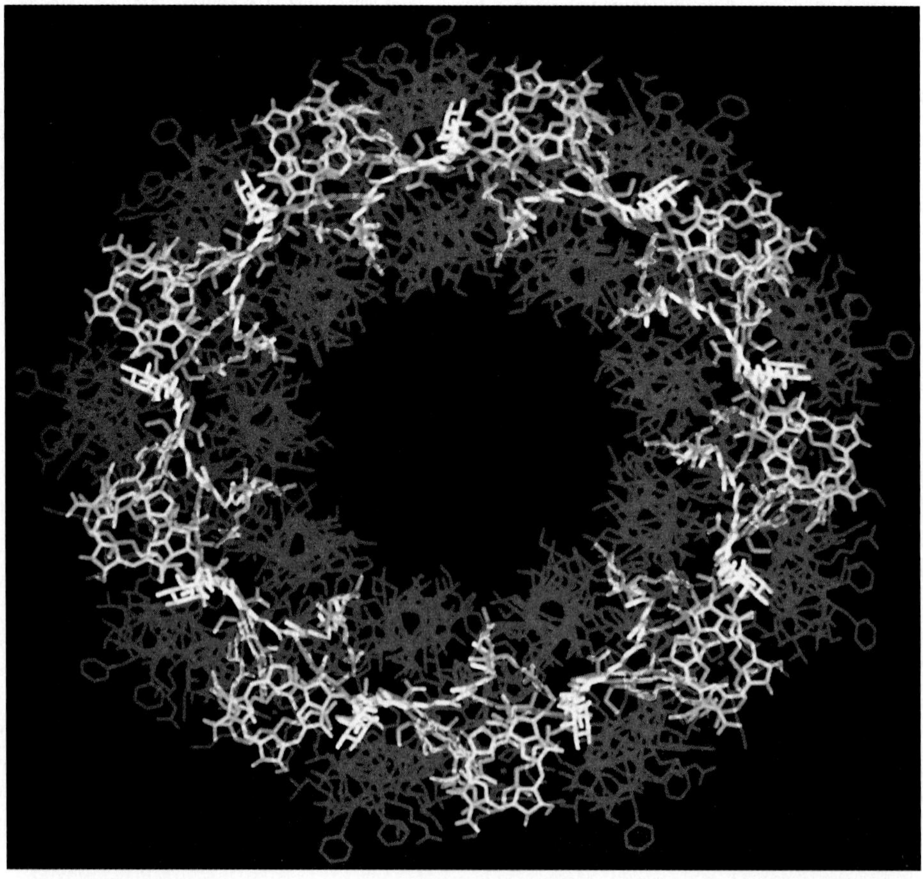

(b)

that are resolved in the crystal structure, six probably are chlorophyll *a* and six are chlorophyll *b*. Light energy absorbed by chlorophyll *b* is transferred rapidly to chlorophyll *a*, and the excitation energy then hops from one LHC-II complex to another until it finds its way to a reaction center of either photosystem I or photosystem II. LHC-II is the most abundant membrane protein in chloroplasts and typically accounts for about half of the chlorophyll in the thylakoid membrane. Additional antenna chlorophyll molecules are bound to other proteins that associate more specifically with one or the other of the two types of reaction center.

R. J. Cogdell, N. W. Isaacs, and their colleagues have determined the structure of a photosynthetic bacterial antenna complex by x-ray crystallography. The bacterial *B800-850 complex*, or *LH2*, contains multiple copies of two small polypeptides (α and β), each of which folds into a single transmembrane α helix. The α polypeptides pack side by side to form a hollow cylinder, which is surrounded by a cylinder of β polypeptides (see fig. 17.17*b*). A ring of 18 molecules of bacteriochlorophyll is sandwiched between the polypeptide helices near the periplasmic face of the membrane. These bacteriochlorophylls interact with each other in a staggered, end-to-end arrangement and give the complex a strong absorption band at 850 nm. The LH2 complex also contains nine molecules of bacteriochlorophyll in more widely separated positions closer to the cytoplasmic face of the membrane. This set of bacteriochlorophylls contributes an absorption band at 800 nm. Light energy moves rapidly from the isolated bacteriochlorophylls to the ring of 18 bacteriochlorophylls and then jumps to another complex on its way to the reaction center. The reaction center itself may reside in the center of a similar but larger cylindrical antenna complex containing a ring of 32 bacteriochlorophylls (*B875*, or *LH1*).

Along with bacteriochlorophyll or chlorophyll, antenna systems of both plants and bacteria contain a variety of carotenoids. These *accessory pigments* fill in the antenna's absorption spectrum in regions where chlorophylls do not absorb well. As shown in Figure 17.5, chlorophyll *a* absorbs red or blue light well, but not green. Carotenoids, which are long, linear polyenes (fig. 17.18), have absorption bands in the green region. The energy they absorb is transferred to the chlorophyll molecules of the antenna and from there to the reaction centers. This transfer is thermodynamically downhill because the carotenoid's excited singlet state has a higher energy than the lowest excited singlet state of chlorophyll. Carotenoid molecules can be seen in the crystal structures of both of the antenna complexes shown in figure 17.17. LHC-II contains two molecules of the carotenoid lutein, which stretch across the membrane making close contacts with some of the chlorophylls. The bacterial B800-850 complex contains nine carotenoid molecules. Carotenoids serve other functions (box 17A).

Figure 17.18

Structures of β-carotene, a major carotenoid in many types of plants, and spheroidene, a common carotenoid of photosynthetic bacteria. About 350 different natural carotenoids exist. These vary widely in color, depending principally on the number of conjugated double bonds.

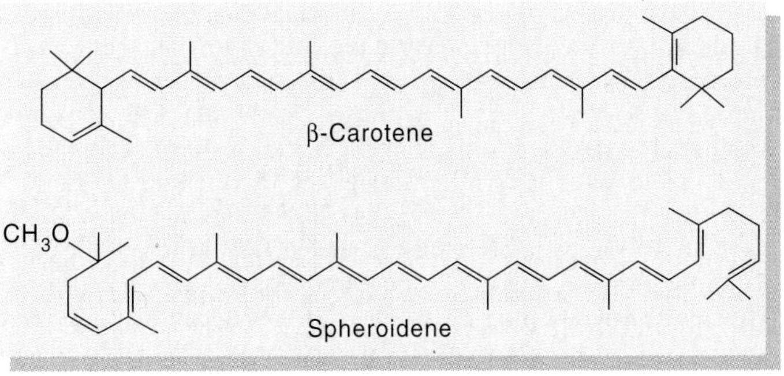

β-Carotene

Spheroidene

Chloroplasts Have Two Photosystems Linked in Series

We mentioned earlier that chloroplasts contain two types of reactive chlorophyll complexes, P700 and P680. Together with their antennae and their initial electron acceptors and donors, the reaction centers that contain P700 or P680 form two distinct assemblies called *photosystem I* and *photosystem II*. Much evidence indicates that photosystems I and II are connected in series, as shown in figure 17.19. Excitation of P700 (photosystem I) generates a strong reductant that transfers electrons to NADP$^+$ by way of several secondary electron carriers. Excitation of P680 (photosystem II) generates a strong oxidant, which oxidizes H_2O to O_2. The reductant formed in photosystem II injects electrons into a chain of carriers that connect the two photosystems. This scheme, which is called the Z scheme, was first suggested by R. Hill and F. Bendall in 1960. Note that the Z scheme differs from the bacterial electron transport chain that we discussed above in being linear rather than cyclic.

Figure 17.19

The Z scheme for the photosynthetic apparatus of plants. Two photochemical reactions are required to drive electrons from H_2O to NADP$^+$. The electron carriers are positioned vertically according to their $E^{\circ\prime}$ values, with the strongest reductants (most negative $E^{\circ\prime}$ values) at the top. Electron flow downward is thermodynamically spontaneous.

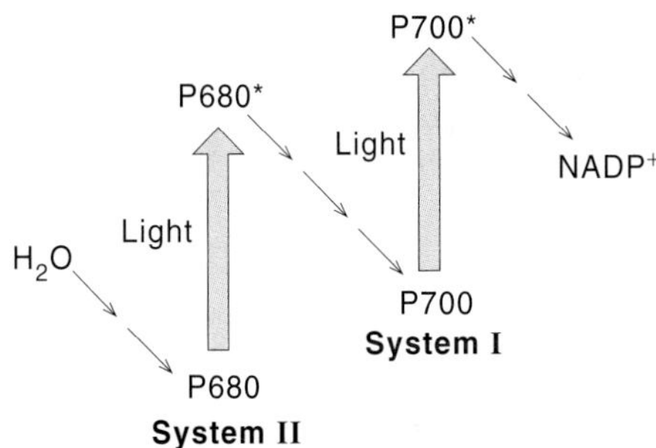

Figure 17.20 gives a more detailed picture of the Z scheme. The electron acceptors on the reducing side of photosystem II resemble those of purple bacterial reaction centers. The acceptor that removes an electron from P680 appears to be a molecule of pheophytin *a*. The second and third acceptors are molecules of plastoquinone, whose structure was shown in figure 17.11. As in the bacterial reaction center, electrons move one at a time from the first plastoquinone to the second. When the second plasto-

Carotenoids Protect Cells against Damage by O_2

In addition to transferring excitation energy to the chlorophylls, carotenoids play an important role in protecting the cell against damage by O_2 at high light intensities. If the antenna is flooded with light too rapidly for the reaction centers to keep pace, the antenna chlorophylls discharge most of the extra energy as fluorescence and heat. However, the excited molecules also have an opportunity to evolve into an excited triplet state, in which the spins of the two unpaired electrons are parallel (fig. 1*a*). Excited triplet states are relatively long-lived. They cannot decay to the ground (singlet) state unless the electronic spin changes again, and this does not happen readily. One way that they do decay is by reacting with molecular O_2, which has a triplet ground state. The reaction returns the chlorophyll to its ground state and promotes the O_2 to an excited singlet state (fig. 1*b*). This change can have lethal consequences for the cell because singlet O_2 is extremely toxic. It reacts irreversibly with a variety of groups in proteins, nucleic acids, and lipids.

Carotenoids intervene to prevent these destructive side reactions by quenching the excited triplet chlorophyll before it has a chance to react with O_2 (fig. 1*c*). In this process, the chlorophyll returns immediately to its ground state, and the carotenoid is elevated to an excited triplet state. The carotenoid triplet state cannot generate singlet O_2, because it lies below singlet O_2 in energy. Instead, it decays harmlessly to the ground state. Carotenoids also can quench singlet O_2 itself.

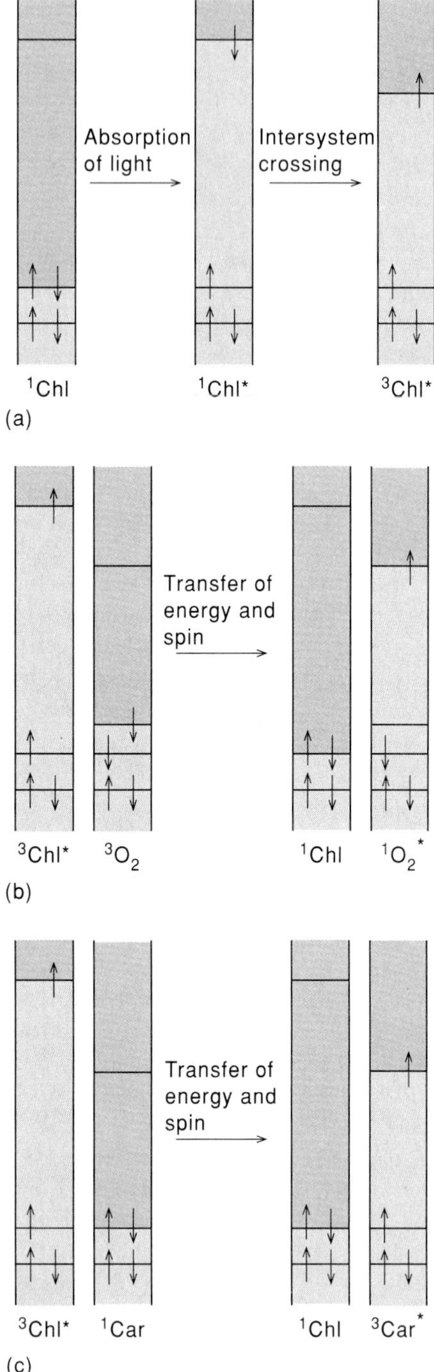

Figure 1

Carotenoids protect the system from the toxic effects of oxygen. (*a*) At high light intensities, chlorophyll has a tendency to be promoted to a relatively long-lived excited triplet state. (*b*) Chlorophyll in the excited triplet state can return to the ground state by reacting with molecular oxygen. This promotes oxygen to an excited singlet state, which is quite toxic. (*c*) Carotenoids return chlorophyll in the excited triplet state to the ground state before it has a chance to react with molecular oxygen.

Figure 17.20

The Z scheme. [$(Mn)_4$ = a complex of four Mn atoms bound to the reaction center of photosystem II; Y_Z = tyrosine side chain; Phe *a* = pheophytin *a*; Q_A and Q_B = two molecules of plastoquinone; Cyt *b*/*f* = cytochrome b_6f complex; PC = plastocyanin; Chl *a* = chlorophyll *a*; Q = phylloquinone (vitamin K_1); Fe–S_X, Fe–S_A, and Fe–S_B = iron–sulfur centers in the reaction center of photosystem I; FD = ferredoxin; FP = flavoprotein (ferredoxin-NADP oxidoreductase).] The sequence of electron transfer through Fe–S_A and Fe–S_B is not yet clear.

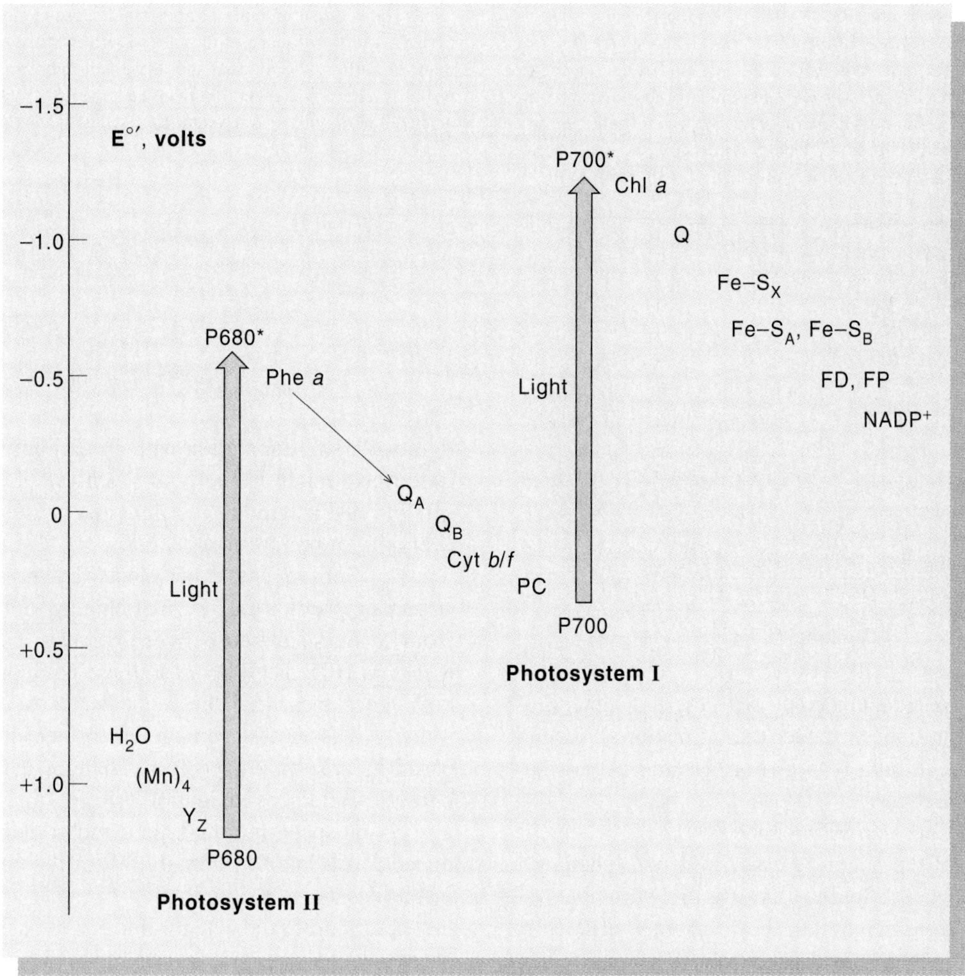

quinone becomes doubly reduced, it picks up protons from the stromal side of the thylakoid membrane and dissociates from the reaction center to join a pool of plastoquinone molecules in the membrane.

The chain of carriers between the two photosystems includes the cytochrome b_6f complex and a copper protein called plastocyanin. The cytochrome b_6f complex is very similar to the mitochondrial and bacterial cytochrome bc_1 complexes. It contains a cytochrome with two *b*-type hemes (cytochrome b_6), an iron–sulfur protein, and the *c*-type cytochrome *f*. As electrons move through the cytochrome b_6f complex from reduced plastoquinone to cytochrome *f*, the plastoquinone probably executes a Q cycle similar to the cycle that was presented for ubiquinone in the mitochondrial complex III and in the bacterial photosynthetic apparatus (see fig. 16.12). The cytochrome b_6f complex provides electrons to plastocyanin, which transfers them to $P700^+$ in the reaction center of photosystem I. Additional details on the electron carriers between P700 and NADP and between H_2O and P680 will be presented later, after we discuss some of the evidence supporting the Z scheme.

Some of the earliest observations that led to the Z scheme came from measurements of how the quantum yield of O_2 evolution in algae depends on the excitation wavelength. The quantum yield of O_2 evolution is the molar ratio of O_2 evolved to photons absorbed. In green algae the quantum yield is relatively independent of wavelength between 400 and 675 nm but falls off drastically in the far-red region near 700 nm (fig. 17.21). This seems odd, because most of the absorbance in the far-red region is due to chlorophyll *a*. How could light absorbed by chlorophyll *a* be used less efficiently than light absorbed by carotenoids or other accessory pigments? A clue came from R. Emerson's finding in 1956 that 700-nm light is used more efficiently if it is superimposed on a background of weak blue-green light (see fig. 17.21). It was found subsequently that the blue-green light actually does

Figure 17.21

The quantum yield of O_2 evolution from a suspension of algae (*Chlorella pyrenoidosa*) as a function of the excitation wavelength. Measurements were made without any supplementary light (lower curve) and with supplementary blue-green light (upper curve). The quantum yield was calculated as moles of O_2 evolved per einstein of light incident on the sample; O_2 evolution caused by the supplementary light alone was subtracted. Without the supplementary light the quantum yield falls off precipitously in the far-red region above 680 nm. The antenna of photosystem I absorbs light well in this region, but that of photosystem II does not. Supplementary blue-green light, which is absorbed well by photosystem II, increases the quantum yield with which the far-red light can be used. (Source: From R. Emerson et al., Some factors influencing the long-wavelength limit of photosynthesis, *Proc. Natl. Acad. Sci. USA* 43:133, 1957.)

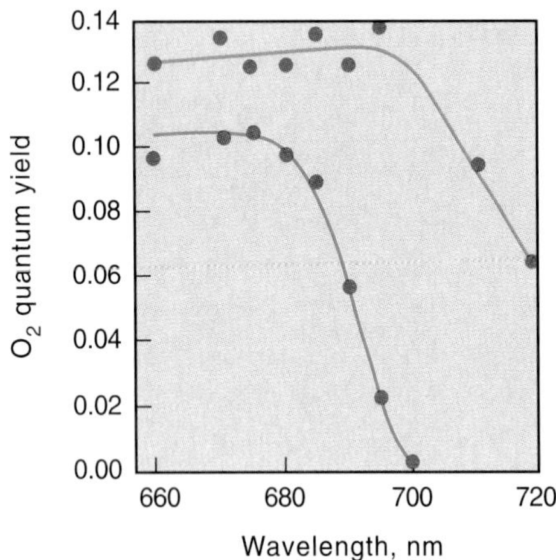

Figure 17.22

Illumination of a suspension of red algae (*Porphyridium cruentum*) with far-red light (680 nm) causes an oxidation of cytochrome *f*, as measured by an absorbance change at 422 nm. (The box below the trace indicates the period when the far-red light was on.) Green (562-nm) light superimposed on top of the far-red light causes the cytochrome to become more reduced. Additional measurements showed that green light also could cause cytochrome oxidation if it was turned on in the absence of the far-red light when the cytochrome was initially reduced. This result indicates that the green light can drive either photosystem I or photosystem II. Far-red light can cause only cytochrome oxidation, because it is absorbed only by photosystem I.

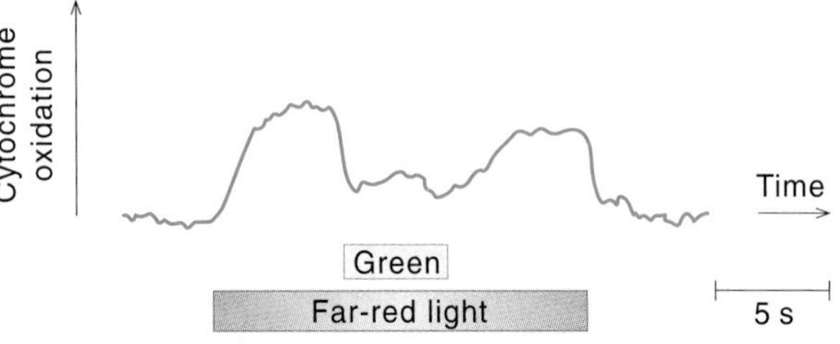

not have to be presented simultaneously with the red light. Illumination with blue-green light improves the utilization of far-red light even if the blue-green light is turned off several seconds before the red light is turned on.

These observations can be explained by the Z scheme if the absorption spectra of the antennae associated with photosystems I and II are somewhat different. Because the two photosystems must operate in series, light will be used most efficiently when the flux of electrons through photosystem II is equal to that through photosystem I. If light of some wavelengths excites one of the photosystems more frequently than the other, some of the light will be wasted.

The effectiveness with which light of different wavelengths excites the two photosystems can be explored by blocking one of the systems. For example, herbicides such as 3-(3,4-dichlorophenyl)-1,1-dimethylurea (DCMU) block electron flow between the two molecules of plastoquinone in the reaction center of photosystem II. Photosystem I continues to function well in the presence of DCMU if a reductant is added to provide electrons to the carriers in the chain connecting the two photosystems. Experiments with various excitation wavelengths showed that photosystem I can absorb and use virtually all wavelengths of light up to about 740 nm. Similar studies of photosystem II showed that it is not excited well by far-red light. The pool of electron carriers between the two photosystems will therefore be drained of electrons if algae are illuminated with far-red light in the absence of added reductants. P700 will remain oxidized and will be unable to respond to the light. Photosystem II does absorb blue-green light well, so illumination with green light will replenish the pool. Electrons remain in the pool when the blue-green light is turned off, and that is why far-red light can be used effectively for a time even after a period of darkness.

Subsequent experiments by Louis Duysens provided strong support for the idea that photosystem I withdraws electrons from carriers situated between the two photosystems while photosystem II feeds electrons to these components. Duysens and his colleagues measured the redox states of several of the carriers directly. The easiest component to measure is cytochrome *f*, because its absorption spectrum changes markedly when the cytochrome undergoes oxidation. (Refer to figure 16.4 to see the absorption spectra of a *c*-type cytochrome in the oxidized and reduced forms and to figure 17.20 for the position of cytochrome *f* in the Z scheme.) Illumination of algae with far-red light causes essentially all of the cytochrome *f* in the cells to become oxidized, as we would expect if far-red light drives photosystem I but not photosystem II (fig. 17.22). When a supplementary green light is turned on, some of the cytochrome returns quickly to the reduced state. This makes sense if green light can drive photosystem II. During the illumination, individual cytochrome molecules will cycle repeatedly between the oxidized and reduced states, so the cytochrome population as a whole is in a steady state of partial oxidation. Green light actually can cause either a net oxidation or a net reduction of cytochrome *f*, depending on the initial conditions, so it must be able to excite either photosystem II or photosystem I. Far-red light, however, can cause only oxidation of the cytochrome.

The electron carriers in photosystem II are located mainly in the stacked, granal regions of the thylakoid membrane, whereas those of photosystem I occur mainly in the unstacked stromal lamellae (fig. 17.23). The cytochrome b_6f complexes that participate in the transport of electrons between the two photosystems are found in both regions of the membrane. The distribution of photons between the two photosystems depends partly on the relative numbers of antenna complexes in the two regions of the

Figure 17.23

The photosystem II complexes are located mainly in the stacked (granal) regions of the thylakoid membrane, whereas photosystem I complexes are most abundant in the unstacked stromal lamellae. The cytochrome b_6f complex is located in both regions of the membrane. Some types of antenna complexes also reside in both regions and can shift from one region to the other in response to a change in conditions. The ATP-synthase occurs mainly in the unstacked regions. These conclusions came from studies in which thylakoid membranes were fragmented gently by ultrasound and then separated by density-gradient centrifugation into fractions representing stacked and unstacked membranes.

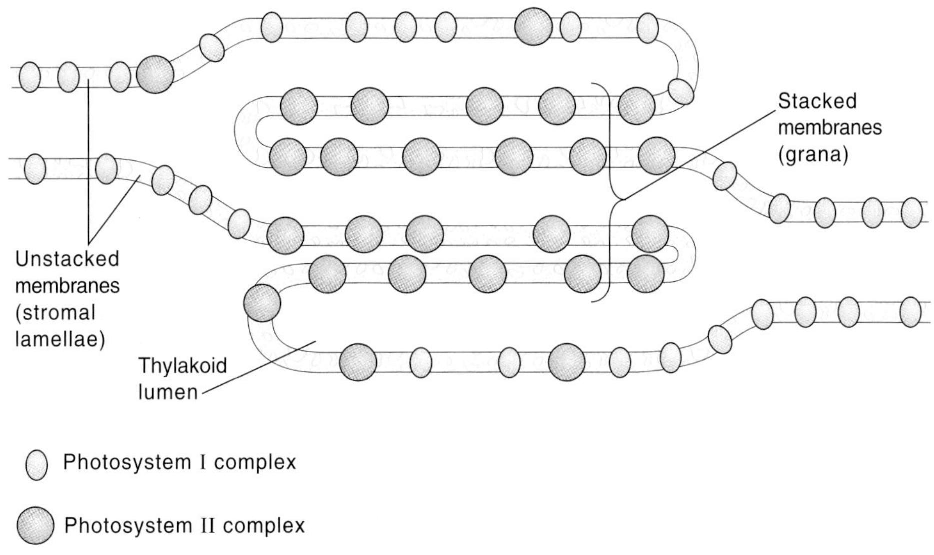

membrane and can change as a result of movements of antenna complexes from one region to another. Movements of the LHC II antenna complex appear to be regulated by phosphorylation of the protein. When chloroplasts are illuminated with light that is absorbed better by photosystem I than by photosystem II, the mobile LHC II complex moves into the stacked regions, and a larger fraction of the excitations then are passed to photosystem II.

If the reaction centers of photosystem I and photosystem II are segregated into separate regions of the thylakoid membrane, how can electrons move readily from photosystem II to photosystem I? The answer appears to be that the plastoquinone that is reduced by photosystem II can diffuse rapidly in the plane of the membrane, just as ubiquinone does in the mitochondrial inner membrane. Plastoquinone thus carries electrons from photosystem II to the cytochrome b_6f complex. Similarly, plastocyanin is a small protein that can diffuse in the thylakoid lumen. Plastocyanin thus can act as a mobile electron carrier from the cytochrome b_6f complex to a reaction center of photosystem I, just as cytochrome *c* carries electrons from the mitochondrial cytochrome bc_1 complex to cytochrome oxidase and as a *c*-type cytochrome provides electrons to the reaction centers of purple bacteria (see figure 17.14).

Photosystem I Reduces NADP⁺ by Way of Iron-Sulfur Proteins

On the reducing side of photosystem I, the earliest electron acceptor appears to be a molecule of chlorophyll *a* (see fig. 17.20). The second acceptor appears to be a quinone, phylloquinone (vitamin K_1) (see fig. 17.11). In these respects, photosystem I resembles photosystem II and the purple photosynthetic bacteria, which use pheophytin *a* and bacteriopheophytin *a* as initial electron acceptors and a quinone as the secondary acceptor. From this point on, however, photosystem I is different. Its next set of electron carriers consists of iron–sulfur proteins instead of additional quinones.

The first of the iron–sulfur proteins to be characterized was ferredoxin, a small, soluble protein found in the chloroplast stroma. Ferredoxin is designated as FD in figure 17.20. It contains two iron atoms and two atoms of inorganic sulfide, which are bound to the sulfur atoms of four cysteine residues in the type of complex that was illustrated in figure 11.21*a*. Photosystem I also contains three additional iron–sulfur clusters that are more firmly associated with the reaction center. These are designated $Fe–S_X$, $Fe–S_A$, and $Fe–S_B$ in figure 17.20. Each of $Fe–S_X$, $Fe–S_A$, and $Fe–S_B$ probably has four iron atoms and four inorganic sulfides held by four cysteines in the cubic structure shown in figure 11.21*b*. The cysteines of $Fe–S_X$ are provided by the two main polypeptides of the reaction center, which also bind P700 and the chlorophyll and quinone that act as the initial electron acceptors. Each polypeptide evidently contributes two cysteines to the $Fe–S_X$ cluster. $Fe–S_A$ and $Fe–S_B$ are both bound to a separate, small polypeptide.

As we discussed in chapter 16, iron–sulfur proteins undergo one-electron oxidation–reduction reactions. The $E^{\circ\prime}$ values for these transitions are about -0.7 V for $Fe–S_X$, -0.59 for $Fe–S_B$, -0.54 for $Fe–S_A$, and -0.40 V for FD. The quinone that is reduced in photosystem I probably transfers an electron to $Fe–S_X$. $Fe–S_X$ reduces $Fe–S_A$ and $Fe–S_B$, which in turn reduce FD (see fig. 17.20). From FD, electrons move to a flavoprotein ferredoxin-NADP oxidoreductase and then to $NADP^+$.

O_2 Evolution Requires the Accumulation of Four Oxidizing Equivalents in the Reaction Center of Photosystem II

The oxidation of H_2O to O_2 requires the removal of four electrons for each O_2 produced. According to the Z scheme each electron must traverse the photochemical reactions of both photosystem I and photosystem II, so at least eight photons (2×4) have to be absorbed for each O_2 that is released. Currently accepted experimental measurements of the number of photons that are needed, the quantum requirement, are indeed on the order of 8 to 12. (The quantum requirement is the reciprocal of the quantum yield of O_2 evolution.) The quantum requirement will exceed 8 if some of the photons that are absorbed are lost as heat or fluorescence from the antenna or if some of the electrons pumped through the photosystems are not removed to $NADP^+$ but instead cycle back into the electron-transport chain between the two photosystems.

If the photosystem II reaction center can transfer only one electron at a time, how does the photosynthetic apparatus assemble the four oxidizing equivalents that are needed for the oxidization of H_2O to O_2? One possibility would be that several different photosystem II reaction centers cooperate, but this seems not to happen. Instead, each reaction center progresses independently through a series of oxidation states, advancing to the next state each time it absorbs a photon. O_2 evolution occurs only when a reaction center has accumulated four oxidizing equivalents. This conclusion comes principally from measurements of the amount of O_2 that is evolved on each flash when algae or chloroplasts are excited with a series of short flashes after a period of darkness. Pierre Joliot found that essentially no O_2 is released on the first and second flashes (fig. 17.24). On the third flash, however, there is a burst of O_2. After this the amount of O_2 released on each flash oscillates, going through a maximum every fourth flash.

Kok pointed out that this pattern can be explained if the photosystem II reaction center cycles through five different oxidation states, S_0 through S_4, as shown in figure 17.25. When the system reaches state S_4, O_2 is given off and the reaction center returns to state S_0. The fact that the first burst of O_2 comes on the third flash instead of the fourth can be explained if the reaction centers relax mainly into S_1 rather than into S_0 during the dark period before the flashes. Added reductants can in fact convert the reaction centers from S_1 to S_0, so the first peak of O_2 occurs on the fourth flash. The gradual damping of the oscillations in figure 17.24 is due to reaction centers that get out of phase, either because they miss being excited on one of the flashes or because they are excited twice and advance two steps.

The component that undergoes oxidation as photosystem II progresses from one of the S states to the next is a complex of four atoms of manganese bound to one of the two central polypeptides of the reaction center. $P680^+$ draws electrons from the Mn complex by way of a tyrosine in this polypeptide (Y_z in fig. 17.20). In the course of this reaction the phenolic side chain of the tyrosine is oxidized transiently to a free radical. Although the detailed structure of the Mn complex is not yet known, the complex appears to be constructed of two dimers of Mn atoms bridged by oxygens as shown in figure 17.26. The Mn dimers probably are linked together by another oxygen bridge and possibly by the carboxylate oxygens of aspartic acid or glutamic acid residues. The other ligands of the Mn atoms probably include a histidine side chain and a chloride ion. A calcium ion that also forms an essential part of the complex could be bound to another carboxylate group that is attached to one of the Mn atoms (see fig. 17.26).

Studies of the x-ray absorption spectrum of the Mn complex indicate that in state S_0, three of the Mn atoms are at the Mn(III) oxidation level and one at Mn(IV). The transitions from state S_0 to S_1 and from S_1 to S_2 appear to represent sequential oxidation of additional Mn atoms from Mn(III) to Mn(IV). The transition from S_2 to S_3 seems not to result in oxidation of any of the

Figure 17.24

O_2 evolution from a suspension of chloroplasts that is excited with a series of 24 flashes after having been kept in the dark for several minutes. Little or no O_2 is evolved on the first two flashes. O_2 evolution peaks on the third flash and on every fourth flash thereafter. The oscillations in O_2 evolution are damped, and after many flashes the yields converge on the level indicated by the horizontal line.

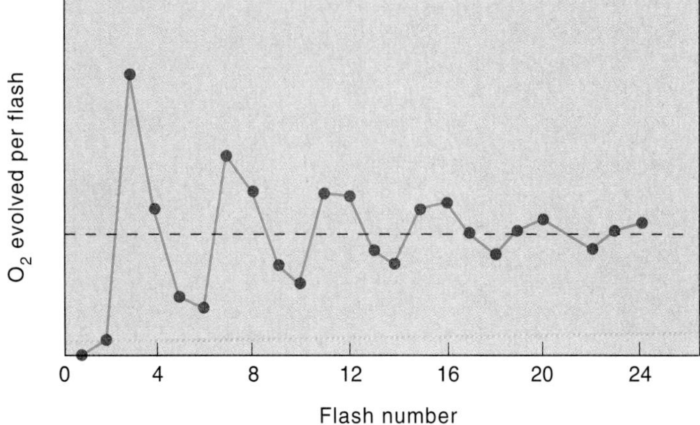

Figure 17.25

The five oxidation states of the O_2-evolving apparatus. One electron (e^-) is removed photochemically in each of the transitions between states S_0 and S_4. S_4 decays spontaneously, releasing O_2. Protons are released at several steps of the cycle.

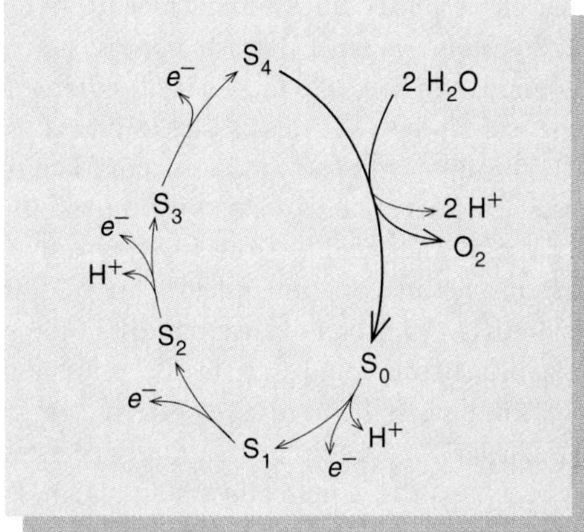

Figure 17.26

Proposed model for the O_2-evolving Mn complex of photosystem II. The ligands of the Mn atoms probably include oxygen (μ-oxo) bridges, the oxygens of carboxylic amino acid side chains, a histidine side-chain N, and a Cl^- ion. This structure is based largely on x-ray absorption fine structure and ESR spectroscopy. (From V. K. Yachandra, V. J. DeRose, M. J. Latimer, I. Mukerji, K. Sauer, and M. P. Klein, Where plants make oxygen: A structural model for the photosynthetic oxygen-evolving manganese cluster, *Science* 260:675, 1993.)

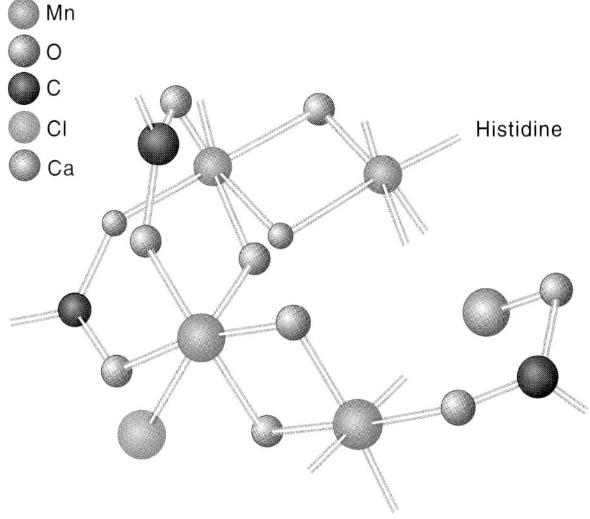

Mn atoms, and there are indications that it involves oxidation of the histidine side chain that is bound to the complex. What happens in the transition to S_4 is currently unclear.

The $E°$ of the $P680^+/P680$ redox couple must be greater than +1 V, because the oxidized manganese complex that $P680^+$ generates is powerful enough to oxidize water. At pH 5, which is approximately the pH of the thylakoid lumen, the O_2/H_2O couple has an $E°$ of +0.94 V. $P700^+/P700$, by contrast, has an $E°'$ of only about +0.50 V. It is puzzling that $P680^+$ is so much stronger an oxidant than $P700^+$ if they both consist of chlorophyll *a*. However, the redox potential of the chlorophyll radical is likely to depend strongly on the polarity of the surroundings in the protein.

Flow of Electrons from H_2O to $NADP^+$ Drives Proton Transport into the Thylakoid Lumen; Protons Return to the Stroma through an ATP-Synthase

As in the mitochondrial inner membrane and in the cell membrane of purple photosynthetic bacteria, the flow of electrons through the chloroplast's cytochrome b_6f complex results in translocation of protons across the thylakoid membrane. When plastoquinone undergoes reduction in photosystem II and in the cytochrome b_6f complex, protons are taken up from the stromal side of the membrane. When the plastoquinone is reoxidized, protons are released into the thylakoid lumen. This lowers the pH in the lumen and makes the inside positively charged with respect to the stroma. In current schemes of the Q cycle catalyzed by the cytochrome b_6f complex, four protons move across the membrane for each pair of electrons that proceed to photosystem I. Two more protons are released in the lumen for each molecule of H_2O that is oxidized to 1/2 O_2, and one proton is taken up from the stroma for each $NADP^+$ that is reduced to NADPH (fig. 17.27).

Proton movement back out through the thylakoid membrane is conducted by an ATP-synthase, and this movement drives the formation of ATP. The chloroplast ATP-synthase is structurally very similar to the mitochondrial ATP-synthase (see fig. 16.23). Its head-piece, which contains the catalytic sites, is generally called CF_1. Its hydrophobic base-piece, CF_o, includes the proton-conducting channel.

Observations on chloroplasts played a key role in the development of the chemiosmotic theory of oxidative phosphorylation, which we discussed in chapter 16. Andre Jagendorf and his colleagues discovered that if they illuminated chloroplasts in the absence of ADP, the chloroplasts developed the capacity to form ATP when ADP was added later, after the light was turned off. The amount of ATP that could be synthesized was much greater than the number of electron-transport assemblies in the thylakoid membranes, so the energy to drive the phosphorylation could not have been stored in an energized form of one of the electron carriers. Protons were taken up from the solution outside the thylakoid membranes during the illumination, and the ability to form ATP was correlated with the magnitude of the pH gradient across the membrane. If a pH gradient in the right direction was set up across the thylakoid membrane by suddenly raising the pH on the outside, chloroplasts could make ATP without any illumination at all (fig. 17.28).

The formation of ATP by photosynthetic systems is often called photophosphorylation to distinguish it from the process that is coupled to respiration in mitochondria. Although the two processes are very similar, they do differ in a few details. The respiratory chain pumps protons outward across the inner membrane, from the mitochondrial matrix to the intermembrane space (see chapter 16). The chloroplast electron-transport chain pumps protons into the thylakoid lumen. The electrochemical potential gradient for protons across the mitochondrial membrane consists mainly of an electrical potential; in chloroplasts it consists mainly of a pH gradient, which can be greater than three pH units during strong illumination. In both cases, however, the F_1 or CF_1 head-piece of the proton-conducting ATP-synthase is on the side of the membrane that becomes more alkaline. The relative importance of the pH gradient or the electrical potential probably depends on the permeability of the membrane to other ions such as Cl^-. Movement of Cl^- into the thylakoid lumen reduces the electrical potential difference across the membrane, allowing the buildup of a larger pH gradient.

In the Z scheme, photosystem II, the cytochrome b_6f complex, and photosystem I operate in series to move electrons from H_2O to $NADP^+$ and to create an electrochemical potential gradient for protons across the thylakoid membrane. In addition to this linear pathway, chloroplasts in some plant species may use a cyclic electron-transfer scheme that includes photosystem I and the cytochrome b_6f complex but not photosystem II. If the concentration of $NADP^+$ is low, electrons ejected by the reaction center of photosystem I can be passed from one of the iron–sulfur

Figure 17.27

Transport of two electrons from photosystem II through the cytochrome b_6f complex to photosystem I results in the movement of four protons from the chloroplast stroma to the thylakoid lumen. The proton translocation probably occurs in a Q cycle resembling that illustrated in figure 16.12. Two more protons are released in the lumen for each molecule of H_2O that is oxidized to O_2, and one additional proton is removed from the stroma for each molecule of $NADP^+$ reduced to NADPH. (Two protons are taken up when the flavoprotein is reduced by photosystem I; one ends up on NADPH, and the other is returned to the solution.) The flow of protons from the thylakoid lumen back to the stroma through an ATP-synthase (CF_0–CF_1) drives the formation of ATP. The abbreviations used in this figure are the same as in figure 17.20.

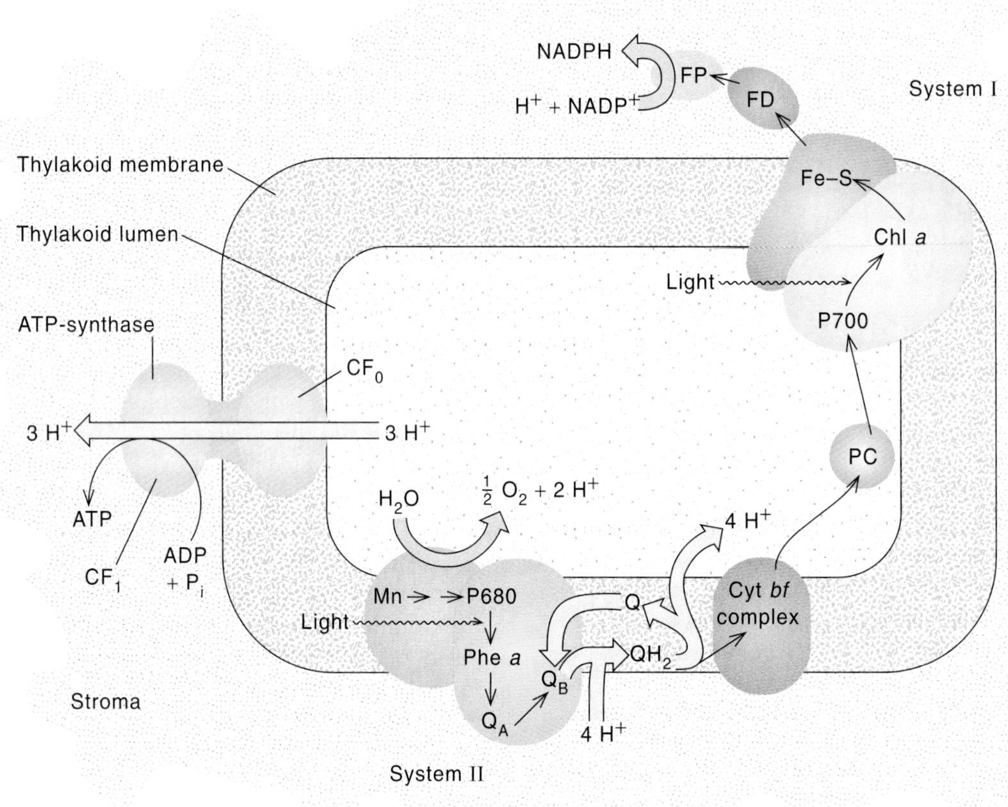

Figure 17.28

Chloroplasts can use an electrochemical potential gradient for protons to form ATP in the dark. In a two-step experiment, spinach chloroplasts first were equilibrated at an acidic pH (pH 4.0). The pH of the solution then was raised quickly to a value between 6.7 and 8.8, as indicated on the abscissa of the graph, and ADP and P_i were added. Because the pH in the thylakoid lumen was buffered at pH 4, the sudden increase in the external pH created a pH gradient of between 2.7 and 4.8 pH units across the thylakoid membrane. Proton efflux through the ATP-synthase can drive the formation of ATP. Under optimal conditions, one molecule of ATP was formed for approximately every five chlorophylls in the sample (see the ordinate scale on the graph). For comparison, chloroplasts contain only about one molecule of P700 and one of P680 per 400 chlorophylls. (Source: From A. Jagendorf and E. Uribe, ATP formation caused by acid-base transition of spinach chloroplasts, *Proc. Natl. Acad. Sci. USA* 55:197, 1966.)

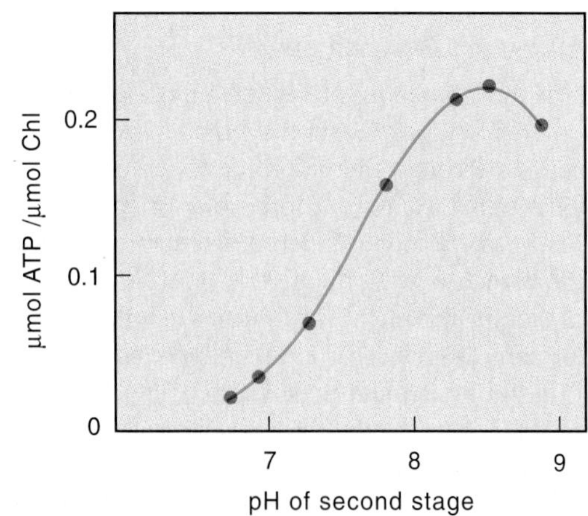

proteins FD_A and FD_B to a *b*-type cytochrome (cytochrome b_{563}) and from there to the plastoquinone pool between the two photosystems (see fig. 17.20). From plastoquinone, electrons can return to P700 via the cytochrome b_6f complex and plastocyanin. The Q cycle that is included within this cyclic electron pathway results in proton translocation across the thylakoid membrane and thus could contribute to the transmembrane electrochemical potential gradient that drives the ATP-synthase. Cyclic electron transport probably does not occur to a significant extent under physiological conditions in cyanobacteria and terrestrial plants but may be important in some algae.

Carbon Fixation: The Reductive Pentose Cycle

The ATP and NADPH that are generated by the photosynthetic electron-transfer reactions are used to drive fixation of CO_2. Reactions in which CO_2 is incorporated into carbohydrates were discovered in the early 1950s by Melvin Calvin and his co-workers in some of the first biochemical studies employing radioactive tracers. A key experiment was to expose algae to a brief period of illumination in the presence of $^{14}CO_2$ and then to disrupt the cells quickly and search for organic molecules that had become labeled with ^{14}C. The material that turned out to be labeled most rapidly was glycerate-3-phosphate, and almost all of its ^{14}C was found to be in the carboxyl group. To identify the precursor of the glycerate-3-phosphate, Calvin's group looked for changes in the steady-state concentrations of other compounds when they turned a continuous light on or off or when they suddenly raised or lowered the concentration of CO_2. These experiments showed that glycerate-3-phosphate is formed from ribulose-1,5-bisphosphate. A molecule of CO_2 is incorporated in the process, so one molecule of the five-carbon sugar ribulose-1,5-bisphosphate gives rise to two molecules of the three-carbon glycerate-3-phosphate (fig. 17.29). The reaction is catalyzed by ribulose bisphosphate carboxylase, which is found in the chloroplast stroma.

Further studies revealed that other enzymes in the stroma can convert glycerate-3-phosphate back to ribulose-1,5-bisphosphate. The reactions of this reductive pentose cycle, or Calvin cycle, are shown in figure 17.29. The 3-phosphoglycerate is first phosphorylated to glycerate-1,3-bisphosphate at the expense of ATP and then reduced to glyceraldehyde-3-phosphate by NADPH. (Note that the nucleotide specificity in the reductive step differs from that of the cytosolic glyceraldehyde-3-phosphate dehydrogenase, which uses NAD^+ and NADH.) Glyceraldehyde-3-phosphate and its isomerization product dihydroxyacetone phosphate can combine to form fructose-1,6-bisphosphate under the influence of aldolase. Fructose-6-phosphate, formed by hydrolysis of the fructose-1,6-bisphosphate, combines with another molecule of glyceraldehyde-3-phosphate, generating xylulose-5-phosphate and erythrose-4-phosphate. The enzyme that catalyzes this reaction is similar to the transketolase of the pentose phosphate pathway (see chapter 14). The erythrose-4-phosphate that is formed goes on to combine with dihydroxyacetone phosphate to give sedoheptulose-1,7-bisphosphate in a second reaction catalyzed by aldolase. After hydrolysis to sedoheptulose-7-phosphate the seven-carbon sugar reacts with glyceraldehyde-3-phosphate in another transketolase reaction, forming ribose-5-phosphate and a second molecule of xylulose-5-phosphate. Xylulose-5-phosphate and ribose-5-phosphate both can be isomerized to ribulose-5-phosphate. Finally, ribulose-5-phosphate is phosphorylated to ribulose-1,5-bisphosphate at the expense of an additional ATP, completing the cycle.

In figure 17.29, note that three molecules of ribulose-1,5-bisphosphate are regenerated for every three that are carboxylated. In the process, three molecules of CO_2 are taken up, and there is a net gain of one molecule of glyceraldehyde-3-phosphate. The expenses are nine molecules of ATP converted to ADP and six molecules of NADPH oxidized to $NADP^+$ or three molecules of ATP and two of NADPH consumed per CO_2 fixed. The glyceraldehyde-3-phosphate that is produced is exported from the chloroplast to the cytosol or converted to hexoses for storage in the chloroplast as starch.

Ribulose Bisphosphate Carboxylase/Oxygenase, Photorespiration, and the C_4 Cycle

Ribulose bisphosphate carboxylase typically accounts for more than half the soluble protein in a leaf. It is surely the world's most abundant enzyme and probably the most abundant protein. The enzyme found in plants contains eight copies each of two types of subunits. The larger subunit, which has a molecular weight of 56,000 and contains the catalytic site, is synthesized within the chloroplast under the direction of chloroplast DNA and ribosomes. The smaller subunit, with a molecular weight of 14,000, is synthesized on cytoplasmic ribosomes under the direction of nuclear DNA and has to come into the chloroplast for assembly of the complete enzyme. The smaller subunit is lacking in the enzyme isolated from some species of bacteria, and what role it plays in plants is still unclear.

In addition to serving as a substrate, CO_2 activates ribulose bisphosphate carboxylase by binding to the ϵ-amino group of a lysyl residue in the large subunit to form a carbamate (lysine—NH—CO_2^-). This process is catalyzed by another enzyme and requires Mg^{2+}, which binds to the carboxyl group of the carbamate. The Mg^{2+} in turn forms part of the binding site for a second molecule of CO_2, which acts as the substrate in the carboxylase reaction. The reactive molecule of CO_2 probably adds to the enolate of ribulose-1,5-bisphosphate to form 2-carboxy-3-ketoarabinitol-1,5-bisphosphate as an intermediate, as shown in figure 17.30.

Studies of ribulose bisphosphate carboxylase took on additional complexity with the discovery that the same active site on the enzyme also catalyzes a competing reaction in which O_2 replaces CO_2 as a substrate. The products of the oxygenase reaction are 3-phosphoglycerate and a two-carbon acid, 2-phosphoglycolate (fig. 17.31). Phosphoglycolate is oxidized to CO_2 by O_2 in a series of additional reactions involving enzymes in the cytosol, mitochondria, and another organelle, the peroxisome. This photorespiration is not coupled to oxidative phosphorylation, and it appears to constitute a severe drain on chloroplast metabolism. In some plants, as much as one-third of the CO_2 that is fixed is re-

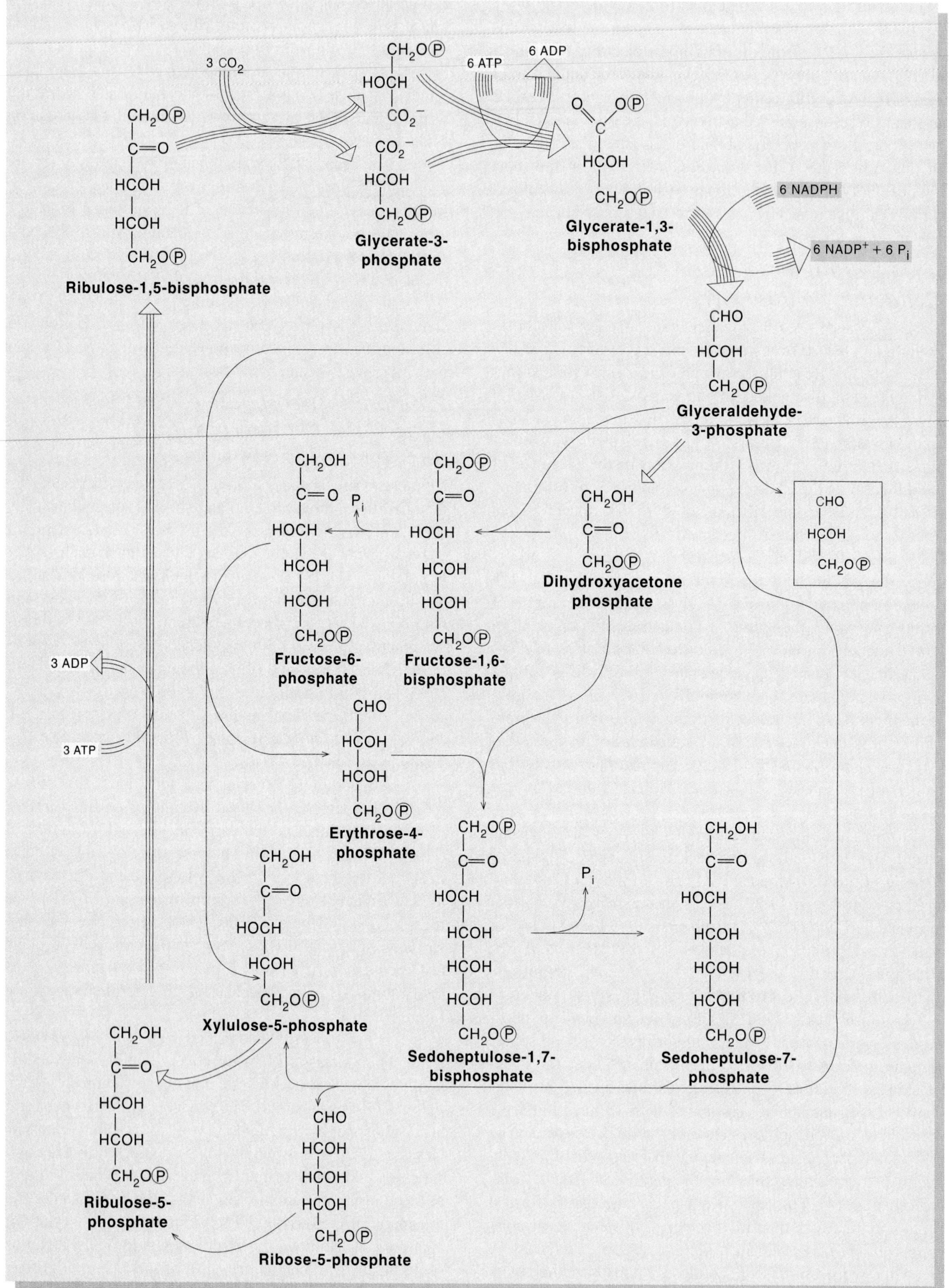

3 CO_2
6 ATP
6 ADP
Ribulose-1,5-bisphosphate
Glycerate-3-phosphate
Glycerate-1,3-bisphosphate
6 NADPH
6 $NADP^+$ + 6 P_i
Glyceraldehyde-3-phosphate
Dihydroxyacetone phosphate
Fructose-1,6-bisphosphate
Fructose-6-phosphate
P_i
Erythrose-4-phosphate
Xylulose-5-phosphate
Sedoheptulose-1,7-bisphosphate
Sedoheptulose-7-phosphate
Ribulose-5-phosphate
Ribose-5-phosphate
3 ADP
3 ATP

Figure 17.29

The reductive pentose cycle, or Calvin cycle. The number of arrows drawn at each step in the diagram indicates the number of molecules proceeding through that step for every three molecules of CO_2 that enter the cycle. The entry of three molecules of CO_2 results in the formation of one molecule of glyceraldehyde-3-phosphate (box on right), and requires the oxidation of six molecules of NADPH to $NADP^+$ and the breakdown of nine molecules of ATP to ADP.

leased again by photorespiration. If photorespiration has any benefit to the plant, it is not obvious.

The outcome of the competition between the carboxylase and oxygenase reactions depends on the concentrations of CO_2 and O_2. The K_m of the activated enzyme for the substrate molecule of CO_2 is about 20 μM, and that for O_2 is about 200 μM, so CO_2 is the preferred substrate. In illuminated chloroplasts, however, the concentration of O_2 is elevated as a result of the photosynthetic splitting of H_2O. The concentrations of the two gases

Figure 17.30

The reaction catalyzed by ribulose bisphosphate carboxylase involves 2-carboxy-3-ketoarabinitol-1,5-bisphosphate as an enzyme-bound intermediate. The intermediate probably forms by the addition of CO_2 to the enolate of ribulose-1,5-bisphosphate. The substrate is known to be CO_2 rather than bicarbonate.

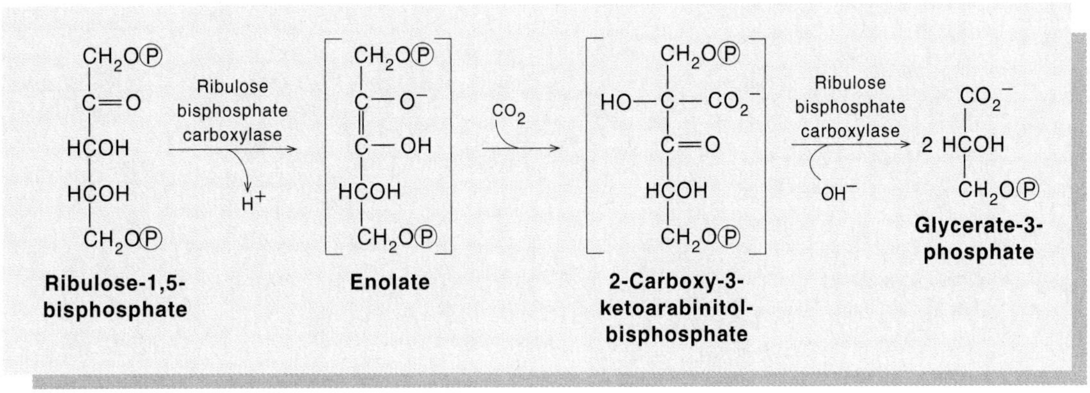

Figure 17.31

Photorespiration results from the oxygenation reaction catalyzed by ribulose bisphosphate carboxylase/oxygenase. 2-Phosphoglycolate generated by the reaction moves from the chloroplast to the cytosol, where other enzymes break it down to CO_2, H_2O, and P_i. The oxygenation reaction, like carboxylation, does not require light directly. It occurs mainly during illumination, however, because the formation of the substrate, ribulose-1,5-bisphosphate, requires ATP and NADPH (see fig. 17.29).

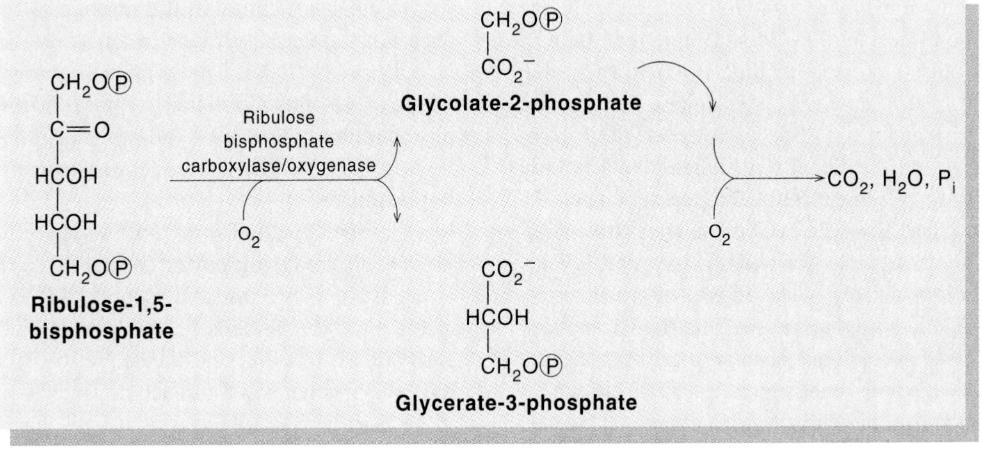

Figure 17.32

The C-4 pathway requires the cooperation of two types of cells. Mesophyll cells (*left*) take up CO_2 from the air and export malate to the bundle sheath cells (*right*). The bundle sheath cells return pyruvate to the mesophyll cells and fix the CO_2 using ribulose bisphosphate carboxylase and the reductive pentose cycle.

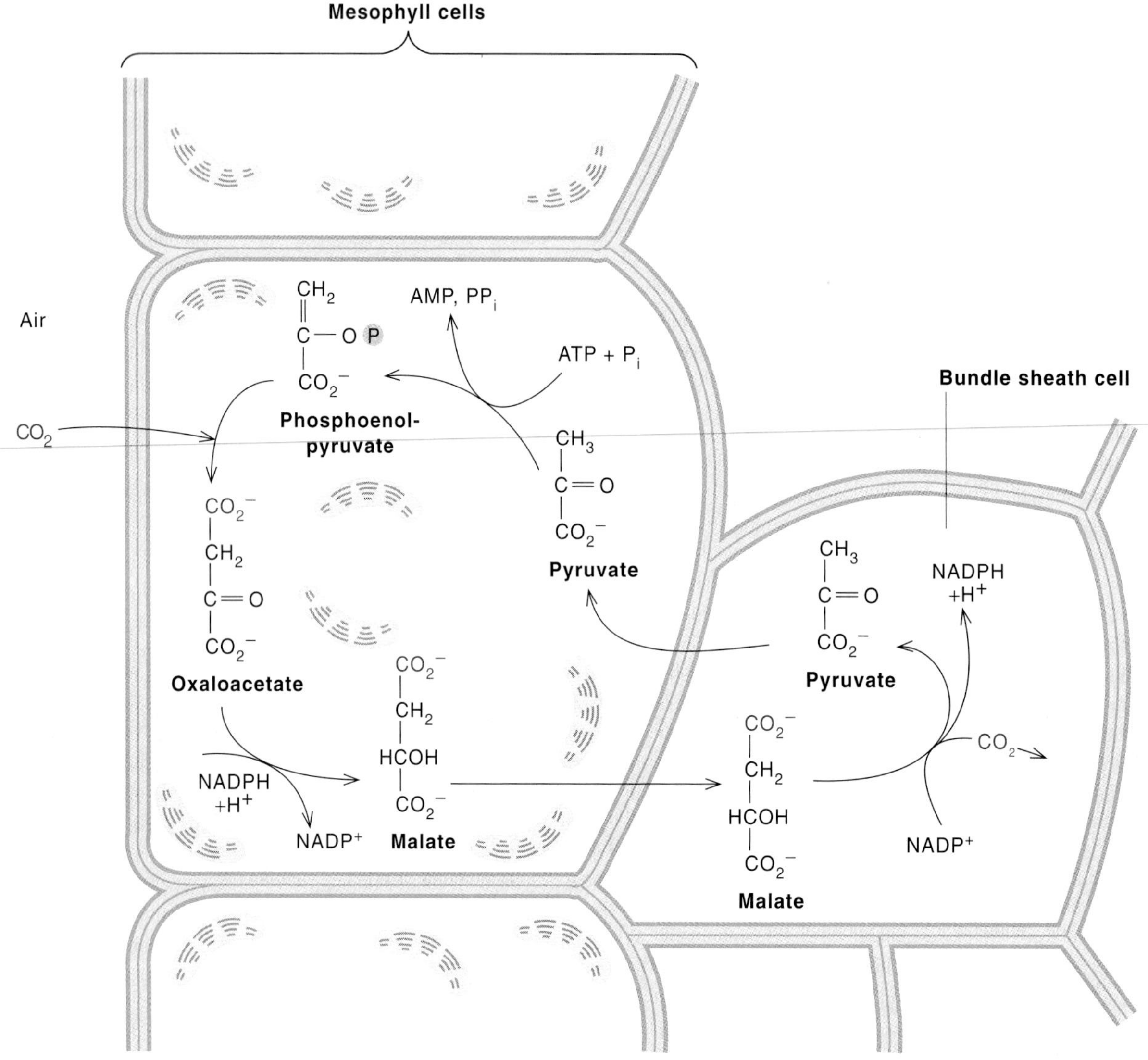

are frequently on the order of the K_m values, making O_2 a serious competitor. Plants that live in dry climates face the additional problem that opening the stomata in the leaves to improve the exchange of CO_2 and O_2 with the atmosphere can result in dehydration.

If photorespiration is only a liability, it would seem that plants should have evolved a ribulose bisphosphate carboxylase that had minimal activity as an oxygenase. However, the oxygenase activity of the enzyme purified from higher plants is only modestly below that of the enzymes from photosynthetic bacteria. Some species of plants have, however, evolved an alternative strategy for favoring the carboxylase reaction over the oxygenase: They increase the concentration of CO_2 in the region of the enzyme. These plants, which include corn, sugar cane, and numerous tropical species, have a layer of specialized mesophyll cells at the outer surface of the leaf. The mesophyll cells take up CO_2 from the air by the carboxylation of phosphoenolpyruvate to obtain oxalacetate (fig. 17.32). Oxalacetate is reduced to malate, which then moves from the mesophyll cells to the bundle sheath cells that surround the vascular structures in the interior of the leaf. Here malate is decarboxylated to pyruvate in an oxidative reaction that reduces $NADP^+$ to NADPH. The pyruvate returns to the mesophyll cell, where it is phosphorylated to phosphoenolpyruvate. The phosphorylation is driven by the splitting of ATP to AMP and pyrophosphate and the subsequent hydrolysis of the pyrophosphate to phosphate.

The result of this cyclic series of reactions is the delivery of CO_2 and reducing power (NADPH) to the bundle sheath cells, at a cost of the hydrolysis of one ATP to AMP (see fig. 17.32). The bundle sheath cells fix the CO_2 by the ribulose-bisphosphate carboxylase reaction and the reductive pentose cycle. Plants that have this mechanism for concentrating CO_2 can grow considerably more rapidly than species that do not, particularly in hot, dry climates. The auxiliary cycle is called the C_4 cycle because it involves the four-carbon acids malate and oxalacetate. In some species, oxalacetate is transaminated to aspartate rather than being reduced to malate, and it is aspartate that moves to the bundle sheath cells. There is currently interest in the possibility of increasing agricultural productivity by using genetic engineering techniques to introduce the C_4 pathway into plant species that lack it.

Other Biochemical Processes Involving Light

Living things depend on light as a source of energy. They also have found a number of other uses for light. Vision is one of these and is taken up separately in chapter 29, where it can be treated in the necessary detail. In the remainder of this chapter we will consider bioluminescence and the role of light in the synchronization of circadian and seasonal rhythms.

Phytochrome Synchronizes Circadian and Seasonal Rhythms in Plants

Most eukaryotic organisms have endogenous rhythms of metabolic activity with periods of about 24 hours. The circadian, or diurnal, rhythms are triggered by light, and under natural conditions they are locked in phase with the 24-h cycle of day and night. Many organisms also have seasonal cycles that are set by the length of the day. In mammals, synchronization of circadian rhythms depends on reactions that occur in the retina of the eye, but organisms that lack nervous systems depend on other types of photoreceptors. In some unicellular organisms, such as paramecia, the photoreceptive pigment appears to be a flavin or flavoprotein. In higher plants the major receptor is a pigment-protein complex called phytochrome.

In addition to synchronization of diurnal and seasonal rhythms, phytochrome has been implicated in literally hundreds of other responses of plants to light. These include germination of seeds, changes in ion transport, and synthesis of numerous enzymes. Some of its effects involve changes in gene expression, including an increase in the rate of transcription of the gene for the small subunit of ribulose bisphosphate carboxylase. Other effects, such as changes in the movement of Ca^{2+} across the cytoplasmic membrane, seem too rapid for this and may reflect a more immediate site of action. Some of the effects appear to be mediated by GTP-binding proteins and the Ca^{2+}-binding protein calmodulin.

The phytochrome protein is a dimer of subunits with molecular weights of about 120,000. The pigment is an open-chain tetrapyrrole (fig. 17.33). Phytochrome has two interconvertible forms (fig. 17.34). When the molecule is in one form, called P_R, phytochrome has an optical absorption maximum near 670 nm. Absorption of light converts phytochrome from P_R to the other form (P_{FR}), which absorbs maximally at longer wavelengths (730 nm). (The subscripts R and FR stand for "red" and "far-red.") This transformation probably involves *cis–trans* isomerization of a double bond in the pigment (see fig. 17.33) followed by conformational changes in the protein. P_{FR} is evidently the biologically active form of phytochrome, but exactly what it does is not clear.

Figure 17.33

The prosthetic group of phytochrome is an open-chain tetrapyrrole, which is bound to the protein as a cysteine thioether. Conversion of P_R to P_{FR} probably involves *cis–trans* isomerization of the double bond indicated in red.

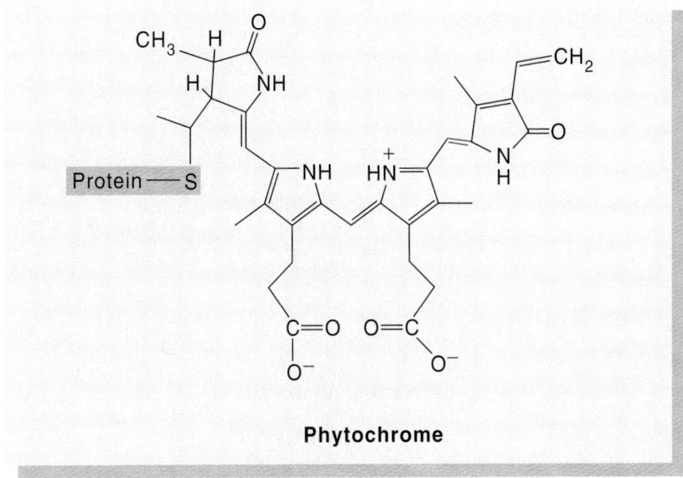

Figure 17.34

Phytochrome exists in two interconvertible forms, P_R and P_{FR}, which have different absorption spectra. The form that is synthesized initially, P_R, has an absorption band at 670 nm. When it is excited with light, P_R is converted to the biologically active species, P_{FR}, which absorbs at 730 mn. P_{FR} decays back to P_R in the dark, and it can be converted to P_R by excitation with light. P_{FR} also is degraded relatively quickly by proteolysis. P_R and P_{FR} may be conformational isomers that share a common excited state.

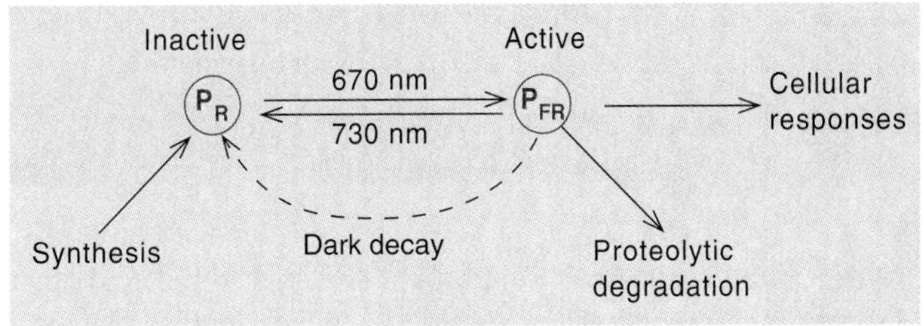

Phytochrome decays slowly from P_{FR} back to P_R in the dark. It also can be returned to P_R immediately by exciting it with light (see fig. 17.34). Because P_R absorbs maximally at 670 nm and P_{FR} at 730 nm, phytochrome can be cycled back and forth between the active and inactive forms by excitation flashes that alternate between the two wavelengths. Continuous illumination sets up a steady-state mixture, in which the amount of P_{FR} depends on the color of the light. Because chlorophyll absorbs red light better than it does far-red light, the color of the light reaching a leaf depends on whether the leaf is shaded by other leaves. The information collected by phytochrome can therefore be useful in controlling the direction and rate of growth of the plant.

Bioluminescence

Conceptually, bioluminescence is the reverse of photosynthesis. A reduced substrate reacts with O_2 and is converted to an oxidized

Figure 17.35

Bioluminescent reactions in the sea pansy *Renilla reniformis*. The structures in brackets are plausible enzyme-bound intermediates but have not been identified conclusively. They are based partly on analogous nonenzymatic reactions of dioxetanones and on the observation that one of the O atoms of the CO_2 that is released comes from the O_2.

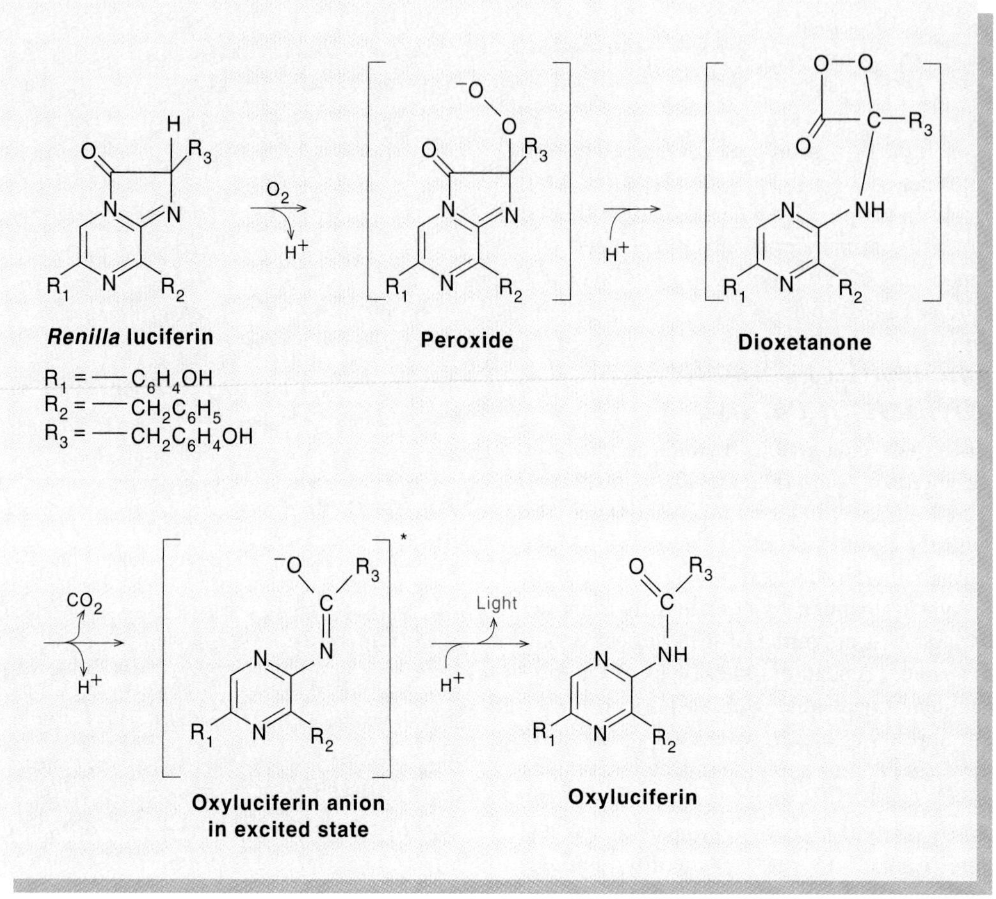

product in an excited electronic state. The excited molecule then decays to the ground state, emitting light. The process occurs in several groups of bacteria and fungi; in marine invertebrates such as sponges, shrimp, and jellyfish; and in a variety of terrestrial creatures, including earthworms, centipedes, and insects. The bacteria that emit light generally live symbiotically with fish in special luminous organs. In some cases the evolutionary benefits of bioluminescence seem clear: Fireflies use it for communication; fish, to attract or locate prey or to confuse predators. In other cases, such as in fungi, the benefits are not obvious, and one is struck by the expense of the process. The energy that is released in the oxidation–reduction reaction could, in principle, be directed into ATP production instead of being emitted as light.

The substrates that undergo oxidation with the emission of light are all called luciferins, although their structures vary among different bioluminescent species. The oxidized products are termed oxyluciferins, and the enzymes that catalyze the oxidations are called luciferases. Figures 17.35 and 17.36 show the structures of the luciferins and oxyluciferins used by two species, the sea pansy (*Renilla reniformis*) and the firefly (*Photinus*). The figures also indicate likely intermediates in the oxidation reactions. Although the luciferins have rather different structures in the two species, the chemistry appears to be much the same. O_2 reacts with a heterocyclic ring, forming a peroxide at a carbon that is bound to a nitrogen and to a carboxylic amide or ester. The peroxide cyclizes to give a cyclic peroxide, or dioxetanone, with the release of the amine or alcoholic group that was bound to the carboxyl. The cyclic peroxide then decomposes with the release of CO_2, forming the oxyluciferin in an excited electronic state.

The decomposition of the dioxetanone intermediate may involve an internal oxidation–reduction reaction in which an electron is transferred to the peroxide from another part of the molecule. Such a reaction could generate a linked pair of free radicals ($D^{+\cdot}\ A^{-\cdot}$) analogous to the species that are formed in the photochemical reactions of photosynthesis. After CO_2 splits off, an electron could return from the reduced component to the oxidized component, putting the product in an excited state.

One difference between fireflies and *Renilla* is that the

Figure 17.36

Bioluminescent reactions of fireflies. The excited oxyluciferin that emits light is believed to be a dianion, in which the phenolic and enolic oxygens are both ionized. The ionization state can be deduced by comparing the emission spectrum of the luminescence with those of model compounds.

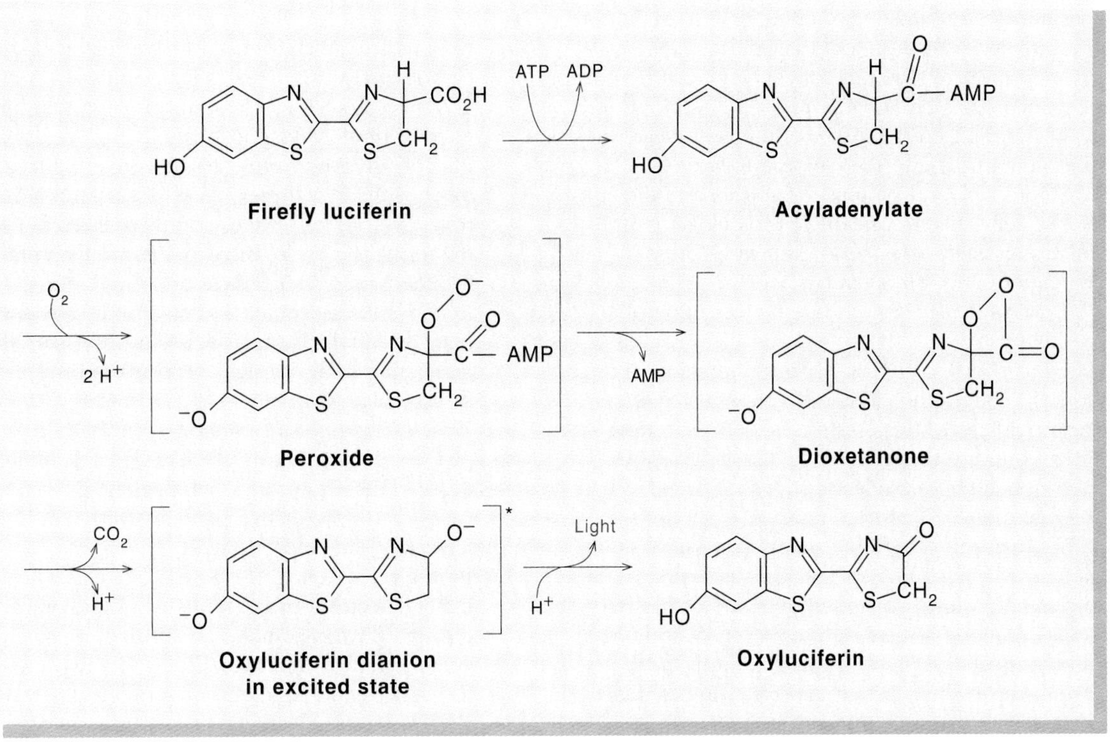

luciferase of fireflies activates the carboxyl group of the luciferin by forming an acyladenylate (see fig. 17.36). This step resembles the activation of amino acids for protein synthesis (chapter 34), but it is unique to fireflies among the bioluminescent systems that have been studied. Since the activation requires ATP, measurements of the light that is emitted by firefly oxyluciferin provide a sensitive assay for ATP.

Luminescent bacteria use a long-chain aliphatic aldehyde such as decanal as a luciferin, oxidizing it to the carboxylic acid. $FMNH_2$ is oxidized simultaneously to FMN. Again, a peroxide appears to be an intermediate in the oxidation. The excited product is probably a hydroxy derivative of FMN, which subsequently loses H_2O to form FMN.

In bioluminescent organisms the molecule that actually emits the light is sometimes not the oxyluciferin itself, but a pigment on another protein, to which the excited oxyluciferin transfers its energy. This is conceptually like a reverse operation of the light-harvesting antenna systems of photosynthesis.

Summary

In this chapter we have examined photochemical reactions, primarily those that occur in photosynthesis. The chief points made in the chapter are as follows.

1. The primary source of energy for biosynthetic reactions is sunlight. Photosynthesis, the process of capturing sunlight and converting it into chemical energy, occurs in numerous species of bacteria and algae, in addition to higher plants. In plants the photochemical reactions of photosynthesis take place in the chloroplast thylakoid membrane. In bacteria they take place in the cell membrane.
2. Almost all photosynthetic organisms take advantage of the fact that chlorophyll or bacteriochlorophyll becomes a strong reductant when it is excited with light. Photooxidation of chlorophyll or bacteriochlorophyll occurs in pigment–protein complexes, the reaction centers. Chloroplasts contain two types of reaction centers: photosystem I, which contains a reactive chlorophyll complex called P700, and photosystem II, which contains P680. The reactive complex in purple photosynthetic bacteria (P870) is a bacteriochlorophyll dimer.
3. The thylakoid membrane and the bacterial cell membrane also contain molecules that serve as antennae. The antenna systems consist of small pigment–protein complexes assembled into large arrays that can contain hundreds of pigment molecules per reaction center. When the antenna absorbs a photon, the excitation energy moves rapidly from complex to

complex by resonance energy transfer until the energy is trapped in a reaction center.

4. The bacterial reaction center contains two molecules of bacteriopheophytin and two of bacteriochlorophyll in addition to the two bacteriochlorophylls of P870. When P870 is excited, it transfers an electron to one of the bacteriopheophytins. The bacteriopheophytin then reduces a quinone, Q_A, which in turn reduces another quinone, Q_B. When P870 is excited a second time, a second electron is sent through the same carriers to Q_B. The electron released by P870 is replaced by one from a *c*-type cytochrome. The doubly reduced Q_B picks up protons from the solution on the inside of the cell. Electrons move from the quinol (QH_2) to the *c*-type cytochrome via a cytochrome bc_1 complex, and protons are released to the solution outside the cell. This movement of protons across the membrane creates an electrochemical potential gradient across the membrane. Proton movement back into the cell is mediated by an ATP-synthase and drives the formation of ATP.
5. Photosystems I and II of chloroplasts operate in series. The photooxidation of P680 in photosystem II generates a strong oxidant that can oxidize H_2O to O_2. In this process, a manganese complex bound to the reaction center protein progresses through five different oxidation states (S_0 to S_4), advancing to the next higher state each time P680 undergoes photooxidation. When the complex reaches state S_4, O_2 is given off. The electron released by P680 goes to a pheophytin, then to a molecule of plastoquinone, and from there to a second plastoquinone. When the second plastoquinone has been doubly reduced, it picks up protons from the chloroplast stromal space and dissociates from the reaction center. Electrons move from the reduced plastoquinone to P700 of photosystem I by means of a cytochrome b_6f complex and a copper protein (plastocyanin).
6. Photooxidation of P700 in photosystem I reduces a chlorophyll, which transfers electrons to a series of membrane-bound iron–sulfur proteins, probably by way of a quinone. From the bound iron–sulfur centers, electrons move to the soluble iron–sulfur protein ferredoxin and then to a flavoprotein that reduces NADP. There also may be some cyclic electron flow, in which electrons go from the iron–sulfur centers to the carriers between the two photosystems.
7. Electron movement through the cytochrome b_6f complex between the chloroplast photosystems results in proton translocation from the stroma to the thylakoid lumen. In addition, protons are released in the lumen when H_2O is oxidized and are taken up in the stromal space when NADP is reduced. Protons move from the thylakoid lumen back to the stroma through an ATP-synthase (CF_0–CF_1), driving the formation of ATP.
8. ATP and NADPH are used for the incorporation of CO_2 into carbohydrates. CO_2 reacts with ribulose-1,5-bisphosphate to give two molecules of 3-phosphoglycerate, under the influence of ribulose bisphosphate carboxylase/oxygenase. 3-Phosphoglycerate can then be converted back to ribulose-1,5-bisphosphate at the expense of ATP and NADPH in the reductive pentose cycle. For every three molecules of CO_2 that enter the cycle, there is a gain of one molecule of glyceraldehyde-3-phosphate.
9. Ribulose bisphosphate carboxylase/oxygenase also catalyzes a reaction in which ribulose-1,5-bisphosphate reacts with O_2, generating 2-phosphoglycolate and 3-phosphoglycerate. Phosphoglycolate is oxidized to CO_2 and H_2O by additional enzymes. This photorespiration appears to be a wasteful process. Some species of plants use an auxiliary cycle of reactions (the C_4 cycle) to concentrate CO_2 in the cells that carry out the reductive pentose cycle. This allows the plants to fix CO_2 at elevated rates.
10. Biological systems have found other uses for light. In plants, light regulates a broad range of physiological processes through phytochrome, a complex of a protein and an open-chain tetrapyrrole. Phytochrome has two interconvertible forms, P_R and P_{FR}. P_R is converted to P_{FR} when it is excited with red light. P_{FR} reverts to P_R upon excitation with far-red light. The formation of P_{FR} initiates numerous responses, but the mechanism of its action is unknown.
11. In bioluminescence, oxidation of a small molecule (a luciferin) by O_2 creates a product in an excited electronic state. The product decays to its ground state by emitting light. Luciferins vary from species to species, but many are heterocyclic compounds that react with O_2 to form a cyclic peroxide that decomposes by releasing CO_2.

Selected Readings

Amesz, J. (ed.), *Photosynthesis*. Amsterdam: Elsevier, 1987. Review articles on many aspects of photosynthesis.

Bauer, C. E., and T. H. Bird, Regulatory Circuits Controlling Photosynthesis Gene Expression. *Cell* 85:5, 1996.

Deisenhofer, J., and H. Michel, Crystallographic refinement at 2.3 Å resolution and refined model of the photosynthetic reaction centre from *Rhodopseudomonas viridis*. *J. Mol. Biol.* 246:429, 1995.

Duysens, L. N. M., and J. Amesz, Function and identification of two functional systems in photosynthesis. *Biochim. Biophys. Acta* 64:243, 1962. Evidence supporting the Z scheme is obtained by measuring oxidation and reduction of cytochrome *f* in algae illuminated with light of various colors.

Forbush, B., B. Kok, and M. McGloin, Cooperation of charges in photosynthetic O_2 evolution: II. Damping of flash yield, oscillation, deactivation. *Photochem. Photobiol.* 14:307, 1971. Oscillations in O_2 yield during a train of flashes provide evidence for the four S states.

Jagendorf, A. T., and E. Uribe, ATP formation caused by acid-base transition of spinach chloroplasts. *Proc. Natl. Acad. Sci. USA* 55:197, 1966. Chloroplasts can form ATP in the dark if an electrochemical potential gradient for protons is set up across the thylakoid membrane.

McDermott, G., S. M. Prince, A. A. Freer, A. M. Hawthornthwaite-Lawless, M. Z. Papiz, R. J. Cogdell, and N. W. Isaacs, Crystal structure of an integral membrane light-harvesting complex from photosynthetic bacteria. *Nature* 374:517, 1995.

Schneider, G., Y. Lindvist, and C. I. Brandèn, Rubisco: structure and mechanism. *Ann. Rev. Biophys. Biomol. Struct.* 21:119, 1992. Structure and activities of ribulose-1,5-bisphosphate carboxylase/oxygenase.

Woodbury, N. W., M. Becker, D. Middendorf, and W. W. Parson, Picosecond kinetics of the initial photochemical electron transfer reaction in bacterial photosynthetic reaction centers. *Biochem.* 24:7516, 1985. Fast spectrophotometric techniques are used to follow the initial steps in reaction centers purified from photosynthetic bacteria.

Yachandra, V. K., V. J. DeRose, M. J. Latimer, I. Mukerje, K. Sauer, and M. P. Klein, Where plants make oxygen: a structural model for the photosynthetic oxygen-evolving manganese cluster. *Science* 260:675, 1993. Fast spectrophotometric techniques are used to follow the initial steps in reaction centers purified from photosynthetic bacteria.

PROBLEMS

1. The word "fixed" is often used to describe what happens to CO_2 in photosynthesis or to N_2 in nitrogen "fixation." How do you explain the meaning of "to fix" a gas or "a fixed gas" to an English major?
2. If the conversion of light energy into chemical energy occurred with 100% efficiency, how many moles of ATP could be produced per mole of photons of blue light (450 nm) and red light (700 nm)?
3. What arguments can you make for or against the idea that the chlorophylls (fig. 17.4) are aromatic compounds?
4. The lowest-energy absorption band of P870 occurs at 870 nm.
   (a) Calculate the energy of an einstein of light at this wavelength.
   (b) Estimate the effective standard redox potential ($E°'$) of P870 in its first excited singlet state, given that the $E°'$ for oxidation in the ground state is +0.45 V.
5. The traces shown here are measurements of optical absorbance changes at 870 and 550 nm when a suspension of membrane vesicles from photosynthetic bacteria was excited with a short flash of light. Downward deflection of the traces represent absorbance decreases. Explain the observations. (Absorption spectra of a *c*-type cytochrome in its reduced and oxidized forms are described in the previous chapter.)

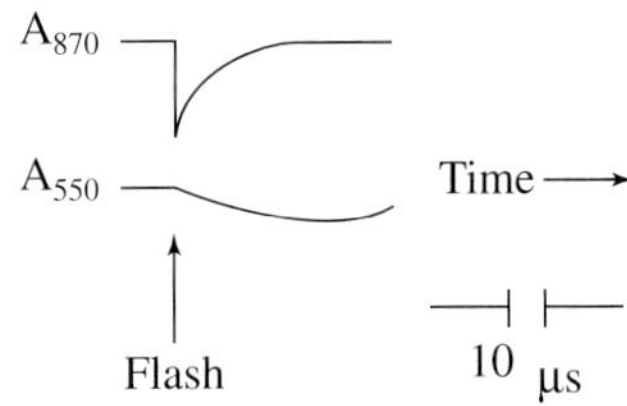

6. You add a nonphysiological electron donor to a suspension of chloroplasts. When you illuminate the chloroplasts, the donor becomes oxidized. How can you determine whether this process involves both photosystems I and II? (In principle, the donor can transfer electrons either to some component on the O_2 side of photosystem II or to a component between the two photosystems.)
7. Ubiquinone has an absorbance band at 275 nm. This band bleaches when the quinone is reduced to either the semiquinone or the dihydroquinone. The anionic semiquinone has an absorption band at 450 nm, but neither the quinone nor the dihydroquinone absorbs at this wavelength. A suspension of purified bacterial reaction centers was supplemented with extra ubiquinone and reduced cytochrome *c* and was then illuminated with a series of short flashes of light. The absorbance at 275 nm decreased on the odd-numbered flashes as shown on the first curve. The absorbance at 450 nm increased on the odd-numbered flashes but returned to the original level on the even-numbered flashes as shown in the second curve.
   (a) Explain the patterns of absorbance changes at the two wavelengths.
   (b) Why is it necessary to have reduced cytochrome *c* present in order to see these effects?

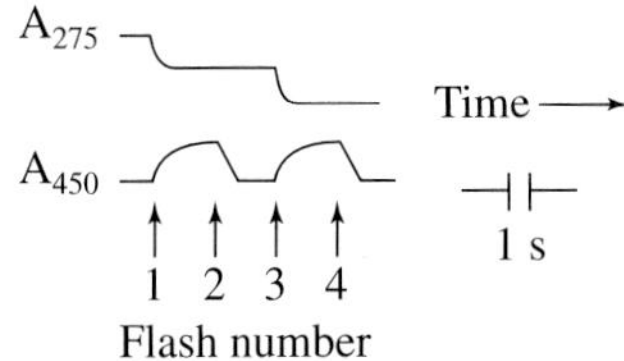

8. Explain why in green plants the quantum efficiency of photosynthesis drops sharply at wavelengths longer than 680 nm.
9. Do you predict that plant cells lacking carotenoids are more or less susceptible to damage by photooxidation? Explain.
10. Molecular oxygen is an alternative substrate for the ribulose biphosphate carboxylase–oxygenase and is also a competitive inhibitor with respect to CO_2 fixation. Explain.
11. In the formation of phosphoenolpyruvate from pyruvate in the mesophyll cells (fig. 17.32), notice the energetics of the reaction. How does this process compare energetically with the formation of phosphoenolpyruvate from pyruvate in the start of gluconeogenesis?
12. Provide a mechanism for the conversion of pyruvate to phosphoenolpyruvate in the mesophyll cells (fig. 17.32), and account for the consumption of an ATP and P_i and the production of AMP, PP_i, and phosphoenolpyruvate.
13. Propose a mechanism for the "oxygenase" activity of ribulose bisphosphate carboxylase.
14. When examining the information described in figure 17.14 or figure 17.20, one is struck by the similarities with the electron-transport scheme (chapter 16). What types of components do these two schemes have in common?
15. What differences can you find between the pentose phosphate pathway (figs. 14.22 through 14.25 and associated text) and the Calvin cycle?
16. Reduction of 3 moles of CO_2 to form 1 mole of triose phosphate requires 9 moles of ATP and 6 moles of NADPH.
   (a) What is the source of NADPH in the reduction of CO_2?
   (b) Account for the ATP consumed in the formation of triose phosphate.
   (c) Assume that the CO_2 initially is added to PEP in the C-4 pathway. What is the additional cost in ATP per mole of CO_2 added?
   (d) Is additional NADPH required for CO_2 fixation in the C-4 plant?
17. O_2 is an alternative substrate for ribulose bisphosphate carboxylase/oxygenase and is also a competitive inhibitor with respect to CO_2 fixation. Explain.
18. Compare and contrast the initial process of CO_2 fixation in C_3 and C_4 plants. In each case, indicate which stable organic molecule initially becomes labeled with ^{14}C labeled CO_2.
19. C_4 plants ultimately use ribulose bisphosphate carboxylase/oxygenase yet have lower rates of photorespiration than do C_3 plants. Explain.

CHAPTER 18

Structures and Metabolism of Polysaccharides and Glycoproteins

Hexoses link together to form straight-chain or branched-chain oligosaccharides in glycoproteins.

We have seen that small carbohydrates such as glucose, fructose, and pyruvate occupy key roles in energy metabolism and supply carbon skeletons for the synthesis of other compounds. Polymeric carbohydrates also play an important role in energy metabolism as short-term energy-storage compounds. We now turn our attention to the extremely varied roles that oligosaccharides and polysaccharides play in other forms of metabolism. Polymeric carbohydrates are important as major structural compounds in plant and bacterial cell walls and in the extracellular matrix. Branched-chain carbohydrates covalently linked to proteins give proteins specific surface characteristics that are frequently exploited to signal protein targeting and in a variety of cell recognition processes.

Since the main building blocks of all polysaccharides are hexoses, we will begin this chapter by considering the structure and synthesis of the hexoses. In the remainder of the chapter we focus mainly on the structure and synthesis of complex oligosaccharides and polysaccharides. Special emphasis is given to glycoproteins, because there have been major advances in this all-important subject in recent years.

Figure 18.1

Monosaccharide interconversions. The following abbreviations are used here and throughout the chapter: Gal = galactose, Xyl = xylose, Glc = glucose, GlcUA = glucuronic acid, GlcNAc = *N*-acetylglucosamine, GalNAc = *N*-acetylgalactosamine, Fuc = fucose, Man = mannose, ManNAc = *N*-acetylmannosamine, NeuNAc = *N*-acetylneuraminic acid (the most common type of sialic acid = Sia). Sialic acid is a more general term including N- and O-substituted derivatives of neuraminic acid. Compounds in capital boldface type are the unmodified sugars or their amine derivatives.

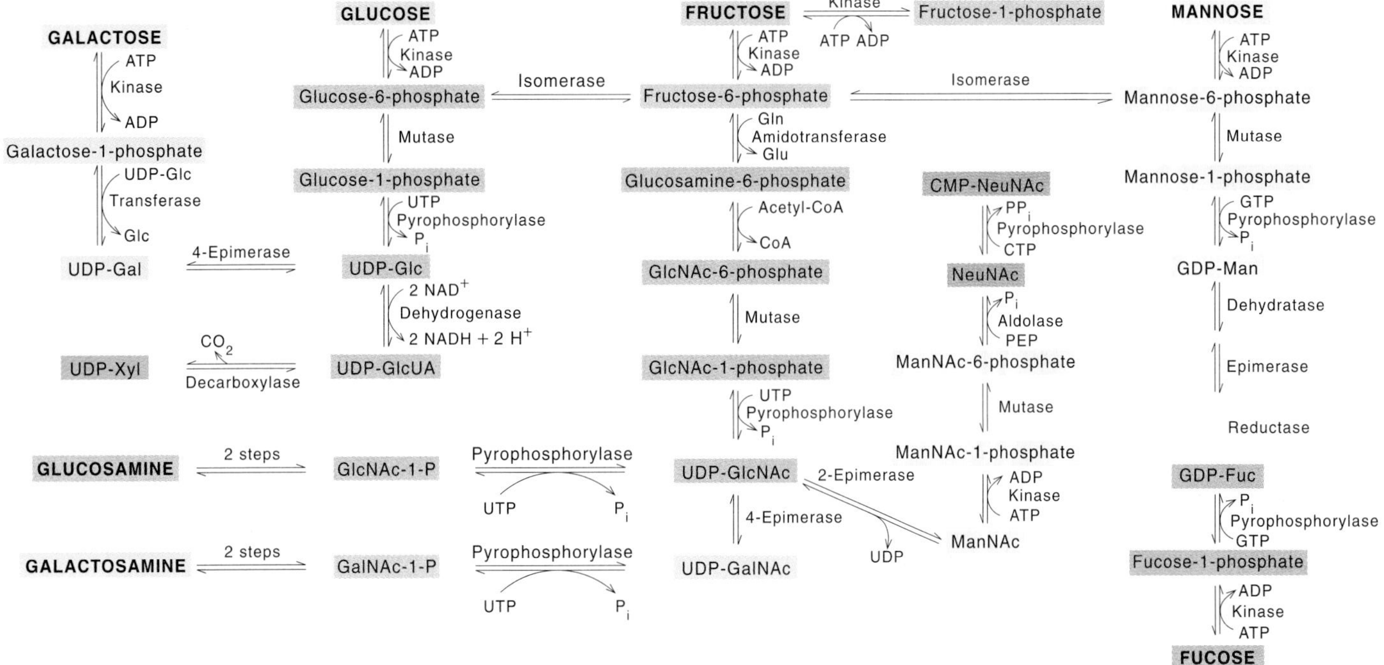

Monosaccharides Are Often Formed by Interconversions between Hexoses

We discussed the biosynthesis of glucose from simpler starting materials in chapter 14. The biosynthesis of other hexoses is linked to glucose by a complex network of single-step reactions (fig. 18.1). In fact, glucose serves as the precursor for the synthesis of many other hexoses without any rearrangement of the central carbon atoms.

Although the list of hexoses of biologic origin is large, we can make certain generalizations about the network of pathways by which they are metabolically connected. Neutral hexoses are never interconverted; some interconversions occur at the level of the monophosphorylated hexose, but the majority of interconversions occur at the level of the nucleotide sugars. Finally, hexose modification (e.g., addition of an *N*-acetyl group or dehydrogenation) generally takes place before polymerization to form a nucleotide sugar.

The Hexose Monophosphate Pool Includes Mannose as Well as Glucose and Fructose

Derivatives of glucose, mannose, fructose, and galactose are the most common sugars found in oligo- and polysaccharides. Three of these hexoses—glucose, mannose, and fructose—belong to the same hexose monophosphate pool (fig. 18.2). They are readily interconverted, with little or no difference in the free energy of the different compounds involved.

We discussed the interconversion of glucose and fructose phosphates in chapter 14, but we omitted mannose from that discussion because it is not centrally involved in either glycolysis or gluconeogenesis. Mannose is, however, a major component of many complex carbohydrates. Mannose-6-phosphate, the 2-epimer of glucose-6-phosphate, can be made directly from fructose-6-phosphate in a reaction analogous to the interconversion of fructose-6-phosphate and glucose-6-phosphate. The phosphomannoisomerase enzyme that catalyzes this isomerization holds the substrate so that a proton can be added to the planar C-2 of the intermediate enediol on the side opposite that on which addition occurs in the phosphoglucoisomerase reaction (see fig. 18.1). Subsequently, mannose-6-phosphate can be converted to mannose-1-phosphate by a mutase specific for phosphomannose. Mannose and mannose derivatives are incorporated into complex polysaccharides by way of GDP-mannose, which is made from mannose-1-phosphate and GTP as shown in Fig. 18.1.

Galactose Is Not a Member of the Hexose Monophosphate Pool

Galactose, the other main hexose found in structural polysaccharides, is not a member of the central hexose monophosphate pool.

Figure 18.2

Members of the hexose monophosphate pool. The sugars shown are all freely interconvertible, with little change in free energy involved in the conversions.

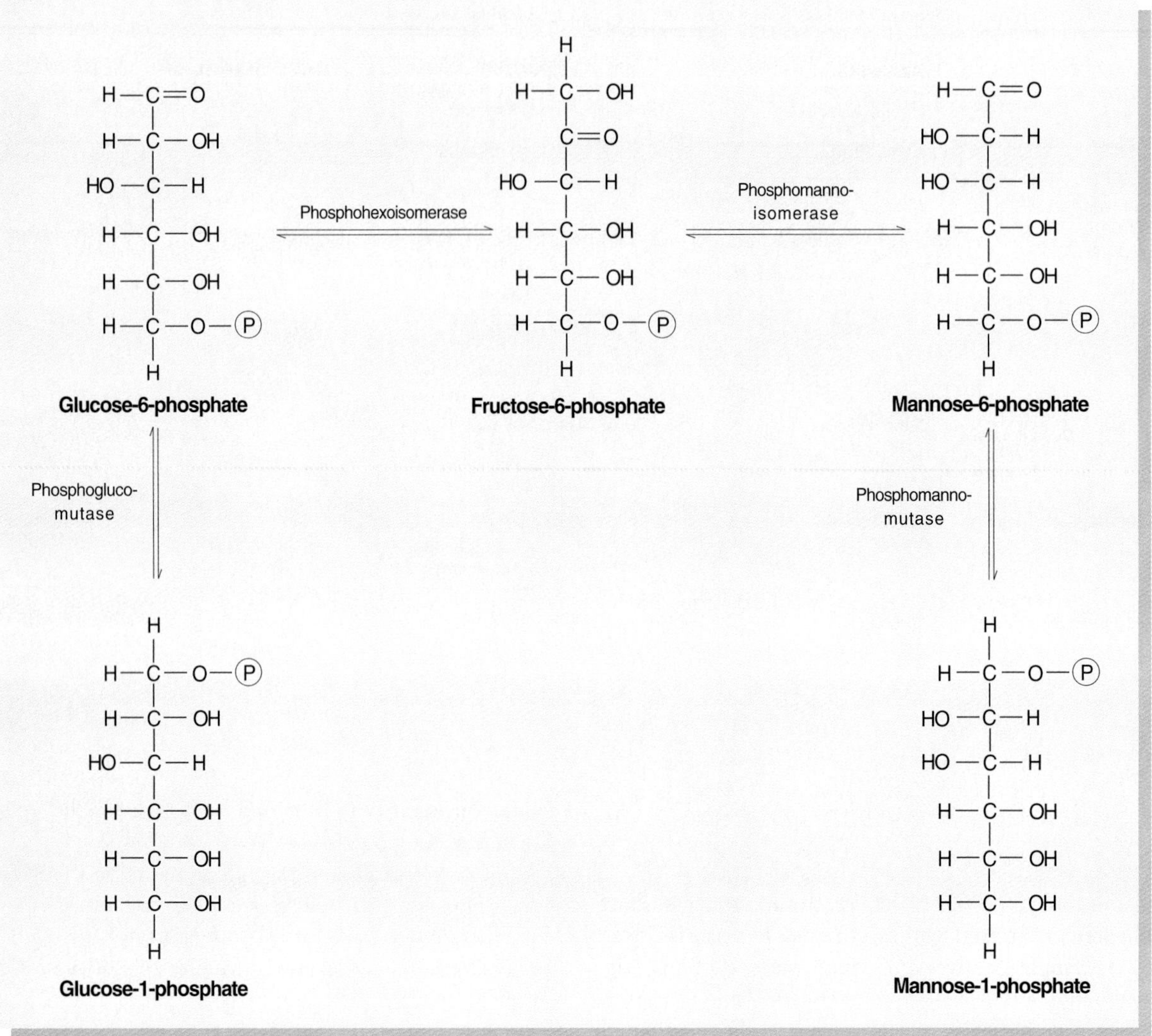

The only route from the main pool to UDP-galactose goes through UDP-activated derivatives of glucose (see fig. 18.1). By the simplest route, UDP-galactose, the 4-epimer of UDP-glucose, is produced by way of UDP-glucose. The reactions are

$$\text{Glucose-1-phosphate} + \text{UTP} \rightleftharpoons \text{UDP-glucose} + 2\text{P}_i$$

$$\text{UDP-glucose} \rightleftharpoons \text{UDP-galactose}$$

UDP-glucose 4-epimerase, which catalyzes the isomerization of UDP-glucose to UDP-galactose, contains a tightly bound molecule of NAD^+ or $NADP^+$, even though no net oxidation is involved. It is believed that the —OH group at C-4 is transiently oxidized to a carbonyl, which can then be reduced by addition of a hydride ion to either side, thus producing either UDP-glucose or UDP-galactose (see fig. 11.7). Two monosaccharides are considered epimers (e.g., galactose is the 4-epimer of glucose) when they differ only by the configuration of a single hydroxyl group other than at C-1; anomers differ in configuration at C-1 (α or β)(see chapter 13).

Hexose Modifications Involve Alterations or Additions of Small Substituents

Thus far we have considered only the isomerization of hexoses. Many modifications of hexoses involve the alteration of existing groups or the addition of small groups to the hexose moiety. (Structures for some modified hexoses are illustrated in fig. 18.3.) Oxidation of carbon 6 to a carboxylate group produces a uronic acid. The dehydrogenases that catalyze those oxidations use NAD^+ as the oxidizing agent. For example, UDP-glucose (UDP-Glc) is converted to UDP-glucuronic acid (UDP-GlcUA) in this way.

Replacement of the —OH group at carbon 2 by $—NH_2$ yields an amino sugar, which is usually modified further. For ex-

Figure 18.3

Some of the sugar building blocks found in polysaccharides.

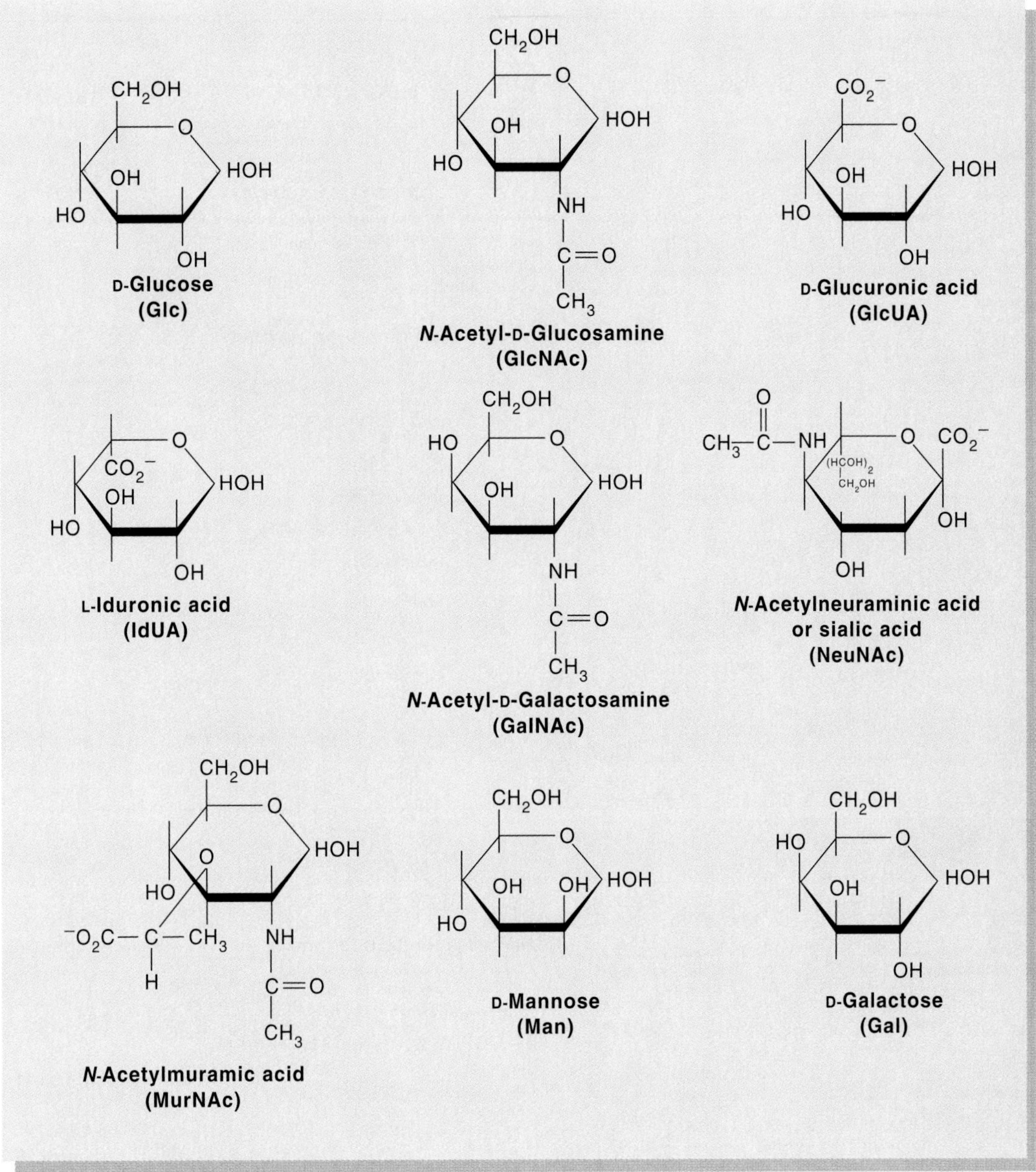

ample, fructose-6-phosphate is the monosaccharide precursor of several important derivatives (fig. 18.4). The process begins with glucosamine formation, a one-step reaction in which the amide nitrogen of glutamine is transferred to the C-2 carbon of fructose. This is the first step in a biosynthetic pathway leading to several different sugar monomers. The amine is acetylated to *N*-acetylglucosamine-6-phosphate, which is activated by reaction with UTP to form UDP-*N*-acetylglucosamine. This derivative may be directly incorporated into a polymer or may be converted to other polymer precursors. In one step, UDP-*N*-acetylglucosamine can be epimerized to UDP-*N*-acetylgalactosamine, or in six steps it can be converted to CMP-*N*-acetylneuraminic acid, also called CMP-sialic acid (see fig. 18.3 for the structure of sialic acid). The CMP-sialic acids are the only nucleotide sugars that occur as nucleoside monophosphate derivatives.

Little is known about the regulation of synthesis of various nucleotide sugar derivatives. If that regulation follows the general scheme suggested in chapter 12 for a biosynthetic pathway, then the first enzyme in a sequence should be negatively regulated by the end product of the sequence. In agreement with this principle it has been found in rat liver that UDP-*N*-acetylglucosamine regulates its own synthesis by inhibiting the amido-

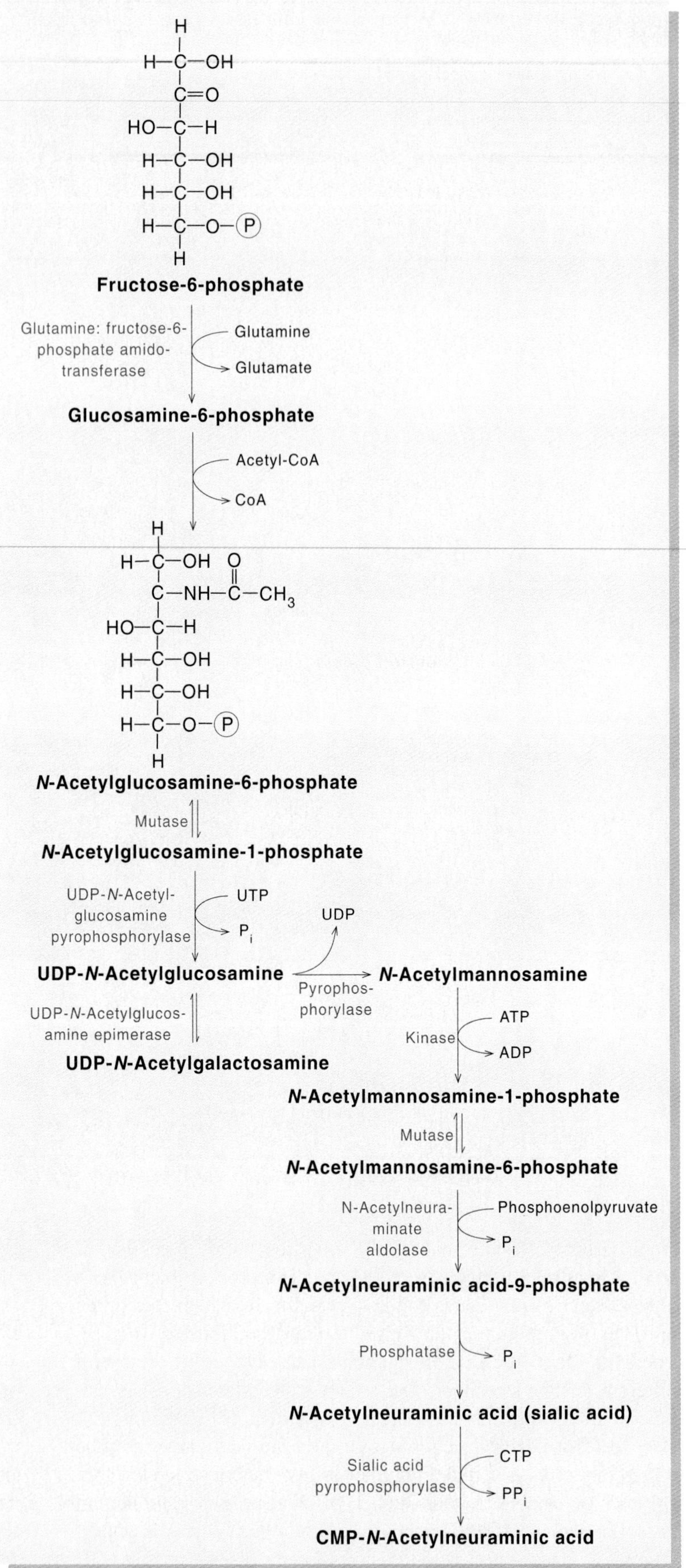

Figure 18.4

Some derivatives formed from fructose-6-phosphate. This figure elaborates on one branch of figure 18.1.

Figure 18.5

The mechanism of lactose formation. The more realistic chair forms are shown for the hexoses to make it easier to appreciate the stereochemistry of the reaction. Note how the configuration at C-1 becomes inverted in this reaction. Lactose synthase is two proteins: $\beta(1,4)$-galactosyltransferase and α-lactalbumin. The mechanism for this reaction differs from that found in polysaccharide synthesis or degradation, neither of which shows inversions; these latter reactions involve an oxonium ion intermediate.

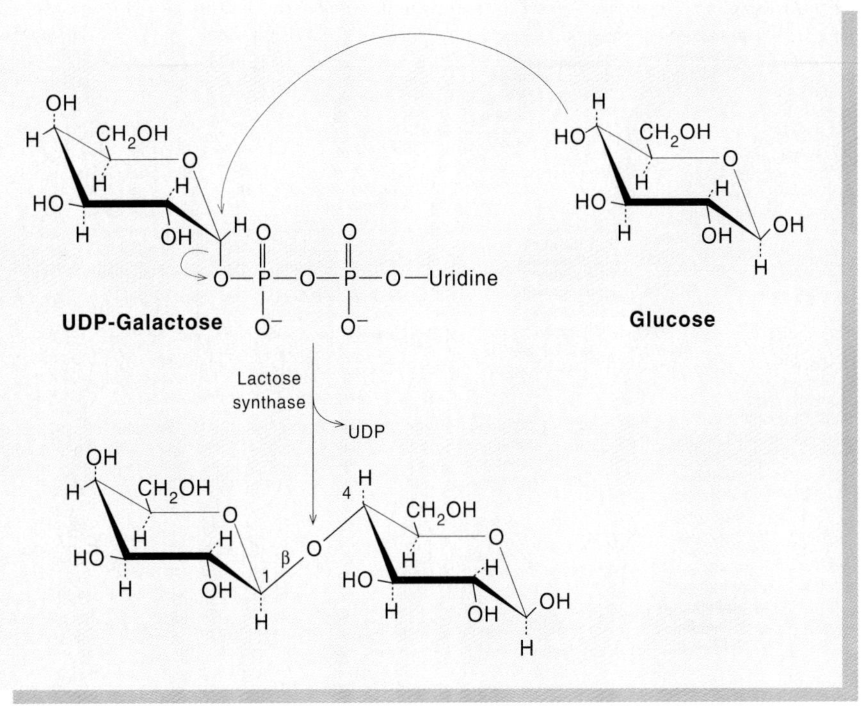

transferase that catalyzes the conversion of fructose-6-phosphate to glucosamine-6-phosphate, the first step that is specific to this pathway (see fig. 18.4).

Disaccharide Biosynthesis

In disaccharide synthesis, only one of the participating hexoses is activated. The disaccharide lactose is formed in the mammary gland from D-glucose and UDP-galactose by the action of lactose synthase (fig. 18.5). The reaction involves nucleophilic displacement of UDP from UDP-galactose by the C-4 hydroxyl group of a free glucose.

Formation of lactose involves an unusual mechanism for controlling enzyme specificity. Lactose synthase is actually a complex of two proteins: (1) Galactosyltransferase is found not only in mammary gland, but in all body tissues. It catalyzes the reaction

$$\text{UDP-galactose} + N\text{-acetyl-D-glucosamine} \longrightarrow \text{UDP} + N\text{-acetyllactosamine}$$

(2) α-Lactalbumin of milk has no catalytic activity of its own. Rather, it alters the specificity of galactosyltransferase so that the latter will utilize D-glucose instead of *N*-acetyl-D-glucosamine as the galactose acceptor. As a result, α-lactalbumin makes lactose instead of *N*-acetyllactosamine:

$$\text{UDP-galactose} + \text{D-glucose} \longrightarrow \text{UDP} + \text{lactose}$$

Sucrose is synthesized from glucose and fructose. First, fructose-6-phosphate is produced from glucose-6-phosphate. The latter is also converted via glucose-1-phosphate to uridine diphosphate glucose (UDP-glucose), which then reacts with fructose-6-phosphate to give UDP and sucrose-6-phosphate. The phosphate is removed by a single enzymatically catalyzed hydrolysis to yield sucrose and inorganic phosphate. In some plants, sucrose is formed simply by the reaction of UDP-glucose with fructose.

Energy-Storage Polysaccharides Are Simple Homopolymers

Most polysaccharides that are used for energy storage are simple homopolymers of glucose linked by $\alpha(1,4)$-glycosidic bonds. Some aspects of the synthesis of these polyglucose molecules were described in chapter 14. In all cases the C-1 atom of the monomer is activated as in disaccharide synthesis (fig. 18.6). Sometimes this activation is supplied by a UDP-derivative, sometimes by an ADP-derivative (table 18.1). The $\Delta G°$ of this reaction is about -3.2 kcal/mole.

The enzyme glycogen synthase requires a primer oligosaccharide with at least four glucose residues, to which it adds successive glucosyl groups. In addition to $\alpha(1,4)$ bonds, glycogen contains $\alpha(1,6)$ bonds. The latter bonds are made by the branching enzyme amylo-α(1,4-1,6)-*trans*-glycosylase. This enzyme transfers a terminal oligosaccharide fragment of six or seven glucosyl residues from the end of the main glycogen chain to the 6-hydroxyl group of a glucose residue somewhere in a glycogen chain. The reaction produces a branched-chain polymer from a straight-chain polymer, as shown in figure 18.7. As you might expect, the free energy change in this reaction is very small, since

Figure 18.6

Elongation step in glycogen synthesis. The gross chemistry of this reaction is quite similar to that of lactose synthesis except that there is no inversion about C-1, since it proceeds by a different mechanism, probably similar to that used by starch synthase (see fig. 14.3).

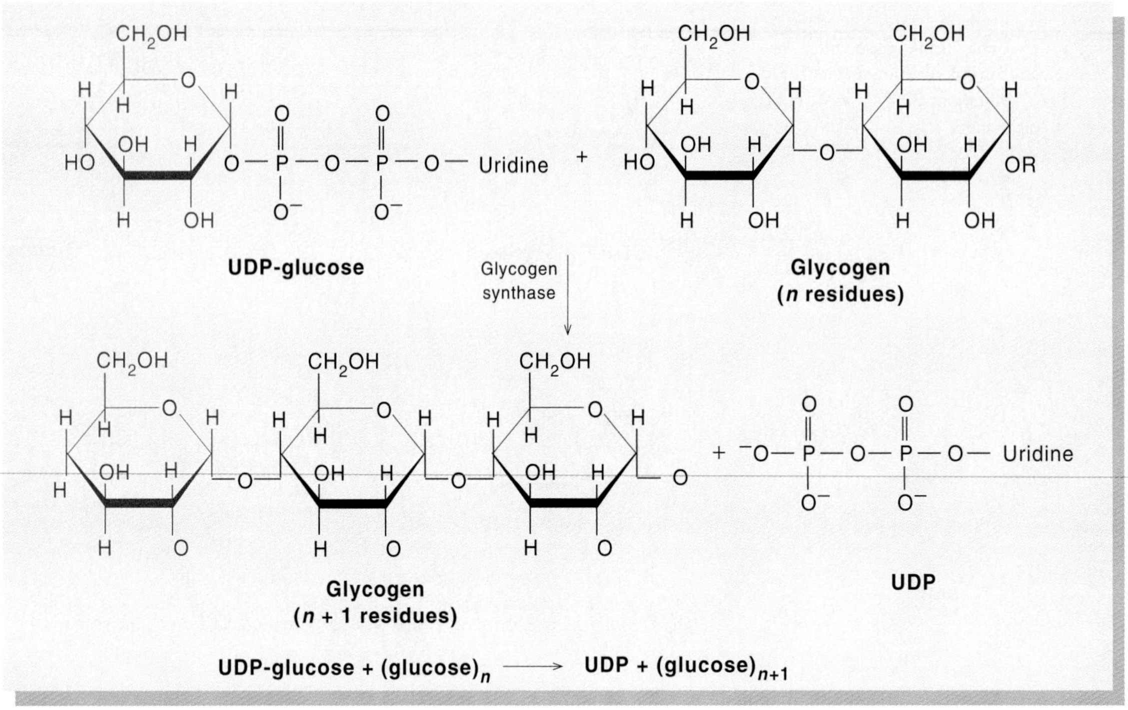

$$\text{UDP-glucose} + (\text{glucose})_n \longrightarrow \text{UDP} + (\text{glucose})_{n+1}$$

Table 18.1
Some Storage Polysaccharides

Source	Polysaccharide	Monosaccharide Component(s)	Glycosyl Donor	Polymer Structure
Primarily muscle and liver cells of animals	Glycogen	Glucose	UDP-glucose	$\alpha(1,4)$ with $\alpha(1,6)$ branchpoints
Bacterial glycogen	Glycogen	Glucose	ADP-glucose	$\alpha(1,4)$ with $\alpha(1,6)$ branchpoints
Green algae	Amylose	Glucose	ADP-glucose	Linear $\alpha(1,4)$
Leaves, stem, roots, and seeds of higher plants	Amylopectin	Glucose	ADP-glucose	Linear $\alpha(1,4)$ with $\alpha(1,6)$ branchpoints
Some bacteria	Dextran	Glucose	Sucrose	Linear $\alpha(1,6)$ with $\alpha(1,2)$, $\alpha(1,3)$ or $\alpha(1,4)$ branchpoints

very similar chemical linkages are involved. In plant tissues, starch synthesis occurs by an analogous pathway that is catalyzed by amylose synthase. ADP-glucose is the preferred glucose donor.

Structural Polysaccharides Include Homopolymers and Heteropolymers

The most abundant structural polysaccharide is plant cellulose, a straight-chain homopolymer of glucose with a $\beta(1,4)$ linkage (see chapter 13). Cellulose is formed by the same general mechanism as glycogen, using nucleotide diphosphate sugars.

Chitin Contains a Different Building Block

Many polysaccharides consist of sugars other than glucose or contain different combinations of sugars. In some cases the new sugar is merely a derivative of glucose. Chitin is an example of a structural polysaccharide that uses a modified derivative of glucose. Chitin is found in the shells of crustaceans and insects and in the

Figure 18.7

Schematic diagram showing the action of the "branching enzyme" in glycogen formation. A terminal hexasaccharide fragment is shown as being transferred from a 1,4 straight-chain linkage to a 1,6 branchpoint. No activation is involved because the energy change on reaction is very small.

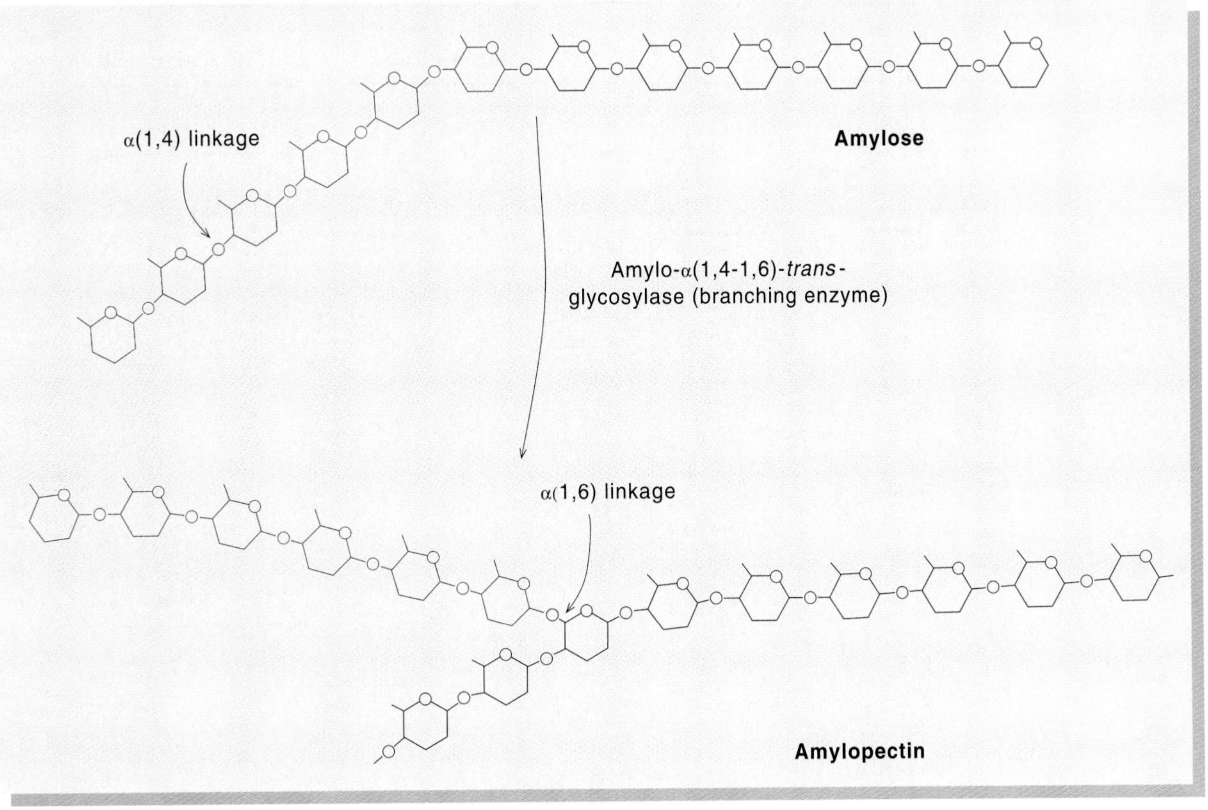

cell walls of fungi; it is a linear $\beta(1,4)$ polymer of *N*-acetyl-D-glucosamine (fig. 18.8) formed from the activated monomer UDP-*N*-acetylglucosamine (UDP-GlcNAc). A major rigid component of bacterial cell walls, the peptidoglycan, which we will discuss later, could be regarded as a substituted chitin.

Whereas most polysaccharide syntheses use nucleoside diphosphate sugars as substrates, in the formation of dextran the disaccharide sucrose serves as the substrate. The energy of the glycosidic bond between glucose and fructose in this disaccharide drives the formation of an $\alpha(1,6)$ polymer of glucose according to the following reaction:

$$n \text{ Sucrose} \xrightarrow[\textbf{Sucrase}]{\textbf{Dextran}} \text{Dextran} + n \text{ Fructose}$$

Dextrans formed by bacteria growing on the surface of teeth are an important component of dental plaque.

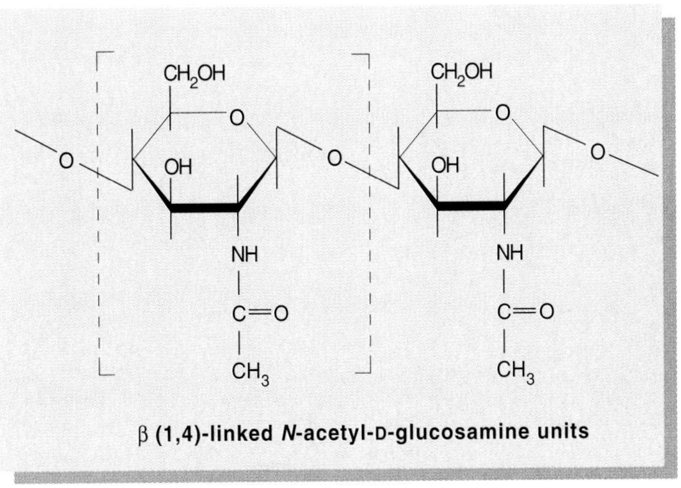

Heteropolysaccharides Contain More Than One Building Block

A large variety of carbohydrates contain either modified sugars, like the *N*-acetylglucosamine found in chitin, or two or more different sugars in straight-chain or branched-chain linkages.

Structurally, the simplest and best known of the heteropolysaccharides are the glycosaminoglycans. These are long, unbranched polysaccharide chains composed of repeating disaccharide subunits in which one of the two sugars is either *N*-acetylglucosamine or *N*-acetylgalactosamine (table 18.2). Glycosaminoglycans are highly negatively charged because of the presence of

Figure 18.8

Structure of the repeating unit of chitin: a $\beta(1,4)$ homopolymer of *N*-acetyl-D-glucosamine.

Table 18.2
Structure of Glycosaminoglycans

Polysaccharide	Monosaccharide Units[a] *A*	Monosaccharide Units[a] *B*	Susbsituents	Repeating Unit
Hyaluronate	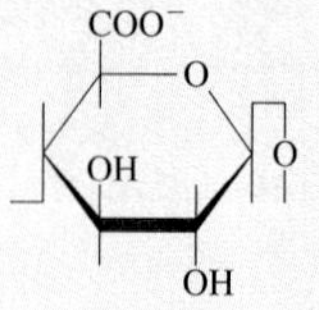 β-D-GlcUA	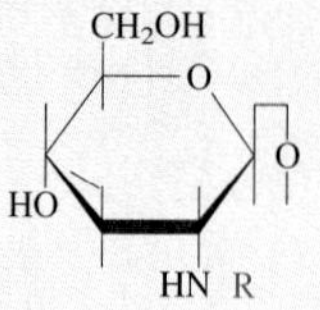 β-D-GlcN	$R = -C(=O)CH_3$	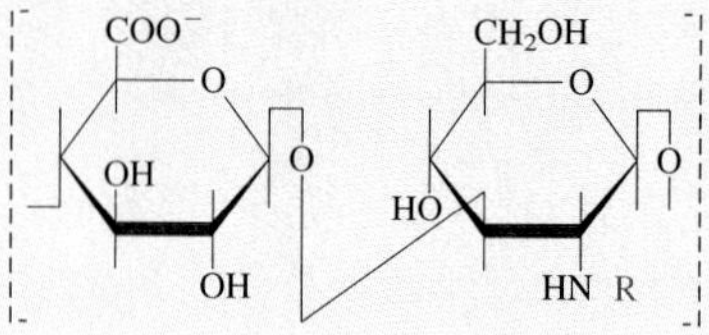
Chondroitin sulfates Dermatan sulfate	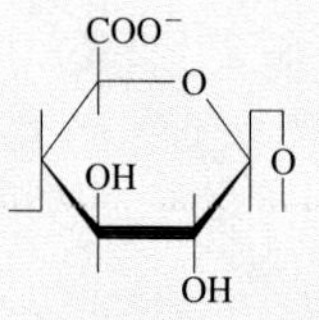 β-D-GlcUA 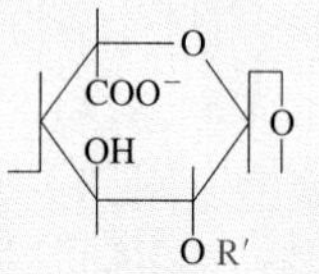 α-L-IdUA	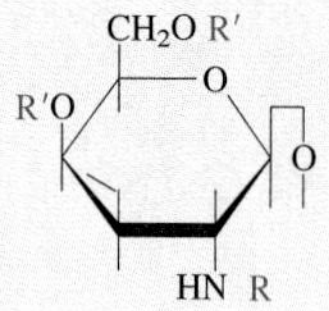 β-D-GalN	$R = -C(=O)CH_3$ $R' = -H$ or $-SO_3^-$	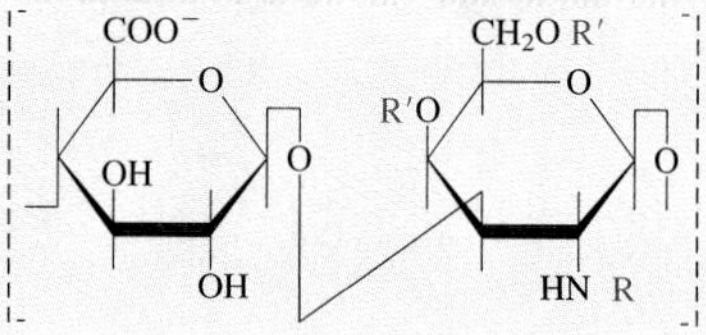
Heparan sulfate and heparin	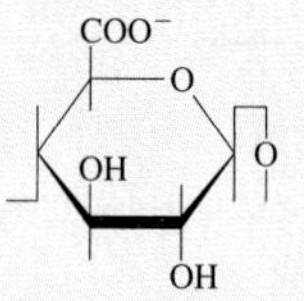 β-D-GlcUA 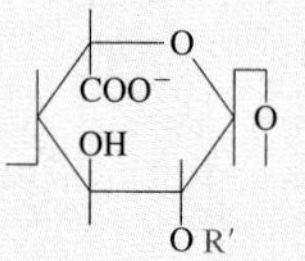 α-L-IdUA	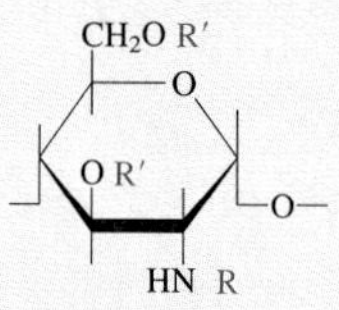 α-D-GlcN	$R = -C(=O)CH_3$ $R' = -H$ or $-SO_3^-$	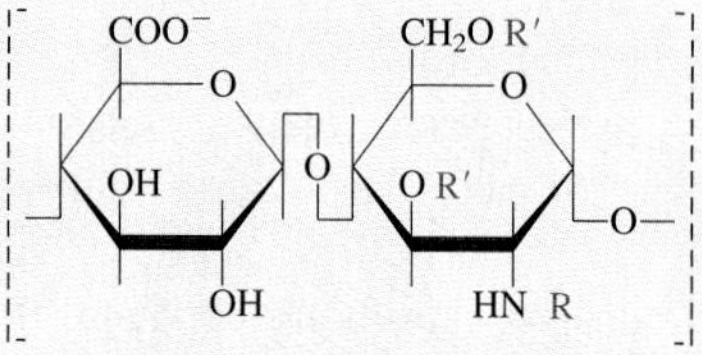
Keratan sulfate	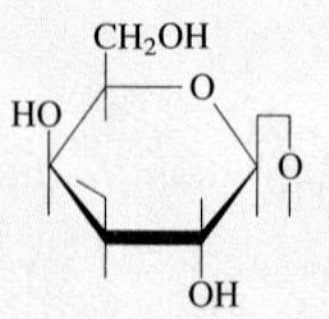 β-D-Gal	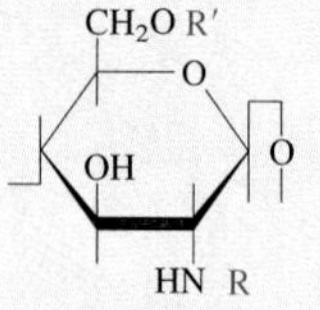 β-D-GlcN	$R = -C(=O)CH_3$ $R' = -H$ or $-SO_3^-$	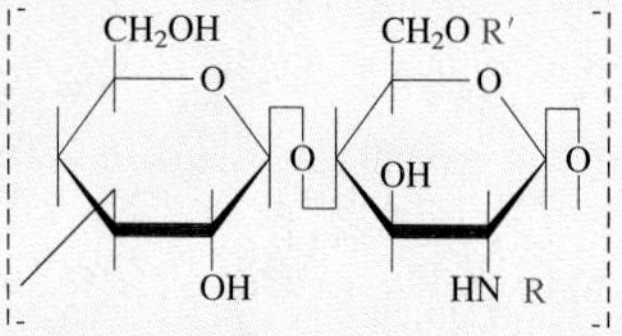

[a]The polysaccharides are depicted as linear polymers of alternating A and B monosaccharide units.
Abbreviations: GlcUA, glucuronic acid; IdUA, iduronic acid; GalN, galactosamine; GlcN, glucosamine; Gal, galactose.

Figure 18.9

Structure of the repeating unit of hyaluronic acid: GlcUAβ(1,3)GlcNAcβ(1,4).

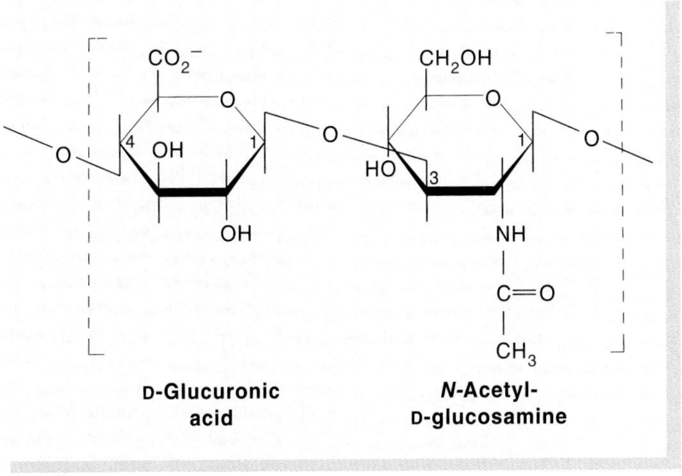

carboxyl or sulfate groups on many of the sugar residues. The high negative charge causes the polymeric chains to adopt a stretched or extended conformation. Their extended structure gives a high viscosity to the surrounding region, even in a dilute solution of the polysaccharide. Glycosaminoglycans are usually found in extracellular space in multicellular organisms, where they produce a viscous extracellular matrix that resists compression. Such an environment can be beneficial to the organism in various ways; it can provide a passageway for cell migration, supply lubrication between joints, or help maintain certain structural shapes such as the ball of the eye.

Hyaluronic acid is a copolymer of D-glucuronic acid and *N*-acetyl-D-glucosamine (fig. 18.9). It is formed by the alternating addition of UDP-glucuronic acid and UDP-*N*-acetylglucosamine to the ends of a growing chain. It is much larger than other glycosaminoglycans, reaching molecular weights in excess of 10^6.

Proteoglycans Are Complexes of Proteins with Glycans

Most glycosaminoglycans are linked to a core protein as lateral extensions, forming a proteoglycan with a highly extended, brush-like structure (fig. 18.10). The linkage between the glycosaminoglycans and the core protein within the proteoglycan is mediated by a specific trisaccharide unit that is linked on one side to the repeating disaccharide unit of the glycosaminoglycan and on the other side to a serine hydroxyl group of the core protein (see fig. 18.10). In the extracellular matrix, hyaluronic acid and the other proteoglycans form aggregates with each other as well as with other macromolecular components, such as serum glycoproteins, growth factors, collagen, elastin, or the outer plasma membrane of cells. Hyaluronic acid, when complexed with other proteoglycans through their core proteins, produces very large complexes (fig. 18.11).

Thus far we have seen that glycosaminoglycans and their proteoglycans form highly extended aggregates that impart a rigid, gel-like structure to the extracellular matrix. It seems likely that many more specific reactions occur involving these compounds. However, it should be emphasized that relatively little is known about the organization or reactions of these molecules in the extracellular matrix. One exception is the case of the glycosaminoglycan heparin. We know that this compound functions as a highly

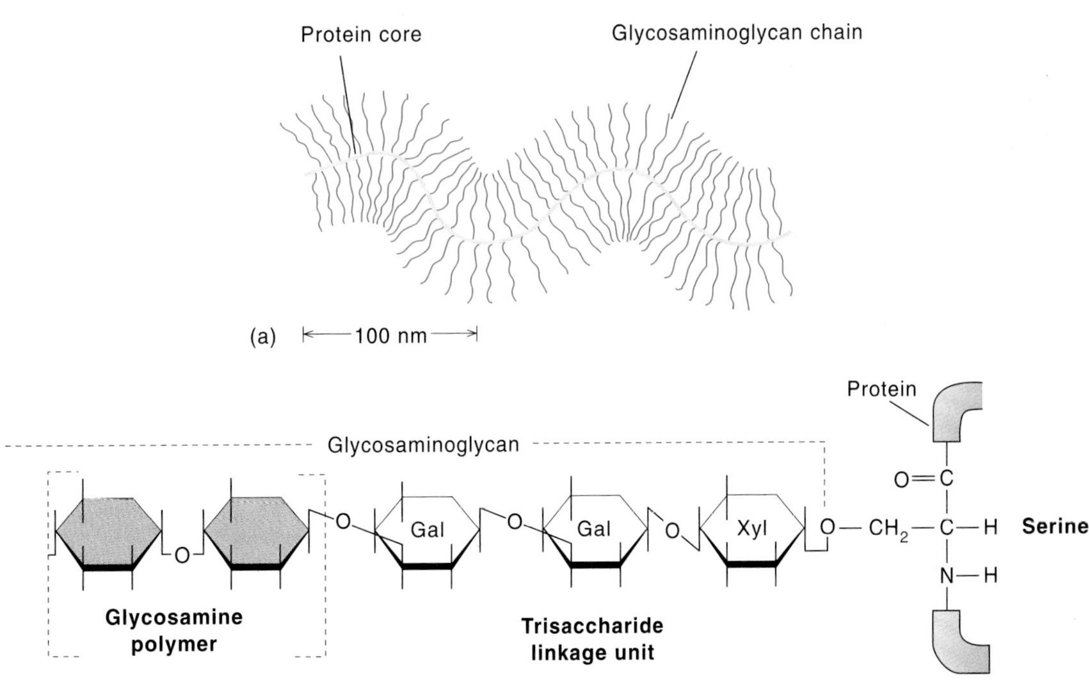

Figure 18.10

(*a*) Structure of a typical proteoglycan. (*b*) The attachment site between a serine in the core protein and the glycosaminoglycan.

Figure 18.11

A hyaluronic acid–proteoglycan complex. The individual proteoglycans contain a core protein by which are linked various glycosaminoglycans as shown in figure 18.10. The core protein is noncovalently associated with a single hyaluronic acid molecule via two link proteins. Electron micrograph below shows a large aggregate (*A*) and a small aggregate (*B*) whose sizes are determined mainly by the length of the hyaluronate central filament of the individual aggregates. (From Buckwalter, J. A. and Rosenberg, L. Coli. Rel. Res. (1983) 3, 489–504.)

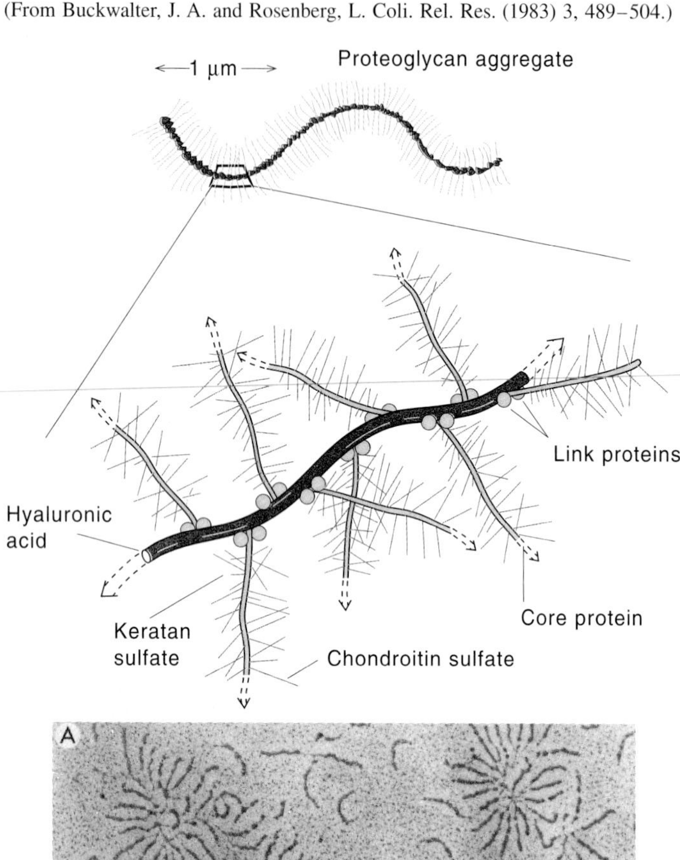

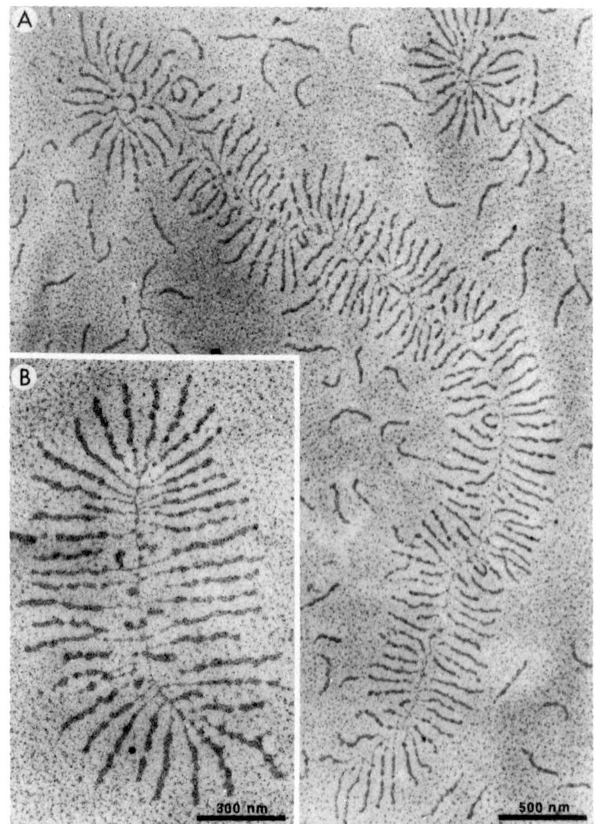

specific anticoagulant. First it forms a complex with the plasma protein antithrombin III. This complex in turn inhibits the serine proteases of the blood-clotting system.

In Glycoproteins, Oligosaccharides Are Covalently Linked to N or O Atoms in Protein Amino Acid Side Chains

Proteins are frequently adorned by straight-chain or branched oligosaccharides, in which case they are called glycoproteins. This type of modification can serve a variety of functions. It can stabilize the protein, facilitate its correct folding, be part of a lipid anchor for attaching the protein to a membrane, or provide the protein with characteristics that allow it to be recognized by a carbohydrate-binding protein.

The carbohydrate moiety in a glycoprotein is attached to the polypeptide chain by a covalent connection to certain amino acid side chains. Two types of glycosidic linkages are commonly found: the O-glycosidic linkage involves attachment of the carbohydrate to the hydroxyl group of serine, threonine, or hydroxylysine; the N-glycosidic linkage involves attachment to the amide group of asparagine (fig. 18.12). Conformational constraints appear to be a major factor in determining the addition, although other factors also play a role. Thus in the case of O-linked sugars, clusters of serines or threonines appear to be a preferred substrate. For N-glycosidic linkage, only asparagine residues of the sequence Asn-X-Ser(Thr) will accept carbohydrates. In this sequence, X may be any amino acid except proline.

Glycoproteins are found in all cellular compartments and are also secreted by the cell. Those found in the cytoplasm have a simple modification, consisting of a single *N*-acetylglucosamine that is O-glycosidically linked to Ser(Thr). Collagen, a secreted glycoprotein of the extracellular matrix, also has simple carbohydrates — the disaccharide Glcβ(1,2)Gal linked to hydroxylysine.

Most secreted and membrane-bound glycoproteins have more complicated oligosaccharides containing between 4 and 30 sugar residues. A typical O-linked carbohydrate of a membrane glycoprotein is shown in figure 18.13. This structure can be further extended and branched by the addition of galactose, *N*-acetylgalactosamine, *N*-acetylglucosamine, fucose, or sialic acid in different linkages. Larger O-linked structures with branching occur on proteins with a high Ser(Thr) content, forming mucins. These molecules are found in body fluids and carry the blood group determinants (see fig. 18.25).

In contrast to the usual O-linked oligosaccharides of mammalian glycoproteins, N-linked carbohydrates always contain mannose and *N*-acetylglucosamine and may also contain galactose, fucose, and sialic acid in combinations that vary as a result of differences in branching and linkage relationships. Three general types of structure form the basis of the variations in N-linked carbohydrates. These are termed oligomannosyl, hybrid, and lactosamine-containing (or complex) structures (see fig. 18.13). They all have a common core of five sugars attached to Asn because they are all synthesized by a common pathway. In fact, the wide

Figure 18.12

Linkages found between oligosaccharides and proteins in glycoproteins. (*a*) An O-glycosidic linkage found in many glycoproteins and mucins; an analogous linkage occurs between *N*-acetylglucosamine and serine in glycoproteins. (*b*) An N-glycosidic linkage found in many glycoproteins. (*c*) An O-glycosidic linkage found only in collagen.

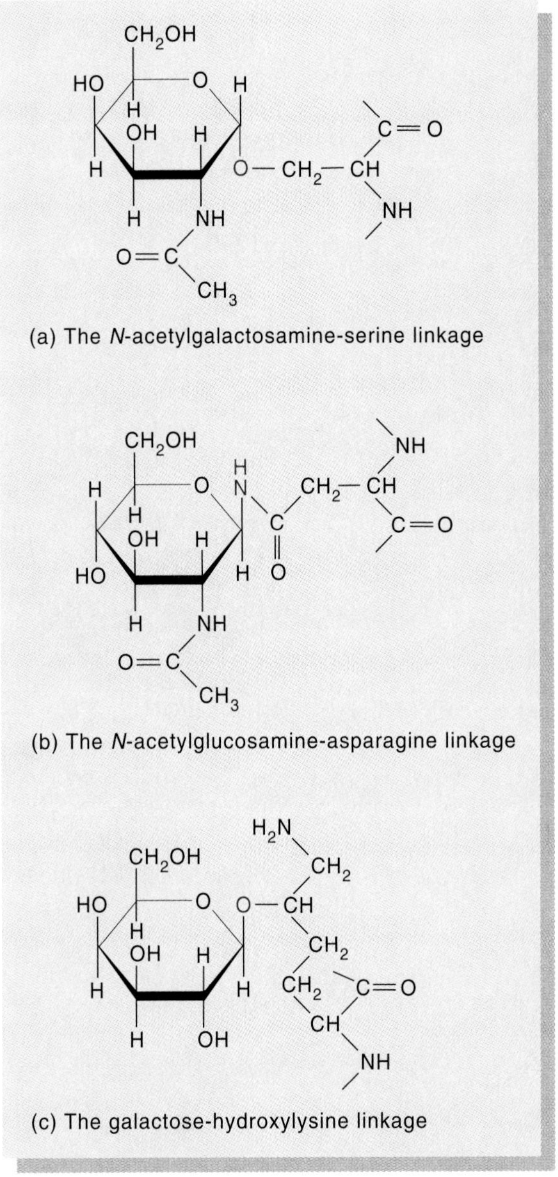

(a) The *N*-acetylgalactosamine-serine linkage

(b) The *N*-acetylglucosamine-asparagine linkage

(c) The galactose-hydroxylysine linkage

variety of N-linked carbohydrates found in glycoproteins reflects intermediates of the biosynthetic pathway. The particular structures associated with a completed glycoprotein are a result of the conformation of that protein during biosynthesis, the availability of specific glycosyltransferase enzymes in the host cell, and the speed with which the glycoprotein travels through the secretory pathway along which glycosylation enzymes are located (see below).

Yeast glycoproteins carry only the oligomannosyl type of N-linked carbohydrates with mature structures that may contain hundreds of mannose residues. Thus far, such structures have been found only in yeasts. Yeasts do not synthesize hybrid or lactosamine-containing structures, although the initial steps of N-linked carbohydrate biosynthesis appear to be the same in all eukaryotes (see below).

Carbohydrate Modification Is Important in Targeting Certain Enzymes to the Lysosomes

Soluble lysosomal hydrolases are targeted to lysosomes by a specific carbohydrate recognition marker that they acquire in the Golgi complex. Oligomannosyl carbohydrates on soluble lysosomal enzymes carry a phosphate residue at the 6-position of one or more mannoses (Man-6-P). These phosphorylated mannose residues are recognized by a glycoprotein called the Man-6-P receptor, which binds and transports lysosomal enzymes via several cellular compartments. In an acidic, prelysosomal compartment the binding between the Man-6-P receptor and the lysosomal enzyme is disrupted. The lysosomal enzyme continues in a vesicle destined to fuse and thereby deliver its contents to the lysosome. Fibroblasts from patients with a lysosomal storage disease called I-cell disease cannot add the carbohydrate recognition marker and consequently their lysosomal hydrolases are largely secreted instead of being targeted to the lysosome. As a result, many molecules that are normally degraded by lysosomal hydrolases accumulate in the lysosomes. Morphologists have termed these dense lysosomes inclusion bodies, hence the name I-cell disease.

Thus far, no other cases of protein targeting by carbohydrate modification are known, although this is a distinct possibility.

A Carbohydrate-Lipid Serves to Anchor Some Glycoproteins to the Cell Surface

Both yeast and higher eukaryotes share an unusual carbohydrate modification that is found at the C-terminal end of several diverse types of glycoproteins located at the extracellular surface. The oligosaccharide is unique in structure and is linked to the protein via phosphatidylethanolamine. A glucosamine residue of the oligosaccharide is, in turn, linked to phosphatidylinositol, which is attached to two fatty acid chains (diacylglycerol) that anchor the whole molecule in the plasma membrane. This complex modification is termed a glycosylphosphatidylinositol (GPI) anchor, and proteins that carry such an anchor are said to be glypiated. The structure of a GPI anchor from a mammalian cell surface glycoprotein is shown in figure 18.14. The presence of this anchor in a glycoprotein may be ascertained by treatment with nitrous acid, which cleaves the glucosamine-*myo*-inositol linkage, or with phospholipases that cleave the link between *myo*-inositol and diacylglycerol. GPI-anchored molecules can also be released from the cell surface by the action of a simple phospholipase. This suggests that a possible function of glypiation is to provide a mechanism for regulating the concentration of the glypiated glycoprotein at the cell surface and in tissue fluids.

In the human disease called paroxysmal nocturnal hemoglobinuria, many molecules that are normally membrane-anchored by glypiation are found free in the blood. One of these molecules is decay-accelerating factor whose action at the cell surface prevents red blood cell lysis by complement. In its absence

Figure 18.13

Representative carbohydrate structures found on mammalian glycoproteins. (*a*) An O-glycosidically linked carbohydrate from a red blood cell membrane glycoprotein. (*b*) An oligomannosyl N-linked carbohydrate. (*c*) A hybrid N-linked carbohydrate. (*d*) A complex, lactosamine-containing carbohydrate. The latter three structures are found on many membrane and secreted glycoproteins of mammalian cells. Structure (*d*) may be much more complex, with the addition on the core mannose residues of extra branches of various lengths terminating with different sugars. It should be noted that all the N-linked carbohydrates (structures *b, c,* and *d*) have a common core of five sugars. (NeuNAc = sialic acid; Gal = galactose; GlcNAc = *N*-acetylglucosamine; Man = mannose; Fuc = fucose; Asn = asparagine; Ser = serine.)

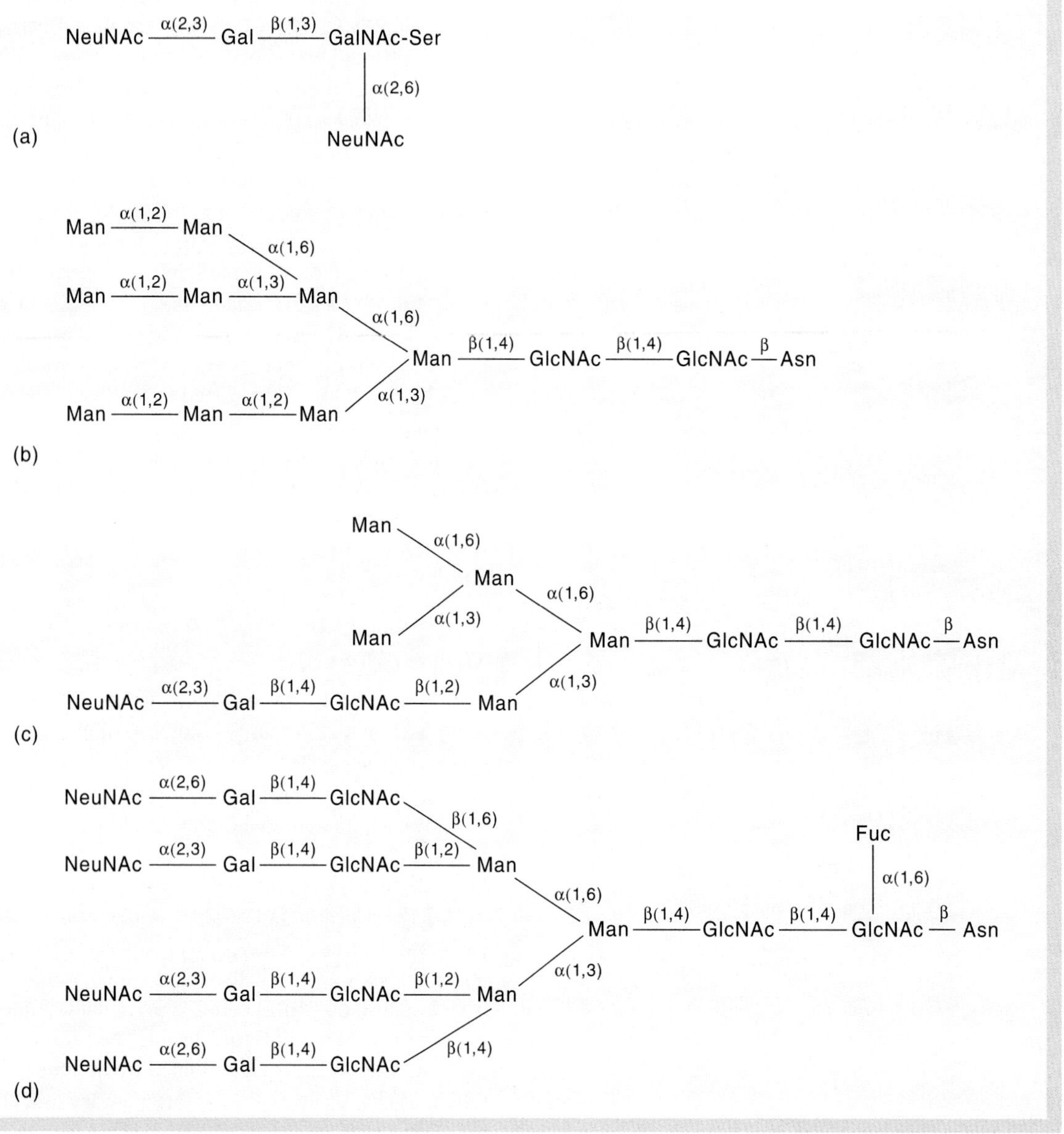

(a)
NeuNAc —$\alpha(2,3)$— Gal —$\beta(1,3)$— GalNAc-Ser
$\alpha(2,6)$
NeuNAc

(b)
Man —$\alpha(1,2)$— Man
$\alpha(1,6)$
Man —$\alpha(1,2)$— Man —$\alpha(1,3)$— Man
$\alpha(1,6)$
Man —$\beta(1,4)$— GlcNAc —$\beta(1,4)$— GlcNAc —β— Asn
$\alpha(1,3)$
Man —$\alpha(1,2)$— Man —$\alpha(1,2)$— Man

(c)
Man
$\alpha(1,6)$
Man
$\alpha(1,6)$
$\alpha(1,3)$
Man
Man —$\beta(1,4)$— GlcNAc —$\beta(1,4)$— GlcNAc —β— Asn
$\alpha(1,3)$
NeuNAc —$\alpha(2,3)$— Gal —$\beta(1,4)$— GlcNAc —$\beta(1,2)$— Man

(d)
NeuNAc —$\alpha(2,6)$— Gal —$\beta(1,4)$— GlcNAc
$\beta(1,6)$
Fuc
NeuNAc —$\alpha(2,3)$— Gal —$\beta(1,4)$— GlcNAc —$\beta(1,2)$— Man
$\alpha(1,6)$
$\alpha(1,6)$
Man —$\beta(1,4)$— GlcNAc —$\beta(1,4)$— GlcNAc —β— Asn
$\alpha(1,3)$
NeuNAc —$\alpha(2,3)$— Gal —$\beta(1,4)$— GlcNAc —$\beta(1,2)$— Man
$\beta(1,4)$
NeuNAc —$\alpha(2,6)$— Gal —$\beta(1,4)$— GlcNAc

from the membrane, much lysis occurs, leading to the presence of hemoglobin in the urine.

The GPI anchor serves to locate glycoproteins on the outer leaflet of the plasma membrane, where they are significantly more mobile than membrane proteins that span the bilayer. These glycoproteins are thus more accessible to other extracellular molecules and are more readily released by a phospholipase.

Carbohydrates of the Plasma Membrane Are Important in Cell Recognition

GPI-anchored proteins are not the only molecules that possess carbohydrates on the outside leaflet of the plasma membrane. Carbohydrates also appear prominently on the outside leaflet of the plasma membrane in the form of N- and O-linked

Figure 18.14

Glycosylphosphatidylinositol anchor of the membrane glycoprotein Thy 1. Many enzymes and receptors at the cell surface are anchored in the membrane via the diacylglycerol (DAG) portion of phosphatidylinositol, which is linked to a glycan that is in turn linked to the carboxyl-terminal amino acid of a protein via phosphoethanolamine. The oligosaccharide structure of the glycan region is quite different from that of O-linked or N-linked oligosaccharides. Most novel is the presence of glucosamine instead of *N*-acetylglucosamine. The bond between the glucosamine and *myo*-inositol can be cleaved by nitrous acid, allowing identification of proteins that carry this modification. In addition, the glycan and protein can be released from diaclyglycerol by cleavage with various phospholipase enzymes.

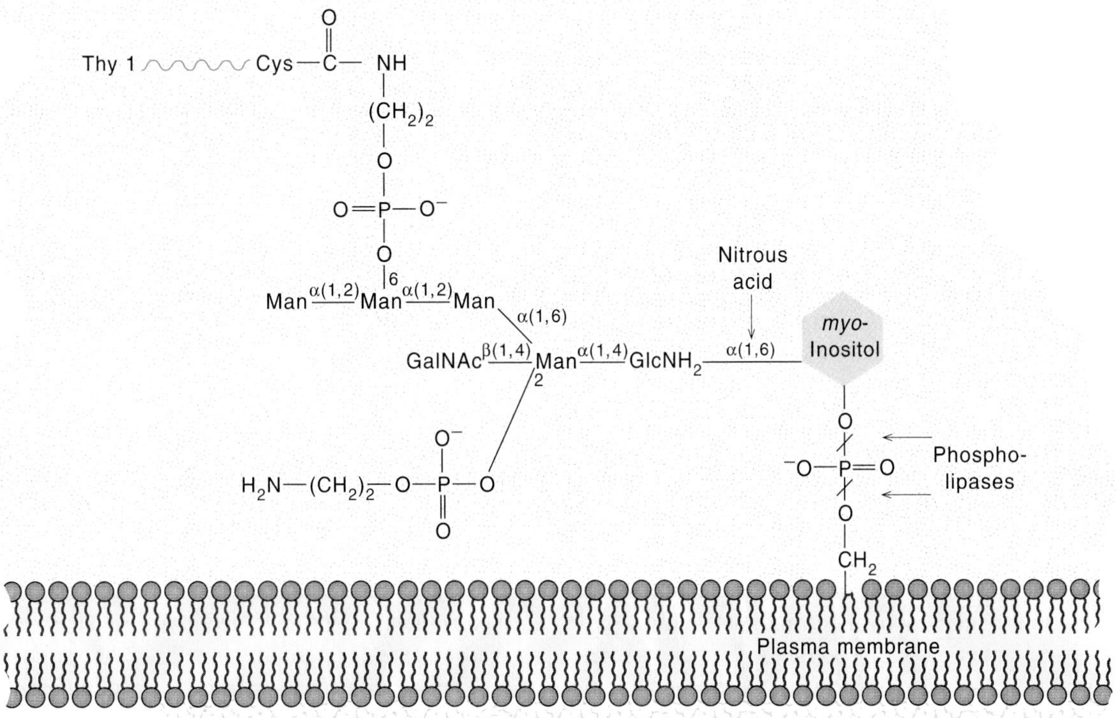

glycoproteins, glycolipids, and proteoglycans (figure 18.15). In addition to being accessible for recognition by carbohydrate-binding proteins, cell-surface carbohydrates also appear to be important in cell–cell interactions. Many infectious agents, such as bacteria, viruses, or parasites, recognize and bind to host cells via specific carbohydrate structures. For example, influenza virus has a glycoprotein termed hemagglutinin (because it causes agglutination or clumping of red blood cells), which binds to cells via the sialic acid residues of cell surface glycoconjugates (glycoproteins or glycolipids). Some hemagglutinins bind only to sialic acid linked $\alpha(2,3)$ to galactose, while others bind only to sialic acid linked $\alpha(2,6)$ to galactose. These binding specificities are due to the presence of a particular amino acid at a single position in the sialic acid binding site of the hemagglutinin. Such exquisite specificity shows the potential for the functional consequences of changes in carbohydrate structure. Other important roles of carbohydrate as cell recognition markers include the binding of cholera toxin to a specific glycolipid, GM_1; the binding of terminal sialylated, fucosylated lactosamine sugar sequences by cellular adhesion molecules; and the recognition of mammalian eggs

Figure 18.15

Diagram of a eukaryotic cell plasma membrane. Oligosaccharides are found on integral, transmembrane glycoproteins, glycolipids, glycophosphatidylinositol (GPI)-anchored glycoproteins, and glycoconjugates adsorbed at the cell surface. All carbohydrates on these molecules face the outside of the cell except O-linked *N*-acetylglucosamine, which may be found on the cytoplasmic portion of glycoproteins. Since many carbohydrates terminate in sialic acid, the external surface of the plasma membrane is negatively charged. The oligosaccharides of membrane glycoconjugates form a sugar coat, "glycocalyx," for the mammalian cell, which can be readily seen in the electron microscope by staining with ruthenium red or other sugar-binding dyes. Proteoglycans are found in the extracellular matrix and provide the support substance for cells in all tissues. Proteoglycan chains have also been found on certain membrane glycoproteins. Many membrane glycoproteins are receptors that bind other glycoproteins (e.g., growth factors), absorbing them to the cell surface.

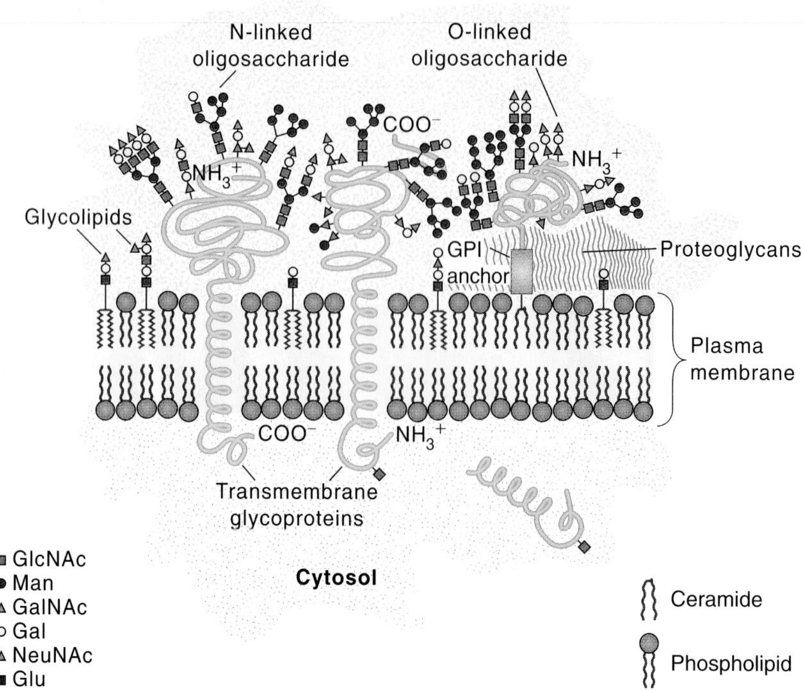

by sperm. It is likely that the species specificity of egg fertilization is due to the inability of sperm from one species to recognize the oligosaccharides of glycoproteins in the zona pellucida that surrounds the eggs of another species.

Proteins that bind specifically to carbohydrates in glycoconjugates are termed lectins. Vertebrate lectins are proteins of nonimmune origin that bind carbohydrates and may cause agglutination of cells or precipitation of glycoproteins. Some vertebrate lectins are integral membrane proteins, while others are soluble glycoproteins that are found in tissue fluids. Correlative evidence suggests several biological functions for vertebrate lectins, including the clearance of aged red blood cells from the circulation, tissue-specific localization of lymphoid cells (box 18A), and formation of specific connections between nerve cells.

Determination of Carbohydrate Primary Structure Requires Purification before Structural Analysis

In order to correlate the structure of carbohydrates with their function, we must determine the precise arrangements of sugars in an oligosaccharide. For glycoproteins, often we must also determine the types of oligosaccharide structures at each glycosylation site. To do this, we first separate protease-generated fragments of the protein each containing one glycosylation site, then release the oligosaccharide from the peptide. Oligosaccharide moieties can be obtained free of protein by chemical or enzymic release from the protein backbone or by exhaustive proteolysis to give glycopeptides. Oligosaccharide mixtures can be separated on the basis of size or charge by standard chromatographic procedures and on the basis of structure by lectin-affinity chromatography (fig. 18.16). Plant seeds are the source of lectins for chromatography. For example, concanavalin A from the jack bean binds oligomannosyl N-linked carbohydrates but does not bind O-linked moieties or branched N-linked structures. Other lectins that are used in lectin-affinity chromatography include wheat germ agglutinin (binds sialic acid and *N*-acetylglucosamine), ricin (binds galactose), and lotus lectin (binds fucose).

Studies using plant lectins as reagents that detect changes in carbohydrate structures have been instrumental in showing that carbohydrate structures at the cell surface change during embryonic development and, later, during differentiation of cells in the maturing organism. In the adult, certain oligosaccharides are synthesized in a tissue-restricted manner. Such developmentally regulated structures are precisely the carbohydrates that are reexpressed when a cell becomes cancerous. It is likely that vertebrate lectins exist that recognize these carbohydrate changes and initiate a biological response. Progress in identifying the biological consequences of carbohydrate structural changes, in relation to certain biological states, should improve in the next few years with the cloning of many vertebrate lectins as well as glycosyltransferases—the enzymes that catalyze the synthesis of mammalian oligosaccharides. Now that clones encoding these molecules are available, it is possible to investigate causal relationships by trying various strategies for abrogating gene activity or causing molecules to be inappropriately expressed.

Once an oligosaccharide has been purified, the sequence of sugars can be determined by sequential digestion with exoglycosidase enzymes (fig. 18.17*a*). These are hydrolases that remove

Selectins Are Cell Adhesion Molecules with a Lectin Domain That Functions in Cell Recognition

Since many different molecules with lectin domains have now been identified, it appears that enormous potential is available to the cell for utilizing carbohydrate–protein interactions for physiological recognition events. For example, consider the well-characterized endothelial cell selectin known as E-selectin. This glycoprotein is synthesized after endothelial cells are activated by a cytokine and is expressed within 4–6 h as a transmembrane molecule on the cell surface. E-selectin functions to bind leukocytes and facilitate their transport from the bloodstream across the endothelial cell layer of postcapillary venules to sites of inflammation or tissue damage. The means by which E-selectin recognizes the correct subset of leukocytes from all the cells in the bloodstream is at least in part via their carbohydrates. The structure that appears to be necessary and sufficient for E-selectin binding in vitro is NeuNAca(2,3)Galβ(1,4)[Fuca(1,3)]GlcNAc(β1) (see fig. 1). However, there is evidence that E-selectin ligands are shed from the leukocyte cell surface; if so, the physiological ligand may be a particular glycoconjugate that carries this carbohydrate structure. This carbohydrate is an example of a developmentally regulated structure that is expressed at different stages of ontogeny and is often reexpressed in cancer cells. Its availability at the cell surface may be controlled by the regulated expression of a specific α(1,3)fucosyltransferase gene.

Two other selectin molecules, termed P-selectin (from platelets) and L-selectin (from lymphocytes), have the same general molecular features as E-selectin. All the selectins recognize the specific terminal group of sugars shown and related structures that contain ducose, sialic acid and/or sulfate. In the animal these sugars are found on O-linked structures that are clustered on mucinlike membrane-associated glycoproteins. Therefore selectins recognize specific carbohydrate structures attached to particular proteins.

Figure 1

Arrangement of functional units in a typical selectin.

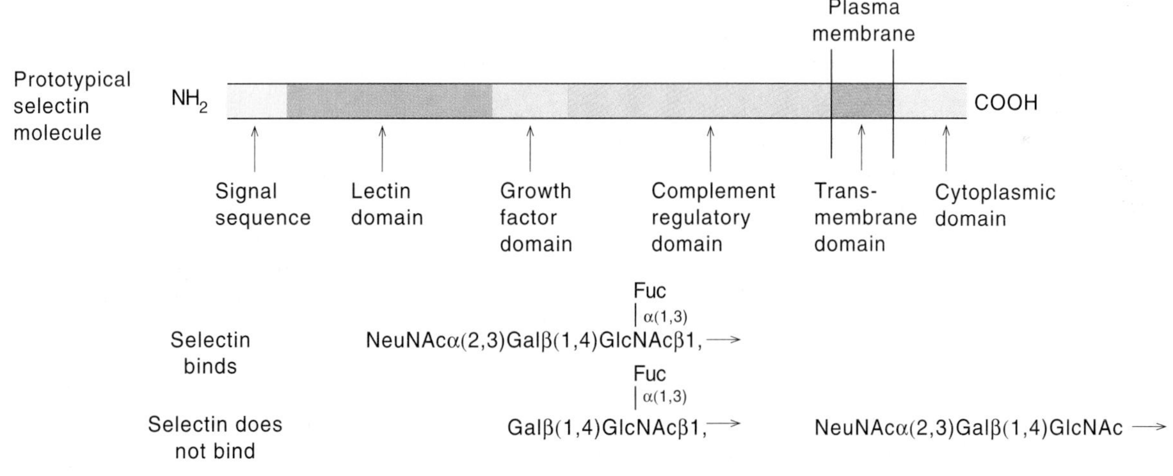

terminal, nonreducing sugars. They are highly specific for the particular sugar and its anomeric linkage (α or β). Some glycosidases are also specific for the type of linkage, cleaving a β(1,2), for example, but not a β(1,4) linkage. Thus by sequential exoglycosidase digestion in combination with methods for detecting sugar removal (conventional or lectin-affinity chromatography), it is possible to determine the general structure of an oligosaccharide. Endoglycosidases are also most useful for carbohydrate structural analysis because they are able to recognize certain structural elements. Different endoglycosidases (D, F, or H) can be used to distinguish between complex, oligomannosyl, or hybrid structures as shown in figure 18.17*b*. The composition of an oligosaccharide can be determined after acid hydrolysis by gas chromatography or by separation of sugar oxyanions formed at high pH. The

Figure 18.16

Separation of N-linked carbohydrates by lectin-affinity chromatography. Radiolabeled carbohydrates are released from a glycoprotein by an endoglycosidase (*N*-glycanase) that cleaves between the Asn of the protein and the first GlcNAc residue of the carbohydrate. The carbohydrates can then be separated into branched or biantennary (two branches) lactosamine-containing species or oligomannosyl species by affinity chromatography on concanavalin A-sepharose. Branched carbohydrates do not bind to the column (nor do O-linked oligosaccharides), biantennary carbohydrates bind and are eluted by 10-mM α-methylglucoside(α-MG), while oligomannosyl carbohydrates bind more strongly to the column and are eluted by increasing concentrations of α-methylmannoside (α-MM; 10 mM and 100 mM). Although several types of oligosaccharides are completely separated by this method, each peak may include a mixture of related structures. For example, hybrid structures that may contain GlcNAc, Gal, and sialic acid attached to the $Man_5GlcNAc_2$ shown are also eluted with α-MM(10).

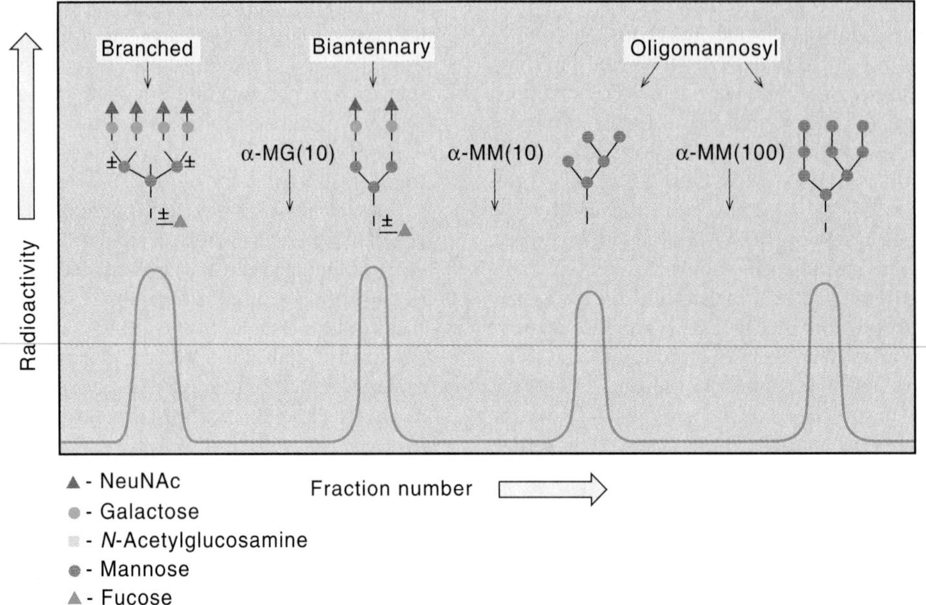

types of linkages between sugars can be determined by methylation analysis as described in figure 13.11. With combined information a complete structure can usually be deduced.

A much more rapid method of structural analysis is possible if $\geq$50 μg of pure oligosaccharide is available. This is high field proton (^{1}H) nuclear magnetic resonance (NMR) spectroscopy. NMR spectroscopy is nondestructive and gives a spectrum in which one to three protons of each sugar—for example, the anomeric carbon (C-1) proton or the C-2 proton or methyl group protons—are resolved. The resolution improves dramatically with two-dimensional (2D) NMR, allowing all protons in small oligosaccharides to be resolved. An example of a partial NMR spectrum of an oligosaccharide is interpreted in box 18B. Basic aspects of the NMR technique were explained in chapter 5.

In the discussion that follows, we will see that the rich variety of oligosaccharides found in glycoconjugates is reflected by the large number of enzymes involved in their synthesis.

Oligosaccharides Are Synthesized in a Concerted Fashion by Specific Glycosyltransferases

All oligosaccharide structures of glycoconjugates are synthesized by glycosyltransferases according to the following general reaction:

Nucleotide-sugar (Donor glycose) + Glycoconjugate $\rightleftharpoons$ Glycose-*O*-glycoconjugate (Product) + Nucleotide diphosphate

The reaction is driven towards the formation of product by the hydrolysis of the released nucleotide diphosphate. Most glycosyltransferase enzymes require a divalent cation (usually manganese), and most are highly specific for the acceptor substrate, nucleotide-sugar, and linkage by which the new sugar is attached to its acceptor. For example, there are two different enzymes that attach GlcNAc in β(1,2) linkage to mannose for the synthesis of a biantennary N-linked carbohydrate (fig. 18.18). This is because the carbohydrate acceptor to which the GlcNAc is added differs markedly in each case. The specificity of glycosyltransferases is generally such that with few exceptions there exists one unique enzyme for each and every type of glycosidic bond known.

In the biosynthesis of oligosaccharides, each sugar is added in a stepwise and orderly fashion to the previous sugar. For example, in the case of O-glycosidically linked carbohydrates attached to Ser or Thr, the first sugar is added directly to the protein by a specific *N*-acetylgalactosaminyl glycosyltransferase. The second sugar is subsequently added by a different glycosyltransferase and so on until a mature structure is obtained. The final

Figure 18.17

Glycosidase enzymes used for structural analysis of oligosaccharides. (*a*) Exoglycosidase enzymes remove terminal, nonreducing sugars with specificity for the sugar and whether it is in α or β linkage. Used sequentially and in conjunction with various separation techniques, exoglycosidases reveal the sequence of sugars in an oligosaccharide. Endoglycosidases may be used to remove the intact oligosaccharide from the protein backbone. The enzyme that cleaves the GlcNAc–Asn bond is an amidase or peptide-*N*-glycosidase. It requires the glycosylated Asn to be substituted with an amino acid on both sides (that is, to be within a peptide) for its action. Upon removal of the oligosaccharide the Asn is converted to aspartic acid in the peptide. (*b*) Each endoglycosidase cleaves between the core GlcNAc residues. However, different endoglycosidases vary in their specificity for cleaving N-linked structures, as noted in the diagram. In every case the GlcNAc residue next to the Asn is left attached. The endoglycosidases act on glycopeptides in which the Asn need not be substituted.

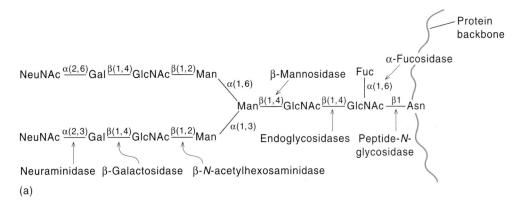

Protein backbone
α-Fucosidase
NeuNAc α(2,6) Gal β(1,4) GlcNAc β(1,2) Man
β-Mannosidase
Fuc
α(1,6)
α(1,6)
Man β(1,4) GlcNAc β(1,4) GlcNAc β1 Asn
α(1,3)
NeuNAc α(2,3) Gal β(1,4) GlcNAc β(1,2) Man
Endoglycosidases
Peptide-*N*-glycosidase
Neuraminidase
β-Galactosidase
β-*N*-acetylhexosaminidase
(a)

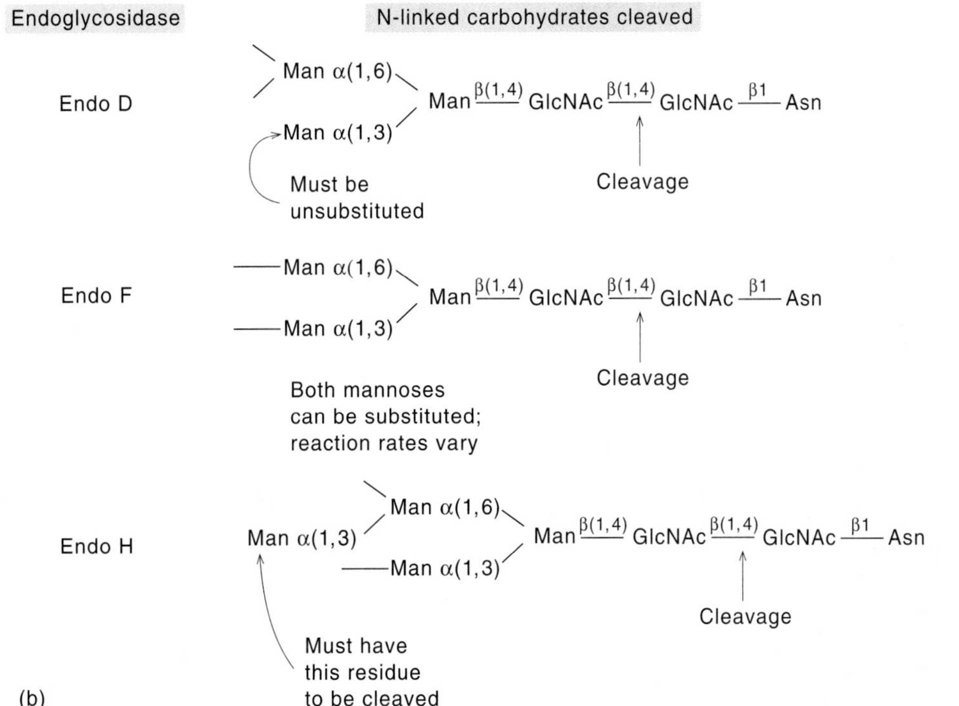

Structural Analysis of an Oligosaccharide by Nuclear Magnetic Resonance (NMR) Spectroscopy

Nuclear magnetic resonance (NMR) spectroscopy exploits the fact that when a spinning, paramagnetic, charged particle (e.g., a proton) is placed in a magnetic field, it aligns mainly with the field and precesses about the field with a frequency (the Lamar frequency) dependent on the particle properties and the strength of the magnetic field (see chapter 5). To obtain an NMR spectrum, a sample of protons is placed in a strong magnetic field (generated by the magnet of the NMR spectrometer) and is irradiated with a range of radiofrequency energies at 90° to the main field. This treatment causes all the protons in the sample to absorb energy at their characteristic frequency, flipping their magnetic orientations with respect to their original state. After the applied pulse field is switched off, the protons gradually relax to precess about the main field. Receiver coils in a probe surrounding the sample detect the frequencies of precessing protons as a set of oscillating electric currents, induced by the precessing magnetic vectors, which constitute the NMR signal. The magnitude of the induced voltage decays exponentially, giving rise to a free induction decay (FID). An FID is actually a mixture of sine waves arising from each of the chemical classes of protons in the sample. The FID is called the time domain of the NMR signal. Fourier transformation decodes the frequencies in the FID so they can be displayed in a plot of amplitude (amount) versus frequency.

The Lamar frequency of a proton is precisely dictated by its chemical environment and is expressed as a chemical shift in parts per million (ppm). For oligosaccharides, therefore, the chemical shift of an anomeric proton of a particular sugar (e.g., galactose) will vary depending on the structure of the oligosaccharide in which the sugar exists (see fig. 1). Powerful NMR spectrometers can resolve many of the protons in a complex oligosaccharide. By identifying the chemical shifts of specific protons for each sugar in oligosaccharides of known structure (determined by other methods), data banks have been acquired that allow complete structures to be deduced.

Figure 1

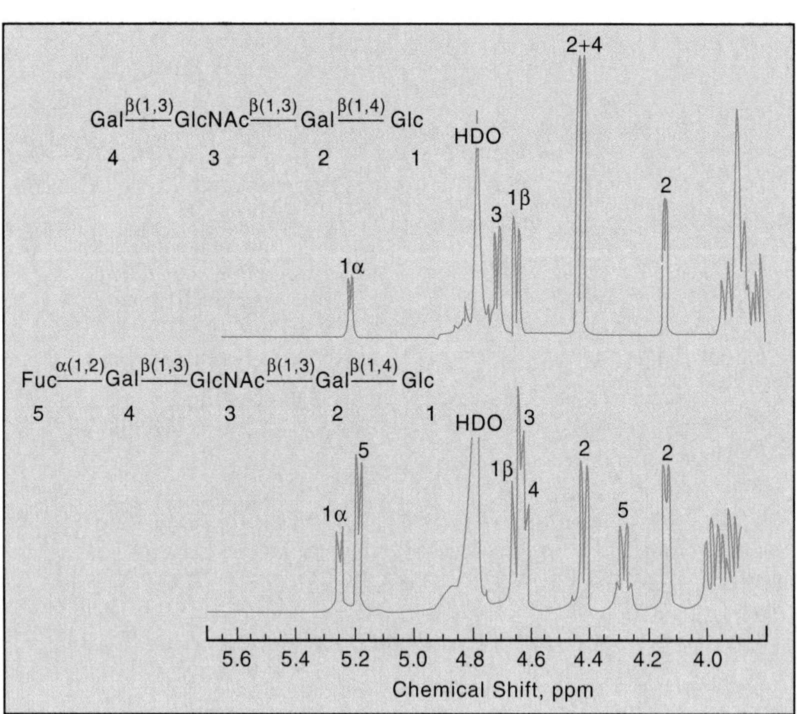

Oligosaccharide structural determination by high field ^{1}H-NMR spectroscopy. These traces are the partial spectra at 500 MHz of two related oligosaccharides from human milk. Several protons attached to one or more carbons of each sugar resonate in this region. The individual proton peaks are numbered to correspond to the sugar from which they are derived. Two peaks are obtained for the monomeric proton (H1) of glucose, depending on whether it derives from the α or β anomeric form (1α, 1β). The two resonances seen for fucose (5) and galactose (2) derive from two protons: the H1 and H5 for fucose, the H1 and H4 for galactose. Other regions of the spectrum (not shown) resolve other reporter protons (e.g., protons of the *N*-acetyl group of GlcNAc and the CH_3 group of fucose). HDO derives from the deuterium oxide solvent. (Source: C. Campbell and P. Stanley, The Chinese hamster ovary glycosylation mutants LEC11 and LEC12 express two novel GDP-fucose: *N*-acetylglucosaminide 3-α-L fucosyltransferase enzymes. (Source: From C. Campbell and P. Stanley, The Chinese hamster ovary glycosylation mutants LEC11 and LEC12 express two novel GDP-fucose: *N*-acetylglucosaminide 3-α-L fucosyltransferase enzymes. *J. Biol. Chem.* 259(18):11208-11284, 1984.)

Figure 18.18

In general, glycosyltransferases are specific for the acceptor, the sugar transferred, and the linkage whose formation they catalyze. The *N*-acetylglucosaminyltransferases termed GlcNAc-TI and GlcNAc-TII both transfer *N*-acetylglucosamine to mannose in $\beta(1,2)$ linkage, but their acceptor substrates are completely different. This sequence of biosynthetic steps was deduced from analyses of a mammalian cell mutant that lacks GlcNAc-TI activity.

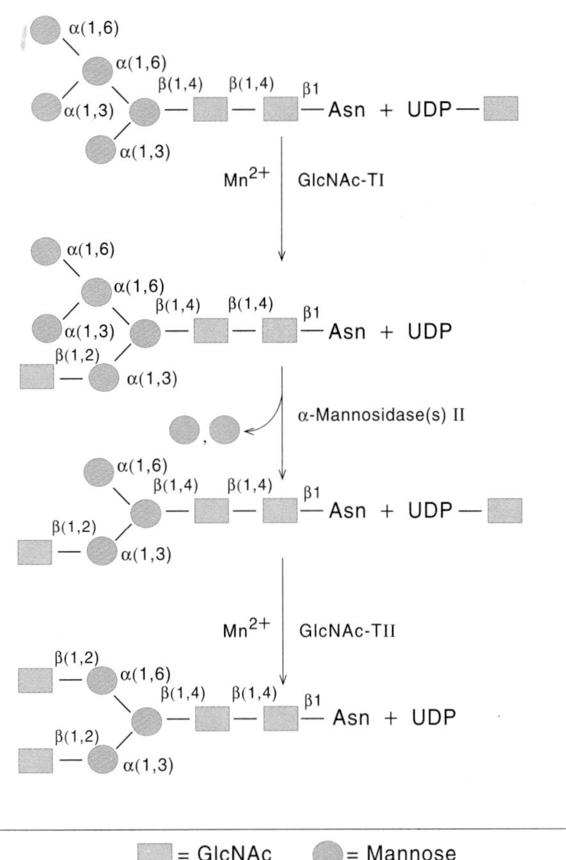

structure of an oligosaccharide will depend on the different glycosyltransferases present in a cell, the conformation of the protein during sugar addition, and the speed with which the protein traverses the secretory pathway. Thus O-glycosidically linked carbohydrates can be as simple as one sugar (GalNAc or GlcNAc) or can contain many sugars in a variety of linkages and branching patterns.

In addition to the high specificity of glycoslytransferases, another major reason for the ordered addition of oligosaccharides is that the glycosyltransferases that catalyze sugar addition are compartmentalized within the cell. Therefore during biosynthesis, glycoproteins come in contact with particular glycosyltransferases when they traverse different cellular compartments. All glycoproteins (with the exception of cytoplasmic and nuclear glycoproteins carrying *O*-GlcNAc residues) have their oligosaccharide portions synthesized in the lumen of membrane-bound compartments.

The early stages of glycoprotein synthesis are further complicated by the fact that oligosaccharides are added and often modified before protein synthesis is completed. We will not discuss the mechanism of protein synthesis in any detail until chapter 34, but there are certain facts that must be presented here if we are to understand how oligosaccharide synthesis is associated with glycoprotein synthesis.

In a eukaryotic cell, protein synthesis takes place on ribosomes (see chapters 1 and 34). All the ribosomes in a cell are identical except that they become transiently associated with different messenger RNA molecules, which determine the specific protein that will be synthesized. Ribosomes in the process of protein synthesis exist as independent bodies in the cell cytosol or as membrane-complexed bodies on the endoplasmic reticulum. The endoplasmic reticulum is a membrane-bounded organelle that is found in the cytoplasm. Ribosomes bind to the endoplasmic reticulum by a complex sequence of events. All proteins destined to be N-glycosylated and most that are O-glycosylated are sequestered from the cytoplasm by translocation during synthesis into the lumen of the endoplasmic reticulum (fig. 18.19). These proteins are marked by the presence of a short ($\sim$20) hydrophobic amino acid stretch termed a signal sequence at their extreme NH_2 terminus. During translation the signal sequence emerges first from the large ribosomal subunit. The signal sequence and ribosome are recognized by a complex termed the signal recognition particle (SRP). The binding of SRP blocks further translation until binding between SRP and the SRP receptor in the endoplasmic reticulum membranes is achieved. Once the ribosome complex is bound via receptor proteins that recognize ribosomes and the SRP receptor, translation resumes and the protein being synthesized is extruded into the endoplasmic reticulum. As a rule the signal sequence is cleaved during this process, giving rise to a new NH_2 terminus for the surviving polypeptide chain.

The biosynthesis of O- and N-linked oligosaccharides follows two different routes. N-linked oligosaccharides are added cotranslationally in the endoplasmic reticulum to certain Asn-X-Ser/Thr sequences as a complex of sugars originally synthesized on a lipid carrier. Some glycoproteins receive O-linked *N*-acetylgalactosamine in a compartment just beyond the endoplasmic reticulum, but most O-linked glycosylation occurs in the Golgi complex. This is a membranous organelle to which all nonresident proteins of the endoplasmic reticulum are transferred by a budding process diagrammed schematically in figure 18.20. In the Golgi complex, oligosaccharide biosynthesis continues with the sequential addition (and removal) of sugar residues. After exiting the *trans* Golgi network by a budding process, glycoproteins are delivered to the plasma membrane as resident integral membrane proteins, or they are secreted from the cell or sent to lysosomes.

Biosynthesis of N-Linked Oligosaccharides

The N-linked oligosaccharides are divided into three major classes: complex or lactosamine-containing (lactosamine is the disaccharide Galβ(1,4)GlcNAc), hybrid, and oligomannosyl. An example of a structure in each class is given in fig. 18.13. All structures include a common core of five sugars linked to Asn: Manα(1,3)[Manα(1,6)]Manβ(1,4)GlcNAcβ(1,4)GlcNAcβ1Asn. This results from the fact that all of these structures are initially

Figure 18.19

Diagram of events involved in the targeting and translocation of proteins destined to be completed in the endoplasmic reticulum. Proteins that include a signal sequence of approximately 20 hydrophobic amino acids at their NH_2-terminus are recognized by a signal recognition particle (SRP). This complex of proteins and 7S RNA binds to the signal sequence of the nascent protein and the large ribosomal subunit. SRP binding causes translation arrest and the targeting of the translocation complex to the endoplasmic reticulum (ER), where it is bound by SRP receptor (SR) as well as a ribosome receptor (RR) protein. These and several other resident membrane proteins form a pore complex called the translocon. As protein synthesis resumes, SRP and its receptor are released from the ribosome, the growing chain traverses the aqueous pore of the translocon and enters the lumen of the endoplasmic reticulum. Signal peptidase (SP) degrades the signal sequence, creating a new NH_2 terminus, and N-linked carbohydrates are added by oligosaccharyltransferase (OST) to glycoprotein. When synthesis is complete, soluble ER proteins are released into the lumen of the ER. Proteins and glycoproteins with hydrophobic regions elsewhere in the molecule span the ER membrane so that their C terminus or N terminus or both are on the cytoplasmic face of the ER membrane.

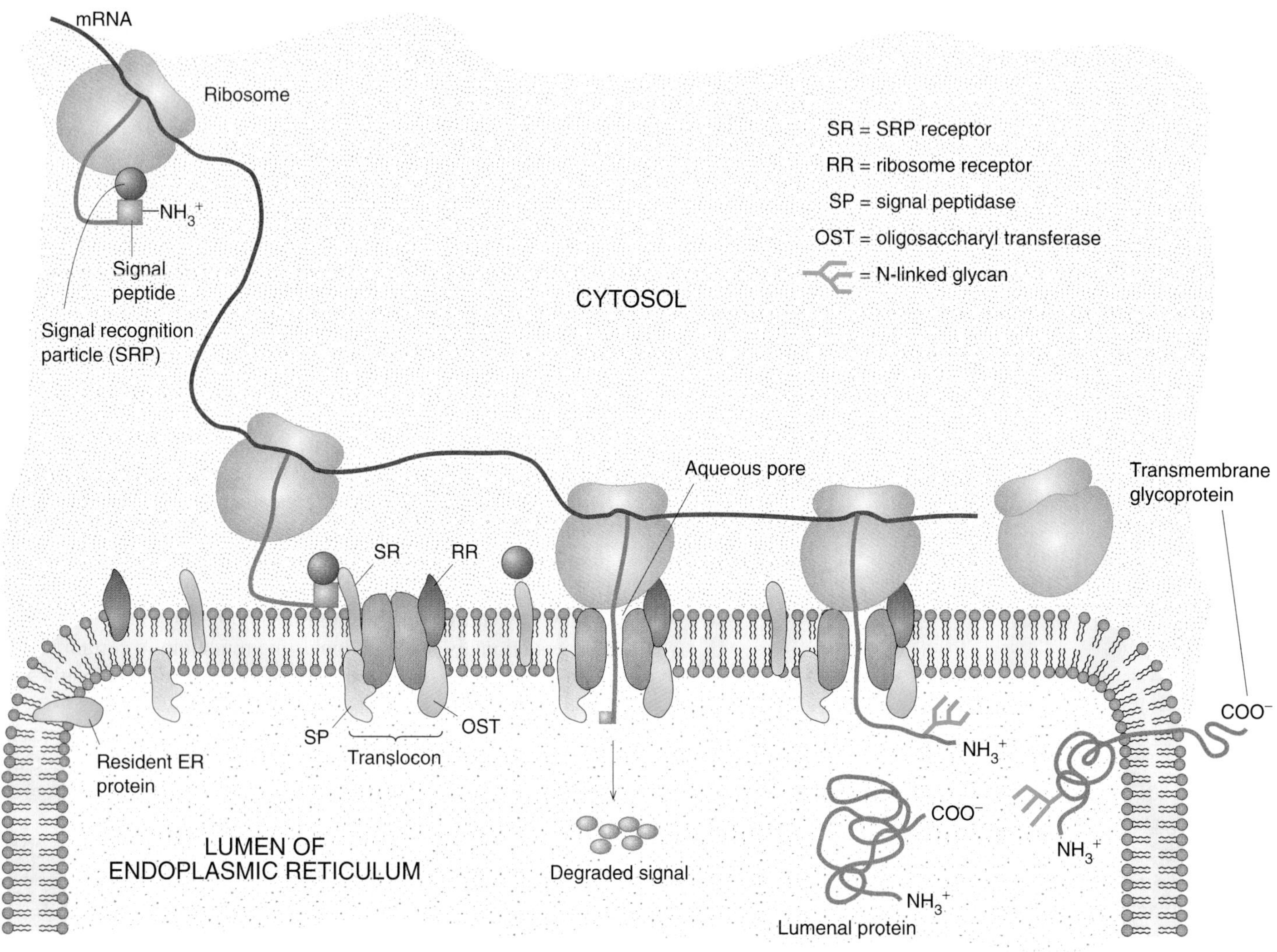

synthesized by a common pathway that exists in all eukaryotes, including yeast. This pathway involves the sequential addition of sugars to dolichol phosphate, a membrane-associated polyprenol lipid (fig. 18.21).

Dolichol phosphate (Dol-P) serves as the substrate for a glycosyltransferase that adds phospho-*N*-acetylglucosamine from UDP-GlcNAc to form Dol-P-P-GlcNAc. The first GlcNAc transfer and the subsequent addition of four more sugar residues occurs on the cytoplasmic side of the endoplasmic reticulum membrane as shown in figure 18.22. The first seven sugars are transferred to the growing oligosaccharide from their appropriate nucleotide-sugar conjugates (UDP-GlcNAc or GDP-Man), which are present in the cytoplasm. At this stage the Dol-P-oligosaccharide portion is translocated to the lumenal side of the endoplasmic reticulum membrane. The next sugars are added from Dol-P-Man and Dol-P-Glc precursors rather than the corresponding nucleotide sugar derivatives. The Dol-P sugars themselves are synthesized by the transfer of mannose from GDP-mannose or of glucose from UDP-glucose to dolichol phosphate to form Dol-P-Man and Dol-P-Glc, respectively. These reactions probably occur on the cytoplasmic face of the endoplasmic reticulum, where Dol-P has access to both nucleotide sugars. The final structure assembled on dolichol contains two *N*-acetylglucosamines, nine mannoses, and three glucoses linked in the manner shown in fig-

Figure 18.20

Schematic diagram of a cross-section through a cell showing the relative locations of nucleus, endoplasmic reticulum (ER), Golgi complex, *trans* Golgi network, and plasma membrane. Glycoproteins synthesized in the lumen of the ER pass to the *cis* cisterna of the Golgi complex by a sequential membrane budding and fusion mechanism. The Golgi cisternae are classified into *cis,* medial, and *trans* in the order of increasing distance from the nucleus. Mature glycoproteins exit from the *trans* Golgi network in membrane-bound secretory vesicles. Depending on the nature of the membranes, the vesicles have one of several different fates. Following fusion with other membranes their contents are delivered outside the cell, to the plasma membrane, or to other organelles inside the cell, such as lysosomes (see fig. 18.24).

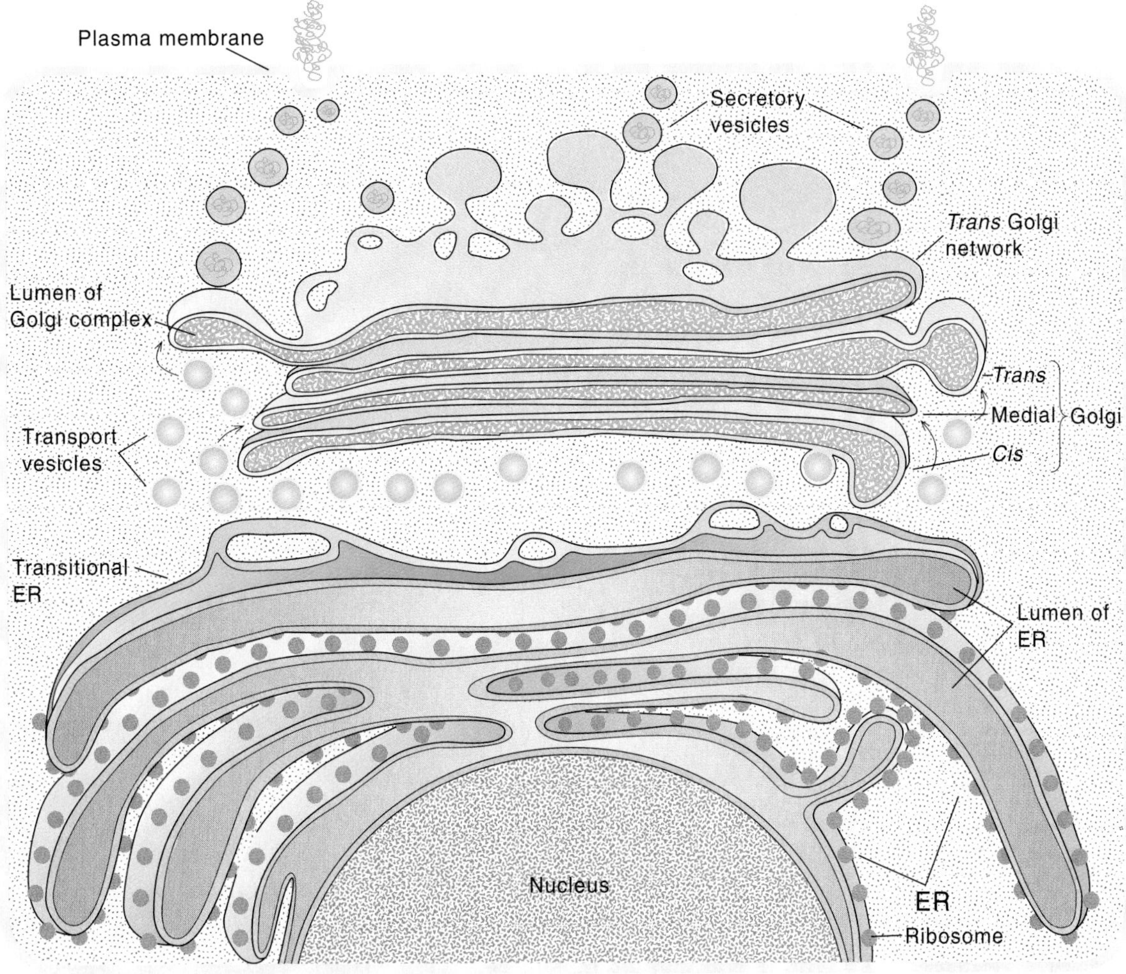

Figure 18.21

The structure of dolichol phosphate. The hydrocarbon portion of the molecule has an affinity for membranes. The phosphate end forms an activated complex with oligosaccharide intermediates.

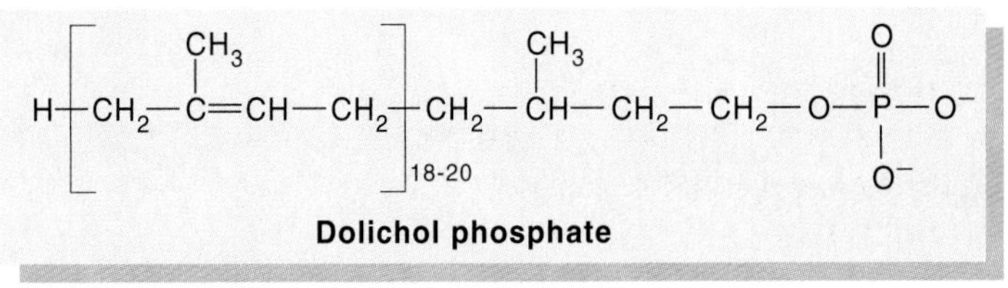

$$H-\left[CH_2-\underset{}{\overset{CH_3}{\overset{|}{C}}}=CH-CH_2\right]_{18\text{-}20}-CH_2-\overset{CH_3}{\overset{|}{CH}}-CH_2-CH_2-O-\underset{O^-}{\underset{|}{\overset{O}{\overset{\|}{P}}}}-O^-$$

Dolichol phosphate

ure 18.22. This oligosaccharide moiety is transferred as a unit by the enzyme oligosaccharyltransferase to the Asn residues in certain Asn-X-Ser(Thr) sequences of growing peptide chains. The transfer and the subsequent biosynthesis of N-linked carbohydrates are shown schematically in figure 18.23.

Once attached to the protein, the oligosaccharide is processed (or trimmed) by specific exoglycosidases that sequentially remove terminal sugars (see fig. 18.23). The glucoses are removed by at least two different α-glucosidases: α-glucosidase I, which removes the first glucose, and α-glucosidase II, which re-

Figure 18.22

Synthesis of the dolichol-linked oligosaccharide. Biosynthesis starts on the cytosolic side of the endoplasmic reticulum. Seven single-step reactions lead to the heptasaccharide that is linked to dolichol pyrophosphate. The heptasaccharide is translocated across the membrane of the endoplasmic reticulum to the lumen, where seven additional one-step reactions take place. The activated sugars for these additions are formed on the cytosolic side and translocated to the lumen as hexose-P-dolichol complexes. The initial step in the pathway is inhibited by tunicamycin. The synthesis of Dol-P-Man is inhibited by amphomycin.

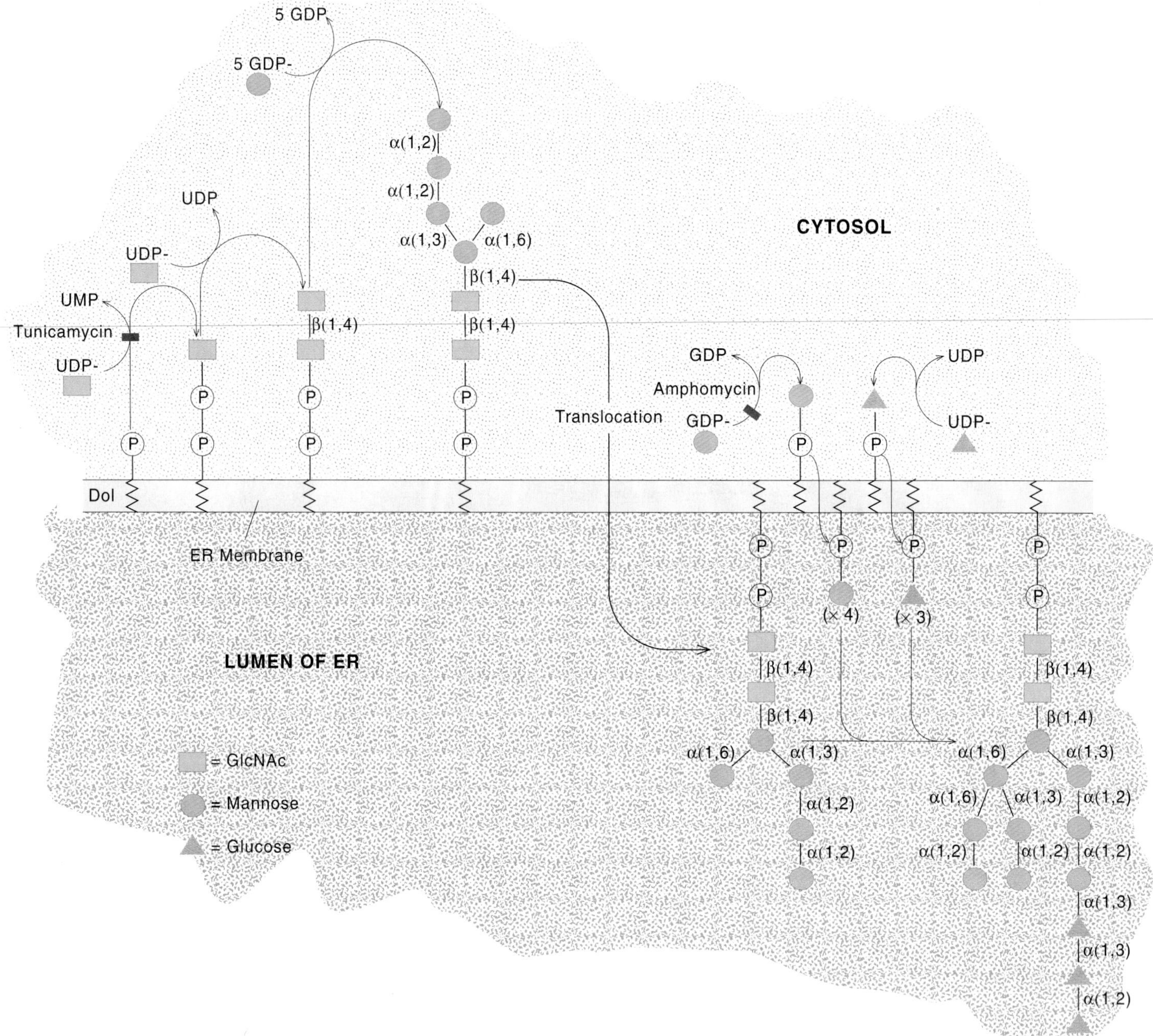

moves the next two glucoses. Following this, the first mannose is removed by a specific endoplasmic reticulum α-mannosidase. Yeast and mammalian cells appear to follow exactly the same pathway to the generation of the Man_8 intermediate. At this point many glycoproteins are translocated by vesicular transport to the lumen of the *cis* membranes of the Golgi complex.

In the *cis* Golgi, either the Man_8 oligosaccharide is further processed or, if it is attached to a lysosomal hydrolase, it is modified by the addition of P-GlcNAc residues that are transferred from UDP-GlcNAc by an α-phospho-*N*-acetylglucosaminyltransferase. This transferase acts only on the oligomannosyl carbohydrates of certain lysosomal enzymes, which it recognizes by a spe-

Figure 18.23

Schematic diagram illustrating the transfer of oligosaccharide to protein in the endoplasmic reticulum and the subsequent processing and maturation of oligosaccharides in the Golgi complex. In the endoplasmic reticulum the oligosaccharide synthesized on Dol-P is transferred to the Asn(N) of an Asn-X-Ser(Thr) sequence in the growing peptide chain. The transfer is catalyzed by oligosaccharyltransferase (OST), which is associated with the translocon (see fig. 18.19). Subsequent trimming of the oligosaccharide occurs by specific exoglycosidases that sequentially remove terminal glucoses and a specific mannose residue. When synthesis of the protein portion is complete, the glycoprotein is transferred to the *cis* compartment of the Golgi complex. For most glycoproteins, further processing of the $Man_8GlcNAc_2$ oligosaccharide occurs in this compartment. However, soluble lysosomal enzymes are recognized by a phospho-*N*-acetyglucosaminyltransferase, which transfers P-GlcNAc to the C-6 of mannose residues. Subsequently an α-*N*-acetylglucosaminidase removes the GlcNAc, exposing Man-6-P which is recognized by the Man-6-P receptor. Lysosomal enzymes may have several N-linked carbohydrates, some of which are converted to typical complex carbohydrates in the later compartments of the Golgi complex.

In the medial Golgi the synthesis of a biantennary structure terminating in GlcNAc is achieved (see fig. 18.18). Also in this compartment, additional GlcNAc residues may be added to the core mannose residues to give as many as seven branches. Fucose, galactose, and sialic acid residues are added in the *trans* Golgi and/or *trans* Golgi network compartments. The membranes of the Golgi compartments contain nucleotide-sugar translocases specific for transferring each nucleotide sugar across the membrane.

Inhibitors can block the pathway at specific steps, causing the accumulation of biosynthetic intermediates. Castanospermine inhibits the α-glucosidase that removes the first glucose; deoxymannojirimycin inhibits the α-mannosidase I that removes mannose from the Man_8 intermediate, and swainsonine blocks the α-mannosidase II that removes mannose from the Man_5 intermediate. In the presence of swainsonine, hybrid N-linked carbohydrates are synthesized.

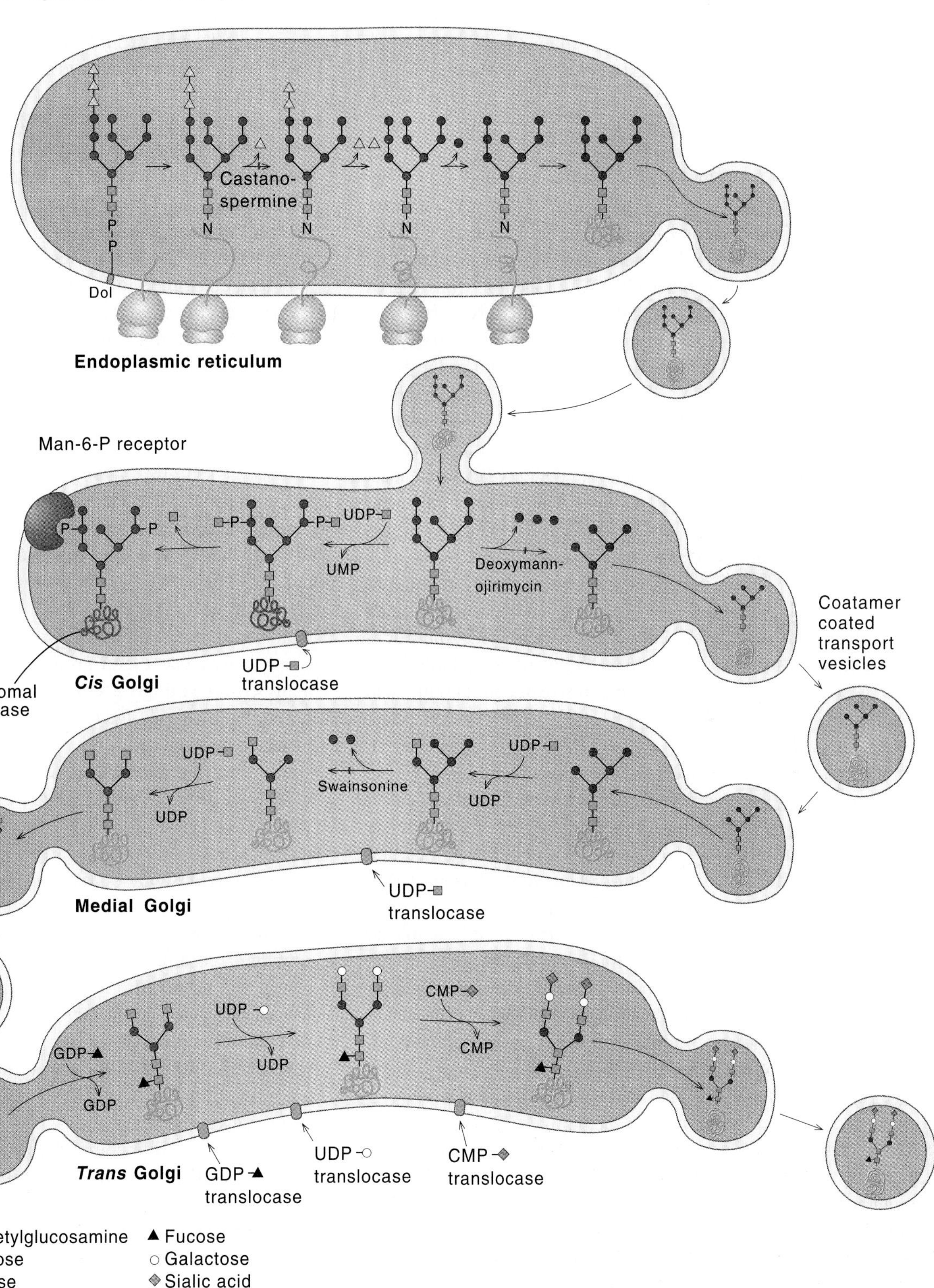

cific protein–protein interaction. Once transferred, the GlcNAc is removed by an α-*N*-acetylglucosaminidase, thereby exposing Man-6-P residues for interaction with the Man-6-P receptor. Lysosomal hydrolases subsequently traverse all compartments of the Golgi and therefore may also contain N-linked carbohydrates of the complex type at other glycosylation sites. In yeast the $Man_8GlcNAc_2$ oligosaccharide is elongated by the addition of many mannose residues in the Golgi complex.

The α-mannosidase I in the *cis* Golgi removes mannose residues from the Man_8 intermediate in mammalian cells. Altogether, three mannoses are removed to generate the Man_5 substrate for *N*-acetylglucosaminyltransferase I (GlcNAc-TI), whose action initiates the conversion of oligomannosyl carbohydrates to hybrid or complex carbohydrates (see fig. 18.18). The addition of the first branch GlcNAc residue (which occurs in medial Golgi cisternae) generates the substrate for the last processing glycosidase, α-mannosidase II. The second, branch GlcNAc residue is added by GlcNAc-TII (see fig. 18.18), generating an intermediate that can be acted on by several other GlcNAc branching transferases and by a fucosyltransferase. The addition of all GlcNAc residues probably occurs in the medial Golgi, while the fucose is probably added in the *trans* Golgi. Also in the *trans* Golgi and in the *trans* Golgi network, branch GlcNAc residues are substituted with galactose, which is in turn substituted with sialic acid. Many other types of terminal sugar additions have been described, including the presence of long polylactosamine (Galβ(1,4)GlcNAc) sequences, fucose on galactose and GlcNAc residues, GalNAc and galactose on galactose residues and so on. The structure of mature complex N-linked carbohydrates may vary enormously as the result of a variety of factors. In addition to a wide range of different sugar combinations, the presence of sulfated and phosphorylated residues has also been described. Most of these latter modifications are thought to occur in the *trans* Golgi or *trans* Golgi network. These compartments are variable in size and potentially also in function in different cell types.

The specific location of the glycosylation reactions of N-linked carbohydrate biosynthesis is a major factor determining final oligosaccharide structure. Glycoproteins that reside in a particular compartment bear oligosaccharides that reflect the pathway to that point. Similarly, if a glycoprotein accumulates in a certain compartment because of a mutation that prevents normal transit (e.g., causes misfolding), it will bear the oligosaccharides typical of its final location (see fig. 18.23). As a result it is possible to tell in which compartment a glycoprotein resides by determining the structure of its N-linked carbohydrates. The compartmentalization of glycosyltransferases has also proved useful for developing assays of the transit of molecules between membrane compartments of the secretory pathway (box 18C).

Not all of the vesicles that bud from the *trans* Golgi network are destined for the plasma membrane (fig. 18.24). Lysosomal hydrolases bound to the Man-6-P receptor are sorted into vesicles coated with a fibrous protein known as clathrin that fuse with an acidic prelysosome. In this environment of acidic pH ($\leq$5) the binding between Man-6-P receptors and lysosomal hydrolases is broken, and they subsequently are sorted independently. Vesicles carrying lysosomal hydrolases ultimately deliver them to a mature lysosome, while the Man-6-P receptor (which is an integral membrane protein) cycles between the plasma membrane, Golgi membranes, and endosomes.

The endoplasmic reticulum–Golgi network also transfers nascent glypiated glycoproteins to the plasma membrane (box 18D).

Biosynthesis of O-Linked Oligosaccharides

Glycoproteins in the secretory pathway receive GalNAc in O-glycosidic linkage from UDP-GalNAc via a transferase that acts directly on Ser or Thr residues of proteins. No dolichol-linked intermediates are involved. Most kinetic labeling studies indicate that GalNAc is added in Golgi membranes, in a compartment just beyond the endoplasmic reticulum and before the *cis* Golgi. There are several GalNAc transferases that may initiate O-linked glycosylation. The second sugar in O-linked oligosaccharides is usually galactose, and that is added in the *trans* Golgi compartment. The remainder of the O-linked oligosaccharide is synthesized in the Golgi by sequential sugar additions from nucleotide-sugars catalyzed by specific glycosyltransferases. There is a certain order to the sequence of sugars added, but no invariant core sequence as is found in N-linked carbohydrates. Cytoplasmic and nuclear proteins with *O*-GlcNAc residues that are not further modified are synthesized by a cytoplasmic glycosyltransferase.

The O-linked oligosaccharides of membrane glycoproteins are usually not very large although they may exhibit branching and various terminal sugar sequences. However, in body fluids very large O-linked structures are found on proteins termed mucins. These oligosaccharides often terminate with sugars that are highly immunogenic. These sugars form the basis of the human blood type and are therefore known as blood group substances. Because blood-group-specific sugars are expressed on many O-linked oligosaccharides of cells and fluids and are highly immunogenic, individuals will mount an immune response against blood that is of a different type. For this reason, individuals of similar blood groups can accept blood from each other, but individuals of different types usually cannot.

The best-characterized blood grouping scheme is the ABO scheme. All people carry a gene that is responsible for generating their ABO blood type. This gene comes in three variations (alleles) that code for related proteins. The A allele codes for a glycosyltransferase that catalyzes the addition of a terminal *N*-acetylgalactosamine (GalNAc) residue onto a particular trisaccharide found mainly in O-linked carbohydrates; the B allele differs in only four amino acids and encodes a glycosyltransferase that catalyzes the transfer of a galactose (Gal) residue to the same trisaccharide; the O allele has a mutation and therefore generates a nonfunctional protein (fig. 18.25). An individual may have two copies of one allele of the ABO gene or a copy of two different alleles. The possible combinations give rise to six different genotypes (AA, BB, OO, AB, AO, BO) and the expression of four different combinations of blood group antigens (type A from AA and AO people, type B from BB and BO people, type AB from AB people, and type O from OO people). When both A and B alleles

Figure 18.24

Sorting of proteins for their final destination occurs in the *trans* Golgi network (TGN). At 20°C, secreted proteins accumulate in the TGN. By raising the temperature to 37°C, the sorting and secretion of synchronized populations of marker proteins can be studied. Two major routes to the plasma membrane have been identified: the constitutive pathway and the regulated pathway. Proteins in the latter pathway (including many prohormones) accumulate in secretory granules that fuse with the plasma membrane on receiving a biochemical stimulatory signal. As the name suggests, passage of integral membrane proteins and soluble proteins through the constitutive pathway occurs constantly, with no requirement for a stimulatory signal. Soluble lysosomal hydrolases bound to the Man-6-P receptor exit from the TGN in clathrin-coated vesicles. Clathrin is a fibrous protein that coats certain vesicles. These vesicles fuse with an acidic, prelysosomal compartment where the lysosomal hydrolases are released from the Man-6-P receptor and are ultimately delivered to lysosomes. The Man-6-P receptor does not go to lysosomes but recycles between the plasma membrane, Golgi membranes, and prelysosomes.

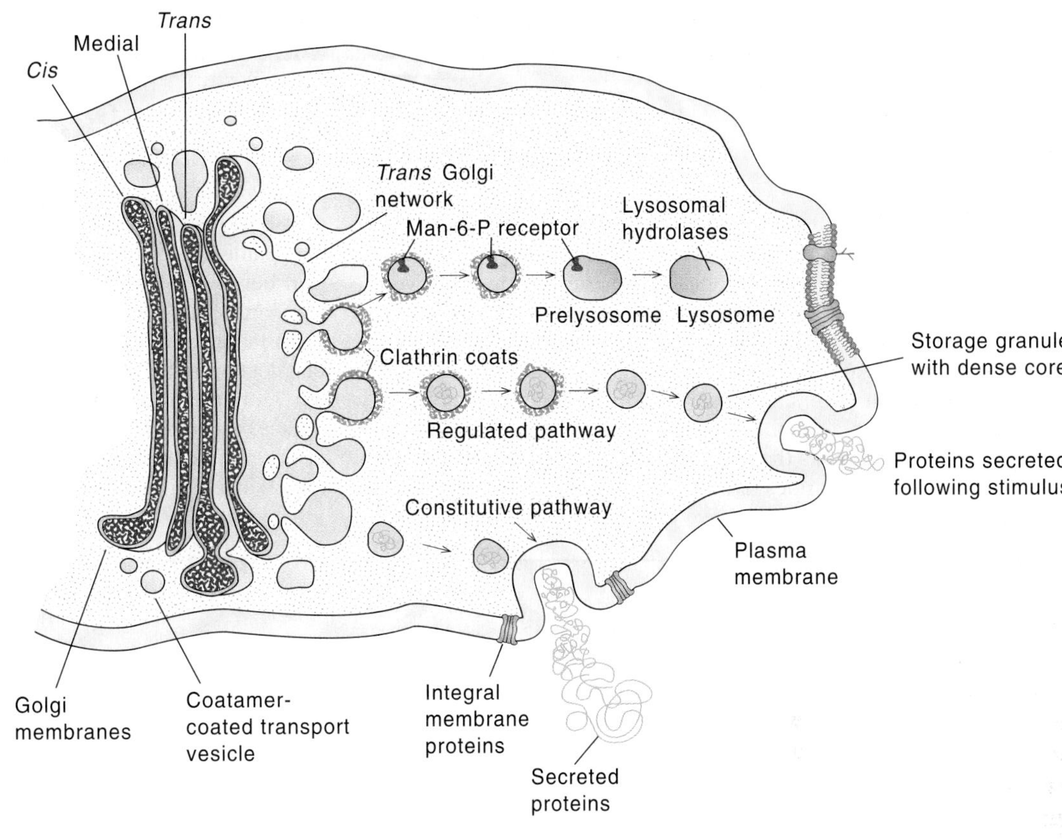

Figure 18.25

The structures and reactions at the oligosaccharide termini of the ABO human blood group antigens. The A gene and the B gene each encode a glycosyltransferase that adds either GalNAc (A gene glycosyltransferase) or Gal (B gene glycosyltransferase), to the termini of oligosaccharides on red blood cell glycoproteins. Individuals may carry two A genes, two B genes, one A and one B gene or one or two genes termed O, at the ABO genetic locus. The O gene produces a non-functional transferase. Individual differences are genetically inherited and are reflected in the structures of their blood group antigens. Therefore, an individual who is AA or AO will express only the A antigen; an individual who is BB or BO will express only the B antigen; an individual who is AB will express both antigens; and an individual who is OO will express neither antigen. OO individuals are able to donate blood to any person because they do not express A or B antigens on their red blood cells and consequently do not induce an immune response.

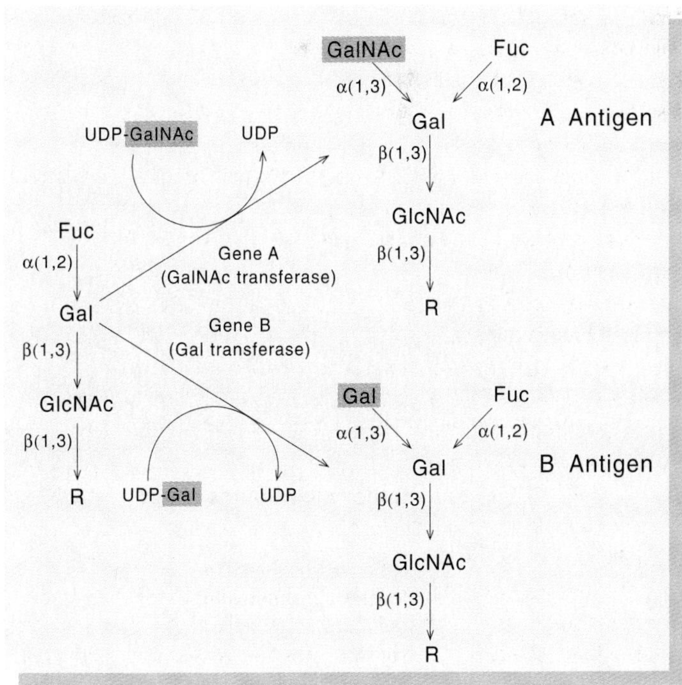

In Vitro Assay for Transit between Golgi Membrane Compartments

The compartmentalization of glycosylation reactions allows us to assay the molecular mechanisms of transit between Golgi membranes. Donor Golgi membranes lacking the GlcNAc-T1 transferase can be obtained from Lec1 or 15B mutant cells (see table 18.4). These membranes cannot further modify the oligosaccharides of a marker glycoprotein even if donor–donor membrane transport occurs. However, when a marker glycoprotein is transported into an acceptor Golgi membrane compartment prepared from wild-type cells that have GlcNAc-T1 activity, GlcNAc is added to the oligosaccharides of the marker glycoprotein. With the help of radiolabeled UDP-GlcNAc we can therefore measure transport between the *cis* and medial Golgi compartments. Similar assays using other glycosylation mutants can be used to follow transport between ER and Golgi or between different medial and *trans* compartments. Such assays have been the key to identifying molecules involved in vesicular transport reactions. These include vesicular coat proteins, membrane-docking proteins, and small GTP-binding proteins that are required for transport. The assay shown in figure 1 assumes forward transport, as would be expected from the vectorial functioning of the secretory pathway. However, this assay could also measure certain retrograde transport steps.

Figure 1

Mutants lacking specific glycosylation enzymes can transfer their partially processed glycoproteins to Golgi from wild-type cells for further processing.

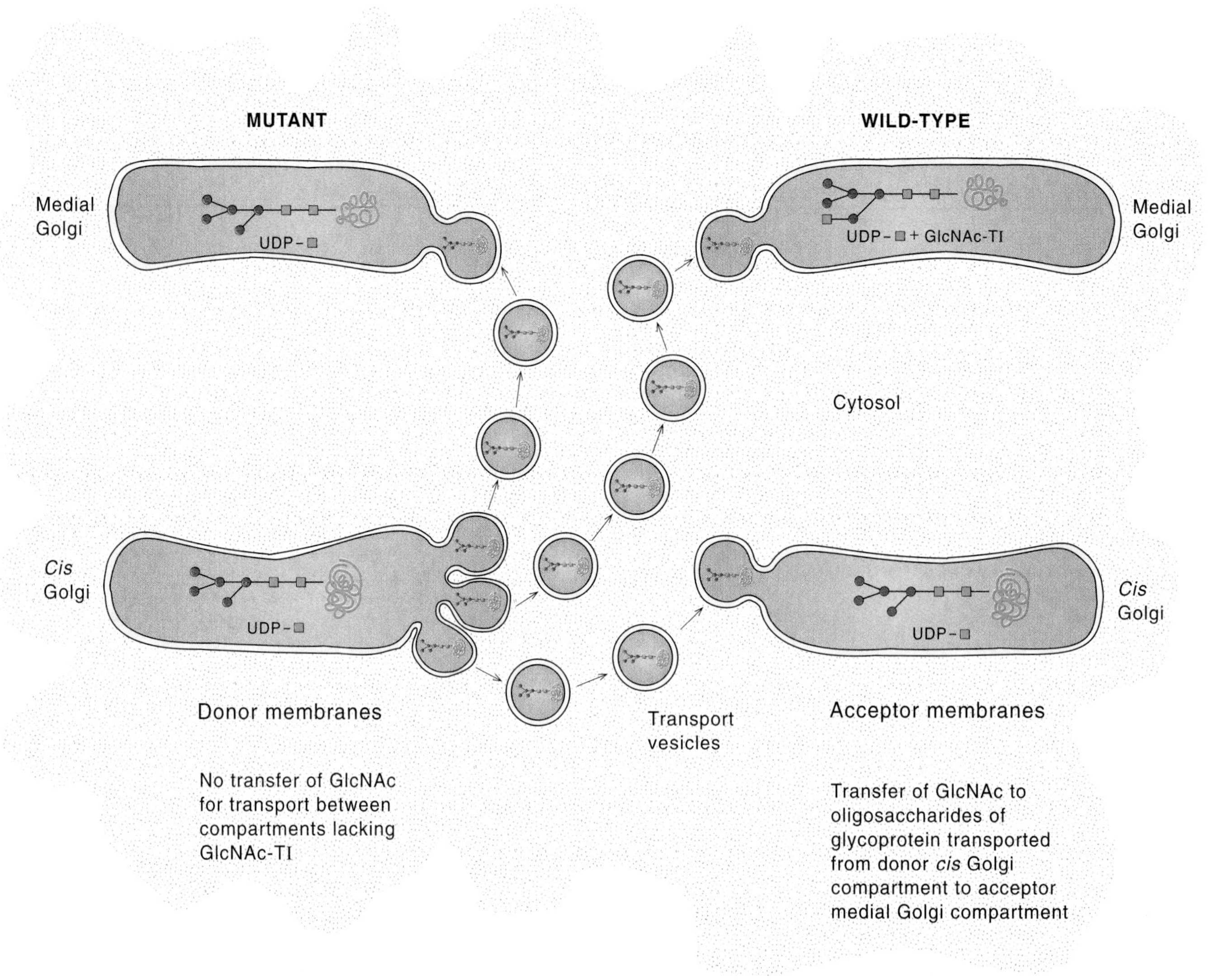

Synthesis of the Glycosylphosphatidylinositol (GPI) Anchor

The core region of the glycan in the GPI anchor that is found at the carboxy terminus of glypiated proteins (see fig. 18.14) is synthesized in the endoplasmic reticulum (ER). The first reaction occurs in the cytoplasmic leaflet of the ER membrane and involves the transfer of GlcNAc from UDP-GlcNAc to the C-6 of *myo*-inositol in a phosphatidylinositol lipid (PI). The GlcNAc is subsequently deacetylated to become glucosamine (GlcN). In mammalian cells and yeast a fatty acid is then added to one of the OH groups in the inositol to generate GlcN-acylPI. Three mannose residues are subsequently added from Dol-P-Man to form Man_3GlcN-acylPI. Each Man residue may be modified by an ethanolaminephosphate (EthN-P) residue that is transferred from phosphatidylethanolamine(PE). It is the EthN-P on the terminal Man that is covalently complexed to the carboxyl amino acid of the glypiated protein. The transfer of the anchor to the protein occurs immediately after synthesis of the protein in a transamidation reaction. Recognition of proteins that are to receive a GPI anchor requires a hydrophobic stretch of amino acids at the C terminus of the protein. This hydrophobic terminus is removed by cleavage at an internal amino acid to generate the new C-terminal amino acid to which the anchor becomes bound. Thus glypiation of a protein involves removal of C-terminal amino acids and transfer of a preformed GPI anchor to the new C terminus immediately after synthesis of the protein is complete. The glycan moiety can be further modified by the addition of other sugar residues such as Gal or GalNAc. These reactions probably take place during transit through the Golgi compartments. Glypiated proteins are finally localized on the outer leaflet of the plasma membrane of the cell.

Patients with the blood disorder paroxysmal nocturnal hemoglobinuria (PNH) do not synthesize GPI anchors because of a block at the first step in biosynthesis (fig. 1). In these patients, proteins that are normally glypiated are either degraded in the ER or secreted from the cell. The loss of certain glypiated proteins from the surface of red blood cells leads to their lysis and thus to the presence of hemoglobin in the urine as well as to other complications. PNH arises owing to a spontaneous mutation in a blood precursor cell and is present only in that cell and its descendants. Therefore PNH patients have both affected and unaffected cells in their blood.

Figure 1

Steps in the synthesis of a GPI anchor.

PI
Blocked in patients with PNH → UDP-GlcNAc
GlcNAc α(1,6)-PI
→ Ac
GlcN α(1,6)-PI
FA
GlcN α(1,6)-acyl PI
Dol-P-Man[a]
Man α(1,2) Man α(1,6) Man α(1,4) GlcN α(1,6)-acyl PI
EthN-P-Man α(1,2) Man α(1,6) Man α(1,4) GlcN α(1,6)-acyl PI
(EthN-P) (EthN-P)
Protein
Protein—C(=O)—NH—CH_2—CH_2—O—P(=O)(O⁻)—O
$Man_3(EthN\text{-}P)_2$GlcN-acyl PI

[a]The modification of each Man residue by EthN-P may occur immediately after its transfer from Dol-P-Man and before the addition of the subsequent Man.

are present in AB people, both the A and B blood group antigens are expressed on blood cells and in mucins. The presence of the A and B glycosyltransferases can be readily detected in the milk of a lactating female. When the O allele is present in two copies (in OO people), the terminus of the trisaccharide is left unsubstituted. Such people are able to donate blood to all others without inducing an immune response and are therefore known as universal blood donors.

Specific Inhibitors and Mutants Are Used to Explore the Roles of Glycoprotein Carbohydrates

The functions of the carbohydrate moieties of glycoproteins are difficult to study because of the intimate physical association between oligosaccharides and the protein backbone. However, there are several ways to produce glycoproteins with carbohydrates that are severely truncated or missing entirely, so we can deduce the normal function of the affected carbohydrates.

For glycoproteins that contain N-linked carbohydrates, we can completely inhibit oligosaccharide addition by synthesizing the glycoprotein in the presence of tunicamycin (see fig. 18.22). The protein portion of many glycoproteins is synthesized and translocated through the secretory pathway essentially normally in the presence of tunicamycin and may be studied by functional assays. Alternatively, we can eliminate each glycosylation sequence by site-directed mutagenesis of the cloned glycoprotein. Following transfection, the protein portion devoid of carbohydrates will be produced. This approach has the advantage that the altered protein is synthesized in a normal cell in contrast to a drug-treated cell, in which all glycoproteins are affected, as happens with tunicamycin treatment. An analogous approach is to introduce cloned glycoprotein genes into *E. coli,* which has none of the transferases required for eukaryotic N- or O-linked glycosylation. However, *E. coli* has other modifying enzymes that may give rise to a protein with modifications not usually associated with it in mammalian cells. Therefore in structure/function studies we usually try to synthesize mammalian proteins in a mammalian host.

Although several glycoproteins have been efficiently produced without their oligosaccharides, many glycoproteins cannot be synthesized devoid of carbohydrates because they require oligosaccharides for proper folding and translocation competence. In such cases we instead aim to produce glycoproteins with truncated carbohydrates, to see the effect of the new structures on the biological activity of the glycoprotein. One way to achieve this aim is to use inhibitors that stop oligosaccharide processing, such as castanospermine, deoxymannojirimycin, or swainsonine (table 18.3). The glycoproteins so produced have immature N-linked carbohydrates that are quite distinct in structure from mature complex oligosaccharides. A complementary approach, which has the advantage of producing glycoproteins with truncated oligosaccharides that are quite homogeneous, is to use mutants of mammalian cells or yeast that are unable to complete the synthesis of N-linked carbohydrates. When cloned glycoproteins are transfected into such mutants, the carbohydrates produced will be intermediates characteristic of the biosynthetic step that is blocked in the mutant.

There are many glycosylation mutants of cultured mammalian cells and yeast. They have been selected as rare survivors of treatments that kill cells that express a particular carbohydrate or glycoprotein at the cell surface. For example, many plant lectins are toxic to mammalian cells and can be used to select for mutants that no longer bind the lectin at the cell surface and are therefore resistant to the toxic effects of the lectin. We can also select glycosylation mutants using toxic reagents directed at glycoproteins that require carbohydrates for their stable cell surface expression. Some glycoproteins require a cluster of O-linked oligosaccharides close to the membrane to protect them from being attacked by a protease. Glycoproteins with a glycosylphosphatidylinositol (GPI) anchor must have Dol-P-Man for their synthesis (see box 18D). In mutants lacking Dol-P-Man synthase a complete GPI anchor is not synthesized, and glycoproteins that are normally localized to the plasma membrane are secreted from the cell. Loss of Dol-P-Man synthase also affects N-linked carbohydrate biosynthesis by prohibiting the addition of the last four mannose residues to the dolichol oligosaccharide (see fig. 18.22). However, the cell compensates by adding three glucose residues to the $Man_5GlcNAc_2PPDol$ intermediate and transfers that to protein. This structure is processed and extended to form the usual range of complex N-linked carbohydrates by an "alternative" pathway, which requires that GlcNAc-TI transfers GlcNAc to $Man_3GlcNAc_2Asn$ instead of $Man_5GlcNAc_2Asn$ (see figs. 18.18 and 18.23). Thus although glycoproteins synthesized in this mutant may carry normal complex N-linked carbohydrates, they will have oligomannosyl residues only up to $Man_5GlcNAc_2Asn$. This $Man_5GlcNAc_2Asn$ intermediate is distinct from the processing intermediate of the medial Golgi and is not susceptible to cleavage by endo H.

Glycosylation mutants that are particularly useful for producing glycoproteins with modified carbohydrates are summarized in table 18.4. In some mutants both N- and O-linked carbohydrates are modified, while in others only N-linked carbohydrates are affected. We could also obtain glycoproteins with altered carbohydrates by treating the glycoproteins with glycosidases to remove particular sugars, but sugars and oligosaccharides are often sterically protected by the protein, and therefore complete deglycosylation would require treatment of unfolded (often denatured) protein. For this reason the use of glycosylation mutants frequently provides a superior approach for making altered carbohydrates.

Bacterial Cell Walls Are Composed of Polysaccharides Cross-Linked by Peptides

As we noted in chapter 1, a unique feature of bacteria is the cell wall that surrounds the plasma membrane and provides the mechanical strength that enables bacteria to resist shear and osmotic shock. The cell wall is composed of a network of linear heteropolysaccharides cross-linked by peptides. A structure of this sort is called a peptidoglycan. Some bacterial cells (Gram nega-

Table 18.3
Effects of Certain Inhibitors on Glycoprotein Synthesis

Glycosylation Inhibitor[a]	Major N-linked Carbohydrates Found on Glycoproteins in Presence of Inhibitor
Castanospermine inhibits α-glucosidase I	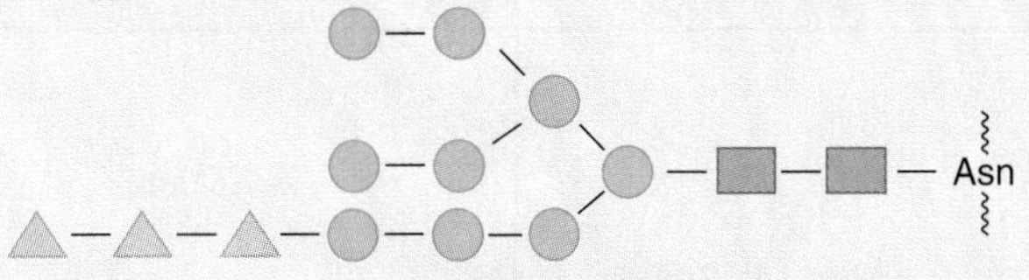 Asn ($Glc_3Man_9GlcNAc_2Asn$) Unblocked mannoses can be cleaved by α-mannosidase I in *cis* Golgi.
Deoxymannojirimycin inhibits α-mannosidase I	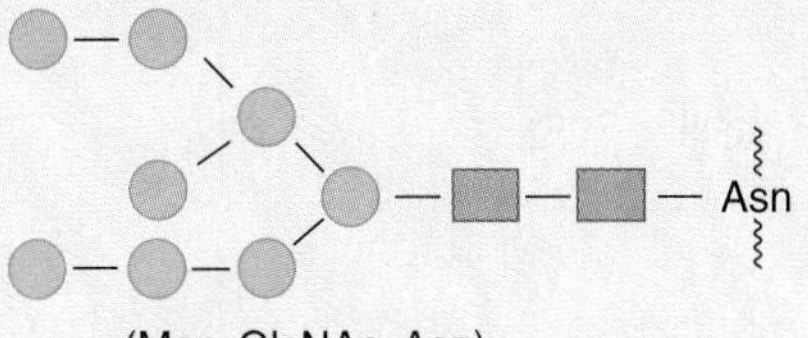 Asn ($Man_8GlcNAc_2Asn$)
Swainsonine inhibits α-mannosidase II	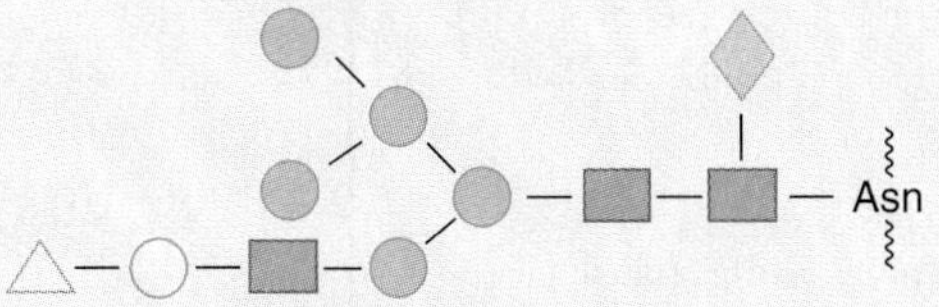 Asn (Hybrid) Lower antennae can have all residues found on antennae of complex N-linked structures.

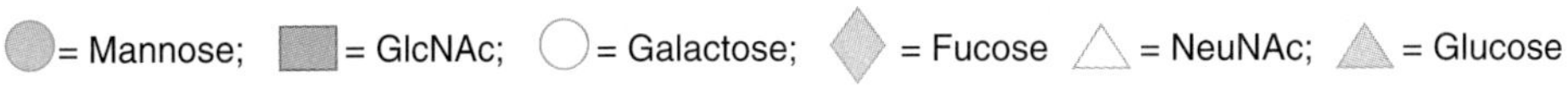

= Mannose; = GlcNAc; = Galactose; = Fucose = NeuNAc; = Glucose

[a]For site of action of each inhibitor, see fig. 18.23.

tive cells) also possess an outer membrane composed of lipids, proteins, and polysaccharides. The main structural features of a Gram negative bacterial cell envelope, which is a composite of the two membranes and the cell wall, are illustrated in figure 18.26.

Here we will discuss the structural aspects of the cell and describe its biosynthesis. The peptidoglycan that constitutes the cell wall is a polymeric structure consisting of a heteropolysaccharide composed of amino sugars in one dimension, cross-linked through branched polypeptides in the other (fig. 18.27). The amino sugars alternate in the polymer, forming the glycan strands (see fig. 18.27). The carboxyl group of the lactic acid moiety of the N-acetylmuramic acid is substituted by a tetrapeptide, which in the Gram positive bacterium *Staphylococcus aureus* has the sequence L-alanyl-D-γ-glutamyl-L-lysyl-D-alanine. In this tetrapeptide the glutamyl residue is attached through its γ-COOH rather than its α-COOH. All the muramic acids are substituted in this way to form peptidoglycan strands. Variations of the same basic structure occur in most bacterial species. The peptidoglycan strands are further linked to each other by means of an interpeptide bridge. In *S. aureus* this bridge is a pentaglycine chain that extends from the terminal carboxyl group of the D-alanine residue of one tetrapeptide to the ε-NH_2 group of the third amino acid, L-lysine, in another tetrapeptide (see fig. 18.27). The third dimension is probably built up by bridges extending in different planes. This gigantic macromolecule has the mechanical stability required for the cell wall.

Bacterial Cell Wall Biosynthesis

A segment of bacterial cell wall has the structure of a two-dimensional network containing a parallel array of linear heteropolysaccharide strands extended in one direction, which are cross-linked with strands of an oligopeptide in the perpendicular direction (see fig. 18.27). Biosynthesis of the cell wall is unusual in two respects: (1) It is an example of the synthesis of a regularly cross-linked polymer, and (2) part of the synthesis takes place inside the plasma membrane and part takes place outside the plasma membrane. For descriptive purposes the synthesis of peptidoglycan can be conveniently broken into three stages, which occur at different locations in the cell: (1) synthesis of UDP-*N*-acetylmuramyl-pentapeptide, (2) polymerization of *N*-acetylglucosamine and *N*-acetylmuramyl-pentapeptide to form the linear peptidoglycan strands, and (3) cross-linking of the peptidoglycan strands.

Table 18.4
Glycosylation Mutants That Are Useful for Producing Glycoproteins with Modified Carbohydrates

Mutant[a] Cell Lines	Glycosylation Deficiency	Carbohydrates on Glycoproteins: N-Linked Complex Sites[b,c]	Carbohydrates on Glycoproteins: O-Linked Sites
Lec1 Clone 15B	GlcNAc-TI	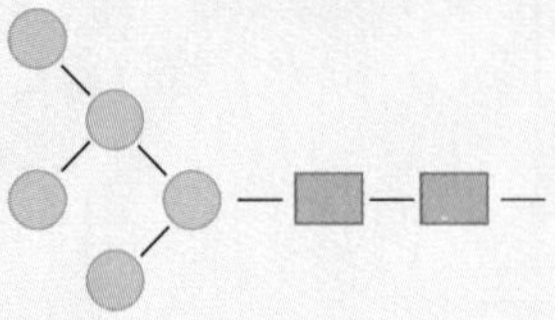 Asn	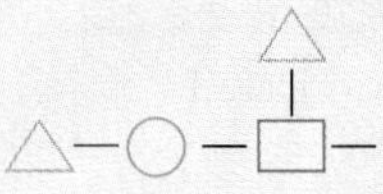 Ser
1d1D	UDP-Glc-4-epimerase	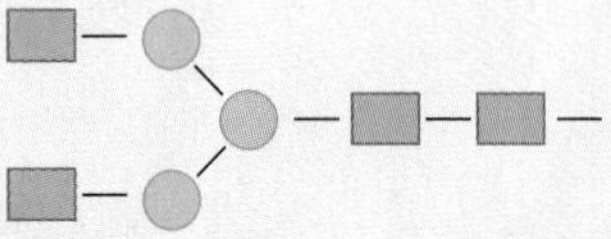 Asn	Ser
Lec8 Clone 13	UDP-Gal Golgi translocase	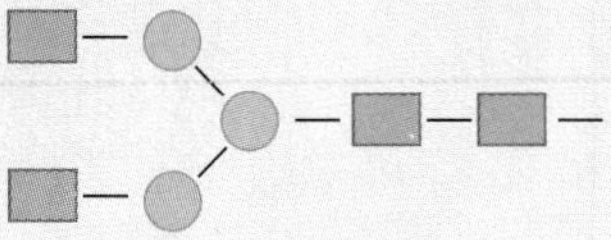 Asn	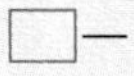 Ser
Lec2 Clone 1021	CMP-NeuNAc Golgi translocase	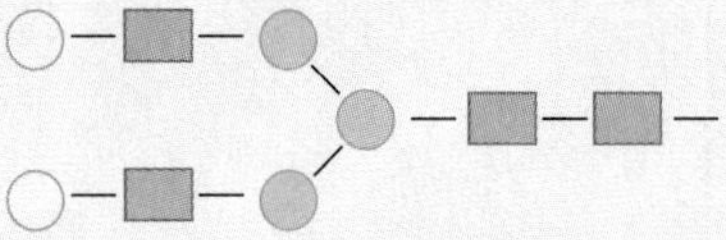 Asn	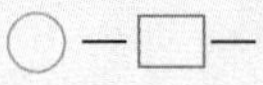 Ser

= Mannose; = GlcNAc; = Galactose; = GalNAc; = NeuNAc

[a]All mutants were derived from Chinese hamster ovary (CHO) cells.

[b]All N-linked complex structures may have fucose on the GlcNAc next to Asn, and more than two antennae.

[c]In all these mutants, sites on glycoproteins that normally carry N-linked oligomannosyl units will continue to do so.

Synthesis of the UDP-N-Acetylmuramyl-Pentapeptide Monomer Occurs in the Cytoplasm

The first stage in cell wall synthesis (fig. 18.28) involves the synthesis of UDP-*N*-acetylmuramyl-pentapeptide. First, the condensation of *N*-acetylglucosamine-1-phosphate with UTP leads to the formation of UDP-*N*-acetylglucosamine. A specific transferase catalyzes a reaction with phosphoenolpyruvate to give the 3-enolpyruvylether of UDP-*N*-acetylglucosamine. The pyruvyl group is then reduced to lactyl by an NADPH-linked reductase, thus forming the 3-*O*-D-lactylether of *N*-acetylglucosamine. This compound is known as UDP-*N*-acetylmuramic acid (see fig. 18.28 for its detailed structure).

Conversion of UDP-*N*-acetylmuramic acid to its pentapeptide form occurs by the sequential addition of the necessary amino acids. Each step requires ATP and a specific enzyme that ensures the addition of amino acids in the proper sequence; L-alanine is added first, followed by D-glutamic acid, L-lysine (attached by its α-amino group to the γ-carboxyl group of the glutamic acid), and finally the dipeptide D-alanyl-D-alanine is added as a unit. The latter dipeptide is formed by two enzymatic reactions: conversion of L-alanine to D-alanine by a racemase, followed by the linking of the two alanine residues in an ATP-requiring reaction to form D-alanyl-D-alanine. All of these reactions occur in the cytoplasm of the bacterial cell.

Formation of Linear Polymers of the Peptidoglycan Is Membrane-Associated

The most complex stage in peptidoglycan synthesis takes place on the plasma membrane. It may be divided into five steps, which are illustrated in figure 18.29. This stage involves the polymerization of *N*-acetylglucosamine and *N*-acetylmuramyl-pentapeptide-containing residues into peptidoglycan strands.

In step 1 (see fig. 18.29), UDP-*N*-acetylmuramyl-pentapeptide reacts with a 55-carbon isoprenyl alcohol known as undecaprenol phosphate. This lipid is similar in structure (fig. 18.30) and function to the dolichol phosphates involved in glycoprotein synthesis in eukaryotes. A pyrophosphate linkage is formed with the lipid, and UMP is released.

In step 2 (see fig. 18.29), *N*-acetylglucosamine is added to the lipid intermediate by means of a typical transglycosylation

Figure 18.26

Diagram of a Gram negative cell envelope. The trimers of matrix protein of the outer membrane are associated with lipoprotein and with lipopolysaccharide (of variable polysaccharide length), and lipoprotein is covalently bound to peptidoglycan. Diagram also illustrates some general properties of membranes. Phospholipid molecules are illustrated with a circle for the polar groups and a line for each fatty acid acyl moiety. (Courtesy M. Inouye.)

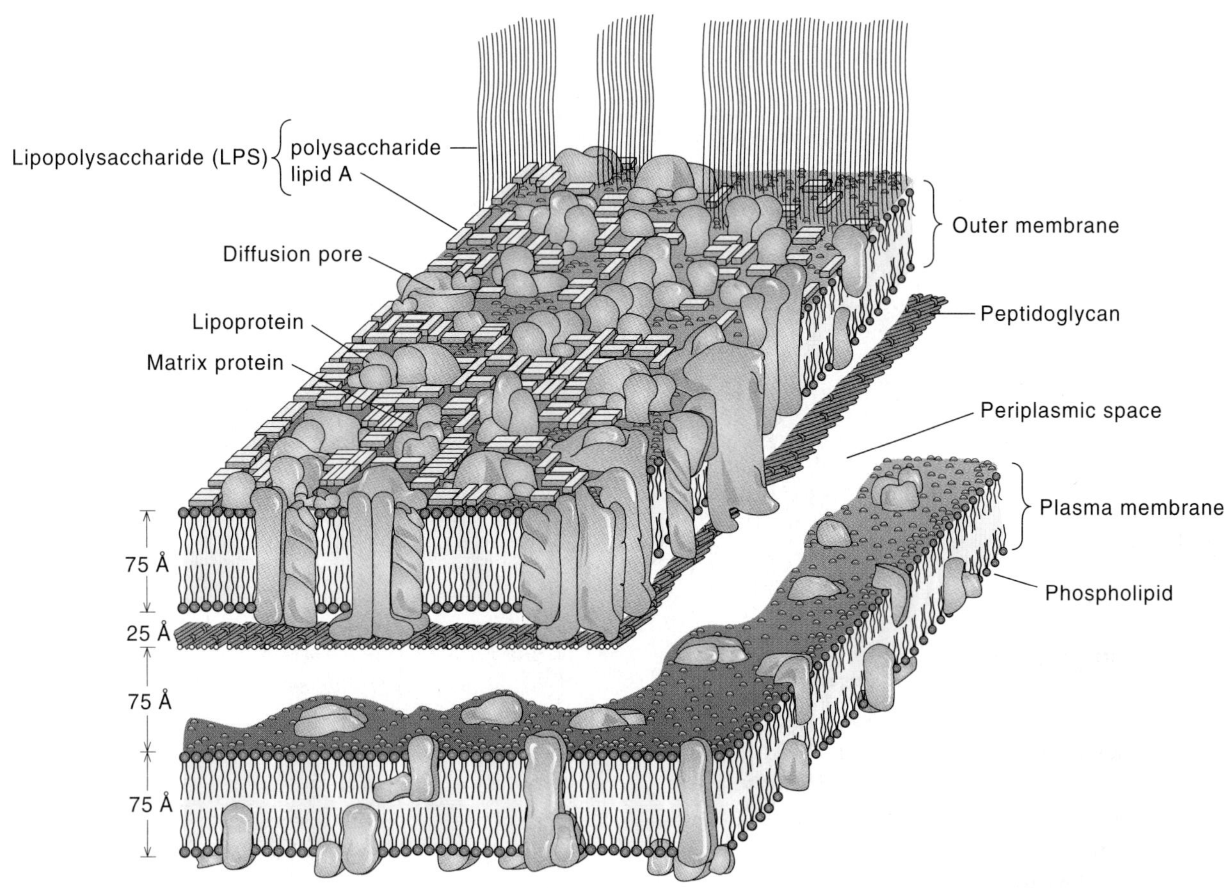

from UDP-*N*-acetylglucosamine, and UDP is released. In step 3, five glycine residues are sequentially added to the ϵ-amino group of lysine. Curiously, these glycines are activated by ester formation to a transfer RNA molecule. This form of amino acid activation is rarely seen except in protein synthesis (see chapter 34). Clearly, the primary function of transfer RNAs is for protein synthesis. The involvement of transfer RNA in cell wall synthesis seems opportunistic or it may indicate a function that tRNAs possessed prior to the advent of translation systems.

In step 4 the disaccharide–oligopeptide unit is transferred from the lipid intermediate to the growing peptidoglycan, and lipid pyrophosphate is generated. It is in this step that we see the dual function of the undecaprenol lipid. The lipid not only serves to activate the monomer for addition to polymer, but also transports the monomer from the cytoplasmic side of the membrane to the extracellular side of the membrane, where cell wall assembly must take place. Little is known about the details of this transport process.

In the fifth and final step, one phosphate is hydrolyzed to regenerate the phospholipid, which then can react once again with UDP-*N*-acetylmuramyl-pentapeptide and participate in another cycle, resulting in the addition of a new unit to the growing peptidoglycan strand. The antibiotic bacitracin is a specific inhibitor of the dephosphorylation of the pyrophosphate form of the lipid.

Cross-Linking of the Peptidoglycan Strands Occurs on the Noncytoplasmic Side of the Plasma Membrane

The cross-linking of peptidoglycan strands takes place outside the cell membrane, at the site of the preexisting wall. Since no ATP or other obvious energy source is available there, a mechanism independent of any external energy source has evolved for this reaction. The reaction is a transpeptidation in which the terminal amino end of an open cross-bridge attacks the terminal peptide bond in an adjustment strand to form a cross-link. The terminal alanine residue from the strand that becomes cross-linked is thus eliminated (fig. 18.31).

Figure 18.27

Structure of the peptidoglycan of the cell wall of *Staphylococcus aureus.* (*a*) In this representation, X (*N*-acetylglucosamine) and Y (*N*-acetylmuramic acid) are the two sugars in the peptidoglycan. Light green circles represent the four amino acids of the tetrapeptide L-alanyl-D-glutamyl-L-lysyl-D-alanine. Dark green circles are pentaglycine bridges that interconnect peptidoglycan strands. The nascent peptidoglycan units bearing open pentaglycine chains are shown at the left of each strand. TA—P is the teichoic acid antigen of the organism, which is attached to the polysaccharide through a phosphodiester linkage. (*b*) The structures of X (*N*-acetylglucosamine) and Y (*N*-acetylmuramic acid) are connected by $\beta(1,4)$ linkages that alternate in the glycan strand. (*c*) The structure of a segment of the peptidoglycan before and after the final cross-linking reaction.

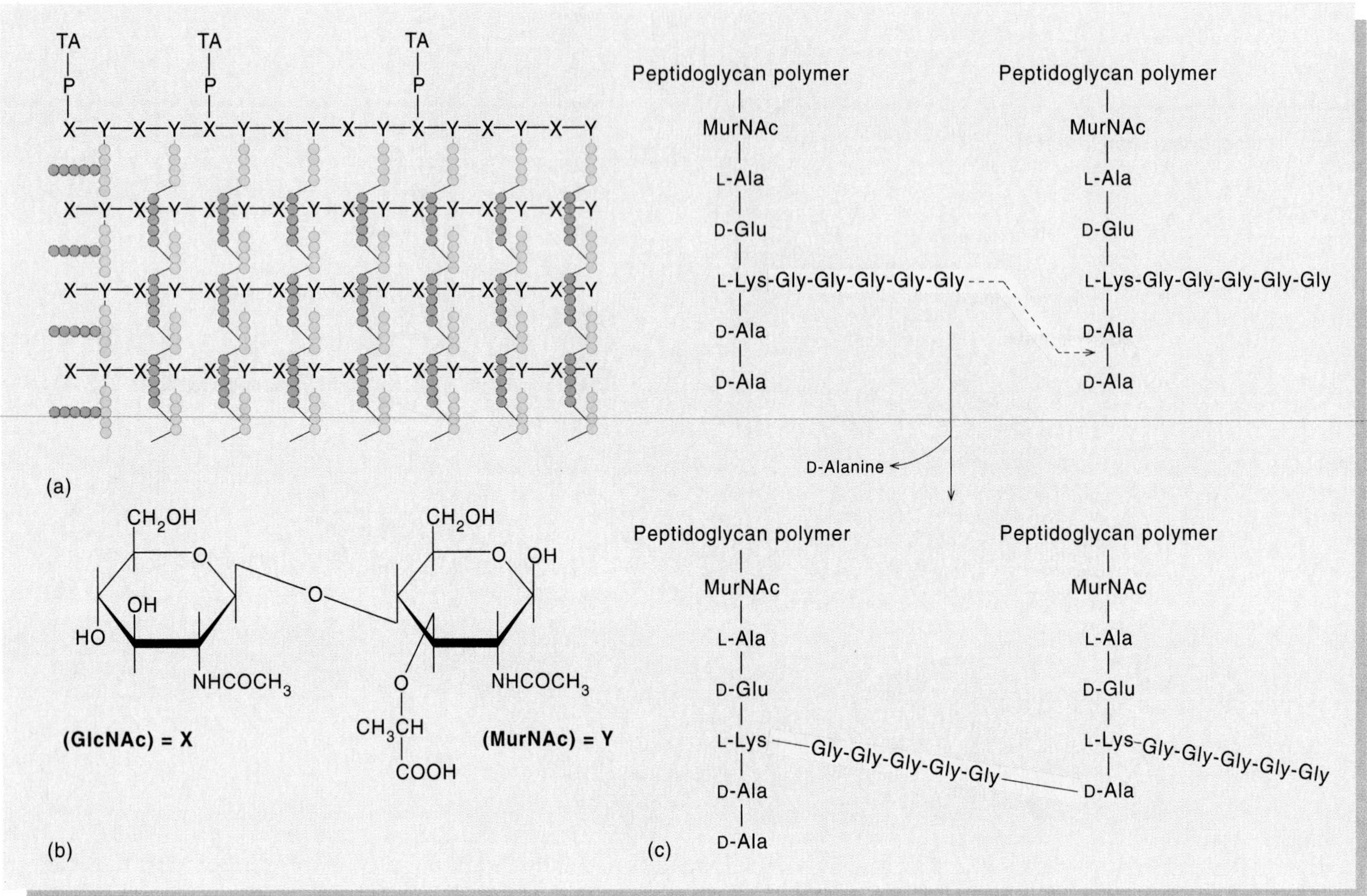

Penicillin Inhibits the Transpeptidation Reaction

Penicillin has been unequaled for usefulness in combatting bacterial diseases and infections. During the 50 years since Fleming brought penicillin to the attention of microbiologists, many biochemists and pharmacologists have been interested in the mechanism by which this potent antibiotic kills bacteria. It is known that penicillin inhibits the cross-linking reaction by acting as a structural analog of the terminal D-alanyl-D-alanine residue of the peptidoglycan strand. The similarity between the conformation of the penicillin molecule and one of the conformations of the dipeptide D-alanyl-D-alanine is shown in figure 18.32.

It is thought that the transpeptidase (TPase) first reacts with the substrate to form an acyl enzyme intermediate, with the elimination of D-alanine, and that this active intermediate then reacts with another strand to form the cross-link and regenerate the enzyme. Because penicillin is an analog of alanyl-alanine, it should fit the substrate-binding site, with the highly reactive CO—N bond in the β-lactam ring in the same position as the bond involved in the transpeptidation. It thus has the potential to acylate the enzyme, forming a penicilloyl enzyme, and thereby inactivate it (see fig. 18.31). In support of this view is the fact that penicilloyl is the piece of the antibiotic found in inhibited enzymes. An acyl group derived from the substrate is used to form an acyl enzyme intermediate. Most important for verifying the proposed mechanism of penicillin action, the antibiotic-derived penicilloyl

Figure 18.28

The first stage of cell wall synthesis: formation of UDP-*N*-acetylmuramyl-pentapeptide (full structure shown at bottom). Points of inhibition by the antibiotics penicillin and phosphonomycin are indicated.

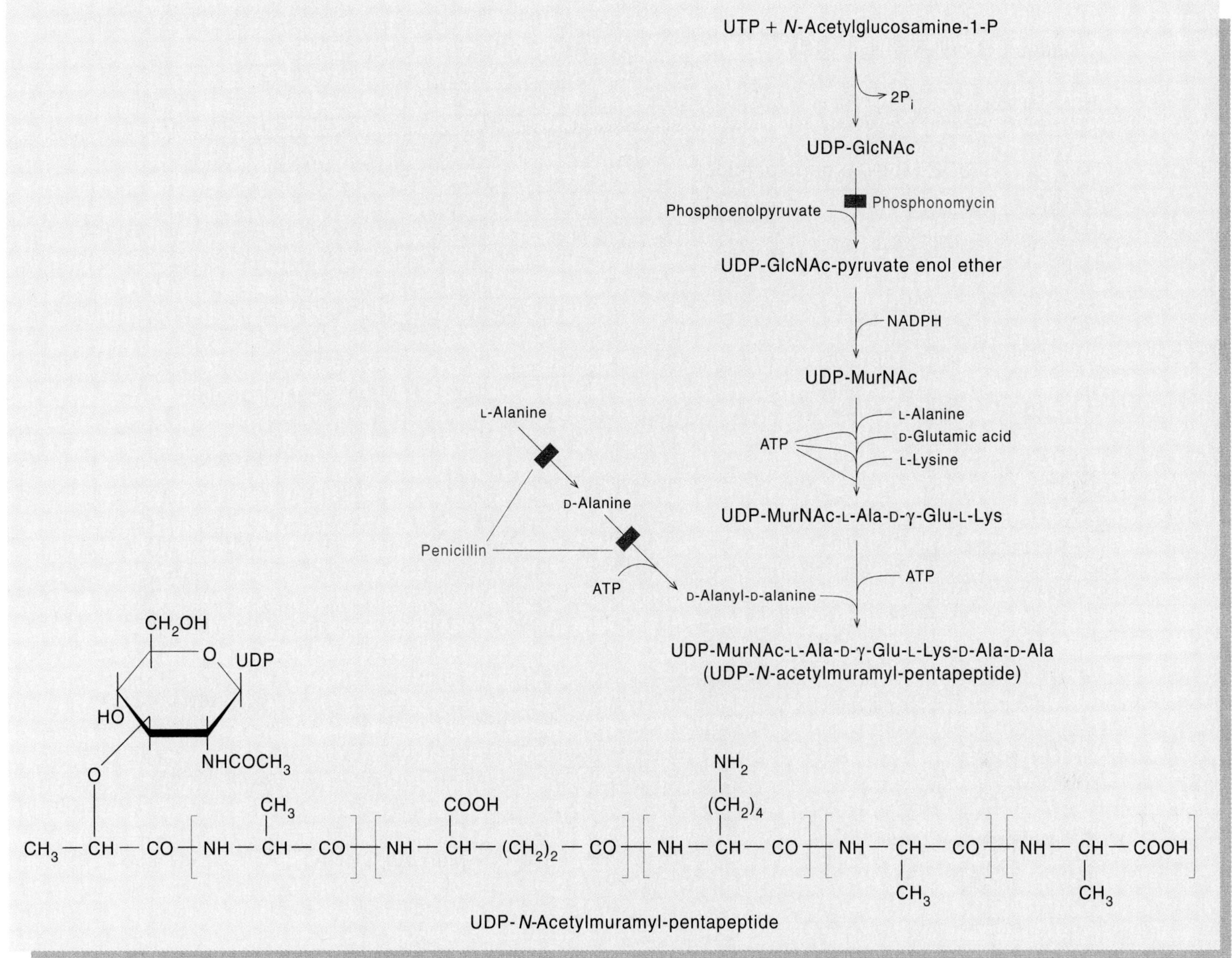

moiety and the substrate-derived acyl moiety are substituted on the same site in the penicillin-sensitive enzymes from a variety of genera of bacteria.

A number of other antibiotics in addition to penicillin (phosphonomycin, bacitracin, and vancomycin) block other stages in cell wall synthesis. Their points of action are indicated in figures 18.28 and 18.29. In addition to their biological and medical importance, these antibiotics have been extremely useful in elucidating the biosynthetic pathway. This is because they permit accumulation of the product before the blocked step; the product can then frequently be isolated and confirmed as a genuine intermediate in the pathway.

Figure 18.29

The second stage of cell wall synthesis. An ATP-requiring amidation of glutamic acid that occurs between steps 2 and 3 has been omitted. Points of action of the antibiotic inhibitors bacitracin and vancomycin are indicated.

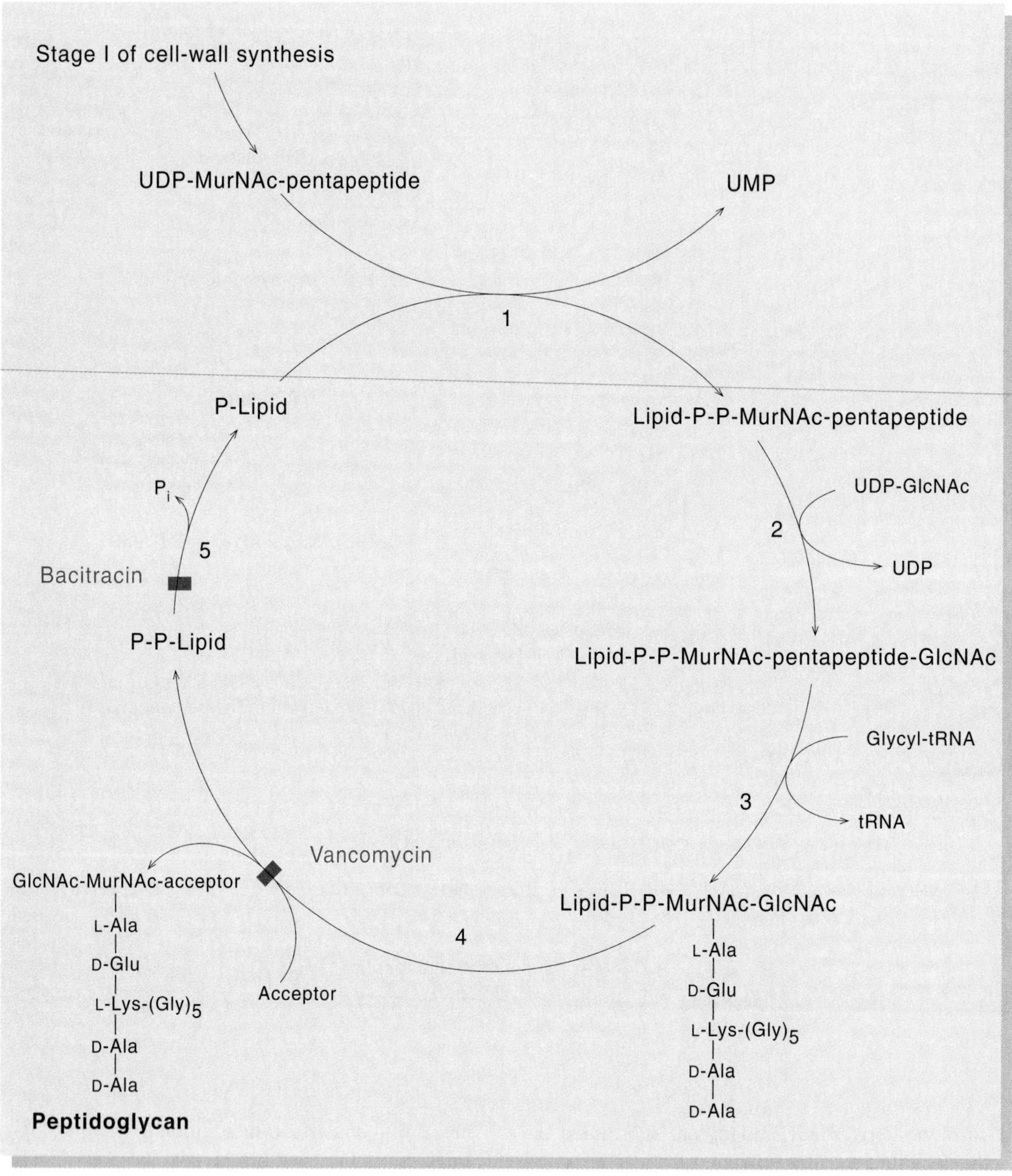

Figure 18.30

The structure of undecaprenol phosphate. Isoprene phosphates such as undecaprenol phosphate are important carriers and activators in the synthesis of oligosaccharides. Note the similarities between undecaprenol phosphate and dolichol phosphate (Fig. 18.21).

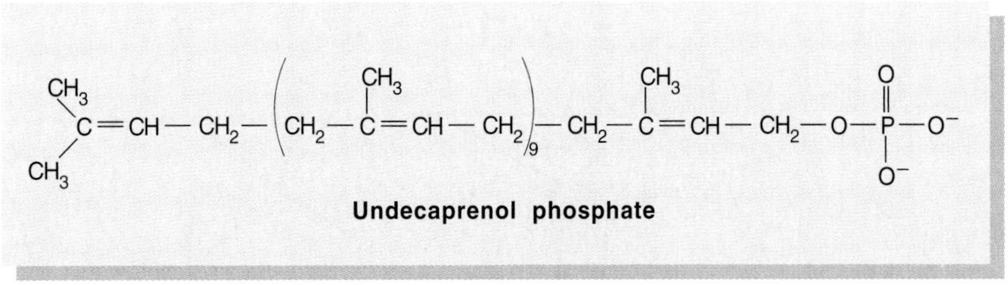

$$(CH_3)_2C{=}CH{-}CH_2{-}\left(CH_2{-}\overset{CH_3}{\overset{|}{C}}{=}CH{-}CH_2\right)_9{-}CH_2{-}\overset{CH_3}{\overset{|}{C}}{=}CH{-}CH_2{-}O{-}\underset{O^-}{\underset{|}{\overset{O}{\overset{\|}{P}}}}{-}O^-$$

Undecaprenol phosphate

Figure 18.31

The third stage of cell wall synthesis. This diagram shows the cross-linking reaction and the mechanism of inhibition by penicillin in the bacterium *Staphylococcus aureus.* (*a*) The end of the peptide side chain of a glycan strand. (*b*) The end of the pentaglycine substituent from an adjacent strand.

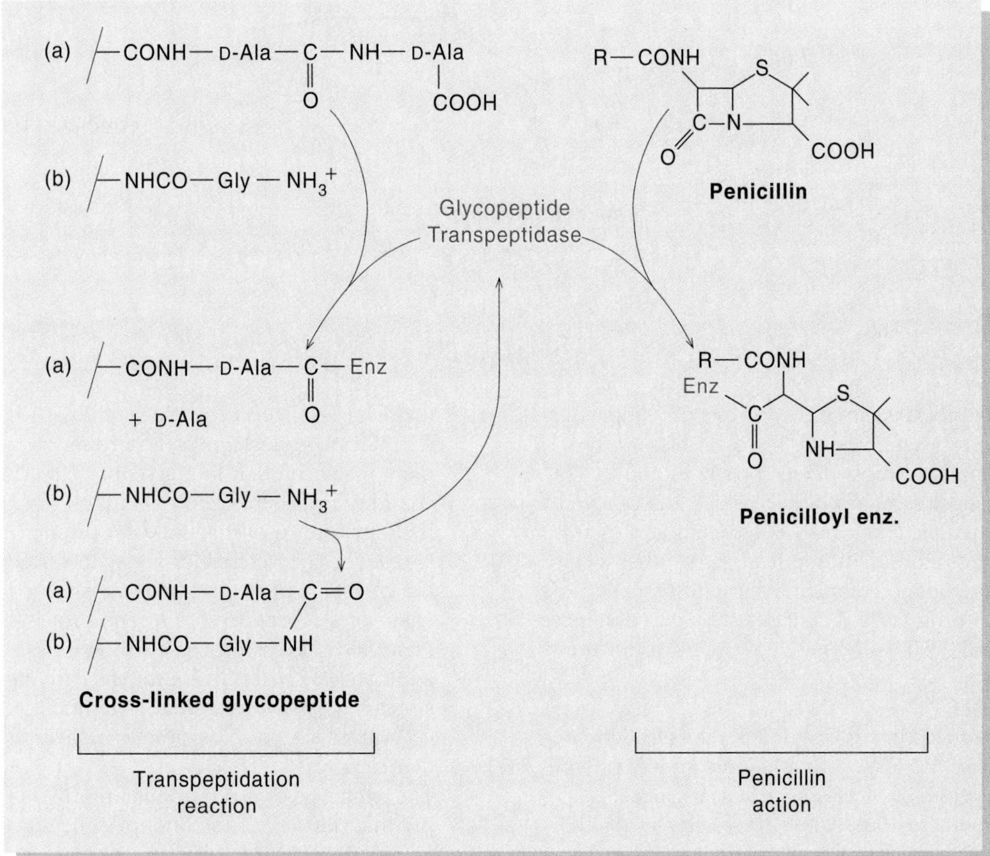

Figure 18.32

Stereomodels of penicillin (middle, left) and of the D-alanyl-D-alanine end of the peptidoglycan strand (middle, right). Arrows indicate the position of the CO—N bond in the β-lactam ring of penicillin and of the CO—N bond in D-alanyl-D-alanine at the end of the peptidoglycan strand.

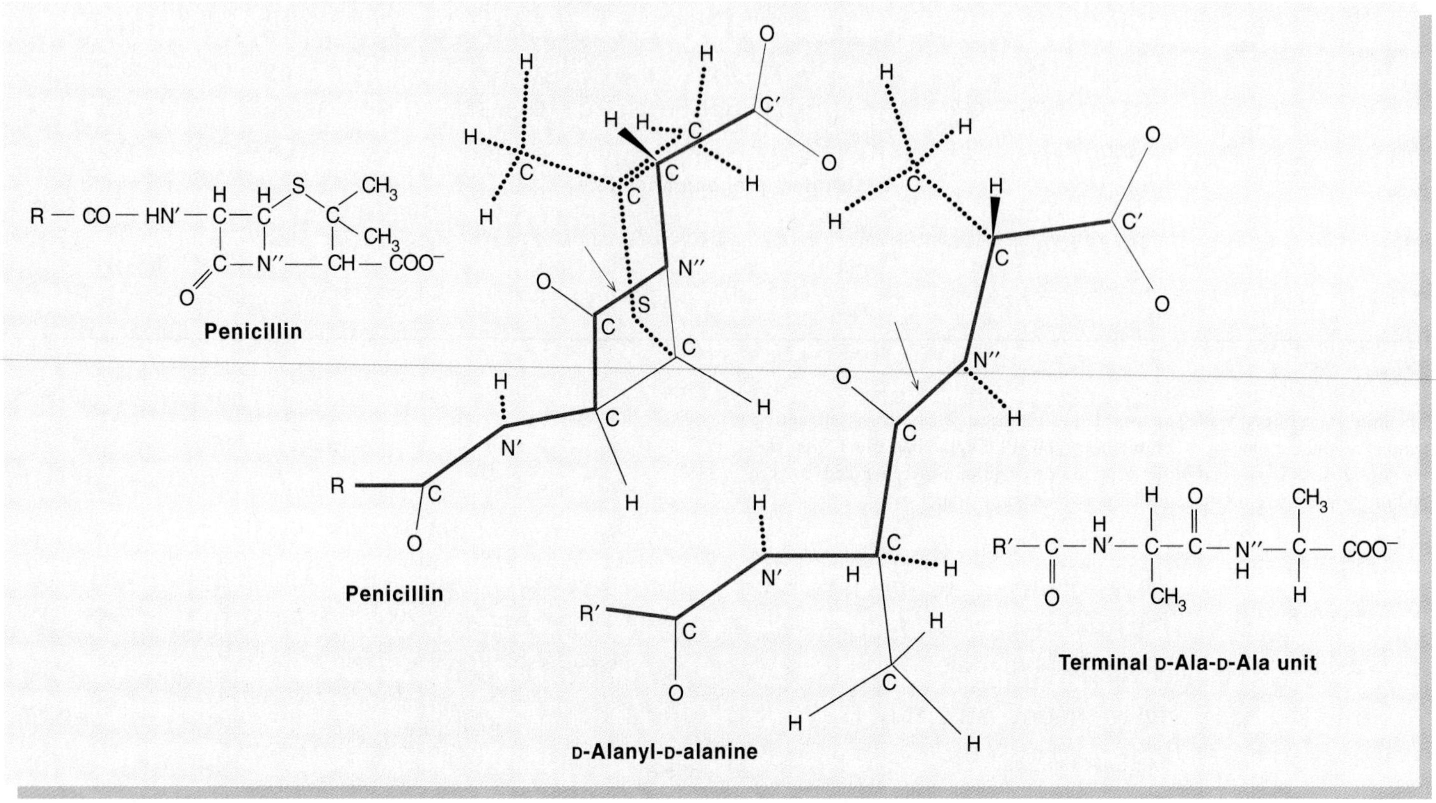

Summary

In this chapter we have focused on the synthesis of complex carbohydrates. We began by examining the hexoses that are the building blocks of complex carbohydrates, briefly turned to synthesis of some of the most common disaccharides, then moved to simple homopolysaccharides and a brief consideration of heteropolymers that contain more than one hexose. We then dealt with glycoproteins that contain complex linear and branched carbohydrates attached to proteins and concluded by describing the synthesis of the bacterial cell wall. The chief points in our presentation are as follows:

1. Hexoses, which are the primary building blocks of oligosaccharides and polysaccharides, come in a large variety of types. All hexoses can be thought of as derivatives of glucose through a series of conversions. These conversions usually occur at the level of the monophosphorylated sugar or the nucleoside diphosphate sugar. The nucleotide-sugar is also the activated substrate for formation of disaccharides, oligosaccharides, and polysaccharides.
2. In higher animals many different branched-chain oligosaccharides are found as conjugates in glycolipids and glycoproteins. A large number of sugars and specific glycosyltransferases are involved in oligosaccharide synthesis. We can distinguish two types of oligosaccharides, according to their mode of attachment to the protein in a glycoprotein. The O-linked oligosaccharides are synthesized directly on the amino acid side-chain hydroxyl group of a serine or a threonine. The N-linked oligosaccharides are synthesized first on a long-chain dolichol phosphate and then transferred to the asparagine side chain of a receptor protein. The protein-attached N-linked oligosaccharide is processed by removal of certain sugars and addition of others. Glycoproteins carrying altered or immature structures can be obtained from mutant or inhibitor-treated cells in which the biosynthetic pathway is blocked at a particular step.
3. The synthesis of glycoproteins mostly takes place in two cytoplasmic organelles: the endoplasmic reticulum and the Golgi apparatus. Glycoproteins move between the compartments of these organelles and from the Golgi complex to the plasma membrane in coatamer-coated vesicles. The oligosaccharide portion of the glycoprotein is usually important in its folding, solubility, and stability and may be required for recognition by carbohydrate-binding proteins. The Man-6-P receptor sorts lysosomal hydrolases to lysosomes because of recognition of

their N-linked oligomannosyl carbohydrates with Man-6-P residues. Cell–cell adhesion events important in lymphoid cell trafficking are mediated by recognition of specific carbohydrates on particular cell surface glycoproteins.

4. The bacterial cell wall contains a heteropolymeric polysaccharide chain that is cross-linked by peptide linkages. The complexity of the resulting peptidoglycan results in part from the complex repeating units and in part from the fact that a cross-linked polymer is being made outside the cell. The partially completed polysaccharide structures are transferred from the cytoplasm to extracellular space by attachment to a long-chain bactoprenol lipid (undecaprenol) that can traverse the cell membrane. The lipid is similar in structure and function to the dolichol phosphate used in oligosaccharide synthesis in animals. Various antibiotics that block specific steps in cell wall synthesis are of great importance. They have also been very helpful in elucidating the biochemical pathway.

Selected Readings

Albeijon, C., and C. B. Hirschberg, Topography of glycosylation reactions in the endoplasmic reticulum. *Trends Biochem. Sci.* 17:32–36, 1992. Summarizes the location in the lumen and membranes of the endoplasmic reticulum of molecules and reactions involved in carbohydrate biosynthesis.

Brandli, A. W., Mammalian glycosylation mutants as tools for the analysis and reconstitution of protein transport. *Biochemistry* 276:1–12, 1991. Describes the uses of glycosylation mutants in developing assays to follow vesicular transport between cellular compartments.

Elbein, A. D., Inhibitors of the biosynthesis and processing of N-linked oligosaccharide chains. *Ann. Rev. Biochem.* 56:497–531, 1987.

Ferguson, M. A. J., and A. F. Williams, Cell surface anchoring of proteins via glycosylphosphatidylinositol structures. *Ann. Rev. Biochem.* 57:285–320, 1988. Excellent review of the discovery and basic features of glypiated proteins.

Fukuda, M. N., Hempas disease: Genetic defect of glycosylation. *Glycobiology* 1:9–15, 1990. Reviews the evidence suggesting that CDAII is a disease arising from a defect in N-linked carbohydrate biosynthesis.

Ginsberg, V., and P. Robbins, eds., *Biology of Carbohydrates.* New York: JAI Press, 1991. Chapters written by different authors on several of the diverse biological functions of carbohydrates.

Hart, G. W., Glycosylation. *Curr. Opinion Cell Biol.* 4:1017–1023, 1992. Summarizes the many types and functions of carbohydrate modification found on eukaryotic glycoproteins.

Hassell, J. R., J. H. Kimina, and L. Cantly, Proteoglycan core protein families. *Ann. Rev. Biochem.* 55:539–568, 1986. Review of protein sequences and properties of core proteins in proteoglycans.

Herscovics, A., and P. Orlean, Glycoprotein biosynthesis in yeast. *FASEB J.* 7:540–550, 1993.

Hillmen, P., M. Bessler, P. J. Mason, W. M. Watkins, and L. Luzzatto, Specific defect in *N*-acetylglucosamine incorporation in the biosynthesis of the glycosylphosphatidylinositol anchor in cloned cell lines from patients with paroxysmal nocturnal hemoglobinuria. *Proc. Natl. Acad. Sci. USA* 90:5272–5276, 1993. The basis of the acquired disease paroxysmal nocturnal hemoglobinuria is identified as the first step in the pathway of GPI anchor biosynthesis.

Kelly, R. J., L. K. Ernst, R. D. Larsen, J. G. Bryant, J. S. Robinson, and J. B. Lowe, Molecular basis for H blood group deficiency kin Bombay (O_h) and para-Bombay individuals. *Proc. Natl. Acad. Sci. USA* 91:5843–5847, 1994. Mutations that occur in the (1,2) fucosyltransferase that makes the H blood group determinant are responsible for the blood groups termed Bombay and para-Bombay.

Kennedy, J. F., *Carbohydrate Chemistry.* Oxford, England: Oxford University Press, 1985. Monograph describing the basics of carbohydrate chemistry including ways in which to detect different types of sugars.

Kleene, R., and E. G. Berger, The molecular and cell biology of glycosyltransferases. *Biochim. Biophys. Acta* 1154:283–325, 1993. Reviews the molecular biology, structure, and functional relationships of mammalian glycosyltransferases.

Kochetkov, N. K., and V. N. Shibaev, Glycosyl esters of nucleoside pyrophosphates, *Adv. Carbohydr. Chem. Biochem.* 28:307–325, 1973. A concise review of the chemistry and biochemistry of nucleoside pyrophosphate sugars and derivatives.

Krieger, M., P. Reddy, K. Kozarsky, D. Kingsley, and M. Penman, Analysis of the synthesis, intracellular sorting and function of glycoproteins using a mammalian cell mutant with reversible glycosylation defects. *Methods Cell Biol.* 32:57–84, 1989. Describes the properties of the CHO glycosylation mutant IdID, which can be used to obtain glycoproteins lacking O-linked carbohydrates.

Kukowsaka-Latallo, J. R., R. D. Larsen, R. P. Nair, and J. B. Lowe, A cloned human cDNA determines expression of a mouse stage specific embryonic antigen and the Lewis blood group $\alpha(1,3/1,4)$ fucosyltransferase. *Genes Dev.* 4:1288–1303, 1990. Presents the first expression cloning of a glycosyltransferase cDNA and shows that this enzyme can transfer fucose in two different linkages in contrast to most glycosyltransferases that conform to the rule "one transferase, one linkage."

Lis, H., and N. Sharon, Lectins as molecules and as tools. *Ann. Rev. Biochem.* 55:35–67, 1986. Review of the nature and sugar-binding properties of plant and animal lectins.

Lis, H., and N. Sharon, Protein glycosylation: Structural and functional aspects. *Eur. J. Biochem.* 218:1–27, 1993. Up-to-date overview of the structures of mammalian carbohydrates and their intramolecular and intermolecular roles in glycoproteins.

McNeil, M., A. G. Darvill, S. C. Fry, and P. Albersheim, Structure and function of the primary cell walls of plants. *Ann. Rev. Biochem.* 53:625–664, 1984. An insight into the complex carbohydrates involved in plant cell walls and their biologically active breakdown products.

Pigman, W., *Carbohydrates: Chemistry and Biochemistry,* Vol. 1B, 2d ed. New York: Academic Press, 1980. Basic text on carbohydrate chemistry and biochemistry by one of the first authors of such a text.

Rothman, J. E., Mechanisms of intracellular protein transport. *Nature* 372:55–63, 1994. Review of the molecules identified initially by in vitro assays of intercompartmental vesicular transport and their roles in the secretory pathways of mammalian cells and yeast.

Ruoslahti, E., Structure and biology of proteoglycans. *Ann. Rev. Cell. Biol.* 4:229–255, 1988.

Sharon, N., and H. Lis, Carbohydrates in cell recognition. *Sci. Am.* January: 82–89, 1993.

Stanley, P., Glycosylation mutants of animal cells. *Ann. Rev. Genet.* 18:5251–5252, 1984. Summary of mammalian mutants affected in specific glycosylation reactions and how they may be used to identify carbohydrate functions.

Stanley, P., Glycosylation engineering. *Glycobiology* 2:99–107, 1992. A review of the uses of mammalian cell glycosylation mutants for producing recombinant glycoproteins with a desired complement of carbohydrate structures that may be advantageous for optimal biological activity or targeting of the glycoprotein.

Sweeley, C. C., and H. Nunez, Structural analysis of glycoconjugates by mass spectrometry and nuclear magnetic resonance spectroscopy. *Ann Rev. Biochem.* 54:765–801, 1985.

Takeda, J., and T. Kinoshita, GPI-anchor biosynthesis. *Trends in Biochem. Sci.* 20:367–371, 1995.

Varki, A., Selectin ligands. *Proc. Natl. Acad. Sci. USA* 91:7390–7397, 1994. A discussion of the criteria that must be applied to identify ligands for selectins that operate in vivo. Summary of selectins and their ligands.

Vliegenthart, J. F. G., L. Dorland, and H. van Halbeek, High-resolution H-nuclear magnetic resonance spectroscopy as a tool in the structural analysis of carbohydrates related to glycoproteins. *Adv. Carbohydr. Chem. Biochem.* 411:209–374, 1983. Review of IH-NMR spectra of several hundred carbohydrates with their chemical shift assignments.

Von Figura, K., and A. Hasilik, Lysosomal enzymes and their receptors. *Ann. Rev. Biochem.* 55:167–193, 1986. Review of the experiments proving the many soluble lysosomal enzymes are sorted to lysosomes through their association with a Man-6-phosphate receptor.

Weis, W. I., K. Drickamer, and W. A. Henderson, Structure of a C-type mannose-binding protein complexed with an oligosaccharide. *Nature* 360:127–134, 1992. First x-ray structure of an animal lectin complexed with a complex carbohydrate.

Yamamoto, F., H. Clausen, T. White, J. Marken, and S. Hakomon, Molecular genetic basis for the histo blood group ABO system. *Nature* 345:229–233, 1990. Cloning of the first blood group generating glycosyltransferases and molecular description of the ABO gene.

Yocum, R. R., J. R. Rasmussen, and J. L. Strominger, The mechanism of action of penicillin: Penicillin acylates the active site of *Bacillus stearothermophilus* D-alanine carboxypeptidase. *J. Biol. Chem.* 255:3977–3986, 1980. Demonstration of the reaction by which penicillin specifically disrupts bacterial cell-wall biosynthesis.

Problems

1. α-Lactalbumin has at least two biological functions: (1) a nutritional protein for young mammals and (2) a component of lactose synthetase that alters the specificity of galactosyl transferase. What problem(s) can you envision that might have occurred in the evolution of this bifunctional protein?
2. In the N-glycosidic linkages in glycoproteins the Asn-AA_x-Ser(Thr) sequence is involved with the Asn amide nitrogen bonded to the sugar residue (fig. 18.12b). A hydrogen bond is thought to exist between the serine (threonine) alcohol group and the β-carbonyl oxygen of the Asn (stabilizing the geometry). Draw the structure of this hydrogen bond.
3. During the cross-linking reaction in peptidoglycan biosynthesis (fig. 18.31) a D-alanine is cleaved from the peptidoglycan polymer. What do you think is the fate of that D-alanine?
4. The synthesis of dextrans does not use nucleotide sugars as seen in the synthesis of other glucose polymers. What is the driving force for this reaction, i.e., how would you analyze the thermodynamics?
5. How can lectins be used to identify complex carbohydrate structures on glycolipids or glycoproteins? (Hint: You could label the lectins with radioactive iodine [^{125}I] or use lectins bound to a column matrix.)
6. A great variety of different oligosaccharides result from a limited number of sugars. Explain.
7. How can two different genes for different glycosyltransferases determine ABO blood group types in humans? Explain why blood group O is considered a universal donor. If you have AB type blood, why can you accept any blood type? A small number of people lack the H antigen (the glycosyltransferase that adds Fuc α1, 2) and have Bombay type blood. What blood type could you give to a person with Bombay type blood and why?
8. There is a lot of interest in expressing human genes in bacteria to produce a useful human protein in large amounts. Why would this approach not be expected to work with some human proteins?
9. A mutant cell line that does not synthesize Dol-P-Man does not express glypiated proteins on its cell surface. Why?
10. The sequence of amino acids in a protein is determined by the base sequence in DNA. What determines the sequence of sugar residues in oligosaccharides and polysaccharides?
11. The interconversion of glucose and galactose occurs at the UDP-hexose level (fig. 18.1) with a 4-epimerase. The epimerase is unusual in that it has a covalently bound NAD^+. What function do you propose for the NAD^+, and what intermediate do you expect in the reaction?
12. Many amino sugars are found in nature, but glucosamine and galactosamine (figs. 18.1 and 18.2) are by far the most common. By examining their biosynthesis, can you explain why the most common amino sugars in nature are 2-amino sugars?
13. Throughout this chapter, notice the occurrence of *N*-acetylglucosamine and *N*-acetylgalactosamine in oligo- or polysaccharides. Why are these sugars present in the *N*-acetyl form?
14. Gluconic acid, as its phosphate derivative, is a metabolite in the pentose phosphate pathway. Why isn't gluconic acid, like glucuronic acid, listed in figure 18.1 as a common component of oligo- or polysaccharides?
15. The hydrophobic amino acid sequence that serves as the binding site for the SRP (fig. 18.19) is on the N terminus of the protein. Why isn't it on the C terminus?
16. Figure 18.29 shows a series of reactions in a cycle. Is this a series of reactions like the tricarboxylic acid cycle, or does something actually "go" around the cycle?
17. Compare the pentapeptide portion of the peptidoglycans (fig. 18.29) with a normal sequence of amino acids in a protein. What differences can you identify?
18. What complications have to be overcome to synthesize complex carbohydrates (such as cell-wall components and O antigens) outside the cell?
19. Bacteria starved for an essential nutrient are not affected by penicillin. Can you explain this?

PART

5

METABOLISM OF LIPIDS

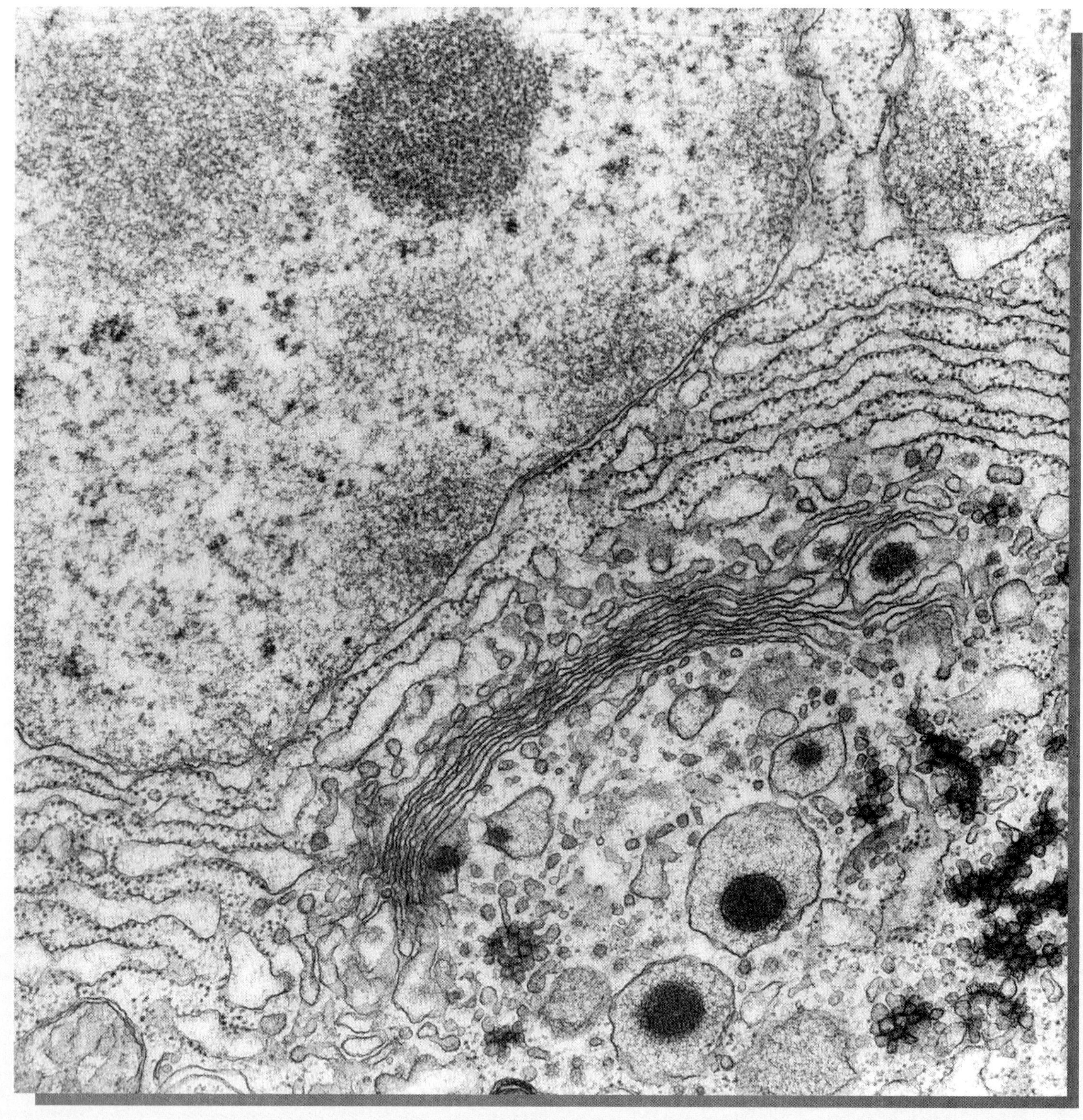

A transmission electron micrograph of an acidic proteoglycan secreting cell from a copepod. The nucleus is at upper left with rough endoplasmic reticulum closest to the nuclear membrane. The smooth endoplasmic reticulum are closest to the Golgi apparatus, which is centrally located in this micrograph. (Courtesy Estate of Dr. Brij L. Gupta, 1964, Ph.D. Thesis, Cambridge University, England.)

C H A P T E R

LIPIDS AND MEMBRANES

19

Membranes contain amphipathic lipid and protein molecules that form bilayers with the hydrophilic groups exposed.

Every cell is surrounded by a plasma membrane that creates a compartment where the functions of life can proceed in relative isolation from the outside world (fig. 19.1). The plasma membrane keeps proteins and other essential materials inside the cell. But the plasma membrane is not simply an inert barrier. Proteins embedded in it work to bring nutrients into the cell and extrude waste products. Other membrane-bound proteins may sense the cell's surroundings, communicate with other cells, or act to move the cell to a new location. In aerobic bacteria, plasma membranes house the electron-transfer reactions that provide the cell with energy.

In addition to their plasma membrane, eukaryotic cells also contain internal membranes that define a variety of organelles (fig. 19.2). Each of these organelles is specialized for particular functions: The nucleus synthesizes nucleic acids, mitochondria oxidize carbohydrates and lipids and make ATP, chloroplasts carry out photosynthesis, the endoplasmic reticulum and the Golgi apparatus synthesize and secrete proteins and lipids, and lysosomes digest proteins. Additional membranes divide mitochondria and chloroplasts into even finer, more specialized subcompartments. Like the plasma membrane, organellar membranes act as barriers to the leakage of proteins, metabolites, and ions; they contain transport systems for import and export of materials, and they are the sites of enzymatic activities as diverse as cholesterol biosynthesis and oxidative phosphorylation.

Although membranes from different organelles or cells may have very different activities, they share the following basic properties:

1. As a rule, biological membranes are impermeable to polar molecules or ions. The nuclear membrane and the outer mitochondrial membrane do have pores that admit relatively large molecules (see fig. 19.2). More commonly, however, an ion or polar molecule can cross a biological membrane only if the membrane has a protein that is a specific transporter for that molecule.
2. Membranes are not rigid but rather adapt flexibly to changes in cellular or organellar shape and size.
3. They are durable. The plasma membrane of an erythrocyte, for example, experiences constant buffeting as the blood courses through the capillaries, and yet it survives for the

Figure 19.1

Cross section of microvilli of cat intestinal epithelial cells, showing the trilaminar (three-layer) structure of the cytoplasmic membranes (165,000×). (Courtesy of Dr. S. Ito.)

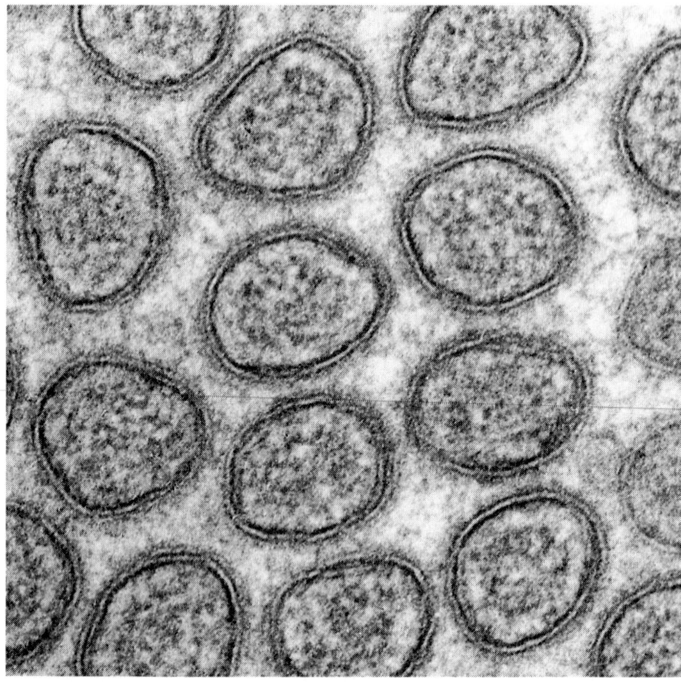

Figure 19.2

Electron micrograph of a cell from the rat pancreas, showing several different intracellular organelles. (PM = plasma membrane; NE = nuclear envelope; Nu = nucleolus; M = mitochondrion; ER = endoplasmic reticulum; Go = Golgi apparatus; arrows show pore complexes in the nuclear envelope; 25,000×.) (From S. L. Wolfe, *Biology of the Cell,* 2d ed., Copyright © 1981, Wadsworth Publishing Co.)

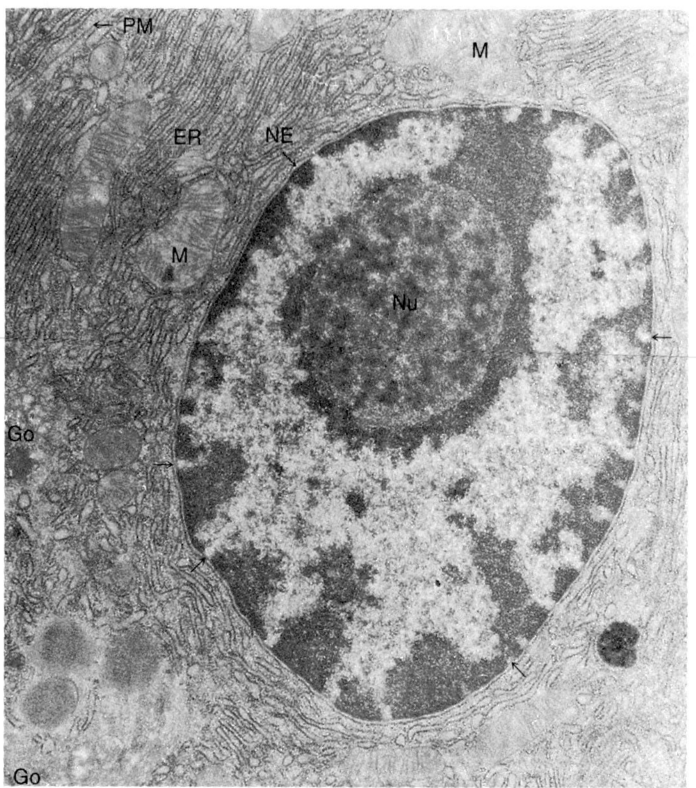

lifetime of the cell, a matter of months. The membrane either must be remarkably resistant to damage or must reseal very quickly if it is breached.

4. When viewed with an electron microscope after thin-sectioning and staining, membranes typically have a trilaminar appearance of two dark lines separated by a lighter space, with a total thickness on the order of 40 Å (see fig. 19.1).
5. Membranes contain proteins that are not simply structural in nature but have a variety of enzymatic activities. The identities and amounts of these proteins vary in different cells and may change with time in response to changing conditions.

Our first goal in this chapter is to explain how biological membranes can be, on the one hand, thin and flexible, and, on the other, durable and functional. We start by examining the constituents of membranes with the aim of developing a general model for membrane structure. In the following chapter we turn to the question of how cells transport materials across membranes.

The Structure of Biological Membranes

Biological membranes consist primarily of protein and lipids. The relative amounts of these materials vary considerably, depending on the source of the membrane. At one extreme the inner mitochondrial membrane is about 80% protein and 20% lipid by weight; at the other the myelin sheath membrane is about 80% lipid and 20% protein. The plasma membrane of human erythrocytes contains about equal amounts of protein and lipid. Many membranes also contain small amounts of carbohydrates. These almost always are covalently attached to either proteins (as glycoproteins) or lipids (as glycolipids or lipopolysaccharides). The mitochondrial inner membrane has little or no carbohydrate, but the myelin membrane has about 3% carbohydrate by weight, and the erythrocyte plasma membrane has about 8%.

Different Membrane Structures Can Be Separated According to Their Density

As with the structure of any biological entity, in order to study the structure of biological membranes we must first isolate them in a more or less intact form from the cell. In eukaryotic cells this problem is complicated by the existence of several different membrane systems in addition to the plasma membrane, each surrounding a specific organelle. To separate membrane fractions, we must first disrupt the plasma membrane under conditions that leave subcellular organelles intact. One common procedure involves mild ho-

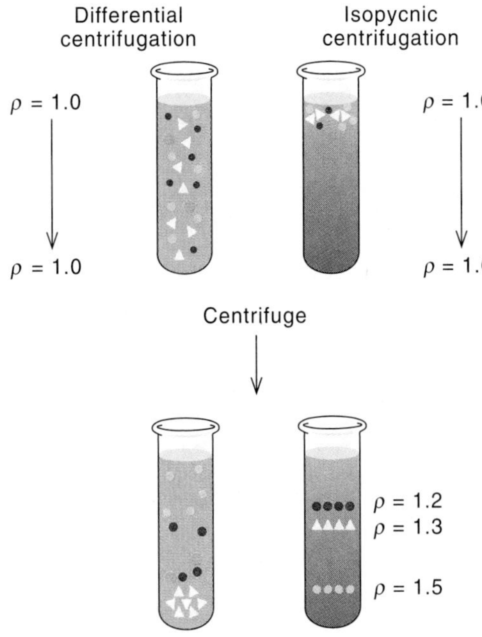

Figure 19.3

Comparison of differential centrifugation (left), which separates on the basis of size, and isopycnic centrifugation (right), which separates on the basis of density. ρ is the density in grams per milliliter.

mogenization in a medium in which the osmolarity is below the normal physiologic value (hypotonic medium). Another method, nitrogen cavitation, involves forcing nitrogen gas into the cells under pressure and then rapidly releasing the pressure to "explode" the cell membrane.

Organelles can be isolated from disrupted cells by differential centrifugation, which separates them on the basis of their size (fig. 19.3). Ruptured plasma membrane fragments can be purified from the same mixture by equilibrium-density-gradient (isopycnic) centrifugation because of their low density (high lipid content) relative to intact organelles (table 19.1, column 4). This technique relies on centrifuging the sample into a preformed gradient of a solute, such as sucrose. When equilibrium is reached, each type of membrane or organelle is found in the region of the gradient corresponding to its own density (see fig. 19.3). Gradients of synthetic sucrose polymers (Ficoll) or colloidal silica particles (Percoll) also are used in these separations because of their inertness, ability to form stable gradients, and impermeability to biological membranes.

The properties of isolated rat liver organelles are summarized in table 19.1. The entries in column 2 provide some idea of the relative proportions of these organelles in the mammalian liver. For example, mitochondria represent 25% of the total cell protein, while lysosomes comprise about 2%. Interestingly, the plasma membrane, which completely surrounds the cell, also represents only 2% of the total protein. In addition to the soluble protein that is membrane-free, there is considerable soluble protein within the organelles, leaving somewhat less than 50% of the total cell protein in the membrane-associated form.

Membranes Contain Complex Mixtures of Lipids

Two main types of lipids occur in biological membranes: phospholipids and sterols. The predominant phospholipids in most membranes are phosphoacylglycerols, which are phosphate esters of the three-carbon alcohol, glycerol. A typical structure is that of

Table 19.1
Properties of Rat Liver Organelles

Organelle	Percent of Cell Protein	Diameter (μm)	Equilibrium Density in Sucrose (g/ml)	Organelle-Specific Enzyme Marker
Liver cell	100	20	1.20	—
Nuclei	15	5–10	1.32	DNA polymerase
Golgi apparatus	2	2	1.10	Glycosyl transferases
Mitochondria	25	1	1.20	Monoamine oxidase (outer membrane); cytochrome *c* (inner membrane)
Lysosomes	2	0.5	1.20	Acid phosphatase
Endoplasmic reticular vesicles	20	1.0	1.15	Cytochrome b_5 reductase and cytochrome b_5; glucose-6-phosphatase
Cytoplasmic membrane	2	—	1.15	Na^+-K^+ ATPase; viral receptors
Soluble protein	30	<0.01	—	—

Source: From M. H. Saier, Jr., and C. D. Stiles, *Molecular Dynamics in Biological Membranes,* Heidelberg Science Library, Vol. 22, Copyright © 1975, Springer-Verlag, New York, N.Y. Reprinted with permission.

Table 19.2
Fatty Acids Frequently Found in Membrane Phospholipids

Name	Structure	Abbreviation[a]
Saturated Fatty Acids[b]		
Myristic acid	$CH_3(CH_2)_{12}CO_2H$	14:0
Palmitic acid	$CH_3(CH_2)_{14}CO_2H$	16:0
Stearic acid	$CH_3(CH_2)_{16}CO_2H$	18:0
Unsaturated Fatty Acids[c]		
Palmitoleic acid	$CH_3(CH_2)_5\overset{H}{C}{=}\overset{H}{C}(CH_2)_7CO_2H$	$16{:}1^{\Delta 9}$
Oleic acid	$CH_3(CH_2)_7\overset{H}{C}{=}\overset{H}{C}(CH_2)_7CO_2H$	$18{:}1^{\Delta 9}$
Linoleic acid	$CH_3(CH_2)_4\overset{H}{C}{=}\overset{H}{C}CH_2\overset{H}{C}{=}\overset{H}{C}(CH_2)_7CO_2H$	$18{:}2^{\Delta 9,12}$
Linolanic acid	$CH_3CH_2{-}\overset{H}{C}{=}\overset{H}{C}{-}CH_2{-}\overset{H}{C}{=}\overset{H}{C}\ CH_2\ \overset{H}{C}{=}\overset{H}{C}(CH_2)_7COO^-$	$18{:}3^{\Delta 9,12,15}$
Arachidonic acid	$CH_3(CH_2)_3(CH_2{-}\overset{H}{C}{=}\overset{H}{C})_4(CH_2)_3CO_2H$	$20{:}4^{\Delta 5,8,11,14}$

[a]The abbreviation indicates that total number of carbon atoms and, following the colon, the number of double bonds and their positions relative to the carboxyl group. The Δ superscript indicates the start location of the double bond(s). Numbering of carbons starts from the carboxyl carbon. Thus $18{:}2^{\Delta 9,12}$ indicates two double bonds, one between carbon 9 and 10 and one between carbons 12 and 13.

[b]A saturated fatty acid usually is esterified at C-1 of the glycerol.

[c]An unsaturated fatty acid usually is present at the C-2 position. Some bacterial phospholipids have a fatty acid with an internal cyclopropane ring here.

phosphatidylcholine (lecithin):

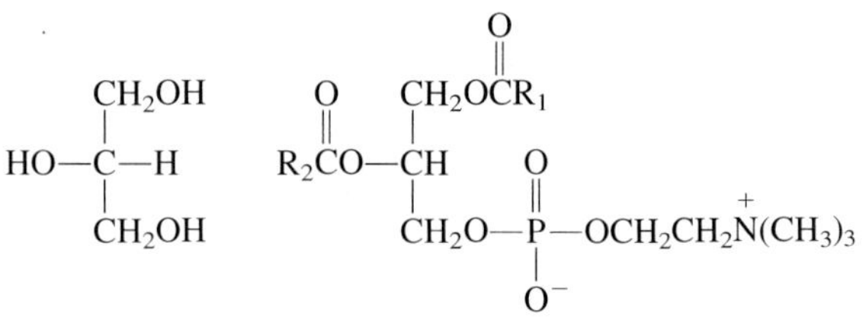

$$\begin{array}{c} CH_2OH \\ | \\ HO{-}C{-}H \\ | \\ CH_2OH \end{array} \qquad \begin{array}{l} \qquad\qquad\quad CH_2O\overset{O}{\overset{\|}{C}}R_1 \\ \qquad\qquad\quad\ | \\ R_2\overset{O}{\overset{\|}{C}}O{-}CH \\ \qquad\qquad\quad\ | \\ \qquad\qquad\quad CH_2O{-}\overset{O}{\overset{\|}{P}}{-}OCH_2CH_2\overset{+}{N}(CH_3)_3 \\ \qquad\qquad\qquad\qquad\ \ | \\ \qquad\qquad\qquad\qquad\ \ O^- \end{array}$$

Glycerol Phosphatidylcholine

Here R_1 and R_2 are long, fatty acid side chains. The parent fatty acids R_1CO_2H and R_2CO_2H usually have an even number of carbon atoms; 16- and 18-carbon acids are the most common. The acid esterified to the hydroxyl group on C-1 of the glycerol (that at the top of phosphatidylcholine is drawn above) usually has a fully saturated chain, whereas the acid attached at C-2 often has one or more double bonds, which are almost always *cis* double bonds. Table 19.2 lists some of the fatty acids commonly found in these positions. A phosphatidylcholine that has palmitic acid esterified at both the C-1 and C-2 positions of the glycerol is known by the name dipalmitoylphosphatidylcholine. One with palmitic acid at C-1 and oleic acid at C-2 is called a 1-palmitoyl-2-oleoylphosphatidylcholine.

The phosphate group in phosphatidylcholine forms ester linkages both with the hydroxyl group on C-3 of the glycerol and also with a second alcohol, choline ($HOCH_2CH_2N(CH_3)_3{}^+$). In other phospholipids a variety of other alcohols occupy this second position, and like choline, these all contain polar or electrically charged substituents. Table 19.3 shows the most common of these alcohols and indicates how the phosphoglycerides containing them are named by appending the name of the alcohol to the prefix "phosphatidyl." Free phosphatidic acid, in which the phosphate group is not esterified in this position (fig. 19.4), is an intermediate in phospholipid biosynthesis and metabolism but is not a major constituent of most biological membranes.

In addition to phosphoglycerides, membranes from animal cells usually contain a second group of polar lipids, the sphingolipids. Sphingomyelin, which is representative of this group, has the structure

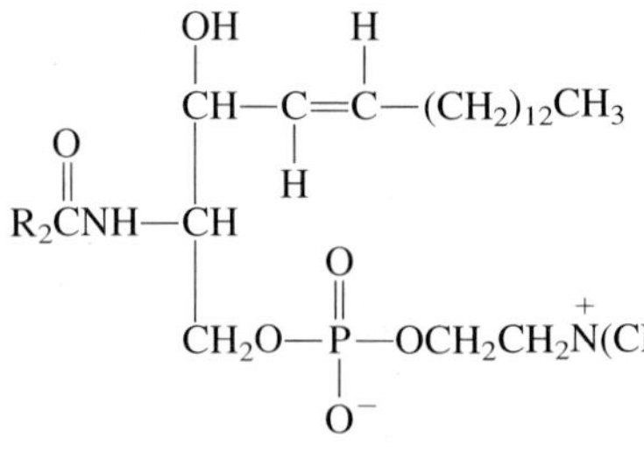

$$\begin{array}{l} \qquad\qquad\quad OH \qquad H \\ \qquad\qquad\quad\ | \qquad\quad | \\ \qquad\qquad\quad CH{-}C{=}C{-}(CH_2)_{12}CH_3 \\ \qquad O \qquad\ | \qquad\ | \\ \qquad \| \qquad\ \ | \qquad H \\ R_2CNH{-}CH \\ \qquad\qquad\quad\ | \qquad\ O \\ \qquad\qquad\quad\ | \qquad\ \| \\ \qquad\qquad\quad CH_2O{-}P{-}OCH_2CH_2\overset{+}{N}(CH_3)_3 \\ \qquad\qquad\qquad\qquad | \\ \qquad\qquad\qquad\qquad O^- \end{array}$$

Sphingomyelin

R_2 is of variable length. Although sphingomyelin does not contain glycerol, its structural similarity to phosphatidylcholine is evident (fig. 19.5). Both molecules have two long hydrocarbon side

Table 19.3
Major Phosphoglycerides

Alcohol (HO—X)	Formula	Phospholipid
Choline	$HO-CH_2CH_2N(CH_3)_3^+$	Phosphatidylcholine (lecithin)
Ethanolamine	$HO-CH_2CH_2NH_3^+$	Phosphatidylethanolamine
Serine	$HO-CH_2CH(CO_2^-)NH_3^+$	Phosphatidylserine
Glycerol	$HO-CH_2CH(OH)CH_2OH$	Phosphatidylglycerol
Phosphatidylglycerol	$R_2C(=O)O-CH(CH_2OC(=O)R_1)-CH_2O-P(=O)(O^-)-O-CH_2CH(OH)CH_2-OH$	Diphosphatidylglycerol (cardiolipin)
myo-Inositol	OH OH H H H OH OH H HO H H OH (cyclohexane ring)	Phosphatidylinositol

Figure 19.4

Structure of phosphatidic acid, a phosphoglyceride. A saturated fatty acid is esterified at carbon 1 of the glycerol, and a *cis*-unsaturated fatty acid at carbon 2. The cluster of polar and charged oxygens makes the head group hydrophilic, in marked contrast to the hydrophobic fatty acid chains.

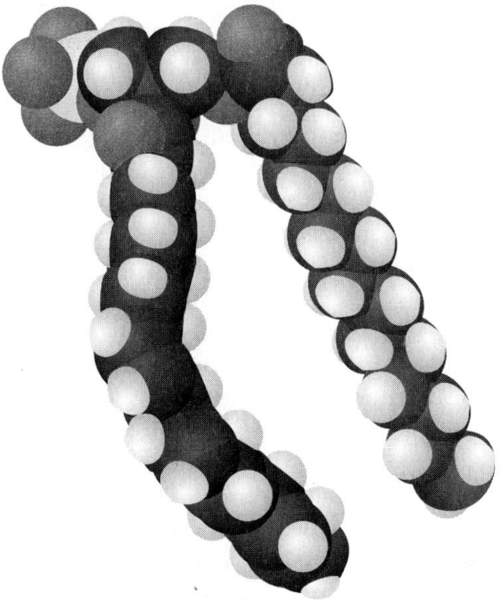

O^- | $O=P-O$ | O^- — CH_2 — CH — CH_2 — O — C(=O) ; CH — O — C(=O)

Figure 19.5

Structure of a sphingomyelin.

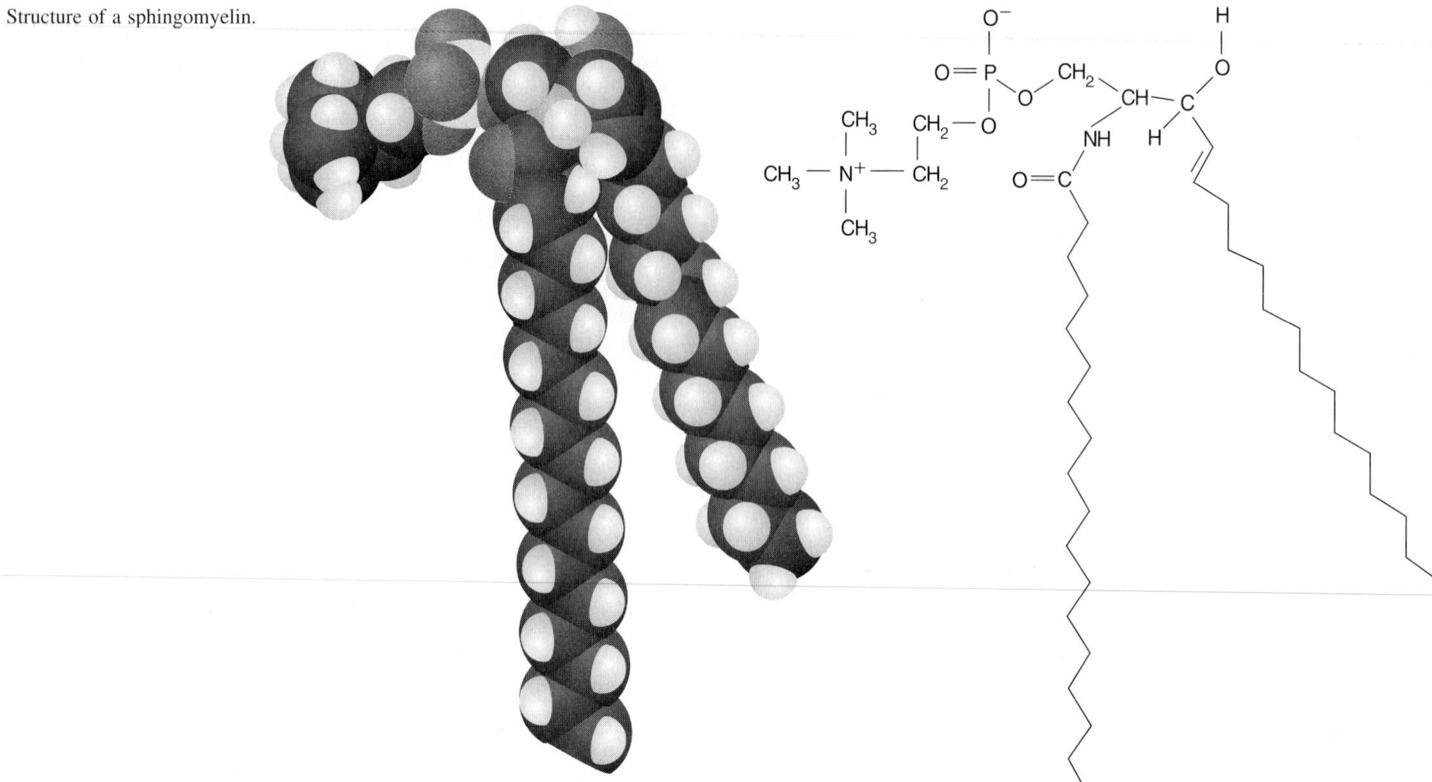

chains and a negatively charged phosphate group esterified to the positively charged choline.

Other sphingolipids have a carbohydrate such as galactose or a short oligosaccharide chain esterified to the hydroxyl in place of the phosphocholine of sphingomyelin. These compounds are called glycosphingolipids or, more generally, glycolipids. The prototype, in which the substituent is a single galactosyl residue, is called galactosylceramide.

The second major type of lipid found in some biological membranes is cholesterol. Cholesterol (fig. 19.6) is an isoprenoid compound with four fused rings, a short aliphatic chain, and a single hydroxyl group. It occurs in membranes both in its free form and esterified with long-chain fatty acids. Table 19.4 compares the lipid compositions of membranes from several biological sources. As a rule, phosphatidylcholine is the major phospholipid of animal cell membranes, whereas phosphatidylethanolamine predominates in bacteria. Most bacterial membranes have no cholesterol or sphingolipids. Cholesterol also is low in some organellar membranes of animal cells, such as the inner mitochondrial membrane, but can account for almost one-third of the total lipid in cytoplasmic membranes. Plant cell membranes have no cholesterol, but some contain relatively large amounts of glycolipids.

Phospholipids Spontaneously Form Ordered Structures in Water

Phospholipid molecules are said to be amphipathic, a term derived from the Greek, meaning having ambivalent feelings. The polar head group of the molecule is intrinsically soluble in water; the fatty acid tails are hydrophobic. The space-filling models shown in figures 19.4 and 19.5 emphasize the segregation of these polar and nonpolar groups into two distinct domains. In this regard, phospholipids, resemble detergents such as sodium dodecylsulfate (SDS), which has an ionic sulfate head and a single, long hydrocarbon tail:

$$CH_3(CH_2)_{11}O{-}\overset{\overset{\displaystyle O}{\|}}{\underset{\underset{\displaystyle O}{\|}}{S}}{-}O^-Na^+$$

Sodium dodecylsulfate

When a detergent is mixed with water, it aggregates spontaneously into spherical or ellipsoidal micelles, in which the polar head groups of the molecules are exposed to water but the hydrophobic tails are sequestered (fig. 19.7*a*). Surprisingly, this process is driven, not by a decrease in energy, but rather by an increase in entropy associated with removing the hydrocarbon chains from water. If a hydrocarbon is dissolved in water, the water molecules surrounding it adopt a netlike structure that is more highly ordered than the structure of pure liquid water (see fig. 3.4). Burying the hydrocarbon tails of the detergent molecules in the center of a micelle frees many water molecules from these nets and increases the overall amount of disorder in the system.

Phospholipids themselves generally do not form micelles, because, instead of having a single hydrophobic tail, they have two. The two tails are too bulky to pack together in a spherical or

Table 19.4
Lipid Compositions of Membranes

Source	Lipid Composition[a] (% of total lipids)								
	Cholesterol	**PC**	**SM**	**PE**	**PI**	**PS**	**PG**	**DPG**	**Glycolipids**
Rat Liver									
Plasma membrane	30	18	14	11	4	9	—	—	—
Rough endoplasmic reticulum	6	55	3	16	8	3	—	—	—
Inner mitochondrial membrane	3	45	3	25	6	1	2	18	—
Nuclear membrane	10	55	3	20	7	3	—	—	—
Golgi	8	40	10	15	6	4	—	—	—
Lysosomes	14	25	24	13	7	—	—	5	—
Rat Brain Myelin	22	11	6	14	—	7	—	—	21
Rat Erythrocyte	24	31	9	15	2	7	—	—	3
***E. coli* Plasma Membrane**	0	0	—	80	—	—	15	5	—

[a]PC = phosphatidylcholine; SM = sphingomyelin; PE = phosphatidylethanolamine; PI = phosphatidylinositol; PS = phosphatidylserine; PG = phosphatidylglycerol; DPG = diphosphatidylglycerol (cardiolipin).

Source: Adapted from M. K. Jain and R. C. Wagner, *Introduction to Biological Membranes,* John Wiley & Sons, New York, 1980.

Figure 19.6

Structure of cholesterol, in three different views. The conventional projection is shown at the top right. The more realistic space-filling and conformational models are shown at the lower left and the lower right, respectively.

ellipsoidal micelle. They hold the individual molecules apart enough that portions of the tails remain in contact with water. There is, however, another solution to the problem: The phospholipids can form a bilayer. In this type of aggregate the hydrocarbon tails of two monolayers of phospholipids pack together while the polar head groups face outward and remain in contact with water on either side (fig. 19.7*b*). A sheetlike planar bilayer removes the hydrophobic tails from water everywhere except around the edges of the sheet, and by curling up into a spherical vesicle, the bilayer can get rid of its edges. Such a vesicle is called a liposome. Multilayered liposomes resembling onions can be formed by enclosing one vesicle inside another (fig. 19.8).

Although other types of aggregates can form under special conditions, when phospholipids are agitated in the presence of excess water, they tend to aggregate spontaneously to form bilayers. Electron micrographs of bilayers (see fig. 19.8) immediately call to mind the trilaminar appearance of biological membranes (see figs. 19.1 and 19.2). Furthermore, the idea that biological membranes might be built of phospholipid bilayers suggests an explanation for the observation that membranes

Figure 19.7

Structures formed by (*a*) detergents and (*b*) phospholipids in aqueous solution. Each molecule is depicted schematically as a polar head group (•) attached to one or two long, nonpolar chains. Most detergents have one nonpolar chain; phospholipids have two. At very low concentrations, detergents or phospholipids form monolayers at the air–water interface. At higher concentrations, when this interface is saturated, additional molecules form micelles or bilayer vesicles (liposomes).

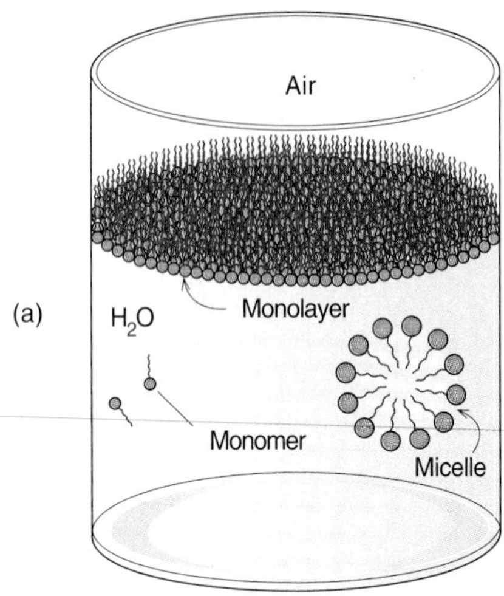

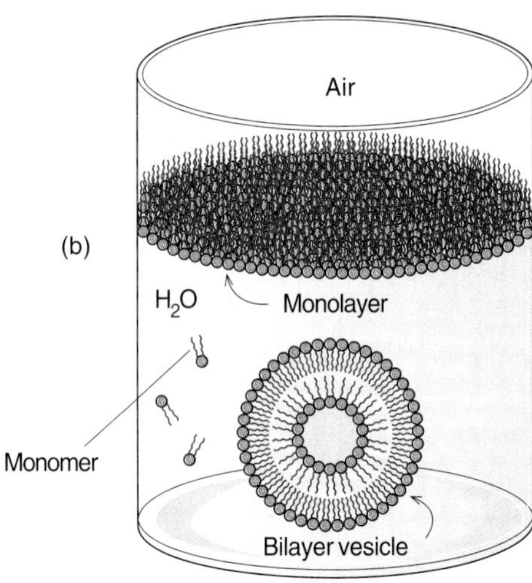

Figure 19.8

Multilayered vesicles (liposomes) formed from sonically dispersed phosphatidylcholine in the presence of 10% diacetylphosphate and 2% potassium phosphotungstate. Each vesicle has a trilaminar structure consisting of two dark layers separated by a light layer. The dark layers contain the electron-dense phosphotungstate ion; the light layer corresponds to the hydrophobic interior of the bilayer.

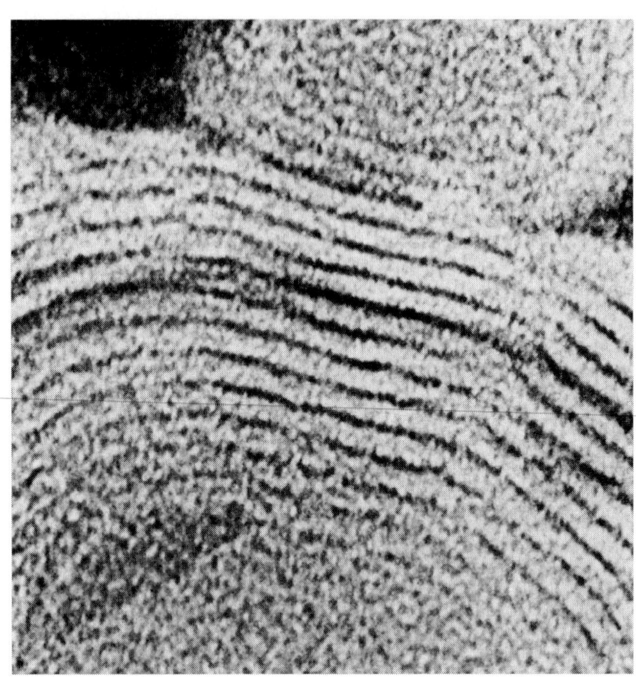

are largely impermeable to ions. To cross a phospholipid bilayer, an ion would have to traverse the apolar region of hydrocarbon tails, where it would not be well solvated. Small, lipid-soluble molecules, on the other hand, are expected to pass across such a membrane relatively easily, and this is indeed observed to occur. Finally, the presence of a phospholipid bilayer explains the flexibility of natural membranes and the ability of membranes to reseal if they are punctured. A bilayer is held together, not by bonds or electrostatic attractions between individual phospholipid molecules, but rather by the entropy increase that results when the hydrocarbon side chains come together and shed their coats of water.

The idea of a lipid bilayer was first proposed by E. Gorter and F. Grendel, who showed in 1925 that the phospholipid content of the erythrocyte plasma membrane is approximately the amount needed to enclose the cell with a bilayer. Subsequent x-ray diffraction measurements confirmed this picture (fig. 19.9).

Membranes Have Both Integral and Peripheral Proteins

In spite of the arguments presented in the previous section, it is reasonable to ask whether biological membranes are held together simply by phospholipid bilayers. Is it not necessary to reinforce their structure by cross-linking the proteins in the membrane with covalent bonds or at least by favorable electrostatic interactions among the proteins? Considering that some membranes contain even more protein than lipids, any model for membrane structure needs to confront the questions of how these proteins are positioned and of what role, if any, proteins play in holding the membrane together. A possible answer to these question is that a layer of protein coats each side of the phospholipid bilayer, as shown in figure 19.10. This model was suggested by H. Davson and J. F. Danielli in 1935 and was elaborated by them and others over the course of the next 30 years. It was suggested that the protein coats could take the form of β-pleated sheets. The Davson-Danielli

Figure 19.9

(*a*) Relative electron densities of myelin membranes from rabbit optic nerve (red line) and sciatic nerve (green line) as a function of distance from the center, measured by x-ray diffraction. The density profile is due mainly to lipids because of the high lipid–protein ratio in myelin. (*b*) Structural interpretation of the electron-density profile, based on the approximate ratios of lipids in mammalian nerve myelin membranes. Side chains from the opposite leaflets of the bilayer do not appear to interdigitate to any great extent.

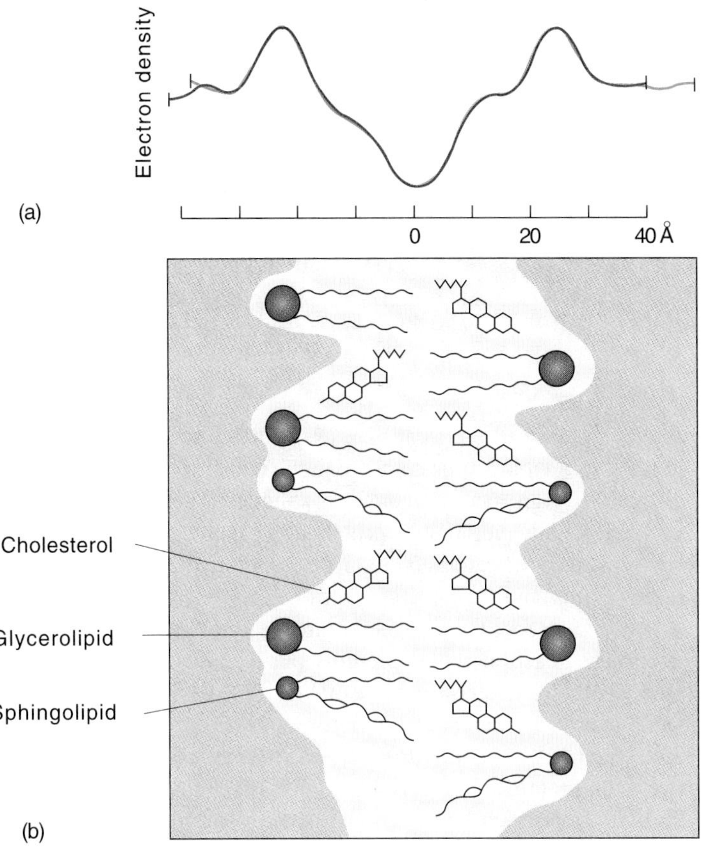

Figure 19.10

In an early inaccurate model for the structure of biological membranes a phospholipid bilayer was coated on both sides by protein in an unfolded or β-pleated sheet conformation. This model reflected the prevailing view of membrane structure from about 1940 until the early 1970s.

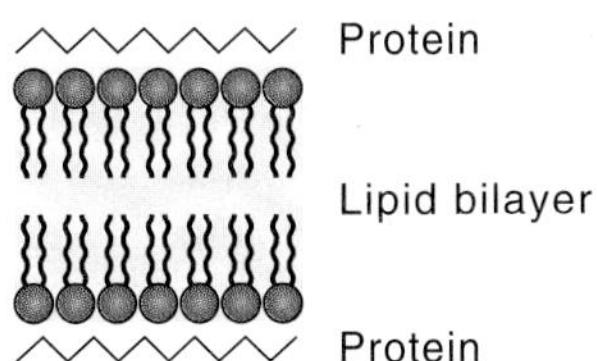

model was attractive because it seemed to account for the trilaminar appearance, impermeability to ions, and durability of membranes. It did not account for the fact that membrane proteins exhibit a variety of enzymatic activities, because the protein in the model played a purely structural role. But at the time the model was proposed, the enzymatic activities of membrane proteins were not recognized as fully as they are today.

One asset of the Davson-Danielli model was its amenability to experimental tests. The model placed all the proteins on the surface of the membrane, where they would be exposed to water and presumably would interact electrostatically with the polar head groups of the phospholipids. It thus made several predictions. First, to have a stable, sheetlike conformation in contact with water, the proteins found in membranes should contain amino acids with predominantly hydrophilic side chains. Second, it should be possible to wash the proteins off the surface of the phospholipid bilayer by treating membranes with solutions of salts at high ionic strengths. Salt solutions would shield electrostatic interactions holding the proteins to the phospholipid heads. Third, physical probes of protein conformation should show that membrane proteins are unfolded or have high complements of β structure and relatively little α helix. Finally, high-resolution electron-microscopic images should show that membranes have smooth surfaces; and if it is possible to examine the middle of the phospholipid bilayer, no protein should be found there. On each of these predictions the model ran into trouble.

Washing preparations of biological membranes with salt solutions usually does remove some proteins, but in most cases they are only a relatively minor fraction of the total protein. Such experiments have led to the realization that biological membranes contain two classes of proteins: peripheral and integral. It is the peripheral proteins that can be removed by washing with salts. In addition to high salt concentrations, EDTA (ethylenediaminetetraacetic acid, a chelator of Ca^{2+} or Mg^{2+} ions) or urea often is used to solubilize these proteins. As a group, peripheral membrane proteins have amino acid compositions similar to those of soluble proteins. On the order of 30% of their residues may be hydrophobic and 70% hydrophilic or neutral. They exhibit the full range of secondary structures and, again, are not remarkable in this regard.

Integral membrane proteins are much more difficult to extract. To solubilize them, it usually is necessary to resort to the use of detergents. Detergents disrupt the phospholipid bilayer of the membrane and incorporate lipids and the hydrophobic portions of proteins into their micelles. Figure 19.11 shows some of the detergents that are used for this purpose. After a membrane has been disrupted, integral membrane proteins can be purified by column chromatography and other conventional techniques, but they almost always require the continued presence of a detergent to remain in solution.

Integral Membrane Proteins Contain Transmembrane α Helices

As we might expect from the fact that they are soluble only in the presence of detergents, integral membrane proteins tend to have comparatively high contents of hydrophobic amino acid residues. Between 40% and 60% of their side chains may be hydrophobic. In addition, hydrophobic residues often show a curious distribution in the amino acid sequence. They

Figure 19.11

Several of the synthetic detergents used for dissolving membranes and solubilizing integral membrane proteins. Triton X-100 and octylglucoside are nonionic detergents; cetyltrimethylammonium bromide and sodium dodecylsulfate (SDS) are ionic. SDS is also an effective denaturant of proteins and is used in polyacrylamide-gel electrophoresis (see chapter 7).

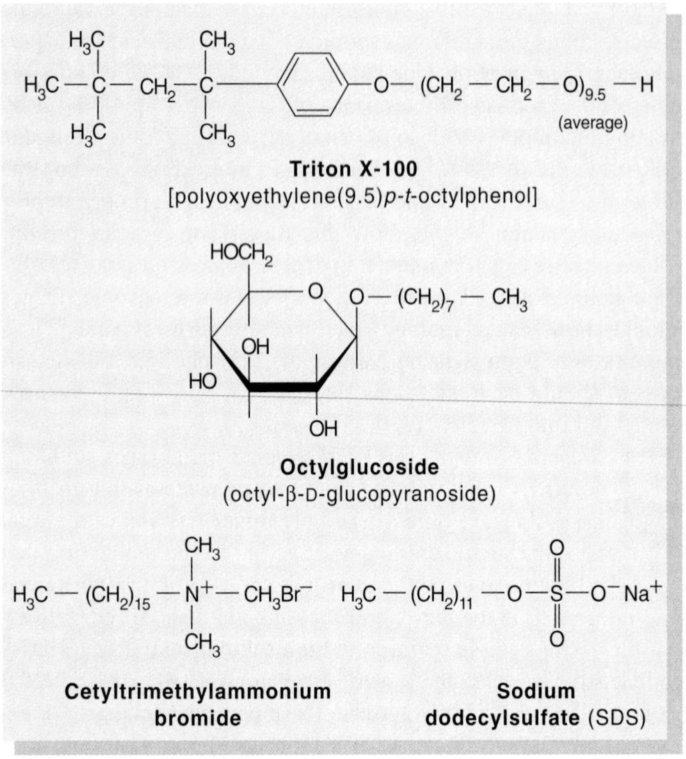

$(H_3C)_3C-CH_2-C(CH_3)_2-C_6H_4-O-(CH_2-CH_2-O)_{9.5}-H$ (average)

Triton X-100
[polyoxyethylene(9.5)*p*-*t*-octylphenol]

$HOCH_2$, OH, HO, OH; $O-(CH_2)_7-CH_3$

Octylglucoside
(octyl-β-D-glucopyranoside)

$H_3C-(CH_2)_{15}-N^+(CH_3)_2-CH_3Br^-$

Cetyltrimethylammonium bromide

$H_3C-(CH_2)_{11}-O-SO_2-O^-Na^+$

Sodium dodecylsulfate (SDS)

Figure 19.12

Hydropathy index for the erythrocyte protein glycophorin as a function of position in the amino acid sequence. The hydropathy index was averaged over a "window" of seven amino acids, and the window was advanced along the sequence. Glycophorin has a single membrane-spanning segment in the region of residues 75–94. (See fig. 19.19 for a model of how the protein sits in the membrane.) (Reprinted from *J. Mol. Biol.*, vol. 157, J. Kyte and R. F. Doolittle, A simple method for displaying the hydropathic character of a protein, page 105, 1982, by permission of the publisher Academic Press Limited, London.)

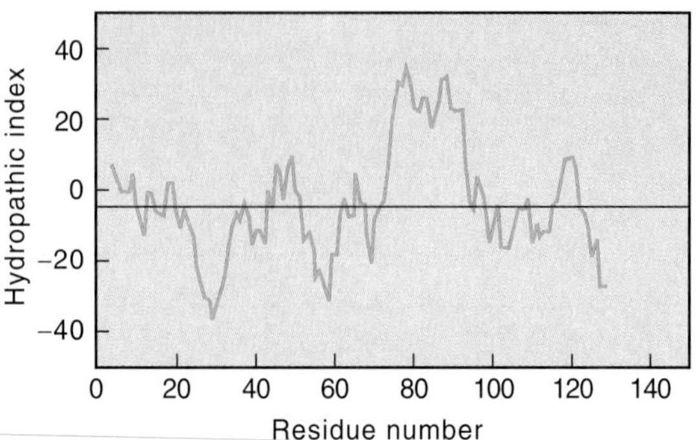

frequently appear in strings of approximately 20 residues, separated by stretches of hydrophilic residues. To describe this aspect of protein structure quantitatively, it is useful to assign each of the 20 natural amino acid residues a number that expresses the side chain's relatively hydrophobicity or hydrophilicity. This number can be obtained by measuring the free energy change associated with transfer from an organic solvent to water. On a hydropathy scale, residues with strongly hydrophobic side chains such as isoleucine usually are given positive numbers, and residues with hydrophilic side chains such as arginine are given negative numbers. Figure 19.12 shows the average hydropathy index of the amino acids in glycophorin, a small protein purified from erythrocyte membranes, as a function of location in the amino acid sequence. This protein has one region of predominantly hydrophobic residues extending for about 20 residues. Such regions are uncommon in water-soluble proteins, but larger integral membrane proteins may contain 10 or more of them.

An α helix of 20 amino acid residues is approximately 30 Å long (1.5 Å per residue). This is about the right length to reach across the hydrocarbon region of a bilayer of phospholipids containing 16- and 18-carbon fatty acids. Hydropathy plots like that of figure 19.12 therefore suggest that integral membrane proteins contain one or more α-helical regions extending completely across the bilayer. Physical measurements of secondary structure in integral membrane proteins, though subject to greater technical difficulties than measurements on soluble proteins, support this view. Integral membrane proteins typically have substantial amounts of α-helical structure oriented perpendicular to the plane of the membrane.

Very few integral membrane proteins have been crystallized. The reaction-center proteins purified from membranes of photosynthetic bacteria are a notable exception. These proteins were discussed in chapter 17. Before their crystal structures were elucidated, analysis of hydropathy plots suggested that each of the two main protein subunits is folded into five transmembrane α helices, and one such helix was predicted to occur in another subunit. The crystal structures provided a beautiful confirmation of these predictions (see fig. 17.12*a*). Successful crystallization of the reaction-center proteins was achieved by including small, amphipathic molecules such as heptane-1,2,3-triol in the solution. These molecules evidently help to fill up the crevices that are occupied by water in crystals of water-soluble proteins.

A three-dimensional structure also has been elucidated for bacteriorhodopsin, an integral membrane protein of the halophilic (salt-loving) bacterium *Halobacterium halobium.* This protein has been studied intensively because of its remarkable activity as a light-driven proton pump (see chapter 16). It forms well-ordered arrays in two-dimensional sheets that can be studied by electron diffraction. Measurements of the diffraction patterns show clearly that bacteriorhodopsin has seven transmembrane helices (fig. 19.13).

Electron microscopy has provided additional evidence that membrane proteins typically have globular structures rather than sheetlike structures as suggested in figure 19.10. The electron micrographs in figures 19.1 and 19.2 show specimens that were sliced into extremely thin sections (≈100 nm) and then stained with an electron-dense reagent that reacted chemically with

Figure 19.13

A model for the structure of bacteriorhodopsin, a membrane protein from *Halobacterium halobium.* The protein has seven membrane-spanning segments connected by shorter stretches of hydrophilic amino acid residues.

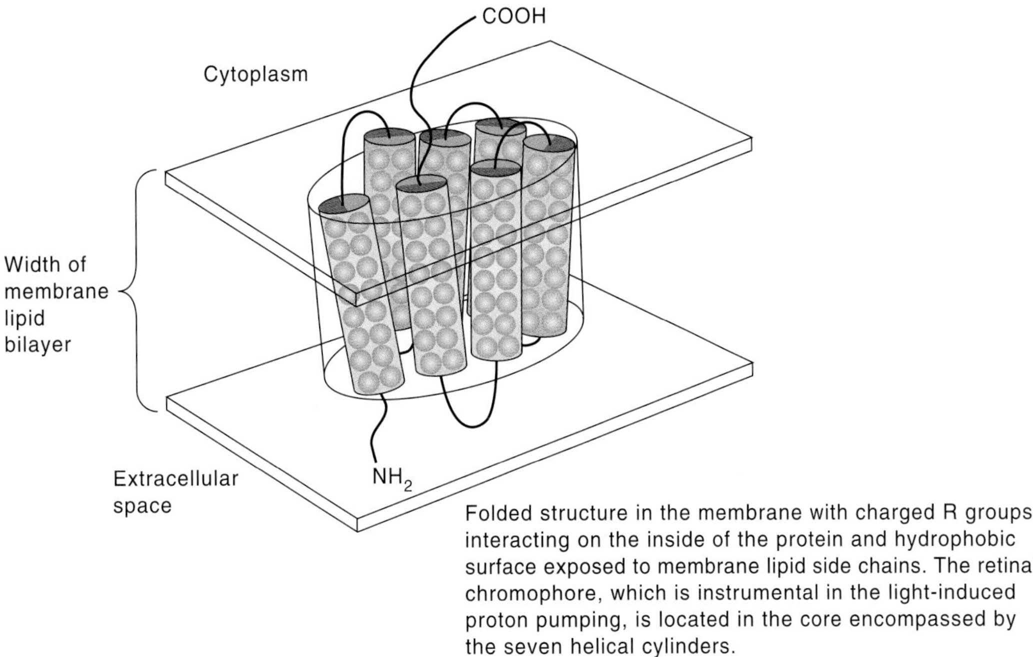
COOH
Cytoplasm
Width of
membrane
lipid
bilayer
Extracellular
space
NH_2
Folded structure in the membrane with charged R groups interacting on the inside of the protein and hydrophobic surface exposed to membrane lipid side chains. The retinal chromophore, which is instrumental in the light-induced proton pumping, is located in the core encompassed by the seven helical cylinders.

double bonds or other functional groups in the membrane. In an alternative technique, called negative staining, a thicker specimen is flooded with a dense, but chemically inert, material that simply fills all the spaces around the membrane. With negative staining, the surfaces of unsliced membranes from some cells and organelles display numerous globular projections that turn out to be proteins. The inner mitochondrial membrane, for example, is decorated with the head pieces of the proton-conducting ATP-synthase (see fig. 16.23).

Another microscopic technique is to freeze the specimen and then fracture it with a knife. A knife cutting through the frozen specimen splits the membrane down the middle, exposing the inside of the bilayer (fig. 19.14*a*). If the Davson-Danielli model for membrane structure were correct, the two exposed surfaces would be featureless. However, electron micrographs of metallic casts of such samples reveal surfaces studded with particles of various sizes (fig. 19.14*b*). Additional studies indicate that these particles are proteins that are deeply embedded in the membrane. The particles seen on the inner and outer leaflets of the bilayer usually differ in size and distribution because of an asymmetrical disposition of the proteins across the bilayer.

Proteins and Lipids Can Move around within Membranes

During the 1960s, various alternatives to the Davson-Danielli model were proposed. Some investigators abandoned the idea of a phospholipid bilayer and suggested instead that membranes consist of aggregates of lipid–protein complexes. However, in 1972, Jon Singer and Garth Nicolson incorporated all of the available information into a model they called the fluid-mosaic model. This model, which is now generally accepted, retains the idea that the phospholipid bilayer is the primary structural element of biological membranes. But unlike the Davson-Danielli model, it proposes the integral membrane proteins are embedded in the bilayer, in some cases only partially but in other cases extending all the way across (fig. 19.15). Peripheral proteins are attached more loosely by ionic interactions with protruding portions of integral proteins or with phospholipid head-groups. A key feature of the fluid-mosaic model is that the integral proteins are, in most cases, not linked together by protein–protein interactions. They are free to diffuse laterally in the bilayer or to rotate about an axis perpendicular to the plane of the membrane. The entire structure thus has the potential of being dynamic rather than static.

Frye and Edidin provided a striking visual demonstration of the dynamic nature of membrane structure. They labeled proteins on the plasma membranes of two samples of cells with fluorescent dyes: human cells with a dye that emitted red light and mouse cells with a dye that emitted green light (fig. 19.16). The two populations of cells then were mixed and treated with Sendai virus, which causes individual cells to fuse. Immediately after the fusion, red fluorescence from the human proteins could be seen on one half of the hybrid membrane, and green fluorescence from the mouse proteins on the other half. But within a few minutes the two types of proteins were intermingled over the entire surface.

Figure 19.14

Freeze-fracture electron microscopy. (*a*) When struck with a sharp knife, membranes embedded in ice usually fracture between the monolayer leaflets of the lipid bilayer. (*b*) Freeze-fracture electron micrograph of the plasma membrane of *Streptococcus faecalis,* showing a large number of protrusions (presumably proteins) on the outer fracture face and the relative lack of such particles on the inner fracture face (inset). (From H. C. Tsien and M. L. Higgins, Effect of temperature on the distribution of membrane particles in *Streptococcus faecalis* as seen by the freeze-fracture technique. *J. Bacteriol.* 118:725, 1974. Reprinted with permission from American Society for Microbiology.)

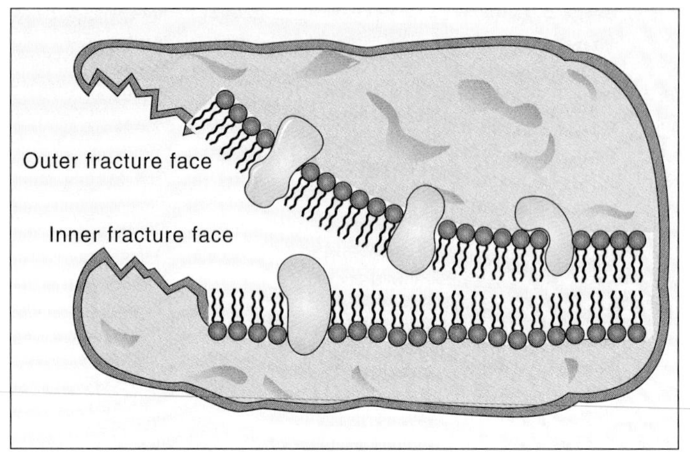

(a)

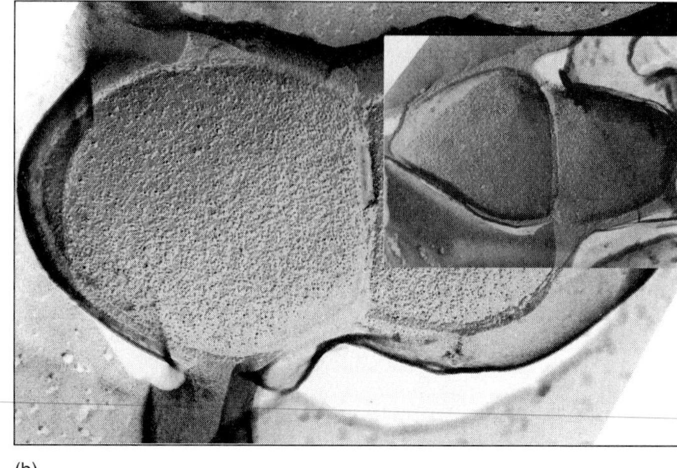

(b)

Figure 19.15

The fluid-mosaic model for biological membranes as envisioned by Singer and Nicolson. Integral membrane proteins are embedded in the lipid bilayer; peripheral proteins are attached more loosely to protruding regions of the integral proteins. The proteins are free to diffuse laterally or to rotate about an axis perpendicular to the plane of the membrane. For further information, see S. J. Singer and G. L. Nicolson. The fluid mosaic model of the structure of cell membranes. *Science* 175:720, 1972.

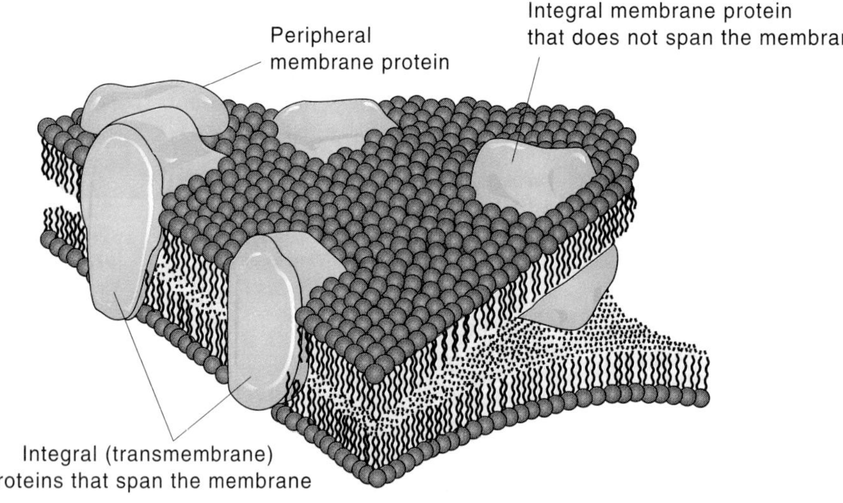

If integral membrane proteins are free to diffuse in the membrane, we expect the same to be true of the individual phospholipid molecules that make up the bilayer. To study the dynamics of these motions, phospholipids were labeled with a fluorescent dye that decomposed irreversibly when it was illuminated by a strong laser. When a laser flash was focused to a small spot on the surface of a cell, the labeled phospholipids in this region abruptly ceased fluorescing (fig. 19.17). Fluorescence then rapidly reappeared as the bleached molecules diffused out of the illuminated region and fresh phospholipids diffused in from outside.

Phospholipid molecules in the plasma membrane diffuse rapidly enough to go from one end of an average-sized animal cell to the other in a few minutes. In a bacterial cell such a trip would take only a few seconds. Integral membrane proteins move more slowly than phospholipids, as we expect in view of their greater mass. Diffusion of membrane proteins plays essential roles in

Figure 19.16

Frye and Edidin's experiment demonstrating lateral diffusion of membrane proteins. (*a*) Proteins on the plasma membranes of human and mouse cells were labeled with dyes that fluoresced at different wavelengths. (*b*) The two populations of cells were mixed and infected with a virus that causes cells to fuse. At short times after mixing, red and green fluorescence from the original cells was seen in separate parts of the membranes of the fused cells. (*c*) Within about 30 min the two populations of proteins had become intermingled over the entire surface.

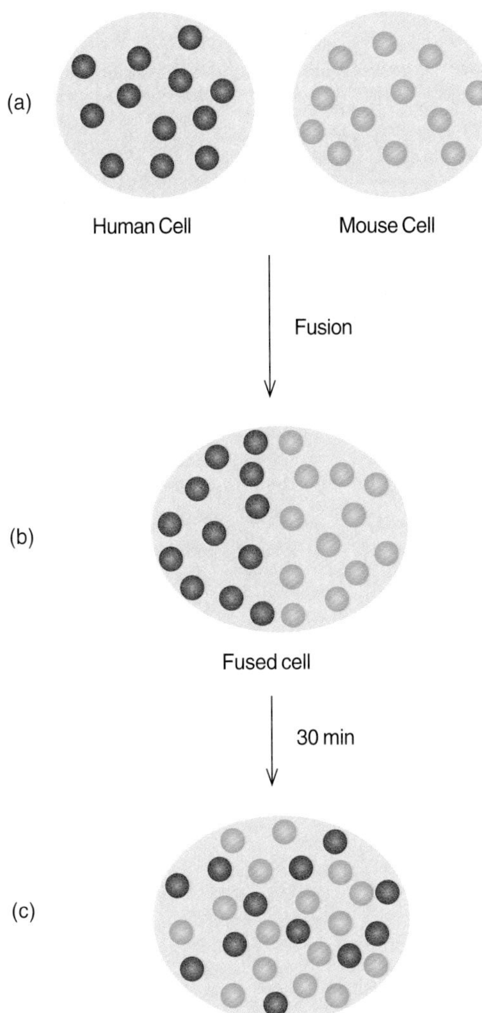

Figure 19.17

Lateral diffusion rates can be studied by measuring the recovery of fluorescence after photobleaching. (*a*) A phospholipid or other component of the membrane is labeled with a fluorescent dye, and the fluorescence from a small region of the membrane is measured through a microscope. The observation region is indicated here by the dashed circle. (*b*) When a laser flash is focused on the observation region, the dye molecules here are destroyed, and the amplitude of the fluorescence decreases. (*c*) The fluorescence signal increases again as fresh molecules diffuse into the observation region. (*d*) The diffusion rate is obtained from a plot of the fluorescence as a function of time.

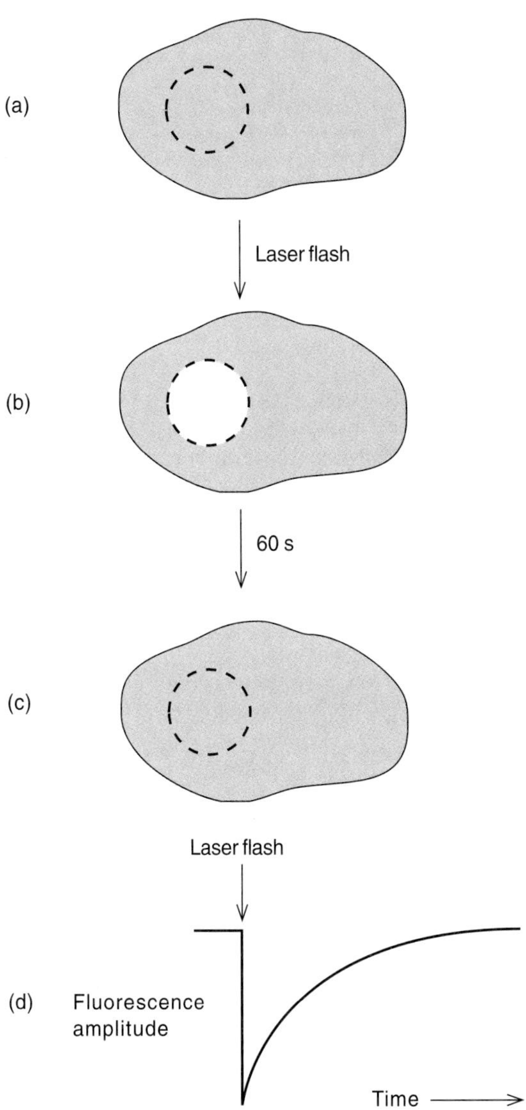

many biochemical processes, including the cellular uptake of lipoproteins (chapter 21), responses of cells to hormones (chapters 20 and 27), immunological reactions (chapter 37), vision (chapter 29), and the transport of nutrients and ions. As we see in a later section, however, some membrane proteins cannot move about rapidly because they are attached to cytoskeletal scaffolds.

Biological Membranes Are Asymmetrical

Although phospholipids diffuse laterally in the plane of the bilayer and rotate more or less freely about an axis perpendicular to this plane, movements from one side of the bilayer to the other are a different matter. Diffusion across the membrane, a transverse, or flip-flop, motion, requires getting the polar head group of the phospholipid through the hydrocarbon region in the center of the bilayer (fig. 19.18). Flip-flop motions of phospholipids do occur, and they can be catalyzed enzymatically, but they are much slower than the other types of motions we have described. The same arguments apply even more forcefully to membrane proteins. For a protein to invert its orientation in the membrane, the hydrophilic domain that sticks out into the solution on one side of the membrane has to pass through the center of the bilayer. This rarely oc-

Figure 19.18

(*a*) Phospholipids can rotate relatively freely about an axis perpendicular to the plane of the bilayer. (*b*) Rotation about an axis parallel to the plane of the bilayer (flip-flop) is much slower because it requires moving the polar head group across the bilayer.

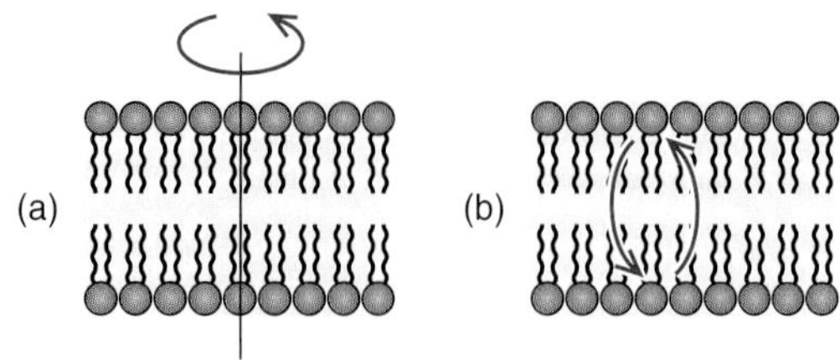

curs. Once a protein has been inserted into a membrane in a particular orientation, it usually retains that orientation indefinitely. An important consequence of these considerations is that biological membranes are both structurally and functionally asymmetrical. An enzyme or receptor embedded in the membrane usually provides a binding site for its substrate or effector only on one side of the membrane.

The orientations of membrane proteins can be probed by examining their reactions with reagents that do not leak across the membrane rapidly and that are introduced only on one side of the membrane or the other. Proteases are useful reagents here. For example, carboxypeptidase, which digests proteins from the C-terminal end, can work on a membrane protein only if this end of the protein is exposed on the side of the membrane on which the enzyme is presented. Erythrocyte membranes are ideal for experiments of this sort, because they can be turned inside out experimentally. In the case of glycophorin, the erythrocyte protein that we mentioned previously in connection with hydropathy plots, the C-terminal end of the protein is inside the cell, and the N-terminal end is outside (fig. 19.19). This agrees with the conclusion drawn from the hydropathy plot (see fig. 19.12) that glycophorin makes one pass through the membrane.

Glycophorin is one of the numerous glycoproteins found in the plasma membranes of erythrocytes and other cells. These proteins provide good illustrations of membrane structural asymmetry, because the oligosaccharides attached to them are almost always on the outside of the cell. Glycophorin has about 100 carbohydrate residues attached to the protein at 16 sites between residues 2 and 50; these are all on the outside of the erythrocyte (see fig. 19.19).

Lipids also show asymmetrical distributions between the inner and outer leaflets of the bilayer. In the erythrocyte plasma membrane most of the phosphatidylethanolamine and phosphatidylserine are in the inner leaflet, whereas the phosphatidylcholine and sphingomyelin are located mainly in the outer leaflet. A similar asymmetry is seen even in artificial liposomes prepared from mixtures of phospholipids. In liposomes containing a mixture of phosphatidylethanolamine and phosphatidylcholine, phosphatidylethanolamine localizes preferentially in the inner leaflet, and phosphatidylcholine in the outer. For the most part, the asymmetrical distributions of lipids probably reflect packing forces de-

Figure 19.19

Topography of glycophorin in the mammalian erythrocyte membrane. Carbohydrate residues (small blue hexagons) are attached to the hydroxyl groups of threonine and serine residues in the N-terminal domain of the protein. The N terminus and all of the carbohydrates are outside the cell; the C-terminal domain of the protein is inside. The hydrophobic, membrane-spanning domain is flanked by charged amino acid residues that may interact electrostatically with the polar head groups of the phospholipids.

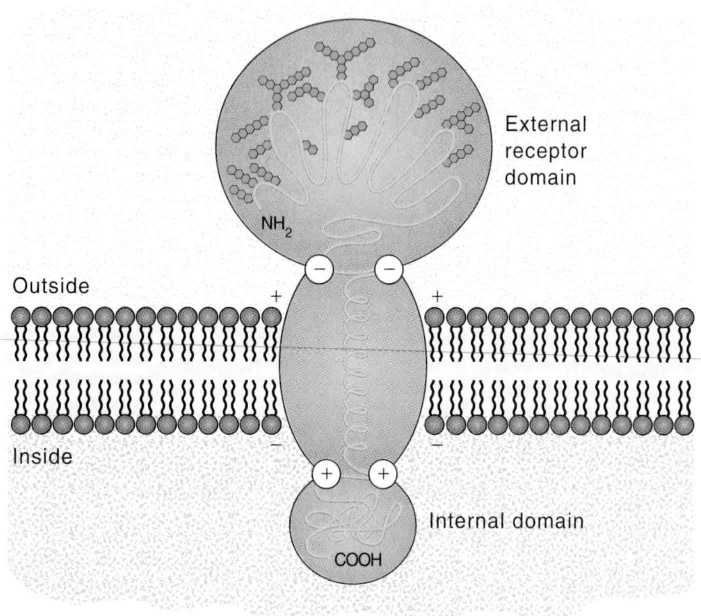

termined by the different curvatures of the inner and outer surfaces of the bilayer. By contrast the disposition of membrane proteins reflects the mechanism of protein synthesis and insertion into the membrane. We return to this topic in chapter 34.

Membrane Fluidity Is Sensitive to Temperature and Lipid Composition

The degree of fluidity of biological membranes is a function of both temperature and lipid composition. Bilayers containing a single type of phospholipid show an abrupt increase in fluidity (melting) as the temperature is raised through a characteristic and narrow range (T_m). The transition can be measured by a variety of physical techniques. One technique that is frequently applied here is differential scanning calorimetry. In this technique a sample of interest and an inert reference material are warmed slowly, side by side. The calorimeter records differences between the amounts of heat that must be applied to the two samples to keep their temperatures the same. If the sample undergoes an endothermic phase transition such as melting at a characteristic temperature, extra heat must be applied. Phospholipid bilayers exhibit such a transition at their T_m (fig. 19.20).

The phase transition of a phospholipid bilayer is not as sharp as the melting of a crystaline solid. It extends over several degrees, rather than just a fraction of a degree. In addition, al-

Figure 19.20

(*Top*) Differential scanning calorimetry of various phospholipids dispersed in water. Heat absorption is indicated by a trough in the plot relating differential heat flow to temperature. The lowest point in the trough is the phase transition temperature (T_m): (*a*) dipalmitoyl phosphatidylethanolamine; (*b*) dimyristoyl lecithin; (*c*) dipalmitoyl lecithin; (*d*) egg lecithin (plus ethylene glycol to prevent freezing). (Reproduced, with permission, from the *Annual Review of Biophysics and Bioengineering,* Volume 5, © 1976, by Annual Reviews, Inc.) (*Bottom*) Molecular interpretation of the heat-absorbing reaction during the phase transition.

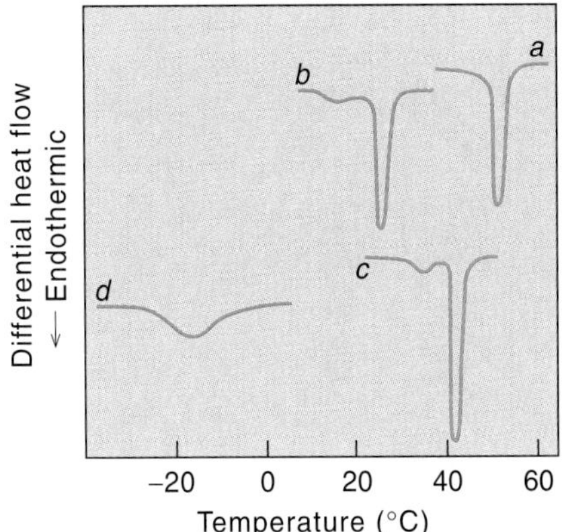

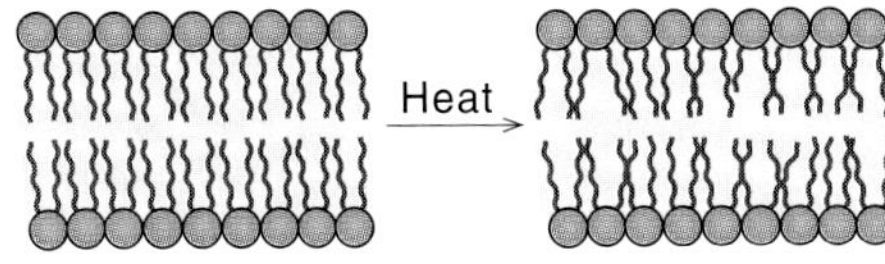

Figure 19.21

The amount of disorder in the fatty acid side chains in a phospholipid bilayer as a function of temperature. The side chains become much more disordered as the temperature increases through the T_m. The *solid curve* represents a bilayer that does not contain cholesterol; the *dotted curve,* a bilayer with the same phospholipid plus about 25% cholesterol. The amount of random, disorderly motion of the fatty acid chains can be measured quantitatively by deuterium- or ^{13}C-NMR. At any given temperature the disorder is greater near the tips of the chains (toward the middle of the bilayer) than it is close to the head groups.

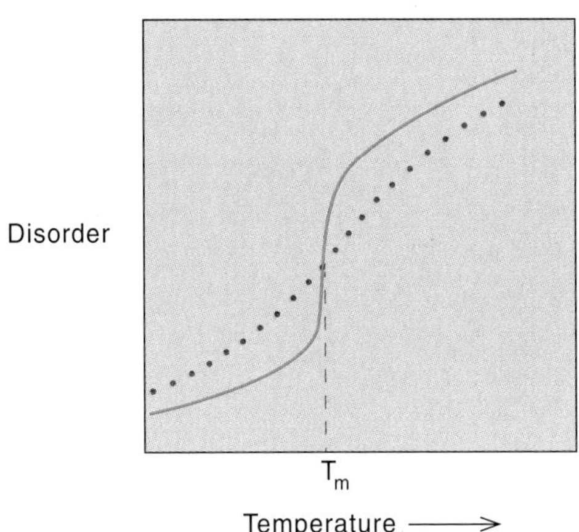

Table 19.5
Transition Temperatures for Aqueous Suspensions of Phospholipids

Phospholipid[a]	T_m (°C)
Di-14:0 phosphatidylcholine	24
Di-16:0 phosphatidylcholine	41
Di-18:0 phosphatidylcholine	58
Di-22:0 phosphatidylcholine	75
Di-18:1$^{\Delta 9}$ phosphatidylcholine	−22
Di-14:0 phosphatidylethanolamine	51
Di-16:0 phosphatidylethanolamine	63
1-18:0, 2-18:1$^{\Delta 9}$ phosphatidylethanolamine	3
Di-14:0 phosphatidylglycerol	23
Di-16:0 phosphatidylglycerol	41

[a]Phospholipid abbreviations are as in tables 19.2 and 19.4; Di-14:0 phosphatidylcholine, for example, refers to dimyristoyl (14 carbons, 0 double bonds) phosphatidylcholine.

Source: Adapted from M. K. Jain and R. C. Wagner, *Introduction to Biological Membranes,* John Wiley & Sons, New York, 1980.

though the fluidity of the bilayer increases substantially at the T_m, it also increases gradually with temperature both below and above this point (fig. 19.21). In this regard, the behavior of a bilayer on warming calls to mind the gradual softening of butter.

The temperature of the melting transition (T_m) depends strongly on the phospholipid composition in the bilayer. Increasing the length of the fatty acid chains increases the T_m; *cis* double bonds decrease the T_m (table 19.5). Increasing the length of the chains makes the formation of kinks more difficult because each kink must disrupt the packing of a large number of atoms. Double bonds introduce rigid kinks that interfere with the packing of neighboring chains even below the T_m. The T_m also depends on the nature of the polar head group of the phospholipid; phosphatidylethanolamines have higher T_m values than phosphatidylcholines or phosphatidylglycerols containing the same fatty acids (see table 19.5). Membranes that contain mixtures of phospholipids may show partial melting transitions at several temperatures, or (depending on their composition) they may have a single, broadened transition (see fig. 19.20).

We have not yet said much about the second major constituent of eukaryotic membrane lipids: cholesterol. Cholesterol broadens the melting transition of the phospholipid bilayer (see fig. 19.21). Below the T_m, cholesterol disorders the membrane because it is too bulky to fit well into the neatly packed arrangement of the fatty acid chains that is favored at low temperatures. Above the T_m, cholesterol restricts further disordering because it is too large and inflexible to join in the rapid fluctuations of the chains. If the amount of cholesterol in a phospholipid bilayer is increased to about 30%, roughly the amount in the plasma membranes of typical animal cells, the melting transition becomes so broad as to be almost undetectable.

Figure 19.22

The structure of spectrin and the location of spectrin in the cytoskeleton. (*a*) An $\alpha\beta$ dimer of spectrin. Both α and β subunits are extended structures consisting of end-to-end domains of 106 aminoacyl residues folded into three α helices; the subunits twist about one another loosely as shown. (*b*) The erythrocyte membrane skeleton. Spectrin tetramers ($\alpha_2\beta_2$), shown in yellow, are linked to the cytoplasmic domain of the anion channel (blue) by the protein ankyrin (red), and to glycophorin and actin filaments by protein 4.1. This structure lends stability to the red cell membrane while maintaining sufficient flexibility to allow erythrocytes to withstand substantial shear forces in the peripheral circulation.

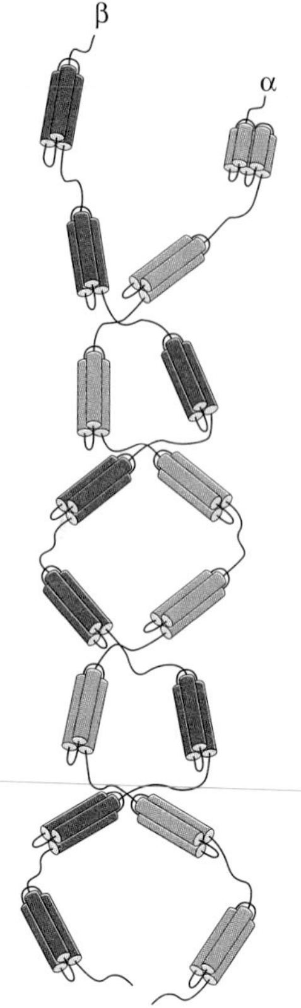

(a)

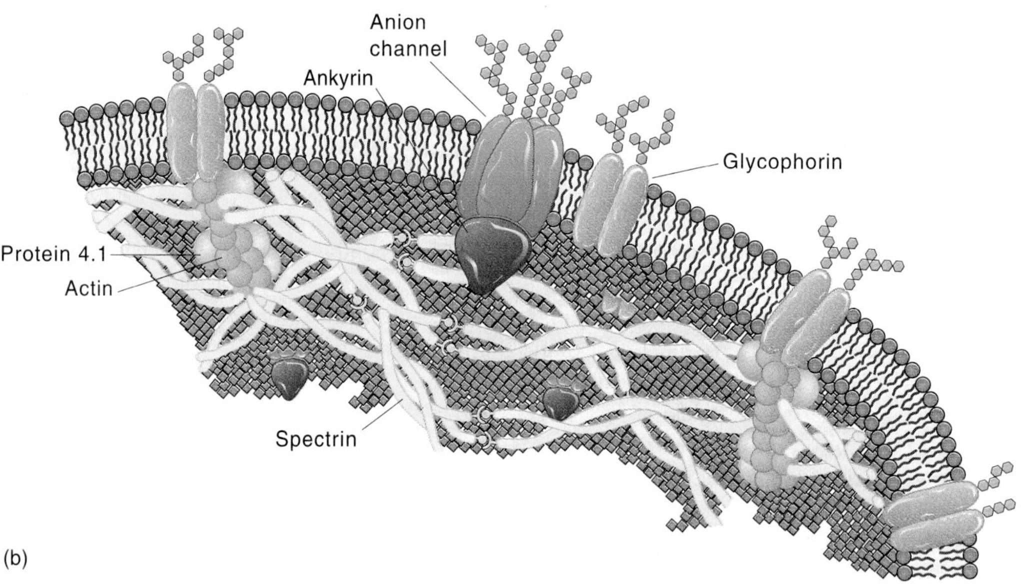

(b)

Cells regulate the lipid compositions of their plasma membrane so that a reasonable membrane fluidity is maintained. They do this by controlling fatty acid biosynthesis so as to vary the lengths of the fatty acid chains and the ratio of unsaturated to saturated fatty acids (see chapter 22). If cells are grown at low temperatures, their phospholipids contain more unsaturated fatty acids or fatty acids with shorter chains or both. These adjustments shift the T_m to lower temperatures, with the result that (to the extent that the melting transition is sharp enough to be measurable) the T_m remains close to or slightly below the ambient temperature. This has been shown in bacteria, in plants, and even in animal cells from different parts of the same organism. For example, the ratio of unsaturated to saturated fatty acids is higher in winter wheat than in summer wheat of the same species and higher in tissue near a reindeer's foot than in the thigh.

Figure 19.23

Membrane-associated cytoskeletal components of cultured mouse cells. (PM = plasma membrane, MF = microfilaments, MT = microtubules; 54,000×.) (From G. L. Nicolson, Transmembrane control of the receptors on normal and tumor cells: I. Cytoplasmic influence over cell surface components, *Biochem. Biophys. Acta* 457:57, © 1976. Reprinted with permission from Elsevier Science Publishers, *Amsterdam The Netherlands.*)

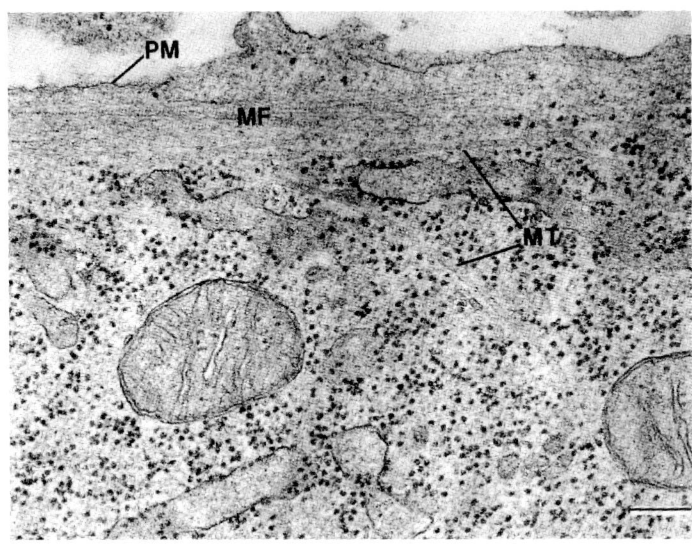

Figure 19.24

Schematic illustration of a so-called focal contact, showing how extracellular fibronectin is believed to be indirectly attached to the intracellular cytoskeleton through a transmembrane fibronectin receptor and several other peripheral membrane proteins.

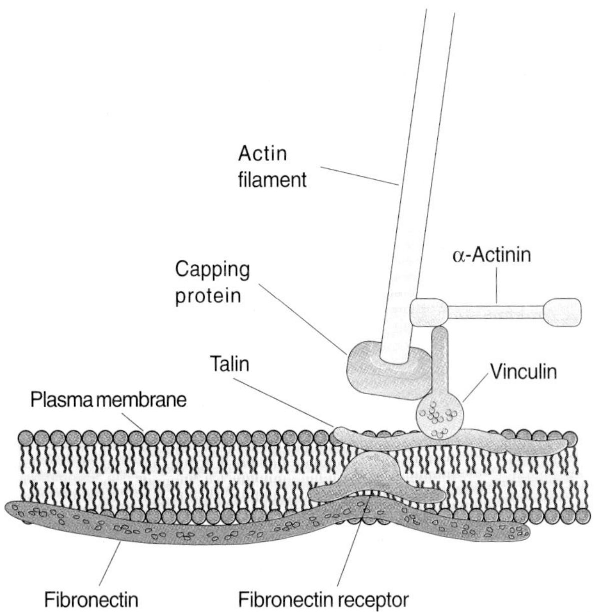

Some Proteins of Eukaryotic Plasma Membranes Are Connected to the Cytoskeleton

Contrary to the general theme that membrane structures are highly dynamic, some of the proteins in eukaryotic plasma membranes are linked to other proteins so that their motions are severely restricted. These linked proteins are attached to the cytoskeleton that is situated in the cytosol of nearly all eukaryotic cells. In the erythrocyte the cytoskeleton is constituted mainly of spectrin, a protein that has two chains, each with approximately 100 subunits (fig. 19.22). The integral membrane protein glycophorin is linked indirectly to the mesh formed by spectrin through another protein called the band-4.1 protein. The spectrin cytoskeleton is important for maintaining the biconcave shape of the erythrocyte. Cells with a genetic deficiency of spectrin round up into spheres and are more fragile than normal erythrocytes.

One of the major integral proteins of the erythrocyte membrane is the anion channel, or band-3 protein, which moves Cl^- and HCO_3^- anions across the membrane. The anion transporter has two identical subunits with molecular weights of about 95,000, and each subunit probably has 10 or 11 transmembrane helices. The band-3 protein is attached to the spectrin cytoskeleton through a smaller protein, ankyrin. The cytosolic domain of the anion transporter also binds the glycolytic enzyme glyceraldehyde-3-phosphate dehydrogenase.

The cytoskeletons of other eukaryotic cells typically include both microtubules and microfilaments, which consist of long, chainlike oligomers of the proteins tubulin and actin, respectively. Bundles of microfilaments often lie just underneath the plasma membrane (fig. 19.23). They participate in processes that require changes in the shape of the cell, such as locomotion and phagocytosis. In some cells, cytoskeletal microfilaments appear to be linked indirectly through the plasma membrane to peripheral proteins on the outer surface of the cell (fig. 19.24). Among the cell surface proteins connected to this network is fibronectin, a glycoprotein believed to play a role in cell–cell interactions. The lateral diffusion of fibronectin is at least 5,000 times slower than that of freely diffusible membrane proteins.

Summary

In this chapter we discussed the structure and function of biological membranes. First we considered their structure, starting with an examination of the constituents of membranes. Then we turned to questions concerning the transport of materials across membranes.

1. Biological membranes consist primarily of proteins and lipids whose relative amounts vary considerably.
2. In order to study the structure of biological membranes, they must first be isolated in a more or less intact form from the cell. A popular technique for doing this involves equilibrium density gradient centrifugation.

3. There are two main types of lipids in biological membranes: Phospholipids and sterols. The predominant phospholipids in most membranes are phosphoglycerides, which are phosphate esters of the three-carbon alcohol glycerol.
4. Owing to their amphipathic nature, phospholipids spontaneously form ordered structures in water. When phospholipids are agitated in the presence of excess water, they tend to aggregate spontaneously to form bilayers, which strongly resemble the types of structures they form in biological membranes.
5. Membranes contain proteins that merely bind to their surface (peripheral proteins) and those that are embedded in the lipid matrix (integral proteins). Integral membrane proteins contain transmembrane α helices.
6. Proteins and lipids have considerable lateral mobility within membranes.
7. Biological membranes are asymmetric. Consistent with this asymmetry, a protein that has been inserted into a membrane in a particular orientation usually retains that orientation indefinitely.
8. On heating, phospholipid bilayers undergo a phase transition (melting) from an ordered to a disordered state. The melting temperature (T_m) depends strongly on the phospholipid composition in the bilayer. Increasing the length of the fatty-acid chains increases the T_m; *cis* double bonds decrease the T_m. Cholesterol broadens the melting transition of the phospholipid bilayer. Cells regulate the lipid compositions of their plasma membrane so that a reasonable membrane fluidity is maintained.
9. Some proteins of eukaryotic plasma membranes are connected to the cytoskeleton; this connection inhibits their lateral mobility with the membrane.

Selected Readings

Bennett, V., The membrane skeleton of human erythrocytes and its implications for more complex cells. *Ann. Rev. Biochem.* 54:273, 1985.

Bennett, V., Ankyrins. *J. Biol. Chem.* 267:8703–8706, 1992.

Bretscher, M., The molecules of the cell membrane. *Sci. Am.* 253(4):100–108, 1985.

Cowan, S. W., Bacterial porins: Lessons from three high-resolution structures. *Curr. Opin. Struct. Biol.* 3:501–507, 1993.

Davies, K. E., and S. E. Lux, Hereditary disorders of the red cell membrane. *Trends Genet.* 5:222–227, 1989.

Deisenhofer, J., O. Epp, N. Miki, R. Huber, and H. Michel, X-ray structure analysis of a membrane protein complex: Electron density map at 3-Å resolution and a model of the chromophores of the photosynthetic reaction center from *Rhodopseudomonas viridis*. *J. Mol. Biol.* 180:385, 1984.

Gennis, R. B., *Biomembranes: Molecular Structure and Function.* New York: Springer-Verlag, 1989.

Henderson, R., J. M. Baldwin, T. A. Ceska, F. Zemlin, E. Beckmann, and K. H. Downing, Model for the structure of bacteriorhodopsin based on high resolution electron cryo-microscopy. *J. Mol. Biol.* 213:899, 1990.

Inesi, G., D. Lewis, D. Nikic, A. Hussain, and M. Kirtley. Long-range intramolecular linked functions in the calcium transport ATPase. *Adv. Enzymol.* 65:182, 1992.

Jain, M. K., *Introduction to Biological Membranes,* 2nd ed. New York: Wiley, 1988.

Jennings, M. L., Topography of membrane proteins. *Ann. Rev. Biochem.* 58:999, 1989.

Kaback, H. R., E. Bibi, and P. D. Roepe, β-Galactoside transport in *E. coli:* A functional dissection of lac permease. *Trends Biochem. Sci.* 15:309, 1990.

Schofield, A. E., R. M. Reardon, and M. J. A. Tanner, Defective anion transport activity of the abnormal band 3 in hereditary ovalocytotic red blood cells. *Nature* 355:836–838, 1992.

Silverman, M., Structure and function of hexose transporters. *Ann. Rev. Biochem.* 60:757, 1991.

Singer, J. J., The structure and insertion of integral proteins in membranes. *Ann. Rev. Cell. Biol.* 6:247–296, 1990.

Singer, S. J., The molecular organization of membranes. *Ann. Rev. Biochem.* 43:805, 1974.

Vance, D. E., and J. E. Vance (eds.), *Biochemistry of Lipids, Lipoproteins and Membranes.* Amsterdam: Elsevier, 1996.

Problems

1. A lecturer who is inclined to personify molecules to assist in describing their properties, might use the word "schizophrenic" in the same sentence as "phosphatidylethanolamine." Explain the use of the word "schizophrenic."
2. Which would you expect to have the fastest nonenzymatic flip-flop motion (fig. 19.18): phosphatidylethanolamine or phosphatidylserine?
3. Glycophorin contains a sequence of amino acid residues that is thought to span a membrane (fig. 19.12). Assume that the region of the peptide with a positive hydropathy index exists as an α helix. Use the dimensions of a helix presented in figure 5.8 and chapter 5 to calculate the thickness of a membrane. If the length of an extended alkane is 1.25 Å per methylene, estimate the thickness of a membrane.
4. Membranes are known to repair themselves if punctured. Can you explain how this might occur?
5. Examine the list of ingredients on a container of salad dressing. Without much searching you should be able to find lecithin listed. Remember that ingredients are ordered by weight of the item. (The materials that contribute the most are listed first.) After identifying the major components of salad dressing, propose a function for the lecithin.
6. Both triacylglycerols and phospholipids have fatty acid ester components, but only one can be considered amphipathic. Indicate which is amphipathic, and explain why.
7. Why does the Davson-Danielli membrane model predict that the exposed inside of the lipid bilayer is featureless?
8. The relative orientation of polar and nonpolar amino acid side chains in integral membrane proteins is "inside-out" relative to that of the amino acid side chains of water-soluble glubular proteins. Explain.
9. What physical properties are conferred on biological membranes by phospholipids? How can the charge characteristics of the phospholipids affect binding of peripheral proteins to the membrane? What role might divalent metal ions play in the interaction of peripheral membrane proteins with phospholipids?
10. Differentiate between peripheral and integral membrane pro-

teins with respect to location, orientation, and interactions that bind the protein to the membrane. What are some strategies used to differentiate between peripheral and integral proteins by means of detergents or chelating agents?

11. Frequently, integral membrane proteins are glycosylated with complex carbohydrate arrays. Explain how glycosylation further enhances the asymmetrical orientation of integral proteins.
12. Integral transmembrane proteins often contain helical segments of the appropriate length to span the membrane. These helices are composed of hydrophobic amino acid residues. In transmembranous proteins with multiple segments that span the membrane, you may find some hydrophilic residue side chains. Why are hydrophilic side chains not favored in single-span membrane proteins? How may the hydrophilic side chains be accommodated in multiple-span proteins?
13. Compare the relative efficiency of extraction of free fatty acid, fatty acid methylester, and triacylglycerol into organic solvent. Which component(s) have a pH dependency of extraction? Why?
14. The term phosphatidylcholine (PC) defines a class of phospholipids. What portion of the phosphatidylcholine molecule is common to all membranes of the class? What portion of the molecule is variable among the PCs?
15. You are characterizing a protein in a membrane fraction that was dissolved in octylglucoside. You have estimated the molecular weight to be approximately 60,000. However, upon exhaustive treatment to remove most of the detergent, the protein elutes from a 100,000-M_r cutoff gel exclusion column in the void (excluded) volume. What can you conclude from these data?

C H A P T E R

Mechanisms of Membrane Transport

While the primary function of a membrane is to seal off a territory, it must provide a means for the transport of substances and signals across the membrane.

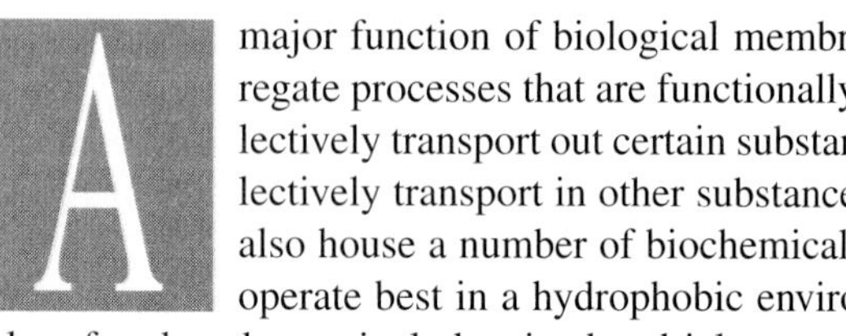

A major function of biological membranes is to segregate processes that are functionally related, to selectively transport out certain substances, and to selectively transport in other substances. Membranes also house a number of biochemical pathways that operate best in a hydrophobic environment; examples of such pathways include mitochondrial transport and the later steps in lipid synthesis. Membranes also permit the development of concentration gradients, which are crucial to the functioning of the mitochondrial and chloroplast ATPases and to neurotransmission. We have already dealt with a number of these phenomena, and we will deal with a good many more. In this chapter we will confine ourselves to a consideration of the mechanisms of membrane transport and a less detailed consideration of the "transport" of signals across membranes.

Transport of Materials across Membranes

Living cells depend on an influx of phosphate and other ions and of nutrients such as carbohydrates and amino acids. They extrude certain ions, such as Na^+, and rid themselves of metabolic end products. How do these ionic or polar species traverse the phospholipid bilayer of the plasma membrane? How do pyruvate, malate, the tricarboxylic acid citrate, and even ATP move between the cytosol and the mitochondrial matrix (see figs. 15.16 and 16.26)? The answer is that biological membranes contain proteins that act as specific transporters, or permeases. These proteins behave much like conventional enzymes; they bind substrates and they release products. Their primary function, however, is not to catalyze chemical reactions but to move materials from one side of a membrane to the other. In this section we discuss the general features of membrane transport and examine the structures and activities of several transport proteins.

Most Solutes Are Transported by Specific Carriers

For the sake of comparison, let us consider movement of glucose across an inert, porous membrane such as a piece of dialysis tubing (fig. 20.1). The pores in dialysis tubing are large enough that glucose and water molecules can diffuse from either side of the membrane to the other with little hindrance. If the solutions on the two sides initially contain different concentrations of glucose, molecules diffuse from the more

Figure 20.1

(*a*) Glucose can diffuse in either direction through the pores in dialysis tubing. (*b*) If the glucose concentration inside the dialysis bag is initially higher than the concentration outside, glucose diffuses out spontaneously until the two concentrations are equal. (*c*) At any given time, the net rate of diffusion is proportional to the difference between the two concentrations: $v_{in \rightarrow out} = k[C_{in}] - k[C_{out}] = k[\Delta C]$. (*d*) The measured rate of transport across a biological membrane usually is not simply proportional to the concentration difference across the membrane but rather approaches a maximum of high values of $[\Delta C]$.

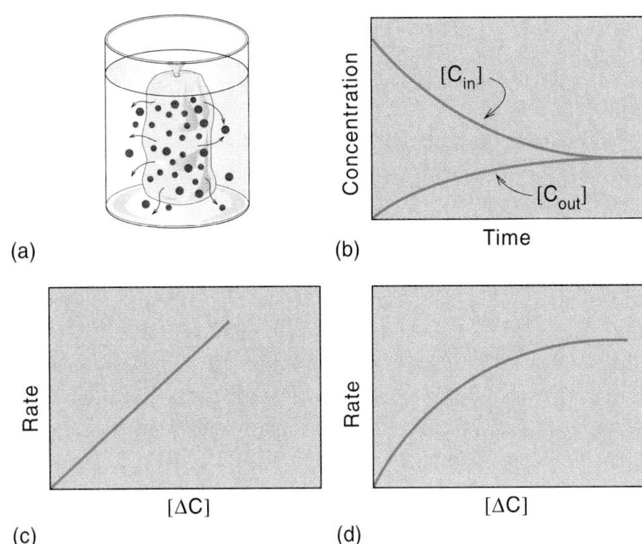

Figure 20.2

The bacterial lactose-transport protein (lactose permease) transports β galactosides, such as lactose, *o*-nitrophenyl-β-galactoside, and isopropyl-β-thiogalactoside. It does not transport galactosides with an α-glycosidic linkage.

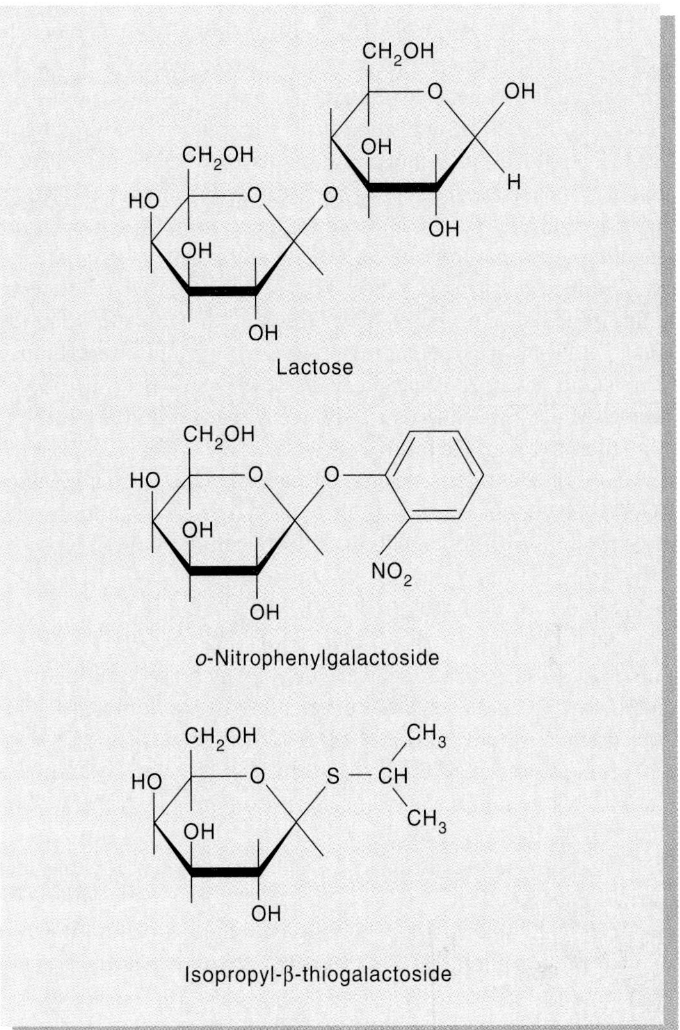

concentrated solution to the more dilute. The net rate of this diffusion is proportional to the difference between the two concentrations $[\Delta C]$, as shown in figure 20.1*c*.

By contrast, the kinetics of transport across a biological membrane usually does not exhibit such a linear dependence on ΔC. Instead, the rate approaches an asymptote at high concentration differences, much as the rate of an enzymatic reaction approaches a maximum rate at high substrate concentrations (fig. 20.1*d*). This behavior suggests that the transport protein has a specific binding site for the material that it transports. The overall rate of transport is limited by the number of these sites in the membrane and thus fits an expression analogous to the Michaelis-Menten equation for enzyme kinetics:

$$v = \frac{V_{max}[\Delta C]}{K_m + [\Delta C]} \quad (1)$$

Just as in enzyme kinetics (see equation (25) in chapter 1), K_m here is an algebraic function of the microscopic rate constants for binding, dissociation, and translocation of the substrate in either direction.

The high specificity of biological transport systems provides compelling support for this interpretation. For example, whereas dialysis tubing is equally permeable to D- and L-glucose, most eukaryotic cells take up D-glucose rapidly but not L-glucose. In bacteria the synthesis of a specific transport system for a particular nutrient often can be induced by adding a small amount of the nutrient to the growth medium. Thus if *E. coli* is grown on medium without lactose, transporters for lactose usually are not found in the plasma membrane. When lactose is added to the culture medium, however, synthesis of a protein that transports lactose is switched on, and the cells soon begin to demonstrate this activity. The lactose-transport system also catalyzes the uptake of a variety of galactosides that resemble lactose having a $\beta(1,4)$-glycosidic linkage (fig. 20.2). It does not transport α-galactosides.

Like conventional enzymes, transport proteins often are sensitive to specific inhibitors. For example, the anion transporter of the erythrocyte plasma membrane, which moves HCO_3^- and Cl^- ions across the membrane, is inhibited by 1,2-cyclohexadione or derivatives of stilbenedisulfonate. Because these inhibitors react covalently with amino acid residues of the protein, they afford useful probes for the location and structure of the anion-binding site.

Some Transporters Facilitate Diffusion of a Solute down an Electrochemical Potential Gradient

The presence of a specific transport system in the plasma membrane does not necessarily imply that a cell can pump the substrate in a direction opposite to the way the substrate spontaneously diffuses. In some cases the net flow always proceeds down the concentration gradient, just as glucose always diffuses out of the more concentrated of two solutions separated by a dialysis membrane. The glucose transporter in the plasma membranes of most animal cells behaves in this way. Muscle cells continue to take up glucose from the blood only because the molecules that enter the cell are quickly modified by metabolic reactions that keep the cytosolic glucose concentration low. And yet, for a given concentration gradient, the uptake of D-glucose is much faster than the uptake of L-glucose or other materials that the cell is not equipped to transport. Transport that proceeds in the same net direction as simple diffusion but that is catalyzed by a specific transport system is termed facilitated diffusion.

If the solute is uncharged, the free energy change associated with moving 1 mol of the material from a solution with concentration $[C_1]$ to a solution with concentration $[C_2]$ is

$$\Delta G_{1\rightarrow 2} = 2.3RT\log\frac{[C_2]}{[C_1]} \qquad (2)$$

This expression, which reflects the entropy increase resulting from distributing the molecules more randomly between the two solutions, holds for both free and facilitated diffusion. If $[C_1] > [C_2]$, $\Delta G_{1\rightarrow 2}$ is negative and molecules diffuse spontaneously from solution 1 to solution 2. Net flux across the membrane ceases when $[C_1] = [C_2]$ and $\Delta G_{1\rightarrow 2}$ is zero.

If the solute carries a net charge, there is an additional thermodynamic effect of moving the charge across any difference in electric potential that exists between the solutions on the two sides of the membrane. The free energy change then is

$$\Delta G_{1\rightarrow 2} = 2.3RT\log\frac{[C_2]}{[C_1]} + z\mathcal{F}\,\Delta\psi \qquad (3)$$

where z is the charge, $\mathcal{F}$ is the Faraday constant, and $\Delta\psi$ is the electric potential difference ($\Delta\psi = \psi_2 - \psi_1$). The equilibrium condition here is

$$\log\frac{[C_2]}{[C_1]} = -\frac{z\mathcal{F}\,\Delta\psi}{2.3RT}$$

We used this expression in chapter 16 (see equation (9) in chapter 16) to evaluate $\Delta\psi$ across the mitochondrial inner membrane.

Among the transporters that facilitate diffusion of charged species down electrochemical potential gradients are the Na^+ and K^+ channels of nerve cells. These channels are closed in the resting cell but open transiently when the cell is stimulated electrically. The action potential that sweeps down the cell reflects a sudden electric depolarization of the membrane caused by Na^+ flowing into the cell, followed by a repolarization as K^+ flows out. Na^+ influx is thermodynamically downhill because the intracellular Na^+ concentration (about 12 mM) is much lower than the extracellular concentration (145 mM) and the internal electric potential is negative relative to the external potential by about 60 mV. These conditions are set up by the membrane Na^+–K^+ pump, which we discuss shortly. K^+ ions, on the other hand, are close to electrochemical equilibrium in the resting cell: The tendency of the negative $\Delta\psi$ to pull K^+ into the cell is nearly balanced by a higher intracellular concentration that favors K^+ efflux. (The K^+ concentrations are about 140 mM inside the cell and 4 mM outside.) When the sudden influx of Na^+ causes $\Delta\psi$ to shoot through zero and even change sign, this balance is upset and K^+ starts flowing out. The Na^+ and K^+ channels both close again after 1 to 2 ms, and the Na^+–K^+ pump soon reestablishes the original conditions.

The anion transporter of the erythrocyte provides another important, though less dramatic, illustration of facilitated diffusion. CO_2 produced in respiring tissues diffuses spontaneously into the red blood cells, where carbonic anhydrase catalyzes its reaction with H_2O to form H_2CO_3. The H_2CO_3 then ionizes to HCO_3^-. Some of the HCO_3^- so formed is bound by hemoglobin in the erythrocytes, but a substantial amount is moved back out to the blood plasma by the anion transporter. These processes are reversed when the blood reaches the lungs and CO_2 is exhaled. Transport of HCO_3^- in either direction across the erythrocyte membrane is coupled to the movement of Cl^- in the opposite direction. Because an exchange of one anion for another results in no net flow of charge across the membrane, the transport of HCO_3^- and Cl^- is independent of $\Delta\psi$. The ΔG for moving 1 mol of HCO_3^- into the cell in exchange for 1 mol of Cl^- is

$$\Delta G_{HCO_3^-/Cl^-} = 2.3RT\log\frac{[HCO_3^-]_{in}}{[HCO_3^-]_{out}} + 2.3RT\log\frac{[Cl^-]_{out}}{[Cl^-]_{in}}$$

$$= 2.3RT\log\left\{\frac{[HCO_3^-]_{in}[Cl^-]_{out}}{[HCO_3^-]_{out}[Cl^-]_{in}}\right\} \qquad (4)$$

Such coupled transport of two solutes in opposite directions is termed antiport.

Active Transport against an Electrochemical Potential Gradient Requires Energy

Cells can transport some materials against gradients of concentration or electric potential or both. As we mentioned in the previous section, eukaryotic cells pump out Na^+ ions even though the external Na^+ concentration is greater than the cytosolic concentration and even though the electric potential inside is more negative than the potential outside. Many cells take up amino acids even though the cytosolic concentration exceeds the external concentration, and cells that line the stomach pump out protons against a concentration gradient of more than a million to one. Transport against such concentration and electric gradients is called active transport.

To drive active transport, a cell must couple transport to another process that is thermodynamically favorable, so that the total ΔG is negative. Cells have developed a variety of schemes for this coupling (fig. 20.3). Some transport systems, including the Na^+–K^+ pump of animal cells,

Figure 20.3

Cells drive active transport in a variety of ways. The plasma-membrane Na^+–K^+ pump of animal cells (*a*) and the plasma-membrane H^+ pump of anaerobic bacteria (*b*) are driven by the hydrolysis of ATP. (*c*) Eukaryotic cells couple the uptake of neutral amino acids to the inward flow of Na^+. (*d*) Uptake of β-galactosides by some bacteria is coupled to inward flow of protons. (*e*) Electron-transfer reactions drive proton extrusion from mitochondria and aerobic bacteria. (*f*) In halophilic bacteria, bacteriorhodopsin uses the energy of sunlight to pump protons. (*g*) *E. coli* and some other bacteria phosphorylate glucose as it moves into the cell and thus couple the transport to hydrolysis of phosphoenolpyruvate.

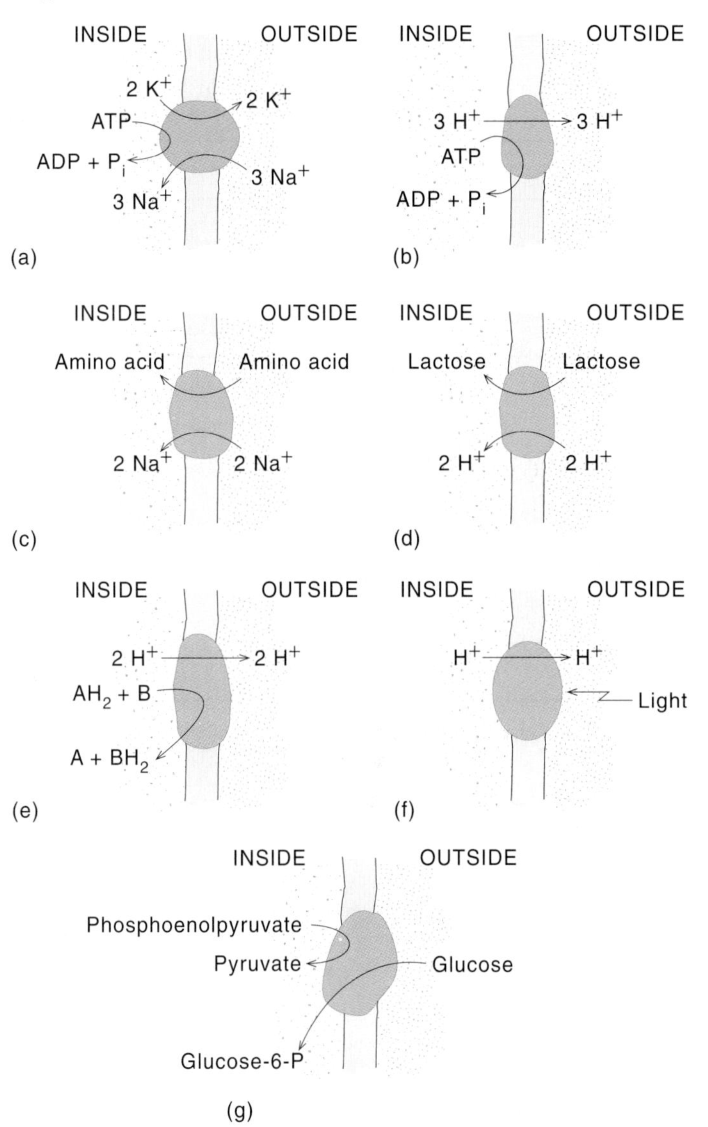

drive active transport by the hydrolysis of ATP (see fig. 20.3*a*). The Na^+–K^+ pump may consume as much as 70% of the ATP that nerve cells synthesize, and even nonexcitable cells spend a substantial portion of their energy resources in this way. Table 20.1 lists some of the other active-transport systems energized by ATP. Many cells contain ATPases that pump protons into lysozomes or intracellular vacuoles. ATPases that pump protons outward are found in the epithelial cells that line the stomach and in the plasma membranes of anaerobic bacteria (see fig. 20.3*b*). The bacterial enzymes are similar to the proton-conducting F_1/F_0 ATP-synthases of mitochondria and chloroplasts (see chapters 16 and 17).

Active transport of a solute against a concentration gradient also can be driven by a flow of an ion down *its* concentration gradient. Table 20.2 lists some of the active-transport systems that operate in this way. In some cases the ion moves across the membrane in the opposite direction to the primary substrate (antiport); in others the two species move in the same direction (symport). Many eukaryotic cells take up neutral amino acids by coupling this uptake to the inward movement of Na^+ (see fig. 20.3*c*). As we discussed previously, Na^+ influx is downhill thermodynamically because the Na^+–K^+ pump keeps the intracellular concentration of Na^+ lower than the extracellular concentration and sets up a favorable electric potential difference across the membrane. Another example is the β-galactoside transport system of *E. coli,* which couples uptake of lactose to the inward flow of protons (see fig. 20.3*d*). Proton influx is downhill because electron-transfer reactions (or, under anaerobic conditions, the proton-conducting ATPase) set up gradients of pH and electric potential across the membrane.

Proton pumps driven by electron-transfer reactions occur in mitochondria, bacteria, and chloroplasts (see fig. 20.3*e* and chapters 16 and 17). Bacteriorhodopsin, which was mentioned in chapter 20 in connection with the structure of membrane proteins, uses light energy to pump protons out of *Halobacterium halobium* (see fig. 20.3*f*). Some bacterial cells can energize sugar transport by still another strategy, group translocation. They modify the sugar by phosphorylation as they transport it (see fig. 20.3*g*). The overall free energy change thus includes the negative ΔG of the phosphorylation reaction, which in this case is provided ultimately by hydrolysis of phosphoenolpyruvate.

Isotopes, Substrate Analogs, Membrane Vesicles, and Bacterial Mutants Are Used to Study Transport

To measure rates of transport, methods must be available for identifying molecules that have crossed the membrane. Radioactively labeled substrates are commonly used for this purpose. In a typical experiment the substrate is added to a suspension of cells, time is allowed for transport to occur, the cells are collected and rinsed rapidly by centrifugation or filtration, and the amount of labeled substrate in the cells is measured. Control measurements usually are made with a labeled substrate that the cells do not transport.

In some cases, rapid spectrophotometric assays can be used to measure transport rates. These assays often employ analogs of the normal, physiological substrate of the transporter and take advantage of metabolic reactions that alter the spectroscopic properties of the solute after it enters the cell. The bacterial β-galactoside transporter, for example, can be studied by using *o*-nitrophenyl-β-galactoside in place of the physiological substrate, lactose (see fig. 20.2). Once *o*-nitrophenyl-β-galactoside is inside the cell, β-galactosidase hydrolyzes it to galactose

Table 20.1
Some Systems of Active Transport Driven by ATP Hydolysis

Substrate	Direction of Pump	Organism, Tissue, or Organelle
Na^+/K^+	Na^+ out, K^+ in	Eukaryotic cell plasma membranes
Ca^{2+}	In	Muscle sarcoplasmic reticulum
Ca^{2+}	Out	Eukaryotic cell plasma membranes
H^+/K^+	H^+ out, K^+ in	Stomach epithelial cell plasma membranes
H^+	In	Lysosomes and secretory vesicles; vacuoles in plants and fungi
H^+	Out	Anaerobic bacteria (F_1/F_0 ATPase)

Table 20.2
Some Systems of Active Transport Driven by Ion Symport or Antiport

Substrate	Cotransported Ion	Organism or Tissue[a]
Neutral amino acids	Na^+ (symport)	Eukaryotic cells and bacteria
Glucose	Na^+ (symport)	Intestine and kidney
Glutamate	Na^+ (symport)	*E. coli*
Lactose	H^+ (symport)	*E. coli*
Xylose	H^+ (symport)	*E. coli*
Glucose	H^+ (symport)	Yeast
Galactose	H^+ (symport)	Yeast
Serotonin, other amines	H^+ (antiport)	Adrenal medulla (secretory vesicles)

[a]Except for the amine transporter of adrenal secretory vesicles, all the transport systems listed here operate in the plasma membrane.

and *o*-nitrophenol, which is measurable by its yellow color:

$$\underset{\text{(colorless)}}{o\text{-nitrophenyl-}\beta\text{-galactoside}} \xrightarrow{\text{transport}} o\text{-nitrophenyl-}\beta\text{-galactoside}$$

$$\xrightarrow[\text{H}_2\text{O}]{\substack{\beta\text{-galactosidase} \\ \text{(inside cell)}}} \text{galactose} + \underset{\text{(yellow)}}{o\text{-nitrophenol}}$$

If the ability of a transporter to work against a concentration gradient is to be determined, a means must be found to prevent metabolic removal of the transported solute. Here, nonmetabolizable analogs of the substrate can be useful. Glucose transport, for example, can be studied with the analog 2-deoxyglucose, which most cells cannot metabolize beyond formation of the hexose phosphate. In bacteria, mutant strains that cannot metabolize the normal substrate are often used. For example, accumulation of lactose against a concentration gradient can be studied in mutants that lack β-galactosidase or by using the

Figure 20.4

Facilitated diffusion carried out by the glucose transporter from human erythrocytes. The solubilized transporter was introduced into phospholipid vesicles, and transport was measured with D-glucose and L-glucose as substrates. The initial transport rate is much greater with D-glucose. L-glucose enter the vesicles slowly by simple diffusion but eventually reaches the same internal concentration as D-glucose. (Source: Adapted from M. Kasahara and P. C. Hinkle. Reconstitution and purification of the D-glucose transporter from human erythrocytes. *J. Biol. Chem.* 252:7384, 1977.)

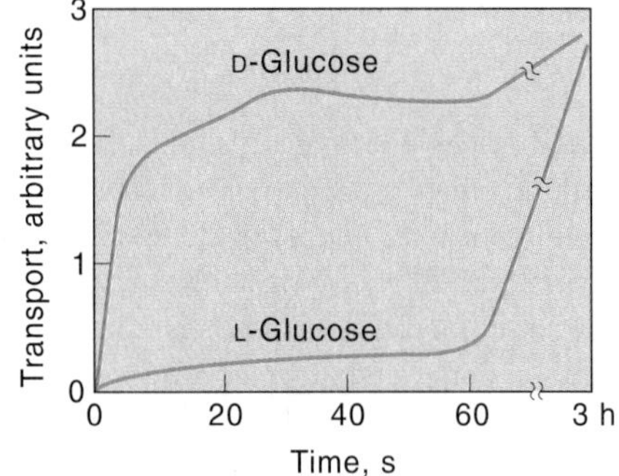

Figure 20.5

Secondary structure of lactose permease, based on the protein's hydropathy profile. Amino acid residues are designated by the one-letter code. The boxes indicate hydrophobic segments that probably form transmembrane α helices. Residues implicated in the binding of β-galactosides or in coupling galactoside transport to proton movements are circled. These residues were identified largely by studies of genetically engineered bacterial mutants. For example, replacing the circled Glu (E) residue in helix X (Glu-325) by Ala renders the protein unable to catalyze lactose/H^+ symport, although the mutant protein still catalyzes facilitated diffusion of lactose. (Source: From H. R. Kaback, Use of site-directed mutagenesis to study the mechanism of a membrane transport protein, *Biochem.* 26:2071, 1987; and J. C. Collins, S. F. Permuth, and R. J. Brooker, Isolation and characterization of lactose permease mutants with an enhanced recognition of maltose and diminished recognition of cellobiose, *J. Biol. Chem.* 264:14698, 1989.)

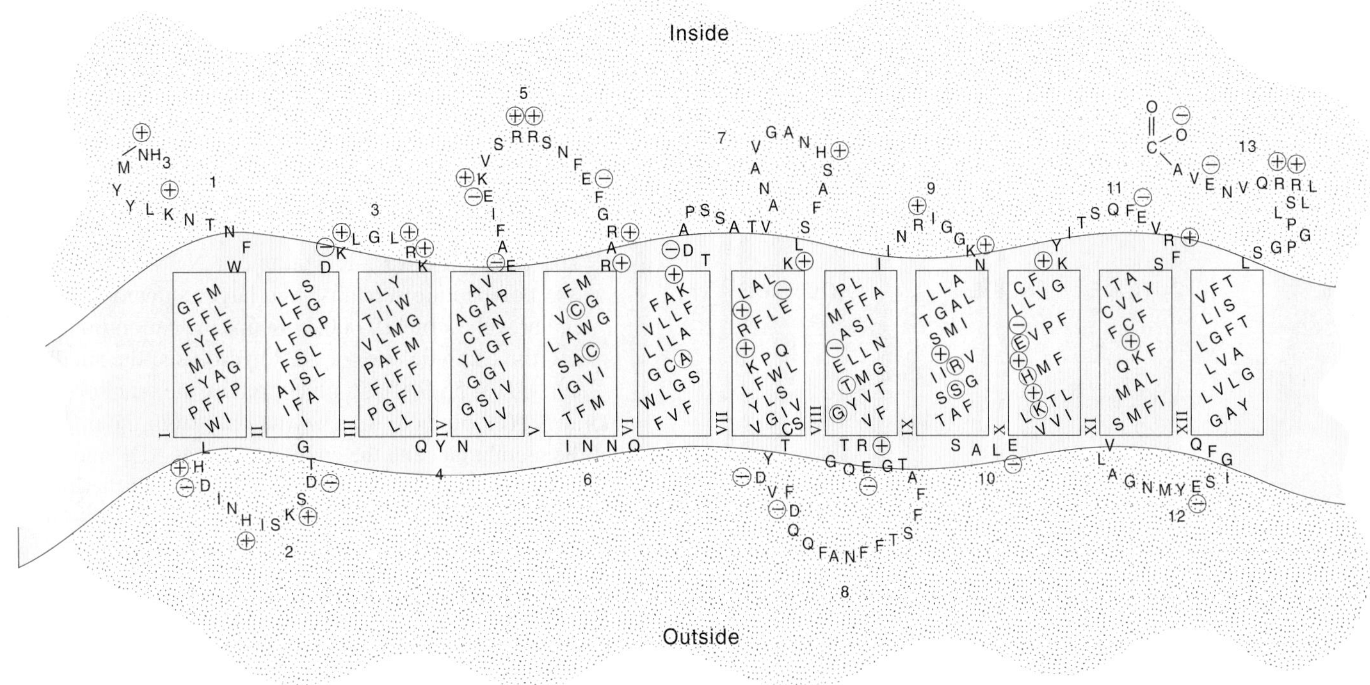

nonmetabolizable substrate analog isopropyl-β-thiogalactoside (see fig. 20.2).

To study transport in the absence of complicating metabolic processes, it often is advantageous to work with isolated membrane vesicles rather than with whole cells. Cytoplasmic membrane vesicles can be obtained from either eukaryotic or bacterial cells after homogenization or osmotic lysis. Transport proteins that have been solubilized with detergents also can be reincorporated into synthetic phospholipid vesicles (fig. 20.4).

The amino acid sequences of numerous transport proteins have been elucidated during the last 10 years. As with conventional enzymes, molecular biological techniques now make it possible to modify individual amino acid residues in the proteins and thus to probe the roles these residues play in catalysis of facilitated diffusion or active transport.

Molecular Models of Transport Mechanisms

Judging from their hydropathy profiles, transport proteins typically contain multiple α helices stretching across the membrane. The sequence of the β-galactoside transporter (lactose permease) of *E. coli* illustrates this point (fig. 20.5). This protein probably has 12 transmembrane α helices linked by short runs of more hydrophilic amino acid residues. Amino acid residues that are implicated in binding of substrates (lactose or other β-galactosides) or in the coupling of lactose transport to proton movements are distributed among five of the helices (see fig. 20.5). Although the three-dimensional structure of the protein is not known, it seems likely that these helices pack together in a parallel bundle. Such a bundle could enclose a tubular channel that stretches across the membrane. However, the channel probably is partly occluded by amino acid side chains, because it does not render the membrane freely permeable to ions. Even protons can negotiate the channel only in the presence of an appropriate galactoside.

Figure 20.6 shows a schematic model of how lactose permease could couple the movement of protons and lactose across the membrane. In this model, the transmembrane channel is never fully open but is always plugged at one end or the other. The lactose-binding site is centrally located and is accessible to either the extracellular solution or the intracellular solution, depending on where the channel is open. Transitions between these two states might occur by relatively minor conformational changes in the protein. The lactose-binding site itself also is presumed to switch between two states with different dissociation constants for lactose, depending on whether or not a nearby amino acid residue is protonated. Binding of a proton from the solution on either side favors the binding of lactose from the same side. The transporter thus would tend to pick up both a proton and lactose from the so-

Figure 20.6

A schematic model for lactose/H^+ symport catalyzed by the lactose permease. In the model, the protein forms a channel with gates that can open to expose lactose- and proton-binding sites to the solution on one side of the membrane or the other. In step 1, binding of a proton from the extracellular solution increases the affinity of the lactose-binding site for lactose. Binding of lactose (L) from the extracellular solution (step 2) results in a conformational change that switches the states of the gates (step 3). Lactose and a proton dissociate to the intracellular solution (steps 4 and 5), and the protein relaxes to a conformation that has a low affinity for both lactose and protons. In step 6 the gates switch back to their original state, exposing the unloaded binding sites to the external solution. In living cells the cycle is driven in the indicated direction by the electrochemical potential gradient for protons (the intracellular solution has a higher pH and a negative electric potential relative to the extracellular solution). The protein can transport a proton into the cell only if a galactoside moves along with it. (Source: From J. C. Collins, S. F. Permuth and R. J. Brooker, Isolation and characterization of lactose permease mutants with an enhanced recognition of maltose and diminished recognition of cellobiose, *J. Biol. Chem.* 264:14698, 1989.)

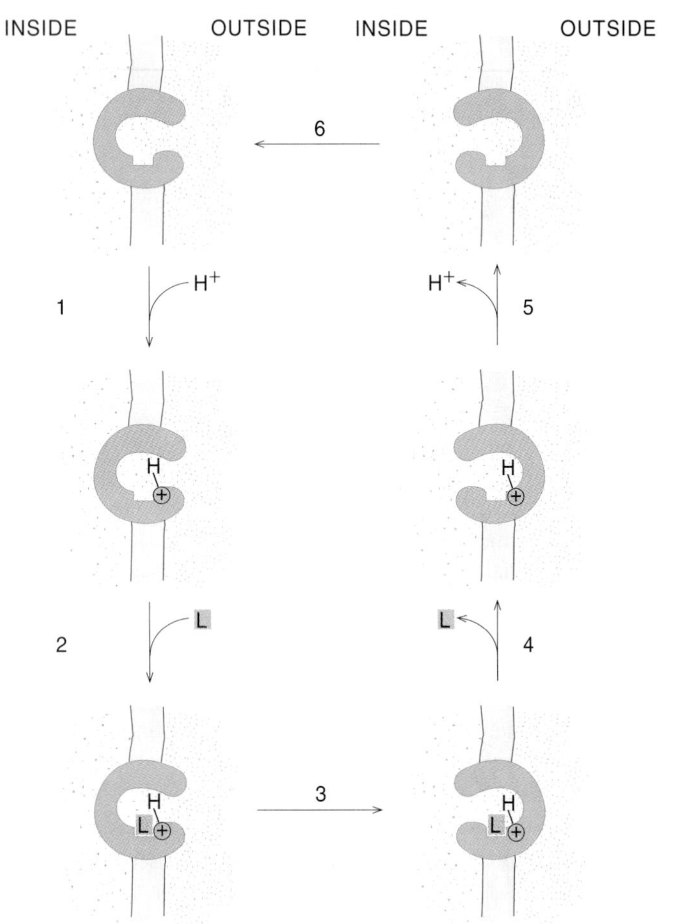

lution with the lower pH or the more positive electrical potential and to release them on the other side. A critical feature of the model is that, if the proton-binding site is occupied, conformational changes that expose this site to the opposite side of the membrane can occur only if the lactose-binding site also is occupied. This restriction prevents the protein from catalyzing proton transport in the absence of galactoside transport.

Similar models can rationalize the mechanisms of transport systems that facilitate the diffusion of a single species such as glucose. Again, although the three-dimensional structures of these proteins are not known, hydropathy profiles of the amino acid sequences indicate the presence of numerous α helices that most likely form partially blocked channels across the membrane. The erythrocyte glucose transporter probably has 12 such helices. Facilitated diffusion could result from conformational changes that open the channel alternately on one side of the binding site for glucose and then on the other side. In agreement with the notion that the protein can switch between two conformational states, certain substrate analogs bind to the glucose site best from the extracellular solution, whereas others bind best from inside the cell.

The Catalytic Cycle of the Na^+-K^+ Pump Includes Two Phosphorylated Forms of the Enzyme

The eukaryotic Na^+–K^+ pump, or Na^+–K^+ ATPase, has two polypeptide subunits. The larger subunit, with a molecular weight of 110,000, contains the cation-binding sites and the catalytic site for ATP hydrolysis; the smaller subunit ($M_r \approx 55{,}000$) is a glycoprotein, the function of which is unknown. ATP binds to the enzyme only from the intracellular side of the membrane, and the enzyme releases ADP and P_i on the same side. Na^+ ions also bind most strongly from the intracellular side, but K^+ ions bind best from the extracellular side. Ouabain, a specific inhibitor long used as an arrow poison, attaches to the enzyme on the extracellular side. The ion-pumping and ATPase activities of the enzyme are tightly coupled in the sense that ATP hydrolysis occurs at an appreciable rate only if Na^+ and K^+ both are present on appropriate sides of the membrane.

For each ATP that it splits, the Na^+–K^+ pump moves three Na^+ ions out of the cell and brings in two K^+ ions. This means that the pump is intrinsically electrogenic: It creates an electric potential gradient across the membrane because it moves more positive charges out than it brings in. However, the $\Delta\psi$ of about −60 mV that is built up across the plasma membrane of a typical nerve cell probably owes more to leakage of K^+ back out of the cell through other types of channels. K^+ tends to leak out because the Na^+–K^+ pump elevates the internal K^+ concentration above the extracellular concentration, and each K^+ that exits carries a positive charge. As we discussed previously, K^+ approaches an equilibrium when

$$\log \frac{[K^+{}_{in}]}{[K^+]_{out}} = -\frac{\mathcal{F}\,\Delta\psi}{2.3RT}$$

In the course of the catalytic cycle of the Na^+–K^+ ATPase, the terminal phosphate group of ATP is transferred to the enzyme, where it forms a carboxylic-phosphoryl anhydride with an aspartyl residue:

$$\text{ATP} + \text{Enzyme}{-}\text{CH}_2\text{CO}_2^- \xrightleftharpoons{Na^+} \text{ADP} + \text{Enzyme}{-}\text{CH}_2\overset{\overset{\text{O}}{\|}}{\text{C}}{-}\text{O}{-}\underset{\underset{\text{O}}{\|}}{\overset{\overset{\text{O}^-}{|}}{\text{P}}}{-}\text{O}^- \quad (5)$$

This step of the reaction is promoted by Na^+ ions. K^+, on the other hand, promotes hydrolysis of the phosphorylated enzyme to yield P_i:

$$\text{Enzyme—CH}_2\overset{\overset{\text{O}}{\|}}{\text{C}}\text{—O—}\overset{\overset{\text{O}^-}{|}}{\underset{\underset{\text{O}}{\|}}{\text{P}}}\text{—O}^- \xrightleftharpoons{K^+} \text{Enzyme—CH}_2\text{CO}_2^- + \text{P}_i \quad \textbf{(6)}$$

The curious thing about these reactions is that both of them are readily reversible. The reversibility of reaction (5) is not remarkable, because the $\Delta G°$ of hydrolysis of a carboxylic-phosphoryl anhydride is comparable to that of ATP. However, we do not ordinarily expect the reversal of reaction (6) to occur readily, because it requires forming a carboxylic-phosphoryl anhydride without using ATP or any other external source of energy. The explanation must be that the enzyme

$$\text{—CH}_2\overset{\overset{\text{O}}{\|}}{\text{C}}\text{—O—}\overset{\overset{\text{O}^-}{|}}{\underset{\underset{\text{O}}{\|}}{\text{P}}}\text{—O}^-$$

species that forms from P_i in the reversal of reaction (6) is not identical with the species that forms from ATP in reaction (5). The conformations of the protein evidently differ significantly in two phosphorylated species, as a result of differences in the interactions of the enzyme with Na^+ and K^+.

These observations suggest a model like that shown in figure 20.7. The enzyme is depicted in two conformations, E_1 and E_2, both of which can exist in phosphorylated states. $E_1 \sim P$ has a "high-energy" phosphoryl bond that is in equilibrium with the β-γ-pyrophosphoryl bond of ATP; E_2—P has a "low-energy" bond in equilibrium with P_i. Binding of Na^+ from inside the cell occurs in conjunction with phosphorylation of the enzyme, in the E_1 conformation, to form $E_1 \sim P$. In the model, it is the spontaneous relaxation from $E_1 \sim P$ to E_2—P that drives the cycle in the direction of Na^+ uptake. K^+ binds preferentially to E_2—P from the extracellular solution and is imported when the dephosphorylated enzyme returns from conformation E_2 to E_1.

The Ca^{2+}-pumping ATPases (see table 20.1) have phosphorylated intermediates similar to those of the Na^+–K^+ ATPase, and so does the H^+–K^+ ATPase of stomach epithelial cells. These enzymes are all closely related structurally and probably all operate by similar mechanisms, although the Ca^{2+} pump moves only one type of ion. (The Ca^{2+} ATPase moves two Ca^{2+} ions across the membrane for each ATP that it splits; the gastric enzyme pushes one H^+ out and brings one K^+ in.) But phosphorylated intermediates are not a hallmark of all ion-pumping ATPases. As we discussed in chapter 16, the F_1–F_0 ATPases of mitochondria and bacterial plasma membranes do not form such intermediates. The H^+-pumping ATPases of lysosomes and vacuoles constitute a third, structurally distinct group of enzymes that remain less thoroughly studied.

Some Membranes Have Relatively Large Pores

Mitochondria, Gram negative bacteria, and chloroplasts have double envelopes of membranes. As we discussed, their inner membranes are impermeable to ions and polar molecules other than those that are transported by specific proteins. The outer membranes, by contrast, are indiscriminately permeable to molecules with molecular weights below about 600. The permeability of the outer membranes reflects the presence of pores with diameters of about 10 Å, which are formed from proteins called porins.

Figure 20.7

A model for operation of the Na^+–K^+ pump. The enzyme exists in two conformational states, E_1 and E_2. In state E_1 the ion-binding site is open to the intracellular solution and preferentially binds Na^+ ions; in E_2 the ion-binding site faces the extracellular solution and preferentially binds K^+. ATP phosphorylates the enzyme in the E_1 conformation to give a "high-energy" intermediate, $E_1 \sim P$. When the enzyme relaxes to the "low-energy" conformation E_2—P, the Na^+ ions dissociate on the extracellular side of the membrane and K^+ ions are bound. Hydrolysis of E_2—P releases P_i inside the cell. This returns the protein to the E_1 conformation so that the K^+ ions are released inside the cell.

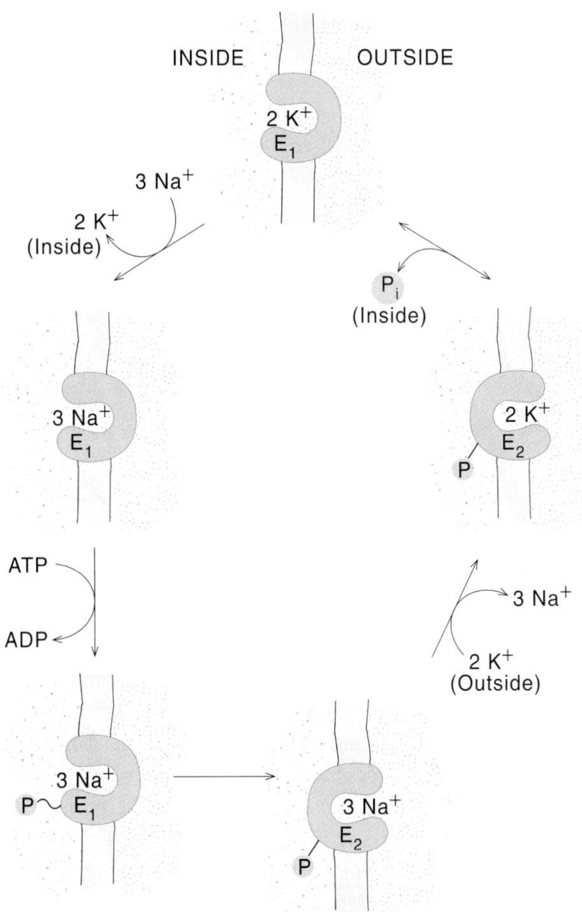

Figure 20.8 shows a model of porin from the outer membrane of the bacterium *Rhodobacter capsulatus*. Unlike most integral membrane proteins, porins have a β-stranded secondary structure. The *R. capsulatus* protein appears to form a 16-stranded β barrel that crosses the membrane as a tube. The tubular molecules aggregate as trimers with three parallel pores.

Pores called gap junctions occur in the plasma membranes of cells in some eukaryotic tissues, including liver, brain, and cardiac muscle. These channels allow molecules with molec-

Figure 20.8

A model of the structure of porin from the outer membrane of *Rhodobacter capsulatus.* Part (*a*) shows the α-carbon backbones of a trimer of porin molecules viewed along an axis approximately perpendicular to the plane of the membrane. Each molecule forms a tube that passes across the membrane. Part (*b*) shows an individual porin monomer, enlarged slightly from (*a*) and viewed along an axis approximately in the plane of the membrane. The molecule folds as a β-barrel with 16 antiparallel β strands. (Reprinted from *FEBS Lett.,* vol. 267, M. S. Weiss et al., The three-dimensional structure of porin from Rhodobacter capsulatus at 3 A resolution, page 268, 1990, with kind permission of Elsevier Science-NL, Sara Burgerhartstraat 25, 1055 KV Amsterdam, The Netherlands.)

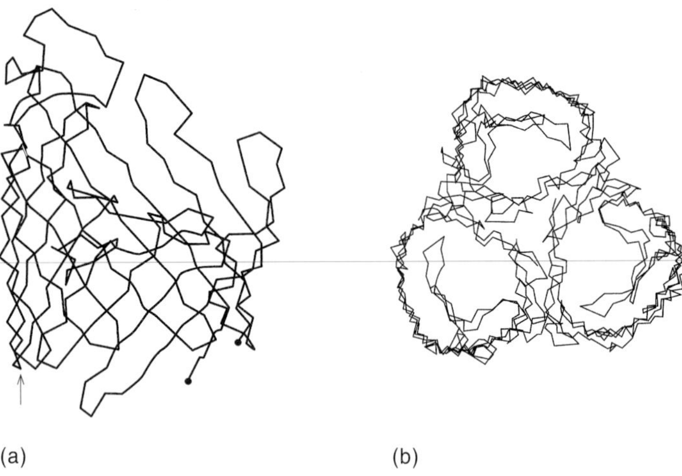

Figure 20.9

Gap junctions are formed from hexagonal arrays of a rod-shaped protein. Arrays in the membranes of adjacent cells line up to form intercellular channels that can transmit small molecules from cell to cell in some tissues. (Source: Adapted from L. Makowski, D. L. D. Caspar, W. C. Phillips, and D. A. Goodenough, Gap junction structures: II. Analysis of the x-ray diffraction data, *J. Cell Biol.* 74:629, 1977.)

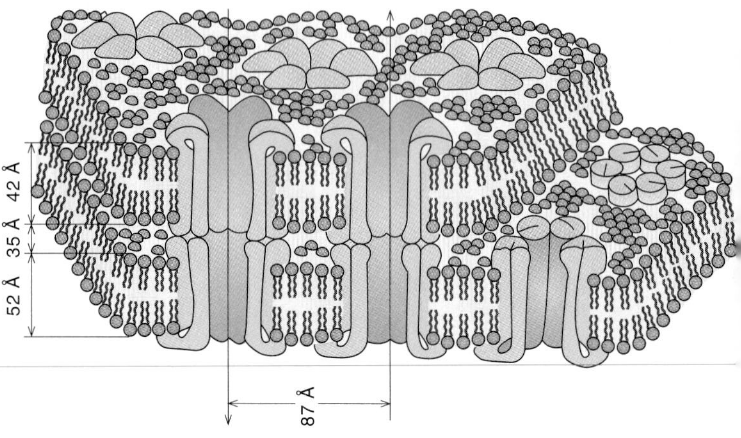

ular weights up to about 1,000 to diffuse between adjacent cells of the same type. Flow of ions through gap junctions provides intercellular communication that enables the heart to beat synchronously. In some tissues, gap junctions participate in distributing nutrients and coordinating embryological development. Gap junctions are formed from hexagonal arrays of a rod-shaped protein that stretches across the plasma membrane. At the center of the hexagon the cluster of rods evidently leaves a channel through the membrane (fig. 20.9). The hexagonal arrays in the membranes of adjacent cells line up so that small molecules can flow through them from the cytosol of one cell to the cytosol of the other without leaking out to the interstitial space.

Gap junctions close if the Ca^{2+} concentration of the cell rises above normal physiological levels. Because the Ca^{2+}-ATPase ordinarily keeps the cytosolic Ca^{2+} concentration below 10^{-7} M, a large increase in $[Ca^{2+}]$ could be a signal that a cell has exhausted its energy supplies, and the closing of gap junctions under these conditions could serve to cut off the drain of materials from living cells to neighboring cells that have died.

Vesicular Transport

Thus far we have considered transport of substances across membranes through channels composed of transmembrane proteins. In terms of amount of substances transported, vesicular transport by endocytotic and exocytotic processes far exceeds transport by discrete protein channels. The topic of vesicular transport is fragmented in this text because it relates to a variety of distinct biochemical functions. It is mentioned briefly in chapter 6 in connection with protein targeting, a topic that is elaborated upon in chapters 34 and 37. It is discussed again in chapter 18 in conjunction with glycoprotein synthesis and transport. Synaptic vesicles that transport neurotransmitters are discussed in chapter 28. Finally, lipid and lipoprotein transport by vesicles is discussed in chapter 23. In this chapter the goal is to give a brief overview of vesicular transport with special reference to basic mechanisms that relate to protein transport.

Vesicle formation and fusion are intrinsic characteristics of structures formed by lipids in water. Thus if a purified preparation of membrane lipids are added to an aqueous solution, they will spontaneously form bilayer vesicles with a range of sizes. If an aqueous solution of spontaneously formed vesicles is vortexed, the average vesicle size will decrease by a fission process. If allowed to sit for some time, the suspension of vortexed vesicles will form somewhat larger vesicles by a fusion process. These observations testify to the ease of breakage and reformation of vesicles. A small but noticeable activation energy appears to be required for the fusion and fission processes to take place. This activation energy most likely arises from the disruption of preexisting membranes. Although this disruption does not require covalent bond breakage, it does necessitate the exchange of energetically favorable lipophilic contacts for unfavorable contacts between the lipid components and water. It seems likely that factors other than the lipids could play a significant role in catalyzing these breakage and reformation processes. Consistent with this notion, there are proteins that are specifically associated with the outer leaflet of the vesicular membranes. These membrane-bound proteins are believed to facilitate the breakage and reformation processes associated with endocytosis and exocytosis.

There are three major types of vesicles involved in protein transport. First there are vesicles that transport proteins between the endoplasmic reticulum and the various Golgi compartments. These vesicles are coated on the cytoplasmic side with a characteristic mix of about five proteins collectively called coatomers. Coatomers concentrate at the budding point before the pinching off process that results in a newly formed vesicle. The coatomers dissociate from the vesicle soon after it has formed, suggesting that the main role of coatomers is to facilitate vesicle formation and perhaps to modulate the size of the vesicle. Another class of proteins called clathrins appear to serve a similar function for vesicles that package lysosomal proteins bound for the lysosome or regulated secretory vesicles carrying proteins destined for extracellular transport. A third class of vesicles that do not appear to contain any protein coat are associated with proteins that are continuously secreted.

In addition to the major proteins that make up the protein coats, smaller amounts of more specific proteins that play a role in directing vesicles to specific locations have been postulated. However, we know little about this subject except for the transport of lysosomal proteins that appear to be under the direction of the mannose-6-phosphate receptor (see chapter 18).

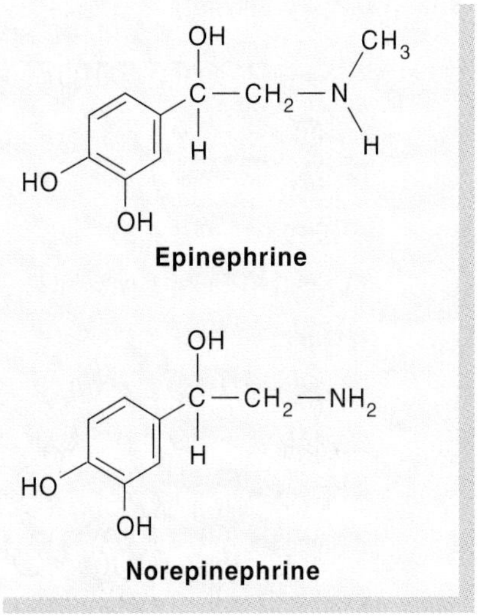

Hormone Receptors and Enzymes in Membranes Transport Signals

In addition to the transport of molecules, membrane-bound proteins are frequently involved in the transmission of signals between cells. This form of communication is carried by chemical messengers. These messengers coordinate the growth, development, and metabolic activities of different cells and tissues, allowing the organism to respond coherently to changing environments. Some messengers act only on neighboring cells; others travel to distant tissues through the bloodstream. The neurotransmitters acetylcholine, norepinephrine, and dopamine are examples of chemicals that convey messages to neighboring cells. They are released at the termini of nerve cells near other nerve or muscle cells. Binding of the transmitter to a receptor in the membrane of the second cell elicits a rapid and often dramatic response, typically a spike of electrical activity in a nerve or a contraction of a muscle. We will discuss the activities of nerve cells further in chapter 28.

The hormones produced by the endocrine glands act on cells that are far removed from the tissues where the hormones are released. As we saw in chapter 14, the adrenal hormone epinephrine and the pancreatic hormone glucagon stimulate glycogenolysis and inhibit glycogen synthesis in muscle and liver. Insulin, which also is produced by the pancreas, has opposing effects. These hormones also regulate the metabolism of carbohydrates and lipids in adipose tissue (see fig. 21.25). The endocrine system also includes the pituitary, thyroid, and parathyroid glands, the adrenal cortex; and the gonads (testis or ovary). The pituitary elaborates a large family of polypeptide hormones that act on other endocrine glands or on the central nervous system: adrenocorticotropic hormone stimulates production of steroids by the adrenal cortex; follicle-stimulating hormone and luteinizing hormone stimulate the ovaries to undergo follicular development, steroid synthesis, and ovulation; thyrotropin stimulates the synthesis

Figure 20.10

The hormones epinephrine (adrenalin) and norepinephrine (noradrenalin) are synthesized and secreted by the adrenal medulla (the outer portion of the adrenal gland). These compounds are often called the catecholamines because they are derivatives of catechol (1,2-dihydroxybenzene).

OH CH3 C—CH2—N H H HO OH

Epinephrine

OH C—CH2—NH2 H HO OH

Norepinephrine

of the thyroid hormones; met-enkephalin, leu-enkephalin, and β-endorphin have effects resembling those of opiates on the brain.

Cells that line the stomach and intestine release peptide hormones that act on other cells of the digestive system. These hormones reach their target cells by way of the digestive juices rather than through the bloodstream.

Coordination of cell proliferation and differentiation in animals depends on messages carried by a series of hormonelike polypeptides called growth factors. These polypeptides, which include epidermal growth factor, platelet-derived growth factor, fibroblast growth factor, nerve growth factor, and several insulin-like growth factors, typically are produced by one type of cell in a tissue and act to stimulate meiosis in other types of cells in the same tissue. Abnormal growth factor metabolism is associated with certain types of cancer (chapter 38).

When a hormone or growth factor reaches its target cell, it binds to a specific receptor that in some manner must trigger a change in enzymatic activity within the cell. The receptors for steroid hormones are located in the nucleus, but those for growth factors and most hormones are integral membrane proteins of the plasma membrane. In many cases the initial changes in enzyme activity also involve enzymes associated with plasma membrane. In the remainder of this chapter we will discuss three types of membrane-bound receptors and the reactions that they control. Other aspects of hormone production and function are discussed in chapter 27.

Many Hormone Receptors Trigger G Proteins to Activate or Inhibit Adenylate Cyclase

The adrenal medulla secretes two closely related hormones: epinephrine and norepinephrine (fig. 20.10). Epinephrine and norepinephrine bind to several types of receptors in their target tissues,

Figure 20.11

Structure of the human β-adrenergic receptor. The protein has seven regions of hydrophobic amino acids that probably form transmembrane α helices. Epinephrine binds to the amino-terminal domain of the protein, which lies outside the cell and is glycosylated; the carboxyl-terminal domain of the receptor is in the cytosol. Open circles represent amino acids that are identical with those in the hamster protein. (Reprinted with permission from H. G. Dohlman, M. G. Caron, and R. J. Lefkowitz, A family of receptors coupled to guanine nucleotide regulatory proteins, *Biochemistry* 26:2657, 1987. Copyright 1987 American Chemical Society.)

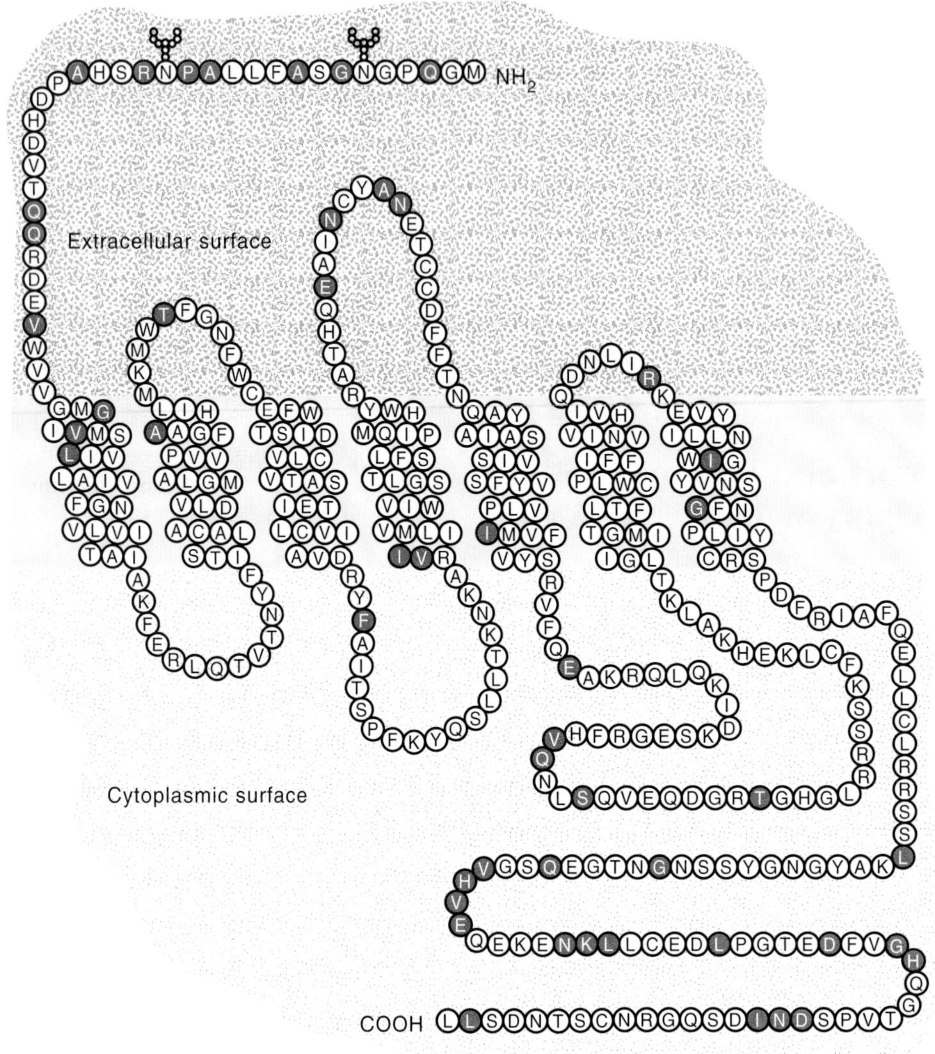

and these adrenergic receptors control a variety of cellular processes, depending on the tissue. Binding of epinephrine to the β-adrenergic receptors of liver or muscle leads to activation of the enzyme adenylate cyclase, which converts ATP to c-AMP (see fig. 14.19). Cyclic AMP, in turn, switches on the c-AMP-dependent protein kinase that converts phosphorylase *b* to the more active phorphorylase *a* and converts glycogen synthase to its less active D form (see figs. 14.18 and 14.20). In adipose cells the c-AMP-dependent protein kinase activates an enzyme that hydrolyzes triacylglycerols (chapter 21). The heart responds with an accelerated beat; blood vessels that supply skeletal muscles respond by dilating, increasing the flow of blood to the muscles; and the bronchial tubes of the lungs expand to increase the flow of air. Binding of norepinephrine to β-adrenergic receptors has similar effects. Surprisingly, however, binding of norepinephrine to another type of receptor, the α_2-adrenergic receptor, has diametrically opposite effects. Formation of the α_2-receptor-hormone complex leads to inhibition of adenylate cyclase. A third type of adrenergic receptor (the α_1 receptor) does not work through adenylate cyclase at all, but by a different mechanism that we will introduce presently.

Judging from its amino acid sequence, the β-adrenergic receptor has seven transmembrane α helices (fig. 20.11). A similar cluster of seven helices occurs in the α_2-adrenergic receptor and, perhaps surprisingly, in the purple-bacterial protein bacteriorhodopsin (see fig. 20.11) and in the light-sensitive protein of the eye, rhodopsin. We will see in chapter 29 that the events that follow excitation of rhodopsin with light are closely related to those resulting from the binding of epinephrine to the β-adrenergic receptor.

Binding of epinephrine to a β-adrenergic receptor on the outer surface of the target-cell membrane evidently changes the structure of the cytosolic or intramembranous parts of the receptor in a way that brings about activation of adenylate cyclase. Adenylate cyclase also resides in the plasma membrane of the cell. However, the epinephrine–receptor complex does not interact directly with this enzyme. Its effect is mediated by an additional membrane-bound protein termed a G protein. As shown in figure 20.12, the G protein has three subunits: α, β, and γ. In its resting state, the G protein's α subunit binds a molecule of GDP. When the G protein interacts with the hormone–receptor complex, the affinity of the guanine-nucleotide site for GDP is decreased, but the affinity for GTP is increased. The G protein therefore releases its bound GDP and picks up a molecule of GTP. It also may dissociate from the $\beta\gamma$ subunit complex. Replacement of the GDP by GTP allows the α subunit to bind transiently to adenylate cyclase, and it is this binding that switches on the cyclase (see fig. 20.12).

The hormone receptor, G protein, and adenylate cyclase probably are not linked together in the absence of the hormone. They find each other by lateral diffusion in the membrane after the hormone binds to the receptor. This was demonstrated by fusing mutant cells that lacked receptors but contained adenylate cyclase with cells that had functional receptors but lacked adenylate cyclase or the G protein. The hybrid cells were able to produce c-AMP in response to hormones shortly after they were fused.

Each hormone–receptor complex can act catalytically to modify the binding site for guanine nucleotides in multiple copies of the G protein. The GTP-α-subunit complex of a G protein also can proceed to activate many molecules of adenylate cyclase. However, these events soon come to a halt because the hormone–receptor complex reverts spontaneously to an inactive form and the molecule of GTP bound to the α subunit of the G protein is hydrolyzed to GDP. Hydrolysis of the bound GTP is catalyzed slowly by the α subunit of the G protein itself.

A nearly identical sequence of events occurs when norepinephrine binds to an α_2-adrenergic receptor, except that the result in this case is inhibition of adenylate cyclase. Different results are obtained because the two types of hormone–receptor complex react specifically with two different types of G proteins. β-Adrenergic receptors interact with G_s proteins, which have a stimulatory α subunit ($G_{\alpha s}$); α_2-adrenergic receptors react with G_i proteins, which have an inhibitory α subunit ($G_{\alpha i}$). The β and γ subunits of the G_s and G_i proteins appear to be identical. Opiates and angiotensin, a peptide hormone from the liver that regulates blood pressure, act through receptors that interact with G_i proteins. Glucagon, parathyroid hormone, and many of the pituitary and gastrointestinal hormones work through G_s proteins.

The reactions initiated by the binding of hormones to their receptors are sensitive to several medically important drugs, and they also play prominent roles in the diseases cholera and pertussis (whooping cough). For example, propanolol, a drug that blocks the binding of epinephrine to β-adrenergic receptors, is used to treat high blood pressure and to control cardiac activity in patients who have experienced a heart attack. In cholera, which is caused by an infestation of the bacterial species *Vibrio cholerae,* the deactivation of the GTP–$G_{\alpha s}$ complex is sidetracked. Cholera bacteria secrete a toxin that binds to the plasma membranes of the epithelial cells that line the gut. A portion of the toxin penetrates the cell and catalyzes covalent attachment of an ADP-ribose group to the $G_{\alpha s}$ subunit of the G_s protein (fig. 20.13). This modification blocks hydrolysis of GTP to GDP on the

Figure 20.12

Binding of epinephrine to a β-adrenergic receptor in the plasma membrane of a cell induces the receptor to react with a membrane-bound G_s-protein that has GDP bound to its α subunit ($G_{\alpha s}$). This causes the $G_{\alpha s}$ subunit to exchange its GDP for GTP. Activation of the GTP/GDP exchange reaction is indicated by ⊕ in the figure. The GTP–$G_{\alpha s}$ complex in turn activates adenylate cyclase, which converts ATP to 3′,5′-cyclic-AMP (c-AMP). Cyclic-AMP activates the cytosolic c-AMP-dependent kinase, which phosphorylates numerous other enzymes, depending on the nature of the cell. Binding of hormone to α_2-adrenergic receptors causes similar events, except that the hormone–receptor complex reacts with a G_i protein instead of with G_s, and the $G_{\alpha i}$ subunit inhibits adenylate cyclase instead of activating it.

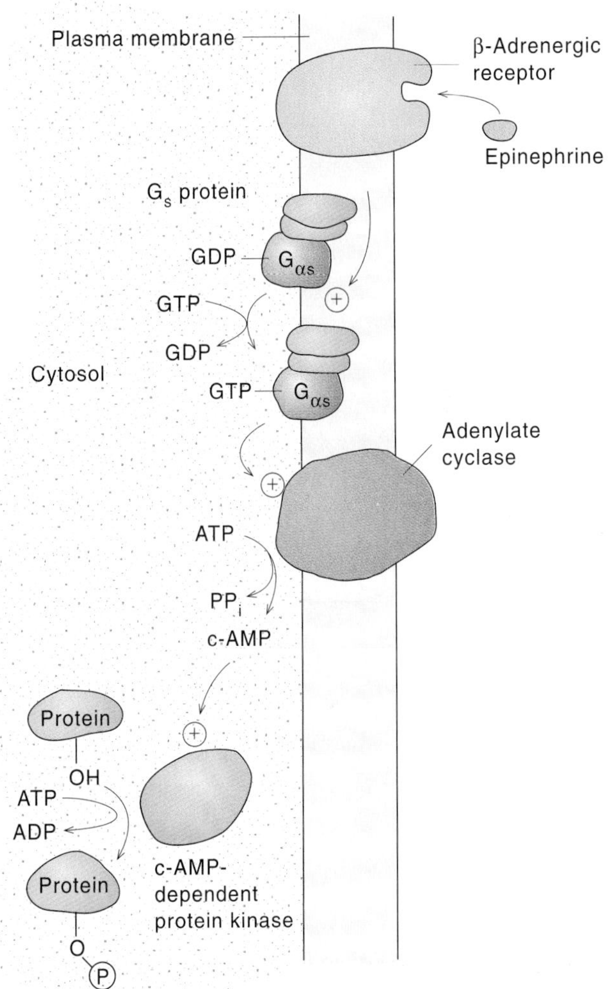

Figure 20.13

(a) Cholera toxin catalyzes transfer of the ADP-ribose moiety of NAD^+ to a specific arginyl residue of the $G_{\alpha s}$ subunit of the G_s-protein. (b) This locks the $G_{\alpha s}$ subunit in its active form by preventing it from hydrolyzing bound GTP to GDP. (c) Pertussis toxin causes an ADP-ribosylation of the G_i protein; this locks $G_{\alpha i}$ subunit in its inactive form by preventing exchange of GTP for the bound GDP.

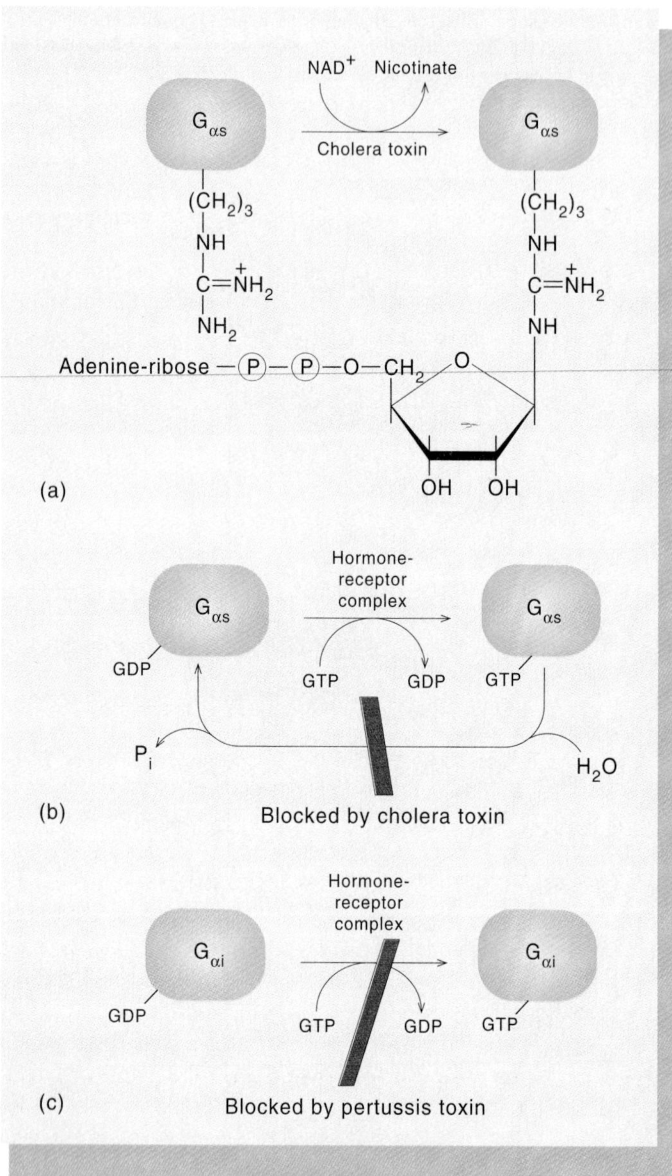

Figure 20.14

The insulin receptor has two membrane-spanning β chains linked by disulfide bonds to two α chains that sit on the outer surface of the plasma membrane. Each β chain has a tyrosine-kinase catalytic site. Binding of insulin activates the tyrosine-kinase sites.

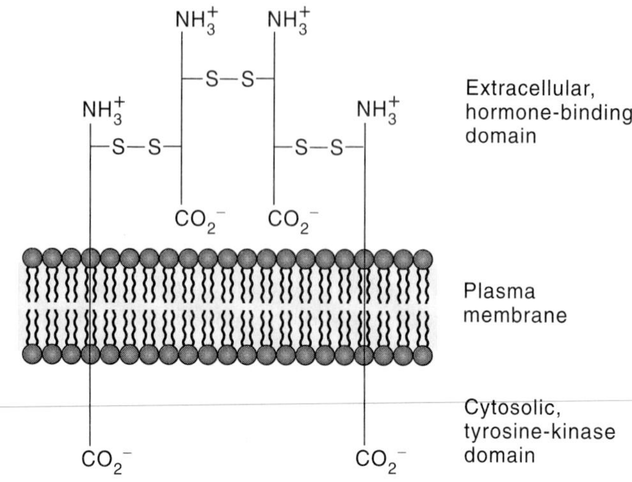

GTP–$G_{\alpha s}$ complex and thus locks the complex permanently in its active form. The resulting persistent activation of adenylate cyclase leads to a continuing loss of fluid from the intestinal cells, which is lethal unless the patient receives adequate liquids to replace the lost water and salts. In pertussis the invading bacteria (*Bordatella pertussis*) secrete a toxin that causes a similar ADP-ribosylation of the $G_{\alpha s}$ subunit in the G_i protein. The result in this case is to prevent the exchange of GTP for the bound GDP and thus to block the normal inhibition of adenylate cyclase (see fig. 20.13).

The Receptors for Insulin and Some Growth Factors Are Tyrosine Kinases

Whereas the β-adrenergic receptor consists of a single polypeptide chain, the receptor for insulin has the more complex $\alpha_2\beta_2$ subunit structure shown in figure 20.14. Each of the β chains probably has a single transmembrane helix. The binding site for insulin is in the extracellular domain, which is made up mainly of the α chains. In the cytosolic domain of each β chain there is a tyrosine-kinase enzymatic site that can catalyze the phosphorylation of protein tyrosine residues:

$$\text{Protein}-C_6H_4-OH + ATP^{4-} \longrightarrow \text{Protein}-C_6H_4-OPO_3^{2-} + ADP^{3-} + H^+$$

Binding of insulin switches on the activity of the insulin receptor as a tyrosine kinase. Curiously, each of the tyrosine-kinase sites first carries out an autophosphorylation of the receptor itself, attaching phosphate residues to two specific tyrosines in its companion β subunit. This modification causes a further stimulation of the activity of the tyrosine kinase, which now begins to operate on other proteins in the cell.

The receptor for epidermal growth factor (EGF) resembles the insulin receptor in having a tyrosine-kinase site in its cytosolic domain. The two proteins also have homologous regions in their extracellular domains. But instead of having separate α and β chains, the EGF receptor in the active form consists of a dimer of identical, membrane-spanning polypeptides, each of which resembles a composite of the insulin receptor's subunits.

Figure 20.15

Phosphatidylinositol-4,5-bisphosphate (PIP_2) and the two second messengers, diacylglycerol and inositol trisphosphate (IP_3), that are derived from it. The highly unsaturated fatty acid esterified at the C-2 position of PIP_2 is arachidonic acid.

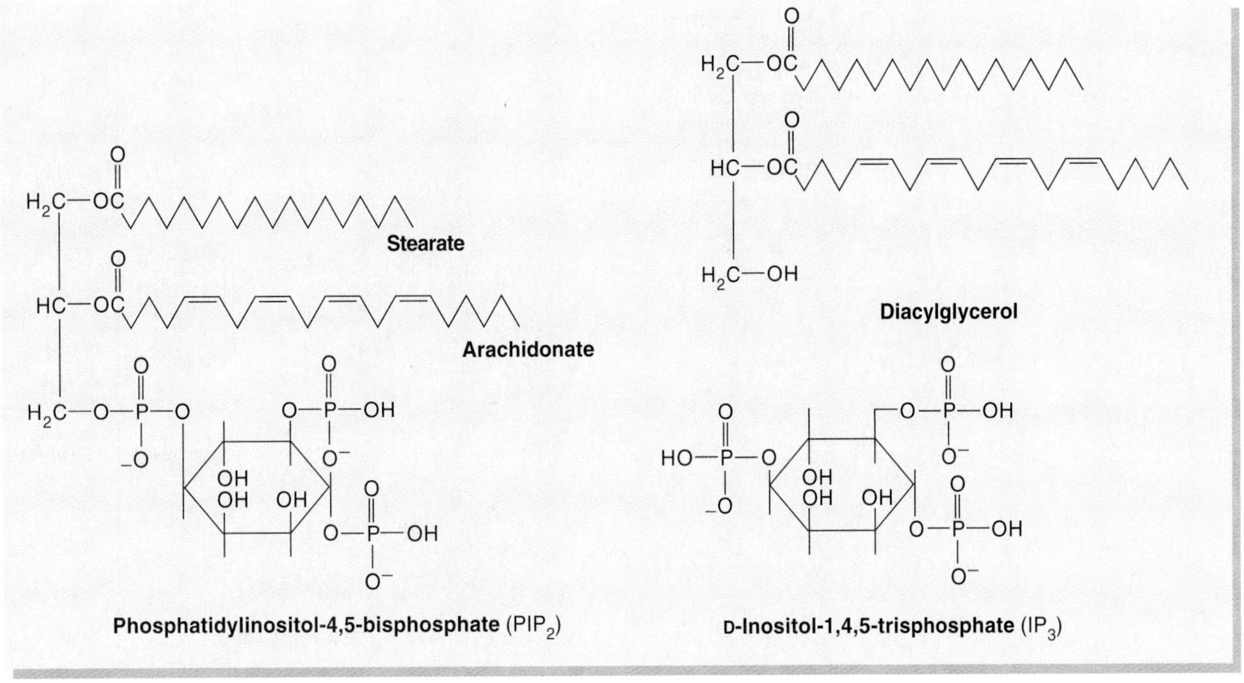

The EGF monomer associates only to form a dimer in the presence of growth factor. Interest in growth-factor receptors has heightened with the discovery that many of the known viral oncogenes, genes that can transform virally infected cells into tumors, encode similar tyrosine-kinase domains. Because growth factors ordinarily promote cell proliferation, it appears that the tyrosine kinase activities introduced from the viral genome may induce the uncontrolled cell growth and division that is characteristic of tumors. However, the key proteins that are the substrates of these kinases remain to be identified.

Other Receptors Trigger Breakdown of Phosphatidylinositol to Inositol Trisphosphate and Diacylglycerol

The α_1-adrenergic receptor is the gateway to still a third pathway of intercellular communications. It resembles the β- and α_2-adrenergic receptors in working through a G protein, but this is still another type of G protein, G_q. Like G_s and G_i, G_q responds to the hormone-receptor complex by exchanging GTP for a molecule of bound GDP on its α subunit. But instead of interacting with adenylate cyclase, G_q activates a phospholipase (phospholipase C) that specifically hydrolyzes an unusual phospholipid, phosphatidylinositol-4,5-bisphosphate (PIP_2, fig. 20.15). Phospholipase C is associated peripherally with the plasma membrane of the cell, and PIP_2 is among the phospholipids that make up this membrane. Hydrolysis of PIP_2 by phospholipase C releases diacylglycerol, which remains in the membrane, and the small, polar molecule inositol-1,4,5-trisphosphate (IP_3), which diffuses out into the cytosol (see fig. 20.15). Both of these products act as secondary messengers to regulate other processes.

Diacylglycerol activates a membrane-bound protein kinase (protein kinase C) that phosphorylates serine and threonine residues. In addition, further hydrolysis of diacylglycerol yields the highly unsaturated fatty acid arachidonic acid, which serves as a precursor for the formation of prostaglandins (see chapter 22). Inositol triphosphate acts at the membrane of the endoplasmic reticulum within the cell, stimulating the release of Ca^{2+} from internal stores (fig. 20.16). The Ca^{2+} that emerges binds to a small, soluble protein, calmodulin, which proceeds to activate several different protein kinases.

In addition to norepinephrine, vasopressin (a polypeptide hormone from the pituitary that controls blood pressure and water balance) and, in some tissues, acetylcholine operate through the phosphatidylinositol pathway. In the endocrine cells of the pancreas, acetylcholine induces secretion of insulin; in smooth muscle it induces contraction.

Figure 20.16

The phosphatidylinositol (PIP_2) pathway. Binding of certain hormones to their membrane receptors leads to activation (⊕) of GTP/GDP exchange on a G_q protein, which then activates phospholipase C. Phospholipase C cleaves inositol trisphosphate (IP_3) from PIP_2, leaving diacylglycerol (DG) in the membrane. Diacylglycerol activates protein kinase C, which phosphorylates several regulatory enzymes, thereby activating or inhibiting them. The polar IP_3 binds to receptors on the endoplasmic reticulum membrane, resulting in the release of Ca^{2+} into the cytosol. Ca^{2+} activates another group of protein kinases. Hormonal stimulation is short-lived because IP_3 and DG are degraded rapidly to components that ultimately are recycled to PIP_2 and Ca^{2+} is pumped back into the endoplasmic reticulum.

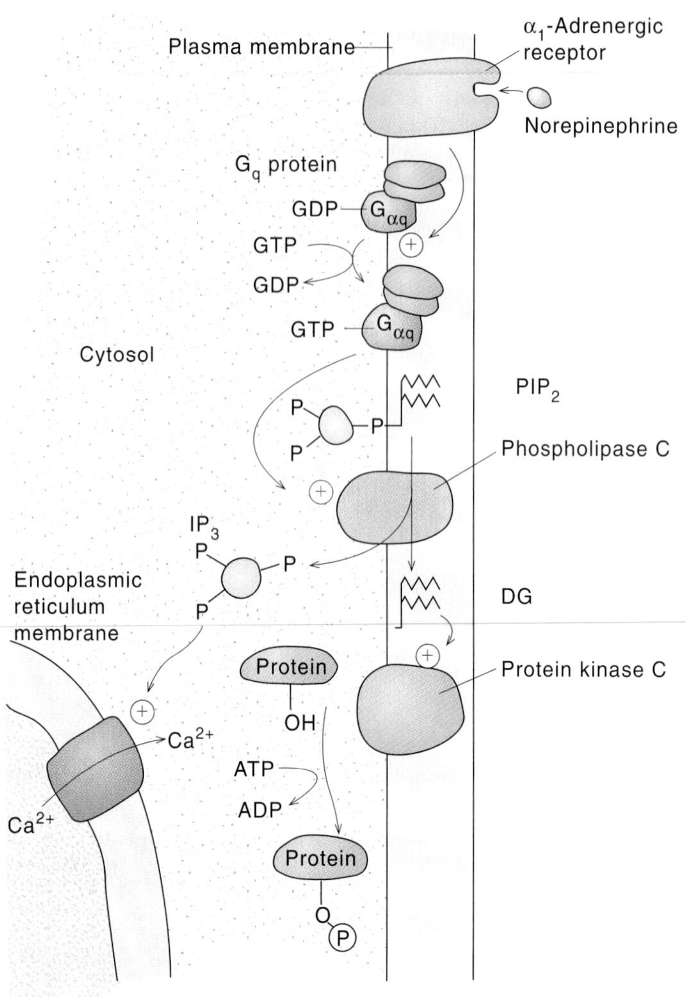

Summary

This chapter is divided into two parts. In the first part we discuss the transport of materials across membranes. In the second part we discuss the transduction of signals across the cell membrane.

1. Biological membranes contain proteins that act as specific transporters of small molecules into and out of the cell. Most solutes are transported by specific carriers that are invariably proteins.
2. Some transporters facilitate diffusion of a solute from a region of relatively high concentration or down a favorable electrochemical potential gradient. Such transporters do not require energy, since the transport is in the thermodynamically favorable direction.
3. Other transporters move solutes against an electrochemical potential gradient and require an energy-producing process to make them functional. Cells drive such active transport processes in a variety of ways. The transport can be coupled to the hydrolysis of a high-energy phosphate, to the cotransport of another molecule down an electrochemical potential gradient, or to the modification of the transported molecule soon after it crosses the membrane.
4. Proteins involved in active transport frequently change their structure during the transport process. For example, the Na^+–K^+ pump includes two phosphorylated forms of the enzyme involved in the transport of these two ions.
5. Some membranes contain relatively large pores, which allow for the free passage of molecules with molecular weights up to 600. For example, the outer membranes of Gram negative bacteria contain pores with diameters of about 10 Å, which are formed from proteins called porins.
6. Pores called gap junctions occur in the plasma membranes of cells in some eukaryotic tissues. These channels allow molecules with molecular weights up to about 1000 to diffuse between adjacent cells of the same types.
7. Hormone receptors and enzymes in membranes mediate communication between cells. The action of a hormone depends on the type of receptors that a cell possesses and the types of protein factors that are associated with those receptors on the cytosolic side of the plasma membrane. In this chapter we discuss three types of membrane-bound receptors: β-adrenergic, α-adrenergic, and α_2-adrenergic receptors in different cell types.

Selected Readings

Ames, G. F.-L., C. S. Mimura, S. R. Holbrook, and V. Shyamala, Traffic ATPase: A superfamily of transport proteins operating from *E. coli* to humans. *Adv. Enzymol.* 65:1–47, 1992. Excellent review of binding-protein-dependent transport systems in bacteria and related eukaryotic transport proteins that are also ATPases.

Bell, G., et al., Structure and function of mammalian facilitative sugar transporters. *J. Biol. Chem.* 268:19161–19164, 1993.

Boyer, P. D., A perspective of the binding change mechanism for ATP synthesis. *FASEB J.* 3:2164–2178, 1989. Reviews evidence for the mechanism of H^+-coupled ATP synthesis by the F_1/F_0 ATPase.

Bronner, F., and A. Kleinzeller (ed.), *Current Topics in Membranes and Transport*. New York: Academic Press. Continuing series reviewing some of the current problems in biological transport.

Dharmavaram, R. M., and J. Konisky, Characterization of a P-type ATPase of the archaebacterium *Methanococcus voltae*. *J. Biol. Chem.* 264: 14085–14089, 1989. Recent discovery of a P-type ATPase in an archaebacterium, which may have significant evolutionary implications.

Friedlander, M., and M. Mueckler (eds.), Molecular biology of receptors and transporters, *Int. Rev. Cyto.* 137A, 1992. This recent volume contains eight review articles on various types of transport systems in both eucaryotic and procaryotic cells.

Gennis, R. B., *Biomembranes: Molecular Structure and Function*. N.Y.: Springer-Verlag, 1989. Excellent book dealing with both membrane protein structure and function.

Higgins, C. F., M. P. Gallagher, M. L. Mimmack, and S. R. Pearce, A family of closely related ATP-binding subunits from prokaryotic and eukaryotic cells. *Bioessays* 8:111–118, 1988. An interesting summary of recent evidence for the structural and functional relatedness of seeming divergent proteins that bind ATP.

Inesi, G., D. Lewis, D. Nikic, A. Hussain, and M. Kirtley, Long-range intramolecular linked functions in the calcium transport ATPase. *Adv. Enzymol.* 65:182–215. Recent review of the structure and mechanism of the Ca^{2+}-ATPase, with an emphasis on how Ca^{2+} transport and ATP hydrolysis may be linked.

Kaback, H. R., E. Bibi, and P. D. Roepe, β-Galactoside transport in *E. coli*: a functional dissection of lac permease. *Trends. Biochem. Sci.* 15:309–314, 1990. Structure-function relationships in the lactose permease.

Meadow, N. O., D. K. Fox, and S. Roseman, The bacterial phosphoenolpyruvate:glycose phosphotransferase system. *Ann. Rev. Biochem.* 59:497–542, 1990. A recent review of this carbohydrate transport system that is commonly found in anaerobic and facultatively anaerobic bacteria.

Rothman, J. E., and L. Orci, Budding vesicles in living cells. *Sci. Am.* 274:70–75, 1996.

Saier, M. H., Jr., *Mechanisms and Regulation of Carbohydrate Transport in Bacteria*. New York: Academic Press, 1985. Reviews the various mechanisms of carbohydrate transport in bacteria, as well as transport regulation.

Silverman, M., Structure and function of hexose transporters. *Ann. Rev. Biochem.* 60:757–794, 1991. Excellent review of recent work on sugar transporters in eukaryotic cells.

Spudich, J. L., and R. A. Bogomolni, Sensory rhodopsins of halobacteria. *Ann. Rev. Biophys. Biophys. Chem.* 17:193–215, 1988. Review of structure and function of bacteriorhodopsin and other related proteins of halobacteria.

Problems

1. Besides the comments made in the text associated with figure 20.3 on the free energy changes associated with phosphorylation during sugar transport, what more can be said on the unidirectional aspect of this type of transport system?
2. Elementary portrayals of transport systems have often utilized "revolving doors" as analogies. After reading this chapter, what objection(s) do you have to the revolving door idea?
3. Membrane vesicles of *E. coli* that possess the lactose permease are preloaded with KCl and are suspended in an equal concentration of NaCl. It is observed that these vesicles actively, although transiently, accumulate lactose if valinomycin is added to the vesicle suspension. No such active uptake is observed if KCl replaces NaCl in the suspending medium. Explain these results in light of what you know about the mechanism of lactose transport and the properties of valinomycin.
4. The Nernst equation relates the electric potential $\Delta\Psi$ resulting from an unequal distribution of a charged solute across a membrane permeable to that solute to the ratio between the concentration of solute on one side and on the other:

$$m\,\Delta\Psi - \frac{-2.3RT}{F}\log\frac{[\mathrm{Sol}]_1}{[\mathrm{Sol}]_2}$$

where m is the charge on the solute, $2.3RT/F$ has a value of about 60 mV at 37°C, and $[\mathrm{Sol}]_1$ and $[\mathrm{Sol}]_2$ refer to the concentrations of solute on either side of the membrane. Consider a planar phospholipid bilayer separating two compartments of equal volume. Side 1 contains 50 mM KCl and 50 mM NaCl, whereas side 2 contains 100 mM KCl.
   (a) If the membrane is made permeable only to K^+, for example, by addition of valinomycin, what is the magnitude of $\Delta\Psi$?
   (b) If the membrane is made permeable to H^+ and K^+, in which direction does H^+ initially flow?
   (c) If the membrane can be made selectively permeable to both K^+ and Cl^-, what is the value of $\Delta\Psi$ and the ion concentrations on both sides of the membrane at equilibrium? (*Hint:* Initially, K^+ diffuses down its concentration gradient accompanied by an equivalent amount of Cl^-. Equilibrium is established when the potentials due to K^+ and Cl^- equal each other and the overall membrane potential.)
5. Predict the effects of the following on the initial rate of glucose transport into vesicles derived from animal cells that accumulate this sugar by means of Na^+ symport. Assume that initially $\Delta\Psi = 0$. $\Delta pH = 0$ (pH = 7), and the outside medium contains 0.2 M Na^+, whereas the vesicle interior contains an equivalent amount of K^+.
   (a) Valinomycin.
   (b) Gramicidin A.
   (c) Nigericin.
   (d) Preparing the membrane vesicles at pH 5 (in 0.2 M KCl), resuspending them at pH 7 (in 0.2 M NaCl), and adding 2,4-dinitrophenol.

6. In *E. coli,* lactose is taken up by means of proton symport, maltose by means of a binding (ABC-type) protein-dependent system, melibiose by means of Na^+ symport, and glucose by means of the phosphotransferase system (PTS). Although this bacterium normally does not transport sucrose, suppose that you isolated a strain that does. How do you determine whether one of the four mechanisms just listed is responsible for sucrose transport in this mutant strain?
7. You are growing some mammalian cells in culture and measure the uptake of D-glucose and L-glucose (see data in the chart). What type of transport is observed with these sugars? (*Hint:* Plot V versus [sugar] and $1/V$ versus 1/[sugar].) Explain the significance of these data.

[Sugar] (mM)	V (mM cm s^{-1}) × 10^7 *D-glucose*	*L-glucose*
0.100	166	4.8
0.167	252	8.0
0.333	408	16
1.000	717	50

8. (a) Using Fick's law, show that the diffusion coefficient D has the dimensions of area per unit time.
   (b) The diameter of a pore channel is about 10^{-9} m, and its length is 4×10^{-9} m. In planar membrane bilayers, glucose traverses this channel at the rate of about 50 molecules per channel per second at room temperature when the concentration of glucose is 3×10^{-6} M on one side of the membrane. Calculate the diffusion coefficient for glucose through the pore channels under these conditions.
9. Intracellular vacuoles in the yeast *Saccharomyces cerevisiae* are membrane-bounded organelles that are known to concentrate within them a variety of basic amino acids, including arginine (net charge = +1). Vesicles prepared from these vacuoles lack an electron-transport chain, and arginine uptake into them is dependent on extravesicular ATP. A membrane potential $\Delta\psi$ has no effect on ATP-dependent arginine uptake in the absence of a proton gradient, while proton ionophores and dicyclohexylcarbodiimide (a known inhibitor of the F_1/F_0 ATPase) greatly inhibit accumulation of arginine by this system. Upon addition of ATP in the absence of arginine, the intravesicular pH of these vesicles drops. Describe a mechanism for the energization of arginine transport in this system, taking into account all these observations.
10. In some instances the efflux of a radioactively labeled transport substrate out of preloaded cells or vesicles is transiently stimulated by addition of the same nonradioactive transport substrate to the outside. This phenomenon is known as trans-stimulation and occurs with transport systems that are reversible (i.e., can operate in either direction). Can you think of an explanation for trans-stimulation in view of what is known about the molecular mechanisms of transmembrane transport?
11. Outline a molecular mechanism by which, and the conditions under which, an H^+ symport system (such as the *E. coli* lactose permease system) might operate to actively accumulate a metabolite such as lactose.
12. If the cells in problem 7 are treated with $HgCl_2$ (mercury reacts with —SH groups in proteins), the rate of transport of D-glucose is the same as that of L-glucose. What is indicated about D-glucose transport?
13. The translocation of K^+ was studied by using an artificial membrane system (this membrane system had a phase transition T_m of 41°C) with valinomycin and gramicidin. The results showed that K^+ translocation with gramicidin was high over a broad temperature range, while valinomycin translocated K^+ at a high rate only above 41°C. What models of translocation are suggested by these data for each of these polypeptide antibiotics?
14. While there are many different and diverse types of transport systems in cells, these systems seem to be evolutionarily related and can be accommodated within a unified model. Outline the overall mechanism that is similar in these transport systems.

C H A P T E R

Metabolism of Fatty Acids

Fatty acid synthesis and breakdown both occur in steps involving two carbon units but by totally different mechanisms.

atty acids, as components of phospholipids, are important structural elements of biological membranes and, like carbohydrates, are important sources of energy. In this chapter we focus on the role of fatty acids in energy metabolism. We see that fatty acids are both assembled and degraded in blocks of two carbon atoms (fig. 21.1). Despite this obvious similarity, the processes of synthesis and breakdown are very different. In contrast to the biosynthesis and catabolism of carbohydrates the two processes for metabolism of fatty acids use completely different sets of enzymes, and they occur in different cellular compartments. These differences facilitate regulatory mechanisms that ensure that biosynthesis and catabolism do not occur simultaneously.

In the first section we deal with reactions associated with fatty acid breakdown. The second section covers the pathway for fatty acid biosynthesis. Finally, we consider the regulatory mechanisms that determine the conditions under which each of these processes occurs.

Fatty Acid Degradation

Fatty Acids Originate from Three Sources: Diet, Adipocytes, and de novo Synthesis

Fatty acids are mainly derived from the diet. In fact, 30%–40% of the calories ingested each day in the average human diet are

Figure 21.1

Synthesis and degradation of fatty acids. Both synthesis and degradation of fatty acids occur by multistep pathways involving totally different enzymes located in different parts of the cell. Both pathways are similar in that two carbon atoms are either added (in synthesis) or deleted (in degradation) in each round of a repetitious, cyclical process.

The acetyl-CoA used as substrate for fatty acid synthases in the cytosol originates from the mitochondrion. This acetyl-CoA condenses with CO_2 to form malonyl-CoA, which eliminates CO_2 after an initial condensation reaction. After three more steps a two-carbon unit is added to a growing fatty acid chain. This cycle repeats itself many times until a fatty acid chain with 16 carbon atoms has been synthesized. Variations of this pathway give rise to unsaturated and polyunsaturated fatty acids.

The degradation of fatty acids also involves a repetitious process yielding a two-carbon unit (acetyl-CoA) in every cycle. This acetyl-CoA can be fed into the TCA cycle, or, alternatively, it can be used to make ketone bodies.

Elaborate controls prevent synthesis and degradation of fatty acids from occurring simultaneously.

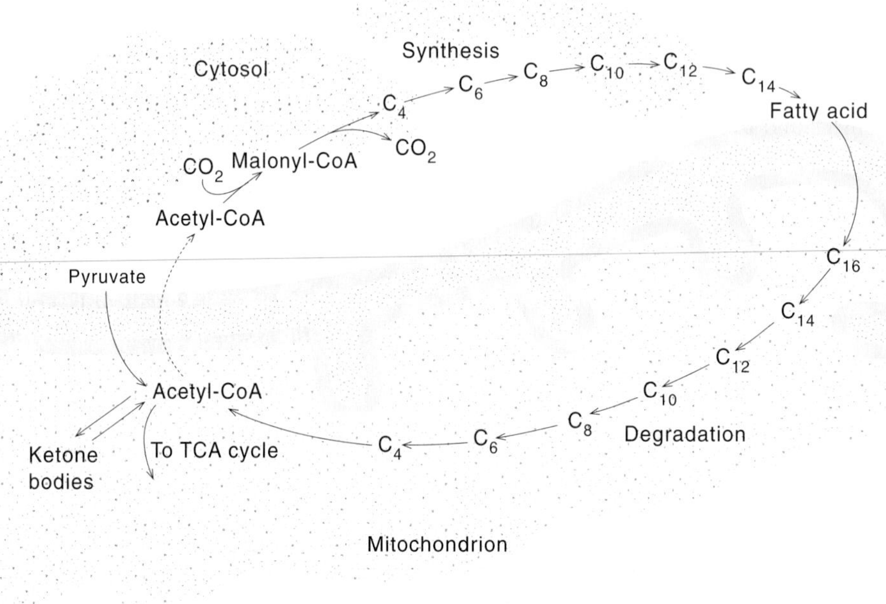

provided by the fatty acid components of triacylglycerols and phospholipids. Because of the relationship between a diet high in fat and heart disease, Heart Associations throughout the world recommend that no more than 30% of the calories in our diet be derived from fatty acids. Dietary lipids are degraded by lipases and phospholipases, which are enzymes secreted into the intestinal lumen. Like other digestive enzymes, many of the lipases and phospholipases are derived from the pancreas (fig. 21.2*a*). The fatty acids released by this process are absorbed by cells in the intestinal wall, and triacylglycerols are resynthesized (see fig. 21.2*b*). These lipids are then packaged into spherical lipoproteins, particles of lipids and proteins, known as chylomicrons, which are secreted into lymphatic vessels through which they eventually reach the bloodstream. Once in the circulatory system, the triacylglycerol components of the chylomicrons are degraded to fatty acids and glycerol by the enzyme lipoprotein lipase, which is attached to the luminal (inner) side of capillary vessels in heart, muscle, adipose (commonly known as fat), and other tissues (fig. 21.3). The released fatty acids are transported into these tissues for generation of energy, storage as an energy reserve in the form of triacylglycerols, or biosynthesis of membrane phospholipids.

If the diet does not provide sufficient lipid to satisfy immediate needs, the fatty acids stored in the adipose tissue in the form of triacylglycerols may be required (discussed later; see fig. 21.25). In addition, most tissues have the capacity to convert carbohydrates and some amino acids into fatty acids.

Fatty Acid Breakdown Occurs in Blocks of Two Carbon Atoms

Recall that fatty acid chains nearly always contain an even number of carbon atoms (see chapter 19). Investigations into the mechanisms of fatty acid catabolism began around the turn of the century when Fritz Knoop reported experiments that indicated fatty acids were degraded by removal of two carbons at a time. The data indicated that carbon 3 of a fatty acid was oxidized with subsequent cleavage between carbons 2 and 3:

$$\underset{\beta}{\overset{3}{R}CH_2}\underset{\alpha}{\overset{2}{CH_2}}\overset{1}{COOH} \longrightarrow R\underset{\beta}{\overset{O}{\overset{\|}{C}}}\underset{\alpha}{\overset{2}{CH_2}}\overset{1}{COOH} \longrightarrow R\overset{O}{\overset{\|}{C}}OH + \underset{\alpha}{\overset{2}{CH_3}}\overset{1}{COOH} \qquad \textbf{(1)}$$

Carbon 2 is also known as the alpha (α) carbon and carbon 3 as the beta (β) carbon. Hence the term β oxidation was coined.

Figure 21.2

(*a*) Triacylglycerols are digested in the intestine by pancreatic lipase. (*b*) The fatty acids and monoacylglycerol are absorbed into the intestinal wall. The cells in the wall, enterocytes, resynthesize triacylglycerol and package the lipid with proteins to give lipoprotein particles called chylomicrons, which are secreted from the enterocytes into the lymph.

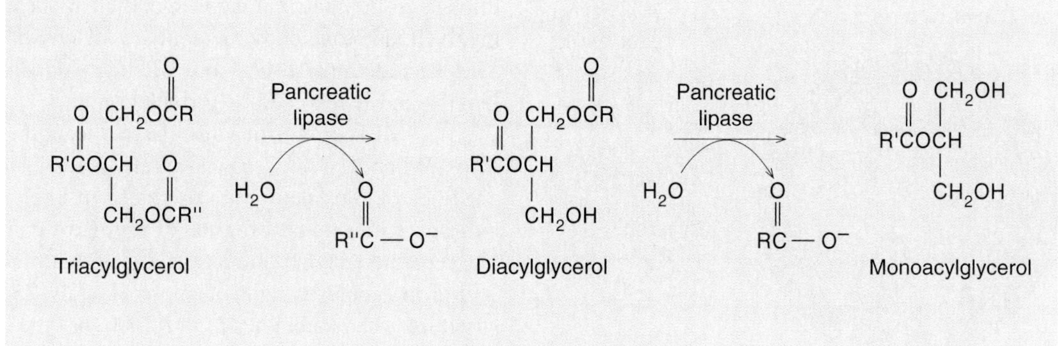

(a)

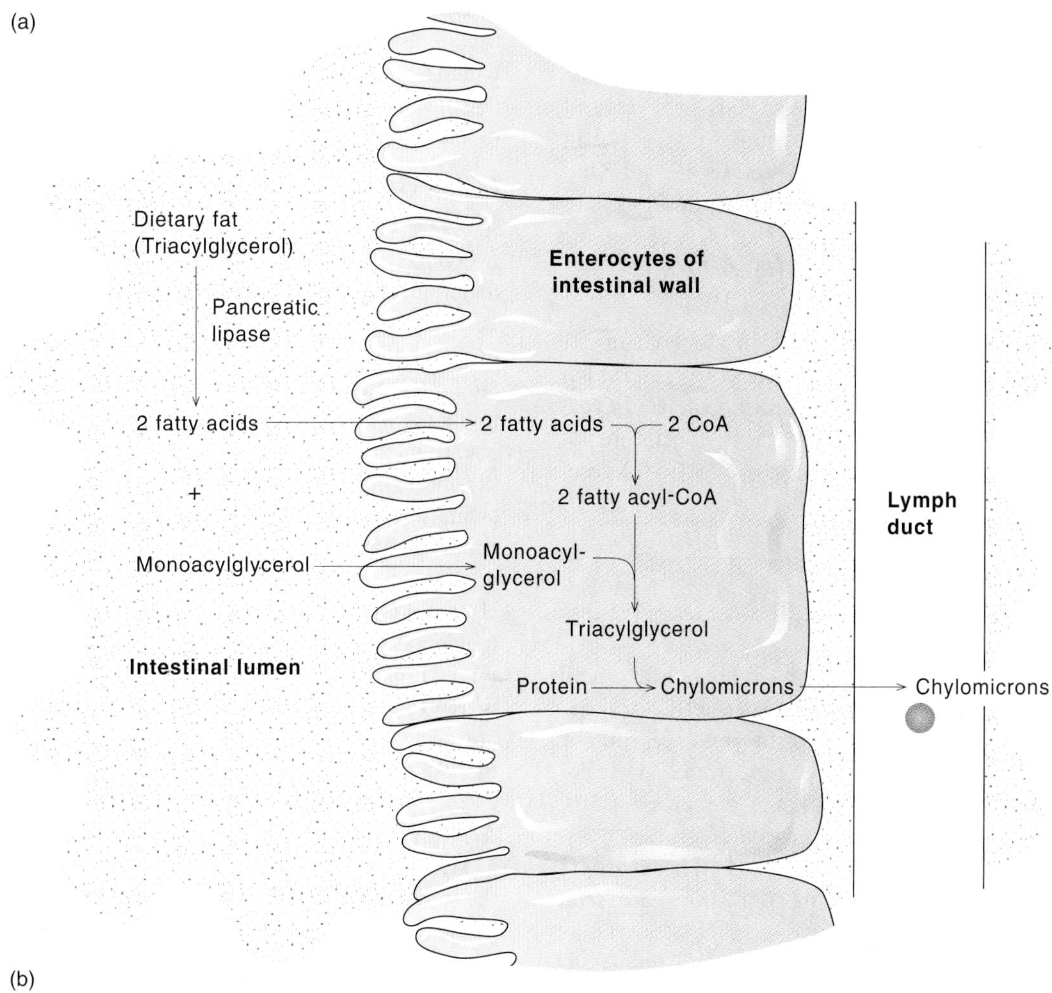

(b)

The next major experimental step was the demonstration in 1943 by J. M. Munoz and Luis Leloir that fatty acids could be oxidized in a cell-free system. This was followed by Albert Lehninger's demonstration that fatty acid oxidation occurred in liver mitochondria and, apparently, involved an "active acetate." Experiments by Fritz Lipmann proved that coenzyme A was involved in the formation of "active acetate":

$$\text{Acetate} + \text{ATP} + \text{CoA} \rightarrow \text{“active acetate”}$$

Subsequently, in 1951, Feodor Lynen, working with yeast, demon-

Figure 21.3

The chylomicrons are delivered to other tissues in the body via the bloodstream. In the tissues the chylomicrons bind to the endothelial cells of the capillaries and are degraded to fatty acids and glycerol by lipoprotein lipase. This enzyme is secreted by cells in the tissue and is bound to the endothelial cells.

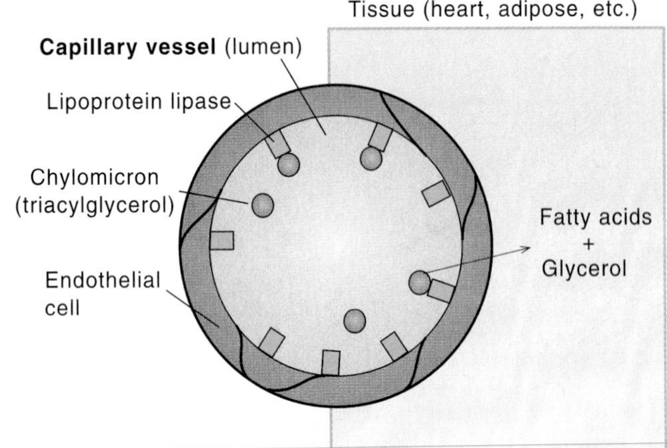

strated that "active acetate" was acetyl-CoA. At this stage, several laboratories conceived of the idea that CoA might play a role in the activation of fatty acids for β oxidation, and by 1954 the basic outline of β oxidation as we know it today was developed.

The Oxidation of Saturated Fatty Acids Occurs in Mitochondria

We now explore the remarkable process by which a long-chain saturated fatty acid is converted into two-carbon units (acetate), which can be oxidized to CO_2 and H_2O via the tricarboxylic acid cycle and the electron-transport chain. Fatty acids that enter cells are activated to their CoA derivatives by the enzyme acyl-CoA ligase,

$$\text{RCOO}^- + \text{ATP} + \text{CoA} \xrightarrow{\text{Mg}^{2+}} \text{R}\overset{\text{O}}{\overset{\|}{\text{C}}}\text{—CoA} + \text{PP}_i + \text{AMP}$$

and transported into the mitochondria for β oxidation as we discuss later in this chapter.

The outline for the β oxidation of a saturated fatty acid is shown in figure 21.4. In the first reaction the acyl-CoA is dehydrogenated by acyl-CoA dehydrogenase to yield the α,β (or 2,3)*trans*-enoyl-CoA and $FADH_2$. The electrons from $FADH_2$ are channeled into the electron-transport chain (see chapter 16). Enoyl-CoA is subsequently hydrated stereospecifically by enoyl-CoA hydrase to yield the 3-L-hydroxyacyl-CoA. The hydroxyl group is oxidized by 3-hydroxyacyl-CoA dehydrogenase with NAD^+ as coenzyme to yield 3-ketoacyl-CoA and NADH. The final step in the sequence is catalyzed by thiolase to form acetyl-CoA and an acyl-CoA that is two carbons shorter than the initial substrate for β oxidation. This acyl-CoA can undergo another cycle of β oxidation to yield $FADH_2$, NADH, acetyl-CoA, and an acyl-CoA with two fewer carbons (see fig. 21.4*b*). The enzymatic steps are cycled until, in the last sequence of reactions, the four-carbon unit,

$$\text{Butyryl-CoA}\ \ (\text{CH}_3\text{CH}_2\text{CH}_2\overset{\text{O}}{\overset{\|}{\text{C}}}\text{-CoA})$$

is degraded to two acetyl CoAs.

As has already been stated, the last reaction in figure 21.4*a* is catalyzed by thiolase. Mitochondria contain two types of thiolase in the matrix. One is specific for acetoacetyl-CoA and is involved in ketone body metabolism (discussed in a later section). The second thiolase acts on 3-ketoacyl-CoAs of various chain lengths and is involved in β oxidation.

Microorganisms also have the ability to oxidize fatty acids. When *E. coli* is grown on fatty acids instead of glucose, the enzymes of β oxidation are produced in large quantities. The enzyme activities (except for that of acyl-CoA dehydrogenase, which is a separate enzyme) are associated with a multienzyme complex with a molecular weight of approximately 250,000 and an $\alpha_2\beta_2$ structure. The α subunit (M_r = 78,000) has the enoyl-CoA hydrase and the 3-hydroxyacyl-CoA dehydrogenase activities, whereas the β subunit (M_r = 42,000) has the thiolase activity. Two other activities associated with the α subunit are enoyl-CoA isomerase, which is involved in oxidation of unsaturated fatty acids (discussed later), and hydroxyacyl-CoA epimerase, which is involved in β oxidation of D-3-hydroxy fatty acids.

Fatty Acid Oxidation Yields Large Amounts of ATP

β Oxidation, in combination with the tricarboxylic acid cycle and respiratory chain, provides more energy per carbon atom than any other energy source, such as glucose and amino acids. The equations for the complete oxidation of palmitoyl-CoA are shown in table 21.1. Equation (1) in the table shows the oxidation of palmitoyl-CoA by the enzymes of β oxidation. Each of the products of equation (1) is further oxidized by the respiratory chain, equations (2) and (3), or by the tricarboxylic acid cycle and the respiratory chain, equation (4). When the reactions of equations (1)–(4) are summed, the result is equation (5). Hence complete oxidation of one molecule of palmitoyl-CoA yields 108 ATP + 16 CO_2 + 123 H_2O and CoA. The water generated by β oxidation seems almost incidental to this process but is crucial in several animal species. For example, the oxidation of fatty acids is used as a major source of H_2O by the killer whale. This animal lives in the sea but cannot drink the salt water. Similarly, the fat stored in the hump of the camel serves as a source of energy and water in the desert.

The oxidation of palmitate to CO_2 and H_2O yields a $\Delta G^{\circ\prime}$ of $-2{,}340$ kcal/mole. Since the formation of 108 ATPs has a $\Delta G^{\circ\prime}$ of $+787$ kcal/mole, approximately 34% (787/2,340) of the standard free energy for the oxidation of palmitate is conserved in the formation of ATP.

The yield of ATP from the oxidation of palmitoyl-CoA can be compared with the yield from glucose. Both are major sources of energy in our bodies. Perhaps the most meaningful comparison is in terms of the number of ATPs produced per carbon atom. For glucose the yield is 30/6 = 5 ATPs per carbon atom, and for palmitoyl-CoA it is 108/16 = almost 7 ATPs per carbon.

RCH$_2$CH$_2$—C(=O)—CoA
Acyl-CoA

Acyl-CoA dehydrogenase (FAD → FADH$_2$)

H
RC=C—C(=O)—CoA
H
Enoyl-CoA

Enoyl-CoA hydrase (H_2O)

OH
RC—CH$_2$—C(=O)—CoA
H
3-L-Hydroxyacyl-CoA

3-L-Hydroxyacyl-CoA dehydrogenase (NAD$^+$ → NADH + H$^+$)

RC(=O)—CH$_2$—C(=O)—CoA
3-Ketoacyl-CoA

Thiolase (CoA)

RC(=O)—CoA + CH$_3$—C(=O)—CoA
Acyl-CoA **Acetyl-CoA**
(a)

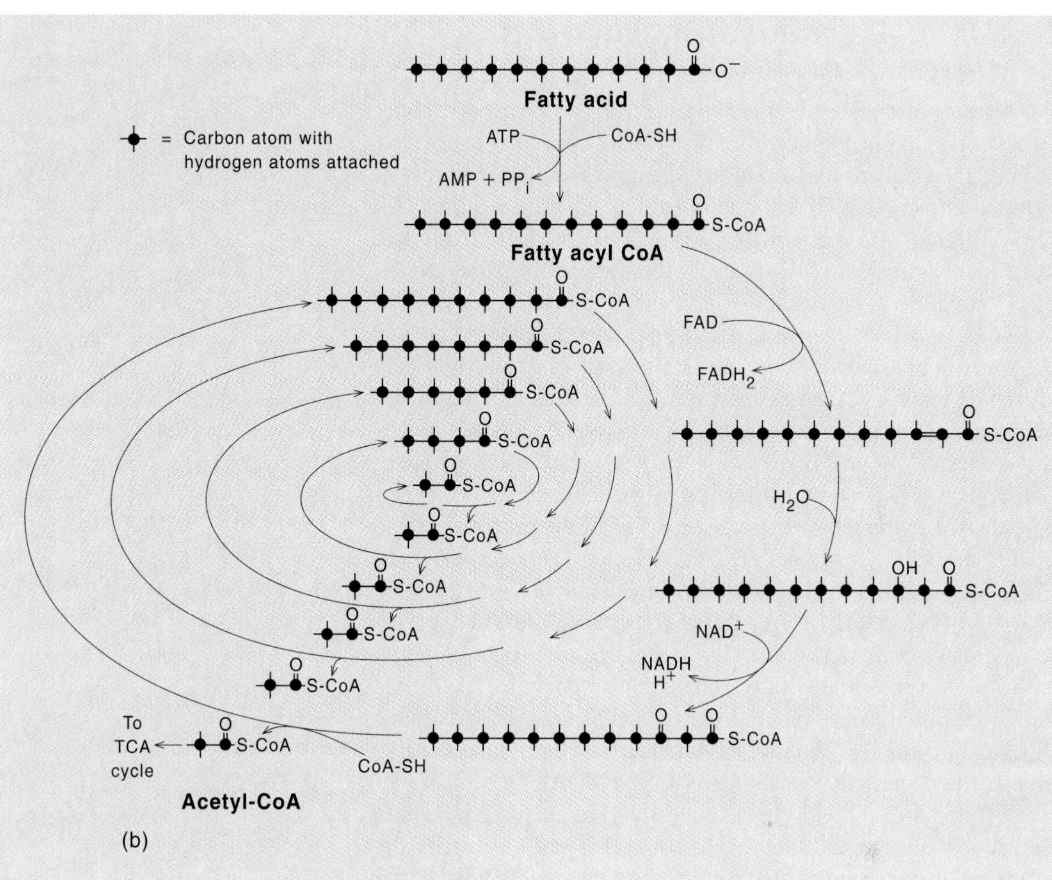

Figure 21.4

(*a*) The β oxidation of fatty acids consists of four reactions in the matrix of the mitochondrion. Each cycle of reactions results in the formation of acetyl-CoA and an acyl-CoA with two fewer carbons. (*b*) The cyclic nature of the β oxidation cycle. In some illustrations the —SH group of CoA is indicated for emphasis. Likewise, in some illustrations the —S— group in an acyl-CoA structure is indicated.

Table 21.1
Equations for the Complete Oxidation of Palmitoyl-CoA to CO_2 and H_2O

1. $CH_3(CH_2)_{14}CO\text{-}CoA + 7\ FAD + 7\ H_2O + 7\ CoA + 7\ NAD^+ \longrightarrow 8\ CH_3CO\text{-}CoA + 7\ FADH_2 + 7\ NADH + 7\ H^+$
2. $7\ FADH_2{}^a + 10\ P_i{}^b + 10\ ADP + 3\tfrac{1}{2}\ O_2 + 10\ H^+ \longrightarrow 7\ FAD + 17\ H_2O + 10\ ATP$
3. $7\ NADH^a + 18\ P_i + 18\ ADP + 3\tfrac{1}{2}\ O_2 + 25\ H^+ \longrightarrow 7\ NAD^+ + 25\ H_2O + 18\ ATP$
4. $8\ CH_3CO\text{-}CoA + 16\ O_2 + 80\ P_i + 80\ ADP + 80\ H^+ \longrightarrow 8\ CoA + 88\ H_2O + 16\ CO_2 + 80\ ATP$
5. $CH_3(CH_2)_{14}CO\text{-}CoA^c + 108\ P_i + 108\ ADP + 23\ O_2 + 108\ H^+ \longrightarrow 108\ ATP + 16\ CO_2 + CoA + 123\ H_2O$

[a]The yield of ATP per oxidation of $FADH_2$ and NADH is assumed to be 1.5 and 2.5, respectively.

[b]Remember that condensation of ADP and P_i yields a molecule of H_2O as well as ATP.

[c]See figure 15.4 when accounting for the production of H_2O from the oxidation of acetyl-CoA because each TCA cycle uses two molecules of H_2O.

Thus the oxidation of palmitoyl-CoA yields more ATP per carbon. The chemical reason for this difference is that 15 of the 16 carbons in palmitate are in the completely reduced state (i.e., no oxygen substitution), whereas all 6 carbons of glucose are partially oxidized (each carbon has a hydroxyl group). The more a carbon atom is substituted with oxygen, the less energy can be harnessed by oxidation to CO_2, which is the derivative of carbon that is completely oxidized.

Additional Enzymes Are Required for Oxidation of Unsaturated Fatty Acids in Mitochondria

We learned in chapter 19 that unsaturated fatty acids are commonly found in triacylglycerols and phospholipids. Nature has also evolved mechanisms for oxidation of these fatty acids. The process is based on β oxidation, but additional enzymes are required. As seen in figure 21.5 oleoyl-CoA (18:1Δ^9) can be degraded in the same manner as stearoyl-CoA (18:0) through the first three cycles of β oxidation. The resulting product, Δ^3-*cis*-dodecenoyl-CoA (12:1Δ^3), is not a substrate for acyl-CoA dehydrogenase because of its *cis* double bond. This step is bypassed by a reaction, catalyzed by enoyl-CoA isomerase, in which the double bond is isomerized to Δ^2-*trans*-dodecenoyl-CoA. This intermediate is a normal substrate for the β oxidation enzyme, enoyl-CoA hydrase. Subsequently, the normal route for β oxidation resumes. Since acyl-CoA dehydrogenase is not used in one of the rounds of β oxidation of oleoyl-CoA, one fewer $FADH_2$ and 1.5 fewer ATPs are made than in the β oxidation of the corresponding saturated fatty acyl-CoA (i.e., stearoyl-CoA). Enoyl-CoA isomerase is active with Δ^3-*cis* or Δ^3-*trans* acyl-CoAs that contain from 6 to 16 carbons.

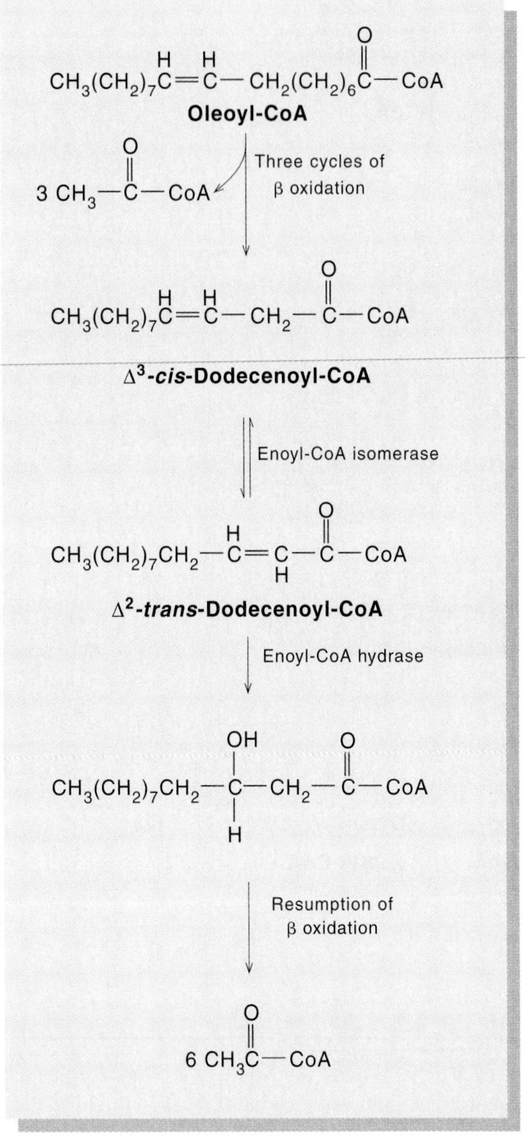

Polyunsaturated fatty acids are also degraded by β oxidation, but the process requires enoyl-CoA isomerase and an additional enzyme, 2,4-dienoyl-CoA reductase (fig. 21.6). For example, the degradation of linoleoyl-CoA (18:2$\Delta^{9,12}$) begins, like that of oleoyl-CoA, with three rounds of β oxidation and results in a Δ^3-*cis* unsaturated fatty acyl-CoA that is not a substrate for acyl-CoA dehydrogenase. Isomerization of the double bond to the Δ^2-*trans* position by enoyl-CoA isomerase allows the resumption of one cycle of β oxidation and generation of a Δ^4-*cis*-enoyl-CoA. Subsequently, acyl-CoA dehydrogenase produces Δ^2-*trans*,Δ^4-*cis*-dienoyl-CoA (see fig. 21.6), which has two conjugated double bonds that are not interrupted by a methylene group. This derivative is not a substrate for what would normally be the next enzyme in β oxidation, enoyl-CoA hydrase. Instead, the intermediate is converted by 2,4-dienoyl-CoA reductase to Δ^3-*trans* enoyl-CoA, then to the Δ^2-*trans*-isomer by enoyl-CoA isomerase (see fig. 21.6). With the double bond in the Δ^2-*trans*-configuration, β oxidation can resume. Since acyl-CoA dehydrogenase is bypassed twice during the β oxidation of linoleoyl-CoA, two fewer $FADH_2$ and three fewer ATPs are generated compared with the β oxidation of stearoyl-CoA (18:0).

Fatty Acids with an Odd Number of Carbons Are Oxidized to Propionyl-CoA

Fatty acids with an odd number of carbons are rare in many mammalian tissues. However, in ruminant mammals such as cows and sheep, the oxidation of odd-chain-length fatty acids can account for as much as 25% of their energy requirements. Consequently, straight-chain fatty acids with seventeen carbons will be oxidized by the normal β oxidation sequence and give rise to seven acetyl-CoAs and one propionyl-CoA:

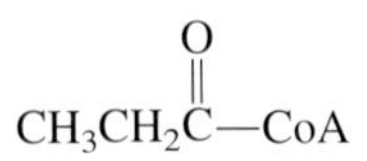

$$CH_3CH_2\overset{O}{\overset{\|}{C}}{-}CoA$$

Propionyl-CoA

This three-carbon acyl-CoA also is a product of degradation of

Figure 21.5

Unsaturated fatty acids such as oleoyl-CoA can also be degraded to acetyl-CoA in mitochondria. Three cycles of β oxidation result in an acyl-CoA (Δ^3-*cis*-dodecenoyl-CoA) that is not a substrate for acyl-CoA dehydrogenase. This problem is circumvented by isomerization of the double bond to the Δ^2-*trans* position by enoyl-CoA isomerase. Complete oxidation of the remainder of the molecule by the enzymes of β oxidation is now possible.

$$CH_3(CH_2)_7\overset{H}{C}{=}\overset{H}{C}{-}CH_2(CH_2)_6\overset{O}{\overset{\|}{C}}{-}CoA$$

Oleoyl-CoA

Three cycles of β oxidation → $3\ CH_3{-}\overset{O}{\overset{\|}{C}}{-}CoA$

$$CH_3(CH_2)_7\overset{H}{C}{=}\overset{H}{C}{-}CH_2{-}\overset{O}{\overset{\|}{C}}{-}CoA$$

Δ^3-*cis*-Dodecenoyl-CoA

⇅ Enoyl-CoA isomerase

$$CH_3(CH_2)_7CH_2{-}\overset{H}{C}{=}\underset{H}{C}{-}\overset{O}{\overset{\|}{C}}{-}CoA$$

Δ^2-*trans*-Dodecenoyl-CoA

↓ Enoyl-CoA hydrase

$$CH_3(CH_2)_7CH_2{-}\overset{OH}{\underset{H}{C}}{-}CH_2{-}\overset{O}{\overset{\|}{C}}{-}CoA$$

↓ Resumption of β oxidation

$$6\ CH_3\overset{O}{\overset{\|}{C}}{-}CoA$$

Figure 21.6

Polyunsaturated fatty acids such as linoleoyl-CoA are also oxidized in the mitochondria. In addition to the enzymes of β oxidation and enoyl-CoA isomerase, the process requires 2,4-dienoyl-CoA reductase.

$$CH_3(CH_2)_4CH{=}CH{-}CH_2{-}CH{=}CH{-}CH_2(CH_2)_6C(=O){-}CoA$$

Δ^9-***cis***, Δ^{12}-***cis***

↓ Three cycles of β oxidation

$$CH_3(CH_2)_4CH{=}CH{-}CH_2{-}CH{=}CH{-}CH_2{-}C(=O){-}CoA + 3\ CH_3{-}C(=O){-}CoA$$

Δ^3-***cis***, Δ^6-***cis***

↓ Enoyl-CoA isomerase

$$CH_3(CH_2)_4CH{=}CH{-}CH_2{-}CH_2{-}CH{=}CH{-}C(=O){-}CoA$$

Δ^2-***trans***, Δ^6-***cis***

↓ One cycle of β oxidation

$$CH_3(CH_2)_4CH{=}CH{-}CH_2{-}CH_2{-}C(=O){-}CoA + CH_3{-}C(=O){-}CoA$$

Δ^4-***cis***

↓ Acyl-CoA dehydrogenase

$$CH_3(CH_2)_4CH{=}CH{-}CH{=}CH{-}C(=O){-}CoA$$

Δ^2-***trans***, Δ^4-***cis***

↓ 2,4-Dienoyl-CoA reductase ($NADPH + H^+ \rightarrow NADP^+$)

$$CH_3(CH_2)_4CH_2{-}CH{=}CH{-}CH_2{-}C(=O){-}CoA$$

Δ^3-***trans***

↓ Enoyl-CoA isomerase

$$CH_3(CH_2)_4CH_2{-}CH_2{-}CH{=}CH{-}C(=O){-}CoA$$

Δ^2-***trans***

↓ Four cycles of β oxidation

$$5\ CH_3{-}C(=O){-}CoA$$

Acetyl-CoA

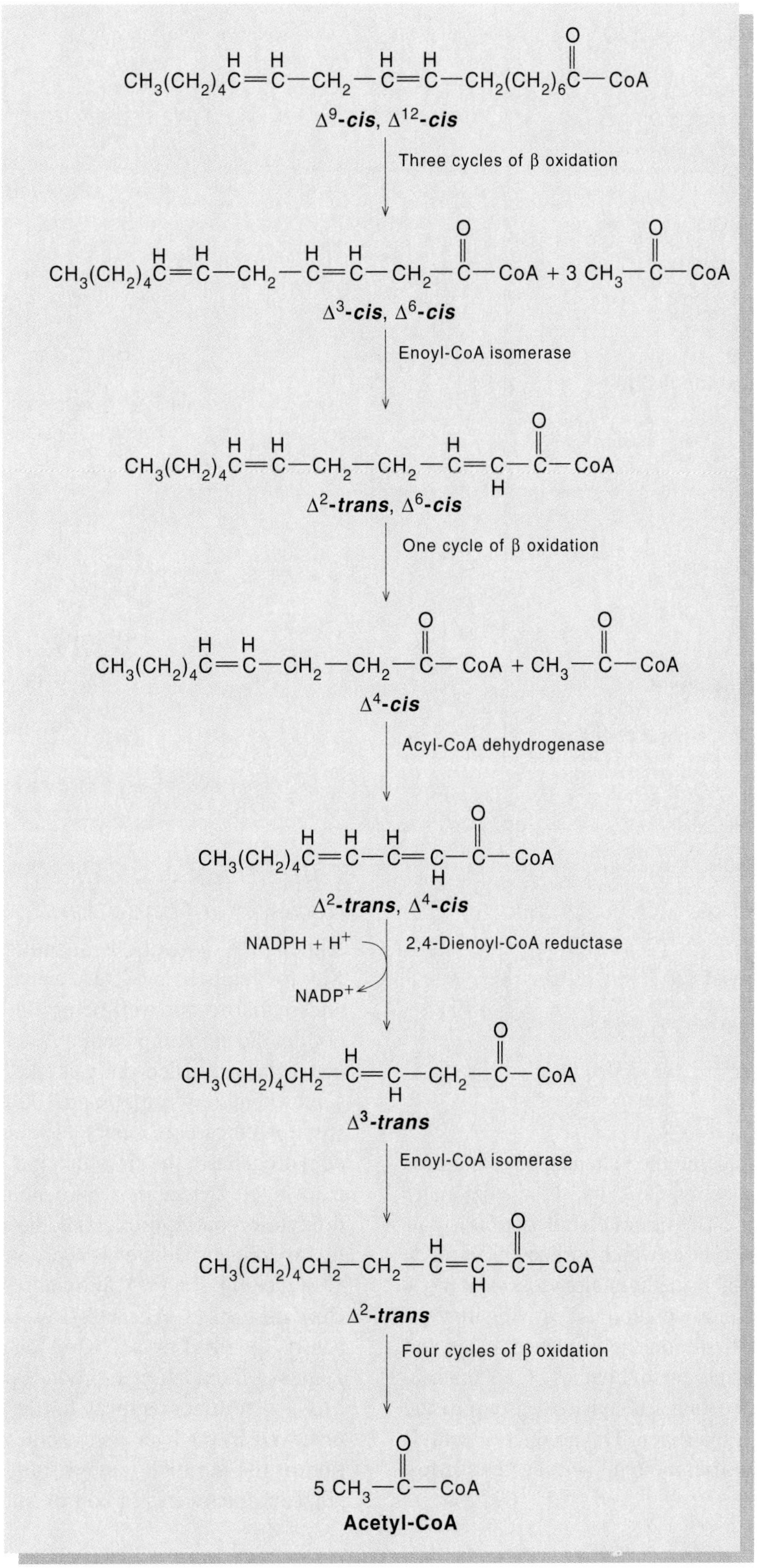

Figure 21.7

Propionyl-CoA generated by the β oxidation of odd-chain fatty acids is converted to succinyl-CoA, an intermediate of the tricarboxylic acid cycle.

$CH_3CH_2C(=O)—CoA + ATP + CO_2 + H_2O$
Propionyl-CoA

Propionyl-CoA carboxylase

COO^- / $H—C—CH_3$ / $C(=O)—CoA$
D-Methylmalonyl-CoA

Methylmalonyl-CoA racemase

COO^- / $H_3C—C—H$ / $C(=O)—CoA$
L-Methylmalonyl-CoA

Methylmalonyl-CoA mutase

COO^- / $H_2C—CH_2$ / $C(=O)—CoA$
Succinyl-CoA

Figure 21.8

The oxidation of phytanic acid involves oxidation of the α carbon and decarboxylation. The product, pristanic acid, can be degraded by β oxidation after conversion to the CoA derivative.

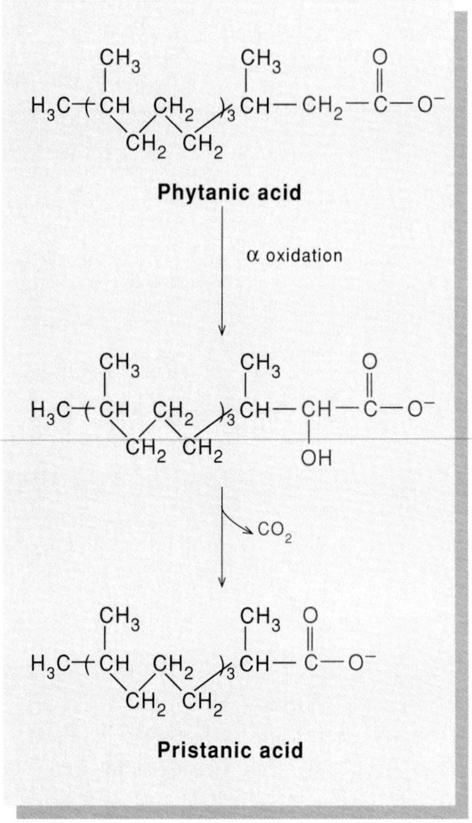

$H_3C(CH(CH_3)CH_2CH_2CH_2)_3CH(CH_3)—CH_2—C(=O)—O^-$
Phytanic acid

α oxidation

$H_3C(CH(CH_3)CH_2CH_2CH_2)_3CH(CH_3)—CH(OH)—C(=O)—O^-$

CO_2

$H_3C(CH(CH_3)CH_2CH_2CH_2)_3CH(CH_3)—C(=O)—O^-$
Pristanic acid

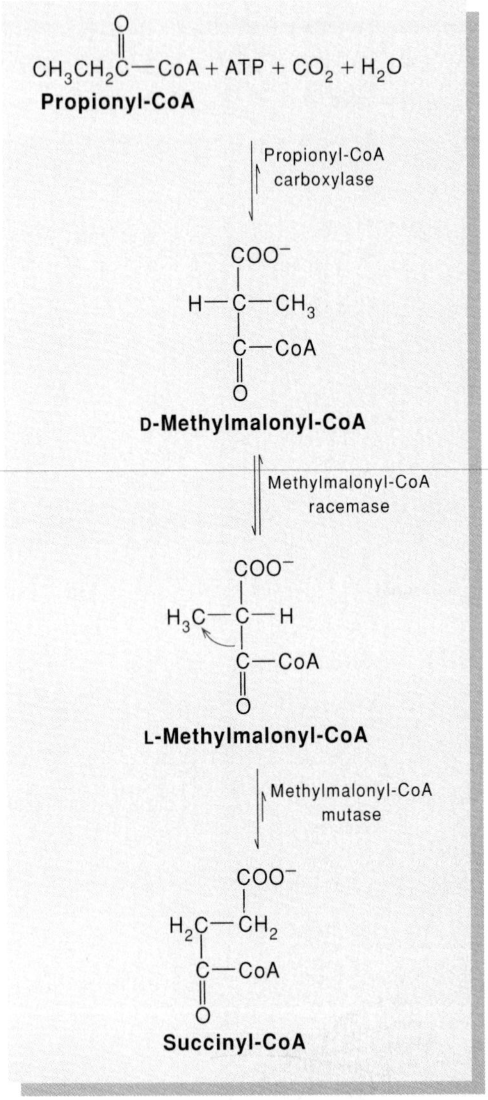

the amino acids valine and isoleucine (see chapter 25). The propionyl-CoA is converted to succinyl-CoA by three enzymatic steps, as indicated in figure 21.7. The initial carboxylation is catalyzed by propionyl-CoA carboxylase, which utilizes biotin as a cofactor. In the second reaction, D-methylmalonyl-CoA is converted to its optical isomer, L-methylmalonyl-CoA, by methylmalonyl-CoA racemase. The last step in the sequence, catalyzed by the cobalamin-requiring enzyme, methylmalonyl-CoA mutase, involves an unusual migration of the carbonyl-CoA group to the methyl group in an exchange for hydrogen. The product, succinyl-CoA, can be metabolized in the tricarboxylic acid cycle (see fig. 15.4).

Fatty Acids Can Also Be Oxidized by α or ω Oxidation

Although β oxidation is quantitatively the most significant pathway for catabolism of fatty acids, α oxidation of some fatty acids is essential to our well-being. Phytanic acid (fig. 21.8) is an important dietary component, present in ruminant fat and dairy products. The estimated daily intake of phytanic acid by humans is somewhere between 50 and 100 mg. Because of the methyl substitution on carbon 3, phytanic acid is not a substrate for acyl-CoA dehydrogenase, the first enzyme in β oxidation. This step is circumvented by another mitochondrial enzyme, fatty acid α-hydroxylase, that hydroxylates the α carbon of phytanic acid. The hydroxyl intermediate is decarboxylated to yield pristanic acid and CO_2 (see fig. 21.8). A thiokinase reaction activates the acid to its CoA derivative. Pristanyl-CoA is unsubstituted at carbon 3 and can be oxidized by acyl-CoA dehydrogenase and the normal enzymes of β oxidation to produce propionyl-CoA and an acyl-CoA.

A minor pathway for the oxidation of fatty acids has been observed in rat liver microsomes. This pathway involves oxidation of the terminal methyl, called the ω (omega) carbon, or the adjacent methylene carbon of fatty acids by NADPH and molec-

Figure 21.9

The oxidation of the methyl group (ω oxidation) of certain fatty acids is known to occur. However, the enzymes involved have not been well described.

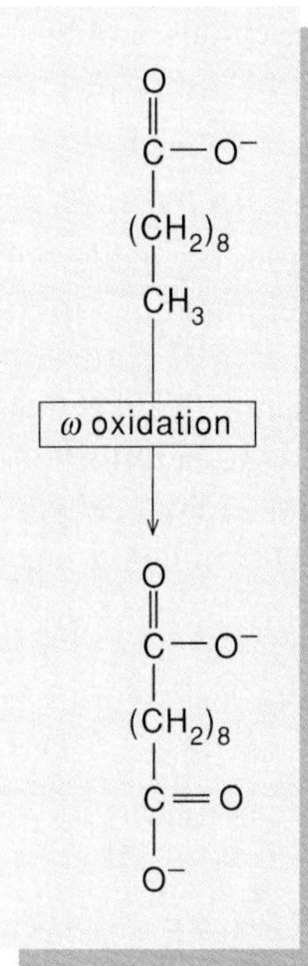

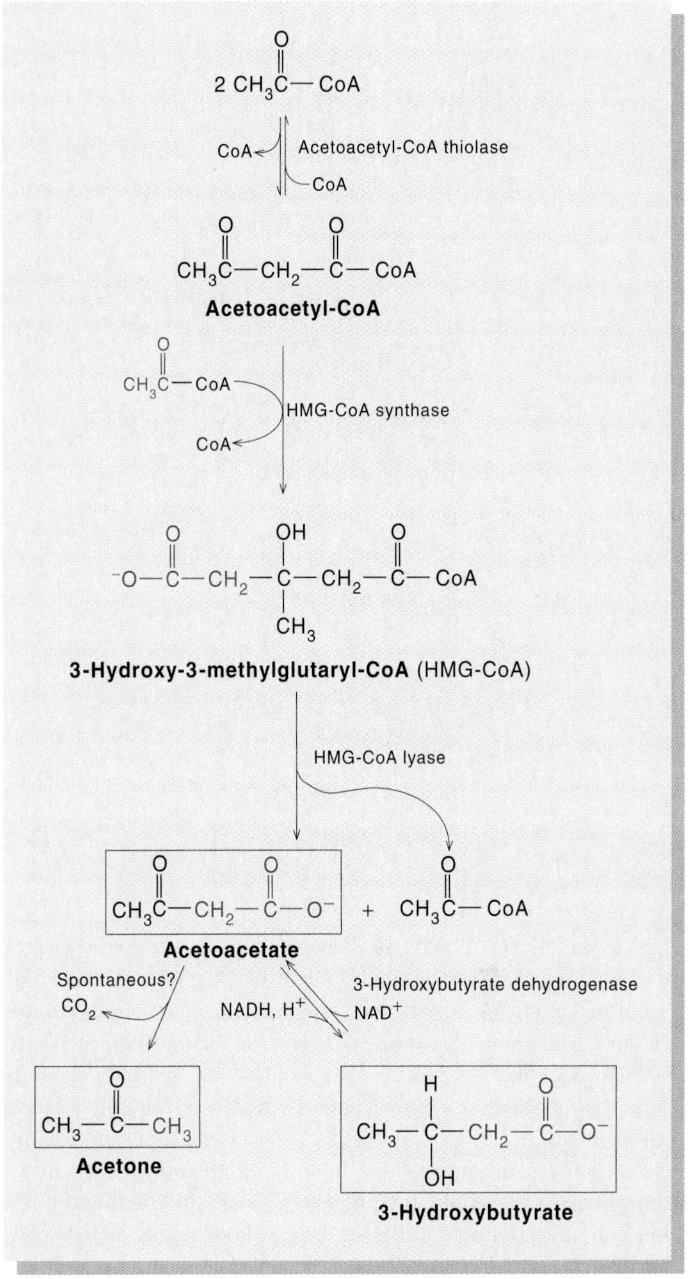

ular oxygen (fig. 21.9). The pathway is probably not quantitatively significant for the oxidation of long-chain fatty acids, and the enzymes involved have not been characterized. However, ω oxidation may be important for the metabolism of fatty acids with 6 to 10 carbons.

Ketone Bodies Formed in the Liver Are Used for Energy in Other Tissues

Ketone bodies (acetoacetate, β-hydroxybutyrate, and acetone; structures are presented in fig. 21.10) are made in the liver when β oxidation of fatty acids is in excess of that required by the liver. These water-soluble, energy-rich compounds are transported to other tissues for generation of energy. As we discuss later on, excess production of ketone bodies, which occurs during starvation or untreated diabetes, can be harmful.

Once fatty acids are degraded in liver mitochondria, the resulting acetyl-CoA can undergo a number of metabolic fates. As we learned in chapter 15, the utilization of acetyl-CoA is of central importance in the tricarboxylic acid cycle. Alternatively, this coenzyme is involved in the synthesis of ketone bodies, which takes place only in the mitochondria (see fig. 21.10). In the first reaction, catalyzed by acetoacetyl-CoA thiolase, two acetyl-CoAs condense to form acetoacetyl-CoA. This reaction is a reversal of the last reaction in β oxidation and is thermodynamically unfavorable in that the equilibrium favors thiolytic cleavage of acetoacetyl-CoA. Hence this compound is formed when the levels of acetyl-CoA rise, which pushes the reaction toward acetoacetyl-CoA synthesis. A third molecule of acetyl-CoA reacts with acetoacetyl-CoA to yield 3-hydroxy-3-methylglutaryl-

Figure 21.10

The biosynthesis of ketone bodies (acetoacetate, hydroxybutyrate, and acetone) occurs in the mitochondria of the liver.

$2\ CH_3C(=O)—CoA$

CoA ← Acetoacetyl-CoA thiolase → CoA

$CH_3C(=O)—CH_2—C(=O)—CoA$

Acetoacetyl-CoA

$CH_3C(=O)—CoA$ → HMG-CoA synthase → CoA

$^-O—C(=O)—CH_2—C(OH)(CH_3)—CH_2—C(=O)—CoA$

3-Hydroxy-3-methylglutaryl-CoA (HMG-CoA)

HMG-CoA lyase

$CH_3C(=O)—CH_2—C(=O)—O^-$ + $CH_3C(=O)—CoA$

Acetoacetate

Spontaneous? CO_2

NADH, H^+ — 3-Hydroxybutyrate dehydrogenase — NAD^+

$CH_3—C(=O)—CH_3$

Acetone

$CH_3—C(H)(OH)—CH_2—C(=O)—O^-$

3-Hydroxybutyrate

CoA (HMG-CoA) in a reaction catalyzed by HMG-CoA synthase. (As we see in chapter 23, the same two reactions are also the first steps in cholesterol biosynthesis, which take place in the cytosol.) In the formation of ketone bodies the next reaction is catalyzed by HMG-CoA lyase and yields acetoacetate and acetyl-CoA. The acetoacetate can be reduced to 3-hydroxybutyrate by an enzyme on the inner membrane of the mitochondrion, 3-hydroxybutyrate dehydrogenase. Although acetoacetate can also be decarboxylated to form acetone, this is normally of minor importance. However, pa-

Figure 21.11

Ketone bodies are converted back to acetyl-CoA in the mitochondria of nonhepatic tissues and used as a source of energy in these tissues.

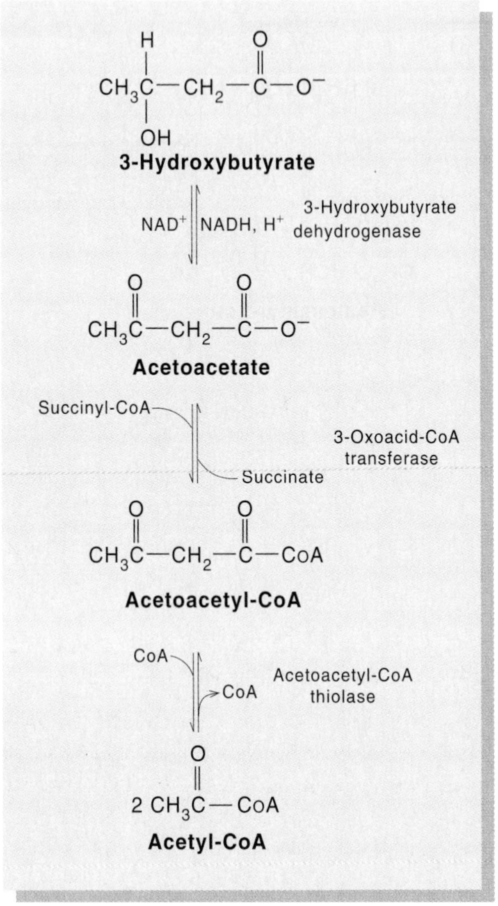

$$CH_3\overset{H}{\underset{OH}{C}}-CH_2-\overset{O}{\overset{\|}{C}}-O^-$$

3-Hydroxybutyrate

$\updownarrow$ NAD^+ / $NADH, H^+$ — 3-Hydroxybutyrate dehydrogenase

$$CH_3\overset{O}{\overset{\|}{C}}-CH_2-\overset{O}{\overset{\|}{C}}-O^-$$

Acetoacetate

$\updownarrow$ Succinyl-CoA / Succinate — 3-Oxoacid-CoA transferase

$$CH_3\overset{O}{\overset{\|}{C}}-CH_2-\overset{O}{\overset{\|}{C}}-CoA$$

Acetoacetyl-CoA

$\updownarrow$ CoA / CoA — Acetoacetyl-CoA thiolase

$$2\ CH_3\overset{O}{\overset{\|}{C}}-CoA$$

Acetyl-CoA

tients with uncontrolled type I diabetes (a disease caused by a lack of production of insulin in the islet cells of the pancreas) have high levels of ketone bodies in their plasma, and their breath has the characteristic odor of acetone. The rise in ketone bodies occurs because cells and tissues are glucose starved. As a result, fatty acids are mobilized to provide energy, which results in an excess production of acetyl-CoA in the liver. Hence the liver converts this excess acetyl-CoA into ketone bodies, which can be used in extrahepatic tissues as a source of energy. An additional complication is that a large rise in the serum levels of ketone bodies lowers the pH of blood because of the increased concentration of these acids. Among other reasons, this condition can be hazardous because of the effect of lower pH on oxygen binding to hemoglobin.

Ketone body synthesis is primarily a liver function, since HMG-CoA synthase is present in large quantities only in this tissue. Acetoacetate and 3-hydroxybutyrate are secreted into the blood and carried to other tissues, where they are converted into acetyl-CoA as described in figure 21.11. The reactions catalyzed by β-hydroxybutyrate dehydrogenase and thiolase are common to both the synthesis and degradation of the ketone bodies. However, the second enzyme in the sequence for degradation shown in figure 21.11, 3-oxoacid-CoA transferase, is present in all tissues but liver. Hence ketone bodies are made in the liver and metabolized to CO_2 and energy in nonhepatic (nonliver) tissues. Ketone bodies can also be used to supply these tissues with acetyl-CoA for fatty acid and cholesterol biosynthesis. Notably, ketone bodies are important sources of energy for the brain during starvation. Normally, glucose is the major source of energy in the brain, and the brain does not use fatty acids as a major source of energy.

β Oxidation Also Occurs in Peroxisomes

Peroxisomes are subcellular organelles whose main function is to process substances for elimination. In 1976, Paul Lazarow and Christian de Duve showed that peroxisomes from rat liver contained a β-oxidation system. The reactions of β oxidation in peroxisomes are similar to those in mitochondria (see fig. 21.4), with the notable exception of the initial dehydrogenation. This reaction is catalyzed by the FAD-containing enzyme acyl-CoA oxidase:

$$E{-}FAD + RCH_2CH_2\overset{O}{\overset{\|}{C}}{-}CoA \longrightarrow RC\overset{H}{=}\underset{H}{C}{-}\overset{O}{\overset{\|}{C}}{-}CoA + E{-}FADH_2 \quad (2)$$

$$E{-}FADH_2 + O_2 \longrightarrow H_2O_2 + E{-}FAD \quad (3)$$

Oxidized flavin is regenerated by reaction with oxygen, which is reduced to H_2O_2, and catalase reduces the H_2O_2 to H_2O. Since the electrons from the β oxidation are not shuttled into the respiratory chain, the peroxisomal pathway is not involved in energy production. Acyl-CoA oxidase is inactive with low-molecular-weight acyl-CoAs such as hexanoyl-CoA or butyryl-CoA. It is suspected that these acyl-CoAs are further catabolized in the mitochondria.

In view of all these limitations you may well wonder what function β oxidation serves in the peroxisomes. It appears that peroxisomes are uniquely involved in the oxidation of very-long-chain fatty acids such as 24:0 and rare fatty acids found in the brain (C_{26}–C_{38} polyenoic fatty acids). The importance of this fact is underscored by the occurrence of rare metabolic disorders where there is an accumulation of very-long-chain fatty acids. In all such cases these disorders have been correlated with defects in the peroxisomal oxidation system.

Summary of Fatty Acid Degradation

In this section we have seen that fatty acids are oxidized in units of two carbon atoms. The immediate end products of this oxidation are $FADH_2$ and NADH, which supply energy through the respiratory chain, and acetyl-CoA, which has multiple possible uses in addition to the generation of energy via the tricarboxylic acid cycle and respiratory chain. Unsaturated fatty acids can also be oxidized in the mitochondria with the help of auxiliary enzymes. Ketone body synthesis from acetyl-CoA is an important liver function for transfer of energy to other tissues, especially brain, when glucose levels are decreased as in diabetes or starvation.

Additional enzymes are available in animals for oxidation of fatty acid with odd numbers of carbons, with methyl substitutions or very-long-chain fatty acids.

Figure 21.12

Reactions catalyzed by acetyl-CoA carboxylase. In *E. coli,* BCCP and the two enzymatic activities (biotin carboxylase and carboxyltransferase) can be separated from each other. In contrast, in the liver all three components exist on a single multifunctional polypeptide.

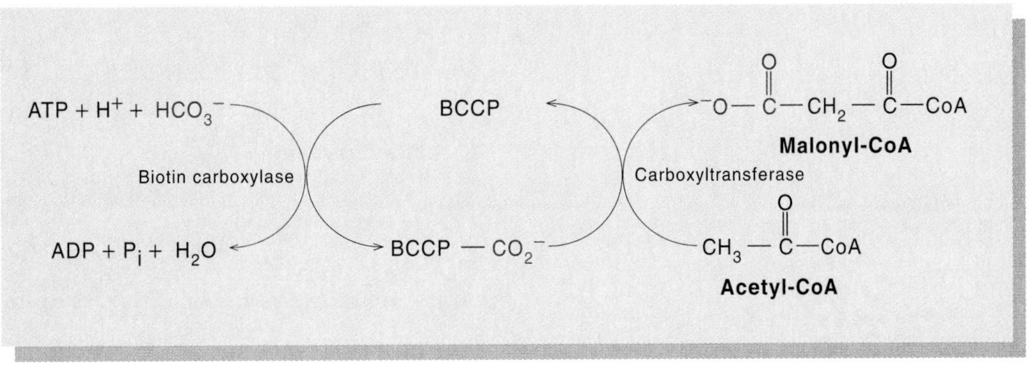

$$\mathrm{ATP} + \mathrm{H^+} + \mathrm{HCO_3^-} \xrightarrow{\text{Biotin carboxylase}} \mathrm{ADP} + \mathrm{P_i} + \mathrm{H_2O}$$

$$\mathrm{BCCP} \longrightarrow \mathrm{BCCP{-}CO_2^-}$$

$$\underset{\text{Acetyl-CoA}}{\mathrm{CH_3{-}\overset{O}{\overset{\|}{C}}{-}CoA}} \xrightarrow{\text{Carboxyltransferase}} \underset{\text{Malonyl-CoA}}{\mathrm{{}^-O{-}\overset{O}{\overset{\|}{C}}{-}CH_2{-}\overset{O}{\overset{\|}{C}}{-}CoA}}$$

Biosynthesis of Saturated Fatty Acids

When energy occurs in excess, for example after a meal, organisms make fatty acids to store this energy as components of triacylglycerol. Research on the mechanism of fatty acid biosynthesis goes back to the very early years of biochemical investigations. The finding that most fatty acids contain an even number of carbon atoms led Rapier to postulate in 1907 that fatty acids were produced by condensation of an activated two-carbon compound. In some of the first isotopic tracer experiments performed in the late 1930s and early 1940s, David Rittenberg and Konrad Bloch used the newly developed heavy isotopes of carbon (carbon 13) and hydrogen (deuterium) and implicated acetate as the two-carbon compound. Subsequently, Feodor Lynen showed a central role for acetyl-CoA in fatty acid biosynthesis. Precisely how acetyl-CoA was converted into fatty acids eluded workers until the late 1950s, when Salih Wakil discovered the involvement of malonyl-CoA. Subsequent progress was rapid, and the scheme for fatty acid biosynthesis as we know it today was elucidated.

The synthesis of saturated fatty acids is very similar in all organisms. The overall reaction for the formation of palmitic acid is

$$\underset{\text{Acetyl-CoA}}{\mathrm{CH_3\overset{O}{\overset{\|}{C}}CoA}} + 7\ \underset{\text{Malonyl-CoA}}{\mathrm{{}^-O\overset{O}{\overset{\|}{C}}CH_2\overset{O}{\overset{\|}{C}}CoA}} + 14\ \mathrm{NADPH} + 14\ \mathrm{H^+} \longrightarrow$$

$$\underset{\text{Palmitic acid}}{\mathrm{CH_3(CH_2)_{14}COO^-}} + 7\ \mathrm{CO_2} + 8\ \mathrm{CoA} + 14\ \mathrm{NADP^+} + 7\ \mathrm{H_2O} \quad \mathbf{(4)}$$

Comparing this equation with the equation for the complete oxidation of palmitoyl-CoA (see table 21.1, equation 1), we find major differences in carriers and intermediates. The principal electron carrier in the anabolic pathway is the NADPH–$NADP^+$ system; in the catabolic pathway, β oxidation, the principal electron carriers are FAD–$FADH_2$ and NAD^+–NADH. The second striking difference between the two pathways is that malonyl-CoA is the principal substrate in the anabolic pathway but plays no role in the catabolic pathway. These differences reflect the fact that the two pathways do not share common enzymes. Indeed, in animal cells the reactions occur in separate cell compartments; biosynthesis takes place in the cytosol, whereas catabolism occurs in the mitochondria (see fig. 21.1).

The First Step in Fatty Acid Synthesis Is Catalyzed by Acetyl-CoA Carboxylase

Eight enzyme-catalyzed reactions are involved in the conversion of acetyl-CoA into fatty acids. The first reaction is catalyzed by acetyl-CoA carboxylase and requires ATP. This is the reaction that supplies the energy that drives the biosynthesis of fatty acids. The properties of acetyl-CoA carboxylase are similar to those of pyruvate carboxylase, which is important in the gluconeogenesis pathway (see chapter 14). Both enzymes contain the coenzyme biotin covalently linked to a lysine residue of the protein via its ϵ-amino group. In the last section of this chapter we show that the activity of acetyl-CoA carboxylase plays an important role in the control of fatty acid biosynthesis in animals. Regulation of the first enzyme in a biosynthetic pathway is a strategy widely used in metabolism.

Acetyl-CoA carboxylase of *E. coli* is a multienzyme complex that consists of three protein components that can be isolated individually: biotin carboxyl carrier protein (BCCP), biotin carboxylase, and carboxyltransferase (fig. 21.12). The reaction sequence involves an initial carboxylation of BCCP, catalyzed by biotin carboxylase. The CO_2 is covalently linked to one of the nitrogen atoms of biotin (fig. 21.13). Subsequently, the CO_2 is transferred from BCCP to acetyl-CoA in a reaction catalyzed by carboxyltransferase, which yields malonyl-CoA. A likely mechanism for the reaction is shown in figure 21.14.

A distinctly different form of acetyl-CoA carboxylase is found in the cytosol of animal tissues. The rat liver enzyme is a dimer composed of two identical subunits (M_r of each = 265,000) with one biotin per subunit. In contrast to the multienzyme complex in *E. coli,* the three functional parts of acetyl-CoA carboxylase in rat liver occur in a single multifunc-

Figure 21.13

Structure of N'-carboxybiotin linked to biotin carboxyl carrier protein (BCCP). BCCP is one of the components of acetyl-CoA carboxylase isolated from *E. coli.*

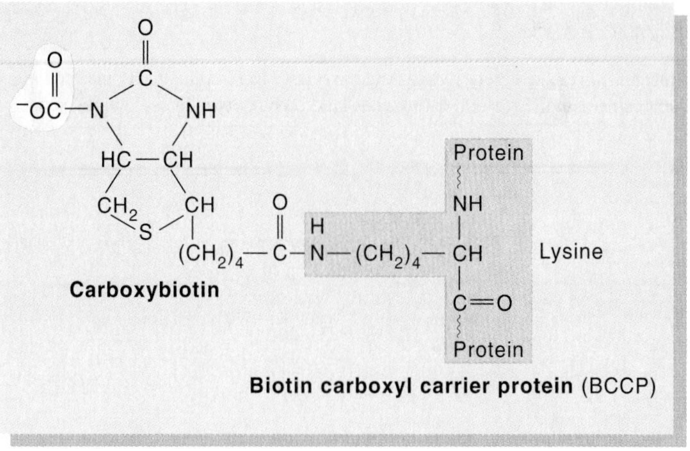

Figure 21.14

Mechanism for the carboxylation of acetyl-CoA. Acetyl-CoA forms a carbanion by proton loss. This carbanion attacks the carbon in carboxybiotin to give malonyl-CoA and biotinate. The biotinate anion is returned to the neutral form by addition of a proton. The R group attached to carboxybiotin is the carboxyl carrier protein.

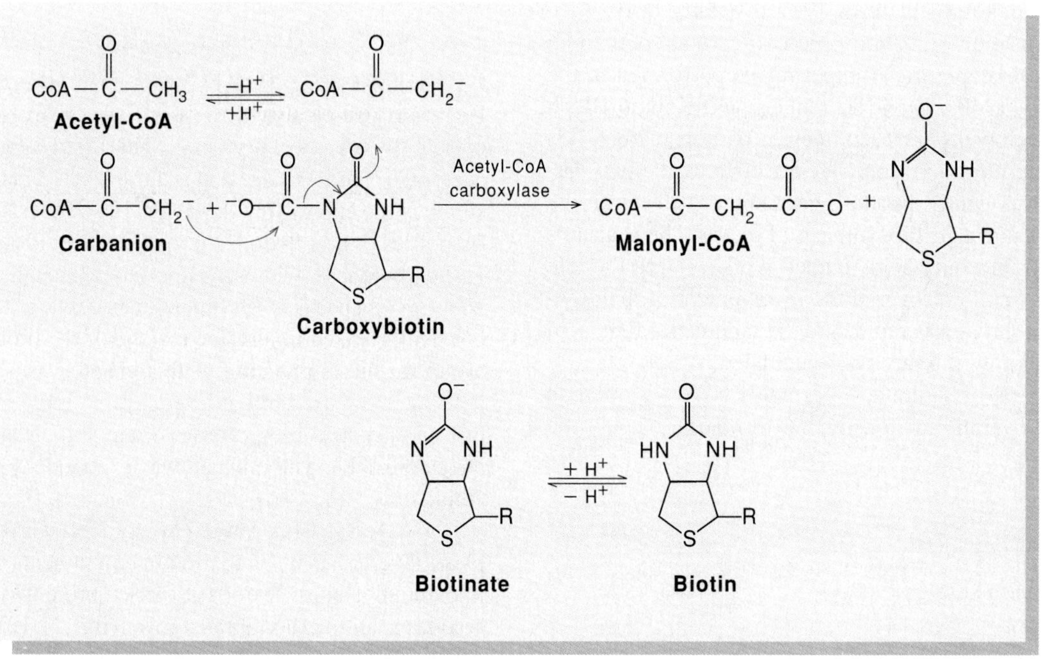

tional polypeptide. The enzyme, as a dimer, has very low activity. However, in the presence of citrate the enzyme oligomerizes to an active form, a polymer with a molecular weight between 4 and 8 million (fig. 21.15). The use of citrate to activate acetyl-CoA carboxylase makes good biochemical sense. When there is an abundance of carbohydrate, production of citrate in the mitochondrion occurs in excess. This citrate is transported to the cytosol, where it can activate acetyl-CoA carboxylase. Acetyl-CoA carboxylase becomes deactivated and depolymerized when incubated with malonyl-CoA or palmitoyl-CoA. The significance of the concentration of palmitoyl-CoA on the rate of fatty acid biosynthesis is discussed when we look at the regulation of fatty acid metabolism later in the chapter. Incidentally, citrate does not activate or polymerize the *E. coli* enzyme.

Seven Reactions Are Catalyzed by the Fatty Acid Synthase

After malonyl-CoA synthesis the remaining steps in fatty acid synthesis occur on fatty acid synthase, which exists as a multienzyme complex. In the initial reactions, acetyl-CoA and malonyl-

Figure 21.15

Activation of acetyl-CoA carboxylase from rat liver in vitro. Citrate activates the enzyme and converts it to a polymer. Either palmitoyl-CoA or malonyl-CoA can reverse this process. Citrate does not activate or polymerize the *E. coli* enzyme.

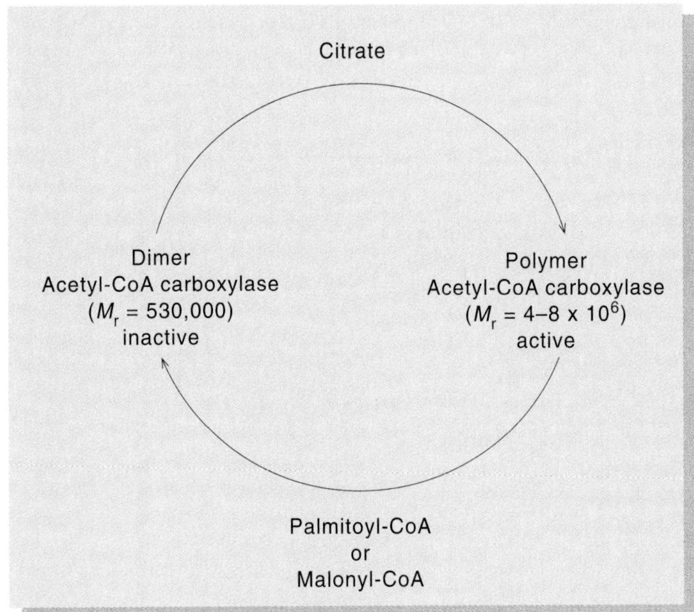

CoA are transferred onto the protein complex by acetyl-CoA transacylase and malonyl-CoA transacylase (step 1 and step 2 in fig. 21.16*a*). The acceptor for the acetyl and malonyl groups is acyl carrier protein (ACP). ACP also carries all of the intermediates during fatty acid biosynthesis. The prosthetic group that binds these intermediates is the coenzyme phosphopantetheine that is bound to ACP by an ester linkage between the phosphate of the coenzyme and a serine hydroxyl side-chain of the protein (fig. 21.17). The sulfhydryl group of the phosphopantetheine is the attachment site for the intermediates during fatty acid synthesis. Note that CoA also contains this phosphopantetheine moiety, which also utilizes its sulfhydryl group to bind fatty acyl residues.

The Condensation Reaction. In the condensation reaction the acetyl group is initially transferred from ACP on to a SH group of 3-ketoacyl-ACP synthase. This acetyl moiety then reacts with malonyl-ACP (step 3 in fig. 21.16*a*) so that the acetyl component becomes the methyl-terminal two-carbon unit of the acetoacetyl-ACP. The release of CO_2 in this condensation reaction provides the extra thermodynamic push to make the reaction highly favorable. A likely mechanism for the condensation reaction is shown in figure 21.18.

The Reduction Reactions. The object of the next three reactions (steps 4 to 6 in fig. 21.16*a*) is to reduce the 3-carbonyl group to a methylene group. The carbonyl is first reduced to a hydroxyl by 3-ketoacyl-ACP reductase. Next, the hydroxyl is removed by a dehydration reaction catalyzed by 3-hydroxyacyl-ACP dehydrase with the formation of a *trans* double bond. This double bond is reduced by NADPH catalyzed by 2,3-*trans*-enoyl-ACP reductase. Chemically, these reactions are nearly the same as the reverse of three steps in the β-oxidation pathway except that the hydroxyl group is in the D-configuration for fatty acid synthesis and in the L-configuration for β oxidation (compare figs. 21.4*a* and 21.16*a*). Also remember that different cofactors, enzymes, and cellular compartments are used in the reactions of fatty acid biosynthesis and degradation.

Continuation Reactions. At this point we have seen one full round of reactions catalyzed by the fatty acid synthase. Each enzyme activity of the complex that we have discussed has been used precisely once. The resulting acyl group on ACP (formed during step 6, fig. 21.16*a*) is transferred to the SH group of the active site of 3-ketoacyl-ACP synthase (the condensing enzyme activity) as was the acetyl group in the first cycle. The acyl group then reacts with another molecule of malonyl-ACP catalyzed by 3-ketoacyl-ACP synthase to yield a six-carbon intermediate with a ketone on the 3 carbon (see fig. 21.16*b*). The reduction reactions (steps 4–6 in fig. 21.16*a*) occur again and a six-carbon intermediate, hexanoyl-ACP, is formed (see fig. 21.16*b*). The hexanoyl group is transferred to the 3-ketoacyl-ACP synthase and condenses with another malonyl-ACP followed by the three reduction reactions (see fig. 21.16*b*). The biosynthetic process continues to recycle until palmitoyl-ACP is made. At this point, in animal cells, palmitate is hydrolyzed from the phosphopantetheine of ACP by the activity of a thioesterase whereas in *E. coli,* palmitoyl-ACP is used directly for phospholipid biosynthesis, as will be discussed in chapter 22. Assay for fatty acid synthase from liver is described in box 21A.

The Organization of the Fatty Acid Synthase Is Different in E. coli *and Animals*

In *E. coli* and plants the enzymes of fatty acid synthase are believed to occur as a multienzyme complex composed of the individual enzyme activities and ACP grouped together as an aggregate. These proteins are not covalently linked and can be isolated from each other. In contrast the enzyme activities and ACP of the fatty acid synthase of animals are covalently linked in a giant multifunctional polypeptide with a molecular weight of 272,000 (2,505 amino acids). The active form of the enzyme consists of a dimer of this peptide as illustrated in figure 21.19. The legend to figure 21.19 explains how this multifunctional protein makes fatty acids. The evolution of a multifunctional protein facilitates the channeling of the intermediates of fatty acid synthesis. For each step in the pathway the next enzyme is always near at hand, and the dilution of intermediates is minimized. In addition, the expression of the seven enzymatic activities of fatty acid biosynthesis is coordinated because only one gene product, as opposed to seven, is required.

Biosynthesis of Monounsaturated Fatty Acids Follows Distinct Routes in E. coli *and Animal Cells*

As mentioned in chapter 19, unsaturated fatty acids are abundant in all living organisms. Alternative mechanisms for the biosyn-

Figure 21.16

Outline of the reactions for fatty acid biosynthesis. Fatty acids grow in steps of two-carbon units and take place on a multienzyme complex. (*a*) The initial reactions of fatty acid biosynthesis are shown. In the first reaction, acetyl-CoA reacts with ACP (acyl carrier protein) to form acetyl-ACP (step 1). ACP is shown with its SH group emphasized (see fig. 21.17) to remind readers that the acyl derivatives are linked to ACP via a thioester bond. Malonyl-CoA, derived from the carboxylation of acetyl-CoA (see fig. 21.12), reacts with ACP to yield malonyl-ACP (step 2). These two ACP derivatives then condense to form 3-ketoacyl-ACP with the release of ACP and CO_2 (step 3). Step 4 involves the reduction of the 3-keto group with NADPH and step 5 the dehydration of 3-hydroxyacyl-ACP to form 2,3-*trans*-enoyl-ACP. The final reaction in the cycle is reduction of this double bond with NADPH (step 6) to give an acyl-ACP with 4 carbons. Screens permit tracking of different molecular groupings. (*b*) The ini-

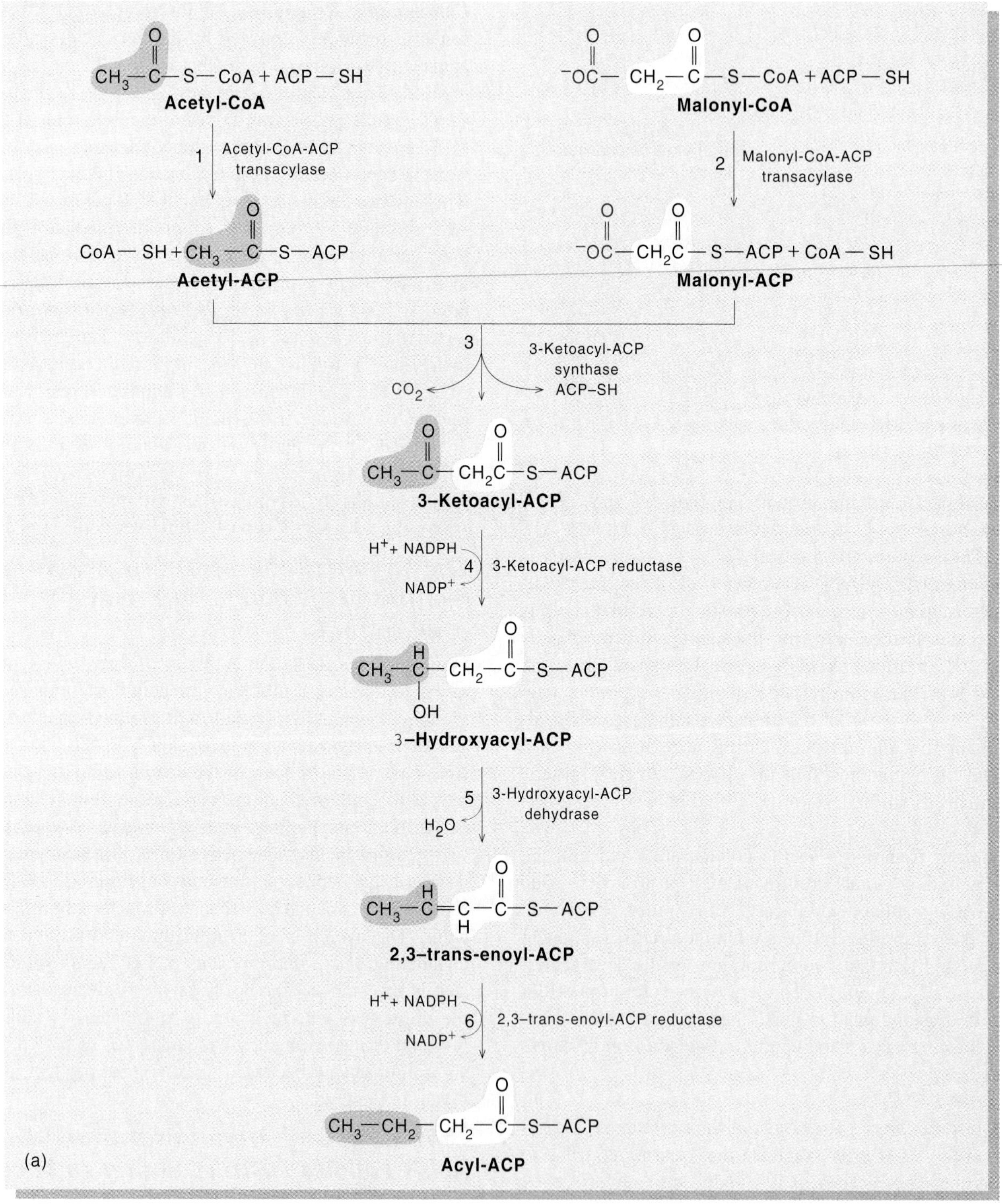

$CH_3-C(=O)-S-CoA + ACP-SH$

Acetyl-CoA

1 Acetyl-CoA-ACP transacylase

$CoA-SH + CH_3-C(=O)-S-ACP$

Acetyl-ACP

$^{-}OC(=O)-CH_2-C(=O)-S-CoA + ACP-SH$

Malonyl-CoA

2 Malonyl-CoA-ACP transacylase

$^{-}OC(=O)-CH_2C(=O)-S-ACP + CoA-SH$

Malonyl-ACP

3 3-Ketoacyl-ACP synthase

CO_2 ACP–SH

$CH_3-C(=O)-CH_2C(=O)-S-ACP$

3-Ketoacyl-ACP

H^++ NADPH 4 3-Ketoacyl-ACP reductase $NADP^+$

$CH_3-CH(OH)-CH_2-C(=O)-S-ACP$

3-**Hydroxyacyl-ACP**

5 3-Hydroxyacyl-ACP dehydrase H_2O

$CH_3-CH=CH-C(=O)-S-ACP$

2,3-trans-enoyl-ACP

H^++ NADPH 6 2,3-trans-enoyl-ACP reductase $NADP^+$

$CH_3-CH_2-CH_2-C(=O)-S-ACP$

(a) **Acyl-ACP**

tial and continuation reactions of fatty acid synthesis are depicted in this part of the figure. The acyl-ACP generated from the first cycle of fatty acid synthesis is condensed with another molecule of malonyl-ACP to yield the 3-keto, 6-carbon intermediate, which is reduced, dehydrated, and reduced to give an acyl-ACP with six carbons. The cycle is then repeated five times with the final yield of an acyl-ACP that contains 16 carbons (palmitoyl-ACP).

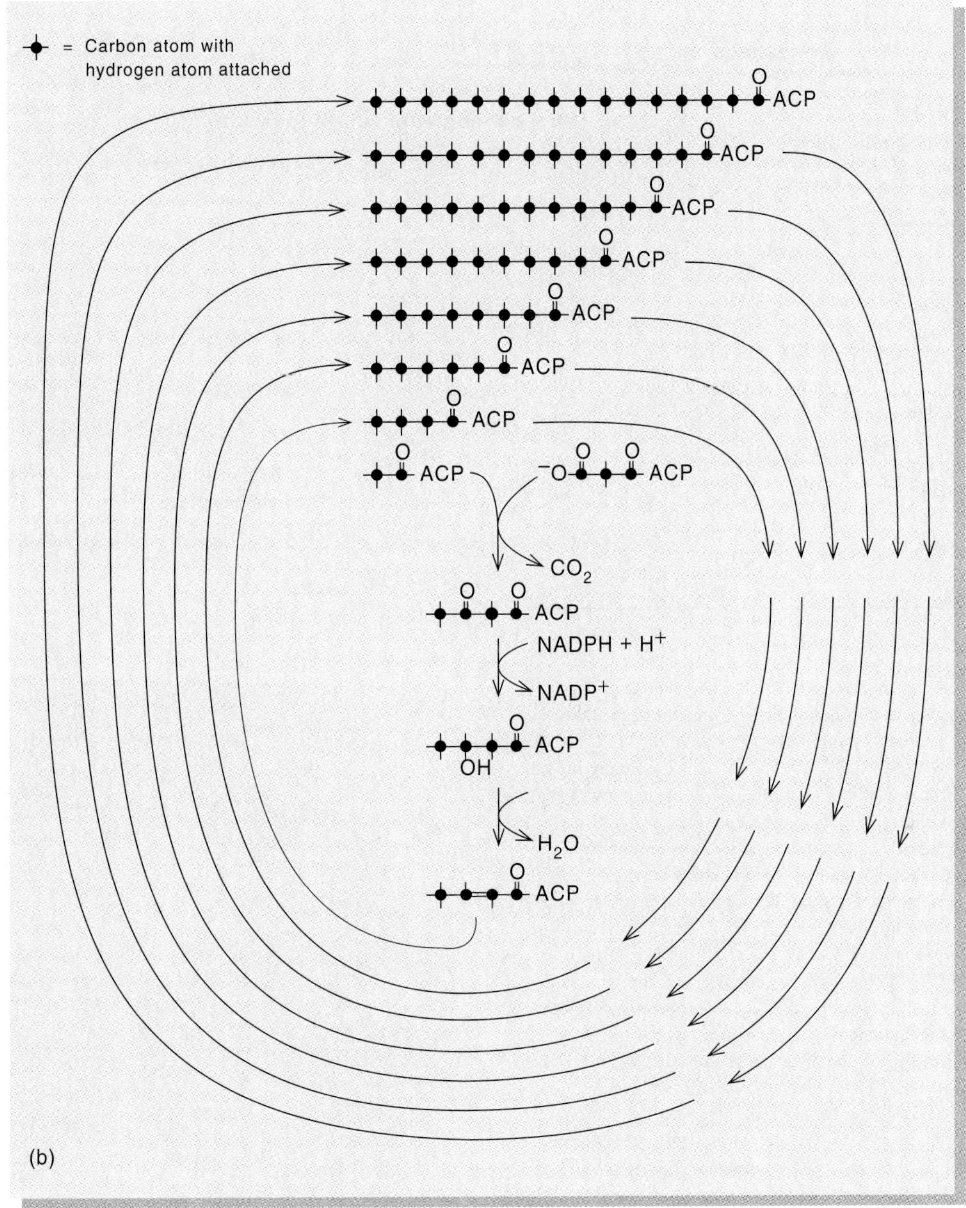

Figure 21.17

The phosphopantetheine group in acyl carrier protein (ACP) and in CoA.

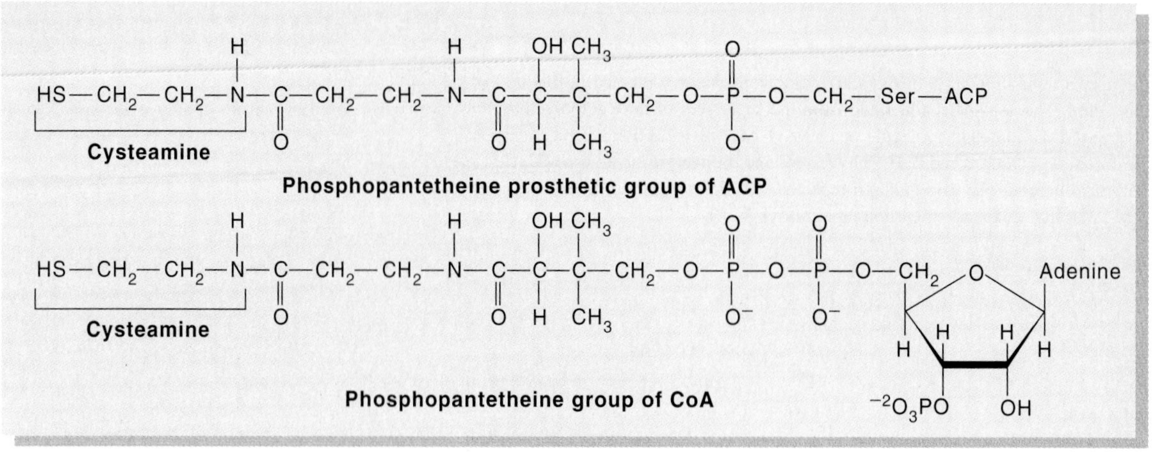

$HS{-}CH_2{-}CH_2{-}NH{-}C(=O){-}CH_2{-}CH_2{-}NH{-}C(=O){-}CH(OH){-}C(CH_3)_2{-}CH_2{-}O{-}P(=O)(O^-){-}O{-}CH_2{-}Ser{-}ACP$

Cysteamine

Phosphopantetheine prosthetic group of ACP

$HS{-}CH_2{-}CH_2{-}NH{-}C(=O){-}CH_2{-}CH_2{-}NH{-}C(=O){-}CH(OH){-}C(CH_3)_2{-}CH_2{-}O{-}P(=O)(O^-){-}O{-}P(=O)(O^-){-}O{-}CH_2$ – ribose ($^{-2}O_3PO$, OH) – Adenine

Cysteamine

Phosphopantetheine group of CoA

Figure 21.18

Formation of acetoacetyl-ACP, catalyzed by 3-ketoacyl-ACP synthase (KSase). The acetyl group is bound to the cysteine residue of KSase (labeled enzyme in the figure). The carbonyl group of the enzyme-bound acetyl is attacked by the central carbon on the malonyl bound to ACP. Finally, the C—S linkage is broken, resulting in acetoacetyl-ACP.

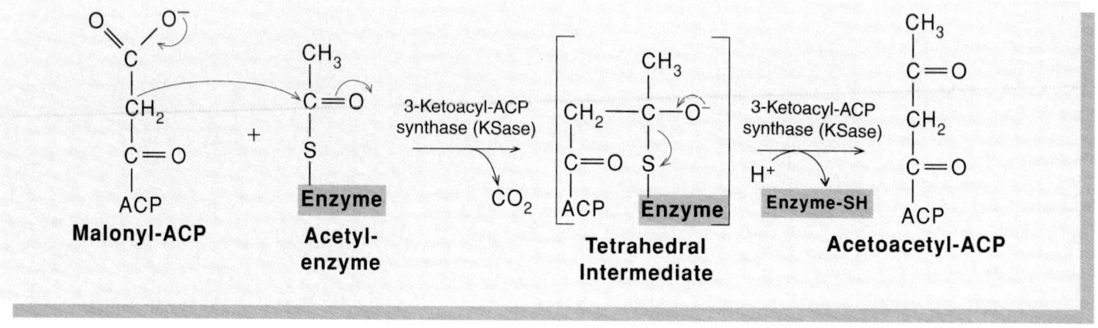

Malonyl-ACP + **Acetyl-enzyme** → (3-Ketoacyl-ACP synthase (KSase); CO_2) → **Tetrahedral Intermediate** → (3-Ketoacyl-ACP synthase (KSase); H^+; Enzyme-SH) → **Acetoacetyl-ACP**

Figure 21.19

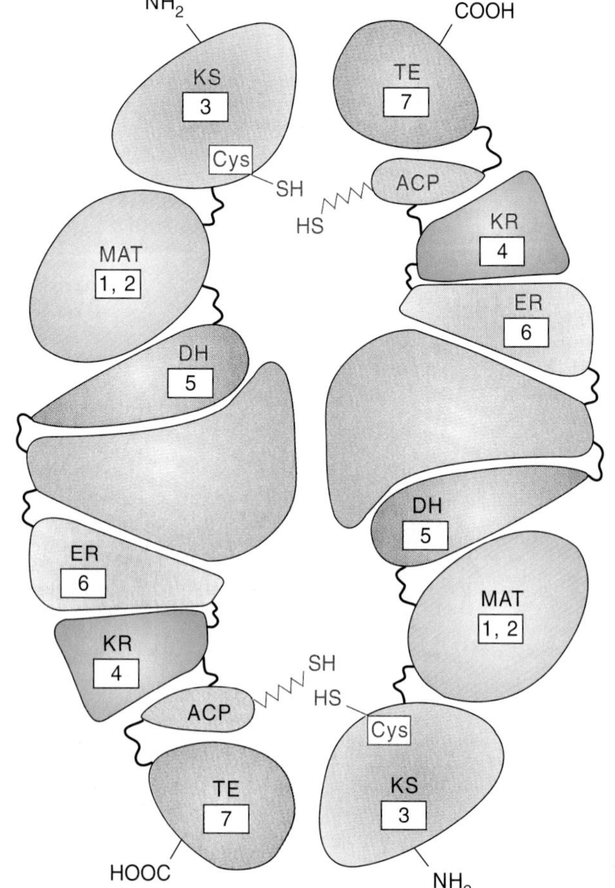

Proposed organization of the enzymatic activities of fatty acid synthase from animal liver. Fatty acid synthase exists as a dimer of two giant identical peptides (M_r = 272,000). Each subunit has one copy of acyl carrier protein (ACP), and each of the enzyme activities involved in fatty acid synthesis is covalently linked. The two peptides are organized in a head-to-tail configuration in such a way that it is possible to make two fatty acid molecules at the same time.

Fatty acid synthesis begins when the substrates, acetyl-CoA and malonyl-CoA, are transferred onto the protein by malonyl-CoA: acetyl-CoA-ACP transacylase (MAT, steps 1 and 2 in fig. 21.16*a*). The numbers in parentheses below the abbreviation of the enzyme in this figure refer to the reactions shown in fig. 21.16. (Whereas *E. coli* has separate enzymes that catalyze the transfer of acetyl- and malonyl-CoA to ACP, both reactions are catalyzed by the same enzymatic activity (MAT) on the animal fatty acid synthase.) Subsequently, 3-ketobutyryl-ACP and CO_2 are formed in a condensation reaction catalyzed by 3-keto-acyl-ACP synthase (KS, step 3 in fig. 21.16*a*). The active site cysteine of KS is represented in the figure by Cys—SH. The 3-ketobutyryl-ACP is reduced to the 3-hydroxy derivative by 3-ketoacyl-ACP reductase (KR, step 4 in fig. 21.16*a*), dehydrated to enoyl-ACP by 3-hydroxylacyl-ACP dehydrase (DH, step 5 in fig. 21.16*a*) and reduced to butyryl-ACP by enoyl-ACP reductase (ER, step 6 in fig. 21.16*a*). The butyryl group is then transferred to the cysteine residue of KS and another malonyl group is transferred to ACP by MAT. Another condensation occurs (step 3 in fig. 21.16*a*) and the 3-ketohexanoyl group is converted to hexanoyl-ACP by the enzymes KR, DH and ER (steps 4, 5, 6 in fig. 21.16*a*). This enzymatic process continues to recycle until palmitoyl-ACP is made. The palmitate is released from the complex by a thioesterase (TE, reaction 7 in fig. 21.16*a*). (From A. K. Joshi and S. Smith, Mutagenesis of the dehydrase domain of fatty acid synthase, *J. Biol. Chem.* 268:22508–22513, 1993. Reprinted by permission.)

21A
BOX

Assaying the Activity of Fatty Acid Synthase

The activity of fatty acid synthase can be assayed by the incorporation of [2-^{14}C]malonyl-CoA or [^{3}H]acetyl-CoA into fatty acid. This scheme illustrates a common principle often utilized in the assay of lipid biosynthetic enzymes. The radioactive substrate is a water-soluble molecule that can easily be separated from the lipid product by extraction of the reaction mixture with an organic solvent such as hexane.

Step 1. Prepare the incubation mixture, which contains:

0.3 mM NADPH
0.1 M phosphate buffer, pH 6.8
5 μM mercaptoethanol
3 μM EDTA
50 μM malonyl-CoA
12 μM [^{3}H]acetyl-CoA (specific radioactivity = 2.0×10^6 dpm/μmole)
Enough distilled H_2O to bring the final volume to 1 ml

Mercaptoethanol is used to keep the SH residues of the enzyme in the reduced state. EDTA chelates divalent cations such as Mg^{2+}, which might inhibit the reaction. All of the materials can be purchased from companies that sell chemical compounds. Radioactive compounds (e.g., [^{3}H]acetyl-CoA) are sold by companies that specialize in the manufacture of radioisotopes.

Step 2. Equilibrate the mixture for 5 min at 37°C in a shaking water bath.

Step 3. Add enzyme (e.g., 2 mg protein), mix thoroughly, and incubate at 37°C for 5 min.

Step 4. Stop the reaction by the addition of 0.1 ml 18% perchloric acid. The perchloric acid lowers the pH to ~1 and thus denatures and inactivates the enzyme.

Step 5. Add 1 ml ethanol and 2 ml hexane, mix thoroughly, and allow the phases to separate. Transfer the upper ether layer to a tube and extract the lower aqueous phase two more times with hexane. The ethanol and water separate from the hexane layer, which floats on the aqueous layer. Because the pH is approximately 1, the fatty acid product is protonated (RCOOH) and can therefore be easily extracted into the ether phase.

Step 6. Evaporate the combined hexane extracts, add liquid scintillation fluid, and determine the radioactivity by liquid scintillation spectrophotometry.

Step 7. Calculate the specific activity of the enzyme. To do this, divide the dpm incorporated into the fatty acid by the specific radioactivity of the acetyl-CoA, the time of the incubation, and the milligrams of protein added to the assay. For example:

$$\text{Specific activity} = \frac{\text{dpm in fatty acid}}{\text{Specific radioactivity of acetyl-CoA} \cdot \text{min} \cdot \text{mg protein}}$$
$$= \frac{50{,}000 \text{ dpm}}{2 \times 10^6 \text{ dpm}/\mu\text{mole} \cdot 5 \text{ min} \cdot 2 \text{ mg protein}}$$
$$= 2.5 \times 10^{-3}\ \mu\text{mole fatty acid formed/min} \cdot \text{mg protein}$$

The specific activity is a measure of the activity of an enzyme as a function of time and amount of protein. When fatty acid synthesis is reduced, for example during a fast, the specific activity will be much lower (about tenfold) than after a carbohydrate-rich meal. The specific activity can also be a measure of the purity of an enzyme preparation. Thus fatty acid synthase in rat liver cytosol might have an activity of 1×10^{-3} μmole fatty acid formed per minute per milligram of protein. By contrast, a pure enzyme, free of all other cytosolic proteins, might have 1000-fold higher specific activity (1 μmole fatty acid formed per minute per milligram of protein).

When the fatty acid synthase is highly purified, it also can be assayed by a spectrophotometric method in which the oxidation of NADPH is followed. As $NADP^+$ is formed, there is a decrease in the absorbance at 340 nm.

---

thesis of unsaturated fatty acids have evolved. Two chemically distinct pathways exist for the introduction of a *cis* double bond into saturated fatty acids: The anaerobic pathway as typified in *E. coli* and the aerobic pathway found in eukaryotes.

As the name "anaerobic" implies, the double bond of the fatty acid is inserted in the absence of oxygen. Biosynthesis of monounsaturated fatty acids follows the pathway described previously for saturated fatty acids until the intermediate β-hydroxydecanoyl-ACP is reached (fig. 21.20). At this point a new enzyme, 3-hydroxydecanoyl-ACP dehydrase, becomes involved. This dehydrase can form the α–β *trans* double bond, and saturated fatty acid synthesis can occur as previously discussed. In addition, this dehydrase is capable of isomerization of the double bond to a *cis* β–γ (or 2,3) double bond as shown in figure 21.20. The β–γ (or 2,3) unsaturated fatty acyl-ACP is subsequently elongated by the normal enzymes of fatty acid synthesis to yield palmitoleoyl-ACP ($16{:}1^{\Delta 9}$). The conversion of this compound to the major unsaturated fatty acid of *E. coli, cis*-vaccenic acid ($18{:}1^{\Delta 11}$), requires a condensing enzyme that we have not previously discussed, 3-ketoacyl-ACP synthase II, which shows a preference for palmitoleoyl-ACP as a substrate. The subsequent conversion to vaccenyl-ACP is catalyzed by the usual enzymes of saturated fatty acid biosynthesis.

In contrast to the anaerobic pathway found in *E. coli,* the aerobic pathway in eukaryotic cells introduces double bonds after the saturated fatty acid has been synthesized. Stearoyl-CoA (18:0) is the major substrate for desaturation. Stearic acid is made by the fatty acid synthase as a minor product, the major product being

Figure 21.20

Anaerobic pathway for biosynthesis of monounsaturated fatty acids in *E. coli.* Synthesis of monounsaturated fatty acids follows the pathway described previously for saturated fatty acids until the intermediate 3-hydroxydecanoyl-ACP is reached. At this point an apparent competition arises between the enzymes involved in saturated and unsaturated fatty acid synthesis. If the β-hydroxydecanoyl-ACP dehydrase isomerizes the double bond to the 2,3 (or β, γ) position, then this product is elongated as shown to give rise to the monounsaturated fatty acids $16:1^{\Delta 9}$ and $18:1^{\Delta 11}$.

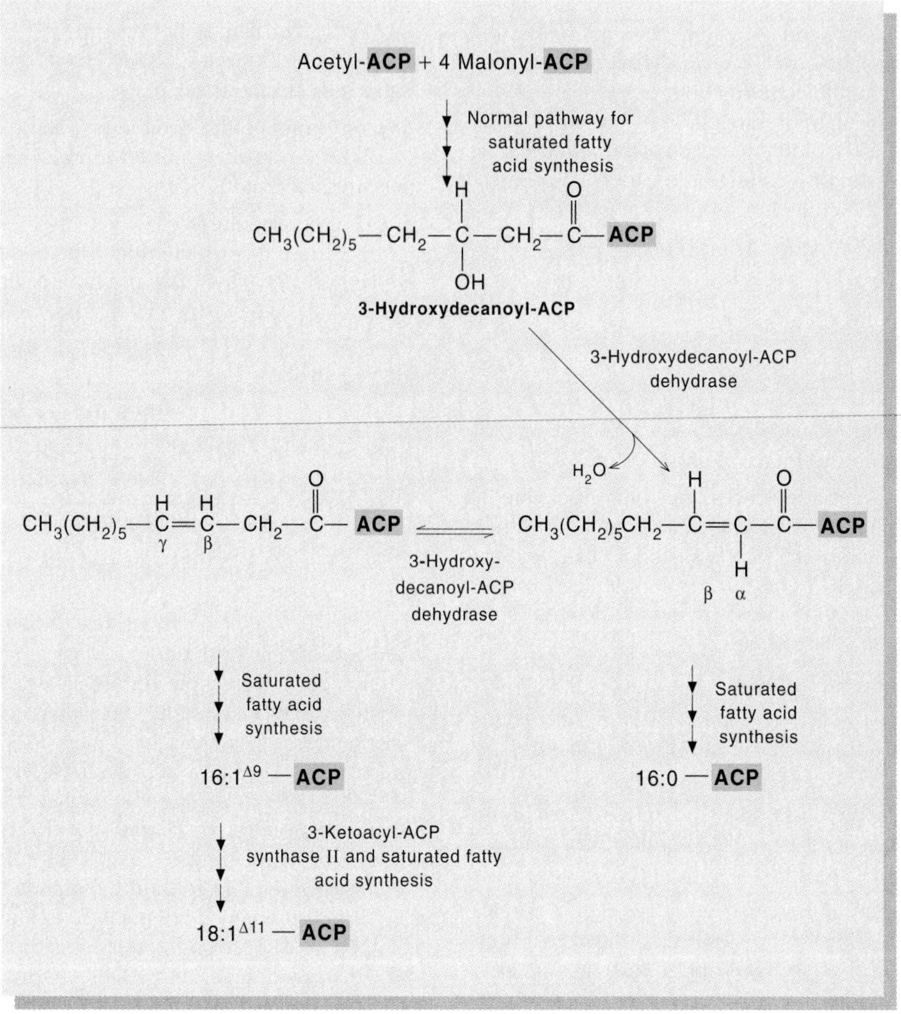

palmitic acid, and is activated to its CoA derivative by acyl-CoA synthase. In eukaryotic cells an enzyme complex associated with the endoplasmic reticulum desaturates stearoyl-CoA to oleoyl-CoA ($18:1^{\Delta 9}$) (fig. 21.21). This remarkable reaction requires NADH and O_2 and results in the formation of a double bond in the middle of an acyl chain with no activating groups nearby. The chemical mechanism for desaturation of long-chain acyl-CoAs remains unclear.

Desaturation requires the cooperative action of two enzymes: cytochrome b_5 reductase and stearoyl-CoA desaturase, in addition to an electron carrier protein, cytochrome b_5. A scheme for this set of reactions is shown in figure 21.21. Cytochrome b_5 reductase is a flavoprotein that transfers electrons from NADH by means of flavin (F) to cytochrome b_5, a heme-containing protein in which Fe^{3+} can be reduced to Fe^{2+}. Stearoyl-CoA desaturase utilizes two electrons from cytochrome b_5 coupled with an atom of oxygen to form a *cis* double bond in the Δ^9 position of stearoyl-CoA.

Biosynthesis of Polyunsaturated Fatty Acids Occurs Mainly in Eukaryotes

E. coli does not have polyunsaturated fatty acids, whereas eukaryotes produce a large variety of polyunsaturated fatty acids. Mammals can desaturate only between the Δ^9 position and the carboxyl end of an acyl chain, whereas plants have the enzymes to desaturate at positions Δ^9, Δ^{12}, and Δ^{15}. Mammals need linoleic acid ($18:2^{\Delta 9,12}$) and linolenic acid ($18:3^{\Delta 9,12,15}$) and are unable to make them. Hence these are considered essential fatty acids in our diet. One derivative of linoleic acid, arachidonic acid ($20:4^{\Delta 5,8,11,14}$), is the biosynthetic precursor of the eicosanoids, which are discussed in chapter 22. A derivative of linolenic acid that has six double bonds, docosahexaenoic acid ($22:6^{\Delta 4,7,11,14,17,20}$), is found in abundance in the phospholipids of the retinal membrane of the eye. The function of this highly unsaturated fatty acid in the retina is unknown. Both arachidonic acid and docosahex-

Figure 21.21

The aerobic pathway for formation of oleoyl-CoA in eukaryotes. In eukaryotes the double bonds are introduced after the C_{16} and C_{18} saturated fatty acid have been synthesized. F stands for flavin.

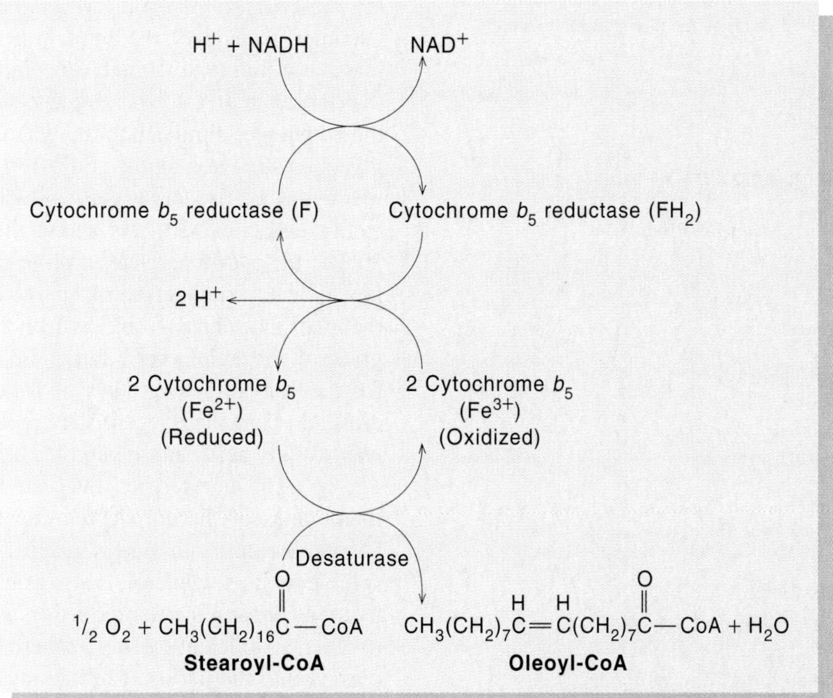

aenoic acid are made in the liver from the respective essential fatty acid precursors, linoleic ($18:20^{\Delta 9,12}$) and linolenic ($18:20^{\Delta 9,12,15}$) acids.

Enzyme complexes occur in the endoplasmic reticulum of animal cells that desaturate at Δ^5 if there is a double bond at the Δ^8 position or at Δ^6 if there is a double bond at the Δ^9 position. These enzymes are different from each other and from the Δ^9-desaturase discussed in the previous section, but the Δ^5 and Δ^6 desaturases do appear to utilize the same cytochrome b_5 reductase and cytochrome b_5 mentioned previously. Also present in the endoplasmic reticulum are enzymes that elongate saturated and unsaturated fatty acids by two carbons. As in the biosynthesis of palmitic acid, the fatty acid elongation system uses malonyl-CoA as a donor of the two-carbon unit. A combination of the desaturation and elongation enzymes allows for the biosynthesis of arachidonic acid and docosahexaenoic acid in the mammalian liver. As an example, the pathway by which linoleic acid is converted to arachidonic acid is shown in figure 21.22. Interestingly, cats are unable to synthesize arachidonic acid from linoleic acid. This may be why cats are carnivores and depend on other animals to make arachidonic acid for them. Also note that the elongation system in the endoplasmic reticulum is important for the conversion of palmitoyl-CoA to stearoyl-CoA.

Summary of the Pathways for Synthesis and Degradation

Before discussing the specific aspects of regulation of fatty acid metabolism, let us review the main steps in fatty acid synthesis and degradation. Figure 21.23 illustrates these processes in a way that emphasizes the parallels and differences. In both cases, two-carbon units are involved. However, different enzymes and coenzymes are utilized in the biosynthetic and degradative processes. Moreover, the processes take place in different compartments of the cell. The differences in the location of the two processes and in the enzymes used make it possible to regulate the two pathways independently.

Regulation of Fatty Acid Metabolism

In light of the principles of pathway regulation discussed in chapter 12, you can anticipate that controls must exist to ensure that fatty acid synthesis and breakdown do not occur at the same time. When an organism has satisfied its immediate energy needs and the limited storage space for glycogen has been filled, most nutrients are directed toward fatty acid synthesis. A virtually unlimited amount of fat can be stored in an animal's body. The last section of this chapter considers the mechanisms employed by animals to regulate the utilization of fatty acids as an energy source.

The Release of Fatty Acids from Adipose Tissue Is Regulated

There are a number of circumstances in which animals call on their energy reserve in adipose (fat) tissue. For example, during fasting or starvation, mobilization of fatty acids from adipose tis-

Figure 21.22

Synthesis in mammalian tissues of arachidonic acid from linoleic acid. The Δ^5 and Δ^6 desaturases are separate enzymes and are also different from the Δ^9 desaturase (fig. 21.21). The mechanisms, however, seem to be the same, involving cytochrome b_5 and cytochrome b_5 reductase. The enzymes for elongation of unsaturated fatty acid such as 18:3 to 20:3 occur on the endoplasmic reticulum.

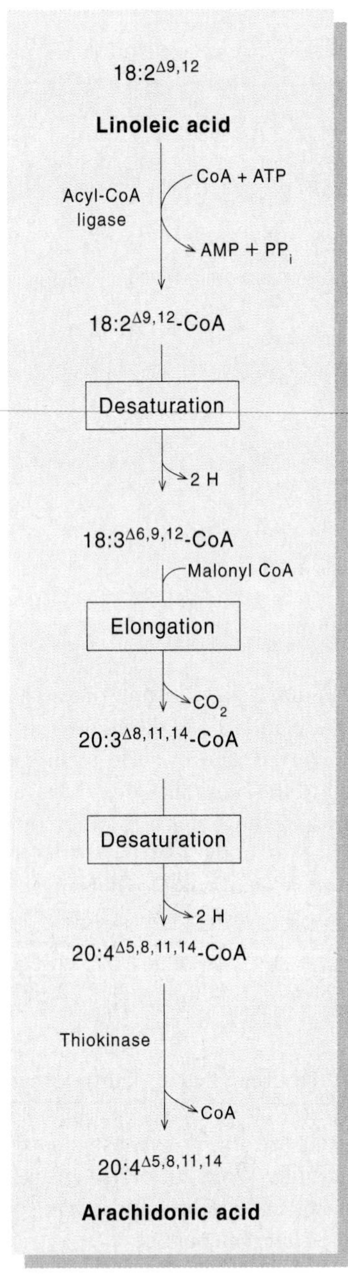

sue is an important source of energy. Prolonged work or exercise also promotes the release of fatty acids from fat tissue. The fatty acids are stored in adipose tissue as components of triacylglycerols. Although triacylglycerols are found in liver, intestine, and other tissues, they are primarily found in adipose (fat) tissue, which functions as the main storage depot of triacylglycerols. In a 70-kg human, 135,000 kcal of energy is stored as triacylglycerols in adipose compared with 450 kcal stored in the liver. The specialized cell in adipose tissue for lipid storage is the adipocyte (fig. 21.24). The cytoplasm of the adipocyte is packed with vesicles that are rich in triacylglycerols and that serve as the long-term energy reserves in mammals. When the energy supply from the diet becomes inadequate, the animal responds to the deficiency by release of a hormonal signal, especially epinephrine and glucagon, that is transmitted to the target tissue. These hormones work on the adipocytes similarly to the way they work on cells containing glycogen storage granules. First, the hormones bind to the plasma membrane of the target cells, which stimulates the synthesis of cyclic AMP (cAMP). As shown in figure 21.25 (also see fig. 20.12), the cAMP activates a protein kinase that phosphorylates a key enzyme, triacylglycerol lipase. The lipase, which is active in the phosphorylated form, hydrolyzes triacylglycerol to diacylglycerol with release of a fatty acid. This reaction is rate-limiting for the complete hydrolysis of triacylglycerols. The diacylglycerols and monoacylglycerols are rapidly hydrolyzed the rest of the way to fatty acids and glycerol.

The unesterified fatty acids move through the plasma membranes of the adipocytes into the bloodstream, where they become bound to the major plasma protein, albumin. The water-soluble product, glycerol, is also released into plasma and removed by the liver for glucose production. Albumin carries the fatty acids to energy-deficient tissues, where fatty acids move from the plasma into the tissues. Cardiac muscle utilizes fatty acids as the major oxidative source of energy for ATP synthesis and therefore removes large amounts of fatty acids from the circulation. At the other extreme the brain does not use fatty acids as a major source of energy but depends mostly on glucose and, to a lesser extent, ketone bodies.

Fatty Acid-Binding Proteins and Acyl-CoA-Binding Protein May Be Important in the Intracellular Trafficking of Fatty Acids

Because of their hydrophobic tail, fatty acids preferentially associate with membrane lipids. How do the fatty acids move from the plasma membrane to the mitochondria and other organelles in the cell? The answer is not known, but fatty acid–binding proteins are major candidates for this job. These are small proteins of 127–132 amino acids that bind long-chain (16–20 carbons) saturated and unsaturated fatty acids to a single binding site with a K_d of 1–4 μM. The rat liver fatty acid binding protein ($M_r = 14{,}273$) is an abundant cytosolic protein and accounts for 3–5% of the mass of cytosolic proteins. Approximately 60% of the long-chain fatty acids found in liver cytosol are bound to this protein.

Although the fatty acid–binding proteins also bind acyl-CoAs, another cytosolic protein has recently been identified that binds acyl-CoAs with greater avidity. The acyl-CoA–binding protein ($M_r = 9{,}938$) does not bind unesterified fatty acids. This protein is present at lower concentrations (approximately 0.5% of cytosolic protein) than fatty acid–binding protein and is a candidate for the transfer of acyl-CoAs between organelles in cells.

Figure 21.23

Comparison between synthesis and degradation of fatty acids in the liver. Both processes involve two carbons at a time and very similar intermediates, even though they go in opposite directions. CoA is also heavily involved in both processes. Here the similarities end. The enzymes used in the two processes are totally different, and the coenzymes are also different. In the degradative direction, FAD and NAD^+ are used, whereas in the synthetic direction the coenzyme NADPH is used. Degradation occurs in the mitochondrial matrix, and synthesis occurs in the cytosol.

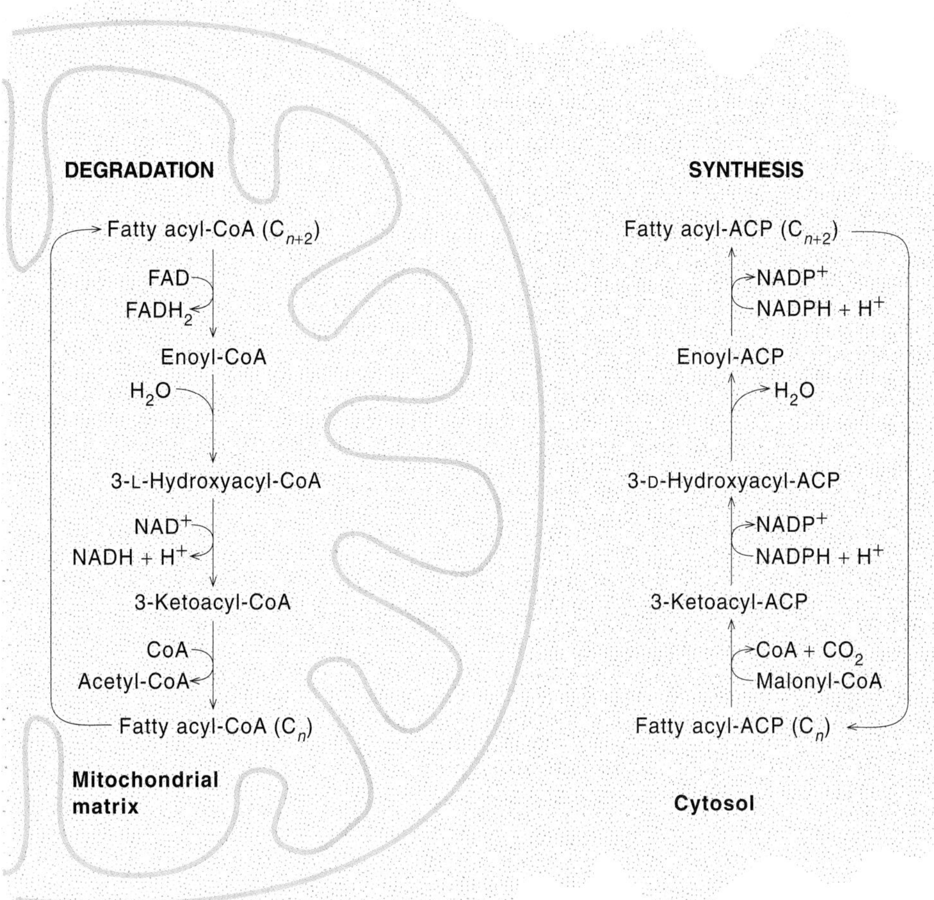

Figure 21.24

Scanning electron micrograph of white adipocytes from rat adipose tissue (600×). (Courtesy of Dr. A. Angel, University of Manitoba, and Dr. M. J. Hollenberg of the University of British Columbia.)

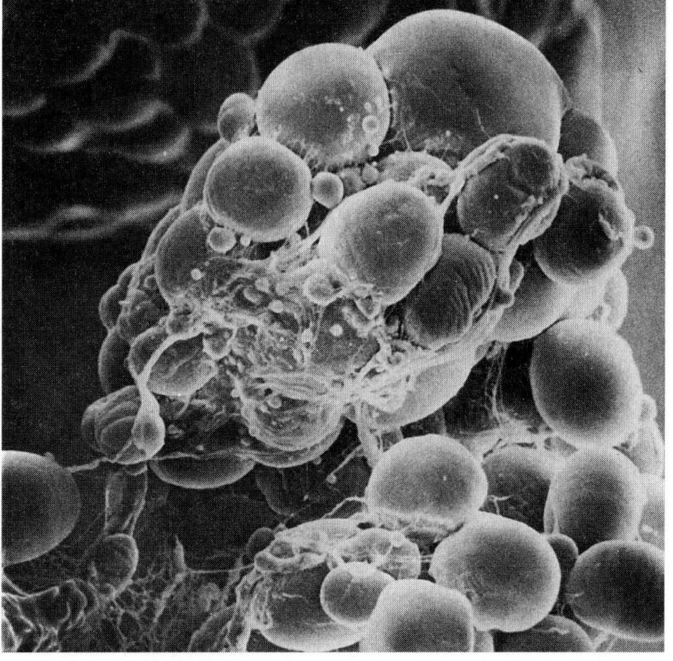

Figure 21.25

When certain hormones (e.g., epinephrine) bind to their receptors in adipose tissue, adenylate cyclase is activated. The cAMP that is formed activates protein kinase A, which phosphorylates triacylglycerol lipase. The phosphorylated form of this enzyme is the active species, and triacylglycerols are degraded to fatty acids. The fatty acids are released into the bloodstream, bound by albumin, and delivered to energy-deprived tissues.

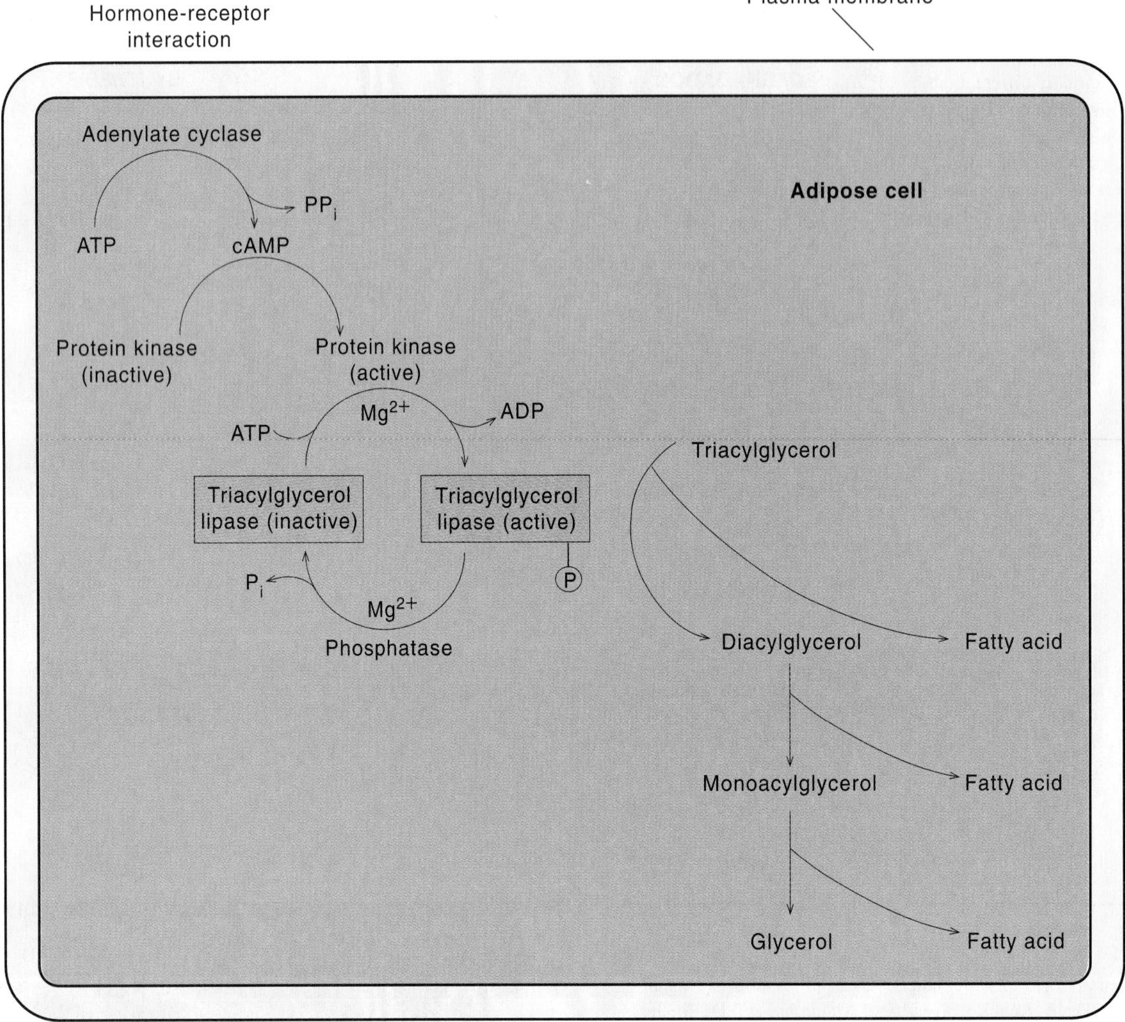

Transport of Fatty Acids into Mitochondria Is Regulated

The level of fatty acids in the bloodstream is one factor that dictates how much fatty acid is delivered to a tissue. Further controls are necessary in the cell to regulate the utilization of the fatty acid, and this depends on the needs of a particular cell type. The heart requires a great deal of energy, so the fatty acids delivered to this organ are largely directed to oxidation, whereas the liver may require fatty acids primarily for membrane lipid and triacylglycerol synthesis. A major control point, therefore, is entry of the fatty acids into the mitochondrion where degradation occurs.

Fatty acids taken into cells are first activated in the cytosol by reaction with CoA and ATP to yield fatty acyl-CoA in a reaction catalyzed by acyl-CoA ligase:

$$\text{RCOO}^- + \text{ATP} + \text{CoA} \xrightarrow{\text{Mg}^{2+}} \text{RCOCoA} + \text{PP}_i + \text{AMP} \quad \textbf{(5)}$$

The acyl-CoAs can penetrate the outer membrane of the mitochondria through a pore but cannot be transported into mitochondria, where β oxidation occurs. They must first be converted to their carnitine derivatives, which can be transported across the inner membrane of the mitochondria:

$$\text{RCOCoA} + \underset{\text{Carnitine}}{(\text{CH}_3)_3\text{N}^+\text{—CH}_2\underset{\substack{|\\ \text{OH}}}{\text{C}}\text{HCH}_2\text{COO}^-} \rightleftharpoons$$

$$\underset{\text{Acyl-carnitine}}{(\text{CH}_3)_3\text{N}^+\text{—CH}_2\underset{\substack{|\\ \text{O}\\ |\\ \text{RC=O}}}{\text{C}}\text{HCH}_2\text{COO}^-} + \text{CoA} \quad \textbf{(6)}$$

This reaction is catalyzed by carnitine acyltransferase I on the

Figure 21.26

Acyl-CoA is not transported across the inner membrane of the mitochondrion. Instead, the acyl-CoA reacts with carnitine to yield the acyl-carnitine derivative. This reaction is catalyzed by carnitine acyltransferase I, which is located on the outer mitochondrial membrane. The acyl-carnitine is transported across the inner membrane by a specific carrier protein. Once inside the matrix of the mitochondrion, the acyl-carnitine is converted back to its acyl-CoA derivative, the substrate for the start of β oxidation. This reaction is catalyzed by carnitine acyltransferase II, which is located on the mitochondrial inner membrane. Note that acyltransferases I and II are oriented in their respective membranes so that the reactions they catalyze occur in the intermembrane space and the mitochondrial matrix, respectively. The carnitine is also transferred by the carrier protein.

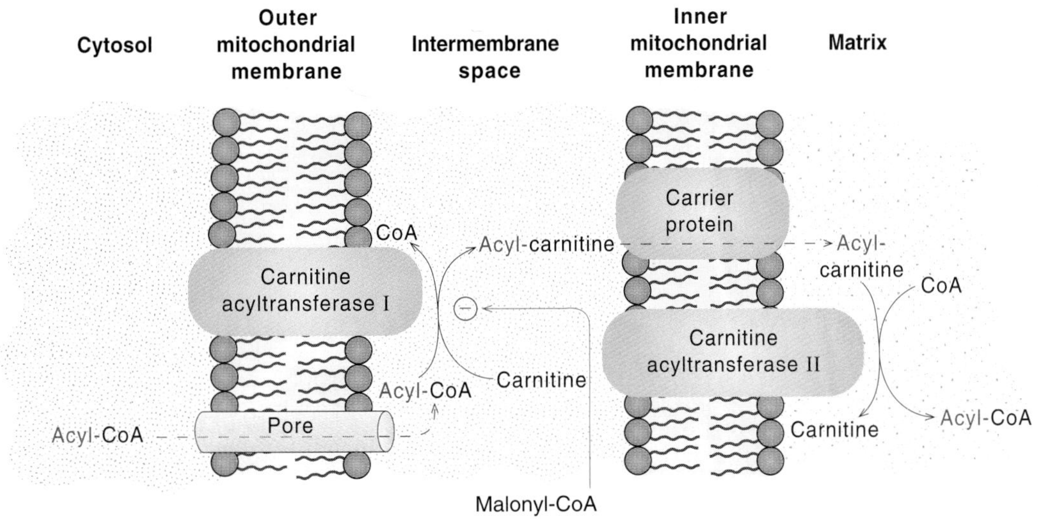

outer membrane (fig. 21.26). A protein carrier in the inner mitochondrial membrane transfers the acyl-carnitine derivatives across the membrane. Once inside the mitochondria, the reaction is reversed by carnitine acyltransferase II to yield a fatty acyl-CoA (see fig. 21.26). Thus at least two distinct pools of acyl-CoA occur in the cell, one in the cytosol and the other in the mitochondrion.

This elaborate chain of reactions provides a number of possible points for regulation of the supply of acyl-CoAs for oxidation. The major control point is carnitine acyltransferase I, which is strongly inhibited by malonyl-CoA. Recall that malonyl-CoA is a major substrate for fatty acid biosynthesis. Thus high levels of malonyl-CoA, which indicate fatty acid synthesis is in progress, prevent fatty acid catabolism, thus avoiding a futile cycle.

Fatty Acid Biosynthesis Is Limited by Substrate Supply

Just as fatty acid oxidation is limited by substrate supply, so is fatty acid synthesis. The overall equation for fatty acid biosynthesis indicates a need for acetyl-CoA, malonyl-CoA, and NADPH. Since malonyl-CoA is derived from acetyl-CoA catalyzed by the acetyl-CoA carboxylase reaction, we can think of acetyl-CoA as the main substrate for fatty acid synthesis. Acetyl-CoA and NADPH are generated from substrates that originate in the glycolytic pathway and TCA cycle. Thus glycolysis generates pyruvate, and in the mitochondria the pyruvate is converted to acetyl-CoA as well as oxaloacetate (fig. 21.27). The mitochondrial membrane is impermeable to acetyl-CoA, which is therefore converted to citrate and transported into the cytosol. There citrate is converted to acetyl-CoA and oxaloacetate by ATP:citrate lyase. The end result is that an intermediate in the glycolytic degradation of glucose, pyruvate, is converted to acetyl-CoA in the cytosol and can be used for fatty acid biosynthesis.

Fatty Acid Synthesis Is Regulated by the First Step in the Pathway

An abundance of acetyl-CoA, which would be produced after a carbohydrate-rich meal, creates the possibility for fatty acid synthesis. Since acetyl-CoA can be used in a variety of ways (e.g., ketone body synthesis and, as we examine in chapter 23, cholesterol biosynthesis), it is essential that specific controls exist to regulate the synthesis of fatty acids. In fact, multiple types of control have been discovered for the initial step in fatty acid biosynthesis catalyzed by acetyl-CoA carboxylase.

We already learned that citrate activates the liver acetyl-CoA carboxylase and converts it to a high-molecular-weight polymer (see fig. 21.15). The activation by citrate is appropriate because excess cytosolic citrate is a good indication that the energy needs of the cell are satisfied. Furthermore, citrate is the most important source of cytosolic acetyl-CoA as described above and in figure 21.27.

The acetyl-CoA carboxylase is inhibited by palmitoyl-CoA (see figs. 21.15 and 21.27). This control is a classic example of end-product inhibition since palmitoyl-CoA is the final product in fatty acid synthesis that is initiated by acetyl-CoA carboxylase.

Acetyl-CoA carboxylase is also regulated by a chain of reactions promoted by the hormones glucagon and epinephrine. Remember that high levels of these hormones serve as a general signal that energy is in short supply or soon may be needed in large amounts (see chapters 12 and 27). Under these conditions it

Figure 21.27

Overview of the conversion of carbohydrate to lipid in rat liver cells and its regulation. Red arrows with pluses and minuses indicate points of activation and inhibition, respectively. Glucagon and epinephrine are the main hormones involved in regulation. Citrate and palmitoyl-CoA are the main substrates involved in regulation. The dashed black arrow indicates compound that passes across the inner mitochondrial membrane.

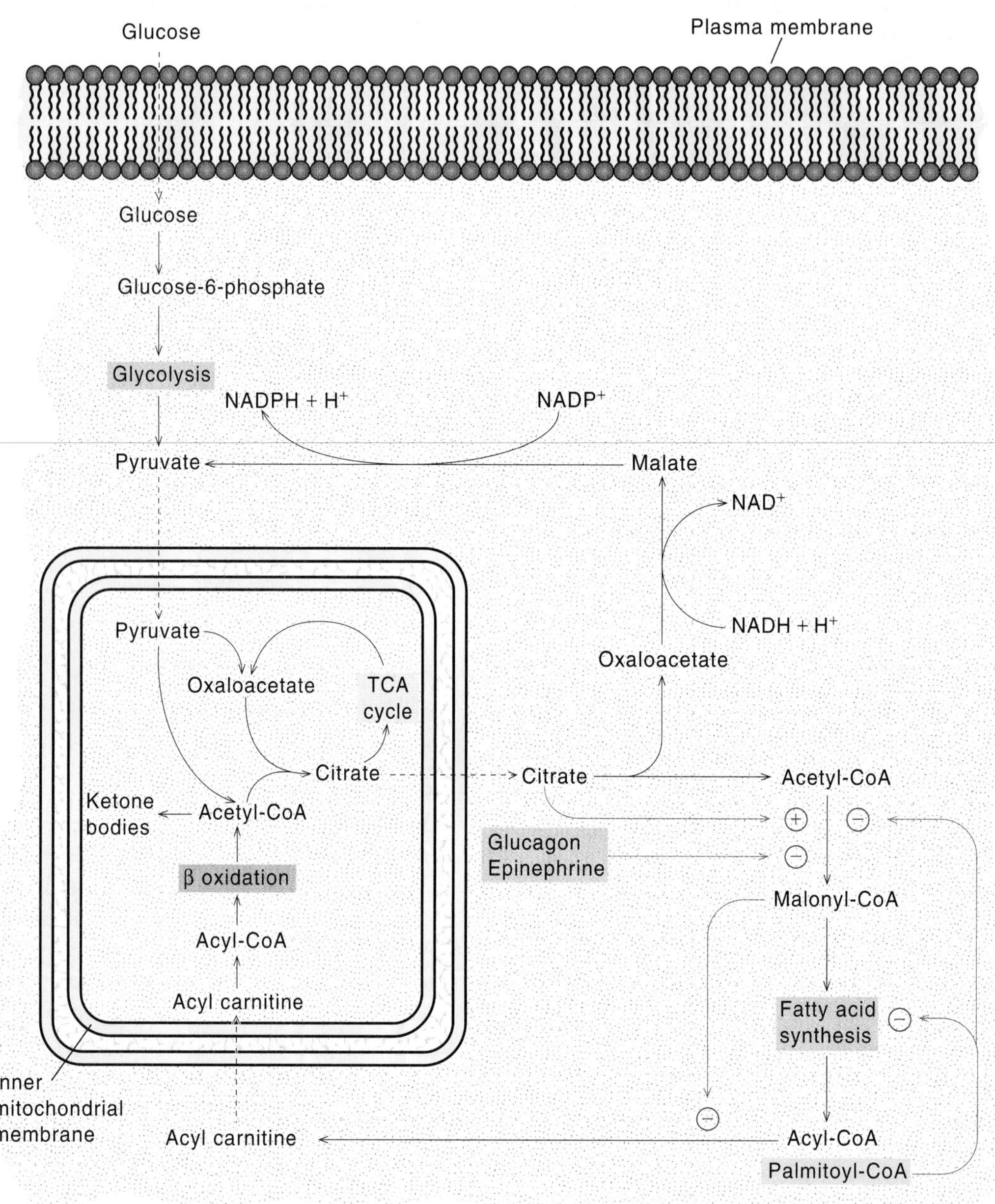

would be inappropriate to divert energy to the synthesis of fatty acids. These hormones bind to receptors on the plasma membrane and stimulate a kinase that phosphorylates and inactivates acetyl-CoA carboxylase (fig. 21.28).

Initially, it seemed that the enzyme that phosphorylates and inactivates acetyl-CoA carboxylase was cAMP-dependent protein kinase. It is now clear that the kinase that phosphorylates and inactivates acetyl-CoA carboxylase is AMP-activated protein kinase (fig. 21.28), and this kinase is regulated by a kinase kinase. The cAMP-dependent kinase is involved, since it appears to phosphorylate and activate the kinase kinase. Interestingly, this kinase kinase also appears to be activated by long-chain acyl-CoA. Thus palmitoyl-CoA, for example, inactivates acetyl-CoA carboxylase not only directly (see fig. 21.15) but also indirectly through the kinase kinase (see fig. 21.28). Insulin, which often opposes the action of glucagon, activates the carboxylase. Recent results suggest that the insulin effect might be due to an inhibition of the AMP-activated kinase by an unknown mechanism. Two phosphatases

Figure 21.28

Regulation of acetyl-CoA carboxylase by phosphorylation/dephosphorylation. In its inactive form, acetyl-CoA carboxylase is phosphorylated on several sites. The action of protein phosphatase 1C or 2A can dephosphorylate the enzyme to an active form. The active form can be phosphorylated back to its inactive form by AMP-activated protein kinase. The cell can regulate the state of enzyme phosphorylation by control of the activity of AMP-activated kinase, which is regulated via the action of a kinase kinase. Note that, in contrast to the carboxylase, the active form of the AMP-activated kinase is phosphorylated. Incubation of cells with glucagon or epinephrine promotes an activation of the kinase kinase, and there is evidence that this is mediated by the phosphorylation of the kinase kinase by cAMP-dependent protein kinase. On the other hand, insulin promotes the activation of the carboxylase by decreasing the AMP-activated kinase. It is also known that the kinase kinase can be activated by fatty acyl-CoA. Thus when there is a requirement for energy (signaled by glucagon or epinephrine) or an excess of fatty acids (e.g., after a fatty meal), the regulated enzyme in fatty acid biosynthesis, acetyl-CoA carboxylase, is inactivated.

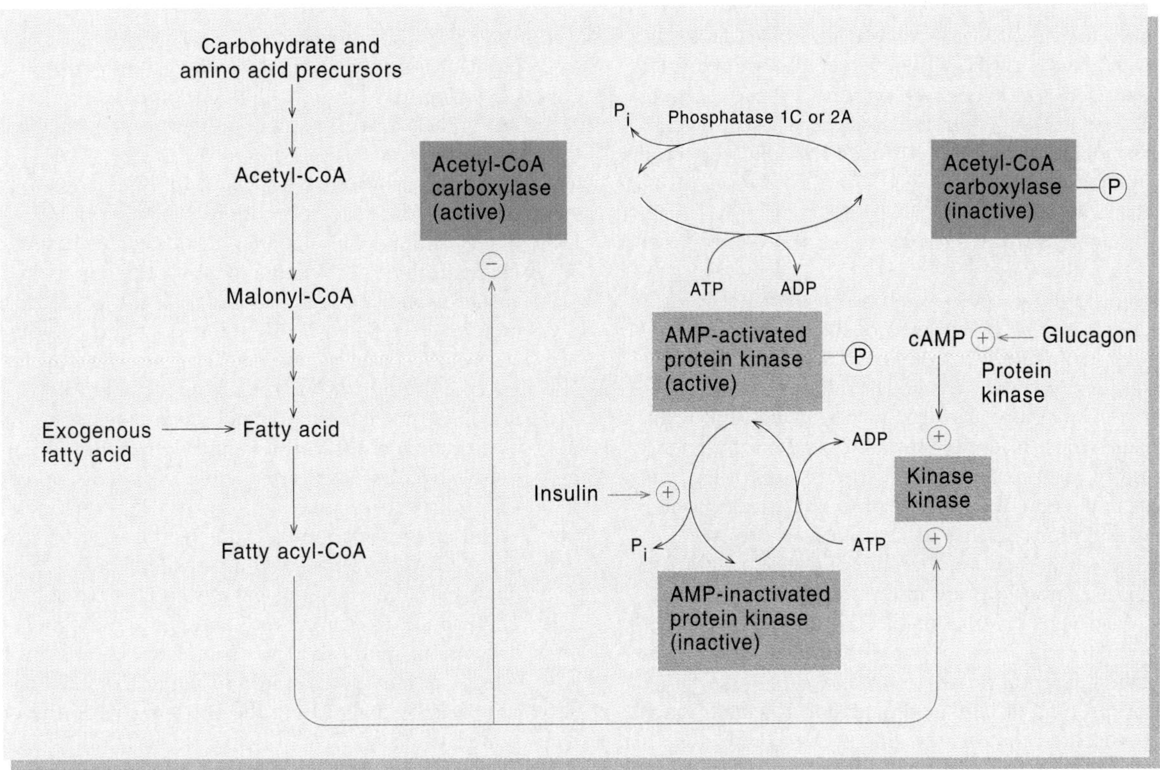

have been implicated in the dephosphorylation of acetyl-CoA carboxylase (see fig. 21.28).

The Controls for Fatty Acid Metabolism Discourage Simultaneous Synthesis and Breakdown

The controls for fatty acid metabolism satisfy most of our expectations for what control systems should do. One of the most important aspects of control is that simultaneous synthesis and breakdown should be discouraged; otherwise, energy would be wasted. Two major hormones are implicated in fatty acid breakdown: glucagon and epinephrine. They encourage mobilization of fatty acids from adipose tissue and simultaneously discourage fatty acid synthesis by inhibiting the formation of malonyl-CoA from acetyl-CoA (see fig. 21.28). Hence the main effect of glucagon and epinephrine is to stimulate breakdown while inhibiting synthesis.

Similarly, factors that stimulate acetyl-CoA carboxylase, the first enzyme in the pathway for fatty acid synthesis, also discourage fatty acid catabolism. This dual effect occurs because the first enzyme in the pathway leads to the formation of malonyl-CoA, which is a potent inhibitor of carnitine acyltransferase I. This inhibition prevents the transport of fatty acids into the mitochondrion, thereby preventing fatty acid breakdown.

Another point to note about the controls of fatty acid metabolism is that they are designed to meet the metabolic needs of the organism. Thus an excess of palmitoyl-CoA in the cytosol signals a shutdown of fatty acid synthesis by inhibiting acetyl-CoA carboxylase. Similarly, an abundance of citrate indicates that the energy needs of the organism are being met and that it is an appropriate time to synthesize fatty acids for energy storage. Citrate accomplishes this by supplying the substrate for fatty acid synthesis, acetyl-CoA, and activates the first enzyme in the pathway, acetyl-CoA carboxylase.

Long-Term Dietary Changes Lead to Adjustments in the Level of Enzymes

Before closing we should point out that, over an extended period, dietary conditions can alter the levels of enzymes involved in fatty acid metabolism. For example, the concentrations of fatty acid synthase and acetyl-CoA carboxylase in rat liver are reduced four- to fivefold after fasting. When a rat is fed a fat-free diet, the concentration of fatty acid synthase is 14-fold higher than in a rat maintained on standard rat chow diet. Current evidence indicates that the levels of these enzymes are governed by the rate of enzyme synthesis, not degradation. It appears that synthesis of the enzyme, in turn, is controlled by the rate of transcription of DNA into mRNA. A question of current interest is how this transcription of DNA is regulated. Several laboratories have identified sequences of DNA in the promoter region of genes involved in fatty acid synthesis and proteins (called transcription factors) that modulate the expression of these genes.

Summary

In this chapter we focused on the synthesis and degradation of long-chain fatty acids and on how these processes are regulated. Most of our discussion was concerned with how these reactions take place in the mammalian liver, although occasionally we referred to other animal tissues and to *E. coli*. The following points are the most important.

1. Fatty acids originate from two sources: diet and de novo synthesis.
2. The degradation of fatty acids occurs by an oxidation process in the mitochondria. The breakdown of the 16-carbon saturated fatty acid, palmitate, occurs in blocks of two carbon atoms by a cyclical process. The active substrate is the acyl-CoA derivative of the fatty acid. Each cycle involves four discrete enzymatic steps. In the process of oxidation the energy is sequestered in the form of reduced coenzymes of FAD and NAD^+. These reduced coenzymes lead to ATP production through the respiratory chain. The oxidation of fatty acids yields more energy per carbon than the oxidation of glucose because saturated fatty acids are in the fully reduced state.
3. Unsaturated fatty acids are also oxidized in mitochondria with the help of certain additional enzymes that facilitate a continuous flow of the oxidation process.
4. The main end product of fatty acid oxidation is acetyl-CoA, which can be used by the tricarboxylic acid cycle for generation of energy. Alternatively, ketone bodies may be formed from condensation of acetyl-CoAs. Ketone bodies are made in the liver and subsequently diffuse into the blood to be carried to other tissues, where they are converted into acetyl-CoA for various metabolic purposes.
5. Fatty acid biosynthesis also occurs in steps of two carbon atoms. Biosynthesis takes place in the cytosol. In addition to occurring in a different cellular compartment from degradation, biosynthesis involves totally different enzymes and different coenzymes.
6. Fatty acid synthesis takes place in eight steps. All except the first step take place on a multienzyme complex. The intermediates on this complex are carried by attachment of the acid group in thioester linkage to phosphopantetheine of the acyl carrier protein (ACP). The multienzyme complex greatly increases the efficiency of fatty acid synthesis, because for each step in the pathway the next enzyme is always near at hand, and the dilution of intermediates is minimized.
7. Regulation of fatty acid metabolism takes place in such a way that simultaneous synthesis and degradation are minimized. Control factors ensure that synthesis occurs primarily when energy is in excess and degradation when energy is needed. When energy is abundant, the synthesis of malonyl-CoA by acetyl-CoA carboxylase stimulates fatty acid synthesis and inhibits carnitine acyltransferase I and, as a result, β oxidation. The hormones epinephrine and glucagon stimulate degradation and inhibit synthesis. They stimulate degradation by facilitating release of fatty acids stored in the adipocytes. They inhibit synthesis by inactivating the first enzyme in the pathway for synthesis, acetyl-CoA carboxylase. Most other control factors interfere with the supply of substrate for either of the two processes.

Selected Readings

Cook, H. W., Fatty acid denaturation and chain elongation in eucaryotes. In D. E. Vance and J. E. Vance (eds.), *Biochemistry of Lipids, Lipoproteins and Membranes*. Amsterdam: Elsevier Science Publishers, 1996. Provides an advanced and current treatment of fatty acid desaturation and its regulation, and cites other key references to this field.

Deuel, H. J., *The Lipids, Biochemistry*, vol. 3. New York: Interscience, 1957. A comprehensive and classical treatise on the biochemistry of lipids until the mid-1950s.

McGarry, J. D., and D. W. Foster, Regulation of hepatic fatty acid oxidation and ketone body production. *Ann. Rev. Biochem.* 49:395, 1980. A now classic review article that summarizes the evidence for the regulation of β oxidation by malonyl-CoA.

Rock, C. O., S Jackowski, and J. E. Cronan, Lipid metabolism in procaryotes. In D. E. Vance and J. E. Vance (eds.). *Biochemistry of Lipids, Lipoproteins and Membranes*. Amsterdam: Elsevier Science Publishers, 1996. Contains current and advanced information on the metabolism of fatty acids in *E. coli* and other prokaryotes.

Salati, L. M., and A. G. Goodridge, Fatty acid synthesis in eucaryotes. In D. E. Vance and J. E. Vance (eds.), *Biochemistry of Lipids, Lipoproteins and Membranes*. Amsterdam: Elsevier Science Publishers, 1996. Provides an advanced treatment of the regulation of fatty acid synthesis and cites other key references related to this topic.

Schulz, H., Oxidation of fatty acids. In D. E. Vance and J. E. Vance (eds.), *Biochemistry of Lipids, Lipoproteins and Membranes*. Amsterdam: Elsevier Science Publishers, 1996. Provides an advanced and current summary of fatty acid oxidation in prokaryotes and eukaryotes.

Wakil, S. J., Fatty acid synthase, a proficient multifunctional enzyme. *Biochemistry* 28:4523, 1989. Reviews the evidence for the current model of the mammalian fatty acid synthase as depicted in figure 21.19.

Problems

1. Examine the catabolism of propionyl-CoA (fig. 21.7). Why didn't nature just carboxylate propionyl-CoA on carbon 3, producing succinoyl-CoA directly, and save herself the effort required to produce two additional enzymes?
2. The consensus is that fats containing unsaturated fatty acids are "better for you" than the corresponding saturated forms. Can this statement be explained by the ATP yield that results on complete oxidation (which in turn reflects the caloric content)? Calculate the number of ATPs produced for the complete oxidation of arachidic ($C_{20:0}$) and arachidonic ($C_{20:4}$) acids to assess any differences in energy value of saturated versus polyunsaturated fatty acids.
3. Examine the chemistry of the reactions presented in figure 21.4*a*. Where else in metabolism have you seen a similar sequence of chemical events? What feature(s) is (are) identical, what different?
4. (a) Calculate the number of moles of ATP produced during the catabolism of a mole of glucose.
   (b) Calculate the number of moles of ATP produced during the catabolism of a mole of decanoic acid, [$CH_3(CH_2)_8CO_2H$]. Although this is not a typical fatty acid, it was selected because it has a molecular mass comparable to that of glucose.
   (c) If the heat of combustion of glucose is −669.9 kcal/mole and that of decanoic acid is −1452 kcal/mole, and if −7.3 kcal are preserved per mole of ATP, what fraction (%) of the energy available is preserved in each case?
   (d) Consider glucose and decanoic acid to represent a typical carbohydrate and fat, respectively. A typical carbohydrate contains 4 kcal/g, and fats contain 9 kcal/g. Calculate the ratio of the nutritional calories per gram for fats to carbohydrates and compare this number with the ratio of the ATPs produced for fats to carbohydrates. Are the numbers consistent?
5. Order the following substances from the least to most oxidized carbon: formaldehyde, carbon dioxide, methane, formic acid, methanol. Use the open-chain form of glucose and the structural formula for decanoic acid and this scheme to determine the "average" oxidation state of the carbons in glucose and decanoic acid. How does this oxidation state compare with the data obtained in problems 24.4(a) and 24.4(b) above? Can you explain your observation?
6. Explain the role of carnitine acyltransferases in fatty acid oxidation.
7. Late-night TV ads are often quite educational. One advertisement touted the merits of eating grapefruit to lose weight, but taking grapefruit in your lunch presents problems. At that point an individual in a lab coat standing in front of a blackboard covered with molecular formulas proceeds to explain how it is the citric acid in the grapefruit that burns up the fat and for only $XX you can get a month's supply of their pills. Use what you know about the citric acid cycle and fatty acid metabolism to deduce whether added citric acid causes weight loss.
8. After working through problem 7 and having read the chapter, you want to cash in on a get-rich-quick diet scam. You start thinking that if only the body could be tricked into converting some of its fatty acids into acetyl-CoA and then resynthesizing fatty acids, there would be a weight loss. If you could actually make this work, would there be a weight loss?
9. Is β oxidation (fig. 21.4) best described as a spiral or cyclic process? Why?
10. Acetoacetate (fig. 21.10) is shown to give rise to acetone by a spontaneous reaction. Can you explain how this might occur?
11. β-Oxoacyl-CoA transferase (see fig. 21.11) is involved in the transfer of a CoASH from succinyl-CoA to acetoacetate to produce succinate and acetoacetyl-CoA. A cursory examination of this reaction suggests a simple transfer of the CoA moiety. However, it is soon realized that the loss of an oxygen by the acetoacetate and the gain of an oxygen by the succinyl group present a dilemma. Produce a rational mechanism that explains the preservation of the thioester energy and solves this dilemma. (*Hint:* Consider a succinyl phosphate intermediate.)
12. Carnitine deficiency in liver is correlated with hypoglycemia. Suggest a plausible explanation for hypoglycemia in the carnitine-deficient human.
13. (a) Liver mitochondria convert long-chain fatty acids to ketone bodies (acetoacetate and β-hydroxybutyrate) that are subsequently transported in the plasma to nonhepatic tissues. Suggest some metabolic advantages of supplying ketone bodies to nonhepatic tissues.
    (b) In what way is β-hydroxybutyrate a better energy source than acetoacetate for nonhepatic tissues?
    (c) Outline the oxidation of β-hydroxybutyrate to acetyl-CoA in heart mitochondria.
14. Predict the effect on oxidation of ketone bodies and of glucose in nonhepatic tissue of individuals with markedly diminished β-oxyacid-CoA-transferase activity. Predict the effect if the activity were absent.
15. (a) For an in vitro synthesis of fatty acids with purified fatty acid synthase, the acetyl-CoA was supplied as the ^{14}C-labeled derivative

    $$^{14}CH_3-\overset{\overset{\displaystyle O}{\|}}{C}-S-CoA$$

    The other reactants, including the malonyl-CoA, were not radioactive. Where is the ^{14}C-label found in palmitic acid?
    (b) If the malonyl-CoA were supplied as the only labeled compound deuterated as shown in the structure below, how many deuterium atoms would be incorporated in palmitate? On which carbon(s) would these deuterium atoms reside?

    $$^{-}O-\overset{\overset{\displaystyle O}{\|}}{C}-CD_2-\overset{\overset{\displaystyle O}{\|}}{C}-S-CoA$$

    (c) If [3-^{14}C]malonyl-CoA (shown below) were used in the reaction, which atoms in palmitate would be labeled? Why?

    $$^{-}O-{}^{14}\overset{\overset{\displaystyle O}{\|}}{C}-CH_2-\overset{\overset{\displaystyle O}{\|}}{C}-S-CoA$$

16. Except for malonyl-CoA formation, all the individual reactions

for palmitate synthesis reside on a single multifunctional protein (fatty acid synthase) in animal cells. It has been shown that a dimer of the multifunctional protein is required to catalyze palmitate synthesis. Explain the molecular basis of this observation.

17. What are the metabolic sources of NADPH used in fatty acid biosynthesis? How many moles of NADPH are required for the synthesis of 1 mole of palmitic acid from acetyl-CoA?
18. Citrate is both a lipogenic substrate and a regulatory molecule in mammalian fatty acid synthesis.
    (a) Explain each function of citrate in fatty acid synthesis.
    (b) Write reactions (including structures) outlining the role of citrate as a lipogenic substrate.
19. Which catalytic activity of the mammalian fatty acid synthase determines the chain length of the fatty acid product?
20. (a) Why is the location of biosynthesis and β oxidation of fatty acids in separate metabolic compartments essential to regulation of fatty acid metabolism in the hepatocyte?
    (b) Would you expect an inhibitor of the extramitochondrial carnitine acyltransferase to mimic the effect of malonyl-CoA on β oxidation? (Assume that the inhibitor can penetrate the cell membrane.) Explain the rationale for your answer.

C H A P T E R

Biosynthesis of Membrane Lipids

Most of the metabolism of lipids occurs on the surfaces of membranes because of the lipophilic nature of the substrates and products.

Thus far we have been concerned with the metabolism of fatty acids in relationship to storage and release of energy (chapter 21). In this chapter we focus on the metabolism of lipids that serve other roles (fig. 22.1). Most fatty acids that are not utilized for energy storage perform important structural roles as integral components of phospholipids and sphingolipids in membranes. The phospholipids of membranes also serve as a reservoir for cellular second messengers. One fatty acyl component of phospholipids, arachidonic acid, is the precursor of eicosanoids, which trigger a wide range of responses. In this chapter we describe the biosynthesis of the main membrane phospholipids and sphingolipids and explain how the second messengers that are stored as components of phospholipids are released in response to extracellular stimuli.

Phospholipids

Phospholipids are ideal compounds for making membranes because of their amphipathic nature (see chapter 19). The polar head groups of phospholipids prefer an aqueous environment, whereas the nonpolar acyl substituents do not. As a result, phospholipids spontaneously form bilayer structures (see fig. 19.7), which are a dominant feature of most membranes. The phospholipid bilayer is the barrier of the cell membrane that prevents the unrestricted transport of most molecules other than water into the cell. Entry of other molecules is allowed if a specific transport protein is present in the cell membrane. Similarly, the phospholipid bilayer prevents leakage of metabolites from the cell. The amphipathic nature of phospholipids has a great influence on the mode of their biosynthesis. Thus most of the reactions involved in lipid synthesis occur on the surface of membrane structures catalyzed by enzymes that are themselves amphipathic.

As in other areas of biochemistry, research on phospholipid biosynthesis was initiated only after the structures of the major phospholipids were elucidated. Beginning in 1927 the structure of phosphatidylcholine was confirmed by chemical synthesis. Subsequently, in 1932, Charles Best (a codiscoverer of insulin) demonstrated that choline was an essential dietary component. Experiments by Don Zilversmit and Irving Chaikoff in the 1940s

Figure 22.1

Outline of pathways for the biosynthesis of major cellular lipids (other than cholesterol) in a mammalian cell. Most of the metabolism of these lipids occurs on membrane surfaces because of the insoluble nature of the substrates and products. These lipids play three major roles: (1) They act as a storehouse of chemical energy, as with triacylglycerols; (2) they are structural components of membranes (boxed compounds); and (3) they act as regulatory compounds (underlined), either as eicosanoids, which act as local hormones, or as phosphorylated inositols and diacylglycerols, which function as second messengers.

Synthesis of most phospholipids starts from glycerol-3-phosphate, which is formed in one step from the central metabolic pathways, and acyl-CoA, which arises in one step from activation of a fatty acid. In two acylation steps the key compound phosphatidic acid is formed. This can be converted to many other lipid compounds as well as CDP-diacylglycerol, which is a key branchpoint intermediate that can be converted to other lipids. Distinct routes to phosphatidylethanolamine and phosphatidylcholine are found in prokaryotes and eukaryotes. The pathway found in eukaryotes starts with transport across the plasma membrane of ethanolamine and/or choline. The modified derivatives of these compounds are directly condensed with diacylglycerol to form the corresponding membrane lipids. Modification of the head groups or tail groups on preformed lipids is a common reaction. For example, the ethanolamine of the head group in phosphatidylethanolamine can be replaced in one step by serine or modified in 3 steps to choline.

Phospholipids containing arachidonic acid serve as the precursor, through liberation of arachidonic acid, of a wide variety of eicosanoids. Similarly, the membrane lipid phosphatidylinositol-4,5-P_2 can undergo hydrolysis to inositol-P_3 and diacylglycerol. Both of these hydrolysis products are regulatory molecules that serve as second messengers (so called because they are formed in response to hormone binding to the plasma membrane).

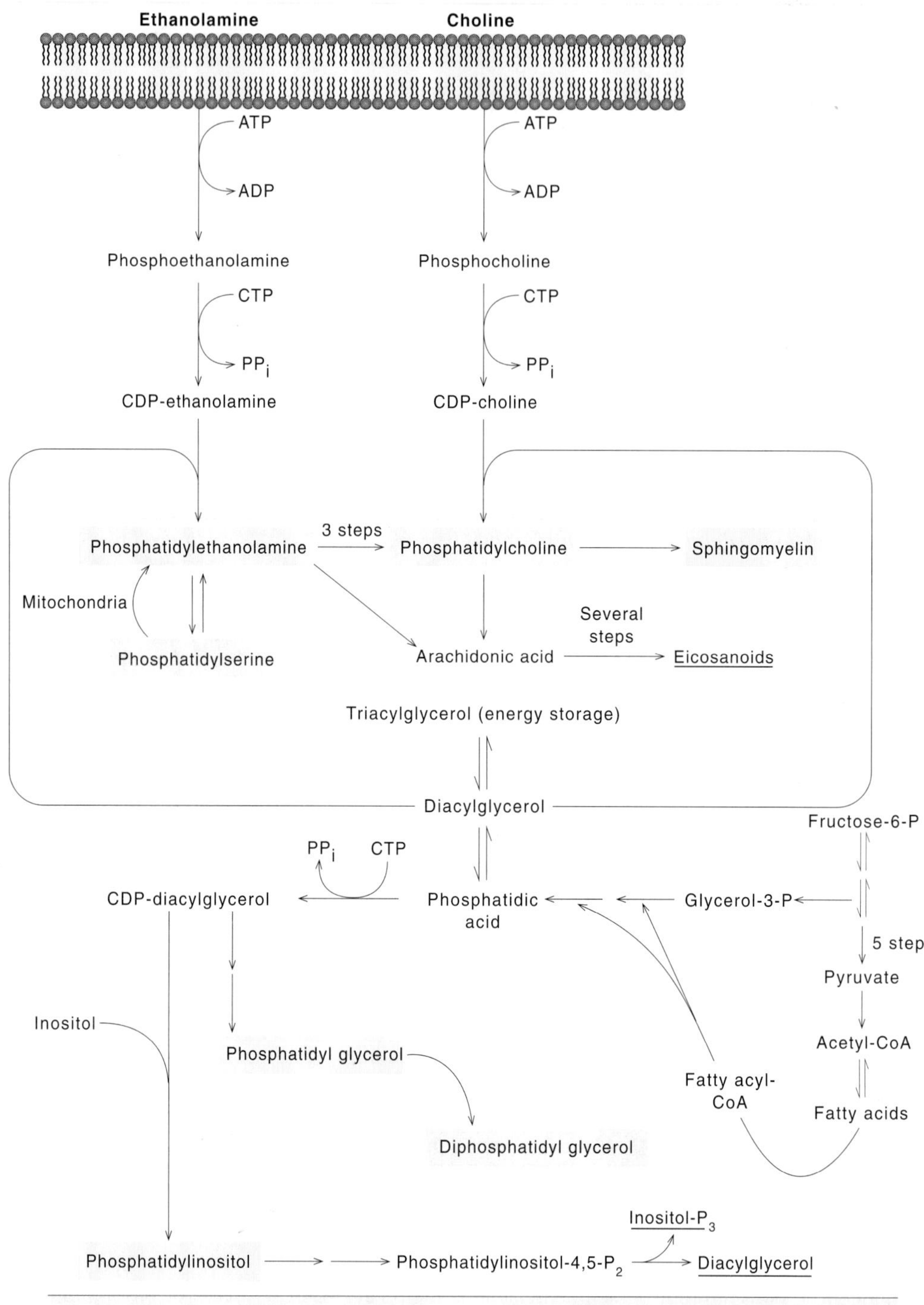

with radioactive phosphorus (^{32}P) provided insight into the metabolism of phospholipids in intact animals. In the 1950s, Eugene Kennedy and co-workers established the role of glycerol-3-phosphate as a precursor of phospholipids and defined the CDP-choline and CDP-ethanolamine pathways. The groundwork laid by these researchers has led to more recent studies that have provided information on the enzymes involved in phospholipid metabolism and the regulation of these pathways.

In E. coli, *Phospholipid Synthesis Generates Phosphatidylethanolamine, Phosphatidylglycerol, and Diphosphatidylglycerol*

Escherichia coli contains three important classes of phospholipids: phosphatidylethanolamine (75%–85%), phosphatidylglycerol (10%–20%), and diphosphatidylglycerol (5%–15%). All three of these

Figure 22.2

The first phase of phospholipid synthesis in *E. coli* and eukaryotes. Additional routes to and from phosphatidic acid, found predominantly in eukaryotes, are shown in brackets.

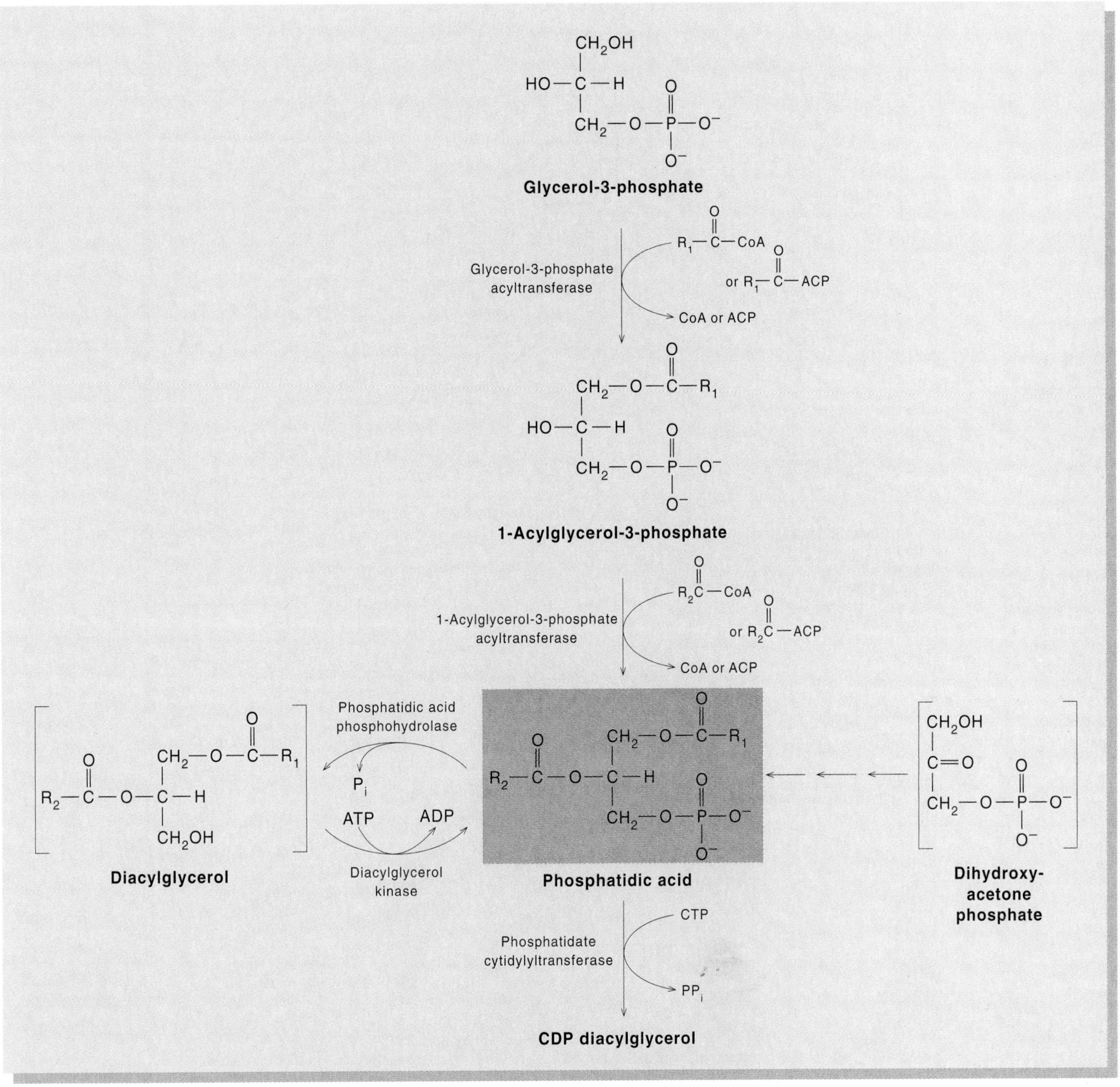

phospholipids share the same biosynthetic pathway up to the formation of CDP-diacylglycerol (fig. 22.2), after which the pathways branch (fig. 22.3).

Most of the enzymes for phospholipid synthesis are located on the inner plasma membrane of *E. coli.* Glycerol-3-phosphate acyltransferase, the first enzyme in the pathway, preferentially utilizes saturated fatty acyl derivatives (palmitoyl-CoA or palmitoyl-ACP) for the initial acylation of glycerol-3-phosphate. (The mechanism for acyltransferase-catalyzed reactions is described in box 22A.) The second enzyme (see fig. 22.2) 1-acylglycerol-3-phosphate acyltransferase, catalyzes phosphatidic acid formation; this enzyme shows a preference for acyl residues with

Figure 22.3

The second phase of phospholipid synthesis in *E. coli,* from CDP-diacylglycerol to the end products.

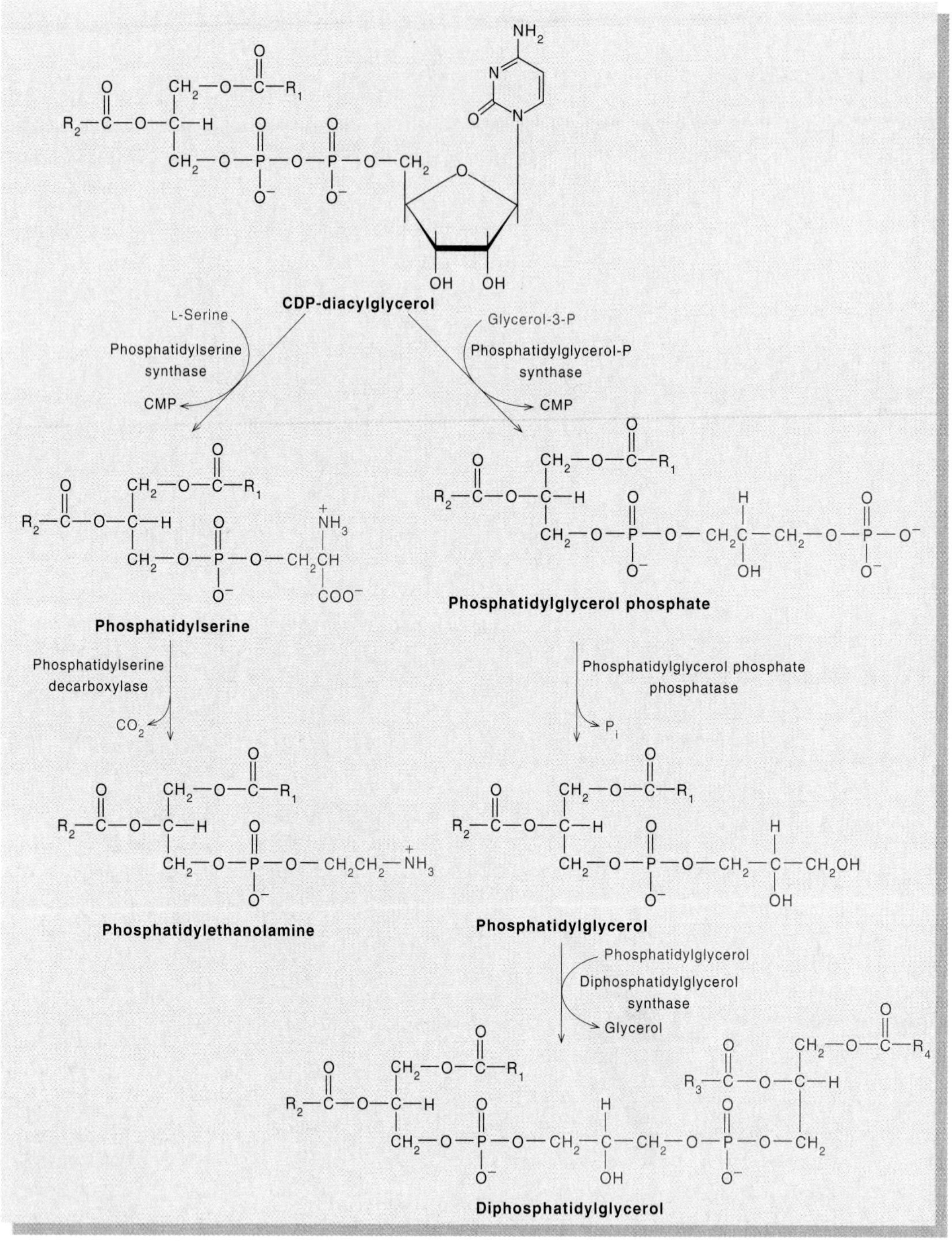

a double bond. The substrate specificities of these two acyltransferases account for there usually being saturated fatty acids in the SN-1 position and unsaturated fatty acids in the SN-2 position of phospholipids. A third transferase reaction converts phosphatidic acid to CDP-diacylglycerol with CTP as a cosubstrate. Interestingly, CTP is the only high-energy nucleotide used for the synthesis of phospholipids in all organisms. CDP-diacylglycerol is a branchpoint intermediate. It is converted either to phosphatidylserine en route to phosphatidylethanolamine or to phosphatidylglycerol phosphate enroute to phosphatidylglycerol and diphos-

Mechanism of Action of Acyltransferases

Nucleophilic addition to the neutral activated acyl group is a favored process and coenzyme A is a good leaving group from the tetrahedral intermediate. This point was discussed in chapter 12 in connection with the formation of acetoacetyl-CoA, and we see it again here in the acyltransferase reactions involving glycerol or glycerol derivatives and acyl-CoA compounds. The reaction is initiated by a nucleophilic attack of acyl-CoA by a hydroxyl, giving rise to a tetrahedral intermediate. This is followed by expulsion of the CoA.

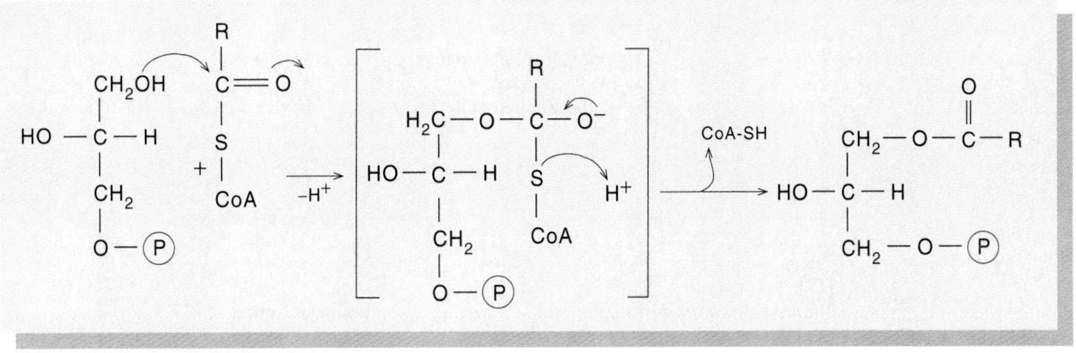

phatidylglycerol (see fig. 22.3). The hydroxyl group of glycerol-3-phosphate or serine reacts with the high-energy pyrophosphate bond of CDP-diacylglycerol.

Phosphatidylglycerol and diphosphatidylglycerol are also synthesized in the mitochondria of eukaryotes by a pathway similar to that in prokaryotes. The only difference is that diphosphatidylglycerol in eukaryotes is made by the reaction of CDP-diacylglycerol with phosphatidylglycerol rather than the condensation of two molecules of phosphatidylglycerol as occurs in *E. coli.*

Phospholipid Synthesis in Eukaryotes Is More Complex

Phospholipid synthesis in eukaryotes is more complex than in *E. coli.* This relates to the other roles phospholipids play in membranes aside from their structural role. Eukaryotes have organelle membranes as well as plasma membranes. In addition, as we will see, the regulation of phospholipid biosynthesis is more complicated than in prokaryotes, partly because a central intermediate in phospholipid biosynthesis, diacylglycerol, is also a precursor of triacylglycerol, the main energy storage lipid.

In the first phase of phospholipid synthesis from glycerol-3-phosphate to phosphatidic acid, the pathways in *E. coli* and eukaryotes are very similar (see fig. 22.2). The major difference is that one additional pathway exists for generation of phosphatidic acid from dihydroxyacetone phosphate, an intermediate in glycolysis. Once phosphatidic acid is made, it is rapidly converted to diacylglycerol or CDP-diacylglycerol (see fig. 22.2), both of which are intermediates for the biosynthesis of eukaryotic phospholipids.

Diacylglycerol Is the Key Intermediate in the Biosynthesis of Phosphatidylcholine and Phosphatidylethanolamine

Phosphatidylcholine and phosphatidylethanolamine, which are quantitatively the most important phospholipids in eukaryotic cells (see table 19.4), are derived from diacylglycerol as shown in figure 22.4 (also see fig. 22.1). Alternatively, when there is an abundance of fatty acids, diacylglycerol is converted into triacylglycerol by diacylglycerol acyltransferase (see fig. 22.4). The biosynthesis of phosphatidylcholine begins with choline that is transported into the cell. It is noteworthy that choline is an essential ingredient in the human diet. Once inside the cell, the choline is rapidly phosphorylated to phosphocholine by choline (ethanolamine) kinase (see fig. 22.4). Phosphocholine reacts with CTP to form CDP-choline in a reaction catalyzed by CTP:phosphocholine cytidylyltransferase. The activity of this enzyme is usually rate-limiting for phosphatidylcholine biosynthesis. In the last reaction, diacylglycerol reacts with CDP-choline to yield phosphatidylcholine, catalyzed by CDP-choline:1,2-diacylglycerol phosphocholine-transferase. Similar to the reactions involving CDP-diacylglycerol, the hydroxyl group of diacylglycerol reacts with the pyrophosphate bond in CDP-choline. The biosynthesis of phosphatidylethanolamine proceeds from ethanolamine in a comparable series of reactions (see fig. 22.4).

Figure 22.4

The second phase of phospholipid synthesis in eukaryotes. Choline or ethanolamine enters the cell via active transport mechanisms and is immediately phosphorylated by the enzyme choline (ethanolamine) kinase. The phosphorylated derivatives of choline and ethanolamine are activated to their CDP derivatives by separate enzymes. The last reaction occurs on the endoplasmic reticulum. The diacylglycerol used as a substrate in this reaction may alternatively be converted to storage lipid (triacylglycerol).

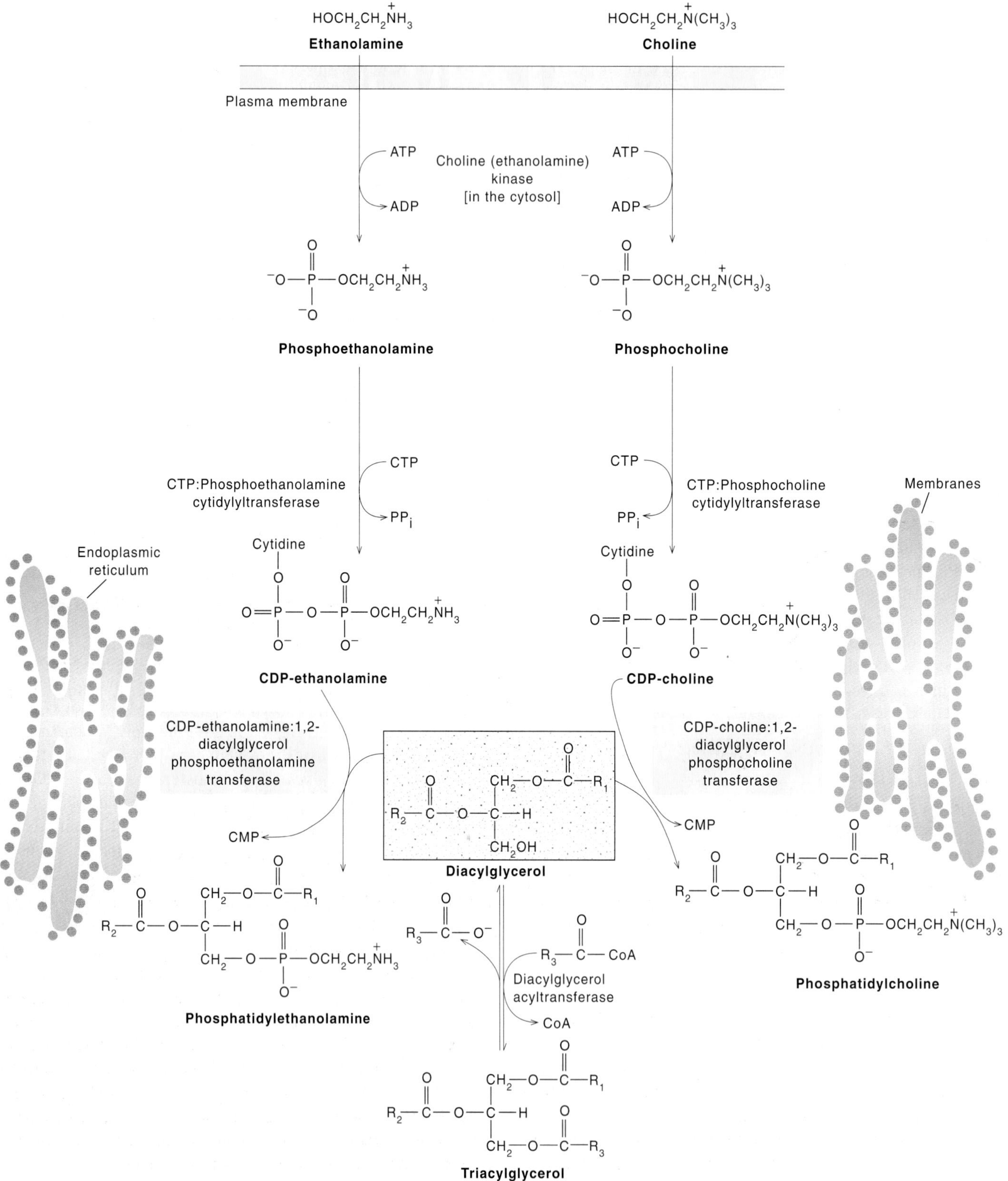

Pathway for Methylation by S-*Adenosylmethionine*

S-Adenosylmethionine (Ado Met) is the primary methylating agent in the cell. It is not considered a coenzyme, so we did not discuss it in chapter 12. It is formed and functioning in the degradative pathway from methionine to propionyl-CoA (see fig. 25.11; also see fig. 24.18). Ado Met also methylates many other compounds, including the ϵ-amino group of lysine, the guanidino group of arginine, the nitrogens of the imidazole side chain of histidine, and the bases of DNA and RNA. Ado Met is discussed here because of its role in the remarkable trimethylation of phosphatidylethanolamine to form phosphatidylcholine.

The positive charge on the sulfur atom of *S*-adenosylmethionine makes Ado Met a powerful alkylating agent. This is because the plus charge on the sulfur converts the *S*-adenosylhomocysteine moiety into an excellent leaving group. Alkylation occurs as a bimolecular substitution reaction.

$^{-}OOC-CH(NH_3^+)-CH_2-CH_2-S^+(CH_3)-CH_2$–ribose(OH, OH)–Adenine

***S*-Adenosylmethionine**

Methyl transfer: X → X–CH_3

$^{-}OOC-CH(NH_3^+)-CH_2-CH_2-S-CH_2$–ribose(OH, OH)–Adenine

***S*-Adenosylhomocysteine**

The liver, fungi, plants, and a few classes of bacteria have the capacity to convert phosphatidylethanolamine to phosphatidylcholine (fig. 22.5 and box 22B). The methyl donor is *S*-adenosylmethionine, which is also important for the methylation of certain proteins, RNA, DNA, and some small water-soluble molecules. The methylation of phosphatidylethanolamine, together with the subsequent degradation of phosphatidylcholine, is the major mechanism by which organisms make choline. Since choline is required and available in our diet, we are not sure why the phosphatidylethanolamine methyltransferase in liver has survived in evolution. By the use of recombinant DNA techniques, it should soon be possible to produce a mouse in which the gene has been deleted. Such an experiment may show an unexpected function for this methylating enzyme

Fatty Acid Substituents at SN-*1 and* SN-*2 Positions Are Replaceable*

Lung tissue manufactures a specialized species of phosphatidylcholine, dipalmitoyl-phosphatidylcholine, in which palmitic acid is attached to both positions 1 and 2 of the glycerol backbone. This species of phosphatidylcholine is the major component of lung surfactant, which functions to maintain surface tension in the lung alveoli so that they do not collapse when air is expelled. Premature babies do not secrete sufficient dipalmitoyl-phosphatidylcholine-enriched surfactant and therefore have major difficulties in breathing. Understanding the biochemistry of lung surfactant phospholipids has led to the development of synthetic surfactant mixtures, which can be sprayed into the lung. This treatment decreases the need for respirators prior to maturation of the surfactant biosynthetic machinery of the infant.

The pathway for the synthesis of dipalmitoylphosphatidylcholine is illustrated in figure 22.6. The starting species of phosphatidylcholine is made by the CDP-choline pathway (see fig. 22.4). The fatty acid at the SN-2 position, which is usually unsaturated, is hydrolyzed by phospholipase A_2, and the lysophosphatidylcholine is reacylated with palmitoyl-CoA. This modification permits alteration of the properties of the phospholipid without resynthesis of the entire molecule, a strategy called remodeling. Deacylation–reacylation of phosphatidylcholine occurs in other tissues and provides an important route for alteration of the fatty acid substituents at both the SN-1 and SN-2 positions. For example, fatty acids at the SN-2 position can be replaced by arachidonic acid, which is stored there until needed for eicosanoid biosynthesis, as we discuss later in this chapter.

Figure 22.5

Conversion of phosphatidylethanolamine to phosphatidylcholine by phosphatidylethanolamine-*N*-methyltransferase. (AdoMet is a standard abbreviation for *S*-adenosyl-L-methionine. Recall that SAM may also be used. AdoHcy is a standard abbreviation for *S*-adenosyl-L-homocysteine.) The structures of AdoMet and AdoHcy and the general mechanism of the methylation reaction are presented in box 22B.

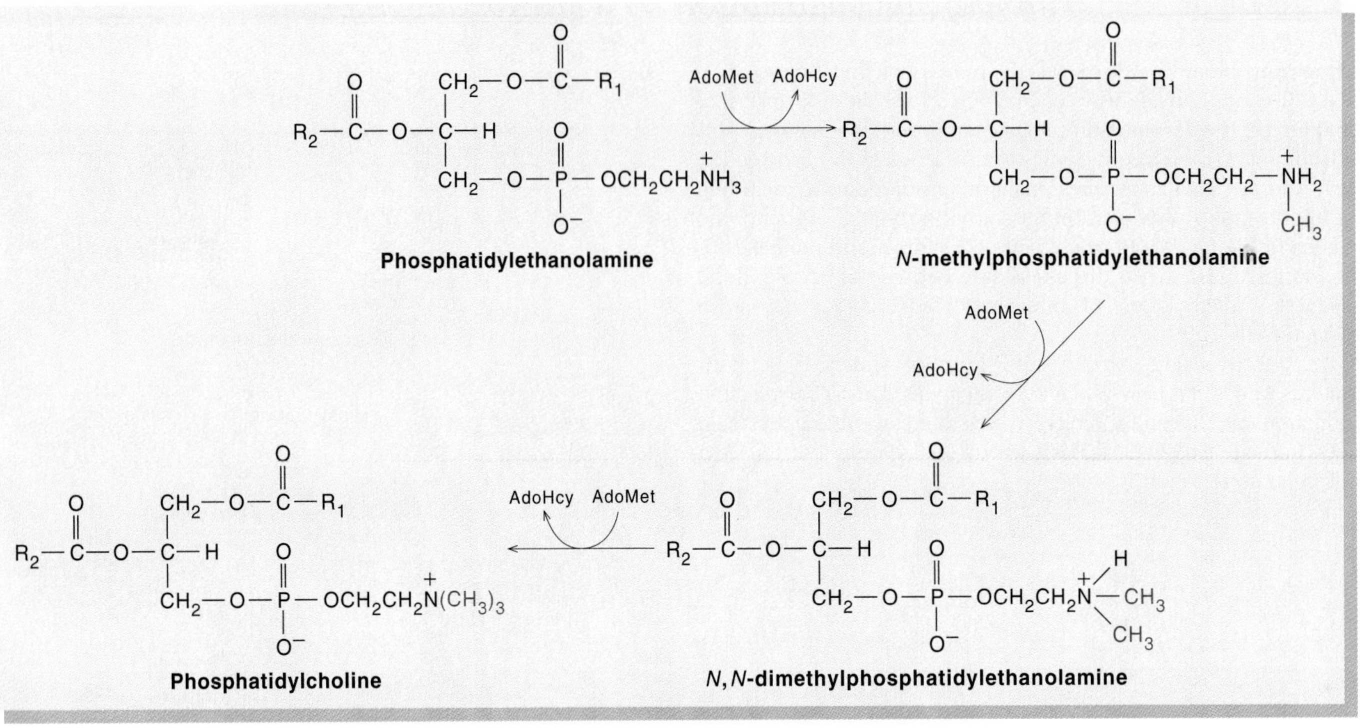

Phosphatidylethanolamine → (AdoMet → AdoHcy) → *N*-methylphosphatidylethanolamine → (AdoMet → AdoHcy) → *N,N*-dimethylphosphatidylethanolamine → (AdoMet → AdoHcy) → Phosphatidylcholine

Figure 22.6

Biosynthesis of dipalmitoylphosphatidylcholine. R_2 is usually an unsaturated fatty acid. Thus this two-step reaction results in the replacement of an unsaturated by a saturated fatty acid at the C-2 position on the glycerol backbone. Dipalmitoylphosphatidylcholine is the major component in lung surfactant, a substance that maintains surface tension in the lung alveoli so that they do not collapse when air is expelled.

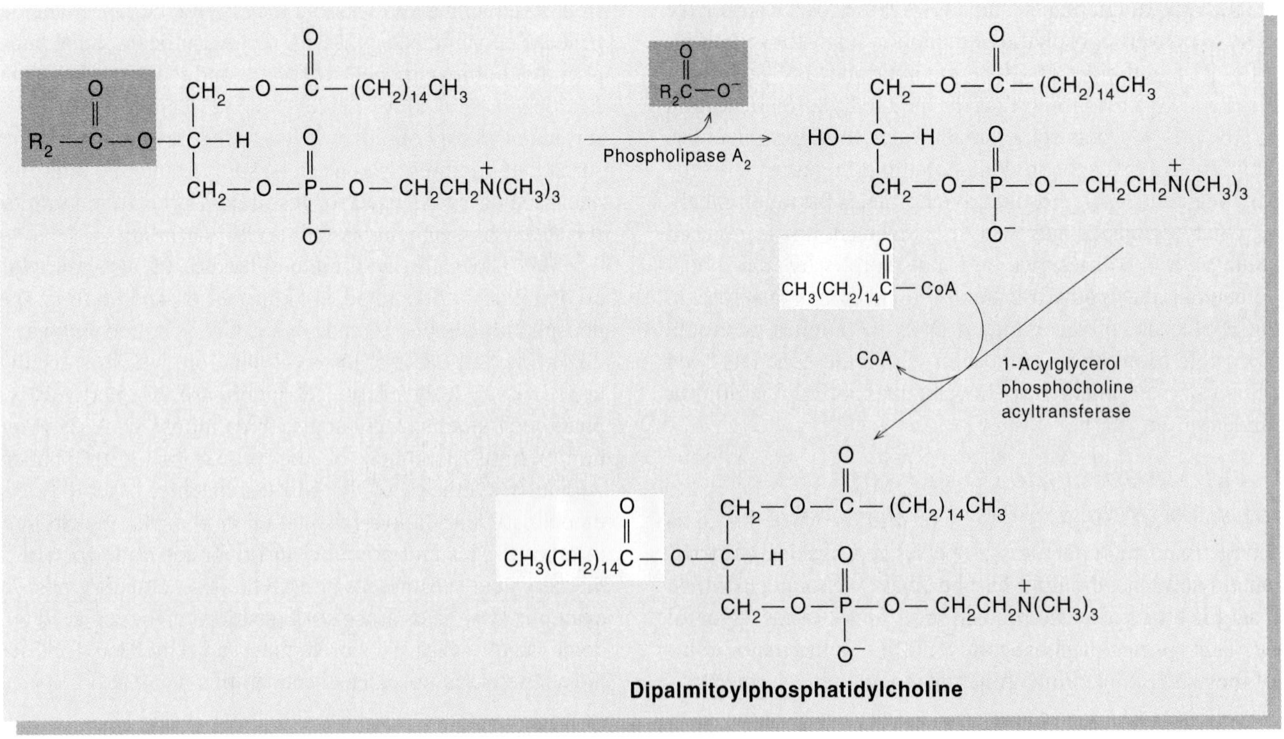

Phosphatidylcholine → (Phospholipase A_2; releases $R_2C(=O)O^-$) → lysophosphatidylcholine → ($CH_3(CH_2)_{14}C(=O)$—CoA → CoA; 1-Acylglycerol phosphocholine acyltransferase) → Dipalmitoylphosphatidylcholine

Phosphatidylinositol-4,5-Bisphosphate, a Precursor of Second Messengers, Is Synthesized via CDP-Diacylglycerol

CDP-diacylglycerol is also a precursor of phosphatidylinositol (fig. 22.7), a lipid that is unique to eukaryotes. Phosphatidylinositol accounts for approximately 5% of the lipids present in animal cell membranes (see table 19.4). Also present, at much lower concentrations, are phosphatidylinositol-4-phosphate and phosphatidylinositol-4,5-bisphosphate. These two derivatives are made from phosphatidylinositol by separate kinases that use ATP as the phosphate donor. When certain hormones bind to the cell surface (e.g., vasopressin binds to hepatocytes), the phosphatidylinositol-4,5-bisphosphate, which is localized to the plasma membrane, is attacked by phospholipase C between the glycerol and phosphate moieties. This reaction yields two cellular second messengers: diacylglycerol and inositol-1,4,5-P_3 (fig. 22.8). Each of these compounds has important regulatory functions. The inositol-1,4,5-P_3 mobilizes calcium from intracellular stores (endoplasmic reticulum); the rise in cytosolic calcium activates a host of different enzymes (see chapter 27). The rise in diacylglycerol activates protein kinase C, which plays a major regulatory role as a phosphorylating agent of an extremely diverse group of proteins (see chapters 20 and 27 and fig. 20.16). Phosphatidylcholine can also serve as a precursor of diacylglycerol without giving rise to inositol-1,4,5-P_3.

The Metabolism of Phosphatidylserine and Phosphatidylethanolamine Is Closely Linked

In prokaryotes, phosphatidylserine is made from CDP-diacylglycerol (see fig. 22.3). The enzyme for this reaction is absent in animal cells, which rely on a base exchange reaction in which serine and ethanolamine are interchanged (fig. 22.9). Although the reaction is reversible, it usually proceeds in the direction of phosphatidylserine synthesis. Phosphatidylserine can be converted back to phosphatidylethanolamine by a decarboxylation reaction in the mitochondria. This may be the preferred route for phosphatidylethanolamine biosynthesis in some animal cells. Furthermore, these two reactions (see fig. 22.9) establish a cycle that has the net effect of converting serine into ethanolamine. This is the main route for ethanolamine synthesis in animal cells. Ethanolamine also enters our bodies as a result of digestion of dietary phosphatidylethanolamine. The primary fate of ethanolamine is for use in the biosynthesis of phosphatidylethanolamine via the CDP-ethanolamine pathway (see fig. 22.4).

Biosynthesis of Alkyl and Alkenyl Ethers

Some phospholipids contain either an *O*-alkyl or an *O*-alkenyl ether species at the SN-1 position, instead of the more common acylester linkage (see chapter 19). The biosynthetic pathway for the alkyl ether species of phosphatidic acid in eukaryotes is indicated in figure 22.10. The initial step involves the acylation of dihydroxyacetone phosphate (DHAP), a reaction discussed in connection with phosphatidic acid biosynthesis. Subsequently, an exchange reaction replaces the 1-acyl group with an alkyl group derived from an alcohol. This alcohol is formed by reduction of an acyl-CoA by NADPH or NADH (fig. 22.11). Following the exchange reaction, reduction of the ketone and acylation of the 2-hydroxyl group occurs. Once formed, the 1-alkyl ether derivative of phosphatidic acid is used for the synthesis of other phospholipids.

Figure 22.7

Reaction that converts CDP-diacylglycerol to phosphatidylinositol in eukaryotic cells.

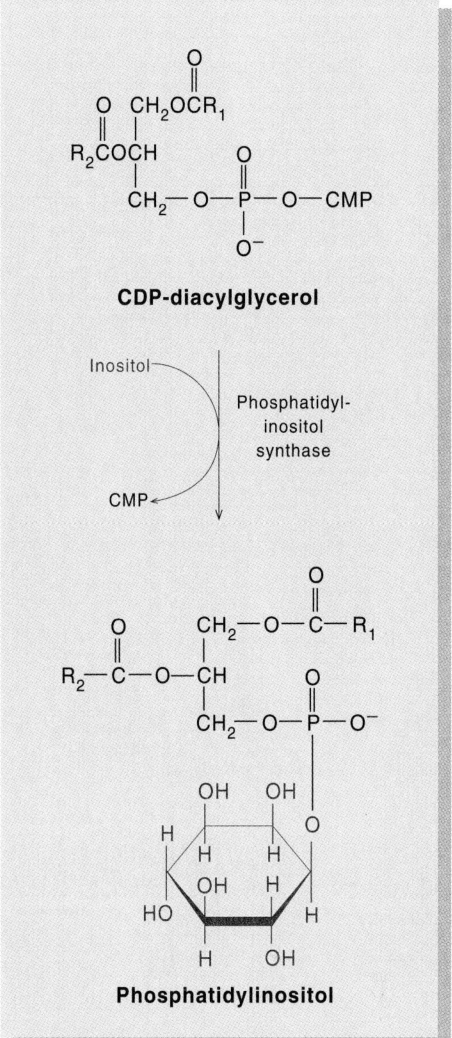

Certain alkyl ether species of phosphatidylcholine possess potent biological activity. For example, 1-alkyl-2-acetylglycerophosphocholine (fig. 22.12), also known as platelet-activating factor, reduces blood pressure in hypertensive rats and causes blood platelets to aggregate at very low hormone levels (10^{-10} M).

In many tissues, 1-alkyl-2-acylphosphatidylethanolamine can be desaturated by an endoplasmic reticulum enzyme, 1-alkyl-2-acylglycerophosphoethanolamine desaturase, to yield the corresponding unsaturated derivative called a plasmalogen (fig. 22.13). This enzyme requires O_2, NADH, and cytochrome b_5, the same cofactors required for the desaturation of stearoyl-CoA (see fig.

Figure 22.8

Phospholipase C degradation of phosphatidylinositol-4,5-P_2.

Phosphatidylinositol-4,5-P_2

Phospholipase C

H_2O

Inositol-1,4,5-P_3

Diacylglycerol

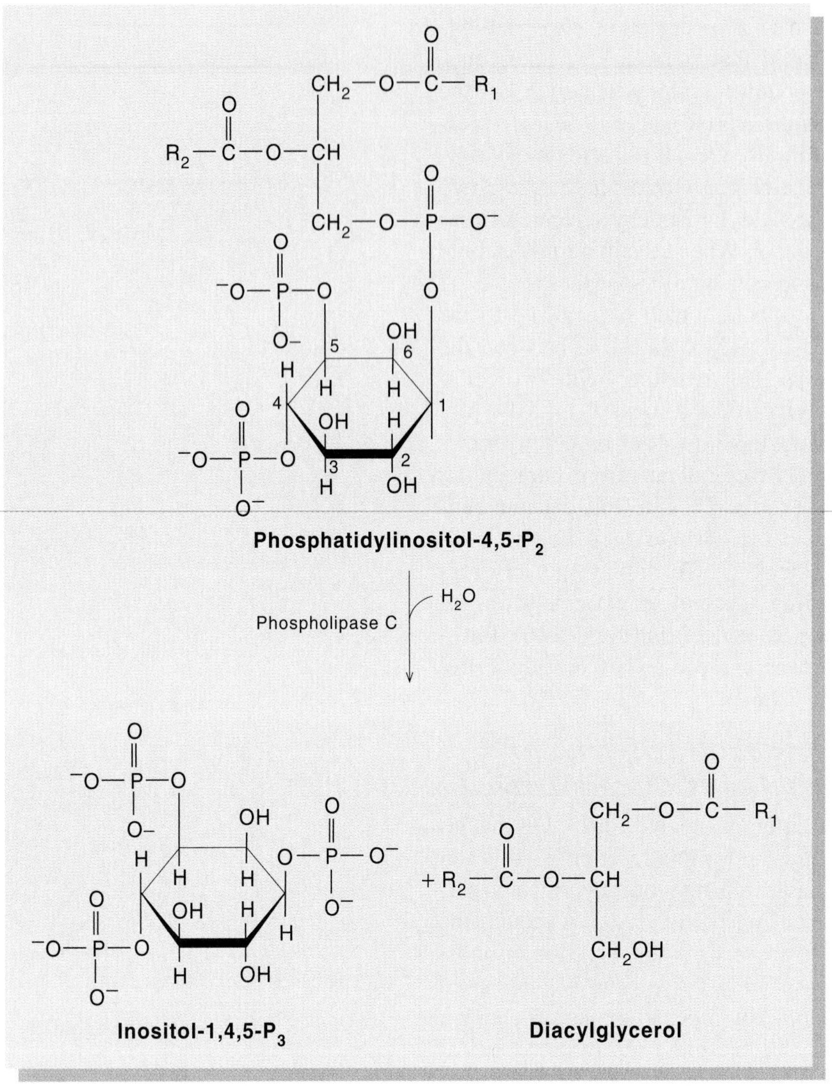

21.21). In many tissues, plasmalogens are minor constituents, but in heart tissue, nearly 50% of phosphatidylethanolamine contains the alkenyl ether at position SN-1. The function of plasmalogens in animals is unknown. However, alkenyl ether–containing phospholipids can protect cells against the deleterious effects of singlet oxygen, which at high concentrations can kill cells.

Ether lipids are also found in microorganisms, notably the eukaryotic protozoans and archaebacteria. Alkyl ether bonds are more stable to hydrolysis than alkyl ester bonds. This greater stability probably accounts for the omnipresence of ether lipids in the membranes of archaebacteria, which frequently experience extremes of pH, salt, and temperature.

The Final Reactions for Phospholipid Biosynthesis Occur on the Cytosolic Surface of the Endoplasmic Reticulum

The final reactions for the biosynthesis of phosphatidylcholine, phosphatidylethanolamine, phosphatidylserine, and phosphatidylinositol all occur on the cytosolic surface of the endoplasmic reticulum and Golgi apparatus (fig. 22.14). By contrast, phosphatidylglycerol and diphosphatidylglycerol are synthesized on the mitochondrial membrane where they remain for the most part.

Two questions concerning the distribution of lipids in membranes are of current interest.

Figure 22.9

Phosphatidylserine biosynthesis in animals is catalyzed by a base exchange enzyme on the endoplasmic reticulum. Decarboxylation of phosphatidylserine occurs in mitochondria. The cyclic process of phosphatidylserine formation from phosphatidylethanolamine and the reformation of phosphatidylethanolamine by decarboxylation has the net effect of converting serine to ethanolamine. This is a major mechanism for the synthesis of ethanolamine in many eukaryotes.

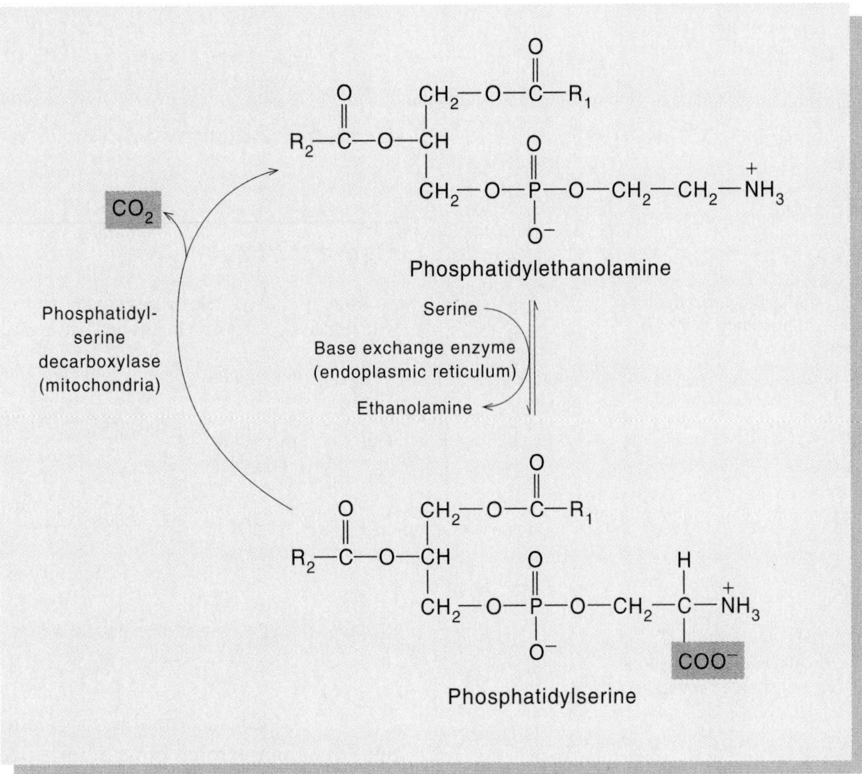

1. Because phospholipid synthesis occurs on the cytosolic side of the endoplasmic reticulum and phospholipids are found on both leaflets of the bilayer, how do these lipids reach the inner leaflet of the bilayer? Possibly proteins called flipases catalyze the movement of phospholipids from one side of the bilayer to the other. Such a flipase has been described in the plasma membrane of erythrocytes. More research is required to determine how important these flipases are in movement of lipids between the two leaflets of the bilayer in the endoplasmic reticulum.
2. How does the cell sort and transport phospholipids from the site of synthesis to the other membranes in the cell? One view is that phospholipid vesicles that bud from the endoplasmic reticulum are targeted to another membrane where the vesicles fuse with the membrane. Alternatively, phospholipid transfer proteins may be involved. Proteins that transfer phospholipids between membranes in vitro have been known for over 25 years, but it has not been demonstrated that they function in this way in vivo.

Recent studies with a yeast mutant deficient in a phosphatidylinositol/phosphatidylcholine transfer protein have localized this protein to Golgi and shown that the transfer protein is required for secretion of proteins from yeast and even for the survival of the yeast. It is also interesting that even though the phosphatidylinositol/phosphatidylcholine transfer proteins from yeast and brain catalyze the same transfer reaction in vitro, there is no homology in structure between the two proteins.

In the Liver, Regulation Gives Priority to Formation of Structural Lipids over Energy-Storage Lipids

The energy state of the cell dictates the relative rates of phosphatidylcholine, phosphatidylethanolamine, and triacylglycerol biosynthesis. When energy is in short supply, the level of cAMP rises, leading to inhibition of fatty acid biosynthesis (see chapter 21). This in turn decreases the supply of diacylglycerol, which limits the synthesis of phosphatidylcholine,

Figure 22.10

Biosynthesis of the alkyl ether species of phosphatidic acid. The first two enzymes in this sequence are restricted to peroxisomes.

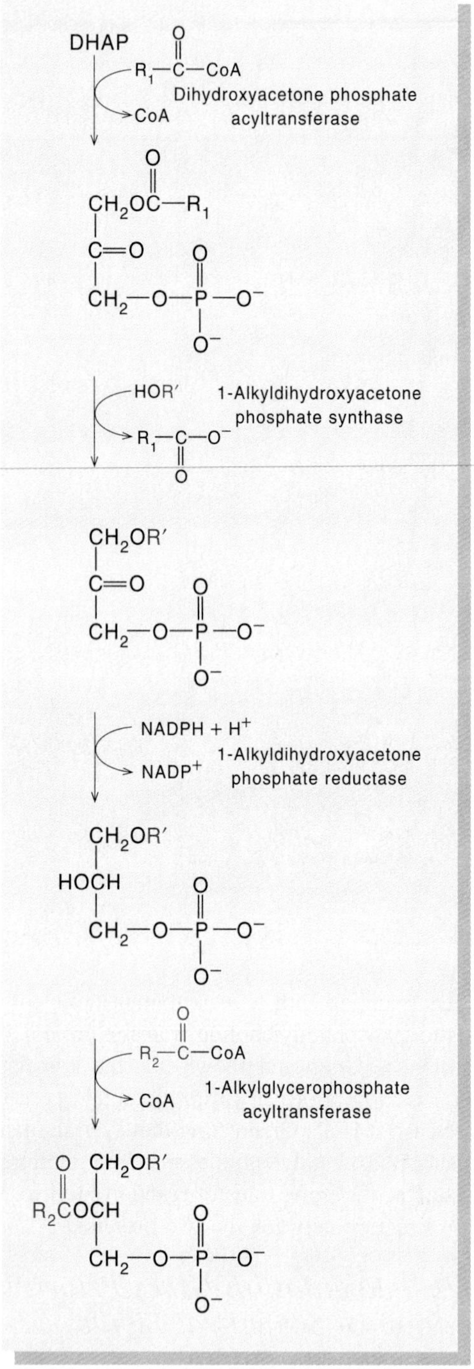

Figure 22.11

Acyl-CoA reductase catalyzes the formation of long-chain alcohols from fatty acyl-CoA.

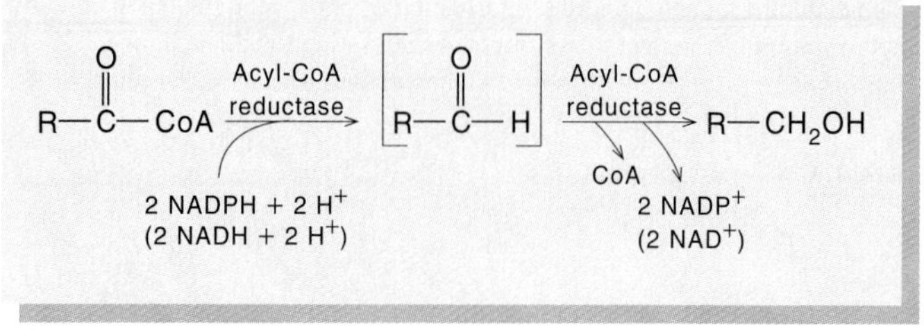

Figure 22.12

Structure of 1-alkyl-2-acetylglycerophosphocholine, a potent phospholipid that will reduce hypertension in rats and cause platelets to aggregate.

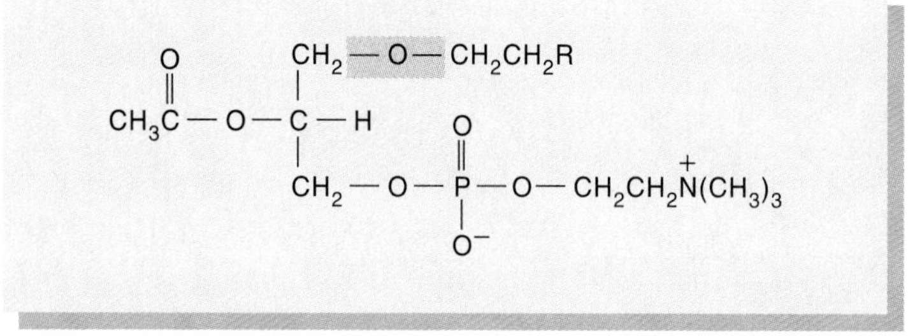

phosphatidylethanolamine, and triacylglycerol. When sufficient diacylglycerol is present, the requirements for the synthesis of the essential membrane components, phosphatidylethanolamine and phosphatidylcholine, are met before an appreciable amount of energy-storage lipid (triacylglycerol) is made. Any excess diacylglycerol and fatty acyl-CoA is funneled into triacylglycerol.

Additional regulation of phosphatidylcholine and phosphatidylethanolamine biosynthesis occurs at the second step in the biosynthetic sequence (see fig. 22.4), where either CDP-choline or CDP-ethanolamine is made. For phosphatidylcholine biosynthesis the activity of CTP:phosphocholine cytidylyltransferase (which makes CDP-choline) is governed by an unusual mechanism. The enzyme exists in a soluble form as an inactive reservoir and is translocated to the cell membranes, where it is activated by membrane lipids (fig. 22.15). This facilitates a rapid response to a sudden requirement for biosynthesis of phosphatidylcholine, which could be vital to maintaining the integrity of the cell membrane. The binding of the cytidylyltransferase to membranes is enhanced by a decrease in phosphatidylcholine, dephosphorylation of the enzyme, or a rise in the concentration of fatty acids or diacylglycerol in the membrane (see fig. 22.15). The regulation by phosphatidylcholine and diacylglycerol levels are good examples of metabolic regulation by feedback and feedforward mechanisms, respectively.

Unlike many other regulatory enzymes involved in lipid metabolism (e.g., acetyl-CoA carboxylase, discussed in chapter 18, and HMG-CoA reductase, the regulatory enzyme in cholesterol biosynthesis, discussed in chapter 23), alterations of the expression of the cytidylyltransferase gene has been reported in only a few examples. The main reason for this appears to be that either fatty acids or cholesterol can be supplied to animals and cells without modification, whereas phosphatidylcholine is catabolized to

Figure 22.13

Formation of plasmalogens. A specific enzyme found in the endoplasmic reticulum desaturates the alkyl group and is specific for the ethanolamine phospholipid. The product of this reaction is called plasmalogen and is abundant in heart tissue and myelin.

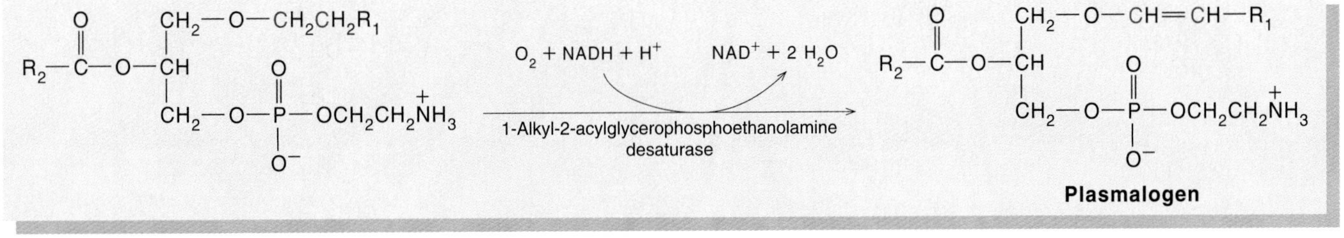

$$\begin{array}{l} CH_2{-}O{-}CH_2CH_2R_1 \\ R_2{-}\overset{O}{\overset{\|}{C}}{-}O{-}CH \\ CH_2{-}O{-}\underset{O^-}{\underset{|}{\overset{O}{\overset{\|}{P}}}}{-}OCH_2CH_2\overset{+}{N}H_3 \end{array} \xrightarrow[\text{1-Alkyl-2-acylglycerophosphoethanolamine desaturase}]{O_2 + NADH + H^+ \quad NAD^+ + 2\,H_2O} \begin{array}{l} CH_2{-}O{-}CH{=}CH{-}R_1 \\ R_2{-}\overset{O}{\overset{\|}{C}}{-}O{-}CH \\ CH_2{-}O{-}\underset{O^-}{\underset{|}{\overset{O}{\overset{\|}{P}}}}{-}OCH_2CH_2\overset{+}{N}H_3 \end{array}$$

Plasmalogen

Figure 22.14

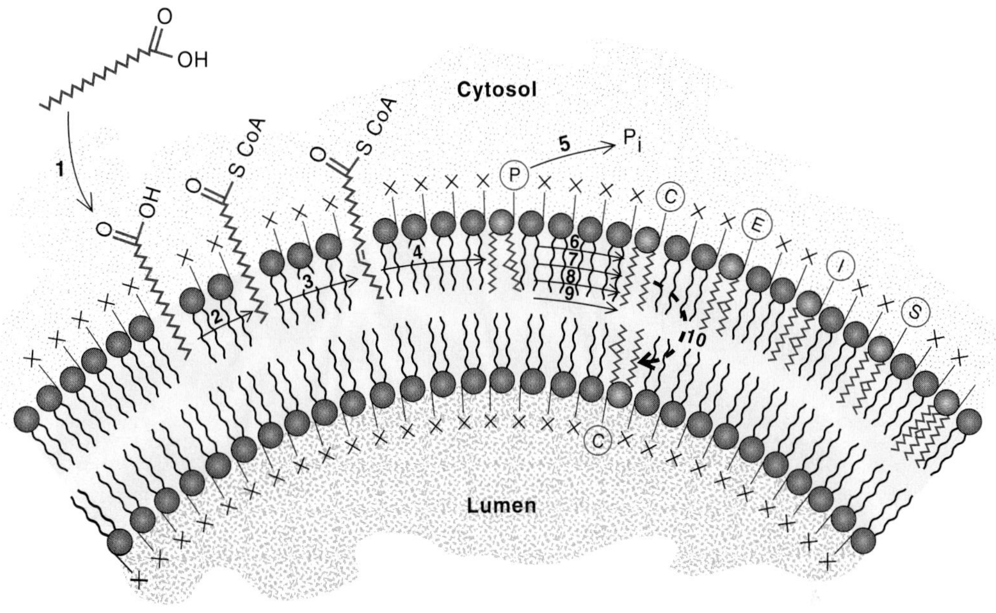

Glycerolipid synthesis on the endoplasmic reticulum from rat liver. Fatty acids are inserted into the cytoplasmic surface of the endoplasmic reticulum (1) and activated to form acyl-CoA thioesters (2). The acyl chains may be elongated or desaturated or both (3). Glycerol-phosphate undergoes acyl-CoA-dependent esterification to form phosphatidic acid (4). The action of phosphatidic acid phosphohydrolase (5) forms diacylglycerols that are converted to phosphatidylcholine and phosphatidylethanolamine by acquisition of phosphocholine and phosphoethanolamine polar head groups (6). Phosphatidylserine synthesis occurs by base exchange (7). Triacylglycerol synthesis occurs by esterification of diacylglycerol (8). CDP-diacylglycerol is an intermediate in the synthesis of phosphatidylinositol (9). Once formed, the glycerolipids may move to the lumenal surface of the endoplasmic reticulum (10). (C = choline; E = ethanolamine; I = inositol; S = serine; X = polar head group C, E, I, or S; P = PO_4.) (Source: From R. M. Bell, L. M. Ballas, and R. A. Coleman, Lipid topogenesis, *J. Lipid Res.* 22:391, 1981.)

Figure 22.15

Proposed mechanisms for regulation of CTP:phosphocholine cytidylyltransferase (CT) in rat liver. The enzyme is found in an inactive form in the cytosol and in an active form bound to the endoplasmic reticulum. The distribution between these two subcellular fractions appears to be primarily regulated by (1) the concentration of phosphatidylcholine, (2) the state of phosphorylation of the enzyme, (3) the concentration of fatty acids, and (4) the concentration of diacylglycerol. The enzyme is also found in the nucleus. How this location relates to the enzyme in the cytoplasm is under investigation at this time.

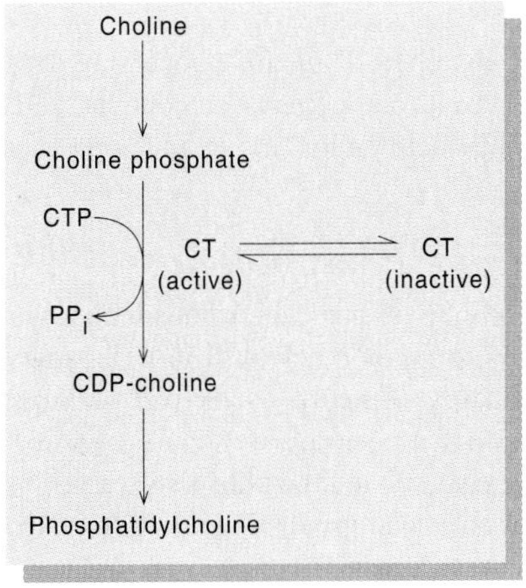

Choline
↓
Choline phosphate
↓ CTP → PP_i (CT (active) ⇌ CT (inactive))
CDP-choline
↓
Phosphatidylcholine

Figure 22.16

Reactions catalyzed by phospholipases. X can be any of the head groups: choline, ethanolamine, serine, glycerol, or inositol.

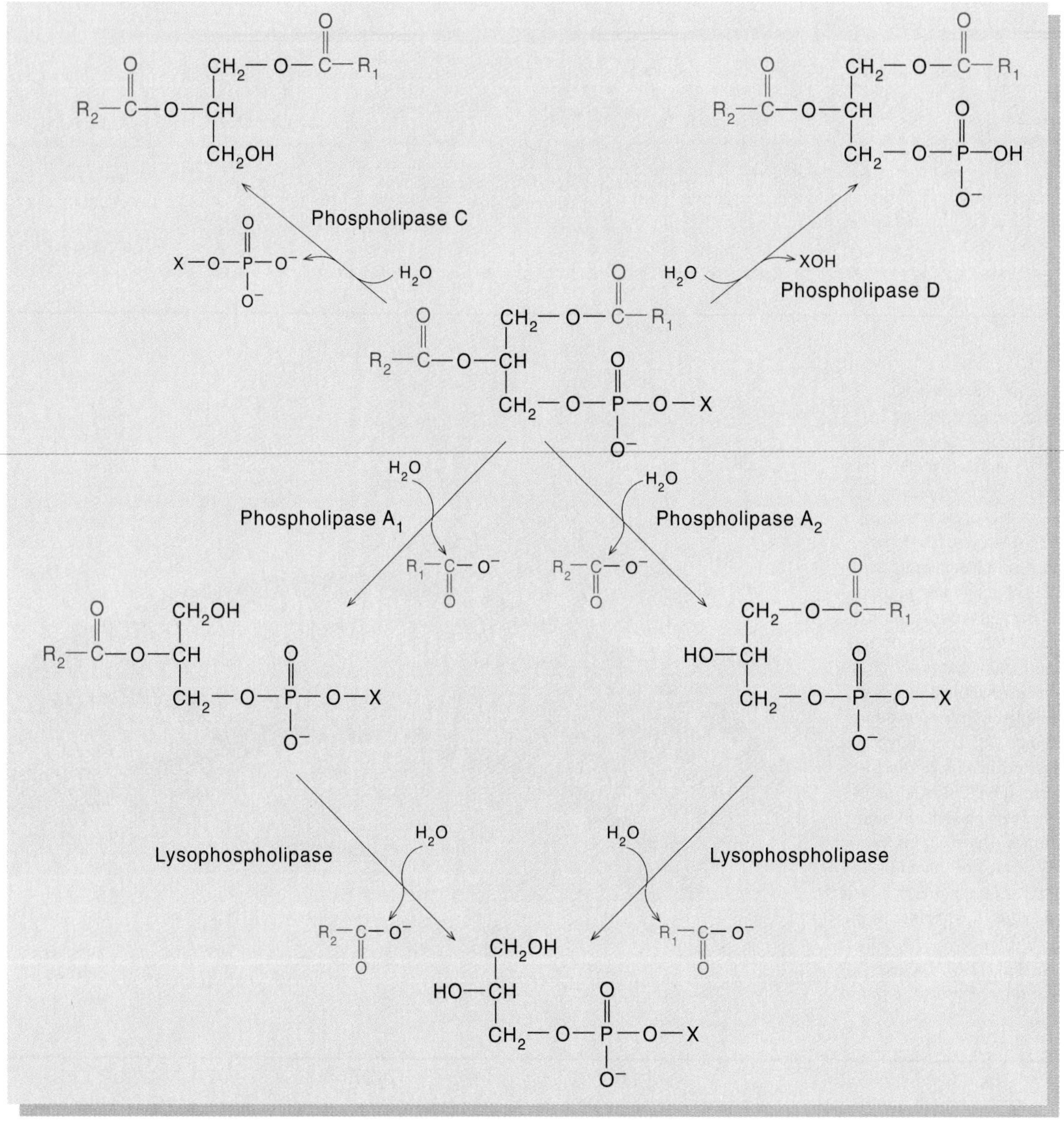

its constituents before being absorbed. Thus this phospholipid always has to be made again from its components. In contrast, cells supplied with sufficient palmitic acid no longer need the enzymes of fatty acid biosynthesis.

Phospholipases Degrade Phospholipids

Enzymes that degrade phospholipids are called phospholipases. They are classified according to the bond cleaved in a phospholipid (fig. 22.16). Phospholipases A_1 and A_2 selectively remove fatty acids from the SN-1 and SN-2 positions, respectively. Phospholipase C cleaves between glycerol and the phosphate moieties; phospholipase D hydrolyzes the head-group moiety X from the phospholipid. Lysophospholipids, which lack a fatty acid at the SN-1 or SN-2 position, are degraded by lysophospholipases.

Phospholipases are found in all types of cells and in various subcellular locations within eukaryotic cells. Some of these enzymes are specific for particular polar head groups; others are nonspecific. Phospholipase A_2 is a major component of snake venom (cobra and rattlesnake) and is partially involved in the deadly effects of these venoms. Because of the high concentration of phospholipase A_2 in these venoms, this enzyme has been studied intensively. The pancreas is also rich in phospholipase A_2, which is secreted into the intestine for digestion of dietary phospholipids.

Figure 22.19

Outline of biosynthesis of some glycosphingolipids. (Glc = glucose, Gal = galactose, GalNAc = *N*-acetylgalactosamine, NeuAc = *N*-acetylneuraminic acid.)

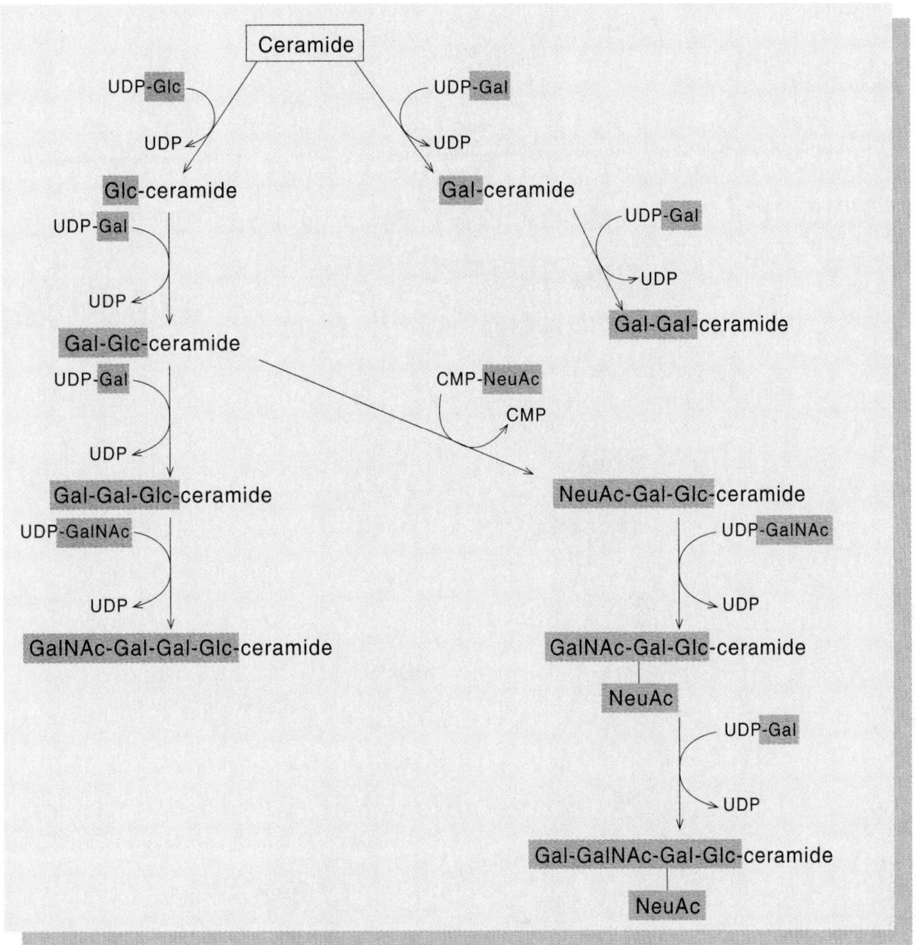

each reaction. Most of the glycosyltransferases involved in the biosynthesis of the glycosphingolipids are located on the lumenal side of the Golgi apparatus. The nucleotide sugars used in biosynthesis are transferred across the membrane into the lumen of the Golgi apparatus by a transporter located on the apparatus' membrane. The biosynthesis of certain key glycosphingolipids is shown in figure 22.19.

Sphingolipids Function as Structural Components, as Specific Cell Receptors, and as Second-Messenger Precursors

The glycosphingolipids are found largely in the plasma membrane and are oriented asymmetrically in the bilayer, with the carbohydrate moieties facing exclusively toward the outside of the cell. The glycosphingolipids appear to have a structural role particularly in the myelin sheath and the membrane of the red cell. However, many observations indicate that the glycosphingolipids are more than just structural lipids and have specific cell surface recognition properties that are similar to plasma-membrane-bound glycoproteins (see chapter 18). This conclusion is consistent with the finding that many cell surface recognition properties are due to the protruding carbohydrate moieties, which may be attached to either lipids or proteins. For example, some glycosphingolipids are blood group antigens. A person who displays the B blood type has the same B antigen oligosaccharide as a component of both a membrane-bound glycosphingolipid and glycoprotein. Moreover, the carbohydrate moieties displayed by the glycosphingolipids and glycoproteins differ according to the organism's stage of development. Cells transformed by tumor viruses also display altered glycosphingolipids, just as is the case with membrane-bound glycoproteins.

Sphingomyelin also has a structural role in cells and myelin, from which it was first isolated, hence the name. Sphingomyelin is particularly enriched in the plasma membrane, but the reason for this is not understood. The catabolic products of sphingomyelin degraded by a phospholipase C–like enzyme (sphingomyelinase) are ceramide and phosphocholine. When a certain polypeptide factor (tumor necrosis factor) binds to the

Figure 22.20

Catabolism of some glycosphingolipids. The glycosyl hydrolase enzymes involved in these reactions are localized in the lysosomes.

Gal-GalNAc-Gal-Glc-ceramide (GM_1)
NeuAc
β-Galactosidase
Gal
GalNAc-Gal-Glc-ceramide (GM_2)
NeuAc
β-*N*-acetylhexosaminidase A
(defective in Tay-Sachs disease)
GalNAc
NeuAc-Gal-Glc-ceramide (GM_3)
GalNAc-Gal-Gal-Glc-ceramide
β-*N*-acetylhexosaminidase
GalNAc
Gal-Gal-Glc-ceramide
Neuraminidase
NeuAc
α-Galactosidase A
(defective in Fabry's disease)
Gal
Gal-Glc-ceramide
β-Galactosidase
Gal
Glc -ceramide
β-Glucosidase
Glc
Gal-Gal-ceramide
α-Galactosidase
Gal
Gal -Ceramide
β-Galactosidase
Gal
Ceramide
Ceramidase
Long-chain base
Fatty acid

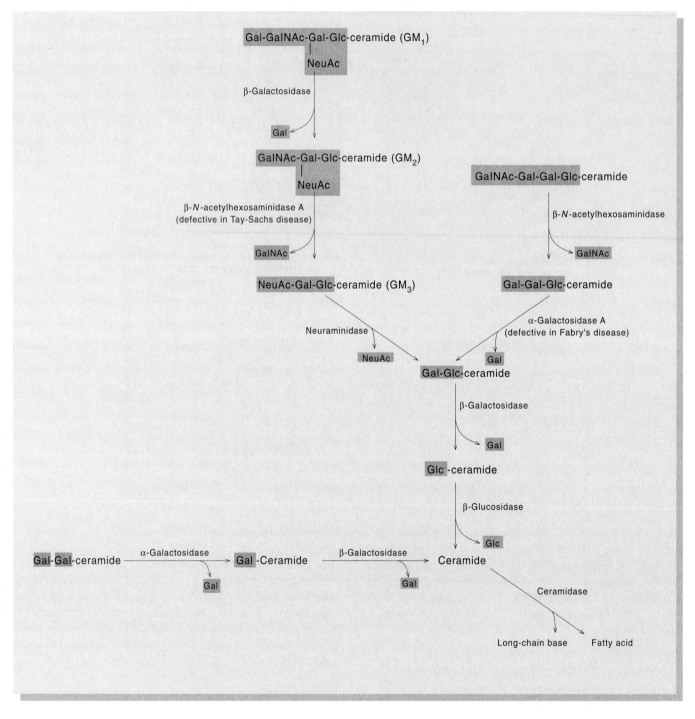

surface of monocyte precursor cells, a plasma membrane sphingomyelinase is activated and generates ceramide, which activates both a protein kinase and a protein phosphate phosphatase, which appear to be involved in the differentiation of these cells into monocytes (white blood cells important in the immune system). Ceramide generated from sphingomyelin has also been implicated in several other signaling pathways. Thus we must also think of sphingomyelin as a precursor of ceramide that can act as a second messenger in the plasma membrane of some cells.

Defects in Sphingolipid Catabolism Are Associated with Metabolic Diseases

The degradation of glycosphingolipids occurs in a stepwise fashion in lysosomes (fig. 22.20). Thus the sphin-

Table 22.1
Inherited Diseases of Sphingolipid Catabolism

Disease	Enzyme Activity That Is Deficient	Reaction
1. Ceramidase deficiency: Farber's lipogranulomatosis	Ceramidase	Ceramide $\longrightarrow$ Fatty acid + Long-chain base
2. Sphingomyelin lipidosis: Niemann-Pick disease	Sphingomyelinase	Sphingomyelin $\longrightarrow$ Ceramide + Phosphocholine
3. Glucosylceramide lipidosis: Gaucher's disease	β-Glucosidase	Glc-ceramide $\longrightarrow$ Glc + Ceramide
4. Galactosylceramide lipidosis: globoid cell leukodystrophy	β-Galactosidase	Gal-ceramide $\longrightarrow$ Gal + Ceramide
5. Sulfatide lipidosis: metachromatic leukodystrophy	Arylsulfatase A	3′-SO_3^--Gal-ceramide $\longrightarrow$ Gal-ceramide + SO_4^{2-}
6. Fabry's disease	α-Galactosidase A	Gal-Gal-Glc-ceramide $\longrightarrow$ Gal + Gal-Glc-ceramide
7. GM_1[a] gangliosidosis	GM_1-β-galactosidase	$GM_1 \longrightarrow$ Gal + GM_2
8. Tay-Sachs disease (GM_2 gangliosidosis)	Hexosaminidase A	$GM_2 \longrightarrow GM_3$ + GalNAc
9. Sandhoff's disease	Hexosaminidases A + B	$GM_2 \longrightarrow GM_3$ + GalNAc

[a]GM is the general abbreviation for a ganglioside. Subscripts are added to distinguish different members of this group of the glycosphingolipids.

golipids, like all the components in the cell, are constantly being biosynthesized and degraded (i.e., the cell constituents turn over). Investigations on sphingolipid degradation have been stimulated by the occurrence of a number of human genetic diseases, the sphingolipidoses, each of which results from the accumulation of one of these lipids. Each of these diseases is caused by a deficiency in the activity of an enzyme involved in the catabolism of one of the sphingolipids. While there is a human disease that results from a metabolic defect in each one of the degradation reactions shown in figure 22.20 and listed in table 22.1, herein we will limit the discussion to two of the better-known disorders.

Fabry's disease was described independently in 1898 by J. Fabry and W. Anderson. Characteristic symptoms are skin rash, pain in the extremities, and renal impairment accompanied by hypertension. Patients lead a reasonably normal life until their fourth decade, when the kidneys usually fail. Little progress was made in understanding this disease until 1963, when Charles Sweeley and Bernard Klionsky described the structure of Gal-Gal-Glc-ceramide as the major lipid that accumulates in the Fabry-affected kidneys. The accumulation of this lipid throughout the body results in the symptoms described above. In 1967, Roscoe Brady demonstrated that the enzymatic defect was a deficiency of the enzyme that degrades this trihexosyl-ceramide. This enzyme was later shown to be α-galactosidase A (see fig. 22.20). The cDNA for the enzyme and the gene that encodes the enzyme have recently been cloned and sequenced. It is now possible to diagnose the disease with biochemical tests based on these discoveries. For example, in prenatal diagnosis, cells are obtained from the amniotic fluid by amniocentesis and culturing. It is then possible to screen the DNA of these cells to see whether a normal gene is present or, alternatively, assay the activity of α-galactosidase A. There is currently no effective treatment for Fabry's disease other than such drastic procedures as a kidney transplant. Fortunately, Fabry's disease is rare, only a few hundred cases being reported throughout the world.

The disease described by Warren Tay in 1881, now known as Tay-Sachs disease, occurs much more frequently. It is estimated that 30–50 children with Tay-Sachs disease are conceived each year in the United States. This disease is devastating, and the children usually do not survive beyond the age of 3. In Tay-Sachs disease the glycosphingolipid abbreviated as GM_2 (see fig. 22.20) accumulates, especially in the brain, as a result of the deficiency of β-*N*-acetylhexosaminidase A.

There is no way of treating Tay-Sachs disease. Enzyme replacement is not considered a likely therapy because infused enzyme cannot penetrate the blood–brain barrier. However, the incidence of the disease has been dramatically decreased by prenatal diagnosis. Tay-Sachs disease is an autosomal recessive disease and so can arise only if both parents are carriers, i.e., if each parent carries a single defective gene for the hexosaminidase A enzyme. In that case there is a 25% chance that a child of these parents will have the disease.

Eicosanoids Are Hormones Derived from Arachidonic Acid

Eicosanoids are a diverse group of hormones, most of which are derived from the C_{20} polyunsaturated fatty acid, arachidonic acid ($20.4^{\Delta 5,8,11,14}$). Most prominent among the group is a series of cyclopentanoic acids known as prostaglandins (PG)

Figure 22.21

Structure of prostaglandin E_2. All prostaglandins are derived from C_{20} fatty acids. The PG stands for prostaglandin. The E specifies the position of oxygen substituents, and the subscript refers to the number of double bonds. From the structure it can be seen that the double bonds of PGE_2 are *trans*Δ^{13} and *cis*Δ^5.

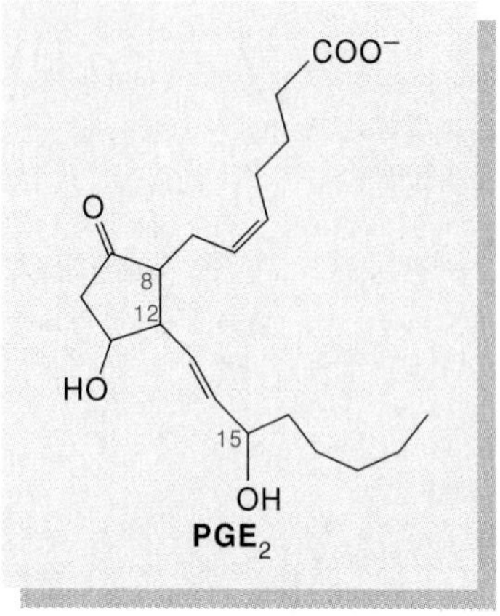

PGE_2

Figure 22.22

The structure of the thromboxane A_2. Thromboxanes differ from prostaglandins in having a cyclic ether (oxane) ring structure.

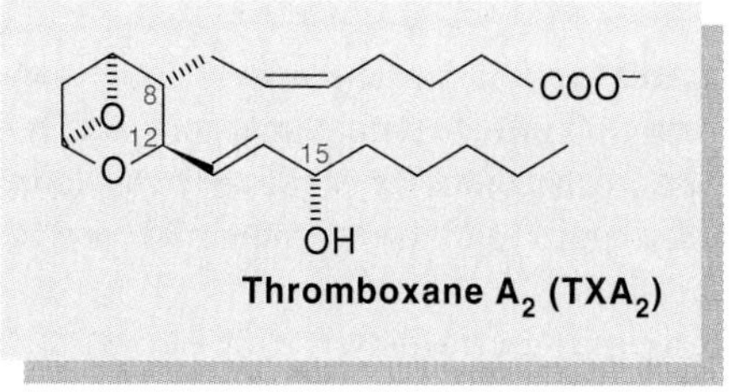

Thromboxane A_2 (TXA_2)

Figure 22.23

Reaction catalyzed by prostaglandin endoperoxide synthase. This enzyme has two catalytic activities, as indicated next to the reaction arrows.

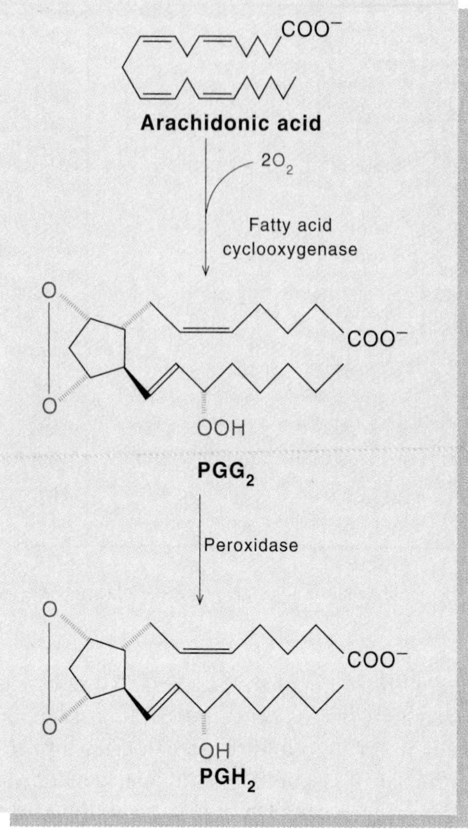

(fig. 22.21). Closely related to the prostaglandins are the eicosanoids known as thromboxanes (TX) (fig. 22.22). They differ from prostaglandins in the ring structure, which for thromboxanes is a cyclic ether (oxane ring) rather than the cyclopentane ring of the prostaglandins.

Oxygenated eicosanoids appear to be unique to animal cells. The first oxygenated eicosanoids to be discovered, the prostaglandins, were extracted from human semen in the 1930s by Ulf von Euler in Sweden. These compounds, which he believed to arise from the prostate gland (hence the name), were injected into animals and caused the uterus to contract and lowered blood pressure. It is now known that Euler's prostaglandins were from the seminal gland, not the prostate, and that they are present in most tissues of both male and female animals. Since the prostaglandins are present only transiently and at very low concentrations, the structures were not elucidated until the 1960s and later, after the development of sensitive instruments such as the gas chromatograph–mass spectrometer. Sune Bergström and Bengt Samuelsson, also from Sweden, were pioneers in studies on the structure and metabolism of the eicosanoids. Pharmacological interest in prostaglandins was awakened in 1971 by John Vane's discovery that aspirin blocked the synthesis of prostaglandins. Bergström, Samuelsson, and Vane were awarded the 1982 Nobel Prize in Medicine for their important discoveries.

Eicosanoid Biosynthesis

Arachidonic acid is not present in significant amounts in tissues as the free acid but is stored as a fatty acid at the SN-2 position of phospholipids. Prostaglandin biosynthesis is initiated by the interaction of a stimulus with the cell surface. Depending on the cell type, the stimulus can take the form of a hormone, such as angiotensin II or antidiuretic hormone, or a protease such as thrombin (involved in blood clotting), or both hormone and protease. These agents bind to a specific receptor that activates a phospholipase A_2, discussed earlier in this chapter, that specifically releases the arachidonic acid from a phospholipid such as phosphatidylcholine. The release of arachidonic acid by phospholipase A_2 is believed to be the rate-limiting step for the biosynthesis of eicosanoids.

The initial steps for the synthesis of prostaglandins and thromboxanes are the oxidation and cyclization of arachidonic acid to yield the prostaglandins PGG_2 and PGH_2 (fig. 22.23). (In the abbreviation, PG stands for prostaglandin. The third letter, G or H, refers to the particular structure, and the subscript 2 notes the

Figure 22.24

Aspirin inactivates cyclooxygenase by acetylation of a serine, probably at or near the active site.

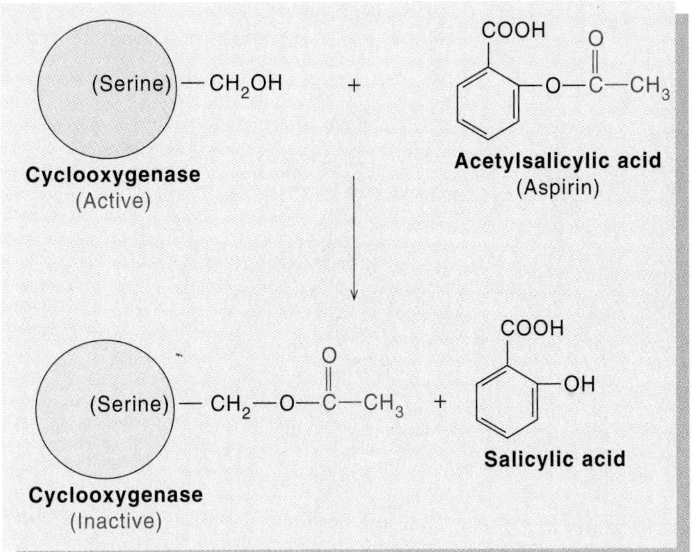

number of double bonds.) This reaction is catalyzed by a single bifunctional enzyme, prostaglandin endoperoxide synthase, that is present on the endoplasmic reticulum. The formation of PGG_2 is catalyzed by the cyclooxygenase component of the enzyme (see fig. 22.23), and the subsequent formation of PGH_2 is catalyzed by the peroxidase component. The cyclooxygenase activity has the unusual property of catalyzing its own destruction. Approximately once in every 1,400 substrate turnovers, the cyclooxygenase activity is irreversibly inactivated. This "suicide reaction" occurs both in vivo and in vitro. The suicide process is an unusual regulatory mechanism, which places an upper limit on cellular prostaglandin biosynthetic activity.

The anti-inflammatory effect of aspirin appears to be due to the acetylation of a serine residue at the active site of the cyclooxygenase, which irreversibly inactivates the enzyme (fig. 22.24). Aspirin has no effect on the peroxidase activity of the enzyme.

PGH_2 is the precursor of various prostaglandins and thromboxanes. Each reaction is catalyzed by a separate tissue-specific enzyme as indicated in figure 22.25. The tissue localization of the enzyme relates to the role of the eicosanoid; for example, thromboxane A_2 (TXA_2) is made in platelets and causes them to aggregate, as occurs in blood clotting, whereas prostaglandin I_2 (PGI_2) is made in arterial walls and inhibits platelet aggregation.

In addition to the formation of prostaglandins and thromboxanes, arachidonic acid is metabolized to several other compounds. The hydroxyeicosatetraenoic acids, or HETEs (fig. 22.26), are formed from arachidonic acid by the enzyme lipoxygenase. Mammalian lipoxygenases found in white blood cells catalyze the insertion of oxygen in the 5, 12, or 15 position of various eicosanoic acids. The reaction mechanism for these insertions involves addition of oxygen to a double bond with the formation of a conjugated *cis–trans* diene (fig. 22.27). Subsequently, the hydroperoxy group is reduced to an alcohol to form the corresponding hydroxy-eicosanoic acid. The physiological functions of hydroxy-eicosanoic acids are not well understood.

The rate of synthesis of lipoxygenase-derived compounds is controlled by both the release of arachidonic acid and the activation of the lipoxygenase. Certain peptides, which promote chemically stimulated movement (chemotaxis) of cells, cause the release of arachidonic acid and the activation of the 5-lipoxygenase.

If the eicosanoids are to serve a useful hormonal function, there must be a rapid means for their removal when their action is no longer required. Consistent with this principle, it has been found that eicosanoids have very short half-lives in vivo; injected prostaglandins do not survive a single pass through the circulatory system.

Eicosanoids Exert Their Action Locally

The diversity of eicosanoid structures is paralleled by a diversity of biological effects. They are generally considered to be hormones that exert their main effect locally. Their target sites include both the cells in which they are formed and neighboring, different cell types. In all cases the effects of prostaglandins are mediated through specific cell surface receptors, as is the case for other hormones that bind to the outside of the plasma membrane.

Specific binding of PGE_2 to a variety of cells correlates with an activation of adenylate cyclase and the accumulation of cAMP. In human adipocytes, PGE_2 causes a 15-fold increase in the concentration of cAMP. In platelets, PGI_2 rather than PGE_2 appears to mediate the increase in cAMP.

The eicosanoids have been implicated as mediators of tissue inflammation, since aspirin's anti-inflammatory effect was shown to be the result of its inactivation of cyclooxygenase. How eicosanoids cause tissue inflammation is the focus of much current research.

The role of prostaglandins in blood clotting is also stimulating a great deal of research. PGI_2 relaxes coronary arteries and inhibits platelet aggregation. TXA_2 has the opposite effects (fig. 22.28). PGI_2 is made in endothelial cells that line blood vessels and inhibits platelet aggregation by binding to a receptor on the plasma membrane of platelets, which causes an increase in cAMP. TXA_2 suppresses the PGI_2-mediated increase in cAMP. It has been speculated that the synthesis of PGI_2 may prevent platelets from binding to arterial walls. In damaged areas of arteries (caused by the disease atherosclerosis, which involves cholesterol metabolism as discussed in chapter 23) the synthesis of PGI_2 may be decreased, and the presence of TXA_2 would cause platelets to aggregate, leading to the formation of a blood clot (thrombosis, a major cause of heart attack and stroke). Interestingly, besides the anti-inflammatory effect of aspirin, a low-dose aspirin (one "baby" aspirin daily or one regular aspirin every three days) has proved useful in the prevention of heart attacks. The low dose of aspirin leads to selective inhibition of platelet thromboxane formation (thus decreased platelet aggregation) without affecting the synthesis of other eicosanoids in other cells.

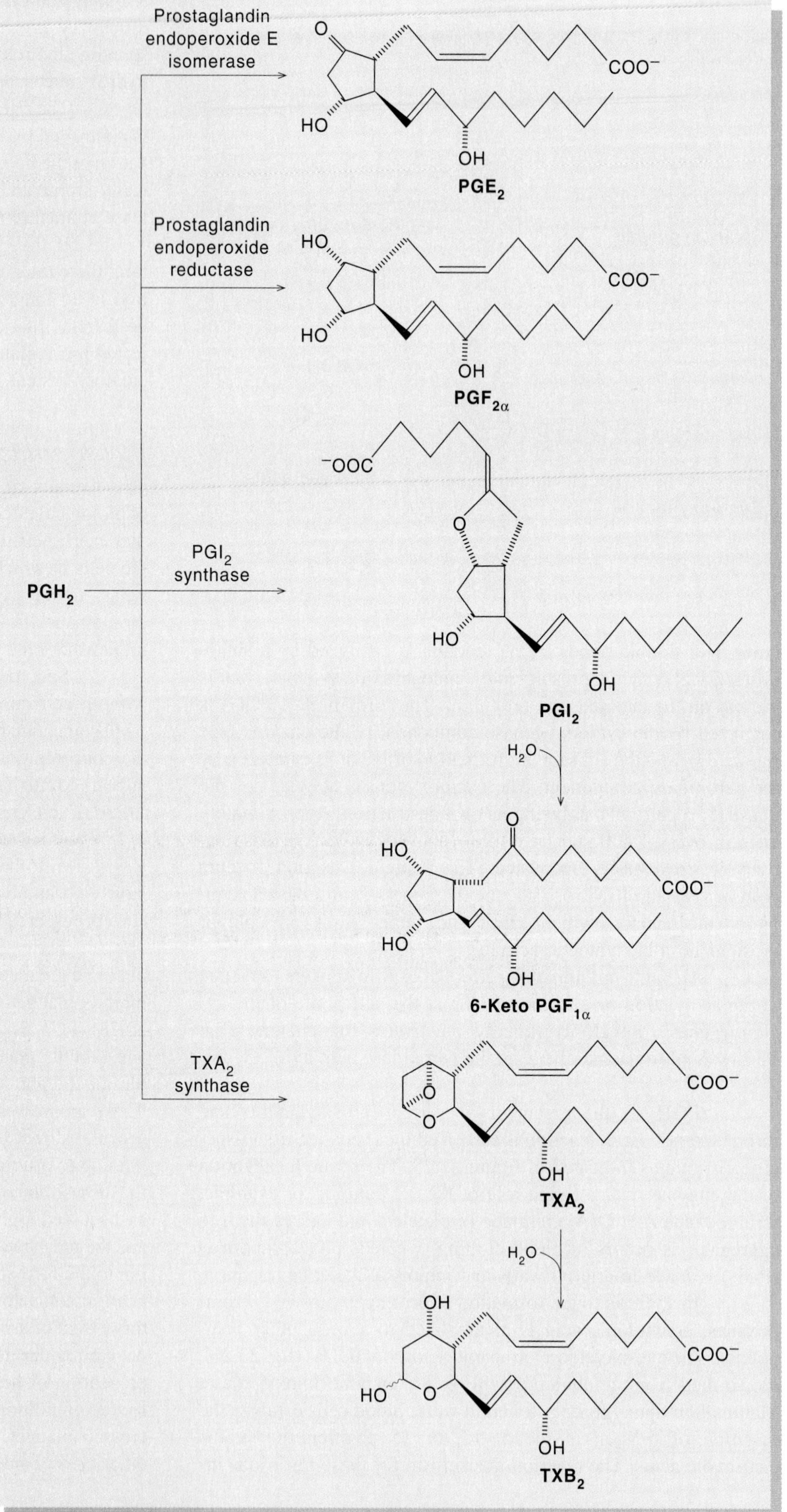

Figure 22.25

Formation of prostaglandins and thromboxanes.

Figure 22.26

Structures of three common hydroxyeicosatetraenoic acids (HETEs). These compounds are all formed directly from arachidonic acid by the action of a specific lipoxygenase, which inserts a hydroxyl group at the indicated location.

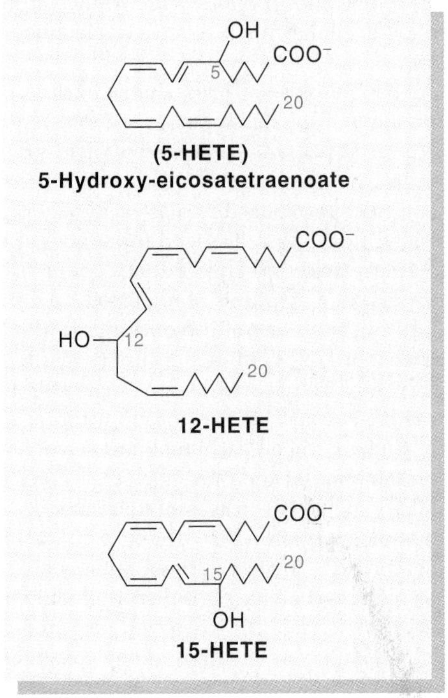

(5-HETE)
5-Hydroxy-eicosatetraenoate

12-HETE

15-HETE

Figure 22.27

The mechanism of the reaction catalyzed by lipoxygenases. This is a simple dioxygenase reaction; there is no net oxidation–reduction of either the fatty acid or the oxygen. A *cis–trans* conjugated diene is formed in the reaction.

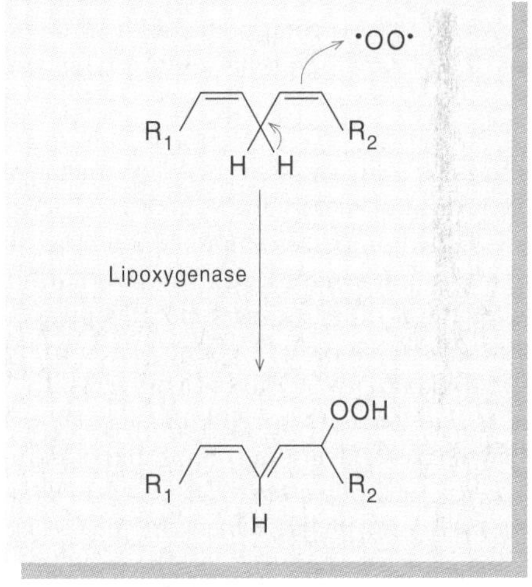

Figure 22.28

Possible actions of PGI_2 and TXA_2 in blood clotting and thrombosis are shown. As indicated in the figure, blood platelets make TXA_2, which promotes constriction of coronary arteries and platelet aggregation. Endothelial cells that line blood vessels make PGI_2, which opposes the action of TXA_2; PGI_2 relaxes coronary arteries and inhibits platelet aggregation. Atherosclerotic lesions commonly develop in humans, often due to hypercholesterolemia (see chapter 23), and lack endothelial cells. There is an increased risk for adherence of platelets to the surface of these lesions. A rise in the concentration of TXA_2 that might occur in platelets attached to these lesions cannot be opposed by PGI_2 because of the absence of endothelial cells. Thus there is an increased risk for formation of a blood clot (thrombosis). If the blood passage is blocked, oxygen and energy supply to the tissue will be decreased leading to tissue death. If a major artery is involved, heart attack or stroke can occur.

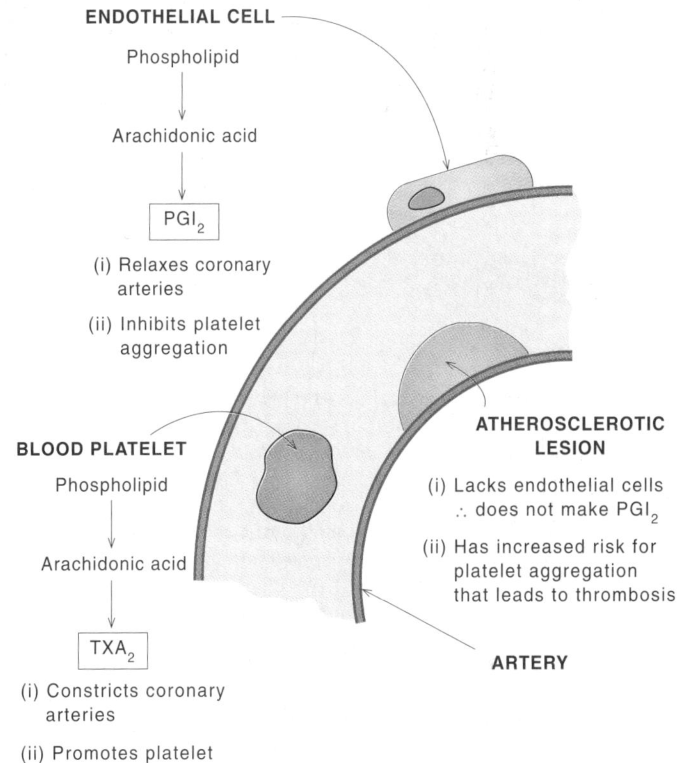

Summary

Fatty acids are the components of a rich variety of complex lipid molecules that play critical structural roles in membranes. Some of these lipids are also the precursors of compounds with hormone or second-messenger activities. In this chapter we focused on the metabolism of these compounds with some mention of their functions. The following points are the highlights of our discussion.

1. Lipid synthesis is unique in that it is almost exclusively localized to the surface of membrane structures. The reason for this restriction is the amphipathic nature of the lipid molecules. Phospholipids are biosynthesized by acylation of either glycerol-3-phosphate or dihydroxyacetone phosphate to form phosphatidic acid. This central intermediate can be converted into phospholipids by two different pathways. In one of these, phosphatidic acid reacts with CTP to yield CDP-diacylglycerol, which in bacteria is converted to phosphatidylserine, phosphatidylglycerol, or diphosphatidylglycerol. In *E. coli,* the major phospholipid, phosphatidylethanolamine, is synthesized by means of this route through the decarboxylation of phosphatidylserine. In the second pathway, found in eukaryotes, phosphatidic acid is hydrolyzed to diacylglycerol, which reacts with CDP-ethanolamine or CDP-choline to yield phosphatidylethanolamine or phosphatidylcholine, respectively. Alternatively, the diacylglycerol may react with acyl-CoA to form triacylglycerol.
2. There are numerous reactions by which the acyl groups or polar head groups of phospholipids might be modified or exchanged. Phospholipids are degraded by specific phospholipases. How phospholipids move from their site of synthesis to other organelles (e.g., phosphatidylcholine is made on the endoplasmic reticulum and is delivered to mitochondria) is a topic of current research.
3. Phospholipids also serve as a reservoir for cellular second messengers. Catabolism of phosphatidylinositol-4,5-bisphosphate by a specific phospholipase C yields inositol-1,4,5-P_3, which mobilizes calcium and diacylglycerol, which activates protein kinase C. A specific phospholipase A_2 is utilized for the release of arachidonic acid, which is converted into eicosanoids. Other phospholipases are found throughout the cell and in extracellular fluids such as pancreatic juice and snake venom.
4. In *E. coli,* phospholipids are used almost exclusively as structural components of the cell membranes, and regulation is known to occur at an early stage in fatty acid synthesis. In the mammalian liver, fatty acids are important precursors of both the structural phospholipids and also of the energy-storage lipid, triacylglycerol. The cell's need for structural lipids is satisfied before fatty acids are shunted into energy storage. The rate of triacylglycerol synthesis is regulated by the availability of fatty acid from both diet and biosynthesis. The regulation of phosphatidylcholine biosynthesis occurs primarily at the reaction catalyzed by CTP:phosphocholine cytidylyltransferase. This enzyme is activated by translocation from a soluble form, where it is inactive, to cellular membranes, where it is activated. The regulation of the biosynthesis of other phospholipids in eukaryotic cells is less well understood.
5. The sphingolipids are important structural lipids found in eukaryotic membranes. The acylation of sphingenine produces ceramide, which reacts with phosphatidylcholine to give sphingomyelin. Ceramide also can react with activated carbohydrates (e.g., UDP-glucose) to form the glycosphingolipids. Studies on the catabolism of the sphingolipids have revolved around inherited diseases, sphingolipidoses, that result from a defect in lysosomal enzymes that degrade sphingolipids.
6. Prostaglandins and thromboxanes are hormonelike substances that affect the function of a cell by binding to a receptor on the cell surface. Prostaglandins and thromboxanes are biosynthesized from C_{20} polyunsaturated fatty acids, primarily arachidonic acid. In the initial reaction, arachidonic acid is converted to PGH_2 by prostaglandin endoperoxide synthase. PGH_2 is then converted to PGE_2, PGF_2, TXA_2, and PGI_2.

Selected Readings

Dennis, E. A., Diversity of group types, regulation, and function of phospholipase A_2. *J. Biol. Chem.* 269:13057–13060, 1994. An article that provides a recent update on developments in the phospholipase field.

Jackowski, S., J. E. Cronan, and C. O. Rock, Lipid metabolism in procaryotes. In D. E. Vance and J. E. Vance (eds.), *Biochemistry of Lipids, Lipoproteins and Membranes.* Amsterdam: Elsevier Science Publishers, 1996. This chapter (2) provides an advanced treatment of genetics and metabolism of phospholipids in *E. coli.*

Kolesnick, R., and D. W. Golde, The sphingomyelin pathway in tumor necrosis factor and interleukin-1 signaling. *Cell* 77:325–328, 1994. An article that provides recent information on sphingomyelin as a precursor of ceramide, a recently identified cellular second messenger.

Merrill, A. H., and C. C. Sweeley, Sphingolipids: metabolism and cell signalling. In D. E. Vance and J. E. Vance (eds.), *Biochemistry of Lipids, Lipoproteins and Membranes.* Amsterdam: Elsevier Science Publishers, 1996. This is an advanced chapter (12) on the chemistry, metabolism, and function of the sphingolipids.

Scriver, C. R., A. I. Beaudet, W. S. Sly, and D. Valle (eds.), *The Metabolic Basis of Inherited Disease,* 6th ed., vol. 2, New York: McGraw Hill, 1989. This book contains indepth chapters on sphingolipid metabolism and many of the sphingolipidoses.

Smith, W. I., and F. A. Fitzpatrick, The eicosanoids: Cyclooxygenase, lipoxygenase and epoxygenase pathways. In D. E. Vance and J. E. Vance (eds.), *Biochemistry of Lipids, Lipoproteins and Membranes.* Amsterdam: Elsevier Science Publishers, 1996. This chapter (11) provides advanced information on the biochemistry, metabolism, and functions of the eicosanoids.

Vance, D. E. (ed.), *Phosphatidylcholine Metabolism,* Boca Raton, Fla.: CRC Press, 1989. This advanced monograph contains 13 chapters on all aspects of phosphatidylcholine biosynthesis and catabolism.

Vance, D. E., Glycerolipid biosynthesis in eucaryotes. In D. E. Vance and J. E. Vance (eds.), *Biochemistry of Lipids, Lipoproteins and Membranes.* Amsterdam: Elsevier Science Publishers, 1996. This chapter (6) covers phospholipid metabolism at an advanced level and discusses the role of phosphatidylinositol and other phospholipids in the generation of second messengers in the cell.

Waite, M., *The Phospholipases.* New York: Plenum Press, 1987. This book provides a complete coverage of phospholipases from the technicalities of assay of phospholipases to the proposed mechanism of catalysis.

Waite, M., Phospholipases. In D. E. Vance and J. E. Vance (eds.), *Biochemistry of Lipids, Lipoproteins and Membranes.* Amsterdam: Elsevier Science Publishers, 1996. This chapter (8) provides advanced knowledge about phospholipases and the hydrolysis of phospholipids.

Weissmann, G., Aspirin. *Sci. Am.* 264:84–90, 1991. An interesting article that discusses the mechanism of action of aspirin in connection with eicosanoid metabolism.

PROBLEMS

1. A logic question: Inositol (fig. 22.7) is a component of several molecules in this chapter. If inositol is thought of as 1,2,3,4,5,6-hexahydroxycyclohexane, how many geometric isomers are there and how many are optically active?
2. During the biosynthesis of 3-ketosphinganine (fig. 22.18) palmitoyl-CoA and serine combine with the loss of CoASH and CO_2 to produce 3-ketosphinganine. Which carbon of serine is lost as CO_2 and what cofactor might you expect to be associated with the enzyme involved in the reaction?
3. The structures of a wide variety of lipids are shown in this chapter. Which items, if any, have chiral carbons?
4. The α hydrogen of serine is preserved during the biosynthesis of 3-ketosphinganine (fig. 22.18). How does this observation contribute to the understanding of the biosynthesis of this substance?
5. Consider that inositol exists in the chair conformation. What conclusions can you reach about the geometry of inositol-1,4,5-trisphosphate (fig. 22.8)?
6. Would you expect phosphatidylserine decarboxylase (fig. 22.3) to be a pyridoxal phosphate enzyme?
7. Sphingolipids as well as other membrane components are constantly degraded and resynthesized. Why does the cell waste so much energy for what appears to be a futile effort?
8. If both parents carry a single defective gene for the hexosaminidase A enzyme, there is a 25% chance that their child will have Tay-Sachs disease. What is the chance that their child will be a carrier of a defective gene?
9. Why don't carriers of a defective gene for the hexosaminidase A enzyme suffer from Tay-Sachs disease (or at least some of the symptoms)?
10. One of the pathways for the biosynthesis of phosphatidic acid in eukaryotes originates with dihydroxyacetone phosphate (from glycolysis). This pathway is shown only as three unlabeled arrows in figure 22.2. Given that the three enzymes starting with dihydroxyacetone phosphate are (1) an acyl transferase, (2) a dehydrogenase, and (3) a second acyltransferase, complete the pathway including any missing substrates.
11. This chapter discusses many of the biosynthetic interconversions that are known to occur with the different phosphatidyl components (e.g., phosphatidylethanolamine, phosphatidylcholine, and phosphatidylinositol) of membranes. Considering only the bioenergetics, what is the major difference between the biosynthetic pathways shown in figures 22.4 and 22.7?
12. Why are human genetic defects in phospholipid biosynthesis not observed?
13. Phosphatidylcholine biosynthesis appears to be regulated principally at the step catalyzed by CTP phosphocholine cytidylyltransferase. Does the type of regulation observed make biochemical sense? Draw a chemical reaction mechanism for this enzyme.
14. The majority of phospholipids contain an unsaturated fatty acid at the C-2 position. Give an example of a phospholipid that has a saturated fatty acid in this position. How is it synthesized?
15. Explain why the reactions from choline to phosphatidylcholine in eukaryotic cells are thermodynamically feasible.
16. Where are ether-linked phospholipids found and what are some their functions?
17. Arachidonic acid is the major precursor of prostaglandins and thromboxanes. If a person were unable to absorb arachidonic acid from the diet but could absorb linoleic acid, could that person still make PGE_2?
18. Low doses of aspirin (one aspirin every other day) are recommended to prevent heart attacks and strokes. Why would three or four tablets per day not work better? (*Hint:* Remember that TXA_2 is made in platelets and that PGI_2 is made in the arterial walls.)
19. Snake venoms contain many types of lipases, including phospholipase A_2. Why would small amounts of this enzyme contribute to some of the toxic effects of snake venom? (Bee venom contains a protein that stimulates phospholipase A_2.)

C H A P T E R

Metabolism of Cholesterol

The tetracyclic molecule of cholesterol synthesized in blocks of five carbon units is a major component of membranes as well as a precursor of steroids and bile acids.

Steroids are tetracyclic hydrocarbons that can be considered derivatives of perhydrocyclopentanophenanthrene. They differ from one another in the degree of saturation of each of the four hydrocarbon rings and in the side-chain substituents attached to these rings.

Steroids are much more common in eukaryotes than in prokaryotes. Cholesterol, the most prominent member of the steroid family, is an important component of many eukaryotic membranes (see chapter 19). In addition, it is the precursor of the other two major classes of steroids: the steroid hormones and the bile acids (fig. 23.1).

Steroid hormones play a key role in the regulation of metabolism. These hormones come in a rich variety, each interacting in a highly specific manner with a receptor protein to effect gene expression in the appropriate target tissue (see chapter 27). Bile acids are the primary degradation product of cholesterol. The bile acids are made in the liver, stored in the gall bladder, and secreted into the small intestine. There they aid in the solubilization of lipids, facilitating their digestion by intestinal lipases.

All of these biologic roles of the steroids figure prominently in human well-being. Defects in cholesterol metabolism are a major cause of cardiovascular disease. It is no wonder that steroids are a central concern in medical biochemistry. In this chapter we will discuss the metabolism of these complex lipids and the plasma lipoproteins in which they and other complex lipids are transported to various tissues.

Biosynthesis of Cholesterol

Early in the 1930s, the structure of cholesterol was finally determined, an achievement that concluded a brilliant chapter in structural organic chemistry. However, the solution to that problem led to the formulation of many new ones. In particular, it was not clear how such a complex structure could be assembled from small molecules.

Work on the biosynthesis of cholesterol began in earnest after Rudolf Schoenheimer and David Rittenberg, at Columbia University, developed isotopic tracer techniques for the analysis of biochemical pathways. In 1941, Rittenberg and Konrad Bloch were able to show that deuterium-labeled acetate ($C^2H_3COO^-$) was a precursor of cholesterol in rats and mice. Subsequently, in collaboration with Edward Tatum and others, Bloch proved that the carbon skeleton of the sterol ergosterol of *Neurospora crassa*

Figure 23.1

Formation of cholesterol and some of its derivatives. All of the carbon atoms of cholesterol are derived from acetyl-CoA by way of mevalonate in a pathway with 33 reaction steps. From cholesterol a wide variety of steroids and bile acids and bile salts are formed. Many of the reactions leading to cholesterol derivatives are organ-specific. Numbers associated with the arrows indicate the approximate number of steps in the sequence, where that is relevant. Most of the reactions are unidirectional. Excess cholesterol is disposed of by conversion to bile acids and excreted as bile salts (as shown). Most of the bile salts are recycled through the liver and intestine.

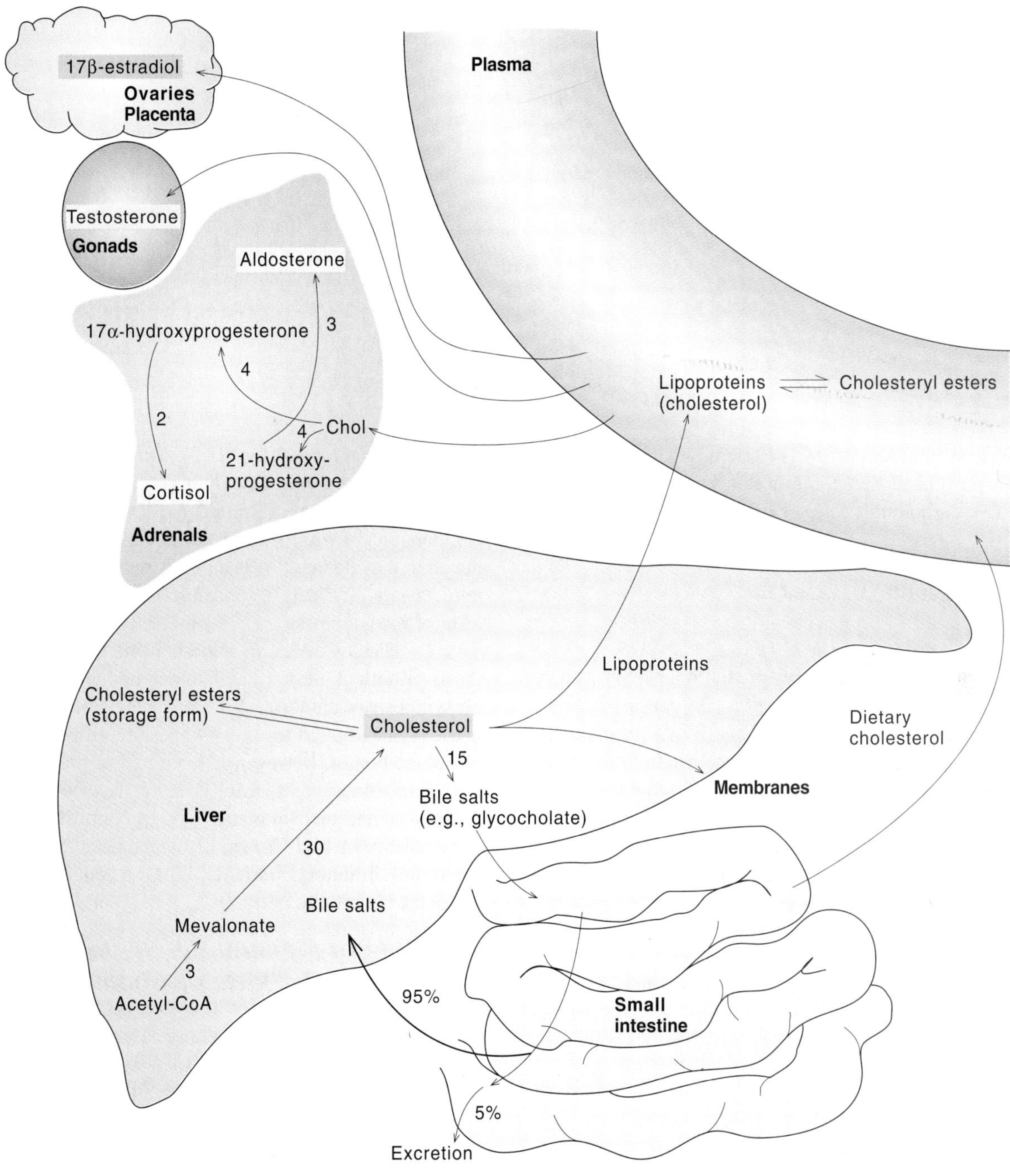

Figure 23.2

Basic scheme for cholesterol biosynthesis proposed by Bloch in 1952 (*left*) and scheme proposed for the cyclization of squalene by Woodward and Bloch in 1953 (*right*).

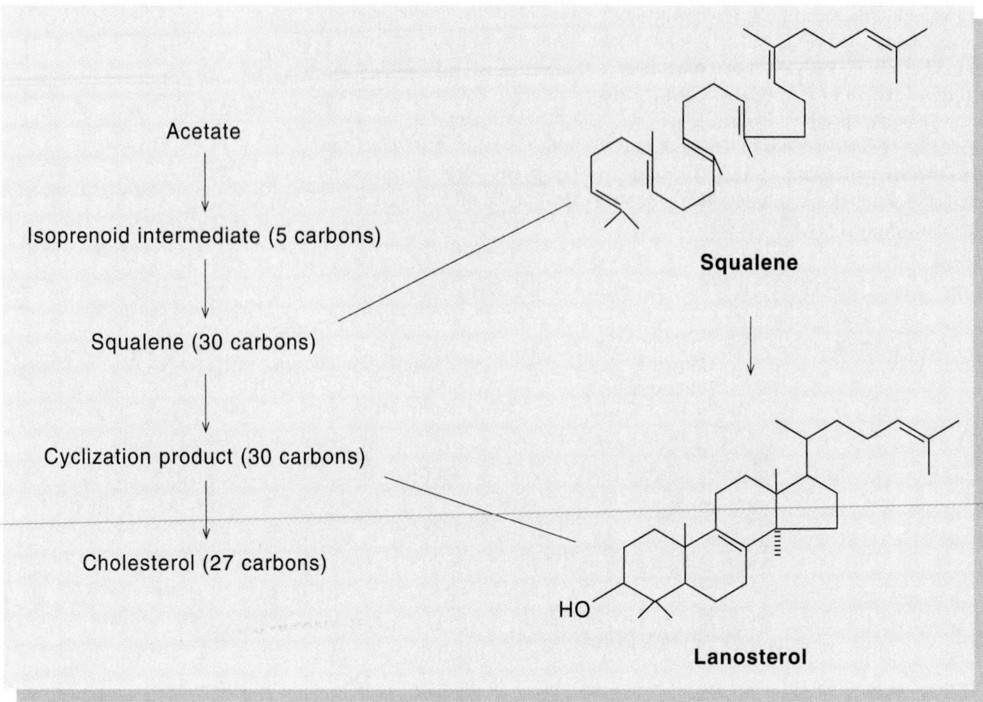

was entirely derived from acetate. In 1949, James Bonner and Barbarin Arreguin postulated that three acetates could combine to form a single five-carbon unit called isoprene:

$$CH_2{=}\underset{\displaystyle H}{\overset{\displaystyle CH_3}{C}}{-}C{=}CH_2$$

This proposal agreed with an earlier prediction of Sir Robert Robinson, that cholesterol was a cyclization product of squalene, a 30-carbon polymer of isoprene units. Thus Bloch postulated a scheme for the biosynthesis of cholesterol as shown in figure 23.2. And in 1952, Bloch and Robert Langdon demonstrated that squalene could readily be converted into cholesterol.

Two difficult problems remained. What was the structure of the isoprenoid intermediate, and how did squalene cyclize to form cholesterol? In 1953, R. B. Woodward and Bloch postulated a cyclization scheme for squalene (see fig. 23.2) that was later shown to be correct. In 1956, the unknown isoprenoid precursor was identified as mevalonic acid by Karl Folkers and others at Merck, Sharpe and Dohme Laboratories. The discovery of mevalonate provided the missing link in the basic outline of cholesterol biosynthesis. Since that time, the sequence and the stereochemical course for the biosynthesis of cholesterol have been defined in detail.

Mevalonate Is a Key Intermediate in Cholesterol Biosynthesis

The sequence of cholesterol biosynthesis begins with a condensation in the cytosol of two molecules of acetyl-CoA, a reaction catalyzed by thiolase (fig. 23.3). The next step requires the enzyme β-hydroxy-β-methylglutaryl-CoA (HMG-CoA) synthase. This enzyme catalyzes the condensation of a third acetyl-CoA with β-ketobutyryl-CoA to yield HMG-CoA. HMG-CoA is then reduced to mevalonate by HMG-CoA reductase. The activity of this reductase is primarily responsible for control of the rate of cholesterol biosynthesis.

HMG-CoA is an important intermediate for the biosynthesis of both cholesterol and ketone bodies (see chapter 20). The biosynthesis of cholesterol is catalyzed by enzymes in the cytosol and enzymes bound to the endoplasmic reticulum. The synthesis of ketone bodies, however, is restricted to the mitochondrial matrix. Thus thiolase and HMG-CoA synthase are found in both mitochondria and cytosol of rat liver. In contrast, HMG-CoA lyase, which cleaves HMG-CoA to ketone bodies (see chapter 20), is located only in mitochondria. HMG-CoA reductase is bound to the endoplasmic reticulum.

The Rate of Mevalonate Synthesis Determines the Rate of Cholesterol Biosynthesis

The thiolase and HMG-CoA synthase exhibit some regulatory properties in rat liver (cholesterol feeding causes a decrease in these enzyme activities in the cytosol but not in the mitochondria). However, primary regulation of cholesterol biosynthesis appears to be centered on the HMG-CoA reductase reaction. HMG-CoA reductase is found on the endoplasmic reticulum, has a molecular weight of 97,092, and consists of 887 amino acids in a single polypeptide chain. The structure of the enzyme

Figure 23.3

Formation of mevalonate. The first two enzymes, thiolase and synthase, are found in both cytosol and mitochondria. The lyase that catalyzes ketone body formation is found only in the mitochondria. The reductase that catalyzes mevalonate formation is found in the endoplasmic reticulum. β-Ketobutyryl-CoA is also known as acetoacetyl-CoA.

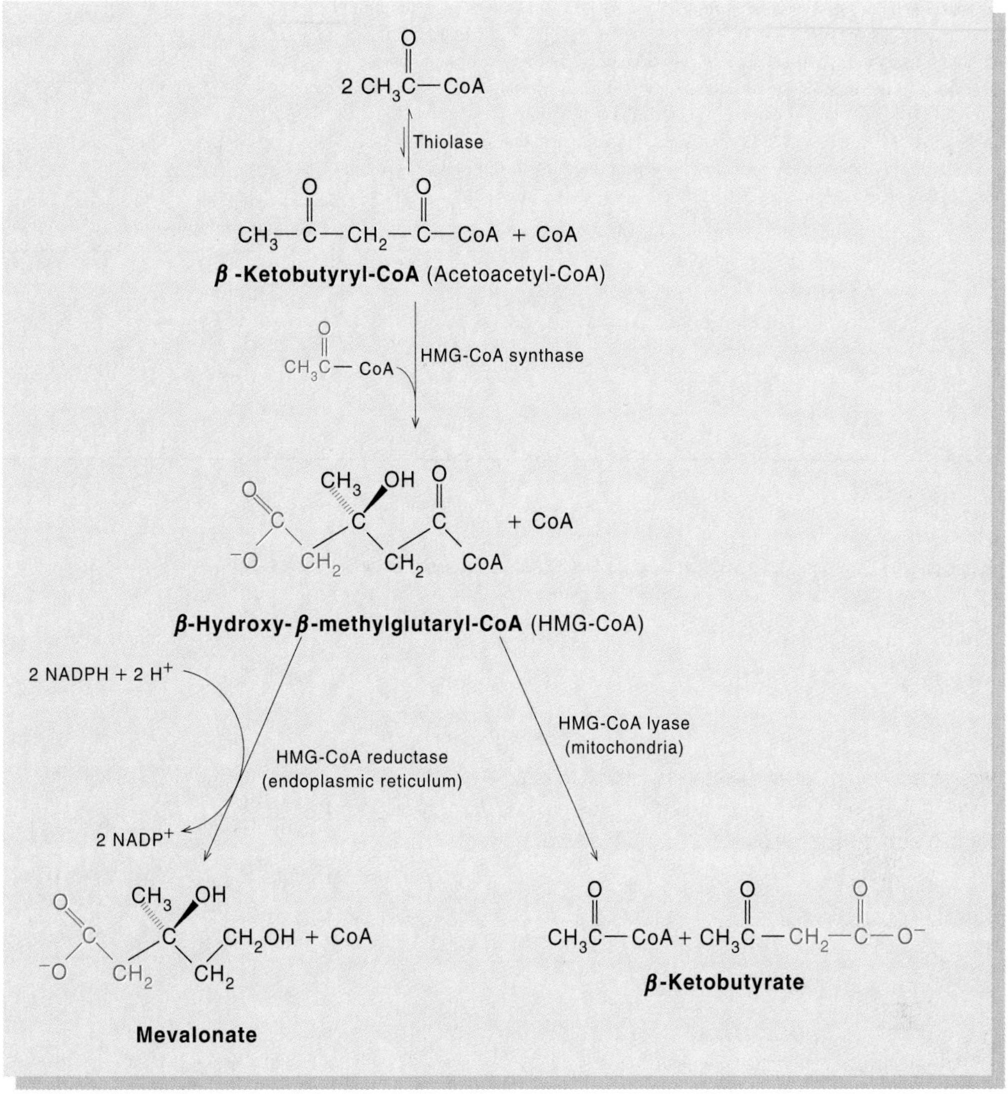

was deduced by Michael Brown and Joseph Goldstein from the sequence of a piece of DNA (cDNA) derived from mRNA that codes for the reductase. The enzyme has two domains (fig. 23.4). The amino-terminal domain has a molecular weight of 35,000, with seven or possibly eight hydrophobic segments that are thought to cross the membrane as shown in figure 23.4. The carboxyl-terminal domain has a molecular weight of 62,000, contains the catalytic site of the enzyme, and is thought to protrude into the cytosol.

The activity of the reductase is regulated by three distinct mechanisms. The first control point is at the level of gene expression. The amount of mRNA produced is modulated by the supply of cholesterol. When cholesterol is in excess, the amount of mRNA for HMG-CoA reductase is reduced, hence less enzyme is made. Depletion of cholesterol enhances the synthesis of reductase mRNA.

The second regulatory mechanism involves the rate of degradation of HMG-CoA reductase. As stated in chapter 21, the amount of an enzyme in a cell is determined by both its rate of synthesis and its rate of degradation. It has been known for some time that the half-life of HMG-CoA reductase is between 2 and 4 h, about 10-fold lower than that of many other proteins on the endoplasmic reticulum. In other words, HMG-CoA reductase is rapidly degraded within the cell. The rate of degradation of the reductase appears to be modulated by the supply of cholesterol. Thus when cholesterol is abundant, the rate of enzyme degradation is twice as fast as when there is a limited supply of cholesterol. The effect of cholesterol on enzyme degradation is mediated by the membrane domain of the enzyme. In support of this point, a mutant enzyme lacking the membrane domain was found to be active in the synthesis of mevalonic acid even though free in the cytosol, and the mutant enzyme had an extended half-life of five times the normal. Moreover, the supply of cholesterol did not affect the half-life. More work needs to be done to determine the exact mechanism of regulating the half-life of this enzyme.

The third regulatory mechanism is phosphorylation–

Figure 23.4

Proposed structure of HMG-CoA reductase derived from studies of recombinant DNA that codes for the enzyme. The enzyme is attached to the endoplasmic reticulum membrane and consists of two domains: the hydrophobic domain, embedded in the membrane, and the catalytic domain, which protrudes into the cytosol. (Source: Adapted from L. Liscum, J. Finer-Moore, R. M. Stroud, K. L. Luskey, M. S. Brown, and J. L. Goldstein, Domain structure of 3-hydroxy-3-methylglutaryl coenzyme A reductase, a glycoprotein of the endoplasmic reticulum, *J. Biol. Chem.* 260:522–530, 1985.)

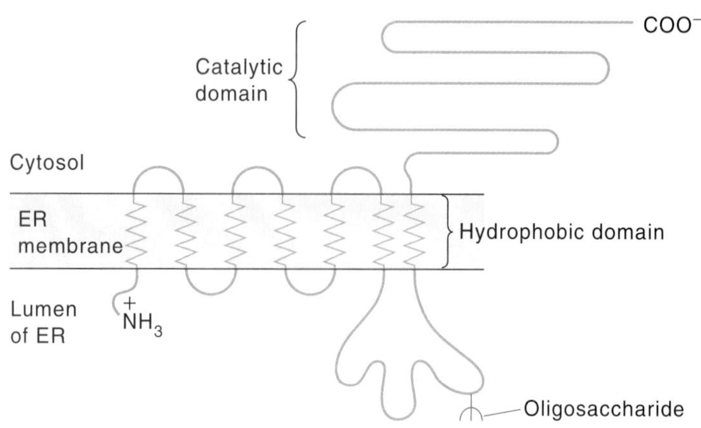

Figure 23.5

Scheme for modulation of the activity of HMG-CoA reductase by phosphorylation–dephosphorylation. The AMP-activated protein kinase is the same enzyme that catalyzes the phosphorylation of acetyl-CoA carboxylase (see fig. 21.28).

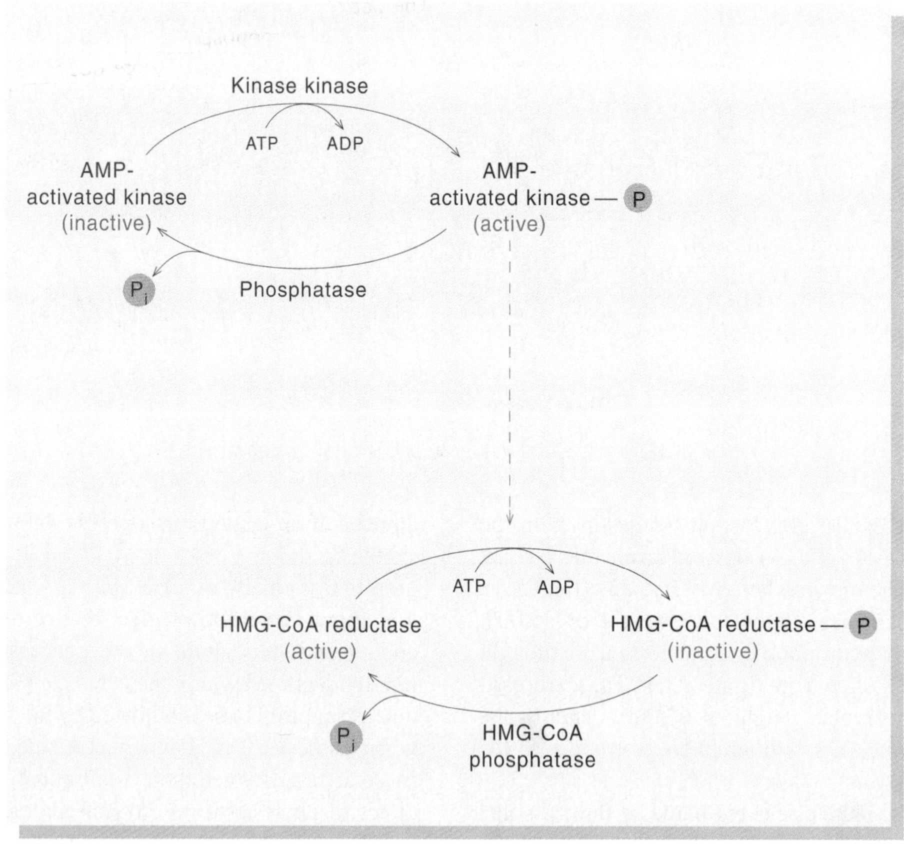

dephosphorylation of the reductase, which causes inactivation and activation as outlined in figure 23.5. The phosphorylation is catalyzed by AMP-activated protein kinase and occurs on serine 871 of the reductase. Recall that the AMP-activated protein kinase also acts on acetyl-CoA carboxylase (see fig. 21.28). AMP levels in the cell rise when there is a depletion of ATP. Thus the inactivation of HMG-CoA reductase by this phosphorylation probably occurs only under conditions in cells and tissues where a serious lack of energy causes ATP levels to decrease. Unlike the first two mechanisms for regulation of the reductase mentioned above, variations in the supply of cholesterol have no effect on the phosphorylation of the enzyme by AMP-activated protein kinase. It is not known, but likely, that the kinase kinase implicated for regulation of cholesterol biosynthesis in fig. 23.5 is the same kinase kinase discussed in fig. 21.28. There is evidence that the phosphatase indicated in fig. 23.5 is protein phosphatase 2C.

Cholesterol biosynthesis is also controlled by plasma low-density lipoproteins, which we will discuss in the context of lipoprotein metabolism later in this chapter.

It has long been recognized that a correlation exists between a high level of serum cholesterol and cardiovascular disease (e.g., most heart diseases, stroke). Most serum cholesterol originates from the liver; thus a drug that would specifically reduce cholesterol biosynthesis has been sought. It would be logical for such a drug to inactivate HMG-CoA reductase, since this enzyme catalyzes the key regulatory step in the pathway. Several fungal metabolites have been isolated that are competitive inhibitors of HMG-CoA reductase. One of the most active compounds is lovastatin (fig. 23.6), which competes favorably (K_I = 0.6 nM) with HMG-CoA for the reductase. Small doses of this drug (20 mg/kg of body weight) lower the levels of plasma cholesterol in humans by 20%. This drug has been approved for the treatment of patients with hypercholesterolemia.

Figure 23.6

Structure of lovastatin acid, a potent competitive inhibitor of HMG-CoA reductase. Note the similarity in structure of the red portion of the molecule with mevalonate, the product of the HMG-CoA reductase reaction.

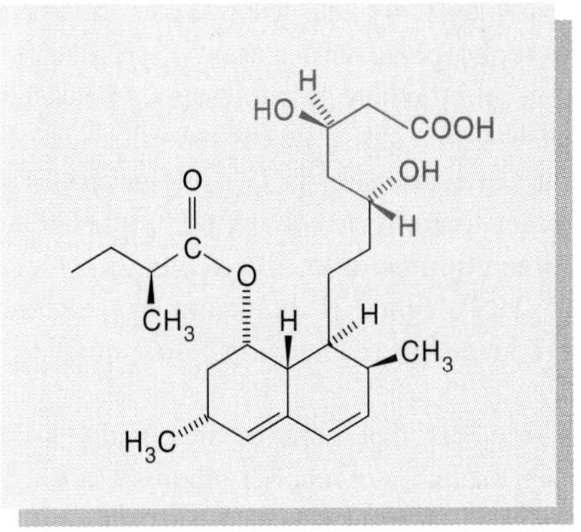

It Takes Six Mevalonates and Ten Steps to Make Lanosterol, the First Tetracyclic Intermediate

In the next segment of the pathway, mevalonate is converted to squalene, which is cyclized to form lanosterol. The first stage in this sequence of reactions is the synthesis of the five-carbon isoprenoid intermediates isopentenyl pyrophosphate and dimethylallyl pyrophosphate. The synthesis involves four enzymes (fig. 23.7). The action of mevalonate kinase and phosphomevalonate kinase produces 5-pyrophosphomevalonate. The third enzyme (see fig. 23.7), pyrophosphomevalonate decarboxylase, catalyzes a decarboxylation and elimination of the 3-hydroxyl group. The decarboxylase probably acts by an initial phosphorylation of the 3-hydroxyl group with ATP, followed by the *trans* elimination of

Figure 23.7

The conversion of mevalonate to isopentenyl pyrophosphate and dimethylallyl pyrophosphate. Mevalonate is converted to isopentenyl pyrophosphate in three steps. Each of these steps requires one ATP cleavage. The conversion of isopentenyl pyrophosphate to dimethylallyl pyrophosphate is readily reversible and does not require any further expenditure of ATP.

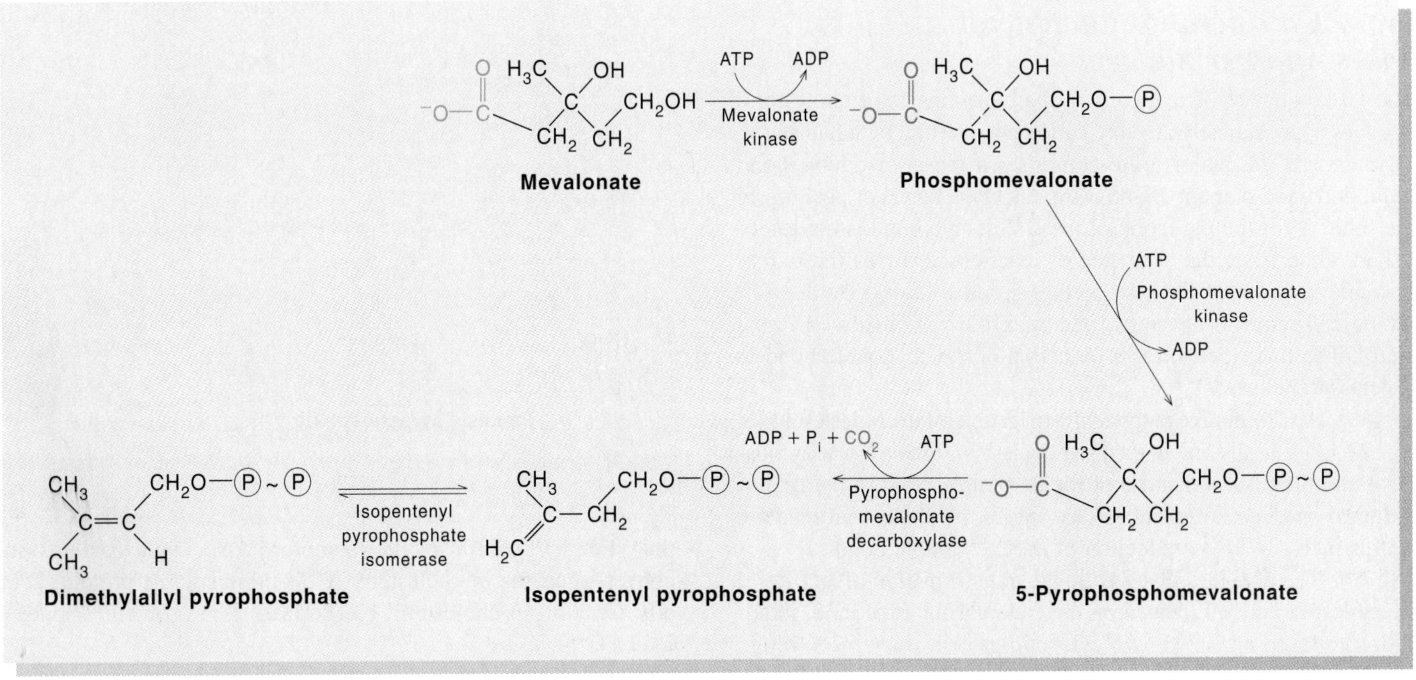

the carboxyl and phosphate to give isopentenyl pyrophosphate. This intermediate can be enzymatically isomerized to 3,3-dimethylallyl pyrophosphate. These two C_5 isoprenoid pyrophosphates react to produce the C_{10} intermediate geranyl pyrophosphate (fig. 23.8). Subsequently, a C_{15} intermediate, farnesyl pyrophosphate, is formed by the reaction of another molecule of isopentenyl pyrophosphate with geranyl pyrophosphate. In addition to being an intermediate in cholesterol biosynthesis, farnesyl pyrophosphate is also a substrate for modification of certain proteins, as discussed in box 23A. Two molecules of farnesyl pyrophosphate react to form presqualene pyrophosphate, which rearranges with the elimination of PP_i to yield the C_{30} intermediate squalene (fig. 23.9). Some of the complex stereochemistry involved in the conversion of mevalonate to squalene is discussed in box 23B.

The enzymes that convert mevalonate to farnesyl pyrophosphate are probably cytosolic, whereas farnesyl transferase (squalene synthase) is tightly associated with the endoplasmic reticulum (see fig. 23.9). Even though such membrane-bound enzymes are difficult to solubilize, this enzyme has been isolated from yeast.

The two remaining reactions in the biosynthesis of lanosterol are shown in figure 23.10. In the first of these reactions, squalene-2,3-oxide is formed from squalene by squalene monooxygenase, an endoplasmic-reticulum-bound enzyme, in a reaction that requires O_2, NADPH, FAD, phospholipid, and a cytosolic protein. As can be seen in figure 23.9, squalene is a symmetrical molecule; hence the formation of squalene oxide can be initiated from either end of the molecule. The oxide is converted into lanosterol by another endoplasmic-reticulum-bound enzyme, 2,3-oxidosqualene lanosterol cyclase. The reaction can be formulated as proceeding by means of a protonated intermediate that undergoes a concerted series of *trans*-1,2 shifts of methyl groups and hydride ions to produce lanosterol (see fig. 23.10).

From Lanosterol to Cholesterol Takes Another 20 Steps

The last sequence of reactions in the biosynthesis of cholesterol involves approximately 20 enzymatic steps, starting with lanosterol. In mammals the major route comprises a series of double-bond reductions and demethylations (fig. 23.11). The exact position in the scheme for the reduction of the Δ^{24} double bond is not established. Otherwise, the sequence of reactions involves the oxidation and removal of the 14α-methyl group followed by the oxidation and removal of the two methyl groups at position 4 in the sterol. The final reaction is a reduction of the Δ^7 double bond in 7-dehydrocholesterol.

An alternative pathway from lanosterol to cholesterol (see fig. 23.11) initially involves three demethylations to give zymosterol and then isomerization of the Δ^8 double bond to the Δ^5 position to produce desmosterol (see fig. 23.11). The final reaction in this pathway is the reduction of the Δ^{24} double bond.

The enzymes involved in the transformation of lanosterol to cholesterol are all located on the endoplasmic reticulum. In addition to these enzymes, two cytosolic proteins have been found that stimulate several of the membrane-associated reactions that convert squalene to cholesterol. How these soluble proteins actually function in cholesterol biosynthesis is a problem of current interest.

Figure 23.8

Biosynthesis of farnesyl pyrophosphate. Farnesyl pyrophosphate is a C_{15} intermediate containing three C_5 isoprenoid subunits. The two transferase reactions involved in the formation of farnesyl pyrophosphate occur by virtually identical mechanisms as shown.

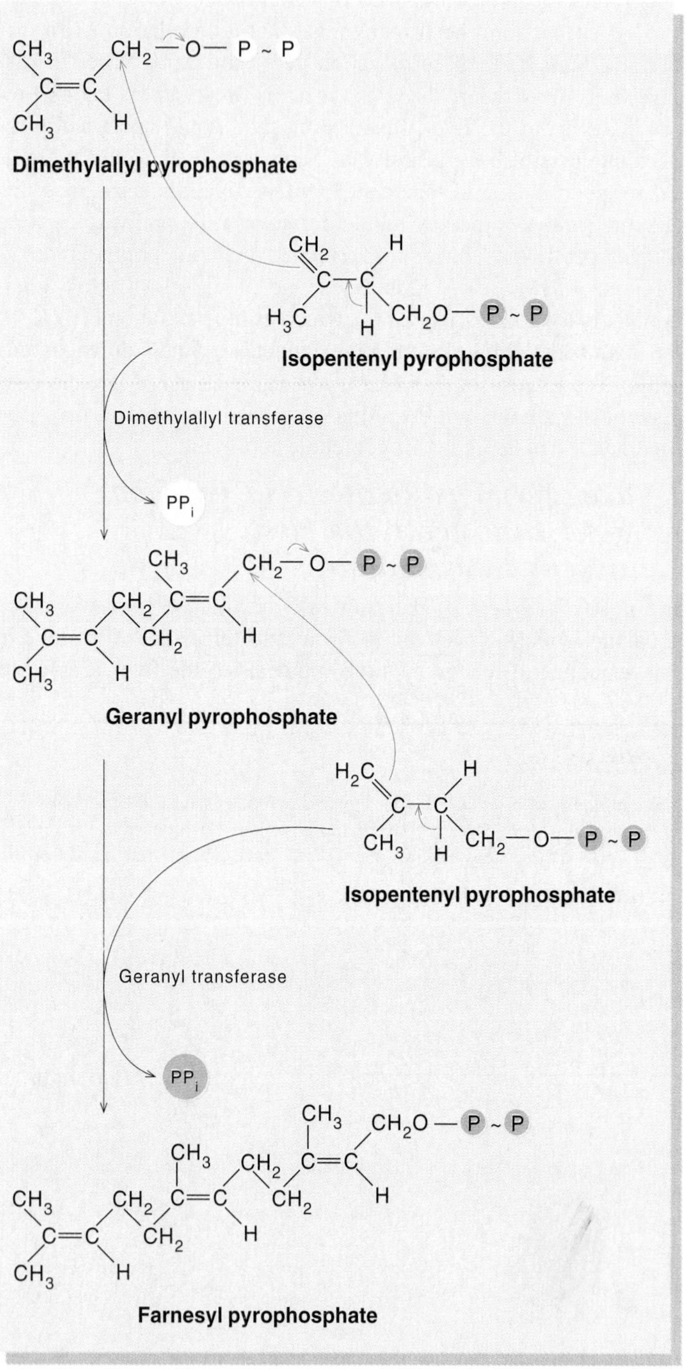

Figure 23.9

Formation of squalene from farnesyl pyrophosphate. Farnesyl transferase is tightly complexed to the endoplasmic reticulum. Three carbons are labeled ●, ▲, ★ in different structures for purposes of tracking them as the reaction proceeds.

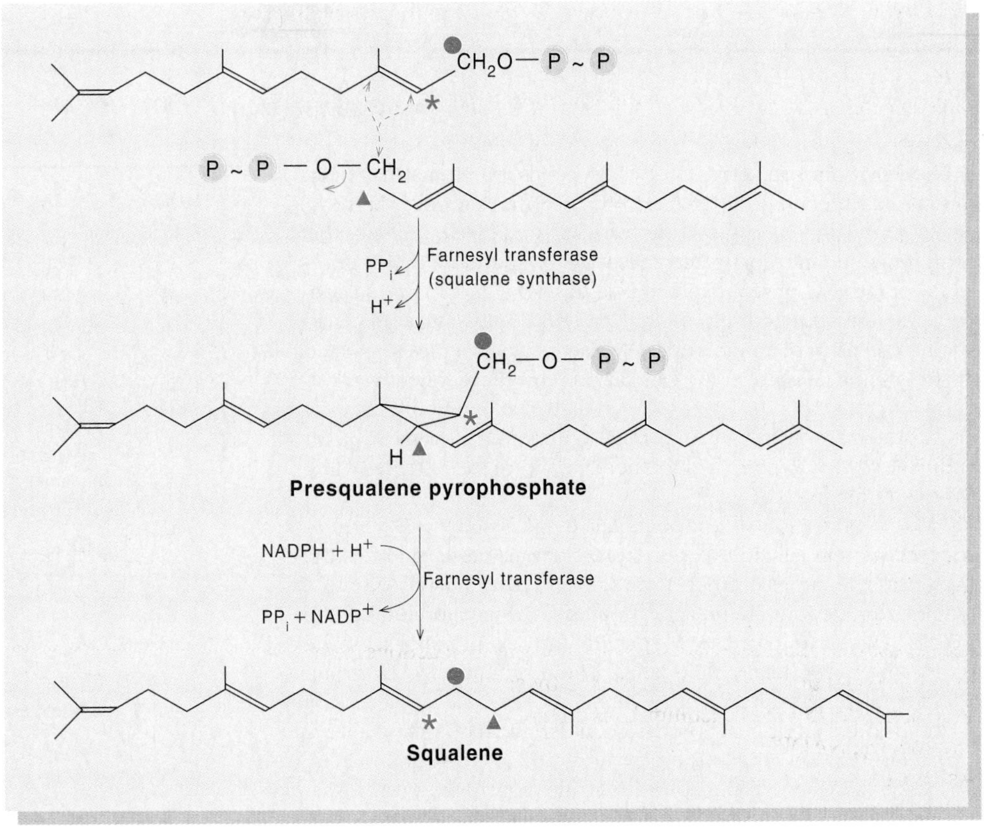

Figure 23.10

The transformation of squalene into lanosterol. The squalene monooxygenase reaction requires O_2, NADPH, FAD, phospholipid, and a cytosolic protein. The cyclase reaction has no known cofactor requirements. The reaction proceeds by means of a protonated intermediate that undergoes a concerted series of *trans*-1,2 shifts of methyl groups and hydride ions to produce lanosterol.

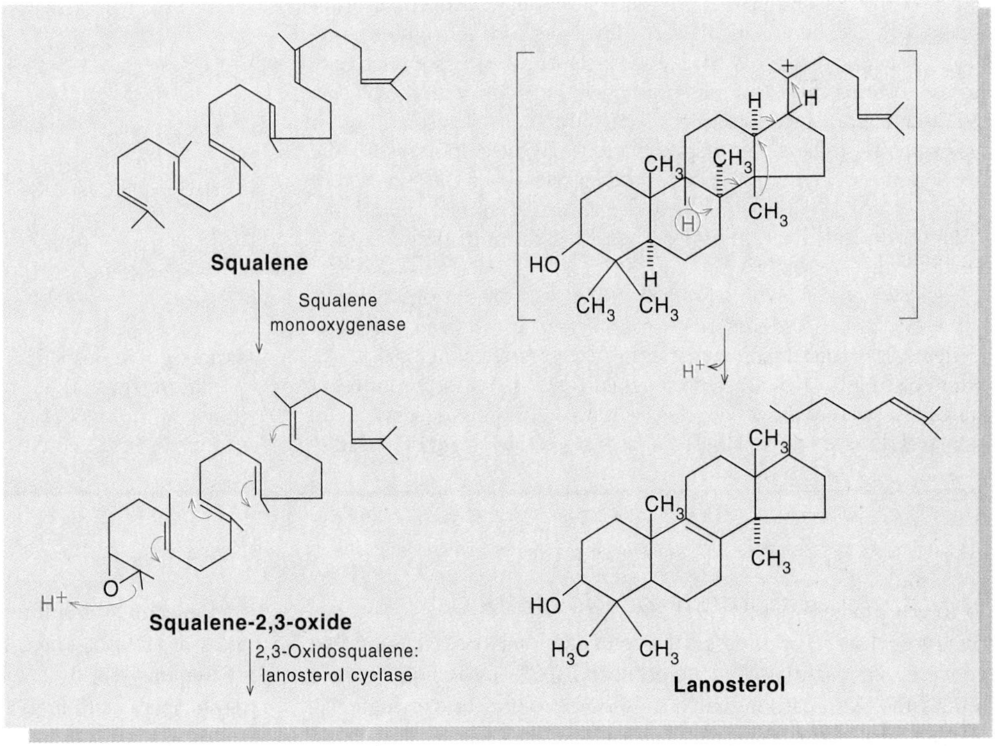

23A
BOX

Isoprenylation of Proteins

Farnesyl pyrophosphate is a 15-carbon polyisoprene that has been known as a precursor of cholesterol for several decades. In 1984, John Glomset and colleagues discovered an unexpected role for this isoprene. When they incubated cells with ^{14}C-mevalonic acid, a precursor of farnesyl pyrophosphate (see figs. 23.7 and 23.8), radioactivity was incorporated into several proteins. Since mevalonic acid is not a precursor of amino acids, Glomset and colleagues suspected a novel modification of some cellular proteins. Subsequent work in many different laboratories has shown that covalent modification of some proteins occurs with two polyisoprenoids, farnesyl (C_{15}) or geranylgeranyl (C_{20}), via a thioether linkage on a cysteine at the carboxy terminus.

The modification of a protein called *ras* is shown in figure 1. After the protein is made, farnesyl: protein transferase catalyzes the transfer of the farnesyl group from farnesyl pyrophosphate onto a cysteine residue four amino acids from the carboxyl terminal of the protein. This is followed by a specific proteolytic removal of the three carboxyl-terminal amino acids. The resulting carboxyl group of cysteine is methylated. The sequence of amino acids necessary for the isoprenylation reaction has been defined as CAAX, where C is cysteine, A is any aliphatic amino acid, and X is any amino acid.

Since farnesyl pyrophosphate is made from HMG-CoA, lovastatin (see fig. 23.6) and related HMG-CoA reductase inhibitors also inhibit the biosynthesis of the isoprenoids and, as a result, inhibit the isoprenylation of proteins. The protein *ras* is normally associated with the plasma membrane of cells and has the property of transforming cells. Cells that are transformed have cancerlike properties. When cells are incubated with lovastatin and farnesyl pyrophosphate synthesis is blocked, *ras* is no longer isoprenylated. Because of this, *ras* is soluble, no longer associated with the plasma membrane, and loses its ability to transform cells. These and other experiments suggest that isoprenylation may be important for the membrane association of certain proteins. However, the function of isoprenylation is more complicated than this example, since some prenylated proteins are found on other cellular membranes and in the cytosol.

The cause of an inherited metabolic disease, choroideremia, has recently been identified as a defect in an enzyme, Rab geranylgeranyl transferase, which transfers geranylgeranyl (C_{20} isoprenoid) to a GTP-binding protein (Rab) that is important in regulating the movement of vesicles within cells. Choroideremia is an X-linked disease (carried by females in a recessive manner and fully expressed in males) characterized by tunnel vision or complete blindness by middle age. This is caused by degeneration of the retina in the eye. How a defect in the isoprenylation of Rab leads to this disease is unknown.

Figure 1

Steps in protein isoprenylation.

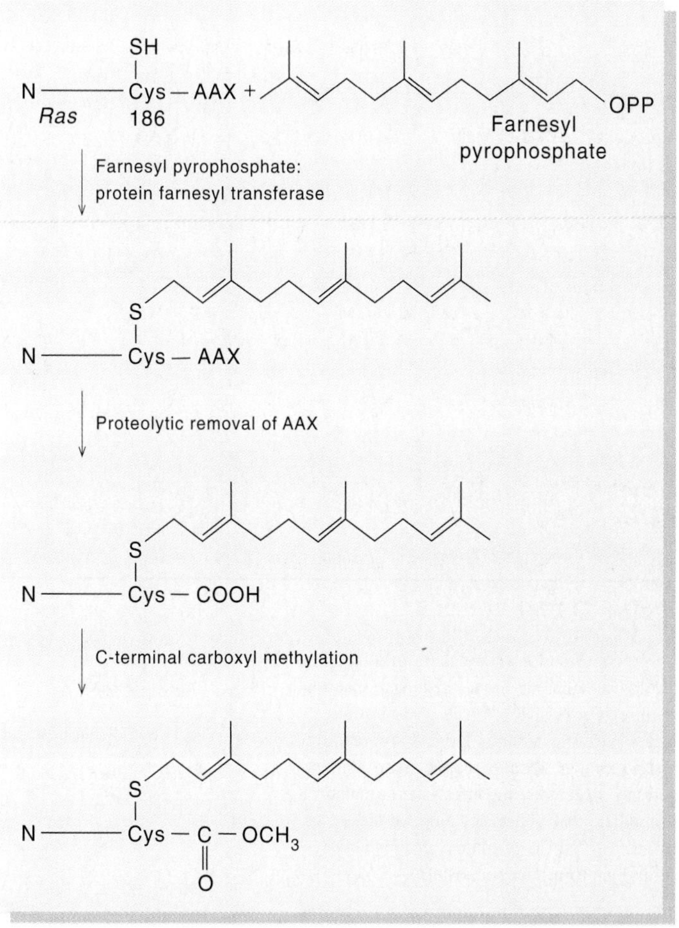

Summary of Cholesterol Biosynthesis

In this section we discussed the remarkable set of reactions that converts a two-carbon precursor, acetyl-CoA, into cholesterol, a tetracyclic hydrocarbon with a rigid ring structure and a single hydroxyl substituent. The reactions occur in the cytosol and on the endoplasmic reticulum. The key enzyme in the sequence is HMG-CoA reductase, which is regulated by the level of gene expression of the enzyme, by enzyme degradation, and by phosphorylation–dephosphorylation. We now examine how cholesterol and other lipids, particularly triacylglycerols, are transported in the plasma.

23B
BOX

Stereochemistry of the Conversion of Mevalonate to Squalene

As we have seen, the biosynthesis of cholesterol from acetyl-CoA is accomplished by means of a complicated route that involves more than 30 different enzymes, numerous cofactors, and at least two cytosolic proteins. Although this certainly represents enough complexity for most people, George Popjak and John Cornforth, in Britain, recognized that it was possible to define the conversion of mevalonate to squalene in a more precise and elegant manner. These two scientists observed in the 1960s that there were 14 "stereochemical ambiguities" in the conversion of pyrophosphomevalonate to squalene. In other words, there were 2^{14}, or 16,384, theoretically possible stereochemical pathways by which mevalonate could be transformed into squalene. This in itself was a remarkable observation. More remarkably, these two men and their collaborators subsequently were able to define precisely which one of these 16,384 possible stereochemical pathways actually occurred.

It is beyond the scope of this introductory text for us to examine each of the 14 stereochemical ambiguities and their resolution. Two examples should demonstrate the principles involved. The reaction catalyzed by pyrophosphomevalonate decarboxylase could involve either a *cis* or a *trans* elimination of the carboxyl and hydroxyl groups to produce isopentenyl pyrophosphate (see fig. 23.7). Cornforth and colleagues solved this stereochemical ambiguity by the synthesis of a stereospecifically deuterium-labeled pyrophosphomevalonate (fig. 1) that was incubated with the decarboxylase. The product of the reaction was isolated, and after several chemical transformations, Cornforth and co-workers were able to distinguish which of the two possible isomers of isopentenyl pyrophosphate was formed. The product was solely the result of a *trans* elimination (fig. 1). Thus the first stereochemical ambiguity was resolved.

Another stereochemical problem was to determine which of the two hydrogens from C-2 of isopentenyl pyrophosphate was lost in its isomerization to dimethylallyl pyrophosphate (fig. 2). Isopentenyl pyrophosphate was chemically labeled with deuterium on the 2-R or 2-S position. Incubation of this substrate with the isomerase and subsequent characterization of the product demonstrated that the pro R hydrogen (H_R) was specifically removed during the isomerization reaction (fig. 2). Further information on the 14 stereochemical ambiguities and their resolution can be found in the book by Ronald Bentley (see Selected Readings).

Figure 1

Synthesis of a stereospecifically deuterium-labeled pyrophosphomevalonate.

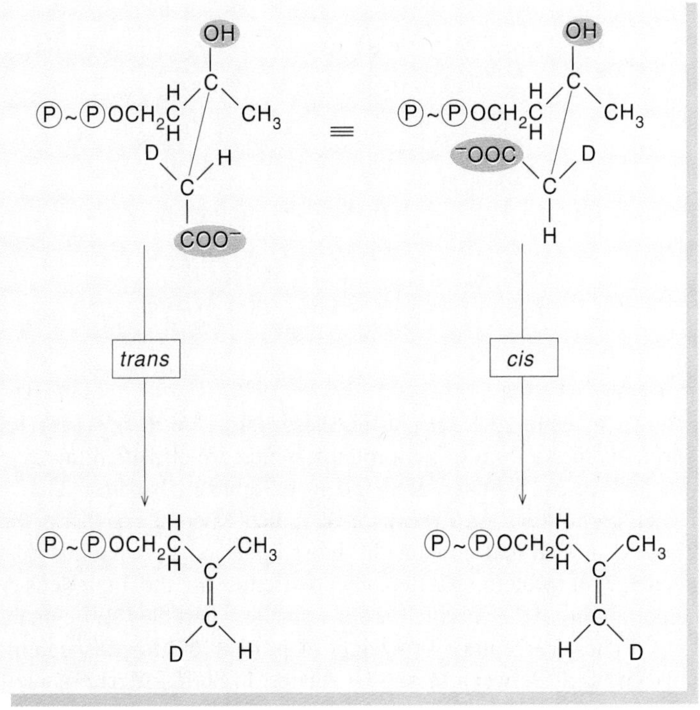

Figure 2

Diagram indicating loss of the H_R hydrogen in the isomerization of isopentenyl pyrophosphate to dimethylallyl pyrophosphate.

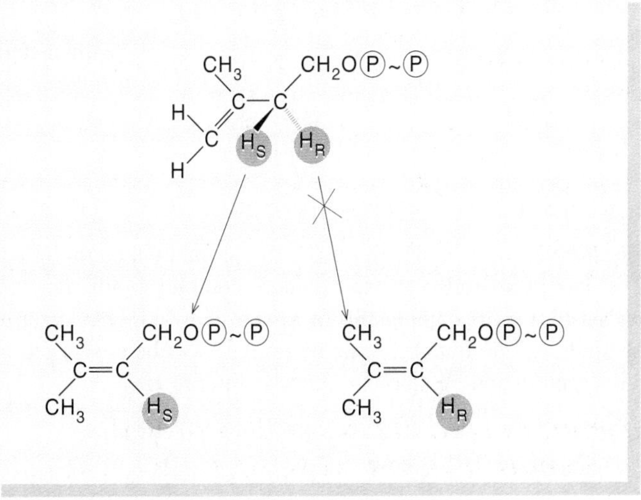

Figure 23.11

Two pathways for the conversion of lanosterol to cholesterol. The major route in mammals proceeds through 7-dehydrocholesterol.

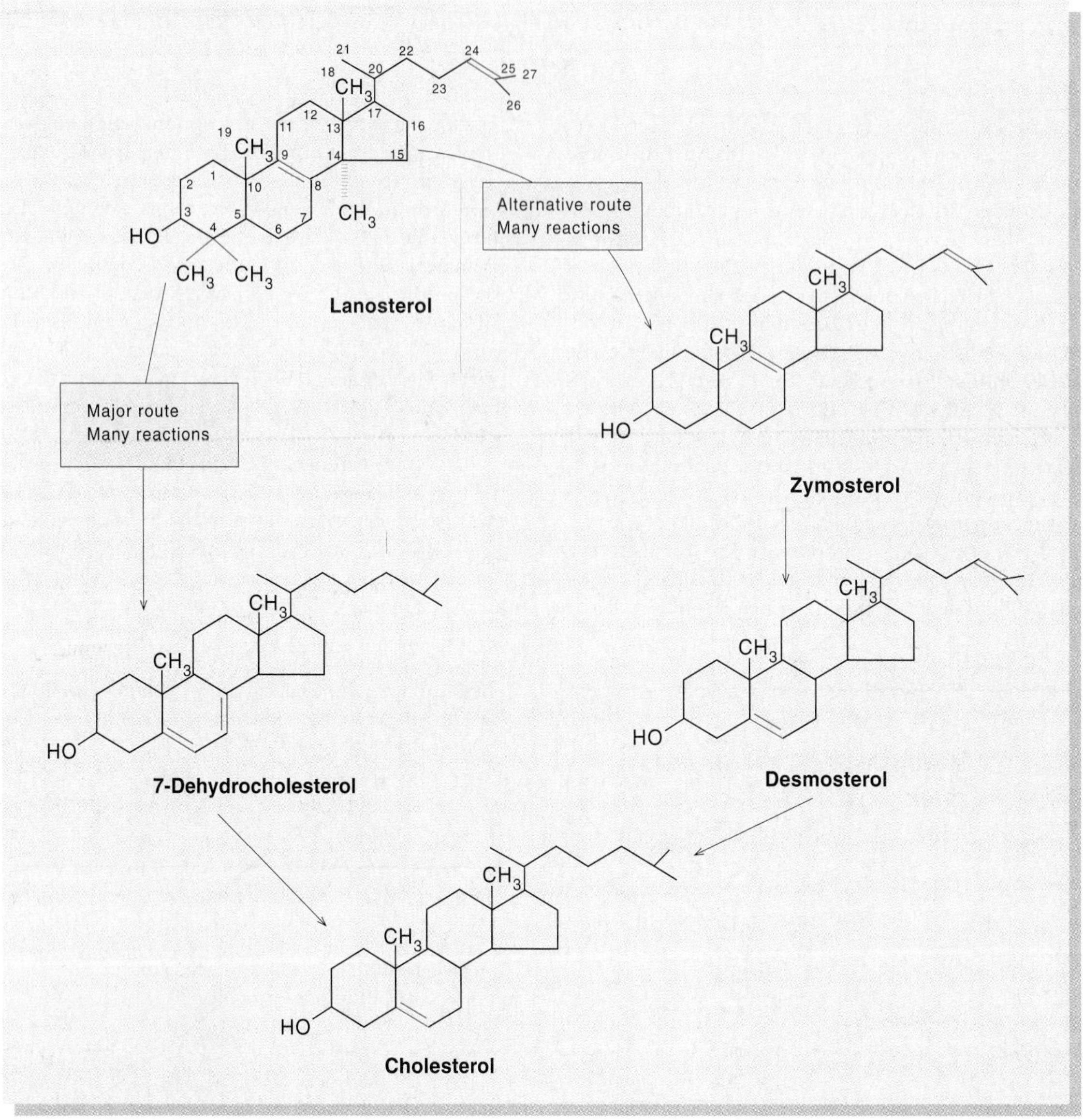

Lanosterol

Alternative route
Many reactions

Zymosterol

Major route
Many reactions

7-Dehydrocholesterol

Desmosterol

Cholesterol

Lipoprotein Metabolism

Unesterified fatty acids are carried in plasma by albumin (chapter 21). The plasma also transports more complex lipids among the various tissues as components of lipoproteins (particles composed of lipids and proteins). In this section we will be concerned with the structure and metabolism of these lipoproteins.

There Are Five Classes of Lipoproteins in Human Plasma

The amounts and types of lipids found in human plasma fluctuate according to the dietary habits and metabolic states of the individual. The normal ranges for the lipid levels in plasma are shown in table 23.1. In plasma these lipids are associated with proteins in the form of lipoproteins, which are classified into five major types on the basis of their density (table 23.2 and fig. 23.12). The lipoproteins of lowest density, the chylomicrons, are the largest in size and contain the most lipid and the smallest percentage of protein. At the other extreme are the high-density lipoproteins (HDL), which are the smallest particles and contain the highest percentage by weight of protein and lowest percentage of lipid. Between these two classes, in both size and composition, are the low-density lipoproteins (LDL), the intermediate-density lipoproteins (IDL), and the very-low-density lipoproteins (VLDL).

Table 23.1
Normal Concentrations of the Major Lipid Classes in Plasma in Humans

Lipid	Concentration (g/l)
Total lipid	3.6–6.8
Cholesterol and cholesteryl ester	1.3–2.6
Triacylglycerol	0.8–2.4
Phospholipid	1.5–2.5

The structures of the various lipoproteins appear to be similar (fig. 23.13). Each of the lipoprotein classes contains a neutral lipid core composed of triacylglycerol and/or a cholesteryl ester. Around this core is a layer of protein, phospholipid, and cholesterol oriented with the polar portions exposed to the surface of the lipoprotein.

There are at least nine apoproteins (see table 23.2) associated with the lipoproteins, as well as several enzymes and a cholesteryl ester transfer protein. The structure and function of these apoproteins have been intensely studied in the past decade, and some of the properties of these apoproteins are summarized in table 23.3. Most of the apoproteins have been sequenced and contain regions that are rich in hydrophobic amino acids, which facilitate binding of phospholipid.

Lipoproteins Are Made in the Endoplasmic Reticulum of the Liver and Intestine

Of the various lipid components of the lipoproteins, only the biosynthesis of cholesteryl esters has not been mentioned. Cholesterol ester is the storage form of cholesteryl in cells. It is synthesized from cholesterol and acyl-CoA by acyl-CoA:cholesterol acyltransferase (ACAT) (fig. 23.14), which is located on the cytosolic surface of liver endoplasmic reticulum. The synthesis of the apoproteins takes place on ribosomes that are bound to the endoplasmic reticulum. As we mentioned previously, the biosynthesis of cholesterol, triacylglycerols, and phospholipids also occurs on the endoplasmic reticulum.

How the various components of the lipoproteins are assembled and secreted into the plasma is not known. Current ideas for this process suggest the transfer of the components from the endoplasmic reticulum to the Golgi apparatus, where secretory (membrane-encapsulated) vesicles are formed. These vesicles would subsequently fuse with the plasma membrane and release their lipoprotein contents into plasma (fig. 23.15).

Patients with the inherited disease abetalipoproteinemia have no chylomicrons, VLDL, or LDL in their bloodstream. As was recently discovered, this is due to a defect in the assembly of VLDL in the endoplasmic reticulum. Patients who have this disease lack a protein (microsomal triacylglycerol transfer protein) found in the lumen of the endoplasmic reticulum that, in vitro, has the property of transferring triacylglycerol and phospholipids between lipoprotein particles. From studies on patients with this rare disease it has become clear that this transfer protein is required for the proper assembly of VLDL and chylomicrons. The exact mechanism by which this protein functions in lipoprotein assembly has not been clarified. Presumably, the protein is involved in the assembly of lipids into the forming particle (step 2 in fig. 23.15).

The plasma lipoproteins appear to be made mainly in the liver and intestine. In the rat, approximately 80% of the plasma apoproteins originate from the liver; the rest come from the intestine. Most of the components of chylomicrons, including apoprotein A, apoprotein B-48, phospholipid, cholesterol, cholesteryl ester, and triacylglycerols, are products of the intestinal cells.

Table 23.2
Composition and Density of Human Lipoproteins

	Chylomicron	VLDL	IDL	LDL	HDL
Density (g/ml)	<0.95	0.95–1.006	1.006–1.019	1.019–1.063	1.063–1.210
Diameter (nm)	75–1,200	30–80	25–35	18–25	5–12
Components (% dry weight)					
Protein	1–2	10	18	25	33
Triacylglycerol	83	50	31	10	8
Cholesterol and cholesteryl esters	8	22	29	46	30
Phospholipids	7	18	22	22	29
Apoprotein composition	A-I, A-II B-48 C-I, C-II, C-III E	B-100 C-I, C-II, C-III E	B-100 C-I, C-II, C-III E	B-100	A-I, A-II C-I, C-II, C-III D E
Classification by electrophoresis	Omega	Pre-beta	Between beta and pre-beta	Beta	Alpha

Figure 23.12

Electron micrographs of low-density lipoproteins (LDL), very-low-density lipoproteins (VLDL), chylomicrons (Chylo), and high-density lipoproteins (HDL). Lipids are transported in plasma as components of these particles. The larger particles contain a higher percentage of lipid and therefore are less dense, whereas the smaller particles have less lipid and a higher percentage of protein, and are more dense. (Electron micrographs courtesy of Dr. Robert Hamilton, University of California, San Francisco.)

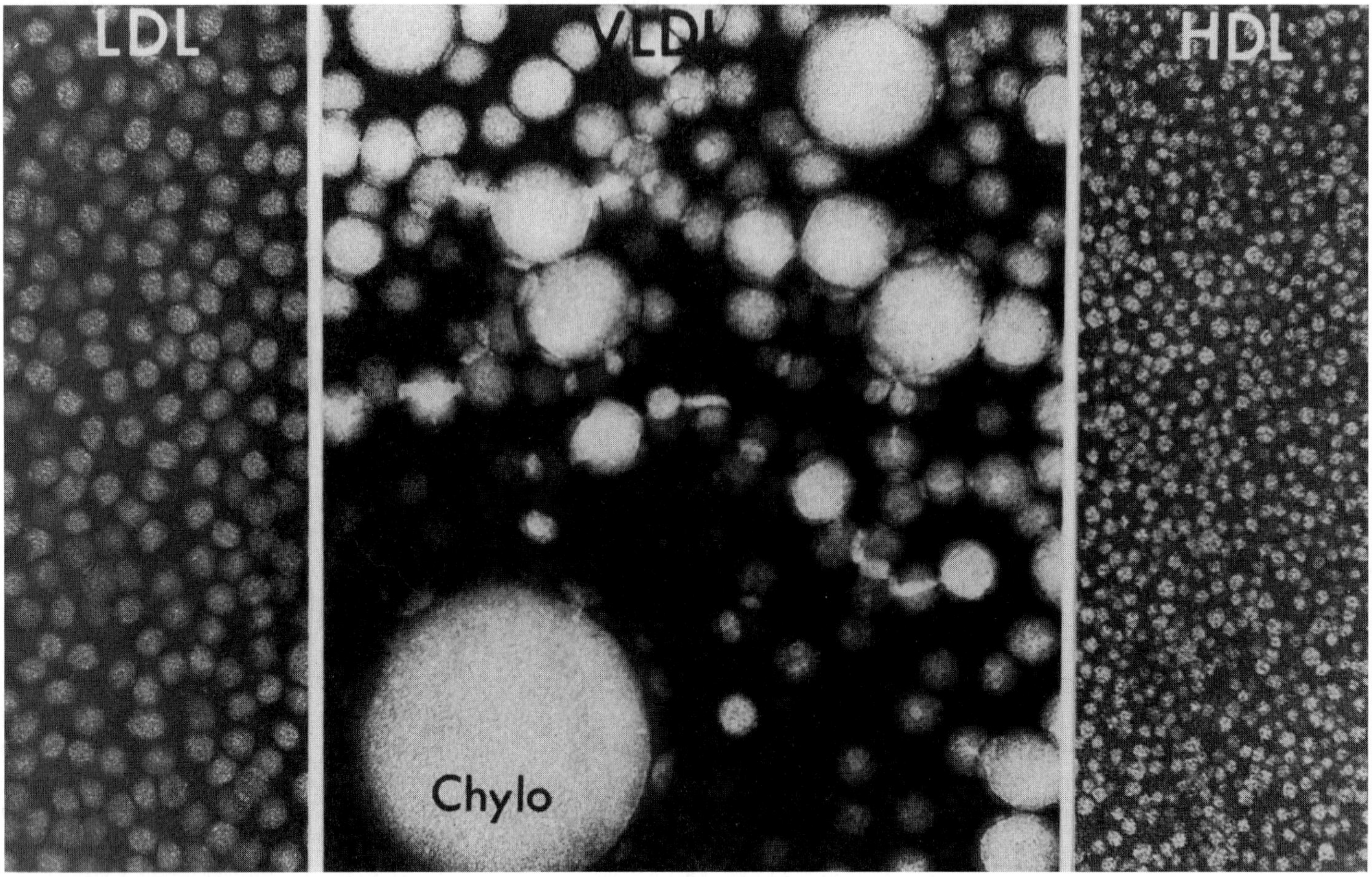

Figure 23.13

Generalized structure of human lipoproteins. The lipoproteins are spherical and vary in diameter from 10 nm to as much as 1,000 nm, depending on the particular proteins and lipids. Each of the lipoprotein classes contains a neutral lipid core composed of triacylglycerol or cholesteryl ester or both. Around the core is a layer of protein, phospholipid, and cholesterol that is oriented with the polar portions exposed to the surface of the lipoprotein. Note the micellar-like construction.

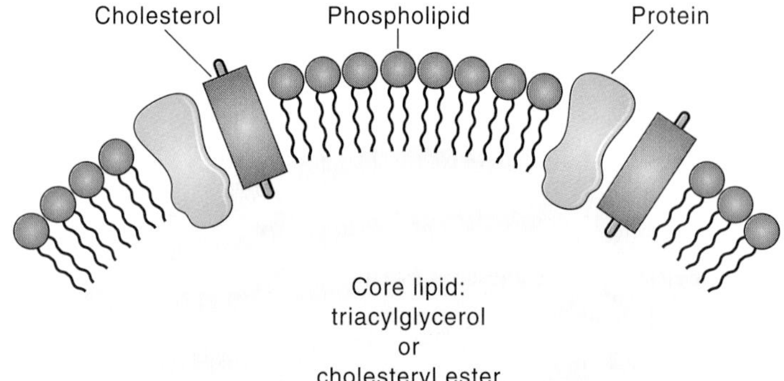

Table 23.3
Properties of the Apoproteins of the Major Human Lipoprotein Classes

Apoprotein	M_r	Plasma Concentration (mg/100 ml)	Miscellaneous
A-I	29,016	90–120	Major protein in HDL (64%); contains 245 amino acids and no carbohydrate; activates LCAT*
A-II	17,400	30–50	Mainly in HDL (20% of dry mass); two identical chains with 77 amino acids each, joined by a disulfide at residue 6
B-100	513,000	80–100	Major protein in LDL; very difficult to solubilize in detergents, is made in the liver and binds to the LDL receptor (see fig. 23.21)
B-48	241,000	<5	A protein found exclusively in chylomicrons
C-I	7,000	4–7	Contains 57 amino acids
C-II	9,000	3–8	Contains 80–85 amino acids; activates lipoprotein lipase
C-III	9,300	8–15	Contains 79 amino acids and inhibits lipoprotein lipase
D	19,000	8–10	Associated with HDL; function unknown
E	33,000	3–6	Found on VLDL and chylomicrons and binds to the LDL receptor (see fig. 23.21)

*LCAT = Lecithin: cholesterol acyltransferase.

Figure 23.14

Biosynthesis of cholesteryl esters. The acyl-CoA: cholesterol acyltransferase involved in cholesteryl ester synthesis is located on the cytosolic surface of liver endoplasmic reticulum.

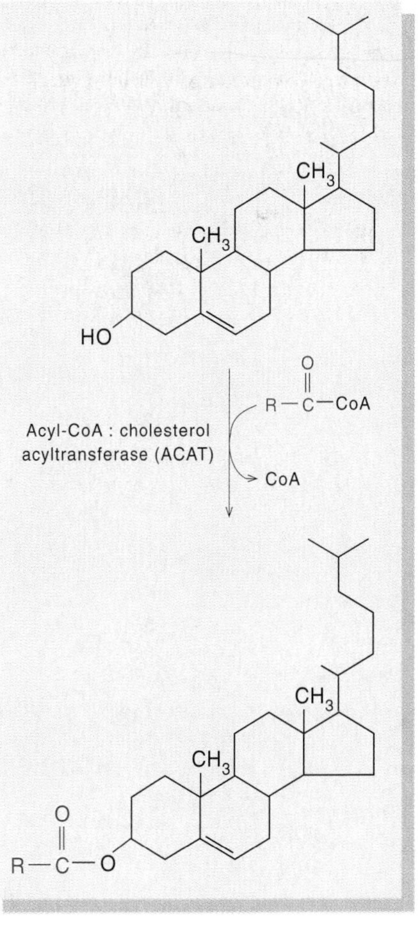

Figure 23.15

Postulated scheme for the synthesis, assembly, and secretion of VLDL by a hepatocyte (liver cell). (1) Synthesis: The apoproteins, phospholipid, triacylglycerol, cholesterol, and cholesteryl esters are synthesized in the endoplasmic reticulum. (2) Assembly: These components are assembled into a prelipoprotein particle in the lumen of the endoplasmic reticulum. (3) Processing: The particle moves to the Golgi apparatus, where modification of the apoproteins occurs. (4) Vesicle formation: A secretory vesicle containing the lipoprotein particles is formed and fuses with the plasma membrane. (5) Secretion: The VLDL is released into the circulation.

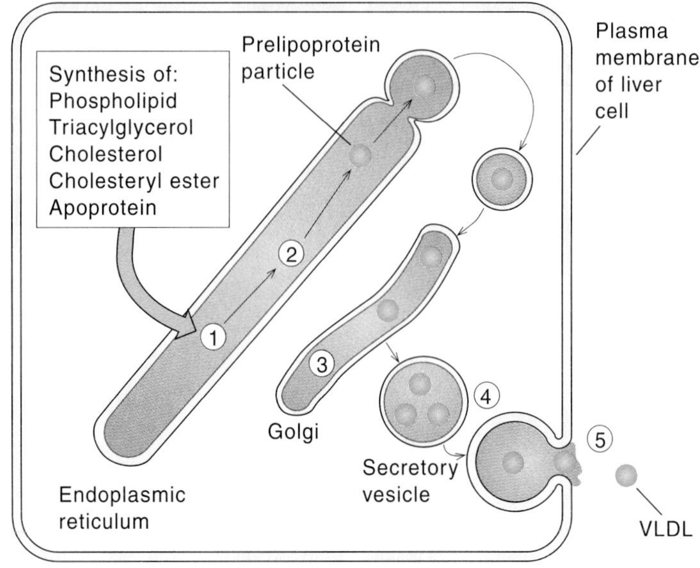

Figure 23.16

Secretion of chylomicrons and their metabolism within the secretory system. Dietary triacylglycerol is degraded in the intestine, the fatty acids are absorbed, and triacylglycerols are resynthesized and packaged into chylomicrons, which are secreted into the lymphatic system (also see fig. 21.2). From there they pass into the capillaries, where the triacylglycerols are degraded to fatty acids by the enzyme lipoprotein lipase. This enzyme is made in various tissues (e.g., heart, adipose tissue, and muscle) and secreted into the lumen of the capillaries, where it attaches to the wall of the vessel. The chylomicron binds to the lipoprotein lipase, and the apo C-II on the lipoprotein activates this lipase, which degrades the triacylglycerol to fatty acids and glycerol (also see fig. 21.3). The fatty acids are absorbed by the tissue, and a smaller lipoprotein particle (chylomicron remnant) is removed by the liver. (Figure kindly supplied by Dr. Alan Attie and Mr. Adam Steinberg, University of Wisconsin.)

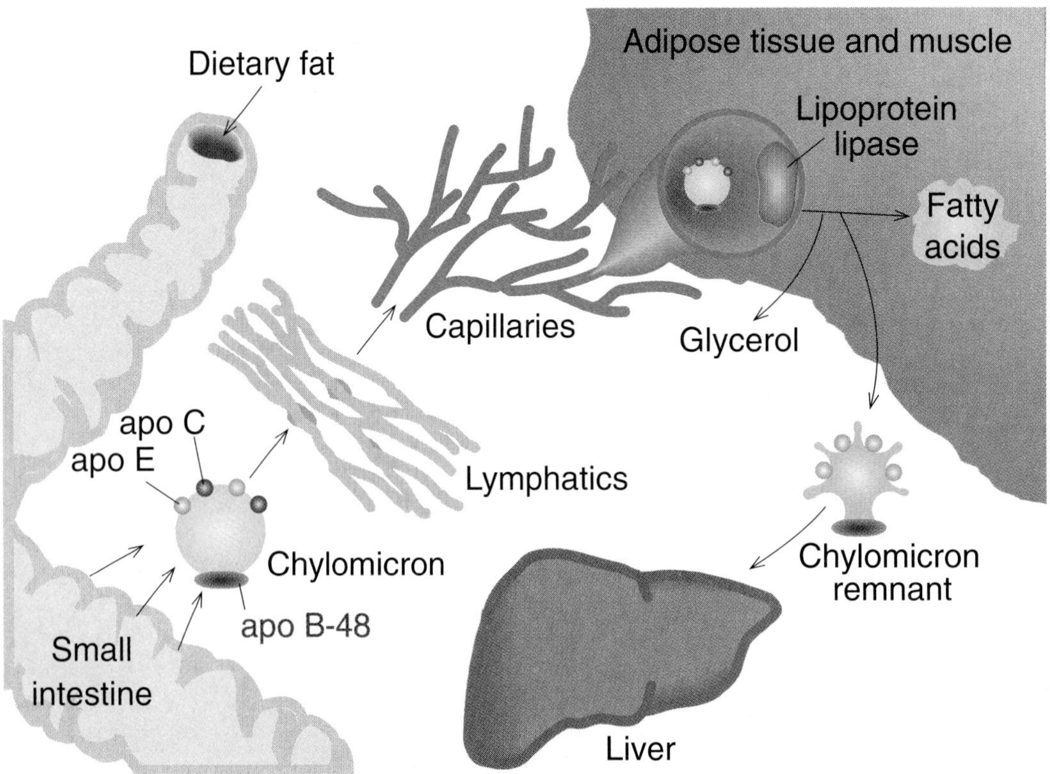

Figure 23.17

Secretion and metabolism of VLDL to form LDL. Many of the steps are similar to those for chylomicron secretion and metabolism (fig. 23.16). In this case, triacylglycerols (TG) and, to a lesser extent, cholesteryl esters (CE) are made in the liver, and a VLDL particle is assembled and secreted into the bloodstream (fig. 23.15). Lipoprotein lipase is again involved in the catabolism of the triacylglycerols to fatty acids to yield a smaller particle, IDL. The IDL can be cleared by the liver or can be degraded further to LDL, which is rich in cholesteryl esters and has apolipoprotein B as the sole apoprotein. LDL delivers cholesterol to tissues in the body or is also removed by the liver. (Figure kindly supplied by Dr. Alan Attie and Mr. Adam Steinberg, University of Wisconsin.)

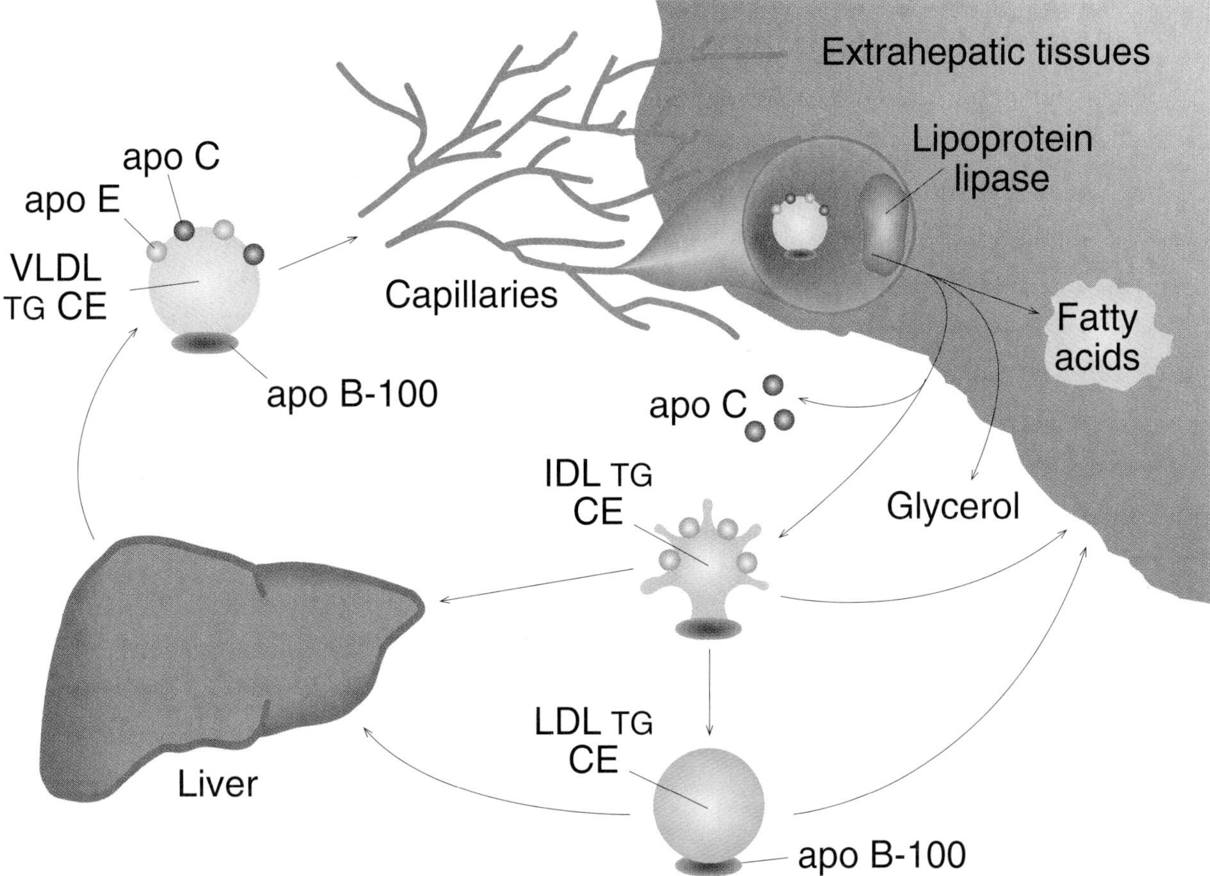

The chylomicrons are secreted into lymphatic capillaries, which eventually enter the bloodstream at the large subclavian vein and therefore bypass the liver. The liver appears to be the major source of VLDL and HDL, which include apo A-I, A-II, B-100, C-I, C-II, C-III, and E and the lipid components of these lipoproteins. Low-density lipoprotein is produced from VLDL, as we will see in a moment.

Chylomicrons and Very-Low-Density Lipoproteins (VLDL) Transport Cholesterol and Triacylglycerol to Other Tissues

Chylomicrons serve as the mode of transport of triacylglycerol and cholesteryl ester from the intestine to other tissues in the body. Chylomicrons are made in the intestine and secreted into the lymphatic system as described in fig. 23.16. Very-low-density lipoprotein functions in a similar manner for the transport of lipid from the liver to other tissues. VLDL are secreted from the liver directly into the bloodstream (fig. 23.17). These two types of triacylglycerol-rich particles are initially degraded by the action of lipoprotein lipase, an extracellular enzyme that is most active within the capillaries of adipose tissue, cardiac and skeletal muscle, and lactating mammary gland. Lipoprotein lipase catalyzes the hydrolysis of triacylglycerols (see fig. 21.2). The enzyme is specifically activated by apoprotein C-II, which is associated with chylomicrons and VLDL (see table 23.2). As a result, this lipase supplies the heart, muscle, adipose, and other tissues with fatty acids, derived from these lipoproteins in the plasma. In both the heart and adipose tissue, the fatty acids produced by lipoprotein lipase can be used for energy or stored as a component of triacylglycerols. Alternatively, the fatty acids can be bound by albumin and transported to other tissues.

As the lipoproteins are depleted of triacylglycerol, the particles become smaller. Some of the surface molecules (apoproteins, phospholipids) are transferred to HDL. In the rat, "remnants" that result from chylomicrons and VLDL catabolism are taken up by the liver. In humans the uptake of remnant VLDL also occurs, but much of the triacylglycerol is further degraded by lipoprotein lipase to give the intermediate-density lipoprotein IDL (see fig. 23.17). This particle is converted into LDL via the action of lipoprotein lipase and enriched in cholesteryl ester via transfer

Figure 23.18

Receptor-mediated uptake of LDL by human skin fibroblasts. Specific LDL receptors are located in coated regions of the plasma membrane. LDL binding results in uptake by endocytosis and formation of a coated vesicle. This vesicle fuses with a lysosome containing many hydrolytic enzymes that degrade the lipoprotein, releasing cholesterol.

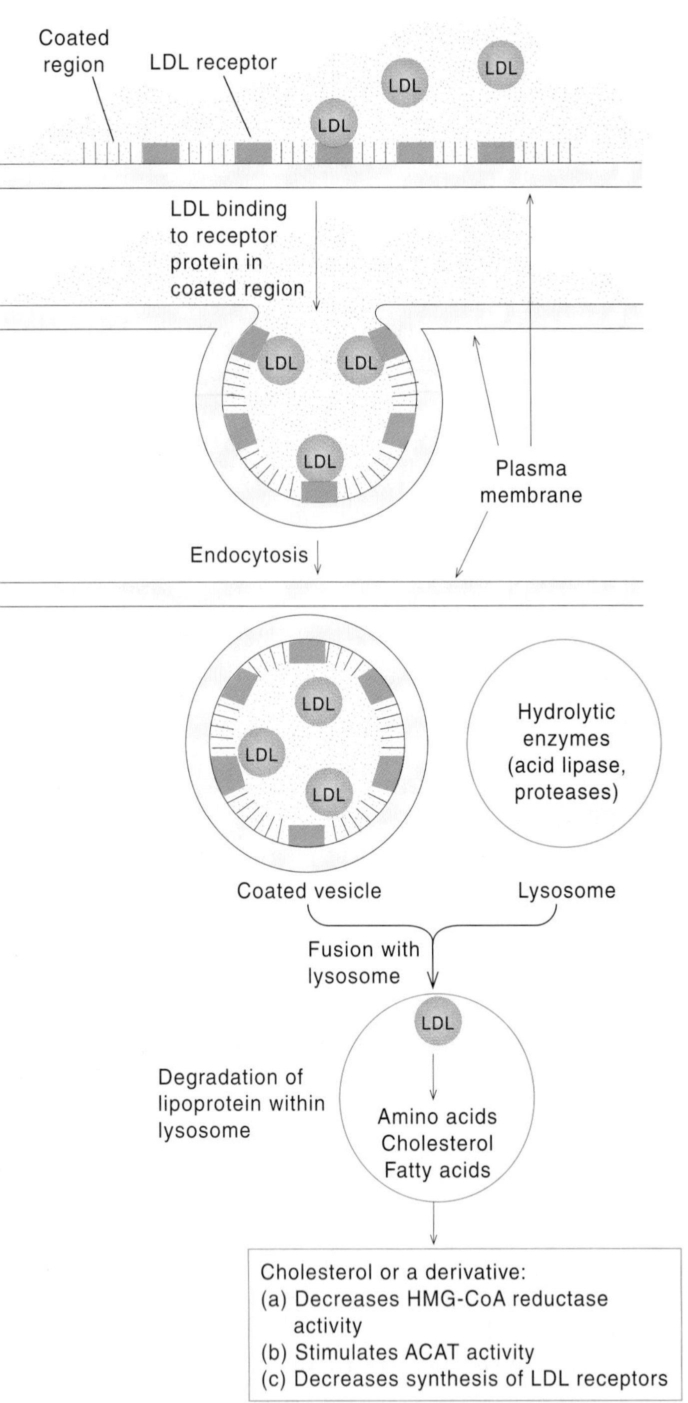

from HDL by the cholesteryl ester transfer protein. The half-life for clearance of chylomicrons and remnants from plasma of humans is 4 to 5 min. The clearance value for VLDL is 1 to 3 h.

Figure 23.19

Electron micrograph of LDL particles (made electron-dense with covalently bound ferritin) bound to coated regions of a human skin fibroblast (97,000×). (From R. G. W. Anderson, M. S. Brown, and J. L. Goldstein, Role of the coated and endocytic vesicle in the uptake of receptor-bound low-density lipoprotein in human fibroblasts. *Cell* 10:351, 1977. © Cell Press.)

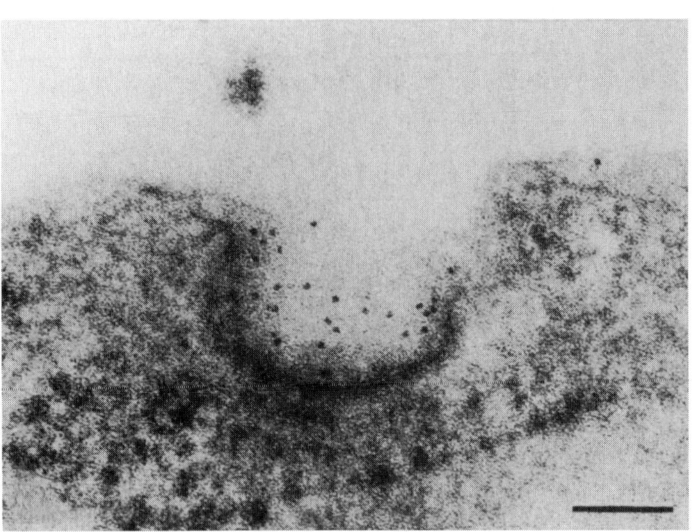

Low-Density Lipoproteins (LDL) Are Removed from the Plasma by the Liver, Adrenals, and Adipose Tissue

Each day approximately 45% of the plasma pool of low-density lipoprotein is removed from human plasma by both the liver and extrahepatic tissues (particularly the adrenals and adipose tissue). The mechanism for the uptake of LDL in extrahepatic tissue has been extensively described by Michael Brown and Joseph Goldstein. They studied the uptake of LDL by human skin fibroblasts grown in cultures in Petri dishes. LDL particles bind to the cell surface by specific receptors that congregate in areas of the plasma membrane called coated regions (figs. 23.18 and 23.19). These areas of plasma membranes engulf the LDL particles (in a process called endocytosis) to form "coated vesicles," which are somehow directed toward and fuse with lysosomes. The LDL particles are degraded within the lysosomes by the action of proteases and lysosomal acid lipases (lipid degradative enzymes). The cholesterol, or a derivative of cholesterol, diffuses from the lysosomes, suppresses the activity of HMG-CoA reductase, and stimulates the activity of acyl-CoA:cholesterol acyltransferase (ACAT). ACAT catalyzes the synthesis of cholesteryl esters (see fig. 23.14), which are then stored within the cell. The cholesterol (or its derivative) also suppresses the synthesis of the LDL receptors and thereby limits the uptake of LDL.

How cholesterol or a derivative suppresses the synthesis of the LDL receptors has recently been clarified by Brown and Goldstein's laboratory. Under conditions in which the level of cholesterol in the cell is low, two proteases cleave a 125-kD protein called sterol regulatory element–binding protein (SREBP) precursor, that is on the endoplasmic reticulum, into a 68-kD protein

Figure 23.20

Regulation of the expression of LDL receptors by cholesterol or a derivative. When the supply of cholesterol is low, a protease cleaves a sterol regulatory element binding protein (SREBP) precursor to give SREBP, which migrates to the nucleus and is a positive effector for the expression of the mRNA for the LDL receptor. This mRNA moves to ribosomes bound to the endoplasmic reticulum, where LDL receptors are made and shuttled to the cell surface (see fig. 23.22). When cholesterol is supplied to a cell, the cholesterol, or more likely a derivative, inhibits the protease with the result that the expression of the LDL receptor gene is down-regulated.

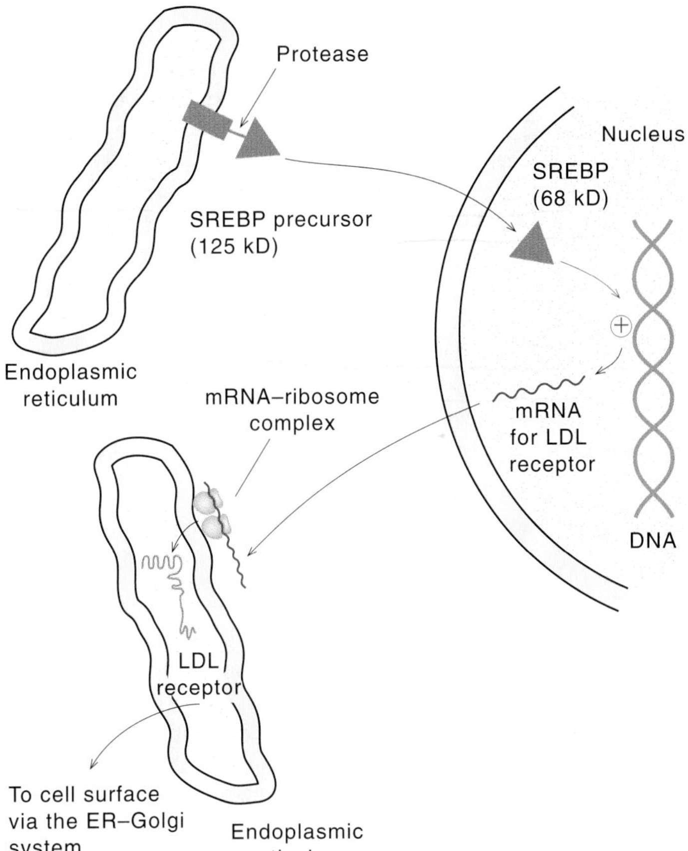

Figure 23.21

The LDL receptor: a single protein with five domains. The cytosolic portion (domain 5) is required for congregation of the LDL receptors in the coated regions of the plasma membrane (see fig. 23.18). Once an LDL molecule is bound to the receptor, both are rapidly internalized by endocytosis.

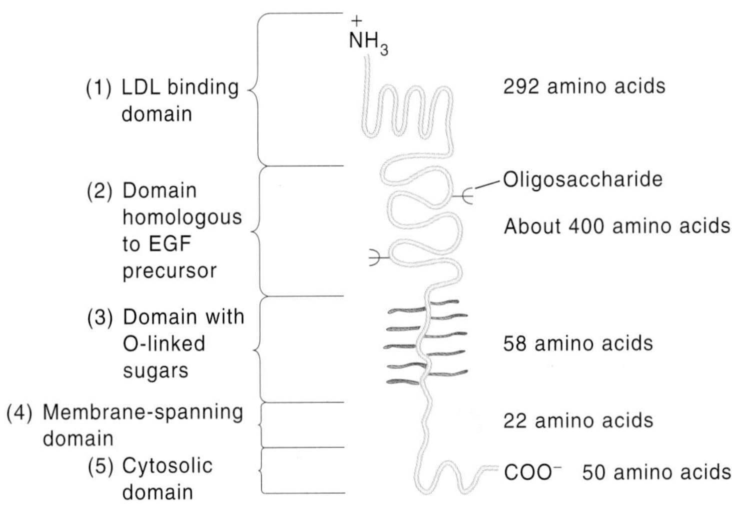

(SREBP) as depicted in fig. 23.20. SREBP migrates to the nucleus, where it activates the LDL receptor gene for the synthesis of mRNA, which is translated into receptors on ribosomes bound to the endoplasmic reticulum. Cholesterol or a derivative down-regulates this process by inhibiting the protease that acts on the SREBP precursor. As a result the SREBP precursor is not cleaved; hence the LDL receptor gene is not activated.

The LDL receptor is a glycoprotein with 839 amino acids and consists of five domains (fig. 23.21). Domain 1 is the binding site for lipoproteins that contain apo B-100 and apo E. Domain 2 is 35% homologous to part of the extracellular domain of the *precursor* to epidermal growth factor (growth factors are discussed in chapter 27). The function of this domain is not clear. The cytosolic portion (domain 5) is required for congregation of the LDL receptors in the coated regions of the plasma membrane (see fig. 23.18). The number of LDL receptors per cell can vary between 15,000 and 70,000, depending on the cell's requirement for cholesterol. Once an LDL molecule is bound to the receptor, both are rapidly internalized by endocytosis ($t_{1/2}$ = 3 min).

Serious Diseases Result from Cholesterol Deposits

Brown and Goldstein were able to deduce the pathway for the uptake and catabolism of LDL largely as a result of their studies on the inherited disease familial hypercholesterolemia. Patients with the homozygous (two defective genes) form of this disease have grossly elevated levels of plasma cholesterol (650–1,000 mg/100 ml) (see table 23.1 for normal values), which is largely carried by an elevated concentration of LDL. One result is the formation of cholesterol deposits in the skin (xanthomas) in various areas of the body. Of greater consequence is the deposit of cholesterol in arteries, which results in atherosclerosis, a condition that is the underlying cause of most cardiovascular diseases. In fact, patients with homozygous familial hypercholesterolemia have symptoms of heart disease by the early teens and usually die from cardiovascular disease before the age of 20. The heterozygotes (individuals with one normal and one defective gene) manifest similar but less severe symptoms. Their plasma cholesterol is in the range of 250–550 mg/100 ml of plasma, and they generally do not have a heart attack before the age of 40. The frequency of the heterozygous form of familial hypercholesterolemia has been estimated at 1 in 500, and that of the homozygous form is, in all likelihood, about 1 in 1 million.

Four different classes of biochemical mutations have been shown to cause familial hypercholesterolemia (fig. 23.22). The most common defect (class 1) is in the synthesis of the receptor. The other classes of mutations are defects in the transport of the receptor to the Golgi (class 2), defects in the binding of LDL (class 3), and inability of the receptors to cluster in coated pits (class 4).

Figure 23.22

Four classes of mutations that disrupt the structure and function of the LDL receptor. Each class of mutation interferes with a different step in the process by which the receptor is synthesized. Class 1 mutations result in defective receptor synthesis in the endoplasmic reticulum. Class 2 mutations lead to defective receptor processing in the Golgi. Class 3 mutations result in defective receptor binding sites for LDL particles. Finally, class 4 mutations result in the inability of a receptor to cluster in coated pits.

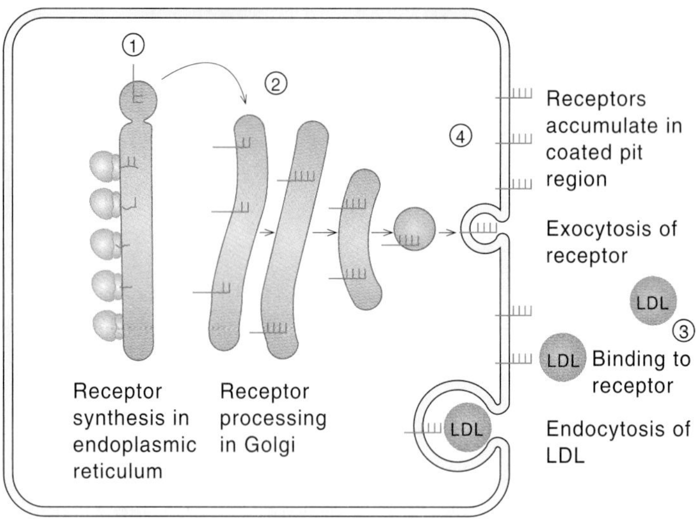

A related disorder, Wolman's disease, has provided further evidence for the receptor-mediated pathway of LDL uptake (see fig. 23.18). Wolman's disease is a very rare inborn error of metabolism (approximately 25 cases diagnosed since 1956) that is characterized by the accumulation of cholesteryl esters and triacylglycerols in various tissues. The disease can be diagnosed within several weeks of birth but is fatal, usually within 6 months. It is caused by a complete lack of a lysosomal acid lipase, which is responsible for the normal catabolism of cholesteryl esters and triacylglycerols in lysosomes. Cholesteryl ester storage disease is a related disorder, caused by a substantial reduction in the activity of lysosomal acid lipase (1 to 20% of normal). The symptoms are far less severe than in Wolman's disease, and patients have survived to the age of 40.

The importance of the work done on the LDL receptor and familial hypercholesterolemia was recognized in 1985 when Joseph Goldstein and Michael Brown were awarded the Nobel Prize in Physiology or Medicine.

High-Density Lipoproteins (HDL) May Reduce Cholesterol Deposits

The catabolism of high-density lipoproteins is a complex process that is currently under investigation. The half-life of HDL in human plasma (5 to 6 days) is much longer than for the other lipoproteins. When HDL is secreted into plasma from liver, it has a discoid shape and is almost devoid of cholesterol ester (fig. 23.23). These newly formed HDL particles are converted into spherical particles by the accumulation of cholesteryl ester. The cholesteryl ester is derived from cholesterol and phosphatidylcholine on the surface of the HDL particle in a reaction catalyzed by lecithin:cholesterol acyltransferase (LCAT) (figs. 23.23 and 23.24). LCAT is a glycoprotein (24% carbohydrate by weight) with a molecular weight of 59,000. This enzyme is associated with HDL in plasma and is activated by apoprotein A-I, a component of HDL (see table 23.3). Associated with the LCAT-HDL complex is cholesteryl ester transfer protein, which catalyzes the transfer of cholesteryl esters from HDL to VLDL or LDL. In the steady state, cholesteryl esters that are synthesized by LCAT would be transferred to these other lipoproteins and catabolized as noted earlier. The HDL particles themselves turn over, but how they are degraded is not firmly established.

Although elevated levels of cholesterol and LDL in human plasma are linked with an increased incidence of cardiovascular disease, recent data have shown that an increase in concentration of HDL in plasma is correlated with a lowered risk of coronary artery disease. Why does an elevated HDL level in plasma appear to protect against cardiovascular disease, whereas an elevated LDL level seems to cause this disease? The answer to this question is not known. An explanation currently favored is that HDL functions in the return of cholesterol to the liver, where it is metabolized and secreted. The net effect would be a decrease in the amount of plasma cholesterol available for deposit in arteries (see box 23C for more information on cholesterol and heart disease).

Bile Acid Metabolism

The conversion of cholesterol to bile acids is quantitatively the most important mechanism for degradation of cholesterol. In a normal human adult, approximately 0.5 g of cholesterol is converted to bile acids each day. The regulation of this process operates at the initial biosynthetic step catalyzed by the endoplasmic reticulum enzyme 7α-hydroxylase (fig. 23.25). The 7α-hydroxylase is one of a group of enzymes called mixed function oxidases, which are involved in the hydroxylation of the sterol molecule at numerous specific sites. A mixed function oxidase is an enzyme complex that catalyzes hydroxylation of a substrate and production of H_2O from a single molecule of O_2. The 7α-hydroxylase is one of several enzymes referred to as cytochrome P450. The hydroxylation of cholesterol also requires NADPH:cytochrome P450 reductase (fig. 23.26).

The subsequent conversion of 7α-hydroxycholesterol to cholic acid is outlined in figure 23.27. These reactions involve oxidation of the 3β-hydroxyl group, isomerization of the double bond, 12α-hydroxylation, reduction of the double bond, and reduction of the 3-keto group to a 3α-hydroxyl group. Additional hydroxylations and oxidation reactions on the side chain lead to cholic acid, one of the two major human bile acids (see chapter 19).

The bile acids are mostly converted to the corresponding bile salts, as shown for the formation of glycocholate in figure 23.28. The bile salts are critically important for the solubilization of lipids in the intestine, as shown in fig. 23.29.

Figure 23.23

Conversion of HDL disk particles into spherical particles. Apoprotein A-I is secreted from liver and intestine either as the free apoprotein or as the apoprotein in association with phospholipid. These newly secreted particles are small and disk shaped. Cholesterol is transferred from the plasma membrane of cells to these disks. Subsequently, lecithin:cholesterol acyltransferase (LCAT), an enzyme present on the HDL, catalyzes the formation of cholesteryl ester from cholesterol and phosphatidylcholine (lecithin). Since the cholesteryl ester is very hydrophobic, it prefers the apolar environment in the core of the HDL particle. As a result, after many enzymatic reactions, the HDL has a large core and is spherical in shape. (Figure kindly supplied by Dr. Haydn Pritchard, University of British Columbia.)

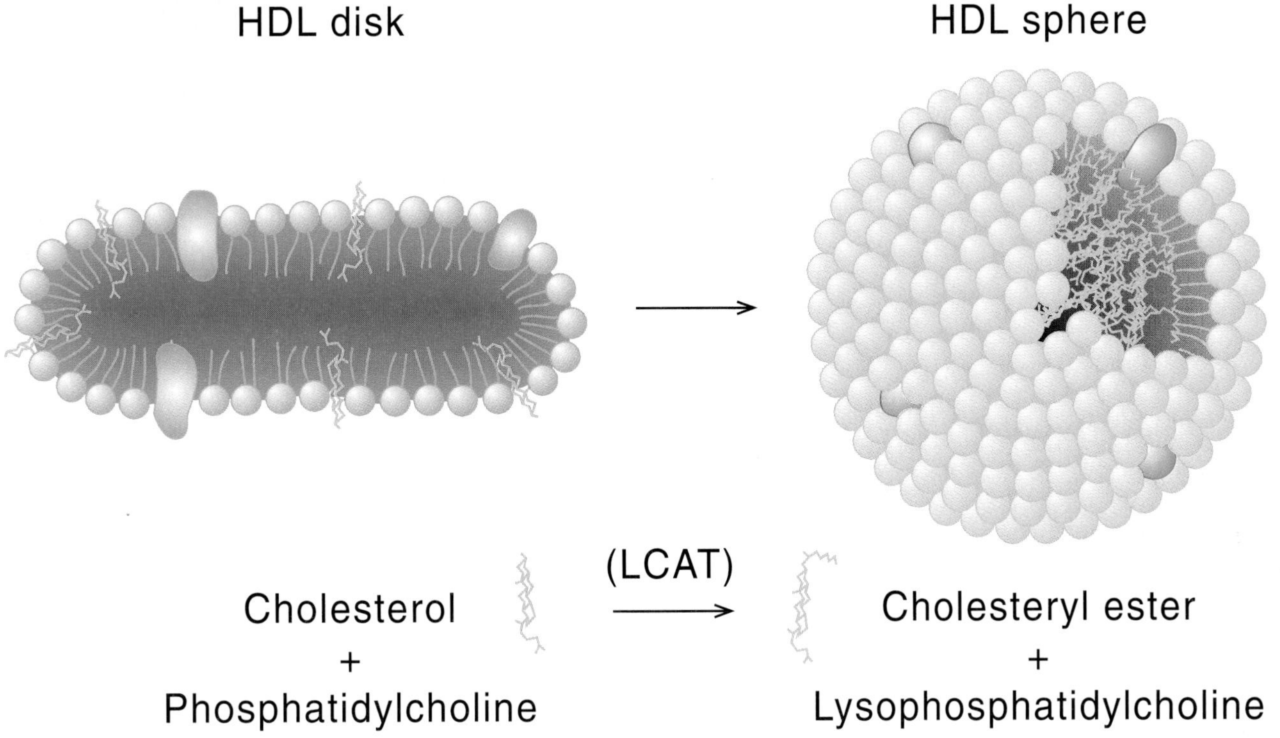

Metabolism of Steroid Hormones

Information on the hormonal functions of steroids and the mechanism by which steroids work is presented in the chapter on hormone action (chapter 27). Here we will focus on the enzymatic processes by which cholesterol is converted to the major steroid hormones.

The initial reaction in steroid hormone biosynthesis is catalyzed by desmolase (side-chain cleavage complex), which is found in the mitochondria of steroid-producing tissues (e.g., adrenals, gonads). The reaction is shown in figure 23.30. Desmolase appears to consist of two hydroxylases containing cytochrome P450, and a lyase. The product, pregnenolone, is subsequently transferred to the endoplasmic reticulum, where an oxidation of the hydroxyl group and isomerization of the double bond produces progesterone (see fig. 23.30). Progesterone is a steroid hormone, and it or pregnenolone is the biosynthetic precursor of all other sterorid hormones.

The initial step in the conversion of progesterone to aldosterone (fig. 23.31) is catalyzed by a 21-hydroxylase, present on endoplasmic reticulum from the adrenal cortex but absent in gonads and placenta. This enzyme ($M_r = 47{,}000$) is also a cytochrome P450 protein and has been purified from adrenocortical endoplasmic reticulum. It is distinct from the 11β-hydroxylase and 18-hydroxylase, two mitochondrial enzymes also involved in aldosterone synthesis (see fig. 23.31). It is curious that the biosynthesis of aldosterone from cholesterol begins in mitochondria with desmolase. Then the next reactions are catalyzed by enzymes on the endoplasmic reticulum, and finally the 21-hydroxy-progesterone is carried back to the mitochondria for the last enzymatic steps. The reason for this subcellular compartmentation is not obvious, and the mechanism by which the cell directs the intermediates from one subcellular site to another is not known.

The 17α-hydroxylase (fig. 23.32) is a mixed function oxidase found on the endoplasmic reticulum in all steroid-secreting

Figure 23.24

Reaction catalyzed by lecithin:cholesterol acyltransferase (LCAT). The resulting cholesteryl ester is transferred to VLDL and LDL particles by a lipid transfer protein.

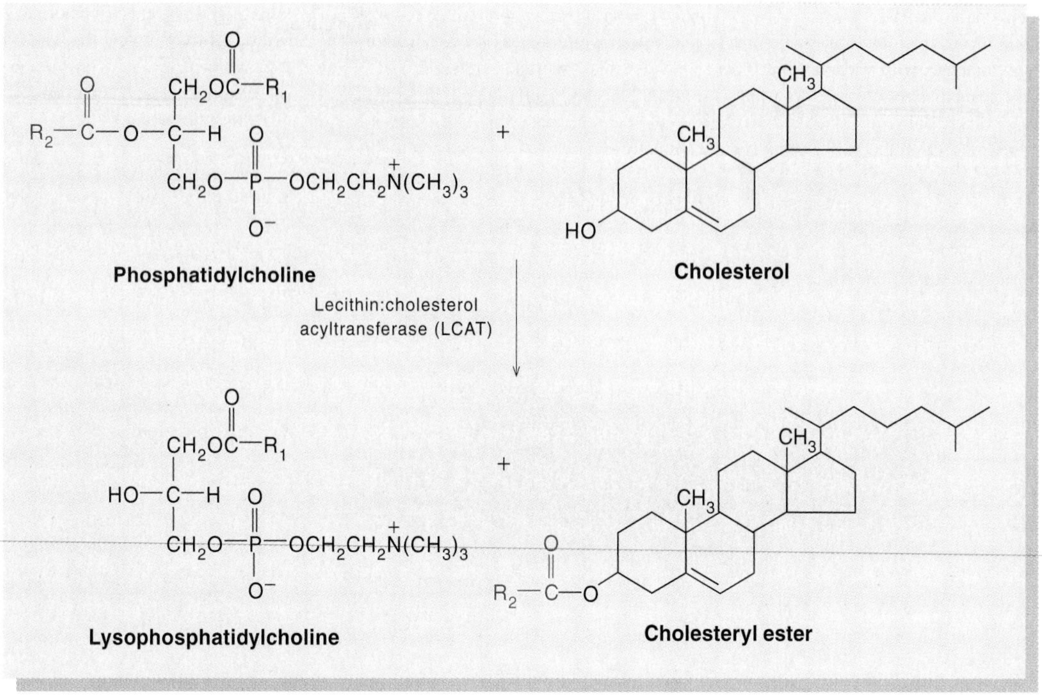

Phosphatidylcholine + **Cholesterol**

Lecithin:cholesterol acyltransferase (LCAT)

Lysophosphatidylcholine + **Cholesteryl ester**

Figure 23.25

Formation of 7-hydroxycholesterol. The committed and rate-limiting reaction for bile acid synthesis is catalyzed by the endoplasmic reticulum enzyme 7α-hydroxylase.

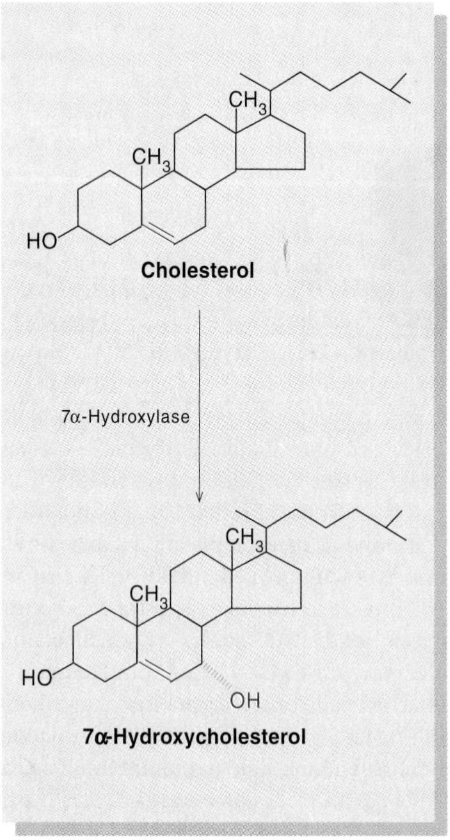

Cholesterol

7α-Hydroxylase

7α-Hydroxycholesterol

Figure 23.26

The reaction for the mixed function oxidase activity of 7α-hydroxylase.

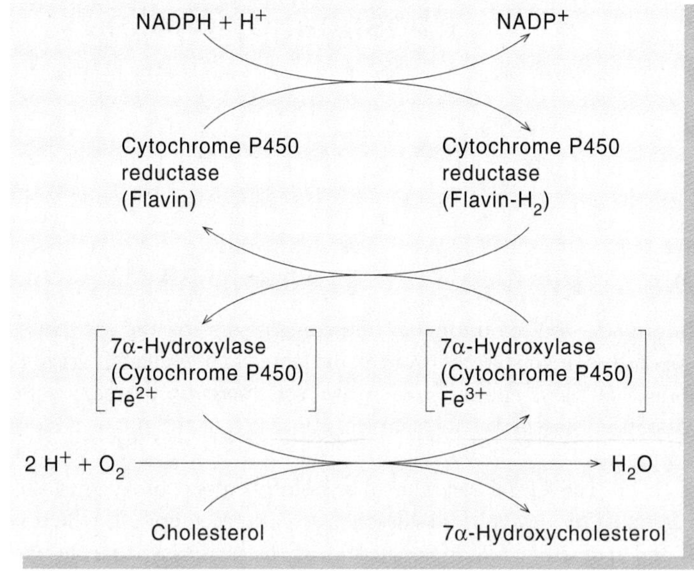

Cholesterol Metabolism and Heart Disease

There is considerable discussion in the lay press about cholesterol and its link to cardiovascular disease because there is a direct relationship between elevated levels of cholesterol in the plasma and the incidence of heart disease. Experts generally agree that people with levels of total cholesterol in plasma above 240 mg/dl (6.2 millimoles/l) for many years are at increased risk of having a heart attack compared with people whose plasma cholesterol level is below 200 mg/dl (5.2 millimoles/l). It is generally recommended that adults should endeavor to achieve levels of total cholesterol (includes both free cholesterol and cholesteryl ester) in plasma of 200 mg/dl (5.2 millimoles/l) or less. As discussed in the text, plasma cholesterol is largely carried in LDL as cholesteryl ester. The cholesteryl ester carried by LDL is sometimes referred to in the lay press as "bad cholesterol."

However, the relationship between cholesterol levels in plasma and cardiovascular disease is not so simple, since high levels of HDL (which contains 30% cholesterol by weight—see table 23.2) appear to protect against heart attack. Not surprisingly, the cholesterol carried by HDL is referred to as "good cholesterol" in the lay press. In addition, there are many other factors that have been linked to increased incidences of heart attack by epidemiological studies. Among these factors are smoking, high blood pressure, genetic background (recall our discussion of familial hypercholesterolemia), diabetes, and obesity.

How can people achieve total cholesterol levels of 200 mg/dl or less? If they were lucky and chose their parents carefully, they will have genes that protect them from high cholesterol levels. Such people can eat a diet that is high in fat and cholesterol and still remain well below the 200 mg/dl threshold. Most adults are not in this category and therefore have to use dietary or drug treatments to reduce plasma cholesterol levels. Cholesterol is enriched in animal meats and dairy products and is absent in vegetables. Thus many people can reduce their plasma cholesterol levels by eating smaller amounts of meat and dairy products. Another complication is that the serum levels of cholesterol are affected by the amount of saturated and polyunsaturated triacyglycerols in the diet. Saturated fats tend to increase plasma cholesterol levels, whereas polyunsaturated fats tend to protect against increased cholesterol in the plasma. Again, meat and dairy products are enriched in saturated fatty acids. In contrast, many fish have high levels of polyunsaturated oils, particularly $C_{22:5}$ and $C_{22:6}$. People in countries such as Japan, who have diets enriched in fish, have lower cholesterol levels in plasma and a reduced incidence of heart disease compared with people in North America and many European countries. Thus reducing the intake of lipid from 40 to 30% of dietary calories by eating more carbohydrate, vegetables, and fish will often result in a 15 to 20% decrease in serum cholesterol. However, there is a wide variation among individuals in how successful the dietary approach is, since genetic factors again come into play.

A drug called lovastatin (see fig. 23.6) has become available by prescription for treatment of hypercholesterolemia. This compound, a competitive inhibitor of HMG-CoA reductase, sharply reduces the rate of cholesterol biosynthesis. Equally important, in response to the reduction of cellular cholesterol levels there is increased expression of LDL receptors, which in turn allows more LDL to be cleared from the bloodstream, particularly by the liver. Thus the combination of this drug plus dietary restriction can result in striking reductions in plasma cholesterol levels (e.g., from 300 to 200 mg/dl). The drug is used effectively by heterozygotes with familial hypercholesterolemia. Interestingly, lovastatin has no significant effect on the serum cholesterol levels in homozygotes with familial hypercholesterolemia, since these patients do not have functional LDL receptors.

High levels of plasma cholesterol do not directly cause heart attacks. Rather, high levels of serum cholesterol over long periods are somehow involved in the development of a disease of the arteries called atherosclerosis. Atherosclerotic plaques are complex lesions in arterial walls that contain abnormal deposits of cholesteryl esters. Precisely how high cholesterol levels in the plasma relate to the development of atherosclerosis is not understood and is a major frontier of medical research today. The actual sudden onset of heart attack appears to result from the adherence of platelets to the atherosclerotic lesion and the subsequent formation of a clot that occludes the blood flow in a coronary artery. The heart tissue is therefore deprived of blood supply, a condition that leads to tissue death.

organs. It is the key enzyme for directing steroids into the synthesis of glucocorticoids, androgens, and estrogens. Hydroxylations at the 11β-position and the 21-position direct the 17-OH progesterone into cortisol, a glucocorticoid made in the adrenals (see fig. 23.32). Alternatively, the gonads will direct 17-OH progesterone to the synthesis of *testosterone* as the result of the action of 17,20-lyase and a reduction of the 17-keto group (see fig. 23.32). Testosterone can be converted to dihydrotestosterone as discussed in chapter 27.

The female sex hormones arise from testosterone via the 19-hydroxylated intermediate (fig. 23.33). The enzyme complex responsible for this conversion is called the aromatase system and is one of the few reactions by which mammals can make an aromatic ring. This complex is associated with the endoplasmic reticulum of cells in the ovary and the placenta.

Thus the second major degradative fate of cholesterol in mammals is the formation of steroid hormones. The enzymes involved are found in mitochondria and endoplasmic reticulum of

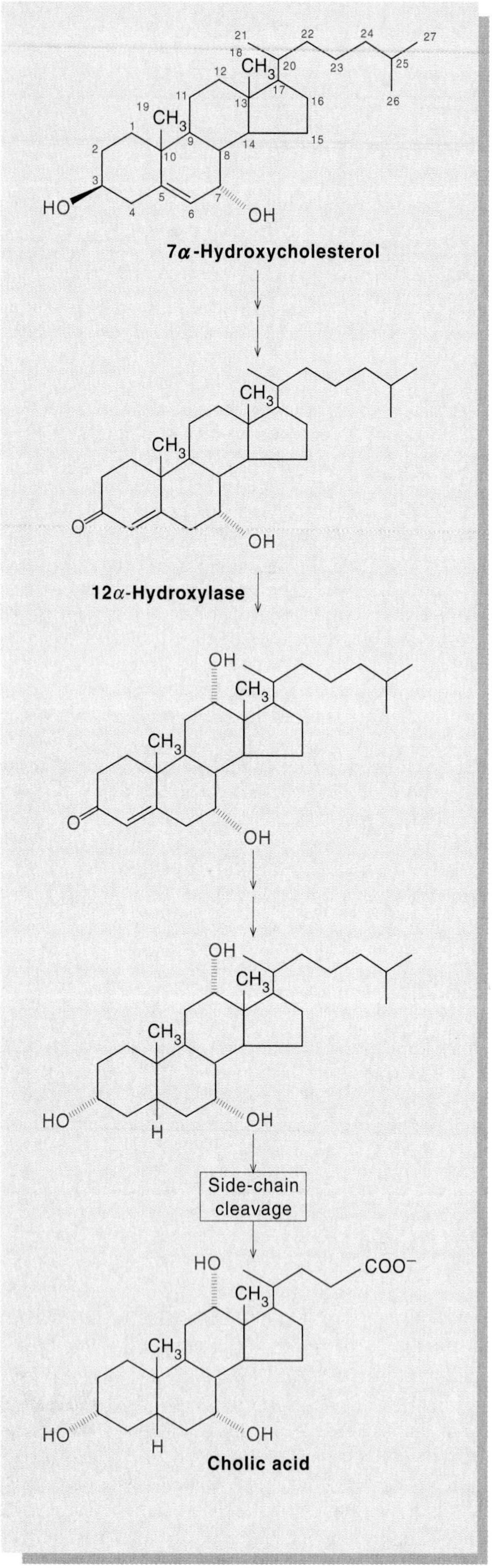

Figure 23.27

Conversion of 7α-hydroxycholesterol to cholic acid. The increase in the number of polar groups in the conversion to cholic acid increases the water solubility. Bile acids possess a 5 hydrogen. As a result, the A and B rings are no longer coplanar but have an A/B *cis*-configuration. This configuration improves the detergent properties of the bile acids so that they are better able to solubilize lipids in the intestine and aid digestion.

Figure 23.28

Conversion of a cholic acid into glycocholate. Glycocholate is an example of a bile salt. The bile salts solubilize lipids in the small intestine so they can be degraded by lipases (see fig. 23.29).

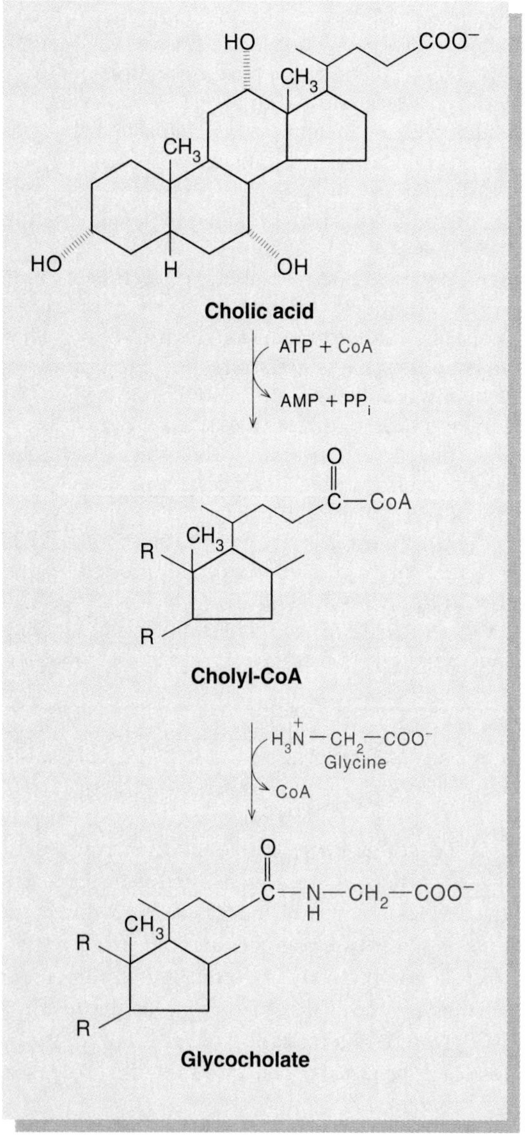

Figure 23.29

Bile salts emulsify fats in the intestine. The hydrophobic side or surface of the bile salt associates with triacylglycerols to form a complex. These complexes aggregate to form a micelle, with the hydrophilic side of the bile salt facing outward. Pancreatic lipase also associates with this micelle. The action of the lipase releases free fatty acids that form a much smaller micelle, which can be absorbed by the intestine.

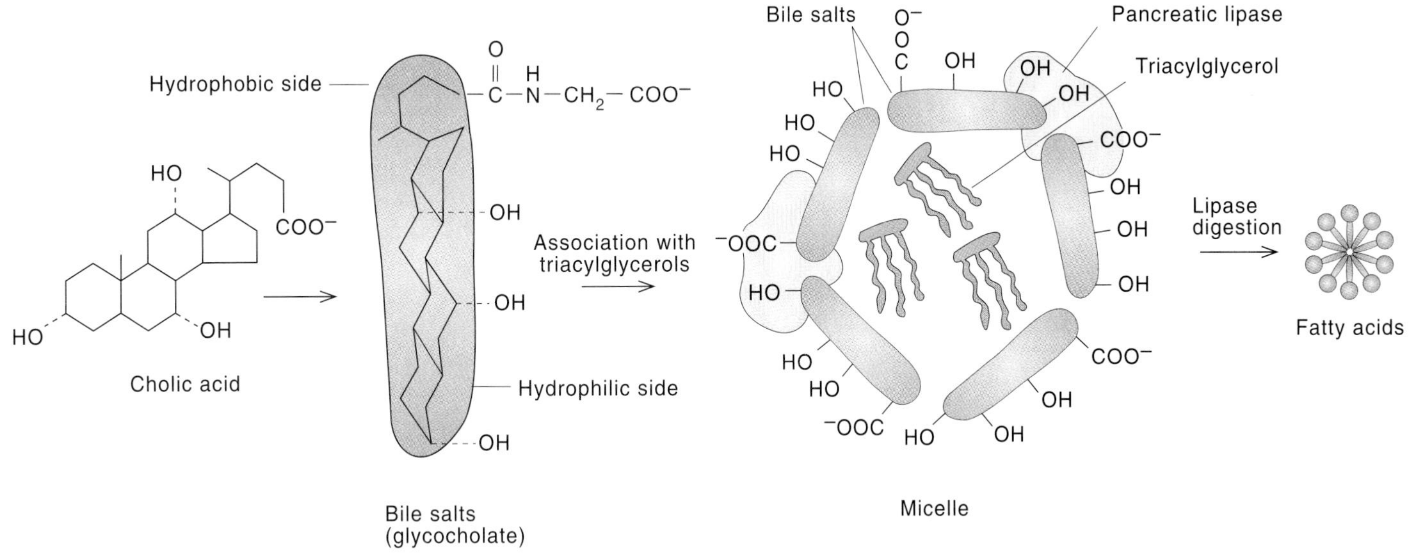

steroid-producing tissues. The hydroxylases are in each case mixed function oxidases that have cytochrome P450 at the active site. We will discuss some of the remarkable effects of these potent hormones in chapter 27.

There is no known pathway in mammals by which the steroid ring nucleus can be degraded to smaller molecules such as acetate. The carbon atoms of the ring system cannot, therefore, be used as a source of metabolic energy. However, many reactions occur, particularly in the liver, for the partial catabolism and inactivation of the steroid hormones. Frequently, these reactions involve reduction of ketone groups or double bonds. These inactive steroids are conjugated to glucuronic acid (fig. 23.34) or sulfate. The excretion of such derivatives results in a very large number of steroid metabolites in the urine. For example, 20 different metabolites of estrogen, conjugated to sulfate or glucuronic acid, have been identified in human urine.

Overview of Mammalian Cholesterol Metabolism

The metabolism of cholesterol in mammals is extremely complex. A summary sketch (fig. 23.35) helps to draw the major metabolic interrelationships together. Cholesterol is biosynthesized largely in the liver (fig. 23.35*a*) or taken in through the diet (fig. 23.35*b*). From the intestine, cholesterol is secreted into the plasma mainly as a component of chylomicrons. These particles are quickly degraded by lipoprotein lipase, and the remnants are removed by the liver. Apoproteins and lipid components of the chylomicrons and remnants appear to exchange with HDL. Cholesterol made in the liver (see fig. 23.35*a*) has several alternative fates. It can be (1) secreted into plasma as a component of HDL and VLDL, (2) stored in droplets as cholesteryl ester, (3) used as a structural component of cell membranes, or (4) converted into bile salts. In plasma, VLDL is degraded to IDL and LDL by the action of lipoprotein lipase and through exchange reactions with HDL. The LDL serves as a major carrier of cholesterol to extrahepatic cells, which include the adrenals and gonads. LDL and HDL are also returned to the liver. The steroid hormones made in the adrenals and gonads are delivered to various target tissues and promote a wide range of metabolic effects. The steroid hormones are eventually excreted as glycosyl conjugates in the urine. The bile salts made in the liver are delivered to the upper intestine, where they aid in the solubilization of dietary lipid. Most of the bile salts are resorbed in the lower intestine and returned to the liver by the portal vein.

Figure 23.30

Biosynthesis of progesterone. Desmolase is found in the mitochondria of steroid-producing tissues. Cholesterol is converted to pregnenolone in the mitochondria and then transferred to the endoplasmic reticulum, where it is converted to progesterone. Progesterone is a hormone as well as the precursor of several other hormones.

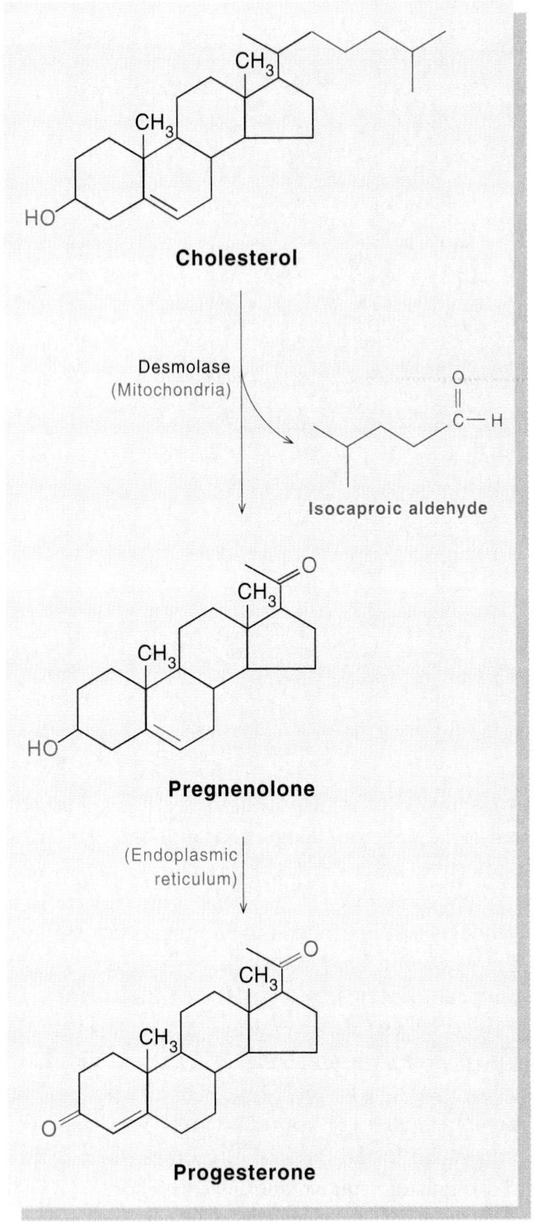

Figure 23.31

Conversion of progesterone to aldosterone. The initial reaction, involving the 21-hydroxylase enzyme, occurs on the endoplasmic reticulum of the adrenal cortex. Three mitochondrial enzymes are involved in the next steps, leading to aldosterone.

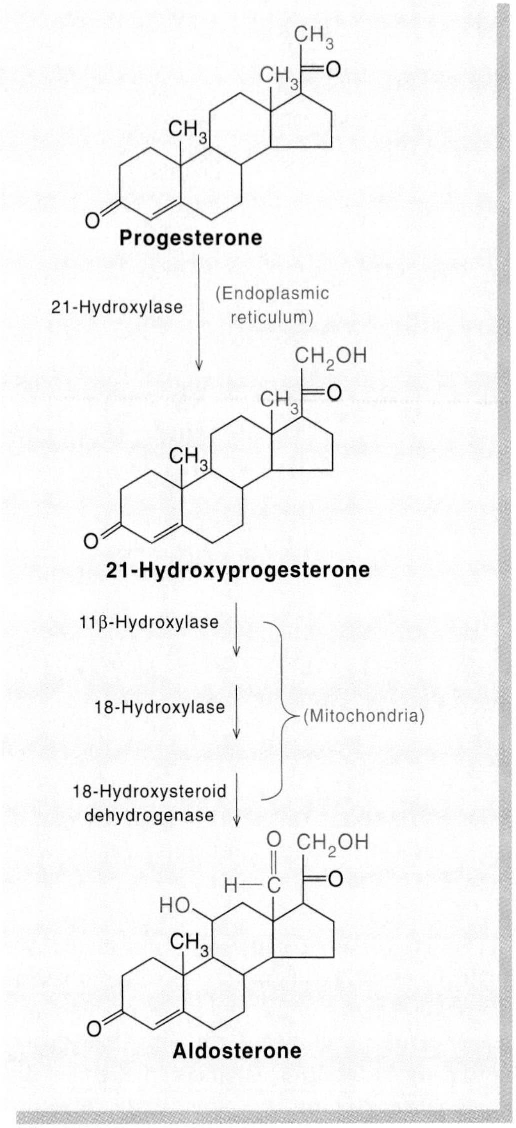

Figure 23.32

Two pathways for the conversion of progesterone to other steroid hormones. The first reaction, catalyzed by 17α-hydroxylase, is found in all steroid-secreting organs. Subsequent conversions to cortisol and testosterone are organ-specific as shown.

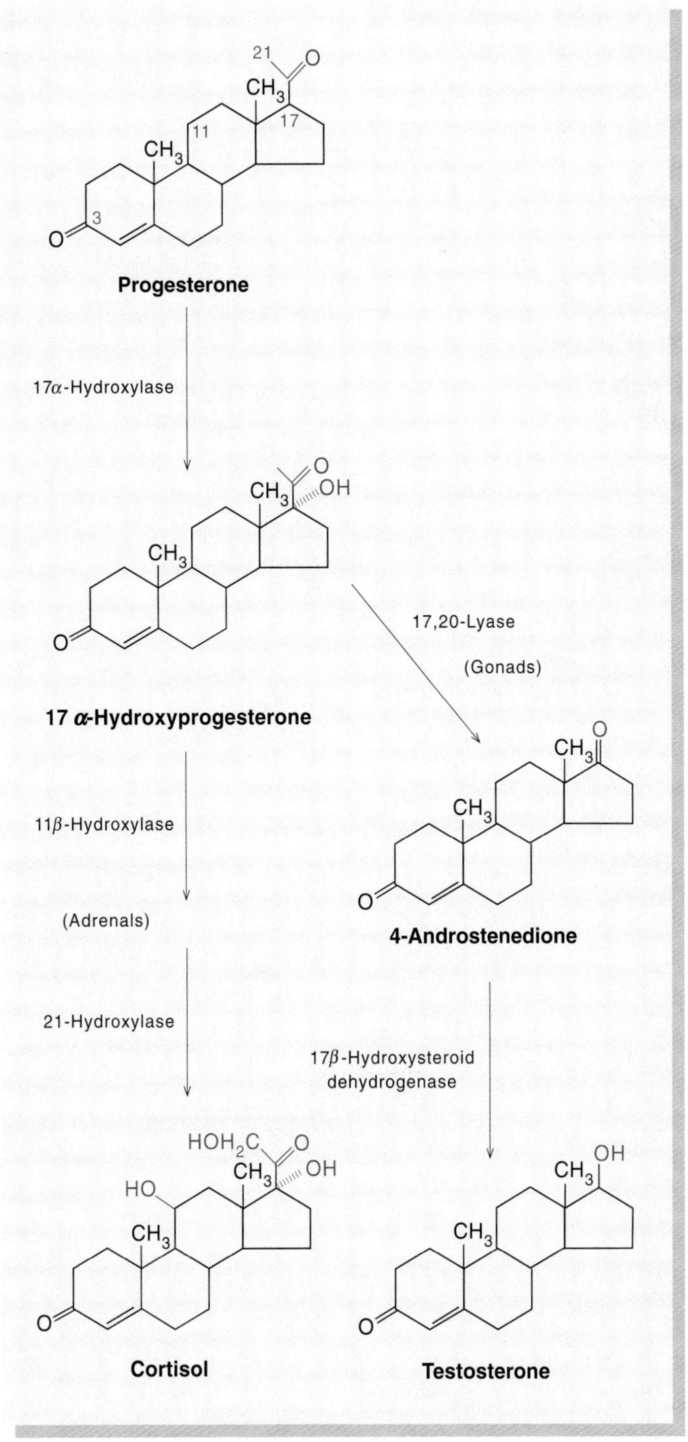

Progesterone → (17α-Hydroxylase) → 17 α-Hydroxyprogesterone

17 α-Hydroxyprogesterone → (11β-Hydroxylase, Adrenals) → (21-Hydroxylase) → Cortisol

17 α-Hydroxyprogesterone → (17,20-Lyase, Gonads) → 4-Androstenedione → (17β-Hydroxysteroid dehydrogenase) → Testosterone

Figure 23.33

The conversion of testosterone into estradiol. These reactions are catalyzed by a complex of enzymes called the aromatase system, which consists of three enzyme activities located on the endoplasmic reticulum. The aromatase system is found in the ovaries and the placenta.

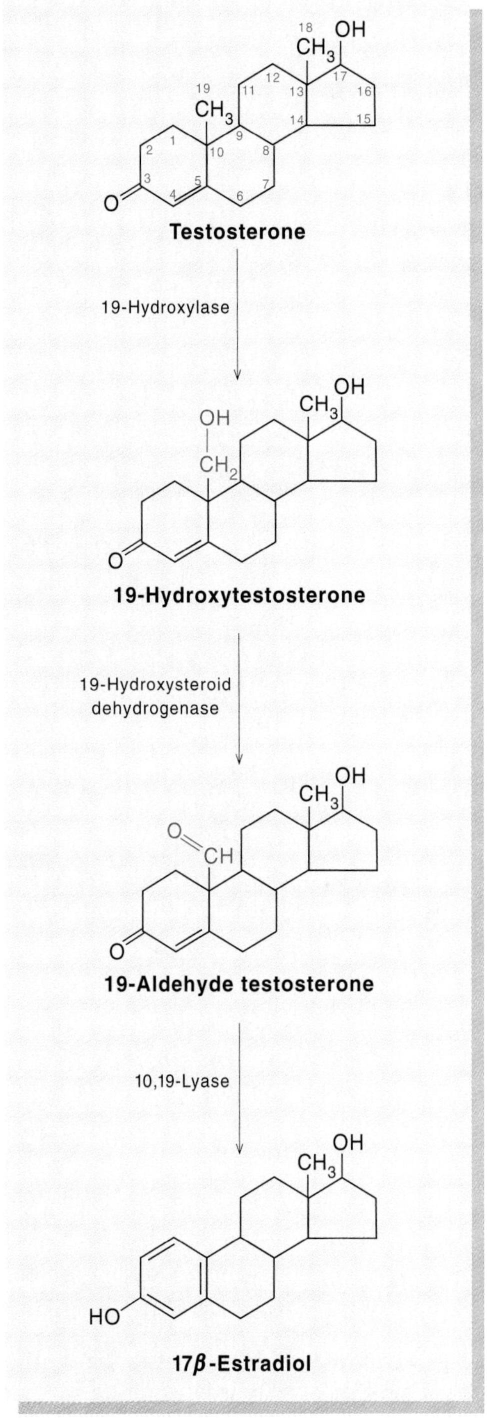

Testosterone → (19-Hydroxylase) → 19-Hydroxytestosterone → (19-Hydroxysteroid dehydrogenase) → 19-Aldehyde testosterone → (10,19-Lyase) → 17β-Estradiol

Figure 23.34

Inactivated steroid hormones are conjugated to glucuronic acid prior to excretion in the urine.

UDP — Glucuronic acid

+

"Inactive steroid" — OH

↓ Glucuronyl transferase

COOH

O O— "Inactive steroid"

OH

OH H

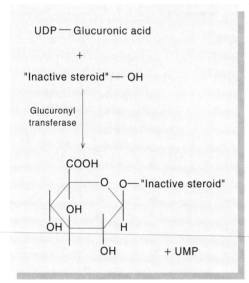

OH + UMP

Figure 23.35

Fate of cholesterol. (*a*) Cholesterol biosynthesized in the liver has several alternative fates. (*b*) Cholesterol obtained from the diet can enter the plasma and subsequently the liver.

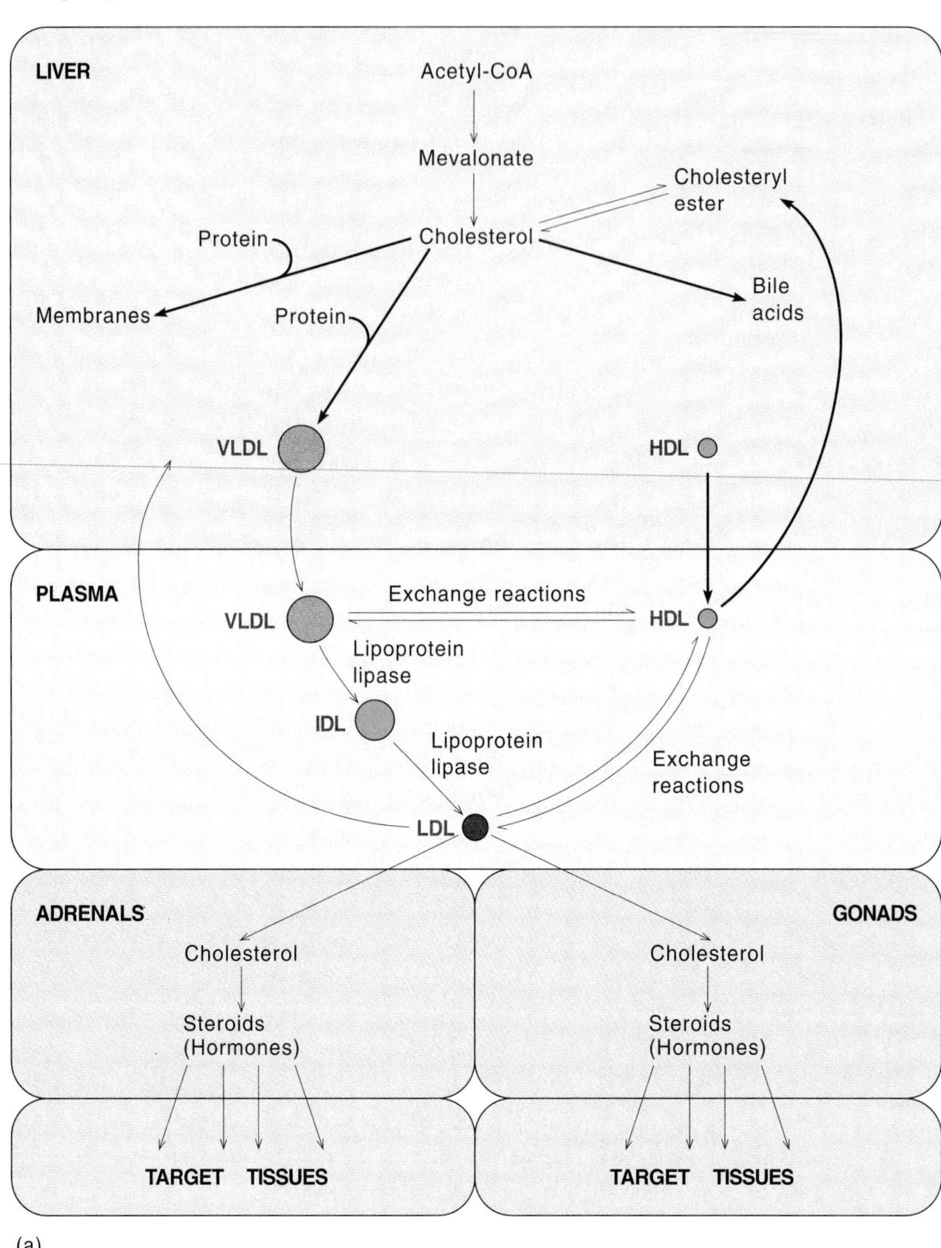

Summary

In this chapter we have dealt primarily with the metabolism of cholesterol, the most prominent member of the steroid family of lipids, and with the associated plasma lipoproteins. The chief points in our discussion are as follows.

1. Steroids are derivatives of the tetracyclic hydrocarbon perhydrocyclopentanophenanthrene. The biosynthesis of steroids begins with the conversion of three molecules of acetyl-CoA into mevalonate, the decarboxylation of mevalonate, and its conversion to isopentenyl pyrophosphate. Six molecules of isopentenyl pyrophosphate are polymerized into squalene, which is cyclized to yield lanosterol. Lanosterol is converted to cholesterol, which is the precursor of bile acids and steroid hormones.

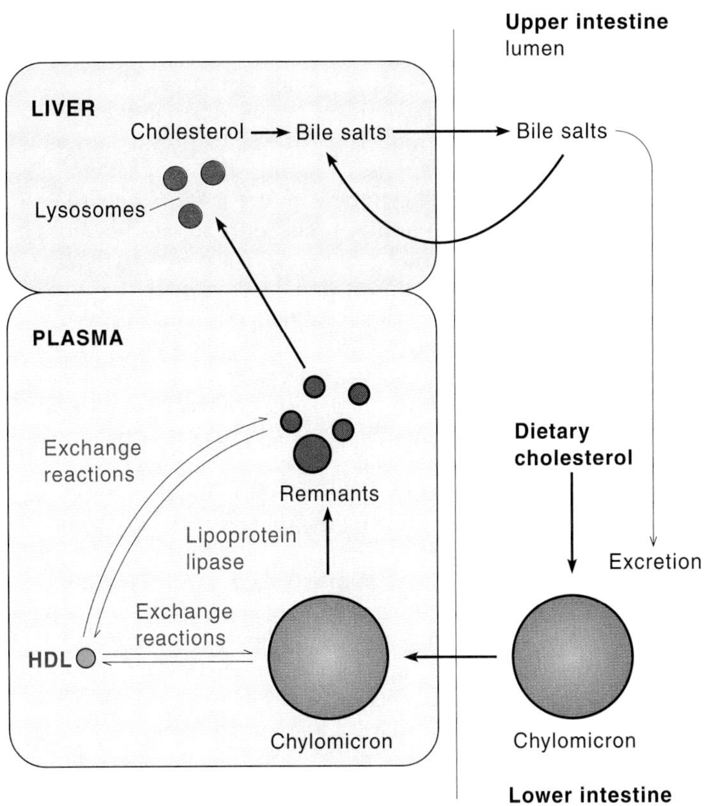

(b)

2. The rate of cholesterol biosynthesis appears to be regulated primarily by the activity of HMG-CoA reductase. This key enzyme is controlled by the rate of enzyme synthesis and degradation and by phosphorylation–dephosphorylation reactions, and synthesis of the reductase is inhibited by cholesterol delivered to cells by means of low-density lipoproteins (LDL).
3. Cholesterol and phospholipids are carried in plasma by lipoproteins, which are synthesized and secreted by the intestine and liver. The major lipoproteins are chylomicrons, very-low-density lipoproteins (VLDL), low-density lipoproteins (LDL), and high-density lipoproteins (HDL). The triacylglycerols in chylomicrons and VLDLs are degraded in plasma by lipoprotein lipase, and the fatty acids and monoacylglycerols are absorbed primarily by heart, skeletal, and adipose tissue. LDLs are removed from plasma by an endocytotic process after binding to specific LDL receptors on the plasma membrane. The LDLs are enzymatically degraded in the lysosomes. In familial hypercholesterolemia, the specific receptors for LDL uptake are inactive. High levels of LDL are associated with an increased risk of cardiovascular disease, whereas high levels of HDL seem to protect against this disease.
4. Bile acids are C_{24} carboxylic acids that are biosynthetically derived from cholesterol. The 7α-hydroxylation of cholesterol is the committed and rate-limiting reaction in the synthesis of bile acids. Salts formed from the bile acids are secreted into the small intestine and aid the solubilization and digestion of lipids.
5. Steroid hormones are biosynthesized from cholesterol in the adrenal cortex, gonads, and placenta. These steroids are important hormones for many specific physiological processes.

Selected Readings

Bentley, R., *Molecular Asymmetry in Biology,* vol. 2. New York: Academic Press, 1970. This book contains a very lucid explanation of the stereochemistry of cholesterol biosynthesis.

Bloch, K., Cholesterol: evolution of structure and function. Chapter 12 in D. E. Vance and J. E. Vance (eds.), *Biochemistry of Lipids, Lipoproteins and Membranes.* Amsterdam: Elsevier Science Publishers, 1991. This article provides an interesting view of how the structure of cholesterol evolved to optimize its function in cells.

Breslow, J. L., Insights into lipoprotein metabolism from studies in transgenic mice. *Ann. Rev. Physiol.* 56:797–810, 1994. This article describes how experiments utilizing transgenic mice have provided new knowledge on lipoprotein metabolism, particularly in reference to atherosclerosis.

Brown, M. S., and J. L. Goldstein, A receptor-mediated pathway for cholesterol homeostasis. *Science* 232:34–47, 1986. An article describing their Nobel Prize–winning research on the LDL receptor and familial hypercholesterolemia.

Davis, R. A., and J. E. Vance, Structure, assembly, and secretion of lipoproteins. Chapter 17 in D. E. Vance and J. E. Vance (eds.), *Biochemistry of Lipids, Lipoproteins and Membranes.* Amsterdam: Elsevier Science Publishers, 1996. This chapter provides an advanced discussion on the assembly and secretion of very-low-density lipoproteins.

Edwards, P. A., and Davis, R. A., Isoprenoids, sterols, and bile acids. Chapter 13 in D. E. Vance and J. E. Vance (eds.), *Biochemistry of Lipids, Lipoproteins and Membranes.* Amsterdam: Elsevier Science Publishers, 1996. The complexities of the regulation of cholesterol biosynthesis are explained in this chapter.

Fielding, P. E., and C. J. Fielding, Dynamics of lipoprotein transport in the human circulatory system. Chapter 18 in D. E. Vance and J. E. Vance (eds.), *Biochemistry of Lipids, Lipoproteins and Membranes.* Amsterdam: Elsevier Science Publishers, 1996. This article reviews the current literature on the intricacies of lipoprotein metabolism in the circulatory system.

Hardie, D. G., Regulation of fatty acid and cholesterol metabolism by the AMP-activated protein kinase. *Biochim. Biophys. Acta* 1123:231–238, 1992. A review article that presents an overview of the AMP-activated kinase and its involvement in the regulation of fatty acid and cholesterol metabolism.

Makin, H. L. J., *Biochemistry of Steroid Hormones,* 2d ed. Oxford: Blackwell, 1984. An advanced and comprehensive treatment of steroid hormones.

Schneider, W. J., Removal of lipoproteins from plasma. Chapter 19 in D. E. Vance and J. E. Vance (eds.), *Biochemistry of Lipids, Lipoproteins and Membranes.* Amsterdam: Elsevier Science Publishers, 1996.

This article provides a clear explanation of the current literature on the uptake of lipoproteins into cells and tissues.

Scriver, C. R., A. L. Beaudet, W. S. Sly, and D. Valle, *The Metabolic Basis of Inherited Disease,* 6th ed., vol. I. New York: McGraw-Hill, 1989.

This book has an introductory chapter on lipoprotein structure and metabolism followed by many excellent chapters on disorders of cholesterol and lipoprotein metabolism.

Problems

1. How many of cholesterol's carbons are chiral?
2. How many of cholesterol's carbons originate as the carboxyl and the methyl carbon of acetate?
3. Why aren't the origins of the carbons in cholesterol closer to 50% from the carboxyl and 50% from the methyl carbons of acetate (see problem 23.2)?
4. Can you comment on the nature of the fatty acid you would expect to find on cholesteryl ester produced by lecithin cholesterol acyltransferase?
5. The frequency of the heterozygous and homozygous forms of a familial hypercholesterolemia are given in the text (page 549). Can you explain the frequencies indicated for the heterozygous and homozygous forms?
6. The conversion of cholyl-CoA to glycocholate (see fig. 23.28) is an example of amide bond formation utilizing a CoASH (thioester) derivative. Are all amide bonds made from CoASH derivatives?
7. The conversion of cholic acid to cholyl-CoA (fig. 23.28) also involves the conversion of ATP to AMP and PP_i. What intermediate would you expect to be involved in this reaction?
8. Do you expect the equilibrium position of the reaction catalyzed by Acyl CoA:cholesterol acyltransferase (fig. 23.14) to be any different from that of lecithin:cholesterol acyltransferase (fig. 23.24)?
9. A patient homozygous for familial hypercholesterolemia (FH) was treated with lovastatin to lower LDL levels in the blood. This treatment did not have any effect on LDL levels. Why? After a number of heart attacks, a heart and liver transplant were done, and LDL levels were dramatically lowered. Why were both organs replaced?
10. During routine investigations the plasma from a family of rats (group 1) was found to have very low concentrations of cholesterol. When the microsomal HMG-CoA reductase from liver was assayed, extremely low activities were found. When the cytosol from normal rats (group 2) was added to the microsomal fraction from group 1 rats, the HMG-CoA reductase activity was gradually restored to normal values. What enzyme activity (or activities) might be deficient in the group 1 rats?
11. How would a dietary resin that absorbs bile salts reduce plasma cholesterol levels?
12. Notice that the metabolic sequences described in this and previous chapters often involve multiple locations. These locations can be at the organ (fig. 23.1) or the subcellular level (fig. 23.30). This is in contrast, for example, with glycolysis, which occurs in the cytosol, or the tricarboxylic acid cycle, which is completely mitochondrial. Can you provide a possible explanation for this multiple location phenomenon?
13. How can elevated levels of cholesterol in the liver lower cholesterol biosynthesis? What effect will elevated cholesterol have on LDL receptors?
14. A person with diabetes is found to show no signs of ketone bodies in plasma even when in diabetic shock. Which enzyme(s) of ketone body synthesis might be deficient? If cholesterol synthesis were normal, would this be a clue as to which enzyme(s) might be deficient?
15. What is the hereditary defect in Wolman's disease? Would you expect HMG-CoA reductase activity to be high or low in skin fibroblasts cultured from patients with this disorder? Would the number of LDL receptors be high or low in these fibroblasts?
16. There is an inherited disease in which lecithin:cholesterol acyltransferase (LCAT) is deficient. What effect would you expect this deficiency to have on the composition of HDL and other lipoproteins in plasma?

PART 6

METABOLISM OF NITROGEN-CONTAINING COMPOUNDS

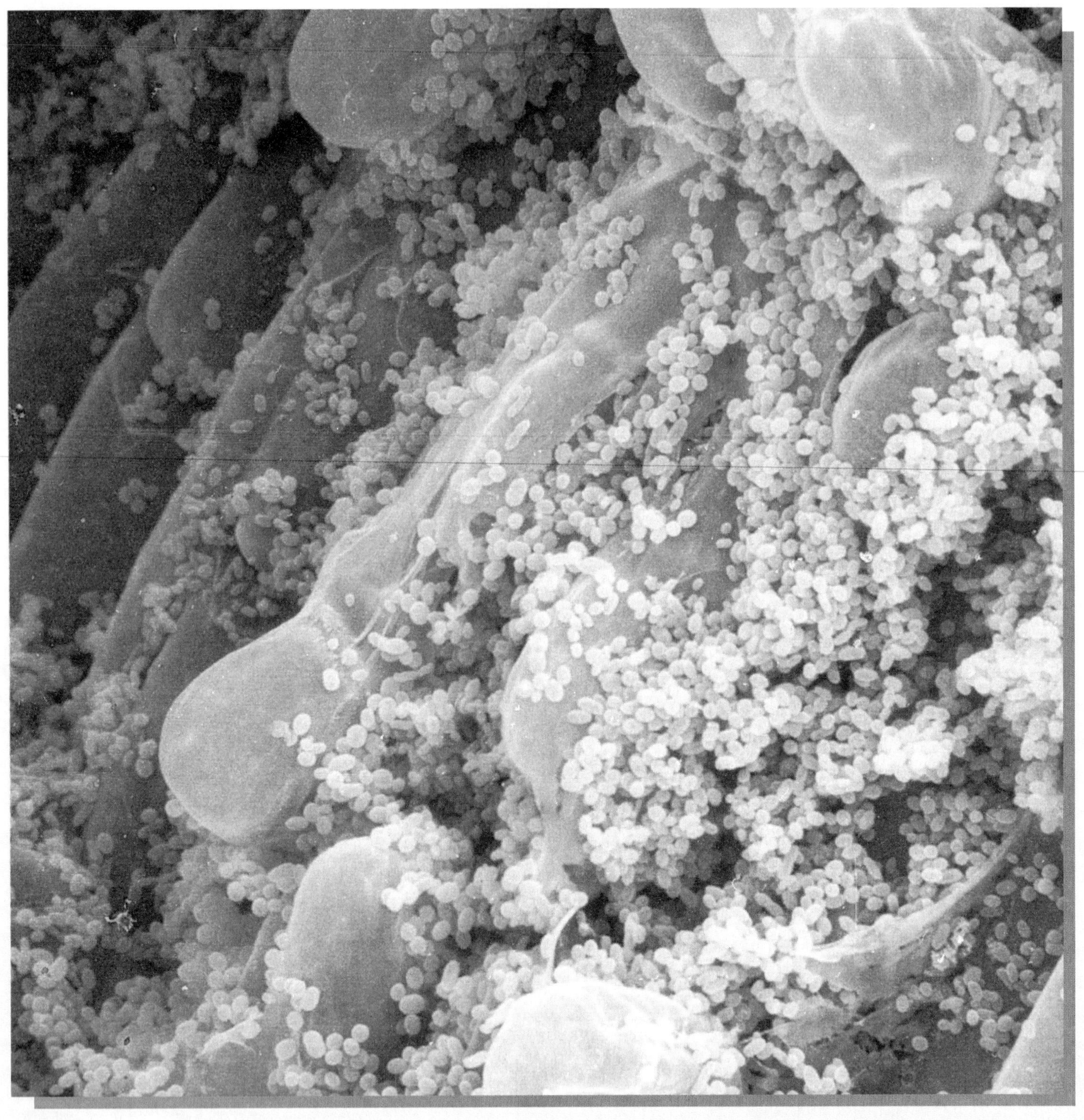

Scanning electron micrograph of part of a sorghum root covered with nitrogen-fixing bacterial cells (Azospirillum brasilense). The magnification is 1540×. Courtesy of Dr. R. Howard Berg, University of Memphis.

CHAPTER 24

Amino Acid Biosynthesis and Nitrogen Fixation in Plants and Microorganisms

The carbon skeletons for amino acid biosynthesis originate from the central metabolic pathways; the first step in amino acid biosynthesis usually involves addition of the α-amino group.

Amino acids are best known as the building blocks of protein, and indeed that is a main function of the 20 L-amino acids that are commonly found in proteins (see chapter 4). In addition to this role, amino acids serve as precursors to many important small molecule compounds, including nucleotides, porphyrins, parts of lipid molecules, and several coenzymes. Amino acids are end-products and intermediates in the conversion of nitrogen and sulfur from inorganic to organic forms. As an alternative energy source, amino acids can be degraded in ways that regenerate ATP from ADP or AMP.

In this chapter we focus on the biosynthesis of amino acids (fig. 24.1) and their role in bringing inorganic nitrogen and sulfur into the biological world. In addition, nonprotein amino acids are discussed briefly.

The Pathways to Amino Acids Are Branchpoints from the Central Metabolic Pathways

Inspection of the amino acid biosynthetic pathways shows that all amino acids arise from a few intermediates of central metabolism (see fig. 24.1). Amino acids derived from a common intermediate are said to be in the same family. For example, the serine family of amino acids, which includes serine, glycine, tryptophan, and cysteine, all contain carbon atoms from glycerate-3-phosphate (see fig. 24.1). Carbon flow from the central metabolic pathways to amino acids is a regulated process that provides amino acids in the amounts needed for maintenance and growth. The flow is regulated mostly by end-product inhibition; thus amino acids (end product) generally inhibit the first enzyme in the pathway for its synthesis.

Our Understanding of Amino Acid Biosynthesis Has Resulted from Genetic and Biochemical Investigations

Escherichia coli served as the organism of choice for dissecting the pathways of amino acid biosynthesis, for two reasons: (1) *E. coli* synthesizes all 20 amino acids commonly found in proteins, and (2) *E. coli* is an organism ideally suited for genetic and biochemical studies.

The Number of Proteins Participating in a Pathway Is Known through Genetic Complementation Analysis

The investigation of an amino acid biosynthetic pathway in *E. coli* begins with the isolation of mutants that are deficient in the capacity to synthesize that amino acid. Mutants of this sort are known as auxotrophs. Each auxotroph bears a mutation in one of the genes that encodes an enzyme required for a step in the biosynthetic pathway of the amino acid. It is possible to determine how many steps are in a particular pathway by the process of complementation analysis (see fig. 12.14).

Biochemists Use the Auxotrophs Isolated by Geneticists

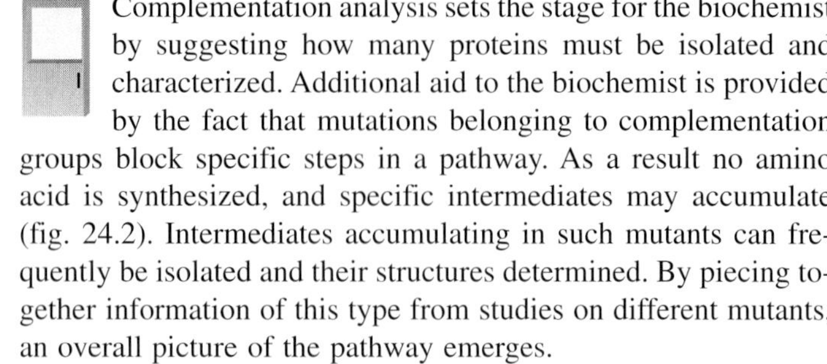

Complementation analysis sets the stage for the biochemist by suggesting how many proteins must be isolated and characterized. Additional aid to the biochemist is provided by the fact that mutations belonging to complementation groups block specific steps in a pathway. As a result no amino acid is synthesized, and specific intermediates may accumulate (fig. 24.2). Intermediates accumulating in such mutants can frequently be isolated and their structures determined. By piecing together information of this type from studies on different mutants, an overall picture of the pathway emerges.

Identifying the intermediates in a pathway is only half the problem. The biochemist must also isolate and characterize each enzyme of the pathway. To this end, extracts from either normal or mutant cells are tested in conjunction with different intermediates to see whether the conversion of one intermediate to the next in the pathway can occur. This conversion is used as an assay for isolating the enzyme by fractionation of the extract until the enzyme of interest is reasonably pure (see chapter 7 for protein purification techniques). In the final stages of biochemical analysis the structure and catalytic mechanism of the enzyme are determined.

Sometimes the elucidation of a pathway is complicated by the fact that common intermediates are used in the synthesis of more than one amino acid. Such "branchpoint" intermediates may be converted into more than one product by different enzymes, each product serving as an intermediate for a particular amino acid. Mutants containing defects in enzymes required for the synthesis of more than one amino acid usually require the addition of two or more amino acids for cell growth.

The Glutamate Family of Amino Acids and Nitrogen Fixation

A common element in all protein-bound amino acids is the α-amino group. Directly or indirectly, the α-amino groups of most amino acids are derived from ammonia by way of the amino groups of L-glutamine. Ammonia itself can be incorporated directly into α-ketoglutarate to make glutamate or into glutamate to make glutamine. The reaction of NH_3 with glutamate represents a key step in the process whereby reduced nitrogen becomes incorporated into organic molecules.

Glutamate also serves as the precursor of glutamine, proline, and arginine (see fig. 24.1). Appropriately, these four amino acids are described as belonging to the glutamate family. In fungi, lysine is also included in this family (see fig. 24.1).

In this section we discuss the pathways for glutamate and glutamine synthesis. We also discuss some other aspects of the nitrogen cycle and proline and arginine synthesis.

Figure 24.1

Outline of the biosynthesis of the 20 amino acids found in proteins. The de novo biosynthesis of amino acids starts with carbon compounds found in the central metabolic pathways. The central metabolic pathways are drawn in black, and the additional pathways are drawn in red. Some key intermediates are illustrated, and the number of steps in each pathway is indicated alongside the conversion arrow. All common amino acids are emphasized by boxes. Dashed arrows from pyruvate to both diaminopimelate and isoleucine reflect the fact that pyruvate contributes some of the side-chain carbon atoms for each of these amino acids. Note that lysine is unique in that two completely different pathways exist for its biosynthesis. The six amino acid families are screened.

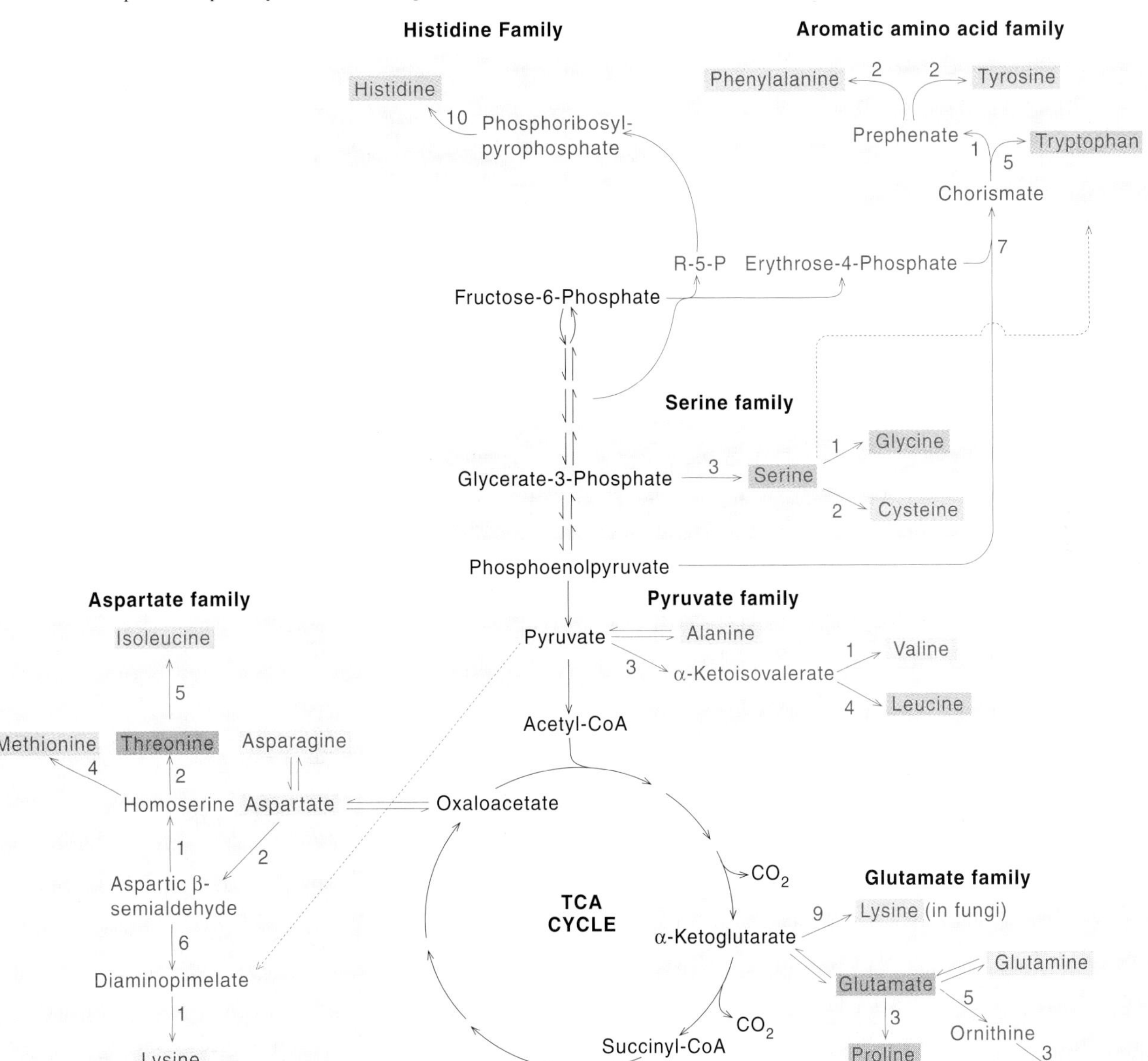

The Direct Amination of α-Ketoglutarate Leads to Glutamate

The most direct route to glutamate (and therefore amino group formation) is that exhibited by many bacteria when grown in a medium containing plentiful amounts of ammonium ion as the sole nitrogen source. The reaction entails a reductive amination catalyzed by glutamate dehydrogenase (fig. 24.3). In *E. coli* this enzyme is specific for NADPH as the hydrogen donor, as might be expected for a biosynthetic reaction involving a reductive step.

The glutamate dehydrogenases from green plants require a very high concentration of NH_3 to be effective. In fact an alternative pathway is responsible for glutamate biosynthesis in most

Figure 24.2

Immediate consequences of a mutation in a biosynthetic pathway. When the mutation leads to an inactive enzyme, the chain of reactions leading to the end product in the pathway is broken. Frequently large amounts of intermediate are produced, accumulate in the cell, and may leak to the environment. If the intermediate is phosphorylated, sometimes dephosphorylation precedes excretion.

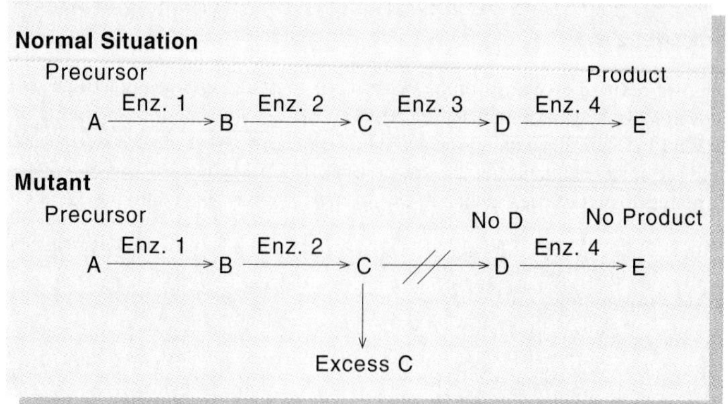

Figure 24.3

The conversion of ammonia into the α-amino group of glutamate and into the amide group of glutamine. The direct amination of α-ketoglutarate by NH_4^+ occurs in many bacteria. In green plants, fungi, and some bacteria the preferred nitrogen source is glutamine. Glutamine itself is formed by the ATP-dependent amidation of glutamic acid.

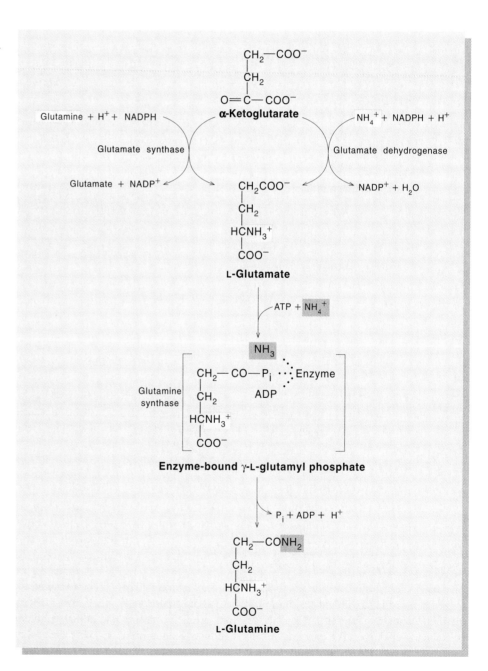

$CH_2—COO^-$
CH_2
$O=C—COO^-$
α-Ketoglutarate

Glutamine + H^+ + NADPH
Glutamate synthase
Glutamate + $NADP^+$

NH_4^+ + NADPH + H^+
Glutamate dehydrogenase
$NADP^+$ + H_2O

CH_2COO^-
CH_2
$HCNH_3^+$
COO^-
L-Glutamate

ATP + NH_4^+

Glutamine synthase

NH_3
$CH_2—CO—P_i$ ⋯ Enzyme
CH_2
ADP
$HCNH_3^+$
COO^-
Enzyme-bound γ-L-glutamyl phosphate

P_i + ADP + H^+

$CH_2—CONH_2$
CH_2
$HCNH_3^+$
COO^-
L-Glutamine

Figure 24.4

The structure of glutamine synthase from *Salmonella typhimurium.* The enzyme consists of 12 identical subunits arranged like a hexagonal prism. (*a*) View down the sixfold axis of symmetry. The monomers in the top ring are alternately colored light and dark blue and the monomers in the bottom ring light and dark red. The active sites of each monomer are marked by pairs of Mn^{2+} ions (white spheres). (*b*) Side view of the enzyme along one of the twofold axes of symmetry, showing only the six nearest subunits. (Copyright 1994 by the Scripps Research Institute/Molecular Graphics Images by Michael Pique using software by Yng Chen, Michael Connolly, Michael Carson, Alex Shah, and AVS, Inc. Visualization advice by Holly Miller, Wake Forest University Medical Center.)

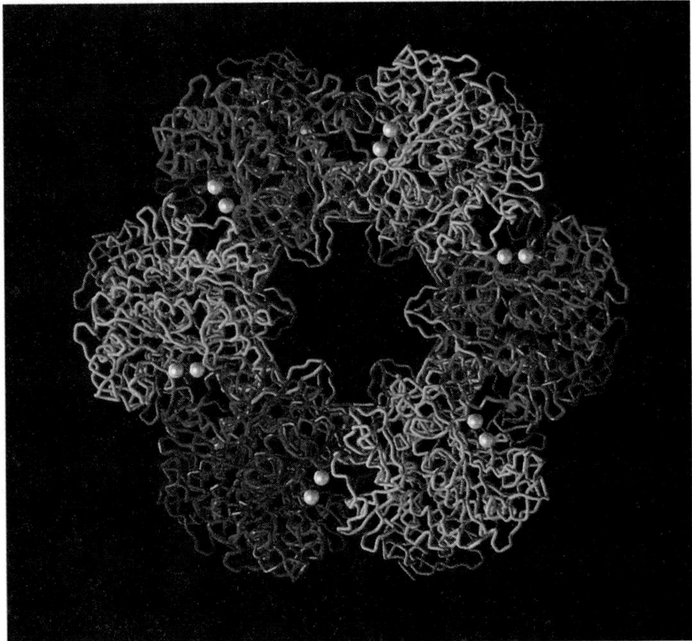

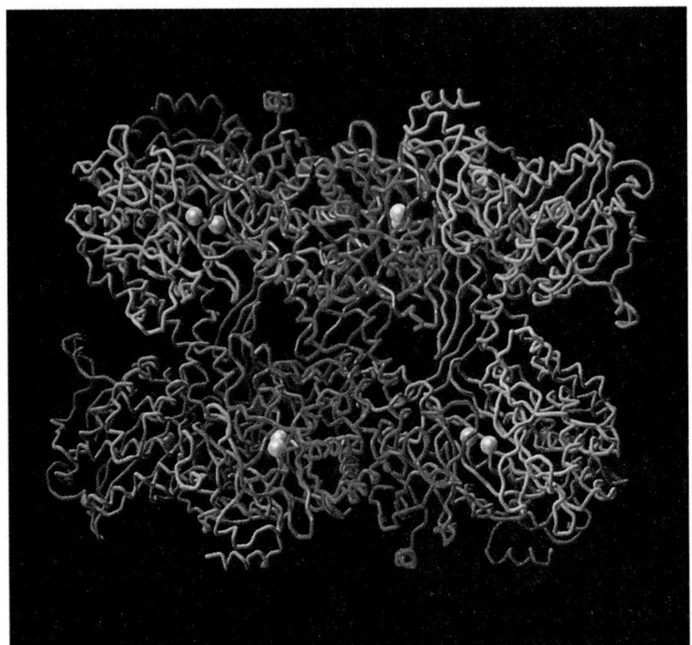

plants. Thus in green plants, as well as many bacteria and fungi, the amino group of glutamate usually originates from the amide group of glutamine. The reaction is catalyzed by glutamate synthase (see fig. 24.3).

Amidation of Glutamate to Glutamine Is an Elaborately Regulated Process

Glutamine, which is so commonly used in amination reactions, is formed by a reaction in which ammonia is directly added to glutamate; the reaction is catalyzed by glutamine synthase. This enzyme is encoded by a gene named glnA. This two-step reaction starts with activation of the γ-carboxyl group of glutamate and yields a γ-glutamyl–enzyme complex (see fig. 24.3); this is followed by γ-glutamyl transfer to NH_3.

Since glutamine synthetase is the main enzyme initiating the flow of ammonia nitrogen into organic compounds, it is not surprising that it is a highly regulated enzyme. The *E. coli* enzyme, studied extensively by Stadtman and Ginsberg, is composed of 12 identical subunits (M_r 50,000) arranged in two hexameric rings (fig. 24.4). The enzyme is negatively regulated by eight nitrogenous compounds: carbamyl phosphate, glucosamine-6-phosphate, tryptophan, alanine, glycine, histidine, cytidine triphosphate, and AMP. With the exception of glycine and alanine these compounds receive an amide nitrogen directly from glutamine during their biosynthesis. Therefore most of these compounds can be considered to be glutamine metabolites. Although glycine and alanine are not direct end products, they are "indicators" of the sufficiency of the nitrogen supply of the cell. Thus the regulation of glutamine synthase can be regarded as an example of end-product inhibition.

Even more important in the regulation of *E. coli* glutamine synthetase activity is the reversible ATP-dependent adenylylation of a specific tyrosyl residue on each subunit. As the enzyme becomes progressively more adenylylated (up to the fully adenylylated form, which contains 12 AMP groups per enzyme molecule), the enzyme becomes progressively less active.

The adenylylation reaction and its reversal by deadenylylation are regulated by the nitrogen status of the cell. The immediate small-molecule effectors of this regulatory system, glutamine and α-ketoglutarate, work in concert with ATP and two protein molecules, PII, encoded by glnB, and a bifunctional uridylyltransferase/uridyl-releasing enzyme. A high glutamine concentration or a high glutamine/α-ketoglutarate ratio signals nitrogen excess. Conversely, a high α-ketoglutarate/glutamine ratio signals nitrogen limitation. Nitrogen excess leads to inactivation of glutamine synthase, whereas nitrogen limitation leads to activation (fig. 24.5). The activation of glutamine synthesis favors the conversion of ammonia and glutamate to glutamine. Although the en-

Figure 24.5

Regulation of glutamine synthase in *E. coli*. The activity of glutamine synthase is inhibited by adenylylation. Both adenylylation and deadenylylation are regulated by a cascade of controls that are responsive to the concentrations of glutamine and α-ketoglutarate. The concentrations of these two compounds are an indication of the nitrogen supply. Thus under conditions of nitrogen excess (*top part of figure*) the concentration of glutamine is high, and the ratio of glutamine to α-ketoglutarate is high. This leads to inactivation of the glutamine synthase. Under conditions of nitrogen limitation (*bottom part of figure*) the α-ketoglutarate concentration is high, and the ratio of α-ketoglutarate to glutamine is high, so the glutamine synthase is activated.

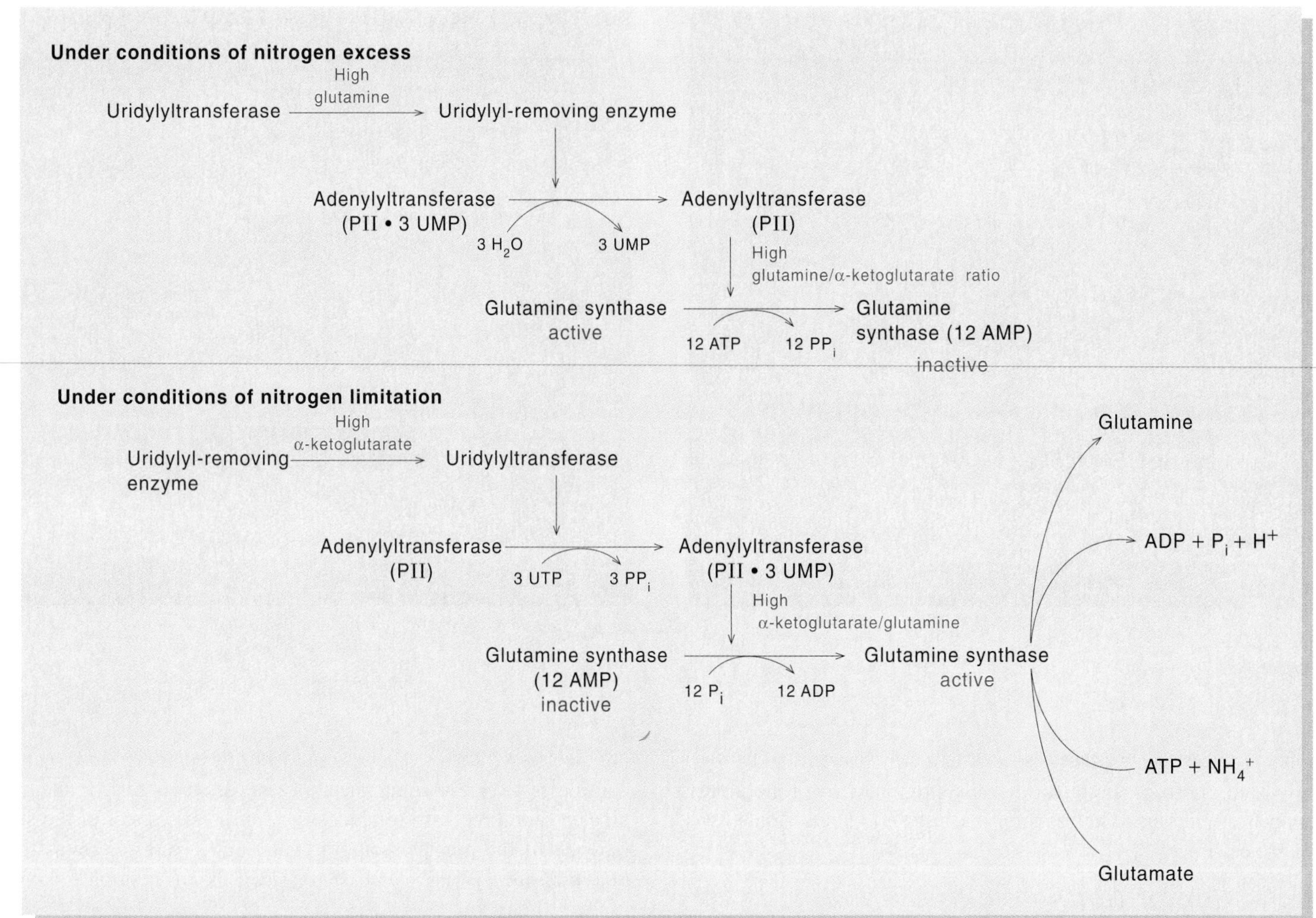

zyme catalyzes the reverse reaction (like any enzyme), the equilibrium is very much in favor of glutamine formation. Both the activation (deadenylylation) and the inactivation (adenylylation) are controlled by a cascade of regulatory interactions illustrated in figure 24.5. As a rule, metabolic signals are greatly amplified by cascade arrangements. The glutamine synthase system is unusual in that the same two regulatory proteins function in both cascades. Both regulatory proteins have two enzymatically distinct sites whose activities are mutually exclusive. Which site in the two regulatory proteins is active depends on the relative concentrations of glutamine and α-ketoglutarate, as already indicated.

Let us consider the situation under conditions of nitrogen excess (see fig. 24.5). The first regulatory protein in the cascade at high glutamine is converted into a uridylyl-removing enzyme. This enzyme hydrolyzes UMP from a trimeric PII protein that acts in concert with adenylyltransferase, at a high glutamine/α-ketoglutarate ratio, to adenylate glutamine synthase. The resulting adenylated enzyme is inactive.

Under conditions of nitrogen limitation the first regulatory enzyme in the cascade is converted into a uridylyltransferase. The uridylyltransferase is activated by high concentrations of α-ketoglutarate, leading to uridylylation of the PII protein. The PII • UMP protein and adenylyltransferase deadenylates the glutamine synthase • 12 AMP when the α-ketoglutarate/glutamine ratio is high. This returns glutamine synthase to a catalytically active form.

The uridylylation state of the PII protein also influences the covalent modification of a transcriptional regulator (NtrC) of glnA, whose ability to act is diminished by dephosphorylation. The unmodified PII protein stimulates the dephosphorylation of

NtrC, which reduces the rate of formation of glnA messenger RNA.

The Nitrogen Cycle Encompasses a Series of Reactions in Which Nitrogen Passes through Many Forms

Glutamine synthase is one of the links in the nitrogen cycle whereby nitrogen passes through several chemical forms. The passage of nitrogen from one form to another involves a chain of widely distributed organisms (fig. 24.6).

Although NH_4^+ is the form in which nitrogen is ultimately incorporated into organic materials, it is often less available to plants or bacteria for biosynthesis than other forms of nitrogen. When present for any length of time in the free state NH_4^+ is likely to be oxidized by nitrifying bacteria (such as *Nitrosomonas* and *Nitrobacter*) to nitrite (NO_2^-) and nitrate (NO_3^-). Reduction of nitrate by plants and bacteria to NH_4^+ is vital to maintaining the nitrogen cycle. Nitrogen fixation whereby atmospheric nitrogen is reduced to NH_3 occurs in a limited number of microorganisms.

Nitrogen Fixation Involves an Enzyme Complex Called Nitrogenase. The Haber-Bosch industrial chemical process for ammonia (NH_3) production, which uses elevated temperatures and pressures, accounts for about 10^8 tons per year of N_2 converted to NH_3. Approximately an equal amount of N_2 is converted to NH_3 by biological nitrogen fixation, which is carried out at ambient temperatures and 1 atm by a variety of free-living bacteria, cyanobacteria, and symbiotic bacteria. The biological reaction is catalyzed by the nitrogenase enzyme system. This enzyme consists of two metalloproteins: the iron–moblydenum protein (FeMo-protein) and the iron protein (Fe-protein). In addition to the nitrogenase proteins a source of reducing equivalents (ferrodoxin or flavodoxin in vivo), ATP, and protons are required for nitrogen fixation. The overall stoichiometry of nitrogen fixation is represented by the equation

$$N_2 + 8\,H^+ + 8\,e^- + 16\,ATP \longrightarrow 2\,NH_3 + H_2 + 16\,ADP + 16\,P_i \quad (1)$$

The nitrogenase MoFe-protein from *Clostridium pasteurianum* is an $\alpha_2\beta\gamma$ tetramer with a molecular mass of 220,000, and the corresponding Fe-protein is a γ_2 dimer with a molecular mass of about 60,000. The FeMo-protein contains two copies of FeMo-cofactors, which are believed to be the substrate-binding and reduction sites (fig. 24.7), and two copies of the iron–sulfur P-cluster pair, which may serve to mediate electron transfer between the Fe-protein and the FeMo-cofactor (see fig. 24.7*c*). The Fe-protein has a single 4Fe • 4S cluster, which transfers electrons to the FeMo-protein in an ATP-dependent manner. Electron transfer from the Fe-protein to the FeMo-protein involves a cycle of association and dissociation of the protein complex concomitant with ATP hydrolysis; the dissociation has been identified as the rate-determining step.

Although it might be expected that only six reducing equivalents would be required for each mole of N_2 (valence state 0) reduced to 2 NH_3 (valence state −3), in fact, the reaction catalyzed by nitrogenase converts two protons (H^+) to hydrogen (H_2) for each molecule of N_2 reduced (see equation 1). As a result the overall equation for the conversion of nitrogen to ammonia requires eight reducing equivalents. Why is hydrogen production a necessary step in the reduction of dinitrogen? It appears that it is necessary for the binding of nitrogen. The reduction of nitrogen by nitrogenase can be viewed as a cycle of eight electron transfers, each of which results in the cyclic transfer of a single electron from reduced ferrodeoxin or flavodoxin to the Fe-protein and the regeneration of the reduced electron donor by cell metabolism (e.g., pyruvate oxidation) (fig. 24.8). On reduction the Fe-protein binds two molecules of ATP, after which it associates with the FeMo-protein. The transfer of an electron to the FeMo-protein is accompanied by the hydrolysis of the two bound ATP molecules and the dissociation of the two proteins. Overall, the sequential functioning of eight Fe-protein reduction–oxidation cycles is re-

Figure 24.6

The nitrogen cycle depicts the flow of nitrogen in the biological world. The proteins of animals and plants are cleaved by many microorganisms to free amino acids from which ammonia (or ammonium ion) is released by deamination. Urea, the main nitrogen excretion product of animals, is hydrolyzed to NH_3 and CO_2. *Nitrosomonas* soil bacteria obtain their energy by oxidizing NH_3 to nitrite, NO_2^-. *Nitrobacter* obtain their energy by oxidizing nitrite, NO_2^-, to nitrate, NO_3^-. Plants and many microorganisms reduce nitrate for incorporation into amino acids, completing the cycle. Other microorganisms reduce nitrate partly to NH_3 and partly to N_2, which is lost to the atmosphere. Atmospheric nitrogen, N_2, can be recaptured, reduced, and converted into organic substances by a limited number of nitrogen-fixing bacteria and algae.

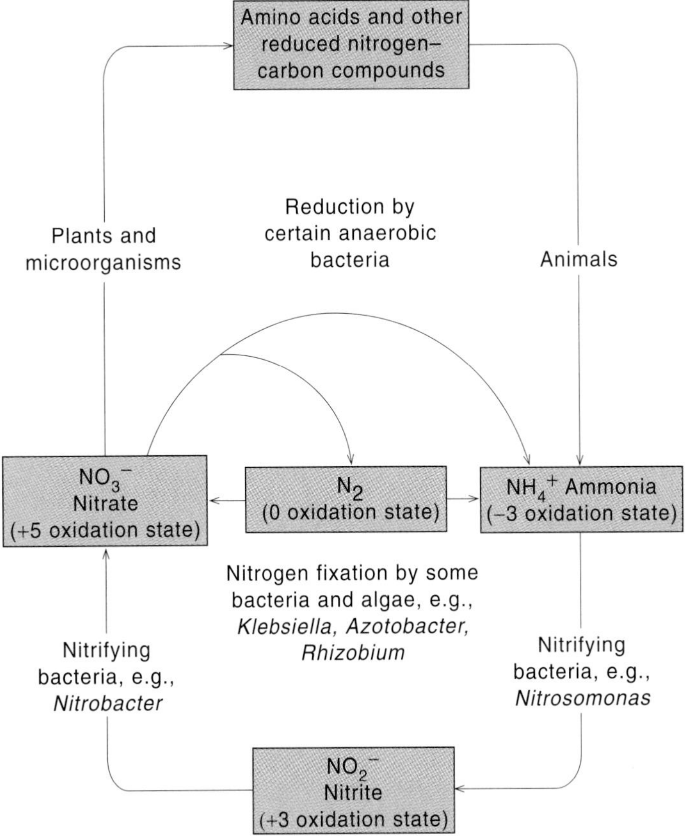

Figure 24.7

The iron–molybdenum cofactor (FeMo-cofactor) and the P-cluster in the nitrogenase of *Azotobacter vinelandii.* The structures of these cofactors are believed to be the same in *Clostridium pasteurianum.* (*a*) The FeMo-cofactor structure determined by X-ray crystallography. (*b*) A possible binding site for dinitrogen in the cavity of the cofactor. Y is an undesignated ligand. (*c*) The structure of the P-cluster, which may serve to mediate electron transfer between the Fe-protein and the FeMo-cofactor in the FeMo-protein.

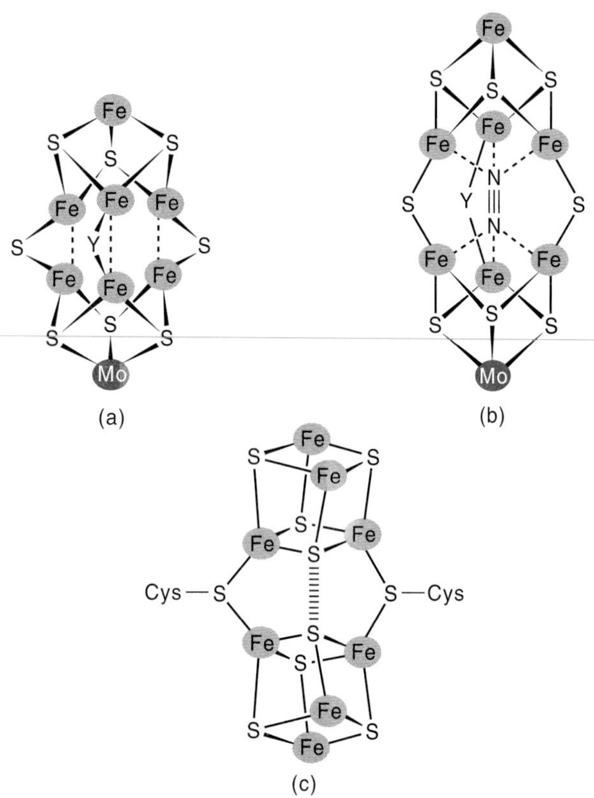

quired for one FeMo-protein cycle. The first three transfers, which result in the production of one H_2 molecule, somehow pave the way for the binding of dinitrogen at the active site (see fig. 24.7*b*). The remaining transfers are necessary to reduce a single dinitrogen to two ammonias.

To reduce dinitrogen to ammonia, a supply of protons is essential in addition to dinitrogen. Whereas proposals have been made for how this occurs as well as for the precise pathways for electron transfer and substrate entry and product release, no definite answers are known.

Nitrate and Nitrite Reduction Play Two Roles. As was pointed out above, NH_3 tends to be oxidized to nitrites or nitrates, if it is not immediately incorporated into organic matter. Nitrate and nitrite reduction play two physiological roles. One, exhibited primarily by bacteria, is a dissimilatory role in which nitrate and nitrite serve as terminal electron acceptors. The reduction serves to oxidize reducing equivalents such as NADH that are generated during oxidation of substrates. The other role played by these reductions is an assimilatory one in which nitrate is first converted to nitrite and thence to NH_3. These reactions are observed in bacteria, fungi, and plants.

Generally, the assimilatory nitrate and nitrite reductases are soluble enzymes that utilize reduced pyridine nucleotides or reduced ferrodoxin. In contrast, the dissimilatory nitrate reductases are membrane-bound terminal electron acceptors that are tightly linked to cytochrome b_1 pigments. Such complexes allow one or more sites of energy conservation (ATP generation) coupled with electron transport.

The ammonium ion (NH_4^+), produced by fermentative bacteria that use nitrite as an oxidant or produced during the decomposition of organic materials, is an important source of nitrogen for many plants and bacteria. Nevertheless, under vigorous aerobic conditions much of the NH_4^+ so produced is converted back to nitrite and nitrate by nitrifying bacteria.

Three Enzymes Convert Glutamate to Proline

The conversion of glutamate to proline involves the activation and reduction of the γ-carboxyl group to yield glutamic-γ-semialdehyde, which spontaneously cyclizes to yield a five-membered ring compound Δ^1-pyrroline-5-carboxylate. A second reduction then yields proline (fig. 24.9). Hydroxyproline found in collagen is formed from proline (see fig. 11.16; also see fig. 34.22) as a post-translational modification.

Arginine Biosynthesis Uses Some Reactions Seen in the Urea Cycle

Another pathway that involves an activation and reduction of the γ carboxyl of glutamate is that leading to L-arginine by way of L-ornithine (fig. 24.10). For ornithine biosynthesis, however, the α-amino group is protected by acetylation prior to carboxyl activation and reduction. Ring closure is thereby prevented, and glutamyl residues destined for arginine biosynthesis are effectively sequestered from those destined for proline biosynthesis.

Ornithine is converted to arginine by means of a series of reactions that we will encounter again when we discuss urea formation in chapter 25.

The Biosynthesis of Amino Acids of the Serine Family (L-Serine, Glycine, and L-Cysteine) and the Fixation of Sulfur

The diversion of 3-phosphoglycerate from the glycolytic pathway into the serine biosynthetic pathway is important, not only for the formation of L-serine, L-cysteine, L-tryptophan, and glycine needed for protein synthesis, but also for other functions these amino acids serve. For example, the conversion of serine to glycine generates one-carbon units for purine, thymine, and methionine biosynthesis and replenishes methyl groups transferred from methionine in many methylation reactions. The carbons of glycine contribute to purine and heme-containing compounds, to glu-

Figure 24.8

The eight-step nitrogenase cycle. Nitrogenase contains two complex protein components: component I and component II. The nitrogenase cycle starts when component II (an Fe-protein) binds to component I (usually an FeMo-protein). The binding requires two ATP molecules. During the bound state, component II transfers an electron to component I. This is followed by the hydrolysis of the two ATPs and the separation of the two components that begin a new cycle. In each step of the cycle the Fe-protein component II picks up a single electron from reduced ferrodoxin or flavodoxin and binds to two ATP molecules. After three single electron (e^-) transfers, an H_2 molecule is released from component II, and a N_2 molecule binds in its place. Following five more electron transfers, the N_2 molecule is fully reduced to 2 NH_3's, which are released.

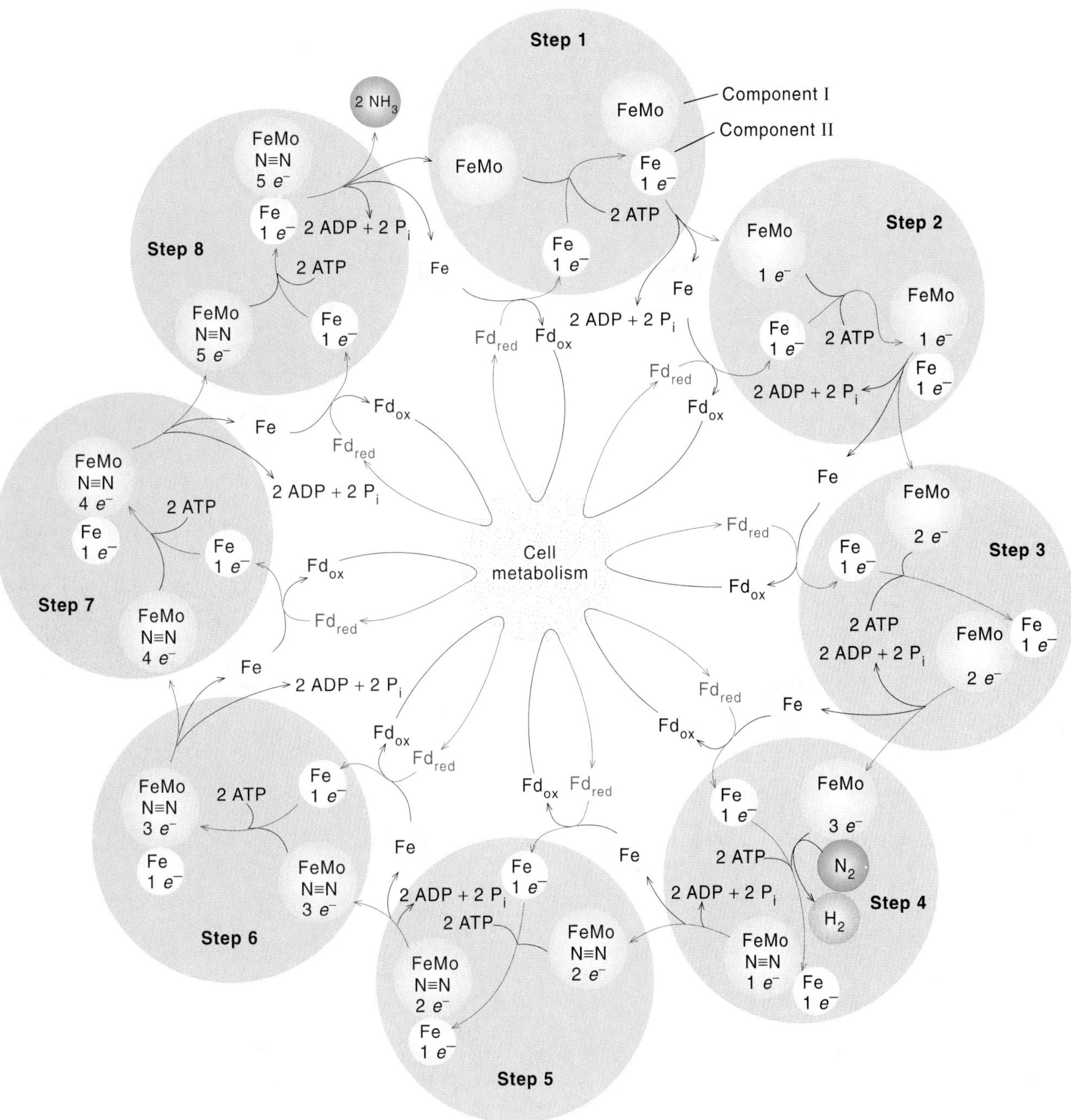

Figure 24.9

The biosynthesis of proline. Proline is synthesized in three steps from L-glutamate. The middle step involves a cyclization that occurs spontaneously.

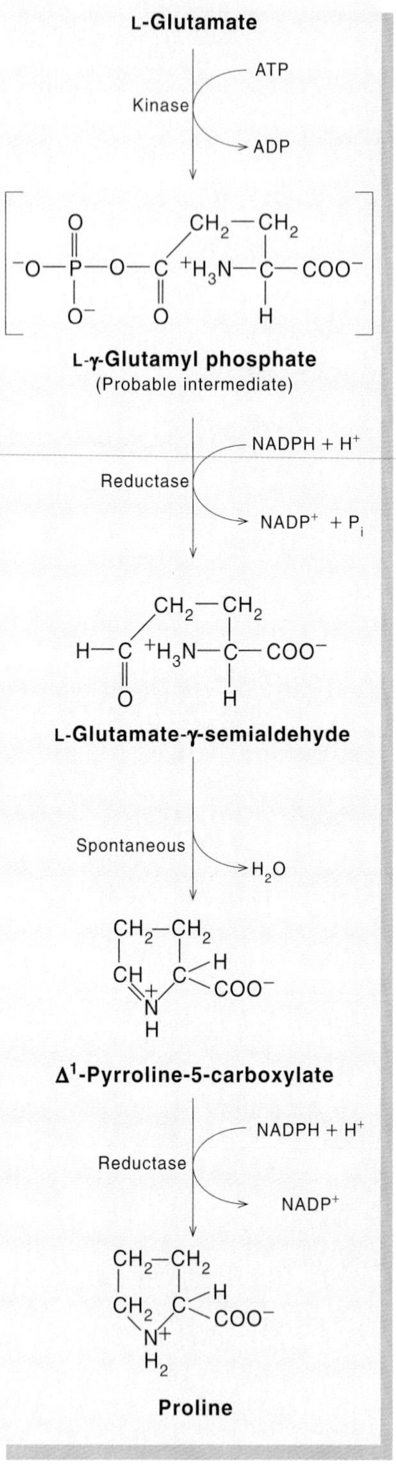

L-Glutamate

Kinase: ATP → ADP

L-γ-Glutamyl phosphate (Probable intermediate)

Reductase: NADPH + H^+ → $NADP^+$ + P_i

L-Glutamate-γ-semialdehyde

Spontaneous: → H_2O

Δ^1-Pyrroline-5-carboxylate

Reductase: NADPH + H^+ → $NADP^+$

Proline

tathione (as do those of cysteine), and, in animals, to certain detoxification products. The oxidative cleavage of glycine provides an additional source of one-carbon units from the α carbon. In many plants and microorganisms, sulfur in the form of sulfide is first incorporated into cysteine and then later is transferred to methionine and other sulfur-containing compounds. In the process, the carbons of cysteine are returned to the glycolytic pathway in the form of pyruvate. Serine itself is incorporated directly into phospholipids and into tryptophan, which therefore also might be considered a member of the serine family of amino acids. However, it will be more appropriate to consider tryptophan biosynthesis later, along with the formation of the other aromatic amino acids.

Three Enzymes Convert 3-Phospho-D-Glycerate to Serine

The initial reaction of L-serine biosynthesis is the oxidation of 3-phosphoglycerate by an NAD^+-linked dehydrogenase (fig. 24.11). This redox reaction is freely reversible. The enzyme activity is regulated by the end product of the biosynthetic sequence, serine. The product of the reaction, 3-phosphohydroxypyruvate, is converted to *O*-phosphoserine by a specific phosphoserine-glutamate transaminase. Removal of the phosphate group by a specific phosphoserine phosphatase yields serine. Phosphoserine is also found as a protein constituent, but in such cases it arises via the post-translational phosphorylation of a seryl residue by an ATP-dependent protein kinase. Such kinases serve an important function in regulating the activity of some enzymes (e.g., see chapters 10 and 14).

Two More Enzymes Convert L-Serine to Glycine

The pathway from serine to glycine consists of a single complex step catalyzed by serine hydroxymethyltransferase (fig. 24.12). The reaction is a pyridoxal-phosphate-dependent aldol cleavage to yield glycine and an "active" formaldehyde unit that is transferred to a tetrahydrofolate cofactor. This reaction is an important one in supplying one-carbon units for other biosynthetic reactions. It has been calculated, however, that the amount of methylene tetrahydrofolate generated during the biosynthesis of the glycine in bacterial protein is not quite sufficient to account for the synthesis of those cellular constituents that require one-transfers of carbon for their formation. (You may wish to review the reactions involving the folic acid coenzymes in chapter 11.) The most likely source of these required extra one-carbon units is the α carbon of glycine; glycine is converted to NH_3, CO_2, and a one-carbon unit as 5,10-methylene tetrahydrofolate by a four-protein complex, the glycine cleavage enzyme. This system has been found in bacteria, plants, and animals.

Cysteine Biosynthesis Involves Sulfhydryl Transfer to Activated Serine

The biosynthesis of L-cysteine results from a sulfhydryl transfer to an activated form of serine. The form of the sulfhydryl group used in cells is unclear. For most plants and microorganisms the sulfur source is sulfate, which must be reduced to the level of sulfide. This reduction is a complex process that we will discuss briefly in the following subsection. For some organisms it appears that the initial sulfhydryl transfer is made to an activated form of

Figure 24.10

The biosynthesis of arginine. This synthesis begins by the *N*-acetylation of the α-amino group. The γ-carboxyl group is phosphorylated before reduction to an aldehyde that is subsequently transaminated. In some organisms the acetyl group, whose purpose is to protect the α-amino group from reacting during intermediate steps, is removed as shown. In others it is preserved by transfer to another glutamate as the initial step in arginine biosynthesis. In the latter the acetyl-CoA-dependent formation of *N*-acetylglutamate serves only an anaplerotic role. The ornithine formed after the deacetylation is converted to arginine by means of a series of reactions that are encountered in urea formation (see fig. 25.6).

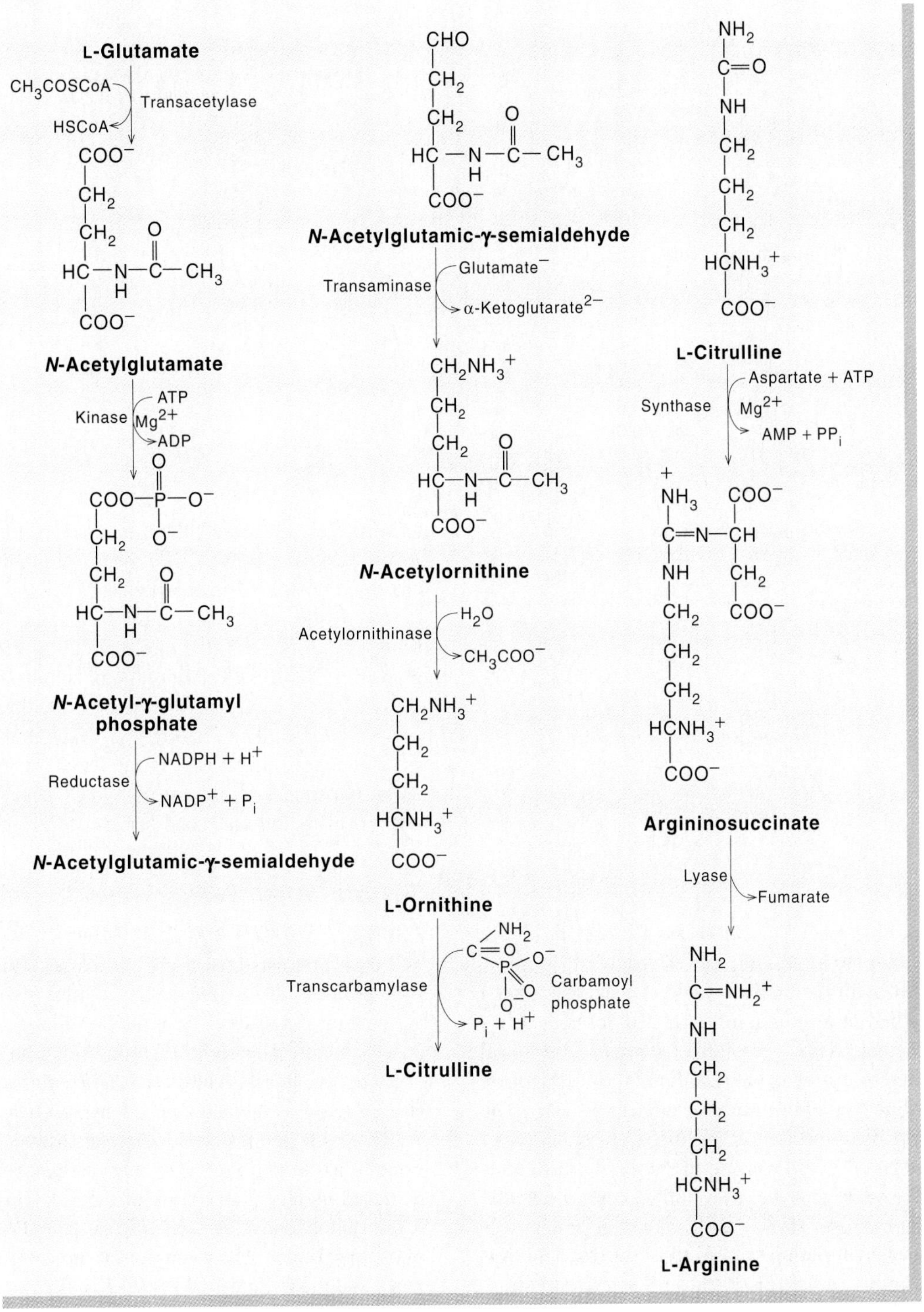

L-Glutamate

$CH_3COSCoA$ → $HSCoA$ (Transacetylase)

COO^-–CH_2–CH_2–HC(–N(H)–C(=O)–CH_3)–COO^-

***N*-Acetylglutamate**

Kinase: ATP, Mg^{2+} → ADP

COO–P(=O)(O^-)–O^- / CH_2 / CH_2 / HC–N(H)–C(=O)–CH_3 / COO^-

***N*-Acetyl-γ-glutamyl phosphate**

Reductase: NADPH + H^+ → $NADP^+$ + P_i

***N*-Acetylglutamic-γ-semialdehyde**

CHO / CH_2 / CH_2 / HC–N(H)–C(=O)–CH_3 / COO^-

***N*-Acetylglutamic-γ-semialdehyde**

Transaminase: $Glutamate^-$ → α-$Ketoglutarate^{2-}$

$CH_2NH_3^+$ / CH_2 / CH_2 / HC–N(H)–C(=O)–CH_3 / COO^-

***N*-Acetylornithine**

Acetylornithinase: H_2O → CH_3COO^-

$CH_2NH_3^+$ / CH_2 / CH_2 / $HCNH_3^+$ / COO^-

L-Ornithine

Transcarbamylase: Carbamoyl phosphate (NH_2–C(=O)–O–P(=O)(O^-)–O^-) → P_i + H^+

L-Citrulline

NH_2 / C=O / NH / CH_2 / CH_2 / CH_2 / $HCNH_3^+$ / COO^-

L-Citrulline

Synthase: Aspartate + ATP, Mg^{2+} → AMP + PP_i

$\overset{+}{NH_3}$–C(=N–CH(COO^-)–CH_2–COO^-) / NH / CH_2 / CH_2 / CH_2 / $HCNH_3^+$ / COO^-

Argininosuccinate

Lyase → Fumarate

NH_2 / C=NH_2^+ / NH / CH_2 / CH_2 / CH_2 / $HCNH_3^+$ / COO^-

L-Arginine

Figure 24.11

The biosynthesis of serine. The end product, serine, inhibits 3-phosphoglycerate dehydrogenase, the first enzyme in the pathway.

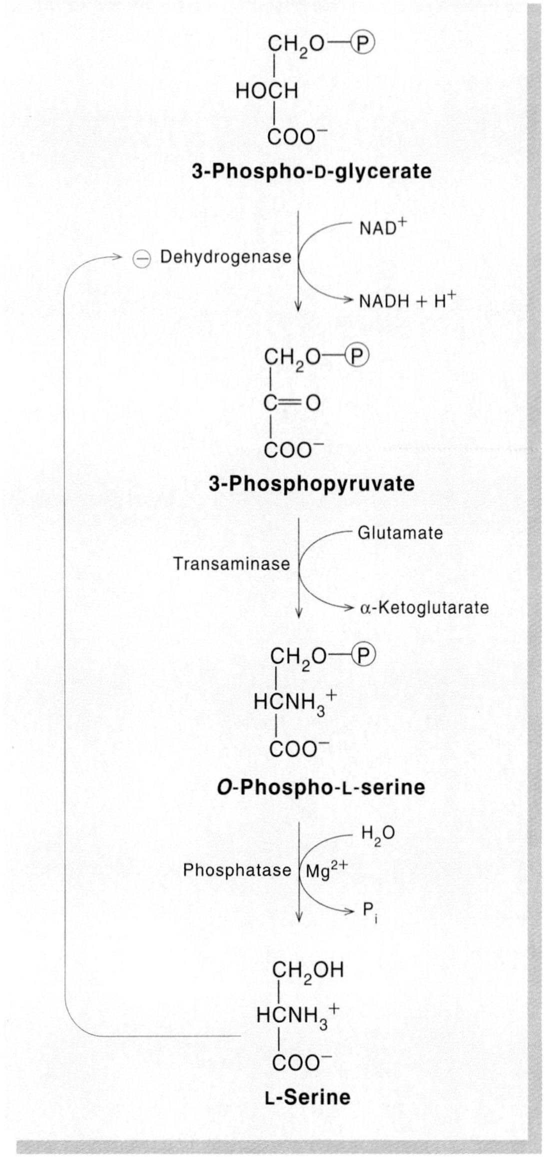

homoserine, a methionine precursor. Subsequently, transfer of the sulfur to cysteine occurs by a transsulfuration pathway.

The direct sulfhydrylation pathway to L-cysteine has been most thoroughly studied in *E. coli* and the related *Salmonella typhimurium* and in the eukaryotic green alga *Chlorella.* The initial step, which is inhibited by cysteine, is a transfer of the acetyl group of acetyl-CoA to serine to yield *O*-acetylserine (fig 24.13*a*). The reaction is catalyzed by serine transacetylase. The formation of cysteine itself is catalyzed by *O*-acetylserine sulfhydrylase, a reaction in which *O*-acetylserine serves as a β-alanyl donor (alanine activated at the β carbon) and H_2S as the β-alanyl acceptor.

The direct sulfhydrylation of *O*-acetylserine to yield cysteine is the most common mechanism for sulfide incorporation.

Figure 24.12

Biosynthesis and oxidation of glycine. Both of these reactions are important for supplying one-carbon units for metabolism by way of the tetrahydrofolate coenzyme.

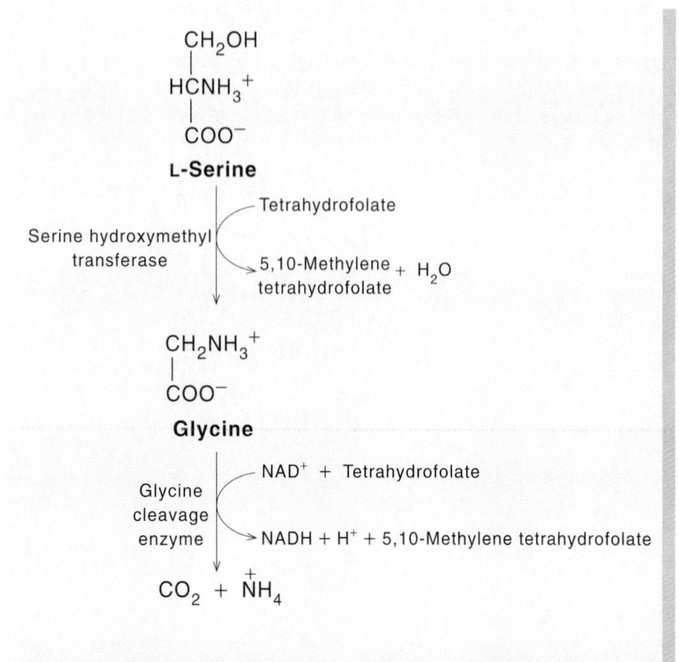

Nevertheless, in some cells the major, if not the sole, mechanism for sulfur incorporation is by means of homocysteine synthase, an enzyme considered later, when we discuss L-methionine biosynthesis. Under such conditions, cysteine formation occurs by transsulfuration, with the intermediate formation of L,L-cystathionine (fig. 24.13*b*). Cystathionine biosynthesis from serine and homocysteine is catalyzed as a simple condensation by cystathionine-β-synthase. The cleavage of cystathionine to yield cysteine, α-ketobutyrate, and NH_4^+ is catalyzed by γ-cystathionase, a pyridoxal-phosphate-containing enzyme. This transsulfuration pathway from methionine also serves as a route for methionine catabolism.

Sulfate Must Be Reduced to Sulfide before Incorporation into Amino Acids

Most sulfur exists in form of inorganic sulfate ion, SO_4^{2-}. To be incorporated into amino acids, it must first be reduced to H_2S. This requires a system for reduction found only in plants and microorganisms. The eight-electron reduction of SO_4^{2-} to H_2S occurs in two stages. The first stage requires activation of sulfate (fig. 24.14) catalyzed by adenylylsulfate pyrophosphorylase, which yields adenosine-5′-phosphosulfate (APS). Further activation is required in *E. coli* and certain other bacteria. This is achieved by phosphorylation of the 3′-OH by APS kinase to yield 3′-phosphoadenosine-5′-phosphosulfate (PAPS) with ATP as the phosphate donor. The reduction of the sulfonyl moiety of APS (in yeast and plants) or of PAPS (in *E. coli*) to sulfite occurs by trans-

Figure 24.13

The biosynthesis of cysteine by direct sulfhydrylation and by a transsulfuration route in which the sulfur is derived from homocysteine. The direct sulfhydrylation pathway (*a*) is indicated as occurring with H_2S as the source of sulfur. The transsulfuration pathway (*b*) passes through homocysteine. Two routes leading to homocysteine are shown. One starts with L-methionine. This is the sole route in animals. The second route involves the homocysteine-synthase-catalyzed conversion of *O*-acetyl-L-homoserine as shown.

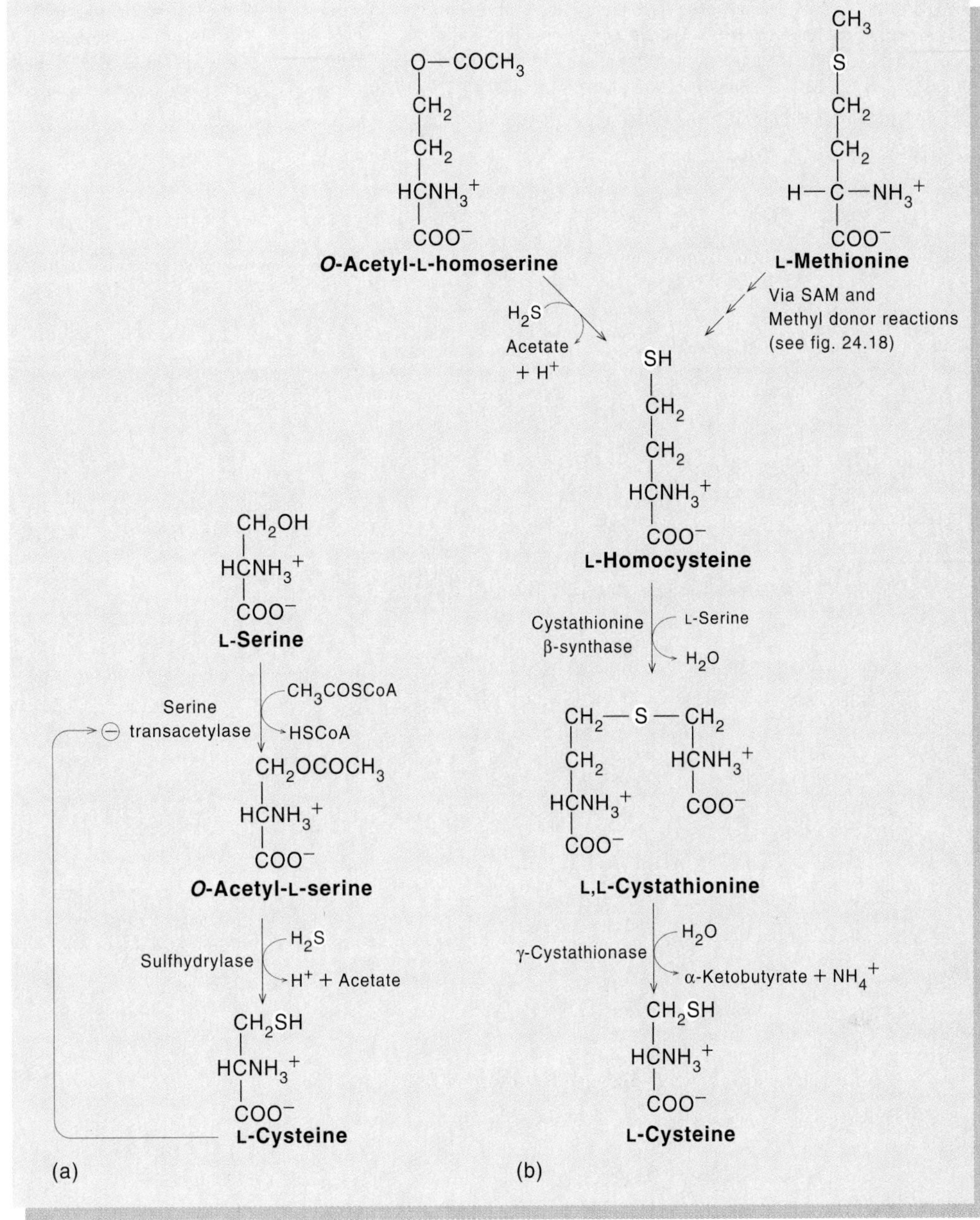

fer of the sulfonyl group to a thiol acceptor, such as thioredoxin (a vicinal dithiol protein), to yield an $—S—SO_3^-$ derivative. Reaction with a second thiol group (on thioredoxin or on a second molecule of glutathione) yields sulfite and an oxidized acceptor.

The six-electron reduction of sulfite to sulfide is catalyzed by sulfite reductase without the release of any free intermediates. Sulfite reductase is a multisubunit complex composed of a flavoprotein and a protein containing both heme and iron–sulfur centers.

Inhibition by cysteine of the active transport of sulfate into the cell may control of the conversion of sulfate to sulfide in *E. coli*.

The Biosynthesis of Some Amino Acids of the Aspartate Family: L-Aspartate, L-Asparagine, L-Methionine, and L-Threonine

The formation of the aspartate family of amino acids (asparagine, methionine, threonine, isoleucine, and, in plants and bacteria, lysine) and the conversion of aspartate to the pyrimidine nucleotides causes a significant drain of carbon from the tricarboxylic acid cycle in organisms such as *E. coli*. The nitrogen of aspartate is used

Figure 24.14

Formation of 3′-phosphoadenosine-5′-phosphosulfate (PAPS), an active intermediate involved in sulfate reduction. The eight-electron reduction of SO_4^{2-} to H_2S is poorly understood except for the initial steps in the activation of sulfate (shown in yellow). Reduction in yeast and plants involves the APS derivative shown. In *E. coli* a PAPS derivative is used.

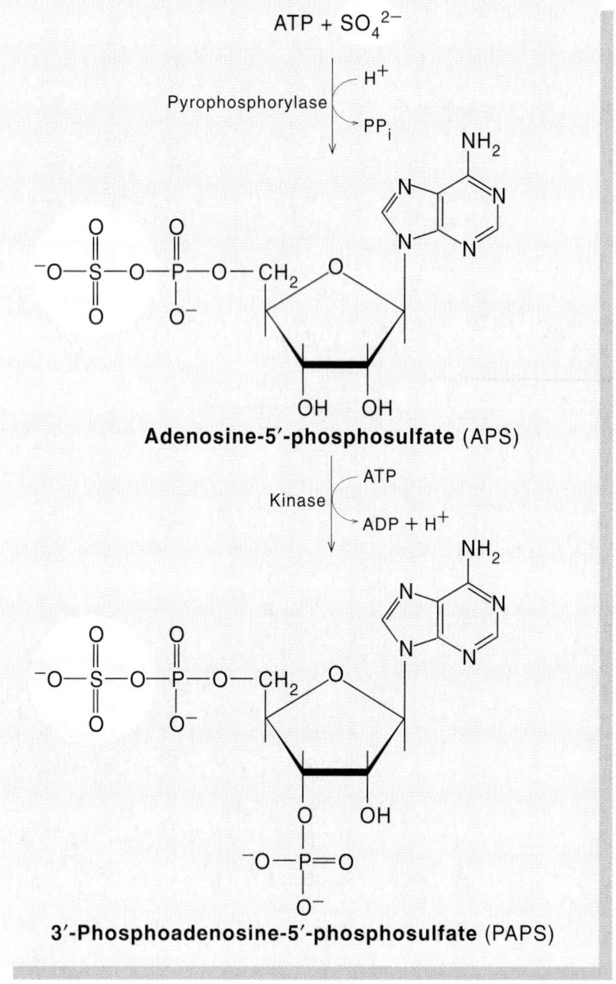

ATP + SO_4^{2-}

Pyrophosphorylase

H^+

PP_i

Adenosine-5′-phosphosulfate (APS)

Kinase

ATP

ADP + H^+

3′-Phosphoadenosine-5′-phosphosulfate (PAPS)

Figure 24.15

The biosynthesis of aspartate and asparagine. The more prevalent transamination reaction is shown. In a limited number of microorganisms an ammonia-dependent asparagine synthetase has been found. The asparagine synthetase involved in the typical amidation reaction shown in the figure also catalyzes a pyrophosphate cleavage of ATP.

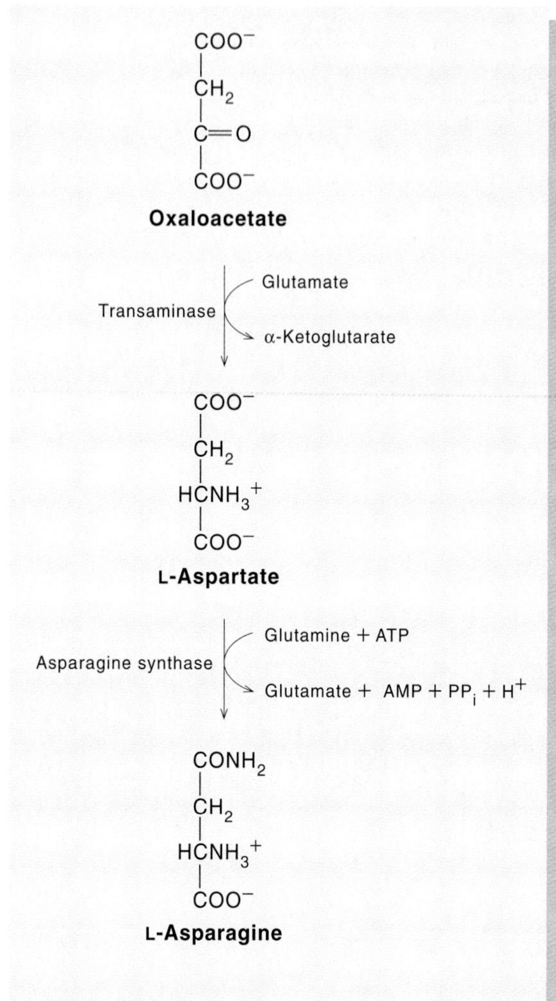

both in the formation of inosinate and its conversion to adenylate and in the conversion of citrulline to arginine.

Aspartate Is Formed from Oxaloacetate in a Transamination Reaction

L-Aspartate is formed from oxaloacetate in a transamination reaction with glutamate as the amino donor (fig. 24.15). In *E. coli* the major protein exhibiting aspartate-glutamate transaminase activity is transaminase A, an enzyme that also exhibits activity with the aromatic acids but not with leucine.

Asparagine Biosynthesis Requires ATP and Glutamine

In most organisms the formation of L-asparagine occurs by an ATP-dependent transfer of the amide group of glutamine to the β carboxyl of aspartate in a reaction catalyzed by asparagine synthetase (see fig. 24.15). The basic mechanism of amidation is different from that catalyzed by glutamine synthase in that it involves β-aspartyladenylate (in contrast to an acylphosphate) as an enzyme-bound intermediate.

L-Aspartic-β-Semialdehyde Is a Common Intermediate in L-Lysine, L-Methionine, and L-Threonine Synthesis

In the conversion of L-aspartate to L-lysine, L-methionine, and L-threonine, a reduction of the β-carboxyl group is necessary. As in the case of the reduction of the glutamate α-carboxyl group the reduction is preceded by a phosphorylation (fig. 24.16).

The NADPH-dependent reduction of β-aspartyl phosphate by aspartic-β-semialdehyde dehydrogenase yields aspartic-

Figure 24.16

The common aspartate family pathway. In some green plants, *O*-phosphohomoserine rather than homoserine itself is the point at which methionine biosynthesis and threonine biosynthesis diverge.

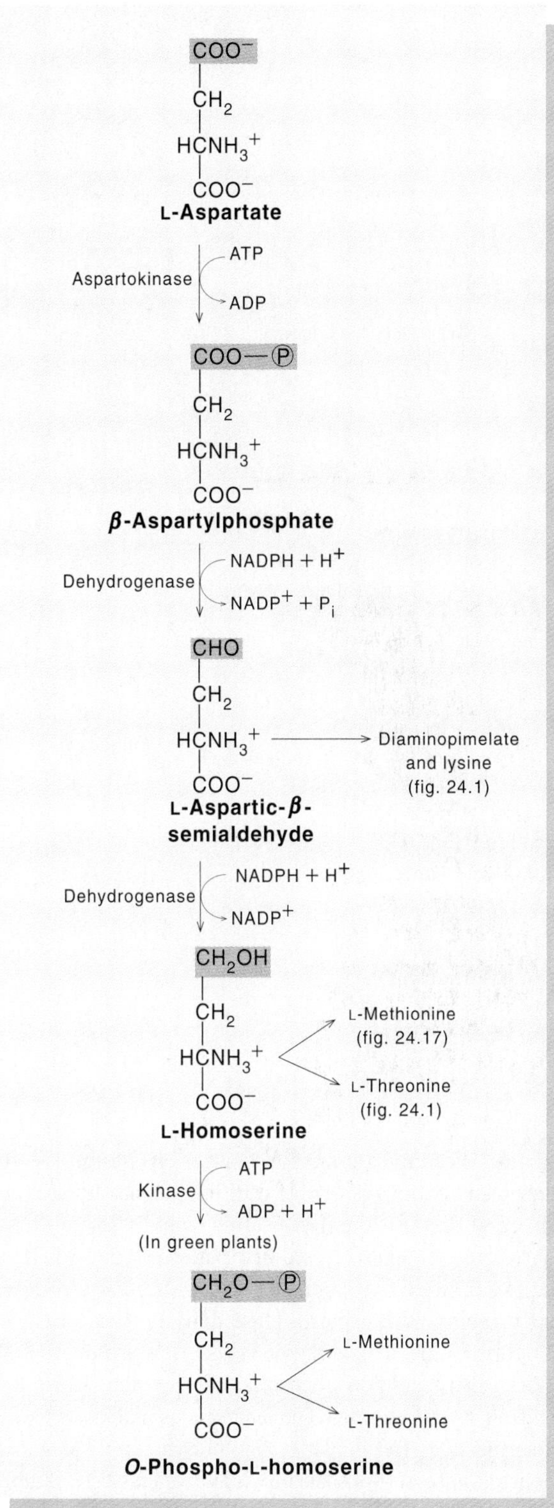

β-semialdehyde, a branchpoint compound from which lysine biosynthesis in plants and bacteria proceeds (see fig. 24.16). The common pathway is longer for methionine and threonine biosynthesis. In most cells, homoserine is a branchpoint compound. Homoserine itself is formed by the reduction of aspartic-β-semialdehyde by homoserine dehydrogenase.

We will consider the intricacies of the control of carbon flow over the common aspartate family pathway after we have described the specific branches.

Methionine Is Important as a Protein Constituent and as a Precursor of Other Cell Components via S-Adenosylmethionine

The conversion of L-homoserine to L-methionine occurs in more than one way. One route, used in some bacteria, consists of acylation of the hydroxyl group of homoserine, a condensation with cysteine to yield cystathionine, cleavage to homocysteine, and methylation to yield methionine (fig. 24.17). The second route, found in yeast and fungi, was referred to earlier and involves a sulfhydryl transfer to homoserine activated by an acyl group to yield homocysteine directly.

Although methionine is itself an end product of the pathway and is incorporated into protein, another important biosynthetic intermediate derived from methionine is *S*-adenosylmethionine (SAM). This intermediate serves as a methyl donor in many reactions and is a precursor of the propylamine groups in spermidine and spermine and of the hormone ethylene in plants. In some organisms, SAM participates in the control of the pathway leading from homoserine to methionine.

An adenosylation of methionine by SAM synthase activates methionine for either methyl group or, with a decarboxylase, propylamine transfers. As figure 24.18 shows, the activation is most unusual in that all three phosphate groups of ATP are cleaved in the activation reaction. In the methyl donor reaction, *S*-adenosylhomocysteine is liberated by one of several methyltransferases and then cleaved by *S*-adenosylhomocysteine hydrolase to homocysteine and adenosine, which can be recycled to methionine and ATP, respectively. In reactions where a propylamine group is transferred to spermidine, methylthioadenosine is liberated. The latter is converted to free adenine and 5′-methylthioribose-1-phosphate by either of two mechanisms. One, demonstrated in rat liver, is the direct conversion via a phosphorylase. The second, demonstrated in plants, protozoans, and bacteria, is via a nucleosidase (hydrolytic) to yield the nonphosphorylated thiomethyl sugar, which is phosphorylated in a subsequent kinase reaction. The adenine can be returned to the adenylate pool by one of the pyrophosphorylases described in chapter 26, and 5′-methylthioribose-1-phosphate is converted to methionine and formate by a series of reactions in which carbons 2 to 5 become carbons 1 to 4 of methionine.

The Carbon Flow in the Aspartate Family Is Regulated at the Aspartokinase Step

The flow of carbon into and through the aspartate family pathway must be regulated in a way that provides ample amounts of the branchpoint compounds aspartic-β-semi-

Figure 24.17

The biosynthesis of methionine. The acyl group employed to activate homoserine varies among different organisms. When a phosphoryl group is used as in green plants, the intermediate, *O*-phosphohomoserine is a branchpoint compound that is converted to either threonine or methionine. The direct sulfhydrylation of activated homoserine (by either H_2S or carrier-bound sulfide) that is found in some organisms is indicated by a broken line. Finally, the methylation of homocysteine may occur by either a tetrahydrofolate-dependent route (shown) or a cobalamin-dependent route (not shown).

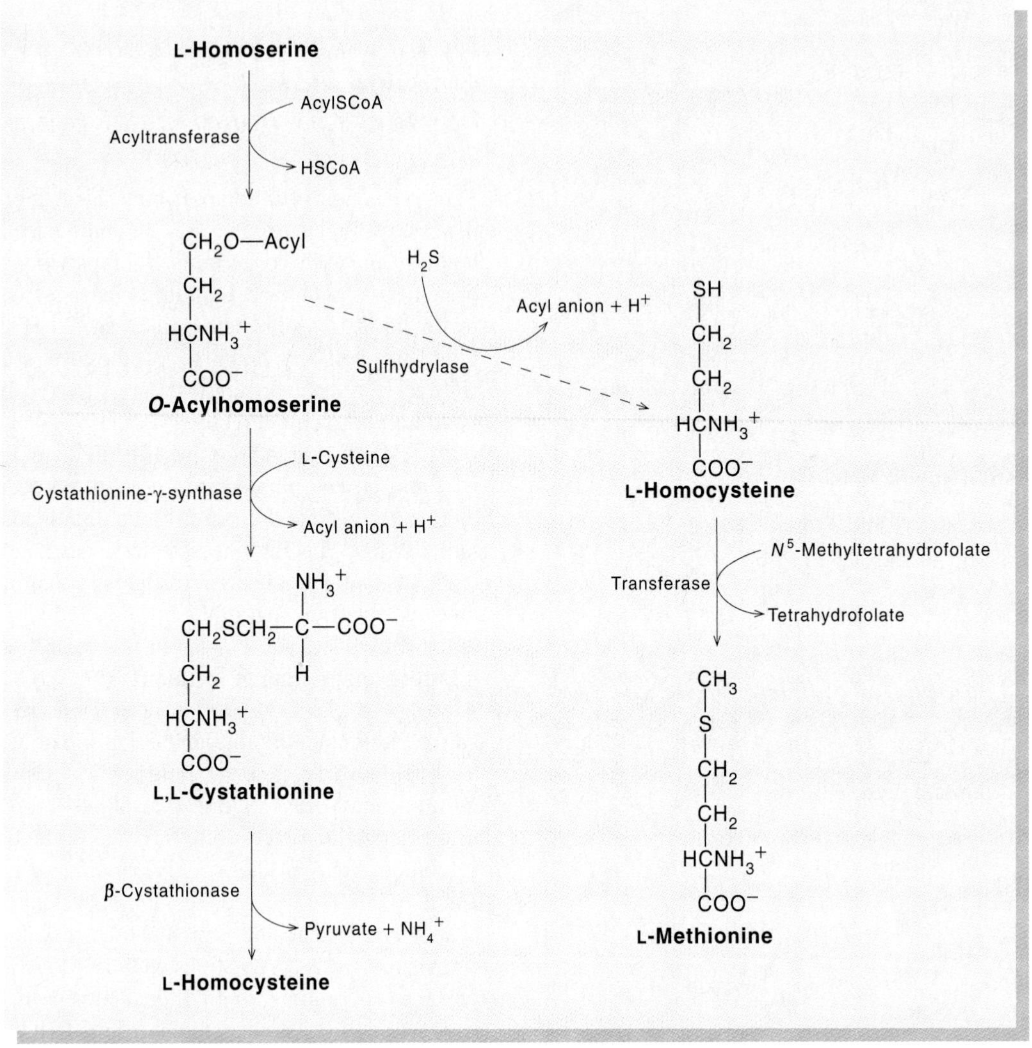

aldehyde (except in fungi) and homoserine (or *O*-phosphohomoserine in most green plants) but does not lead to oversynthesis of either. In most organisms there appear to be negligible pools of either branchpoint compound—perhaps only those amounts bound to product or substrate sites of the respective enzymes. Therefore feedback control cannot be exerted by these intermediates. Rather, in one way or another, control must be exerted by the end products (lysine, methionine, threonine, and isoleucine), which accumulate in measurable amounts.

Although an essentially common pathway has evolved for formation of the aspartate family of precursors, the patterns of feedback control vary considerably. They are of two general kinds. In one, there are single aspartokinases, but the control of the enzyme is multivalent. By multivalent we mean that more than one of the amino acid end products is required to inhibit the enzyme. The other basic pattern is one of multiple aspartokinases, each of which is controlled differently.

A most extensively studied pattern is that of the enteric bacteria, exemplified by *E. coli* (fig. 24.19). *E. coli* possesses three different proteins with aspartokinase activity. One enzyme is inhibited by threonine. This enzyme is part of a protein that also exhibits homoserine dehydrogenase activity, which is also sensitive to inhibition by threonine. The enzyme is called aspartokinase I–homoserine dehydrogenase I. Another distinctly different protein, aspartokinase II–homoserine dehydrogenase II, is not inhibited by any of the multiple end products of the aspartate family pathway, but its synthesis is repressed by methionine (as are other methionine biosynthetic enzymes). The synthesis of aspartokinase

Figure 24.18

The formation of "active" methionine, or S-adenosylmethionine (SAM), and some of its reactions. S-adenosylmethionine is synthesized in one step from L-methionine and ATP. The SAM so formed can serve as a propylamine donor or a methyl donor as shown. In the methyl donor reaction the S-adenosylhomocysteine formed is cleaved to homocysteine and adenosine, which can be used to regenerate methionine and ATP, respectively. The methylthioadenosine liberated in propylamine donor reactions is converted to 5′-methylthioribose-1-phosphate, which is converted to methionine and formate by a series of reactions.

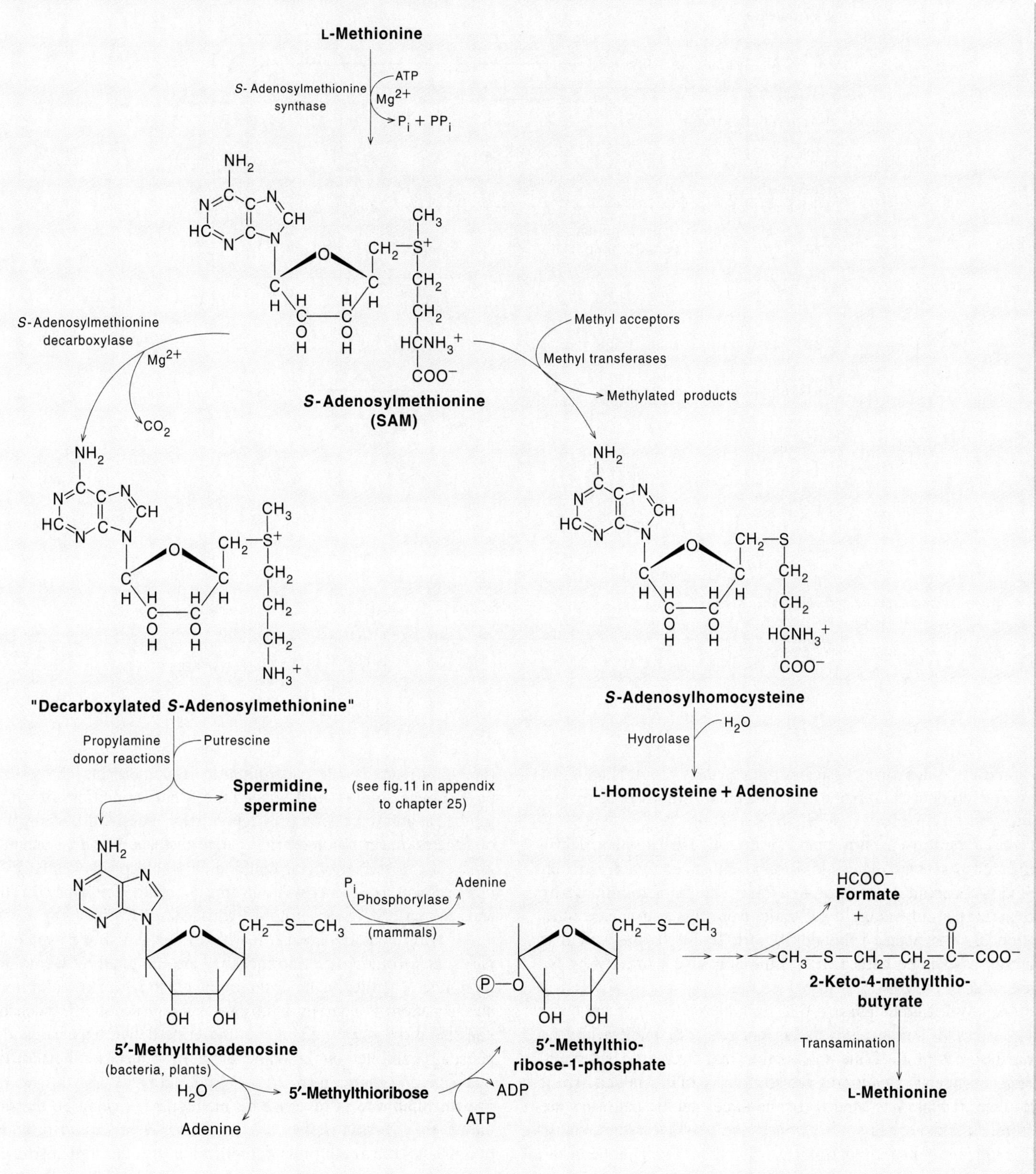

Figure 24.19

Regulation of the aspartokinases of *E. coli*. These enzymes is regulated at the level of enzyme synthesis and at the level of enzyme activity. In *E. coli* there are three aspartokinases that catalyze the conversion of L-aspartate to β-aspartylphosphate. The formation of these aspartokinases is repressed by different amino acids that are end products of this highly branched pathway. In addition, the activities of aspartokinase I and homoserine dehydrogenase I are inhibited by threonine, and that of aspartokinase III is inhibited by lysine (arrows for inhibited reactions are not shown).

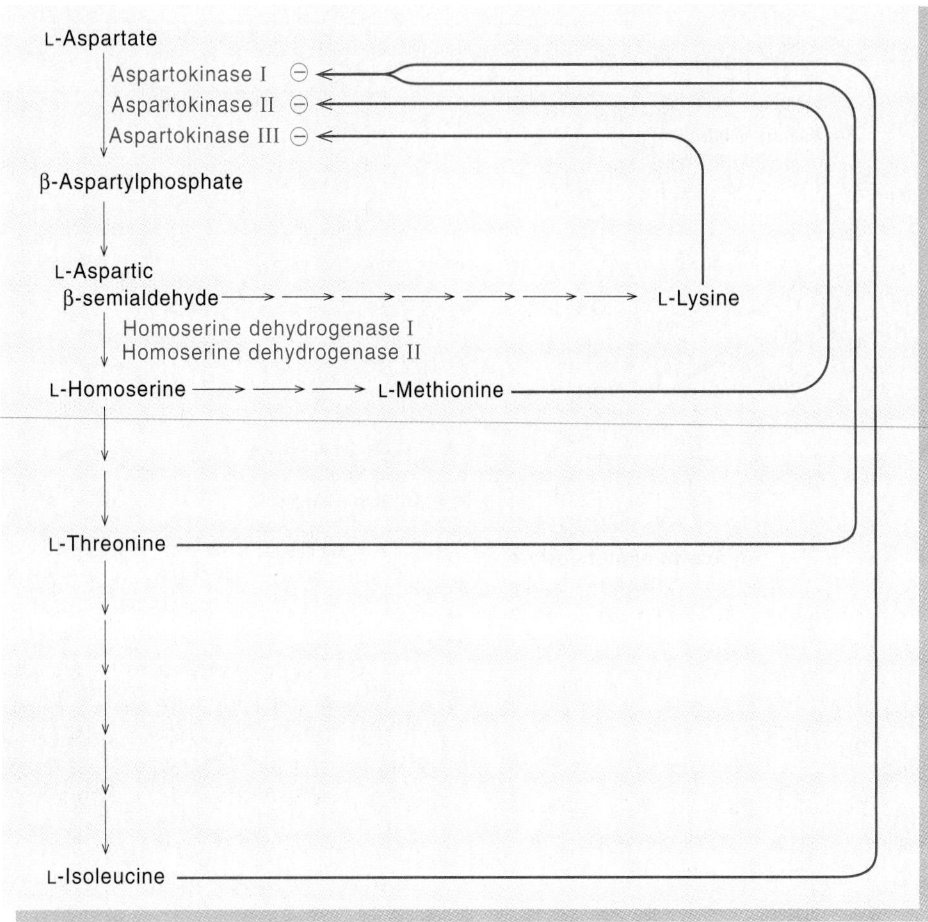

I–homoserine dehydrogenase I is repressed by a combination of two end products, threonine and isoleucine (multivalent repression).

The third aspartokinase in *E. coli,* aspartokinase III, is inhibited by lysine. The inhibition is enhanced synergistically by phenylalanine, leucine, or, to a lesser extent, methionine. Any physiological advantage that results from this synergistic inhibition has not been explained. Aspartokinase III has no other activity associated with it. Its synthesis is also repressed by lysine.

We must emphasize that the intermediates of the common aspartate family pathway are not being channeled into the branches leading to lysine, methionine, and threonine, but rather, there is probably a common pool of intermediates (albeit small) from which materials needed for the three specific pathways are drawn. In some strains of *E. coli,* lysine-sensitive aspartokinase III is the predominant aspartokinase, and there is very little of the methionine-repressible aspartokinase II–homoserine dehydrogenase II in an amino-acid-free medium. Under conditions of strong inhibition and repression (i.e., repressed synthesis of the enzyme) of aspartokinase I–homoserine dehydrogenase I and of aspartokinase III, a starvation for methionine would be prevented by a derepression (i.e., increased synthesis of the enzyme) of the aspartokinase II–homoserine dehydrogenase II.

In those organisms in which there are single aspartokinases, it is common to find that they are inhibited by lysine and threonine. Usually, lysine or threonine alone is weakly inhibitory, so the pattern is actually a strongly synergistic one. In such organisms there is only a single homoserine dehydrogenase, and it is usually inhibited by threonine alone. In some cases the addition of lysine and threonine is strongly inhibitory to growth, and the growth inhibition is reversed by methionine. It may be that aspartokinase II–homoserine dehydrogenase II of *E. coli* provides one means of avoiding this complication. In other organisms the inhibition is less severe, and the proteins are relatively insensitive to the feedback inhibitors after growth in the presence of the inhibitory amino acids. The physical basis of the desensitization is unknown.

The Biosynthesis of Amino Acids of the Pyruvate Family: L-Alanine, L-Valine, and L-Leucine

The pyruvate family of amino acids consists of L-alanine, L-valine, and L-leucine. In addition, pyruvate contributes two carbons to isoleucine and, on the average, two and one-half carbons to lysine. As we mentioned earlier, the biosynthesis of isoleucine, a member of the aspartate family, proceeds via a pathway that parallels that of valine. For this reason we will also consider the biosynthesis of isoleucine here.

L-Alanine Is Formed from Pyruvate in a Transamination Reaction

The formation of L-alanine occurs by a transamination reaction, with glutamate as the amino donor and pyruvate as the acceptor (fig. 24.20). There is no feedback control over alanine formation, and in many forms of bacteria, large intracellular pools of alanine are present unless the nitrogen supply is restricted. However, since the transaminases catalyze completely reversible reactions, this accumulation of alanine does not effect a drain on the supply of pyruvate.

Figure 24.20

The biosynthesis of alanine. This reaction is not regulated. It is readily reversible and therefore constitutes no major drain on the pyruvate supply.

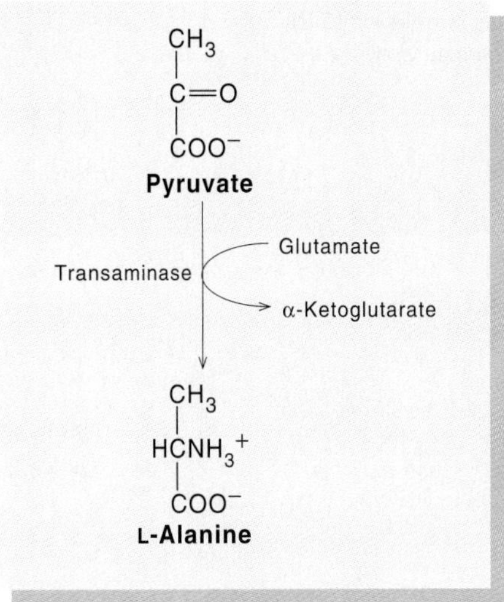

Isoleucine and Valine Biosynthesis Share Four Enzymes

The four enzymes required for valine biosynthesis are also required for the last four, parallel steps in isoleucine biosynthesis (fig. 24.21). The first step in valine biosynthesis is a condensation between pyruvate and "active" acetaldehyde (probably hydroxyethyl thiamine pyrophosphate) to yield α-acetolactate. The enzyme usually has a requirement for FAD, which, in contrast to most flavoproteins, is rather loosely bound to the protein. The same enzyme transfers the acetaldehyde group to α-ketobutyrate, yielding α-aceto-α-hydroxybutyrate, the isoleucine precursor. The α-ketobutyrate, unlike pyruvate, is not one of the key intermediates in many of the central metabolic routes; therefore a specific pathway to α-ketobutyrate must be present.

For nearly all plants, fungi, and bacteria the normal route to α-ketobutyrate is that from aspartate by way of threonine, which in turn is deaminated to α-ketobutyrate. The enzyme threonine deaminase contains pyridoxal phosphate and functions as a dehydratase (reactions not shown), presumably liberating α-aminocrotonate, which upon rearrangement to α-iminobutyrate spontaneously yields α-ketobutyrate and ammonia. There are two additional ways to form α-ketobutyrate, but we will not consider these in this section.

Conversion of the acetohydroxy acids to the β-dihydroxy acid precursors of valine and isoleucine is a complex reaction catalyzed by acetohydroxy acid isomeroreductase. The α,β-dihydroxy acids are both converted to the α-keto acid precursors of valine and isoleucine by a dihydroxy acid dehydrase. Finally, the two amino acids are formed in transamination reactions in which glutamate is the amino donor (branched-chain amino acid-glutamate transaminase).

The pathways to isoleucine and valine are particularly suitable for illustrating the way studies with nutritionally deficient mutants (auxotrophs), isotope incorporation experiments, and enzymatic analysis have been used to decipher the biosynthetic pathways to the amino acid (box 24A).

L-Leucine Is Formed from α-Ketoisovalerate in Four Steps

The biosynthesis of L-leucine involves the lengthening of the carbon chain of α-ketoisovalerate, an intermediate in valine biosynthesis (fig. 24.22).

Amino Acid Pathways Absent in Mammals Offer Targets for Safe Herbicides

In recent years, agribusiness firms have developed empirically several compounds that inhibit essential steps in the biosynthesis of amino acids found in plants but missing in animals. One of these compounds, glyphosate, is a highly specific inhibitor of 5-enol pyruvylshikimate-3-phosphate synthase (an enzyme needed for aromatic amino acid biosynthesis). Glyphosate is the active ingredient in the widely used herbicide Roundup:

$$^{-}O-\overset{\overset{O}{\|}}{\underset{\underset{O^{-}}{|}}{P}}-CH_2-NH-CH_2-C\overset{\nearrow O}{\underset{\searrow O^{-}}{}}$$

Glyphosate

Three other classes of compounds, although quite different from each other, are all inhibitors of acetohydroxy acid syn-

Figure 24.21

The biosynthesis of isoleucine and valine. The reactions leading to valine are catalyzed by the same enzymes that catalyze the corresponding reactions in isoleucine biosynthesis. Common enzymes are screened in yellow.

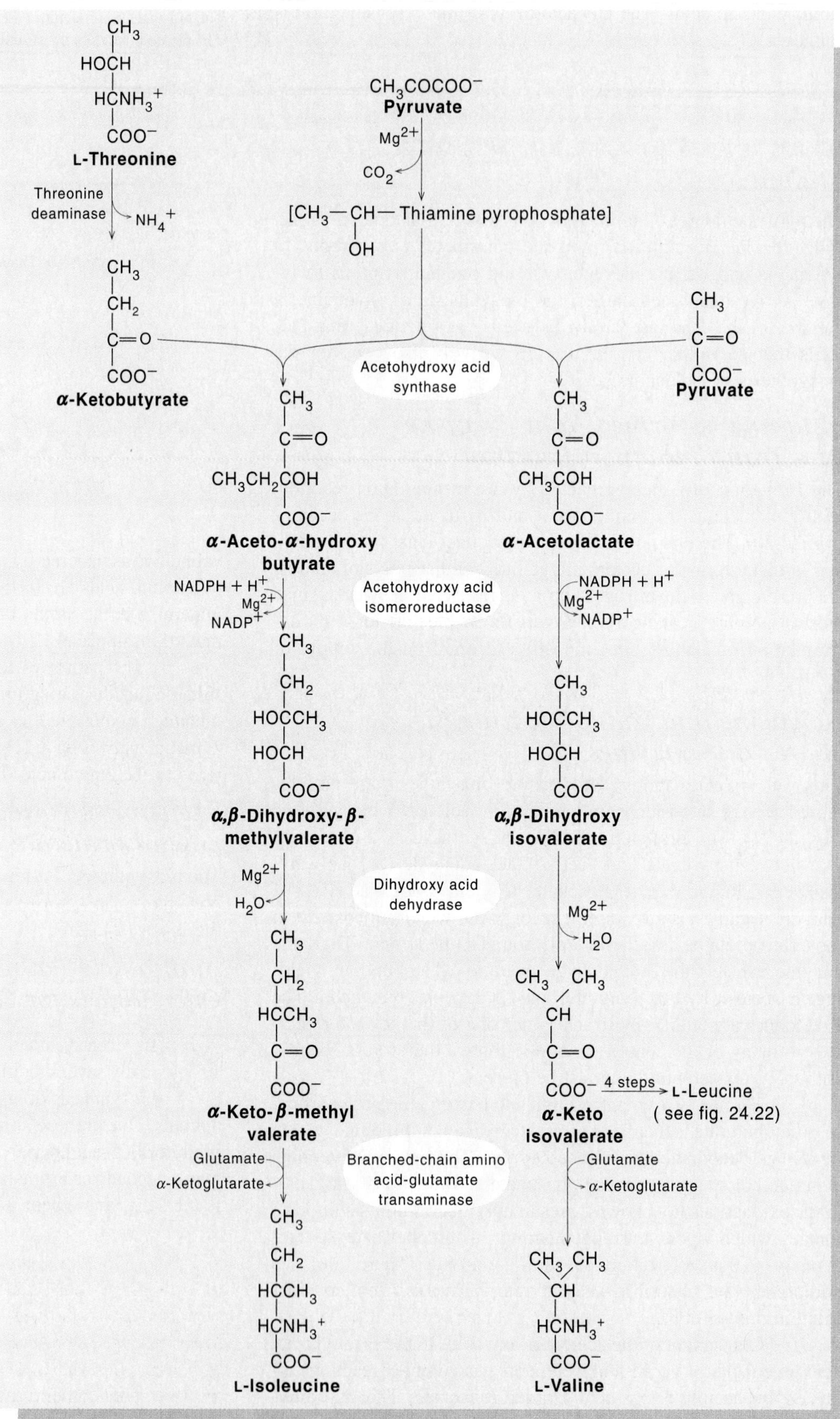

24A BOX

Use of Isotopes as Tracers to Delineate a Complex Pathway

Early studies with mutants of both *Neurospora* and *E. coli* revealed that certain mutants, presumably altered in but a single gene, required not one, but two amino acids: isoleucine and valine. This finding was an apparent contradiction to the one gene, one enzyme concept. When they were examined genetically and nutritionally, it became clear that there were several classes of these doubly auxotrophic mutants. One class was found to accumulate, in the culture fluids, material that fed mutants of several other classes. Analysis revealed that the active material consisted of the α-keto acid precursors of valine and isoleucine (fig. 1). Furthermore, among the mutants that responded to the keto acids, one class was found to accumulate material that fed other isoleucine and valine auxotrophs. This accumulated material was identified as α,β-dihydroxyisovalerate and α,β-dihydroxy-β-methylvalerate. Both these findings were followed by the demonstration of a lack of the branched-chain amino acid-glutamate transaminase in the class accumulating the α-keto acids and the lack of what is now called dihydroxy acid dehydrase in the dihydroxy acid accumulators. It was this loss of a single enzyme that catalyzes the corresponding step in both pathways that accounted for the unexpected double auxotrophy.

The valine and isoleucine auxotrophs that responded to the dihydroxy acids did not feed any other class except one, which appeared to be blocked only in isoleucine biosynthesis and which responded as well to α-ketobutyrate as to α-aminobutyrate. These single blocked auxotrophs were later shown to lack threonine deaminase, the enzyme required only for isoleucine biosynthesis.

Figure 1

Incorporation of lactate and acetate carbon into valine and isoleucine. The colored symbols next to the carbon atoms in lactate, acetate, and the other compounds indicate where the carbons are situated in the precursors and products. Radioactive labeling patterns of this type provide important information for pathway analysis. The isotope studies indicate that no sequence of three carbons in valine could have arisen from lactate (or pyruvate) directly, nor is there any four-carbon sequence in isoleucine that is labeled with acetate carbon. The labeling paradox was readily explained by predicting and later demonstrating that the precursors for valine and isoleucine were acetolactate and acetohydroxybutyrate, respectively, both of which underwent an alkyl group migration to yield the carbon skeletons of valine and isoleucine.

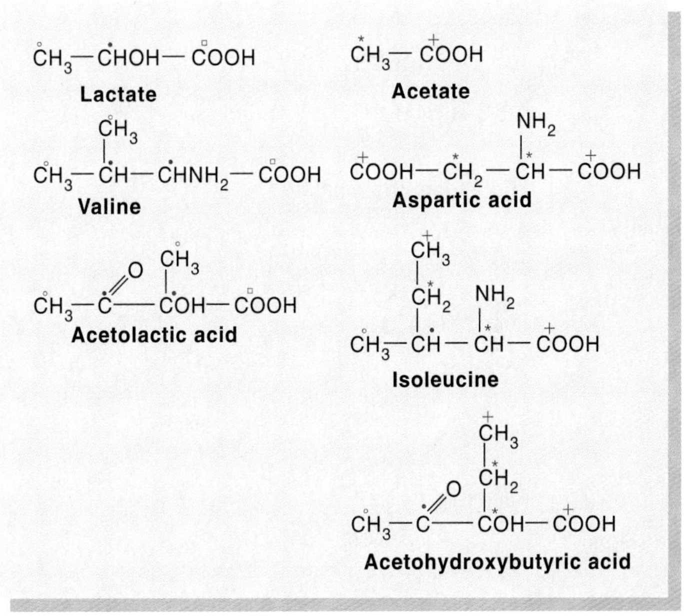

thase (an enzyme required for branched-chain amino acid biosynthesis (see fig. 24.21). These three classes are sulfonylureas, imidazolinones, and triazolopyrimidines, which are the active ingredients in, respectively, Oust, Sceptor, and Broadstrike (fig. 24.23).

Because animals do not synthesize either the aromatic or branched-chain amino acids, these materials can be applied to kill unwanted vegetation without causing harm to domestic animals or humans. Certain derivatives can often be selectively applied to combat noxious plants without appreciable harm to crops. More promising, however, is the prospect of using biotechnology to incorporate man-made genes for enzymes specifically resistant to one of the herbicides into agronomically important species.

Figure 24.22

The biosynthesis of leucine from α-ketoisovalerate, the branchpoint intermediate also used in L-valine synthesis. In the first step the carbon chain of α-ketoisovalerate is lengthened by two carbon atoms. The steps to leucine are completed by specific isomerase, dehydrogenase, and transaminase reactions.

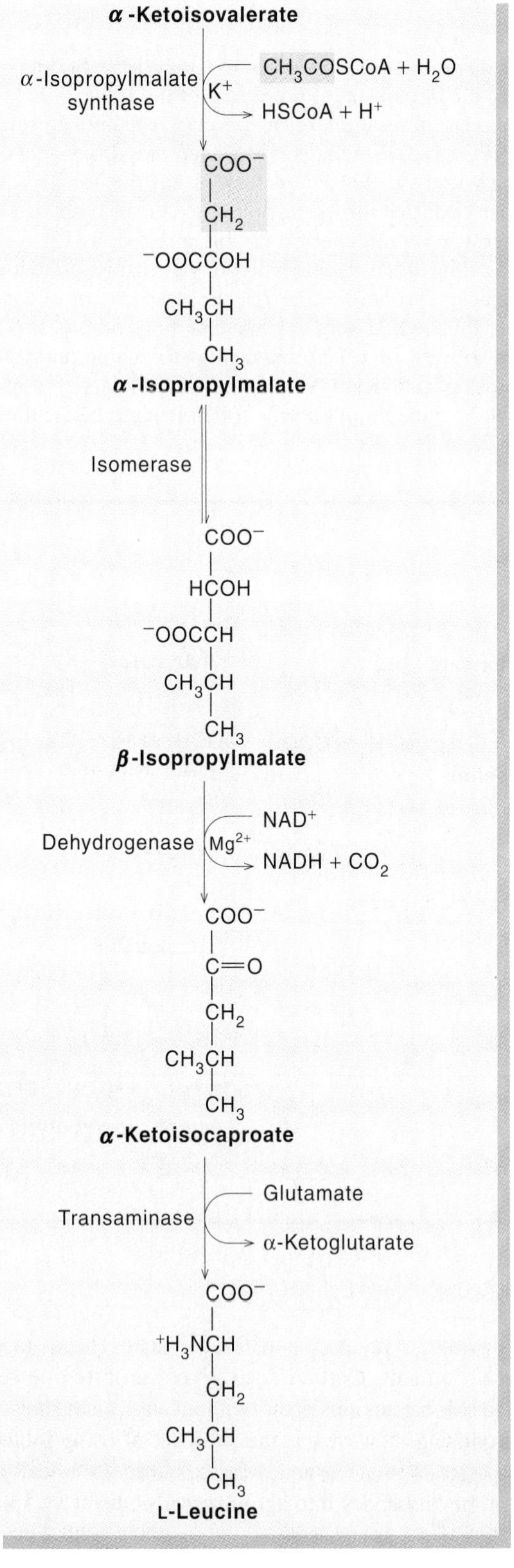

Figure 24.23

Three herbicides considered to be "safe." These compounds all act on an enzyme required for branched amino acid biosynthesis. Since this enzyme is not present in mammals, the herbicides are considered to be harmless to humans.

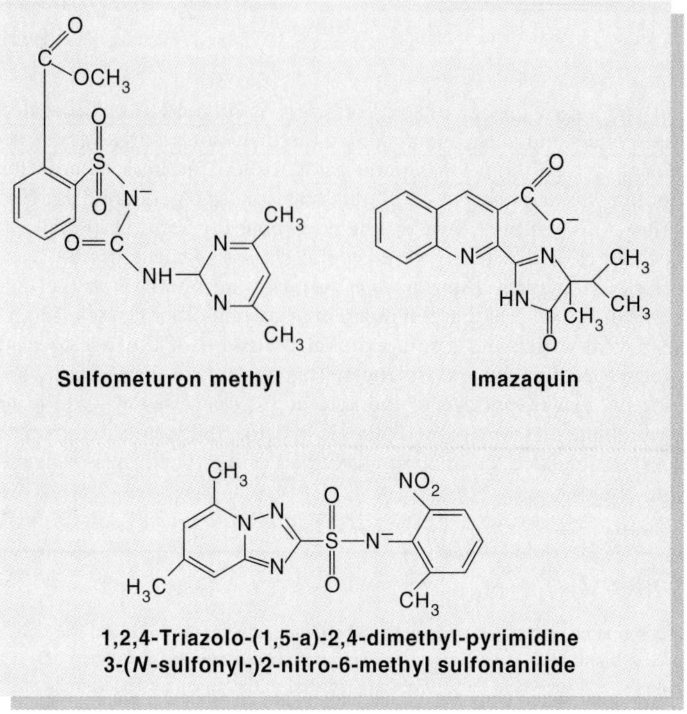

In the Biosynthesis of the Aromatic Family of Amino Acids (L-Tryptophan, L-Phenylalanine, and L-Tyrosine), Chorismate Is a Key Intermediate

The aromatic amino acids phenylalanine, tyrosine, and tryptophan are all formed by means of the shikimate pathway. This pathway is also important for the formation of the aromatic nuclei in or the aromatic precursor of vitamins E and K, folic acid, ubiquinone, and plastoquinone and certain metal chelators, such as enterochelin. The branchpoint compound for all these diverse products is chorismate, which has a prearomatic cyclohexadiene nucleus (fig. 24.24).

The overall route of chorismate synthesis, beginning with the condensation of two intermediates from the central metabolic routes, phosphoenolpyruvate and erythrose-4-phosphate, is illustrated in figure 24.24. Chorismate is the final product of the common pathway to all aromatic compounds, including phenylalanine, tyrosine, and tryptophan.

Prephenate Is a Common Intermediate in L-Phenylalanine and L-Tyrosine Synthesis

In some organisms the pathways from chorismate to L-phenylalanine and L-tyrosine, shown in figure 24.25, are truly separate pathways, even though the first step from chorismate is the same. For

Figure 24.24

The common aromatic (shikimate) pathway leading to chorismate biosynthesis. Note that NAD^+ is required for the conversion of 3-deoxy-D-*arabino*-heptulosonate-7-phosphate to 3-dehydroquinate, but there is no net change in the redox state during the conversion of substrate to product. This was also seen in the epimerizaton of fructose (see chapter 14) and is indicative of an oxidized intermediate. Chorismate formed by this series of reactions is a common intermediate for phenylalanine, tyrosine and tryptophan biosynthesis.

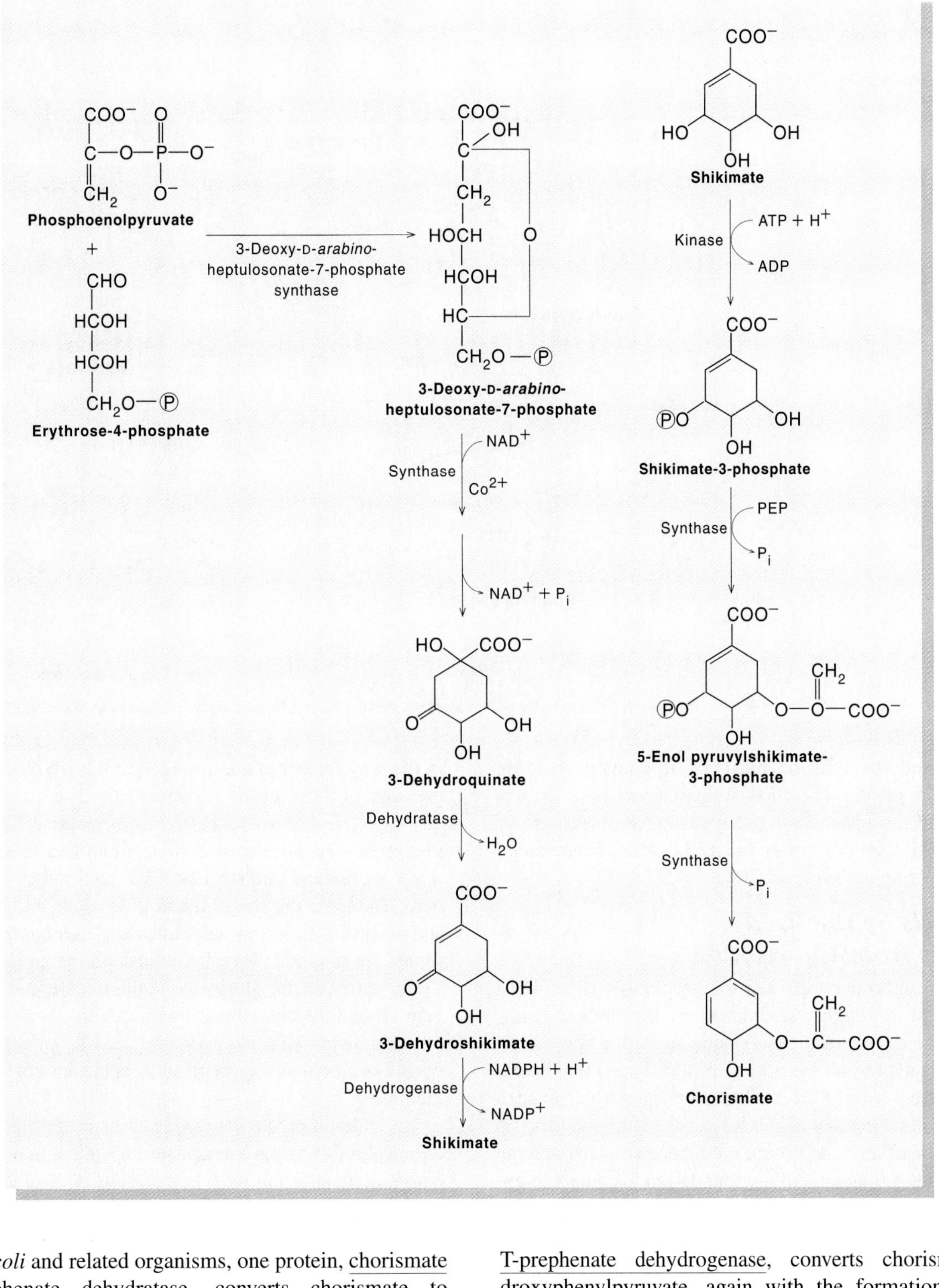

example, in *E. coli* and related organisms, one protein, chorismate mutase P-prephenate dehydratase, converts chorismate to phenylpyruvate, with the intermediate, prephenate, being enzyme-bound. A second protein, the NAD-dependent chorismate mutase T-prephenate dehydrogenase, converts chorismate to 4-hydroxyphenylpyruvate, again with the formation of prephenate occurring as an enzyme-bound intermediate. Both enzymes utilize prephenate, the phenylalanine biosynthetic enzyme yielding

Figure 24.25

The biosynthesis of phenylalanine and tyrosine from the branchpoint compound chorismate. The lines between the mutase and the dehydratase and between the mutase and the dehydrogenase indicate that in both pathways we are dealing with bifunctional enzymes that are exclusive to the indicated pathways. Thus although prephenate is the first intermediate in both pathways, it is not a branchpoint intermediate in the usual sense.

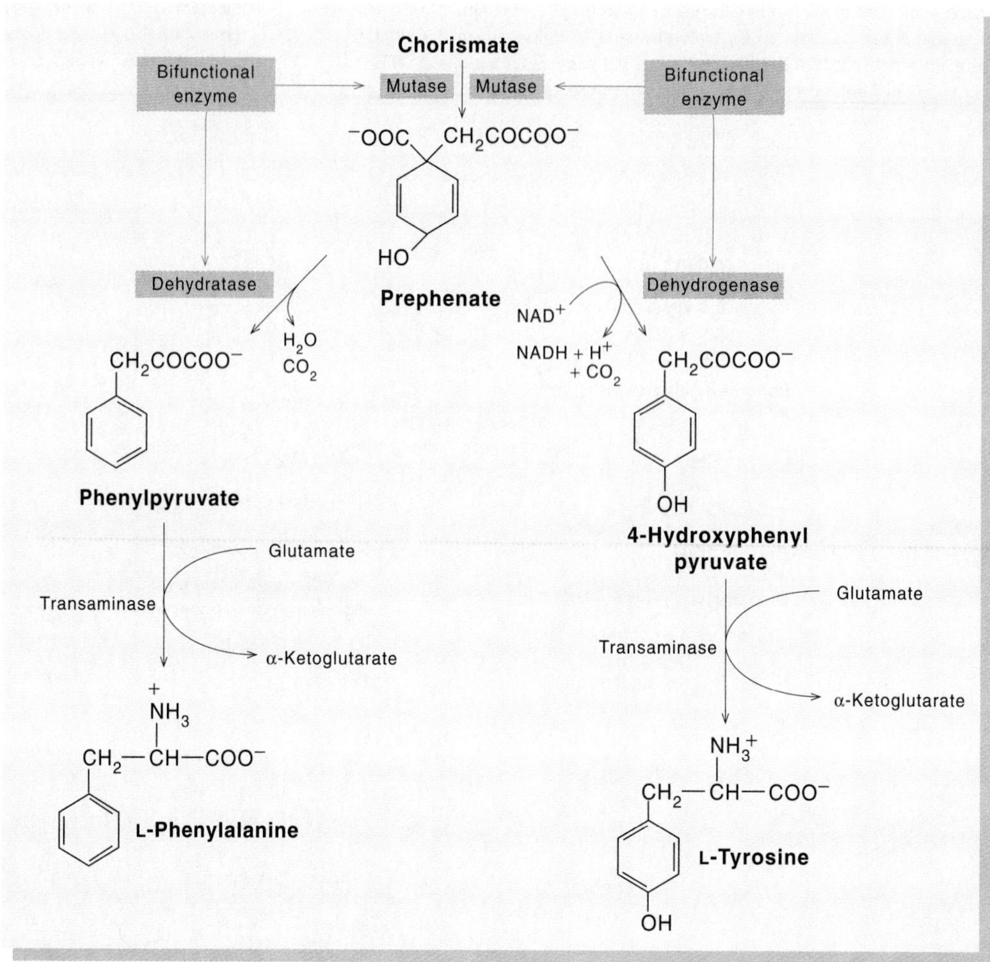

phenylpyruvate and the tyrosine biosynthetic enzyme yielding 4-hydroxyphenylpyruvate. The final aromatization step, the removal of water from prephenate in phenylalanine biosynthesis, or the removal of hydrogen in tyrosine biosynthesis, is accompanied by the loss of the ring carboxyl as CO_2.

Tryptophan Is Synthesized in Five Steps from Chorismate

The pathway leading from chorismate to L-tryptophan (fig. 24.26) is among the most thoroughly studied of any biosynthetic pathway. In *E. coli* the details of the enzymatic steps, the correlation between DNA sequence and the protein products, and the factors controlling the transcription of the structural genes far exceed those known for any other set of related genes. Although this has not been the work of any one group, the extensive gene-enzyme analysis of Charles Yanofsky laid the foundation for others to explore details of some of the enzymatic steps by physical and kinetic approaches. Comparative studies in other bacteria and in fungi have revealed variations upon the themes found in *E. coli,* particularly with respect to the distribution on one protein or another of the sequence of enzyme activities, which are identical in all forms. In addition, these studies have also revealed differences in the way the genes are arranged in the DNA and in the way expression of those genes is controlled.

The first specific step in tryptophan biosynthesis is the glutamine-dependent conversion of chorismate to the simple aromatic compound anthranilate. Like most other glutamine-dependent reactions, the reaction can also occur with ammonia as the source of the amino group. However, high concentrations of ammonia are required. Thus far, almost all the anthranilate synthases examined have the glutamine amidotransferase activity (component II) and the chorismate-to-anthranilate activity (component I) on separate proteins. Component I actually catalyzes two reactions with 2-amino-4-deoxychorismate being an enzyme-bound intermediate.

Anthranilate is transferred to a ribose phosphate chain in a phosphoribosyl-pyrophosphate-dependent reaction catalyzed by anthranilate phosphoribosyltransferase. An isomerase catalyzes an Amadori rearrangement, in which the ribosyl moiety of phosphoribosylanthranilate becomes a ribulosyl moiety. The product, 1-(*o*-carboxyphenylamino)-1-deoxyribulose-5′-phosphate, is cyclized to indoleglycerol phosphate by the removal of water and loss of the ring carboxyl by indoleglycerol phosphate synthase. The final step in tryptophan biosynthesis is a replacement reaction, cat-

Figure 24.26

The biosynthesis of tryptophan from the branchpoint compound, chorismate in *E. coli*. The first step involves the conversion of chorismate to the aromatic compound anthranilate. The anthranilate is transferred to a ribose phosphate chain. The product is cyclized to indoleglycerol phosphate by the removal of water and loss of the ring carboxyl by indoleglycerol phosphate synthase. Finally, in a replacement reaction catalyzed by tryptophan synthase, glyceraldehyde-3-phosphate is removed from indoleglycerol phosphate, and the enzyme-bound indole is condensed with serine. The structure of phosphoribosyl pyrophosphate is shown in figure 24.27. The red arrow indicates that the end product of this pathway inhibits the first enzyme in the pathway.

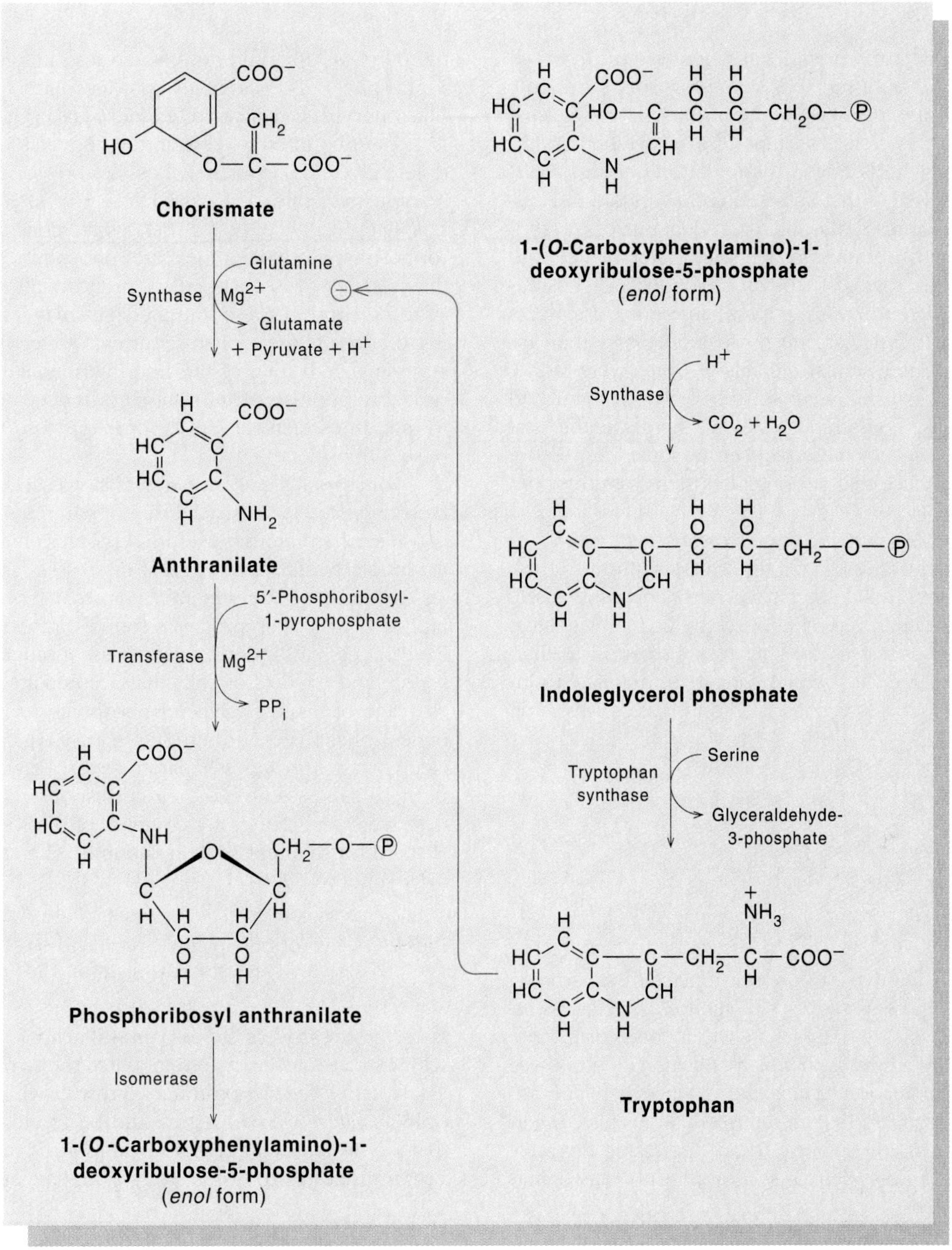

alyzed by tryptophan synthase, in which glyceraldehyde-3-phosphate is removed from indoleglycerol phosphate and the enzyme-bound indole so formed is condensed with serine.

At one time it was thought that indole itself was a free intermediate in tryptophan synthesis, but this notion was dispelled by a combination of biochemical and genetic studies (box 24B).

Among different organisms these five enzyme activities are distributed on different proteins (table 24.1). For example, in *E. coli,* indoleglycerol phosphate synthase catalyzes both the isomerization of phosphoribosylanthranilate and the cyclization step. Of particular interest is the occurrence on a single protein of the catalytic activities for nonconsecutive reactions in some cases. If

24B
BOX

Demonstration That Indole Is Not a Free Intermediate in the Tryptophan Biosynthetic Pathway

The tryptophan pathway provides another example in which nutritional studies with mutants of *Neurospora* and *E. coli*, isotope incorporation studies, and enzymatic analyses have been exploited to reveal the steps in a biosynthetic pathway. For example, early studies with tryptophan-requiring organisms found in nature revealed that some could use indole (a compound known to be formed by the microbial degradation of tryptophan and others could use anthranilate. Later, after Beadle and Tatum introduced the approach of studying metabolism with mutants of the bread mold *Neurospora*, tryptophan-requiring mutants of this organism were found that could use indole or either anthranilate or indole. Still later, similar mutants of *E. coli* were found, and mutants of both organisms were described that accumulated one or the other of these compounds. Clearly, those mutants that grew on anthranilate or indole were blocked in some step before these compounds, and those that accumulated them were blocked in the step after them.

Incorporation studies with isotopes showed that when anthranilate was converted to tryptophan, the carboxyl group of anthranilate was lost as carbon dioxide, but the nitrogen was retained. Because the enzymes in the tryptophan biosynthetic pathway have only a limited specificity, it was possible to substitute 4-methylanthranilate in *E. coli* extracts that could convert anthranilate to indole. This "nonisotope" label was conserved during the conversion to yield 6-methyl indole:

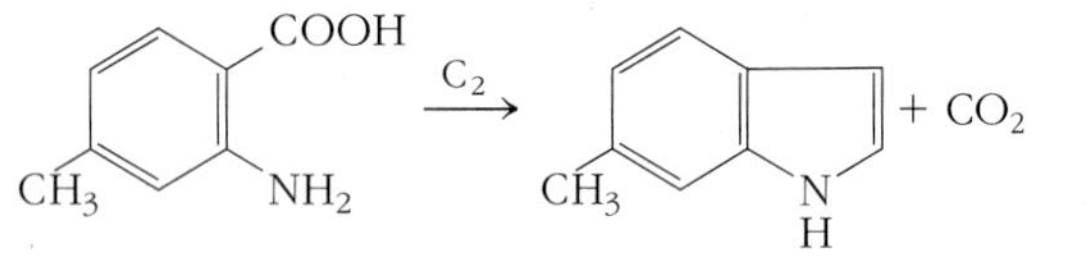

COOH; CH_3; NH_2 $\xrightarrow{C_2}$ CH_3; N H $+ CO_2$

4-Methylanthranilate   6-Methyl indole

It was thus clear that some two-carbon unit replaced the carboxyl carbon of anthranilate. Further studies with such *E. coli* extracts indicated that phosphoribosyl pyrophosphate was a good cosubstrate for the formation of indole from anthranilate. Fractionation of these extracts, as well as examination of mutants blocked between anthranilate and indole, revealed that an intermediate in this conversion was indole-3-glycerol phosphate. Extracts from one group of such mutants could not form indole-3-glycerol phosphate, whereas the other group could not convert it to indole and glyceraldehyde-3-phosphate. The latter group was found to accumulate the dephosphorylated derivative, indole-3-glycerol, in culture fluids.

The two intermediates in the conversion of anthranilate to indole-3-glycerol phosphate, phosphoribosylanthranilate and 1-(*O*-carboxyphenylamino)-1-deoxyribulose-5′-phosphate, were originally postulated to account for the involvement of phosphoribosyl pyrophosphate in indole-3-glycerol phosphate formation. Support for the postulate was obtained when the dephosphorylated derivative of the second of these intermediates was found in the culture fluids of certain tryptophan-requiring bacterial mutants. The corresponding derivative of the first intermediate has not been found, probably because of its instability. Indeed, this compound, when formed in extracts, is rapidly broken down to anthranilate and ribose-5-phosphate.

For several years, indole, which was accumulated by some mutants and used to satisfy the tryptophan requirement by others, was considered an intermediate in tryptophan biosynthesis. Such a role for indole would have been of interest, because it appeared to be an exception to the generalization that biosynthetic intermediates had to bear a charge. It was found that extracts of cells that utilized indole did indeed catalyze the condensation of indole with serine, and extracts of cells that accumulated indole catalyzed the cleavage of indole-3-glycerol phosphate to indole and glyceraldehyde-3-phosphate. Furthermore, *E. coli* mutants of these two classes clearly were affected in separate genes. However, the two products of these genes catalyzed their corresponding reactions faster when they were associated in a complex of the form $\alpha_2\beta_2$, where α and β stand for different protein subunits. The complex itself catalyzes the overall reaction

Indole-3-glycerol phosphate + Serine $\longrightarrow$
Tryptophan + Glyceraldehyde-3-phosphate

faster than either of the separate reactions. The same was found with extracts of *Neurospora* in which the two partial reactions were catalyzed by the same protein and with which no evidence for indole as a free intermediate could be found. Thus it became clear that indole, although historically important in deciphering the pathway to tryptophan, occurs only as an enzyme-bound intermediate.

---

in such cases the proteins were separate from each other in the cell, this arrangement, for example, in *Neurospora*, would necessitate the product of one reaction leaving the product site of one enzyme to be acted upon by another enzyme and then returning to the substrate site of a third enzyme on the same protein that exhibited the first enzyme activity. The persistence of this arrangement during evolution makes attractive the idea that all the tryptophan biosynthetic enzymes exist in the cell as a single multienzyme (and multiprotein) complex. However, if so, the complex must be quite labile, since individual gene products are so readily separated.

The evolution of the gene fusion that resulted in the conversion of the α and β peptides of tryptophan synthase as found in *E. coli* to the single peptide with β and α domains as found in

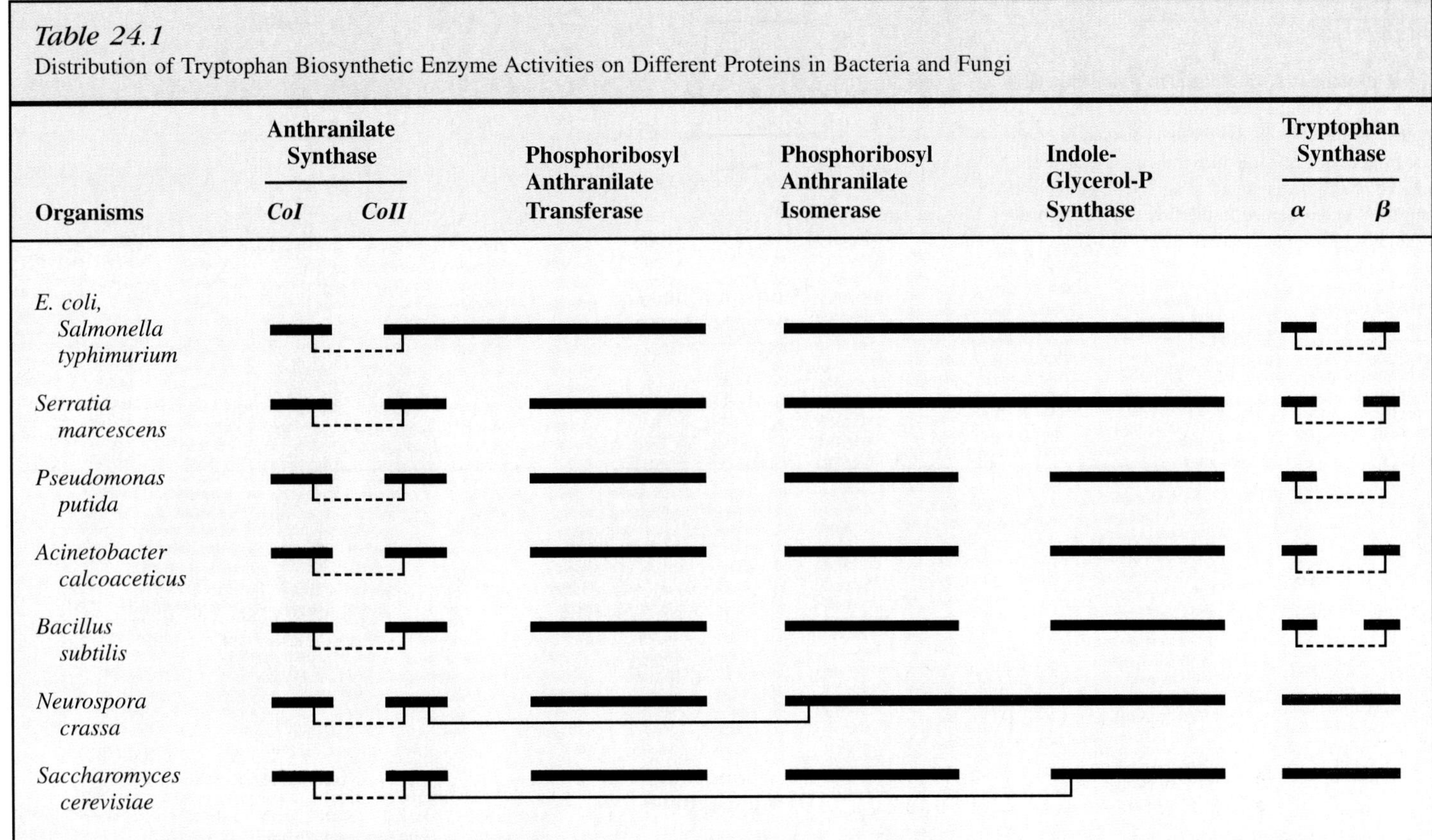

Organisms
Anthranilate Synthase
CoI
CoII
Phosphoribosyl Anthranilate Transferase
Phosphoribosyl Anthranilate Isomerase
Indole-Glycerol-P Synthase
Tryptophan Synthase
α
β
E. coli, Salmonella typhimurium
Serratia marcescens
Pseudomonas putida
Acinetobacter calcoaceticus
Bacillus subtilis
Neurospora crassa
Saccharomyces cerevisiae

Table 24.1
Distribution of Tryptophan Biosynthetic Enzyme Activities on Different Proteins in Bacteria and Fungi

Note: └──┘ = covalent linkage; └------┘ = obligatory association required for full activity.
A single band covering two activities indicates a single polypeptide.

fungi and yeast has been studied experimentally. Fusion of the *trpB* and *trpA* genes in the order in which they occur on the chromosome to yield a single β-α chain results in a protein of only limited activity. In contrast, when the gene fusion is done in the reverse of the genetic order (i.e., *A* to *B*), the resulting α-β chain yields a protein that is much more active. In addition, a short interdomain peptide region is required for full activity in a fusion protein.

Carbon Flow in the Biosynthesis of Aromatic Amino Acids Is Regulated at Branchpoints

Metabolite flow to tryptophan is controlled by inhibition of anthranilate synthase by tryptophan. Regulation of metabolite flow in phenylalanine and tyrosine biosynthesis varies from organism to organism, owing to the variety of enzyme patterns in the conversion of chorismate to the two amino acids. In *E. coli* and related organisms, phenylalanine inhibits both activities of chorismate mutase P-prephenate dehydratase, whereas tyrosine inhibits only the mutase activity of chorismate mutase T-prephenate dehydrogenase.

There are two general patterns of control over the common aromatic pathway. One is that found in *E. coli* and related organisms. The pattern is similar to that of the common aspartate family pathway of the same organism in that there are three isozymic deoxy-*arabino*-heptulosonate-7-phosphate synthases. Each is inhibited by one of the three aromatic amino acids. (There is, in addition, a tryptophan-specific repression of the tryptophan-sensitive enzyme, a tyrosine-specific repression of the tyrosine-sensitive enzyme, and a tryptophan plus phenylalanine-specific multivalent repression of the phenylalanine-sensitive enzyme.) As in the synthesis of the intermediates in the aspartate family common pathway, the three enzymes contribute to a common pool of deoxy-*arabino*-heptulosonate-7-phosphate that is drawn upon for all the compounds formed from chorismate. Indeed, in some strains the phenylalanine-sensitive enzyme is predominant, whereas in others the tyrosine-sensitive enzyme is predominant.

Histidine Constitutes a Family of One

Histidine is unusual in two respects: It is in a family by itself, and both its structure and its pathway show a strong interplay with the purine pathway. The starting point for histidine biosynthesis is phosphoribosyl pyrophosphate (PRPP) as in the purine pathway (see chapter 26). The first specific step in histidine biosynthesis entails a condensation reaction between PRPP and ATP leading to phosphoribosyl ATP (see fig. 24.27). In the fifth step most of the purine nucleotide donated in the first step is returned to the purine pathway while the histidine precursor, imidazole glycerol phos-

Figure 24.27

The biosynthesis of histidine. The 5-aminoimidazole-4-carboxamide ribotide formed during the course of histidine biosynthesis is also an intermediate in purine nucleotide biosynthesis. Therefore it can be readily regenerated to an ATP, thus replenishing the ATP consumed in the first step in the histidine biosynthetic pathway (see fig. 26.14).

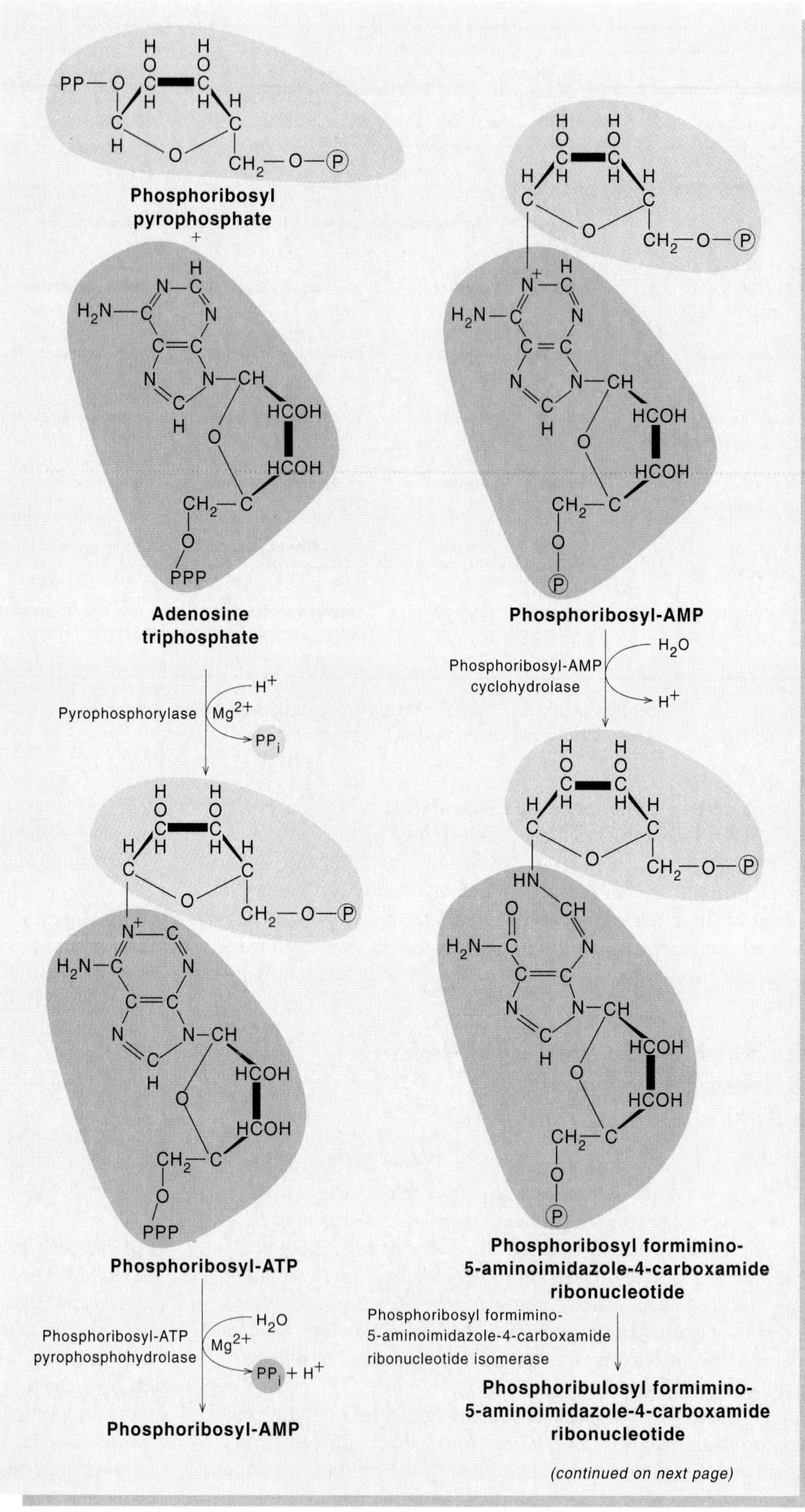

(continued on next page)

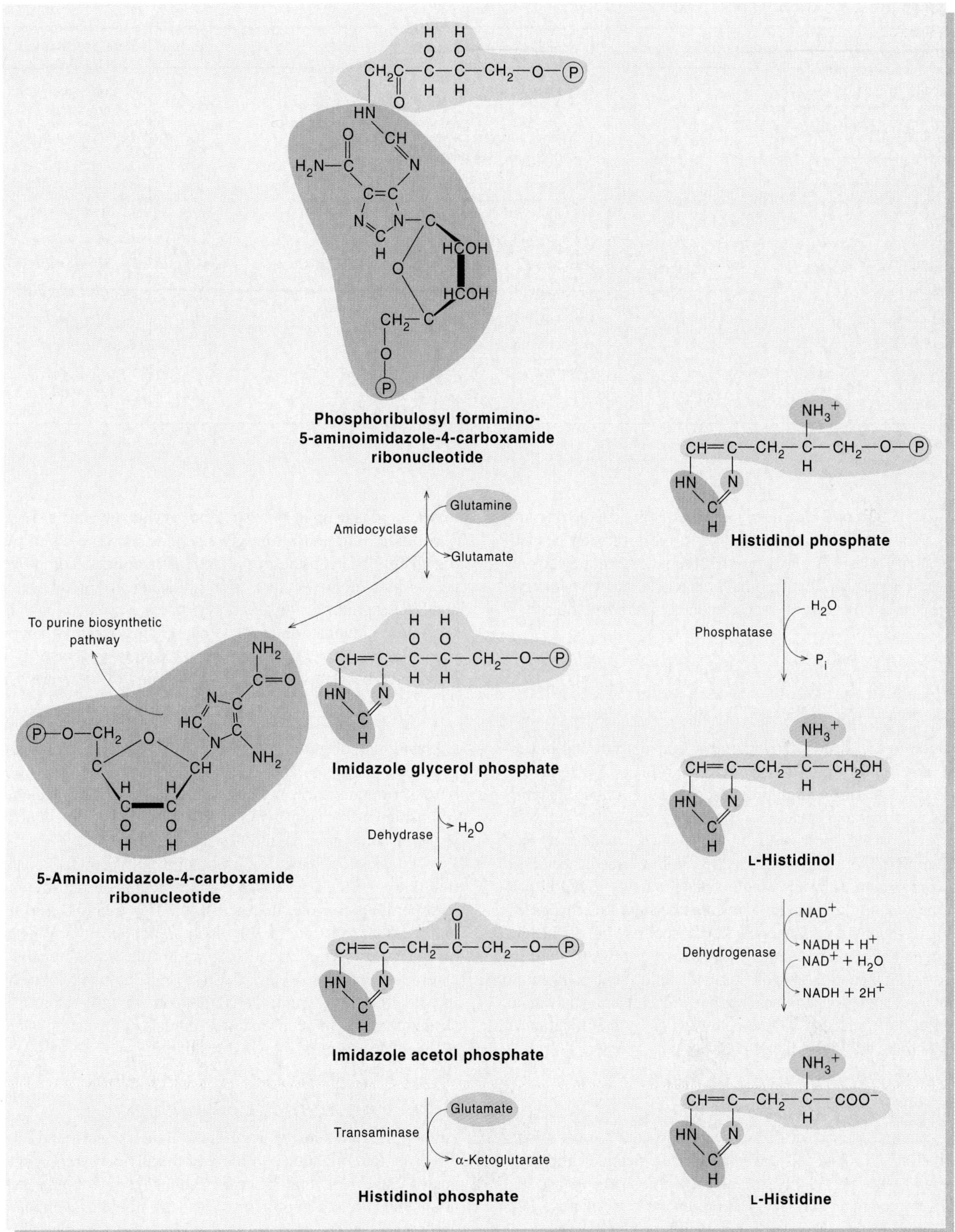

Phosphoribulosyl formimino-5-aminoimidazole-4-carboxamide ribonucleotide
Amidocyclase
Glutamine
Glutamate
To purine biosynthetic pathway
5-Aminoimidazole-4-carboxamide ribonucleotide
Imidazole glycerol phosphate
Dehydrase
H_2O
Imidazole acetol phosphate
Transaminase
Glutamate
α-Ketoglutarate
Histidinol phosphate
Histidinol phosphate
Phosphatase
H_2O
P_i
L-Histidinol
Dehydrogenase
NAD^+
$NADH + H^+$
$NAD^+ + H_2O$
$NADH + 2H^+$
L-Histidine

Table 24.2
Some D-Amino Acids Found in Peptide Antibiotics

Antibiotic	D-Amino Acids Present	Produced By
Actinomycin C_1 (D)	D-Valine	*Streptomyces parralus* and others
Bacitracin A	D-Asparagine, D-glutamate, D-ornithine, D-phenylalanine	*Bacillus subtilis*
Circulin A	D-Leucine	*Bacillus circulans*
Fungisporin	D-Phenylalanine, D-valine	*Penicillium spp.*
Gramicidin S	D-Phenylalanine	*Bacillus brevis*
Malformin A_1, C	D-Cysteine, D-leucine	*Aspergillus niger*
Mycobacillin	D-Aspartate, D-glutamate	*Bacillus subtilis*
Polymixin B_1	D-Phenylalanine	*Bacillus polymyxa*
Tyrocidine A, B	D-Phenylalanine	*Bacillus brevis*
Valinomycin	D-Valine	*Streptomyces fulrissimus*

phate, which now contains a newly synthesized imidazole ring, undergoes additional conversions leading to the formation of histidine. Not only does interplay occur between the histidine pathway and the purine pathway, but the final products of the two pathways, histidine and purine nucleotide, both contain imidazole rings.

Nonprotein Amino Acids Are Derived from Protein Amino Acids

In addition to the 20 amino acids most frequently found in proteins a large group of amino acids occur in plants, bacteria, and animals that are not found in proteins. Some are found in peptide linkages in compounds that are important as cell wall or capsular structures in bacteria or as antibiotic substances produced by bacteria and fungi. Others are found as free amino acids in seeds and other plant structures. Some amino acids are never found in proteins. These nonprotein amino acids, numbering in the hundreds, include precursors of normal amino acids, such as homoserine and diaminopimelate; intermediates in catabolic pathways, such as pipecolic acid; D enantiomers of "normal" amino acids; and amino acid analogs, such as azetidine-2-carboxylic acid and canavanine, that might be formed by unique pathways or by modification of normal amino acid biosynthetic pathways.

A Wide Variety of D-Amino Acids Are Found in Microbes

Certain D-amino acids with some L enantiomers are commonly found both in microbial cell walls and in many peptide antibiotics. For example, the peptidoglycans of bacteria contain both D-alanine and D-glutamate. The latter is present in a γ-glutamyl linkage. In some forms, the α carboxyl of the D-glutamyl residue is either amidated or in peptide linkage with glycine. D-Lysine or D-ornithine is found in the glycopeptide of some Gram positive organisms. The capsule of the anthrax bacillus is composed of a nearly pure homopolymer of D-glutamate in γ linkage. Other bacilli also produce γ-linked polyglutamates, some of which form separate D-glutamate and L-glutamate chains, whereas others form a copolymer of D- and L-glutamate. A wide variety of D-amino acids have been found in antibiotics; some of these are listed in table 24.2.

D-Alanine is found in bacterial cell wall peptidoglycan. L-Alanine is converted to D-alanine by a racemase that contains pyridoxal phosphate as a cofactor. The racemization is followed by the formation of a D-alanyl-D-alanine dipeptide, which is accompanied by the conversion of ATP to ADP. The dipeptide is subsequently incorporated into the glycopeptide (see fig. 18.23).

In most cases of formation of peptides containing D-amino acid, the L form of the amino acid is the substrate for the incorporating enzyme. In contrast, the free D-amino acid is ordinarily a poor substrate for the incorporation reaction. Whether the racemization occurs on the enzyme or afterwards remains to be determined in most cases. In the case of the D-valyl residue formed in penicillin, a tripeptide derivative containing L-valine is an intermediate, and the conversion is thought to occur by way of an α,β-dehydro form of the valyl residue.

There Are Hundreds of Naturally Occurring Amino Acid Analogs

Among the hundreds of nonprotein amino acids found in nature are many that might be considered naturally occurring amino acid analogs and many that are toxic and antagonistic to the usual 20 amino acids found in proteins. Some are found as components of antibiotics; others have been identified as antibiotic substances

Table 24.3
Some Naturally Occurring Nonprotein Amino Acids

Compound	Occurrence	Remarks
Branched-chain and cyclopropane amino acids		
$CH_3—CH_2—CH(CH_3)—CH_2—CHNH_2—COOH$ 2-Amino-4-methylcaproic acid (homoisoleucine)	California buckeye	Leucine antagonizes toxicity
$(CH_3)_2—NCH_2—CHNH_2—COOH$ 2-Amino-3-dimethylaminopropionic acid (azaleucine)	*Streptomyces neocaliberis*	Leucine antagonizes toxicity
$CH_2=C—CHCHNH_2—COOH$ (cyclopropane ring: CH_2 bridging C and CH) 2-(Methylenecyclopropyl)glycine	Lychee seeds	Leucine antagonizes toxicity
Sulfur-containing amino acids		
$CH_3—SCH_2—CHNH_2—COOH$ *S*-Methylcysteine	Broad bean (*Phaseolus vulgaris*)	—
$CH_3—CH=CH—S—CH_2—CHNH_2—COOH$ *S*-(Prop-1-enyl)cysteine	Garlic	—
Aromatic and heterocyclic amino acids		
3-(3-Carboxyphenyl)alanine	Iris (*Iris pseudoacoras*)	—
β-*N*-(3-Hydroxy-4-pyridone)alanine (mimosine)	Mimosa tree, leucaena	Tyrosine; toxic to nonruminants and ruminants lacking deterifying bacteria in their rumens
Pipecolic acid	Widely distributed in plants	Probably not toxic; an intermediate in lysine catabolism
Acidic amino acids		
$HOOC—CH(CH_3)—CH_2—CHNH_2—COOH$ 4-Methylglutamic acid	Sweet pea	—
Basic amino acids		
$NH—O—CH_2—CH_2—CHNH_2—COOH$ (NH bonded to $C=NH$, bonded to NH_2) Canavanine	Jack bean and other legumes	Arginine antagonizes toxicity
$NH_2—CH_2—CHNH_2—COOH$ 2,3-Diaminopropionic acid	Seeds of acacia and mimosa	As the oxalyl derivative, acts as a neurotoxin

themselves. The frequent occurrence of toxic amino acids in the seeds of plants suggests that they might play a role in the protection of the seeds from insects or other predators.

Some typical examples of these naturally occurring analogs are given in table 24.3 along with their sources and the antagonistic L-amino acid, where such an antagonism is known. It should be pointed out, however, that not all the "analogs" are toxic. For example, pipecolic acid, the next higher homolog of proline and an intermediate in lysine degradation, does not interfere in any demonstrable way with proline metabolism.

Summary

In this chapter we have discussed the biosynthesis of amino acids and the roles that certain amino acids play in bringing inorganic nitrogen and sulfur into bioorganic compounds. The following points are the highlights of this discussion.

1. Amino acid biosynthesis is best studied in microorganisms such as *E. coli,* in which all 20 of the amino acids found in proteins are synthesized. Microorganisms are also ideal for such studies because both genetic and biochemical techniques can be harnessed to analyze the pathways. Typically, research begins by isolating mutants defective in single steps in the pathway for a particular amino acid and analyzing the consequences of the mutation.
2. The pathways to amino acids arise as branchpoints from a few key intermediates in the central metabolic pathways.
3. Inorganic nitrogen for amino acid biosynthesis must be derived from nitrate, nitrite, or ammonia in the environment. However, it is incorporated into organic form only as ammonia. For most cells and under most conditions this conversion occurs by means of the amidation of glutamate to yield glutamine. This amide group is then used as a donor of "active" ammonia in numerous reactions, including the reductive amination of α-ketoglutarate. The amino group of glutamate thus formed then serves as the source of amino groups in the biosynthetic pathways of all the amino acids found in proteins. In only some cell types, growing in the presence of a high concentration of ammonia, does the amination of α-ketoglutarate and the amidation of aspartate (to yield asparagine) occur directly by free ammonia.
4. NH_3 is the form in which nitrogen is incorporated into organic materials. Nitrogen exists in the -3 valence state in NH_3. Nitrogen itself actually passes through various forms and valence states as a result of its interactions with different living forms. The valence states range from $+5$ in nitrates to -3 in ammonia or organic materials. In the 0 valence state, nitrogen is a gas. The passage of nitrogen from one form to another involves a chain of reactions performed by widely distributed organisms. The biological fixation of gaseous nitrogen by both free-living and symbiotic nitrogen-fixing bacteria involves an enzyme complex called nitrogenase.
5. The pathway for synthesizing the α-ketoglutarate family of amino acids has two main branches. One of these involves a chain-lengthening process that yields lysine. Glutamate is also a precursor of proline and arginine. Both routes require activation and reduction of a carboxyl group. In the route to arginine, protection of the α-amino group is required to prevent the cyclization reaction essential for proline biosynthesis. The formation of the guanidine group of arginine requires carbamoyl phosphate, which is also a precursor of the pyrimidines.
6. The serine family includes L-serine, glycine, and L-cysteine. Three enzymes convert 3-phospho-D-glycerate to serine and two more enzymes convert L-serine to glycine. Cysteine biosynthesis involves sulhydryl transfer to activated serine. Inorganic sulfate must be reduced to sulfide before it is incorporated into amino acids.
7. The biosynthetic route for the aspartate family is a highly branched pathway with one leading to its amide, asparagine, a second to lysine following condensation with pyruvate, a third to methionine following sulfur and methyl group transfer, and a fourth to threonine. In these pathway branches the α-amino group of aspartate is preserved. The amino group is lost, however, in the route to isoleucine, by which threonine contributes four carbons to isoleucine. Carbon flow in the aspartate family is regulated in the aspartokinase step.
8. The pyruvate family consists of its α-amino analog, alanine, valine (derived from the condensation of two pyruvate molecules), and leucine (made by lengthening the carbon skeleton of valine, much like the chain-lengthening reaction in the fungal lysine pathway, which we have not shown). The steps to valine are paralleled by those to isoleucine and, indeed, are catalyzed by the same enzymes—the difference being that α-ketobutyrate rather than pyruvate is the acceptor of the two-carbon fragment in the condensation step in isoleucine biosynthesis.
9. The aromatic amino acids tyrosine, phenylalanine, and tryptophan derive their aromatic rings from the shikimate pathway, with final aromatization of the ring occurring only in the specific branch pathways leading to the final products. The ring itself arises from a condensation between erythrose-4-phosphate and phosphoenolpyruvate, followed by cyclization. The α-amino group of tryptophan arises only indirectly by transamination with glutamate, since there is an exchange of serine for three carbons that had originated from ribose-5-phosphate.
10. The histidine pathway is a complex one in which a —C—N— unit of adenine serves as a nucleus for condensation with a ribosylphosphate moiety and another nitrogen derived from glutamine. The residue from adenine is, in fact, an intermediate in the purine nucleotide biosynthetic pathway and can thus be recycled by replenishing the lost —C—N— unit. The phosphate group is retained until after the α-amino group is incorporated, an example of the principle that metabolic intermediates bear charged groups.
11. In addition to the 20 amino acids commonly found in proteins, there are many amino acids and amino acid analogs that serve other functions. (Some nonprotein compounds formed from amino acids are discussed.)

Selected Readings

Barker, H. A., Amino acid degradation by anaerobic bacteria. *Ann. Rev. Biochem.* 50:23, 1981. A review of an important group of fermentation pathways of amino acid breakdown that occur in nature and could not be covered in this chapter.

Bishop, P. E., and R. D. Joerger, Genetics and molecular biology of alternative nitrogen fixation systems. *Ann. Rev. Plant Physiol. Plant Mol. Biol.* 41:109–125, 1990.

Chan, M. K., J. Kim, and D. C. Rees, The nitrogenase FeMo-cofactor and P-cluster pair: 2.2 Å resolution structures. *Science* 260:792–794, 1993.

Christen, P., and D. E. Metzler (eds.), *Transaminases.* New York: John

Wiley and Sons, 1985. A series of review chapters describing in detail the scope and mechanisms of transamination reactions.

Fowden, L., P. J. Lea, and E. A. Bell, The nonprotein amino acids of plants. *Adv. Enzymol.* 50:117, 1979. A discussion of the occurrence and biosynthesis of naturally occurring amino acid analogs in plants.

Howard, J. B., and D. C. Rees, Nitrogenase: A nucleotide-dependent molecular switch. *Ann. Rev. Biochem.* 63:235–264, 1994.

Kamberov, E. S., M. R. Atkinson, and A. J. Ninfa, The *Escherichia coli* PII signal transduction protein is activated upon binding α-ketoglutarate and ATP. *J. Biol. Chem.* 270:17797–17807, 1995.

Katz, E., and A. L. Demain, The peptide antibiotics of *Bacillus:* Chemistry, biogenesis and possible functions. *Bacteriol. Rev.* 41:449, 1977. A description of several peptide antibiotics showing the distribution of D-amino acid in these compounds.

Kim, J., D. Woo, and D. C. Rees, X-ray crystal structure of the nitrogenase molydenum-iron protein from *Clostridium pasteurianum* at 3.0 Å resolution. *Biochemistry* 32:7104–7115, 1993.

Ledley, F. D., H. E. Grenett, M. McGinnis-Shelnutt, and S. L. C. Woo, Retroviral-mediated gene transfer of human phenylalanine hydroxylase into NIH 3T3 and hepatoma cells. *Proc. Natl. Acad. Sci. USA* 83:409, 1986.

Mazelis, M., Amino acid catabolism. In B. J. Mifflin (ed.), *The Biochemistry of Plants,* vol. 5. New York: Academic Press, 1980, pp. 541–567. A survey of some of the amino acid catabolic pathways that have been found in plants.

Meister, A., Glutathione metabolism and its selective modification. *J. Biol. Chem.* 263: 17205, 1988. A minireview describing the many important metabolic roles for glutathione.

Miflin, B. J. (ed.), *The Biochemistry of Plants: A Comprehensive Treatise,* vol. 5, *Amino Acids and Derivatives.* New York: Acadmic Press, 1980. This volume contains 10 chapters by several authors detailing amino acid biosynthesis pathways in plants.

Neidhardt, F. C., R. Curtiss, III, J. L. Ingraham, E. C. C. Lin, K. B. Low, Jr., B. Magasanik, W. Reznikoff, M. Riley, M. Schaechter, and H. E. Umbarger (eds.), *Escherichia coli and Salmonella typhimurium: Cellular and Molecular Biology,* vol. 1, 2nd ed. Washington, D.C.: American Society for Microbiology, 1996. This volume contains eleven chapters by several authors describing in detail the pathways of amino acid biosynthesis with particular emphasis on enzymatic and genetic control mechanisms.

Tyerman, S. D., L. F. Whitehead, and D. A. Day, A channel-like transporter for NH_4^+ on the symbiotic interface of N_2-fixing plants. *Nature* 378:629–632, 1995.

Warren, M. J., and A. I. Scott, Tetrapyrrole assembly and modification into the ligands of biologically functional cofactors. *Trends Biol. Sci.* 51:486–491, 1990.

Wellner, D., and A. Meister, A Survey of inborn errors of metabolism and transport in man. *Ann. Rev. Biochem.* 50:911, 1981. This review documents the important of the pathways that break down amino acids in humans.

Yamada, K. S., T. Kinoshita, T. Tsunoda, and K. Aida (eds.), *The Microbial Production of Amino Acids.* New York: John Wiley and Sons, 1972. A collection of essays describing microbial processes used in Japanese industry for the production of amino acids. Includes examples in which the regulatory mechanisms functioning in most cells have been modified or bypassed.

PROBLEMS

1. The two enzymes γ-cystathionase (fig. 24.13) and β-cystathionase (fig. 24.17) are both pyridoxal phosphate enzymes. Which amino group of cystathionine forms a Schiff base (imine) with the pyridoxal phosphate associated with each enzyme?
2. If aspartate with ^{18}O-labeled γ-carboxylate oxygens was supplied to asparagine synthetase, where would the isotopic oxygens reappear?
3. Threonine biosynthesis (fig. 24.16) is shown to proceed from O-phospho-L-homoserine. The process actually involves two enzymatic steps in which an alkene intermediary compound is formed. Complete these two steps and provide names that would describe the two enzymes involved.
4. Notice in the biosynthesis of the two hydroxy amino acids serine (fig. 24.11) and threonine (fig. 24.16 and problem 24.3) that O-phosphate derivatives are involved in both pathways. Examine the methods used to remove the phosphate group in each pathway. What fundamental chemical difference can you deduce between the two pathways?
5. In the reaction promoted by β-isopropylmalate dehydrogenase (fig. 24.22) there is both an oxidation and a decarboxylation. Do you think one reaction precedes the other or do they occur at the same time? Why?
6. During the biosynthesis of histidine (fig. 24.27) the α-amino group is added before the formation of the α-carboxylate group. Is this a typical pattern during the biosynthesis of amino acids?
7. Why did nature "waste" an ATP in glutamine biosynthesis? The lone pair of electrons on an ammonia could have attacked the γ-carbonyl group of a glutamate. Subsequent elimination of an oxide ion and release of a proton from the nitrogen would produce a glutamine without the consumption of an ATP.
8. Does it appear to be a paradox that L-glutamate is both the product and an initial reactant in the glutamine biosynthetic pathway? Assuming that glutamate synthase (fig. 24.3) is utilized, how can you explain this paradox?
9. Examine the bioenergetics of the synthesis of glutamine synthesis from α-ketoglutarate via glutamate synthase or glutamate dehydrogenase (fig. 24.3). Is there a difference?
10. What is the function of NADPH in the reactions catalyzed by glutamate dehydrogenase and glutamate synthase?
11. *E. coli* strains have been isolated that are unable to grow in a medium containing L-valine but lacking L-isoleucine and L-leucine. The same organism can grow on a medium lacking all three amino acids. Provide an explanation.
12. The amino acid 2-aminobutanoate is a product of some bacteria (not a protein component). Predict how the bacteria produce this amino acid.
13. A certain bacterium that was a tryptophan auxotroph was observed to grow well when it was supplied with tryptophan, but as soon as the tryptophan in the environment was exhausted, the bacteria started to excrete a metabolite on the tryptophan biosynthetic pathway. Why didn't the bacteria ex-

crete the metabolite before exhausting the environmental tryptophan?

14. This chapter categorizes the amino acids into families based on the origin of their carbon skeleton. Is this an absolute pattern? Take a closer look at the information in this chapter to produce an answer.
15. What is the function of the acetylation of serine with acetyl-CoA during the biosynthesis of cysteine (fig. 24.13)?
16. Which ribose carbons are incorporated into tryptophan?
17. Molecules with structures as diverse as carbamoylphosphate, tryptophan, and cytidine triphosphate are feedback inhibitors of the *E. coli* glutamine synthase. The feedback inhibition is cumulative, each metabolite exerting a partial inhibition on the enzyme. Why would complete inhibition of the glutamine synthase by a single metabolite be metabolically unsound?
18. Given the structural diversity of the compounds that feedback-inhibit glutamine synthase, would you predict that they interact at a common regulatory site? Why or why not?
19. How does increased synthesis of aspartate and glutamate affect the TCA cycle? How does the cell accommodate this effect?
20. In what sense may indole be viewed as an "intermediate" in L-tryptophan biosynthesis?
21. When ^{14}C-labeled 4-hydroxyproline was administered to rats, the 4-hydroxyproline in newly synthesized collagen was not radiolabeled. Explain.
22. The accumulation of biosynthetic intermediates, or of metabolites derived from these intermediates, has proven to be valuable in the analysis of biosynthetic pathways in microorganisms. It was found that these accumulations occurred only after the required amino acid had been consumed and growth had stopped. How might you account for this observation?

C H A P T E R

Amino Acid Metabolism in Vertebrates

Vertebrates obtain most of their amino acids by nutrition. Amino acid degradation in vertebrates usually starts with the loss of the α amino nitrogen and finishes with the return of the carbon skeletons to the central metabolic pathways.

Amino acid metabolism in vertebrates contrasts sharply with amino acid metabolism in plants and many microorganisms. Most striking is the fact that plants and microorganisms can synthesize all 20 amino acids required for protein synthesis, whereas vertebrates can synthesize only about half this number. This inability leads to complex nutritional needs for vertebrates. We discuss these needs in light of the pathways for biosynthesis that still exist.

In contrast with the biosynthetic pathways, degradation pathways for all the commonly occurring amino acids are found in vertebrates. This makes it possible for vertebrates to disposc of potentially harmful excesses of amino acids and their partial degradation products. The need for efficient degradation pathways is highlighted by the existence of many pathological states that result from deficiencies in the degradative pathways. We discuss some of these pathological states and the mutational deficiencies that lead to them in light of the degradation pathways.

In addition to serving in their role as precursors of proteins, amino acids are essential precursors of other biomolecules, of which we discuss two examples: the synthesis of porphyrin and the synthesis of glutathione.

Humans and Rodents Synthesize Less Than Half of the Amino Acids They Need for Protein Synthesis

In view of the central importance of amino acids in proteins, we might expect that all organisms would possess the necessary enzymes to synthesize the protein amino acids. Surprisingly, only eight protein amino acids can be synthesized by the standard de novo pathways (fig. 25.1). The remainder are synthesized by alternative pathways (sometimes referred to as salvage pathways) or supplied by nutrients.

Figure 25.1

Amino acids that are synthesized de novo in mammals. All such amino acids are related by a small number of steps to glycolysis or TCA cycle intermediates. (Also see fig. 24.1).

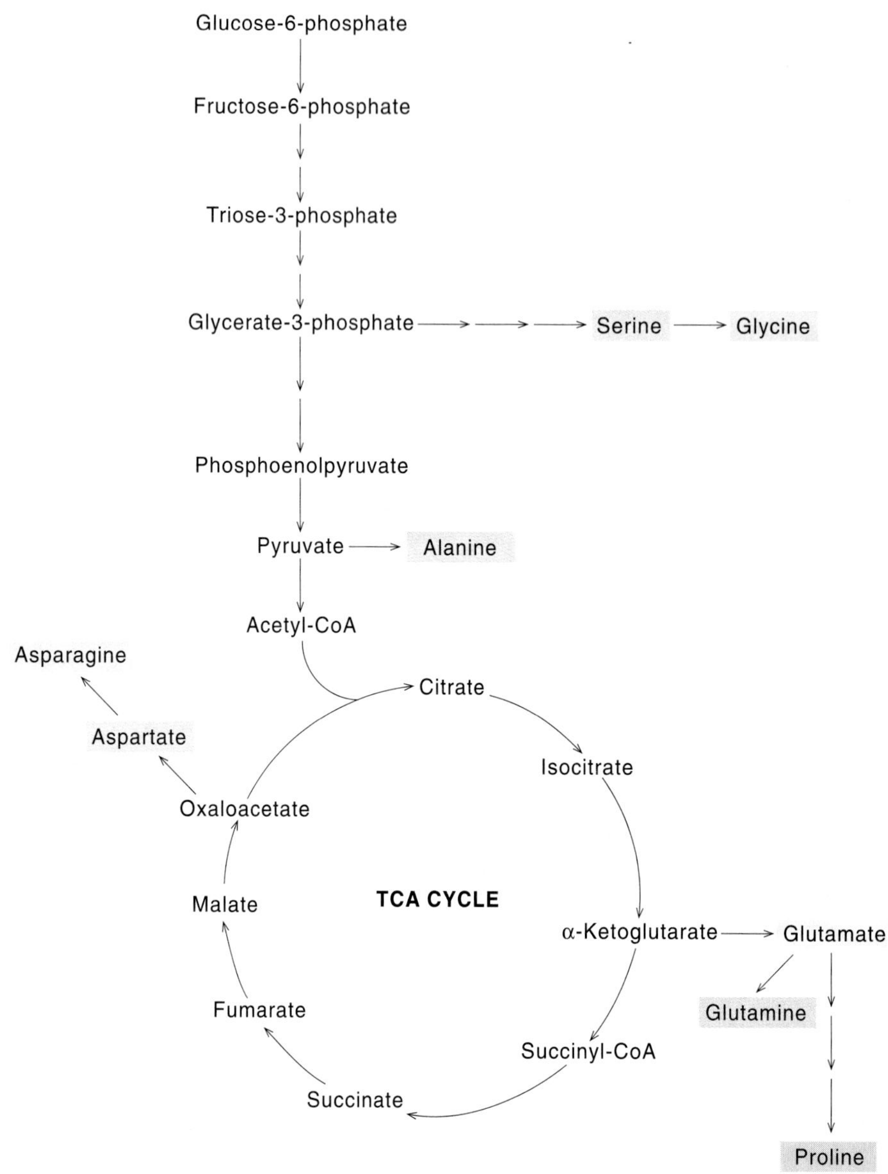

Why did the evolutionary process not favor the preservation of pathways for synthesizing all of the amino acids needed in higher animals? Can this be a case of too great a genetic load? Not likely, since these mammals are believed to have in excess of 50,000 genes and only 84 genes are required to encode all of the enzymes needed to make the 20 major amino acids found in proteins. A more likely reason for this loss of major pathways is that some of the intermediates in amino acid biosynthesis may have toxic effects on other biochemical processes occurring in higher eukaryotes. Perhaps the most probable reason of all is that these amino acids were not needed because these vertebrates eat organisms that contain an abundance of amino acids.

Many Amino Acids Are Required in the Diet for Good Nutrition

The inability of mammals to synthesize all of the amino acids they require has led to the classification of amino acids as essential and nonessential. Conceptually, this distinction seems clear. Practically speaking, it turns out to be more complex to designate essential amino acids. An "essential" amino acid, in this classification, means one that must be supplied in the diet if the organism is to maintain a positive nitrogen balance. As we will see, the absence of a de novo pathway for the biosynthesis of an amino acid does not necessarily mean that the amino acid must be obtained from

Table 25.1
The Essential Amino Acids

For Weight Gain in Protein-Starved Adult Rats		For Positive Nitrogen Balance in Adult Humans		For Mouse L Cells in Culture	
Essential	***Nonessential***	***Essential***	***Nonessential***	***Essential***[a]	***Nonessential***
Histidine	Alanine	Isoleucine	Alanine	Arginine	Alanine
Isoleucine	Arginine[b]	Leucine	Arginine	Cysteine	Asparagine
Leucine	Asparagine	Lysine	Asparagine	Glutamine	Aspartate
Lysine	Aspartate	Methionine	Aspartate	Histidine	Glutamate
Methionine	Cysteine	Phenylalanine	Cysteine	Isoleucine	Glycine
Phenylalanine	Glutamate	Threonine	Glutamate	Leucine	Proline
Threonine	Glutamine	Tryptophan	Glutamine	Lysine	Serine
Tryptophan	Glycine	Valine	Glycine	Methionine	
Valine	Proline		Histidine[c]	Phenylalanine	
	Serine		Proline	Threonine	
	Tyrosine		Serine	Tryptophan	
			Tyrosine	Tyrosine	
				Valine	

[a]The medium also contained 0.25–1% dialyzed horse serum.
[b]Arginine is required in the diet of young rats.
[c]Histidine is required in infant humans.

the diet. For one thing, it is frequently possible for the organism to make one amino acid from another. Furthermore, an alternative pathway for synthesis may supply sufficient amino acid for maintenance if not for growth.

The classical differentiation between essential and nonessential amino acids was made by W. C. Rose. His identification scheme was based on the weight gain of growing white rats that were fed diets containing 19 of the 20 amino acids found in proteins. These results are shown in the first two columns of table 25.1. For humans the results were based not on weight gain or loss but on short-term maintenance of a positive nitrogen balance (see table 25.1, middle columns). If the nitrogen balance was negative (total nitrogen excretion exceeding total nitrogen intake) during the period when a single amino acid was excluded from the diet, it was concluded that tissue protein was being degraded to supply the missing amino acid. For comparison, the amino acids found to be essential and nonessential for a strain (L) of mouse fibroblasts grown in cell culture are also included in table 25.1 (last two columns).

The results in table 25.1 indicate that given 19 other amino acids, the particular amino acid can be formed either by the de novo pathway used by plants or by a "salvage" pathway at the expense of some other amino acid.

As an example, arginine can be synthesized by a circuitous route that starts from the amino acid proline (fig. 25.2). The enzymes needed for de novo synthesis of proline are still formed (or found) in animal tissues. In addition, animals contain a proline oxidase that yields Δ'-pyrroline-5-carboxylic acid, which is in equilibrium with glutamic-γ-semialdehyde. A transaminase converts the semialdehyde to ornithine. From ornithine to arginine, enzymes of the urea cycle are used (see fig. 25.7). Thus although the pathway by which bacteria and plants form arginine is not present in animal tissues, arginine can be formed, but in some organisms (e.g., rats) at a rate insufficient for growth. Two other amino acids normally present in the diet also can be formed from other amino acids: Cysteine can be formed from dietary methionine, and tyrosine can be formed by hydroxylation of phenylalanine. These routes and the corresponding enzymes are listed in table 25.2.

Essential Amino Acids Must Be Obtained by Degradation of Ingested Proteins

Amino acids that originate from catabolism are derived from three sources: dietary proteins, storage proteins, and metabolic turnover of endogenous proteins. Catabolism of dietary proteins is a characteristic of higher animals, whereas the catabolism of storage protein is best illustrated by the germination of protein-storing seeds, such as beans or peas. Turnover of endogenous proteins occurs in all cells. The amino acids to which they are degraded can be re-

Figure 25.2

Synthesis of arginine by the salvage pathway found in vertebrates and by the de novo pathway found in plants and bacteria. The final steps from ornithine to arginine are also part of the urea cycle (see fig. 25.7).

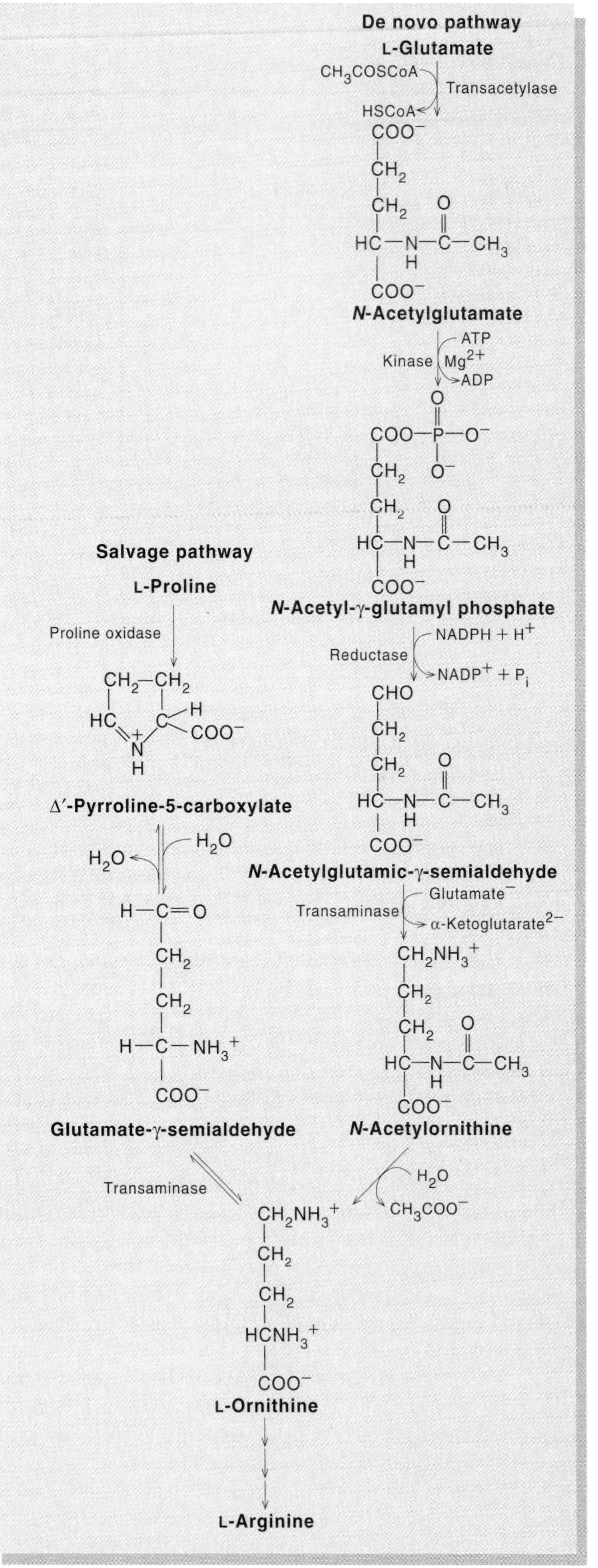

cycled to make new proteins or derivatives that involve amino acids as precursors.

Protein catabolism begins with hydrolysis of the covalent peptide bonds that link successive amino acid residues in a polypeptide chain (fig. 25.3). This process is termed proteolysis, and the enzymes responsible for the action are called proteases. In humans and many other animals, proteolysis occurs in the gastrointestinal tract; this type of proteolysis results from proteases secreted by the stomach, pancreas, and small intestine.

The initial products of proteolytic digestion are free amino acids and small peptides. Further digestion of peptides results from the action of peptidases secreted (or formed) by the intestinal mucosa. Some peptidases act on their substrates by hydrolyzing internal peptide bonds (endopeptidases); others act by removing amino acids one at a time from the end of the peptide (exopeptidases). A sign of the competitiveness of microorganism for the available food supply is that they make proteases and peptidases, which can be secreted into the surrounding medium so as to break down potential nutrient proteins to a suitable size for absorption.

Amino Acids May Be Reutilized or They May Be Degraded When Present in Excess

Degradative pathways for most amino acids begin by removal of the α-amino nitrogen. There are two major routes of deamination: transamination and oxidative deamination.

Transamination Is the Most Widespread Form of Nitrogen Transfer

The process of transamination is illustrated in figure 25.4 for an undesignated amino acid donating its amino group to the TCA cycle intermediate α-ketoglutarate. The reaction leads to an α-keto acid and glutamate. Most transaminases involved in amino acid catabolism exhibit a fairly broad specificity for the α-amino acid. The amino group acceptor is usually α-ketoglutarate. The mechanism for this reaction has been discussed (see fig. 11.5).

Net Deamination via Transamination Requires Oxidative Deamination

Transamination does not result in net deamination because one amino acid is replaced by another amino acid. The main function of transamination in catabolism is to funnel the amino nitrogen into one or a few amino acids. For glutamate to play a role in the net conversion of amino groups to ammonia, a mechanism for glutamate deamination is needed so that α-ketoglutarate can be regenerated for further transamination. Most often this entails the oxidative deamination of glutamate in a reaction catalyzed by an

Table 25.2

Salvage Pathways Allowing the Formation of Certain Nonessential Amino Acids from Other Amino Acids

Amino Acid Formed	Formed From	Enzymes Required
Arginine	Proline	Proline oxidase Ornithine-glutamate transaminase Ornithine transcarbamoylase Argininosuccinate synthase Argininosuccinate lyase
Cysteine	Methionine	*S*-Adenosylmethionine synthase α-Methyltransferase *S*-Adenosylhomocysteinase Cystathionine-β-synthase Cystathionine-γ-lyase
Tyrosine	Phenylalanine	Phenylalanine-4-monooxygenase

Figure 25.3

A protease hydrolyzes a peptide bond. Proteases have varying degrees of specificity, depending on the chemical nature of the R group and the location of the peptide linkage. Exopeptidases attack one or both ends of a polypeptide chain, and endopeptidases attack interior linkages.

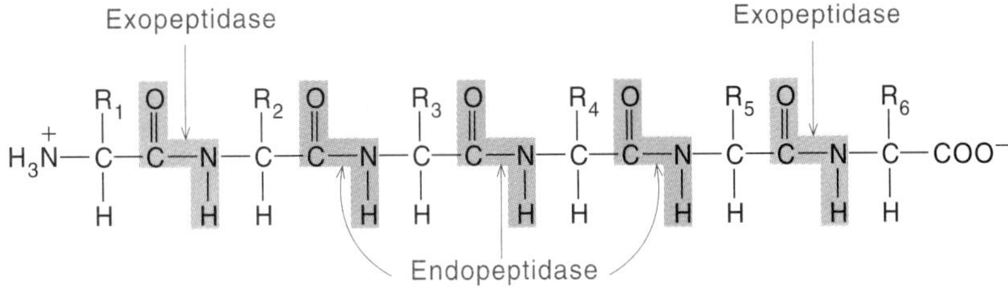

Figure 25.4

Transamination and deamination. Glutamate transaminase catalyzes the transfer of the α-amino group of an amino acid to α-ketoglutarate. The reaction is highly reversible because the reacting functional groups of the products are identical to those of the reactants. Transamination is not deamination. Transamination yields ammonia only if it is linked to another type of deamination process. Here net deamination results from the combined action of glutamate transaminase and glutamate dehydrogenase. In this process the α-ketoglutarate is recycled.

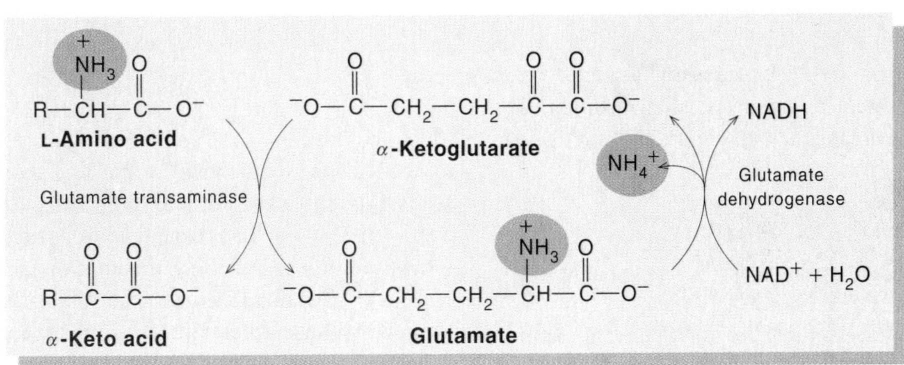

NAD$^+$-linked enzyme, glutamate dehydrogenase. This broadly distributed enzyme is located in the mitochondria of eukaryotic cells. It catalyzes release of the amino group of glutamate, leading to the regeneration of α-ketoglutarate:

$$\text{Glutamate} + \text{NAD}^+ + \text{H}_2\text{O} \longrightarrow \alpha\text{-Ketoglutarate} + \text{NH}_4^+ + \text{NADH} \quad \textbf{(1)}$$

The overall process of transamination of α-ketoglutarate and regeneration of the α-ketoglutarate is shown in figure 25.4.

Figure 25.5

The glutamate-dehydrogenase-catalyzed reaction involves hydride transfer from glutamate to NAD^+, leading to α-iminoglutarate imine, followed by hydrolysis to α-ketoglutarate. The amine group in glutamate is written in the uncharged form (—NH_2) because this is believed to be the reactive species.

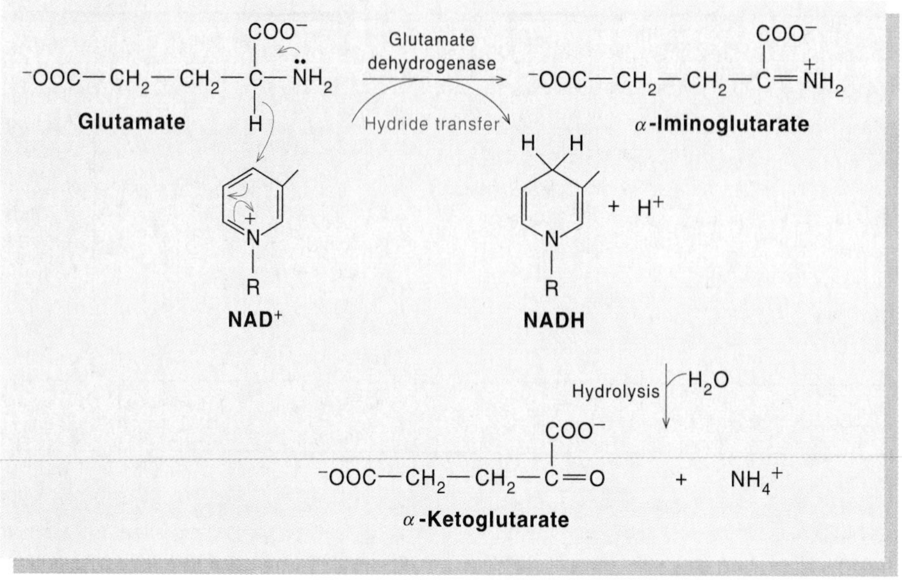

Catalysis by glutamate dehydrogenase starts with a hydride transfer from the α carbon of the amino acid to NAD^+ (fig. 25.5). The resulting α-iminoglutarate hydrolyzes to α-ketoglutarate and ammonia.

We discussed a glutamate dehydrogenase reaction in the previous chapter in conjunction with the biosynthesis of glutamate (see fig. 24.3). The deamination of glutamate to α-ketoglutarate involves the same compounds, but the reaction is reversed. This reaction is close enough to equilibrium that it can occur in either direction, but the glutamate dehydrogenases involved in catalyzing the forward and backward reactions are usually different and located in different places. The glutamate dehydrogenase involved in deamination is located in the mitochondria, and it uses NAD^+ as a cosubstrate. The amination reaction occurs in the cytosol, and it usually is specific for NADPH as cosubstrate.

Some other amino acids also undergo deamination directly by oxidative reactions. Whether transamination or direct deamination is more important as an initial step in amino acid breakdown depends on the organism or tissue under investigation.

In Many Vertebrates, Ammonia Resulting from Deamination Must Be Detoxified Prior to Elimination

The NH_4^+ resulting from deamination of amino acids is converted to ammonia either directly or indirectly (e.g., by means of a transamination to yield a readily deaminated product such as glutamate). In microorganisms using a single amino acid as a nitrogen source, the ammonia so liberated is assimilated and used to form other nitrogen-containing cellular components. When the amino acid is a carbon source, much more ammonia is liberated than is needed for biosynthesis, and it is disposed of by excretion to the surrounding medium. This simple disposal mechanism is adequate for free-living microorganisms because the ammonia is carried away in the surrounding medium or escapes into the atmosphere.

Figure 25.6

Excretory forms of nitrogen in different organisms. NH_3 is the most common end product of nitrogen metabolism. In many organisms, NH_3 is toxic. To prevent the harmful excess of ammonia, it is converted to urea or uric acid before excretion.

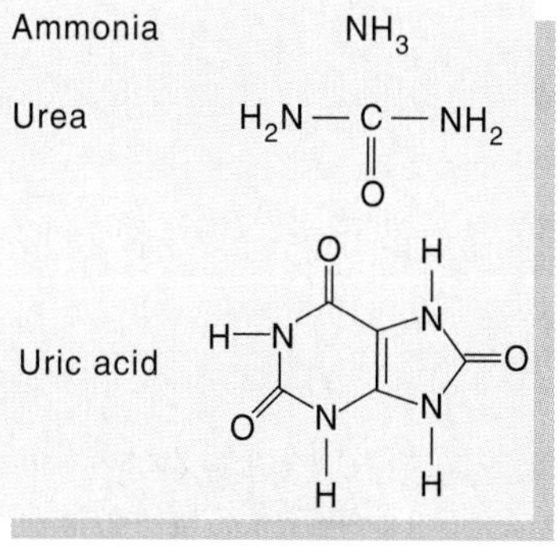

Ammonia is also the major nitrogenous end product in some of the simpler aquatic and marine animal forms, such as protozoa, nematodes, and even bony fishes, aquatic amphibia, and amphibian larvae. Such animals are called ammonotelic. But in many animals, NH_3 is toxic, and its removal by simple diffusion is difficult. Thus in terrestrial snails and amphibia, as well as in other animals living in environments in which water is limited, urea is the principal end product (fig. 25.6). Urea formation also helps to maintain osmotic balance with seawater in cartilagenous fishes. In such animals, most of the urea secreted by the kidney glomerulus is reabsorbed by the tubules. Indeed, the amount of nitrogen excreted by the kidneys of fishes is small compared with that excreted by the gills, and in most fishes, ammonia is the major form of excreted nitrogen.

Another form of "detoxified" ammonia that is used in nitrogen excretion is uric acid. Uric acid is the predominant nitrogen excretory product in birds and terrestrial reptiles (turtles excrete urea, whereas alligators excrete ammonia unless they are dehydrated, in which case they, too, excrete uric acid). Uric acid formed as a product of amino acid catabolism involves the de novo pathway of purine biosynthesis; therefore its formation from NH_3 liberated in amino acid catabolism is described elsewhere (see chapter 26). In mammals, uric acid is exclusively an intermediate in purine catabolism, and in most mammals (primates excluded) it is further converted by uricase to allantoin.

Urea Formation Is a Complex and Costly Mode of Ammonia Detoxification

Urea formation in the liver starts with the multistep conversion of ornithine to arginine (fig. 25.7). This is followed by the breakdown of arginine into ornithine and urea. The cyclic nature of this pathway was first appreciated by Hans Krebs and Paul Henseleit in 1932. In subsequent years the important details of the pathway were determined by many workers, including P. P. Cohen, S. Grisolia, and S. Ratner.

The complete urea cycle as it occurs in the mammalian liver requires five enzymes: argininosuccinate synthase, arginase, and argininosuccinate lyase (which function in the cytosol) and ornithine transcarbamoylase and carbamoyl phosphate synthase (which function in the mitochondria). Additional specific transport proteins are required for the mitochondrial uptake of L-ornithine, NH_3, and HCO_3^- and for the release of L-citrulline.

The free ammonia formed by oxidative deamination of glutamate is converted into carbamoyl phosphate in a three-step reaction requiring two ATP molecules (fig. 25.8):

$$NH_4^+ + HCO_3^- + 2\ ATP \longrightarrow \text{Carbamoyl phosphate} + HPO_4^{2-} + 2\ ADP + 2\ H^+ \quad (2)$$

First the bicarbonate ion is activated (step 1). The activated carbon is subject to nucleophilic attack by ammonia, leading to a carbamate intermediate (step 2). Finally, in a reaction similar to step 1, a second phosphoryl group is transferred to carbamate to form carbamoyl phosphate (step 3).

The carbamoyl group of carbamoyl phosphate has a high group-transfer potential, which is displayed by its transfer to the terminal amino group of ornithine to form L-citrulline (see fig. 25.7). In the process, inorganic phosphate is released. Before further reaction can occur, the citrulline must be transported across the mitochondrial membrane to the cytosol, where the remaining reactions leading to urea formation occur. Citrulline reacts with L-aspartate in an ATP-dependent reaction to form argininosuccinate, AMP, and PP_i. The PP_i is subsequently hydrolyzed to inorganic phosphate, so in effect the cost of this step is two ATP molecules. Argininosuccinate is cleaved to fumarate and L-arginine. The fumarate returns to the pool of TCA cycle intermediates, whereas the arginine becomes hydrolyzed to urea and ornithine. The ornithine is reutilized in further rounds of the urea cycle. Urea diffuses into the bloodstream and is ultimately eliminated through the kidneys in the urine. The stoichiometry for the urea cycle is

$$CO_2 + NH_4^+ + 3\ ATP + \text{Aspartate} + 2\ H_2O \longrightarrow \text{Urea} + 2\ ADP + 2\ P_i + AMP + PP_i + \text{Fumarate} + 6\ H^+ \quad (3)$$

In each turning of the urea cycle, two nitrogens are eliminated, one originating from the oxidative deamination of glutamate and the other coming from the α-amino group of aspartate. Because the PP_i produced in the urea cycle is subsequently hydrolyzed, it takes the equivalent of four high-energy phosphates to form a single molecule of urea. Thus the cost of this form of detoxification of ammonia is quite high.

The Urea Cycle and the TCA Cycle Are Linked by the Krebs Bicycle

The fumarate released in the urea cycle links the urea cycle with the TCA cycle. This fumarate is hydrated to malate, which is oxidized to oxaloacetate. The carbons of oxaloacetate can stay in the TCA cycle by condensation with acetyl-CoA to form citrate, or they can leave the TCA cycle either by gluconeogenesis to form glucose or by transamination to form aspartate as shown in figure 25.9. Because Krebs was involved in the discoveries of both the urea cycle and the TCA cycle, the interaction between the two cycles shown in figure 25.9 is sometimes referred to as the Krebs bicycle.

More Than One Carrier Exists for Transporting Ammonia from the Muscle to the Liver

The urea cycle is a unique function of the liver. Excess ammonia formed in other tissues must be carried in a nontoxic form to the liver. In many tissues, glutamine serves as the carrier of excess nitrogen. The glutamine is formed in the tissues in a reaction, catalyzed by glutamine synthase, that combines NH_3 with glutamate:

$$ATP + NH_4^+ + \text{Glutamate} \xrightleftharpoons{\text{Glutamine synthase}} ADP + P_i + \text{Glutamine} + H^+ \quad (4)$$

This reaction involves activation of the γ-carboxyl group of glutamate to yield a γ-glutamyl enzyme complex, together with the cleavage of ATP to ADP and P_i. In a second step the γ-glutamyl group is transferred to NH_4^+.

After the glutamine reaches the liver, the enzyme glutaminase releases the ammonia from the glutamine by the reaction

$$\text{Glutamine} + H_2O \longrightarrow \text{Glutamate} + NH_4^+ \quad (5)$$

Ammonia is also transported from skeletal muscle to the liver in the form of the amino acid alanine (fig. 25.10). This alanine is formed in the muscle tissue by a transamination reaction between pyruvate and glutamate. Then the alanine is transported by the bloodstream to the liver, where it reacts with α-ketoglutarate to reform pyruvate and glutamate. This reaction is catalyzed by alanine transaminase. The nitrogen originating from the gluta-

Figure 25.7

The urea cycle is a mechanism for removing unwanted nitrogen. Sources of nitrogens involved in urea formation are shown in red. Five enzymes are used in the urea cycle. Three of these function in the cytosol, and two, as shown, function in the mitochondrial matrix. Specific carriers in the inner mitochondrial membrane transport ornithine, citrulline, ammonium ion, and HCO_3^- (CO_2) into and out of the mitochondrial matrix.

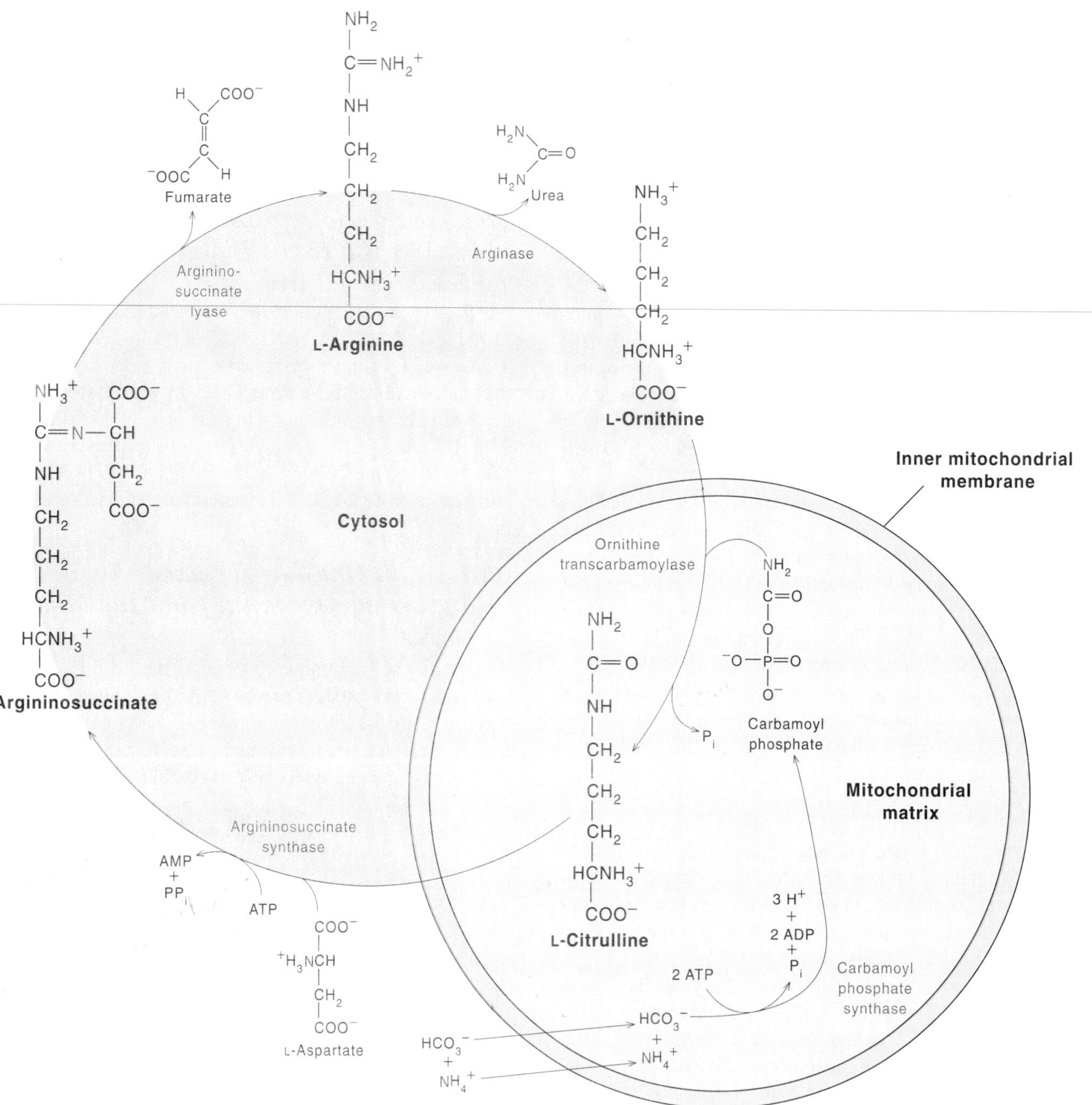

mate is processed by the urea cycle. When the blood glucose concentration is low, the pyruvate resulting from alanine transamination is used to make glucose via the gluconeogenesis pathway. The glucose can be returned to the skeletal muscle to supply immediately available energy to keep the muscle going. Thus the transport of alanine from muscle to liver results in a reciprocal transfer of glucose to muscle. The entire cyclical process is referred to as the glucose–alanine cycle (see fig. 25.10).

Figure 25.8

The mechanism of formation of carbamoyl phosphate. The reaction involves three steps, all of which take place on the same enzyme, carbamoyl phosphate synthase.

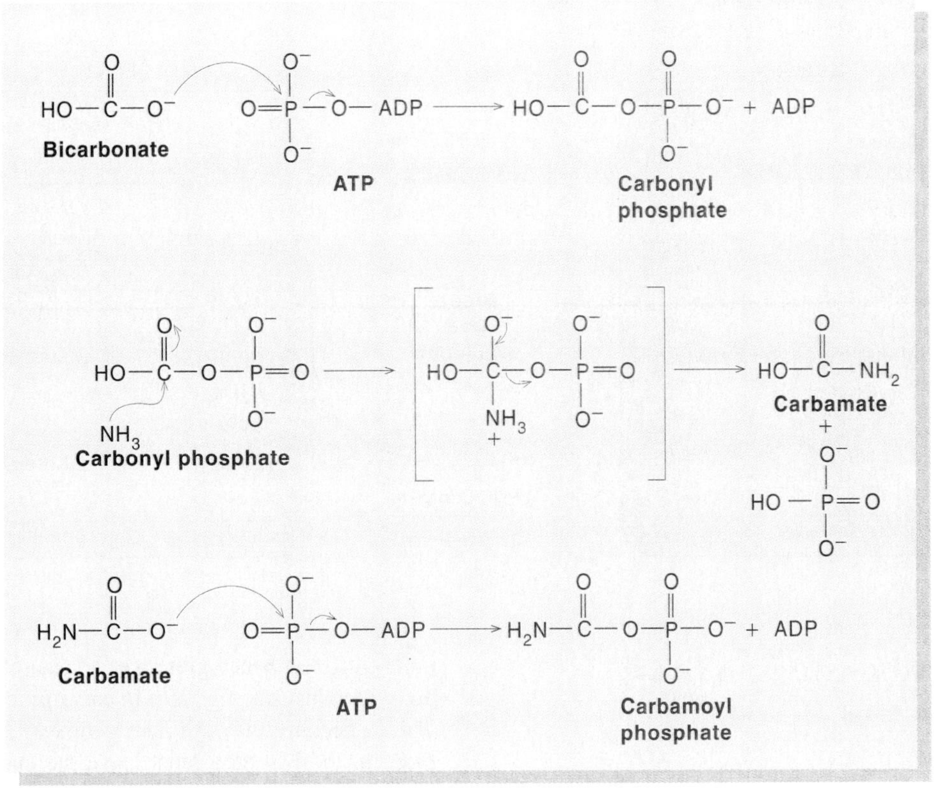

$$HO{-}C(=O){-}O^- \;+\; O{=}P(O^-)_2{-}O{-}ADP \longrightarrow HO{-}C(=O){-}O{-}P(=O)(O^-){-}O^- + ADP$$

Bicarbonate + **ATP** → **Carbonyl phosphate**

$$HO{-}C(=O){-}O{-}P(O^-)_2{=}O \;(+\;NH_3) \longrightarrow \left[HO{-}C(O^-)(\overset{+}{N}H_3){-}O{-}P(O^-)_2{=}O\right] \longrightarrow HO{-}C(=O){-}NH_2 + HO{-}P(O^-)_2{=}O$$

Carbonyl phosphate → **Carbamate** +

$$H_2N{-}C(=O){-}O^- \;+\; O{=}P(O^-)_2{-}O{-}ADP \longrightarrow H_2N{-}C(=O){-}O{-}P(=O)(O^-){-}O^- + ADP$$

Carbamate + **ATP** → **Carbamoyl phosphate**

Figure 25.9

The "Krebs bicycle" involves interaction between components of the TCA cycle (on the left) and the urea cycle (on the right). This interaction explains the origin of the amino group contributed by aspartate to urea formation. The amino group originates from a transamination reaction involving oxaloacetate. The resulting aspartate is deaminated to fumarate, which can be recycled to oxaloacetate.

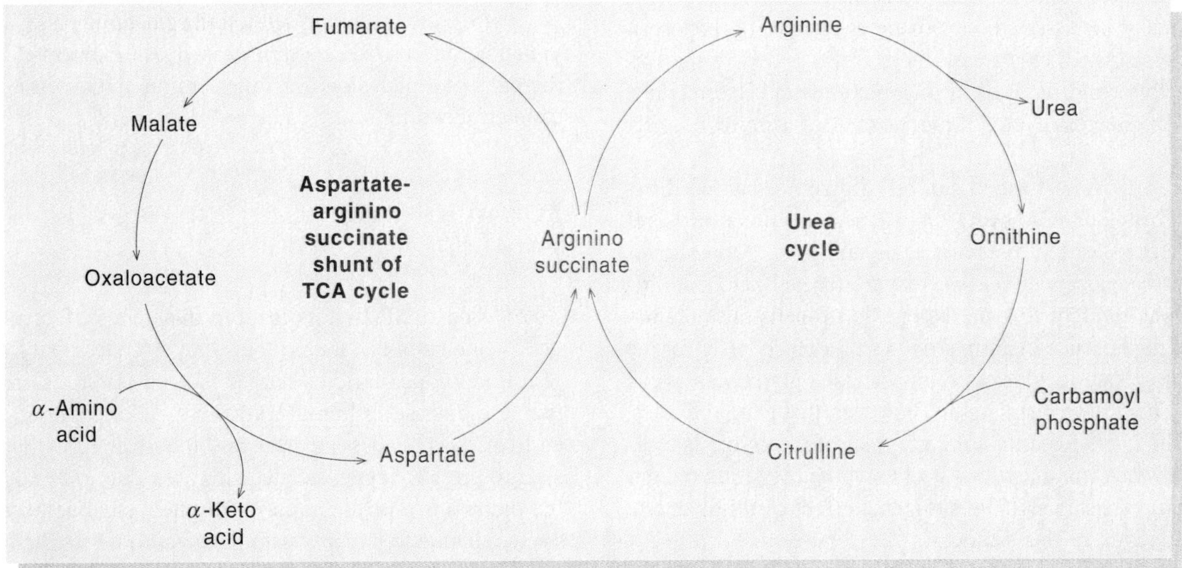

Figure 25.10

The glucose–alanine cycle. Active muscle functions anaerobically and synthesizes alanine by a transamination reaction between glutamate and pyruvate. The alanine is transported to the liver, where the pyruvate is regenerated and converted to glucose by gluconeogenesis. The glucose then is transported back to the muscle tissue, where it is used for energy production in glycolysis.

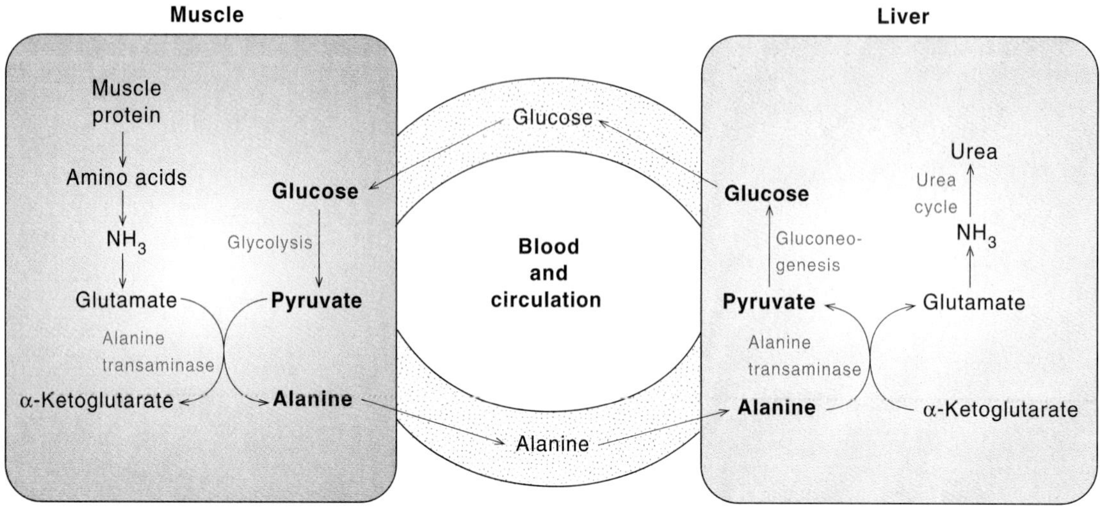

Amino Acid Catabolism Can Serve as a Major Source of Carbon Skeletons and Energy

Thus far we have considered the deamination of amino acids and the fate of the resulting ammonium ion. The carbon skeleton remaining after deamination can be used in various biosynthetic pathways, or it can be degraded by reactions coupled to the production of energy.

Catabolism of amino acids usually entails their conversion to intermediates in the central metabolic pathways. All amino acids can be degraded to carbon dioxide and water by appropriate enzyme systems. In every case the pathways involve the formation, directly or indirectly, of a dicarboxylic acid intermediate of the tricarboxylic acid cycle, of pyruvate, or of acetyl-CoA (fig. 25.11).

Acetyl-CoA so formed can be oxidized to carbon dioxide by means of the TCA cycle or, when cycle function is restricted, can be converted to acetoacetate and lipid. Amino acids metabolized to acetoacetate and acetate are termed ketogenic. At one time it was thought that the ketogenic property was readily explained by the absence in animal tissues of a mechanism for net conversion of acetate residues into glucose (see glyoxylate cycle, chapter 15). There have been claims, however, that animal tissues, particularly the liver, do contain the enzymes of the glyoxylate cycle, although we do not know the level to which they function under different circumstances. The ketogenic effect could be due to the limited function of this pathway.

In contrast to amino acids that lead to C-2 carbon units such as acetate, amino acids that give rise to C-4 or C-5 carbon units such as α-ketoglutarate or any of the four-carbon dicarboxylic acids (see fig. 25.11) can stimulate TCA cycle function. This is because they are intermediates in the cycle. For their further metabolism they must leave the cycle by one of two routes (see fig. 15.1). By one route the conversion of oxaloacetate to phosphoenolpyruvate results in gluconeogenesis when carbohydrate utilization is restricted. For this reason such amino acids are considered glycogenic. By the other route, pyruvate is formed and, after conversion of the latter to acetyl-CoA, can be oxidized completely to carbon dioxide and water, provided that there is ample TCA cycle function.

In the following section the catabolic pathways for phenylalanine and tyrosine are discussed. An extensive discussion of the catabolic pathways for other amino acids is presented in the chapter appendix.

For Many Genetic Diseases the Defect Is in Amino Acid Catabolism

One of the first indications that genes affect phenotypes by the nature of the proteins they encode came from the work of a London pediatrician named Garrod (1908). Garrod was analyzing a disease in humans known as alkaptonuria, in which the cartilaginous tissues are dark and the urine turns black on exposure to air. He suggested that this was due to an abnormality in the metabolism of the amino acid phenylalanine because feeding phenylalanine to patients with this syndrome resulted in increased secretions of homogentisic acid; it is the homogentisic acid that turns black on air oxidation. Long before the full explanation for

Figure 25.11

Pathways for the degradation of the 20 amino acids (highlighted in colored boxes) found in proteins. The strategy followed for amino acid degradation is similar in gross respects (except for direction) to the strategy for amino acid biosynthesis. Thus the α-amino groups are usually removed at an early stage in degradation, and the carbon skeletons filter into the central metabolic pathways. However, enormous differences exist in the specific pathways used in the two processes. Only a handful of amino acids involve similar sequences in both directions, and most of these involve highly reversible reactions, indicated by double arrows. The differences can be seen by comparing figures 25.11 and 24.1. In degradation pathways the carbon skeletons are funneled almost exclusively to intermediates in the TCA cycle. The dashed arrows in this figure associated with tyrosine and isoleucine reflect the fact that the carbon skeletons of these amino acids are split into two components, which are processed separately.

Amino acid degradation serves three purposes: (1) supplying energy, (2) supplying intermediates for the synthesis of other compounds, and (3) removing harmful excesses of certain amino acids. The number of steps in each pathway is indicated by the number alongside the conversion arrow.

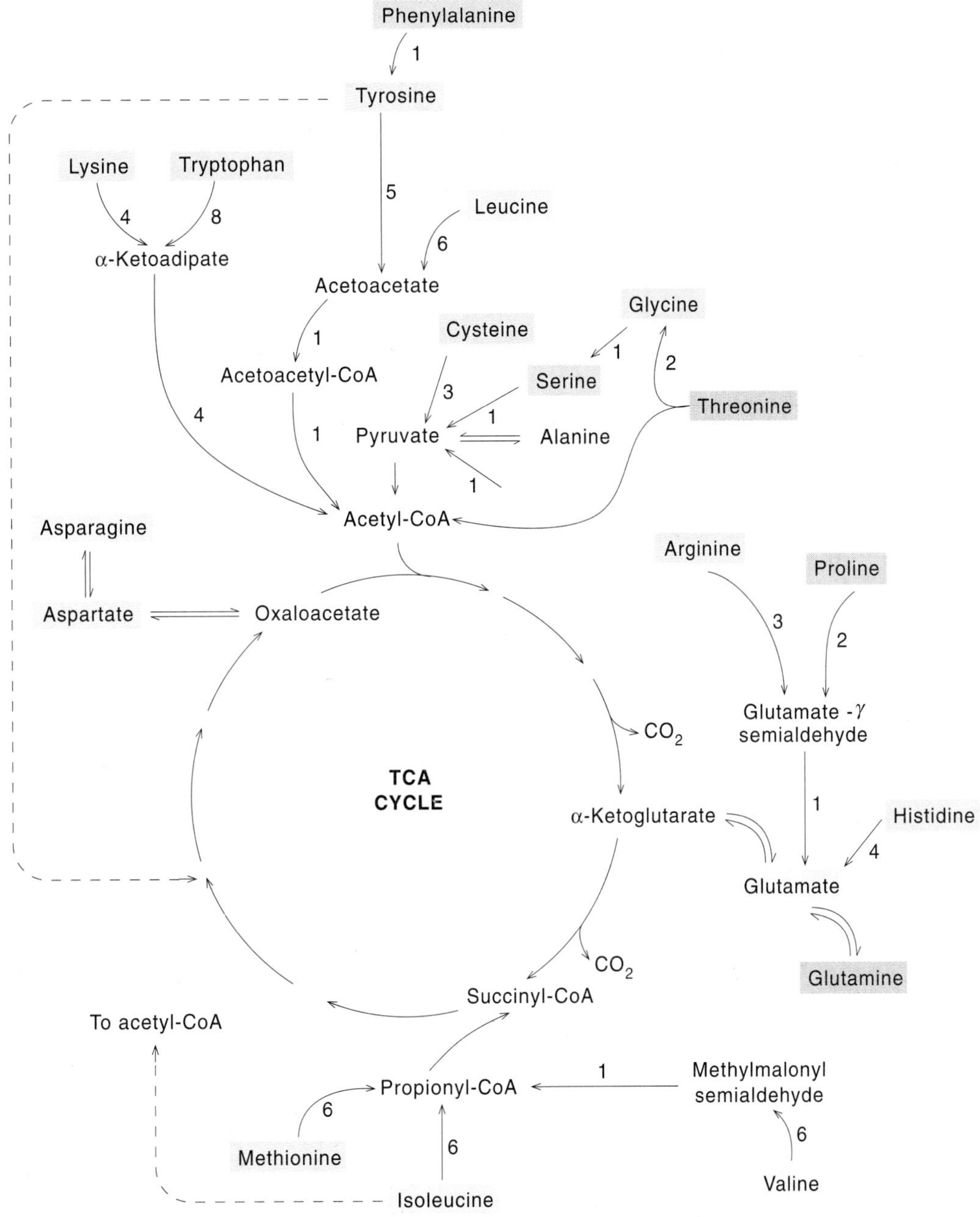

alkaptonuria was known, it had been found that it followed a recessive pattern of inheritance, and Garrod proposed that the condition was due to a defective enzyme. It was not until 1958 that the full biochemistry leading to this problem was appreciated. Humans have developed a multistep metabolic pathway for disposing of the excess phenylalanine (or tyrosine), and one of the intermediates in this pathway is homogentisate (fig. 25.12). Alkaptonuria is due to a deficiency of the homogentisic acid oxidase

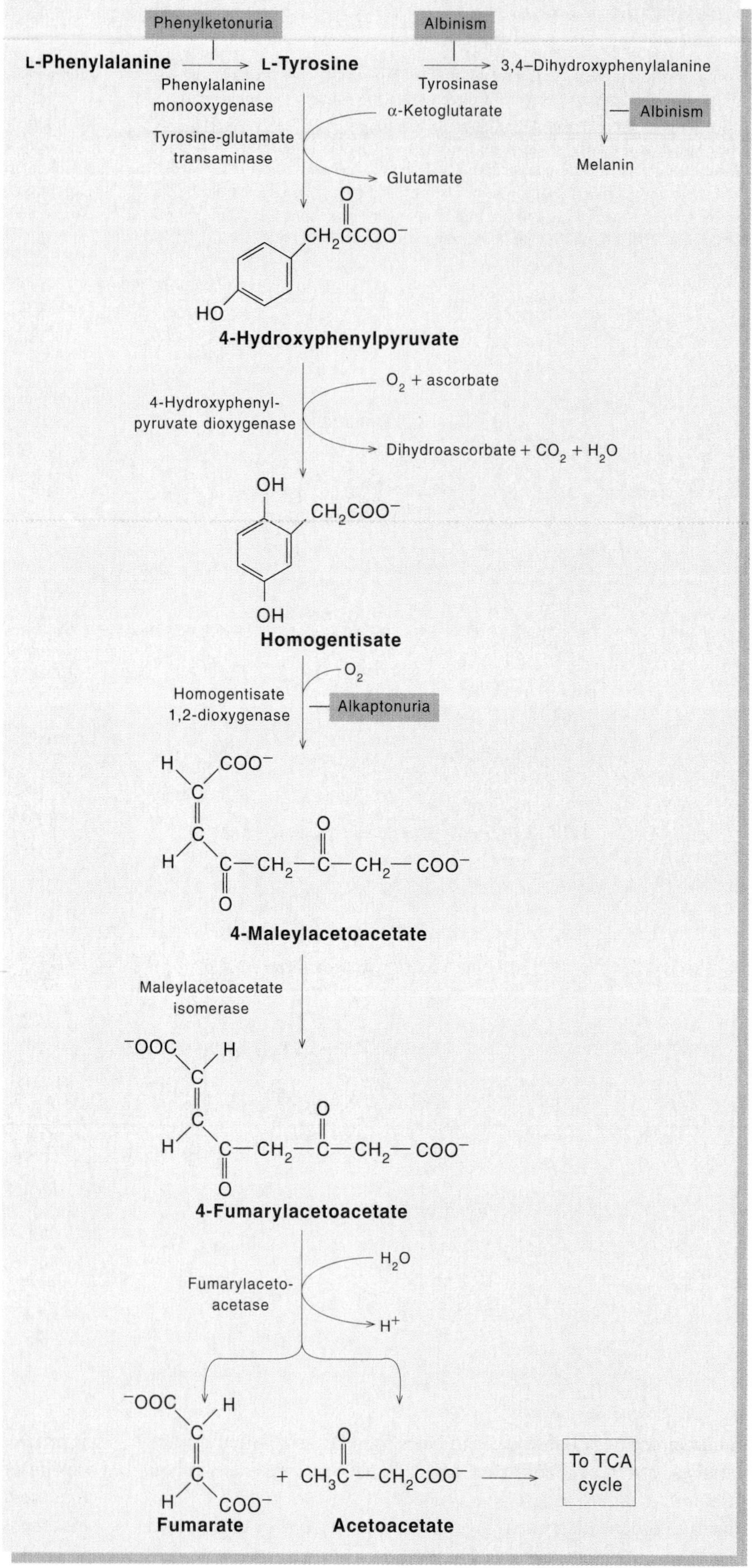

Figure 25.12

The conversion of phenylalanine and tyrosine to fumarate and acetoacetate. Blue boxes indicate a diseased state that results from a defective enzyme.

enzyme. When this enzyme is not present in sufficient amounts, homogentisate accumulates, resulting in the darkening of the urine and other problems associated with this condition.

Garrod was too far ahead of his time to be fully appreciated when he made his great discovery; his incredible insight was noted by George Beadle, who shared the 1958 Nobel Prize with E. L. Tatum in Physiology and Medicine for their contributions to biochemical genetics. Beadle pointed out that the one-gene–one-enzyme hypothesis was implicit in Garrod's work and was actually formulated by Garrod in almost the same terms.

Most Human Genetic Diseases Associated with Amino Acid Metabolism Are Due to Defects in Their Catabolism

Since Garrod's work, researchers have described many metabolic diseases that are due to the inability of the affected individual to dispose of specific dietary components. The diseases may be difficult to treat in the cases of errors in amino acid catabolism, because the culprit amino acid is one of the normal constituents of protein and is required for growth and development as well as for replacement of those body proteins that undergo rapid turnover. Since the affected fetus is usually carried by a mother who is heterozygous for the deficiency (carrying one normal and one defective gene) and whose own metabolism is essentially normal, the development of the fetus is essentially normal. Thus management of these diseases is possible, but it is dependent on prenatal diagnosis or diagnosis soon after birth. Treatment consists of a low-protein diet, carefully selected to supply enough of the culprit amino acid for protein formation but not enough to allow high plasma levels of it or of the offending metabolites. Supplements of nonoffending amino acids, prepared by synthesis or by fermentation processes, could be employed to compensate in part for the low-protein diet. Indeed, the chemical industries have made such preparations available.

Some of the diseases of amino acid catabolism are listed in table 25.3. These naturally occurring defects have been invaluable in demonstrating the obligatory nature of some of the steps in amino acid breakdown. A mutation that causes a defective enzyme usually leads to (1) a substantial accumulation of the intermediate that is a substrate for that enzyme and (2) a drastic lowering of all the intermediates below that step in the pathway.

The fact that these genetic diseases are due to single (recessive) gene mutations gives hope on two fronts. It should soon be possible to identify carriers of any of these traits so that, with the aid of genetic counseling, it will be possible for parents to avoid giving birth to homozygous children that carry two defective genes. For those rare individuals affected, it may one day be possible to provide, through transplant, cells capable of metabolizing the offending amino acids. It has already been demonstrated, for example, that fibroblasts from patients with human "maple syrup" urine disease (see table 25.3) can be transfected with a cDNA for the missing component of branched-chain α-keto acid dehydrogenase to yield cells in which the missing activity is restored.

Amino Acids Serve as the Precursors for Compounds Other Than Proteins

As stated at the outset of this chapter, the primary fate of amino acids is their incorporation into protein. However, amino acids also serve as precursors for a number of other important molecules. These include processes leading to the formation of the porphyrin nucleus found in many oxygen- and electron-carrying proteins, of biologically active amines, and of glutathione. These two topics are examined below.

Porphyrin Biosynthesis Starts with the Condensation of Glycine and Succinyl-CoA

Early isotope tracer experiments by David Shemin permitted the elucidation of the formation of the immediate precursor of the porphyrin needed for the cytochromes and for hemoglobin. These studies indicated that the glycine methylene carbon and nitrogen were incorporated along with both carbons of acetate. Subsequent enzymatic studies in both bacteria and animals revealed a condensation reaction between succinyl-CoA and glycine to yield δ-aminolevulinate and CO_2 (presumably by way of an enzyme-bound β-keto acid, α-amino-β-ketoadipate) (fig. 25.13).

δ-Aminolevulinate is also the precursor to porphobilinogen in plants, blue-green algae, and most eubacteria, but it is not formed by a condensation of glycine and succinyl-CoA. Rather, it is formed from glutamate by reduction of the α-carboxyl group to yield α-glutamyl semialdehyde. As has been emphasized in describing reduction of other carboxyl groups, an "activation" of the carboxyl group is required. The reaction in the δ-aminolevulinate pathway is unique in that the glutamate is transferred to a tRNA acceptor. Hence the reaction requires the expenditure of two high-energy phosphate bonds. The glutamyl tRNA is the substrate for a specific reductase. The reduced product, α-glutamyl semialdehyde, undergoes an unusual intramolecular transamination reaction in which the amino group on C-2 of the semialdehyde is transferred to the C-1 position (which becomes C-5 in δ-aminolevulinate).

The pyrrole monomer porphobilinogen arises from the condensation of two molecules of δ-aminolevulinate with the ions of two water molecules. This reaction is catalyzed by δ-aminolevulinate dehydrase. Condensation of four porphobilinogen molecules yields the branchpoint compound in tetrapyrrole synthesis, uroporphyrinogen III. This is a complex reaction requiring two enzymes: uroporphyrinogen I synthase, which catalyzes a head-to-tail condensation of four porphobilinogen molecules (fig. 25.14), and uroporphyrinogen cosynthase, which inverts one of the units and closes the ring.

Derivatives of porphyrins are coenzymes in a number of oxidation–reduction reactions (see chapters 16 and 17).

Glutathione Is a Multipurpose Reducing Agent

The tripeptide γ-glutamylcysteinylglycine, or glutathione, is found in nearly all cells and plays a variety of roles. It is formed in two steps, each requiring ATP. In the first step, glutamate condenses with a cysteine. The γ- rather than the α-carboxyl of glutamate makes a peptide bond with the α-amino group of cysteine:

$$\text{Glutamate} + \text{Cysteine} + \text{ATP} \xrightarrow[\text{synthase}]{\gamma\text{-Glutamylcysteine}} \gamma\text{-Glutamylcysteine} + \text{ADP} + \text{P}_i \quad \textbf{(6)}$$

In the second step the condensation of the dipeptide with glycine, a normal peptide linkage is made:

$$\gamma\text{-Glutamylcysteine} + \text{Glycine} + \text{ATP} \xrightarrow[\text{synthase}]{\text{Glutathione}} \text{Glutathione} + \text{ADP} + \text{P}_i \quad \textbf{(7)}$$

Figure 25.13

Tetrapyrrole biosynthesis. The sequence by which four porphobilinogen residues are converted to uroporphyrinogen III is the sequential head-to-tail condensation of the four residues by uroporphyrinogen I synthase (porphobilinogen deaminase) to yield the unrearranged hydroxymethylbilane. This unstable intermediate is rearranged and cyclized by uroporphyrinogen III cosynthase to yield uroporphyrinogen III. Hydroxymethylbilane in the absence of cosynthase spontaneously cyclizes to yield the unrearranged urobilinogen I, which is not an intermediate in the pathway (hence the name uroporphyrinogen I synthase for the deaminase).

Table 25.3
Some Inborn Errors of Amino Acid Metabolism in Humans

Amino Acid Catabolic Pathway Involved	Condition	Distinctive Clinical Manifestation	Enzymatic Block or Deficiency
Arginine and the urea cycle	Argininemia and hyperammonemia	Mental retardation	Arginase
	Hyperammonemia	Neonatal death, lethargy, convulsions	Carbamoyl phosphate synthase
	Ornithinemia	Mental retardation	Ornithine decarboxylase
Glycine	Hyperglycinemia	Severe mental retardation	Glycine-cleavage system
Histidine	Histidinemia	Speech defects, mental retardation in some; in others, none	Histidase
Isoleucine, leucine, and valine	Branched-chain ketoaciduria ("maple syrup" urine disease)	Neonatal vomiting, convulsions, and death; mental retardation in survivors	Branched-chain keto acid dehydrogenase complex
Isoleucine, methionine, threonine, and valine	Methylmalonic acidemia	Similar to the preceding except that methylmalonate accumulates	Methylmalonyl-CoA mutase (some patients respond to vitamin B_{12} therapy)
Leucine	Isovaleric acidemia	Neonatal vomiting, acidosis, lethargy, and coma; survivors mentally retarded	Isovaleryl-CoA dehydrogenase
Lysine	Hyperlysinemia	Mental retardation and some noncentral nervous system abnormalities	Lysine-ketoglutarate reductase
Methionine	Homocystinuria	Mental retardation common: several eye diseases and thromboembolism common; osteoporosis and faulty bone structures	Cystathionine-β-synthase
Phenylalanine	Phenylketonuria and hyperphenylalaninemia	Vomiting is an early neonatal symptom, but mental retardation and other neurologic disorders develop in the absence of dietary treatment	Phenylalanine-L-monoxygenase
Proline	Hyperprolinemia, type I	Probably not etiologically associated with any disease; proline excreted	Proline oxidase
Tyrosine	Alkaptonuria	Homogentisic acid in urine darkens on standing; in adult years, pigment deposits cause darkening of skin, cartilage; arthritis develops	Homogentisic acid oxidase
	Albinism	The most common type, oculocutaneous albinism, results in white hair, pink skin, and an extreme photophobia owing to lack of pigment in the eye	Tyrosinase of the melanocyte is absent

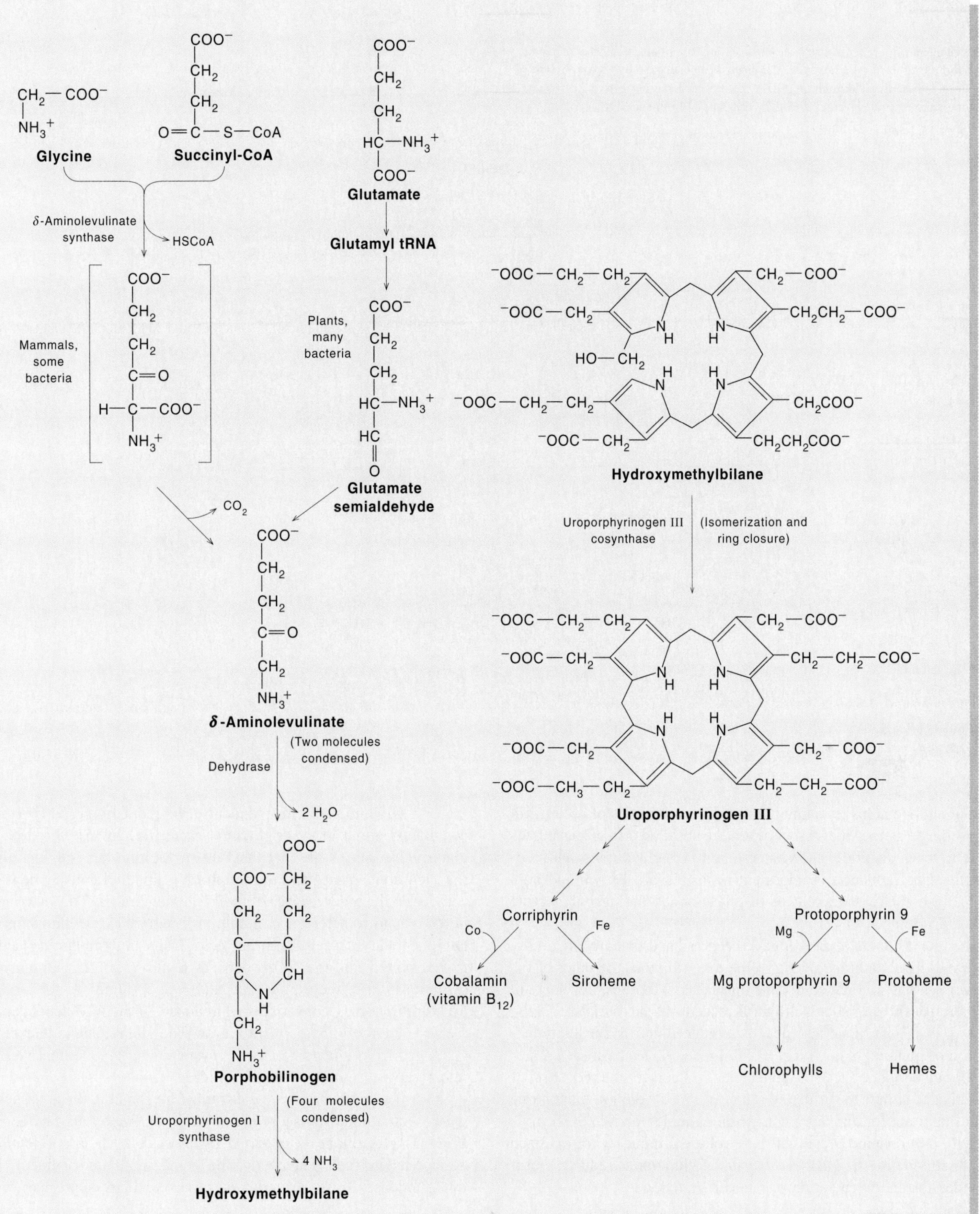

Glycine
Succinyl-CoA
Glutamate
δ-Aminolevulinate synthase
HSCoA
Glutamyl tRNA
Mammals, some bacteria
Plants, many bacteria
Glutamate semialdehyde
CO_2
δ-Aminolevulinate
Dehydrase
(Two molecules condensed)
2 H_2O
Porphobilinogen
Uroporphyrinogen I synthase
(Four molecules condensed)
4 NH_3
Hydroxymethylbilane
Hydroxymethylbilane
Uroporphyrinogen III cosynthase
(Isomerization and ring closure)
Uroporphyrinogen III
Corriphyrin
Co
Fe
Cobalamin (vitamin B_{12})
Siroheme
Protoporphyrin 9
Mg
Fe
Mg protoporphyrin 9
Protoheme
Chlorophylls
Hemes

Figure 25.14

Mechanism of polymerization in a linear tetrapyrrole. Four molecules of porphobilinogen undergo a head-to-tail condensation catalyzed by uroporphyrinogen I synthase to yield a tetrapyrrole. Asterisks indicate nitrogen and carbon atoms derived from glycine; the others are derived from succinyl-CoA.

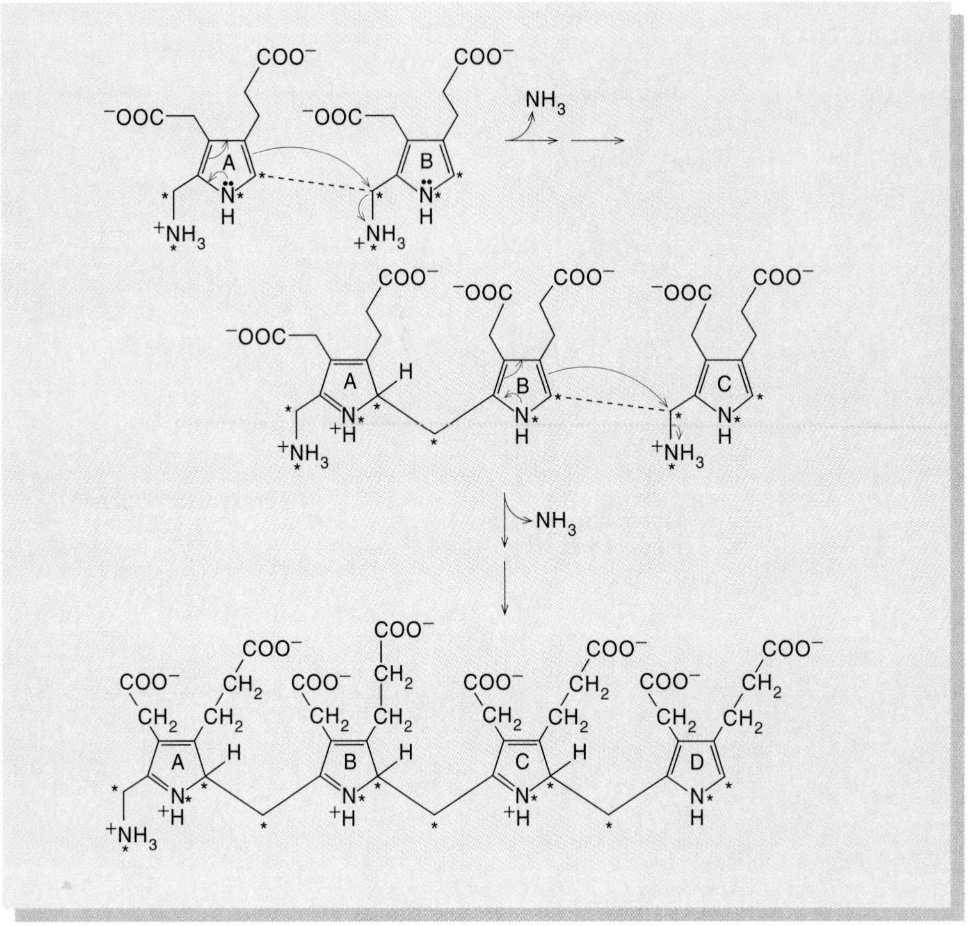

Glutathione helps to maintain the sulfhydryl groups of proteins in a reduced state. An enzyme, protein-disulfide reductase, catalyzes sulfhydryl disulfide interchanges between glutathione and proteins. The reductase is important in insulin breakdown and may catalyze the reassortment of disulfide bonds during polypeptide chain folding.

Mutants of *E. coli* have been isolated that are essentially devoid of glutathione owing to the loss of one or the other of the two synthases. Such cells have normal growth rates, but they are more sensitive to such sulfhydryl reagents as mercurials.

Glutathione also acts as a reduced carrier for the reduction of glutaredoxin, which, like thioredoxin, is a hydrogen donor for nucleotide reductase (see chapter 26), and for the reduction of activated sulfate to sulfite (see chapter 25). In addition glutathione is important for maintaining the iron of hemoglobin in the ferrous state (see chapter 14). In all these roles the reduction of oxidized glutathione by the NADPH-dependent glutathione reductase provides for the regeneration of reduced glutathione.

Glutathione is independent of its reducing property as a γ-glutamyl donor in the γ-glutamyl cycle (fig. 25.15). Of particular significance is the fact that cells exhibiting the activities of the cycle contain substantial amounts of γ-glutamyl transpeptidase activity on the outer surface of their cell membranes. This enzyme is thought to transfer the γ-glutamyl group of extracellular glutathione to an extracellular amino acid. The γ-glutamylamino acid is transported into the cell, where, as a substrate for γ-glutamyl cyclotransferase, the amino acid and 5-oxoproline are released. 5-Oxoprolinase is converted to glutamate in an ATP-dependent cleavage catalyzed by 5-oxoprolinase. The glutathione is then regenerated from glutamate and the cysteine and glycine that were released by cysteinylglycine dipeptidase.

The γ-glutamyl cycle enzymes are found in those tissues where glutathione transport into cells is an important function. Whereas glutathione is exported by most cells, it is efficiently transported into cells that contain the membrane-bound γ-glutamyl transpeptidase. Thus the γ-glutamyl transpeptidase appears to fa-

Figure 25.15

The γ-glutamyl cycle proposed by Alton Meister. The cycle involves enzymes forming glutathione, the excretion of glutathione (1), and, in cells that import glutathione, the γ-glutamyl transpeptidase–dependent transport (2) of another amino acid (or peptide), the cleavage of the intracellular γ-glutamyl amino acid, and the ATP-dependent conversion of 5-oxoproline to glutamate. The cysteinylglycine (Cys-Gly) formed in the reaction is transported by an uncharacterized transport system (3) and cleaved by an intracellular protease (4) or cleaved by a membrane-bound protease and the free amino acids transported. The γ-glutamyl cycle enzymes are found in those tissues for which the transport of glutathione into cells is an important function. This includes liver and kidney cells in animals.

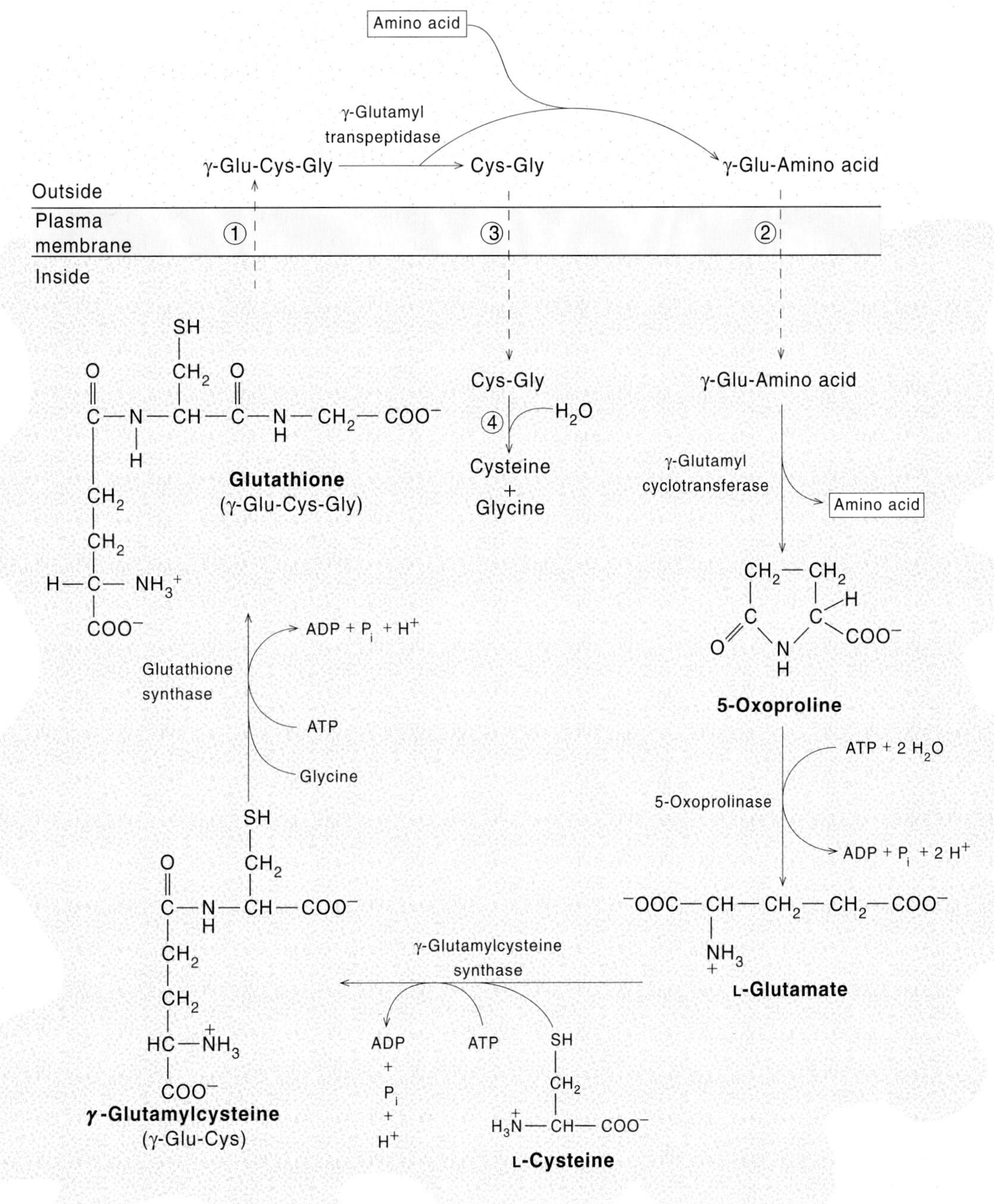

cilitate salvage of glutathione secreted by some tissues into the bloodstream, as well as to permit an energy-driven transport system for amino acids. Either oxidized or reduced glutathione is a substrate for the transpeptidase. The enzyme also catalyzes the hydrolysis of glutathione to glutamate and cysteinylglycine and glutamine to glutamate and ammonia. Many amino acids can serve as acceptors for the γ-glutamyl group, including γ-glutamyl amino acids (to yield γ-glutamyl-γ-glutamyl amino acids) and even glutathione itself. Such a transport across cell membranes is probably especially important for cysteine and methionine, as well as for glutathione.

Summary

1. Only eight of the de novo pathways for amino acid biosynthesis can be found in humans. These amino acids are all related by a small number of steps to glycolytic or TCA cycle intermediates. A number of additional amino acids can be formed from these amino acids. Essential amino acids are those that must be supplied in the diet.
2. For most amino acids the α-amino group is removed at an early stage in catabolism, usually in the first step. Transaminases are specific for different amino acids. Frequently, α-ketoglutarate is the acceptor for the amino group, in which case it is converted into glutamate. The α-ketoglutarate can be regenerated from the glutamate by oxidative deamination.
3. A great deal of excess NH_3 frequently results from amino acid catabolism. This excess ammonia must be eliminated. In bacteria and lower eukaryotes the ammonia can usually be removed by simple diffusion, but in higher eukaryotes this is not feasible. Since the ammonia is frequently quite toxic, it is detoxified before removal by conversion to urea or uric acid. An intricate pathway resulting in the conversion of ammonia into urea involves five enzymes: three located in the cytoplasm and the remaining two in the mitochondrial matrix.
4. All amino acids can be degraded to CO_2 and water via the TCA cycle by the appropriate enzymes; the pathways often contain branchpoints to useful biosynthetic products. In every case the pathways involve the formation of a dicarboxylic acid intermediate of the TCA cycle, of pyruvate, or of acetyl-CoA.
5. The discussion of amino acid catabolism is organized according to the common intermediates formed during degradation. Alanine, glycine, threonine, serine, and cysteine are degraded to acetyl-CoA by way of pyruvate. Threonine is also degraded to acetyl-CoA via pyruvate, but it also yields acetyl-CoA directly. Phenylalanine, tyrosine, tryptophan, lysine, and leucine also lead to acetyl-CoA, but they go by way of acetoacetyl-CoA rather than pyruvate. Arginine, histidine, proline, glutamic acid, and glutamine are all degraded to α-ketoglutarate. Catabolism of methionine, valine, and isoleucine leads to succinyl-CoA. Aspartate and asparagine are converted to oxaloacetate on degradation.
6. The importance of catabolic pathways is underscored by a broad spectrum of human metabolic diseases in each of which one enzyme for normal amino acid catabolism is either missing or defective.
7. Many biologically important routes of amino acid utilization, other than those leading to incorporation into proteins, are known. Some of these routes are distinctly anabolic pathways in which the amino acids serve as an initial substrate in an independent biosynthetic pathway. Other simple pathways involve the conversion of one amino acid to another, such as the formation of tyrosine from phenylalanine. The utilization of glycine in the formation of porphyrin derivatives occurs by very complex highly branched pathways. Some other biologically important pathways lead to the biosynthesis of small peptides, as in the biosynthesis of glutathione.

Selected Readings

Battersby, A. R., C. J. R. Fookes, G. W. J. Matcham, and E. McDonald, Biosynthesis of the pigments of life: Formation of the macrocycle. *Nature* 285:17, 1980. This paper discusses the steps in tetrapyrrole biosynthesis and the pathways diverting this nucleus to chlorophylls, hemes, cytochromes, and other macrocyclic pigments.

Bender, D. A., *Amino Acid Metabolism*. New York: John Wiley and Sons, 1985.

Fowden, L., P. J. Lea, and E. A. Bell, The nonprotein amino acids of plants. *Adv. Enzymol.* 50:117, 1979. A discussion of the occurrence and biosynthesis of naturally occurring amino acid analogs in plants.

Katz, E., and A. L. Demain, The peptide antibiotics of *Bacillus:* Chemistry, biogenesis and possible functions. *Bacteriol. Rev.* 41:449, 1977. A description of several peptide antibiotics showing the distribution of D-amino acid in these compounds.

Kishore, G. M., and D. M. Shah, Amino acid biosynthesis inhibitors as herbicides. *Ann. Rev. Biochem.* 57:627–663, 1988. The focus is on the biosynthesis of essential amino acids.

Meister, A., *Biochemistry of the Amino Acids,* vols. 1 and 2. New York: Academic Press, 1965. The two-volume classic provides a thorough discussion of amino acid literature, occurrence, properties, and metabolism of amino acids up to that time.

Torchinsky, Y. M., Transamination, its discovery, biological and chemical aspects (1937–1987). *Trends Biochem. Sci.* 12:115–117, 1987.

Yamada, K., S. Kinoshita, T. Tsunoda, and K. Aida (eds.), *The Microbial Production of Amino Acids*. New York: John Wiley and Sons, 1972. A collection of essays describing microbial processes used in Japanese industry for the production of amino acids. Includes examples in which the regulatory mechanisms functioning in most cells have been modified or bypassed.

Also see readings at the end of chapter 24.

Problems

1. How does a negative nitrogen balance, when a vertebrate is supplied a diet of 19 amino acids, prove that a particular amino acid is essential?
2. Superficially examine the amino acid biosynthetic pathways in chapter 24 for the essential versus the nonessential amino acids (table 25.1). What is the most obvious statement that you can make? How might you explain this observation considering the statements given on page 598, regarding numbers of genes.
3. Why isn't there an oxidative deaminase specific for each amino acid?
4. In describing the mode of synthesis of δ-aminolevulinate in plants and many bacteria, the statement "an unusual intramolecular transamination reaction (page 609) is made. What is unusual about this transamination and what intermediate might you expect in this reaction?
5. In the description of porphyrin biosynthesis the phrase "head-to-tail condensation" is used (page 609). What is meant by "head-to-tail"?
6. Propose a mechanism for 5-oxoprolinase (fig. 25.15). Why did nature elect to use an ATP for this reaction? Why not just a simple hydrolysis reaction?
7. Examine the reactions promoted by γ-glutamylcysteine synthase and glutathione synthase (fig. 25.15). What reaction intermediate would you expect these reactions to utilize?
8. The pathway for the biosynthesis of arginine (fig. 25.2) has several intermediates that have an *N*-acetyl group. Compare these intermediates with those in the salvage pathway (see fig. 25.2) and propose a reason for the acetyl groups.
9. In the conversion of the *N*-acetyl-γ-glutamyl phosphate to *N*-acetylglutamic-γ-semialdehyde (see fig. 25.2), two processes occur: an elimination of a phosphate and a reduction. Which step occurs first? Also, can you propose a reason for the use of the phosphate group in the first place?
10. Often ammonia is portrayed as "feeding into" the urea cycle; for example, see figure 25.10. Do ammonia molecules "feed into" the urea cycle?
11. What effect does deprivation of dietary pyridoxal phosphate have on the capacity to metabolize amino acids?
12. The de novo biosynthetic pathway for the biosynthesis of arginine has been lost by higher animals. However, the biosynthesis of arginine via a salvage pathway from proline can occur. What arguments can you make that it is unlikely that mammals would completely lose all capabilities to produce arginine?
13. Reduce the names "ketogenic" and "glucogenic" to the simplest possible terms.
14. Use figure 25.11 to deduce which amino acids are glucogenic, which are ketogenic, and which are both.
15. Many of the inborn errors in amino acid metabolism appear to result in mental retardation (see table 25.3). The majority of these individuals appear normal at birth, but their mental capabilities fail to develop. Can you provide an explanation for this observation?
16. Why have many inborn errors been found in the metabolism of amino acids in humans but none in glycolysis, the citric acid cycle, or electron transport?
17. Fumarate is a product of both argininosuccinate lyase (fig. 25.7) and fumarylacetoacetase (see fig. 25.12), even though the reactions are quite different. If you were describing these two reactions to an organic chemistry student using one word per reaction, what words would you pick?
18. The biosynthesis of δ-aminolevulinate is known to occur with the loss of one glycine carbon-bound hydrogen, producing the intermediate shown in figure 25.13. What coenzyme would you expect to participate in this process? How does the same coenzyme stabilize the carbanion formed in the decarboxylation part of the reaction?
19. What are the functions of the two water molecules that are utilized by the enzyme 5-oxoprolinase (see fig. 25.15)?
20. Would you anticipate elevated arginase activity in the liver of an untreated diabetic animal? Why or why not?
21. The concentration of phenylalanine in the blood of neonates is used to screen for phenylketonuria (PKU). Explain the biochemical basis for the correlation of elevated blood phenylalanine concentration and PKU. Explain why restriction of dietary phenylalanine is critically important for youngsters with PKU.
22. (a) L-Glutathione is not a primary gene product, as are proteins. What "information" is used to direct the synthesis of L-glutathione?
    (b) Predict the effect of a glutathione synthase inhibitor on cells exposed to oxidative stress.

Appendix A The Catabolic Pathways for Amino Acids

The discussion of the catabolic pathways is organized into groups of amino acids that give rise to the same main pathway intermediates.

Five Amino Acids Degrade to Acetyl-CoA by Way of Pyruvate

The carbon skeletons of 10 amino acids yield acetyl-CoA. Five of these, alanine, glycine, threonine, serine, and cysteine, are degraded to acetyl-CoA by way of pyruvate (fig. 1). Another five, phenylalanine, tyrosine, tryptophan, lysine, and leucine, go by way of acetoacetyl-CoA. We will discuss first the amino acids that are converted to pyruvate.

Alanine. Alanine undergoes a reversible transamination directly to pyruvate. Recall that this reaction is part of the glucose–alanine cycle (see fig. 25.10).

Threonine. Threonine is degraded in more than one way. In the pathway shown in figure 1, threonine dehydrogenase converts threonine via two enzymes: threonine dehydrogenase and the acetyl-CoA-dependent α-amino-β-ketobutyrate lyase.

Figure 1

Outline of the catabolism of threonine, serine, cysteine, alanine, and glycine to acetyl-CoA by way of pyruvate. When a single enzyme is involved in the transition, the enzyme name is indicated next to the reaction arrow. Otherwise, the number of steps is indicated.

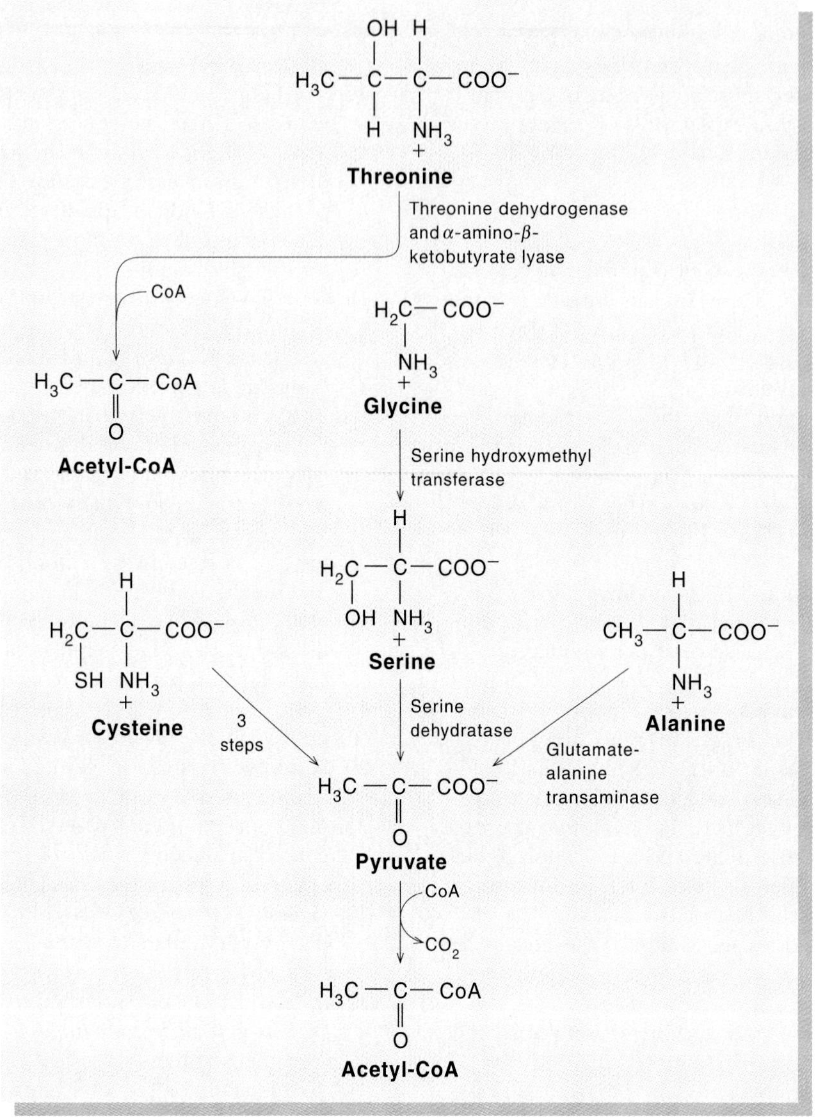

Glycine and Serine. Glycine itself is degraded in two ways, only one of which leads to pyruvate. The pathway to pyruvate involves the conversion of glycine to serine by addition of a hydroxymethyl group carried by N^5,N^{10}-methylenetetrahydrofolate (see fig. 11.14). Subsequently, the serine is converted to pyruvate by a specific serine dehydratase, unless, of course, there is a shortage of serine for biosynthesis.

The major pathway for the catabolism of glycine involves the oxidative cleavage of glycine to CO_2, NH_4^+, and a one carbon unit at the oxidation level of formaldehyde, which is accepted by tetrahydrofolate in a reversible reaction catalyzed by glycine cleavage complex:

$$\text{Glycine} + FH_4 + NAD^+ \rightleftharpoons N^5,N^{10}\text{-Methylene } FH_4' + CO_2 + NADH + NH_4^+$$

Thus even though glycine does not enter the TCA cycle by this mode of degradation, its degradation products are not wasted: the methyl group is donated to the coenzyme for use in one-carbon metabolism, and the NADH also produced in this process can be used directly to yield energy via the electron-transport system (since the glycine cleavage system is mitochondrial) or indirectly via a transhydrogenase to yield reducing power for biosynthesis.

Cysteine. There are several pathways for the catabolism of cysteine. All of these ultimately lead to the formation of pyruvate. The main pathway in animal cells occurs in the three steps, shown in figure 2. Cysteine sulfinate, an intermediate in this pathway, is also a biosynthetic intermediate; upon decarboxylation and oxidation it produces taurine (2-amino-ethanesulfonate), a component of certain bile acids (see chapter 23). Frequently, two cysteines

Figure 2

The main pathway for the conversion of cysteine to pyruvate in animals takes place in three steps. The first intermediate in this pathway, cysteinesulfinate, is a branchpoint that can also lead to taurine. Taurine is a component of certain bile acids (see chapter 23).

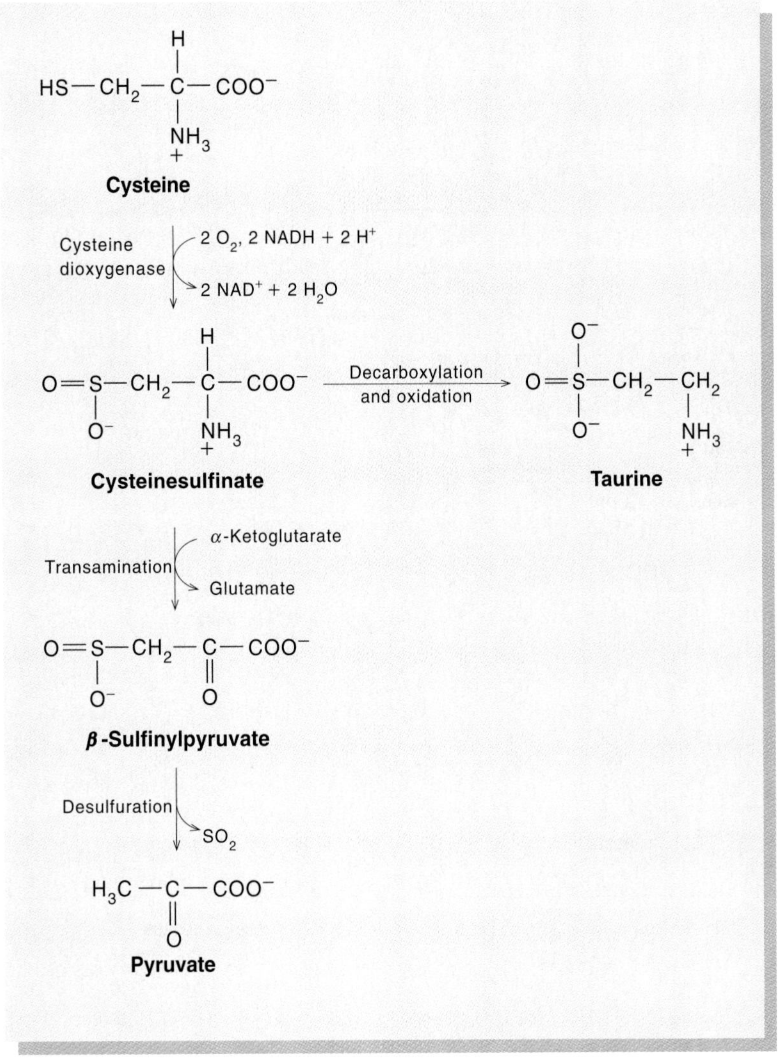

are disulfide-linked into a cystine. On such occasions the cystine is first reduced by an NADH-linked cystine reductase.

Five Amino Acids Degrade to Acetyl-CoA by Way of Acetoacetyl-CoA

The pathways for the degradation of phenylalanine, tyrosine, tryptophan, lysine, and leucine also lead to acetyl-CoA but they go by way of acetoacetyl-CoA rather than pyruvate (fig. 3). All of the pathways contain many steps, as indicated by the numbers next to the reaction arrows.

Lysine and Leucine. The pathways for leucine, lysine, and tryptophan are similar in the final steps and resemble the steps in the β oxidation of fatty acids (see chapter 21). No more will be said here about the catabolism of lysine and leucine.

Tyrosine. The pathways for the aromatic amino acids phenylalanine, tyrosine, and tryptophan are noteworthy not only because their side chains create special complications for degradation, but also because they provide numerous intermediates for the biosynthesis of useful compounds.

The oxidation of tyrosine (and phenylalanine) by the liver proceeds by way of acetoacetate and the dicarboxylic acid fumarate. Thus tyrosine and phenylalanine are both ketogenic and glycogenic.

The first step in tyrosine catabolism involves its conversion to 4-hydroxyphenylpyruvate by a tyrosine-glutamate transaminase of rather broad specificity (see fig. 25.12). The next step is catalyzed by 4-hydroxyphenylpyruvate dioxygenase, a copper-containing enzyme that is stimulated by ascorbate. The enzyme is called a dioxygenase because both atoms of the oxygen become incorporated into the product. The product, homogentisate, results from oxidation of the aromatic ring and an oxidative decarboxylation and migration of the side chain. The aromatic ring is fur-

Figure 3

Outline of the catabolism of lysine, tryptophan, phenylalanine, tyrosine, and leucine. The numbers of enzyme-catalyzed steps are indicated next to the reaction arrows.

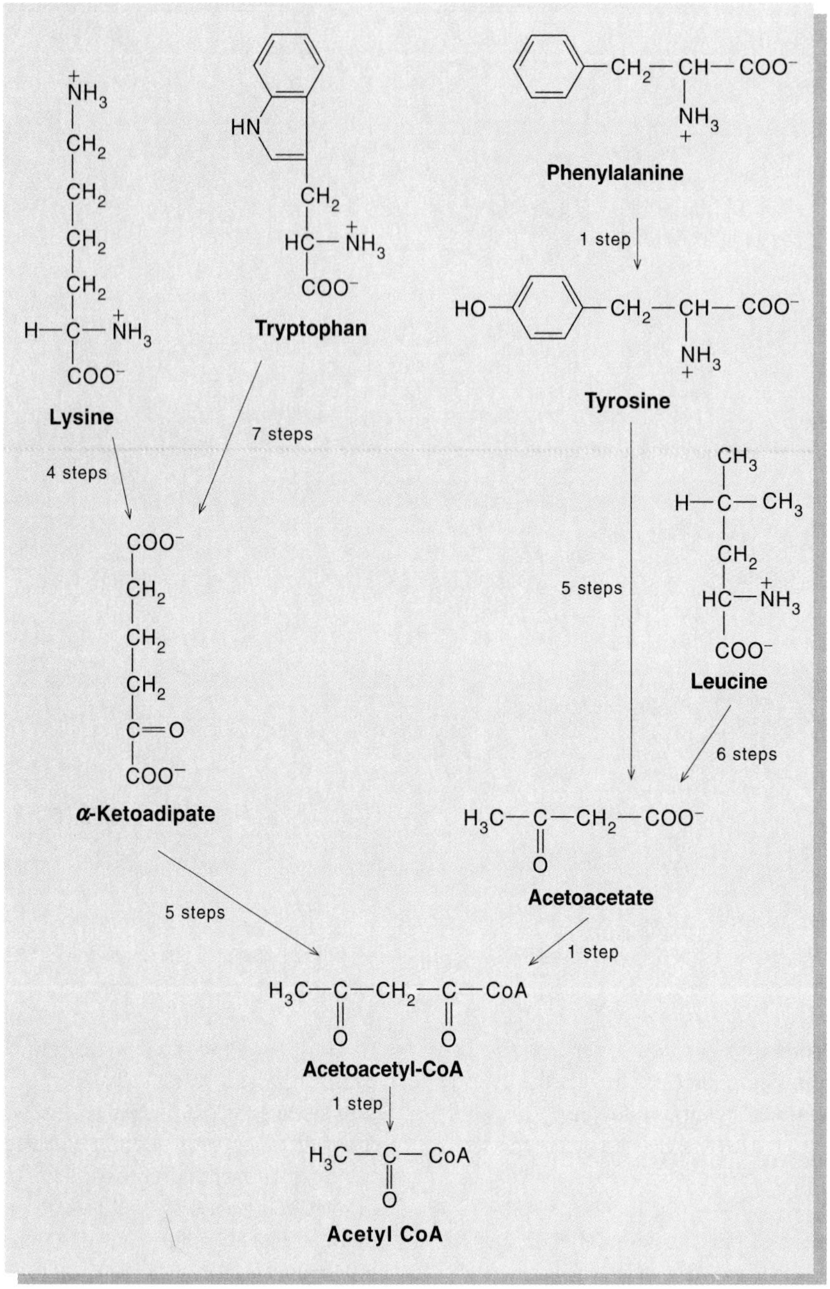

ther oxidized and cleaved by homogentisate-1,2-dioxygenase to 4-maleylacetoacetate. As is apparent from the name, this enzyme is also a dioxygenase. Nearly all cleavages of aromatic rings in biological systems are catalyzed by dioxygenases. The enzyme requires ferrous iron and is also stimulated by ascorbate. An isomerase, maleylacetoacetate isomerase, yields the *trans* compound 4-fumarylacetoacetate, which is hydrolytically cleaved to fumarate and acetoacetate by fumarylacetoacetate hydrolase.

Another route of tyrosine metabolism is that leading to melanin, which results from a two-stage attack on tyrosine by tyrosinase (see fig. 25.12), yielding first dihydroxyphenylalanine (DOPA). The latter is oxidized as a cosubstrate by tyrosinase to yield the 3,4-quinone. The quinone is unstable and undergoes a series of spontaneous reactions that ultimately lead to melanin, a black pigment. In animals, tyrosinase is found only in the organelles known as melanosomes, which are present in specialized

Figure 4

The formation of tyrosine from phenylalanine. This reaction occurs in one step. The enzyme requires tetrahydrobiopterin, a folic-acid-like compound, as a cosubstrate. Tetrahydrobiopterin is occasionally used as an electron carrier coenzyme.

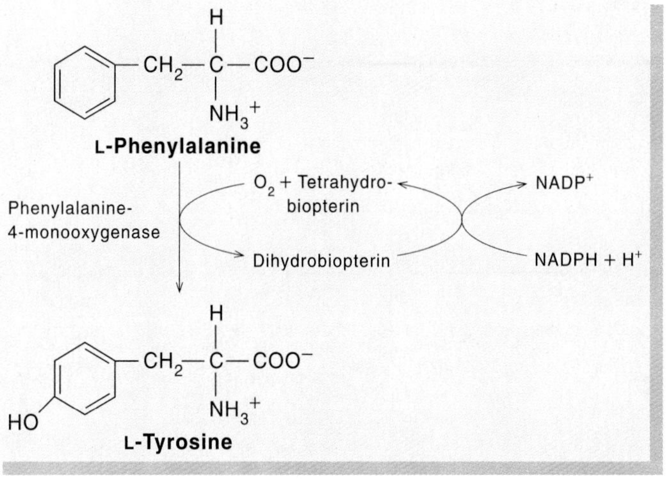

Figure 5

The structures of dihydrobiopterin and tetrahydrobiopterin. Dihydrobiopterin is the oxidized form of the coenzyme.

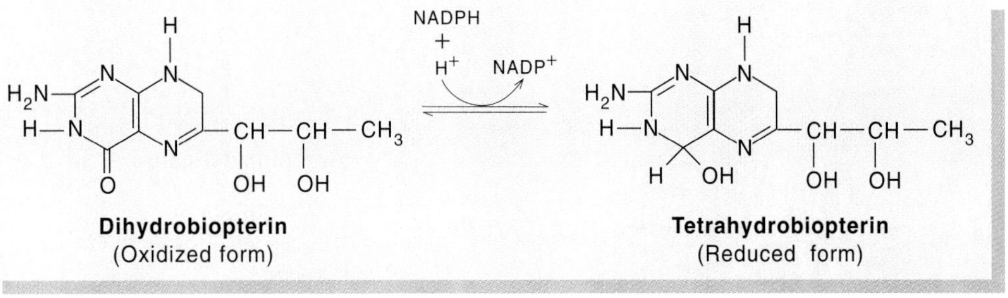

pigment-producing melanocytes found in the epidermis and certain other tissues.

Dopa is also synthesized from tyrosine by tyrosine hydroxylase as a precursor to certain neurohormones (norephinephrine and epinephrine). This reaction occurs exclusively in the adrenal glands (see chapter 27). Between tyrosinase and tyrosine hydroxylase we have an example of two enzymes that carry out the same reaction but for totally different purposes.

Phenylalanine. Phenylalanine is broken down normally by way of tyrosine through the action of phenylalanine-4-monooxygenase, as indicated in figure 4 (see also fig. 25.12). The enzyme requires tetrahydrobiopterin, a folic-acid-like compound, as a cosubstrate. Tetrahydrobiopterin is an infrequently used electron carrier coenzyme (fig. 5). Dihydrobiopterin is its oxidized form. The biopterin is kept in the reduced form by NADPH, the ultimate hydrogen donor in the hydroxylation reaction. The presence of phenylalanine-4-monooxygenase accounts for the fact that tyrosine is not an essential amino acid in mammals, provided that the dietary supply of phenylalanine is sufficient.

Minor pathways for phenylalanine breakdown in animals involve transamination to yield phenylpyruvate. Although phenylpyruvate can be reduced to phenyllactate and metabolized to other phenyl derivatives, the disposal of dietary phenylalanine by these routes is insufficient, so in the inherited absence of the hydroxylation to tyrosine, high blood levels of phenylalanine and phenylpyruvate result.

Tryptophan. The major pathways for tryptophan catabolism in the mammalian liver and for many microorganisms proceed by way of kynurenine (fig. 6). Kynurenine itself can be metabolized in the liver by way of α-ketoadipate, which is also an intermediate in lysine degradation. An interesting variant of the kynurenine pathway allows for the synthesis of the coenzyme nicotinamide.

The first step in the breakdown of tryptophan is catalyzed by tryptophan oxygenase, which yields *N*-formylkynurenine (see fig. 6). The enzyme is a dioxygenase and cleaves the indole ring by incorporating an oxygen atom on both C-2 and C-3 of the indole ring. Kynurenine itself is formed by the liberation of formate by kynurenine formamidase.

Kynurenine is converted to 3-hydroxykynurenine by the NADPH-dependent kynurenine-3-monooxygenase (see fig. 6). Kynureninase, a pyridoxal phosphate enzyme, catalyzes a hydrolytic cleavage of the alanine side chain to yield 3-hydroxyanthranilate.

Figure 6

The main route of tryptophan degradation in mammals leads to α-ketoadipate. The intermediate 2-amino-3-carboxymuconate-6-semialdehyde (ACS) is a branch-point metabolite that can also lead to nicotinamide via quinolate as indicated.

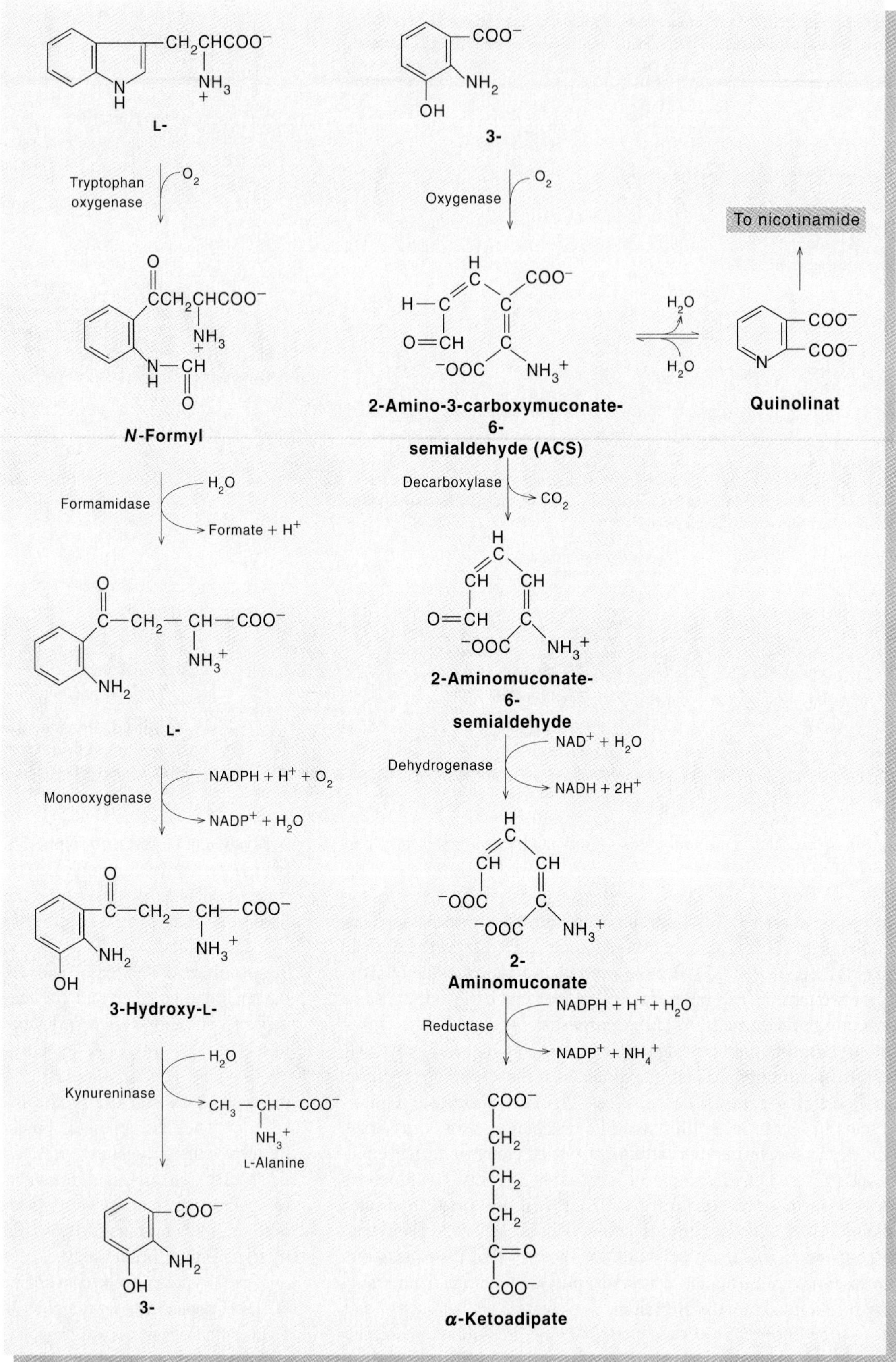

Figure 7

The conversion of α-ketoadipate to acetyl-CoA. The sulfur in the CoA-containing compounds is indicated. α-Ketoadipate is also formed in the liver by the breakdown of lysine (see fig. 3). Thus the degradative pathways for tryptophan and lysine converge at the level of α-ketoadipate. Two molecules of CO_2 are released and two molecules of NADH are produced in the process of converting α-ketoadipate into two molecules of acetyl-CoA.

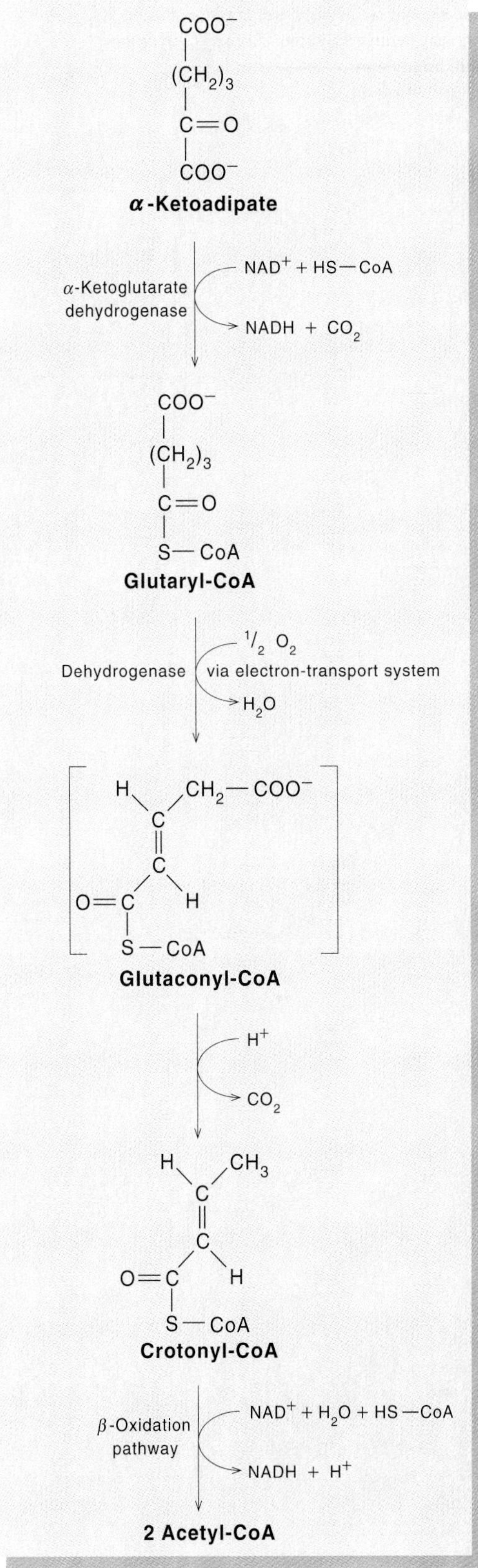

The aromatic ring is cleaved to 2-amino-3-carboxymuconate-6-semialdehyde (ACS) by 3-hydroxyanthranilate oxygenase. Again, this enzyme is a dioxygenase, and oxygen atoms are incorporated on both the carbons at the site of ring cleavage. Ferrous ions are required by the enzyme. ACS can be spontaneously cyclized to quinolinate with the liberation of a molecule of H_2O. Quinolinate is an intermediate in the biosynthesis of nicotinamide, a synthesis that many animals have a limited capacity to perform. The ACS is decarboxylated by a specific decarboxylase to yield 2-aminomuconate-6-semialdehyde. 2-Aminomuconate-6-semialdehyde is oxidized by an NAD-dependent aminomuconate semialdehyde dehydrogenase. The resulting 2-amino muconate is reduced to α-ketoadipate by an NAD(P)H-dependent reductase.

The further catabolism of α-ketoadipate results in the liberation of two molecules of CO_2 and two of acetyl-CoA (fig. 7). The first step is a coenzyme-A-dependent oxidative decarboxylation by an enzyme probably identical to α-ketoglutarate dehydrogenase. The product, glutaryl-CoA, is oxidized by a flavin-linked dehydrogenase to an intermediate common to the oxidation of fatty acids, crotonyl-CoA. (The α-keto derivative glutaconyl coenzyme is probably an enzyme-bound intermediate in this reaction.) Finally, two molecules of acetyl coenzyme are formed by the action of the fatty-acid-oxidizing enzymes (see chapter 21). It should be remembered that α-ketoadipate is also formed in the liver by the breakdown of lysine. Thus the degradative pathways for lysine and tryptophan converge at the level of α-ketoadipate.

Five Amino Acids Degrade to α-Ketoglutarate

α-Ketoglutarate is the endpoint for degradation of five amino acids: arginine, histidine, proline, glutamic acid, and glutamine (fig. 8). As we saw (chapter 24), it is also the starting point for the synthesis of these same five amino acids. Hence these are reversible conversions, like those we saw in the glycolytic–gluconeogenic interconversions (see chapter 14), but fewer enzymes are used in common.

Glutamine and Glutamate. Reactions involving the interconversions of glutamine, glutamate, and α-ketoglutarate were discussed earlier in this chapter when we were considering deamination.

Proline. Proline is converted into Δ^1-pyrolline-5-carboxylate in a reaction catalyzed by proline oxidase. This compound is in equilibrium with glutamate-γ-semialdehyde, which is also an intermediate in arginine catabolism (fig. 9).

Arginine. Arginine is one of the few amino acids for which there is no transaminase reaction. In addition to serving as a source of

Figure 8

Outline of the catabolism of histidine, arginine, proline, and glutamine to glutamate and then to the TCA component α-ketoglutarate.

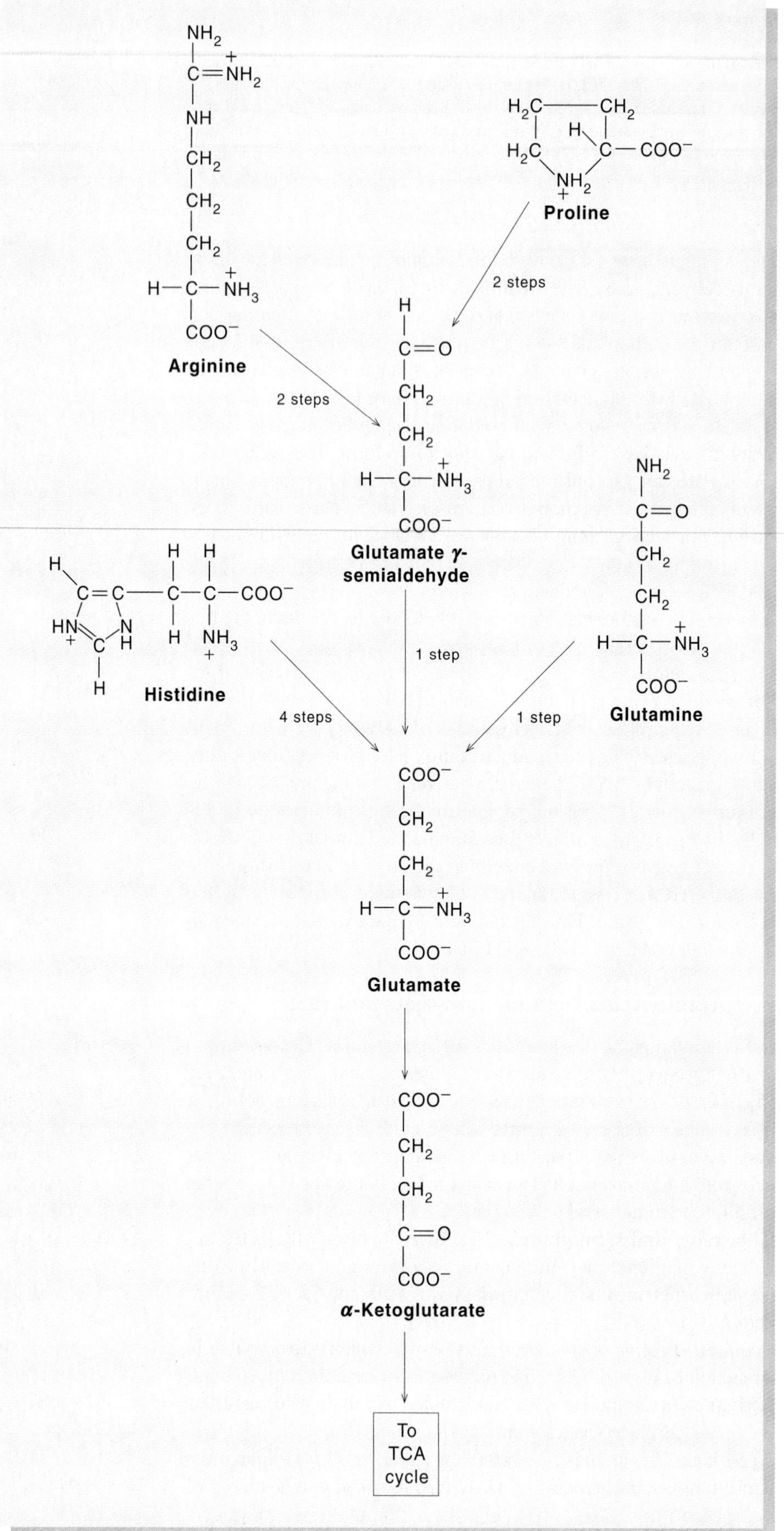

Figure 9

Degradation of proline to glutamate-γ-semialdehyde. This can be thought of as a two-step reaction. The first step is enzyme-catalyzed. The second step is spontaneous and reversible.

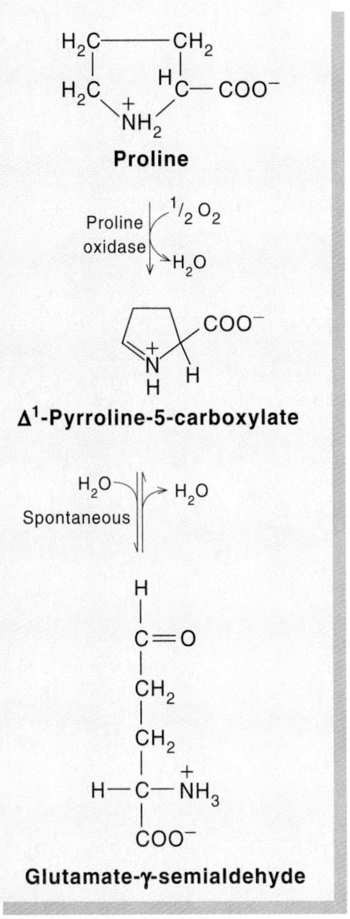

Proline

Proline oxidase ($\frac{1}{2}O_2$ → H_2O)

Δ^1-Pyrroline-5-carboxylate

Spontaneous (H_2O ⇌ H_2O)

Glutamate-γ-semialdehyde

carbon and energy, arginine is a precursor of various essential polyamines and, as we have seen, is required in the urea cycle (see fig. 25.7).

Arginine is converted to ornithine by two different routes (fig. 10). We have discussed one, which provides a mechanism for urea formation and is important in the nitrogen metabolism in many animal species. The other route is one that provides a source of energy for many microorganisms. It is called the arginine dihydrolase pathway. The first enzyme, arginine deiminase, converts arginine to citrulline, with the liberation of NH_3 (fig. 10). Citrulline can be cleaved by a degradative ornithine transcarbamoylase to yield ornithine and carbamoyl phosphate. The carbamoyl phosphate so formed serves as a high-energy phosphate donor for ATP formation in a reaction catalyzed by carbamate kinase. In some animal tissues it appears that the same reaction can be catalyzed by an acetate kinase.

Ornithine, whether formed by the arginase of the urea cycle or arginine dihydrolase pathway, is broken down in most organisms by a transaminase to yield glutamate-γ-semialdehyde or its cyclized derivative, Δ^1-pyrroline-5-carboxylate. This compound can be further oxidized to glutamate by Δ^1-pyrroline-5-carboxylate dehydrogenase, an NAD^+-linked enzyme, or it can be reduced to proline by the normal proline biosynthetic enzyme Δ^1-pyrroline-5-carboxylate reductase in an NADPH-requiring reaction. This route to proline from ornithine accounts for the interconvertibility of ornithine and proline that is seen in many cells and tissues.

Amines Produced from Arginine. Putrescine, the decarboxylated product of ornithine, is important as an intermediate in spermidine and spermine formation. These two polyamines of unknown function are generally found in association with nucleic acids. The decarboxylation is catalyzed by ornithine decarboxylase, and in most bacteria this route is the sole or primary route of putrescine formation (see fig. 10).

The synthesis of the omnipresent polyamines spermidine and spermine requires not only the formation of putrescine, but also the generation of a propylamine group. The propylamine donor is a product derived from *S*-adenosylmethionine by a specific decarboxylase (fig. 11). The same enzyme transfers an aminopropyl group to spermidine to yield spermine.

Histidine. The major route for histidine catabolism in mammals involves the conversion of histidine to glutamate. We shall see that in the process of breakdown, histidine contributes a carbon atom to one-carbon metabolism (fig. 12). In the first step, histidase catalyzes the removal of NH_3 with the formation of urocanate. This step is followed by an internal oxidation and reduction involving addition of the elements of water in a reaction catalyzed by urocanase. The resulting intermediate, 4-imidazolone-3-propionate, contains an imidazolone ring that is opened by a hydrolytic reaction. Cleavage of the product *N*-formino-L-glutamate leads to formimino group transfer to the N^5 position of tetrahydrofolate and free glutamate. Animals that are deficient in folic acid excrete large amounts of formiminoglutamate. Vitamin B_{12} deficiency leads to the same syndrome, a fact suggesting that the one-carbon unit on N^5-formiminotetrahydrofolate is normally transferred to this vitamin (see chapter 11 for further discussion on vitamin B_{12}).

Catabolism of Methionine, Isoleucine, and Valine Leads to Succinyl-CoA

The carbon skeletons of methionine, isoleucine, and valine are degraded by pathways that lead to succinyl-CoA (fig. 13). Although these pathways are rather long, there are some simplifying features: Isoleucine and valine undergo identical reactions in the first four steps of degradation (fig. 14); methionine and isoleucine are reduced to propionyl-CoA. Propionyl-CoA is converted into methylmalonyl-CoA in two steps. Methylmalonyl-CoA, a common intermediate in all three pathways, is converted into succinyl-CoA in one step. The reactions between propionyl-CoA and succinyl-CoA have already been discussed in chapter 18. Once again we see the strong overlap of reactions in lipid and amino acid catabolism, where the carbon skeletons contain several saturated carbon–carbon linkages.

The three keto acids derived by deamination of valine, isoleucine, and leucine are decarboxylated by the same enzyme complex. This enzyme complex also acts on pyruvate and α-ke-

Figure 10

Some of the reactions involved in arginine catabolism. Arginine is converted to ornithine by two different routes. One is involved in urea formation (see fig. 25.7). The other route, known as the dihydrolase pathway, provides a source of energy for many microorganisms. The first enzyme in this pathway converts arginine to citrulline. The citrulline is cleaved to ornithine and carbamoyl phosphate. The carbamoyl phosphate serves as a high-energy phosphate donor for ATP formation. Ornithine can be converted to proline, glutamate, and several polyamines, including putrescine, spermidine, and spermine.

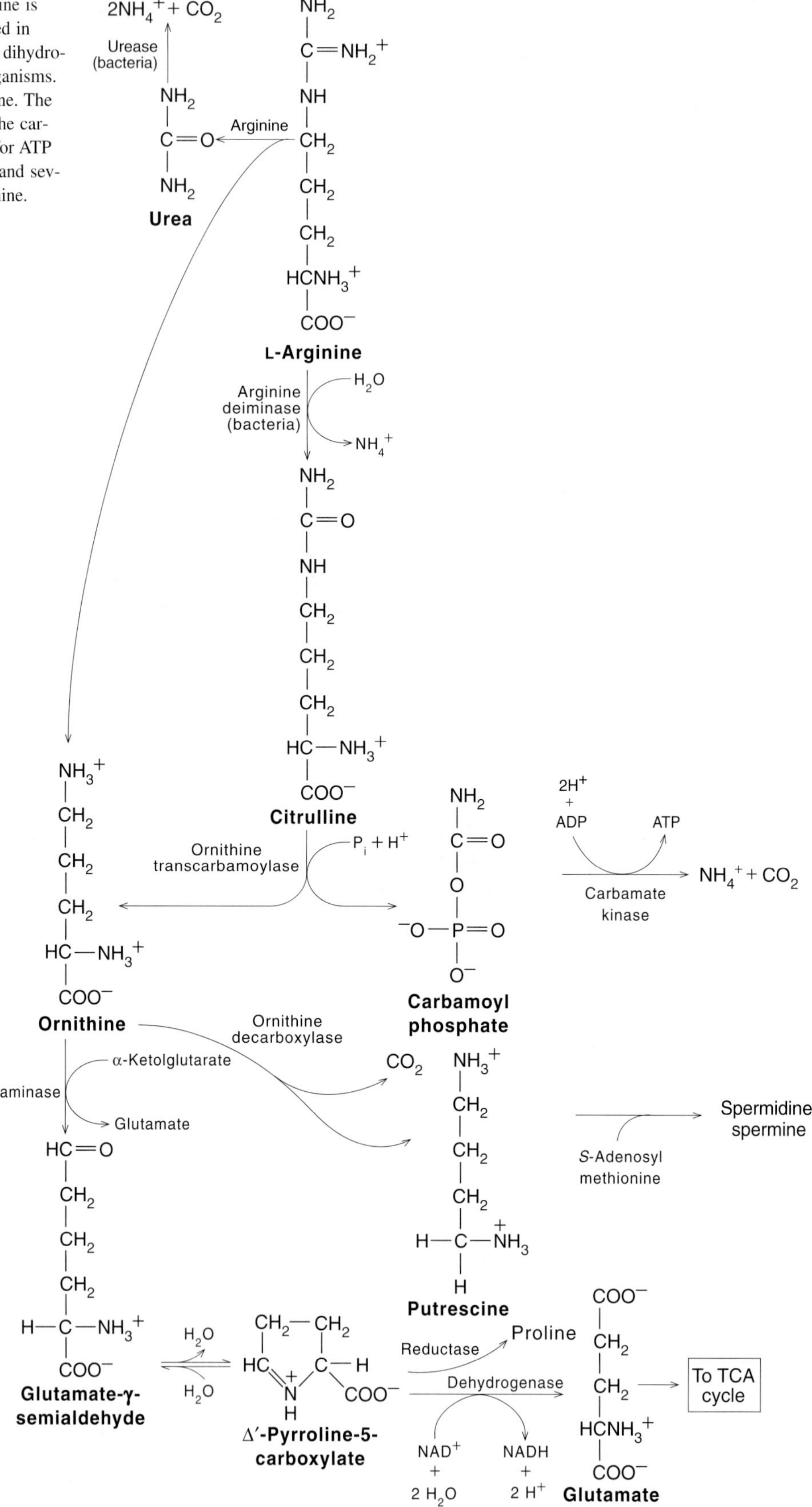

Figure 11

The utilization of putrescine and *S*-adenosylmethionine for the formation of spermidine and spermine. Polyamines are thought to have many functions; they are invariably found complexed with nucleic acids, both DNAs and RNAs.

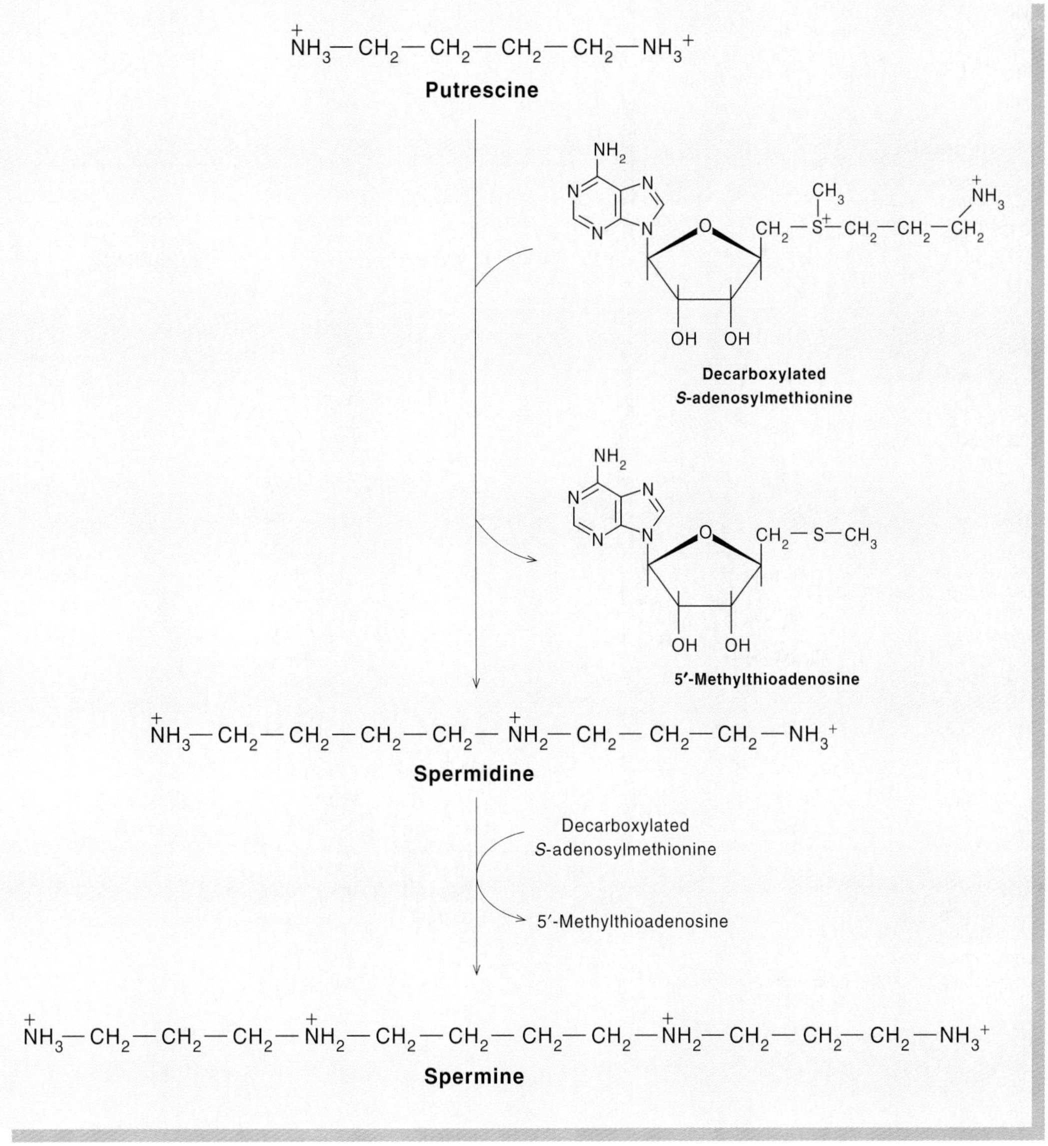

$\overset{+}{N}H_3—CH_2—CH_2—CH_2—CH_2—NH_3^+$

Putrescine

Decarboxylated *S*-adenosylmethionine

5′-Methylthioadenosine

$\overset{+}{N}H_3—CH_2—CH_2—CH_2—CH_2—\overset{+}{N}H_2—CH_2—CH_2—CH_2—NH_3^+$

Spermidine

Decarboxylated *S*-adenosylmethionine

5′-Methylthioadenosine

$\overset{+}{N}H_3—CH_2—CH_2—CH_2—\overset{+}{N}H_2—CH_2—CH_2—CH_2—CH_2—\overset{+}{N}H_2—CH_2—CH_2—CH_2—NH_3^+$

Spermine

tobutyrate, which are products of both threonine and methionine metabolism, respectively. Some people are genetically defective for the presence of this enzyme complex. They accumulate substantial amounts of branched chain α-keto acids in the urine and suffer from a variety of disorders (see table 25.3). Strict control of the diet, allowing only low amounts of these three amino acids, alleviates the immediate symptoms of this disease but, unless begun early after birth, cannot prevent or reverse the mental retardation.

Methionine. The catabolism of methionine involves nine steps leading to succinyl-CoA. The first three steps of this pathway are discussed in chapter 24 (see fig. 24.18). The last three steps, from propionyl-CoA, have already been discussed in chapter 21 (see fig. 21.7).

In the first step, methionine is adenylated to S-adenosylmethionine (SAM). SAM is probably the most used transmethylating agent in the cell. Transfer of the methyl group from SAM to an appropriate receptor leads to *S*-adenosylhomocysteine. This is hydrolyzed to adenosine and homocysteine. The homocysteine is condensed with serine to yield cystathionine (fig. 15), which in one more step is converted to cysteine and α-ketobutyrate. Cysteine, if it is present in excess, can be catabolized by the three-step pathway described in figure 2. The α-ketobutyrate is converted in one step into propionyl-CoA.

Figure 12

The catabolism of histidine. In the final step of histidine breakdown to glutamate, tetrahydrofolate is converted to N^5-formiminotetrahydrofolate. In this way, histidine breakdown contributes a carbon atom to C-1 metabolism.

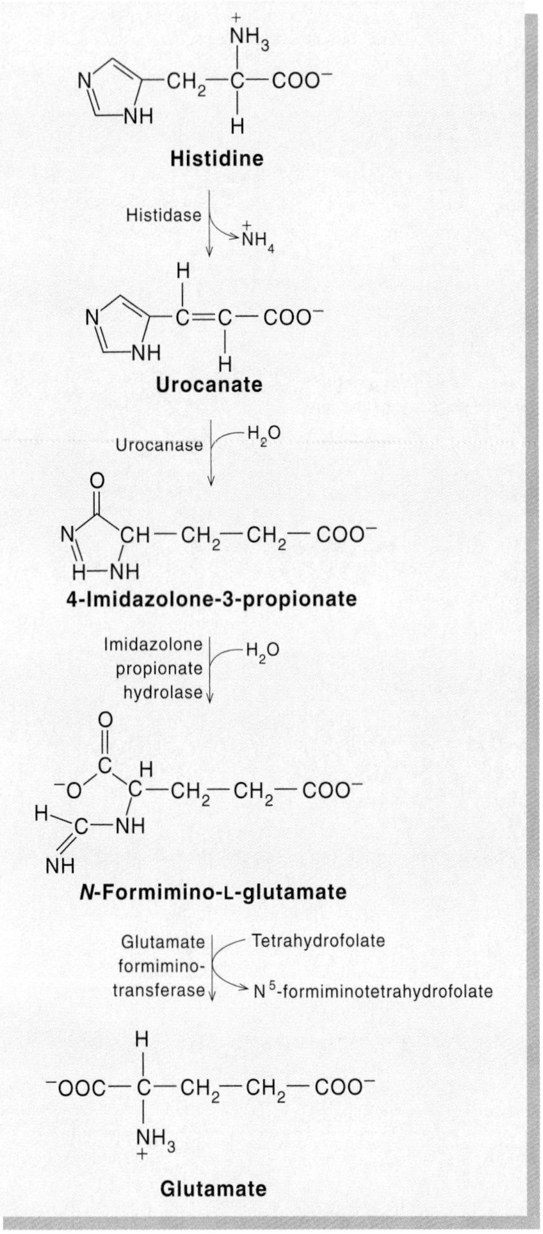

Histidine

Histidase → NH_4^+

Urocanate

Urocanase ← H_2O

4-Imidazolone-3-propionate

Imidazolone propionate hydrolase ← H_2O

***N*-Formimino-L-glutamate**

Glutamate formiminotransferase: Tetrahydrofolate → N^5-formiminotetrahydrofolate

Glutamate

Figure 13

Outline of the catabolism of methionine, isoleucine, and valine to succinyl-CoA.

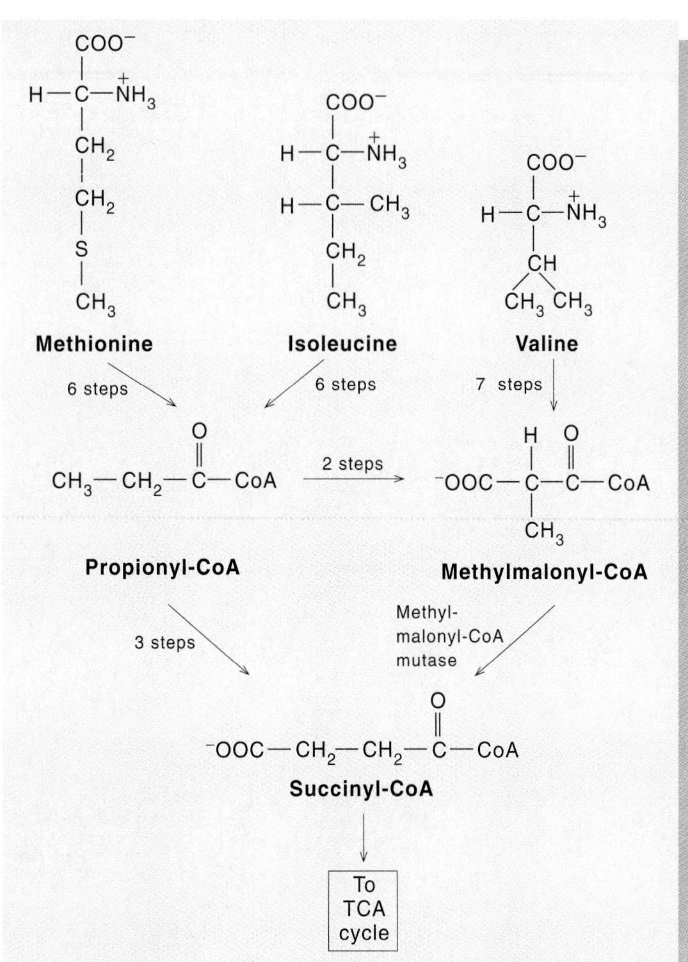

Aspartate and Asparagine Are Deaminated to Oxaloacetate

The last two amino acids to be considered, to complete our discussion of amino acid catabolic pathways, are aspartate and asparagine. The entry of these amino acids into the TCA pool via oxaloacetate involves only two enzymes. Asparaginase converts asparagine to aspartate, and aspartate is reversibly converted into oxaloacetate in a typical transamination reaction with glutamate:

$$\text{Asparagine} + H_2O \longrightarrow \text{Aspartate}^- + NH_4^+$$

$$\text{Aspartate}^- + \alpha\text{-Ketoglutarate} \rightleftharpoons \text{Oxaloacetate} + \text{Glutamate}^-$$

Recall that aspartate is also converted into fumarate in the urea cycle.

Figure 14

Isoleucine and valine undergo identical reactions in the first four steps of degradation.

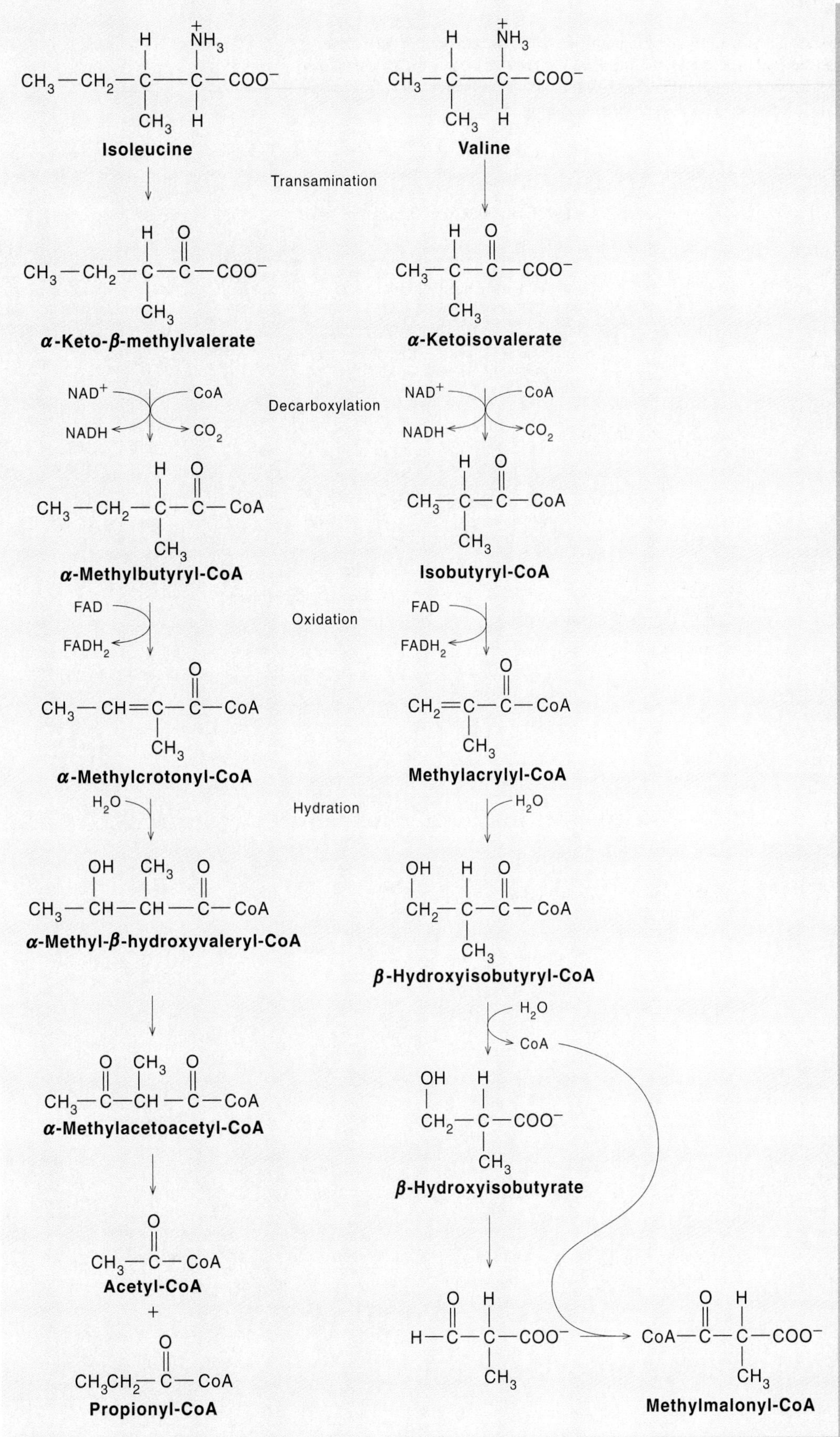

Figure 15

Degradation of homocysteine to propionyl-CoA. These represent the middle three steps in the catabolism of methionine. The first three steps leading to homocysteine are shown in fig. 24.18, and the last three steps leading from propionyl-CoA to succinyl-CoA are illustrated in fig. 21.7.

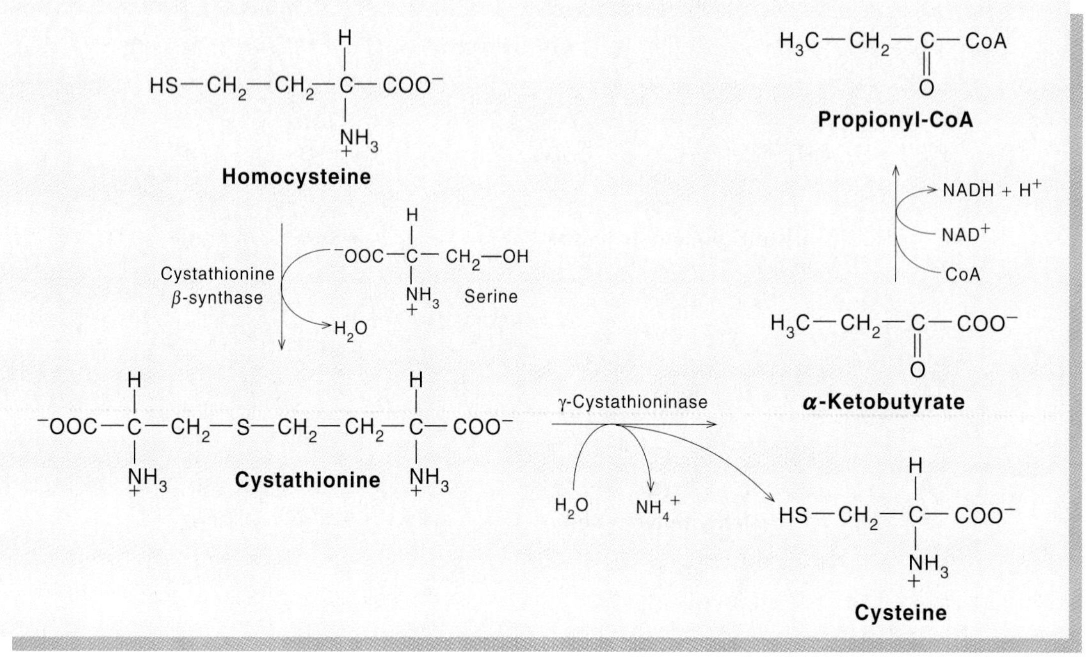

C H A P T E R

NUCLEOTIDES

26

Nucleotides are synthesized from very simple precursors. They function as energy carriers, as regulatory molecules, and as precursors of coenzymes and nucleic acids.

This chapter deals with the biosynthesis of ribonucleotides and deoxyribonucleotides, their role in metabolic processes, and the pathways for their degradation (fig. 26.1). The biosynthesis of nucleotides is a vital process, since these compounds are indispensable precursors for the synthesis of both RNA and DNA. Without RNA synthesis, protein synthesis is halted; and unless cells can synthesize DNA, they cannot divide. Nucleotides are also necessary for constant repair of DNA, a process necessary for cell survival.

It is not surprising that inhibitors of nucleotide biosynthesis are very toxic to cells. As we will see, their toxicity has been used to advantage in the treatment of cancer as well as in the treatment of certain diseases resulting from infections by viruses, bacteria, or protozoans.

Nucleotides play important roles in all major aspects of metabolism. ATP, an adenine nucleotide, is the major substance used by all organisms for the transfer of chemical energy from energy-yielding reactions to energy-requiring reactions such as biosynthesis. Other nucleotides are activated intermediates in the

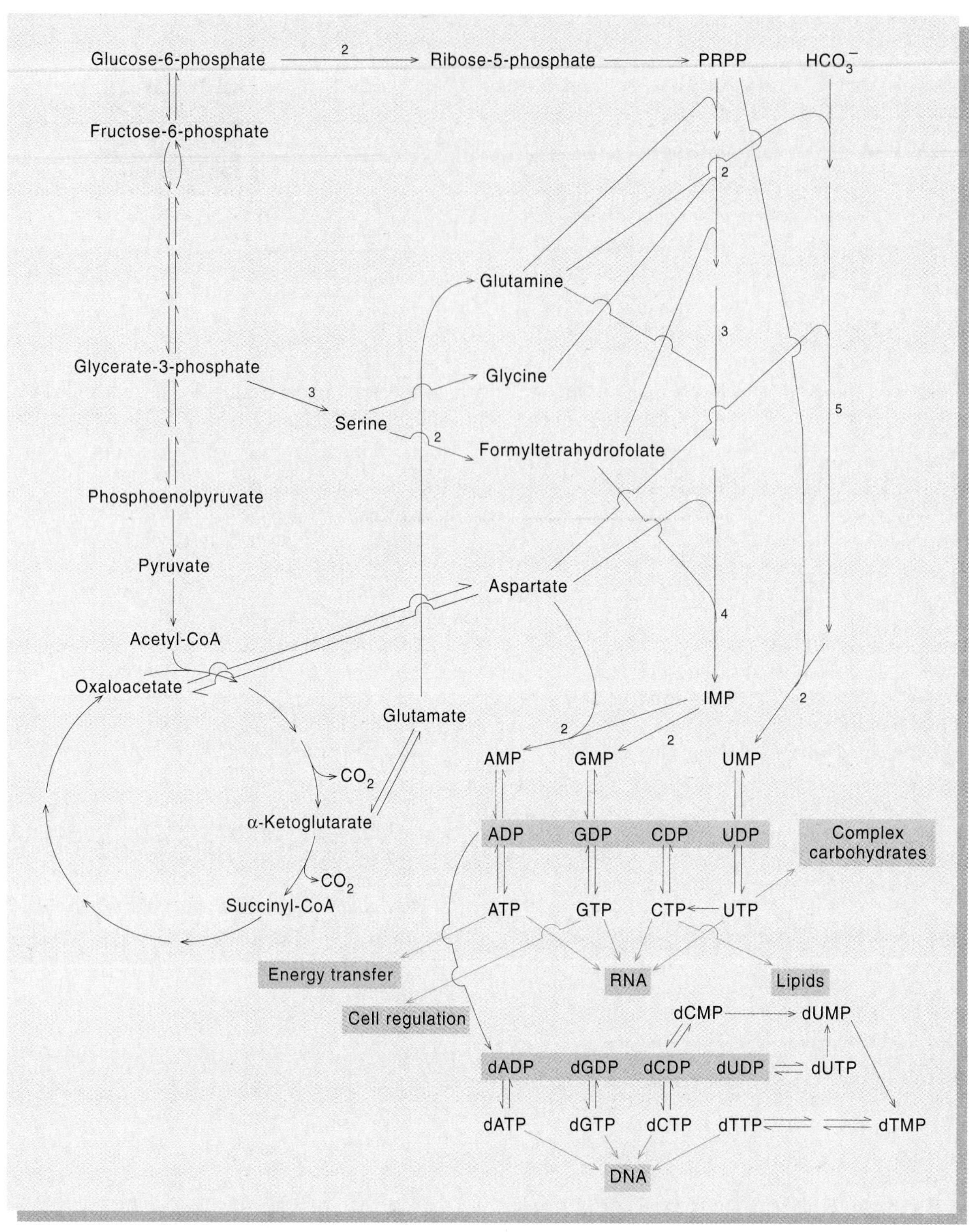

synthesis of carbohydrates, lipids, proteins, and nucleic acids. Adenine nucleotides are components of many major coenzymes, such as NAD^+, $NADP^+$, FAD, and CoA. The critical role played by nucleotides as regulators of metabolism in both prokaryotic and eukaryotic organisms and in signal transduction is described in other chapters.

Pathways for the metabolic degradation of nucleotides are also very important to the organism, as demonstrated by the fact that several genetic defects causing blocks in these pathways have serious consequences for the health of humans and other animals.

The physiological importance of nucleotides is reflected in the careful regulation of their intracellular levels as well as intra- and extracellular levels of nucleosides and nucleobases. Levels of ATP, ADP, and AMP are tightly controlled in a variety of cells, and intracellular levels of dATP, dCTP, dGTP, and dTTP are also carefully regulated under normal conditions. This regulation

Figure 26.1

Synthesis and utilization of nucleotides. Numbers associated with arrows indicate the number of enzymatic steps involved; where no number is shown there is one step. Reactions from the central metabolic pathways are shown in black; the remaining reactions discussed in this chapter are shown in red.

It is remarkable how few starting materials are required for the biosynthesis of nucleotides: ribose-5-phosphate, glutamine, glycine, formyltetrahydrofolate, and aspartate. Serine also provides methylenetetrahydrofolate for converting dUMP to dTMP. The sugar, ribose, is derived as its 5-phosphate from the pentose phosphate pathway and activated by conversion to phosphoribosylpyrophosphate (PRPP). In purine nucleotide synthesis the purine base is built onto PRPP to give IMP, the precursor of AMP and IMP. In contrast to that strategy, the pyrimidine ring is constructed first from HCO_3^-, glutamine, and aspartate before condensation with PRPP to give OMP, the precursor of IMP, and ultimately of all other pyrimidine nucleotides. Aspartate contributes three carbons and a nitrogen to the pyrimidine ring, whereas glycine contributes two carbons and a nitrogen to the purine ring. Other carbon and nitrogen atoms are added one at a time in single steps. Phosphorylation of the ribonucleoside monophosphates (NMPs) gives ribonucleoside diphosphates (NDPs), and further phosphorylation gives ribonucleoside triphosphates (NTPs). A variety of phosphatases can reverse this process to re-form NDPs and NMPs from NTPs. Most other conversions occur at a specific level of phosphorylation. For example, deoxyribose is formed from ribose by reduction of NDPs to deoxyribonucleoside diphosphates (dNDPs), uracil is methylated to thymine at the monophosphate level (dUMP to dTMP), and uracil is converted to cytosine at the triphosphate level (UTP to CTP). The important roles of nucleotides in the cell are shown by the boxes to which the green arrows point.

The diagram does not show the pathways for nucleotide catabolism, which involve conversion of nucleotides to the monophosphate level, then to nucleosides, and finally to bases. Certain of the nucleosides and bases can be carried to other cells, where they are used for reconstruction of nucleotides, the so-called salvage pathway, or the bases can be degraded to excretion products. Unlike the catabolism of carbohydrates, fats, and proteins, nucleotide degradation, except for the recovery of the ribose, makes no contribution to the central metabolic pathways.

is determined by the concentration and location of the enzymes of nucleotide metabolism and by levels of substrates, products, and effectors.

Nucleotide Components: A Phosphoryl Group, a Pentose, and a Base

Each nucleotide is composed of three parts: (1) a heterocyclic nitrogenous base, (2) a pentose, and (3) a phosphoryl group (fig. 26.2). Nucleotides differ from one another through differences in each of these components.

Nucleotide bases (or nucleobases) belong to two classes: pyrimidines and purines (table 26.1). The former have a single six-membered ring containing two nitrogen atoms. In pyrimidine nucleotides, N-1 of the pyrimidine base is attached to the pentose C-1′. The common pyrimidine bases in nucleotides are uracil, cytosine, and thymine (fig. 26.3). Purine bases have a six-membered pyrimidine ring fused to a five-membered imidazole ring, the fused system containing four nitrogen atoms. In purine nucleotides the glycosidic linkage is between N-9 of the purine and C-1′ of the pentose. The common purine bases in nucleotides are adenine, guanine, and hypoxanthine (see fig. 26.3). Nucleotide catabolism leads to the purines xanthine and uric acid (see fig. 26.26), and

Figure 26.2

Structure of a nucleotide. A nucleotide has three components: a phosphoryl group, a pentose, and a nitrogenous base. The carbon atoms of pentose (ribose in the example shown) are given prime designations. The deoxyribonucleotides have deoxyribose as the pentose. This sugar has a hydrogen instead of a hydroxyl group at C-2′ of the pentose. The phosphoryl group is attached as an ester of one of the pentose hydroxyls, most commonly at C-5′. The base (adenine in the example shown) is attached by a glycosidic bond from a ring nitrogen to C-1′ of the pentose. When the base is a purine (e.g., adenine, guanine, or hypoxanthine) and the pentose is deoxyribose, this glycosidic bond is acid-labile. A nucleoside has no phosphoryl group, only a hydroxyl group at C-5′.

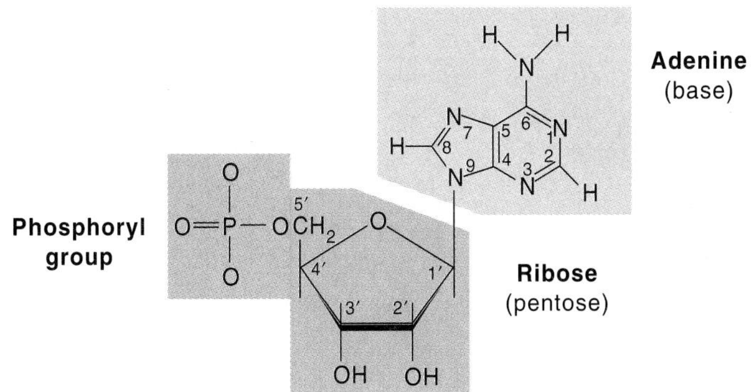

Adenosine-5′-monophosphate
(Adenylic acid)

Table 26.1
Names of Common Bases, Nucleosides, and Nucleotides

Base	Nucleoside[a]	Nucleotide
Purines		
Adenine	Adenosine (A)	AMP[b] or adenylate
Guanine	Guanosine (G)	GMP or guanylate
Hypoxanthine	Inosine (I)	IMP or inosinate
Pyrimidines		
Uracil	Uridine (U)	UMP or uridylate
Thymine	Thymidine (T)	TMP or thymidylate
Cytosine	Cytidine (C)	CMP or cytidylate

[a]With the exception of thymidine these are the names of the ribonucleosides, and deoxyribonucleosides are indicated by the prefix *deoxy* in front of the nucleoside name; thus deoxyadenosine contains deoxyribose instead of ribose. Thymidine indicates the deoxyribose derivative of thymine. Standard one-letter abbreviations are shown in parentheses.

[b]AMP is the abbreviation of the most commonly used name for this nucleotide, adenosine-5′-monophosphate or adenosine-5′-phosphate. Similarly, ADP and ATP designate the 5′-diphosphate and triphosphate of adenosine, respective (see fig. 26.4). Adenylate (or adenylic acid) is an alternative nomenclature.

Uridine-5′-monophosphate (UMP)

Guanosine-5′-monophosphate (GMP)

Inosine-5′-monophosphate (IMP)

(a)

Deoxyadenosine-5′-phosphate (dAMP)

Thymidine-5′-phosphate (dTMP)

Deoxyguanosine-5′-phosphate (dGMP)

Deoxycytidine-5′-phosphate (dCMP)

(b)

Figure 26.3

Structures of (*a*) three common ribonucleotides and (*b*) four common deoxyribonucleotides. See table 26.1 for alternative names and for names of the corresponding bases and nucleosides.

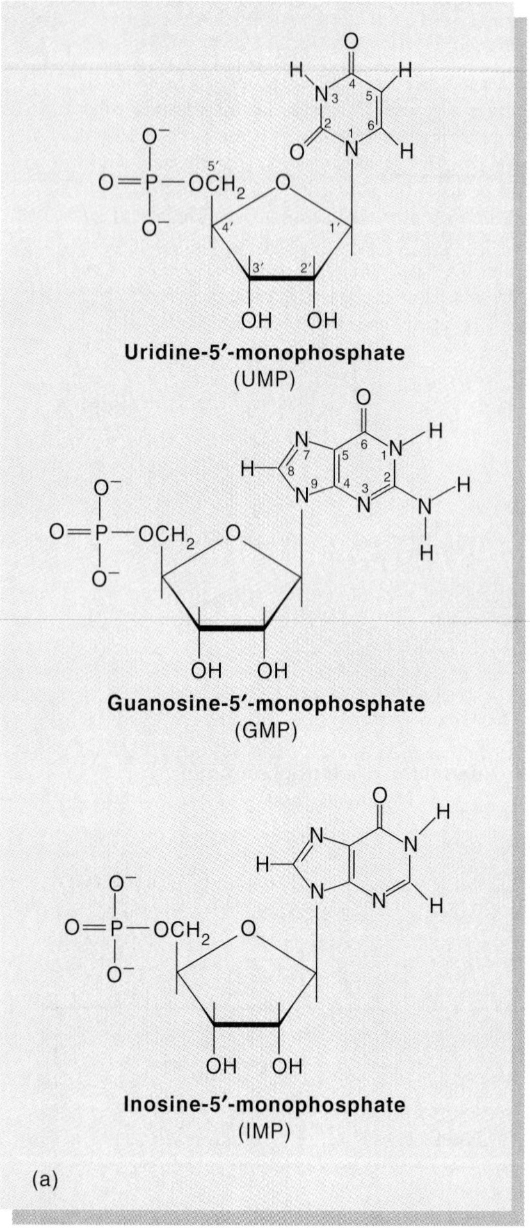

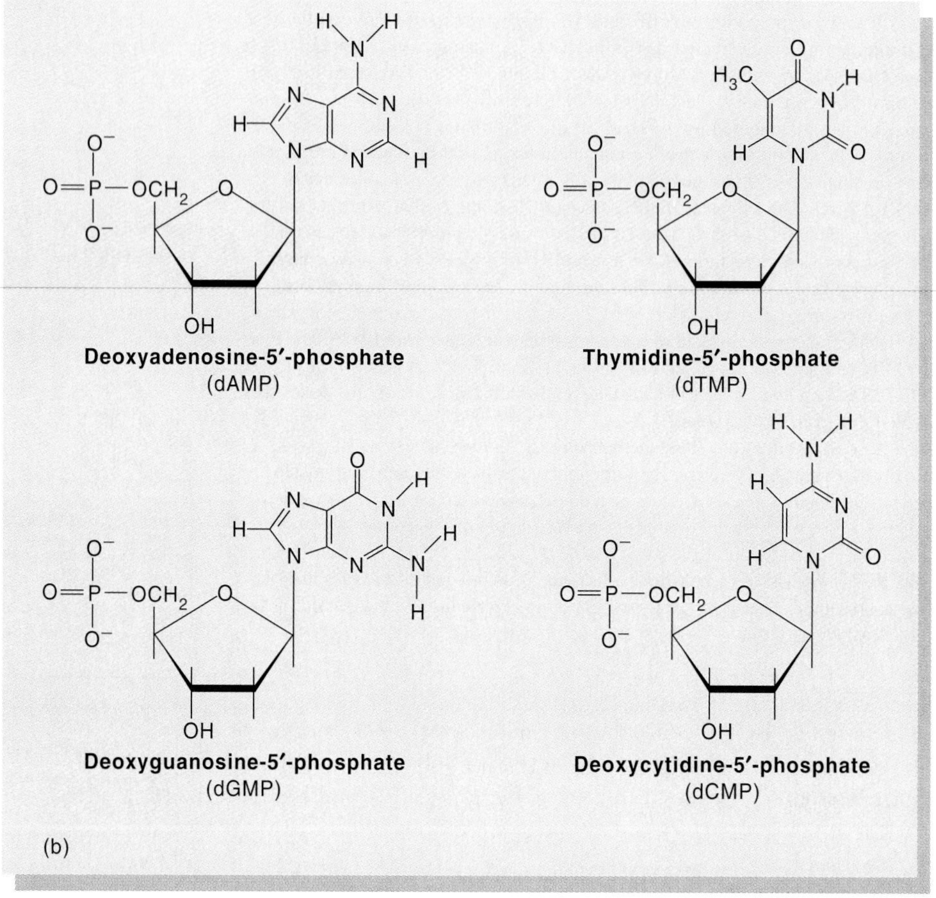

xanthosine monophosphate is a nucleotide intermediate of metabolism (see fig. 26.16).

The pentose component of naturally occurring nucleotides is ribose or 2-deoxyribose (i.e., ribose with a hydrogen instead of a C-2′—OH). In nucleotides the purine or pyrimidine is attached to C-1′ of the pentose in the β configuration. This means that the base is *cis* relative to C-5′ and *trans* relative to the C-3′—OH. The major function of deoxyribonucleotides (those that have 2-deoxyribose as the pentose) is to serve as building blocks for DNA. Although ribonucleotides similarly serve as the units for RNA synthesis, they also have a multitude of other functions in cell metabolism. In some synthetic nucleotides with therapeutic properties, other pentose components, such as arabinose, are present. (See fig. 13.2 for the structure of arabinose.)

The phosphoryl group of nucleotides is most commonly substituted on the C-5′—OH of the pentose. However, in cyclic nucleotides a single phosphoryl group is esterified to both the C-5′—OH and the C-3′—OH (e.g., see fig. 14.19). In nucleic acids each nucleotide unit has one phosphoryl group esterified to the C-5′—OH and another at the C-3′—OH, and these phosphoryl groups link the nucleotide units.

A nucleoside has no phosphoryl groups; it consists of a purine or pyrimidine linked to ribose or deoxyribose. The simplest nucleotides are therefore nucleoside-5′-monophosphates. However, nucleoside-5′-diphosphates and nucleoside-5′-triphosphates (fig. 26.4) are nucleotide forms that are also extremely important in cell metabolism. In diphosphates a second phosphoryl group is added to the first in acid anhydride linkage (also called pyrophosphate linkage), and in triphosphates a second anhydride bond links another phosphoryl group. The nucleoside diphosphates (NDPs) and the nucleoside triphosphates (NTPs) dissociate three and four protons, respectively, from their phosphate groups. In the

Figure 26.4

The general structure of a nucleoside monophosphate, diphosphate, and triphosphate.

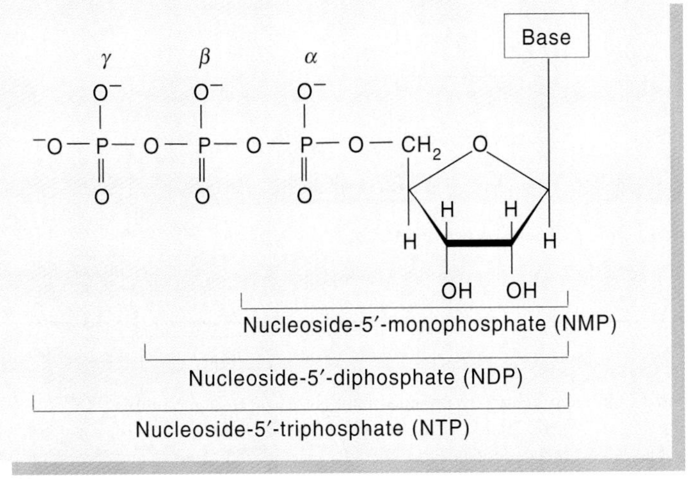

Figure 26.5

Tautomeric equilibrium of guanine and adenine. The keto and amino forms are strongly favored for guanine and adenine, respectively. Comparable tautomeric equilibria exist for thymine, uracil, and cytosine.

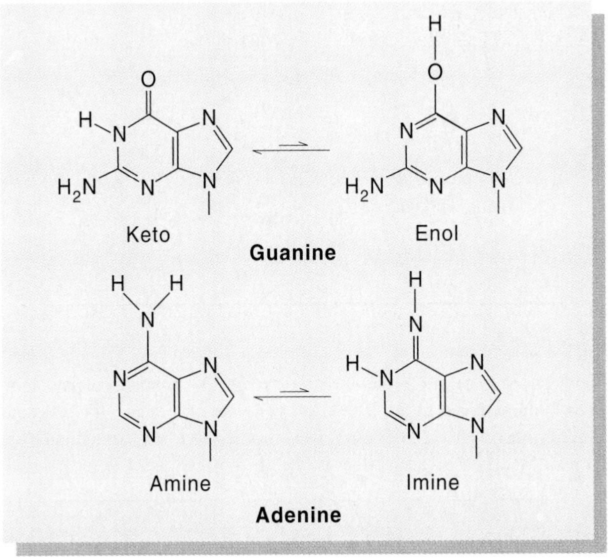

triphosphates the phosphate immediately attached in ester linkage to the 5′-carbon is designated α; the middle phosphate, in pyrophosphate linkage, is called β; and the terminal phosphate, also in pyrophosphate linkage, is called γ. The NDPs and NTPs can form complexes with Mg^{2+} or Ca^{2+} and probably exist in these complexes in the cell. The NDPs and NTPs have a number of important metabolic functions in the cell: They serve as energy-carrying enzyme cofactors (e.g., see chapters 14–17) and as substrates for the biosynthesis of nucleic acids (see chapters 31 and 33).

All of the commonly occurring bases in nucleotides are capable of existing in two tautomeric forms, which differ by the placement of a proton and some electrons. For example, guanosine can undergo a change from a keto form to an enol form as shown in figure 26.5. The keto form is so strongly favored that it is difficult to detect even trace amounts of the enol form at equilibrium. Similarly, the keto forms of thymidine and uridine are strongly preferred. Adenosine and cytidine can isomerize to imino forms, but the amino forms are strongly preferred (see fig. 26.5). Even though the unusual tautomers are present in very small amounts, it is conceivable that, when present in DNA, they contribute to the mutation process.

Some nucleotides undergo protonation in acid, and some undergo deprotonation in base; the relevant p*K* values are listed in table 26.2. At neutrality there is no charge on any of the bases. Three of the bases, A, C, and G, undergo protonation as the pH is lowered. The adenine moiety in AMP (adenosine monophosphate) protonates on the N-1 position of the purine rather than on the amino group (fig. 26.6). The charged form is stabilized by the resonance hybrids shown. In CMP the proton adds to the comparable N-3 ring nitrogen. In guanosine a proton adds to N-7 rather than the amino group (fig. 26.7), again indicating the unusually low basicity of the amino groups on the nucleotides compared with primary aliphatic amines. On the basic side of neutrality both UMP and GMP lose a proton from the imino nitrogens at positions 3

Table 26.2

Ionization Constants of the Ribonucleotides (Presented as p*K* Values)

	Base	Secondary Phosphate	Primary Phosphate
Adenosine-5′-phosphate (AMP)	3.8	6.1	0.9
Uridine-5′-phosphate (UMP)	9.4	6.4	1.0
Cytidine-5′-phosphate (CMP)	4.5	6.3	0.8
Guanosine-5′-phosphate (GMP)	2.4, 9.4	6.1	0.7

Figure 26.6

Uncharged and protonated forms of adenosine. The charged base resonates between the two structures shown on the right. Cytosine protonates in a similar way.

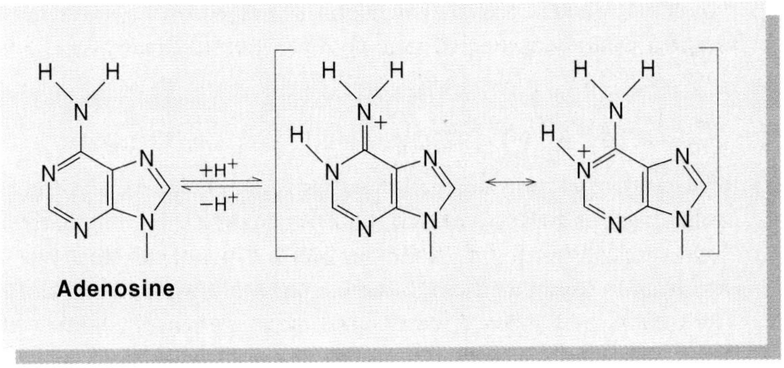

Figure 26.7

Uncharged and protonated forms of guanosine.

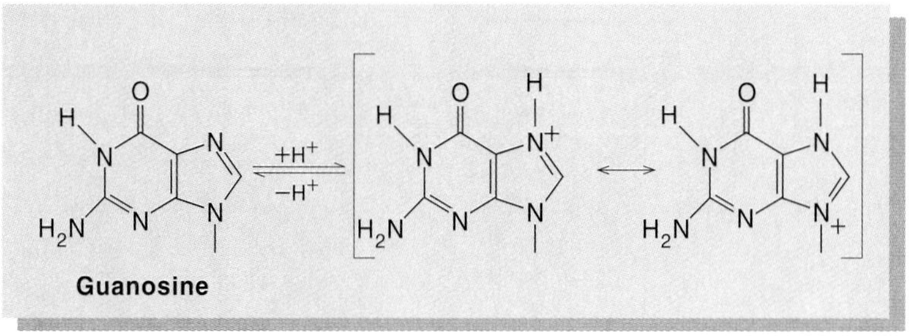

Guanosine

Figure 26.8

Absorption spectra of the common ribonucleoside-5′-monophosphates in protonated and unprotonated forms. The lower pH gives absorbance for the protonated form. Absorbance values are relative to the value at the maximum for each compound. Molar absorbances (that is, the absorbance of a 1-M solution) at the wavelength and pH giving the maximum absorbance are as follows: AMP, 15.4×10^3; UMP, 10.0×10^3; CMP, 13.2×10^3; GMP, 13.7×10^3. The comparable deoxyribonucleotides and the nucleoside di- and triphosphates have similar absorbance curves and molar absorbances.

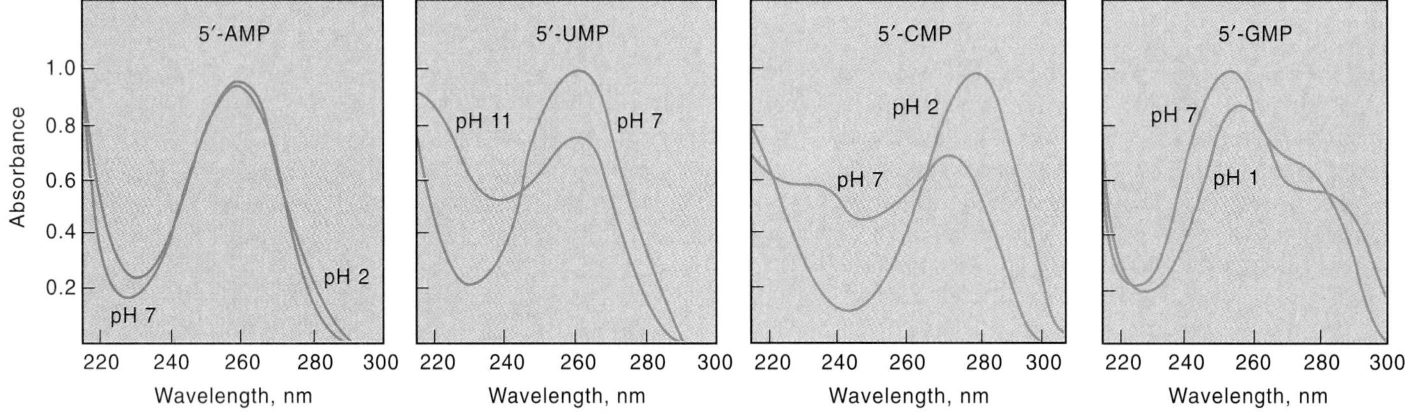

and 1, respectively. As you might expect, the ionization constants for the primary and secondary dissociations of the phosphate group do not differ appreciably for the various nucleosides (see table 26.2).

Owing to the large number of conjugated double bonds in their nitrogen bases, all nucleotides show absorption maxima in the near-ultraviolet range (fig. 26.8). The spectrum is pH-dependent, since protonation or deprotonation changes the electronic distribution in the base rings. The ultraviolet absorption of the nucleotides has been useful in many ways for the study of mononucleotides and polynucleotides. Indeed, it has been most useful in studying nucleic acid conformation (chapter 30).

Overview of Nucleotide Metabolism

The pathways by which cells synthesize, interconvert, and catabolize various purine and pyrimidine nucleotides are summarized schematically in figures 26.9 and 26.10. Cells of different types, or even the same cells in different stages of development, differ greatly in their ability to carry out some of the reactions involved, with some cells favoring one set of reactions and others another. In the rest of the chapter we will deal with the details of these pathways.

Synthesis of Purine Ribonucleotides de Novo

Purine nucleotides can be synthesized in three ways: by de novo synthesis, by reconstruction from purine bases through the addition of the ribose phosphate moiety, or by phosphorylation of nucleosides. The first two pathways are the more important quantitatively, and in both of them, phosphoribosylpyrophosphate (PRPP; α-D-ribofuranose 1-pyrophosphate 5-phosphate) is an essential precursor. It has a similar important role in pathways for pyrimidine nucleotide biosynthesis. PRPP is synthesized from ribose-5-phosphate, which cells synthesize either from glucose-6-phosphate by an oxidative pathway or from intermediates of glycolysis by a nonoxidative pathway (see chapter 14).

The formation of PRPP from ribose-5-phosphate and ATP is catalyzed by ribose-5-phosphate pyrophosphokinase (fig. 26.11). This is an unusual kinase because the pyrophosphoryl

Figure 26.9

Pathways of purine metabolism. Double-headed arrows indicate reversible enzymatic reactions. Separate arrows in opposite directions between metabolites indicate a different enzyme participating in each direction. The diagram is arranged in tiers: purines at the bottom, nucleosides at the next level, then nucleoside mono-, di-, and triphosphates (in ascending order).

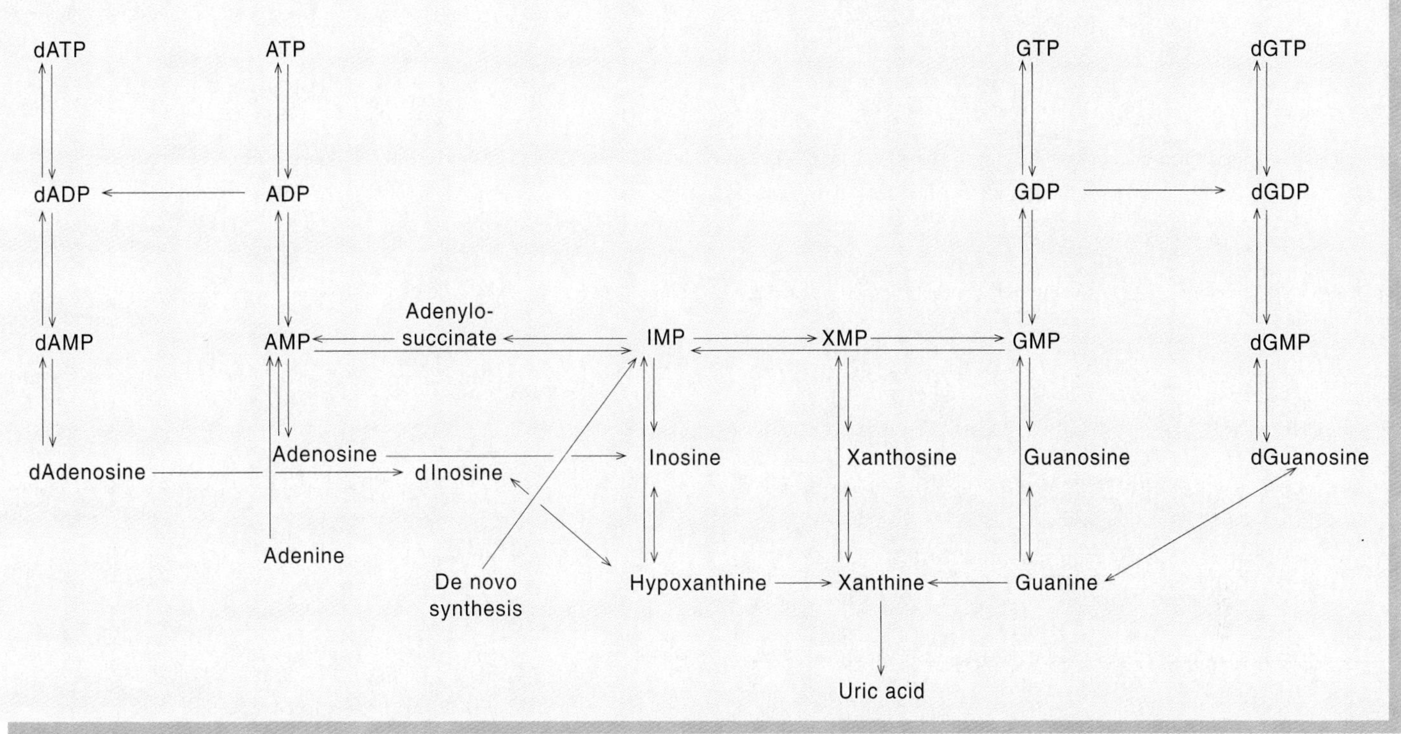

Figure 26.10

Pathways of pyrimidine metabolism. Arrows and layout have a similar significance as in figure 26.9.

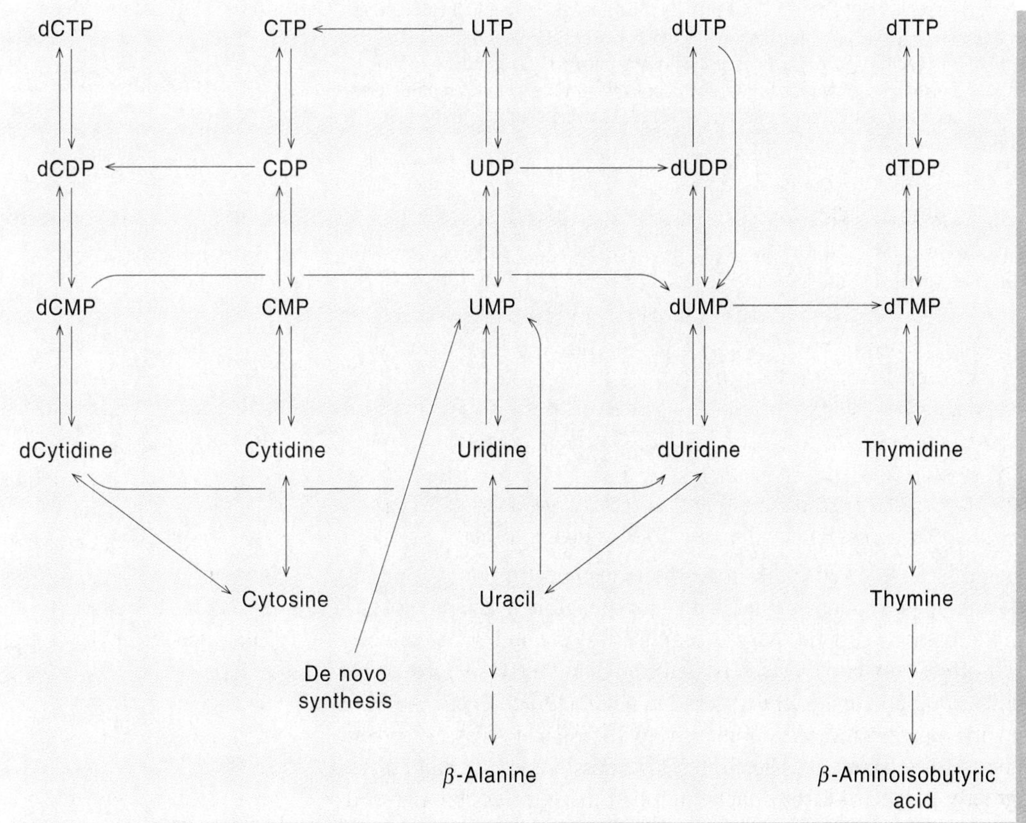

Figure 26.11

Synthesis of phosphoribosylpyrophosphate (PRPP). This is an unusual kinase-catalyzed reaction because the group transferred is the pyrophosphate group rather than the phosphate group.

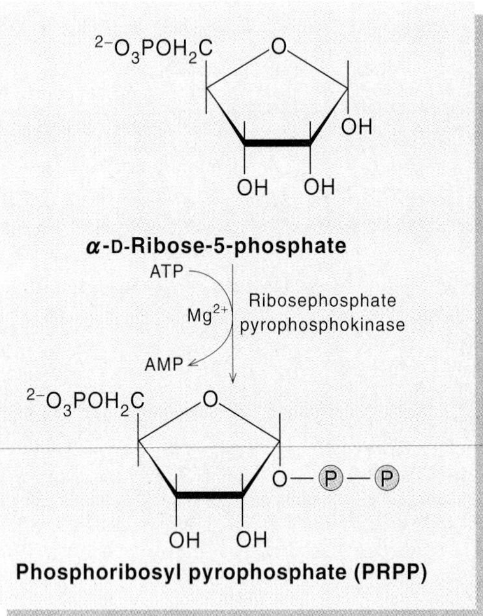

Figure 26.12

Precursors of the purine ring of uric acid in pigeons as determined by isotope labeling experiments. The indicated precursor substances were administered one at a time to pigeons. Each precursor was labeled with isotopic nitrogen or carbon, and in each case the excreted uric acid was purified and degraded chemically. The isotope content of the various degradation products indicated that precursors contributed the specific atoms indicated. Glycine contributed the two bridge carbons (4 and 5) as well as N7.

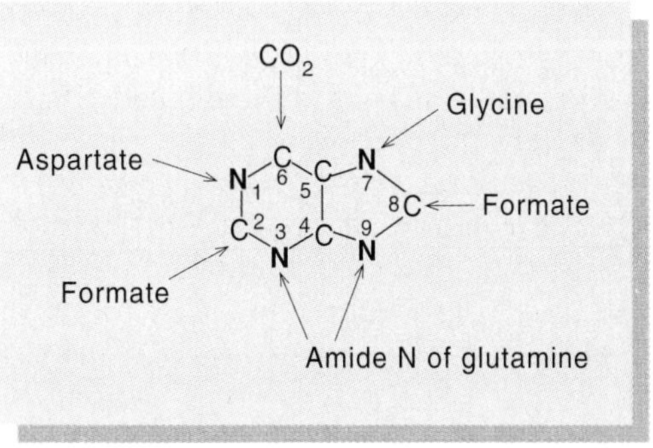

group is transferred rather than the phosphoryl group. As might be anticipated for an enzyme with a key position in several biosynthetic pathways, its activity is regulated by a number of metabolites. Inorganic phosphate is an activator, and the Mg^{2+} ion is both cofactor and activator. Inhibitors include ADP and glycerate-2,3-bisphosphate, which are competitive with respect to ribose-5-phosphate. By contrast, AMP and GDP are noncompetitive inhibitors. At any specific time the concentration of these metabolites as well

Figure 26.13

Summary of incorporation of precursors into the purine ring of IMP. Formate as well as other indirect donors of one carbon, such as serine, donate their atoms by means of the formyl group of a folate derivative in steps 3 and 9.

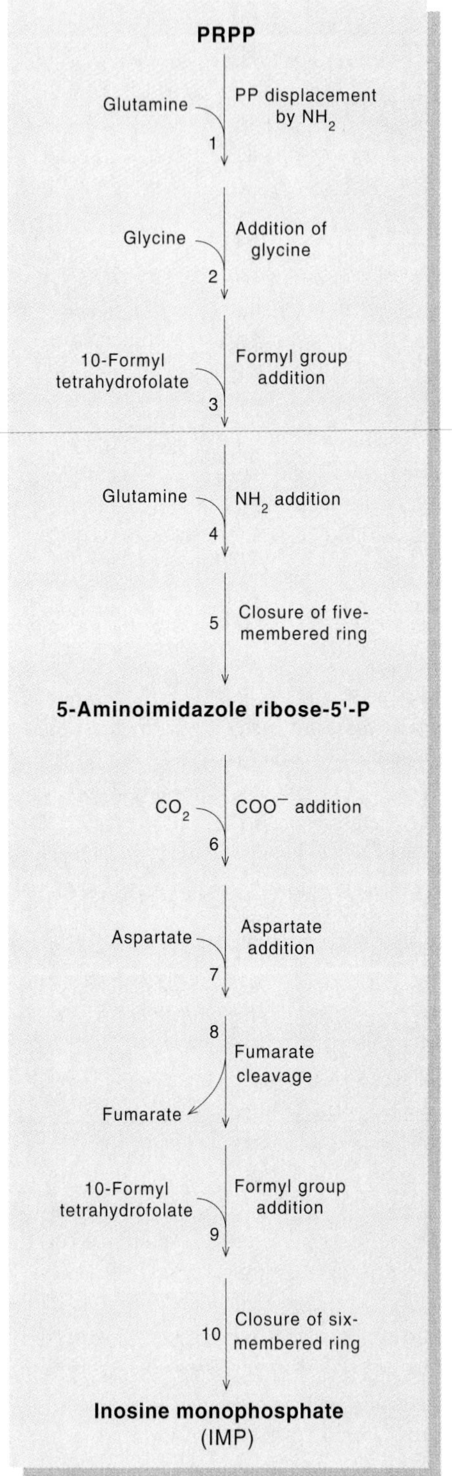

as those of the substrates will determine the activity of the kinase. In particular, the nucleotide inhibitors will curtail synthesis of PRPP when the energy stores of the cell are low.

The ultimate precursors of the purine ring were established by administering isotopically labeled compounds to pigeons and tracing the incorporation of labeled atoms into the purine ring of uric acid. Birds were used in these experiments because they excrete waste nitrogen largely as uric acid, a purine derivative that is easily isolated in pure form. Chemical degradation of the uric acid revealed the origins of the atoms as depicted in figure 26.12. A flow scheme showing the successive incorporation of atoms from these precursors is shown in figure 26.13.

Inosine Monophosphate (IMP) Is the First Purine Nucleotide Formed

The pathway from PRPP to the first complete purine nucleotide, inosine monophosphate, involves ten steps and is shown in figure 26.14. It would seem logical that the purine should be built up first, followed by addition of ribose-5-phosphate, but this is not the case. The starting point is PRPP, to which the imidazole ring is added; the six-membered ring is built up afterward.

The first step, in which phosphoribosylamine is formed, is catalyzed by glutamine phosphoribosylpyrosphosphate amidotransferase, an enzyme containing nonheme iron. The reaction involves inversion of the configuration at C-1 of the ribose and leads to the β configuration that is characteristic of naturally occurring nucleotides. This step involves commitment of PRPP to the purine biosynthetic pathway and, as you might expect, is subject to important feedback inhibitory effects by purine nucleotides. We will examine these effects a little later.

In step 2 the synthase forms an amide bond between the carboxyl group of glycine and the amino group of phosphoribosylamine, with ATP supplying energy and being hydrolyzed to ADP and inorganic phosphate. After these two steps have introduced atoms 4, 5, 7, and 9 of the purine ring, the remaining atoms are introduced one by one (steps 3, 4, 6, 7, and 9).

In step 3, carbon 8 is introduced as a formyl group that is transferred from 10-formyltetrahydrofolate (10-formyl-H_4 folate). (To review the way tetrahydrofolate derivatives accept a one-carbon unit from donors such as serine, glycine, or formate and transfer it to a suitable acceptor in biosynthetic reactions, see chapters 11 and 24; in chapters 11 and 24 10-formyl H_4folate is written as N^{10}-formyl H_4folate, an alternative terminology that is also still in use.) Now the five components that will constitute the imidazole part of the purine are present, but before ring closure occurs, N-3 of the purine ring is introduced (step 4) by transfer of another amino group from glutamine to phosphoribosylformylglycinamide. ATP provides energy for the amido group transfer, being itself hydrolyzed to ADP and phosphate.

The imidazole ring is closed in an essentially irreversible cyclization requiring the presence of Mg^{2+} and K^+ (step 5). Then C-6 of the purine ring is introduced by addition of bicarbonate in the presence of a specific carboxylase (step 6). This carboxylation is unusual in that it does not seem to involve biotin and is not coupled with any energy-yielding process such as ATP hydrolysis. The equilibrium is unfavorable for the formation of the carboxylate. In vivo the reaction proceeds at physiological concentrations of bicarbonate because of coupling with subsequent steps that are thermodynamically favorable.

Next, in steps 7 and 8, N-1 of the purine ring is contributed by aspartate. Aspartate forms an amide with the 4-carboxyl group, and the succinocarboxamide so formed is then cleaved with release of fumarate. Energy for carboxamide formation is provided by ATP hydrolysis to ADP and phosphate. These reactions resemble the conversion of citrulline to arginine in the urea cycle (chapter 25) and the conversion of IMP to AMP (see fig. 26.16). It is worth noting that fumarate released in all these synthetic pathways can be converted to oxaloacetate by fumarase and malate dehydrogenase, and in the process, NAD^+ is reduced to NADH (by malate dehydrogenase). Reoxidation of the NADH by the electron-transport chain then generates three ATPs from ADP and P_i. The ATPs supply some of the energy required for purine synthesis, and the oxaloacetate formed can be used to replenish the supply of aspartate by transamination with glutamate (fig. 26.15).

The final atom of the purine ring is provided in step 9 by donation of a formyl group from 10-formyltetrahydrofolate to the 5-amino group of the almost completed ribonucleotide. In the final step, ring closure is effected by elimination of water to form IMP (inosine monophosphate), the first product with a complete purine ring. Although this final ring closure does not require energy from ATP, closure of the imidazole ring does, and the synthesis of IMP from ribose-5-phosphate requires a total of six high-energy phosphate groups from ATP (assuming hydrolysis of pyrophosphate released during phosphoribosylamine synthesis, step 1 of figure 26.14).

IMP Is Converted into AMP and GMP

IMP does not accumulate in the cell but is converted to AMP, GMP, and the corresponding diphosphates and triphosphates. The two steps of the pathway from IMP to AMP (fig. 26.16) are typical reactions by which the amino group from aspartate is introduced into a product. The 6-hydroxyl group of IMP (tautomeric with the 6-keto group) is first displaced by the amino of aspartate to give adenylosuccinate, and the latter is then cleaved nonhydrolytically by adenylosuccinate lyase to yield fumarate and AMP. In the condensation of aspartate with IMP, cleavage of GTP to GDP and phosphate provides energy to drive the reaction.

Conversion of IMP to GMP also proceeds by a two-step pathway (see fig. 26.16): first, dehydrogenation of IMP to xanthosine-5′-phosphate (XMP), and second, transfer of an amino group from glutamine to C-2 of the xanthine ring to yield GMP. The second reaction also involves the cleavage of ATP to AMP and inorganic pyrophosphate. The latter is in turn hydrolyzed to inorganic phosphate by the ubiquitous inorganic pyrophosphatase in a reaction with a very favorable equilibrium. This hydrolysis is coupled with the GMP synthase reaction because pyrophosphate is a product of the latter and a substrate of the pyrophosphatase. The net result is that the release of two high-energy phosphate groups is used to drive the GMP synthase reaction to completion.

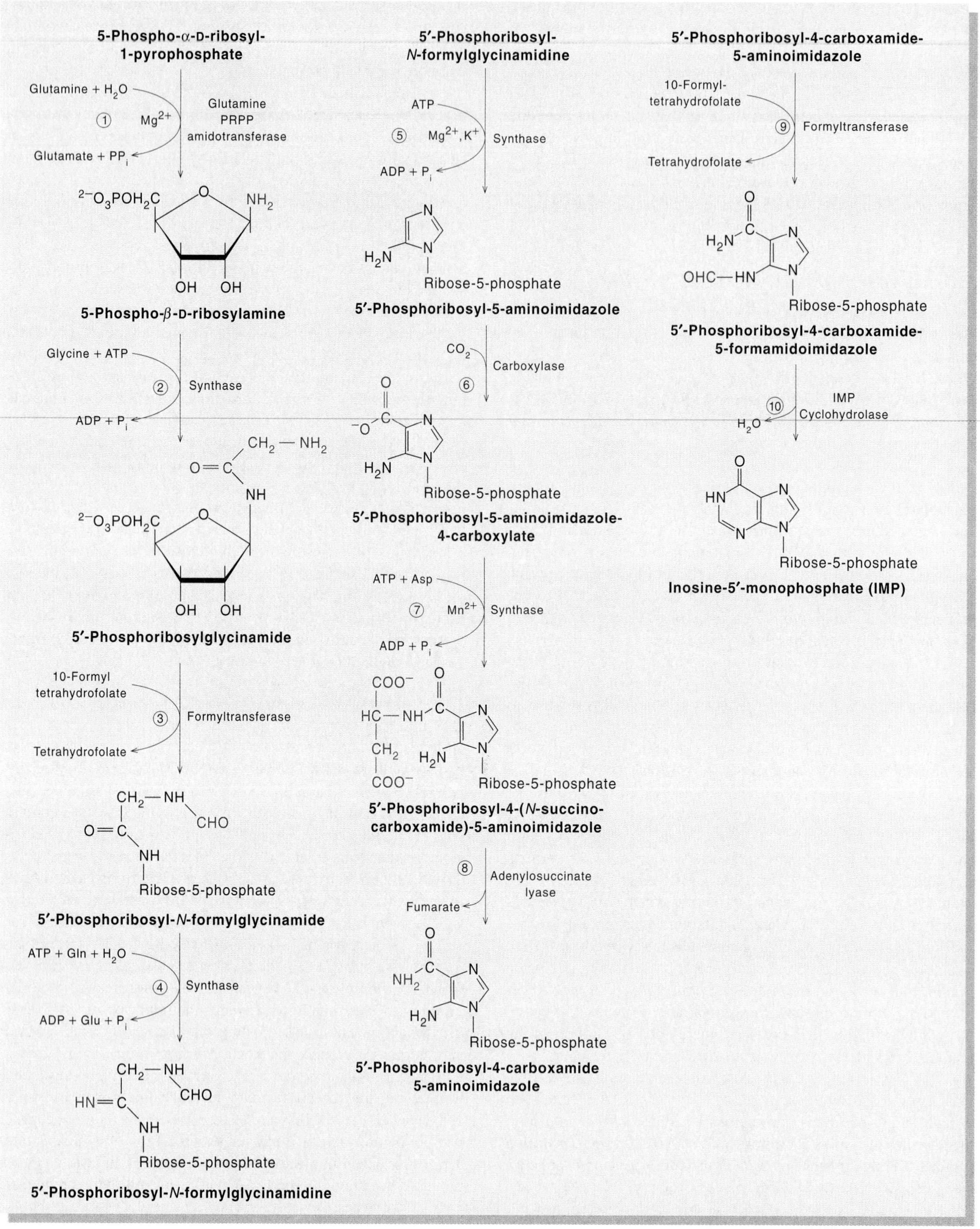

5-Phospho-α-D-ribosyl-1-pyrophosphate

Glutamine + H_2O → Glutamate + PP_i
① Mg^{2+} Glutamine PRPP amidotransferase

$^{2-}O_3POH_2C$ ... NH_2 ... OH OH

5-Phospho-β-D-ribosylamine

Glycine + ATP → ADP + P_i
② Synthase

CH_2 — NH_2
O═C
NH
$^{2-}O_3POH_2C$ ... OH OH

5′-Phosphoribosylglycinamide

10-Formyl tetrahydrofolate → Tetrahydrofolate
③ Formyltransferase

CH_2—NH
CHO
O═C
NH
Ribose-5-phosphate

5′-Phosphoribosyl-*N*-formylglycinamide

ATP + Gln + H_2O → ADP + Glu + P_i
④ Synthase

CH_2—NH
CHO
HN═C
NH
Ribose-5-phosphate

5′-Phosphoribosyl-*N*-formylglycinamidine

5′-Phosphoribosyl-*N*-formylglycinamidine

ATP → ADP + P_i
⑤ Mg^{2+}, K^+ Synthase

N
H_2N
N
Ribose-5-phosphate

5′-Phosphoribosyl-5-aminoimidazole

CO_2
⑥ Carboxylase

O
^-O C
N
H_2N
N
Ribose-5-phosphate

5′-Phosphoribosyl-5-aminoimidazole-4-carboxylate

ATP + Asp → ADP + P_i
⑦ Mn^{2+} Synthase

COO^-
O
HC — NH — C
N
CH_2
H_2N
N
COO^-
Ribose-5-phosphate

5′-Phosphoribosyl-4-(*N*-succinocarboxamide)-5-aminoimidazole

⑧ Adenylosuccinate lyase
→ Fumarate

O
NH_2 C
N
H_2N
N
Ribose-5-phosphate

5′-Phosphoribosyl-4-carboxamide 5-aminoimidazole

5′-Phosphoribosyl-4-carboxamide-5-aminoimidazole

10-Formyl-tetrahydrofolate → Tetrahydrofolate
⑨ Formyltransferase

O
H_2N C
N
OHC—HN
N
Ribose-5-phosphate

5′-Phosphoribosyl-4-carboxamide-5-formamidoimidazole

⑩ IMP Cyclohydrolase
→ H_2O

O
HN
N
N
N
Ribose-5-phosphate

Inosine-5′-monophosphate (IMP)

Figure 26.14

Biosynthetic pathway to inosine monophosphate. Red indicates the group or atom introduced at each step. The first step, in which phosphoribosylamine is formed, involves inversion of the configuration at C-1 of the ribose. In step 2 an amide bond is formed by the synthase between the carboxyl group of glycine and the amino group of phosphoribosylamine. An additional formyl group is added to the glycine amino group in the next step. An additional amide group is donated by a glutamine, and next the five-membered imidazole ring is formed. Further groups are introduced through successive attachments to the imidazole, and finally the purine ring is completed in inosine-5′-monophosphate.

Figure 26.15

Utilization of fumarate formed in reactions in which aspartate acts as donor of an amino group to generate ATP and regenerate aspartate (ETC = electron-transport chain).

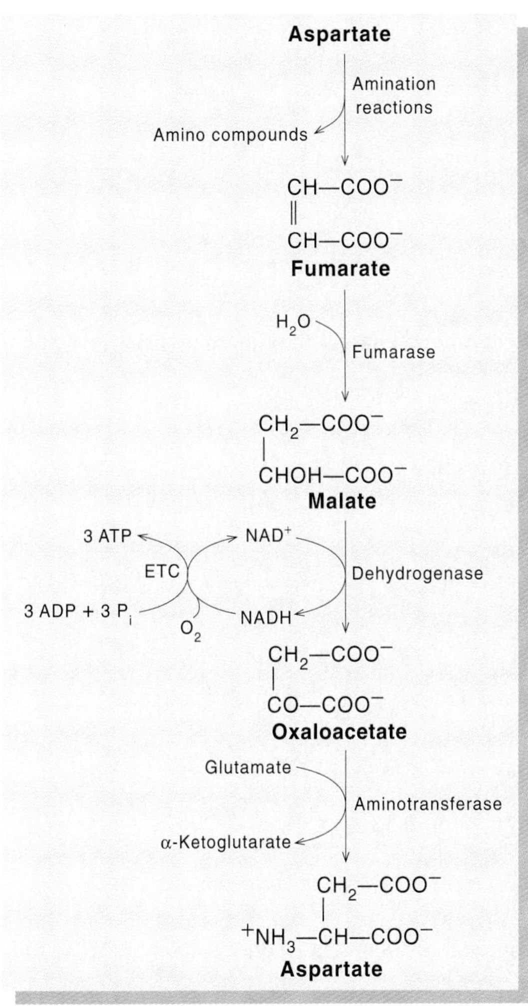

Synthesis of Pyrimidine Ribonucleotides de Novo

Like purine nucleotides, pyrimidine nucleotides can be synthesized either de novo or by the "salvage pathways" from nucleobases or nucleosides (see fig. 26.10). However, salvage is less efficient because, except in the case of utilization of uracil, by bacteria, and to some extent by mammalian cells, nucleobases are not converted to nucleotides directly but only via nucleosides.

The biosynthetic pathway to pyrimidine nucleotides is simpler than that for purine nucleotides, reflecting the simpler structure of the base. In contrast to the biosynthetic pathway for purine nucleotides, in the pyrimidine pathway the pyrimidine ring is constructed before ribose-5-phosphate is incorporated into the nucleotide. The first pyrimidine mononucleotide to be synthesized is orotidine-5′-monophosphate (OMP), and from this compound, pathways lead to nucleotides of uracil, cytosine, and thymine. OMP thus occupies a central role in pyrimidine nucleotide biosynthesis, somewhat analogous to the position of IMP in purine nucleotide biosynthesis. Like IMP, OMP is found only in low concentrations in cells and is not a constituent of RNA.

Early clues to the nature of the pyrimidine pathway were provided by the observations that orotic acid (6-carboxyuracil, fig. 26.17) can satisfy the growth requirement of mutants of the fungus *Neurospora* that are unable to make pyrimidines and that isotopically labeled orotate is an immediate precursor of pyrimidines in *Neurospora* and a number of bacteria.

UMP Is a Precursor of Other Pyrimidine Mononucleotides

The pathway for UMP synthesis is shown in figure 26.18. It starts with the synthesis of carbamoyl phosphate, catalyzed by carbamoyl phosphate synthase. This enzyme is present in microorganisms and in the cytosol of all eukaryotic cells that are capable of forming pyrimidine nucleotides. Eukaryotes also have another carbamoyl phosphate synthase, a distinct enzyme that uses ammonia as a substrate instead of glutamine. It is associated with citrulline formation in the pathway for arginine biosynthesis, and in mammals it is a mitochondrial enzyme that is present predominantly in the liver, where it catalyzes a step in the urea cycle (chapter 25).

Carbamoyl phosphate synthase does not contain biotin and is not activated by it. The enzyme product, carbamoyl phosphate, next reacts with aspartate to form carbamoyl aspartate. The reaction is catalyzed by aspartate carbamoyltransferase (aspartate transcarbamoylase or aspartate transcarbamylase), and the equilibrium greatly favors carbamoyl aspartate synthesis.

In the third step the pyrimidine ring is closed by dihydroorotase to form L-dihydroorotate. Dihydroorotate is then oxidized to orotate by dihydroorotate dehydrogenase. This flavoprotein in some organisms contains FMN and in others both FMN and FAD. It also contains nonheme iron and sulfur. In eukaryotes it is a lipoprotein associated with the inner membrane of the mitochondria. In the final two steps of the pathway, orotate phosphoribosyltransferase yields orotidine-5′-phosphate (OMP), and a specific decarboxylase then produces UMP.

Low activities of orotidine phosphate decarboxylase and (usually) orotate phosphoribosyltransferase are associated with a genetic disease in children that is characterized by abnormal growth, megaloblastic anemia, and the excretion of large amounts of orotate. When affected children are fed a pyrimidine nucleoside, usually uridine, the anemia decreases and the excretion of orotate diminishes. A likely explanation for the improvement is

Figure 26.16

Conversion of IMP to AMP and GMP. In both cases, two steps are required. Note that the formation of AMP requires GTP and the formation of GMP requires ATP. This tends to balance the flow of the IMP down the two pathways.

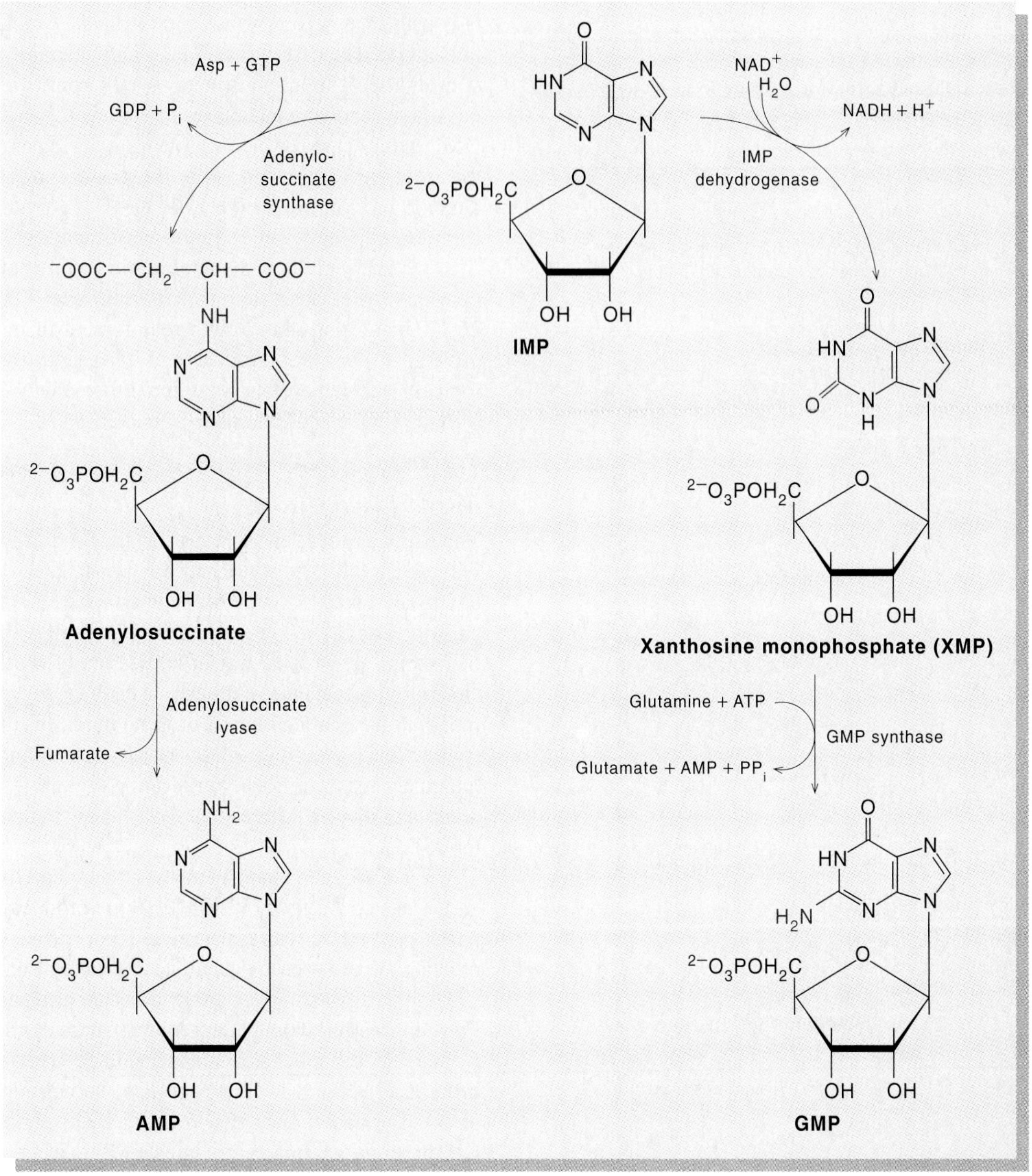

Figure 26.17

The structure of orotic acid. Loss of proton leads to orotate. Deposits of sodium orotate cause a painful condition.

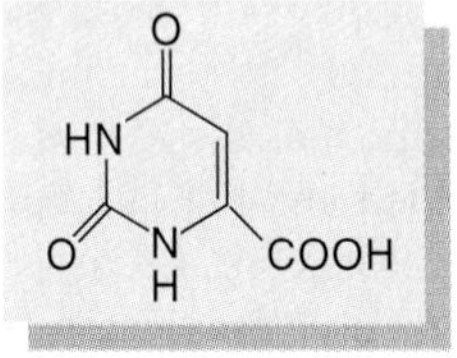

that the ingested uridine is phosphorylated to UMP, which is then converted to other pyrimidine nucleotides so that nucleic acid and protein synthesis can resume. In addition, the increased intracellular concentrations of pyrimidine nucleotides inhibit carbamoyl phosphate synthase, the first enzyme in the pathway of orotate synthesis.

CTP Is Formed from UTP

After UMP is synthesized, it is phosphorylated to UTP. Then CTP is formed by reaction of UTP with glutamine in a reaction driven by the concomitant hydrolysis of ATP to ADP and inorganic phos-

Figure 26.18

Biosynthesis of UMP. Red indicates the parts of the intermediates derived from aspartate. Bold type indicates atoms derived from carbamoyl phosphate. In contrast to purine nucleotide synthesis, where ring formation starts on the sugar, in pyrimidine biosynthesis the pyrimidine ring is completed before being attached to the ribose.

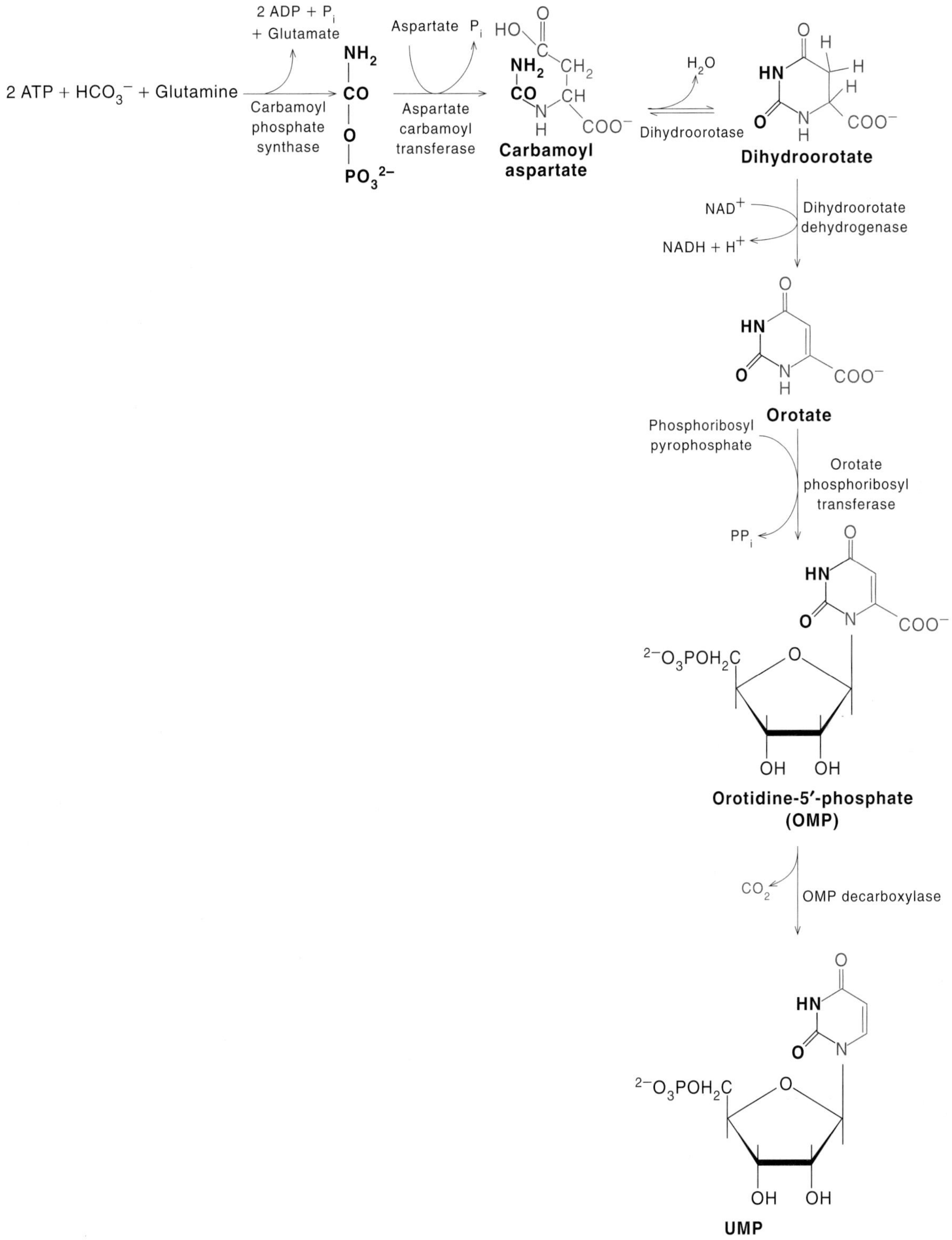

Figure 26.19

The production of CTP by amination of UTP. The conversion of the pyrimidine ring of uracil to cytosine occurs at the level of the nucleoside triphosphate. The enzyme responsible for this conversion is known as cytidine triphosphate synthase.

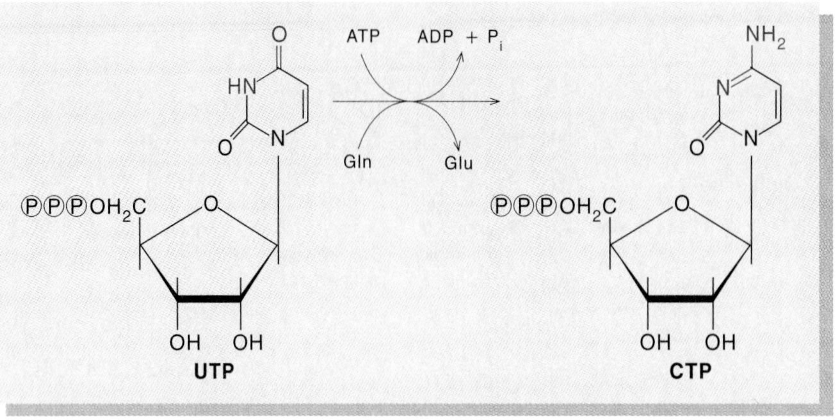

phate (fig. 26.19). Both the mammalian and bacterial cytidine triphosphate synthases can use ammonia as a donor in place of glutamine, but this reaction is of no physiological significance because the K_m for ammonia is very high and the reaction rate is low. With glutamine as the amino donor, GTP is an allosteric activator for CTP synthase from *E. coli* and probably for the mammalian enzyme as well (fig. 26.20). However, there is no stimulation of CTP synthesis from ammonia, a result interpreted to mean that the allosteric effect of GTP is specifically on the release of ammonia from glutamine. This ammonia is then channeled into reaction with UTP in the enzyme active site.

Formation of Deoxyribonucleotides by Reduction of Ribonucleotides

Tracer studies with isotopically labeled precursors have shown that both in mammalian tissues and in microorganisms, deoxyribonucleotides are formed from corresponding ribonucleotides by replacement of the 2′—OH group with hydrogen.

There are three types of ribonucleotide reductase that catalyze this reduction of the ribose ring. The first is widely distributed in nature, occurring in mammalian and plant cells, in yeast, and in some prokaryotes. This type of reductase contains a tyrosyl radical closely associated with nonheme iron, and the prototype is the reductase from *E. coli.* The latter consists of two nonidentical subunits, both contributing to the active site, and is specific for the reduction of diphosphates (ADP, GDP, CDP, and UDP).

The second type of ribonucleotide reductase is restricted to certain microorganisms, including some bacteria, several species of algae, and at least one fungus. Ribonucleotide reductases of this type, for which the enzyme from *Lactobacillus leichmannii* is the prototype, requires adenosyl cobalamin (coenzyme B_{12}) as an obligatory coenzyme. It does not contain nonheme iron; it uses either nucleoside diphosphates or triphosphates, depending on the source of the enzyme; and it consists of only one type of polypeptide chain (although this may form oligomers in some cases). A third type of reductase present in some prokaryotes contains manganese.

Figure 26.20

Cytidine triphosphate synthase is activated by GTP.

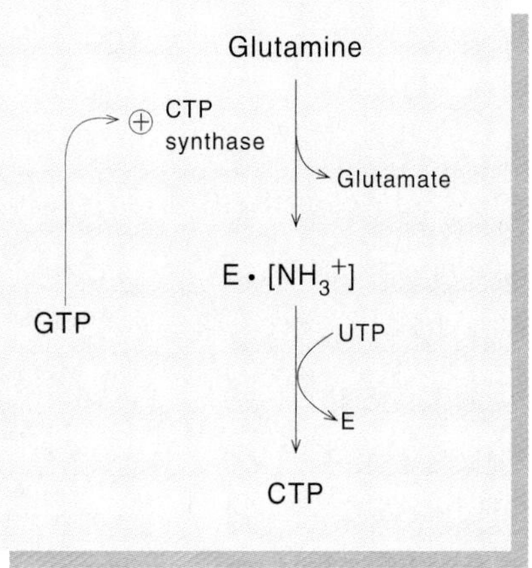

An unusual feature of ribonucleotide reductase is that the reaction it catalyzes involves a radical mechanism. The *E. coli* type of reductase initiates this reaction by the tyrosyl radical-nonheme iron, whereas the *L. leichmannii* reductase initiates radical formation by homolytic cleavage of the C—Co bond of the adenosylcobalamin coenzyme (coenzyme B_{12}). Hydroxyurea and related compounds inhibit the mammalian reductase by abolishing the radical state of the tyrosine residue. Inhibition of DNA synthesis by such compounds is secondary to this effect.

For both major types of reductase the physiological reducing substrate is a low-molecular-weight (13,000) electron-transport protein, thioredoxin. Thioredoxin has two half-cystine residues that are separated in the polypeptide chain by two other residues. The oxidized form of thioredoxin, with a disulfide bridge between the half-cystines, is reduced by NADPH in the presence of a flavoprotein, thioredoxin reductase. The reduced form of

Figure 26.21

Ribonucleotide reductase and the thioredoxin system. In some lactobacilli, vitamin B_{12} is involved in the reduction of ribonucleotide tridiphosphate to deoxyribonucleotide.

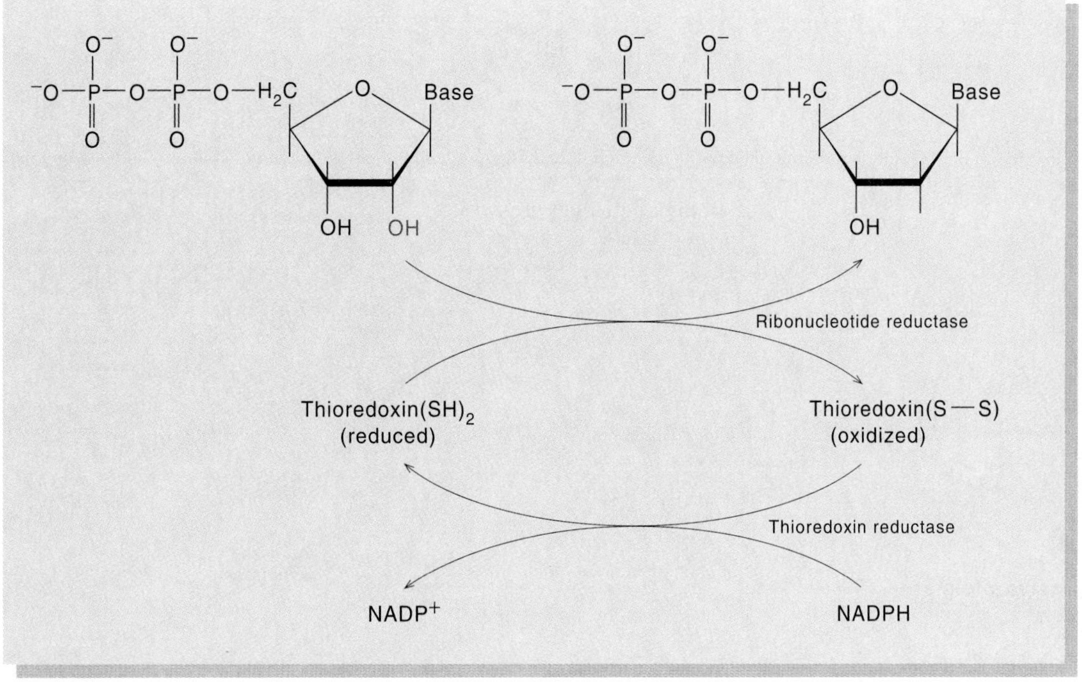

thioredoxin, with two cysteine residues present, is the reducing substrate for ribonucleotide reduction. The flow of electrons from NADPH to ribose is shown in figure 26.21.

An alternative electron-transport system for ribonucleotide reduction has been discovered in *E. coli.* In this case the ultimate source of electrons is again NADPH, but they are passed to glutathione (in a reaction catalyzed by glutathione reductase), and the reduced glutathione, in turn, reduces a small protein called glutaredoxin. It is the reduced glutaredoxin that acts as the reducing substrate in ribonucleotide reduction. The distribution of glutaredoxin in nature remains to be determined. Neither thioredoxin nor glutaredoxin is essential to *E. coli,* but the bacteria cannot survive when both proteins are lost.

Thymidylate Is Formed from dUMP

Since DNA contains thymine (5-methyluracil) as a major base instead of uracil, the synthesis of thymidine monophosphate (dTMP or thymidylate) is essential to provide dTTP (thymidine triphosphate), needed for DNA replication together with dATP, dGTP, and dCTP.

Thymidylate is synthesized from dUMP, and there are two pathways by which the latter may be formed in cells. The major precursor of dUMP is dCMP, which is converted to dUMP by deoxycytidylate deaminase:

$$\text{dCMP} + H_2O \longrightarrow \text{dUMP} + NH_3 \qquad \textbf{(1)}$$

This enzyme is widely distributed in animal tissues, and the enzyme produced by yeast, by bacteriophage T2-infected *E. coli,* and by animal cells has been studied extensively. Deamination of 5-methyldeoxycytidylate and 5-hydroxymethyldeoxycytidylate is also catalyzed by this enzyme

Another route to dUMP is the reduction of UDP to dUDP, followed by phosphorylation of dUDP to dUTP (or direct reduction of UTP to dUTP in some microorganisms). The dUTP is then hydrolyzed to dUMP. This circuitous route to dUMP is dictated by two considerations. First, the ribonucleotide reductase in most cells acts only on ribonucleoside diphosphates, probably because this permits better regulation of its activity. Second, cells contain a powerful deoxyuridine triphosphate diphosphohydrolase (dUTPase). It prevents the incorporation of dUTP into DNA by keeping intracellular levels of dUTP low by means of the reaction

$$\text{dUTP} + H_2O \longrightarrow \text{dUMP} + PP_i \qquad \textbf{(2)}$$

Methylation of dUMP to give thymidylate is catalyzed by thymidylate synthase and utilizes 5,10-methylenetetrahydrofolate as the source of the methyl group. This reaction is unique in the metabolism of folate derivatives in that the folate derivative acts both as a donor of the one-carbon group and also as its reductant, using the reduced pteridine ring as the source of reducing potential. Consequently, in this reaction, unlike any other in folate metabolism, dihydrofolate is a product (fig. 26.22). Since folate derivatives are present in cells at very low concentrations, continued synthesis of thymidylate requires regeneration of 5,10-methylenetetrahydrofolate from dihydrofolate. As shown in figure

Figure 26.22

Thymidylate biosynthesis. (R = *p*-aminobenzoyl-L-glutamate.) Colored symbols indicate atoms that are precursors of the methyl group of thymidylate. The dTMP is formed from dUMP. 5,10-methylenetetrahydrofolate donates the methyl group in this reaction. This methyl group donor is regenerated by a two-step process involving first the reduction of 7,8-dihydrofolate to tetrahydrofolate and then condensation of the hydroxymethyl group of serine with the latter to form 5,10-methylenetetrahydrofolate.

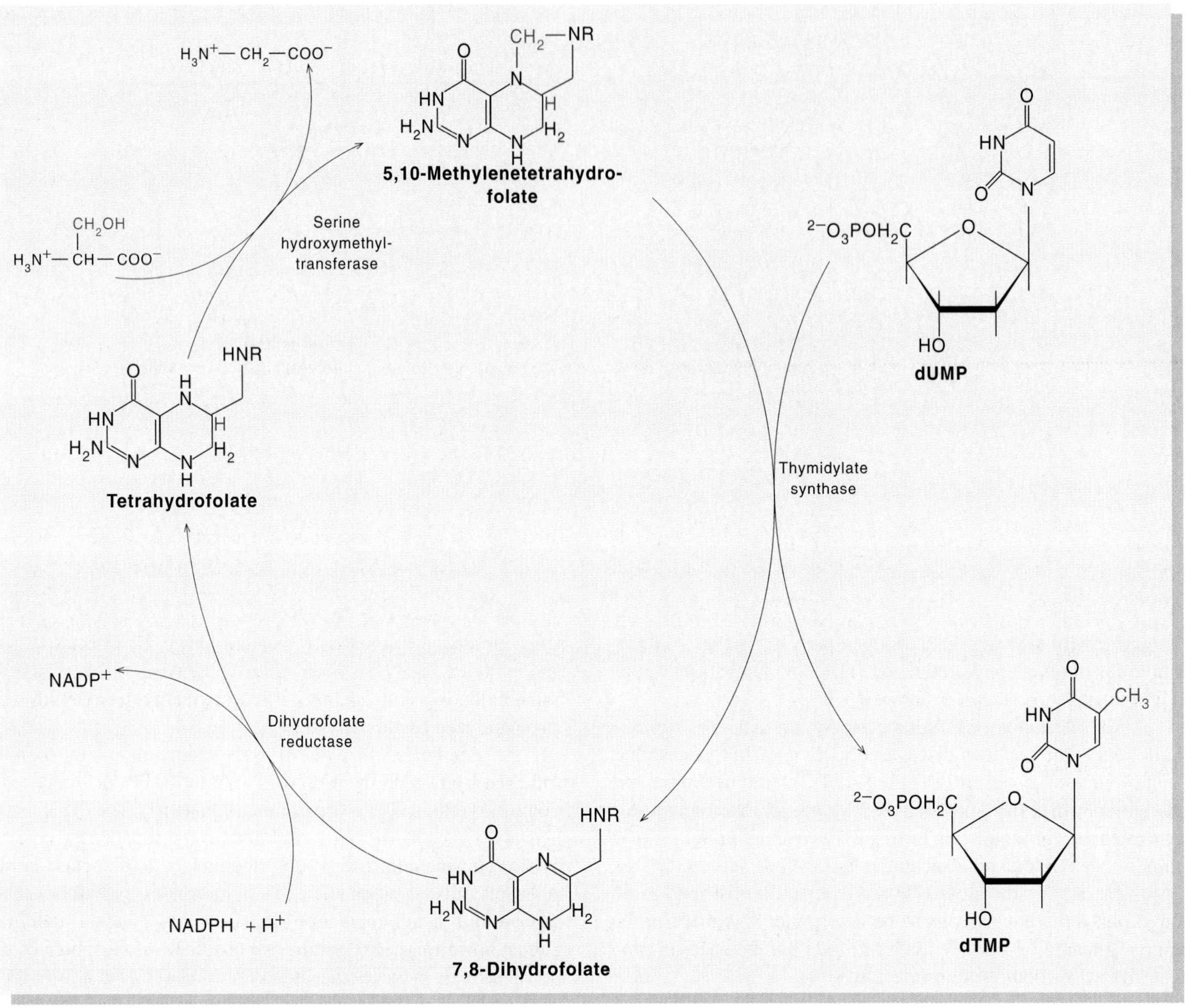

26.22, this occurs in two steps catalyzed by the enzymes dihydrofolate reductase, an $NADP^+$-linked dehydrogenase, and serine hydroxymethyltransferase. Thymidylate synthesis can be interrupted, and consequently the synthesis of DNA arrested, by the inhibition of either thymidylate synthase or dihydrofolate reductase. Many potent inhibitors are known for each of these enzymes, the best known being 5-fluoro-dUMP for thymidylate synthase and methotrexate, trimethoprim, and related compounds for dihydrofolate reductase. We will discuss these and other inhibitors of nucleotide synthesis in a later section.

You should note that reduced folates are present in cells as polyglutamate forms, with up to five additional glutamate residues attached to the terminal carboxyl of the folates. The glutamate residues are attached to each other in γ-peptide linkage. These polyglutamate forms of methylenetetrahydrofolate are the true substrates for thymidylate synthase; they have much lower K_m values and give higher maximum velocity than the monoglutamate. Analogous polyglutamate forms of 10-formyltetrahydrofolate are the true cofactors for purine synthesis. Human thymidylate synthase, as well as the enzyme from bacteria, has been crystallized and the three-dimensional structure deduced from x-ray diffraction results.

Formation of Nucleotides from Bases and Nucleosides (Salvage Pathways)

In addition to the pathways for synthesis de novo, mammalian cells and microorganisms can readily form mononucleotides from purine and, to a lesser extent, pyrimidine bases and their nucleosides. In this way, bases and nucleosides formed by constant breakdown of mRNA and other nucleic acids can be reconverted to useful nucleotides, and the energy expended by the cell in synthesizing the bases is retained.

Transporter systems are present in the membranes of mammalian cells and of some microorganisms for the efficient uptake of bases and nucleosides. The carriers for bases are quite specific and mediate facilitated diffusion (chapter 20). Two types of carrier for facilitated diffusion of nucleosides are widely distributed in membranes of mammalian cells, and concentrative uptake of nucleosides occurs in cells of the intestinal epithelium and of the kidney tubule. Once inside the cell, the bases and nucleosides are converted to nucleotides by enzymes that are widely distributed in mammalian tissues and in microorganisms. The same transport systems and salvage enzymes are responsible for converting base or nucleoside analogs to therapeutically active nucleotide forms.

An ecto-5′-nucleotidase, which is attached to the outer surface of the plasma membrane of many types of mammalian cell, dephosphorylates purine and pyrimidine ribo- and deoxyribonucleoside monophosphates to the corresponding nucleosides. An important function of the ecto-nucleotidase may be the assimilation of nucleotides arising from the dissolution of dying cells. The nucleosides formed by the ecto-5′-nucleotidase are then transported into the cell and are reconverted to nucleotides by nucleoside kinases, as described in the following section.

Purine Phosphoribosyltransferases Convert Purines to Nucleotides

In mammals, specific enzymes for converting purine bases to nucleotides are present in many organs, and in heart muscle this may be the main source of purine nucleotides. The most important of these enzymes is hypoxanthine-guanine phosphoribosyltransferase, which catalyzes the formation of IMP from hypoxanthine and GMP from guanine:

$$\text{Hypoxanthine} + \text{PRPP} \rightleftharpoons \text{IMP} + \text{PP}_i \quad (3)$$

$$\text{Guanine} + \text{PRPP} \rightleftharpoons \text{GMP} + \text{PP}_i \quad (4)$$

The enzyme may also form some XMP from xanthine, but affinity of the enzyme for the latter is nearly 1,000-fold less than for guanine and hypoxanthine. This enzyme therefore returns the major products of purine nucleotide catabolism to nucleotide forms (see fig. 26.9). Some microorganisms contain a separate phosphoribosyltransferase for xanthine in addition to those for adenine and hypoxanthine-guanine.

The equilibria in these phosphoribosyltransferase reactions favor nucleotide synthesis, and since the inorganic pyrophosphate released is rapidly hydrolyzed by inorganic pyrophosphatase, the coupling of these reactions makes the synthesis of nucleotide irreversible. However, the efficiency of salvage is heavily dependent on the intracellular concentration of PRPP.

The role of hypoxanthine-guanine phosphoribosyltransferase in purine salvage has been confirmed by the abnormally high excretion of purines (as uric acid) in humans who lack hypoxanthine-guanine phosphoribosyltransferase. Studies of purine metabolism in cultures of cells from patients with this hereditary disorder also support this conclusion, as did a study involving patients who lack the enzyme xanthine oxidase (to be discussed later). The amounts of xanthine and hypoxanthine excreted by these patients were compared with estimates of the amounts synthesized daily. The results indicated that about 90% of the purines formed were reutilized. The high recovery rate is presumably possible because of the restricted distribution, in tissues of vertebrates, of catabolic enzymes acting on free purines.

Besides this salvage role, however, hypoxanthine-guanine phosphoribosyltransferase is probably important also for the transfer of purines from liver to other tissues. Purine biosynthesis de novo is especially active in the liver, and there is evidence to suggest that extrahepatic cells that have a low capacity for the synthesis of purines de novo, such as erythrocytes and bone marrow cells, depend on uptake of hypoxanthine and xanthine from the blood to fulfill their needs for purine nucleotides. It seems likely that blood levels of xanthine and hypoxanthine, which are normally about 0.40 mM, are maintained by release of these bases from the liver. Some evidence suggests that the bases released by the liver are largely taken up by red blood cells and converted to purine nucleotides that are later broken down again with release of bases to tissues. If this is the case, the factors regulating their release and breakdown are unknown. The uptake of purine bases by extrahepatic tissues, as well as by bacteria, appears to be closely linked to the activity of the purine phosphoribosyltransferases. At least in some cases this is so because the transferases are membrane proteins.

The neurologic disorder of children called the Lesch-Nyhan syndrome is due to a congenital lack of hypoxanthine-guanine phosphoribosyltransferase. The disorder is characterized by aggressive behavior, mental retardation, spastic cerebral palsy, and self-mutilation. Purine metabolism is profoundly disturbed, with greatly increased de novo biosynthesis of purines (200 times normal), overproduction of uric acid (6 times normal), and elevated blood levels of uric acid. Severe gout is caused by the latter in some individuals. The increased rate of purine biosynthesis is probably due to several factors (fig. 26.23). Because nucleotide production is depressed by the lack of phosphoribosyltransferase, the normal feedback inhibition that controls the production of PRPP amidotransferase is lifted. At the same time, decreased utilization of PRPP by phosphoribosyltransferase makes a higher intracellular concentration of PRPP available for PRPP amidotransferase. Finally, decreased nucleotide pools also would increase the PRPP level by deregulating the activity of ribose-5-phosphate pyrophosphokinase.

Adenine phosphoribosyltransferase catalyzes the conversion of adenine to AMP in many tissues by a reaction similar to that of hypoxanthine-guanine phosphoribosyltransferase but is quite distinct from the latter. It plays a minor role in purine sal-

Figure 26.23

Mechanism of overproduction of purine nucleotides in the congenital deficiency of hypoxanthine-guanine phosphoribosyltransferase. The loss of the transferase prevents the recycling of hypoxanthine and guanine. This increases uric acid production as well as the de novo synthesis of purine nucleotides.

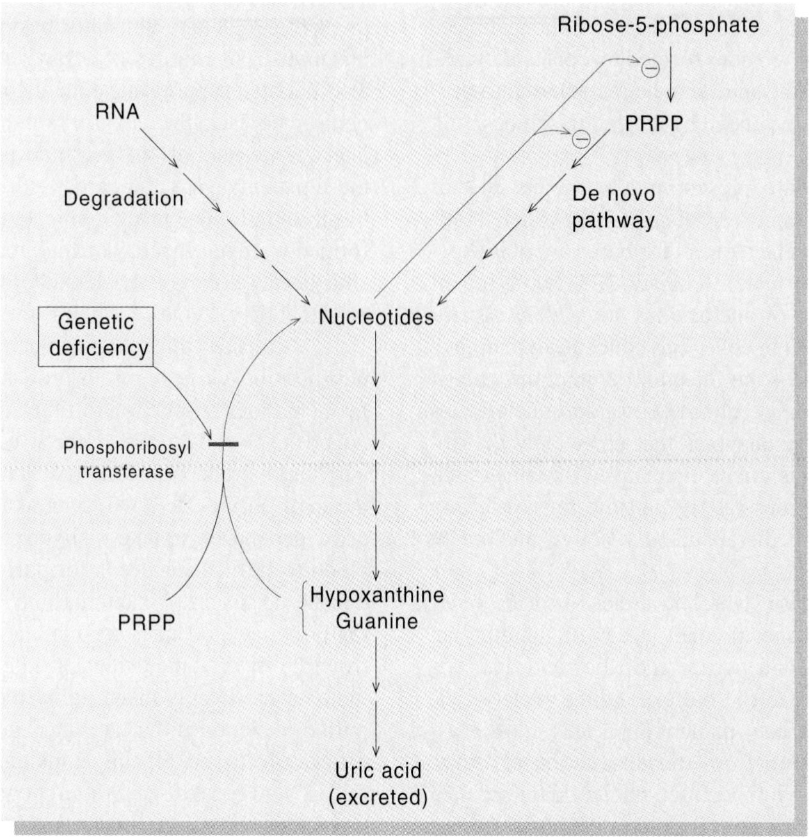

vage, since adenine is not a significant product of purine nucleotide catabolism (see figs. 26.9 and 26.27). The function of this enzyme seems to be to scavenge small amounts of adenine that are produced during intestinal digestion of nucleic acids or in the metabolism of 5′-deoxy-5′-methylthioadenosine, a product of polyamine synthesis.

In theory, hypoxanthine and guanine could also be salvaged by widely distributed purine nucleoside phosphorylase according to the reaction:

$$\text{Purine} + \text{Ribose-1-phosphate} \rightleftharpoons \text{Purine nucleoside} + P_i \quad (5)$$

However, this does not seem to be a quantitatively important salvage pathway and the role of the purine nucleoside phosphorylase seems to be in catabolism, presumably because of the low intracellular concentration of ribose-1-phosphate relative to P_i.

Although purine nucleosides are intermediates in the catabolism of nucleotides and nucleic acids in higher animals and humans, these nucleosides do not accumulate and are normally present in blood and tissues only in trace amounts. Nevertheless, these low concentrations are of great physiological significance in the case of some purine nucleosides. Thus adenosine has a number of effects that are probably mediated through changes in cyclic AMP levels. These effects include actions as a vasodilator, as an inhibitor of platelet aggregation and lipolysis, and as both an inhibitor and stimulator of histamine release. Cells of many vertebrate tissues contain kinases capable of converting purine nucleosides to nucleotides. Typical of these is adenosine kinase, which catalyzes the reaction

$$\text{Adenosine} + \text{ATP} \longrightarrow \text{AMP} + \text{ADP} \quad (6)$$

2′-Deoxyadenosine is also a substrate, though a relatively poor one. In some mammalian cells, deoxycytidine kinase also phosphorylates deoxyadenosine and deoxyguanosine (see the next section). In addition, indirect evidence suggests the existence of a specific deoxyadenosine kinase in human cells. Deoxyadenosine has received special attention as a substrate for kinases because congenital inability to catabolize this nucleoside results in immune deficiency disease. A mitochondrial kinase specific for deoxyguanosine has been reported. None of these kinases has been studied extensively, and their role in various tissues and organs is uncertain.

Salvage of Pyrimidines Is Less Important for Mammals and Goes through Nucleosides

In contrast to the extensive salvage and reutilization of purine bases, pyrimidine bases are poorly utilized by mammalian tissues.

Many bacteria take up pyrimidine bases efficiently for nucleotide synthesis, however, and a phosphoribosyltransferase for uracil has been identified in a few species of bacteria. In some mammalian cells uracil and 5-fluorouracil can be converted directly to UMP and 5-fluoro UMP, respectively, by the action of orotate phosphoribosyl transferase, the enzyme that normally uses orotate as substrate (fig. 26.18).

An alternative route for conversion of uracil to UMP, both in bacteria and in higher animals, is by successive reactions catalyzed by uridine phosphorylase and uridine kinase, respectively:

$$\text{Uracil} + \text{Ribose-1-phosphate} \rightleftharpoons \text{Uridine} + P_i \quad \textbf{(7)}$$

$$\text{Uridine} + \text{ATP} \xrightarrow{Mg^{2+}} \text{UMP} + \text{ADP} \quad \textbf{(8)}$$

Mammalian uridine phosphorylase will also accept deoxyuridine as a substrate, and thymidine is slowly attacked, but other nucleosides are not substrates. Uridine kinase activity has been demonstrated in a variety of bacteria and animal cells, including tumors, and is especially high in cells of high growth rate. Cytidine is the only other physiological nucleoside to act as a substrate for bacterial uridine kinase, but a variety of nucleoside triphosphates can act as phosphoryl donors.

Thymine is similarly converted to dTMP according to the following reactions:

$$\text{Thymine} + \text{Deoxyribose-1-phosphate} \rightleftharpoons \text{Thymidine} + P_i \quad \textbf{(9)}$$

$$\text{Thymidine} + \text{ATP} \longrightarrow \text{dTMP} + \text{ADP} \quad \textbf{(10)}$$

Thymine phosphorylase has been purified from bacterial and mammalian tissues. Thymidine is the preferred substrate, but deoxyuridine and 5-substituted deoxyuridines are also cleaved. Thymidine kinase is widely distributed, and the cytosolic enzyme from mammalian tissues is the most extensively studied of the deoxyribonucleoside kinases. It also accepts deoxyuridine as substrate, is inhibited by dTTP, and has complex kinetics. The activity of thymidine kinase in cells increases dramatically during rapid growth and DNA synthesis. Infection of cells with any of several viruses also induces thymidine kinase activity, together with certain other enzymes concerned with deoxyribonucleotide synthesis. The viral thymidine kinase is quite different from the host enzyme. Significantly, the viral enzymes are subject to none of the allosteric controls of the host enzymes.

A deoxycytidine kinase has been purified from certain bacteria, from thymus, and from tumor cells. It catalyzes the reaction

$$\text{Deoxycytidine} + \text{ATP} \rightleftharpoons \text{dCMP} + \text{ADP} \quad \textbf{(11)}$$

Deoxyadenosine and deoxyguanosine are also substrates for deoxycytidine kinase, but the purine deoxyribonucleosides have much higher K_m values than that of deoxycytidine.

As in the case of adenosine kinase, the main function of pyrimidine nucleoside kinases in mammalian cells is to maintain a balance between intracellular concentrations of nucleosides and nucleoside monophosphates. Together with nucleotidases that hydrolyze the monophosphates to the nucleosides, they constitute an important mechanism by which the cell regulates levels of nucleotides. The nucleoside phosphorylase–nucleoside kinase route for synthesis of pyrimidine nucleoside monophosphates is relatively inefficient for salvage of pyrimidine bases because of their very low concentration in plasma and tissues and because of the relatively low intracellular concentration of ribose-1-phosphate compared with P_i. Uptake of thymidine from the serum and its phosphorylation is an important mechanism for making dTMP in many mammalian cells, including those in the bone marrow.

Conversion of Nucleoside Monophosphates to Triphosphates Goes through Diphosphates

The products of biosynthetic pathways discussed in the preceding sections are, in most cases, mononucleotides. In cells a series of kinases (phosphotransferases) converts these mononucleotides to their metabolically active diphosphate and triphosphate forms.

Nucleoside Monophosphate Kinases. Bacteria and other microorganisms, as well as animal cells, contain a variety of kinases that catalyze reactions of the general type

$$\text{(d)NMP} + \text{ATP} \rightleftharpoons \text{(d)NDP} + \text{ADP} \quad \textbf{(12)}$$

Four types of nucleoside monophosphate kinases are known. These catalyze the phosphorylation of (1) GMP and dGMP, (2) AMP and dAMP, (3) dCMP, CMP, and UMP, and (4) dTMP.

The second of the kinases mentioned, which uses AMP as substrate, is referred to as adenylate kinase. Its activity is high in tissues where the turnover of energy from adenine nucleotides is great, for example, in liver and muscle, and its activity is also high in mitochondria. In these tissues its function is to make more energy available as ATP bond energy. When ADP is formed from ATP in energy-consuming reactions, more ATP can be formed from ADP according to the reaction

$$2\ \text{ADP} \rightleftharpoons \text{AMP} + \text{ATP} \quad \textbf{(13)}$$

Under conditions where energy-generating reactions convert intracellular ADP to ATP, AMP will be phosphorylated by running the preceding reaction from right to left. Adenylate kinase is therefore important in biological systems for maintaining equilibrium among adenine nucleotides as they are depleted or formed by energy transfers.

Nucleoside Diphosphate Kinases. Enzymes of this type have been found in many tissues of animals, plants, and microorganisms. They catalyze reactions of the following general type:

$$N_1TP + N_2DP \rightleftharpoons N_1DP + N_2TP \quad \textbf{(14)}$$

where N_1 and N_2 are purine or pyrimidine ribonucleosides or deoxyribonucleosides. The activity of NDP kinases is relatively high, usually 10- to 100-fold greater than the activity of the monophosphate kinases. As a result, intracellular concentrations of triphosphates are normally much higher than those of diphosphates, which in turn are often higher than those of monophosphates. Unlike the monophosphate kinases, which are substrate-specific, NDP kinases from all sources are active with a wide range of nucleoside diphosphates and triphosphates. They require a divalent cation for activity, and although many metal ions can satisfy this requirement, Mg^{2+} is the physiological cofactor.

Nucleoside diphosphate kinases from many sources have been shown to function by forming a phosphoryl–enzyme intermediate:

$$E + N_1TP \rightleftharpoons E \sim P + N_1DP \quad (15)$$

$$E \sim P + N_2DP \rightleftharpoons E + N_2TP \quad (16)$$

In several cases it has been shown that the phosphoryl group is attached to N-1 of a histidine side chain.

In addition to nucleoside diphosphokinases, a number of other cellular enzymes are capable of converting nucleoside diphosphates to triphosphates. Thus within many cells, GDP and dGDP are converted to their triphosphates by phosphoglycerate kinase and pyruvate kinase at rates comparable to that caused by nucleoside diphosphate kinase.

Inhibitors of Nucleotide Synthesis and Their Role in Chemotherapy

There are several distinct types of inhibitors of nucleotide biosynthesis, each type acting at different points in the pathways to purine or pyrimidine nucleotides. All these inhibitors are very toxic to cells, especially rapidly growing cells such as those of tumors or bacteria, because interruption of the supply of nucleotides seriously limits the cell's capacity to synthesize the nucleic acids necessary for protein synthesis and cell replication. In some cases the toxic effect of such inhibitors makes them useful in cancer chemotherapy or in the treatment of bacterial infections. However, these agents can also damage the replicating cells of the intestinal tract and bone marrow. This danger imposes limits on the doses that can be used safely and in some cases makes the inhibitor unsuitable for chemotherapy.

6-Mercaptopurine (table 26.3) and related thiopurines are potent inhibitors of purine nucleotide biosynthesis, but they are inactive until they are converted to the corresponding ribonucleoside 5′-phosphates. 6-Mercaptopurine is converted by the action of hypoxanthine-guanine phosphoribosyltransferase to the nucleotide 6-thioinosine-5′-monophosphate (T-IMP) (fig. 26.24). The latter inhibits several enzymes of purine biosynthesis, and it is uncertain which effect is primarily responsible for the toxicity. It blocks conversion of IMP to adenylosuccinate and to XMP, key reactions in the formation of AMP and GMP (see fig. 26.16). T-IMP is also capable of "pseudofeedback" inhibition of glutamine PRPP amidotransferase (see fig. 26.14), the first committed step in the purine nucleotide pathway. This enzyme is highly responsive to intracellular concentrations of both normal ribonucleoside 5′-monophosphates and analogs. 6-Mercaptopurine is used clinically in the treatment of leukemia.

Tiazofurin is a thiazole-C-nucleoside that is metabolized to tiazofurin-5′-monophosphate and then to thiazole-4-carboxamide adenine dinucleotide (TAD), an analog of NAD. TAD inhibits IMP dehydrogenase (see fig. 26.16), with consequent depletion of the pools of GMP, GDP, and GTP, but other dehydrogenases are inhibited also.

Figure 26.24

Activated forms of some inhibitors of nucleotide metabolism.

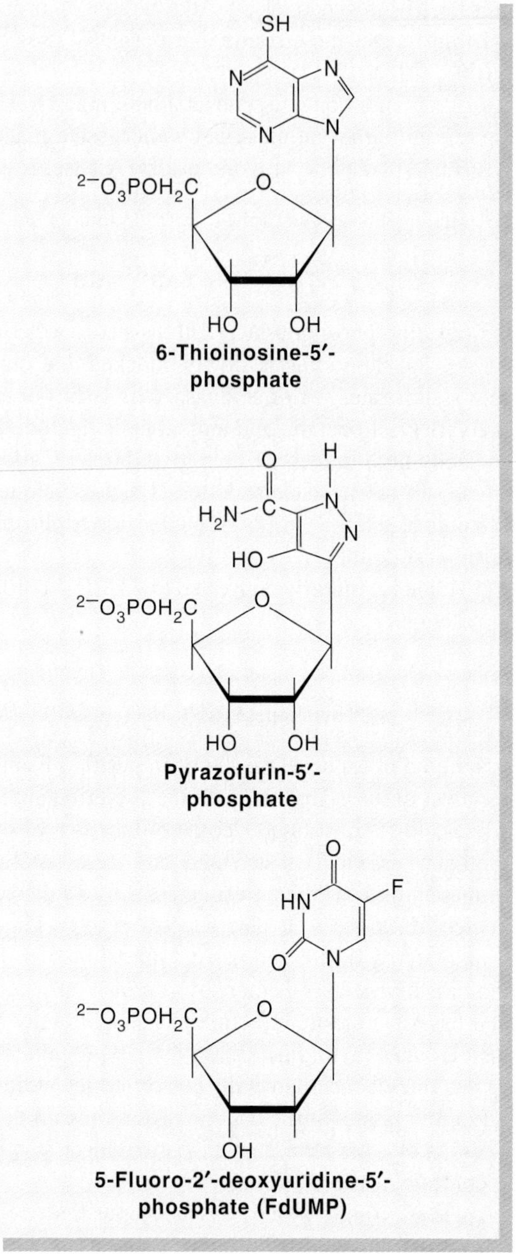

Three inhibitors are shown in table 26.3 that specifically interfere with steps in pyrimidine nucleotide biosynthesis. *N*-(Phosphonacetyl)-L-aspartate (PALA) is a powerful inhibitor of the carbamoyl transferase reaction (see fig. 26.18). PALA was synthesized to act as an analog of the transition state intermediate (fig. 26.25) postulated to be formed in the aspartate carbamoyltransferase reaction, and its tight binding to the enzyme is probably

Table 26.3
Some Inhibitors of Nucleotide Metabolism Together with Enzymes Inhibited

Inhibitor	Enzymes Inhibited	Inhibitor	Enzymes Inhibited
Inhibitors of purine nucleotide biosynthesis		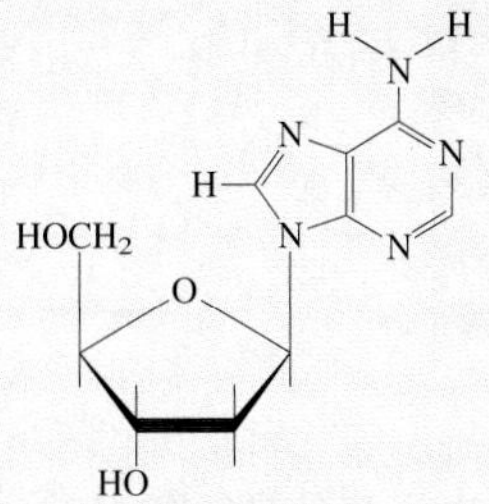 **2-Chlorodeoxyadenosine (2-CdA)**	Figure 26.21, ribonucleotide reductase
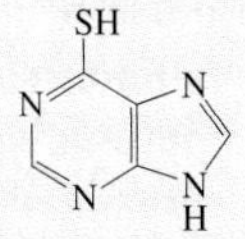 **6-Mercaptopurine**	Figure 26.16, adenylosuccinate synthase, IMP dehydrogenase; figure 26.14, reaction 1		
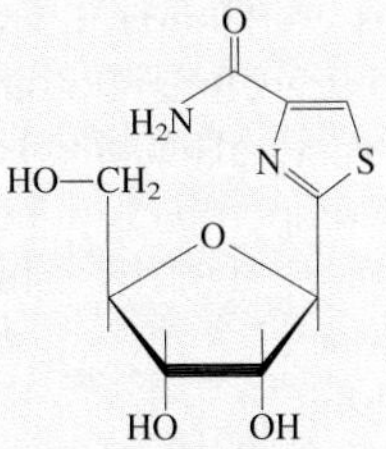 **Tiazofurin**	Figure 26.16, IMP dehydrogenase	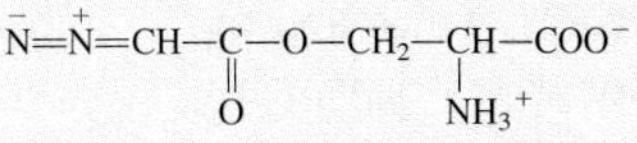 $\bar{N}=\overset{+}{N}=CH-C(=O)-O-CH_2-CH(NH_3^+)-COO^-$ **Azaserine (*O*-diazoacetyl-L-serine)**	Figure 26.14, reactions 1 and 4; figure 26.18, reaction 1; figure 26.20
		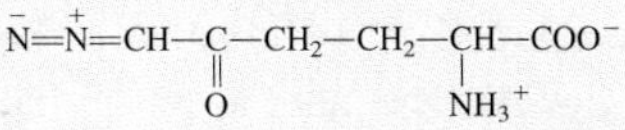 $\bar{N}=\overset{+}{N}=CH-C(=O)-CH_2-CH_2-CH(NH_3^+)-COO^-$ **6-Diazo-5-oxo-L-2-aminohexanoic acid (6-diazo-5-oxo-norleucine, DON)**	Figure 26.14, reactions 1 and 4; figure 26.18, reaction 1; figure 26.20
Inhibitors of pyrimidine nucleotide biosynthesis			
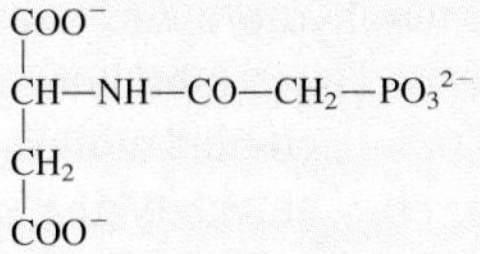 COO^-–CH(–NH–CO–CH_2–PO_3^{2-})–CH_2–COO^- ***N*-(Phosphonacetyl)-L-aspartate (PALA)**	Figure 26.18, aspartate carbamoyltransferase	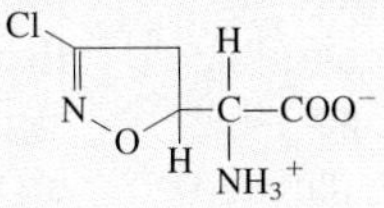 **Acivicin**	Figure 26.14, reaction 1 and 4; figure 26.16, GMP synthase; figure 26.18, reaction 1
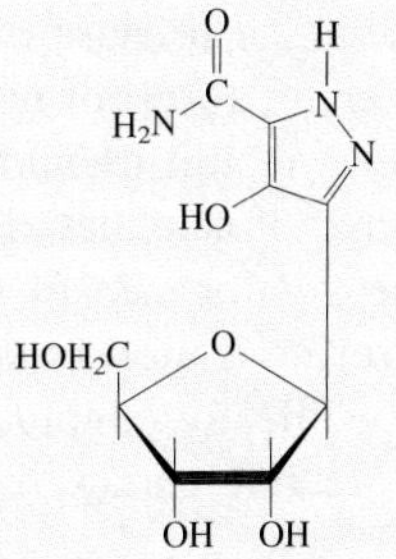 **Pyrazofurin (pyrazomycin)**	Figure 26.18, orotidylate decarboxylase	Indirect inhibitors of nucleotide biosynthesis	
		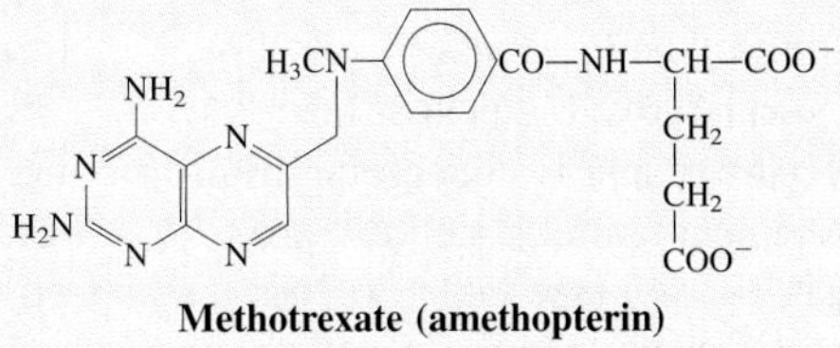 **Methotrexate (amethopterin)**	Figure 26.22, dihydrofolate reductase
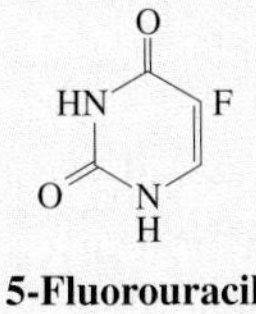 **5-Fluorouracil**	Figure 26.22, thymidylate synthase	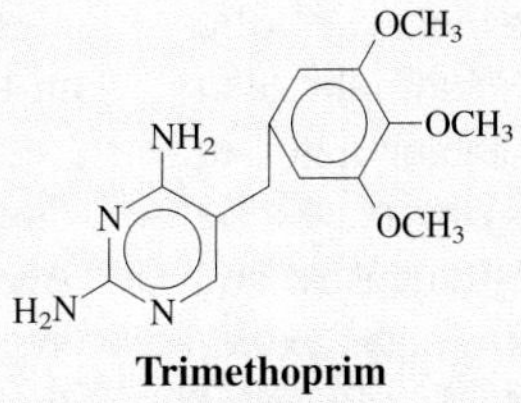 **Trimethoprim**	Figure 26.22, dihydrofolate reductase
Inhibitors of both purine and pyrimidine nucleotide biosynthesis			
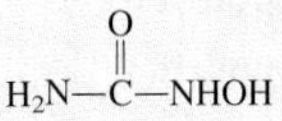 $H_2N-C(=O)-NHOH$ **Hydroxyurea**	Figure 26.21, ribonucleotide reductase	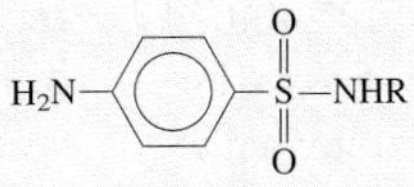 $H_2N-C_6H_4-SO_2-NHR$ **Sulfonamide**	Dihydropteroate synthase

Figure 26.25

Postulated transition state intermediate in the aspartate carbamoyltransferase (fig. 26.18).

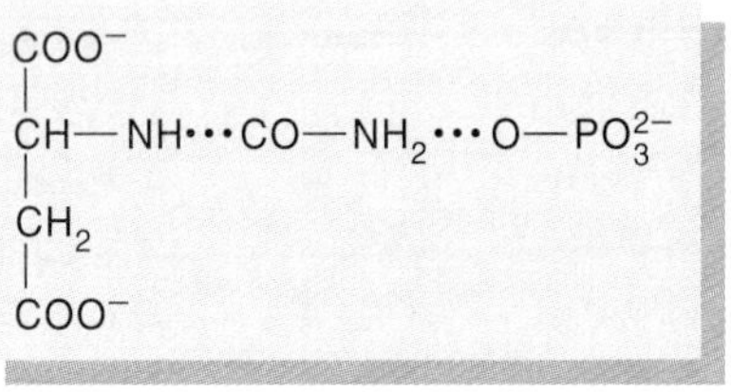

$$\begin{array}{l} COO^- \\ | \\ CH—NH\cdots CO—NH_2\cdots O—PO_3^{2-} \\ | \\ CH_2 \\ | \\ COO^- \end{array}$$

due to its resemblance to the intermediate. PALA is undergoing trials for treatment of colorectal and pancreatic cancer.

Pyrazofurin is a fermentation product. It is converted by kinase action to the 5′-phosphate (see fig. 26.24), which appears to mimic the substrate of orotidylate decarboxylase, for which it is a powerful inhibitor.

5-Fluorouracil, an agent used in treating solid tumors, interferes with thymidylate synthesis (see fig. 26.22). Its major inhibitory effect occurs after its conversion in the cell to 5-fluoro-2′-deoxyuridine-5′-monophosphate (see fig. 26.24). The latter acts as an analog of dUMP and binds very tightly to thymidylate synthase (see fig. 26.22) in the presence of methylenetetrahydrofolate, forming a covalent complex with the enzyme that is unable to undergo the normal catalytic reaction. 5-Fluorouracil may also exert a cytotoxic effect through incorporation of 5-fluorouridine phosphate into RNA. Recently, improved therapeutic results have been obtained with 5-fluorouracil by administering a reduced folate with the inhibitor. This increases the intracellular level of 5,10-methylenetetrahydrofolate, which increases the binding of FdUMP to thymidylate synthase.

Among inhibitors that interfere with the synthesis of both pyrimidine and purine nucleotides (see table 26.3) hydroxyurea interferes with the synthesis of deoxyribonucleotides by inhibiting ribonucleotide reductase of mammalian cells, an enzyme that is crucial and probably rate-limiting in the biosynthesis of DNA. It probably acts by disrupting the iron–tyrosyl radical structure at the active site of the reductase. Hydroxyurea is undergoing trials as an anticancer agent and for treatment of sickle-cell disease.

2-Chlorodeoxyadenosine (2-CdA) (see table 26.3) is an analog of deoxyadenosine and, like the latter, is very cytotoxic for cells, but unlike deoxyadenosine it is resistant to adenosine deaminase (adenosine aminohydrolase), the first enzyme in its catabolic pathway (fig. 26.26). Consequently, when 2-CdA is administered to leukemic patients, the triphosphate of 2-CdA accumulates in leukemic cells. Like dATP, the triphosphate inhibits ribonucleotide reductase. It is also a good substrate for DNA polymerases, and when it is incorporated into the DNA chain, it terminates extension. 2-CdA is under development for the treatment of some forms of leukemia.

Reactions involving glutamine as a substrate are inhibited by the glutamine analogs azaserine and 6-diaza-5-oxo-1-aminohexanoic acid (see table 26.3). Inhibition is irreversible, with formation of a covalent bond between the inhibitor and an amino acid side chain at the catalytic site. The specific reactions inhibited are those catalyzed by glutamine PRPP amidotransferase and phosphoribosyl-*N*-formylglycinamidine synthase in the de novo purine pathway and carbamoyl phosphate synthase and CTP synthase in the pyrimidine pathway.

Acivicin is an analog with a more distant resemblance to glutamine. It is, nevertheless, a potent inhibitor of several steps in purine nucleotide biosynthesis that utilize glutamine. The enzymes it inhibits are glutamine PRPP amidotransferase (step 1, fig. 26.14), phosphoribosyl-*N*-formylglycinamidine synthase (step 4, fig. 26.14), and GMP synthase (see fig. 26.16). In pyrimidine nucleotide biosynthesis the enzymes inhibited are carbamoyl synthase (step 1, fig. 26.18) and CTP synthase (see fig. 26.19). Acivicin is under trial for the treatment of some forms of cancer.

Another group of inhibitors prevents nucleotide biosynthesis indirectly by depleting the level of intracellular tetrahydrofolate derivatives. Sulfonamides are structural analogs of *p*-aminobenzoic acid (fig. 26.27), and they competitively inhibit the bacterial biosynthesis of folic acid at a step in which *p*-aminobenzoic acid is incorporated into folic acid. Sulfonamides are widely used in medicine because they inhibit growth of many bacteria. When cultures of susceptible bacteria are treated with sulfonamides, they accumulate 4-carboxamide-5-aminoimidazole in the medium, because of a lack of 10-formyltetrahydrofolate for the penultimate step in the pathway to IMP (see fig. 26.14). Methotrexate, trimethoprim, and a number of related compounds inhibit the reduction of dihydrofolate to tetrahydrofolate, a reaction catalyzed by dihydrofolate reductase. These inhibitors are structural analogs of folic acid (see fig. 26.27) and bind at the catalytic site of dihydrofolate reductase, an enzyme catalyzing one of the steps in the cycle of reactions involved in thymidylate synthesis (see fig. 26.22). These inhibitors therefore prevent synthesis of thymidylate in replicating cells, and as a secondary effect, synthesis of purine nucleotides is also decreased because of 10-formyltetrahydrofolate depletion. Polyglutamate forms of methotrexate that accumulate in cells also directly inhibit thymidylate synthase, as do polyglutamate forms of dihydrofolate that methotrexate may cause to accumulate, but the significance of these effects in cell killing is uncertain. Methotrexate is used as an anticancer drug, and trimethoprim, which specifically inhibits bacterial dihydrofolate reductase, is used for treating certain bacterial infections.

A number of drugs that are of great importance for the treatment of viral infections and of certain types of cancer are nucleosides or have closely related structures. Examples are shown in table 26.4.

3′-Azido-3′-deoxythymidine (AZT) has received much attention because it is one of the few drugs with proven effectiveness in arresting the course of acquired immune deficiency syndrome (AIDS). The virus that causes this disease, human immunodeficiency virus (HIV), contains an RNA-directed DNA polymerase. Once the virus enters a host cell, this reverse transcriptase catalyzes the synthesis of a double-stranded DNA copy of the viral RNA, as the first step in virus replication. AZT is converted to its monophosphate (AZTMP) by thymidine kinase; thymidylate kinase then converts the monophosphate to the

Figure 26.26

Major pathways of purine degradation in animals. Primates excrete uric acid. Mammals other than primates catabolize uric acid to other end products. In contrast to the catabolism of carbohydrates, lipids, or amino acids, the catabolism of nucleotides results in no energy production in the form of ATP. In both GMP and AMP catabolism, ribose-1-phosphate is released.

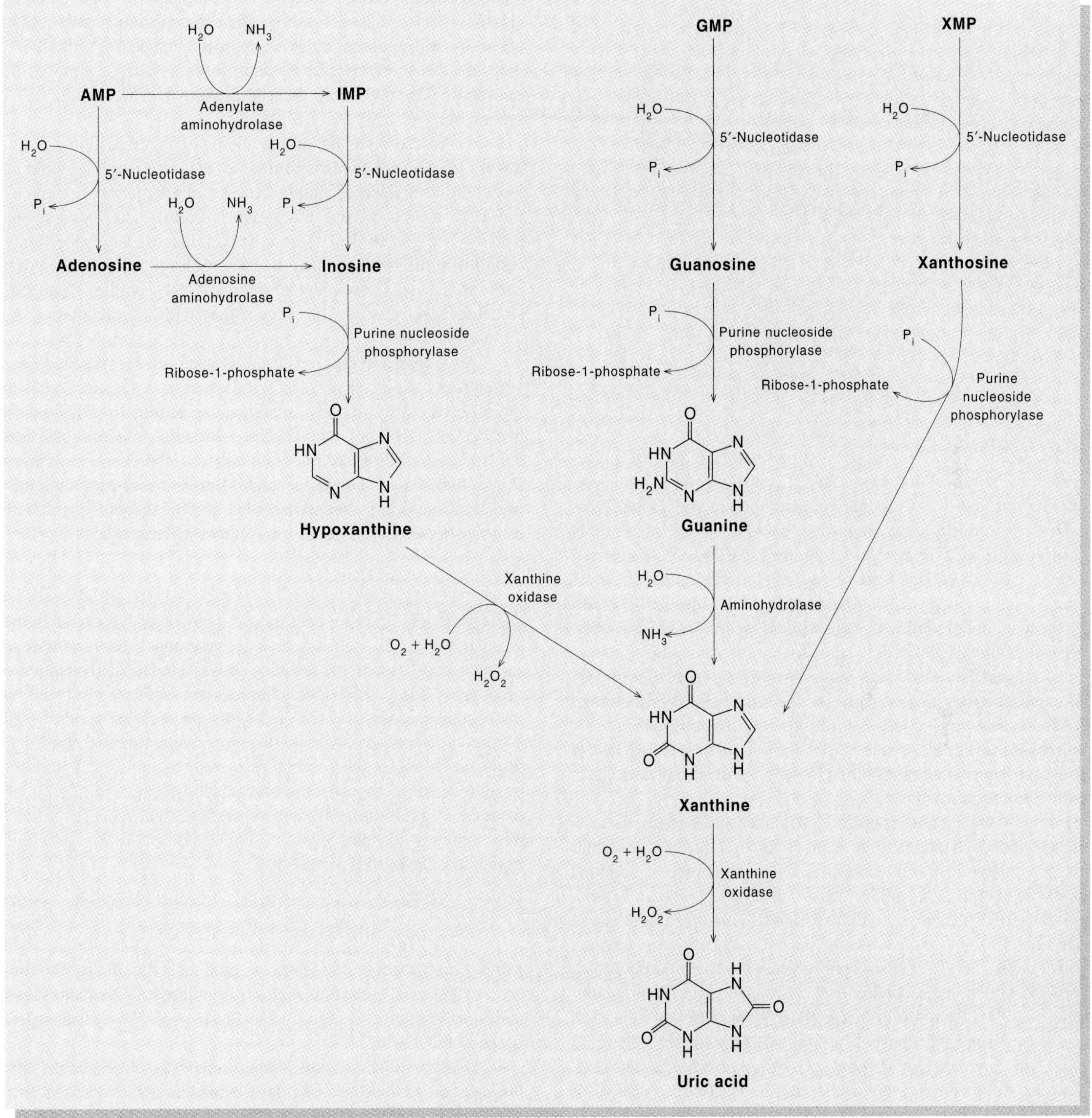

Figure 26.27

Normal metabolites with which indirect inhibitors of nucleotide metabolism compete. Methotrexate and trimethoprim compete with folate and, more importantly, with dihydrofolate (fig. 26.22); sulfonamides (sulfa drugs) complete with *p*-aminobenzoate.

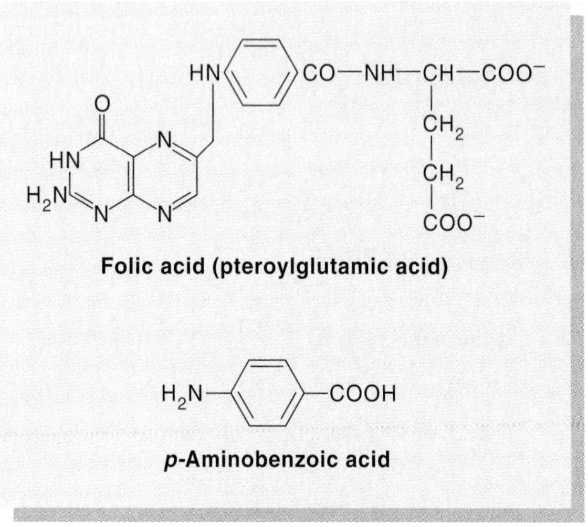

Folic acid (pteroylglutamic acid)

p-Aminobenzoic acid

triphosphate (AZTTP). AZTTP is a powerful inhibitor of the reverse transcriptase. More important, it is also a good substrate for this enzyme and, when incorporated into the growing DNA chain, promptly terminates chain extension, because there is no 3′—OH group on the AZT residue to which the next nucleotide unit can become attached. The effect of AZTTP on DNA polymerase of the host cell is much weaker. The effect of AZT is further increased by the accumulation of considerable amounts of AZTMP in cells. This monophosphate inhibits thymidylate kinase, with a consequent decrease in the concentration of dTTP in the cell. With less competition from dTTP there is more efficient incorporation of AZTTP into viral DNA by the reverse transcriptase. 2′,3′-dideoxyinosine (ddI) is under trial for use against AIDS and is similarly incorporated into viral DNA by the reverse transcriptase with chain termination.

Acyclovir (ACV) is not a true nucleoside, because the guanine residue is attached to an open-chain structure, but the latter mimics deoxyribose well enough for the compound to be accepted as a substrate by a thymidine kinase specified by certain herpes-type viruses. The normal thymidine kinase in mammalian cells will not recognize ACV as a substrate, however, so only virus-infected cells convert ACV to its monophosphate. Once the first phosphate has been added, the second phosphate is added by cellular guanylate kinase, while several other cellular kinases can add the third phosphate. The triphosphate is a more potent inhibitor of the viral DNA polymerases than of cellular DNA polymerases and also inactivates the former but not the latter. The net result is that ACV has been an effective treatment of, and prophylaxis for, genital herpes; can result in dramatic relief of pain associated with "shingles" caused by reactivation of latent varicella-zoster virus; and has been successful in many patients with herpes encephalitis.

Gancyclovir (GCV), like ACV, is an open-chain analog of deoxyguanosine. It is more effective than ACV in the treatment of infections by human cytomegalovirus (HCMV). These infections can cause birth defects in newborns and can be fatal or cause blindness in immunosuppressed individuals such as transplant recipients, cancer patients receiving chemotherapy, or AIDS patients. HCMV-infected cells are more efficient at forming the triphosphate of the drug, though the mechanism is unclear. Furthermore, the viral DNA polymerase is much more sensitive to inhibition (and chain termination) by the triphosphate than are cellular DNA polymerases.

An experimental strategy being explored for tumor eradication in cancer patients is to transfer the gene for the thymidine kinase from the Herpes simplex virus to the target tumor cells by means of a viral carrier, and then administer gancyclovir to the patient. The presence of the viral thymidine kinase in the tumor cells will result in the formation of the triphosphate of gancyclovir in these cells and consequent inhibition of DNA synthesis, whereas since the host cells lack this kinase they will be little affected by the gancyclovir.

Cytosine arabinose (araC) is taken up by cells and converted to its triphosphate (araCTP), which is a substrate for cell DNA polymerases, and is a relative chain terminator. This means that the cell DNA polymerases have difficulty in adding the next nucleotide after araCMP has been added to the chain. Incorporation of two or more successive araCMP residues causes chain termination. araC is a component of a combination of drugs with proven effectiveness against some forms of leukemia.

Catabolism of Nucleotides

Dietary nucleic acids are unaffected by gastric enzymes, but in the small intestine, ribonuclease and deoxyribonuclease I, which are secreted in the pancreatic juice, hydrolyze nucleic acids mainly to oligonucleotides. The oligonucleotides are further hydrolyzed by phosphodiesterases, also secreted by the pancreas, to yield 5′- and 3′-mononucleotides. Most of the mononucleotides are then hydrolyzed to nucleosides by various group-specific nucleotidases or by a variety of nonspecific phosphatases. The resulting nucleosides may be absorbed intact by the intestinal mucosa, or they may undergo phosphorolysis by nucleoside phosphorylases and by nucleosidases to free bases:

$$\text{Nucleoside} + P_i \rightleftharpoons \text{Base} + \text{Ribose-1-phosphate} \qquad \textbf{(17)}$$

$$\text{Nucleoside} + H_2O \rightleftharpoons \text{Base} + \text{Ribose} \qquad \textbf{(18)}$$

Little is known about these enzymes and their specificity. The mucosa of the small intestine are rich in nucleoside phosphorylase, and most of the remaining nucleoside is probably hydrolyzed to bases in this tissue.

Experiments with labeled nucleic acid indicate that purines and pyrimidines of ingested nucleic acids are used only to a small extent for synthesis of tissue nucleic acids, and in the case of purines most of the bases were shown to be catabolized. This finding is consistent with the presence in intestinal mucosa of a high level of xanthine oxidase, a catabolic enzyme.

Table 26.4
Other Nucleoside Analogs Used in Therapy

Analog	Enzyme Inhibited by Active Form	Clinical Condition Treated	Analog	Enzyme Inhibited by Active Form	Clinical Condition Treated
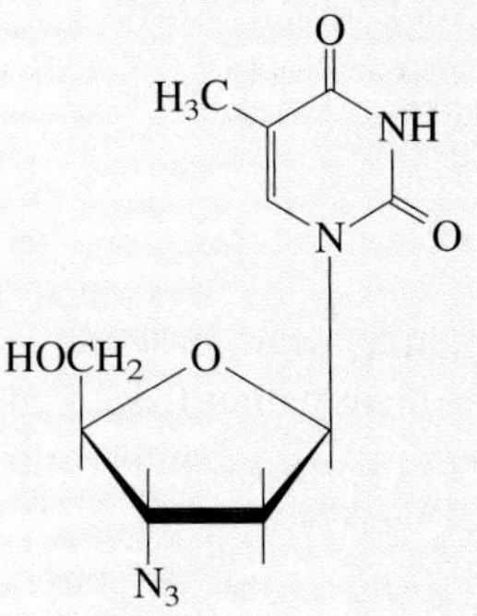 H_3C, O, NH, N, O; $HOCH_2$, O; N_3 Azidothymidine (AZT)	Viral DNA polymerase	AIDS (human immunodeficiency virus infection)	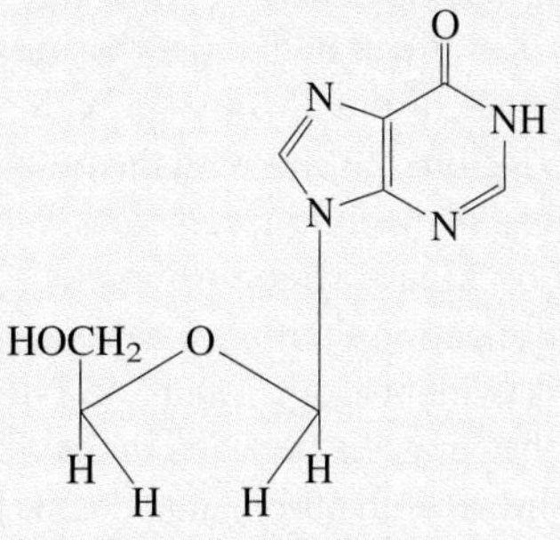 O, N, NH, N, N; $HOCH_2$, O; H, H, H, H Acyclovir (ACV)	Viral DNA polymerase	Herpes simplex virus infections
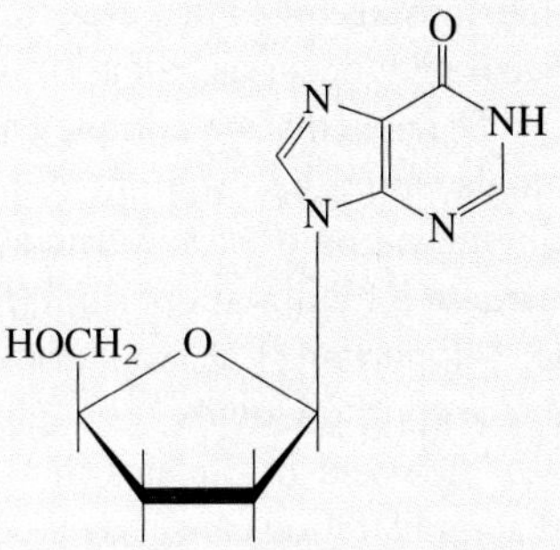 O, N, NH, N, N; $HOCH_2$, O 2′,3′-Dideoxyinosine (ddI)	Viral DNA polymerase	AIDS (human immunodeficiency virus infection)	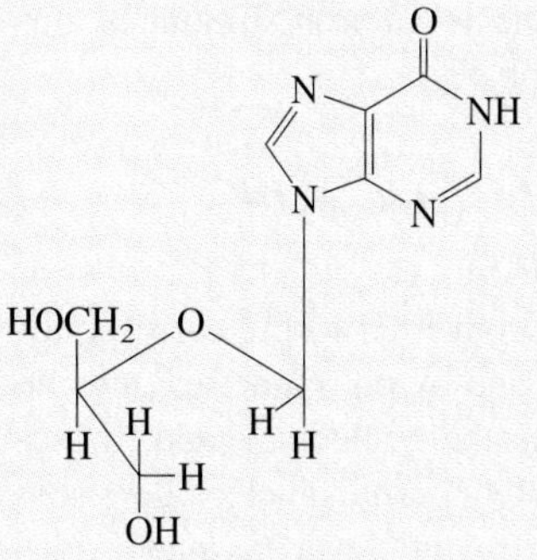 O, N, NH, N, N; $HOCH_2$, O; H, H, H, H, H; OH Gancyclovir (GCV)	Viral DNA polymerase	Human cytomegalovirus infections
			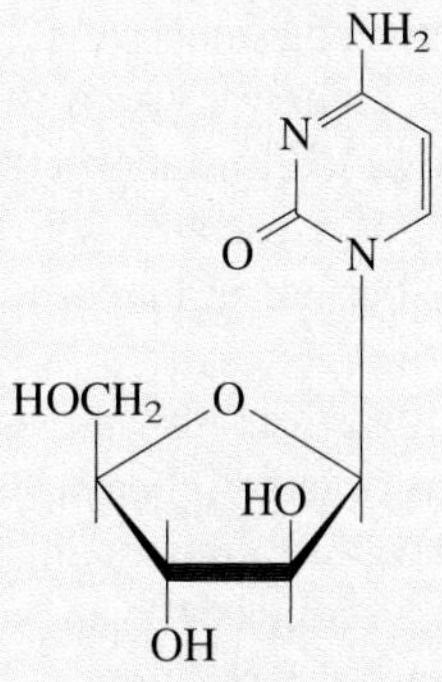 NH_2, N, O, N; $HOCH_2$, O; HO; OH Cytosine arabinoside (araC)	DNA polymerase	Leukemia

Intracellular Catabolism of Nucleotides Is Highly Regulated

Nucleotides are also catabolized within cells by several types of intracellular nucleotidase that hydrolyze the phosphate ester groups to release inorganic phosphate. One of the 5′-nucleotidases that act on monophosphates in mammalian cells is probably located in lysosomes. It has a pH optimum of 5.0 and a broad substrate specificity. In lymphocytes there are two cytoplasmic purine nucleotidases, one acting on deoxyribonucleotides, the other on ribonucleotides. In these cells, three types of cytoplasmic nucleotidase act on pyrimidine nucleotides; one acts on ribonucleotides, another acts on deoxyribonucleotides, and a third is specific for thymidylate.

Although allosteric regulation of 5′-nucleotidases has not been demonstrated, attack of these enzymes on nucleotides is limited under normal circumstances, either by their intracellular localization or by the effects of nucleotide concentrations. Nevertheless, they appear to be involved in cycling of nucleotides along the pathways shown in figures 26.9 and 26.10. The balance between the activity of nucleoside kinases and 5′-nucleotidases serves to regulate intracellular nucleoside and nucleotide levels. An increase in nucleotide levels will result in increased nucleotidase-catalyzed hydrolysis of nucleotides to nucleosides, which are then transported out of the cell or broken down by the nucleoside phosphatases that we described in connection with salvage pathways. The nucleobases formed can exit from the cell with the help of transport proteins. Low intracellular nucleotide levels will result in net formation of nucleotides by kinase action on nucleosides and by phosphoribosyltransferase action on purine bases as well as by de novo synthesis. This is evident from results of experiments with inhibitors of some of the enzymes involved in these pathways, such as adenosine deaminase and purine nucleoside phosphorylase. In certain tissues, specialized nucleotidases have a specific role. For example, an acid phosphatase that serves to provide inorganic phosphate in bone is an iron-stimulated nucleotidase acting on di- and triphosphates.

Purines Are Catabolized to Uric Acid and Then to Other Products

After purine nucleotides have been converted to the corresponding nucleosides by 5′-nucleotidases and by phosphatases, inosine and guanosine are readily cleaved to the nucleobase and ribose-1-phosphate by the widely distributed purine nucleoside phosphorylase (see fig. 26.26). The corresponding deoxynucleosides yield deoxyribose-1-phosphate and base with the phosphorylase from most sources. Adenosine and deoxyadenosine are not attacked by the phosphorylase of mammalian tissue, but much AMP is converted to IMP by an aminohydrolase (deaminase), which is very active in muscle and other tissues (see fig. 26.26). Inherited deficiency of purine nucleoside phosphorylase is associated with a deficiency in the cellular type of immunity but not in humoral immunity.

An adenosine aminohydrolase (deaminase) is also present in many mammalian tissues. This enzyme is of interest because hereditary deficiency of the enzyme is linked to a severe (usually fatal) defect in the immune system, marked by a serious deficiency in lymphocytes and consequent inability to combat infections.

Inosine formed by either route is then phosphorolyzed to yield hypoxanthine. Although, as we have seen, much of the hypoxanthine and guanine produced in the mammalian body is converted to IMP and GMP by a phosphoribosyltransferase, about 10% is catabolized. Xanthine oxidase, an enzyme present in large amounts in liver and intestinal mucosa and in traces in other tissues, oxidizes hypoxanthine to xanthine, and xanthine to uric acid (see fig. 26.26). Xanthine oxidase contains FAD, molybdenum, iron, and acid-labile sulfur in the ratio 1:1:4:4, and in addition to forming hydrogen peroxide, it is also a strong producer of the superoxide anion $\bullet O_2$, a very reactive species. The enzyme oxidizes a wide variety of purines, aldehydes, and pteridines.

Figure 26.28

The structure of allopurinol, an analog of hypoxanthine. Allopurinol inhibits xanthine oxidase and is used in the treatment of gout.

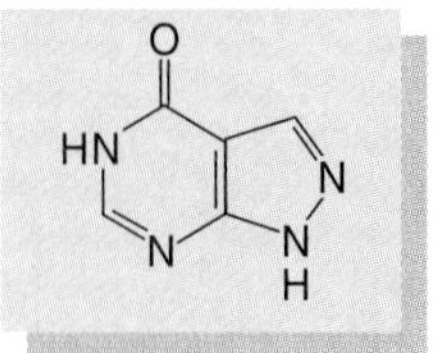

Guanine aminohydrolase (guanine deaminase or guanase), present in liver, brain, and other mammalian tissues, provides another pathway to xanthine, this time for guanine. Subsequent oxidation of xanthine to uric acid then occurs.

Gout is a relatively common ($\approx$3 per 1,000 persons) derangement of purine metabolism that is associated with elevated plasma levels of uric acid. The excessive uric acid leads to painful deposits of monosodium urate in the cartilage of joints, especially of the big toe. Uric acid deposits also may occur as calculi in the kidney, with resultant renal damage. The genetics are complex and incompletely understood. Individuals suffering from gout and other metabolic disorders producing elevation of serum uric acid may be treated with the xanthine oxidase inhibitor allopurinol (fig. 26.28), an analog of hypoxanthine. Allopurinol is also a substrate of xanthine oxidase, but the product, oxypurinol, binds very tightly to the reduced form of xanthine oxidase and inactivates the enzyme. Allopurinol is nontoxic, and administration causes a marked decrease in the uric acid concentration in serum and in urinary excretion of uric acid. During allopurinol treatment, serum uric acid and xanthine are prevented from accumulating by operation of the salvage pathway. Furthermore, the consequent accumulation of IMP, GMP, and AMP causes feedback inhibition of purine biosynthesis (see fig. 26.31).

Mammals other than primates further oxidize urate by a liver enzyme, urate oxidase, which is a copper protein. The product, allantoin, is excreted. Humans and other primates, as well as birds, lack urate oxidase and hence excrete uric acid as the final product of purine catabolism. In many animals other than mammals, allantoin is metabolized further to other products that are excreted: allantoic acid (some teleost fish), urea (most fishes, amphibians, some mollusks), and ammonia (some marine invertebrates, crustaceans, etc.). This pathway of further purine breakdown is shown in figure 26.29.

Pyrimidines Are Catabolized to β-Alanine, NH_3, and CO_2

A number of deaminases present in many cells are able to deaminate cytosine or its nucleosides or nucleotides to the corresponding uracil derivatives. Cytosine aminohydrolase (deaminase) appears to occur only in microorganisms (yeast and bacteria), but cytidine aminohydrolase is widely distributed in bacteria, plants, and mammalian tissues. A distinct deoxycytidine aminohydrolase

Figure 26.29

Degradation of uric acid to excretory products. Mammals other than primates oxidize uric acid further to allantoin. Humans and other primates as well as birds lack urate oxidase and hence excrete uric acid as the final product of purine catabolism. In many animals other than mammals, allantoin is metabolized further to urea or ammonia and CO_2 as shown.

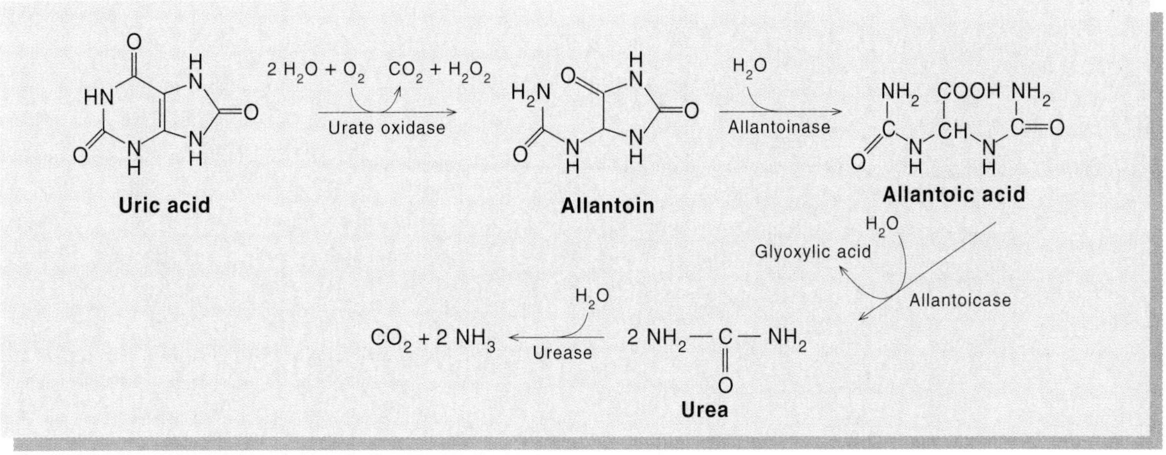

is present in various mammalian tissues and tumors, in plants, and in bacteria. A deoxycytidylate aminohydrolase that is similarly distributed produces dUMP, which is susceptible to attack by 5′-nucleotidase to give deoxyuridine. Although the physiological function of these aminohydrolases is not completely understood, the uridine and deoxyuridine formed can be further degraded by uridine phosphorylase to uracil as previously discussed, so these reactions provide a pathway for converting nucleotides of uracil and cytosine to uracil and ribose-1-phosphate or deoxyribose-1-phosphate (fig. 26.30). Similarly, thymine nucleosides and nucleotides can be converted by 5′-nucleotidase and phosphorylase to thymine.

Enzymes present in mammalian liver are capable of the catabolism of both uracil and thymine. The first reduces uracil and thymine to the corresponding 5,6-dihydro derivatives. This hepatic enzyme uses NADPH as the reductant, whereas a similar bacterial enzyme is specific for NADH. Similar enzymes are apparently present in yeast and plants. Hydropyrimidine hydrase then opens the reduced pyrimidine ring, and finally the carbamoyl group is hydrolyzed off from the product to yield β-alanine or β-aminoisobutyric acid, respectively, from uracil and thymine (see fig. 26.30).

Regulation of Nucleotide Metabolism

Among the reaction pathways that we have described, there exist many possibilities for futile cycles, in which nucleotides built up in the biosynthetic pathways are broken down in catabolic pathways to products closely related to the starting materials. As an example, AMP synthesized from IMP by adenylosuccinate synthase and adenylosuccinate lyase may be hydrolyzed back to IMP by adenylate aminohydrolase (see fig. 26.26). The net result is the conversion of aspartate to fumarate and ammonia and the hydrolysis of GTP to GDP and P_i. To avoid such futile cycles, both biosynthetic and catabolic processes are under tight regulatory controls. The efficiency of these controls is demonstrated by the increased activity of many enzymes involved in nucleotide biosynthesis when cells are proliferating. Evidently, regulatory mechanisms increase nucleotide biosynthesis as intracellular nucleotides are used for the synthesis of RNA and DNA. As we have seen, drastic consequences can attend impairment of the control machinery, as in the Lesch-Nyhan syndrome or intervention with drugs such as 6-mercaptopurine.

Although much remains to be discovered about the details of the regulation of nucleotide metabolism, a number of important control points are rather well understood. We will discuss these here, together with their known effects on intracellular nucleotide pools.

Purine Biosynthesis Is Regulated at Two Levels

Many lines of evidence indicate that the first committed step in de novo purine nucleotide biosynthesis, production of 5-phosphoribosylamine via PRPP amidotransferase, is rate-limiting for the entire sequence. Consequently, regulation of this enzyme is probably the most important factor in control of purine synthesis de novo (fig. 26.31). The enzyme is inhibited by purine-5′-nucleotides, and the nucleotides that are most inhibitory vary with the source of the enzyme. Inhibition constants (K_I) are usually in the range 10^{-3} to 10^{-5} M. The maximum effect of this end-product inhibition is produced by certain combinations of nucleotides (e.g., AMP and GMP) in optimum concentrations and ratios, indicating two kinds of inhibitor-binding sites. This is an example of a concerted feedback inhibition.

Figure 26.30

Degradation of pyrimidine bases. Parts of this pathway are widely distributed in nature. The entire pathway is found in mammalian liver. As in purine nucleotide catabolism, no ATP results from catabolism, and the ribose-1-phosphate is released during catabolism before destruction of the base.

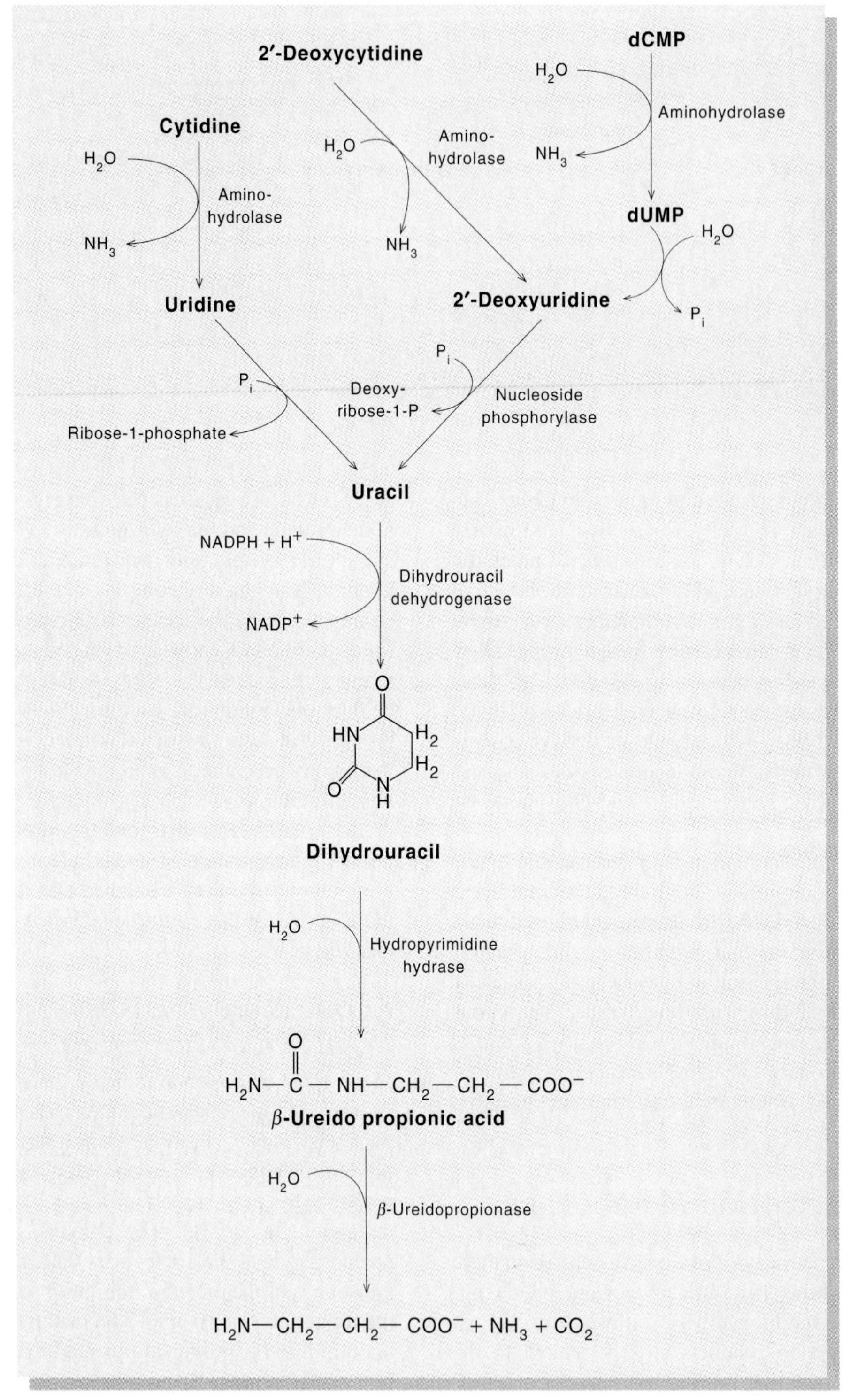

Figure 26.31

Regulations of purine biosynthesis. Red arrows show points of feedback inhibition. In addition to the feedback inhibition, GTP stimulates ATP synthesis and ATP stimulates GTP synthesis. This helps to ensure a balance between the pools of the two nucleoside triphosphates. The full biosynthetic pathways are shown in figures 26.14 and 26.16.

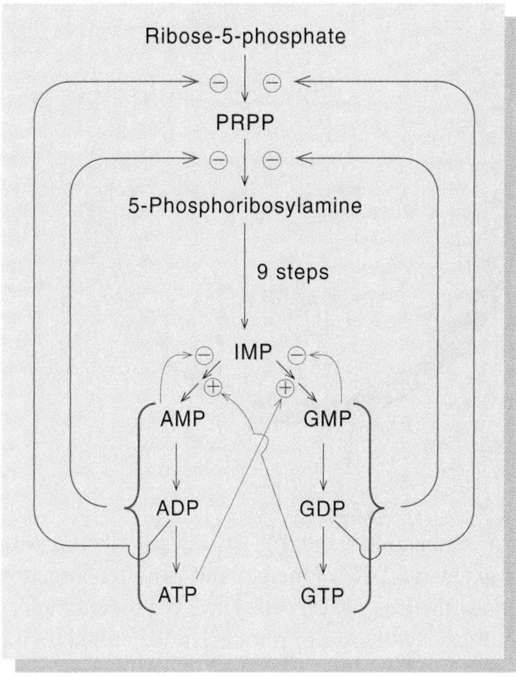

Figure 26.32

Proposed scheme for the regulation of deoxyribonucleotide synthesis in *E. coli* and mammalian cells. Red arrows indicate points of activation and inhibition, respectively. (*Source:* L. Thelander and P. Reichard, "Reduction of ribonucleotides," in *Annual Review of Biochemistry* 48:133, 1979. Copyright © 1979 Annual Reviews Inc. Palo Alto, Calif.)

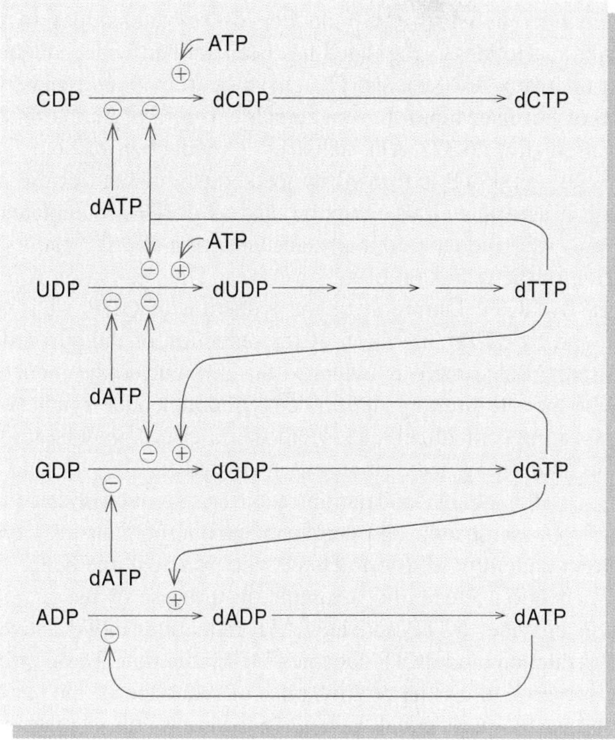

The rate of the amidotransferase reaction is also governed by intracellular concentrations of the substrates L-glutamine and PRPP. Competing metabolic reactions, or drugs that alter the supply of these substrates, also affect the rate of IMP synthesis.

The second important level of regulation of purine nucleotide synthesis is in the branch pathways from IMP to AMP and to GMP (see fig. 26.16). The first of these two reactions leading from IMP to AMP is the irreversible synthesis of adenylosuccinate. This requires GTP as a source of energy and is inhibited by AMP. Of the two reactions required to convert IMP to GMP, the first is irreversible and is inhibited by GMP, while the second requires ATP as a source of energy. Thus there are two types of regulation at this level of purine nucleotide synthesis: (1) a "forward" control, by which increased GTP accelerates AMP synthesis and increased ATP accelerates GMP synthesis, and (2) feedback inhibition, by which AMP and GMP each regulate their own synthesis. Excess AMP also may be converted to IMP by adenylate aminohydrolase and thus can serve as a source of GMP. Adenylate aminohydrolase is activated by ATP and inhibited by GTP, which may serve to control this potential conversion of adenine nucleotides to guanine nucleotides. Finally, when the energy reserves of the cell are low, feedback inhibition of ribose-5-phosphate pyrophosphokinase by ADP and GDP restricts the synthesis of PRPP.

Pyrimidine Biosynthesis Is Regulated at the Level of Formation of Carbamoyl Phosphate (Eukaryotes) or Carbamoyl Aspartate (Bacteria)

In bacteria the first committed step in pyrimidine nucleotide biosynthesis is the formation of carbamoyl aspartate from carbamoyl phosphate and aspartate. In *E. coli* the enzyme catalyzing this step, aspartate carbamoyltransferase, is powerfully inhibited by CTP, which acts chiefly by decreasing the affinity of the enzyme for aspartate (see chapter 10). ATP has the opposite effect, activating the enzyme by increasing its affinity for aspartate. Concentrations of ATP and CTP in *E. coli* are high enough for these nucleotides to influence the intracellular activity of aspartate carbamoyltransferase. However, this is not a regulatory enzyme in all bacterial species and is not involved in regulation of pyrimidine nucleotide synthesis in animal cells.

In eukaryotes, carbamoyl phosphate synthase is inhibited by pyrimidine nucleotides and stimulated by purine nucleotides;

it appears to be the most important site of feedback inhibition of pyrimidine nucleotide biosynthesis in mammalian tissues. However, it has been suggested that under some conditions, orotate phosphoribosyltransferase may be a regulatory site as well.

Deoxyribonucleotide Synthesis Is Regulated by Both Activators and Inhibitors

The manner in which the reduction of ribonucleotides to deoxyribonucleotides is regulated has been studied with reductases from relatively few species. The enzymes from *E. coli* and from Novikoff rat liver tumor have a complex pattern of inhibition and activation (fig. 26.32). ATP activates the reduction of both CDP and UDP. As dTTP is formed by metabolism of both dCDP and dUDP, it activates GDP reduction, and as dGTP accumulates, it activates ADP reduction. Finally, accumulation of dATP causes inhibition of the reduction of all substrates. This regulation is reinforced by dGTP inhibition of the reduction of GDP, UDP, and CDP and by dTTP inhibition of the reduction of the pyrimidine substrates. Since there is evidence that ribonucleotide reductase may be the rate-limiting step in deoxyribonucleotide synthesis in at least some animal cells, these allosteric effects may be important in controlling deoxyribonucleotide synthesis.

The adenosylcobalamin-requiring ribonucleotide reductases from lactobacilli and certain other microorganisms have a different pattern of allosteric effects, the principal one being specific activation effects. For example, in the case of the *L. leichmannii* enzyme, dGTP activates ATP reduction, dATP activates CTP reduction, and dCTP activates UTP reduction. These effects may serve to adjust the relative rates of reduction of the various substrates to more equal values. In addition, the synthesis of *L. leichmannii* enzyme is repressed by the presence in the growth medium of an excess of vitamin B_{12} (cyanocobalamin) or of a deoxyribonucleoside such as thymidine. The repressor for enzyme synthesis is probably dTTP or a closely related nucleotide, which accumulates in the cell when rapid ribonucleotide reduction occurs as a result of an ample cobalamin supply or when deoxynucleoside is supplied. Further deoxyribonucleotide synthesis is then slowed by the decreased level of reductase activity.

A second enzyme on the pathway to dTTP that is subject to allosteric control is deoxycytidylate deaminase, which supplies dUMP for thymidylate synthesis. The enzyme has been studied in depth in mammalian cells, yeast, and bacteriophage T2-infected *E. coli*. In each case the enzyme is allosterically activated by dCTP (hydroxymethyl dCTP for the phage enzyme) and inhibited by dTTP.

Enzyme Synthesis Also Contributes to Regulation of Deoxyribonucleotides during the Cell Cycle

Many of the enzymes participating in de novo synthesis of deoxyribonucleotide triphosphates, as well as those responsible for interconversion of deoxyribonucleotides, increase in activity when cells prepare for DNA synthesis. The need for increased DNA synthesis occurs under three circumstances: (1) when the cell proceeds from the G_0 or resting stage of the cell cycle to the S or synthetic or replication stage (fig. 26.33); (2) when it performs repair after extensive DNA damage; and (3) after infection of quiescent cells with virus. When cells leave G_0, for example, enzymes like thymidylate synthase and ribonucleotide reductase increase as well as the corresponding mRNAs. These increases in enzyme amount supplement allosteric controls that increase the activity of each enzyme molecule. Corresponding decreases in amounts of these enzymes and their mRNAs occur when DNA synthesis is completed.

Figure 26.33

Phases in the life cycle of a typical eukaryotic cell. The cell cycle is divided into the resting stage (G_0), the prereplication stage (G_1), the synthesis or replication stage (S), the postreplication stage (G_2), and the mitotic stage (M).

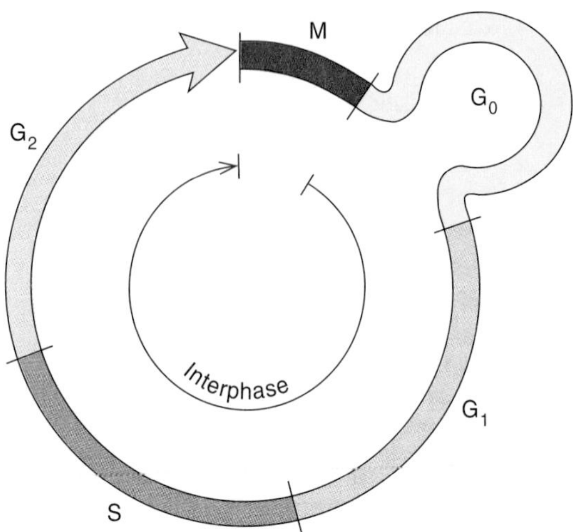

Metabolites Are Channeled along the Nucleotide Biosynthesis Pathways

In addition to the regulatory controls described in the preceding sections, which are mainly allosteric feedback mechanisms, evidence is accumulating that nucleotide biosynthetic pathways are closely controlled through the phenomenon of channeling. This involves an assembly of enzymes catalyzing successive steps in the biosynthetic pathway so that metabolic intermediates pass directly from one enzyme to the next. Thus the metabolites are channeled along the metabolic pathway, with restricted opportunity for their diffusion into the medium or entry into the general metabolic pool in the cell.

The most clearly established way in which channeling of nucleotide precursors is achieved involves multifunctional enzymes, in which several catalytic activities occur on a single polypeptide chain. Another possibility is the noncovalent association of pathway enzymes in complexes that may sometimes be concentrated in a particular intracellular location.

Examples of multifunctional enzymes are provided by the pyrimidine biosynthetic pathway. Although in most prokaryotes, six structural genes code for the six enzymes involved in the de novo synthesis of UMP, in eukaryotes the number of genes is re-

Figure 26.34

Formation of formyl group donors for purine biosynthesis (R = *p*-aminobenzoyl-L-glutamate). A trifunctional enzyme catalyzes the three reactions shown, which generate the formyl donors for steps in purine nucleotide synthesis (see steps 3 and 9 in fig. 26.14).

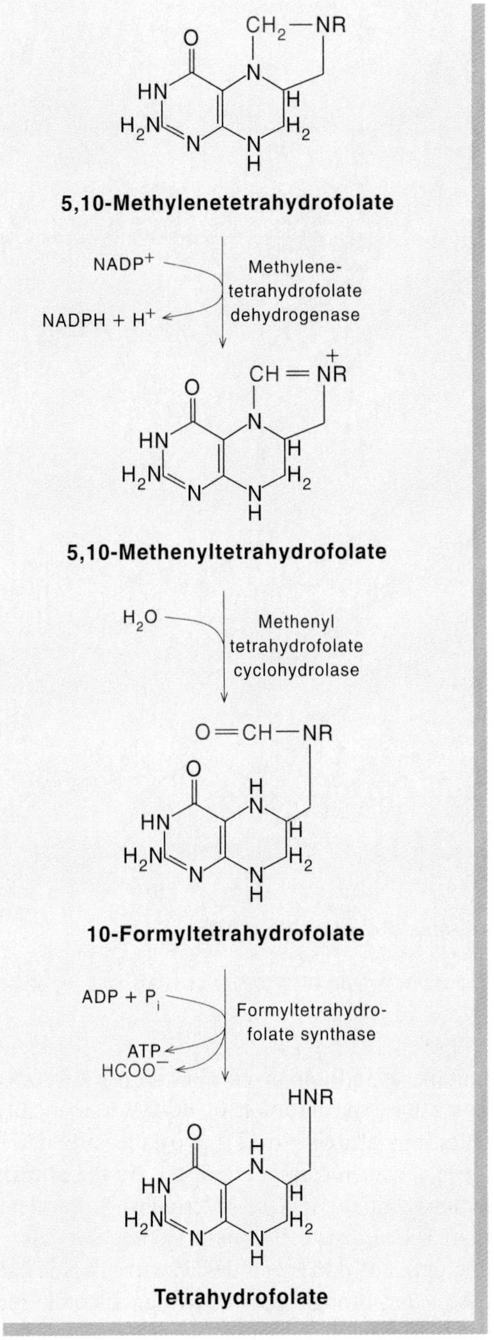

5,10-Methylenetetrahydrofolate

$NADP^+$ → $NADPH + H^+$ — Methylene-tetrahydrofolate dehydrogenase

5,10-Methenyltetrahydrofolate

H_2O — Methenyl tetrahydrofolate cyclohydrolase

10-Formyltetrahydrofolate

$ADP + P_i$ → ATP, $HCOO^-$ — Formyltetrahydrofolate synthase

Tetrahydrofolate

duced because of the production of multifunctional proteins. In the fungus *Neurospora* a single protein has both carbamoyl phosphate synthase activity and aspartate carbamoyltransferase activity, but in mammalian cells a single protein not only has both these activities, but also has dihydroorotase activity. The latter protein is an oligomer (probably a trimer) of a large polypeptide (M_r = 200,000), and there is evidence to indicate that the multifunctional polypeptide is a single gene product. This multifunctional enzyme channels carbamoyl phosphate and carbamoyl aspartate, provided that dihydroorotate is rapidly removed, which is the case in the normal cell.

In mammalian cells the last two steps of the pathway to UMP (see fig. 26.18) are catalyzed by the bifunctional protein that is the product of a single gene. This protein therefore has both orotate phosphoribosyltransferase and OMP decarboxylase activities. Although added OMP is accepted as substrate for the decarboxylase, OMP formed from orotate and PRPP is not released but is preferentially utilized at the decarboxylase site for UMP formation.

In the purine pathway a trifunctional enzyme catalyzes three reactions concerned with generation of the formyl donors for steps 3 and 9 (see fig. 26.14). The enzymatic reactions catalyzed by this protein are shown in figure 26.34. Since 10-formyltetrahydrofolate inhibits the cyclohydrolase, this may serve to regulate the amount of the formyl donor that is available. Channeling is probably further enhanced by noncovalent association of other enzymes of the pathway with the dehydrogenase-cyclohydrolase-synthase enzyme. Some evidence suggests that this loose complex contains serine hydroxymethyltransferase (which generates methylenetetrahydrofolate), the transformylases catalyzing steps 3 and 9 of the purine pathway, and probably all the other enzymes of the pathway. This association of pathway enzymes would permit the efficient generation and use of unstable intermediates like 10-formyltetrahydrofolate and phosphoribosylamine (half-life 38 seconds at 37°C).

Intracellular Concentrations of Ribonucleotides Are Much Higher Than Those of Deoxyribonucleotides

Methods are available for analysis of nucleotides in eukaryotic cells and in mammalian tissues. Concentrations are frequently expressed in terms of picomoles per 10^6 cells or picomoles per microgram of DNA, since it is easier to express analyses on this basis than as intracellular molar concentration. Some estimates in molar terms are available, but the values depend on a number of factors. The presence of nucleosides or bases in the extracellular fluid not only causes increases in the intracellular concentration of the corresponding nucleotides, but through allosteric effects may also affect the size of other nucleotide pools. Transfer of a cell culture to fresh medium may also cause perturbation of intracellular nucleotide levels. Finally, as the cell progresses through its life cycle (see fig. 26.33), there are marked changes in concentration of some nucleotides, particularly an increase of the deoxyribonucleoside triphosphates as the cell enters S phase. Conditions that damage cells, such as starvation, lack of oxygen, or presence of toxic materials, also affect nucleotide levels. Consequently, there are no "correct" values of nucleotide levels, even for a specific type of cell. Some general statements are possible, however. ATP is generally present in normal cells at a concentration of 2–10 millimolar, but other ribonucleoside triphosphates are at lower concentrations (0.05 to 2 mM), depending on conditions

Figure 26.35

Biosynthesis of flavin mononucleotide (FMN) and flavin adenine dinucleotide (FAD) from riboflavin. In the first reaction a kinase transfers a single phosphate to the terminal hydroxyl of the ribose. In the second reaction the AMP moiety is transferred to the phosphate.

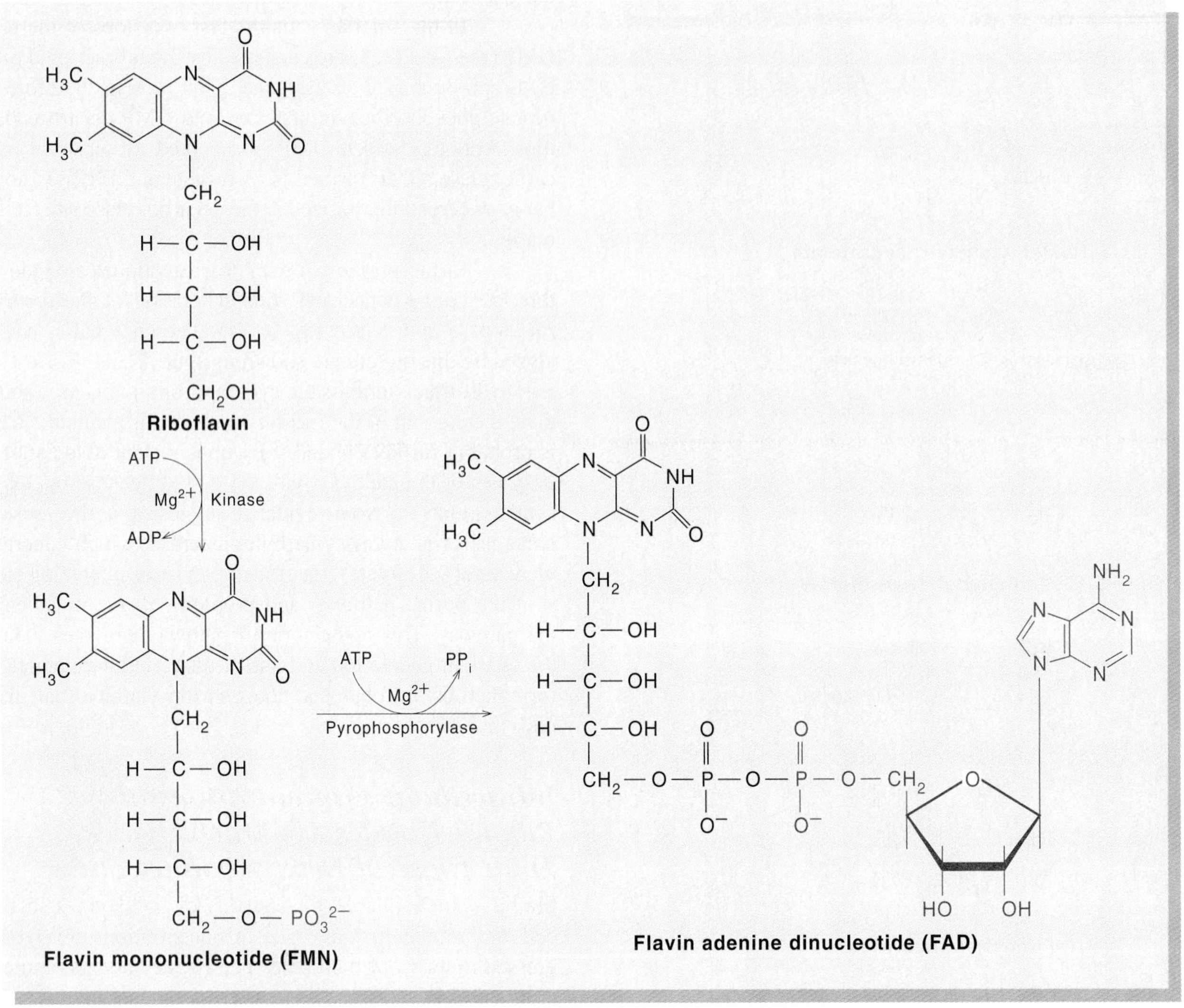

and the cell type. The ribonucleoside mono- and diphosphates have lower concentrations than the corresponding triphosphates, but ADP is typically comparable with or even higher than GTP, CTP, or UTP. Deoxyribonucleoside triphosphates (dNTPs) are present in much lower concentrations, 2–60 μM. The concentration of the dNTP present at lowest concentration, dCTP in some cells and dGTP in others, is sufficient for only a few minutes of DNA synthesis, so during DNA replication, synthesis of dNTPs must keep pace with utilization for DNA synthesis.

In prokaryotes the levels of deoxyribonucleotides vary from undetectable to 200 picomoles per 10^6 cells (compared with 3 to 30 picomoles of dATP, dCTP, and dGTP per 10^6 eukaryotic cells). As in eukaryotic cells, dTTP is usually present at higher concentrations than the other deoxyribonucleoside triphosphates.

The arrest of cell growth by many agents is associated with depletion of one or more of the deoxyribonucleotide pools. Thus thymidine at millimolar concentrations arrests cell growth and decreases the concentration of dCTP dramatically, whereas the concentrations of dATP, dGTP, and especially dTTP increase. This effect is considered to be mediated by the allosteric inhibition of ribonucleotide reductase, referred to earlier. Hydroxyurea, another agent that arrests cell growth by blocking DNA synthesis, depletes the pools of dATP and dGTP, and in some cells it is the effect on the latter, brought about by ribonucleotide reductase inhibition, that is probably critical.

T4 Bacteriophage Infection Stimulates Nucleotide Metabolism

Infection of *E. coli* by T4 phage results in the induction of nearly 30 proteins. Many of these are enzymes that ensure an abundant supply of deoxynucleotides above those produced by the host. As a result, deoxynucleotide concentrations are increased many times.

Figure 26.36

Biosynthesis of NAD^+. In animal tissues, tryptophan degradation leads to quinolinate, which is converted to deamido-nicotinamide mononucleotide (deamido-NMN). Deamido-NMN can also be formed from nicotinate in some mammalian tissues and in many microorganisms. The deamido-NMN is subsequently converted into NAD^+. Another route to NAD^+ starts from nicotinamide. The phosphoribosyltransferase required for this pathway is found in the cytosol of animal tissues. These transferases are responsible for utilization of nicotinate and nicotinamide in the diet.

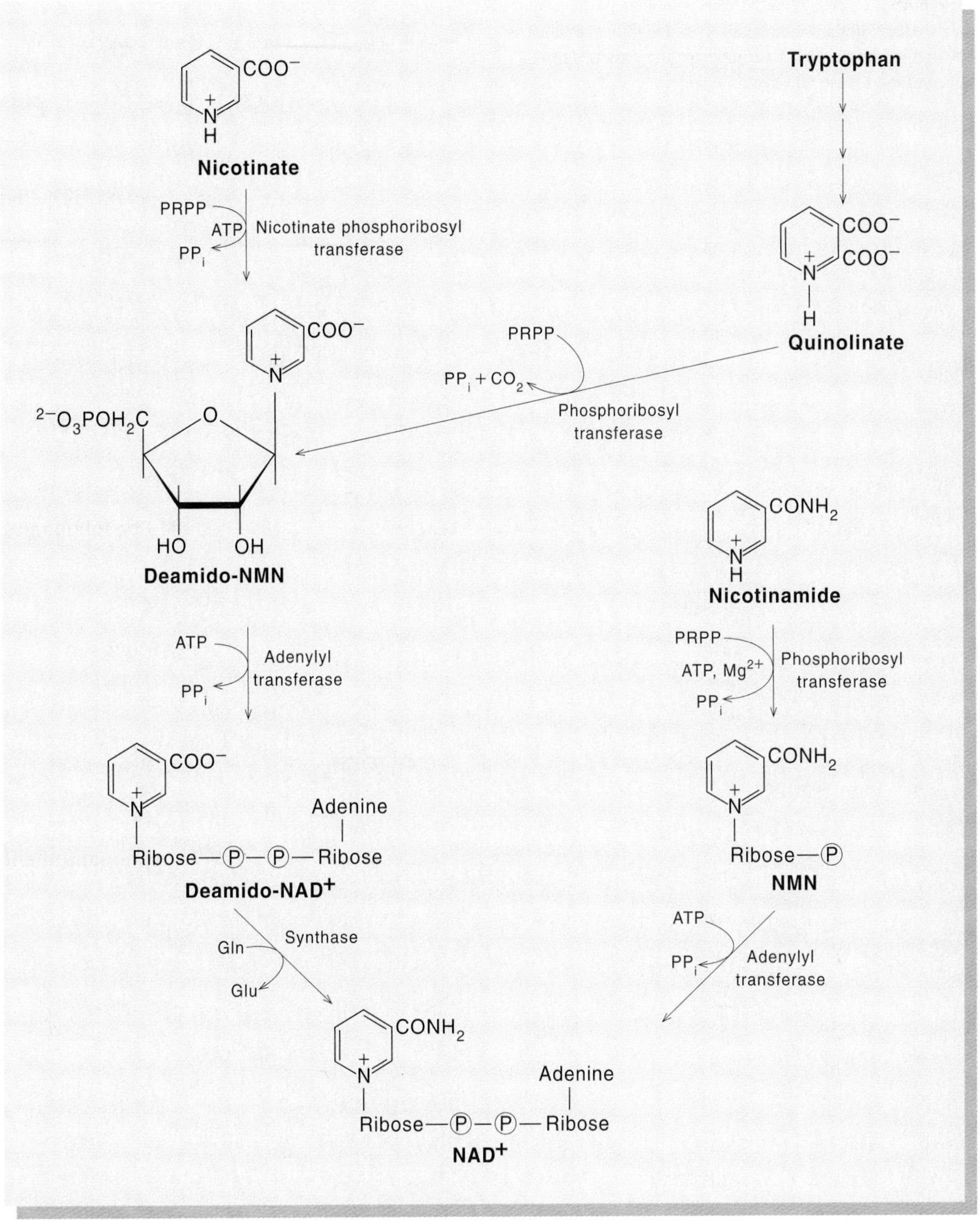

Phage-coded enzymes and related proteins include thioredoxin, ribonucleotide reductase, dihydrofolate reductase, dCMP deaminase, thymidylate synthase, and deoxyribonucleotide kinase. However, the phage relies completely on some host enzymes that are normally present at high levels. Examples of such enzymes are adenylate kinase and nucleoside diphosphate kinase. Completely novel enzymes coded by the phage are endonucleases II and IV, which supply nucleotides directly by degrading host DNA.

The phage DNA contains no cytosine; instead, hydroxymethylcytosine is incorporated. To accomplish this, the phage induces enzymes that hydrolyze dCTP and dCDP, synthesize 5-hydroxymethyl dCMP from dCMP and methylenetetrahydrofolate and phosphorylate hydroxymethyl dCMP. All these enzymes help to ensure rapid and specific synthesis of phage DNA while preventing synthesis of host DNA.

Biosynthesis of Nucleotide Coenzymes

Many of the nucleotides considered thus far play important roles in metabolism and are discussed in other chapters. However, nucleotide coenzymes such as flavin nucleotides, NAD^+, $NADP^+$,

Figure 26.37

Biosynthesis of coenzyme A from pantothenate. This synthesis occurs in the mammalian liver. Pantothenate must be supplied in the diet. Red indicates the groups introduced at the kinase and synthase steps.

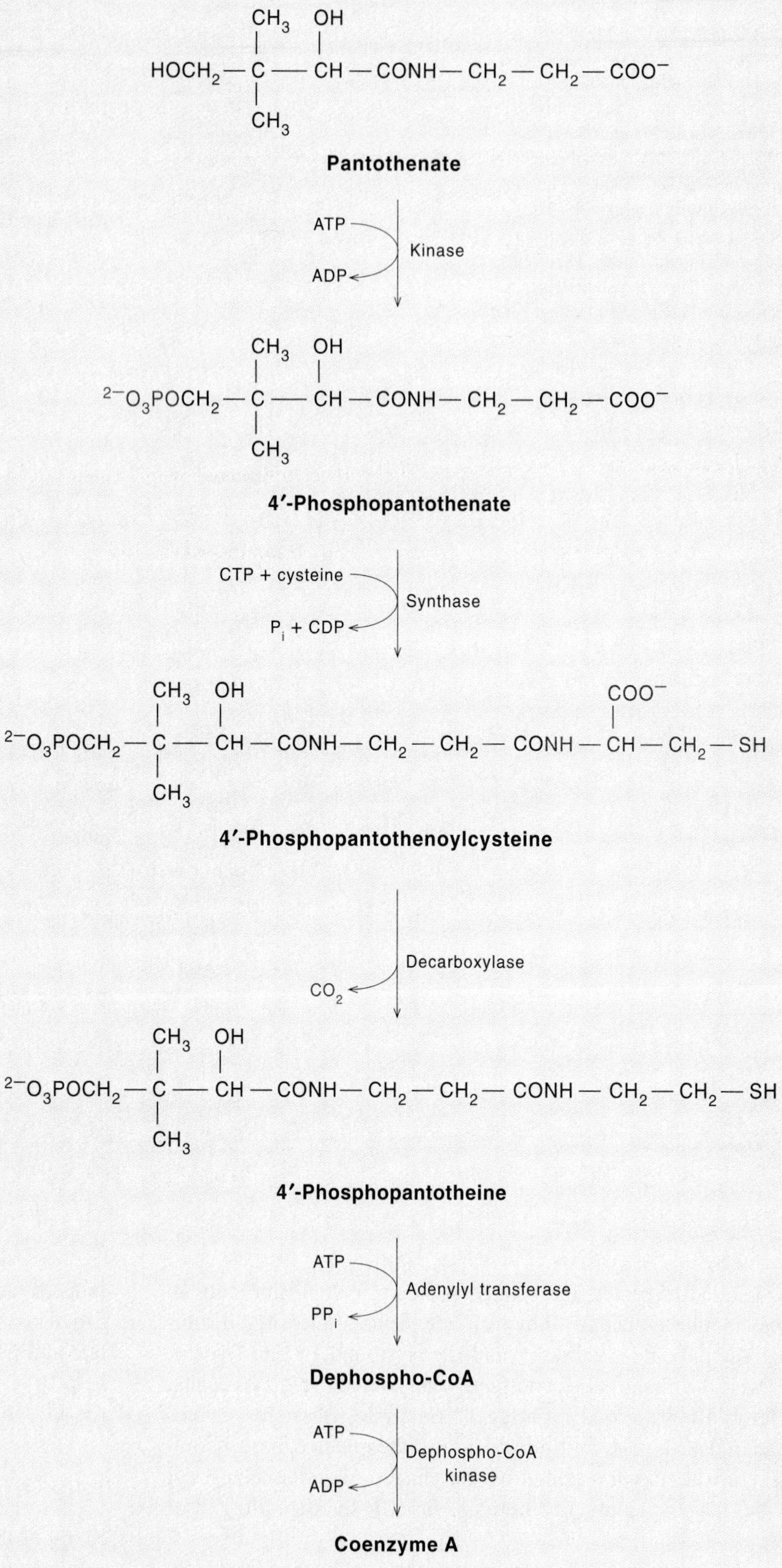

and coenzyme A are also extremely important in metabolism. In this section we will discuss the pathway for the completion of each coenzyme.

Riboflavin, that is, 7,8-dimethyl-10(1′-D-ribityl)isoalloxazine, is synthesized by microorganisms such as the fungus *Eremothecium* and mutants of the yeast *Saccharomyces* in a pathway that starts from GTP. Riboflavin is an essential dietary constituent for mammals and is converted in the body to the mononucleotide or dinucleotide forms that function as the prosthetic groups of many enzymes. Riboflavin is converted to riboflavin-5′-phosphate, more commonly called flavin mononucleotide (FMN), by flavokinase (ATP:riboflavin phosphotransferase), as shown in figure 26.35. The enzyme has been purified from yeast, plants, and liver. It is also present in a variety of other animal tissues (kidney, brain, spleen, and heart).

The other nucleotide form of riboflavin, flavin adenine dinucleotide (FAD), is formed from FMN in a reversible reaction catalyzed by flavin nucleotide pyrophosphorylase (see fig. 26.35). This enzyme is also widely distributed in nature and has been observed in plants, yeast, lactobacilli, and many animal tissues.

The nicotinamide moiety of the coenzymes nicotinamide adenine dinucleotide (NAD^+) and nicotinamide adenine dinucleotide phosphate ($NADP^+$) is synthesized by several routes (fig. 26.36). In liver and other animal tissues, tryptophan degradation forms, among other products, quinolinic acid (chapter 24), which is converted to nicotinate mononucleotide (deamidonicotinamide mononucleotide, deamido-NMN) by quinolinate phosphoribosyltransferase. In the cytosol of cells of many mammalian tissues, and in yeast and other microorganisms, there is present a nicotinate phosphoribosyltransferase that also forms deamido-NMN (see fig. 26.36). A very similar phosphoribosyltransferase present in the cytosol of all animal tissues investigated acts on nicotinamide. These transferases are responsible for utilization of nicotinate and nicotinamide in the diet. The role of ATP in these reactions is unclear. Some transferases do not require it, for others it seems to be an allosteric regulator, and in yet other cases, ATP seems to be hydrolyzed to yield ADP and P_i in equimolar amounts with deamido-NMN formation.

The mononucleotides so formed are converted to the corresponding dinucleotides by NMN adenylyltransferase (see fig. 26.36). In mammalian cells it appears to be a single enzyme that catalyzes both reactions, but an adenylyltransferase acting only on NMN has been isolated from some bacteria (*Lactobacillus fructosus*). A cytoplasmic NAD^+ synthase present in yeast, liver, and other tissues transfers the amino group from glutamine at the expense of ATP hydrolysis (see fig. 26.36).

A cytoplasmic kinase present in liver, mammary gland, and brain is responsible for the formation of $NADP^+$ from NAD^+:

$$NAD^+ + ATP \longrightarrow NADP^+ + ADP \qquad \textbf{(19)}$$

NADH is not a substrate and inhibits competitively with respect to NAD^+.

Coenzyme A is synthesized in the mammalian liver from pantothenic acid (pantoyl-β-alanine), which is required in the mammalian diet. The five steps in the synthesis are shown in figure 26.37. In the last step a specific kinase transfers a phosphoryl group to the 3′-hydroxyl of the adenylate portion of the molecule.

Summary

Nucleotides are the building blocks for nucleic acids; they are also involved in a wide variety of metabolic processes. They serve as the carriers of high-energy phosphate and as the precursors of several coenzymes and regulatory small molecules. Nucleotides can be synthesized de novo from small-molecule precursors or, through salvage pathways, from the partial breakdown products of nucleic acids. The highlights of our discussion in this chapter are as follows:

1. The ribose for nucleotide synthesis comes from glucose, either by means of the pentose phosphate pathway or from glycolytic intermediates through transketolase–transaldolase reactions. Ribose-5-phosphate is converted to phosphoribosylpyrophosphate (PRPP), the starting point for purine synthesis. This pathway also incorporates into purines atoms from glycine, aspartate, glutamate, CO_2, and one-carbon fragments carried by folates. IMP synthesized by this route is converted by two-step pathways to AMP and GMP, respectively.
2. The biosynthetic pathway to UMP starts from carbamoyl phosphate and results in the synthesis of the pyrimidine orotate, to which ribose phosphate is subsequently attached. CTP is subsequently formed from UTP.
3. Deoxyribonucleotides are formed by reduction of ribonucleotides (diphosphates in most cells). Thymidylate is formed from dUMP.
4. All biosynthetic pathways are under regulatory control by key allosteric enzymes that are influenced by the end products of the pathways. For example, the first step in the pathway for purine biosynthesis is inhibited in a concerted fashion by nucleotides of either adenine or guanine. In addition, the nucleoside monophosphate of each of these bases inhibits its own formation from inosine monophosphate (IMP). On the other hand, adenine nucleotides stimulate the conversion of IMP into GMP, and GTP is needed for AMP formation.
5. Inhibitors of nucleotide biosynthesis are toxic to cells, especially to rapidly growing cells, where the need for nucleic acid synthesis is greatest. In limited amounts, some of the inhibitors have chemotherapeutic value in the treatment of cancer and other illnesses. Some analogs of normal nucleosides are proving to be useful in the treatment of AIDS and certain other viral infections.
6. Nucleic acids and nucleotides in food are degraded to nucleosides or free bases before they are ingested, but most of the products are catabolized and excreted. Nucleotides or their partial degradation products from tissues may be reutilized for nucleic acid synthesis. Purine bases from nucleotide break-

down may be reutilized for nucleotide synthesis, or, together with pyrimidine bases, they may be further catabolized for excretion. Purine nucleotides are degraded via guanine, hypoxanthine, and xanthine to uric acid, which in some species is degraded further before excretion. Inherited deficiencies in some of the enzymes involved in nucleotide degradation and salvage cause severe impairment of health, a fact testifying to the importance of the degradative pathways. Nucleotides of uracil and cytosine are degraded via uridine and uracil to simpler substances such as β-alanine.

7. In addition to the nucleotides used as substrates in nucleic acid synthesis, there are a number of other nucleotide-containing molecules that serve various purposes in the cell. These include the coenzymes NAD^+, $NADP^+$, FAD, and CoA.

Selected Readings

Christopherson, R. I., and Szabados, E., Nucleotide biosynthesis in mammals. In L. Agius and H. S. A. Sherratt (eds.), *Channeling in Intermediary Metabolism*. London: Portland Press, 1995. A summary of postulated channeling and a good critical review of the current evidence.

Eaton, W. A., and J. Hofrichter, The biophysics of sickle cell hydroxyurea therapy. *Science* 268:1142–1143, 1995.

Johnston, M. I., and D. F. Hoth, Present status and future prospects for HIV therapies. *Science* 260:1286–1292, 1993.

Jones, M. E., Pyrimidine nucleotide biosynthesis in animals: Genes, enzymes and regulation of UMP biosynthesis. *Ann. Rev. Biochem.* 49:253–279, 1980. Authoritative outline of the regulatory properties of the two multifunctional proteins responsible for pyrimidine nucleotide synthesis in animals.

Kornberg, A., *DNA Replication,* 2d ed. San Francisco: Freeman, 1989. See especially chapter 1.

Manfredi, J. P., and E. W. Holmes, Purine salvage pathways in myocardium. *Ann. Rev. Physiol.* 47:691–705, 1985. Although this review applies specifically to salvage in heart muscle, it is an excellent summary of purine salvage pathways and the metabolism of the purine nucleotides formed.

Mathews, C. K., L. K. Moen, and R. G. Sargent, Enzyme interactions in deoxyribonucleotide synthesis. *Trends Biochem. Sci.* 13:394–397, 1988.

Nordlund, P., B.-M. Sjoberg, and H. Eklund, Three-dimensional structure of the free radical protein of ribonucleotide reductase. *Nature* 345:593–598, 1990.

Plageman, P. G. W., R. M. Woehlheuter, and C. Woffendin, Nucleoside and nucleobase transport in animal cells. *Biochem. Biophys. Acta* 947:405–443, 1988. Reviews the field in detail.

Reichard, P., Interactions between deoxyribonucleotide and DNA synthesis. *Ann. Rev. Biochem.* 57:349–374, 1988. Comprehensive review of deoxyribonucleotide synthesis and its relation to DNA synthesis.

St. Georgiev, V., and J. J. McGowan (eds.), AIDS: Anti-HIV agents, therapies, and vaccines. *Ann. NY Acad. Sci.* 616, 1990. This volume, reporting papers from a conference on the title topic, is introduced by a readable summary of the life cycle of the virus and the stages at which intervention is possible by available therapy.

Stadel, J. M., A. D. Lean, and R. J. Lefkowitz, Molecular mechanisms of coupling in hormone receptor adenylate cyclase systems. *Adv. Enzymol.* 53:1–43, 1982. A current account by a major contributing group, emphasizing the more biochemical aspects of this important system.

Weber, G., Enzyme pattern-targeted chemotherapy. *Adv. Enzyme Regul.* 24:118, 1985. This volume contains several chapters about inhibitors of nucleotide metabolism and their mechanisms of action.

Problems

1. Examine figure 26.21 and determine the fate of the C-2′ oxygen upon conversion of the nucleotide to the 2-deoxynucleotide.
2. Propose a mechanism for formyltetrahydrofolate synthase (fig. 26.34). (Hint: This reaction is often written as an equilibrium; consider the reverse reaction.)
3. Would you expect an ATP to be involved in the synthesis of NAD^+ from deamido-NAD^+?
4. Would you expect the decarboxylation of the cysteine moiety in the synthesis of CoASH (fig. 26.37) to be a pyridoxal phosphate enzyme?
5. When phosphoribosylpyrophosphate (PRPP) is treated with base, 5-phosphoribose-1,2-cyclic phosphate is formed. Produce a mechanism for this reaction. What is the second product in this reaction?
6. How could you prove that if PRPP is formed by

$$\text{Ribose-5-P} + \text{ATP} \longrightarrow \text{PRPP} + \text{AMP} \qquad \textbf{(A)}$$

rather than

$$\text{Ribose-5-P} + \text{ATP} \longrightarrow \text{Ribose-1,5-bisP} + \text{ADP} \qquad \textbf{(B)}$$

then

$$\text{Ribose-1,5-bisP} + \text{ATP} \longrightarrow \text{PRPP} + \text{ADP}$$

by using ATP that is labeled with ^{32}P in either the α (proximal to ribose) or γ (distal to ribose) position? What results would you expect and how would the information from problem 5 help you decide which method is used by nature?

7. How do you explain the observation that pyrimidine biosynthesis in bacteria is regulated at the level of aspartate carbamoyltransferase, whereas most of the regulation in humans is at the level of carbamoyl phosphate synthase?
8. Compare and contrast the pathways for the biosynthesis of IMP (fig. 26.14) with that of UMP (fig. 26.18). What are the two most striking features of this comparison?

9. The first enzyme in the biosynthesis of IMP (fig. 26.14) utilizes a glutamine to provide an amino group. We saw the same phenomenon in the two previous chapters. Why doesn't nature use an ammonium ion rather than the more expensive (in the expenditures of ATP) glutamine as an amine donor?
10. Examine the reactants and products of the second enzyme on the IMP biosynthetic pathway. What reaction intermediate would you expect?
11. What function can you propose for the ATP in the fourth reaction in figure 26.14?
12. What function can you give ATP (in a mechanistic sense) in the fifth reaction (fig. 26.14) in the formation of IMP?
13. What happens if you give allopurinol to a chicken?
14. Explain why patients with Lesch-Nyhan syndrome suffer from severe gout. Although these patients can be treated with allopurinol to relieve the symptoms of gout, this treatment has no effect on the severe mental retardation. Suggest a possible explanation.
15. Why can the toxic effect of sulfanilamide on bacteria be reversed by *p*-aminobenzoate? Why are sulfa drugs not very toxic to humans?
16. Explain how antifolates like methotrexate selectively kill cancer cells. Why do cancer patients lose their hair, intestinal mucosa, cells of the immune system, and so forth when treated with antifolates?
17. When phosphoribosylpyrophosphate (PRPP) is incubated in alkali, 5-phosphoribose-1,2-cyclic phosphate is formed with the release of phosphate. Draw a chemical reaction mechanism for this reaction. (*Hint:* The reaction is similar to the mechanism for the base hydrolysis of RNA.)
18. Using the information in problem 5, design an experiment to prove that ribose-5-phosphate pyrophosphokinase (rib-5-P + ATP $\rightarrow$ PRPP + AMP) transfers a pyrophosphate group, making it an unusual kinase. (In most cases a pyrophosphate group is constructed in two steps. For example, in the formation of mevalonate pyrophosphate two phosphates are *separately* transferred to form the pyrophsophate group.) Also, could you prove that the pyrophosphate group in PRPP is in the α position? (*Hint:* Use [γ-^{32}P]ATP and [α-^{32}P]ATP.) Draw a chemical reaction mechanism for this pyrophosphokinase.
19. Genetic defects in adenosine deaminase (ADA) in humans lead to severe defects in the immune system. This disease has been treated by injecting the enzyme adenosine deaminase and more recently by gene therapy. Discuss how this defect could lead to toxic effects on lymphocytes. (*Hint:* Deoxyadenosine is metabolized by ADA.)
20. Propose a chemical reaction mechanism for the second enzyme in purine biosynthesis, GAR synthase (phosphoribosylamine + ATP + glycine $\rightarrow$ GAR + ADP + P_i).
21. Allopurinol administered together with 6-mercaptopurine under certain conditions enhances the anticancer effectiveness of the latter. How can the known site of action of allopurinol explain this effect?

C H A P T E R

27

Integration of Metabolism in Vertebrates

The metabolic processes of vertebrates are regulated by intercellular and intracellular signaling.

In chapter 12 we described strategies used to organize biochemical conversions so that all essential conversions are thermodynamically feasible and kinetically regulated. In the 14 chapters that followed, we presented numerous examples employing these basic strategies. This is an appropriate time to review how various organizational strategies are integrated in the metabolism of a multicellular organism. In this chapter we review and extend the discussion on the integration of metabolism with a focus on the tissues in vertebrates. First we review how energy metabolism is manipulated between these tissues. Then we expand the discussion to a consideration of other aspects of the metabolism.

Tissues Store Biochemical Energy in Three Major Forms

In addition to relying on nutrition to supply the chemical fuel for their energy needs, vertebrates maintain fuel reserves in various tissues. These reserves are of three types: glycogen, triacylglycerols, and proteins. More than 90% of the fuel reserves in an av-

Table 27.1
Fuel Reserves in a Normal 70-kg Human (kcal)

Organ	Glucose or Glycogen	Triacylglycerols	Mobilizable Proteins
Liver	400	450	400
Brain	8	0	0
Muscle	1,200	450	24,000
Adipose tissue	80	135,000	40

Source: Data from G. T. Cahill, *Clin. Endrocrinol. Metab.* 5:598, 1976.

erage human exist in the form of triacylglycerols stored in the adipose tissue (table 27.1). In obese individuals this fuel reserve can be severalfold higher. Despite their abundance, triacylglycerols in large measure are held in check until the more readily utilizable glycogen reserves located in liver and muscle tissues are close to exhaustion. Generally, glycogen reserves of the liver play the most widely useful role, despite the greater abundance of glycogen often found in muscle tissue. This is because, of the two tissues, only the liver is capable of degrading glycogen to (neutral) glucose that can be secreted into the bloodstream. Once the glycogen reserves have been depleted from the liver, the energy-hungry organism turns to the lipid (triacylglycerol) reserves stored in adipose tissue. The first step in mobilizing lipid for energy consumption involves the hydrolysis of triacylglycerols to fatty acids and glycerol. The fatty acids so produced are transported to other tissues, where they are degraded to the activated two-carbon unit, acetyl-CoA, in a repetitious multistep process (see chapter 21). A third major fuel reserve, protein, mostly from muscle tissues, can be mobilized by breakdown to amino acids that are transported to the liver. In the liver the amino acids are deaminated and converted into TCA cycle intermediates. The use of muscle tissue proteins to satisfy energy needs is bound to physically weaken the organism, so it comes as no surprise that muscle tissue is used for energy purposes only as a last resort. The best-known exception to this rule occurs during periods of prolonged muscle inactivity, as in the case of a temporarily incapacitated limb.

There must be a way in which the energy-requiring tissues can send signals to the energy-producing tissues that reflect their needs. The signals are supplied by circulating hormones, which regulate the metabolic activities of different tissues in the organism (fig. 27.1).

Each Tissue Makes Characteristic Demands and Contributions to the Energy Pool

Some tissues are mainly energy suppliers; others are mainly energy consumers. Still other tissues are important both as consumers and as suppliers. In this section we survey energy metabolism in five well-characterized vertebrate tissues: liver, adipocytes, striated muscle, smooth heart muscle, and brain.

Brain Tissue Makes No Contributions to the Fuel Needs of the Organism

Brain tissue does not contribute to the energy needs of other tissues. Rather, it makes major demands on the glucose supply, accounting for upwards of 50% of the glucose consumption in the resting human. To satisfy its needs, the brain normally absorbs glucose from the bloodstream. It metabolizes the glucose to CO_2 and H_2O by a combination of glycolysis and the TCA cycle. Energy consumption by the brain is approximately constant: A sleeping brain and a wide-awake, mentally alert brain burn about the same amount of glucose. Because of its vital importance and delicate constitution, brain tissue is somehow given priority in meeting demands on the available supplies of blood glucose in times of starvation or other forms of stress, when the glucose supply is limited. Under conditions of prolonged starvation the blood glucose level drops somewhat. As a consequence, the energy needs of brain tissue are met by a combination of glucose and ketone bodies, which are metabolized via the TCA cycle.

Heart Muscle Utilizes Fatty Acids in Preference to Glucose to Fulfill Its Energy Needs

Continuous operation is an obvious necessity for the heart muscle. Its energy needs vary considerably, depending on the physical activity of the organism. For reasons that are unclear, the heart relies mainly on fatty acids supplied from the bloodstream to fulfill its energy needs. Perhaps this is because the fatty acid supply is more reliable than the fluctuating carbohydrate supply. The fatty acids are metabolized by β oxidation in the densely packed mitochondria of the heart muscle cells. Most organisms have a very extensive supply of fatty acids; thus the functioning of the heart muscle is ensured. Under conditions of vigorous activity the heart muscle also uses glucose as a source of energy. Under starvation conditions the heart muscle can switch to using ketone bodies supplied from the liver via the bloodstream. The ketone bodies are degraded in the mitochondria of the heart muscle by way of the TCA cycle. Like the brain, heart muscle makes no energy contributions to other tissues of the organism.

Skeletal Muscle Can Function Aerobically or Anaerobically

Whereas the heart muscle operates under strictly aerobic conditions, skeletal muscle can function either aerobically or anaerobically. The energy consumption of skeletal muscle varies enormously with the extent of muscle activity. During periods of mild exertion, muscle tissue requires moderate levels of glucose, which can be supplied by the blood glucose or the breakdown of the glycogen reserves present in the muscle tissue. This glucose is metabolized aerobically all the way to CO_2 and H_2O. During times of great exertion, muscle tissue uses oxygen faster than it can be supplied by the bloodstream, so that the muscle must operate

Figure 27.1

Some aspects of the regulation of energy metabolism of cells in a typical mammal. Each cell type serves a particular function and has specific requirements for maintenance and growth. Some cells are primarily energy producers, and others are primarily energy consumers. The activities of different cell types are regulated by an intricate hierarchy of hormone-secreting cells. To be susceptible to a specific hormone, a cell must possess a specific hormone receptor. On contact with the receptor the hormone causes a structural change in the hormone receptor, which in turn triggers a chain of reactions in the cell.

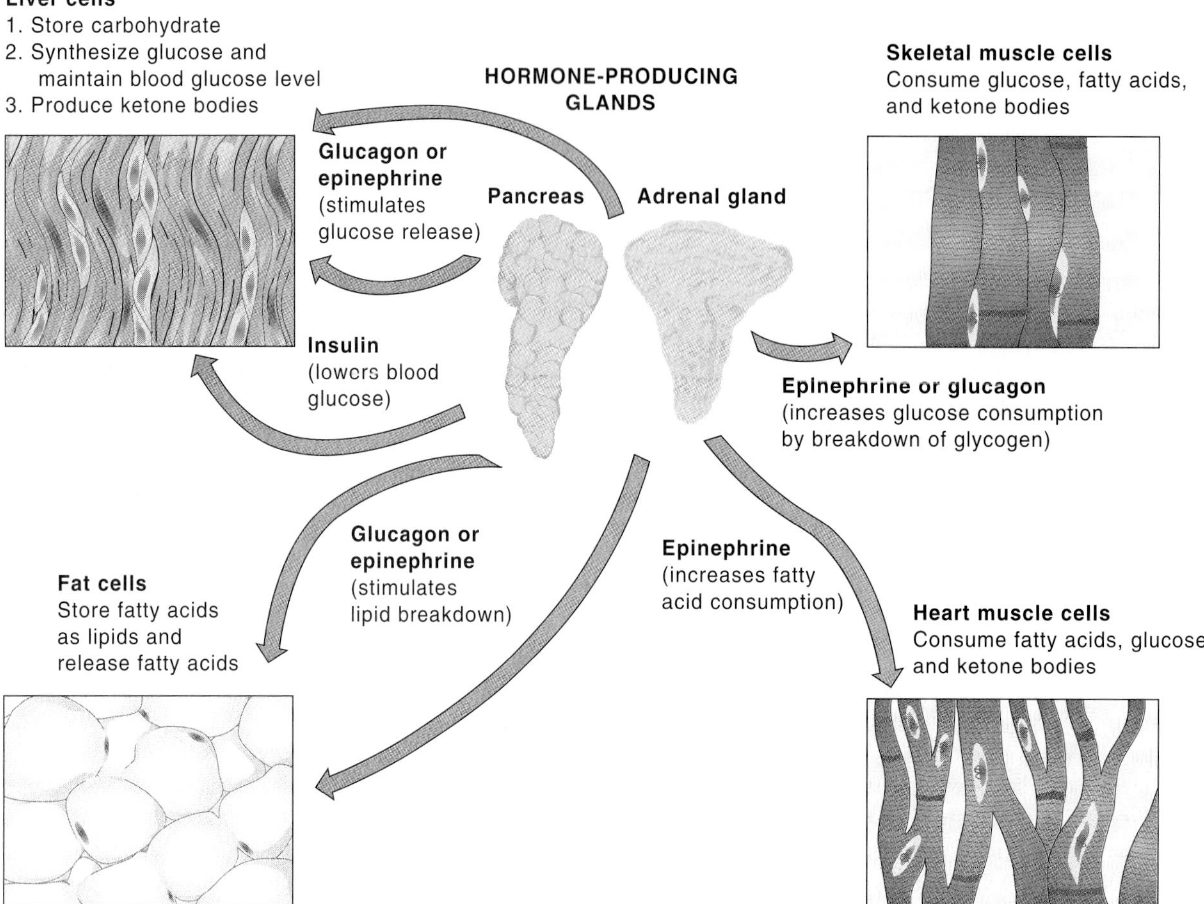

anaerobically. Under these conditions, glucose breakdown stops at the three-carbon acid, lactate. The lactate so produced is secreted into the bloodstream and picked up by the liver, which converts the lactate back to glucose. This glucose can be returned to the muscle for further glycolysis. The cycling of glucose and lactate between skeletal muscle and liver is known as the Cori cycle (fig. 27.2). When the blood glucose level is low, muscle tissue can utilize fatty acids as an alternative supply of energy. The fatty acids are mobilized via the β-oxidation pathway in the muscle mitochondria. Muscle lactate can also be harnessed by heart muscle, which catabolizes it for energy purposes (box 27A).

In times of starvation, muscle tissue is usually quite inactive, so its energy demands are low. In fact, under such conditions the muscles tend to supply energy to the rest of the organism via degradation of muscle protein to amino acids. These amino acids are transported to the liver, where they are converted to glucose or used for biosynthesis of other proteins essential for maintenance of the organism's vital processes.

Adipose Tissue Maintains Vast Fuel Reserves in the Form of Triacylglycerols

Under conditions of nutritional excess, fatty acids are absorbed by adipose tissue, where they are converted to storage lipids in the form of triacylglycerols. The triacylglycerols can be mobilized at a later time, when the carbohydrate energy reserves are low. The process of mobilization starts with conversion of triacylglycerols to fatty acids, which are transported by the bloodstream to the liver for further processing (fig. 27.3).

The Liver Is the Central Clearing House for All Energy-Related Metabolism

Glucose is the most readily utilizable energy source of the organism, and it is a primary function of the liver to maintain blood glucose at a reasonable level for absorption into most tissues. Thus in times of glucose excess the liver absorbs glucose and converts it to glycogen, which it stores for future energy needs. At other

Figure 27.2

The cycling of lactate and glucose between muscle and liver in the Cori cycle. Under conditions of intense activity the muscle operates anaerobically, so the end product of glucose breakdown is lactate. This product can pass through the bloodstream to the liver and be converted to glucose, which can be returned to the muscle.

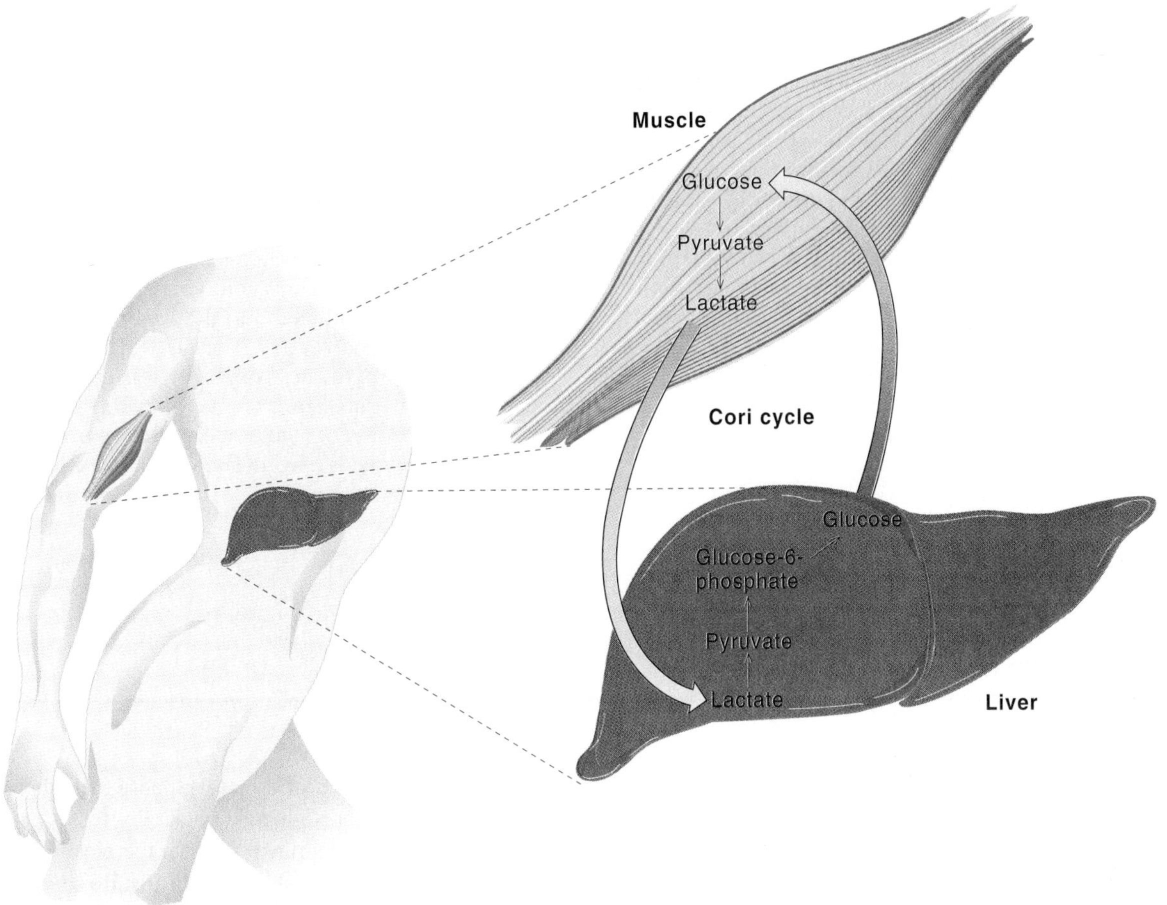

times, when the glucose concentration in the bloodstream drops, the liver converts glycogen into glucose, which it secretes into the bloodstream. When the blood glucose level falls and the liver's glycogen reserves are also exhausted, the liver still has the capacity to synthesize glucose via gluconeogenesis from amino acids that are supplied from protein breakdown. Under starvation conditions the liver forms increasing amounts of ketone bodies (see fig. 21.10). This is due to elevated concentrations of acetyl-CoA, which favor the formation of ketone bodies. The ketone bodies are secreted and used as a source of energy by other tissues, especially those tissues such as the brain that cannot catabolize fatty acids directly.

The major events in fuel storage and energy utilization are summarized in table 27.2. In this table, energy metabolism is considered under conditions of glucose excess, such as just after a meal; under conditions when the glucose supply is scant, as after prolonged periods between meals; and under starvation conditions, when the glucose supply is severely limited.

Most often the metabolic state of an organism is reflected by the small molecules present in plasma. Figure 27.4 illustrates the plasma levels of glucose, ketone bodies, and fatty acids as a function of the number of days of starvation. As you can see, the glucose level drops about 30% as the period of starvation becomes prolonged; the fatty acid level rises about twofold, while the level of ketone bodies rises severalfold.

Pancreatic Hormones Play a Major Role in Maintaining Blood Glucose Levels

Thus far we have argued that it is the liver's job to maintain normal blood glucose levels. But the liver cannot by itself sense the blood glucose levels. It relies on hormonal signals transmitted by the pancreas (fig. 27.5). The pancreas sends two quite different signals. When the blood glucose level is high, the pancreas secretes insulin, which binds to specific receptors located on the outer plasma membranes of insulin-responsive cells (target cells) (see fig. 20.14). The effect of insulin on liver cells

27A
BOX

The Lactate Dehydrogenase of Heart Muscle

Most vertebrates possess at least two genes for lactate dehydrogenase (LDH) that make similar but nonidentical polypeptides called M and H. In embryonic tissue the two genes are equally active, resulting in equimolar amounts of the two gene products and a statistical array of tetramers (M_4, M_3H, M_2H_2, H_3M_1, and H_4 in the ratios of 1:4:6:4:1). These so-called isoenzymes, or isozymes, can usually be detected by their differing electrophoretic mobilities. As embryonic tissue multiplies and differentiates, the relative amounts of the M and H forms change. In pure heart tissue the H_4 tetramer predominates. In skeletal muscle, which functions anaerobically under stress, the M_4 isozyme predominates. It seems likely that the M and H forms were designed to serve different functions. A clue to these functions may be revealed by the inhibiting effect of pyruvate on the dehydrogenase. Pyruvate can form a covalent complex with NAD^+ at the active site of the enzyme according to the following reaction:

The H_4 tetramer shows a much greater inhibition by this compound than the M_4 tetramer. Active muscle tissue is anaerobic and produces a good deal of pyruvate. Inhibition of LDH under anaerobic conditions would shrink the supply of NAD^+ and shut down the glycolytic pathway (see chapter 14) with disastrous consequences. In fact, active muscle tissue has augmented levels of pyruvate but converts this readily to lactate. This conversion is possible because the predominant form of LDH in muscle is M_4, which is only poorly inhibited by excess pyruvate. The function of LDH in aerobic tissue is clear. Heart muscle is aerobic tissue, and consequently, most of its pyruvate is funneled into the Krebs cycle for greater energy production (see chapter 15). The LDH of heart muscle, H_4, might be inhibited by pyruvate to prevent waste of this potential high-energy carbon source or excessive buildup of pyruvate resulting from the conversion of incoming lactate to pyruvate. It seems likely that the H_4 enzyme is used in such tissues to convert absorbed lactate into pyruvate.

$$\text{ADPR—N}^+\text{(pyridinium, CONH}_2) + \underset{\text{Pyruvate}}{CH_3COCOO^-} \longrightarrow \text{ADPR—N(dihydropyridine; H, } CH_2COCOO^-\text{; CONH}_2) + H^+$$

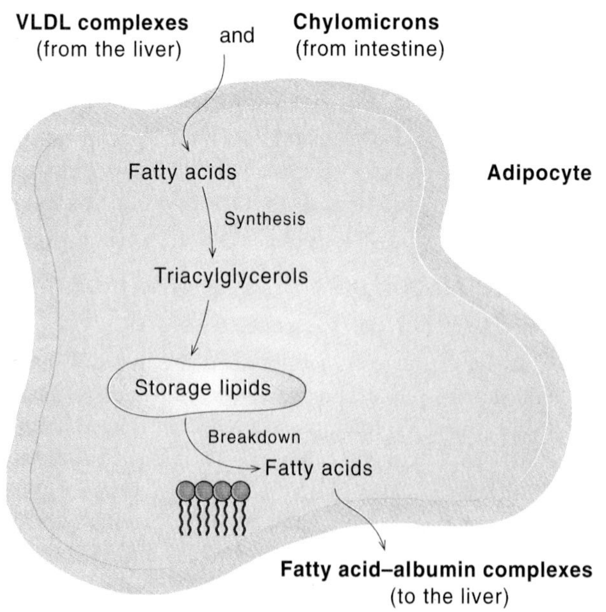

Figure 27.3

Synthesis and degradation of triacylglycerols in adipose tissue. Fatty acids are delivered to adipose tissue. In times of energy excess these are converted to triacylglycerols and stored until needed, at which point the triacylglycerols are converted back to fatty acids.

is to stimulate uptake of excess glucose, which is converted to glycogen. It stimulates glucose uptake by fostering an increase in the number of glucose transporters in the plasma membrane. This effect is reversible so that when the insulin level drops, the glucose transporters are removed from the plasma membrane. When the blood glucose level is low, the pancreas secretes the protein hormone glucagon, which binds to different receptors on liver cells. This hormone stimulates the pathway leading to glycogen breakdown and subsequent secretion of glucose by the liver.

Glucagon and insulin bind to specific receptors on the outer plasma membrane of a target cell. In the case of glucagon this binding indirectly stimulates the enzyme adenylate cyclase, on the inner surface of the membrane, to catalyze the production of cyclic AMP. Depending on the cell type, the cAMP exerts different effects inside the cell. In liver cells, cAMP sets off a chain of reactions that results in the breakdown of glycogen to glucose-1-phosphate and subsequent conversion of glucose-1-phosphate to glucose (see fig. 14.20). In adipocytes the main effect of cAMP is to activate the first enzyme in the pathway for lipid breakdown, triacylglycerol lipase. This enzyme hydrolyzes the triacylglycerol to diacylglycerol with release of fatty acid, the rate-limiting step in the complete hydrolysis of the triacylglycerols. Fatty acids are transported to other tissues (see chapter 23), where they can be metabolized as a source of energy.

Glucagon and insulin are only two of the many hormones

Table 27.2
Main Energy-Related Reactions in Five Vertebrate Tissues*

Tissue	Just after Eating	When Hungry	Starvation Conditions
Brain	Glucose → $CO_2 + H_2O$	Glucose → $CO_2 + H_2O$	Ketone bodies → $CO_2 + H_2O$ Glucose → $CO_2 + H_2O$
Heart muscle	Fatty acids → $CO_2 + H_2O$ Glucose → $CO_2 + H_2O$	Fatty acids → $CO_2 + H_2O$ Glucose → $CO_2 + H_2O$	Ketone bodies → $CO_2 + H_2O$
Skeletal muscle	Glucose → $CO_2 + H_2O$ Glucose → Glycogen Glucose → Lactate	Glycogen → $CO_2 + H_2O$ Glycogen → Lactate Fatty acids → $CO_2 + H_2O$	Fatty acids → $CO_2 + H_2O$ Ketone bodies → $CO_2 + H_2O$ Proteins → Amino acids
Adipose tissue	Fatty acids → Triacylglycerols	Triacylglycerols → Fatty acids	Triacylglycerols → Fatty acids
Liver	Glucose → Glycogen Fatty acids → Triacylglycerols	Fatty acids → $CO_2 + H_2O$ Glycogen → Glucose	Amino acids → $CO_2 + H_2O$ Amino acids → Glucose Fatty acids → $CO_2 + H_2O$ Fatty acids → Ketone bodies

*Light blue indicates imports; dark blue indicates exports.

Figure 27.4

The plasma levels of fatty acids and ketone bodies increase during starvation, whereas the glucose levels decrease. Owing to the dramatic rise in ketone bodies, there are more ATP equivalents in the plasma of the starving animal than in the normal animal.

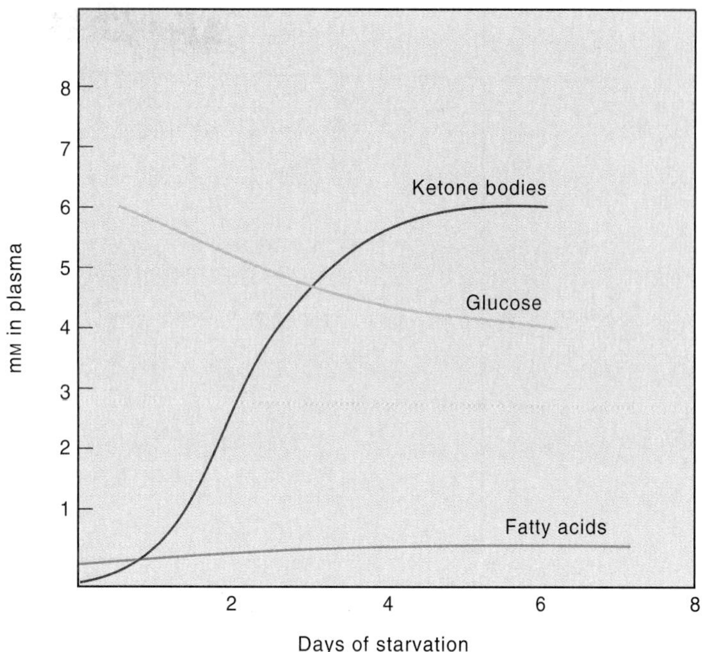

Figure 27.5

The opposing effects of insulin and glucagon on the blood glucose level. Insulin and glucagon are both secreted by the pancreas. In many cases they act on the same tissues, but their effects are opposite. Insulin promotes the storage of glucose as glycogen, the use of glucose as an energy source, and the synthesis of proteins and fats. As a result, insulin tends to lower the blood glucose level. Glucagon acts in the opposite direction, and its action therefore tends to raise the blood glucose level. The secretion of these hormones is regulated by the blood glucose level. A low blood glucose level favors the secretion of glucagon, and a high blood glucose level favors the secretion of insulin.

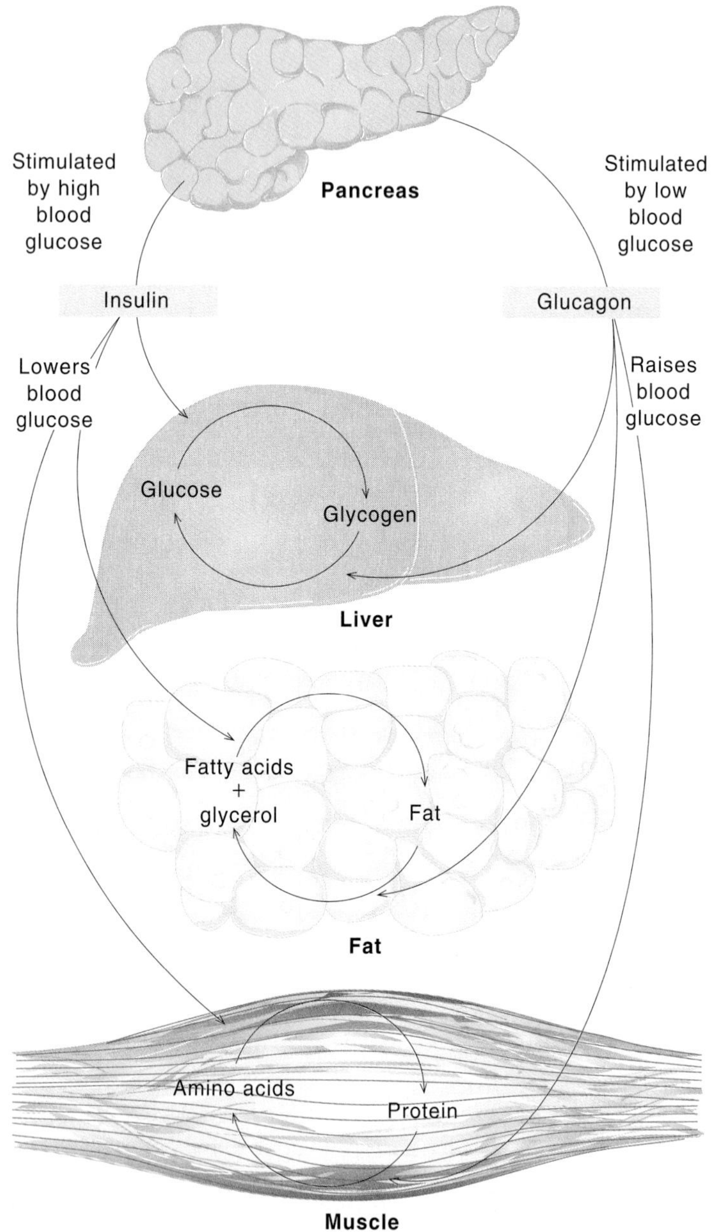

that regulate vertebrate metabolism, and cAMP is only one of several so-called second messengers that transduce the hormone signal to the inside of the cell. In the remainder of this chapter we take a comprehensive look at the hormone and hormonelike systems that provide the main mechanisms for regulating energy metabolism and other metabolic activities.

General Aspects of Cell Signaling

Intercellular communication is a result of signals sent and signals received. Most cells produce signals in the form of either specific cell surface receptors or secreted molecules. Some signals, such as the signals that specify specific cell contacts, are effective only if the signaling cell is in direct contact with the signaled target cell (e.g., see chapter 37). Other signals such as eicosanoids are effective only on cells that are nearby (see chapter 22).

The best-studied signals are the hormones. Hormones are produced in specialized endocrine glands (fig. 27.6); they are effective over a long range. Each cell is programmed to respond in a specific way to a select group of hormonal signals. We have already seen examples of this in chapters 14 and 20 with respect to how different cells respond to glucagon and epinephrine. The specificity of the response to a particular signal is a function of two factors. First, a responding cell must have a receptor that interacts with the signaling molecule; otherwise, the signal will be ignored. Second the type of response evoked by a signal-sensitive cell depends upon the way in which the signal-binding receptor is hooked up to other signal-relay chains in the cell. In this chapter we will also discuss hormones that function by binding to soluble receptors in the cytoplasm or the nucleus.

Figure 27.6

Location of major endocrine glands in humans. The hypothalamus regulates the anterior pituitary, which regulates the hormonal secretions of the thyroid, adrenals, and gonads (ovary in the female and testis in the male).

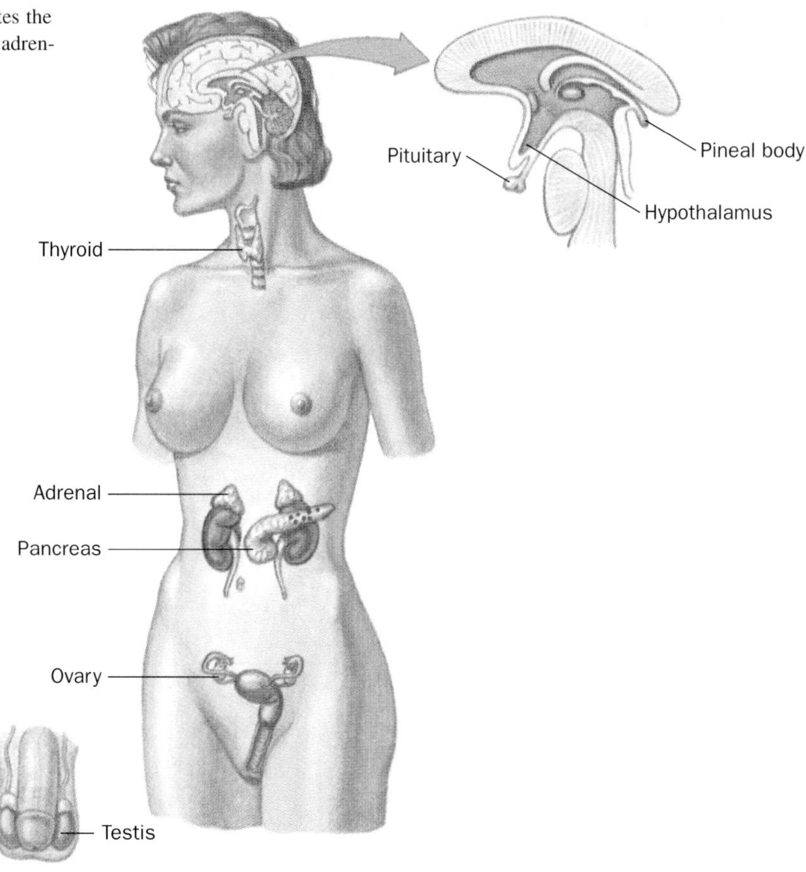

Hormones Are Major Vehicles for Intercellular Communication

In the following sections we discuss how hormones are synthesized, transported, and degraded. Then we explore basic aspects of hormone–receptor interaction and the more direct biochemical consequences of these interactions. We also indicate some of the ways in which the hormonal circuits themselves are regulated.

Hormones Are Synthesized and Secreted by Specialized Endocrine Glands

Table 27.3 lists many of the better-known hormones of vertebrates. Although many different classes of compounds are used as hormones by animals and plants, most vertebrate hormones fall into one of three classes: polypeptides, amino acid derivatives, and steroids.

Polypeptide Hormones Are Stored in Secretory Granules after Synthesis

Insulin, the first polypeptide hormone to be identified, was discovered by Frederick G. Banting and Charles Best in 1922. They found that this substance, which they isolated from the pancreas, restored normal glucose utilization in experimental animals lacking a pancreas. Insulin was also the first protein to be sequenced, a landmark accomplishment achieved by Fred Sanger in 1955. About 20 years later, Steiner discovered that the two polypeptide chains of insulin are synthesized as a single polypeptide, proinsulin, which folds and is cross-linked by disulfide bonds (see fig. 34.20). An internal peptide is then removed by the concerted action of specific proteases. All of these events occur within the pancreatic β cells that synthesize insulin.

With the advent of techniques to isolate and translate mRNA in cell-free systems, it was discovered that the primary translation product of insulin mRNA—called preproinsulin—is even larger than proinsulin (see fig. 34.20). Like all secreted polypeptide hormones, proinsulin is synthesized with a hydrophobic signal sequence at the amino terminus that directs the nascent polypeptide into the endoplasmic reticulum and is then removed. Surprisingly, in some cases several different peptide hormones can be liberated from the same precursor by proteolytic processing. A striking example is preproopiomelanocortin (fig. 27.7), which is a precursor for corticotropin (ACTH), β-lipotropin, three melanocyte-stimulating hormones (MSH), endorphin, and an enkephalin. When several different polypeptide hormones are cleaved from a common precursor, the cleavage pattern can vary to yield a different spectrum of peptides, depending on the cell type. Proteolytic processing of these precursors often occurs at the site of dibasic amino acids.

Table 27.3
Vertebrate Hormones[a]

Hormone	Structure	Function
Pineal		
Melatonin	*N*-Acetyl-5-methyoxytryptamine	Regulates circadian rhythms
Hypothalamus[b]		
Corticotropin-releasing factor (CRF or CRH)	Polypeptide (41 residues)	Stimulates ACTH and β-endorphin secretion
Gonadotropin-releasing factor (GnRF) or (GnRH)	Polypeptide (10 residues)	Stimulates LH and FSH secretion
Prolactin-releasing factor (PRF)	(May be TRH)	Stimulates prolactin secretion
Prolactin-release inhibiting factor (PIF)	(May be 56-residue peptide from GnRH precursor)	Inhibits prolactin secretion
Growth hormone-releasing factor (GRF or GRH)	Polypeptide (40 and 44 residues)	Stimulates GH secretion
Somatostatin (Growth hormone-release inhibiting factor, SIF)	Polypeptide (14 and 28 residues)	Inhibits GH and TSH secretion
Thyrotropin-releasing factor (TRF or TRH)	Polypeptide (3 residues)	Stimulates TSH and prolactin secretion
Pituitary		
Oxytocin (oxytocin)	Polypeptide (9 residues)	Uterine contraction, milk ejection
Vasopressin (antidiuretic hormone, ADH)	Polypeptide (9 residues)	Blood pressure, water balance
Melanocyte-stimulating hormones (MSH)	α Polypeptide (13 residues) β Polypeptide (18 residues) γ Polypeptide (12 residues)	Pigmentation
Lipotropin (LPH)	β Polypeptide (93 residues) γ Polypeptide (60 residues)	Fatty acid release from adipocytes
Corticotropin (adrenocorticotropic hormone, ACTH)	Polypeptide (39 residues)	Stimulates adrenal steroid synthesis
Thyrotropin (thyroid-stimulating hormone, TSH)	2 Polypeptides (α, 96 residues; β, 112 residues)	Stimulates thyroid hormone synthesis
Growth hormone (GH) or somatotropin	Polypeptide (191 residues)	General anabolic effects; stimulates release of insulinlike growth factor-I
Prolactin	Polypeptide (197 residues)	Stimulates milk synthesis
Luteinizing hormone (LH)	2 Polypeptides (α, 96 residues; β, 121 residues)	Ovary: luteinization, progesterone synthesis; testis: interstitial cell development, androgen synthesis
Follicle-stimulating hormone (FSH)	2 Polypeptides (α, 96 residues; β, 120 residues)	Ovary: follicle development, ovulation, estrogen synthesis; testis: spermatogenesis
Thyroid		
Thyroxine and triiodothyronine	Iodinated dityrosine derivatives (see fig. 27.9)	General stimulation of many cellular reactions
Calcitonin	Polypeptide (32 residues)	Ca^{2+} and P_i metabolism
Calcitonin gene-related peptide (CGRP)	Polypeptide (37 residues)	Vasodilator
Parathyroid		
Parathyroid hormone (PTH)	Polypeptide (84 residues)	Ca^{2+} and P_i metabolism

[a]Only the more common hormones of known structure are listed.

[b]Most of the hypothalamic releasing factors are also called hypothalamic regulatory hormones.

[c]Many of these peptides are also found in the brain, where they may modulate neural activity.

[d]The liver secretes α_2-globulin, which is cleaved by renin, a kidney enzyme, to give a decapeptide, proangiotensin, from which the carboxyl-terminal dipeptide is removed to give angiotensin.

Table 27.3
Vertebrate Hormones[a] (continued)

Hormone	Structure	Function
Alimentary tract[c]		
Gastrin	Polypeptide (17 residues)	Stimulates acid secretion from stomach and pancreatic secretion
Secretin	Polypeptide (27 residues)	Regulates pancreas secretion of water and bicarbonate
Cholecystokinin	Polypeptide (33 residues)	Secretion of digestive enzymes
Motilin	Polypeptide (22 residues)	Controls gastrointestinal muscles
Vasoactive intestinal peptide (VIP)	Polypeptide (28 residues)	Gastrointestinal relaxation; inhibits acid and pepsin secretion
Gastric inhibitory peptide (GIP)	Polypeptide (43 residues)	Inhibits gastrin secretion
Somatostatin	Polypeptide (14 residues)	Inhibits gastrin secretion; inhibits glucagon secretion
Heart		
Atrial natriuretic peptide (ANP)	Several active peptides cleaved from precursor polypeptide of 126 residues	Smooth muscle relaxation; diuretic activity
Pancreas		
Insulin	2 Polypeptides (21 and 30 residues)	Glucose uptake, lipogenesis, general anabolic effects
Glucagon	Polypeptide (29 residues)	Glycogenolysis, release of lipid
Pancreatic polypeptide	Polypeptide (36 residues)	Glycogenolysis, gastrointestinal regulation
Somatostatin	Polypeptide (14 residues)	Inhibition of somatotropin and glucagon release
Adrenal cortex		
Glucocorticoids	Steroids (cortisol, corticosterone)	Many diverse effects on protein synthesis and inflammation
Mineralocorticoids	Steroids (aldosterone)	Maintains salt balance
Adrenal medulla		
Epinephrine (adrenalin)	Tyrosine derivative (see fig. 27.10)	Smooth muscle contraction, heart function, glycogenolysis, lipid release
Norepinephrine (noradrenalin)	Tyrosine derivative (see fig. 27.10)	Arteriole contraction, lipid release
Gonads		
Estrogens (ovary)	Steroids (estradiol, estrone)	Maturation and function of secondary sex organs
Progestins (ovary)	Steroids (progesterone)	Ovum implantation, maintenance of pregnancy
Androgens (testes)	Steroids (testosterone)	Maturation and function of secondary sex organs
Inhibins A and B	1 Polypeptide (α, 134 residues; β, 115 and 116 residues)	Inhibit FSH secretion
Placenta		
Estrogens	Steroids	Maintenance of pregnancy
Progestins	Steroids	
Choriogonadotropin	2 Polypeptides (α, 96 residues; β, 147 residues)	Similar to LH
Placental lactogen	Polypeptide (191 residues)	Similar to prolactin
Relaxin	2 Polypeptides (22 and 32 residues)	Muscle tone
Liver		
Angiotensin[d]	Polypeptide (8 residues)	Responsible for essential hypertension
Kidney		
1,25-dihydroxyvitamin D_3	Steroid	Calcium uptake, bone formation

Figure 27.7

Processing pathway of preproopiomelanocortin. This precursor polypeptide is cleaved into a variety of active peptides. With the exception of the signal peptidase cleavage site, the cleavage sites are generally pairs of basic amino acids, although one site contains four. Which active peptides are produced depends on the processing pathway, which varies in different cell types. Thus in the anterior and intermediate lobes of the pituitary gland the precursor polypeptide is cleaved to yield corticotropin and β-lipotropin. In the intermediate lobe only, these polypeptide hormones are further cleaved to yield the melanocyte-stimulating hormones, γ-MSH and α-MSH; γ-lipotropin; and β-endorphin. Numbers refer to number of amino acids.

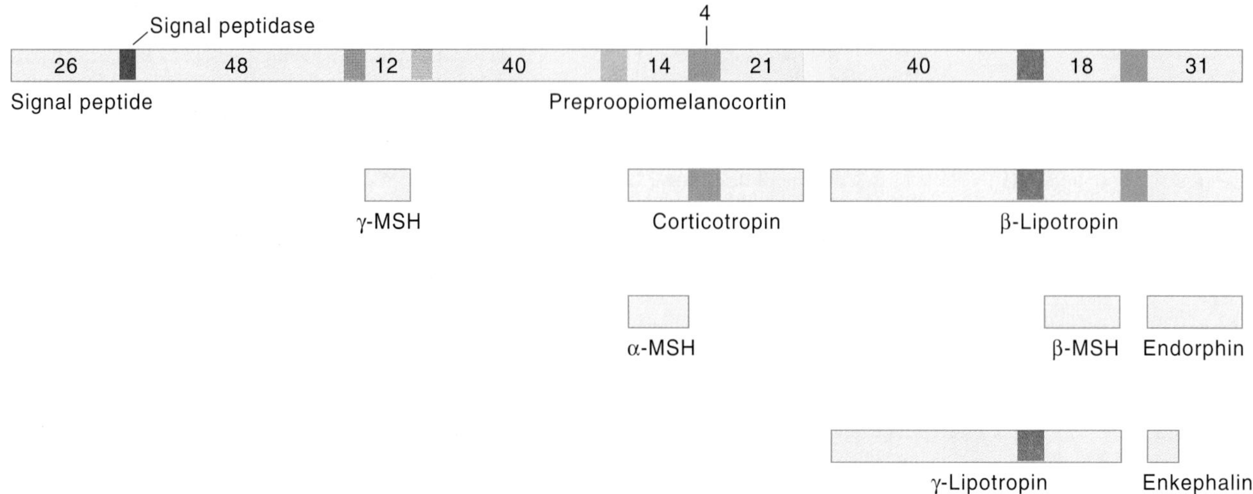

All polypeptide hormones are synthesized from mRNA as precursors, which contain signal peptides that direct them into the lumen of the endoplasmic reticulum. In a few cases (e.g., growth hormone and prolactin) no further processing occurs. However, in most cases, further processing is required. The peptides synthesized by neurosecretory cells of the hypothalamus, for example, are often much larger than the final hormones. Oxytocin and vasopressin, each nine residues long, represent the amino terminals of precursors called proneurophysins, which are 160 and 215 amino acids long, respectively. After oxytocin or vasopressin is cleaved from proneurophysin in the hypothalamus, where it is synthesized, the hormone remains associated with the neurophysin as it passes down the axons to the posterior pituitary, where it is secreted. The neurophysin may serve to protect the hormone from degradation prior to secretion. Somatostatin, another hypothalamic hormone, is a 14-amino-acid product cleaved from the carboxyl terminus of a precursor that contains 121 amino acids. Thyrotropin-releasing hormone (TRH) is the smallest known polypeptide hormone (pyroGlu-His-ProNH_2). In the synthesis of TRH the glutamyl residue is cyclized, and the carboxyl-terminal amide originates from an adjacent glycine. The precursor polypeptide (255 amino acids) contains five copies of the sequence Lys-Arg-Gln-His-Pro-Gly-Arg-Arg within it (fig. 27.8).

Polypeptide hormones are usually stored in secretory granules after their passage through the endoplasmic reticulum and Golgi apparatus. Release of these hormones into the bloodstream is accomplished by fusing the secretory granule membranes with the plasma membrane. This event is often regulated by other hormones.

Thyroid Hormones and Epinephrine Are Amino Acid Derivatives

Thyroxine (T_4) and the more potent triiodothyronine (T_3) are cleaved from a large precursor protein called thyroglobulin. Thyroglobulin exists as a dimer of two identical polypeptides ($M_r \approx$ 330,000). It is a storage protein for iodine and can be considered a prohormone of the circulating thyroid hormones. Thyroglobulin is secreted into the lumen of the thyroid gland, where specific residues are iodinated in one or two positions by a special peroxidase; then two iodinated residues condense as shown in figure 27.9.

The secretion of thyroid hormones starts with endocytosis of the modified thyroglobulin, followed by fusion of the endocytotic vesicles with lysosomes. The lysosomal enzymes then degrade the thyroglobulin, liberating triiodothyronine and thyroxine into the circulation. Only about five molecules of T_3 and T_4 are generated from each molecule of thyroglobulin. Thyroid hormone secretion is stimulated by thyrotropin (TSH), a pituitary hormone that activates adenylate cyclase in its target cells.

Epinephrine, sometimes called adrenalin, was the first hormone to be isolated, characterized, and synthesized. Epinephrine and its precursor, norepinephrine, are synthesized from tyrosine in the chromaffin cells of the adrenal medulla. They are also synthesized by neurons of the central and peripheral nervous system. The biosynthetic pathway is shown in figure 27.10. The first step, which involves oxidation of tyrosine to 3,4-dihydroxyphenylalanine (dopa), is catalyzed by tyrosine hydroxylase. This is the rate-limiting enzyme in the pathway. The amount and

Figure 27.8

Biosynthesis of thyrotropin-releasing hormone (TRH). Five copies of TRH are contained within a 255-amino-acid precursor polypeptide (pre-pro-TRH). The precursor has a hydrophobic signal peptide (dark green) and five copies of the sequence Lys-Arg-Gln-His-Pro-Gly-Arg-Arg (light green). The dibasic amino acids are recognized by specific proteases to liberate Gln-His-Pro-Gly, which is subsequently converted into TRH. The amino-terminal glutamine is converted into pyroglutamine, and the carboxyl-terminal amide is derived from the neighboring glycine, which is removed by a specific enzyme.

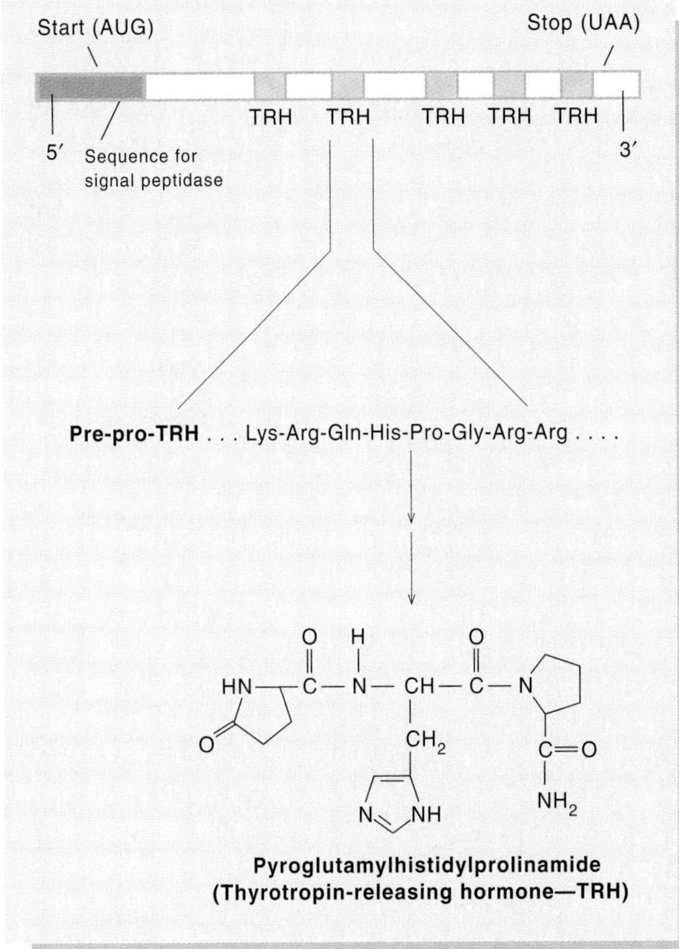

activity of tyrosine hydroxylase are regulated by cAMP-dependent mechanisms that are responsive to the neurotransmitter acetylcholine. This neurotransmitter is liberated by special neurons that impinge on the chromaffin cells. Second, dopa is decarboxylated to dopamine, which is then β-hydroxylated to produce norepinephrine. Finally, the N-methylation of norepinephrine (*S*-adenosylmethionine is the methyl donor) produces epinephrine. Both of these catecholamines (norepinephrine and epinephrine) are stored in chromaffin granules, where they are complexed with ATP and proteins called chromogranins. Neural stimulation of the medulla is mediated by acetylcholine, which binds to receptors on the membranes of medullary cells (see chapter 28). This event leads to a local depolarization and an influx of calcium. As a result, the chromaffin granules fuse with the cell membrane, and a packet of catecholamines, ATP, and protein is extruded into the extracellular fluid.

Steroid Hormones Are Derived from Cholesterol

Steroid hormones are derived from cholesterol by a stepwise removal of carbon atoms and hydroxylation. The steroid hormones are synthesized by cells of the adrenal cortex (in the case of glucocorticoids and mineralocorticoids) and the gonads (in the case of estrogens, progestins, and androgens). The hormone names just mentioned are generic names for entire classes of compounds that interact with specific receptors; for example, estrogen is the generic name for a family that includes 17β-estradiol, estrone, and diethylstilbestrol (DES), a synthetic nonsteroidal estrogen.

The step-by-step synthesis of the steroid hormones pregnenolone and progesterone from cholesterol (C_{27}) was presented in chapter 23 (see fig. 23.30). Note that pregnenolone (C_{21}) and progesterone (table 27.4) (C_{21}) are intermediates in the biosynthesis of all of the major adrenal steroids, including cortisol (C_{21}), corticosterone (C_{21}), and aldosterone (C_{21}) (see figs. 23.31 and 23.32). The same two compounds are intermediates in the synthesis of the gonadal steroid hormones, testosterone (C_{19}) and 17β-estradiol (C_{18}) (see figs. 23.32 and 23.33). Because the synthesis of all these hormones follows a common pathway, a defect in the activity or amount of an enzyme along that pathway can lead to both a deficiency in the hormones beyond the affected step and an excess of the hormones, or metabolites, prior to that step.

Deficiencies in each of the six enzymes involved in the conversion of cholesterol to aldosterone have been observed in humans. Each deficiency gives rise to a characteristic steroid hormone imbalance, with telling clinical consequences. For example, a deficiency in 17-hydroxylase gives rise to inadequate levels of cortisol as well as to inadequate levels of androgens and estrogens, with severe effects on sexual maturation. A deficiency in the next enzyme along the pathway, 21-hydroxylase, blocks the synthesis of adrenal glucocorticoids and mineralocorticoids and leads to an overproduction of testosterone by the adrenals. This overproduction of testosterone is due to metabolic shunting of progesterone into the sex steroid pathway. The synthesis of androgens is exacerbated by the lack of feedback inhibition of cortisol on the hypothalamus; the result is chronic production of CRF and ACTH and perpetual activation of adrenal steroid biosynthesis. In females, excessive androgen production, due to a defect in 21-hydroxylase, leads to masculinization, or, to use the clinical term, female pseudohermaphrodism—that is, a genetic female and male appearance. This reversal of sexual phenotype is explained by the fact that during embryonic development of mammals, the external genitalia develop from common precursor cells. In the absence of hormonal stimulation, they develop into female structures, but androgens direct their development into male structures.

Testosterone is both a hormone and a prohormone. The high levels of testosterone normally produced in the male by the testes play a major role in the growth and function of many tissues in addition to reproductive organs. Essentially all the sexual

Figure 27.9

Pathway of thyroxine (T_4) and triiodothyronine (T_3) synthesis. Thyroid cells actively transport iodine (I^-), which is incorporated into a few tyrosine residues of thyroglobulin by the enzyme iodoperoxidase. After condensation of iodinated tyrosine residues the thyroglobulin is proteolytically degraded, liberating thyroxine and triiodothyronine.

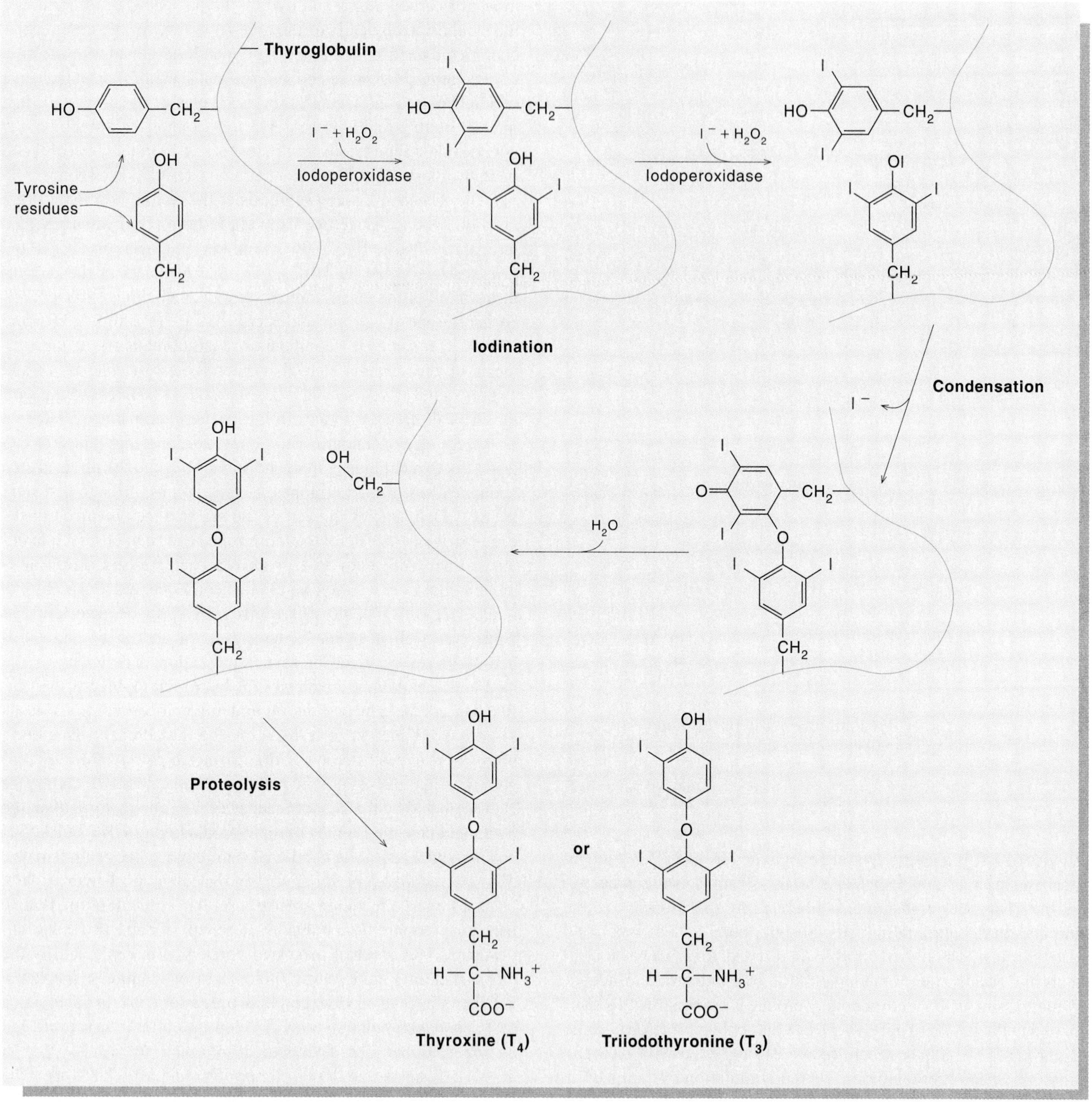

Figure 27.10

Pathway of epinephrine synthesis. Epinephrine and its precursor, norepinephrine, are synthesized from tyrosine. The synthesis occurs in the chromaffin cells of the adrenal medulla and in neurons of the central and peripheral nervous system. The first step, which is catalyzed by tyrosine hydroxylase, is the rate-limiting step in the pathway.

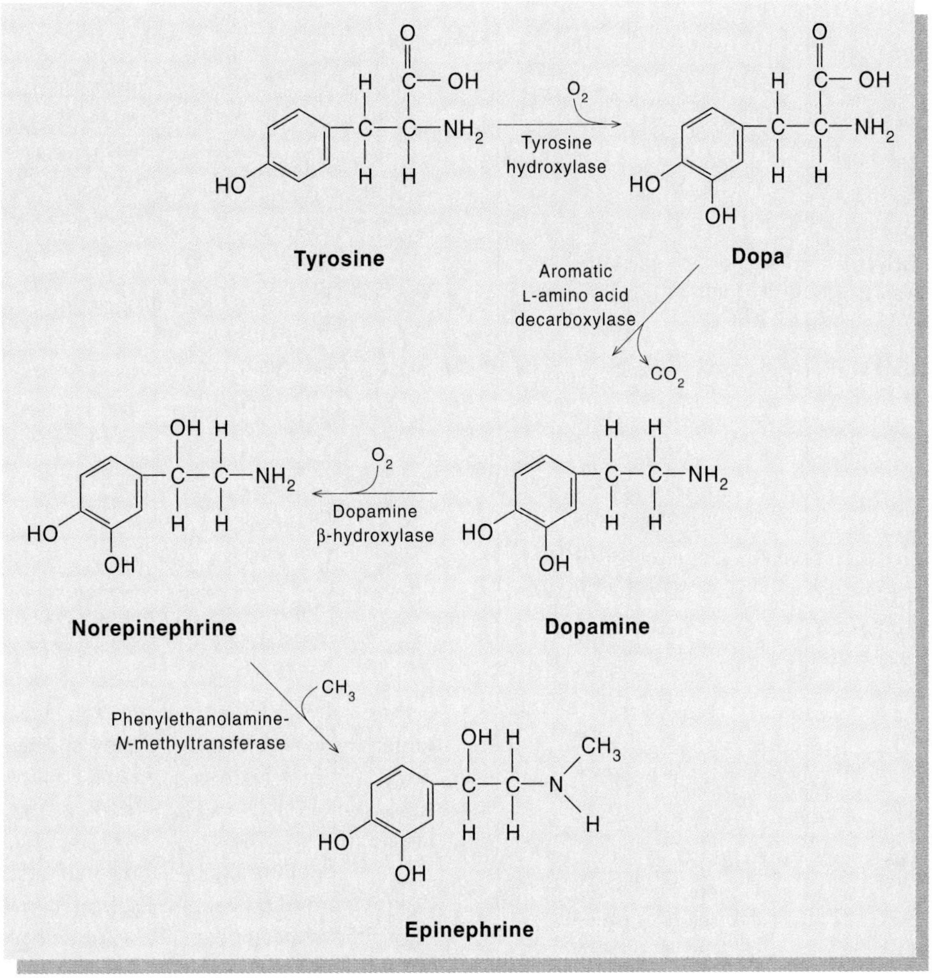

differences in nonreproductive tissues, such as muscle, liver, and brain, are a consequence of androgen action. Although testosterone is the major circulating androgen, many target cells reduce this steroid to 5α-dihydrotestosterone, a steroid that binds to the androgen receptor with higher affinity than testosterone (fig. 27.11). When 5α-reductase is defective, the androgen receptors are only partially activated, and a full androgen response is not obtained. A deficiency in this enzyme therefore leads to abnormal development of male genitalia—that is, they are of female phenotype (a clinical condition referred to as male pseudohermaphrodism, type 2). In this example, testosterone can be considered a prohormone of a more active androgen.

Testosterone is also a prohormone of metabolites that bind to different receptors. For example, some of the effects of androgens on the production of red blood cells (erythropoiesis) are due to the reduction of testosterone to 5β-dihydrotestosterone within precursor cells. 5β-Dihydrotestosterone binds to a receptor that is distinct from the one that binds testosterone and 5α-dihydrotestosterone (see fig. 27.11). Testosterone also influences erythropoiesis by stimulating the kidney to produce a specific growth factor, erythropoietin. Another striking example of testosterone as a prohormone occurs in the brain, where testosterone influences neural development and activity (e.g., male-specific mating behavior and bird songs) by being converted into 17β-estradiol, which interacts with estrogen receptors. Testosterone is metabolized to 17β-estradiol by aromatase, the same enzyme that is involved in 17β-estradiol synthesis in the ovary (fig. 27.12).

Cell specificity of mineralocorticoid action is achieved in a different manner. Aldosterone, cortisol, and corticosterone bind with similar affinities to mineralocorticoid and glucocorticoid re-

Table 27.4
Major Steroid Hormones

Steroid	Function
Progesterone	Precursor of other steroids; prepares uterus for implantation of an egg; prevents ovulation during pregnancy
Aldosterone (a mineralocorticoid)	Increases retention of sodium ions by the renal tubules
Cortisol (a glucocorticoid)	Promotes gluconeogenesis; suppress inflammatory reactions
Testosterone (an androgen)	Promotes male sexual development; promotes and maintains male sex characteristics
Estradiol (an estrogen)	Responsible for sexual development in the female; promotes and maintains female sex characteristics

Figure 27.11

Conversion of testosterone to 5α-dihydrotestosterone (5α-DHT). Receptor-binding studies indicate that 5α-DHT has a higher affinity for the androgen receptor than testosterone does.

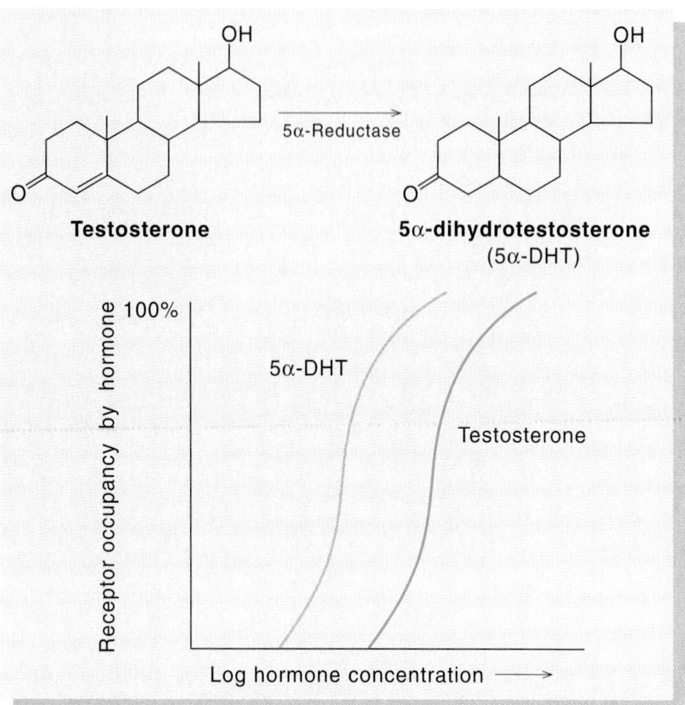

ceptors. However, aldosterone activates only its own receptor in target tissues such as the kidney because of an enzyme, 11β-hydroxysteroid dehydrogenase, that converts the prevalent glucocorticoids into inactive 11-keto derivatives but does not affect aldosterone.

Vitamin D_3 is a precursor of the hormone 1,25-dihydroxyvitamin D_3. Vitamin D_3 is essential for normal calcium and phosphorus metabolism. It is formed from 7-dehydrocholesterol by ultraviolet photolysis in the skin. Insufficient exposure to sunlight and absence of vitamin D_3 in the diet leads to rickets, a condition characterized by weak, malformed bones. Vitamin D_3 is inactive, but it is converted into an active compound by two hydroxylation reactions that occur in different organs. The first hydroxylation occurs in the liver, which produces 25-hydroxyvitamin D_3, abbreviated $25(OH)D_3$; the second hydroxylation occurs in the kidney and gives rise to the active product 1,25-dihydroxyvitamin D_3 $24,25(OH)_2D_3$ (fig. 27.13). The hydroxylation at position 1 that occurs in the kidney is stimulated by parathyroid hormone (PTH), which is secreted from the parathyroid gland in response to low circulating levels of calcium. In the presence of adequate calcium $25(OH)D_3$ is converted into an inactive metabolite, $24,25(OH)_2D_3$. The active derivative of vitamin D_3 is considered a hormone because it is transported from the kidneys to target cells, where it binds to nuclear receptors that are analogous to those of typical steroid hormones. $1,25(OH)_2D_3$ stimulates calcium transport by intestinal cells and increases calcium uptake by osteoblasts (precursors of bone cells).

Figure 27.12

Metabolic conversion of testosterone by target cells. Testosterone (T) is the predominant androgen in the bloodstream. When testosterone enters target cells, it can be metabolized in a variety of different ways. It can either (*a*) bind directly to androgen receptors (R^a) or (*b*) be reduced to 5α-dihydrotestosterone (5α-DHT), which then binds to R^a with higher affinity. (*c*) Other target cells reduce testosterone to 5β-dihydrotestosterone (and other 5β metabolites), which bind to a distinct receptor (R^β). (*d*) Yet other cells convert testosterone into an estrogen, 17β-estradiol, which binds to estrogen receptors (R^e).

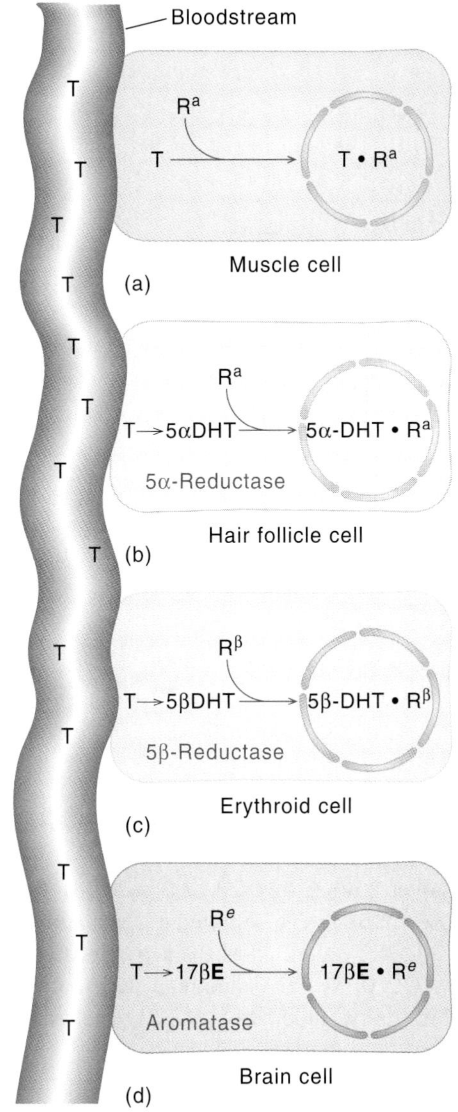

The Circulating Hormone Concentration Is Regulated

The occupancy of hormone receptors can fluctuate greatly and is ultimately determined by the concentration of "free" hormone in the blood. The major determinants of hormone concentrations are (1) the rate of hormone secretion from endocrine cells and (2) the rate of hormone removal by clearance or metabolic inactivation. As we have seen, most hormones (with the exception of steroids) are stored in secretory granules. When the hormone is needed, the granule membranes fuse with the plasma membrane to liberate their contents into the bloodstream. This event is triggered by signals from other hormones or by neural signals. Stimulation of hormonal secretion is usually coupled with an increase of hormone synthesis, so that hormonal stores are replenished.

Most hormones have a half-life in the blood of only a few minutes because they are cleared or metabolized very rapidly. The rapid degradation of hormones allows target cells to respond transiently. Polypeptide hormones are removed from the circulation by serum and cell surface proteases, by endocytosis followed by lysosomal degradation, and by glomerular filtration in the kidney. Steroid hormones are taken up by the liver and metabolized to inactive forms, which are excreted into the bile duct or back into the blood for removal by the kidneys. Catecholamines are metabolically inactivated by O-methylation, by deamination, and by conjugation with sulfate or glucuronic acid.

Thyroid hormones and most steroid hormones are associated with carrier proteins in the serum. The carrier proteins are called, appropriately, thyroxine-binding globulin, transcortin (for cortisol), and sex-steroid-binding protein. These proteins have a high affinity ($K_d \approx 10^{-9}$–10^{-8} M) for their respective hormones. They buffer the concentration of "free" hormone and retard hormone degradation and excretion. The carrier proteins are distinguishable from the intracellular receptors for these hormones.

Hormone Action Is Mediated by Receptors

Hormone action begins with the binding of the hormone to a receptor on (or in) a target cell. Binding of hormone molecule induces a conformational change in its receptor, and this change is detected by other macromolecules. Hence the hormone–receptor interaction can be transduced from one molecule to another. In the simplest scheme a hormone receptor might be a rate-limiting enzyme (fig. 27.14*a*) or be coupled to a rate-limiting enzyme (fig. 27.14*b*). In cases in which the membrane-bound receptors regulate adenylate cyclase, another protein is interposed between the receptor and the adenylate cyclase (fig. 27.14*c*). The protein directly activated by hormone–receptor interaction is sometimes referred to as an acceptor protein. GTP-binding proteins (G proteins) are acceptors for all receptors that activate or inhibit adenylate cyclase (see fig. 20.12). For many membrane receptors the molecular intermediates are still unknown.

Not all hormone receptors are membrane bound. Receptors for steroid hormones are soluble proteins located within the cell and bind to specific DNA sequences when they are activated by the appropriate hormones. The binding of these receptors to DNA in the promoter region of a gene activates transcription, in most cases by helping to assemble an efficient initiation complex. The acceptor proteins in these cases may be other transcription factors (see chapter 37).

Overall, it appears that all hormones act by binding to receptors whether they are located in the cell membrane or inside the cell. Binding of hormones induces a conformational change in the receptor that is transmitted to its active site (if it is an enzyme)

Figure 27.13

The conversion of vitamin D_3 to an active compound. Vitamin D_3 is formed from 7-dehydrocholesterol by ultraviolet photolysis in the skin. Vitamin D_3 is inactive, but it is converted into an active compound by two hydroxylation reactions that occur in different organs. The first reaction occurs in the liver and results in 25-hydroxyvitamin D_3. The second hydroxylation occurs in the kidney and results in 1,25-dihydroxyvitamin D_3.

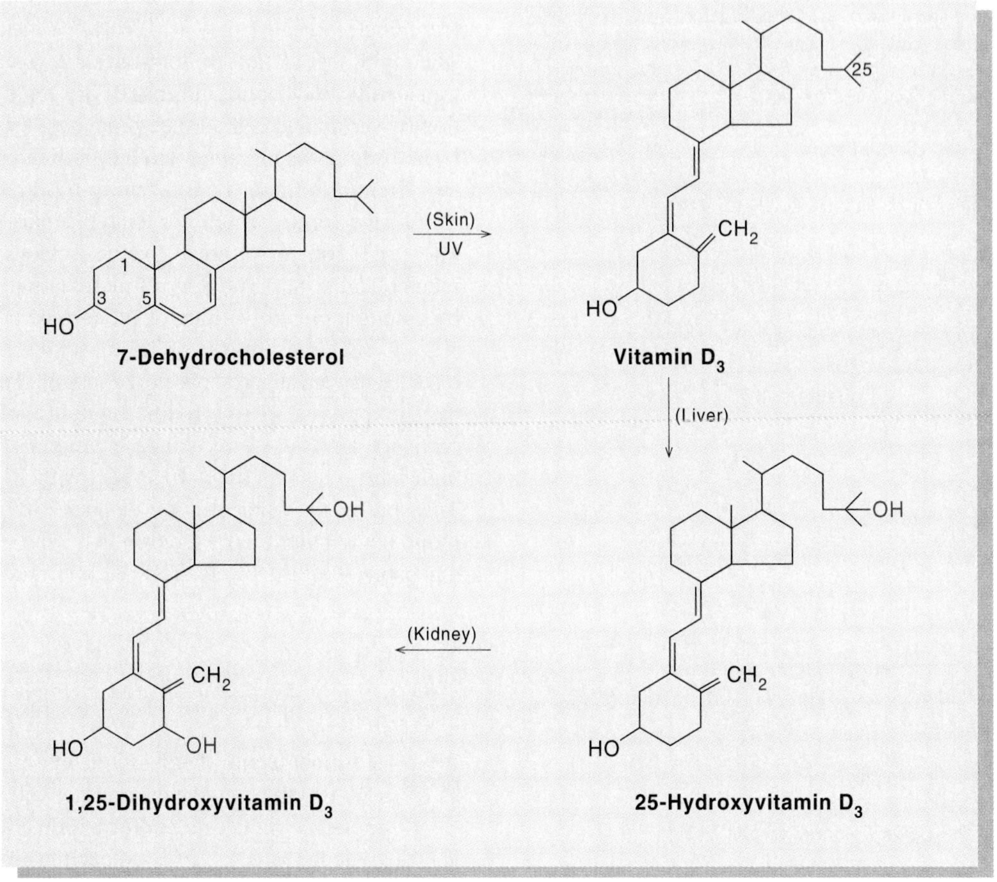

or to other macromolecules. In either case a chain of events is elicited that ultimately affects an array of metabolic processes, ranging from alterations in enzyme activities to changes in gene expression. These changes may lead to major alterations in cell growth, morphology, and function.

Many Plasma Membrane Receptors Generate a Diffusible Intracellular Signal

Activation of many membrane-associated hormone receptors generates a diffusible intracellular signal called a second messenger. Five intracellular messengers are currently known: cyclic AMP, cyclic GMP, inositol triphosphate, diacylglycerol, and calcium. Some aspects of this subject were discussed in chapter 20.

The Adenylate Cyclase Pathway Is Triggered by a Membrane-Bound Receptor

All hormones that trigger release of the cyclic AMP intracellular messenger act through three-component systems in which the first component is the hormone receptor, the second component is a G protein, and the third component is a membrane-bound adenylate cyclase (table 27.5). Some aspects of this subject were discussed in chapter 20 (see figs. 20.12 and 20.13). Cyclic AMP so formed diffuses into the cytosol, where it can activate protein kinase A. The effect of protein kinase A activation depends upon the cell type (see fig. 20.12). For example, if we look at the adipocyte, the chain of reactions initiated by cAMP results in the release of fatty acids from storage lipids. Inspection of table 27.5 shows that there are more than a dozen hormones that trigger adenylate cyclase synthesis in different types of cells. In each case the ultimate effect of the cAMP is to trigger a metabolic response that is most appropriate to the cell type. This can be seen by comparing the hormone referred to in table 27.5 with its function in table 27.3.

Protein Phosphorylation Is the Most Common Way in Which Regulatory Proteins Respond to Hormonal Signals

In chapter 10 we talked about various ways in which regulatory proteins could be activated. We saw that phosphorylation was one way. In the case of hormone signal transduction there are often

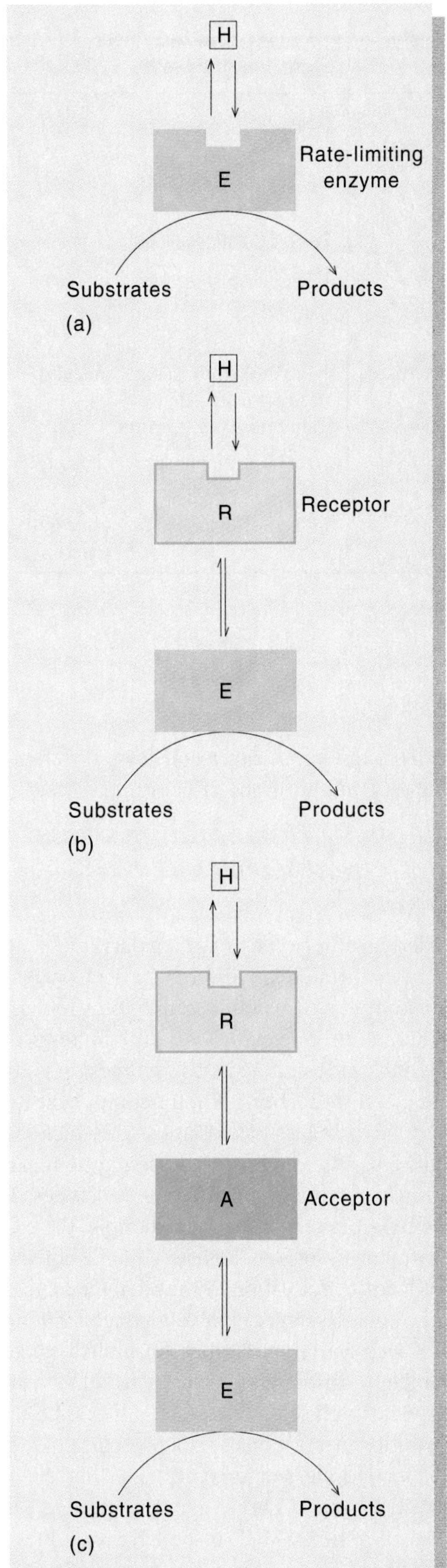

Figure 27.14

Possible mechanisms of hormone action. (*a*) The hormone (H) theoretically activates an enzyme (E) directly as an allosteric effector. (*b*) Alternatively, a separate binding protein for the hormone, called a receptor (R), may then activate an enzyme. (*c*) Another possibility interposes an acceptor protein (A) between the receptor and the enzyme. Each interaction is reversible.

Table 27.5
Hormones That Activate or Inhibit Adenylate Cyclase

Activators
Corticotropin (ACTH)
Calcitonin
Catecholamines (acting on β_1 and β_2 receptors)
Choriogonadotropin
Follicle-stimulating hormone (FSH)
Glucagon
Gonadotropin-releasing hormone (GnRH)
Growth hormone-releasing hormone (GRH)
Luteinizing hormone (LH)
Lipotropin (LPH)
Melanocyte-stimulating hormones (MSH)
Parathormone (PTH)
Secretin
Thyrotropin regulatory hormone (TRH)
Thyrotropin (TSH)
Vasoactive intestinal peptide (VIP)
Vasopressin
Inhibitors
Angiotensin
Catecholamines (acting on α_2 receptors)

long branching chains of reactions that occur in the responding target cell, but almost invariably, protein phosphorylation is the major way in which the business enzymes at the end of the chains are altered. Why is phosphorylation such a popular way for regulating enzymes in the case of signal transduction? First, it is relatively simple to phosphorylate a protein or to dephosphorylate it. Second, the addition of a phosphate group to an exposed amino acid side chain hydroxyl group changes the protein surface dramatically by replacing a mild polar group by a bulky doubly negatively charged phosphate group. Such a change could have a major impact on the activity of an enzyme if the group were strategically placed.

Table 27.6
Some G Protein Receptors and Effectors

G Protein	Receptors for	Effectors	Signaling pathways
G_s	Epinephrine, norepinephrine, histamine glucagon, ACTH, luteinizing hormone, follicle-stimulating hormone, thyroid-stimulating hormone, and others	Adenylate cyclase Ca^{2+} channels	↑ cAMP ↑ Ca^{2+} influx
G_{olf}	Odorants	Adenylate cyclase	↑ cAMP (olfaction)
G_{t1} (rods)	Photons	cGMP phosphodiesterase	↓ cGMP (vision)
G_{t2} (cones)	Photons	cGMP phosphodiesterase	↓ cGMP (color vision)
G_{i1}, G_{i2}, G_{i3}	Norepinephrine, prostaglandins, opiates, angiotensin, many peptides	Adenylate cyclase Phospholipase C Phospholipase A_2 K^+ channels	↓ cAMP ↑ Inositol triphosphate, diacylglycerol, Ca^{2+} Arachidonate release Membrane polarization
G_0	Probably many, but not yet defined	Phospholipase C Ca^{2+} channels	↑ Inositol trisphosphate, diacylglycerol, Ca^{2+} ↓ Ca^{2+} influx
Gq	Norepinephrine, vasopressin	Phospholipase C	↑ Inositol bisphosphate, diacylglycerol, Ca^{2+}

Source: Reprinted with permission from *Nature* (H. R. Bourne et al., The GTPase superfamily: A conserved switch of diverse cell functions, *Nature,* 348:126, 1990). Copyright © 1990 Macmillan Magazines Limited.

Variability in G Proteins Adds to the Variability of the Hormone-Triggered Response

The middle component in the three-component system described in figure 20.12*c* is most frequently a G protein, so named because it binds a molecule of GTP in its active form. Most G proteins are heterotrimers of α, β, and γ subunits that are associated with the membrane that can make contact with both the membrane receptor and an effector molecule such as adenylate cyclase as shown in figure 20.12. In the absence of hormone activation the α subunit of the G protein binds GDP at an allosteric site. On binding of hormone the conformation of the α subunit changes so that the GDP is displaced and replaced by GTP. The GTP-bound α subunit may dissociate from the $\beta\gamma$ subunits, allowing it to activate an effector molecule.* In the case of receptors coupled with G_s, the α_s–GTP complex activates adenylate cyclase, which proceeds to synthesize a burst of cAMP (see fig. 20.12). The period of activation is brief, as the α_s subunit also contains a GTPase activity that hydrolyzes a phosphate residue from the GTP, thereby returning the α_s subunit to its original inactive state.

There are over 200 known G proteins. Inspection of table 27.6 shows that G proteins vary according to the hormone receptor they associate with as well as the effector response that they trigger. This variability in G protein specificities greatly increases the variety of responses that can result from different families of hormone receptors and hormone effector enzymes.

*Recent discoveries indicate that the $\beta\gamma$ complex of the G protein plays a more active role in signal transduction (see Clapham reference).

Multicomponent Hormonal Systems Facilitate a Great Variety of Responses

Before considering other types of secondary messengers associated with hormonal systems, we might reflect on why such systems are often complex. It would certainly be simpler to have just one transmembrane protein with a site for binding the hormone on the extracellular side and a site for synthesizing cAMP on the cytosolic side (see fig. 27.14*a*). A multicomponent system, however, has many advantages, the foremost of which is that it presents many more points for control. Such a system can be used in various ways in the same or in different cell types, leading to a rich variety of possible responses. For example, there is more than one type of hormone system in the liver cell that activates the same G protein that leads to activation of the adenylate cyclase enzyme. (These two hormonal systems involve glucagon and epinephrine.) In addition, a rich variety of other hormonal systems operate through completely different G proteins to either activate or inhibit the adenylate cyclase enzyme (see tables 27.5 and 27.6).

Variability in hormonal response patterns does not stop at the level of second-messenger synthesis. Thus cyclic AMP can activate the well-known cAMP-dependent protein kinase A, but the possibility of other cAMP-responsive enzymes or cAMP-activated regulatory proteins should not be ruled out. The protein kinase activated by cAMP can activate a number of other enzymes. For example, in the liver, phosphorylase kinase is activated and catalyzes the breakdown of glycogen. In adipocytes, triacylglycerol lipase is activated and catalyzes the breakdown of triacylglycerols.

One should think of the individual proteins in a multicomponent hormonal response system as being like the parts of a machine, which may play similar roles in different machines where the overall function is quite different. Each cell type may contain some but not all elements, permitting cells to respond differently to the same hormonal signal or in a similar way to different hormonal signals.

The Guanylate Cyclase Pathway

Another second-messenger system involves receptors that are either coupled to guanylate cyclase or guanylate cyclase itself. One class of receptors, those that are activated by atrial, cardiac, or brain natriuretic factors (a natriuretic factor increases the urinary secretion of sodium), is bifunctional in that the external domain of these membrane receptors binds the peptide hormones, resulting in the activation of an internal cytoplasmic domain that has guanylate cyclase activity and produces cGMP from GTP. This is the best example of the simplest receptor–effector coupling scheme illustrated in figure 27.14*a*. Another class of receptors activates intracellular guanylate cyclases by an unknown mechanism. The cGMP that is produced can either act directly on target molecules such as ion channels or activate a cGMP protein kinase G that can phosphorylate many proteins, resulting in a variety of metabolic changes, depending on the cell type. For example, a cAMP phosphodiesterase can be activated by cGMP-dependent phosphorylation, resulting in lowering of cAMP levels; this effect helps explain why cAMP and cGMP sometimes have opposite effects on cells. The cGMP kinases differ from the cAMP kinases described previously in that the regulatory and catalytic domains are part of the same molecule rather than being encoded by separate genes.

In chapter 29 we will describe the role played by cGMP in the visual response.

Guanylyl Cyclase Can Be Activated by a Gas

Recently, the gas nitric oxide (NO) has been recognized as a signaling molecule in vertebrates. NO synthesis is triggered when acetylcholine is released by autonomic nerves in the walls of a blood vessel. The acetylcholine induces the endothelial cells to synthesize NO. NO is made by the enzyme nitric oxide synthase by the deamination of arginine (fig. 27.15). Because it diffuses readily across membranes, the NO diffuses out of the cell where it is produced and passes directly into neighboring cells. Its action is local because of a short half-life of only 5 to 10 seconds. In many target cells, such as endothelial cells, NO reacts with iron in the active site of the enzyme guanylyl cyclase, stimulating the production of cGMP.

The effect of NO on blood vessels provides an explanation for the effectiveness of nitroglycerine, which has long been used to treat angina (pain due to inadequate blood flow to the heart muscle). The nitroglycerine is converted to NO, which relaxes the blood vessels in the heart, thereby increasing the blood flow to the heart muscle. NO is also produced as a local mediator by activated macrophages to help kill invading microorganisms (see chapter 37).

Figure 27.15

Formation of nitric oxide (NO) from arginine catalyzed by nitric oxide synthase (NOS).

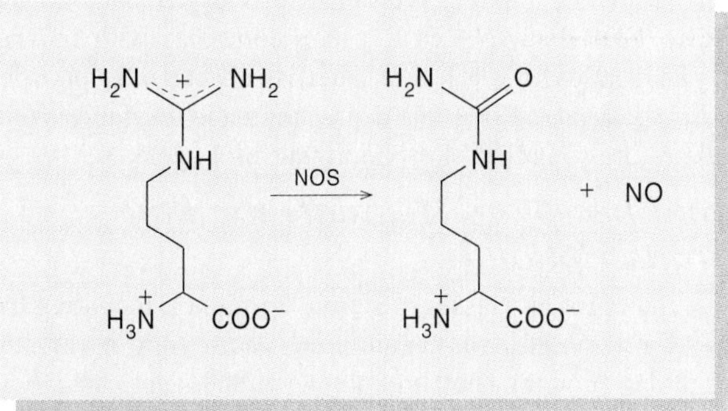

Calcium and the Inositol Trisphosphate Pathway

The outline of another important second-messenger system was elucidated during the last ten years. Chemical messengers that act via this system include a variety of hormones (e.g., catecholamines, vasopressin, and angiotensin) as well as some neurotransmitters (e.g., acetylcholine acting on pancreatic acinar cells to stimulate secretion of digestive enzymes or acting on pancreatic β cells to stimulate insulin secretion). The receptors for these hormones and neurotransmitters are membrane-bound proteins with their hormone-binding sites facing the outside of the cell. Their activation by an appropriate signal is transmitted by a G protein (G_0, or Gq, table 27.6), which then activates a phosphodiesterase (phospholipase C) that cleaves the polar inositol trisphosphate (IP_3) from phosphatidylinositol-4,5-bisphosphate (PIP_2). As a result, IP_3 enters the cytoplasm, whereas the diacylglycerol moiety remains in the membrane (see fig. 20.15). Both breakdown products of PIP_2 play important second-messenger roles (see fig. 20.16). IP_3 stimulates the release of calcium from intracellular stores (residing in the endoplasmic reticulum) into the cytoplasm. The calcium is bound by the protein calmodulin, which then activates one or more calcium-dependent protein kinases. Meanwhile, the diacylglycerol, along with phosphatidylserine, activates a membrane-associated protein kinase C that phosphorylates serine and threonine residues (see fig. 20.16). Thus several kinases are activated by this complex system, and they, like the cAMP-dependent protein kinases, modify the function of rate-limiting enzymes and regulatory proteins involved in a variety of metabolic pathways. In addition to intracellular messengers, the breakdown of PIP_2 also stimulates the production of extracellular modulators of hormone activity. Thus, following the breakdown of PIP_2, arachidonic acid (one of the main fatty acids found in the diacylglycerol moiety of PIP_2) is metabolically converted into eicosanoids of different sorts. (Formation of eicosanoids, as well as their function as local hormones were discussed in chapter 22.)

IP_3 is rapidly degraded to inactive IP_2 and then on to inositol. Meanwhile, diacylglycerol is phosphorylated and then converted to CDP-diacylglycerol, which combines with inositol to form phosphatidylinositol. The latter is subsequently phosphorylated in two steps to PIP_2. The degradation and resynthesis of PIP_2 completes the so-called phosphatidylinositol cycle.

Steroid Receptors Modulate the Rate of Transcription

The receptors for all classes of steroid hormones, including the receptors for 1,25-dihydroxyvitamin D_3, and thyroid hormones are intracellular proteins that are not very abundant, usually only 10^2–10^5 molecules per cell.

All of these classes of receptors have a similar structure composed of several functional domains, including regions that bind the hormone, bind to DNA, activate transcription and allow dimerization. In the absence of hormone these receptors are usually sequestered by other proteins located either in the cytoplasm or nucleus. For example, glucocorticoid receptors are sequestered by hsp90 in the cytoplasm. Binding of a hormone such as corticosterone to the COOH-terminal domain of the receptor causes a conformational change that liberates the receptor and allows it to interact with another activated receptor, forming a homodimer (fig. 27.16). The receptor dimers recognize specific DNA sequences that are located in the promoter region of the genes that these receptors regulate (table 27.7). The positioning of receptors at the promoter allows their activation domains to stimulate transcription. Transcriptional regulation is discussed in chapter 36.

Thyroid hormone receptors differ from steroid hormone receptors in that they are always bound to DNA. In the absence of thyroid hormones these receptors inhibit the expression of genes to which they bind; the addition of hormone converts them into transcriptional activators (see chapter 36).

Hormones Are Organized into a Hierarchy

The synthesis of many hormones is regulated by a cascade of hormones (fig. 27.17). Frequently, a hypothalamic hormone impinges on the pituitary to stimulate synthesis of a hormone that activates hormone synthesis in yet another organ. The end products of these cascades generally feedback-inhibit the production of hormones at the beginning of the cascade (fig. 27.18).

Another common mechanism for modulating hormonal response involves two (or more) hormonal inputs with both positive and negative effects (see fig. 27.18). The hypothalamic peptides, somatostatin and GRF, have opposite effects on GH synthesis and secretion. Similarly, glucagon and insulin have opposite effects on gluconeogenesis in the liver (see the discussion earlier in this chapter), and some of the effects of ecdysone on gene expression in insects are blocked by juvenile hormone (a terpene derivative; fig. 27.19).

Considerable "cross-talk" also occurs between different hormones at the level of receptor function. For instance, the diacylglycerol-activated protein kinase C phosphorylates and thereby inhibits the activity of insulin and epidermal growth factor receptor kinases (epidermal growth factor is discussed in a later section). Likewise, when cAMP-dependent protein kinases are active, they can inhibit receptors for epinephrine by phosphorylating them.

Cells also have mechanisms that tend to prevent chronic stimulation. Exposure of cells to epinephrine leads to an initial sharp rise in cAMP levels; however, cAMP levels fall nearly to basal levels within an hour or so, despite the continuous presence of saturating amounts of epinephrine. Furthermore, if the hormone is removed and the cells are challenged within a few hours, the secondary response is lower. This phenomenon, called desensitization, is associated with both an uncoupling of receptors from adenylate cyclase activation and a decrease in the number of receptors accessible to hormone binding. It occurs only after productive receptor function. The main consequence of desensitization is that exposure to hormone results in transient activation of cellular events, rather than chronic activation.

Figure 27.16

Activation of steroid hormone receptors by the hormone. In the absence of the hormone the steroid receptors are complexed through the hormone-binding domain to another protein known as heat shock protein 90 (hsp90). Both the hormone-binding domain and the hsp90 prevent functional interaction of the receptor with DNA. Binding of the hormone frees the receptor from hsp90 and promotes dimerization of the receptor, which can then bind to the palindromic hormone response element (HRE) and activate transcription.

Without hormone

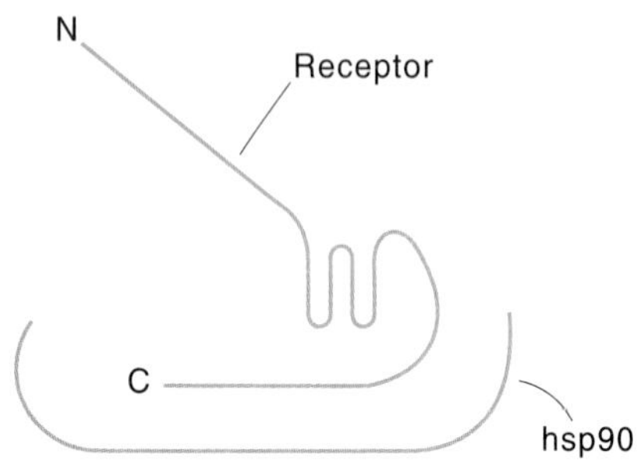

With hormone

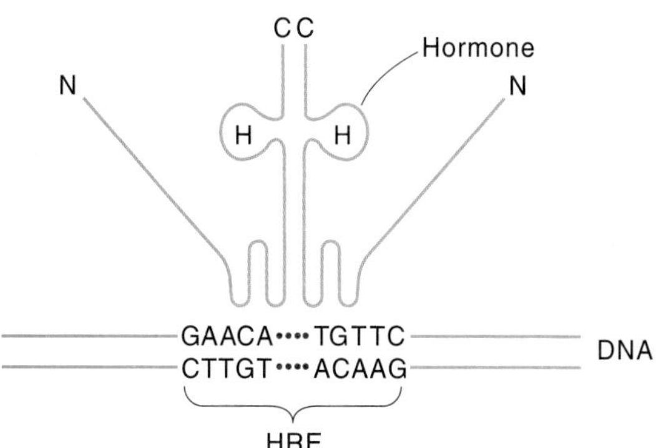

Table 27.7
Regulation of Specific Genes by Steroid and Thyroid Hormones

Hormone / Gene	Tissue	Hormone / Gene	Tissue
Glucocorticoids		**Androgens**	
Tyrosine aminotransferase	Liver	β-Glucuronidase	Kidney
Tryptophan oxygenase	Liver	Aldolase	Prostate
Glutamine synthase	Liver, retina	Prostate-binding proteins	Prostate
Phosphoenolpyruvate carboxykinase	Kidney	Ovomucoid	Oviduct
Ovalbumin	Oviduct	Ovalbumin	Oviduct
Conalbumin	Oviduct (liver)[a]	**1,25-Dihydroxyvitamin D_3**	
α-Fetoprotein (↓)	Liver	Calcium-binding protein	Intestine
α_2-Globulin	Liver	Ostercalcin	Bone
Metallothionein	Liver	**Ecdysone**	
Proopiomelanocortin (↓)	Pituitary	Dopa-decarboxylase	Epidermis
Mammary tumor virus	Mammary gland	Vitellogenin	Fat body
Estrogens		Larval scrum protein I	Fat body
Ovalbumin	Oviduct	**Thyroid Hormones**	
Conalbumin	Oviduct (liver)	Carbamyl phosphate synthase	Liver
Ovomucoid	Oviduct	Growth hormone	Pituitary
Lysozyme	Oviduct	Prolactin (↓)	Pituitary
Vitellogenin	Liver	α-Glycerophosphate dehydrogenase	Liver (mitochondria)
apo-VLDL	Liver	Malic enzyme	Liver
Glucose-6-P-dehydrogenase	Uterus		
Progestins			
Avidin	Oviduct		
Ovalbumin	Oviduct		
Conalbumin	Oviduct		
Uteroglobin	Uterus		

[a]In the liver the product of the conalbumin gene is called *transferrin.*

Note: (↓) means that mRNA levels are decreased by hormone.

Source: Reprinted with permission from *Nature* (H. R. Bourne et al., The GTPase superfamily: A conserved switch of diverse cell functions, *Nature,* 348:126, 1990).
Copyright © 1990 Macmillan Magazines Limited.

The response to many polypeptide hormones also diminishes with chronic stimulation, but the time course is much longer (many hours to days) and the mechanism is different from that involved in desensitization. Binding of insulin, calcitonin, LH, TRH, and EGF to their respective receptors promotes a physiological response but ultimately leads to the clearance of these receptors from the surface and a blunting of that response. The hormone-mediated loss of receptors is often referred to as down-regulation. In this case the receptors are internalized and degraded. Hormone binding leads to a clustering of receptors, as if they were cross-linked. These clusters, or patches, aggregate in membrane structures called coated pits (the intracellular side of the pits is coated with a scaffolding protein called clathrin). Small endosome vesicles that engulf the receptors and their ligands bud off from the coated pits, migrate within the cell, and associate with other membranous structures known collectively as GERL (Golgi–endoplasmic reticulum lysosomes). After a few hours they fuse with lysosomes, at which point the lysosomal enzymes degrade both the receptor and the hormone (fig. 27.20). Replacement of the hormone receptors requires protein synthesis.

Most hormones are released in a cyclic manner (examples include insulin, glucagon, growth hormone, and many of the hypothalamic releasing hormones). In some cases the cycles appear to be autonomously regulated, but in others they are clearly entrained by neural and hormonal signals that may vary, depending on the developmental stage or physiological condition. The

Figure 27.17

Growth hormone cascade. Neurosecretory cells in the hypothalamic region of the brain are activated by neurotransmitters from other neurons to secrete growth hormone releasing factor (GRF), a 44-amino-acid peptide that travels through the portal circulation to the anterior pituitary, where it binds to membrane receptors on somatotroph cells. GRF binding stimulates the production of cAMP, which activates growth hormone (GH) synthesis and secretion. GH, a 191-amino-acid polypeptide, passes into the bloodstream and travels to the liver, where it binds to membrane receptors that probably produce a second messenger (as yet unknown) that activates transcription of insulinlike growth factor-1 (IGF-1) gene, which codes for a 71-amino-acid peptide. IGF-1 is secreted from the liver into the bloodstream and ultimately binds to membrane receptors (which are tyrosine kinases similar to the insulin receptor) located on many peripheral cells. Activation of these receptors leads to cellular proliferation under appropriate conditions.

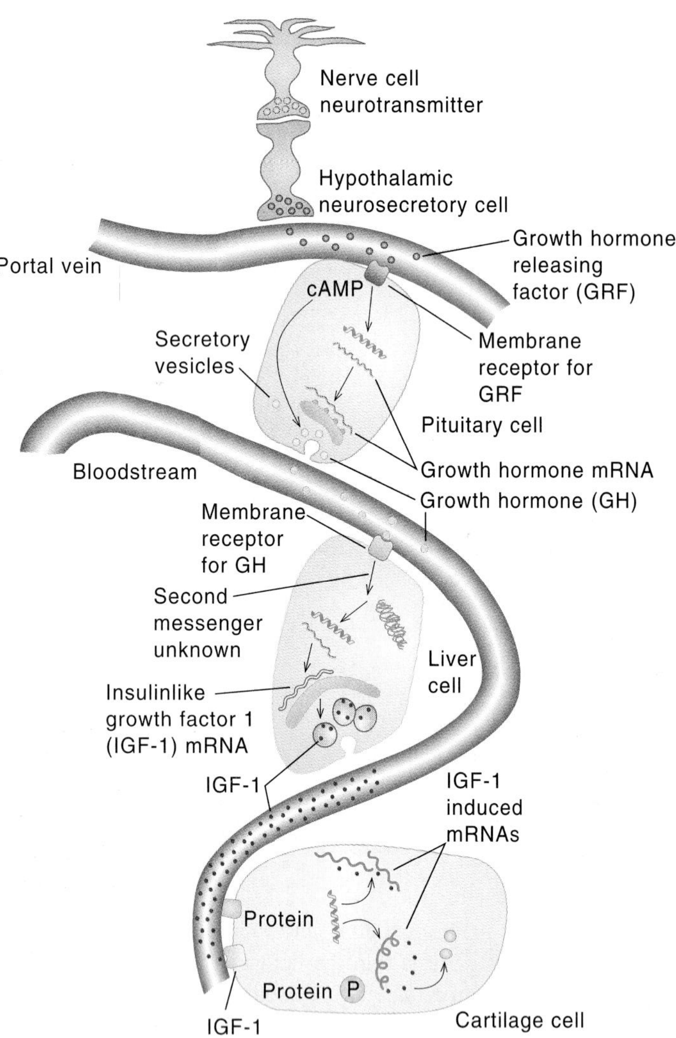

Figure 27.18

Control of hormone synthesis and secretion in the anterior pituitary. Neurosecretory neurons in the hypothalamus liberate polypeptides that either stimulate ⊕ or inhibit ⊖ hormone synthesis by specialized pituitary cells containing the appropriate receptors. For example, GnRH stimulates gonadotroph cells in the pituitary to synthesize and secrete LH and FSH. These pituitary hormones then impinge on target cells, typically stimulating them to make other low-molecular-weight hormones. The end products of these cascades feedback-inhibit hormone production at either or both hypothalamic and pituitary levels.

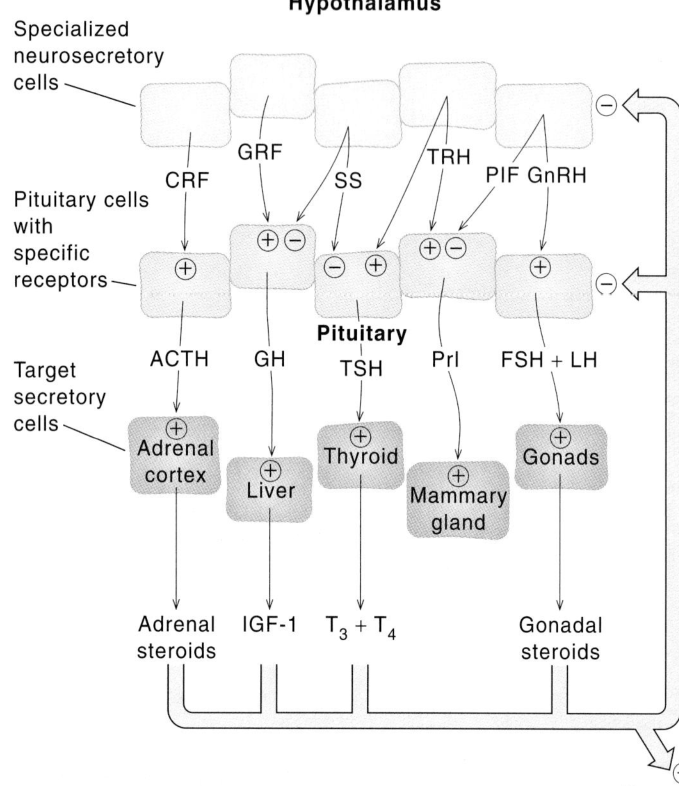

CRF = corticotropin regulatory factor
GRF = growth hormone releasing factor
SS = somatostatin
TRH = thyrotropin releasing hormone
PIF = prolactin inhibitory factor
GnRH = gonadotropin regulatory hormone
IGF-1 = insulinlike growth factor-1
ACTH = corticotropin
GH = growth hormone
TSH = thyrotropin
Prl = prolactin
FSH = follicle-stimulating hormone
LH = luteinizing hormone
T_3, T_4 = thyroid hormones

Figure 27.19

Structure of β-ecdysone and juvenile hormone. These hormones play major roles in the growth and maturation of insects by controlling the timing for molting of the insect exoskeleton.

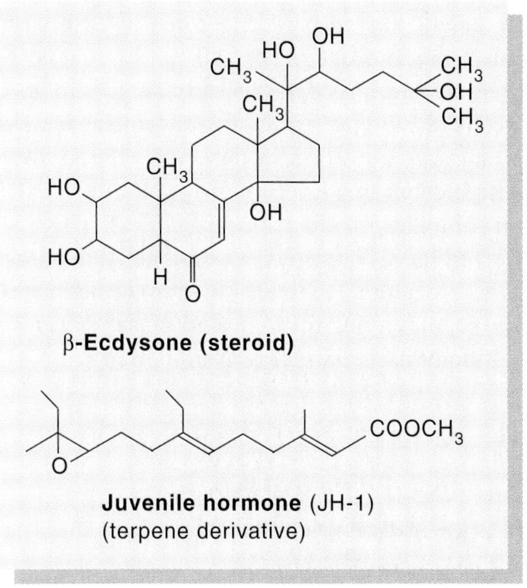

β-Ecdysone (steroid)

Juvenile hormone (JH-1)
(terpene derivative)

Figure 27.20

The down-regulation of receptors by endocytosis. The hormone-mediated loss of receptors is often referred to as down-regulation. Continuous activation of receptors by hormone often leads to patching, a clustering of receptors as if they were cross-linked. Endocytosis of the patches removes them from the cell surface. The endocytotic vesicles, sometimes called receptosomes, fuse with lysosomes, where the contents are degraded by the lysosomal enzymes. Receptosomes also may allow entry of receptors into other cell compartments, such as the nucleus, by fusing with these organelles.

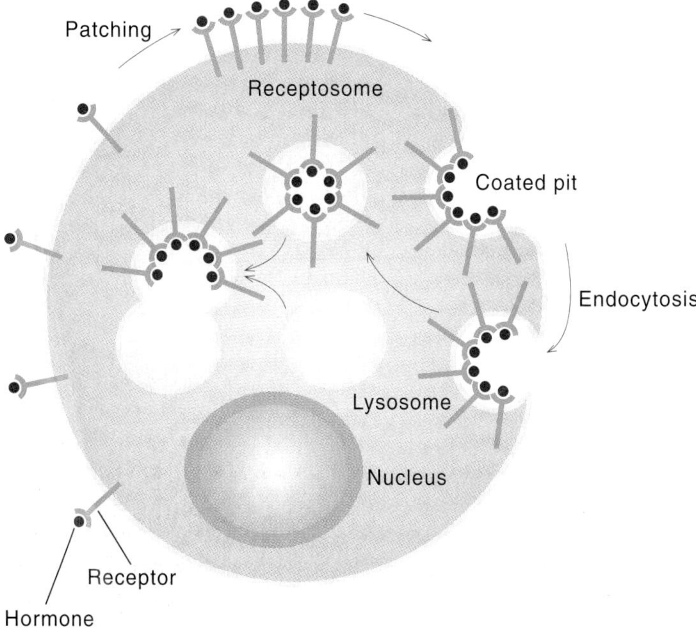

periodic release of peptide hormone can promote distinctly different responses than would be achieved by chronic release.

Diseases Associated with the Endocrine System

Human diseases related to endocrine dysfunction can be broadly grouped into (1) overproduction of a particular hormone, (2) underproduction of a hormone, and (3) target-cell insensitivity to a hormone.

Overproduction of Hormones Is Commonly Caused by Tumor Formation

Most cases of hormonal overproduction are associated with enlargement of the normal endocrine organ, frequently owing to a tumor. Pituitary neoplasms usually affect the production of only one pituitary hormone as a result of the cancerous proliferation of the cell type that normally synthesizes that hormone. Examples include giantism (acromegaly), which is associated with proliferation of the somatotroph cells that synthesize growth hormone, and Cushing's syndrome, which is usually due to overproduction of ACTH. Adrenal and parathyroid tumors that lead to the overproduction of various adrenal steroids and parathyroid hormones have also been described.

Occasionally, tumors of organs that do not usually produce a given hormone (ectopic tumors) synthesize and secrete peptide hormones, as in ACTH synthesis by certain lung tumors and GRH production from a pancreatic islet tumor. Expression of hormones by ectopic tumors represents a curious activation of gene expression in an inappropriate tissue. The same result can be obtained experimentally by producing animals in which a structural gene codes for a hormone that is under the transcriptional control of a promoter from a gene expressed in another tissue. This type of overproduction is illustrated in experiments with metallothionein-growth hormone (fig. 27.21). In this situation the growth hormone gene has been engineered so that it is not under normal feedback control.

The excessive production of thyroid hormone in Graves' disease is associated with an enlarged thyroid gland, but in this case a circulating immunoglobulin that mimics the activity of thyroid stimulating hormone (TSH) is implicated. Finally, we saw in an earlier section how an inappropriate steroid hormone may be secreted in excessive amounts when specific enzymes in the adrenal steroid biosynthetic pathway are present at inadequate levels.

Underproduction of Hormones Has Multiple Causes

A wide variety of defects can lead to inadequate production of hormones. In some conditions, not enough cells produce a given hormone. For example, in juvenile-onset diabetes the number of

Figure 27.21

A transgenic mouse that synthesizes rat growth hormone ectopically and a normal littermate. A chimeric gene with the mouse metallothionein promoter fused to the structural gene of rat growth hormone (rGH) was introduced into the germline of mice by microinjection of cloned DNA into fertilized eggs. Because of the metallothionein promoter, these genes were expressed in many large organs (such as liver, kidney, heart, intestine) that do not normally make GH. As a consequence, the circulating GH level was elevated several hundredfold, an effect that led to increased growth of those mice that inherited the gene. (From R. D. Palmiter, R. L. Brinster, R. E. Hammer, and R. M. Evans. Reprinted by permission from *Nature* 300:611–615. © 1982 Macmillan Magazines Limited. Photo courtesy of Ralph Brinster.)

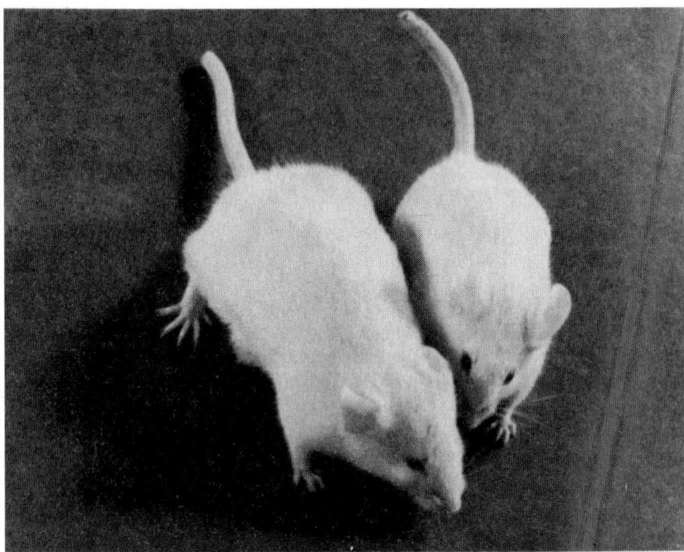

pancreatic islet cells that synthesize insulin decreases. In other, more extreme cases the gene coding for the hormone may be missing or defective; for example, some forms of dwarfism are due to a lack of the growth hormone gene. In most cases, however, the gene coding for the hormone is present, but inadequate amounts of active hormone are produced. Such defects can have many possible explanations. Some of them are genetic. For example, a mutation may affect the rate of synthesis of mRNA coding for the hormone precursor, or a mutation may affect the processing of the mRNA, or an amino acid substitution may decrease the activity or processing of the mRNA, or an amino acid substitution may decrease the activity or processing of a polypeptide hormone. The techniques of molecular biology are being used to discover the causes of many of these genetic disorders. Other defects in hormone synthesis are the consequence of an inadequate supply of precursors. For example, iodine is essential for thyroid hormone synthesis, and its absence leads to goiter. This condition, involving enlargement of the thyroid, is caused by high concentrations of TSH that result from lack of normal feedback by T_3 and T_4 on the hypothalamus. Likewise, synthesis of vitamin D requires ultraviolet irradiation of 7-dehydrocholesterol, without which the formation of $1{,}25(OH)_2D_3$ cannot occur, so that rickets ensues. A rare cause of rickets involves a defect in the enzyme that converts $25(OH)D_3$ to $1{,}25(OH)_2D_3$, the active form of vitamin D. Similarly, defects in enzymes involved in adrenal steroid biosynthesis lead to Addison's disease, and a defect in 5α-reductase leads to one form of testicular feminization, the result of inadequate production of dihydrotestosterone.

Target-Cell Insensitivity Results from a Lack of Functional Receptors

The most dramatic examples of target-cell insensitivity are those in which the correct receptors are lacking. In complete testicular feminization, androgen receptors are missing from all cells. The consequence is phenotypic expression of female characteristics in genotypic males. A rare form of dwarfism (Laron dwarfs) is associated with high plasma levels of growth hormone (GH) but low levels of insulinlike growth factor-1 (IGF-1); these individuals have a defect in the GH receptor. Occasionally, antibodies are directed against receptors and interfere with normal hormone binding, such as occurs in one form of adult-onset diabetes. Remarkably, the antibody itself promotes insulin effects.

Growth Factors Are Proteins That Behave Like Hormones

In addition to the many hormones that have been discussed, a large number of growth factors exist that carry on hormonelike activities. As their name implies, growth factors often stimulate the proliferation of particular cells. They also stimulate differentiation. Sometimes they stimulate both proliferation and differentiation together. All growth factors characterized so far are proteins (table 27.8). Unlike hormones, most growth factors are synthesized by a variety of cell types rather than in specialized endocrine glands. However, the mechanisms of action of growth factors and hormones are very similar.

Many growth factors are cleaved from larger precursors. For example, the 53 amino acid epidermal growth factor (EGF) is cleaved from a precursor of 1,168 amino acids. This precursor is a membrane-spanning protein, with the EGF moiety and nine related sequences in the extracellular domain. EGF is homologous in amino acid sequence to many other growth factors.

All growth factors act by binding to specific receptors on the outer plasma membrane of target cells. The receptors are usually transmembrane proteins that contain the growth factor–binding site on the outer side and an enzyme effector on the cytosolic side of the plasma membrane. In this respect, growth factor receptors resemble the simplest type of hormone-response element (see fig. 27.14*a*). In most cases the enzyme effector is a tyrosine kinase. This is interesting in view of the fact that most soluble cellular kinases are serine or threonine specific. The tyrosine kinases are largely but not entirely confined to membrane proteins.

Activated tyrosine kinase receptors frequently phosphorylate themselves, and this is often crucial to their function. The binding of a growth factor is the trigger that leads to this autophosphorylation. In the case of the EGF receptor it appears that the EGF growth factor upon binding stimulates dimerization, which leads to cross phosphorylation of the two linked receptor monomers. The phosphorylated dimer has a high affinity for proteins that initiate a cascade of reactions associated with the mitogenic response.

Table 27.8
Vertebrate Growth Factors

Factor[a]	M_r	Cell Types Affected
Epidermal growth factor (EGF)[b]	6,400	Epithelial and mesodermal cells
Tumor growth factor-α (TGF-α)	7,000	Same
Insulinlike growth factor-1 (IGF-1)	7,000	Same
Insulinlike growth factor-2 (IGF-2)	7,000	Same
Fibroblast growth factor (FGF)[c]	26,000 (dimer)	Same
Platelet-derived growth factor (PDGF)[b]	31,000	Same
Nerve growth factor (NGF)[c]	13,000	Sensory and sympathetic neurons
Erythropoietin	23,000	Erythroid cell precursors
Macrophage colony-stimulating factor (M-CSF or CSF-1)[b]	70,000 (dimer)	Macrophage precursors
Granulocyte colony-stimulating factor (G-CSF)	25,000	Granulocyte precursors
Granulocyte–macrophage colony-stimulating factor (GM-CSF)	23,000	Granulocytes and macrophage precursors
Multicolony-stimulating factor (multi-CSF) or Interleukin-3 (IL-3)	25,000	Precursors of most hematopoietic cells
Interleukin-2 (IL-2)[c]	13,000	T lymphocytes

[a]The genes for most of these growth factors and many of their receptors have been cloned.

[b]The EGF receptor is homologous to the *erb*B and *neu* oncogenes, and the M-CSF receptor is homologous to the *fms* oncogene. PDGF is homologous to the *sis* oncogene.

[c]There are many members of the FGF, NGF, and interleukin families of growth factors.

The connection between activated growth factor receptors and proliferation has been a subject of intense study because of the close relationship between growth abnormalities and cancer. Normal cells require growth factors (mitogens) for proliferation; in the absence of these factors they reversibly withdraw from the cell cycle and become arrested. By contrast transformed cells (tumor cells) have a relaxed cell cycle control and can traverse the cell cycle in the absence of added growth factors. Cell transformation can be achieved by activation or inappropriate expression of a variety of cellular genes or the obtrusive expression of oncogenes in cancer-causing viruses (see chapter 38). The isolation and characterization of many of these transforming genes, which are better known as oncogenes, have revealed striking homology in their coding sequence with growth factors or their receptors. For example, the oncogene product or oncoprotein *sis* resembles PDGF in its sequence and its affinity for specific receptors, and the EGF receptor is homologous to the erbB oncoprotein. From telling observations of this sort the prevalent idea has evolved that many oncogenes lead to abnormal cell proliferation by directly or indirectly providing a rate-limiting factor, be it a growth factor, its receptor, or a downstream intermediate in a mitogenic pathway.

We might expect that certain downstream elements that transmit the mitogenic signals from the tyrosine kinases of the growth factor receptors could be altered so as to short-circuit the need for the cell-surface signals. It seems probable that this would lead to a transformed state. A likely candidate for such an element has been found in the form of a 21-kD plasma membrane bound protein called Ras which is mutated to a hyperactive state in about one-third of all human cancers. Developmental studies on *C. elegans* and *Drosophila* have identified Ras as a critical determinant of cellular differentiation. Moreover, microinjection of antibodies against Ras into mammalian cells blocked the DNA synthesis induced by receptor tyrosine kinases (RTKs) and most nonnuclear oncoproteins. These results established that across a wide phylogenetic span, Ras normally relays mitogenic and developmental signals initiated by cell-surface receptors into the cytoplasm and the nucleus.

Ras is a G protein of sorts that binds GTP in the active form and GDP in the inactive form. Two classes of signaling proteins regulate Ras activity in influencing the transition between the active and inactive states. GTPase-activating proteins (GAPs) increase the rate of hydrolysis of bound GTP by Ras, thereby inactivating it. These negative regulators are counteracted by guanine nucleotide–releasing proteins (GNRPs), which promote the exchange of bound nucleotide by stimulating the loss of GDP and the subsequent uptake of GTP from the cytosol; they therefore tend to activate Ras. Activated receptor tyrosine kinases bind GAPs directly and GNRPs indirectly. It is the indirect binding to GNRPs that appears to be crucial in driving Ras into its active GTP-bound state.

Once activated, Ras relays a signal downstream by activating a serine–threonine phosphorylation cascade that is highly conserved in eukaryotic cells from yeasts to humans. In stepwise

Figure 27.22

Ras is at the hub of mitogenic signaling. The mitogenic process begins with the binding of a mitogenic effector to a transmembrane receptor, which usually bears a tyrosine kinase on the cytosolic side. Activation of the kinase leads to a chain reaction involving activation of proteins A and B and then Ras. Ras activates Raf-1, which activates MEK, then MAPK and other cytoplasmic proteins and nuclear transcription factors (TFs).

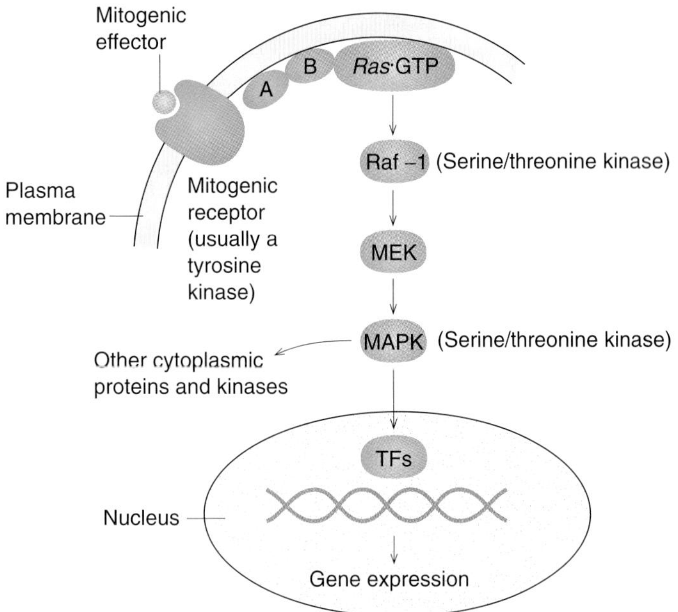

fashion, Ras activates Raf protein, which activates MEK protein, which activates a group of mitogen-activated protein kinases (MAPKs). The MAPKs phosphorylate serine and threonine residues on a wide variety of regulatory proteins that include transcription factors and other serine and threonine kinases (fig. 27.22).

One may wonder why Ras uses a cascade of protein kinases to signal mitogenesis and cell differentiation. If activation of MAP kinase is both necessary and sufficient for Ras action, why not activate the MAP kinase directly through Ras or, better yet, directly through the RTK? The answer to this is not known, but evidently, this cascade confers significant advantages. One possible benefit of the cascade is an increase in the number of targets for positive and negative cross-regulation by other signaling pathways. A cascade also increases the number of steps at which mitogenic signals can be diversified. Indeed, current indications are that Ras is situated at the hub of mitogenic signaling with signals coming in from various sources and being transmitted to different mitogen-activated protein kinases. This is a breaking story with much progress expected in the near future.

Plant Hormones

Plant hormones (also called growth regulators) coordinate the growth and development of plants. The major hormones discovered to date fall into six classes: auxins, cytokinins, gibberellins, abscisic acid, ethylene, and oligosaccharides. The structures of representative members of each class are shown in figure 27.23. All of these hormones are low-molecular-weight compounds; indeed, one hormone, ethylene, is a gas. The polypeptide hormone of higher plants, systemin, is an 18-amino-acid peptide that is released on wounding and activates the plant defense system. Polypeptide and steroid hormones have not been described for higher plants, although some yeasts and fungi use these compounds as mating factors. Each class of compounds elicits many diverse responses, and considerable interaction occurs among different plant hormones in the control of physiological processes. Furthermore, the same process is controlled by different hormones (or combinations of hormones) in different species. These considerations make it difficult to analyze or generalize about the mechanism of plant hormone action.

Auxins are synthesized in the apical buds of growing shoots. They stimulate growth of the main shoot but inhibit the development of lateral shoots; this effect has led to the horticultural practice of pinching off the apical buds to stimulate the formation of bushy plants. The curvature of plants toward the light (phototropism) is thought to be due to transport of auxins away from the light, which stimulates more rapid growth of cells on the darker side of the shoot. Auxins bind to specific membrane proteins. The affinity of auxins for these proteins, coupled with their location and abundance on target cells, supports the view that the membrane proteins may be the receptors that mediate auxin action.

Auxin stimulation of growth occurs in two phases. The earliest response (called the rapid response) is an increase in proton transport out of the cell, which occurs after a lag of a few minutes. This hydrogen ion pump is thought to be coupled with a membrane ATPase, but it is not clear whether the receptor interacts directly with the ATPase or whether other intermediates are involved. It is thought that lowering the extracellular pH activates enzymes that partially degrade the cell wall, thereby loosening it and allowing for cell expansion. A subsequent effect of auxin (the slow response) is to increase the synthesis of proteins and nucleic acids, resulting in sustained growth. Auxin has been shown, for example, to increase the amount of cellulase mRNA during pea cell expansion. A 20-kD membrane protein that binds auxins has been characterized that appears to be the auxin receptor in that an antibody against it blocks auxin-stimulated events.

Cytokinins are adenine derivatives that are produced in the roots and that promote growth and differentiation of numerous tissues. Cytokinins can overcome the auxin-mediated inhibition of lateral shoots; in other tissues both hormones act synergistically. Plant cells can be grown in culture if auxins and cytokinins are provided. With relatively balanced concentrations of both hormones, cells proliferate but remain unorganized. If the ratio of auxins to cytokinins is high, shoots develop; if the ratio is low, root development is favored. Intermediate concentrations of both hormones promote undifferentiated growth, with neither roots nor shoots. As these results show, plant cells can display a wide range of physiological responses. They are much more plastic in their developmental potential than animal cells. Indeed, normal plants can be grown from a single tissue culture cell. Although this diversity of responses is fascinating, it has also been difficult to define the mechanism of action of plant hormones.

Figure 27.23

Some common plant hormones. All of these hormones are low-molecular-weight compounds. One hormone, ethylene, is a gas.

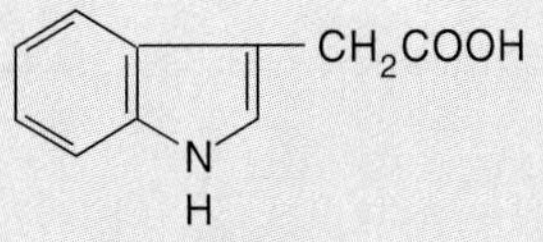

Auxin
(indole acetic acid)

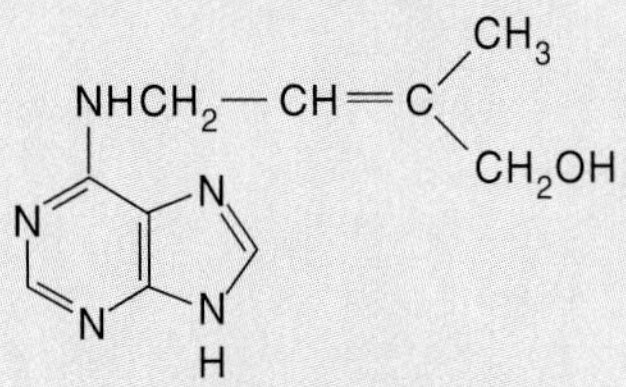

Cytokinin
(zeatin)

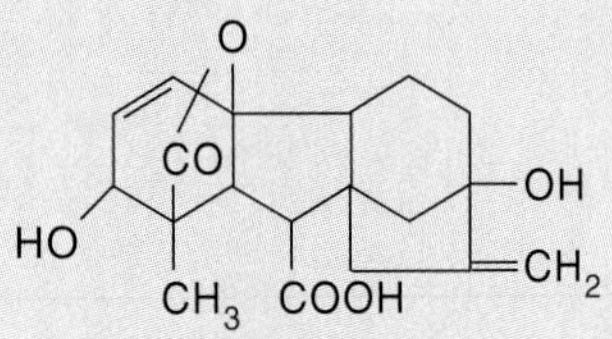

Gibberellin
(gibberellic acid)

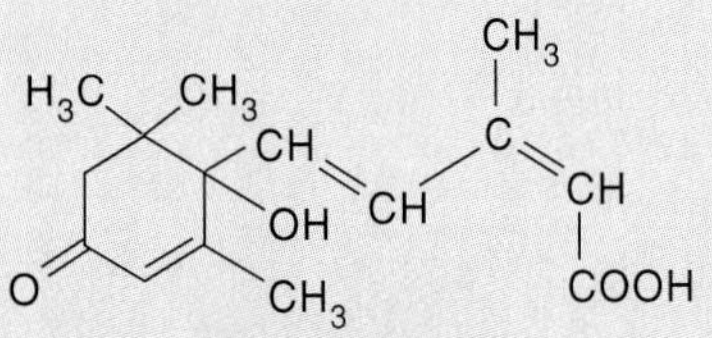

Abscisic acid

$H_2C{=}CH_2$

Ethylene

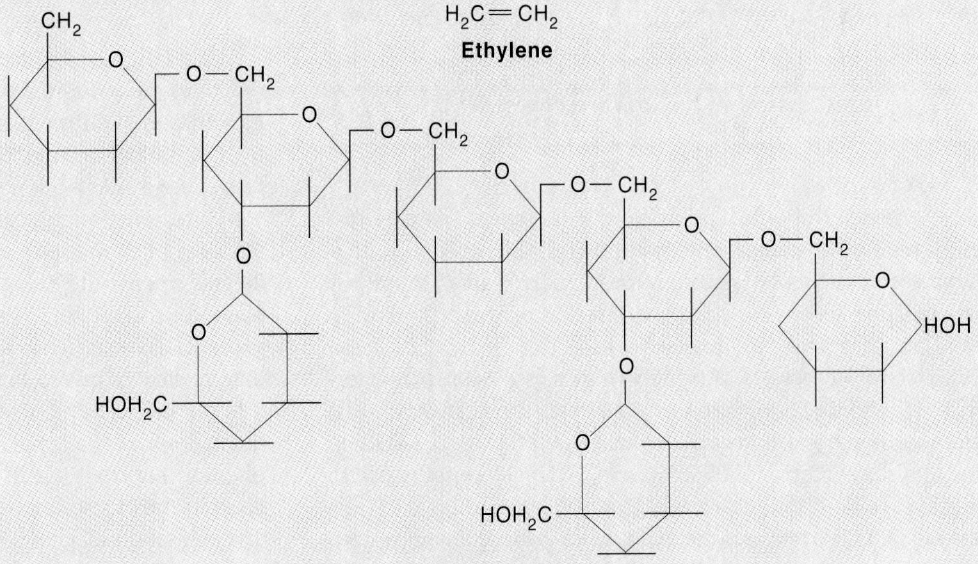

Oligosaccharin
(a heptaglucoside)

Figure 27.24

Synthesis of ethylene, a gaseous plant hormone involved in fruit ripening and flower senescence.

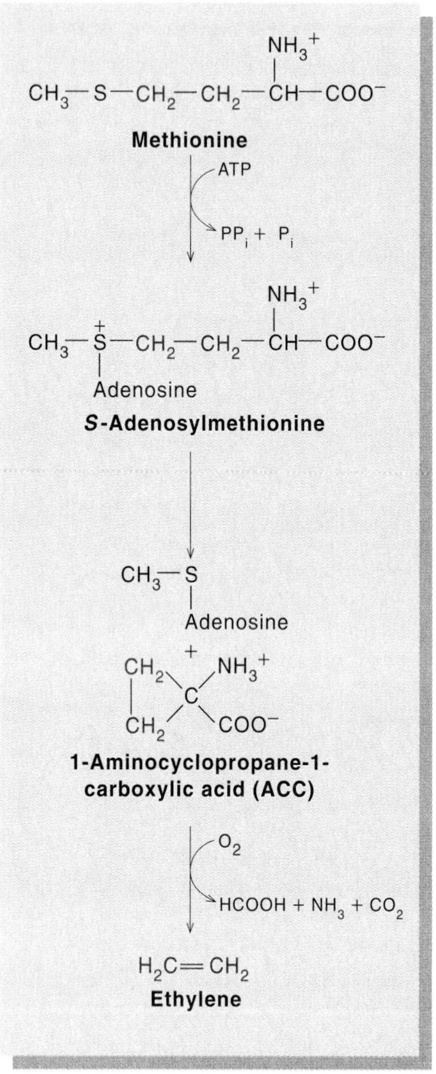

$$CH_3{-}S{-}CH_2{-}CH_2{-}\underset{}{CH}(NH_3^+){-}COO^-$$

Methionine

$\downarrow$ ATP $\rightarrow$ $PP_i + P_i$

$$CH_3{-}\overset{+}{S}(\text{Adenosine}){-}CH_2{-}CH_2{-}CH(NH_3^+){-}COO^-$$

***S*-Adenosylmethionine**

$\downarrow$

$$CH_3{-}S{-}\text{Adenosine}$$

$$+$$

$$\text{(cyclo-}CH_2{-}CH_2{-})C(NH_3^+)(COO^-)$$

1-Aminocyclopropane-1-carboxylic acid (ACC)

$\downarrow$ O_2 $\rightarrow$ $HCOOH + NH_3 + CO_2$

$$H_2C{=}CH_2$$

Ethylene

Gibberellins also promote shoot elongation, and frequently they act synergistically with auxins. Gibberellins stimulate the accumulation of specific mRNAs such as amylase mRNA in germinating seeds. This fact suggests that their receptors (or a second messenger) act at the genetic level.

Abscisic acid is antagonistic to many other plant hormones. It inhibits germination of seeds and shoot growth while promoting resting bud formation and leaf senescence. Wilting stimulates the synthesis of abscisic acid in the chloroplasts within mesophyll cells of the leaves. The abscisic acid in this case is a stress signal that stimulates the guard cells to close and thus minimize water loss through the stomata. Abscisic acid stimulates K^+ efflux from guard cells and into adjacent cells; thus it may act by means of membrane receptors to modulate ion pumps, as was suggested for auxins.

Figure 27.25

Crown gall tumors of plants are caused by certain strains of the bacterium *Agrobacterium* that can infect plant cells and introduce new genetic information coding for plant hormones. The transforming DNA of these bacterial strains is carried on a large plasmid; it contains genes coding for rate-limiting enzymes in hormone biosynthesis. (*a*) The transforming DNA of wild-type bacterial plasmid codes for rate-limiting enzymes involved in both auxin and cytokinin biosynthesis, resulting in undifferentiated growth (called callus). (*b*) Mutations that disrupt the gene involved in auxin biosynthesis give rise to tumors that produce only cytokinins, which leads to a proliferation of leaves and shoots. (*c*) Alternatively, mutations that disrupt the gene involved in cytokinin synthesis give rise to tumors that produce only auxin. In these tumors, only roots differentiate.

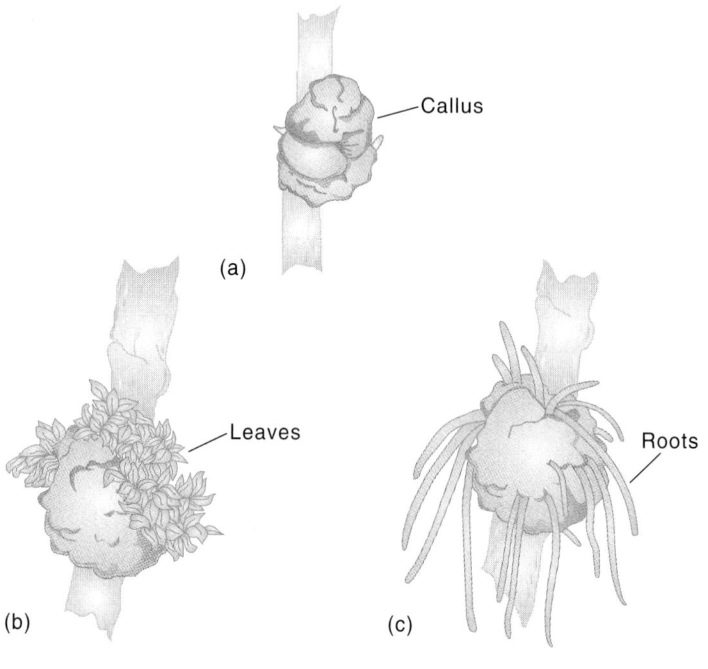

Oligosaccharin is one member of complex oligosaccharides that function in defense against disease, in control of plant growth, and in differentiation. They are released from the cell wall by specific degradative enzymes. For example, when bacteria or fungi infect plants, they produce enzymes that degrade the cell wall, liberating oligosaccharides. Some of these compounds stimulate the plant to make antibiotics or proteases that help the plant defend against the pathogen or insect. There they stimulate the plant cells to produce antibiotics that inhibit the growth of the pathogen. When certain plant cells are damaged, they can produce the enzymes required to liberate oligosaccharides, thus allowing a hormonal response even when the pathogen does not produce the appropriate enzymes. Auxins also stimulate the production of oligosaccharides, which then counteract the auxin-stimulated growth; thus these oligosaccharides serve as feedback inhibitors. The inhibition of lateral shoot growth that has been attributed to auxin may actually be due to the production of oligosaccharides.

Addition of oligosaccharides to combinations of auxin plus cytokinins also influences the differentiation of shoots and roots in tissue culture. Indeed, many of the pleiotropic effects originally attributed to auxins and other plant hormones may actually be mediated by oligosaccharins.

Ethylene, although gaseous, has effects that are comparable to those of other hormones. It plays an important role in transverse rather than longitudinal growth of cells. It also stimulates fruit ripening and flower senescence, and it inhibits seedling growth. Ethylene is synthesized from *S*-adenosylmethionine (SAM), as shown in figure 27.24. The conversion of SAM to 1-aminocyclopropane-1-carboxylic acid, the immediate precursor of ethylene, is stimulated by auxins, wounding, and anaerobiosis. Once again we see the interplay of hormones in the regulation of cell activity.

Plant tumors result from uncontrolled hormone production. Crown gall tumors, for example, are due to the infection of plant wounds by certain strains of *Agrobacterium.* These bacteria carry a large plasmid, the tumor-inducing, or T_1 plasmid, part of which is incorporated into the plant genome. This DNA encodes several genes that stimulate cell proliferation. One gene product is a rate-limiting enzyme involved in auxin biosynthesis, and another controls cytokinin biosynthesis. Together they promote the rapid but undifferentiated growth that is the crown gall. Interestingly, inactivation of one of these genes promotes the growth of shoots at the site of infection, whereas inactivation of the other promotes the growth of roots (fig. 27.25). These effects are reminiscent of the action of auxins and cytokinins in tissue culture.

Summary

In this chapter we focused on the ways in which various metabolic activities are integrated, with special attention to the nature and functioning of hormones. The following points are the highlights of our discussion.

1. Tissues store biochemically useful energy in three major forms: carbohydrates, lipids, and proteins. Each tissue makes characteristic demands on and contributions to the energy supply of the organism.
2. Hormones are chemical messengers formed in specific tissues. They circulate between tissues of multicellular organisms and serve to coordinate metabolic activities, maintain homeostasis of essential nutrients, and prepare the organism for reproduction.
3. Most hormones fall into three classes: polypeptides, steroids, and amino acid derivatives. Polypeptide hormones are synthesized from large precursors. Steroid hormones are derivatives of cholesterol. Thyroid hormones and epinephrine are amino acid derivatives.
4. A number of factors—synthesis, rate of release, and rate of elimination—determine the concentration of circulating hormone.
5. Hormones act by reversibly binding to proteins called receptors, an event that results in a conformational change that is detected by other macromolecules (acceptors) and that eventually leads to activation of rate-limiting enzymes. Each class of hormones binds to specific receptors that activate other membrane proteins.
6. Most membrane receptors generate a diffusible intracellular signal called a second messenger. Five intracellular messengers are currently known: cyclic AMP, cyclic GMP, inositol triphosphate, diacylglycerol, and calcium. Second messengers usually activate or inhibit the action of one or more enzymes.
7. Steroid hormones penetrate the cell and bind to receptors in the nucleus, and they activate (or sometimes repress) transcription of specific genes. Thyroid hormones act similarly.
8. A large number of diseases are due either to overproduction or underproduction of hormones or to insensitivity of target tissues to circulating hormones. Knowledge of hormone biosynthesis, secretion, and interaction with target cells is essential to an understanding of the biochemical basis of these disorders.
9. In addition to classical hormones, other chemical messengers called growth factors serve to coordinate growth of tissues during development. Most growth factors interact with specific membrane-bound receptors that carry tyrosine kinases on the cytosolic side of the membrane. The binding of growth factors to the receptors triggers tyrosine kinase activity, which activates the receptors by autophosphorylation. Receptor activation initiates a chain of reactions involving regulatory proteins. At the business end of this chain, regulatory enzymes that are involved in either cell proliferation or differentiation or both are activated by phosphorylation.
10. Plants make several hormones that regulate growth and differentiation.

Selected Readings

Annual Reviews of Biochemistry and *Annual Reviews of Physiology.* Over 40 volumes in each series with many relevant reviews in the area of hormone receptors and hormone action. Provides good access to primary literature.

Avruch, J., X.-F. Zhang, and J. M. Kyriakis, Raf meets Ras: Completing the framework of a signal transduction pathway. *TIBS* 19:279–284, 1994.

Baxter, J. D., and K. M. MacLoed, Molecular basis for hormone action. In P. K. Bondy and L. E. Rosenberg (eds.), *Metabolic Control and Disease.* Philadelphia: Saunders, 1980. A useful summary.

Berridge, M. J., Inositol trisphosphate and calcium signalling. *Nature* 361:315–325, 1993.

Bredt, D. S., and S. H. Snyder, Nitric oxide: A physiologic messenger molecule. *Ann. Rev. Biochem.* 63:175–196, 1994.

Carafoli, E., and J. T. Penniston, The calcium signal. *Sci. Amer.* 253(5):70–78, 1985.

Carpenter, G., Receptors for epidermal growth factor and other polypeptide mitogens. *Ann. Rev. Biochem.* 56:881–914, 1987.

Clapham, E. D., and E. J. Neer, New roles for G-protein $\beta\gamma$-dimers in transmembrane signalling. *Nature* 365:403–406, 1993.

Clark, E. A., and J. S. Brugge, Integrins and signal transduction pathways: The road taken. *Science* 268:233–240, 1995.

Clark, E. A., S. J. Shattil, and J. S. Brugge, Regulation of protein tyrosine kinases in platelets. *TIBS* 19:464–469, 1994.

Collins, S., M. G. Caron, and R. J. Lefkowitz, From ligand binding to gene expression: New insights into the regulation of G-protein-coupled receptors. *Trends Biochem. Sci.* 17:37–39, 1992.

Cooper, D. M. F., N. Mons, and J. W. Karpen, Adenylyl cyclases and the interaction between calcium and cAMP signalling. *Nature* 374:421–424, 1995.

Czech, M. P., J. K. Klarlund, K. A. Yagaloff, A. P. Bradford, and R. E. Lewis, Insulin receptor signaling. *J. Biol. Chem.* 263:11017–11020, 1988.

Daum, G., et al., The ins and outs of Raf kinases. *TIBS* 19:474–479, 1994.

deGroot, L. J., et al. (eds.), *Endocrinology*, 3 vols. New York: Grune and Stratton, 1979. Over 2,000 pages of comprehensive treatment, primarily from a medical point of view.

DeVos, A. M., M. Ultsch, and A. A. Kossiakoff, Human growth hormone and extracellular domain of its receptor: Crystal structure of the complex. *Science* 255:306–312, 1992.

Egan, S. E., B. W. Giddings, M. W. Brooks, L. Buday, A. M. Sizeland, and R. A. Weinberg, Association of Sos Ras exchange protein with Grb2 is implicated in tyrosine kinase signal transduction and transformation. *Nature* 363:45–51, 1993.

Fantl, W. J., D. E. Johnson, and L. T. Williams, Signalling by receptor tyrosine kinases. *Ann. Rev. Biochem.* 62:453–481, 1993.

Feig, L. A., The many roads that lead to RAS. *Science* 260:767–768, 1993.

Funder, J. W., Mineralcorticoids, glucocorticoids, receptors and response elements. *Science* 259:1132–1133, 1993.

Gerisch, G., Cyclic AMP and other signals controlling cell development and differentiation in *Dictyostelium*. *Ann. Rev. Biochem.* 56:853–879, 1987.

Guillemin, R., Peptides in the brain: The new endocrinology of the neuron. *Science* 202:390, 1978. Nobel laureate speech.

Hubbard, S. R., L. Wei, L. Ellis, and W. A. Hendrickson, Crystal structure of the tyrosine kinase domain of the human insulin receptor. *Nature* 372:746–754, 1994.

Iniguez-Lluli, J., C. Kleuss, and A. G. Gillman, The importance of G-protein $\beta\gamma$ subunits. *Trends Cell Biol.* 3:230–235, 1993.

Jones, A. M., Surprising signals in plant cells. *Science* 263:183–184, 1994. An update on signal transduction pathway in plants.

Kikkawa, U., A. Kishimoto, and Y. Nishizuku, The protein kinase C family: Heterogeneity and its implications. *Ann. Rev. Biochem.* 58:31–44, 1989.

Lambright, D. G., J. P. Noel, and H. E. Hamm, Structural determinants for activation of the α-subunit of a heterotrimeric G protein. *Nature* 369:621–628, 1994.

Lefkowitz, R. J., G-protein-coupled receptor kinases. *Cell* 74:409–412, 1993.

Lehmann, J. M., L. Jong, A. Fanjul, J. F. Cameron, X. P. Lu, P. Haefner, M. I. Dawson, and M. Pfahl, Retinoids selective for retinoid X receptor response pathways. *Science* 258:1944–1946, 1992.

Lienhard, G. E., Life without the IRS. *Nature* 372:128–129, 1994.

Lowy, D. R., and B. M. Willumsen, Function and regulation of RAS. *Ann. Rev. Biochem.* 62:851–891, 1993.

Marx, J., Two major signal pathways linked. *Science* 262:988–990, 1993.

Mayer, B. J., and D. Baltimore, Signalling through SH2 and SH3 domains. *Trends Cell Biol.* 3:8–13, 1993.

Miyajima, A., T. Hara, and T. Kitamura, Common subunits of cytokine receptors and the functional redundancy of cytokines. *TIBS* 17:378–382, 1992.

Mochly-Rosen, D., Localization of protein kinases by anchoring proteins: A theme in signal transduction. *Science* 268:247–251, 1995.

Mustelin, T., and P. Burn, Regulation of src family tyrosine kinases in lymphocytes. *TIBS* 18:215–220, 1993.

Napier, R. M., and M. A. Venis, From auxin-binding protein to plant hormone receptor? *Trends Biochem. Sci.* 16:72–75, 1991.

Nishida, E., and Y. Gotoh, The MAP kinase cascade is essential for diverse signal transduction pathways. *TIBS* 18:128–130, 1993.

O'Malley, B. W., and L. Birnbaumer, *Receptors and Hormone Action*, 3 vols. New York: Academic Press, 1977. Review articles by many authors cover most aspects of hormone action. Somewhat dated.

Parkinson, J. S., Signal transduction schemes of bacteria. *Cell* 73:857–871, 1993.

Pawson, T., Protein modules and signalling networks. *Nature* 373:573–580, 1995.

Pelech, S. L., and D. E. Vance, Signal transduction via phosphatidylcholine cycles. *Trends Biochem. Sci.* 14:28–30, 1989.

Pifkis, S. J., M. R. El-Maghrabi, and T. H. Claus, Hormonal regulation of hepatic gluconeogenesis and glycolysis. *Ann. Rev. Biochem.* 57:755–784, 1987.

Rasmussen, H., The cycling of calcium as an intracellular messenger. *Sci. Am.* 261(4):66–73, 1989.

Recent Progress in Hormone Research. New York: Academic Press. An annual publication with nearly 40 volumes. A good place to find a recent summary.

Riddiford, L. M., and J. W. Truman, Biochemistry of insect hormones and insect growth regulators. In *Biochemistry of Insects*. New York: Academic Press, 1978.

Schmidt, H., and U. Walter, NO at work. *Cell* 78:919–925, 1994.

Simon, M. I., P. Strathmann, and N. Gautan, Diversity of G proteins in signal transduction. *Science* 252:802–808, 1991.

Simpson, I. A., and S. W. Cushman, Hormonal regulation of mammalian glucose transport. *Ann. Rev. Biochem.* 55:1059–1089, 1986.

Somiyo, A. P., and A. V. Somiyo, Signal transduction and regulation in smooth muscle. *Nature* 372:231–236, 1994.

Stahl, N., and G. D. Yancopoulos, The alphas, betas and kinases of cytokine receptor complexes. *Cell* 74:587–590, 1993.

Sutherland, E. W., Studies on the mechanism of hormone action. *Science* 177:401, 1972. Nobel laureate speech related to discovery of cAMP as second messenger.

Tamiguchi, T., Cytokine signaling through nonreceptor protein tyrosine kinases. *Science* 268:251–255, 1995.

Taylor, S. S., D. R. Knighton, J. Zheng, J. M. Sowadski, C. S. Gibbs, and M. J. Zoller, A template for the protein kinase family. *TIBS* 18:84–89, 1993.

Wagner, R. L., Aprilelli, J. W., McGrath, M. E., West, B. L., Baxter, J. D., and Fletterick, R. J. A structural role for hormone in the thyroid hormone receptor. *Nature* 378:690–697, 1995.

Wittinghofer, A., and E. F. Pai, The structure of Ras protein: a model for a universal molecular switch. *Trends Biochem. Sci.* 16:382–387, 1991.

Yarden, Y., and A. Ullrich, Growth factor receptor tyrosine kinases. *Ann. Rev. Biochem.* 57:443–478, 1988.

PROBLEMS

1. What differences/similarities are there between (1) an allosteric effector and its corresponding enzyme and (2) a hormone and its receptor protein?

2. What, if anything, actually goes around the cycle in the phosphatidylinositol cycle?

3. Notice the term "cross-talk" in quotations in the text (p. 686). Why is this expression in quotes?
4. If the M and H polypeptide monomers of lactate dehydrogenase are produced in a 1:4 mole ratio in a particular tissue, what will the M_4:M_3H:M_2H_2:MH_3:H_4 ratio be? Assume a statistical assembly of the monomers.
5. Examine the reaction that forms ethylene from 1-aminocyclopropane-1-carboxylic acid (fig. 27.24). Do you see a problem with this reaction? Can you suggest one or more solutions to the problem?
6. What cofactor would you expect to be associated with aromatic L-amino acid decarboxylase (fig. 27.10)?
7. By comparisons with metabolic reactions presented elsewhere in the text, propose a pathway for the formation of auxin from one of the protein amino acids.
8. Most metabolic conversions can occur in either direction by using different pathways. Eukaryotic cells frequently take advantage of subcellular compartments to separate oppositely directed pathways. Use fatty acid synthesis and degradation as an example, and discuss the design of the paths from the point of view that they are thermodynamically favorable and kinetically regulated. (Be sure to consider subcellular compartments.) Discuss the design of glycolysis and gluconeogenesis in the liver.
9. If a starving person (one who has gone a number of weeks with no food) is given a shot of insulin, what happens? Explain your answer.
10. Why are all known hormone receptors proteins? Can other macromolecules serve as receptors?
11. What is the advantage of the fact that muscle tissue can use an anaerobic metabolism?
12. List several reasons why polypeptide hormones are synthesized as precursors.
13. A patient has a hypothyroid condition. He has low serum T_3 and T_4 levels and elevated levels of serum thyroid-stimulating hormone (TSH). After injection of thyrotropin-releasing hormone (TRH), his serum TSH goes even higher. Is his defect primary (thyroid), secondary (pituitary), or tertiary (hypothalamus)?
14. Thyrotropin-releasing hormone (TRH) contains an unusual amino acid, pyroglutamate (fig. 27.8). Have we encountered this amino acid before under a different name? Where do you think the name "pyroglutamate" came from?
15. Activation of most membrane-associated hormone receptors generates a second messenger. What is a second messenger and what are the five second messengers currently known? What role do G proteins play in second-messenger formation?
16. Is vitamin D a hormone or a vitamin? Explain your answer.
17. Inhibitors of protein synthesis have been shown to block both the rapid and slow auxin-mediated growth responses. How do you explain these observations?

CHAPTER

NEUROTRANSMISSION

Nerve-impulse propagation is a form of intracellular communication in which the communicating cells are linked through complementary membranes that either secrete or respond to small-molecule neurotransmitters.

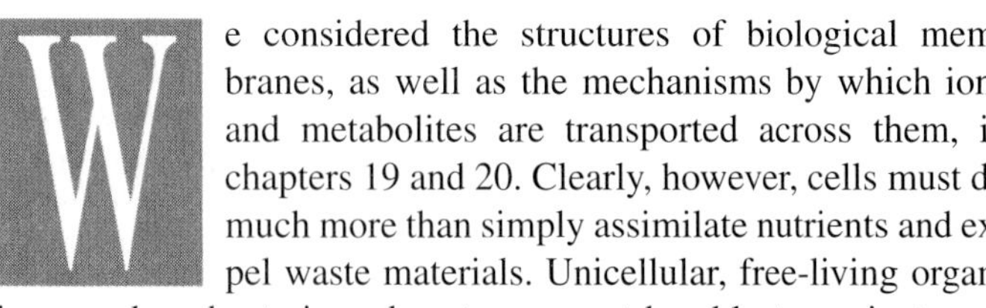

We considered the structures of biological membranes, as well as the mechanisms by which ions and metabolites are transported across them, in chapters 19 and 20. Clearly, however, cells must do much more than simply assimilate nutrients and expel waste materials. Unicellular, free-living organisms such as bacteria and protozoa must be able to navigate and successfully compete in environments that often contain many perils to their livelihood. Multicellular organisms have the additional problem of coordination and signaling among many diverse, differentiated cell types to ensure the efficient functioning of the organism as a whole. The necessary interactions of a cell with its environment and with other cells must obviously involve processes associated with the cytoplasmic membrane. In many cases, transmembrane transport is part of the signaling processes between cells and their environment.

In chapter 27 we dealt with one type of communication mechanism in animals. Hormones released in one part of the body interact with receptors, which are sometimes membrane-bound, to affect processes occurring in other cells that are often quite distant from the cells releasing the signal. This mechanism allows for efficient coordination of the wide variety of metabolic events continually taking place in higher organisms. Another, quite different phenomenon takes place in the cells of sensory organs—for example, in the retina cells of the eye of vertebrate animals. Light energy is converted by these cells into nerve impulses, which are translated by the brain into an image of our surroundings. This interconversion involves transmembrane movements of ions and will be considered in chapter 29.

In this chapter we will examine the mechanisms by which nerve impulses are propagated, to illustrate how excitable tissues function and interact in higher organisms. As we will show, the fundamental mechanisms revealed by studies of membrane transport in both prokaryotic and eukaryotic cells (chapter 20) can account in large part for the seemingly more complex phenomena of nerve-impulse propagation and transmission.

Nerve-Impulse Propagation

As early as the late eighteenth century, experiments by Galvani and Volta suggested that the transmission of nerve impulses and muscular contraction involved electrical signals. In 1902, J. Bernstein first proposed that the unequal distribution of K^+ across the nerve-cell membrane and the selective permeability of this membrane for the potassium ion were responsible for a resting potential known to exist in nerve and muscle fibers. He further believed that excitation of a nerve cell involved a transient collapse in this selective permeability such that other ions were able to penetrate the nerve-cell membrane and abolish the resting potential. If these changes in ion permeability could move down the axon of a nerve cell and be transmitted to other cells, they could provide the basis for propagation of nerve impulses.

In the 1930s the isolated squid giant nerve axon became available for experimentation, and its size made it especially amenable to electrophysiological measurements. Experiments pioneered by A. L. Hodgkin and A. F. Huxley soon established the essential ionic movements associated with impulse propagation, and the resultant local changes in membrane potential could be measured. It became clear that the transmembrane transport of ions was important in nerve cells for their signaling function and for the actual signal conductance itself. As we will see, two basic types of transmembrane transport proteins are intimately involved in these processes: ion channels, which participate in the propagation of the impulse down the axon, and neurotransmitter receptors (also ion channels), which are involved in the transmission of signals between nerve cells and their target cells.

An Unequal Distribution of Ionic Species Results in a Resting Transmembrane Potential

To understand how nerve impulses are generated along the axon of a nerve cell (fig. 28.1), we must first understand the basis for the electric potentials that exist across the neuronal membrane. An unequal distribution of ionic species across a biological membrane that is permeable to these molecules can result in a transmembrane electric potential, $\Delta\Psi$ (chapter 20). For a membrane system permeable to several ionic species, the numerical value of $\Delta\Psi$ can be approximated by the Goldman equation, derived by D. E. Goldman in 1943:

$$\Delta\Psi \text{ (``in'' relative to ``out'')} = \frac{2.3\,RT}{F}\log_{10}\left(\frac{\Sigma P_c[\mathrm{C}]_{\mathrm{out}} + \Sigma P_a[\mathrm{A}]_{\mathrm{in}}}{\Sigma P_c[\mathrm{C}]_{\mathrm{in}} + \Sigma P_a[\mathrm{A}]_{\mathrm{out}}}\right) \quad \textbf{(1)}$$

where C and A are univalent cations and anions, respectively, and P_c and P_a refer to their permeability coefficients* across the membrane of interest. Since multivalent ions are generally not quantitatively significant in contributing to $\Delta\Psi$ in resting neuronal membranes, we usually ignore them in calculating $\Delta\Psi$. If the membrane is selectively permeable to one ion only, for example C, equation (1) reduces to the familiar Nernst equation:

$$\Delta\Psi = E_c = \frac{2.3\,RT}{F}\log_{10}\frac{[\mathrm{C}]_{\mathrm{out}}}{[\mathrm{C}]_{\mathrm{in}}} \quad \textbf{(2)}$$

where E_c refers to the equilibrium electric potential of C.

It is the unequal distribution of protons and other cations that gives rise to transmembrane potentials that can drive ATP synthesis and secondary active transport in many cells (chapter 20). In nerve cells a similar situation exists as summarized in table 28.1. Thus the external environment of the nerve cell contains a high concentration of sodium ions and a low concentration of K^+, while the reverse is true for the cytoplasm of the nerve cell, called the axoplasm. Furthermore, the extracellular concentration of Cl^- is 5 to 10 times that of the axoplasm. A number of other impermeant anions—largely organic molecules, proteins, and nucleic acids—maintain an approximate charge neutrality in the axoplasm. Resting nerve cells are much more permeable to K^+ than to Na^+, and it is this selective permeability that allows an electric potential to develop across the membrane in the presence of a K^+ concentration gradient. It is, of course, the role of Na^+–K^+ ATPase to maintain these unequal distributions of cations across the neuronal membrane, as we discussed in chapter 20.

*The permeability coefficient is equal to the diffusion coefficient D divided by the width of the membrane l.

Figure 28.1

Schematic diagram of a typical motor neuron (nerve cell conducting impulses to muscle cells). All neuronal types, regardless of their shape, size, location, and function, exhibit four anatomically distinct compartments, each of which has a distinctive role in signaling: the cell body, dendrites, an axon, and the nerve terminals. The dendrites and cell body make up the input component of the neuron; the axon is the long-range signaling component, and the presynaptic terminal is the output component.

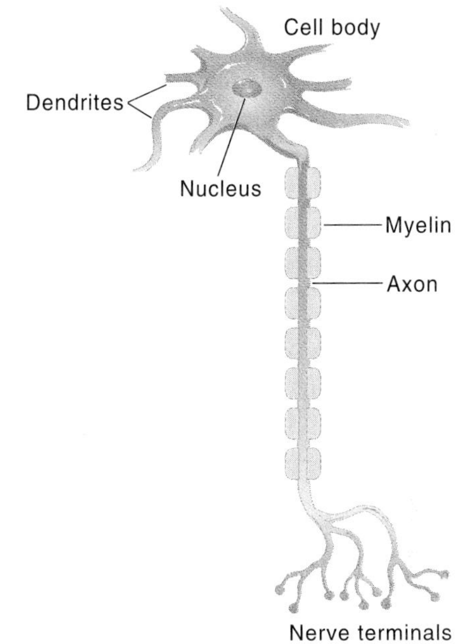

Table 28.1
Ionic Concentrations Inside (Axoplasm) and Outside (Blood) the Squid Giant Axon

Ion	Inside (mM)	Outside (mM)
Na^+	50	440
K^+	400	20
Cl^-	40–150	560

Source: Data from S. W. Kuffler and J. G. Nicholls, *From Neuron to Brain,* Sinauer Associates, Sunderland, Mass., 1976.

To calculate the resting membrane potential $\Delta\Psi$ across the axonal membrane from equation (1), we can use the values in table 28.1 and the permeabilities of Na^+ and Cl^- relative to K^+ (0.04 and 0.45, respectively), assuming an intracellular Cl^- concentration of 50 mM:

$$\Delta\Psi = 60 \log \left(\frac{20 + 0.04(440) + 0.45(50)}{400 + 0.04(50) + 0.45(560)} \right) = -62 \text{ mV} \quad \textbf{(3)}$$

This value is close to the experimentally measured resting potential across a squid axonal membrane.

An Action Potential Is the Transient Change in Membrane Potential Occurring during Nerve Stimulation

The use of giant axons from squid nerve cells in electrophysiological experiments has greatly aided our understanding of the electrical events that take place during nerve stimulation. An experimental apparatus for measuring changes in the potential across the membrane of such an axon, which has a diameter of approximately 0.5 mm, is schematically illustrated in figure 28.2. The apparatus consists of a pair of stimulating electrodes connected to a current source and a second pair of recording electrodes located slightly farther down the axonal segment. The latter electrodes are connected to a sensitive recording device, such as an oscilloscope, and can be used to measure time-dependent changes in the membrane potential.

Initially, the potential measured in the system shown in figure 28.2 is about −60 mV, that is, the resting membrane potential (fig. 28.3). If a brief current is applied at the stimulating electrodes, a time-dependent change in the membrane potential may be recorded on the oscilloscope, as shown in figure 28.3. This so-called action potential occurs only if the stimulus is sufficient to depolarize the membrane by about 20 mV (i.e., to about −40 mV). Weaker stimuli give small local potential changes, while current pulses greater than this threshold value give a curve similar in shape and height to that shown in figure 28.3, independent of the magnitude of the stimulus. During the development of the action potential the value of $\Delta\Psi$ across the axonal membrane rises in about 1 ms to nearly +40 mV. This is followed by a somewhat slower return to the resting potential, during which time the membrane potential drops transiently below the resting value, to about −75 mV. This value, which is close to the one predicted by the Nernst equation if the membrane is permeable only to K^+ (the potassium equilibrium potential), marks the state referred to as hyperpolarization.

Classic experiments by A. L. Hodgkin and A. F. Huxley established that the changes in membrane potential occurring during nerve stimulation are due to transient changes in the permeability of the membrane to Na^+ and K^+ ions. As illustrated in figure 28.3, the rapid rise in $\Delta\Psi$ to a positive value is accompanied by a large increase in the relative permeability of Na^+, while the return to resting potential is correlated with the inactivation of Na^+ permeability and a transient increase in K^+ permeability. An important conclusion from these observations is that the permeabilities of the axonal membrane to Na^+ and K^+ depend on the mem-

Figure 28.2

A device for eliciting and recording action potentials along the squid giant nerve axon. Brief closure of the switch connected to the stimulating electrode causes a current pulse into the axoplasm. If an impulse is generated, resultant potential changes can be detected by the recording electrode, which is connected to an oscilloscope or other recording device.

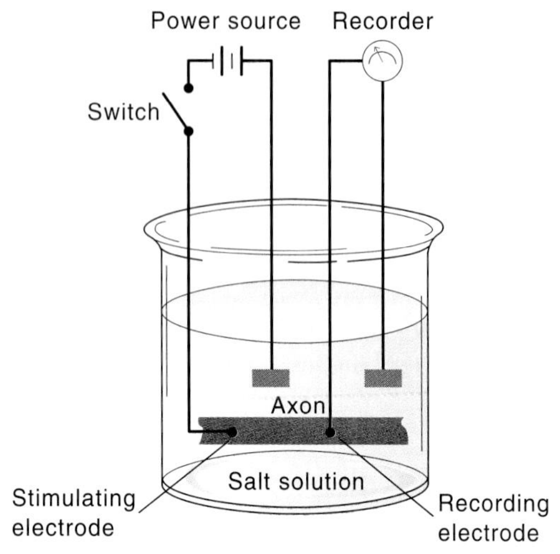

Figure 28.3

A typical action potential V (blue) that might be recorded by the instrument in figure 28.2 if the stimulating current is sufficient to depolarize the membrane by at least 20 mV. The membrane potential returns to its resting value of about −60 mV within about 5 ms. Accompanying changes in relative Na^+ permeability, g_{Na+} (red), and K^+ permeability, g_{K+} (green), are also shown. Depolarization is correlated with an increase in g_{Na+}, while repolarization is accompanied by a decrease in g_{Na+} and a transient increase in g_{K+}. Note that the membrane potential transiently becomes more negative than the resting value (hyperpolarizes) until g_{K+} returns to its normal value. The time of appearance of the action potential after stimulation at $T = 0$ depends on the distance between the recording and stimulating electrodes. (*Source:* S. W. Kuffler and J. G. Nicholls, *From Neuron to Brain.* Copyright © 1976 Sinauer Associates, Sunderland, Mass.)

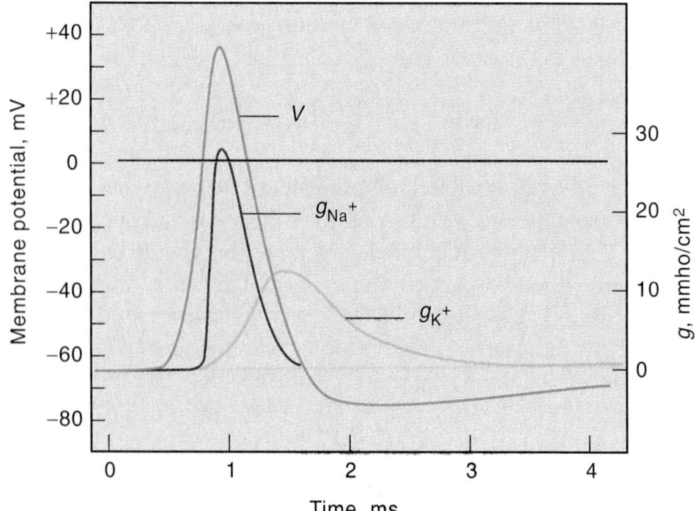

brane potential. Thus depolarization above the threshold, leading to a more positive $\Delta\Psi$, first leads to an increased permeability of Na^+ followed by inactivation of this phenomenon and an increase in the membrane permeability of K^+. The latter events tend to hyperpolarize the membrane, increasing the negative $\Delta\Psi$ and decreasing the Na^+ permeability.

As we mentioned in the preceding section, the membrane potential depends on the relative permeabilities and concentration gradients of electrolytes across the membrane [equation (1)]. The resting potential is largely dependent on the K^+ gradient, since unstimulated nerve membranes have a high permeability only to this cation. At the height of depolarization, however, the membrane is much more permeable to Na^+ than to K^+. Because the Na^+ gradient is opposite to that of K^+, Na^+ influx rapidly causes the membrane potential to become positive, approaching but never reaching the value it would have if the membrane were permeable only to Na^+. You can see this if you substitute the values of $[Na^+]_{in}$ and $[Na^+]_{out}$ into equation (2). You will get a value of 57 mV for $\Delta\Psi$, assuming that the membrane is permeable only to Na^+. In reality, this value is never attained because an increased permeability of the membrane to K^+ follows the change in Na^+ permeability, which itself is only transient (see fig. 28.3).

The ionic movements leading to an action potential can therefore be summarized as follows:

1. Stimulation leads to an influx of Na^+ into the axoplasm, down its electrochemical gradient, owing to an increased permeability of the membrane to this cation.
2. The change in $\Delta\Psi$ resulting from this flow increases the membrane permeability of K^+, which flows out of the cell and down its electrochemical gradient; the result is the reestablishment of a negative $\Delta\Psi$, accompanied by an inactivation of the influx of Na^+.
3. The membrane potential eventually becomes sufficiently negative to return both K^+ and Na^+ permeabilities to their normal values, and $\Delta\Psi$ resumes its resting level.

The events shown in figure 28.3 record fluctuations in $\Delta\Psi$ and ionic currents at one point on the axonal membrane as a function of time during passage of an action potential through this point. The action potential, however, is conducted down the axon as a wave of depolarization–repolarization events through the following mechanism: Depolarization of a given area of the membrane causes current to flow in the axoplasm from the more positive (depolarized) region to neighboring regions. This flow, in turn, triggers an action potential across the neighboring section of the membrane, and so forth (fig. 28.4). Thus the action potential provides a mechanism whereby the transmitted signal is constantly amplified to maintain a constant amplitude. In a regular cable without amplification the propagated pulse decreases with distance as a result of resistance and leakage. Without the action potential a current pulse would therefore be reduced to an insignificant level after traveling a very short length along the axon. In nerve cells of invertebrate animals, as well as in many cells in vertebrates, nerve impulses are therefore conducted along axons and dendrites by these local currents and action potentials.

Figure 28.4

Nerve impulses in unmyelinated nerves are conducted through local current movements that propagate the action potential. A portion of the resting axonal membrane is shown at the top. Arrival of an action potential causes local depolarization of the membrane (*middle*), which is propagated from left to right by local currents shown by the arrows. (*Source:* From A. L. Hodgkin, *Proceedings. Royal Society of London. Series B. Biological Sciences* 148:1, 1957.)

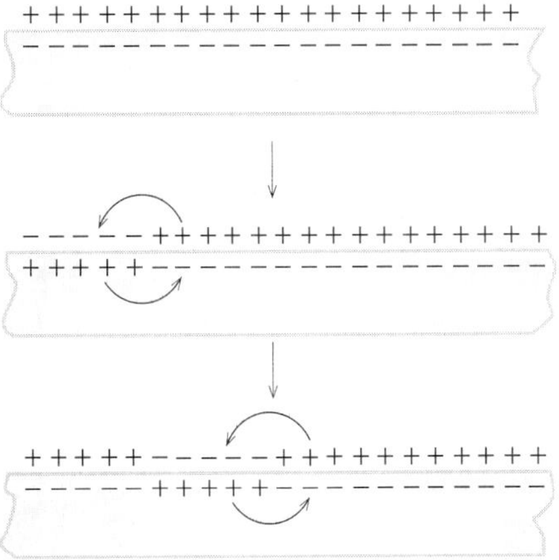

The axons of many nerve cells of higher animals, however, are also surrounded by a multilayered myelin sheath, each layer consisting of a typical lipid bilayer membrane (fig. 28.5). At intervals, the spacing of which depends on the fiber diameter, this myelin insulation is interrupted by the so-called nodes of Ranvier (fig. 28.6). Nerve impulses in these types of nerve fibers are conducted in a saltatory manner, that is, with action potentials "jumping" from node to node, where the axonal membranes are in direct contact with the extracellular fluid (see fig. 28.6). The insulation provided by the myelin sheath allows for efficient current conduction within the axoplasm by preventing signal loss, and as a result, this type of impulse propagation can be much more rapid than that observed in unmyelinated fibers of similar diameter. Indeed, impulse propagation velocities of over 100 m/s have been recorded in myelinated nerve fibers.

Gated Ion Channels

By a slight modification of the experimental setup shown in figure 28.2, it is possible to hold the membrane potential across an axonal membrane constant at any predetermined value. This is accomplished by connecting the recording electrodes to a device called a feedback amplifier. When any potential change is sensed by the electrodes, the feedback amplifier compensates for it by applying a current to keep the voltage constant. In such a voltage-clamped situation, ionic movements can be inferred from the amount of current necessary to hold $\Delta\Psi$ at its predetermined value. It was the use of a voltage clamp by Hodgkin and Huxley that allowed them to deduce the movements of Na^+ and K^+ that accompany the appearance of the action potential. Because the changes in membrane permeability to Na^+ and K^+ were not superimposable in time, they tentatively

Figure 28.5

Cross section of a myelinated nerve axon from the superior cervical ganglion of the rabbit. Note the multilayered membrane of the myelin sheath that serves as an electric insulator. (Micrograph courtesy of Dr. T. Lentz.)

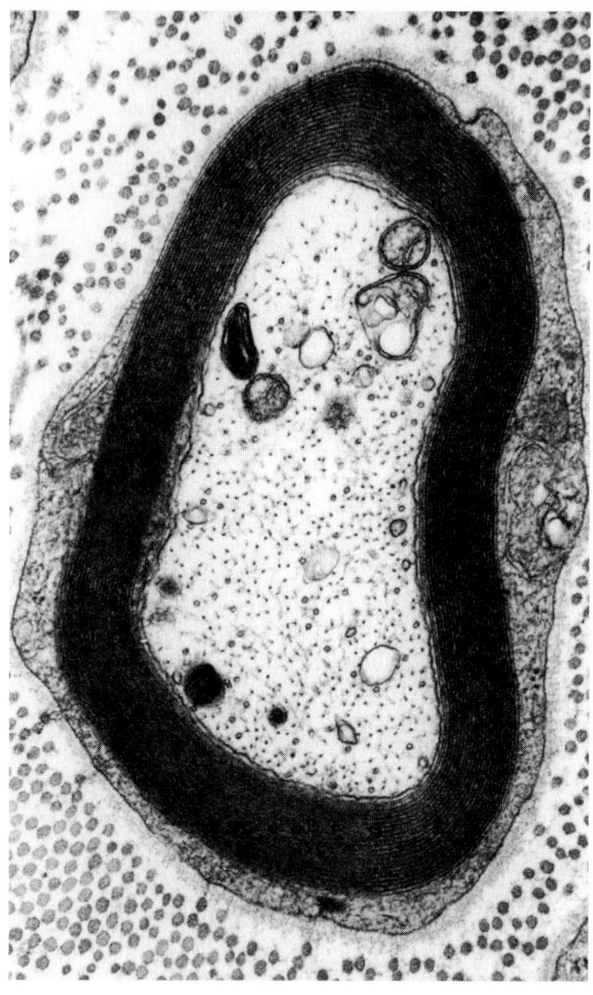

Figure 28.6

Schematic illustration of a longitudinal cross section of a myelinated axon. The sheath is interrupted at various intervals by the nodes of Ranvier, where the cytoplasmic membrane is exposed to the surrounding fluid. Nerve impulses are conducted in a saltatory manner in myelinated axons, as illustrated in this schematic diagram. Impulse propagation is from left to right, and the arrows show the accompanying local current movements.

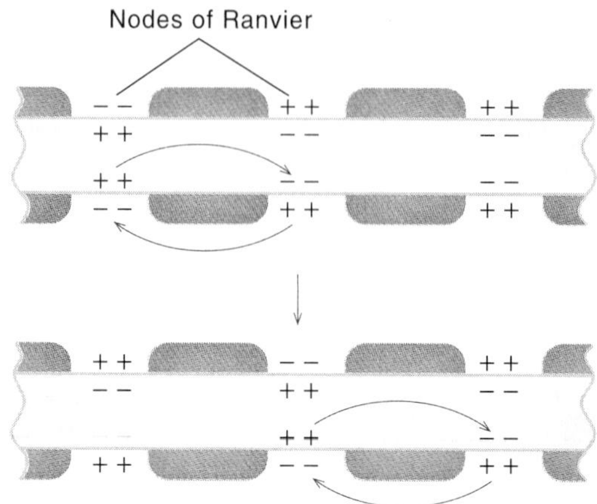

concluded that two different "channels" were involved in the transmembrane movements of these ions.

Separate Channels for Na^+ and K^+ Have Been Found in Excitable Cell Membranes

Compounds that specifically block the conductance of the nerve membrane to either Na^+ or K^+ provided the first direct evidence for separate Na^+ and K^+ channels in the nerve-cell membrane. Tetrodotoxin and saxitoxin (fig. 28.7) are both potent nerve poisons that specifically inhibit the transmembrane movement of Na^+ by binding to the outside surface of nerve-cell membranes without affecting K^+ permeability, as measured in voltage-clamp experiments. Similarly, tetraethylammonium ions (fig. 28.8) have been shown to bind to the axoplasmic membrane surface and to specifically block the outward flow of K^+.

A large body of evidence has accumulated that shows that K^+ and Na^+ movements through axonal membranes are mediated by separate channels that act as gated pores (chapter 20). During a typical action potential lasting about 5 ms, it has been calculated that at least 60 Na^+ ions pass through a single Na^+-specific channel (i.e., about 12,000 per second). Since the neuronal Na^+–K^+ ATPase normally pumps less than 200 Na^+ ions per second out of the cell, this pump cannot be directly involved in the development of the action potential. Instead, a process involving passive diffusion through a transmembrane pore best explains these rapid Na^+ fluxes. Furthermore, voltage-clamp studies by B. Hille established that molecules such as the K^+–monohydrate complex, with dimensions larger than about 0.4 nm^2 in cross section (the size of a monohydrated Na^+ ion), do not pass readily through the Na^+ channel. However, smaller molecules, such as monohydrated Li^+ ions, and other small cations such as hydroxylamine and hydrazine, do. This mechanism, that is, exclusion on the basis of size rather than chemical structure, is more characteristic of a pore than of a specific membrane permease (chapter 20).

Additional properties of the Na^+ channel have been inferred from many other experiments. For example, the pH dependence of Na^+ permeation, as well as other experimental observations, suggest that a negatively charged carboxylate group (—COO^-) is present in the pore and is involved in the interacting with the cationic molecules that can traverse the Na^+ channel. It is presumably this interaction that allows the positively charged guanidinium groups of tetrodotoxin and saxitoxin (see fig. 28.7) to bind to the channel, but the sizes of these poisons prevent their passage through the membrane, with the result that Na^+ conduction is blocked.

Estimates of the number of Na^+ channels in a variety of nerve types have been obtained by the use of radioactively labeled tetrodotoxin or saxitoxin molecules. From these studies, it is apparent that in unmyelinated nerve fibers, the density of Na^+ channels in the membrane is quite low. For example, the olfactory nerve of the garfish has only about 30 to 40 Na^+ channels per square

Figure 28.7

(*a*) The structure of tetrodotoxin, a compound that specifically blocks Na^+ channels in nerve cell membranes. This extremely toxic compound is found in the liver and ovaries of the Japanese puffer fish (*Spheroides rubripes*). (*b*) The structure of saxitoxin, a compound found in certain marine dinoflagellates ("plankton"), which are constituents of the so-called red tide. Mussels and clams that have fed upon these organisms are therefore extremely poisonous, and commercial shell fishing is banned in areas where these dinoflagellates appear. Saxitoxin is also a Na^+-channel blocking agent, and both compounds most probably interact with the channel through their positively charged guanidino groups.

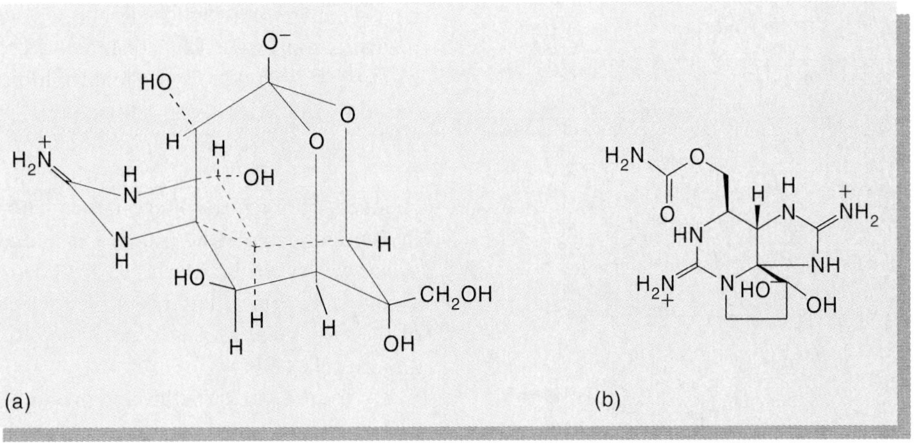

(a)

(b)

Figure 28.8

Tetraethylammonium ion, a compound that specifically inhibits K^+ channels in nerve cell membranes. The ethyl groups presumably sterically hinder passage of this cation through the channel, "plugging" the channel and thereby blocking the transport of K^+.

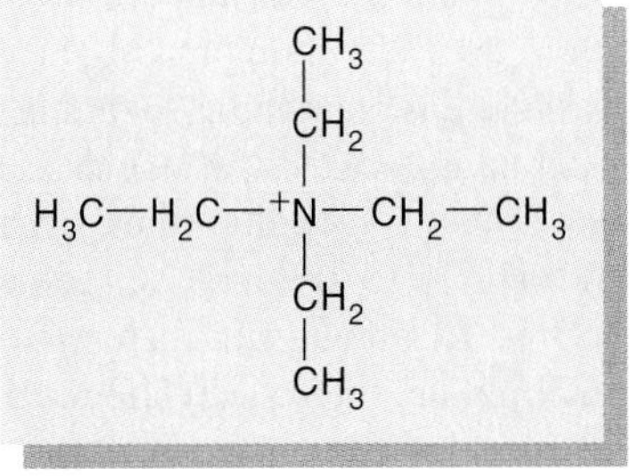

micrometer of membrane, which corresponds to only about 0.2% of the total surface area of the phospholipid bilayer. In the squid giant axon, which has a fiber diameter some 2,500 times that of the garfish olfactory nerve, this value is increased only to several hundred Na^+ channels per square micrometer, or about 2% of the surface area. The situation is quite different, however, in myelinated nerves of vertebrate animals. In this case, Na^+ channels are found in significant numbers only at the nodes of Ranvier, as we might anticipate from the mechanism of impulse propagation in these types of nerve cells (see fig. 28.6). Here, the channel density is on the order of 10^4 channels per square micrometer, or about 60% of the total membrane surface area. Indeed, Na^+ fluxes at the nodes of Ranvier have been estimated to be 10 to 100 times larger than those in unmyelinated axons during propagation of the action potential.

In contrast to the situation with Na^+ channels, there are many fewer K^+ channels in the axonal membrane (about one-tenth) the number in the squid giant axon). Nevertheless, it has been demonstrated that the K^+ pore is also quite specific, barring the passage of cations both smaller (Na^+) and larger (Cs^+; tetramethylammonium ions) than K^+. It seems likely, therefore, that both the diameter of the channel and specific interactions of channel components with the K^+–H_2O complex confer cation selectivity on this channel as well.

The Gating Properties of Ion Channels

How do channels selective for Na^+ and K^+ ions "sense" the membrane potential and react by opening or closing during various stages of action-potential propagation? Although this question is still far from being answered in molecular detail, a number of clues have emerged from work on the gating properties of these channels, their purification and reconstitution into artificial membrane systems, and the recent cloning and sequencing of genes encoding Na^+ and K^+ channels. The results of these investigations show that ion channels in nerve and muscle cell membranes can assume multiple conformations, one of which is more permeable to the ion in question than the others.

An important observation that bears on this question is the discovery of gating currents, first predicted by Hodgkin and Huxley. Under appropriate conditions a small current opposite in direction to that carried by Na^+ during the opening of the Na^+ channels can be detected just before the development of the action potential. This current is short in duration (0.1 ms) and precedes the inward movement of Na^+ (fig. 28.9). One widely accepted proposal is that the Na^+ channel has a "built-in" voltage sensor that has a large dipole moment. Depolarization of the membrane, which elicits the action potential, would displace or rearrange the dipole in response to the change in the electric field, and this process would be detected as the gating current. The change in the orientation of the dipole could then be the trigger

Figure 28.9

Illustration of the gating current that precedes the inward-directed Na^+ current associated with depolarization during the action potential. The gating current is believed to be related to the voltage-dependent opening of the Na^+ channels and perhaps may reflect rearrangement of a dipolar "voltage sensor" associated with the channel gate. (Source: Adapted from C. M. Armstrong and F. Bezanilla, Charge movement associated with the opening and closing of the activation gates of the Na channels, *J. Gen. Physiol.* 63:533, 1974.)

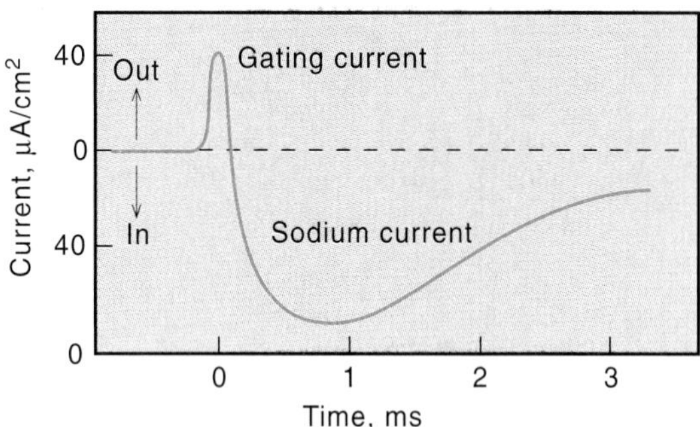

by which the Na^+ channel is opened, presumably by means of a conformational change in this protein. Indeed, a candidate for such a voltage sensor in both Na^+ and K^+ channels has recently been recognized from the primary structures of these proteins (see the following discussion).

Additional evidence for voltage-dependent conformational changes that allow the gating of the Na^+ channel has been obtained by using nerve poisons from North African scorpions and sea anemone. These toxins, which are basic polypeptides, bind to sites on the Na^+ channel that are distinct from those that bind the channel-blocking agents tetrodotoxin and saxitoxin. They appear to exert their physiological effects by slowing inactivation of the Na^+ channel after its initial opening during the action potential. Binding of scorpion toxin to Na^+ channels in nerve and muscle-cell membranes has been shown to be voltage-dependent, a fact suggesting that this toxin recognizes a conformation of the channel that also depends on the membrane potential (the "open" conformation). Toxin binding would thus reflect the conformational state of the channel, which, in turn, depends on $\Delta\Psi$. These and other experiments have led to the conclusion that both the Na^+ and K^+ channels can exist in at least three conformations: closed (resting $\Delta\Psi$), open, and inactive. The latter two states are responsible for the transient increase and then decrease in Na^+ and K^+ permeabilities, in sequence, observed during the action potential.

Further Evidence Concerning the Structure and Function of Ion Channels

One major approach to understanding structure–function relationships in the ion channels of nerve-cell membranes has been to purify these proteins and to study their properties in artificial membrane systems. The ability of Na^+ channels to tightly bind specific neurotoxins has been used as an assay in such purifications. Na^+ channels have been purified from the electric organ of the electric eel (*Electrophorus electricus*), which is rich in such channels, as well as from a number of higher vertebrate sources. A major component of all these preparations was a protein with a polypeptide-chain molecular weight of about 260,000, while variable amounts of a second polypeptide (M_r = ca. 37,000) were also found in some preparations. These two subunits have been referred to as the α and β subunits of the Na^+ channel, respectively.

Several highly purified preparations of the Na^+ channel have been reconstituted into proteoliposomes. In these experiments, reconstituted Na^+ channels were activated by channel activators and blocked by channel inhibitors such as saxitoxin and tetrodotoxin. Moreover, the cation specificity of reconstituted Na^+ channels was the same as that for the native channel: $Li^+ = Na^+ > K^+ > Rb^+ > Cs^+$. Further, in their purified form these channels retained a voltage dependency similar to that observed in the cells from which they were obtained. Since most of these properties could be attributed to the α subunit in many of these systems, it appears that this polypeptide alone is sufficient for most of the biological functions attributed to the Na^+ channel in nerve and muscle cells.

A major breakthrough in attempts to understand the structure and function of ion channels of excitable cells came with the cloning and sequencing of the cDNA that encodes the α subunit of the Na^+ channel from *Electrophorus electricus* by S. Numa and collaborators in 1984. The deduced amino acid sequence showed that this protein consists of 1,820 amino acid residues and has several remarkable features. First, the protein consists of four highly homologous domains, each of which may span the membrane six times, with both the N- and C-terminal regions on the cytoplasmic surface of the membrane (fig. 28.10). The membrane-spanning regions include both potential hydrophobic α helices (regions 1, 2, 5, and 6 in figure 28.10) and amphipathic α helices (regions 3 and 4). Strikingly, potential membrane-spanning region 4 (S4) of all four domains contains a stretch in which every third aminoacyl residue is basic (either lysine or arginine). A number of workers have proposed that this region may be the positively charged voltage sensor that moves outward in response to depolarization of the nerve-cell membrane, thus producing the gating current. In contrast, proposed membrane-spanning region 3 (S3) can form a highly amphipathic helix with a net negative charge. Synthetic peptides containing the amino acid sequence of S3 have been shown to form ion-selective channels in phospholipid bilayer membranes. Moreover, computer-assisted three-dimensional structural predictions have shown that four S3 peptides could form a tetrameric transmembrane pore with hydrophilic side chains lining this channel (fig. 28.11). The dimensions of the channel are strikingly similar to those deduced for the native Na^+ channel on the basis of its size selectivity. The negatively charged, acidic residues inside and at the entrances to the pore could account for its cation selectivity and also for the results of the pH-dependence studies of Na^+-channel function mentioned earlier. Additionally, the loop between S5 and S6 has recently been shown to have a role in channel function and may therefore comprise part of the pore. However, whether each of the four homologous domains of the Na^+ channel, or a complex including, for example, the S3 region and the S5/S6 loop from each of these domains, forms the

Figure 28.10

A functional map of the Na^+ channel α subunit. The transmembrane folding model of the α subunit is depicted with experimentally demonstrated sites of cAMP-dependent phosphorylation (P), interaction of site-directed antibodies that define transmembrane orientation (≻), covalent attachment of α-scorpion toxins (ScTx), glycosylation (ΨΨ), and modulation of channel inactivation (h). (Reprinted with permission from W. A. Catterall, Structure and function of voltage-sensitive ion channels, *Science* 242:50, Oct. 7, 1988. Copyright 1988 American Association for the Advancement of Science.)

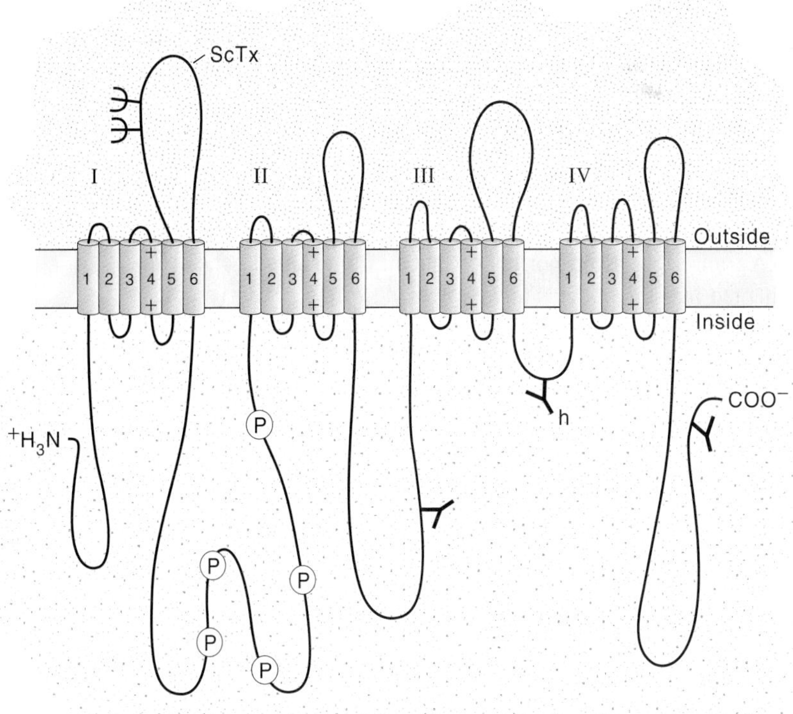

Figure 28.11

End view of energy-optimized parallel tetramer of the sodium channel S3 segment. Colors: light blue, α-carbon backbone; red, acidic residues; blue, basic residues; yellow, polar and neutral residues; and purple, lipophilic residues. This view is from the intracellular face of the membrane, at which the N-terminus of each segment is predicted to be located. From *Proteins: Structure, Function and Genetics* 8:226–236, ©1990. Reprinted by permission of John Wiley & Sons. Photograph courtesy of M. Montal.

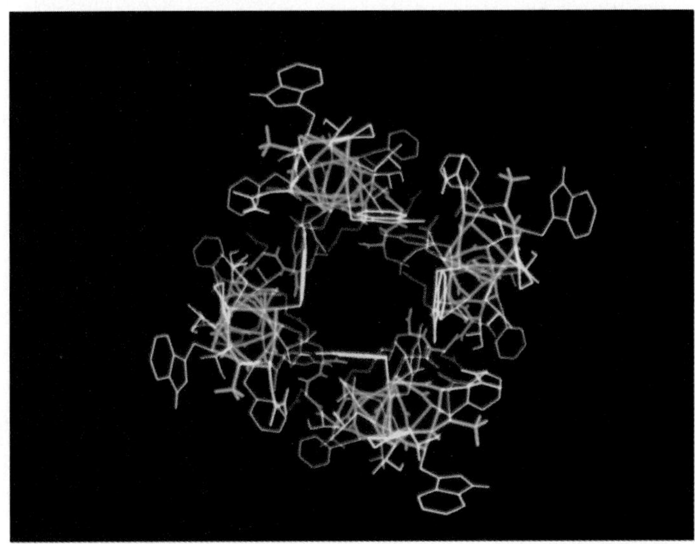

functional channel remains to be determined. Finally, a region between homologous domains I and II that is exposed in the cytoplasm has been found to be phosphorylated in a cyclic-AMP-dependent manner (see fig. 28.10). Phosphorylation of Na^+ channels inhibits Na^+ influx and therefore may have a regulatory role in excitable cells.

Genes encoding Na^+ channels from higher organisms so been cloned and sequenced, and they have a structure that is highly homologous to the one we have just described. More recently, the gene corresponding to the *Shaker* locus of *Drosophila* has also been cloned and sequenced. True to its name, a mutation in this locus causes severe neurological malfunction in the affected fruit fly, which is due to a defect in one type of K^+ channel. The deduced amino acid sequence of this K^+ channel revealed a 616-residue polypeptide that showed significant primary and secondary structural homology to each of the four domains of the Na^+ channel. Furthermore, models of the intramembrane structure of this protein were very similar to that of a single Na^+ channel domain, complete with approximately six membrane-spanning regions, an S4-like voltage sensor domain, and an S3-like domain with an enrichment of negatively charged residues. These properties of the K^+ channel from *Drosophila,* and the fact that even unicellular eukaryotes, such as *Paramecium,* and bacteria, such as *E. coli,* have voltage-sensitive K^+-selective channels, have led to the suggestion that all such channels may have had a common evolutionary ancestor that was similar in structure to the present-day K^+ channels. Presumably, tandem duplication events involving such an ancestral gene could have given rise to the present-day Na^+ channel gene, consisting of four tandem segments that are highly homologous to each other and to the K^+-channel gene.

Very recently, the powerful tool of site-directed mutagenesis has been used to study structure–function relationships in ion channels. Such studies have been aided by the development of techniques to study specific mutant proteins expressed in a model system, oocytes from South American frogs of the genus *Xenopus*. These relatively large cells are easily manipulable, and mRNA transcribed from normal or mutant ion channel genes can be injected into the oocytes and expressed. Voltage-clamp studies of *Xenopus* oocytes, into which mRNAs encoding various ion channels have been injected, have shown that the channel proteins are expressed in a functional form in the oocyte membrane. Thus far, use of this technique has yielded its greatest benefits in the area of understanding the mechanism of ion channel inactivation.

Experiments by R. W. Aldrich and co-workers have provided strong evidence that ion channel inactivation may operate by a relatively simple "ball-and-chain" mechanism (fig. 28.12). These workers studied a number of site-directed mutants of the *Drosophila Shaker* gene for their inactivation properties after injecting the corresponding mRNA for the gene into *Xenopus* oocytes. Previous experiments had suggested that the extreme N terminus of the *Shaker* gene product (K^+ channel) may be involved in inactivation, so the focus was on mutations in this region. The results of the experiments can be summarized as follows:

1. Mutations within the first 11 amino acid residues of this protein that resulted in hydrophobic to polar substitutions gave a K^+ channel in which inactivation was slowed relative to the wild-type protein. Furthermore, mutations that neutralized or eliminated positively charged residues at positions 14, 16, 17, 18, and/or 19, immediately following the hydrophobic N terminus, also slowed the inactivation rate of the channel.
2. A region immediately C-terminal to the positively charged cluster of amino acids (residues 23–37) was also studied. This region is rich in hydrophilic amino acids, and when deletion mutations were made in this region, inactivation was faster than in the wild-type protein. Conversely, insertion mutations in this region slowed inactivation of the channel.
3. Finally, a deletion mutant that lacked most of the first 20 aminoacyl residues failed to inactivate, as we would expect from the previous results. However, inactivation could be restored by adding back a synthetic peptide that had the same sequence as the amino-terminal 20 amino acid residues of the native *Shaker* channel. The degree of restoration of inactivation was directly proportional to the amount of peptide added.

The simplest interpretation of these results is that the first 20 amino acid residues of this K^+ channel comprise a "ball," with the positively charged residues on the surface and the hydrophobic residues forming the core. It is this structure that is responsible for K^+ channel inactivation by virtue of its binding to the inside surface of the pore, presumably at least partially by interacting with negatively charged residues near the pore opening. Mutations that disrupt this structure disrupt inactivation. The next 20 residues or so are hypothesized to be a "chain" that can "swing" the ball into or out of the pore opening. Shortening the chain by deletion mutation brings the ball closer to the opening and thus speeds up inactivation, while the reverse is true for insertion mutations. However, the free ball can inactivate a mutant protein lacking an attached ball, as long as it is provided in a high enough concentration. Perhaps the blocking effect of the triethylammonium cation, which we mentioned earlier, is simply due to its mimicking the positively charged region of the inactivation "ball." Figure 28.12 is thus a reasonable schematic model for how inactivation may work, at least in the case of one ion channel. Application of similar molecular biological techniques to the cloned genes of other ion channels will no doubt cast further light on structure–function relationships in these ubiquitous proteins.

Figure 28.12

"Ball-and-chain" model for inactivation of the K^+ channel. (*a*) Open conformation, (*b*) inactive conformation. (*Source:* From "Research News," *Science* 250:507, 1990. Reprinted by permission of Suzanne Black.)

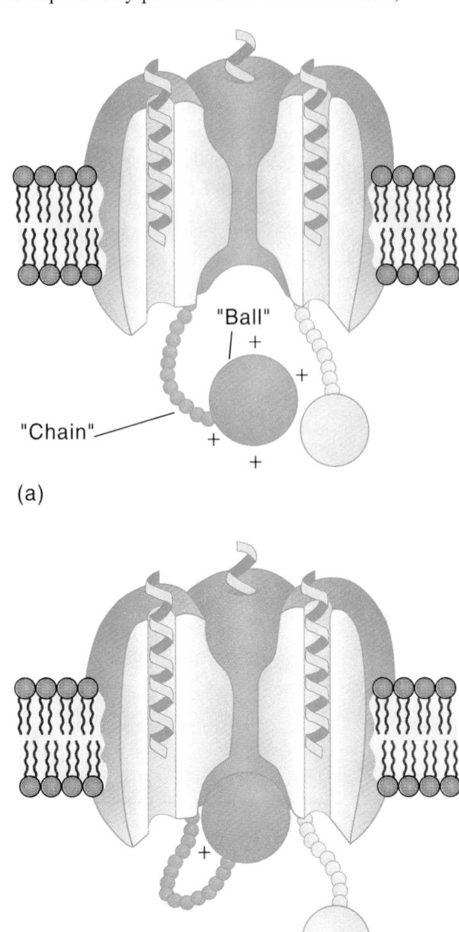

Figure 28.13

Schematic diagram of a synaptic junction in which acetylcholine is the chemical transmitter. Arrival of an action potential at the terminus of the presynaptic cell (*top*) stimulates Ca^{2+} uptake, which triggers release of acetylcholine (ACh) from vesicles near the terminus of the presynaptic cell. Release is accomplished by vesicle fusion with the plasma membrane. The interaction of acetylcholine with its receptors in the postsynaptic membrane triggers depolarization, thus initiating an action potential in the postsynaptic cell (*bottom*).

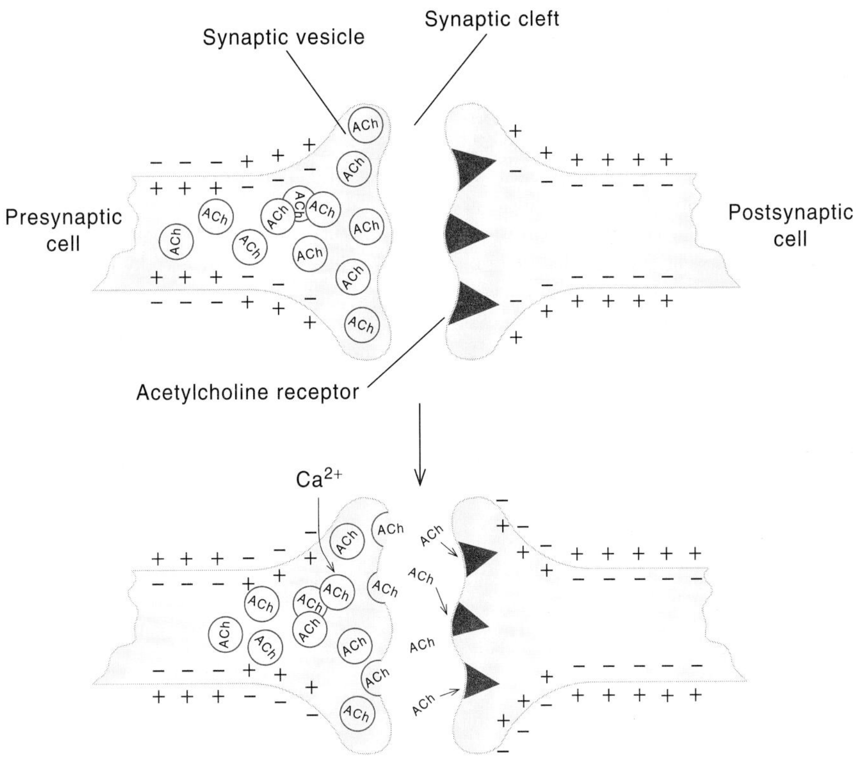

Synaptic Transmission: A Chemical Mechanism for Communication between Nerve Cell and Target Cell

Nerve impulses, propagated along the axon by mechanisms outlined in the preceding sections, must be transmitted between nerve cells or from nerve cells to muscle or glandular tissues in order for their effects (e.g., contraction or secretion) to take place. This type of intercellular communication can occur either by an electrical mechanism of coupling or by chemical transmission across specific connections, called synapses, between nerve cells and target cells. An example of electrical transmission was given in chapter 20. Direct cell-to-cell contact by means of gap junctions can allow action potentials to be transmitted between cells by ionic mechanisms. In another mechanism, specific chemicals called neurotransmitters are released at the presynaptic membrane of one cell in response to membrane depolarization. These chemicals then diffuse to receptors in the postsynaptic membrane of the recipient cell, where a new action potential may be generated.

Acetylcholine Is a Common Chemical Neurotransmitter

The process of synaptic transmission is depicted schematically in figure 28.13 for acetylcholine, the excitatory neurotransmitter at vertebrate neuromuscular junctions (motor end plates). Acetylcholine is synthesized in nerve cells by the enzyme cholineacetyltransferase,

$$H_3C-\overset{\overset{CH_3}{|}}{\underset{\underset{CH_3}{|}}{N^+}}-CH_2CH_2-OH + \text{Acetyl-CoA} \xrightarrow{\text{Cholineacetyltransferase}}$$

$$H_3C-\overset{\overset{CH_3}{|}}{\underset{\underset{CH_3}{|}}{N^+}}-CH_2CH_2-O-\overset{\overset{O}{\|}}{C}-CH_3 + \text{CoA}-\text{SH} \qquad \textbf{(4)}$$

and is packaged in units of 10^3 to 10^4 molecules within synaptic vesicles, which are abundant near the cytoplasmic membrane of the presynaptic axon. The arrival of an action potential triggers a large increase in the permeability of the presynaptic membrane to Ca^{2+}, and this ion flows into the axoplasm down its chemical gradient. Fusion of the synaptic vesicles with the plasma membrane and the concomitant release of acetylcholine into the synaptic cleft are promoted by this increase in intracellular Ca^{2+} (box 28A). In a typical neuromuscular synaptic junction, several hundred synaptic vesicles empty their contents into the synaptic cleft by this mechanism in response to a single action potential. The resultant large increase in the local concentration of acetylcholine is

The Synaptic Vesicle Leads to Regeneration of Neurotransmitter-Filled Vesicles

Once the synaptic vesicle has released its neurotransmitter, a new cycle is begun to regenerate fresh vesicles. According to T. C. Südhof, the cycle can be divided into nine steps, as shown in figure 1: (1) docking of the vesicle at the plasma membrane, (2) priming so that the vesicle is competent for fast Ca^{2+}-triggered fusion and exocytosis, (3) fusion and exocytosis, (4) endocytosis of the empty synaptic vesicle membranes, (5) translocation of the vesicles, (6) endosome fusion, (7) regeneration of synaptic vesicles by budding from endosomes, (8) accumulation of neurotransmitters by active transport, and (9) translocation of synaptic vesicles filled with neurotransmitters back to the plasma membrane.

There are many processes that involve the shuttling of membranes between the cell surface and endosomes. Of these the synaptic vesicle cycle is unique in its speed and tight regulation. Exocytosis and fusion occur in less than 1 ms, endocytosis in less than 5 s, and the remaining steps of the cycle in about 55 s. The vesicles are associated with two functional classes of proteins: transport proteins involved in the uptake of neurotransmitters and trafficking proteins that mediate synaptic vesicle–membrane docking, fusion, and budding.

Figure 1

The nine steps of the synaptic vesicle (SV) cycle in the presynaptic nerve terminal.

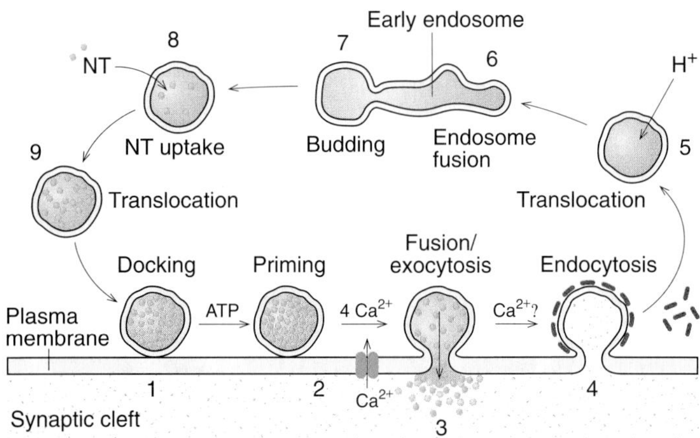

"sensed" by a protein, the acetylcholine receptor, located in the cytoplasmic membrane of the postsynaptic cell. Binding of the neurotransmitter to many receptor molecules triggers an action potential in the recipient cell. The acetylcholine is then rapidly degraded into acetate and choline by an enzyme in the synaptic cleft called acetylcholinesterase

$$H_3C-\overset{CH_3}{\underset{CH_3}{\overset{|}{\underset{|}{N^+}}}}-CH_2CH_2-O-\overset{O}{\overset{\|}{C}}-CH_3 + OH^- \xrightarrow{\text{Acetylcholinesterase}}$$

$$H_3C-\overset{CH_3}{\underset{CH_3}{\overset{|}{\underset{|}{N^+}}}}-CH_2CH_2-OH + CH_3-\overset{O}{\overset{\|}{C}}-O^- \quad \textbf{(5)}$$

and the resting potential of the postsynaptic membrane is soon restored.

How does the binding of a neurotransmitter to its receptor promote depolarization of the postsynaptic membrane? One of the first clues came from studies by Bernard Katz and collaborators, who showed that acetylcholine increases the ionic permeability of the postsynaptic membrane. Subsequent studies with radioactive tracers and by the voltage-clamp technique established that the membrane permeabilities to both Na^+ and K^+ were increased simultaneously by the action of acetylcholine. Because in the resting cell K^+ permeability is already quite high relative to Na^+ permeability, the major effect of acetylcholine binding and receptor channel opening is a substantial influx of Na^+ down its concentration gradient. The inward Na^+ current tends to collapse (or depolarize) the negative resting potential toward zero. This lo-

Figure 28.14

Acetylcholinesterase inhibitors. (*a*) Eserine, or *physostigmine,* is an alkaloid that forms a relatively stable covalent carbamoyl intermediate with an active-site serine on the enzyme that is hydrolyzed only very slowly. (*b*) Parathion and malathion are organophosphorus compounds that also form stable covalent complexes with the active-site serine of acetylcholinesterase. They are widely used agriculturally as insecticides.

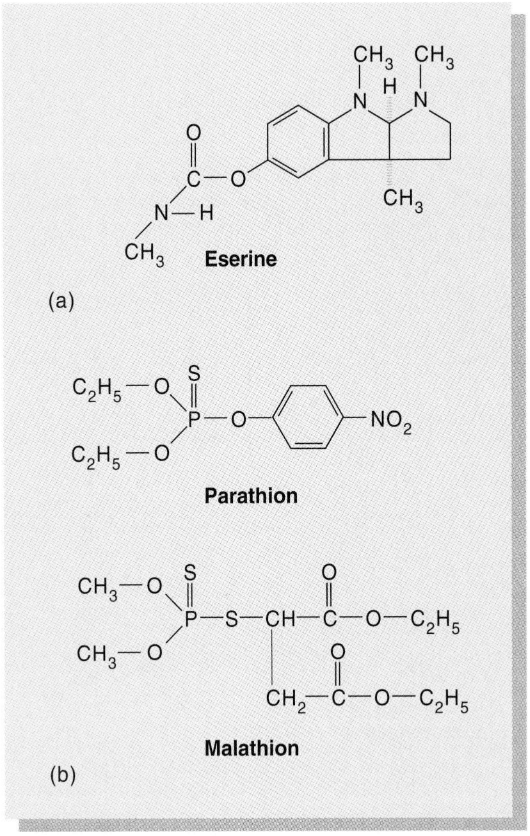

(a) **Eserine**

(b) **Parathion** **Malathion**

Figure 28.15

The structure of *d*(+)-tubocurarine, an active component of the neurotoxin curare. This compound binds to the acetylcholine-binding site on its receptor, preventing synaptic transmission and subsequent depolarization of the postsynaptic cell membrane.

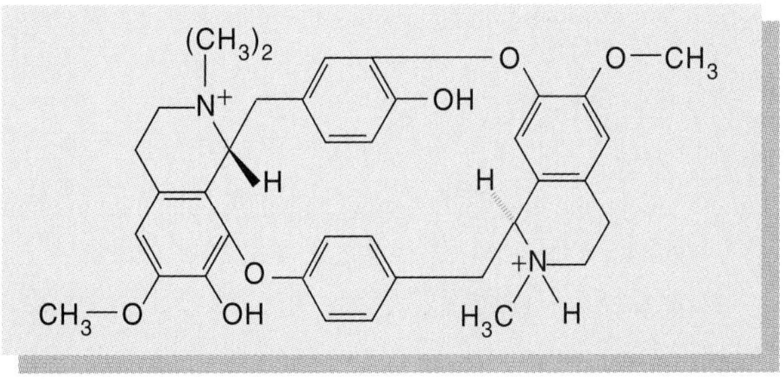

Figure 28.16

Decamethonium ion, an agonist of cholinergic systems. This compound binds to the acetylcholine receptor, but because it cannot be degraded, it causes persistent depolarization of the postsynaptic membrane.

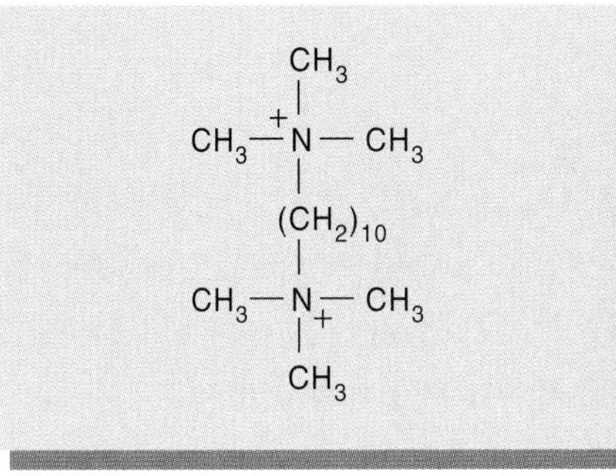

cal perturbation in $\Delta\Psi$ is enough to initiate a new action potential in the recipient neuronal or muscular membrane as long as a sufficient number of receptor molecules bind the neurotransmitter. In fact, the receptor itself contains the ion channel through which both Na^+ and K^+ can flow. The number of occupied receptors at a given moment thus dictates the magnitude of the inward flow of Na^+ and the resulting magnitude of the change in the membrane potential. Rapid hydrolysis of acetylcholine by acetylcholinesterase bound to the postsynaptic membrane then quickly reduces the number of transmitter–receptor complexes and repolarizes the membrane until a new action potential triggers the release of more neurotransmitter from the presynaptic membrane.

To elucidate the steps just outlined, researchers have made use of specific inhibitors of both acetylcholinesterase and the acetylcholine receptor. Compounds such as eserine (fig. 28.14*a*) block the hydrolytic enzyme and thus can be used to study the effects of acetylcholine in cholinergic systems (those using acetylcholine as a transmitter) under conditions where it cannot be hydrolyzed. Likewise, certain organophosphorus compounds efficiently inhibit acetylcholinesterase by forming stable covalent intermediates with an active-site serine in the enzyme. The widely used insecticides parathion and malathion are examples of this class of nerve poisons (fig. 28.14*b*). Neuromuscular junctions exposed to acetylcholinesterase inhibitors are paralyzed because the persistent presence of acetylcholine prevents repolarization of the postsynaptic membrane to restore its excitability. In fact, in such a situation the acetylcholine receptor eventually becomes desensitized, that is, it remains closed to ion flow for long intervals even in the presence of the neurotransmitter.

Specific blocking agents of the acetylcholine receptor include *d*-tubocurarine, an active component of neurotoxin curare (fig. 28.15), and the snake venom poisons α-bungarotoxin (from snakes of the genus *Bungarus*) and cobratoxin. The latter are small basic proteins with masses around 7,000 daltons. All three of these substances interact noncovalently with the receptor and interfere with acetylcholine binding, thus blocking depolarization of the postsynaptic membrane. They are referred to as antagonists of the cholinergic systems. Another type of acetylcholine receptor inhibitor is exemplified by the divalent cation decamethonium (fig. 28.16), which "locks" the ion channel of the receptor in the open state and thus leads to a constant depolarization of the recipient cell membrane. Such compounds, referred to as agonists, mimic

Figure 28.17

Biosynthesis from L-tyrosine and inactivation by monoamine oxidase of catecholamine neurotransmitters. Norepinephrine, epinephrine, and dopamine are confirmed neurotransmitters in various systems, and L-DOPA is a probable neurochemical messenger (see table 28.2).

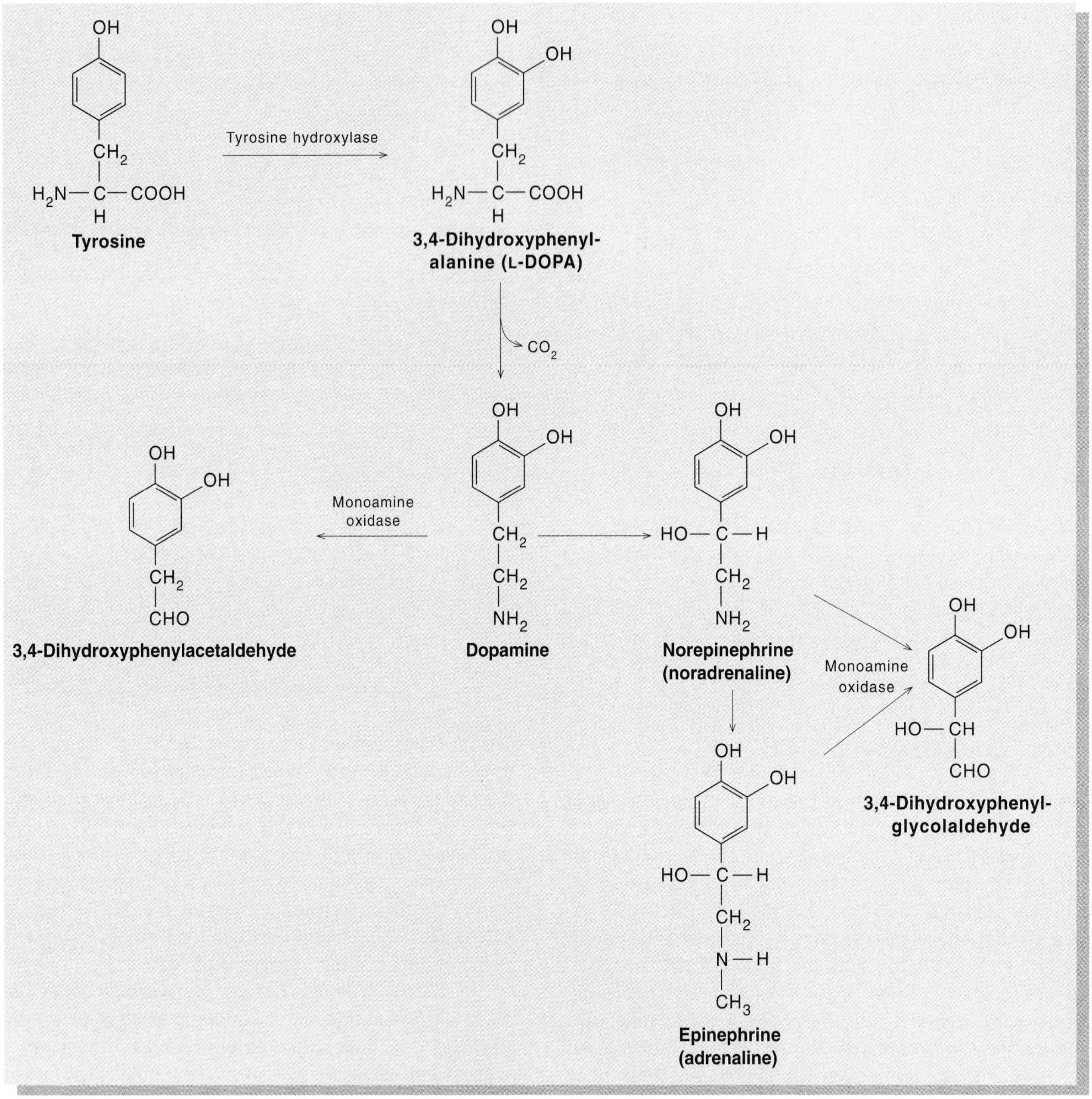

the effect of acetylcholine but cannot be rapidly inactivated, so they block resensitization of the postsynaptic membrane. With agonists and antagonists as tools, workers have been able to investigate properties of the acetylcholine receptor in the open and closed states and to study the effects of these compounds on the permeability of the postsynaptic membrane. Agonists and antagonists have also been useful in the purification of acetylcholine receptors, as we will describe shortly.

A Number of Other Compounds Also Serve as Neurotransmitters

The best-documented examples of chemical neuromessengers other than acetylcholine are the catecholamines, certain amino acids (and derivatives), and a variety of peptides. The catecholamines are all derived biosynthetically from L-tyrosine (fig. 28.17) and include the hormones norepinephrine and epinephrine (adren-

Figure 28.18

The structure of chlorpromazine, a drug that blocks dopamine receptors and has been used in the treatment of psychological disorders such as schizophrenia.

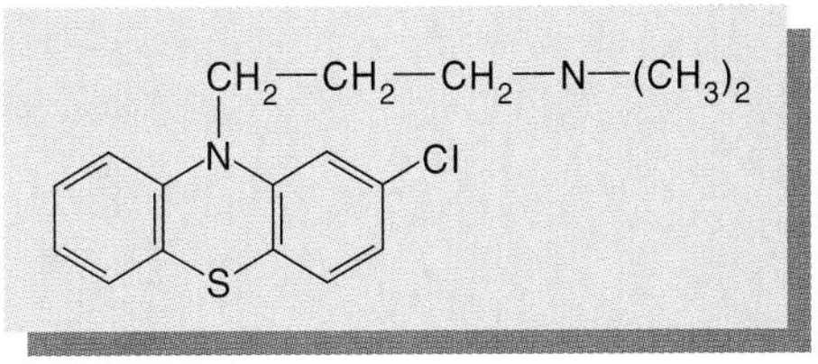

aline). Because these compounds are also synthesized in the adrenal gland, neurons that use these substances as chemical transmitters are said to be adrenergic. Sympathetic nerve fibers that innervate smooth-muscle cells in internal organs such as the heart, spleen, and gut have been shown to release norepinephrine at their terminals by mechanisms similar to those used by cholinergic neurons. Norepinephrine and related amines also have been shown to serve as neurotransmitters in a number of nerve pathways in the brain. Like acetylcholine, catecholamines may be inactivated by chemical modifications. Inactivation may be effected by a methylation reaction or by an oxidation reaction catalyzed by the enzyme monoamine oxidase (see fig. 28.17). In some cases they can also be resorbed through the presynaptic membrane after their release, providing an additional mechanism for removal from the synaptic cleft.

The hydroxylation of tyrosine, which is the first unique step in the biosynthesis of catecholamines, is catalyzed by tyrosine hydroxylase and yields the compound 3,4-dihydroxyphenylalanine (L-DOPA). The neurological disorder Parkinson's disease is associated with an underproduction in the human brain of the catecholamine transmitter dopamine, which is derived from L-DOPA by a decarboxylation reaction (see fig. 28.18). L-DOPA has therefore been found to be an effective drug in some instances in the treatment of Parkinson's disease. Interestingly, overproduction of dopamine in the brain also occasionally occurs and appears to be associated with psychological disorders such as schizophrenia. In these cases, dopamine-receptor-blocking drugs, such as chlorpromazine (fig. 28.18), have been found to be useful therapeutic agents.

Amino acids that are believed to have roles as neurotransmitters include glutamic acid and glycine. The amino acid derivatives histamine (synthesized by the decarboxylation of histidine), 5-hydroxytryptamine (or serotonin, derived from tryptophan), and gamma-aminobutyric acid (or GABA, a decarboxylation product of glutamic acid) have all been shown to be transmitters in various systems as well. The reactions involved in the biosynthesis of these compounds are diagrammed in figure 28.19. GABA is used most often as an inhibitory transmitter. Interaction of this compound with its receptor on many postsynaptic membranes results in a large increase in the membrane permeability to Cl^- and/or K^+ ions. This change inhibits the postsynaptic cell, often by hyperpolarizing its membrane (recall, for example, that the equilibrium potential of K^+ is more negative than the resting potential). Most target tissues, in fact, are innervated by more than one type of nerve fiber, each using a different neurotransmitter. Thus different signals can be relayed to the recipient cells, some stimulatory and others inhibitory. Given the complexity of the nervous systems of higher animals, additional neurotransmitters are continually being discovered. Some compounds currently thought to be chemical transmitters are listed in table 28.2.

Table 28.2
Neurochemical Messengers

Compounds	Status[a]
Acetylcholine	C
Catecholamines	
Norepinephrine (noradrenaline)	C
Epinephrine (adrenaline)	C
L-DOPA	P
Dopamine	C
Octopamine	C
Amino acids (and derivatives)	
Glutamate	C
Aspartate	P
Glycine	C
Proline	Pos
Gamma-aminobutyrate (GABA)	C
Tyrosine	Pos
Taurine	P
Alanine	Pos
Cystathionine	Pos
Histamine	C
Serotonin (5-hydroxytryptamine)	C
Peptides	
Substance P	P
Cholecystokinin	P
Neurotensin	P
Enkephalins	P
Somatostatin	P

[a]C = confirmed neurotransmitter; P = probable; Pos = possible.

The Acetylcholine Receptor Is the Best-Understood Neurotransmitter Receptor

The structure and properties of the acetylcholine receptor are by far the best understood among all neurotransmitter receptors. One reason for this is their abundance in postsynaptic membranes found in the electric organs of the electric eel (*Electrophorus*) and the electric ray (*Torpedo*). These organs contain stacks of cells (electroplaxes), each cell of which receives nerve endings on one side but not on the other. Release of acetylcholine by the presynaptic

Figure 28.19

Biosynthesis of the neurotransmitters histamine, gamma-aminobutyrate (GABA), and serotonin (5-hydroxytryptamine).

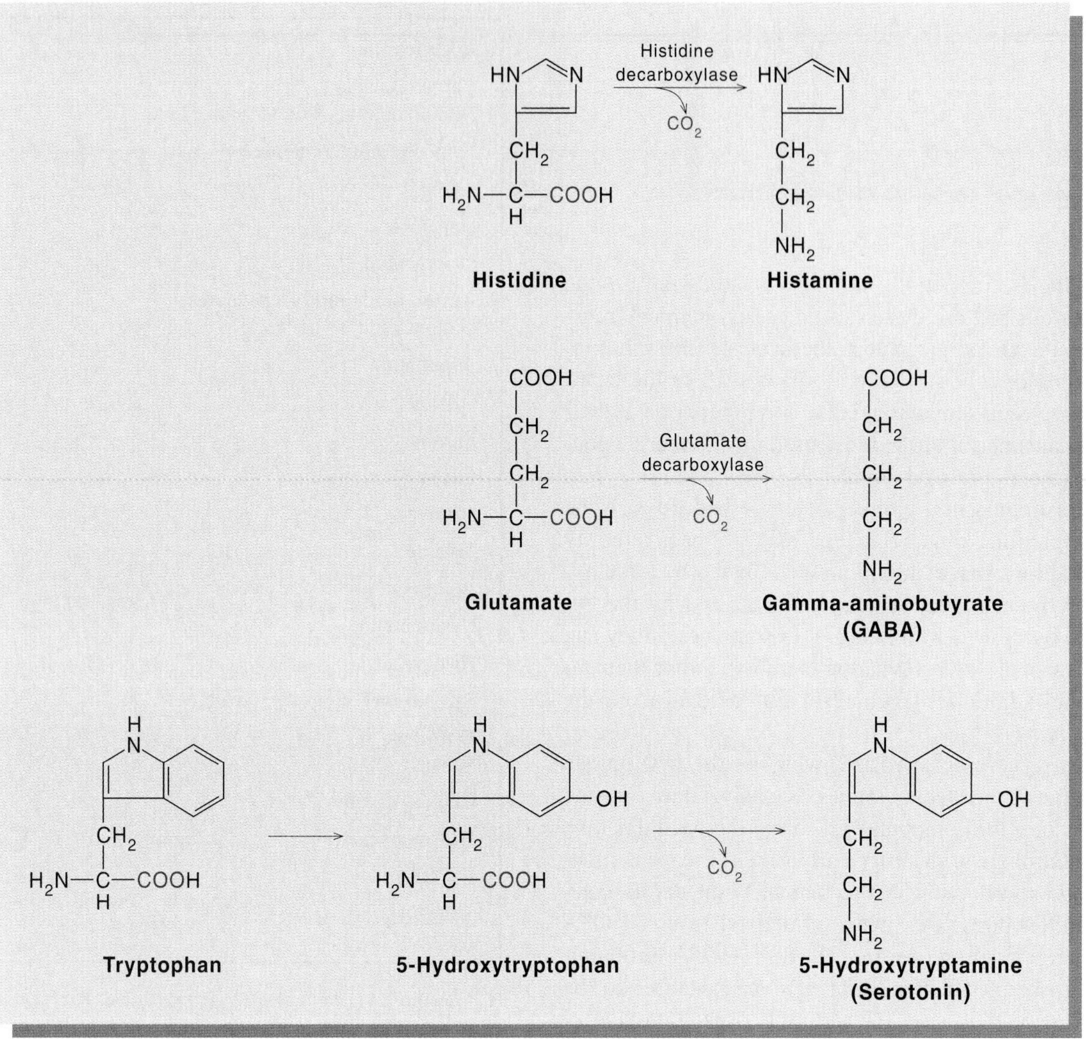

Histidine → (Histidine decarboxylase, $-CO_2$) → Histamine

Glutamate → (Glutamate decarboxylase, $-CO_2$) → Gamma-aminobutyrate (GABA)

Tryptophan → 5-Hydroxytryptophan → ($-CO_2$) → 5-Hydroxytryptamine (Serotonin)

nerve endings thus depolarizes the postsynaptic membrane by virtue of the binding of the neurotransmitter to the membrane-bound receptors, while the other face of the cell remains at its resting potential. In this way each cell, when stimulated, can attain a potential difference of over 100 mV between its two faces, and thousands of such stacked cells, present in the electric organ, can consequently emit an electric discharge of several hundred volts.

The acetylcholine receptor was purified for the first time from membranes of the electric organ of the electric ray fish. When the purified receptor was incorporated into phospholipid vesicles, the vesicles became permeable to Na^+ in the presence of the acetylcholine analogs. This permeability was blocked by agents that were known to inhibit synaptic transmission, including protein neurotoxins from snake venoms (α-bungarotoxin, from snakes of the genus *Bungarus,* or cobratoxin from cobras).

The acetylcholine receptor has four subunits (α, β, γ, and δ, with $M_r \approx 52$, 56, 63, and 66×10^3, respectively). The complete receptor includes two copies of the α subunit, and one of each of the others. The overall structure, as visualized by electron microscopy, resembles a cylindrical bundle of five approximately parallel rods, with a water-filled channel along the axis of the cylinder (fig. 28.20). This assembly projects about 70 Å into the synaptic cleft on one side of the membrane and about 40 Å into the intracellular space on the other (fig. 28.21). The binding site for acetylcholine is in the extracellular region. The central channel has an opening with a diameter of about 25 Å in the synaptic cleft but narrows to only a few angstroms near the center of the membrane.

The individual subunits of the acetylcholine receptor have homologous amino acid sequences. Each of them has four regions of hydrophobic amino acid residues that probably are embedded in the membrane (M1, M2, M3, and M4), although only one transmembrane helix (the M2 helix) from each subunit is clearly resolved in the structural maps currently provided by electron mi-

Figure 28.20

Images of the acetylcholine receptor from a view looking down on the plane of the membrane. This picture was made by mathematically filtering electron micrographs of ordered, two-dimensional arrays of the receptor. Each of the four pentameric structures shown represents an individual acetylcholine receptor with its central transmembrane pore. The dark areas indicate the five subunits ($\alpha_2\beta\gamma\delta$). (Reprinted with permission from R. M. Stroud, M. P. McCarthy, and M. Schuster, Nicotinic acetylcholine receptor superfamily of ligand-gated ion channels, *Biochemistry* 29:11009, 1990. Copyright 1990 American Chemical Society.)

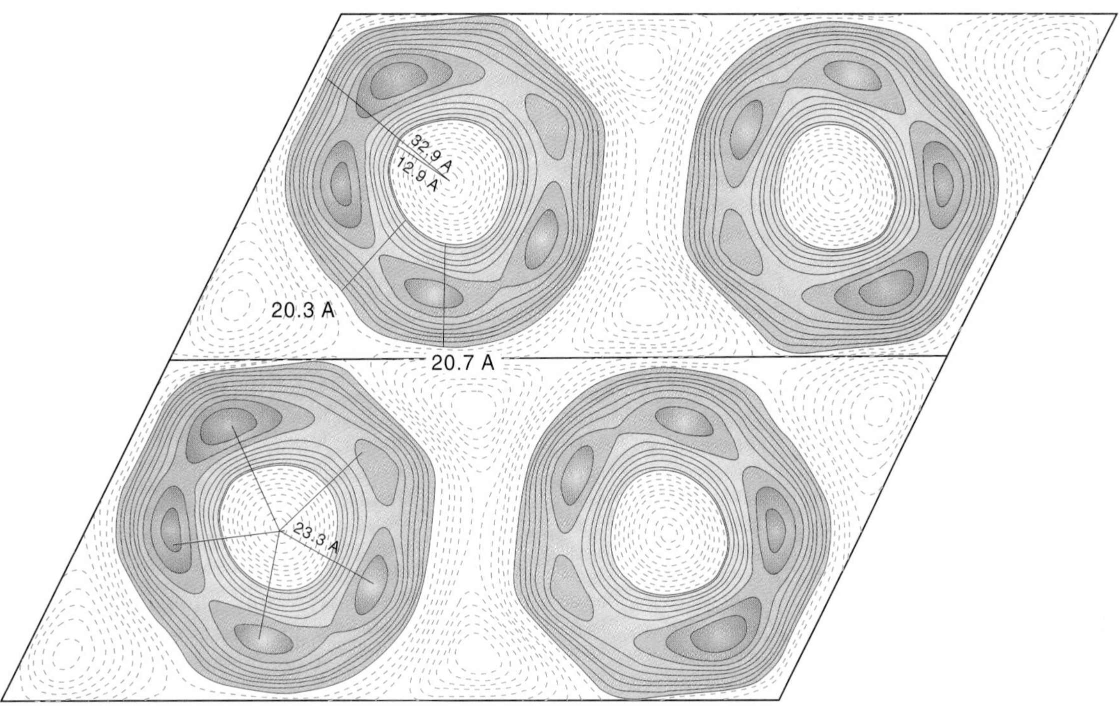

croscopy. These helices also include many uncharged polar residues, notably serines and threonines, which probably line the receptor's central channel (fig. 28.22). The M2 helices are bent near the center of the phospholipid bilayer, at the site of a leucine residue that is conserved in the amino acid sequences of all the subunits (fig. 28.23). The leucine side chains appear to form a tight, hydrophobic ring that blocks the flow of hydrated ions at this point.

Since the binding sites of the receptor are located in a domain of the protein that is remote from the region that forms the gate of the channel (see fig. 28.21), the binding of ACh must bring about a long-range conformational change in the receptor. It appears that the binding causes one subunit to shift relative to another, thereby propagating a localized distortion of the binding pocket to the gate of the channel.

The structure of the open pore is quite different from that of the closed pore. The open pore association involves side-to-side interactions around a barrel, whereas the closed pore association involves interactions between exposed side chains forming a tight central ring (see fig. 28.23). These changes in configuration of the helices are linked to a change in orientation of the kinks. Instead of pointing inward, as in the closed configuration, they have rotated over to the side, opening up the pore in the region where the gate was located. The kink acts as a region of flexure between two α-helical segments accommodating an extended conformational

Figure 28.21

Cutaway diagram of an acetylcholine-gated ion channel.

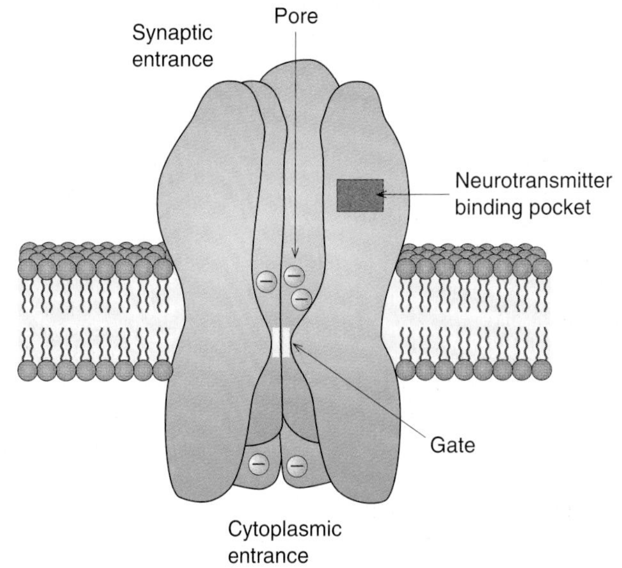

Figure 28.22

A model of the channel formed by the M2 regions of the five subunits of the acetylcholine receptor. Polar amino acid side chains, shown in red, line the channel. (Model of the channel formed by the M2 regions of the five subunits of the acetylcholine receptor, modeled by A. Kamb, J. Newdoll and R. M. Stroud, using the software MidasPlus of the UCSF Computer Graphics Lab.)

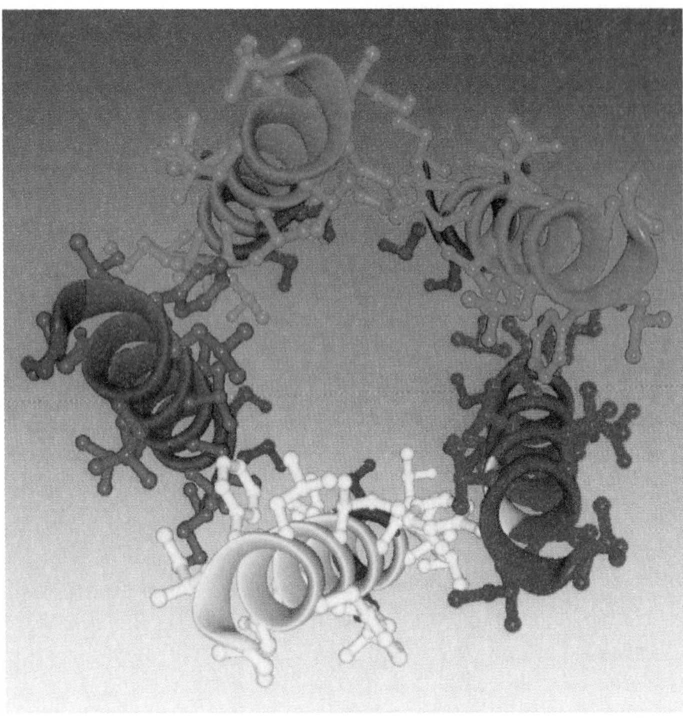

Figure 28.23

A model of the gate in the closed acetylcholine channel. The M2 helices are bent about halfway across the phospholipid bilayer. A leucine side chain (L) from each of the five helices projects into the channel here, forming a hydrophobic barrier to the diffusion of ions. (Reprinted from *J. Mol. Biol.*, vol. 229, N. Unwin, Nicotinic acetylcholine receptor at 9 Å resolution, page 1101, 1993, by permission of the publisher Academic Press Limited, London.)

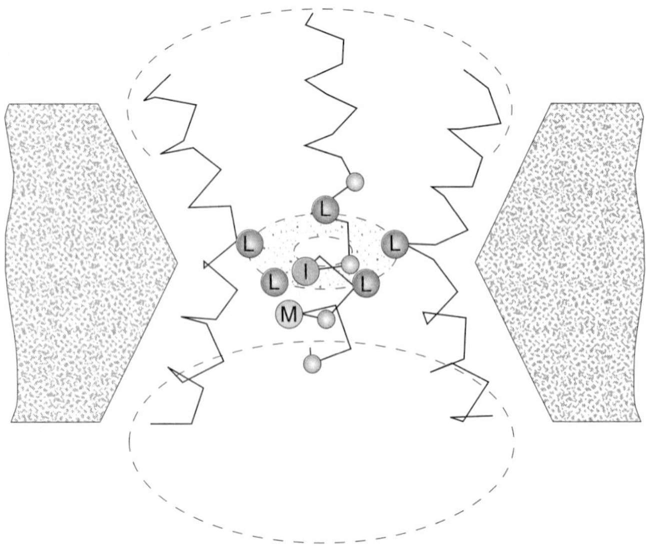

change. Kinks between α-helical segments have been found to have analogous roles in other proteins.

Synaptic Receptors Coupled to G Proteins Produce Slow Synaptic Responses

In the 1970s, evidence emerged that virtually all of the neurotransmitters that activate ligand-gated channels to produce rapid synaptic potentials lasting only milliseconds also interact with a second, even larger class of receptors that produce synaptic responses that persist for seconds or minutes. Thus a single cell releasing a single transmitter can produce a wide variety of actions on different target cells by acting on distinct receptors.

Slow synaptic responses are transduced by members of the family of receptors with transmembrane-spanning domains that do not couple to ion channels directly but do so indirectly by means of GTP-binding (G) proteins. G proteins couple this class of receptors to effectors, that is, enzymes that produce intracellular second messengers such as cAMP and cGMP (see chapter 27). These second messengers can activate channels directly, but more commonly, these messengers activate a protein kinase that regulates channel function by phosphorylating the channel protein or an associated regulatory protein.

The study of slow synaptic potentials mediated by second messengers has added new features to our picture of chemical transmission. By acting through second messengers, neurotransmitters can modify proteins other than channel proteins, thereby activating a coordinated molecular response within the postsynaptic cell. In addition, second messengers can modify transcriptional regulatory proteins and in this way control gene expression. Thus second messengers can both produce a covalent modification of preexisting proteins and regulate the synthesis of new proteins. This latter class of synaptic action can lead to long-lasting structural changes at synapses such as are likely to be involved in the establishment of long-term memory.

Synaptic Plasticity and Learning

Research has shown that learning consists of at least two distinct families of mental processes. First, learning about people, places, and things is called explicit learning because it involves a conscious awareness of what is learned and recalled. Second, learning of motor skills, perceptual strategies, habits, classical conditioning, sensitization, and habituation is called implicit learning because it is carried out without recourse to consciousness. These two major forms of learning have been localized to different neural systems in the brain. Explicit learning involves regions deep within the temporal lobe of the cerebral cortex, including the hippocampus. Implicit learning does not involve the hippocampus but involves the specific sensory and motor systems used during the learning process.

Aspects of both types of learning are represented at the level of individual synapses as changes in the strength of synaptic connections. In the few cases so far analyzed, the storage of short-term memory involves the strengthening of existing synaptic connections through covalent modification of preexisting proteins by means of one or another second-messenger cascade. The storage of long-term memory seems to involve the growth of new synaptic connections through the activation of gene expression and new proteins. The mechanisms that may be involved in these processes are one of the major interests in modern neurobiology.

Excitability Is Found in Many Different Cell Types

Although nerve cells are among the most extensively studied excitable cells, excitability has been demonstrated not only in other cell types in higher animals but also in organisms as simple as unicellular protozoans and in algae. In ciliated protozoa, for example, membrane depolarization events have been shown to regulate swimming behavior, while in certain algae, localized transmembrane movements of ions appear to play a role in early development. Membrane excitability thus appears to be a common phenomenon in many different types of cells and to play a role in a variety of physiological processes. In the following chapter we will examine one such process: vision. In this case, photons—impinging upon specialized nerve cells containing photoreceptors—initiate nerve impulses that are sent to the brain, which translates these signals into an image of our environment.

SUMMARY

In this chapter we have examined the way in which nerve impulses are propagated along the neurons and are transmitted to receptor cells. The following points were central to our discussion.

1. Nerve cells are highly specialized for the reception of external signals and the transmission of those signals to other cells in the organism. Signal transmission within a neuron involves waves of membrane depolarization–repolarization events that travel the length of the axon to the nerve endings. The arrival of such a wave at a single region of the axon can be recorded as an action potential, which is generated by a transient influx of Na^+, followed by a transient efflux of K^+ and then a return of the membrane potential to its resting state.
2. Separate, voltage-gated channels for Na^+ and K^+ have been demonstrated in many types of excitable cells, and have been shown to be responsible for the ionic movements and changes in membrane potential associated with the action potential. Na^+ channels from a number of sources consist of an α subunit (M_r = 260,000) and variable amounts of a β subunit (M_r = 37,000). The best-characterized K^+ channel is that encoded by the *Drosophila Shaker* locus. It is a 616-residue polypeptide that has significant primary and secondary structural homology to each of the four domains of the Na^+ channel.
3. Transmission of the action potential from a neuron to another cell occurs most commonly by the release of chemical neurotransmitters at the synapse, triggered by the arrival of the action potential at the terminus of the axon. Neurotransmitter compounds include acetylcholine, the catecholamines such as epinephrine and norepinephrine, certain amino acids, and a variety of neuropeptides. Neurotransmitters bind to receptors in the postsynaptic membrane, initiating a response in the target cell that is, again, triggered by the influx of cations.
4. The best-studied neurotransmitter receptor is that for acetylcholine. It consists of four different polypeptide subunits (α, β, γ, and δ) in a molar ratio of 2:1:1:1 with a total mass of about 255,000 daltons. Biochemical and electron-microscopic techniques have established that all five subunits span the postsynaptic membrane. These subunits comprise a pentagonally symmetrical shell, which delineates a water-filled channel that is undoubtedly important in the translocation of cations initiated by ligand binding.
5. Molecular biological techniques and modeling studies have shown how both the ion channel proteins and neurotransmitter receptors may be arranged in the membrane, and have identified regions in these proteins that are important for ion translocation and ligand binding. Site-directed mutagenesis studies of a *Drosophila* K^+ channel also provide evidence as to how the opening and closing of such channels may occur. Further application of these techniques will probably continue to increase our understanding of structure–function relationships in these proteins.

SELECTED READINGS

Brisson, A., and P. N. T. Unwin, Quaternary structure of the acetylcholine receptor. *Nature* 315:457, 1985. This article describes the three-dimensional electron image analysis of tubular crystals of the acetylcholine receptor grown from native membrane vesicles.

Catterall, W., Structure and function of voltage-sensitive ion channels. *Science* 242:60, 1988. Excellent review of the subject through 1987.

Galzi, J. L., A. Devillers-Thiery, N. Hussy, S. Bertrand, J.-P. Changeux, and D. Bertrand, Mutations in the channel domain of a neuronal nicotinic receptor convert ion selectivity from cationic to anionic. *Nature* 359:500–505, 1992. In this article, site-directed mutagenesis is used to probe amino acid residues in the acetylcholine receptor that may be important for ion selectivity.

Hille, B., *Ionic Channels in Excitable Membranes,* 2nd edition. Sunderland, Mass.: Sinauer, 1991. This book covers recent research on a variety of nerve and muscle ion channels. Theory of ion channel function and techniques of analysis are also reviewed.

Hooper, N. M., and A. J. Turner, Specificity of the Alzheimer's amyloid precursor protein secretase. *TIBS* 20:15–16, 1995.

Hoshi, T., W. N. Zagotta, and R. W. Aldrich, Biophysical and molecular mechanisms of *Shaker* potassium channel inactivation. *Science* 250:533, 1990. Evidence for the "ball-and-chain" mechanism of ion channel inactivation is presented in this article.

Jessell, T. M., and E. R. Kandel, Synaptic transmission: A bidirectional and self-modifiable form of cell-cell communication. *Cell* 72(Suppl. 1)1–30, 1993.

Katz, B., *Nerve, Muscle and Synapse.* New York: McGraw-Hill, 1966. Excellent overview of early and classical work on neurotransmission.

Kuffler, S. W., and J. G. Nicholls, *From Neuron to Brain: A Cellular Approach to the Function of the Nervous System.* Sunderland, Mass.: Sinauer Associates, 1984. This book covers aspects of neurotransmission and neurophysiology at the cellular level in an easily readable form.

Olivera, B. M., G. Miljanich, J. Ramachandran, and M. E. Adams, Calcium channel diversity and neurotransmitter release: The ω-conotoxins and ω-agatoxins. *Ann. Rev. Biochem.* 63:823–868, 1994.

Putney, J. W., Excitement about calcium signaling in inexcitable cells. *Science* 262:676–678, 1993.

Raffioni, S., R. A. Bradshaw, and S. E. Buxser, The receptors for nerve growth factor and other neurotrophins. *Ann. Rev. Biochem.* 62:823–850, 1993.

Rehm, H., and B. L. Temple, Voltage-gated K^+ channels of the mammalian brain. *FASEB J.* 5:164, 1991. Excellent synopsis of recent work on structure–function relationships in the K^+ channel, including molecular genetic approaches.

Sherrington, R., et al., Cloning a gene bearing missense mutations in early-onset familial Alzheimer's disease. *Nature* 375:754–762, 1995.

Südhof, T. C., The synaptic vesicle cycle: a cascade of protein-protein interactions. *Nature* 375:645–651, 1995.

Unwin, N., Neurotransmitter action: Opening of ligand gated ion channels. *Cell* 72(Suppl. 1):31–41, 1993.

Unwin, N., Acetylcholine receptor channel imaged in the open state. *Nature* 373:37–43, 1995. Describes the mechanism of acetylcholine-induced channel opening.

Yan, S. D., et al., Rage and amyloid-β peptide neurotoxicity in Alzheimer's disease. *Nature* 382:685–691, 1996.

PROBLEMS

1. Where have you previously encountered choline? How is choline biosynthesized in humans, and in nature in general?
2. Why does nature have acetylcholine esterase? What would be the consequences of a lack of this enzyme?
3. Many of the illicit psychedelic drugs contain the indol ring system. How might you explain their drug action?
4. Examine the nonpeptide neurotransmitters listed in table 28.2. What chemical feature(s) can you find in common? Does the pattern change if you consider only the confirmed neurotransmitters?
5. Monoamine oxidase catalyzes three reactions listed in figure 28.17. What are the missing reactant(s)/product(s) for these reactions?
6. What is the source of the aldehyde oxygen in the product of monoamine oxidase (fig. 28.17)?
7. Given that ascorbic acid is a cosubstrate in the conversion of dopamine to norepinephrine (fig. 28.17), produce a balanced reaction for this process.
8. What fundamental biochemical difference can you see in the reactions associated with problems 4–6?
9. $HONH_3^+$ and $H_2NNH_3^+$ ions have been shown to readily pass through the Na^+ pores, but the $CH_3NH_3^+$ ion does not. How might you explain this observation?
10. How can you explain that Na^+ pores readily transport Li^+ but not Rb^+ while the K^+ pores readily transport Rb^+ but not Li^+?
11. Explain why a nerve cell membrane exhibits an "all-or-none" response (i.e., action potential) independent of the magnitude of an electric or chemical stimulus (above a threshold value).
12. List the criteria for demonstrating that a particular compound acts as a neurotransmitter in a given system, assuming that the mechanism of synaptic transmission in most instances is analogous to that found in cholinergic systems.
13. In reconstituted transport systems it is often important to demonstrate that the rate of transport is similar to that observed in vitro, that is, that the transport protein is fully functional in the reconstituted state. For systems in which in vivo fluxes are very rapid (e.g., cation flux through the acetylcholine receptor), it is often difficult to measure these rates directly in reconstituted vesicles. Thallous ion (Tl^+) is known to pass readily through the "open" state of the acetylcholine receptor. It also very efficiently quenches the fluorescence emission of the fluorophore 8-aminonaphthalene-1,3,6-trisulfonate (ANTS), which is relatively impermeable to phospholipid bilayers.
    (a) Using this information, outline a series of experiments to measure Tl^+ fluxes in reconstituted proteoliposomes containing purified acetylcholine receptors.
    (b) Actual Tl^+ fluxes into proteoliposomes containing an average of two acetylcholine receptor channels per vesicle in the presence of agonists have been measured to be 200 moles/(liter • s). If the average inner diameter of such vesicles is 400 Å, what is the number of Tl^+ ions transported per second by each activated acetylcholine receptor channel? How does this value compare with the rate of Na^+ flux measured in vivo?
14. Describe in your own words, and critique, the evidence that voltage-sensitive K^+ channels may be regulated by a "ball-and-chain" mechanism.
15. Patients with Parkinson's disease are treated with DOPA and inhibitors of dopa decarboxylase. This treatment will raise levels of dopamine in certain areas of the brain. However, dopamine given as a drug has no effect! How does this treatment of Parkinson's disease work? (*Hint:* Think about the blood–brain barrier and membrane permeability.)
16. Draw a chemical reaction mechanism for the reaction of the nerve poison parathion with acetylcholinesterase (see fig. 28.14*b*). How do these types of inhibitors prevent repolarization of the postsynaptic membrane?
17. What are the data that support a gated-pore-type model for ion permeability conferred by the acetylcholine receptor, as opposed to a carrier-mediated mechanism?

C H A P T E R

VISION

Vision, like photosynthesis, starts with the excitation of an electron from one molecular orbital to another orbital of higher energy.

The photochemistry of vision is different from that of photosynthesis. Perhaps this is not surprising. Animals that can see use light to obtain information; photosynthetic organisms use light as a source of energy. However, vision and photosynthesis both start with the excitation of an electron from one molecular orbital to another orbital of higher energy. The excited molecule must then undergo a transformation to a metastable product. Since chlorophyll can undergo such a transformation with a quantum yield near 1.0, it is not hard to imagine an eye that uses electron-transfer reactions like those that work so well in photosynthesis. In fact, there are synthetic eyes that work very much that way: diode-array television camera tubes. But the eyes of multicellular animals are different. Instead of chlorophyll, their light-sensitive cells contain a complex called rhodopsin, which consists of a protein, opsin, and a linear polyene, 11-*cis*-retinal. Instead of undergoing oxidation when it is excited with light, the retinal isomerizes.

In this chapter we will examine the biochemistry involved in this reaction to light. We will focus primarily on vertebrates, with only occasional references to invertebrate photochemistry, but we will conclude with a short section on the light-reactive substance found in one species of bacterium.

The Visual Pigments Found in Rod and Cone Cells

The eyes of vertebrates are marvelously complex organs, with many different types of specialized cells (fig. 29.1). Light rays entering the eye are refracted by the cornea, the clear tissue at the front of the eye. The light traverses an aqueous chamber and reaches the lens, which is densely packed with proteins called crystallins. Adjustments in the shape of the lens focus a sharp optical image onto the retina, a thin layer of tissue that lines the back of the eye. The retina is a neural tissue with several different layers of cells. Some of these cells, the rod and cone cells, contain the visual pigments. Other cells make synaptic connections to the rods or cones and to additional neural cells that carry impulses to the brain.

In both rods and cones the light-sensitive molecules are collected in a layered system of membranes at one end of the cell (figs. 29.2 and 29.3). The membranes form by invaginations of the cytoplasmic (plasma) membrane near the middle of the cell. In cone cells the membranes remain contiguous with the cytoplasmic membrane. In rods they pinch off to form a stack of autonomous flattened vesicles, or disks, in the outer segment of the cell. The outer segment of each rod contains from 500 to 2,000 of

Figure 29.1

Structure of the human eye.

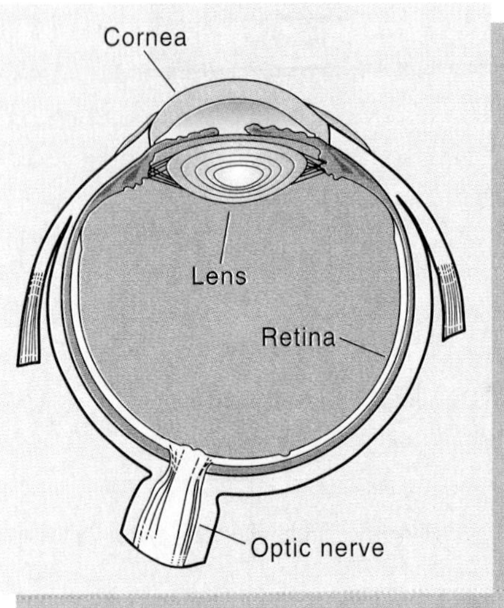

Figure 29.2

Thin-section electron micrograph (59,200×) of a portion of a rod cell in a rabbit retina. Part of the outer segment is shown at the top and part of the inner segment at the bottom. D = disk, C = cilium, M = mitochondrion. (Courtesy Dr. Ann H. Milam.)

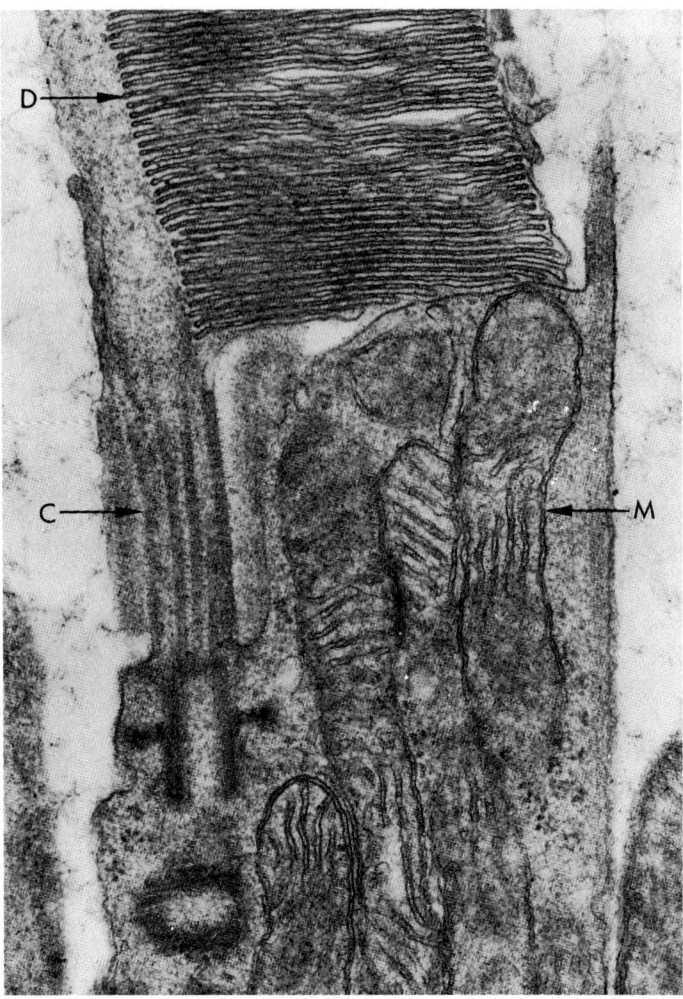

these disks. This region of the cell is connected by a thin cilium to the inner segment, which is packed with mitochondria and ribosomes. The basal part of the cell ends in a synaptic junction with another neural cell called a bipolar cell. In a living rod cell, disks are constantly forming at the base of the outer segment and moving in a file to the tip of the outer segment, where they are sloughed off and phagocytosed by the underlying epithelial cells. It takes about 10 days for a disk to make this journey. Exactly why the disks need to be replaced so rapidly is unclear.

Rod cells are specially adapted for vision in dim light. Cones provide visual acuity in bright light and also serve for perception of color. Animals such as owls, which have very high visual sensitivity in dim light but cannot distinguish colors, have only rod cells. Some animals, such as pigeons, have only cones and are skilled at distinguishing colors in bright light but inept at night vision. Primates have both rods and cones, with rods considerably outnumbering cones in all but the central portion of the retina.

*Rhodopsin Consists of 11-*cis*-Retinal Bound to a Protein, Opsin*

Figure 29.4 shows the structures of 11-*cis*-retinal and its more stable isomer all-*trans*-retinal. The retinals are related to the alcohol retinol, or vitamin A_1. These compounds cannot be synthesized de novo by mammals, but they can be formed from carotenoids, such as β-carotene, which are abundant in carrots and some other vegetables. A deficiency of vitamin A causes night blindness, along with a serious deterioration of the eyes and a number of other tissues.

There are animals with eyes in several different phyla, including Mollusca, Arthropoda, and Annelida, in addition to Chordata. The eyes of arthropods differ substantially from those of molluscs and chordates anatomically, and they apparently originated independently, after the phyla had separated in evolution. Some animals, such as sea turtles and amphibians in certain stages of development, use a retinal that has an additional double bond in the ring. The parent molecule in this case is vitamin A_2, or 3,4-dehydroretinal (see fig. 29.4). In all cases, however, the photochemically active protein complex is made from the 11-*cis* isomer of the aldehyde. 11-*cis*-Retinals must somehow be particularly fit for the task of responding to light. (Some unicellular organisms, such as *Euglena,* have light-sensitive organelles that contain a carotenoid instead of a retinal, but little is known of the biochemistry of these primitive receptors.)

Opsin is an integral membrane protein with a molecular weight of about 38,000. It accounts for about 95% of the protein in the disk membranes. Although a high-resolution structure of the rhodopsin has not yet been obtained, analyses of the amino acid sequence and electron microscopic studies of two-dimensional crystals have shown that the protein has seven transmembrane α

Figure 29.3

Schematic diagram of a rod cell. The orientation of the cell is the same as in figure 29.2. In the retina many rod cells are stacked side by side, with the outer segments all pointing out to the periphery of the eye. Light enters the cells end-on through the inner segment after passing through several layers of other neural cells. In cone cells the outer segments are shorter and are conical rather than cylindrical in shape.

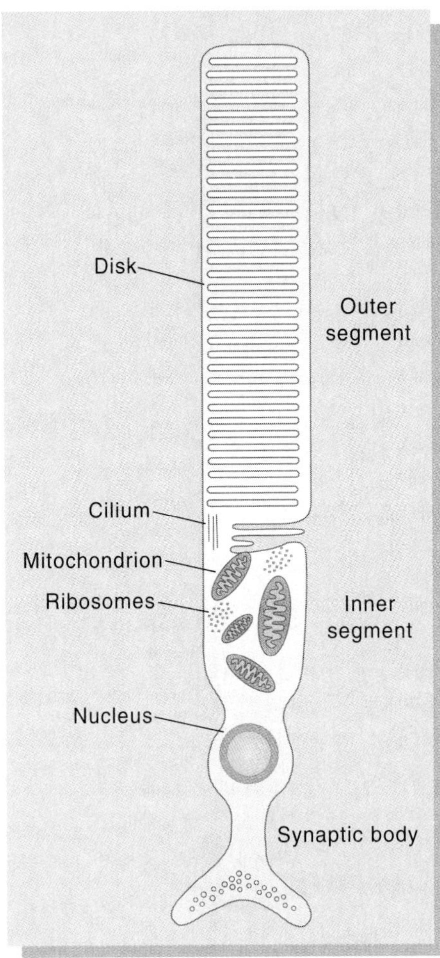

helices. Figure 29.5 shows the likely arrangement of the helices in the membrane. The carboxyl-terminal end of the protein is in the cytoplasmic space outside the disk, and the amino-terminal end is inside the disk. The 11-*cis*-retinal is bound by a Schiff base, or aldimine, linkage (see fig. 29.4) to the ϵ-amino group of lysine 296 in helix G and probably is nestled between helices C and F (see fig. 29.5).

In solution, 11-*cis*-retinal absorbs maximally near 380 nm, but in rhodopsin the peak is at 500 nm (fig. 29.6). The absorption spectrum of rhodopsin is essentially identical to the spectrum of the sensitivity of the rod cells (after correction for absorption in the cornea and lens). The perception of color by cones depends on the fact that there are three different types of cones with different absorption spectra. One of these absorbs blue light (440 nm) maximally, the second absorbs green (530 nm), and the third absorbs yellow (750 nm). In some species of birds the third component absorbs maximally at 630 nm. However, all cones contain 11-*cis*-retinal bound to proteins that are similar to opsin. (Some fish use another strategy altogether for color vision. Their eyes have carotenoid-containing oil droplets of three colors, and these act as filters in front of the receptor cells.)

Light Isomerizes the Retinal of Rhodopsin to All-trans

The discovery that rhodopsin contains the 11-*cis* isomer of retinal came as a surprise. In the early 1950s, Ruth Hubbard, George Wald, and their co-workers found that rhodopsin decomposes into retinal and the apoprotein opsin when the retina is exposed to light. The retinal that was released was the all-*trans* isomer. When opsin was mixed with a crude preparation of retinal, rhodopsin was regenerated. Crystalline all-*trans*-retinal, however, did not bind to opsin. The regeneration achieved with the crude preparation proved to be due to a small amount of the 11-*cis* isomer in the mixture. Since rhodopsin binds 11-*cis*-retinal in the dark but releases all-*trans*-retinal after illumination, Hubbard and Kropf concluded that the action of light in vision is to isomerize the chromophore about its C-11=C-12 double bond. This change in structure must somehow be translated into an electrophysiological signal that can be transmitted to the brain.

Transformations of Rhodopsin Can Be Detected by Changes in Its Absorption Spectrum

The release of all-*trans*-retinal from opsin takes several minutes and is too slow to be an obligatory step in visual perception. To explore the changes in rhodopsin that precede the release of all-*trans*-retinal from the protein, Toru Yoshizawa and Wald measured the optical absorbance changes that occurred when they illuminated rhodopsin at low temperatures. Illumination at liquid N_2 temperature (77 K) caused the absorption band of the rhodopsin to shift from 500 to 543 nm. The product of this transformation is now called bathorhodopsin. Bathorhodopsin is stable indefinitely in the dark at 77 K, but if it is warmed above about 130 K, it decays spontaneously to a species that absorbs maximally at 497 nm. This species is called lumirhodopsin. If the sample is warmed further, to about 230 K, lumirhodopsin decays to metarhodopsin I, which absorbs at 478 nm. Above about 255 K, metarhodopsin I decays to metarhodopsin II, which absorbs at 380 nm. These transformations are outlined in figure 29.7.

Subsequent kinetic measurements by other investigators have shown that rhodopsin passes through the same series of states if it is excited with a short flash at physiological temperatures. The numbers on the right side of figure 29.7 give approximate half-times for the transformations at 37°C. Agreement between the kinetic studies and low-temperature experiments may seem routine, but it might not have turned out this way at all. The trapping of a metastable state at low temperature can depend on the thermodynamic properties of the state, as well as on its position in the kinetic sequence. For example, side products that normally are unimportant could accumulate when the normal pathway is blocked by lowering the temperature.

How can the absorption spectrum of rhodopsin go

Figure 29.4

Structures of retinals, retinols, and β-carotene. The structure of 11-*cis*-retinal *(top)* indicates the numbering system used for the carbons. In rhodopsin, 11-*cis*-retinal is bound by a protonated Schiff's base linkage to a lysine of opsin.

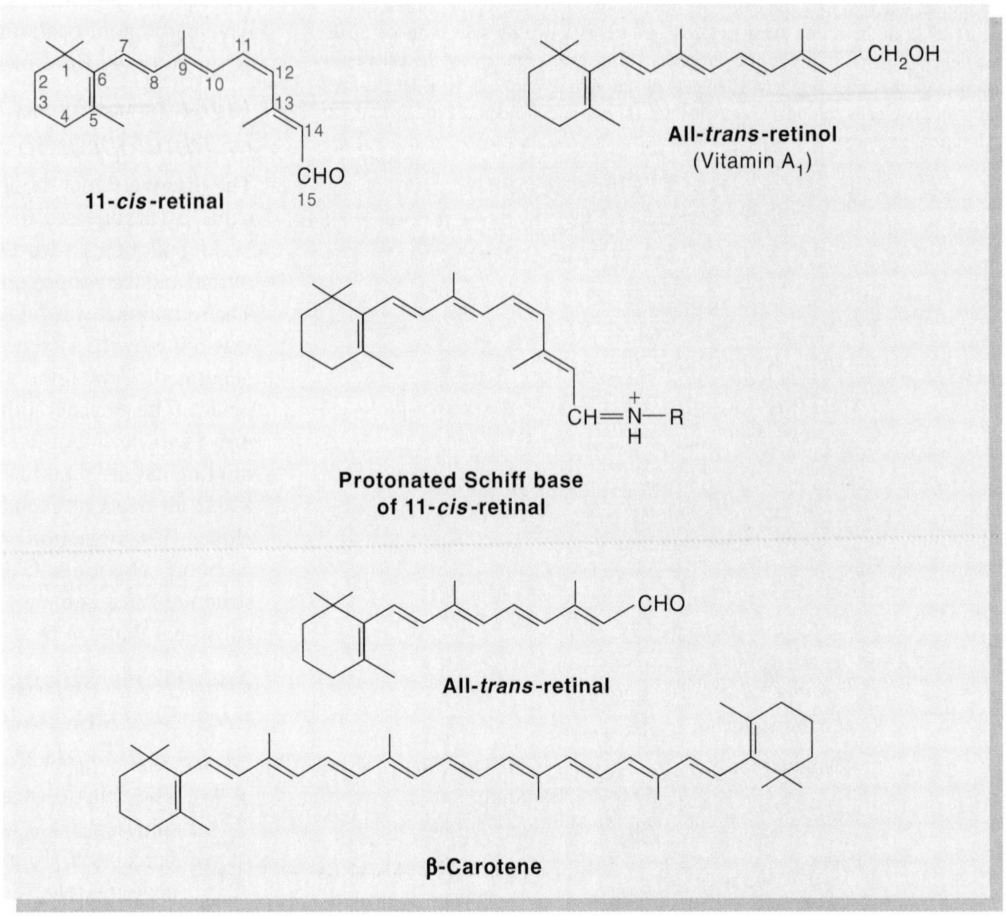

11-*cis*-retinal

All-*trans*-retinol
(Vitamin A_1)

Protonated Schiff base of 11-*cis*-retinal

All-*trans*-retinal

β-Carotene

Figure 29.5

A model for the folding of rhodopsin. The cytoplasmic side of the disk membrane is at the top, and the disk lumen is at the bottom. Rhodopsin's seven transmembrane α helices are represented as columns labeled A–F. 11-*cis*-Retinal is attached to lysine 296 in helix G by a Schiff base linkage. Helix F has been cut away in the drawing to show the approximate location of the retinyl chain. The negatively charged carboxylate group of glutamic acid 113 in helix C acts as a counterion for the protonated Schiff base. Studies of site-directed mutations have identified seven residues in helices C and F as close neighbors of the retinyl group. The protein also has two hexasaccharides bound to asparagines 2 and 15 and palmitoyl groups bound to cysteines 322 and 323. Cysteines 110 and 187 are linked by a disulfide bond. Transducin interacts with rhodopsin on the cytoplasmic side of the membrane, probably in the loops between helices C and D and helices E and F. Near the carboxyl-terminal end, rhodopsin has four threonine and two serine residues (labeled OH in the figure). Phosphorylation of these residues by rhodopsin kinase terminates interactions with transducin. Rhodopsin is structurally homologous to a large family of hormone receptors and neuroreceptors that interact with G proteins and transmit signals across membranes.

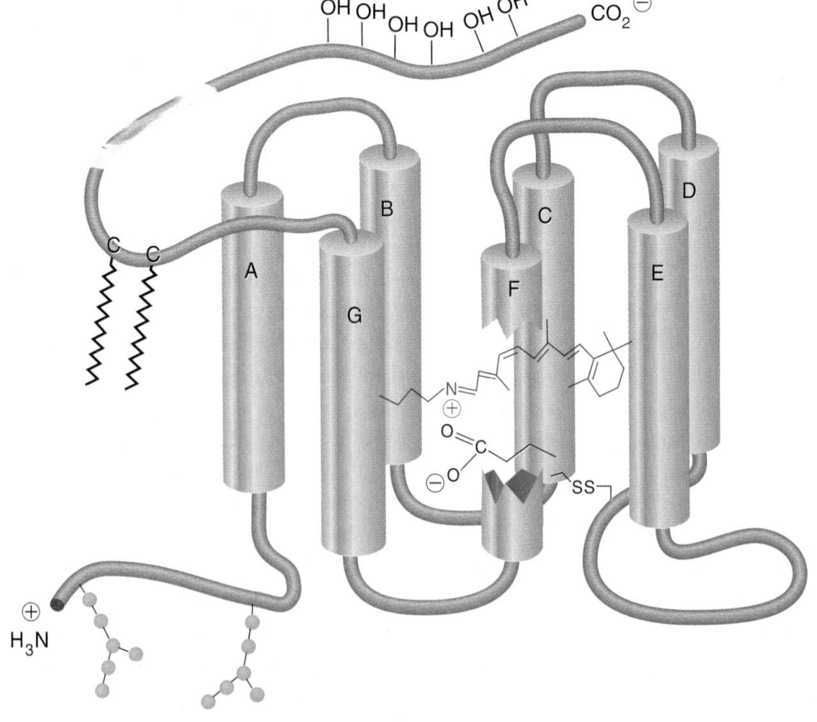

Figure 29.6

Absorption spectra of rhodopsin and of 11-*cis*-retinal in hexane solution. The absorption of rhodopsin in the 280-nm region is due mainly to the opsin.

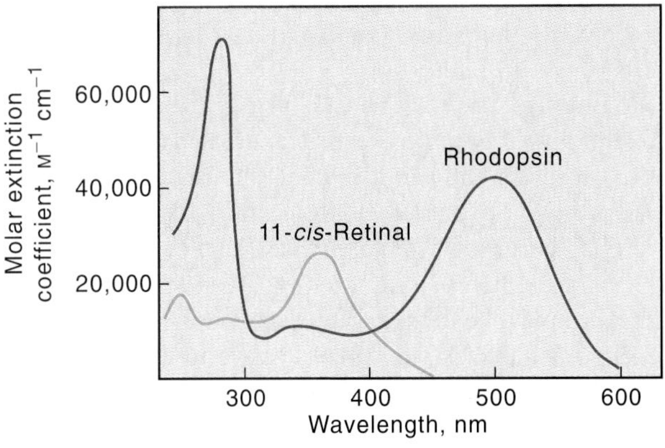

Figure 29.7

Photochemical transformations of rhodopsin. The numbers in parentheses indicate the optical absorption maxima of the intermediates. The numbers in red on the right are approximate half-times for the conversions in rod outer segment membranes near 37°C. The steps following the formation of bathorhodopsin have progressively higher thermal activation energies. They can be blocked by lowering the temperature below the temperatures indicated on the left. The asterisk indicates the excited state of rhodopsin.

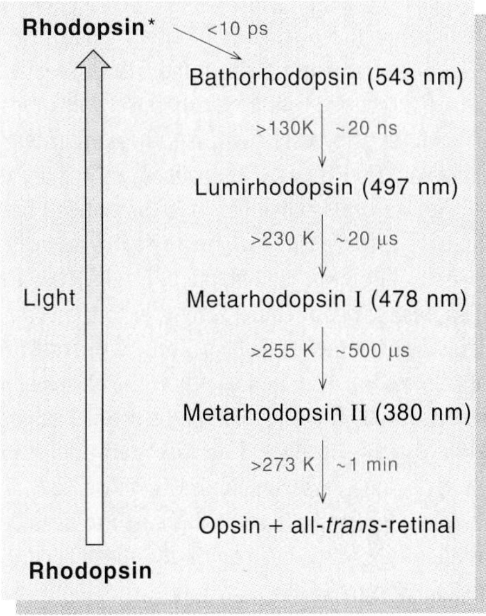

Figure 29.8

Excitation of the protonated Schiff's base causes a movement of positive charge from the N toward the ring. The covalent linkage between C11 and C12 loses much of its double-bond character. The valence bond diagrams shown here should not be taken too literally, but they give a good qualitative picture of the redistribution of electrons that occurs when the molecule is excited.

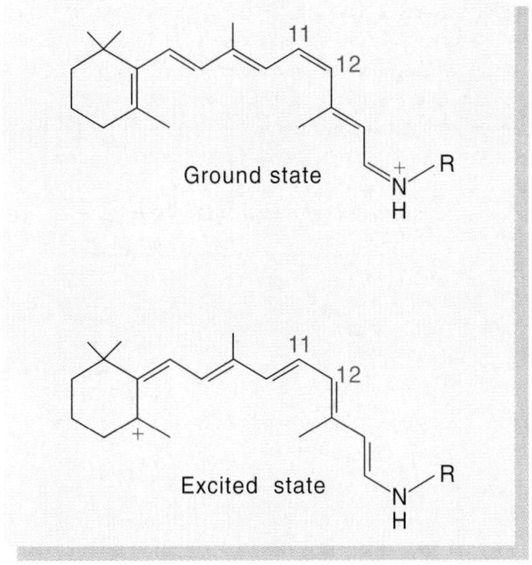

through such wild changes from 500 to 543 nm, and eventually to 380 nm? In thinking about this, we are reminded that rhodopsin's initial absorption spectrum is already shifted by 120 nm compared with the spectrum of free 11-*cis*-retinal (see fig. 29.6). Further, the absorption maxima of the cone pigments vary from 450 to 630 nm, in spite of the fact that these pigments all contain 11-*cis*-retinal. The explanation for these spectral differences depends partly on the fact that the nitrogen atom of the retinylidine Schiff base linkage in rhodopsin is protonated and is therefore positively charged (fig. 29.8). The protonation state of the nitrogen has been demonstrated clearly by NMR and resonance Raman spectroscopy. When the retinal absorbs light and is raised to an excited state, the electron density on the nitrogen increases, and the positive charge moves to the opposite end of the molecule, as represented roughly in figure 29.8. Because of the redistribution of charge, the relative energies of the excited and ground states are extremely sensitive to the positions of other charged, dipolar, or polarizable groups nearby. An arrangement of charged groups that stabilizes the excited state relative to the ground state will shift the absorption spectrum to longer wavelengths.

There must be at least one charged group near the retinylidine Schiff base in rhodopsin: the anionic counterion that is needed to balance the positive charge on the nitrogen. The anion is probably the carboxylate of the side chain of a glutamate residue, Glu 113, because mutating this residue to glutamine causes a dramatic change in the pK_a of the Schiff base (see fig. 29.5). In the mutant rhodopsin the Schiff base is not protonated at physiological pH, and the absorption maximum is at 380 nm instead of 500 nm. The structure of the protein thus can affect the absorption spectrum of the pigment profoundly. Moving the carboxylate group of Glu 113 farther away from the retinylidine Schiff base would destabilize the ground state relative to the excited state, shifting the absorption spectrum to longer wavelengths.

Isomerization of the Retinal Causes Other Structural Changes in the Protein

Let us now consider bathorhodopsin, the first metastable product of the photochemical reaction. If bathorhodopsin is excited with long-wavelength light at 77 K, it can be converted back to

Figure 29.9

Potential energy surfaces of ground and excited states in rhodopsin as functions of the angle of rotation of the bond between carbons 11 and 12. A rotational angle of 0° means that the retinyl group is 11-*cis*-(rhodopsin); 180° means that it is all-*trans* (bathorhodopsin). In the ground state (lower curve), rhodopsin is stabilized with respect to bathorhodopsin, possibly because the positive charge of the protonated Schiff's base interacts more favorably with a negatively charged group of the protein (see fig. 29.10). The energy of the excited state (upper curve) is less sensitive to the position of the Schiff's base than the energy of the ground state is, because the positive charge has moved to the opposite end of the retinal molecule (see fig. 29.8). The energy of the excited state is minimal when the rotation angle is about 90°. Rhodopsin is raised from the ground state to the excited state by light (vertical arrow). It relaxes back to the ground state along a path like that indicated by the dashed arrow, usually ending up as bathorhodopsin but sometimes returning to the original 11-*cis*-isomer.

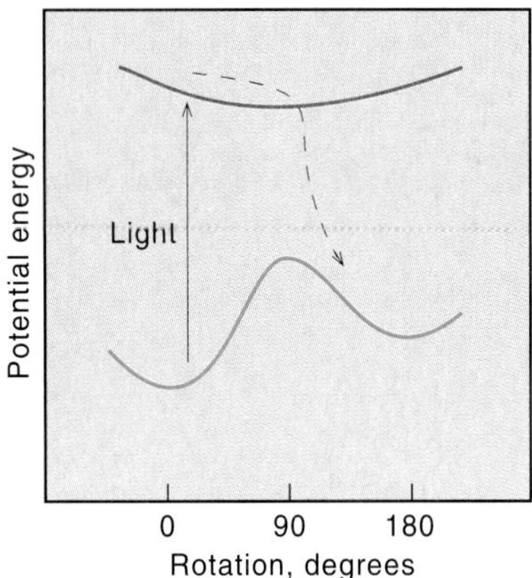

rhodopsin. Resonance Raman measurements support the view that the retinal in bathorhodopsin has isomerized to the all-*trans* form, but that it continues to be held as a protonated Schiff base. Additional evidence that bathorhodopsin contains all-*trans*-retinal has been obtained by studying a modified form of rhodopsin, isorhodopsin, which contains the 9-*cis* isomer of retinal. When isorhodopsin is illuminated, it gives rise to bathorhodopsin with 10^{-11} s, just as rhodopsin (11-*cis*) does. The only common product that could form directly from both the 9-*cis* and 11-*cis* isomers is the all-*trans* isomer.

Isomerization of the retinal Schiff base can occur when the molecule is excited with light, because the C-11=C-12 bond loses much of its double-bond character in the excited state. The valence bond diagrams of figure 29.8 illustrate this point qualitatively. In the ground state of rhodopsin the potential energy barrier to rotation about the C-11=C-12 bond is probably on the order of 30 kcal/mole. This barrier essentially vanishes in the excited state. In fact, molecular orbital calculations suggest that the energy of the excited molecule is minimal when the C-11=C-12 bond is twisted by about 90° (fig. 29.9). The 11-*cis* and all-*trans* molecules thus have a common excited state. When the molecule decays from the excited state to a ground state, it can end up in either of these isomeric forms. It turns out that the excited state decays to bathorhodopsin (all-*trans*) about 67% of the time and to rhodopsin (11-*cis*) about 33% of the time. Isorhodopsin (9-*cis*) and other isomers are formed only in small amounts (about 1%). Similar ratios of products hold no matter whether we start by exciting rhodopsin or bathorhodopsin.

The high energy barrier to the isomerization of rhodopsin in the dark is physiologically important because it limits the "noise" in our perception of light. If the barrier were low enough to be overcome thermally, the discrimination of light from dark would be more difficult.

Because opsin binds 11-*cis*-retinal, but not all-*trans*-retinal, the all-*trans* molecule that is created in bathorhodopsin must find itself initially in a binding site that is tailored for the 11-*cis* structure. In solution, 11-*cis*-retinal is less stable than all-*trans*-retinal, because of steric repulsion between the methyl group on C-13 and the hydrogen atom on C-10 (see fig. 29.4). On the protein there must be compensating interactions that decrease the free energy of the 11-*cis* complex relative to that of the all-*trans* complex. These interactions will be disrupted when the pigment is isomerized. The pronounced shift of the absorption spectrum to longer wavelengths in bathorhodopsin could be explained if the isomerization resulted in a movement of the Schiff base nitrogen, increasing the distance between the positively charged nitrogen and its counterion (fig. 29.10). Such a movement would destabilize the ground state of bathorhodopsin relative to the excited state, as we discussed earlier for rhodopsin. A repositioning of the Schiff base also is likely to lead to changes in the structure of the protein surrounding the binding site. If the positively charged nitrogen moves closer to a second anionic group, as illustrated in figure 29.10, a new hydrogen bond could form between this group and the Schiff base nitrogen, and we would expect the protein to respond by adjusting the positions of other nuclei nearby.

The reorganization of rhodopsin that is set in motion by isomerization of the retinylidine Schiff base continues as the system relaxes through the states lumirhodopsin, metarhodopsin I, and metarhodopsin II. Physical measurements indicate that purified rhodopsin undergoes particularly substantial changes in protein conformation during the transition from metarhodopsin I to metarhodopsin II. The extent of these changes and the kinetics of formation of metarhodopsin II are sensitive to the type and number of phospholipids surrounding the protein. Conformation changes also undoubtedly occur in this step when the rhodopsin is in place in the disk membrane, but they are subtler and may be confined to small regions of the protein. The nitrogen of the retinyl Schiff base loses its proton as metarhodopsin II is formed, and a second, unidentified group takes up a proton from the solution. The loss of the positive charge on the nitrogen accounts for the shift of the absorption maximum to 380 nm, where unprotonated Schiff bases of all-*trans*-retinal absorb. The Schiff base linkage also becomes accessible to reagents in the aqueous solution during or shortly after the formation of metarhodopsin II.

The formation of metarhodopsin II is fast enough to be an obligatory step in visual transduction. It clearly is associated with changes in the interactions between rhodopsin and its surroundings. A reasonable hypothesis, therefore, is that the changes in protein structure allow metarhodopsin II to initiate an interaction with some other component of the disk membrane. We will explore the nature of this component in the following sections.

Figure 29.10

A model for the photochemical conversion of rhodopsin *(top)* to bathorhodopsin *(bottom)*. In rhodopsin the proton of the Schiff's base nitrogen is hydrogen bonded to a counterion, A_1^-. The retinal is twisted about the C-12—C-13 and C-6—C-7 single bonds. (As noted in figure 29.4, free retinal is twisted in this way, but the amount of twisting in rhodopsin is still uncertain.) When rhodopsin is converted to bathorhodopsin *(bottom)*, the ring end of the retinal is presumed to be locked in position. Isomerization about the C-11═C-12 double bond flips the protonated nitrogen and its positive charge away from the counterion A_1^-. The separation of electrical charge could account for the shift of the absorption spectrum to longer wavelengths. The isomerization also can lead to proton movements. The proton of the Schiff's base could be passed to another basic group (A_2^-). Other groups of the protein, including the lysine attached to the retinal, also must adjust to the new geometry of the retinyl Schiff's base. (For additional discussion, see B. Honig et al., An external point-charge model for wavelength regulation in visual pigments, *J. Am. Chem. Soc.* 101:7084, 1979, and Photoisomerization, energy storage, and charge separation: A model for light-energy transduction in visual pigments and bacteriorhodopsin, *Proc. Natl. Acad. Sci. USA* 76:2503, 1979.)

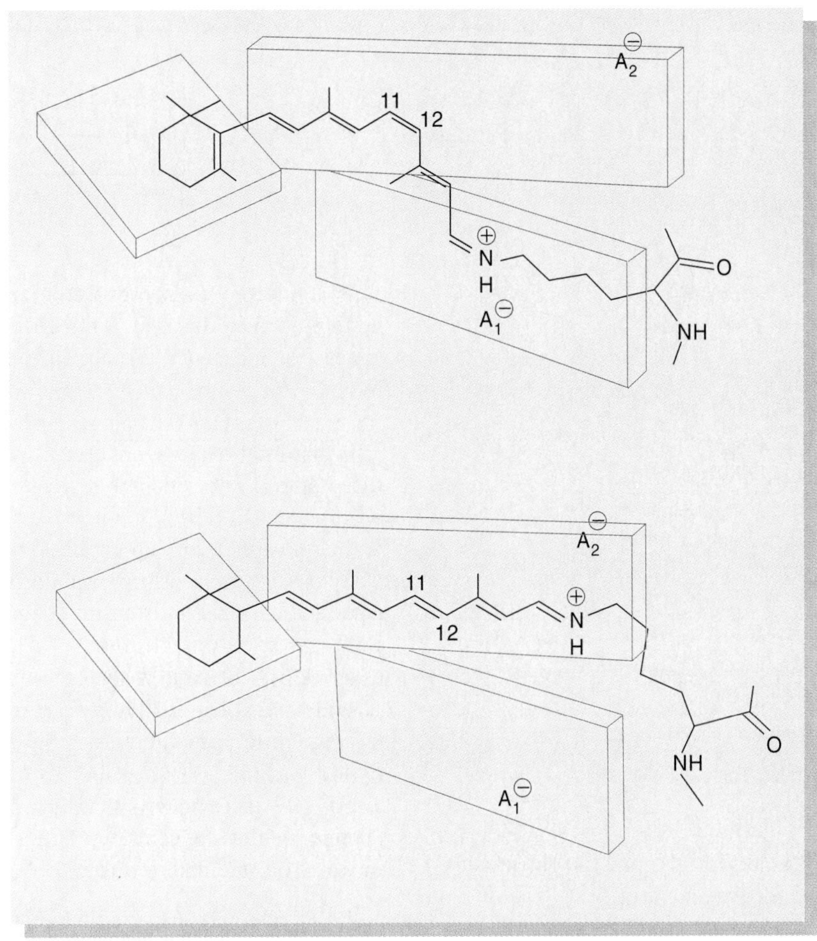

The Conductivity Change That Results from Absorption of a Photon

The human eye is amazingly sensitive. After being in the dark for a time, we can perceive continuous light that is so weak that an individual rod cell absorbs a photon, on the average, only once every 38 minutes. A flash of light is detectable if approximately six rods each absorb one photon. This means that the absorption of a single photon by any one of the approximately 3×10^7 molecules of rhodopsin in a rod must be sufficient to excite the cell and trigger a neuronal response. The combined responses of six rods can elicit the sensation of seeing.

The electrophysiological response of a rod or cone to light involves a change in the permeability of the cytoplasmic membrane to cations. In the inner segment and basal parts of the cell, the cytoplasmic membrane contains a Na^+–K^+ pump, which uses ATP to move Na^+ out of the cell and K^+ in. (Recall our discussion of ion pumps in chapter 20.) K^+ can diffuse back out of the cell relatively freely, and its efflux causes the membrane to become negatively charged by about 20 mV on the inside relative to the outside. Na^+, on the other hand, cannot readily pass back across the membrane into the inner segment. Electrophysiological measurements by William Hagins showed that Na^+ flows outside the cell to the region of the outer segment and that it reenters the outer segment through channels that are selectively permeable to cations (fig. 29.11). Once back inside, the Na^+ flows to the inner segment to be pumped out again. The round trip takes on the order of a minute. When a rod cell of a vertebrate's retina is excited

Figure 29.11

Na^+ is pumped out of the inner segment (solid arrows) and diffuses back into the cell through channels in the outer segment (dashed arrows). In rods of vertebrates the Na^+ channels are held open in the dark, and they close in the light. (In invertebrates the channels open in the light.)

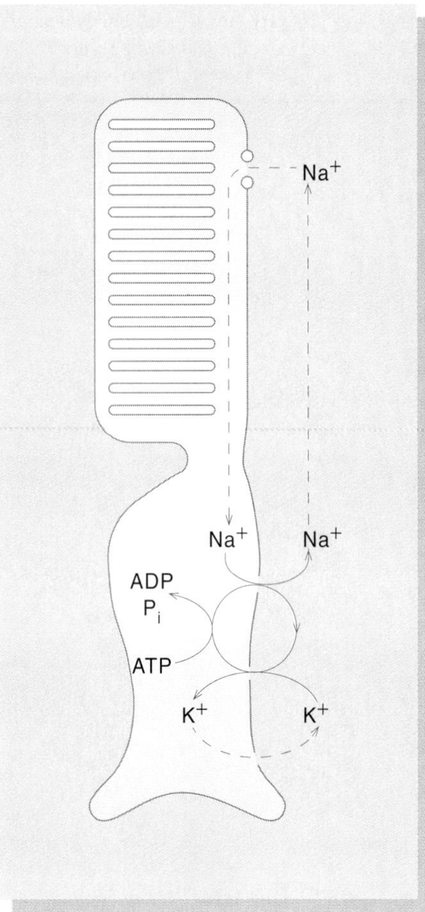

Figure 29.12

cGMP increases the electric conductivity of the rod cell cytoplasmic membrane. The conductivity was measured with a patch of membrane held at the tip of a micropipet. Experiments with solutions containing various ions showed that cGMP affects the movement of Na^+ and other cations across the membrane. Note the sigmoidal concentration dependence: the Hill coefficient is 1.6. (For more details, see E. E. Fesenko et al. *Nature* 313:310, 1985.)

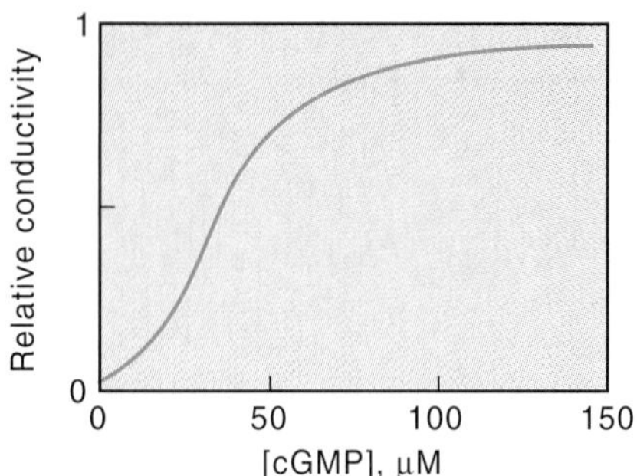

with light, some of the cation channels in the outer segment suddenly close, decreasing the inward movement of Na^+. The interruption of the Na^+ current causes an increase in the electrical potential across the cytoplasmic membrane. This hyperpolarization causes a change in the movement of a neurotransmitter to the bipolar cell that makes a synaptic junction with the rod. The bipolar cell then sends a signal to a ganglion cell, the third component in the hierarchy of retinal neurons.

The hyperpolarization of the rod and the response of the bipolar cell are graded effects. Absorption of a single photon causes the membrane potential to increase by about 1 mV, with the effect peaking about 1 s after the excitation. Up to about 100 photons per rod, the more light the cell absorbs, the larger the amplitude of the hyperpolarization and the faster the hyperpolarization occurs. The reaction of the ganglion cell, however, is all-or-none. When the ganglion cell is triggered, it responds with an action potential that proceeds to the brain. Sensitivity to weak light is enhanced (with a sacrifice in spatial resolution) by having multiple rod cells connected to each bipolar cell and multiple bipolar cells connected to each ganglion cell. The output of cone cells is not summed in this way. This difference partly explains why the cones provide better visual acuity but lower sensitivity than the rods.

Analogous events occur in the eyes of invertebrates, except that light causes an increase in the cation permeability of the receptor cell rather than a decrease. In the rods of vertebrates the absorption of a single photon decreases the current of Na^+ into the outer segment by about 3%. The inflow of approximately 10^6 Na^+ ions is transiently prevented. Exactly how many Na^+ channels have to close in order to achieve this effect is uncertain, but estimates have ranged from 25 to 1,000. Since only one molecule of rhodopsin is excited, the cell must have a mechanism for amplifying the effect of light by a substantial factor. Recall also that the cytoplasmic membrane of the rod outer segment is not contiguous with the disk membranes that hold the rhodopsin (see fig. 29.3). This fact suggests that the excitation of rhodopsin causes a change in the concentration of a diffusible transmitter, which moves from the disk to the cytoplasmic membrane.

The Effect of Light Is Mediated by Guanine Nucleotides

There is strong evidence that the diffusible component that moves between the disk and the cytoplasmic membrane is 3′,5′-cyclic-GMP (cGMP). Electrophysiological measurements have shown that cGMP causes an increase in the permeability of the cytoplasmic membrane to Na^+ (fig. 29.12). This appears to be a direct effect of cGMP on the Na^+ channels, rather than an indirect effect mediated by a kinase, because it can be seen in the absence of ATP.

The disk membranes of the rod outer segment contain a phosphodiesterase, which hydrolyzes 3′-5′-cyclic-GMP (cGMP) to 5′-GMP. If the disk membranes are kept in the dark, the phosphodiesterase remains in a relatively inactive state. When the disks are illuminated and rhodopsin is converted to metarhodopsin II, the activity of the phosphodiesterase increases. For each molecule of rhodopsin that is excited, on the order of 500 molecules of the

Figure 29.13

The large arrows show the main steps in the activation of the cGMP phosphodiesterase by light. Light converts rhodopsin (R) to a metastable form (M_{II}, probably metarhodopsin II). M_{II} reacts catalytically with transducin, which contains GDP bound to one of its three subunits (α). The interaction with M_{II} changes the specificity of the nucleotide binding site so that GDP is released and GTP is bound. This causes the α subunit to dissociate from the other two subunits of transducin (β and γ). In the dark the phosphodiesterase (PDE) is inactive because of the presence of an inhibitory polypeptide (I). The I polypeptide is removed by binding to the complex of the transducin α subunit and GTP. The active phosphodiesterase then can hydrolyze cGMP to 5′-GMP. M_{II} can go on to activate several hundred additional molecules of transducin before slower enzymatic reactions (thin arrows) gradually convert it to an inactive form. The inactivation involves conversion of the rhodopsin to a phosphorylated form (R-P) and an interaction with another protein (arrestin A). The GTP bound to transducin's α subunit eventually is hydrolyzed to GDP. When this happens, I dissociates and returns to the phosphodiesterase, the three subunits of transducin reassemble, and the system returns to its resting state. (For details, see M. Chabre, Trigger and amplification mechanisms in visual phototransduction, *Ann. Rev. Biophys. Chem.* 14:331, 1985.)

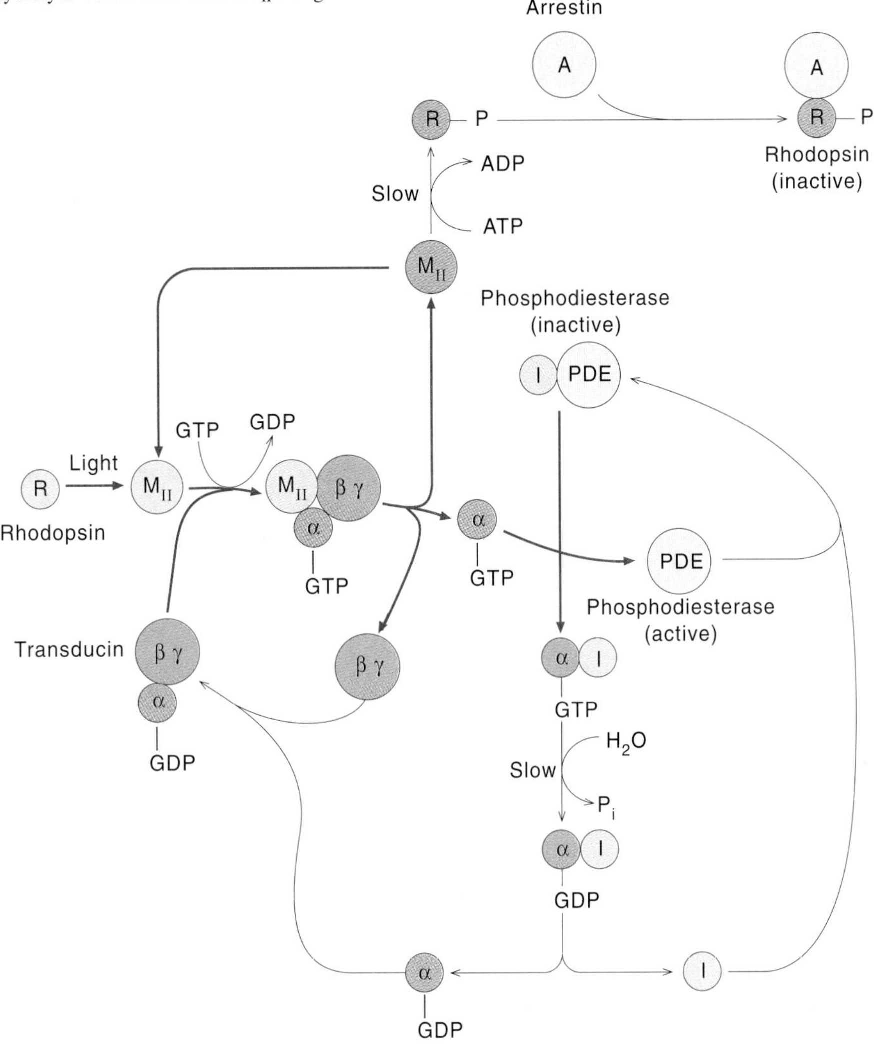

phosphodiesterase are activated. Illumination thus results in a decrease in the cGMP content of the cell. A drop in the cGMP concentration is in the right direction to cause a decrease in Na^+ permeability of the cell membrane (see fig. 29.12) and thus to cause a hyperpolarization of the membrane. In agreement with this scheme, injecting extra cGMP into rod outer segments temporarily inhibits the hyperpolarization caused by light.

Unlike rhodopsin, the phosphodiesterase is a peripheral membrane protein that can be readily solubilized. Its activation by light requires the presence of GTP and is associated with the binding of GTP to a second peripheral membrane protein called transducin. Transducin is a member of the family of G proteins that participate in the activation or inhibition of adenylate cyclase by hormones in other tissues (see chapter 27). Like other G proteins, transducin consists of three subunits, α, β, and γ (fig. 29.13). In the resting state, the α subunit contains a molecule of bound GDP. When rhodopsin is transformed to metarhodopsin II by light, it interacts with transducin, causing GTP to displace the bound GDP. Once GTP is attached, the α subunit separates from the β and γ subunits and binds to an inhibitory subunit of the phosphodiesterase. The removal of the inhibitory component activates the phosphodiesterase.

Each molecule of metarhodopsin II that is generated by light appears to be able to trigger about 500 molecules of transducin to bind GTP in place of GDP. This number agrees with the number of phosphodiesterase molecules that are activated, and accounts for much of the amplification in the response to light. Additional amplification results from the enzymatic action of each phosphodiesterase on many molecules of cGMP. These effects can be demonstrated with purified preparations of rhodopsin, transducin, and the phosphodiesterase. Purified, illuminated rhodopsin that has been incorporated into phospholipid vesicles is capable of activating transducin, and the complex of transducin's α subunit with GTP is capable of activating the phosphodiesterase in the absence of rhodopsin. The interaction of rhodopsin with transducin appears to involve the loops connecting helix C to D and helix E to F on the cytoplasmic side of the disk membrane (see fig. 29.5).

In order for the eye to respond rapidly to changing light intensities, the activation of the phosphodiesterase must be a transient response that quickly switches off again. The activation of transducin's α subunit is reversed by hydrolysis of the bound GTP to form bound GDP and free inorganic phosphate (see fig 29.13). When this happens, the α subunit separates from the inhibitory polypeptide of the phosphodiesterase and recombines with the β and γ subunits of transducin. The inhibitory polypeptide then recombines with the phosphodiesterase, returning the phosphodiesterase to its resting, inactive state. All of these processes can be blocked by replacing the nucleotide bound to transducin with a nonhydrolyzable analog of GTP.

To switch off the response to light, two additional things must happen: Metarhodopsin II must decay to a form that is incapable of activating additional molecules of transducin, and the cGMP that was hydrolyzed must be resynthesized in order to reopen the cation channels in the plasma membrane. The inactivation of metarhodopsin involves phosphorylation of the protein at multiple serine and threonine residues near the carbonyl-terminal end (see fig. 29.5) and binding of the phosphorylated rhodopsin to another protein called arrestin. The regeneration of the cGMP is carried out by a guanylate cyclase, which catalyzes the synthesis of cGMP from GTP. This enzyme is stimulated by still another regulatory protein but is strongly inhibited by Ca^{2+}. Ca^{2+} enters the rod outer segment through the same cation channels that admit Na^{+}, and it is exported by a transport protein that exchanges Ca^{2+} for Na^{+}. The Ca^{2+} concentration in the cell thus decreases upon illumination, when the cation channels close. The drop in the Ca^{2+} concentration results in an activation of the guanylate cyclase. Changes in the intracellular Ca^{2+} concentration also appear to underlie the ability of the eye to adapt to large changes in illumination intensity.

Regeneration of 11-*cis*-Retinal by Way of a Retinyl Ester

The transformations of rhodopsin begin with 11-*cis*-retinal bound to opsin and culminate in the release of all-*trans*-retinal from opsin. How is the 11-*cis*-isomer regenerated? The regeneration turns out to be more complex than we might have expected (fig. 29.14). The all-*trans*-retinal first is reduced to all-*trans*-retinol by NADH. The all-*trans*-retinol then leaves the retina and is taken up by the pigment epithelium, a separate tissue at the back of the eye, where it is esterified by an acyltransferase. This enzyme transfers a fatty acid from phosphatidylcholine to the retinol, forming an all-*trans*-retinyl ester. The retinyl ester then breaks down in an unusual, concerted reaction that couples isomerization of the retinol and the release of the free fatty acid. 11-*cis*-Retinal is regenerated by oxidation of the 11-*cis*-retinol in the pigment epithelium. It then returns to the retina and is taken up again by opsin. The free energy that is needed to drive the conversion of retinol from all-*trans* to the less stable 11-*cis* isomer thus is provided by hydrolysis of a phospholipid.

Rhodopsin Movement in the Disk Membrane

Each molecule of rhodopsin that is converted to metarhodopsin II is capable of causing some 500 molecules of transducin to bind GTP. Since this happens within about 0.5 s after the absorption of a photon, either rhodopsin or transducin, or both, must move about rapidly enough to encounter thousands of reaction partners per second. Rhodopsin is firmly embedded in the disk membrane and transducin is attached to the surface of the membrane, so they need to diffuse only in two dimensions in the plane of the membrane. But is it likely that they could move this rapidly? Actually, evidence that rhodopsin can diffuse rapidly within the membrane was obtained prior to the discovery of transducin and the phosphodiesterase.

The initial indications that rhodopsin moves around in the membrane came from studies of linear dichroism of rod outer segment disks. A material is said to exhibit linear dichroism if the strength of its optical absorption, when measured with polarized light, depends on the orientation of the polarizer. The polarization of a light beam is defined by the orientation of the light's electric field, as indicated in figure 17.6. Molecules in solution usually do not exhibit linear dichroism because they are free to tumble about. At any given time, the solution contains molecules with all possible orientations relative to the polarizer, so the absorbance of the solution will not change if we rotate the polarizer. But if all the molecules are fixed in position with their molecular axes aligned, the system generally will exhibit linear dichroism. In the case of retinal or its Schiff base, light is absorbed best when the polarization makes the electric vector of the light parallel to the long axis of the molecule. You can get a feeling for this by examining the diagrams in figure 29.8. In order to convert the molecule from the ground state to the excited state, the electric field of the light must move electron density away from the ring end of the molecule and in the direction of the nitrogen.

The vertebrate retina is well suited for measurements of linear dichroism, because the rod cells are neatly aligned with their long axes perpendicular to the plane of the tissue. Suppose that we send a weak beam of polarized light through the rods from the side (fig. 29.15*a* and *b*). The polarization of the beam then could be made either parallel or perpendicular to the long axis of the cells. The planes of the disk membranes in the rod outer segments are aligned perpendicular to the cell axis, so the polarization of the light will be either perpendicular or parallel to the membrane

Figure 29.14

11-*cis*-retinal is regenerated from all-*trans*-retinal in a series of reactions that involve the formation of a retinyl ester as an intermediate. Most of these reactions occur in the pigment epithelium. (R_1 and R_2 = fatty acid side chains; R_3 = choline.)

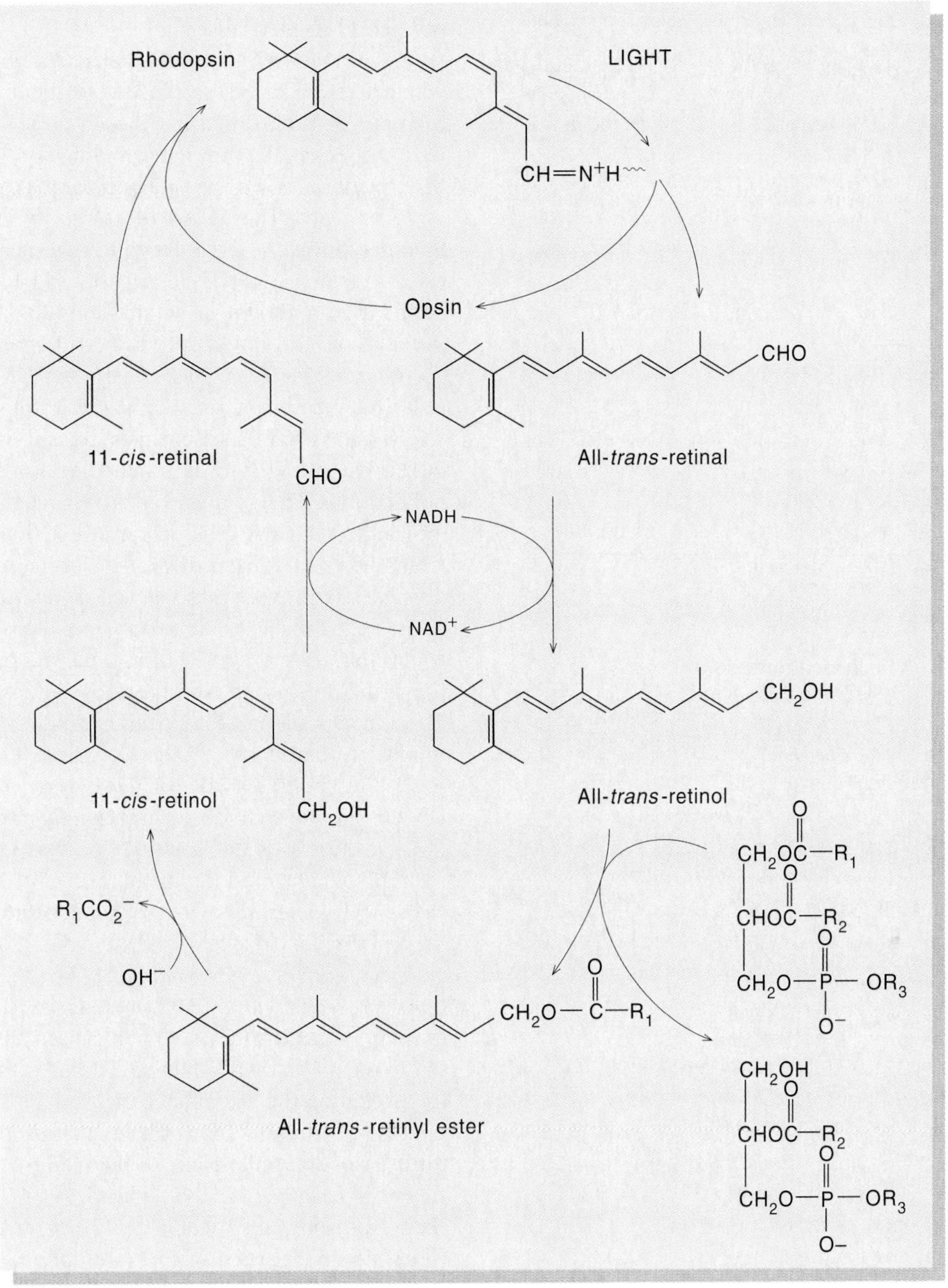

surface. It turns out that the rhodopsin absorbs much more strongly when the light is polarized parallel to the membranes than it does when the polarization is perpendicular. This means that the long axes of the retinyl groups must be held more or less parallel to the plane of the membrane. Now suppose that we send the light beam through the cells end-on (see fig. 29.15*c* and *d*). The polarization then must be parallel to the plane of the disk membranes, no matter how the polarizer is turned. With light coming through the cells end-on, the absorbance of the rhodopsin is found to be independent of the direction of the polarization. This means that the retinyl groups are not aligned in any particular direction within the plane of the membrane, but instead can take on all possible orientations in this plane.

It is interesting to note in passing that the orientation of the retinyl groups in the retinas of vertebrates optimizes our visual sensitivity for unpolarized light. Most of the light reaching the retinas passes through the rods end-on and is absorbed well, whatever its polarization (see fig. 29.15*c* and *d*). Insects, however,

Figure 29.15

When rod cells are illuminated from the side, the light can be polarized either perpendicular (*a*) or parallel (*b*) to the planes of the disk membranes. The absorbance of the rhodopsin is much greater for parallel polarization than it is for perpendicular. When the cells are illuminated end-on, the light's electric field has to be parallel to the membranes, no matter how the light is polarized (*c* and *d*). With end-on illumination the absorbance of the rhodopsin is independent of the polarization. These observations show that the 11-*cis*-retinyl groups (small arrows in disks) are held parallel to the plane of the membrane but can point in any direction in this plane.

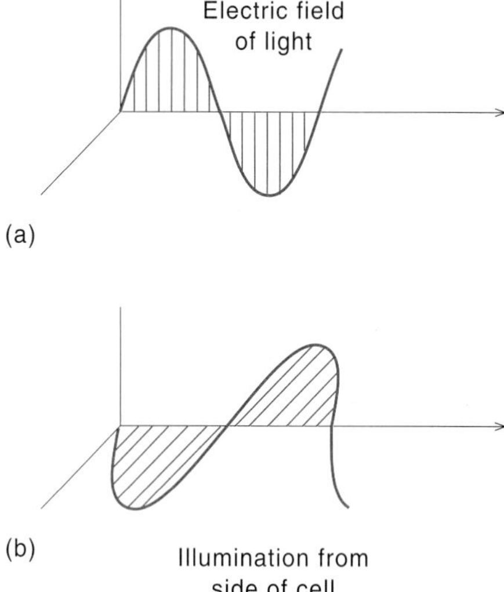

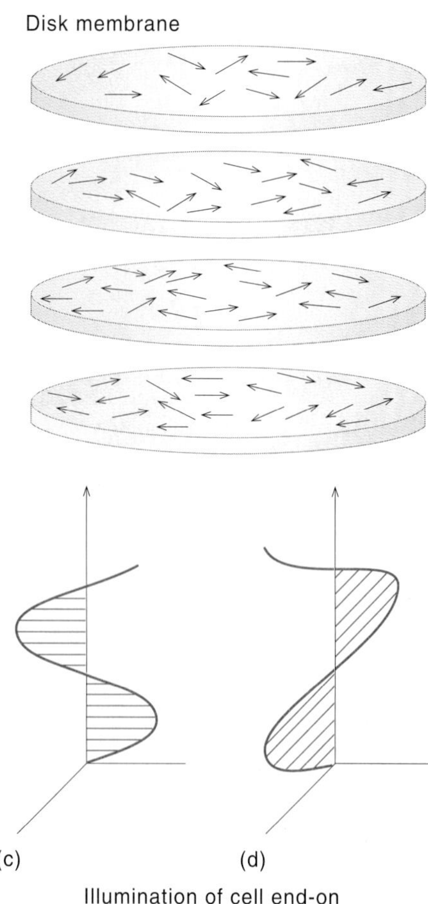

can distinguish between horizontally and vertically polarized light. This ability can be advantageous, because light that is reflected by smooth surfaces of leaves or water is partially polarized, whereas light coming directly from the sky is unpolarized. Insects may make this distinction by having separate sets of cells with differently aligned retinyl groups.

How can measurements of linear dichroism reveal whether rhodopsin molecules move about within the membrane? Suppose that the retina is exposed to a weak flash of polarized light that enters the rods end-on. Some of the retinyl groups in the disk membranes will be oriented parallel to the polarization, and some will not. The light will *selectively* excite those molecules that are parallel to the polarization, because these have the highest absorbance. Most of the molecules that are excited will be converted to bathorhodopsin and the series of states that descend from it, and their absorption spectrum will change. If we measure the absorbance changes a short time after the excitation flash, using polarized measuring light that passes through the cells end-on, we see that the absorbance changes depend on how the polarization of the measuring light is oriented with respect to the excitation polarization. The polarized measuring light selectively measures molecules that have a particular orientation. Before the excitation flash, there is no linear dichroism for measurement made end-on; after the flash, there is. However, if the rhodopsin molecules can rotate in the plane of the membrane, the orientations of the molecules that were excited will in time become randomized, and the linear dichroism that is induced by the excitation will disappear. The kinetics of this disappearance will depend on how rapidly rhodopsin rotates.

Experiments of this sort were done by Richard Cone. Rhodopsin proved to rotate surprisingly rapidly. The time required for a rotation is about 2×10^{-5} s. This means that the environment of rhodopsin in the membrane must be highly fluid.

To determine whether rhodopsin can diffuse laterally in the membrane, Cone and Paul Leibman independently used a microspectrophotometric technique. They excited individual rod outer segments end-on near one edge of the disks through a microscope. Initially, the excitation caused absorbance changes only in the region that was illuminated, but within a few seconds after the excitation the absorbance changes became uniformly distributed across the disk. This means that rhodopsin must be able to diffuse rapidly in the plane of the membrane. The diffusion coefficient is calculated to be about 5×10^{-9} cm^2/s. At this speed, rhodopsin molecules would collide with each other 10^5 to 10^6 times a second, and it would take only about a second for a single rhodopsin to collide with most of the molecules of transducin on the surface of the disk.

Bacteriorhodopsin: A Bacterial Pigment-Protein Complex That Resembles Rhodopsin

Halobacterium halobium is a red-colored, halophilic (salt-loving) bacterium that thrives in tidal salt flats. Its red color comes from two components of its cytoplasmic membrane: carotenoids and a purple pigment–protein complex that bears a striking resemblance to rhodopsin. The purple complex, bacteriorhodopsin, was first described by Walter Stoeckenius and his colleagues. Bacteriorhodopsin is not distributed randomly throughout the cell membrane, but rather is collected in patches that are held together by strong interactions among the bacteriorhodopsins. The patches, or purple membranes, can be isolated relatively simply if the cells are disrupted by being placed in distilled water. Bacteriorhosopsin is the only protein in the purple membrane. It forms a highly ordered two-dimensional array that is almost crystalline in nature. Electron diffraction studies have shown that the individual proteins are folded into seven α-helical stretches extending from one side of the membrane to the other (see fig. 19.13).

Like rhodopsin, bacteriorhodopsin contains retinal bound to a lysine as a protonated Schiff base. The retinal isomerizes when it is excited with light. However, bacteriorhosopsin differs from rhodopsin in that the retinal starts out in the all-*trans* form rather than as the 11-*cis* isomer. The main product of the isomerization is the 13-*cis* isomer. Like rhodopsin, the excited bacteriorhodopsin progresses through a series of metastable states with different absorption spectra. The first of these states resembles bathorhodospin in having an absorption spectrum that is shifted markedly to longer wavelengths. A later state resembles metarhodopsin II in that the retinyl Schiff base has lost its proton. A major difference, however, is that the transformations of bacteriorhodopsin are cyclic. Rather than dissociating from the protein, the retinal returns to the all-*trans* form, and the complex relaxes back to its original state and is ready to operate again.

The role of bacteriohodopsin in *H. halobium* is to pump protons across the cytoplasmic membrane. We discussed some aspects of this pumping in chapter 20. As the excited bacteriorhodopsin undergoes its cyclic series of transformations, a proton is taken up from the solution inside the cell and released on the extracellular side of the membrane. This probably is the proton that is initially on the nitrogen atom of the Schiff base, because the Schiff base is deprotonated and then reprotonated in the course of the cycle. Isomerization of the retinal could move the proton with the nitrogen from a position where it equilibrates indirectly with the cytosol to a position where it equilibrates with the extracellular solution. Figure 29.10 will serve to illustrate this idea. The figure was drawn to show the isomerization of 11-*cis*-retinal to all-*trans*-retinal that occurs in rhodopsin, but the principles here are conceptually the same. The isomerization of the retinyl chain pushes the proton on the Schiff base nitrogen from the region of one functional group (A_1^- in the figure) to the region of another (A_2^-). From A_2H a proton could be conducted to the solution on the extracellular side of the membrane. When the retinal then returns to its original state, the proton lost by the nitrogen could be replaced by one from a third group on the cytoplasmic side of the membrane.

As in other types of photosynthetic bacteria, the protons that are pumped across the membrane by bacteriorhodopsin in the light can reenter the cell by way of a proton-conducting ATP-synthase, generating ATP. *H. halobium* also contains a transport system that moves Na^+ ions out of the cell in exchange for protons coming in. Removal of Na^+ is a major chore for an organism that lives in water containing 4-M NaCl. The cytoplasmic membrane also has two other pigment–protein complexes that execute photochemical cycles similar to the transformations of bacteriorhodopsin. One of these, halorhodopsin, acts as a light-driven pump for Cl^-. The other, sensory rhodopsin, is involved in phototaxis, the tendency of the cells to swim in the direction of a source of light.

Summary

In this chapter we have described the biochemical mechanisms involved in vision. The main points of our discussion are as follows.

1. Light rays entering the eye of a vertebrate are refracted by the cornea and focused to form an image on the retina. The retina contains several layers of cells. Some of these cells, the rods and cones, contain the visual pigments that are responsible for the initial response to light; other cells make synaptic connections to rods or cones and to additional neural cells that carry impulses to the brain.
2. The light-sensitive protein complex in the rods, rhodopsin, consists of 11-*cis*-retinal bound as a Schiff base to a protein, opsin. Rhodopsin is an integral constituent of membranes that form a stack of disks at one end of the cell. Cone cells, which are responsible for the perception of color, contain similar complexes in infoldings of the plasma membrane.
3. When rhodopsin absorbs light, the retinal isomerizes to the all-*trans*-isomer. This change initiates a series of transformations of the pigment–protein complex that result in an interaction with another protein, transducin. The interaction with rhodopsin causes one of transducin's subunits to take up a molecule of GTP in exchange for bound GDP, to dissociate from the other subunits, and to react with a phosphodiesterase that hydrolyzes cGMP. This reaction activates the phosphodiesterase. The resulting drop in cGMP concentration leads to a decrease in the Na^+ permeability of the cytoplasmic membrane. A hyperpolarization of the membrane then triggers the transmission of a signal at the rod's synaptic junction to the adjacent neural cell.
4. Bacteriorhodopsin, a bacterial pigment–protein complex, resembles rhodopsin (it contains all-*trans*-retinal) but serves a different function. Like rhodopsin, bacteriorhodopsin progresses through a series of metastable states when it is excited with light. In the course of these transformations, protons are taken up from the solution inside the cell and are released on the extracellular side of the membrane, generating an electrochemical potential gradient for protons across the membrane. Protons reenter the cell by way of an ATP-synthase, energizing the formation of ATP.

Selected Readings

Cone, R., Rotational diffusion of rhodopsin in visual photoreceptor membrane. *Nature* 236:39, 1972. Excitation of the retina with polarized light creates a transient linear dichroism. The decay of the dichroism is explained by rotation of rhodopsin.

Fesenko, E. E., S. S. Kolesnikov, and A. L. Lymbarsky, Induction by cyclic GMP of cation conductance in plasma membrane of retinal rod outer segment. *Nature* 313:310, 1985. cGMP increases the Na^+ conductance of the cytoplasmic membrane.

Khorana, H. G., Rhodopsin, photoreceptor of the rod cell. *J. Biol. Chem.* 267:1, 1992. A review covering the structure of rhodopsin and the roles of cGMP, GTP, the phosphodiesterase, and transducin in the response of the rod cell to light.

Nathans, J., Rhodopsin: Structure, function, and genetics. *Biochem.* 31:4923–4932, 1992. A review of recent work on the structure and reactions of rhodopsin.

Renaud, J.-P., N. Rochel, M. Ruff, V. Vivat, P. Chambon, H. Gronemeyer, and D. Moras, Crystal structure of the RAR-γ ligand binding domain to all-*trans* retinoic acid. *Nature* 378:681–689, 1995.

Wald, G., The molecular basis of visual excitation. *Nature* 219:800, 1968. Wald's Nobel Prize address, describing early work on the isomerization of retinal in rhodopsin.

Problems

1. Is the bond type between the palmitoyl moieties and rhodopsin typically found in proteins?
2. What functions would you surmise for the two hexasaccharides and two palmitoyl residues in rhodopsin?
3. What role, if any, does the disk lumen have in the processes involved in vision?
4. What chemical characteristics would you look for in a compound that absorbs visible light?
5. The spectrophotometers in biochemistry laboratories are very sensitive because they use a photomultiplier tube as a detector and convert photons of light into an elecrical signal. The signal subsequently is multiplied to produce a strong signal. How do our eyes multiply the signal derived from light to increase sensitivity?
6. Draw traces showing how the optical absorbance of a fresh suspension of rod outer segment disks might change as a function of time when the suspension is excited with a short flash of light at 37°C. Show the absorbance at (a) 545 nm and (b) 480 nm. Select the time scale for each trace judiciously so that the traces illustrate the kinetics of the major absorbance changes that occur at the two wavelengths. (You may need to use two traces with diffferent time scales at each wavelength to show both the initial absorbance change and its decay.)
7. If rhodopsin is illuminated at 500 nm at 77 K, the absorbance of the sample at 500 nm decreases. If the sample is then illuminated at 550 nm (still at 77 K), the absorbance at 500 nm increases again. Explain.
8. Why is the absorption spectrum of metarhodopsin II so different from that of metarhodopsin I?
9. Provide evidence and a model for the involvement of cGMP as a mediator in visual transduction.
10. In solution, 11-*cis*-retinal maximally absorbs light near 380 nm, but in rhodopsin the peak is at 500 nm. Explain the spectral shift in the retinal bound to the protein.
11. How do our eyes detect color? Why do we not see color in dim light?

PART

7

Storage and Utilization of Genetic Information

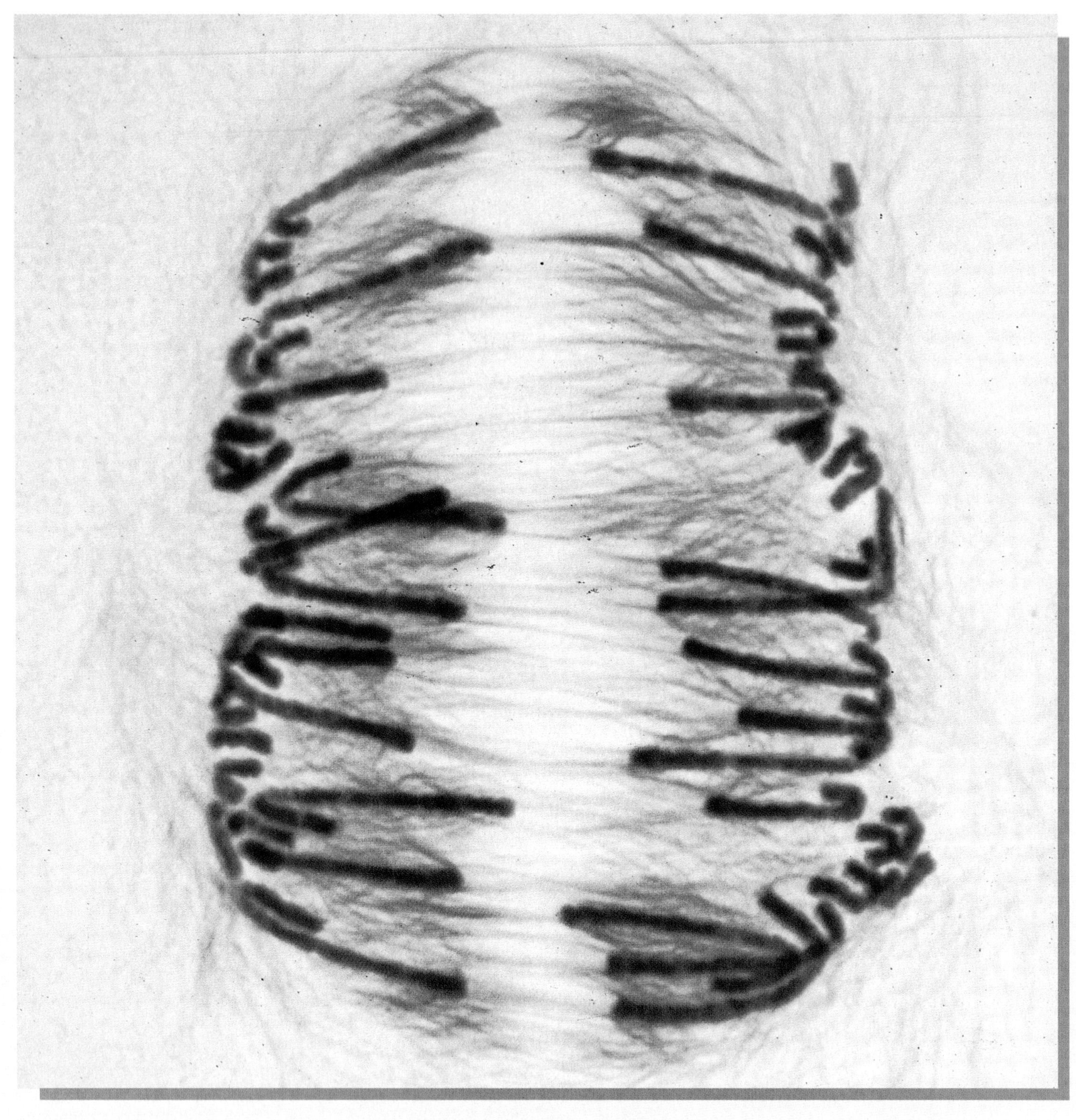

Light micrograph of late mitosis in a plant. Microtubules are stained red and chromosomes are counterstained blue. (Courtesy of Dr. Andrew Bajer, University of Oregon.)

C H A P T E R

Structures of Nucleic Acids and Nucleoproteins

Most DNAs form a double-stranded helical structure in which nitrogen bases from opposing chains form complementary structures that are hydrogen bonded with one another.

Nucleic acids occupy a unique position in the biochemical world. Not only are they involved in many important reactions, but they carry genetic information, which must be faithfully duplicated so that it can be passed from one cell generation to the next and from one organism to the next. Nonetheless, this information must be mutable to produce the variability on which the evolutionary selection process feeds. Finally, the DNA must be selectively transcribed so that each cell can synthesize the proteins it needs.

In the following five chapters we focus on the biochemistry of DNA, RNA, and protein, with occasional reminders of the genetic significance of these closely related processes. Appropriately, this section on nucleic acids and protein metabolism begins with a description of key experiments that demonstrated the genetic significance of nucleic acids. Following this we turn to a consideration of the structural properties of DNA and chromosomes. Next, reactions involving DNA are discussed, first purely biochemical reactions (chapter 31) and then in vitro and in vivo reactions in which DNA is manipulated (chapter 32). Chapter 33 deals with information transfer from DNA to RNA, and finally, chapter 34 covers the mechanism of information transfer from RNA to protein.

The Genetic Significance of Nucleic Acids

After the discovery around the turn of the century that genes are carried by chromosomes, a great deal of effort went into characterizing the sizes and shapes of chromosomes. But it was not until much later that significant progress was made in elucidating the chemical nature of the gene. Biologists were aware, as early as 1900, that chromosomes are composed of both nucleic acids and proteins. However, the seemingly simple chemical composition of nucleic acids misled early investigators into believing that nucleic acids were a purely structural component of the chromosome. The favored theory was that the arrangement of specific proteins along the chromosome accounted for gene specificity. This notion was dispelled in 1944 when Oswald T. Avery and his colleagues at the Rockefeller Institute (now University) demonstrated that purified deoxyribonucleic acid (DNA) contains the genetic determinants of the bacterium *Diplococcus pneumoniae* (now called *Strepto-*

coccus pneumoniae). In this section we discuss Avery's results, as well as some other historically important observations that led up to his experiments.

Transformation Is DNA-Mediated

In 1928, Fred Griffith was experimenting with two different strains of pneumococcus. Type S bacteria (S for smooth, from the appearance of bacterial colonies on agar plates) are encapsulated by polysaccharide. The capsules protect them from the host immune system, making the S bacteria pathogenic; even when small numbers of S bacteria are injected into mice, death results. By contrast, R bacteria (R for rough colonies) are nonencapsulated; they are readily attacked by the mouse's immune system and consequently are nonpathogenic. Although heat-killed S bacteria by themselves are nonvirulent, Griffith found that when heat-killed S bacteria were mixed with live R bacteria and introduced into a susceptible laboratory mouse, death of the animal frequently occurred (fig. 30.1*a*). S bacteria could then be detected in the blood. Apparently, the genetic factor required for encapsulation was transferred from killed S cells to live R cells.

Griffith's result was duplicated outside of the animal by Dawson and Sia in 1930. Both bacterial strains were grown in liquid growth medium and distinguished by the distinctive appearance of their colonies on plates (see fig. 30.1*b*). In parallel with the results obtained in the mouse, heat-treated extracts of S cells transformed R cells into S cells.

More than 10 years later, Avery provided convincing proof that the active agent in the S cell extracts was the cellular DNA. He and his co-workers did this by purifying the DNA from S cells and showing that it had the capacity to transform R cells into S cells in vitro (see fig. 30.1*b*). The transforming activity in the purified extract was destroyed if the extract was first incubated with the enzyme DNase, which specifically degrades DNA. By contrast the transforming activity was not affected by RNase, proteases, or enzymes that degrade capsular polysaccharides. It was subsequently found that DNA could be used to transfer many other genetic traits between the appropriate pairs of donor and recipient bacterial strains. For example, resistance to the antibiotics streptomycin or penicillin could be transferred, with the DNA of resistant cells, to sensitive cells. The transformation studies with different traits conferred by the donor DNA showed that the active genetic material being transferred faithfully reflected the genetic patterns of the donor strains. The procedure of altering the genetic composition of one cell strain by exposing it to DNA of another strain is termed transformation.

Transformation occurs naturally in only a small number of bacterial species other than *Streptococcus pneumoniae*. In organisms that are not transformed naturally, laboratory procedures are available to make the cell envelope partially permeable (e.g., of *E. coli*) and permit uptake of DNA. This is termed artificial transformation. It can be brought about by Ca^{2+} treatment and temperature shock, but the most efficient method found to be useful with most organisms is electroporation, or electric shock. For instance, by exposing yeast to electric field pulses it is possible to obtain transformation efficiencies that are 10^4 times greater than those obtained by conventional methods. Transformation is an essential step in the cloning of genes (see chapter 32). Transformation with self-replicating plasmids that may be carrying different genes has made it possible to select cells carrying the gene(s) of interest from a large population of cells.

Studies on Viruses Confirm the Genetic Nature of Nucleic Acid

For some time, doubts lingered about whether DNA, protein, or capsular polysaccharide was the transforming factor. During this period, other investigators performed some elegant experiments showing that nucleic acid was the genetic substance of bacterial viruses (bacteriophages) and plant viruses.

In 1952, A. D. Hershey and Martha Chase demonstrated the independent functions of viral protein and viral nucleic acid in the *E. coli* bacteriophage T2. This bacterial virus is composed of about equal weights of DNA and protein. The linear, duplex DNA is encompassed by a protein shell in a polyhedral "head" or capsid, which is connected to a protein tail that facilitates infection of cells. When a single bacteriophage particle infects a bacterium, it causes cell death and cell disruption (lysis), accompanied by the release of several hundred viruses within 30 min after infection. The fate of the DNA and the protein components of the infecting viruses can be followed by labeling with an appropriate radioactive isotope (fig. 30.2).

In an initial round of infection, Hershey and Chase used inorganic ^{32}P-labeled phosphate to label the DNA component of the virus and ^{35}S-labeled inorganic sulfate to label the protein component. The phosphorus became incorporated into the phosphoryl groups of the viral DNA and the sulfate into the sulfur-containing amino acids of the viral protein. In a second round of infection they added the radioactively labeled bacteriophages to cells in an unlabeled medium. Shortly after the bacterial host was infected, they subjected the phage-bacterium complex to vigorous agitation in a Waring blender. This treatment sheared the phage from the attachment sites on the bacterial cell wall. The bacteria were then sedimented by centrifugation, leaving the DNA-free viral protein in the supernatant. Analysis for radioactivity showed that most of the ^{32}P-labeled viral DNA sedimented with the bacteria, whereas most of the ^{35}S-labeled viral protein remained in the supernatant. That result demonstrated that most of the viral DNA entered the cells (see fig. 30.2*b*) during infection. Quantitative isotopic analysis of the progeny phage particles indicated that they retained less than 1% of the original infecting viral protein but about two-thirds of the original infecting DNA. In other words, most of the original DNA that entered the cell on infection also became an integral part of the progeny phage, but the protein did not. Since the genetic material should survive replication, the results strongly implicated DNA as the genetic component of bacteriophage T2. Long after these experiments were performed, it was shown that purified, deproteinized phage DNA could itself produce phage after being taken up by host cells. Infection of bacteria with phage-derived DNA is referred to as transfection.

Figure 30.1

In vivo and in vitro evidence that DNA causes transformation of *Streptococcus pneumoniae*. (*a*) Transformation experiment by Griffith. R bacteria are nonvirulent. S bacteria are virulent. A mixture of R bacteria and heat-killed S bacteria is also virulent if transformation has occurred. (*b*) Transformation experiment by Avery and co-workers. When bacteria from a liquid culture are spread on a semisolid medium, each cell adheres to the medium at random. As time passes, the cells and their offspring grow and divide, leading to visible clones or colonies, each of which arose from a single cell. R and S cells each have a distinct clonal morphology. Whereas R cells produce small rough colonies, S colonies are smooth. R cells exposed to DNA from S cells produce a mixture of both types of colonies: untransformed R colonies and transformed S colonies.

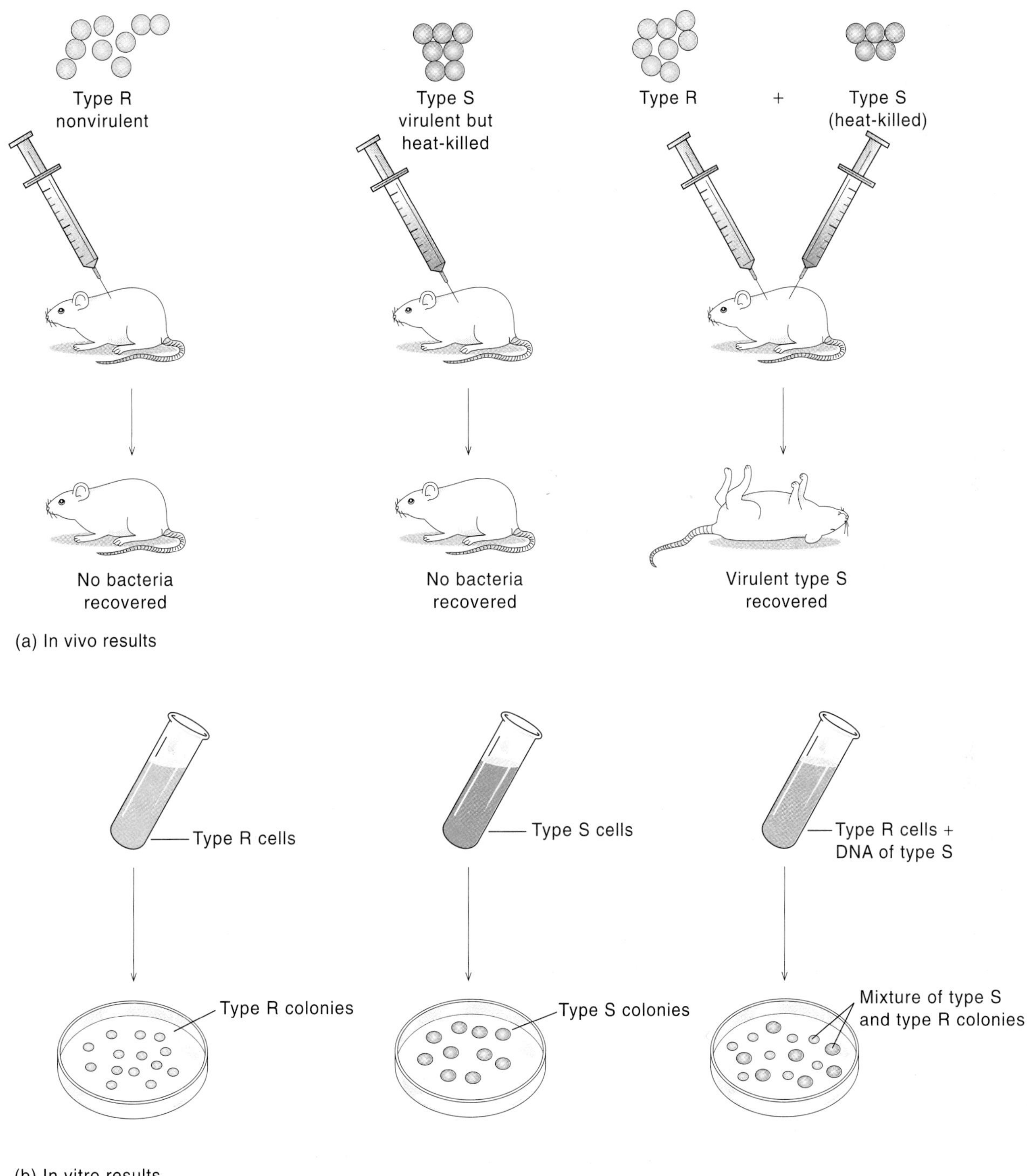

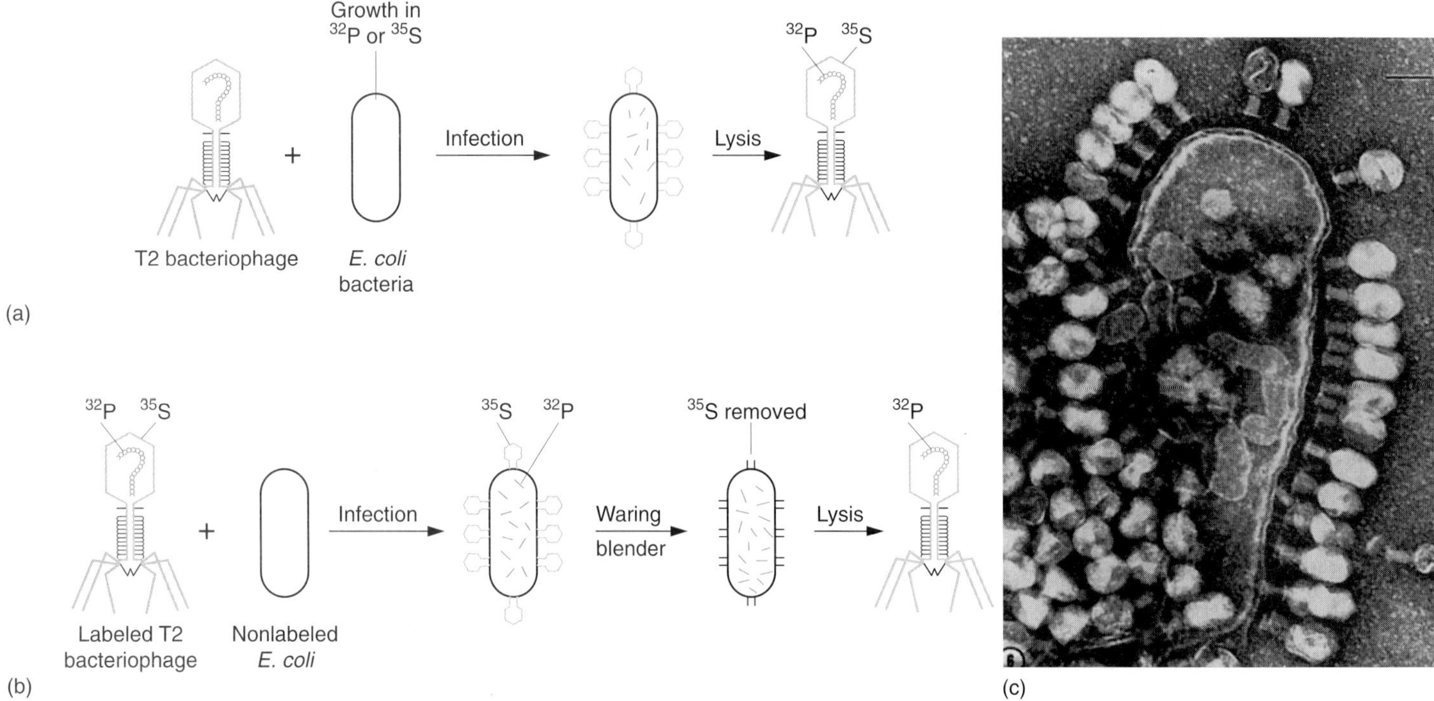

Figure 30.2

Hershey-Chase experiment demonstrating that the DNA but not the protein of the T2 bacteriophage is passed from parent to progeny phage particles. (*a*) Viruses with ^{32}P-labeled DNA or ^{35}S-labeled protein were prepared by growing the viruses on bacteria in growth medium containing $^{32}P\text{-}PO_4^{3-}$ or $^{35}S\text{-}SO_4^{2-}$, respectively. (*b*) When labeled phage particles infect an unlabeled *E. coli* cell, only the ^{32}P label appears in the progeny phage particles, showing that the DNA structure but not the protein structure is preserved during phage replication. (*c*) T4 phage (closely related to phage T2) lysing an *E. coli* cell. (Courtesy of Lee D. Simon, Waksman Institute, Rutgers State, University of New Jersey.)

Early experiments on plant viruses provided even more convincing evidence identifying nucleic acid as the genetic substance of a virus. This work was done on tobacco mosaic virus (TMV), the first virus to be crystallized. TMV is a plant virus with a molecular weight of 40×10^6. The virion contains one molecule of single-stranded RNA with a molecular weight of 2×10^6 and 2,130 identical protein subunits, each with a molecular weight of 18,000. In 1956, Gierer and Schramm showed that the deproteinized viral RNA produced lesions on tobacco plant leaves similar in appearance to the lesions produced by the intact virus.

Shortly thereafter (1957), Fraenkel-Conrat and Singer obtained similar evidence that RNA was the genetic substance of TMV (fig. 30.3). Using two different strains of the virus, HR and TMV, each with readily identifiable protein components, they generated reconstituted viruses in vitro in all possible combinations:

HR-protein + HR-RNA
HR-protein + TMV-RNA
TMV-protein + HR-RNA
TMV-protein + TMV-RNA

The reconstituted virus made from purified RNA and protein components had infectious activities comparable to those of the normal viruses. After infection the protein components were examined. For all combinations the type of viral protein found in progeny virus was determined by the type of RNA in the reconstituted particles, a very elegant proof that the nucleic acid carries the genetic determinants of the viral protein.

The studies that we have described in this section show that the genetic information required for replication of a cell (e.g., pneumococcus) or a virus (T2, TMV) is carried by nucleic acids. In the case of a cell the genetic information is always carried by DNA. In the case of viruses the genetic information can be carried by either single- or double-stranded DNA or by RNA, depending on the virus. Sometimes the bacterial or viral DNAs are linear overall; sometimes they are circular. The only case in which circular RNAs have been encountered is in viroids—low-molecular-weight, single-stranded RNAs that cause several important diseases in cultivated plants but are not encapsidated. Viroids are the smallest known agents that cause infectious disease.

Structural Properties of DNA

The early work equating genetic material with DNA plunged genetics into an entirely new vocabulary of chemical terms. The genetic consequences of these early studies could be understood, but the chemistry was new. The remainder of this chapter focuses on the chemical and structural properties of DNA.

The amount of DNA per cell differs widely among different organisms (fig. 30.4). Mammalian cells contain about 1,000-fold more DNA than bacterial cells. Bacterial viruses such

Figure 30.3

Reconstitution experiment with tobacco mosaic virus. The RNA and protein components of TMV may be separated and reconstituted. If two strains are used and if the RNA and protein parts from different viruses are used in the reconstitution experiment, the progeny virus will always reflect the parental type from which the RNA was taken.

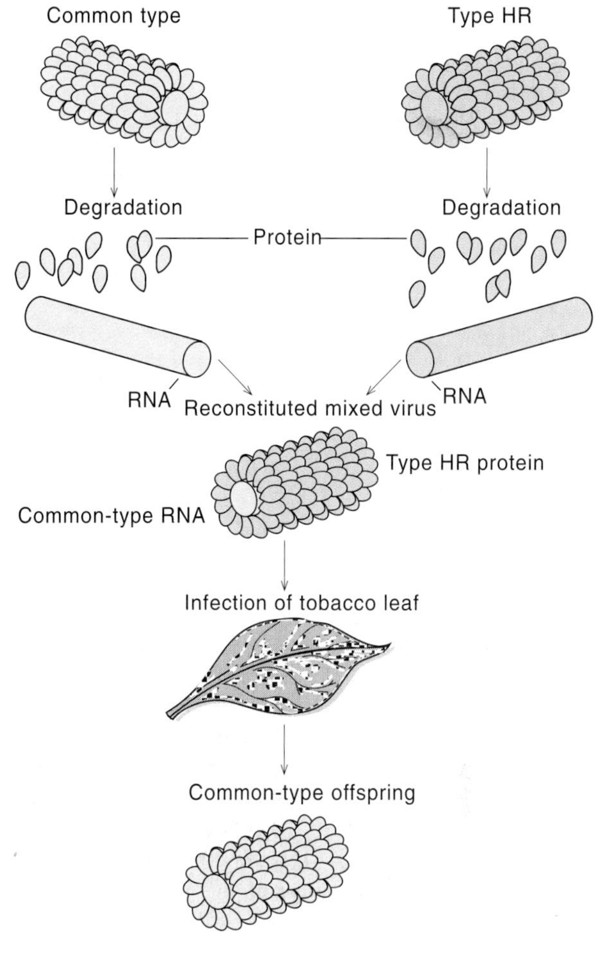

Figure 30.4

Genome size in different cells, viruses, and plasmids. Plasmids are small circular DNA molecules that replicate autonomously in cells harboring them. Unlike viruses, they do not form any complex nucleoprotein structures. In the case of plasmids, most viruses, and bacteria, the genome size is equivalent to the size of the chromosomal DNA because there is only one chromosome. For all of the remaining eukaryotic organisms listed here, the genome is subdivided into two or more chromosomes. Some organisms contain more than 100 chromosomes. All chromosomes are believed to contain a single DNA molecule.

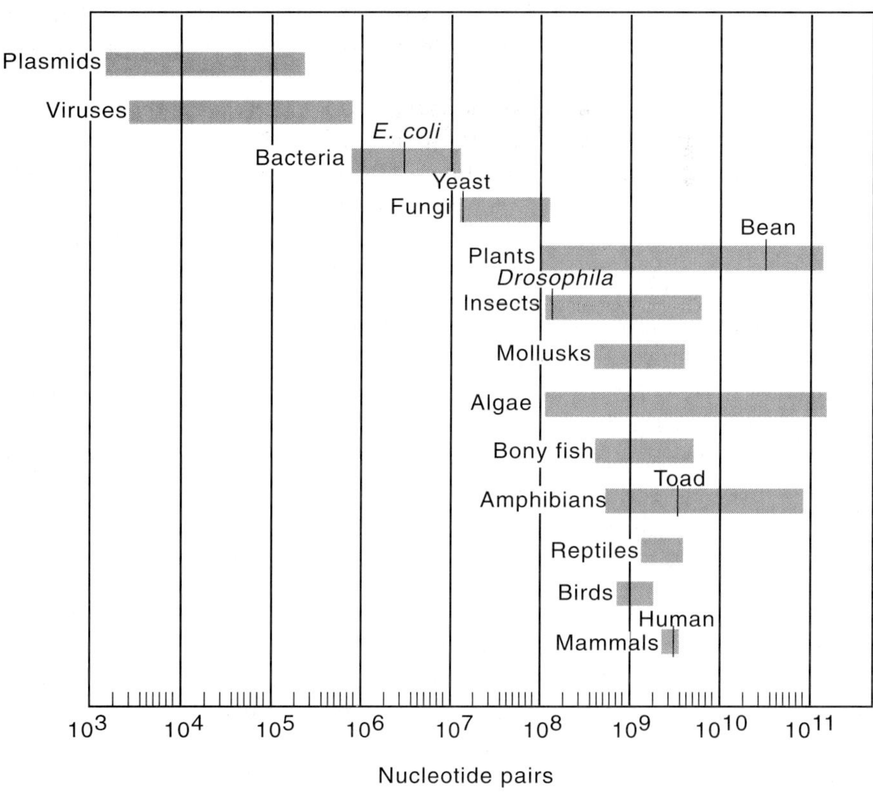

Figure 30.5

The structure of a deoxyribonucleotide. Drawn in abbreviated form at lower left. The illustrated structure is written pTpApCpG.

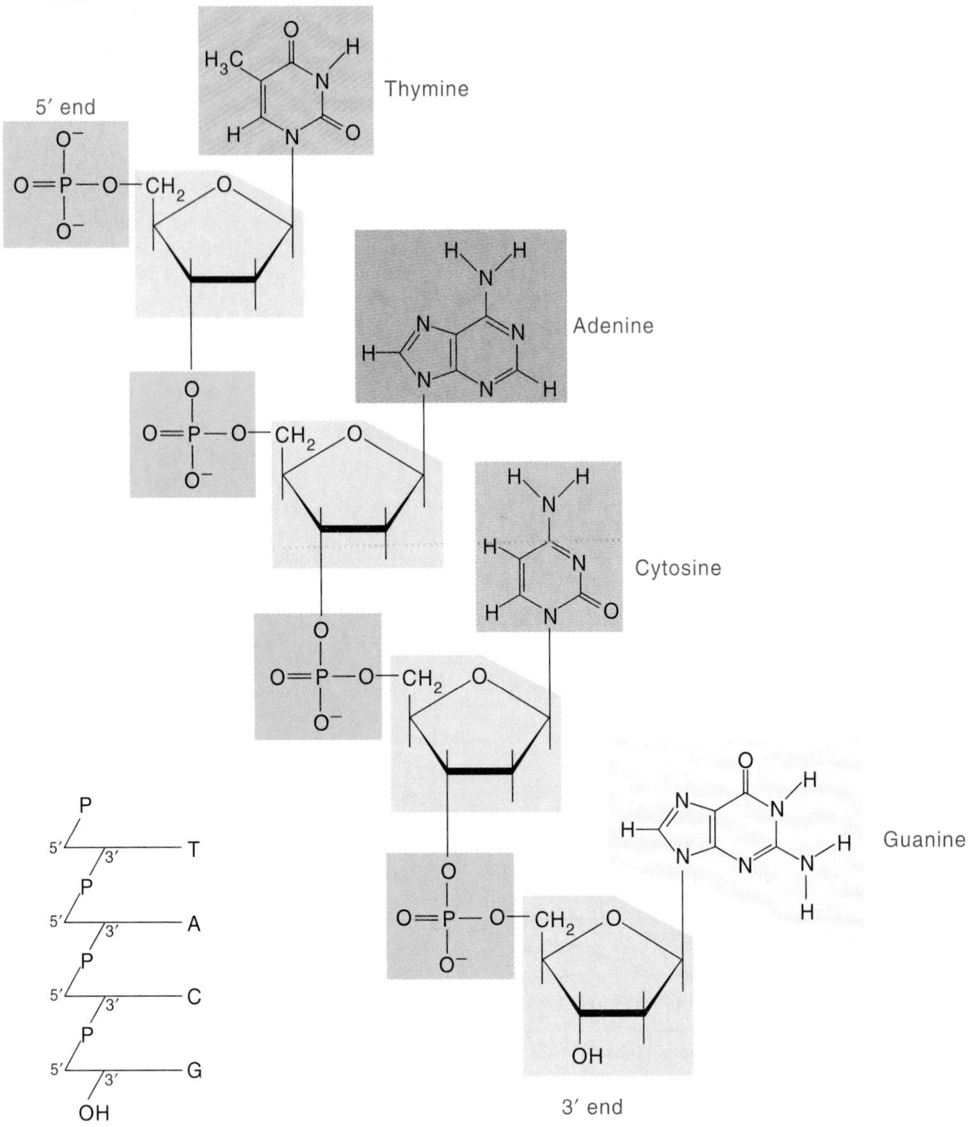

as the T type (T1–T7) bacteriophages (phages) that infect *E. coli* contain 10- to 20-fold less DNA than the bacterial host chromosome. The DNA of the smallest viruses is about 1/10 the size of the smallest T phage DNA, containing barely enough genetic material to accommodate about 10 genes. This finding is consistent with the fact that viruses do not contain sufficient genetic information for independent growth but can grow only parasitically in the host cells they infect. On the other hand, the amount of DNA per cell is not always directly proportional to the amount of genetic information an organism carries. This is because complex eukaryotes contain a great deal of noninformational DNA in their chromosomes.

In addition to the main DNA associated with the cell or the virus, informational DNA is found in special organelles, such as chloroplasts and mitochondria. This DNA carries genes whose products are exclusively associated with organelle function.

The Polynucleotide Chain Contains Mononucleotides Linked by Phosphodiester Bonds

Nucleotides are the building blocks of nucleic acids; their structures and biochemistry were discussed in chapter 26. When a 5′-phosphomononucleotide is joined by a phosphodiester bond to the 3′-OH group of another mononucleotide, a dinucleotide is formed. The 3′-5′-linked phosphodiester internucleotide structure of nucleic acids was firmly established by Alexander Todd in 1951. Repetition of this linkage leads to the formation of polydeoxyribonucleotides in DNA or polyribonucleotides in RNA. The structure of a short polydeoxyribonucleotide is shown in figure 30.5. The polymeric structure consists of a sugar phosphate diester backbone with bases attached as distinctive side chains to the sugars.

The polynucleotide chain has a directional sense with 5′

Table 30.1
Base Composition of DNAs from Different Sources

	(A) **Adenine**	(G) **Guanine**	(C) **Cytosine**	(5-MC) **5-Methyl-cytosine**	(T) **Thymine**	$\frac{A + T}{G + C + 5\text{-}MC}$
Human	30.4	19.6	19.9	0.7	30.1	1.53
Sheep	29.3	21.1	20.9	1.0	28.7	1.38
Ox	29.0	21.2	21.2	1.3	28.7	1.36
Rat	28.6	21.4	20.4	1.1	28.4	1.33
Hen	28.0	22.0	21.6		28.4	1.29
Turtle	28.7	22.0	21.3		27.9	1.31
Trout	29.7	22.2	20.5		27.5	1.34
Salmon	28.9	22.4	21.6		27.1	1.27
Locust	29.3	20.5	20.7	0.2	29.3	1.41
Sea urchin	28.4	19.5	19.3		32.8	1.58
Carrot	26.7	23.1	17.3	5.9	26.9	1.16
Clover	29.9	21.0	15.6	4.8	28.6	1.41
Neurospora crassa	23.0	27.1	26.6		23.3	0.86
Escherichia coli	24.7	26.0	25.7		23.6	0.93
T4 bacteriophage	32.3	17.6		16.7[a]	33.4	1.91

[a]In the T4 bacteriophage all of the cytosine exists in the 5-hydroxymethyl form 5-HMC.

and 3′ ends. Either of these ends may contain a free hydroxyl group or a phosphorylated hydroxyl group. The structure shown in figure 30.5 contains a phosphate group on the 5′ end but none on the 3′ end. By convention, one writes a nucleic acid sequence from the 5′ to the 3′ end, so a comparable structure is written pTpApCpG. With no phosphate on the 5′ end, the structure is designated TpApCpG; alternatively, if the terminal phosphate is on the 3′ end rather than the 5′ end, the structure is written TpApCpGp. When the phosphates are not indicated, the structure is indicated by dashes on either end: –TACG–. The letter "d" or "r" sometimes precedes the capital letter of the nucleotide to indicate a deoxyribo- or a riboderivative.

Most DNAs Exist as Double-Helix (Duplex) Structures

Like most other types of biological macromolecules, nucleic acids adopt highly organized three-dimensional structures. The dominant factors that determine nucleic acid conformations are the limitations imposed by the stereochemistry of the polynucleotide chains, the high negative charge resulting from the regularly repeating phosphate groups, and the noncovalent affinities between purine and pyrimidine bases.

A body of chemical information that proved vital to understanding DNA structure came from Erwin Chargaff's analyses of the nucleotide composition of duplex DNAs from various sources (table 30.1). Although the base compositions varied over a wide range, Chargaff found that within the DNA of each source that he examined, the amount of A was very nearly equal to the amount of T, and the amount of G was very nearly equal to the amount of C. The C is present as both unmodified C and, to a lesser extent, 5-methyl-cytosine, which results from postreplicative modification. The two equalities were the first indication that regular complexes occur between A and T and between G and C in DNA.

While searching for the significance of these equalities, James Watson noted that hydrogen-bonded base pairs with the same overall dimensions could be formed between A and T and between G and C (fig. 30.6). The A-T base-paired structure has two hydrogen bonds, whereas the G-C base pair has three. The hydrogen-bonded pairs are formed between bases of opposing strands and can arise only if the directional senses of the two interacting chains are opposite or antiparallel (fig. 30.7). With this notion in mind, Francis Crick took a closer look at the x-ray diffraction pattern produced by DNA and was able to interpret the diffraction pattern in terms of a helix (see box 30A) composed of two polynucleotide strands. In this structure, the planes of the base pairs are perpendicular to the helix axis, and the distance between adjacent pairs along the helix axis is 3.4 Å, bringing them into close contact (fig. 30.8). The structure repeats itself after 10 residues, or once every 34 Å along the helix axis; the repeating distance is referred to as the pitch length or just the pitch. An average-sized bacterial gene, which encodes the information to make a single protein, is about 1,000 bp (base pairs) in length, equivalent to 100 helical turns. As we see (chapters 31 and 33), the com-

Figure 30.6

Dimensions and hydrogen bonding of (*a*) thymine to adenine and (*b*) cytosine to guanine. Note that two hydrogen bonds are formed in the A-T base pair and three in the G-C base pair. The overall dimensions of the base pairs are the same. Consequently, they fit at any position in an otherwise regular polymeric structure. (*Source:* Adapted from S. Arnott, M. H. F. Wilkins, L. D. Hamilton, and R. Langridge, Fourier synthesis studies of lithium DNA, part III: Hoogsteen models, *J. Mol. Biol.* 11:391, 1965.)

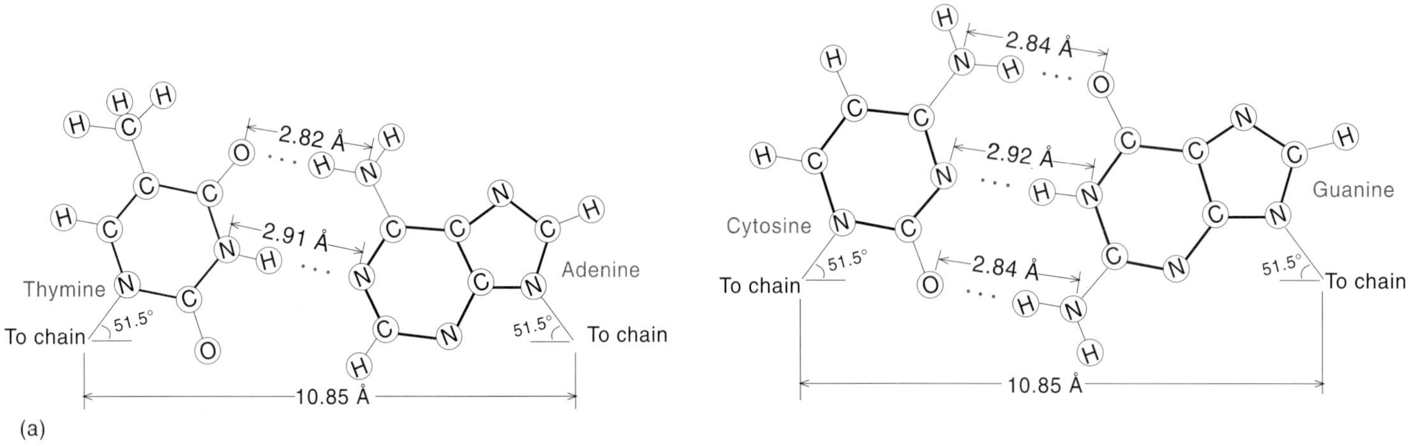

Figure 30.7

Segment of DNA, drawn to emphasize the hydrogen bonds formed between opposing chains. Each type of base is represented by a different color, with the sugar–phosphate backbones in blue and yellow. Note the three hydrogen bonds in the G-C pairs and the two in the A-T pairs (A, red; T, green; G, yellow; C, blue). The two strands are antiparallel: One strand (left side) runs 5′ to 3′ from top to bottom, and the other strand (right side) runs 5′ to 3′ from bottom to top. The planes of the base pairs are turned 90° to show the hydrogen bonds between the base pairs. (*Source:* Adapted from A. Kornberg, The Synthesis of DNA, *Scientific American,* October 1968.)

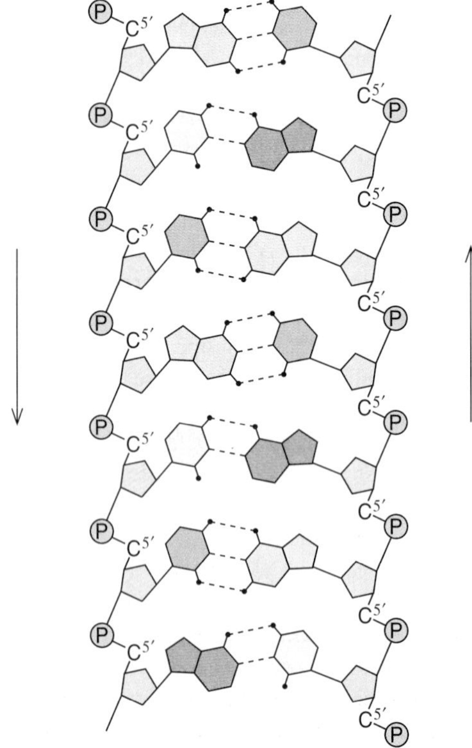

plementary structure of duplex DNA hints at how the genetic material is faithfully replicated as well as how it is expressed.

An important feature of the helical structure is the grooved nature of the surface resulting from the helical twist. Alternating wide (major) and narrow (minor) grooves are displayed in a side view of the helix structure (see fig. 30.8*a* and *b*). Different sections of the purine and pyrimidine bases are exposed in these two grooves as indicated in figure 30.8*c*. Many different proteins interact with DNA; most of the interactions occur with the phosphoryl groups on the outer surface of the structure and with the purine and pyrimidine bases in the major groove because of its greater accessibility. Specific instances of DNA-protein interactions are considered in later chapters (chapters 35 and 36).

Hydrogen Bonds and Stacking Forces Stabilize the Double Helix

Several factors account for the stability of the double-helix structure. The negatively charged phosphoryl groups are all located on the outer surface, where they have a minimum effect on one another. The repulsive electrostatic interactions generated by these charged groups are often partly neutralized by interaction with cations such as Mg^{2+}, basic polyamines (such as putrescine and spermidine), and the positively charged side chains of chromosomal proteins. The core of the helix is composed of the base pairs held together by the specific hydrogen bonds and also by favorable stacking interactions between the planes of adjacent base pairs. These stacking interactions are complex, involving dipole–dipole interactions and van der Waals forces; this results in a stacking energy comparable in magnitude to the stabilizing energy generated by the hydrogen bonds between the base pairs. The result is that stacking is maximized in most nucleic acid structures. In this connection it is noteworthy that two fully extended polynucleotide strands can form a hydrogen-bonded base-paired complex, leading to a stepladderlike structure (fig. 30.9). In this struc-

30A

BOX

X-Ray Diffraction of DNA

An x-ray pattern of DNA (fig. 1) is obtained by holding a stretched fiber containing many DNA molecules in a vertical direction and exposing it to a collimated monochromatic beam of x-rays. Only a small percentage of the x-ray beam is diffracted. Most of the beam travels through the specimen with no change in direction. A photographic film is held in back of the specimen; a hole in the center of the film allows the incident undiffracted beam to pass through (fig. 2). Coherent diffraction occurs only in certain directions, specified by Bragg's law: $2d \sin \theta = n\lambda$. Here d is the distance between identical repeating structural elements: θ is the angle between the incident beam and the regularly spaced diffracting planes, λ is the wavelength of x-rays used, and n is the order of diffraction, which may equal any integer but is usually strongest for $n = 1$. The most important point is that $\sin \theta \approx \theta$ and $d \approx 1/\theta$, so a spot far out on the photographic film is indicative of a repeating element of small dimension and vice versa.

Watson and Crick were the first to appreciate the significance of strong 3.4-Å and 34-Å spacings and the central crosslike pattern, which reflects a helix structure in the x-ray diffraction pattern of DNA. They interpreted this as arising from the hydrogen-bonded antiparallel double-helix structure.

Figure 1

Diffraction pattern of a fibrous sample of DNA. (© M. H. F. Wilkins.)

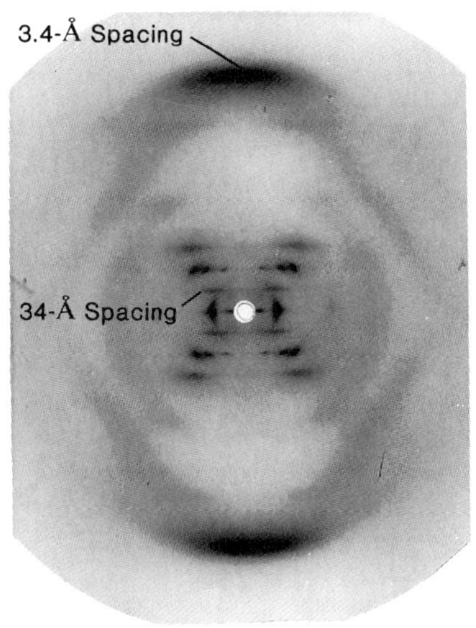

Figure 2

Camera setup for obtaining DNA diffraction pattern.

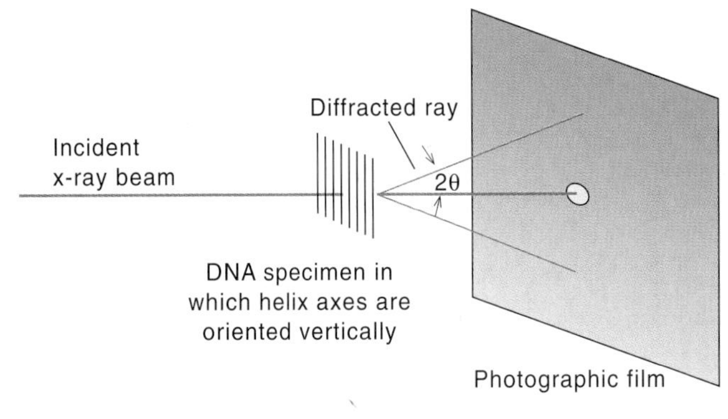

ture, the chains do not form a helix but lie straight, with a distance of 6.8 Å between adjacent nucleotides in the direction of the long axis. This 6.8-Å distance between adjacent base pairs produces a gap that would presumably be filled by water. Such a conformation is unstable because the planes of the bases prefer close contacts with one another, as does water. The stepladder structure is related to the helix structure by a simple right-handed (clockwise) twist (see fig. 30.9*a*). Following this operation the distance between base pairs decreases until they are in close contact, with a spacing of 3.4 Å.

Conformational Variants of the Double-Helix Structure

The same base-pairing arrangement is found in all naturally occurring double-helix structures. However, the inherent flexibility in the furanose ring of the sugar and the degrees of freedom generated by several rotatable single bonds per residue—six in the sugar phosphate backbone and one in the C-1′ N-glycosidic linkage (fig. 30.10)—lead to considerable variation in the conformations adopted by double-helix structures. Four puckered confor-

Figure 30.8

The most common form of the double-helix DNA. The base-paired structure shown in figure 30.7 forms the helix structure shown in (*a*) and (*b*) by a right-handed twist. The two strands are antiparallel as indicated by the curved arrows in (*a*). (Reprinted with permission from *Nature* (171:737, 1953) Copyright 1953 Macmillan Magazines Limited.) In (*b*) a space-filling model depicts the sugar–phosphate backbones as strings of mostly gray, red, white, and yellow spheres, and the base pairs are rendered as dark blue spheres arranged in horizontal plates composed of dark blue spheres. (Reprinted with permission from *Nature* (175:834, 1955) Copyright 1955 Macmillan Magazines Limited.) In (*c*) the orientation of the groups in the base pairs with respect to the major and minor grooves is indicated.

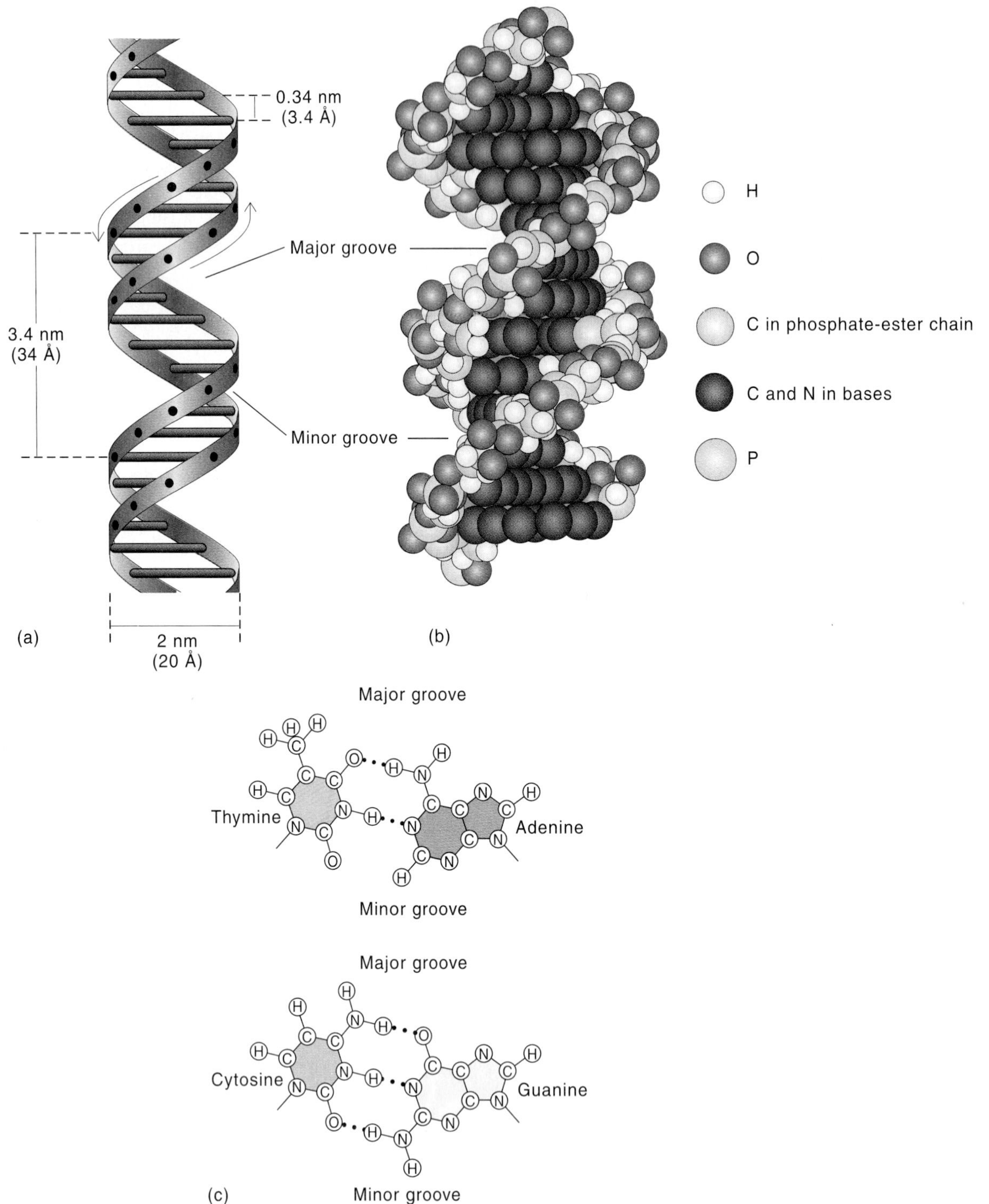

Figure 30.9

Different conformations of base-paired DNA: (*a*) the untwisted straight ladder, (*b*) the normal spiral ladder. The stepladder structure is unstable; it can be converted into a spiral ladder by a right-handed twist, a change that permits the planes of the base pairs to come into close contact.

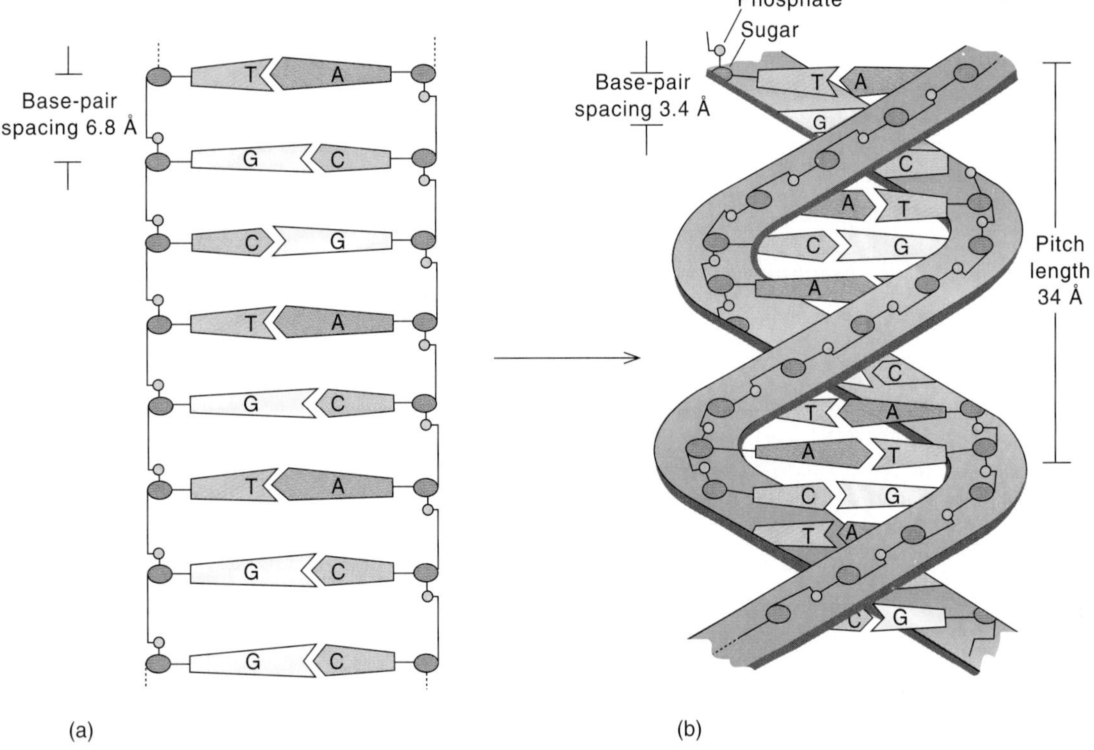

Figure 30.10

A segment of polynucleotide chain with rotatable bonds indicated by curved arrows. The presence of so many rotatable bonds permits the backbone of a polynucleotide chain to adopt many different conformations. (*Source:* From W. K. Olson and P. J. Flory, Different conformations of deoxyribose in DNA, *Biopolymers*, 11:1, 1972.)

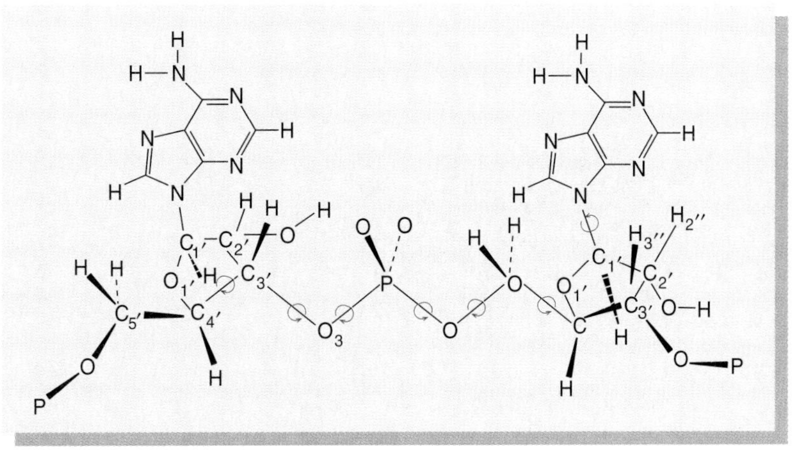

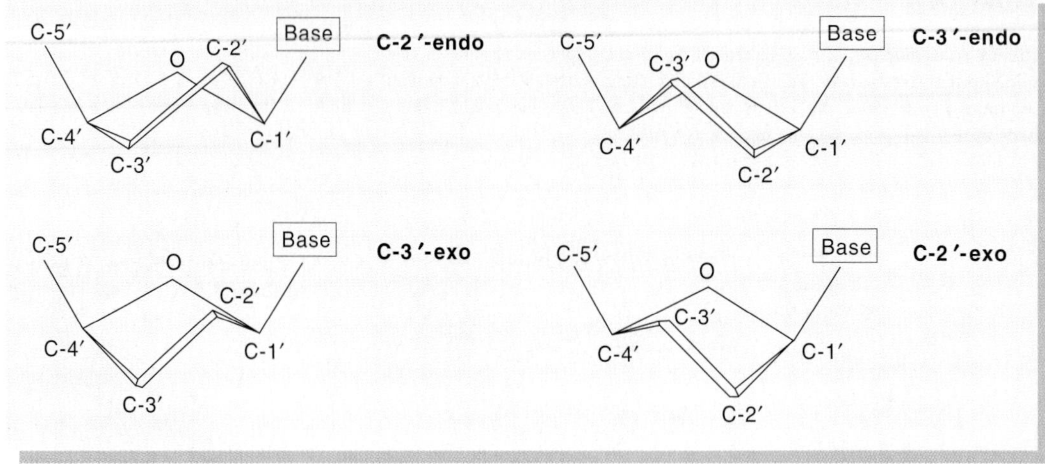

Figure 30.11

Four pucker conformations of the furanose rings of ribose and deoxyribose that are deemed energetically feasible. The flexibility of these furanose rings contributes to the possible variations in the conformation of the double helix.

Narrow groove

Wide groove

© IRVING GEIS

B DNA

A DNA

36°

Base pairs

Backbone

Base pairs

32.7°

Backbone

Figure 30.12

The A and B forms of DNA. (*Top*) A view perpendicular to the helix axis. (*Bottom*) A view of two adjacent base pairs, looking down the helix axis. A single turning is shown for each duplex. DNA usually is found in the B form. The A form has been observed in microcrystals of DNA from which a significant amount of the water has been removed. Dimensions of the two structures are compared in table 30.2. (Illustration copyright by Irving Geis. Reprinted by permission.)

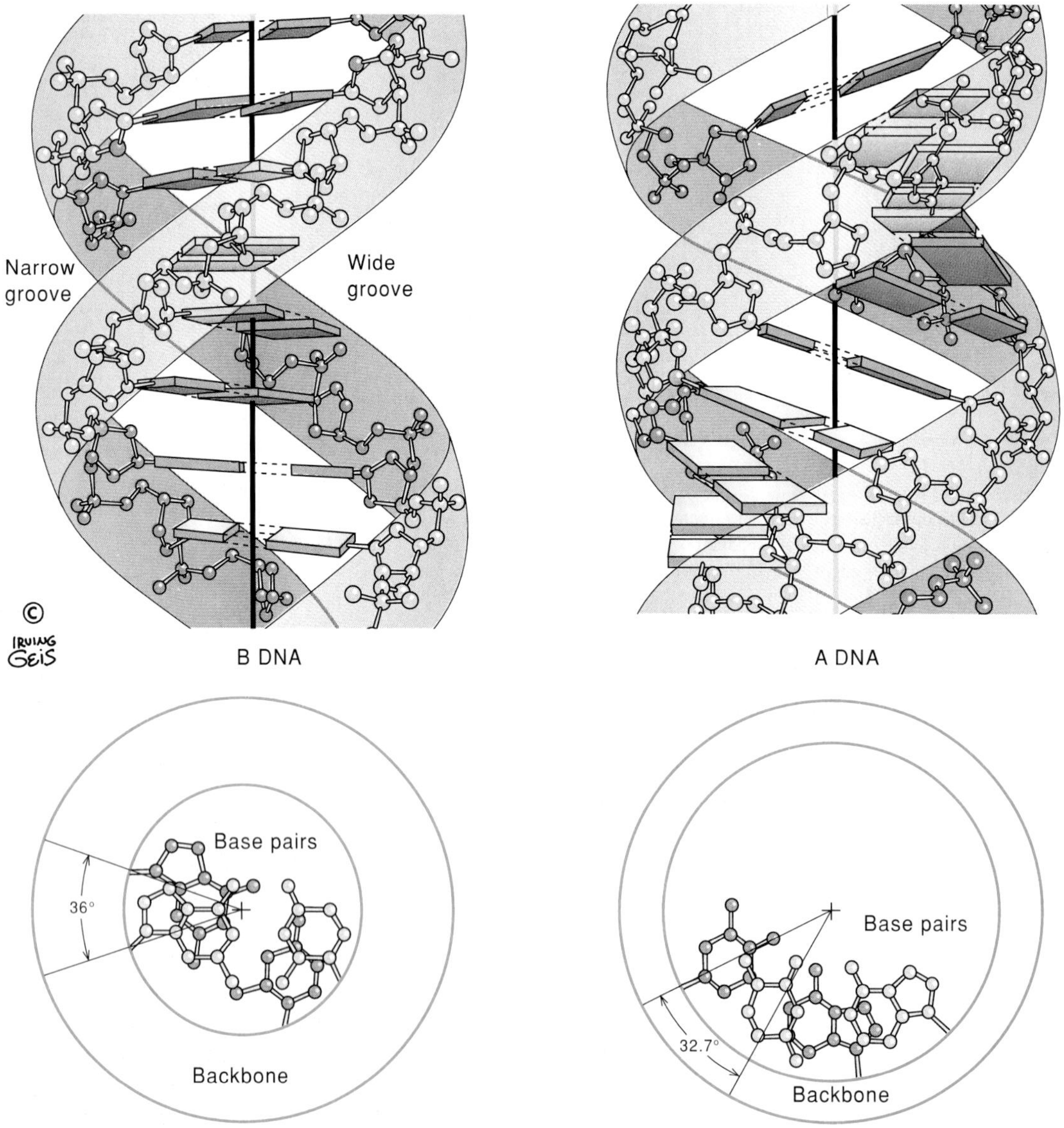

Figure 30.13

Space-filling models of (*a*) B DNA and (*b*) Z DNA. The irregularity of the Z DNA backbone is illustrated by the heavy lines that go from phosphate to phosphate residue along the chain. In contrast, B DNA has a smooth line that connects the phosphate groups. The space-filling model is excellent for displaying the volume occupied by molecular constituents and the shape of the outer surface. (Reprinted with permission from A. Wang et al., Left-handed double helical DNA: Variations in the backbone conformation, *Science,* 211, 1981. Copyright 1981 American Association for the Advancement of Science.)

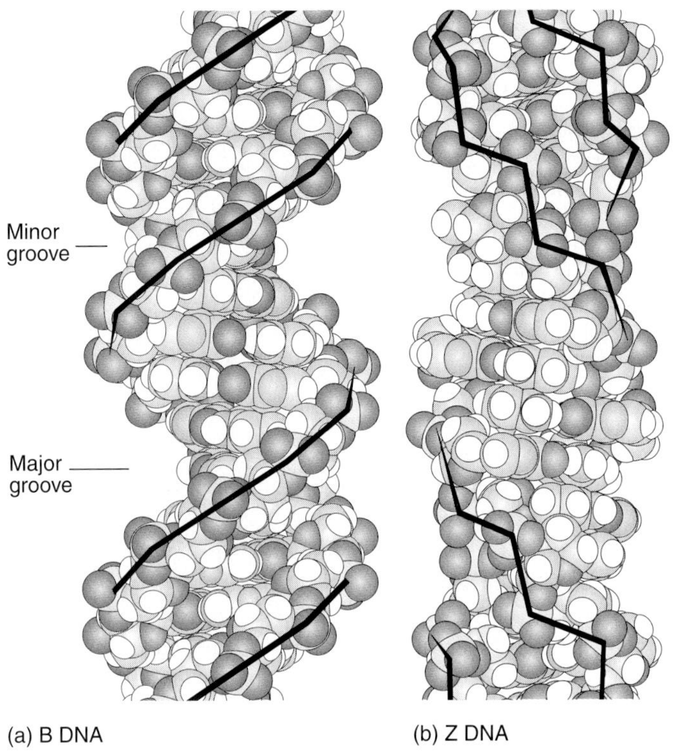

Figure 30.14

The change in topological relationship if a four-base-pair segment of B DNA is converted into Z DNA. Such a conversion could be accomplished by rotation of the bases relative to those in B DNA. This rotation is shown diagrammatically by coloring one surface of the bases. All of the colored areas are at the bottom in B DNA. In the segment of Z DNA, however, four of them are turned upward. The turning is indicated by the curved arrows.

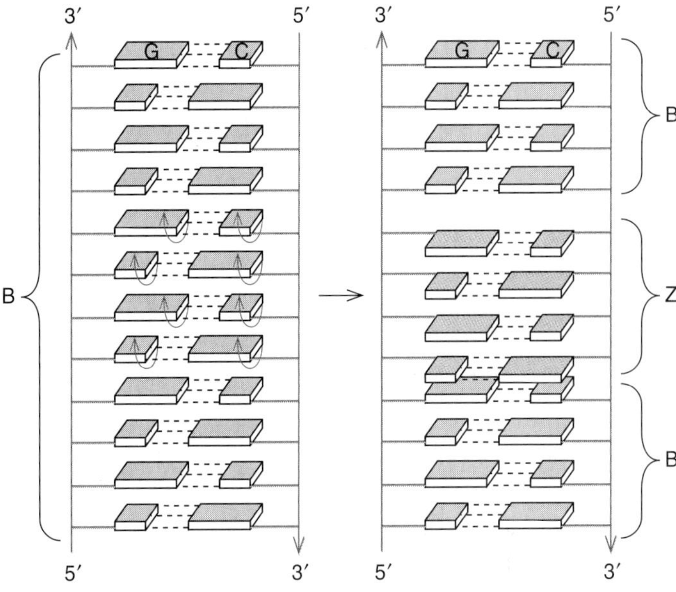

mations for the sugar, with small differences in stability, are shown in figure 30.11. In the double-helix structure shown in figure 30.8, the furanose rings are in the C-2′-endo conformation. This is believed to be the major conformation adopted by the sugars in DNA, both when free in aqueous solution and in chromosomes. It is known as the B form of DNA. When some of the water is removed from the hydrated DNA fibers, the double helices push even closer together, and the structure changes to the so-called A form, which has about 11 bases per turn and base pairs that are tilted about 20° with respect to the helix axis (fig. 30.12*b*). In the A form, the furanose rings have changed their pucker to the C-3′-endo conformation. The furanose rings in RNA have a stronger preference for the C-3′-endo conformation, with the result that RNA double helices adopt a structure similar to the DNA A form, even at high degrees of hydration.

The most striking conformational variant observed for a DNA double helix with Watson-Crick base pairing is referred to as the Z form. This structure was first detected by Alex Rich and his co-workers for the deoxyoligonucleotide d(CpGpCpGpCpG), which crystallizes into an antiparallel double helix with a left-handed rather than a right-handed twist (fig. 30.13). The Z form is a considerably slimmer helix than the B form and contains 12 base pairs per turn rather than ten. In the Z form, the planes of the base pairs are rotated approximately 180° with respect to the helix axis from their orientation in the B form (fig. 30.14). The flipping of the base pairs involves different conformational changes in the G and C residues in the alternating GC structure. In the case of the G residues the base is rotated by 180° about the glycosidic bond, a change resulting in a transition from the anti conformation found in B DNA to the syn conformation (fig. 30.15). Model-building studies indicate that the anti conformation of the nucleotide (found in B DNA) has less steric crowding than the syn conformation but that it is far easier for a purine nucleotide to adopt the syn conformation than for a pyrimidine nucleotide. In fact, cytidine remains in the anti conformation in Z DNA. The flipping of the cytidine base in going from the B to Z conformation involves rotation of the entire cytidine residue while maintaining the anti conformation. The effect is to make the sugar–phosphate backbone follow a zigzag course (see fig. 30.13). Thus the name Z DNA is an appropriate descriptive designation for this structure. A number of the structural parameters associated with the A, B, and Z helices are summarized in table 30.2.

Because of the different orientations of the G and C residues in Z DNA, this DNA conformation requires that the sequence of purine and pyrimidine bases be strictly alternating. Many other arrangements that involve alternating purine and pyrimidine residues can adopt the Z conformation. For example, a duplex containing alternating T and G residues in one strand and

Figure 30.15

The syn and anti conformations of deoxyadenosine and deoxycytidine. Purine nucleosides can readily adapt to either conformation by a simple rotation about C-1—N-9 glycosidic bond. Pyrimidine nucleosides are considerably less stable in the syn conformation because of steric hindrance between the sugar and the C-2 carbonyl groups.

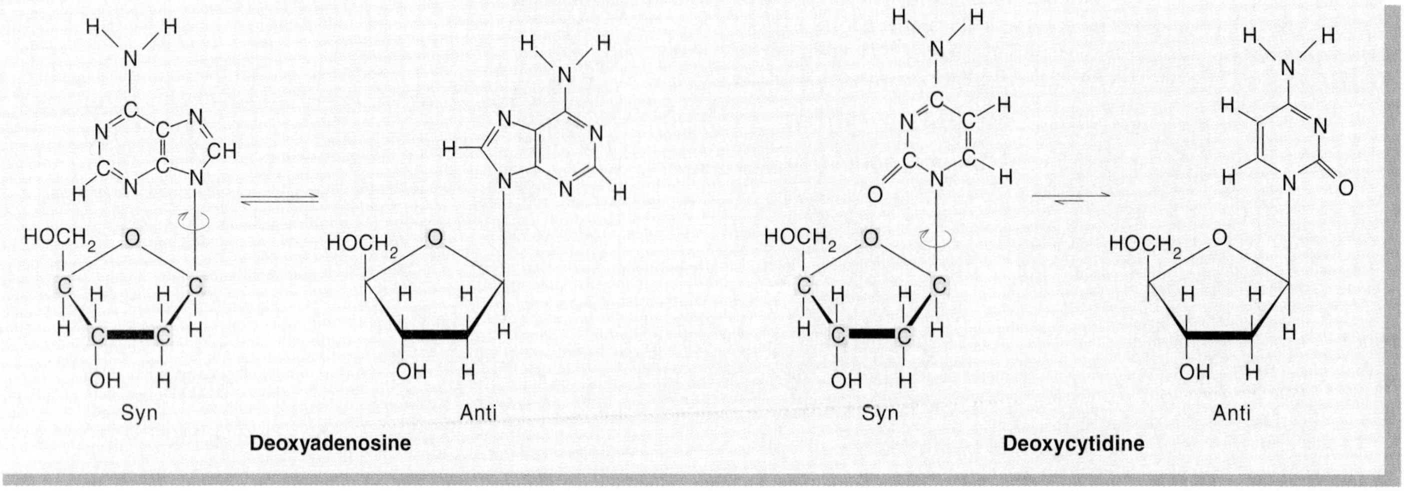

Table 30.2
Some Structural Parameters of A, B, and Z DNA

	A DNA	B DNA	Z DNA
Helix sense	Right-handed	Right-handed	Left-handed
Residues per turn	11	10	12 (6 dimers)
Rise per residue	2.55 Å	3.4 Å	3.7 Å
Helix pitch	28 Å	34 Å	45 Å
Base-pair tilt	20°	6°	7°
Rotation per residue	33°	36°	−60° (per dimer)
Glycosidic conformation			
Deoxycytidine	Anti	Anti	Anti
Deoxyguanosine	Anti	Anti	Syn
Sugar pucker			
Deoxycytidine	C-3′-endo	C-2′-endo	C-2′-endo
Deoxyguanosine	C-3′-endo	C-2′-endo	C-3′-endo

the complementary A and C residues in the other strand can adopt a Z conformation. An alternating A-and-T DNA sequence has never been observed in a Z conformation. The reason is believed to be the way in which water molecules orient around an A-T base pair in the Z helix. The arrangement is not as satisfactory as that observed for the G-C base pair, creating a less stable structure. Other factors that favor the stability of Z DNA are methylation of the 5 position of C residues and negative supercoiling of the DNA. We will discuss negative supercoiling in the next section.

The biological significance of Z DNA is currently unclear. However, several cellular proteins that bind specifically to Z DNA have been isolated from the nuclei of *Drosophila* fruit flies. The mere existence of such proteins suggests that they may function in some specific role when they encounter stretches of DNA that can adopt a Z conformation. A mutation-genetic analysis of such Z-DNA-binding proteins would be helpful to ascertain their role(s). Current speculation on the biological significance of the Z conformation centers on the general notion that the conformation plays a regulatory role in gene expression and possible genetic recombination.

It is appropriate at this point to say a few words about RNA conformations even though the main discussion of this will be presented in chapters 33 and 34, where it is more relevant. Double-helical RNA is unable to assume a B-DNA-like conformation because of steric clashes involving its 2′-OH groups. Rather, it usually assumes a conformation resembling A DNA, which has 11 base pairs per helical turn, a pitch of 30 Å, and its base pairs inclined to the helix axis by 14°. Many RNAs, for example, transfer RNAs and ribosomal RNAs (whose structures are detailed in chapter 33) and contain complementary sequences that form double-helical stems. Hybrid double helices, which consist of one strand each of RNA and DNA, are also thought to have an A-DNA-like conformation.

Helical Structures That Use Additional Kinds of Hydrogen Bonding

In addition to the A, B, and Z helices observed for DNA many other conformations have been observed by Alex Rich and others for synthetic polynucleotides. Polyriboadenylic acid (A) and

Figure 30.16

Four-stranded structures formed by oligonucleotides contain clusters of Gs. (*a*) Short oligonucleotides containing runs of four Gs tend to form four-stranded structures in which the four chains are oriented in the same direction. (*b*) A single polynucleotide chain containing regularly spaced clusters of four Gs could form a four-stranded structure in which two segments of the chain are oriented in one direction and two segments are oriented in the opposite direction. This type of structure may exist at the ends of eukaryotic chromosomes.

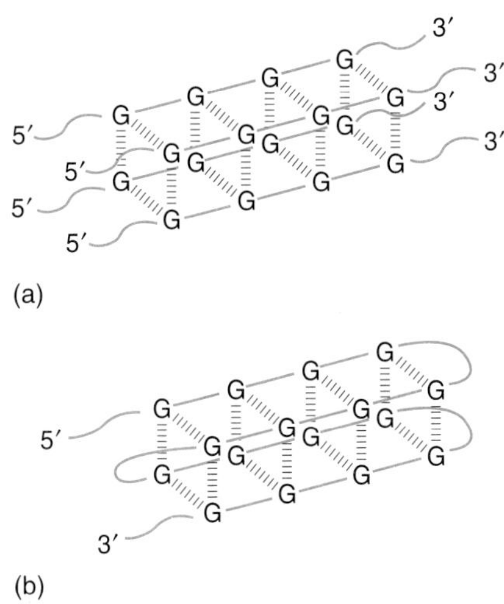

Figure 30.17

Topology of negative and positive supercoil.

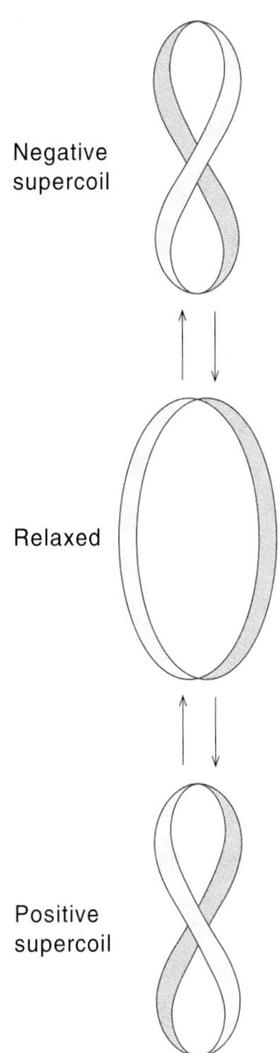

polyribouridylic acid (U) in 1 : 1 base-pair ratios form a duplex resembling the A form of DNA. In 1 : 2 ratios of one A chain to two U chains they form a triplex that displays additional kinds of hydrogen bonding. Two chains of A make a double helix.

Four chains of G can form a quadruple helix. This has been known for a long time but has become more interesting in recent years because it may have a counterpart in biological systems. The ends of eukaryotic chromosomes known as telomeres commonly have repetitious sequences that contain clusters of G residues. For example, the ciliates Oxytricha and Euplotes contain many repeats of the octameric sequence TTTTGGGG at the 3′ ends of each chromosomal DNA strand. For the most part, the opposing strand has complementary base pairs as expected, but at the very 3′ ends, usually 12 to 16 nucleotides protrude beyond the complementary C-rich strand. In vitro, oligonucleotides containing the TTTTGGGG repeats have been found to interact in various ways to form four-stranded helical segments held together by a hydrogen-bonded network. The precise hydrogen-bonding pattern observed depends on the relative orientation of the interacting chains. Four separate chains containing sequences of Gs are most likely to interact in parallel fashion (fig. 30.16*a*). A single polynucleotide strand containing repetitious sequences could form only a four-strand complex by folding back on itself so that two segments of the chain are oriented in one direction and two in the other direction (see fig. 30.16*b*). Guowei Fang and Thomas Cech found that the telomere-binding protein of Oxytricha promotes four-stranded complex formation by telomeric DNA. This has led them to believe that such structures are likely to exist in vivo at the chromosome telomeres.

Duplex Structures Can Form Supercoils

The detection of different conformations of DNA underscores the inherent flexibility built into the DNA duplex. All the conformations discussed thus far involve regular linear duplexes. Energetically favorable interactions with other molecules, particularly proteins, can induce additional conformations that do not result in major changes in either base pairing or stacking. Several conformations are believed to play important roles in different situations. Bends are known to be important in structures formed by chromosomes (discussed later in this chapter). Cruciforms, in which a single chain folds back on itself into a hairpinlike duplex, are important as intermediates in DNA and RNA synthesis. Supercoiled DNA is a very common type of tertiary structure in which the double-helix segments twist around each other. Supercoiled DNA is topologically constrained by being covalently closed and circular or by being complexed to proteins so that the ends of the DNA cannot rotate freely.

DNA can form right-handed (negatively supercoiled) or left-handed (positively supercoiled) supercoils (fig. 30.17). Nega-

Figure 30.18

Ethidium bromide and DNA intercalated with ethidium bromide. Molecules such as ethidium bromide can intercalate DNA because they are flat rings of the same thickness as DNA base pairs. In order to accommodate the ethidium bromide molecule the duplex must untwist in the region of intercalation. This increases the separation between the planes of the base pairs. We saw the effect of complete untwisting in figure 30.9.

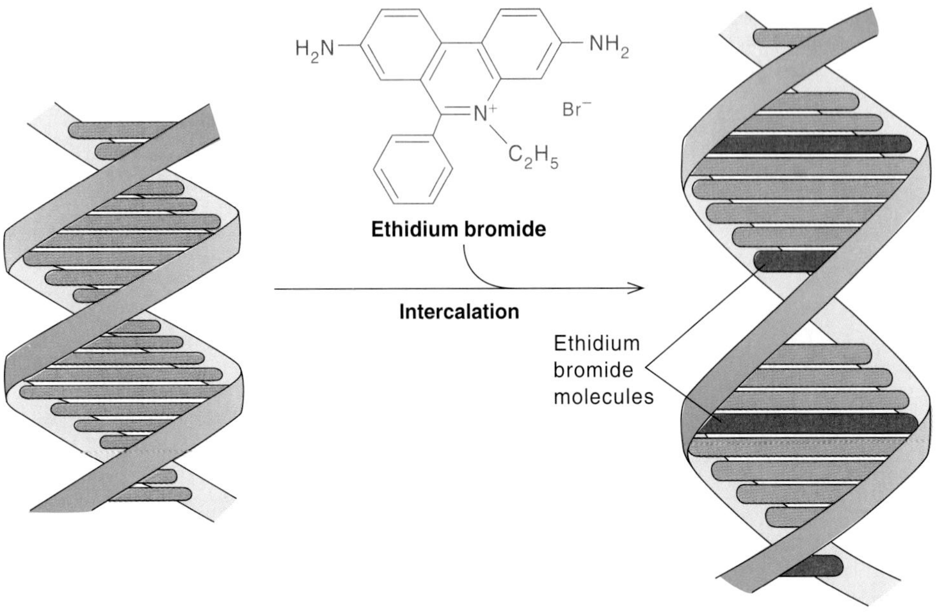

tive supercoiling imparts a torsional stress to the DNA that favors unwinding, whereas positive supercoiling favors tighter winding of the double helix. Supercoiling imparts a more compact structure to a circular duplex, which makes it sediment more rapidly in a centrifuge. Thus either a positively or a negatively supercoiled DNA will sediment more rapidly than a circular duplex with no supercoiling. Jerome Vinograd demonstrated this tendency by adding ethidium bromide (fig. 30.18) to circular DNA of the small phage PM2. This DNA has about 40 negatively supercoiled turns when it is isolated from cells. Addition to ethidium bromide to duplex DNA leads to the binding of ethidium between adjacent base pairs, a type of binding known as intercalation (see fig. 30.18). Normally, the base pairs are nearly closely packed, so such binding would not be possible without an alteration in the DNA structure. In order to accommodate the ethidium, the duplex unwinds by about −27° per base pair. This unwinding reduces the negative supercoiling in naturally occurring circular duplexes such as PM2 DNA. As more ethidium is added, enough unwinding takes place to eliminate all the supercoiling. At this point, the DNA is in the most extended state and thus has its slowest sedimentation rate (fig. 30.19). Addition of more ethidium causes the DNA to adopt a positively supercoiled form that again increases its sedimentation rate.

Supercoiling of circular duplex DNA is quantitatively considered in terms of the linking number (L), an integer that specifies the number of complete turns made by one strand around the other. The linking number can change only if a covalent linkage in the DNA backbone is broken. If a molecule of DNA is projected onto a two-dimensional surface, the linking number is de-

Figure 30.19

The sedimentation coefficients of closed circular phage PM2 DNA in 2.85-M CsCl containing varying amounts of ethidium bromide. The more compact the structure, the faster it will sediment. Supercoiled DNA is more compact than relaxed circular DNA. Thus when ethidium bromide is added to negatively supercoiled DNA the sedimentation constant decreases, indicating that the DNA is becoming more relaxed. A minimum sedimentation constant is reached when the fully relaxed structure is obtained. Addition of ethidium bromide beyond this point results in an increase in the sedimentation constant again, as the DNA becomes positively supercoiled. (*Source:* From B. M. J. Revet et al., Direct determination of the superhelix density of closed circular DNA by isometric titration, *Nature (New Biol.)* 229:10, 1971.)

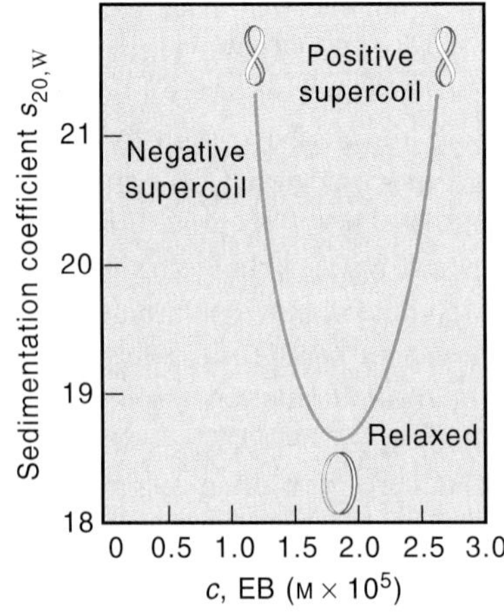

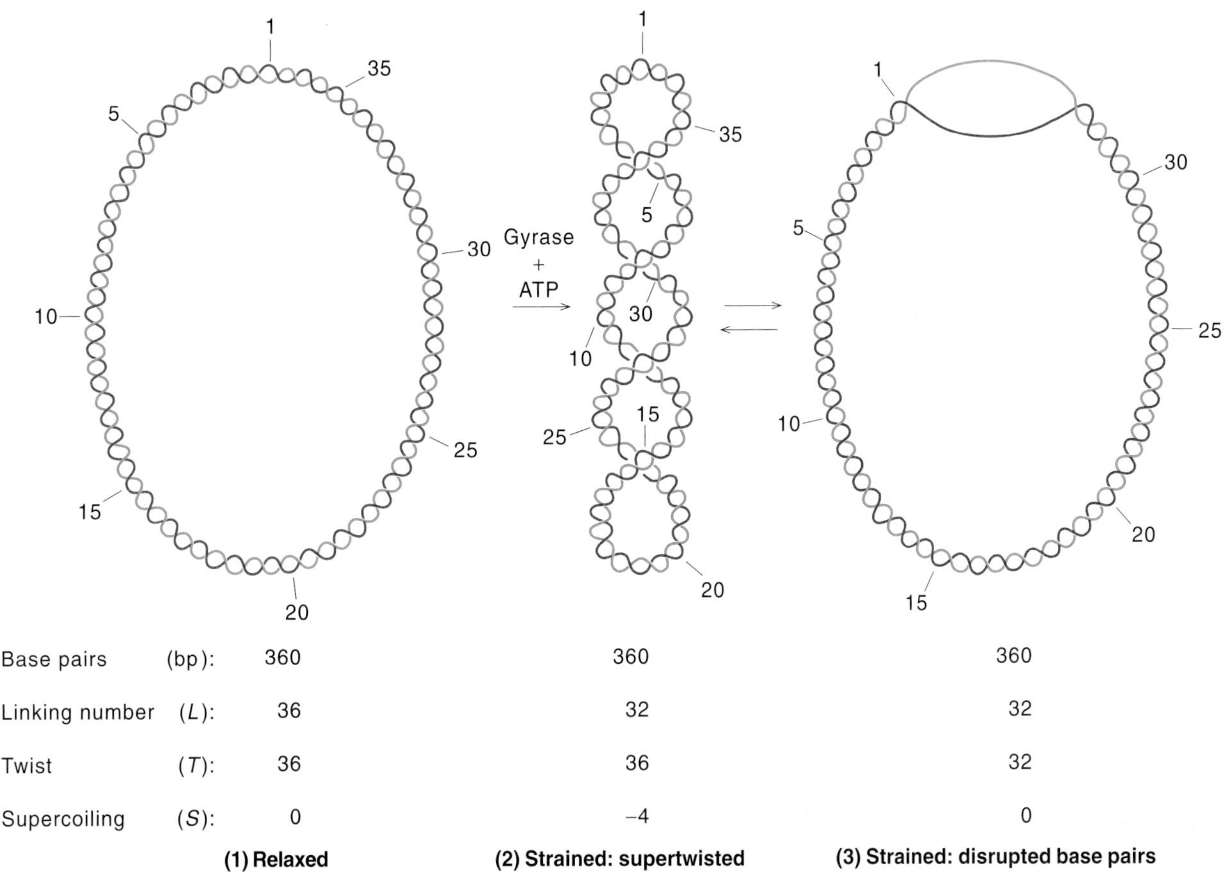

Figure 30.20

A circular duplex molecule in different topological states. (Adapted from a diagram supplied by M. Gellert.) The linking number can be changed only by breakage and re-formation of the phosphodiester linkages, as shown in the conversion of the relaxed circular form (1) to the strained negatively supercoiled form (2). The strain in the negatively supercoiled form can be partitioned in different ways between twist (T) and supercoiling (S) as shown in the interconversion between (2) and (3). No phosphodiester linkages are broken in making this interconversion, and consequently, no change occurs in the linking number (L) as indicated.

fined as the excess of right-handed over left-handed crossings of one strand over the other. Linear duplex DNA with free ends adopts a conformation in solution close to the B form, with about 10 base pairs per turn. Therefore a closed circular duplex with this extent of twist is presumed to be under no torsional strain and is said to be relaxed. Because B DNA is a right-handed helix, the linking number is normally positive by the sign convention we have adopted. The values of the linking number of relaxed DNA, $L°$, will be distributed over a narrow range of integral values centered around 1 per 10 base pairs. DNA with a mean linking number smaller than this is termed negatively supercoiled, or underwound; DNA with a larger linking number is termed positively supercoiled or overwound. The deviation of the linking number from its relaxed value, $\Delta L = L - L°$, can be partitioned between twist (altered double-helix coiling) and supercoiling:

$$L = \text{Twist } (T) + \text{Supercoiling } (S)^*$$

At the present time we do not know precisely how L will partition between twist and supercoiling for helices under torsional stress. For example, consider the hypothetical situation illustrated in figure 30.20. A 360-base-pair structure in the circular relaxed form ($L = +36$, $T = +36$, $S = 0$) is indicated to the left. Exposure to the bacterial enzyme DNA gyrase will introduce negative supercoils into such a structure in a reaction that requires ATP (see chapter 31). If four negative supertwists are introduced, L will be reduced to $+32$. Barring other changes, T will remain fixed and S will become -4. In fact, the torsional strain introduced by the four negative supertwists will tend to reduce T, causing a partial unwinding of the duplex. At one extreme this effect could lead to the unwinding of four helical turns, in which case the supercoiling would disappear entirely ($L = +32$, $T = +32$, $S = 0$). The actual situation would probably lead to a reduction in the negative value of S and a concomitant reduction in T that would be spread over the entire duplex without any localized total unwinding of the double helix as pictured. Note that the linkage number L changes only when covalent bonds are broken, as in the case of gyrase treatment.

For DNA with a molecular weight of less than 10^7, agarose gel electrophoresis is a most effective method for assessing the extent of supercoiling. (For analysis of higher-molecular-

Application of Gel Electrophoresis for Chromosome Analysis

A variety of methods, both absolute (light scattering) and empirical (sedimentation, viscosity, gel electrophoresis), exist to determine the size of nucleic acids. The most accurate method is to determine the base sequence of the nucleic acid. Then sequenced restriction fragments can be used as standards for comparison to samples whose molecular weight is being studied.

To analyze heterogeneous DNA populations, gel electrophoresis offers the best approach. However, very large DNA molecules —in the range of 150 kb or greater—tend to migrate with size-independent mobilities. (The abbreviation kb stands for "kilobase pair," or 1000 base pairs.) A greatly improved, new technique called pulsed field gel electrophoresis can resolve large DNA molecules, the size of the *E. coli* genome and intact, individual yeast chromosomal DNA, on agarose gels. The gently isolated DNA samples are subjected alternately to two approximately orthogonal electric fields. First, one field is applied and turned off. Then the other field is applied and turned off. This process is repeated with a pulse time of many seconds over a period of several hours. The precise pulse time depends on the size of the molecules being separated, being longer for larger molecules.

When large DNA molecules enter a gel in response to an electric field the molecule must elongate parallel to the field. When the field is shut off and a new field is applied, perpendicular to the long axis of the DNA, the molecule must reorient. The reorientation time should be quite sensitive to the size or molecular weight of the DNA. Since the length of the stretched-out DNA molecule is generally larger than the pores in an agarose gel, this reorientation is absolutely essential if the DNA is to undergo any net migration in response to the new field. Therefore the smaller molecules have a mobility advantage over the larger molecules, which results in a size-based separation.

Another development is the electrophoretic analysis of partially denatured DNA, a technique that is very useful in the detection of single base substitutions of cloned DNA. Fragments of DNA that are wild type or that contain a single base mismatch migrate into a polyacrylamide gel containing an ascending denaturing gradient of urea and formamide in the same direction as the electrical field. At a critical depth the mobility of partially denatured DNA slows down abruptly. The distance moved is a measure of the stability of the sequence, rather than the overall base composition or size of the remaining unmelted duplex. Differences in the partial denaturation of wild type and mutant DNA molecules allow one to separate and detect them. This method provides a quick method to detect differences—even a single base-pair change—in closed homologous fragments from wild-type and mutant sources.

---

weight DNAs, see box 30B). DNA isomers differing by 1 in linking number form separate bands in the gel (fig. 30.21). The more highly supercoiled molecule will migrate more rapidly through the gel as a result of its more compact structure.

In chapter 31 we will discuss enzymes called topoisomerases, which relax positively or negatively supercoiled DNA, as well as topoisomerases such as DNA gyrase (mentioned earlier), which only generate negatively supercoiled DNA from relaxed DNA. Gyrases that catalyze the formation of negatively supercoiled DNA have been found only in bacteria. Consistent with this fact, all double-helix DNA found in bacteria that is topologically constrained by not having free ends, such as circular DNA, is negatively supercoiled. Circular DNA isolated from virus-infected eukaryotic cells (for example, simian virus 40, or SV40 DNA) is frequently found to be negatively supercoiled, but only after deproteinization. In such instances the DNA is not supercoiled in cells. Instead, the supercoiling results from removal of the proteins that normally are bound to the DNA and cause it to be underwound in its native state.

The biological importance of supercoiling has been clearly established only for DNA in bacteria. Drugs, such as novobiocin, that specifically inhibit DNA gyrase have a lethal effect except in strains that have a novobiocin-resistant DNA gyrase. On the other hand, mutants containing an altered topiosomerase, relaxing enzyme, which relaxes negatively supercoiled DNA, have been isolated from *E. coli.* Careful analysis of such mutants has shown that they are double mutants; that is, they have an altered DNA gyrase that is less active, in addition to the defective relaxing enzyme. This observation suggests that in a normal bacterium the extent of negative supercoiling is carefully adjusted by the relative activities of the gyrase and relaxing enzymes. Negative supercoiling imparts a torsional stress or tension to the DNA structure that can be relieved to some extent by unwinding of the double helix. Both the initiation of DNA synthesis (chapter 31) and the transcription of certain DNA sequences or genes (chapter 32) are strongly dependent on negative supercoiling, probably because of the necessity to unwind at least one turn of the double helix during the initiation process. This same torsional stress explains why Z helix formation is encouraged by negative supercoiling; since in Z DNA the twists are left-handed, the transition from the B to Z form will tend to relax negatively supercoiled DNA.

DNA Denaturation Involves Separation of Complementary Strands

In the laboratory, double-stranded DNA can be separated into single strands. The process of separating the polynucleotide strands of duplex nucleic acid structures is called denaturation. Denaturation disrupts the secondary binding forces that hold the strands together. Recall that the secondary binding forces include the edge-to-edge hydrogen bonds between the base pairs of opposing

Figure 30.21

Electrophoretic patterns of highly supercoiled or partially supercoiled DNA. Strip A represents a sample of circular duplex DNA obtained by deproteinization of the animal virus SV40. In strips B and C the DNA has been exposed for increasing times to an enzyme (topoisomerase) that catalyzes relaxation. Adjacent bands differ by 1 in linking number. (From W. Keller, Characterization of purified DNA-relaxing enzyme from human tissue culture cells, *Proc. Natl. Acad. Sci. USA* 72:2553, 1975.)

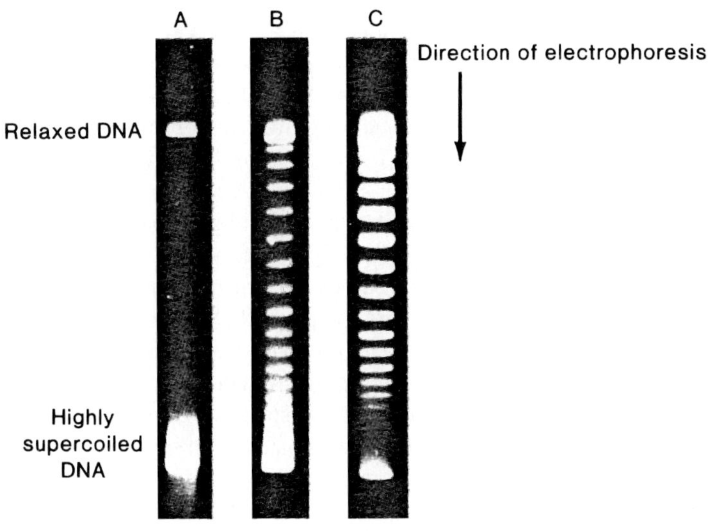

Figure 30.22

Effect of temperature on the relative absorbance of native, renatured, and denatured DNA. When native DNA is heated in aqueous solution, its absorbance does not change until a temperature of about 80°C is reached, after which the absorbance rises sharply, by about 40% (curve *a*). On cooling the absorbance falls, but along a different curve, and it does not return to its original value (curve *b*). Renatured DNA, in which the two strands have been brought back into perfect register, shows a sharp melting curve similar to native DNA (curve *c*). Renatured DNA is prepared from denatured DNA by holding the temperature at about 25°C below the denaturation temperature for an extended time. This subject is discussed in detail later in the text. The temperature at which the native DNA is half denatured is labeled T_m.

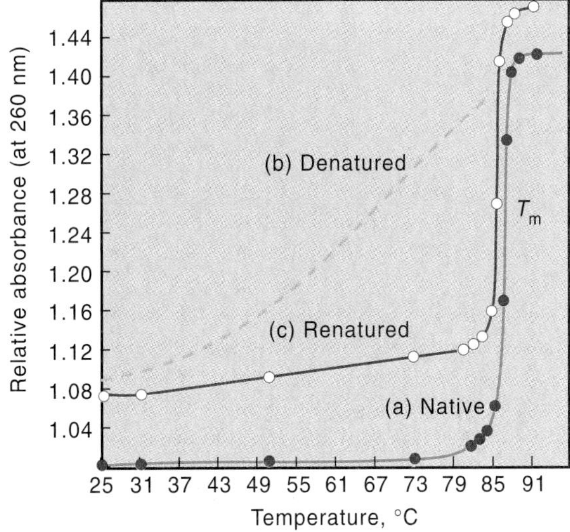

strands and the face-to-face stacking forces between the planes of adjacent base pairs. Individually, these secondary forces are weak, but when they act cooperatively they give rise to a DNA duplex that is highly stable in aqueous solution. The conditions required to denature DNA provide us with a measure of the strength of these interactions.

One of the simplest ways to denature DNA is by heating. The extent of denaturation at any temperature can be measured by the change in ultraviolet absorbance of a solution of DNA. A rise in absorbance coincides with the disruption of the regular base-paired structure, and the separation of the two polynucleotide strands from one another. The sharpness of the disruption of the regularly hydrogen-bonded base-paired native structure may be likened to the melting of a pure organic compound. It is customary to refer to this ultraviolet absorption temperature profile as a melting curve (curve *a* in fig. 30.22).

The melting temperature, T_m, of DNA is defined as the temperature at the midpoint of the absorption increase. This is about 85°C for the example shown in figure 30.22. Rapid cooling of the denatured DNA solution leads to re-formation of intrastrand hydrogen bonds but in a nonspecific, irregular manner. The absorbance decreases, but only by about three-fourths of the total original increase, and the decrease occurs over a much broader range of temperatures, as shown by curve *b* in figure 30.22. On subsequent reheating and cooling, the absorbance follows this cooling curve in the appropriate direction, indicating that denaturation results in an irreversible change.

The melting curve is an excellent tool for detecting DNAs that occur naturally in the single-stranded conformation. For example, in certain bacteriophages that infect *E. coli,* such as ϕX174, fd, or M13, the DNA exists as a single, circular strand. The ultraviolet absorption temperature curve for the DNA of these phages is similar in shape to that observed for denatured DNA (curve *b* in fig. 30.22). Broad melting curves also are characteristic of most RNAs, which rarely have regions of regular base pairing that extend for more than 10 or 20 residues.

Melting curves also have provided evidence that the stability of the double-helix structure is a function of its base composition. The midpoint of denaturation (T_m) of naturally occurring DNAs is precisely correlated with the average base composition of the DNA: The higher the mole percent of G-C base pairs, the higher the T_m (fig. 30.23). This seems reasonable, because the G-C base pair contains three hydrogen bonds, whereas the A-T base pair contains only two (see fig. 30.6); thus DNA with a greater G-C content is expected to be more stable. As indicated above, base stacking is also believed to contribute to the stability of the duplex structure. In general the interaction energy gained by stacking between adjacent G-C base pairs is greater than that gained by interaction between A-T base pairs.

Other factors present in aqueous solution can affect the stability of the double-helix structure in a positive or a negative way. For example, salt has a stabilizing effect, which is mainly due to the repulsive electrostatic interactions between the nega-

Figure 30.23

Dependence of the temperature midpoint (T_m) of DNA on the content of guanine and cytosine. As the percentage of G + C increases, the T_m increases. Two curves are shown to illustrate the point that the denaturation temperature is shifted to lower values when the ionic strength is lowered.

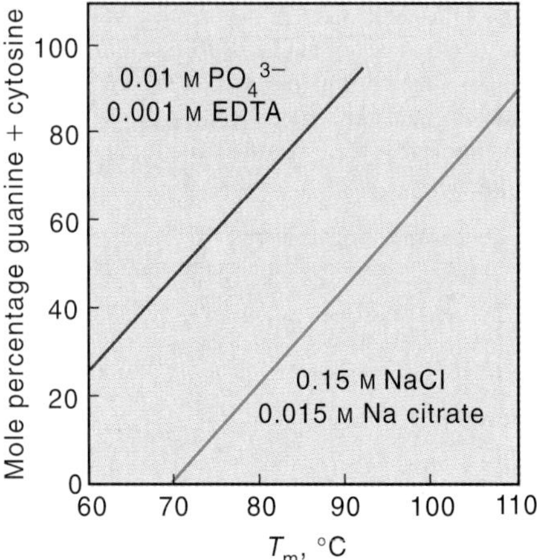

tively charged phosphate groups. Salt shields this charge interaction and therefore stabilizes the duplex structure. Thus DNA in 0.15 M NaCl denatures at a T_m about 20°C higher than DNA in 0.01 M phosphate. In pure water (no salt present), DNA denatures at room temperature. Extremes of pH also have a destabilizing effect on the double-helix structure. When the pH is above 11.5 or below 2.3, extensive deprotonization or protonization, respectively, occurs in the hydrogen-bonding groups of the bases, which in turn disrupts the hydrogen-bonded structure. Alkali is an excellent DNA denaturant; it permits rapid separation of the strands without degradation. Alkali both denatures and degrades RNA to 2′(3′)-mononucleotides.

Many solutes that can form hydrogen bonds also lower the melting temperature (decrease the stability) of double-helix structures. The organic compounds formamide and urea are frequently used to lower the denaturation temperature as well as to prevent reaggregation in DNA manipulations when it is important to avoid nonspecific aggregation. Reagents that increase the solubility of the DNA bases (e.g., methanol) or disrupt the water shell around them (e.g., trifluoroacetate) reduce the hydrophobic interactions between the bases and lower the T_m. Most proteins that bind to DNA inhibit denaturation. However, some DNA-binding proteins destabilize the native state. Proteins of this class usually bind preferentially to single-stranded DNA, thereby favoring separation of the double strands. We discuss so-called single-strand binding proteins in the next chapter because they play an important role in DNA synthesis.

DNA Renaturation Involves Duplex Formation from Single Strands

We have seen that when a solution of heat-denatured DNA is allowed to cool rapidly, the regularly hydrogen-bonded structure does not reform. However, reassembly of the two separated polynucleotide strands into the native structure, called renaturation, is possible under certain specialized conditions.

The first indication that renaturation was possible came from Paul Doty's laboratory in experiments performed by Julius Marmur. They observed that when transforming DNA was heated and rapidly cooled, it was biologically inactive; however, when denatured DNA was slowly cooled, a small percentage of the initial transforming activity was recovered. Further experiments showed that the optimal temperature for this recovery of activity, or renaturation, was about 25°C below the T_m. Similar to denaturation, renaturation can be followed spectrophotometrically. If the temperature of a denatured DNA solution is maintained at $T_m - 25$°C for a long time, the absorbance of the solution gradually decreases until it approaches a value close to that of native DNA (see curve *c* in fig. 30.22). The optimum temperature for renaturation is called the annealing temperature. At the annealing temperature, irregularly hydrogen-bonded structures are unstable, but regularly hydrogen-bonded structures are stable. Consequently, prolonged exposure of denatured DNA at this temperature favors the formation of regularly base-paired structures.

Kinetic analysis indicates that renaturation is a two-step process. In the slow step, effective contact is made between two complementary regions of DNA originated from separate strands. This rate-limiting step called nucleation is a function of the concentration of complementary strands. Nucleation is followed by a relatively rapid zippering up of adjoining base residues into a duplex structure. The main steps involved in denaturation and renaturation are depicted in figure 30.24.

Since nucleation involves interaction between two molecules, it should occur at a rate proportional to the square of the concentration of single strands. If c is the concentration of single-stranded DNA at time t, then the rate equation for the loss of single-stranded DNA is

$$-\frac{dc}{dt} = k_2c^2 \qquad (1)$$

where k_2 is the rate constant for a second-order reaction. Starting with a concentration c_0 of completely denatured DNA, the amount of single-stranded DNA left after renaturation for time t is given by

$$\frac{c}{c_0} = \frac{1}{1 + k_2c_0t} \qquad (2)$$

At time $t_{1/2}$, when half of the DNA is renatured, $c/c_0 = 0.5$ and $t = t_{1/2}$, from which it follows that

$$c_0t_{1/2} = \frac{1}{k_2} \qquad (3)$$

The rate of renaturation is also a function of chain length, but this effect is usually eliminated as a variable by shearing the starting DNA down to a uniform size. For a typical renaturation experiment, the values of c/c_0 are plotted as a function of c_0t, and the resulting curve is referred to as a "cot" curve.

In figure 30.25, cot curves for four DNA samples and one synthetic polyribonucleotide sample are presented. The mouse

Figure 30.24

Steps in denaturation and renaturation of a DNA duplex. In step 1 the temperature is raised to the point where the two strands of the duplex separate. If denatured DNA is slowly cooled, the events depicted as steps 2 and 3 follow. In step 2 a second-order reaction occurs in which two complementary strands of DNA must collide and form interstrand hydrogen bonds over a limited region. Step 3 is a first-order reaction in which additional hydrogen bonds form between the complementary strands that are partially hydrogen-bonded (zippering). Once complementary strands are partially bonded, the zippering reaction occurs rapidly. In the overall process, step 2 is rate-limiting.

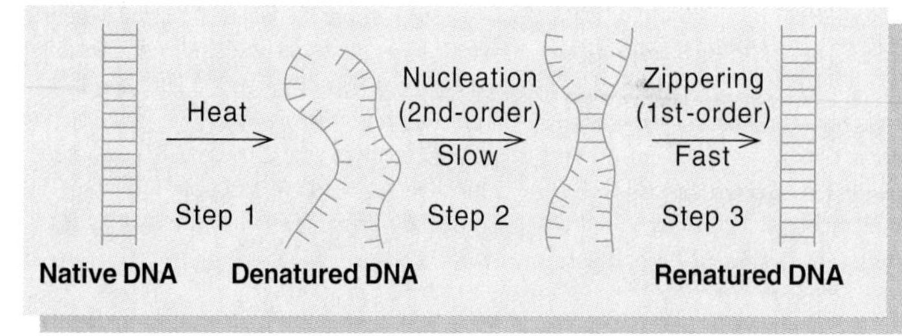

Figure 30.25

Reassociation of double-stranded nucleic acids from various sources. The genome size is indicated by arrows near the upper nomographic scale. Over a factor of 10^{10} this value is proportional to the c_0t (the "cot") required for half-reaction. All DNAs were sheared so that they have approximately the same fragment size (about 400 nucleotides, single-stranded). Correction has been made to give the rate that would be observed at 0.18 M sodium ion concentration. No correction for temperature has been applied because it was approximately optimum in all cases. The labels for the different DNAs should not concern the average reader. Mouse satellite and calf (nonrepetitive fraction) are fractions of the genome obtained from the indicated animals. (*Source:* Adapted from R. J. Britten and D. E. Kohne, Repeated sequences in DNA, *Science* 161:529, 1968.)

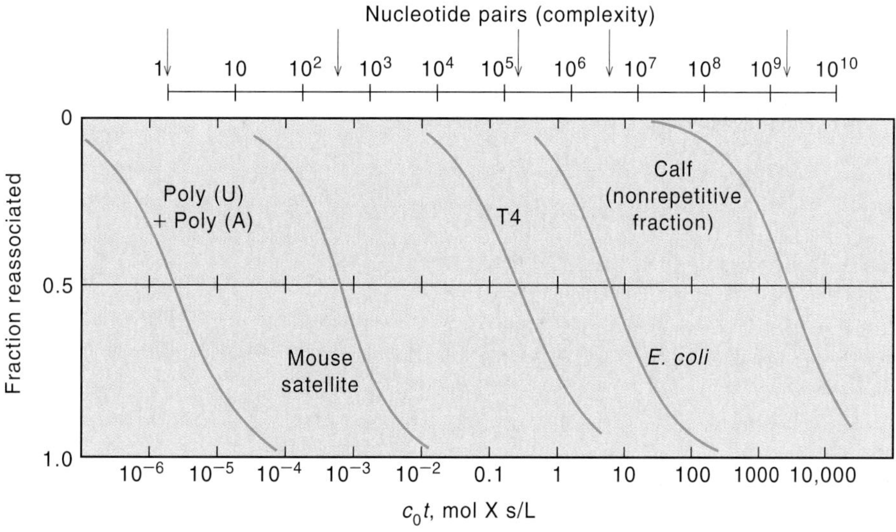

satellite DNA is a fraction of the DNA from the mouse that contains highly repetitious sequences. The T4 DNA is the total DNA isolated from the bacteriophage T4. Similarly the *E. coli* DNA is the total DNA isolated from *E. coli* bacteria. The calf nonrepetitive fraction represents a fraction of calf nuclear DNA that contains mostly sequences that are represented only one time per haploid genome. At the top of this semilogarithmic plot is an additional scale indicating the nucleotide complexity, *N*, which is defined as the number of nucleotides in a nonrepeating sequence. If no sequences in the cellular DNA repeat, then *N* is equal to the number of nucleotides in the genome. It can be seen that $c_0t_{1/2}$ is proportional to *N* for these samples.

This family of curves has been used to calibrate more complex situations in which the test DNA is a mixture of unique sequence and repetitive DNA. In such cases the repetitive DNA fraction tends to anneal more rapidly. Results for several species are presented in figure 30.26. Bacterial DNA (*E. coli*) contains very little repetitive DNA (0.3%). This is mainly accounted for by the eight genes for *E. coli* ribosomal RNA that have nearly identical sequences. In complex eukaryotes, single-copy DNA (i.e., nonrepetitive DNA) accounts for 40%–70% of the DNA, most of the remainder being roughly divided between middle repetitive ($<10^4$ copies/genome) and highly repetitive ($>5 \times 10^4$ copies/genome). Further analyses of eukaryotic gene structure by a variety of other techniques have shown that most genes encode proteins that belong to the unique (single-copy) class. The middle repetitive class includes transfer RNA genes and ribosomal RNA genes that form part of the biochemical machinery for protein synthesis (chapter 34), as well as the genes encoding histones, the main chromosomal proteins. Some highly repetitive sequences occur in tandem (see mouse satellite DNA in fig. 30.25), and still other repetitive DNA elements are distributed at random through-

Figure 30.26

Distribution of single-copy DNA and repetitive-sequence DNA in various organisms. The width of bands and the number below the bands indicate the fraction of total cellular DNA in each class. For example, in calf about 55% of the DNA sequences are single copy, about 38% are present in somewhat fewer than 10^5 copies, and about 3% are present in somewhat fewer than 10^6 copies. (Adapted from R. J. Britten and D. E. Kohne, Repeated segments of DNA, *Sci. Amer.* 222(4):24–31, 1970. Copyright © 1970 by Scientific American, Inc. All rights reserved. Reprinted by permission.)

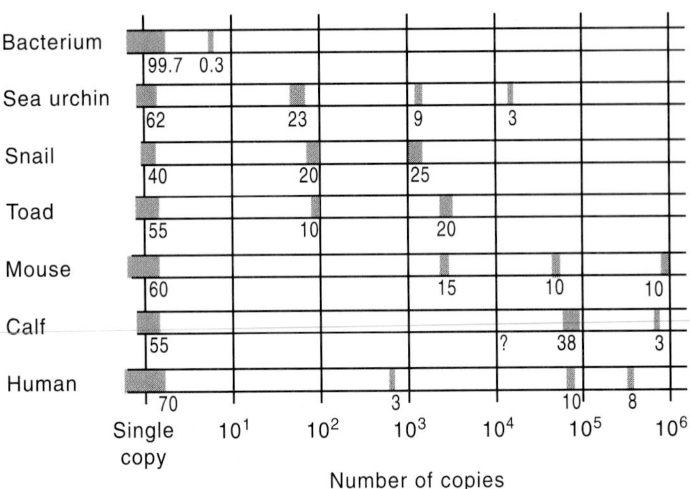

Figure 30.27

The *E. coli* chromosome exists as a circular, folded, supercoiled duplex (*a*). This can be converted to a partially unfolded structure by brief treatment with RNase (*b*). There are 50–100 loops in the structure; supercoiling may be selectively eliminated from individual loops by single-strand nicking of the DNA within the loop (*c*).

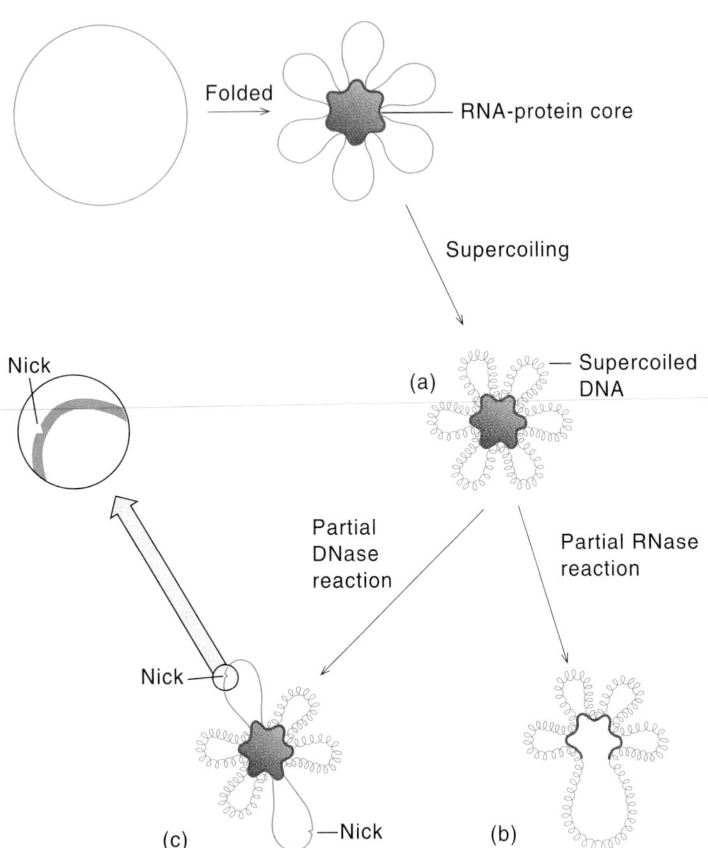

out the genome. It has been argued that some families of repetitive DNA, referred to as "selfish DNA," represent "parasitic" sequences that replicate together with the genome without conferring any positive or negative characteristics to the host cells that have these sequences.

Chromosome Structure

All types of nucleic acids interact with proteins. Chromosomal DNA forms stable nonspecific complexes with structural proteins that stabilize their tertiary structure; it also forms transient complexes with enzymes and regulatory proteins that modulate DNA and RNA metabolism. The gross tertiary structure of DNA in *E. coli* and a typical eukaryotic chromosome is described in the next section.

Physical Structure of the Bacterial Chromosome

A single chromosome of *E. coli* contains about 3×10^9 daltons or about 4.5×10^6 bp of DNA. If all of this DNA were in a duplex structure stretched end to end, it would be 1.5 mm long, which is about 75 cell diameters. In fact, the chromosome is circular, centrally located in the cell, and highly folded, so it is only about 2 μm across. No dramatic change in chromosome morphology is seen prior to cell division as in eukaryotes. Clearly, the degree of compaction observed throughout the cell cycle does not interfere with transcription to any great extent because all the genes in *E. coli* are readily expressible.

Electron micrographs of the *E. coli* chromosome suggest a folded circular structure containing about 40–100 supercoiled loops (diagrammatically indicated in fig. 30.27). It is believed that the folded structure is held together by an RNA-protein core, although the manner in which this is done is not well understood. The structure is further stabilized because the core forms a complex with positively charged polyamines and certain basic proteins. D. E. Pettijohn and his co-workers have provided evidence of such a core. They first showed that the individual supercoiled loops maintain their supercoiling independently of one another. Thus if a single nick is introduced into one of the loops by limited DNase action, that loop adopts an expanded relaxed conformation, but supercoiling in the other loops is maintained (see fig. 30.27). Limited RNase or protease treatment causes the partial breakdown of the looped structures without interfering with the supercoiling (see fig. 30.27). These results have led to the conclusion that each of the loops is a domain, the lateral motion of which is restricted by an RNA-protein core complex.

The Genetic Map of Escherichia coli

The circular *E. coli* chromosome contains enough base pairs to make about 3,000 average-sized genes. The relative positions of over half of these genes are known. In regions where our understanding is reasonably complete the impression is obtained of a tightly organized genome. Coding regions are interspersed with regulatory regions; based on the limited sequence information that is currently available, stretches of DNA with no apparent function make up a very small percentage of the total genome. Frequently, genes with a related function are clustered. The clustered genes are usually transcribed into single expression units (messenger RNAs) containing the information for the synthesis of several functionally related proteins (see chapters 34 and 35).

Eukaryotic DNA Is Complexed with Histones

DNA in eukaryotic chromosomes exists in a highly compacted form known as chromatin, a complex of DNA with a great variety of proteins. Five proteins called histones are present in large amounts (table 30.3) and are believed to form a regularly repeating structural motif. The remaining proteins are present in smaller amounts and are irregularly distributed.

Electron-microscopic and x-ray diffraction studies on chromatin suggest that the DNA forms a coiled-coil structure with the histones of chromatin. A breakthrough in our understanding of the nucleohistone complex came when D. E. Olins and A. L. Olins observed that chromatin viewed after sudden swelling in water had a beaded structure (fig. 30.28). The beads, called nucleosomes, contain four of the five histones; they are about 10 nm in diameter, and the spacing between the beads is about 14 nm. Brief enzymatic digestion of chromatin with micrococcal nuclease fragments this structure. The DNA–histone fragments from this partial digestion give rise to a banded pattern on agarose gel electrophoresis that suggests nucleoprotein structures containing 200 bp of DNA or multiples thereof (400, 600, 800 bp, etc.). Electron-microscopic examination of the individual fractions isolated after gel electrophoresis confirms the suggested correlation between the size of the DNA estimated on gels and the number of nucleosomes. Thus the most rapidly moving DNA band seen on gels was derived from a structure containing one nucleosome, and the second fastest migrating species contains nucleosome dimers, and so forth. Evidently, the brief treatment with endonuclease preferentially cleaves DNA in the internucleosomal region, where the DNA is least likely to be protected from enzyme attack. More extensive nuclease treatment results in a single band on gels that contain a single nucleosome with 140 bp of DNA. It appears that exhaustive nuclease digestion has removed all of the DNA that is not in direct contact with the nucleosome. From these results it was deduced that nucleosomes contain a core of histone with 140 base pairs of DNA wrapping; an additional 60 bp of more exposed DNA connects adjacent nucleosomes.

The histones present in chromatin are of five major types: H1, H2a, H2b, H3, and H4 (see table 30.3). The lysine-rich histone H1 is not present in the nucleosome core particles, as evidenced by its release on extensive nuclease treatment and the finding that H1 is the only histone that readily exchanges between free and DNA-bound histone. H1 may play a key role in the conversion of chromatin to the highly compacted chromosome that occurs immediately before cell division. The other eight histones, two each of the other four histones, form the protein core of the nucleosome. These protein octamers do not come apart even when chromosomes duplicate.

An illustration of the coiled-coil structure of the nucleosome is presented in figure 30.29. The 140 bp of DNA make about one and three-quarters superhelical turns about the histone octamer. An additional 60 bp of spacer DNA (not shown) connect adjacent nucleosomes. A survey of chromatins from different species and in different tissues of the same species has shown that the spacer DNA actually varies in length from about 20 to 95 bp.

Salt bridges between positively charged basic amino acid side chains of histones and the negatively charged DNA phos-

Table 30.3
Characteristics of Histones

Name	Ratio of Lysine to Arginine	M_r	Copies per Nucleosome
Histone H1[a]	20	21,000	1 (not in bead)
Histone H2a	1.2	14,500	2 (in bead)
Histone H2b	2.5	13,700	2 (in bead)
Histone H3	0.7	15,300	2 (in bead)
Histone H4	0.8	11,300	2 (in bead)

[a]Not found in lower eukaryotes such as yeast.

Figure 30.28

Swollen fibers of chromatin from the nucleus of the chicken red blood cell. The electron micrograph is enlarged about 325,000× and negatively stained with uranyl acetate. (Micrograph courtesy of A. L. Olins and D. E. Olins.)

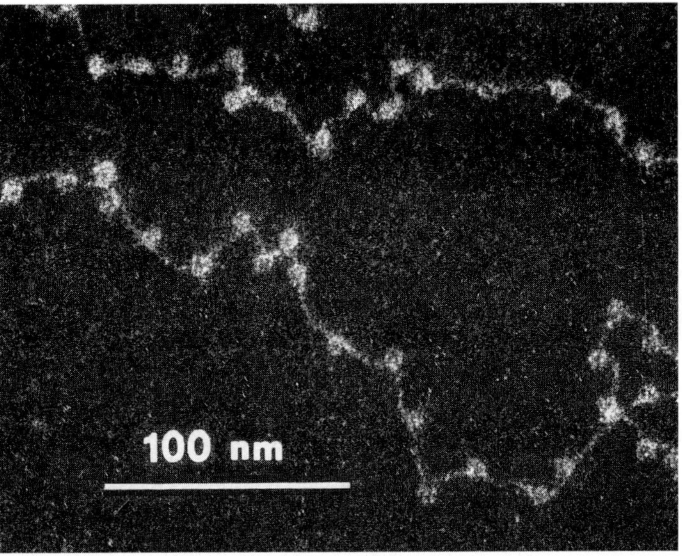

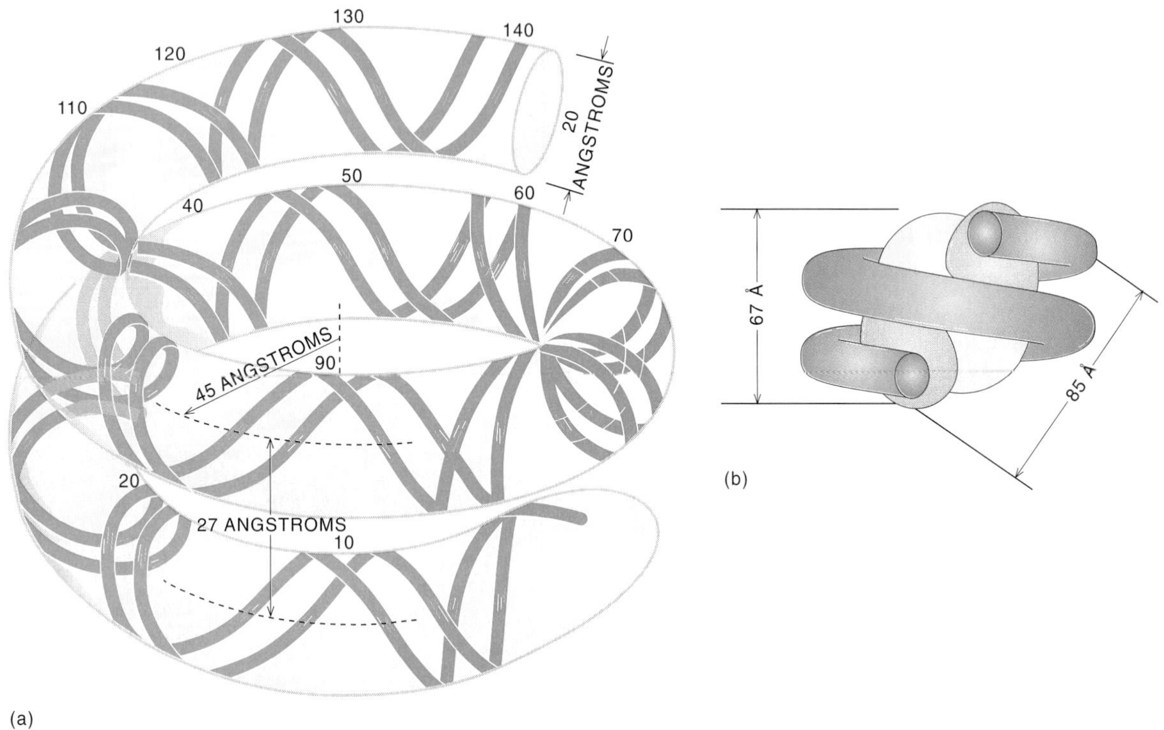

Figure 30.29

(*a*) Path of DNA that can account for the bipartite structure of the nucleosome core is a superhelix with an external diameter of 110 Å and a pitch of 27 Å; the turns of the 20-Å-wide DNA helix are nearly in contact. About 80 bp of DNA occur per turn; the nucleosome core, an enzymatically reduced form of the nucleosome consisting of some 140 bp, has about one and three-quarter turns wrapped on it. The histone octamer complex, containing two each of histones H2a, H2b, H3, and H4, is packed on the inside of the DNA coiled-coil structure. (© George V. Kelvin. Scientific American, Reprinted with permission.) In (*b*) we see this histone octamer inserted into the nucleosome core. The H3-H4 tetramer is shown in yellow, and an H2a-H2b dimer is shown at each end in purple. For further details on the structure of the histone octamer, consult the Selected Readings list.

phates play a major role in stabilizing the DNA–histone complex. Indeed, treatment of chromatin with concentrated NaCl (1–2 M), which is known to disrupt electrostatic bonds, causes a complete dissociation of DNA and histone in the nucleohistone complex.

Higher-order structures beyond that of the nucleosome are less well understood. Electron-microscopic investigations indicate two types of fibers with diameters of 10 nm and 30 nm. To account for the 10-nm fiber, the nucleosomes can be arranged edge to edge in a zigzag fashion to produce a fibril that is 10 nm wide. When the ionic strength is raised on an isolated preparation of chromatin, the fibrils reversibly condense into an irregularly supercoiled fiber about 30 nm in diameter. The nucleosome particles are thought to have their cylindrical axes approximately perpendicular to the long axis of the 30-nm fiber with six to seven nucleosomes per turn (fig. 30.30). Further coiling of these structures is necessary to explain the much larger structures seen in mitotic chromosomes.

Organization of Genes within Eukaryotic Chromosomes

The organization of genes within a typical eukaryotic chromosome is far more complex and less well understood than in prokaryotes. It is highly likely that a much lower percentage of the DNA is informational in complex eukaryotes than in prokaryotes. *E. coli* contains about 3,000 genes; the human genome is 1,000 times larger, but it is very unlikely to contain 1,000 times the number of functional genes. Most estimates hover around 50,000. Even if there were 200,000 human genes, this would still leave approximately 90% of the DNA with no coding function. Measurements on the fruit fly *Drosophila melanogaster* indicate that the coding information for proteins in most genes probably accounts for no more than one-tenth of the base pairs within the gene. The discovery that noncoding regions (introns) occur between coding regions (exons) helps to explain the excessive amounts of DNA that appear to be present in eukaryotes, although it is probably not the whole story. Noncoding regions serving as control loci may be larger on the average in eukaryotes. Nonfunctioning genes that have lost their initiation sites for being expressed (promoters) and repetitive genetic elements with no apparent coding function also help to account for the large amount of noncoding DNA.

As we have already discussed, not all repetitive DNA in eukaryotes is noncoding. Thus histone genes, for example, are typically reiterated many times. The five different histone genes are usually clustered, and this cluster is then tandemly repeated many (up to 100 or more) times. Ribosomal RNA genes are also tandemly clustered. Other nonidentical but functionally related genes that show clustering include the globin genes and the immunoglobulin genes.

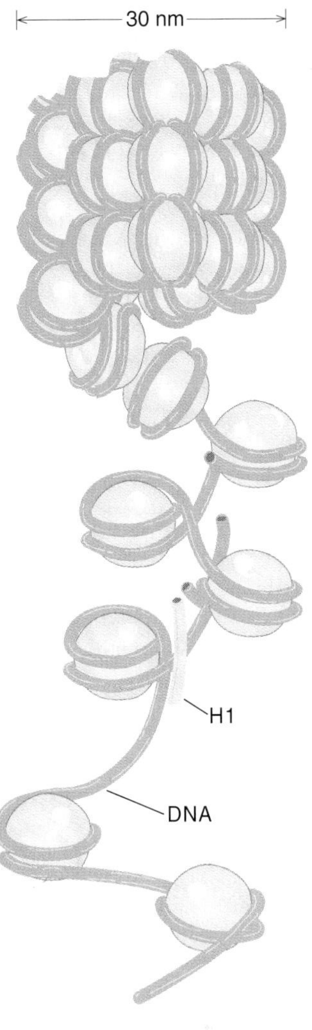

Figure 30.30

Helical superstructures might be formed with increasing salt concentration (bottom to top) as is suggested here. The zigzag pattern of nucleosome (1, 2, 3, 4) closes up, eventually to form a solenoid, a helix with about six nucleosomes per turn. (The helix is probably more irregular than it is in this drawing.) Cross-linking data indicate that H1 molecules on adjacent nucleosomes make contact. Extrapolation from the zigzag form to the solenoid suggests (but does not prove) that the aggregation of H1 at higher ionic strengths gives rise to a helical H1 polymer (not shown) running down the center of the solenoid. In the absence of H1 (bottom) no ordered structures are formed. The details of H1 associations are not known at this time; the drawing is meant to indicate only that H1 molecules contact one another and link DNA. (© George V. Kelvin. Scientific American, Reprinted with permission.)

SUMMARY

1. The genetic material of cells and viruses consists of DNA or RNA. That DNA bears genetic information was first shown when the heritable transfer of various traits from one bacterial strain to another was found to be mediated by purified DNA.
2. All nucleic acids consist of covalently linked nucleotides. Each nucleotide has three characteristic components: (1) a purine or pyrimidine base, (2) a pentose, and (3) a phosphate group. The purine or pyrimidine bases are linked to the C-1′ carbon of a deoxyribose sugar in DNA or a ribose sugar in RNA. The phosphate groups are linked to the sugar at the C-5′ and C-3′ positions. The purine bases in both DNA and RNA are always adenine (A) and guanine (G). The pyrimidine bases in DNA are thymine (T) and cytosine (C); in RNA they are uracil (U) and cytosine. The bases may be postreplicatively or posttranscriptionally modified by methylation or other reactions in certain circumstances.
3. DNA exists most typically as a double-stranded molecule, but in rare instances it exists (in some phages and viruses) in a single-stranded form. The continuity of the strands is maintained by repeating 3′,5′-phosphodiester linkages formed between the sugar and the phosphate groups; they constitute the covalent backbone of the macromolecule. The side chains of the covalent backbone consist of the purine or pyrimidine bases. In double-stranded, or duplex, DNA the two chains are held together in an antiparallel arrangement. The base composition of DNA varies characteristically from one species to another in the range of 25%–75% guanine plus cytosine. Specific pairing occurs between bases on one strand and bases on the other strand. The complementary base pairs are either A and T, which can form two hydrogen bonds, or G and C, which can form three hydrogen bonds. The duplex is stabilized by the edge-to-edge hydrogen bonds formed between these planar base pairs and face-to-face interactions (stacking) between adjacent base pairs. Twisting of the duplex structure into a helix makes stacking interactions possible.
4. The right-handed helical structure, known as B DNA, is the

most commonly occurring conformation of linear duplex DNA in nature. In this structure, the distance between stacked base pairs is 3.4 Å, with approximately 10 base pairs per helical turn. The inherent flexibility of the structure, however, makes a variety of conformations possible under different conditions. In some instances, nucleotide sequence and degree of hydration dictate which conformations are favored. DNA interacts with a variety of proteins inside the cell, and these proteins can also have a significant influence on its secondary and tertiary structure.

5. Circular DNA molecules, which are topologically confined so that their ends are not free to rotate, can form supercoils that are either right-handed (negative) or left-handed (positive). Negative supercoiling exerts a torsional tension favoring the untwisting of the primary right-handed double helix, whereas positive supercoiling has the opposite effect. Negatively supercoiled DNAs are most commonly observed in prokaryotes, which contain an enzyme that generates the supercoiled structure.
6. When duplex DNA (or RNA) is heated, it dissociates (denatures) into single strands. The temperature at which denaturation occurs (the melting temperature) is a measure of the stability of the duplex and is a function of the G-C content of the DNA. A preparation of denatured DNA may be renatured in the native duplex structure by maintaining the temperature about 25°C below the melting temperature. The rate of renaturation is a measure of the sequence complexity of the DNA. In prokaryotes, which consist predominantly of unique sequences, the complexity (the number of base pairs) is approximately equal to the genome size. However, eukaryotic cells contain DNAs of varying sequence complexity that renature at quite different rates. The fastest renaturing fractions are present in many copies per nucleus, whereas the slowest renaturing fractions are present in single copies. Analyses by other techniques have shown that some of the repetitive DNA sequences exist as tandemly repeated structures, while other types of repetitive sequences are dispersed throughout the genome.
7. In the chromatin of eukaryotic cells DNA forms a coiled-coil structure with an approximately equal weight of a mixture of five basic proteins known as histones. Four of these histones in pairs form an octamer around which the DNA duplex occurs in a left-handed helix. The DNA octamer is called a nucleosome. Each nucleosome contains about 140 base pairs of DNA in a nuclease-resistant "nucleosome core" and approximately 60 base pairs of spacer between core particles. Histone H1 binds to the chromatin independently of the octamer and is the first histone to dissociate from the chromatin when the ionic strength is raised. Beyond the nucleosome the higher-order structure of the chromosome involves coiled-coil structures with varying degrees of regularity.

Selected Readings

Arents, G., R. W. Burlingame, B.-C. Wang, W. E. Love, and E. N. Moudrianakis, The nucleosomal core histone octamer at 3.1 Å resolution: A tripartite protein assembly and a left-handed superhelix. *Proc. Natl. Acad. Sci. USA* 88:10148–10152, 1991.

Arents, G., and N. Moudrianakis, Topography of the histone octamer surface: Repeating structural motifs utilized in the docking of nucleosomal DNA. *Proc. Natl. Acad. Sci. USA* 90:10489–10493, 1993.

Chu, G., D. Vollrath, and R. W. Davis, Electrophoresis of yeast chromosomes. *Science* 234:1583, 1986.

Daniels, D. L., G. Plunkett III, V. Burland, and F. R. Blattner, Analysis of the *Escherichia coli* genome: DNA sequence of the region from 84.5 to 86.5 minutes. *Science* 257:771–778, 1992.

Dervan, P. B., Reagents for the site-specific cleavage of megabase DNA. *Nature* 359:87–88, 1992.

Dickerson, R. E., The DNA helix and how it is read. *Sci. Am.* 249(6)d:94–111, 1983.

Fang, G., and T. R. Cech, The β subunit of Oxytricha telomere-binding protein promotes G-quartet formation by telomeric DNA. *Cell* 74:875–885, 1993.

Hershey, A. D., and M. Chase, Independent functions of viral proteins and nucleic acid in growth of bacteriophage. *J. Gen. Physiol.* 36:39–56, 1952.

Hillary, C. M., J. T. Finch, B. F. Luisi, and A. Klug, The structure of an oligo(dA), oligo(dT) tract and its biological implications. *Nature* 330:221–236, 1987.

Kang, C., X. Zhang, R. Ratliff, R. Moyzis, and A. Rich, Crystal structure of four-stranded Oxytricha telomeric DNA. *Nature* 356:126–131, 1992.

Kim, S. H., Three-dimensional structure of transfer RNA. *Prog. Nuc. Acid Res. Mol. Biol.* 17:181–216, 1973.

Kornberg, R. D., and A. Klug, The nucleosome. *Sci. Am.* 244(2):52–64, 1981.

Kornberg, R. D., and Y. Lorch, Chromatin structure and transcription. *Ann. Rev. Cell. Biol.* 8:563–587, 1992.

Laughlan, G., A. I. H. Murchie, D. G. Norman, M. H. Moore, P. C. E. Moody, D. M. J. Lillely, and B. Luisi, The high-resolution crystal structure of a parallel-stranded guanine tetraplex. *Science* 265:520–524, 1994.

Lerman, L. S., S. G. Fischer, I. Hurley, K. Silverstein, and N. Lumelsky, Sequence-determined DNA separations. *Ann. Rev. Biophys. Bioeng.* 13:399–423, 1983.

Meervelt, L., D. Vlieghe, A. Dautant, B. Gallois, G. Precigoux, and O. Kennard, High-resolution structure of a DNA helix forming (C · G)*G base triplets. *Nature* 274:742–744, 1995. A structure that may play a role in homologous recombination.

Morse, R. H., and R. T. Simpson, DNA in the nucleosome. *Cell* 54:285–287, 1988.

Nadeau, J. G., and D. M. Crothers, Structural basis for DNA bending, *Proc. Natl. Acad. Sci.* 86:2622–2626, 1989.

Ramakrishnan, V., J. T. Finch, V. Graziano, P. L. Lee, and R. M. Sweet, Crystal structure of globular domain of histone H5 and its implications for nucleosome binding. *Nature* 362:219–223, 1993.

Rich, A., A. Nordheim, and A. H.-J. Wang, The chemistry and biology of left-handed Z DNA. *Ann. Rev. Biochem.* 53:791–846, 1984.

Roberts, R. W., and D. M. Crothers, Stability and properties of double and triple helices: Dramatic effects of RNA or DNA backbone composition. *Science* 258:1463–1466, 1992.

Saenger, W., *Principles of Nucleic Acid Structure.* New York: Springer-Verlag, 1984.

Schmid, M. B., Structure and function of the bacterial chromosome. *Trends Biochem. Sci.* 13:131–135, 1988.

Schwartz, D. C., and C. R. Cantor, Separation of yeast chromosome-sized DNAs by pulsed field gradient gel electrophoresis. *Cell* 37:67–75, 1984.

Strobel, S. A., L. A. Doucette-Stamm, L. Riba, D. E. Housman, P. B. Dervan, Site-specific cleavage of human chromosome 4 mediated by triple-helix formation. *Science* 254:1639–1642, 1991.

Structures of DNA, *Cold Spring Harbor Symp. Quant. Biol.* 47, 1983.

van Holde, K. E., *Chromatin.* New York: Springer-Verlag, 1988.

Watson, J. D., and F. H. C. Crick, Molecular structure of nucleic acids. *Nature* 171:737–738, 1953.

PROBLEMS

1. When Erwin Chargaff began studying the composition of nucleic acids from different tissues and different organisms, there was already strong evidence for DNA as the genetic material, despite some chemists' belief that the structure of DNA consisted of a simple tetranucleotide (or a repeating tetranucleotide). What were the hypotheses that were tested in Chargaff's work, and how did the results support his hypothesis?
2. Why is the bond that holds nucleotides together in nucleic acids called a "phosphodiester" bond?
3. How many different base-paired structures with two hydrogen bonds can be made by using guanine (G) and thymine (T)? Which one of these is most similar to the standard Watson-Crick (G-C) base pair?
4. Draw the structure of the enol form of thymine. Can this molecule form a base pair that is similar in size and shape to a standard Watson-Crick base pair?
5. How can the DNA double helix have a constant diameter even though each strand consists of purines and pyrimidines of different sizes?
6. (a) List the hydrogen bond donors and acceptors that are available in the major and minor grooves of B-form DNA.
   (b) The intermolecular forces that stabilize DNA–protein complexes often involve hydrogen bonds between specific amino acids and the exposed surfaces of the bases. Explain why interaction with the major groove is more common among DNA-binding proteins than is interaction with the minor groove.
   (c) Most DNA-binding proteins (histones, for example) are rich in the basic amino acids arginine and lysine. Explain this observation.
7. Why are RNA duplexes and RNA–DNA hybrids unlikely to adopt the B conformation?
8. (a) How many base pairs would be found in the DNA pictured at the right in figure 30.20 if two negative supercoils were introduced without breaking the backbone?
   (b) Describe the effects of a short stretch of Z-DNA on the overall conformation of this DNA.
9. Standard conditions for hydrolyzing RNA to nucleotides are 0.3 N NaOH at 37°C for 16 h. Draw a chemical reaction mechanism for this hydrolysis. Why is DNA not hydrolyzed under these conditions?
10. Why is DNA denatured at either low pH (pH 2) or high pH (pH 11)? (*Hint:* See the pK values in table 26.2.)
11. You are given a sample of nucleic acid extracted from a virus. How would you determine whether the virus has an RNA or DNA genome and whether it is single- or double-stranded?
12. In the early 1980s, Stanley Prusiner and co-workers found evidence that a protein is the causative agent of scrapie, an infectious disease in sheep. Why was this finding initially met with considerable skepticism? If you were reviewing this work, what experiments would you suggest to confirm that the infectious agent in scrapie is protein and not nucleic acid?
13. Explain the differences in structure between the following nucleotides: AUG and GTA.
14. The enzyme S1 nuclease specifically digests single-stranded nucleic acids to mononucleotides. The enzyme can also cleave double-stranded DNA to a limited extent, primarily at stretches that are rich in A and T, suggesting that these regions have some single-stranded character. What does this observation reveal about the structure of double-stranded DNA?
15. Suggest a structure for the base triplet holding together the triplex formed between one chain of polyriboadenylic acid and two chains of polyribouridylic acid. (*Hint:* Examine carefully *all* possible hydrogen-bonded interactions between A and U.) Why might this triplex structure be stabilized by the addition of $MgCl_2$?
16. The following equation was derived empirically to describe the melting temperature, T_m, of a double-stranded DNA of length L base pairs in the presence of different concentrations (M) of monovalent cations and different percentages (% form) of formamide. The composition of the DNA duplex is given by % G + C.

    $$T_m = 81.5°\text{C} + 16.6[\log M] + 0.41[\%\ \text{G} + \text{C}] - 0.61[\%\ \text{form}] - 500/L$$

    How does this equation incorporate information about the forces stabilizing the DNA double helix?
17. Your supervisor shows you two test tubes and announces, "The labels fell off these tubes in the freezer! One was supposed to contain DNA from *E. coli* and one was DNA from *Mycobacterium tuberculosis.* I can't figure out which is which, so you will just have to grow cells and prepare more DNA." Because *M. tuberculosis* is a dangerous pathogen, you wish to avoid culturing the cells. Can you devise a simple strategy to determine which sample is from which organism, based on the fact that *E. coli* DNA contains 25% G + C and *M. tuberculosis* DNA is 70% G + C? Give an example of the expected results.
18. A column packed with the material hydroxyapatite preferentially binds double-stranded DNA over single-stranded DNA. The double-stranded DNA can be eluted by changing the salt concentration. Describe a procedure that could be used to isolate a fraction of eukaryotic DNA enriched for highly repetitive DNA.
19. When histone proteins are isolated from chromatin, their mass is equal to that of the DNA, and the molar ratio of four of the histones is 1:1:1:1 (H2a:H2b:H3:H4), while H1 is found in half the amount (0.5). Discuss whether or not these data fit the bead-and-string model for nucleosomes.
20. One of the intriguing questions still to be answered about the structure of the eukaryotic genome is how the information encoded in the DNA is retrieved. How might the clustering of functionally related genes facilitate the job of rapidly and specifically altering the pattern of genes that are expressed during different stages of an organisms's development?

C H A P T E R

31

DNA Replication, Repair, and Recombination

The genetic message carried by the sequence of bases in the DNA is replicated by a template mechanism.

From the complementary duplex structure of DNA it is a short intuitive hop to a model for replication that satisfies the requirement for one round of DNA duplication for every cell division. Such a proposal was made by Watson and Crick when they proposed the duplex structure for DNA (fig. 31.1). First, the double helix unwinds; next, mononucleotides are absorbed into complementary sites on each polynucleotide strand; and finally these mononucleotides become linked to yield two identical daughter DNA duplexes. What could be simpler! Subsequent biochemical investigations showed that in many respects this model for DNA replication was correct, but they also indicated a much greater complexity than was initially suspected. Part of the reason for the complications is that replication must be very fast to keep up with the cell division rate, and it must be very accurate to ensure faithful transfer of information from one cell generation to the next.

In addition to the replication process there are two other major areas of normal DNA metabolism. One is concerned with repair of damaged DNA, and the other is concerned with recombination between DNA molecules. In this chapter we deal mainly with the replication process, but we also consider repair and recombination.

The Universality of Semiconservative Replication

The Watson-Crick model for DNA replication is called semiconservative because the daughter duplexes arising from replication each contain one old (conserved) strand and one new strand.

Figure 31.1

Watson-Crick model for DNA replication. The double helix unwinds at one end. New strand synthesis begins by absorption of mononucleotides to complementary bases on the old strands. These ordered nucleotides are then covalently linked into a polynucleotide chain, a process resulting ultimately in two daughter DNA duplexes.

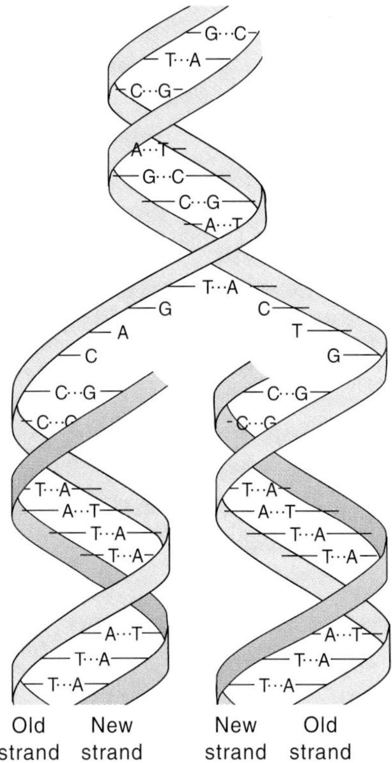

Figure 31.2

The Meselson-Stahl experiment demonstrating semiconservative replication for *E. coli* chromosomal DNA. Cesium chloride (CsCl) density-gradient centrifugation is used to discriminate between DNAs of different densities. When a concentrated solution of CsCl is centrifuged at high speed (50,000–100,000 times gravity), a stable concentration gradient of CsCl develops, with the concentration increasing along the direction of the centrifugal force. Since CsCl is much denser than water, this concentration gradient produces a density gradient. Macromolecules of DNA present in the solution are driven by the centrifugal field into the region where the solution density is equal to their own density. If the DNA has a uniform density, then after many hours of centrifugation (about 36 h), equilibrium is established, with all of the DNA concentrated in a single band. *E. coli* DNA has different densities when cells are grown in ^{14}N or ^{15}N medium (*a* and *b*). When cells containing pure heavy DNA (^{15}N–^{15}N DNA) are grown in ^{14}N medium for one generation, all of the DNA is of intermediate density (^{14}N–^{15}N). After two generations of growth in ^{14}N medium, the cells contain equal amounts of light (^{14}N–^{14}N) DNA and intermediate density DNA (*e*). In subsequent generations the hybrid DNA reappears in constant amounts, but the amount of light DNA increases. These results support the model of a semiconservative mode of DNA replication.

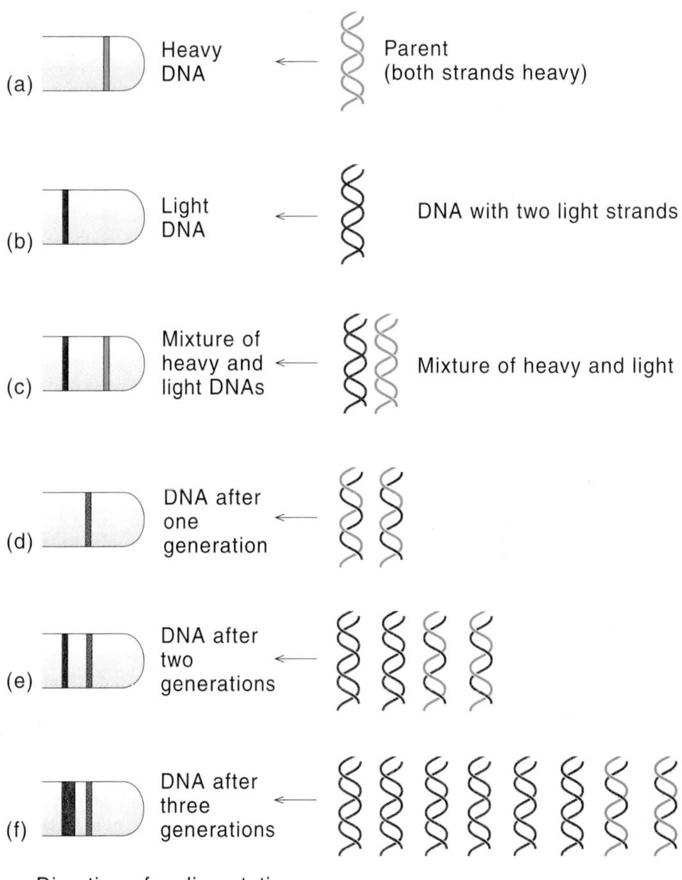

Matthew Meselson and Frank Stahl conceived a way of demonstrating the semiconservative mode of replication that involved the use of isotopes that result in DNAs of different densities after replication. For this purpose, *E. coli* bacteria were grown for several generations on a medium in which all the nitrogen was of the heavy ^{15}N isotope type (normal nitrogen is ^{14}N). The resulting DNA had a greater than normal density because the ^{15}N was incorporated into the bases of the DNA. Following this step, the bacteria were transferred to a growth medium containing normal (^{14}N) nitrogen, and the cells were allowed to go through one or more rounds of duplication. The DNA from these cells was then isolated and analyzed by density-gradient centrifugation. Pure ^{15}N DNA produces a single band of DNA (fig. 31.2*a*). The same is true for ^{14}N DNA (see fig. 31.2*b*). The only difference is that denser DNA produces a band farther down the tube. Thus the location of the DNA in the centrifuge tube can be used as a measure of the density of the DNA. Cells containing pure ^{15}N DNA were allowed to grow in ^{14}N medium for one or more generations, and the DNA from them was similarly analyzed. After precisely one generation in ^{14}N medium the only band visible in the isolated DNA was that corresponding to ^{15}N–^{14}N hybrid DNA (see fig. 31.2*d*). These data argue strongly against the conservative mode of replication, but they do not discriminate between the semiconservative and other modes of replication in which both strands might be labeled approximately equally. For this purpose the results in further generations were examined. Only in the semiconservative mode would equal amounts of DNA with the intermediate hybrid density and the light density be expected after two generations. This result is shown in figure 31.2*e*. In the third generation we still see a band with the hybrid density, but the amount of pure light DNA has increased (see fig. 31.2*f*). These

Figure 31.3

Autoradiographs of *Vicia faba* chromosomes labeled with [^{3}H]thymidine. The labeled thymidine becomes incorporated into the chromosomal DNA. A suitably labeled preparation is flattened and subjected to film exposure. Small dots indicate radioactive disintegration in the exposed film. (*a*) The first metaphase after replication in the presence of [^{3}H]thymidine. (*b*) The second metaphase after an additional replication in nonradioactive medium. (*c*) A diagrammatic interpretation of the results shown in (*a*) and (*b*). Radioactive single strands of DNA are shown in color. Radioactive chromatids at metaphase are also indicated in color. Colchicine has been used to inhibit spindle fiber formation and thus the anaphase separation of sister chromatids. Under these "C-metaphase" conditions, separation of sister chromatids is delayed. In (*a*) both sister chromatids are labeled uniformly. In (*b*) the sister chromatids are not labeled uniformly. The large chromosome at the top has one chromatid labeled and one virtually unlabeled. The homolog to its right has two exchanges (a labeled segment moved into the lower chromatid). The two small chromosomes to the lower left of it are lying one on top of the other. Both have one sister chromatid exchange. The small chromosome to the upper left has one sister chromatid exchange and the one at the lower left is lightly labeled but probably has two exchanges. (Autoradiographs courtesy of J. H. Taylor.)

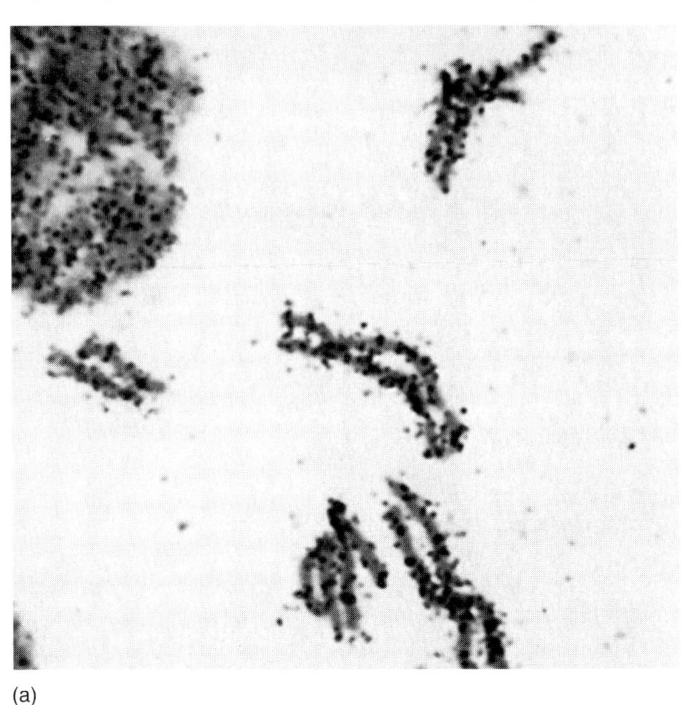

(a)

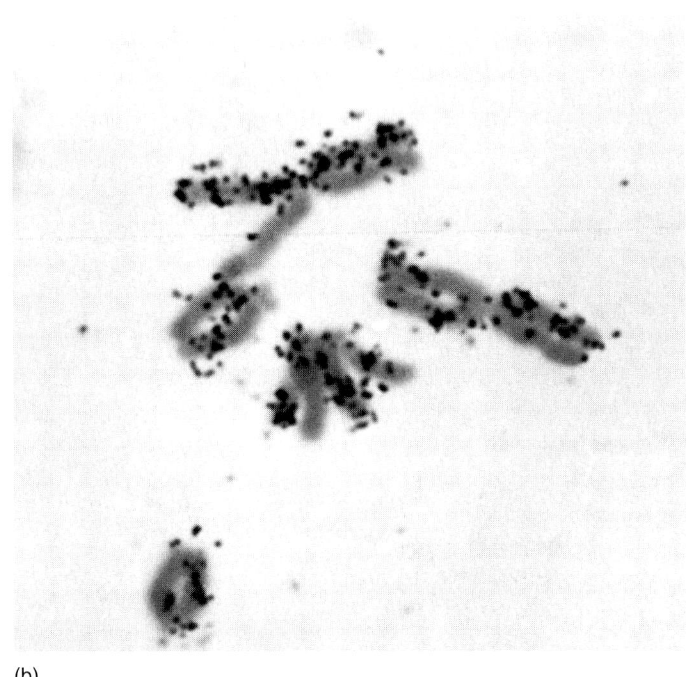

(b)

In presence of colchicine

Duplication with labeled thymidine

First C-metaphase after labeling; (a) above

Duplication without labeled thymidine

Second C-metaphase after labeling; (b) above

(c)

results strongly support the semiconservative mode for DNA replication. Indeed, it is hard to think of another mode of replication that could give rise to these results.

Similar experiments have been performed on mammalian cells grown in tissue culture, using bromouracil as the density label. Bromouracil contains a bromine atom instead of a methyl group on the 5 position of thymine. Bromouracil, when incorporated into DNA, can substitute for most of the thymine, leading to DNA with a substantially higher than normal density. The results with eukaryotic DNA were found to parallel those in *E. coli.* Thus it appears that semiconservative replication of cellular DNA is a general phenomenon.

Taylor, Woods, and Hughes devised another method for labeling DNA to follow its replication. They selectively labeled chromosomal DNA with tritiated (^{3}H-labeled) thymidine and then visualized the distribution of the radioactive label directly in autoradiographs of chromosomes (fig. 31.3*a*); the same pattern of labeling was visible after the cells subsequently replicated in the absence of tritiated thymidine (fig. 31.3*b*). Bean seedlings were used for those experiments because cell division is very rapid in their growth tips. A sufficient period of time was allowed for some of the cells in the seedlings to undergo one round of DNA duplication (incorporating the radioactive label) and cell division. After this the seedlings were transferred to a fresh solution containing colchicine without the ^{3}H label. Colchicine is a plant alkaloid that inhibits normal mitosis by interacting with microtubule proteins necessary for mitotic spindle formation; it does not inhibit DNA synthesis nor the replication of chromosomes but delays the formation of daughter cells by inhibiting chromatid segregation. The advantage of colchicine treatment in this experiment is that in tissue so treated, many cells become arrested in the state where the sister chromatids are paired. Cells that were examined shortly after the transfer from the [^{3}H] thymidine medium were examined at mitosis, and all the chromosomes appeared to be labeled uniformly (see fig. 31.3*a*). When the cells were allowed to duplicate their chromosomes in unlabeled medium, only one of the two chromatids in each chromosome pair was labeled (see fig. 31.3*b*). That is exactly the result that would be expected if each chromatid is composed of a single linear duplex of DNA that replicates semiconservatively (see fig. 31.3*c*). Similar results have been obtained for other eukaryotic cell types.

The universality of the semiconservative mode of DNA replication follows from the complementary nature of the DNA duplex. As the duplex unwinds, it presents two templates for the binding of complementary nucleotides. Subsequently, polymerization results in two duplexes, each with one old strand and one new strand. Thus the genetic information contained in the base sequence is directly transferred from one generation of DNA to the next by the capacity of DNA single strands to serve as templates for the assembly of complementary mononucleotides.

Overview of DNA Replication in Bacteria

More is known about the replication of DNA in *E. coli* than in any other system. The *E. coli* bacterium contains a single circular chromosome with about 4.5×10^4 bp. Within this chromosome there are about 3,000 genes, each gene being represented by a unique sequence of bases and encoding the genetic information essential for the synthesis of a specific protein. Exactly how many of these genes are required for DNA synthesis is not known; so far about 30 have been found, and in many cases the gene-encoded proteins have been extensively characterized. In figure 31.4 the locations of some of these genes are indicated, together with the locations of unique initiation (*oriC*) and termination points for DNA replication. First we consider the overall strategy for bacterial DNA replication; then we discuss some of the proteins involved in the synthesis.

Figure 31.4

A diagram of *E. coli* chromosome. The origin and approximate region of termination of replication are indicated. Locations of some genes involved in DNA replication are also indicated. The functions of many of these genes are described in table 31.1.

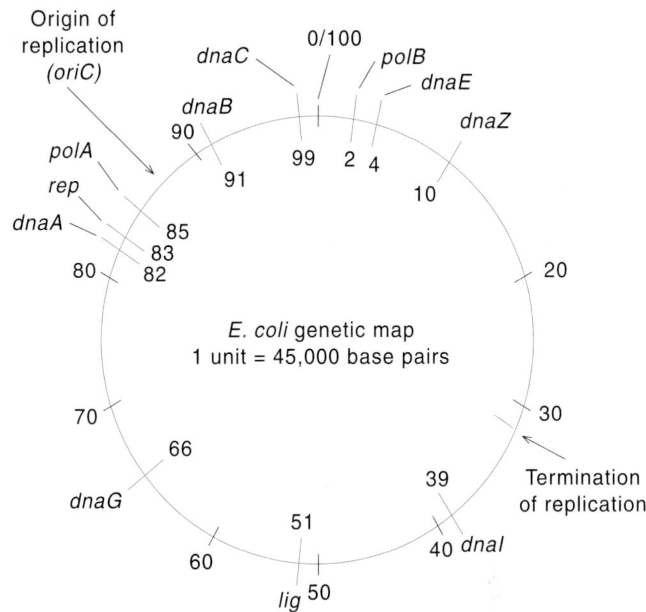

Growth during Replication Is Bidirectional

Replication of the *E. coli* chromosome can be made visible by very delicate techniques developed by John Cairns and Ric Davern for isolating intact ^{3}H-labeled chromosomes and subjecting them to autoradiography. After one round of replication in labeled medium, chromosomes appear circular and uniformly labeled.

Initiation of a second round of replication leads to a replication "eye" at the initiation site of replication (fig. 31.5). As synthesis proceeds, the size of the replication eye becomes larger; at this stage the replicating chromosome is referred to as a theta structure because it has the appearance of the Greek letter θ. Semiconservative replication is consistent with the density of the autoradiographic tracks made by different parts of the chromosome

Figure 31.5

Simulated autoradiographs of the *E. coli* chromosome after one or more replications in the presence of [^{3}H]thymidine. After one round of replication the autoradiograph shows a circular structure that is uniformly labeled. The second round of replication begins with the formation of a replication "eye." One branch in the replication eye is twice as strongly labeled as the remainder of the chromosome, indicating that this branch contains two labeled strands. This structure is consistent with semiconservative replication for the *E. coli* chromosome.

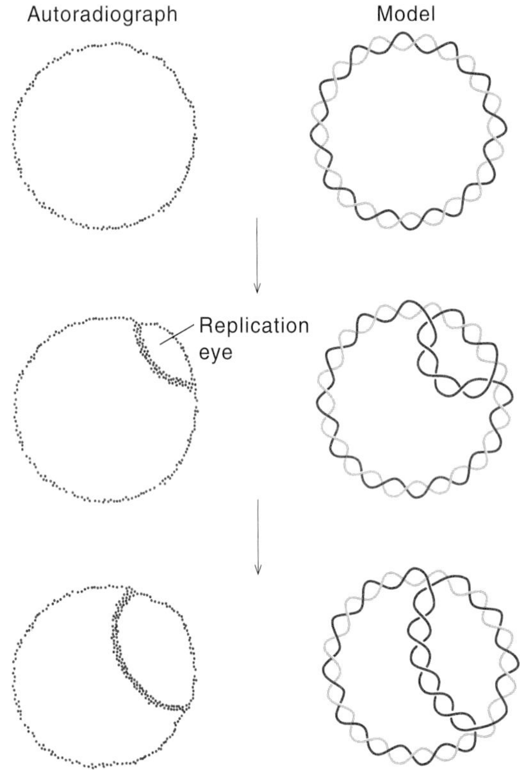

Figure 31.6

Schematic diagrams of two different modes of DNA synthesis at the growth fork(s). (*a*) In unidirectional replication, one growth fork occurs; (*b*) in bidirectional replication, two occur. Red indicates regions containing newly synthesized DNA.

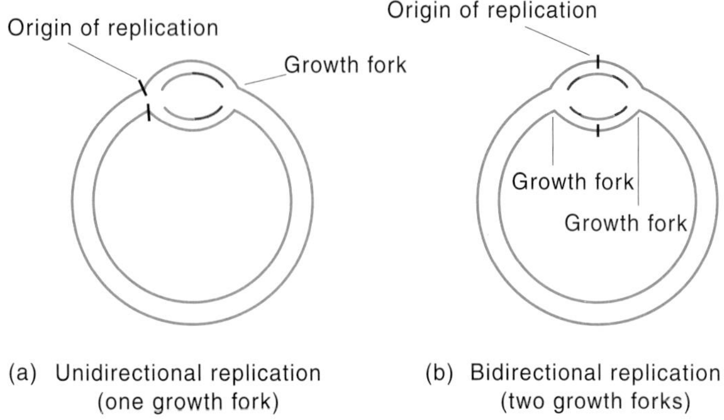

after one and two rounds of replication in [^{3}H]thymidine (see fig. 31.5).

It seems clear that the replication eye must contain two partially separated parental DNA strands that are base-paired with short strands of newly synthesized DNA. Not resolved by these autoradiographs is the question of whether replication occurs in one or both directions about the origin of replication. If growth is unidirectional, we expect one growth point (fig. 31.6*a*), called a growth fork; if growth is bidirectional, we expect two growth points or two growth forks (see fig. 31.6*b*). Although examples of both types of replication occur, it is believed that most bacterial chromosomes, including *E. coli,* replicate bidirectionally.

Evidence for bidirectional replication is supported by additional experiments. Growing bacteria were briefly exposed to radioactive thymidine, after which the replicating chromosomes were examined by autoradiography; both of the forks in the replicating structures were intensely labeled (see fig. 31.6*b*). These results show that both forks must be active in replication, a finding consistent with bidirectional growth.

Growth at the Replication Forks Is Discontinuous

Continuous synthesis on both strands of the replication fork would require synthesis in the $5' \rightarrow 3'$ direction on one strand and in the $3' \rightarrow 5'$ direction on the other because of the antiparallel nature of the DNA duplex (fig. 31.7*a*). Continuous synthesis on both strands seems unlikely, since the only known enzymes that catalyze DNA synthesis add bases to the growing chain in the $5' \rightarrow 3'$ direction (see below). For this reason it was hypothesized that replication is discontinuous on one of the branches at the replication fork (see fig. 31.7*a*). Meticulous electron-microscopic examination of replication forks in bacterial viruses did in fact show that transient gaps sometimes are apparent on one of the DNA strands near the replicating fork. Observations such as this led to the notion of leading-strand and lagging-strand synthesis (see fig. 31.7*b*). The $5' \rightarrow 3'$ synthesis of the leading strand can occur continuously in the direction of unwinding at the replication fork. But synthesis of the lagging strand in the $5' \rightarrow 3'$ direction occurs only in discontinuous spurts in the opposite direction.

This concept of discontinuous synthesis was supported by the finding that about half of the newly synthesized DNA is first made in small pieces that subsequently become incorporated into larger units of DNA. Small replication fragments were first detected in the laboratory of Okazaki. This was done by exposing growing cells to ^{3}H-labeled thymidine for a very short time (2–10 s), followed by rapid isolation of the radioactively labeled DNA. After longer labeling times, 1–2 min, most of the labeled DNA was found in much larger segments of DNA (fig. 31.8).

A closer examination of the Okazaki fragments led to detection of short stretches of ribonucleotides at the 5′ ends. From this and many other observations on different systems it was determined that a new DNA chain can be initiated only by attaching the first deoxynucleotide through its 5′-phosphate to the

Figure 31.7

Models for synthesis at the replication fork. (*a*) Continuous synthesis on both strands. Note that both growth arrows are pointing in the same direction, which would require growth in the $5' \rightarrow 3'$ direction on one strand and in the $3' \rightarrow 5'$ direction on the other. If growth occurs only in the $5' \rightarrow 3'$ direction, synthesis would have to be discontinuous on one strand, as in (*b*). Alternatively, it could be discontinuous on both strands (*c*).

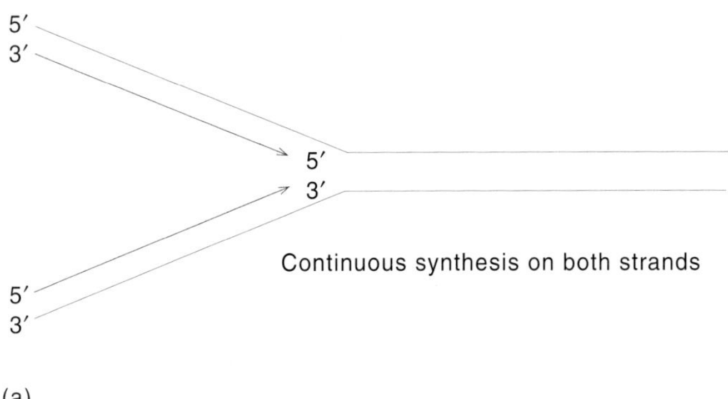

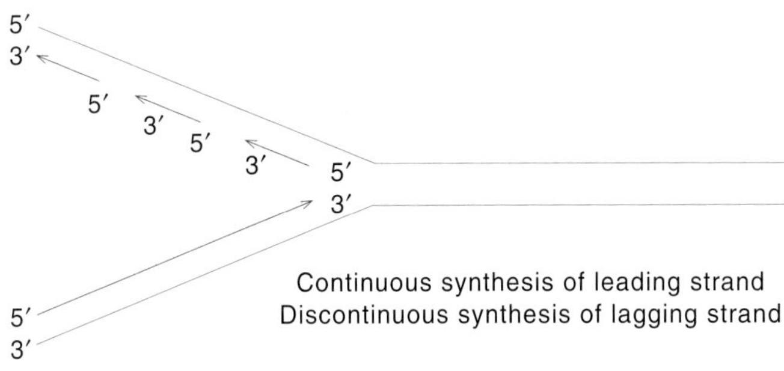

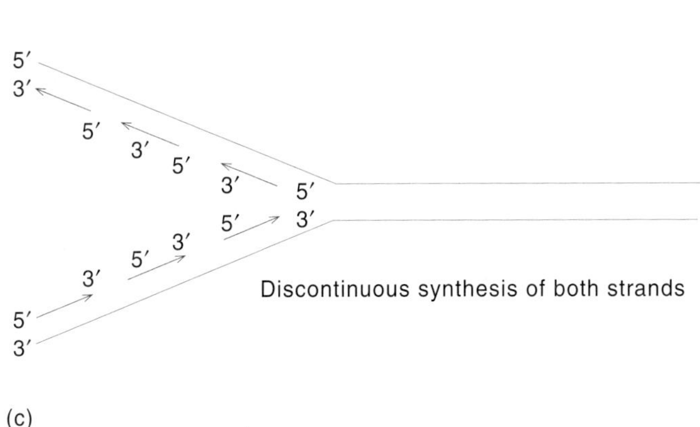

3′-OH of a short RNA polynucleotide. An RNA strand that functions in this capacity is called a primer. Primers are formed at points along the chromosome; they base-pair with the single-stranded template DNA in the regions where they are formed.

A detailed model for discontinuous synthesis could now be proposed (fig. 31.9). First, RNA primers are made on the sin-

Figure 31.8

Sedimentation analysis of *E. coli* DNA from cells labeled with [^{3}H]thymidine for different lengths of time. Sedimentation analysis is done on an alkaline sucrose gradient. The alkali denatures the DNA so that it becomes single-stranded. Short-term labeling (2 to 10 s) preferentially labels the most slowly sedimenting DNA, in the size range of 1,000–2,000 base pairs. In long-term labeling (1 to 2 min), most of the labeled DNA is of much higher molecular weight.

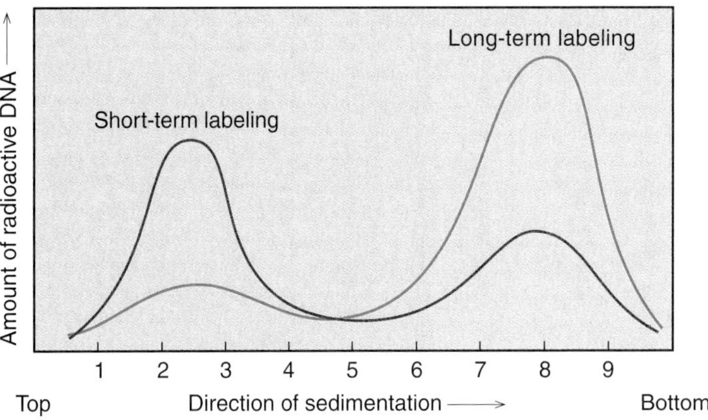

Figure 31.9

A model for discontinuous DNA synthesis. Synthesis occurs in a region that has been partially single-stranded. First, RNA primers are formed at various points on the single-stranded region (1). DNA synthesis starts at the 3′ ends of the primers (2). Primers are removed (3). Gaps between DNA fragments are filled in by further DNA synthesis (4). Fragments are ligated to make one long, continuous piece of DNA (5). Newly synthesized RNA and DNA are indicated in red.

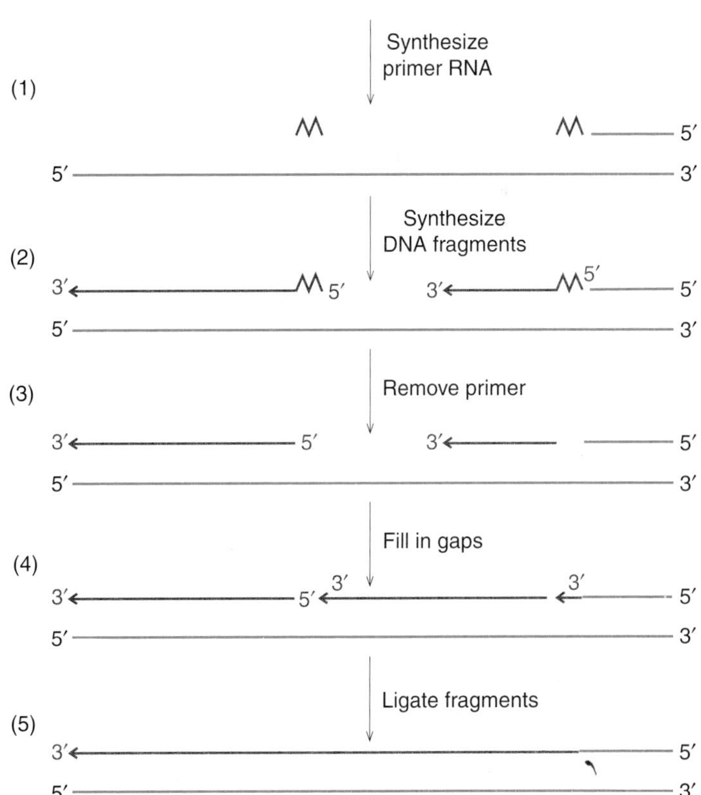

Table 31.1
Proteins Involved in DNA Replication

Protein	Gene(s)	Function
DnaA	*dnaA*	Initiator protein: Binds *oriC;* promote double-helix opening: DnaB loading
DnaC	*dnaC*	Complexes with DnaB; delivers DnaB to DNA
DNA polymerase III holoenzyme		
α	*dnaE*	Polymerase
ϵ	*dnaQ*	3′ to 5′ exonuclease
θ	*holE*	Unknown
τ	*dnaX*	
γ	*dnaX*	
δ	*holA*	
δ'	*holB*	
χ	*holC*	
ψ	*holD*	
β	*dnaN*	Processivity factor: "Sliding clamp"
PolI	*polA*	Prime removal and gap filling; Initial leading strand synthesis on ColE1
Ligase	*lig*	Joins nascent DNA fragments
Gyrase	*gyrA*	Type II topoisomerase; replication swivel, DNA supercoiling, decatenation
	gyrB	
DnaB	*dnaB*	5′ to 3′ helicase and activator of primase
Primase	*dnaG*	Primer synthesis
SSB	*ssb*	Binds single-stranded DNA

Source: Adapted from T. A. Baker and S. H. Wickner, Genetics and enzymology of DNA replication in *Escherichia coli. Ann. Rev. Genetics,* 26:447, 1992.

gle-strand region of the template; then DNA is synthesized. Finally, the RNA is removed from the fragments, and the gaps are filled in and ligated. An understanding of the detailed steps of this process has come largely from analysis of simpler viral replicating systems and more recently from examination of a system intended to resemble the *E. coli* chromosome (see below).

Proteins Involved in DNA Replication

Table 31.1 contains a list of some proteins involved in DNA replication and their functions. How does one go about analyzing such a complex situation? Historically, three general methods have been used for the identification and characterization of the proteins involved in DNA replication: purification, reconstitution, and mutation. Insofar as possible, all three methods are used together

The first method involves isolation of proteins with enzymatic activities that are logically related to the replication process, such as DNA polymerases and ligases. This is the classical biochemical approach and can be applied to any biological system. After isolation and characterization, several approaches may be used to demonstrate that the purified enzyme is active in the replication process in vivo. Sometimes this can be done by using inhibitors that act on both the purified protein and the cellular process. The concentration of inhibitor required to inhibit DNA replication in vivo should be approximately the same as that required to inhibit the purified enzyme in vitro. In prokaryotes, mutations have been very useful for confirming the functions of isolated proteins in the replication process. The induction of a new enzyme activity associated with a biological process, such as virus infection or cell proliferation, also provides useful evidence.

A second method used to identify proteins needed for replication involves reconstitution. Whole-cell lysates containing all of the components necessary for replication are fractionated, and the DNA replication process is then reconstituted with various combinations of the purified or partially purified proteins. Components of the replication system are recognized on the basis of their ability to restore overall activity in vitro. This procedure can be applied to any organism, even when relevant genetic mutants are not available.

The third method uses genetic mutation as the primary tool. This method requires the isolation of conditional mutants, that is, mutants that behave normally under one set of conditions but abnormally under another. Most commonly, temperature-sensitive DNA replication mutants are used. Such mutants have been isolated in *E. coli,* and they grow normally at a low (permissive) temperature (33°C) but poorly or not at all at a high (nonpermissive) temperature (41°C). Preliminary analysis of the temperature-sensitive step provides clues to the stage of replication affected. For example, the length of time required for DNA synthesis to stop after cells are shifted from permissive to nonpermissive temperatures can indicate whether the mutation occurs in a protein involved in initiation or elongation. If the mutation is in the gene for a protein required for elongation, most DNA synthesis stops immediately at the nonpermissive temperature because the majority of cells are usually at some stage in the elongation process. If the mutation is in a gene required only for initiation of replication, most DNA synthesis continues for some time and stops when the rounds of replication in progress are completed. In vitro assays are then used to aid in purifying the corresponding proteins from wild-type cells. Extracts from cells with the temperature-dependent defect are not active in DNA synthesis at the elevated temperature, but activity can be restored by adding the corresponding protein from normal wild-type cells. This complementation assay can be used to aid in the purification of particular replication proteins. To prove that the correct protein has been purified from wild-type cells, the proteins from the temperature-sensitive mutant also must be purified and shown to be abnormal, frequently exhibiting unusual instability at elevated temperatures.

Characterization of DNA Polymerase I in Vitro

We look at *E. coli* DNA polymerase I as an example of how biochemical and genetic studies are used to characterize a DNA replication enzyme. Then we consider the other major proteins involved in DNA replication in *E. coli* before examining the proteins that participate in the synthesis of DNA in different types of organisms.

The Watson-Crick proposal of a complementary duplex structure for DNA stimulated a search for a DNA polymerase enzyme with certain implied properties. The enzyme should require an intact DNA chain to serve as a template for the absorption of complementary bases, and the newly synthesized DNA should be a complement of one of the template DNA chains. Arthur Kornberg and his co-workers isolated a DNA-synthesizing enzyme from cells of *E. coli* that satisfied these requirements and named it DNA polymerase; it is now known as DNA polymerase I (PolI), or the Kornberg enzyme. This enzyme requires a DNA template, the four commonly occurring deoxynucleotide triphosphates, and Mg^{2+} ions for making DNA. The enzyme catalyzes the addition of mononucleotides to the 3′-OH end of a growing chain (fig. 31.10). Simultaneously with the formation of this linkage, the linkage between the two phosphates is broken, releasing a pyrophosphate group. The energy produced by the cleavage provides the energy necessary for linking the mononucleotide to the growing DNA chain. Subsequent cleavage of the pyrophosphate to orthophosphates ensures the irreversibility of the reaction. Bases added to the growing chain are determined by the sequence of bases in the DNA template. As nucleotides complementary to those on the template are added, the single-stranded DNA template gradually becomes converted to a double helix. Structure studies have shown that the enzyme has a complex surface with specific attachment sites for the template chain, the growing chain, and monomer nucleoside triphosphate. The enzyme is highly selective, because the only nucleotides it links to the growing chain are those that form Watson-Crick base pairs with the template strand. As the new chain lengthens by synthesis, the enzyme moves along the template one base at a time.

In addition to the characteristic template-directed polymerization activity, PolI contains an activity that results in the removal of mononucleotides from the 3′ end of a polynucleotide strand. This enzyme activity leads to cleavage of the bond between the 5′-phosphate of the terminal residue and the 3′-OH group of the penultimate residue. Since this is the same linkage that is made during synthesis, the degradation reaction may be thought of as a reversal of the polymerization process. For net chain elongation beyond the 3′-OH end of the DNA strand, the polymerization rate must exceed the depolymerization rate. Polymerization is much faster than depolymerization if the correct base-paired nucleotide is inserted into the growing chain. If a mismatched base is accidentally inserted, the opposite is true, and the mismatched base is usually removed. The combined polymerization–depolymerization reaction has been viewed as a "proofreading," error-reducing mechanism in DNA synthesis.

E. coli DNA polymerase I has a second associated activity catalyzing $5' \rightarrow 3'$ degradation of DNA. Whereas $3' \rightarrow 5'$ degradation activity of the enzyme is much more effective on unpaired or mispaired bases, the $5' \rightarrow 3'$ activity cleaves preferentially at base-paired regions. The ability of PolI to degrade DNA (or RNA) in the $5' \rightarrow 3'$ direction as well as to carry out a polymerization reaction suggests a role in removing RNA primer during gap filling.

Crystallography Combined with Genetics to Produce a Detailed Picture of DNA PolI Function

The question arises as to whether the three activities of PolI just described all originate from a single active site or from more than one site. Cleavage of PolI by the protease subtilisin leads to a small fragment ($M_r = 30{,}000$) with $5' \rightarrow 3'$ nuclease activity and a large fragment ($M_r = 70{,}000$), called the Klenow fragment, exhibiting the polymerization and $3' \rightarrow 5'$ depolymerization activities. A brilliant series of investigations by Tom Steitz and his co-workers led to a detailed understanding of the Klenow fragment. These studies were done on wild-type and mutant enzyme lacking the nuclease activity. From their studies Steitz and his colleagues concluded that within the Klenow fragment the polymerase and $3' \rightarrow 5'$ exonuclease activities were located in different regions. Thus the polymerase has three enzyme activities, all resulting from different parts of the same protein molecule. The approximate locations of these activities are indicated relative to a chain undergoing synthesis in figure 31.11. Recent investigations on cocrystals of DNA and PolI indicate that the DNA makes a sharp bend between the $3' \rightarrow 5'$ exo and the polymerase active sites. This somewhat clumsy-looking structure may facilitate the detection of imperfections in newly synthesized DNA and the backing up process that must accompany proofreading excision by the $3' \rightarrow 5'$ exo.

Figure 31.10

Template and growing strands of DNA. (*a*) Nucleotides are added one at a time to the 3′-OH end of the growing chain. Only residues that form Watson-Crick H-bonded base pairs with the template strand are added. (*b*) Covalent bond formation between the 3′-OH end of the growing chain and the 5′-phosphate of the mononucleotide is accompanied by pyrophosphate removal from the substrate nucleoside triphosphate. One or two Mg^{2+} ions are also involved in the growth step.

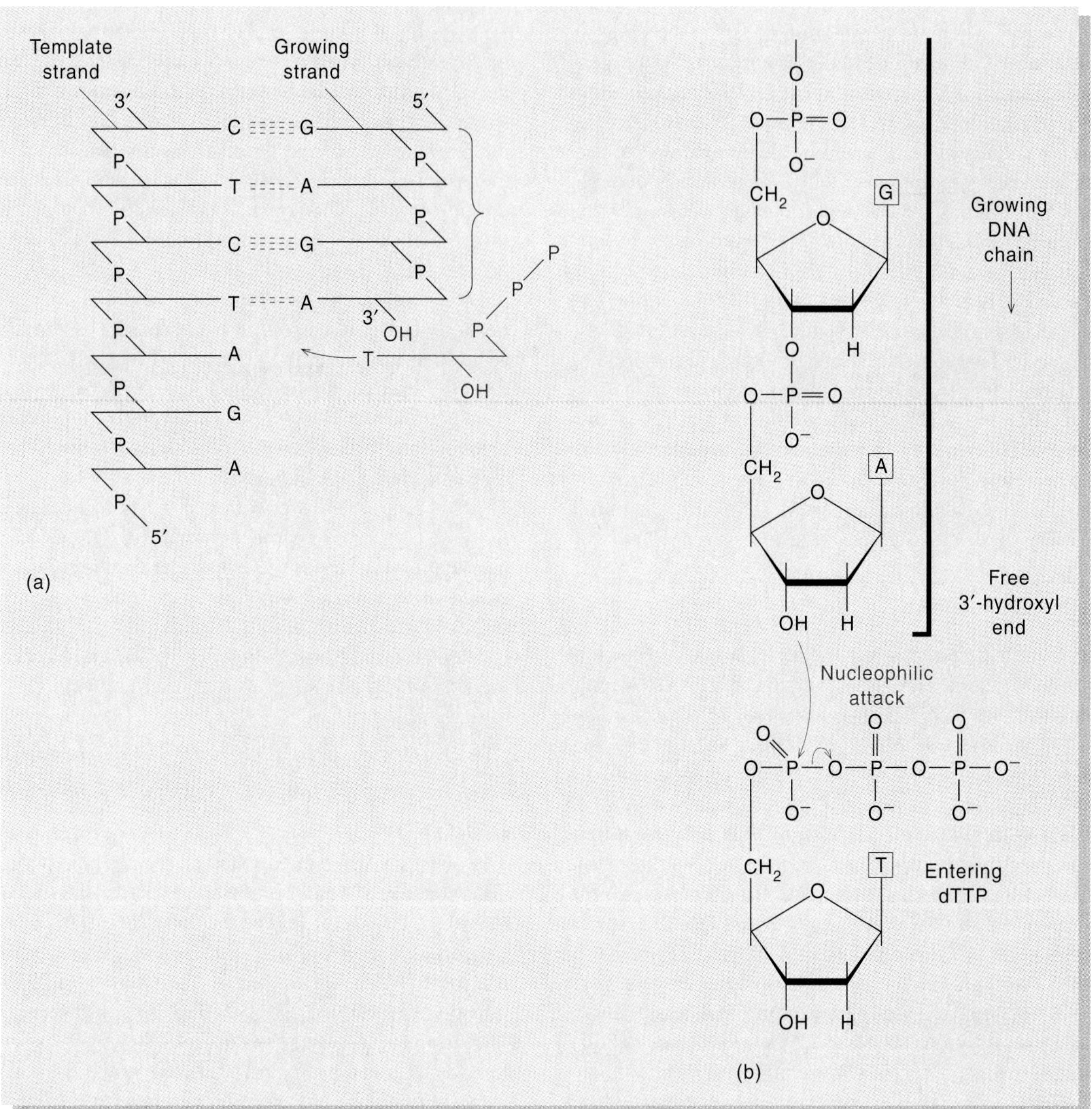

Establishing the Normal Roles of DNA Polymerases I and III

All of Kornberg's early work was done on DNA polymerase in cell-free systems. The fact that the enzyme had so many of the properties expected for a DNA-replicating enzyme led most observers to believe that it was the DNA-replicating enzyme. Final proof, however, that an enzyme functions in the same capacity in vivo requires the isolation of mutants that affect its behavior. Our current understanding of the role of PolI exemplifies the importance of correlating a given biochemical behavior with knowledge gained by studying mutants.

From the time of its initial discovery it took about 20 years to reach our current understanding of the physiological role of PolI. A major step in this direction was taken by Cairns and DeLucia, who laboriously scanned several thousand strains of mutagenized *E. coli* to find one that contained almost no PolI polymerizing activity (1%–2% of normal). This mutant grew well under normal conditions, suggesting that PolI was not

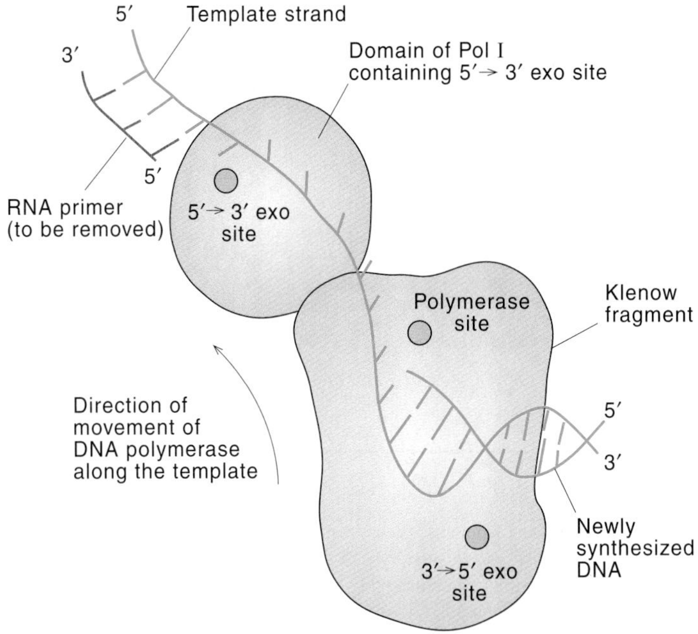

Figure 31.11

A schematic drawing of the possible structure of the DNA–PolI complex in an elongating complex of DNA. The Klenow fragment contains the $5' \rightarrow 3'$ polymerase site and the error-correcting $3' \rightarrow 5'$ exonuclease site. The fragment cleaved from the PolI enzyme by limited proteolysis contains the $5' \rightarrow 3'$ exonuclease site. The $5' \rightarrow 3'$ exonuclease removes RNA leaders from Okazaki fragments to make way for polymerization. Any mismatches formed by the polymerase result in the elongating strand being displaced and hydrolyzed at the $3' \rightarrow 5'$ exonuclease site. (*Source:* This drawing is primarily based on the structural studies of L. S. Beese, V. Derbyshire, and T. A. Steitz, Structure of DNA polymerase I Klenow fragment bound to duplex DNA, *Science* 260:352–355, 1993.)

Table 31.2
Properties of Polymerases I, II, and III of *E. coli*

	PolI	PolII	PolIII
Molecules per cell	400	—	15
Turnover number[a]	600	30	9,000
Structural gene[b]	*polA*	*polB*	*polC* (*dnaE*)
$5' \rightarrow 3'$ polymerizing activity	+	+	+
$3' \rightarrow 5'$ exonuclease activity	+	+	+
$5' \rightarrow 3'$ exonuclease activity	+	−	−
Mutant loci	*polA*	*polB*	*polC, dnaN, dnaX, dnaZ, dnaQ*
Lethality	Viability reduced only when $5' \rightarrow 3'$ exonuclease affected	No effect	Conditional lethality

[a]Nucleotides polymerized/min/molecule of enzyme at 37°C.

[b]Only the structural gene for the largest protein subunit in the enzyme is recorded.

involved in replication. However, the mutant in question was hypersensitive to ultraviolet irradiation. Since ultraviolet irradiation was known to damage DNA, the hypersensitivity of the mutant suggested that PolI might be involved in repairing chromosome damage. The discovery of the polymerase mutant had a profound effect on thought and experimental design in further studies on DNA biosynthesis. An important general principle was underscored by this unexpected finding—that a function should not be assigned to an enzyme on the basis of its in vitro properties alone. Only genetic mutants make meaningful in vivo correlates possible. In cell-free extracts from a mutant *E. coli* strain that did not contain PolI polymerizing activity, it was subsequently possible to detect two additional DNA polymerizing enzymes. These were named DNA polymerases II (PolII) and III (PolIII). The behavior of conditional lethal mutants of PolIII led to the conclusion that this is the main replication enzyme. Some of the properties of the three enzymes are summarized in table 31.2.

For a few years following the observations of DeLucia and Cairns it was assumed that PolI was not important in replication but only in repair. However, further genetic and biochemical studies provided convincing evidence that PolI is a multifunctional enzyme possessing, in addition to $5' \rightarrow 3'$ polymerizing activity, the $3' \rightarrow 5'$ and $5' \rightarrow 3'$ degradation activities described earlier. The original mutant of PolI isolated by Cairns and DeLucia was inactivated only in its polymerizing function. Subsequently, mutants in PolI that affect the $5' \rightarrow 3'$ degradation function were found to be conditionally lethal and did not permit elongation of DNA synthesis under nonpermissive conditions; thereby, it was demonstrated that this activity of the PolI enzyme is indispensable for chromosome replication. In this regard it should be mentioned that of the three known *E. coli* DNA polymerases, only PolI has a $5' \rightarrow 3'$ exonuclease activity.

Other Proteins Required for DNA Synthesis in Escherichia coli

As mentioned earlier, discontinuous DNA synthesis necessitates the existence of an enzyme for joining the newly synthesized segments (see fig. 31.9). Such an enzyme has been found in a variety of cell types and is called polynucleotide ligase (fig. 31.12).

Topoisomerases constitute a family of enzymes that can introduce negative supercoils into DNA or relax supercoiled DNA (fig. 31.13). They can also catalyze the linking together (catena-

Figure 31.12

Steps in the sealing of a DNA nick, catalyzed by DNA ligase. The bacterial ligase uses NAD^+ to make an enzyme–AMP intermediate. Mammalian DNA ligases and bacteriophage T4 ligase use ATP for the same purpose.

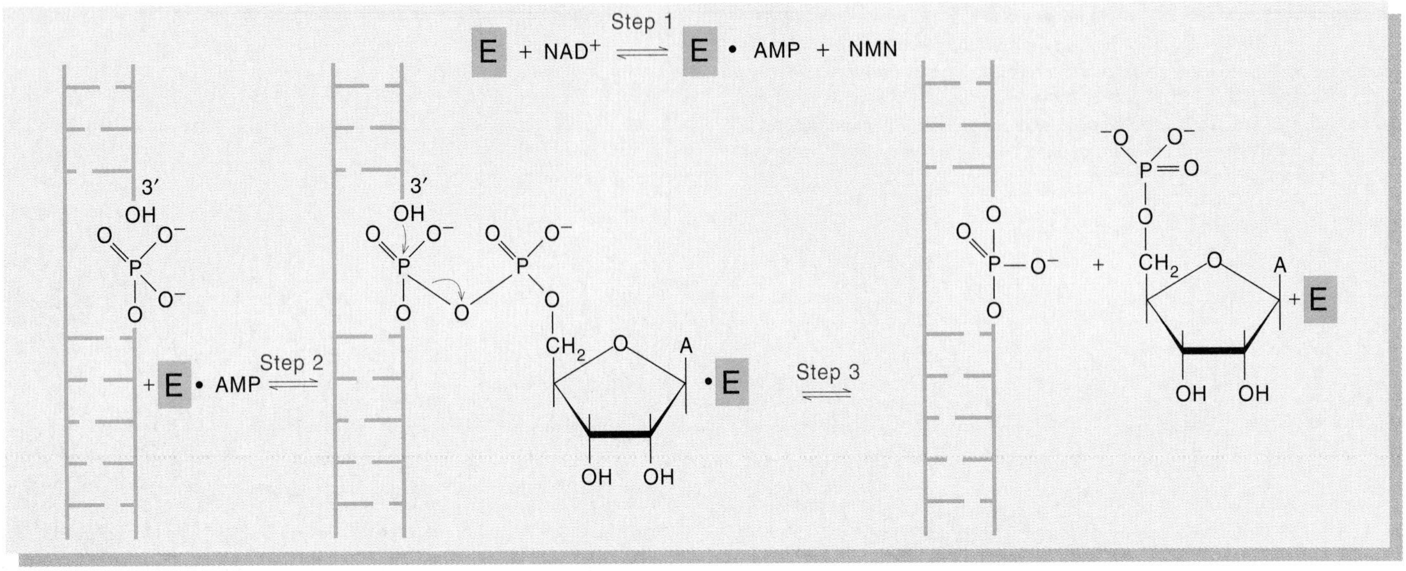

Step 1: $E + NAD^+ \rightleftharpoons E \cdot AMP + NMN$

tion) of double-stranded circular DNAs or the decatenation of linked circular molecules (fig. 31.14). It is clear that these enzymes serve vital functions because mutants carrying defective topoisomerase of one sort or another show severely impaired growth properties in both prokaryotic and eukaryotic cells. Topoisomerases are classified as type I or type II according to whether they change the linking number in steps of one or steps of two, respectively. Type I enzymes produce transient single-strand breaks in the double helix, whereas type II enzymes produce transient double-strand breaks (see fig. 31.13). Type I topoisomerase of *E. coli* shows a preference for highly negatively supercoiled DNA and is inactive on positively supercoiled DNA. This enzyme can also catalyze the catenation (interlocking) of double-stranded circular DNA or the separation of catenanes into simple circles provided that at least one circle contains a single-stranded break (see fig. 31.14). Type II topoisomerases can carry out the catenation reaction as well; in this case no single-stranded breaks are required. Type I topoisomerases that have been isolated from eukaryotic cells differ in two important respects from the *E. coli* enzymes: They do not require Mg^{2+} ions, and they can relax positively as well as negatively supercoiled DNA. Type II topoisomerases from bacteria but not from eukaryotes are able to catalyze the conversion of relaxed duplex DNA into a high-energy negatively superhelical form in a reaction that requires ATP.

At the growth fork it is necessary that the parental double helix be unwound to present further single-stranded regions to serve as templates for continued replication. The dnaB protein of *E. coli* is believed to be a helicase that directly catalyzes this process.

Figure 31.13

Type I and type II topoisomerases relax negatively supercoiled DNA in steps of one and steps of two, respectively. Type II topoisomerases can also add additional negative supercoils (as indicated by the double arrow). The latter reaction requires energy input, which is provided by ATP cleavage.

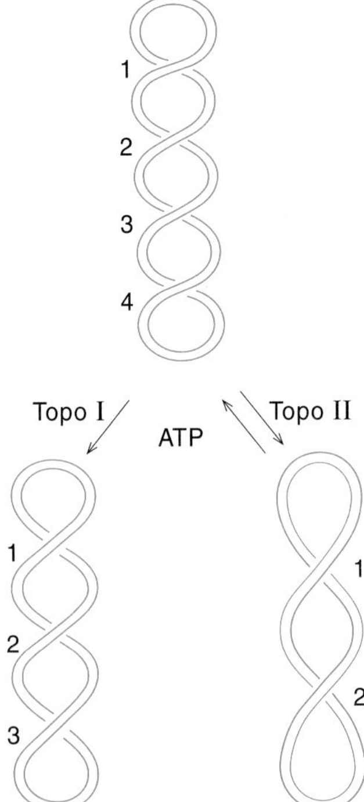

Figure 31.14

Catenation by topoisomerases. (*a*) Two circular DNAs can be catenated by type I topoisomerase only if one of the DNAs is nicked. This is not necessary in using a type II topoisomerase. (*b*) Electron micrographs of catenated DNA before (i) and after (ii) incubation with DNA gyrase. The catenate contains one large circular DNA and one small circular pBNP66 plasmid DNA. (*Source:* Adapted from M. Gellert, L. M. Fisher, H. Ohmori, M. H. O'Dea, and K. Mizuchi, DNA gyrase: Site-specific interactions and transient double-strand breakage of DNA, *Cold Spring Harbor Symp. Quant. Biol.* 45:301, 1981.)

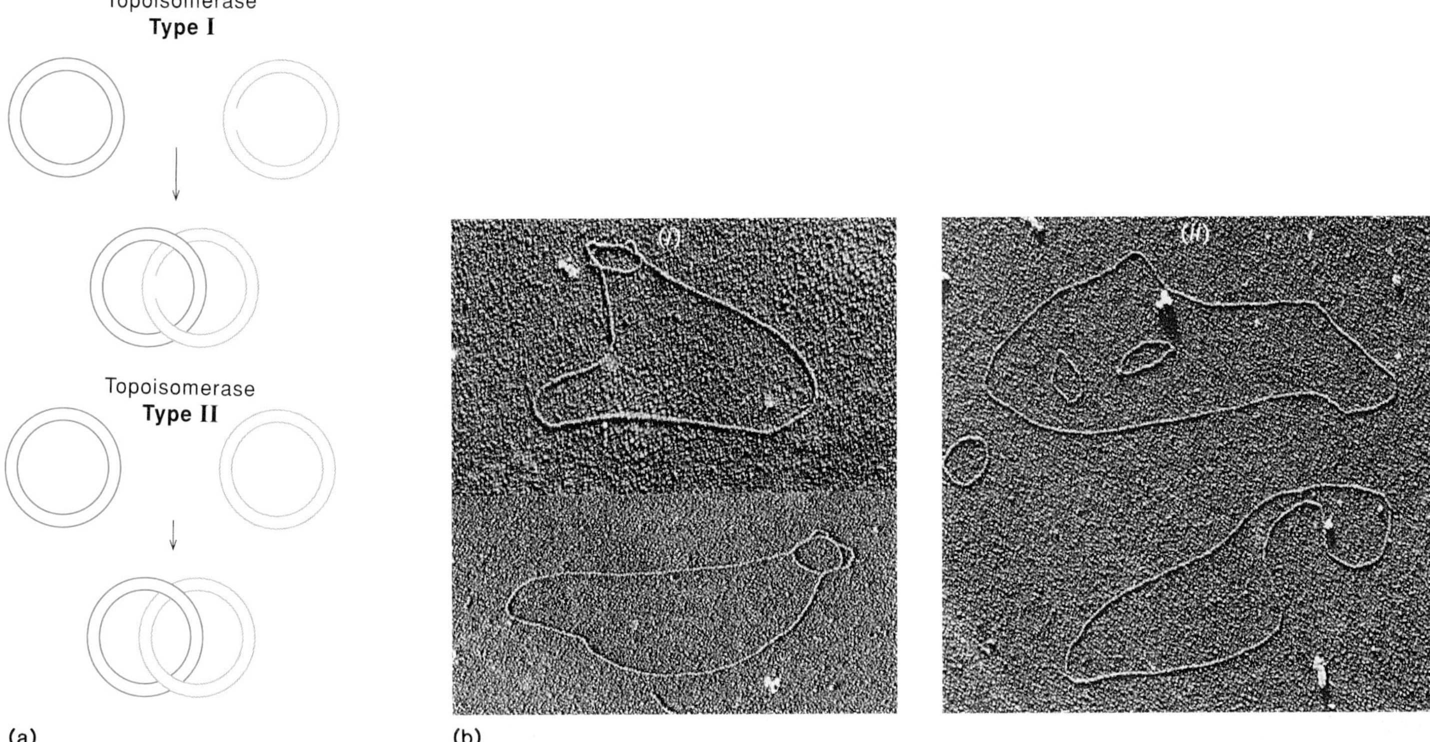

Single-strand binding protein (SSB) is found in abundance in *E. coli*, and it is believed to be bound in mass at the replication fork. This fact and other evidence described later on indicate that it plays an important role at the growth fork. Two functions can be suggested, both inferred from its preference for binding to single-stranded DNA. It could protect single-stranded DNA from nucleases, or it could facilitate unwinding by inhibiting rewinding.

There are two enzymes that can catalyze the synthesis of RNA with the help of a DNA template. One of these, called primase, is encoded by the *dnaG* gene. Primase catalyzes the synthesis of small primer RNAs that are required for DNA synthesis. The other enzyme, called RNA polymerase, catalyzes the synthesis of all the other RNAs found in *E. coli.*

Replication of the Escherichia coli *Chromosome*

The elongation step in *E. coli* chromosomal DNA synthesis is depicted in figure 31.15. As the DNA gradually unwinds, new synthesis takes place on the two strands. At the heart of the unwinding process we find a DnaB helicase, and just behind it we find primase. Single-strand binding protein (SSB) coats the newly unwound DNA strands. The primase travels in the same direction as the DnaB helicase, making periodic pauses to synthesize short RNA primers on the lagging strand. PolIII extends the leading strand and also extends the lagging strand. In the latter case the first DNA bases are added to the primers. Following the action of PolIII on the lagging strand, PolI replaces the RNA bases with DNA bases, and DNA ligase then links the short DNAs on the lagging strand.

Initiation and Termination of Escherichia coli *Chromosomal Replication*

Two aspects of *E. coli* chromosomal replication still to be considered are initiation and termination. From what has been said, we conjecture that replication initiates at a unique site, proceeds bidirectionally, and terminates at a point where the two oppositely advancing growth forks meet. To study initiation, it was first necessary to isolate that segment of the chromosome that carries the unique origin of replication. This was done by chopping the bacterial chromosome into small pieces, circularizing the small pieces, and then transfecting them into cells to see which pieces could sustain autonomous replication. In this way a unique segment of chromosome was discovered that contains 245 bp required to initiate replication.

Figure 31.15

Model for the replication fork in the *E. coli* chromosome. The DnaB protein unwinds the duplex while the primase directly behind it synthesizes RNA primer on the lagging strand. SSB protein binds to single-stranded regions wherever they are. DNA polymerase III progressively adds nucleotides to the leading strand and also adds nucleotides to the lagging strand starting at the RNA primers. After this, DNA polymerase I replaces the RNA primer (blue wiggly line) with the DNA equivalent. Finally, DNA ligase knits the short DNA segments together on the lagging strand. (*Source:* Adapted from K. Arai, R. L. Low, and A. Kornberg, Movement and site selection for priming by the primosome in phage ϕX174 DNA replication, *Proc. Natl. Acad. Sci. USA* 78:711, 1981.)

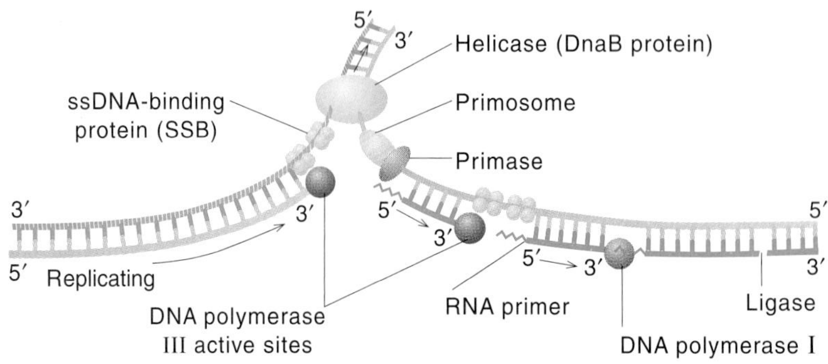

With circular minichromosomes containing this unique segment termed oriC, it has been possible to study the initiation and replication process in vitro. First, in crude extracts it was possible to show that replication initiates within or near the *oriC* sequence and proceeds bidirectionally. About 13 different proteins participate in the *oriC*-directed DNA replication as judged by their activity in a reconstituted replication reaction.

Initial complex formation entails binding of the dnaA protein to several sites (probably four) of the *oriC* region (fig. 31.16). This is followed by the cooperative binding of more dnaA proteins. In all, 20–40 dnaA proteins bind to a region of about 200 bp. This binding is climaxed by a structural change in the dnaA–*oriC* complex, which results in localized melting of the DNA at one end of the *oriC* region. The melting reaction, which reproduces the local opening of the duplex requires that the DNA be negatively supercoiled. As we have already discussed, negative supercoiling favors unwinding of the duplex (see chapter 30).

The dnaB protein is transferred to this complex from a dnaB–dnaC complex. The dnaC dissociates, leaving the dnaB protein bound to the template. Melting of the DNA duplex proceeds bidirectionally from *oriC* as the dnaB helicase migrates from the dnaA–*oriC* complex to provide a template for the priming and replication enzymes. Addition of the priming and replication enzymes results in immediate initiation and elongation. Continued elongation requires gyrase and SSB. Gyrase provides a swivel to permit continued unwinding of parental strands while SSB stabilizes the transiently single-stranded regions. DNA gyrase proves useful in the late stages of replication as well, because many circular dimers are formed that require decatenation and negative supercoils so they can function again as active templates.

Synthesis May Take Place Concurrently on Both Strands

Concurrent replication of both strands, rather than the jerky sequence of synthesis of one strand and then the other, might be achieved if priming of nascent fragments of the lagging strand were integrated with continuous synthesis of the leading strand.

Figure 31.16

Model of the initiation complex at the *E. coli* replication origin (*oriC*). R1-R4 indicates four 9-bp sequences that are primary binding sites for the DnaA protein. The consensus sequence for these binding sites is TTATCCACA, and their relative orientation is indicated by the arrows. After the binding of about 20 dnaA proteins at *oriC*, a small region of the DNA melts. This permits the binding of the dnaB helicase. The binding of dnaB is assisted by dnaC, which does not bind itself. The helicase unwinds a significant region around *oriC*, creating the necessary room for assembly of the replication apparatus. (*Source:* Adapted from T. A. Baker and S. H. Wickner, Genetics and enzymology of DNA replication in *Escherichia coli*, *Ann. Rev. Genetics* 26:447, 1992).

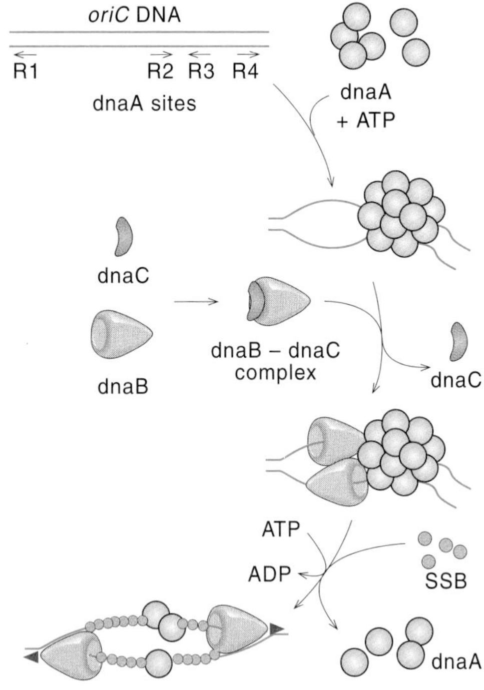

Concurrent replication would also require a more complex holoenzyme possessing primase and twin active sites for polymerization (fig. 31.17). Evidence for such a structure includes (1) twin poly-

Figure 31.17

Hypothetical asymmetric dimeric structure suggested by Kornberg for the DNA polymerase III holoenzyme of *E. coli*. The holoenzyme appears to be organized as an asymmetric dimer; a pair of core subassemblies, each with a potential for polymerase action, has an asymmetric distribution of auxiliary subunits that may endow each one with different properties, one suited to the continuous synthesis of one strand and the other to the discontinuous synthesis of the other strand. The α subunit contains the polymerase activity. The ϵ subunit contains the 3′ to 5′ exonuclease proofreading activity, which is active only on attachment to the α subunit. Several of the subunits are essential to processivity. Among them, the β subunit is held loosely in the holoenzyme and dissociates during the cycling of the polymerase from one template to another. (*Source:* Adapted from A. Kornberg, DNA replication, *J. Biol. Chem.* 263:1–4, 1988.)

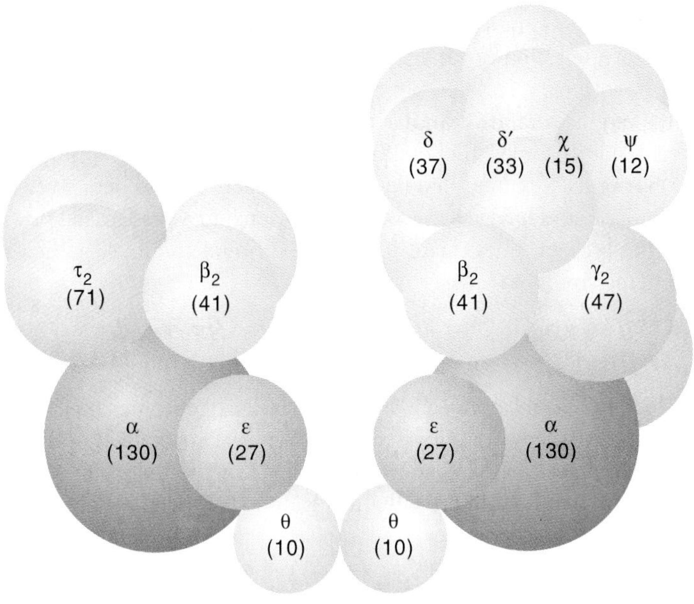

merase subassemblies in the PolIII holoenzyme and (2) complexing of primase by some polymerases. In this hypothetical scheme for concurrent replication, looping of the lagging strand by 180° gives it the same orientation as the leading strand at the replication fork (fig. 31.18). A primer generated by primase is extended by polymerase as the lagging strand template is drawn past it. When synthesis reaches the 5′ end of the previous nascent fragment, the lagging strand template is released. Helicase action and continuous synthesis of the leading strand periodically expose lengths of template for priming of nascent fragments on the lagging strand.

DNA Replication in Eukaryotic Cells

Mutant studies have not been that helpful in finding proteins with known functions in vertebrate DNA replication. The major approach in vertebrate investigations has been isolation of proteins having activities logically related to DNA replication and isolation of replicative intermediates. Different forms of DNA polymerases, DNA ligases, topoisomerases, single-strand binding proteins, and unwinding enzymes have all been found.

Figure 31.18

Hypothetical scheme for concurrent replication of leading and lagging strands by an asymmetric, dimeric polymerase associated with a primosome and a helicase in a replisome. In this hypothetical scheme for concurrent replication, looping of the lagging strand by 180° gives it the same orientation as the leading strand at the fork. A primer generated by primase is extended by polymerase as the lagging strand template is drawn past it. When synthesis reaches the 5′ end of the previous nascent fragment, the lagging strand template is released and unlooped. Helicase action and continuous synthesis of the leading strand periodically expose lengths of template for priming of nascent fragments. (*Source:* Adapted from A. Kornberg, DNA replication, *J. Biol. Chem.* 263:1–4, 1988.)

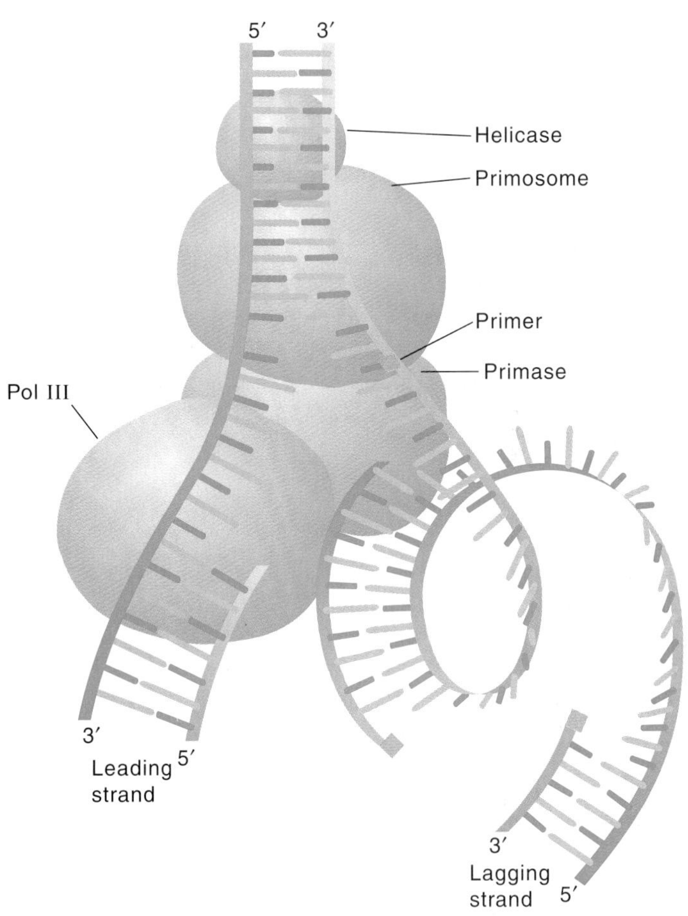

Eukaryotic Chromosomal DNA

Although general features of DNA replication in eukaryotes are thought to be similar to those of prokaryotes, there are some interesting differences. The chromosomes of higher eukaryotic organisms are quite large, in some cases 1,000 times larger than their bacterial counterparts. For these larger DNA molecules to replicate in a reasonable time, they have multiple origins of replication. The simultaneous synthesis of DNA at several points along the chromosome has been demonstrated by incorporating radioactive nucleotides into the replicating chromosomes for a short period and then observing the distribution of radioactive DNA by autoradiography (fig. 31.19). Multiple regions

Figure 31.19

Multiple-origin model for eukaryotic chromosomal DNA replication. (*a*) Autoradiograph of short-term labeling of a eukaryotic chromosome during replication and its interpretation. (*b*) Overall replication scheme for a eukaryotic chromosome. Only a short region of the chromosome is shown. It is believed that replication origins are occupied by a multiprotein replication complex most of the time.

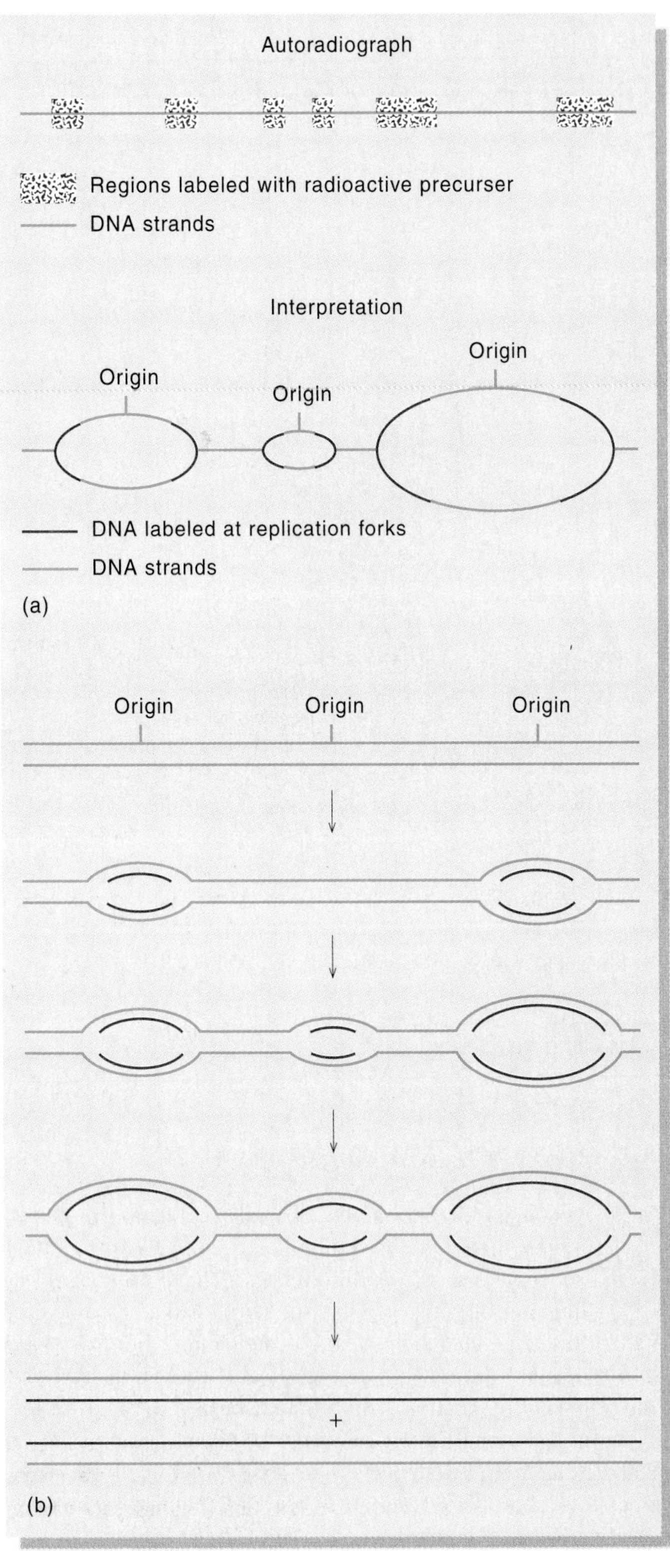

of incorporated label are observed, and replication proceeds bidirectionally from these regions. Termination of replication occurs at the point where the growth forks from two adjacent replication units meet (see fig. 31.19). DNA on the lagging strand of a fork is made discontinuously. The Okazaki fragments are much shorter than those found in prokaryotes, averaging only between 100 and 200 nucleotides in length. Synthesis of these DNA fragments is initiated on primers that are found covalently attached to the 5′ ends of newly synthesized fragments.

Remember that DNA in eukaryotic chromosomes is associated with histones in complexes called nucleosomes (see fig. 30.29). The disassembly of the DNA–histone complex presumably occurs directly in front of the replication fork. Two observations suggest that nucleosome disassembly may be a rate-limiting step in the migration of the replication fork in chromatin: (1) The rate of migration of replication forks in eukaryotes is slower than in prokaryotes, and (2) the length of the replication fragments on the lagging strand is similar to the length of DNA between adjacent nucleosomes (about 200 bp). Before the newly replicated DNA is reassembled into nucleosomes, the RNA primers must be removed, and the gaps must be filled by enzymes that are unknown at the present time. Finally, replication fragments must be linked together by DNA ligase.

SV40 Is Similar to Its Host in Its Mode of Replication

Much has been learned about the replication of chromosomal DNA in eukaryotes by studying a simple monkey virus known as SV40. The viral genome of SV40 consists of a circular duplex DNA molecule of about 5,200 bp with one origin for replication. The SV40 viral genome is complexed with histones to form a nucleoprotein structure very similar to that observed for chromatin. Since SV40 encodes only a single replication protein (T antigen), the virus must make extensive use of the host's cellular replication machinery. As a result many similarities exist between viral and cellular DNA replication. In both cases, initiation of DNA synthesis involves two nascent strands. The leading strand grows continuously while the lagging strand grows discontinuously by joining together small (about 200 bp) segments of DNA that are independently initiated with RNA primers. Completion of replication occurs when two oppositely moving forks meet. In linear cellular chromosomes the two merging forks originate from adjacent origins, but in circular SV40 chromosomes they have a single origin.

The development of an efficient cell-free replication system has greatly accelerated progress in understanding the molecular mechanisms involved in SV40 DNA replication. The origin of replication is recognized by the viral T antigen. In addition to its specific binding activity, the T antigen has helicase activity. Once it is bound to the origin, T antigen enters the duplex and catalyzes the unwinding of the two DNA strands. Unwinding appears to be a critical step that establishes the replication forks and generates the single-stranded DNA regions required for the priming and elongation of nascent strands.

In addition to specific nucleotide sequence elements, the T-antigen-mediated unwinding reaction requires accessory proteins contributed by the host cell. For example, a single-stranded

DNA-binding protein is required to prevent reassociation of the single strands exposed during unwinding. Such a protein has been found, and it binds specifically to single-stranded DNA.

Of four distinguishable DNA polymerase activities α, β, γ, and δ, it appears that α and δ are required for SV40 DNA replication. DNA polymerase α is composed of four subunits. The largest subunit contains the polymerase active site. The smallest subunit of DNA polymerase α contains a primase capable of synthesizing short RNA transcripts that can serve as primers for subsequent DNA chain elongation by the polymerase-carrying subunit. Each time a polymerase binds to the template primer, it adds a certain number of nucleotides to the growing chain before it dissociates. A highly processive enzyme adds a great number of residues before it dissociates. DNA polymerase α is not a highly processive enzyme because fewer than 100 nucleotides are polymerized per binding event.

The properties of DNA polymerase δ, which is also required for SV40 DNA replication, contrast with those of DNA polymerase α. First, δ has no primase activity; second, it is a highly processive enzyme capable of catalyzing the polymerization of more than 1,000 nucleotides per binding event. Third, it has a $3' \rightarrow 5'$ exonuclease that serves a proofreading function during polymerization.

These differences between the two polymerases initially led to the suggestion that DNA polymerase α (polα) might be best suited to serve as the lagging-strand polymerase and that DNA polymerase δ (polδ) would be best suited to serve as the leading-strand polymerase. However, recent evidence suggests that DNA polδ is involved in replication of both strands, while polα and primase are involved in the synthesis of RNA–DNA primers. This evidence was obtained by Shou Waga and Bruce Stillman using a cell-free system for the complete replication of SV40 DNA. This system contains SV40 DNA, SV40 T antigen, and several purified cellular proteins.

The process of SV40 replication starts when a double hexamer of T antigen formed in the presence of ATP binds to the unique origin for replication (*ori*) and causes structural distortion of the DNA. This stable initiator protein–DNA complex then binds the cellular replication protein A (RPA), which leads to a more extensive unwinding of the DNA mediated by the DNA helicase function of SV40 T antigen. The T antigen–RPA complex binds polα/primase, and a short RNA–DNA chain is synthesized at the *ori*. Continued elongation of the DNA requires a switch from polα to polδ. The cellular replication factor C (RFC) protein binds to the 3′ end of the nascent DNA strand and loads proliferating cell nuclear antigen (PCNA) and polδ onto the template, thereby replacing polα. The processive RFC/PCNA/polδ complex then extends the nascent DNA strands from the continuously synthesized leading-strand complex. The polα/primase continues to prime and synthesize DNA discontinuously on the lagging strand template. On the lagging strand, polα to δ switching is required for the complex synthesis of every Okazaki fragment (fig. 31.20). A $5' \rightarrow 3'$ exonuclease (MF1) and DNA ligase I are required together with RNase H to complete DNA replication to yield covalently closed duplex DNA. These proteins are required for removal of the primer RNA from Okazaki fragments and for DNA ligation to yield intact lagging-strand DNA. Complete synthesis of both lagging and leading strands at the replication fork is a highly coordinated process (fig. 31.21).

Figure 31.20

Model for DNA polymerase switching mechanism during lagging-strand replication. RPA is a cellular DNA-binding protein. PCNA is the proliferating nuclear antigen. RFC is a cellular replication factor. RNaseH is an RNAse that specifically degrades RNA that is hydrogen bonded by Watson-Crick pairing to DNA. (Reprinted with permission from S. Waga and B. Stillman, Anatomy of a DNA replication fork revealed by reconstitution of SV40 DNA replication in vitro, *Nature* 369:207–212, 1994. Copyright 1994 Macmillan Magazines Limited.)

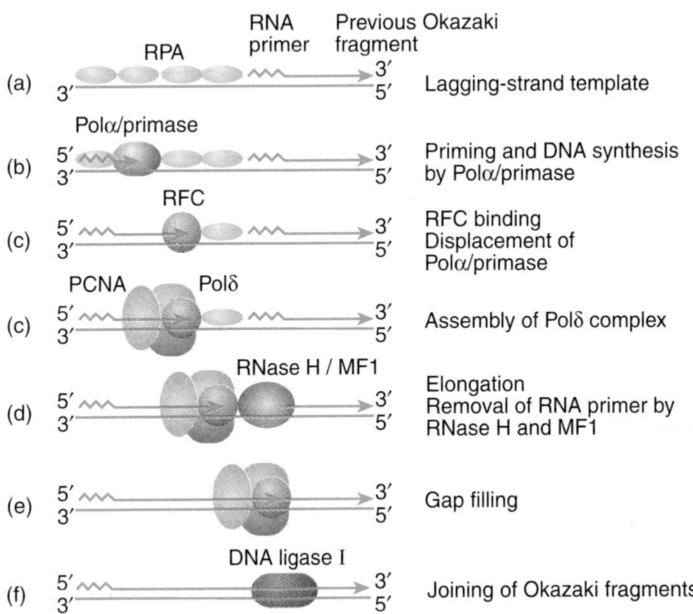

Characterization of the SV40 DNA replication reaction has provided a foundation for future studies on the mechanism and regulation of chromosomal DNA replication in eukaryotes. In this system, however, SV40 T antigen functions as an initiator, DNA helicase, and primase-loading protein. It will therefore be necessary to investigate the biochemistry of DNA replication from cellular origins of DNA replication to achieve a full appreciation of eukaryotic cell DNA replication.

Initiation of Chromosomal Replication in Eukaryotes

The question as to what triggers initiation of chromosomal replication is a complex topic that is being intensively studied. Despite this, the subject is not that well understood, and we shall limit ourselves to a few remarks about the origins of replication. The most precise information about initiation of replication comes from studies on the budding yeast *Saccharomyces cerevisiae*. Functionally, origins in yeast have been identified genetically as autonomously replicating sequence (ARS) elements; these are short DNA sequences that cause plasmids containing them to replicate autonomously once during S phase. There are about 400 ARS elements distributed over the 17 chromosomes of *S. cerevisiae*. Each ARS contains an essential 11-bp consensus sequence, (A/T)TTTAAT(A/G)TTT(A/T), known as the A element or core element. The core element and the adjacent 3′ sequence are bound by a complex of six proteins that comprise the origin recognition

: 31.21

el for elongation at the eukaryotic DNA replication fork. The polymerase primase complex is about to synthesize an initiator RNA–DNA on the lagging-strand template. The complex contains two molecules of polymerase δ: One of them, on the lagging-strand template, is elongating the DNA strand from the initiator DNA previously synthesized by polymerase α–primase; the other, on the leading-strand template, is continuously elongating the leading strand. In SV40 DNA replication the DNA helicase at the fork is SV40 T antigen. (Reprinted with permission from S. Waga and B. Stillman, Anatomy of a DNA replication fork revealed by reconstitution of SV40 DNA replication in vitro, *Nature* 369:207–212, 1994. Copyright 1994 Macmillan Magazines Limited.)

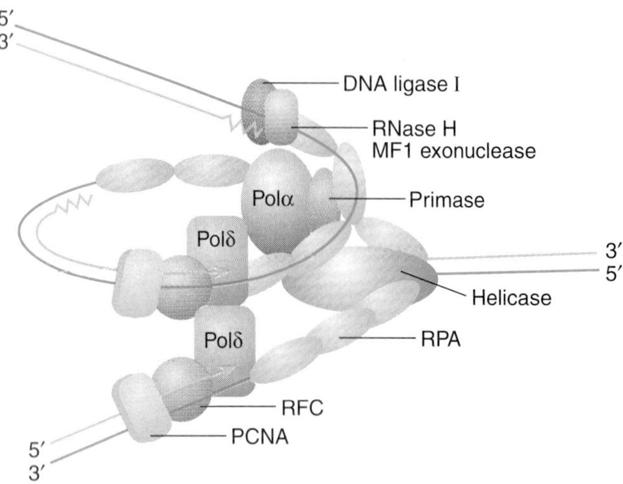

complex (ORC), at least one of which, ORC2, is essential for normal origin function in vivo. ORC binds to ARS elements throughout the cell cycle, implying that newly replicated ARS elements are rapidly bound by the ORC complex during S phase. The ORC–ARS complex could therefore serve as a target whereby cell-cycle-regulated proteins can activate DNA synthesis, either by binding to the complex or by posttranslational modification of ORC subunits in late G1.

In mammalian cells, cytological staining reveals a reproducible temporal pattern of early- and late-replicating chromosomal bands, implying that mammalian replication initiates at specific locations. However, there is no known consensus sequence for eukaryotic chromosomal origins other than those in *S. cerevisiae*. Indeed, initiation of chromosomal replication in mammals can occur anywhere in a region many kilobases long. For example, there is a 55-kb region downstream of the dihydrofolate reductase gene in Chinese hamster ovary cells that acts as an origin of replication. It may be that, instead of using unique essential sequences, mammalian cells use dispersed, redundant *cis*-acting elements.

Mitochondrial DNA Replicates Continuously on Both Strands

Mitochondria in mammalian cells contain circular, supercoiled DNA (mtDNA) about 5 μ in size ($M_r = 10 \times 10^6$; about 15,000 bp). It appears to be replicated by DNA polymerase γ, since this is the only DNA polymerase present in mammalian mitochondria. Furthermore, mtDNA replicates even in cells where DNA polymerase α is absent or inhibited by aphidicoline. (fig. 31.22).

Figure 31.22

Structure of the DNA polymerase α inhibitor aphidicolin.

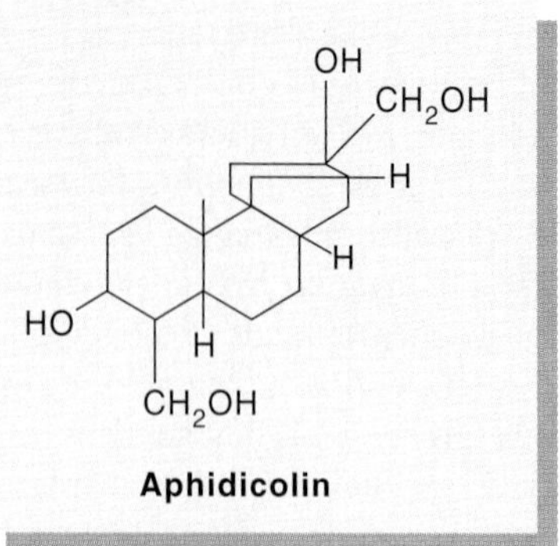

The replication of animal mtDNA has been studied most extensively in mouse L cells (fig. 31.23), which have two origins of replication, one for each strand (referred to as H and L). DNA replication is initiated at the first site on the H strand (O_H) to form a displacement loop (D loop). It is a triple-stranded structure that includes the newly synthesized short daughter H strand. Replication continues unidirectionally until completion. When H strands synthesis is two-thirds complete, L strand synthesis is initiated (O_L) and elongated in the direction opposite to that of H strand synthesis. The daughter molecules segregate, and synthesis proceeds to completion prior to closure with the introduction of about 100 negative superhelical turns. The overall rate of mtDNA synthesis is about 270 nucleotides/min, about 0.5% the rate at which *E. coli* DNA replicates.

An interesting observation is that mammalian mtDNA contains a small percentage of ribonucleotides, which have been monitored by measuring susceptibility to alkali and to ribonuclease H. It has been found that mouse L-cell mtDNA has ribosubstitutions in the two replication origin regions. Their function is unknown.

Several Systems Exist for DNA Repair

We have seen that chromosomes are usually formed by a single DNA molecule, regardless of their size. Such a large molecule makes an easy target to attack and damage. In fact, limited damage occurs quite frequently. A single lesion in a DNA molecule left unrepaired could interefere with the replication process and probably would cause cell death. Clearly, it is highly advantageous to have a repair system; indeed, most cells have more than one.

DNA damage is caused by a variety of physical, chemical, and biological agents. Physical agents include ultraviolet light and ionizing radiation. Damage can be in the form of a missing, incorrect, or modified base or an alteration in the structural integrity of the DNA strands by breaks, cross-links, or dimerization of bases, usually pyrimidines.

Adjacent pyrimidine bases in a DNA strand form dimers with high efficiency after absorbing ultraviolet light (fig. 31.24). By contrast, purines are quite resistant to damage by ultraviolet.

Figure 31.23

Replication model for mouse mitochondrial DNA. (Blue solid line: parental heavy (H) strands. Black solid lines, parental light (L) strands. Blue dashed lines: daughter H strands. Black dashed lines: daughter L strands.) The order of replication is clockwise, starting at D mtDNA. O_H and O_L are the origins of H- and L-strand synthesis, respectively. The double arrows reflect the metabolic instability of D-loop strands and consequent equilibrium between D mtDNA and C mtDNA. Expanded D-loop replicative intermediates are termed Exp-D prior to initiation of L-strand synthesis and Exp-D(l) after initiation of L-strand synthesis. The caret marks the interruption of at least one phosphodiester bond in the H strand of the daughter molecule. (β Gpc = gapped circular daughter molecule.) Each replicative form is discussed in order in the text.

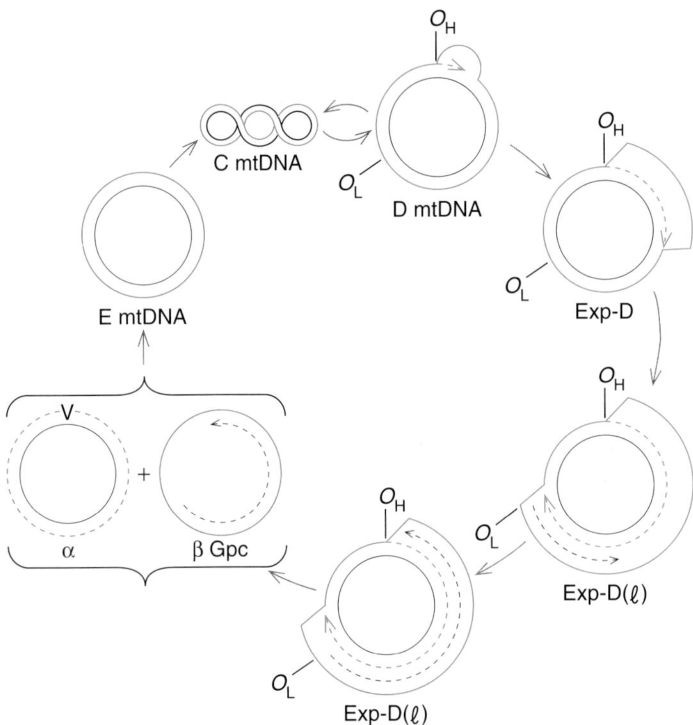

Figure 31.24

Structure of a thymine dimer formed in DNA by exposure to short-wavelength ultraviolet light.

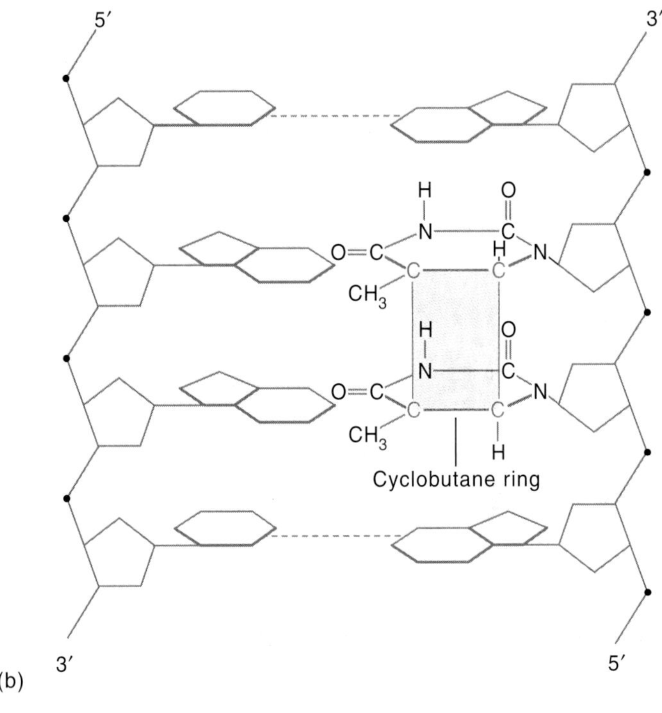

Figure 31.25

Thymine dimers may be monomerized from DNA by enzymatic photoreactivation. In this case no nucleotides are removed in the repair reaction.

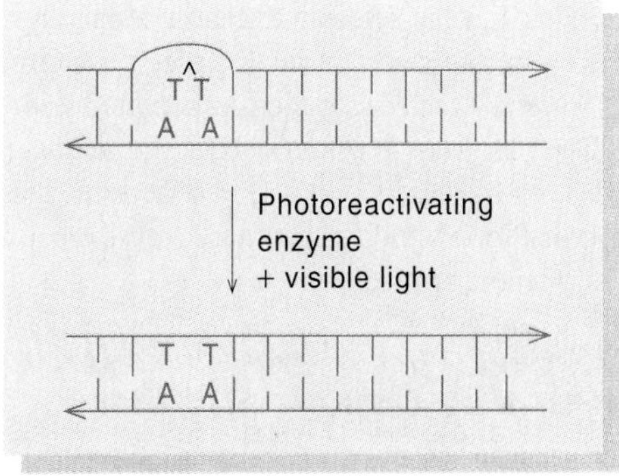

Pyrimidine dimers formed within an otherwise intact DNA duplex have provided a useful substrate to assay for DNA repair. These dimers can be repaired directly by enzymatic photoreactivation (fig. 31.25). The photoreactivation enzyme binds to the DNA containing the pyrimidine dimer and uses visible light to cleave the dimer without breaking any phosphodiester bonds.

Systems that function with the help of visible light are quite common in microorganisms. This relates to the fact that sunlight is a mixture of ultraviolet and visible light. The potentially harmful effects of moderate doses of ultraviolet are partially overcome by the repair processes triggered by the visible light.

Many species, including humans, cannot carry out light repair. In fact all species have alternative methods for removing pyrimidine dimers and other forms of DNA damage (fig. 31.26). This type of repair, known as excision repair, entails removal of the damaged region and resynthesis of the excised region. Special glycosylases that recognize abnormal or incorrectly paired bases can cleave N-glycosidic bonds to generate an apurinic or apyrimidinic site (see fig. 31.26*a*). Alternatively, a double incision is made in the strand that carries the lesion by a complex of three proteins encoded jointly by the *uvrA, uvrB,* and *uvrC* genes (see fig. 31.26*b*). In the case of pyrimidine dimers an incision is made of seven nucleotides 5′ to the pyrimidine dimer, followed by a second incision of three or four nucleotides 3′ to the same dimer. The *uvrD* gene product, a helicase, together with PolI and possibly a

Figure 31.26

Pyrimidine dimers and other forms of DNA damage can be removed by a general excision repair mechanism. The first reaction in this form of repair involves forming nicks about the damaged region of the DNA. In (*a*) we see the mode of incision of UV-irradiated DNA by the pyrimidine-dimer-specific glycosylase and AP endonuclease activities of *Micrococcus luteus* and bacteriophage T4. In (*b*) we see the mode of incision of the uvrABC endonuclease of *E. coli.* (*Source:* Adapted from G. Walker, Inducible DNA repair systems, *Ann. Rev. Biochem.* 54:425, 1985.)

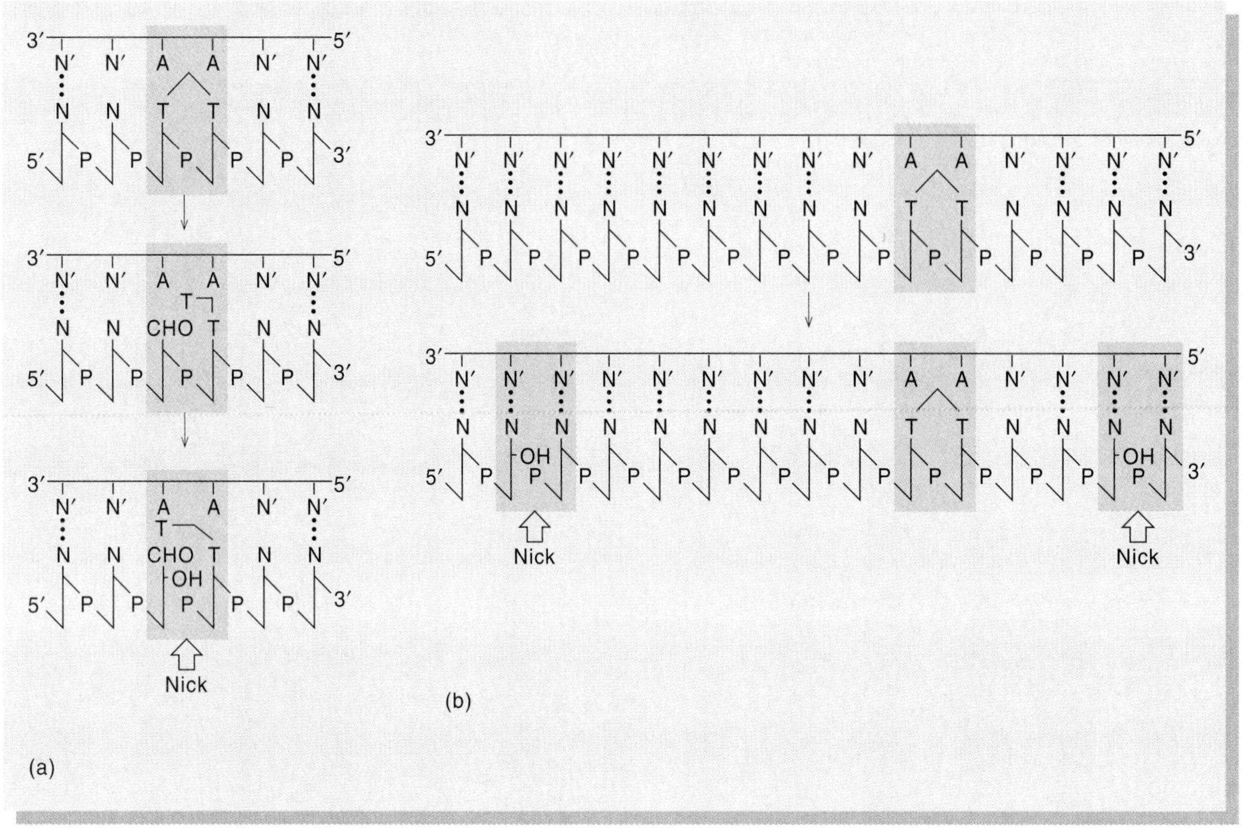

single-strand binding protein, releases the 12- to 13-nucleotide oligomer generated by the incision of the *uvrABC*–enzyme complex. After release of this oligomer, PolI and DNA ligase resynthesize the excised region and ligate the nicks.

Humans have a similar excision repair system. In this case the excised regions are 29 nucleotides in length. In humans the inability to repair DNA damage is associated with rare genetic syndromes. The best known is xeroderma pigmentosum. There are several variations of this condition. People with the disease are unable to repair the DNA damage caused by exposure to ultraviolet light and some chemicals.

The Mismatch Repair System Is Important for Maintaining Genetic Stability

The excision repair system can remove all sorts of DNA damage as long as it is confined to one strand. It also removes mismatches. However, since mismatches contain normal mispaired bases that result primarily if not exclusively from replication errors, it is important to remove the base that was incorrectly added to the growing chain during synthesis and not the one that is in the template strand. In fact the excision-repair system cannot distinguish between the two chains, so its action on a mismatch is just as likely to fix a mutation as to eliminate it. Fortunately, the excision repair system is relatively ineffective in acting on mismatches compared to the mismatch repair system, which can discriminate between old and new chains.

Because mismatches consist of normal bases, mismatch repair systems rely on secondary signals within the helix to identify the newly synthesized DNA strand. The requisite strand specificity for processing of replication errors in *E. coli* is provided by patterns of adenine methylation at GATC sequences. Because GATC modification occurs after DNA strand synthesis, newly synthesized DNA exists briefly in an unmethylated state, and it is this transient absence of modification that targets repair to the new DNA strand. The mechanism of replication error correction by the methyl-directed pathway depends on several proteins, some of which are primarily involved in mismatch repair and designated by Mut and others that serve a variety of functions (fig. 31.27). Repair is initiated by binding of MutS to the mismatch, followed by the addition of MutL. Assembly of this complex leads to activation of a latent GATC endonuclease associated with the MutH protein, which incises the unmodified strand at a hemimethylated d(GATC) sequence. The resulting strand break can occur on either side of the mismatch. The ensuing excision reaction, which

Figure 31.27

Steps in the correction of a mismatch by the mismatch repair system in *E. coli.* The mismatch repair system discriminates between the two DNA chains by lack of methylation in the newly synthesized chain. In repairing the mismatch, only the unmethylated strand is attacked. (From M. Grilley, J. Griffith, and P. Modrick, *J. Biol. Chem.* 268:11830, 1993. Reprinted with permission.)

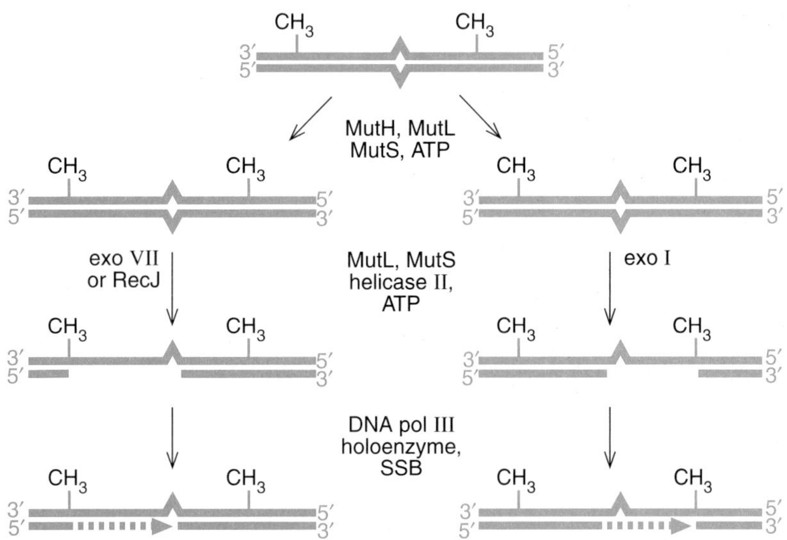

depends on MutS, MutL, and the cooperative action of DNA helicase II (the MutU protein) with an appropriate exonuclease, removes that portion of the unmodified strand spanning the GATC site and the mismatch. This reaction is strictly exonucleolytic, initiating at the strand break and proceeding toward the mismatch without regard to the location of the strand break. The unusual bidirectional excision capability implies that the methyl-directed system keeps track of which side of the mismatch the strand signal is located on.

Mismatch repair systems have been found in eukaryotes, including yeasts and humans. Lesions in the mismatch repair system in humans are believed to be responsible for certain forms of cancer.

Synthesis of Repair Proteins Is Regulated

In *E. coli* the synthesis of many enzymes involved in repair is regulated by the so-called SOS system. Two proteins, lexA and recA, form the working machinery of this regulatory system (fig. 31.28). Under normal conditions the lexA protein inhibits the expression of about 17 genes (the *din* genes), the encoded proteins of which are exclusively involved in DNA repair. The lexA protein does this by binding tightly to the control regions of these genes.

The recA protein has two functions: One is to catalyze DNA recombination (see next section); the other is to facilitate expression of the *din* genes during times of chromosomal disorder. It does the latter by stimulating self-cleavage of the lexA protein. The loss of lexA repressor triggers the transient expression of the *din* genes. The recA protein is active in this capacity only when bound to a single-stranded DNA fragment. Once the chromosome damage has been repaired, the concentration of DNA fragments diminishes, and recA returns to its normally quiescent state. This mechanism of control ensures that the *din* genes will be expressed only under conditions where there is chromosome damage.

DNA Recombination

Individuals belonging to the same species carry approximately the same number of genes positioned in the same relative locations on their homologous chromosomes; each gene is represented by more than one variant, which frequently results in different visible characteristics (phenotypes) in different individuals. The different representations of a gene are referred to as alleles for that gene (see box 5B). Alleles that give rise to different phenotypes serve as useful markers for detecting recombination between homologous chromosomes. Most cells (diploid cells) of an organism contain pairs of homologous chromosomes, one from each parent. During formation of the sex cells (haploid cells) a reduction takes place so that each sex cell contains only one chromosome of each type. The process of chromosome segregation, which takes place during formation of the sex cells, is called meiosis (fig. 31.29). Homologous chromosomes segregate at random during the first phase of meiosis (see fig. 31.29*a*). This results in the production of sex cells with a very large number of possible combinations of chromosomes. Some combinations of alleles undoubtedly have selective advantages over others. Indeed, the primary advantage of mating seems to be to produce new combinations of genes. But what of the alleles on the same chromosomes? Nature has provided another mechanism for reshuffling alleles on the same chromosome, a process called recombination. Although recombination can take place in most cells of the organism, meaningful genetic recombination that can be passed from one generation to the next takes place only during the first phase of meiosis (see fig. 31.29*c*).

In the 1960s, Matthew Meselson demonstrated that homologous recombination involves breakage and rejoining between

Figure 31.28

Model for the SOS regulatory system. In normally growing cells the SOS functions associated with DNA repair are not expressed. This is because lexA repressor inhibits their transcription. LexA repressor also inhibits its own expression and that of *recA*. SOS functions are turned on by a series of reactions that starts with DNA damage. DNA damage results in an inducing signal that activates a self-cleavage function of lexA. This leads to cleavage of the lexA protein, so that all genes that were formerly inhibited by lexA are expressed. Once the damage is repaired, the level of lexA repressor builds up again, and the SOS genes return to their usual repressed state.

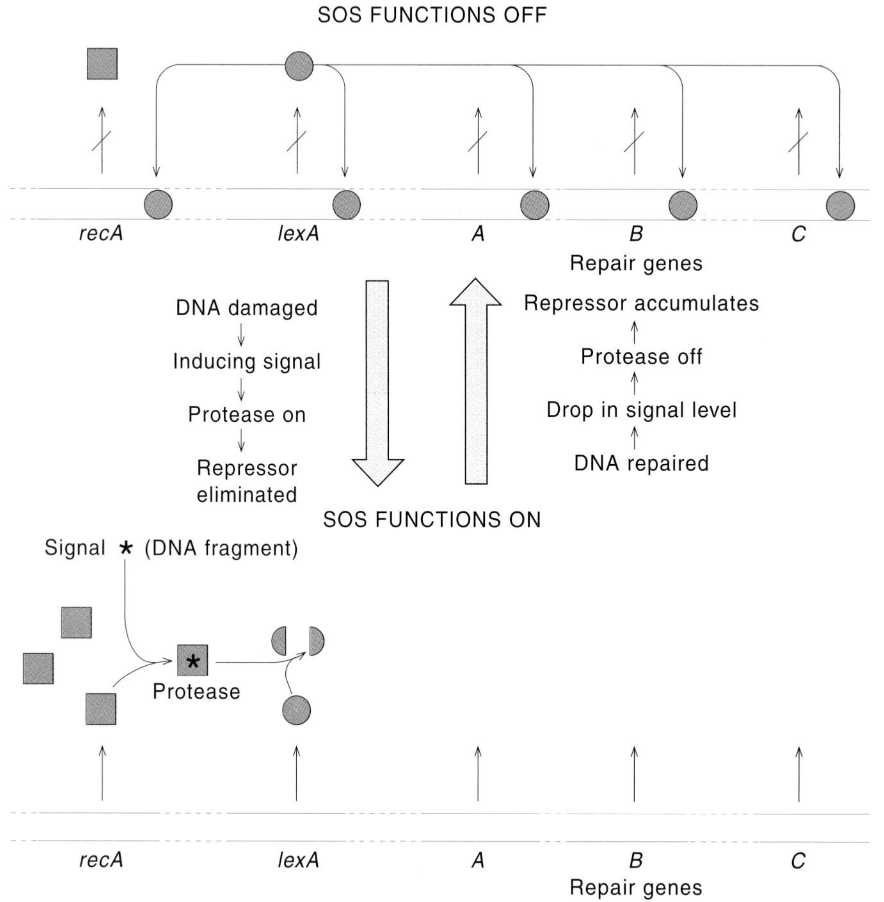

existing chromosomes. He made this demonstration on bacteriophages carrying different alleles that infect the same cell. With the help of isotopic labeling he showed that recombinant chromosomes can form without DNA replication, thus demonstrating the breakage–rejoining mechanism. Only a small fraction of new DNA appears in the recombinant chromosome, and this can be accounted for by the repair processes that accompany breakage and rejoining. From his studies, Meselson proposed a three-step model for homologous recombination (fig. 31.30): (1) formation of staggered breaks in two parental DNAs, (2) base pairing between single-stranded regions of the two parental types, and (3) repair synthesis. The most important aspect of the Meselson model is that it explained why homologous recombination is so common—a key step in the process involves interaction between complementary single-strand regions originating from homologous chromosomes.

Meselson's results spearheaded investigations of the mechanism of recombination in eukaryotes. Here also the mechanism of recombination can be explained by breakage and rejoining with only a small amount of new DNA synthesis being required in the region of the join. In eukaryotes it was found that recombination is invariably reciprocal. Thus if chromosome AB recombines with chromosome a,b so that Ab forms as a recombinant product, then aB also forms.

Enzymes Have Been Found in Escherichia coli *That Mediate the Recombination Process*

We will focus on the biochemistry of homologous recombination in *E. coli,* since this is one of the best-understood systems. Cursory inspection of the Meselson model suggests the need for special enzymes to mediate the recombination process. First an endonuclease is needed to make the initial strand breaks that are required for strand invasion. Second an enzyme is required to mediate formation of the recombination complex. Finally, an enzyme is probably required to resolve the recombinant duplexes. Genetic and biochemical studies on *E. coli* have led to the discovery of these three types of enzymes.

Figure 31.29

Meiosis. The process of meiosis involves two cell divisions with two segregation cycles, meiosis I and meiosis II. These cycles are pictured for a hypothetical cell containing four chromosomes. (*a*) During the first cycle, homologous chromosomes pair and then segregate. (*b*) During the second cycle the sister chromatids from each chromosome segregate. The second cycle is very much like mitosis (see fig. 1.17). Each diploid cell that enters meiosis ultimately yields four haploid cells. These haploid cells are the sex cells for the next round of mating. (*c*) Recombination between homologous chromosomes during meiosis I leads to different combinations of alleles. In this illustration, alleles for the same gene are represented by corresponding capital or lowercase letters.

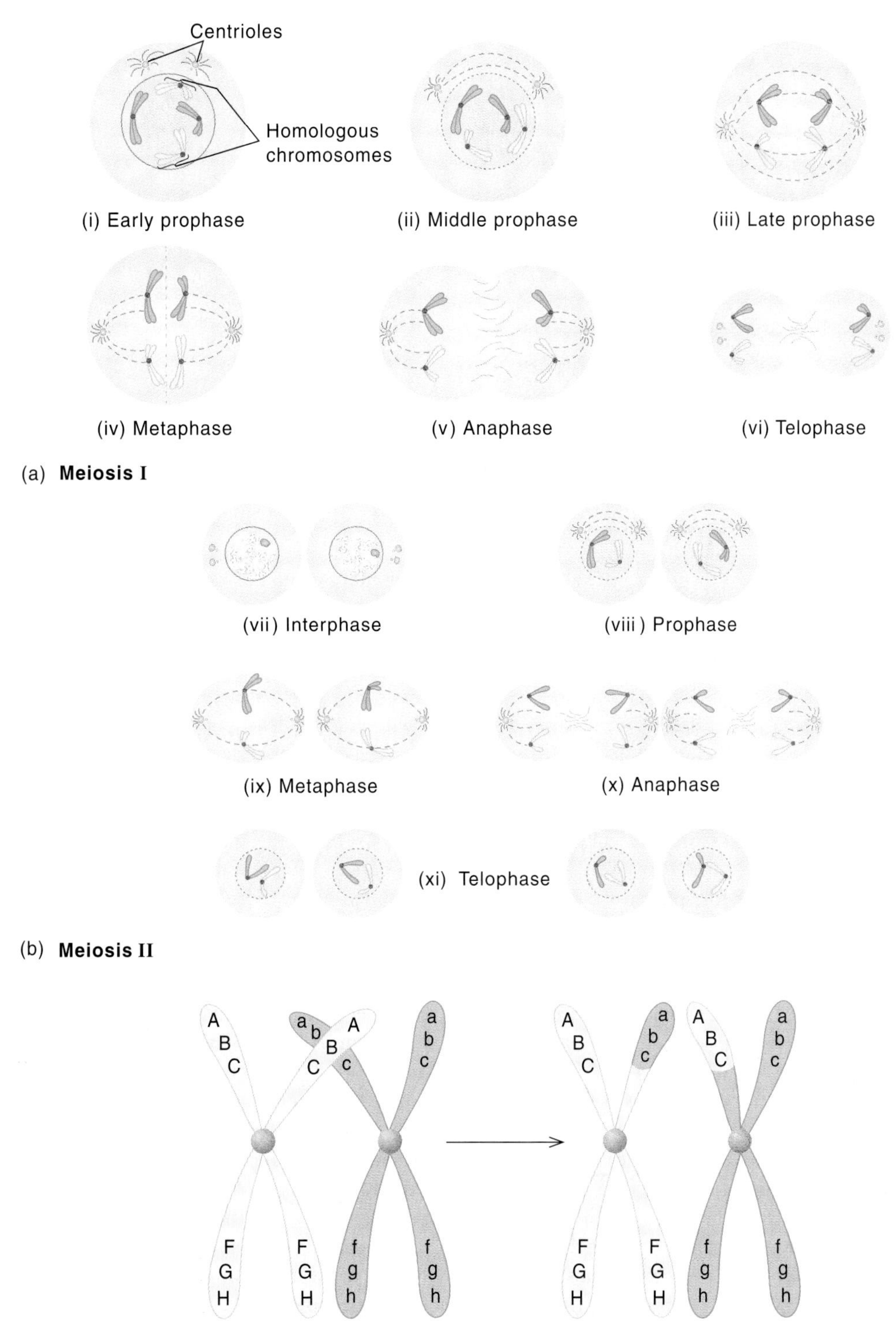

Figure 31.30

Meselson model for phage recombination (circa 1964). Phage form staggered breaks. Base pairing leads to an annealed complex with gaps. Gaps are mended by repair synthesis.

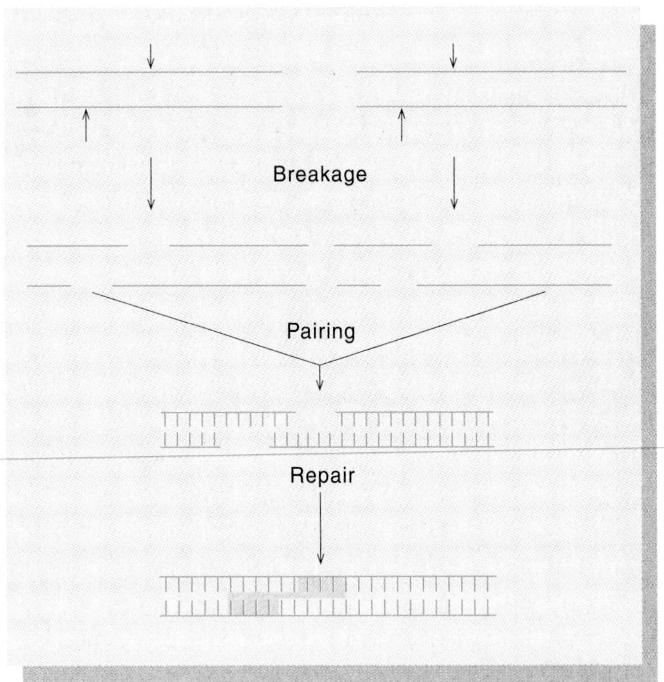

Figure 31.31

Reactions catalyzed by purified recA protein in vitro. RecA catalyzes a number of different reactions between DNA strands, all of them involving the unwinding and winding of base-paired structures. (*a*) D-loop formation by interaction between supercoiled circular duplex DNA and single-stranded DNA. (*b*) Strand exchange between a gapped circular duplex structure and a linear duplex structure. (*c*) Complex formation between two helices, one of which is gapped.

(a) recA enzyme — D-loop formation

(b) recA enzyme — Strand exchange and linear duplex formation

(c) recA enzyme — Complex formation between two helices, one of which is gapped

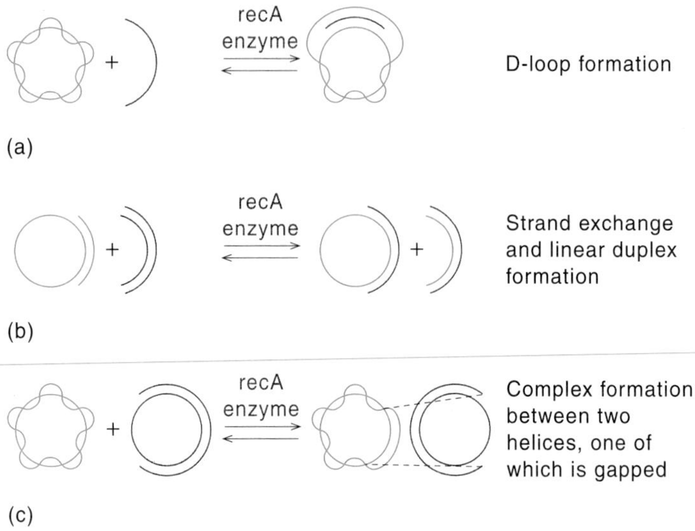

The search for enzyme activity associated with recombination has been based on analysis of organisms carrying mutations that affect general recombination. Such mutants, known as *rec* mutants, were first discovered more than 20 years ago, and new ones are still being found. *E. coli rec* mutations, *recA, recB, recC,* and *recD* reduce recombination efficiency to different extents. The *recA* mutants are totally deficient in homologous recombination, whereas *recB* or *recC* mutants can recombine, but only about 10^{-4} times as well as wild-type cells.

We have discussed the repair role played by the coprotease function of *recA* (see fig. 31.28). The second class of reactions catalyzed by recA protein indicates a prominent role in the initiation of homologous pairing during genetic recombination. Thus the purified recA protein catalyzes various forms of complex formation and exchange between duplex and single-stranded DNAs in reactions that entail the cleavage of ATP (fig. 31.31). D-loop formation can result when the recipient is a circular duplex and the donor is a single strand (see fig. 31.31*a*). Strand exchange can occur when a single-stranded fragment is removed in the presence of a double-stranded fragment (see fig. 31.31*b*). RecA protein also causes a complex to form between two circular helices, provided that at least one input helix is gapped on one strand (see fig. 31.31*c*). Interestingly, while the two helices must contain homologous sequences in order to form a four-stranded structure, the gap may lie in a nonhomologous region.

The fact that mutations in *recA* completely eliminate recombination, as well as the finding that the recA-related protein catalyzes strand exchange reactions like those required in the recombination models we have been discussing, makes it highly likely that *recA* plays a key role in these processes in vivo. If this is so, then recA protein is clearly situated at the hub of activities in the recombination complex.

The recA enzyme does not catalyze any recombination event unless at least one free end of a single strand or a DNA duplex is available. Thus it seems unlikely that the recA protein could be involved in making the initial incision(s) in the duplex structure that is (are) required to initiate recombination. There are reasons for believing that the *recB* and *recC* genes are involved in this capacity. First, a mutation in either of these genes reduces recombination by a factor of about 10^{-4}. Second, these two genes together with *recD* encode the subunits of a nuclease known as exoV. ExoV is both a nuclease and a helicase. The helicase activity is best demonstrated in vitro under conditions where the nuclease activity is blocked. This can be done by adding Ca^{2+}. When linear duplex DNA is incubated with exoV in the presence of ATP and Ca^{2+}, it begins to unwind at one end. As the unwinding progresses, the single-stranded regions collapse back on each other to reform a duplex. A double-loop structure is maintained in the vicinity of the migrating enzyme (fig. 31.32). What is the point of this migrating behavior? Further experiments indicate that exoV scans the duplex, looking for preferred sequences, where it makes single-strand incisions.

The recombination process triggered by the initial scissions made by exoV and mediated by the recA enzyme are not sufficient for recombination between interacting chromosomes. An additional enzyme, resolvase, is required for efficient separation of the recombining chromosomes in the recA–chromosome complex. This protein is encoded by the *ruvC* gene in *E. coli* and is required

Figure 31.32

DNA unwinding by exoV enzymes. This is best observed in vitro in the presence of ATP and Ca^{2+}. The Ca^{2+} inhibits the exonucleolytic activity of the enzyme. The ATP provides the energy that drives the enzyme in a concerted manner into the duplex structure. The duplex unwinds in front of the path of the enzyme and rewinds in back of the enzyme.

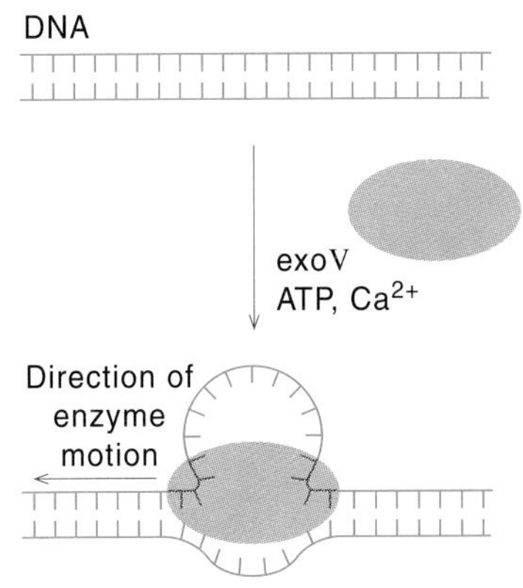

for homologous recombination in the bacterium. Its effectiveness has also been demonstrated in a cell-free in vitro system.

Incidentally, the fact that the recA enzyme carries two activities, one for homologous recombination and one for regulating the concentration of DNA repair genes, including itself, leads to the suspicion that recA is also more directly involved in DNA repair. Recombination between two homologous chromosomes badly damaged in different regions could lead to one normal chromosome. Although technically a haploid organism, *E. coli* carries two copies of the bacterial chromosome for a considerable time prior to cell division. Thus the necessary identical pairs of chromosomes are present for homologous recombination.

Other Types of Recombination

There are two other types of recombination, on which we will not elaborate. Site-specific recombination is limited to highly select regions of the genome and to very specific functions. One of the best-understood examples of site-specific recombination is the integration of bacteriophage λ DNA into the *E. coli* host chromosome (see chapter 35). Finally, nonspecific recombination occurs between nonhomologous regions of chromosomes with little or no sequence homology. In the next section we see that at one stage in its life cycle a retrovirus recombines in this way.

RNA-Directed DNA Polymerases

It is a commonly held belief that RNA preceded DNA in the early evolution of living systems. If this was the case, then the first DNA polymerases must have been capable of transferring sequence information from RNA to DNA. Enzymes of this sort are called reverse transcriptases because they do the reverse of common transcriptases (see chapter 33). Reverse transcriptases no longer play the central role in genetic information transfer, but they are still found in all species and function in a number of capacities in both cellular and viral metabolism.

Retroviruses Are RNA Viruses That Replicate through a DNA Intermediate

RNA viruses that cause tumors (oncogenic RNA viruses) are called retroviruses because their life cycle involves a DNA intermediate. The ability of retroviruses to use such a route for replication hinges on a viral-encoded enzyme called reverse transcriptase, an enzyme with three discrete activities: (1) It catalyzes the synthesis of DNA from the viral plus-strand; (2) it catalyzes the synthesis of DNA plus-strand from the viral minus-strand DNA; and (3) it catalyzes the degradation of the viral RNA from an RNA–DNA heteroduplex.

Following viral infection the viral plus-strand first functions as an mRNA for the synthesis of the viral proteins. Once the viral reverse transcriptase has been synthesized, it is used in conjunction with the viral RNA template and a tRNA primer for the synthesis of minus-strand DNA. Next the RNase H activity of the viral enzyme removes extensive sections of RNA from the DNA–RNA heteroduplex so that the newly synthesized minus-strand DNA can be used as a template for the synthesis of viral plus-strand DNA. The complete process of duplex DNA synthesis is depicted in figure 31.33. The duplex DNA is subsequently integrated more or less indiscriminately at many sites in the host genome. There it can remain indefinitely or until the cell dies. The integrated viral DNA serves as a template for the synthesis of viral plus strand, completing the nucleic acid cycle that began with virus infection. Additional properties of retroviruses, including their association with malignant transformation, are taken up in chapter 38.

Hepatitis B Virus Is a DNA Virus That Replicates through an RNA Intermediate

Hepatitis B is an animal virus with a small circular DNA genome. A gap in one strand is bridged by an incomplete complementary strand. The longer strand has a protein bound at its 5′ end. A DNA polymerase present in the mature virus particle can elongate the 3′ ends of the incomplete strands. An unusual feature of this virus is that replication involves an RNA intermediate that must be reverse transcribed.

Some Transposable Genetic Elements Encode a Reverse Transcriptase That Is Crucial to the Transposition Process

A widely distributed group of genetic elements exist that are capable of integrating at random sites in the host chromosome. These transposable genetic elements usually carry genes to facilitate their own transposition in addition to conferring specific properties on the host cells in which they are located. Many of these transposable elements encode a reverse transcriptase that functions in the transposition process. The first step in transposition of elements of this type involves transcription of the genetic element. The resulting RNA, like the retroviral RNA, has two functions: one as a messenger RNA for the synthesis of element-encoded proteins and one as a template for the synthesis of the transposable element DNA. The transposable element integrates at other sites on the host chromosome without any apparent regard for sequence homology at the sites of integration. Transposable genetic elements

Figure 31.33

A model for the generation of double-stranded DNA carrying two copies of LTR. The sequence X is a marker for the plus-strand; the complementary sequence X′ occurs on the minus-strand. (*a–c*) Synthesis of minus-strand DNA from the genomic RNA template using a tRNA primer (represented by inverted J in step *a*). (*d*) Degradation of the RNA portion of the resulting DNA:RNA hybrid by RNAseH. (*e*) Bridge between the newly synthesized segment of minus-strand and repeated sequences at the 3′ end of genomic RNA. (*f*) Extension of minus-strand DNA (leftward) and initiation of plus-strand DNA (heavy arrow moving rightward). (*g*) Completion of 300-nucleotide fragment of plus-strand DNA. (*h*) Bridge between sequences repeated in minus-strand DNA and in the 300-nucleotide fragment. (*i*) Completion of synthesis and tidying up: = LTR, with sequences from the 5′ and 3′ ends of the viral genome separated by the terminal repeat. In this figure, red stands for RNA and blue stands for DNA.

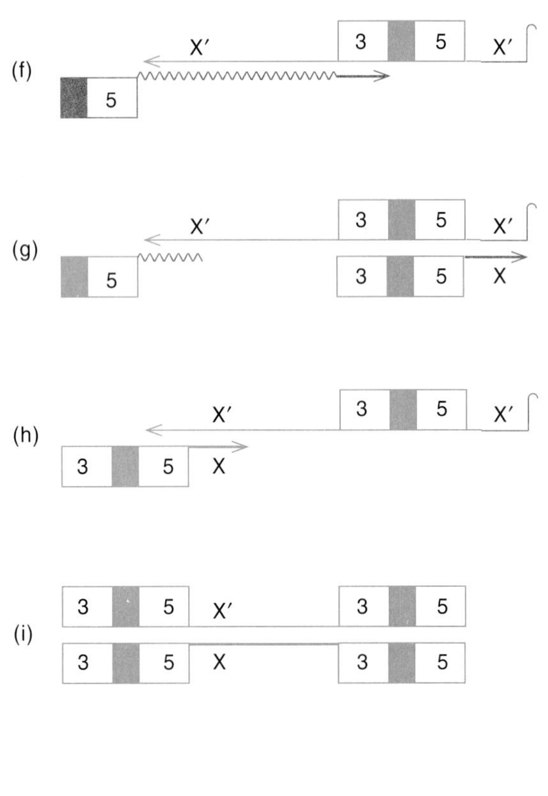

that use a reverse transcriptase in this way have been found in yeast, *Drosophila,* and mice, and they are presumed to be present in a very wide range of organisms.

Bacterial Reverse Transcriptase Catalyzes Synthesis of a DNA-RNA Molecule

Except as intermediates, molecules in which DNA and RNA are covalently linked to each other have been found only in certain bacteria. For example, some strains of *E. coli* carry a genetic element that encodes a reverse transcriptase. This genetic element produces one long transcript that folds into a complex three-dimensional structure. Part of this molecule serves as a template for the synthesis of a reverse transcriptase, which recognizes a specific G residue on the folded RNA as a primer start site for DNA synthesis. Synthesis starts from the 2′-OH group of this G residue using another part of the RNA as a template. DNA synthesis arrests at a specific point on the RNA template. An RNase H removes that region of the RNA in which DNA–RNA heteroduplex has been formed, leaving a covalently linked DNA–RNA molecule. No function has been found for this unusual nucleic acid (see the Linn and Maas reference in the Selected Readings).

Telomerase Facilitates Replication at the Ends of Eukaryotic Chromosomes

Circular bacterial chromosomes are initiated by RNA primers. At some stage the RNA primers must be eliminated and replaced by DNA. Owing to the circular nature of the chromosome, an upstream DNA molecule can always serve as a primer for regions from which RNA primers are eventually removed. This guarantees that the primer requirement does not interfere with complete replication of the chromosome.

Since eukaryotic chromosomes are linear, the ends of these chromosomes require a special solution to ensure complete replication. This can be seen in figure 31.34. At the very end of a

Figure 31.34

Synthesis at the ends of a eukaryotic chromosome. One end of the linear DNA of a eukaryotic chromosome is diagrammed. (*a*) A flush-ended DNA duplex presents a problem for completing synthesis at the 5′ end. This is because of the RNA primer requirement for DNA synthesis. When the primer at the 5′ end is removed, there is no conventional way to fill the gap. (*b*) A solution to this problem. The ends of eukaryotic chromosomal DNAs consist of highly repetitious tandem repeats (telomeres). These repeats on the 3′ end serve as both primer and template for extending the 3′ end. The extended 3′ end can accommodate a primer RNA, so after chromosomal DNA replication no loss occurs from the 5′ end of the DNA. Another process is needed to remove the extension from the 3′ end. New synthesis is indicated in red. The zigzag represents primer.

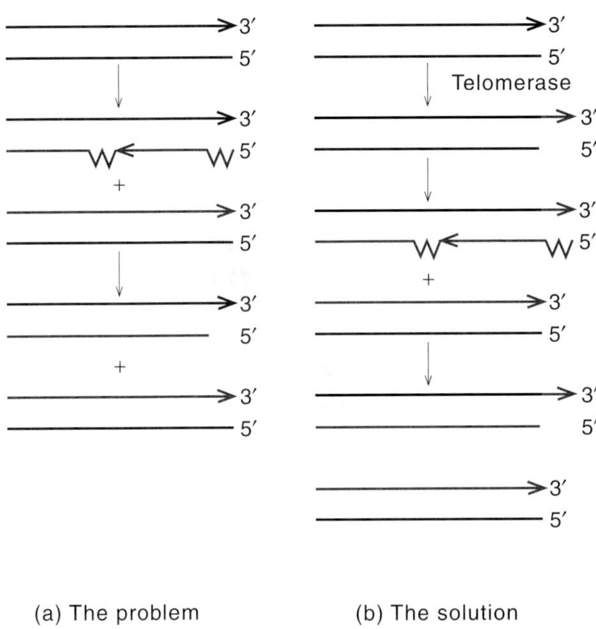

Figure 31.35

Telomerase extends the 3′ end of the eukaryotic chromosomal DNA by a cyclic repeat of three steps. The sequence shown is for the telomeres and telomerase found in the ciliated protozoa tetrahymena. (Reproduced, with permission, from the *Annual Review of Biochemistry*, Volume 61, © 1992, by Annual Reviews Inc.)

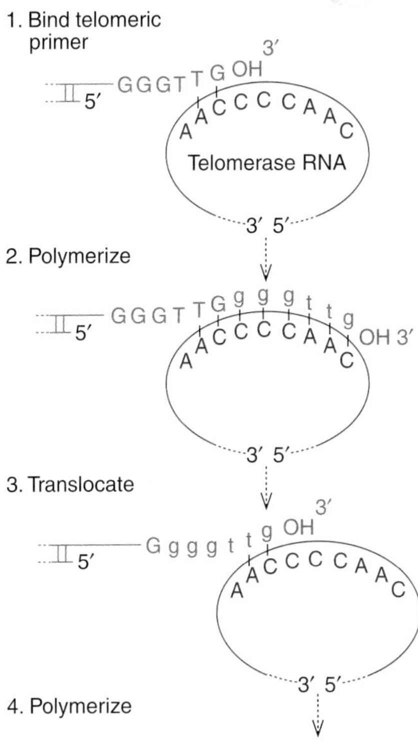

linear duplex a primer is necessary to initiate DNA replication. After RNA primer removal there is bound to be a gap at the 5′ end of the newly synthesized DNA chains. Since DNA synthesis always requires a primer, the usual way of filling this gap is not going to solve the problem. This dilemma is overcome by a special structure at the ends (telomeres) of eukaryotic chromosomes and a special type of reverse transcriptase (telomerase) that synthesizes telomeric DNA. In many eukaryotes the telomeres contain short sequences (frequently hexamers or octamers) that are tandemly repeated many times (see fig. 30.16). Telomerase contains an RNA that binds to the 3′ ends and also serves as a template for the extension of these ends (fig. 31.35). Prior to replication the 3′ ends of the chromosome are extended with additional tandemly repeated hexamers. The 3′ ends are extended sufficiently that there is room to accommodate an RNA primer. In this way there is no net loss of DNA from the 5′ ends as a result of replication. After replication the 3′ end is somewhat longer than it was before because of the extra added tandem repeats. An additional mechanism is necessary to keep the telomeres from growing indefinitely by this effect. Possibly related to this requirement, it has been observed that the telomeres become shorter in somatic tissues with aging. This is believed to be due to a deficiency of telomerase in most somatic tissues.

Recently, many interesting correlations between cell division potential and telomerases have been observed in different types of tissues. Thus telomerases appear to be lacking in normal human somatic tissues but reactivated in cancers where immortal cells are likely to be required to maintain tumor growth. A parallel phenomenon has been observed in cultured cells from many different tissues. Cells with a limited capacity for continued growth and duplication lack telomerase, but immortal cell lines contain readily detectable telomerase.

Other Enzymes That Act on DNA

Many other enzymes modify DNA and degrade it in various ways. In the next chapter we discuss some of those enzymes that have been particularly useful in DNA manipulations. In chapter 37 we discuss DNA splicing operations that are essential for assembling complex antibody genes prior to their expression.

SUMMARY

This chapter deals with reactions involved in DNA synthesis, degradation, repair, and recombination. The chief points to remember are as follows.

1. DNA replication proceeds by the synthesis of one new strand on each of the parental strands. This mode of replication is called semiconservative, and it appears to be universal. DNA synthesis initiates from a primer at a unique point on a prokaryotic template such as the *E. coli* chromosome. From the initiation point, DNA synthesis proceeds bidirectionally on the circular bacterial chromosome. The bidirectional mode of synthesis is not followed by all chromosomes. For some chromosomes, usually small in size, replication is unidirectional.
2. In eukaryotic systems, replication can start at several points (still not well defined) along the chromosome. Replication is usually bidirectional about each initiation site. The termination points of replication are interspersed between initiation sites. In most cases of unidirectional or bidirectional replication, synthesis occurs nearly (but not exactly) simultaneously on both strands of the parent DNA template. Because synthesis can occur only in the $5' \rightarrow 3'$ direction on the growing chain, and the two strands in the parent duplex are oriented in opposite directions, synthesis can occur continuously on only the leading strand. On the other (lagging) strand it must pause for the template to unwind. Synthesis on the lagging strand does not occur continuously but rather in small discontinuous spurts, generating Okazaki fragments.
3. Many proteins are required for DNA synthesis and chromosomal replication. These include polymerases; helicases, which unwind the parental duplex; enzymes that fill in the gaps and join the ends in the case of lagging-strand synthesis; enzymes that synthesize RNA primers at various points along the DNA template; topoisomerases, which permit rotation and supercoiling; and single-strand DNA-binding proteins, which stabilize single-stranded regions that are transiently formed during replication. Most of these proteins have been isolated from whole cells and studied in cell-free systems.
4. In *E. coli,* mutations have been isolated in the genes encoding a number of these enzymes. Many of these mutations are conditional because the functional enzymes involved are required for DNA synthesis and cell viability. Mutants carrying mutationally altered proteins have been important in confirming their roles predicted from cell-free studies.
5. In eukaryotes, considerable progress has been made in studying the in vitro replication of animal viruses, such as SV40. The importance ascribed to the enzymes that have been characterized is largely based on a comparison of their properties with similar prokaryotic enzymes whose functions are better understood.
6. Many enzymes that act on DNA are involved in processes other than DNA synthesis. They include DNA repair enzymes, DNA degradation enzymes, and DNA recombination enzymes.
7. Enzymes that catalyze the synthesis of DNA using an RNA template are known as reverse transcriptases. The first reverse transcriptase discovered was encoded by an RNA retrovirus. This enzyme is needed in the virus replication cycle. Some animal viruses pass through an RNA intermediate and also require a reverse transcriptase to replicate the viral DNA. Similarly, a number of transposable elements found in cellular chromosomes replicate through RNA intermediates; they usually encode a reverse transcriptase. A unique reverse transcriptase called telomerase is used to synthesize the DNA at the ends of linear eukaryotic chromosomes.

SELECTED READINGS

Baker, T. A., and S. H. Wickner, Genetics and enzymology of DNA replication in *Escherichia coli. Ann. Rev. Genet.* 26:447–469, 1992.

Bauer, W. R., F. H. C. Crick, and J. H. White, Supercoiled DNA. *Sci. Am.* 243(4):118–133, 1980.

Beese, L. S., V. Derbyshire, and T. A. Steitz, Structure of DNA polymerase I Klenow fragment bound to duplex DNA. *Science* 260:352–355, 1993.

Beese, L. S., and T. A. Steitz, Structural basis for the $3' \rightarrow 5'$-exonuclease activity of *E. coli* DNA polymerase I: A two metal ion mechanism. *EMBO J.* 10:25–33, 1991.

Bell, S. P., and B. Stillman, ATP-dependent recognition of eukaryotic origins of DNA replication by a multiprotein complex. *Nature* 357:128–134, 1992.

Berger, J. M., S. J. Gamblin, S. C. Harrison, and J. C. Wang, Structure and mechanism of DNA topoisomerase II. *Nature* 379:225–232, 1996.

Blackburn, E. H., Telomerases, *Ann. Rev. Biochem.* 61:113–129, 1992.

Bohr, V. A., and K. Wasserman, DNA repair at the level of the gene. *Trends Biochem. Sci.* 13:429–432, 1988.

Bramhill, D., and A. Kornberg, A model for initiation at origins of DNA replication. *Cell* 54:915–918, 1988.

Campbell, J. L., Eukaryotic DNA replication. *Ann. Rev. Biochem.* 55:733, 1986.

Challberg, M. D., and T. J. Kelly, Animal viruses and DNA replication. *Ann. Rev. Biochem.* 58:671–717, 1989.

Coverley, D., and R. A. Laskey, Regulation of eukaryotic DNA replication. *Ann. Rev. Biochem.* 63:745–776, 1994.

Cox, M. M., and I. R. Lehman, Enzymes of general recombination. *Ann. Rev. Biochem.* 56:229–262, 1987.

DeBondt, H. L., J. Rosenblatt, J. Jancarik, H. D. Jones, D. O. Morgan, and S.-H. Kim, Crystal structure of cyclin-dependent kinase 2. *Nature* 363:595–602, 1993.

Demple, B., and P. Karran, Death of an enzyme: Suicide repair of DNA. *Trends Biochem. Sci.* 8:137–139, 1983.

Diller, J. D., and M. K. Raghuraman, Eukaryotic replication origins: Control in space and time. *TIBS* 19:320–325, 1994.

Dunderdale, H. J., F. E. Benson, C. A. Parsons, G. J. Sharples, R. G. Hoyd, and S. C. West, Formation and resolution of recombination intermediates by *E. coli* RecA and RunC protein. *Nature* 354:506–510, 1991.

Fangman, W. F., and B. J. Brewer, Activation of replication origins with yeast chromosomes. *Ann. Rev. Cell. Biol.* 7:375–402, 1991.

Fink, G. R., J. D. Boeke, and D. J. Garfinkel, The mechanisms and consequences of retrotransposition. *Trends Genet.* 2:118–123, 1986.

Gavin, K. A., Hidaka, M., and Stillman, B., Conserved initiator proteins in eukaryotes, *Science* 270:1667–1670, 1995.

Guzder, S. N., P. Sung, V. Bailly, L. Prakash, and S. Prakash, RAD25 is a DNA helicase required for DNA repair and RNA polymerase II transcription. *Nature* 369:578–582, 1994.

Holliday, R., A different kind of inheritance. *Sci. Am.* 260(4):60–73, 1989. On DNA methylation.

Itoh, T., and J. Tomizawa, Antisense RNA. *Ann. Rev. Biochem.* 60:631–652, 1991. Includes an excellent description of the initiation of DNA synthesis for Col E1 plasmid.

Kim, B., and J. W. Little, LexA and CI repressors as enzymes: Specific cleavage in an intermolecular reaction. *Cell* 73:1165–1173, 1993.

Kohlstaedt, L. A., J. Wang, J. M. Friedman, P. A. Rice, and T. A. Steitz, Crystal structure at 3.5 Å resolution of HIV-1 reverse transcriptase complexed with an inhibitor. *Science* 256:1783–1790, 1992.

Kong, X-P, R. Onrust, M. O'Donnell, and J. Kuriyan, Three-dimensional structure of the β subunit of *E. coli* DNA polymerase III holoenzyme: A sliding clamp. *Cell* 69:425–437, 1992.

Kornberg, A., and T. A. Baker, *DNA replication,* 2d ed. New York: Freeman, 1991. A magnificent up-to-date, clearly written and thorough treatment.

Landy, A., Dynamic, structural and regulatory aspects of site-specific recombination. *Ann. Rev. Biochem.* 58:913–950, 1989.

Linn, D., and W. Maas, Reverse transcriptase-dependent synthesis of a covalently linked branched DNA–RNA compound in *E. coli* B. *Cell* 56:891–904, 1989.

Lohman, T. M., W. Bujalowski, and L. B. Overman, *E. coli* single strand binding protein. *Trends Biochem. Sci.* 13:250–255, 1988.

Maxwell, A., and M. Gellert, Mechanistic aspects of DNA topoisomerases, *Adv. Prot. Chem.* 38:69–107, 1986.

McHenry, C. S., DNA polymerase III holoenzyme of *E. coli. Ann. Rev. Biochem.* 57:519–550, 1988.

Meselson, M., and F. W. Stahl, The replication of DNA in *Escherichia coli. Proc. Natl. Acad. Sci. USA* 44:671–682, 1958. A classic paper.

Messer, W., and W. Noyer-Weidner, Timing and targeting: The biological function of Dam methylation in *E. coli. Cell* 54:734–737, 1988.

Modrich, P., Mismatch repair, genetic stability and cancer. *Science* 266:1959–1960, 1994.

Morgan, D. O., Principles of CDK regulation. *Nature* 374:131–134, 1995.

Newlin, C. S., Yeast chromosome replication and segregation. *Microbiol. Rev.* 52:568–601, 1988.

Ogawa, T., and T. Okazaki, Discontinuous DNA replication. *Ann. Rev. Biochem.* 57:519–550, 1988.

Radman, M., and R. Wagner, The high fidelity of DNA replication. *Sci. Am.* 259(2):40–46, 1988.

Rao, B. J., S.-K. Chiu, L. R. Bazemore, G. Reddy, and C. M. Radding, How specific is the first recognition step of homologous recombination? *TIBS* 20:109–113, 1995.

Sancar, A., Mechanisms of DNA excision repair. *Science* 266:1954–1956, 1994.

Savva, R., K. McAuley-Hecht, T. Brown, and L. Pearl, The structural basis of specific base-excision repair by uracil-DNA glycosylase. *Nature* 373: 487–493, 1995.

Shapiro, J. A., *Mobile Genetic Elements.* New York: Academic Press, 1983.

Smith, G. R., Homologous recombination in *E. coli:* Multiple pathways for multiple reasons. *Cell* 58:807–809, 1989.

Stahl, F. W., Genetic recombination. *Sci. Am.* 256(2):90–101, 1987.

Toyn, J. H., W. M. Toone, B. A. Morgan, and L. H. Johnston, The activation of DNA replication in yeast. *TIBS* 20:70–73, 1995.

Varmus, H., Reverse transcription. *Sci. Am.* 257(3):56–64, 1987.

Waga, S., and B. Stillman, Anatomy of a DNA replication fork revealed by reconstitution of SV40 DNA replication in vitro. *Nature.* 369:207–212, 1994.

Wang, T. S. F., Eukaryotic DNA polymerases. *Ann. Rev. Biochem.* 60:513–553, 1991.

Problems

1. In the Meselson-Stahl experiment illustrated in figure 31.2, a sample of the DNA shown in tube 4 (labeled with ^{15}N followed by one generation in ^{14}N) was heat-denatured prior to being subjected to centrifugation in a CsCl density gradient. This gradient showed two peaks of single-stranded DNAs of different densities. How did this experiment further confirm the process of semiconservative replication?
2. Explain why Cairns and co-workers used [^3H]-thymidine to label replicating *E. coli* DNA in the experiments shown in figure 31.5.
3. Draw the chemical reaction mechanism for the formation of a phosphodiester linkage during DNA synthesis. Discuss the significance of the pyrophosphate product that is formed. What is the significance of the Mg^{2+} requirement?
4. The genetic map represents the distribution of eight genes (*a–h*) as they occur on a bacterial chromosome. The plot is a graphic representation of the average number of copies for each of the eight genes for a cell that grows with a doubling time equal to the time required for a complete round of DNA replication.
   (a) Estimate the location of the origin of replication.
   (b) Infer whether replication is bidirectional or unidirectional.

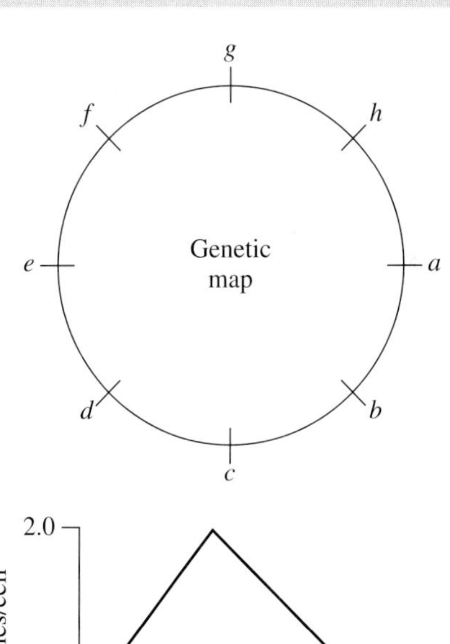

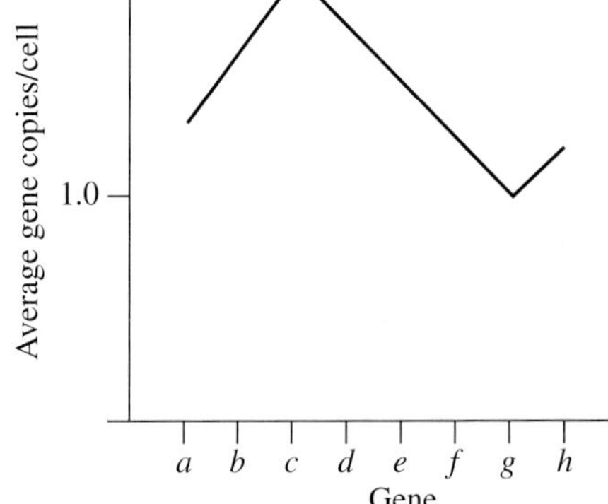

5. In the graph, *E. coli* cells were labeled with radioactive thymidine for a short pulse (10 s), followed by a chase with an excess of nonradioactive thymidine. The DNA was extracted and centrifuged in alkaline sucrose gradients (under high pH conditions the DNA denatures). Explain what these data imply, and interpret these results in light of our current model for DNA replication.

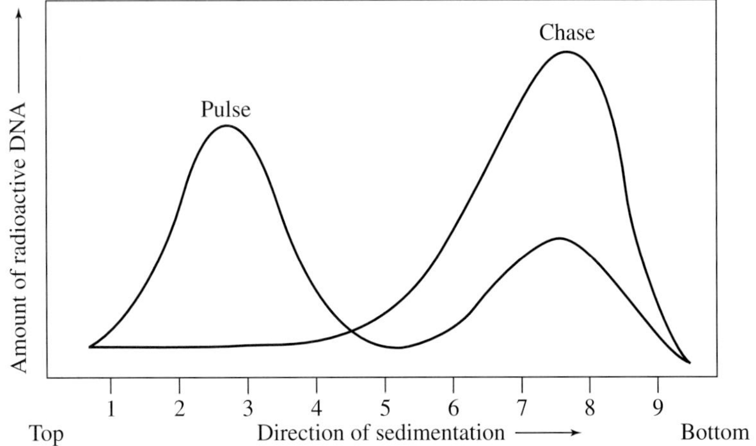

6. Almost all enzymes that make DNA in a template-dependent fashion require a primer. How does the use of a primer increase the fidelity of DNA synthesis, and why is this primer usually RNA?
7. Which of the DNA molecules illustrated can stimulate incorporation of dNTPs by the Klenow fragment of *E. coli* DNA polymerase I?

(a)

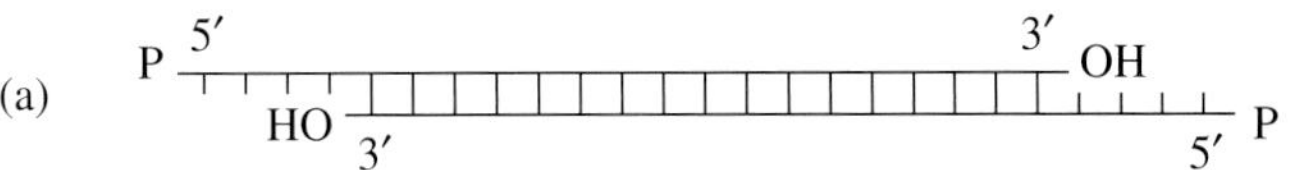

(b)

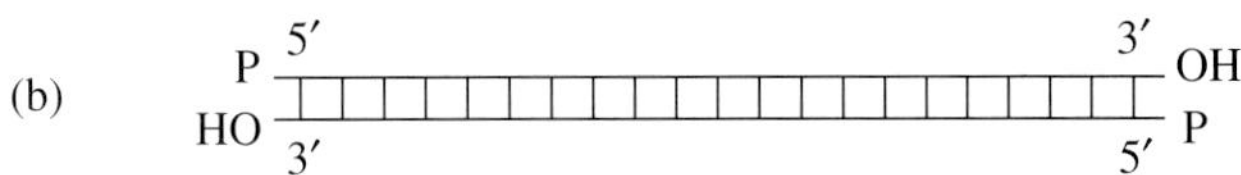

(c)

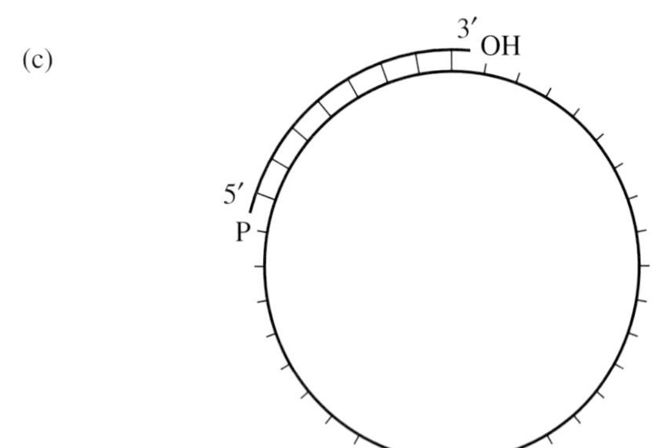

(d)

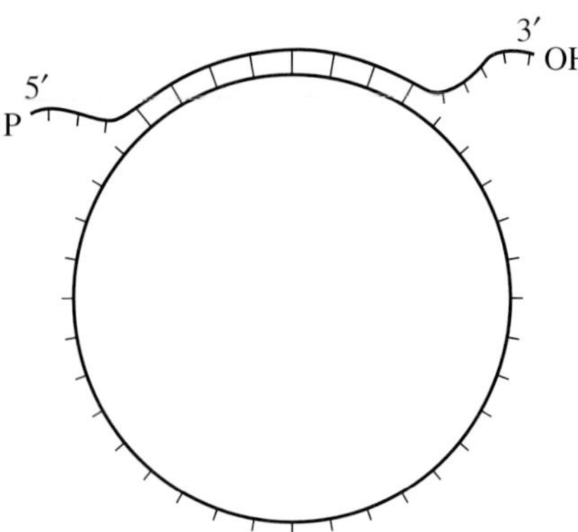

8. Even with its proofreading activity, *E. coli* DNA polymerase III still exhibits a measurable rate of nucleotide misincorporation (about one mistake per 10^{10} nucleotides incorporated). Mutants of *E. coli* DNA polymerase III can be isolated that have a lower than normal rate of misincorporation. Why might such mutants, which can be said to have hyperaccurate DNA replication, be selected against during the course of evolution?
9. *E. coli* has a genomic complexity of about 4×10^6 base pairs, and each replication fork can move at a rate of about 10^3 bp/s (base pairs per second). How long does it take to replicate the *E. coli* chromosome? With an ample carbon source and ideal growth conditions, cells of *E. coli* can divide in about 20 min. How can this shorter division time occur if the rate of fork migration remains constant at 10^3 bp/s?
10. Cairns and de Lucia isolated a mutant strain of *E. coli* that had only about 1% of the DNA polymerase I activity found in wild-type cells, yet the strain replicated its DNA at a normal rate. Explain how this discovery was important in understanding the role of the different DNA polymerases in replication and repair.
11. Would it be theoretically possible for a DNA polymerase to exist that could add nucleotides to the growing DNA chain in a 3′ to 5′ direction? Why would such an enzyme be unable to proofread the DNA it has synthesized? (*Hint:* Think of the source of the pyrophosphate that is released as each nucleotide is incorporated.)
12. Why does the reaction catalyzed by the *dnaB* gene product require ATP hydrolysis? Does this protein alter the linking number of DNA? Explain.
13. Predict whether DNA synthesis would stop immediately or whether a longer time at the nonpermissive temperature would be required for complete cessation of DNA synthesis to occur in the following temperature-sensitive mutants (refer to tables 31.1 and 31.2):
    (a) *dnaA*^{ts}
    (b) *dnaE*^{ts}
    (c) *dnaG*^{ts}
    (d) *dnaN*^{ts}
    (e) *lig*^{ts}
14. List the steps in DNA replication that require ATP, and explain why these steps exhibit an energy requirement. Are there any steps that require energy but do not use ATP for this purpose?
15. Compare and contrast the excision repair and photoreactivation mechanisms for correction of ultraviolet-induced thymine

dimers. How would you be able to discriminate between these mechanisms by using a [^{3}H]thymidine label?

16. Normal human fibroblasts are grown in culture and then exposed to UV light. A short time later, the DNA is extracted and applied to an alkaline sucrose gradient, and the data in graph (*a*) are observed. Another sample of cells is also exposed to UV light, but about 12 h are allowed to pass before the DNA is extracted and applied to an alkaline sucrose gradient (graph *b*). Explain these results on the basis of what you know about DNA repair. (*Hint:* See fig. 31.26.) Another sample of fibroblast cells, isolated from a patient with xeroderma pigmentosum (a disease resulting from the inability to repair DNA damage caused by UV light), was exposed to UV light and then applied to a gradient after a short time. Would you expect the data to resemble those in graph *a* or graph *b*? Why?

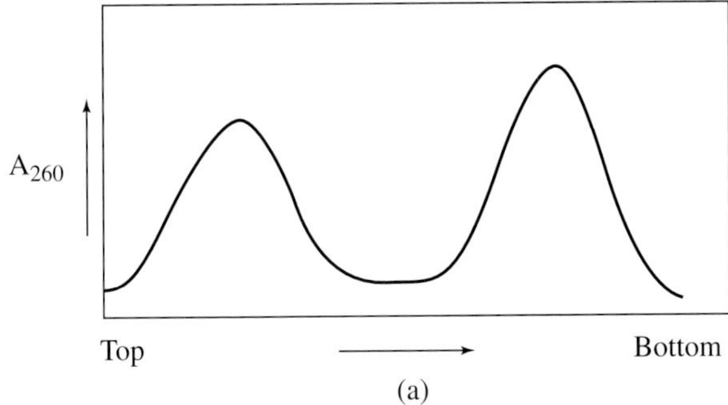

(a)

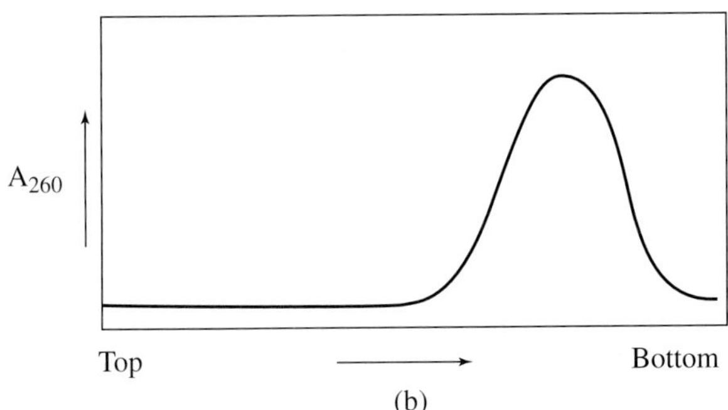

(b)

17. Draw the products of oxidative deamination of adenine, guanine, and cytosine. Propose a general scheme for the correction of this type of DNA damage. Why is the uracil-DNA glycohydrolase very important in DNA repair? Why is thymidine-containing DNA and not uracil-containing DNA the product of modern evolution?
18. Explain how genetic exchange between sister chromatids can be inferred from autoradiograph results such as those in figure 31.3b.
19. What alteration in recA or lexA proteins would explain the phenotypes of the following mutants?
    (a) *recA441:* Dominant, SOS response occurs at 42°C even in the absence of DNA damage
    (b) *recA430:* Recessive, recombination-proficient, but no SOS response
    (c) *lexA*(Ts): Recessive, SOS response occurs at 42°C even in the absence of DNA damage
20. Outline in general terms how a retrovirus replicates its RNA. What unique viral protein is used in this replication process? What serves as the primer for viral replication?
21. A telomerase RNA contains the sequence CAACCCCAA, which is thought to serve as the template in the replication of telomeres with the sequence $(TTGGGG)_n$. What do you predict would happen to the telomeres if the telomerase RNA sequence shown above were mutated to CGACCCCAA?
22. If you were to design an artificial chromosome that could replicate in yeast cells, what would be the minimal sequences required, and why?

C H A P T E R

32

DNA Manipulation and Its Applications

DNA from different sources can be knitted together and thereafter integrated into the genome of living cells.

A broad array of techniques has made it possible to investigate the fine structure of DNA as well as to redesign existing genes. The impact of these techniques cannot be overestimated. It is possible to map and isolate genes from complex eukaryotes—feats that could never have been accomplished by classical genetic methods. Genes can be redesigned to achieve a variety of practical goals. Newly designed genes may be inserted into the same species from which the original unmodified genes came or into other species.

In this chapter we first consider DNA sequencing. Then we explore the different approaches for amplifying and isolating specific genes or gene segments. Following this, we examine the methods currently available for restructuring existing DNA sequences. Then we look at some of the major advances in our understanding of gene structure and location that have resulted from the new technology. This part of the discussion focuses on two examples: the mapping of the human globin gene family and the mapping of the gene responsible for the genetically inherited disease cystic fibrosis. Finally, we consider possible uses of recombinant technology in the production of genetic medicines.

Sequencing DNA

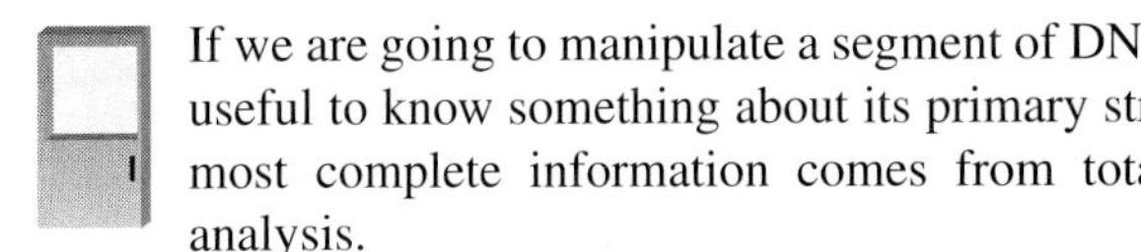

If we are going to manipulate a segment of DNA, it is most useful to know something about its primary structure. The most complete information comes from total sequence analysis.

Initial efforts at sequencing nucleic acids were confined to RNA molecules that could be readily isolated in pure form. The first sequence to be determined was that for tyrosine tRNA from yeast. From roughly 100 pounds of yeast, Robert Holley was able to isolate enough of the tyrosine tRNA to carry out a sequence analysis. This historically significant effort required a number of enzymes and chromatographic techniques. We shall not elaborate on this accomplishment because sequencing of RNA is no longer done this way. In fact, the sequencing of RNA is usually done by sequencing of "cDNA," which results from the reverse transcription of RNA into DNA.

Two quite different methods have been developed for sequencing DNA. One method, developed by Walter Gilbert and Alan Maxam and involving cleavage of preexisting DNA, uses a chemical approach. A second method, involving premature termination of newly synthesized DNA, uses an enzymatic approach that was developed by Fred Sanger.

For pure sequencing, Sanger's is the method of choice. It employs chain-terminating dideoxynucleoside triphosphates to produce a continuous series of fragments in reactions catalyzed by DNA polymerase. Dideoxynucleoside triphosphates (ddXTPs) resemble deoxynucleoside triphosphates except that they lack a 3′-OH group. They can add to a growing chain during polymerization, but they cannot be added onto, and as a result they act as chain terminators.

The DNA being sequenced is mixed with a suitable primer, radioactive dXTPs, DNA polymerase I (PolI), and a small amount of one ddXTP. The primers determine where DNA synthesis starts, and the ddXTP determines the base type where elongation stops. The products of four separate reaction mixtures, each differing only by the ddXTP it contains, are analyzed in figure 32.1. As depicted, reaction 1, using ddATP, contains all fragments with an A terminus; reaction 2, using ddCTP, contains all C terminations; and so on. After the newly synthesized oligonucleotides are separated from the template by denaturation, they are fractionated by electrophoresis for a limited time on polyacrylamide gels. The positions of the fragments on the gel are detected by autoradiography. The sequence is read directly from the composite autoradiogram, starting with the fastest-moving (smallest) band at the bottom of the gel and moving up. If the first band is in reaction 3, it is a G residue; the next band up, appearing from reaction 4, would be T; and so on. Up to 800 residues can be read from a single gel.

Methods for Amplification of Select Segments of DNA

Cellular genomes are very large; even *E. coli* contains more than a million base pairs, and eukaryotic genomes frequently contain a billion or more base pairs in one complete genome. Because of their large size it is impractical to fractionate the cellular genome and expect to obtain enough of a particular DNA segment for sequencing or other investigations. Two methods of amplifying defined segments of DNA have been developed. The first of these takes the desired segment and amplifies it in vitro with DNA PolI using DNA primers that bind to the ends of the region of interest. This method involves repeated cycles of synthesis and is appropriately named the polymerase chain reaction (PCR) method. The second method inserts the DNA segment of interest into a plasmid or virus that can be amplified in vivo. We discuss both of these methods because they are both useful in many ways.

Amplification by the Polymerase Chain Reaction

PCR entails enzymatic amplification of specific DNA sequences using two oligonucleotide primers that flank the DNA segment to be amplified (fig. 32.2). The primers must complement opposite strands so that after annealing, their 3′ ends in effect face each other (see fig. 32.2*b*).

The PCR procedure has three steps, which are usually repeated many times in a cyclical manner:

1. Denaturation of the original double-stranded DNA sample at high temperature
2. Annealing of the oligonucleotide primers to the DNA template at low temperature (37°C)
3. Extension of the primers using DNA polymerase

These steps are illustrated in figure 32.2. Each set of three steps comprises a cycle. The extension products of one primer provide a template for the other primer in a subsequent cycle, so each successive cycle essentially doubles the amount of DNA. This results in the exponential accumulation of the specific target fragment by approximately 2^n, where n is the number of cycles. The specific target fragment is also referred to as the "short product" and is defined as the region between the 5′ ends of the extension primers. Each primer is physically incorporated into one strand of the short product.

Other products are also synthesized during the succession of cycles, such as the "long product" of indefinite length, which is derived from the template molecules. However, the amount of long product increases only arithmetically during each cycle of the amplification process because the quantity of original template remains constant.

At the end of the PCR process the short product is so overwhelmingly abundant compared with the long product that its purification is usually not required.

DNA Cloning

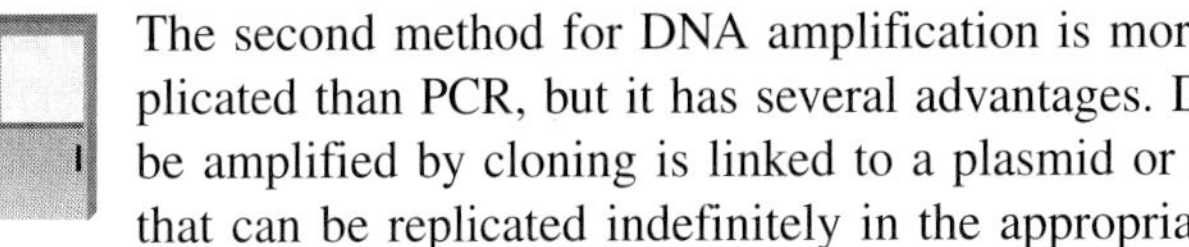

The second method for DNA amplification is more complicated than PCR, but it has several advantages. DNA to be amplified by cloning is linked to a plasmid or a virus that can be replicated indefinitely in the appropriate host

Figure 32.1

The Sanger dideoxynucleoside method of sequencing DNA. (*a*) A suitable template is chosen, and the primer is chosen so that DNA synthesis begins at the point of interest. The primer is radioactively labeled. In addition to the template–primer complex the reaction mixture contains all four radioactive deoxyribonucleoside triphosphates and small amounts of a single dideoxynucleoside triphosphate. The dideoxy compound serves as a chain terminator. (*b*) After synthesis in the presence of DNA polymerase I the products of the reaction mixture are separated by gel electrophoresis and analyzed by autoradiography. The gel is run under denaturing conditions in warm urea so that single-stranded fragments separate strictly according to size. For a given dideoxy compound all fragments terminating with that particular base should give rise to bands on the gel. The interpretation of the gel pattern is given in (*c*). The smallest labeled fragment moves the fastest and appears at the bottom of the gel. (*d*) A typical sequencing film. The sequence begins CAAAAAACGG. (*Source:* Reproduced with the permission of Life Technologies, Inc., Gaithersburg, Md.)

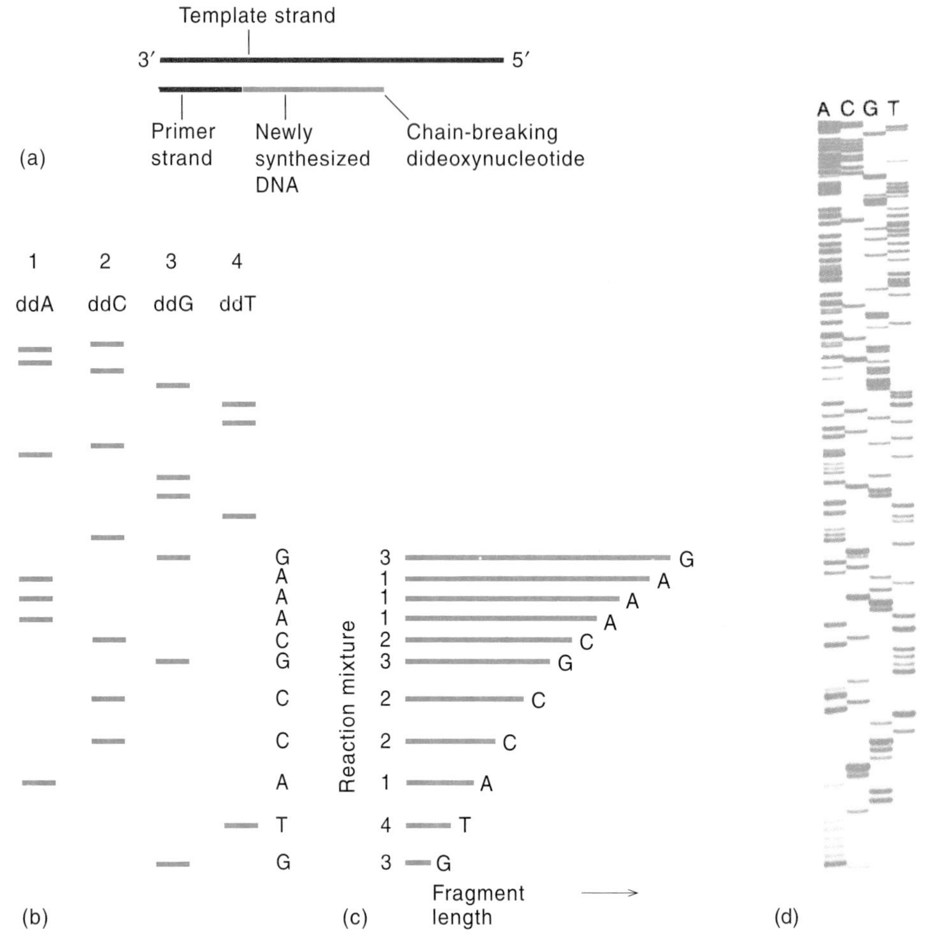

cell. After amplification the DNA of interest can be cut from the plasmid or virus and reisolated by gel electrophoresis. Cloning is not only useful for amplifying a segment of DNA, it can be adapted to the isolation of a DNA segment of interest from a large mixture such as is obtained from the isolation of the entire genome.

Restriction Enzymes Are Used to Cut DNA into Well-Defined Fragments

Most of the enzymes that are absolutely essential for cloning were discussed in the previous chapter. The most important enzymes that have not been discussed yet are the restriction enzymes. Systematic cleavages of duplex DNA at specific sites require restriction enzymes. Each species of bacteria harbors a unique restriction enzyme, and hundreds of restriction enzymes with different specificities have been isolated, giving researchers a great deal of choice as to how and where DNA is cut. Some of the most commonly used restriction enzymes and their recognition sites are indicated in table 32.1. Most of these enzymes recognize a sequence of either four or six contiguous base pairs. The cleavage sites are situated so that a blunt-ended or staggered-ended DNA results from the cleavage reaction. As a rule the recognition sites are located on an axis of symmetry so that the freshly cleaved segments have identical structures at their ends.

A viral genome cleaved exhaustively with a particular restriction enzyme usually yields several fragments. Some restriction enzyme cleavage sites for the 5,300-bp (5.3-kb) SV40 virus genome are shown in figure 32.3. The duplex fragments obtained after cleavage can be separated according to size by gel electrophoresis. Nondenaturing conditions are used so that the duplex strands stay together. The larger a fragment is, the more slowly it migrates on the gel. After electrophoresis for a time sufficient to separate the fragments, the gel is stained with a fluorescent dye

Figure 32.2

Steps in the polymerase chain reaction (PCR). The DNA to be amplified is denatured and annealed with two oligonucleotides that flank the region of interest. These oligonucleotides (or primers) are extended. Extension continues to the ends of the DNA strands. The products are again denatured and annealed to primers for a second round of extension. This process of denaturation, annealing, and primer extension is repeated many times. The primary product of the reaction is duplex DNA, bounded by the sequences of the primers. (Reprinted with permission from J. L. Marx, Multiplying genes by leaps and bounds, *Science* 240:1408–1410, June 10, 1988. Copyright 1988 American Association for the Advancement of Science.)

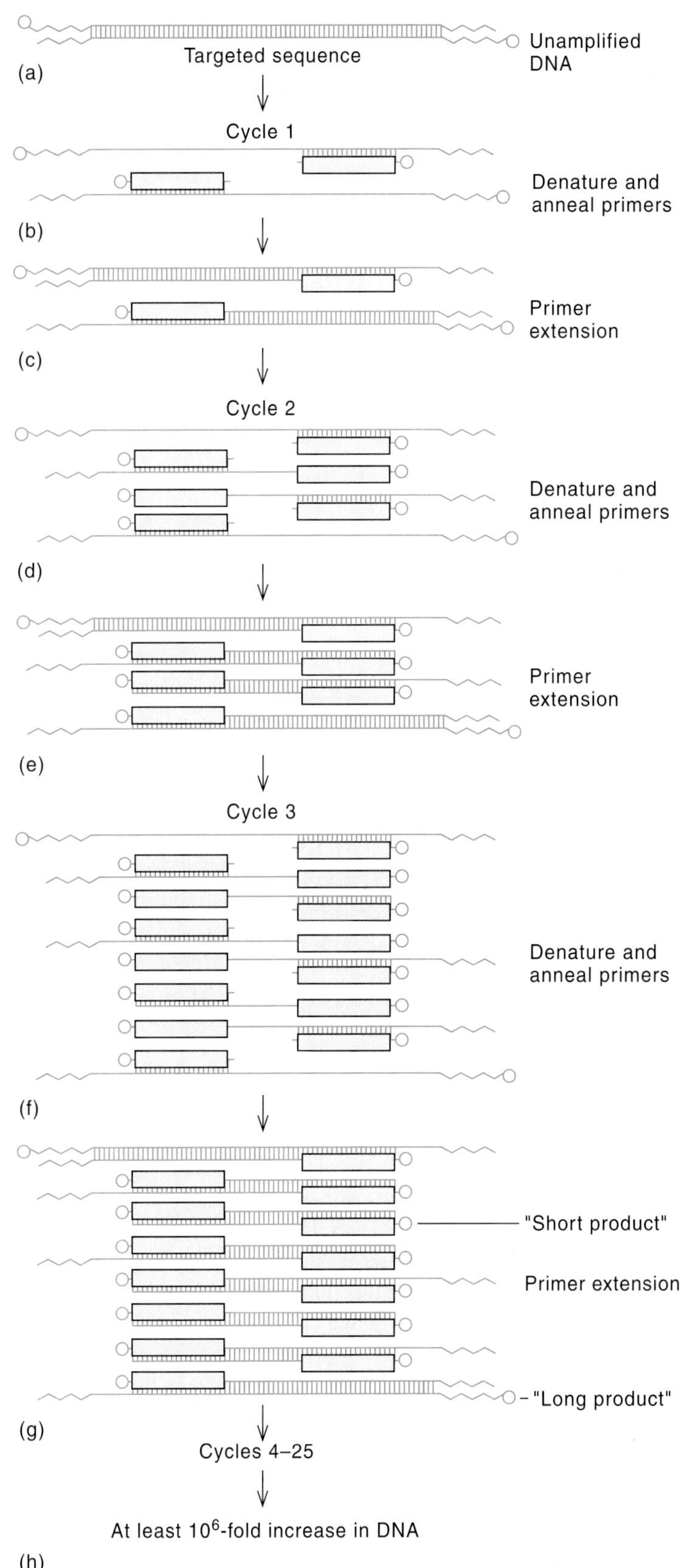

such as ethidium bromide and viewed under long-wavelength ultraviolet light (long-wavelength UV is used because it does not damage the DNA). Individual fragments may be extracted from the gel for sequencing, PCR amplification, or cloning (described later on).

The problem of determining how a set of restriction fragments are normally connected is resolved by determining the sequences by a second set of fragments cut with a different restriction enzyme. The overlapping information obtained from the two sets of fragments permits a determination of the complete sequence of the intact genome. The strategy of sequencing overlapping fragments is identical to that used in primary structure determination of proteins (see chapter 4).

Plasmids Are Used as Vectors to Clone Small Pieces of DNA

In a simple procedure for "DNA cloning," an autonomously replicating plasmid and insert DNA are cut with a restriction enzyme, and then the pieces are annealed and covalently joined by the action of DNA ligase. The resulting recombinant molecules are then transfected into *E. coli,* where they replicate. When plasmid vectors are used, a population of permeabilized cells is bathed in the plasmid DNA containing the inserted DNA. Because only a small number of cells become transfected by this procedure, a way to select cells that carry the desired hybrid plasmids is needed.

A particularly useful plasmid vector for selecting transfected cells called pBR322 is itself a hybrid plasmid (fig. 32.4). This plasmid contains two genes, amp^r and tet^r, which confer resistance to penicillin and tetracycline, respectively. *Pst*I restriction fragments of foreign DNA may be inserted into the unique *Pst*I restriction site on pBR322 (see fig. 32.4). This is done by digesting pBR322 with *Pst*I, mixing with the restriction fragments to be cloned at low temperatures to permit annealing to take place between the two DNAs, and finally ligating the annealed fragments with DNA ligase. The product contains some of the original pBR322 and some pBR322 with inserted foreign DNA. When this mixture is used in transfection, most cells are not transfected, some are transfected with pBR322, and some are transfected with the desired hybrid plasmid. The three types of cells may be readily distinguished by their drug-resistant properties. Normal cells do not grow in the presence of tetracycline or penicillin. Transfected cells with the DNA inserted in the plasmid are tetracycline-resistant but penicillin-sensitive, because the insert has disrupted the amp^r gene. Cells containing the desired plasmids can be dis-

Table 32.1
Recognition Sequences and Cutting Sites of Selected Restriction Enzymes

Enzyme	Recognition Sequences	Enzyme	Recognition Sequences
AluI	↓ AGCT TCGA ↑	HpaII	↓ CCGG GGCC ↑
BamHI	↓ GGATCC CCTAGG ↑	KpaI	↓ GGTACC CCATGG ↑
BglII	↓ AGATCT TCTAGA ↑	MboI	↓ GATC CTAG ↑
ClaI	↓ ATCGAT TAGCTA ↑	PstI	↓ CTGCAG GACGTC ↑
EcoRI	↓ GAATTC CTTAAG ↑	PvuI	↓ CGATCG GCTAGC ↑
HaeII	↓ GGCC CCGG ↑	SalI	↓ GTCGAC CAGCTG ↑
HindII	↓ GTPyPuAC CAPuPyTG ↑	SmaI	↓ CCCGGG GGGCCC ↑
HindIII	↓ AAGCTT TTCGAA ↑	XmaI	↓ CCCGGG GGGCCC ↑

tinguished from those containing pBR322 by "replica plating" (fig. 32.5). The first step in using this approach is to spread a large population of treated bacteria on an agarose plate containing growth medium. Within 12 h a seemingly homogeneous "lawn" of cells develops on the surface of the agarose. Actually, the lawn results from the growth of many microcolonies to the point of confluency. At this point a piece of velvet is lightly pressed against the surface of the plate, and this impression is transferred to other agarose plates containing growth medium with tetracycline or penicillin plus tetracycline. Only the transfected cells produce colonies on the plates containing the antibiotics, and because of their small number, each of these gives rise to readily detectable clones. The clones present on the tetracycline-containing plates, which are missing on the penicillin-plus-tetracycline plates, most likely contain the desired hybrid plasmids (see fig. 32.5). These clones are usually plucked from the tetracycline plates and retested to eliminate any uncertainty about the original drug testing. Once this is confirmed, the appropriate plasmid-containing cells are grown in liquid culture. After a moderate density of growth is achieved, the plasmid DNA is selectively amplified by overnight growth. Plasmid DNA replication continues for several hours, until each cell contains 1,000 to 2,000 copies of the small circular plasmid DNA. This DNA is readily separable from the host DNA and can be characterized by its rate of migration on gel electrophoresis (fig. 32.6) or other more specific tests to determine whether it contains the inserted DNA sequence. If desired, the inserted sequence may be removed from the plasmid vector by digestion with *Pst*I, the restriction enzyme used in the initial construction of the hybrid plasmid. The cleaved fragments can be separated readily by gel electrophoresis. Many plasmid vectors, other than pBR322, have specific advantages for other purposes.

Bacteriophage λ Vectors Are Useful for Cloning DNA Segments of up to 24 kb

Bacteriophage λ possesses a number of advantages as a cloning vector. DNA fragments as large as 24 kb can be propagated by using such vectors. The primary pool of clones can be amplified by limited phage growth as plaques, and the entire collection of phage

Figure 32.3

Cleavage map of the SV40 genome. The zero point of the map is the unique *Eco*R1 site. For clarity the circular genome is shown opened at the R1 site, and the cleavage sites (and resulting fragments) for each restriction enzyme are indicated on a separate line.

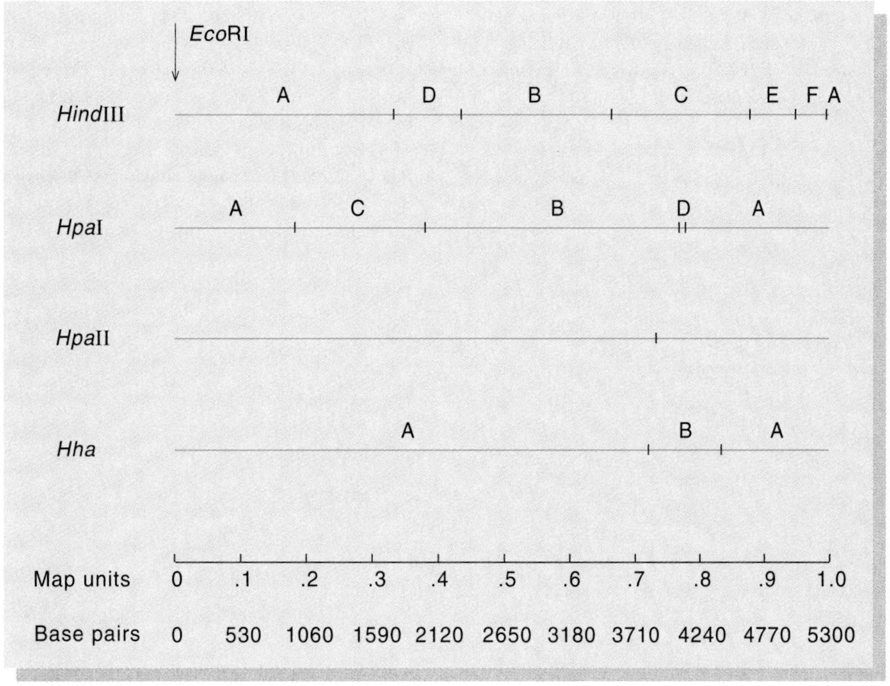

clones (recognized as clear plaques) can be stored for long periods in a small volume (fig. 32.7).

Because it does not accommodate molecules of DNA that are much longer than the viral genome, the use of λ as a vector for cloning substantial DNA fragments requires the removal of a significant portion of the viral DNA beforehand. Fortunately, the central third of the genome contains genes that are not essential for phage production and can therefore be deleted without loss of infectivity.

Bacteriophage λ vectors that accommodate foreign DNA fragments generated by a variety of restriction endonucleases have been constructed. The recombinant DNA molecules that incorporate some of these vectors can be introduced directly into *E. coli* by transfection. Alternatively, recombinant DNA molecules can be packaged into phage particles and subsequently infected into suitable host cells.

Cosmids Are Used to Clone Segments of DNA between 25 kb and 50 kb in Length

Although plasmids and bacteriophage λ are both highly useful vectors, the size of the DNA fragments that can be cloned in them is limited. With plasmids the larger the fragment of foreign DNA inserted, the lower the efficiency of ligation and transfection, making the cloning of DNA fragments larger than 15 kp experimentally difficult. In λ vectors the length of the nonessential region of λ DNA limits fragment size to 24 kb or less. Also, the original λ vectors do not allow propagation of viable bacterial cells that carry the inserted DNA fragment; the insert is propagated as part of a virus that lyses the cell.

Cosmids were developed as vectors for cloning very large DNA fragments (box 32A). The first part of their name, "cos," comes from the fact that cosmids contain the cohesive ends, or *cos,* sites of normal λ. These ends are essential for packaging the DNA into λ phage heads. The last part of their name, "mid," indicates that cosmids carry a plasmid origin of replication like the one found in the pBR322 plasmid. Such cosmids can be used for cloning in the same way as any other plasmid vector. However, because cosmids also contain the *cos* sites, cosmid DNA along with an inserted DNA fragment can be packaged as a λ phage. The result after packaging is a defective but nevertheless infectious phage particle. Once the cosmid and the inserted DNA fragment are introduced by infection into a λ-sensitive cell, the plasmid replicates. Since cosmids lack the entire bacteriophage genome except for the region adjacent to the *cos* sites, these vectors can propagate exogenously derived DNA fragments of up to 40–50 kb in length.

Yeast artificial chromosomes have been developed that can clone hundreds of kilobases of DNA. These are discussed later when we consider cloning in yeast.

Shuttle Vectors Can Be Cloned into Cells of Different Species

Vectors that include replication systems derived from more than one host species are known as shuttle vectors. Such vectors commonly include a replication system able to function in *E. coli* and one that works in a second host, which may be bacterial or eukaryotic. Initial cloning and amplification of the DNA segment to

Figure 32.4

(*a*) Structure of the pBR322 plasmid and (*b*) construction of a hybrid plasmid containing the pBR322 vector and a segment of foreign DNA. For pBR322 the unique sites for various restriction enzymes are indicated. Also indicated are the locations of the tetracycline (*tet*^r^) and the ampicillin (*amp*^r^) resistance genes and the origin for DNA replication. The hybrid plasmid is constructed by treating the plasmid and the foreign DNA with the *Pst*I restriction enzyme and mixing the two DNAs together in the presence of DNA ligase.

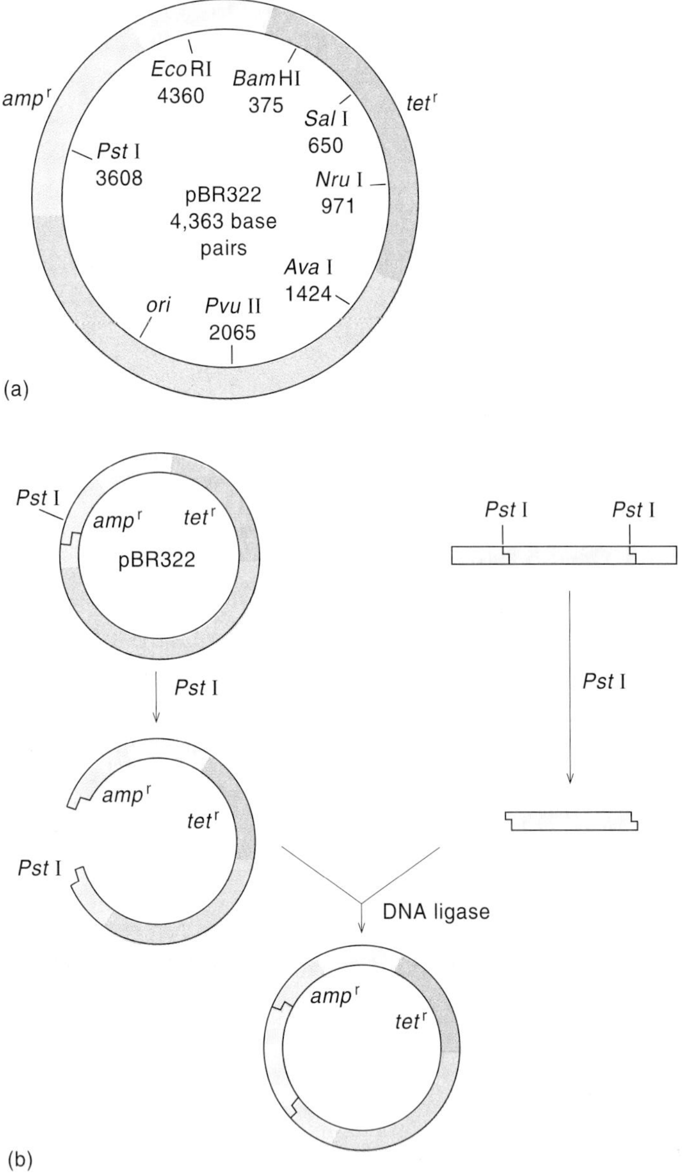

Figure 32.5

Application of the replica-plating technique to the detection of hybrid plasmid-containing cells. About 10^7 bacteria are spread on a plate. After overnight growth the plate (master plate) appears as a uniform "lawn" of bacteria, but in reality it consists of very small colonies that have merged to give a uniform appearance. A piece of velvet is lightly pressed against the surface of this lawn, and some cells stick to the velvet. Several essentially identical impressions of the lawn are transferred to fresh plates, which contain normal medium or normal medium supplemented with antibiotics. The results on the replica plates after overnight growth are indicated. The plate in normal medium again gives rise to a lawn of cells as virtually all of the cells transferred grow into colonies. The plates containing antibiotics give rise to only a few colonies, each of which is derived from a single cell that carries the plasmid-conferred drug resistance(s).

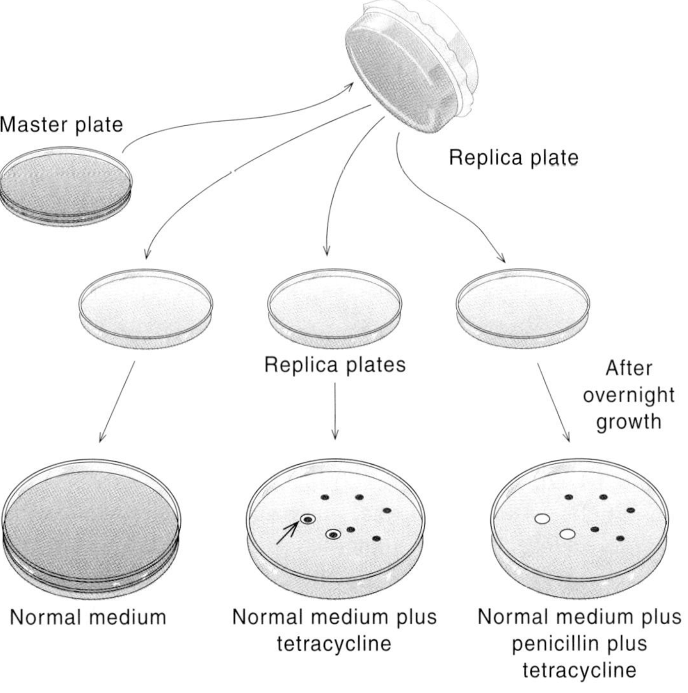

be studied are often carried out in *E. coli* because it is easier to make large quantities when culturing in *E. coli.* The recombinant DNA molecule, consisting of the "bifunctional vector" plus the cloned segment of DNA, is then introduced into the second host, where the purpose is usually to measure the expression of the genes carried by the vector. Shuttle vectors that can replicate in both *E. coli* and yeast are the most common (see fig. 32.10).

Constructing a "Library"

Cloning can involve a single vector-linked DNA fragment or a collection of independently isolated vector-linked DNA fragments derived from a single organism. Such a collection is termed a "library" and may serve as the source of well-defined sequences from a given organism. Each clone of a library harbors a particular DNA segment from the desired organism. Within the entire library a sequence may be repeated, but other sequences may be missing. The ideal library, which can only be approached, represents all of the sequences with the smallest possible number of clones.

A library from the same cell or organism can be prepared in two ways. The genome may be fragmented and ligated to the appropriate vector to produce a genomic DNA library. An alternative approach is to construct a cDNA library in which the DNA fragments to be cloned are obtained by reverse transcription from the cellular RNA. Each of these libraries has advantages and dis-

Figure 32.6

Electrophoretogram of restriction enzyme digests of pBR322 and pBR322 with a DNA insert at the *Pst*I site. The insert is assumed to have no internal *Bam*HI restriction sites. In channels A and B the pBR322 is predigested with *Pst*I and *Bam*HI, respectively. The resulting DNA migrates with the same mobility because the plasmid has one site for each of these enzymes and therefore has the same molecular weight. In C and D the hybrid plasmid containing a DNA insert is treated with *Bam*HI and *Pst*I, respectively. In C the hybrid plasmid has been linearized by one cut at the *Bam*HI site in the tet^r gene. It runs more slowly than the pBR322 because it is larger owing to the DNA insert. In D the plasmid cuts at two *Pst*I sites located between the pBR322 sequences and the insert sequences. Consequently, one segment migrates at the rate of a linearized pBR322 plasmid. The other segment, also linearized, migrates at a rate characteristic of the size of the DNA insert. The electrophoresis is run from left to right; fragments are stained with ethidium bromide and photographed with UV light.

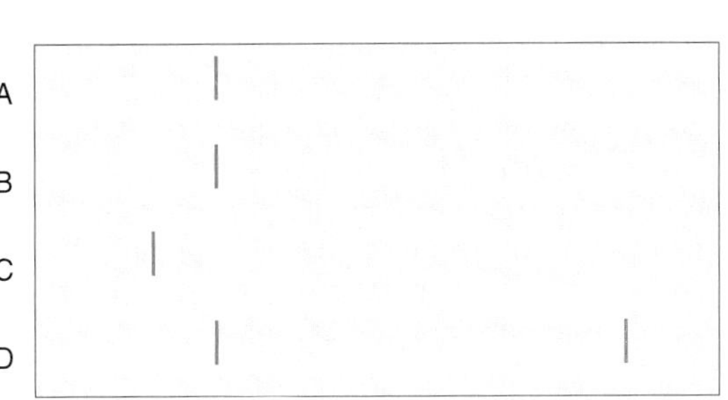

Figure 32.7

The nutrient agar plate contains a continuous lawn of *E. coli* bacteria except for circular clearings that represent phage plaques. Each plaque was originally derived from a single phase infecting a single *E. coli* bacterium. After infection the phage multiplies, ultimately producing about 100 mature viruses. The phages also produce an enzyme that causes the harboring cell to lyse. When this happens the phages are released, and each of them infects a neighboring cell and goes through the same infectious cycle. The process continues. Each cycle takes about 30 min. Eventually, a visible clearing can be seen on the plate.

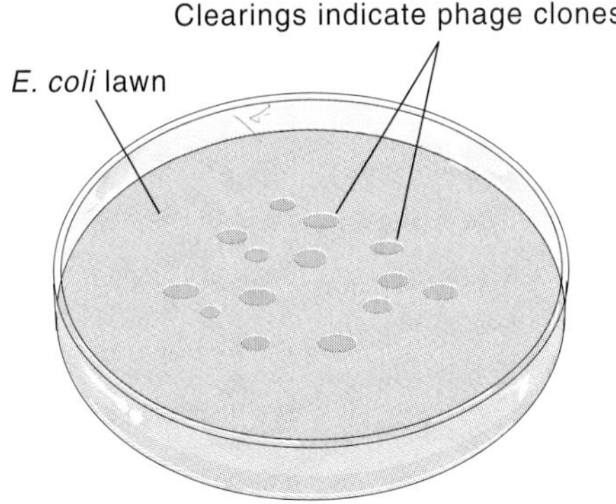

advantages, and for a specific purpose, one library is usually preferred over the other.

The vast majority of DNAs within a library are uncharacterized. As a rule, the task of finding the desired genes or sequences within a library greatly exceeds the task of constructing the library.

A Genomic DNA Library Contains Clones with Different Genomic Fragments

A major concern in constructing a genomic DNA library is to maximize the probability that all segments of the genome are represented. If the genomic DNA is prepared by cutting with a restriction enzyme, an added concern is that the enzyme cleaves genes of interest at one or more sites. To increase the likelihood of isolating desired genes in one piece, different restriction enzymes can be used on different parallel preparations. But even if the genes of interest are not cut by the enzyme(s) chosen, the DNA fragments produced may be inconveniently small to work with. An enzyme that recognizes a sequence of six bases (a six-cutter) gives an average fragment size of 4,096 bp ($(\frac{1}{4})^6 = 1/4{,}096$), which is a reasonable size for making a plasmid library but much smaller than the desirable size for cloning in λ or a cosmid vector. Therefore when large, randomly generated fragments are desired, the method of choice usually entails making an incomplete digest with a four-cutter restriction enzyme, which produces overlapping ends that can be readily cloned into the chosen vector as described earlier. The extent of digestion is controlled so that cleavage occurs at only some of the restriction enzyme recognition sites and the average size of the fragments produced is in the desired range. The conditions used thus depend on whether the product is going to be cloned in a plasmid, a λ phage, or a cosmid vector. Table 32.2 gives the minimum number of clones (i.e., the size of the library) required to fully represent the entire genome in a genomic DNA library, as a function of the average size of the cloned fragments and the size of the genome. Since DNA fragments in a population are cloned on a random basis, the chance of finding a given single-copy gene in a library of the indicated size is 50%. A clone bank should be 3 to 10 times the minimum size to give a high probability that a particular segment is represented.

A cDNA Library Contains Clones Reflecting the mRNA Sequences

A cDNA library consists of a collection of clones that contain DNA copies of the cellular or organismic RNA. If the RNA is ob-

32A
BOX

Construction of a Cosmid

The pJB8 vector is a 5.1-kb plasmid containing a λ cos site, an amp^r selectable marker, an origin of replication, and several possible cloning sites. To construct it, first one of the staggered-end ligation sites of the plasmid was opened with the appropriate restriction enzyme (step 1). In this case the *Bam*HI site was chosen. The linearized vector was then treated with alkaline phosphatase (step 2) to prevent recircularization of the vector in the subsequent ligation. The remaining procedures were dictated by the goal of this protocol, which was to produce very large cloned segments of DNA. Thus ligation had to yield large DNA fragments that were produced by random cutting (and thus were representative of the entire genome) and that could be inserted into the *Bam*HI site of the cosmid vector. On a random basis any very large DNA fragment is likely to be cut several times by even a six-cutter restriction enzyme under conditions of complete digestion. In order to avoid this problem and to obtain maximum randomness of fragments of the appropriate size, the *Mbo*I restriction enzyme was used under conditions of very limited digestion. This enzyme recognizes the four-base sequence of GATC, whereas *Bam*HI recognizes a six-base sequence, GGATCC. The two enzymes produce identical overlapping fragments (see fig. 1, upper right). Thus the eukaryotic fragments should ligate efficiently with the *Bam*HI restricted cosmid vector (step 3). Because limited digestion conditions were used, only some of the *Mbo*I (and *Bam*HI) sites were cleaved, and these varied from molecule to molecule on a random basis. Thus some molecules in the population were likely to contain an uncleaved enzyme recognition site at any given position. After ligation the resulting concatamers were packaged in vitro into λ particles and introduced into a suitable *E. coli* strain (step 4). Transformants were then selected with ampicillin.

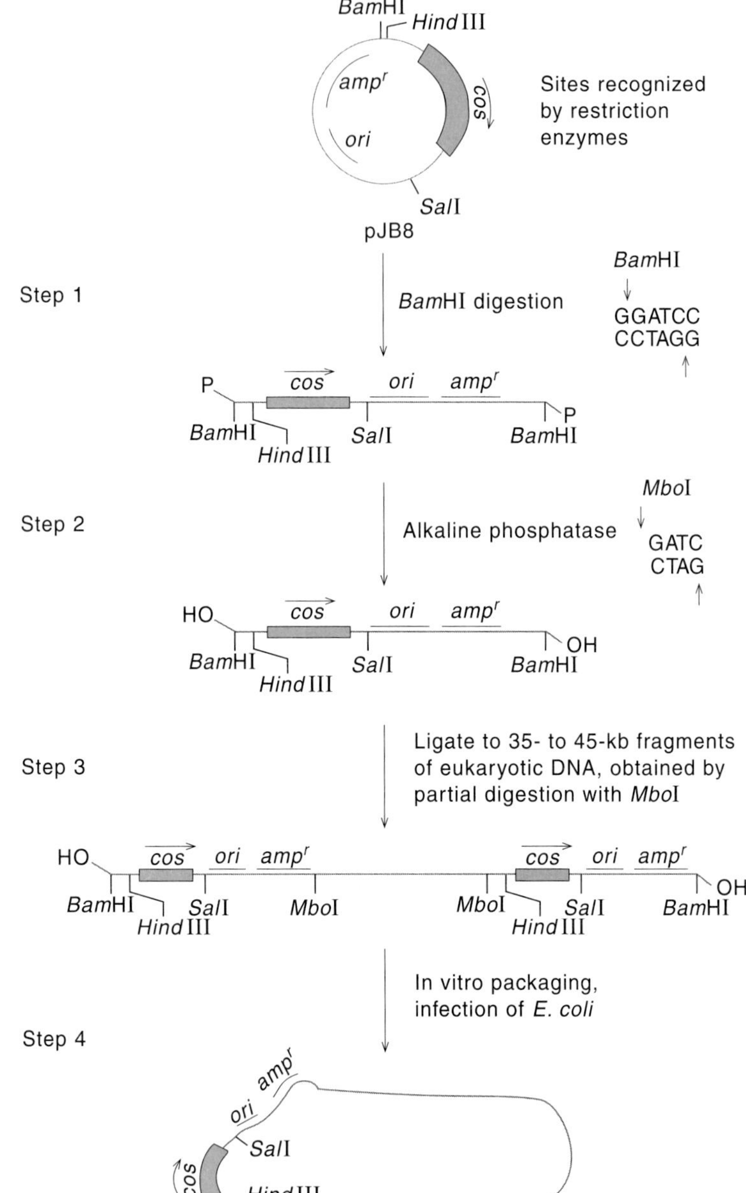

Figure 1

A plasmid containing a λ *cos* site is linearized by treatment with the *Bam*HI restriction enzyme (step 1). The linear structure is treated with alkaline phosphatase to remove the 5′ terminal phosphates (step 2). Eukaryotic restriction fragments are ligated to the linearized plasmid in the presence of a large excess of the latter (step 3). The concatamer shown, which contains two *cos* sites, packages sufficiently into a λ phage that can be introduced by infection into an *E. coli* cell. The transfected structure readily circularizes (step 4). (*Source:* From Geoffrey Zubay, *Genetics.* Copyright © 1987 Benjamin/Cummings Publishing Company Inc., Menlo Park, Calif. Reprinted by permission of the author.)

tained from a differentiated multicellular organism, then the library varies in composition according to the type of cell used as the RNA source and to the physiological state of the cell. This variation is a reflection of the relative abundances of particular mRNAs made by different cell types. If a cDNA species corresponding to a particular gene product is desired, it is often possible to select a cell type that is suspected to synthesize a large amount of the corresponding mRNA or mRNA-related protein. Thus pituitary cells can be used if cDNA encoding growth hormone is desired, whereas liver cells can be used if a serum albumin cDNA is the goal. The mRNAs present in low amounts clearly require the screening of a larger library than the mRNAs present in medium or high abundance.

Once the crude mRNA fraction has been isolated from

Size of Cloned DNA Fragment (bp)	Genome Size (bp) 2 × 10^6 (e.g., bacteria)	2 × 10^7 (e.g., fungi)	3 × 10^9 (e.g., mammals)
5 × 10^3	400	4,000	600,000
10 × 10^3	200	2,000	300,000
20 × 10^3	100	1,000	150,000
40 × 10^3	50	500	75,000

Table 32.2

Theoretical Number of Clones Required to Fully Represent the Entire Genome of Various Organisms

the chosen cells or tissue, it is converted to duplex DNA molecules with the help of reverse transcriptase. This duplex DNA does not have "sticky ends" for insertion into a vector. For this purpose, DNA linkers are attached to the ends. Linkers are synthetic single-stranded oligonucleotide segments (6, 8, 10, or 12 bases in length) that self-associate to form symmetrical, blunt-ended, double-stranded molecules containing the recognition sequence for a particular restriction enzyme. Figure 32.8 shows an eight-base linker (CCTGCAGG) containing a *Pst*I recognition site. This linker self-associates to produce an eight-base, blunt-ended, duplex structure that adds to the double-stranded cDNA in the presence of T4 ligase. The resulting product is treated with *Pst*I to produce the characteristic 3′ overhang. The plasmid, linearized with *Pst*I, and the two DNAs are mixed and reacted with ligase to produce plasmid with the insert.

Figure 32.8

Insertion of cDNA into pBR322 plasmid by the linker method. The strategy here is to open up the plasmid with a restriction enzyme that makes staggered cuts and to attach linkers that contain the same recognition site to the cDNA. After the linkers are attached to the cDNA, the duplex is treated with the same restriction enzyme (*Pst*I) to expose the overhangs. The two DNAs are mixed together and ligated. After transfection, cells containing the hybrid plasmids are recognized by tetracycline resistance and ampicillin sensitivity. Identification of the insert is discussed in the text.

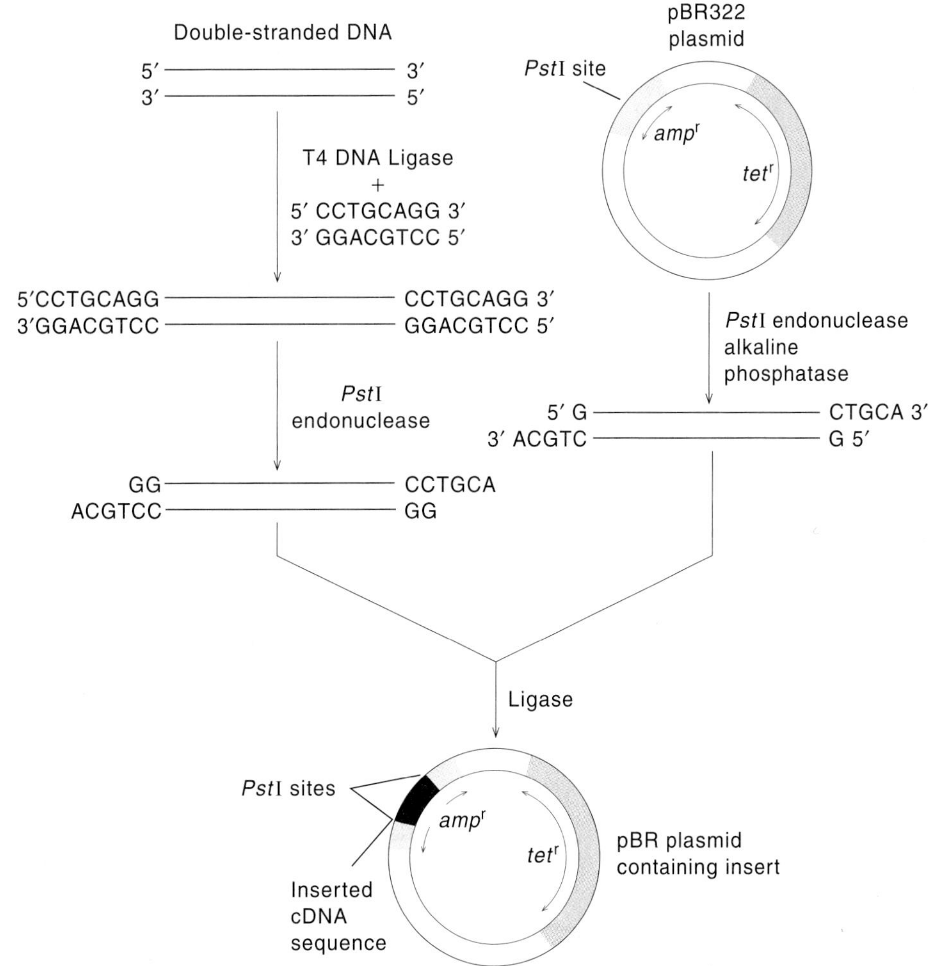

Numerous Approaches Can Be Used to Pick the Correct Clone from a Library

A library can contain thousands or even tens of thousands of different kinds of clones (see table 32.2), making it a challenging endeavor to isolate a clone with the DNA of interest. Most currently used procedures for screening large numbers of colonies for plasmids or phage that contain specific DNA inserts are variants of the colony hybridization method developed by Grunstein and Hogness. This procedure makes use of a specific radioactive probe that contains some sequences complementary to those in the DNA of interest. The colonies to be screened are first grown on agar petri plates (fig. 32.9). A replica of each plate is made on another agar plate, which is stored for reference. A replica is also made on a nitrocellulose filter. The colonies formed on the filter are lysed and the contents are denatured simultaneously by treatment with sodium hydroxide. After heating, the denatured DNA is fixed on the filter at each site where a colony was located. The DNA on the filter is then hybridized with a radioactively labeled nucleic acid probe complementary to the specific DNA sequence to be selected. The presence of hybridized probe at sites occupied by DNA derived from colonies that include the DNA fragment of interest is detected by autoradiography. The colony whose DNA hybridizes with the nucleic acid probe can then be picked from the reference plate, which contains a viable bacterial colony at a corresponding location.

Figure 32.9

Colony hybridization procedure used to identify bacterial clones harboring a plasmid containing a specific DNA. Step 1: Replica-plate the colonies containing plasmids onto nitrocellulose paper. Step 2: Lyse cells with NaOH and fix denatured DNA to paper. Step 3: Hybridize to ^{32}P-labeled DNA carrying the desired sequence and autoradiograph the product. Locations of desired DNA should be emphasized in the autoradiograph. Clones carrying desired plasmids (circled) may then be isolated from a corresponding agar replica plate carrying untreated colonies.

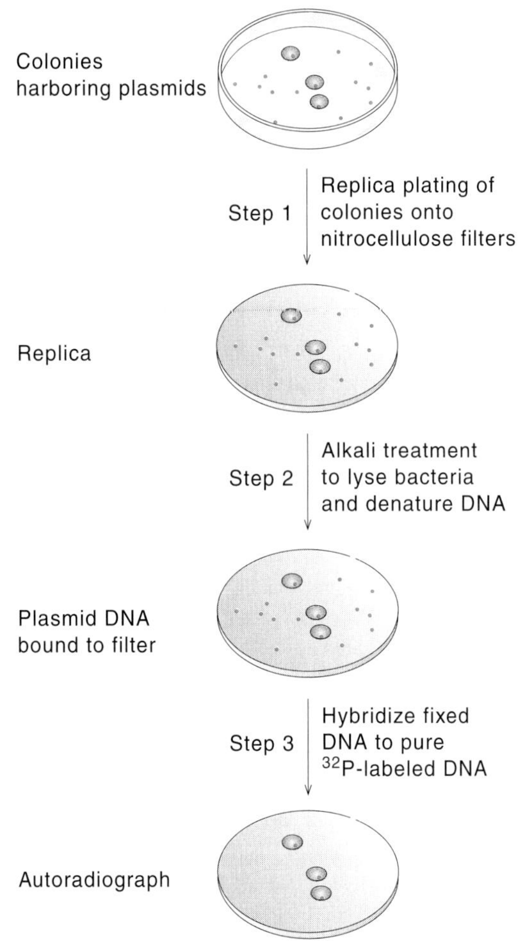

Cloning in Systems Other Than *Escherichia coli*

Despite the success and broad applications of *E. coli* cloning systems, gene products cannot always be made in this bacterium. Sometimes they are not synthesized in their entirety, or they are rapidly broken down after synthesis. In addition, for the study of certain processes indigenous to other species (e.g., photosynthesis, antibiotic production) it is often necessary to use a host bacterial species that carries out the process naturally, rather than *E. coli.*

Effective gene cloning systems are available for a variety of bacterial hosts, including *Bacillus subtilis, Streptomyces* species, and *Agrobacter tumefaciens.* Cloning systems have also been developed for eukaryotic hosts. In this section we will consider examples of cloning in three eukaryotic systems: the yeast *Saccharomyces cerevisiae,* mammalian cells in tissue culture, and plant cells.

Yeast Artificial Chromosomes Are Used for Cloning DNA Fragments as Large as 500 kb in Length

Yeast artificial chromosomes (YACs) have been developed for cloning huge DNA fragments. A YAC vector contains all the essential elements for making a chromosome: a centromere, a chromosomal origin for replication, and two telomeres (fig. 32.10). In addition it carries two selectable markers for replication in yeast and a cloning site. For purposes of vector amplification it also carries an *ori* for replication in *E. coli* and a marker suitable for selection in *E. coli.* These sequences are assembled so that when the vector is cut with the appropriate restriction enzymes, it separates into two segments, each of which carries a selectable marker for growth on yeast and a telomere at one end. Genomic DNA to be cloned is partially digested with a restriction enzyme. The resulting fragments may be fractionated according to size by pulse field gel electrophoresis (see box 30B). Fragments in the desired size range are plucked from the gel and ligated to the YAC arms. Transfection into yeast cells is carried out under selective growth conditions so that every transfected cell contains both arms of the newly created YAC chromosomes. The YACs replicate and segregate in the same way as ordinary chromosomes. Screening of the YAC clones follows the Grunstein-Hogness procedure (see fig. 32.9).

Figure 32.10

Procedure for making a yeast artificial chromosome containing a large genomic DNA insert. First a YAC shuttle vector is constructed that has the capacity for replication and selection in both *E. coli* and yeast. For this purpose it contains *ori,* a replication origin that works in *E. coli,* and ARS, a replication origin that works in yeast. It contains Amp^r for selection in *E. coli* and *TRP*1 and *URA*3 for selection in yeast. It also contains a centromere, so replicated chromosomes will segregate regularly in yeast, and two telomeres, which will be needed for the chromosome ends. The *Eco*RI restriction site is the cloning site, and the two *Bam*HI restriction sites are used to linearize the circular vector for chromosome construction. After treatment of the vector with *Eco*RI and *Bam*HI the two elements of the vector are ready to receive the genomic DNA insert. The genomic DNA insert is prepared by partial digestion of the genomic DNA with *Eco*RI and size fractionation by pulse field gel electrophoresis. Genomic DNA fragments of the appropriate size are mixed with the vector fragments and ligated.

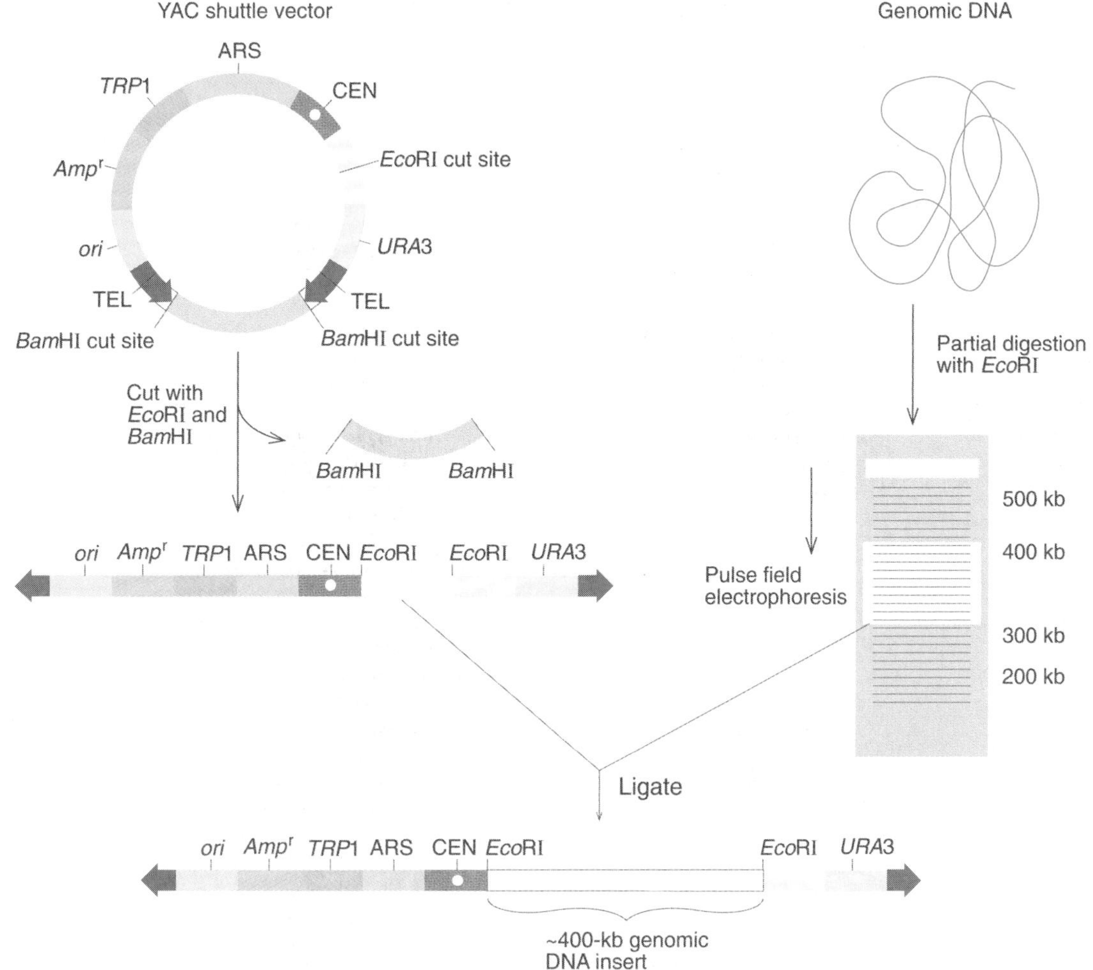

Studies on Cloned Genes in Mammals Start with Tissue Culture Cells

Mammalian cells from various sources can be adapted for growth as single cells in liquid culture or on plates. Such cells can be formally treated like bacteria or yeast; single cells give rise to genetically homogeneous colonies. Cultured cells have proved to be effective recipients of cloned DNAs.

The simplest procedure for transfection of mammalian cells utilizes purified DNA. The DNA is mixed with calcium chloride and sodium phosphate to give a finely divided calcium phosphate–DNA precipitate. Treated cells appear to take up the DNA by endocytosis. Transfection efficiencies, measured as the fraction of treated cells that become transfected, vary from 10^{-4} to several percent by this method.

Cultured mammalian cells permit many types of studies that would be impractical to carry out on whole animals. However, only a few mutants provide readily selectable genetic markers; this limitation creates a problem in selecting for transformed colonies. One of the best systems available for selection involves the thymidine kinase (*tk*) gene of herpes simplex virus. Mutants that are tk^- have been isolated for a number of cell types, including mouse, rat, and human. Such cells cannot grow on the selective medium known as HAT. The tk^- cells can be transformed to tk^+ cells by plasmid DNA that contains the herpes simplex virus *tk* gene. Other genes of interest can be inserted into the same plasmid vector so that selection of tk^+ cells after transformation leads to a high probability of coselection of adjacent genes.

A versatile experimental system using DNA transfection involves the mouse teratocarcinoma cell (TCC). With this system

Figure 32.11

Manipulation of mouse teratoma cells. The cells from a teratocarcinoma may be dispersed and grown in tissue culture. These cells can be injected into an embryo (*a*), in which case the resulting animal is a chimera in which some cells come from the original parents and others arise from the cells injected into the blastocyst. (*b*) Alternatively, these cells may be implanted subcutaneously, in which case the animal develops a tumor at the site of implantation. (*Source:* From G. Zubay, *Genetics,* Benjamin/Cummings, Menlo Park, Calif., 1987, p. 301.)

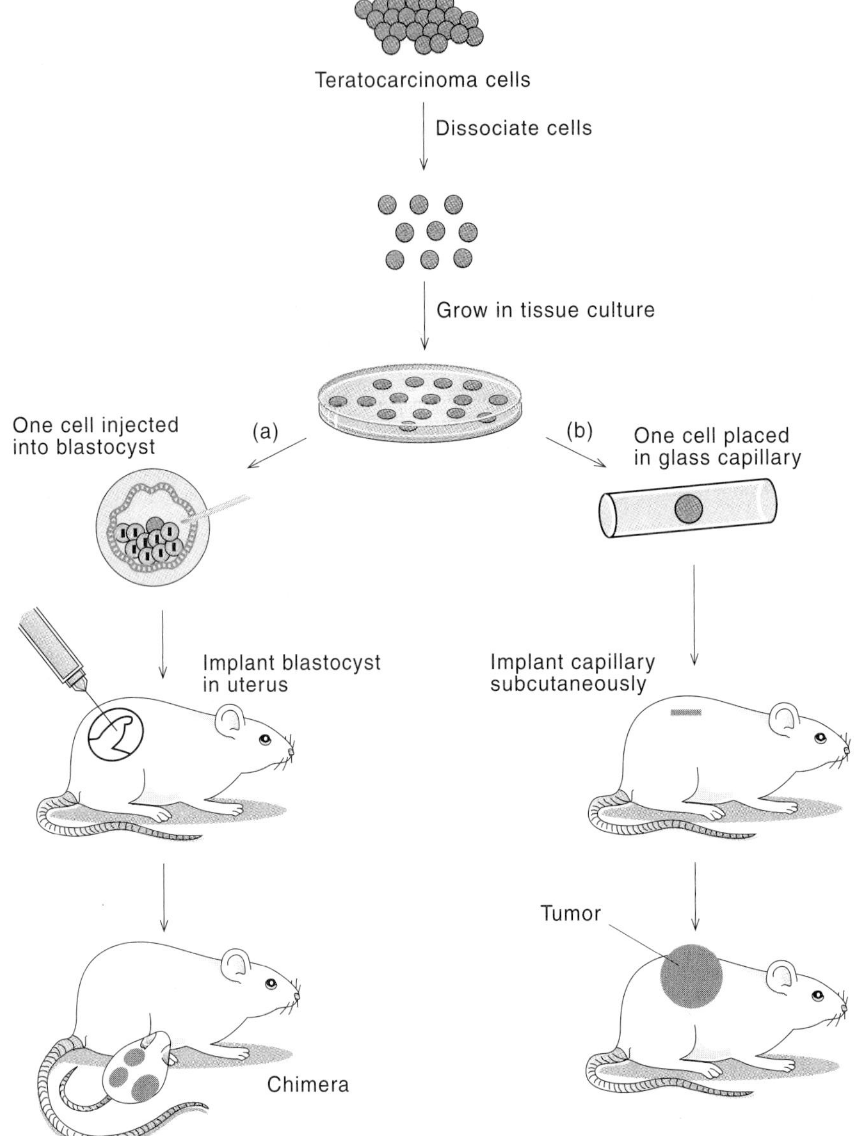

it is possible to transfect cells in tissue culture as in other tissue culture systems and then to implant the cells into a growing embryo so that the cells become part of the animal. By this procedure we can study the effects of cells altered by transfection in any organ of the whole mammalian organism.

The teratoma is a unique type of tumor found in many kinds of mammals. It is composed both of neoplastic cells, such as occur in other tumors, and also of many kinds of differentiated cells. A typical teratoma may include nerve cells, muscle cells, blood cells, skin cells, and other differentiated cells all mixed together with neoplastic stem cells. Only the stem cells are neoplastic, producing more stem cells or more differentiated cells, usually both. In many ways the stem cells behave like embryonic cells. They can be dissociated into single cells and cultured in vitro like bacterial cells. Cells grown from the culture can be reintroduced into the animal by subcutaneous implantation, in which case they produce a tumor. Most remarkably, they can be introduced into an early embryo (blastocyst) to produce a hybrid chimera where, in favorable cases, all of the tissues possess some cells from the tumor parent (fig. 32.11). Even egg cells have been isolated that are derived from the tumor parent. These egg cells, when fertilized by normal sperm, result in progeny that could truly be said to have a tumor for a mother. This example shows that the neoplastic stem cells are capable of reverting to completely normal behavior when subjected to the embryonic environment.

Taking advantage of this quite remarkable result, Bea Mintz and her co-workers have shown that teratocarcinoma (TCC) stem cells can function as vehicles for the introduction of specific recombinant genes into mice. Mice that carry a foreign gene are referred to as transgenic, and the foreign DNA is termed a transgene. For this purpose, TCC cells were first grown in single-cell tissue culture, and cells with a thymidine kinase deficiency were selected and treated with DNA containing the human β-globin gene and the thymidine kinase (*tk*) gene. Then tk^+-transformed cells were selected with the help of HAT medium, as explained earlier. Further tests revealed that the majority of the transformants also had copies of the human β-globin gene. Thus although there is no facile selection procedure for the human β-globin gene, it can be successfully cotransfected by linking it to the *tk* gene. Such altered TCC cells can then be introduced into early embryos to engineer new mouse strains carrying the human β-globin gene. This experiment shows that virtually any gene can be transfected into the whole animal by this two-step procedure. In current research it has been found convenient to use embryonal stem cells (ES cells) in place of the stem cells derived from teratomas. ES cells are obtained by culturing the inner cell mass of mouse blastocysts.

Oncogenes Can Be Selected from a Genomic Library by Subculture Cloning

Sometimes a gene can be identified when it confers an obvious phenotype on a mammalian cell. However, even in such cases, indirect selection procedures often must be used because of the limitations in the cloning vehicles currently available for mammalian cells. In a strategy called subculture cloning, a genomic library of mammalian DNA is made in *E. coli,* and then clones from the library are tested in subgroups in mammalian cells until one exhibiting the desired properties is identified. The DNA of interest is then propagated in *E. coli.* This selection method is illustrated in figure 32.12 for the isolation of an oncogene from a chicken lymphoma. A library containing 200,000 phage clones was divided into sublibraries, and the DNA of the clones in each sublibrary was introduced by transfection into mouse fibroblast cells growing in tissue culture. Cellular transformation was assayed by the appearance of a clone or clones of rapidly dividing cells (foci) superimposed on a background of slowly dividing, nontransformed cells. The sublibrary that assayed positively was further subdivided and retested. This procedure was repeated until a clone was identified that conferred the rapid growth properties associated with oncogene transformation.

Subculture cloning depends on the sensitivity of the assay used for detecting a small number of positive clones in a heterogeneous population. Alternatives to subculture cloning have been devised, but these methods still clone genes in bacteria and then detect them in mammalian cells. However, papilloma viruses, which can function as true plasmid vectors in mammalian cells, may permit the direct cloning and selection of genes in mammalian cells.

Cloning in Plants Has Been Accomplished with a Bacterial Plasmid

Tissue culture systems for plants are hard to manipulate and grow for extended periods of time. The difficulties of cloning DNA in plants are compounded by the fact that there is no plasmid that can be introduced directly into plant cells. Much plant DNA cloning uses a naturally occurring host–vector system involving *Agrobacter tumefaciens,* a pathogenic soil bacterium that causes crown gall tumors in a wide variety of dicotyledonous plants. (It is noteworthy that this is the only known case of naturally occurring nucleic acid transfer between a prokaryote and a eukaryote.) These plant tumors are caused by tumor-inducing (T1) plasmids in some agrobacteria; related bacteria that lack the plasmid are not tumorigenic.

The T1 plasmids are large circular DNA duplexes ranging in size from 160 to 240 kb. Bacterial infection of the plant leads to development of a tumor, usually near the junction of root and stem, as a consequence of a plasmid segment called T-DNA. T-DNA is integrated into the nuclear genome of the plant cells, where it alters the metabolism of the tumor cells so that small molecules known as opines are synthesized in large quantities. Different T1 plasmids encode genes that lead to the synthesis of different types of opines. While these compounds are of no apparent use to the plant's cells, they provide a growth substance for the agrobacteria that contain the T1 plasmid. Other genes carried by the T1 plasmid confer upon the host bacteria the ability to metabolize the T-DNA-encoded opine and use it as a substrate to promote bacterial growth.

It is impractical to use the entire T1 plasmid as a cloning vector because of its size. However, foreign genes can be transferred into plants by introducing them into the T-DNA segment of T1, reinserting the T-DNA into T1, and then letting the natural biological process of T-DNA transfer to the plants cells do the rest (box 32B). Some plant cells carrying part or all of the T1 plasmid can regenerate whole plants, so it is possible to measure the effects of transfected genes on any tissues of the whole plant.

Site-Directed Mutagenesis Permits the Restructuring of Existing Genes

By combining different procedures of manipulation, it is possible to make discrete changes in genes. This technique, called site-directed mutagenesis, is one of the most important in modern genetics and biochemistry. The first site-directed mutagenesis studies were carried out by David Shortle and Daniel Nathans in 1978 with the help of the mutagen sodium bisulfite, which deaminates C residues so that they become converted into U residues.

Directed mutagenesis as it is practiced today is based on the chemical synthesis of a deoxyoligonucleotide that contains discrete changes in its sequence from that normally observed in the genome under investigation. These changes may be single-base or multibase; they may involve base changes, base deletions, or base additions.

Figure 32.12

Subculture cloning procedure. Subculture cloning is used in successive steps to isolate the clone that contains the gene of interest. In this case the gene of interest was an oncogene originating from a chicken lymphoma and identified by the rapid growth characteristics it confers on transformed cells. A λ library was made with the chicken lymphoma nuclear DNA. The library was subdivided into ten approximately equal lots (sublibraries). From each lot the DNA was isolated and used in a transformation assay. Lot 2 scored positively. This lot was further subdivided, and the procedure was repeated again and again until a pure plaque carrying the oncogene was isolated.

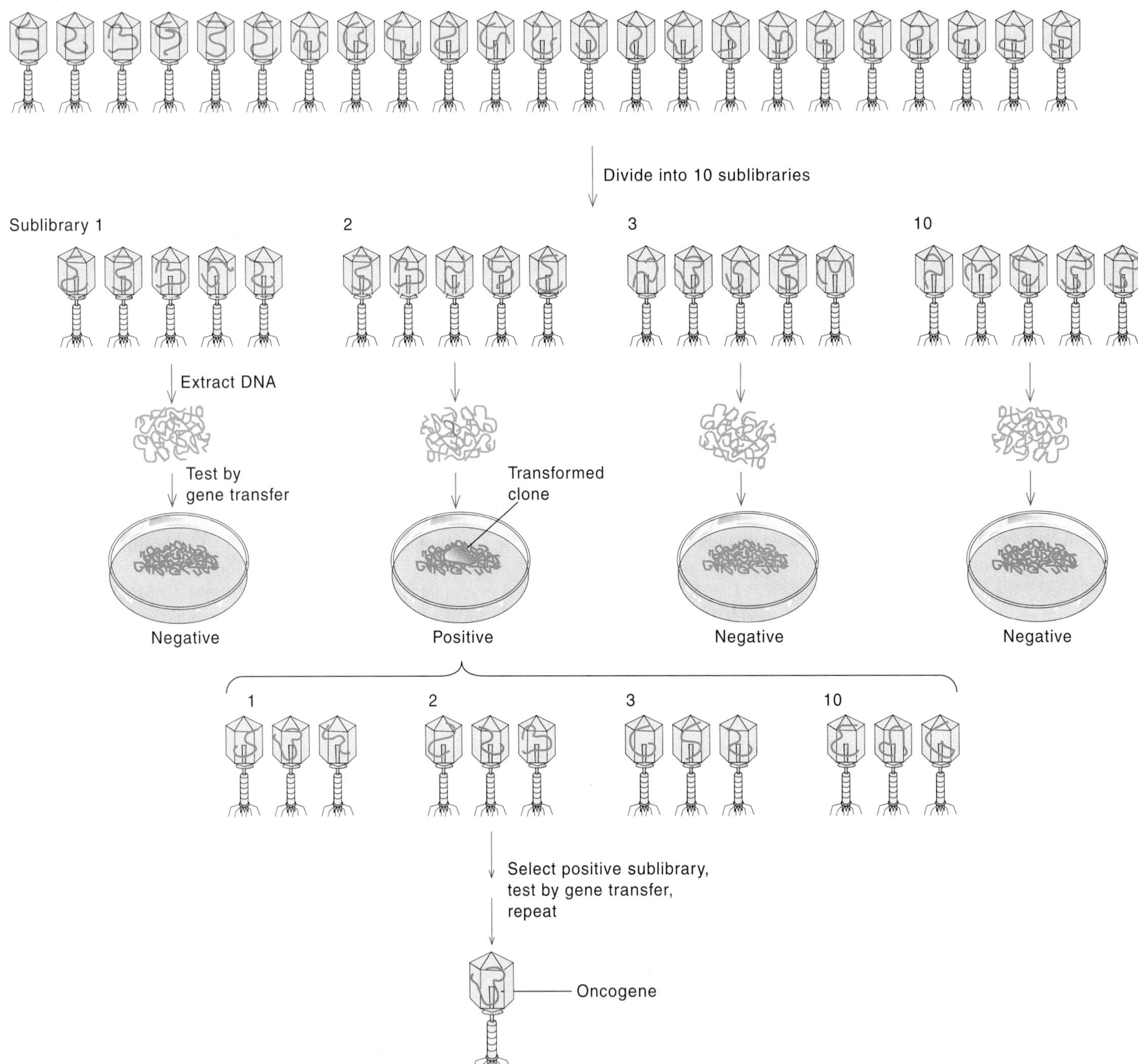

Many variations of site-directed mutagenesis exist. One can start out with a circular, single-stranded DNA and anneal it to a synthetic primer DNA carrying the desired changes (fig. 32.13). This primer can be extended, and the resulting product can be transfected. Finally, one selects clones of cells containing the plasmid with the desired changes.

The polymerase chain reaction (PCR) may become the method of choice in the future for carrying out site-directed mu-

32B
BOX

Construction of a T1 Plasmid Derivative Suitable for Cloning

The following procedure is quite complex, but it introduces some new and very worthwhile principles of cloning. First, a segment of T-DNA that included a preselected restriction enzyme cleavage site was cloned into pBR322 for replication in *E. coli* (step 1 in figure 1). The plant gene to be cloned was covalently linked to a selective genetic marker, the kanamycin resistance (*kan*^{r}) gene (step 2). The segment containing both the *kan*^{r} gene and the plant gene was spliced into a T-DNA segment cloned in the pBR322 plasmid (step 3), which was used for amplification of the plant gene in *E. coli*. However, since pBR322 does not replicate in agrobacteria, the segment containing the T-DNA and the *kan*^{r} gene was inserted into the plasmid pRK290, which replicates in either bacterial host (step 4). The modified T-DNA fragment, now attached to pRK290, was introduced by transformation into an agrobacterium that contained a T1 plasmid (step 5).

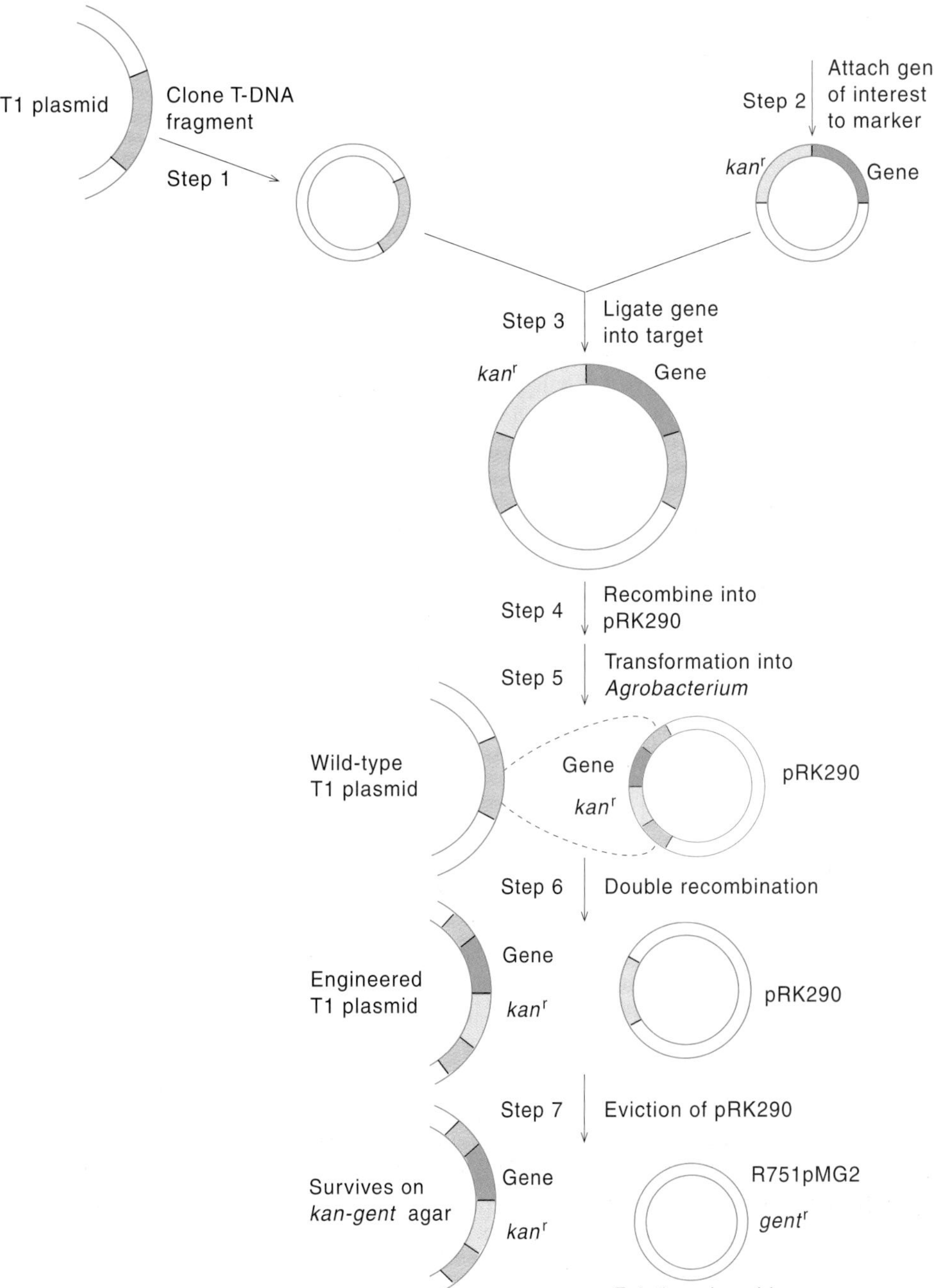

Figure 1

Construction of T1 plasmid for cloning. (*Source:* From Geoffrey Zubay, *Genetics*. Copyright © 1987 Benjamin/Cummings Publishing Company Inc., Menlo Park, Calif. Reprinted by permission of the author.)

Figure 32.13

Scheme for oligonucleotide-directed mutagenesis of double-stranded circular plasmid DNA. Supercoiled plasmid circles are nicked in one strand and rendered partially single-stranded by treatment with exonuclease. The gapped circles are hybridized with a homologous oligodeoxynucleotide carrying, by design, some mismatches. In vitro DNA synthesis, primed in part by the oligodeoxynucleotide, leads to heteroduplex plasmid circles. (*Source:* After G. Dalbadie-McFarland, L. W. Cohen, A. D. Riggs, C. Morin, K. Itakura, and J. H. Richards, Oligonucleotide-directed mutagenesis as a general and powerful method for studies of protein function, *Proc. Natl. Acad. Sci. USA* 79:6408–6412, 1982.)

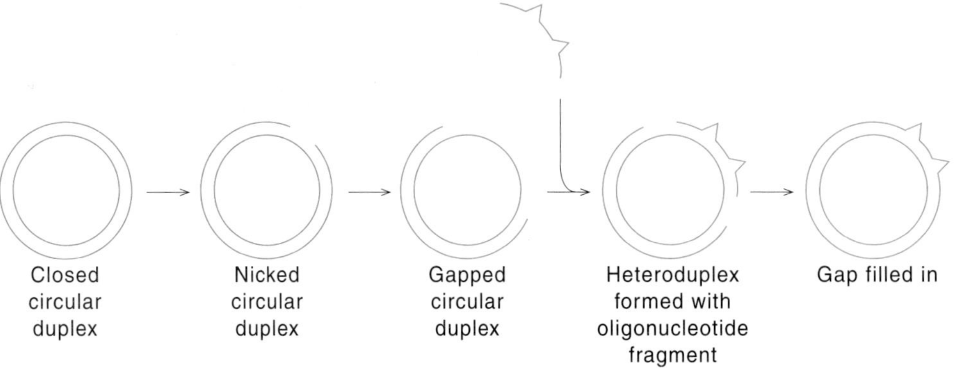

tagenesis. In its most general form, the use of PCR for this purpose requires a piece of duplex starting DNA, two outside flanking primers that are perfect complements to segments of opposing strands, and two complementary primers with the desired changes in their sequence as diagrammed in figure 32.14. The steps followed parallel the steps of PCR amplification.

PCR amplification can be coupled with classical cloning methods using cloning vectors. For this purpose the PCR product should contain restriction sites at its ends that are suitable for cloning. Thus PCR amplification and cloning need not be thought of as alternatives for particular purposes but as complementing each other to give a greater variety of approaches.

Targeted Gene Replacement in Mammalian Cells

Originally, when it was discovered that DNA could be transfected into mammalian cells and integrated into the chromosomes, there was cause for jubilation. However, jubilation gave way to despair when it was realized that genes do not always integrate at homologous locations. This creates a problem when the goal is to replace the wild-type allele with a mutant allele. It turns out that homologous recombination does occur but that it is difficult to detect because it is rare. Remember that we are dealing with a diploid organism, so even if homologous integration does occur, a single mutant allele will not change the phenotype unless it is dominant to the wild type, which is rarely the case. Therefore there is no way of selecting for the rare homologous recombinant.

Mario Capacchi devised a selection technique for the rare transformants where homologous integration has occurred (fig. 32.15). In the first step the gene of interest is inactivated by insertion of the *neo*^{r} gene into its protein-coding region (see fig. 32.15*a*). The *neo*^{r} gene will serve later as a marker to indicate that the vector DNA took up residence somewhere in the genome. The vector has also been engineered to carry a second marker at one end, the

Figure 32.14

Illustration of a general method of mutagenesis using PCR. Primers are represented as short lines with arrowheads pointing in the 3′ direction. The bump in primers 2 and 3 and their products represent a mismatched base, a deliberate alteration in base sequence from that present in the starting DNA. Of the four major products resulting from step 3, only D is extendable by DNA polymerase.

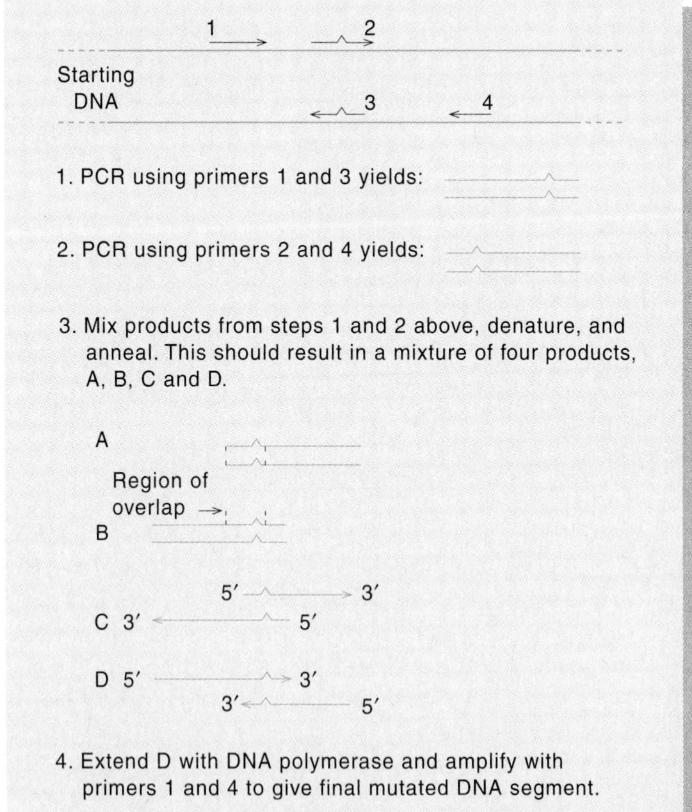

1
2
Starting DNA
3
4
1. PCR using primers 1 and 3 yields:
2. PCR using primers 2 and 4 yields:
3. Mix products from steps 1 and 2 above, denature, and anneal. This should result in a mixture of four products, A, B, C and D.
A
Region of overlap
B
5′
3′
C 3′
5′
D 5′
3′
3′
5′
4. Extend D with DNA polymerase and amplify with primers 1 and 4 to give final mutated DNA segment.

Figure 32.15

Targeted gene replacement in cultured mouse cells. (*a*) The cloned gene which is to replace the targeted gene in the cell is altered by insertion of the neomycin resistance gene (*neo*^{r}). Near the cloned gene on one side the *Herpes simplex* virus, thymidine kinase gene is attached. This targeting vector is transfected into mouse embryonal stem cells. (*b*) The targeting vector has three possible fates after transfection. It could replace the normal gene by homologous recombination (left); it could integrate at random (middle); or it could fail to integrate, in which case there is no change (right). Treatment of cells with neomycin and gancyclovir should select for homologous recombinants. This is because homologous recombinants are resistant to neomycin because they have the *neo*^{r} gene and they are unaffected by ganciclovir. Cells that have integrated the targeting vector by nonhomologous recombination will usually integrate the *tk* gene as well as the split gene containing *neo*^{r}. The *tk* gene makes the cells sensitive to ganciclovir so that such cells will die. Cells that have not integrated the vector will be killed by the neomycin.

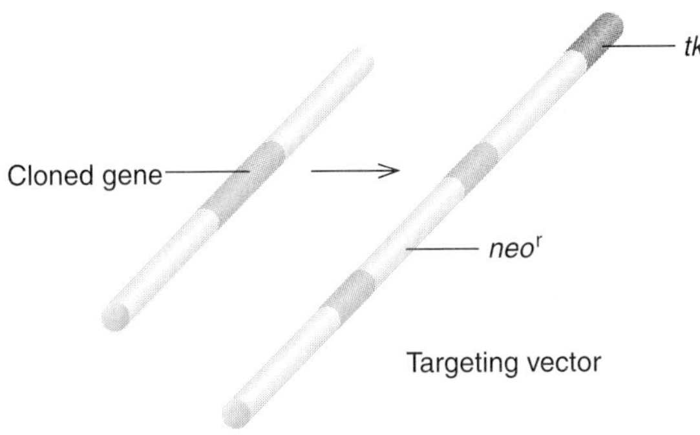

(a) Cloned gene is altered to produce targeting vector

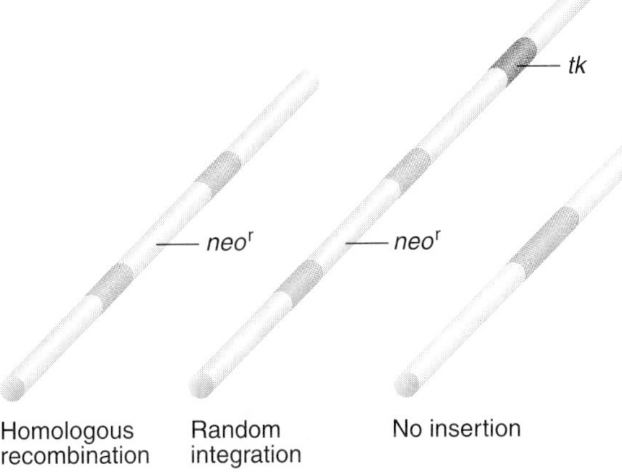

(b) Three possible fates of targeted gene after treatment of cells with targeting vector

herpes thymidine kinase (*tk*) gene. This marker will be used to indicate that the integrated vector has not integrated at the homologous location. Once a vector with its dual markers has been made, it is introduced into the cells isolated from a mouse embryo.

When all goes well, homologous recombination occurs (see fig. 32.15*b*, left); the vector lines up next to the normal gene (the target) on a chromosome in a cell, so identical regions are aligned; then those regions on the vector (together with any DNA in between) take the place of the original gene, excluding the *tk* marker at the tip. In many cells, though, the full vector (complete with the *tk* marker) integrates randomly into the genome (see fig. 31.15*b*, middle).

To isolate cells carrying a targeted mutation, all cells are put into a medium containing neomycin and ganciclovir. Neomycin is lethal to cells unless they carry a *neo*^{r} gene, so it eliminates cells where no integration has occurred. Meanwhile, ganciclovir kills cells that harbor a *tk* gene (see chapter 26), thereby eliminating cells bearing a randomly integrated vector. Consequently, the only cells that survive the double drug treatment and proliferate are those harboring the homologously integrated gene from the targeting vector.

To go from this situation to the whole animal, a process similar to that described in figure 32.11 is followed. Once two animals of opposite mating types have been isolated that are heterozygous for the targeted mutated gene, they can be mated to produce an offspring that should be homozygous for the double mutant. Such offspring should result with a .25 probability.

Recombinant DNA Techniques Were Used to Characterize the Globin Gene Family

The human globin family is a paradigm for studying differential gene activity during development and the molecular basis of genetic disorders in gene expression. Hemoglobin is a tetramer containing two α-like and two β-like subunits (see chapter 6). These proteins are encoded by a small number of genes that are expressed sequentially during development. The information summarized in figure 32.16 indicates that the α-like and β-like globin gene families have coordinated programs for expression: Two switches exist for the β-like genes, whereas a single switch results in activation of adult α-globin production early in fetal life. A combination of classical and recombinant DNA techniques has been used to show that the α-like genes are located in a single cluster on chromosome 16 and the β-genes are located in a single cluster on chromosome 11 (fig. 32.17). We focus on the contributions to our understanding of the globin genes that have resulted from investigations using the recombinant DNA approach.

DNA Sequence Differences Were Used to Detect Defective Hemoglobin Genes

All of the hemoglobin genes are represented by two or more alleles within the human population; the genes are said to be polymorphic. This polymorphism frequently shows up in readily detectable phenotypes when the differences occur in vital areas of the polypeptide chains. Polymorphisms show up in the DNA even more frequently because the DNA contains sequence differences in both the coding and the noncoding regions of a gene, and these differences can be detected even when no visible effect is apparent in the organism. Because of their frequency and ease of detection by recombinant DNA methods, DNA polymorphisms have become extremely useful in mapping the human genome.

Figure 32.16

Changes in types of hemoglobin observed in early development. A single switch in gene expression is observed for α-like chains. Two switches in gene expression are observed for β-like chains. The corresponding tetrameric hemoglobin molecules observed at different stages in development are also indicated.

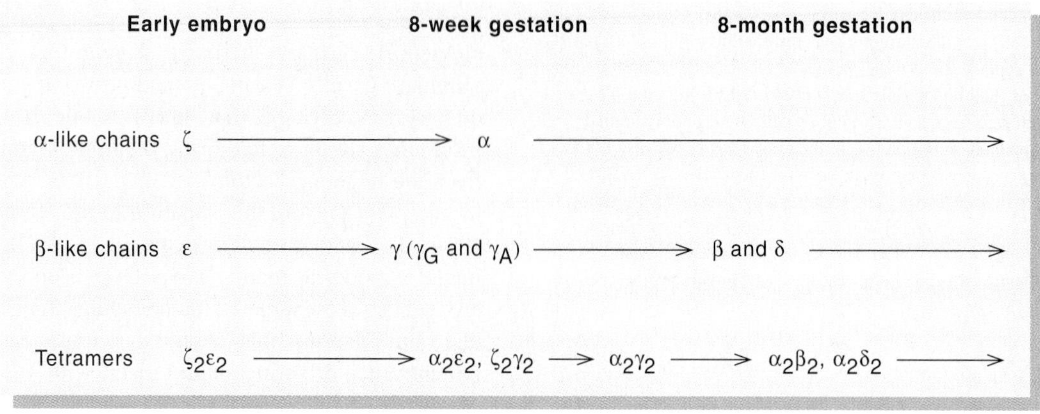

Figure 32.17

The chromosomal localization and genomic organization of the human globin genes. The α- and β-globin gene complexes are positioned on chromosomes 16 and 11, respectively. For each complex the arrangement of genes on the chromosome is depicted above, and the general structure of the major gene is shown below, together with the location of the intervening sequences, or introns (IVS), and codon numbers. Coding regions are shown by solid boxes and IVS regions by open boxes. Genes with the ψ symbol in front are called pseudogenes because they are sequence-related but not expressed.

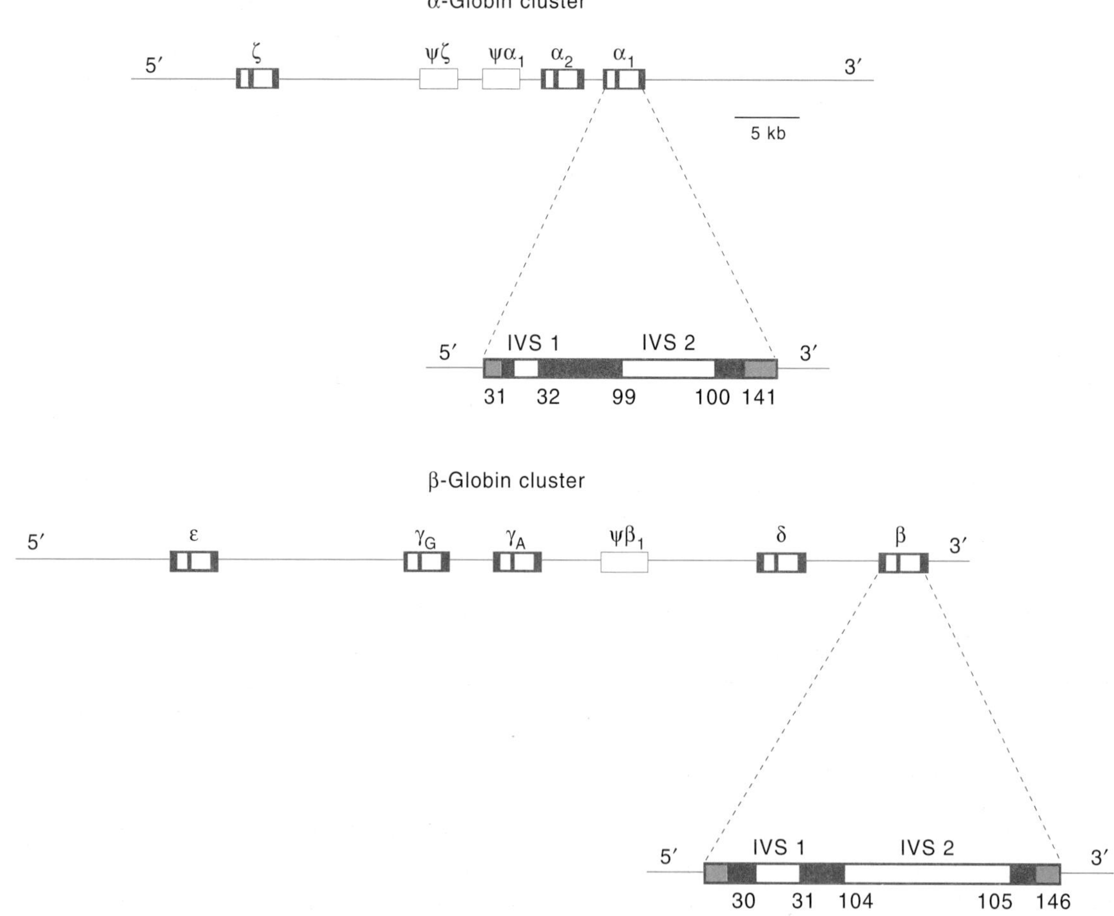

Whereas DNA polymorphisms should be recognizable by sequence differences, it is usually more convenient to detect these polymorphisms by differences in the size of DNA fragments obtained with restriction enzymes. Differences observed in this way are called restriction fragment length polymorphisms (RFLPs).

Kan and Dozy were the first to discover an allele-linked DNA polymorphism in the globin genes. With it they predicted which fetuses carried normal and which carried abnormal sickle-cell genes for β-globin. To analyze the DNA for these differences, they used a technique called Southern blotting. The steps involved in Southern blotting are illustrated in figure 32.18. First the genomic DNA from the test subject is digested with a restriction enzyme to yield specific DNA fragments. These fragments are separated according to size by agarose gel electrophoresis. Next the DNA is denatured and transferred from the agarose gel to a cellulose nitrate sheet. The DNA firmly bound to the sheet is hybridized with a radioactively labeled DNA probe, which carries some of the sequences of interest. The radiolabel, which hybridizes to specific regions of the sheet, is detected by autoradiography. By comparing the results obtained from the DNA of different individuals, one can see whether the labeled DNAs move with the same or a different mobility. If they move differently, there must be a RFLP difference between the individuals. Detection of an RFLP by this means usually depends on the restriction enzyme used in the initial digestion. Some enzymes show a difference; others do not.

To apply this technique to their hemoglobin studies, Kan and Dozy first had to prepare a DNA probe that carried specific sequences in the β-globin gene. This was a relatively simple task because reticulocytes, which contain vast quantities of hemoglobin, are also greatly enriched in the β-globin mRNA. From the purified β-globin mRNA isolated from reticulocytes, a cDNA probe was made with the help of reverse transcriptase. Using a *Hpa*I restriction enzyme digest of the total genomic DNA in conjunction with Southern blotting, Kan and Dozy showed that the normal β-globin gene was contained within a 7.6-kb *Hpa*I restriction fragment, whereas the β-globin gene of sickle-cell anemia, Hb^s, was contained within a 13-kb fragment (fig. 32.19). Further analysis showed that the RFLP resulted from a *Hpa*I restriction site 5 kb to the 3′ side of the β-globin gene that was present in the normal case and absent in Hb^s. Subsequent analysis showed sequence differences within the coding regions. One may wonder why the RFLP outside the coding region was so commonly associated with the abnormal gene. A possible explanation for this is that in the distant past a mutation occurred that resulted in the 7.6-kb type and a few 13-kb types before introduction of the sickle-cell gene mutation. After the Hb^s mutation was introduced into the 13-kb type, it became greatly expanded because of the selective advantages of this gene in heterozygotes. In this connection it should be noted that heterozygotes carrying one normal gene and one sickle-cell gene fare far better when infected by malaria. As a result, in areas where malaria is prevalent the heterozygote has a selective advantage over the normal homozygote.

Knowledge of this and other polymorphisms has been used for pre- or postnatal diagnosis of the sickle-cell gene. Such information can be of great practical value in genetic counseling.

Figure 32.18

The steps involved in assaying by Southern blotting. The DNA to be analyzed is digested with a restriction enzyme (1). The resulting fragments are electrophoresed on an agarose gel (2). The DNA fragments on the gel are transferred to a cellulose nitrate sheet by placing the cellulose nitrate sheet next to the gel and passing solvent through the gel into the sheet. Flow of the solvent is maintained by blotting the far side of the cellulose nitrate sheet with paper towels. The DNA, first denatured with alkali, flows with the solvent but gets stuck in the sheet (3). The sheet is hybridized to radioactively labeled DNA containing the gene sequence of interest (4). The hybridized sheet is autoradiographed to determine the location of the labeled restriction fragment on the gel (5).

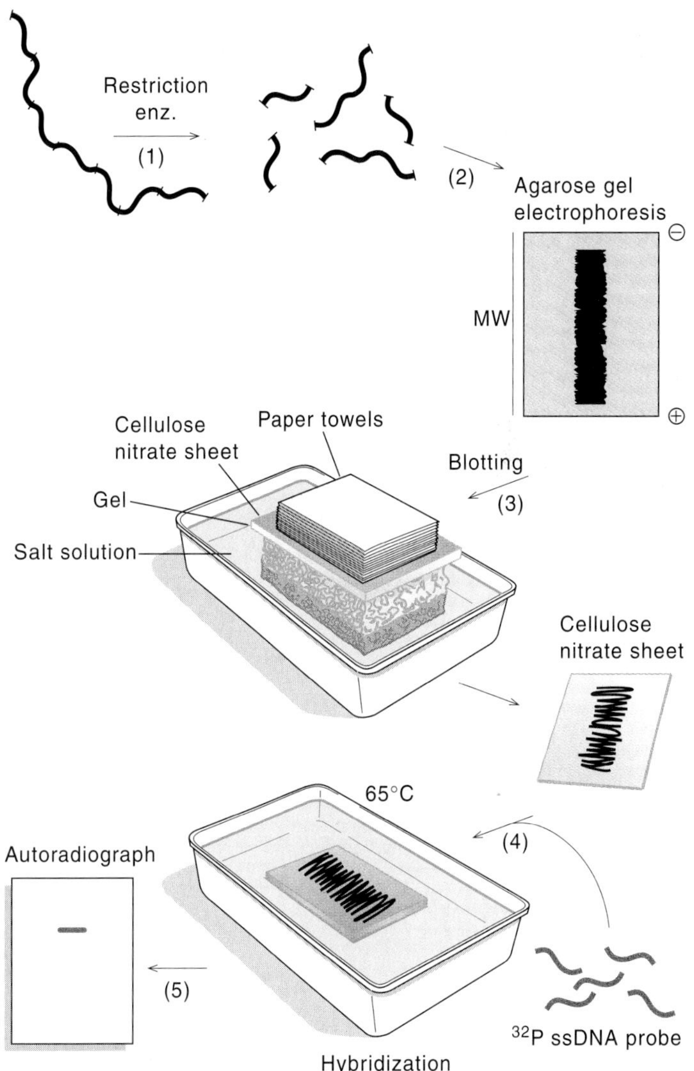

Incidentally, it is now possible to diagnose sickle-cell disease (which occurs in individuals who are homozygous for the sickle-cell gene) with greater certainty because the point mutation leading to the defect in the coding region itself produces a recognizable RFLP.

The β-Globin cDNA Probe Was Used to Characterize the Normal β-Globin Gene

Detailed mapping with DNA probes was first successfully executed on the human β-globin gene. All members of a human ge-

Figure 32.19

Inheritance pattern of an RFLP associated with sickle-cell disease. Humans carry two alleles for the same gene, and each offspring inherits one allele from each of its parents in an entirely random fashion. Normal individuals are homozygous for normal *Hb* alleles; individuals with sickle-cell trait are heterozygous, with the one normal *Hb* allele and one Hb^s allele; and individuals with sickle-cell disease are homozygous for the Hb^s allele. At the top (*a*) we see a three-generation pedigree analysis for a family that carries both the normal and the sickle-cell gene for β-globin. Males are represented by squares and females by circles. A purple circle or square indicates an individual who is homozygous normal. A half-filled circle or square (red/purple) indicates a heterozygous individual with sickle-cell trait. A filled circle or square (red) indicates a homozygous individual with sickle-cell disease. In (*a*) both sets of grandparents produce a heterozygous individual with sickle-cell trait. Because one of the grandparents is homozygous normal and the other is heterozygous, there is a 50% chance that the grandparent mating will give rise to a heterozygous offspring as shown and also a 50% chance that they will have normal offspring (not shown). The two heterozygous parents have an increased chance of having abnormal offspring because in this mating, each parent carries one abnormal gene or allele. There is a 25% chance of a homozygous sickle-cell anemic offspring, a 50% chance of an abnormal offspring with sickle-cell trait, and a 25% chance of a normal offspring. Below the pedigree chart is the electrophoretic pattern of a *Hpa*I digest probed with β-globin cDNA by the Southern blotting technique (*b*). At the bottom we see an interpretation of the normal and abnormal DNAs (*c*).

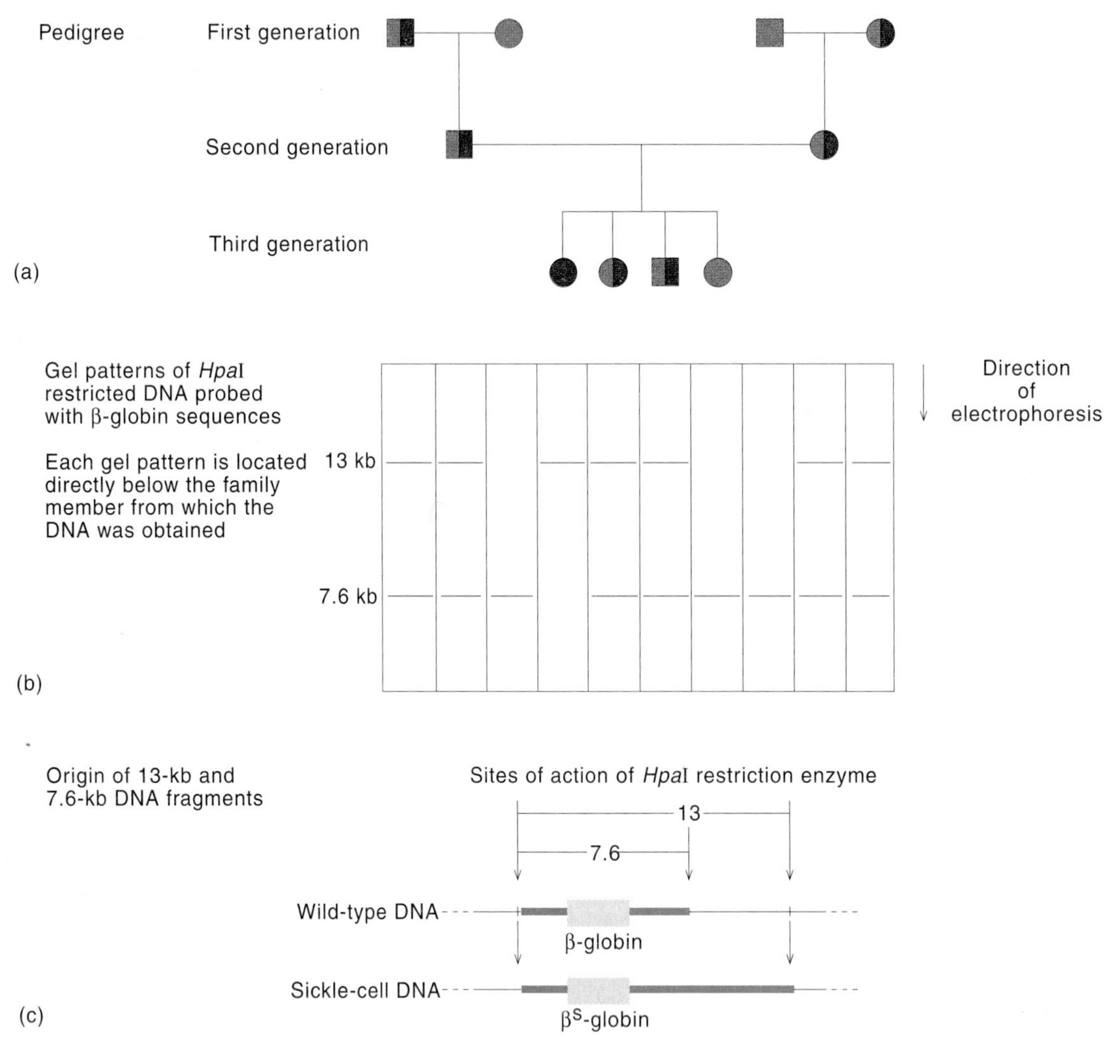

nomic library that annealed to the radioactive cDNA probe for the β-globin gene were isolated, and each of these was sequenced. This analysis resulted in a complete description of the β-globin gene (see fig. 32.17). The β-globin gene is appreciably longer than the β-globin mRNA; in addition to containing regions that are present in the final mRNA the gene contains two "intervening" regions not represented in the mRNA sequences. We have more to say about the significance of these intervening sequences (IVS, also called introns) in the next chapter.

Chromosome Walking Permitted Identification and Isolation of the Regions around the Adult β-Globin Genes

The original cDNA probe carrying the β-globin mRNA sequences could detect only members in the genomic library that contained sequences homologous to those present in the probe. To explore the region flanking the β-globin gene, the genomic library was

Figure 32.20

The linkage map of the human β-globin gene locus as shown by the structural analysis of overlapping λ genomic clones. Both $\lambda H\beta G1$ and $\lambda H\beta G3$ clones contained the entire β-globin gene. Other clones detected by "walking" led to the discovery of other β-globinlike genes. These included four genes that are expressed and two pseudogenes that are not expressed. The genomic segments of the clones isolated are shown together with the cleavage sites for the enzyme *Eco*R1. The numbers on the top line indicate the size of the fragments in kilobase pairs.

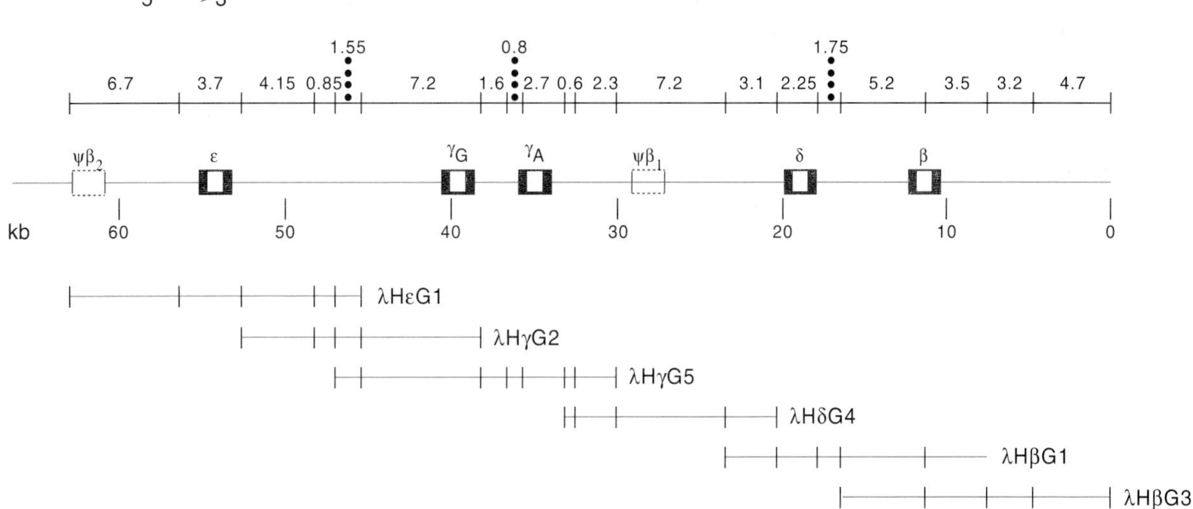

probed further. For this purpose, members of the genomic library that hybridized with the original cDNA were themselves converted into radioactive probes, and these were used to locate additional members of the library that contained sequences flanking the β-globin gene. A cyclic repetition of this process resulted in a gradual extension of sequence information in and around the β-globin gene. Using the library in this manner to extend the map is known as chromosome walking (fig. 32.20). A parallel approach was used to extend the map around the adult α-globin gene (see fig. 32.17). It can be seen that in both cases several genes occur in a cluster for each of the protein types. Most of these can be correlated with the genes that are expressed at different times during development. In addition, genes, called pseudogenes, occur that have strong sequence similarities to known genes but are never expressed. It is not clear whether these pseudogenes have a function or simply represent evolutionary "junk" that has not been removed.

Walking and Jumping Were Both Used to Map the Cystic Fibrosis Gene

By combining the linkage information obtained from RFLP mapping with other DNA manipulation techniques, it has been possible to locate genes causing serious genetic disorders even when these genes are known only from their inheritance patterns. The list of serious disorders that can be linked to single genetic loci is growing. It includes Huntington's disease, Duchenne's muscular dystrophy, polycystic kidney disease, cystic fibrosis, chronic granulomatous disease, peripheral neurofibromatosis, central neurofibromatosis, familial polyposis coli, and multiple endocrine neoplasia. One of the most spectacular achievements has been identification of the gene causing cystic fibrosis (CF).

Cystic fibrosis is the most common serious genetic disorder in Caucasian populations. The major clinical symptoms include chronic pulmonary disease, pancreatic exocrine insufficiency, and an increase in the concentration of sweat electrolyte. Bearers of this disease, who are readily diagnosed, often die of congestive lung complications before age 30. Pedigree analysis shows that a single gene inherited in autosomal recessive fashion results in the disease syndrome. The frequency of the disease is 1 in 2,000, from which it may be calculated that the carrier frequency is about 1 in 20 (the frequency of the heterozygote for a rare allele is twice the square root of the frequency of the homozygote). By classical genetic analysis the *CF* locus has been assigned to the long arm of chromosome 7 near the *Met* locus. A map of the region containing the *Met* locus and the *CF* gene is shown in figure 32.21.

In many genetic disorders the analysis must begin before the responsible gene and its protein product are known. Therefore it is not possible to locate the gene directly, as in the case of the β-globin gene, and then determine its approximate location in subsequent analysis. For genes such as cystic fibrosis the approximate location is determined by conventional recombination analysis, and then researchers attempt to close in on the gene by the process called reverse genetics.

Detailed genetic mapping by recombination frequency is not practical with human genes that are separated by map distances of less than 1 centimorgan (cM)* because of the small number of test recombinant crosses that are ordinarily available for observation. Unfortunately, 1 cM on the human genome is still equivalent to a physical distance of about 10^6 (1 million) bp. One

*Genetic loci 1 cM apart recombine 1% of the time at meiosis.

Figure 32.21

Map of restriction fragment length polymorphisms (RFLPs) closely linked to the cystic fibrosis (CF) gene. The inverted triangle near the right-hand end indicates the location of the ΔF_{508} mutation characteristic of most persons with cystic fibrosis disease. (*Source:* Adapted from B-S. Kerem, J. M. Rommens, J. A. Buchanan, D. Markiewicz, T. K. Cox, A. Chakravarti, M. Buchwald, and L-C. Tsui, Identification of the cystic fibrosis gene: Genetic analysis, *Science* 245:1075, 1989.)

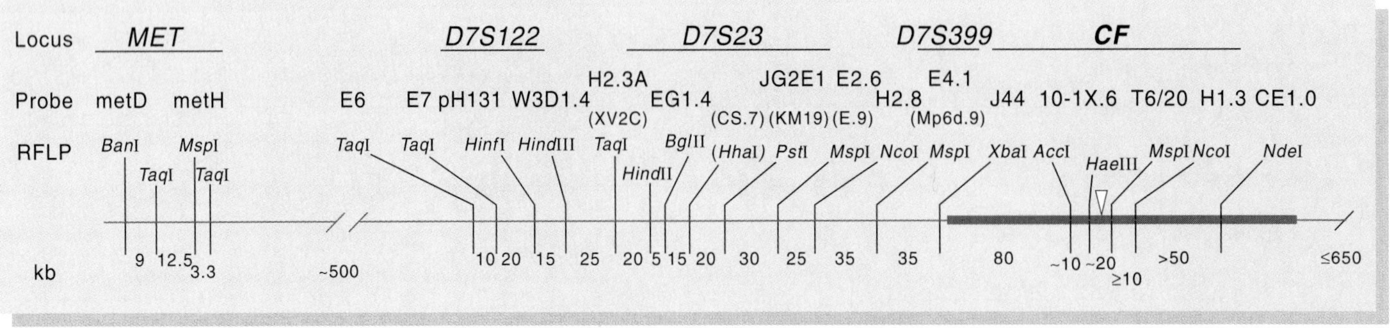

obstacle to the use of chromosome walking over such long distances is the size limitation of probes. Even cosmids, which provide the largest probes, cannot harbor probes larger than 40 kb, so it would take 25 cosmids end-to-end to span 10^6 base pairs. A walk involving this many probes would take a very long time, if it were possible at all. Another hindrance to such a long walk is that the human genome is sprinkled with segments of repetitive DNA. Those regions interrupt the walk because upon encountering such a region the probe will anneal to many members of the library, which could be situated almost anywhere in the genome. Therefore walking must be replaced by a procedure called chromosome jumping. Probes for jumping carry segments of DNA that are not ordinarily next to one another and that will therefore anneal to two regions that are a considerable distance apart. This distance is determined by how the probes are made.

Suitable jumping probes can be constructed in several ways. In one particularly elegant approach, the genome is first digested with a restriction enzyme that recognizes rare sites in the DNA (fig. 32.22). *Not*I, for example, is a "rare cutting enzyme" that recognizes a sequence of eight bases,

↓
GCGGCCGC

The fragments produced by complete digestion of the human genome with *Not*I are, on the average, about 500 kb long. These fragments are circularized in the presence of a small marker DNA, so thereafter the marker DNA is always located between the ends of the original large fragments. The circular DNA is recut with a restriction enzyme that produces fragments suitable in size for cloning in λ. The marker DNA used contains a gene essential for λ multiplication on a suitable *E. coli* host. In this way the only λ clones that develop are those that carry the marker DNA and hence the ends of the original restriction fragments.

A jumping library is of very limited value on its own because it permits only one jump if used by itself. This is because the members of such a library would not be expected to cross-anneal. Therefore a jumping library is always used in conjunction with a complementary linking library (see fig. 32.22). Starting from the same DNA sample from which the jumping library was made, a partial digest is made with a restriction enzyme that gives considerably smaller restriction fragments, about 20 kb in size. These fragments are circularized in the presence of the same marker DNA that was used to construct the jumping library, and the resulting circular pieces are recut by complete digestion with the *Not*I enzyme. Only the circular pieces containing a *Not*I site are linearized by this treatment, so only these fragments become part of the linking library. The linearization step is followed by ligation of the linear fragments into the λ vector and selection on the same *E. coli* strain that requires the marker DNA for λ replication.

Each member of the linking library carries sequences on both sides of *Not*I restriction sites in the original DNA. In contrast, in the jumping library each member carries sequences from one side of two adjacent *Not*I sites. The jumping library and the complementary linking library are used in strict alternation as illustrated in figure 32.23. Since there are two directions in which jumping can occur initially, the direction is determined by the first annealing. After the first annealing, the direction of future jumps is fixed. By using other markers it is possible to determine whether the direction of the first jump was the one of interest. Sometimes several jumps have to be made before the direction can be known.

In the cystic fibrosis investigation, starting from each *Not*I site determined by the jumping-linking library analysis, the sequences surrounding the *Not*I sites were scrutinized with a walking library made from the same starting DNA. The purpose of these local walks was to get as much information as possible about the region in the hope that by trial and error, one of the probes would be the lucky one that overlapped the cystic fibrosis gene. The joint procedure of jumping and walking is illustrated in figure 32.24.

All of this analysis would have been in vain in the absence of some criterion for knowing when the goal of finding the cystic fibrosis gene had been reached. This is where clues from the physiological nature of the condition became useful. In a bril-

Figure 32.22

Construction of a jumping library (left) and a complementary linking library (right). The jumping library is constructed by exhaustive digestion of total nuclear DNA with a rare cutting *Not*I restriction enzyme. The cut fragments are circularized in the presence of a small gene required for amplification in a special *E. coli* host. The circularized molecules are recut with *Bam*HI restriction enzyme and ligated into a λ vector. The linking library is constructed from the same starting DNA. In this case the DNA is partially digested with *Sau*3A restriction enzyme. The resulting fragments are circularized in the presence of the same small gene required for amplification and the circularized fragments are recut with *Not*I. The linear fragments are ligated into a suitable λ vector and amplified in the special *E. coli* host.

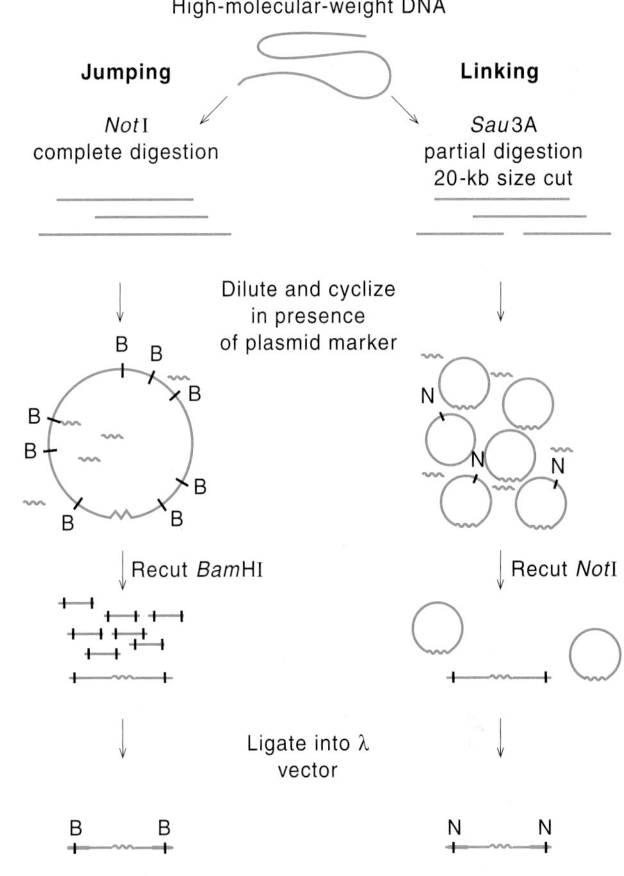

Figure 32.23

Directional jumping by alternating between a jumping library and a complementary linking library. Because of the way in which these two libraries were constructed (see fig. 32.22), each member of the linking library should contain sequences on both sides of a single *Not*I site, whereas each member of the jumping library should contain sequences on one side of adjacent *Not*I sites. By alternating between the two libraries, it should be possible to traverse large regions on the genome relatively quickly and in a unique direction. The process starts with a member of either library that originates from the general region of the cystic fibrosis gene. If the search starts with a member of the jumping library as suggested in the figure, then this member is used to probe the linking library. The member of the linking library found in this way is then used to probe the jumping library, and the process is repeated as many times as necessary.

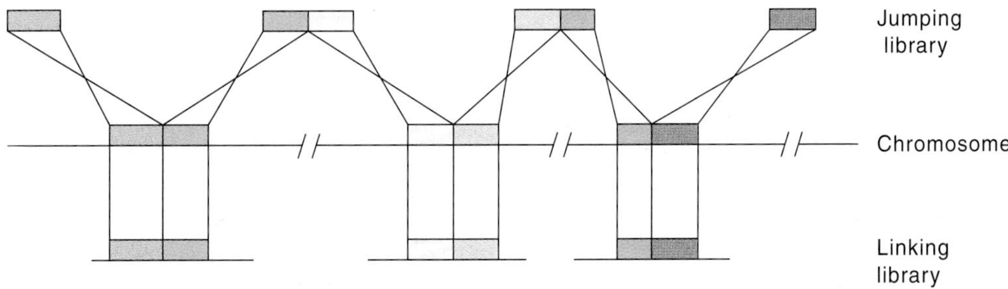

liant strategy a cDNA library was prepared from the messenger RNA fraction of sweat gland tissue. Recall that in cystic fibrosis there is sweat gland malfunction. Therefore it seemed that the mRNA for the cystic fibrosis gene might be well represented in the messenger RNA fraction of the sweat gland cells.

While the walking and jumping process was in progress, each new segment mapping in the general region of interest was tested against the sweat gland cDNA library. Finally, a member of the walking library was found that annealed with a member of the sweat gland cDNA library. Was this match fortuitous or did it mean

Figure 32.24

Jumping and walking to find the cystic fibrosis gene. Following each jump, the locus defined by the jump was used as a starting point for a chromosome walk. Each DNA segment so found was hybridized to a sweat gland cDNA library until a match was found. It seemed likely that the sweat gland cDNA library would have a good representation of the cystic fibrosis gene transcript because the disease involves the sweat glands.

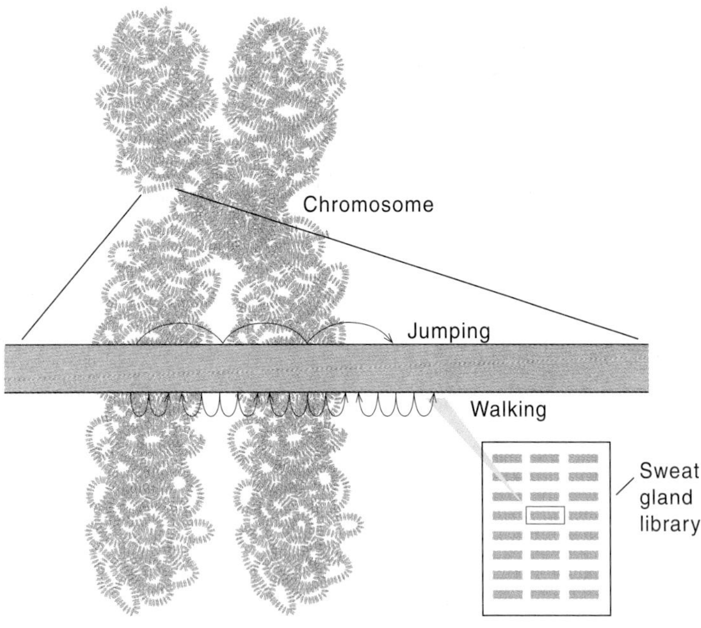

Figure 32.25

Predicted structure of the CFTR protein. Cylinders represent membrane-spanning helical segments. The cytoplasmically oriented nucleotide binding folds (NBFs) are shown as blue spheres with slots to indicate the points of entry of nucleotides. R represents the large polar domain, which is linked to two halves of the protein molecule. Charged amino acids are shown as small circles with the charge sign. Net charges on the internal and external loops joining the membrane cylinders and on regions of the NBFs are contained in open squares. Potential sites for phosphorylation by protein kinases A or C (PKA or PKC) and N-glycosylation (N-linked CHO) are indicated. (K = Lys; R = Arg; H = His; D = Asp; E = Glu.) (Reprinted with permission from J. R. Riordan et al., Identification of the cystic fibrosis gene, *Science* 245:1066, Sept. 8, 1989. Copyright 1989 American Association for the Advancement of Science.)

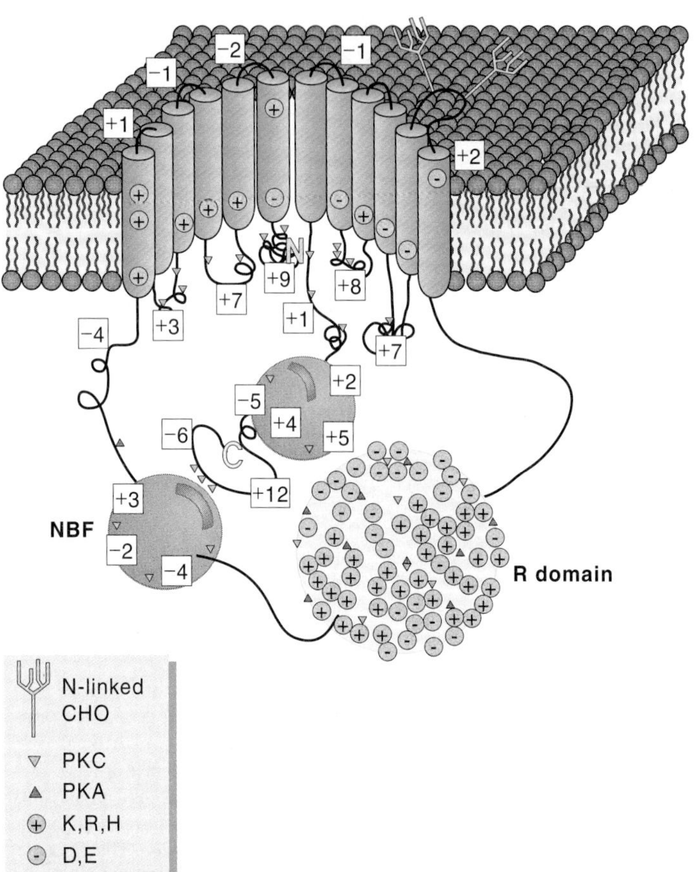

the cystic fibrosis gene was in hand? To answer this question, the researchers used the cDNA discovered in this way to probe the genomic library and thereby were able to map a gene that extended over a region of about 250 kb with 23 intervening sequences. In extracts of sweat gland tissue they detected a unique transcript, approximately 6,500 nucleotides in size, that matched the transcript size expected from this gene. The protein predicted from a sequence analysis of this transcript consists of two similar motifs, each with (1) a domain having properties consistent with membrane association and (2) a domain believed to be involved in ATP binding (fig. 32.25). Finally, the researchers discovered that most CF patients carry a three-base deletion in this transcript, eliminating a phenylalanine residue from the protein. Therefore they reasoned that the protein in cystic fibrosis patients should bear a defect. This defect correlated with the notion that CF patients have a faulty membrane protein that leads to the secretion problems characteristic of the disease. The fact that the abnormal gene is located as close as can be detected by classical genetics to the CF locus added additional support to the notion that the CF gene had been found.

The protein encoded by the cystic fibrosis gene has now been characterized in considerable detail. This protein, known as the cystic fibrosis transmembrane conductance regulator (CFTR), is a 168-kD membrane-spanning glycoprotein containing regulatory and nucleotide-binding domains. CFTR functions as a cAMP-stimulated channel for halides, water, and various small solutes.

Finding the disease gene has not, of course, meant an end to the problem of cystic fibrosis. However, the characterization of the disease gene will be a tremendous aid in diagnosing carriers and fetuses that are homozygous for the disease gene. It also should be a help in focusing approaches to finding a cure for the disease.

Will Nucleic Acids Ever Become Useful Therapeutic Agents?

Medical advances in the treatment of human disease have been punctuated by three breakthroughs in approach: the employment of vaccines to bolster the immune system, the use of antibiotics to defend against bacterial infection, and the use of antiviral drugs to interfere with the metabolism of certain viral pathogens. Our knowledge of nucleic acids and how to manipulate them leads me to believe that we are on the verge of another breakthrough, this time one in which nucleic acids will be the magic bullet.

There are three basically different ways in which nucleic acids could be used in therapy. New genes could be introduced that encode proteins that might be useful to the organism. New genes could be introduced that have been engineered to produce a ribozyme that is designed to attack a specific messenger. Finally, new genes could be introduced that encode an inhibitory RNA that might be expected to inhibit transcription of target genes or inhibit the translation of their mRNA.

Whereas nucleic acids will probably make ideal "medicines" someday because they can be engineered to interact strongly and specifically with a predesignated target nucleic acid, there is still a major problem of finding a way to introduce the nucleic acids into the organism. Recently, P. Felgner and his coworkers found that mice injected with DNA containing a gene from influenza A virus were thereafter immune to the influenza. The DNA is believed to have transfected the muscle cells and thereafter resulted in the production of sufficient influenza protein to immunize the animal. Assuming the correctness of this report, direct DNA injection has the advantage that it is simple, but it has a disadvantage in the majority of cases where it would be desirable to target a specific cell type.

For the purpose of targeting specific cell types, viruses that carry specific cell surface receptors would appear to be well suited. Retroviruses that convert their RNA genome to DNA and insert it permanently into the chromosomes of the cells it infects are attracting a great deal of attention by those interested in gene therapy. One problem here is that the retrovirus may make an ideal carrier but not have the desired specificity for cell attachment. In this regard, N. Kasahara and A. Dozy have recently reported that the envelope protein of a retrovirus can be modified so that it specifically infects cultured human cells that carry the receptor for the blood protein erythropoietin. This is a very exciting observation because it should be possible to modify the envelope protein in other ways so as to direct the virus to a predesignated cell type.

Summary

1. Sequencing DNA uses chemical methods to cleave specific bases in preexisting DNA or carries out the synthesis under conditions where synthesis is interrupted at specific bases.
2. A specific segment of DNA can be synthesized in vitro by the polymerase chain reaction. Short segments of DNA bordering the segment of interest are added to a mixture containing the segment of interest, a DNA polymerase, and the deoxyribotriphosphate substrates. The DNAs are first denatured, then annealed, and then synthesized. This cycle is repeated 20 or more times by raising the temperature to stop synthesis and lowering the temperature for annealing and synthesis. The outcome is a mixture in which the vast majority of the DNA is newly synthesized DNA bounded by the sequences of the added primers.
3. Another procedure for amplification cuts DNA containing the segment of interest into small pieces with a restriction enzyme. The cut pieces are incorporated into a plasmid or virus "vector" to be amplified in a suitable host. After growth the mixture is plated to produce a mixture of bacterial or viral clones. The clone or clones of interest are identified, often by hybridization of the clones after replica plating with a radioactive probe, followed by autoradiography to find the clone of interest.
4. Most cloning has been done in *E. coli*. Yeast is the most used eukaryotic host. Cloning is also possible in a number of plant and animal cells.
5. Mapping with recombinant DNA probes was first applied to the human globin genes. Starting probes were obtained by isolating the globin messenger from reticulocytes and converting it into a cDNA, which was used to scan a human genomic library for cross-hybridizing members. Once detected and purified, these cross-hybridizing members carrying globin messenger sequences were themselves converted to radioactive probes and used to further scan the genomic library for nearby sequences. By repeating this cycle several times, a process known as chromosome walking revealed a region around the adult hemoglobin gene that contained several closely related genes associated with hemoglobin.
6. Frequently, alleles of the same gene can be distinguished by restriction site differences in the genes themselves or in nearby locations. Alleles identified in this way are said to show restriction fragment length polymorphism. The allele responsible for sickle-cell disease was identified in this way.
7. The cystic fibrosis gene has been mapped by chromosome walking and jumping, a newer approach in which the relevant probes contain segments of the genome that are normally located about 500 kbp from one another. A cDNA library was made from normal sweat gland tissue, chosen because of the disease's association with abnormal release of sweat salt suggested that the sweat gland would contain an abundance of the messenger associated with the gene. By hybridizing the genomic DNA probes with the cDNA sweat gland library, a segment of genome was identified as a candidate for the cystic fibrosis gene. This gene was characterized in detail and found to encode a complex transmembrane protein that carries a specific amino acid change in over half of the persons with cystic fibrosis. This correlation is overwhelming support that the gene responsible for cystic fibrosis has been mapped and characterized.

Selected Readings

Aldhous, P., Fast tracks to disease genes. *Science* 265:2008–2009, 1994. Two new techniques for scanning the genome promise great advances in tracking the roots of disorders caused by multiple genes.

Barinaga, M., Step taken toward improved vectors for gene transfer. *Science* 266:1326–1327, 1994.

Boerrigter, M. E. T. I., Dolié, M. E. T., Martus, H.-J., Gossen, J. A., and Vijg, J., Plasmid-based transgenic mouse model for studying in vivo mutations. *Nature* 377:657–658, 1995.

Burke, D. T., G. F. Carle, and M. V. Olsen, Cloning of large segments of exogenous DNA into yeast by means of artificial chromosome vectors. *Science* 236:806–812, 1987.

Capacchi, M. R., Targeted gene replacement. *Scientific American* March: 52–59, 1994. Researchers can now create mice bearing any chosen mutations in any known gene. The technology is revolutionizing the study of mammalian biology.

Caruthers, M. H., A. D. Barone, S. L. Beaucage, D. R. Dodds, E. F. Fisher, L. J. McBride, M. Matteucci, Z. Stabinsky, and J. Y. Tang, Chemical synthesis of deoxyoligonucleotides. *Methods Enzymol.* 154:287–313, 1987.

Cohen, J. S., and M. E. Hogan, The new genetic medicines. *Scientific American* December:76–82, 1994. Synthetic strands of DNA are being developed as drugs. They can potentially attack viruses and cancers without harming healthy tissue.

Cohen, S. N., A. Change, H. Boyer, and R. Helling, Construction of biologically functional bacterial plasmids *in vitro*. *Proc. Natl. Acad. Sci. USA* 70:3240–3244, 1973. Of historical interest.

Dervan, P. B., Reagents for the site-specific cleavage of megabase DNA. *Nature* 359:87–88, 1992.

Gusella, J. F., DNA polymorphism and human disease. *Ann. Rev. Biochem.* 55:831–854, 1986.

Hasegawa, H., W. Skach, O. Baker, M. C. Calayag, V. Lingappa, and A. S. Verkman, A multifunctional aqueous channel formed by CFTR. *Science* 258:1477–1449, 1992. The cystic fibrosis gene product (CFTR) is a complex protein that functions as an adenosine 3,5-monophosphate (cAMP)-stimulated ion channel and possibly as a regulator of intracellular processes.

Hunkapiller, T., R. J. Kaiser, B. F. Koop, and L. Hood, Large-scale and automated DNA sequence determination. *Science* 254:59–67, 1991. State of the art on the mammoth project to sequence the human genome.

Jackson, D. A., R. H. Symons, and P. Berg, Biochemical method for inserting new genetic information into DNA of Simian Virus 40: Circular SV40 DNA molecules containing lambda phage genes and the galactose operon of *E. coli*. *Proc. Natl. Acad. Sci. USA* 69:2904–2909, 1972. Of historical interest.

Jaenisch, R., Transgenic animals. *Science* 240:1468–1474, 1989.

Jayawardhana, R., Naked DNA points way to vaccines. *Science* 259:1691–1694, 1993. Direct injections of a gene from the influenza A virus can immunize mice, sparking hopes that this simple and cheap approach can trip up other cunning pathogens.

Johnston, M., et al., Complete nucleotide sequence of *saccharomyces cerevisiae* chromosome VIII. *Science* 265:2077–2082, 1994. Only 22% of the genes were previously identified. On the average, there is one gene for every two kilobases.

Kan, Y. W., and A. M. Dozy, Antenatal diagnosis of sickle cell anaemia by DNA analysis of amniotic-fluid cells. *Lancet* II:910–912, 1978. Of historical interest.

Kerem, B., J. M. Rommens, J. A. Buchanan, D. Markiewicz, T. K. Cox, A. Chakravarti, M. Buchwald, and L. C. Tsui, Identification of the cystic fibrosis gene: Genetic analysis. *Science* 245:1073–1079, 1989.

Mansour, S. L., K. R. Thomas, and M. R. Capecchi, Disruption of the proto-oncogene *int*-2 in mouse embryo-derived stem cells: A general strategy for targeting mutations to non-selectable genes. *Nature* 336:348–352, 1988.

Maxam, A. M., and W. Gilbert, A new method of sequencing DNA. *Proc. Natl. Acad. Sci. USA* 74:560–564, 1977. Of historical interest.

Morell, V., Huntington's gene finally found. *Science* 260:28–30, 1993. The long hunt for the gene defective in Huntington's disease is over, raising hopes that researchers will be able to figure out what causes the condition.

Mullis, K. B., The unusual origin of the polymerase chain reaction. *Scientific American* 262:56–65, 1990. A surprisingly simple method for making unlimited copies of DNA fragments.

Riordan, J. R., J. M. Rommens, B. S. Kerem, N. Alon, R. Rozmahel, Z. Grezelczak, J. Zielenski, S. Lok, N. Plasvsic, J.-L. Chou, M. T. Drumm, M. C. Iannuzzi, F. S. Collins, and L.-C. Tsui, Identification of the cystic fibrosis gene: Cloning and characterization of complementary DNA. *Science* 245:1066–1073, 1989.

Saiki, R. K., S. Scharf, F. Faloona, K. B. Mullis, G. T. Horn, H. A. Erlich, and N. Arnheim, Enzymatic amplification of globin genomic sequences and restriction site analysis for diagnosis of sickle cell anemia. *Science* 230:1350–1354, 1985.

Sambrook, J., E. F. Fritsch, and T. Maniatis, *Molecular Cloning: A Laboratory Manual,* 2d ed., Cold Spring Harbor Laboratory Press, Cold Spring Harbor, N.Y., 1989. A three-volume collection that is thorough but somewhat dated.

Sanger, F., Sequences, sequences, and sequences. *Ann. Rev. Biochem.* 57:1–28, 1988. A scientific memoir. Of historical interest.

Sanger, F., and A. R. Coulson, A rapid method for determining sequences in DNA by primed synthesis with DNA polymerase. *J. Mol. Biol.* 94:444–448, 1975.

Thomas, K. R., and M. R. Capacchi, Site-directed mutagenesis by gene targeting in mouse-embryo-derived stem cells. *Cell* 51:503–512, 1987. Of historical interest.

Wagner, R. W., Gene inhibition using antisense oligodeoxynucleotides. *Nature* 372:333–335, 1994.

Watson, J. D., M. Gilman, J. Witkowski, and M. Zoller, *Recombinant DNA,* 2d ed. Scientific American Books. New York: W. H. Freeman Company, 1992. This text is an excellent elementary text on the subject of recombinant DNA. It contains many exciting chapters on specific applications and is extremely well referenced.

Welsh, M. J., and A. E. Smith, Cystic fibrosis. *Scientific American* December: 52–59, 1995.

Problems

1. Read the rest of the sequence in the autoradiogram in figure 32.1 as far as possible. Why do bands appear closer together near the top of the autoradiogram? Where would the primer be with respect to the sequence you have read from the autoradiogram?
2. A procedure similar to Sanger's dideoxy method of DNA sequencing, except for the substitution of reverse transcriptase for DNA polymerase, can be used to sequence RNA. Briefly describe this method.
3. What are the major advantages of the polymerase chain reac-

tion (PCR) method for amplifying defined segments of DNA as opposed to the use of conventional cloning methods? How might the PCR method be used to test for infection with the AIDS virus, and how would this be an improvement over the antibody test that is currently used? (The current ELISA test is an indirect test for the presence of antibodies against the HIV proteins.)

4. The first step in designing a PCR-based strategy for the study of a gene is the choice of primers. Since oligonucleotide synthesis is, in some cases, a significant cost, it may be tempting to use only short primers (<20 nucleotides). What is the disadvantage of using such short oligonucleotides? What considerations other than size might be important in primer design?
5. Calculate the expected frequency of occurrence of restriction sites for *PstI* and *HindIII* (see table 32.1) in the DNA from a thermophile (89% G + C) and from *E. coli* (52% G + C).
6. You just isolated a novel recombinant clone and purified the desired insert (a 10,000-bp linear duplex DNA) from the vector. Now you wish to map the recognition sequences for restriction endonucleases A and B. You cleave the DNA with these enzymes and fractionate the digestion products according to size by agarose gel electrophoresis. Comparison of the pattern of DNA fragments with marker DNAs of known sizes yields the following results:
   (a) Digestion with A alone gives two fragments, of lengths 3,000 and 7,000 bp.
   (b) Digestion with B alone generates three fragments, of lengths 500, 1,000, and 8,500 bp.
   (c) Digestion with A and B together gives four fragments, of lengths 500, 1,000, 2,000, and 6,500 bp.
   Draw a restriction map of the insert, showing the relative positions of the cleavage sites with respect to one another.
7. Draw the ends of a DNA fragment digested with the restriction endonuclease *BamHI*. How do these ends differ from those generated by *MboI*? If *MboI* and *BamHI* ends were to be ligated together, would the resulting junction be cleavable by *BamHI* or *MboI*?
8. If a DNA preparation were subjected to digestion with *ClaI* and then incubated with the Klenow fragment of DNA polymerase I and [α-^{32}P]dCTP, what would be the structure at the ends of the resulting radiolabeled fragments? Could the same strategy be used for radioactive labeling of fragments generated by *KpnI*?
9. Which of the methods listed below would serve as the best starting point to accomplish each objective?
   Methods
   (a) Polymerase chain reaction
   (b) Isolating a clone from a cDNA library
   (c) Isolating a clone from a genomic library
   Objectives
   (1) Characterizing sequences controlling expression of a gene
   (2) Characterizing a mutation in a known gene
   (3) Deducing a protein sequence from its coding sequence
10. A transgenic mouse line carrying a mutant CFTR gene has been prepared. How might the study of these mice benefit humans suffering from cystic fibrosis?
11. One of the problems with the use of Ti plasmids in plants is that the bacterial host, *Agrobacterium tumefaciens,* can sometimes survive prolonged antibiotic treatment in regenerating plantlets. How could you verify that the transfer of a DNA segment of interest from bacterial T-DNA to plant genomic DNA had occurred and that all of the bacterial DNA was gone?
12. The Southern blot technique is often used to compare genes from different organisms. For example, one could use the human globin gene probe described in the text to determine the extent of homology between globin genes from different primates. How could one reduce the stringency of hybridization conditions (step 4 of fig. 32.18) to permit such a "heterologous" hybridization?
13. Why is the Klenow fragment of *E. coli* DNA polymerase I preferable to the unmodified form of DNA polymerase I in site-directed mutagenesis procedures?
14. An unusual feature of the sickle-cell variant of the β-globin gene is that it directly alters a cleavage site for restriction endonuclease *MstII*. *MstII* recognizes the sequence CCTGAGG, which is mutated to CCTG<u>T</u>GG in the sickle-cell gene. How would you use this information and the Southern blot method to analyze fetal cells in amniotic fluid to determine whether the fetus carries sickle-cell anemia? What problems might you encounter in using this method?
15. Describe a procedure using the PCR technique that could be used to determine whether a normal individual is a carrier of the cystic fibrosis ΔF508 mutation. What problems could you anticipate with this method?
16. Transformation of the yeast *Saccharomyces cerevisiae* with linear fragments of genes results in a high frequency of homologous recombination. What would happen if you transformed *S. cerevisiae* with a fragment of a yeast gene into the middle of which was inserted an antibiotic resistance gene and then selected for resistance to that antibiotic? Why would such "knock-out" mutations be useful?
17. Your dog has excavated a remarkably well-preserved dinosaur bone (its identity was confirmed by your paleontologist friend), from which you have extracted a tiny bit of DNA. Using primers complementary to highly conserved regions of ribosomal RNA genes, you have used the polymerase chain reaction to amplify what may be a segment of the dinosaur ribosomal RNA gene. You decide to compare this sequence with other ribosomal RNA sequences in the sequence databases. Unexpectedly, your sequence turns out to be nearly identical to canine rRNA genes but more distantly related to those of birds and crocodiles (which are presumed to be dinosaurs' closest living relatives). What do you suppose happened?
18. Why would telomeric DNA be underrepresented in a genomic library that was constructed from DNA digested with a restriction enzyme in a vector treated with a compatible restriction enzyme?
19. Why, in a cDNA library prepared in a vector that has appropriately positioned transcription and translation start signals, would only one in six clones containing the desired DNA also be able to express the corresponding protein?

C H A P T E R

RNA Synthesis and Processing

Single-stranded RNA molecules are transcribed from select regions of the DNA genome.

In prokaryotes DNA, RNA, and protein synthesis all take place in the same cellular compartment. In eukaryotes the DNA is compartmentalized in the cell nucleus, and it became clear long before the biochemistry of these three processes was understood that DNA synthesis takes place in the nucleus, whereas the bulk of protein synthesis takes place in the cytoplasm. From these observations on eukaryotes it was self-evident that DNA cannot be directly involved in the synthesis of protein but must somehow transmit its genetic information for protein synthesis to the cytoplasm. Careful experiments with radioactive labels were used to demonstrate that RNA synthesis takes place in the nucleus; much of this RNA is degraded quickly; most of the portion that survives is transferred to the cytoplasm (fig. 33.1). These observations supported the notion that RNA is the carrier of sequence information for the synthesis of proteins.

In this chapter we focus on the structure and metabolism of the major classes of cellular RNA.

The First RNA Polymerase to Be Discovered Did Not Require a DNA Template

The first enzyme discovered that could catalyze polynucleotide synthesis was a bacterial enzyme called polynucleotide phosphorylase (PNPase). This enzyme, isolated by Severo Ochoa and Marianne Grunberg-Manago in 1955, could make long chains of 5′-3′-linked polyribonucleotides starting from nucleoside diphosphates. However, there was no template requirement for this synthesis, and the sequence was uncontrollable except in a crude way by adjusting the relative concentrations of different nucleotides in the starting materials.

Shortly after Ochoa's studies, Sam Weiss began a search for a DNA-directed RNA polymerase. His experimental design was influenced by the theory that RNA must be made on a DNA template if it is to carry the genetic message from DNA to protein. In his experimental design he was influenced by Kornberg's discovery that DNA synthesis required nucleoside triphosphates for substrates rather than nucleoside diphosphates. With crude liver extracts, Weiss was able to demonstrate a capacity for RNA synthesis that was severely inhibited by DNase. Weiss's results touched off systematic investigations of RNA metabolism in many laboratories. A continuous expansion of research effort from that time on has yielded a wealth of understanding about the transcription process and related aspects of RNA metabolism.

DNA–RNA Hybrid Duplexes Suggest That RNA Carries the DNA Sequences

Sol Spiegelman reasoned that if RNA was made on a DNA template, it should be complementary to one of the DNA chains. In this case it should be possible to make a DNA–RNA duplex by annealing, as was done for complementary DNA chains (see chapter 31). Using ^{3}H-labeled T2 bacteriophage DNA and ^{32}P-labeled RNA that was synthesized after T2 infection of *E. coli* cells, he was able to show that the newly synthesized RNA forms a specific complex with the viral DNA (fig. 33.2). It was presumed that the specific complex must involve Watson-Crick-like base pairing between an RNA and a DNA chain, with uracil playing the role of thymine in the RNA. Complementary interaction was a strong indication that the RNA was a product of DNA-directed synthesis. Although most RNAs are synthesized as a result of transcription from a DNA template, different strategies are used in their synthesis and in their posttranscriptional modification. The strategy used in a given case is strongly related to the RNA species, of which there are several types. For this reason it is important that we discuss the different classes of RNA found in the cell before we consider their mode of synthesis or postsynthesis modification.

There Are Three Major Classes of RNA

There are three major types of RNA that are transcribed from the DNA template: messenger RNA (mRNA), ribosomal RNA (rRNA), and transfer RNA (tRNA). These three RNAs work together in protein synthesis. Some of the properties of the three RNAs as they are found in *E. coli* are summarized in table 33.1.

Figure 33.1

In eukaryotic cells the nuclear membrane separates the processes of RNA and protein synthesis. This can be demonstrated with radioactive substrates that are precursors of RNA and protein. Immediately after exposure of cells to labeled precursors, the RNA label becomes fixed in the nucleus, and the protein label becomes fixed in the cytoplasm. Eventually, most of the labeled RNA becomes transferred to the cytoplasm, and a fraction of the labeled protein becomes transferred to the nucleus.

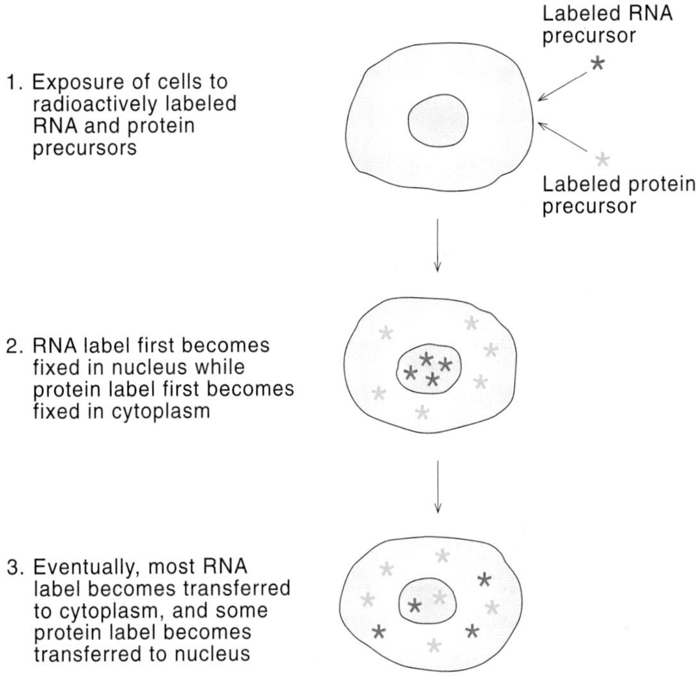

Labeled RNA precursor
1. Exposure of cells to radioactively labeled RNA and protein precursors
Labeled protein precursor
2. RNA label first becomes fixed in nucleus while protein label first becomes fixed in cytoplasm
3. Eventually, most RNA label becomes transferred to cytoplasm, and some protein label becomes transferred to nucleus

Messenger RNA Carries the Information for Polypeptide Synthesis

Messenger RNA carries the message from the DNA for the synthesis of a polypeptide chain with a specific sequence of amino acids. Each protein is unique, resulting in an enormous variety of mRNAs. It should be noted (see table 33.1) that although 40%–50% of the RNA synthesized in *E. coli* is mRNA, at any given time mRNA accounts for only about 3% of the cellular RNA. This is because bacterial mRNA is very unstable, with an average half-life of only 1–3 min. Messenger RNAs are heterogeneous in both size and sequence, reflecting the fact that each mRNA encodes the information for the synthesis of a different protein(s).

Transfer RNA Carries Amino Acids to the Template for Protein Synthesis

The function of transfer RNA is to carry amino acids to the mRNA template, where the amino acids are linked into a specific order. For this purpose each tRNA molecule contains a site for the attachment of an amino acid and a site, the anticodon, that recognizes the corresponding three-base codon on the mRNA (see chap-

Figure 33.2

Demonstration that phage RNA has sequences complementary to phage DNA. S. Spiegelman and B. Hall used ^{3}H-labeled T2 DNA and ^{32}P-labeled RNA, the latter made after T2 infection of *E. coli* cells. The two nucleic acids were mixed together, annealed, and then centrifuged to equilibrium in a CsCl density gradient. In (*a*) the DNA was first heat-denatured and then mixed with RNA and annealed at 65°C prior to centrifugation. In (*b*) the DNA denaturation step was left out. Most of the RNA goes to the bottom of the centrifuge tube because of its

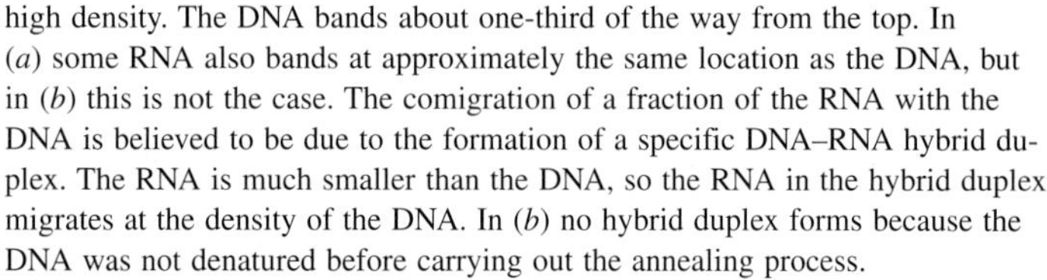

high density. The DNA bands about one-third of the way from the top. In (*a*) some RNA also bands at approximately the same location as the DNA, but in (*b*) this is not the case. The comigration of a fraction of the RNA with the DNA is believed to be due to the formation of a specific DNA–RNA hybrid duplex. The RNA is much smaller than the DNA, so the RNA in the hybrid duplex migrates at the density of the DNA. In (*b*) no hybrid duplex forms because the DNA was not denatured before carrying out the annealing process.

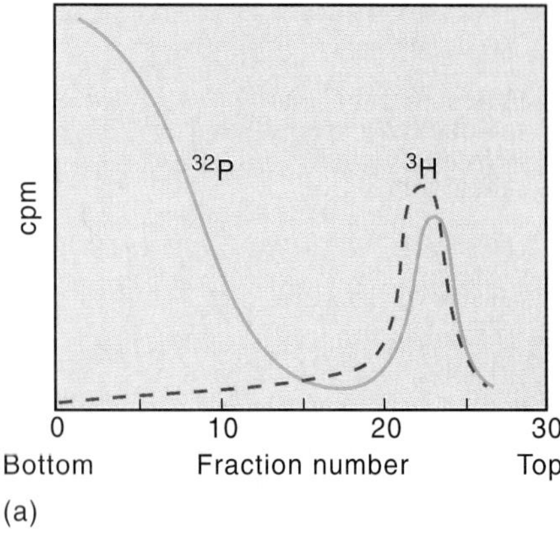

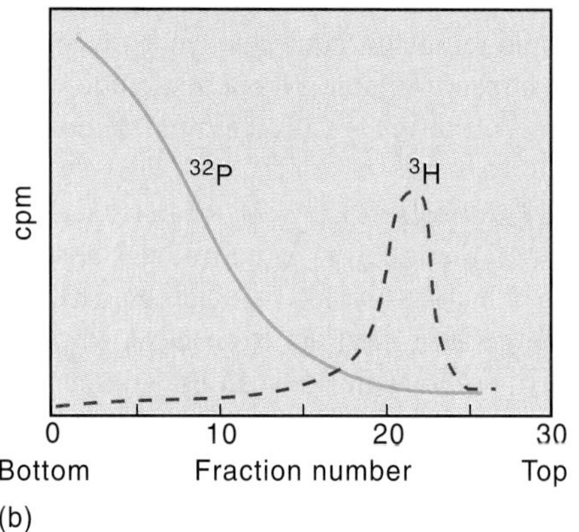

Table 33.1
Types of RNA in *Escherichia coli*

Type	Function	Number of Different Kinds	Number of Nucleotides	Percent of Synthesis	Percent of Total RNA in Cell	Stability
mRNA	Messenger	Thousands	500–6000	40–50	3	Unstable ($t_{1/2}$ = 1–3 min)
rRNA	Structure and function of ribosomes	3 {23S, 16S, 5S}	2800 1540 120	50	90	Stable
tRNA	Adapter	50–60	75–90	3	7	Stable

Source: G. Zubay, *Biochemistry,* 2d ed., Macmillan, New York, 1988, p. 893.

ter 34). Since there are only 20 amino acids that are incorporated into proteins and more than twice this number of tRNAs in most cells, several amino acids must be represented by more than one tRNA. Each amino acid is enzymatically attached to the 3′ end of one or more tRNAs by a specific aminoacyl-tRNA synthase that recognizes both the amino acid and the tRNA.

The primary structure is known for all *E. coli* tRNAs, and the three-dimensional structures of some of them have been determined by x-ray crystallography. The three-dimensional structure of phenylalanine tRNA of yeast is shown in figure 33.3. All tRNAs show similarities in their folded structures with four loops. Except for the variable loop, the loops are usually the same size in different tRNAs. The $tRNA^{Phe}$ molecule contains 20 bp that are hydrogen-bonded in Watson-Crick fashion. G-U can also form a hydrogen-bonded base pair in folded RNA structures that is slightly distorted compared to the standard Watson-Crick base pairs (see fig. 34.8). An additional 40 or so hydrogen bonds are formed in other ways (fig. 33.4). These additional hydrogen bonds and the accompanying base stacking stabilize the $tRNA^{Phe}$ in the complex folded structure shown in figure 33.3*a*.

All tRNAs contain several ribonucleotides that differ from the usual four (12 in the case of $tRNA^{Phe}$). The structures for some of these are shown in figure 33.5. Only four ribonucleotides are incorporated into RNA in the transcription process. All of the rare bases found in the mature tRNA result from posttranscriptional modification.

The complex folded structures adopted by tRNAs illustrate the fact that nucleic acids with a properly adjusted primary

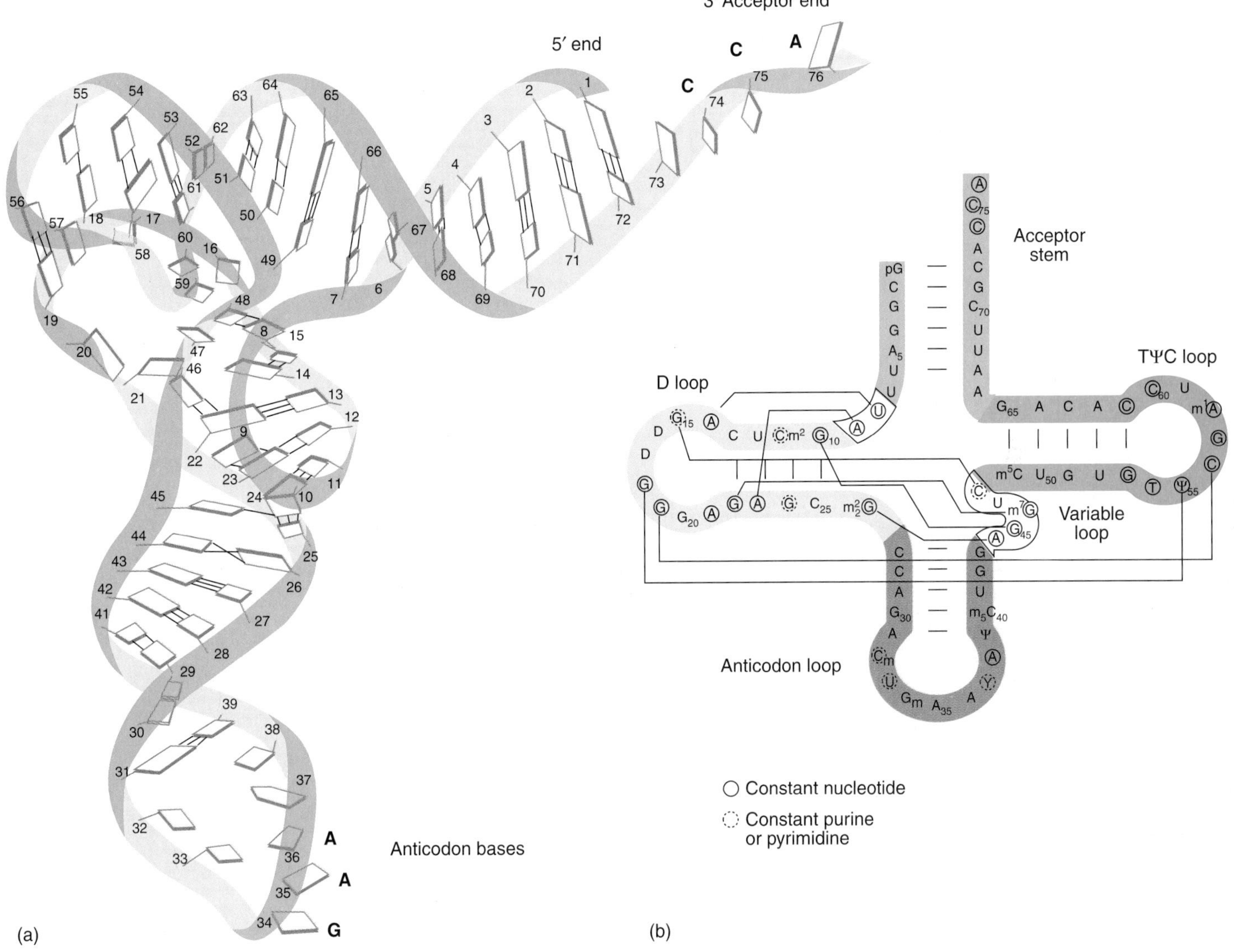

Figure 33.3

The tertiary structure of yeast phenylalanine tRNA. (*a*) The full tertiary structure. Purines are shown as rectangular slabs, pyrimidines as square slabs, and hydrogen bonds as lines between slabs. (*Source:* From G. J. Quigley and A. Rich, Structural domains of transfer RNA molecules, *Science,* 194:796, 1976.) (*b*) Nucleotide sequence. Residues that appear in most of the yeast tRNAs and residues that appear to be constantly a purine or a pyrimidine are indicated. Residues involved in tertiary base pairing are shown connected by solid lines. Several of the nucleotides are methylated. These are indicated by a lowercase m. (Illustration copyright by Irving Geis. Reprinted by permission.)

sequence can adopt complex secondary and tertiary structures. Apropos of this, Francis Crick once said that transfer RNA is an RNA molecule trying to look like a protein.

Ribosomal RNA Is an Integral Part of the Ribosome

The bulk of the cellular RNA is ribosomal RNA. Although seven genes exist in *E. coli* for rRNA, they all lead to essentially the same three ribosomal RNA molecules (see table 33.1), which differ substantially in size. The three rRNAs are always found in a complex with proteins in a functional component known as the ribosome. The ribosome is the site where mRNA and tRNAs meet to engage in protein synthesis. In *E. coli,* ribosomes are referred to as 70S particles, a measure of their rate of sedimentation and hence their size (S refers to Svedberg units, which as defined in chapter 7). A 70S ribosome consists of two dissociable subunits: a 50S subunit and a 30S subunit. Each of these contains both RNA and protein. The 50S subunit contains 23S and 5S rRNAs. The 30S subunit contains a single 16S rRNA (fig. 33.6). Eukaryotic ribosomes are similar in structure, although they are somewhat larger (80S) and contain mostly larger rRNAs (25–28S, 18S, 5S, and an additional 5.5–5.8S; this additional rRNA corresponds in sequence to the first 150 nucleotides of the prokaryotic large rRNA subunit). Chloroplasts and mitochondria have ribosomes and

Figure 33.4

Some of the tertiary hydrogen-bonded interactions found in yeast phenylalanine tRNA. Superscripts refer to base number in the tRNA molecule. It can be seen that there are many hydrogen-bonding arrangements in a tRNA structure that are not found in a duplex DNA structure. (*Source:* G. J. Quigley and A. Rich, "Structural domains of transfer RNA molecules" in *Science,* 194:796, 1976. Copyright © 1976 American Association for the Advancement of Science, Washington, D.C.)

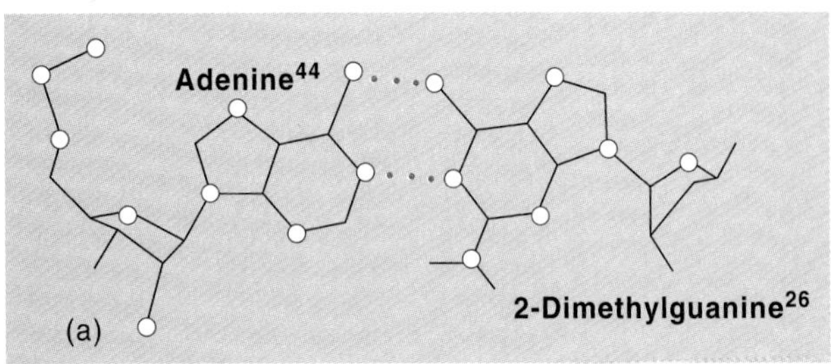

Figure 33.5

The structures of some modified nucleosides found in tRNA. The parent ribonucleosides are shown on the left in yellow screens. The other bases found in RNA result from posttranscriptional modification.

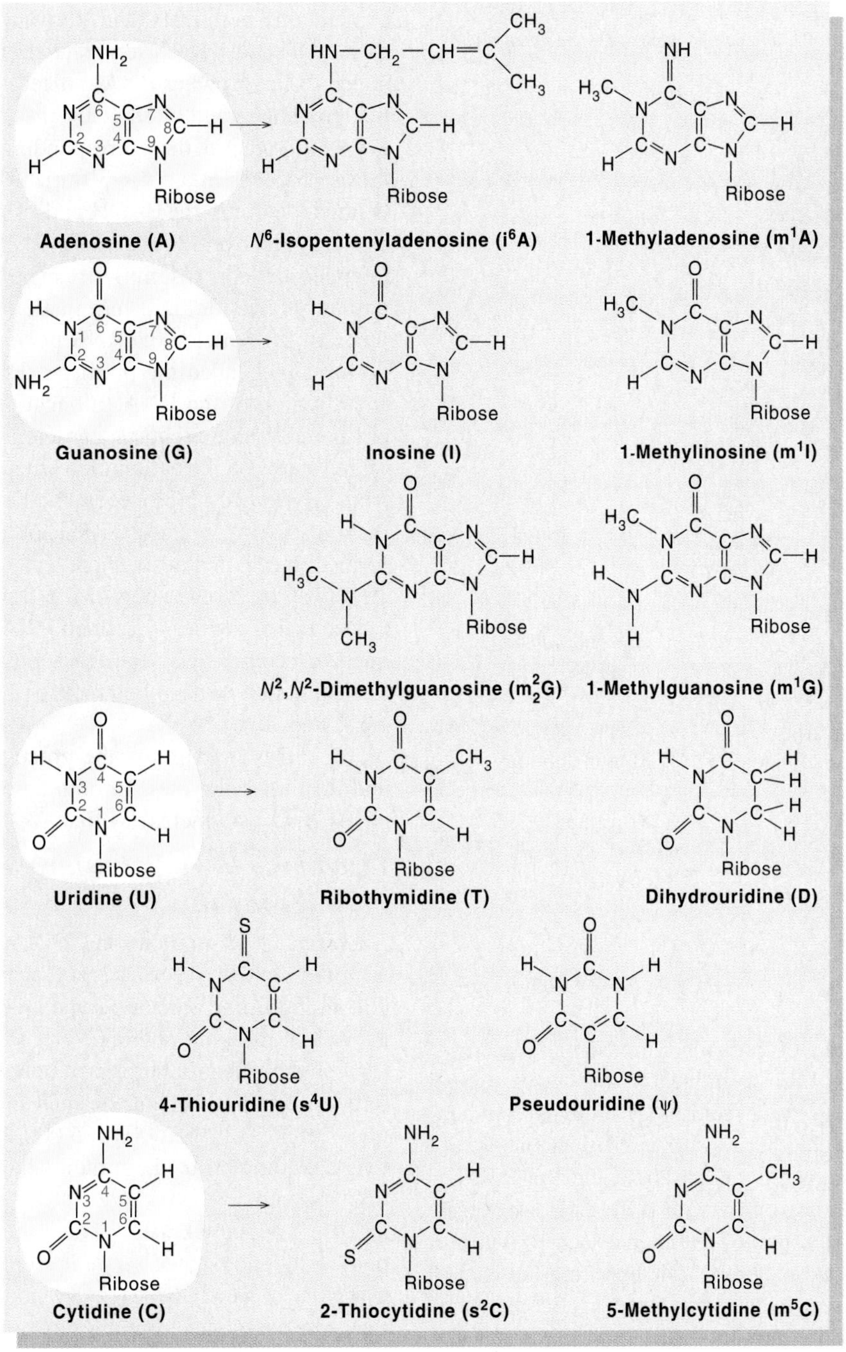

Adenosine (A) → N^6-Isopentenyladenosine (i^6A), 1-Methyladenosine (m^1A)

Guanosine (G) → Inosine (I), 1-Methylinosine (m^1I), N^2,N^2-Dimethylguanosine (m^2_2G), 1-Methylguanosine (m^1G)

Uridine (U) → Ribothymidine (T), Dihydrouridine (D), 4-Thiouridine (s^4U), Pseudouridine (ψ)

Cytidine (C) → 2-Thiocytidine (s^2C), 5-Methylcytidine (m^5C)

rRNA that are distinctly different from those present in the cytoplasm and strongly resemble those of prokaryotes, testifying to their evolutionary origin from bacteria.

The Fine Structure of the Ribosome Is Beginning to Emerge

Results from many different experimental approaches are beginning to coalesce to produce a three-dimensional picture of the ribosome that includes the location of its individual structural components and functional sites. The development of this picture has been especially challenging because the ribosome is large, fragile, and structurally complex and has resisted efforts to produce crystals that are capable of giving high-resolution structural information.

The current view of the overall morphology of ribosomes is based largely on electron-micrographic studies of the subunits

Figure 33.6

Composition of the *E. coli* ribosomes. The 70S ribosome can dissociate into a 50S and a 30S subunit. In vitro this can be done by lowering the Mg ion concentration. The individual subunits can be dissociated into their constituent RNAs and proteins by exposure to urea denaturant. Molecular weights are given for the subunits and the proteins, and the numbers of nucleotides are given for the RNAs.

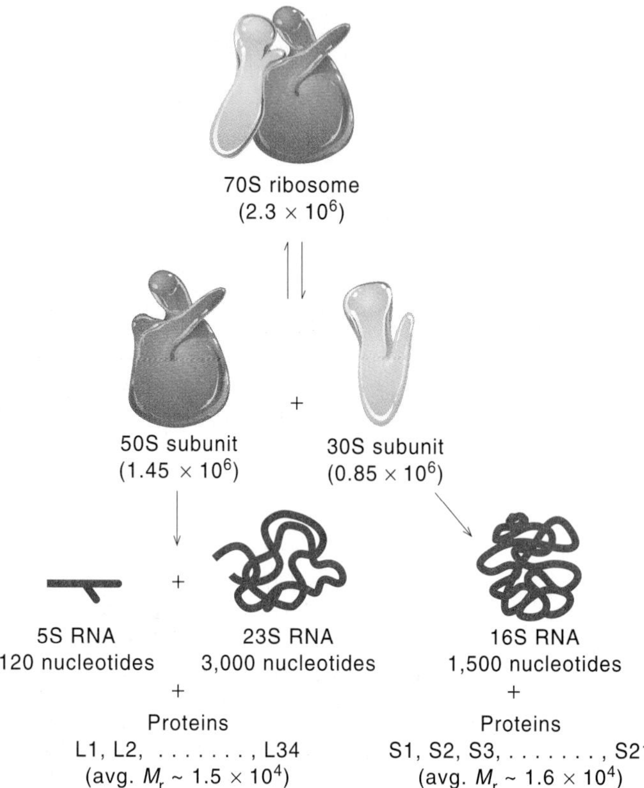

of *E. coli*. From this analysis it appears that both subunits are asymmetrical (see fig. 33.6).

The relative location of individual ribosomal proteins within the two subunits has been examined in two ways. One method involves determining which ribosomal proteins can be chemically cross-linked to each other and has yielded an elaborate grid of spatial relationships based on the frequency of cross-linking. The other method relies on neutron diffraction whereby the individually deuterated ribosomal proteins are located within the ribosomal subunit. The two methods of determining the location of ribosomal proteins have yielded a consistent spatial picture.

Information concerning protein locations within the 30S subunit has been combined with a secondary structure model of 16S rRNA and the location of protein-binding sites in rRNA to generate a partial three-dimensional picture of where the rRNA and proteins are situated in this ribosomal subunit (fig. 33.7).

Overview of the Transcription Process

All DNA-dependent RNA polymerases carry out the following reaction:

$$\text{NTP} + (\text{NMP})_n \xrightarrow[\text{DNA}]{\text{Mg}^{2+}} (\text{NMP})_{n+1} + \text{PP}_i$$

The subsequent breakdown of PP_i ensures the irreversibility of this reaction, as is the case in DNA polymerization reactions; this helps explain why both DNA and RNA polymerase utilize NTPs rather than NDPs. The DNA template strand determines which base is added to the growing RNA molecule. For example, a cytosine in the template strand of DNA means that a complementary guanine is incorporated at the corresponding location of the RNA. Synthesis proceeds in a $5' \rightarrow 3'$ direction, with each new nucleotide being added onto the 3′-OH end of the growing RNA chain.

The overall process for RNA synthesis on a duplex DNA template, called transcription, can be conceptually divided into initiation, elongation, and termination (fig. 33.8). In the initiation phase of the reaction the RNA polymerase binds at a specific site on the DNA called the promoter. Here it unwinds and unpairs a small region of the DNA. Elongation begins by the base pairing of ribonucleotide triphosphates to one strand of the DNA, followed by the stepwise formation of covalent bonds from one base to the next. As elongation proceeds, the DNA unwinds progressively in the direction of synthesis. The short stretch of DNA–RNA hybrid formed during synthesis is prevented from becoming longer than 10–20 bp by a rewinding of the DNA and the simultaneous displacement of the newly formed RNA. Termination occurs at a sequence recognized by the RNA polymerase as a stop signal. At this point the ternary complex of DNA, RNA, and polymerase breaks up.

First we discuss the process of transcription in bacteria and then in eukaryotes. In both cases we consider the properties of the RNA polymerase(s) first.

Bacterial RNA Polymerase Contains Five Subunits

Escherichia coli contains one RNA polymerase, which transcribes all three major types of RNA. The active enzyme is a pentamer containing four different polypeptide chains with a total molecular weight of about 460,000. The subunits of the enzyme can be separated by electrophoresis on polyacrylamide gels. The four different polypeptide chains, termed β', β, σ^{70}, and α, have molecular weights of 165,000, 155,000, 72,000, and 35,000, respectively. Additional σ-like proteins have been identified, which we discuss later.

A complex with the subunit structure $\alpha_2\beta\beta'\sigma^{70}$ can carry out the functions necessary for synthesis of RNA and is referred to as the holoenzyme. Holoenzyme can be reversibly separated into two components by chromatography on a phosphocellulose column:

$$\underset{\text{Holoenzyme}}{\alpha_2\beta\beta'\sigma^{70}} \rightleftharpoons \underset{\text{Core polymerase}}{\alpha_2\beta\beta'} + \underset{\text{Sigma-70}}{\sigma^{70}}$$

The enzyme without the sigma factor, called core polymerase, retains the capability to synthesize RNA, but it is defective in the ability to bind and initiate transcription at true initiation sites on the DNA. In fact, when RNA polymerase was first purified from

Figure 33.7

Arrangement of components in the *E. coli* 30S ribosomal particle. (*a*) The relative locations of the ribosomal proteins, numbered 1–21. (Illustration prepared by Dr. Malcolm Capel from data described in M. S. Capel, M. Kjeldguard, D. M. Engelman, *J. Mol. Biol.* 200:66–87, 1988.) (*b*) The conformation of the rRNA. The location of the proteins is indicated by numbers given in the figure. (Illustration prepared by S. Stern, B. Weiser, and H. F. Noller from data described in *J. Mol. Biol.* 204:447–481, 1988.)

(a)

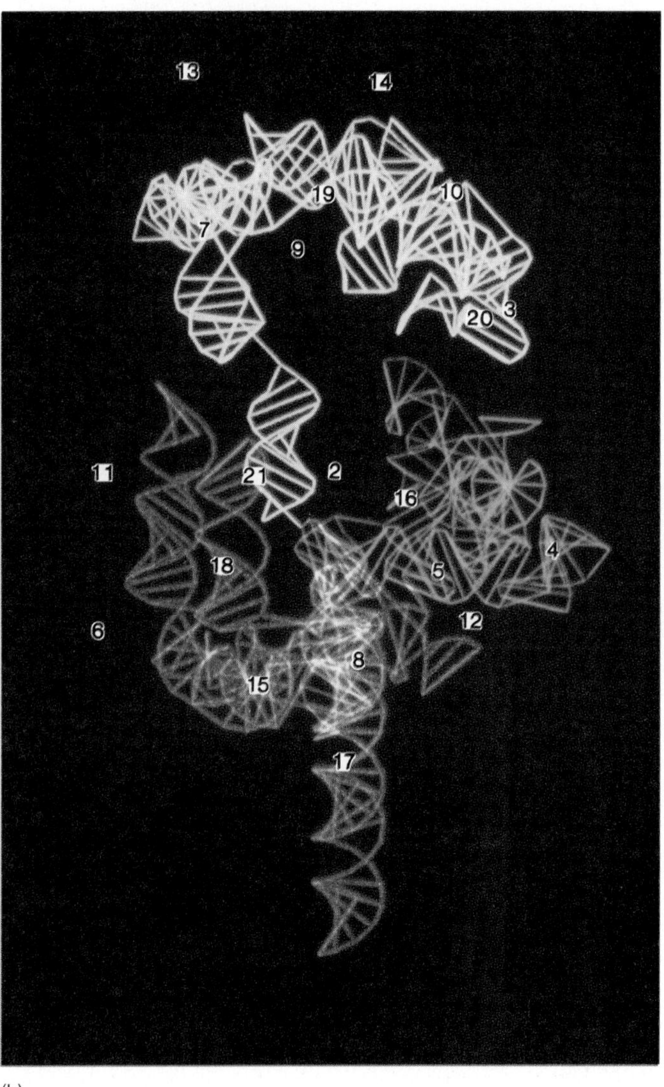

(b)

crude extracts, it was missing the σ factor. The assay for polymerase involved the use of DNA with single-strand nicks. When a DNA template was used that did not have single-strand nicks, this enzyme was not active. This led to a search for a missing factor. When this factor (σ^{70}) was added back to the purified core enzyme and the uncut DNA template, the enzyme was able to bind tightly and selectively initiate RNA chains. Because of its role in binding and initiation, σ^{70} is often referred to as an initiation factor.

The precise functions of the subunits of the core enzyme are not known. β' is a basic (positively charged) polypeptide thought to be involved in DNA binding. The β subunit is the site of binding of several inhibitors of transcription and is thought to contain most or all of the active sites for phosphodiester bond formation. The α subunit is necessary for reconstituting active enzyme from separated subunits.

Binding at Promoters

In the cell, RNA polymerase transcribes only select regions of the DNA; in these regions it synthesizes RNA that is complementary to one of the DNA strands. This selective action is possible because the holoenzyme is able to recognize and form a stable complex with DNA at specific promoters. RNA polymerase is able to form unstable nonspecific complexes at any place on the template, mainly by an interaction with the DNA phosphates, but it either

Figure 33.8

Overview of RNA synthesis.

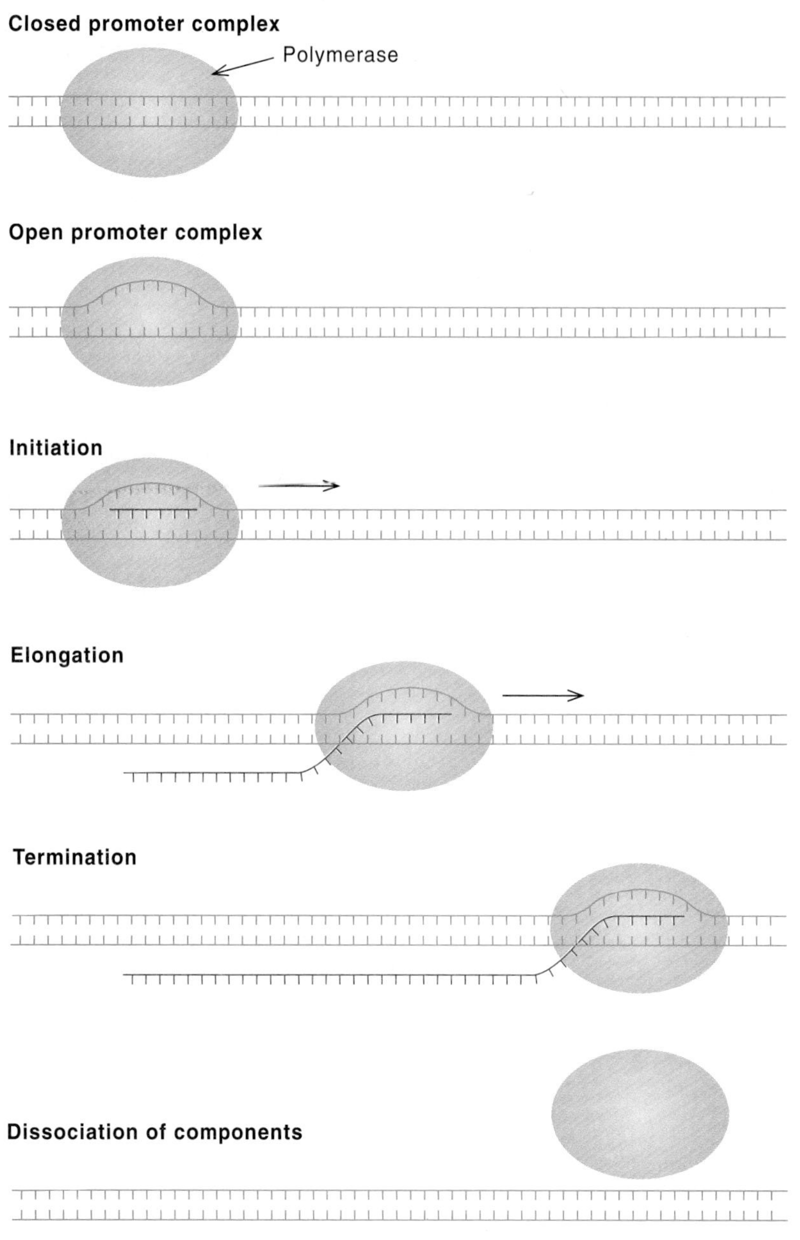

rapidly dissociates and rebinds or slides along the DNA until it reaches a promoter. It then forms a moderately stable complex with the promoter, most likely interacting with particular nucleotides in the −35 and −10 regions of the promoter. Initially, the complex contains DNA in the double-helical or unmelted state; in this state they are referred to as closed promoter complexes (fig. 33.9). The next step involves breaking the internucleotide H bonds (melting) over a stretch of about 10 bp of DNA, from positions −9 to 2, and a conformational change in the polymerase; this complex is termed the open promoter complex.

Common features have been identified in the sequence of over 250 different *E. coli* promoters at two locations: one 6-bp region centered at −35 (35 bp upstream of the initiation site) and one 6-bp region centered at −10 contain sequences that are similar but not identical in all bacterial promoters. The average sequences in both of these regions (TTGACA at −35 and TATAAT at −10) are termed the consensus sequences. A promoter with the consensus sequences in both regions leads to a very high level of transcription. Remarkably, no naturally occurring promoter has the consensus sequence. It seems likely that this is part of a strategy that ensures that regulatory proteins can stimulate promoters and thereby exert an influence over their activity. This issue is taken up in chapters 35 and 36. Promoter strength in the absence of regulatory proteins decreases the more the actual sequences deviate

Figure 33.9

Various types of RNA polymerase–DNA binding complexes. A nonspecific complex is formed at any point along the DNA. The closed promoter complex is formed at a polymerase-binding site. Following the formation of the closed promoter complex, an open promoter complex is formed at the same site.

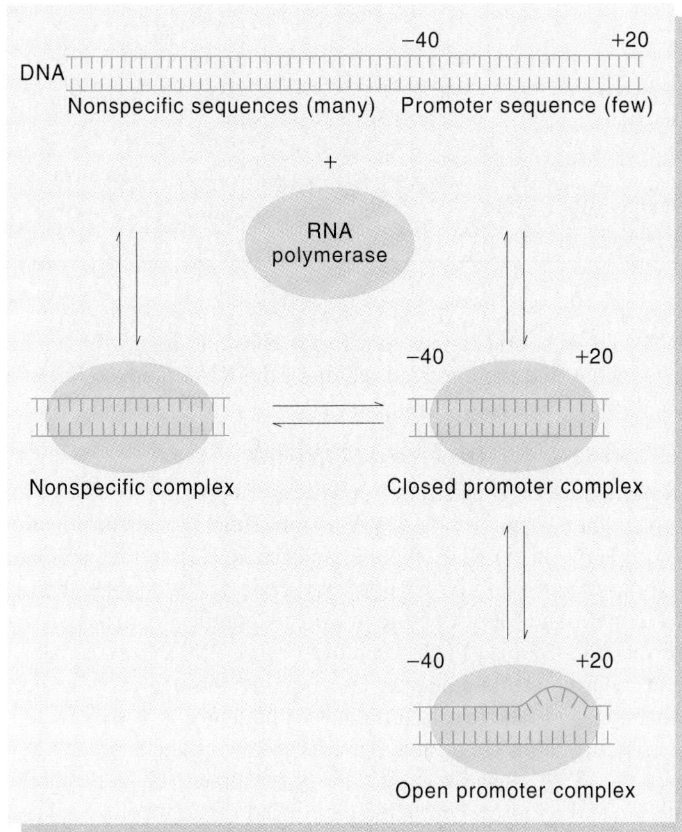

from the consensus sequences, with some bases being more important than others. Naturally occurring promoters differ greatly in strength, which is best defined in terms of the frequency of RNA initiation. Some promoters are very weak, such as that associated with the *lac* repressor gene (discussed in chapter 35); as a result the *lac* repressor gene is transcribed only once in 20–40 min. At the other extreme, one finds very strong promoters for rRNA genes, which are transcribed at the rate of one per second. This 2,000-fold difference in transcription rate is primarily a function of the base sequence of the promoter.

Initiation at Promoters

Once the polymerase binds to the promoter and strand separation occurs, initiation usually proceeds rapidly (1–2 s). The first, or initiating, NTP, which is usually ATP or GTP, binds to the enzyme. The binding is directed by the complementary base in the DNA template strand at the start site. A second NTP binds, and initiation occurs on formation of the first phosphodiester bond by a reaction involving the 3′-hydroxyl group of the initiating NTP with the inner phosphorus atom of the second NTP. Inorganic pyrophosphate derived from the second NTP is a product of the reaction. This process is illustrated in figure 33.10.

Figure 33.10

Details of phosphodiester bond formation. The α, β, and γ phosphates are indicated on the initiating NTP, which in this case is ATP. The colored ovals represent NTP-binding sites on the RNA polymerase. The biochemistry of bond formation in RNA synthesis is very similar to that in DNA synthesis.

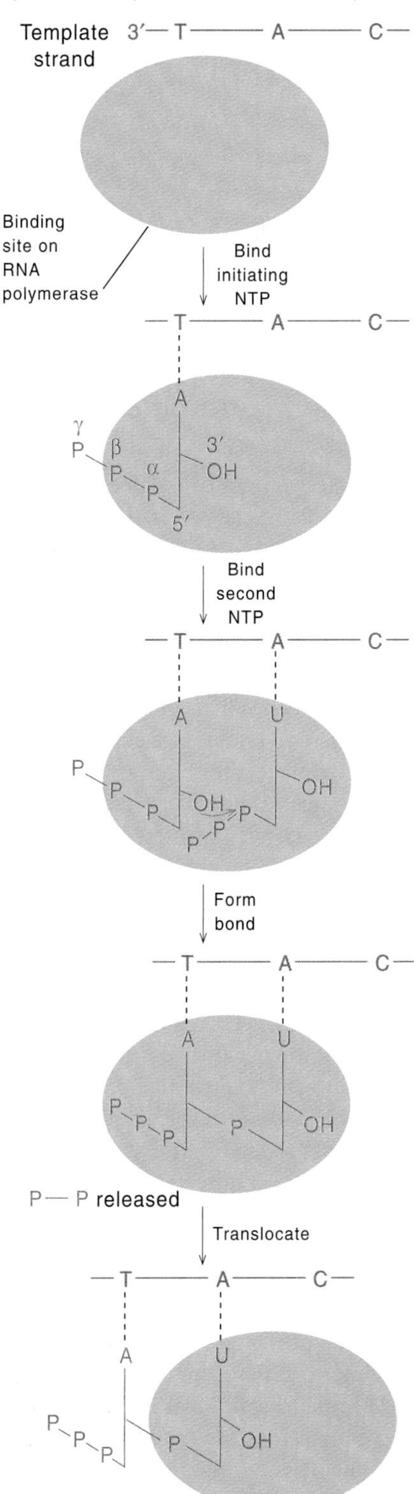

Table 33.2
Summary of *Escherichia coli* Sigma Factors

Sigma Factor	Gene	Consensus Sequence −35 Region	−10 Region	Genes Recognized
σ^{70}	*rpoD*	TTGACA	TATAAT	Most genes
σ^{32}	*rpoH (hptR)*	CTTGAA	CCCCAT-TA	Heat-shock-regulated
σ^{54}	*rpoN (ntrA)*	$CTGGCACN_5TTGCA$		Nitrogen-regulated
σ^{E}	Not identified	GAACTT	TCTGA	*rpoH, htrA*
σ^{F}	*flaI*	TAAA	GCCGATAA	Flagellar, chemotaxis
σ^{K}	*rpoK (katF)*	Unknown		*KatE*

Alternative Sigma Factors Trigger Initiation of Transcription at Promoters with Different Consensus Sequences

Although most promoters in *E. coli* appear to utilize σ^{70} as an initiation factor, several additional sigma-like initiation factors have been discovered that bind to core polymerase to form holoenzymes that recognize different species of promoters. One such factor is the product of the *rpoH* gene, which is involved in heat shock regulation. This factor is called sigma-32 (σ^{32}) because of its molecular weight of 32,000. It becomes bound to core polymerase in *E. coli* cells that have been subjected to heat shock and directs the polymerase to bind to and initiate at a class of promoters (the heat shock promoters) responsible for high-level expression of a dozen or so heat shock genes. Another such factor, σ^{54}, ($M_r = 54{,}000$) encoded by the *rpoN* gene is required for expression of certain genes involved in nitrogen metabolism. The consensus sequences associated with different classes of promoters are given in table 33.2.

Elongation of the Transcript

After initiation has occurred, chain elongation proceeds by the successive binding of the nucleoside triphosphate complementary to the base at the growth point in the template strand, bond formation with pyrophosphate release, and translocation of the polymerase one base farther along on the template strand. Transcription proceeds in the $5' \rightarrow 3'$ direction, antiparallel to the $3' \rightarrow 5'$ strand of the templating DNA strand. Once elongation has produced an RNA chain about 10 bases long, the σ subunit dissociates from the holoenzyme, leaving core polymerase to continue the elongation reaction until a terminator signal is reached. The released σ is available to bind to a free core polymerase and re-form a holoenzyme capable of binding at the same or other promoters, where it initiates additional RNA chains.

As the polymerase traverses the DNA, it must continually cause a melting or strand separation of the DNA so that a single DNA template strand is available at the active site of the enzyme. During elongation, one base pair re-forms behind the active site for every base pair opened in front of it. The short transient RNA–DNA hybrid duplex that forms between the newly synthesized RNA and the unpaired region of the DNA helps to hold the RNA to the elongating complex.

Termination of Transcription

Termination of transcription involves stopping the elongation process at a region on the DNA template that signals termination and release of the RNA product and the RNA polymerase. Most terminators are similar in that they code for a double-stranded RNA stem-and-loop structure just preceding the 3′ end of the transcript (fig. 33.11). Such structures cause RNA polymerase to pause, terminate, and detach. Two types of terminators have been distinguished. The first is sufficient without any accessory factors; it contains about six uridine residues following the stem and loop (see fig. 33.11). The second type of terminator lacks the polyU stretch and requires a protein factor called rho to facilitate release.

Rho binds to the RNA and is able to traverse RNA in an ATP-requiring reaction in the $5' \rightarrow 3'$ direction on the RNA. C residues on the RNA are absolutely essential for this to happen. A helicase activity in rho is essential for its termination activity. If rho catches up with the RNA polymerase at a pause site or a termination site, it is highly likely to catalyze termination. This general picture of rho action is supported by the finding that interjection of a translational block that inhibits ribosome movement encourages premature release of the transcript. In *E. coli,* ribosomes normally traverse the nascent transcript before transcription is completed. These ribosomes make it difficult for rho to bind and reach the elongating polymerase. A translational block facilitates rho binding and migration to the elongating RNA polymerase.

Comparison of Escherichia coli *RNA Polymerase with DNA PolI and PolIII*

In table 33.3 the bacterial RNA polymerase and the bacterial DNA polymerase are compared. Both types of enzymes are DNA-template-directed and require four NTPs and a divalent cation. While RNA polymerase makes single-stranded chains involved directly or indirectly in protein synthesis, the DNA polymerases replicate and repair the duplex DNA. Only RNA polymerase rec-

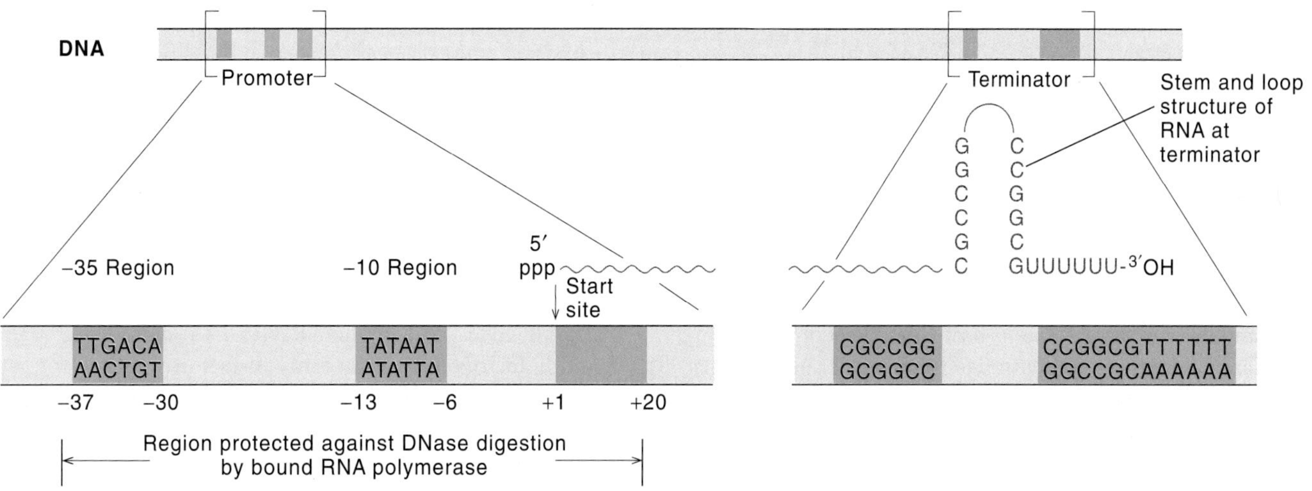

Figure 33.11

Important features of a typical transcription unit. DNA is shown with promoter and terminator regions expanded below. RNA is transcribed starting in the promoter region at +1 and ending after the stem and loop of the terminator.

Table 33.3
Comparison of *Escherichia coli* RNA Polymerase with DNA Polymerases I and III

	RNA Polymerase	**DNA Polymerases I and III**
Similarities		
DNA-template-directed	Yes	Yes
Requires 4 NTPs	Yes (rNTPs)	Yes (dNTPs)
Requires divalent cation	Yes	Yes
Differences		
Function	Transcription	Replication and repair
Initiates chains	Yes	No
Terminates chains	Yes	No
Recognizes sequences	Yes	No
Uses intact duplex template	Yes	No
Product	Single-strand RNAs	Duplex DNA strands
Proofreading	?	Yes

ognizes start-and-stop sequences in the duplex DNA. Thus far, only the DNA polymerases have been shown to have a proofreading function, but the possibility of a proofreading function for RNA polymerase has not been ruled out.

Important Differences Exist between Eukaryotic and Prokaryotic Transcription

Although the basic mechanism by which RNA is synthesized is quite similar in prokaryotes and eukaryotes, several important differences exist. Most parts of the bacterial DNA are readily accessible to RNA polymerase binding and transcription. By contrast, most DNA in eukaryotic cells exists in a condensed form (chromatin), which is less readily accessible to transcription. The fraction of DNA accessible to the RNA polymerase in any given cell type is especially sensitive to cleavage by mild treatment with bovine pancreatic DNase I. These regions of the DNA often contain bound RNA polymerase, modified histones, and additional nonhistone proteins. Active regions are often undermethylated compared with the total DNA. Most of the methylated groups in DNA are on the C residues in the CG sequences.

Whereas histones are likely to interfere with transcription initiation, particularly when they are bound to the promoter, there are many cases where histones are relatively scarce in the region of the promoter, suggesting either that the DNA sequences are not particularly favorable for histone binding or that there are other circumstances that lead to a local scarcity of bound histones. In this regard we shall see that there are certain proteins that may bind to the promoter even when the polymerase is not there. These proteins, which are part of the preinitiation complex, could serve the dual purpose of keeping the promoters relatively histone free.

We have stated that translation begins before transcription is completed in prokaryotes. The situation is quite different in eukaryotes, where transcription and translation occur in different cellular compartments separated by the nuclear membrane. Large precursors of mRNA are synthesized in the nucleus; these

Table 33.4
Comparison of Eukaryotic DNA–Dependent RNA Polymerases

Type	Location	RNAs Synthesized	Sensitivity to α-Amanitin
RNA polymerase I	Nucleolus	Pre-rRNA	Resistant
RNA polymerase II	Nucleoplasm	hnRNA, mRNA	Sensitive
RNA polymerase III	Nucleoplasm	Pre-tRNA, 5S RNA	Sensitive to very high levels
Mitochondrial	Mitochondria	Mitochondrial	Resistant
Chloroplast	Chloroplasts	Chloroplast	Resistant

become complexed with proteins to form ribonucleoprotein particles which are modified and processed to form smaller mRNAs that become transported across the nuclear membrane to the cytoplasm.

Eukaryotes Have Three Nuclear RNA Polymerases

Unlike prokaryotes, in which the major types of RNA are synthesized by one RNA polymerase, eukaryotic cells contain three nuclear RNA polymerases, each responsible for transcribing a different class of RNAs.

Nuclear extracts can be fractionated by chromatography on DEAE-cellulose to give three peaks of RNA polymerase activity. (The use of column chromatography is explained in chapter 7.) These three peaks correspond to three different RNA polymerases (I, II, and III), which differ in relative amount, cellular location, type of RNA synthesized, subunit structure, response to salt and divalent cation concentrations, and sensitivity to the mushroom-derived toxin α-amanitin. The three polymerases and some of their properties are summarized in table 33.4.

RNA polymerase I is localized to the nucleolus and transcribes a large precursor that is later processed to form rRNA. It is completely resistant to inhibition by α-amanitin. RNA polymerase II is located in the nucleoplasm and transcribes large precursor RNAs (sometimes called heterogeneous nuclear RNA, or hnRNA) but are processed to form cytoplasmic mRNAs. It is also responsible for the synthesis of most viral RNA in virus-infected cells. PolII is very sensitive to α-amanitin, being inhibited by 50% to 0.05 μg/ml. RNA polymerase III is also located in the nucleoplasm and transcribes small RNAs, such as 5S RNA and the precursors to tRNAs. This enzyme is somewhat resistant to α-amanitin, requiring about 5 μg/ml to reach 50% inhibition.

Eukaryotic RNA Polymerases Are Not Fully Functional by Themselves

Crude enzyme preparations of RNA polymerases I, II, and III have been shown to be capable of selective transcription of defined DNA templates, initiating at sites known in some cases to be utilized in vivo. Fractionation of these polymerase-containing extracts has revealed many additional protein factors that stimulate in vitro transcription. These factors are divided into two categories. Basal transcription factors are required for the transcription of virtually all genes of a particular class. In addition to basal transcription factors there is an array of factors required for activated transcription. Transcription studies are currently in a stage of rapid development, and the distinction between factors required for basal transcription and activated transcription is not always clear. The situation is much more complex than for prokaryotes.

First we discuss basal transcription factors. With all three polymerases, two or more basal transcription factors are required; some of these factors bind to the promoter before the polymerase can bind.

Messenger RNA Transcription by Polymerase II

Because many eukaryotic genes have been cloned during the last few years, it has become possible to compare the DNA sequences preceding genes that may act as promoter-like signals for RNA polymerase II. One feature that stands out for most polII promoters is a common sequence, TATAAA, called the TATA box, found usually 25–30 bp before the transcription start site.

The first step in the formation of the polymerase II initiation complex involves the binding of transcription factor IID (TFIID) to the TATA box. TFIID itself consists of a universal TATA-binding protein (TBP) and a number of associated factors (TAFs). TBP, the protein at the heart of TFIID, opens the way to transcription (fig. 33.12). One can define the recognition motif for TBP in structural terms as an 8-bp TATA element. The interaction with the TATA element causes the DNA to bend through an angle of about 80°, and in consequence the TATA element unwinds considerably, by about 110°. What implications does this dramatic change in DNA structure hold for the mechanism of transcription? Clearly, the bending at the TATA element could bring other initiation factors closer together than on linear DNA. The unwinding of the TATA element could well be a nucleation step for further unwinding toward the start site and thus for the eventual separation of the two DNA strands required for the initiation of transcription.

In a purified in vitro system it can be shown that the other factors required for formation of the transcription initiation complex are deposited in stepwise fashion once the TFIID protein is in place (fig. 33.13). For an update on transcription factors associated with PolII the reader may wish to see Jacobson and Tjian reference.

Figure 33.12

The structure of the TBP–TATA complex. The TBP is a saddle-shaped protein with an approximately twofold axis of symmetry. In this figure it is shown as a green ribbon that follows the course of its α-carbon atoms. The sugar–phosphate backbone of the DNA is shown in yellow, and the eight base pairs of the TATA element (whose 5′ end is on the left) in red. The remainder of the DNA, shown in blue, is computer generated. The complexed portion of the double helix is unwound to give a ladder-like appearance, and its axis is bent into a right-handed supercoil arc. Overall, the DNA is bent through an angle of 80° at the TATA element. (Courtesy, Dr. Paul B. Sigler, Yale Univ.)

Figure 33.13

Formation of a polymerase II initiation complex. Most PolII promoters contain a TATA box. Transcription factor D (TFIID) binds to the TATA box before the polymerase or other factors bind. The remaining factors bind in the order shown. Each of the protein factors including PolII is a multisubunit protein. The complex shown is for basal transcription. Many promoter complexes also involve activator proteins that usually bind to the complex shown and to an upstream site on the DNA. Although it is not shown in the figure a factor known as TFIIA is believed to enhance the DNA binding affinity of TBP and to mediate efficient activation of various enhancer binding proteins.

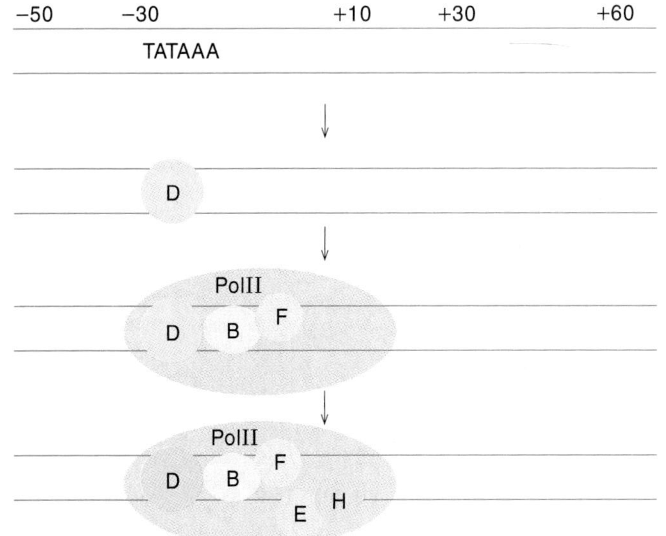

Recent studies by A. Koleske and R. Young on the yeast PolII enzyme suggest an alternative to the gradual stepwise formation of the initiation complex. The key observation is that a holoenzyme complex that accounts for 20–50% of the yeast PolII can be precipitated with the antibodies directed against transcriptional regulatory proteins encoded by SRB genes. These proteins appear to bind tightly to PolII. This has led to the suggestion that much of the holoenzyme is assembled independently and is available for binding directly to the TFIID/TATA complex. In many cases, DNA-bound upstream activators may also be involved in stabilizing the new initiation complex (fig. 33.14). The finding that much of the PolII exists as a holoenzyme does not necessarily exclude the alternative mechanism in which the initiation complex forms by the stepwise accretion of transcription factors. The two processes could occur jointly in different circumstances.

Another suggestion made by Koleske and Young relates to how PolII dissociates from the initiation complex at the start of transcription. The carboxy-terminal domain of the PolII-designated CTD is believed to be important in making contacts with protein factors that are distinct from the basal transcription factors. Since the PolII CTD is generally found in an unphosphorylated form in transcription initiation complexes but is highly phosphorylated in molecules in the midst of transcription elongation, CTD phosphorylation has been proposed to be a regulatory event governing some aspect of the transcription initiation process. Such a phosphorylation event could disrupt multiple interactions between the

Figure 33.14

Formation of an initiation complex by the yeast RNA polymerase II (PolII) holoenzyme. Multiple interactions between the holoenzyme, activator proteins, and transcription factor TFIID stabilize the initiation complex. In addition to the PolII enzyme the yeast holoenzyme depicted contains a number of transcription factors that are believed to interact primarily with activators (especially the SRB factors). It also contains other factors such as TFIIH and TFIIB that are believed to interact primarily at the TFIID-binding site.

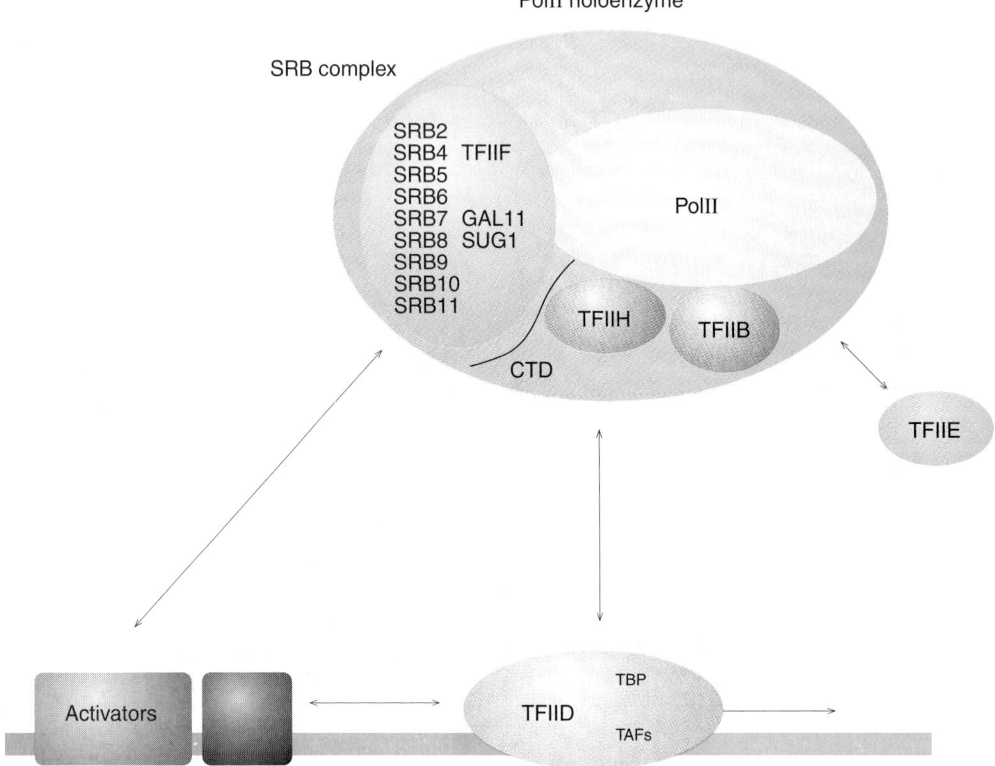

PolII CTD and other components of the transcription initiation complex to allow the enzyme to dissociate from those components and exit the promoter (fig. 33.15).

The PolI Promoter Has Two Elements

For the formation of the RNA polymerase I initiation complex two additional proteins are required (fig. 33.16). Transcription factor UBF binds specifically to sequences in an upstream element and a core element. The second protein, SL1, cannot bind to the DNA by itself but binds cooperatively to the DNA in the presence of UBF. Once these two protein complexes have formed, a single PolI binds to the core element, completing formation of the initiation complex. It is presumed that the upstream element interacts in some way with the core element to stabilize the initiation complex.

For some time it was thought that the TATA-binding protein TBP was involved in PolII promoters only where a TATA box was present. However, it turns out that TBP is a universal transcription factor. In the case of PolI initiation complexes, TBP is found in the SL1 complex along with three TBP-associated factors (TAFs). The TAFs that give rise to the SL1 complex are different from those that give rise to the TBP-containing TFIID complex described above. It seems likely that the TBP in SL1 makes direct contact with the DNA even though PolI promoters do not contain a TATA box.

Some PolIII Promoters Have Downstream Elements

PolIII promoters show a great deal of variability depending upon the types of genes they transcribe. In the case of 5S RNA genes a region necessary for transcription is located 40–80 bp downstream of the transcription start site (fig. 33.17*a*). One of the additional protein complexes needed for selective transcription of the 5S gene (TFIIIA) binds to this site and, in conjunction with two other transcription factors (TFIIIB and TFIIIC), directs RNA polymerase to bind and initiate transcription. The TFIIIB multisubunit factor contains TBP, which is believed to make contact with the DNA and the polymerase in the final initiation complex. The binding site for factor TFIIIA was determined by the DNA "footprinting" technique (box 33A), a technique commonly used to locate binding sites for specific DNA-binding proteins.

PolIII polymerase also transcribes tRNA genes. In this case, TFIIIA does not participate in the transcription process (see fig. 33.17*b*).

In Eukaryotes, Promoter Elements Are Located at a Considerable Distance from the Polymerase-Binding Site

Eukaryotic promoters often contain binding sites adjacent to the polymerase-binding site where additional protein factors may

Figure 33.15

Model for PolII leaving the initiation complex to begin transcription. Dissociation of the initiation complex during promoter clearance may be facilitated by carboxy-terminal domain (CTD) phosphorylation. The change in the state of phosphorylation of the CTD may disrupt interactions among proteins within the RNA polymerase II (PolII) initiation complex.

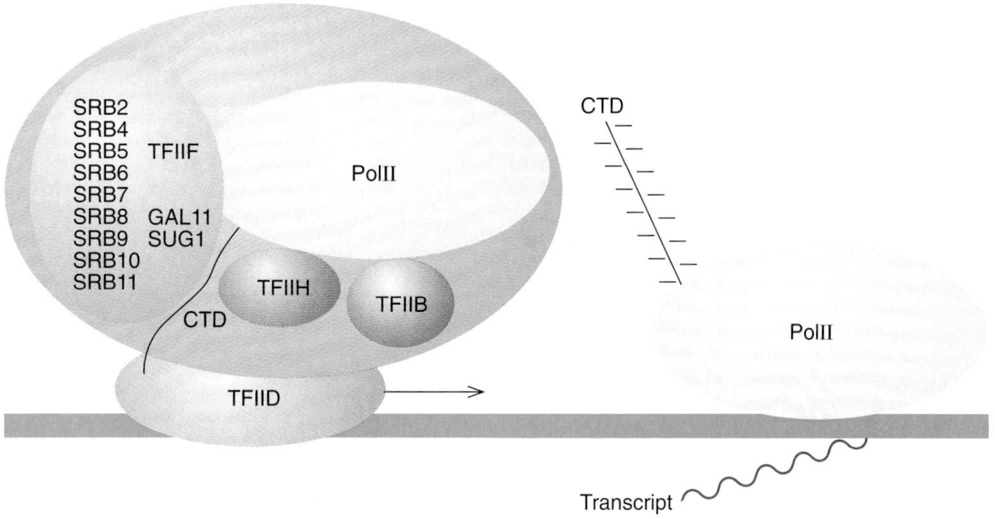

Figure 33.16

Initiation complex for PolI promoters. Binding of PolI is preceded by the binding of UBF and SL1 in that order. UBF and SL1 bind at two locations: the upstream element and the core. PolI binds only at the core. It is not clear at this point why the complex formed at the upstream element facilitates initiation complex formation.

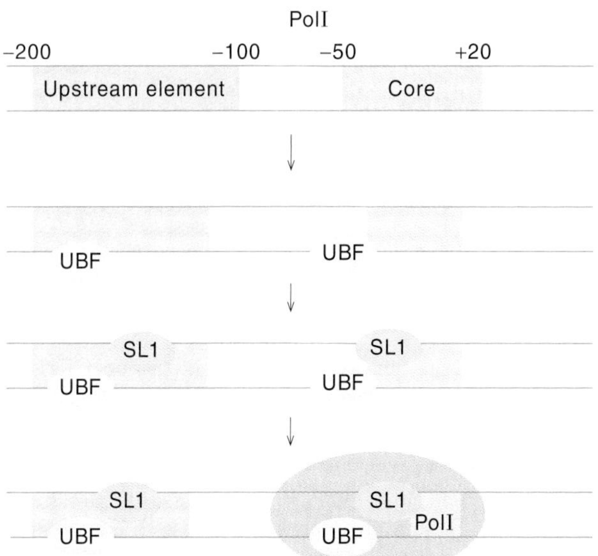

Figure 33.17

Initiation complexes involving PolIII transcription of (*a*) 5S genes and (*b*) tRNA genes.

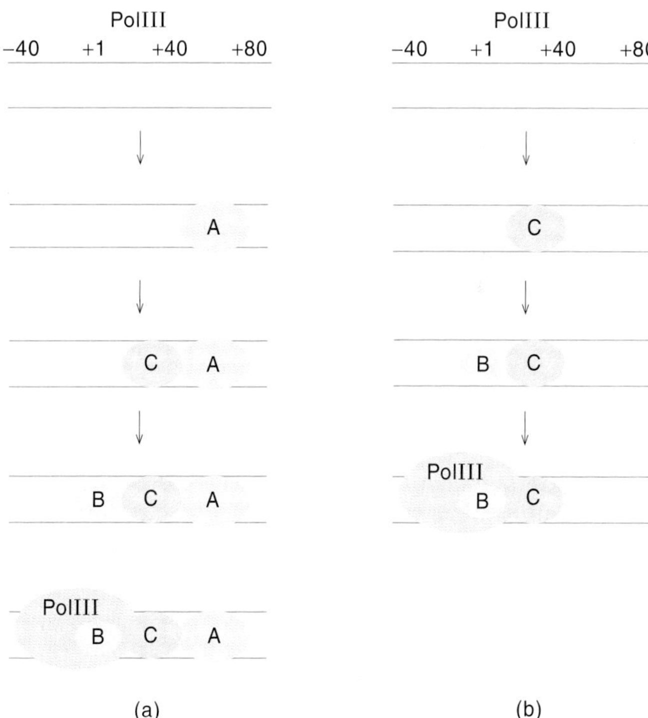

bind. Promoter regions further upstream from the polymerase-binding site are also quite common. The transcription factors that bind to these regions tend to be less general and more gene-specific than the factors that bind next to the polymerase. In the case of yeast, additional promoter elements called upstream activator sequences (UAS) are usually located 40–200 bp upstream from the transcription start site (fig. 33.18). The UAS elements bind additional transcription activation factors, which interact with those that bind at the TATA box. In vertebrates, activator elements called enhancers are found upstream, downstream, and even in the middle of genes. Remarkably, these elements may be located as far

Use of the Footprinting Technique to Determine the Binding Site of a DNA-Binding Protein

To determine the site of action of the transcription factor TFIIIA, the approximate location of its binding site on a 5S gene of *Xenopus borealis* was determined by the "footprinting" technique. DNA fragments containing the 5S gene were labeled at the 5′ end with ^{32}P mixed with TFIIIA and then digested with DNAse I. The resulting DNA fragments were electrophoresed on a polyacrylamide gel that was subsequently autoradiographed (fig. 1). Regions of the DNA that were protected from DNase attack by TFIIIA binding appear as a blank spot (footprint) on the autoradiogram. The footprint shows that the protected region is situated between the 45th and the 90th base pair.

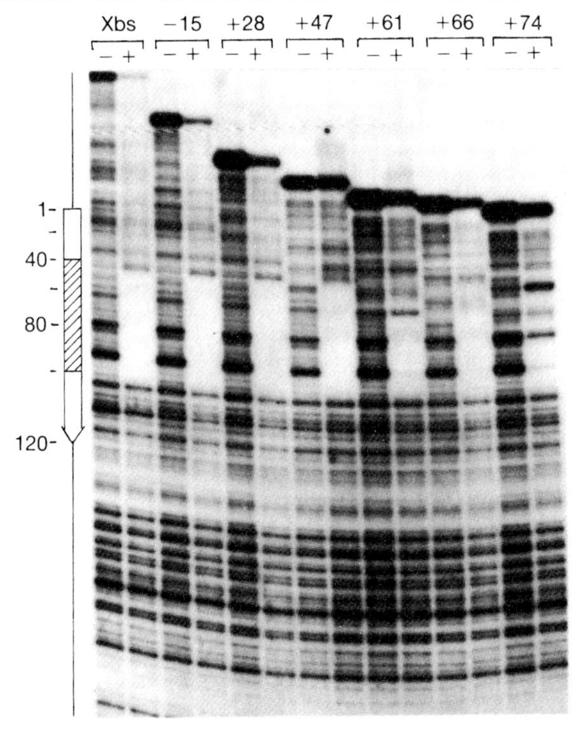

Figure 1

DNase I protection (footprinting) experiment on 5S DNA of *Xenopus borealis*. The diagram on the left indicates the region on the gel that corresponds to the 5S RNA gene. Arrow points in the direction of transcription. Cross-hatched area indicates the region that binds transcription factor protein. Column labeled Xbs refers to an intact gene containing 160 bp of the 5′ flanking sequence, the 5S RNA gene (120 bp), and the 3′ flanking sequence (138 bp). The various deleted 5S DNAs are preceded by 74 bp of the plasmid pBR322 sequence. Numbers in other columns refer to portions of the 5S gene that have been deleted. All samples were subjected to partial digestion with DNAse I, then were electrophoresed and autoradiographed. In + columns, transcription factor protein was added before DNAse I treatment. (*Source:* Courtesy Dr. Donald D. Brown, Carnegie Institute of Washington.)

as 10 kb from the gene they influence. Like UAS elements, enhancers serve as binding sites for additional transcription activators. We will have much more to say about transcription activators in chapter 36, where we discuss regulation of gene expression in eukaryotes.

Many Viruses Encode Their Own RNA Polymerases

Three strategies are used by different DNA viruses to accomplish transcription of viral DNA. The first type utilizes the host RNA polymerase, in some cases modifying it or synthesizing new promoter-specific factors to direct it to read the viral promoters. Examples of such viruses are bacteriophage ϕX174, and T4 (of *E. coli*).

The second type of virus utilizes the host RNA polymerase to transcribe "early" viral genes, including a gene for a new RNA polymerase that transcribes exclusively the remaining "late" viral genes. *E. coli* bacteriophages T7 and T3 are the best known examples of this type. T7 RNA polymerase recognizes specifically the T7 late promoters, all of which contain a nearly identical sequence of 18–22 nucleotides immediately upstream of the 5′-triphosphate terminal GTP start site. T7 RNA polymerase also recognizes specific termination points on the template and ignores the ones normally recognized by *E. coli* RNA polymerase.

A third type of virus, exemplified by bacteriophage N4, carries a virus specific RNA polymerase in its virion. This polymerase enters the cell together with the viral DNA and transcribes some early viral genes. Some of these genes code for specificity factors that direct the host RNA polymerase to transcribe late

Figure 33.18

Cis elements involved in transcription (*a*) in yeast and (*b*) in vertebrates. Upstream activator sequences (UAS) in yeast are similar in function to upstream enhancers in vertebrates. Yeast has no parallel to downstream enhancers found in vertebrates.

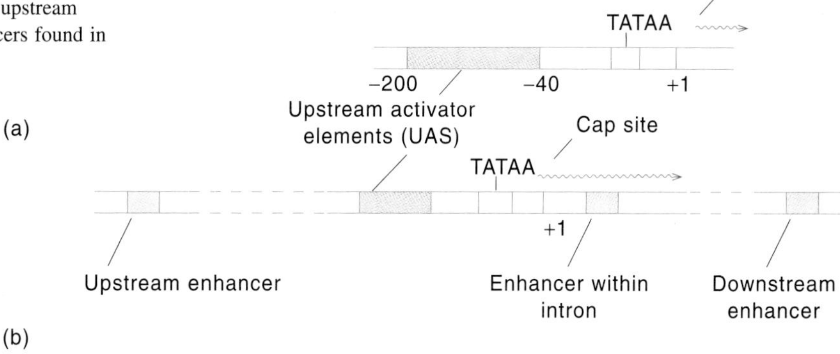

Table 33.5
RNA-Synthesizing Enzymes

	Template	Primer	Molecular Weight of Subunit(s)	Gene Name	Subtrate	Inhibition by Rifampicin
Template-dependent						
Enzymes from bacteria						
Holoenzyme (*E. coli*)	DNA	—	155,000 151,000 70,000 36,500	*rpoC* *rpoB* *rpoD* *rpoA*	4 NTPs	Yes
DNA primase	DNA	—	65,000	*dnaG*	4 NTP, 4 dNTP	No
Enzymes from phage or phage-infected bacteria						
T7 RNA polymerase	T7 DNA	—	99,000	T7 *gene1*	4 NTPs	No
N4 RNA polymerase	N4 DNA	?	350,000	Viral	4 NTPs	No
Qβ replicase	Qβ RNA	—	65,000 55,000 43,000 35,000	*rpsA* Viral *tuf* *tsf*	4 NTPs	No
Template-independent						
CCA enzyme	—	3′ end tRNA	45,000	*cca*	CTP ATP	No
Poly(A) polymerase (eukaryotic)	—	3′ end mRNA			ATP	No
Polynucleotide phosphorylase	—	3′ end RNA	86,000 48,000	*pnp*	4 NDPs	No

Source: G. Zubay, *Biochemistry,* 2d ed., Macmillan, New York, 1988, p. 813.

genes. Vaccinia virus is another example of a virus that contains a virion-encapsulated RNA polymerase.

RNA-Dependent RNA Polymerases of RNA Viruses

The RNA genomes of single-stranded RNA bacterial viruses, such as Qβ, MS2, R17, and f2, are themselves mRNAs. Bacteriophage Qβ codes for a polypeptide that combines with three host proteins to form an RNA-dependent RNA polymerase (replicase). The three host proteins are ribosomal protein S1 and two elongation factors for protein synthesis: EF-Tu and EF-Ts (table 33.5). The Qβ replicase functions exclusively with the Qβ RNA plus-strand template. It first makes a complementary RNA transcript (minus-strand) and ultimately uses the minus-strand as a template to synthesize mul-

Figure 33.19

Comparative processing of major transcripts in prokaryotes and eukaryotes.

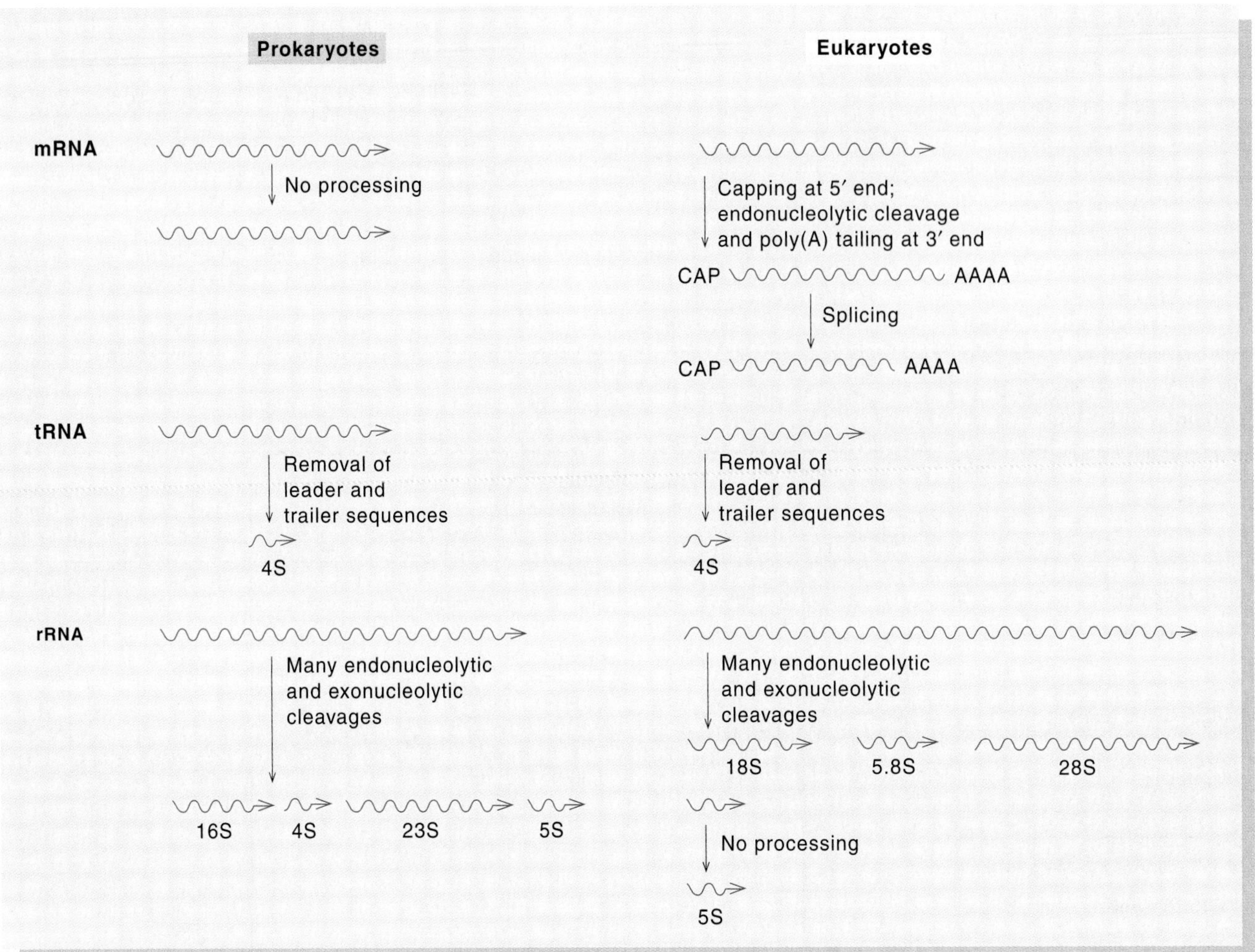

tiple copies of viral RNA plus-strands. Like the DNA-dependent RNA polymerases, the replicase utilizes rNTPs and transcribes in the $5' \rightarrow 3'$ direction. The phage RNA must first act as an mRNA to direct the synthesis of the aforementioned component of the replicase, since uninfected cells do not have an RNA-dependent RNA polymerase or replicase.

RNA tumor viruses (retroviruses) that infect animal cells exhibit a different replication strategy. In their virions they carry an enzyme that uses the viral RNA as a template to synthesize a DNA copy (see chapter 31). This DNA becomes integrated into the host genome. Subsequently, the viral RNA is transcribed from the integrated viral DNA using host cell RNA polymerase.

Other Types of RNA Synthesis

In addition to the cellular enzyme(s) that catalyzes DNA-directed RNA synthesis, cellular enzymes are involved in polyribonucleotide synthesis that do not use a template. Some of the properties of these enzymes are summarized in table 33.5. We have already mentioned polynucleotide phosphorylase in this chapter, and in chapter 31 we discussed the importance of DNA primase to DNA synthesis.

Two enzymes are known to add ribonucleotides posttranscriptionally to the 3′ hydroxyl end of specific RNAs. One adds the CCA sequence found in all tRNAs at their 3′ ends. The 3′ terminal adenine in this sequence serves as the amino acid attachment site. The 3′-CCA is relatively unstable and is continually being rejuvenated by this enzyme, called the CCA enzyme, or tRNA nucleotidyltransferase.

In eukaryotes, 100–200 adenosine residues are added to the 3′ ends of most mRNAs by a poly(A) polymerase. This addition occurs in the nucleus before the mRNA is fully processed and transported to the cytoplasm.

Posttranscriptional Alterations of Transcripts

Most RNAs are not made in their final functional forms as they peel off the DNA template (fig. 33.19). They must undergo backbone phosphodiester bond cleavages into smaller molecules (processing) and individual base changes (modification). The types of alterations that pre-tRNA and pre-rRNA transcripts undergo are very similar in prokaryotes and eukaryotes. We focus on the situation in *E. coli* because it is the best understood.

Processing and Modification of tRNA Require Several Enzymes

Transfer RNAs are processed from larger precursors in both prokaryotic and eukaryotic cells. This processing involves two types of nucleases: endoribonucleases, which cleave at internal sites in the RNA, and exonucleases, which remove nucleotides from the ends of the chains.

The processing steps of *E. coli* tyrosine $tRNA^{Tyr}$ are diagrammed in figure 33.20. The initial transcript has, in addition to the 85 nucleotide residues of the final product, 41 residues at the 5′ end and 225 residues at the 3′ end; it probably folds to form the typical cloverleaf structure of the mature tRNA prior to processing. Processing begins when a specific endonuclease called RNaseF cleaves the precursor at a site three nucleotides beyond what will be the 3′ end of the mature tRNA. Another endonuclease, RNaseP, then cleaves the remaining RNA to produce the mature 5′ end. At the 3′ end, exonuclease RNaseD sequentially removes additional nucleotides, usually until it reaches the 3′ terminal CCA sequence. Some tRNAs encode the CCA terminal sequence; some do not. Following processing, individual bases on the tRNA molecule are modified by a variety of enzymes, including methylases, deaminases, thiolases, pseudourydylating enzymes, and transglycosylases. Some of the modified bases that result from the action of these enzymes are illustrated in figure 33.5.

Processing of Ribosomal Precursor Leads to Three RNAs

Both eukaryotic and prokaryotic cells synthesize large precursors to rRNA that are processed to produce the mature rRNAs. The scheme for processing rRNA in *E. coli* is summarized in figure 33.21.

The initial transcript is over 5,500 nucleotides long and includes, from the 5′ end of the RNA: a 16S rRNA, a spacer region with one or two tRNAs, a 23S rRNA, a 5S rRNA, and in some cases one or two additional tRNAs. Extra bases are found preceding and following each of these RNAs. Primary processing events include the endonucleolytic action by RNaseIII to produce pre-16S and pre-23S RNAs. This is followed by the action of specific ribonucleases to produce the tRNAs and pre-5S rRNA. Secondary processing by endonucleases M16, M23, and M5 results in mature 16S, 23S, and 5S RNAs, respectively. Extra bases on the 3′ end of the tRNAs are removed by exonuclease RNaseD. This processing scheme has been deduced by observing the accumulation of intermediates in mutant strains defective in one or more of the nucleases and by cleaving the intermediates in vitro with purified or partially purified nucleases.

Eukaryotic Pre-mRNA Undergoes Extensive Processing

In prokaryotes most mRNAs function in translation with no prior alterations. Indeed, little opportunity arises for any alterations because translation starts on the nascent transcript before transcription has been completed. By contrast, in eukaryotes, transcription occurs in the nucleus, and a transcript undergoes extensive changes before being transported to the cytoplasm. First, the 5′ end of the message is modified, a process called capping (fig. 33.22). This usually occurs before transcription has been completed. Changes

Figure 33.20

Processing and modification of *E. coli* tyrosine tRNA. (T = ribothymidine; ψ = pseudouridine; i^6A = isopentenyladenosine; mG = methylguanosine; s^4U = thiouridine.) See figure 33.5 for the structures of these modified bases.

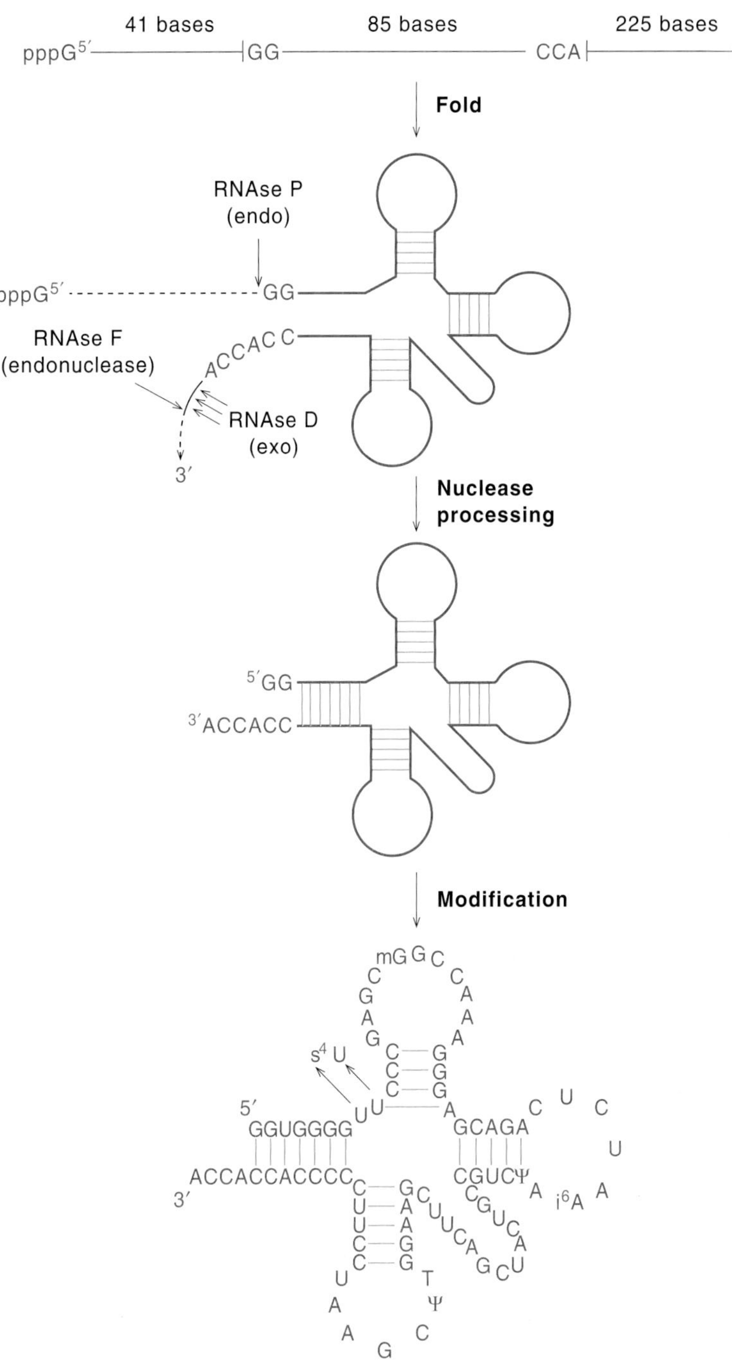

Figure 33.21

Processing of *E. coli* ribosomal RNA. The ribosomal RNA is transcribed as one long RNA molecule, which contains the sequences for the three ribosomal RNAs and one or two tRNA molecules. Many processing sites occur, and many different enzymes are involved in the processing, as indicated by the vertical arrows and the symbols associated with these arrows. The various nucleases are described in table 33.6. (*Source:* Adapted from D. Apirion and P. Gegenheimer, Processing of bacterial RNA, *FEBS Lett.* 125:1, 1981.)

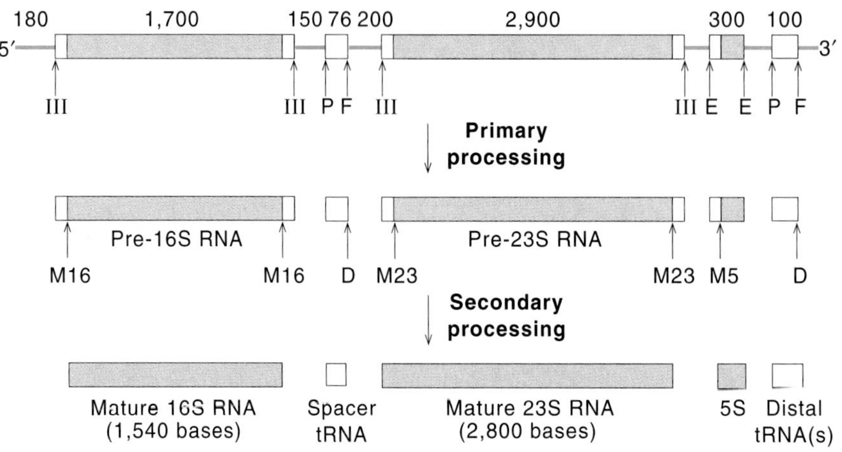

Figure 33.22

Structure of the 5′ methylated cap of eukaryotic mRNA. A 7-methylguanosine (in red) is attached through a triphosphate linkage formed between its 5′-OH and the 5′-OH of the terminal residue in the initial transcript. Note that the 2′-OH groups on the last two bases of the initial transcript have also been modified by methylation (in red). N_1, N_2, and N_3 can be any purine or pyrimidine base.

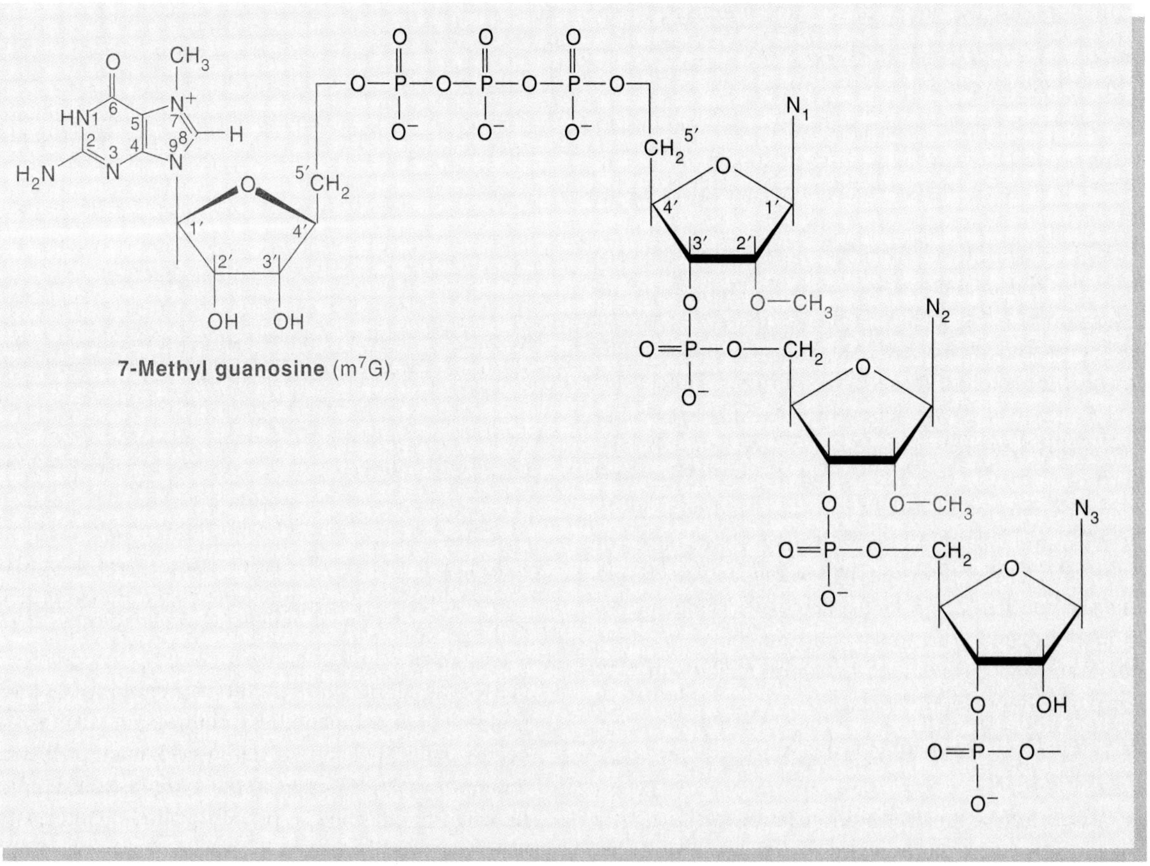

7-Methyl guanosine (m^7G)

at the 3′ end of the transcript are part of the termination mechanism because the PolII enzyme does not appear to recognize any termination signal. The RNA polymerase transcribes well beyond the useful part of the message. These extended transcripts are usually cleaved about 10–30 nucleotides downstream of an AAUAAA sequence, after which poly(A) polymerase adds about 200 adenylate residues to the 3′ end. In many cases, especially in higher eukaryotes, noncoding regions called introns are removed from interior locations, and the remaining message is reunited. This process is called splicing because it is similar to the process of splicing in film editing. We discuss splicing in some detail.

To assist in the modification and processing of mRNAs, eukaryotic cells contain in their nuclei small nuclear RNAs (snRNAs) that are complexed with specific proteins to form small nuclear ribonucleoprotein particles (snRNPs). These RNAs have been named U1, U2, U3, . . . , U13 and range in size from 100 to 220 bases. One, U3, is found in the nucleolus, the site of rRNA synthesis. All are very abundant, and as many as 1 million copies of most of them may occur per nucleus. Their base sequences are highly conserved among organisms, and all seem to contain unusual trimethylguanosine structures at their 5′ ends. Each snRNP contains an snRNA and 6–12 proteins, some of which are common to all snRNPs. The snRNPs appear to be involved in the processing and modification of RNAs, including splicing (U1, U2, U4, U5, U6), formation of the 3′ ends of histone mRNAs (U7), and maturation of rRNA (U3). Polyadenylation is unique in the processing reactions in that the complex machinery involved contains only proteins.

Figure 33.23

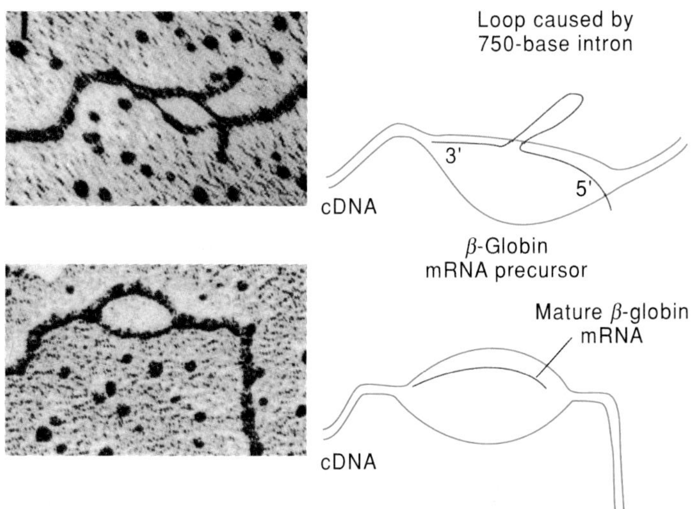

Electron micrograph showing mouse β-globin precursor mRNA (nascent transcript) hybridized with DNA (cDNA) complementary to mature mRNA (upper photo). A control experiment (lower photo) shows that mature mRNA forms a perfect hybrid with DNA complementary to mature mRNA (cDNA) as expected. The intron region in the nascent transcript is indicated by a loop in the upper figure. Schematics indicating the RNA and cDNA are shown on the right. A second small intron is present in the precursor mRNA near the 5′ end and is the reason that the 5′ end of the RNA is not hybridized to the cDNA in the upper photo. (*Source:* From A. Kinniburgh, J. Mertz, and J. Ross, The precursor of mouse β-globin messenger RNA contains two intervening RNA sequences, *Cell* 14:681, 1978. © Cell Press.)

Splicing Entails the Removal of Internal Sequences. It was established in the early 1970s that a great deal of RNA (hnRNA) turns over in the nucleus without ever reaching the cytoplasm. What could this RNA be? Was it a specific type of RNA that never made it to the cytoplasm, or was it evidence for processing of mRNA precursors? The big surprise came in 1977 when Ric Robert's and Phil Sharp's laboratories simultaneously discovered that the mRNAs of adenovirus undergo extensive processing in which internal segments are removed from the mRNA precursors. Very soon thereafter it was found that this phenomenon was general and widespread, especially in higher eukaryotes.

One of the early demonstrations of sequence removal resulted from the finding that mouse β-globin precursor mRNA did not form a perfect hybrid with DNA complementary to mature mRNA. When the hybrid was observed with an electron microscope, a loop in the DNA of the heteroduplex appeared, which suggested that the precursor contained internal sequences not present in the mature mRNA (fig. 33.23). These noncoding intervening sequences (introns) are interspersed with coding sequences (exons).

The presence of introns implies a function, but in many cases the presence of introns may represent no more than a stage in the evolution of a gene. This argument is supported by the finding that introns are far less common in unicellular eukaryotes such as the yeast *Saccharomyces* and they are very rare in prokaryotes such as *E. coli*. Frequently, splice points are correlated with "domains" that define protein structural units (see chapter 5). Similar domains are often seen in different proteins. For example, the exons of hemoglobin encode three structural domains of different types, whereas the heavy-chain immunoglobulin exons encode four domains that are quite similar in structure.

Splicing Is a Two-Step Process. By attaching the promoter for the bacteriophage SP6 RNA polymerase to the β-globin gene, it has become possible to transcribe the gene in vitro with SP6 RNA polymerase to produce abundant amounts of β-globin pre-mRNA. This source of precursor mRNA was used to determine the steps and factors involved in splicing. The process involves cleavage of the pre-mRNA at the 5′ splice site to generate the 5′-proximal exon and an RNA species containing the intron in a "lariat" configuration connected to the distal exon (fig. 33.24). The lariat is formed via a 2′–5′ phosphodiester bond, which joins the 5′ terminal guanosine of the intron to an adenosine residue within the intron at a spot 18–40 nucleotides upstream of the 3′ splice site. These two RNA species are probably held together in a noncovalent complex until the next step (see fig. 33.24, step 2) in the reaction, which entails cleavage at the 3′ splice site to generate the free intron RNA and ligation of the two exons via a 3′–5′ phosphodiester bond. U1, U2, U4, U5, and U6 snRNAs, as well as the associated protein factors, are necessary for the splicing reaction to occur in vitro. These factors are assembled in a large 40–60S ribonucleoprotein called a spliceosome. The snRNA components of the splicing apparatus interact both among themselves and with

Figure 33.24

Splicing scheme for pre-mRNA. In step 1 the 2′-OH on an adenosine attacks a phosphate that is 5′-linked to a guanine residue. This leads to a lariat configuration connected to the distal exon. The lariat is formed via a 2′–5′ phosphodiester bond, which joins the 5′ terminal guanosine of the intron to an adenosine residue within the intron, 18–40 nucleotides upstream of the 3′ splice site. In the next step a cleavage occurs at the 3′ splice site to generate the free intron RNA, and a ligation occurs of the two exons via a 3′–5′ phosphodiester bond. Bases usually found in the region of the splice sites are indicated. The spliceosome is presumed to hold the reacting components in the proper juxtaposition to facilitate the two-step reaction.

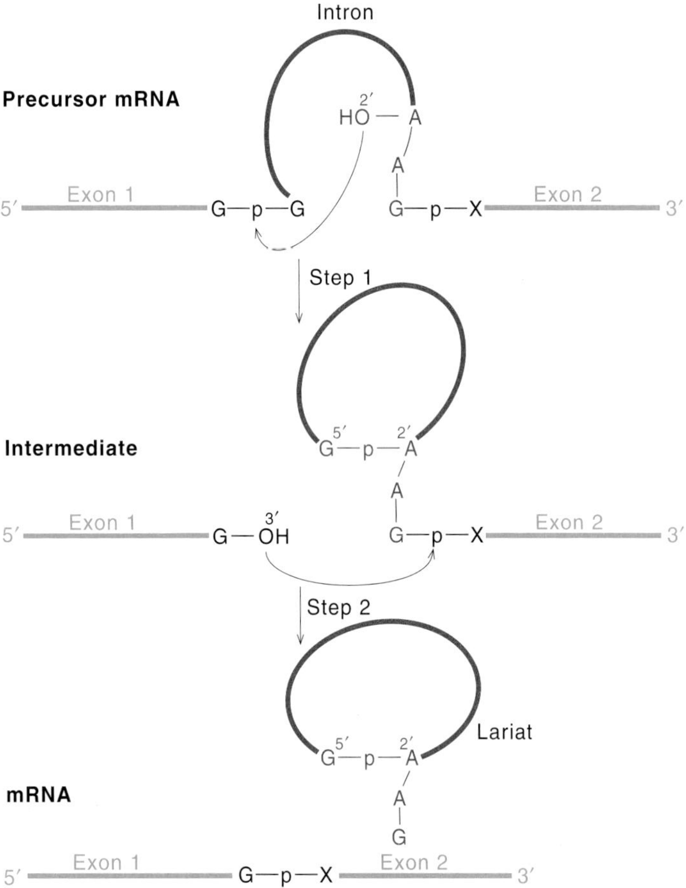

Figure 33.25

Self-splicing of pre-rRNA from the protozoan *Tetrahymena*. The first step is a transesterification reaction in which the 3′ hydroxyl group of a guanosine attacks the phosphodiester bond at the 5′ splice site. The second step involves another transesterification reaction in which the 3′ hydroxyl group of the upstream exon attacks the phosphodiester bond at the 3′ splice site and displaces the 3′ hydroxyl group of the intron.

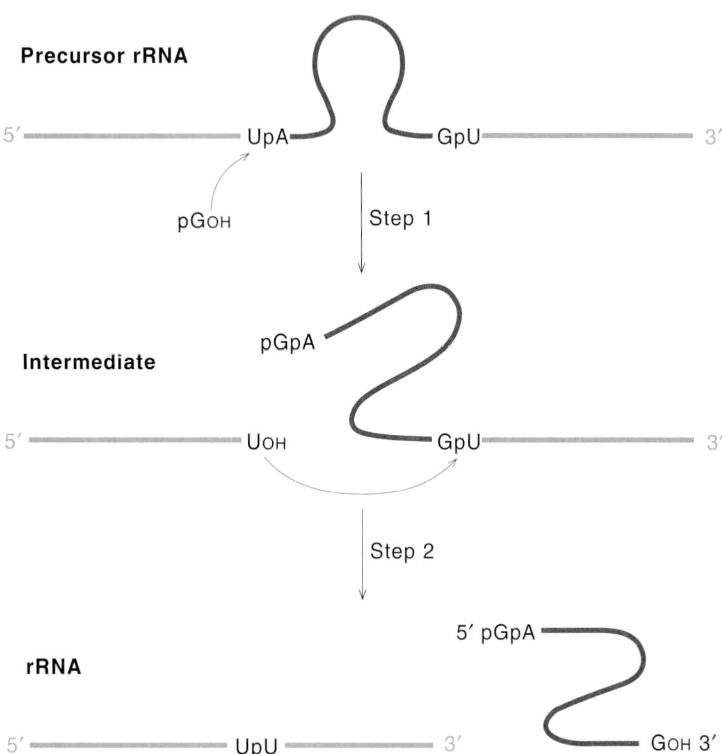

the substrate RNA by means of base pairing interactions, which foster changes in structure that bring reacting groups into apposition and probably create catalytic centers. First U1 binds to the 5′ splice site. Then U2 binds to the branch site. The latter reaction requires an ATP hydrolysis and commits the pre-mRNA to the splicing pathway. Interaction between U1 and U2 brings the splicing sites close together. U5/U4/U6 binds as a trimer with U5 binding the exon at the 5′ site and U6 binding to U2. This requires another ATP hydrolysis. Following this, U5 shifts from exon to intron. U4 release triggers the catalytic reaction of U6 resulting in the first transesterification reaction (see fig. 33.24, step 1). Following this, cleavage occurs at the 3′ splice site, and the exons are simultaneously ligated (see fig. 33.24, step 2). The spliced mRNA is released. U2, U5, and U6 remain bound to the so-called lariat until it becomes debranched.

Removal of internal sequences in eukaryotes is not restricted to mRNA processing. It also occurs in the processing of rRNA and some tRNAs. In tRNAs the mechanism appears to be different in that the signal for splicing originates not from the primary sequence but from the secondary or tertiary structure of the pre-tRNA.

Some RNAs Are Self-Splicing

In 1982 the splicing of rRNA from the protozoan *Tetrahymena* was shown by Tom Cech to involve a startling mechanism. When the pre-rRNA was incubated under the appropriate conditions of salt, Mg^{2+}, and guanosine nucleotide, splicing of the RNA occurred without the help of a protein enzyme! This ability of RNA to act as a catalyst for its own splicing has given rise to the term "ribozyme" and has broadened our definition of enzymes.

The mechanism of the self-splicing reaction is shown in figure 33.25. The first step is a transesterification in which the 3′ hydroxyl group of the guanosine cofactor attacks the phosphodiester bond at the 5′ splice site. The second step is another transesterification reaction in which the 3′ hydroxyl group of the upstream exon attacks the phosphodiester bond at the 3′ splice site and displaces the 3′ hydroxyl group of the intron. The final reaction products are the joined exons and the excised intron.

Since its initial discovery, self-splicing has been found to occur for RNAs from a wide variety of organisms. However, the fraction of unprocessed RNAs that can be shown to undergo self-splicing in vitro is quite small in all cases. Certain precursor RNAs that exhibit self-splicing produce intron lariats, just like those seen in the commonly observed splicing reactions that take place in most precursor RNAs with introns. In this case the guanosine nucleotide is not required. This fact suggests that originally RNAs may have been self-splicing and that the spliceosome components have evolved to improve catalytic efficiency and perhaps to regulate the process.

Self-splicing of RNA shows that in certain cases, RNA has enzymelike activity. However, it does not by itself demonstrate true enzyme behavior. Recall that an enzyme is a substance that acts as a catalyst—it accelerates a reaction without itself being consumed in the reaction. More recently, Zaug and Cech have shown that a fragment of the self-splicing ribosomal RNA intervening sequence of *Tetrahymena thermophila* can act as an enzyme in vitro. This RNA fragment has been shown to catalyze the breakage and rejoining of oligonucleotide substrates in a sequence-dependent manner, with a $K_m = 42\ \mu\text{M}$ and a $k_{cat} = 2\ \text{min}^{-1}$. With pentacytidylic acid as the substrate, successive cleavage and rejoining reactions lead to the synthesis of higher polymers of polycytidylic acid (fig. 33.26).

Cech and his colleagues' discovery that RNA can function as an enzyme raises many possibilities both for biochemical reactions as they are currently understood and for reactions that may have played a major role in prebiotic evolution (see later discussion).

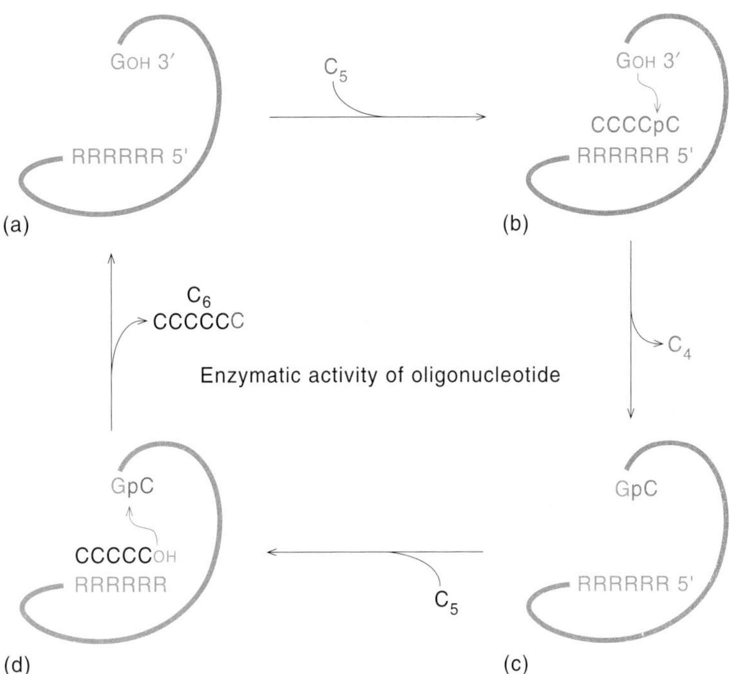

Figure 33.26

Model for the enzymatic mechanism of the 19-base oligonucleotide intervening sequence isolated by Cech from a *Tetrahymena* precursor rRNA. (*a*) The oligonucleotide enzyme (ribozyme) is shown with the oligopyrimidine-binding site (RRRRRR), a sequence of six purines near its 5′ end, and a guanine with a 3′ hydroxyl group at its 3′ end. (*b*) The enzyme binds its substrate, a pentacytidylic acid. (*c*) Nucleophilic attack by the terminal 3′-OH of the guanine residue leads to formation of the covalent intermediate and displacement of the tetracytidylic acid. (*d*) A second pentacytidylic acid binds at the enzyme site, and by transesterification a hexanucleotide is formed.

Degradation of RNA by Ribonucleases

Although a large number of nonspecific exonucleases and endonucleases have been identified in many organisms, the role of most of them is not understood. Some are extracellular or secreted enzymes that presumably function in breaking down RNA to recycle the purine and pyrimidine bases. Intracellular ribonucleases also may be involved in recycling, as well as in some aspects of processing, as described earlier. Some of the better-known enzymes are listed in table 33.6.

E. coli RNase I is an endonuclease that is located in the periplasmic space between the cell membrane and the cell wall. RNase II is an intracellular 3′ exonuclease that rapidly degrades RNA fragments. Mutants defective in these enzymes were very important in the development of in vitro translation systems, since extracts from cells defective in these enzymes allowed mRNA to be translated without being rapidly degraded.

RNase H has the unusual property of degrading the RNA strand from a RNA–DNA hybrid molecule and may be involved in the removal of RNA primers during DNA replication. Some endonucleases have marked preferences for cleavage after certain bases. Pancreatic RNase A cleaves at the 3′ side of pyrimidines, while RNase T1 cleaves only after guanosine residues. Both leave 3′ phosphate termini. Because of their specificity, these enzymes have proven very useful in the analysis of RNA sequences. S1 nuclease digests single-stranded DNA or RNA.

Finally, there are exonucleases that digest in the $3' \rightarrow 5'$ direction and others that digest in the $5' \rightarrow 3'$ direction. All known intracellular exonucleases in *E. coli* degrade from $3' \rightarrow 5'$ and produce 5′-NMPs or 5′ NDPs, which can be readily recycled for use in RNA synthesis.

It is likely that degradation of mRNA proceeds by numerous endonucleolytic cleavages and then $3' \rightarrow 5'$ exonucleolytic digestion of the resulting fragments. Differences in decay rates of bacterial mRNAs result from differences in mRNA sequence and structure. RNase II and PNPase, the major bacterial exonucleases involved in mRNA turnover, rapidly degrade single-stranded RNA from the 3′ end but are impeded by 3′ stem-and-loop structures.

Some Ribonucleases Are RNAs

RNase P, which we mentioned earlier, was the first nucleic acid–containing ribonuclease to be discovered. The enzyme is a ribonucleoprotein composed of both a 20-kD protein and a 377-nucleotide RNA molecule. Sid Altman discovered that the RNA component is catalytically active, whereas the protein component helps to maintain the three-dimensional structure of the RNA. This discovery led to Altman sharing the Nobel prize with Cech who discovered self-splicing.

Table 33.6
Representative Enzymes Involved in Processing and Degradation

Enzyme	For *E. coli* Genes: Gene Name	Map Position	Type	Product	Specificity
Processing					
RNaseIII	*rnc*	55′	endo	3′-OH, 5′-PO_4	Specific, long, double-stranded RNA
RNase D	*rnd*	40′	3′ → 5′ exo	5′-NMPs	Nonspecific, but stops at CCA
RNase E	*rne*	24′	endo		Specific
RNase F	*rnf*	?	endo		Specifically cuts 3′ to tRNA-like structures
RNase P	*rnpA* *rnpB*	83′ 70′	endo	3′-OH, 5′-PO_4	Specifically cuts 5′ to tRNA-like structures
RNase M16			endo(s)		Specific, cuts pre-16S to 16S RNA
RNase M23			endo(s)		Specific, cuts pre-23S to 23S RNA
RNase M5			endo		Specific, cuts pre-5S to 5S RNA
Degradation					
RNase I	*rna*	14′	endo	3′-PO_4 oligos	Nonspecific
RNase II	*rnb*	28′	3′ → 5′ exo	5′-NMP	Nonspecific
Polynucleoside phosphorylase	*pnp*	69′	3′ → 5′ exo	5′-NDP	Nonspecific
RNase H	*rnh*	5′	endo		Nonspecific, digests RNA out of RNA–DNA duplex
Bovine pancreatic RNase A			endo	Py-3′-PO_4	Specific, cuts 3′ to pyrimidines
Aspergillus RNase T1			endo	G-3′-PO_4	Specific, cuts 3′ to guanine
Aspergillus S1 nuclease			endo	5′-NMP	Nonspecific, cuts single-stranded RNA or DNA
Bovine spleen phosphodiesterase			5′ → 3′ exo	3′-NMP	Nonspecific
Snake venom phosphodiesterase			3′ → 5′ exo	5′-NMP	Nonspecific

Certain circular and low-molecular-weight single-stranded plant pathogenic RNAs called viroids can self-cleave at specific sites in the presence of Mg^{2+}. This property is considered important in their replication in vivo by a rolling-circle mechanism. All RNAs produced by the rolling-circle mechanism are polymers of indefinite length. The self-cleavage reaction results in linear monomeric molecules that can subsequently be circularized. The self-cleavage site occurs at a specific locus in a secondary structure that has been described as a hammerhead structure (fig. 33.27). Short RNA molecules that anneal to form a hammerhead structure with a consensus sequence undergo self-cleavage as shown in figure 33.27*b* and *c*.

Catalytic RNA May Have Evolutionary Significance

It seems likely that the range of catalytic activities for RNA will be expanded by future discoveries. For many years prior to the discovery of RNA catalysis, molecular evolutionists struggled with the question of which came first, RNA or protein. Proteins make excellent enzymes, but they lack the template activity required for transmission of information and evolution. On the other hand, nucleic acids (RNA) have excellent template activity, like DNA, but they did not have any known catalytic activities. For a period of time it was impossible to imagine how either of these molecules

Figure 33.27

Hammerhead structures that undergo self-cleavage. Cleavage occurs at site indicated by the arrow. (*a*) The structure formed by a sequence found in a viroid or a virusoid. (*b*) A short circular RNA molecule with the "consensus" sequence that also undergoes cleavage. (*c*) A hybrid molecule, formed by two RNAs, that undergoes a comparable cleavage reaction. In the presence of excess "substrate" the 19-mer functions as an enzyme, since the same 19-mer can cleave many substrates.

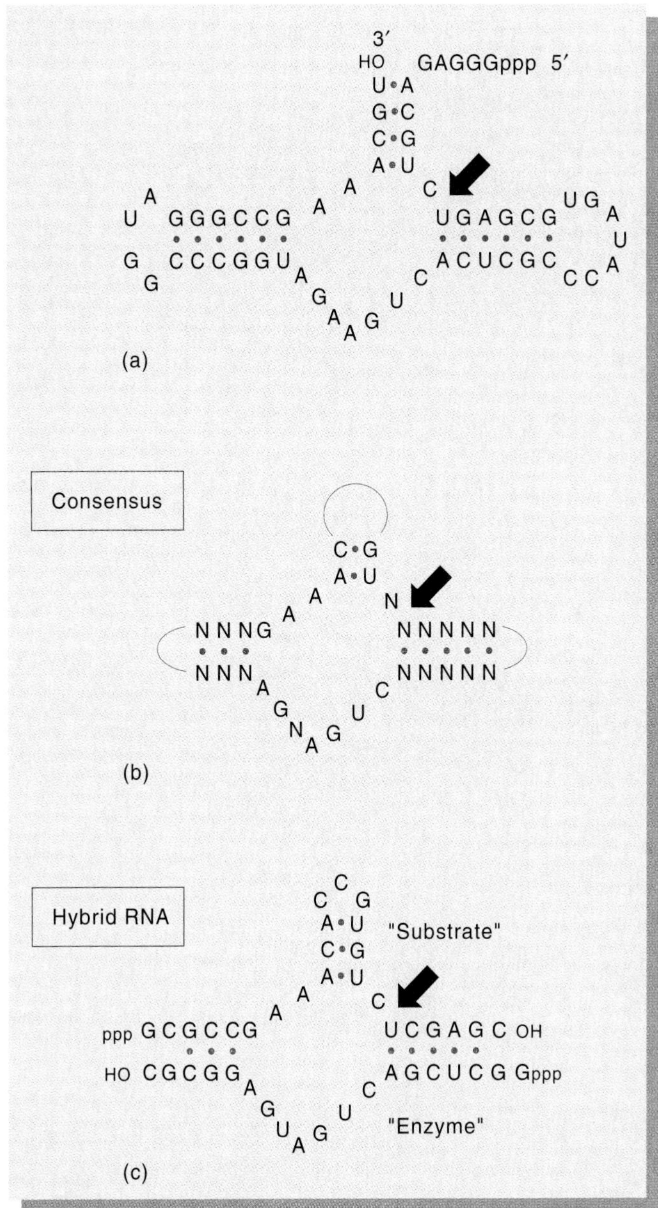

Figure 33.28

Editing of the cytochrome b message in the flagellated protozoan *Leishmania tarentolae*. Editing, which involves the insertion of 11 U residues, creates a translatable RNA where no translation was possible before. Structure shown in mRNA is after editing.

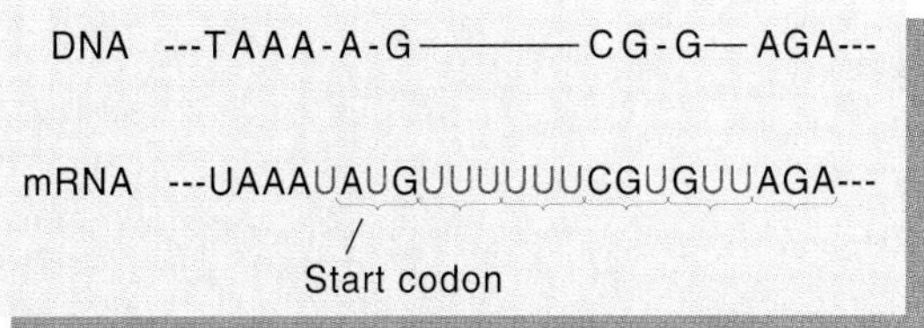

could have evolved first, since proteins cannot exist without templates for ordering amino acids in a polypeptide chain and RNAs cannot exist without enzymes. This dilemma appears to have been resolved by the recent discovery of ribozymes. Now a reasonable argument can be made that the first enzymes were RNAs and the first proteins were made under the direction of ribozymes.

Although this argument gives us a theoretically satisfying resolution to a central dilemma in molecular evolution, it still remains to demonstrate that RNAs have the versatility to function as enzymes for a much wider range of reactions. In this regard it seems unlikely that the full range of ribozymes that probably existed in early evolution are still present. More than likely, most primitive ribozymes have long since been replaced by more efficient protein enzymes.

This creates a dilemma of a different sort. How are we to demonstrate ribozyme activities that no longer exist? Perhaps we must design our own ribozymes to demonstrate the full potential for RNA to act as an enzyme. This approach is being used by Jack Szostak and Gerald Joyce, who independently have developed in vitro systems to select for RNA molecules with predesignated activities. So far, Joyce has selected for a de novo synthesized RNA that can cleave DNA, while Szostak has selected for an RNA that catalyzes primer extension by trinucleotides. Rapid progress using this approach to assess the potential of RNA to function as an enzyme is being made.

RNA Editing Involves Changing Some of the Primary Sequence of a Nascent Transcript

On some occasions the nascent transcript requires alterations in its sequence to convert it into a translatable mRNA. For example, in the mitochondrial message for cytochrome c oxidase from the protozoan *Leishmania tarentolae,* several U residues are added at different points to the nascent transcript (fig. 33.28). To make the editing changes, a so-called guide RNA must form a complex with the nascent transcript. The changes in sequence are dictated by the complementary sequence in the guide RNA in the region where editing occurs. In addition to the guide RNA a special enzyme system is required that is capable of inserting and removing bases at various points in the nascent transcript.

Editing seems like a roundabout way to get a functional message. Why not just make the transcript so that it can be translated in the first place. Possibly, editing serves a regulatory func-

Figure 33.29

Inhibitors of RNA synthesis.

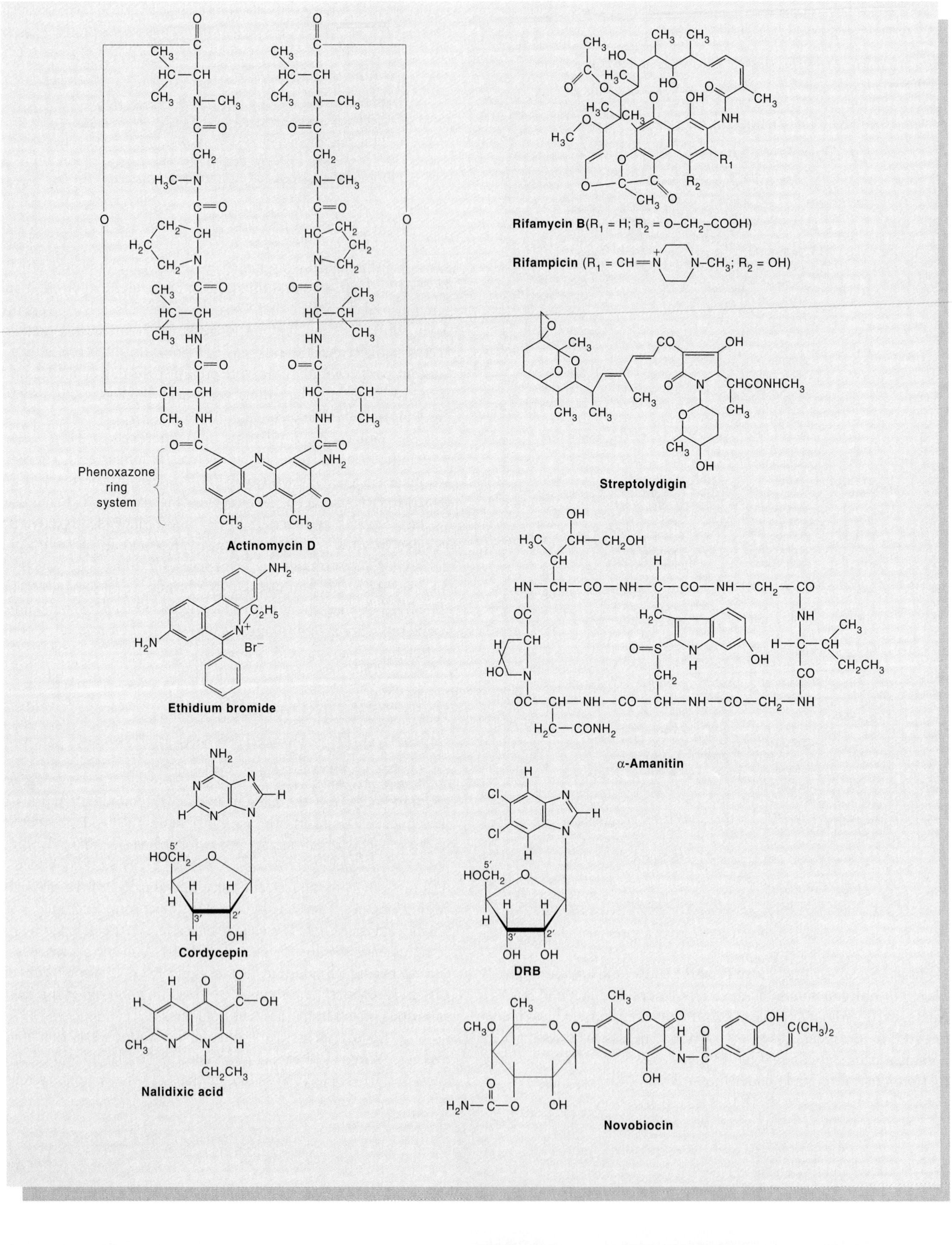

Phenoxazone ring system

Actinomycin D

Rifamycin B (R_1 = H; R_2 = O–CH_2–COOH)

Rifampicin (R_1 = CH=N⁺(piperazine)N–CH_3; R_2 = OH)

Streptolydigin

Ethidium bromide

α-Amanitin

Cordycepin

DRB

Nalidixic acid

Novobiocin

tion. Alternatively, it has been proposed that editing may be a device used to accelerate the evolutionary process for a gene.

Inhibitors of RNA Metabolism

A large variety of inhibitors of RNA synthesis have been identified. Some of these inhibitors have proved useful in elucidating transcription mechanisms, and some have facilitated selection for superior mutant strains with enzymes that are resistant to their inhibition. The inhibitors fall into three classes (fig. 33.29).

Some Inhibitors Act by Binding to DNA

The best-known example of inhibitors that bind to DNA is actinomycin D, an antibiotic produced by *Streptomyces antibioticus*. The inhibition of RNA synthesis is caused by the insertion (intercalation) of its phenoxazone ring between two G-C base pairs, with the side chains projecting into the minor groove of the double helix, hydrogen-bonded to guanine residues. RNA polymerase binding to DNA that contains actinomycin D is only slightly impaired, but RNA chain elongation in both eukaryotes and prokaryotes is blocked. Ethidium bromide also intercalates into DNA and at low concentrations preferentially binds to negatively supercoiled DNA (see chapter 30). It has been used to selectively inhibit transcription in mitochondria, which contains supercoiled DNA.

Some Inhibitors Bind to RNA Polymerase

Rifampicin is a synthetic derivative of a naturally occurring antibiotic, rifamycin, that inhibits bacterial DNA-dependent RNA polymerase but not T7 RNA polymerase or eukaryotic RNA polymerases. It binds tightly to the β subunit. Although it does not prevent promoter binding or formation of the first phosphodiester bond, it effectively prevents synthesis of longer RNA chains. It does not inhibit elongation when added after initiation has occurred. Another antibiotic, streptolydigin, also binds to the β subunit; it inhibits all bond formation.

The most useful inhibitor of eukaryotic transcription has been α-amanitin, a major toxic substance in the poisonous mushroom *Amanita phalloides*. The toxin preferentially binds to and inhibits RNA polymerase II (see table 33.4). At high concentrations it also can inhibit RNA polymerase III but not RNA polymerase I or bacterial, mitochondrial, or chloroplast RNA polymerases.

Some Inhibitors Are Incorporated into the Growing RNA Chain

Cordycepin in its 5′-triphosphorylated form is a substrate analog that is incorporated into growing RNA chains by most RNA polymerases. It causes chain termination after incorporation, since it does not contain the 3′ hydroxyl group necessary for the formation of the next phosphodiester bond.

Summary

In this chapter we described the synthesis, transcription, and posttranscriptional reactions undergone by the three major classes of RNA. The main points we covered are as follows:

1. RNA is synthesized in the $5' \rightarrow 3'$ direction by the formation of 3′–5′ phosphodiester linkages between four ribonucleoside triphosphate substrates, analogous to the process of DNA synthesis. The sequence of bases in RNA transcripts catalyzed by DNA-dependent RNA polymerases is specified by the complementary sequences of the DNA template strand.
2. Some newly synthesized RNA transcripts are the functional species, whereas others must be modified or processed into the mature functional species. Modifying enzymes add nucleotides to the 5′ or 3′ ends or alter bases within the RNA, such as by methylation of specific residues. Specific processing enzymes cleave RNA internally, splice together noncontiguous regions of a transcript, or remove nucleotides from the 5′ or 3′ ends.
3. The major classes of RNA in both prokaryotes and eukaryotes are messenger RNA, ribosomal RNA, and transfer RNA. These distinct classes of RNA play specific functional or structural roles in the translation of genetic information into proteins.
4. DNA-dependent synthesis of RNA in *E. coli* is catalyzed by one enzyme, consisting of five polypeptide subunits. The complete holoenzyme is composed of four polypeptides (the core enzyme) and an additional polypeptide that confers specificity for initiation at promoter sequences in the DNA template.
5. The steps involved in transcription include binding of polymerase at the initiation site, initiation, elongation, and termination.
6. In eukaryotes most transcription takes place in the nucleus. Three nuclear RNA polymerases, I, II and III, are responsible for the synthesis of rRNA, mRNA, and small RNA transcripts, respectively. The polymerases contain more subunits than in *E. coli,* and other proteins must bind at the initiation sites or near the initiation sites before the polymerases can begin transcription.
7. Viruses sometimes use the host RNA polymerase in a modified form and sometimes synthesize their own RNA polymerase.
8. One of the more interesting processing reactions of nascent transcripts involves the removal of internal sequences. This type of processing is referred to as splicing. Most splicing reactions appear to require host proteins; however, some require only RNA, a fact demonstrating that RNA is capable of functioning like an enzyme in making and breaking of phosphodiester linkages in polyribonucleotides. Catalytic RNAs may have evolutionary significance.
9. Inhibitors of RNA synthesis may be classified according to their mechanism of action. Some bind to DNA, some bind to RNA, and some are incorporated into the growing RNA chain during transcription.

Selected Readings

Abelson, J., Recognition of tRNA Precursors: A role for the intron. *Science* 255:1390, 1992.

Altman, S., M. Baer, C. Guerrier-Takada, and A. Vioque, Enzymatic cleavage of RNA by RNA. *Trends Biochem. Sci.* 11:515–518, 1986.

Barinaga, M., Ribozymes: Killing the messenger. *Science* 262:1512–1514, 1993.

Bass, B. L., Splicing: The new edition. *Nature* 352:283–284, 1991.

Bear, D. G., and D. W. Peabody, The *E. coli* rho protein: An ATPase that terminates transcription. *Trends Biochem. Sci.* 13:343–348, 1988.

Beaudry, A. A., and G. R. Joyce, Directed evolution of an RNA enzyme, *Science* 257:635–641, 1992.

Beckmann, H., J.-L. Chen, T. O'Brien, and R. Tjian, Coactivator and promoter-selective properties of RNA polymerase I TAFs. *Science* 270:1506–1509, 1995.

Bjork, G. R., J. U. Ericson, C. E. D. Gustafsson, T. G. Hdagervall, Y. H. Josson, and P. M. Wikstrom, Transfer RNA modification. *Ann. Rev. Biochem.* 56:263–287, 1987.

Blumenthal, T., Trans-splicing and polycistronic transcription in *Caenorhabditis elegans. TIG* 11:132–136, 1995.

Burtis, K. C., and B. X. Baker, *Drosophila* double sex gene controls somatic sexual differentiation by producing alternatively spliced mRNAs encoding related sex-specific polypeptides. *Cell* 56:997–1010, 1989.

Cattaneo, R., RNA editing: In chloroplast and brain. *Trends Biochem. Sci.* 17:4–6, 1992. A recent review dealing with RNA editing.

Cech, T., RNA editing: World's smallest introns? *Cell* 64:667–669, 1991.

Chowrira, B. M., A. Berzal-Herranz, and J. M. Burke, Novel guanosine requirement for catalysis by the hairpin ribozyme. *Nature* 354:320–323, 1991. In some cases the guanine amino group serves a catalytic role in self-splicing.

Comai, L., J. C. B. M. Zomerdijk, H. Beckmann, S. Zhou, A. Admon, and R. Tjian, Reconstitution of transcription factor SL1: Exclusive binding of TBP by SL1 or TFIID subunits. *Science* 266:1966–1972, 1994.

Comai, L., N. Tanese, and R. Tjian, The TATA-binding protein and associated factors are integral components of the RNA polymerase I transcription factor, SL1. *Cell* 68:965–976, 1992.

Crick, F., Central dogma of molecular biology. *Nature* 227:561–563, 1970. Of historical interest.

Decker, C. J., and R. Parker, Mechanisms of mRNA degradation in eukaryotes. *TIBS* 19:336–340, 1994.

Deutscher, M. P., The metabolic role of RNases. *Trends Biochem. Sci.* 13:136–139, 1988.

Dorit, R. L., L. Schoenbach, and W. Gilbert, How big is the universe of exons? *Science* 250:1377–1382, 1990.

Drapkin, R., and D. Reinberg, The essential twist. *Nature* 369:523–524, 1994.

Duratowski, S., The basics of basal transcription by RNA polymerase II. *Cell* 77:1–2, 1994.

Eick, D., A. Wedeland, and H. Heumann, From initiation to elongation: Comparison of transcription by prokaryotic and eukaryotic RNA polymerases. *TIG* 10:292–296, 1994.

Forster, A. C., A. C. Jeffries, C. C. Sheldon, and R. H. Symons, Structural and ionic requirements for self-cleavage of virusoid RNAs and trans self-cleavage of viroid RNA. *Cold Spring Harb. Symp. Quant. Biol.* 52:249–259, 1987.

Geiduschek, E. P., and G. P. Tocchini-Valentini, Transcription by RNA polymerase III. *Ann. Rev. Biochem.* 57:873–914, 1988.

Geiger, J. H., S. Hahn, S. Lee, and P. B. Sigler, Crystal structure of the yeast TFIIA/TBP/DNA complex. *Science* 272:830–836, 1996.

Haas, E. S., D. P. Morse, J. W. Brown, F. J. Schmidt, and N. R. Pace, Long-range structure in ribonuclease P RNA. *Science* 254:853–856, 1991.

Hall, B. D., and S. Spiegelman, Sequence complementarity of T2-DNA and T2 specific RNA. *Proc. Natl. Acad. Sci. USA* 47:137–146, 1964. The first use of RNA–DNA hybridization. Of historical interest.

Helmann, J. D., and M. J. Chamberlin, Structure and function of bacterial sigma factors. *Ann. Rev. Biochem.* 57:839–872, 1988.

Henry, R. W., C. L. Sadowski, R. Kobayashi, and M. A. Hernandez, TBP–TAF complex required for transcription of human snRNA genes by RNA polymerases II and III. *Nature* 374:653–662, 1995.

Hou, Y.-M., and P. Schimmel, A simple structural feature is a major determinant of the identity of a transfer RNA. *Nature* 333:144–145, 1988.

Jacobson, R. H., and R. Tjian, Transcription Factor IIA: A structure with multiple functions. *Science* 272:827–828, 1996.

Khoury, G., and P. Gruss, Enhancer elements. *Cell* 33:313–314, 1983. The first report on enhancers.

Kiages, N., and M. Strubin, Stimulation of RNA polymerase II transcription initiation by recruitment of TBP in vivo. *Nature* 374:822–823, 1995.

Koleske, A. J., and R. A. Young, The RNA polymerase II holoenzyme and its implications for gene regulation. *TIBS* 20:113–115, 1995.

Landick, R., and J. W. Roberts, The shrewd grasp of RNA polymerase. *Science* 273:202–203, 1996.

Landweber, L. R., and W. Gilbert, RNA editing as a source of genetic variation. *Nature* 363:179–182, 1993.

Leff, S. D., M. G. Rosenfeld, and R. M. Evans, Complex transcriptional units: Diversity in gene expression by alternative RNA processing. *Ann. Rev. Biochem.* 55:1091–1117, 1986.

Lewin, B., Chapter 31 in *The Apparatus for Nuclear Splicing in Genes,* Vol. 5. New York: Oxford University Press, 1994. Good review.

Lorsch, J. R., and J. W. Szostak, In vitro evolution of new ribozymes with polynucleotide kinase activity. *Nature* 371:31–36, 1994.

Murphy, F. L., J.-H. Wang, J. L. Griffith, and T. R. Cech, Coaxially stacked RNA helices in the catalytic center of *Tetrahymena* ribozyme. *Science* 265:1700–1712, 1994.

Nikoliu, D. B., S.-H. Hu, J. Lin, A. Gasch, A. Hoffman, M. Horikoshi, N.-H. Chua, R. G. Roeder, and S. K. Burly, Crystal structure of TF IID TATA-box binding protein. *Nature* 360:40–45, 1992.

O'Neill, E. M., and E. K. O'Shea, Cyclins in initiation. *Nature* 374:121–122, 1995.

Padgett, R. A., P. J. Grabowski, M. M. Komarska, S. Seller, and P. A. Sharp, Splicing of messenger RNA precursors. *Ann. Rev. Biochem.* 55:1119–1150, 1988.

Peterson, M. G., J. Inostroza, M. E. Maxon, F. Osvaldo, A. Admon, D. Reinberg, and R. Tjian, Structure and functional properties of human general transcription factor IIE. *Nature* 354:369–373, 1991.

Petska, S., J. A. Langer, K. C. Zoon, and C. E. Samuel, Interferons and their actions. *Ann. Rev. Biochem.* 56:757–777, 1987.

Proudfoot, N. J., How RNA polymerase II terminates transcription in higher eukaryotes. *Trends Biochem. Sci.* 114:105–110, 1989.

Pyle, A. M., F. L. Murphy, and T. R. Cech, RNA substrate binding site in the catalytic core of the *Tetrahymena* ribozyme. *Nature* 358:123–128, 1992.

Simpson, L., and D. A. Maslov, RNA editing and the evolution of parasites. *Science* 264:1870–1871, 1994.

Sousa, R., Y. J. Chung, J. P. Rose, and B.-C. Wang, Crystal structure of bacteriophage T7 RNA polymerase at 3.3 Å resolution. *Nature* 364:593–599, 1993.

Steitz, J. A., "Snurps." *Scientific American* 258(6):56–63, 1988.

Stuart, K., RNA editing in mitochondrial mRNA of trypanosomatids. *Trends Biochem. Sci.* 16:68–72, 1991. RNA editing is processing that involves the removal, addition, and modification of nucleotides in the coding regions of nascent transcripts.

Symons, R. H., Small catalytic RNAs. *Ann. Rev. Biochem.* 61:641–671, 1992.

Tollerney, D., Small nucleolar RNAs guide ribosomal RNA methylation. *Science* 273:1056–1057, 1996.

Wahle, E., and W. Keller, The biochemistry of polyadenylation. *TIBS* 21:247–250, 1996.

Weiner, A. M., mRNA splicing and autocatalytic introns: Distant cousins or the products of chemical determination? *Cell* 72:161–164, 1993.

Wolffe, A. P., Nucleosome positioning and modification: Chromatin structures that potentiate transcription. *TIBS* 19:240–244, 1994.

Xing, Y., C. V. Johnson, P. R. Dobner, and J. B. Lawrence, Higher level organization of individual gene transcription and RNA splicing, *Science* 259:1330, 1993.

PROBLEMS

1. Base-stacking interactions are an important stabilizing force in nucleic acid structures. Describe how base stacking contributes to the tertiary structure of the tRNA molecule.
2. G_{22} of $tRNA^{Phe}$ is involved in a Watson-Crick interaction with C_{13} *and* a non-Watson-Crick interaction with m^7G_{46}. Propose a structure for this base "triplet."
3. The illustration is a schematic drawing representing a portion of Oscar Miller and Barbara Hamkalo's electron micrograph of transcription and translation in progress in *E. coli.*

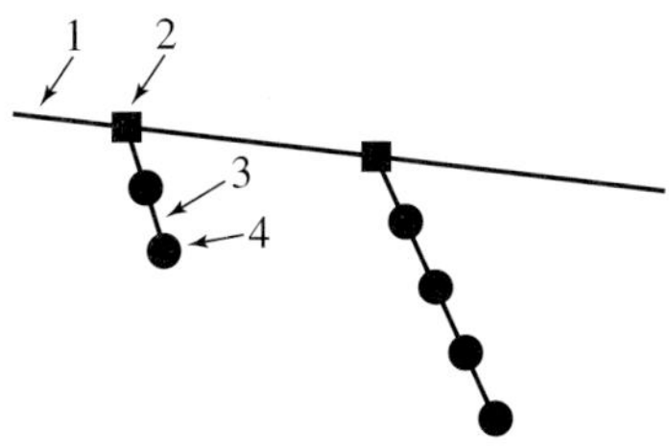

   (a) Identify 1, 2, 3, and 4.
   (b) Indicate the 3′ and 5′ ends of the sense (template) strand of DNA.
   (c) Indicate the 3′ and 5′ ends of the mRNA.
   (d) Draw four peptides in the process of being synthesized, indicating relative lengths.
   (e) Indicate the N- and C-terminal ends of the longest peptide.
   (f) Indicate with an arrow the direction in which RNA polymerase is moving.
   (g) What parts of this diagram would be different in eukaryotes?
4. Why are single-strand binding protein and DNA helicase not required for transcription, as they are for replication?
5. The x-ray crystal structure of T7 RNA polymerase (a small single-subunit bacteriophage-encoded RNA polymerase) has revealed that while this enzyme exhibits amino acid sequence similarity with DNA polymerases, there are striking structural similarities with the Klenow fragment of DNA polymerase I. What features of the two enzymes might lead to similarities in structure? What features of DNA polymerase structure would you expect to be missing from the RNA polymerase structure?
6. Identify the promoter in the *E. coli* sequence shown below. Describe a mutation that would result in more efficient transcription of this operon (i.e., a mutation that would result in a stronger promoter).

GCGGGATCGTTGTATATTTCTTGACACCTTTTCGGCATCGCCCTAA
AATTCGGCGTCCTCATATTG

7. T7 and SP6 bacteriophage RNA polymerases are important tools in molecular biology and are used to make RNA transcripts in vitro. The addition of inorganic pyrophosphatase to these reactions can increase the yield of the RNA product. Explain the increased yield.
8. Although 40–50% of the RNA being synthesized in *E. coli* at any given time is mRNA, only about 3% of the total RNA in the cell is mRNA. Explain.
9. Human RNA polymerase II generates RNA at a rate of approximately 3,000 nucleotides per minute at 37°C. One of the largest mammalian genes known is the 2,000 kbp (1 kbp = one kilobase pair = 1,000 base pairs) gene encoding the muscle protein dystrophin. How long would it take one RNA polymerase II molecule to completely transcribe this gene? How long would it take to transcribe the adult β-globulin gene (1.6 kbp)?
10. The TATA-binding protein, in addition to flattening and widening the DNA by interacting (atypically) with the minor groove, also introduces a sharp bend (>100°) into its recognition sequence upon binding. How might these changes in the structure of the DNA facilitate assembly of the RNA polymerase II transcription initiation complex?
11. In early research with intact eukaryotic mRNA, the RNA appeared to have two 3′ ends and no 5′ terminus. Explain.
12. Compare the ATP requirement for splicing of mRNA introns with that of the T4 DNA ligase reaction.
13. The following hypothetical RNA is a precursor made in a eukaryotic nucleus, and its synthesis is inhibited by low levels of α-amanitin:

pppAUUAUGCCGAUAAGGUAAGUA–(N_{100})–
AUCUCCCUGCAGGGCGUAACCAAUAAACGACGACGACGUCACC

   Indicate the final processed RNA found in the cytoplasm, and point out important features.
14. Predict the effects of the following mutations on mRNA precursors, splicing intermediates, and mRNA derived from an intron-containing gene:
   (a) Changing the A at position 6 of the branch point TACTAAC sequence to a C
   (b) Changing the G at the 5′ end of the intron to a T
   (c) Changing the G at the 3′ end of the intron to an A
15. A particular eukaryotic DNA virus is found to code for two mRNA transcripts, one shorter than the other, from the same region on the DNA. Analysis of the translation products reveals that the two polypeptides share the same amino acid sequence at their amino-terminal ends but are different at their carboxyl-terminal ends. The longer polypeptide is coded by the shorter mRNA! Suggest an explanation.
16. Intestinal mRNA for human apolipoprotein-B was shown to possess a U at one specific position, whereas a C should have been found on the basis of the corresponding gene sequence. (This sequence alteration, which is not found in the

hepatic form of the mRNA, creates a translation stop codon, resulting in production of a truncated form of apolipoprotein-B in the intestine.) On the basis of your knowledge of DNA repair enzymes, suggest several possible mechanisms for this type of RNA editing. Assuming that you were able to prepare a nuclear extract that supported RNA editing in vitro, how might you test some of your suggested mechanisms?

17. Compare the splicing mechanisms of mRNA introns and Group I introns. What aspects of these pathways are similar? How do they differ?

18. A temperature-sensitive mutant of yeast TBP (TATA-binding protein), a subunit of TFIID, was used to study TBP function in vivo. At the restrictive temperature, transcription by all three RNA polymerases, not just RNA polymerase II, stopped almost immediately. What is the simplest explanation of this observation?

C H A P T E R

Protein Synthesis, Targeting, and Turnover

The arrangement of amino acids in polypeptide chains is determined by the arrangement of codons in messenger RNA molecules.

Proteins are informational macromolecules, the ultimate heirs of the genetic information encoded in the sequence of nucleotide bases within the chromosomes. Each protein is composed of one or more polypeptide chains, and each peptide chain is a linear polymer of amino acids. The order of the amino acids commonly found in the polypeptide chain is determined by the order of nucleotides in the corresponding messenger RNA template. In this chapter we examine four aspects of protein metabolism: (1) the process whereby amino acids are ordered and polymerized into polypeptide chains; (2) posttranslational alterations in polypeptides, which occur after they are assembled on the ribosome; (3) the targeting process whereby proteins move from their site of synthesis to their sites of function; and (4) the proteolytic reactions that result in the return of proteins to their starting material, amino acids.

Figure 34.1

Overview of reactions in protein synthesis. (aa_1, aa_2, aa_3 = amino acids 1, 2, 3.) Protein synthesis requires transfer RNAs for each amino acid, ribosomes, messenger RNA, and a number of dissociable protein factors in addition to ATP, GTP, and divalent cations. First the transfer RNAs become charged with amino acids, then the initiation complex is formed. Peptide synthesis does not start until the second aminoacyl tRNA becomes bound to the ribosome. Elongation reactions involve peptide bond formation, dissociation of the discharged tRNA, and translocation. The elongation process is repeated many times until the termination codon is reached. Termination is marked by the dissociation of the messenger RNA from the ribosome and the dissociation of the two ribosomal subunits. The polypeptide chain sometimes folds into its final form without further modifications. Frequently, the folded polypeptide chain is modified by removal of part of the polypeptide chain or by addition of various groups to specific amino acid side chains. The completed polypeptide chain migrates to different locations according to its structure. It may remain in the cytosol or it may be transported into one of the cellular organelles or across the plasma membrane into extracellular space. Proteins have different lifetimes. Eventually, a protein is degraded, usually down to its component amino acids.

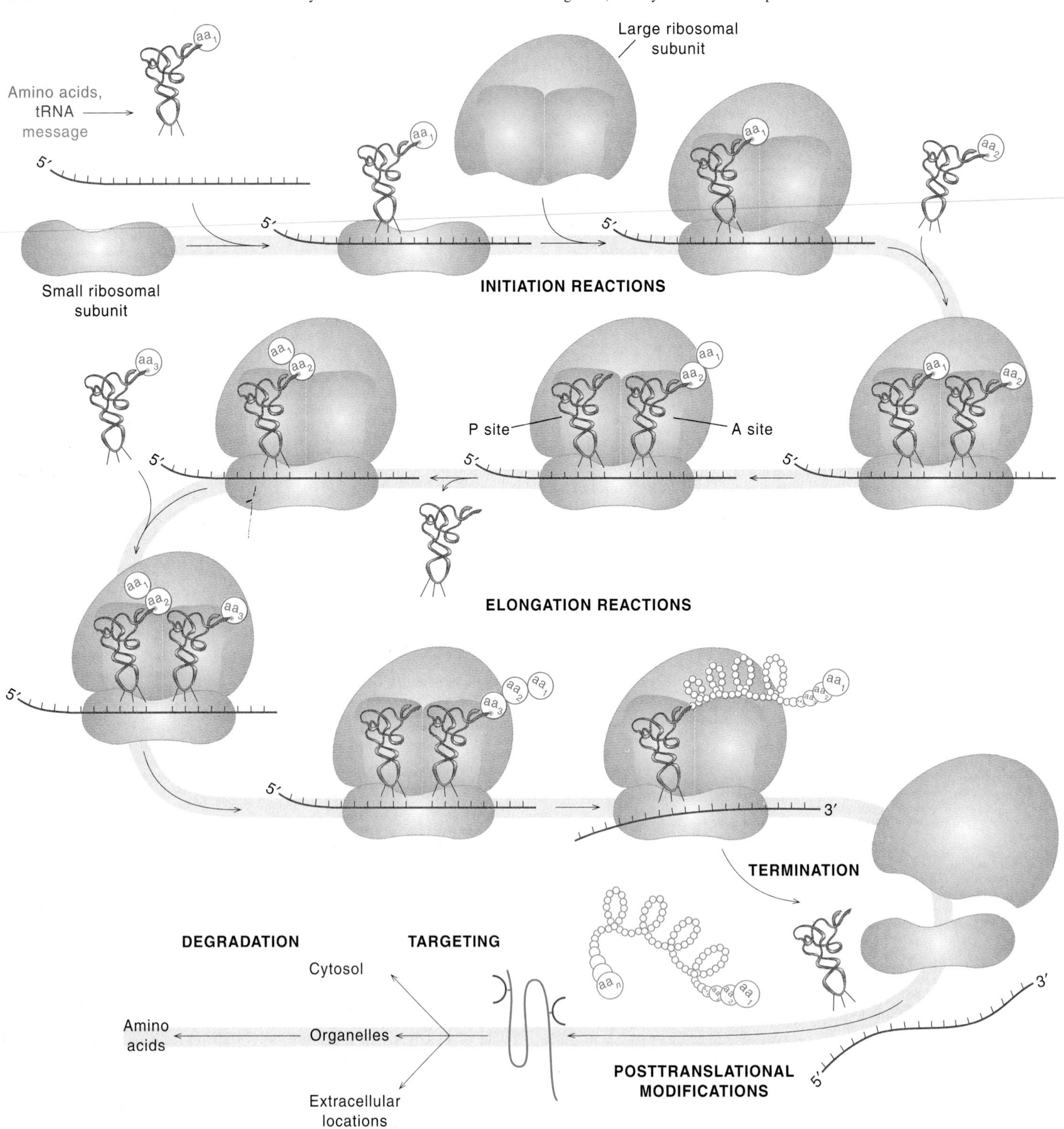

Figure 34.2

(*a*) Electron micrograph of *E. coli* polysomes. Ribosomes are the dark structures connected by the faintly visible mRNA strand. A DNA strand connecting the polysomes from which the mRNA is being transcribed is visible as a horizontal line. (From O. L. Miller, B. A. Hankalo, and C. A. Thomas, "Visualization of bacterial genes in action," *Science* 169:392, 1970, © 1970 by AAAS.) (*b*) Line drawing for clarification. In bacteria, translation usually begins before transcription is completed, resulting in a DNA–mRNA–ribosome complex as shown. As the mRNA grows, the number of ribosomes associated with it increases. Here the mRNA appears to be growing from left to right. It is not possible to see the growing polypeptide chains on the ribosome because of the staining procedure used and the relatively small size of the polypeptide chains.

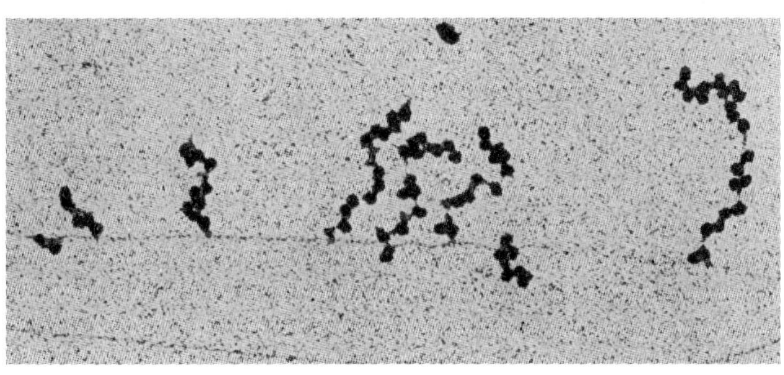

(a)

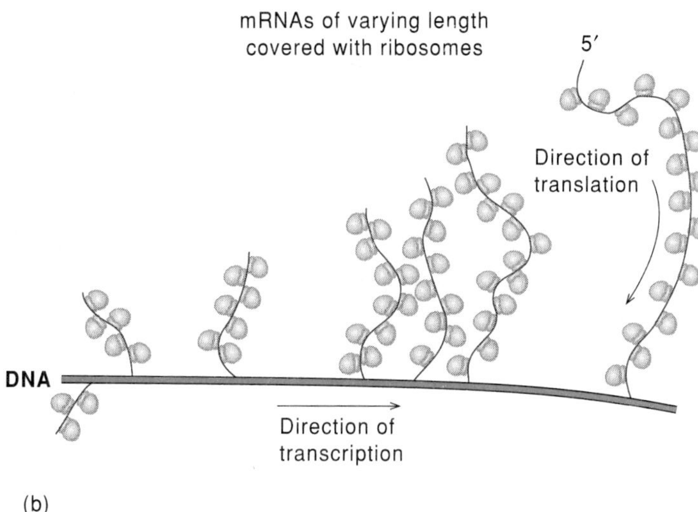

(b)

Figure 34.3

(*a*) Electron micrograph of mammalian rough endoplasmic reticulum. Continuous sheets of membrane create a compartment distinct from the surrounding cytosol. (*Source:* From Dr. S. L. Wolfe, *Biology of the Cell,* 2d ed., 1981, Wadsworth Publishing Co.) (*b*) Clarifying line drawing shows expanded region of endoplasmic reticulum with ribosomes attached to the surface of the membrane facing the cytosol. The term "rough endoplasmic reticulum" arose because at low magnification the attached ribosomes give the endoplasmic reticulum a rough appearance.

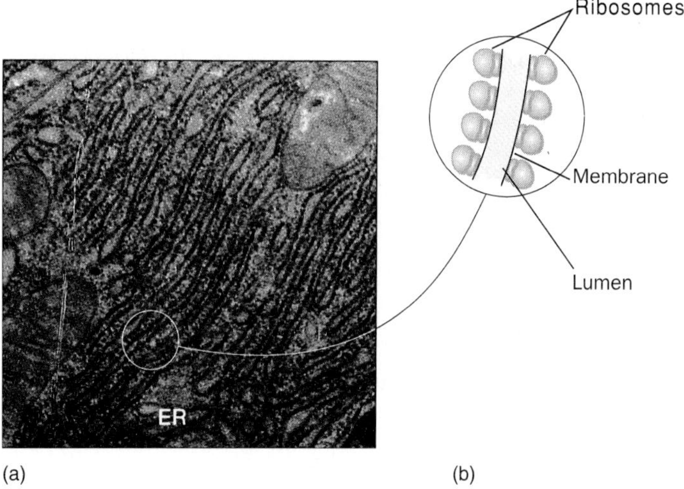

(a) (b)

The Cellular Machinery of Protein Synthesis

Amino acids are assembled into polypeptides on ribosomes (fig. 34.1). Prior to their interaction with messenger RNA (mRNA), amino acids are covalently attached to transfer RNA (tRNA) to form aminoacyl-tRNAs. The aminoacyl-tRNAs attach to specific sites on the mRNA. Messenger RNA contains the instructions for translation in the form of the genetic code that specifies the amino acid sequence of the polypeptide to be synthesized. Each ribosome binds to and moves along the messenger RNA while producing a single polypeptide. The direction of polarity of translation is from the 5′ to the 3′ terminus on the message, whereas the polypeptide is synthesized from the amino to the carboxyl terminus. As a rule, the information in a single mRNA molecule is translated simultaneously and sequentially by a number of ribosomes that form a structure called a polysome. In prokaryotic cells, translation of mRNA begins while it is still being transcribed, so assemblies of nascent polypeptides and mRNAs can be seen on chromosomal DNA (fig. 34.2). In eukaryotic cells, transcription and translation are separate, and polysomes occur either free in the cytosol or bound to membranes in the endoplasmic reticulum (fig. 34.3).

Messenger RNA Is the Template for Protein Synthesis

The mRNA molecule carries the genetic message in the form of a sequence of nucleotides that determines the order of amino acids in the polypeptide chain. Each amino acid is represented in the mRNA by a sequence of three nucleotides called a codon. Codons are arranged in a contiguous reading frame, which is flanked on either side by bases that are not translated. These untranslated regions frequently have roles in regulating the processing and expression of the message. The 5′ end of the reading frame begins with a start codon, usually consisting of the nucleotides AUG, a sequence that codes for the amino acid methionine. Methionine is always used to initiate translation. The 3′ end of the reading frame contains one or more of three stop codons: UAA, UAG, or UGA. Stop codons serve as signals to terminate the polypeptide chain. The 3′ end of the message in eukaryotes usually contains a posttranscriptionally added poly(A) tail. This poly(A) tail increases the lifetime of the message and also plays a role in initiation of translation by facilitating binding of small ribosomal subunits to mRNA, with participation of a poly(A) binding protein.

Figure 34.4

Simplified diagram of mRNA structure. (*a*) Typical eukaryotic mRNA. An AUG start codon is located near the 5′ end of the mRNA. The single reading frame ends with one of the three trinucleotide sequences that represents a stop codon. Frequently, but not always, the 5′ end of the mRNA is capped, and the 3′ end contains a poly(A) tail. The cap structure is described in figure 33.22. (*b*) Typical bacterial mRNA. Some bacterial mRNAs contain more than one reading frame; some contain only one. Each reading frame contains a start and a stop codon. Recognition of the start codons is facilitated by the presence of a ribosome recognition sequence (see table 34.5). The space between any two reading frames in the same transcript varies from one transcript to the next and the reading frames can overlap in some cases.

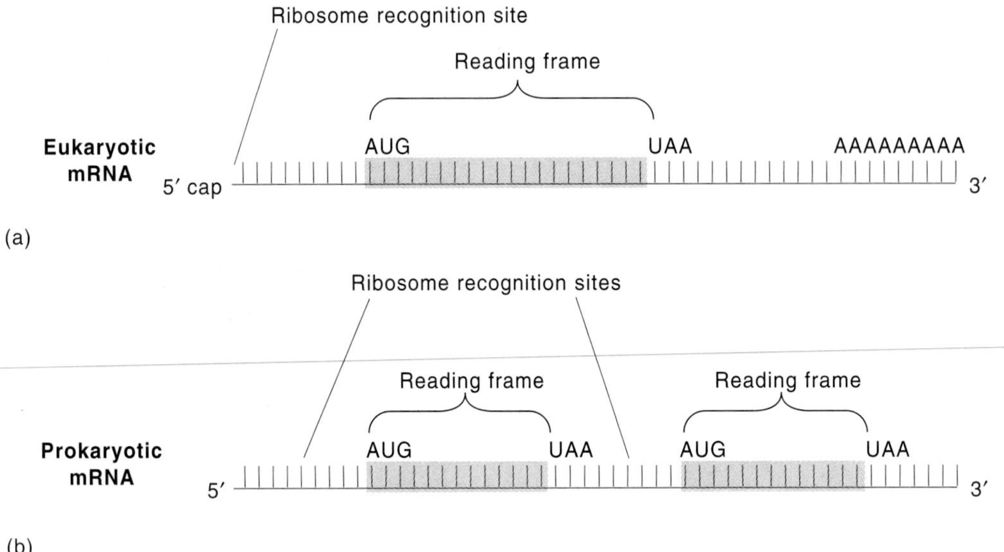

The 5′ end of the mRNA plays a special role in the selection of the start codon. This selection process occurs in fundamentally different ways in prokaryotes and eukaryotes. In prokaryotes (fig. 34.4) the mRNA contains a specific ribosome-binding site upstream of the initiating start codon. This binding site allows the ribosome to identify the correct initiating AUG. Prokaryotic mRNAs are frequently polycistronic, meaning that they encode more than one polypeptide chain. Internal ribosome-binding sites allow for independent initiation at internal reading frames. Eukaryotic mRNAs, on the other hand (see fig. 34.4), are usually monocistronic, encoding only one polypeptide chain. In keeping with this architecture, eukaryotic ribosomes interact with a ribosome entry site at the 5′ terminus of mRNA and move along it by a scanning mechanism to find the start codon. Initiation in eukaryotes usually occurs at the first AUG sequence downstream from the ribosome entry site. In multicellular eukaryotes the initiating AUG sequence is selected by a preferred initiation context of adjacent bases. Recognition of initiation sites for translation is discussed in greater detail in the section entitled "Translation Begins with the Binding of mRNA to the Ribosome."

Transfer RNAs Order Activated Amino Acids on the mRNA Template

Transfer RNAs contain two crucial functional sites: A site for attachment of the amino acids and a site that interacts with the mRNA during translation on the ribosome. At least one type of tRNA corresponds to each of the 20 types of amino acids that are incorporated into proteins. For accurate translation to occur, tRNAs must be distinguishable from one another by the molecules that recognize these specific types while still being recognizable to the molecules that interact with all tRNAs. A major challenge has been to understand how the similarities and differences in the structures of different tRNAs are related to the functions that they play in protein synthesis. Here we summarize the structural features found in all tRNAs that allow them to perform their common functions.

Transfer RNAs that correspond to a single amino acid type are known as cognate tRNAs. In writing, the different cognate tRNAs are designated by a superscript. For example, $tRNA^{Phe}$ and $tRNA^{Ser}$ designate the tRNAs that correspond to the amino acids phenylalanine and serine, respectively. Some amino acids have several different cognate tRNAs, which are called isoacceptor tRNAs. Isoacceptor tRNAs are distinguished from one another by subscripts. A single cell typically contains a total of approximately 50 different tRNAs.

The tRNAs are built on a structural theme so similar that a single figure can describe the common features of their primary, secondary, and tertiary structures (fig. 34.5). All tRNAs are composed of a single polynucleotide chain of 70–95 residues. This chain folds back on itself to form a cloverleaf structure composed of four double-stranded stems and four single-stranded loops. The overall folding is such that 5′ and 3′ ends of the molecule are brought together to create a stem with seven regular Watson-Crick base pairs. This stem is known as the acceptor stem because during protein synthesis the amino acid is transiently attached to the ribose at this terminus. The 3′ end of the molecule is always terminated with the sequence CCA, which identifies the tRNA to the components that interact with all of these molecules during protein synthesis.

The unpaired loops of tRNA are named according to their unique structural features. Loop I varies in size from 7 to 11 unpaired bases and frequently contains the unusual base dihy-

Figure 34.5

The general structure of a tRNA molecule. (*a*) Representation in the form of a cloverleaf, which is the simplest way of observing the secondary structure. (*b*) A more realistic drawing of the three-dimensional folded structure. Color coding shows how the various loops in the cloverleaf structure correspond to the parts of the folded structure. For a more detailed drawing showing the hydrogen bonds and the names of the individual bases, see figure 33.3. (Adapted from A. Rich and S. H. Kim, The three-dimensional structure of transfer RNA, *Sci. Amer.* 238:52–62, January 1978. Copyright © 1978 by Scientific American, Inc. All rights reserved. Reprinted by permission.)

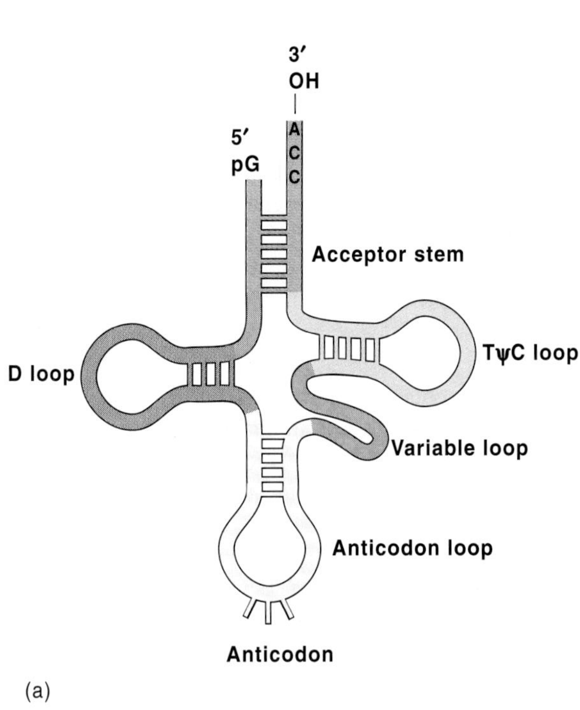

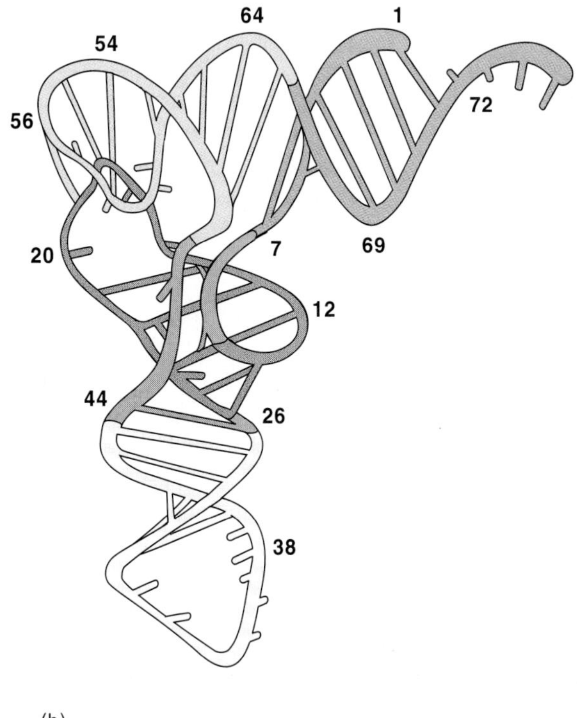

drouracil; it is designated the D loop. Loop II contains the three bases known as the anticodon and is therefore designated the anticodon loop. This portion of the tRNA plays a key role in translation by pairing with the complementary codon in mRNA and thus serves to order amino acids in their appropriate sequence. Loop III, the variable loop, may contain as few as 3 or as many as 21 bases, making it the major site of size variation in tRNA. Loop IV contains the unusual ribothymidine and pseudouridine bases as part of an invariant sequence; for this reason it is known as the TψC loop.

All tRNAs in solution fold into a three-dimensional L-shaped structure like that of $tRNA^{Phe}$ (see fig. 34.5*b*). This structure is composed of two helical arms joined at right angles. The ribose moiety to which the amino acid is joined is at the end of one arm, identifying it as the acceptor arm. The anticodon is at the end of the other arm, identifying it as the anticodon arm. The 3′ terminal adenosine and anticodon are separated by a diagonal distance of about 75 Å.

Ribosomes Are the Site of Protein Synthesis

We considered the structure of the ribosome in some detail in the previous chapter without referring to those sites that are functionally important in protein synthesis. Here we can localize some of the functional sites on the two ribosomal subunits (fig. 34.6). The mRNA binds to the smaller ribosomal subunit. The peptidyl transferase is an integral part of the 50S subunit, and the elongation factor EF-G binds to the 50S subunit. The nascent polypeptide chain exits through a channel in the 50S subunit. Two functional sites occur on the 50S subunits, called the P site (for peptidyl) and the A site (for aminoacyl) where adjacent tRNAs align with the messenger (see fig. 34.1). A third site for tRNA binding has also been identified by Knud Nierhaus and co-workers, called the E site (for exit).

Before we discuss the reactions that occur on the ribosome, it is useful to consider the crucial interaction between the anticodon on the tRNAs and the codons on the mRNAs.

The Genetic Code

The concept of a genetic code grew out of the realization that both nucleic acids and proteins are linear polymers made of a limited number of building blocks, plus the knowledge that the structure of proteins is genetically determined. The genetic code is simply the sequence relationship between nucleotides in genes (or mRNA) and the amino acids in the proteins they encode. The coding ratio is the number of nucleotides in an mRNA required to represent an amino acid. Since there are four different nucleotides in RNA and 20 different amino acids in proteins, the minimum acceptable coding ratio is 3:1. A singlet code, in which one nucleotide represents one amino acid, could code for only four different amino acids because there are only four different nucleotides in DNA or RNA. A doublet code, in which two nucleotides code for a single

Figure 34.6

Functional sites on the prokaryotic ribosome. This figure shows a ribosome in the process of elongating a polypeptide chain. The mRNA and the EF-Tu-aminoacyl-tRNA complex are more closely associated with the 30S subunit. The peptidyl transferase and the EF-G elongation factor are associated with the 50S subunit, as is the nascent polypeptide chain. IF and EF are standard abbreviations for proteins that function as initiation factors and elongation factors, respectively. (*Source:* From C. Bernabeu and J. A. Lake, Nascent polypeptide chains emerge from the exit domain of the large ribosomal subunit: Immune mapping of the nascent chain, *Proc. Nat'l. Acad. Sci. USA* 79:3111–3115, 1982. Copyright © 1982 National Academy of Sciences, Washington, D.C. Reprinted by permission.)

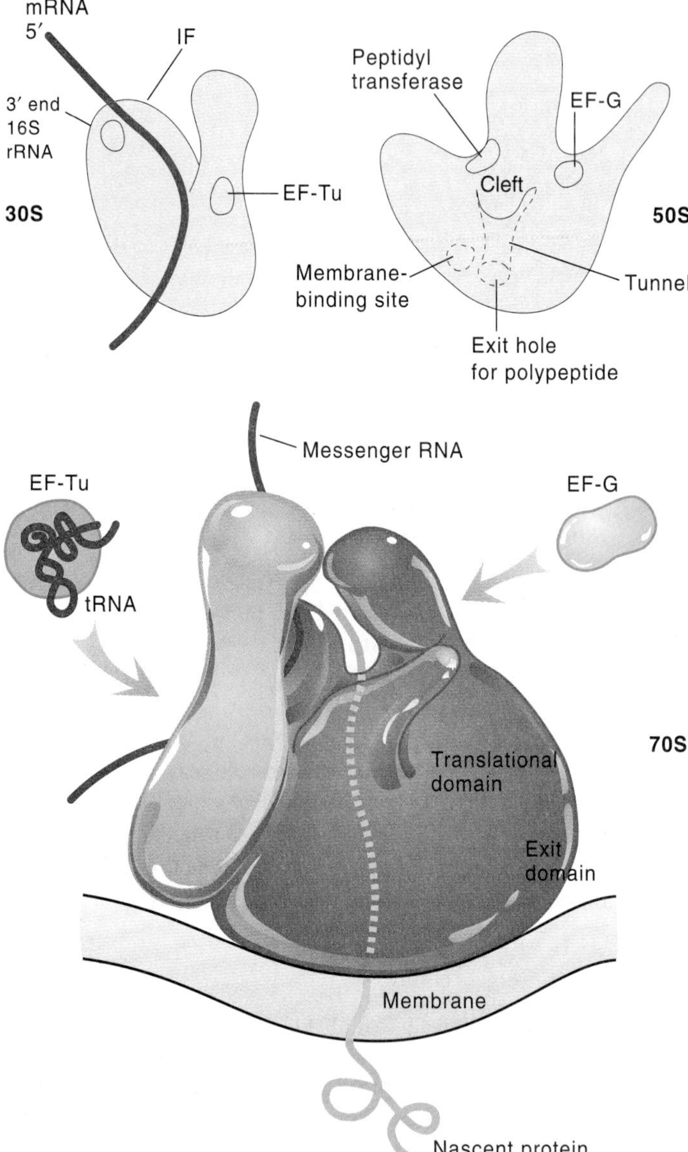

amino acid, could code for 16 amino acids since there are 16 possible doublet sequences in RNA. A triplet code, made from four nucleotides taken three at a time, generates a total of 64 possible triplet sequences. This would be more than enough to represent the 20 amino acids found in proteins.

In fact, a triplet code creates a problem of a different sort. What is the use of the extra trinucleotide sequences? Allowing for the three triplets that represent stop signals, we are still left with 61 possible triplets to code for just 20 amino acids. A nondegenerate code, in which a unique triplet codes for each amino acid, would imply that 41 triplet sequences are never found in translation reading frames. On the other hand, a degenerate code, in which all triplets are used, requires that each amino acid be represented, on the average, by more than one triplet sequence. We return to this point later. For now, we simply note that in most cases, point mutations in which one base pair in the DNA is replaced by another result either in no amino acid change or in a change to another amino acid. This is the result that we expect if most of the triplet sequences do in fact represent amino acids.

Another question relates to the location of codons in the translation reading frame. Codons could be arranged in a close-packed side-by-side arrangement. In this event each nucleotide within the translation reading frame would represent a code letter within one, and only one, codon. Alternatively, codons could be separated by one or more "spacer nucleotides." In this event some nucleotides within the reading frame would represent code letters within codons, while others would serve as spacer nucleotides. Finally, we can imagine an overlapping codon arrangement, in which all nucleotides represent code letters, and some nucleotides represent code letters in more than one codon. However, biochemical experiments that we describe later in this chapter favor a nonoverlapping triplet code without spacers. Further genetic experiments in support of this type of code (not elaborated on here) indicate that when a single base pair is added or deleted, a change occurs in the reading frame of the message downstream of the change. Such an alteration in the DNA is known as a frameshift mutation. A frameshift mutation generates a new termination point because the frameshift changes the reading of the original stop codon, and a new stop codon will come into frame at some other location in the sequence.

The Code Was Deciphered with the Help of Synthetic Messengers

In 1961, Marshall Nirenberg and Heinrich Mattaei were attempting to synthesize proteins in cell-free extracts of *E. coli.* These extracts, containing the essential cellular components, were supplemented with nucleotides, salts, and radioactive amino acids, which were used so that the expected small amounts of protein could be detected. For mRNA they chose to use viral RNA from tobacco mosaic virus (TMV), because their goal was to make viral protein. As a control for these first "incorporation" experiments, they needed an mRNA that would not be expected to code for a protein. For this purpose they chose poly(U), which contained only uridine residues. When Nirenberg and Mattaei added this control RNA to their cell-free system, substantial incorporation was evident. They showed that this synthetic messenger specifies the synthesis of a polypeptide containing only phenylalanine residues (polyphenylalanine).

A wave of excitement was set in motion by this discovery. Soon afterwards it was demonstrated that poly(A) promotes polylysine synthesis and poly(C) promotes polyproline synthesis. From these observations it seemed clear that the code words UUU, AAA, and CCC correspond to the amino acids phenylalanine, lysine, and proline, respectively.

Figure 34.7

Demonstration that poly(UC) codes for a Ser-Leu repeating peptide. When poly(UC) containing a strict alternating sequence of U and C is used as an mRNA, only serine and leucine are incorporated. Analysis of the polypeptide product indicates that it contains an alternating sequence of serine and leucine.

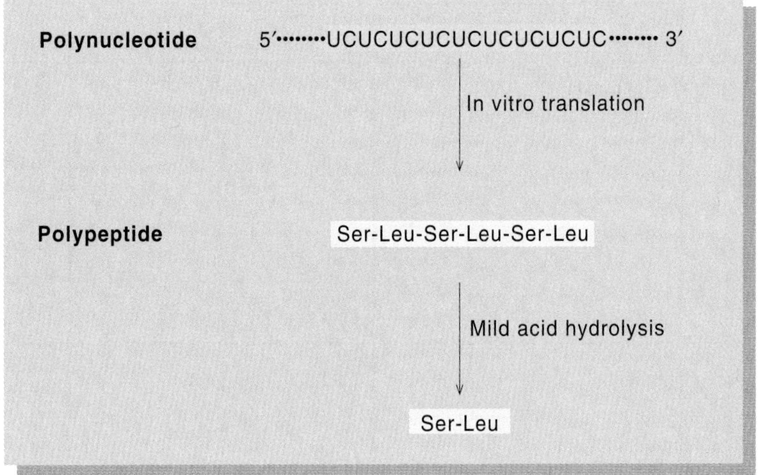

Polynucleotide phosphorylase was used to produce "RNA" of random sequence, the composition of which reflected the mixture of nucleoside diphosphates in the reaction mixture. Mixed polynucleotides containing two bases were used in the incorporation system and shown to incorporate a pattern of amino acids consistent with a triplet code, but the observed incorporation could not define the code sequence.

The chemical synthesis of short oligonucleotides of known sequence provided a way out of this dilemma. First, Philip Leder and Nirenberg showed that nucleotide triplets cause the specific binding of aminoacyl-tRNA to ribosomes. They observed that this binding could be readily measured by simply passing a solution containing radioactive aminoacyl-tRNA, ribosomes, and the trinucleotide through a nitrocellulose filter. Ribosomes, along with bound ligands, were adsorbed to the filter, and the coding specificity of some triplets was quickly established. Thus UUC, UCU, and CUU were shown to specify phenylalanine, serine, and leucine, respectively. However, many triplets did not give clear binding signals, and there was concern that the binding reaction might not display the correct specificity.

Deciphering of the code was ultimately completed by the translation of polynucleotides of repeating sequences. These polynucleotides with long repeat sequences were produced by chemically synthesizing short, defined-sequence oligomers and then amplifying them enzymatically. When repeating dinucleotides were translated (fig. 34.7), they yielded repeating dipeptides, and when repeating trinucleotides (not shown) were translated, they yielded up to three individual peptides, each containing only a single amino acid. Eventually, the sequence of bases from natural messages was correlated with the sequence of amino acids in the proteins they encoded.

The Code Is Highly Degenerate

A triplet code, one made from four nucleotides taken three at a time, generates a total of 64 different triplet sequences or codons. Three of these codons, as we will see, are utilized to terminate translation and are not generally used to specify amino acids. The remaining 61 codons and the 20 amino acids can be neatly summarized by grouping codons with the same first and second bases into a grid (table 34.1). The four horizontal sections are composed of codons with the same first base. The four vertical sections are composed of codons with the same second base. The boxes representing the vertical–horizontal intersections contain codon families, the members of which differ only in their 3′ terminal base. For example, the codons UCU, UCC, UCA, and UCG constitute a family encoding serine. Thus the genetic code is degenerate, that is, one amino acid is generally specified by multiple or synonymous codons.

As we have indicated, the codon AUG is the only one generally used to specify methionine, but it serves a dual function in that it is also used to initiate translation. Occasionally, GUG and UUG are also read as an initiating codon specifying methionine in bacteria, but in internal positions these codons are always read as valine and leucine, respectively. In eukaryotes, initiation at codons other than AUG is much less frequent than in prokaryotes. The UGA triplet also serves a dual function; it is usually recognized as a stop, but on occasion it serves as a codon for selenocysteine (box 34A).

Organisms differ in the frequency with which they utilize synonymous codons. Some synonymous codons are used frequently, and some are used infrequently. For *E. coli* and yeast this frequency of use, known as codon usage, correlates with the abundance in the organism of the tRNAs that recognize particular synonymous codons. Also, for the same two species, proteins that occur in the greatest abundance employ high-usage codons much more frequently and low-usage codons much less frequently than the average proteins. In some differentiated eukaryotic cells the intracellular concentrations of tRNA isoacceptors are adjusted to meet the codon demands of the message being translated. This is notable in the case of silk fibroin protein, approximately 85% of which is comprised of the amino acids

Table 34.1
The Genetic Code

First Position (5′ end)	Second Position								Third Position (3′ end)
	U		**C**		**A**		**G**		
U	UUU	Phe	UCU	Ser	UAU	Tyr	UGU	Cys	**U**
	UUC		UCC		UAC		UGC		**C**
	UUA	Leu	UCA		UAA	STOP	UGA	STOP	**A**
	UUG		UCG		UAG		UGG	Trp	**G**
C	CUU	Leu	CCU	Pro	CAU	His	CGU	Arg	**U**
	CUC		CCC		CAC		CGC		**C**
	CUA		CCA		CAA	Gln	CGA		**A**
	CUG		CCG		CAG		CGG		**G**
A	AUU	Ile	ACU	Thr	AAU	Asn	AGU	Ser	**U**
	AUC		ACC		AAC		AGC		**C**
	AUA		ACA		AAA	Lys	AGA	Arg	**A**
	AUG	Met	ACG		AAG		AGG		**G**
G	GUU	Val	GCU	Ala	GAU	Asp	GGU	Gly	**U**
	GUC		GCC		GAC		GGC		**C**
	GUA		GCA		GAA	Glu	GGA		**A**
	GUG		GCG		GAG		GGG		**G**

Source: Robert F. Weaver, and Philip W. Hedrick, *Basic Genetics.* Copyright © 1991 Times Mirror Higher Education Group, Inc., Dubuque, Iowa. All Rights Reserved. Reprinted by permission.

glycine, alanine, and serine. During differentiation of the silk gland in worms, tRNAs cognate to these codons are amplified and attain levels of about 70% of the total tRNA.

Wobble Introduces Ambiguity into Codon-Anticodon Interactions

For 61 triplets to act as codons, tRNAs must interact specifically with each triplet. Strict Watson-Crick base pairing between codon and anticodon would require 61 different anticodons and, correspondingly, 61 different tRNAs. As the characterization of tRNAs progressed, it became clear that in many cases, individual tRNAs could recognize more than one codon. In all cases the different codons recognized by the same tRNA were found to contain identical nucleotides in the first two positions and a different nucleotide in the third position (in the 3′ position of the codon). This relationship is the reason that the genetic code can be so neatly arranged, as shown in table 34.1, by codon families that differ in their third base.

The 3′ terminal redundancy of the genetic code and its

34A

BOX

Selenocysteine Incorporation During Translation Elongation

The unusual amino acid selenocysteine (a derivative of cysteine in which the sulfur atom is replaced by a selenium atom) in an essential component in a small number of proteins. These proteins occur in prokaryotes and eukaryotes ranging from *E. coli* to humans. In all cases, selenocysteine is incorporated into protein during translation in response to the codon UGA. This codon usually serves as a termination codon but occasionally, in some required but unknown context of bases, is used to specify selenocysteine instead.

In *E. coli* the products of four genes (*selA, selB, selC,* and *selD*) are required for the incorporation of selenocysteine. The product of the *selC* gene is $tRNA^{Ser}$ (a suppressor tRNA the anticodon of which is UCA). The first step in the incorporation of selenocysteine is catalyzed by seryl-tRNA synthase. The products of *selA* and *selD* function in the subsequent conversion of Ser-$tRNA^{Ser}$ to selenocysteyl-$tRNA^{Ser}$. The probable pathway is

$$tRNA^{Ser} \longrightarrow \text{Seryl-}tRNA^{Ser} \longrightarrow \text{Phosphoseryl-}tRNA^{Ser} \longrightarrow \text{Selenocysteinyl-}tRNA^{Ser}$$

Incorporation of selenocysteinyl-$tRNA^{Ser}$ into protein in response to the UGA codon requires SELB (the protein product of the *selB* gene in *E. coli*). SELB is homologous in sequence to EF-Tu and probably replaces it in translation by specifically recognizing selenocysteinyl-tRNA and UGA in the appropriate sequence context. Selenocysteinyl-$tRNA^{Ser}$, in combination with SELB, must be capable of competing with termination factors for the translation of the termination codon when it occurs in the "right" context of bases. A stem-loop structure in the mRNA is required just downstream of the UGA codon specifying selenocysteine. As discussed in the text for programmed translational frameshifts (see section entitled "Ribosomes Can Change Reading Frame during Translation"), this structure is similar to other "stimulators" involved in recoding events.

Other proteins of unknown function that are homologous to EF-Tu are known to exist. Conceivably, these proteins might participate in the incorporation of other rare amino acids into unique positions in proteins. Mutants of EF-Tu have also been found which stimulate ribosomes to read through stop codons.

---

mechanistic basis were first appreciated by Francis Crick in 1966. He proposed that codons and anticodons interact in an antiparallel manner on the ribosome in such a way as to require strict Watson-Crick pairing (that is, A-U and G-C) in the first two positions of the codon but to allow other pairings in its 3′ terminal position. Nonstandard base-pairing between the 3′ terminal position of the codon and the 5′ terminal position of the anticodon alters the geometry between the paired bases; Crick's proposal, labeled the wobble hypothesis, is now viewed as generally describing the codon–anticodon interactions that underlie the translation of the genetic code.

By inspecting the geometry that would result from different wobble pairings (fig. 34.8) and recognizing that inosine, the deaminated form of adenine, frequently occurs in tRNAs in the 5′ position of the anticodon, Crick grasped the relationship between codon and anticodon. According to this relationship (table 34.2), when C or A occurs in the 5′ position of an anticodon, it can pair only with G or U, respectively, in the 3′ position of a codon. Transfer RNAs containing either G or U in the 5′, or wobble position of the anticodon can each pair with two different codons, whereas an inosine (I) in this position produces a tRNA that can pair with three codons differing in the 3′ base. There are instances, however, when pairing in the wobble position is even more flexible than originally proposed by Crick. Subsequent analysis of many tRNAs has confirmed this hypothesis as a feature of the translation of the "universal" genetic code (but see the following discussion).

A careful comparison of the wobble rules with the genetic code indicates that the minimum number of tRNAs required to translate all 61 codons is 31. With the addition of $tRNA_i^{Met}$ the total comes to 32. Most cells contain many more than this minimum number of tRNA types.

The Code Is Not Quite Universal

It was originally believed that exactly the same genetic code is utilized by all systems. Initial experiments supported this conclusion, since it was found that some mammalian mRNAs could be faithfully translated in cell-free bacterial systems. We now know, however, that significant variations in the meaning of specific code words occur in many systems. The exact scope and nature of these variations are still being discovered, but the current view is that the variations in genetic meaning reflect divergences from the standard, or "universal," genetic code described earlier, rather than independent origins of the genetic code.

Many of the known variations in the genetic code are found in genes of mitochondria and chloroplasts. It is easy to see why these systems might be more plastic, since they frequently encode only 10–20 proteins. The remainder of the organellar proteins are derived by importing nuclear gene products.

The tRNAs used to translate mitochondrial mRNAs are entirely derived from mitochondrial chromosomes. The first clue that something was unusual about the mitochondrial genetic code was that only 24 types of tRNA could be found. According to Crick's rules of wobble, 32 tRNAs minimally are required for the translation of all 61 codons. One possible solution to this conundrum was that mitochondrial genes do not utilize all 61 codons. Another possibility was that the wobble rules might be different for mitochondria. In fact, the latter is the case. Crick's original wobble rules stated that at least two different tRNAs are needed to translate four codon families. In all of these cases (with the exception of the codon fam-

Figure 34.8

Examples of standard (*a*) and wobble (*b* and *c*) base pairs formed between the first base in the anticodon and the third base in the codon.

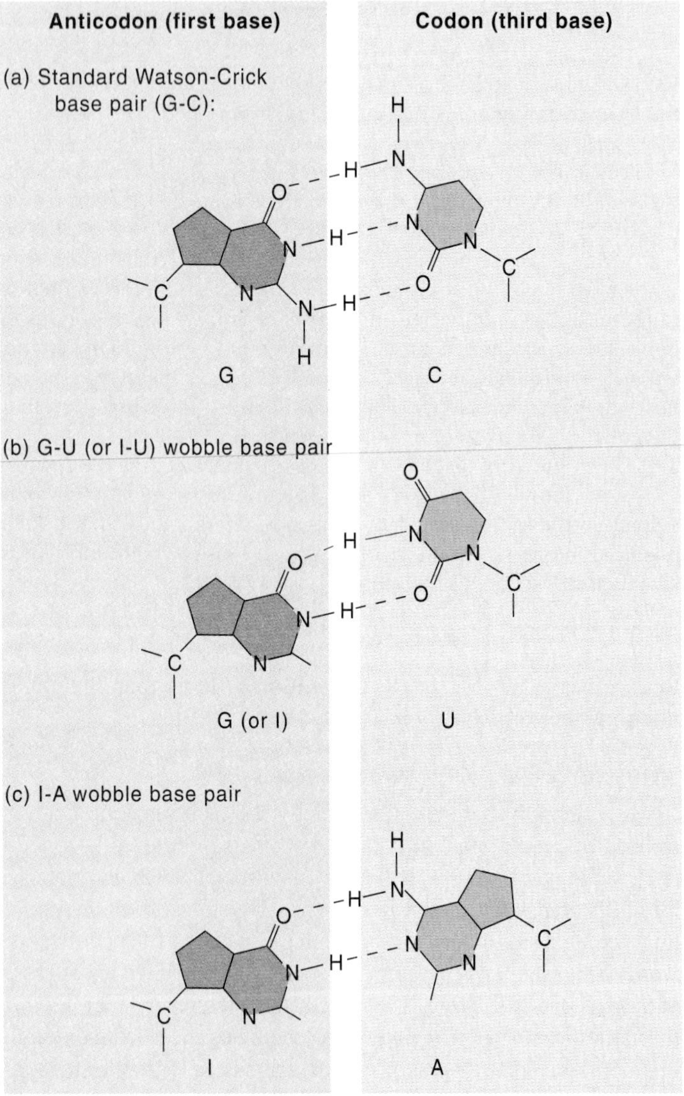

ily specifying Arg; see the footnote to table 34.3), single tRNAs have been found responsible for specifying all four code words, and these tRNAs all contain a U in the "wobble" position of their anticodons. It appears that the mitochondrial ribosome allows these tRNAs to pair with all four members of the codon family. The six mitochondrial tRNAs that pair with the normal two codons contain an altered U in the wobble position, and this modification causes them to conform to the normal "wobble" rules.

Another peculiarity of the mitochondrial code emerged from a study of yeast codon usage. By comparing tables 34.3 and 34.1 you will see that the yeast mitochondrial code has several differences in code word meaning. The codons beginning with CU represent Thr instead of Leu, the AUA codon represents Met instead of Ile, and the UGA codon represents Trp rather than a stop signal.

Table 34.2
The Wobble Rules of Codon–Anticodon Pairing

5′ Base of Anticodon	3′ Base of Codon
C	G
A	U
U	A or G
G	C or U
I	U, C, or A

The Rules Regarding Codon-Anticodon Pairing Are Species-Specific

The genetic code differs very little between species. By contrast, considerable differences occur between species in the anticodon translation system of tRNA, as evidenced by the mitochondrial tRNA system. In all systems the bases in the anticodon–codon complex run antiparallel, as in standard double-helix pairing, and in all cases, only Watson-Crick-like base pairing occurs between the first two bases in the codon and the opposing bases in the anticodon segment of the tRNA. However, for the 3′ base in the codon, the rules for pairing vary with the species and with the base in question. These rules, summarized in table 34.4, are as follows.

When the 5′ base in the anticodon is a G, it can pair with either a U or a C, and this is true in all organisms. When the 5′ base in the anticodon is an A, it can pair only with a U in the codon. However, it is rare that an A is found in this position of the anticodon. An A in this position in eukaryotes is usually deaminated to an inosine (I) base, which has an expanded capacity for pairing. Base A can pair only with U, but I can pair with U, C, or A. In eubacteria, deamination is limited to the conversion of the ACG sequence to an ICG anticodon. When C is the 5′ base in the anticodon, it pairs with G only. An exception to this rule is found in eubacteria, in which the C is covalently modified in the tRNA that recognizes the AUA codon. Thus in this instance the modified C can pair with a 3′ A in the codon. A U base in the 5′ position of the anticodon shows the greatest variability. An unmodified U can pair with any of the four bases in the 3′ position of the codon. This situation is reflected in the U-family box in mitochondria (see table 34.3).

U can also be modified in various ways. In mitochondria, one type of modification permits a U to pair with either an A or a G in the two-codon sets. In eukaryotes, another type of U modification limits U to pairing with an A in the codon, and in eubacteria a third type of U modification permits U pairing with U, A, or G but not C in family boxes.

The Steps in Translation

Protein synthesis involves more than 100 different proteins and more than 30 kinds of RNA molecules. A considerable achievement was the development of prokaryotic in vitro protein synthesis systems in which all of the constituent components in pure forms were added to the reaction mixture, and translation of nat-

Table 34.3
The Genetic Code of Yeast Mitochondria

First Position (5′ end)	Second Position: U			C			A			G			Third Position (3′ end)
	Codon	Amino acid	Anticodon	Codon	Amino acid	Anticodon	Codon	Amino acid	Anticodon	Codon	Amino acid	Anticodon	
U	UUU	Phe	AAG	UCU	Ser	AGU	UAU	Tyr	AUG	UGU	Cys	ACG	U
	UUC			UCC			UAC			UGC			C
	UUA	Leu	AAU*	UCA			UAA	STOP		UGA	Trp	ACU*	A
	UUG			UCG			UAG			UGG			G
C	CUU	Thr	GAU	CCU	Pro	GGU	CAU	His	GUG	CGU	Arg	GCA[b]	U
	CUC			CCC			CAC			CGC			C
	CUA			CCA			CAA	Gln	GUU*	CGA			A
	CUG			CCG			CAG			CGG			G
A	AUU	Ile	UAG	ACU	Thr	UGU	AAU	Asn	UUG	AGU	Ser	UCG	U
	AUC			ACC			AAC			AGC			C
	AUA	Met	UAC[a]	ACA			AAA	Lys	UUU*	AGA	Arg	UCU*	A
	AUG			ACG			AAG			AGG			G
G	GUU	Val	CAU	GCU	Ala	CGU	GAU	Asp	CUG	GGU	Gly	CCU	U
	GUC			GCC			GAC			GGC			C
	GUA			GCA			GAA	Glu	CUU*	GGA			A
	GUG			GCG			GAG			GGG			G

Source: From S. G. Bonitz et al., Codon recognition rules in yeast mitochondria, in *Proc. Natl. Acad. Sci. USA* 77:3167, 1980.

The codons (5′ → 3′) are at the left and the anticodons (3′ → 5′) are at the right in each box. (* designates U in the 5′ position of the anticodon that carries the —$CH_2NH_2CH_2COOH$ grouping on the 5′ position of the pyrimidine.)

[a]Two tRNAs for methionine have been found. One is used in initiation, and one is used for internal methionines.

[b]Although an Arg tRNA has been found in yeast mitochondria, the extent to which the CGN codons are used is not clear.

urally occurring mRNAs ensued. Coupled transcription-translation systems were also developed, in which the mRNA was first synthesized in vitro from a DNA copy of the gene prior to translation. Application of this kind of technology to eukaryotic systems has recently led to the astonishing result that intact, infectious poliovirus particles could be generated from the viral genome by in vitro reactions in the absence of living cells. Much of what we know about the mechanics of translation was derived from in vitro protein synthesis experiments. The process begins by the attachment of amino acids to specific tRNA molecules. Subsequent steps take place on the ribosome; amino acids are transported to the ribosome on their tRNA carriers, and they do not leave the ribosome until they have become an integral part of a polypeptide chain.

Synthases Attach Amino Acids to tRNAs

A unique class of enzymes, called aminoacyl-tRNA synthases, attach amino acids to their cognate tRNAs. This attachment serves two functions: (1) The linkage between amino acid and tRNA activates the amino acid, making the subsequent formation of a peptide bond energetically favorable. (2) The tRNA directs the amino acid to a designated location on a messenger RNA so that the amino acid is incorporated at the appropriate location in the polypeptide chain.

All synthase reactions proceed in two separate steps (fig. 34.9). In the first step the synthase recognizes its corresponding amino acid and its second substrate, ATP, and forms a mixed anhydride bond between the carboxyl group of the amino acid and the phosphate of AMP with the release of PP_i:

$$\text{Amino acid} + \text{ATP} \longrightarrow \text{Aminoacyl-AMP} + PP_i \quad \textbf{(1)}$$

The equilibrium constant for this reaction is about 1, so the energy derived from the cleavage of the phosphate anhydride of ATP is conserved in the mixed anhydride. Aminoacyl-AMP remains tightly bound to the enzyme, and, as we will soon show, this fact has allowed researchers to crystallize this important complex and analyze its structure.

The second reaction catalyzed by the aminoacyl-tRNA synthases results in the attachment of the amino acid through an ester linkage to the 3′ terminal ribose of tRNA:

$$\text{Aminoacyl-AMP} + \text{tRNA} \longrightarrow \text{Aminoacyl-tRNA} + \text{AMP} \quad \textbf{(2)}$$

Table 34.4
Anticodon–Codon Pairing

Anticodon First Base	Codon Third Base	Examples
U	U, C, A, G	Mitochondrial code in family boxes
*U	A, G	Mitochondrial code in two-codon sets
†U	A	Eukaryotes
‡U	U, A, G	Eubacteria in family boxes
C	G	All codes
*C	A	Bacteria, isoleucine codon AUA
G	U, C	All codes
A	U	Rare
I	U, C, A	Eukaryotes, ICG in eubacteria

*, †, ‡ = Various modifications of U (Yokoyama et al., Proc. Natl. Acad. Sci. 82:4905, 1985).

*C = modified C.

Source: From Thomas Jukes et al., *Cold Spring Harbor Symp. Quant. Biol.* 32:775, 1987. Copyright © 1987 Cold Spring Harbor Laboratory Press, Cold Spring Harbor, N.Y. Reprinted by permission.

Synthases differ with respect to their site of attachment to tRNA. Some synthases form the 2′ ester, some form the 3′ ester, and still others produce a mixture of the two. The specificity of the synthases was determined by analyzing their ability to act on tRNA derivatives lacking one or the other terminal hydroxyl group. Once esterified to the terminal ribose, the aminoacyl group can migrate between the vicinal 2′ and 3′ hydroxyl groups. Thus in cells, aminoacyl-tRNAs are mixtures of 2′ and 3′ esters. Only the 3′ derivative is a substrate for the subsequent transpeptidation reaction catalyzed by the ribosome.

The sum of the two reactions catalyzed by aminoacyl-tRNA synthases is

$$\text{Amino acid} + \text{ATP} + \text{tRNA} \longrightarrow \text{Aminoacyl-tRNA} + \text{AMP} + \text{PP}_i \quad (3)$$

This overall reaction is reversible but is driven to completion by the subsequent hydrolysis of PP_i to two equivalents of P_i through the action of ubiquitous pyrophosphatases. Thus the formation of aminoacyl-tRNA consumes two equivalents of ATP. The energy that ultimately drives the formation of the peptide linkage during protein synthesis is derived from the ester linkage that joins amino acids to tRNA.

Each Synthase Recognizes a Specific Amino Acid and Specific Regions on Its Cognate tRNA

Our understanding of synthase reactions and the types of active sites involved in these reactions was advanced substantially by the crystallization and structural solution of tyrosyl-tRNA synthase complexed with the reaction intermediate tyrosyl-adenylate (fig. 34.10). The reaction intermediate is bound in a deep cleft in the enzyme and interacts with it through 11 hydrogen bonds. Six of these bonds are with the AMP moiety, and five are with the tyrosyl moiety of the intermediate. The amino acid selectivity of tyrosyl-tRNA synthase is thus determined primarily by the formation of specific hydrogen bonds with the amino acid.

Most cells contain only one synthase for each of the 20 amino acids specified by the genetic code. Each enzyme must be capable of recognizing its unique amino acid and one or more cognate tRNAs. Solving the puzzle of how synthases recognize tRNAs has been one of the major challenges in understanding the nature of the translation of the genetic code itself. The identity elements in tRNAs that confer specific recognition by their cognate synthases are now known for all 20 tRNA families.

Synthases fall into two categories with respect to the importance of the anticodon as a specificity element that they recognize. Some synthases (17 in *E. coli*) recognize the anticodon and some do not (3 in *E. coli*). This distinction has been demonstrated in several ways. First, three tRNAs can be genetically altered in their anticodon without changing the specificity of their recognition by the cognate synthase. Seventeen synthases apparently require an unaltered anticodon to recognize their cognate tRNAs. The acceptor stem is important for all synthases, and for seven synthases the region in and around the variable loop is also important.

The crystal structure of glutaminyl-tRNA synthase, complexed with tRNA and ATP, has been determined by Tom Steitz and his colleagues (fig. 34.11). This accomplishment provided the first structure of a tRNA–protein complex and thus offered important insight into the general nature of the recognition of tRNAs by proteins.

Glutaminyl-tRNA synthase is known to require the integrity of the anticodon loop of tRNA for recognition and is thus presumed to interact with both "ends" of the tRNA. In keeping with this interpretation the synthase is asymmetrical and longer (≈ 100 Å) than the tRNA. Moreover, the crystal structure demonstrates the occurrence of significant contact between the protein and bases in the anticodon loop. The additional contacts between the synthase and the tRNA appear to occur along the inside of the L-shaped structure. The recognition of the tRNA appears to arise from direct hydrogen bonding between amino acid residues of the protein and bases of the tRNA over a wide region of the tRNA structure.

The acceptor stem of the bound tRNA substrate plunges deeply into the active site pocket created by the dinucleotide fold, also the site of ATP binding. Binding in the active site induces a significant conformational change in the 3′ terminal CCA acceptor sequence. This conformational change appears to cause the melting of the A-U pair at the end of the acceptor stem in a manner that allows these bases to interact with amino acid side chains of the protein.

Aminoacyl-tRNA Synthases Can Correct Acylation Errors

Another important function of aminoacyl-tRNA synthases is to discriminate cognate and noncognate amino acids. In many cases the differences in binding energies of amino acids to aminoacyl-tRNA synthases do not provide the required accuracy of translation. This has necessitated the evolution of a second determinant

Figure 34.9

Formation of aminoacyl-tRNA. This is a two-step process involving a single enzyme that links a specific amino acid to a specific tRNA molecule. In the first step (1) the amino acid is activated by the formation of an aminoacyl-AMP complex. This complex then reacts with a tRNA molecule to form an aminoacyl-tRNA complex (2).

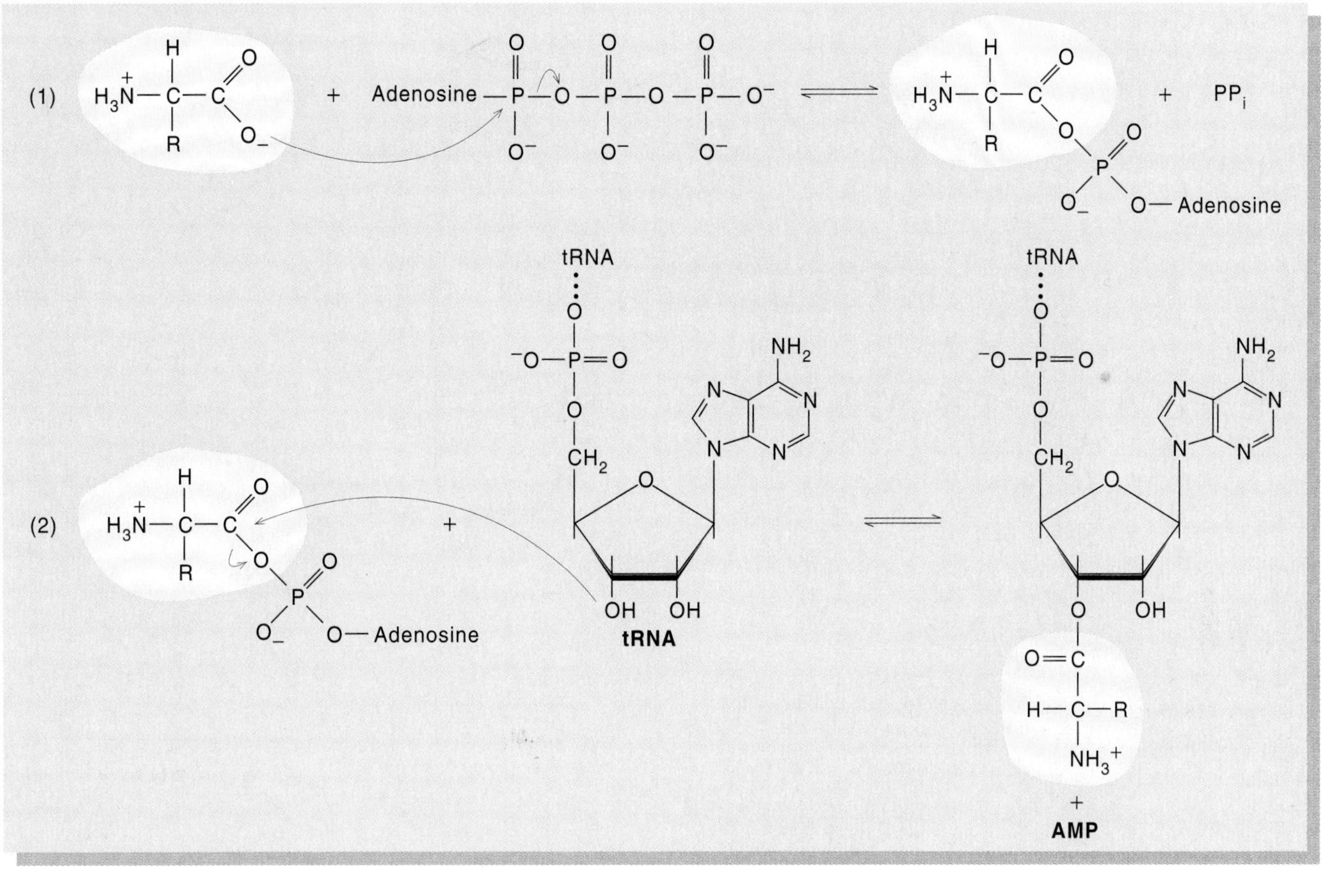

of specificity, proofreading or editing mechanism, that involves the expenditure of energy to remove errors and prevent mischarging. Studies with purified aminoacyl-tRNA synthases indicate that misactivated amino acids can be edited via different pathways, including an "adenylate" pathway (in which the noncognate aminoacyl-adenylate is rejected prior to attachment of amino acid to tRNA) and a "misacylation-deacylation" pathway (in which the noncognate amino acid is removed after attachment to tRNA). One of the best understood examples of proofreading is that of methionyl-tRNA synthase (MetRS), which must discriminate against homocysteine (Hcy), the immediate precursor of methionine in the methionine biosynthetic pathway (see fig. 24.17). This involves misactivation of Hcy to form the enzyme-bound Hcy-AMP, which is then enzymatically converted to the cyclic compound homocysteine thiolactone through the adenylate pathway:

$$\text{MetRS} + \text{Hcy} + \text{ATP} \underset{\text{PP}_i}{\rightleftharpoons} \text{MetRS} \cdot \text{Hcy-AMP} \longrightarrow$$

$$\text{Hcy-thiolactone} + \text{MetRS} + \text{AMP} \qquad (4)$$

As shown in *E. coli,* yeast, and some cultured mammalian cells the incorporation of homocysteine in place of methionine is thereby avoided. The sequential action of initial selectivity and proofreading contributes to an overall error frequency of aminoacylation of less than 1 in 10,000.

A Unique tRNA Initiates Protein Synthesis

The translation of every protein begins with the incorporation of the amino acid methionine. A unique initiator tRNA, $\text{tRNA}_i^{\text{Met}}$, is responsible for the incorporation of this initiating methionine in all protein-synthesizing systems, and it also plays an important role in selecting the appropriate translation start site in mRNA. Generally, two tRNAs that specify methionine occur in cells. We designate the one that is responsible for the incorporation of internal methionines simply as tRNA^{Met}. A single methionyl-tRNA synthase is responsible for activating both methionine isoacceptor tRNAs.

The functional discrimination between the two isoacceptor tRNAs results from specific interactions between the tRNAs and protein initiation and elongation factors. Initiation factors recognize $\text{tRNA}_i^{\text{Met}}$, and elongation factors recognize tRNA^{Met}. Clearly, $\text{tRNA}_i^{\text{Met}}$ and tRNA^{Met} must possess structural features that allow them to be distinguished from all other tRNAs by the single methionyl-tRNA synthase, but they must be sufficiently different that they can be discriminated by the factors.

Figure 34.10

Recognition of an adenylylated amino acid by the proper synthase. Shown is adenyl tyrosine bound to the tyrosyl synthase. A network of H bonds not only stabilizes the reaction but also serves to discriminate between different amino acid residues. MC designates main chain (backbone) carbonyl or amino groups participating in hydrogen bonding.

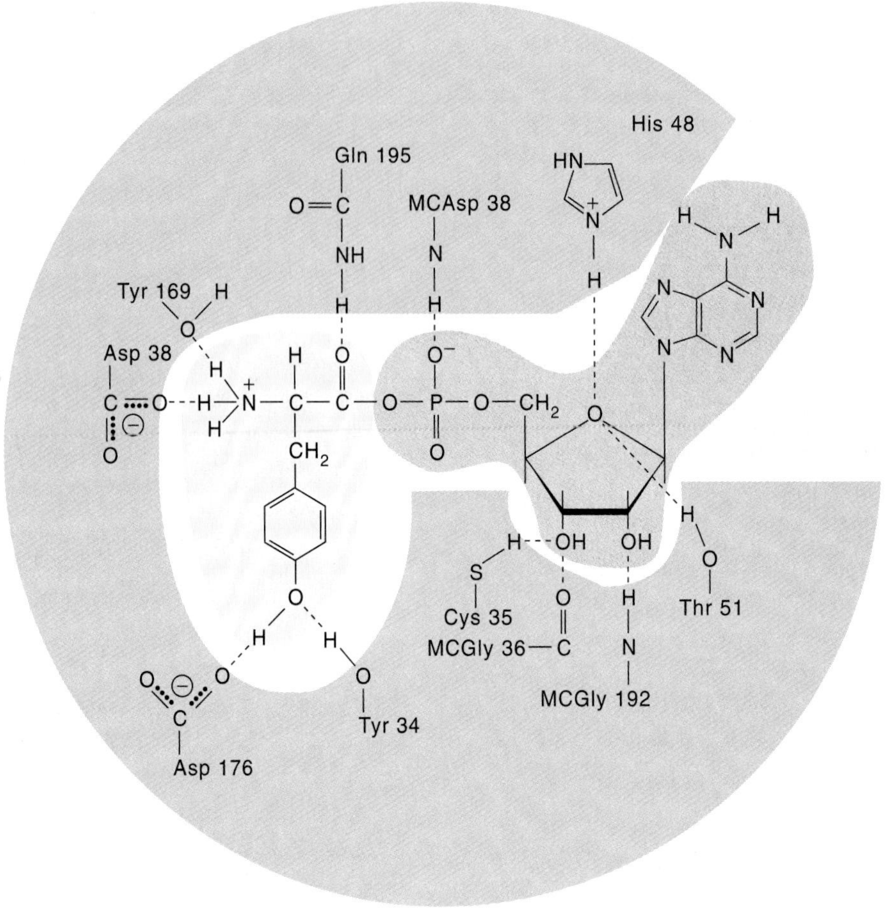

In prokaryotic systems a specific formylating enzyme exists that can recognize $tRNA^{Met}$ and formylate, its amino terminus utilizing N^{10}-formyltetrahydrofolate as the formyl donor. This reaction serves to ensure that the initiator does not participate in the elongation reactions. This recognition step is not possessed by eukaryotic systems. As a result of this formylation reaction, initiator tRNA in prokaryotes is designated $tRNA_f^{Met}$.

Translation Begins with the Binding of mRNA to the Ribosome

One of the major demands of protein synthesis is to select the appropriate initiator codon, generally AUG, for translation. This is accomplished at the level of the ribosome by the binding of the small ribosomal subunit to mRNA. The recognition of the appropriate start codon occurs in different ways in prokaryotes and eukaryotes.

In eukaryotes the AUG sequence closest to the 5′ end of the mRNA usually serves as the start codon for the single protein encoded by each mRNA. The small ribosomal subunit binds at the 5′ end of the mRNA and moves along the strand until it encounters an AUG sequence that is recognized by base pairing with the anticodon of Met-$tRNA_i^{Met}$. In higher eukaryotes, but not in yeast, the recognition of this initiating AUG is facilitated by an initiation context of flanking bases. The preferred initiation context is GCCGCCpurCCAUGG. How this sequence is recognized is unknown, but when the initiation context of the first AUG departs from this sequence, the 40S subunit can bypass it and initiate at the next AUG downstream. There are also some instances where eukaryotic ribosomes find an internal ribosome entry site (IRES), apparently governed by a complex superstructure of the mRNA. In these messages, ribosomes bypass the 5′ end of the message and initiate translation at the first AUG downstream from the IRES.

In prokaryotic systems the initiating AUG codon may occur at any point within the mRNA, and more than one start site may occur within the same mRNA. How do prokaryotic ribosomes select the appropriate initiating codon from the much more abundant internal AUG sequences? The answer was suggested in the early 1970s when Shine and Dalgarno noticed that bacterial mRNAs contain a complementary purine-rich region (which has become known as the Shine-Dalgarno sequence) centered approximately 10 bases toward the 5′ side of the initiating AUG sequence and that *E. coli* 16S RNA contains a seven-base pyrimidine-rich sequence near its 3′ terminus (table 34.5). They proposed that base pairing between

Figure 34.11

Solvent-accessible surface representation of the GlnRS enzyme complexed with tRNA and ATP. The region of contact between tRNA and protein extends across one side of the entire enzyme surface and includes interactions from all four protein domains. The acceptor end of the tRNA and the ATP are seen in the bottom of the deep cleft. Protein is inserted between the 5′ and 3′ ends of the tRNA and disrupts the expected base pair between U1 and A72. (*Source:* From M. G. Rould, J. J. Persona, D. Söll, and T. Steitz, "Structure of *E. coli* glutamyl-tRNA synthetics complexed with $tRNA^{Gln}$ and ATP at 2.8-Å resolution, implications for tRNA discrimination," *Science* 246:1135–1142, 1989, © 1989 by the AAAS.)

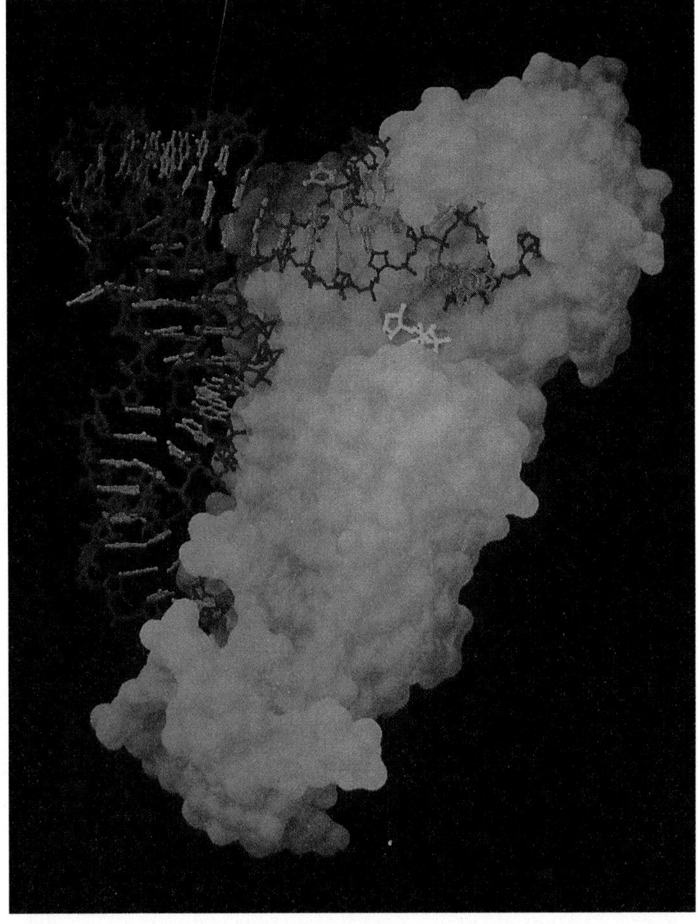

these complementary sequences could serve to align the initiating AUG for decoding. Such base pairing is the major mechanism for codon initiator recognition in *E. coli*. Thus mutations in the Shine-Dalgarno sequence of mRNA that improve pairing enhance translation, and mutations that decrease pairing decrease translation. This pairing has been exploited to develop systems of "specialized ribosomes," in which plasmids bearing genes for 16S rRNA with altered sequences in the complementary region to Shine-Dalgarno are introduced into cells. Ribosomes which subsequently form, containing this mutant 16S rRNA, will selectively translate messages with complementary mutant Shine-Dalgarno sequences, independently introduced into the cells via genes on other plasmids.

Another application of this knowledge has been to design complementary RNAs to the 5′ end of messages, which can pair with the Shine-Dalgarno and AUG initiation region in a double-stranded structure to selectively block ribosome access and prevent translation of that message. Such complementary RNA is called antisense RNA. Antisense RNA is also used for regulation of expression in several naturally occurring systems.

Table 34.5
The Sequence of the 3′ End of *E. coli* 16S RNA and Some Shine-Dalgarno Sequences at the 5′ End of Bacterial mRNAs

	The pyrimidine-rich complement to the Shine-Dalgarno sequence
16S rRNA	3′ ··· HOAUUCCUCCA CUA ··· 5′
lacZ mRNA	5′ ··· ACACAGGAAACAGCUAUG ··· 3′
trpA mRNA	5′ ··· ACGAGGGGAAAUCUGAUG ··· 3′
RNA polymerase β mRNA	5′ ··· GAGCUGAGGAACCCUAUG ··· 3′
r-Protein L10 mRNA	5′ ··· CCAGGAGCAAAGCUAAUG ··· 3′
	The purine-rich Shine-Dalgarno sequence; The initiation codon

The importance of the Shine-Dalgarno sequence is underscored by the action of the bacterial toxin colecin E3. This toxin inactivates the small subunit of the prokaryotic ribosome by cleavage of about 50 residues from the 3′ terminus of 16S rRNA. The cleavage disrupts the sequence that is complementary to the Shine-Dalgarno sequence and thus specifically inhibits the initiation process. Because of the fundamental differences between prokaryotic and eukaryotic initiation just described, colecin E3 does not inhibit the eukaryotic ribosome.

Dissociable Protein Factors Play Key Roles at the Different Stages in Protein Synthesis on the Ribosome

At each stage in protein synthesis on the ribosome—initiation, elongation, and termination—a different set of protein factors is engaged by the ribosome. Why do such protein factors, which are crucial to the translation, exist separately from the ribosome? Why must they cycle on and off the ribosome, and why are their functions not performed by firmly bound ribosomal proteins? The answers to these questions appear to lie in the economy of structure that is afforded by the cycling strategy. Since factors cannot interact simultaneously with the ribosome, it makes sense for the ribosome to interact with them sequentially at a single site.

Protein Factors Aid Initiation

Even though specific differences distinguish the initiation process in eukaryotes and prokaryotes, three things must be accomplished to initiate protein synthesis in all systems: (1) The small ribosomal subunit must bind the initiator tRNA; (2) the appropriate initiating codon on mRNA must be located; and (3) the large ribosomal subunit must associate with the complex of the small subunit, the initiating tRNA, and mRNA. Nonribosomal proteins,

Figure 34.12

Formation of the initiation complex for protein synthesis in prokaryotes. *E. coli* has three initiation factors bound to a pool of 30S ribosomal subunits. One of these factors, IF-3, holds the 30S and 50S subunits apart after termination of a previous round of protein synthesis. The other two factors, IF-1 and IF-2, promote the binding of both fMet-tRNA^{fMet} and mRNA to the 30S subunit. The binding of mRNA occurs so that its Shine-Dalgarno sequence pairs with 16S rRNA and the initiating AUG sequence with the anticodon of the initiator tRNA. The 30S subunit and its associated factors can bind fMet-tRNA^{fMet} and mRNA in either order. Once these ligands are bound, IF-3 dissociates from the 30S subunit, permitting the 50S subunit to join the complex. This releases the remaining initiation factors and hydrolyzes the GTP, which is bound to IF-2.

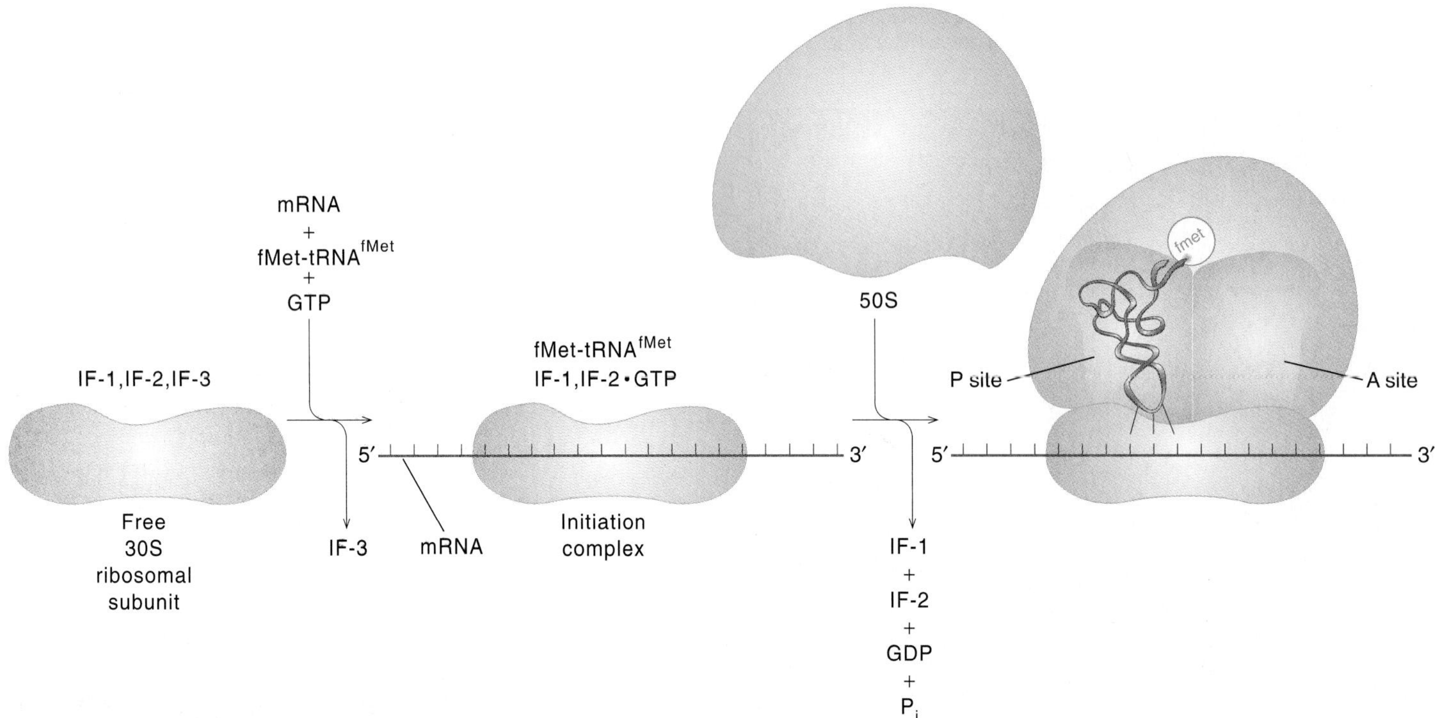

known as initiation factors (IFs), participate in each of these three processes. IFs interact transiently with a ribosome during initiation and thus differ from ribosomal proteins, which remain continuously associated with the same ribosome.

E. coli has three initiation factors (fig. 34.12) bound to a small pool of 30S ribosomal subunits. One of these factors, IF-3, serves to hold the 30S and 50S subunits apart after termination of a previous round of protein synthesis. The other two factors IF-1 and IF-2, promote the binding of both fMet-tRNA$_f^{Met}$ and mRNA to the 30S subunit. As we noted before, the binding of mRNA occurs so that its Shine-Dalgarno sequence pairs with 16S RNA and the initiating AUG sequence with the anticodon of the initiator tRNA. The 30S subunit and its associated factors can bind fMet-tRNA$_i^{Met}$ and mRNA in either order. Once these ligands are found, IF-3 dissociates from the 30S, permitting the 50S to join the complex. This releases the remaining initiation factors and hydrolyzes the GTP that is bound to IF-2. The initiation step in prokaryotes requires the hydrolysis of one equivalent of GTP to GDP and P_i.

Initiation of protein synthesis in eukaryotic cells requires a much more complex spectrum of initiation factors, abbreviated eIF. At least nine separate factors have been identified, some of which are composed of as many as 11 different peptide subunits. The exact function of only a few of these factors is known. The main features of the process are outlined in figure 34.13. As in the prokaryotic system, the reaction begins with the small subunit held apart from the large subunit by an antiassociation factor and ends with the hydrolysis of GTP and joining of the large subunit. The intervening reactions are different in prokaryotes and eukaryotes. The Met-tRNA$_i^{Met}$ first binds to the small subunit as a ternary complex with eIF-2 and GTP. The resulting preinitiation complex then binds to the 5′ end of the mRNA with the aid of several factors, one of which, eIF-4F, contains a subunit that specifically binds to the terminal cap structure of mRNA. Binding to mRNA is followed by a scanning reaction that moves the small subunit along the mRNA, usually to the first AUG, in a reaction driven by the hydrolysis of ATP to ADP and P_i. The mRNA binding and scanning reactions, which have no counterpart in the prokaryotic system, position the ribosome on the initiating AUG sequence in the appropriate flanking nucleotide sequence.

Three Elongation Reactions Are Repeated with the Incorporation of Each Amino Acid

At the conclusion of the initiation process, the ribosome is poised to translate the reading frame associated with the initiator codon. The translation of the contiguous codons in mRNA is accomplished by the sequential repetition of three reactions with each amino acid. These three reactions of elongation are similar in both prokaryotic

Figure 34.13

Formation of the initiation complex for protein synthesis in eukaryotes. The reaction begins with the small subunit held apart from the large subunit by an antiassociation factor and ends with the hydrolysis of GTP and joining of the large subunit as in prokaryotes. The intervening reactions are different. A much more complex spectrum of initiation factors (eIFs) is involved, and the exact function of only a few of these factors is known with certainty. The Met-tRNA$_i^{Met}$ first binds to the small subunit as a ternary complex with the eIF-2 and GTP. This ternary complex then binds to the 5′ end of mRNA with the aid of several factors, one of which, eIF-4F, contains a subunit that specifically binds to the terminal cap structure of the mRNA. Binding to mRNA is followed by a scanning reaction that moves the small subunit along the mRNA, usually to the first AUG, in a reaction driven by the hydrolysis of ATP to ADP and P_i.

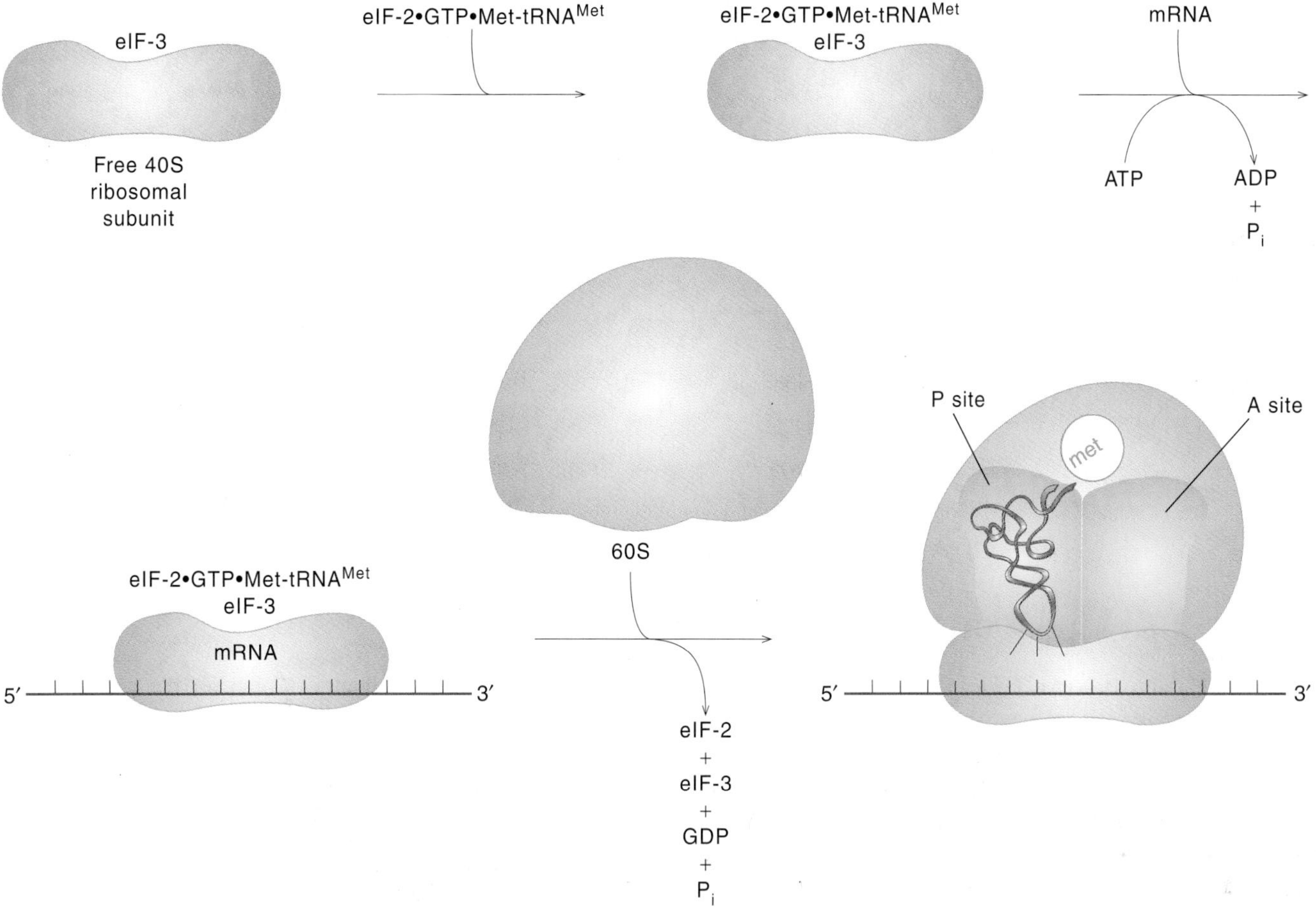

and eukaryotic systems; two of them require nonribosomal proteins known as elongation factors (EF). Interestingly, the actual formation of the peptide bond does not require a factor and is the only reaction of protein synthesis catalyzed by the ribosome itself.

The elongation reactions begin with the binding of the aminoacyl-tRNA specified by the codon immediately adjacent to the initiator codon. The binding of this aminoacyl-tRNA is catalyzed by an aminoacyl-tRNA binding factor, designated EF-Tu in bacteria and EF-1 in eukaryotic systems. The factor interacts with the ribosome as a ternary complex bound to both aminoacyl-tRNA and GTP. The binding of this complex to the ribosome is coupled to GTP hydrolysis. A productive complex forms if the anticodon of the tRNA in the complex is complementary to the codon bound to the A site on the ribosome. Following the binding of aminoacyl-tRNA, EF-Tu is released from the ribosome as a complex with GDP. A second elongation factor, EF-Ts, catalyzes the regeneration of the EF-Tu–GTP complex so that it can again bind aminoacyl-tRNA. The series of reactions involved in this regeneration is depicted in figure 34.14. EF-1 is a multisubunit protein that combines the functional properties of EF-Tu and EF-Ts.

Peptide bond formation occurs immediately following the dissociation of the binding factor from the ribosome. This reaction is known as transpeptidation, and the enzymatic center that catalyzes it is known as peptidyl transferase, although it promotes the conversion of an ester to a peptide bond. Noller and coworkers have found that ribosomes stripped of almost all of their protein can still catalyze the transpeptidation reaction, suggesting that the peptidyl transferase activity is contained in the rRNA. Transpeptidation is generally assumed to proceed by nucleophilic attack by the amino group of the incoming aminoacyl-tRNA on the carbonyl of the ester of peptidyl-tRNA with the formation of a tetrahedral intermediate (fig. 34.15).

The final step in elongation is known as translocation (fig. 34.16). This reaction, like aminoacyl-tRNA binding, is catalyzed by a factor (the translocation factor, known as EF-G in prokaryotic systems and EF-2 in eukaryotic systems) that cycles on and

Figure 34.14

Addition of the second aminoacyl-tRNA to the ribosome complex and the accompanying EF-Tu, EF-Ts cycle in *E. coli*. The purpose of the cycle is to regenerate another protein aminoacyl-tRNA complex suitable for transferring further aminoacyl-tRNAs to the A site on the ribosome.

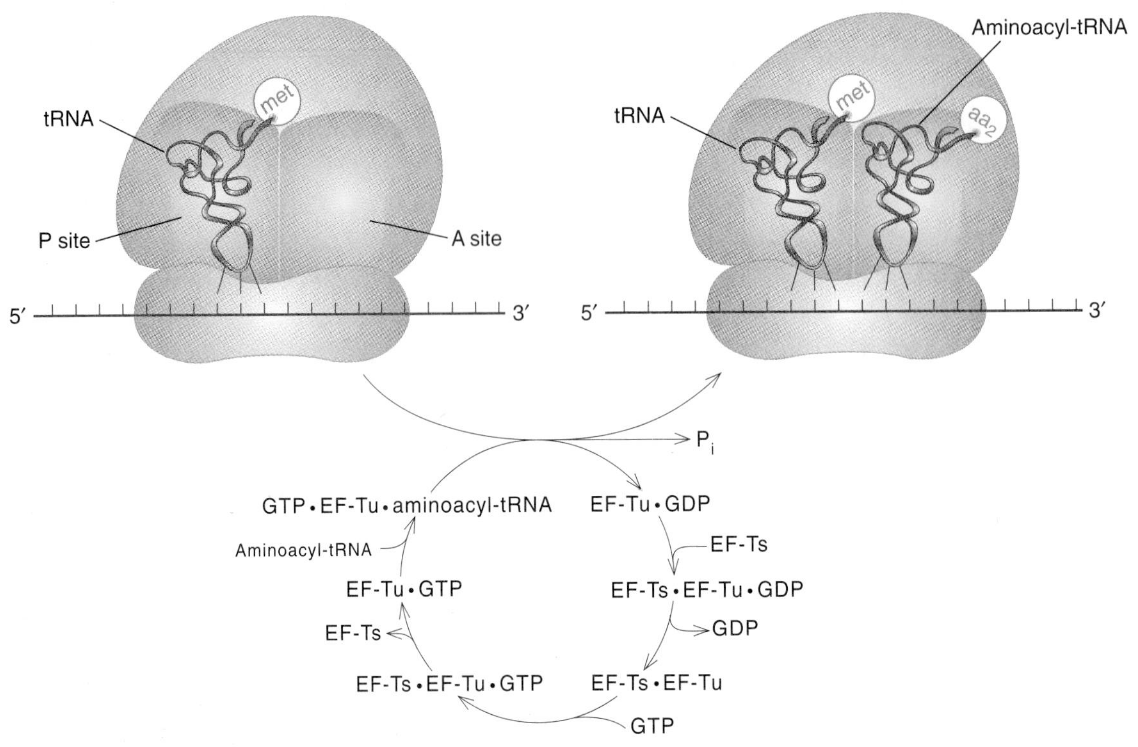

off the ribosome and hydrolyzes GTP in the process. The overall purpose of translocation is to move the ribosome physically along the mRNA to expose the next codon for translation.

The structural differences between bacterial and eukaryotic elongation factors are highlighted by the selective action that diphtheria has on eukaryotic systems (box 34B).

In Addition to the P Site and the A Site for Binding tRNAs the Ribosome May Possess a Third Site, the E Site

Almost 30 years ago it was noticed that 70S *E. coli* ribosomes could bind more than two equivalents of tRNA. For example, in the presence of poly(U) it could be shown that approximately three equivalents of tRNA bind at saturation. Since the P and A sites both must bind aa-tRNA in the first step in protein synthesis, the binding of a third tRNA to the ribosome seemed without purpose, and its significance was discounted. However, more sophisticated binding experiments have raised anew the possibility of a third binding site for tRNA.

Studies with heteropolymeric mRNA done by Knud Nierhaus support the notion of a discrete third site for binding deacylated tRNA; this site has been named the exit site, or the E site. Using a short synthetic messenger containing the sequence AUGAAAACC, which should represent the code words for methionine, lysine, and threonine in that order, the deacylated initiator tRNA for methionine ($tRNA_i^{Met}$) was added to the ribosomes in the presence of the synthetic mRNA. The $tRNA_i^{Met}$ sets the reading frame for the mRNA so that the AUG codon in the mRNA is juxtaposed in the P site. Under these conditions, only the lys-tRNA can bind next because the lysine AAA codon is the adjacent downstream codon. Next, if elongation factor EF-G and GTP are added, translocation appears to occur. This is demonstrated by the fact that the lysine on the lysyl-tRNA can now react with puromycin, whereas prior to the addition of EF-G and GTP it could not. (This puromycin test is a diagnostic test for binding of an aminoacyl-tRNA to the P site because puromycin is a structural analog to the amino acylated 3′-adenosine of aminoacyl tRNA.) The most important observation relevant to the existence of an E site is that the deacylated $tRNA_i^{Met}$ is not discharged by this operation. This experiment suggests that there is no coupling of translocation to release of deacylated tRNA, in contrast to the prediction of the classical two-site model.

Additional experiments done by Nierhaus suggest a possible function for the binding of the deacylated tRNA to the E site. An experiment involving the poly(U)-ribosome system was used for this purpose. When the P site is filled with a peptidyl-tRNA or an aa-tRNA, the A site will bind a phe-tRNA at either 0°C or 37°C. However, when the P site is filled and the E site contains a deacylated tRNA, phe-tRNA can no longer bind to the A site at 0°C, but it will bind at 37°C. It has been argued that filling the E site induces a low-affinity A site, so activation energy is required to convert this low-affinity A site into a high-affinity site. This result suggests a negative cooperativity between tRNA binding at the E and A sites. Such a negative cooperativity would discourage the bind-

Figure 34.15

Formation of the first peptide linkage. The formylmethionine group is transferred from its tRNA at the P site to the amino group of the second aminoacyl-tRNA at the A site of the ribosome. This involves nucleophilic attack by the amino group of the second amino acid on the carboxyl carbon of the methionine. The resulting bond formation attaches both amino acids to the tRNA at the A site.

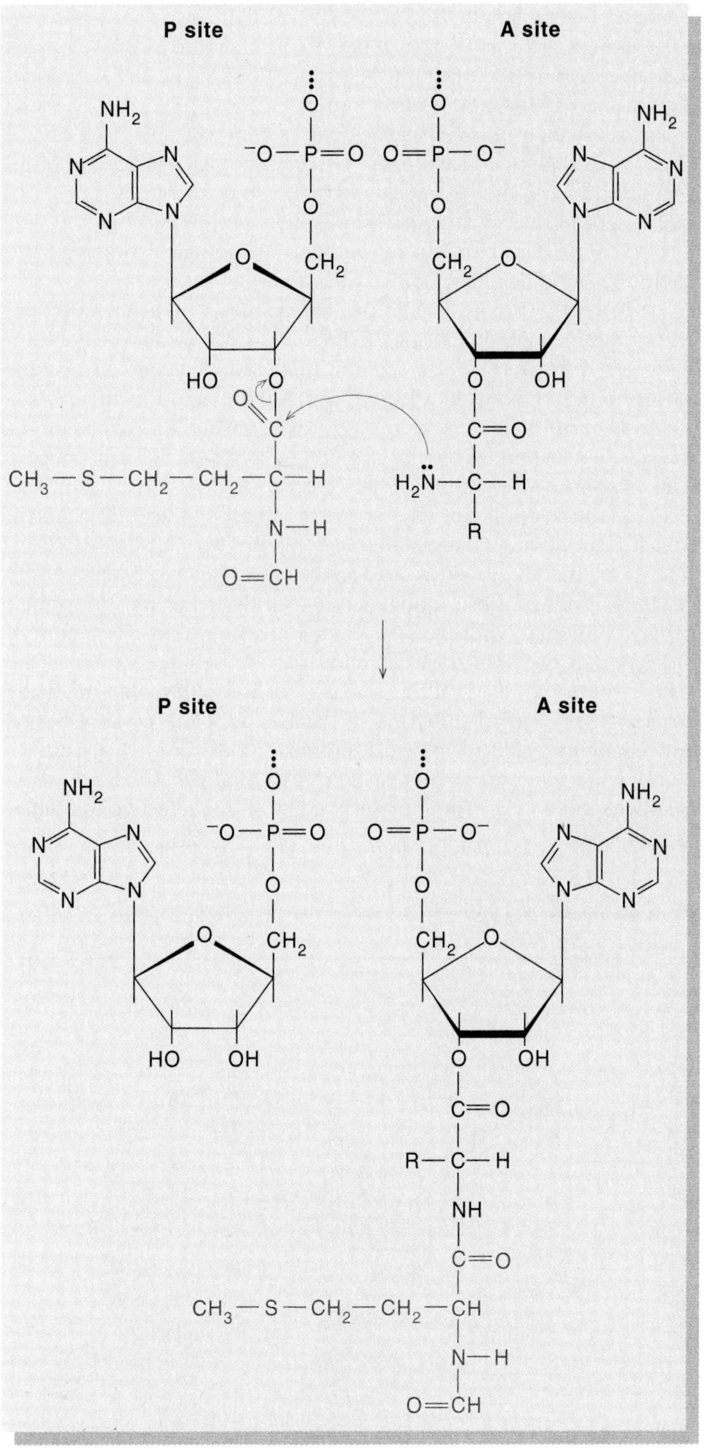

P site A site

NH_2 NH_2

$^-O-P=O$ $O=P-O^-$

CH_2 CH_2

HO OH

$CH_3-S-CH_2-CH_2-C-H$ $H_2\ddot{N}-C-H$

N—H R

O=CH

P site A site

NH_2 NH_2

$^-O-P=O$ $O=P-O^-$

CH_2 CH_2

HO OH OH

C=O

R—C—H

NH

C=O

$CH_3-S-CH_2-CH_2-CH$

N—H

O=CH

ing of a tRNA at the A site unless it contains the appropriate anticodon. This effect should improve the fidelity of translation.

Despite the persuasive nature of these experiments the support for a discrete E site is purely biochemical. Some genetic evidence that relates to the E site is needed to finally establish the existence and functional significance of the E site.

Two (or Three) GTPs Are Required for Each Step in Elongation*

The hydrolysis of GTP plays a conspicuous role in the translation process. Two equivalents of GTP are hydrolyzed during elongation with the incorporation of each amino acid. This hydrolysis accounts for about half of the total energy consumed during protein synthesis. The chemical and functional purposes of GTP hydrolysis are best understood in the case of *E. coli* EF-Tu and EF-G. The sites of GTP binding and hydrolysis are situated on these factors. The interaction of the factor–GTP complex with the ribosome is believed to activate the hydrolytic site. GTP hydrolysis leads to GDP. No covalent intermediates are formed. Rather, the change of bound ligand to GDP that results from hydrolysis and release of P_i is believed to change the conformation of the factor. The factor when bound to GTP is thought to be in the "on" configuration, able to bind to the ribosome and through binding to cause either aminoacyl-tRNA binding or translocation. The binding of GDP to the factor puts it in the "off" configuration and causes it to dissociate from the ribosome. Hydrolysis itself is not required for these changes to occur. This point is demonstrated by the fact that the nonhydrolyzable analog of GTP, guanylyl methylene diphosphonate (GMPPCP), containing a methylene bridge instead of an oxygen between the β and γ phosphorus, can substitute for it in the reactions catalyzed by the elongation factors (fig. 34.17). The use of this analog slows these reactions by delaying the dissociation of the factors from the ribosome.

The translation factors that interact with GTP are members of the so called G protein superfamily, which includes the signal-transducing G proteins that link membrane receptors with their intracellular targets (see chapter 27) and the Ras proteins that function as growth regulators (see chapter 38, Carcinogenesis and Oncogenes). All G proteins bind and hydrolyze GTP, and it is believed that they all act by the same general mechanism we have just described. Thus, when bound to GTP, these proteins are in their "active" configuration, and when bound to GDP as a result of hydrolysis, they are converted to their "inactive" configuration.

Termination of Translation Requires Release Factors and Termination Codons

The last step in translation involves the cleavage of the ester bond that joins the now complete peptide chain to the tRNA corresponding to its C-terminal amino acid (fig. 34.18). Termination requires a termination codon, mRNA, and at least one protein release factor (RF). The freeing of the ribosome from mRNA during this step requires the participation of a protein called ribosome-releasing factor (RRF).

*It has been proposed that two GTPs instead of one may be involved in the EF-Tu reaction (see the Weijland and Parmeggiani paper in the Selected Readings).

34B

BOX

Mechanisms of Damage Produced by Certain Toxins

A single exotoxin, diphtheria toxin, is responsible for the pathogenesis of *Corynebacterium diphtheriae* and the disease diphtheria. The pathogenic consequences can be prevented by immunization with toxoid, an inactivated form of the purified toxin. Curiously, the structural gene for the toxin is carried by a bacterial virus, called β phage, that must infect the bacterium to induce toxin production. The widespread immunization against diphtheria employed in the United States has caused β phage, but not *C. diphtheriae,* largely to disappear. A catalytically identical but structurally very different exotoxin is produced by *Pseudomonas aeruginosa.*

In the cytoplasm of the cell the catalytic portion of diphtheria toxin acts as a very specific protein-modifying enzyme. It catalyzes the ADP-ribosylation and consequent inactivation of EF-2 by the following reaction:

$$\text{EF-2} + \text{NAD}^+ \longrightarrow \text{ADP-ribosyl-EF-2} + \text{Nicotinamide} + \text{H}^+$$

This reaction is reversible when conducted in vitro, but under the conditions of pH and nicotinamide concentration that exist in the cell, it is irreversible. Thus diphtheria toxin kills cells by irreversibly destroying the ability of EF-2 to participate in the translocation step of protein synthesis elongation. A number of other protein toxins have subsequently been found to ADP-ribosylate and inactivate cellular proteins involved in other essential cellular pathways. For example, cholera and pertussis toxins ADP-ribosylate and inactivate proteins important to cAMP metabolism.

The enzymatic specificity of diphtheria toxin deserves special comment. The toxin ADP-ribosylates EF-2 in all eukaryotic cells in vitro whether or not they are sensitive to the toxin in vivo, but it does not modify any other protein, including the bacterial counterpart of EF-2. This narrow enzymatic specificity has called attention to an unusual posttranslational derivative of histidine, diphthamide, that occurs in EF-2 at the site of ADP-ribosylation (see fig. 1). Although the unique occurrence of diphthamide in EF-2 explains the specificity of the toxin, it raises questions about the functional significance of this modification in translocation. Interestingly, some mutants of eukaryotic cells selected for toxin resistance lack one of several enzymes necessary for the posttranslational synthesis of diphthamide in EF-2 that is necessary for toxin recognition, but these cells seem perfectly competent in protein synthesis. Thus the *raison d'être* of diphthamide, as well as the biological origin of the toxin that modifies it, remains a mystery.

Some other toxins inhibit protein synthesis by inactivating the ribosome through alterations of rRNA. A fungal toxin, α-sarcin, inactivates the ribosome by specific nuclease cleavage of a single phosphodiester bond in a purine-rich region found in the 23–28S RNA of all ribosomes. A second group of toxins, known as the ribosome-inactivating proteins, is abundant in plants of many species. The best-known example of this type of toxin is ricin, the toxic agent in the castor bean. These proteins all act by the curious mechanism of removing a single adenine residue, by N-glycolytic cleavage, from the site in rRNA adjacent to the one that is cleaved by α-sarcin. Both of these modifications of rRNA alter the ribosome's ability to interact with factors, a fact suggesting that this region of rRNA is part of the factor interaction site on the large subunit. One of the ribosome-inactivating proteins, trichosanthin, was originally isolated from a plant tuber as the active principle of an Oriental folk medicine used to induce abortions. More recently, trichosanthin has been found to selectively inhibit viral protein synthesis in HIV-infected T cells. The biological basis of this selective inhibition is unknown.

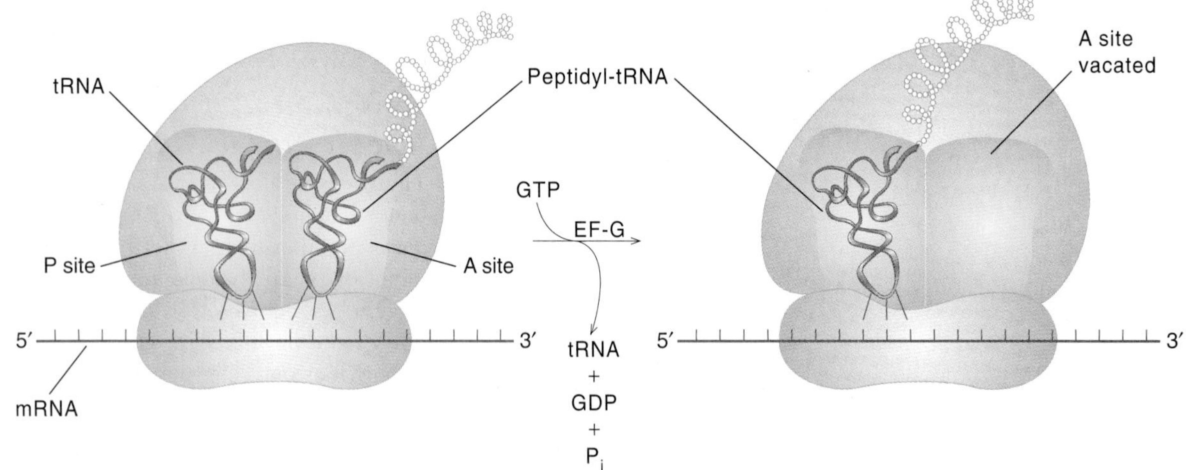

Figure 34.16

The translocation reaction in *E. coli.* The translocation reaction occurs immediately after peptide synthesis. It involves concerted movement of the discharged tRNA from the P site to the E (not shown) site of the peptidyl-tRNA and mRNA from the A site to the P site so that the peptidyl-tRNA is bound in the P site to the same three nucleotides in the mRNA. The A site is vacated and ready for the addition of another aminoacyl-tRNA. Translocation in eukaryotes is similar except that EF-2 is involved instead of the EF-G factor.

Figure 1

Inactivation of the EF-2 factor by diphtheria toxin through the ADP-ribosylation of a modified histidine side chain.

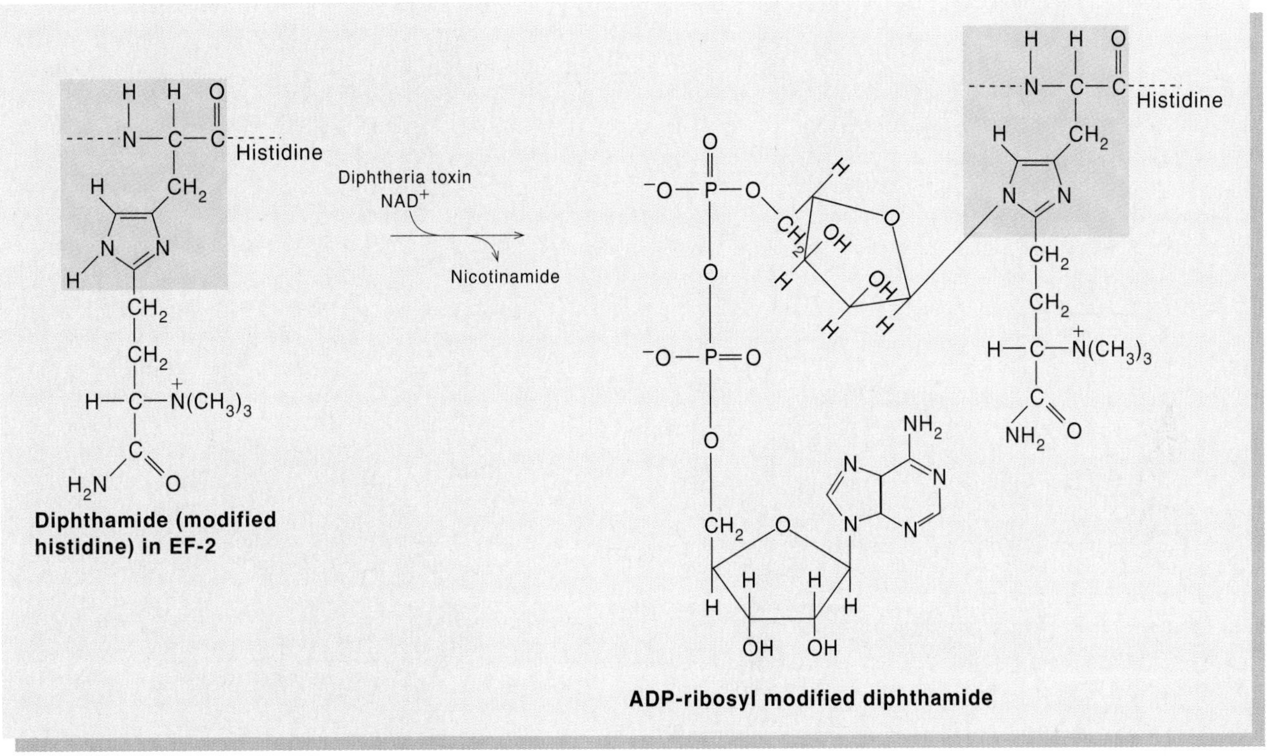

In *E. coli,* termination codons that arrive at the A site on the ribosome are recognized by one of three protein release factors. RF-1 recognizes UAA and UAG, and RF-2 recognizes UAA and UGA. The third release factor, RF-3, does not itself recognize termination codons but stimulates the activity of the other two factors.

The consequence of release factor recognition of a termination codon in the A site is to alter the peptidyl transferase center on the large ribosomal subunit so that it can accept water as the attacking nucleophile rather than requiring the normal substrate, aminoacyl-tRNA (fig. 34.19). In other words, the termination reaction serves to convert the peptidyl transferase into an esterase. This feature of the termination reaction is clearly seen in the simple in vitro reaction that occurs when *E. coli* ribosomes are combined with fMet-$tRNA_i^{Met}$, RF-1, and two separate nucleotide triplets: AUG and UAA (see fig. 34.19). Formylmethionine is produced by hydrolysis, and this reaction is specifically inhibited by antibiotics such as sparsomycin that inhibit peptidyl transferase.

Figure 34.17

The structure of guanylyl methylene diphosphonate (GMPPCP).

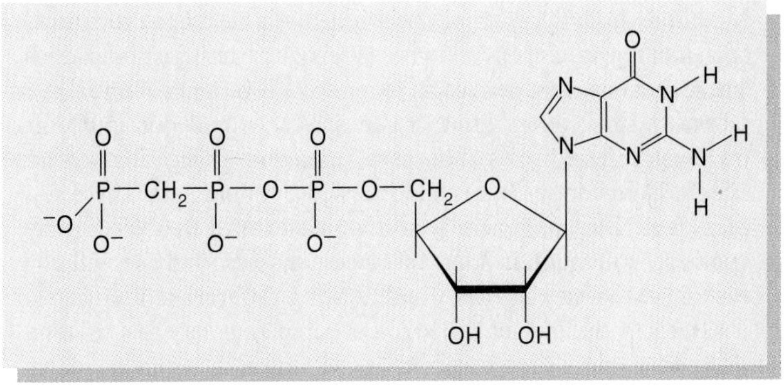

Figure 34.18

The release reaction in *E. coli*. The release reaction occurs when the codon adjacent to the anticodon–codon complex is one of the stop codons, for example, UAA. The stop is recognized by release factor proteins that cause the peptidyl transferase to transfer the nascent polypeptide to water, forming a free polypeptide. Following the release of the polypeptide, the final tRNA and the mRNA dissociate from the ribosome, and the ribosome dissociates into its constituent subunits. RF is an abbreviation for protein release factor.

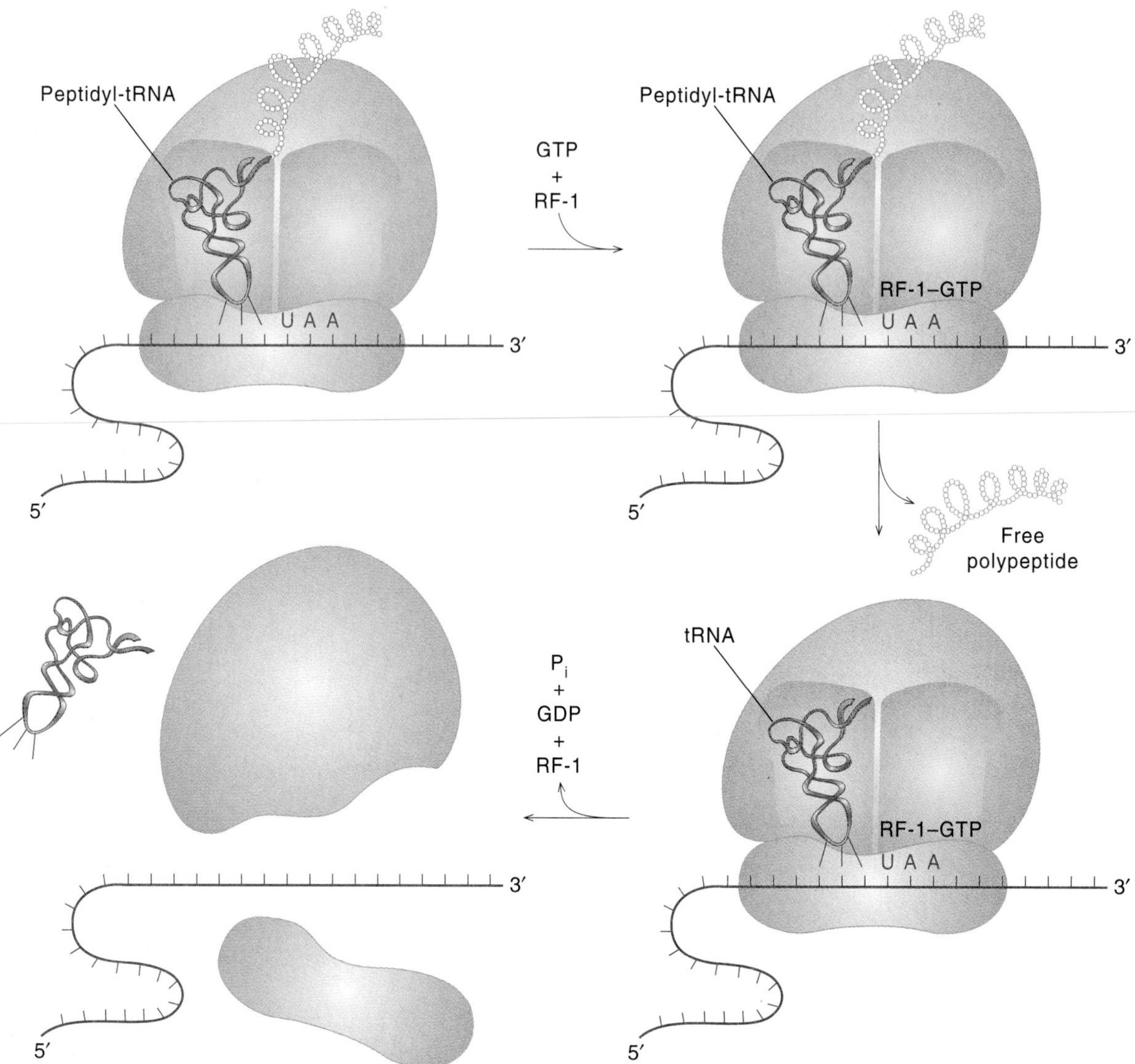

tRNA Mutations Can Suppress Termination

Mutations in anticodons of specific tRNAs have been identified, such that the mutant anticodons can recognize termination codons. These mutants were derived as second-site revertants of initial mutations to stop codons in other genes. Such stop codon mutations had been referred to as "nonsense" mutations, hence the second-site tRNA mutations were called nonsense suppressors. These suppressor tRNAs are generally still aminoacylated by their cognate synthases (although in some instances, another synthase will now recognize the altered tRNA and attach a different amino acid to it). Because the mutant anticodon is complementary to a termination codon, these suppressor tRNAs will insert the amino acid they are carrying into the growing peptide chain in response to the stop codon. When suppressing a nonsense mutation in another gene, activity of the other gene product will be regained by the cell unless the substituted amino acid inserted by the suppressor tRNA adversely affects the conformation or active site of the resulting protein. tRNA anticodon mutants which recognize UAG stop codons are called amber suppressors, while those which recognize both UAG and UAA (due to wobble) stop codons are called ochre suppressors. UGA-reading tRNAs are sometimes called opal suppressors. Redundancy of many tRNA genes permitted such mutations to arise, since cells still retained the capacity to decode normal codons from the other normal copies of the tRNA.

Because cells still contain release factors, which also recognize the stop codons, suppression (or, reading of stop codons by the mutant tRNAs) is not complete. This competition between release factors and suppressor tRNAs gives a characteristic effi-

Figure 34.19

An in vitro assay for release factors.

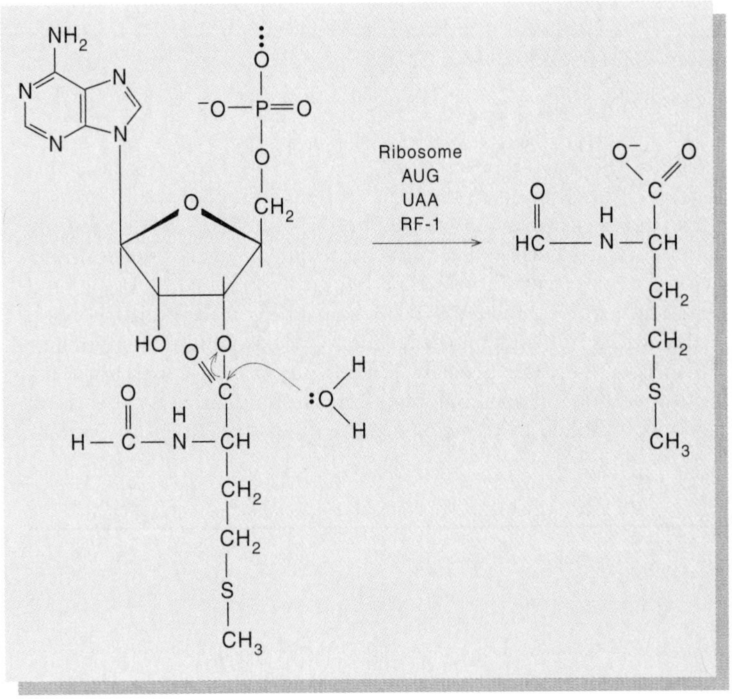

ciency of suppression for any individual suppressor tRNA. Amber suppressors can have efficiencies of suppression up to 50–60%, while ochre suppressors will range from 5–10% efficiency. The reason for this is that highly efficient ochre suppressors would be lethal to the host, since normal chain termination would be suppressed too frequently, and too many non-functional proteins would be generated. The reason why amber suppressors can have high efficiency is that the amber codon (UAG) is infrequently used as a sole termination codon after a normal reading frame.

The nucleotide sequence immediately surrounding a particular codon can have a dramatic effect on the efficiency of reading that codon, first noticed in studies of nonsense suppression. The codon context can affect the efficiency of suppression by several orders of magnitude. This may be related to interactions between adjacent tRNAs on the ribosome.

Ribosomes Can Change Reading Frame during Translation

Normally, ribosomes march in lock step down the mRNA, one codon at a time, beginning at the initiating AUG and ending at a termination codon. In this way, translation is rigidly maintained in a single reading frame. In the last few years a number of specific exceptions to this seemingly rigid rule have been discovered in both prokaryotic and eukaryotic systems. In these cases the correct expression of specific messages requires the ribosome to violate the rule of lock-step translation. One way that this occurs is known as programmed translational frameshifting. Frameshifting generally occurs one base at a time; this change, or slipping, can be in either the +1 or the −1 direction. Another way involves skipping as many as 50 bases downstream in the mRNA. This process is known as translational jumping. From an informational point of view, translational jumping is analogous to mRNA splicing except that the information is not removed but is simply ignored by the translation system. Both frameshifting and jumping are favored at specific sites on the mRNA. In most cases these sites are known to require the presence of specific sequences or specific secondary structures in the mRNA termed stimulators. These structures, which usually occur just downstream of the frameshift site, are frequently stem-loops and/or pseudoknots, in which the loop part of the stem-loop is also base-paired with a downstream portion of the RNA that has folded over. These structures are thought to impede the progress of ribosomes during translation, thereby facilitating the frameshift. The frequency of these events can be as high as 50% or more of all ribosome transits. These kinds of events have been termed recoding. In most instances, these events appear to involve realignment (or slippage) of one or two tRNAs on the ribosome with codons in the message.

A notable instance of frameshifting occurs in the gene for the release factor RF-2, which recognizes the UGA stop codon in *E. coli*. This gene does not make RF-2 unless a frameshift takes place at an in-frame UGA sequence. When there is an abundance of the release factor, the frameshift is unlikely to occur because release occurs when the ribosome reaches this point on the messenger. However, when the release factor is in short supply, the ribosome is likely to pause at this site and then undergo a frameshift to make functional release factor. In this example, frameshifting results in a negative feedback mechanism for regulating the amount of release factor that is synthesized.

Translational fidelity of bacterial ribosomes is influenced in a more general way by specific ribosomal proteins and by the binding of antibiotics to the ribosome (box 34C).

Protein Folding Is Mediated by Protein Chaperones

Protein folding into the final three-dimensional structure usually begins during translation. The discovery of a ubiquitous subgroup of proteins that bind and fold other proteins has sparked a continuing wave of repetitions of the Anfinsen refolding experiment (see Chapter 5) with pure proteins, but modified by the addition of so-called molecular chaperones to the refolding buffer. Such experiments suggest that the protein chaperones improve the yield of correctly folded proteins, not by violating the principle of protein self-assembly, but by suppressing and reversing chain interactions that produce incorrectly folded proteins. These incorrect interactions are more prevalent in the complex and highly concentrated intracellular environment than in a dilute solution of pure refolding protein, and the chaperones presumably evolved to combat them. Plausible models for how they do this appear frequently. But the experiments on which these models are based are still remote from the conditions that occur inside cells, so their conclusions need to be checked in systems that are closer to the in vivo situation.

The folding of polypeptides emerging from eukaryotic ribosomes was analyzed by Hartl and his co-workers in a mammalian translation system using firefly luciferase as a model protein. The growing polypeptide interacts with a specific set of molecular chaperones, including Hsp70, Hsp40, and TRIC. The ordered as-

Antibiotics Inhibit by Binding to Specific Sites on the Ribosome

Many antibiotics (generally, small organic compounds with therapeutic utility) prevent bacterial growth by inhibiting translation (see fig. 1). This is not surprising, because translation is both a complex and metabolically essential process. Also, the bacterial ribosome is structurally distinct from the ribosome in the eukaryotic cytoplasm, and thus specific bacteria inhibitors can be found.

A great many antibiotic inhibitors of ribosome function belong to the class known as aminoglycoside antibiotics. Of these, streptomycin is the best known and the best investigated. Streptomycin binding produces a variety of functional alterations in the ribosome. One of the first to be recognized was the loss of translational fidelity. When bound to the small subunit, streptomycin distorts its structure so as to allow altered codon–anticodon pairing and the consequent incorporation of incorrect amino acids. Indeed, mutations to streptomycin resistance frequently prevent antibiotic binding and involve alterations in ribosomal proteins that are known to play a role in maintaining the fidelity of translation. Such mutants display an increased accuracy phenotype, and are slower in the rate of translational elongation. Streptomycin binding also alters the ribosome's ability to participate properly in the initiation reactions.

Figure 1

The structures of some antibiotic inhibitors of protein synthesis. All of the inhibitors shown function by binding to specific sites on the ribosome.

$R=CH_3NH$

Streptomycin

Chloramphenicol

Tetracycline

Erythromycin

sembly of these three components on the nascent chain forms a large complex that allows the cotranslational formation of protein domains and the completion of folding in an ATP-dependent process once the chain is released from the ribosome.

In *E. coli* the heat shock proteins DnaK, DnaJ, and GrpE (also produced at normal temperatures) can bind to nascent protein chains, with hydrolysis of ATP and ADP, as the proteins are being completed by ribosomes. In the released complex, proper folding of the protein can ensue. Frequently, the released complexes are transferred to supercomplexes in the GroEL and GroES system of *E. coli*. These supercomplexes are ring-shaped hollow cylinders that sequester and protect the folding intermediates and can also renature denatured proteins. These complexes, comparable in size to a ribosome, consume 100–200 ATP molecules in refolding a single polypeptide.

Targeting and Posttranslational Modification of Proteins

Despite the fact that only 20 amino acids (plus selenocysteine and formylmethionine in prokaryotic systems) are known to be directly specified by the genetic code, chemical analysis of mature proteins has revealed hundreds of different amino acids, all of them structural variants on the original 20. This structural diversity, which greatly expands the chemical lexicon of proteins, results from posttranslational modification of the primary products of translation. Our knowledge of the nature and significance of enzymatic reactions that bring about these important alterations is still very incomplete.

In addition to the modification of amino acid side chains many cases occur in which parts of the originally synthesized polypeptide chain are removed during the process of maturation. The types of modification and processing that polypeptide chains undergo is strongly related to the site of protein synthesis and to the mechanisms that are involved in targeting the polypeptide chain to its final destination. In fact, processing frequently begins during polypeptide synthesis.

In bacteria either the formyl group at the amino terminus or the entire formyl methionine residue itself is removed from the nascent peptide chain while the protein is still being assembled. As a general rule, if the next residue after formyl-methionine has a long side chain, only the formyl group is removed (by a specific enzyme), and the final protein product begins with methionine. On the other hand, if the next residue has a short side chain, the entire formylmethionine is removed (by a different enzyme), and the final product will start with the next residue (typically glycine, alanine, or serine). In eukaryotes a major division involving the types of processing reactions that polypeptide chains undergo is related to the site of protein synthesis and the final destination of the protein. Proteins synthesized on free polysomes either remain in the cytosol or are targeted to the mitochondria, the chloroplasts, or the nucleus. Those that remain in the cytosol are often ready to function as soon as they are released from the ribosome. Those that must be transported to another site usually undergo substantial modification during the transport process. Many proteins, however, are synthesized on membrane-bound polysomes rather than on free polysomes. In bacteria such proteins are synthesized on polysomes associated with the inner plasma membrane; in eukaryotes they are synthesized on polysomes associated with the endoplasmic reticulum. Proteins that are synthesized on ribosomes bound to the endoplasmic reticulum (ER) either remain in the ER or are targeted to the Golgi apparatus, secretory granules, the plasma membrane, or lysosomes. As a rule, proteins that are synthesized on membrane-bound polysomes undergo more extensive modification before they reach their final destination and become fully functional (see chapter 18).

Proteins Are Targeted to Their Destination by Signal Sequences

Protein transport in all systems is accomplished by a single underlying mechanism; each polypeptide destined for transport contains an amino acid sequence known as a signal, or leader sequence that identifies the polypeptide to the appropriate transporting system. This generality was first recognized in the middle 1970s by Gunter Blobel, who articulated the underlying signal hypothesis (see fig. 18.19). Frequently, the signal sequence is cleaved from the parent polypeptide during the transport process. A protein containing a signal sequence that is cleaved on transport is known as a preprotein. A protein containing a peptide sequence that must be removed for the protein to be active is known as a proprotein, and a protein that contains both of these sequences is known as a preproprotein.

Many protein hormones are synthesized in a preproprotein form. For example, insulin mRNA is translated into a single polypeptide chain, preproinsulin, that contains 84 residues of proinsulin plus a 23-residue signal peptide (fig. 34.20). Within the islets of Langerhans of the pancreas the N-terminal sequence is removed cotranslationally in targeting proinsulin to the Golgi apparatus, and the two disulfides joining the ends of the molecule are formed. Following this, the C-peptide region of proinsulin is removed to yield the mature circulating form of insulin with 51 residues in two disulfide-linked peptides.

Signal sequences usually occur at the amino terminus of polypeptides to be transported. Signal peptide (SP) sequences are generally characterized by a positively charged amino terminus followed by a hydrophobic region of about 10–12 residues, followed by a cleavage site, which is a sequence recognized by the signal peptidase. The presence of hydrophobic residues reflects the fact that protein transport invariably involves the movement of proteins across lipid membranes. The hydrophobic residues both target the protein to the appropriate compartment and initiate penetration into the membrane. Stop-transfer or signal anchor sequences are similar to SP sequences, but their hydrophobic regions are much longer (about 24 residues and up), and they are not cleaved by signal peptidases. Stop-transfer sequences anchor proteins in the membrane.

Some Mitochondrial Proteins Are Transported after Translation

Some proteins, especially those destined for the eukaryotic mitochondria and chloroplasts, are transported after their synthesis on free polysomes is complete. Such transport is known as posttranslational transport. In the case of posttranslational transport it is believed that the polypeptide to be transported must be unfolded from its native folded configuration by a system of polypeptide-chain-binding proteins (PCBs) before it can pass through the mem-

Figure 34.20

Processing of insulin. Insulin is synthesized by membrane-bound polysomes in the β cells of the pancreas. The primary translation product is preproinsulin, which contains a 24-residue signal peptide preceding the 81-residue proinsulin molecule. The signal peptide is removed by signal peptidase, cutting between Ala (−1) and Phe (+1), as the nascent chain is transported into the lumen of the endoplasmic reticulum. Proinsulin folds and two disulfide bonds cross-link the ends of the molecule as shown. Before secretion a trypsinlike enzyme cleaves after a pair of basic residues 31, 32 and 59, 60; then a carboxypeptidase B-like enzyme removes these basic residues to generate the mature form of insulin.

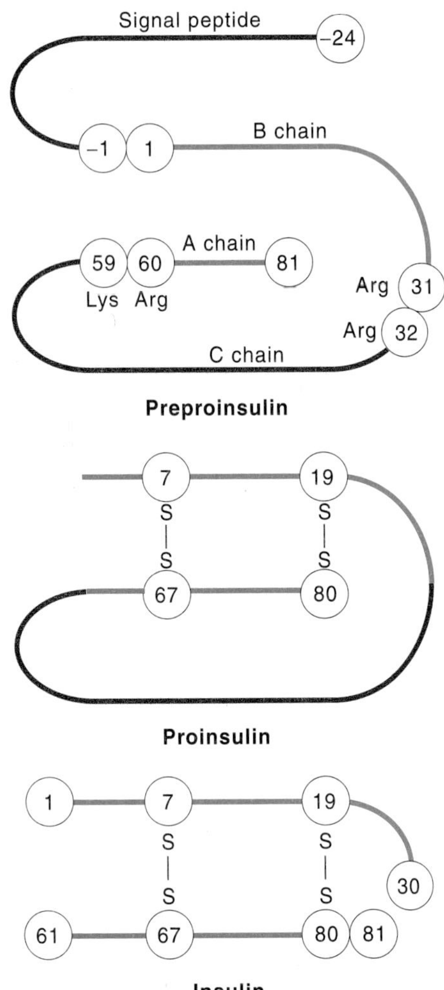

brane. Posttranslational transport into the mitochondrion requires both ATP and the membrane electrical potential. Presumably, the energy from one or both of these sources is used to unfold the protein or separate it from the PCB system so that it can pass through the membrane. ATP is also consumed in the process of pulling the polypeptide chain across the membrane.

The transport of proteins from the cytoplasm to the mitochondrion is further complicated by the fact that mitochondria themselves are compartmentalized. They have both an outer membrane and an inner membrane (see chapter 16). Some proteins are targeted to either of these membranes, some are targeted to the fluid layer enclosed by the inner membrane, called the matrix, and others such as cytochrome c_1 are targeted to the fluid layer bounded by the inner membrane and the outer membrane, known as the intermembrane space (fig. 34.21). Cytochrome c_1 has two signal sequences located in series at the amino-terminal end of the preprotein. The first is recognized by a specific receptor protein attached to the outer membrane of the mitochondrion. This receptor protein guides the polypeptide chain to a transport channel, across which it is transported into the matrix. Once in the matrix, the first signal sequence is removed by a specific protease. A second signal sequence, which is exposed by this cleavage reaction, results in the transport of the polypeptide chain by a similar mechanism across a transport channel into the intermembrane space. In intermembrane space the second signal sequence is removed, and the protein folds into its mature configuration, which includes associating with a heme group.

As in the case of mitochondrial proteins, some of the proteins of chloroplasts are synthesized directly in the organelle, whereas others are synthesized in the cytosol and must be transported. The mechanisms for transport are similar to those observed for transported mitochondrial proteins.

All nuclear proteins are synthesized on free polysomes in the cytosol. These proteins must have some way of making their way back to the nucleus. The nuclear membrane is pockmarked with large pores that are freely permeable to small molecules and medium-sized proteins. Despite this, many proteins and ribonucleoproteins destined for the nucleus contain one or more nuclear localization sequences (NLSs) or other determinants that lead to their being actively and selectively transported into the nucleus in an ATP-dependent process. For example, the large T antigen of the SV40 virus carries a highly positively charged seven-amino-acid NLS, pro-lys-lys-lys-arg-lys-val. Unlike the situation in other organelles, unfolding of the protein for transport is not required for proteins that are bound for the nucleus.

Eukaryotic Proteins Targeted for Secretion Are Synthesized in the Endoplasmic Reticulum

The endoplasmic reticulum (ER) is the largest membrane-bounded organelle in a typical eukaryotic cell. The ER consists of a continuous network of tubules and cisternae extending throughout the cytoplasm, with a total surface area many times that of the plasma membrane. Most of the ER is studded with ribosomes to form the rough endoplasmic reticulum (RER). The ribosomes of the RER are the site of synthesis of membrane and secretory proteins and are the starting point for the protein secretory pathway. The membrane and lumen of the ER contain a characteristic set of proteins that function to process secretory and membrane proteins. After only a short time in the RER, secretory proteins are transported, by a process of vesicle budding and fusion, to the Golgi apparatus and from there to the cell surface, secretory granules, or lysosomes. As you will recall, the Golgi apparatus is a flattened stack of membranes that is the primary site of protein targeting as well as a principal site of carbohydrate addition to form glycoproteins (see chapter 18).

Let us review the way in which proteins that are destined for assembly into membranes, for secretion, or for targeting to other areas of the cell enter the lumen of the ER (see fig. 18.19). The initial targeting of nascent polypeptides to the ER membrane results from the cotranslational recognition of a signal sequence

Figure 34.21

Two successive translocations are required to target proteins such as cytochrome c_1 to the intermembrane space. The precursor of cytochrome c_1 has two uptake-targeting sequences at its N terminus. The first targets the polypeptide to the matrix. In the matrix the first target sequence is cleaved by a specific protease. The second target sequence is thereby exposed and targets the polypeptide to intermembrane space, where the second target sequence is removed by another protease. The molecule folds and adds heme to become fully functional. (*Source:* After F. U. Hartl, J. Ostenmann, B. Guiard, and W. Neupert, *Cell* 53:1021–1027, 1987; and E. C. Hurt and A. P. G. M. van Loon, How proteins find mitochondria and intramitochondrial compartments, *Trends Biochem. Sci.* 11:204–207, 1986.)

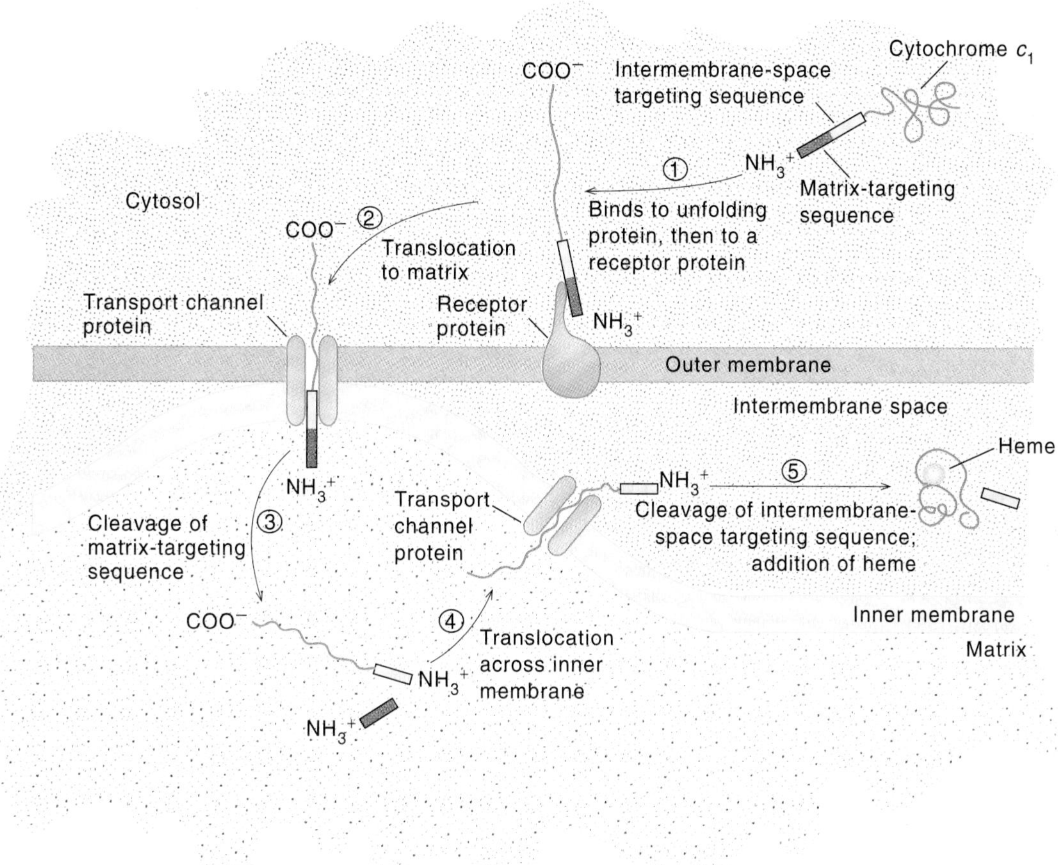

by a ribonucleoprotein complex called the signal recognition particle, or SRP. The SRP is a complex that consists of six different proteins and a single 300-residue RNA molecule designated 7S RNA. This RNA is structurally and functionally related to a 4.5S RNA which is an essential molecule in *E. coli.* This complex is especially adapted to recognize and bind to the signal sequence when the total nascent chain has reached a length of about 90 amino acid residues. On binding, the SRP arrests translation of the nascent chain while it searches for a specific receptor known as the SRP receptor, or "docking protein." The SRP receptor is an integral membrane protein that protrudes on the inner face of the ER membrane. When docking is complete, the SRP dissociates from the ribosome, which is now bound by the signal sequence of its nascent polypeptide to the SRP receptor. After dissociation of the SRP translation resumes, and translocation of the nascent polypeptide across the membrane begins. GTP is essential for the docking maneuver, and its hydrolysis is probably essential for the subsequent release of SRP. Furthermore, sequencing of one of the protein subunits of SRP, SRP54, has revealed that it is homologous to the GTP-binding domain of the G proteins. Thus signal sequence recognition by SRP may operate in a manner analogous to the GTP-dependent functioning of members of the G protein family.

Numerous nonmembranous proteins are retained by and function in the ER. Retention of these proteins is accomplished by a surprisingly simple mechanism. Soluble ER proteins all share the common C-terminal sequence Lys-Asp-Glu-Leu (or KDEL in single-letter language). A protein receptor appears to be bound to the ER membrane that recognizes and binds this sequence. The passive nature of this retention mechanism is illustrated by the fact that if the four C-terminal residues are removed from an ER protein, it is no longer retained by the organelle.

Proteins That Pass through the Golgi Apparatus Become Glycosylated

Glycosylation presents a conspicuous modification of proteins as they pass through the Golgi apparatus (see chapter 18). Many of the cell surface and secretory proteins produced here are glycoproteins. Because of the diversity of the glycosylation reactions that occur in the Golgi and the fact that this organelle is the principal site of protein transport, it seems likely that the carbohydrate moieties attached to proteins are responsible for targeting them to their destinations.

One clear-cut example supporting this conclusion is the role of mannose-6-phosphate residues in targeting proteins to the lysosome. Digestive enzymes destined for the lysosome are identified by the presence of mannose-6-phosphate by binding to a specific mannose-6-phosphate receptor protein. The critical step in targeting glycoproteins from the Golgi to the lysosome is their recognition by the enzymes that catalyze the two-step phosphorylation of terminal mannose residues in their N-linked oligosaccharide side chains. Failure to phosphorylate the mannose residues results in the secretion of lysosomal enzymes. This malfunction, you may remember, is associated with the syndrome known as I-cell disease, a condition that leads to the crowding of lysosomes with damaged proteins that it cannot degrade (see chapter 18).

Processing of Collagen Does Not End with Secretion

Recall that collagen is an extracellular matrix protein that serves as a major constituent of many connective tissues (see figs. 5.13 to 5.16). Collagen fibrils have a distinctive banded pattern with a periodicity of 680 Å. Individual fibrils are composed of three polypeptide chains wound around one another in a right-handed helix with a total length of 3,000 Å. Each of the polypeptide chains in the triple helix has a repetitious tripeptide sequence, Gly-X-Y, where X is frequently a proline and Y is frequently a hydroxyproline. The latter amino acid is not one of the 20 that are specified genetically, so it must be formed posttranslationally by a modification of some of the prolines.

Since collagen is a secreted protein, we know that it must follow the route of synthesis that starts on a ribosome bound to the endoplasmic reticulum (fig. 34.22). This is where the fun begins, as we find that the nascent collagen polypeptide has extensive N and C termini (150 and 250 amino acids, respectively, for type I collagen) that are not found in mature collagen. The function of these extensions appears to be to facilitate the initial interaction of chains in triplets and to stabilize the triple helices once they have been formed. The first modification reaction to take place in the endoplasmic reticulum is hydroxylation of specific proline residues and some lysine residues by two different enzyme systems. Glycosylation begins soon thereafter with O-glycosylation of certain hydroxylysine residues and N-glycosylation of certain asparagine residues. Next, the modified polypeptide chains, in clusters of three, form interchain disulfides near the C termini. This brings the chains in close proximity, thereby facilitating the winding reaction that leads to triple-helix formation. The winding reaction proceeds in the C to N direction. The triple-helix procollagen molecule is packaged into a secretory vesicle. Following exocytosis, the ends of the procollagen molecule are removed by extracellular proteases. Collagen fibrils form by the spontaneous association of the mature collagen molecules.

Table 34.6
Half-Lives of Some Proteins in Mammalian Cells

Enzyme	Half-life (h)
Rapidly degraded	
1. c-myc, c-fos, p53 oncogenes	0.5
2. Ornithine decarboxylase	0.5
3. δ-Aminolevulinate synthase	1.1
4. RNA polymerase I	1.3
5. Tyrosine aminotransferase	2.0
6. Tryptophan oxygenase	2.0
7. β-Hydroxyl-β-methylglutaryl coenzyme A reductase	2.0
8. Deoxythymidine kinase	2.6
9. Phosphoenolpyruvate carboxykinase	5.0
Slowly degraded	
1. Arginase	96
2. Aldolase	118
3. Cytochrome b_5	122
4. Glyceraldehyde-3-phosphate dehydrogenase	130
5. Cytochrome *b*	130
6. Lactic dehydrogenase (isoenzyme 5)	144
7. Cytochrome *c*	150

Bacterial Protein Transport Frequently Occurs during Translation

The problem of protein transport in bacterial systems is relatively simple. Polypeptides synthesized in the cytoplasm may function there, may be inserted into the plasma membrane, or may be secreted by being passed through this membrane.

Most noncytoplasmic bacterial proteins are targeted into or across the plasma membrane while they are being synthesized on the ribosome. This process is known as cotranslational transport. Cotranslational secretion of proteins in *E. coli* involves a group of proteins encoded by the *sec* (for secretion) genes. Some of these proteins are membrane proteins that function by recognizing and binding the signal sequence of nascent polypeptide chains as they emerge from the ribosome. As a consequence of cotranslational transport and the action of the sec proteins, most bacterial ribosomes engaged in the synthesis of secreted proteins are tethered to the cytoplasmic face of the plasma membrane by their nascent peptide chains. A leader peptidase that is an integral membrane protein cleaves the leader sequence from the secreted polypeptide.

Protein Turnover

Cellular proteins are continuously being formed and degraded. At first glance, continuous degradation appears to be wasteful. How-

Figure 34.22

Major events in the posttranslational processing of collagen.

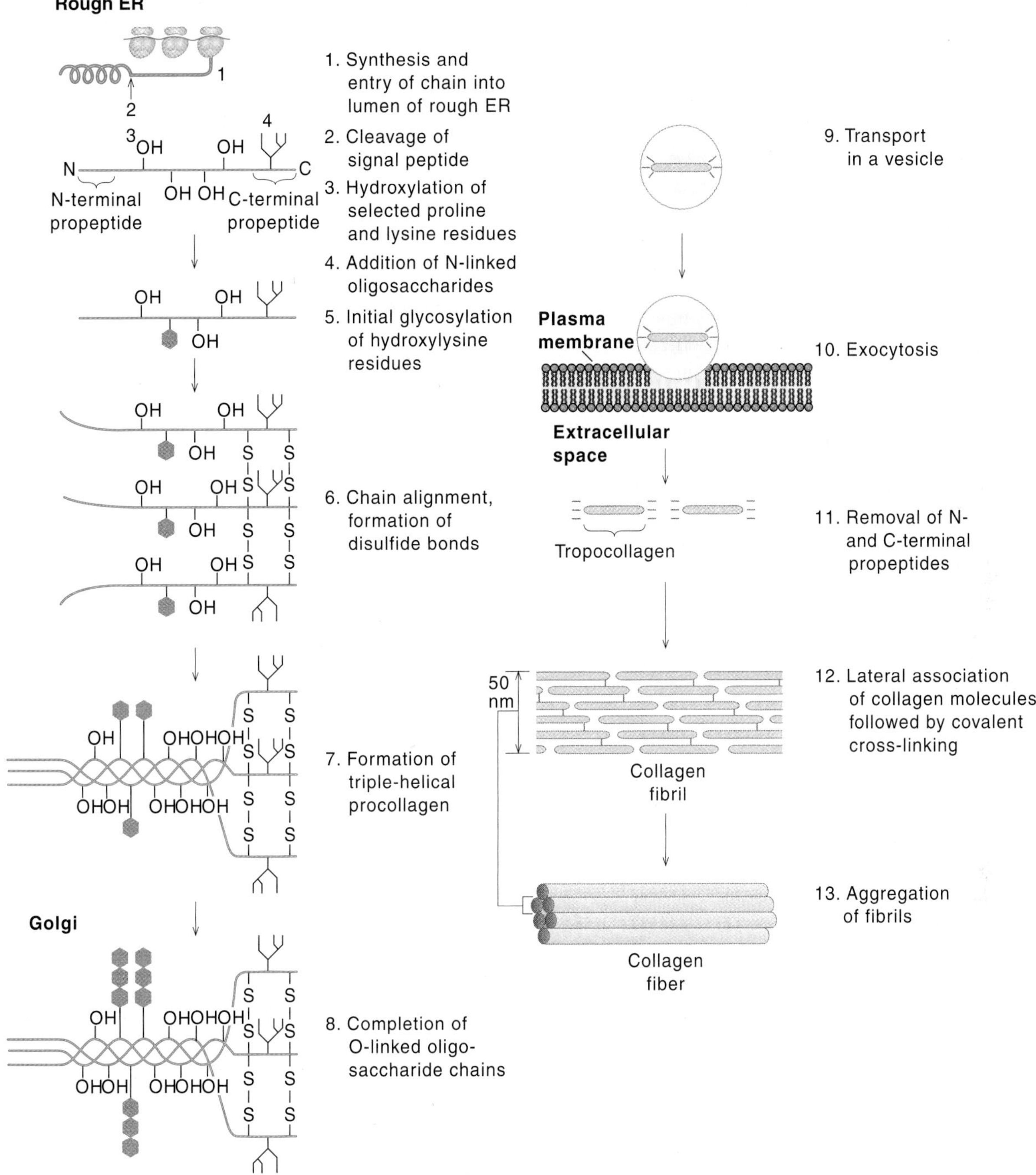

ever, protein degradation is of major biological importance in regulating protein levels, in protecting against the accumulation of abnormal proteins, in controlling growth and development, and in allowing adaptation to changing environmental conditions.

Although many proteolytic enzymes are known to exist, we are just beginning to understand how they operate to govern the levels and types of proteins within cells—to distinguish those proteins that must be degraded from those that must be preserved.

The Lifetimes of Proteins Differ

The level of a protein within a cell is determined by the balance between its rates of synthesis and degradation. As a consequence, changes in protein levels can be brought about by changes either in synthetic or degradative rates. Moreover, a rapid rate of degradation ensures that the concentration of a protein rises or falls rapidly when its synthetic rate changes.

The half-lives of eukaryotic proteins are different and distinct. Representative examples are listed in table 34.6. It is apparent that degradative rates of individual proteins within a single cell can vary over a wide range. At least in the mammalian liver, enzymes that occupy important metabolic control points are degraded most rapidly, whereas especially long-lived proteins are rarely the sites of metabolic control. In addition, it has been established that degradative rates of specific proteins can vary with changes in physiological conditions. Thus the protein degradative mechanism is in some way tuned to metabolic control.

How are individual proteins selected for hydrolysis by the proteolytic machinery? Two sequence characteristics have been identified as correlating with rates of protein degradation. Several years ago, it was observed through sequence analysis that rapidly degraded liver proteins, those with half-lives of less than 2 h, nearly all contain regions of their sequence that are rich in the amino acids proline, glutamate, serine, and threonine. These regions, involving from 10 to 60 amino acid residues, have been designated PEST sequences on the basis of the single-letter designations of their constituent amino acids. Presumably, the PEST sequences create structural domains that are recognized by proteolytic enzymes.

A second structural feature that has been correlated with protein degradative rates is the N-terminal residue of the mature form of cytoplasmic proteins. This relationship, summarized in table 34.7, has become known as the N-terminal rule. Here the presumption is that the amino-terminal residue is at least partly responsible for recognition by the degradative machinery. Proteins that are degraded according to the N-terminal rule are believed to be recognized by the ubiquitin-ATP-dependent pathway, which we describe shortly.

Table 34.7

Correlation between Half-Lives of Cytosolic Proteins and Amino Acid Residue at the N Terminal

Amino-Terminal Residue	Half-Life
Stabilizing	
Methionine	
Glycine	
Alanine	> 20 h
Serine	
Threonine	
Valine	
Destabilizing	
Isoleucine	≈ 30 min
Glutamate	
Tyrosine	
Glutamine	≈ 10 min
Proline	≈ 7 min
Highly Destabilizing	
Leucine	
Phenylalanine	
Aspartate	≈ 3 min
Lysine	
Arginine	≈ 2 min

Source: From A. Bachmjuir et al., *In vivo* half-life of a protein is a function of its amino-terminal residue, *Science* 234:179, 1986.

Abnormal Proteins Are Selectively Degraded

A very important function of protein degradation is to protect the organism against the consequences of intracellular accumulation of abnormal proteins. All cells possess a degradative system that is capable of recognizing "abnormal" proteins.

The features of this degradative system are best seen in *E. coli,* in which altered intracellular proteins can be readily manipulated. For example, it has been known for many years that incomplete chains of β-galactosidase are rapidly degraded even though the completed protein is very stable. In a similar way, many protein alterations that result from miscoding induced by streptomycin, or by the incorporation of amino acid analogs, also fail to accumulate because the protein products that contain them are rapidly degraded. This system presumably limits expression of heterologous proteins in bacteria that are degraded because they cannot form their native structure.

Another manifestation of the defense against abnormal proteins is seen in the heat shock response. This defense reaction, involving programmed changes in gene transcription and translation, is exhibited by essentially all cells under stressful conditions. The cells reduce their overall rates of gene transcription and translation and for a brief time produce a small repertoire of proteins called heat shock proteins (hsp). Some hsp proteins play a role in the folding and transport of proteins during normal cellular function. In the heat shock response it is believed that these proteins protect against the presence of damaged proteins by binding to them and promoting either their refolding or their proteolytic degradation.

Proteolytic Hydrolysis Occurs in Mammalian Lysosomes

The classic studies of Christian DeDuve in the 1960s established that mammalian cells contain a degradative organelle, the lysosome, that is produced in the Golgi apparatus and contains a large number of proteases and other hydrolytic enzymes. A similar vacuole containing hydrolytic enzymes is also present in yeast and higher plants. Together, the hydrolytic enzymes in these organelles are capable of completely degrading many macromolecules and delivering their monomeric units to the cytoplasm for further metabolism. The metabolic importance of the lysosome is demonstrated by the existence of various lysosomal storage diseases, in which one or more hydrolytic enzymes is missing. An accumulation of undegraded molecules results from these genetic diseases, frequently with devastating consequences.

A primary function of the lysosome is to digest protein-containing particles derived from the extracellular space. One mechanism of delivery is the process of endocytosis. Endocytosis is the invagination of a group of occupied receptors on the plasma membrane. Most mammalian cells can also engulf large extracellular particles by the less specific processes of pinocytosis and phagocytosis. The endocytic vesicles formed by these processes fuse with lysosomes to form secondary lysosomes where hydrolysis occurs. Lysosomes also function to degrade intracellular proteins, especially under conditions of nutritional deprivation. Under these circumstances, lysosomes can engulf cytoplasmic contents to form autophagic vacuoles and thus recycle cellular proteins.

Lysosomal proteases are called cathepsins, a name derived from the Greek term meaning "to digest." The interior of the lysosome is acidic, and the cathepsins, like all lysosomal hydrolases, possess acidic pH optima and exhibit little enzymatic activity at neutral pH. This characteristic protects the cell from autolytic breakdown that might result from leakage of lysosomal contents into the neutral cytoplasm.

Ubiquitin Tags Proteins for Proteolysis

A number of different proteolytic systems are thought to be responsible for the degradation of soluble proteins in the cytoplasm of eukaryotic cells. One of the best understood is that which involves ATP and the protein ubiquitin. Ubiquitin is a small protein of only 76 residues. It occurs universally in eukaryotic cells and is highly conserved in sequence; only three residues distinguish the ubiquitin in yeast and humans. The covalent attachment of ubiquitin to proteins is thought to "tag" them for subsequent hydrolysis by a 26S complex, which contains a key proteolytic component, the 20S proteosome. This nonlysosomal pathway was first found to mediate the rapid elimination of highly abnormal proteins that arise by mutation or postsynthetic damage. In addition this pathway is now known to degrade many critical regulatory proteins that must be rapidly destroyed for normal growth and metabolism.

Three enzymes participate in the ATP–ubiquitin system that prepares proteins for proteolysis (fig. 34.23). In the first step of this series of reactions the C-terminal glycine residue of ubiquitin is activated by forming a thiol ester with a specific activating enzyme designated E_1. This reaction is driven by the hydrolysis of ATP to AMP and PP_i. Ubiquitin is then transferred to a sulfhydryl group of a second protein (E_2). In a similar reaction the ubiquitin is transferred to a sulfhydryl group of a third protein, E_3. The E_3 proteins are responsible for identifying proteins for degradation (some on the basis of the identity of their N-terminal residue and some because they are "abnormal") and catalyzing the covalent transfer of ubiquitin from E_3 to these proteins. This attachment is via isopeptide bonds that join the carboxyl group of the previously activated C-terminal glycine residue of ubiquitin with ϵ-amino groups of lysine side chains of the targeted proteins. Proteins tagged in this way are then recognized and degraded by specific proteases that release ubiquitin so that it can recycle.

ATP Plays Multiple Roles in Protein Degradation

ATP plays a conspicuous and rather surprising role in the degradation of proteins. Clearly, the hydrolysis of peptide bonds is a reaction that, in and of itself, does not require the input of energy. Nonetheless, ATP is required for the action of many proteolytic enzymes, independent of its role in ubiquitination that we have just seen.

Figure 34.23

The ubiquitin marking system targets certain proteins for degradation. At least three enzymes, E_1, E_2, and E_3, are involved in addition to ubiquitin-specific proteases.

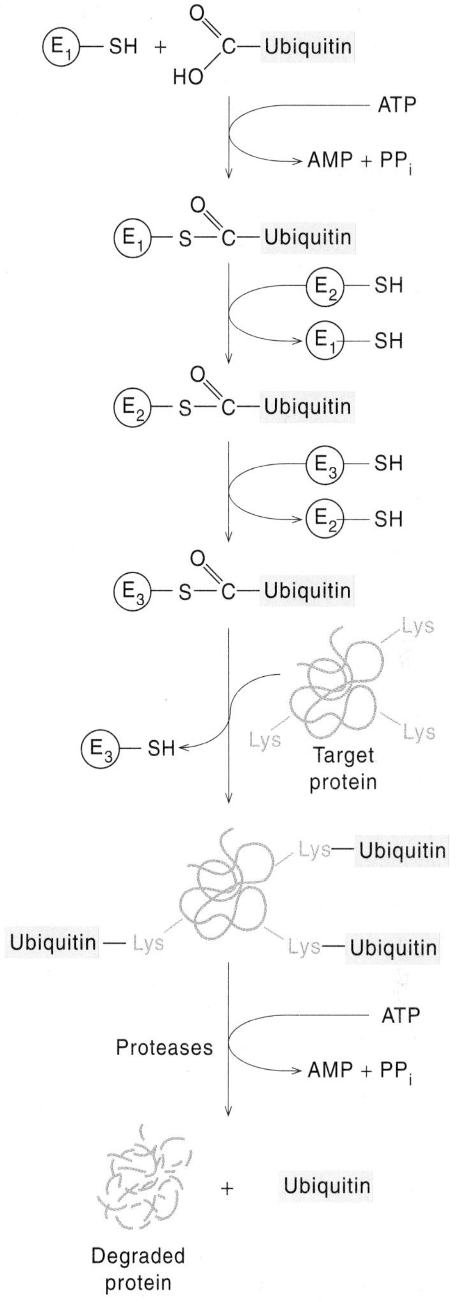

One particularly well-understood ATP-dependent protease is the La protease that is the product of the *lon* gene of *E. coli.* This protease, like many ATP-dependent proteases, is a large protein, and its ability to hydrolyze proteins is tightly coupled to its ability to hydrolyze ATP. Approximately two ATPs are hydrolyzed to ADP and P_i for each peptide bond that is hydrolyzed. It appears that the hydrolysis of ATP is required to activate the proteolytic active site of La. Other proteases seem to require ATP as an allosteric effecter to activate their hydrolytic sites, but this ATP is not hydrolyzed.

Thus ATP serves at least three roles in intracellular proteolysis: (1) It functions in the tagging of proteins through covalent attachment of ubiquitin, (2) it activates proteases such as La through hydrolysis, and (3) it serves as a positive allosteric effecter of other proteases without being hydrolyzed. The presumed function of this energetic requirement is to ensure the fidelity of protein degradation. It is, after all, nearly as important to cellular function to degrade proteins correctly as it is to synthesize them correctly.

Summary

We focused in this chapter on the complex mechanisms of protein synthesis. The following points are central to this subject.

1. Three types of RNA carry out protein synthesis: ribosomal RNA, transfer RNA, and messenger RNA. Ribosomal RNA is invariably complexed with many proteins to form ribosomes, on which amino acids are assembled into polypeptides. The amino acids are brought to the ribosomes attached to transfer RNAs. The messenger RNA contains the instructions for translation in the form of the genetic code. Messenger RNAs form transient complexes with ribosomes. Individual aminoacyl-tRNAs align with specific sites on the messenger RNAs. The interacting site on the messenger is the codon; the interacting site on the tRNA is the anticodon.
2. The part of the messenger that is translated is the reading frame. Eukaryotic messages generally carry only one reading frame, whereas prokaryotic messengers may carry more than one. In prokaryotes the initiation codons are recognized by a ribosome-binding site just upstream of the start codon.
3. Most transfer RNAs have common parts and uncommon parts. The common parts facilitate binding of the aminoacyl-tRNAs to common sites on the ribosome. The uncommon sites permit specific reactions with charging enzymes that covalently attach the correct amino acids to the correct tRNA. Another uncommon site on the tRNAs is the anticodon, which leads to specific complex formation with the complementary codon site on the messenger.
4. Attachment of the amino acid to the tRNA is catalyzed by a specific aminoacyl tRNA synthase, which recognizes all the cognate tRNAs for specific amino acid.
5. A unique methionyl-tRNA recognizes the initiation codon on all messages.
6. The genetic code is the sequence relationship between nucleotides in the messenger RNA and amino acids in the proteins they encode. Triplet codons are arranged on the messenger in a nonoverlapping manner without spacers.
7. The code was deciphered with the help of synthetic messengers with a defined sequence, by analyzing the types of polypeptide chains that were made when these messengers were used in an in vitro protein-synthesizing system.
8. The genetic code is highly degenerate, with most amino acids represented by more than one codon. In many cases the 3′ base in the codon may be altered without changing the amino acid that is encoded.
9. The codon–anticodon interaction is limited to Watson-Crick pairing for the first two bases in the codon but is considerably more flexible in the third position.
10. Translation begins with the binding of the ribosome to mRNA. A number of protein factors transiently associate with the ribosome during different phases of translation: initiation factors, elongation factors, and termination factors.
11. Initiation factors contribute to the ribosome complex with the messenger RNA and the initiator methionyl-tRNA. Elongation factors assist the binding of all the other tRNAs and the translocation reaction that must occur after each peptide bond is made. Termination factors recognize a stop signal and lead to the termination of polypeptide synthesis and the release of the polypeptide chain and the messenger from the ribosome.
12. A large number of antibiotics have been characterized that inhibit protein synthesis. These antibiotics are usually made by a particular microorganism, and they inhibit protein synthesis in a broad family of other organisms, mostly bacterial.
13. Specific enzymes called chaperones catalyze folding after polypeptide synthesis.
14. Proteins are targeted to their destination by signal sequences built into the polypeptide chain. These signals are usually located at the N-terminal end of the protein and are generally cleaved during protein maturation.
15. Posttranslational modifications include many covalent alterations: Polypeptide processing, attachment of carbohydrate or lipid groups to specific side chains, and addition of many other low-molecular-weight ligands to side chains.
16. Intracellular protein degradation is not random. Different proteins have quite different half-lives, which are related to specific structural features. Imperfectly folded proteins and polypeptide fragments are frequently degraded most rapidly. In eukaryotes, lysosomes play a major role in protein degradation.

Selected Readings

Atkins, J. F., R. B. Weiss, S. Thompson, and R. F. Gesteland, Towards a genetic dissection of the basis of triplet decoding, and its natural subversion: Programmed reading frameshifts and hops. *Ann. Rev. Genet.* 25:201–228, 1991. Ribosome "gymnastics."

Bjork, G. R., J. U. Ericson, C. E. D. Gustafsson, T. G. Hagervall, Y. H. Jonsson, and P. M. Wilkstrom, Transfer RNA modification. *Ann. Rev. Biochem.* 56:263–288, 1987.

Böck, A., K. Forchhammer, J. Heider, and C. Baron, Selenoprotein synthesis: An expansion of the genetic code. *Trends Biochem. Sci.* 16:463–467, 1991.

Chamorro, M., N. Parkin, and H. E. Varmus, An RNA pseudoknot and an optimal heptameric shift site are required for highly efficient ribosomal frameshifting on a retroviral messenger RNA. *Proc. Natl. Acad. Sci. USA* 89:713–717, 1992.

Chapeville, F., F. Lipman, G. von Ehrenstein, B. Weisblum, W. J. Ray, and S. Benzer, On the role of soluble ribonucleic acid in coding for amino acids. *Proc. Natl. Acad. Sci. USA* 48:1086–1092, 1962. A classic paper that showed that specificity of amino acid insertion into protein lies with the tRNA.

Cooper, A. A., and T. H. Stevens, Protein splicing: Self-splicing of genetically mobile elements at the protein level. *TIBS* 20:351–356, 1995.

Crick, F. H. C., Codon–anticodon pairing: The wobble hypothesis. *J. Mol. Biol.* 19:548–555, 1966. A classic paper.

Cyr, D. M., T. Langer, and M. G. Douglas, DnaJ-like proteins: Molecular chaperones and specific regulators of Hsp70. *TIBS* 19:176–181, 1994.

Ellis, R. J., Chaperoning nascent proteins. *Nature* 370:96–97, 1994.

Englander, S. W., In pursuit of protein folding. *Science* 262:848–850, 1993.

Frydman, J., E. Nimmesgern, K. Ohtsuka, and U. Hartl, Folding of nascent polypeptide chains in a high molecular mass assembly with molecular chaperones. *Nature* 370:111–117, 1994.

Gilmore, R., Protein translocation across the endoplasmic reticulum: A tunnel with toll booths at entry and exit. *Cell* 75:5889–5892, 1993.

Goldberg, A. L., Functions of the proteasome: The lysis at the end of the tunnel. *Science* 268:522–524, 1995.

Goldman, E., and H. Jakubowski, Uncharged tRNA, protein synthesis, and the bacterial stringent response. *Mol. Microbiol.* 4:2035–2040, 1990. The effects of deacylated tRNA.

Hartl, F.-U., R. Hlodan, and T. Langer, Molecular chaperones in protein folding: The art of avoiding sticky situations. *Trends Biochem. Sci.* 19:20–25, 1994.

Herchtold, H., L. Reshetnikova, C. O. A. Relser, N. K. Schirmer, M. Sprinzl, and R. Hilgenfeld, Crystal structure of active elongation factor Tu reveals major domain rearrangements. *Nature* 365:126–132, 1993.

Hershko, A., and A. Ciechanover, The ubiquitin system for protein degradation. *Ann. Rev. Biochem.* 61:761–807, 1992.

Hoagland, M. B., M. L. Stephenson, J. F. Scott, L. I. Hecht, and P. Zamecnik, A soluble ribonucleic acid intermediate in protein synthesis. *J. Biol. Chem.* 231:241–257, 1958. Describes the pioneering tracer studies that chart the course of amino acid into polypeptide chain.

Jakubowski, H., and E. Goldman, Editing of errors in selection of amino acids for protein synthesis. *Microbiol. Rev.* 56:412–429, 1992. Reviews the role of editing in maintenance of translational accuracy by aminoacyl-tRNA synthetases.

Kawashima, T., C. Berthet-Colominas, M. Wulff, S. Cusack, and R. Leberman, The structure of the Escherichia coli EF-Tu-EF-Ts complex at 2.5Å resolution. *Nature* 379:511–518, 1996.

Kessler, F., G. Blobel, H. A. Patel, and D. J. Schnell, Identification of two GTP-binding proteins in the chloroplast protein import machinery. *Science* 266:1035–1039, 1994.

Khorana, H. G., H. Büchi, H. Ghosh, N. Gupta, T. M. Jacob, H. Kössel, R. Morgan, S. A. Narang, E. Ohtsuka, and R. D. Wells, Polynucleotide synthesis and the genetic code. *Cold Spring Harbor Symp. Quant. Biol.* 31:39–49, 1966. A classic description of cell-free protein synthesis experiments with synthetic polyribonucleotide messages to determine codon assignments.

Kleinkauf, H., and H. Dohren, Nonribosomal polypeptide formation on multifunctional proteins. *Trends Biochem. Sci.* 8:281–283, 1983.

Koazk, M., Structural features in eukaryotic mRNAs that modulate the initiation of translation. *J. Biol. Chem.* 266:19867–19870, 1991. An update of the scanning model for eukaryotic translation initiation, by the originator of the model.

Lithgow, T., B. S. Glick, and G. Schatz, The protein import receptor of mitochondria. *TIBS* 20:98–101, 1995.

Lovett, P. S., and E. J. Rogers, Ribosome regulation by the nascent peptide. *Microbiol. Rev.* 60:366–385, 1996. Includes description of translational attenuation.

Lowe, J., D. Stock, B. Jap, P. Zwicki, W. Baumeister, and R. Huber, Crystal structure of the 20S proteasome from the archaeon *T. acidophilum* at 3.4 Å resolution. *Science* 268:533–539, 1995.

Min Jou, W., G. Haegeman, M. Ysebaert, and W. Fiers, Nucleotide sequence of the gene coding for the bacteriophage MS2 coat protein. *Nature* 237:82–88, 1972. The first natural gene coding for a protein that was sequenced, confirming the genetic code table.

Moore, P. B., The ribosome returns. *Nature* 331:223–227, 1988.

Ng, D. T. W., J. D. Brown, and P. Walter, Signal sequences specify the targeting route to the endoplasmic reticulum. *J. Cell Biol.* 134:269–278, 1996.

Nierhaus, K. H., The allosteric three-site model for the ribosomal elongation cycle: features and future. *Biochemistry* 29:4997–5008, 1990. Reviews the ribosomal E-site and the pre- and post-translocational states.

Nirenberg, M. W., and J. H. Mattaei, The dependence of cell-free protein synthesis in *E. coli* upon naturally occurring or synthetic polyribonucleotides. *Proc. Natl. Acad. Sci. USA* 47:1588–1602, 1961. The landmark paper reporting the finding that poly(U) stimulates the synthesis of polyphenylalanine.

Noller, H. F., V. Hoffarth, and L. Zimniak, Resistance of peptidyl transferase to protein extraction procedures. *Science* 256:1416–1418, 1992. Demonstration that peptidyl transferase is an RNA.

Pain, V. M., Initiation of protein synthesis in mammalian cells. *Biochem. J.* 235:625, 1986. A comprehensive review of the complex process of translational initiation in mammalian systems, emphasizing the mechanism of the process and its regulation.

Pfeffer, S. R., and J. E. Rothman, Biosynthetic protein transport and sorting by the endoplasmic reticulum and Golgi. *Ann. Rev. Biochem.* 56:829, 1987. An excellent overview of the major features of protein targeting in eukaryotic cells.

Randall, L. L., and S. J. S. Hardy, High selectivity with low specificity: How Sec B has solved the paradox of chaperone binding. *TIBS* 20:65–70, 1995.

Rapoport, T. A., Transport of proteins across the endoplasmic reticulum membrane. *Science* 258:931–936, 1992. This review includes description of SRP function; the journal issue contained several articles related to membranes and transport.

Rould, M. A., J. J. Perona, and T. A. Steitz, Structural basis of anticodon loop recognition by glutaminyl-tRNA synthetase. *Nature* 352:213–218, 1991.

Rothman, J. E., and F. T. Wieland, Protein sorting by transport vesicles. *Science* 272:227–234, 1996.

Saks, M. E., J. R. Sampson, and J. N. Abelson, The transfer RNA identity problem: A search for rules. *Science* 263:191–197, 1994.

Schaeffner, M., U. Nuber, and J. M. Hulbregise, Protein ubiquitination involving an E_1-E_2-E_3 enzyme ubiquitin thioester cascade. *Nature* 373:81–83, 1995.

Schnell, D. J., F. Kessler, and G. Blobel, Isolation of components of the chloroplast protein import machinery. *Science* 266:1007–1012, 1994.

Siegel, V., and P. Walter, Each of the activities of signal recognition particle (SRP) is contained within a distinct domain. *Cell* 52:39–49, 1988.

Sprinzl, M. Elongation factor Tu: A regulatory GTPase with an integrated effector. *TIBS* 19:245–250, 1994.

Stansfield, I., K. M. Jones, and M. F. Tuite, The end in sight: terminating translation in eukaryotes. *TIBS* 20:489–491, 1995.

Tobias, J. W., T. E. Shrader, G. Rocap, and A. Varshavsky, The N-end rule in bacteria. *Science* 254:1374–1377, 1991.

Todd, M. J., P. V. Vitanen, and G. H. Lorimer, Dynamics of the chaperonin ATPase cycle: Implications for facilitated protein folding. *Science* 265:659–666, 1994.

Wagner, E. G. H., and R. W. Simons, Antisense RNA control in bacteria, phages, and plasmids. *Ann. Rev. Microbiol.* 48:713–742, 1996.

Weidmann, B., H. Sakai, T. A. Davis, and M. Weidmann, A protein complex required for signal-sequence-specific sorting and translocation. *Nature* 370:434–440, 1994. A description of a nascent-polypeptide associated complex (NAC) in eukaryotes.

Weijland, A., and A. Parmeggiani, Toward a model for the interaction between elongation factor Tu and the ribosome. *Science* 259:1311–1314, 1993. Revises the energetics of protein synthesis: two GTPs (instead of one) for EFTu to bring aminoacyl-tRNA to the A site on the ribosome.

Wickner, W. T., How ATP drives proteins across membranes. *Science* 266:1197–1198, 1994.

Wolin, S. L., From the elephant to *E. coli:* SRP-dependent protein targeting. *Cell* 77:787–790, 1994. This minireview describes the bacterial analog of the SRP system.

Yanofsky, C., Gene structure and protein structure. *Scientific American* 216(5):80–94, 1967. A classic description of colinearity of coding regions of a gene and its specified protein product.

Zhang, S., G. Zubay, and E. Goldman, Low-usage codons in *E. coli,* yeast, fruit fly and primates. *Gene* 105:61–72, 1991. Reviews biased codon usage and possible functions.

Problems

1. Assume that you have a copolymer with a random sequence containing equimolar amounts of A and U. What amino acids would be incorporated, and in what ratio, when this copolymer is used as an mRNA?
2. A single tRNA can insert serine in response to three different codons: UCC, UCU, or UCA. What is the anticodon sequence of this tRNA?
3. The relationship between tRNAs and their synthases is sometimes called the "second genetic code." Explain.
4. For each of the six leucine codons, write down all possible $tRNA^{Leu}$ anticodons.
5. The *su3* mutation of *E. coli* is an altered tRNA that "suppresses" UAG nonsense mutations by inserting tyrosine in response to this termination codon. What is the most likely mutation in this strain? Explain how the cells survive despite this alteration in a tRNA-coding gene.
6. The AUA codon for isoleucine is rarely used in highly expressed genes in *E. coli.* Can you predict from this observation which $tRNA^{Ile}$ isoacceptors might be poorly represented in *E. coli* cells?
7. Draw the structures of the first dipeptides made in bacterial protein synthesis reactions when the initiation codon is followed by codons for either alanine or leucine. Which part of the peptide, and what kind of chemical bond, forms the attachment to tRNA?
8. During protein synthesis, GTP hydrolysis is involved in translocation and in transport of aa-tRNA to the ribosome, but not in peptide bond formation. Where does the energy for peptide bond formation come from?
9. Which of the following mRNA sequences is from a prokaryotic source and which is from a eukaryotic organism? Using what you know about initiation of protein synthesis and the genetic code, deduce the amino acid sequence of the polypeptides encoded by each of these mRNAs.
   (a) m^7GpppGCUUGUGUCCUCAUGGCUUUUACAAAACAGUG-GCGUCCAGAUUAAACCAUUCAGGCUU
   (b) CCUCAACUAUGGAGGUCAGCCAUGGCAUUCACUAAG-CAAUGGAGGCCUGACUGAGCUUACCGUA
10. The effect of single-point mutations on the amino acid sequence of a protein can provide precise identification of the codon used to specify a particular residue. Assuming a single base change for each step, deduce the wild-type codon in each of the following cases.
    (a) Gln → Arg → Trp

    (b) Glu → Lys → Ile

    (c) Leu
↙ ↓ ↘
Ser Val Met

    (d) Thr
↙ ↓ ↘
Ile Pro Lys
11. Researchers often design degenerate oligonucleotides based on a protein sequence for use as hybridization probes to isolate the corresponding gene. (A degenerate oligonucleotide is actually a mixture of oligonucleotides whose sequences differ at positions corresponding to degeneracies in the genetic code.) The N-terminal amino acid sequence of a protein is

    Met-Val-Asp-Ser-Asn-Trp-Ala-Gln-Cys-Asn-Pro-Ala-Thr

    Give the sequence of the least degenerate 20-residue-long oligonucleotide that hybridizes to the gene encoding this protein.
12. Why does tyrosyl-tRNA synthase not bind phenylalanine as effectively as it binds tyrosine? (Refer to figure 34.10 to answer this question.)
13. Although the roles of IF-2, EF-Tu, EF-G, and RF-3 in protein synthesis are quite different, all four of these proteins share a domain with significant amino acid sequence similarity. Suggest a role for this conserved domain.
14. The antibiotic fusidic acid inhibits protein synthesis by preventing EF-G from cycling off of the ribosome. Fusidic acid resistant mutants of EF-G have been isolated. Fusidic acid resistance is recessive to sensitivity. In other words, an *E. coli* cell containing two EF-G genes, one resistant and one sensitive, will still be sensitive to the antibiotic. Why? (*Hint:* Look at figure 34.16.)
15. What are the major differences between the mechanisms of protein import into endoplasmic reticulum and those of protein import into mitochondria?
16. It has been observed that in some cases a single gene encodes closely related proteins found in mitochondria and the cytoplasm. Propose a mechanism utilizing two promoters whereby differently compartmentalized proteins could be produced from a single gene.
17. What are the possible amino acid changes that could result from a single nucleotide change in a GAA codon? Knowing the structures of the amino acids, what would you predict the effects of the altered amino acids to be?
18. Why are PEST sequences more likely to be found in regulatory proteins than in structural proteins?
19. Explain why toxins such as ricin and α-sarcin are such potent toxins, that is, act at such low concentrations (one molecule per cell). What is unusual about this group of toxic molecules?
20. An *E. coli* strain bears amber mutations in both the arabinose and tryptophan operons. Introduction of the genetic locus *su1* (coding for a serine amber suppressor tRNA) leads to a phenotype in which the cells no longer require tryptophan but are still unable to metabolize arabinose. Conversely, introduction of the *su2* locus (which codes for a glutamine amber suppressor tRNA) leads to a phenotype in which the cells no longer require tryptophan but now can metabolize arabinose. How would you account for these observations?

CHAPTER

Regulation of Gene Expression in Prokaryotes

In bacteria the level of expression of a particular messenger RNA is a function of the affinity of the RNA polymerase for a DNA promoter; this affinity is modulated by regulatory proteins that bind to the DNA or the RNA polymerase.

In all biological systems, gene expression is regulated so that gene products are produced either before or as they are needed. In this chapter we examine the mechanisms that ensure efficient regulation in the bacterium *Escherichia coli* and the bacteriophage λ.

E. coli maintains all of its genes in a state where they can be turned on or turned off on short notice. The short messenger lifetime makes it possible to control gene expression from the transcription level. The lack of separate compartments for RNA and protein synthesis has fostered mechanisms where translation actually exerts a direct role on transcription. These are some of the special features that have influenced the evolution of regulatory systems in *E. coli.*

Control of Transcription Is the Dominant Mode of Regulation in *Escherichia coli*

The *E. coli* chromosome contains about 3,000 genes. This system is regulated so that under conditions of active growth, only about 5% of the genome is actively transcribed at any given time. The remainder of the genome is either silent or transcribed at a very low rate. When growth conditions change, some active genes are turned off, and other, inactive genes are turned on. The cell always retains its totipotency, so within a short time (seconds to minutes in most cases) and given appropriate circumstances, any gene can be fully turned on. The fully expressing rRNA gene makes one copy per second, a fully turned-on β-galactosidase gene makes about one copy per minute, and a fully turned-on biotin synthase gene makes about one copy every 10 min. In the maximally repressed state, all these genes express less than one transcript every 10 min.

The level of transcription for any particular gene usually results from a collection of control elements organized into a hierarchy that coordinates all the metabolic activities of the cell. For example, when the rRNA genes are highly active, so are the genes for ribosomal proteins, and the latter are regulated in such a way that stoichiometric amounts of most of the ribosomal proteins are produced. When glucose is abundant, most genes involved in processing more complex carbon sources are turned off by a process called catabolite repression. If the glucose supply is depleted and lactose is present, then the genes involved in lactose catabolism are expressed. In *E. coli* the production of most RNAs and proteins is regulated primarily if not exclusively at the transcriptional level, although notable exceptions occur. Rapid response to changing conditions is ensured partly by a short mRNA lifetime—on the order of 1–3 min for most mRNAs. Some mRNAs have appreciably longer lifetimes (10 min or longer) and the consequent potential for much higher levels of protein synthesis per mRNA subject to translational control. Examples of all these situations are considered. Finally, the fine-level control for any particular enzyme system is subject to regulation by activators or inhibitors (see chapter 10).

Figure 35.1

Schematic diagram of DNA conformation in the rapid-start complex. The consensus sequences for two regions of DNA that are important for polymerase binding are shown.

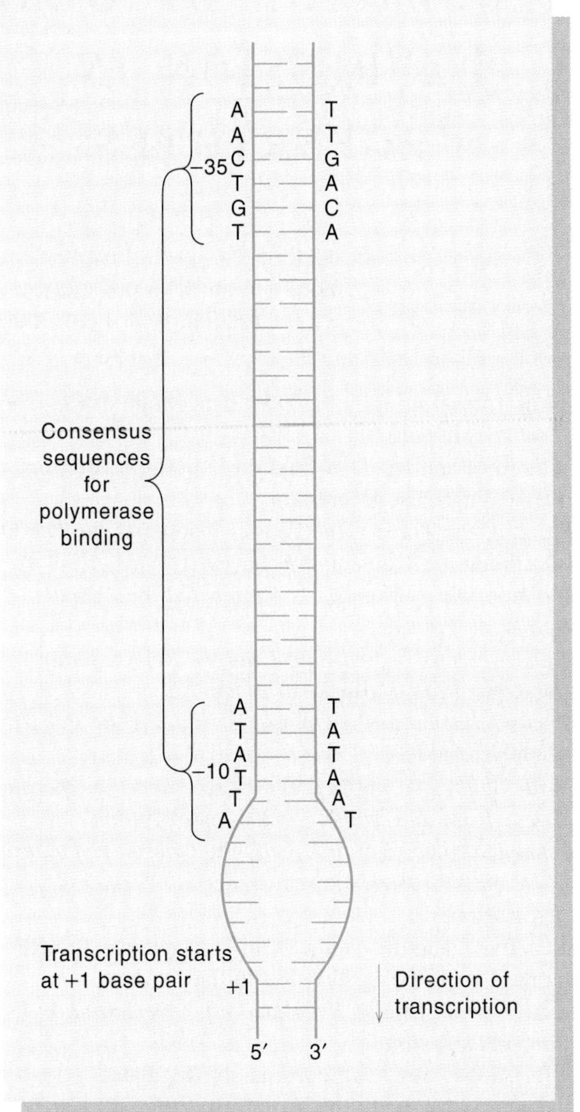

The Initiation Point for Transcription Is a Major Site for Regulating Gene Expression

The rate of initiation of transcription can be regulated in several ways, most of which influence the rate of formation of the RNA polymerase–DNA promoter complex. The primary sequence of nucleotides in the promoter region is the first factor to be considered. The closer this sequence is to the consensus sequence, the greater is the affinity of the polymerase for the promoter (fig. 35.1).

The rate of initiation of transcription also can be altered by changes in the RNA polymerase structure. This can occur by subunit replacement, subunit covalent modification, or small-molecule-induced allosteric transition. During a temperature upshift ($30 \rightarrow 42°C$) the usual σ subunit (σ^{70}) is partially replaced by an alternative σ factor (σ^{32}), changing the types of promoters recognized by the polymerase. In bacteriophage T4 infection the subunits of the polymerase become ribose-adenylated, lowering the affinity of polymerase for bacterial promoters and raising the affinity for phage promoters. Binding of guanosine tetraphosphate (ppGpp) to RNA polymerase changes the structure of the polymerase so that it has a greatly lowered affinity for rRNA, tRNA, and ribosomal protein promoters and at the same time a somewhat greater affinity for some other promoters.

Finally, the rate of initiation of RNA synthesis can be controlled by auxiliary regulatory proteins that affect the rate of formation of the polymerase–promoter complex. According to their positive or negative action on gene expression, regulatory

Figure 35.2

Different genetic elements of the *lac* operon. The operon contains a control region, the promoter–operator region, and three structural genes, *z*, *y*, and *a*. The *i* gene, a repressor, is also shown. It is not part of the operon, but it is located at an adjacent site on the genome with its own promoter.

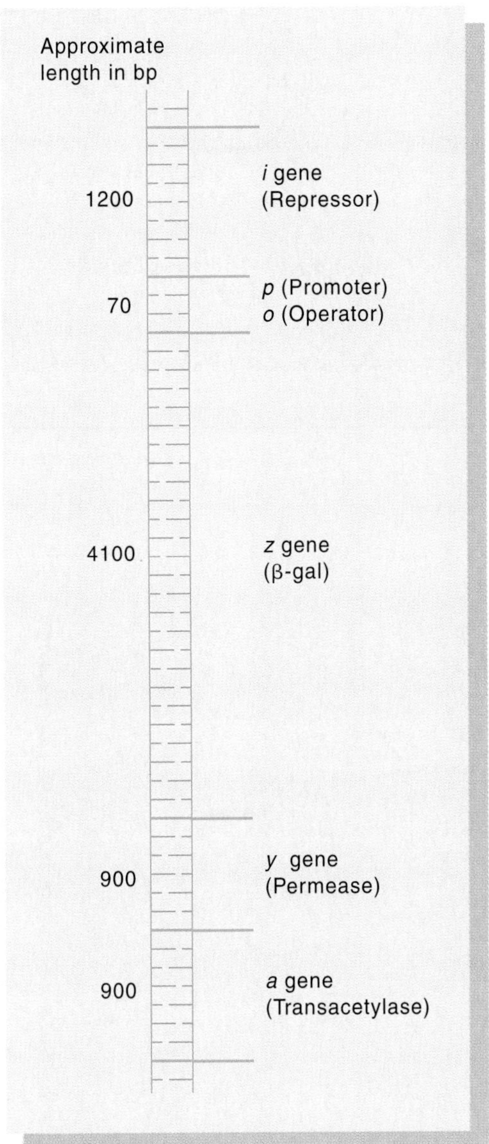

proteins are known as activators or repressors, respectively. Activators augment polymerase binding to the promoter, whereas repressors have the opposite effect.

Regulation of the Three-Gene Cluster Known as the *Lac* Operon Occurs at the Transcription Level

In bacteria it is common to find units of expression that contain clusters of two or more functionally related genes. Such is the case with the *lac* operon, a three-gene cluster associated with the metabolism of the dissacharide lactose (fig. 35.2). The first of these genes, the *z* gene, encodes β-galactosidase, which hydrolyzes β-galactosides, in particular lactose, to produce the monosaccharides glucose and galactose. The middle gene, *y*, encodes lactose permease, which is associated with the active transport of lactose into the cell. The third gene, *a*, encodes thiogalactoside transacetylase. A useful function has yet to be found for this gene.

Figure 35.3

Effect of inducer on β-galactosidase synthesis. Differential plot expressing accumulation of β-galactosidase as a function of increase in mass of cells in a growing culture of *E. coli*. Because the abscissa and ordinate are expressed in the same units (micrograms of protein), the slope of the straight line gives galactosidase as the fraction (*P*) of total protein synthesized in the presence of inducer. (*Source:* After Melvin Cohn, *Bact. Rev.* 21:140, 1957.)

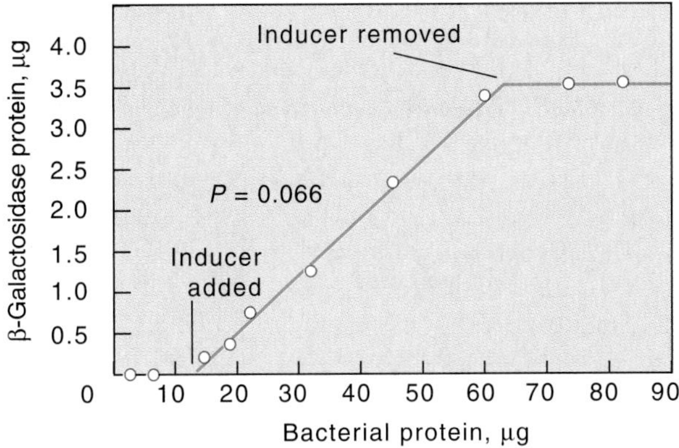

Expression of the *lac* operon is regulated by controlling elements, which are separate from the structural genes. The controlling elements consist of a promoter locus, which is the site where RNA polymerase binds and initiates transcription; the promoter locus also contains sites for the binding of a repressor and an activator. The *i* gene encodes the repressor, and the *crp* gene encodes the activator.

β-Galactosidase Synthesis Is Augmented by a Small-Molecule Inducer

Wild-type *E. coli* cells grown in the absence of lactose contain an average of 0.5–5.0 molecules of β-galactosidase per cell, whereas bacteria grown in the presence of an excess of lactose or certain lactose analogs contain 1,000–10,000 molecules per cell. Radioactive amino acid has been used as a tracer to show that the increase in enzyme activity observed on induction results from de novo protein synthesis. When excess β-galactoside inducer is added, enzyme activity increases at a rate proportional to the increase in total protein within the culture (fig. 35.3). Enzyme formation reaches its maximum rate within 3 min after inducer is added. Removal of inducer leads to cessation of enzyme synthesis in about the same amount of time.

A large number of compounds have been tested for their capacity to induce β-galactosidase. All inducers contain an intact, unsubstituted galactosidic residue (fig. 35.4). Many compounds that are not themselves substrates for β-galactosidase, such as thiogalactosides, are good inducers (such compounds are called gratuitous inducers because they are not substrates for the enzyme).

Figure 35.4

Inducers of the *lac* operon. (*a*) All inducers have the β-galactoside structure shown, in which R can be a variety of substituents. (*b*) Allolactose is the natural inducer when cells are grown on lactose. (*c*) Isopropyl-β-D-thiogalactoside (IPTG) is a synthetic inducer useful in the laboratory; the β-oxygen is replaced by a β-sulfur atom. This change prevents hydrolysis by β-galactosidase.

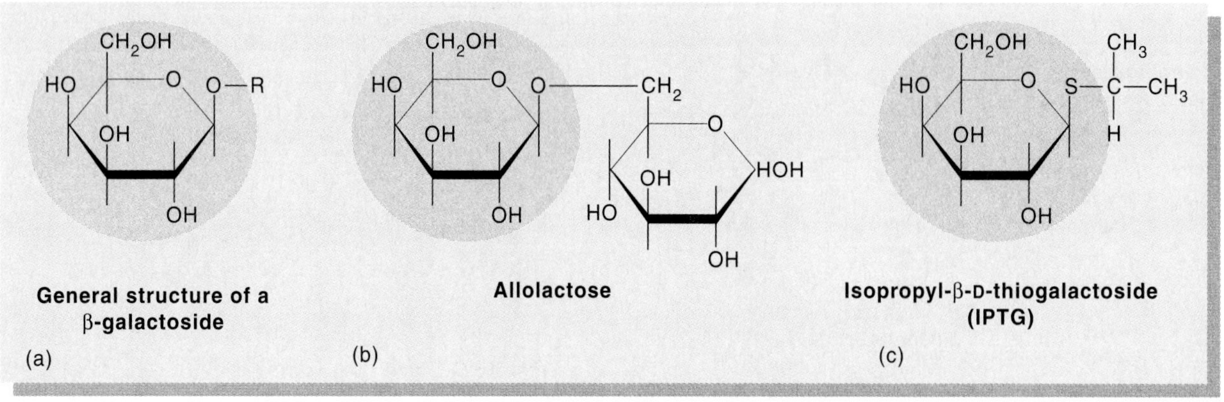

Figure 35.5

Conversion of lactose to allolactose, the natural inducer of the *lac* operon. Ultimately, lactose is broken down to its constituent monosaccharides, galactose and glucose.

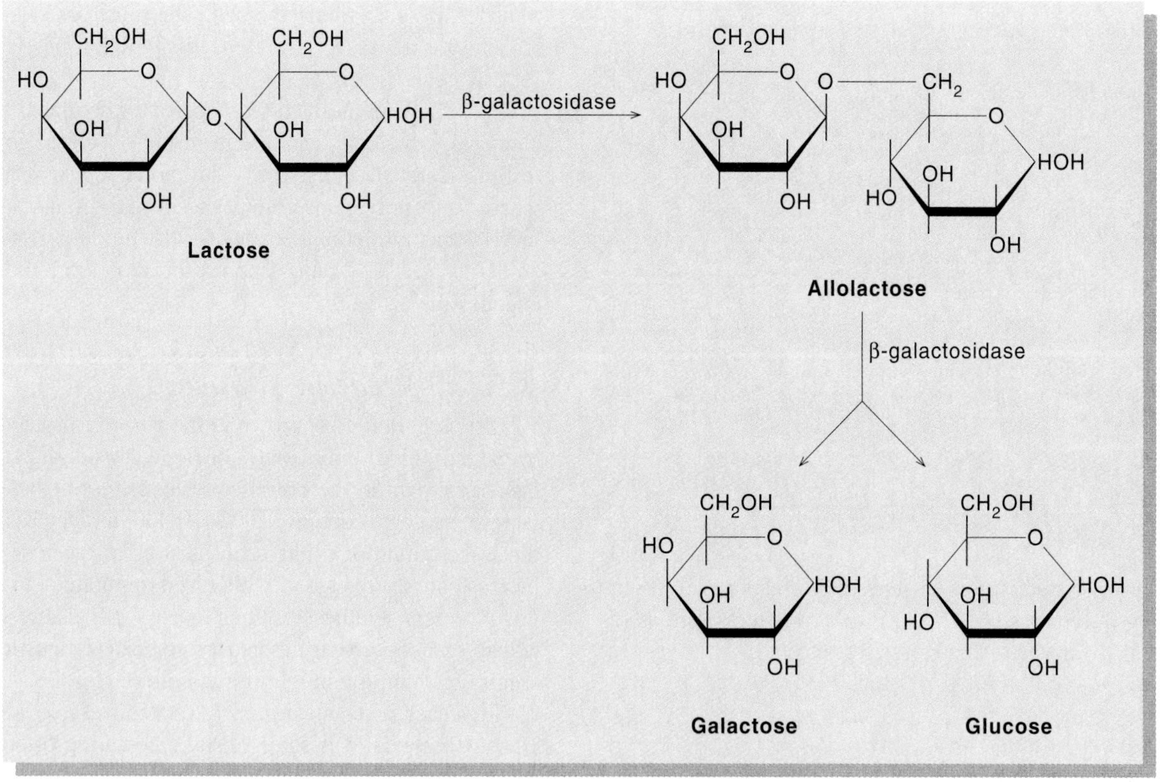

No correlation exists between affinity for β-galactosidase and the capacity to induce. Lactose, the natural substrate of the operon, is not an inducer in vivo. Rather, allolactose, which is formed as an intermediate in lactose metabolism in the presence of the very limited amount of β-galactosidase that exists in uninduced cells, is believed to be the natural inducer (fig. 35.5). The three proteins of the *lac* operon are coordinately induced, that is, they are induced to the same extent by the same inducer. These results suggest that the receptor for the inducer is distinct from the proteins encoded by the operon, and the inducer acts at one site.

35A
BOX

Genetic Concepts and Genetic Notation

Much of the early work on the *lac* operon was purely genetic. It is essential that certain aspects of genetics be understood. The information in this box should be adequate for those with no prior exposure to genetics except for what they have already encountered in this text.

Genes are specified by one or more small letters in italic. Thus z indicates the gene for β-galactosidase, and *lac* indicates the operon. Frequently, a superscript is appended to the genetic symbol. The two most common superscripts are +, indicating a normal (wild-type) gene, and −, indicating a nonfunctioning (mutant) gene. Different representations of the same gene are referred to as alleles. Thus z^+ and z^- are both alleles of the z gene.

Cells that carry a single copy of each gene are referred to as haploids. Cells that carry two copies of each gene are referred to as diploids. Bacteria are haploid cells because they carry a single chromosome with a unique representation for each gene. Bacterial cells that are partial diploids (merodiploids) may occur naturally, or they may be selected for by genetic techniques.

A favored method for constructing merodiploids is to infect the bacterial cell with a virus or a plasmid DNA that carries the extra genes of interest. The F plasmid is commonly used for this purpose. In strict usage the genetic representation for a cell carrying the *lac* operon on the chromosome and the F plasmid would be $z^+y^+a^+//Fz^+y^+a^+$, where the diagonal lines separate the host chromosome to the left and the plasmid chromosome to the right. As a rule, however, only one diagonal line is used to separate the genetic symbols. Also, for convenience, if all the alleles for a given gene are wild type, they may not be shown. Thus $z^+y^+a^+/Fz^-y^+a^+$ and z^+/Fz^- may be taken as representations of the same genetic state in cases in which it is understood that the *lac* operon is present on both the host and the plasmid chromosome.

Two genetic elements located on the same chromosome are said to be in the *cis* orientation. Two genetic elements located on different chromosomes in the same cell are said to be in the *trans* orientation. In the merodiploid $z^-y^-a^+/Fz^+y^+a^+$ the two mutant genes are in the *cis* orientation. In the merodiploid $z^-y^+a^+/Fz^+y^-a^+$ they are in the *trans* orientation.

A major reason for using merodiploids is to study the interaction between different alleles of the same gene. This often tells us a great deal about how a gene or the gene product functions. The two simplest types of interactions are dominant and recessive. A cell that is z^+/Fz^- behaves like a z^+ cell as far as the metabolism of β-galactosidase is concerned. Therefore the z^+ allele is dominant to the z^- allele, or, conversely, the z^- allele is recessive to the z^+ allele.

A Gene Was Discovered That Leads to Repression of Synthesis in the Absence of Inducer

Two distinct types of mutations have been observed in the genes associated with the *lac* operon. One class of mutations includes structural gene mutations: (1) β-galactosidase mutations ($z^+ \rightarrow z^-$), expressed as the loss of the capacity to synthesize active β-galactosidase; (2) permease mutations ($y^+ \rightarrow y^-$), expressed as the loss of the capacity to concentrate lactose; and (3) transacetylase mutations ($a^+ \rightarrow a^-$), expressed as the loss of the capacity to form thiogalactoside transacetylase. The other class of mutations involves controlling elements of the operon such as i gene mutations ($i^+ \rightarrow i^-$), expressed as the capacity to synthesize large amounts of β-galactosidase even in the absence of inducer. This type of i gene mutation is called a constitutive mutation. Structural mutations usually affect only the enzyme in whose gene the mutation occurs. In contrast, constitutive mutations invariably affect the amounts of all three structural gene products but not their structures (at this point you may wish to refer to box 35A for more information on genetic concepts and notation).

The most informative genetic studies were performed on cells containing two copies of the *lac* operon with point mutations in different genes. Partial diploids (merodiploids) of this sort are constructed by mating experiments in which a second copy of the *lac* region is incorporated into the test cell by mating. For this purpose the F factor plasmid is a favorite. Merodiploids of the type $z^+y^-a^-/Fz^-y^+a^+$ or $z^-y^+a^+/Fz^+y^-a^-$ behave like normal wild-type cells with respect to the expression and metabolism of the *lac* operon. This demonstrates that the distribution of normal genes and mutant genes on the two chromosomes does not influence the phenotype as long as there is at least one functional gene of each type. This tells us that each of the structural genes behaves as an independent entity not affected by other genes on the same operon.

The study of merodiploids of the types i^+z^-/Fi^-z^+ gives the same inducible phenotype. This demonstrates that the i^+ inducible allele is dominant to the i^- constitutive allele and that it is active on the same chromosome (*cis*) or on a different chromosome (*trans*) with respect to the structural gene it influences (table 35.1). In fact the i^+ gene could be moved to any location on the chromosome and it would still show dominant behavior. The fact that the influence of the i gene is not sensitive to location suggests that i gene action results from a diffusible gene product. The dominance of the inducible i^+ allele to the constitutive i^- allele suggests that the former corresponds to the active form of the i gene.

Further understanding of i gene function has come from study of rare mutations designated i^s. Mutants bearing this allele are noninducible, meaning that they have lost their capacity to ex-

Table 35.1
Expression of β-Galactosidase as a Function of Genotype

Genotype[a]	Phenotype	
	−Inducer	**+Inducer**
$i^+o^+z^+$	−	+
$i^-o^+z^+$	+	+
$i^+o^+z^-$	−	−
$i^+o^+z^+/i^-o^+z^+$	−	+
$i^so^+z^+/i^+o^+z^+$	−	−
$i^+o^cz^+$	+	+
$i^+o^cz^+/i^+o^+z^+$	+	+
$i^+o^cz^+/i^+o^+z^-$	+	+
$i^+o^cz^-/i^+o^+z^+$	−	+

[a]The diagonal indicates that two *lac* operons, including the i^{1}, are present in the same cell. Under "phenotype," a + indicates a high level expression of β-galactosidase, and a − indicates low level or the absence of expression.

press the structural gene products of the operon. In merodiploids of the constitution i^+/i^s, the i^s allele is dominant, that is, the merodiploids cannot synthesize structural gene products even in the presence of inducer (see table 35.1). The most likely explanation for the i^s mutant is that it is an allele of i in which the repressor is not influenced by the inducer, so it always represses.

A Locus Adjacent to the Operon Is Found to Be Required for Repressor Action

The vast majority of constitutive mutants result from i^- mutations. However, occasional constitutive mutants have been mapped outside of the i gene in the region of the *lac* operon promoter. Rare mutants of this type designated o^c are much easier to isolate by selection for constitutivity in cells diploid for the i^+ gene. This selection procedure minimizes the chance of finding constitutive mutants that result from i gene mutations because both copies of the i^+ gene would have to mutate to i^- simultaneously to give a constitutive phenotype. If the probability of an $i^+ \rightarrow i^-$ mutation is 10^{-6}, the probability of two such events occurring simultaneously in the same cell is 10^{-12}, which is extremely unlikely. As a result, constitutive mutations obtained under these circumstances are almost always of the o^c type. Like i^- mutations, o^c mutations affect the quantity of β-galactosidase synthesized but not its structure.

In merodiploids of the type o^+z^+/o^cz^+, β-galactosidase is constitutively expressed, showing that the o^c allele is dominant to o^+ in this situation. In merodiploids of the type o^cz^+/o^+z^- the o^c-allele is dominant also; but in o^cz^-/o^+z^+ it is recessive. Thus the o^c-allele is dominant only when it is located *cis* to the structural genes it influences. From this result, François Jacob and Jacques Monod inferred that $o^+ \rightarrow o^c$ mutations correspond to a modification of the DNA structure that affects the ability of the repressor to bind.

Figure 35.6

Schematic model illustrating the operon hypothesis. This diagram is modified from the original proposed by Jacob and Monod, who thought i gene repressor was an RNA rather than a protein. (*a*) The i gene encodes a repressor that binds tightly to the operator o locus, thereby preventing transcription of the mRNA from the z, y, and a structural genes. (*b*) When inducer is present, it combines with repressor, changing its structure so it can no longer bind to the operator locus. Inducer also can remove repressor already complexed with the o locus.

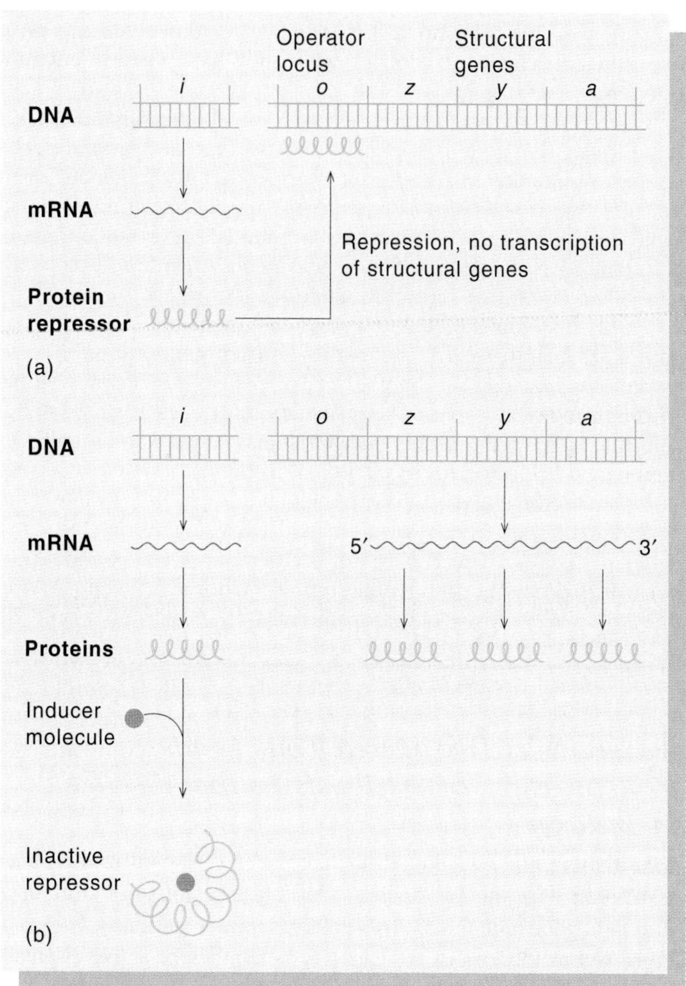

Genetic Studies on the Repressor Gene and the Operator Locus Lead to a Model for Repressor Action

The behavior of the various mutations we have just discussed led Jacob and Monod to propose a model for the regulation of protein synthesis. The genetic elements of this model consist of a structural gene or genes, a regulator gene, and an operator locus (fig. 35.6).

1. The structural gene produces an mRNA that serves as a template for protein synthesis.
2. The regulator gene (not itself part of the operon) produces a repressor that can interact with the operator locus.

Figure 35.7

The operator locus (the presumptive repressor binding site). Bases are numbered +1 for the first base transcribed and −1 for the base before that. Regions showing dyad symmetry are underlined and overlined. Arrows indicate point mutations leading to the constitutive phenotype (o^c). Circled bases are those groups that are strongly protected against reaction with dimethylsulfoxide when *lac* repressor is bound. Shaded circles indicate those groups that become cross-linked to repressor in the presence of ultraviolet light when thymine in the DNA is replaced by 5-bromouracil.

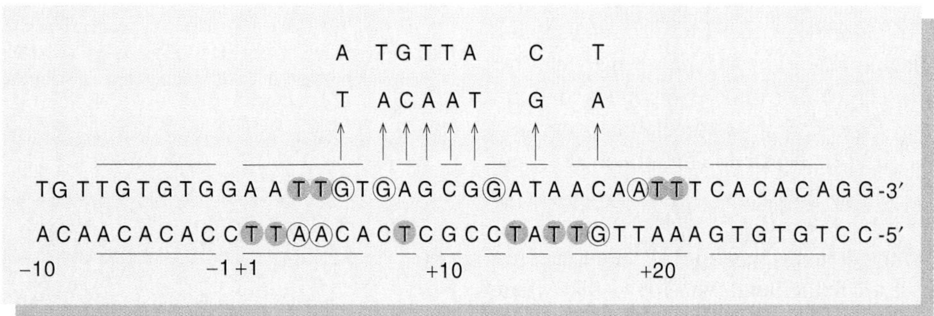

3. The operator is always adjacent to the structural genes it controls.
4. The operator and its associated structural genes are referred to as the operon.
5. The repressor molecule combines with the operator locus to prevent the structural gene(s) from synthesizing mRNA. In induction the inducer combines with the repressor, changing its structure so that it no longer binds to the operator; this region of the genome is then free to combine with RNA polymerase.

The Jacob-Monod operon hypothesis provided a tremendous stimulus for investigations directed toward understanding not only the *lac* system but other genetic regulatory systems as well.

Biochemical Investigations Verify the Operon Hypothesis

Genetic studies led to the operon hypothesis. Biochemical investigations were essential to provide direct evidence for the hypothesized properties of the repressor. The first task was to isolate the repressor.

According to the operon hypothesis, inducer is supposed to bind to repressor. Walter Gilbert and Benno Muller-Hill used ^{14}C-labeled isopropyl-β-D-thiogalactoside (IPTG), the strongest known inducer (see fig. 35.4), to monitor repressor purification from a crude cell extract. IPTG has a further advantage for such studies in that it is completely stable. A crude extract of disrupted cells was fractionated by standard protein purification procedures, and the fraction containing the ^{14}C-IPTG-labeled product was isolated.

Several properties of normal and abnormal repressor were studied. The repressor was found to bind strongly to the *lac* promoter. This binding was disrupted by adding inducer. Promoter DNA containing an o^c mutation was not effective in binding repressor. Only repressor prepared from i^+ cells was effective in binding to DNA.

A more detailed characterization of the operator binding site was made. Eight o^c mutations were found to involve base replacements near the center of this region. The reactivity of wild-type operator to various chemical agents was determined in the presence and in the absence of repressor. Repressor binding substantially decreases the reactivity toward dimethylsulfate of several purine bases in the operator region (fig. 35.7). Dimethylsulfate reacts with N-7 of guanine and the N-3 of adenine in the double helix. Finally, if all the thymines in the DNA are replaced by bromouracil, a number of the bromouracil bases become cross-linked to the bound repressor in the presence of ultraviolet light. Taking into account the normal twist of the double helix, all the groups shown by chemical methods to be in the vicinity of the repressor are situated on one side of the double helix.

The sequence of bases in the operator region shows a remarkable symmetry property. Twenty-eight out of 36 of the base pairs in this region of the promoter are located on a twofold (dyad) axis of symmetry (see fig. 35.7). We shall see that dyad symmetry is a common property for repressor or activator binding sites and that it relates to the way in which DNA and regulatory proteins interact.

Biochemical proof that repressor inhibits operon expression was shown by M. Lederman, J. DeVries, and G. Zubay in a cell-free system in which most of the components were prepared from an i^- extract. Addition of repressor to such an extract inhibited the synthesis of β-galactosidase. This inhibition was reversed by addition of IPTG inducer.

The detailed biochemical studies not only confirmed the Jacob-Monod hypothesis, they gave further details about the nature of repressor interaction and a detailed characterization of the repressor binding site. The repressor binds mainly to one side of the double helix, over a 36-base region covered by the symmetry axis. It inhibits expression because it binds to a site that overlaps the polymerase binding site.

An Activator Protein Is Discovered That Augments Operon Expression

Although the genetic and biochemical studies on the action of repressor on the *lac* operon answered many questions about gene expression of the *lac* operon, they left equally important questions unanswered. It had been known since the turn of the century that the *lac* operon expresses at a greatly reduced level if lactose and glucose are present simultaneously. Either of these sugars can be used by the bacterium as a source of carbon compounds and energy, but the lactose is not utilized to any appreciable extent until the glucose supply has been exhausted. This effect is called catabolite repression. As long as glucose is available, lactose is underutilized.

A turning point in our understanding of catabolite repression was provided by Earl Sutherland, who found that when glucose was added to growing *E. coli* cells, the level of 3′,5′-cAMP (cAMP) was drastically reduced. Could the lack of cAMP be responsible for the poor expression of the *lac* operon in the presence of glucose? In support of this, A. Ullman and J. Monod and I. Pastan and R. Perlman found that large quantities of cAMP added to the growth medium could partially reverse the glucose catabolite repression effect. In a cell-free system containing crude extracts from *E. coli* and DNA containing the lac operon, D. Chambers and G. Zubay found that the low-level expression of the *lac* operon could be greatly increased by addition of cAMP. This provided support for the notion that cAMP was playing a direct role in activating the *lac* operon. Further investigations were facilitated by Jonathan Beckwith's genetic studies and the isolation of key mutants relating to the action of cAMP.

Beckwith and his colleagues isolated a large family of mutants that were permanently catabolite-repressed. These mutants fell into two categories: those that could be phenotypically corrected by growing in the presence of cAMP and those that could not. Mutants in the first class were believed to be defective in the synthesis of cAMP; those in the latter class were presumed to be defective in the protein receptor for cAMP. Cell-free extracts were prepared from both of these mutants. When used for cell-free synthesis of β-galactosidase, mutants of the first type were found to be greatly stimulated by addition of cAMP, confirming the belief that these mutants were defective in the synthesis of cAMP but nothing else. When extracts from mutants of the second type were used, cAMP had no stimulating effect, suggesting that a protein necessary for cAMP action was missing or defective. Further cell-free studies were performed in which mutants of the second type were used in conjunction with partially purified extracts from a normal strain. Addition of small amounts of extracts from a normal strain reestablished the stimulatory effect of the cAMP. The purification of the cAMP receptor protein was monitored with this system. Ultimately, a single protein called CAP was found to be responsible for the effect. Soon afterwards, it was found that CAP is a dimer composed of identical subunits, each with an M_r of 22,000. CAP binds to DNA and this binding is greatly stimulated in the presence of cAMP. The cAMP apparently alters the conformation of CAP so that it can form a strong complex with DNA at the *lac* promoter region.

Figure 35.8

CAP-binding region in the *lac* promoter. Regions showing dyad symmetry that are believed to interact strongly with CAP are overlined and underlined. Circled bases are those that are strongly protected from reaction with dimethylsulfate when CAP is bound. Point mutations L8 or L29 result in a promoter that is not stimulated in transcription by cAMP. Red dots indicate those phosphate positions that, if ethylated by ethylnitrosourea, block CAP binding. If the area of interest exists as a normal DNA double helix when binding CAP, then most of the groups implicated in CAP binding appear on one side of the DNA.

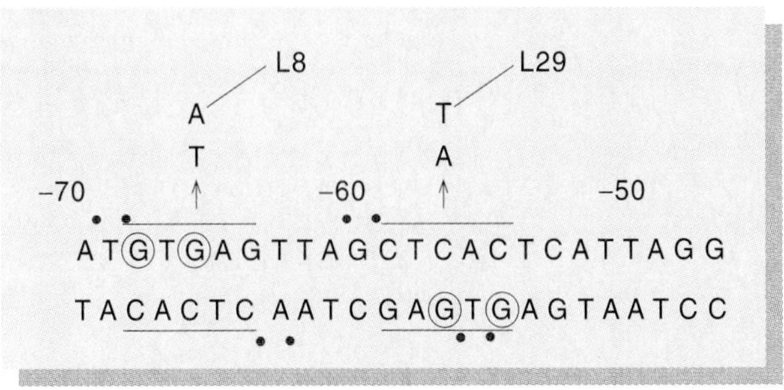

A series of genetic deletions was used by Beckwith to demonstrate that the site necessary for CAP stimulation of the *lac* operon is in the −50−−80 region of the *lac* operon. A 14-bp segment between −53 and −68 shows dyad symmetry for 12 of the 14 bp (fig. 35.8).

In the absence of bound repressor the full sequence of reactions leading to initiation of transcription is summarized by the following set of equations. First, the coactivator cAMP combines with the activator CAP, which then binds in the −60 region of the promoter. This complex stimulates the binding of RNA polymerase to an adjacent site on the promoter. CAP stimulates binding of RNA polymerase by an affinity between CAP and polymerase. Current indications are that CAP interacts with the carboxy terminal domain of an α subunit of the RNA polymerase holoenzyme.

$$\text{cAMP} + \text{CAP} \rightleftharpoons \text{cAMP-CAP}$$

$$\text{cAMP-CAP} + \text{DNA} \rightleftharpoons \text{cAMP-CAP-DNA}$$

$$\text{cAMP-CAP-DNA} + \text{Polymerase} \rightleftharpoons \text{cAMP-CAP-DNA-polymerase}$$

The overall strategy in creating a promoter sequence responsive to cAMP-CAP activation can be summarized as follows. The nucleotide sequence in the RNA polymerase-binding site is adjusted so that polymerase by itself produces a low level of transcription. An adjacent site for binding CAP is created so that when CAP is binding, the additional affinity contributed by favorable contacts between the CAP and the polymerase convert this into a high-level promoter. The crucial importance of the promoter sequence in eliciting the polymerase response to CAP is shown by the fact that certain mutations that result in producing the consensus sequence for the promoter (see fig. 35.1) eliminate the catabolite repression effect for the *lac* operon in vivo as well as the need for cAMP in the cell-free system.

Figure 35.9

The tryptophan operon, indicating the location of the different genes, the polypeptide chains, the resulting enzyme complexes, and the reactions catalyzed by the enzyme complexes.

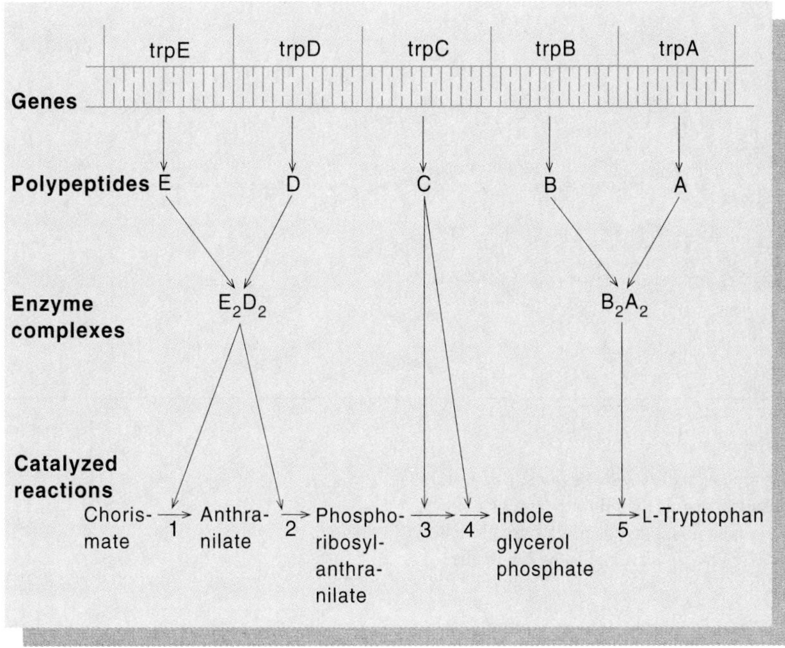

Enzymes That Catalyze Amino Acid Biosynthesis Are Regulated at the Level of Transcription Initiation

Among the genes concerned with biosynthetic processes, those uniquely involved in synthesis of the amino acid tryptophan are possibly the best understood. This is mostly a result of the efforts of Charles Yanofsky and his colleagues, who have used a wide variety of genetic and biochemical techniques to probe the complexities of this system.

Wild-type *E. coli* can synthesize all 20 types of amino acids from simpler substrates, but for most of them it does so only when they are not available in adequate amounts in the growth medium. For example, synthesis of enzymes for tryptophan synthesis is sharply reduced when the external tryptophan supply is high. Lowering available L-tryptophan selectively stimulates synthesis of the mRNA and associated enzymes (fig. 35.9). The five contiguous structural genes are transcribed as a single polycistronic mRNA. Initiation of transcription is regulated in part by the interaction of the tryptophan repressor, the protein product of the *trpR* gene, with its target site on the DNA, the *trp* operator, *trpO* (fig. 35.10). Binding of L-tryptophan to the repressor causes a structural alteration that is essential for strong specific binding to the *trpO* locus. Like *lac* repressor, the *trp* repressor binds at a site that overlaps the RNA polymerase-binding site, and the region where the repressor binds contains a number of bases arranged with dyad symmetry (fig. 35.11).

The most significant difference between the action of the *trp* and *lac* repressors relates to the function of the small-molecule effector. In the case of *lac* the effector molecule allolactose acts as an antirepressor (inducer), causing release of repressor from the operator; in the case of *trp* the effector molecule L-tryptophan acts as a corepressor, stimulating the binding of repressor to the operator. It should be obvious that the difference in action of these small-molecule effectors, the concentrations of which dictate the level of operon activity, is well suited to the different metabolic needs of the cell satisfied by the two operons.

The trp *Operon Is Also Regulated after the Initiation Point for Transcription*

From this point on, the close parallel between the regulation of the *trp* operon and the *lac* operon ends. The *trp* operon has no positive control system like cAMP-CAP, but it does have another means for regulating transcription at a site downstream from the initiation point. The existence of a second signal for regulation was first suspected when it was discovered that a genetic deletion of some of the bases between the initiation site for transcription and the first structural gene (*trpE*) raised the level of expression of the operon eight- to tenfold. This was true even in strains with a defective repressor gene ($trpR^-$ strains). How could this be, and what could be the mechanism of action? An important clue was provided by sequence analysis of the 162 bases in the *trp* leader region, that is, the region between the initiation site for transcription and the initiation site for translation of the first structural gene (fig. 35.12). This leader region contains a potential initiation codon

Figure 35.10

Schematic diagram of the repressor control of *trp* operon expression. The *trp* promoter (*P*) and *trp* operator (*O*) regions overlap. The *trp* aporepressor is encoded by a distantly located *trpR* gene. L-Tryptophan binding converts the aporepressor to the repressor that binds at the operator locus. This complex prevents the formation of the polymerase–promoter complex and transcription of the operon that begins in the leader region (*trpL*). Only a fraction of the transcripts extends beyond the attenuator locus in the leader region. The regulation of this fraction is discussed in the text.

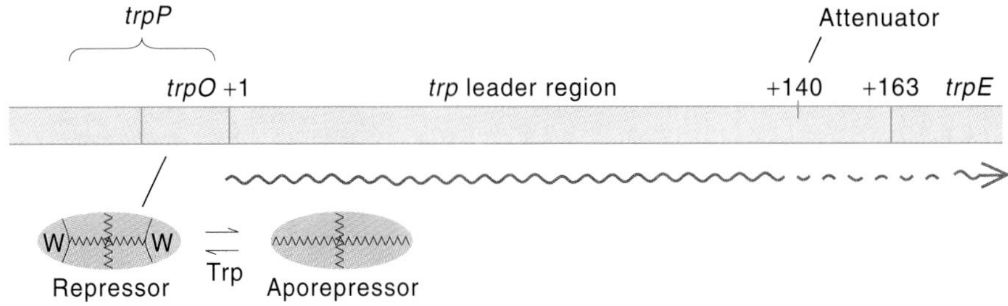

Figure 35.11

The promoter–operator region of the tryptophan operon. Two regions, PBS1 and PBS2, where polymerase binds are bracketed. Regions within the repressor-binding site showing dyadic symmetry are underlined and overlined. Single-base changes that lead to operator constitutive (o^c) mutants are indicated below the duplex.

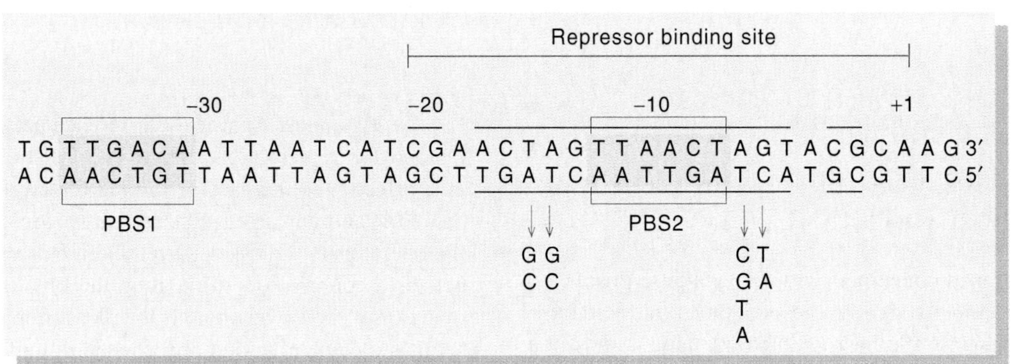

Figure 35.12

The leader region for the tryptophan operon. The region of the leader RNA containing the hypothesized leader polypeptide is shown. The translation start of the trpE protein is also shown.

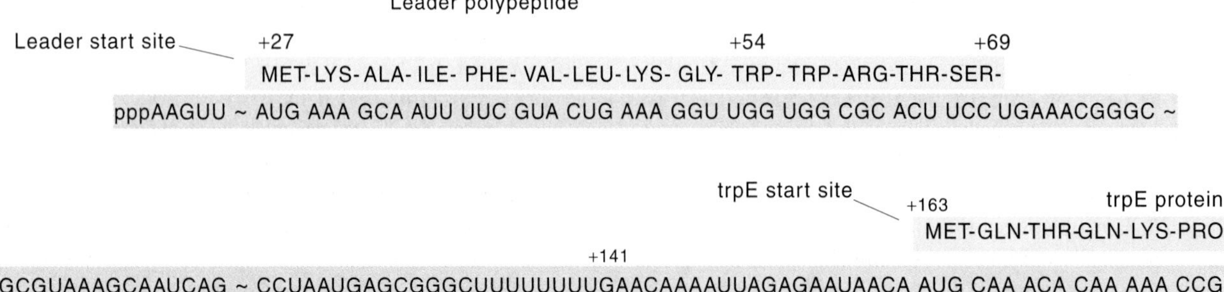

(bases 27–29), two tandem *trp* codons (bases 54–59), and a terminator codon (bases 69–71). A so-called leader peptide of 14 amino acids would result from translation of this region.

There are numerous reasons for believing that translation of the leader peptide up to or through the *trp* codons regulates attenuation of mRNA transcription. First of all, selective starvation of cells for tryptophan relieves attenuation and permits most RNA polymerase molecules to read through the leader region. The only other amino acid that relieves attenuation of the *trp* operon when it is lacking is arginine; arginine starvation is about 80% as effective as tryptophan starvation. It should be noticed that an *arg* codon is located adjacent to the two *trp* codons in the leader region. Most telling of all was the finding that a mutation resulting in the replacement of the AUG start codon by AUA, which should eliminate translation of the leader peptide, also prevents transcription beyond the attenuator.

Other experiments indicated that the fraction of tRNA that is charged with an amino acid is a crucial factor in the attenuation response. This has been examined in vivo by comparing the *trp* operon enzyme levels in *trpR*$^-$ strains that are otherwise normal with strains that are defective in some respect in charged $tRNA^{Trp}$. Such structural defects in $tRNA^{Trp}$ or in the charging enzyme elevates expression, probably by permitting polymerase to transcribe through the attenuator.

These results support the hypothesis that transcription read-through requires partial translation of the leader sequence. However, only if the translation pauses or stops in the region where the *trp* or *arg* codons occur is read-through favored. A careful examination of the secondary-structure possibilities in the attenuator region suggests why this is so. The leader region RNA between bases 50 and 141 has the potential to form a variety of base-paired conformations. Figure 35.13 illustrates the most likely secondary structures that form in terminated *trp* leader RNA. These are based on analysis of regions of the transcript that show resistance to RNase T1 digestion under mild conditions and the base pairing established by studies of defined oligonucleotides. Four regions of base pairing that can form three stem-and-loop structures have been proposed. Region 1, which includes the tandem *trp* codons and the leader peptide translation stop codon (bases 54–68), can base-pair with region 2 (bases 76–91). Although region 2 (bases 74–85) also should be able to base-pair with region 3 (bases 108–119), stem-and-loop 2 · 3 has not been observed in vitro, presumably because stem-and-loop 3 · 4 and stem-and-loop 1 · 2 form preferentially. Region 3 (bases 114–121) can base-pair with region 4 (bases 126–134). The existence of this stem and loop is inferred from T_1 RNase-resistance of the GC-rich region from residue 107 to the 3′ end of the transcript. The stem-and-loop structure formed between regions 3 and 4, followed by a sequence of U residues, is a common structure for a transcription terminator (see chapter 33). Hence it is expected that conditions under which this structure is preserved would favor transcription termination. In support of this a number of single-base replacement mutations have been isolated that lead to mispairing in the 3 · 4 stem; all of these lower the level of transcription termination in the leader to some extent.

Figure 35.13

Proposed secondary structures in the leader RNA. Four regions, labeled 1, 2, 3, and 4 at the left, can base-pair to form three stem-and-loop structures (1.2, 2.3, and 3.4). The arrows in the main figure mark the RNase T1 cleavage sites.

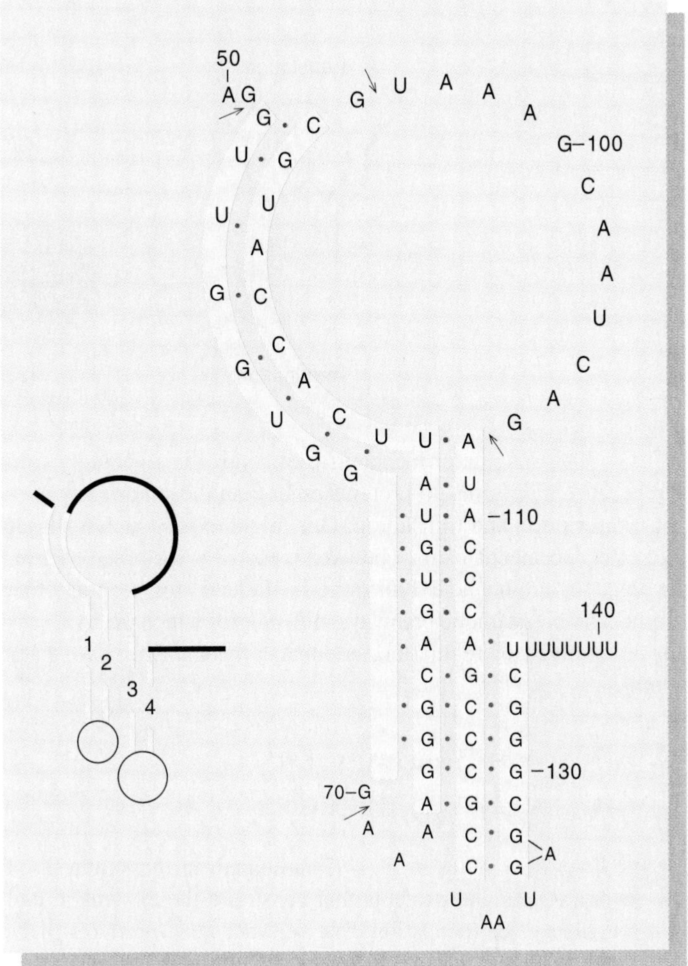

The absence of translation of the leader would not perturb this 1 · 2 and 3 · 4 structure (fig. 35.14*a*). Consistent with this, changing the initiator codon of the leader peptide by a single base prevents read-through, as discussed earlier. On the other hand, selective starvation resulting from either tryptophan or arginine deprivation stimulates transcription read-through. Most likely, this is because the ribosome stalls in the region of the *trp* or *arg* codons (bases 54–62). The resulting rupture of the base-paired 1 · 2 structure would make region 2 available for pairing with region 3. This would encourage disruption of the stem-and-loop 3 · 4 structure, resulting in transcription read-through (see fig. 35.14*b*). In the presence of an adequate supply of all amino acids, translation would proceed beyond this critical region, so region 2 would not be available for base pairing, and the 3 · 4 loop would be maintained, favoring transcription termination at the attenuator (see fig. 35.14*c*).

Figure 35.14

Model for attenuation in the *trp* operon, showing ribosome and leader RNA. (*a*) Where no translation occurs, as when the leader AUG codon is replaced by an AUA codon, stem-and-loop 3.4 is intact, and termination in the leader is favored. (*b*) Cells are selectively starved for tryptophan so that the ribosome stops prematurely at the tandem *trp* codons. Under these conditions, stem-and-loop 2.3 can form, and this is believed to lead to the disruption of stem-and-loop 3.4. (*c*) All amino acids, including excess tryptophan, are present so stem-and-loop 3.4 is present.

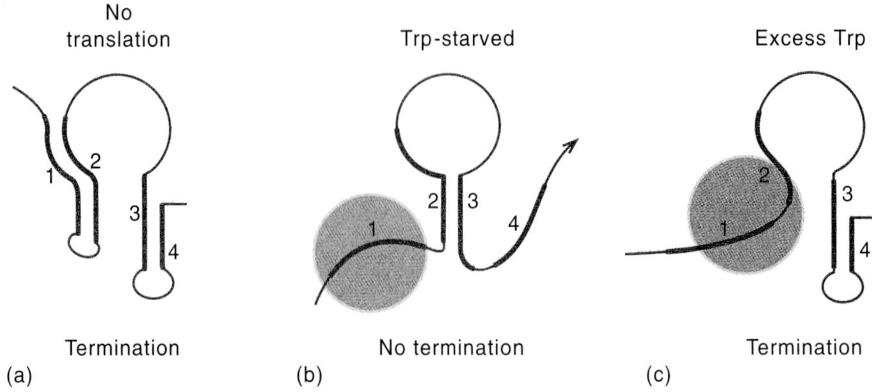

This attenuator mechanism of control is amazingly simple because it requires no proteins other than those normally used for transcription and translation. One might expect such a simple and effective mechanism to be used repeatedly for other operons involved in amino acid biosynthesis. Indeed, for several other amino acid biosynthetic pathways in *E. coli* for which tRNA charging is involved in regulation, attenuator mechanisms have been found.

Genes for Ribosomes Are Coordinately Regulated

Over 100 genes in *E. coli* participate in the synthesis of the RNAs and proteins that constitute the enzymatic machinery for translation. The relevant gene products make up between 20% and 40% of the dry cell mass. In rapidly growing cells, about 85% of the RNA is ribosomal, 10% is tRNA, and most of the remainder is mRNA. The various RNAs and proteins are produced according to need. For ribosomes and tRNAs this results in a synthesis rate that is roughly proportional to the cell growth rate; the relative amounts of the three rRNAs (16S, 23S, and 5S), the 60 or so tRNAs, and the 50 ribosomal proteins, are consistent with the stoichiometric needs for making ribosomes.

Control of rRNA and tRNA Synthesis by the rel *Gene*

Under conditions of rapid growth, *E. coli* cells contain about 10^4 ribosomes. The maximum rate of reinitiation at the ribosomal gene promoter is about one per second. In rapid growth, *E. coli* can duplicate once every 20 min, which would allow for the synthesis of only about 1,200 molecules of rRNA if there were only one gene for ribosomal RNA. In fact, there are seven copies for ribosomal RNA operons in the bacterial chromosome, which makes it possible for rRNA synthesis to maintain the necessary pace under conditions of rapid growth. These rRNA (*rrn*) operons are dispersed at seven locations around the circular *E. coli* chromosome.

Figure 35.15

A typical rRNA (*rrn*) operon contains two promoters and genes for 16S, 23S, and 5S rRNA and a single 4S tRNA gene. The four fully processed RNAs are derived from a single intact 30S primary transcript.

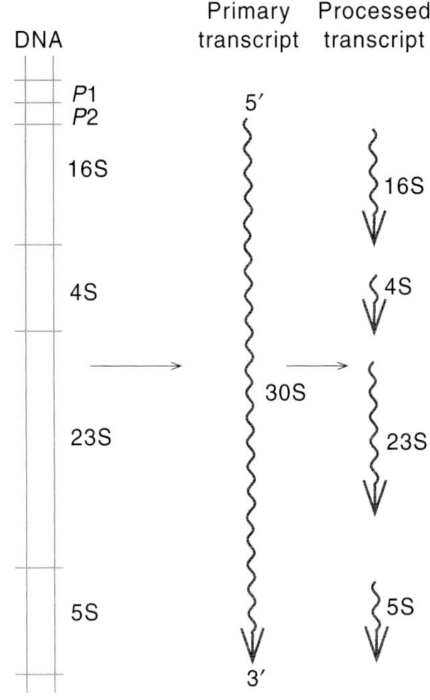

Each operon is transcribed into a single transcript, which is processed into four or five RNAs (fig. 35.15). The order of RNAs in the original transcript starting from the 5′ end is 16S, 4S, 23S, and 5S. In some operons, additional 4S genes for tRNA are located downstream from the 5S gene.

Given the enormous demands that are placed on the *rrn* operons during times of rapid growth, it would not be surprising to find that these genes have been especially designed so that they can function very efficiently. All seven *rrn* operons are designed

Figure 35.16

A typical rRNA promoter from *E. coli* (the rrnB P1 promoter). The promoter includes elements recognized directly by the RNA polymerase at −10 and −35 and an AT-rich element that is believed to make contact with the α subunit of the RNA polymerase. Further upstream, between −60 and −150 bp, there are three binding sites for Fis, a protein that activates transcription from promoters containing these sites by 10- to 20-fold in vitro. About 120 bp downstream of the P1 promoter there is another promoter (P2). The actual sequence present in the mature 16S rRNA begins about 300 bp downstream of the P1 start site. Immediately downstream of the P2 start site there are sequences that have been found to be essential for preventing premature transcription termination (not shown). (Reprinted with permission from W. Ross et al., A third recognition element in bacterial promoters: DNA binding by the α subunit of RNA polymerase, *Science* 262:1407–1413, 1993. Copyright 1993 American Association for the Advancement of Science.)

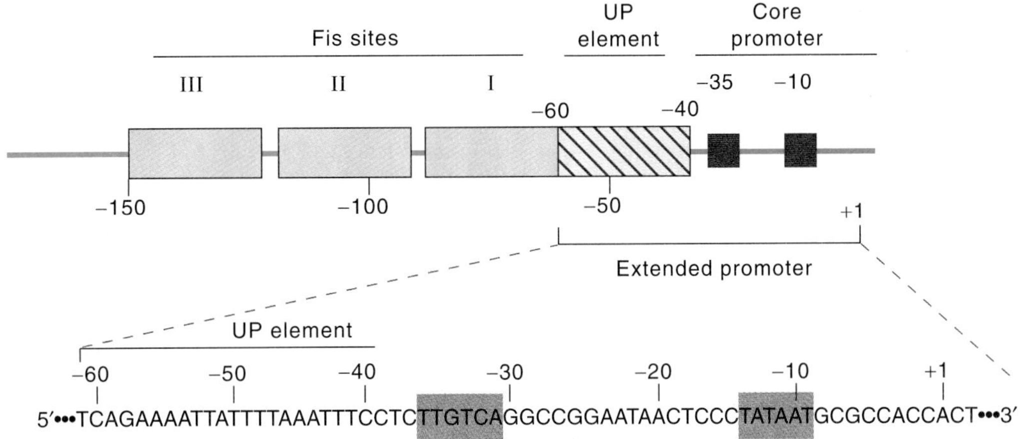

similarly as far as crucial control sequences are concerned (fig. 35.16). First, they all possess two promoters in tandem, which probably permit a more rapid rate of reinitiation for the gene than would be possible with only one promoter. But the main reason for the high activity of these operons is the structure of the promoters themselves. At the −10 and −35 regions they both possess sequences close to the consensus sequence. Upstream of the upper promoter (P1) there are two key control regions (see fig. 35.16). The first, located between −40 and −60 bp, is an AT rich region, which boosts transcription by about 30-fold. Mutation and binding studies of W. Ross and her co-workers indicate that the RNA polymerase α subunit(s) interacts with this so-called UP region. Above this, in the −60 to −150 region, there are three binding sites for a DNA-binding protein called Fis that appears to act as an activator. Binding of Fis at these sites increases promoter activity by another tenfold.

The *rrn* promoters are special in another way. It should be recalled from our discussion of the *trp* operon that when the polymerase pauses just beyond a stem-and-loop structure, there is a possibility of premature termination unless a translating ribosome disrupts the stem-and-loop structure. Since there are no ribosomes involved in translating the *rrn* transcripts but frequent regions where such stem-and-loop structures would be expected to form, there must be another way to prevent premature termination. It has been found that the sequences just downstream of the second promoter (P2) are crucial for this purpose. The exact way in which these sequences function is under study; however, there is good reason to believe that they function in a way that closely resembles a similar antitermination device that prevents termination of early transcripts in λ (discussed later).

As stated earlier, rRNA synthesis is usually maintained at a rate proportional to the gross rate of protein synthesis. In a normal wild-type cell, when protein synthesis is limited (e.g., by amino acid availability), M. Cashel and J. Gallant have shown that the ppGpp concentration rises rapidly from about 50 μM to 500 μM (fig. 35.17). Concomitantly, rRNA synthesis ceases abruptly. This is part of the phenotype known as the stringent response: First, amino acid deprivation or other factors that slow down protein synthesis provoke an increased rate of accumulation of ppGpp; this in turn leads to an inhibition of rRNA synthesis.

Figure 35.17

Guanosine tetraphosphate (ppGpp) concentration under normal conditions, after amino acid starvation, and after readdition of amino acids (● = wild-type cells; ▲ = *relA* cells; and ■ = *spoT* cells).

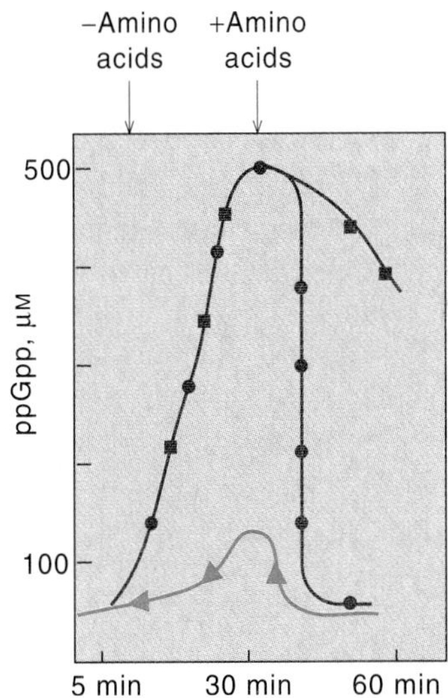

Observations on different mutants indicate that the concentration of ppGpp is regulated by a careful balance between its rate of synthesis (controlled by the *rel* gene product) and rate of breakdown (controlled by the *spoT* gene product). Correlated observations indicate that the synthesis of rRNA is inversely proportional to the ppGpp concentration. In a *relA* mutant cell, neither the rapid rise in ppGpp concentration nor the cessation of rRNA synthesis is seen when amino acids are removed. The *relA* gene encodes a protein that is required for ppGpp synthesis. If amino acids are reintroduced into the growth medium of a wild-type culture, the ppGpp concentration falls rapidly (half-life about 20 s), and the rate of rRNA synthesis rises rapidly. In a mutant called $spoT^-$ the normal rise in ppGpp level is observed on amino acid starvation, but the level of ppGpp falls much more slowly on readdition of amino acids to the growth medium. Correlated with this, the rate of rRNA synthesis in a $spoT^-$ mutant also increases very slowly on readdition of amino acids.

The synthesis of ppGpp has been studied both in crude cell-free extracts of *E. coli* and in a partially purified system to determine what factors influence its rate of synthesis. It was found that ppGpp is synthesized on the ribosome from GTP in the presence of the protein encoded by the wild-type *relA* gene. Maximum ppGpp synthesis occurs in the presence of ribosomes associated with mRNA and uncharged tRNA with anticodons specified by the mRNA. If the uncharged tRNA bound to the ribosome acceptor site (A site) is replaced by charged tRNA, the rate of ppGpp synthesis is greatly lowered. If uncharged tRNA anticodons are not complementary to the mRNA codons exposed on the ribosome for protein synthesis, ppGpp synthesis does not occur.

Cell-free synthesis studies also strongly support the notion that ppGpp directly inhibits RNA synthesis. Thus in a cell-free system the DNA-directed synthesis of rRNA with *E. coli* RNA polymerase is strongly inhibited by 100–200 μM ppGpp. Such experiments have led to the hypothesis that ppGpp, by binding to RNA polymerase, alters its structure so that it has a lowered affinity for rRNA promoters. The in vivo and in vitro studies on ppGpp and rRNA have resulted in a model for how the level of amino acid charging of tRNA controls the rate of rRNA synthesis (fig. 35.18). First, uncharged tRNA that is codon-specific for the exposed codons on the mRNA becomes bound to the ribosome acceptor site, creating a situation unfavorable for protein synthesis but favorable for ppGpp formation. Second, the ppGpp diffuses and binds to RNA polymerase, thereby lowering its affinity for rRNA promoters.

This alteration of the RNA polymerase by ppGpp affects the ability of RNA polymerase to interact with promoters in a differential way. For the promoters of the rRNA operons the polymerase–promoter interaction is strongly inhibited by ppGpp. For some other promoters the effect of ppGpp is actually stimulating. As first observed in the cell-free system, ppGpp stimulates expression from the *lac* and *trp* operons. In more recent in vivo studies, Cashel has found that the biosynthetic operons associated with arg, gly, his, leu, lys, phe, ser, thr, and val biosynthetic enzymes have an absolute requirement for ppGpp. This is an appropriate response because the absence of an amino acid leads to ppGpp buildup, which then stimulates the genes required for biosynthesis of the amino acid.

The observations on ppGpp's role in rRNA synthesis show that this nucleotide is an important control factor regulating rRNA synthesis, but it does not eliminate the possibility that other factors also affect the level of rRNA. In vivo and in vitro evidence indicates that the inhibitory effect of ppGpp on transcription extends to most tRNA and ribosomal protein genes. Ribosomal protein gene expression also appears to be regulated at the translational level.

Translational Control of Ribosomal Protein Synthesis

In exponentially growing *E. coli* cells, synthesis rates of most ribosomal proteins are nearly identical and coordinately regulated. M. Nomura and his co-workers have suggested that free ribosomal proteins inhibit the translation of their own mRNA and that as long as the assembly of ribosomes removes ribosomal proteins, the corresponding mRNA escapes this feedback inhibition. This hypothesis has been tested in vitro by using a protein-synthesizing system with various template DNA molecules carrying ribosomal protein genes; it has been tested in vivo by examining the effect of overproduction of certain ribosomal proteins on the synthesis of other ribosomal proteins using various recombinant plasmids. By these means, it was found that certain ribosomal proteins selectively inhibit the synthesis of other ribosomal proteins whose genes are part of the same operon; this autogenous, or self-imposed, inhibition occurs at the level of the translation of the mRNA rather than at the level of the transcription of mRNA. Figure 35.19 typifies how this scheme works for the regulatory protein L1. L1 and L11 are encoded by the P_{L11} operon. L1 can form a complex either with the 5′ end of its own mRNA or with 23S rRNA. It binds more strongly to the 23S rRNA. However, if an excess of L1 occurs over the available 23S rRNA, then L1 binds to the 5′ end of the mRNA, thereby inhibiting the synthesis of both L1 and L11. In this way the amounts of L1 and L11 proteins synthesized are kept in register with the amounts of rRNA synthesized.

Other operons carrying ribosomal protein genes are regulated in a similar manner. Most ribosomal protein genes exist in operons with promoters at one end. In some cases, individual operons are regulated by one of the encoded ribosomal proteins; in other cases, operons appear to be subdivided into units of regulation; and individual units of regulation are regulated at the translation level by one of the translation products.

Regulation of Gene Expression in Bacterial Viruses

Bacterial viruses rarely coexist indefinitely with the host cell. Temperate viruses can adopt either an active, replicating state or a dormant, prophage state. In the prophage state, the viral genome exists at a low copy number, sometimes in a host-integrated form as with λ bacteriophage, and other times in an independent plasmidlike chromosome, as with P1 bacteriophage. In the active, replicating state, temperate viruses duplicate

Figure 35.18

Schematic diagram of ppGpp synthesis and the hypothesized mechanism for its action. ppGpp is synthesized on the ribosome when there is a peptidyl-tRNA on the P site of the ribosome and uncharged tRNA on the A site. The ppGpp probably inhibits rRNA synthesis by complexing with the RNA polymerase.

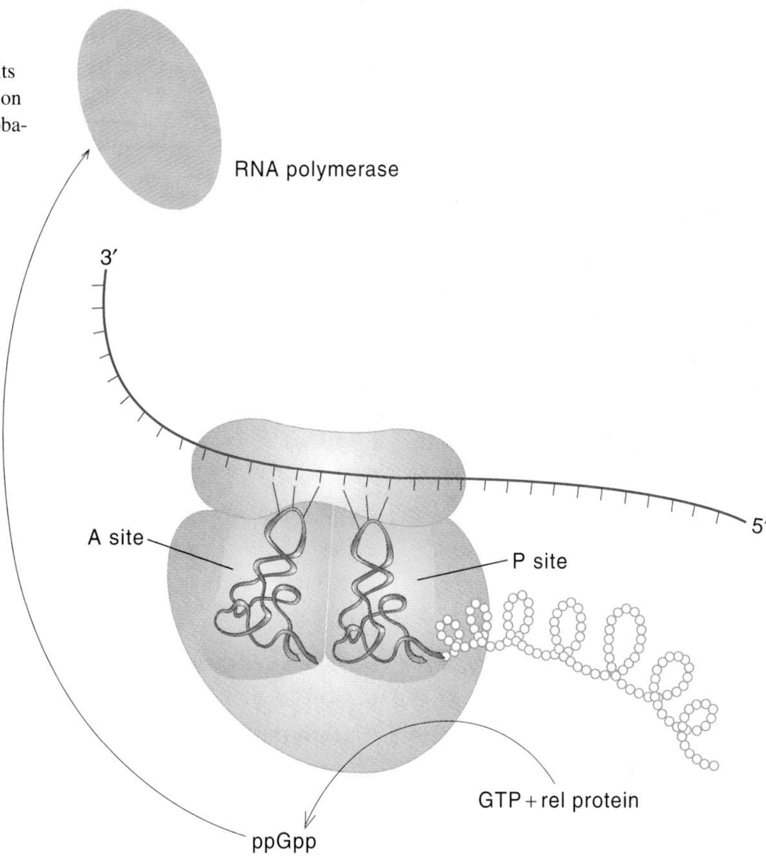

Figure 35.19

Schematic diagram explaining autogenous inhibition by L1. L1 can form a complex either with the 5′ end of its own mRNA or with 23S rRNA. If there is an excess of L1 over the available 23S rRNA, then L1 binds to the 5′ end of the mRNA and inhibits both L1 and L11 synthesis. The inhibition of L1 translation by this binding depends on the fact that the translation of L1 and L11 is somehow coupled.

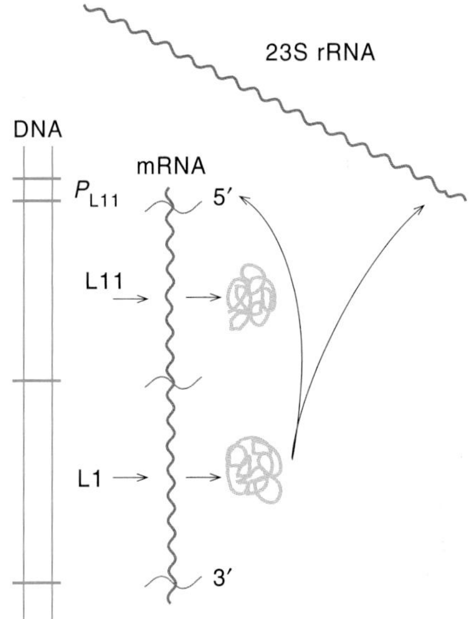

rapidly, so within about an hour they kill the host cell and release infectious particles. Most temperate viruses in the prophage state utilize the unmodified host RNA polymerase. But once they enter the lytic cycle, whether they are temperate or virulent, they either modify or replace the host polymerase to enhance their own transcription. The lytic life cycle of viruses employs multiple regulatory proteins that function in a cascade. The pattern of regulation is unidirectional and irreversible. A small percentage of virus genes designated as early genes are expressed first. Subsequently, late genes are turned on, and early genes are sometimes turned off, until mature virus particles are assembled and exit the cell. The pattern of gene expression observed for a bacteriophage has been likened to the pattern of expression observed in the development of differentiated cells in multicellular organisms (see chapter 36).

In this chapter we consider the patterns of gene expression found for λ, the best understood of the temperate phages.

λ Metabolism Is Directed by Six Regulatory Proteins

λ encodes six regulatory protein: cI, cII, cIII, cro, N, and Q (table 35.2). The developmental processes accompanying infection center on the roles of these six regulatory proteins and the factors that govern the level of their expression.

Table 35.2
Regulatory Proteins of λ

Protein	Function
cI	At low concentrations represses P_R and P_L and activates P_{RM}; at high concentrations represses P_R, P_L, and P_{RM}
cII	Activator for P_{int} and P_{RE}
cIII	Stabilizes cII
cro	At low concentrations represses P_{RM}; at high concentrations represses P_{RM}, P_R, and P_L
N	An antiterminator at t_{L1}, t_{R1}, and t_{R2}
Q	An antiterminator at t_{6S}

The Dormant Prophage State of λ Is Maintained by a Phage-Encoded Repressor

About half of the time when λ infects a cell, it adopts a dormant lysogenic state in which the virus is linearly integrated into the host genome. This state is maintained by moderate amounts of the λ-encoded cI repressor. The cI repressor prevents the lytic cycle from developing by inhibiting two promoters: the P_L promoter for early leftward transcription and P_R the promoter for early rightward transcription (fig. 35.20). It does this by binding to sites that prevent the polymerase from binding, just like the bacterial repressor proteins we have already discussed. The repressor binding site closest to the P_R promoter is actually a composite of three adjacent repressor binding sites O_{R1}, O_{R2}, and O_{R3}, which have different affinities for repressor. At moderate concentrations (about 100 copies of cI repressor per cell), O_{R1} and O_{R2} are occupied. At higher concentrations of cI the O_{R3} site is also occupied.

Binding of repressor at O_{R1} and O_{R2} has two effects: It inhibits transcription from the P_R promoter while it stimulates the promoter for repressor maintenance P_{RM}. Thus at moderate concentrations the cI repressor functions simultaneously as a repressor and as an activator. The stimulation of leftward transcription from the P_{RM} promoter is due to favorable contacts made between the repressor bound at O_{R2} and the polymerase bound to the P_{RM} promoter. At more elevated concentrations of cI when the O_{R3} binding site is also occupied by repressor, the P_{RM} promoter is inhibited. This prevents overproduction of cI.

As long as a moderate concentration of the cI repressor is present, λ remains dormant. Exposure of lysogenic cells to ultraviolet light or DNA-damaging drugs such as mitomycin C leads to wholesale destruction of the cI protein. This is because DNA damage activates the host recA protein in a way that leads to destruction of the cI repressor. With *cI* gone the prophage spontaneously enters the lytic cycle. This mechanism of prophage activation seems to be an instruction to the prophage that when conditions are unfavorable in the cell, it is time to replicate and find a new host cell to infect or lysogenize.

Events That Follow Infection of Escherichia coli *by Bacteriophage λ Can Lead to Lysis or Lysogeny*

Infection of *E. coli* by λ starts by attachment of the virus to the bacterial membrane and injection of the viral genome (fig. 35.21). This is followed by circularization catalyzed by the cellular DNA ligase enzyme. Only three genes are expressed at this time: the regulatory gene *N* from the P_L promoter and the regulatory genes *cro* and *cII* from the P_R promoter. The N protein combines with the host RNA polymerase, permitting extension of the early right and early left transcripts. This leads to expression of the other regulatory genes: *cII, cIII,* and *Q* and replication and recombination genes. At this point the infection process can proceed in either of two mutually exclusive directions: the lytic direction or the lyso-

Figure 35.20

A segment of the λ genome containing the P_{RM} and nearby P_L and P_R promoters. Transcription from P_{RM} results in cI synthesis. An expanded view is shown of the three cI binding sites flanking the P_{RM} and P_R promoters. Transcripts that originate from the three promoters are shown by the wavy red arrows. The O_{R1}, O_{R2}, and O_{R3} sites are binding sites for the cI repressor.

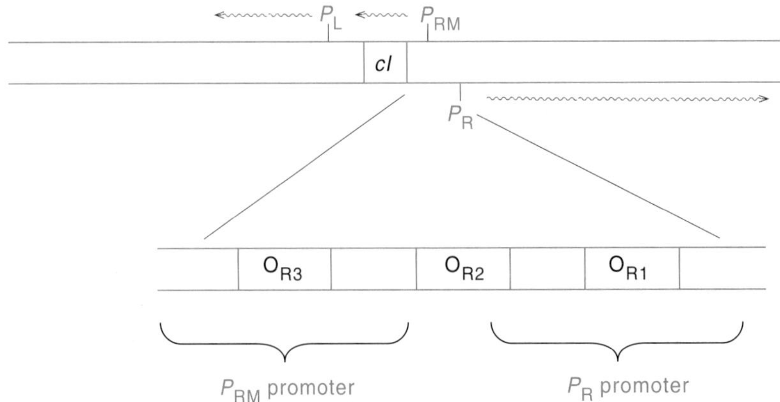

Figure 35.21

Overview of λ development.

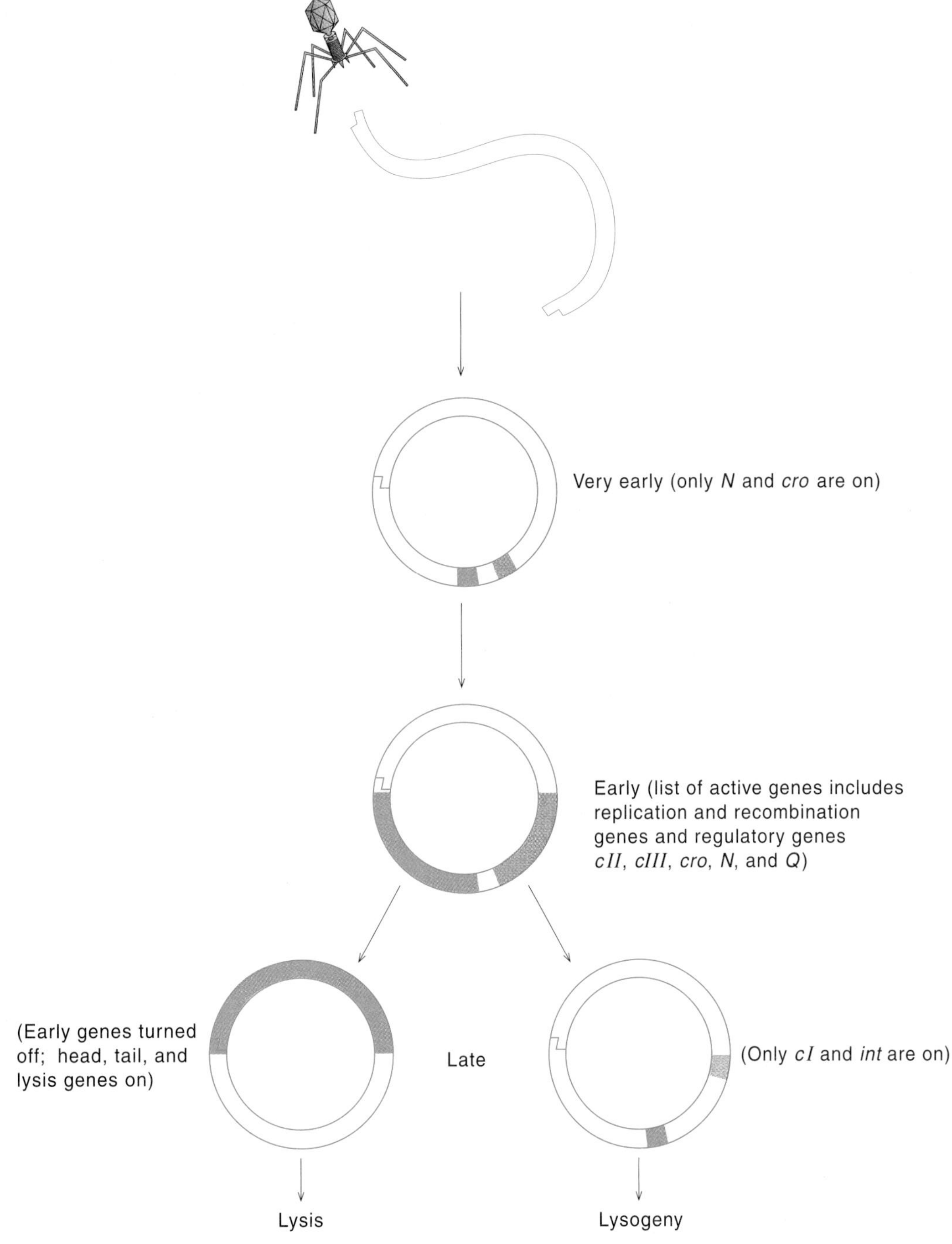

genic direction. First we consider the events that take place in the lytic direction.

The N Protein Is an Antiterminator That Results in Extension of Early Transcripts

The N protein produced by early leftward transcription influences the RNA polymerase by causing it to ignore certain termination signals. For N to act, it must first bind to the polymerase. Remarkably, N can attach to a transcribing polymerase only at two polymerase pause sites, which are designated nut_L and nut_R (fig. 35.22). At these points during transcription the polymerase pauses and can pick up the N protein. The polymerase structure is altered after binding N, so it no longer recognizes the early termination sites, t_{RI}, t_{R2}, and t_{L1}, as stop signals. (See box 35B for the proposed structure of the antitermination complexes.) As a result, polymerase transcribes through these termination sites, producing

Figure 35.22

Early transcripts. The early left transcript is initiated from the P_L promoter. In the absence of N protein this transcript terminates at t_{L1}. In the presence of N protein the polymerase picks up an N protein at the N utilization site, nut_L. This makes it possible for the polymerase to transcribe through t_{L1}. The early right transcript tends to terminate at t_{R1} unless N protein is present. In this case the N protein becomes bound to the polymerase at the nut_R site.

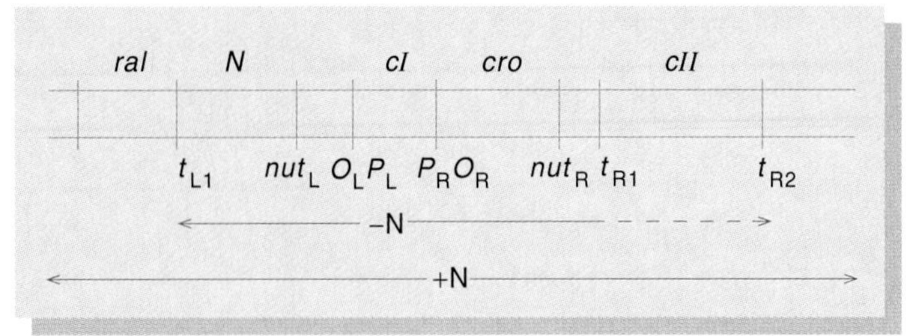

longer transcripts from which additional gene products are expressed (see figs. 35.22 and 35.23).

Another Antiterminator, the Q Protein, Is the Key to Late Transcription

Extension of the rightward transcript stimulated by the N protein leads to expression of the *Q* gene protein, a key regulatory protein for late transcription. The Q protein also functions as an antiterminator, but the way in which it acts is believed to be quite different; in this case, antitermination occurs at the t_{6S} terminator (see fig. 35.23). This leads to transcription of all of the late genes, including the lysis gene and the genes for the head and tail proteins of the mature phage. At late times, other factors encourage rightward transcription from the $P_{R'}$ promoter.

Cro Protein Prevents Buildup of cI Protein during the Lytic Cycle

As the lytic cycle progresses, the phage DNA replicates. The increase in the number of gene copies can result in overexpression of the cI, which could shut down the lytic cycle. This effect is

Figure 35.23

Bacteriophage λ DNA. *Top:* Circular map indicating locations of main control genes and early and late functions. *Bottom:* Expanded region containing control genes and main promoters and terminators. Arrows indicate main transcripts and conditions under which they are active. In the absence of N protein, about half the transcription beginning at P_R reads through t_{R1} to t_{R2} (indicated by the dashed portion of the arrow). More information on regulatory proteins and promoters is given in table 35.2.

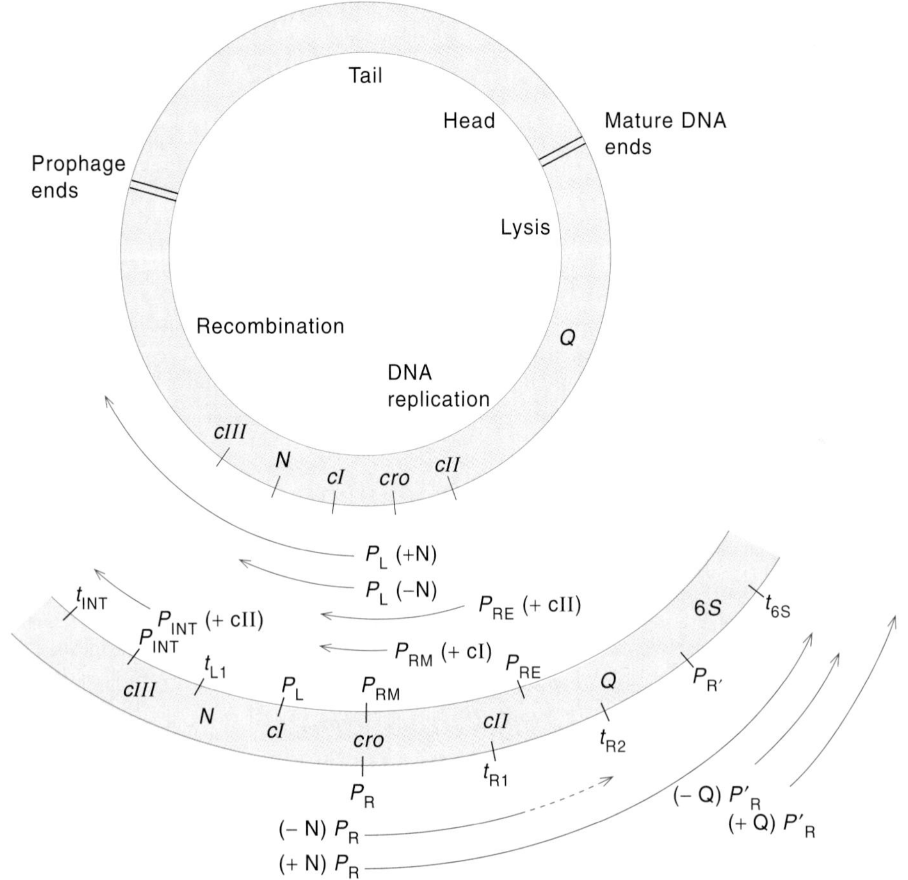

35B

BOX

Antitermination by the λN Protein

Jack Greenblatt and his colleagues have proposed that antitermination by the λN protein results from either of two closely related complexes (fig. 1). A key feature of these systems is that the control signals are located in the nascent RNA and are connected to the elongating RNA polymerase by RNA looping and multiple protein–protein interactions. The elongation factors are bound to the RNA polymerase as it transcribes DNA downstream from the control signals.

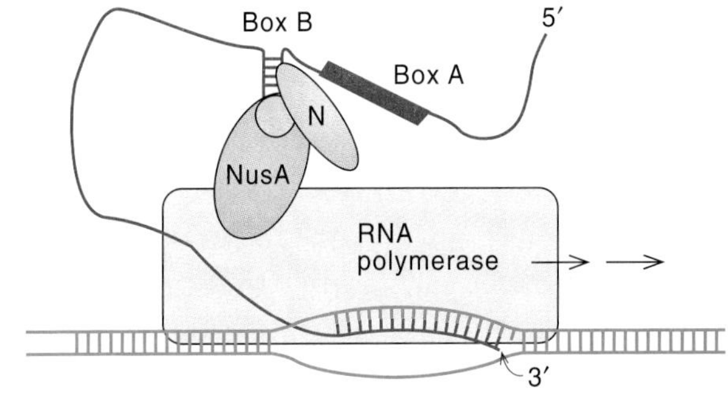

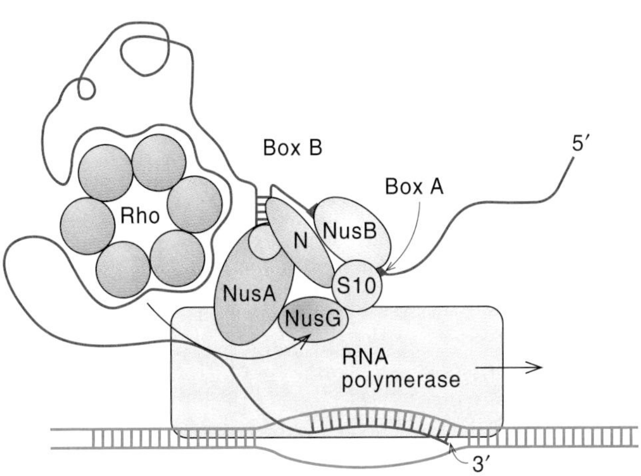

Figure 1

Antitermination by the λN protein. Two types of termination complexes are shown. (*a*) In nonprocessive antitermination, a complex is formed between the λN protein, the nascent RNA transcript, the host NusA protein and RNA polymerase. This complex inhibits pausing by the RNA polymerase. (*b*) In processive termination a complex is formed in which additional host factors are involved. This includes NusB, NusG, and S10. Interaction between NusG and the host Rho hexamer is believed to prevent Rho from releasing the nascent transcript. (Reprinted with permission from J. Greenblatt, J. R. Nodell, and S. W. Mason, Transcription antitermination, *Nature* 364:402, 1993. Copyright 1993 Macmillan Magazines Limited.)

overcome by the regulatory protein cro, which is specifically designed to inhibit cI synthesis during late infection.

The cro protein and the cI protein bind to exactly the same sites on the DNA. Despite this fact, the cI protein is required for lysogeny, whereas the cro protein is required for lysis. These requirements can be shown with mutants. A cI^- mutant invariably undergoes lysis, whereas a cro^- mutant can lysogenize but cannot complete the lytic cycle. This remarkable difference in the behavior of cro and cI results from the fact that although they bind to the same sites, they do so with totally different relative affinities (fig. 35.24). Cro binds preferentially to O_{R3} and less strongly to O_{R1} and O_{R2}.

The binding of cro to O_{R3} turns off cI expression originating from the P_{RM} promoter. At high concentrations, cro binds to one or more of the other sites, turning off the P_R promoter. The P_L promoter is also turned off at high concentrations of cro. The most important physiologic effect of cro is the turning off of the P_{RM} promoter. In the absence of cro, cI expression increases because of the increase of the number of cI gene copies resulting from early replication of viral DNA during the lytic cycle. This buildup shuts down transcription from the P_R and P_L promoters before sufficient transcripts have been made to ensure the lytic pathway. The binding properties of cI and cro to the tripartite operator shared by P_{RM} and P_R are shown in fig. 35.24.

Late Expression along the Lysogenic Pathway Requires a Rapid Buildup of the cII Regulatory Protein

Thus far, we have considered the events that take place after infection or after activation of the prophage that lead to phage replication and ultimately lysis. As already indicated, when λ infects a cell, the cells can follow this lytic pathway, or, alternatively, they

Figure 35.24

Segment of the λ genome showing the three operators O_{R1}, O_{R2}, and O_{R3} around the P_{RM} and the P_R promoters. The cI and cro regulatory proteins bind to these operators with different relative affinities. The net result of these differing affinities is that cI is required for lysogeny and cro is required for the lytic cycle.

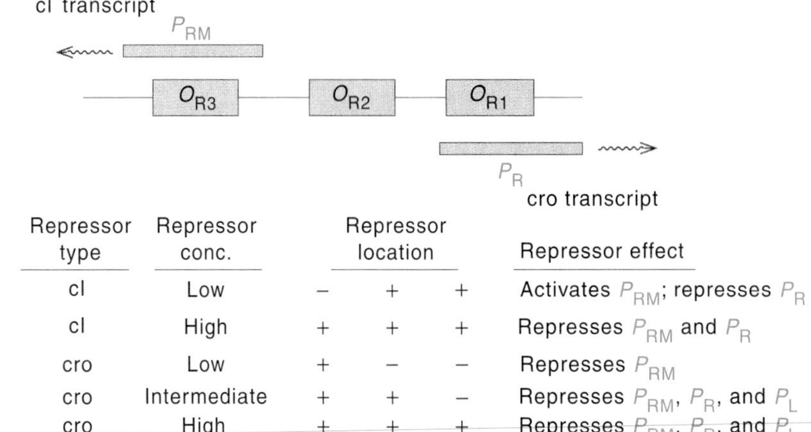

Repressor type	Repressor conc.	Repressor location			Repressor effect
cI	Low	−	+	+	Activates P_{RM}; represses P_R
cI	High	+	+	+	Represses P_{RM} and P_R
cro	Low	+	−	−	Represses P_{RM}
cro	Intermediate	+	+	−	Represses P_{RM}, P_R, and P_L
cro	High	+	+	+	Represses P_{RM}, P_R, and P_L

Figure 35.25

A segment of the λ genome containing the P_{RM} and nearby P_L, P_R, and P_{RE} promoters. Transcripts that can originate from the three promoters are shown by the wavy arrows. The P_{RE} and P_{RM} require different activators for expression. The transcript from P_{RE} is 10 times more effective in cI expression because it has a Shine-Dalgarno sequence for ribosome attachment.

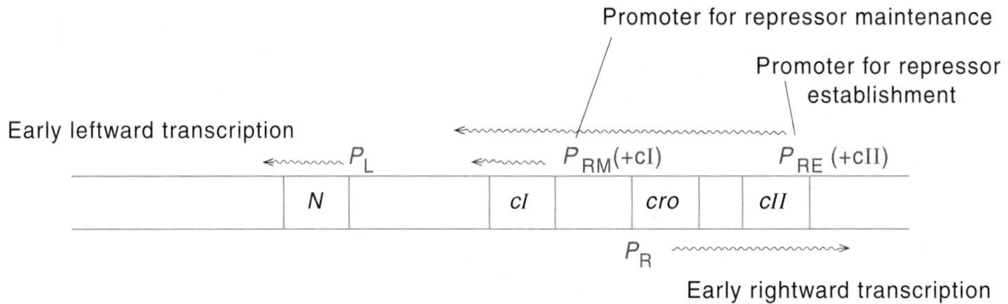

can follow the lysogenic pathway, in which case the λ genome becomes dormant and integrates into the host genome. The choice appears to depend primarily on the level of cII protein formed during the early stages of infection. If this level is sufficiently high, the lysogenic pathway is followed; otherwise, the lytic pathway is followed. Although we do not understand precisely what controls the level of cII expression, it is easy to appreciate the importance of cII to elaboration of the lysogenic pathway. Thus cII protein is an activator for two promoters: P_{RE} and P_{int} (see fig. 35.23). The P_{RE} promoter leads to transcripts that are very efficient in producing high levels of the cI repressor (fig. 35.25). High levels of cI can shut down the P_L and P_R promoters before the lytic cycle gets underway. The other promoter activated by cII is P_{int}, which leads to int protein synthesis. This protein is required for the integration of the phage genome into the host genome, a process that completes the lysogenic process. After integration, moderate levels of cI resulting from transcription initiated at the P_{RM} promoter (as described above) keep both the P_L and the P_R promoters in the repressed state.

Interaction between DNA and DNA-Binding Proteins

So far we have focused on where regulatory proteins bind to the DNA and under what conditions they bind. The ability to isolate and sequence DNA promoters and to isolate and characterize regulatory proteins has greatly increased our appreciation of how specific complexes are formed between DNA and regulatory proteins.

The three-dimensional structures of about 20 regulatory proteins have been determined; in four cases the structures of the specific DNA–regulatory protein complexes have also been determined.

Recognizing Specific Regions in the DNA Duplex

The problem of regulation in prokaryotes such as *E. coli* is far simpler than in complex eukaryotes because of the smaller number of genes. It seems likely that a system of this complexity (about 3,000 genes) should contain no more than 100–300 regulatory proteins because genes with related functions are often under the control of the same regulatory proteins.

Each member of this diverse family of regulatory proteins must be able to scan the DNA structure and recognize control regions with a high degree of specificity. How is this possible, particularly as the specific regions of DNA are embedded in a core of base pairs that are already hydrogen-bonded to each other? Present indications are that the secondary structure of DNA is left intact when these regions interact with their regulatory proteins. Thus the base pairs must be identifiable by the groups that are exposed in the core of an intact duplex structure. As in most specific interactions between nucleic acids and proteins, it seems likely that the specificity of interactions between DNA and regulatory proteins must result primarily from hydrogen bond interactions. Inspection of the DNA duplex in the major and minor grooves reveals that many of the hydrogen-bonding groups of the base pairs are available for such interactions (see fig. 30.8). In the major groove the GC and AT base pairs both have three hydrogen-bonding groups accessible for interaction with other molecules. In the minor groove the GC base pair also has three, but the AT base pair has only two. There are other considerations. The major groove can accommodate a larger protein element than the minor groove. An α helix or a two-chain β structure can fit comfortably into the major groove so that their polar side chains have no difficulty making hydrogen bonds with a core of base pairs. Such structures cannot be fitted into the minor groove, the contacts of which are probably limited to those that can be made with an extended polypeptide chain. On the basis of size considerations alone, it seems likely that regulatory proteins make most of their specific contacts with DNA in the major groove. Quite remarkably, almost all regulatory proteins involve α-helical elements interacting in the major groove.

The Helix-Turn-Helix Is the Most Common Motif Found in Prokaryotic Regulatory Proteins

In prokaryotes such as *E. coli* the helix segment recognized by the DNA is part of a larger domain known as the helix-turn-helix motif. A protruding recognition helix is supported by a second segment of helix, which stabilizes the recognition helix and fixes its orientation with respect to the remainder of the regulatory protein.

The primary sequence for the helix-turn-helix motif consists of a 20-amino-acid sequence in which six of the amino acids tend to be conserved (residues 4, 5, 8, 9, 10, and 15 in fig. 35.26). Four of these residues, 4, 5, 10, and 15, usually have hydrophobic side chains; they make stabilizing contacts between the two helices, ensuring the preservation of their mutual orientation. Two other residues, 8 and 9, are important for making the bend between the two segments of α helix. Residue 9 is frequently a glycine because its small side chain (—H) is often necessary to make a bend. The remaining amino acid residues vary considerably in different proteins. These are the residues that face either the DNA or the remaining portion of the regulatory protein.

Figure 35.26

Helix-turn-helix motif. Helix 2 is the recognition helix. Individual amino acids are numbered. Residues 4, 5, 8, 9, 10, and 15 tend to be conserved in different regulatory proteins. (*Source:* From Carl Branden and John Tooze, *Introduction to Protein Structure,* Garland Publishing, Inc., New York, 1991, p. 102.)

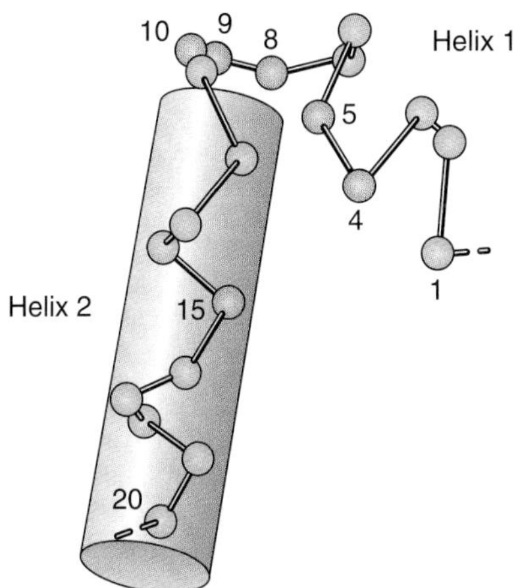

The simplest way for a segment of protein helix to make contact in the major groove is for it to run parallel to the groove. Although it adopts this orientation in some cases, it frequently orients itself in other ways in different regulatory proteins. In all known orientations, ample opportunity arises for many of the amino acid side chains to make contact with the hydrogen-bonding groups in the DNA base pairs.

Helix-Turn-Helix Regulatory Proteins Are Symmetrical

Three helix-turn-helix regulatory proteins are depicted in figure 35.27: cro repressor, λ repressor, and CAP activator. Of these, only in cro is the recognition helix oriented so that it runs parallel to the major groove when bound to the DNA. In all three cases the regulatory proteins are homodimers containing monomers with recognition helices spaced precisely 34 Å apart along the direction of the DNA helix's axis so that they can make identical contacts with adjacent major grooves of the DNA duplex. The strategy seems clear; the regulatory protein contains two identical half-sites for interaction with two virtually identical half-sites in the DNA.

Beyond helix 2 in the helix-turn-helix protein we see a great deal of variability in overall structure in the different regulatory proteins. This variability is not surprising because the remainder of the structure serves quite different functions in different regulatory proteins.

The symmetrical aspect of the regulatory protein-binding sites is mirrored in the DNA sequence of the binding site. Recall that the sequence in this region for the *lac* repressor–binding site (see fig. 35.7), the CAP-binding site (see fig. 35.8), and the *trp* repressor–binding site (see fig. 35.11) all show a dyad axis of symmetry with respect to the arrangement of base pairs.

Figure 35.27

The structures of three regulatory proteins. They all possess twofold axes of symmetry, and the protruding helical cylinders that interact with adjacent major grooves on the DNA (red) are spaced about 34 Å apart. N and C labels indicate the N and C termini of the polypeptide chains.

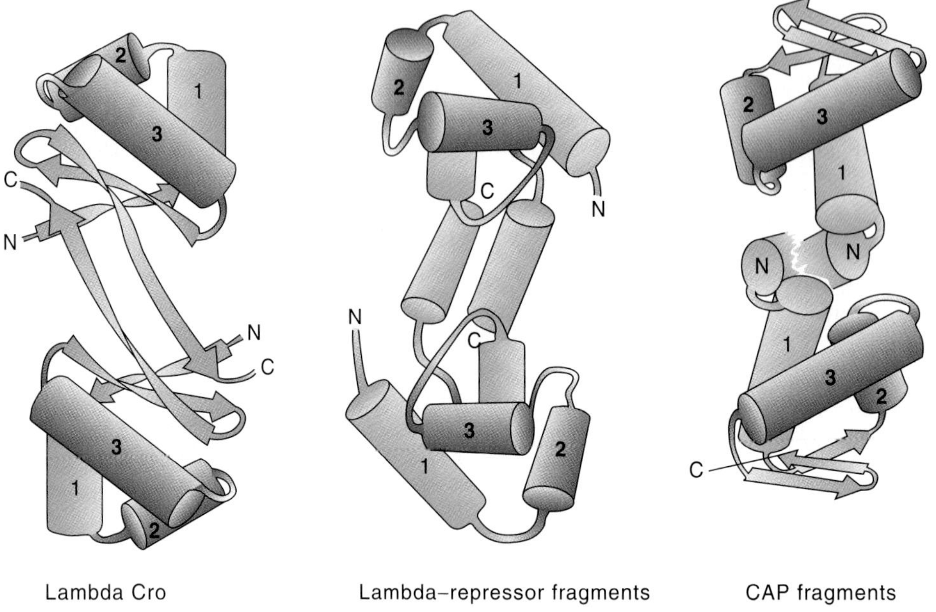

Figure 35.28

The x-ray structure of the CAP–cAMP dimer in complex with a self-complementary 30-bp duplex DNA. The CAP dimer's two helix-turn-helix (HTH) motifs bind in successive major grooves of the DNA. The HTH motif's N-terminal helix is blue, and its C-terminal recognition helix is red. It can be seen that the binding of the CAP produces two kinks in the DNA structure, leading to an overall change in direction of the double helix of about 90°. (*Source:* Courtesy, Dr. Thomas Steitz, Yale University.)

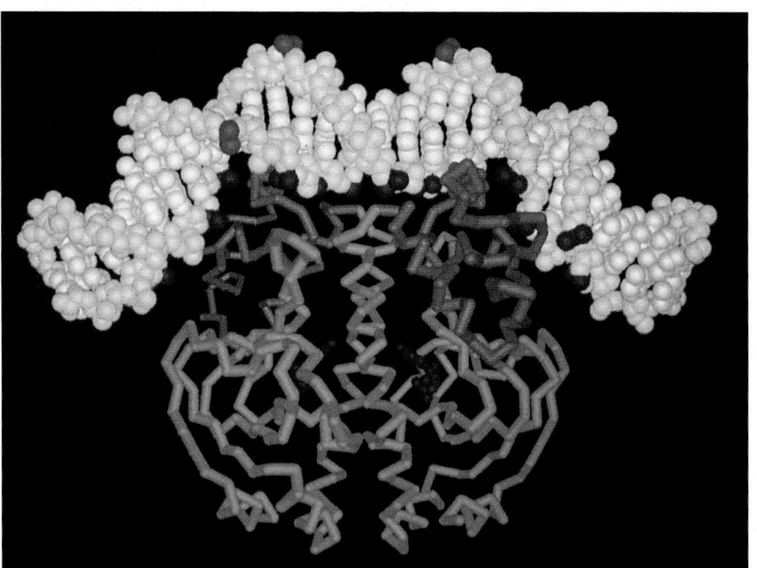

DNA-Protein Cocrystals Reveal Gross Features of the Complex

Crystallographic studies of regulatory protein–DNA complexes reveal many interesting features. The presumption as to where the bulk of the protein is binding has been confirmed in this way. Also some rather bizarre features of the complex that could not have been predicted from the protein and DNA structures alone become apparent. In the case of the CAP–DNA complex the DNA duplex structure is distorted so that there is an overall change in direction of the duplex of 90° (fig. 35.28). It is not clear what advantage this has for the function of the regulatory protein. One cannot help but be reminded of the TBP–TATA complex (see fig. 33.12), in which a severe bend is also produced by the DNA–protein interaction. It should be noted that in the TBP–TATA complex the protein exerts its effect by a pushing action rather than a pulling action as in the case of CAP.

Figure 35.29

Specific interactions between three different repressors and their operator binding sites. Only half the operator binding site is shown because identical contacts are made with the other half. The numbers associated with the amino acid side chains refer to the distance of amino acids from the amino-terminal end of the protein. Nucleotides are numbered from the central dyad at the operator. (*a*) The 434 phage repressor. (*b*) The λ cI repressor. (*c*) The *trp* repressor. (*Source:* Adapted from T. Steitz, *Q. Rev. Biophys.* 23:236, 1990.)

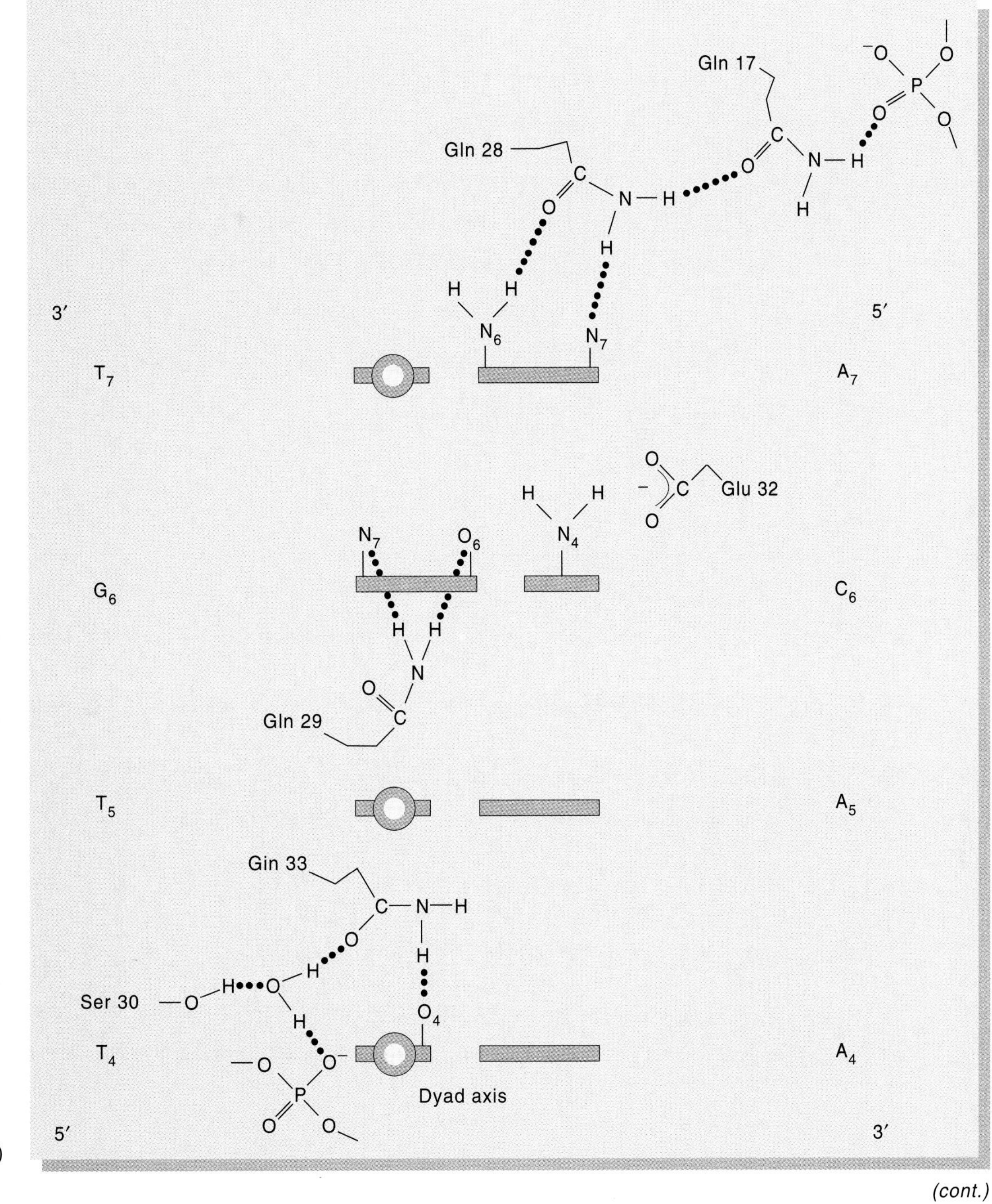

(cont.)

From Cocrystal Studies, the Specific Contacts between Base Pairs and Amino Acid Side Chains May Be Determined

Examination of DNA–regulatory protein complexes have permitted reasonable guesses to be made about the precise nature of the contacts between amino acid side chains and DNA in many cases. Figure 35.29 illustrates three examples: one for the 434 phage repressor (fig. 35.29*a*), one for the λ cI repressor (fig. 35.29*b*), and one for the *trp* repressor (fig. 35.29*c*). In all cases, only half-sites are depicted because symmetry considerations dictate that the two half-sites should have virtually identical structures.

In the case of the two phage repressors, several contacts are observed between individual amino acid side chains and individual base pairs. In addition, one or two hydrogen bonds are ob-

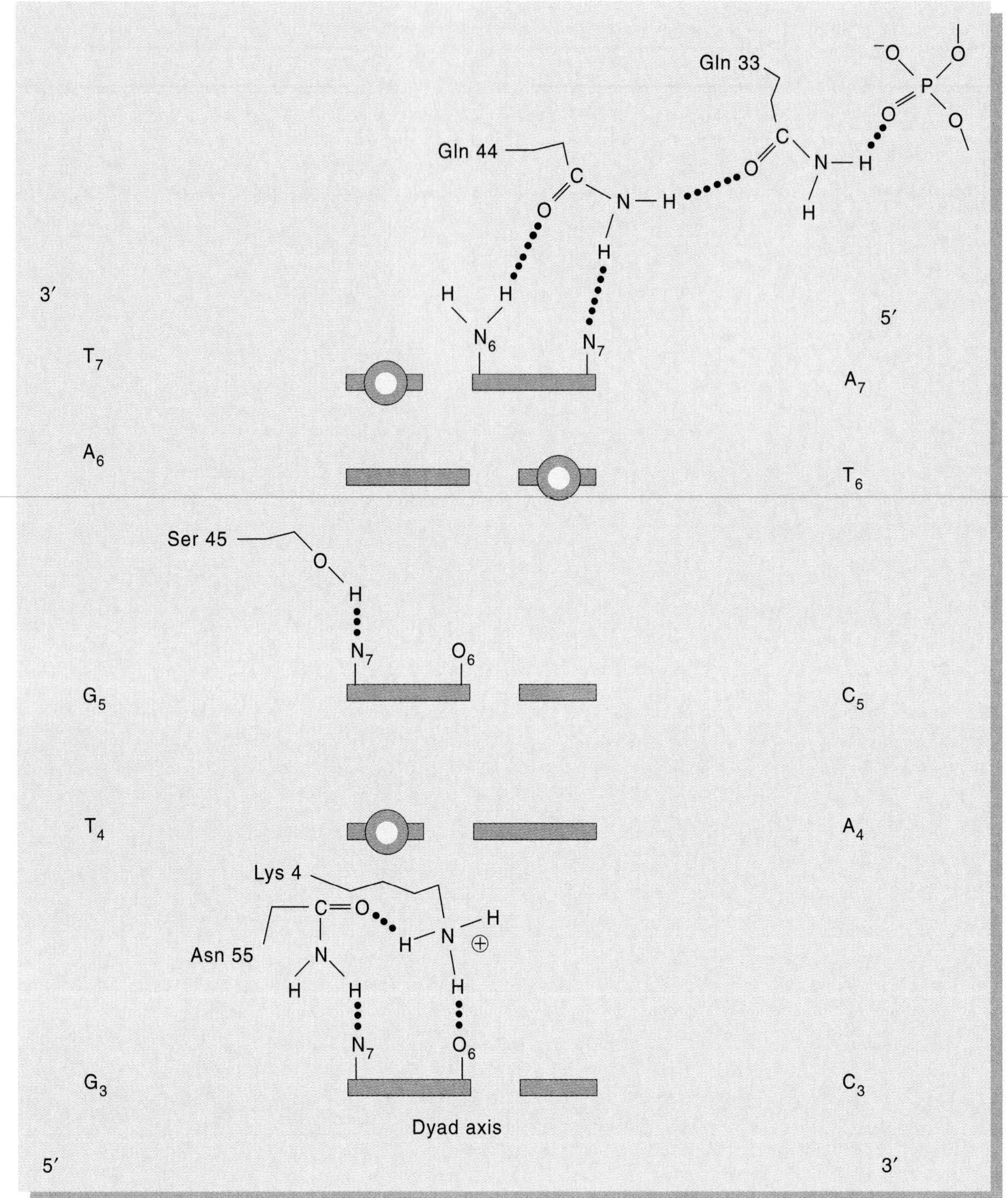

Gln 33
Gln 44
Ser 45
Lys 4
Asn 55
3′
5′
5′
3′
T_7
A_7
A_6
T_6
G_5
C_5
T_4
A_4
G_3
C_3
N_6
N_7
O_6
Dyad axis

(b)

(cont.)

Figure 35.29 (cont.)

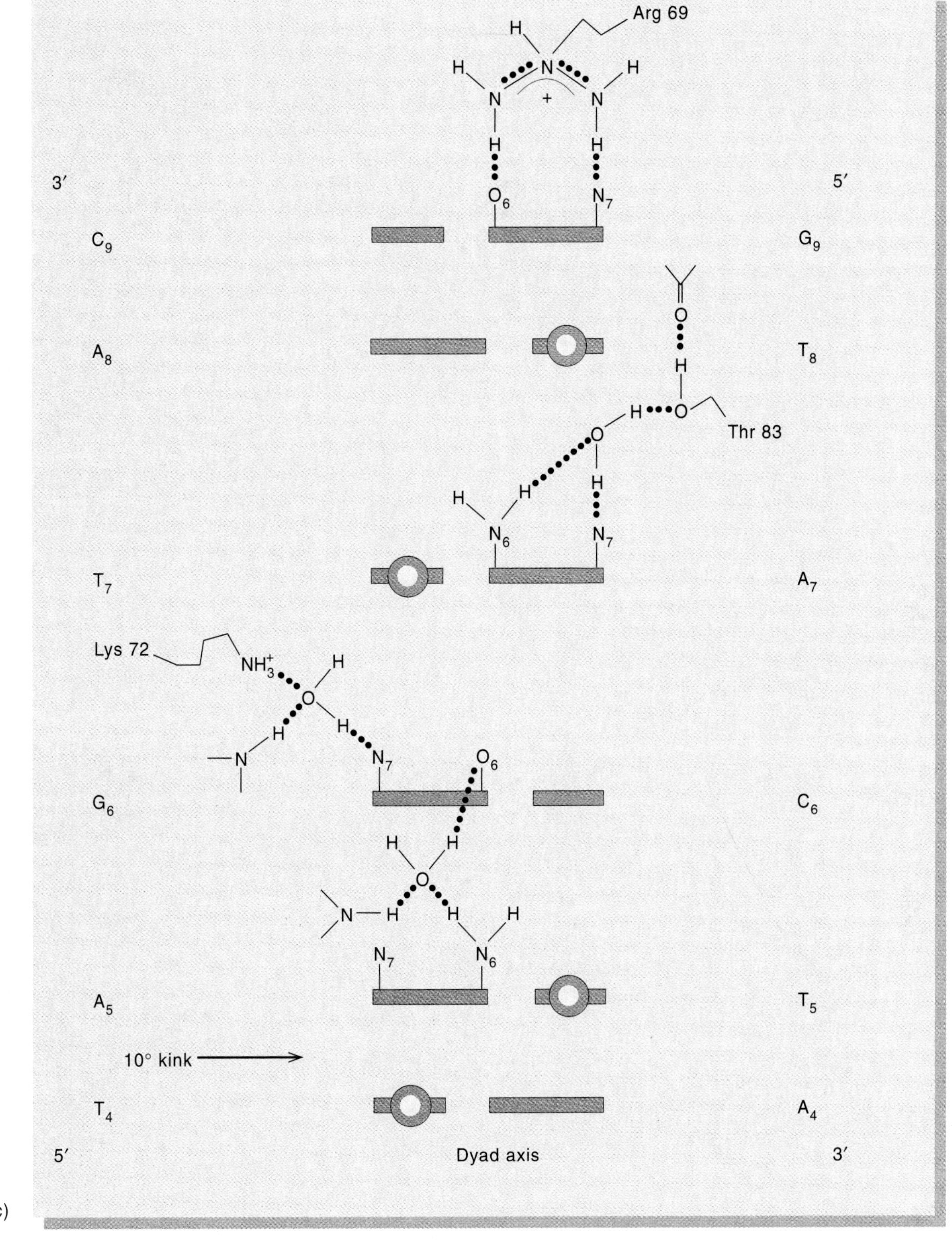

Figure 35.30

Diagram of the *met* repressor–DNA structure illustrating the regions of the dimeric *met* repressor that contact DNA. The two-stranded β sheet of the repressor is bound in the major groove of DNA, where it forms the sequence-specific interactions.

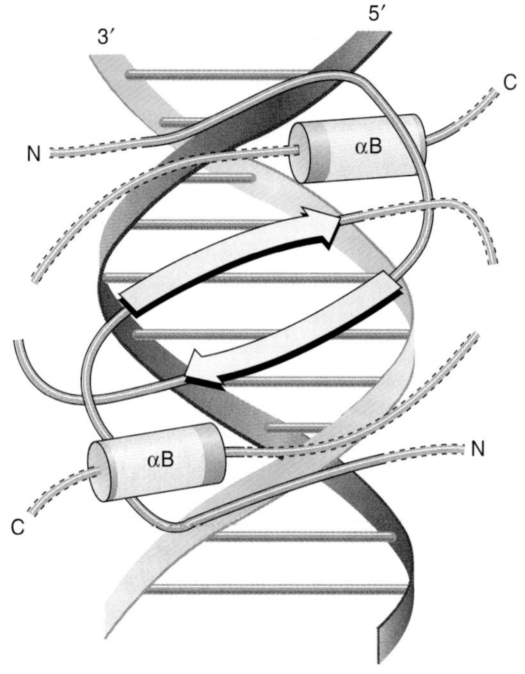

Figure 35.31

Docking of the *trp* repressor to the *trp* operator. The repressor does not bind unless it first complexes with tryptophan. Note how the angle of the recognition helices changes when tryptophan is bound. With regard to the *trp*R/*O* interaction, the bound tryptophan not only causes a readjustment of the so-called recognition helix, but also shapes the side chains in its immediate environment in a very special way.

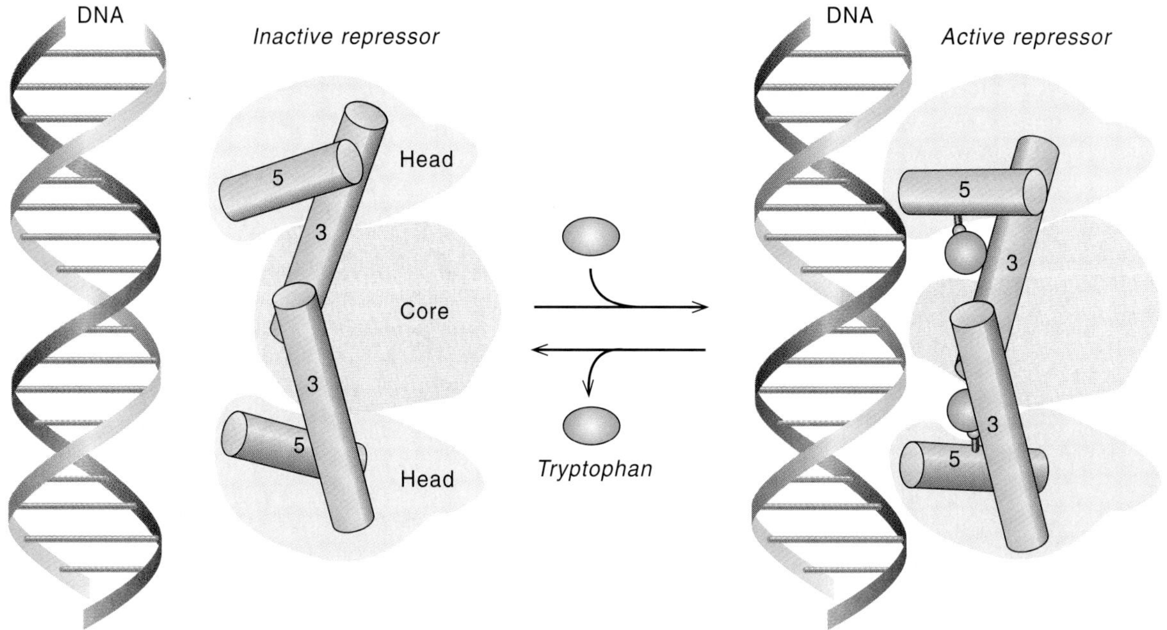

served to nearby phosphate groups. In both of these cases the side chain that is contacting a base participates in the hydrogen bond to a phosphate group, either directly or indirectly.

In these two examples we can see that the amino acids participating in direct interaction are frequently glutamine or asparagine, the two amino acids with amide side chains. The presence of both a hydrogen bond donor and a hydrogen bond acceptor gives the amide side chain the advantage of greater versatility than most amino acid side chains for forming hydrogen bonds. Indeed, in one case we can see that glutamine makes two hydrogen bonds with an adenine. In four of these five examples, glutamine is used. The greater length of the glutamine side chain gives it more "reaching power" for making contact with the base pairs.

Despite the apparent advantages in the amide-containing amino acids, we do not always find amides used as part of the recognition sequence. For example, the *trp* repressor does not use any amides. The basic amino acids, lysine and arginine, and the hydroxylic amino acid, threonine, are the main interacting amino acids in the *trp* repressor. The side chain in arginine can make a pair of hydrogen bonds with a guanine, which should result in a particularly strong interaction.

The *trp* repressor presents an unusual variation for interaction with the base pairs in its operator. Except for the arginine–guanine interaction, the interactions are mediated by a water molecule (see fig. 35.29*c*). For example, a threonine hydroxyl group makes a hydrogen bond with a water molecule, which in turn makes two hydrogen bonds with an adenine.

Some Regulatory Proteins Use the β-Sheet Motif

Despite the dominance of the α-helical motif as a recognition unit, the β-sheet motif has been used in a limited number of cases. In these cases (three are known), two chains in the antiparallel orientation interact with half-sites in adjacent large grooves of the DNA (see fig. 35.30 for an example). (The β-sheet motif obviously works in these cases, despite the fact that it is more difficult to stabilize than a segment of α helix.)

Small Molecules Often Modulate Regulatory Protein Interaction

A common but not universal feature of regulatory proteins is their sensitivity to small-molecule effectors. The lambdoid phage repressors are examples of molecules that almost always function as repressors; their action is controlled merely by their concentration. When they are needed, they are synthesized; when they are no longer needed or their presence is undesirable, they are selectively removed by degradation.

For many other bacterial regulatory proteins the action of the regulatory protein depends on specific small-molecule effectors. As we have seen in the case of the *lac* repressor, the binding of allolactose results in release of the repressor from the operator. By contrast, the binding of cAMP is necessary for the binding of the CAP protein to its DNA-binding site. The small molecule signals a particular metabolic state, and the response of the regulatory protein is appropriate to the metabolic needs of the cell.

It is often difficult to see the change in regulatory protein structure as a result of the binding of its small-molecule effector. In spite of this the theory is that the small-molecule effector changes the structure of the regulatory protein sufficiently to cause release or binding to the appropriate DNA-binding site as the case may be.

In the case of the *trp* repressor the binding of tryptophan to the repressor converts the inactive repressor (aporepressor) into an active repressor by a slight shift in the orientation of the recognition helices (fig. 35.31).

RNA Sometimes Acts as a Repressor

In chapter 32 we discussed the possibility that oligonucleotides complementary to select regions of naturally occurring DNAs or RNAs could inhibit expression at either the transcription or translation levels and that this could have therapeutic value. But what of naturally occurring situations where the interaction between complementary nucleic acids serves a regulatory function? In this regard it should be recalled that in their initial formulation of the operon model of control, Jacob and Monod proposed that repressors were RNAs rather than proteins. Of course, we now know that protein repressors have taken center stage as regulators of gene expression, but this does not mean that RNAs in some cases could not function as repressors. Currently, there are numerous examples that have been cited. In fact the activator protein CAP provides an example of this in which an RNA repressor is part of a negative feedback system for inhibiting transcription of the *crp* gene that encodes CAP. The *crp* gene contains a second promoter downstream from the main promoter; this second promoter is strongly activated by the cAMP–CAP complex. Transcription from this second promoter is initiated 12 nucleotides downstream and on the opposite strand from the initiation site of the *crp* gene. The initial nucleotides of the divergent RNA are complementary to 10 of the first 11 nucleotides of the CAP mRNA (fig. 35.32). The next 11 nucleotides of CAP mRNA and the 5′ end of CAP mRNA could produce a structure similar to a ρ-independent terminator (see chapter 33), a change that could lead to abortive transcription of the CAP mRNA.

Figure 35.32

Model for the inhibition of *crp* transcription by divergent RNA. The proposed RNA–RNA hybrid between the 5′ end of *crp* mRNA and the initial 14 nucleotides of the divergent RNA is shown. The complete structure of the proposed ρ-independent terminator is shown below. The structure consists of the RNA–RNA duplex followed by an A•U-rich RNA•DNA hybrid of 11 bp. Okamoto and Freundlich proposed that the RNA•RNA hybrid causes the polymerase to pause, and as suggested for ρ-independent terminators, the instability of the A•U-rich RNA•DNA hybrid leads to the release of the transcript and termination of transcription. (*Source:* Adapted from K. Okamoto and M. Freundlich, Mechanism for the autogenous control of the *crp* operon: Transcriptional inhibition by a divergent RNA transcript, *Proc. Natl. Acad. Sci. USA* 83:5000–5004, 1986.)

+2 +11
A U G C U A C A G U crp mRNA
U A C G A U G U C A Divergent RNA
+14 +2
U U
U

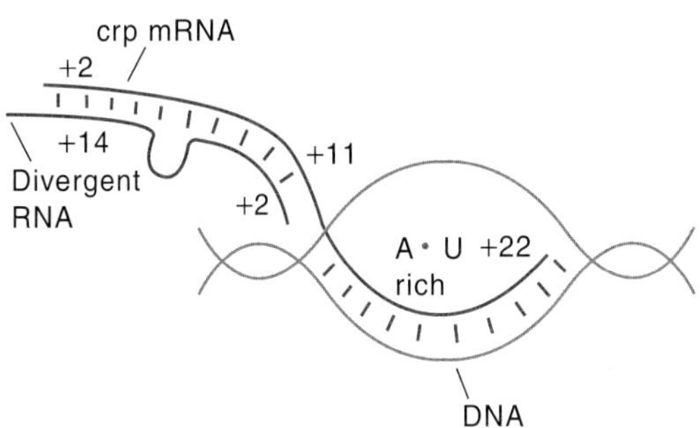

Summary

In this chapter we discussed the regulatory systems of the *E. coli* bacterium and the λ bacteriophage. The main points in our presentation are as follows.

1. *Escherichia coli* carries about 3,000 genes. Only a small fraction of the genome is actively transcribed at any given time. But all of the genes are in a state where they can be readily turned on or turned off in a reversible fashion. The level of transcription is regulated by a complex hierarchy of control elements.
2. In the most common form of control, expression is regulated at the initiation site of transcription. There are several ways of doing this, all of them revolving around protein or small-molecule factors that influence the binding of RNA polymerase at the transcription start site.
3. The *lac* operon, a cluster of three genes involved in the catabolism of lactose, exemplifies both positive and negative forms of control that influence the rate of initiation of transcription. Jacob and Monod identified the repressor as a negative control element that is *trans*-dominant and the operator as a *cis*-dominant site for binding the repressor. Transcription is initiated by RNA polymerase at the promoter, which overlaps the operator site on one side of the three structural genes of the *lac* operon. The tight complex between repressor and operator prevents initiation, and it is broken when lactose is present. The lactose is readily converted to allolactose, which binds to the lac repressor. This changes the structure of the repressor so that it dissociates from the DNA.
4. Initiation of transcription proceeds at a greatly increased rate when cyclic AMP is present. This is because cyclic AMP forms a complex with the CAP activator protein, which then binds at a site adjacent to the polymerase-binding site. The CAP protein enhances polymerase binding at the adjacent site by cooperative binding.
5. Lactose is the substrate of the enzymes of the *lac* operon. In the absence of lactose there is no use for enzymes of the *lac* operon.
6. CAP and cAMP activate a large number of genes in *E. coli* that are concerned with catabolism. When glucose is present, the cAMP is greatly lowered, and the *lac* operon is expressed at a very low level, even when lactose is present. This is because glucose is a more readily metabolizable carbon source than lactose.
7. The *trp* operon contains a cluster of five structural genes associated with tryptophan biosynthesis. Initiation of transcription of the *trp* operon is regulated by a repressor protein that functions similarly to the lac repressor. The main difference is that the trp repressor action is subject to control by the small-molecule effector, tryptophan. When tryptophan binds the repressor, the repressor binds to the *trp* operator. Thus the effect of the small-molecule effector here is opposite to its effect on the *lac* operon. When tryptophan is present, there is no need for the enzymes that synthesize tryptophan.
8. The *trp* operon has a control locus called an attenuator about 150 bases after the transcription initiation site. The attenuator is regulated by the level of charged tryptophan tRNA, so between 10% and 90% of the elongating RNA polymerases transcribe through this site to the end of the operon. Low levels of *trp* tRNA encourage transcription through the attenuator.
9. Ribosomal RNA and protein synthesis are both controlled at the level of initiation of transcription. This is a result of the direct binding of guanosine tetraphosphate, ppGpp, to the RNA polymerase. This binding decreases the affinity of RNA polymerase for the initiation sites of transcription. Guanosine tetraphosphate is synthesized when the general level of amino-acid-charged tRNA is low.
10. The synthesis of ribosomal proteins is regulated at the level of translation. Certain ribosomal proteins bind to specific sites on the ribosomal RNAs or their own mRNAs. In the absence of the ribosomal RNAs, they bind to their own mRNAs, which inhibits their translation. This form of translational control regulates the rate of synthesis of ribosomal proteins so that it does not exceed the rate of ribosomal RNA synthesis.
11. Viruses borrow heavily on the host enzymatic machinery to obtain energy for synthesis, as well as for replication, transcription, and translation. The virus infective cycle is strongly irreversible. Virus infection is followed by the gradual turning on of viral genes. Viral enzymes are the first viral gene products; in late infection the virus structural proteins are favored. The irreversible lytic cycle of the virus is directed by a cascade of controls.
12. In λ the host RNA polymerase is used throughout. Regulation is achieved through a series of repressors and activators, as well as two viral proteins that bind directly to the RNA polymerase. The viral proteins that bind to the polymerase modify it so that it can transcribe through provisional stop signals.
13. When λ phage infects an *E. coli* cell, it does not always produce viral progeny. Sometimes it integrates its genome into the host genome and replicates only as the host genome replicates. This so-called lysogenic state can be disrupted by DNA-damaging conditions such as exposure to UV radiation. Under these conditions the dormant viral genome enters the active replication cycle.
14. Bacterial regulatory proteins are controlled by small-molecule effectors; viral regulatory proteins are not. Bacterial genes are regulated in a highly reversible manner; viral genes are usually turned on only once.
15. Proteins that regulate transcription usually bind to specific sites on the DNA. The recognition process involves specific hydrogen bonds formed between amino acid side chains of the protein and the base pairs of the DNA. Most of this interaction takes place in the major groove of the DNA, which is more accessible to the protein. The vast majority of regulatory proteins interact with the DNA from the side chains of a segment of α helix that fits snugly into the major groove. The binding site on the DNA usually consists of two half-sites, which are arranged on a dyad axis of symmetry that matches two half-sites on the regulatory protein. The two half-sites are situated in adjacent major grooves on one side of the DNA.

Selected Readings

Anderson, J. E., M. Ptashne, and S. C. Harrison, Structure of the repressor-operator complex of bacteriophage 434. *Nature* 306:846–852, 1987.

Ansari, A. Z., J. E. Bradner, and T. V. O'Halloran, DNA-bend modulation in a repressor-to-activator switching mechanism. *Nature* 374:371–375, 1995.

Antson, A. A., et al., The structure of trp RNA-binding attenuation protein. *Nature* 374:693–700, 1995.

Brennan, R. G., and B. W. Matthews, The helix-turn-helix DNA binding motif. *J. Biol. Chem.* 264:1903–1906, 1989. A mini review.

Busby, S., and R. H. Ebright, Promoter structure, promoter recognition, and transcription activation in prokaryotes. *Cell* 79:743–746, 1994.

Dardel, F., M. Panvert, and G. Gayat, Transcription and regulation of expression of the *E. coli* methionyl-tRNA synthetase gene. *Mol. Gen. Genet. 223*:121–133, 1990.

Das, A., Control of transcription termination by RNA-binding proteins. *Ann. Rev. Biochem.* 62:893–930, 1993.

Egushi, Y., T. Itoh, and J.-I. Tomizawa, Antisense RNA. *Ann. Rev. Biochem.* 60:631–652, 1991.

Friedman, A. M., T. O. Fischmann, and T. A. Steitz, Crystal structure of the lac repressor core tetramer and its implications for DNA looping. *Science* 268:1721–1727, 1995.

Gilbert, W., and B. Muller-Hill, Isolation of the lac repressor. *Proc. Natl. Acad. Sci. USA* 56:1891–1898, 1966. First isolation of a repressor protein.

Gold, L., Posttranscriptional regulatory mechanisms in *Escherichia coli*. *Ann. Rev. Biochem.* 56:199–234, 1988.

Goodrich, J. A., and W. R. McClure, Competing promoters in prokaryotic transcription. *Trends Biochem. Sci.* 16:394–396, 1991. Two or more bacterial promoters are often found in close proximity and may compete for the binding of RNA polymerase.

Gralla, J. D., Transcriptional control: Lessons from an *E. coli* promoter data base. *Cell* 66:415–418, 1991.

Greenblatt, J., J. R. Nodwell, and S. W. Mason, Transcriptional antitermination. *Nature* 364:401–406, 1993. Contains an up-to-date treatment on how antitermination works in bacteriophage λ.

Harrison, S. C., A structural taxonomy of DNA-binding domains. *Nature* 353:715–719, 1991.

Helman, J. D., and M. J. Chamberlain, Structure and function of bacterial sigma factors. *Ann. Rev. Biochem.* 57:839–872, 1988.

Jacob, F., and J. Monod, Genetic regulatory mechanisms in the synthesis of proteins. *J. Mol. Biol.* 3:318–356, 1961. A classic paper.

Jeng, J.-A., R. C. Johnson, and R. E. Dickerson, Hin recombinase bound to DNA: The origin of specificity in major and minor groove interactions. *Science* 263:348–355, 1994.

Kaiser, D., and R. Losick, How and why bacteria talk to each other. *Cell* 73:873–886, 1993.

Kang, C. H., R. Chan, I. Berger, C. Lockshin, L. Green, L. Gold, and A. Rich, Crystal structure of the T4 regA translational regulator protein at 1.9 Å resolution. *Science* 268:1170–1173, 1995.

Kustu, S., A. K. North, and D. S. Weiss, Prokaryotic transcriptional enhancers and enhancer-binding proteins. *Trends Biochem. Sci.* 16:397–401, 1991. First discovered in eukaryotes, enhancers have now been found to exist for a number of prokaryotic genes.

Lewis, M., G. Chang, N. C. Horton, M. A. Kercher, H. C. Pace, M. A. Schumacher, R. G. Brennan, P. Lu, Crystal structure of the lactose operon repressor and its complexes with DNA and inducer. *Science* 271:1247–1254, 1966.

Losick, R., and P. Stragier, Crisscross regulation of cell-type-specific gene expression during development in *B. subtilis*. *Nature* 355:601–604, 1992.

Nomura, M., R. Gourse, and G. Baughman, Regulation of the synthesis of ribosomes and ribosomal components. *Ann. Rev. Biochem.* 53:75–117, 1984.

Pabo, C. O., and R. T. Sauer, Transcription factors: Structural families and principles of DNA recognition. *Ann. Rev. Biochem.* 61:1053–1095, 1992.

Parkinson, J. S., Signal transduction schemes of bacteria. *Cell* 73:857–872, 1993.

Ptashne, M., *A Genetic Switch: Gene Control and Phage λ*. Cambridge, Mass.: Cell Press, and Palo Alto, Calif.: Blackwell Scientific, 1987.

Roberts, J. W., RNA and protein elements of *E. coli* and λ transcription antitermination complexes. *Cell* 72:653–656, 1993.

Ross, W., K. K. Gosink, J. Salomon, K. Igarashi, C. Zou, A. Ishihama, K. Severinov, and R. L. Gourse, A third recognition element in bacterial promoters: DNA binding by the α-subunit of RNA polymerase. *Science* 262:1407–1413, 1993.

Shakked, Z., G. Guzikevich-Guerstein, F. Frolow, D. Rabinovich, A. Joachimiak, and P. B. Sigler, Determinants of repressor/operator recognition from the structure of the trp operator binding site. *Nature* 368:469–473, 1994.

Steitz, T. A., Structural studies of protein–nucleic acid interaction: The sources of sequence-specific binding. *Quar. Rev. Biophys.* 23:205–280, 1990. A very readable and very thorough review of the subject with excellent illustrations, written by one of the pioneers in the field.

Storz, G., L. A. Tartaglia, and B. N. Ames, Transcriptional regulator of oxidative stress-inducible genes: Direct activation by oxidation. *Science* 248:189–194, 1990.

Wagner, R. W., Gene inhibition using antisense oligodeoxynucleotides. *Nature* 372:333–335, 1994.

Weintraub, H., Antisense RNA and DNA. *Sci. Am.* 262(1):40–46, 1990.

Wolberger, C., Y. Dong, M. Ptashne, and S. C. Harrison, Structure of phage 434 Cro/DNA complex. *Nature* 335:789–795, 1988.

Yanofsky, C., Operon-specific control by transcription attenuation. *Trends Genet.* 3:356–360, 1987.

Zubay, G., M. Lederman, and J. DeVries, DNA-directed peptide synthesis. III: Repression of β-galactosidase synthesis and inhibition of repressor by inducer in a cell-free system. *Proc. Natl. Acad. Sci. USA* 58:1669–1675, 1967.

Zubay, G., D. Schwartz, and J. Beckwith, Mechanism of activation of catabolite-sensitive genes: A positive control system. *Proc. Natl. Acad. Sci. USA* 66:104–110, 1970. First isolation of an activator protein.

Problems

1. The *lac* promoter/operator region is found in many vectors used by molecular biologists to clone genes. Not infrequently, high levels of transcription driven by the *lac* promoter generate toxic levels of the cloned gene product, causing *E. coli* cells to grow poorly. Faced with such a situation, how could you minimize expression of the *lac* promoter?
2. What would be the phenotype, with respect to β-galactosidase production in the presence and absence of inducer, of the following *E. coli* mutants:
   (a) $i^s o^+ z^+$
   (b) $i^s o^c z^+$
   (c) $i^s o^c z^- / i^+ o^+ z^+$
   (d) $i^+ o^c z^- / i^- o^+ z^+$
3. Compare the mechanisms of action of *lac* and *trp* repressors (include similarities and differences).
4. What effect do you predict the following mutations would

have on transcription of the *lac* operon in *E. coli*? Explain your reasoning.

(a) A mutation in $lacO_1$ (the operator) that alters one of the bases recognized by *lac* repressor.

(b) A mutation in *lacI* (the repressor gene) that abolishes its ability to bind allolactose but does not affect DNA binding.

(c) A mutation in the *lacI* Shine-Dalgarno sequence that improves its complementarity to the 3′ end of 16S rRNA.

(d) A mutation that generates a nonsense codon in the *lacA* gene

5. Explain why, when *E. coli* is grown in the presence of *both* glucose and IPTG, β-galactosidase protein levels are lower than when it is grown only in the presence of IPTG.
6. Although *E. coli* promoters generally conform to a rather well-defined consensus sequence, no perfect match to this consensus has ever been observed in a naturally occurring promoter. Suggest an explanation.
7. A deficiency in phenylalanyl-tRNA does not have the same effect on transcription of the *trp* operon as does a deficiency in tryptophanyl-tRNA. Why not?
8. The *ilv* operon in *E. coli* codes for the enzymes threonine deaminase, acetohydroxy acid synthase, acetohydroxy acid isomeroreductase, dihydroxy acid dehydrase, and branched-chain amino acid glutamate transaminase. (Refer to figure 24.21 for the roles of these enzymes in amino acid biosynthesis.) Near the 5′ end of the transcribed region of the *ilv* operon is an open reading frame encoding the following amino acid sequence:

fMet-Thr-Ala-Leu-Leu-Arg-Val-Ile-Ser-Leu-Val-Val-Ile-Ser-Val-Val-Val-Ile-Ile-Ile-Pro-Pro-Cys-Gly-Ala-Ala-Leu-Gly-Arg-Gly-Lys-Ala-COOH

Suggest why and how this operon is regulated by the levels of available isoleucine, leucine, and valine.

9. A mutation in the *trp* leader region is found to result in a reduction in the level of *trp* operon expression when the mutant is grown in a rich medium. However, when the mutant is expressed in a medium lacking glycine, a stimulation in the level of *trp* enzymes is observed. Explain these observations. What do you anticipate is the effect of growing the mutant in a medium lacking both glycine and tryptophan?
10. Draw a graph of rRNA gene transcription levels under the conditions shown in figure 35.16.
11. The rRNA genes of *E. coli* are present in multiple copies to facilitate the production of many copies of rRNA during periods of rapid growth. If ribosomal proteins and RNAs need to be assembled in a 1:1 ratio, then why are single-copy genes adequate for ribosomal protein expression?
12. Referring to figure 35.29c, draw a detailed structure of the interaction of residue Arg69 of the *trp* repressor with G9 of its recognition sequence. Explain why binding of repressor proteins to DNA does not disrupt the DNA double helix.
13. A gene encoding an enzyme that breaks down polychlorinated organic pollutants has been isolated from a bacterium. The gene was cloned in *E. coli* and sequenced. The 5′ end of the gene has the sequence given below (spaces every 10 nucleotides are used for clarity). The 5′ end of the corresponding mRNA has the sequence ACUAUUGACA, and the sequence at the N-terminus of the protein is fMet-Leu-Glu-Ile-Ala. You would like to study the properties of this enzyme but can obtain it only in small quantities, even after transforming *E. coli* with the cloned gene. What changes would you consider making in the gene sequence (using site-directed mutagenesis) to increase expression of the protein in *E. coli*?

. . . AGCTCGACTT GCAGACTATC GATAGCGCTA
TAGCGTGTGG ACTATTGACA
. . . TCGAGCTGAA CGTCTGATAG CTATCGCGAT ATCGCACACC
TGATAACTGT
GCACTGGATT ATAGCATGCT GGAAATCGCT . . .
CGTGACCTAA TATCGTACGA CCTTTAGCGA . . .

14. A single operon in *E. coli* contains the following genes in the following order: ribosomal protein S21 (*rpsU*), the primase involved in DNA replication (*dnaG*), and the σ subunit of RNA polymerase (*rpoD*). Suggest mechanisms other than transcription of RNA degradation that might account for the observation that the expression of DNA primase is 60-fold lower than that of the σ subunit.
15. How does the clustering of genes on the bacteriophage lambda genome (fig. 35.23) facilitate its genetic regulation?
16. A variant of bacteriophage λ that is cI^- can undergo the lytic cycle, but it cannot lysogenize. A cro^- mutant can lysogenize but cannot undergo the lytic cycle. What sort of behavior would you expect from a $cI^- cro^-$ double mutant?

C H A P T E R

Regulation of Gene Expression in Eukaryotes

Gene expression in eukaryotes is regulated at all levels from gene to functional protein.

The structural and metabolic differences between prokaryotes and eukaryotes are reflected by many differences in the modes used to regulate gene expression (fig. 36.1). The physical separation of transcription and translation in eukaryotes permits more elaborate processing of messenger RNA while eliminating the possibility of regulatory processes that require strict coupling of the transcription and translation processes. The much greater size of the eukaryotic genome necessitates the existence of many more DNA-binding proteins for regulating transcription, and we will see that basically new types of regulatory systems have developed to fulfill this need. Finally, two kinds of regulatory phenomena occur, which are almost unique to multicellular eukaryotes. First, some regulatory processes facilitate communication (including metabolic regulation) between cells; second, other regulatory devices trigger changes during the course of development that lead to differentiated cells.

It was much easier to discuss regulation in prokaryotes, in which so much of what is known comes from work on *E. coli* and related bacteriophages. In eukaryotes the important information comes from many quarters, and the array of processes is wider by far. We therefore needed to be brief and had to leave out a great

Figure 36.1

Overview of some of the unique features associated with regulation of gene expression in higher organisms. Eukaryotes may be unicellular, as in the case of yeast, or multicellular, as in the case of vertebrates. Regulation of gene expression in all eukaryotes bears a close resemblance to regulation of gene expression in simple unicellular prokaryotes. Yet some striking differences exist. Thus multicellular eukaryotes are composed of differentiated cells that can express only a limited number of the total genes in their chromatin. And most of the chromatin in any particular cell type is highly condensed (heterochromatic), but a small fraction of the chromatin is expanded, or swollen (euchromatic). It is the latter portion of the chromatin that has the potential to be expressed. The regions of the chromatin that are euchromatic are different in different cell types. To convert a region of the genome from the potentially active to the active state requires the binding of regulatory proteins to the promoter. The role of these proteins is to create an environment favorable for the binding of RNA polymerase. Here we come to one of the most striking differences between prokaryotes and eukaryotes. In prokaryotes the regulatory proteins usually bind in the immediate vicinity of the RNA polymerase. In eukaryotes the regulatory proteins may be bound to the DNA at several places, ranging from a point adjacent to the polymerase-binding site to points 1000 or more bases away. Although the explanation for how distantly bound regulatory proteins can influence gene expression is not totally clear, the consensus is that chromosome folding is the most likely way of bringing such proteins into the immediate proximity of the RNA polymerase-binding site. The function of all of the regulatory proteins binding to a single promoter of the DNA is to promote formation of the transcription initiation complex.

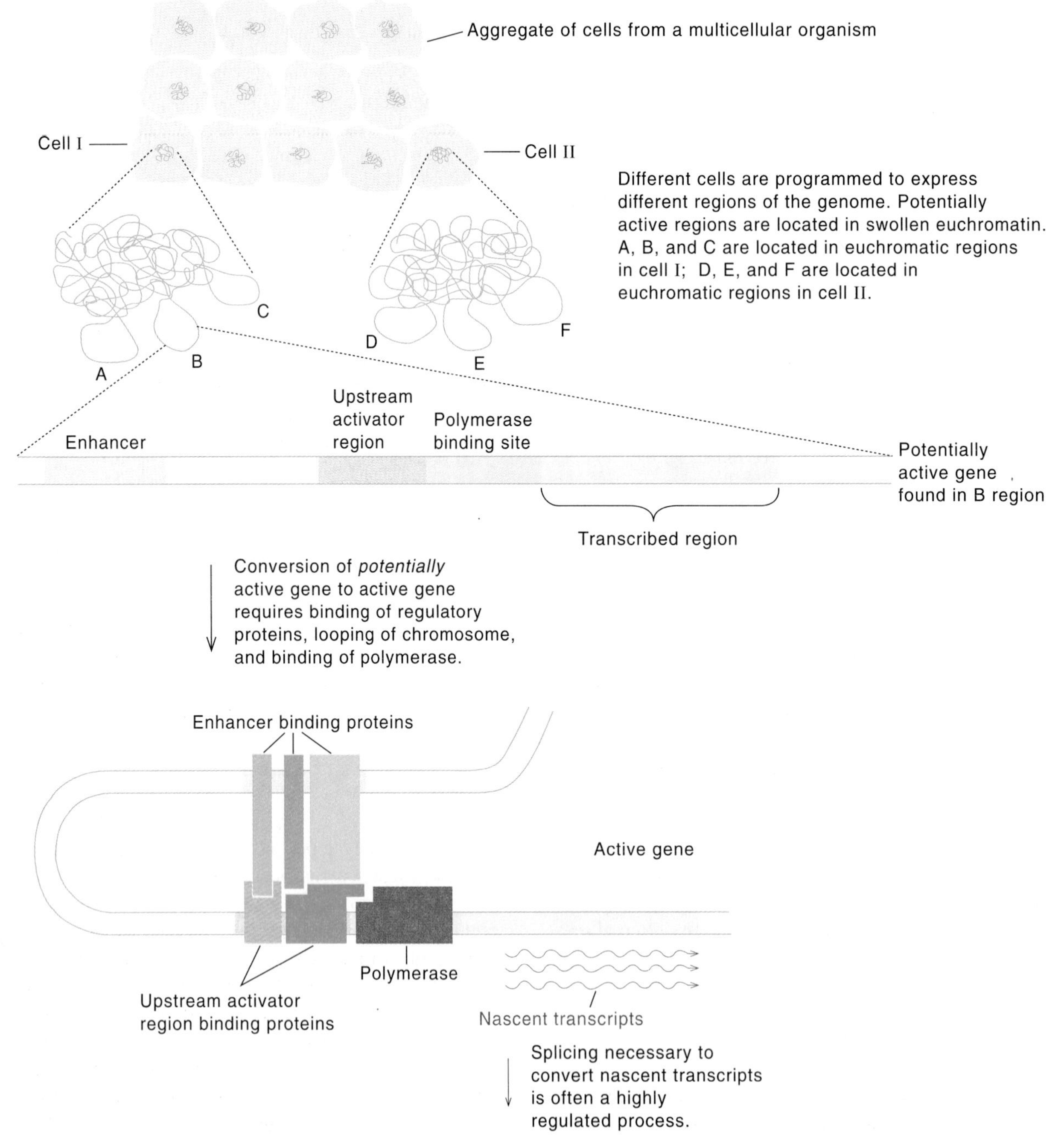
Aggregate of cells from a multicellular organism
Cell I
Cell II
Different cells are programmed to express different regions of the genome. Potentially active regions are located in swollen euchromatin. A, B, and C are located in euchromatic regions in cell I; D, E, and F are located in euchromatic regions in cell II.
A
B
C
D
E
F
Enhancer
Upstream activator region
Polymerase binding site
Potentially active gene found in B region
Transcribed region
Conversion of *potentially* active gene to active gene requires binding of regulatory proteins, looping of chromosome, and binding of polymerase.
Enhancer binding proteins
Active gene
Polymerase
Upstream activator region binding proteins
Nascent transcripts
Splicing necessary to convert nascent transcripts is often a highly regulated process.

Figure 36.2

Diagram of a haploid yeast cell. A yeast cell contains many of the organelles characteristic of a typical eukaryotic cell. Duplication occurs by a budding process. The bud gradually grows until, just before pinching off, it contains a nuclear equivalent of chromosomes, as well as some mitochondria and other elements present in the cytoplasm. Mechanical strength of the yeast cell is guaranteed by a thick cell wall.

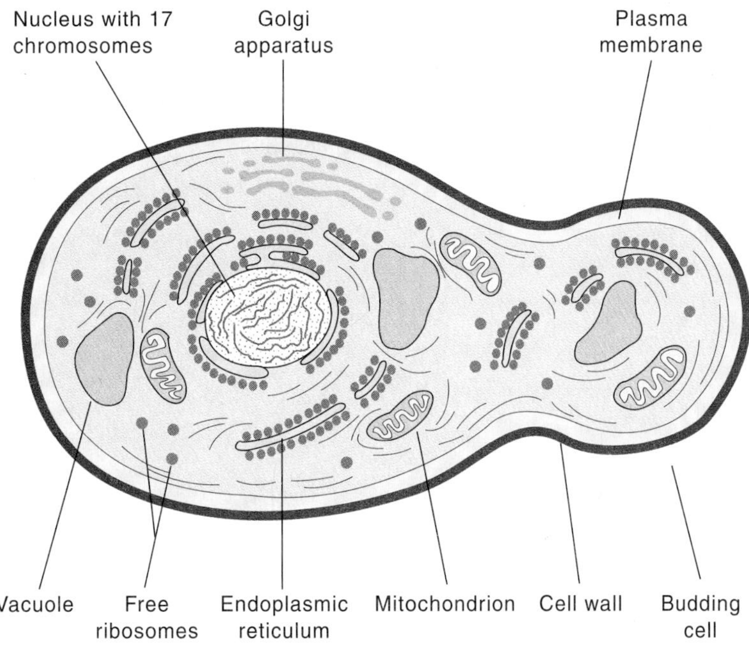

deal. In so doing, we pondered what was most important for students to know. To compensate for the necessary omissions, we included an extensive reference list. Consult this for topics we left out or discussed only briefly. The following chapters on immunobiology and carcinogenesis are also of value on those subjects.

First we look at a unicellular organism, and then we examine mechanisms prevalent in multicellular eukaryotes.

Gene Regulation in Yeast: A Unicellular Eukaryote

The yeast *Saccharomyces cerevisiae* has a cell volume ten times larger and a genome about four times larger than those of *E. coli*. Yeast cells are nearly spherical in shape, divide by budding, and are bounded by a cytoplasmic membrane that is surrounded by a thick polysaccharide cell wall (fig. 36.2). Structures visible in the electron microscope include a nucleus, mitochondria, and endoplasmic reticulum, as well as ribosomes, a Golgi apparatus with secretory vesicles, and several types of granular and vesicular inclusions. The number of mitochondria, microsomes, and ribosomes fluctuates widely with growth conditions, reflecting the presence of regulatory devices that control their numbers.

Galactose Metabolism Is Regulated by Specific Positive and Negative Control Factors in Yeast

Yeast mRNAs are usually translated so that only the 5′ proximal AUG is recognized as a translation start. This minimizes the value of polycistronic mRNAs. Even when functionally related genes are clustered, they usually give rise to separate transcripts. Three of the four genes (*GAL7, GAL10,* and *GAL1*) associated with galactose utilization are clustered on chromosome XI, whereas the fourth, for galactose transport, is specified by a gene (*GAL2*) located on chromosome XII. (*Note:* In yeast the normal wild-type genotype is capitalized, and the mutant genotype is written in lowercase.) Expression of the four structural genes is regulated by specific positive and negative controls. Transcription of the *GAL1, GAL7,* and *GAL10* genes is increased over 1,000-fold when galactose is present, suggesting that galactose is an inducer.

Each of the structural genes is associated with a distinct mRNA. The *GAL7* and *GAL10* genes are transcribed from the same DNA strand, whereas the *GAL1* gene, approximately 600 bp from *GAL10,* is transcribed from the complementary DNA strand (fig. 36.3).

Yeast is ideally suited for genetic analysis because it can be grown and examined in either the haploid or diploid state. Genetic investigations of the GAL system in diploids indicates two *trans*-acting gene products that regulate expression: *GAL4* and *GAL80.* Recall from observations on *E. coli* that *trans*-acting genetic loci usually signify that diffusible gene products are the active agent. Most *gal4* mutants are uninducible for the *GAL* structural genes in the haploid state but inducible in the diploid state when paired with a *GAL4* wild-type gene. This suggests that the active *GAL4* gene product is a positive control protein (an activator) like CAP in the *lac* operon. Rare *GAL4*^{c} mutants result in constitutive expression of the *GAL* structural genes, implying that the structure of the GAL4^c protein in these unusual mutants is modified so that it is no longer repressible. Most *gal80* mutants also

Figure 36.3

Model for the regulation of enzymes of the *GAL* system. (*a*) Three structural genes synthesize distinct mRNAs and enzymes (transferase, epimerase, and kinase). Arrows next to the genes indicate the direction of transcription. Synthesis requires the GAL4 protein. However, it is inactive in the absence of inducer because of complex formation with the GAL80 protein. The GAL80 protein does not prevent the GAL4 protein from binding to specific sites on the DNA, but it does prevent GAL4 protein from activating transcription once bound. (*b*) GAL4 protein becomes active when the GAL80 protein is removed by adding inducer, which binds to the GAL80 protein. Recent evidence suggests that GAL80 protein may not dissociate from GAL4 on induction.

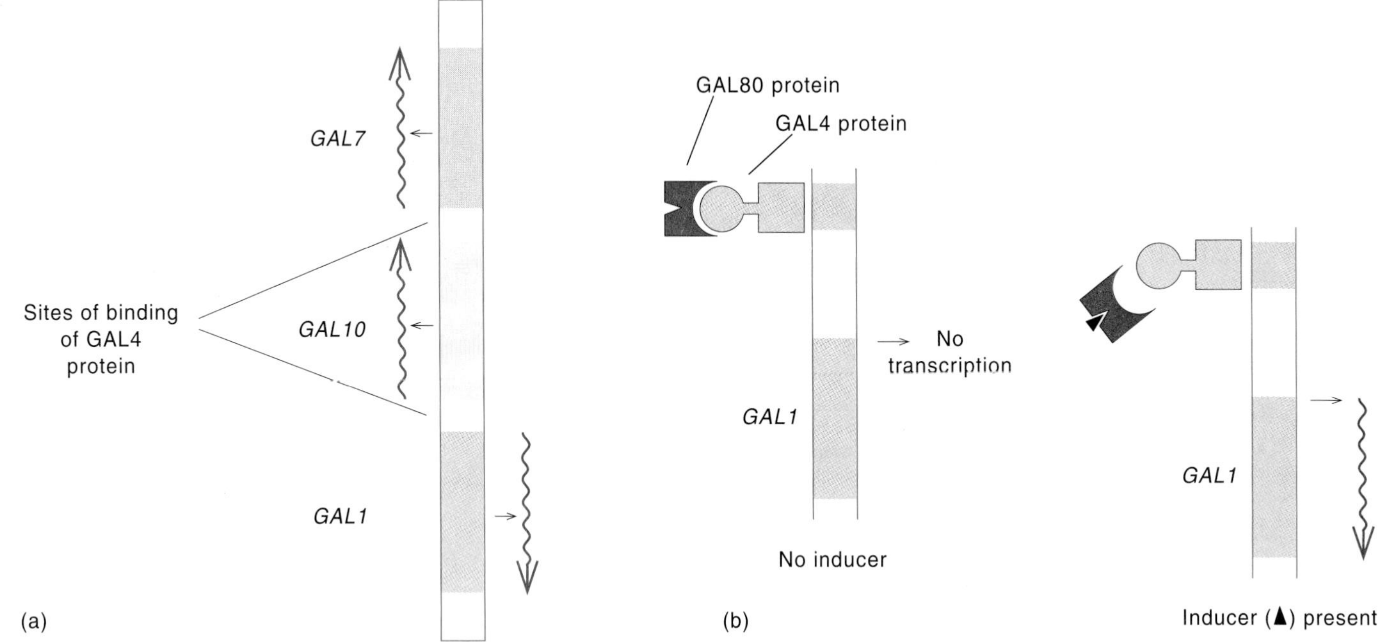

give rise to constitutive expression of the *GAL* genes, which suggests that GAL80 normally acts as a negative control protein (a repressor) of *GAL* gene expression. Rare *GAL80*^{s} mutants are uninducible in the haploid or diploid state where they are paired with wild-type *GAL80.* On the basis of these results it was proposed that the wild-type GAL80 protein binds the inducer galactose and that this binding converts the protein to an inactive form. The GAL80^s protein appears to have lost the site for galactose binding. As a result, it functions as a repressor even in the presence of galactose. This is similar to the situation for i^s mutants in the *lac* operon of *E. coli* (see chapter 35). The phenotypes for different regulatory gene mutants are summarized in table 36.1.

Based on these genetic studies a model was proposed for the mechanism of regulation of the *GAL* system (see fig. 36.3*b*). In this model the GAL4 protein binds to a site upstream of the gene(s) it regulates and promotes RNA polymerase II–dependent transcription of these genes. The GAL80 protein prevents GAL4 protein from functioning as an activator by binding to it. Induction by galactose results from the binding of the sugar to the GAL80 protein, which changes its structure so that the GAL4–GAL80 complex is altered. It is not clear if actual breakup of the complex is required to relieve the repression.

The proposed model postulates *cis*-acting sequences upstream of the target genes for GAL4 protein. The GAL4 protein binds to four related 17-bp sequences located between 150 and 200 bases upstream of the *GAL10* gene. The specific DNA sequences to which the GAL4 protein binds were determined by reacting the DNA with dimethylsulfate and then following the methylation pattern by nucleotide sequencing to determine which guanine bases are protected by the regulatory protein. The sequences that showed protection in a wild-type *GAL4* strain were not pro-

Table 36.1
Phenotypes of Haploid and Diploid Yeast with Regulatory Gene Mutations

Genotype	GAL10 (epimerase)[a] Noninduced	GAL10 (epimerase)[a] Induced	GAL1 (kinase) Noninduced	GAL1 (kinase) Induced
Wild type	–	+	–	+
gal4	–	–	–	–
gal4/GAL4	–	+	–	+
gal80	+	+	+	+
GAL4^c	+	+	+	+
GAL80^s	–	–	–	–
GAL80^s/*GAL80*	–	–	–	–

[a]In the induced state, galactose is added to the growth medium. It should be noted that GAL10 and GAL1 respond in identical fashion to the different mutations.

Figure 36.4

Ptashne's domain-swap experiment. The GAL4 protein normally binds at a site upstream from the *GAL1* gene. In the absence of GAL80 this activates transcription from the *GAL1* gene. If the DNA-binding domain of GAL4 is replaced by the DNA-binding domain of the lexA protein, transcription from the *GAL1* gene does not occur because the hybrid regulatory protein does not bind at the appropriate site. However, if the DNA-binding site is also changed to that of the DNA-binding site for the lexA protein, then the *GAL1* gene becomes active again.

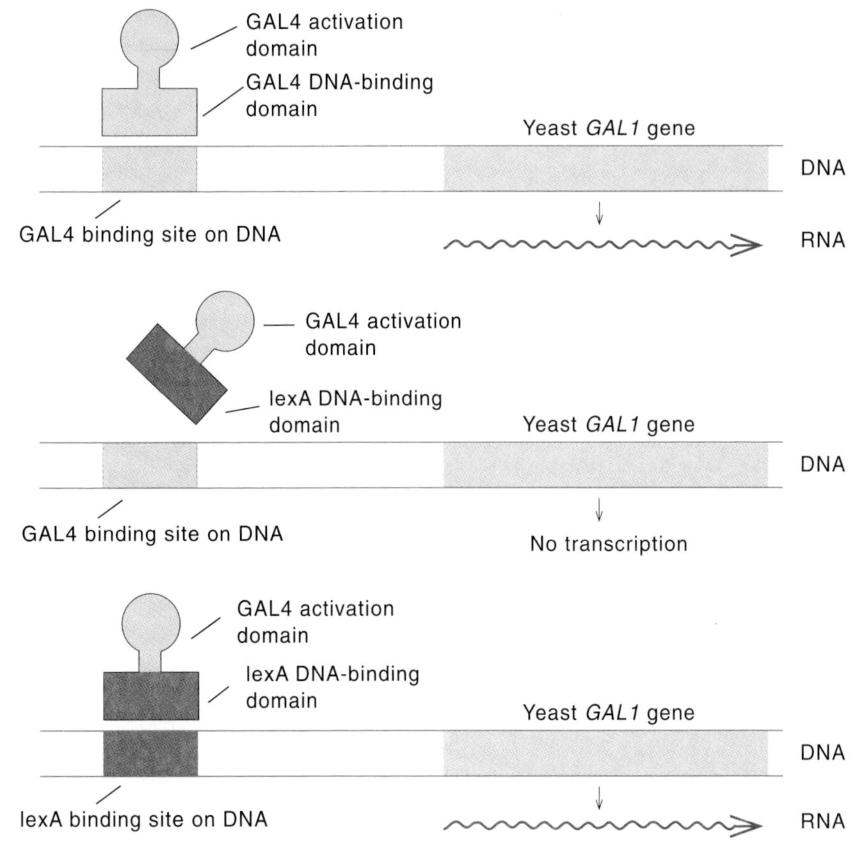

tected in the *gal4* mutant derivative. Further manipulations of the promoter region showed that only one of the 17-bp sequences is needed for normal regulation.

The GAL4 Protein Is Separated into Domains with Different Functions

The model for GAL4 protein specifies a multifunctional protein with binding sites for DNA and GAL80 protein as well as a site that somehow activates transcription. Deletion studies were performed to see if these different sites could be allocated to different parts of the 881-amino-acid GAL4 protein. A mutant protein containing only the first 98 N-terminal amino acids still binds DNA but cannot bind the GAL80 protein or activate transcription. Additional mutant studies show that amino acids 148–196 and 768–881 are required to activate transcription and that amino acids 851–881 are required for GAL80 protein building.

Further experiments indicate that different parts of the GAL4 protein and the lexA bacterial repressor protein (see chapter 31) are functionally interchangeable. In a "domain swap" experiment, the DNA-binding domain of the lexA protein was inserted in place of the GAL4 DNA-binding domain (fig. 36.4). As might be expected, this hybrid protein could not activate the wild-type *GAL1* gene for transcription unless other changes were made in the DNA. If the DNA-binding site for the GAL4 protein was replaced by the DNA-binding site for the lexA protein, then the same hybrid protein was able to activate transcription of the *GAL1* gene.

The *GAL* system shows both similarities and differences to typical bacterial regulatory systems. In both systems, DNA-binding regulatory proteins play a major role. However, in the yeast system the binding site for the regulatory proteins is often located some distance upstream from the RNA polymerase-binding site. In yeast such distant sites required for activation are referred to as upstream activator sequences (UASs). Another difference between yeast and *E. coli* is with respect to the localization of the regulatory protein on the DNA. The GAL4 regulatory protein is bound to its promoter site in both the inactive and active states. In *E. coli,* regulatory protein binding at the promoter is invariably associated with the protein in its active state, whether that be as a repressor or as an activator.

Mating Type Is Determined by Transposable Elements in Yeast

Yeast has two haploid cell mating types: *MAT*α (or simply α) and *MATa* (or *a*); on contact, haploid cells of opposite mating types fuse to form a single diploid (a/α) cell. Diploid cells can grow and divide indefinitely as diploid cells, or they can sporulate, a process in which they undergo meiosis and give rise to two α cells and two *a* cells for each diploid cell. The haploid mating type is determined by specific sequences at the *MAT* locus (fig. 36.5). These

Figure 36.5

Yeast chromosome III, showing the *HML*, *MAT*, and *HMR* loci. *W* is a region common to *MAT* and *HML* but not *HMR* (≈ 750 bp). *X* is a region found at *MAT*, *HMLα*, and *HMRa* (≈ 700 bp). Y_a is a specific substitution found at *MATa* and *HMRa* (≈ 600 bp). Y_α is a specific substitution found at *HMLα* and *MATα* (750 bp). Z_1 is a region found at *MAT*, *HMLα*, and *HMRa* (≈ 250 bp). Z_2 is a region found at *MAT* and *HMLα* (≈ 70 bp).

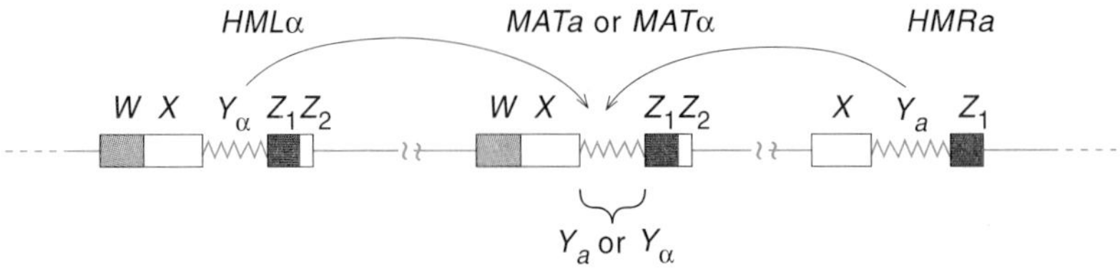

Figure 36.6

Structure of mating loci determinants in yeast. The genetic regions *HMLα* and *HMRa* are normally silent, whereas *MATa* or *MATα* are active. *MATa*, which contains the Y_a segment, expresses transcript *a*1. *MATα*, which contains the Y_α segment, expresses two transcripts: α_1 and α_2. The structures in and around the Y_a and Y_α segments are the same at the storage locus and the expression locus. The inactivity at the storage loci results from far upstream *cis*-acting elements not present at *MAT*. These elements, in conjunction with the four *SIR* gene products, negatively regulate expression of mating type genes at the storage loci.

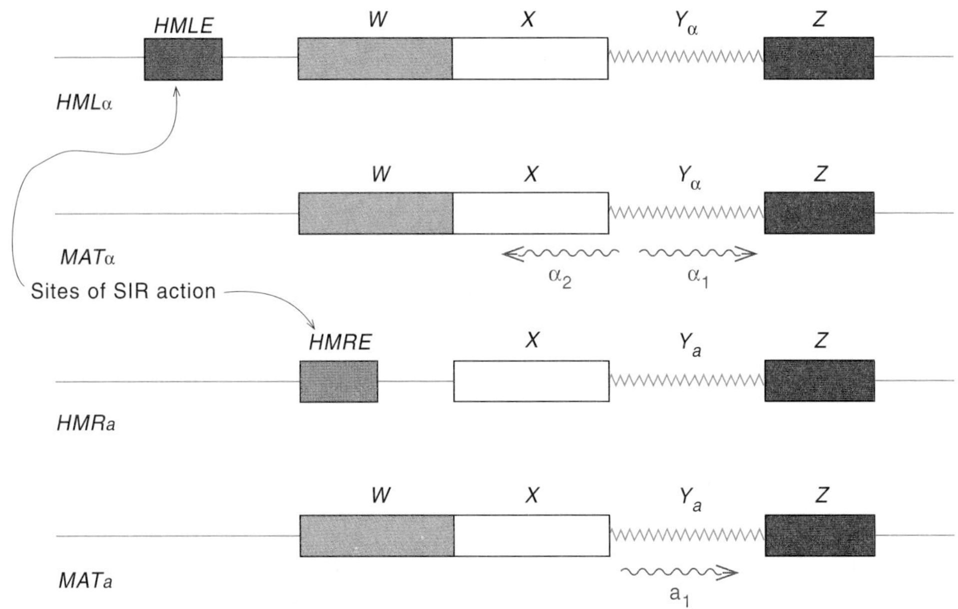

sequences are found at the *HMLα* or *HMRa* locus, where they are usually not expressed. When the sequences stored at the *HMLα* locus are transposed to *MAT*, they express; similarly, when sequences are transposed from *HMRa* to *MAT*, they express.

Why are the *α* and *a* sequences expressed only when they are present at the *MAT* locus? The explanation for this was suggested by the finding of four unlinked genes: *SIR1–SIR4*. If any of these genes are mutated, the *a* and *α* sequences at *HMLα* and *HMRa* express just as they do at *MAT*. Further experiments showed that *trans*-acting regulatory factors, encoded by the *SIR* genes act as repressors at "silencer" sites (*HMLE* and *HMRE*) adjacent to *HMRa* and *HMLα*. Remarkably, these silencer sites are located more than a kilobase away from the regions they control (fig. 36.6). The silencer sites are not found at or near the mating type locus (*MAT*), explaining why the *a* and *α* coding sequences are expressed when transposed to *MAT*.

An important question remaining is how a single genetic locus determines the haploid yeast cell mating type. Haploid cells that carry *MATα* behave as *α* cells; similarly, cells that carry *MATa* behave as *a* cells; finally, diploid cells that carry both *MATα* and *MATa* do not express any haploid-specific genes. To explain the key role of *MAT*, it was proposed that the *MAT* locus encodes regulatory proteins that control unlinked genes that determine mating type. Subsequently, it was shown that the *MATα* locus contains two genes: *MATα*1 and *MATα*2, which encode the *α*1 and *α*2 regulators, respectively. Ira Herskowitz and his co-workers found that *α*1 turns on expression of the *α*-specific genes and *α*2 turns off expression of *a*-specific genes (fig. 36.7).

Figure 36.7

Regulation of *a*-specific genes (*a*sg) and α-specific genes (αsg) in α and *a* haploid cells. In an α cell, $\alpha2$ inhibits *a*sg and $\alpha1$ activates αsg. In an *a* haploid cell the *a*1 transcript made at *MATa* does not appear to exert any regulatory function. ($\oplus$ = positively regulated; $\ominus$ = negatively regulated.)

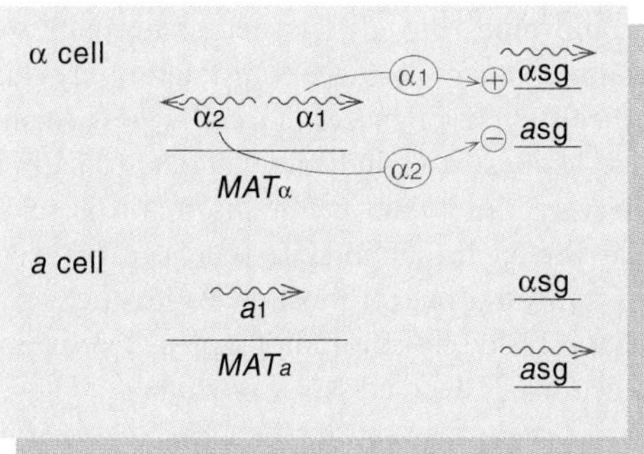

In addition to understanding the regulatory mechanisms that give the haploid cells their specific properties, considerable progress has been made in determining the mechanism that regulates meiosis in diploid cells. Diploid *a*/α cells undergo sporulation and meiosis when they are subjected to nutritional starvation. It is believed that the reactions leading to meiosis are triggered by a complex of the $\alpha2$ and *a*1 proteins, which repress the haploid specific gene *RME*1 (regulator of meiosis), allowing the cells to initiate meiosis (fig. 36.8).

The control of mating type in yeast illustrates a specialized role of mobile or transposable genetic elements in which mobile elements hop from one site to another. Mobile genetic elements are commonly found in both prokaryotes and eukaryotes, but most of them show little site specificity. Indeed, their almost random insertion behavior has led to the proposal that their major function in most cases is to promote evolutionary change.

Mating type in yeast is much better understood than mating type of multicellular organisms such as vertebrates, where sex type is determined by gene dosage. For example, in humans, males carry one X chromosome, whereas females carry two X chromosomes. The mechanism used in yeast and other fungi is more responsive to factors in the external environment, which may be more suitable for a unicellular organism that is directly exposed to its surroundings.

This concludes our discussion of regulatory phenomena in yeast. Most of the regulatory processes in yeast are highly reversible, just as they are in *E. coli*. In multicellular eukaryotes, additional regulatory mechanisms often show an irreversible character.

Figure 36.8

Regulation of meiosis in yeast. Haploid cells are not able to initiate meiosis because they express *RME*1, which is a negative regulator of meiosis. Diploid cells are able to initiate meiosis because they make a complex of $\alpha2$ and *a*1 that inhibits the synthesis of the *RME*1 product.

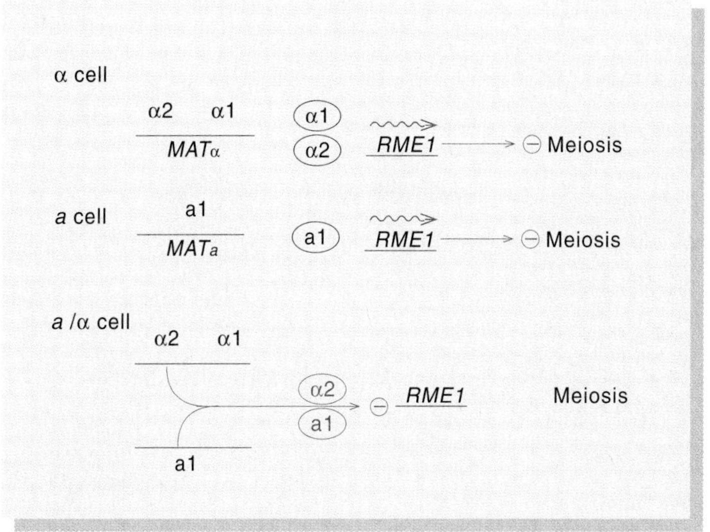

Gene Regulation in Multicellular Eukaryotes

We have seen that yeast differentiates into three cell types: the *a* and α haploid types and the *a*/α diploids. In multicellular eukaryotes we must deal with a much wider range of cell types and a much larger genome. First we consider some aspects of the changing chromosome architecture, which reflects changes in gene activity. Then we examine the underlying changes in the way that expression is controlled at the transcriptional, posttranscriptional, and translational levels. Finally, we consider some developmental processes that are triggered by changes in gene expression.

Nuclear Differentiation Starts in Early Development

In the mature adult state, a multicellular eukaryote contains many cells, each with a potential for the constitutive or inducible expression of a unique subset of genes belonging to the total genome. Whereas most differentiated cells in the adult contain the normal amount of genomic DNA, the expression of only a fraction of the genome in any particular cell type suggests that the average nucleus may have lost its totipotency. The first direct evidence that nuclei irreversibly differentiate during development came from R. Briggs and T. J. King in 1952. Working with frog eggs, they showed that the original nucleus from an unfertilized egg could be replaced by the diploid nucleus from a developing animal (fig. 36.9). If the inserted nucleus came from a frog very early in development (blastula), a mature frog developed. If, however, the nucleus came from a later stage of development, no growth or only limited growth ensued. These nuclear transplantation experiments demonstrated that during development the nucleus assumes a pattern of expression that is difficult or impossible to reverse even when the partially differentiated nucleus is returned to the environment of the egg cytoplasm.

In 1962, J. Gurdon took this type of experiment one step further by showing that adult frogs could be developed by inject-

Figure 36.9

Nuclear transplantation technique. The nucleus of an unfertilized egg is mechanically removed with a micropipette (*left*), and a nucleus from a blastula cell is removed and microinjected into the enucleated egg.

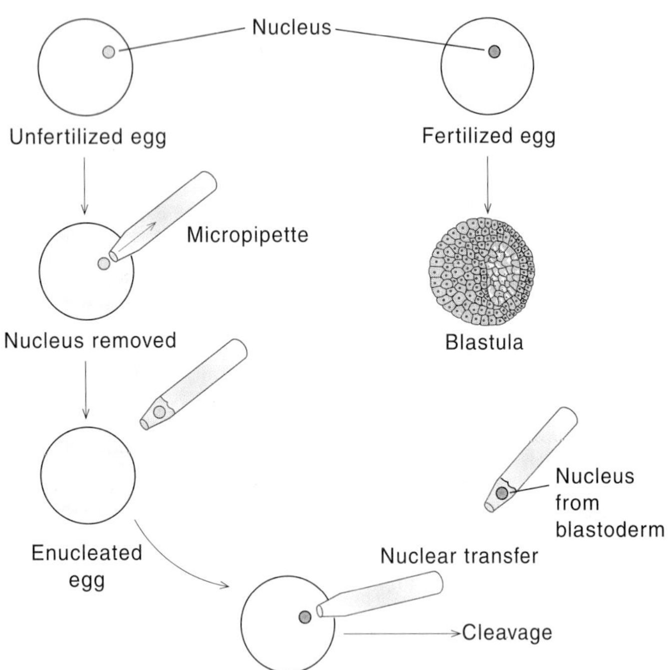

ing enucleated eggs with nuclei from the intestinal epithelium of tadpoles. The success frequency was much lower than when nuclei from earlier stages of development were used. Nevertheless, such results showed the influence of the cytoplasm on nuclear expression; they also demonstrated that in certain cases, nuclei already tentatively committed to a specific pathway of development can be reprogrammed by placing them in a different environment. The most important lesson to be gained from such experiments is that the nucleus tends to assume a stable pattern of expression that is ultimately dictated by its surroundings.

A most convincing piece of evidence for the importance of cytoplasm on nuclear development comes from studies on teratoma, a unique type of tumor found in many kinds of mammals. It is composed both of neoplastic cells, like other tumors, and of many kinds of differentiated cells. Only the stem cells are neoplastic, producing more stem cells or more differentiated cells, usually both. In many ways the stem cells behave like embryonic cells. They can be dissociated into single cells and cultured in vitro like bacterial cells. Most remarkably, they can be implanted in an early embryo (blastocyst) to produce a hybrid chimera, in which, in favorable cases, all of the tissues possess some cells from the tumor parent (see fig. 32.11). This result shows that the neoplastic stem cells are capable of reverting to completely normal behavior when subjected to the embryonic environment.

These pioneering studies demonstrating nuclear differentiation and dedifferentiation are supported by a broad range of genetic and biochemical findings that indicate that chromosomal structural differences often can be correlated with changes in the potential for gene expression.

Chromosome Structure Varies with Gene Activity

In interphase cells, chromosomes are in a dispersed form called chromatin. Chromatin exists in a highly condensed form known as heterochromatin or a swollen form known as euchromatin. All nuclei contain both types of chromatin. Some chromosomes or parts of chromosomes are always heterochromatic (constitutive heterochromatin); others are heterochromatic only during certain times of the cell cycle or in certain cell types (facultative heterochromatin). Many studies indicate that euchromatin is relatively active in RNA synthesis and heterochromatin is relatively inactive. Whereas the euchromatic state seems to be necessary for a high level of transcription, for most genes it is not sufficient. This has been demonstrated by autoradiography of cells grown briefly in the presence of ^{3}H-labeled RNA precursors. Those regions of the genome that are transcriptionally active incorporate the label; the extent of labeling is proportional to the amount of RNA synthetic activity. Some euchromatic regions appear active by this test, whereas others do not.

Giant Chromosomes Permit Direct Visualization of Active Genes

Polytene chromosomes are unusually large chromosomes found in the salivary glands of certain insects. Because of their size, they are convenient vehicles for studying the relationship between chromosome structure and activity. Polytene chromosomes are produced by the repeated replication of interphase chromosomes without separation, resulting in a large number of chromosomes that remain laterally aligned. For example, the DNA content in the giant salivary gland cells of *Drosophila melanogaster* may be as much as a thousand times that of other cells in the fruit fly. Microscopically, the stained chromosomes appear as cross-banded extended bodies (fig. 36.10). About 5,000 bands are in *Drosophila,* and it is tempting to associate single bands with single genes. However, the average DNA content found in each band is 30,000 bp, which is considerably more DNA than necessary to encode the average protein (about 1,000 bp).

Transcription in polytene chromosomes usually is associated with local swellings of the bands, called puffs (see fig. 36.10). Sometimes puffing results in a broadening and lengthening, and sometimes the extended DNA projects laterally into loops that combine to form a large ringlike structure. As a rule, a puff originates from a single band, but it can result from swelling of one or more bands. The extent of incorporation of labeled RNA precursors into a puff is approximately proportional to the size of the puff. A detailed investigation of the puffing patterns as a function of the physiological state of the organism shows that different bands become activated, swollen, and transcriptionally active in a sequentially related, tissue-specific fashion.

The correlation of puffs with genetic functions has focused on two types of gene products: Secretory proteins and proteins formed in response to heat shock. Certain puffs in *Drosophila* can be artificially induced by the insect hormone ecdysone, which is instrumental in regulating development. Some puffs are induced directly by exposure to the hormone, whereas others are induced

Figure 36.10

A segment of giant chromosome from the midge larva of *Rhyncosciara* at different stages during development. The arrows and connecting lines indicate comparable bands. Changes in the extent of swelling of different regions reflect the activity of those regions in transcription. (*Source:* Drawing based on the results of M. F. Breuer and C. Pavin, *Chromosoma* 7:257–280, 1946.)

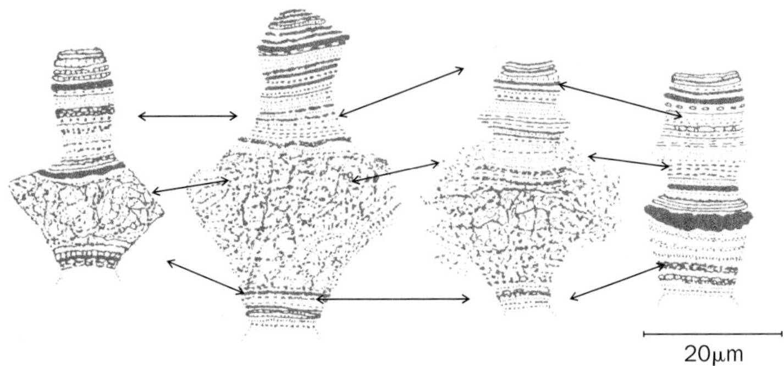

Figure 36.11

Diagram of nuclei obtained from cells in the mucous membrane of the human mouth. (*a*) Nucleus from a female, showing one Barr body (arrow). (*b*) Nucleus from a male, with no Barr body. (*c*) Nucleus from an XXX female, showing two Barr bodies. (*d*) Nucleus from an XXXX female, showing three Barr bodies.

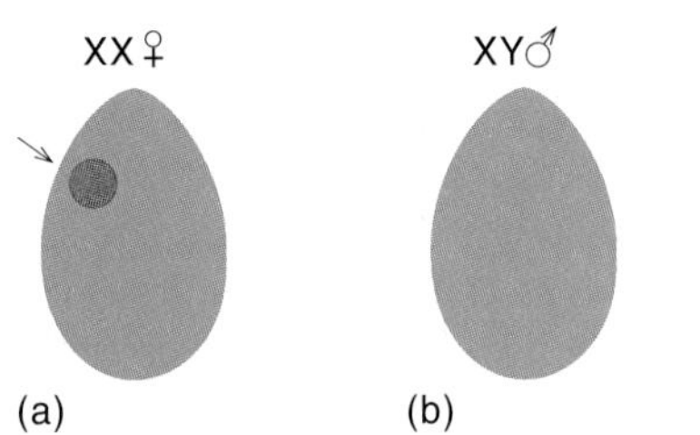

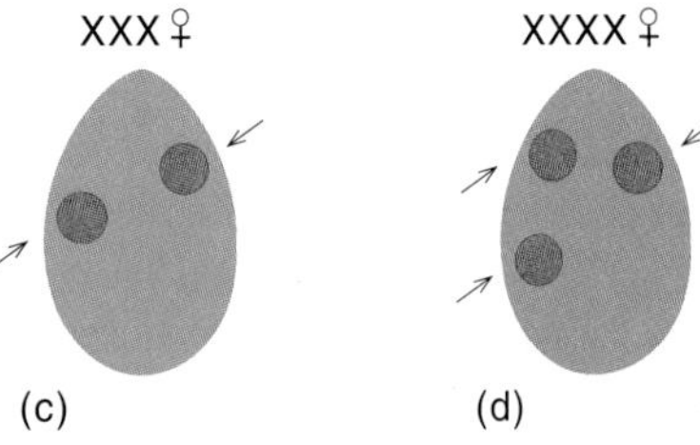

indirectly, depending for puffing on the gene products of the more directly induced puffs. This dependence is shown by adding drugs that inhibit protein synthesis, with the result that secondary puffs do not form.

In Some Cases, Entire Chromosomes Are Heterochromatic

In some cases, entire chromosomes can be heterochromatized and stay that way through successive cell divisions. Male and female mammals are distinguished by the fact that females carry two X chromosomes, and males carry only one. Invariably, one of the X chromosomes in the somatic tissue of females is inactivated and condensed into a heterochromatic state known as a Barr body (fig. 36.11). The consequences of X-chromosome inactivation are readily observed in females that are heterozygous for an X-linked mutation.

For example, the enzyme glucose-6-phosphate dehydrogenase (G-6-PD) is encoded by an X-linked gene. A female that is heterozygous for this gene may carry two alleles that produce electrophoretically distinct forms of the same enzyme (A and B). When isolated cells from a skin biopsy are cloned, each clone contains either the A or the B form of the G-6-PD enzyme but never both. If every skin cell of a single organism were to be analyzed for the enzyme, large homogenous patches expressing one or the other of the G-6-PD alleles would be found. This pattern indicates that the decision as to which X chromosome of the female is inactivated is made at the multicellular stage in the embryo. Once the decision is made, that X chromosome remains inactive through successive cell divisions. There is an equal chance that the X chromosome from either parent is inactivated. In genetically abnormal cells that contain three or four X chromosomes, two or three Barr bodies, respectively, can be found (see fig. 36.11).

X-chromosome inactivation provides a simple means of maintaining equal amounts of active X-linked genes in both males and females. This form of so-called dosage compensation is not found in all species. For example, in *D. melanogaster,* neither of the X chromosomes in the female cells is condensed into a Barr body. In this species, dosage compensation, which regulates the activity of specific alleles, operates by a different mechanism, so many alleles in the female X are about 50% as active as the corresponding alleles in the male. However, not all alleles in the *Drosophila* X chromosome are regulated in this way, so for certain genes, twice as much gene product is produced in females as in males.

Biochemical Differences between Active and Inactive Chromatin

Much of the preceding discussion on chromatin structure indicates that active chromatin exists in a more swollen state than inactive chromatin. Several biochemical changes accompany the transition from condensed to swollen chromatin. These changes include a redistribution of nucleosomes along the DNA duplex, chemical modification of histones, alteration in the pattern of nonhistone chromosomal protein binding, and chemical modification of the DNA. Currently, most of these changes are discussed in a general, descriptive manner because their causes and consequences are not known.

DNA in Active Chromatin Is More Susceptible to DNase Degradation. The greater accessibility of transcriptionally active chromatin has been elegantly demonstrated by Groudine and Weintraub for the hemoglobin and ovalbumin genes in the chicken. Chromatin was isolated from chicken erythrocytes, in which the hemoglobin genes were recently very active, and from the oviduct, in which the ovalbumin gene is very active. These two chromatins were degraded by DNaseI using an assay that measured gross DNA degradation as well as that of specific genes. In both cases the rate of DNA degradation from the active gene was much greater than that from average DNA. Thus in erythrocyte chromatin the globin gene DNA was rapidly degraded, whereas in oviduct chromatin the ovalbumin gene DNA was rapidly degraded. Extensive regions surrounding the active genes were also quite susceptible to DNaseI hydrolysis.

DNA Methylation Is Correlated with Inactive Chromatin. Methylation of cytosine in CpG sequences may play a regulatory role in the gene expression. In chapter 30 it was noted that 5-methylcytosine (5-MC) is the main modified base in vertebrate DNA. Most of the 5-MC occurs in the dinucleotide CpG. Indeed, in mammals and birds, approximately 50%–70% of all such dinucleotide sequences are modified. In this connection it is noteworthy that CpG sequences occur much less frequently than would be expected statistically. The mechanism of methylation has been explored by DNA transfection of certain tissue culture grown cells. If the DNA used in transfection is methylated, the pattern of methylation is maintained through many cell duplications. Likewise, when unmethylated DNA is used in transfection, a nonmethylated pattern is maintained. These results indicate that under many circumstances, methylation is passively maintained by a signal that recognizes hemimethylated DNA. If the C residue in a CpG dinucleotide is methylated, immediately after semiconservative DNA replication, only the parental DNA chain is methylated; the passive methylation system signals methylation of the corresponding C residue in the new DNA chain (fig. 36.12).

The extent of methylation of a gene is correlated with its ability to transcribe. Given that DNA methylation usually reduces transcription, two important, closely related questions remain unanswered: How is methylation regulated in vivo? How does methylation interfere with transcription? Since methylation is known not to interfere with the elongation phase of RNA synthesis, it seems likely that methylation blocks initiation. The binding of polymerase and other regulatory proteins at the initiation locus is sensitive to modification of these nucleotides. The precise inhibition mechanisms, however, await further elucidation.

Before turning to other questions, we note that methylation or something like methylation could be at the root of X-chromosome inactivation. In this connection it may be relevant that *Drosophila,* which does not show the methylation phenomenon, does not show X-chromosome inactivation either.

Figure 36.12

Hemimethylated DNA. The unmethylated C residue in the indicated sequence is destined to be methylated shortly after DNA replication.

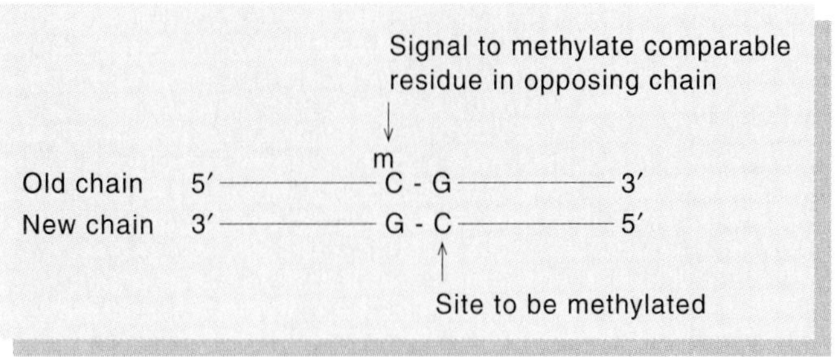

Until recently, the significance of DNA methylation to gene expression has been hampered by the lack of genetic correlates. This situation has changed because it is now possible to breed mouse embryos that are genetically deficient in DNA methyltransferase activity. Genes inherited from paternal and maternal genomes are not equivalent, and both are required for mammalian development. The difference between the parental genomes is believed to be due to gamete-specific differential modification, a process known as genomic imprinting. Methylation patterns can be inherited in a parent-of-origin-specific manner, suggesting that DNA methylation may play a role in genomic imprinting. The functional significance of DNA methylation in genomic imprinting was strengthened by the finding that CpG islands in three imprinted genes, H19, insulinlike growth factor (Igf-2), and insulinlike growth factor 2 receptor (Igf-2r), are differentially methylated depending on their parental origin. Rudy Jaenisch and his coworkers examined the expression of these imprinted genes in mutant mice that are deficient in the DNA methyltransferase activity. The expression of all three genes was abnormal in mutant embryos; the normally silent paternal allele of the H19 gene was activated, whereas the normally active paternal allele of the Igf-2 gene and the normally active maternal allele of the Igf-2r gene were both repressed. These results demonstrate that DNA methylation is required to control differential expression of the paternal and maternal alleles of imprinted genes.

Histones May Play an Active Role in Transcription

The role of the nucleosome in the folding of DNA has often been thought of as purely a packaging one. However, in some instances, nucleosomes are precisely positioned around regulatory sequences of genes, suggesting a more active role for histone in transcription. A case in point has emerged from the studies of H. Richard-Foy and G. Hager on the chromatin organization in the vicinity of the major promoter for the integrated DNA copy of the mouse mammary tumor virus (MMTV) in tissue culture cells. This promoter is known to be regulated by steroid hormones. Nucleosome positioning was determined by localization of sites sensitive to

Figure 36.13

Region in and around the early transcription start sites of the SV40 genome. The base pair number on the circular genome is indicated. There are several transcription start sites. One cluster of start sites (early), used initially after infection, is located about 27 bp downstream from the TATA box. The other cluster of start sites (late early) is used at later times. This cluster of start sites is located upstream of the TATA box. One of the products of early transcription is the protein known as large T antigen. This protein binds in and around the core *ori.* T antigen inhibits early transcription. The upstream positive control *cis* elements include the tandem 72-bp enhancer sequences and three tandem 21-bp sites. The latter function in conjunction with host-encoded protein known as SP1.

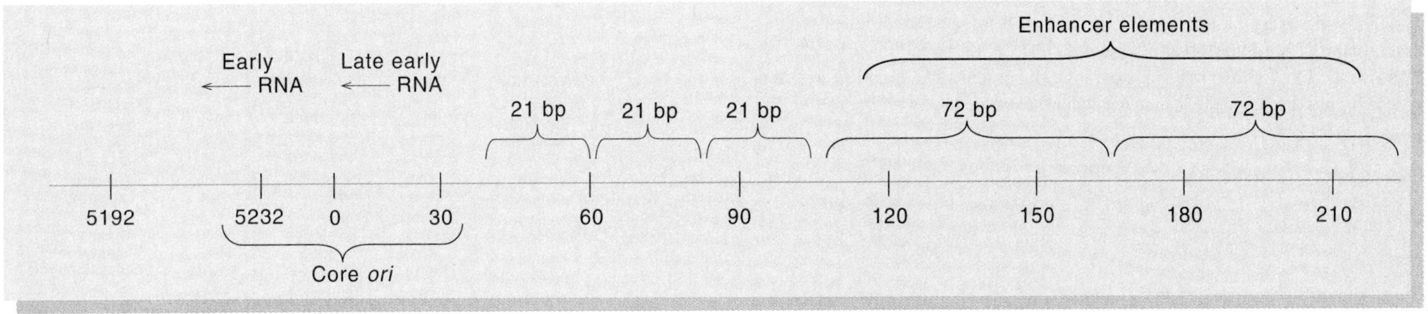

cleavage by micrococcal nuclease. In the absence of hormone a regular cutting pattern is obtained with cleavage sites at +136, −60, −250, −444, −651, −826, and −1019 reactive to the Cap site of the transcript. In the presence of hormone the cutting pattern is unchanged, except for a region between −60 and −250 that becomes hypersensitive to cleavage. This region contains the DNA sequences to which steroid receptor complexes bind during transcriptional activation.

The precise localization of nucleosomes in and around the MMTV major promoter suggests that there is something about the DNA sequences in this region that favors nucleosome formation in certain regions and disfavors it in others. It seems likely that this feature of the promoter is designed to promote access of the transcription apparatus to this promoter. Thus the possibility that histones actually facilitate transcription in instances such as this must be seriously considered.

Enhancers Are Promoter Elements That Operate over Great Distances

In chapter 33 (see fig. 33.13) we saw that several proteins in addition to RNA polymerase II are required for the initiation of transcription at typical eukaryotic promoters. In addition to signals for regulatory protein binding in the immediate vicinity of the polymerase-binding site, other signals function at more distant locations. In yeast DNA, sequences involved in this type of regulation have been called UAS sequences because their action is usually confined to regions located a few hundred bases upstream from the promoters they influence. In vertebrates, *cis*-active sequences operating over much greater distances have been discovered. These sequences have been named enhancers because they enhance expression of genes located in their vicinity. Enhancer signals are effective downstream as well as upstream of the promoters they influence. Moreover, many enhancers are equally effective in either orientation on the DNA.

The first enhancer was discovered by George Khoury in the SV40 virus. It should be recalled that the SV40 genome is a circular DNA duplex containing about 5,300 bp. In SV40-infected cells, viral RNA synthesis is divided into early and late phases. Early transcription originates from a block of sites illustrated in figure 36.13. A TATA box, typical of eukaryotic gene promoters (see chapter 33) is situated about 27 bp upstream from the transcription start site. Immediately upstream of the TATA box the SV40 promoter contains three tandemly repeated GC-rich segments that bind a transcription activator protein known as SP1. Further upstream from the transcription start site a tandemly repeated 72-bp sequence occurs (−116 to −188 and −189 to −261 from the 5′ end of the messenger). Removal of one of these sequences has no effect on transcription, but removal of both 72-bp sequences greatly lowers early transcription.

Surprisingly, the precise location or orientation of the 72-bp segment is not critical to the stimulating effect on transcription. Thus the 72-bp segment remains effective after inversion or after translocation further upstream or downstream from the transcription start site. Foreign genes inserted into DNA containing the SV40 enhancer are frequently stimulated in the same way as the SV40 early region, demonstrating the general stimulating effect of this enhancer.

A more detailed analysis of the SV40 enhancer shows that it is divisible into smaller elements 15–20 bps in length that bind one or more protein factors specifically. These elements, sometimes called enhansons, are ineffective when separated from one another. It is believed that different enhancers are composed of different combinations of enhansons (fig. 36.14). The emerging picture of the enhancer is of a complex, multicomponent segment of DNA that can bind different combinations of protein activators. The use of protein activators in combinations appears to be a major strategy used in multicellular eukaryotes to produce a greater variety of responses. It has the advantage that, given a finite set of regulatory proteins, a much greater variety of combinations can be produced. The use of complexes containing combinations of regulatory proteins has implications for the structures of regulatory proteins that are discussed later on.

Figure 36.14

Enhancers are frequently composed of two or more components, or enhansons. Each enhanson contains binding sites for two or more proteins, which can interact with each other when bound to the enhanson if not before. An enhancer that contains two enhansons, each of which can bind two different proteins, can be arranged in 2^4, or 16, ways with respect to binding sites. From considerations such as this, it is clear that a great deal of variety is possible in enhancer construction with a very limited number of DNA-binding sites for different regulatory proteins. The proteins that bind to the enhancer do not all make direct contact with the DNA. Often one finds pyramids of proteins that bind to each other in which only the base of the pyramid makes direct contact with the DNA.

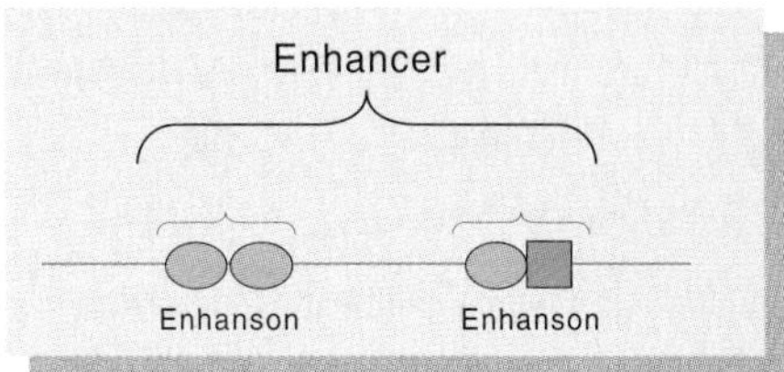

Figure 36.15

Possible mechanisms for enhancer action. (1) The enhancer draws the associated gene to the nuclear matrix, where it is more accessible to the transcription apparatus. (2) The enhancer is the initial binding site for an element that subsequently moves. (3) The enhancer folds or loops, depending on its polarity, to bind with other promoter elements.

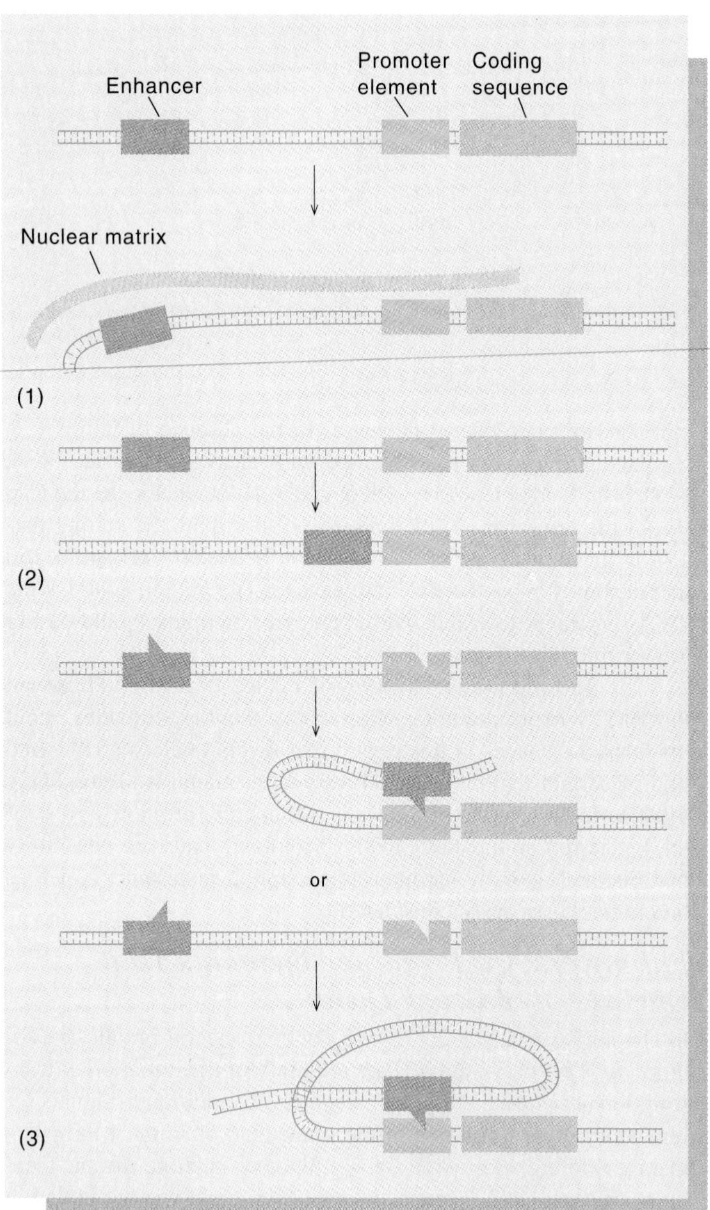

Several proposals have been made for how enhancers influence transcription over long distances: (1) The enhancer element may function as an attachment point to a structural component of the nucleus to stimulate transcription; (2) it may serve as an initial binding site for some factor required for transcription that must subsequently move along the duplex to the initiation point for transcription; or (3) folding or looping of the chromosome may bring the *cis*-bound enhancer proteins into close contact with the gene it stimulates (fig. 36.15). Currently, the third possibility is strongly favored, but this does not exclude the other two mechanisms.

DNA-Binding Proteins That Regulate Transcription in Eukaryotes Are Often Asymmetrical

In the previous chapter we discussed DNA-binding proteins that regulate transcription in prokaryotes. The principles that govern recognition between proteins in eukaryotes show some similarities and some differences. In both cases, specific recognition is dominated by interactions that take place in the major groove of the DNA. The specific interactions usually involve H-bond formation between the base pairs in the DNA and the amino acid side chains in the proteins. In both cases the α helix is the most common element used for DNA recognition. The most striking difference between DNA-binding proteins in prokaryotes and eukaryotes has to do with the symmetry of the interaction. In prokaryotes, DNA-binding proteins almost always are composed of an equal number of identical subunits that interact in a symmetrical fashion with the DNA. Half-sites in the protein bind to half-sites in the DNA, which are usually spaced one helix turn apart. In eukaryotes many cases are known where proteins interact in an asymmetrical fashion with the DNA. In some cases we find eukaryotic proteins that contain multiple sites for interaction within one polypeptide chain, and in other cases we find structurally distinct proteins forming heterodimers that interact with the DNA as well as with other protein components of the RNA polymerase. Let's take a look at some of the main types of gene regulatory proteins that are found in eukaryotes.

The Homeodomain

The homeodomain has a motif that has been recognized in a large family of eukaryotic regulatory proteins. The name "homeodomain" derives from the fact that many mutations that affect the body plan in a developing *Drosophila* embryo are referred to as

Figure 36.16

The homeobox sequence found in three *Drosophila* regulatory proteins. The sequence of amino acids along the main, continuous line is that found in the *Antp* homeobox. At points where the sequence of ftz proteins is different, the differences are shown above the corresponding Antp proteins, and at points where the sequence of ubx proteins is different they are shown below the corresponding Antp proteins. Certain proteins isolated from the human and mouse embryos have segments with similar sequences. Note the high basicity of the sequence, which should favor electrostatic binding to nucleic acid.

				5					10					15					20
			Thr																
Arg	Lys	Arg	Gly	Arg	Gln	Thr	Tyr	Thr	Arg	Tyr	Gln	Thr	Leu	Glu	Leu	Glu	Lys	Glu	Phe
Ser			Gly																

				25					30					35					40
					Ile							Asp			Asn			Ser	
His	Phe	Asn	Arg	Tyr	Leu	Thr	Arg	Arg	Arg	Arg	Ile	Glu	Ile	Ala	His	Ala	Leu	Cys	Leu
	Thr		His										Met		Tyr				

				45					50					55					60
Ser															Ser			Asp	Arg
Thr	Glu	Arg	Gln	Ile	Lys	Ile	Trp	Phe	Gln	Asn	Arg	Arg	Met	Lys	Trp	Lys	Lys	Glu	Asn
															Leu				Ile

homeotic mutations. (This is taken up later.) Walter Gehring found that many of these homeotic genes and other genes that regulate development in *Drosphila* encode regulatory proteins that possess a common 180-base segment in the 3′ exon. The base homology ranges between 60% and 80%, depending on the specific genes being compared, and the amino acid homology is even higher (up to 87%, fig. 36.16). The high basicity of the amino acid sequence in this region—called the homeobox—and other structural features suggest that this region encodes a DNA-binding protein that regulates gene expression by binding to specific sites on the DNA. Indeed, sufficient amino acid homology exists between the homeobox and the helix-turn-helix motif seen in bacterial regulatory proteins to suggest that the gross structures of these very distantly related regulatory proteins are quite similar.

One of the most exciting features relating to the homeobox is that very similar sequences have been identified in many other animals, including frogs, mice, and even humans. This finding suggests that homeobox proteins occur in a wide range of organisms, possibly playing similar roles in regulating developmental processes.

The amino acid sequence of the homeodomain suggests a helix-turn-helix motif. The recognition helix in the homeodomain (helix 3) is longer and makes more contacts with the DNA core than the recognition helices from bacterial regulatory proteins (fig. 36.17).

Although an isolated homeodomain can fold correctly and bind DNA with a specificity similar to that of the intact homeodomain-containing protein, it is believed that the precise DNA-binding specificity is modulated by other regions of the protein. Protein–protein interactions may also have a role in modulating many homeodomain–DNA interactions. For example, the yeast $\alpha 2$ protein forms homodimers, but it also forms a heterodimer with a related homeodomain protein $a1$ (see fig. 36.8). Each of these complexes interacts preferentially at different sites on the DNA.

We tentatively conclude that the asymmetrical interaction of the homeodomain protein with DNA gives it a versatility not displayed by the symmetrical helix-turn-helix homodimers found in bacteria. We will have more to say about homeobox proteins later in this chapter when we discuss *Drosophila* development.

Zinc Fingers

Another DNA-binding motif that is very common in eukaryotic regulatory proteins is the zinc finger. The hallmark of the zinc finger is a tetrahedrally bound zinc linked to the side chains of cysteine and histidine in varying combinations. The zinc stabilizes the conformation of the regulator protein so that the DNA recognition domain and the transcription activation domain are fixed relative to one another. Zinc fingers may constitute the largest class of eukaryotic regulatory proteins; over 200 proteins with this motif have been recognized. It was first identified as the DNA-binding structure in the RNA polymerase III transcription factor TFIIIA that binds to the internal control region of the 5S rRNA gene (discussed in chapter 33). Zinc fingers that resemble TFIIIA are present in the mammalian transcription factor SP1 and in a variety of other regulatory proteins found in eukaryotes. This type of zinc finger motif consists of about 30 amino acids with two cysteine and two histidine residues that stabilize the domain by tetrahedrally coordinating a Zn^{2+} ion (fig. 36.18). A region of about 12 amino acids between the cysteine–histidine pairs is characterized by scattered basic residues and several conserved hydrophobic residues. Proteins in this family usually contain tandem repeats of the 30-residue zinc finger. The crystal structure of a zinc finger–DNA complex containing three fingers from zif268 and a consensus zif-binding site shows that the zinc fingers bind as α helices in the major groove and wrap partway around the double helix (fig. 36.19). Each finger docks against the DNA in a similar way and makes base contacts with a 3-bp subsite arranged

Figure 36.17

(*a*) Complex formed between the *Drosophila* homeobox protein engrailed and DNA seen from two angles. In addition to the main contacts made by the recognition helix in the major groove, additional contacts are made in the minor groove, which can be seen in (*b*). The three helical cylinders are numbered 1, 2, and 3. The other numbers correspond to the numbering system used for the homeobox sequence (see fig. 36.16). (*Source:* From C. R. Kissinger, B. Liu, E. Martin-Blanco, T. B. Kornberg, and C. O. Pabo, Crystal structure of an engrailed homeodomain-DNA complex at 2.8 Å resolution: A framework for understanding homeodomain-DNA interactions, *Cell* 63:579–590, November 2, 1990. Copyright © Cell Press. Reprinted by permission.)

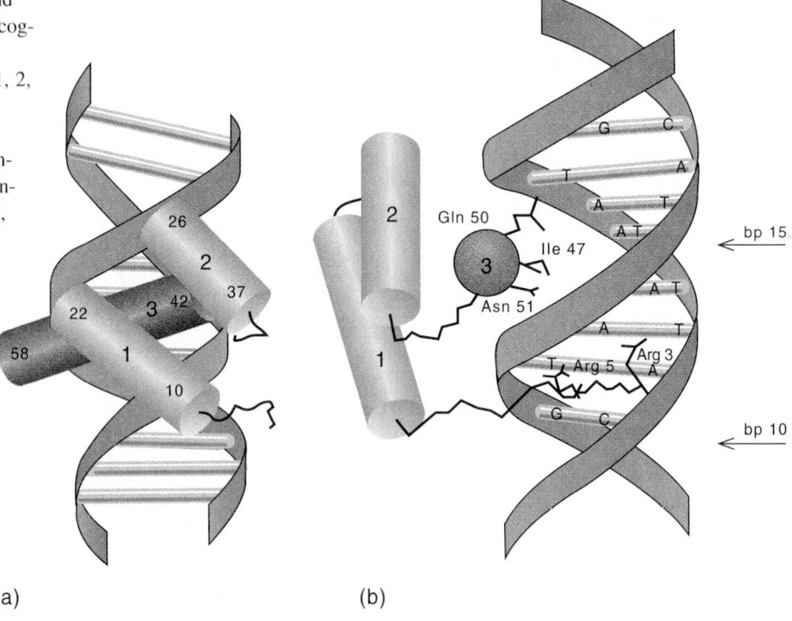

Figure 36.18

Schematic representation of the C_2-H_2 zinc finger found in Xfin from *Xenopus laevis.* (*Source:* Adapted from M. S. Lee et al., *Science* 245:645, 1989.) The recognition helix is stabilized by a complex involving zinc. Cysteine sulfurs are yellow, and histidine nitrogens are blue.

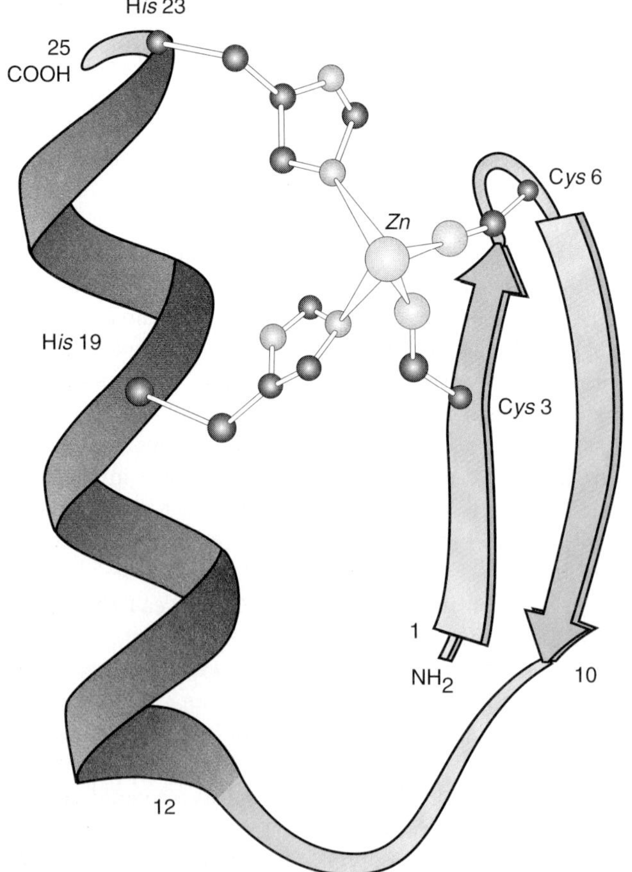

Figure 36.19

The complex formed between the zif268 zinc finger protein and DNA. The three fingers fit into the major groove and wrap partway around the duplex. The zinc is not shown in this figure. (Reprinted with permission from N. P. Pavletich and C. O. Pabo, Zinc finger-DNA recognition: Crystal structure of a Zif268 DNA complex at 2.1 Å, *Science* 252:809–817, 10 May 1991. Copyright 1991 American Association for the Advancement of Science.)

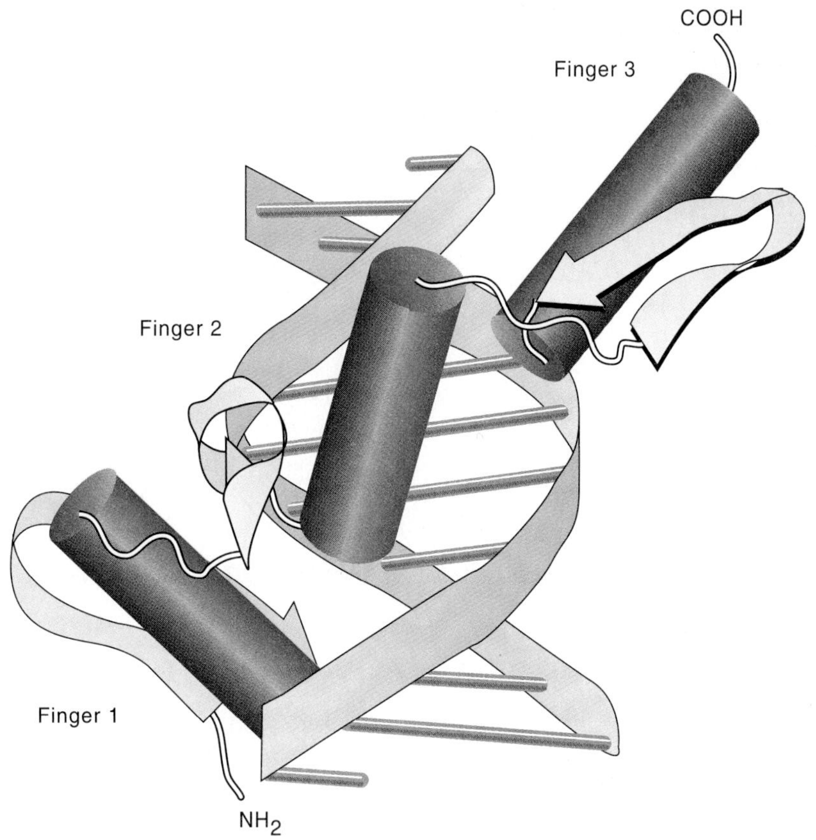

in a way that reflects the helical pitch of the DNA and the 3-bp periodicity of the subsites.

Steroid Hormone Receptors Constitute a Special Class of Zinc Finger Regulatory Proteins

In chapter 27 we discussed roles of steroid hormones in the control of a wide variety of cellular processes in higher eukaryotes, including development, differentiation, and stimulus response. We saw that the steroid hormones are made in specialized endocrine glands from which they are released to circulate in the bloodstream. They pass freely through the plasma membrane of target cells, where they form complexes with specific hormone receptors (HRs). This induces a structural alteration in the receptor favoring binding of the complex to a hormone receptor element (HRE) on the DNA.

Structurally, the steroid hormone receptor constitutes a superfamily that includes receptors for the steroids, estrogen (ER), progesterone (PR), glucocorticoid (GR), mineral glucocorticoid (MR), and androgen (AR). In addition, it includes receptors for thyroid hormone (TR), vitamin D (VDR), retinoic acid (RAR), 9-*cis* retinoic acid (RXR), and ecdysone (EcR). A variety of isoforms of TR, RAR, RXR, ER, PR, and EcR have been identified. These isoforms may be expressed in distinct cell types and developmental stages, suggesting that they play specific physiological roles.

The hormone receptor proteins contain discrete functional domains. The carboxy-terminal domain contains the hormone-binding site and a dimerization region (the receptors bind to DNA as dimers). This domain also interacts with components of the transcriptional machinery and thereby modulates gene expression. The amino-terminal domain also interacts with the transcriptional machinery, whereas the central domain interacts with a specific DNA-binding site (the HRE). The highly conserved DNA-binding domain of the steroid hormone receptors contains two zinc fingerlike sequence motifs (fig. 36.20*a*). In hormone receptors, zinc interacts exclusively with cysteine residues; but the spacing between these is similar to that of the histidine and cysteine residues in the TFIIIA-type zinc fingers. The three-dimensional structure in solution has been determined by using nuclear magnetic resonance spectroscopy and shows that the two zinc finger motifs fold to form a single structural domain (fig. 36.20*b*).

Complementary studies of the regions upstream of target genes have resulted in the identification of the HREs recognized by the hormone receptor proteins. The consensus sequences for these proteins are remarkably similar (fig. 36.21). These consen-

Figure 36.20

The structure of the central DNA-binding domain of the estrogen receptor. (*a*) The amino acid sequence and the linkages to the two zincs. Functional regions involved in DNA binding and dimerization are also indicated. (*b*) The three-dimensional folded structure of the same region. (From D. Rhodes and A. Klug, Zinc fingers, *Sci. Amer.,* February 1993. Copyright © 1993 by Scientific American, Inc. All rights reserved. Reprinted by permission.)

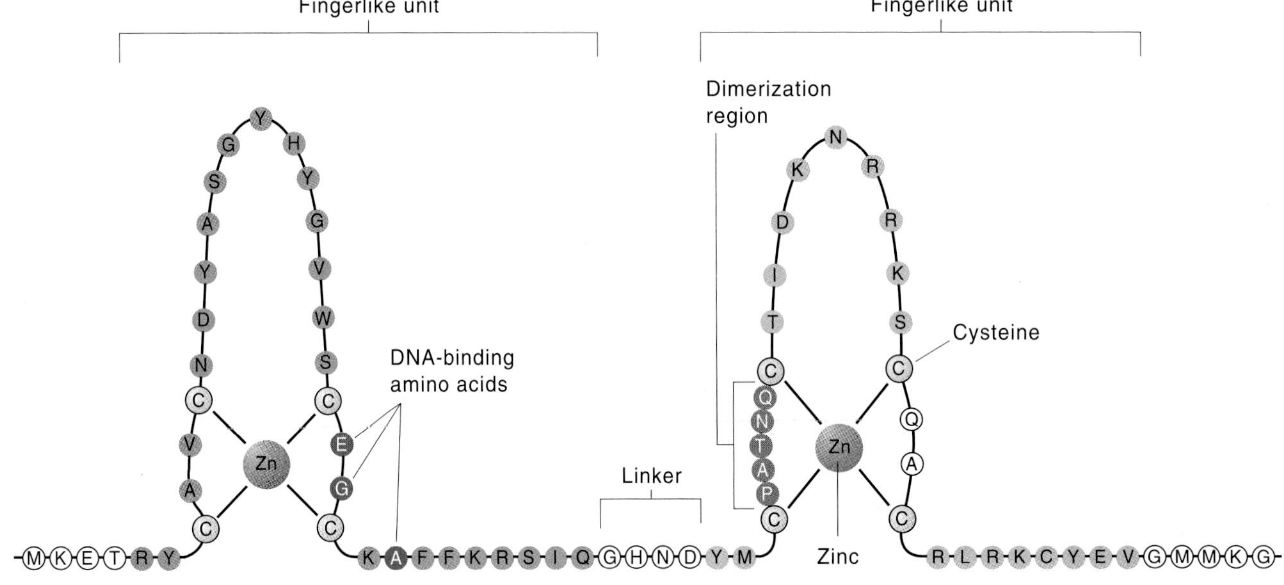

(a)

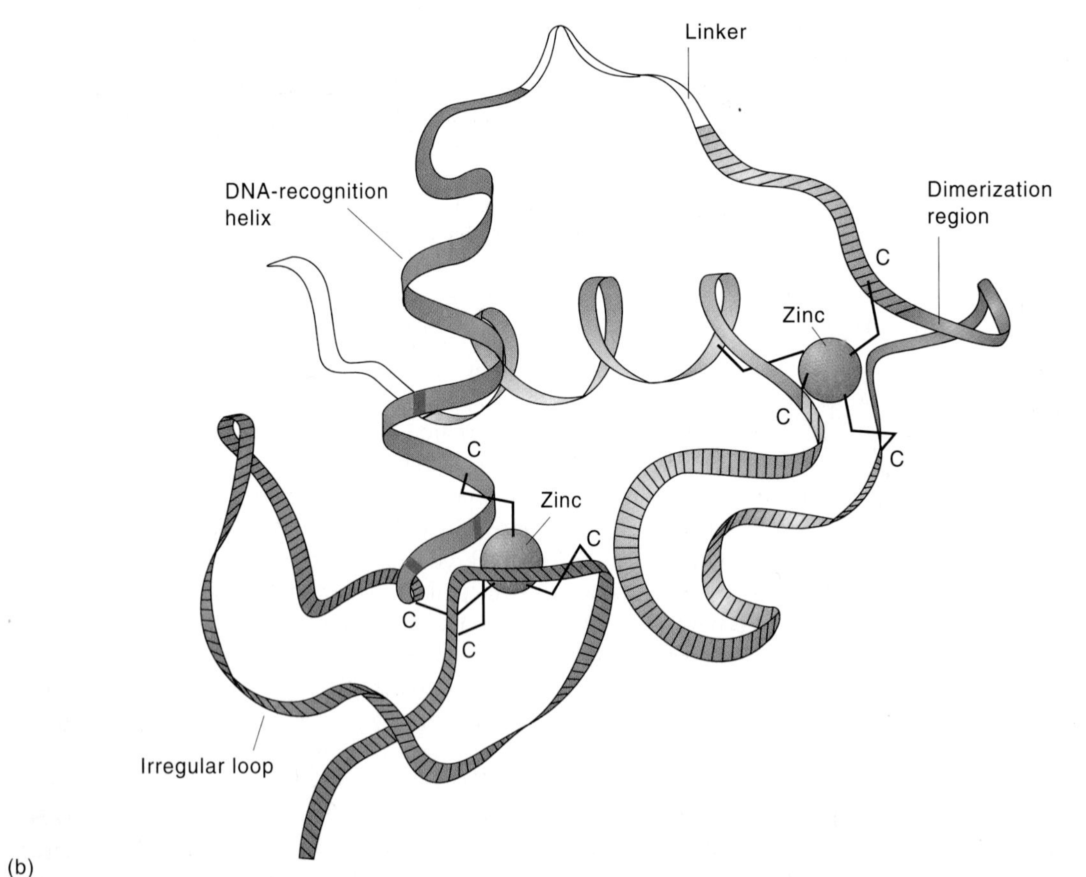

(b)

Figure 36.21

Consensus sequences for a variety of hormone receptor elements.

GRE (glucocorticoids, androgens, mineralocorticoids, progestins)	AGAACA NNN TGTTCT
ERE (estrogen)	AGGTCA NNN TGACCT
TRE (thyroid hormone, retinoic acid)	AGGTCA - - - TGACCT

sus sequences show perfect palindromes, as would be expected if the receptors were organized as dimers with a dyad axis of symmetry. Figure 36.22 suggests how a typical receptor dimer is organized so the DNA binding domains make identical contacts with adjacent large grooves on the DNA.

Leucine Zipper

A third type of DNA-binding domain was first described for the mammalian enhancer-binding protein C-EBP. Proteins of this class show a primary sequence similarity consisting of a highly conserved stretch of about 30 amino acids with a substantial net basic charge immediately followed by a region containing four leucine residues positioned at intervals of seven amino acids. The latter segment, named the leucine zipper by Steve McKnight, is required for dimerization and for DNA binding. It is believed that dimerization of proteins in this group is stabilized by hydrophobic interactions between closely apposed α-helical leucine repeat regions of the two proteins. The two helical cylinders are believed to be oriented in parallel in a coiled-coil fashion (fig. 36.23; also see chapter 5 for an expanded discussion of coiled-coil structures). The main function of the leucine motifs is to bring two proteins together so they can form homodimers or heterodimers that bind to dissimilar half-sites on the DNA.

Helix-Loop-Helix

The helix-loop-helix motif appears to be another way of creating heterodimers that can bind to asymmetric sites on the DNA. Like the leucine zipper proteins, the helix-loop-helix proteins have a basic region that contacts the DNA and a neighboring region that mediates dimer formation. Based on sequence patterns, it has been proposed that this dimerization region forms an α helix, a loop, and a second α helix. Like the leucine zipper protein, the activity of the helix-loop-helix proteins is modulated and usually augmented by heterodimer formation. For example, the MyoD protein, which appears to be the primary signal for differentiation of muscle cells, binds DNA most tightly when it forms a heterodimer with the ubiquitously expressed E2A protein.

This brief description of types of regulatory proteins found primarily if not exclusively in eukaryotes by no means includes all of the known types of DNA-binding proteins. Furthermore, the progress in discovering new regulatory proteins is so rapid that anything we say here is bound to need supplementation if the reader wants to be up to date on this subject.

Figure 36.22

Schematic model for the protein–DNA complex formed between a hormone receptor dimer and its HRE. The small round spheres represent the zinc ions. (From Schwabe and Rhodes, Beyond zinc fingers, *TIBS* 16:292, 1991.)

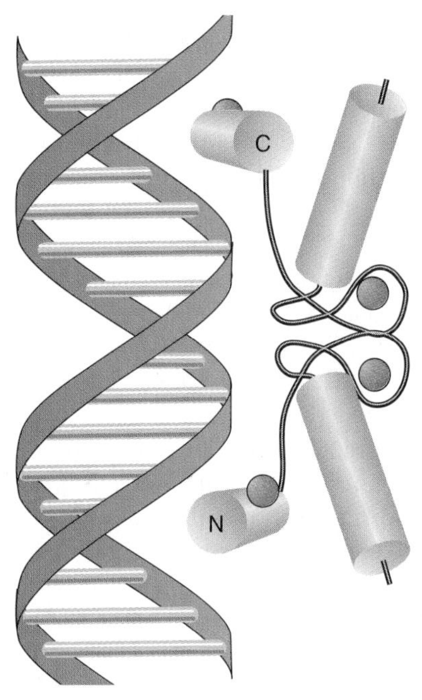

Figure 36.23

Diagrammatic sketch of a leucine zipper dimer. The protein monomers are held together by interaction between leucine side chains (green knobs in the upper part of the structure). The part of the protein monomers that interacts with the major groove of the DNA is shown (in red). (*Source:* Adapted from C. R. Vinson et al., *Science* 246:911, 1989.)

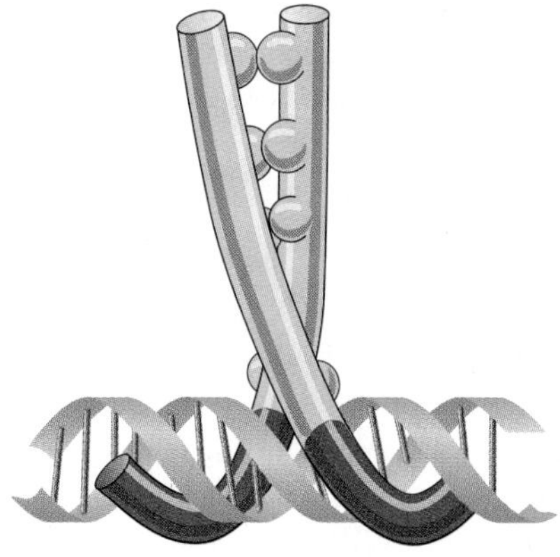

Transcription Activation Domains of Transcription Factors

Thus far we have focused on the DNA-binding domains of regulatory proteins. Many regulatory proteins have additional domains that are involved in transcription activation. Our understanding of transcription activation domains and the factors with which they interact is far less complete. The amino acid sequences of mammalian DNA regulatory proteins suggest the existence of at least three different types of activation domains: acidic, glutamine-rich, and proline-rich.

Transcriptional activation functions of DNA-binding factors depend on regions of 30 to 100 amino acids that are separate from the DNA-binding domains. Factors often have more than one activation domain, and several apparently unrelated structural motifs have been identified that confer these functions. The first defined activation regions in eukaryotic transcription factors were identified by studies of the yeast factor, GAL4 (see fig. 36.4). The activation domains of these factors consist of relatively short stretches of amino acids with significant negative charge that can form amphipathic α-helical structures.

Deletion analysis of the transcription factor of SP1 has revealed four separate regions that contribute to transcriptional activation; all lie outside the zinc-finger-binding domain. The two strongest activation domains contain about 25% glutamine and very few charged amino acid residues. Several other transcription factors show glutamine-rich regions in the domain required for transcription activation.

These are but two of the types of domains found to be involved in transcription activation. Undoubtedly, this list will grow in the near future, as will our understanding about how these transcription activation domains work. Thus far they are likely to represent regions that function by contacting other regulatory proteins, transcription factors, and the RNA polymerase itself.

The recent results of Koleske and Young cited in chapter 33 for yeast PolII support a model in which the assembled holoenzyme is recruited to promoters at which TFIID is already bound. If this model is correct, then two of the major regulatory steps in transcription initiation are formation of a TFIID-bound promoter and association of the holoenzyme with this "landing pad." Activator proteins probably facilitate formation of one or both of these steps, most likely by interactions of the activators with TFIID or different sites on the holoenzyme.

Alternative Modes of mRNA Splicing Present a Potent Mechanism for Posttranscriptional Regulation

We turn the discussion to modes of regulating gene expression that work at the posttranscriptional level. The basic mechanisms of RNA splicing were discussed in chapter 33. RNA splicing was first discovered for transcription of adenovirus where different reading frames are connected to the same 5′ end. Thus we have known about the phenomenon of alternative splicing as long as we have known about splicing itself. Alternative splicing occurs for eukaryotic viruses such as SV40 and polyoma, as well, and it is also a common phenomenon in eukaryotic genes that contain multiple exons in their nascent transcripts. It is clearly a regulatory phenomenon in viruses, since we see a shift in the types of splicing as virus infection progresses. For eukaryotic genes it is also a regulatory phenomenon, since different modes of splicing are found in different tissues of the same multicellular organism. The type of splicing found usually involves segments from the same transcript, but on occasion, transplicing occurs where two independent transcripts participate in a common splicing operation. In the most common splicing situation a promoter is present at one end of the transcript and the combination of exons that is used in the mature mRNA varies according to the pattern of splicing. In some cases, one or more exons are excluded from the message by selective splicing. In other cases, part of an intron is fused to one of the exons to make the final message. Splicing can also be influenced by the choice of the promoter or the choice of the polyadenylation site. In the latter two situations, parts of the upstream or downstream regions of the gene are excluded from the nascent transcript, and so the splicing of the transcript is limited to the RNA remaining.

One particularly elaborate example of alternative splicing is seen in the gene for tropomyosin in vertebrates. Recall that tropomyosin is a key component of vertebrate striated muscle (see table 6.2). The mRNA for tropomyosin found in striated muscle undergoes nine splices in the process of maturation (fig. 36.24). Variants of tropomyosin resulting from alternative splicing are found in other tissues of the same organism. It seems likely that the types of splicing lead to tropomyosin variants that are best suited to the needs of the tissues.

P. Bingham and his co-workers have uncovered another regulatory role for splicing in *Drosophila.* In three different genes they observed that the final splices are subject to regulation. These final splicing operations appear to be regulated by the protein encoded by the mature mRNA. If that protein is present in sufficient concentrations, then it inhibits the final splices, thereby preventing the mRNA from leaving the nucleus and becoming active in translation. This type of splicing is believed to be common, and it is considered highly likely that it provides the key to sexual differentiation in *Drosophila.*

Gene Expression Is Also Regulated at the Levels of Translation and Polypeptide Processing

Following messenger formation, the amount and types of proteins can be modulated in additional ways. The initial polypeptide can be processed in various ways so that different polypeptides or proteins are expressed in different tissues. Such a situation exists for processing the precursor polypeptide preproopiomelanocortin (see fig. 27.7). This polypeptide is processed in different ways in the anterior and intermediate lobes of the pituitary gland to give rise to different hormones in these two tissues.

Because of the longer lifetime of eukaryotic messengers, it seems likely that translation level controls should play a greater role in regulation of eukaryotic gene expression. Following a slow start, more and more mechanisms for translational control are being discovered. Three of the better-understood ones are discussed here.

Figure 36.24

Alternative modes of splicing of the tropomyosin gene transcript in different tissues. (*Source:* Adapted from R. E. Breitbart, A. Andreadis, and B. Nadal-Ginard, Alternative splicing: A ubiquitous mechanism for the generation of multiple protein isoforms from single genes, *Ann. Rev. Biochem.* 56:467–495, 1987.)

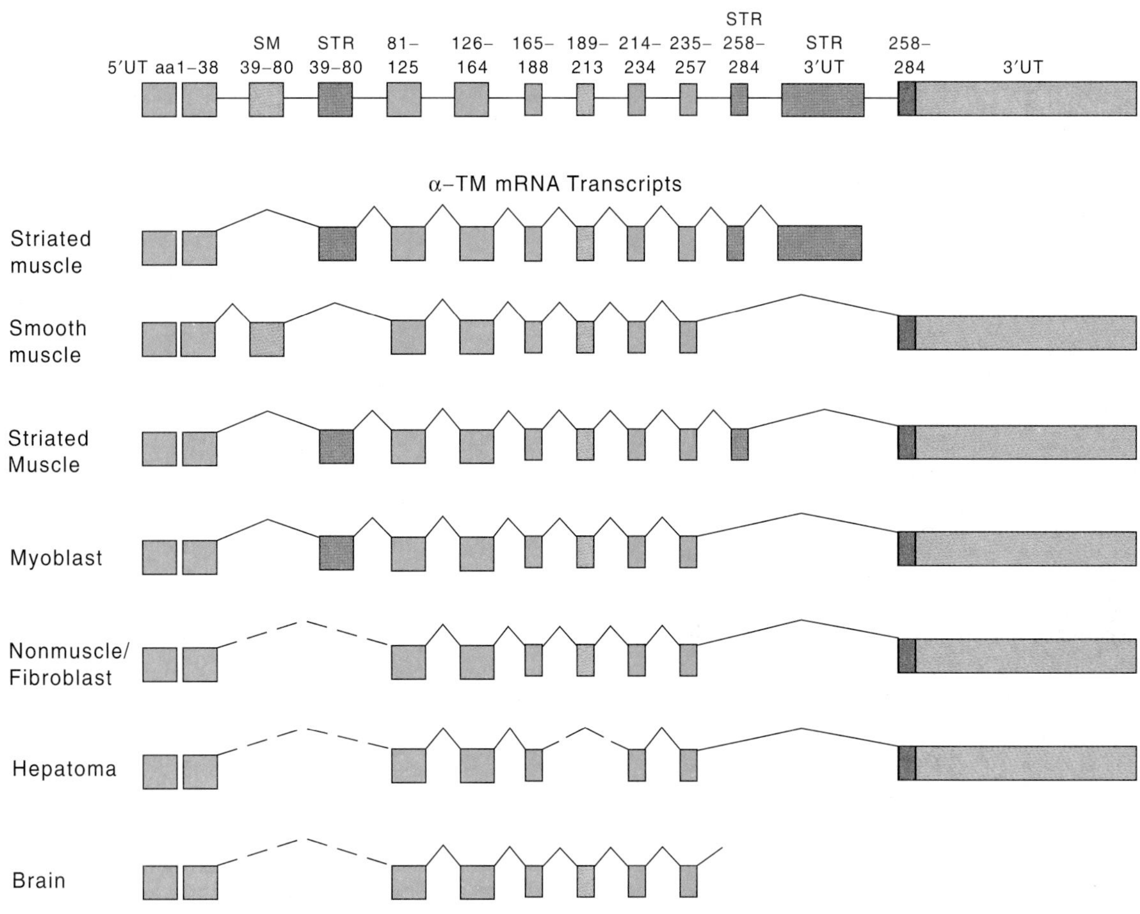

Inactivation of eukaryotic translation factors by covalent modification is one of the few mechanisms known to regulate the rate of translation. Specific protein kinases have been identified that phosphorylate and inactivate both eIF-1 and eIF-2. The significance of the phosphorylation of eIF-2 as a regulatory mechanism of the elongation rate is still not clear, but the phosphorylation of eIF-2 appears to be a general mechanism for controlling translation initiation in many cells.

The regulation of translation through the phosphorylation of eIF-2 is best understood as it operates in the rabbit reticulocyte. Two protein kinases specific for the a subunit of eIF-2 have been purified from reticulocytes. One of these kinases, termed the heme-regulated inhibitor (HRI), serves to coordinate the rate of hemoglobin synthesis (more than 90% of the total protein synthesized in the reticulocyte is hemoglobin) with the availability of hemin (the precursor of the heme group in hemoglobin). Hemin binds to and inhibits the activity of HRI, thereby enhancing the rate of globin synthesis (fig. 36.25).

The second eIF-2-specific kinase appears to be present at low levels in most mammalian cells. This kinase is activated by double-stranded RNA and for this reason has been named the double-stranded RNA-activated inhibitor (DAI). DAI may play a role in defending cells against invasion by viruses, as double-stranded RNAs are frequently found after viral infection.

HRI and DAI are different proteins, but both phosphorylate the same amino acid residue in a subunit of eIF-2, and, in consequence, both kinases inhibit protein synthesis by the same mechanism. Phosphorylated eIF-2 is capable of catalyzing the binding of Met-tRNA to the ribosome, but it does so in a stoichiometric rather than a catalytic manner. This mechanism of inhibition results from the fact that eIF-2 dissociates from the ribosome as a complex with GDP, and in order to recycle, the bound GDP must exchange for GTP. Phosphorylated eIF-2 binds to, but is unable to dissociate from, the guanine nucleotide exchange factor (GEF) that catalyzes the exchange of GTP for GDP. Because cells contain fewer copies of GEF than eIF-2, all of the GEF can be sequestered by partial phosphorylation of eIF-2, and protein synthesis initiation ceases for lack of GEF. This makes the rate of protein synthesis initiation exquisitely sensitive to the state of phosphorylation of eIF-2.

Figure 36.25

Regulation of protein synthesis in the rabbit reticulocyte. The vast majority of the protein synthesized in the rabbit reticulocyte is hemoglobin. The gross rate of protein synthesis in the reticulocyte is controlled indirectly by the concentration of heme. Heme inactivates a kinase that would otherwise inactivate the initiation complex involving eIF-2 and eIF-2B. The kinase phosphorylates the eIF-2 factor, making it impossible for the eIF-2–eIF-2B complex to exchange GDP for GTP.

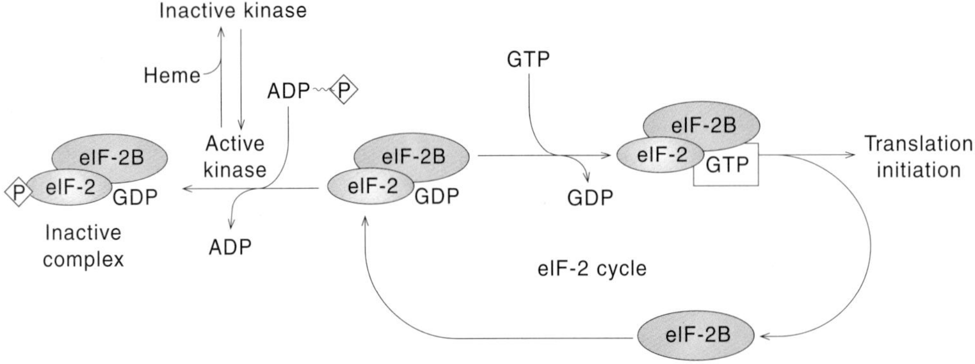

Figure 36.26

Model for regulation of GCN4 translation by differential ribosome scanning. The transcript containing the GCN4-ORF also contains four short upstream open reading frames. (*a*) Under conditions of amino acid abundance, reading of ORF1 is usually followed by reading of ORF3 or ORF4, after which the ribosome drops off the mRNA. (*b*) Under starvation conditions, reading of ORF1 is usually followed by reading of the GCN4-ORF. The explanation for this differential recognition of ORFs appears to reside in the replacement of one initiation factor by another. (*Source:* M. Altmann and H. Trachsel, Regulation of translation initiation and modulation of cellular physiology. *TIBS* 18:430, 1993.)

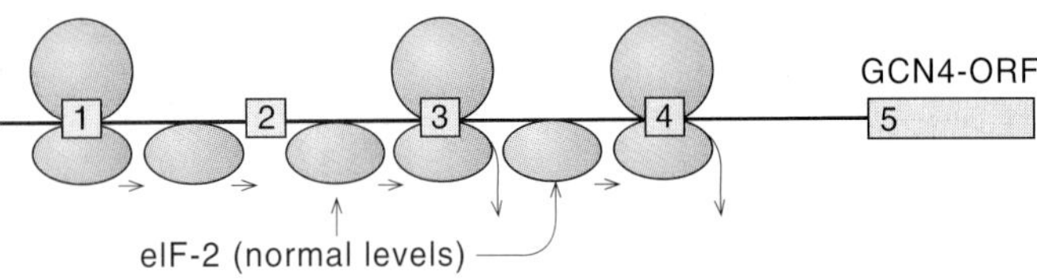

(a) Nonstarvation

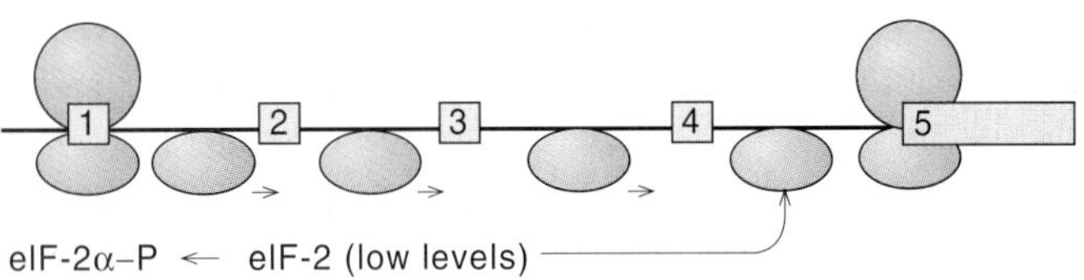

(b) Starvation

How Translation Controls Transcription in Eukaryotes

Transcription and translation are not directly coupled in eukaryotes as they are in bacteria. This eliminates the direct control of transcription by attenuator mechanisms, as we saw in the case of the *trp* operon (see fig. 35.14). However, this has been replaced by an indirect mechanism whereby the translation of the mRNA for a regulatory protein is regulated. An increasing number of open reading frames are being shown to function as *cis*-acting regulatory signals able to moderate expression of the downstream reading frame.

One of the best-understood examples of this mode of regulation is found for the yeast *GCN4* gene, which encodes a transcription factor that regulates transcription of more than 30 biosynthetic genes in response to amino acid starvation. Production of GCN4 protein is enhanced by one to two orders of magnitude in response to amino acid limitation, and its synthesis is regulated primarily at the translation level. Translational control of GCN4 synthesis entails the differential recognition of four short upstream open reading frames (ORFs) that precede the reading frame for GCN4 (fig. 36.26). Under normal conditions, where there is an abundance of amino acids, the level of active translation initiation factor eIF-2 is high. In this situation, ORF1 is translated, and this is followed by the translation of one of the downstream ORFs (usually ORF3 or ORF4). After this the ribosome usually is released from the GCN4 mRNA, so no active GCN4 protein is synthesized. Under amino acid starvation conditions the level of active eIF-2 is low. As a result, after translation of ORF1, ribosomes bypass the downstream ORFs and resume translation at the start codon for GCN4. The resulting synthesis of GCN4 activates the genes for amino acid biosynthesis.

Patterns of Regulation Associated with Developmental Processes

As cells proliferate within the embryo, they assume different properties; changes are evidenced by the genes they express and the proteins they synthesize. Developmental differences can be maintained or altered during subsequent cell duplication. During embryonic growth, cells that are initially capable of following any pathway of development become committed to a particular pathway. As a rule, pathways have many branchpoints, so when progeny cells of partially committed cells reach a branchpoint, they may differentiate further down a more specialized pathway. This process of gradual commitment, repeating itself many times, gives rise to a complex pattern of cell lineages, all arising from the initially fertilized egg.

It seems likely that the pivotal events in the evolution of a differentiated cell reflect changes in gene expression that result from a complex hierarchy of controls. The key to understanding differentiation will be to identify the regulatory factors responsible for the controls involved in differentiation and to explain how they act.

Early Development in *Drosophila* Leads to a Segmented Structure That Is Preserved to Adulthood

Two organisms that have become most popular for studying development are mice and flies. Mice are popular because their biochemistry is very close to that of humans, but they are much easier than humans to manipulate both genetically and biochemically. Much of what we know about immunobiology has resulted from studies on mice (see chapter 37). *Drosophila* fruit flies became popular in the early part of the twentieth century because for a complex organism they were very suitable for genetic studies. They multiply rapidly and can be handled in large numbers, so it is relatively easy to obtain mutants of *Drosophila.* In the study of development, as in so many areas of biochemistry, it is best to start with relevant mutants and then look for biochemical differences. Biochemical and genetic studies on *Drosophila* have been greatly facilitated in the last 20 years by general advances in molecular biology and specific advances in recombinant DNA methodology. In the remainder of this chapter we will focus on some aspects of early development in *Drosophila.*

Some steps in the development of *Drosophila* are shown in figure 36.27. Beginning shortly after fertilization, the new diploid nucleus undergoes a rapid series of divisions with no segregation of nuclei into separate cells. After the eighth division, when there are 256 nuclei (see fig. 36.27*c*), the nuclei begin to migrate to the periphery of the egg cytoplasm. After another nuclear doubling, cell membranes form around a group of cells at the posterior end of the egg; these cells are progenitors of germ cells for the subsequent generation. The remaining nuclei continue to divide until there are about 6,000 nuclei at the periphery (see fig. 36.27*e*). At this point, about two hours after fertilization, membranes are formed, separating the nuclei into a monolayer of cells. The resulting structure, known as the blastoderm, is essentially a cell monolayer enclosing the yolk.

The blastoderm divides into fourteen discrete segments: Md, Mx, Lb, T1–T3, and A1–A8 (see fig. 36.27*f*). The first three segments, Md, Mx, and Lb, become part of the head structure. The remaining segments become subdivided into two compartments, with anterior and posterior parts. During gastrulation there is a continued cell duplication, folding of sheets of cells, and mass migration of segments (see fig. 36.27*g*). The embryo eventually hatches into the first larval stage (see fig. 36.27*h*). Cells within the larva are of two types. About 80% of them are fully functional in the larva. The remaining 20% are embryonic precursors to adult tissues rather than larval tissues. These latter cells form packets within the larva, called imaginal disks, that are arrested in development until pupation (fig. 36.28), when a single hormone, ecdysone, triggers their differentiation into specific adult structures.

Early Development in Drosophila *Involves a Cascade of Regulatory Events*

There are two distinct phases in *Drosophila* embryogenesis; the first precedes cellularization of the blastoderm and is associated with a cascade of interacting regulators; the second occurs after cellularization and depends on intercellular signals that must be carried, at least in part, by membrane-bound proteins. We will focus on events occurring in the first phase because they are better understood.

Recall that in the infectious cycle of the λ bacteriophage, each stage is characterized by the expression of specific gene products under the control of one or more regulatory proteins. A new phase in the development of the bacteriophage results from the synthesis of one or more new regulatory proteins. A hierarchy determines the order in which the different regulatory proteins make their appearance. Early development in *Drosophila* is very similar except for its greater complexity. In fact, some of the earliest regulatory proteins in the developing oocyte are supplied by surrounding cells. Another difference in *Drosophila* development is that the large size of the oocyte permits the establishment of concentration gradients within a single cell. The regulatory protein encoded by a messenger RNA injected at the anterior end of the oocyte is likely to exist in highest concentration at the anterior end and at lowest concentration at the posterior end. Similarly, mRNAs injected into the oocyte at the posterior end establish regulatory-protein gradients in the opposite direction. These gradients are crucial for regulating the expression of genes involved in early development. Some gradients are also established in the perpendicular direction, on the dorsal-ventral axis, but we will overlook these for purposes of discussion.

Three Types of Regulatory Genes Are Involved in Early Segmentation Development in Drosophila

The first insight into the genetic system directing *Drosophila* development came from the discovery of bizarre mutations that affect the body plan (fig. 36.29). For example, the mutation *Antennapedia* results in a pair of extra legs sprouting from the head in place of antennae. Another mutation, *Bithorax,* results in an extra pair of wings appearing where normally there should be much smaller appendages called halteres. These mutations occur in regulatory genes known as homeotic genes.

In addition to homeotic genes there are two other types of genes that influence the segmentation pattern; these are called maternal-effect genes and segmentation genes (table 36.2). Maternal-effect genes are so called because they affect the phenotype only according to the information present in the female parent. Maternal-effect mutations occur in genes responsible for establishing the anterior–posterior axes in the young embryo. Segmen-

Figure 36.27

Steps in the development of *Drosophila melanogaster.* (*a*) The fertilized egg contains a single zygotic nucleus. (*b*) This divides every 10 min. (*c*) After eight divisions, when there are 256 nuclei, nuclear migration toward the outer cortex structure begins. (*d*) Eventually, all the nuclei form a monolayer on the cortex surface. The first nuclei to become enclosed are the pole cells, which become germ cells in the adult organism. (*e*) The fully formed blastoderm contains about 6,000 cells, which form a monolayer around the cortex. (*f*) Even before a cell membrane has formed around the nuclei, the embryo has become functionally divided into a segmented structure. (*g*) During gastrulation there is continued cell duplication, a folding of sheets of cells, and mass migration of segments. (*h*) Eventually, this structure hatches into the first larval stage. (*i*) The remaining stages between the larva and the adult fly are not illustrated. In (*f*) through (*i*), various segments are labeled. Three segments, Md, Mx, and Lb, fuse to make the head structure. Thoracic segments T1–T3 and abdominal segments A1–A8 retain their segmental appearance in the adult.

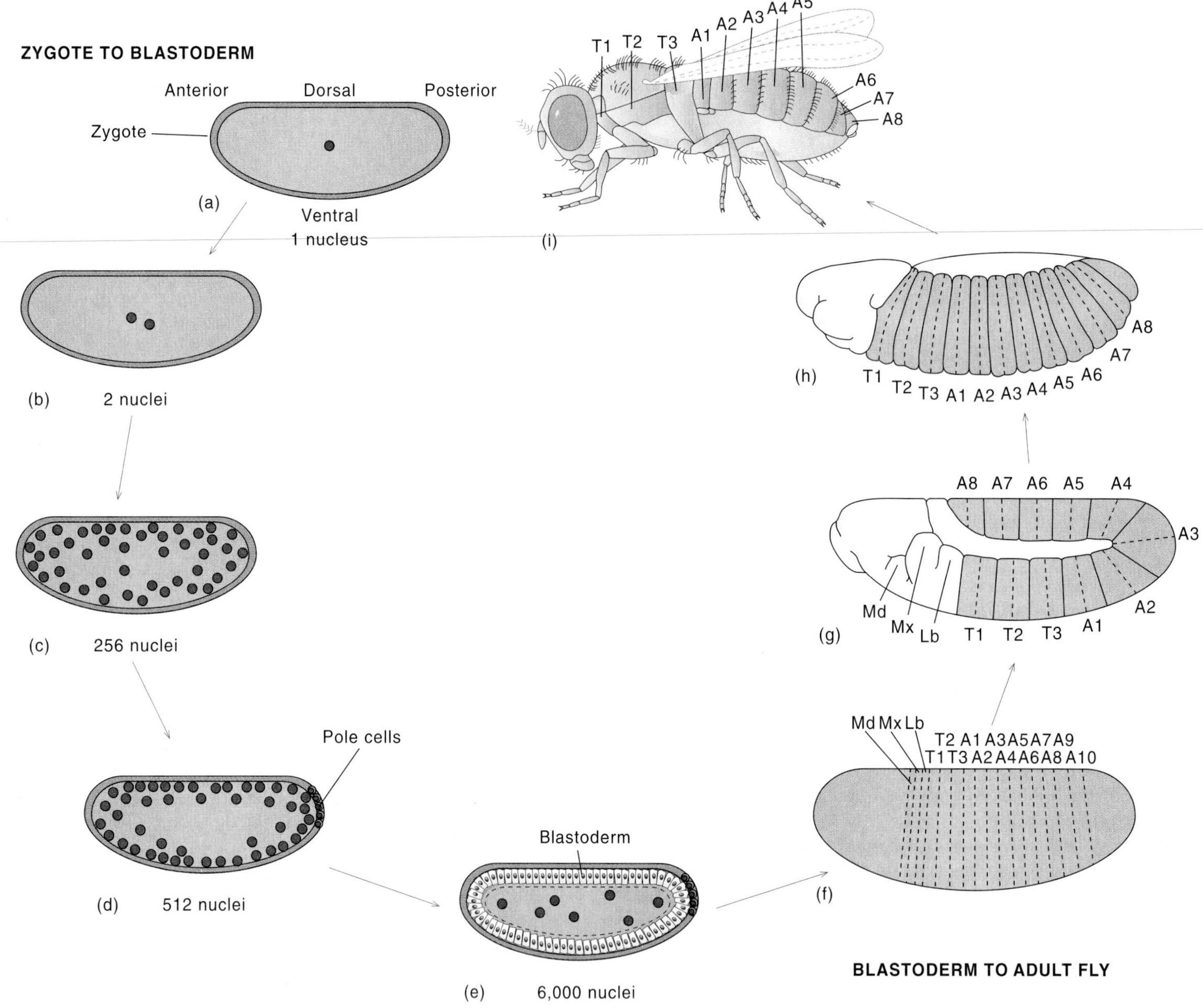

tation mutations affect the number and polarity of the body segments. Homeotic mutations are more specific than segmentation mutations; they result in changes in structures that are uniquely associated with individual segments or subsegments.

Analysis of the Genes That Control the Early Events of Drosophila *Embryogenesis*

Once a mutation that affects development has been detected, a way must be found to assess the presence or absence of the related

Figure 36.28

Imaginal disks in a mature larva and the structures they lead to in the adult fly.

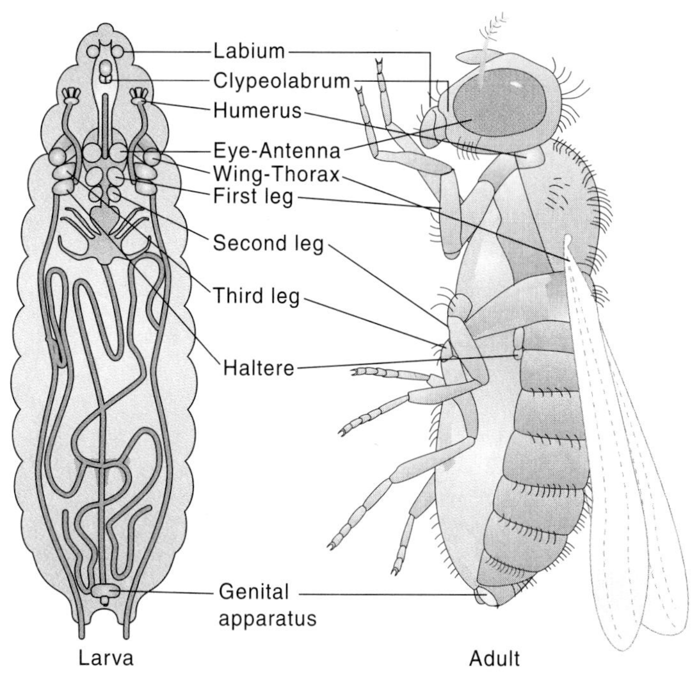

Figure 36.29

Abnormal phenotypes resulting from homeotic mutations. *Antennapedia* results in a pair of extra legs sprouting from the head in place of antennae. *Bithorax* results in an extra pair of wings appearing where halteres normally appear.

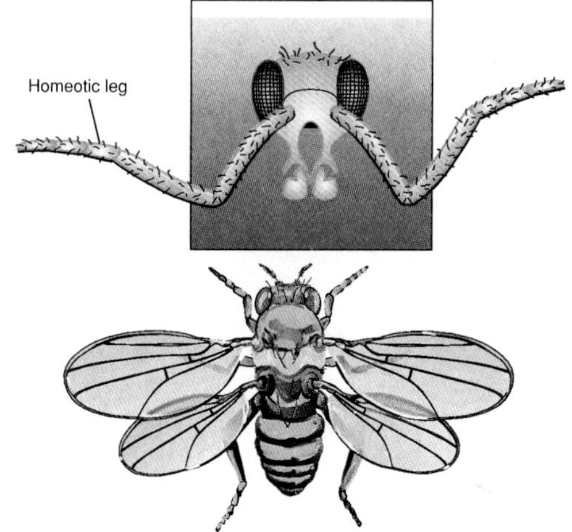

gene product in the developing embryo. A principal tool for doing this is autoradiography using ^{32}P-labeled specific DNA probes. The embryo is labeled by in situ hybridization with the radioactive probe specific for the transcript of a gene. Both the location and the intensity of the labeling give an indication of the level of expression of the gene. In some cases the protein product relating to a specific transcript can also be detected by use of a specific antibody-labeling technique. By using this approach, the appearance of some of the major regulatory genes that are involved in early development has been traced.

Table 36.2

Some of the Regulatory Genes Involved in Segmentation Development

Maternal-effect genes—establish gradients
Anterior
Bicoid (*bcd*)
Posterior
Oskar (*osk*)
Dorsal–ventral
Dorsal (*dl*)
Segmentation genes
Gap genes—define four broad regions in the egg
Hunchback (*hb*)
Krüppel (*Kr*)
Knirps (*kni*)
Pair-rule genes—define seven bands
Runt (*runt*)
Hairy (*h*)
Fushi tarazu (*ftz*)
Even skipped (*eve*)
Paired (*prd*)
Odd paired (*opa*)
Segment-polarity genes—define fourteen bands
Engrailed (*en*)
Wingless (*wg*)
Gooseberry (*gb*)
Homeotic genes—specify structures associated with individual segments
BX-C locus
ANT-C locus

Maternal-Effect Gene Products for Oocytes Are Frequently Made in Helper Cells. The first regulatory gene products to play a significant role in *Drosophila* development are active before fertilization. The bicoid (*bcd*) gene is a major determinant of the anterior–posterior pattern. Like many other maternally expressed genes, *bcd* is transcribed in the ovary in specialized cells that form a cluster around the future anterior pole of the developing oocyte. The bcd RNA synthesized in these so-called nurse cells passes into the oocyte through cytoplasmic canals and becomes localized at the anterior pole of the oocyte (fig. 36.30). Similarly, the oskar (*osk*) group genes, such as *nos,* are involved in the synthesis and deposition of products that become localized at the posterior pole of the egg (see fig. 36.30*a*). Females lacking a functional copy of the *bcd* gene produce eggs that develop into embryos with no head or thorax. The same effect can be achieved by removal of material from the anterior pole of a normal egg. Embryos produced by female mutants for *nos* develop normal head and thoracic segments, but lack the entire abdomen.

Figure 36.30

Key events in the expression of developmental regulatory genes in the blastoderm embryo. Events take place in the order shown over a period of a few hours. Measurements are based on the use of probes for detecting the transcripts of various regulatory genes. Description of genes is given in table 36.2. (*Source:* Adapted from P. W. Ingham, The molecular genetics of embryonic pattern formation in Drosophila, *Nature* 335:25–34, 1988.)

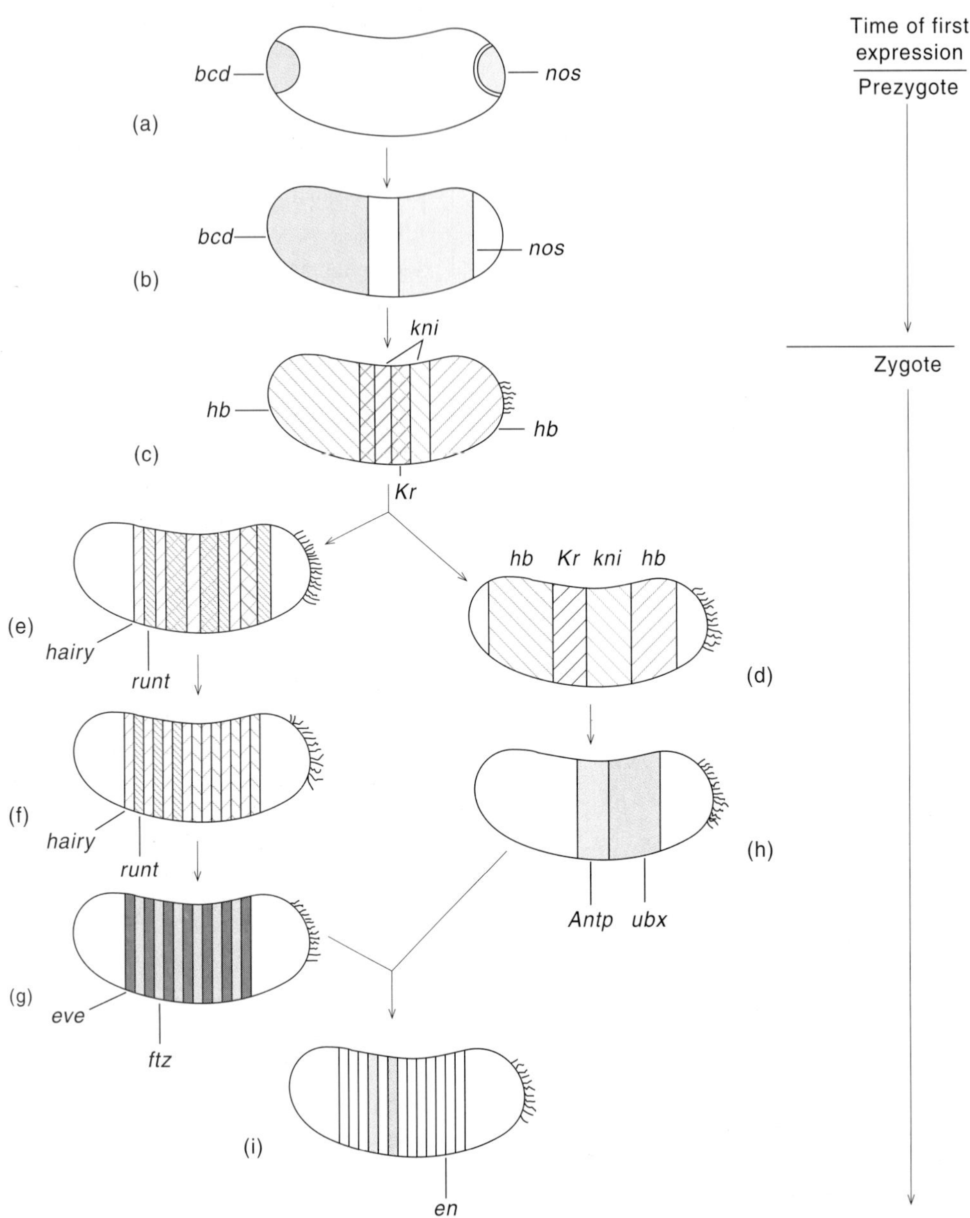

The pattern-forming process continues upon fertilization; as the *bcd* RNA is translated, the protein diffuses from the anterior pole so that it becomes distributed over about half of the length of the egg (see fig. 36.30*b*). Simultaneously, information encoded and localized at the posterior pole during oogenesis by members of the *osk* group begins to move forward.

By the beginning of the precellular blastoderm stage there are gradients for at least two different maternally encoded products along the anterior–posterior axis of the embryo. This quantitative information is now transformed into qualitative differences in the form of region-specific gene expression, by a process requiring interaction between the maternally derived products and the zygotic genome that resulted from fertilization.

Gap Genes Are the First Segmentation Genes to Become Active. The first segmentation genes to become active in the zygote are members of the gap class, so-called because their mutants lack major regions of the body, thereby creating a gap in the anterior–posterior pattern. The three known members of this class are hunchback (*hb*), Krüppel (*Kr*), and knirps (*kni*). These genes are expressed two division cycles prior to cellularization. Expression of *hb* is restricted to two regions: one extending from the anterior pole to 50% of the egg length, the other from the posterior pole to about 25% of the egg length (see fig. 36.30*c*). Initially, the *Kr* gene is expressed in a single broad band in the middle of the embryo, whereas *kni* is expressed in two distinct domains, one anterior and one posterior to the *Kr* band. These transcriptional domains for *hb, Kr,* and *kni* are influenced by the maternally derived information encoded by the *bcd* and *osk* group genes. Thus mutants that lack *bcd* activity do not express *hb,* whereas *Kr* is extended anteriorly in an unusually broad domain. The absence of the posterior maternal determinants results in the extension of the *Kr* domain posteriorly. These observations suggest that the *bcd* and *osk* group genes both act to repress transcription of *Kr* in the anterior and posterior regions, respectively, thereby restricting its region of expression to the central portion of the embryo. In contrast, *bcd* appears to act as a positive regulator of *hb,* and *osk* appears to be a positive regulator of *kni.* In normal embryos the transcriptional domains of *hb, Kr,* and *kni* narrow with time, giving rise to sharp boundaries of expression. This process is driven by negative effects between the gap genes (see fig. 36.30*d*), *hb* and *Kr* mutually repressing one another and *kni* acting as a negative regulator of *Kr.* Thus the establishment of stable domains of gap gene expression is a two-step process: first, a differential response to graded levels of maternal determinants, and second, a mutual repression effect, leading to the generation of stable boundaries between adjacent domains.

The gap gene products regulate the position-specific expression of other genes, those belonging to the pair-rule class of segmentation genes.

Periodic Gene Expression Is Initiated by Pair-Rule Genes. A common feature of pair-rule genes is their transient expression in seven bands during the period of cellularization of the blastoderm. Despite this similarity, each pair-rule gene is unique in its pattern of expression. *Runt* and *hairy* are initially expressed throughout the embryo (see fig. 36.30*e*); a restricted pattern of expression of these two genes begins earlier than for other pair-rule genes.

By the beginning of the last interphase before cellularization, transcripts of both *hairy* and *runt* localize to two complementary series of seven bands that encircle the embryo (see fig. 36.30*f*). The generation of these banded patterns is a key event in the pattern-forming process. The sorts of interactions taking place probably involve the local enhancement of expression of one or another gene in response to local levels of one or more gap gene products. Such interactions result in a patchy pattern of *hairy* and *runt,* which is further refined and stabilized into a regular banded pattern by their mutual repression (see fig. 36.30*f*), in a manner analogous to the stabilization of the gap gene expression domains.

The complementary bands of *runt* and *hairy* are instrumental in the refinement of the spatial patterns of expression of other pair-rule genes.

Hairy and Runt Proteins Function as Negative Regulators for the Expression of Other Pair-Rule Genes. *Hairy* and *runt* act as negative regulators for the expression of the pair-rule genes *fushi tarazu* (*ftz*) and *even skipped* (*eve*), respectively (see fig. 36.30*g*). Both *ftz* and *eve* encode homeodomain proteins, a strong indication that they act as transcriptional regulators of genes of the homeotic and segment-polarity class. The interactions between *hairy* and *runt* and between *ftz* and *eve* generate complementary sets of seven bands expressing one or the other gene (the locations of *ftz* and *eve* expression are shown in figure 36.30). When they first become visible, each band is four nuclei wide. At this stage, each *ftz* and *eve* band coincides with the boundaries of a so-called parasegment. A parasegment is a unit that is out of register with a morphological segment but coincident with the domains of activity and function of the homeotic genes (fig. 36.31). At the same time as the *ftz* and *eve* patterns are resolved, the pattern of expression of a third homeobox-containing pair-rule gene called *paired* (*prd*) is also changing (see fig. 36.31 for the pattern of *prd* expression). Initially, *prd* is expressed in seven broad bands also, but in this case the bands are each about six nuclei wide. This pattern is then transformed into one of fourteen bands by the elimination of expression from the central region of each of the original bands. As with *ftz* and *eve,* the generation of this pattern depends on the activity of *hairy* and *runt.* Thus in a process resembling the establishment of the gap gene domains, information encoded by two components is transformed into a more complex form by the differential response of the various secondary pair-rule genes (*ftz, eve, prd*) to this information.

Figure 36.31

A region of the blastoderm showing three of the fourteen parasegments and the initial and ultimate spheres of expression for certain pair-rule and segment-polarity genes (all 14 parasegments are shown in figure 36.33). The dashed regions for *prd, opa, ftz,* and *eve* show a region of expression that occurs only at early stages, before the narrowing process. The circles represent cells in the early stage in the blastoderm, when the parasegments are only four cells in width.

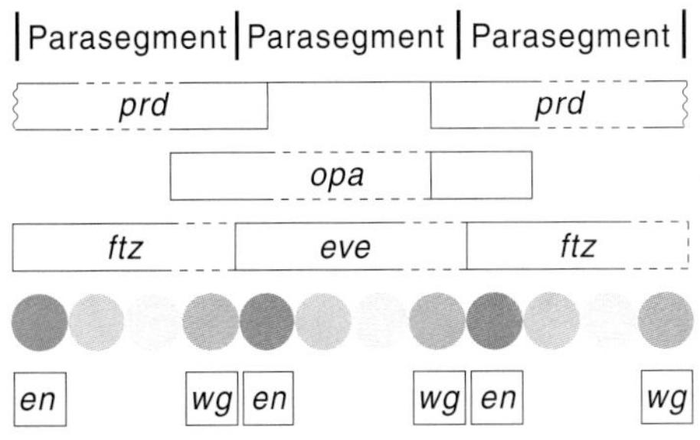

***The Boundaries of the Parasegments Are Set by Wingless* (Wg) *and Engrailed* (en).** The combinations of pair-rule gene activities define the states of particular blastoderm cells. These cells subsequently serve as reference points for the elaboration of the pattern within individual parasegments during the period of limited cell proliferation that follows. In particular, the pair-rule genes have been shown to define the domains of expression of the segment polarity genes *en* and *wg.*

At the end of the cellularization process and the onset of gastrulation, transcripts of both the *en* and *wg* genes accumulate in fourteen narrow bands along the anterior–posterior axis of the embryo, patterns that represent the first evidence of its segmental organization. At this time each band of *en* expression is only one

Figure 36.32

Influence of developmental regulatory genes on one another. An arrow with a plus sign ⊕ at the arrowhead indicates a positive effect on expression, whereas a minus sign ⊖ at the arrowhead indicates a negative effect on expression. The *gap* genes are essential for the establishment of the *hairy* and *runt* banded patterns, but the precise way in which they do this is unclear.

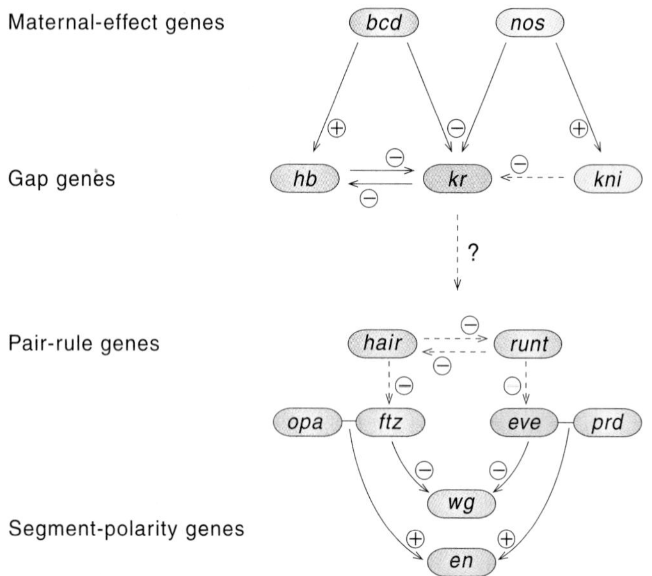

cell wide. Each of the *en* bands represents the anterior boundary of a parasegment. The establishment of alternate *en* bands requires the combined activities of different sets of pair-rule genes. In the case of the odd-numbered bands the *en* expression cells are those that express both *eve* and *prd,* whereas the even-numbered bands require the expression of both *ftz* and another pair-rule gene, *odd-paired* (*opa*).

In contrast to this positive control of *en, wg* expression is repressed by the combination of eve and ftz proteins. At the end of blastoderm the domains of eve and ftz narrow, so single-cell-wide bands come to separate them, and it is these cells that initiate expression of *wg.* The net result is the generation of two sets of cells that are adjacent to one another and mark the anterior and posterior limits of each of the parasegments (see fig. 36.31). These sets of cells, one expressing *en* and the other expressing *wg,* subsequently serve as reference points for the specification of position in each developing parasegment.

The positive and negative effects established between maternal-effect and pair-rule genes are summarized in figure 36.32. This diagram tells an incomplete story. The most serious gap in our knowledge is how the gradients of maternal-effect and gap genes help to deliver the signal for the establishment of the strict periodic patterns of expression seen for most of the segmentation genes. It seems unlikely that the signals in the form of gradients could ever institute such a pattern without other factors being involved. Possibly, some of the signals for expression of pair-rule genes are established in the oocyte before fertilization.

Homeotic Genes Specify Parasegment Character. By the time cellularization of the blastoderm is complete, different cells express different combinations of pair-rule genes (fig. 36.33). At this stage, the embryo is subdivided into a series of repeating units. Despite their similarity, each parasegment is programmed to follow a unique pathway of differentiation, as evidenced by subsequent events. Most probably, the expression of the homeotic genes is influenced by the periodic information generated by the pair-rule and the segment-polarity genes, superimposed on the gradients of information deposited by the maternal genes and the gap genes. We do not know what information turns on specific homeotic genes in specific parasegments, but we do know a most exciting beginning of a story that should unfold in the near future.

Most of the homeotic genes are clustered in two giant loci, the *Antennapedia* (*ANT-C*) and *bithorax* (*BX-C*) complexes. The *ANT-C* locus occupies about 100 kb of the genome, and the *BX-C* locus occupies about 300 kb of the genome. Both of these loci are situated on chromosome 3. The *BX-C* locus is subdivided into three sections, *Ubx, Abd-A,* and *Abd-B,* according to which body parts are affected by different mutations. The *Ubx* region affects thoracic segments T2 and T3 and the anterior portion of the A1 segment. The *Abd-A* region affects the posterior part of A1 and abdominal segments A2 to A4. Finally, the *Abd-B* region affects the abdominal region from A5 to A8. The regions affected by different parts of the *ANT-C* locus include the head and the three thoracic segments (fig. 36.34). It is fascinating that the arrangement of the genes in the two homeotic loci is colinear with the body plan; this colinearity may be related to the temporal mode of expression of the different genes in the two loci. Thus transcription of a region the size of *ANT-C* would take about two hours; the temporal expression of promoter proximal and promoter distal regions of the loci could be crucial in ensuring the correct specification of body parts during development. Another aspect of the gene arrangement and the body plan that is probably related to this factor is that a mutation at a given locus usually results in the conversion of the segment in question to the phenotype of the adjacent segment in the anterior direction. It is as though proceeding from the anterior to the posterior direction results in a stepwise increase in the amount of the locus that gets expressed.

The *ANT-C* and *BX-C* loci both contain vary large introns and very few reading frames. These genes first become transcribed just prior to cellularization of the blastoderm, at about the same time as the pair-rule genes. Initially, they are expressed at rather uniform levels in broad, overlapping domains. The boundaries of these domains are defined both by the maternal-effect organizing activities and by the gap genes. For example, transcripts from both of the known promoters of the *ANT-C* locus accumulate throughout the *Kr* domain. The longer transcript, P1, depends absolutely on *Kr* expression, whereas the shorter transcript, P2, appears to be defined by a combination of *bcd, osk,* and *hb* expression. Similarly, the initial *Ubx* domain is dependent on activation by the *osk* group activity and repression by *hb* activity. These broad regional differences are modulated by interactions with the early pair-rule pattern to generate a series of unique parasegmental states. Expression of *Scr, Antp,* and *Ubx* is specifically elevated in the cells of parasegments 2, 4, and 6, respectively, where the ftz protein is active. These enhanced levels of transcription are subsequently maintained throughout embryogenesis.

Figure 36.33

Pattern of expression of some pair-rule and segment-polarity genes. When the pattern of expression of maternal-effect genes and gap genes is superimposed on this regularly repeating pattern, it gives each region of the blastoderm a unique mixture of developmental regulatory genes.

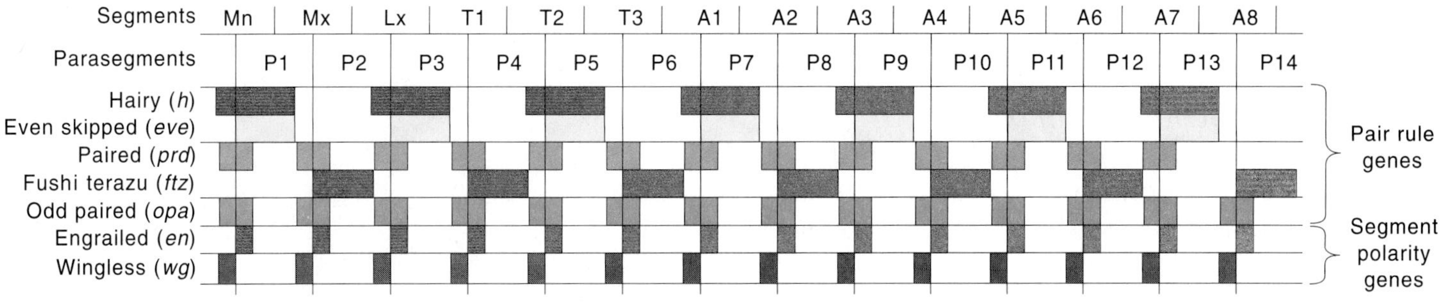

Figure 36.34

Genetic maps of the antennapedia complex (*ANT-C*) and the bithorax (*BX-C*). The red dashed lines specify the regions affected by mutations in a given locus. The bithorax complex is shown in greater detail. By lethal complementation assay, Sanchez-Herrero has shown that the *BX-C* is divided into three complementation groups: *Ubx, Abd-A,* and *Abd-B.* Within each of these complementation groups there are nonlethal point mutations (such as *abx, bx, bxd,* and *pbx* in *Ubx*), which cover specific phenotypes in the regions encompassed by the dashed lines. *Ubx* is required for the development of the posterior compartment of the mesothorax (T2p) through the anterior compartment of the first abdominal segment (A1a). The *Abd-A* function is required for morphogenesis of A1p through A4, and *Abd-B* is required for A5 through A8. At least three essential homeotic functions have been assigned to the antennapedia complex: *Dfd, Scr,* and *Antp.* These are shown on the genetic map below the schematic of the fly. The primary domains of *ANT-C* function are indicated. *Antp* function is required for proper segment morphogenesis of the thorax. Analysis of Scr^- and Dfd^- mutant embryos suggests that these genes are required for the differentiation of the prothorax and posterior head regions.

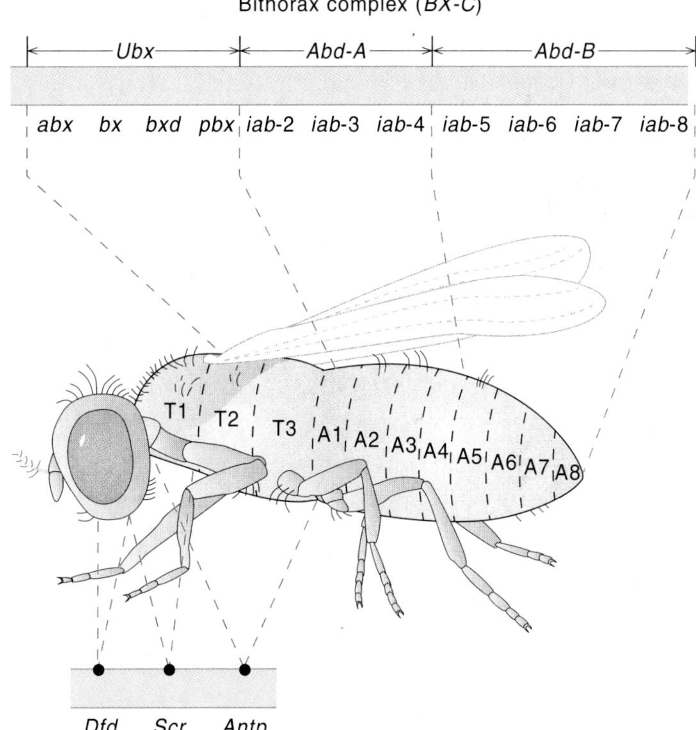

The activity of the homeotic gene domains is dependent in part on interactions between the genes themselves. In general, the most posteriorly expressed genes (such as *AbdA* and *AbdB* of the *BX-C*) repress the expression of the more anteriorly expressed genes (such as *Ubx* and *Antp*). In addition to generating differences between parasegments, homeotic genes are also expressed differentially between germ layers. The differential expression serves to modulate the basic pattern that underlies the organization of each segment.

Cell-Cell Interaction Is Important in the Elaboration of the Developmental Pattern in Parasegments

We have seen how the cells at the boundaries of each parasegment are specified, which is reflected by their expressing particular segment-polarity genes. Although it is possible that the intervening blastoderm cells likewise become determined in response to pair-rule gene signals, genetic studies of specific mutants suggest that they remain unspecified at this stage. Specification of cells within each parasegment would then occur as the cells divide in response to signals mediated by cell–cell interaction. For example, as cells divide, they would assume states dependent on their apposition to *wg*-expressing (or *en*-expressing) cells. Further divisions would allow for the intercalation of additional cell states. Such a process requires the existence of signaling molecules and receptors capable of transducing and receiving information between cells. Although we do not know too much about the specific apparatus used for the transmission of such signals in *Drosophila,* it presumably employs the types of structures that are used in the transmission of hormonal signals (see chapter 27) as well as the type of signals that are involved in B-T cell interaction in antibody formation (see chapter 37).

Early Development in Drosophila and Vertebrates Shows Striking Similarities

Drosophila fruit flies and other invertebrates develop along a very different pathway from that of vertebrates. Despite this, W. McGinnis has found that at the earliest stages of development, flies and vertebrates share a common pattern of expression of the so-called homeobox genes. Genes with very similar homeobox sequences have been identified in flies and many other animals, including frogs, mice, and even humans. This discovery suggests that despite the differences in the final appearance of the animals, they use closely related genes to specify the basic body plan along the anterior–posterior or head–tail axis. A comparison of homeobox genes in mice and humans that are believed to be comparable in both structure and function with *Drosophila* homeobox genes is shown in figure 36.35. Currently, McGinnis is attempting to test this correlation by transplanting human homeobox genes into *Drosophila* to see how they affect the phenotype.

Figure 36.35

Spatial and functional arrangement of homeobox genes in *Drosophila* and comparable homeobox genes that have been found in mice and humans. (From W. McGinnis and M. Kuziora, The molecular architects of body design, *Sci. Amer.,* February 1994. Copyright © 1994 by Scientific American, Inc. All rights reserved. Reprinted by Permission.)

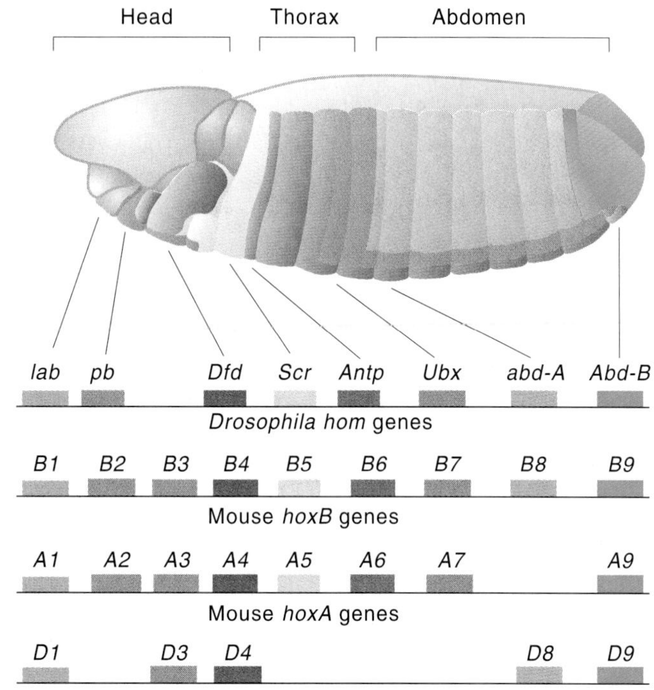

Summary

Some mechanisms of gene expression that are found in eukaryotes are rarely, if ever, seen in prokaryotes. Still, *trans*-acting regulatory proteins that bind to *cis* effector sites on the genome are present in eukaryotic systems as well as in *E. coli.* The best-understood unicellular eukaryote is the budding yeast *Saccharomyces cerevisiae.* Gene regulation, particularly of development, can be quite complex in multicellular eukaryotes. Our discussion in this chapter focused on the following points.

1. In eukaryotes, individual genes encode transcripts for a single polypeptide chain. Although functionally related genes are clustered, each gene has its own promoter.
2. In yeast, genes of the *GAL* system are under the joint control of two *trans*-acting genes that encode regulatory proteins: *GAL4* and *GAL80.* GAL4 protein is an activator that binds at an upstream site, and GAL80 is a repressor that inhibits GAL4 action by binding to it. The ability of the GAL80 protein to

bind to GAL4 is lost in the presence of galactose, which binds to the GAL80 protein, causing a reversible allosteric change in its structure.

3. Yeast has two haploid cells of opposite mating types and one nonmating diploid cell that results from the fusion of haploid cells of opposite mating type. The mating type is determined by the *MAT* locus. Information for the mating type is stored at other, silent loci and is expressed only if it is transposed to the *MAT* locus. Mating-type information is not expressed at the storage loci because of a complex repressor system of proteins interacting in *cis* fashion over a considerable distance at the control centers of the storage loci.
4. The greatest difference between the regulatory systems in yeast and *E. coli* is that yeast regulatory proteins can bind at a long distance from the RNA polymerase-binding site and still be effective.
5. Complex multicellular eukaryotes differentiate irreversibly so that different cell types express a different profile of genes. Genes that are expressed are usually associated with swollen chromatin. Proteins found in active regions of the genome show characteristic modifications.
6. Enhancers are elements of the genome that generally stimulate transcription. They resemble yeast UAS sequences in yeast in that they can function over long distances. In fact, they can function over even greater distances than UAS sequences, and they are effective in either orientation: either upstream or downstream from the promoter. A possible mechanism for enhancer and UAS function over long distances is that the chromosome folds to bring the proteins bound at the enhancer site in close proximity to other regulatory proteins or RNA polymerase bound at the promoter.
7. A typical enhancer is composed of a cluster (usually two or three) of *cis* sites called enhansons; each enhanson contains binding sites for a unique combination of regulatory proteins. Enhansons must be clustered to be effective.
8. In chapter 34 we discussed DNA-binding proteins that regulate transcription in prokaryotes. In prokaryotes and eukaryotes, specific recognition is dominated by H bond interactions that take place in the major groove of the DNA. In both cases the α helix is the most common element used for DNA recognition. The most striking difference between DNA-binding proteins in prokaryotes and eukaryotes has to do with the symmetry of the interaction. In prokaryotes the binding proteins almost always interact in a symmetrical fashion with the DNA. In eukaryotes many of the cases that have been examined so far involve proteins that interact in an asymmetrical fashion with the DNA. In many cases the regulatory proteins interact in multisubunit complexes that contain nonidentical subunits. Four different types of structural motifs are discussed: the homeodomain, the zinc finger, the leucine zipper, and the helix-loop-helix.
9. Posttranscriptional regulation is an important mode of regulation in eukaryotes. Examples are given of three types of posttranscriptional regulation: alternative modes of mRNA splicing, regulation at the translation level, and alternative modes of polypeptide processing.
10. The existence of many kinds of regulatory mutants has helped to advance our understanding of early development, especially in the fruit fly *Drosophila melanogaster*. Regulatory gene products are proteins that activate or repress other genes. Early development in *Drosophila* is a sequence of events in which different regulatory proteins gradually come into play in cascade fashion, controlling a wide range of enzymes and structural proteins and also influencing each other. Until the blastoderm stage the nuclei in a developing *Drosophila* embryo are not separated by cellular membranes. As a result, the regulatory proteins and other gene products may diffuse freely from their site of synthesis to other nuclei in the embryo. At the late blastoderm stage the nuclei become cellularized. From this point on, the influence of regulatory proteins made in one cell must be exerted on another cell at the level of the cell membrane.

Selected Readings

Achneider, R. J., and T. Shenk, Impact of virus infection on host cell protein synthesis. *Ann. Rev. Biochem.* 56:317–332, 1987.

Altmann, M., and H. Trachsel, Regulation of translation initiation and modulation of cellular physiology. *TIBS* 18:429–432, 1993.

Bingham, P. M., T. Chou, I. Mims, and Z. Zachari, On/off regulation of gene expression at the level of splicing. *Trends Genet.* 4:134, 1988.

Blau, H. M., How cells know their place. *Nature* 358:284–286, 1992.

Brietbart, R. E., A. Andreadis, and B. Nadal-Ginard, Alternative splicing: A ubiquitous mechanism for the generation of multiple protein isoforms from single genes. *Ann. Rev. Biochem.* 56:467–495, 1987.

Brown, D. D., How a simple animal gene works. In *The Harvey Lectures,* Series 76, pp. 27–44. New York: Academic Press, 1982.

Chen, H.-Z., T. Hoey, and G. Zubay, Purification and properties of the *Drosophila* zen protein. *Mol. Cell. Biochem.* 79:181–189, 1988. First evidence that a homeobox protein binds to DNA as a monomer.

Chen, J.-J., and I. M. London, Regulation of protein synthesis by heme-regulated eIF-2 α kinase. *TIBS* 20:105–108, 1995.

Cowell, I. G., Repression versus activation in the control of gene transcription. *TIBS* 19:38–42, 1994.

Duboule, D., and G. Morata, Colinearity and functional hierarchy among genes of the homeotic complexes. *TIG* 10:358–364, 1994.

Evans, R. M., The steroid and thyroid hormone receptor superfamily. *Science* 240:889–895, 1988.

Fastinejad, F., T. Perlmann, R. M. Evans, and P. B. Sigler, Structural determinants of nuclear receptor assembly on DNA direct repeats. *Nature* 375:203–211, 1995.

Ferre-D'Amare, A. R., G. C. Prendergast, E. B. Ziff, and S. K. Burley, Recognition by Max of its cognate DNA through a dimeric b/HLH/Z domain. *Nature* 363:38–44, 1993.

Forsburg, S. L., and L. Guarente, Communication between mitochondria and the nucleus in regulation of cytochrome genes in the yeast *Saccharomyces cerevisiae*. *Ann. Rev. Cell. Biol.* 5:153–180, 1989.

Funder, J. W., Mineralocorticoids, glucocorticoids, receptors and response elements. *Science* 259:1132–1133, 1993.

Gabrielsen, O. S., and A. Sentenac, RNA polymerase III (C) and its transcription factors. *Trends Biochem. Sci.* 16:412–416, 1991.

Geballe, A. P., and D. R. Morris, Initiation codons within 5′-leaders of mRNAs as regulators of translation. *TIBS* 19:159–164, 1994.

Gehring, W. J., The molecular basis of development. *Scientific American* 253(4): 152–162, 1985.

Gehring, W., Homeo boxes in the study of development. *Science* 236: 1245–1252, 1987.

Gimeno, C. J., and G. R. Fink, The logic of cell division in the life cycle of yeast. *Science* 257:626, 1992.

Gruenberg, D. A., S. Natesan, C. Alexandrc, M. Z. Gilman. Human and *Drosophila* homeo domain proteins that enhance the DNA-binding activity of serum response factor. *Science* 257:1089–1095, 1992.

Grunstein, M., Histones as regulators of genes. *Scientific American* 267:68–74, 1992.

Guarente, L. P., Regulatory proteins in yeast. *Ann. Rev. Genet.* 21:425–452, 1987.

Gurdon, J., Egg cytoplasm and gene control in development. The Croonian Lecture, 1976. *Proc. R. Lond. B.* 198:211–247, 1977. A classic paper.

Hahn, S., Structure and function of acidic transcription activators. *Cell* 72:481–483, 1993.

Hanes, S. D., and R. Brent, A genetic model for interaction of the homeodomain recognition helix with DNA. *Science* 251:426–430, 1991.

Hanna-Rose, W., and U. Hansen, Active repression mechanisms of eukaryotic transcription repressors. *TIG* 12:229–234, 1996.

Harrison, S. C., A structural taxonomy of DNA-binding domains. *Nature* 353:715–719, 1991.

Hinnebusch, A. G., Involvement of an initiation factor and protein phosphorylation in translational control of GCN4 mRNA. *Trends Biochem. Sci.* 15:148–152, 1990.

Karin, M., and T. Smeal, Control of transcription factors by signal transduction pathways: The beginning of the end. *Trends Biochem. Sci.* 17:418–422, 1992.

Karlsson, S., and A. W. Nienhius, Developmental regulation of human globin genes. *Ann. Rev. Biochem.* 54:1071–1108, 1985.

King, T., and R. Briggs, Serial transplantation of embryonic nuclei. *Cold Spring Harb. Symp. Quant. Biol.* 21:271–290, 1956. A classic paper.

Koleske, J., and R. A. Young, An RNA polymerase II holoenzyme responsive to activators. *Nature* 368:466–469, 1994. At last we are seeing the use of genetics to determine essential components of RNA polymerase II.

Lai, E., and J. E. Darnell, Jr., Transcriptional control in hepatocytes: A window on development. *Trends Biochem. Sci.* 16:427–429, 1991.

Laybourn, P. J., and J. T. Kadonaga, Role of nucleosomal cores and histone H1 in regulation of transcription by RNA polymerase II. *Science* 254:238–245, 1991.

Lee, M. S., S. A. Kliewer, J. Provencal, P. E. Wright, and R. M. Evans, Structure of the retinoid X receptor α DNA binding domain: A helix required for homodimeric DNA binding. *Science* 260:1117–1121, 1993.

Leuther, K. K., and S. A. Johnston, Nondissociation of GAL4 and GAL80 *in vivo* after galactose induction. *Science* 256:1333–1336, 1992.

Lewis, E. B., A gene complex controlling segmentation in *Drosophila*. *Nature* 276:565–510, 1978. A classic paper.

Li, E., C. B. Beard, and R. Jaenisch, Role for DNA methylation in genomic imprinting. *Nature* 366:362–365, 1993. A genetic study demonstrating that DNA methylation is required for the differential expression of inherited paternal and maternal alleles.

Lumb, K. J., and P. S. Kim, Measurement of interhelical electrostatic interactions in the GCN4 leucine zipper. *Science* 268:436–440, 1995.

Marmorstein, R., M. Carey, M. Ptashne, and S. C. Harrison, DNA recognition by GAL4: Structure of a protein-DNA complex. *Nature* 356:408–414, 1992.

Marzluff, W. F., and N. B. Pandey, Multiple regulatory steps control histone messenger-RNA concentrations. *Trends Biochem. Sci.* 12:49–51, 1988.

McClintock, B., Controlling elements and the gene. *Cold Spring Harb. Symp. Quant. Biol.* 21:197–216, 1956. A classic paper.

McGinnis, W., and M. Kuziora, The molecular architects of body design. *Scientific American* February: 58–66, 1994.

Nevins, J. R., Transcriptional activation by viral regulatory proteins. *Trends Biochem. Sci.* 16:435–439, 1991.

Nusslein-Volhard, C., and E. Wiechaus, Mutations affecting segment numbers and polarity in *Drosophila*. *Nature* 287:795–801, 1980. A classic paper.

O'Shea, E. K., J. D. Klemm, P. S. Kim, and T. Alber, X-ray structure of the GCN4 leucine zipper, a two-stranded parallel coiled coil. *Science* 254:539–544, 1991.

Parkhurst, S. M., D. Bopp, and D. Ish-Horowicz, X:A ratio, the primary sex-determining signal in *Drosophila*, is transduced by helix-loop-helix proteins. *Cell* 63:1179–1191, 1990.

Pearce, D., and K. R. Yamamoto, Mineralocorticoid and glucocorticoid receptor activities distinguished by nonreceptor factors at a composite response element. *Science* 259:1161–1164, 1993.

Pomerantz, J. L., and P. A. Sharp, Homeodomain determinants of major groove recognition. *Biochemistry* 33:10851–10852, 1994.

Raghow, R., Regulation of messenger RNA turnover in eukaryotes. *Trends Biochem. Sci.* 12:3358–3360, 1987.

Rhodes, D., and A. Klug, Zinc fingers. *Scientific American* February: 56–65, 1993.

Richard-Foy, H., and G. L. Hager, Sequence-specific positioning of nucleosomes over the steroid-inducible MMTV promoter. *EMBO* 6:2321–2328, 1987.

Ronne, H., Glucose repression in fungi. *TIG* 11:12–17, 1995.

Ruvkun, G., and M. Finney, Regulation of transcription and cell identity by POU domain proteins. *Cell* 64:475–478, 1991.

Sauer, F., S. K. Hansen, and R. Tjian, Multiple TAFs directing synergistic activation of transcription. *Science* 270:1783–1788, 1995.

Sauer, F., S. K. Hansen, and R. Tjian, Multiple TAF_{II}s directing synergistic activation of transcription. *Science* 270:1783–1788, 1995.

Schler, A. F., and W. L. Gehring, Direct-homeo domain-DNA interaction in the autoregulation of the *fushi tarazu* gene. *Nature* 356:804–806, 1992.

Schwabe, J. W. R., and D. Rhodes, Beyond zinc fingers: Steroid hormone receptors have a novel structural motif for DNA recognition. *Trends Biochem. Sci.* 16:291–296, 1991.

Serfling, E., M. Jasin, and W. Schaffner, Enhancers and eukaryotic gene transcription. *TIG* August: 224–230, 1985.

Sherman, A., M. Shefer, S. Sagee, and Y. Kassir, Post-transcriptional regulation of IME1 determines initiation of meiosis in *Saccharomyces cerevisiae*. *Mol. Gen. Genet.* 237:375–384, 1993.

Singer, S. J., Intercellular communication and cell-cell adhesion. *Science* 255:1671–1677, 1992. Considers important aspects of regulation we didn't have time to cover.

Stark, G., M. Debatisse, E. Giulotto, and G. M. Wahl, Recent progress in understanding mechanisms of mammalian gene amplification. *Cell* 57:901–908, 1989.

Thompson, C. C., and S. I. McKnight, Anatomy of an enhancer. *Trends Genet.* 8:232–236, 1992.

Tjian, M., and T. Maniatis, Positive control of pre-mRNA splicing in vitro. *Science* 256:237–240, 1992.

Tsai, M. J., and B. W. O'Malley, Molecular mechanisms of action of steroid/thyroid receptor superfamily members. *Ann. Rev. Biochem.* 63:451–486, 1994.

Weinzierl, R. O. J., B. D. Dynlacht, and R. Tjian, Largest subunit of *Drosophila* transcription factor IID directs assembly of a complex containing TBP and a coactivator. *Nature* 362:511–517, 1993.

Wolffe, A. P., Nucleosome positioning and modification: Chromatin structures that potentiate transcription. *TIBS* 19:240–244, 1994.

PROBLEMS

1. Why is attenuation control in eukaryotes unlikely?
2. The domain-swap experiment illustrated in figure 31.4 demonstrated that the *lexA* DNA binding domain from *E. coli* is functionally interchangeable with the equivalent domain from the GAL4 protein if and only if the *lexA* DNA recognition site replaces the GAL4 equivalent. Would the GAL1 gene with a *lexA* binding site upstream be transcribed in the presence of intact *lexA* protein (with no GAL4 activation domain)? Why or why not?
3. A GAL4 mutation ($GAL4^C$) leads to constitutive synthesis of the GAL1 gene product in haploid yeast. Propose an explanation for the effect of this mutation.
4. The effects of nested deletions of upstream regions on the expression of a gene have been important in illuminating the importance of these *cis*-acting control regions. Describe a series of such deletions that could be used to demonstrate the importance of the GAL4 binding site in the *GAL1* upstream region.
5. Explain how starvation of yeast for any one of 10 amino acids derepresses the synthesis of more than 30 enzymes in nine different amino acid biosynthetic pathways (general amino acid control). What would be the effect on general amino acid control of deleting the four short upstream reading frames preceding the GCN-4 coding sequences in its mRNA?
6. How do you expect a deletion of *HMLE* to affect the expression of mating-type genes in yeast? Compare this effect with the deletion of the α_2 gene from MAT_α. Consider both homothallic and heterothallic backgrounds.
7. Color blindness is X-chromosome linked. Bearing in mind the phenomenon of X-chromosome inactivation, suggest an explanation for the observation that females who are heterozygous for the defective gene can discriminate colors.
8. Expression of some genes in eukaryotic cells can be induced by treatment with 5-azacytidine. Explain how this happens.
9. High salt concentrations weaken the interaction of histones with DNA but have little effect on the binding of many regulatory proteins. Explain these observations in terms of how these molecules interact with DNA.
10. List some characteristics that distinguish active chromatin from inactive chromatin.
11. The restriction endonuclease HpaII cleaves the sequence CCGG only if the second C is unmethylated. The enzyme MspI cleaves the same sequence, whether or not it is methylated. How do the globin-specific sequences in erythroblast DNA (erythroblasts are red blood cell precursors) differ from other tissues in their susceptibility to these two enzymes?
12. Which of the exons shown in figure 36.24 are found in all splice variants of tropomyosin? What would be the significance of such constant exons? What might be the purpose of exons that are present in some mRNAs but not in others?
13. Explain how a DNA sequence (enhancer sequence) located 5,000 bp from a gene transcription start site can stimulate transcription even if its orientation is reversed.
14. When mammalian cells in culture are treated with the antifolate methotrexate (see chapter 26), cells can be selected that are resistant to high levels of this toxic compound. The enzyme dihydrofolate reductase becomes elevated about 1,000-fold in these resistant cells. What mechanism can account for such a large amount of this enzyme being made, and how does this protect the cell from the toxic effects of methotrexate? How could you test for the molecular mechanism of this drug resistance?
15. What kind of changes would have to be made in a typical eukaryotic structural gene for its protein product to be expressed in bacteria?
16. In the *Xenopus oocyte* a large number of ribosomes are made in a short time to handle the rapid demand for cell growth during cleavage stages. How is this large amount of rRNA made in such a short time?
17. Based only on the definition of maternal-effect genes, segmentation genes, and homeotic genes, which would you predict would act earliest in development of the *Drosophila* embryo, and which would act latest?
18. Given a cloned fragment of a *Drosophila* gene, how could you determine which chromosomal band(s) contain the gene?

C H A P T E R

IMMUNOBIOLOGY

37

A broad arsenal of weapons enables the immune system to roust and destroy foreign invaders.

When vertebrates are invaded by foreign agents, they can mobilize a versatile set of adaptive immune responses to form specifically reactive cells and proteins. These responses constitute the principal means of defense against pathogenic microorganisms and viruses and probably also against host cells that undergo transformation into cancer cells.

Throughout most of this century the subject of immunology has attracted some of the keenest minds in biology. As a result, the intricacies of this fascinating subject have been largely unraveled, and its understanding is providing a strong bridge between the fields of biochemistry and physiology. In this chapter we give an introduction to some of the major findings in immunobiology and the experiments that have led to our current level of understanding.

Overview of the Immune System

The immune system was first studied in humans, but mice became a popular subject for immune system studies when researchers began to appreciate how close the mouse and human systems were in their organization and action. Currently, both systems are under study in many different laboratories. Although we focus here on mice, we often refer to parallel observations on humans.

The immune system is an example of a developmental process that takes place in the mature organism. In the new cells that are constantly being generated, changes arise in the genes of the immune system. This variability results from DNA splicing and point mutations. It benefits the organism by providing the cells of the immune system with the widest possible range of specificities. As soon as the organism is invaded by foreign agents, usually viruses or bacteria, the immune system is activated. Those immune system cells that carry the specific immune receptors for interacting with the foreign agent are stimulated to proliferate. In a matter of days, clones of immune system cells with the appropriate specificity have been produced, and the organism fends off the invader with the specific immunologic tools provided by those cells. The clones of cells tend to persist for some time, usually months to years, which accounts for the fact that the second time the same foreign invader makes its presence known, it is usually rejected promptly and without crisis.

Two different classes of white cells, or lymphocytes, are

Figure 37.1

B cells and T cells follow different pathways of development. In mammals, B lymphocytes mature in the bone marrow and then migrate to secondary lymphoid organs. On exposure to foreign substances known as antigens, they proliferate to produce immunoglobulins. T lymphocytes mature in the thymus gland. They also can be stimulated to proliferate by exposure to an appropriate antigen.

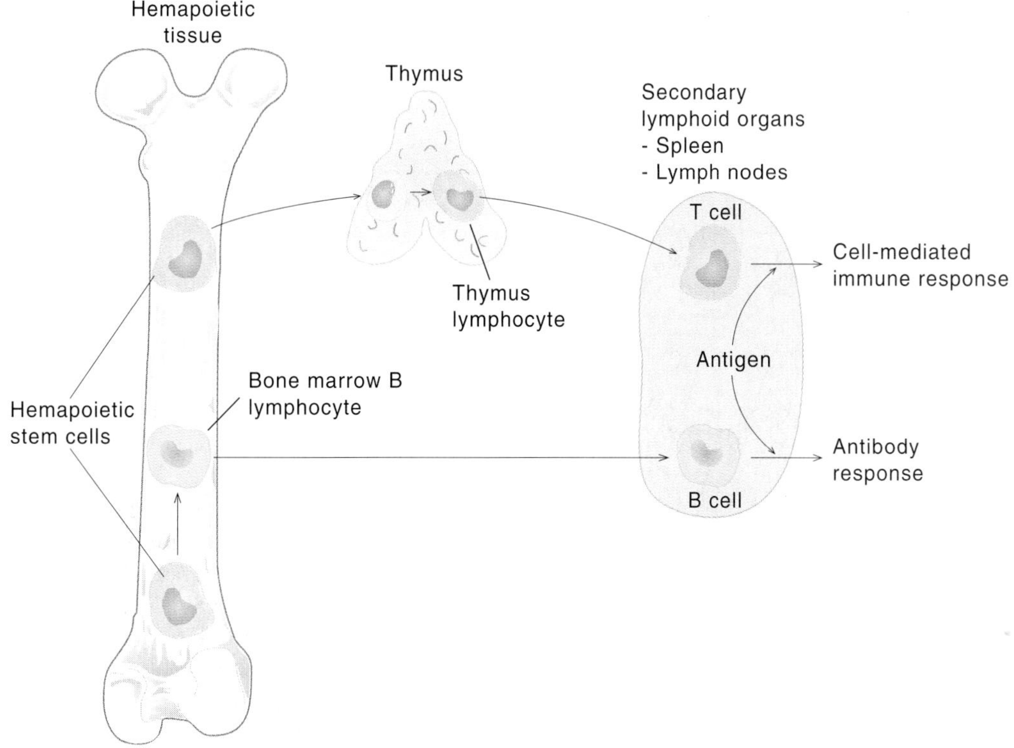

associated with the immune response: the *B cells* and the *T cells.* In mammals, B lymphocytes mature in the bone marrow and then migrate to secondary lymphoid organs (fig. 37.1). On exposure to foreign substances known as antigens, they proliferate and produce immunoglobulin proteins known as antibodies, which they secrete into the bloodstream. T lymphocytes, by contrast, mature in the thymus gland before migrating to the secondary lymphoid organs. They also can be stimulated to proliferate by exposure to an appropriate antigen, but their specific effector molecules remain firmly bound to the cellular membrane, as opposed to being secreted. Three main types of T cells have been recognized: killer T cells, which specifically destroy target cells; helper T cells, which promote the maturation of antigen-stimulated B and T cells; and suppressor T cells, which block the effects of T helper cells.

The Humoral Response: B Cells and T Cells Working Together

There are two basically different types of immune response. The first, called the humoral, or antibody, response, involves the concerted action of both B cells and T cells, and the active agents are the antibody, or immunoglobulin, proteins secreted by the B cells into the bloodstream. The combined B–T immune response is characteristic of most vertebrates. The second type of immune response, called the cell-mediated response, involves only T cells, and the active agent is often the circulating T cell itself, which attacks the foreign agent. This type of immune response is limited to certain groups of vertebrates, including mammals. We first discuss the B–T cell response mediated by antibodies.

Immunoglobulins Are Extremely Varied in Their Specificities

A detailed investigation of immunoglobulin structure provided the first leads about how immunoglobulins of such a wide variety are synthesized. The most common type of immunoglobulin is the 7S molecule, known as immunoglobulin G (IgG). This immunoglobulin has a molecular weight of about 150,000 daltons. It is composed of four polypeptide chains: two heavy (H) chains with molecular weights of about 50,000 and two light (L) chains with molecular weights of about 25,000 (fig. 37.2).

To obtain a detailed understanding of immunoglobulin structure, it was necessary to isolate pure antibody proteins for sequencing. Fortunately, it was discovered that certain plasma cell tumors known as myelomas produce enormous amounts of pure immunoglobulins, which have the same gross structure as the mixed immunoglobulins isolated from serum. Each myeloma appears to originate from a single cell turned cancerous. As a result, each myeloma serves as the source of a pure antibody protein, the

Figure 37.2

Structure of immunoglobulin G (IgG). The light (L) and heavy (H) chains have repeating domains, each with about 110 amino acid residues and an approximately 60-member S—S bonded loop. The C and V refer to regions of the sequence that are relatively constant or quite variable, respectively, in different IgG species. (Illustration copyright by Irving Geis. Reprinted by permission.)

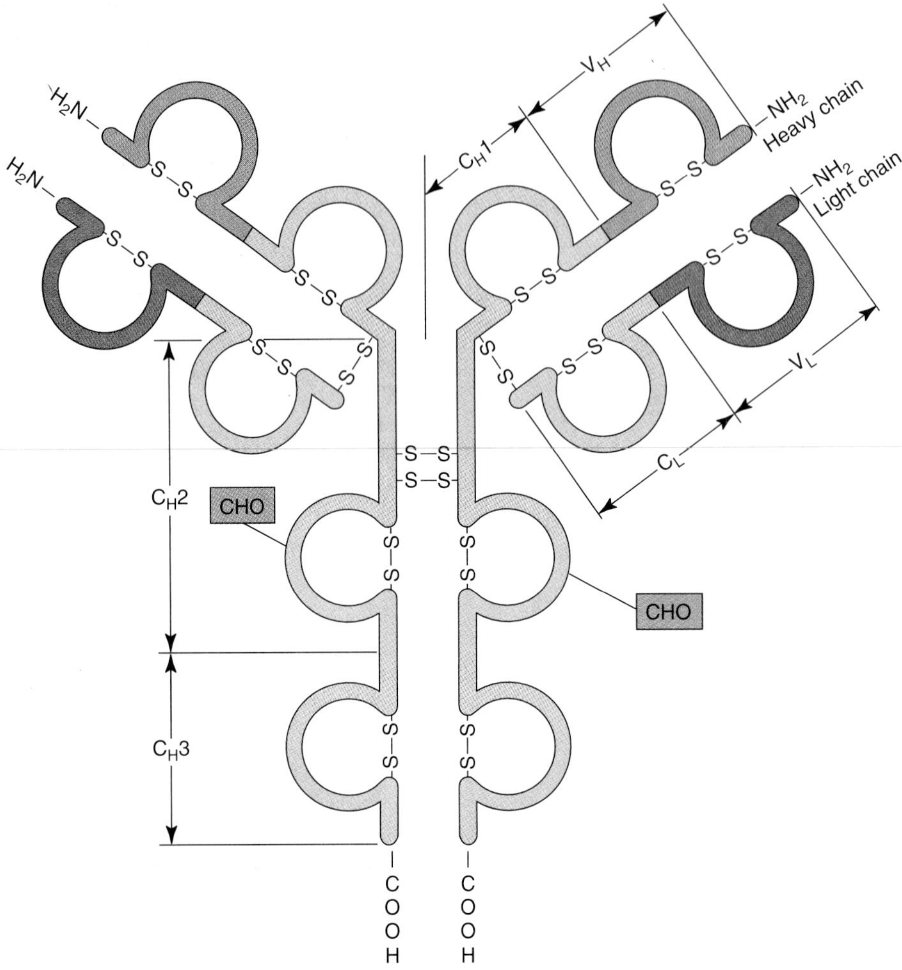

sequence and other properties of which can be determined. In some cases myelomas synthesize both heavy and light chains, like normal antibody-forming cells, but in other cases they synthesize only one or the other type of chain. Comparison of the sequences from a number of different immunoglobulins derived from myelomas has shown that both the heavy and the light chains for immunoglobulins of the same class are divided into regions of relatively constant sequence (the *C* segments) and regions of relatively variable sequence (the *V* segments). In the intact immunoglobulins the antigen-binding domain is composed exclusively of *V* segments originating from the H and L chains (see fig. 37.2), each tetramer containing two equivalent sites for the binding of a specific antigen.

As information on antibody sequences has accumulated, it has become increasingly clear that within the *V* regions of both the heavy and light chains, three segments account for most of the variability (fig. 37.3). These regions are called the hypervariable regions, and they are known to be the parts of the antibody molecule in most direct contact with the antigen in the antigen–antibody complex.

Immunoglobulin G (IgG), the tetrameric species we have been discussing thus far, is not the only type of immunoglobulin found (table 37.1). Most of the other known antibodies—IgA, IgM, IgD, and IgE—also involve a closely related tetrameric structure, sometimes forming larger aggregates and always associated with different functions. For instance, IgM is a 19S antibody accounting for 5%–10% of the serum Ig. It is an aggregate of five tetramers that is formed early during the immune reaction, soon to be diminished in quantity and overshadowed by large amounts of IgG. IgA occurs in various polymeric forms. It normally accounts for about 15% of the total immunoglobulin found in serum, and in addition it is the principal immunoglobulin in exocrine secretions. Each of the different classes of immunoglobulin possesses distinct, heavy chains. Indeed, even within the human IgG class, antisera tests reveal four different types of IgG (IgG1 through IgG4), each with a distinctive H chain, comprising

Figure 37.3

Approximate locations of the hypervariable regions in the heavy and light chains of IgG. Each hypervariable segment is believed to make up part of the site that binds to antigen, known as the complementarity-determining region (CDR). The CDR regions are located in the loop regions of the variable domains where they can make close contact with antigen.

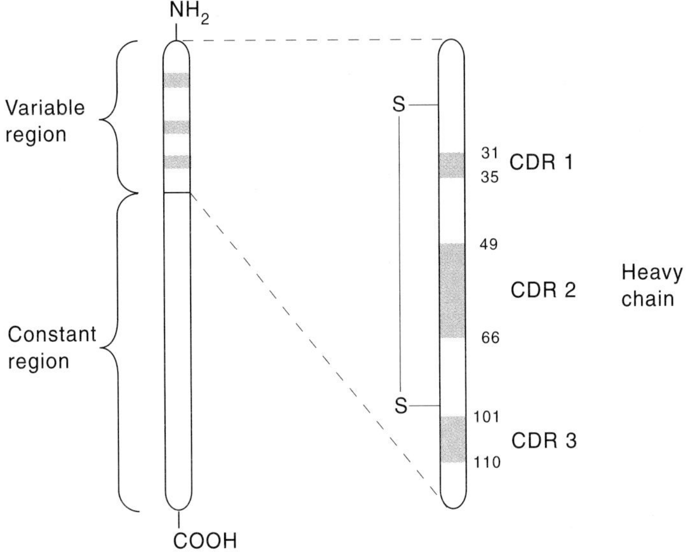

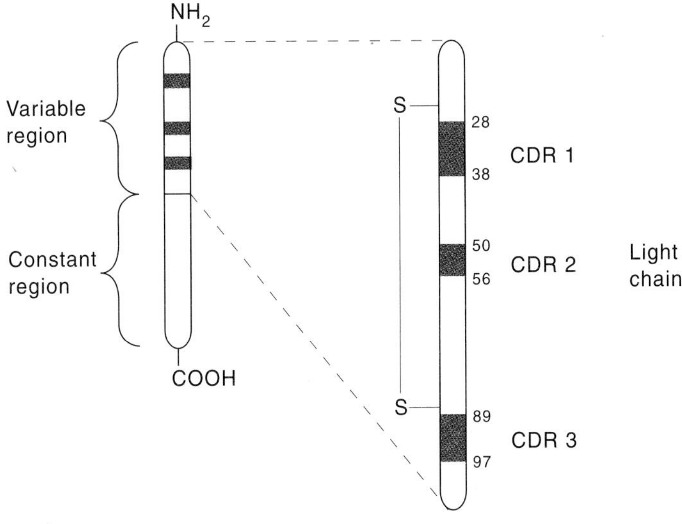

about 70%, 19%, 8%, and 3%, respectively, of the total IgG proteins. The light chains, of which there are two types, are common to all classes of immunoglobulins.

Immunoglobulins that are produced in response to antigens are themselves highly immunogenic (that is, they stimulate antibody formation) when injected into genetically nonidentical organisms. The serological responses induced by using immunoglobulins as antigens or immunogens have been useful in classifying them according to their antigenic determinants.

So-called isotypic determinants are shared by all immunoglobulin molecules of a given class. For example, the human IgG molecules are classified into four isotypes because they are recognized by heterologous antisera produced in one species against the immunoglobulins of another species. Within a given isotype the different immunoglobulins can usually be separated into sets, called allotypes, that are distinguished by minor antigenic differences. Allotypic differences usually reflect alternative amino acid substitutions within otherwise quite similar amino acid sequences of the *C* region of the antibody. Allotypes show typical Mendelian inheritance patterns. Antisera that are useful for discriminating between allotypes are usually made by injecting an individual who lacks a specific allotype with immunoglobulins from an individual who carries the allotype. Finally, idiotype refers to the specific antigenic determinant of the antibody. Some idiotypic determinants are limited to a single immunoglobulin; others are shared by a small number of immunoglobulins. Whereas isotypic and allotypic differences usually result from differences in the constant portion, or *C* segments, of the immunoglobulin polypeptide chain, idiotypic differences usually result from differences in the variable portion, or *V* segment. As we know already, this is the region that contains the binding site for the foreign agent or antigen.

Antibody Diversity Is Augmented by Unique Genetic Mechanisms

Antibodies have been studied most extensively in the mouse, an ideal vertebrate for both genetic and biochemical manipulations. It is believed that mice can synthesize more than a million antibodies with different antigenic specificities. This enormous diversity of proteins is generated from a limited amount of genetic information with the help of two mechanisms: somatic recombination and somatic mutation.

DNA Splicing Brings Different Parts of the Antibody Gene Together. The involvement of somatic recombination was first demonstrated by examining the structure of a specific antibody gene in embryonic and adult immunoglobulin-forming tissue. The adult tissues favored for many studies are myelomas. These tumorous tissues, which we mentioned earlier, produce a homogeneous population of polypeptide chains. They have served as a convenient source of pure immunoglobulin polypeptide chains as well as their mRNAs.

Recall that the generalized structure of the predominant serum antibody consists of two identical heavy chains and two identical light chains (see fig. 37.2). In the mouse there are two classes of light chains, which differ appreciably in the constant regions of the polypeptide chains. These are referred to as the kappa (κ) and the lambda (λ) light chains. Using highly inbred, genetically identical (isogenic) strains of mice, S. Tonegawa and his co-workers isolated the DNA from embryonic cells and two different myeloma tumor cells: one that produces homogeneous λ light chains (strain H2020) and one that produces κ light chains (strain MOPC321). These DNAs were digested with the *Eco*R1 restriction enzyme and electrophoresed on an agarose gel. The gels contained an enormous variety of restriction fragments representing total nuclear DNA, and consequently no discrete pattern of bands could be seen with a stain for nucleic acid.

Table 37.1
Different Isotypes Found in Humans

Class	Heavy Chain	Light Chain	Molecular Formula	Molecular Weight (daltons)	Physiological Functions
IgG	γ	κ or λ	$\gamma_2\kappa_2$ $\gamma_2\lambda_2$	150,000	Complement fixation; placental transfer; stimulation of ingestion by macrophages
IgA	α	κ or λ	$\alpha_2\kappa_2$ $\alpha_2\lambda_2$	160,000 320,000	Localized protection of external secretions
IgM	μ	κ or λ	$\mu_2\kappa_2$ $\mu_2\lambda_2$	900,000	Complement fixation; early immune response; stimulation of ingestion by macrophages
IgD	δ	κ or λ	$\delta_2\kappa_2$ $\delta_2\lambda_2$	185,000	Found on cell surfaces; function unknown
IgE	ϵ	κ or λ	$\epsilon_2\kappa_2$ $\epsilon_2\lambda_2$	200,000	Stimulates mast cells to release histamines

However, when the electrophoresed gels containing the DNA were denatured and hybridized with ^{32}P-labeled DNA containing the sequences found in the RNA for the λ chain, a specific pattern of bands showed up in autoradiographs (fig. 37.4). The R1 digest of DNA from the λ-containing myeloma (H2020) showed four bands; the DNAs of the embryo and the κ-containing myeloma (MOPC321) showed three bands in common with the first DNA but were missing the fourth band.

These results strongly suggested that at some point during development from the embryonic state, a rearrangement had taken place in the H2020 myeloma cells on one of the homologous pairs of chromosomes carrying the λ-chain gene. Further analyses with more specific radioactive probes, containing sequences from either C_λ (the constant region of the λ chain) or V_λ (the variable region of the λ chain), were done. Only the 7.4-kb fragment originating from the λ myeloma cell hybridized to both probes. Therefore the *Eco*R1 fragment must contain both V_λ and C_λ DNA. The 8.6-kb fragment hybridized to the C_λ but not the V_λ probe. Therefore this fragment must contain C_λ sequences but no V_λ sequences. Both the 3.5-kb fragment and the 4.8-kb fragment hybridized exclusively to the V_λ probe, so they must contain only V_λ sequences. Still further analyses indicated that the 7.4-kb fragment arose from a recombinational event between the 3.5-kb fragment and the 8.6-kb fragment (fig. 37.5). The 4.8-kb fragment originates from another V_λ sequence located in the same chromosome. The mouse has only two V_λ sequences in the embryo. As a rule, only one of the pairs of homologous chromosomes recombines to yield a productive antibody gene. This explains why the 3.5-kb and 8.6-kb bands are still visible in the H2020 myeloma DNA preparation. The κ myeloma cell producer shows the same pattern as the embryonic cell. Had it been probed with a DNA carrying the V_κ antibody sequence, it would have shown differences from the embryonic pattern.

Figure 37.4

Analysis of DNA fragments containing λ_1 gene sequences from mouse embryo and myeloma cells. High-molecular-weight DNAs extracted from myeloma H2020, a λ-chain producer (A), from a 13-day-old BALB/c embryo (B), and from myeloma MOPC321, a κ-chain producer (C) were digested to completion with *Eco*R1, electrophoresed on agarose gel, and transferred to nitrocellulose membrane filters and hybridized with a nick-translated *Hha*I fragment of the plasmid B1 DNA. (*Source:* After C. Brack, M. Hirama, R. Lemhard-Schuller, and S. Tonegawa, A complete immunoglobulin gene is created by somatic recombination, *Cell* 15:1–14, 1978.)

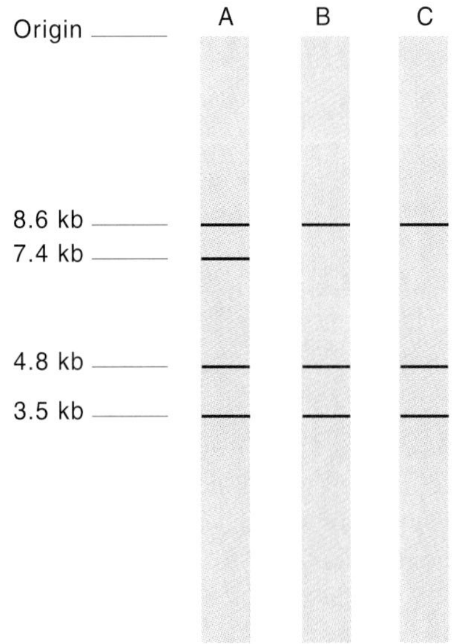

The success achieved by Tonegawa and others in this type of sequence detection of the immunoglobulin genes led to a massive research effort and a general picture of antibody gene organization. All antibody polypeptide chains are derived from split

Figure 37.5

Arrangement of mouse λ_1 gene sequences in embryos and λ_1 chain-producing plasma cells. The vertical arrows point to *Eco*R1 restriction sites. The blue dashed diagonal lines point to hypothesized splice points that explain the difference in structure of the region in the two cell types. The boxed regions represent coding regions, and I_1 and I_2 between boxed regions indicate first and second introns in the spliced gene.

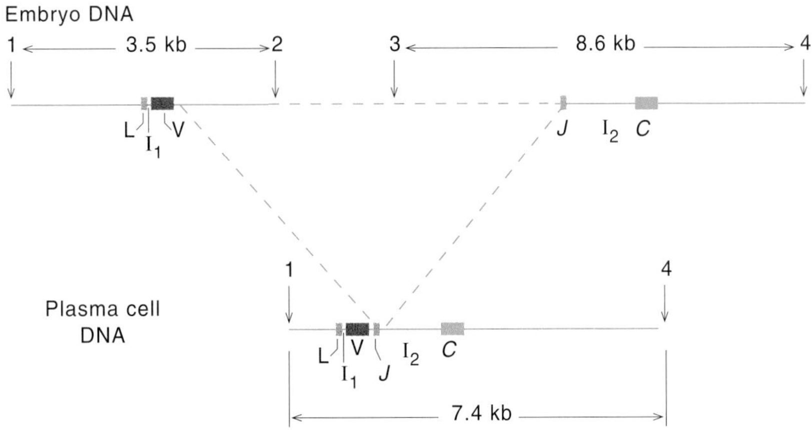

genes. Antibody genes are encoded by three unlinked gene families located on different chromosomes. In the human, chromosomes 2, 22, and 14 encode those gene clusters associated with the kappa (κ) light chains, the lambda (λ) light chains, and the heavy chains, respectively.

The light chain is encoded by three distinct DNA elements—a variable (V_L), a joining (J_L), and a constant (C_L) element (fig. 37.6). During the differentiation of a B cell a V_L element (encoding the first 95 amino-terminal residues) and a J_L segment (encoding about 13 amino acids) are joined by DNA splicing. A complete light-chain gene transcript consists of three exons and two introns, arranged in the following order: 5′L, I_1, *VJ*, I_2, *C*, 3′. Here L refers to sequences in the leader, or signal, peptide, which is removed in the mature immunoglobulin, and I_1 and I_2 refer to the first and second introns, respectively, which are removed by RNA splicing. The heavy chain is encoded by four distinct DNA elements: a V_H, a *D*, a J_H, and a C_H element. In heavy-chain variable region formation a V_H element (encoding about 99 amino acids), a *D* element (encoding 1–15 amino acids), and a J_H element (encoding about 15 amino acids) are joined in two DNA-splicing steps. The heavy-chain gene family has several closely linked C_H genes, which determine the immunoglobulin class.

Class switching, involving an additional DNA-splicing operation, frequently occurs with heavy-chain genes. Thus an immature B cell, which initially expresses a μ chain, results in IgM antibodies. Subsequently, the same cell can be induced to differentiate further so that the same V_H region of the genome becomes relocated next to a C_γ gene, causing the cell to produce an IgG antibody with the same V_H region associated with a different C_H region. In cases of class switching, the second DNA splicing involves removal of the intervening genes. Thus in the case of C_μ to C_γ switching, the region containing C_δ must be removed by the second splice reaction (fig. 37.7). In the case of C_μ to C_α class switching, all of the *C* genes between C_μ and C_α must be removed by the second splice reaction.

Figure 37.6

Organization and DNA splicing of the mouse immunoglobulin genes. The organization is depicted before and after somatic cell rearrangement. For each class, only one example of a rearranged chromosome is shown. A light chain is encoded by three distinct DNA elements, a variable (*V*), a joining (*J*), and a constant (*C*) element. Boxes indicate exons; lines connecting boxes indicate introns. Somatic cell arrangement involves the joining of a *V* and a *J* segment on the same chromosome. The heavy chain is encoded by four distinct types of DNA elements: a *V*, a *D*, a *J*, and a *C* element. In heavy-chain splicing, a *V* element, a *D* element, and a *J* element are joined in two DNA-splicing steps. The splice between *D* and *J* elements occurs first. The splices shown provide only one example of many that might occur in this system. (*Source:* From G. Zubay, *Genetics*, Benjamin/Cummings, Menlo Park, Calif., 1987, p. 808.)

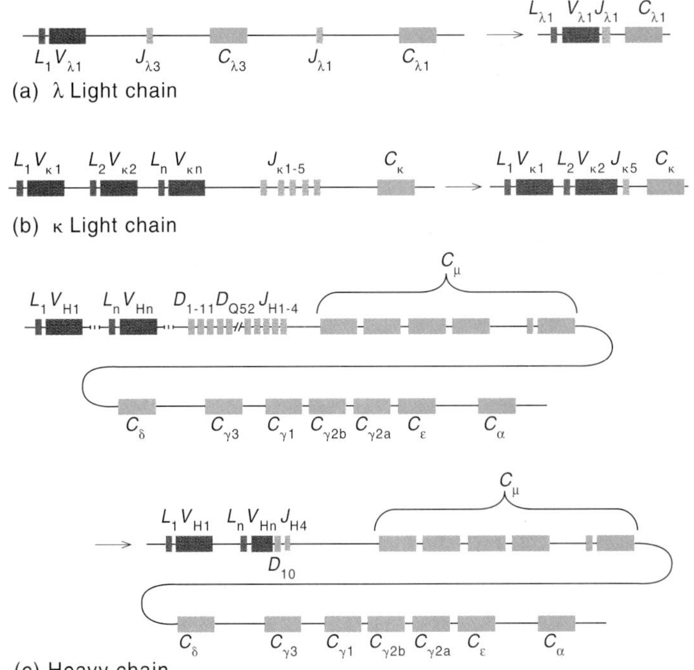

Figure 37.7

Class switching of heavy-chain genes involves additional DNA splicing. Switching from C_μ to $C_{\gamma 1}$ is illustrated. The additional splicing involves removal of the continuous segment carrying the C_μ through the $C_{\gamma 3}$ genes. (*Source:* From G. Zubay, *Genetics,* Benjamin/Cummings, Menlo Park, Calif., 1987, p. 809.)

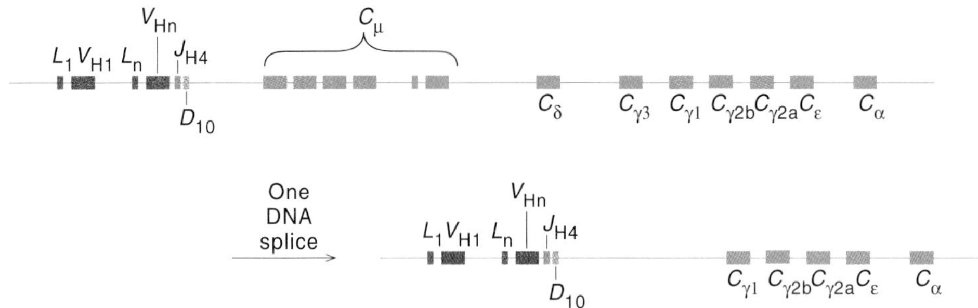

Class switching of the heavy-chain genes illustrates a temporally regulated process associated with the DNA rearrangements that can occur in the antibody-forming genes. Because any given cell normally synthesizes only one type of antibody, a special type of regulatory mechanism must respond in a systematic way to antigenic stimulation. In B cell development it is known that the V_H–D–J_H joining occurs first, leading to the formation of μ heavy chains. Subsequently, V_L–J_L joining occurs, with the production of functional light chains.

As a rule, the DNA-splicing reactions leading to the joining process occur in only one of the alleles for each of these chains. The mechanism for this phenomenon, known by the name allelic exclusion, is not understood. Another type of exclusion process results in the production of only κ or λ light chains in a given B cell. This process is called isotypic exclusion. Indirect evidence that the κ gene family is expressed before the λ gene family comes from the examination of human B-cell lines that are active κ- or λ-chain producers. In all B cells producing λ chains, the C_κ genes are either deleted or rearranged. By contrast, in all cells producing κ chains, the C_λ genes are in the germ-line configuration. The obvious inference is that a B cell becomes a κ producer first and then becomes a λ producer. Other evidence exists for regulatory signals that may control isotypic exclusion. When a pre-B cell is fused with a myeloma cell that has been producing a functional heavy chain but no functional light chain, the hybrid cell is capable of expressing a new light chain as a result of a V–J joining of gene segments in the pre-B cell. If, however, the myeloma cell was already producing a functional light chain, no new light chains or DNA rearrangements result from the cell fusion process.

Alternative Pathways Exist for RNA Splicing of Heavy-Chain mRNAs. The heavy-chain transcript contains numerous exons and introns. In addition to the introns between the L_H and V_H exons and between the V_H–J_H and C_H exons, each C_H region contains several introns. The hinge regions of γ chains (see fig. 37.2) and the intramembrane and cytoplasmic portions of all C_H chains are encoded by independent exons. Most of the RNA-splicing operations are presumed to occur by standard mechanisms similar to those described in chapter 33. In the case of the heavy-chain genes, it has been shown that alternative RNA-splicing patterns involving the C_H region can yield molecules that function as membrane-bound or secreted forms of the antibody. Thus two membrane exons have been detected in the 3′ flanking regions of both the C_μ and C_γ genes. Alternative pathways of RNA splicing give rise to two mRNAs, which encode separately the membrane-bound and secreted forms of the heavy chains. It seems likely that the alternative splicing pattern is temporally regulated, because the membrane-bound forms are favored before and the secreted forms are favored after antigenic stimulation. Amino acid as well as nucleotide sequencing of the μ_M chain (μ_M stands for the membrane-bound form of the μ chain) has revealed a 26-residue hydrophobic segment at the COOH terminus, which anchors in the lipid bilayer of the membrane.

Table 37.2
Estimated Number of Germ-Line Gene Segments for Different Mouse Antibody Genes

	Light Chains		**Heavy Chains**
	κ	λ	H
C	1	4	8
V	90–300	2	>1000
J	5	4	4
D	—	—	12

Source: Geoffrey Zubay, *Genetics,* Benjamin/Cummings, Menlo Park, Calif., 1987. Reprinted by permission of the author.

Somatic Mutation Contributes to Antibody Diversity. Various estimates, some exceeding a million, have been put forward for the number of different antibodies an organism can make. A good deal of this diversity results from the different combinations of heavy and light chains that can be generated by making alternative splicings between the different gene segments that make up the light and heavy chains (table 37.2). Superimposed on the variation from this source, additional diversity is provided by somatic

Figure 37.8

DNA splicing of an antibody gene brings an enhancer element (*E*) close to the promoter region (*P*) of the gene. This step is depicted for a heavy-chain gene after the initial two splices (see fig. 37.6).

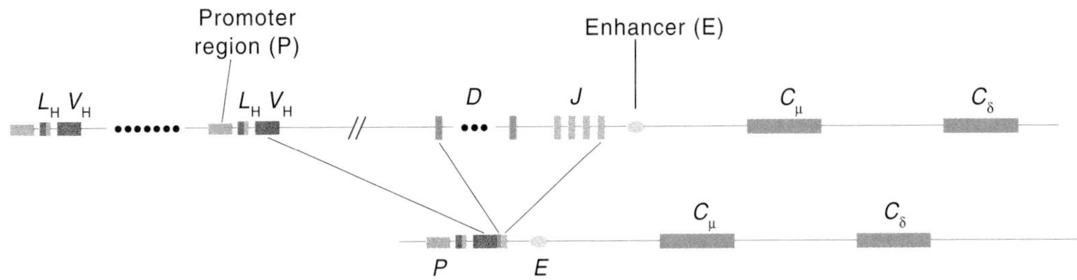

mutation within the coding regions. Several subsets of κ and heavy chains and their germ-line *V* segments have been analyzed by cloning and sequencing. The results confirm that somatic mutations amplify the diversity encoded in the germ-line genome.

A V_κ probe prepared for the κ genes expressed in a particular myeloma, MOPC167, detected only one major band in the Southern gel blot analysis of total cellular DNA. This unusual situation permitted a relatively simple sequence comparison to be made between this segment, found in the germ-line, and its rearranged counterparts in two myeloma lines: MOPC167 and MOPC511. Four and five nucleotide differences were found, respectively, in the V_κ regions of these two lines. The changes were completely different in both lines. This investigation and other similarly directed investigations show that the regions in which mutations are introduced during somatic differentiation are highly restricted to the V_l–J_l or V_H–D–J_H segments. The density of mutations observed is about three times higher in the complementarity-determining regions (CDRs) of the variable segments than in the connecting framework regions (FRs) of the variable segments. (See fig. 37.3 for the approximate location of the CDR regions.) The CDRs are the regions in the immunoglobulin that play a critical role in the interaction with antigen. Mutations produced in these regions must occur either during or after DNA splicing, as they are not found in unrearranged (unspliced) DNA. Somatic mutation clearly increases the number of possible antibodies almost without limit. Most important, it leads to selectable changes in just those regions of the immunoglobulins that are responsible for antigen recognition. Several mechanisms for hypermutation have been proposed; none have been verified.

DNA Splicing Brings an Enhancer Element Close to the Promoter. DNA sequences derived from the germ-line J_H–C_μ region are required for accurate and efficient transcription from a functionally rearranged V_H promoter (fig. 37.8). Similar to viral transcriptional enhancer elements (see chapter 36), these cellular sequences stimulate transcription from either the homologous V_H gene segment promoter or a heterologous SV40 promoter. They are active when placed on either the 5′ or the 3′ side of the rearranged V_H gene segment, and they function when their orientation is reversed. However, unlike viral enhancers, the immunoglobulin gene enhancer appears to act in a tissue-specific manner because it is active in mouse B cells but not in mouse fibroblasts.

The discovery of this enhancer suggests another benefit that results from splicing: It brings an enhancer in close proximity to the promoter just when that promoter activity becomes an important part of the cell's function.

Interaction of B Cells and T Cells Is Required for Antibody Formation

Thus far, our discussion of antibody formation has centered on those reactions that occur at the gene level and on the structure of the antibody. However, since the immune response in vivo involves reactions between whole cells, we must also consider what happens at the level of cell-to-cell interactions. Some of the pluripotent stem cells of the bone marrow replicate in an undifferentiated state, maintaining their pluripotency, whereas others replicate and differentiate at the same time. The latter cells can differentiate along various pathways (fig. 37.9). Some of those that become committed to the lymphocyte pathway differentiate further into B cell and T cell lineages. T cells mature in the thymus, whereas B cells mature in the marrow in most vertebrates (see fig. 37.2). Subsequently, both cell types relocate in secondary lymphatic organs, the spleen and the lymph nodes, and also more diffusely in all tissues. Here they remain dormant until they encounter specific antigens that trigger them to complete their differentiation into fully mature effector B cells and effector and regulator T cells. Each B cell contains immunoglobulins of a specific idiotype bound to its surface, usually of the IgM type and to a lesser extent of the IgD and IgG types.

Antigens Stimulate the Formation of B Cell Clones. The first time a foreign antigen is injected into the bloodstream, there is a long delay before specific antibody appears, and the response is fairly weak. The second time the same antigen is injected, the response is both more rapid and more intense (fig. 37.10). A great deal of research effort has gone into seeking an explanation for this difference between primary and secondary responses. Specific labeling studies show that secondary lymphatic organs contain a very heterogeneous mixture of B cells, carrying membrane-bound antibodies of different idiotypes. When sufficient amounts of anti-

Figure 37.9

Pluripotent stem cells in the marrow give rise to more stem cells and to various progenitors. The lymphocyte progenitors give rise to B cell and T cell lineages. The final differentiation step in both B cell and T cell development requires antigenic stimulation.

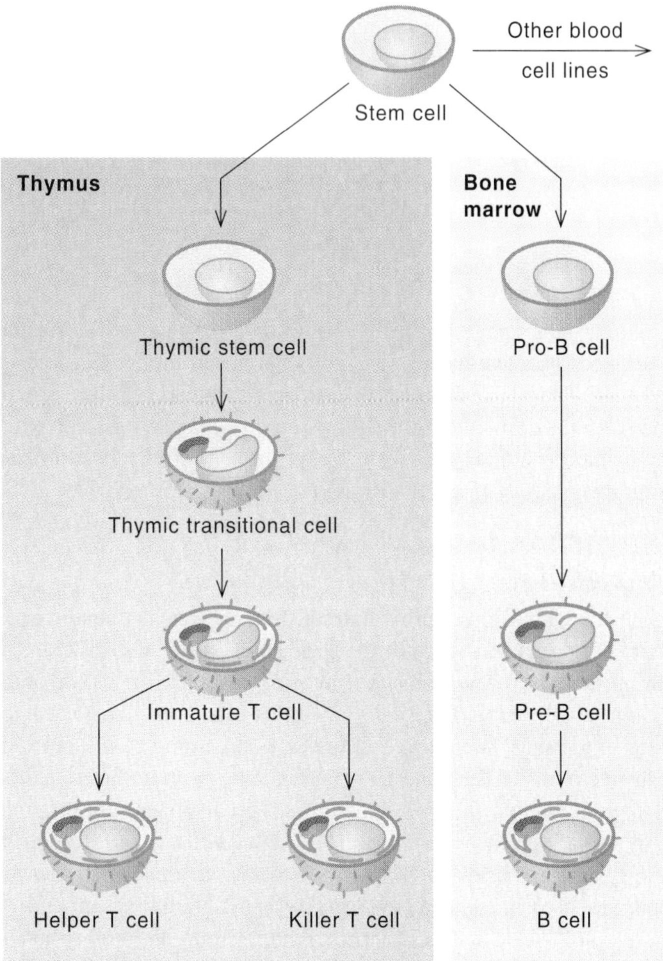

Figure 37.10

Kinetics of IgM and IgG appearance in the serum following a first and a second exposure to the same antigen. On first exposure a delay of several days occurs before any antibody appears in the serum. IgM appears before IgG. After many days the serum level falls. If the system is exposed to a second equivalent dose of antigen, the appearance of IgM and IgG is much faster, and the maximum level of IgG is much higher. (*Source:* From G. Zubay, *Genetics,* Benjamin/Cummings, Menlo Park, Calif., 1987, p. 812.)

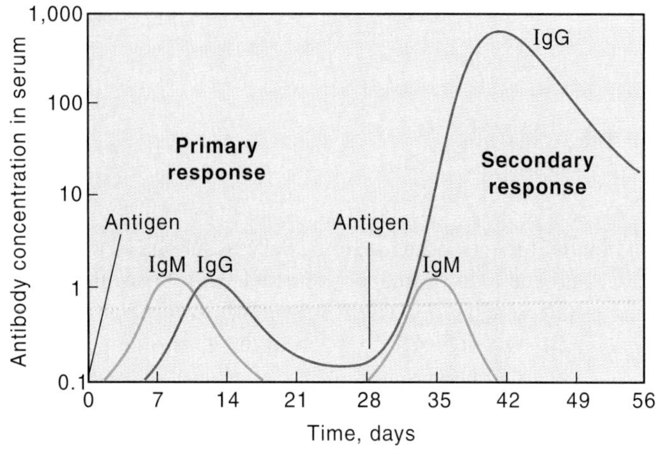

Figure 37.11

(*a*) An immature B cell and (*b*) a fully developed antibody-producing B cell. Most notable in (*b*) is the expanded cytoplasm with densely packed endoplasmic reticulum. (*Source:* Electron micrographs courtesy of Dr. Dorthea Zucker-Franklin, New York University Medical Center.)

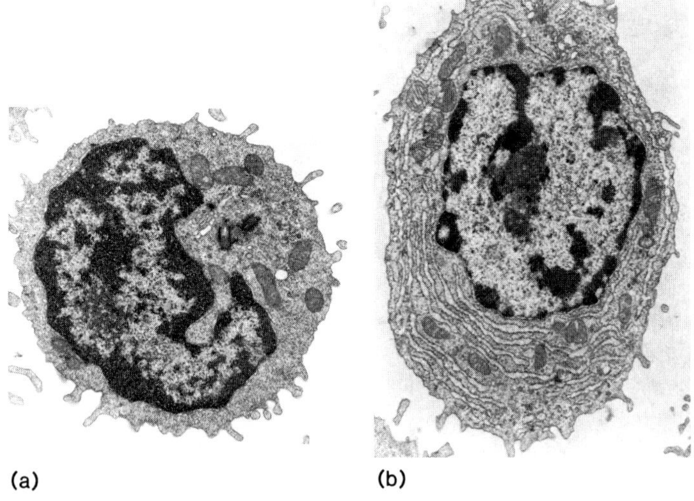

gen bind to the surface antibody, the B cell becomes triggered to divide and make more cells of the same type. Some of these continue to divide, making clones of memory cells in the lymphatic organs, whereas others terminally differentiate into plasma cells richly laced with endoplasmic reticulum that is geared to the production of the antibody of the same idiotype as that which is membrane-bound (fig. 37.11). Whereas the idiotype and consequently the V_L and V_H parts of the antibody are the same as in the original, unstimulated B cell, the *C* regions change. The first secreted antibody is of the secretory IgM type. This is gradually replaced by the secretory IgG type, which dominates in the secondary response. As we have noted (see fig. 37.7), the replacement of IgM by IgG is indicative of an additional DNA splicing.

Helper T Cells Trigger B Cell Division and Differentiation. The importance of T cells in mediating the B cell response to most antigens is demonstrated by the fact that thymectomized animals, which are depleted in T cells, usually show a great attenuated reaction to foreign antigen. A normal response can be restored by transplantation of thymus tissue from a genetically identical animal. Activated helper T cells interact with B cells whose antibodies can recognize the same antigen that stimulated the helper T cell. The complex between the T and B cell results in the clustering of a large number of transmembrane proteins, causing a process known as CAP formation. CAP formation by some unknown

Figure 37.12

Stimulation of antibody-forming cells involves three types of cells. Typically, antigen is phagocytosed by a macrophage. The phagocyte partially digests the antigen and "presents" the processed antigen on its outer plasma membrane. A specific helper T cell binds the antigen to a receptor bound to its outer plasma membrane. The T cell usually has many of the same types of receptors and binds many copies of the antigen in a similar way. This stimulates T cell proliferation. These T cells interact with B cells that display similar processed antigen. Several B cell receptors involved in a similar way become focused in a local region of the outer plasma membrane, a process known as CAP formation. CAP formation triggers the proliferation and differentiation of the B cells involved.

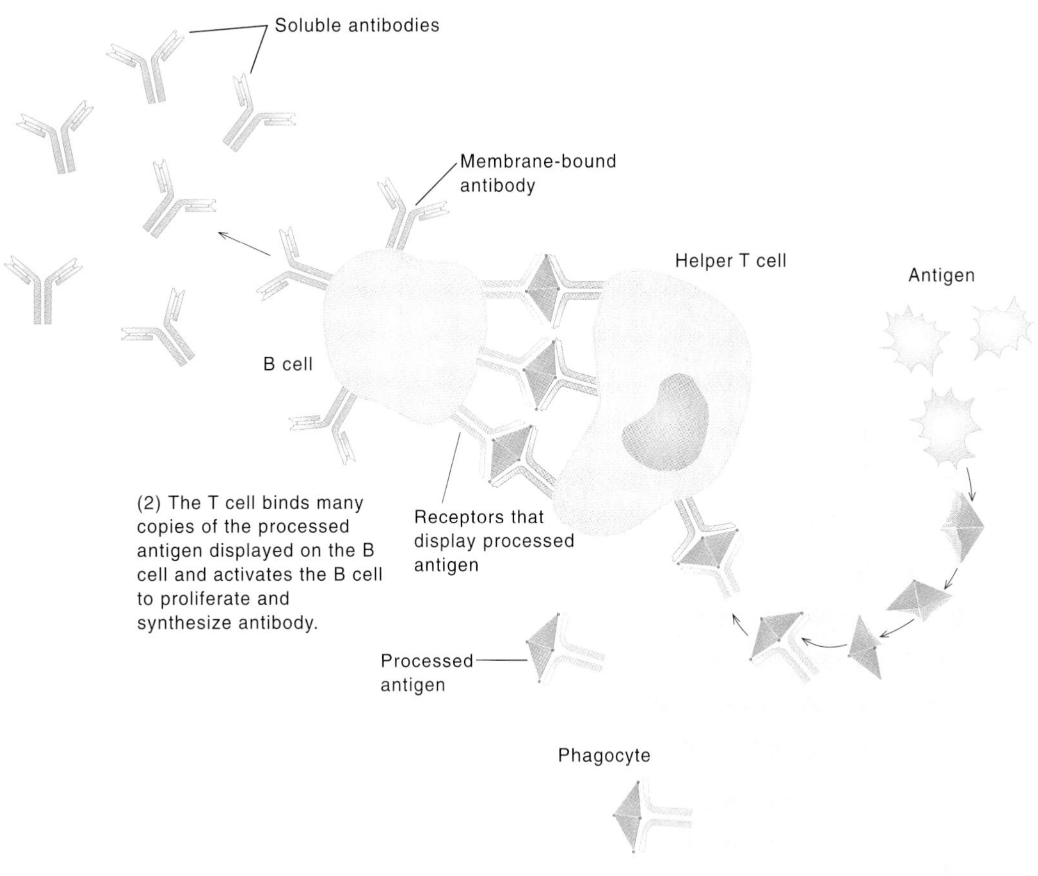
Soluble antibodies
Membrane-bound antibody
Helper T cell
Antigen
B cell
(2) The T cell binds many copies of the processed antigen displayed on the B cell and activates the B cell to proliferate and synthesize antibody.
Receptors that display processed antigen
Processed antigen
Phagocyte
(1) The phagocyte partially digests the antigen and presents the processed antigen on its outer plasma membrane. T cells that interact with these membrane-bound processed antigens are stimulated to proliferate.

means triggers the proliferation and differentiation of the B cells involved.

Support for the role of helper T cells in focusing the antigen comes from the observation that T-independent stimulation of B cells is frequently elicited by polymeric antigens. Such polymeric antigens constitute another way of presenting a large amount of antigen at one location on the B cell membrane, thereby focusing the B cell receptors.

Before they can help other lymphocytes respond to antigen, helper T cells must be activated themselves. This activation occurs when a helper T cell recognizes a foreign antigen bound on the surface of a specialized antigen-presenting cell. The latter cells are found in most tissues; they are derived from bone marrow and constitute a heterogeneous group, including dendritic cells in lymphoid organs and certain types of macrophages. Together with B cells, which can present antigen to helper T cells, these are the main cell types that react with helper T cells. The sequence of events involving antigen-processing cells, helper T cells, and antibody-forming B cells is depicted in figure 37.12.

Note that the mechanism proposed for helper T cell function requires that T cells have surface receptors that recognize antigen-processing cells, antigen itself, and the appropriate B cell. Later in this chapter we discuss the nature of T cell receptors, as well as the surface structures they recognize, in a broader context.

T Cell Action Is Frequently Augmented by the Secretion of Hormone-Like Proteins Called Interleukins

The interactions between immune and inflammatory cells are mediated in large part by certain proteins, called lymphokines, or interleukins (IL), that are able to promote cell growth, differentiation, and functional activation. Interleukins resemble hormones, which also function as intercellular messengers. In contrast to hormones, however, they are secreted by isolated cells rather than dis-

Table 37.3
Properties of Some Interleukins

Interleukin (IL)	Approximate Molecular Weight	Source	Target	Action
IL-1	15,000	Antigen-presenting cells	Helper T cells	Helps activate
IL-2	15,000	Some helper T cells	All activated T cells and B cells	Stimulates proliferation
IL-3	25,000	Some helper T cells	Various hemopoietic cells	Stimulates proliferation
IL-4	20,000	Some helper T cells	B cells	Stimulates proliferation, maturation, and class switching to IgE and IgG1
IL-5	20,000	Helper T cells that make IL-4	B cells, eosinophils	Promotes proliferation and maturation
IL-6	25,000	Some helper T cells and macrophages	Activated B cells, T cells	Promotes B cell maturation to Ig-secreting cells; helps activate T cells
IL-7		Stromal cells of the bone marrow	Hematopoietic precursors	Enhances development

crete glands. Several interleukins have been described; each has unique biological activities as well as some that overlap with the others (table 37.3). For example, macrophages produce IL-1 and IL-6, whereas T cells produce IL-2–IL-6, and bone marrow stromal cells produce IL-7. IL-1 and IL-6 not only play important roles in immune cell function, they also stimulate a spectrum of inflammatory cell types and induce fever. IL-2 is a potent proliferative signal for T cells. IL-1, IL-3, IL-4, and IL-7 enhance the development of a variety of hematopoietic precursors. IL-4–IL-6 also serve to enhance B cell proliferation and antibody production.

The Complement System Facilitates Removal of Microorganisms and Antigen-Antibody Complexes

Many antigen–antibody complexes are eliminated by phagocytosis. Others are attacked by complement, a group of serum proteins that aids in the defense against microorganisms. Individuals with a deficiency in their complement system are subject to repeated bacterial infections, just as are individuals who are deficient in antibodies themselves.

About 20 different proteins are included in the complement system: proteins C1–C9, factors B and D, and a series of regulatory proteins. All these proteins are made in the liver, and they circulate freely in the blood and extracellular fluid. Activation of the complement system involves a cascade of proteolytic reactions. In addition to forming membrane attack complexes, the proteolytic fragments released during the activation process promote dilation of blood vessels and the accumulation of phagocytes at the site of infection.

Figure 37.13

The principal stages in complement activation. Complement activation occurs exclusively on the microbial cell membrane, where it is triggered by bound antibody or microbial envelope polysaccharides, both of which activate early complement components. Two sets of early components belong to two distinct pathways of complement activation. Activation of each complement system involves a cascade of proteolytic reactions. Each component of the complement system is a proenzyme that is activated by the preceding component of the chain by a limited proteolytic cleavage. The ultimate result of this chain reaction is the development of a complex that attacks the cell membrane.

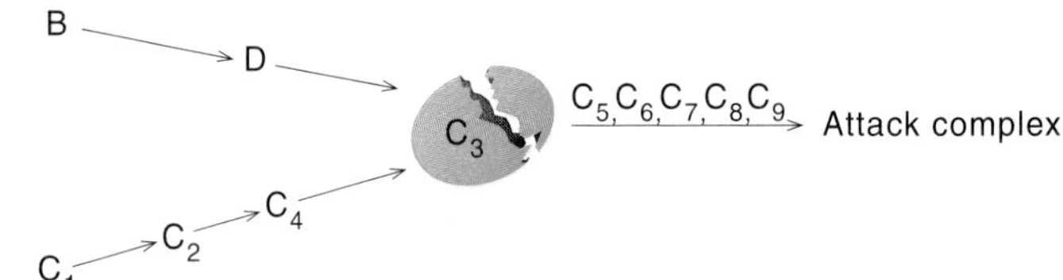

The activation of complement occurs exclusively on the microbial cell membrane, where it is triggered either by bound antibody or microbial envelope polysaccharides, both of which activate early complement components. Two sets of early components belong to two distinct pathways of complement activation: C1, C2, and C4 belong to the pathway that is triggered by antibody binding; factors B and D belong to the alternative pathway that is triggered by microbial polysaccharides (fig. 37.13). The early components of both pathways ultimately act on C3. The early components and C3 are proenzymes that are activated sequentially by limited proteolytic cleavage. As each proenzyme in the sequence is cleaved, it is activated to generate a serine protease, which cleaves the next proenzyme in the sequence. Many of these cleavages liberate small peptide fragments and expose a membrane-binding site on the larger fragment. The larger fragment binds tightly to the target cell membrane by its newly exposed membrane-binding site and helps to carry out the next reaction in the sequence. In this way, complement activation is confined largely to the cell surface, where it began.

The components of the complement attack complex have a very short half-life if released from the complex. This limitation confines their destructive action to the point of assembly, the surface membrane of the microorganism.

The Cell-Mediated Response: A Separate Response by T Cells

Most B cells are fixed in the secondary lymphatic organs: the spleen and the lymph nodes. They are stimulated by antigen only if the antigen is transported to the lymph nodes. In many cases, however, antigens are poorly transported. To rid the organism of antigens that are not readily transported, a second type of immune system exists that is independent of the B cells. This second type of immunity is not found in all vertebrates; it is restricted to mammals, birds, certain amphibians, and bony fishes. Long before the distinction between T and B cells was appreciated, it was known that certain types of immunity, which manifest a delayed-type hypersensitivity response, could be transferred from immunized animals to nonimmunized animals with leukocytes but not with antibody-containing serum. Subsequently, other forms of leukocyte-mediated immunity were discovered that involved increased phagocytic and bacteriocidal activity of macrophages, lysis of virus-infected cells and tumor cells, and the rejection of skin grafts from genetically dissimilar organisms.

The two types of cell-mediated immunity involve basically different types of T cells. In the delayed-type hypersensitivity response, the T cell, T_D, that reacts specifically with antigens secretes interleukins that attract and activate macrophages or other leukocytes, thereby causing a slowly developing inflammatory response. A second type of T cell, known as the killer T cell, T_K, reacts specifically with antigen that is bound to target cells, causing their lysis.

Tolerance Prevents the Immune System from Attacking Self-Antigens

A thorough understanding of the immune system requires an explanation for how the immune system is able to discriminate between foreign antigens and its own antigens (self-antigens). The ability to recognize self-antigens and thus to avoid making an antagonistic response to them is called tolerance. Many serious diseases are believed to be due to a breakdown in the self-tolerance mechanism. Conversely, conditions occur under which a foreign antigen is able to establish itself so that it becomes tolerated, for either a short or a long time. Tolerance and rejection are so closely related that it seems likely that a full understanding of one entails a full understanding of the other.

Figure 37.14

Establishment of tolerance to tissue grafts. If a newborn X mouse is injected with Y cells when it is young, it is tolerant to Y tissue transplants when it becomes an adult. If the X mouse is not exposed to Y cells at an early age, it readily rejects the tissue graft.

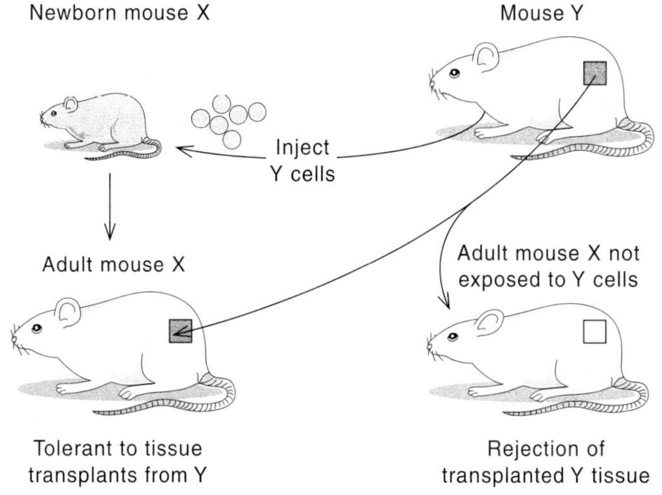

Several conditions favor the establishment of tolerance:

1. Tolerance is much easier to establish in the fetus or the newborn than in the adult. Thus if an organism is exposed as a fetus to a soluble foreign antigen or a foreign tissue graft, it may become indefinitely tolerant to the antigen. This situation is illustrated for two genetically nonidentical strains of mice in figure 37.14. Cells from the Y mouse are injected into a newborn X mouse. The same X mouse as an adult is the recipient of a skin graft (transplant) from a Y-type mouse. Normally, such a graft would be rapidly rejected. However, because of the early exposure to the Y cells, the graft is tolerated indefinitely. This result indicates that the mouse that was exposed to the Y cells just after birth recognizes Y cells later in life as self-antigens.
2. Very high levels of antigen frequently favor the development of tolerance, whereas intermediate levels of antigen favor the development of an immune system.
3. Frequently, the route of antigen administration is critical in the development of tolerance. An intravenously injected antigen is more likely to promote a tolerant condition than is antigen injected subcutaneously.

Two cellular mechanisms may account for most forms of tolerance. In one of these the clones of T or B cells that could otherwise be activated by the antigen are destroyed or inactivated. In the other mechanism the clones in question are still present but do not respond because they are blocked by T_S cells. In general, B cells are more difficult to make tolerant than T cells. This difference is evident from observations that longer and higher doses of antigen are required to establish tolerance to the B-dependent system. Moreover, the period of tolerance is generally shorter when B cells are involved.

Adult mice may be made tolerant to foreign antigens by destroying their immune system with whole-body irradiation and then supplying them with transplants of bone marrow and thymus from a foreign donor, thus refurbishing their system with competent B and T cells, respectively. With the new immune system the animal is now tolerant to any tissue from a mouse that is isogenic with the mouse used to supply the marrow and the thymus.

The irradiated host organism also supplies a useful means of testing the state of B and T cells from a nonirradiated isogenic mouse. B cells may be effectively transferred by bone marrow transplants, and T cells may be transferred by thymus tissue transplants. The donor cells may come from a normal, untreated isogenic strain. In this event the irradiated recipient shows normal immunological behavior. Alternatively, the donor cells may come from a donor that has been made either sensitive or tolerant to a particular antigen. By doing various combinations of experiments using different isogenic donors, it is possible to determine the relative importance of B cell and T cell involvement with different antigens.

T Cells Recognize a Combination of Self and Nonself

All T cells, whether they fight infection directly or regulate the activity of other effector cells, recognize antigen only in combination with a cell membrane surface. Of greatest significance the antigen must be associated with a cell from an organism that is genetically similar or identical to that of the T cell. This requirement was first demonstrated in an experiment where T cells were removed from a mouse that had been immunized with a virus. Immunization meant that the extracted T cells had been activated to direct themselves against cells infected with this particular virus. Thus these T cells could kill fibroblasts infected with the virus in tissue culture. However, they could not kill fibroblasts infected with the virus if the fibroblasts came from a mouse with a different genetic background. Thus sensitized killer T cells do not recognize foreign antigens alone; they recognize foreign antigens only in combination with determinants that are present on their host's own cells (fig. 37.15). These common determinants are a subset of the proteins encoded by the so-called major histocompatibility complex (MHC).

MHC Molecules Account for Graft Rejection

MHC molecules were recognized long before their major biological function was appreciated. They were initially defined as the main target antigens in transplantation reactions, and because of this they became known as major histocompatibility antigens. Experiments on mice demonstrated that graft rejection is an immune response to the surface antigens of the grafted cells. Subsequently, it was shown that these reactions are mediated by T cells and that they are directed against a family of glycoproteins encoded by a group of genes called the major histocompatibility complex. MHC molecules are transmembrane proteins found on the cell membranes of all higher vertebrates.

A vertebrate does not normally need to be protected against invasion by foreign vertebrate cells, so the antagonistic reaction of its T cells to foreign MHC molecules was both an obstacle to transplant operations and an enigma to immunologists. The enigma was solved when it was discovered that MHC molecules contributed to the binding of T lymphocytes on those host cells that have foreign antigen on their surface, as in the case of the virus-infected cells we have just discussed (see fig. 37.15).

There Are Two Major Types of MHC Proteins: Class I and Class II

There are two major classes of MHC proteins that have similar types of structures (fig. 37.16). Class I molecules contain three external domains, each about 90 residues in length; a transmembrane region; and a cytoplasmic domain. The third external domain is noncovalently associated with a small polypeptide known as the β_2 microglobulin. Class II molecules are composed of two noncovalently associated polypeptide chains: α and β, whose overall structure resembles that of the class I complex.

Several genes encode each type of MHC protein, and several alleles represent different MHC proteins for each of these genes. This gives rise to a tremendous variety of MHC proteins, with the consequence that it is extremely unlikely that any two members of the same species will possess the same assortment of MHC proteins. Thus two humans are not likely to carry the same MHC proteins on their cell membranes unless they are identical twins. This is the reason why transplants between two individuals are almost always rejected—it is usually just a matter of time. Highly inbred strains of mice, which are genetically identical, display the same MHC proteins and accordingly take grafts from each other without any rejection.

The two classes of MHC proteins are displayed on different cell types. Class I MHC proteins are found on almost all nucleated cells, including killer T cells. Class II MHC proteins are found mainly on cells involved in the immune response, including antigen-presenting cells, B cells, and helper T cells, but not killer T cells.

T Cell Receptors Resemble Membrane-Bound Antibodies

The specific receptors found on the membranes of T cells resemble the highly specific antibodies found on the membranes of B cells (fig. 37.17). Specific T cell receptors (TCRs) are composed of two disulfide-linked peptide chains (called α and β). Each of these chains shares with antibodies the distinctive property of a variable amino-terminal region and a constant carboxyl-terminal region.

With one exception, all the mechanisms used by B cells to generate antibody diversity are also used by T cells to generate T cell receptor diversity. The one mechanism that does not appear to operate in T cell receptor diversification is somatic hypermutation. This is presumably because mutation would be likely to generate killer T cells that would wantonly attack self-molecules. This is much less of a problem for B cells, since most self-reactive B cells could not be activated without the aid of specific helper T cells.

Figure 37.15

Activated T cells bind and kill only those cells that display a foreign antigen and a familiar cell surface antigen. This illustration is very schematic. We will shortly describe the structures responsible for cell–cell interaction in more precise detail.

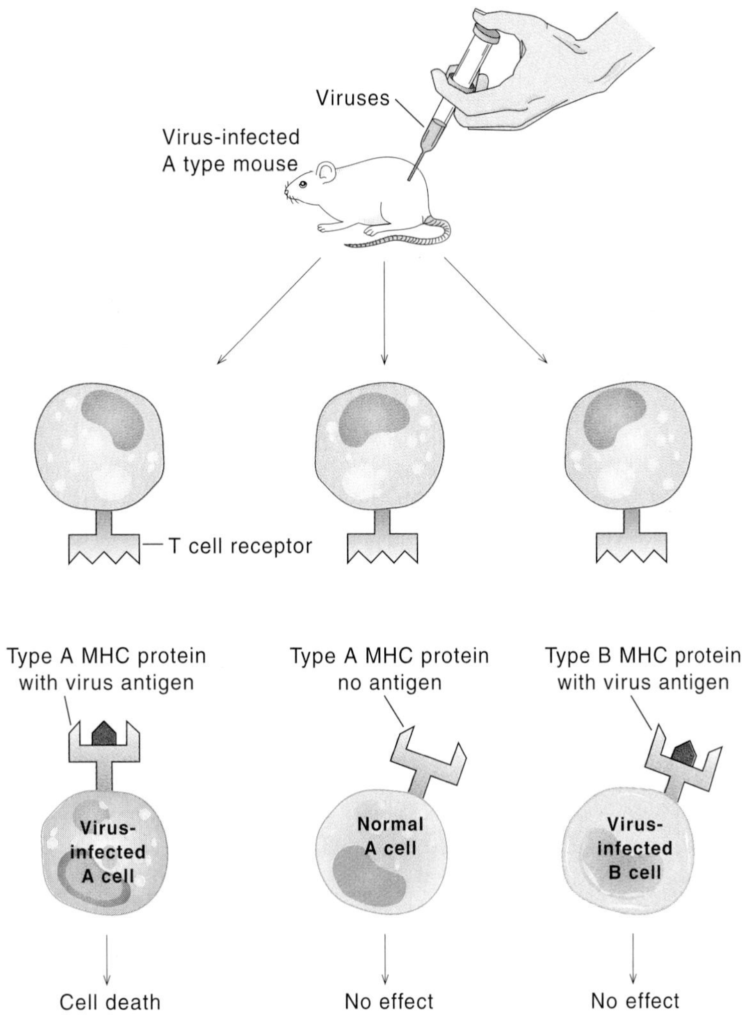

Additional Cell Adhesion Proteins Are Required to Mediate the Immune Response

In addition to membrane-bound antibodies, TCR proteins and MHC proteins, other transmembrane proteins are required to obtain the specific cell–cell interactions necessary to mediate the immune response (table 37.4). These additional proteins are also important in a number of nonimmune reactions in which cell–cell adhesion is important to the organism. We consider only a limited number of these additional proteins because of the enormous complexity of this subject. Three transmembrane proteins directly involved in the T cell reaction with other cells are members of a family of cell–cell adhesion proteins (the CD family of proteins). The CD3 protein is found on both killer and helper T cells; it forms a complex with the membrane-bound TCR proteins. The CD8 and CD4 proteins interact with this complex, but unlike the CD3 proteins, CD8 is found only on killer T cells, whereas CD4 is found only on helper T cells. The presence of CD4 and CD8 proteins limits the range of binary complexes formed by T cells with other cells. This is because CD4 and CD8 have selective affinities for MHC class II and MHC class I proteins, respectively. As a result, killer T cells form complexes only with cells that display a processed foreign antigen in a complex with the MHC I protein; this includes most nucleated cells other than those that belong to the immune system. Similarly, helper T cells are limited to reactions with cells that display the appropriate processed antigen in a complex with the MHC II protein; this includes B cells or phagocytes (fig. 37.18).

From this discussion we can see that a productive interaction between T cells and other cells requires a highly specific reaction between a TCR protein and a processed antigen that is supplemented by less specific interactions between other transmembrane proteins on the two cell types.

Left out of this discussion is an explanation for the source of the processed antigen on the antigen-presenting B cell. Imma-

Figure 37.16

Structures of class I and class II molecules as they would appear in the lipid bilayer of the cell membrane. Models show striking similarities between the two types of molecules. S—S indicates a disulfide bridge in the class I and class II molecules and the β_2 microglobulin.

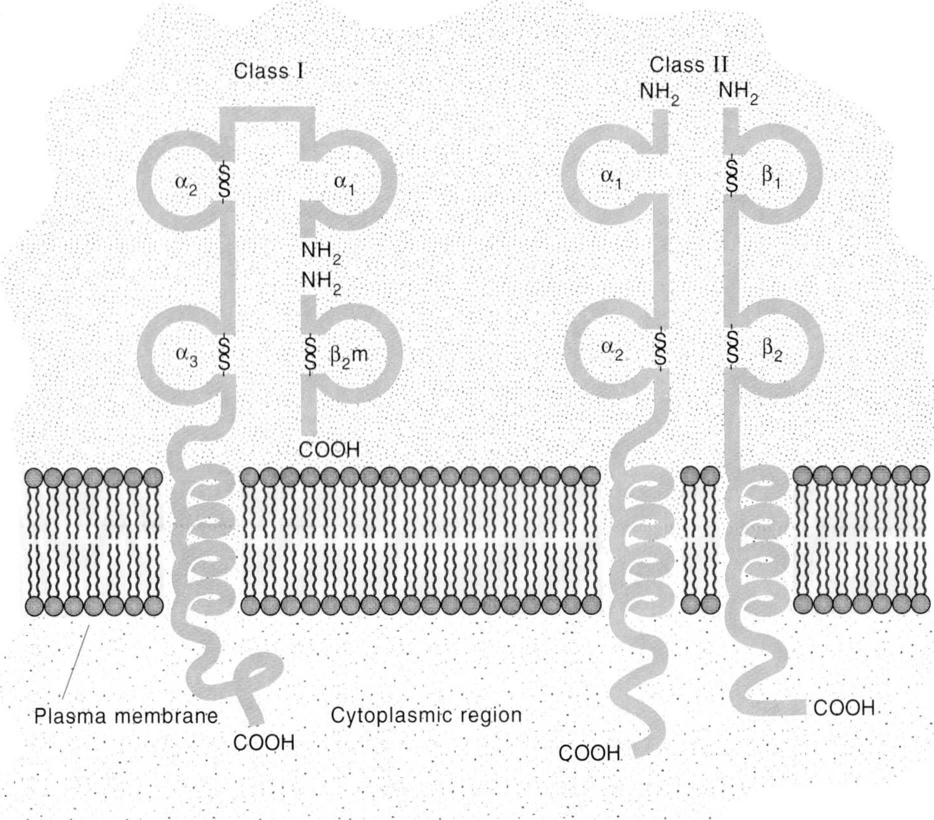

Figure 37.17

A T cell receptor heterodimer. T cell receptors are transmembrane proteins.

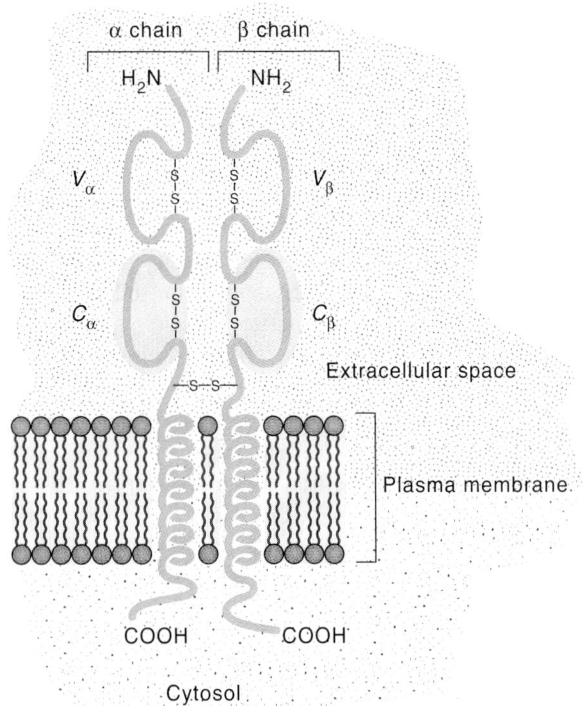

Table 37.4
Key Accessory Proteins Found on the Surface of T Cells

Protein	Molecular Weight	Location	Function
CD3	γ chain = 25,000 δ chain = 20,000 ϵ chain = 20,000 ζ chain = 16,000	All T cells	Helps transduce signal when antigen–MHC complex binds to T cell receptors
CD4	55,000	Helper T cells	Promotes adhesion to antigen-presenting cells and to B cells; signals T cell
CD8	70,000 (homodimer or heterodimer)	Cytotoxic T cells	Promotes adhesion to infected target cells; signals T cell

Figure 37.18

Antigen-specific reactions involving T cells. (*a*) Killer T cells interact and destroy target cells that display complexes of processed antigen with an MHC I protein. Effective interaction requires that the TCR protein interact specifically with the processed antigen. All of the remaining reactions between the two cells are general reactions that occur between other killer T cells and other target cells. (*b*) Helper T cells interact specifically with antigen-presenting B cells, provided that a specific complex is made between the TCR protein and the processed antigen complexed to the MHC II protein on the B cell. The unlabeled membrane proteins in this figure represent proteins that interact nonspecifically to strengthen the binary cell reaction. Processed antigen is indicated in red.

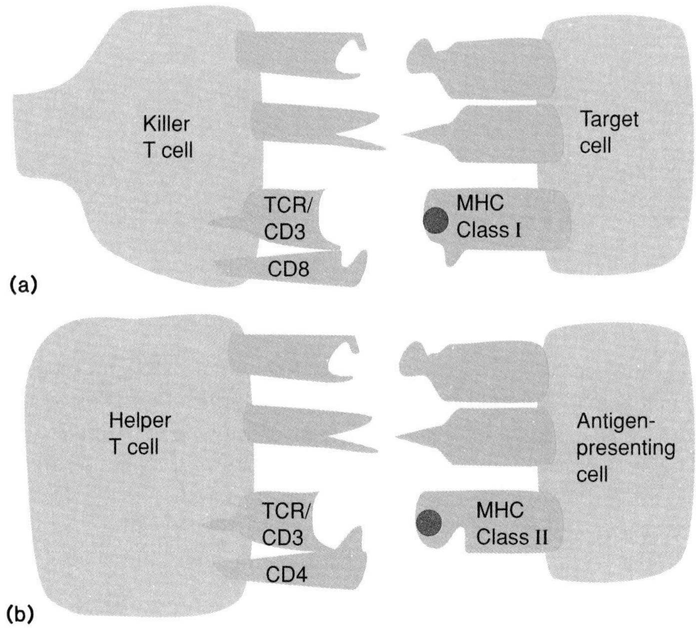

ture B cells display a membrane-bound antibody, which interacts with circulating complementary antigen. Once bound to the antibody, the antigen is internalized, processed, and ultimately displayed by the MHC II protein on the cell surface of the same B cell. It is the interaction between such a B cell and the appropriate helper T cell that triggers the B cell to divide and differentiate.

The Immune System in Action: A Broad Arsenal of Weapons Enables the Immune System to Roust and Destroy Foreign Invaders

When *Streptococcus pneumoniae* infects the lungs, it colonizes the alveoli. There it multiplies, causing tissue damage, inflammation, and sometimes death. Because these bacteria live outside of cells and near the bloodstream, they would seem to make easy prey for the macrophages and other phagocytic cells of the immune system. However, the pneumococcal bacteria escape direct attack by phagocytes because they possess a protective polysaccharide to which phagocytes cannot bind. The capsular polysaccharides offer a more attractive target for the antibody-producing B cells. The surface membranes of B cells bristle with specific receptors for foreign antigens. Each B cell carries receptors for only one kind of antigenic determinant. Because the molecular structures of capsular polysaccharides are highly repetitious, the same capsular antigens appear repeatedly on the surface. As a result, when the B cell with the appropriate surface receptors makes contact with the bacterium, many receptors on a single B cell latch onto the capsule simultaneously. The resulting clustering of B cell receptors causes capping, which leads to B cell activation. The B cell divides, and some of the daughter cells produce soluble antibodies specific for the capsular polysaccharides. These antibodies coat the bacterium, making it attractive to complement and direct ingestion by macrophages.

The antibody response is the simplest in the immune system's arsenal. Survival following infection involves a straightforward race between antibody production and pathogen replication. The resistance to infection in people who have been vaccinated or previously exposed to an infectious agent usually depends on the antibody response.

Many microbes establish infections inside cells, where antibodies and complement cannot reach them. The T cells provide the main defense against such infections. The means by which the T cells fight such intracellular infections is amply illustrated by the way in which T cells ward off infections by the flagellated protozoan Leishmania. Leishmania can infect the liver, spleen, or skin. One usually recovers from the skin infection after a prolonged period, but liver or spleen infections are often fatal if left untreated.

The primary cell target of Leishmania is the macrophage. During routine scavenging in the bloodstream, macrophages engulf Leishmania organisms and package them into vacuoles. These vacuoles fuse with lysosomes that contain proteolytic enzymes that digest and kill most microbes. Leishmania can endure this proteolytic attack and even thrive during it. As a result, the parasite can multiply inside the vacuole until the infected macrophage is overwhelmed and dies.

The body's way of eliminating intracellular parasites sequestered in this way involves sending a distress call to the circulatory T cells. The MHC II molecules of the macrophage are imported into the vacuoles containing the Leishmania organisms. Here they become loaded with peptides shed by the parasites or cleaved from them by partial proteolysis. Once displayed on the cell surface, these complexes alert passing CD4 T cells. The diversity of receptors made by the T cell population ensures that a match can be found for virtually any peptide–MHC combination. This recognition event develops into an immunologic response if and only if the macrophage also provides an additional signal to the T cell. One surface molecule that can provide this signal is B7, which macrophages often express when they become infected. B7 is recognized by a separate protein, CD28, on the T cell's surface. When a CD4 T cell receives the dual signal, it releases various cytokines, the most potent of which is γ-interferon. Once inside the macrophage, the interferon prompts the macrophage to produce agents such as tumor necrosis factor and nitric oxide, which lead to the microbe's destruction.

Like Leishmania, viruses establish infections inside cells of the body that are beyond the reach of antibodies. But unlike Leishmania, viruses live in the cytosol of the cell rather than inside a vacuole. Here they use the synthetic machinery of the cell to manufacture their own proteins. As a result, the viral proteins intermingle with other cellular proteins and so present a less easily isolated target for the molecules of the immune system. Despite this confusing arrangement, MHC molecules can still find and display peptide fragments from viruses. This process is quite similar to the one that reveals Leishmania infection, but since it occurs in most tissue cells, MHC I rather than MHC II is involved in the display of the peptide fragments from the viruses. Complexes of the peptide–MHC I molecules attract select CD8 cells that bear receptors for class I MHC complexes. When CD8 T lymphocytes detect antigenic peptides, they usually secrete substances that result in death of the target cell. T cells destroy cells either by disrupting the integrity of the cellular membrane or by triggering programmed cell death.

Figure 37.19

Some of the proteins belonging to the immunoglobulin superfamily. The antigen-binding domains are shaded in blue. (*Source:* Bruce Alberts, Dennis Bray, Julian Lewis, Martin Raff, Keith Roberts, and James D. Watson, *Molecular Biology of the Cell,* 3rd ed., Garland Publishing Co., 1994.

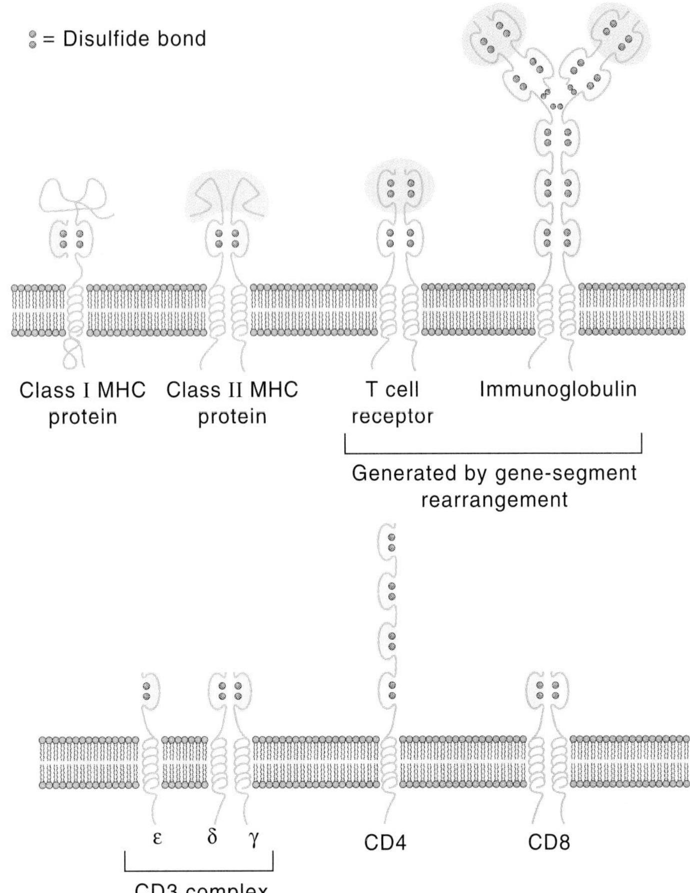

Immune Recognition Molecules Are Evolutionarily Related

The striking structural similarity between immune recognition proteins almost certainly reflects their evolution from a common ancestor. All of them contain one or more immunoglobulinlike domains, each with about 90 amino acids stabilized by a conserved disulfide linkage (fig. 37.19). The simplest member of this family of genes is β_2-microglobulin (see fig. 37.16). Possibly, this or a similar protein was the ancestor for the remaining immune recognition proteins.

Summary

In this chapter we considered the major findings that led to our understanding of the immune system in vertebrates, especially mice and humans. The following points are the highlights of our discussion.

1. Two different classes of white cells (lymphocytes) are associated with the immune response: the B cells and the T cells. B cells produce antibodies that are secreted and aggregated foreign substances known as antigens. T cells are of different

types: Some of them assist B cells to become antibody-forming cells; others mount an attack on antigens of their own.

2. The most common type of antibody is a tetramer composed of two identical heavy (H) chains and two identical light (L) chains. Each chain is divided into a relatively constant sequence (the *C* segments) and a relatively variable sequence (the *V* segments). Within the variable regions there are subsegments known as the hypervariable regions. These subsegments are the regions that specifically bind different antigens.
3. According to some estimates, each organism is capable of making more than a million different antibodies from a limited amount of genetic information. This feat is accomplished with the help of somatic recombination and somatic mutation. Somatic recombination involves splicing at the DNA level. It brings specific *C* and *V* regions together to make a unit that can be transcribed. Somatic mutation is focused on those parts of the gene that encode the hypervariable regions.
4. T helper cells focus the antigen on an immature B cell. In response, some B cells turn into antibody-forming cells, and some B cells turn into memory cells for production of specific antibodies at a later date. The first time an organism is exposed to a specific antigen, it takes longer to mount an immunological response. That is because of the lack of memory cells for making a specific antibody.
5. Complement is a group of serum proteins that aids in the defense against microorganisms and removal of antibody–antigen complexes.
6. T cells give rise to an immune response of their own. The two types of cell-mediated immunity involve basically different types of T cells. In the delayed-type response, the T cell reacts specifically with antigens and secretes lymphokines. These are substances that attract macrophages or other leukocytes, thus producing a slowly developing inflammatory response. A second type of T cell reacts specifically with antigen bound to target cells and causes their lysis.
7. Tolerance prevents the immune system from attacking self-antigens. To understand tolerance, we must appreciate how T cells work. T cells recognize a combination of self and nonself. The cell surface antigens recognized by T cells are known as the major histocompatibility complex (MHC). If two organisms carry the same histocompatibility antigens, the tissues from the two organisms are completely compatible. The histocompatibility antigens resemble antibodies in structure.

Selected Readings

Ada, G. L., and G. Nossal, The clonal selection theory. *Sci. Am.* 257(2): 62–69, 1987.

Alberts, B., et al., The immune system, Chapter 23 in *Molecular Biology of the Cell,* 3rd ed. New York and London: Garland, 1994.

Berke, G., The CTL's kiss of death. *Cell* 81:9–12, 1995. Describes two mechanisms that are used by cytotoxic lymphocytes to kill target cells.

Boon, T., Teaching the immune system to fight cancer. *Scientific American* March: 82–90, 1993.

Cheng, G., A. M. Cleary, Zheng-Sheng Ye, D. I. Hong, S. Lederman, and D. Baltimore, Involvement of CRAFI, a relative of TRAF in CD40 signaling. *Science* 267:1494–1998, 1995.

Ghosh, G., G. V. Duyne, S. Ghosh, and P. B. Sigler, Structure of NF-B p50 homodimer bound to a B site. *Nature* 373:303–317, 1995.

Honjo, T., and S. Habu, Origin of immune diversity: Genetic variation and selection. *Ann. Rev. Biochem.* 54:803–830, 1985.

Kaappes, D., and J. L. Strominger, Human class II major histocompatibility genes and proteins. *Ann. Rev. Biochem.* 57:991–1028, 1988.

Kersh, G. J., and P. M. Allen, Essential flexibility in the T-cell recognition of antigen. *Nature* 380:495–498, 1996.

Klein, J., *Immunology.* London: Blackwell Scientific Publications, 1990.

Krieg, A. M., et al., CpG motifs in bacterial DNA trigger direct B-cell activation. *Nature* 374:546–549, 1995.

Leder, P., The genetics of antibody diversity. *Scientific American* 246(5):102–115, 1982.

Lieber, M. R., Site-specific recombination in the immune system. *FASEB J.* 5:2934–2944, 1991.

Life, Death and the Immune System, *Scientific American* September 1993. This issue is a marvelous collection of articles on various aspects of the immune system. Each article is written by experts in the field: Life, death and the immune system, G. J. V. Nossal; How the immune system develops, I. L. Weissman and M. D. Cooper; How the immune system recognizes invaders, C. A. Janway, Jr.; How the immune system recognizes the body, P. Marrack and J. W. Kappler; Infectious diseases and the immune system, W. E. Paul; AIDS and the immune system, W. C. Greene; Autoimmune disease, L. Steinman; Allergy and the immune system, L. M. Lichtenstein; The immune system as a therapeutic agent, H. Wigzell; Will we survive?, A. Mitchison.

Milstein, C., Monoclonal antibodies, *Sci. Am.* 243(4):66–74, 1980. An exciting, tissue-culture technique for obtaining pure antibodies specific for an antigen of choice.

Mizel, S. B., The interleukins. *FASEB J.* 3:2379–2388, 1989.

Nowak, M. A., R. M. Anderson, A. R. McLean, T. F. W. Wolks, J. Goudsmit, and R. M. May, Antigenic diversity thresholds and the development of AIDS. *Science* 254:963–969, 1991.

Pennisi, E., Teetering on the brink of danger. *Science* 271:1665–1667, 1996.

Rajewsky, K. Clonal selection and learning in the antibody system. *Nature* 381:751–757, 1996.

Rini, J. M., U. Schulze-Gahmen, and I. A. Wilson, Structural evidence for induced fit as a mechanism for antibody–antigen recognition. *Science* 255:959–965, 1992.

Service, R. F., Triggering the first line of defense. *Science* 265:1522–1524, 1994. Vaccines that activate mucosal immunity are often the body's first chance to ward off infection.

Silver, M. L., H.-C. Guo, J. L. Strominger, and D. C. Wiley, Atomic structure of a human MHC molecule presenting an influenza virus particle. *Nature* 360:367–372, 1992.

Singer, S. J., Intercellular communication and cell–cell adhesion. *Science* 255:1671–1677, 1992. A comprehensive review of the major cell–cell interactions important in immune reactions.

Springer, T. A., Adhesion receptors of the immune system. *Nature* 346:425–434, 1990. An excellent review.

Sutton, B. J., and H. J. Gould, The human IgE network. *Nature* 366:421–428, 1993. An understanding of the IgE system holds the key to understanding and intervening in the aetiology of allergic diseases.

Tonegawa, S., Somatic generation of antibody diversity. *Nature* 302:801–803, 1983. A classic paper.

Tonegawa, S., The molecules of the immune system. *Scientific American* December 1985.

Weiss, R. A., How does HIV causes AIDS? *Science* 260:1273–1278, 1993. An exciting up-to-date account of the various routes whereby the HIV virus attacks and destroys the immune system.

PROBLEMS

1. Where do T cells and B cells *mature* in the body? Where do the mature T and B cells *reside* in the body?
2. Why does humoral immunity require both T and B cells?
3. What are the two types of cell-mediated immunity?
4. Anti-idiotypic antibodies are produced upon injection of an individual with antibodies produced in response to a specific antigen in a different individual. Explain why anti-idiotypic antibodies *sometimes* mimic the binding properties of the original antigen.
5. How can DNA splicing help to stimulate a higher level of transcription of an immunoglobulin gene?
6. Compare the results you would obtain if you were to perform a polymerase chain reaction with oligonucleotides from the L and C portions of the light chain gene using embryonic and rearranged genomic DNAs as template.
7. Describe a PCR-based method that could be used to demonstrate whether B-cell mRNA encodes the membrane-bound or secreted form of the heavy chain.
8. You have two strains of genetically nonidentical mice (strain A and strain B). How could an adult mouse from strain A be made to tolerate a transplant from strain B?
9. How could all the immune recognition proteins have evolved from a single common ancestor protein?
10. Recently, antibodies with catalytic activity have been produced. How would you develop a catalytic antibody for a specific reaction? (*Hint:* How does an enzyme work?)
11. If you inject a pure protein into a rabbit and raise antibodies to a high titer, the antibodies produced will be a mixed population with various binding constants. Why would a single protein produce many different antibodies?
12. If you added a thiol compound such as 2-mercaptoethanol at a high concentration (1 mM) to an antibody and antigen mixture, no precipitation complex would form even if the concentration of antibody and antigen were optimal. Explain this observation. (*Hint:* How do thiol compounds work?)

CHAPTER

Cancer and Carcinogenesis

Cancer results from a breakdown in the regulation of cell growth.

In the complex developmental process that leads to a mature human being from a fertilized egg, each cell takes up a role in a specialized organ or tissue. Each group of cells proliferates to a certain point and then takes up a post in a well-organized community of cells. In the case of a solid organ such as the liver, cells of several different types grow until the organ takes on a fixed size and shape, and then growth of the organ ceases. Normally, further growth occurs at a slow rate at which time new cells appear as old cells are replaced. However, if part of the liver is amputated, a rapid regenerative process ensues so that the liver regrows to approximately the same size it had before the amputation. Other organs or tissues may show more or less growth than the liver, but as a rule, new cells appear only as old and damaged cells are replaced. Furthermore, as a rule, the cells of normal tissues do not mix.

Therefore you do not find liver cells mixing with intestinal cells or any other types of cells, a feature suggesting that cells have surface properties that favor interaction of like cells. This was dramatically demonstrated by a classic experiment in the 1940s in which embryonic tissues were dissociated into single cells and allowed to reassociate in the presence of other cell types; they usually reassociated with cells of a similar type.

All of these rules with regard to growth control and cell recognition are violated by cancer cells. Cancer cells are relatively easy to recognize when they occur in groups in the organism (fig. 38.1). A cluster of cancer cells divides more rapidly, and it usually clashes with the architecture of normal cells in its immediate surroundings. From cytological inspection alone it can be speculated that cancer cells are breaking the rules. It appears that the processes that regulate cells have been disrupted; cell growth is uncontrolled, and patches of morphologically distinct cells show by their invasive character that they do not recognize normal territorial boundaries between cells of different types. Cancer cells are deadly because ultimately, they invade and disrupt a vital area. In this chapter we present a brief overview of some aspects of carcinogenesis with an emphasis on recent achievements and future prospects.

Figure 38.1

A benign glandular adenoma (*a*) and a malignant glandular adenocarcinoma (*b*). In both cases the tumor cells differ morphologically from the surrounding tissue and show a pattern of disorganized growth.

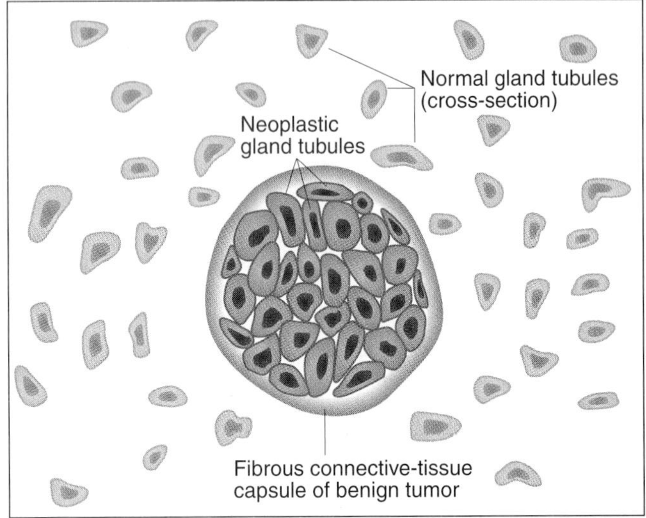

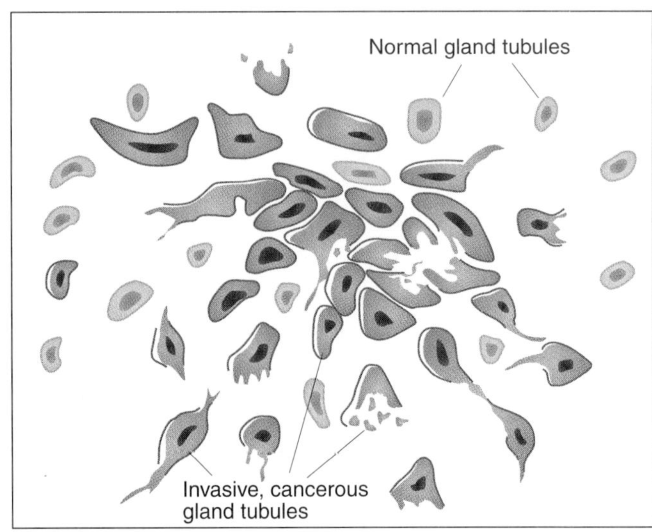

Cancers Are Cells out of Control

Cancer is not a stage in the orderly evolution of a complex organism; it is an organismic catastrophe. Perhaps it is only reasonable to expect that multicellular eukaryotes, with their many dividing cells, occasionally produce a mutant cell that gets out of control and destroys the system. The notion that cancer cells are out of control was proposed more than half a century ago. In the intervening time we have learned a great deal about the biochemistry of growth regulation. This knowledge has helped us to understand the molecular basis of cancer in many cases. Conversely, the study of cancer has added to our knowledge of normal growth regulation.

Environmental Factors Influence the Incidence of Cancers

It is very unlikely that we will find cures for most types of cancer in the foreseeable future. Therefore it is a good idea to put considerable effort into cancer prevention, especially because there are reasons for believing that the vast majority of cancers are preventable. The frequency of different kinds of cancer varies enormously in different human populations (table 38.1). For example, the incidence of breast and colon cancer is much higher in the United States than in Japan, whereas that of stomach cancer is much higher in Japan. In these cases, diet is most likely the culprit. The liver cancer rate is highest in third world countries. Once again, diet seems the most likely cause. Lung cancer is more prevalent in countries where smoking is popular, and indeed, it is well documented that heavy smokers are much more likely to get lung cancer than nonsmokers. Overall, it is clear that much of the variation in cancer incidence is environmental rather than genetic, since these differences tend to disappear from one generation to the next in migrant populations. For example, the high incidence of stomach cancer in Japan is not matched by a similar figure among Japanese-Americans in the United States.

The conviction is growing that the carcinogenic agents in the environment are active as cancer-causing agents because they produce mutations. The hunt for carcinogens has been facilitated by Bruce Ames, who developed a test for carcinogens based on the mutagenic action of a compound on bacteria. Ames tests have shown that many carcinogens originate from food and chemical pollutants. In addition to chemicals, ionizing radiation is carcinogenic. For example, the indications are strong that the ionizing radiation we get from the sun is the major cause of skin cancer. Therefore it is understandable that the incidence of skin cancer in the United States is much higher in the South than in the North. High dosages of radioactive materials or x-rays are also strongly correlated with the incidence of numerous forms of cancer.

In many cases, exposure to certain viruses is correlated with specific types of cancers (table 38.2). For example, liver cancer is 300 times more common in individuals who have previously been infected with hepatitis B virus. Although we can only speculate on the precise reasons for this, such numbers indicate that liver disease resulting from hepatitis B virus infection is almost certainly the major cause of human liver cancer. Likewise, cancer of the B lymphocytes known as Burkitt's lymphoma is strongly correlated with infection by the Epstein-Barr virus. Once again, we do not know the precise chain of events that lead to cancer. Hepatitis B and Epstein-Barr are DNA viruses.

Certain RNA viruses have also been shown to be correlated with cancer. The best known is the human immunodeficiency virus (HIV-1, the virus that causes AIDS). The AIDS virus is be-

Table 38.1
Variation in the Incidence of Cancers by Country

Type of Cancer	Country of High Incidence	Country of Low Incidence	Ratio (High/Low)
Skin	Australia	India	200
Esophagus	Iran	Nigeria	200
Lung	England	Nigeria	35
Prostate	United States	Japan	15
Liver	Mozambique	England	40
Breast	Canada	Israel	7
Rectum	Denmark	Nigeria	20
Colon	United States	Nigeria	10
Ovary	Denmark	Japan	6
Stomach	Japan	Uganda	25

Table 38.2
Human Cancers for Which Viruses Are the Causal Agent

Virus	Type of Cancer
Hepatitis B virus (HBV)	Liver cancer
Epstein-Barr virus (EBV)	Burkitt's lymphoma (cancer of B lymphocytes)
Human T cell lymphotropic virus (HTLV-1)	Adult T cell leukemia
Human immunodeficiency virus (HIV)	Kaposi's sarcoma
Human papillomavirus (HPV)	Warts and carcinoma of the uterine cervix

lieved to cause cancer indirectly by incapacitating the immune system. Various types of cancer follow infection by the AIDS virus, a fact testifying to the key role the immune system plays in protecting the body from cancer. The RNA virus known as human T cell leukemia virus type I (HTLV-I) causes T cell leukemia. It is believed that this virus carries a cancer-causing gene (an oncogene), the continuous expression of which upsets the normal metabolism, causing infected cells to become transformed to cancer cells.

Cancerous Cells Are Almost Always Genetically Abnormal

In the previous section we stated that a strong correlation exists between the mutagenic properties of a chemical and its potency as a carcinogen. This is consistent with other evidence that mutation often plays a causal role in carcinogenesis.

The most graphic illustrations of the correlation between genetic abnormality and cancer come from the frequent association of certain types of tumors with highly specific chromosomal translocations. Chromosomal translocations involve the movement of a segment of chromosomes from one location to another. Frequently, translocations occur between different chromosomes in a reciprocal fashion. This is the case for translocations between human chromosome 8 and the chromosomes that carry the antibody genes: 14, 2, and 22.

A reciprocal exchange of segments of chromosome from the ends of chromosomes 8 and 14 is frequently found in the genome of cancer cells of the Burkitt's lymphoma (fig. 38.2). Burkitt's lymphoma is a tumor of B lymphocytes; for unknown reasons its occurrence is most common in Africa. The resulting tumor cells usually secrete antibodies. The fact that antibody genes, which are so active in B cells, map to the same chromosomal regions involved in the specific translocations in Burkitt's tumors led to the suggestion that a specific oncogene on chromosome 8 might be activated if it were placed near one of these antibody genes. This suggestion proved to be correct; the oncogene involved is *c-myc*. We have more to say about *c-myc* later.

The first chromosome abnormality found to be associated with a cancer was the Philadelphia chromosome, which arises as a result of a reciprocal translocation between chromosomes 9 and 22 (fig. 38.3). The tumor associated with this translocation is chronic myelogenous leukemia and results from the activation of the *c-abl* oncogene, which is normally located on chromosome 9.

Whereas it is a common belief that tumor formation involves mutation, certain properties associated with the etiology of teratocarcinoma do not fit with this concept. Teratocarcinoma is a unique type of tumor in which the tumor is associated with a cluster of cells of different types, including stem cells and differentiated cells of various types (see chapter 32). Only the stem cells are oncogenic. These stem cells may be replicated in vitro. On subcutaneous injection these cells produce tumors, but on injec-

Figure 38.2

Translocation associated with Burkitt's lymphoma. The protooncogene *c-myc,* on the end of chromosome 8q, is translocated to the end of chromosome 14q, adjacent to the immunoglobulin constant gene, *Ig-Cμ.* (From J. Yunis, The chromosomal basis of human neoplasia. *Science* 221:227–236, Jan. 1, 1983. Copyright 1983 by the AAAS. Reprinted by permission.)

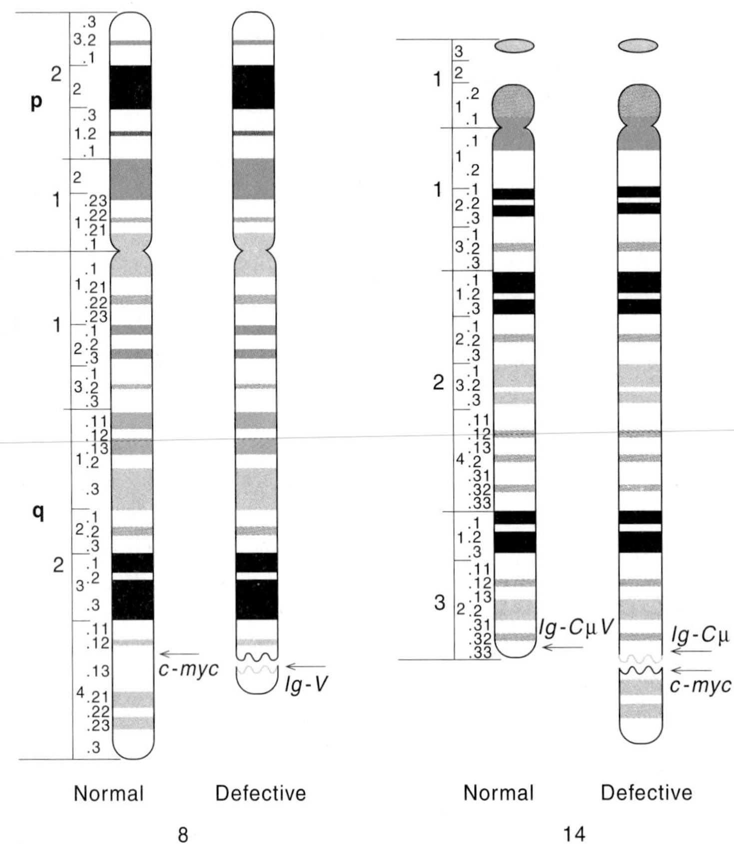

tion into a blastocyst embryo these stem cells revert to normal cells, which can be identified in the adult animal by the genetic markers they carry. If these stem cells bear a mutation that accounts for their oncogenicity, then it is hard to see how this mutation could revert to wild type by placing it in the cytoplasmic environment of the embryo. Possibly, the teratoma is an exceptional situation in which the tumor is caused by abnormal development not necessitating a mutation.

Transformed Tissue Culture Cells Are Closely Related to Cancer Cells

In the 1950s, systematic methods were developed for taking cells directly from the tissues of whole animals and inducing them to grow as single cells in liquid culture.

An outgrowth of this technology was the development of pure cell lines—cells that divide indefinitely. Primary cultures of dispersed single cells from tissue fragments usually die out after 50 or so duplications. However, during the growth of a primary culture, occasional altered cells appear that possess a somewhat different morphology; they frequently take over the culture, because they are capable of initiating new cultures with far fewer

Figure 38.3

The Philadelphia chromosome arises by a reciprocal translocation between chromosomes 9 and 22. Note that the *abl* protooncogene is moved in the translocation process.

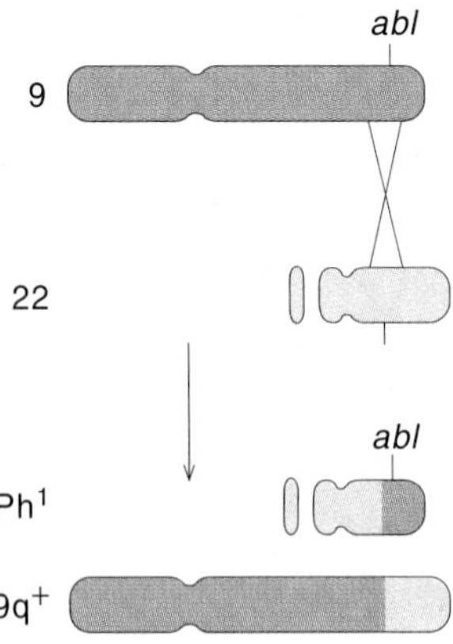

cells. A clone that is derived from one of these indefinitely dividing variants—which has in effect become immortal—is designated a cell line. The transition of primary culture cells to cells that can grow indefinitely is considered to be a step in the direction of becoming cancerous.

Typically, continuous cell lines grow in culture only on a solid support or "anchor" (the surface of a petri dish, for example) and in the presence of relatively high concentrations of nutrients. Even then, they divide only as long as the culture is sparse. When the cell density increases beyond a critical point, the growth rate decreases sharply; in fact, the cells of some continuous lines stop dividing altogether once they have formed a confluent monolayer.

At the density that normally halts growth, rare transformed cells continue to multiply and may reach cell densities up to 20 times higher than those of untransformed cells. By picking and subculturing such transformed cells, it is possible to establish clonal lines of transformants and to ask in what ways such cells differ from their untransformed progenitors.

The differences between transformed and untransformed cells can be classified in three ways: changes in cell growth, changes in cell surface properties, and genetic alterations. These changes are too great in number and too diverse in quality to be caused directly by one or a few mutations. Clearly, then, most of the observed alterations must be secondary and tertiary events that have occurred as a consequence of some primary event. It is the aim of much current work with tumor-causing agents, especially certain viruses, to discover the molecular nature of this primary event.

Transformed cells sometimes, but not always, give rise to tumors when injected into an isogenic animal. There may be numerous reasons for the failure to produce tumors on all occasions. One of the best-known reasons for lack of tumor production is immunological rejection. The transformed cell resulting from exposure to a tumor virus frequently displays viral antigens that are recognized as foreign to the animal host. This recognition excites a positive immune response that can lead to rejection of the transformed cells.

Just as transformed cells frequently do not develop into tumors when injected into whole animals, so it is that tumor cells do not always grow well when dispersed into tissue culture. Nevertheless, tumor cells usually adapt more readily to growth in tissue culture than do normal cells taken from the whole animal.

Many Tumors Arise by Mutational Events in Cellular Protooncogenes

A growing number of cellular genes have been implicated in a wide variety of cancers. These genes do not lead to cancers in their normal state but only after a mutational event has somehow resulted in their altered expression. For that reason, such genes are called protooncogenes in their normal state and oncogenes in their mutated, cancer-causing state. Most protooncogenes, including *c-abl* and *c-myc,* were not known until very similar genes were found as a result of studies of tumor-causing viruses.

Oncogenes Are Frequently Associated with Tumor-Causing Viruses

The first virus linked to tumor production was discovered in 1908 by Ellerman and Bang. Their finding was followed in 1911 by the better-known discovery by Peyton Rous of a virus that produces sarcomas in chickens (now known as Rous sarcoma virus, or RSV). Later (1932), Shope showed that a papilloma virus produced cutaneous tumors in rabbits, and Lucke (1934) showed that adenoviruses produced renal adenocarcinoma in the frog. These and other discoveries made in the 1940s and 1950s indicated that exposure to certain viruses could result in tumor production. However, viral oncogenes had not yet been identified, and skeptics still classified cancer-causing viruses along with other carcinogens.

The development of tissue culture techniques in the 1950s permitted a systematic investigation of the causal link between tumor viruses and cellular transformation in culture. In many ways the properties of cells transformed in culture resembled those of tumor cells in vivo, and so the tissue culture system was considered to be a convenient way to explore the oncogenic properties of viruses. It was noted with some tumor viruses that the response of cells of different species was quite different. On cells that were normally permissive for virus replication, no tumors arose. This was hardly surprising, in view of the fact that the virus replication cycle, particularly for DNA viruses, usually results in cell death. What was surprising was the finding that the same virus on another cell type, in which the virus could not replicate, sometimes resulted in cell transformation. Observations of this sort led Renato Dulbecco to hypothesize that a parallel existed between lysogeny of *Escherichia coli* by λ and transformation of animal cells by oncogenic viruses. In λ the viral genome can exist either as an independent entity, leading to virus replication and cell death, or as an integrated passive state, leading to a lysogenic cell (see chapter 35). With λ the two states—lytic, or active, and lysogenic, or passive—are attainable in the same cell type. Tumor-causing viruses are capable of existing in two states also. However, in this case the two states, actively replicating and passively integrated, are not normally attainable in the same cell type. Usually, two cell types of different species are required. In one cell type the viral genome invariably exists in the actively replicating form, causing immediate cell death; in another cell type the viral genome can exist only in a passive integrated form. Dulbecco suggested that in the integrated state the viral genome expresses a limited number of gene products that lead to cellular transformation.

So far, so good. Dulbecco's suggestion appeared to constitute a reasonable hypothesis for explaining the oncogenicity of certain DNA viruses. But it was also known that certain RNA viruses cause tumors. Indeed, one of the first known cancer-causing viruses, the Rous sarcoma virus, was an RNA virus. To explain the oncogenicity of RNA viruses, Dulbecco proposed that cancer-causing RNA viruses form a DNA copy of their genome that can be integrated into the host cellular genome. This bold proposal—that an RNA genome could be converted into a DNA copy—was confirmed by experiments in both David Baltimore's and Howard Temin's laboratories. Both groups found an enzyme, re-

Table 38.3
Properties of DNA Virus Oncogenes

Viral Gene and/or Gene Product	Function in Lytic Infection	Function in Transformation	Other Properties
Adenovirus			
E1A 9 kd, 26 kd, 32 kd	Required for expression of other early genes	Immortalizes cells; partial transformation	General transcription activator
E1B 9 kd, 55 kd	Required for lytic infection	Required for full transformation and tumorigenic cells Either 19 kd or 55 kd suffices	Associates with p53 55-kd protein found primarily in nucleus; 19-kd protein in endoplasmic reticulum and in nucleus
SV40			
Big T 90 kd	Binds to specific sites; necessary for DNA synthesis; inhibition of early RNA synthesis and activation of late RNA synthesis	Immortalizes and transforms	Stimulates host DNA synthesis on microinjection; associates with p53
Small t	None (?)	Required for full transformation on some cell lines	—
Polyoma			
Big T 100 kd	Binds to DNA; necessary for DNA synthesis and regulation of early and late RNA synthesis	Immortalizes cells; only amino terminal 40% fragment is required	Associates with p53; ATPase activity; gyrase activity
Middle T 56 kd	None (?)	Transforms cells and makes them tumorigenic	Associates with $pp60^{c\text{-}src}$
Small t	None (?)	Transformed cells adhere less firmly when small t is present	Found in nucleus

verse transcriptase, uniquely associated with RNA tumor-causing viruses that could synthesize DNA and an RNA template.

The finding that many viruses cause cancer only after inserting their genome into the host genome has spurred new efforts in cancer research. More and more attention has been paid to viruses. With the help of genetic manipulation the cancer-causing genes within the viruses have been identified and their gene products have been studied. The results of these studies are providing a broad platform of understanding about the kinds of biochemical processes that result in the conversion of normal cells to cancer cells.

The Role of DNA Viral Genes in Transformation Reflects Their Role in the Permissive Infectious Cycle

Usually, tumor-causing DNA viruses possess oncogenes that bear no sequence relationships to any host genes. We can grasp the role of DNA viral oncogenes in cellular transformation by appreciating the role these genes play in the virus lytic cycle (table 38.3). First of all, the oncogenes we have found in the adenovirus, polyoma virus, and SV40 virus all are expressed early in virus infection. These genes also trigger a transient transformation process in most cells they infect, where a permissive virus replication cycle is not possible. Thus it seems likely that these genes are primarily regulatory. In SV40 and polyoma the so-called big-T genes are clearly regulatory in productive infection because they are involved in regulating both viral DNA and RNA synthesis. In adenovirus there is no comparable gene, but the early genes *E1A* and *E1B* may play a similar, albeit indirect role. *E1A* functions as a transcription activator.

Another lead as to how a DNA virus transforms cells is to look at the cellular proteins they influence. There is a striking parallel here between the functions of adenovirus *E1B* and SV40 big T in this regard. These two genes are required for immortalization or transformation, and they both form a tight association with the cellular protein p53, which has been implicated in normal growth regulation and will be discussed at length below. Finally, the polyoma middle-T protein, which is required for transformation by polyoma virus, forms a strong complex with the cellular protein encoded by the protooncogene *c-src*. It seems likely that these associations between viral oncogenes and host proteins result in changes in regulatory functions and that if we knew what regulatory functions were being affected we would be on the threshold of understanding the biochemical mechanism(s) involved in transformation. We will have more to say about this after we have considered the retroviruses.

Figure 38.4

Hypothesis for how a retrovirus picks up a cellular protooncogene. The RNA–DNA cycle for retroviruses is shown in figure 31.33. The proviral DNA is believed to recombine the cellular DNA in such a way that it picks up a region of the cellular DNA containing a protooncogene. Red segments represent coding regions of c-*onc*.

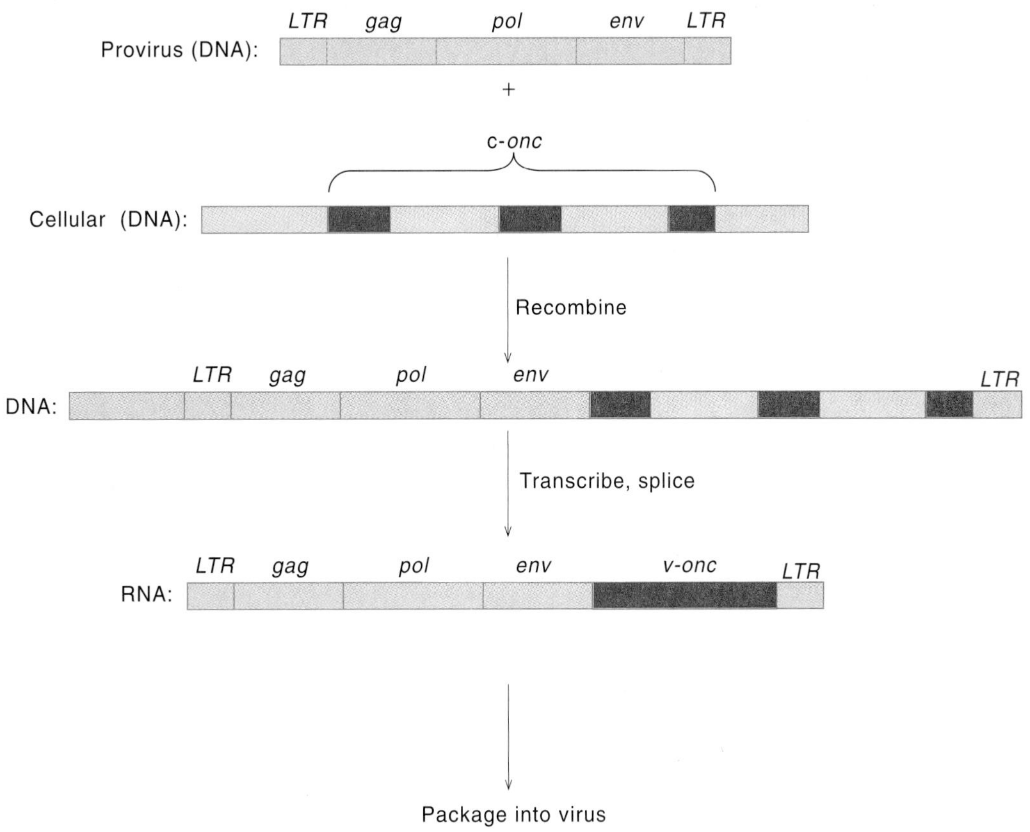

Retroviral-Associated Oncogenes That Are Involved in Growth Regulation

The first intensively studied oncogene associated with a retrovirus was the RSV *src* gene. It is clear that the *src* gene found in RSV, *v-src,* is structurally quite similar to a normal host gene found in the chicken genome called *c-src.* Subsequent work has shown that at least 30 other animal retroviruses exist that have acquired a cellular oncogene during their evolution. This uncanny association of retroviruses with cancer-causing genes has made retroviruses useful devices for scanning the cellular genome for the presence of protooncogenes. It seems likely that retroviruses become associated with oncogenes as the result of a series of recombination events between the DNA copy of the retroviral genome and the host protooncogene (fig. 38.4). The frequent occurrence of such associations is probably due to the survival advantage that oncogenes seem to confer on the virus.

A listing of some of the better characterized retrovirus-associated cellular oncogenes appears in table 38.4. In this table the name of the oncogene is indicated in the leftmost column. Proceeding to the right in the table are indicated (1) the virus of origin, (2) the viral gene product, (3) the cellular homolog of the viral product, (4) the activity associated with the viral gene product, and (5) the subcellular location of the viral gene product, which in most cases is similar to the subcellular location of the cellular homolog.

The src *Gene Product*

The protein encoded by the *src* gene has a molecular weight of 60 kD. Like many other oncogenic proteins, it is bound to the inner plasma membrane, and it possesses a tyrosine phosphokinase activity; that is, it catalyzes the addition of a phosphate group to the tyrosine hydroxyl groups of various proteins. It is considered highly likely that the kinase activity of the src protein is associated with its transforming activity because its loss in mutant *src* genes leads to loss of transforming activity. However, since many different proteins are phosphorylated by the src kinase, it is not clear which phosphorylations are crucial to the complex physiological response that triggers transformation.

The sis *Gene Product*

Many oncogenes are associated with tumors produced from a restricted class of differentiated cells. Such is the case with the *sis*

Table 38..4
Some Retroviral Oncogenes and Their Cellular Homologs

Oncogene	Viral Origin[a]	Viral Gene Product	Cellular Homolog	Activity	Subcellular Location
sis	Simian sarcoma virus	$p28^{sis}$	PDGF B-chain	PDGF agonist	Cytoplasm
src	Rous sarcoma virus	$p60^{v\text{-}src}$	$p60^{c\text{-}src}$	Tyrosine kinase	Plasma membrane
abl	Abelson murine virus	$P120^{gag\text{-}abl}$	$p150^{c\text{-}abl}$	Tyrosine kinase	Plasma membrane
erbB	Avian crythroblastosis virus	$gp65^{erbB}$	Truncated EGF receptor	Tyrosine kinase	Plasma membrane
myc	Avian myelocytomatosis virus MC29	$P110^{gag\text{-}myc}$	$p49^{c\text{-}myc}$	Binds DNA	Nucleus
H-ras	Harvey murine sarcoma virus	$p21^{v\text{-}Hras}$	$p21^{c\text{-}H\text{-}ras}$	Threonine kinase binds GDP or GTP	Plasma membrane
K-ras	Kirsten murine sarcoma virus	$p21^{v\text{-}Kras}$	$p21^{v\text{-}Kras}$	Threonine kinase binds GDP or GTP	Plasma membrane
fos	FBJ murine osteo-sarcoma virus			Binds DNA	Nucleus
jun	Avian sarcoma virus (ASV 17)			Binds DNA	Nucleus

[a]Only one example of a virus is given for each oncogene.

oncogene, found in the simian sarcoma virus. This oncogene is believed to be closely related to the gene for platelet-derived growth factor (PDGF). PDGF is a small protein, synthesized in platelets, that stimulates the growth and division of target cells that carry a membrane-bound specific receptor for PDGF—in particular, the mesenchymal cells involved in wound healing. Tumors originate when the target cells carrying the PDGF receptor mutate to cells capable of synthesizing their own PDGF. This situation is believed to lead to uncontrolled growth because the factor continuously stimulates cell proliferation in the very cells in which it is synthesized (fig. 38.5).

The PDGF protein contains two polypeptide chains: A and B, with molecular weights of 28,000 and 32,000, respectively. The amino acid sequence of the PDGF protein is very similar to that of the predicated sequence of the transforming p28 sis protein of the SSV virus, indicating a common ancestral origin. Furthermore, it is believed that the SSV virus causes tumors by synthesizing large amounts of the viral p28 sis protein, which stimulates the unregulated proliferation of target cells that carry the PDGF receptor.

The erbB *Gene Product*

Epidermal growth factor (EGF) is a small mitogenic protein that simulates the proliferation of cells carrying specific membrane-associated EGF receptors. The EGF receptor has a strong amino acid sequence homology with $gp65^{erbB}$, the transforming protein of avian erythroblastosis virus (AEV). The EGF receptor has tyrosine kinase activity, like the src protein. In addition, the EGF receptor contains an extracellular domain that binds the EGF growth factor.

The *v-erbB* oncogene acts to expand a pool of highly mitotic, undifferentiated erythroid precursor cells, but these are poorly tumorigenic, because they differentiate at high rates into postmitotic, end-stage red cells. Another potential oncogene, *v-erbA,* blocks differentiation of erythroid precursors but creates no tumors because it is unable to provide the mitogenic impetus needed to expand the pool of stem cells. The two genes, *erbA* and *erbB,* carried into erythroid precursors together by avian erythroblastosis virus, act in concert to create an aggressive erythroleukemia; *v-erbB* drives expansion of the pool of undifferentiated precursor cells, whereas *v-erb4* blocks their conversion by the differentiation pathway. The *v-erbA* allele that participates in formation of chicken erythroleukemias is a mutant version of a transcriptional regulatory protein, the chicken thyroid hormone (triiodothyronine) receptor. Function of the wild-type receptor protein is blocked in the presence of *v-erbA,* because the latter occupies critical DNA-binding sites in a way that precludes association by the wild-type receptor protein and inhibits transcription.

The ras *Gene Product*

One of the most studied protooncogenes is *ras.* The *ras* gene found in normal mammalian cells is closely related to the *ras* oncogenes of Harvey and Kirsten murine sarcoma viruses, *H-ras* and *K-ras,* respectively. *Ras* oncogenes have also been isolated with the help of DNA transfection techniques. Activated *ras* genes have been isolated from a wide variety of dissimilar neoplasms, including carcinomas, sarcomas, neuroblastomas, lymphomas, and leukemias. All members of the *ras* gene family encode closely related proteins of approximately 21 kd, which have been designated p21. Unlike the oncogenes associated with many retroviruses, the level

Figure 38.5

Mechanism of mitogenesis in normal and transformed cells. (*a*) Schematic representation of the growth-factor-related mitogenic pathway in normal cells. Here (1) represents the growth factor, (2) the growth factor receptor, and (3) the intracellular messenger system that transmits the mitogenic signal from the receptor to the nucleus. (*b*) Schematic representation of a possible perturbation of the growth-factor-related mitogenic pathway in transformed cells. Here, (1) represents endogenous production of growth factor that may stimulate the cell. The endogenously produced factor may be secreted and interact with growth factor receptors at the cell surface (as shown) or, alternatively, activate the receptor in an intracellular compartment.

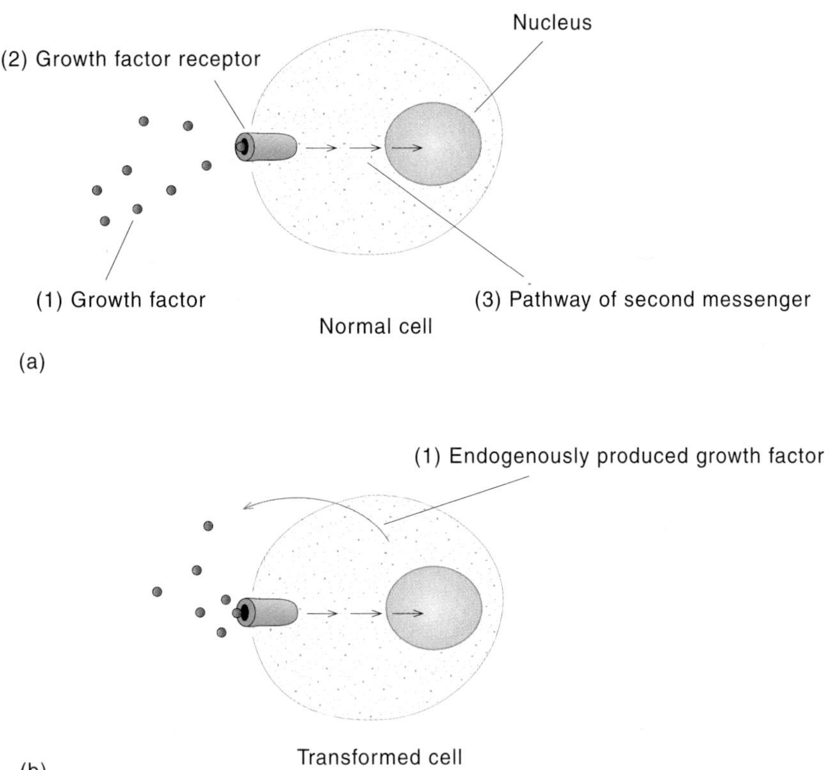

of p21 expression is similar in normal cells and in many different human tumor-cell lines. Nucleotide sequence analysis of the *H-ras* transforming gene isolated from a human bladder carcinoma cell line has shown that the transforming activity of this gene is the consequence of a point mutation altering amino acid 12 of wild-type p21 from glycine to valine (fig. 38.6).

Subcellular fractionation and immunofluorescence of normal cellular and viral Ras proteins have indicated that p21 is localized on the inner face of the plasma membrane. Both the normal and the mutant Ras proteins bind GTP specifically and strongly, but only the normal protein has GTPase activity.

The function of Ras was discussed in chapter 27 (see fig. 27.22). Ras normally relays mitogenic and developmental signals initiated by cell surface receptors into the cytoplasm and the nucleus. The immediate action of Ras in mammalian cells that have been studied is to bind a protein called Raf, a serine/threonine kinase that proceeds to phosphorylate MEK, another downstream regulatory protein in the mitogenic pathway.

The myc *Gene Product*

The *c-myc* gene, identified originally as the cellular homolog of the transforming determinant carried by avian myelocytomatosis virus, is altered in association with a broad spectrum of neoplasms. Consistent with the observation that altered *c-myc* is associated with tumors of diverse origin, it has been observed that normal *c-myc* is expressed in a variety of tissues. Thus *c-myc* appears to encode a function associated with a ubiquitous metabolic pathway.

The *c-myc* gene encodes a 49-kD protein that is highly concentrated in the nucleus of the cell. The concentration of c-myc protein normally varies appreciably with the metabolic state of the cell, increasing by more than an order of magnitude in cells just prior to chromosome duplication and cell division.

In most human tumors associated with the *c-myc* gene, the concentration of c-myc protein is greatly amplified. Burkitt's lymphoma, which we discussed earlier, is a notable exception. In this case the c-myc transcript is marginally, and in some cases not at all, increased by comparison with control lymphoblastoid cell lines. Recall that in Burkitt's lymphoma, reciprocal chromosomal translocations are found that involve a chromosome carrying *c-myc* protooncogene (chromosome 8) and one of the chromosome segments (usually chromosome 14) bearing immunoglobulin genes.

In at least one Burkitt's lymphoma cell line, the structure of the amino acid–coding portion of the translocated *c-myc* gene, and hence its predicted protein product, has not been altered, a

Figure 38.6

Interaction between the *ras* protooncogene and GDP. The protein is shown as a green ribbon, interacting with GDP (dark purple, yellow, and orange). The dark purple rectangle represents the guanosine; the dark yellow pentagon, the ribose; and the orange circles, the phosphates of GDP. The domains of the protein interacting with each of these parts of GDP are represented by sleeves on the protein ribbon, color-coded to match the corresponding part of GDP. The P domain of the protein contains glycine 12: this amino acid is in a critical position next to the phosphates of the nucleotide. The N and C termini of the protein are labeled. (From L. Tong et al., Structure of the ras protein. *Science* 245:243, July 21, 1989. Copyright 1989 by the AAAS. Reprinted by permission.)

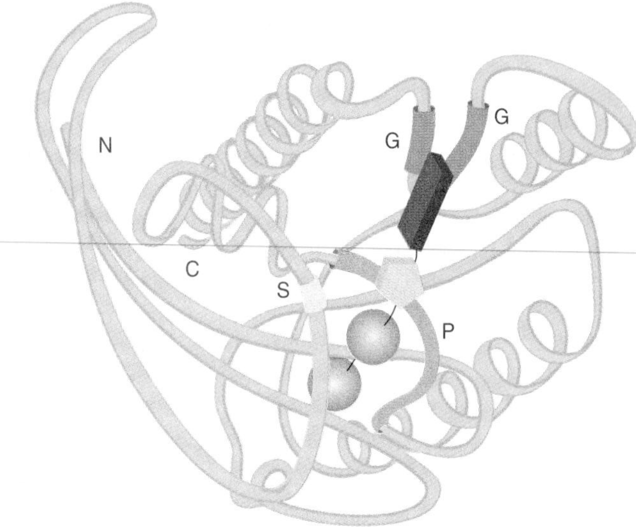

fact suggesting that activation of *c-myc* must be mediated via a regulatory disturbance. Normally, c-myc protein synthesis is strongly regulated with respect to the cell cycle and tightly repressed in quiescent cells. The translocated *c-myc* gene appears to be somewhat deregulated with respect to its expression during various phases of the cell cycle.

Members of the *myc* gene family have been implicated in the control of normal cell proliferation as well as in neoplasia. A more direct role for *myc* genes in transformation is indicated by their ability to transform primary rat embryo fibroblasts in association with the *c-H-ras* oncogene.

An important clue to *c-myc* function was the discovery in the conserved carboxy-terminal regions of three structural motifs: the leucine zipper (LZ), helix-loop-helix (HLH), and basic region (B). These motifs were originally defined in a number of other sequence-specific DNA-binding proteins but had not previously been found within a single protein.

The comparatively weak homooligomerization efficiency of the c-myc HLH/LZ suggested that a partner protein(s) might exist that heterooligomerizes with c-myc to form a specific DNA-binding complex. A protein termed max (for myc-associated "X" factor) was shown to interact with c-myc in a manner that required the integrity of the c-myc HLH/LZ motif. Max was also demonstrated to associate with N- and L-myc proteins, but not with the HLH/LZ proteins USF or AP-4, nor with several other HLH or leucine zipper proteins. In DNA-binding assays, c-myc-max specifically recognized a c-myc-binding site (CACGTG) in a manner that required both an intact max basic region and the HLH/LZ motifs.

In yeast model systems it has been observed that myc is a transcription activator but only when present in a heterodimer with max. Max appears to be essential for DNA binding. Max dimer can bind to DNA on its own, but it does not activate transcription on its own. Mammalian genes that are normally activated by myc are still not known.

It has been reported that the myc amino-terminal and central regions can specifically interact with the retinoblastoma (RB) tumor suppressor protein in vitro. This finding (discussed later in the chapter) prompts the suggestion that RB may directly facilitate myc function, but it is not yet known whether the observed interaction reflects a physiologically relevant association in vivo.

The jun *and* fos *Gene Products*

The *v-jun* oncogene was discovered as a 0.93-kb insert in the genome of a replication-defective retrovirus, avian sarcoma virus 17 (ASV17), isolated from chicken sarcoma. ASV17 causes fibrosarcomas in chickens and oncogenic transformation in cultured avian embryonic fibroblasts. The oncogenic potential of ASV17 is due to the presence of the *v-jun* gene, which is derived from the cellular *c-jun* gene. The gene is believed to be oncogenic in the virus because it is expressed there in higher amounts.

A great deal of excitement was generated by the discovery that the jun protein can substitute for the yeast transcription activator GCN4. Comparative analysis of the sequences of the two proteins showed that a strong homology occurs over about one-third of their length in the DNA-binding parts. Both of these proteins bind to the same consensus sequence, TGACTCA.

The oncogene of the FBJ murine osteosarcoma virus (*fos*) codes for a related nuclear protein that participates in transcriptional regulation. In human fibroblasts the fos protein is mostly associated with *c-jun.* The fos–jun complex, known as AP-1, binds specifically to DNA. Since fos alone does not show specific DNA binding, it is believed that jun is responsible for this affinity. Although jun can form homodimers that bind to DNA, the AP-1 heterodimers formed between fos and jun show a greater affinity. The heterodimers are also more effective in transcription activation; therefore the heterodimer is probably the functionally relevant state of the jun and fos proteins.

Structural analysis indicates that the fos and jun proteins belong to a class of DNA-binding proteins that share the conserved structural motif known as the leucine zipper (see fig. 36.23). Thus the dimerization of these two proteins is mediated by hydrophobic interaction between the leucine side chains of two leucine zipper domains.

The normal mechanism of activation of AP-1 illustrates how a signal received at the cell surface is transduced into a signal for cell division. For example, PDGF binding to its specific plasma membrane receptor on a target cell can trigger a chain of reactions that leads to activation of protein kinase C by a well-known pathway that we discussed in chapter 20 and again in chapter 27 (see fig. 20.16). The activation of protein kinase C results in the phosphorylation and concomitant activation of a number of proteins, including AP-1. As a result of activation, AP-1 stimulates transcription of several genes whose expression favors cell division. Given this central role in regulating cell proliferation, it is not surprising to find that the deregulated activity of AP-1 contributes to tumorigenesis.

The Transition from Protooncogene to Oncogene

All of the oncogenes thus far discovered are associated with a cellular homolog that is required for normal growth. The transition of the protooncogene to an oncogene is accompanied by abnormal expression of the gene products. In many cases, such as *src, jun,* or *sis,* excessive amounts of the gene product are synthesized. In some other cases, such as *H-ras* or *K-ras,* normal amounts of the oncogenic product are synthesized, but the protein encoded by the oncogene is altered so that it behaves differently. In still other cases, such as *myc,* the oncogene product is similar to the cellular homolog in amount as well as structure, but the time during the cell cycle when it is produced is altered.

Oncogene products assume specific locations within the cell. Usually, they are associated with one of two locations: the plasma membrane or the nucleus. This specificity is consistent with the hypothesis that oncogene products are associated with normal cellular metabolism relating to regulation of cell proliferation; it seems likely that many of the elements regulating cell proliferation would be found at the membrane and nucleus of the cell.

It is not surprising to find oncogenes with varying specificity for producing tumors. Some components of the cell proliferation regulatory apparatus are probably quite general. An oncogene like *myc* is probably associated with one of these and is therefore associated with tumors of widely varying origins. By contrast, an oncogene like *sis* is specifically associated with cells that are designed to be triggered into proliferation by PDGF. Hence tumors associated with this oncogene are limited to cell types possessing the PDGF membrane receptor.

Tumor Suppressor Genes Are Genes Whose Presence Is Needed to Block Transformation

Thus far we have discussed oncogenes whose unregulated activities foster the transformed state. An oncogene is usually dominant to the wild-type unaltered protooncogene to which it is related by one or more mutational events. There is another class of genes whose normal activity suppresses tumorigenesis. As a rule, tumor suppressor genes are dominant to their defective alleles. Thus a single copy of a wild-type suppressor gene can mask the presence of a defective allele. The notion that one or more genetic elements could inhibit the proliferation of tumor cells was demonstrated in a series of experiments using somatic cell genetics. In these experiments a cancerous cell, able to form tumors in animals, was fused with a nontumorigenic normal cell. The hybrid cell line grew well in cell culture, retaining many of the properties of the cancerous cells, but it no longer produced tumors in animals.

The Retinoblastoma Gene

The retinoblastoma (*Rb*) gene is one of the better-known tumor suppressor genes. This gene was originally spotted in the human malady known as bilateral retinoblastoma, in which tumors develop in the retinas of both eyes. Humans with this condition usually inherit a genetically recessive mutation on chromosome 13. Sometimes the mutation is microscopically visible as a small chromosomal deletion. In general, recessive mutations do not affect the phenotype unless they are present in the homozygous state. However, in the course of the very extensive cell duplication that ensues in retinal development, there is a high probability that at least one cell in each retina will mutate in the wild-type allele of the genetic locus in question. This change results in a homozygous state and in the conversion of the normal cell to a tumorous cell that proliferates rapidly. As a result of this somatic mutation, both eyes usually develop tumors.

In view of its broad tissue distribution, it might be expected that *Rb* would be implicated in a variety of tumor types. However, *Rb* gene inactivation seems restricted to a narrow subset of tumors. In fact, mice that are heterozygous for the *Rb* gene are not predisposed to retinoblastoma, but some display pituitary tumors arising from cells in which the wild-type *Rb* allele is absent. Transgenic mice that lack the *Rb* gene entirely progress almost normally through the first half of embryonic development but then die.

The 108-kb *Rb* gene encodes a 105-kD nuclear phosphoprotein. In normal cells the Rb protein is permanently present, no matter whether the cells are in Go or cycling, but its state of phosphorylation changes. The nuclear localization of pRb and its DNA-binding ability suggest a role in transcriptional regulation. The dephosphorylated (active) Rb protein is thought to function in G1 as part of the braking mechanism to inhibit passage past Start, a point in late G1. In the Go cell, Rb contains little phosphate and appears to hinder the transcription of genes, such as *fos* and *myc,* which are required for proliferation. These genes are transcribed at a high level in mutant cells that lack a functional copy of the *Rb* gene and at a much lower level in cells with a functional copy of the *Rb* gene.

p53 Is the Most Common Gene Associated with Human Cancers

There are many indications that the oncogenes of the DNA tumor viruses target the major cellular tumor suppressors. This appears to be part of a plan to eliminate the cellular blocks to proliferation. In fact, both Rb and p53 are targeted by the SV40 T antigen. Indeed, p53 was first detected through its association with the SV40 big T oncoprotein in virus-transformed cells. The viral oncoproteins sequester p53 protein in large inactive complexes.

While most mutants of p53 go unnoticed in the heterozygous state, many mutant p53 alleles favor growth and trans-

Figure 38.7

Crystal structure of a p53-DNA. The core domain structure consists of a β sandwich that serves as a scaffold for two large loops and a loop-sheet-helix motif. The two loops, which are held together in part by a tetrahedrally coordinated zinc atom (in red), and the loop-sheet-helix motif form the DNA binding surface of p53. Residues from the loop-sheet-helix motif interact in the major groove of the DNA while an arginine from one of the two large loops interacts in the minor groove. The six most frequently mutated amino acids are shown in yellow. (From *Crystal structure of a p53 tumor suppressor-DNA complex: understanding tumorigenic mutations,* by Dr. Nikola Pavletich, SKI, *Science* 265:346. © 1994 by AAAS.)

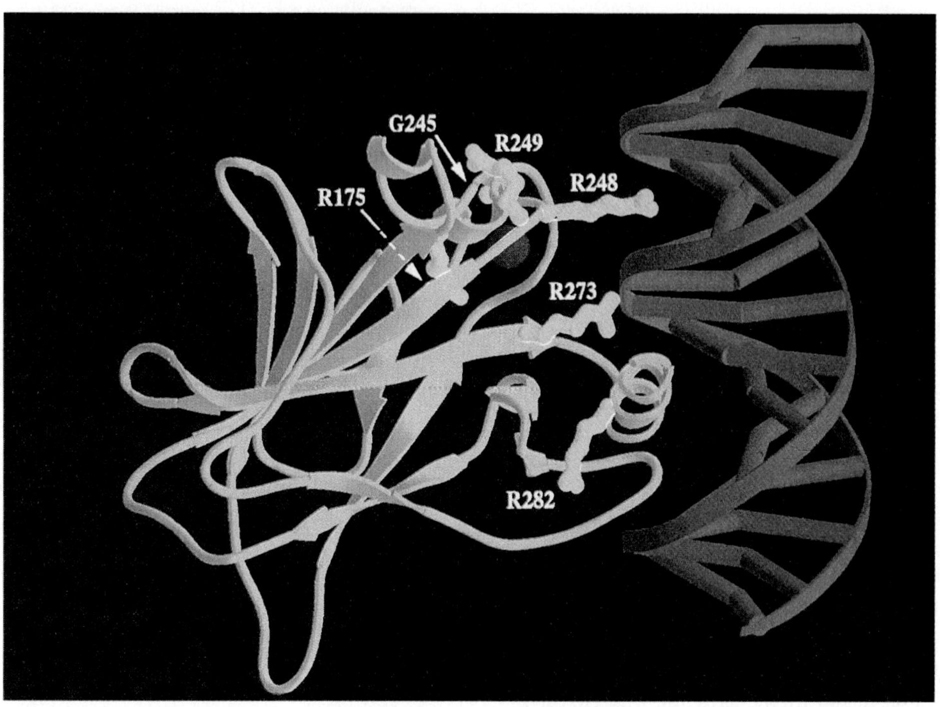

formation in cells that continue to carry intact, wild-type p53 gene copies. Accordingly, such mutant p53 alleles act in a dominant fashion. This created some confusion in determining the role of p53 because in the case of a typical suppressor gene the defective mutant type is recessive to the wild type. Two explanations for this behavior have been offered. First, since p53 acts in the form of an oligomer, it is possible that an oligomer that contains even a single defective subunit might not function properly. Another possible explanation for the dominance of certain p53 mutants is even more intriguing. The defective mutant might drive the cotranslated wild-type protein into the mutant conformation. It should be noted that this is precisely the currently favored explanation for the transmission of prion-related diseases (see chapter 5).

The best demonstration that the loss of active normal p53 explains the oncogenic behavior of mutant p53 comes from studies in which a null mutation was introduced into the gene by homologous recombination in murine embryonic stem cells. Transgenic mice homozygous for the null allele appear to be normal but are prone to the development of a variety of neoplasms by six months of age. Thus normal P53 appears to be dispensable during embryonic development, but its absence predisposes the animal to neoplastic disease.

Inactivation of the p53 tumor suppressor gene occurs in over half of all human tumors, implying that loss of the gene represents a fundamentally important step in cancer pathogenesis. The p53 protein is known to bind to DNA (fig. 38.7) and at least in part functions as a transcription activator through sequence-specific interactions with DNA containing the sequence 5′-PuPuPuC(A/T)(T/A)GPyPy-3′, typically in the context of two such sequence motifs separated by 0–13 bp.

At least two important events are regulated by p53 in connection with its function as a tumor suppressor. First, p53 has been shown to induce cell cycle arrest at the G1/S border. Second, p53 can induce programmed cell death or apoptosis. Thus expression of wild-type p53 in some p53-deficient tumor cell lines results in spontaneous cell death. In other p53-deficient tumor cell lines, however, the mere restoration of p53 activity is insufficient to trigger apoptosis but does render cells relatively more sensitive to induction of apoptosis after exposure to radiation or DNA-damaging chemotherapeutic drugs. This finding suggests that p53 may be involved in genome surveillance and DNA repair. The specific roles played by p53 in this process presumably include the arrest of cycling cells before S-phase to allow for repair of damaged DNA prior to DNA replication and the induction of apoptosis for cases in which the DNA damage is too severe to be properly repaired. Loss of p53 may thus contribute to the genomic instability common in tumor cells by allowing tumor cells to replicate damaged DNA and thereby propagate genetic defects.

In search of specific gene targets for p53, T. Miyashita and J. Reed made the observation that restoration of p53 in a murine leukemia cell was associated with increases in *bax* mRNA and bax protein. Since *bax* was known to be an accelerator of apoptosis, this suggested that the *bax* gene might be a specific target of p53. Subsequent isolation of the *bax* gene and in vitro studies showed that this was the case, thus providing the first example of a proapoptotic gene whose expression is directly regulated by p53. It seems likely that there must be others.

Understanding Cell Growth and Cell Death Is Crucial to Our Understanding of the Transition between Normal Cells and Cancer Cells

It seems likely that an in-depth study of cell cycle control will reveal causal links to the oncogenes and tumor suppressor genes we have been discussing. Clear indications for some of these links already exist. It behooves us to consider the basic aspects of this subject before delving briefly into possible linkages. As we have already described (see chapter 26), the cell cycle is divided into five phases: Go, G1, S, G2, and M. The key events in this cycle include the start of DNA replication and the start of mitosis. Starts are believed to be triggered by the interaction of two types of proteins (fig. 38.8). The first are the cyclin-dependent protein kinases (Cdks), which induce processes by phosphorylating selected proteins on critical serine or threonine residues. The second is a family of specialized activating proteins, called cyclins, that bind to Cdk molecules and control their ability to phosphorylate appropriate target proteins. We may think of the activation and disassembly of cyclin–Cdk complexes as the pivotal events in the cell cycle. Cyclins undergo a fresh round of synthesis and degradation in each cell division cycle.

In multicellular organisms, homeostasis requires that a balance between cell proliferation and cell death be maintained. Much more is known about proliferation vis-à-vis the cell cycle than cell death. Nevertheless, recent evidence suggests that alterations in cell survival contribute to the pathogenesis leading to cancer. Hence we must consider apoptosis as well as proliferation if we are to have a full understanding of this subject. Furthermore, since we started this chapter by describing cancer cells as cells that are out of control, it would not be surprising to find that all of the proteins involved in the pathogenic state are regulatory. So far this appears to be the case.

Is it possible that we are already very close to an understanding of the key proteins involved in the pathogenic state? Let's take a second look at the properties of the two major tumor suppressor genes we have been discussing: *Rb* and *p53*. The loss of function of the retinoblastoma (*Rb*) tumor-suppressor gene releases a G1/S barrier to growth. Are we to assume that the alteration in both alleles of this one gene are sufficient to result in unlimited growth? Most likely not. For one thing, human tumors that have been thoroughly analyzed have often lost both *p53* and *Rb*. Second, DNA-virus transforming proteins target both *Rb* and *p53* for inactivation as though targeting just one of these genes may not be enough to create a cellular environment suitable for unrestricted growth. This may mean that the loss of one tumor-suppressor gene (*Rb*) may somehow be compensated by the activity of another tumor suppressor gene that serves as a backup to prevent neoplastic growth. How could this be? Simple enough, based on what we already know. If *Rb* is defective, it could trigger a higher level of *p53* activity and thereby a higher level of apoptosis, great enough to prevent net growth. On the other hand, if *p53* is also inactivated, then growth might be expected to get out of control, which it appears to do in the limited number of cases that have been studied.

Figure 38.8

Key events in control of the cell cycle. Two key families of proteins play major roles in cell cycle control. These are the cyclin-dependent protein kinases (Cdks), which trigger processes by phosphorylating the proteins involved and the cyclins which activate the Cdks by binding to them.

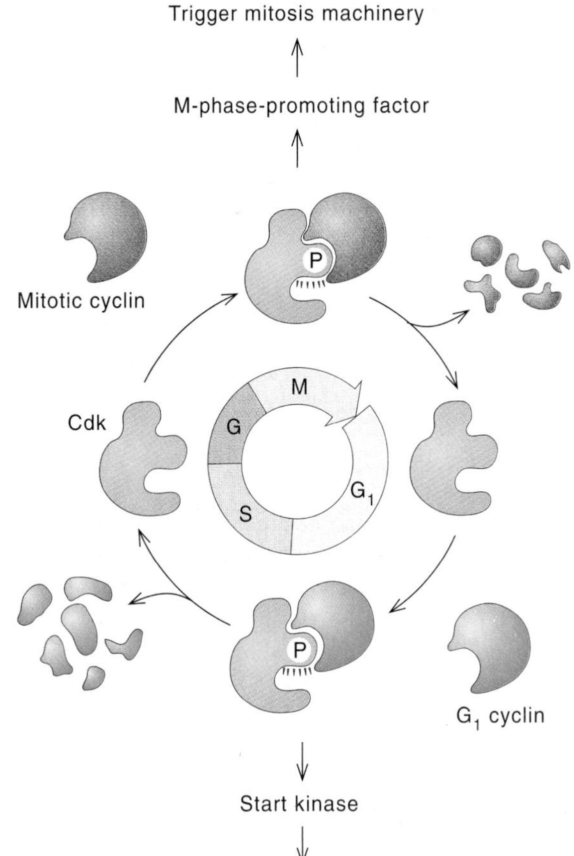

How Close Are We to Understanding the Multistep Process That Leads to Cancer?

A wide range of observations indicate that tumorigenesis is a multistep process involving several mutations, each of which results in discrete changes in the cellular metabolism. If this is the case, then we might expect that any particular oncogene would have the capacity for affecting only one step in the overall process. This may be true as far as the oncogenes are concerned, but the tumor

Figure 38.9

Rat embryo fibroblasts after several days of growth on plates. Normally, these cells stop growing as they approach confluency (top frame). Overgrowth of the monolayer by rounded cells is indicative of cellular transformation. Two transformed foci are seen in cells pretreated with both *ras*- and *myc*-containing DNA (bottom frame). The transformed cells produced similar transformed cells when replated and, when injected into whole animals, produce tumors with a much higher frequency than normal cells.

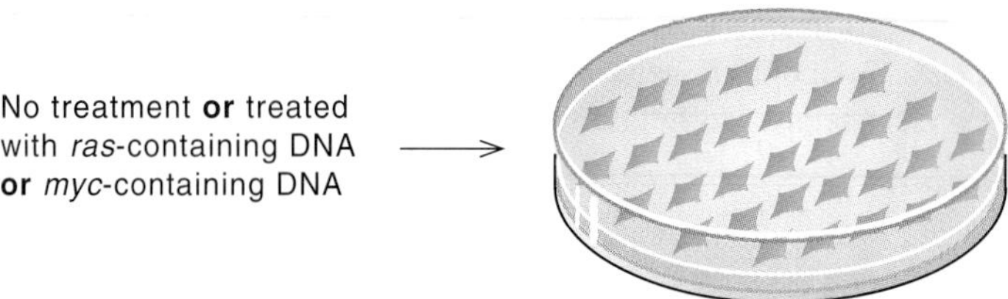

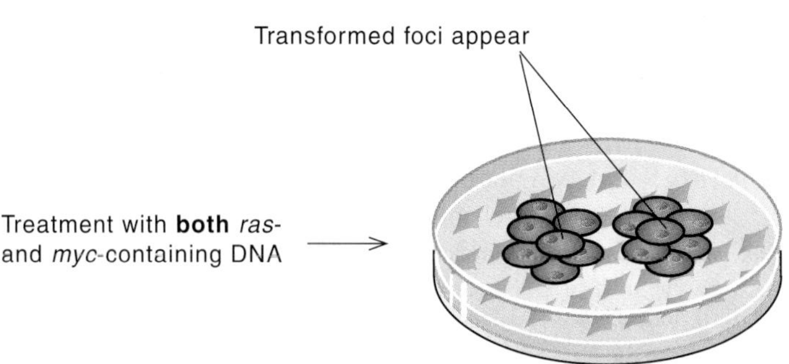

suppressor genes seem to be master genes affecting a cascade of regulatory processes, so they could be very special. Let's ignore this for the sake of argument. We have seen that a multistep biochemical pathway ordinarily requires a separate enzyme for each step. Enzymes that function in the different steps of a pathway are sometimes said to complement one another. The question arises, In carcinogenesis, do different oncogenes show a similar pattern of complementation? For example, can oncogenes that are cellular in origin, such as N-*ras,* complement one or the other of the polyoma oncogenes? In fact, middle-T antigen of polyoma and N-*ras* oncogenes produce no new phenotypes when they are cotransfected into rat embryo fibroblasts (REFs). But large-T antigen and *ras* together achieved dramatic results, producing rapidly expanding foci. This study shows that the conversion of a normal cell to a tumor cell can be achieved by the complementary action of two distinct oncogenes; in this case, one is cellular and one is viral. *Ras* functions at the plasma membrane, while large T antigen probably affects events in the nucleus at the level of transcription.

In parallel studies, H-*ras* and *myc* oncogenes have been shown to cooperate to produce dense foci of morphologically transformed cells from REF cultures (fig. 38.9). Thus two cellularly derived oncogenes have been shown to complement each other to produce a fully transformed phenotype. Parallel experiments have been done on whole mice to show that the combination of *ras* and *myc* is highly tumorigenic. Experiments of this sort suggest that at a minimum, two different oncogene functions are required to convert a normal cell into a tumorigenic one.

Recent experiments of T. C. Thompson and his colleagues have taken the mouse system to another level. By combining *ras* and *myc* in cells heterozygous for a defective p53 mutation, it was found that almost all tumors would progress to the metastic state. In all cases the metastic cells had suffered a complete loss of functional p53.

Experiments of this sort are being performed in various combinations, and conversely, naturally occurring tumorous tissue is being analyzed to see how many oncogenes and altered tumor suppressor genes it contains. This will probably not give us a full picture, since there are still relevant genes that have not yet been characterized. Nevertheless, we may expect rapid progress in the characterization of tumorous tissues and our understanding of the causal relationships that exist between different oncogenes and tumor suppressor genes in the near future. This is a period of exponential growth in this area of cancer research.

Is There a Cure for Cancer?

At the outset of this chapter we stated that a cure for cancer(s) was not likely to be forthcoming in the near future. So environmental causal factors should be addressed. Since environmental factors probably account for about 80% of the life-threatening cancers, this is no small item. Next we might ask how important early detection is for successful treatment. This is very unclear. It should be appreciated that most reports on such matters are written by people with vested interests.

Now let's face up to the enormity of the problem. The more we learn about cancer, the more we become aware that the disease is the result of subtle alterations in proteins that the cell already possesses. This may eliminate the immunological approach, although not necessarily.

Optimistically, within the next few years we will have a nearly complete picture of the regulatory proteins involved in some kinds of cancers and will be able to diagnose a number of the defects that characterize some of the most commonly occurring cancers. Even if diagnosis leads to an accurate description of the tumor, this does not indicate how the condition should be treated. I am reminded of the fact that the molecular biological explanation for sickle-cell anemia was fully determined in the late 1940s, but no cure or reasonable treatment is yet in sight. In the case of cancer the situation is compounded by the fact that cancer comprises dozens of diseases. We must hope that ingenuity and the biotechnology will lead us to workable cures. I myself am most intrigued by the possibility of solutions that are based on gene therapy.

SUMMARY

From the time of their discovery, cancer cells have always appeared to be unruly. We have reached that stage in our understanding of cancer cells when we can point to specific aberrations in the genome. In this chapter we took the view that an understanding of cancer can be achieved by analyzing the regulatory pathways involved in growth control because most cancers appear to originate from mutations in specific genes involved in growth regulation. The following points are the highlights of our discussion.

1. To judge by the frequency of occurrence of cancers in different countries, environmental factors have more influence on the incidence of cancers than inherited genetic factors do. This conclusion is reinforced by studies of migrant populations.
2. Chromosomal translocations are frequently associated with specific types of cancer. This is direct evidence that genetic abnormalities can lead to cancer.
3. Many properties of transformed cells grown in tissue culture resemble cancer cells. The factors that lead to uncontrolled growth in vivo can be studied in vitro by the effects they have on tissue culture cells.
4. A number of tumors arise from recessive mutations in which the mutations appear to be in growth-control genes.
5. Growth-control genes that lead to cancers when they are altered in some way are referred to as protooncogenes. They become oncogenes, that is, cancer-causing genes, by mutation. Host protooncogenes are frequently very similar in structure to oncogenes carried by tumor causing viruses.
6. Dulbecco proposed that cancer-causing viruses insert their oncogenes into the host genome. It appears that cancer-causing viruses are associated with a limited number of DNA viruses and RNA viruses known as retroviruses, which replicate through a DNA intermediate. It is this DNA intermediate that usually gets inserted into the host genome.
7. Abnormal expression accompanies the transition from protooncogene to oncogene. Three types of abnormalities occur: (1) excessive production of the gene product; (2) altered behavior of the gene product, such as a change in its regulatory properties; and (3) expression at a time during the cell cycle when the gene is not normally expressed.
8. Tumor suppressor genes regulate growth. When they are defective or deleted, growth is likely to get out of hand. Thus they differ from protooncogenes, which cause problems only when they become modified to assume a hyperactive form. Some tumor suppressor genes are designed to regulate proliferation. Other tumor suppressor genes are designed to encourage programmed cell death or at the minimum to delay the cell cycle to permit time for DNA repair. In the hierarchy of control it is possible, but not proven, that cellular protooncogens are normally under the control of tumor suppressor genes.
9. A fully developed cancer requires mutations in several genes; it is not clear whether there is a preferred progression in which these mutations are likely to occur.
10. Finding a cure for cancer may require totally different therapies from ones that have been used in medicine thus far.

SELECTED READINGS

Aaronson, S. A., Growth factors and cancer. *Science* 254:1146–1152, 1991.

Amati, B., S. Dalton, M. W. Brooks, T. D. Littlewood, G. L. Evans, and H. Land, Transcriptional activation by the human c-myc oncoprotein in yeast requires interaction with max. *Nature* 359:423–426, 1992.

Arany, Z., D. Newsome, E. Oldread, D. M. Livingston, and R. Eckner, A family of transcriptional adaptor proteins targeted by the EIAA oncoprotein. *Nature* 374:81–88, 1995.

Barinaga, M., Forging a path to cell death. *Science* 273:735–737, 1996.

Baserga, R., Oncogenes and the strategy of growth factors. *Cell* 79:927–930, 1994.

Bischoff, J. R., D. H. Kirn, A. Williams, C. Heise, S. Horn, M. Muna, L. Ng, J. A. Nye, A. Sampson-Johannes, A. Fattaey, and F. McCormick, An Adenovirus Mutant That Replicates Selectively in p53-Deficient Human Tumor Cells *Science* 274:373–376, 1996.

Cho, Y., S. Gorina, P. D. Jeffrey, and N. P. Pavletich, Crystal structure of a p53 tumor suppressor-DNA complex: Understanding tumorigenic mutations. *Science* 265:346–358, 1994.

Cobrinik, E., S. F. Dowdy, P. W. Hinds, S. Mittnacht, and T. A. Weinberg, The retinoblastoma protein and the regulation of cell cycling. *Trends Biochem. Sci.* 17:312–315, 1992.

Cooper, G. M., *Elements of Human Cancer.* Boston: Jones and Bartlett, 1991.

Dobner, T., N. Horikoshi, S. Rubenwolf, and T. Shenk, Blockage by adenovirus E4 or f6 of transcriptional activation by the p53 tumor suppressor. *Science* 272:1470–1473, 1996.

Donehower, L. A., M. Harvey, B. L. Slagle, M. J. McArthur, C. A. Montgomery, J. S. Butel, and A. Bradley, Mice deficient for p53 are developmentally normal but susceptible to spontaneous tumours. *Nature* 356:215–221, 1992. A remarkable new technique for obtaining null mutations demonstrates that embryogenesis is normal in the absence of p53. However, animals develop a variety of neoplasms in the first 6 months when p53 is lacking.

Downward, J., The ras superfamily of small GTP-binding proteins. *Trends Biochem. Sci.* 15:469–472, 1990.

Eck, M. J., S. K. Atwell, S. E. Shoelson, and S. C. Harrison, Structure of the regulatory domains of the Src-family tyrosine kinase Lck. *Nature* 368:764–768, 1994.

Eva, Y. H., P. Lee, N. Chang, Y.-C. Hu, J. Wang, C.-C. Lai, K. Herrup, W.-H. Lee, and A. Bradley. Mice deficient for Rb are nonviable and show defects in neurogenesis and haematopoiesis. *Nature* 359:288–294, 1992.

Feig, L. A., The many roads that lead to ras. *Science* 260:757–758, 1993. Ras can be activated by a number of different transduction pathways.

Friend, S. H., Genetic models for studying cancer susceptibility. *Science* 259:774–775, 1993.

Gallo, R. C., The AIDS virus, *Scientific American* 256(1):46–56, 1987.

Halauska, F. G., Y. Tsujimoto, and C. M. Croce, Oncogene activation by chromosome translocation in human malignancy. *Ann Rev. Genet.* 21:321–345, 1987.

Hamel, P. A., B. I. Gallie, and R. A. Phillips, The retinoblastoma protein and cell cycle regulation. *TIBS* 8:1880–1885, 1992.

Hartwell, L. H., and M. B. Kastan, Cell cycle control and cancer. *Science* 266:1821–1827, 1994.

Hausen, H., Viruses in human cancers. *Science* 254:1167–1173, 1991.

Henderson, B. E., R. K. Ross, and M. C. Pike, Toward the primary prevention of cancer. *Science* 254:1131–1144, 1991.

Hotomana, N., et al., Massive cell death of immature hematopoietic cells and neurons in Bcl-2-deficient mice. *Science* 267:1506–1508, 1995.

Jacks, T., A. Fazeli, E. M. Schmitt, R. T. Bronson, M. A. Goodell, and R. A. Weinberg, Effects of an Rb mutation in the mouse. *Nature* 359:295–300, 1992.

Kamb, A., Cell-cycle regulators and cancer. *TIBS* 11:136–140, 1995.

Krumm, A., and M. Groudine, Tumor suppression and transcription elongation: The dire consequences of changing partners. *Science* 269:1400–1401, 1995.

Lane, D. P., p53, guardian of the genome. *Nature* 358:15–16, 1992. Proposes that p53 monitors the integrity of the genome.

Levine, A. J., The tumor suppressor genes. *Ann. Rev. Biochem.* 62:623–651, 1993. Comprehensive but somewhat dated review because of rapid advances being made in this area of cancer research.

Linzer, D. I. H., The marriage of oncogenes and anti-oncogenes. *Trends Genet.* 4:245–247, 1988.

Liotta, L. A., Cancer cell invasion and metastasis. *Sci. Am.* 266:54–63, 1992. A most important aspect of carcinogenesis that we did not deal with in our chapter.

Mack, D. H., J. Vartikar, J. M. Pipas, and L. A. Laimins, Specific repression of TATA-mediated but not initiator-mediated transcription by wild-type p53. *Nature* 363:281–283, 1993. The p53 protein may repress the activity of certain promotors by direct interaction with TATA box-dependent transcription machinery.

Makela, T. P., P. J. Koskinen, I. Vastrik, and K. Alitalo, Alternative forms of max as enhancers or suppressors of myc-ras cotransformation. *Science* 256:373–376, 1992.

Malkin, D., F. P. Li, L. C. Strong, J. F. Fraumeni, C. E. Nelson, D. H. Kim, J. Kassel, M. A. Gyrka, F. Z. Bischoff, M. A. Tainsky, and S. H. Friend, Germ line p53 mutations in a familial syndrome of breast cancer, sarcomas, and other neoplasms. *Science* 250:1233–1238, 1990.

Marx, J., Two major signal pathways linked. *Science* 262:988–990, 1993. Cyclic AMP blocks transmission of signals from Ras to Raf-1.

Milburn, M. V., L. Tong, A. M. deVos, A. Brunger, Z. Yamaizumi, S. Nishimura, and S. H. Kim, Molecular switch for signal transduction: Structural differences between active and inactive forms of protooncogenic ras proteins. *Science* 247:939–945, 1990.

Milner, J., Flexibility the key to p53 function? *TIBS* 20:49–51, 1995.

Miyashita, J., and J. C. Reed, Tumor suppressor p53 is a direct transcriptional activator of the human *bas* gene. *Cell* 80:293–299, 1995.

Ottral, Z. N., C. L. Milliman, and S. J. Korsmeyer, Bcl-2 heterodimerizes in vivo with a conserved homolog, bax, that accelerates programmed cell death. *Cell* 74:609–619, 1993.

Paparassiliou, A. G., M. Trier, C. Chavrier, and D. Bohmann, Targeted degradation of *c-fos,* but not *v-fos,* by a phosphorylation-dependent signal on *c-jun*. *Science* 258:1941–1949, 1992.

Rabbits, T. H., Chromosomal translocations in human cancer. *Nature* 372:143–149, 1994.

Solomon, E., J. Borrow, and A. D. Goddard, Chromosome aberrations and cancer. *Science* 254:1153–1160, 1991.

Thangue, N. B., DRTF1/E2F: An expanding family of heterodimeric transcription factors implicated in cell-cycle control. *TIBS* 19:108–114, 1994.

Thompson, C. B., Apoptosis in the pathogenesis and treatment of disease. *Science* 267:1456–1461, 1995. Alterations in cell survival contribute to the pathogenesis of a number of human diseases, including cancer.

Thompson, T. C., S. H. Park, T. L. Timme, C. Ren, J. A. Eastham, L. A. Donehower, A. Bradley, D. Kadmon, and G. Yong. Loss of p53 function leads to metastasis in *ras* plus *myc*-initiated mouse prostate cancer. *Oncogene* 10:869–879, 1995.

Varmus, H. E., Reserve transcription. *Sci Am.* 257(3):56–64, 1987.

Weinberg, R. A., Finding the anti-oncogene. *Scientific American* 259(3):44–51, 1988.

Weinberg, R. A., The retinoblastoma gene and cell growth control. *Trends Biochem. Sci.* 15:199–202, 1990.

Weinberg, R. A., Tumor suppressor genes. *Science* 254:1138–1146, 1991. Provides an update on the mechanisms of action of a number of better known oncogenes.

Willinghofer, A., and E. F. Pai, The structure of ras protein: A model for a universal molecular switch. *Trends Biochem. Sci.* 16:382–387, 1991.

PROBLEMS

1. What are some of the differences between a cancer cell and a normal cell?
2. The methionine analog L-ethionine (which has an ethyl group instead of a methyl group attached to the sulfur) causes liver cancer in rats. This analog is a nonmutagenic carcinogen that inhibits cellular methylation. How might this analog cause cancer? (*Hint:* What role does DNA methylation play in gene expression?)
3. Renato Dulbecco suggested that DNA viruses that cause cancer do so by a mechanism similar to lysogeny of *E. coli* by λ virus, that is, by integration of the viral genome into host cell DNA. How can RNA viruses cause cancer?
4. Why is an oncogene like *myc* found associated with many different types of cancer, while the oncogene *sis* is found only in cancers with the PDGF cell surface receptor?
5. Explain why mutations of *ras* often result in a dominant oncogene (one that has transforming activity even when a wild-type copy of the gene is present), while oncogenic mutations in p53 are usually recessive.
6. What are some of the ways a cellular proto-oncogene can be converted into an active oncogene?
7. How does overexpression of the *src* gene cause cancer?
8. The AIDS virus (HIV) can cause many types of cancers (Kaposi's sarcoma, B-cell lymphoma, etc.), yet HIV does not have a viral oncogene. Explain how HIV can cause cancer.

39

THE HUMAN IMMUNODEFICIENCY VIRUS (HIV) AND ACQUIRED IMMUNODEFICIENCY SYNDROME (AIDS)

Successful therapies for the prevention or treatment of AIDS may require a breakthrough in the technology.

For thousands of years, viruses have been the scourge of humankind, causing diseases that have taken their steady toll in suffering and life. Toward the middle of this century it seemed as though we were reaching a point where we could handle almost everything that nature had in its arsenal and could focus medical research on degenerative maladies such as cancer and heart disease. Then the AIDS pandemic erupted in 1981. This slowly growing pandemic promises to take many human sacrifices before it is brought under control. It makes us realize that viruses of all sorts are not just standing still; they are constantly evolving to seek out new hosts on which they can thrive. Now that humankind is the dominant animal species on the planet, we may safely assume that we are the prime target of these predators. Furthermore, the adaptability of viruses, measured by their replication and mutation rates, far exceeds ours. By these measures it would seem that we are very likely to be overwhelmed. Yet there is one way in which we are very superior. Our intellect gives us a means of defending ourselves. Without it the AIDS attack might wipe out over half the human race or perhaps even eliminate the species as more virulent forms develop. With our intellect we will very likely bring this pandemic under control within the next 10 years, though not without considerable effort and loss of life. This experience highlights the importance of biochemistry and associated disciplines to our very survival.

This chapter is divided into three main sections: the discovery and incidence of AIDS, the molecular biology of the HIV virus, and the present status and future prospects for HIV therapies.

Discovery and Incidence of AIDS

In early 1981 a handful of cases of bacterial pneumonia were reported to the National Centers for Disease Control (CDC) in Atlanta. These cases were all in young homosexual men from the same area of Los Angeles. Following this, similar reports began to trickle in involving homosexual men living in New York City, San Francisco, and Los Angeles. In addition to bacterial pneumonia these victims had other opportunistic infections that would not be commonly observed in normal young men. Usually, such ailments are limited to individuals with impaired

Figure 39.1

Estimated distribution of HIV-infected adults in different parts of the world as of mid-1993. (*Source:* M. H. Merson, Slowing the spread of HIV: Agenda for the 1990s, *Science* 260:1266, 1993.)

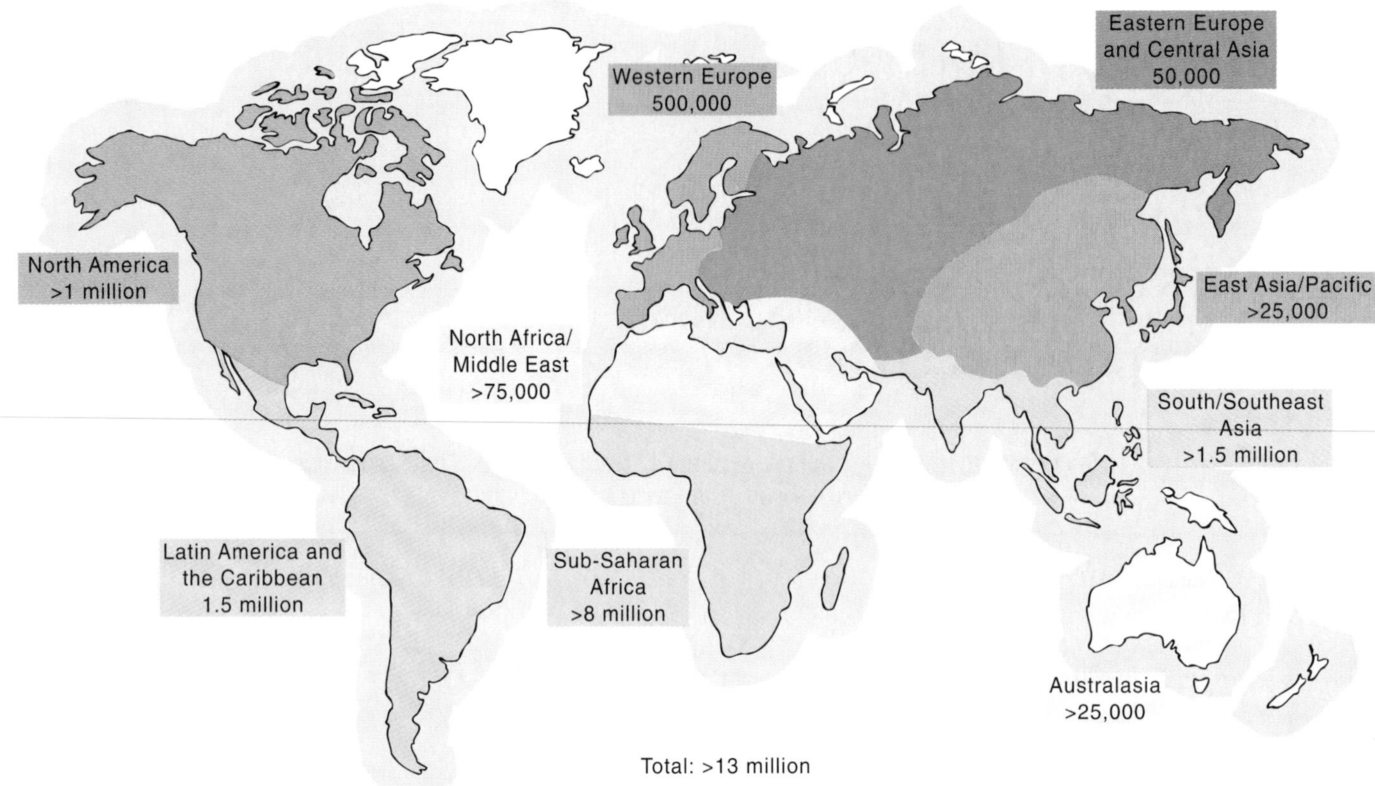

immune systems. This would include individuals born with immune deficiencies, transplant patients receiving immunosuppressive drugs, and cancer patients receiving chemotherapy. In late 1981 a report appeared in the *New England Journal of Medicine* that correlated this condition with a decreased cell count of circulating $CD4^+$ T cells. Soon thereafter, the CDC proposed that this newly recognized disorder be called acquired immunodeficiency syndrome (AIDS).

Although AIDS was first identified in homosexual men in the United States, it soon began to be reported in other groups, including users of intravenous drugs, recipients of blood transfusions, and sexual partners of people with AIDS. Collectively, these findings suggested that AIDS is transmissible through intimate contact.

In a few years, similar reports of AIDS began to appear in many other sectors of the United States and the world. By 1993, more than a million AIDS cases had been reported in the United States, and over 10 million cases had been reported worldwide (fig. 39.1). At the present time (1996) it is estimated that more than 2% of the world's population may have acquired this usually fatal disease.

AIDS Is Associated with a Retrovirus

It should be recalled that retroviruses are RNA viruses whose life cycle involves a DNA intermediate that is integrated into the host genome (see chapter 31). Retroviruses gained prominence in the 1970s when they were discovered and found to constitute a major class of cancer-causing viruses (see chapter 38). In researching AIDS, it was not long before a new retrovirus was identified as being associated with AIDS. The first identification was made at the Pasteur Institute by Luc Montagnier and his co-workers, who isolated a retrovirus from an AIDS patient in 1983. Soon thereafter, Robert Gallo's group at the National Cancer Institute in Bethesda, Maryland, identified a retrovirus in tissue cultured from an AIDS patient. Subsequent studies revealed that the two viral isolates were representatives of the same virus (fig. 39.2). In 1986 an international committee named this virus the human immunodeficiency virus, or HIV. Following the discovery of an antigenic variant in 1986, the originally isolated viruses were designated HIV-1, and the variant was designated HIV-2. Several closely related simian viruses (SIVs) were discovered in monkeys and chimpanzees.

Figure 39.2

AIDS virion structure. Numbers associated with proteins are kilodalton masses (e.g., gp120 is a 120-kd protein). (From W. C. Greene, Mechanisms of disease: The molecular biology of human immunodeficiency virus type 1 infection, *New England Journal of Medicine* 324(5):309, 1991. Copyright 1991 Massachusetts Medical Society. Reprinted by permission of The New England Journal of Medicine.)

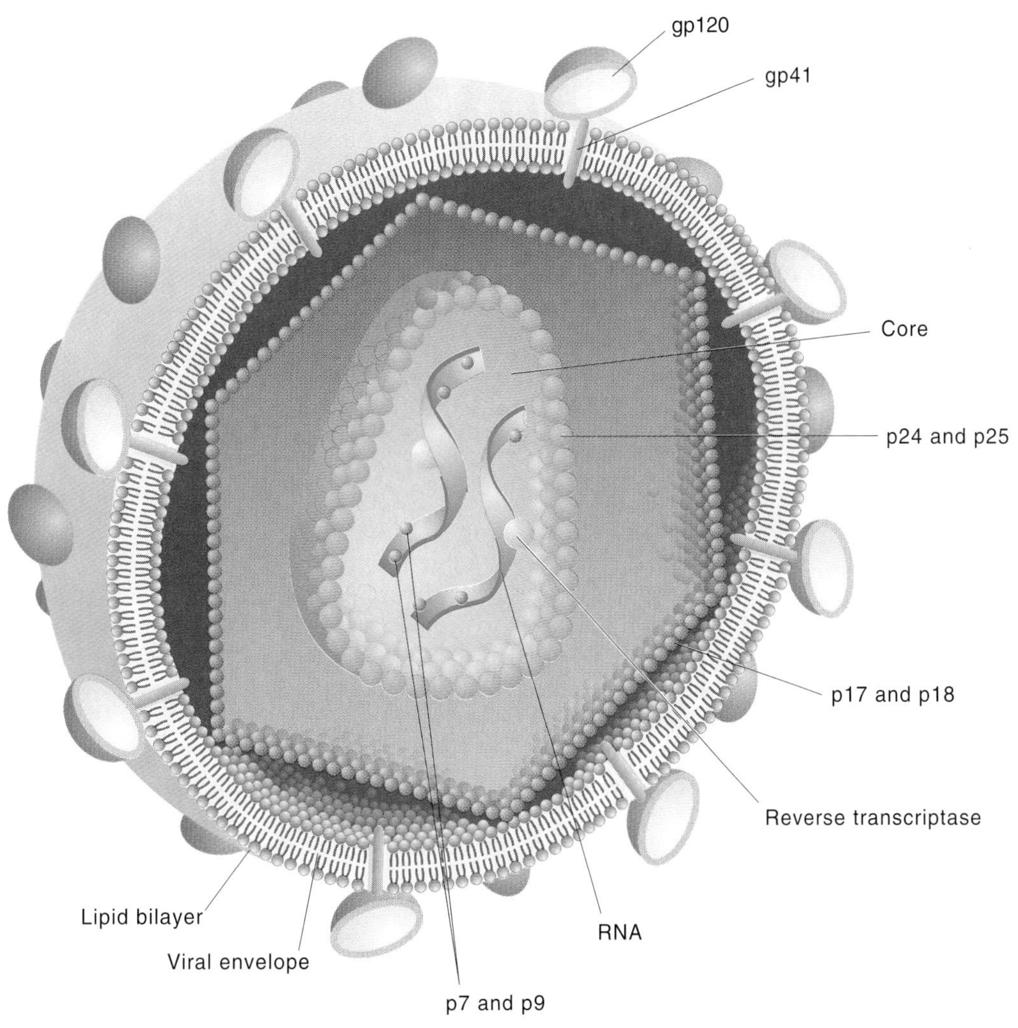

Clinical Diagnosis of AIDS

The finding that the AIDS syndrome is invariably associated with a well-defined virus greatly aided in the clinical diagnosis. The distribution of the HIV virus mirrors the spread of AIDS (see Fig. 39.1). Although many more cases of AIDS have been reported in Africa than elsewhere, it is not clear that AIDS started in Africa; it appeared in many parts of the world almost simultaneously. This no doubt attests to the mobility of the human population. Notwithstanding the question of origin, it is remarkable that as of 1991 the proportions of men and women infected with HIV in sub-Saharan Africa and the Indian subcontinent were approximately equal, while in other locations the disease is four to ten times more common in men than in women (table 39.1). Whether this is due to different life styles or to a longer presence of AIDS in Africa and India is not known.

Table 39.1
Estimated Proportion of Men and Women Infected with HIV[a]

	Men	Women
North America	1 in 75	1 in 700
South America	1 in 125	1 in 500
Western Europe	1 in 200	1 in 1400
Sub-Saharan Africa	1 in 40	1 in 40

[a]Incidence by sex is for adults aged 15 to 49.

Source: WHO Communicable Disease Scotland Weekly Report, 25/8/90, and *Science,* 25:372, 1991.

Early manifestations of HIV infection differ from individual to individual. HIV infection sometimes causes an acute mononucleosis-like illness, which is generally followed by an asymptomatic latency period that usually lasts for several years. Often, however, individuals manifest no obvious symptoms upon initial HIV infection and pass without notice into the asymptomatic period of latency. Finally, a significant number of individuals who have been infected with HIV show a persistent enlargement of multiple lymph nodes but no concurrent illness. In all three cases the individual is infected with HIV but is not yet diagnosed as having AIDS.

The initial clinical manifestations of HIV infection pass in a matter of weeks. This is followed by a prolonged asymptomatic period, which may last for 8 to 12 years. During this period, very little virus is present in the peripheral blood. Following the asymptomatic period, virus reappears in the blood, accompanied by a general collapse of the immune system. As a result, the victim becomes susceptible to any one of a number of infectious diseases, and death usually follows from complications in one to three years.

Although the HIV virus is not associated with any specific oncogenes like many other retroviruses (see chapter 38), certain cancers are known to be associated with HIV infection. This is believed to be due to the deficiency in cell-mediated immunity. Kaposi's sarcoma stands out in this category, appearing in approximately half of all AIDS patients. Other cancers that are associated with AIDS include non-Hodgkins lymphomas, Burkitt's lymphoma, and immunoblastic sarcoma. Incidentally, the high frequency of cancer associated with AIDS is one of the best indications that the immune system normally protects us from many types of cancers.

Is HIV Sufficient to Cause AIDS?

There are several reasons why some investigators have questioned whether HIV infection causes AIDS. First, there is no experimental animal in which the course of the disease can be followed after deliberate virus infection. Second, there is a very extensive latency period after initial infection, during which time the HIV virus is often difficult to detect in the peripheral blood system. Third, we do not know precisely how HIV causes AIDS, although tremendous progress has been made in that direction in the last few years. Finally, some cases of severe immune deficiency have come to light in individuals who have tested negative for HIV-1 and HIV-2 even by the most sensitive PCR tests.

It seems likely that rare cases that look like AIDS but in which no HIV virus can be detected at any stage of the disease must have a different cause. Nevertheless, this does not negate a causal relationship between HIV and the global AIDS pandemic. The epidemiological evidence for a causal relationship is most compelling. Likewise, the fact that more than 50% of those who have been infected with HIV for 10 years or more have progressed to AIDS strongly supports a causal relationship (fig. 39.3). By comparison the only other known human retroviral pathogen, human T cell leukemia virus (HTLV), causes adult T cell leukemia in fewer than 6% of the infected people during lifelong infection. In spite of this low percentage, few doubt the role of the virus because T cell leukemia is much more likely among people who have been infected with HTLV than the general population.

Figure 39.3

Proportion of individuals surviving without AIDS plotted with data combined from several U.S. and United Kingdom cohorts (*5, 62*). ●, HIV^+ homosexual men; ▼, HIV^+ hemophilic men who were > 20–25 years of age at time of infection; ▲, HIV^+ hemophilic men who were < 20 years of age at time of infection; and □, HIV^- hemophiliacs and homosexuals. (Reprinted with permission from R. A. Weiss, How does HIV cause AIDS? *Science* 260:1274, 1993. Copyright 1993 American Association for the Advancement of Science.)

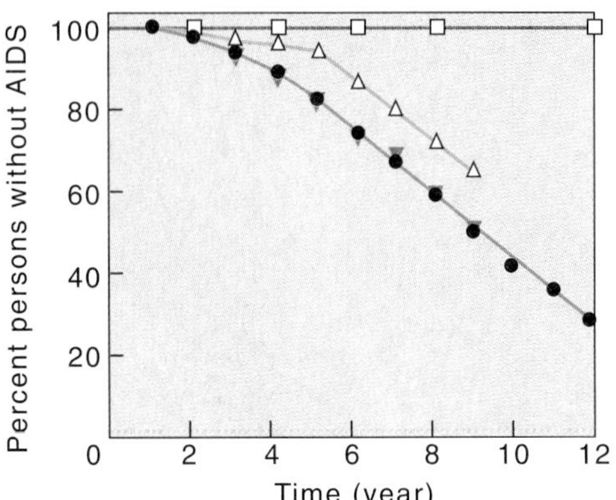

HIV Belongs to the Cytopathic Subgroup of the Retrovirus Family

Retroviruses fall into two groups: transforming and cytopathic. The transforming retroviruses carry oncogenes and induce changes in cell growth that lead to cancer. The best-studied human viruses of this group are HTLV-1 and HTLV-2, which cause T cell leukemia. Infection of T lymphocytes with either of these viruses causes the cells to express a receptor for interleukin 2 (IL-2), so these cells, which secrete IL-2 in the activated state, stimulate their own division in an unregulated way.

The cytopathic retroviruses are members of the lentivirus family. One branch of this group includes human immunodeficiency virus (HIV-1 and HIV-2) and simian immunodeficiency virus (SIV). HIV-1, which is epidemic in Central Africa, the Americas, Europe, Asia, and Haiti, infects both humans and chimpanzees but causes immune suppression only in humans. HIV-2 was originally isolated in West Africa; few cases have been reported outside of Africa. Most HIV-2 strains appear to spread more slowly than HIV-1 and to be less pathogenic. HIV-2 infects humans, chimpanzees, macaque monkeys, and baboons and thus has a broader host range than HIV-1. The infection of macaque monkeys with some strains of HIV-2 results in symptoms of immune suppression and holds promise as an animal model for studying AIDS.

HIV-1 and HIV-2 exhibit only 40–50% sequence homology. The HIV-2 sequence appears to be more closely related to SIV, showing 75% homology with most known SIV strains. From sequence data it seems likely that SIV arose first as a non-

Figure 39.4

Time course of events following HIV infection. The upper graph indicates the cell counts for CD4$^+$ peripheral blood lymphocytes (PBL). The lower graph indicates the new factors that appear as a result of the infection. CTL = cytotoxic T lymphocytes. (Reprinted with permission from R. A. Weiss, How does HIV cause AIDS? *Science* 260:1274, 1993. Copyright 1993 American Association for the Advancement of Science.)

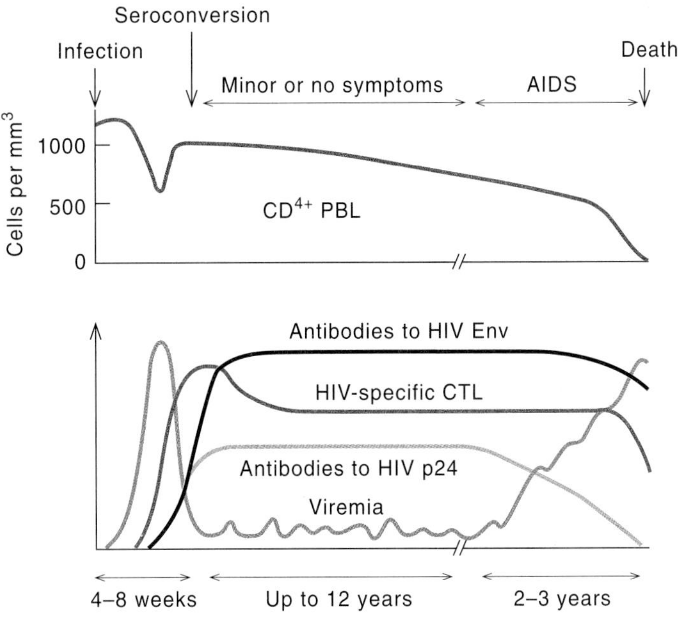

pathogenic lentivirus in African nonhuman primates and that transmission of SIV to humans gave rise to HIV-2.

Molecular Biology of the HIV Virus

Before the discovery of the HIV virus, many viruses, both bacterial and eukaryotic, had been studied and characterized in molecular detail. This information and the techniques used to obtain this information have been immensely useful in speeding up the characterization of the HIV virus. Currently, we probably know more about the HIV virus than any other virus. However, because of its complex life style and the nature of the disease it causes, a great deal of information is still lacking.

Tissue Specificity of HIV

When we speak of the host range of an animal virus, we are concerned with three things: the species range that it infects, the types of cells within a given species that it infects, and the consequences of the infection. We have touched on most of these topics, but there are still major points to raise that relate to the types of cells that the HIV virus infects in humans. The main cells infected are CD4$^+$ helper T lymphocytes and macrophages. T lymphocytes and macrophages become infected via the CD4 surface receptor, as shown by the ability of monoclonal antibodies to CD4 to block HIV entry into these cells in culture.

Several CD4$^-$ cell types can be infected by HIV in vitro, including epithelial and endothelial cells as well as astroglial, oligodendroglial, and neuronal cells from the central nervous system, in which the galactocerebroside molecule acts as an HIV receptor. Some of these CD4$^-$ cell types may play a critical role in the transmission of the HIV virus from one human to the next.

Electron microscope and immunofluorescence studies indicate that HIV is found in massive amounts in the lymph nodes, even in the asymptomatic phase of infection. In addition to helper T lymphocytes and macrophages, virus particles are associated with follicular dendritic cells. Infected macrophages could be important reservoirs outside the blood and as carriers of HIV to different organs. These nonproliferating mature cells can sustain HIV production in vitro. Proliferating mature cells can also sustain HIV production in vitro for a considerable time without being killed by the virus.

Course of the HIV Infection Leading to AIDS

Figure 39.4 depicts the progression of HIV infection from initial infection to AIDS, focusing on the condition of the peripheral blood. At the time of seroconversion, IgM antibody to HIV antigens can be detected. Within a few weeks, antibody of the IgG class appears that is specific for many of the viral structural proteins, including gp160, gp120, gp41, p24, and p17 (table 39.2). The appearance of antibodies to p24, a core protein, is most useful because its presence correlates with the period of viral latency. As the antibody to p24 begins to decline, which can take many years, there is a corresponding increase in the appearance of p24 antigen in the serum. The decline in antibody to p24 and

Table 39.2
HIV Genes and Their Encoded Proteins and Functions

Gene	Immediate Protein Product (kD)	Functions of Final Processed Protein Products
gag	53	Nucleocapsid core products: p17, p24, p9, and p7
env	160	Envelope glycoproteins: gp120 and gp41
pol	Precursor	Enzymes:
		p66 reverse transcriptase and RNaseH
		p51 reverse transcriptase
		p10 protease
		p32 integrase
vif	23	Promotes infectivity of cell-free virions
vpr	15	Weak transcription activator
tat	14	Strong transcription activator
rev	19	Controls splicing
nef	27	Down-regulates transcription
vpu	16	Required for efficient viral assembly and budding

increase in p24 antigen are associated with the progression from latency into lytic infection and have been used to clinically predict the onset of the terminal phase of the disease.

Is there a true latency state of HIV infection? It is likely that some T cells harbor genuinely latent HIV genomes as proviruses that do not express viral RNA. In vivo, latently infected lymphocytes can be detected that carry complete, integrated provirus. These are probably memory cells that may have become infected while in an active state. Notwithstanding the true latency of HIV in individual resting lymphocytes, the infection is not latent during the asymptomatic phase. Actively expressed HIV is found in lymph nodes and other lymphoid organs and in tissue macrophages at all stages of infection, indicating a much higher level of virus activity than in circulating T cells.

The final phases of the disease, in which the viral antigens reemerge in the bloodstream, signals the collapse of the lymph node defense system. In this phase of the disease the number of circulating $CD4^+$ T cells is very low, sometimes below detection limits.

The HIV Genome

Most retroviruses carry three genes designated *gag, pol,* and *env* that encode the viral core proteins, the enzymes required for replication, and the envelope proteins, respectively. Each of these genes encodes a polyprotein precursor that becomes cleaved to render the final gene products. The pol polyprotein is cleaved to generate three enzymes: reverse transcriptase, protease, and integrase. The *gag* gene encodes a precursor polyprotein that is cleaved by the viral protease to yield p24, p7, p9, and p17; p17 and p24 constitute the protein core of the viral particle. The *env* gene encodes a glycosylated precursor polyprotein (gp160), which becomes cleaved to yield gp120 and gp41.

The HIV genome is more complicated than most retroviruses (fig. 39.5). In addition to the basic set of genes, *gag, env,* and *pol,* it harbors six additional genes: virion infectivity factor (*vif*), viral protein R (*vpr*), transactivator (*tat*), regulator of expression of virion proteins (*rev*), negative expression factor (*nef*), and viral protein V (*vpu*) (see table 39.2). *Tat* and *rev* are overlapping genes that are translated in different reading frames. The *tat, rev,* and *nef* genes encode regulatory protein that control the expression of the structural genes *gag, pol,* and *env.* The protein encoded by *tat* up-regulates transcription, while the protein encoded by *nef* down-regulates transcription. The protein encoded by *rev* promotes transcription of the viral structural proteins that are necessary for viral assembly in the lytic state. The *vif* and *vpu* genes encode proteins required for virion maturation. Finally, the *vpr* gene encodes a regulatory protein that is a weak transcriptional activator. The proviral genes are flanked by tandem repeats (LTRs) that contain the sequences for controlling viral expression and integration.

The HIV Virus Life Cycle

The virus life cycle includes events from the time of contact with a target cell to the time at which the virus progeny leave that cell to infect another cell. In the case of the HIV virus this is complicated by the fact that the virus can adopt a dormant state as a provirus and by the fact that the virus can infect a variety of cell types with quite different outcomes. It seems likely that the regulatory factors that control the life cycle may function differently in different cell types. We will not attempt to cover all of these topics but will focus on some of the main ones.

Virus Entry. When HIV encounters a cell with a high-affinity CD4 receptor, a specific binding reaction occurs between the virus envelope glycoprotein, gp120, and the receptor molecule. CD4 is a surface glycoprotein found on a variety of cells of hematopoietic origin (see chapter 37). The CD4 protein is present in low concentrations on macrophages and antigen-presenting dendritic cells. It is present at high concentrations on the surface of immature T lymphocytes as well as on the surface of mature lymphocytes of the $CD4^+$ circulating T helper cells. HIV-1 also infects some cells that appear to lack CD4 via a low-efficiency reaction that is dependent on fusion activity of the virus envelope glycoprotein. This may be quite important in the transmission of the disease from one individual to another or in the transmission of the virus from one cell type to another in the same individual.

Viral entry occurs by fusion of virus and cell membranes. The fusion reaction is mediated by the envelope glycoprotein of the virus. The exterior subunit of the virus (gp120) specifies binding to CD4, whereas the transmembrane protein of the envelope glycoprotein (gp41) mediates the fusion event.

Selective Killing of Infected Cells. The envelope glycoprotein–mediated reactions are often toxic for cells that have an abun-

Figure 39.5

Location of genes in the proviral DNA of the HIV virus. (From W. C. Greene, AIDS and the immune system, *Sci. Amer.,* September 1993, page 105. Copyright © 1993 by Scientific American, Inc. All rights reserved. Reprinted by permission.)

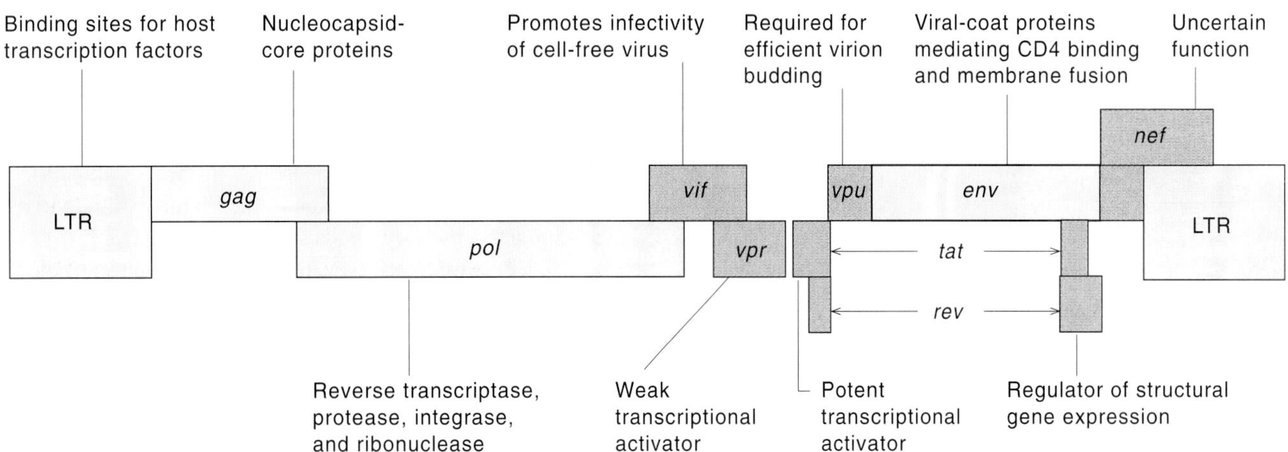

dance of surface CD4 proteins. Two types of killing occur. If a single cell produces large amounts of virus and has a high concentration of surface CD4, a fusion may occur between the budding virus and the surface CD4. This fusion reaction can destroy the integrity of the cell membrane and result in cell death. Cell death can also result from a virus-producing cell fusing via an envelope glycoprotein–mediated reaction to a $CD4^+$ cell. This fusion reaction of the infected cell mimics the early envelope glycoprotein–mediated virus-to-cell-membrane fusion that occurs early in infection. One infected cell may fuse successively with up to several hundred uninfected $CD4^+$ T cells, resulting in a large multinucleated giant cell that has a short half-life. This is one way in which uninfected T helper cells get eliminated in an HIV-infected organism, but it is probably not the main one.

Synthesis and Integration of Viral DNA. The single-stranded viral RNA is converted into a double-stranded DNA copy by the concerted action of the viral reverse transcriptase and the viral ribonuclease H. The DNA copy of the viral genome is inserted into the host genomic DNA by a process that is mediated by the virus enzyme integrase. Integrase is packaged within the virus particle. Since integrase is needed prior to the synthesis of new viral proteins, it is clear why this enzyme is packaged along with the other enzymes involved in viral DNA synthesis.

Activity of the Viral Promoter Is Linked to the State of Activation of the Host Cell. Virus expression begins with synthesis of a complete RNA copy of proviral DNA. The site of transcription initiation is located in the terminally redundant viral sequences, the long-terminal repeats (LTRs) that flank the virus sequences that code for virus proteins. The LTR upstream of the coding sequences is the promoter for the virus. The LTR downstream of the coding sequences plays some indeterminant role depending upon where it is integrated into the host genome.

The HIV-1 LTR promoter is a weak promoter in most cell types. Therefore very little RNA is made, and what is made is translated poorly. The HIV-1 LTR can become an active promoter capable of synthesizing large amounts of viral RNA in activated T cells. This results partly from a change in the cellular factors that recognize sequences of the promoter. In addition, the virus specifies a protein called the transactivator (tat), which prevents RNA polymerases from falling off during the course of transcription. The transactivator may also increase the translation efficiency of viral RNAs that are made.

The promoter region and the proteins it binds in the active state are indicated in figure 39.6. The core promoter and enhancer, containing the TATA box and binding sites for Sp1 and NF-kB, span a region of 250 nucleotides. Sp1 is a transcription activator found in a wide variety of cell types. NF-kB is more specialized; the NF-kB enhancer was originally discovered as a κ-immunoglobulin enhancer that correlated with κ gene transcription. Since then, NF-kB has been shown to activate a number of genes in several tissues. NF-kB is situated in the cytoplasm in most cells in the inactive form, complexed to an inhibitor named IkB. Stimulation by a number of agents results in the dissociation of the IkB–NF-kB complex and correlated transcription of target genes. This activation is thought to lead to release of IkB from the complex, with subsequent translocation of NF-kB to the nucleus.

In addition to playing a major role in HIV transcription, NF-kB takes part in the activation of several T cell genes required for T cell proliferation, including the genes encoding IL-2 and the high-affinity IL-2 receptor. This dual action links the state of HIV activation to the state of T cell activation.

Although sequences upstream of the NF-kB sites contribute only marginally to HIV-1 promoter activity in vitro or in transfected cells, indications are that the upstream sequences between positions -130 and -201 are nevertheless important for viral replication (or proviral promoter activity) in peripheral blood lymphocytes (PBLs) and in some T cell lines. This region of the promoter includes the binding sites for two activators that are highly enriched in T cells: LEF, a lymphocyte-specific protein that is found in immature B and T cells and in mature T cells, and the

Figure 39.6

Structure of the HIV promoter. Labels designate different transcription factors. (Reproduced, with permission, from the *Annual Review of Biochemistry*, Volume 63, © 1994, by Annual Reviews, Inc.)

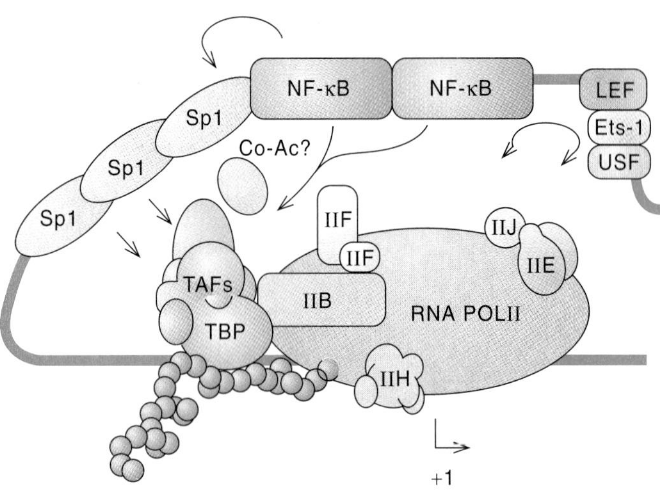

thymocyte-enriched Ets-1 protein. In addition, this region contains a DNA sequence for binding the cellular protein USF. LEF generates a strong (130°) bend in the DNA that may influence the local structure of the promoter. The LEF activation domain is preferentially active in T cells and is strongly influenced by the context of its binding site, indicating that it may act in concert with other T cell–specific proteins such as Ets-1. Transient expression assays indicate that overexpression of LEF strongly activates the HIV-1 promoter. This part of the HIV-1 promoter is similar to the enhancer for the α chain of the T cell receptor, which also binds LEF and Ets-1.

Regulation of Splicing Regulates Translation. The initial full-length RNA that is made sustains multiple splicing events. These multiply spliced viral mRNAs are exported to the cytoplasm from the nucleus where they are translated. The translation products of these small mRNAs include the tat protein and two other regulatory proteins called rev and nef. The tat protein accelerates the rate of viral RNA transcription throughout the course of the infection.

The regulatory protein rev has a profound effect on the fate of the primary RNA transcripts within the nucleus. In the absence of rev, only multiply spliced RNAs accumulate in the cytoplasm. In the presence of rev, unspliced and singly spliced viral messenger RNAs can accumulate in the cytoplasm. These unspliced and singly spliced RNAs are essential for the translation of the capsid and envelope glycoproteins of the virus.

Full-length viral RNA is used as a template to make the capsid protein and the replicative enzymes. The capsid proteins are made as a polyprotein precursor. The replicative enzymes are made as a fusion protein with a capsid protein precursor. Different but overlapping reading frames specify the capsid and replicative enzymes.

The viral envelope glycoprotein is made from singly spliced viral messenger RNA. A leader sequence on the newly synthesized envelope glycoprotein directs the ribosomes to the rough endoplasmic reticulum. The newly synthesized envelope glycoprotein precursor assembles into a dimer on the rough endoplasmic reticulum. Complex carbohydrates are added as the glycoprotein traverses the Golgi apparatus. The envelope glycoprotein is eventually deposited on the surface of the cell, where it concentrates on the membrane in the vicinity of the virus particle; subsequently, the virus buds from the cell surface.

Late maturation events include activation of the viral protease, which cleaves the polyproteins for capsid proteins and the capsid replicative enzyme precursor. Finally, there are two viral genes, *vif* and *vpu,* that act late in the virus life cycle to facilitate virus release and to increase virus particle infectivity. It is not quite clear how these associated virus-encoded proteins work.

Present Status and Future Prospects for the Prevention and Treatment of AIDS

The total collapse of the immune system in AIDS reflects the central role of $CD4^+$ helper T cells in both humoral and cell-mediated responses and in the regulation of both responses. We discussed the importance of helper T cells to humoral immunity in chapter 37. Another class of T helper cells that secrete IL-2 is essential for the conversion of $CD8^+$ Tc cells into cytotoxic lymphocytes. IL-2 also stimulates the proliferation of natural killer (NK) cells, which are important in nonspecific killing of tumor cells. Because of their decreased level of IL-2, AIDS patients have diminished NK-cell activity. This probably explains why various types of tumors are frequently observed in the terminal phase of AIDS.

The precise mechanism by which HIV infection manages to kill off virtually all of the $CD4^+$ cells may not matter in understanding how HIV causes AIDs, but it may be important for developing a therapeutic strategy. Direct killing could result from immune attack on virus-infected cells. Indirect killing of uninfected cells could start with adsorption of gp120 virus envelope protein produced in great excess to uninfected cells bearing the CD4 receptors. Thus the eventual destruction of most or all of the $CD4^+$ T cells need not require infection of more than a small fraction of them. If this explanation is correct, the virus system has clearly outwitted the immune system, so we must intercede to outwit the virus.

We *have* interceded; more funds have probably been allocated to research on AIDS than to research on any other disease with a single cause. As a result, more is known about the virus that causes AIDS than about any other virus. In spite of this we do not know for sure how this virus is transmitted from one individual to another. Nor do we understand the chain of events that ultimately leads to the destruction of the infected individual's immune system.

So far, what we have learned about disease prevention concerns social habits more than anything else. Treatment of infected victims has also been meager. The disease has become much easier to diagnose since the discovery of the HIV virus. No vaccines have thus far proved to be effective. Drug therapies have shown very limited effectiveness with one exception. In cases in which a pregnant mother has AIDS, AZT therapy is approximately 80% effective in protecting the unborn fetus. Many vaccines and drug therapies are undergoing clinical trials. Many more drugs are being screened in tissue culture systems.

In this section we discuss some of the problems, some of the proposals for prevention or cures, and what we consider to be the most promising approaches. The discussion is divided into three sections: immunotherapy, drug therapy, and gene therapy.

Immunotherapy

If one were designing a disease that would be most difficult to guard against or cure, AIDS would come close to perfection. It attacks the immune system in such a way that it pits one part of the immune system against the other and ends up by destroying the entire system. Since transmission of the HIV virus requires intimate contact, it should be possible to minimize transmission if people's habits could be changed. Unfortunately, this is most difficult to do. An individual who may be spreading the disease may pass through a seemingly normal period in which he or she has many opportunities to transmit the disease. Indeed, the people who are spreading the disease may not even know that they are infected. A vaccine that would give immunity would provide the ideal method for preventing spread of the disease, but the development of an effective vaccine against AIDS has proved to be anything but routine.

There are several problems with the vaccine approach. Most vaccines prepared thus far have not passed clinical trials. A great deal of the problem here may be that the virus mutates so rapidly that it inevitably escapes from antibodies directed against the initial antigen(s) used to develop the vaccine. Another problem may be that the initial contact with HIV is more likely to be with cells that carry the virus than with naked virus.

Since the HIV virus is intimately involved with the immune system, it seems likely that the longer we investigate the HIV virus, the more we will learn about the immune system. The situation is analogous to progress in cancer research. After 40 years of intensive investigations we still have no effective means for treating most solid tumors, but the research in search of a cure has been very illuminating as regards understanding how cell growth is regulated.

In the 1980s, Tim Mosmann and Robert Coffman found that they could clone two distinct subtypes of helper T cells in mice that were identifiable because each pumped out a different set of cytokines. One subtype of helper T cells secreted the cytokines interleukin-2 (IL-2) and γ-interferon. This is called the Th1 class. The Th2 subset secreted IL-4 and IL-5. The cytokines secreted by Th1 cells trigger cell-mediated responses that clear the body of infected cells. The Th2 cytokines activate the other arm of the immune system, leading to the production of antibodies that can prevent cells from becoming infected in the first place. It was also reported that a strong Th1 response inhibits a strong Th2 response and vice versa (fig. 39.7). It is not clear that these observations made on the mouse immune system are completely relevant to the human system. Nevertheless, the late Jonas Salk and his co-workers suggested a strategy of prophylactic vaccination against HIV that builds on this concept and additional observations.

Figure 39.7

The two types of helper T cells, the proteins they secrete, and the processes that are affected positively or negatively by these protein secretions. The negative cross-regulation between the Th1 and Th2 cells is speculative.

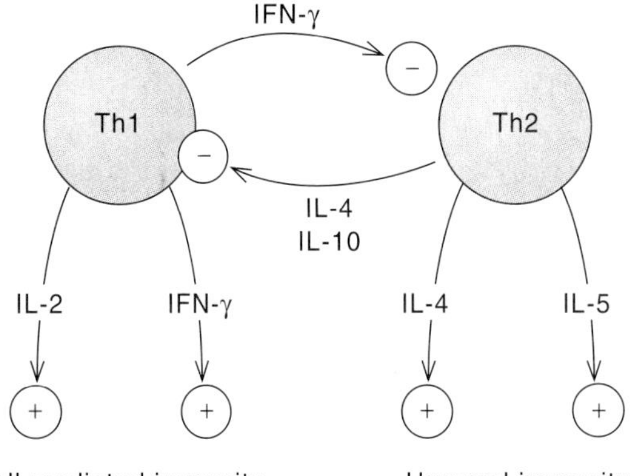

Some individuals who have been exposed to the virus and are therefore at high risk for HIV infection remain apparently uninfected; they do not have antibodies to HIV in their blood. Nevertheless, in one study of 97 such individuals who were seronegative for HIV, 49% exhibited cell-mediated immunity to HIV as evidenced by the fact that their T cells responded to HIV peptides in vitro. Such HIV-specific cell-mediated responses have been seen in homosexual men with known sexual exposure, intravenous drug users, health care workers exposed by accidental needle stick, and newborn infants of HIV-positive mothers. In a control group in which exposure to HIV seemed much less likely, only 2% of 163 individuals exhibited responses to these peptides.

Since infection with HIV under natural circumstances is thought to be transferred primarily by infected cells that are more likely to be susceptible to a cell-mediated defense, we must ask how a strong and stable cell-mediated Th1-type response against HIV could be mounted. One possibility is that low doses of antigen may promote a Th1 response. In general, small quantities of nonreplicating antigens, subimmunogenic for induction of antibody, induce only cell-mediated immunity.

Salk and his co-workers suggested that a low-dose vaccination regimen might generate cellular immunity exclusively, while a high-dose vaccination is likely to develop cellular and antibody immunity simultaneously. In the former case a subsequent challenge by the HIV virus is likely to be thwarted by the cellular immune system, minimizing any buildup of an antibody response. However, if the subject has been exposed to a high-dose vaccination, a subsequent challenge by the HIV virus is likely to strengthen the antibody response at the expense of the cellular immune system response.

Since cell-mediated responses against HIV-1 are directed, at least in part, against epitopes of viral proteins that are relatively invariant among different isolates, the approach suggested here could be protective against most or all HIV-1 variants.

Many other approaches to developing a vaccine suitable to conferring protection against the virus are being considered. A vaccine that contains multiple antigens, such as might be conferred by a live-attenuated virus vaccine, might be effective even if the virus mutates rapidly. One would not expect it to mutate so rapidly that it would escape from a mixture of antibodies to many different antigens of the virus.

While we are still on the subject of vaccines, it should be mentioned that M. A. Nowak and A. J. McMichael have suggested that the antigenic diversity generated by the HIV virus may be pathogenic because it eventually overloads the immune system. This notion seems unlikely, since the genetic diversity and load of SIV in asymptomatic African monkeys appears to be similar to those in humans, yet only humans develop AIDS. Therefore it seems improbable that the imaginary threshold of antigenic diversity has overloaded the human system in the case of HIV infection, especially since there is direct evidence that the lymph nodes are being systematically destroyed during the asymptomatic period of HIV infection.

Drug Therapies

A massive effort has been launched to find drugs that will selectively inhibit the HIV virus. The systematic way to go about this would be to determine the structure of the virus proteins intended to be targeted (which has been done by the crystallographers) and then design a drug that would bind to a site on the target protein and thereby inhibit its function. This approach is being pursued, but most of the effort that goes into this work is less systematic. Inevitably, drug testing proceeds through several stages before a drug is licensed. First, the candidate drug is tested in a tissue culture system to see whether it is effective in inhibiting HIV replication. If it passes this test, the drug may next be tested in an animal model system. If the drug proves to be effective against HIV in animals, it may be approved for testing on small, select groups of humans in clinical trials. Finally, if testing at this level shows promise, the drug may eventually become licensed for general use.

There is a great deal of debate among scientists and other interested parties regarding the testing of drugs for AIDS therapy. It has become a political football, and the stakes are high. The decisions that are made are not always the best ones scientifically. In general it would not be unreasonable to say that too much effort and hope have been pinned on finding a miracle drug that has not been adequately tested in preclinical trials. This has probably slowed progress.

In view of the rapid mutation rate of the virus, it seems likely that no one drug will do the job but rather that a combination of two or more drugs will be required. Although progress on this type of therapy has been slow, I am optimistic that drug treatments will become a dominant therapy once effective drugs or combinations of drugs administered simultaneously have been found.

Within the retroviruses, HIV is one of the most complex viruses; this is good for drug therapy, as it means that there are more potential targets. Currently, drugs that act on the viral reverse transcriptase have received the most attention. A drug that inhibits the reverse transcriptase could be used as a preventive drug as well as on people who have contracted the disease. As a preventive drug it might preclude the integration of the proviral DNA in the host genome. Once the proviral DNA has become established in the human host, it is hard to see how it could ever be totally eliminated unless there were a way of eliminating all of the cells into which it has become integrated.

Currently, all drugs that have been approved for use in HIV disease are nucleoside analogs that are targeted to the viral reverse transcriptase. This includes 3′-azidothymidine (AZT), 2′,3′-dideoxyinosine (ddI), and 2′,3′-dideoxycytidine (ddC). Following phosphorylation by a cellular kinase, these nucleotide analogs become incorporated into the DNA, which leads to premature termination. AZT has the longest history in AIDS treatment. It was identified in 1984 as active against HIV in cell culture. Clinical trials began shortly thereafter. Trials that were purported to show a survival advantage for individuals with advanced disease were completed in 1986, and the drug was licensed for general use shortly thereafter. Unfortunately, more recent expanded data indicate that AZT shows no difference in survival after three years of treatment.

The current status of AZT received a boost when it was discovered that AZT protects the unborn fetus of a diseased mother. The drug is approximately 80% effective in this capacity.

The combination of AZT and ddI has shown some promise. ddC, on a molar basis, is the most potent of the three approved drugs in cell culture assays and has been approved for use with AZT in adults with a reduced $CD4^+$ T cell count. None of the licensed reverse transcriptase inhibitors singly or in combination have proved to be entirely successful. One possible explanation for their poor showing is a failure to maintain adequate drug levels at the site of viral replication over extended periods. Another possible explanation for drug failure is the emergence of drug-resistant strains.

My own feeling is that properly designed drugs that bind to the reverse transcriptase will ultimately be found that are very effective. This belief stems from the fact that the reverse transcriptase is a virtually unique enzyme associated with the retrovirus. It should be possible to design potent drugs that would selectively block this enzyme.

Other protein targets that are currently receiving some attention in the drug design department include *tat* and *rev,* which are both key regulatory proteins affecting expression of the integrated proviral DNA.

Gene Therapy

The subject of gene therapy was briefly discussed in chapter 32. In general, gene therapy is potentially useful for the purpose of introducing a normal gene to replace a defective host gene or to introduce an antigene to inhibit the activity of a viral gene. In the case of therapy directed against the HIV virus the introduction of an antigene that would produce a segment of RNA that is complementary to one of the RNAs of the virus might interfere with the translation of that RNA and thereby create a block in the virus life cycle. Another approach would be to introduce a "gene" that would transcribe an RNA that would bind to the viral promoter to form a region of triple helix that would inhibit viral RNA transcription. It is possible to introduce natural or synthetic genes into cells in tissue culture and get them to express. The major stumbling block to gene therapy for humans is that one must hit most or all the target cells but in a highly selective manner. Delivery systems that will do this will probably take many years to develop. Despite this gene therapy is an approach with a tremendous potential, and research relating to it should be strongly supported.

No area of biological investigation has ever moved with more intensity than the investigations on the HIV virus in the past few years. It stands to reason that a good deal of current information on the behavior of this virus will be lacking from this text. The interested reader should stay abreast by referring to the journals *Science* and *Nature* which are particularly good at highlighting recent exciting discoveries.

I would like to conclude this brief chapter by citing some recent discoveries which give us cause for optimism that an effective treatment for AIDS may not be too far away. These discoveries grew out of observations made by Paul Maddon in the late 1980s. He and his colleagues showed that mouse cells genetically engineered to express human CD4 on their surfaces could not be infected with HIV. This finding launched a decade-long hunt for a "second receptor" that HIV uses in conjunction with CD4. In early 1996 W. A. Paxton and his coworkers described several individuals who had remained free from infection by HIV-1 despite repeated exposure through sexual intercourse with HIV-positive partners. When HIV-1 was applied to leukocytes from these exposed-uninfected subjects, it was found that cells from two of them were highly resistant to infection with strains of the virus that are believed to be important in person-to-person transmission. Following this Rong Liu and his colleagues showed that the two exposed uninfected subjects harbor identical mutations on both chromosomal copies of the gene that encodes chemokine receptor 5 (CCR-5), a member of the seven-transmembrane G-protein-coupled family. This observation indicates the importance of CCR-5 in HIV-1 transmission and suggests that targeting the HIV-1-CCR-5 interaction may provide a means of preventing or slowing disease progression.

Summary

AIDS is the most devastating disease to appear in the twentieth century. Normal biological defense mechanisms seem to be totally inadequate to cope with the HIV virus that causes AIDS. In this chapter we have considered the following: (1) the discovery of AIDS and its association with the HIV virus, (2) the molecular biology of the HIV virus, and (3) the present status and future prospects for the prevention and treatment of AIDS.

1. We first became aware of AIDS in 1981 as a result of numerous reports filtering into the Centers for Disease Control that involved opportunistic infections in young homosexual men. Soon thereafter, the syndrome was recognized in many countries around the world. In most countries the disease is far more common among men than among women.
2. Studies of diseased tissues quickly revealed the presence of a retrovirus that was named the HIV virus. Two variants of this virus have been found: HIV-1 and HIV-2. HIV-2 is closely related to a similar virus found in monkeys and chimpanzees.
3. The early diagnosis of AIDS would be difficult if it were not for the association with the virus. Any diagnosis that positively confirms the presence of HIV would probably be accepted at the present time. In the late stages of the disease many types of infections are likely because of the destruction of the immune system. Blood analysis should indicate a drop in the $CD4^+$ T cell count.
4. There may be other diseases in which there is an immune deficiency but no evidence of the HIV virus. In such rare cases it seems unlikely that one is dealing with the same disease. As far as we know, HIV infection inevitably leads to death, despite a long latency period. HIV virus infects a very wide range of cell types. However, the cells that it infects most readily are cells of the hematopoietic cell line that display the CD4 protein on their cell membranes. This membrane-bound protein is a receptor with a high affinity for the gp120 protein in the HIV virus envelope. Following entry the HIV virus replicates with different efficiencies in different cell types. Helper T cells exemplify a cell type in which the virus replicates most efficiently.
5. Following initial exposure to the HIV virus and after a few weeks, the HIV virus all but disappears from the peripheral blood, probably as a result of the immune system's defensive responses. During this period the virus is very active in the lymph nodes. After a number of years it finally becomes visible in the blood again, signaling the destruction of the lymph node defense system. At this stage the helper T cell population is usually drastically reduced.
6. The HIV genome carries the three genes *gag*, *pol*, and *env*, which are characteristic of most retroviruses. It also carries an additional six genes whose functions are primarily regulatory.
7. The life cycle of the HIV virus is typical of that observed for most retroviruses involving a DNA intermediate that is incorporated into the host genome prior to transcription of any of the viral genes.
8. The host regulatory proteins that activate the T cell for replication are in many cases identical to the regulatory proteins that activate the virus promoter for transcription. In this way, replication of the virus is linked to the replication of the host T cell.
9. No successful therapies for the prevention or treatment of AIDS exist. Currently, a great deal of effort is being exerted on three fronts: immunotherapy, drug therapy, and gene therapy.

SELECTED READINGS

Cohen, J., AIDS research: The mood is uncertain. *Science* 260:1254–1255, 1993.

Cohen, J., Controversy: Is KS really caused by new herpesvirus? *Science* 268:1847–1848, 1995.

Dyda, F., A. B. Hickman, T. M. Jenkins, A. Engelman, R. Craigie, and D. R. Davies, Crystal structure of the catalytic domain of HIV-1 integrase: Similarity to other polynucleotidyl transferases. *Science* 266:1981–1986, 1994.

Embretson, J., M. Zupancic, J. L. Ribas, A. Burke, P. Racz, K. Tenner-Racz, and A. T. Haase, Massive covert infection of helper T lymphocytes and macrophages by HIV during the incubation period of AIDS. *Nature* 362:359–362, 1993.

Fantini, J., N. Yahl, and J.-C. Chermann, Human immunodeficiency virus can infect the apical and basolateral surfaces of human colonic epithelial cells. *Proc. Natl. Acad. Sci. USA* 88:9297–9301, 1991.

Fenouillet, E., J. C. Gluckman, and I. M. Jones, Functions of HIV envelope glycans, *TIBS* 19:65–70, 1994.

Gougeon, M. L., and L. Montagnier, Apoptosis in AIDS. *Science* 260:1269–1270, 1993.

Greene, W. C., AIDS and the immune system. *Scientific American* September 3:99–105, 1993.

Haseltine, W. A., Molecular biology of the human immunodeficiency virus type I. *Faseb* 5:2349–2360, 1991.

Haynes, B. F., Scientific and social issues of human immunodeficiency virus vaccine development. *Science* 260:1279–1285, 1993.

Hill, C. M., and D. R. Littman, Natural resistance to HIV? *Nature* 382:668, 1996.

Johnston, M. I., and D. F. Hoth, Present status and future prospects for HIV therapies. *Science* 260:1286–1293, 1993.

Jones, K. A., and B. M. Peterlin, Control of RNA initiation of elongation at the HIV-1 promoter. *Ann. Rev. Biochem.* 63:717–743, 1994.

Ki, C. J., D. J. Friechman, C. Wang, V. Metelev, and A. B. Pardee, Induction of apoptosis in uninfected lymphocytes by HIV-1 tat protein. *Science* 268:429–431, 1995.

Kohlstaedt, L. A., J. Wang, J. M. Friedman, P. A. Rice, and T. A. Steitz, Crystal structure of 3.5 Å resolution of HIV-1 reverse transcriptase complexed with an inhibitor. *Science* 256:1783–1790, 1992.

Levy, J. A., Comprehensive review of HIV pathogenesis. *Microbiol. Rev.* 57:183, 1993.

Litvak, S., et al., Priming of HIV replication by tRNA: Role of reverse transcriptase. *TIBS* 19:114–118, 1994.

Maddox, J., Where the AIDS virus hides away. *Nature* 362:287, 1993.

Maddox, J., Duesberg and the new view of HIV. *Nature* 373:189, 1995. A distinguished virologist disputes the causal relationship of HIV to AIDS.

McClure, M. O., et al., HIV clearance in an infant? *Nature* 375:637, 1995.

Merson, M. H., Slowing the spread of HIV: Agenda for the 1990s. *Science* 260:1266–1268, 1993.

Moore, J., and R. Anderson, The who and why of HIV vaccine trials. *Nature* 372:313–314, 1994.

Nowak, M. A., and A. J. McMichael, How HIV defeats the immune system. *Scientific American* August:58–65, 1995.

O'Brien, C., HIV integrase structure catalyzes drug search. *Science* 266:1946, 1994.

Pantaleo, G., C. Graziosi, J. F. Demarest, L. Butuni, M. Montroni, C. H. Fox, J. M. Orenstein, D. P. Kotler, and A. S. Fauci, HIV infection is active and progressive in lymphoid tissue during the clinically latent stages of disease. *Nature* 362:355–358, 1993.

Peliska, J. A., and S. J. Benkovic, Mechanism of DNA strand transfer reactions catalyzed by HIV-1 reverse transcriptase. *Science* 258:1112–1118, 1992.

St. Clair, M. H., et al., Resistance to ddI and sensitivity to AZT induced by a mutation in HIV-1 reverse transcriptase. *Science* 253:1557–1559, 1991.

Salk, J., P. A. Bretscher, P. L. Salk, M. Clerici, and G. M. Shearer, A strategy for prophylactic vaccination against HIV. *Science* 260:1270–1272, 1993.

Wei, X., et al., Viral dynamics in human immunodeficiency virus type 1 infection. *Nature* 373:117–126, 1995.

Weiss, R. A., How does HIV cause AIDS? *Science* 260:1273–1278, 1993.

PROBLEMS

1. Diagram the life cycle of HIV.
2. The fundamental similarities among all mammalian retroviruses include their genomic structure (*gag, pol,* and *env* genes, flanked by LTRs). Why, then, do these retroviruses exhibit such dramatic differences in life cycle?
3. Explain the role of RNA splicing in the production of HIV particles.
4. The structure of AZT, one of the drugs used in the treatment of AIDS, is shown in table 26.4. Sketch the pathway for the conversion of this drug to its triphosphate derivative.
5. AZT-resistant mutations of the HIV reverse transcriptase map to the polymerase active site, to residues that are thought to be involved in DNA-template interactions. What is surprising about this finding?
6. Explain why retroviruses, such as HIV, exhibit a higher mutation frequency than the DNA genomes of their host cells.
7. Assuming that the cleavage site for the HIV protease is known, suggest how a peptide analog could be designed to use as an inhibitor of this protease. Why might such an inhibitor be effective in preventing the spread of HIV?
8. The HIV-1 promotor directs high-level transcription in frog oocytes, and the transcripts appear to be processed. However, the mRNA is not translated. The translational block is due to a nontranscribed sequence in the upstream region of the promoter. What is surprising about these results? Can you speculate about a possible explanation for this finding?

Appendix A
Some Landmark Discoveries in Biochemistry

In this appendix we list, in chronological order, some of the most important discoveries made in biochemistry during the past two centuries. It is impossible, for reasons of space, to give credit to every worker who has made a significant contribution, but it is possible to identify certain events as milestones and thus to show how progress in this field has accelerated with the passage of time.

1770–1774
Priestley showed that oxygen is produced by plants and consumed by animals.

1773
Rouelle isolated urea from urine.

1828
Wöhler synthesized the first organic compound, urea, from inorganic components.

1838
Schleiden and Schwann proposed that all living things are composed of cells.

1854–1864
Pasteur proved that fermentation is caused by microorganisms.

1864
Hoppe-Seyler crystallized hemoglobin.

1866
Mendel demonstrated the segregation and independent assortment of alleles in pea plants.

1893
Ostwald showed that enzymes are catalysts.

1898
Camillo Golgi described the Golgi apparatus.

1905
Knoop deduced the β oxidation mechanisms for fatty acid degradation.

1907
Fletcher and Hopkins showed that lactic acid is formed quantitatively from glucose during anaerobic muscle contraction.

1910
Morgan discovered sex-limited inheritance in *Drosophila.*

1912
Warburg postulated a respiratory enzyme for the activation of oxygen.

1913
Michaelis and Menten developed a kinetic theory of enzyme action.

1922
McCollum showed that lack of vitamin D causes rickets.

1926
Sumner crystallized the first enzyme urease.

1926
Jansen and Donath isolated vitamin B_1 (thiamine) from rice polishings.

1926–1930
Svedberg invented the ultracentrifuge and used it to demonstrate the existence of macromolecules.

1928
Levene showed that nucleotides are the building blocks of nucleic acids.

1928
Szent-Györgyi isolated ascorbic acid (Vitamin C).

1928–1933
Warburg deduced the iron–porphyrin presence in the respiratory enzyme.

1929
Burr and Burr discovered that linoleic acid is an essential fatty acid for animals.

1931
Englehardt discovered that phosphorylation is coupled to respiration.

1932
Warburg and Christian discovered the "yellow enzyme," a flavoprotein.

1933
Krebs and Henseleit discovered the urea cycle.

1933
Embden and Meyerhof demonstrated the intermediates in the glycolytic pathway.

1935
Schoenheimer and Rittenberg first used isotopes as tracers in the study of intermediary metabolism.

1935
Stanley first crystallized a virus, tobacco mosaic virus.

1937
Krebs discovered the citric acid cycle.

1937
Warburg showed how ATP formation is coupled to the dehydrogenation of glyceraldehyde-3-phosphate.

1938
Hill found that cell-free suspensions of chloroplasts yield oxygen when illuminated in the presence of an electron acceptor.

1939
C. Cori and G. Cori demonstrated the reversible action of glycogen phosphorylase.

1939
Lipmann postulated the central role of ATP in the energy-transfer cycle.

1939–1946
Szent-Györgyi discovered actin and the actin–myosin complex.

1940
Beadle and Tatum deduced the one gene–one enzyme relationship.

1942
Bloch and Rittenberg discovered that acetate is the precursor of cholesterol.

1943
Chance applied spectrophotometric methods to the study of enzyme–substrate interactions.

1943
Martin and Synge developed partition chromatography.

1944
Avery, MacLeod, and McCarty demonstrated that bacterial transformation is caused by DNA.

1947–1950
Lipmann and Kaplan isolated and characterized coenzyme A.

1948
Leloir discovered the role of uridine nucleotides in carbohydrate metabolism.

1948
Hogeboom, Schneider, and Palade refined the differential centrifugation method for fractionation of cell parts.

1948
Kennedy and Lehninger discovered that the tricarboxylic acid cycle, fatty acid oxidation, and oxidative phosphorylation all take place in mitochondria.

1949
Christian deDuve discovered lysosomes.

1950–1953
Chargaff discovered the base equivalences in DNA.

1951
Pauling and Corey proposed the α-helix structure for α-keratins.

1951
Lynen postulated the role of coenzyme A in fatty acid oxidation.

1952
Palade, Porter, and Sjostrand perfected thin sectioning and fixation methods for electron microscopy of intracellular structures.

1952–1954
Zamecnik and his colleagues developed the first cell-free systems for the study of protein synthesis.

1953
Vincent du Vigneaud synthesized the first biologically active peptide hormone, ocytocin.

1953
Woodward and Bloch postulated a cyclization scheme for squalene, leading to cholesterol.

1953
Sanger and Thompson determined the complete amino acid sequence of insulin.

1953
Hokin and Hokin showed that acetylcholine induces the rapid biosynthesis of phosphatidylinositol in pigeon pancreas.

1953
Horecker, Dickens, and Racker elucidated the 6-phosphogluconate pathway of glucose catabolism.

1953
Watson and Crick and Wilkins determined the double-helix structure of DNA.

1954
Hugh Huxley proposed the sliding filament model for muscular contraction.

1955
Ochoa and Grunberg-Manago discovered polynucleotide phosphorylase.

1955
Kennedy and Weiss described the role of CTP in the biosynthesis of phosphatidylcholine.

1956
Kornberg discovered the first DNA polymerase.

1956
Umbarger reported that the end product isoleucine inhibits the first enzyme in its biosynthesis from threonine.

1956
Dorothy Crawfoot Hodgkin determined the structure of coenzyme B12.

1956
Ingram showed that normal and sickle-cell hemoglobin differ in a single amino acid residue.

1956
Anfinsen and White concluded that the three-dimensional conformation of proteins is specified by their amino acid sequence.

1956
Leloir determined the pathway to uridine diphosphate glucose (UDPG).

1957
Hoagland, Zamecnik, and Stephenson isolated tRNA and determined its function.

1957
Sutherland discovered cyclic AMP.

1958
Weiss, Hurwitz, and Stevens discovered DNA-directed RNA polymerase.

1958
Meselson and Stahl demonstrated that DNA is replicated by a semiconservative mechanism.

1959
Wakil and Ganguly reported that malonyl-CoA is a key intermediate in fatty acid biosynthesis.

1959
Krebs and Fischer discovered protein kinases.

1960
Kendrew reported the x-ray analysis of the structure of myoglobin.

1961
Jacob and Monod proposed the operon hypothesis.

1961
Jacob, Monod, and Changeux proposed a theory of the function and action of allosteric enzymes.

1961
Mitchell postulated the chemiosmotic mechanism of oxidative phosphorylation.

1961
Nirenberg and Matthaei reported that polyuridylic acid codes for polyphenylalanine.

1961
Marmur and Doty discovered DNA renaturation.

1962
Racker isolated F_1 ATPase from mitochondria and reconstituted oxidative phosphorylation in submitochondrial vesicles.

1966
Maizel introduced the use of sodium dodecylsulfate (SDS) for high-resolution electrophoresis of protein mixtures.

1966
Crick proposed the wobble hypothesis.

1966
Gilbert and Muller-Hill isolated the lac repressor.

1968
Glomset proposed the theory of reverse cholesterol transport in which HDL is involved in the return of cholesterol to the liver.

1968
Meselson and Yuan discovered the first DNA restriction enzyme. Shortly thereafter, Smith and Wilcox discovered the first restriction enzyme that cuts DNA at a specific sequence.

1969
Zubay and Lederman developed the first cell-free system for studying the regulation of gene expression.

1970
Howard Temin and David Baltimore discovered reverse transcriptase.

1971
Vane discovered that aspirin blocks the biosynthesis of prostaglandins.

1972
Jon Singer and Garth Nicolson proposed the fluid mosaic model for membrane structure.

1973
Cohen, Chang, Boyer, and Helling reported the first DNA cloning experiments.

1975
Brown and Goldstein described the low-density lipoprotein receptor pathway.

1975
Sanger and Barrell developed rapid DNA-sequencing methods.

1976
Michael Bishop and Harold Varmus discovered the *c-src* gene in uninfected cells, which is homologous to the *v-src* gene in the Rous sarcoma virus.

1977
Starlinger discovered the first DNA insertion element.

1977
McGarry, Mannaerts, and Foster discovered that malonyl-CoA is a potent inhibitor of β oxidation.

1977
Splicing of RNA simultaneously discovered in Robert's and Sharp's laboratories.

1977
Nishizuka and co-workers reported the existence of protein kinase C.

1978
Shortles and Nathans did the first experiments in directed mutagenesis.

1978
Tonegawa demonstrated DNA splicing for an immunoglobulin gene.

1981
Cech discovered RNA self-splicing.

1981
Steitz determined the structure of CAP protein.

1981–1982
Palmiter and Brinster produced transgenic mice.

1983
Mullis amplified DNA by the polymerase chain reaction (PCR) method.

1984
Schwartz and Cantor developed pulsed field gel electrophoresis for the separation of very large DNA molecules.

1984
Michel, Deisenhofer, and Huber determined the structure of the photosynthetic reaction center.

1984
Blobel discovered the mechanism for protein translocation across the endoplasmic reticulum membrane: the signal hypothesis.

1988
Elion and Hitchings shared the Nobel Prize for design and synthesis of therapeutic purines and pyrimidines.

1989
Synder and colleagues purified and reconstituted the inositol-1,3,4-P_3 receptor.

1994
Gilman and Rodbell shared the Nobel Prize for their discovery of G-proteins and the role of these proteins in signal transduction in cells.

1995
Lewis, Nusslein-Volhard, and Wieschaus shared the Nobel Prize for their discoveries concerning the genetic control of early embryonic development in Drosophila.

Appendix B
Answers to Odd-Numbered Problems

Chapter 2

1. Elevated temperature, increased pressure, large concentrations of reactants, and extremes of pH may be applied to chemical reactions carried out in vitro. For example, the synthesis of urea from ammonia and CO_2 requires application of several hundred atmospheres of pressure at elevated temperature. Reactions within the cell occur under a much more restricted range of temperature, pressure, pH, and reactant concentration than is possible in in vitro reactions. Temperatures of 25+/−15°C, near neutral pH, and micro- to millimolar concentrations of reactant are conditions typically found in the cell. Urea is formed in the liver using ammonium ions provided by amino acids and CO_2 (HCO_3^-) provided by oxidative metabolism. The process utilizes energy from metabolism carried out at normobaric pressure and comparatively mild temperature. Urea synthesis is possible in vivo because of the presence of enzymes that efficiently catalyze the reactions.
3. Intensive thermodynamic properties are independent of the amount of material in the state (e.g., temperature, density), whereas extensive properties depend on the amount of material (energy, mass).
5. The water molecules surrounding a hydrophobic molecule form a clathrate structure and are more highly ordered than water molecules in bulk solution. Removal of the hydrophobic molecule to a nonaqueous environment may increase the order of the hydrophobic groups, but that unfavorable entropic contribution is more than compensated by the increased disorder of the water molecules that were previously in the more highly ordered structure (favorable entropic contribution).
7. The reaction shown describes the formation of oxaloacetate from the oxidation of malate in the TCA or Krebs cycle. Given the large (+) $\Delta G^{\circ\prime}$ value for the reaction, the reaction lies strongly in favor of the reactants at standard state equilibrium. However, conditions in the cell impose a steady state whose conditions are far removed from the thermodynamic standard state. The reaction will proceed toward oxaloacetate formation in the cell if the concentration of products is kept small. This is accomplished in the mitochondria in two ways:
   (a) NADH is oxidized by the mitochondrial electron transport system, thereby diminishing the product (NADH) concentration and replenishing the reactant (NAD^+) concentration.
   (b) Oxaloacetate is condensed with acetyl-CoA in the citrate synthase-catalyzed formation of citrate. This condensation is thermodynamically favorable (approximately −8 kcal/mole). Coupling citrate synthase and malate dehydrogenase through the common intermediate oxaloacetate is an example of an exergonic (thermodynamicmally favorable) reaction driving an endergonic (thermodynamically unfavorable) reaction.
9. (a)

$$\text{Lactose} + \text{HOH} \rightarrow \text{D-Galactose} + \text{D-Glucose} \quad \Delta G^{\circ\prime} = -4.0 \text{ kcal/mole}$$

The equilibrium constant is related to the free energy change by the expression

$$\Delta G^{\circ\prime} = -2.3RT \log K'_{eq}$$

$$\text{Log } K'_{eq} = \frac{(-4 \text{ kcal/mole})}{(-1.36 \text{ kcal/mole})} = 2.94$$

$$K'_{eq} = 8.7 \times 10^2$$

(b) The free energy for the synthesis of lactose from D-galactose plus D-glucose is +4 kcal/mole because the reaction considered is the reverse of that examined in 2.9a:

$$\text{D-Galactose} + \text{D-Glucose} \rightarrow \text{Lactose} + \text{HOH}$$

K'_{eq} for the synthesis of lactose from the component parts is

$$\text{Log } K'_{eq} = \frac{(4 \text{ kcal/mole})}{(-1.36 \text{ kcal/mole})}$$

$$K'_{eq} = 1.1 \times 10^{-3} \text{ (inverse of 870)}$$

(c) Consider the reactions

$$\begin{array}{r} \text{UDP-Galactose} + \text{HOH} \rightarrow \text{UDP} + \text{Galactose} \ \Delta G^{\circ\prime} = -7.3 \text{ kcal/mole} \\ \text{D-Galactose} + \text{D-Glucose} \rightarrow \text{Lactose} + \text{HOH} \ \Delta G^{\circ\prime} = +4.0 \text{ kcal/mole} \\ \hline \text{UDP-Galactose} + \text{D-Glucose} - \text{Lactose} + \text{UDP} \ \Delta G^{\circ\prime} = -3.3 \text{ kcal/mole} \end{array}$$

The equilibrium constant of the final reaction shown is calculated from the overall free energy change of the coupled reactions:

$$K'_{eq} = 2.7 \times 10^2$$

In this example a thermodynamically unfavorable reaction (formation of a glycosidic bond between D-galactose and D-glucose) is coupled to a thermodynamically favorable reaction (hydrolysis of UDP-galactose).
11. In the absence of pyrophosphatase the reactions are

$$\text{H}^+ + \text{R—COO}^- + \text{CoASH} \rightarrow \text{R—CO—SCoA} + H_2O \ \Delta G^{\circ\prime} = +10 \text{ kcal/mole}$$

$$H_2O + \text{ATP} \rightarrow \text{AMP} + \text{PP}_i + H_2O \ \Delta G^{\circ\prime} = -7.5 \text{ kcal/mole}$$

$$\text{R—COO}^- + \text{ATP} + \text{CoASH} \rightarrow \text{R—CO—SCoA} + \text{AMP} + \text{PP}_i$$

$$\Delta G^{\circ\prime} = +2.5 \text{ kcal/mole}$$

$$\Delta G^{\circ\prime} = -2.3RT\log K'_{eq}$$

$$K'_{eq} = 1.5 \times 10^{-2} \qquad +2.5/-1.36 = \log K'_{eq}$$

In the presence of pyrophosphatase the reactions are

$$\text{H}^+ + \text{R—COO}^- + \text{CoASH} \rightarrow \text{R—CO—SCoA} + H_2O \quad \Delta G^{\circ\prime} = +10 \text{ kcal/mole}$$

$H_2O + ATP \rightarrow AMP + PP_i + H—^+$ $\Delta G^{\circ\prime} = -7.5$ kcal/mole

$H_2O + PP_i \rightarrow 2\ P_i + H^+$ $\Delta G^{\circ\prime} = -7.5$ kcal/mole

$R—COO^- + CoASH + ATP + H_2O \rightarrow R—CO—SCoA + AMP + 2\ P_i + H^+$ $\Delta G^{\circ\prime} = -5.0$ kcal/mole

$$K'_{eq} = 4.8 \times 10^3$$

Pyrophosphatase removes the pyrophosphate from the reaction and shifts the equilibrium toward R—CO—SCoA formation. The shift is energetically equivalent to the hydrolysis of a second mole of ATP.

Some of the values for $\Delta G^{\circ\prime}$ were based on data in *Handbook of Biochemistry,* 2d ed., CRC Press, Boca Raton, Fla., 1970.

13. In the first case, where the repressor protein is cut in half, the binding enthalpy for each part would be essentially half the enthalpy value for the intact repressor. The entropy would be less favorable (less positive) because of the chelation effect. As a result, the free energy would be less favorable (less negative).

    In the second case, where one of the binding sites on the DNA is eliminated, the binding enthalpy for the repressor would be approximately half that for repressor binding to the unmodified DNA sequence. The entropy would depend on the extent of hydration and the extent of mobility of the unbound portion of the repressor. Again, one would expect the free energy for binding to be less favorable.

Chapter 3

1. Both 1-butanol and 1-aminobutane can participate in the formation of three hydrogen bonds with water molecules while a given molecule of butanoic acid could participate in up to five hydrogen bonds with water. As a weak acid, butanoic acid is slightly ionized, while 1-aminobutane is protonated in an equilibrium reaction with water, which result in hydrated ions and an increase in entropy. The limited solubility of 1-butanol indicates van der Waals interaction between the alkyl portions of 1-butanol molecules is sufficiently strong that only a limited amount of the alcohol is dispersed in solution supported by hydrogen bonding. The differences in the solubility of these substances indicates the contribution of the ionization, hydrated ions, and entropy associated with the ionization to the solubility of a substance.
3. The extrapolated value (HOH), even though it is water in formula, can participate in only three hydrogen bonds because one of the hydrogens on HOH is extrapolated from carbon-bound hydrogens. An individual water molecule can participate in up to four hydrogen bonds, producing a partial molar volume of 18 ml/mole. The fourth hydrogen bond allows the water molecules to pack together more compactly than HOH and results in a smaller partial molar volume.
5. $$K_a = \frac{[H^+][A^-]}{[HA]} \qquad [HA] + [A^-] = 0.0084\ \text{M}$$

$$K_a = \frac{X^2}{0.0084 - X} \qquad 10^{-3.15} = 7.079 \times 10^{-4}\ \text{M} = X$$

$$K_a = 6.52 \times 10^{-5} \qquad pK_a = 4.18 \quad \text{(benzoic acid)}$$

7. The buffer should have a pH of 6.87. Examine the Henderson-Hasselbalch equation; the pH results from the ratio of the conjugate acid–base pair and does not actually depend upon the concentration of the acid–base pair. The actual concentration is important in determining the buffer capacity.
9. $$\frac{3.71 \text{ g citric acid}}{192 \text{ g citric acid/mole}} = 0.0193 \text{ mole citric acid}$$

$$\frac{2.91 \text{ g KOH}}{56.1 \text{ g KOH/mole}} = 0.0519 \text{ mole KOH}$$

After reaction:

0.0060 mole $HCit^{2-}$
0.0133 mole Cit^{3-}

Using the Henderson-Hasselbalch equation:

$$pH = 6.39 + \log \frac{0.0133 \text{ mole}/0.25 \text{ l}}{0.0060 \text{ mole}/0.25 \text{ l}} = 6.39 + 0.35 = 6.74$$

$$[H^+] = 10^{-6.74} = 1.82 \times 10^{-7}\ \text{M}$$

11. $$[H_2PO_4^{1-}] + [HPO_4^{2-}] = 0.035\ \text{M}$$

$$6.87 = 7.21 + \log \frac{[HPO_4^{2-}]}{[H_2PO_4^{1-}]}$$

Solving for both unknowns:

$$[HPO_4^{2-}] = 0.011\ \text{M}$$
$$[H_2PO_4^{1-}] = 0.024\ \text{M}$$

Considering 100 ml prior to KOH addition:

$$HPO_4^{2-} = 0.0011 \text{ mole}$$
$$H_2PO_4^{1-} = 0.0024 \text{ mole}$$
$$(0.015 \text{ l})(0.075 \text{ M/l}) = 0.0011 \text{ mole of } OH^-$$

After KOH addition:

$$pH = 7.21 + \log \frac{(0.0022 \text{ mole } HPO_4^{2-}/115 \text{ ml})}{(0.0013 \text{ mole } H_2PO_4^{1-}/115 \text{ ml})} = 7.44$$

13. $$[H^+] = 10^{-7.71} = 1.9 \times 10^{-8}\ \text{M}$$

$$[H^+][OH^-] = 1 \times 10^{-14}$$
$$[OH^-] = 5.3 \times 10^{-7}\ \text{M}$$

Chapter 4

1. The α-amino and α-carboxyl groups in peptides are separated by a greater number of atoms than in a single amino acid. The electrostatic and inductive interactions between these two groups are diminished with increasing distance, which results in a higher pK_a for the α-carboxyl and lower pK_a for the α-amino of a peptide.
3. The amino acid has no net charge at the isoelectric pH. The pI value of amino acids with side chains that are not ionizable are isoelectric at pH that is the arithmetic mean of the pK values of the carboxyl and the α-amino groups. For ex-

ample, the pI of glycine is (2.35 + 9.78)/2 = 6.07. Amino acids whose R groups are ionizable may be divided between those whose side chains contribute to the positive charge and those contributing to the negative charge. If the R group contributes positive charge (basic amino acid), the fractional charge residing on the α-amino group and on the R group must total 1. This is achieved when the pH is the arithmetic mean of pK_{amino} plus pK_R. Similarly, the isoelectric pH of an acidic amino acid is equal to the arithmetic mean of $pK_{carboxyl}$ and pK_R:

$$\text{pI histidine} = [(pK_{amino}) + (pK_{amidazole})]/2$$
$$\text{pI} = 7.69$$

$$\text{pI aspartic acid} = [(pK_{carboxyl}) + (pK_R)]/2$$
$$\text{pI} = 2.95$$

$$\text{pI arginine} = (pK_{amino} + pK_R)/2$$
$$\text{pI} = 10.74$$

Apply the Henderson-Hasselbalch equation to calculate the ratio of unprotonated to protonated groups on aspartate, then calculate the fractional charge on each group. At pH = pI (2.95) the α-amino group will be virtually fully protonated and will contribute one positive charge. The fractional charge on the carboxyl groups are

$$\alpha\text{—COOH: } pK_a = 1.99$$

$$\text{pH} = pK_a + \log\left[\frac{A^-}{HA}\right]$$

$$2.95 = 1.99 + \log\left[\frac{A^-}{HA}\right]$$

$$\log\left[\frac{A^-}{HA}\right] = 0.96$$

$$\left[\frac{A^-}{HA}\right] = 9.1, \text{ and } [A^-] + [HA] = 1$$

$[A^-] = 0.9$ fractional negative charge (90% of the α-carboxylate will be unprotonated at any time)

A similar calculation will reveal that the fractional negative charge on the β-carboxylate will be significantly less because the pH (2.95) is below the pK_a. Hence

$$[A^-] = 0.1 \text{ negative charge on the } \beta\text{-carboxylate}$$

Sum positive and negative charge contributions are

α-amino group	α-carboxyl	β-carboxyl
(+1)	(−0.9)	(−0.1) = 0

These data demonstrate that the net charge on aspartic acid is zero at pH 2.95 and thus verify 2.95 as the isoelectric pH.

5. Assuming a substance is an effective buffer + / − 1 pH unit of the pK_a, then this mixture would buffer between pH (ca.) 1 to 7 and 8.5 to 10.5.
7. The only ionizable group on this peptide that has a pK_a near 4.4 is the 4.0 of the aspartyl side chain. Using the Henderson-Hasselbalch equation:

$$4.4 = 4.0 + \log\frac{[\text{pep}^0]}{[\text{pep}^{+1}]}$$

$$2.5 = \frac{[\text{pep}^0]}{[\text{pep}^{+1}]} \text{ or } \frac{2.5}{1.0}$$

or +1 charge per 3.5 molecules or +0.3/peptide molecule.

9. At pH 6.2 the following amino acids have the indicated charges: Val, 0; Asp, −1; His +0.4; and Lys, +1. The elution order would be Val and Asp in the hold-up volume followed by His then Lys. In order to separate Val and Asp (and the rest of the amino acids from each other) a buffer with a pH near 3 must be chosen. At a pH of 3 the elution order would be Asp, Val, His, Lys.
11. There are no alkenyl, alkynyl, ethers, aldehydes, ketones, esters, carboxylic acid anhydrides, etc.
13. The approximate charges on the fragments are CT_1 (0), CT_2 (−1), and CT_3 (−2). Upon anion-exchange chromatography they are expected to elute in the following order: CT_1, CT_2, and CT_3. What would you expect to be the source of the names for these peptides?
15. Upon exposure to concentrated base, arginine decomposes to produce ornithine, ammonia, and carbonate. Why aren't similar products produced from arginine by concentrated acid?
17. DNP analysis indicates an N-terminus Val and an internal or C-terminus Lys in the heptapeptide. Analysis after hydrolysis with trypsin indicates a tetrapeptide (T2) that contains Lys (cation at pH 6), an amide amino acid, and Phe (260 nm) and a tripeptide (T1) that contains Cys, Tyr (275 nm), and Glu or Asp (anion at pH 6). At this point, only Val has not been assigned to a tryptic fragment. Because only three of the four amino acids of T2 have been assigned, the T2 must contain Val. Because T2 contains the Lys (and Val), the order of the tryptic fragments must be T2 then T1 in the heptapeptide. Analysis with chymotrypsin indicates that CT1 contains Cys, an aromatic amino acid, and Lys (cation). CT2 can be deduced to contain an aromatid amino acid, and because a free Asp is produced by chymotrypsin, CT2 must contain Gln. Because these are chymotrypsin-produced fragments, the two tripeptides must precede the Asp in the heptapeptide. When combined, the above information allows for the determination of the heptapeptide sequence:

Val Gln Phe Lys Cys Tyr Asp

Chapter 5

1. Structures such as the α helix or pleated sheet require that the amino acids all have the same geometry. Any type of life analogous to ours (i.e., DNA → RNA → AA sequence → shape → function) that had "specific" amino acids as D- or L-geometries would have difficulties handling any mutations (or evolving), because any L- to D-geometry change (or the reverse) would have major impacts on α-helix or pleated-sheet structures. If all of the amino acids were of the "same" geometry, then any mutation would change only the amino acid and not the geometry, which would allow for the preservation of the basic shape of the protein.
3. Heat is synonymous with energy, and with sufficient energy

the *trans* nature of peptide bonds can be isomerized. If a significant number of peptide bonds are isomerized in a protein molecule, the likelihood that they all return to the *trans* form upon cooling is very small. Any *cis*-peptide bonds would have major impacts on the protein's structure.

5. Nothing can be said for certain. What happens depends upon the total of all of the other interactions within the protein that determines the protein's structure. Glutamic acid residues have the highest probability of being in a helix, while glycyl groups have the lowest probability (see fig. 5.41). A glycine allows greater flexibility than is typically permitted within a helix.
7. The α helix is a rather rigid, rodlike secondary structural element that cannot easily change direction in space without breaking the helical arrangement. Frequently, β bends or loops, or in some instances segments of random coil structure, allow the helical elements to change direction and pack into a more compact globular structure. Myoglobin has a high helical content, but the segments of helix fold back on one another and pack into a globular protein.
9. The sequence given can assume an α-helical arrangement with the hydrophobic side chains located along the outside of the helix and exposed to solvent. The α helix will be distorted at the proline residue and may enter a β turn. The arrangement of the hydrophobic residues would likely limit water solubility, because the increased organization of water structure surrounding the hydrophobic side chains decreases the entropic contribution to the stability of the system. However, the segment of hydrophobic helix would be stabilized by insertion into the hydrophobic environment of the membrane. Moving the hydrophobic side chains on the helix from aqueous contact into the lipid bilayer would increase the entropic contribution of the water molecules to the system. Hydrophobic segments are found in proteins that are bound to biological membranes.
11. Metals are held in proteins by interaction with specific amino acid residues (ligands). The metal is bound to the protein in one of a few specific three-dimensional arrangements of the ligands. Binding of the metal to the protein may require the ligands to be arranged roughly in a square planar, a tetrahedral, or an octahedral orientation. The precise geometry of the metal–ligand complex is required for biological activity. Ligands to the metal do not necessarily arise from the same segment of the protein. For example, the ligands to Fe in the iron-containing superoxide dismutase from *E. coli* are histidyl side chains (His-26, His-73, and His-160) and an aspartyl residue (Asp-156). The numbers represent the position of the amino acid in the primary sequence with the N-terminal as number 1. Within the same family of proteins the ligands are apparently highly conserved. (See Carlios, A., et al., Iron superoxide dismutase: Nucleotide sequence of the gene from *Escherichia coli* K_{12} and correlation with crystal structure, *J. Biol. Chem.* 263:1555–1562, 1988; and Barra, D., et al., The primary structure of iron superoxide dismutase from *Photobacterium leiognathi, J. Biol. Chem.* 262:1001–1009, 1987.)

   Changing one or more of the ligands may abolish metal binding to the protein or may alter the spatial arrangement around the metal so that the biological function is abolished. Conservation of residues around the ligand side chains preserves the local environment and conformation of the ligands, allowing the metal to be bound. Nonconservative replacements at sites distant from the metal-binding site may have little influence on the metal binding region.
13. Each peptide bond may be considered to be an approximately planar structure that rotates as a single unit around the peptide C_α—N and the C_α—carbonyl bonds. The planar structure of the peptide bond removes a number of degrees of freedom of rotation. Moreover, rotation of the peptide bond around the C_α—N and C_α—carbonyl carbon bonds is further restricted because of unfavorable interactions between atoms comprising the peptide bonds or between atoms on the peptide and the side chains. The sterically allowed rotational angles that minimize energetically unfavorable interactions between peptide bond substituents accommodate only a limited number of regular structures. (Examine the Ramachandran plot in the text to identify these allowed conformations.)
15. The detergent replaces the membrane, producing a soluble enzyme, and allows the enzymes to be purified free of the membrane.

Chapter 6

1. The early chemists had no way of knowing that each hemoglobin (Hb) contains four Fe^{2+}:

$$\frac{1 \text{ g Hb}}{} \left| \frac{1 \text{ mole Hb}}{64{,}500 \text{ g Hb}} \right| \frac{4 \text{ mole Fe}}{1 \text{ mole Hb}} \left| \frac{1 \text{ mole } Fe_2O_3}{2 \text{ mole Fe}} \right| \frac{160 \text{ g } Fe_2O_3}{1 \text{ mole } Fe_2O_3}$$

$$\frac{1000 \text{ mg}}{\text{g}} = 4.9 \text{ mg } Fe_2O_3$$

3.

(a) $CO_2 + H_2N\text{-hemoglobin} \longrightarrow {}^{-}O\overset{\overset{\displaystyle O}{\|}}{C}HN\text{-hemoglobin} + H^+$

(b) $HOCO_2^- + H_2N\text{-hemoglobin} \longrightarrow$

$${}^{-}O\overset{\overset{\displaystyle O}{\|}}{C}HN\text{-hemoglobin} + H_2O$$

5. During a sickle-cell crisis the red blood cells collapse from their normal "disk" shape into a sickle (or quarter moon) shape. This collapsed shape has a smaller cross sectional area than the original "disk" shape. In the capillaries the red blood cells normally pass in a single file. If the cells are sickled, it is possible for them to wedge together, clogging the capillary, stopping the flow of red blood cells, and depriving the adjacent tissues of oxygen. The oxygen deprivation produces the pain during a crisis.
7. Individuals with sickle-cell anemia have a functional gene for the gamma chain. If the production of fetal hemoglobin could be "turned back on," the affected individuals could function normally except in pregnancy. During pregnancy, net transfer of oxygen from the mother to the fetus would

be inhibited because both mother and fetus would have the same fetal hemoglobin with equal affinity for oxygen.

9. The interaction between the histidyl F8 residue and heme iron is a ligand–metal ion interaction. In this complexation reaction the ligand provides a lone pair of electrons. Both the protonated and unprotonated imidizol ring of histidine have a lone pair(s) of electrons. The protonated form of histidyl F8 is positively charged, which is less likely to interact favorably with the Fe^{2+} of the heme. Also note that during coordinate covalent bond formation the imidizol nitrogen takes on a formal positive charge, which would be unfavorable in a protonated (already positive) histidyl group. If the unprotonated form is utilized, then the lone pair which complexes the iron is in an sp^2 orbital. The coordinate covalent bond is then in conjugation with the imidazol ring, allowing resonance.

11. $$\frac{32{,}000}{} \cdot \frac{\text{AA}}{105\ \text{D}} \cdot \frac{\text{turn}}{3.6\text{AA}} \cdot \frac{5.4\ \text{Å}}{\text{turn}} = 457\ \text{Å}$$

Because the two 457-Å-long strands are twisted into a helix, the resulting TM is shorter than 457 Å.

13. Many explanations are possible. One involves a DNA sequence suitable for a β structure and an α structure that underwent a sequence of gene duplications:

$$\beta\text{-loop-}\alpha \rightarrow (\beta\text{-loop-}\alpha)_2 \rightarrow (\beta\text{-loop-}\alpha)_4 \rightarrow (\beta\text{-loop-}\alpha)_8$$

15. The protein complex sediments similarly to the 64,000 M_r marker protein hemoglobin but sediments similarly to the 17,000 M_r myoglobin marker when 2 M NaCl is included in the buffer. The structure predicted for the complex is a tetramer of approximately 16,000 M_r subunits associated through ionic or electrostatic interactions but not by covalent bonds. The high ionic strength of the 2 M NaCl disrupts the ionic interactions, causing dissociation of the tetramer to the monomers.

Chapter 7

1.

Step	Volume (ml)	Total Protein (mg)	Total Units	Specific Activity (U/mg)	Yield (%)	Purification (*n*-fold)
Cell extract	2,800	70,000	2,700	0.039	100	1
$((NH_4)_2SO_4)$ fractionation	3,000	25,400	2,300	0.091	85	2.3
Heat treatment	3,000	16,500	1,980	0.12	73	3.1
DEAE chromatography	80	390	1,680	4.3	62	110
CM-cellulose	50	47	1,350	29	50	740
Bio-Gel A	7	35	1,120	32	41	820

Sample calculation (DEAE-cellulose step)
Specific activity is defined as units of enzymatic activity per mg protein:

$$\text{Specific activity} = \frac{1{,}680\ \text{U}}{390\ \text{mg}}$$

$$\text{Specific activity} = 4.3\ \text{U/mg}$$

Yield is calculated as the percentage of the initial activity units remaining after the specified isolation step:

$$\text{Yield} = \frac{1{,}680\ \text{U}}{2{,}700\ \text{U}} \times 100$$

$$\text{Yield} = 62\%$$

N-fold purification is the ratio of the specific activity at a given step in the purification to the specific activity of the enzyme in the cell extract:

$$n\text{-fold purification} = \frac{4.3\ \text{U mg}^{-1}}{3.9 \times 10^{-2}\ \text{U mg}^{-1}}$$

$$n\text{-fold purification} = 110$$

The DEAE chromatography yields the greatest purification. The specific activity of the enzyme following the DEAE chromatography was increased 36-fold compared with the specific activity of the previous step. The ammonium sulfate fractionation caused a 2.3-fold increase; the heat step, 1.4-fold; CM-cellulose, 6.6-fold; and gel exclusion chromatography, 1.1-fold.

The specific activity of the pure 6-phosphogluconate dehydrogenase was increased 820-fold compared to the cell extract. The enzyme must have been present as 1/820 of the total protein. The initial percentage was therefore

$$\frac{1}{820} = 1.2 \times 10^{-3} = 0.12$$

3. The protein is predicted to adhere to DEAE if the net charge on the protein is negative and that of the diethylaminoethyl exchange group is positive. The DEAE exchanger group is protonated and positively charged at pH up to approximately 9 but is unprotonated and neutral above pH 9, assuming a pK_a of about 8.5 for the DEAE exchanger. The protein will be negatively charged at pH more basic than the isoelectric point (pl = 6.0). Thus buffer in the pH range just above 6 to just below 9 will ensure that the exchanger group is positively charged and the protein is negatively charged, and ionic interaction between the two would likely occur.

5. Gel exclusion chromatography is generally of limited value when large volumes of protein solution are used. The most effective separations occur when the smallest possible volume of protein solution is loaded onto the gel exclusion column, avoiding the broad, diffuse, overlapping protein bands that usually result when large volumes are loaded onto the column. The 3-liter protein pool after heat treatment is hardly conducive to gel exclusion chromatography on less than industrial scale. For example, the volume loaded onto gel exclusion columns is ideally 5% of the column volume. Were this criterion applied to chromatography of the 3-liter sample, a column containing 60 liters of gel exclusion resin would be required. In addition, the less highly purified sample is more likely to have a contaminating protein that would coelute with the desired protein.

7. Contaminant B (75,000 M_r) can be easily separated from both ribonuclease and contaminant A by gel exclusion chromatography but not by ion-exchange chromatography. The similarity of pI values for the ribonuclease and contaminant B suggests that separation based on charge likely would be of little value. Contaminant A may be separated from ribonuclease by ion-exchange chromatography (pI values differ markedly) but not by gel exclusion chromatography (molecular weights are similar).

   Ion-exchange chromatography is suited not only for separation of proteins but also for concentration of the proteins in dilute solution. Assume that the volume of the contaminated ribonuclease solution is too large for gel exlusion chromatography. The solution is dialyzed in buffer whose pH is more acidic than the pI of ribonuclease but more basic than the pI of contaminant A. Under these conditions the ribonuclease and contaminant B will be positively charged, but contaminant A will be negatively charged. A CM-cellulose ion-exchange column in the same buffer as the proteins will bear a negative charge on the carboxymethyl groups. Contaminant A will elute in the low-ionic-strength buffer used to load the column, while ribonuclease and contaminant B are predicted to adhere to the column but will likely be eluted at approximately the same point in a linear KCl gradient. The pooled fraction from the CM-cellulose (containing ribonuclease and contaminant B) in a smaller volume than the original sample may now be chromatographed on a gel exlusion resin that excludes 50,000 M_r glubular proteins. Contaminant B will elute in the void volume, but ribonuclease will be retarded in the included volume.

9. Calculate an R_M value for each molecular weight standard based on the distance the protein standard migrated divided by the distance the tracking dye migrated. These values graphed versus the logarithm of the respective molecular weights establish a standard curve from which the subunit molecular weight of the pyrophosphatase can be estimated.

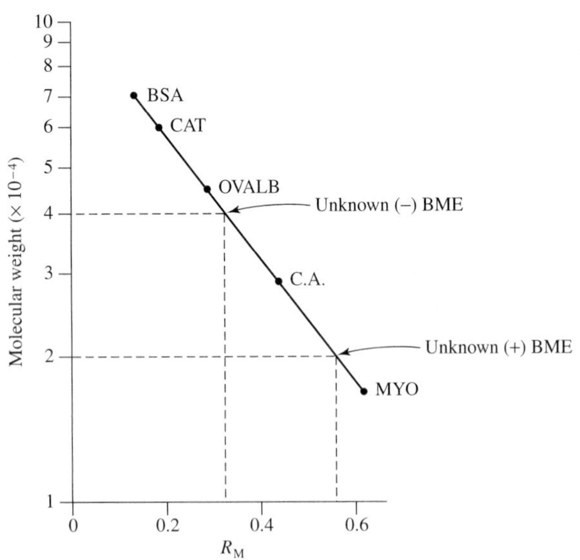

Protein	Subunit Molecular Weight	R_f
Serum albumin	69,000	0.13
Catalase	60,000	0.18
Ovalbumin	43,000	0.29
Carbonic anhydrase	29,000	0.43
Myoglobin	17,000	0.62

Pyrophosphatase (+ 2-mercaptoethanol)	0.56 (20,000 M_r)
Pyrophosphatase (− 2-mercaptoethanol)	0.32 (40,000 M_r)

Native (nondenatured) molecular weight determined with the calibrated gel exclusion chromatography is 39,000. SDS-PAGE of pyrophosphatase that had been denatured and reduced with 2-mercaptoethanol yielded a 20,000 M_r peptide, whereas omission of the reductant yielded a 40,000 M_r peptide. The pyrophosphatase is thus a dimer of 20,000 molecular weight subunits joined by intersubunit disulfide bonds.

11. Protein denaturation is always a concern, but a pH of 8.5 would not have a deleterious effect on most proteins. The major concern is the pK_a of the DEAE cellulose, which is near pH 10. The loss of the protonated amine form of DEAE renders the DEAE cellulose useless as an ion-exchange material.

13. Affinity chromatography is an elegant, rather specific method to separate a protein or enzyme from a mixture. The technique is based on the specific, strong binding of a substrate, substrate analog, inhibitor, or antibody to a protein. An initial strategy to isolate the ATP-binding protein involves constructing an affinity resin by linking ATP to an insoluble support through a spacer side chain of appropriate length. The spacer allows the protein to bind the affinity probe with minimal steric or chemical interference from the insoluble support. However, the protein of interest, or another protein in the crude extract, may hydrolyze ATP and destroy the affinity probe. As an alternative approach, construct the affinity probe using a nonhydrolyzable analog of ATP (e.g., imido- or methylene analog). The ATP-binding protein(s) would be predicted to bind to the affinity probe. The contaminating proteins are washed from the column and the bound protein eluted with ATP, increased ionic strength, pH alteration, or an empirically determined combination of these.

15. The KCl provides alternative counter ions to compete with the salt bonds between the protein and exchange resin on the column. The increasing KCl concentration (a gradient) causes proteins with the weakest salt bonds to the column to elute first.

17.

Pure. Step	Volume (ml)	Protein (mg)	Total Units	Spec. Act. (U/mg)	Yield (%)	Fold Purified
Extract	1500	4500	19,500	4.33	100	1
DNase/ dialysis	1496	2693	18,100	6.72	92.8	1.6
Alumina gel	800	600	12,300	20.5	63.1	4.7
Concen- tration	90	441	9,900	22.4	50.8	5.2
Ammonium sulfate	9	75.6	6,030	79.8	30.9	18.4
DEAE Cellulose	30	18	3,600	200	18.5	46.2

Chapter 8

1. Reaction order is the power to which a reactant concentration is raised in defining the rate equation. The rate equation describing the forward reaction in the example is

$$v_r = k(A)(B)$$

where (A) and (B) are expressed as molar concentration and k has the units $\text{M}^{-1}\ \text{s}^{-1}$. The units of v_r are M s^{-1}. The concentration of substrates A and B are each raised to the first power in the rate equation. The reaction is first-order in A and in B. The overall reaction order is the sum of the order contributions of each reactant and is second-order.

The rate equation for the reverse reaction is

$$v_r = k_2(C)^2$$

where (C) is expressed as M and k has the dimensions $\text{M}^{-1}\ \text{s}^{-1}$. The concentration term of reactant (C) in the rate equation is raised to the second power, and the reaction is second-order.

3. The Michaelis-Menten equation for the reaction

$$E + S \underset{k_2}{\overset{k_1}{\rightleftharpoons}} ES \overset{k_3}{\rightarrow} E + P$$

defines K_m in terms of the individual rate constants: $K_m = (k_2 + k_3)/k_1$.

K_s is the dissociation constant for ES → E + S and in terms of the individual rate constants is k_2/k_1. K_m is equal to the [ES] dissociation constant [K_s] when the value of k_3 is much smaller than the value of k_2.

5. The steady-state approximation is based on the concept that the formation of [ES] complex by binding of substrate to free enzyme and breakdown of [ES] to form product plus free enzyme occur at equal rates. A graphical representation of the relative concentrations of free enzyme, substrate, enzyme–substrate complex, and product is shown in figure 8.8 in the text. Note that [S] is initially much larger than [E_r]. If we assume that

$$E + S \underset{k_2}{\overset{k_1}{\rightleftharpoons}} ES \overset{k_3}{\rightarrow} E + P$$

and that [P] is zero, the rate of formation of [ES] is expressed as

$$v_f = k_1\ [E][S]$$

and the rate of breakdown is expressed as

$$v_r = k_2\ [ES] + k_3\ [ES]$$

$$v_r = (k_2 + k_3)\ [ES]$$

At steady state the two velocities v_f and v_r are equal. One can apply the distribution (or conservation) expression (E_t = [E] + [ES]) and the kinetic constants to arrive at the Michaelis-Menten expression, an expression derived using the steady-state assumption:

$$v = \frac{V_m[S]}{K_m + [S]}$$

where $V_m = k_3[E_t]$ and $K_m = (k_2 + k_3)/(k_1)$.
Steady-state approximation may be assumed until the substrate concentration is depleted with a concomitant decrease in the concentration of [ES].

7. The rate of a chemical reaction increases with temperature as defined by the Arrhenius expression (see chapter 8 in the text). Since an enzyme is a catalyst for a chemical reaction, the rate of an enzyme catalyzed reaction increases with increased temperature. However, the catalyst, a protein, is structurally labile and is inactivated (denatured) at elevated temperatures. The precise temperature at which the enzyme is inactivated varies with the specific enzyme. The figure in problem 7 illustrates the expected increase in reaction rate with increased temperature until the temperature at point A is reached. The temperature at point A roughly approximates the maximum temperature at which the catalyst (enzyme) is stable. Denaturation or inactivation removes the catalyst from the reaction, and the reaction rate decreases because the observed velocity is dependent on the concentration of enzyme. There is no "temperature optimum" for a catalyst (enzyme).

9. Normally, the chemical substance that is the easiest to quantify is chosen, and it does not matter whether it is the appearance of a product or disappearance of a reactant that is determined.

11. The kinetic data are analyzed by graphing the reciprocal of the velocity as a function of the reciprocal of the substrate concentration. The kinetic constants V_{max} and K_m are obtained from the resulting Lineweaver-Burk plot.

Substrate (mM^{-1})	Velocity (sec $M^{-1} \times 10^{-7}$)
10	1.04
8	0.89
6	0.74
4	0.60
2	0.45
1	0.38

The intercept on the negative x-axis is the value of $(-1/K_m)$:

$$-(K_m)^{-1} = -4.17 \text{ mM}^{-1}$$

$$K_m = 0.24 \text{ mM} = (2.4 \times 10^{-4} \text{ M})$$

The y-intercept is $(V_{max})^{-1}$:

$$(V_{max})^{-1} = 0.305$$

$$V_{max} = 3.3 \times 10^{-7} \text{ M s}^{-1}$$

Turnover number (k_{cat} or k_3) $= V_{max}/[E_t]$

$$k_{cat} = (3.3 \times 10^{-7} \text{ M s}^{-1})/(10^{-11} \text{ M}) = 3.3 \times 10^4 \text{ s}^{-1}$$

Specificity constant $= k_{cat}/K_m$:

$$= (3.3 \times 10^4)/(2.4 \times 10^{-4}) = 1.4 \times 10^8 \text{ M}^{-1} \text{ s}^{-1}$$

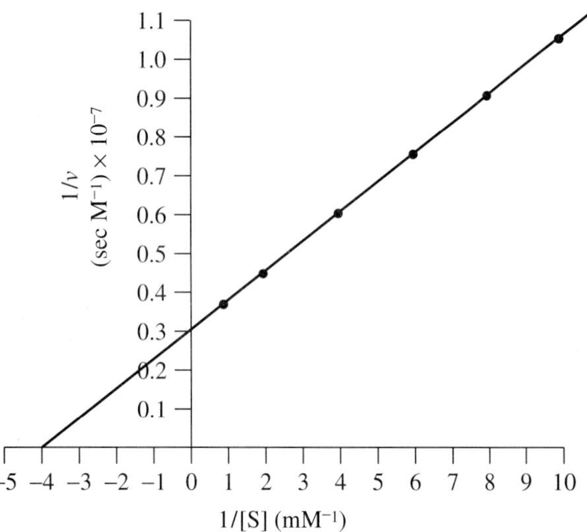

13. Often the lowest substrate concentration samples contain the greatest errors in concentration (because of problems delivering very small volumes). These latter samples then give the highest 1/[S] values. In drawing a line, it is human nature to place a greater emphasis on the larger 1/[S] values (i.e., the line ends up being primarily determined by the data points containing the greatest errors).
15. RNA was not expected to have a sufficient variety of functional groups and possible types of intramolecular interactions to form a structure capable of a specific interaction with a substrate. Also, before the mentioned discovery, all enzymes investigated had been found to be proteins, so all enzymes discovered in the future were expected to be proteins.
17. Of the substances listed in table 8.1, glucose is the preferred substrate for hexokinase and *N*-benzoyltyrosinamide is preferred by chymotrypsin. The preference is assessed by considering the K_m. The substance having the smallest K_m (lowest concentration of substrate that gives half-maximal velocity) is the preferred substrate.
19.

[S], mM	$\frac{1}{[S]}$	Glyoxylate Formed after 10 min, μM	$\frac{1}{V_0}$	Glyoxylate Formed after 10 min with 0.5 mM Oxaloacetate, μM	+I $\frac{1}{V_0}$
0.0318	31.45	0.0420	23.81	0.0040	250
0.0464	21.55	0.0583	17.15	0.0055	182
0.0593	16.86	0.0700	14.29	0.0075	133
0.1185	8.44	0.0955	10.47	0.0131	76.3
0.2222	4.50	0.1167	8.57	0.0233	42.9

Via a Lineweaver-Burk plot:

	V_{max}	K_m
−I	0.14 μM/10 min	0.077 mM
+I	0.14 μM/10 min	1.0 mM

Oxaloacetate is a competitive inhibitor.

Chapter 9

1. In general the amino acids with polar side chains participate in the catalytic process.
3. Only ser 195 has sufficient nucleophilic character to react with diisopropylfluorophosphate. This nucleophilicity is produced by the "charge relay system" described in the text.
5. Consider that the volume of a sphere = $4.18r^3$ and 10^{21} $nm^3 = 1 \text{ cm}^3$.

$$\frac{1 \text{ OH}^-}{4.18(0.15)^3 \text{ nm}^3} \cdot \frac{1000 \text{ cm}^3}{1 \text{ L}} \cdot \frac{10^{21} \text{ nm}^3}{1 \text{ cm}^3} \cdot \frac{1 \text{ mole}}{6 \times 10^{23} \text{ ions}} = 118 \text{ M}$$

This is a pH of 16.1! It should be realized that in terms of a solution this is not a realistic concentration; however, this number is still valid in considering volumes comparable to those of molecular volumes. If this minute volume happens to contain a hydroxide ion, then this is a valid pH for that volume. This gives us another way to visualize enzyme mechanisms; in a minute space an enzyme can produce pH extremes.
7. (a) The reaction catalyzed by lactate dehydrogenase

$$CH_3—CHOH—COO^- + NAD^+ \rightarrow CH_3—CO—COO^- + NADH + H^+$$

is shifted toward product formation, as is any chemical reaction, either by decreasing product concentration, increasing reactant concentration, or both. A proton is released as a product of the lactate dehydrogenase reaction, and increasing pH lowers the H^+ concentration with a shift of the reaction toward pyruvate formation.

An increase of 1 pH unit lowers proton concentration by a factor of 10, but there are limits to the amount that pH can effectively be increased in an enzyme-catalyzed reaction, because many enzymes are inactivated at high pH.

(b) Chemically trapping the pyruvate will have the same effect as increasing the pH of the lactate dehydrogenase reaction. Including a chemical trap to react with pyruvate removes the product, resulting in a shift toward product production. However, NADH, the other product, will continue to accumulate.

(c) Reoxidation of the NADH by the nonenzymatic reaction

$$\text{NADH} + \text{Tetrazolium}_{\text{ox}}\ (\text{yellow}) \rightarrow \text{NAD}^+ + \text{Formazan (purple)}$$

has the advantages of:

(1) removing the product NADH from the equilibrium;

(2) replenishing the reactant NAD^+, providing a longer linear initial rate; and

(3) increasing the sensitivity of the lactate dehydrogenase activity measurement. NADH absorbs light at 340 nm with an extinction coefficient of 6.22 $\text{mM}^{-1}\ \text{cm}^{-1}$. The formazan absorbs at 578 nm with an extinction decrease of 15 $\text{mM}^{-1}\ \text{cm}^{-1}$. The increased sensitivity of the formazan chromophore thus decreases the amount of enzyme required in each assay. (See Abdallah, M. A., and J.-F. Biellmann, *Eur. J. Biochem. 112*:331–333, 1980, for details of the tetrazolium-coupled assay.)

9. The active site could retain the arginine residue to provide electrostatic interaction and stabilization of the substrate carboxyl terminus as is the case in carboxypeptidase A. The substrate binding site in the enzyme would be predicted to contain acidic and polar side chains to bind the basic side chain of the residue to be cleaved. Mechanistic features including polarization of the carbonyl by metal (possibly zinc) and the electrostatic stabilization of the tetrahedral anionic intermediate by the active site transition metal may be similar to those of carboxypeptidase A.

11. (a) The active center amino acid residues shown in the diagram form a catalytic triad similar to the Asp-His-Ser arrangement found in the trypsin family of proteases, except for the substitution of the thiol (—SH) for the alcohol (—OH).

The nucleophilicity of the cysteine sulfhydryl is enhanced by polarization of the sulfur–hydrogen bond by the neighboring imidazole group from histidine. The thiolate adds to the carbonyl, forming an anionic tetrahedral intermediate presumably stabilized by electrostatic interactions with adjacent groups of peptide bond hydrogens. The release of the C-terminal portion of the peptide substrate leaves a thioester as the covalent adduct at the active site. Subsequent hydrolysis of the thioester thus releases the N-terminal portion of the substrate. A mechanism illustrating these fundamentals is shown.

(b) The sulfhydryl group of cysteine adds to the activated double bond of *N*-ethylmaleimide to form the S-(*N*-ethyl) succinimido thioether adduct whose structure is shown.

Cysteine

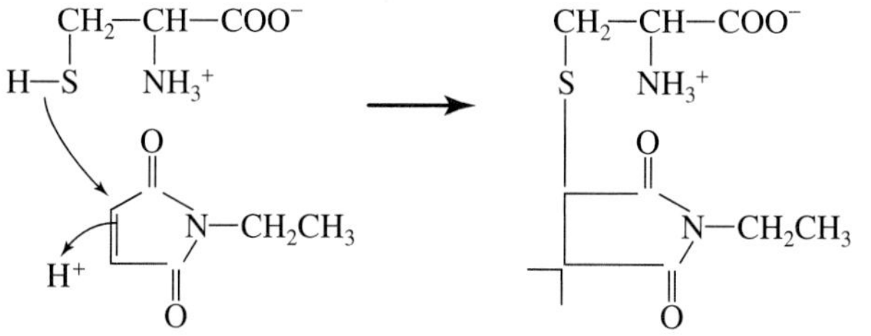

N-ethylmaleimide → S-(*N*-ethyl) succinimido-L-cysteine

One would predict the reactivity of the cysteine SH group with *N*-ethylmaleimide (NEM) to increase with increasing pH if the thiolate anion reacts more rapidly with NEM than does the protonated SH group. The pK_a of the cysteine —SH group is 8.3. The fraction of the cysteine existing as the thiolate anion can be calculated by using the Henderson-Hasselbalch expression:

$$\text{pH} = \text{p}K_a + \log\ (\text{RS}^-)/(\text{RSH})$$

where RS^- and RSH represent the fraction of sulfhydryl group in the thiolate and protonated forms, respectively.

At ph 5 the ratio of $(RS^-)/(RSH)$ is 5.0×10^{-4}. However, at pH 7.5 the ratio $(RS^-)/(RSH)$ is 1.6×10^{-1}, a value about 320-fold larger than the ratio at pH 5. The reaction of cysteine with NEM at pH 7.5 is therefore faster than at pH 5.5 because of the larger fraction of the thiolate anion.

(c) The cysteinyl SH group in the active center of papain would be predicted to react more rapidly with NEM than would any of the other cysteines in the protein because of the increased nucleophilic character of the active site sulfhydryl group. The increased nucleophilicity is a result of the polarization of the proton from the SH group afforded by the adjacent imidazole.

13. The transition-state intermediate is bound more tightly to the enzyme active site than are either substrates or products and presumably fits well into the three-dimensional structure of the active site. Transition-state analogs are chemically similar to the transition-state intermediate and should also be tightly bound by the enzyme active site. The transition-state analogs, used as antigens, should elicit amino acid residues in the antibody combining site that are structurally complementary to the enzyme active site. Hence rate enhancement and substrate specificity similar to the enzyme could be mimicked. Antibodies catalyzing reactions requiring a cofactor, or specific coenzyme functional residue, pose the added difficulty of choosing and synthesizing the appropriate transition-state analog or attachment of a cofactor by chemical means after the catalytic antibody is formed. (See Schultz, P. G., R. A. Lerner, and S. J. Benkovic, *Chem Eng. News.* May 28, 1990, 26–40; and Benkoric, S. J., *Annu. Rev. Biochem.* 61:29–54, 1992, for reviews.)

Chapter 10

1. Factor VIII is a protease zymogen, and it must be hydrolyzed to be activated.
3. Yes. If the interaction weren't an equilibrium, then it would be either a covalent modification or indistinguishable from a covalent modification.
5. An intra-protein disulfide bond has a negligible impact on the molecular mass of a protein, while an inter-protein disulfide essentially doubles the molecular mass of the protein. A method such as gel filtration (chapter 7) that separates on the basis of size can be used to distinguish between intra- and inter-molecular disulfide bonds in proteins.
7. The Ile in chymotrypsinogen is the sixteenth amino acid. During the conversion to chymotrypsin a 15-amino-acid peptide is cleaved from chymotrypsinogen. The original amino acid number system was retained in chymotrypsin, making the N-terminus amino acid number 16.
9. In principle, futile cycles may occur if two enzymes (or two metabolic pathways) catalyze reactions that are the reverse of each other. The result of a futile cycle is generally the hydrolysis of high-energy intermediates, nonproductive use of reductant, or both. Futile cycling may be minimized in the cell by regulation of key enzymes in the opposing pathways or by physically separating the opposing pathways in different subcellular compartments. Each of those strategies has been demonstrated to occur in the cell.

    Protein kinase and phosphoprotein phosphatase catalyze opposing reactions. Consider the reactions

    $$\text{Protein (Ser)} + \text{ATP} \rightarrow \text{Protein (Ser-P}_i) + \text{ADP} \qquad \textbf{(1)}$$

    $$\text{Protein (Ser-P}_i) + H_2O \rightarrow \text{Protein (Ser)} + P_i \qquad \textbf{(2)}$$

    $$\text{Sum:} \quad \text{ATP} + H_2O \rightarrow \text{ADP} + P_i$$

    The enzymes are (1) protein kinase and (2) phosphoprotein phosphatase. The (Ser) represents a serine residue, in this example, that is reversibly phosphorylated. However, threonine and tyrosine side chains are also phosphorylated by protein kinases.
11. Metabolic pathways are frequently regulated at the step that commits a metabolite to the pathway, at branches in the pathway, and at the enzyme that catalyzes the final step in a branch of the pathway. In the example, assume that products G, H, and J are essential end products. Each would likely inhibit its own production without inhibiting production of the others. Thus product J could feedback-inhibit enzyme 7 and inhibit enzyme 8 by product inhibition, G inhibits enzyme 6, and H inhibits enzyme 9. Products G and H could cumulatively inhibit enzyme 4, assuming that it were the committed step for that branch of the pathway. The committed step (enzyme 1) may be cumulatively inhibited by each of the products. Each product alone may cause only minimal inhibition.
13. Phosphofructokinase (PFK) catalyzes the transfer of phosphate from ATP to C-1 of fructose-6-phosphate. Hence ATP is a substrate for the enzyme. PFK is the committed step in the glycolytic pathway whose product is pyruvate, a metabolite whose oxidation drives ADP phosphorylation in the mitochondria. Activity of PFK is therefore responsive to the relative amount of ATP in the cytosol. Two possible mechanisms by which PFK is inhibited by APT include (1) substrate inhibition by ATP and (2) interaction of ATP at a regulatory site different from the active site. Substrate inhibition occurs upon binding of the substrate (ATP) to the enzyme-product complex, forming a dead-end (catalytically inactive) ternary complex. Thus the complex (PFK-fructose-1,6-bisphosphate-ATP) would be catalytically inactive. PFK is inhibited upon binding ATP to a regulatory site that is different from the active site. ATP thus plays a dual role as a substrate and as a ligand that binds to a regulatory site (see fig. 10.6 in the text).
15. Binding of O_2 to the iron of hemoglobin causes the iron to shift to a low-spin *d*-electron configuration from a high-spin configuration. The low-spin Fe^{2+} has a smaller ionic radius compared to the high-spin Fe^{2+}, allowing the iron to move approximately 0.7 Å into the plane of the heme ring. The imidazole group liganded to the heme iron also moves. This motion is transmitted to the other subunits in the tetrameric hemoglobin through the interaction of the two α and two β subunits. Separation of the tetramer into two $(\alpha + \beta)$ dimers would destroy the cooperativity exhibited by the tetramer. There could be some cooperativity exhibited by the $(\alpha + \beta)$ dimer, but not likely on the scale exhibited by the hemoglobin tetramer. Thus there should be a greater degree of saturation with O_2 at lower O_2 tensions to the dimer than to the tetramer. The O_2-binding curve for the (α and β) dimer would likely resemble the O_2-binding curve of myoglobin.

    The 2,3-bisphosphoglycerate (2,3-BPG) binds to deoxyhemoglobin and stabilizes the conformation of the deoxyhemoglobin relative to the oxyform. The 2,3-BPG binds in a crevice formed at the intersection of the subunits of the tetramer. Were the tetramer dissociated to dimers, the crevice would be destroyed, precluding the binding and subsequent stabilizing effect of 2,3-BPG. Thererore 2,3-BPG would not be expected to alter the O_2 affinity of the $(\alpha + \beta)$ dimers.

Chapter 11

1. An early step in the transaminase mechanism is the loss of the α hydrogen to the surface of the enzyme. Presumably, this hydrogen on the surface of the enzyme is not readily exchangeable with the solvent, and this general acid serves as the source of the α hydrogen on amino acid 2.
3. Typical metabolic biological redox reactions involve two electrons and two hydrogen ions. In the case of $FAD/FADH_2$ the two electrons and two hydrogen ions are "nicely contained" in the reduced form. The $NAD^+/NADH$ situation is another story. The two electrons, but only one of the hydrogen ions, are incorporated into the NADH. The second hydrogen ion is released as a free ion. The $NADH_2$, although incorrect, is a simplification. The disturbing aspect comes when it is realized that the $NADH_2$ users may not know the actual chemistry involved.

5. (The mechanism shown in parts (a) and (b) below is based on structures shown in figure 11.5 in text.)

(a) Refer to figure 11.5*a* in the text for the structures of each intermediate step.

PLP

Schiff base

Steps 1. & 2. Form Schiff base
3. Remove proton α-C
4. Protonate C-4'
5. & 6. Hydrolyze Schiff base
Release α-keto acid and pyridoxamine phosphate

Transfer of amino group from pyridoxamine phosphate to pyruvate, forming alanine occurs by reversal of the steps shown above. Other transaminases use other α-amino acids and α-keto acids.

(b) Figure 11.5*b* in the text gives the structures of intermediates shown in the proposed mechanism. Alternative mechanisms are described by Walsh and Boeker and Snell. (See the Selected Readings at the end of the chapter.)

α-Decarboxylation occurs from Schiff base form reactions 1–2, in part (a):

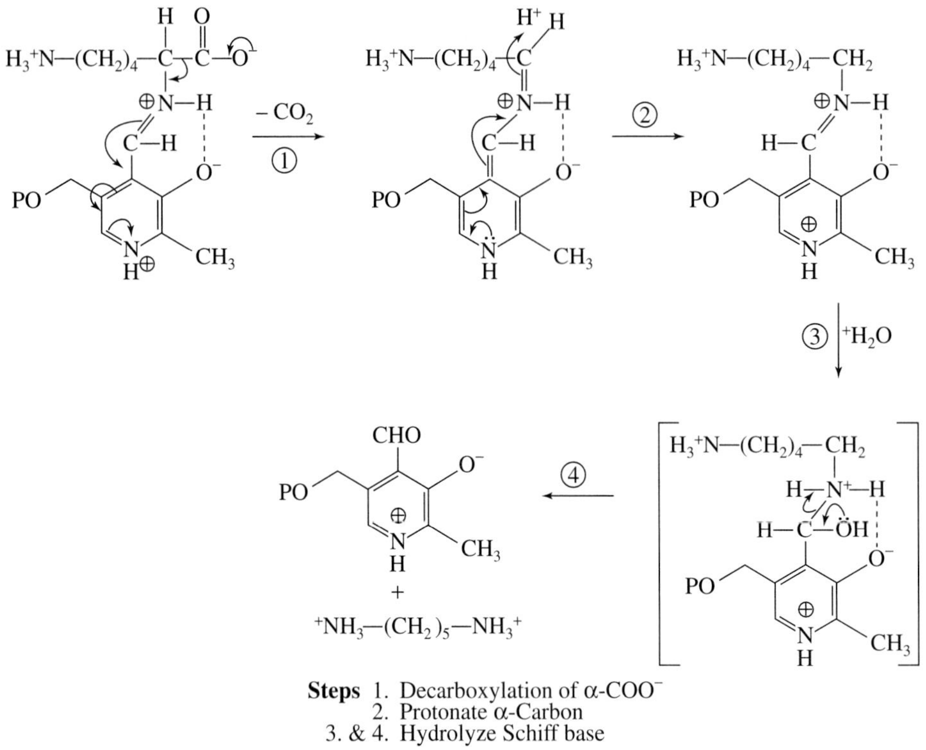

Steps 1. Decarboxylation of α-COO^-
2. Protonate α-Carbon
3. & 4. Hydrolyze Schiff base

(c) β-Decarboxylation. Form Schiff base steps 1–3, from part (a):

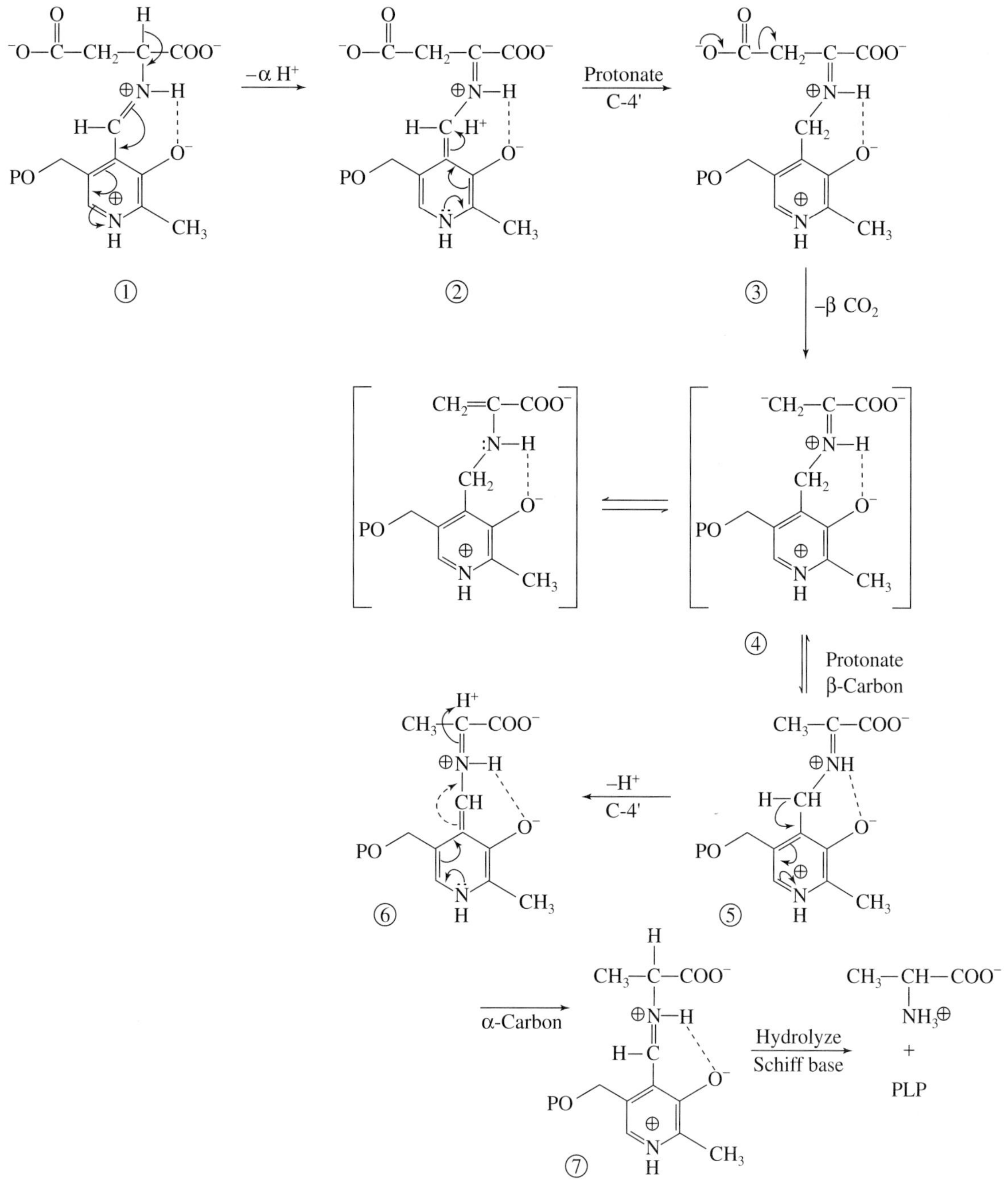

−α H^+

Protonate C-4'

① ② ③

−β CO_2

④

Protonate β-Carbon

−H^+ C-4'

⑥ ⑤

α-Carbon

Hydrolyze Schiff base

⑦

$CH_3-CH-COO^-$ $NH_3^{\oplus}$ + PLP

7.

(a)

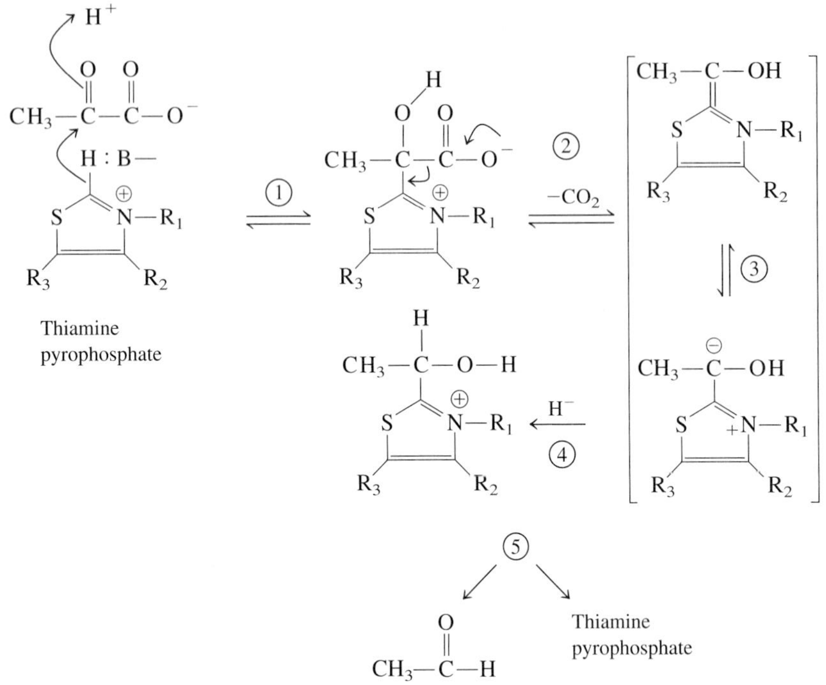

H^+

$CH_3-\overset{O}{\overset{\|}{C}}-\overset{O}{\overset{\|}{C}}-O^-$

H : B—

S N—R_1 ⊕

R_3 R_2

Thiamine pyrophosphate

① ⇌

$CH_3-C(OH)-\overset{O}{\overset{\|}{C}}-O^-$

S N—R_1 ⊕

R_3 R_2

② $-CO_2$ ⇌

CH_3-C-OH

S N—R_1

R_3 R_2

⇅ ③

$CH_3-\overset{\ominus}{C}-OH$

S +N—R_1

R_3 R_2

H^- ← ④

$CH_3-C(H)-O-H$

S N—R_1 ⊕

R_3 R_2

⑤

$CH_3-\overset{O}{\overset{\|}{C}}-H$

Thiamine pyrophosphate

Steps 1. Deprotonation of TPP to ylid form, nucleophilic addition of ylid to α-ketogroup
2. Decarboxylation
3. Resonance stabilization
4. Protonation, elimination of TPP

(b)

Fructose-6-P

H^+

$HO-CH_2-C(=O)-CH(OH)-CH(OH)-HC(OH)-CH_2OP$ ① $+H^+$ → $HO-CH_2-C(OH)(TPP)-C(H)-(CHOH)_2CH_2OP$

TPP

② → Erythrose-4-P ($CHO-(CHOH)_2-CH_2OP$)

③ Resonance forms: $H-C(OH)(H)-C^{\ominus}(OH)-TPP^{\oplus}$ ⇌ $H-C(OH)(H)-C(OH)=TPP$

④ OH^-

$HC{=}C-OH$ (TPP) → $H_3C-C(=O)$ (TPP)

Acetyl TPP ⑤ P_i → $CH_3-C(=O)-OP$ (Acetyl phosphate) + TPP

Steps 1. Deprotonation to ylid form, nucleophilic addition of ylid to fructose-6-phosphate
2. Oxidation of C-3, release of erythrose-4-phosphate
3. Resonance stabilization of intermediate
4. Elimination of -OH from C-2
5. Transfer of acyl group to phosphate

9. (a) Malate dehydrogenase

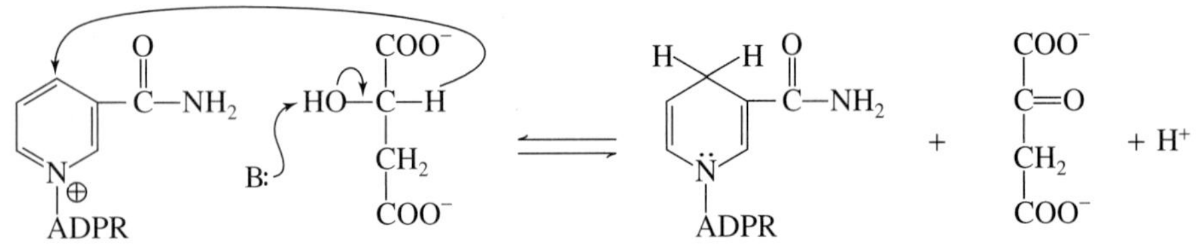

NAD^+ + L-malate ⇌ NADH + Oxaloacetate + H^+

(NAD⁺) (L-malate) (NADH) (Oxaloacetate)

Hydride transfer from C-2 of malate to nicotinamide ring, loss of H^+ from C-2—OH

(b) Malic enzyme

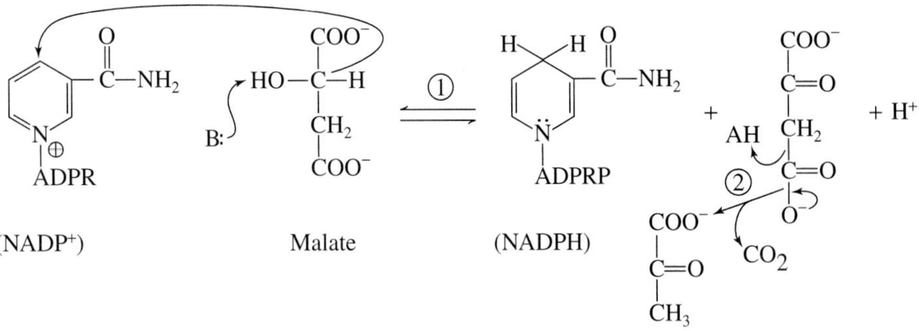

(NADP⁺) Malate (NADPH) Pyruvate

Steps 1. Transfer of hydride from C-2 of malate to $NADP^+$.
2. β-Decarboxylation of oxaloacetate.

11. (a) Reduced flavins in solution are rapidly reoxidized by molecular oxygen forming superoxide radical and H_2O_2, metabolites of oxygen that are toxic to the cell. Were free FAD or FMN used in biological transfer, for example in the mitochondrial electron system, oxygen could compete with other oxidants for the reduced flavin and disrupt energy transduction. If the flavin component of mitochondrial succinate dehydrogenase were directly reoxidized by molecular oxygen, rather than by ubiquinone (see chapter 16 in the text), oxidative phosphorylation driven by succinate oxidation would be abolished. However, sequential transfer of the reducing equivalents via the electron transport chain to cytochrome oxidase drives phosphorylation of approximately 1.5 moles ADP per mole of succinate with the reduction of O_2 to H_2O rather than H_2O_2. Enzyme-bound flavins are usually shielded from uncontrolled oxidation by O_2.

(b) Tightly bound $NAD(P)^+$ is an advantage to enzymes catalyzing rapid H:—removal and readdition to a substrate in a stereospecific fashion. Freely diffusing NAD(P)H is an advantage in transferring reducing equivalents among catalytic sites of various dehydrogenases. NAD(P)H in solution is not rapidly oxidized by molecular oxygen, as is soluble FADH or FMNH, and is suited in its role as a freely diffusing redox reagent in the cell.

13. The flavoprotein (amino acid oxidase) oxidizes the α-amino acid to an α-imino acid with reduction of the enzyme-bound flavin. In aqueous solution the imino acid hydrolyzes with the release of the ammonium ion and formation of the α-keto acid. If not excreted, the ammonium ion may accumulate to toxic levels within the cell. The reduced flavoprotein is oxidized by molecular oxygen, forming H_2O_2. Although most cells have antioxidant enzymes (catalase, glutathione peroxidase) to scavenge the hydrogen peroxide, these defenses may be breached if the production of H_2O_2 is elevated.

In PLP-dependent transamination the amino group is retained in organic molecules of low toxicity. The α-amino group is transferred from the amino acid to the pyridoxal phosphate at the active site of the amino transferase, and the α-keto acid is released. Rather than being hydrolyzed and released into solution, the amino group of pyridoxamine is transferred to an acceptor α-keto acid, in some cases with ultimate transfer to a nontoxic excretory product (urea).

15. Polycyclic aromatic hydrocarbons, derived primarily from the pyrolysis of organic materials, are hydrophobic and dissolve in the lipid fraction of the cell. Their hydrophobicity in general makes their excretion from the body difficult. In addition, their lack of reactive functional groups precludes attachment of hydrophilic substituents to form water-soluble adducts. The liver microsomal cytochrome P-450 (polysub-

strate monoxygenase) system oxidizes the polycyclic aromatic hydrocarbon (PAH) by addition of an oxygen atom, derived from O_2, to a double bond, to form an epoxide. Hydrolysis of the epoxide, catalyzed by epoxide hydrolase, yields hydroxyl groups (dihydrodiol), thus generating functional groups that accept glucuronic acid from UDP-glucuronic acid. The glucuronide adduct is significantly more hydrophilic than either the unmetabolized PAH or the hydroxylated PAH and is subsequently excreted. (See Grover, P. L., *Xenobiotica,* 16:915–931, 1986.)

17. The following coenzymes contain the AMP moiety: NAD^+, NADH, $NADP^+$, NADPH, FAD, $FADH_2$, and CoASH.
19. Most definitions for nucleotides and nucleosides include the work "sugar," and most definitions for sugars require either an actual or a potential aldehyde or ketone. The substance present in riboflavin and flavin mononucleotide is a reduced sugar or a sugar alcohol (ribitol); therefore they are not a nucleoside or a nucleotide, but for simplicity and historical reasons they are included in this category.
21. α-Keto acid dehydrogenases contain bound thiamine pyrophosphate, lipoyl, and FAD. NAD and CoASH are "substratelike" coenzymes. Counting the pantotheinate in the CoASH, here are five vitamin components in this reaction/enzyme system.
23. A solution of an iron–sulfur protein is colored and, when acidified, produces H_2S, which becomes obvious to most individuals.

Chapter 12

1. for ATP + AMP $\rightleftharpoons$ 2 ADP

   (a) ATP $\rightarrow$ ADP + P_i − 8.4 kcal/mole

   (b) ATP $\rightarrow$ AMP + PP_i − 8.4 kcal/mole

   (c) $PP_i \rightarrow 2\ P_i$ − 7.9 kcal/mole

   2 × reaction a − reaction b − reaction c $\rightarrow$ ATP + AMP $\rightarrow$ 2 ADP 0.5 kcal/reaction

   0.5 kcal/reaction $\rightarrow K_{eq}$ 0.43 (This is less than the value given in text.)

   Via energy charge,

   $$0.9 = \frac{[ATP] + 0.5[ADP]}{[ATP] + [ADP] + [AMP]}$$

   $$0.43 = \frac{[ADP]^2}{[ATP][AMP]}$$

   Letting [ATP] = 1 and solving, we have [ATP]:[ADP]:[AMP] = 1.00:0.14:0.05.

3. Some common isotopes used by biochemists include the following:

   ^2H: Stable; detected by NMR, MS; few compounds available.

   ^3H: Radioactive; many compounds available; detected by liquid scintillation.

   ^{13}C: Stable; detected by NMR, MS; few compounds available.

   ^{14}C: Radioactive; many compounds available; detected by liquid scintillation.

   ^{15}N: Stable; detected by NMR, MS; few compounds available.

   ^{18}O: Stable; detected by NMR, MS; few compounds available.

   ^{32}P: Radioactive; limited number of compounds available; detected by liquid scintillation; 14-day half-life can be a problem.

   ^{33}P: Radioactive; limited number of compounds available; detected by liquid scintillation; 25-day half-life can be a problem.

   ^{35}S: Radioactive; few compounds available; detected by liquid scintillation; 88-day half-life can be a problem.

5. If nature were to try to reverse a reaction promoted by a kinase, she would face a major equilibrium position problem. ATP is being synthesized from compounds with a relatively low energy content! This could happen only if the substrate for the kinase were in very high concentration (i.e., a previous reaction that was very exothermic) and if the reaction following the kinase were very exothermic and could assist by "pulling" the reaction forward.
7. A metabolic pathway is a sequence of enzymes and chemical intermediates. The chemical intermediates are interconverted by the enzymes. There is no specific arrangement of pathways; the sequence of steps can be linear, branched, or cyclic.
9. NAD^+ is typically reduced to NADH during the oxidation of food stuffs. The oxidation of NADH to NAD^+ by the electron transport scheme produces the majority of the ATP utilized by each cell. NADPH, produced mainly in the pentose phosphate pathway (see chapter 14), is primarily used as a reducing agent during biosynthesis processes.
11. The conversion of glucose to pyruvic acid represents an oxidation of carbons originally present in the glucose. When written as an overall equation, it becomes

    $$\underset{\text{Glucose}}{C_6H_{12}O_6} \rightarrow \underset{\text{Pyruvic acid}}{2\ C_3H_4O_3}$$

    To balance this reaction, 4 H^+ + 4 e^- must be added to the right side. The fact that glucose loses electrons indicates that an oxidation has occurred.
13. Glycolytic catabolism of glucose is a thermodynamically favorable process that results in the formation of lactate and net phosphorylation of ADP to form ATP. If gluconeogenesis were simply the reversal of glycolysis, the process would be thermodynamically unfavorable. Although glycolysis and gluconeogenesis have some reactions in common, there are key steps in gluconeogenesis that allow the input of metabolic energy and in this way overcome the thermodynamic barriers. ATP (4 moles) and GTP (2 moles) are required per mole of glucose synthesized from lactate in the liver. The gluconeogenic pathway becomes thermodynamically favorable because of the ATP and GTP energy input.
15. Futile cycles are series of metabolic reactions or pathways whose combined reaction is that of hydrolyzing high-energy metabolic intermediates for no apparent metabolic advantage. The reactions

$$\text{Glucose} + \text{ATP} \rightarrow \text{Glc-6-P}_i + \text{ADP (glucokinase)} \quad (1)$$

$$\text{Glc-6-P}_i + \text{HOH} \rightarrow \text{Glucose} + \text{P}_i \text{ (glucose-6-phosphatase)}$$
in sum catalyze the hydrolysis of ATP: (2)

$$\text{ATP} + \text{HOH} \rightarrow \text{ADP} + \text{P}_i \text{ (ersatz ATPase)} \quad (3)$$

The net standard-state free energy change ($\Delta G^{\circ\prime}$) for reaction (3) is about -7.5 kcal/mole, and the equilibrium will strongly favor product formation.

If the glucokinase and glucose-6-phosphatase reactions were unregulated, rapid hydrolysis of ATP with release of heat energy would ensue. However, glucokinase, a liver cytosolic enzyme, and glucose-6-phosphatase, an enzyme located in the liver endoplasmic reticulum, are separated by a membrane that imposes a permeability barrier to Glc-6-P, a common metabolite. Hence regulating the transport of Glc-6-P from cytosol to the endoplasmic reticulum diminishes futile cycling.

Chapter 13

1. The geometry around the highest numbered chiral carbon, which is typically the next-to-last carbon in a simple sugar, determines the D- or L- nature of that monosaccharide. (Also, this is a good word to know for the GRE.)
3. The number of sugars is 2^n, where n is the number of chiral centers but does not include the chiral carbon involved in producing the D-series of sugars. Because carbon 2 of ketoses is a carbonyl group, only the geometry of carbons 3 and 4 remain to produce the four D-ketohexoses ($2^2 = 4$). The carbonyl group on carbon 1 of the aldoses allow carbons 2, 3, and 4 to produce eight D-aldohexoses ($2^3 = 8$).
5. Often, pyranoses and furanoses are portrayed by using Fischer projections containing "bent" bonds (which should raise the blood pressure of any properly trained biochemist). The use of the projectional Haworth is considered to be superior and may be surpassed only by the perspective drawings used to show substituted cyclohexanes in the chair conformation.
7. The fraction of each anomeric form may be calculated by using two linear equations. Let X be the fraction of β anomer and Y the fraction of α anomer. The fractions equal 1 (or 100%):

$$X + Y = 1$$

The specific rotation contributed to the mixture is a function of the amount of each anomer:

$$18.7X + 112.2Y = 52.7$$

Solving the two linear equations reveals that the mixture contains 0.364 (36.4%) α anomer and 0.636 (63.6%) β anomer.
9. Cellulose is a polymer of glucose linked in a linear chain through β(1,4) glycosidic bonds. Exhaustive methylation and mild acid hydrolysis should yield a preponderance of 2,3,6-tri-*O*-methylglucose, derived from the glucosyl residues in the chain, and a small fraction of 2,3,4,6-tetra-*O*-methylglucose from the nonreducing terminal residues.

Glycogen is a highly branched polymer of glucose molecules linked via α(1,4) glycosidic bonds with α(1,6)-linked branches. Methylation and acid hydrolysis will yield the following mixture: 2,3,4,6-tetra-*O*-methylglucose, derived from the nonreducing terminal glucose residues; 2,3,6-tri-*O*-methylglucose, from glucosyl residues in the chain; and 2,3-di-*O*-methylglucose from the glucose residues at the branchpoint. One can differentiate glycogen from cellulose by the presence of the 2,3-di-*O*-methylglucose. One would also expect the ratio of the 2,3,4,6-tetra-*O*-methylglucose to 2,3,6-tri-*O*-methylglucose to be larger in glycogen than in cellulose because of the larger number of branches in glycogen, each contributing a nonreducing glucosyl residue.
11. Trehalose (A) is two α-D-glucoses linked through carbon 1 of each sugar. However, the correct answer to this question would also include (B) two β-D-glucoses linked through carbon 1 and (C) one α-D-glucose linked through carbon 1 to carbon 1 of a β-D-glucose.

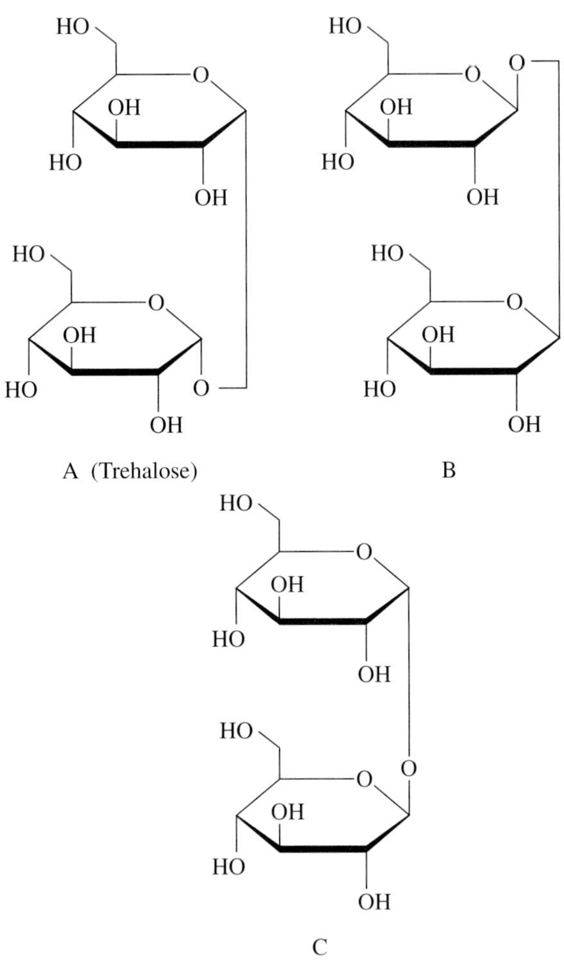

13. Maltose, lactose, and cellobiose are reducing sugars; sucrose and trehalose are nonreducing sugars.

Chapter 14

1. Isolation excludes oxygen, producing an anaerobic environment and enticing the yeast to make ethanol. The bubbles are carbon dioxide. When the bubbles cease, it is the indication that the yeast has exhausted the fermentable carbon

source. The syrup provides an additional carbon source for fermentation, but the primary purpose at this point is carbonation (driven into solution by the confinement).

3. Carbon 1 of glucose is lost as CO_2 in an early reaction in the pentose phosphate pathway. Carbons 3 and 4 of glucose are lost as CO_2 with the formation of acetyl-CoA after glycolysis. Carbons 1, 2, 5, and 6 are lost in the various turns of the citric acid cycle. By assessing the rate of $^{14}CO_2$ production from ^{14}C-1-glucose versus ^{14}C-6-glucose, the relative significance of the different pathways in a tissue can be determined.

5. Glycerate-2,3-bisphosphate was encountered as an allosteric effector of hemoglobin (see chapter 6).

7. In the first example the actual chemistry is identical: an aldose-ketose interconversion. Notice that the geometry of the alcohol on carbon 2 is identical. The second example has essentially the same chemistry. It is logical that nature could have stumbled upon the same chemical mechanism twice or, more likely, that the mechanism evolved once, and following gene duplication, the segments of the gene controlling the substrate specificity changed on one copy of the gene.

9. (a) Hydrolysis rather than phosphorolysis of the Ga3PDH-bound thioester releases 3-phosphoglycerate rather than 1,3-bisphosphoglycerate:

E—SH + H—C(=O)—CHOH—CH_2OP_i (E bound to NAD^+) —Thiohemiacetal formation→ E—S—C(O^-)(H)—R (NAD^+) → NADH / NAD^+ → E—S—C(=O)—R (NAD^+), ^-OH —Hydrolysis→ E(SH)(NAD^+) + COO^-—CHOH—CH_2OP_i

3-Phosphoglycerate

E—SH represents glyceraldehyde-3-phosphate dehydrogenase

(b) There would be no net yield of ATP if glycolysis began with glucose. Conversion of glucose to fructose-1,6-bisphosphate (FBP) requires 2 moles of ATP per mole of FBP formed. Cleavage of FBP by aldolase yields the equivalent of 2 moles of glyceraldehyde-3-phosphate ($Ga3P_i$). The oxidation and phosphorolysis of $Ga3P_i$ by Ga3PDH is one of the two phosphate activation steps in glycolysis. The dehydration of 2-phosphoglycerate by enolase to form PEP is the second phosphate-activation step. Metabolism of the two moles of $Ga3P_i$ to pyruvate yields 4 moles of ATP, a net 2 moles per glucose. Hydrolysis rather than phosphorolysis of the enzyme-bound thioester at the Ga3PDH step therefore abolishes net ATP yield from glycolysis.

(c) Considerably more free energy, and greater potential for ATP formation, is available from the aerobic metabolism of glucose compared with the anaerobic glycolysis of the sugar. Oxidation of the pyruvate formed in glycolysis should yield sufficient ATP for growth of the aerobe, even if the net ATP yield from glycolysis was zero.

11. Arsenate is structurally similar to inorganic phosphate and substitutes for phosphate in the Ga3PDH reaction. The enzyme-bound thioester is cleaved by arsenolysis rather than phosphorolysis to form a mixed anhydride of the phosphoglycerate and arsenate. The mixed anhydride spontaneously hydrolyzes, regenerating the arsenate and forming 3-phosphoglycerate. Arsenate acts catalytically rather than stoichiometrically in dispelling the energy of the thioester bond, and the substrate-level formation of ATP is bypassed because the phosphoanhydride (1,3-bisphosphoglycerate) is not formed. Arsenate does not uncouple phosphorylation at the pyruvate kinase step because there is no labile organic intermediate with which arsenate can react to displace or replace the phosphate. The activated phosphate at C-2 of 2-phosphoglycerate was derived initially from ATP, not orthophosphate.

13. The free energy available from a reaction depends on the energies of the products compared with the substrates. Dehydration of 2-phosphoglycerate "traps" phospho-(enol)pyruvate in the enolate form. The hydrolysis products of PEP are phosphate and the enol form of pyruvate, but (enol) pyruvate is significantly less stable than (keto) pyruvate and rapidly tautomerizes to the more stable keto form. The tautomerization drives the reaction strongly toward products, resulting in a larger free energy difference between substrate and product.

15. (a) **(1)** Pyruvate carboxylase (Review biotin-dependent carboxylation in chapter 11 of text.)

ATP + HCO_3^- + E—biotin → $^-$O—C(=O)—N(biotin ring, S, R) + ADP + P_i

Carboxybiotin

B: H—CH_2—C(=O)—C(=O)—O^- + $^-$O—C(=O)—N(biotin) → $^-$OC(=O)—CH_2—C(=O)—CO^-

Oxaloacetate
+
E-biotin

(2) GTP-dependent PEPcarboxykinase (See the review by Utter and Kollenbrander in *The Enzymes,* vol. 6 (ed. P. Boger), Academic Press, NY, pp. 136–154, 1972.)

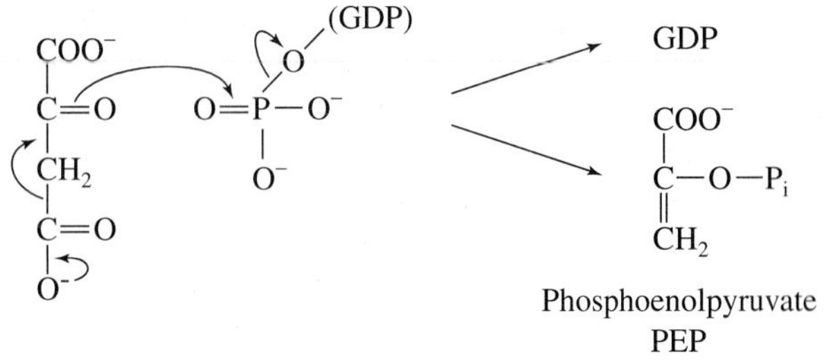

COO^-—C=O—CH_2—C=O—O^- + $O=P(O^-)_2$—O (GDP) → GDP + COO^-—C(—O—P_i)=CH_2

Phosphoenolpyruvate
PEP

(b) $\Delta G^{\circ\prime}$ overall is approximately -1.0 kcal/mole if the hydrolysis of PEP releases 14 kcal/mole and if the hydrolysis of ATP (GTP) releases 7.5 kcal/mole. If the overall free energy change for the coupled reactions is as calculated, the formation of PEP from pyruvate is metabolically feasible. Consider the following reactions and remember that the overall $\Delta G^{\circ\prime}$ is calculated from the sum of the component reactions.

$$\text{Pyr} + \text{ATP} + CO_2 \rightarrow \text{OAA} + \text{ADP} + P_i$$

$$\text{OAA} + \text{GTP} \rightarrow \text{PEP} + \text{GDP} + CO_2$$

$$\text{Pyr} + \text{ATP} + \text{GTP} \rightarrow \text{PEP} + \text{ADP} + \text{GDP} + P_i$$

but

$$\text{ATP} \rightarrow \text{ADP} + P_i \qquad \Delta G^{\circ\prime} = -7.5 \text{ kcal/mole}$$

$$\text{GTP} \rightarrow \text{GDP} + P_i \qquad \Delta G^{\circ\prime} = -7.5 \text{ kcal/mole}$$

$$\text{Pyr} + P_i \rightarrow \text{PEP} \qquad \Delta G^{\circ\prime} = +14 \text{ kcal/mole}$$

$$\text{Pyr} + \text{ATP} + \text{GTP} \rightarrow \text{PEP} + \text{ADP} + \text{GDP} + P_i \; \Delta G^{\circ\prime} = -1 \text{ kcal/mole}$$

17. (a) PFK-2 is active when the insulin/glucagon ratio is high (for example, when the animal is fed a carbohydrate-rich meal). FBPase-2 will be active when the insulin/glucagon ratio is low (blood glucose is low). Thus the level of fructose-2,6-bisphosphate, the activator of PFK-1 and inhibitor of FBPase-1, is increased when the insulin/glucagon ratio is high. The rate of glycolysis is increased, whereas the rate of gluconeogenesis is diminished. When the insulin/glucagon ratio is low, PFK-2 is inactive but FBPase-2 is active. The level of fructose-2,6-bisphosphate decreases, and PFK-1 is much less active because the level of its activator (fructose-2,6-bisphosphate) is low. The inhibition of FBPase-1, imposed by fructose-2,6-bisphosphate, is relieved and gluconeogenesis occurs.

(b) If fructose-2,6-bisphosphate were still present, although not being actively synthesized, it would continue to stimulate PFK-1 and inhibit FBPase-1. Thus simply halting the synthesis of the PFK-1 activator and FBPase-1 inhibitor may not be sufficient to inhibit glycolysis and activate gluconeogenesis.

(c) Activation of FBPase-2 causes the hydrolysis of fructose-2,6-bisphosphate and alleviates inhibition of FBPase-1 and prevents activation of PFK-1. Glycolysis is effectively inhibited because of inhibition of the rate-controlling enzyme, PFK-1, whereas gluconeogenesis is activated by alleviation of the inhibition of FBPase-2.

(d) The insulin level will markedly decline, so the insulin/glucagon ratio will decline. As indicated in part (a), FBPase-2, FBPase-1, pyruvate carboxylase, and PEP-carboxykinase activities should increase. PFK-2 and PFK-1 activities will decline.

(e) The administered insulin will restore the insulin/glucagon ratio. The cAMP level in the liver cell is predicted to decrease, as are the activities of FBPase-2, FBPase-1, pyruvate carboxylase, and PEP-carboxykinase as discussed in part (a).0

Chapter 15

1. The sequence

   Acetaldehyde + Acetyl-CoA → 3-Hydroxybutanoate → 2-Hydroxybutanoate → 2-Ketobutanoate (+NADH) → CO_2 + Propionyl-CoA (+NADH) → Propionate (+ATP) → Acrylate (+$FADH_2$) → 3-Hydroxypropionate → Acetaldehyde (+NADH)

   will produce the same amount of NADH, ATP, and FAD per acetyl-CoA as the "real cycle." There are several other possible sequences that could answer this question.

3. To return the four-carbon dicarboxylic acid to the oxidation state that was borrowed to start the cycle.

5. *cis*-Aconitate is a relatively stable compound that historically came from monkshood, a plant known since the Middle Ages for its antipyretic properties. *cis*-Aconitate is a substrate for aconitase (observed by Krebs); but when provided with citrate, aconitase produces isocitrate without the release of aconitate (observed later). *cis*-Aconitate is not a free intermediate with a different enzyme producing and consuming it within the cycle; therefore it is not listed as a free intermediate in the tricarboxylic acid cycle.

7. Two sequential oxidations involving NAD^+ produce 2-hydroxyacetic acid, which is then oxidized by another NAD^+-dependent dehydrogenase to produce glyoxylate. The latter item then condenses with acetyl-CoA to make malate. Actually, as the question was worded, any sequence of three oxidations involving NAD^+ to produce glyoxylate would be an acceptable answer.

9. The energy available in citryl-CoA could easily be captured in an ATP by a mechanism analogous to succinyl-CoA synthetase. However, because malate dehydrogenase is so unfavorable and is "pulled" by the equilibrium position of the next enzyme, citrate synthetase, any change could be critical. If the majority of the -7.7 kcal/mole (table 15.1) released by the reaction catalyzed by citrate synthetase were captured by the synthesis of ATP, there would be a major impact on the overall thermodynamics of the cycle.

11. The reactions catalyzed by isocitrate dehydrogenase and α-ketoglutarate dehydrogenase are both oxidative decarboxylation reactions. But the similarity ends at that point. In one case a secondary alcohol is oxidized to a ketone; in the other a ketone is oxidized to the carboxylic acid level of oxidation (a thioester). With isocitrate, dehydrogenase oxidation clearly occurs prior to decarboxylation, while with α-ketoglutarate dehydrogenase, decarboxylation occurs before oxidation. The energy released during the oxidative de-

carboxylations (table 15.1) is substantially different. This is particularly apparent when the energy content of the thioester produced is considered (recovered by succinyl-CoA synthetase).

13. In general, only pyruvate, citrate, and malate readily cross the mitochondrial membrane (fig. 15.16).
15. (a) There will be stoichiometric conversion of 3 μmoles pyruvate to acetyl-CoA with release of 3 μmoles CO_2 and production of 3 μmoles of NADH. However, there will be no further metabolism of the acetyl-CoA because the TCA cycle intermediates are lacking. The amounts of GDP and P_i will remain unchanged.
    (b) Upon addition of the TCA cycle intermediates, acetyl-CoA will condense with oxaloacetate to form citrate. Aconitase will catalyze the formation of isocitrate, in equilibrium with citrate. However, isocitrate dehydrogenase will be inactive because little, if any, NAD^+ will be available. The NAD^+ initially added to the mixture was reduced stoichiometrically by the pyruvate dehydrogenase activity. Hence there should be no detectable increase in CO_2 released from the TCA cycle.
    (c) Reoxidation of the NADH in the absence of TCA cycle intermediate(s) or in the absence of pyruvate carboxylase would have no affect on CO_2 evolution. The TCA cycle intermediates are lacking, and there is no mechanism to replenish them in the system described in the problem.
    (d) Reoxidation of the NADH will regenerate NAD^+ and alleviate the inhibition of the TCA cycle, and CO_2 will be evolved: 6 additional μmoles from the 3 μmoles of acetyl-CoA and some CO_2 from the citrate, isocitrate, and α-ketoglutarate that were added. These intermediates will be oxidized. GDP will be phosphorylated to GTP by the succinate thiokinase reaction.
17. (a) Hydroxypyruvate $\rightarrow$ B $\rightarrow$ D $\rightarrow$ C $\rightarrow$ A $\rightarrow$ pyruvate
    (b) Hydroxypyruvate + NADH + H^+ $\rightarrow$ Pyruvate + NAD^+ + HOH
    (c) (A) Phosphoenolpyruvate; (B) glycerate; (C) glycerate-2-phosphate; (D) glycerate-3-phosphate. A, C, and D are glycolytic intermediates.
    (d) ATP is used by the glycerate kinase to phosphorylate glycerate (intermediate B); that step provides the phosphorylated substrate required by the phosphoglyceromutase. In the phosphoglyceromutase reaction there is a net transfer of phosphate from C-3 to C-2 of the glycerate moiety. The 2-phosphoglycerate is dehydrated by the catalytic action of enolase, and the product, PEP, is used by pyruvate kinase to phosphorylate ADP. Hence the ATP used to phosphorylate glycerate is regenerated when pyruvate is formed from PEP.
19. Glyceraldehyde-3-phosphate is bound to glyceraldehyde-3-phosphate dehydrogenase in a thiohemiacetal bond that is subsequently oxidized to an enzyme-bound thioester. Phosphorolytic cleavage of the thioester forms free enzyme, and the product, 1,3-bisphosphoglycerate (a mixed anhydride) diffuses from the enzyme. 1,3-Bisphosphoglycerate is substrate for the ADP-dependent phosphoglycerokinase whose products are 3-phosphoglycerate and ATP. In the reaction sequence described, the inorganic phosphate is activated by addition to an organic substrate.

    The formation and transfer of "activated phosphate" by the succinate thiokinase follow a route similar to that described except that an enzyme-bound phosphoryl group is involved. Succinyl-CoA is formed by α-ketoglutarate dehydrogenase from the oxidative decarboxylation of α-ketoglutarate. (a) The succinate thiokinase catalyzes phosphorolysis of the succinyl-CoA to succinylphosphate, a mixed anhydride. (b) The phosphate from the mixed anhydride is transferred to an imidazole side chain of an active site histidine, forming a phosphoramide. The phosphate is subsequently transferred from the phosphoramide on the enzyme to GDP. Each of the enzyme pairs (glyceraldehyde-3-phosphate dehydrogenase, phosphoglycerokinase and α-ketoglutarate dehydrogenase, succinate thiokinase) catalyzes substrate-level phosphorylation.
21. In mammalian liver, acetyl-CoA provides reducing equivalents by its oxidation in the TCA cycle and subsequent oxidative phosphorylation to provide ATP. The complete oxidation of acetyl-CoA provides approximately 10 moles of ATP per mole of acetyl-CoA oxidized by the mitochondrial TCA and electron transport system. Hence the ATP required to resynthesize glucose from lactate, pyruvate, or their equivalent is provided by the net oxidation of acetyl-CoA. The carbon source for gluconeogenesis, pyruvate (lactate), must be converted to phosphoenolpyruvate for reentry into the gluconeogenic pathway. Reversal of the pyruvate kinase reaction is not metabolically feasible, so a series of reactions bypassing the metabolic block is used. In these reactions, pyruvate is carboxylated to oxaloacetate, the oxaloacetate is reduced to L-malate, and the malate is transported to the cytosol. Reoxidation of the L-malate to oxaloacetate in the cytosol provides substrate for the GTP-dependent PEP carboxykinase.

    Acetyl-CoA is also an obligatory activator of the pyruvate carboxylase that catalyzes carboxylation of pyruvate to oxaloacetate. The oxaloacetate is used either to replenish the TCA cycle or for reduction to malate and subsequent export to the cytosol for gluconeogenesis.

Chapter 16

1. In chemistry, "coupled" often refers to two reactions that share a chemical component. The equilibrium position of one (endergonic) reaction is affected by the exergonic nature of the second reaction. In this chapter, consisting of a series of redox reactions, the shared component are electrons and their potentials. The term "uncoupled" refers to the lack of synthesis of ATP from phosphate and ADP brought about by a collapse of a hydrogen ion gradient. Any substance that facilitates this collapse of the hydrogen ion gradient is referred to as an uncoupler.
3. (a) The chemical nature of the functional group that is oxidized and reduced is as follows: (1) Iron–sulfur protein: Iron–sulfur clusters, whose molecular composition may be 1 iron, no inorganic sulfide; 2 irons, 2 inorganic sul-

fides; 4 irons, 4 inorganic sulfides. (2) Flavoproteins use either FAD or FMN, whose redox center is an isoalloxazine ring. (3) Quinone: The benzoquinone ring (quinone ring) is fully reduced to the hydroquinone upon addition of 2 electrons.

(b) The number of electrons required to reduce the oxidized form are as follows: (1) Iron–sulfur proteins: 1 electron. (2) Flavoproteins: 2 electrons. A stable semiquinone (1-electron reduction product) may be produced. (3) Quinone: 2 electrons. Stable semiquinone may be formed.

(c) Stoichiometry of protons taken up per electron: (1) Iron–sulfur proteins: no proton taken up. (2) Flavoprotein: 1 proton per electron. (3) Quinone: 1 proton per electron.

The flavin and the quinone redox centers may accept or donate either 1 or 2 electrons and therefore are positioned as electron-transfer agents between obligatory 2-electron donors (NADH) and obligatory electron acceptors (cytochromes or iron–sulfur centers).

5. (a) The oxidation–reduction reactions being considered are

$$\text{cyt } a_3\ (Fe^{3+}) + 1\ e^- \rightarrow \text{cyt } a_3\ (Fe^{2+}) \qquad E^{\circ\prime} = +350 \text{ mV}$$

$$\text{cyt } c\ (Fe^{2+}) \rightarrow \text{cyt } c\ (Fe^{3+}) + 1\ e^- \qquad E^{\circ\prime} = -240 \text{ mV}$$

$$\text{cyt } a_3\ (Fe^{3+}) + \text{cyt } c\ (Fe^{2+}) \rightarrow \text{cyt } a_3\ (Fe^{2+}) + \text{cyt } c\ (Fe^{3+}) \qquad \Delta E^{\circ\prime} = +110 \text{ mV}$$

$\Delta G^{\circ\prime} = -nF\Delta E^{\circ\prime}$, where n is electron equivalents per mole and F is the Faraday constant (23.06 kcal $eq^{-1}\ V^{-1}$). In this reaction, n = 1 eq per mole.

$$\Delta G^{\circ\prime} = -(1 \text{ eq/mole})\ (23.06 \text{ kcal } V^{-1}\ eq^{-1})\ (+0.11V)$$
$$\Delta G^{\circ\prime} = -2.5 \text{ kcal/mole}$$

(b)

$$O_2 + 4e^- + 4\ H^+ \rightarrow 2\ H_2O \qquad E^{\circ\prime} = +820 \text{ mV}$$

$$4 \text{ cyt } c\ (Fe^{2+}) \rightarrow 4 \text{ cyt } c\ (Fe^{3+}) + 4\ e^- \qquad E^{\circ\prime} = -240 \text{ mV}$$

$$O_2 + 4 \text{ cyt } c\ (Fe^{2+}) + 4\ H^+ \rightarrow 2\ H_2O + 4 \text{ cyt } c\ (Fe^{3+})$$

Therefore using the approach shown in part (a), $\Delta E^{\circ\prime}$ = +580 mV, $\Delta G^{\circ\prime}$ = −53 kcal/mole O_2 reduced (n = 4 equivalents).

(c)

$$\text{Succinate} \rightarrow \text{Fumarate} + 2\ e + 2\ H^+ \qquad E^{\circ\prime} = +30 \text{ mV}$$

$$2 \text{ cyt } c\ (Fe^{+3}) + 2\ e^- \rightarrow 2 \text{ cyt } c\ (Fe^{2+}) \qquad E^{\circ\prime} = +240 \text{ mV}$$

$$\text{Succinate} + 2 \text{ cyt } c\ (Fe^{+3}) \rightarrow \text{Fumarate} + 2 \text{ cyt } c\ (Fe^{2+}) + 2\ H^+$$

Therefore $\Delta E^{\circ\prime}$ = +270 mV, and $\Delta G^{\circ\prime}$ = −12.4 kcal/mole (n = 2 equivalents).

(d) Consider the reactions

$$NAD^+ + 2\ e^- + H^+ \rightarrow NADH \qquad E^{\circ\prime} = -320 \text{ mV}$$

$$UQH_2 \rightarrow UQ + 2\ e^- + 2\ H^+ \qquad E^{\circ\prime} = -110 \text{ mV}$$

$$UQH_2 + NAD^+ \rightarrow UQ + NADH + H^+$$

Therefore $\Delta E^{\circ\prime}$ = −430 mV, and $\Delta G^{\circ\prime}$ = +19.8 kcal/mole (n = 2 equivalents). Note that $\Delta E^{\circ\prime}$ values are positive for exergonic reactions and are negative for endergonic reactions.

7. The ratio of reduced to oxidized cytochrome may be calculated using the Nernst expression:

$$E = E^{\circ\prime}\ \frac{-2.3RT}{nF} \log \text{(Red)/(Ox)}$$

where 2.3 RT/F is appropriately 60 mV/n. (Red) and (Ox) represent the concentration (or fraction) of the reduced component and the oxidized component of the couple, respectively:

$$E = E^{\circ\prime} - 60 \text{ mV/1} \log [\text{(Red)/(Ox)}]$$

where

$$E = +300 \text{ mV}, \qquad E^{\circ\prime} = +240 \text{ mV}$$

$$+300 \text{ mV} = 240 \text{ mV} - 60 \text{ mV} \log [\text{(Red)/(Ox)}]$$

$$60 \text{ mV} = -60 \text{ mV} \log [\text{(Red)/(Ox)}]$$

Therefore (Red)/(Ox) = 10^{-1}, but (Red) + (Ox) = 1.

$$(\text{Red}) = 0.1\ (\text{Ox})$$
$$0.1\ (\text{Ox}) + (\text{Ox}) = 1$$
$$(\text{Ox}) = .9$$

Thus at +300 mV the cytochrome *c* half-cell will be 90% oxidized and 10% reduced.

9. (a) Cytochrome *c* accepts an electron from complex III and in turn transfers the electron to cytochrome oxidase for the reduction of O_2 to water. The cytochrome *c* may be tightly associated with the complexes (the linear chain model for electron-transfer system) or may diffuse between complex III and cytochrome oxidase during electron transfer. If the cytochrome *c* were tightly associated in the chain, we would predict that the electron acceptor site, associated with complex III, and the electron donor site, associated with cytochrome oxidase, would be at different sites on the cytochrome *c* protein. Acetylation of specific residues on the cytochrome *c* would be predicted to alter interaction with either but not likely both complexes. In fact, acetylation of lysine residues around the crevice of the heme ring alters electron transfer from complex III as well as transfer to cytochrome oxidase. Thus the data are consistent with diffusion of cytochrome *c* between complex III and the cytochrome oxidase. Electrons are likely donated and accepted from the same area on the cytochrome *c* structure.

   (b) The lysines surrounding the heme crevice on cytochrome *c* should interact electrostatically with anionic amino acid side chains (aspartic acid and glutamic acid) spatially located on complex III and on cytochrome oxidase to facilitate interaction with the cytochrome *c*. If the residues on the complexes at the site of interaction with cytochrome *c* were positively charged (lysine or arginine residues), the electrostatic repulsion would hamper the approach of cytochrome *c* and reduce the probability of electron transfer.

11. (a) Tightly coupled mitochondria exhibit an obligatory codependence of electron transfer, as measured by O_2 uptake and phosphorylation of ADP. Mitochondria are said to be in state 3 when respiration is measured with substrate and ADP, phosphate, and O_2 present. Under these conditions the substrate is oxidized, O_2 is reduced and ADP is phosphorylated to ATP. In state 4 the conditions are identical to state 3 except that ADP is omitted. In perfectly coupled mitochondria there should be no reduction of O_2 in state 4. However, perfect coupling has not been observed experimentally, and the degree of coupling is estimated by the Respiratory Control Ratio, the ratio of rate of O_2 reduction by mitochondria in state 3 divided by the rate of O_2 reduction in state 4. Larger values of the ratio are consistent with a greater codependence on ADP phosphorylation and electron transfer.

   (b) Substrates are oxidized in the mitochondria to provide ATP for cellular processes. Unregulated oxidation of substrate wastes energy resources and eliminates the response of the mitochondrial production of ATP to cellular need. Respiratory control reflects the inhibition of mitochondrial respiration when cellular ATP levels are high and the stimulation of respiration when ATP levels are low (ADP levels are high). Respiratory control is a function of coupled mitochondria. Oxidation of NADH and succinate is blocked when ATP levels are high. Glycolysis is slowed by the lack of ADP (or AMP) stimulation of phosphofructokinase, and the TCA cycle is slowed by the increased NADH to NAD^+ ratio. When ATP demand is high, ADP levels will stimulate the oxidation of succinate and NADH. Elevated ADP concentration and high NAD^+/NADH ratio will stimulate glycolysis and the TCA cycle.

   (c) Uncoupled mitochondria oxidize substrate rapidly but fail to phosphorylate ADP. The free energy from the oxidation of NADH (approximately 52 kcal/mole) and succinate (approximately 39 kcal/mole) is released as heat. For animals that hibernate, a mechanism to elevate body temperature upon emerging from hibernation is important. This mechanism, nonshivering thermogenesis, involves brown fat mitochondria that contain the uncoupler thermogenin. The electron-transport system is partially uncoupled, and heat released by the oxidation of fat provides the increase in body temperature.

Chapter 17

1. The word "fixed" refers to converting a gas into a liquid or solid. Much of the early chemistry (18th century) was done with gases because they could be produced in a pure form by various reactions. The only reliable method for quantifying materials in this time period was by weighing. To weigh a gas, it was converted into a solid or liquid by a chemical reaction, and the resulting solid or liquid was weighed. Because the gas was "no longer moving about," it was "fixed."
3. Chlorophyll molecules are flat, completely conjugated ring compounds that obviously have a resonance form. Chlorophylls have a ring containing 9 pi bonds or 18 pi electrons, which fits the $n = 4$ situation in the Huckel $(4n + 2)$ rule. The (*cis, trans, trans*)$_3$ double-bond pattern of the chlorophylls also fits that of [18]annulene, which is known to be aromatic.

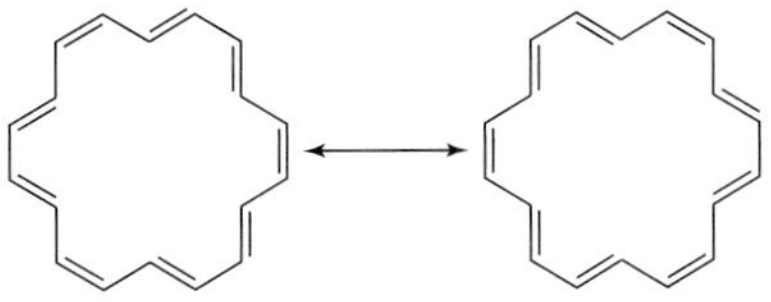

[18]Annulene

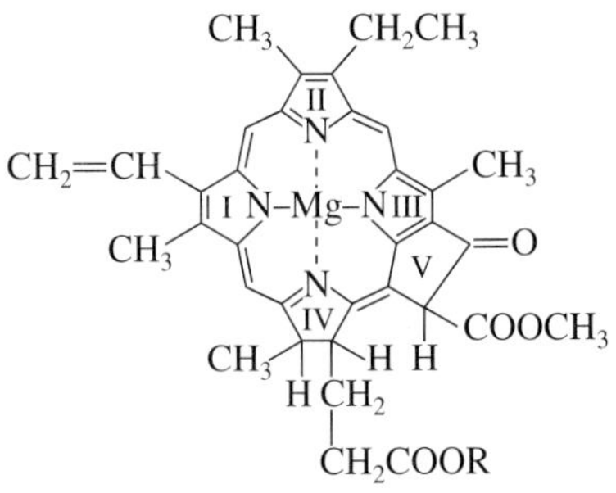

CH_3 CH_2CH_3 II $CH_2{=}CH$ CH_3 I N–Mg–N III CH_3 V =O IV CH_3 $COOCH_3$ H H H CH_2 CH_2COOR

5. Upon illumination the chromatophore P_{870} is activated by absorption of a photon of light. The absorbance at 870 nm decreases because the pi-cation radical of the oxidized chromatophore has a lower absorbance at that wavelength. Thus the trace monitoring 870 nm decreases upon illumination of the chromatophore. The *c*-type cytochrome is added initially in the reduced form (cyt c^{2+}) and absorbs at 550 nm. Oxidized cytochrome *c* (cyt c^{3+}) has only a small absorbance at 550 nm. Electron transfer from reduced cytochrome *c* to the pi-cation radical (P_{870^+}) regenerates the ground state P_{870} and forms oxidized cytochrome *c*. The absorbance of the P_{870} increases to the initial level, and the absorbance at 550 nm of the cytochrome *c* pool decreases. The explanation is consistent with the observed upward trace at 870 nm and the downward trace at 550 nm over 30 μs.
7. (a) Illumination of the bacterial photocenter generates an active reductant that transfers an electron to ubiquinone, forming the semiquinone. Some, but not all, of the ubiquinone is reduced by a single flash. The absorbance at 275 nm decreases because the concentration of ubiquinone is diminished and because the semiquinone radical does not absorb 275-nm light. The absorbance increase at 450 nm is consistent with the formation of semiquinone radical that absorbs 450-nm light. The second flash activates transfer of a second electron from the photocenter to reduce the bound semiquinone to dihydroquinone. The dihydroquinone absorbs at neither 275 nm nor 450 nm. Reduction of the semiquinone form abolishes the absorbance at 450 nm. The decrease in absorbance at 275 nm is consistent with the decrease in oxidized (ubiquinone) concentration. The sequence of events is repeated with the third and fourth flashes.
   (b) Transfer of an electron from the activated photocenter (P_{870}*) to the ubiquinone leaves the oxidized pi-cation radical form of the reaction center (P_{870+}), which must be reduced to allow a second light-activation cycle. In this experiment, reduced cytochrome *c* is the electron donor to P_{870+}. Were the reductant omitted, the P_{870+} might oxidize the reduced quinone, or the activated reaction center P_{870}* might return to ground state by emission of fluorescence.
9. Oxygen, a product of photosynthesis, may damage the chloroplast if the ground-state (triplet) oxygen is activated to the singlet state. Single oxygen may be formed by activated (singlet) chlorophyll that decays to the triplet state. Whereas energy transfer between singlet-state chlorophyll and triplet-state oxygen does not occur, triplet-state chlorophyll may activate triplet-state oxygen to the reactive singlet state. Singlet-state oxygen causes cumulative oxidative damage to the chloroplast. Carotenoids compete with the oxygen for the triplet-state chlorophyll and inhibit singlet oxygen production. Plants deficient in carotenoids risk photooxidative damage because of an increased flux of singlet oxygen. Although there are enzymes that scavenge the reduced, activated oxygen intermediates (superoxide, hydrogen peroxide), there is no enzymatic activity thus far identified that catalytically scavenges singlet oxygen.
11. The mechanisms are different, but the energetic outcome is the same. In the start of gluconeogenesis, pyruvate is converted to phosphoenolpyruvate with the consumption of two equivalents of "ATP" (ADP = product). In the phosphoenolpyruvate production in mesophyll cells the conversion involves the production of an AMP from an ATP. Considering the ubiquitous pyrophosphatase, this is the energetic equivalent of the conversion of 2 ATP to 2 ADP.
13. The enolate intermediate (fig. 17.30) reacts with oxygen to produce a cycloperoxide intermediate, which decomposes into glycerate-3-phosphate and glycolate-2-phosphate (fig. 17.31).

Ribulose-1,5-bisphosphate

Glycolate-2-phosphate + Glycerate-3-phosphate

15. The Calvin cycle uses many of the metabolites and enzymes found in the pentose phosphate pathway and glycolysis. Some of the many differences noted include the following: The involvement of erythrose-4-phosphate and dihydroxyacetone phosphate in the production of sedoheptulose-1,7-bisphosphate, phosphorylation of ribulose-5-phosphate to ribulose-1,5-bisphosphate, the addition of CO_2 to ribulose-1,5-bisphosphate and the production of two glyceraldehyde-3-phosphates. Also notice that the coenzyme specificity for the glyceraldehyde-3-phosphate dehydrogenase is $NADP^+$, while the corresponding enzyme in glycolysis is specific for NAD^+.
17. Ribulose bisphosphate carboxylase/oxygenase (RuBisCO) catalyzes the addition of CO_2 to ribulose-1,5-bisphosphate, yielding two moles of 3-phosphoglycerate per mole of CO_2 added. Alternatively, RuBisCO catalyzes the addition of O_2 to ribulose-1,5-bisphosphate, yielding one mole each of 3-phosphoglycerate and phosphoglycolate. Either CO_2 or O_2 binds to the catalytic site of the ribulose bisphosphate carboxylase/oxygenase (RuBisCO) and reacts with the ribulose-1,5-bisphosphate that is bound to the enzyme. Molecules that bind to the same form of the enzyme (active site) are competitive inhibitors. Thus O_2 competitively inhibits carboxylase activity, and CO_2 competitively inhibits oxygenase activity.
19. The absorption of CO_2 from the air and the addition of CO_2 to ribulose bisphosphate occur in separate compartments in

the C_4 plant. In the mesophyll cell of the C_4 plant, CO_2 is added initially to PEP, yielding oxaloacetate and subsequently malate and/or aspartate. O_2 does not compete with CO_2 (HCO_3) for PEP carboxylation. The four-carbon organic acids are translocated to the bundle sheath cells, where aspartate is converted to oxaloacetate by transamination and the oxaloacetate is reduced to malate. CO_2 is released by oxidative decarboxylation of malate in the bundle sheath cells of C_4 plants, causing the CO_2/HCO_3^- concentration to exceed the CO_2/HCO_3^- concentration in the mesophyll cell. The larger concentration of CO_2 increases the fraction of ribulose-1,5-bisphosphate that is carboxylated rather than oxygenated. Net photorespiration is lower in C_4 plants than in C_3 plants.

Chapter 18

1. As a nutritional item, α-lactalbumin must denature readily, be rapidly hydrolyzed by digestive enzymes, and contain the essential amino acids in their required ratio. As a component of lactose synthetase it has to be able to bind to galactosyl transferase, altering its specificity site. During evolution, α-lactalbumin appears to have answered to two drummers.
3. Cell-wall assembly and cross-linking of the peptidoglycan takes place outside of the cell membrane. The D-alanine released by the cross-linking process is no longer "inside" of the bacteria, and it is expected that the majority of the D-alanine will be lost to the surrounding environment.
5. Lectins are proteins that bind specific carbohydrate structures (for example, concanavalin A binds oligomannosyl N-linked sugars but not O-linked or branched mannose). Glycolipids could be separated on thin layer chromatography (TLC), and a specific radioactive lectin could be used to locate specific structures. The location of these structures could be detected by exposing the TLC to x-ray film (radioautography). The lectin could also be linked to a column matrix, and the compounds of interest could be applied to the column. Only structures recognized by the lectin would bind. Lectin affinity columns are very useful in purifying complex carbohydrates (see fig. 18.16).
7. Only two genes determine the ABO blood group scheme (fig. 18.25). The A gene encodes a glycosyltransferase that adds a terminal *N*-acetylgalactosamine residue onto a core oligosaccharide, and the B gene encodes a glycosyltransferase that adds a galactose residue (fig. 18.25). If both genes are present, then both antigens are present, resulting in blood type AB. If both genes are missing, then the core oligosaccharide is not substituted, and the blood type is O. People with O type blood are considered universal donors because they have only the core oligosaccharide, which will not generate antibodies when their blood is given to people with A or B type blood. If you have the AB blood type, all types of blood are compatible because you already have all the different blood antigens; antigens in the transfused blood would not be considered foreign. People who lack the H antigen cannot add the fucose residue to the core oligosaccharide. These people can receive blood only from another person with Bombay type blood (lacking the H antigen) but can donate blood to people with Bombay type, A, B, or O.
9. In mutant cells lacking Dol-P-mannose synthase a complete glycophosphitadyl inositol (GPI) is not made, and the glycoproteins with GPI anchors, which are normally found in the plasma membrane, are secreted from the cell because they are no longer attached by their anchors to the membrane.
11. The enzyme bound NAD^+ oxidizes the secondary alcohol on carbon 4 of the UDP-hexose to a 4-keto group. The resulting bound NADH then reduces the 4-keto group to a secondary alcohol. Alcohol geometry can either be oxidized to the keto intermediate or be produced by the reduction step.

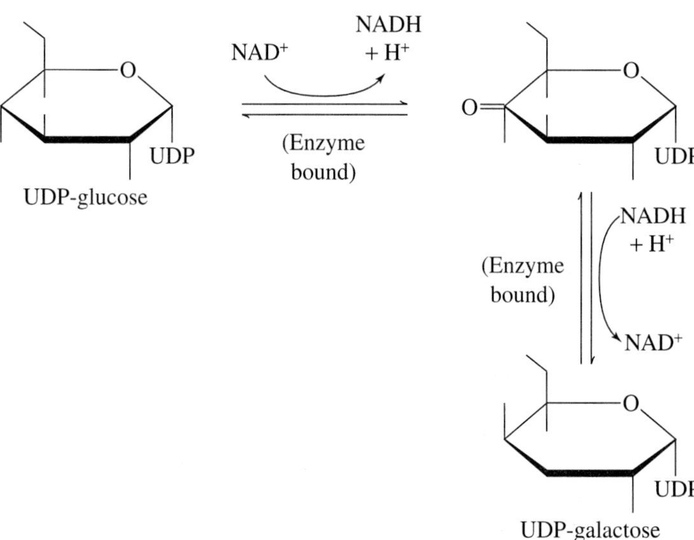

13. By acetylating the amino group, nature has eliminated the possibility of a protonated amine. Why the charge needs to be eliminated is curious, particularly since glucuronic acid residues are common components of oligo- and polysaccharides.
15. Proteins are synthesized from the N to C direction. The 20 hydrophobic amino acid sequence that serves as a signal is synthesized first so that translocation into the lumen of the endoplasmic reticulum can be initiated while translation is in progress. If the signal were on the C terminus, translocation could not occur during translation.
17. The pentapeptide portion of the peptidoglycans contain D-amino acids and peptide linkages involving the γ carboxyl group of glutamate, neither of which are found in normal proteins.
19. Penicillin inhibits the cross-linking reaction in cell-wall biosynthesis. Cell-wall biosynthesis is needed only when bacteria are growing (multiplying). A starving cell is not growing.

Chapter 19

1. The word "schizophrenic" is often used to mean "two personalities." In this case, one end of the molecule, the acyl ends, are hydrophobic and the ester, phosphate, and ethanolamine portions are hydrophilic. What term should the instructor use instead of schizophrenic? (The answer is amphipathic.)

3. The helix is 30 Å long. A membrane based on a 30 CH_2 is 37.5 Å thick.

Helix length:

$$\frac{20\ \text{AA} \mid \text{turn} \mid 5.4\ \text{Å}}{\mid 3.6\ \text{Å} \mid \text{turn}} = 30\text{Å}$$

Length of alkyl portion of two fatty acids (assuming palmityl):

$$\frac{2\ \text{palmityl} \mid 15\ \text{CH}_2 \mid 1.25\ \text{Å}}{\text{thick} \mid \text{Palmityl} \mid \text{CH}_2} = 37.5\ \text{Å}$$

5. The major ingredients are vegetable oil, water, and vinegar. The lecithin serves as an emulsifying agent that allows the aqueous and lipid phases to be dispersed in each other and increases the time required for phase separation.
7. The Davson-Danielli model of a membrane has the fatty acyl portions of two phospholipid bilayers in contact in the center of the membrane. When this model is "split," the new surfaces created should consist of the ends of the fatty alkyl side chains of the phospholipids. The electron micrographs of these exposed surfaces would be expected to be rather uniform (featureless).
9. Phospholipids are amphipathic molecules that associate as a lipid bilayer to form membranes. The hydrophobic side chains of the esterified fatty acyl groups are oriented toward the center of the membrane, whereas the phosphate and any substituent esterified to the phosphate face the aqueous phase. The phospholipids impose a charged hydrophilic surface to the membrane while maintaining a hydrophobic bilayer that is, in general, impermeable to ionic species. In addition, the nature of the esterified fatty acyl group affects the fluidity of the membrane.

   Peripheral proteins are associated with the exterior or interior face of the phospholipid bilayer, whereas the integral proteins are partially dissolved in the hydrophobic membrane. Peripheral proteins bind to the membrane primarily through weak ionic interactions, either directly or through metal ions. In the latter case, divalent metal ions may interact with the negatively charged phosphate or the carboxyl group of phosphatidyl serine of the phospholipid and with negatively charged amino acid side chains of the peripheral protein.
11. Integral membrane proteins have negligible rates of transverse motion (flip) across the membrane, presumably because there is a high energy barrier to the movement of the solvent-exposed hydrophilic portion of the protein through the hydrophobic interior of the membrane. The highly hydrated oligosaccharide associated with some integral protein contributes considerable hydrophilic character to the protein at the aqueous interface and contributes to the energy barrier to transverse motion of the protein across the membrane.
13. Extraction of solute from an aqueous to a nonaqueous solution depends on its relative solubility in each of the solvents. The solubility will depend to a large degree on the relative hydrophobicity or hydrophilicity of the solute. Hydrophilic molecules readily dissolve in aqueous media, whereas hydrophobic molecules do not. Triacylglycerides and the fatty acid methyl ester are hydrophobic and will readily partition into the nonaqueous solvent. Neither of these molecules has an ionizable group; their extraction into a nonpolar solvent is therefore not expected to be pH-dependent. The free (unesterified) fatty acid is an alkyl carboxylic acid whose alkyl chain is hydrophobic but whose carboxylate is hydrophilic.

    The pK_a value for these carboxylic acids is approximately 5, and the carboxyl group will exist as an anion at $pH > 5$. Hence extraction of the fatty acid anion into nonaqueous media is retarded by the energetically unfavorable transfer of a charged group into a nonpolar environment where solvation and H bonding are precluded. Protonation of the carboxylate ($pH < pK_a$) abolishes the charge on the molecule, diminishes the hydrophilicity, and increases the ease of extraction into organic solvent.
15. The protein dissolved in octylglucoside is most likely an integral membrane protein whose surface amino acid side chains are hydrophobic. The detergent must be present to prevent aggregation of protein molecules. The aggregate is more stable in the absence of detergent because aggregation excludes the water that would surround the hydrophobic amino acid side chains. The data are consistent with aggregation of the protein upon removal of detergent. The aggregate is at least a dimer (>100,000 M_r).

Chapter 20

1. The phosphorylated sugar is charged, which assists in keeping it inside the cell. (Charged molecules are less likely to diffuse across cell membranes.)
3. Valinomycin will cause an electrogenic flow of K^+ out of the vesicles, down its concentration gradient. Thus a $\Delta\Psi$ will be induced across the membrane, interior negative. This $\Delta\Psi$ then drives uptake of lactose via the well-known H^+-symport system. As H^+ is accumulated in the vesicles along with lactose, $\Delta\Psi$ is collapsed, and lactose can no longer be actively accumulated. This explains the transience of the phenomenon.
5. (a) Stimulation of glucose transport because a $\Delta\Psi$ (interior negative) would result from the electrogenic flow of K^+ out of the vesicles.
   (b) Inhibition, because gramicidin would collapse the Na^+ gradient present initially.
   (c) No effect (neither $\Delta\Psi$ nor the Na^+ gradient should be affected).
   (d) Stimulation (H^+ would flow down the pH gradient out of the vesicles, creating a $\Delta\Psi$, interior negative).
7.

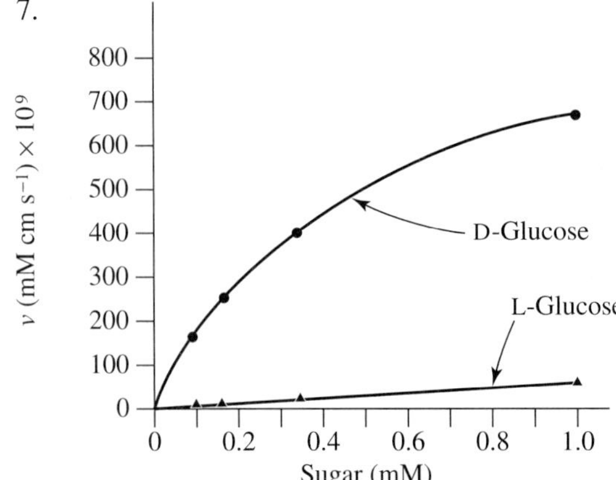

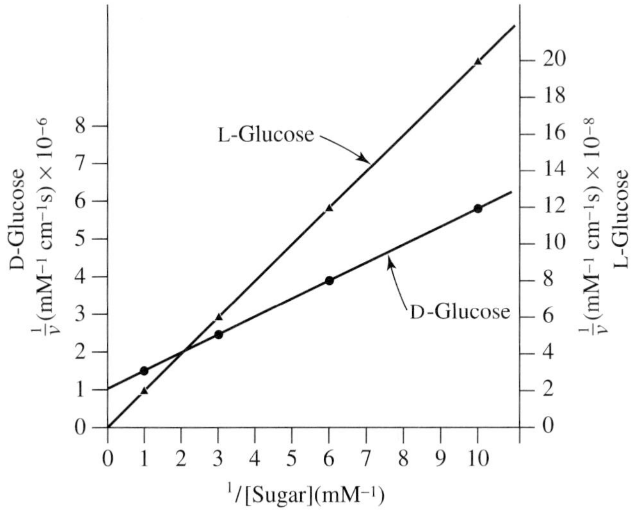

It is clear from the data that the D-glucose is taken up by the cells by facilitated transport (permease), while L-glucose is moving into the cell by passive transport. The D-glucose has a maximum velocity of uptake of about 1×10^9 mM s^{-1} (Y intercept in the plot; analogous to V_{max}) and is saturable. L-glucose is not saturable and is moving through the membrane by passive diffusion. Actually, the data on L-glucose could be used as a control for passive diffusion and could be subtracted from the D-glucose data to obtain the true values for the transport system.

9. The observations suggest that these vesicles contains F_1/F_0 ATPase with the F_1 portion oriented on the outside. Since a $\Delta\Psi$ alone has no effect on arginine transport, the process must be electroneutral. ATP hydrolysis by the ATPase would pump protons into the vesicles, and this stimulates arginine uptake. Proton ionophores prevent proton accumulation, and thus arginine transport is dependent on a ΔpH, interior acidic. The simplest mechanism is therefore an H^+-arginine exchanger (antiporter).
11. The lactose permease transports both lactose and H^+ concomitantly. The binding of H^+ and lactose to the carrier on the outside of the membrane presumably triggers a conformational change that deposits both molecules on the inside. Accumulation of lactose should continue as long as there is a ΔP (interior negative) or until the inside concentration of lactose becomes high enough that the efflux rate balances the influx rate.
13. The transport of K^+ by valinomycin occurred only at a high rate above 41°C, which is the "melting temperature" of this membrane system. At the higher temperature the membrane is more fluid and allows the carrier to diffuse through the bilayer (mobile carrier). Since the antibiotic gramicidin transports K^+ well at all temperatures studied and is insensitive to the physical state of the phospholipid bilayer, this ionophore must form a static pore through the bilayer.

Chapter 21

1. The biotin-facilitated addition of CO_2 to a substrate requires a carbanion on the substrate to approach the carboxylate carbon on the carboxy-biotin. Hydrogens on carbons α to carbonyls are more acidic than those on carbons β to carbonyls because of the formers' ability to form a resonance-stabilized carbanion. The third carbon was not "just" carboxylated because of the difficulty (energy) involved in forming the required carbanion.
3. The citric acid cycle contains the same metabolic sequence: "alkane oxidized to an alkene, which is then hydrated to form an alcohol, which is then oxidized to a ketone." The geometry of the "alcohol and alkene" are the same and the coenzymes utilized are identical. The major difference in β oxidation is performed on the CoA derivatives.
5. Ordering from the most reduced to most oxidized produces: methane, methanol, formaldehyde, formic acid, carbon dioxide. Five of the carbons in glucose correspond to methanol, and one is analogous to formaldehyde in terms of oxidation state. Nine of the carbons in decanoic acid correspond to the oxidation level of methane, and one matches that of formic acid. Glucose and decanoic acid are essentially an alcohol and an alkane, respectively. Because glucose is "already partially oxidized," it yields less energy (ATP) than decanoic acid when completely oxidized to carbon dioxide.
7. The availability of citrate has no relationship to the flow of metabolites through the tricarboxylic acid cycle. Increased citrate concentrations result in increased cytoplasmic acetyl-CoA concentrations, which in turn increases fatty acid biosynthesis.
9. Viewed from a reaction sequence viewpoint, it is a cyclic process. From a substrate viewpoint it is a decreasing spiral because the substrate's length decreases with each turn around the spiral.
11. Many mechanisms are conceivable. One possibility is: (1) approach of a phosphate on succinyl-CoA to produce succinyl phosphate (a mixed anhydride), (2) approach of an acetoacetate oxygen on the mixed anhydride to transfer a phosphoryl (PO_3) group and produce acetoacetyl phosphate, (3) approach by CoASH on the C-1 carbonyl of acetoacetyl phosphate to eliminate a phosphate and produce acetoacetyl-CoA.

$$^-O_2CCH_2CH_2-\overset{O}{\overset{\|}{C}}-S-CoA \xrightarrow[P_i]{CoASH} {}^-O_2CCH_2CH_2-\overset{O}{\overset{\|}{C}}-O-\overset{O}{\overset{\|}{P}}(O^-)-O^-$$

$$CH_3-\overset{O}{\overset{\|}{C}}-CH_2-\overset{O}{\overset{\|}{C}}-O^- \quad \longrightarrow \text{Succinate}$$

$$CH_3\overset{O}{\overset{\|}{C}}CH_2\overset{O}{\overset{\|}{C}}-S-CoA \xleftarrow[P_i]{} CH_3-\overset{O}{\overset{\|}{C}}-CH_2-\overset{O}{\overset{\|}{C}}-O-\overset{O}{\overset{\|}{P}}(O^-)-O^- \quad (\uparrow \text{CoASH})$$

13. (a) Nonhepatic tissues use ketone bodies as a rich energy source supplied by the liver. The ketone bodies are water soluble and easily transported in the blood. Oxidation of ketone bodies by heart muscle, for example, spares the oxidation of glucose, allowing the carbohydrate to be metabolized by erythrocytes and the brain. Oxidation of 1 mole acetoacetate yields a net 19 moles ATP.
(b) β-hydroxybutyrate supplies an additional reducing equivalent per mole compared to acetoacetate. Oxidation of 1 mole β-hydroxybutyrate yields 1 mole acetoacetate plus 1 mole NADH. Subsequently, oxidation of the NADH provides 2.5 additional moles ATP compared to the oxidation of acetoacetate.

(c)

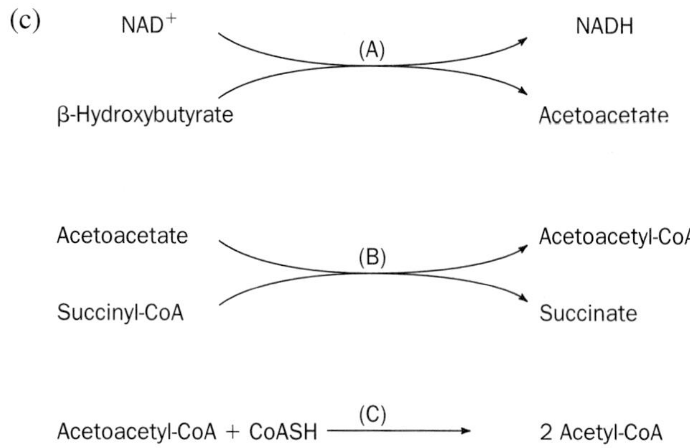

(A): NAD$^+$-dependent β-hydroxybutyrate dehydrogenase
(B): 3-Ketoacyl-CoA transferase
(C): Thiolase

15. (a) Acetyl-CoA is used directly only in the first cycle of condensation and reduction in palmitate synthesis. The methyl group of acetyl-CoA will become C-16 of palmitate. If 2-^{14}C acetyl-CoA were the only labeled metabolite, only C-16 of palmitate would be labeled.
(b) Synthesis of palmitate requires condensation and reduction of one acetyl and seven malonyl units. Each malonyl-CoA unit added undergoes two reduction steps with an intervening dehydration. The dehydration step removes one of the two protons from the methylene group derived from malonyl-CoA. Thus the product will retain only one of the two deuterium labels as shown.

$$CH_3-\overset{O}{\overset{\|}{C}}-CD_2-\overset{O}{\overset{\|}{C}}-S-ACP \xrightarrow[+H^+]{NADPH \;\; NADP^+} CH_3-\underset{H}{\underset{|}{\overset{OH}{\overset{|}{C}}}}-CD_2-\overset{O}{\overset{\|}{C}}-S-ACP$$

$$CH_3-\overset{OH}{\overset{|}{CH}}-CD_2-\overset{O}{\overset{\|}{C}}-S-ACP \xrightarrow{-D^+,\, OH^-} CH_3-CH{=}CD_2-\overset{O}{\overset{\|}{C}}-S-ACP$$

$$CH_3-\overset{OH}{\overset{|}{CH}}{=}CD_2-\overset{O}{\overset{\|}{C}}-S-ACP \xrightarrow[+H^+]{NADPH \;\; NADP^+} CH_3-CH_2-CHD-\overset{O}{\overset{\|}{C}}-S-ACP$$

There will be seven deuterium atoms per palmitate molecule, one residing on each of the carbons 2, 4, 6, 8, 10, 12, and 14. Carbon 16 originates directly from acetyl-CoA and is not labeled.
(c) Condensation of two-carbon units to the growing fatty acyl chain is driven in part by decarboxylation of the malonyl unit. The carbon shown labeled in the problem is lost as CO_2 during decarboxylation and is not incorporated into palmitate.

$$CH_3-\overset{O}{\overset{\|}{C}}-S-\text{Acyl carrier protein} + {}^-O-{}^{14}C-CH_2-\overset{O}{\overset{\|}{C}}-S-ACP \longrightarrow CH_3-\overset{O}{\overset{\|}{C}}-CH_2-\overset{O}{\overset{\|}{C}}-S-ACP + {}^{14}CO_2 + ACP-SH$$

17. NADPH for fatty acid biosynthesis is supplied by the glucose-6-phosphate and 6-phosphogluconate dehydrogenases in the pentose pathway and by the NADP$^+$-dependent-malic enzyme. Oxaloacetate released by the ATP-citrate lyase may be reduced to L-malate by the cytoplasmic NAD$^+$-dependent malate dehydrogenase. Malic enzyme oxidatively decarboxylates L-malate to pyruvate and reduces NADP$^+$ to NADPH. Pyruvate enters the mitochondrial matrix for resynthesis of oxaloacetate by carboxylation.
Palmitate formation requires seven cycles of malonyl-CoA condensation, each of which requires 1 mole NADPH for reduction of the ketoacyl group to β-hydroxyacyl-ACP and another to reduce the enoyl-ACP. Reduction of 1 mole of acetyl-CoA and 7 moles of malonyl-CoA to form 1 mole of palmitate requires 14 moles NADPH.
19. Thiolesterase activity for the fatty acid synthase is more active with palmityl-ACP than with shorter fatty acyl ACPs. Palmitate, the 16-carbon fatty acid, is thus preferentially released.

Chapter 22

1. There are eight isomers, one of which has two optical forms.
3. In general the following are chiral: carbon 2 of lipids that contain phosphatidic acid or its analogs, several carbons in the inositol derivatives, α carbon in the serine residue of phosphatidylserine, carbons 2 and 3 in sphingolipids, and a multitude of carbons in the glycosyl residues of glycolipids. Could a triacyl glyceride ever have a chiral carbon?
5. The phosphates on carbons 1, 4, and 5 and the hydroxys on carbons 2 and 3 will all be in the equatorial positions, minimizing any steric constraints. Only the hydroxy on carbon 6 will be in an axial position.
7. One possibility is the removal of fatty acyl groups that have been damaged by oxidation with molecular oxygen (or other oxygen species). In addition these processes allow the adjustment of the fluidity of the membrane to any environmental changes.
9. Carriers of a defective gene for the hexosaminidase A enzyme still have a functional copy of the gene. In this case,

one functional gene is capable of producing enzyme amounts sufficient to prevent the disease. This produces a situation known as a recessive disease.

11. Figure 22.4 involves a CDP-ethanolamine or CDP-choline, which reacts with diacylglycerol to produce phosphatidylethanolamine or phosphatidylcholine. Figure 22.7 has a CDP-diacylglycerol that reacts with an inositol to produce a phosphatidylinositol. The activated component is different; in the first case (fig. 22.4) the minor component is activated, while in the production of phosphatidylinositol (fig. 22.7) the diacylglycerol is activated.
13. The CTP:phosphocholine cytidylyltransferase is found free in the cytosol and membrane-bound in the endoplasmic reticulum (ER); however, only the membrane-bound fraction is active. When phosphatidylcholine levels decrease in the membrane, there is increased binding of cytidylyltransferase to the ER, where the enzyme is active. After phosphatidylcholine levels increase, the cytidylyltransferase is released in the cytosol, where it is inactive. Phosphorylation of the enzyme also releases it from the membrane. Fatty acids and diacylglycerol will increase binding of the cytidylyltransferase. When fatty acid levels are high, diacylglycerol levels will increase, providing substrate for phosphatidylcholine biosynthesis, and activate the rate-limiting enzyme CTP:phosphocholine cytidylyltransferase by increasing its binding to the ER membrane.

    The chemical reaction mechanism for the cytidylyltransferase is

CTP + Phosphocholine → CDP-choline + Pyrophosphate

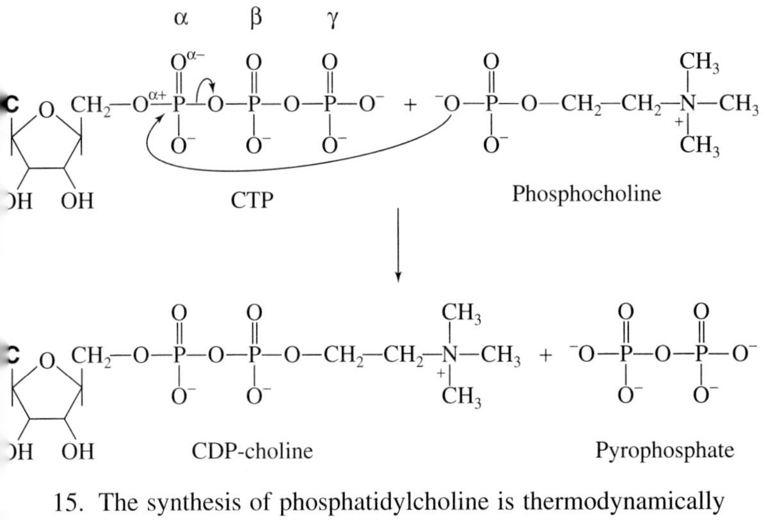

15. The synthesis of phosphatidylcholine is thermodynamically feasible because of the ATP used to phosphorylate choline and the CTP used to form CDP-choline (see the mechanism in the answer to problem 13). The cytidylyltransferase reacts phosphocholine with CTP to form CDP-choline and pyrophosphate. The hydrolysis of this pyrophosphate by pyrophosphatase is the major driving force for these reactions.
17. In animal cells the C_{20} polyunsaturated fatty acid arachidonic acid, which is the precursor for synthesis of PG_2 and TX_2 prostaglandins and thromboxanes, may be derived from the diet or from desaturation and elongation of linoleic acid. The scheme for the synthesis of arachidonic acid ($20{:}4^{\Delta 5,8,11,14}$) from linoleic acid ($18{:}2^{\Delta 9,11}$) is

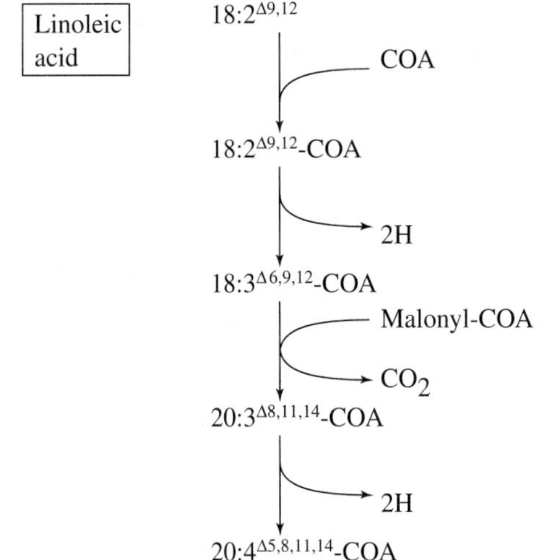

19. Phospholipase A_2 selectively removes fatty acids from the SN-2 position of phospholipids. As discussed in solution 3, arachidonic acid is stored in membranes by linkage at the SN-2 position in phospholipids. If a lot of arachidonic acid were released, it would stimulate the synthesis of prostaglandins, which then would induce inflammation. The release of arachidonic acid from phospholipids is believed to be the rate-limiting step in eicosanoid biosynthesis.

Chapter 23

1. Eight of cholesterol's 27 carbons are chiral: carbons number 3, 8, 9, 10, 13, 14, 17, and 20. See figure 19.6 for carbon-numbering schemes.
3. Each isoprene unit represents two carboxyl and three methyl carbons of acetate (see fig. 23.7). Six isoprene units are assembled into squalene (12 carboxyl, 18 methyl carbons), and three methyl carbons are lost in the conversion of lanosterol to cholesterol to give a 12:15 ratio.
5. Let the gene frequency of familial hypercholesterolemia be B and the "normal" gene frequency be A; then assuming only two alleles, $A + B = 1$ and $A^2 + 2AB + B^2$ describes the population where A^2 are "normal," 2AB are heterozygotic, and B^2 are homozygotic for familial hypercholesterolemia. If $2AB = 1/500$ (see page 549), then $AB = 0.001$, A is approximately 0.999, and B is 0.001. Then $B^2 = (0.001)^2 = 0.000001$, which is the one in a million indicated on page 549.
7. The adenylate derivative (a mixed anhydride) of cholic acid is the most likely intermediate:

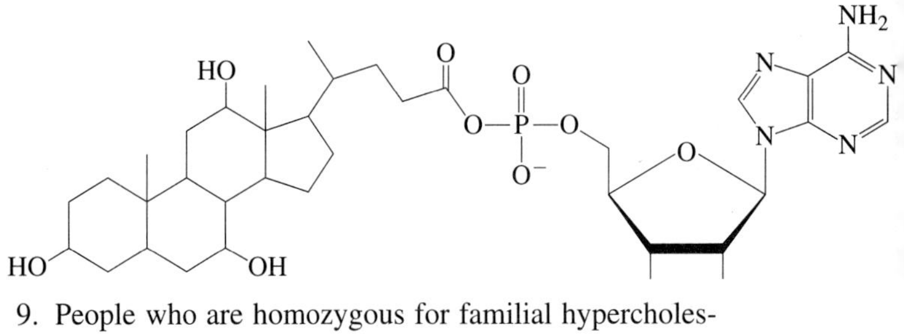

9. People who are homozygous for familial hypercholes-

terolemia (FH) do not make any receptors for LDL and cannot remove LDL from their blood. This leads to very high levels of serum LDL, which results in cardiovascular disease. (Some of the highest levels of serum cholesterol observed have been found in FH patients.) If a normal person is given an inhibitor for HMG-CoA reductase, cholesterol synthesis is inhibited in the liver. Lower levels of cholesterol then signal the synthesis of increased levels of LDL receptors. This increases the uptake of LDL into the liver and reduces serum LDL. In a patient with FH this has little effect, since there are no LDL receptors. The only effect is that the liver does not make as much cholesterol and does not contribute as much to serum LDL levels. The patient in question was given a heart transplant to replace the damaged heart and, at the same time, a liver transplant. The new liver will make normal amounts of LDL receptors and have normal uptake of LDL from the blood. This will dramatically lower serum LDL levels and prevent the new heart from developing coronary artery disease. If the liver transplant had not been done, the heart transplant would have been to no avail.

11. Bile salts lead a cyclic life with more than 95% of the molecules excreted into the small intestine with the bile being reabsorbed. By reducing the amount of absorption, the supply of bile salts is decreased. Because bile salts are biosynthesized from cholesterol (fig. 23.1), the resulting increased bile salt biosynthesis reduces the cholesterol levels.
13. The major regulation of cholesterol biosynthesis occurs with the first committed step in the pathway, HMG-CoA reductase. This enzyme is located on the outer surface of the endoplasmic reticulum (ER). It has two structural domains: One region of the protein has seven hydrophobic segments, which cross the membrane of the ER, while the other region protrudes into the cytosol and has the catalytic site. The activity of the reductase is regulated by three mechanisms. When cholesterol is high, the gene for the reductase is turned off and less mRNA for HMG-CoA reductase is produced. High cholesterol levels will also reduce the half-life by affecting the membrane domain of the enzyme by dissolving into the ER membrane. The third mechanism involves the phosphorylation–dephosphorylation of the reductase. When the enzyme is phosphorylated, the reductase is less active. High levels of cholesterol will also reduce the number of LDL receptors synthesized, thus reducing the amount of cholesterol brought into the liver from the blood.
15. Wolman's disease is caused by a complete lack of lysosomal acid lipase, the activity responsible for normal catabolism of cholesterol esters and triacylglycerols in lysosomes. In individuals suffering from this condition, the uptake and accumulation of these materials within extrahepatic cells is severely affected.

    The major part of cholesterol transport in plasma is achieved through the medium of low-density lipoprotein (LDL) particles. These particles ferry cholesterol esters to extrahepatic cells, such as skin fibroblasts, from a variety of sources, including the liver and the extrahepatic cells themselves. In the extrahepatic tissues, LDLs are first taken up in an endocytotic process mediated by specific LDL receptors; they are then exposed to lysosomal degradation by fusion of endocytotic and lysosomal vesicles. In normal individuals this degradation includes removal of the acyl group from cholesterol esters to give cholesterol, a reaction catalyzed by lysosomal acid lipase. This free cholesterol released into the cytosol provokes reductions in both HMG-CoA reductase activity, reducing the cell's own ability to synthesize cholesterol, and LDL receptor synthesis, reducing further uptake of LDL.

    In individuals suffering from Wolman's disease, the enzyme catalyzing the hydrolysis of cholesterol esters is defective. As a result, the activity of HMG-CoA reductase and the synthesis of LDL receptors remain high, leading to accumulation in extrahepatic tissues of cholesterol esters and triacylglycerols. (LDL particles contain approximately 10% triacylglycerol and 46% cholesterol ester by weight.)

Chapter 24

1. The pyridoxal phosphate attaches to β-cystathionase (fig. 24.17) to the amino group on the three-carbon moiety (the same portion that becomes pyruvate). With γ-cystathionase (fig. 24.13) the pyridoxal phosphate forms a Schiff base with the amino group on the four-carbon moiety (the same portion that becomes α-ketobutyrate).
3. An O-phosphohomoserine phosphate lyase would convert O-phosphohomoserine to 2-amino-3-butenoic acid followed by a 2-amino-3-butenoic acid hydratase that would produce threonine.
5. Most likely the oxidation occurs first, creating a carbonyl group, followed by the decarboxylation. The carbonyl group allows the carbanion formed by the decarboxylation to be resonance stabilized (which in turn facilitated the decarboxylation step).
7. The oxide ion is a very poor leaving group. Notice that the ATP actually contributes a phosphoryl group (PO_3 moiety), which is then eliminated as an HPO_4^{2-} upon the approach of an $:NH_3$. The function of the ATP is to assist in the removal of an oxygen from the carboxylate group:

$$\underset{\text{Glutamate}}{R{-}\overset{O}{\overset{\|}{C}}{-}O^-} \quad {}^-O{-}\overset{O}{\overset{\|}{\underset{O^-}{\underset{|}{P}}}}{-}O{-}\overset{O}{\overset{\|}{\underset{O^-}{\underset{|}{P}}}}{-}AMP \xrightarrow{\text{ADP}}$$

$$\underset{H_3\ddot{N}}{R{-}\overset{O}{\overset{\|}{C}}}{-}O{-}\overset{O}{\overset{\|}{\underset{O^-}{\underset{|}{P}}}}{-}O^- \xrightarrow{HPO_4^{2-}} \underset{\text{Glutamane}}{R{-}\overset{O}{\overset{\|}{C}}{-}NH_2}$$

9. Using glutamate synthetase the overall reaction is

$$\alpha\text{-Ketoglutarate} + \text{Gln} + 2\ \text{ATP} + \text{NADPH} + 2\ \text{NH}_3 \rightarrow 2\ \text{Gln} + 2\ \text{ADP} + 2\ \text{P}_i + \text{NADP}^+$$

or net

$$\alpha\text{-Ketoglutarate} + 2\ \text{ATP} + \text{NADPH} + 2\ \text{NH}_3 \rightarrow \text{Gln} + 2\ \text{ADP} + 2\ \text{P}_i + \text{NADP}^+$$

With glutamate dehydrogenase the overall reaction is

$$\alpha\text{-Ketoglutarate} + \text{ATP} + \text{NADPH} + 2\ \text{NH}_3 \rightarrow \text{Gln} + \text{ADP} + \text{P}_i + \text{NADP}^+$$

The two sequences differ by an ATP.

11. The availability of valine inhibited an enzyme in the early stages of valine biosynthesis (such as acetohydroxyacid synthetase, fig. 24.21). Because this inhibited enzyme is common to the biosynthetic pathways of isoleucine, valine, and leucine, the *E. coli* strain did not grow because of a lack of leucine and isoleucine.
13. Why make more when you already have enough? The environmentally derived tryptophan inhibits the first enzyme in the pathway, preventing flow of material through the pathway. Only when the environmental tryptophan is depleted will material flow through the pathway, or in this case, because of the auxotroph, only partway through the pathway.
15. In this two-step pathway the acetyl group is added in the first reaction and an acetate leaves, while an HS^- is added in the second step. The function of the acetyl group is for the removal of the serine oxygen. The OH group is a poor leaving group; acetate is better. Also notice the O^- acetyl homoserine in figure 24.13. Did nature have to use an acetyl group for the removal of oxygen? See the answer to problem 1 of this chapter for an idea to answer this latter question.
17. Glutamine synthase catalyzes the formation of glutamine, the organically bound nitrogen of which is used in the biosynthesis of a number of structurally unrelated nitrogen-containing compounds. Glutamine synthase activity is well regulated, as you would predict for an enzyme at the beginning of a multibranched biosynthetic system. The *E. coli* glutamine synthase is regulated by feedback inhibition exerted by products of pathways dependent on glutamine and by covalent modification.

    The metabolites carbamoylphosphate, tryptophan, and cytidine triphosphate, as well as glucosamine-6-phosphate, histidine, and AMPs, are synthesized by pathways that incorporate nitrogen directly from glutamine. The supply of glutamine must respond to the demand generated by the several pathways. Hence total inhibition of the glutamine synthase by a single product could inhibit biosynthesis of critically needed end products from other pathways. Cumulative inhibition of the glutamine synthase by end products of the various pathways modulates the supply of glutamine in direct response to the demand.
19. Synthesis of aspartate by transamination of oxaloacetate and of the glutamate by transamination of α-ketoglutarate depletes the concentration of oxaloacetate and α-ketoglutarate in the TCA cycle. If these TCA cycle intermediates were not replenished, the rate of acetyl-CoA oxidation and subsequent ATP production would be markedly diminished. Pyruvate, derived from the glycolytic metabolism of glucose, is used to replenish the concentration of each of the cycle intermediates. As was noted in the solution to the previous problem, oxaloacetate is formed from the carboxylation of pyruvate, catalyzed by pyruvate carboxylase. Synthesis of citrate, from the condensation of oxaloacetate and acetyl-CoA, in turn replenishes the concentration of the other intermediates.
21. Hydroxylation of proline in collagen is a posttranslational modification. Proline, rather than 4-hydroxyproline, is incorporated into collagen precursors and is subsequently hydroxylated by proline-4 hydroxylase. Neither free proline nor proline bound to the prolyl-t-RNA is hydroxylated. The hydroxylated peptides are then assembled and processed into collagen.

Chapter 25

1. If a diet lacks an essential amino acid, then that amino acid must be obtained by the breakdown of body proteins. The metabolism of the amino acids produced, in addition to the deficient amino acid by this process, increases the amount of nitrogen excreted (urea) and produces a negative nitrogen balance.
3. Because α-ketoglutarate is a "universal" substrate for transaminases, the combination of a transaminase for each amino acid and glutamate dehydrogenase produces the same effect of having an oxidative deaminase for each amino acid.
5. The head-to-tail expression refers to the two- and three-carbon side chains on porphobillinogen. Notice that on the linear tetrapyrrole (see fig. 25.14) the side chain sequence is 2, 3, 2, 3, 2, 3, 2, 3, while on uroporphyrinogen III (see fig. 25.13) if one goes clockwise from the 10 o'clock position, the sequence is 2, 3, 2, 3, 2, 3, 3, 2.
7. From the ADP and P_i products it can be deduced that a mixed anhydride (carboxylate from the incoming amino acid) is most likely involved.
9. The reduction occurs first, producing an O-phosphoryl hemiacetal followed by an elimination of the phosphate to form an aldehyde. The purpose of the phosphate is for the elimination of an oxygen. It was added by the previous enzyme as a phosphoryl group (PO_3) and is eliminated as a phosphate.
11. Many enzymes that act upon amino acids have pyridoxyl phosphate as a cofactor. The transaminases mentioned throughout this chapter utilize pyridoxal phosphate. If pyridoxal phosphate were limited by diet, the metabolism of amino acids would be impaired significantly.
13. The ketogenic amino acids are ones that, by catabolism, produce acetyl-CoA (or acetoacetate), which then elevate serum levels of "ketone bodies." The glycogenic amino

acids produce C_3 or C_4 (not acetoacetate) metabolites that increase serum levels of glucose.

15. Many inborn errors in amino acid metabolism result in elevated (or depressed) concentrations of one or more individual metabolites. In a fetus there is the possibility that the elevated concentrations can "spill over" the placenta and be disposed of by the mother. At birth the child appears normal, but with time any accumulation of metabolites can affect brain development.
17. Argininosuccinate lyase produces fumarate by an elimination reaction, while fumarylacetoacetase produces fumarate by an hydrolysis reaction. (*Note:* A tautomer different from that shown in figure 25.7 is utilized in this mechanism.)
19. One of the water molecules is involved in the hydrolysis of the amide; the second water molecule is consumed in the hydrolysis of the ATP.
21. The phenylalanine supplied by the diet is used in the biosynthesis of proteins or is catabolized in a pathway dependent on the formation of tyrosine. The formation of tyrosine is catalyzed by phenylalanine-4-monooxygenase, an enzyme whose activity is deficient in phenylketonurics. As observed in other metabolic pathways, concentrations of metabolites on the substrate (supply) side of a metabolic block will accumulate to above normal levels, whereas concentrations of metabolites on the product side will be markedly diminished. Hence phenylalanine, the substrate of phenylalanine monooxygenase, will accumulate, an event reflected in the blood concentration of phenylalanine. Tourian and Sidbury report that blood levels in excess of 20 mg phenylalanine per 100 ml blood of individuals on a nonrestricted diet are considered diagnostic for phenylketonuria or PKU-like disorders (see A. Y. Tourian and J. B. Sidbury, in *The Metabolic Basis of Inherited Disease,* eds. J. B. Stanbury, J. B. Wyngaarden, and D. S. Fredrickson. New York: McGraw-Hill, 1978, pp. 240–255).

    Accumulation of phenylalanine or products of the minor secondary pathway for phenylalanine metabolism (phenylpyruvate, phenyllactate) apparently adversely affects neuronal development, leading to severe mental retardation. Accumulation of the products may be prevented by management of dietary intake of phenylalanine.

Chapter 26

1. One of the products of ribonucleotide reductase is water with the oxygen originating as the C-2′ oxygen of the ribonucleotide.
3. It is anticipated that the amide would be made in a manner analogous to that of glutamine, that is, the production of a mixed anhydride with the conversion of ATP to ADP. The phosphate is then displaced by the approach of an NH_3 on the mixed anhydride's carbonyl to produce NAD^+.
5. When PRPP is treated with base, an alkoxy ion is formed by the removal of a H^+ from the alcohol group on C-2. The resulting alkoxy ion approaches the phosphorus on the α phosphate of the PP_i moiety and, with the elimination of the phosphate, forms 5-phosphoribose-1,2-cyclophosphate. The second product of the reaction is phosphate.
7. Carbamoyl phosphate synthetase contributes to two processes: (a) the initial enzyme in the biosynthesis of pyrimidines and (b) a component in the synthesis of arginine biosynthesis (fig. 25.2) or the urea cycle (fig. 25.7). In bacteria both of these processes occur within the same compartment. If carbamoyl phosphate were controlled by the product(s) of the pyrimidine biosynthetic pathway, then control of arginine biosynthesis would be lost. Therefore in bacteria the second enzyme in the pathway (aspartate carbamoyltransferase) is controlled. In humans the carbamoyl phosphate synthetase involved in the urea cycle is contained in the mitochondria (fig. 25.7), isolated from the cytosol counter part that is involved in the biosynthesis of pyrimidines. Because the two carbamoyl phosphate synthetases are in separate cellular compartments in humans, there is no impact on the urea cycle by control of the cytosol carbamoyl phosphate synthetase by pyrimidine pathway products.
9. Typically, the mechanism for amine addition involves a nucleophilic approach by nitrogen. A lone pair of electrons is required on the nitrogen. Ammonium ions are protonated at physiological pH and do not have a lone pair. The amide of glutamine is not protonated and carries a lone pair of electrons.
11. At some point during the mechanism a phosphoryl group is transferred from the ATP to the carbonyl carbon on the glycyl moiety of the 5'-phosphoribosyl-N-formylglycinamide. Later during the mechanism, P_i is eliminated, taking with it what was the carbonyl oxygen. The phosphorylation converts the oxygen into a much better leaving group.
13. Allopurinol inhibits xanthine oxidase (fig. 26.27). Because uric acid is the major nitrogen excretion substance in birds, it is expected that allopurinol would have a catastrophic effect on chickens.
15. Sulfa drugs compete with the normal substrate (para-amino benzoic acid) during the biosynthesis of folic acid in bacteria (see table 26.3). Increased concentrations of para-aminobenzoic acid reverse this competition. Because humans do not synthesize folic acid, sulfa drugs are "nontoxic" to humans.

17.

PRPP

$+P_i$

Ribose-1,5-bisphosphate

19. If a person lacked the enzyme adenosine deaminase (ADA), the nucleoside deoxyadenosine would accumulate to high levels in some cells. The deoxyadenosine would then be phosphorylated to dATP, which inhibits the enzyme ribonucleotide reductase, leading to the inhibition of DNA replication (see the illustration). This genetic defect has been treated by injecting patients with the enzyme adenosine deaminase, which is apparently taken up by the cells, reducing deoxyadenosine levels. The newest approach to treatment, however, is to use gene therapy to insert a gene for ADA into lymphocytes. This is the first genetic defect to be treated by gene therapy.

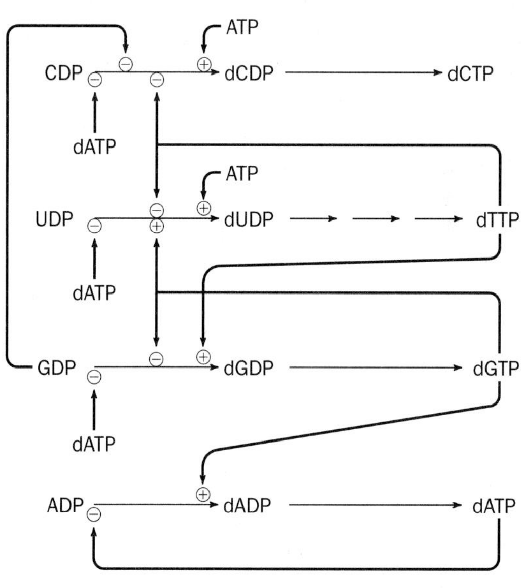

Source: L. Thelander and P. Reichard, "Reduction of Ribonucleotides," *Annual Review of Biochemistry,* 48:133, 1979, Annual Reviews Inc., Palo Alto, CA.

21. Allopurinol is an analog of hypoxanthine that inhibits the activity of xanthine oxidase. Xanthine oxidase, which catalyzes the conversion of xanthine to uric acid (see fig. 26.27), plays an important role in the degradative pathway for 6-mercaptopurine, oxidizing 6-thioxanthine to 6-thiouric

acid, which is then excreted. Inhibition of xanthine oxidase would therefore lead to elevated intracellular levels of 6-thio IMP (6-mercaptopurine is converted to 6-thio IMP by HGPRTase). 6-Thio IMP inhibits key steps in the conversion of IMP to AMP and GMP, and the formation of phosphoribosylamine from PRPP and glutamine, which is the first committed step in the de novo pathway for purine nucleotide synthesis. The resulting depletion of purine nucleotides would therefore kill fast-growing cells, such as cancer cells, far more effectively than 6-mercaptopurine alone.

Chapter 27

1. Both interactions involve shape changes in proteins (i.e., allosteric interactions). Allosteric effectors and hormones are typically small molecules. Allosteric enzymes are enzymes, while hormone receptors may be enzymes. Allosteric effectors/enzymes are typically within a cell, while hormones and receptors typically involve different locations in multicellular organisms.
3. When one speaks, the tones, gestures, and facial expressions often carry part of the message. It is human nature in general, or maybe specifically the nature of biochemists, to personify the inanimate components of processes that make up living systems. This personification now allows the use of significant vocabulary that was previously unavailable. In this specific situation, "cross-talk" refers to the results of a multitude of interactions among the components of the system.
5. As written, there are two electrons and two hydrogen ions in the products that are not in the reactants. Possible solutions include the addition of NADH + H^+ and NAD^+ as additional reactants/product (NADPH or $FADH_2$ could substitute). Alternatively, the production of a more oxidized product (CO_2 instead of formic acid) would solve the problem.
7. Auxin could be formed by transamination of tryptophan with α-ketoglutarate followed by the action of an α ketoacid dehydrogenase (comparable to pyruvate or α-ketoglutarate dehydrogenases), then a reaction analogous to that of the succinate dehydrogenase. (Alternatively, the latter reaction could be a simple hydrolysis of the thioester.)
9. A starving person is metabolizing fats for energy and generating large amounts of acetyl-CoA, which is converted to ketone bodies. The brain has switched from depending entirely on glucose for energy to depending on ketone bodies. When this starving person is given a dose of insulin that normally would be large enough to generate insulin shock (a rapid drop in blood glucose levels that causes a person to pass out because the brain is not getting enough glucose), nothing happens because the brain is getting its energy from ketone bodies, not glucose.
11. The energy requirements of muscle tissues are typically met by aerobic processes. During significant and continuous exertions, energy requirements can be greater than can be met by the circulatory system's capacity to deliver oxygen. If additional energy (ATP) is needed, it can be obtained by anaerobic processes.
13. This patient most likely has a primary defect, that is, of the thyroid. Since his thyroid cannot produce T_3 and T_4, his TSH levels are elevated owing to lack of feedback inhibition. His TRH (from the hypothalamus) cannot be the cause, since under such conditions, TSH (from the pituitary) would be reduced or eliminated instead of elevated.
15. Most membrane-associated hormone receptors generate a diffusible intracellular signal called a second messenger. The five currently known second messengers are cyclic AMP, cyclic GMP, inositol triphosphate, diacylglycerol, and calcium. Part of the transmembrane signaling that occurs when a ligand binds to the membrane-associated receptor is the activation of a membrane-bound G protein (the receptor interaction stimulates the exchange of GTP for GDP) and the resulting G protein-GTP complex, which can activate or inhibit an intracellular enzyme associated with the inner cell membrane. For example, adenylate cyclase can be activated to produce cAMP (a second messenger that activates protein kinases). Another example would be the activation of phospholipase C by a G protein-GTP complex, which leads to the generation of inositol triphosphate (the inositol triphosphate causes the release of calcium from the lumen of the endoplasmic reticulum into the cytosol) and diacylglycerol, both of which stimulate the phosphorylation of proteins leading to a cellular response.
17. Both the rapid and slow growth responses initiated by auxin binding are mediated by protein synthesis. The rapid response is due to an increase in proton transport out of the cell by a pump coupled with a membrane ATPase. Synthesis of polypeptide factors that stimulate proton transport (and the associated ATPase) may occur in the first few minutes after auxin binding. These polypeptides and the events they trigger probably modulate mechanisms controlling cellular growth. The slow response is due to an increase in the synthesis of proteins and nucleic acids necessary for sustained growth. This dependency of both rapid and slow auxin-mediated growth responses upon de novo synthesis of proteins accounts for their sensitivity to the effects of inhibitors of protein synthesis.

Chapter 28

1. Choline is a component of the phospholipid phosphatidyl choline, which is produced by the sequential methylation (SAM) of phosphatidyl ethanolamine. Humans require the choline moiety preformed in their diet.
3. The indol derivatives could act as analogs of the neurotransmitter serotonin or related compounds. Analogs could inhibit normal binding and/or inactivation of serotonin, causing a change in neurotransmission and a corresponding alteration in the perception of reality.
5. $H_2O + R'CH_2NHR + O_2 \rightarrow R'CHO + H_2O_2 + RNH_2$

   Does the destruction of the amine nature of these neurotransmitters cause you to rethink your answer to problem 4? Because monoamine oxidase destroys the amine nature of the neurotransmitters, it is one way they can be inactivated after they have performed their function.

7. $$\text{Dopamine} + O_2 + \text{Ascorbic acid} \rightarrow \text{Water} + \text{Norepinephrine} + \text{Dehydroascorbic acid}$$

9. All of the ions mentioned in this problem have about the same diameter as the monohydrated sodium that is actually transported by the sodium pore. The fact that $CH_3NH_3^+$ is not readily transported suggests that hydrogen bonding between the pore-lining hydroxy groups and the cation to be transported may be rather important. $CH_3NH_3^+$ can participate in only three hydrogen bonds, while the other two ions mentioned can participate in as many as six hydrogen bonds.
11. The conformational state of an Na^+ channel is dependent on $\Delta\Psi$ across the membrane. Current pulses or chemical stimuli that depolarize the membrane by about 20 mV lead to an abrupt opening of all Na^+ channels. Depolarizations below this value are not sufficient to open the channels, whereas no more channels can be opened by greater current pulses than are activated at the threshold depolarization value. This results in an "all-or-none" action potential response, since the magnitude of the potential spike is related to the number of open Na^+ channels, which abruptly changes at values of $\Delta\Psi$ around -40 mV.
13. (a) Since ANTS is membrane-impermeable, it can be trapped in reconstituted proteoliposomes during their preparation. If Tl^+ is added to the outside of such vesicles loaded with the fluorophore ANTS, a decrease in fluorescence emission of ANTS would accompany Tl^+ flow down its concentration gradient into the vesicles. The rate of fluorescence emission decay in the presence of agonists such as acetylcholine should then be proportional to the rate of Tl^+ entry into vesicles through acetylcholine receptor channels (when corrected for the same value obtained from vesicles devoid of receptors). Since those rates should have half-times on the order of milliseconds (if they resemble physiological rates), the experiments would have to be conducted in an instrument designed to measure the kinetics of rapid processes, such as a stopped-flow device.
    (b) The average volume of one vesicle is about 3.3×10^{-20} liter from the information given. This yields 6.6×10^{-18} mole Tl^+ transported per vesicle per second. If each vesicle has an average of two functional channels, 2×10^6 ions per second per channel are transported. Fluxes measured for Na^+ in vivo are on the order of 10^7 ions per channel per second (see the text). Thus the reconstituted system approaches the rates of ion flux measured in undisrupted cells.*

**Source:* This experimental approach is described in W. C.-S.Wu, H.-P. H. Moore, and M. A. Rafter, *Proc. Natl. Acad. Sci. (USA)*78:775, 1981.

15. DOPA is given to Parkinson's disease patients to raise levels of dopamine in the brain. If dopamine were given directly, it would not pass the blood–brain barrier. DOPA, the precursor, is a zwitterion that can pass the blood–brain barrier. When DOPA is given, one of the dopa decarboxylase inhibitors is given to prevent the DOPA from being converted to dopamine in the blood and tissues in the body. This allows serum levels of DOPA to remain high and pass the blood–brain barrier. The dopa decarboxylase inhibitors also do not pass the blood–brain barrier, allowing decarboxylation of DOPA to dopamine in the brain. Thus by using the decarboxylation inhibitors, much lower doses of DOPA can be given, reducing the side effects of this treatment.
17. The acetylcholine receptor appears to be acting by a gated-pore-type mechanism rather than a carrier-mediated one. One of the main types of data pointing to this gated-pore mechanism is the high Na^+ flow seen in the open channel. The carrier-mediated model could not account for such a high flux of Na^+. Also, the detailed structure of the polar channel found in the receptor fits the pore model as seen in Na^+ and K^+ channels.

Chapter 29

1. The thioester linkage between the palmitoyl residue and rhodopsin is not a typical bond type found in proteins. When thioesters are found in proteins, they are often intermediates in reaction mechanisms.
3. All of the processes described in the text occur in the cell membrane, cytoplasm, or disk membrane. The disk lumen has a minimal volume that produces disk "surfaces" (membranes) that are essentially parallel. The "stacked" disks are aligned perpendicularly to the incident light, increasing the likelihood that a photon will interact with a rhodopsin molecule.
5. The absorption of light by rhodopsin can be amplified by a factor of almost one million by using an enzymatic cascade using the G protein transducin. Photoactivated rhodopsin interacts with transducin, allowing the binding of GTP to one of the subunits in transducin. The GTP-transducin subunit complex then activates a phosphodiesterase that hydrolyzes cyclic GMP. Cyclic GMP keeps the Na^+ channel open. One rhodopsin can activate many phosphodiesterases, which can quickly lower cyclic GMP levels. This again is an example of the importance of cascade mechanisms in biochemistry.
7. Rhodopsin has an absorption maximum at 500 nm. Illumination at this wavelength converts rhodopsin to bathorhodopsin, which absorbs maximally at longer wavelengths (543 nm). At 77° K, bathorhodopsin does not decay to lumirhodopsin. Illumination at 500 nm thus causes the absorbance at 500 nm to decrease and the absorbance at 540 to 550 nm to increase. Light at 550 nm is absorbed mainly by bathorhodopsin and can convert the bathorhodopsin back to rhodopsin. (The same excited state can be generated by exciting either rhodopsin or bathorhodopsin. No matter how it is generated, the excited

state decays to bathorhodopsin about 70% of the time and to rhodopsin about 30% of the time. Continuous illumination at any given wavelength will set up a photostationary state in which the mixture of rhodopsin and bathorhodopsin depends on the wavelength. Longer wavelengths are more likely to be absorbed by the bathorhodopsin in the mixture and so favor the net formation of rhodopsin.)

9. $$C + ATP \xrightarrow[\text{cGMP}]{\text{Kinase}} C\text{-}P + ADP$$

Phosphatase: $C\text{-}P + H_2O \rightarrow C + P_i$

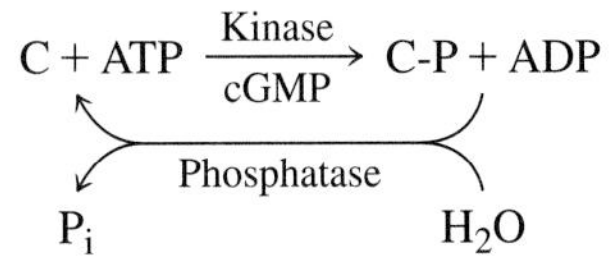

C = inactive Na channel in outer segment plasma membrane; C-P = channel activated by phosphorylation. The phosphorylation is catalyzed by a kinase that requires cGMP. Inactivation of the channel is catalyzed by a phosphatase that is independent of cGMP or is stimulated by cGMP. (This scheme is speculative; you may think of a better one. See the text for evidence.)

11. There are three types of cone cells that absorb blue, green, and yellow light, thereby allowing us to see color. These cone cells are not connected to bipolar cells as the rod cells are. Rod cells are connected to bipolar cells that are connected to each ganglion cell, allowing summing of the signals from the rods. Since the cone cells are not connected, they are not sensitive to dim light as the rod cells are. Therefore we do not see color in dim light.

Chapter 30

1. Erwin Chargaff's hypothesis was as follows: If nucleic acids serve a genetic function, then there should be chemically different nucleic acids in different organisms. He assumed that these chemical differences would most likely be reflected in more-or-less subtle differences in the nucleotide composition of DNA isolated from different species. Chargaff's laboratory developed methods for the precise analysis of nucleic acid hydrolysates, demonstrating that the base composition of DNA varies, depending on the species (but not the tissue) from which it was isolated. This observation, that the DNA of any given species is chemically distinct from that of any other species, is consistent with a possible role of DNA in the transmission of hereditary information.
3. The three base pairs shown in the figure can be formed between guanosine and thymine. The one shown in (a) is most similar to a standard Watson-Crick base pair. Structures like those in (b) and (c) may help hold together the three-dimensional configuration of some nucleic acids; for example, by replacing T with U, similar non-Watson-Crick base pairs help stabilize tRNA and rRNA.

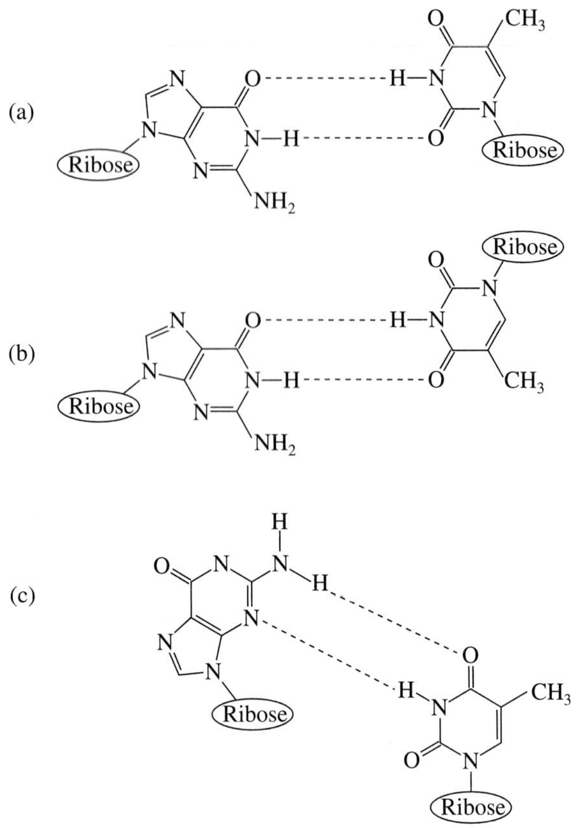

(a) Ribose N N NH$_2$ O N—H H—N O CH$_3$ N Ribose

(b) Ribose N N NH$_2$ O N—H H—N O CH$_3$ N Ribose

(c) H N H O N N N Ribose O H N CH$_3$ O N Ribose

5. The diameter of the DNA double helix is established by the outside dimensions of each purine–pyrimidine base pair and the angle of the bases with respect to the phosphodiester backbone. Since AT and GC base pairs have nearly identical widths, and since the base pairs are positioned in exactly the same way between deoxyriboses of the backbone, then the diameter of the double helix must remain constant.
7. RNA duplexes and RNA–DNA hybrids are typically found in the A conformation. The 2′-OH group of the ribose in RNA sterically prevents the B duplex from forming. In the C2′-endo ribose conformation, found in B-DNA, a 2′-OH would be in very close proximity with the neighboring base pair.
9. The chemical reaction mechanism for the base hydrolysis of RNA is shown in the figure. Obviously, DNA could not be hydrolyzed by this mechanism because it lacks a 2′-OH group.

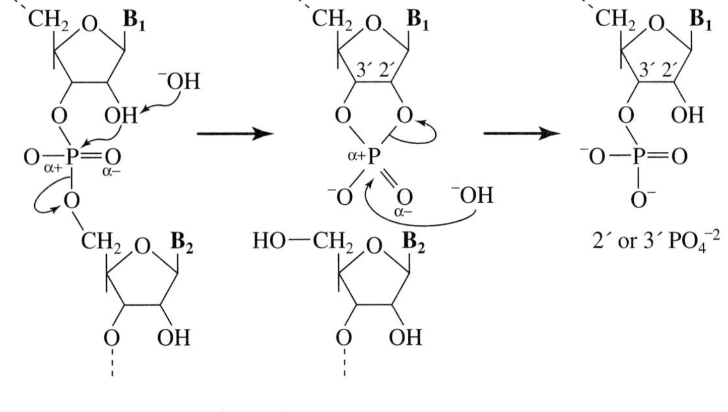

CH_2 O B_1 ^-OH O OH O—P=O α+ α− O CH_2 O B_2 O OH → CH_2 O B_1 3′ 2′ O O α+P ^-O O α− ^-OH HO—CH_2 O B_2 O OH → CH_2 O B_1 3′ 2′ O OH ^-O—P=O O^- 2′ or 3′ PO_4^{-2}

11. A simple approach would be to take small aliquots of the sample and treat them with the enzymes ribonuclease (RNase) or deoxyribonuclease (DNase). Digestion of the sample by one of these enzymes would indicate whether the sample is RNA or DNA. Another method is to treat a small sample with alkali, which degrades RNA to mononucleotides but only denatures DNA to the single-stranded form. One way to detect whether these treatments had any effect on the nucleic acid is to subject it to electrophoresis on an agarose gel. The original viral nucleic acid should yield one (or more) discrete high-molecular-weight bands, which would disappear after the appropriate treatment.

    To determine whether the nucleic acid is single-stranded or double-stranded, it could be heated. A very sharp increase in absorbance at 260 nm would indicate a double-stranded RNA or DNA, while a broader melting curve would suggest a single-stranded nucleic acid. One could also analyze the nucleotide composition of the nucleic acid. Equivalence between A and T and between G and C would strongly suggest that the nucleic acid is double-stranded.

13. In the absence of any other information, one would have to assume that AUG is an RNA trinucleotide and that GTA is DNA (because the presence of U normally indicates RNA and T indicates DNA). The other major difference between these two trimers is the orientation of the purine bases: AUG has an adenine attached to its 5′ ribose, followed by a uracil and then a guanine at the 3′ end. GTA has a guanine at its 5′ end, followed by a T and then an A at its 3′ end.

15. The structure of one U-A-U base triplet is illustrated in the figure. It is possible to draw a similar triplet with the second uracil turned in the other direction, but this would position two of the phosphodiester backbones too close to each other. The triple helical structure would be held together by hydrogen bonding and stacking interactions between bases. The addition of $MgCl_2$ would help stabilize the triple helix by offsetting the electrostatic repulsion between the charged phosphodiester backbones of the three DNA chains.

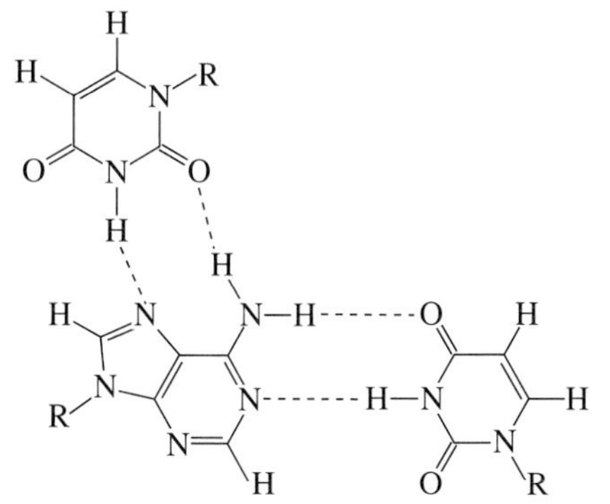

17. The two DNAs will have different melting temperatures because of their differing G + C contents. Take a small sample and measure the absorbance at 260 nm while slowly raising the temperature. You should see the graph shown in the figure. Viscosity can also be used to determine DNA melting points. In chapter 32, the Southern blotting technique is described, which could be used to distinguish these two DNAs in a procedure that is much more cumbersome than the melting point determination described above.

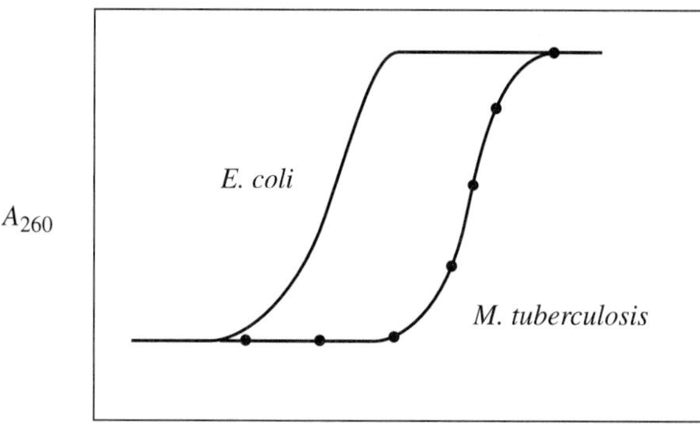

19. The ratio of the histone in chromatin supports the model proposed for nucleosome structure. The core of the nucleosome is made up of an octamer of two molecules each of H2A, H2B, H3, and H4. H1 histone seals off the nucleosome; that is, there is only one H1 per nucleosome. The observed 2:1 molar ratio of core histones to H1 fits the model proposed for the nucleosome.

Chapter 31

1. If DNA replication were dispersive, then each strand of each daughter molecule would have an intermediate density in the experiment. The fact that after one generation, one strand of the DNA still had the same density as the parental (^{15}N) DNA served to further corroborate the conclusion that DNA replication is semiconservative.

3. The reaction for phosphodiester bond formation is as shown in the figure. In this reaction, two phosphate bonds are hydrolyzed (when the pyrophosphate hydrolyzed by pyrophosphatase is included) for each phosphodiester bond formed. The excess energy released apparently is required because the assembly of an ordered nucleic acid with a defined sequence requires additional energy beyond that required to drive phosphodiester bond formation. Mg^{2+} can bind to the transition state intermediate (trigonal bipyramid intermediate) and stabilize it, lowering the energy of activation. The metal ion can also promote the reaction by charge shielding: The Mg^{2+}–NTP complex is the actual substrate, with the metal reducing the negative charge on the phosphate groups so as not to repel the electron pair of the attacking nucleophile.

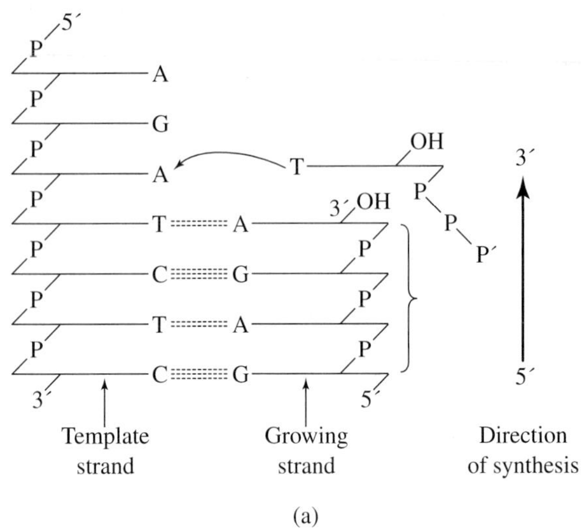

(a)

(b)

5. These data support the model in which one strand (the leading strand) is made in a continuous mode, while the other strand (lagging strand) is synthesized in discontinuous fashion. Small pieces (Okazaki fragments) are made first and then become incorporated into large segments of DNA. When the DNA is applied to the alkaline sucrose gradient, the DNA is denatured and becomes single-stranded. Small single-stranded fragments would not sediment as rapidly in the sucrose gradient as would the large segments of DNA. During the pulse, label is incorporated into the small fragments of the lagging strand as well as the leading strand. The chase allows one to observe the eventual product, in which the small fragments of the lagging strand are joined to form longer segments.
7. (a) Yes, this DNA can stimulate incorporation of dNTPs by Klenow enzyme. This double-stranded DNA has single-stranded 5′ overhangs much like those generated by many restriction enzymes (chapter 32). The recessed 3′ ends would act as primers and be extended by the Klenow fragment's DNA polymerase activity, using the 5′ ends as templates, to generate a fully base-paired double-stranded DNA.
   (b) This DNA is unlikely to stimulate dNTP incorporation, since it lacks single-stranded regions to serve as replication templates. (The 3′ to 5′ proofreading exonuclease activity of the Klenow fragment would probably not chew back the 3′ ends of this fragment, since it is fully base-paired.)
   (c) This molecule may stimulate dNTP incorporation, depending on whether the 3′ to 5′ exonuclease activity of Klenow fragment were able to trim back the 3′ end of the linear oligonucleotide, enabling it to serve as a primer for replication of the circular DNA.
   (d) The circular DNA in this molecule can serve as a replication template, with the linear oligonucleotide serving as a primer. The 3′ end of the linear fragment would be elongated to form a fully double-stranded circular molecule containing a single nick where the newly synthesized DNA meets the 5′ end of the primer. Of the molecules illustrated, this one is likely to be the most efficient at stimulating incorporation of dNTPs into DNA.
9. Since a replication fork moves at about 10^3 bp/s and because there are two replication forks, the DNA is being replicated at about 2×10^3 bp/s. At this rate, it would take *E. coli* about 33 min to replicate its genome:

$$\frac{4 \times 10^6}{2 \times 10^3} = 2 \times 10^3 \text{ s, or about 33 min}$$

Cells divide faster when one round of replication starts before the previous one is finished, generating multiple replication forks (multifork replication) and leading to division times of about 20 min.

11. It would be possible for an enzyme to synthesize DNA (or RNA) in a 3′ to 5′ direction (see the figure). However, because the available nucleotides in present-day organisms have 5′-triphosphates, pyrophosphate would be released from the 5′ end of the nascent strand, rather than from the incoming nucleotide. Unfortunately, such an enzyme would be incapable of proofreading because if it were to excise a misincorporated nucleotide, there would be no β- and γ-phosphates available at the 5′ end of the chain to drive the incorporation of the next nucleotide.

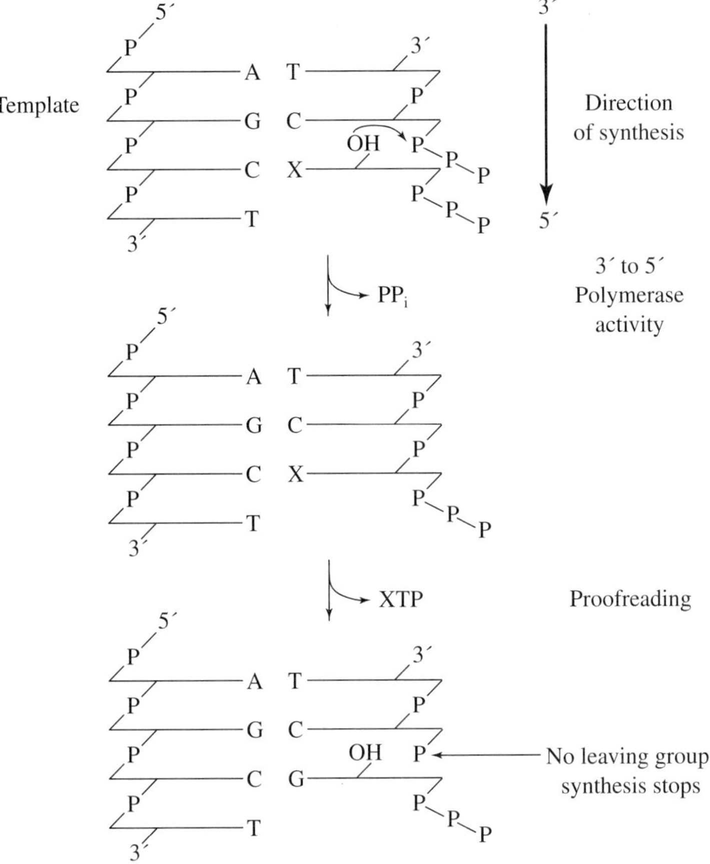

13. (a) $dnaA^{ts}$: DNA synthesis would stop after a time lag, since the *dnaA* gene product is involved in initiation of DNA replication.
(b) $dnaE^{ts}$: DNA synthesis would stop immediately upon a shift to the nonpermissive temperature, since the *dnaE* gene encodes a subunit of DNA polymerase III.
(c) $dnaG^{ts}$: DNA synthesis would stop immediately upon a shift to the nonpermissive temperature, since the *dnaG* gene encodes primase, which is required almost continuously for lagging strand replication. Leading strand replication cannot continue without lagging strand synthesis.
(d) $dnaN^{ts}$: DNA synthesis would stop immediately upon a shift to the nonpermissive temperature, since the *dnaN* gene encodes the processivity factor for DNA polymerase III.
(e) lig^{ts}: DNA synthesis would stop after a time lag, since the product of the *lig* gene, DNA ligase, is not directly required for DNA replication, but only for joining lagging strand fragments.

15. In photoreactivation, the thymine dimer is repaired by using light energy. There is no breakage of the phosphodiester backbone or even of the N-glycosidic linkage with the base. In excision repair, the phosphodiester backbone is cleaved at two positions near the thymine dimer. A helicase unwinds the damaged DNA, which is then discarded. A patch of new DNA is synthesized by DNA polymerase I, using the undamaged strand as template, and the upstream cleavage site as primer. In this case, the energy for repair is supplied by the cleavage of multiple high-energy phosphate compounds.

The two mechanisms are easily distinguished by allowing repair to take place in the presence of ^{3}H-thymidine. (The ^{3}H label would be incorporated into TTP via salvage pathways.) In photoreactivation, there would be no ^{3}H incorporated into the repaired thymine(s), while excision repair would result in the incorporation of ^{3}H.

17. The products of oxidative deamination of adenine, guanine, and cytosine are hypoxanthine, xanthine, and uracil, respectively (see the text for structures). Often, these deaminated bases are recognized by specific glycohydrolases and cleaved, leaving an AP site suitable for excision repair. Uracil-DNA glycohydrolase is very important in DNA repair because it excises uracils from DNA. These uracils arise from deaminated cytosine, the most frequent spontaneous DNA modification. DNA may have evolved to contain thymidine instead of uracil to provide a mechanism to distinguish deaminated cytosines from legitimate thymines.

19. (a) *recA441* has a mutation in the RecA protease activity that causes it to become active at 42°C in the absence of DNA damage. This mutant is dominant because its protease will inactivate LexA protein even in the presence of wild-type RecA.
(b) *recA430* is a mutant with an inactive RecA protease, which cannot activate LexA by itself and therefore cannot trigger the SOS response. The rest of the protein has normal activities, so the mutation does not affect recombination.
(c) *lexA*(Ts) has a mutation that weakens the LexA–DNA interaction at 42°C, perhaps because the protein becomes unstable at that temperature. When this mutant protein dissociates from the DNA, it allows the induction of the SOS response even in the absence of DNA damage. This mutation is recessive because wild-type copies of LexA would remain bound at 42°C.

21. The telomerase RNA mutation, C<u>G</u>ACCCCAA, would, over time, lead to an alteration in the sequence of the telomeres. The telomeric DNA sequence would change from $(TTGGGG)_n$ to $(TCGGGG)_n$ (see fig. 31.35). In addition,

telomere replication would most likely be less efficient than with the wild-type telomerase, because the pairing between telomere and telomerase RNA would be impaired.

Chapter 32

1. With spaces introduced for clarity, the rest of the sequence in the autoradiogram in figure 32.1 reads as follows:

   5′-CAAAAAACGG ACCGGGTGTA CAACTTTTAC TATGGCGTGA CACCTAAATT ATAGGCAGAA ATAAGTACAT GACTATTGGG AGGAGCAGGA ACAAGTAGG-3′

   The more slowly migrating the fragment, the larger it is and the smaller the difference between it and its neighbors. Consequently, the separation between bands on the gel becomes smaller and smaller as one gets farther away from the sequencing primer. The sequence indicated at the 5′ end would be closest to the sequencing primer. The primer sequence itself cannot be read from the sequencing gel, because it is not labeled. By convention (and because it makes it easier to read), the DNA sequence deduced from a sequencing gel is that of the DNA synthesized during the sequencing reactions. Some bands on the sequencing gel autoradiogram are less intense than others. These variations are most likely due to substrate and template-dependent variations in the efficiency of replication by the DNA polymerase used.

3. The primary advantage of the PCR method over conventional recombinant DNA cloning is the speed at which a specific sequence of DNA can be cloned in vitro. (Instead of several days to clone a specific DNA fragment from a complex mixture, it can now be done in an afternoon.) It is possible to start with an mRNA population, convert it to cDNA with reverse transcriptase, and then amplify the cDNA sequence of interest, dramatically reducing the number of steps and the time required for cloning.

   The PCR method has been applied to the detection of many genetic defects and has become the method of choice for detecting many types of infections by both bacteria and viruses. One of the problems with the Western blotting test for HIV infection (infection with the AIDS virus) is that the test is indirect, testing for the presence of antibodies produced by the patient against the AIDS virus. It may take months after infection for a person to produce antibodies at high enough levels to be detected. The PCR method, however, could detect infection within five days. The DNA could be isolated from white blood cells and subjected to PCR amplification, using specific probes for HIV. The presence of the provirus could be detected. This method could detect one infected lymphocyte out of a million.

5. Because DNA is not a regularly repeating polymer, it is impossible to predict *exactly* where and how often a given restriction enzyme cleavage site will occur. However, assuming that DNA is a completely random polymer, we can predict the probability of occurrence of a certain sequence of bases and can therefore predict the average frequency of occurrence of the enzyme's cleavage site in any particular DNA.

   With a G + C content of 80%, and since [G] = [C] and [A] = [T], the base composition of the thermophilic DNA is 40% G, 40% C, 10% A, and 10% T. To calculate the probability of occurrence of a given sequence, simply multiply the probability of occurrence of each individual nucleotide in the sequence. For example, the probability of the sequence -CTGCAG- (a *PstI* site) occurring in DNA that is 80% G + C is $(0.4)^4 \times (0.1)^2$ = once every 3906 bp. An -AAGCTT- (*HindIII*) site should occur on average every $(0.4)^2 \times (0.1)^4$ = once every 62,500 bp. Thus, this very G + C–rich DNA is predicted to be cleaved more often by the restriction enzyme with the more G + C–rich recognition sequence (*PstI*). *PstI* would, on average, give shorter fragments than would *HindIII* with this DNA.

   The *E. coli* genome composition is approximately 26% G, 26% C, 24% A, and 24% T. *PstI* would cleave this genome $(0.26)^4 \times (0.24)^2$ = once every 3,800 bp and *HindIII* $(0.26)^2 \times (0.24)^4$ = once every 4,500 bp. Note that because the *E. coli* nucleotide composition is not as strongly biased as that of the thermophile, *PstI* and *HindIII* sites are expected to occur with similar frequencies in *E. coli.*

   Typically, of course, there is considerable variability in the sizes of restriction fragments produced by any one enzyme from a given DNA. The above estimates represent likely averages, and the range of fragment sizes can still vary from a few base pairs to millions of base pairs.

7. There is no difference between the single-stranded overhangs generated by *BamHI* and *MboI.* These ends are complementary (they are termed "compatible" ends), and they can anneal and be ligated to each other by DNA ligase as shown in the figure. The resulting ligation product contains an *MboI* site and can be cleaved by that enzyme. It only has a 25% probability of restoring a *BamHI* site, however, depending on the nucleotide located adjacent to the *MboI* site. In other words, if the nucleotide 5′ to the *MboI* site is a G, ligation of *MboI* to *BamHI* would regenerate a sequence cleavable by *BamHI,* but not if that nucleotide is an A, C, or T.

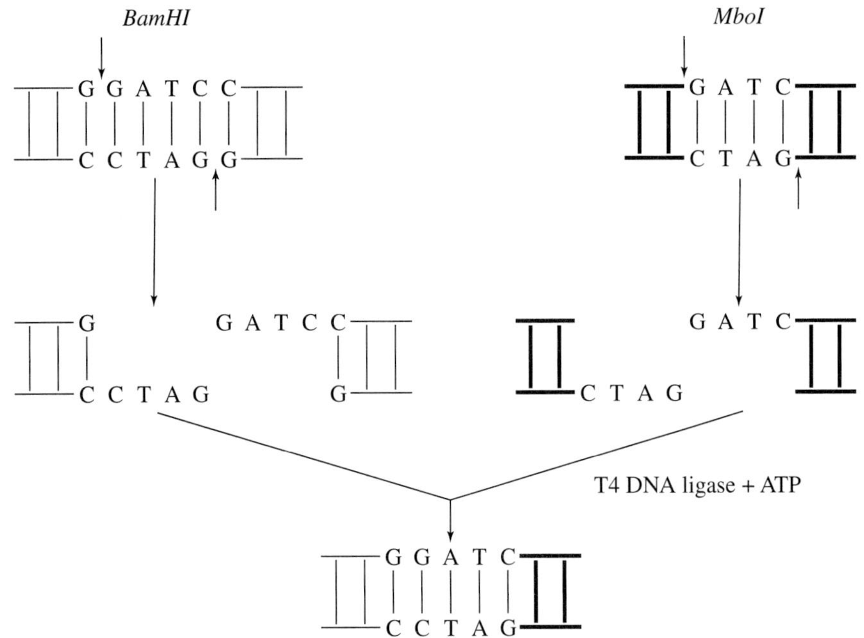

9. (a) Polymerase chain reaction:
   (2) Characterizing a mutation in a known gene. (Since the gene sequence is known, it should be easy to design primers flanking known mutation sites and to use the PCR to amplify the corresponding DNA for sequence analysis.)

   (b) Isolating a clone from a cDNA library:
   (3) Deducing a protein sequence from its coding sequence. (Since the goal in this case is to study only the coding sequence, it would be best to start with cDNA, which lacks regulatory sequences and introns.)

   (c) Isolating a clone from a genomic library:
   (1) Characterizing sequences controlling expression of a gene. (The sequences regulating expression of a gene are often in the nontranscribed part of the gene and therefore would not be represented in a cDNA library.)

11. Verification that bacterial T-DNA has been transferred into the plant genome is easily accomplished by Southern hybridization. DNA would be prepared from a small amount of tissue from the regenerating plantlets and digested with a restriction enzyme that is known to cut at least once within the T-DNA. The pattern of restriction fragments hybridizing to a T-DNA probe would be different, depending on whether the T-DNA is still present on the Ti plasmid or whether it has integrated into the host plant genome.

13. The unmodified form of DNA polymerase I has 5′ to 3′ exonuclease activity, which has the potential to destroy the mutagenic primer in site-directed mutagenesis procedures. Once the Klenow fragment of *E. coli* DNA polymerase I has completely replicated the circular DNA and reached the 5′ end of the mutagenic primer, it must stop because it lacks 5′ to 3′ exonuclease.

15. Two oligonucleotide primers would be designed to correspond to sequences on either side of the ΔF508 mutation in the cystic fibrosis gene, as illustrated in the figure. PCR with these two primers and genomic DNA from the tested individual would yield a product that was 3 bp shorter for the mutation than the wild-type gene. Differences of 3 bp can be detected on some types of agarose gels or on denaturing polyacrylamide gels such as those used for DNA sequence analysis. Southern blotting, using oligonucleotide probes designed to hybridize only to the mutant or to the wild-type CFTR gene, would be even better for identifying mutant and wild-type PCR products. The main problem with this procedure, as with any PCR-based method, is the potential for contamination. A minute sample of DNA contaminating any component of the PCR reaction could give an erroneous result.

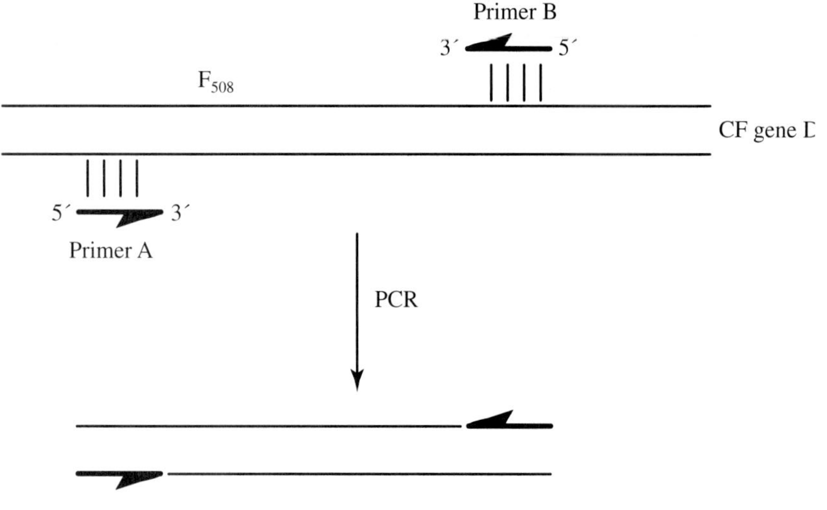

17. Your dog's DNA has contaminated the experiment. Because the dinosaur DNA (if any is present) is highly fragmented, it is likely to be exceedingly difficult to amplify. The DNA from epithelial cells in the dog's saliva, though present in small amounts, was intact and therefore more easily amplified. Because this is a common problem, scientists who isolate DNA from ancient specimens try to use an internal fragment of the sample that has not been exposed to outside sources of contamination.
19. When cDNA is incorporated into a cloning vector, there is a 50% probability that it will be incorporated in the wrong orientation. Of those cDNAs positioned with the correct strand attached to the transcription and translation signals, many will not have the codons encoding the amino acids of the protein positioned correctly for translation. In fact, since there are three nucleotides in each codon, there is a 1 in 3 chance of correctly positioning the cDNA with respect to the translational control signals. There is a $1/2 \times 1/3 = 1/6$ overall probability of expressing the protein encoded by the cDNA.

Chapter 33

1. The bases within the base-paired region of each arm of the tRNA cloverleaf stack in a fashion similar to the base stacking described in chapter 30 for DNA. (There is an important difference, however: Since the 2′-OH of RNA cannot fit in a B-form helix, the base-paired regions in RNA molecules adopt the A-form helix, in which the base pairs are tilted with respect to the helical axis.) In addition to the base stacking within base-paired regions, there is also stacking of one helix on top of another in the tRNA molecule. For example, the acceptor stem stacks with the TΨC stem to form one nearly continuous stacked double helix. The anticodon stem and the D stem also stack on top of one another.
3. The answers are summarized in the figure.

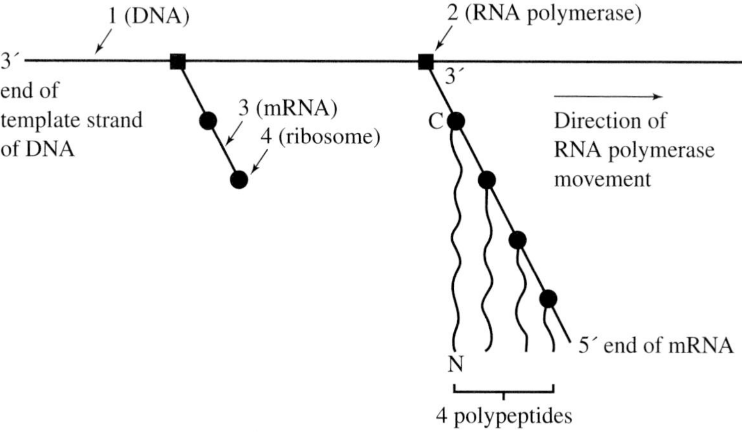

(a) 1, 2, 3, and 4 are the DNA, RNA polymerase, mRNA, and ribosome, respectively.
(b) The 5′ end of the template strand of DNA would be farthest from the transcription start and therefore would be the end closest to the longer RNA strand(s).
(c) The 3′ ends of the mRNAs would be the last to be synthesized and would therefore be the closest to the DNA.
(d) Peptides in the process of being synthesized would remain attached to the ribosomes; and since the ribosomes are moving from the 5′ end of the mRNA towards its 3′ end, the longest nascent polypeptides would be attached to the ribosomes closest to the 3′ end of the mRNA.
(e) Since proteins are synthesized from N-terminus to C-terminus, the N-terminus of any nascent polypeptide would be the part farthest from the ribosome.
(f) RNA polymerase is moving the 3′ end of the template strand of the DNA.
(g) In eukaryotes, RNA synthesis takes place in the nucleus, where the nascent transcripts are first processed into mature mRNAs. The mature mRNAs are then transported into the cytoplasm, where they become associated with ribosomes and are translated.

5. Both RNA polymerase and DNA polymerase have an active site in which are found the single-stranded DNA template, the 3′ end of the newly synthesized strand base-paired with its template, and a binding site for the incoming nucleoside triphosphate. In both x-ray crystal structures, a handlike conformation was observed, in which this active site is found in the palm. A major difference between the two structures is that RNA polymerase lacks the proofreading 3′ to 5′ exonuclease active site adjacent to its polymerase active site.
7. Pyrophosphate is one of the products of the reaction occurring during transcription with any of the RNA polymerases (—NpNpN + pppX → —NpNpNpX + PP_i). Pyrophosphatase would cleave the pyrophosphate to orthophosphate ($PP_i \rightarrow 2\ P_i$), preventing product inhibition and helping to drive the reaction to produce more full-length transcripts. The cleavage of pyrophosphate is a common theme in many biochemical reactions generating a negative Gibbs free energy [$-\Delta G$].
9. The amount of time to completely transcribe the dystrophin gene is

   (2×10^6 nucleotides)/(3×10^3 nucleotides per min) = 670 min, or more than 11 hours!

   The amount of time to generate one transcript from the β-globin gene is

   (1,600 nucleotides)/(3×10^3 nucleotides per min) = about 0.5 min
11. One common way to determine the 3′ end of an RNA is to treat the RNA with Na-periodate and oxidize the *cis*-diols to dialdehydes, which can then be reduced to dialcohols with $^3H\text{-}BH_4$. Normally, the only nucleotide in the sequence that has *cis*-diols is the 3′ terminus, which has the 2′ and 3′ hydroxyls. However, since eukaryotic mRNAs are capped at the 5′ end, they have a guanosine at the 5′ terminus as well as the adenosine at the 3′ end (due to the poly(A) tail) with *cis*-diols. These eukaryotic mRNAs would appear to lack a normal 5′ end by many types of 5′ end analysis, leading to the paradox of an RNA with no 5′ end and two 3′ ends.
13. Since the synthesis of this RNA is sensitive to low levels of α-amanitin, it must be transcribed by RNA polymerase II and processed into a mature RNA before being transported to the cytoplasm for translation. The processing steps include cap formation, removal of intron sequences, and poly(A) tail addition. The consensus sequences that direct these processing events are shown in part (*a*) of the figure. The final processed mRNA and its important features are shown in part (*b*) of the figure.

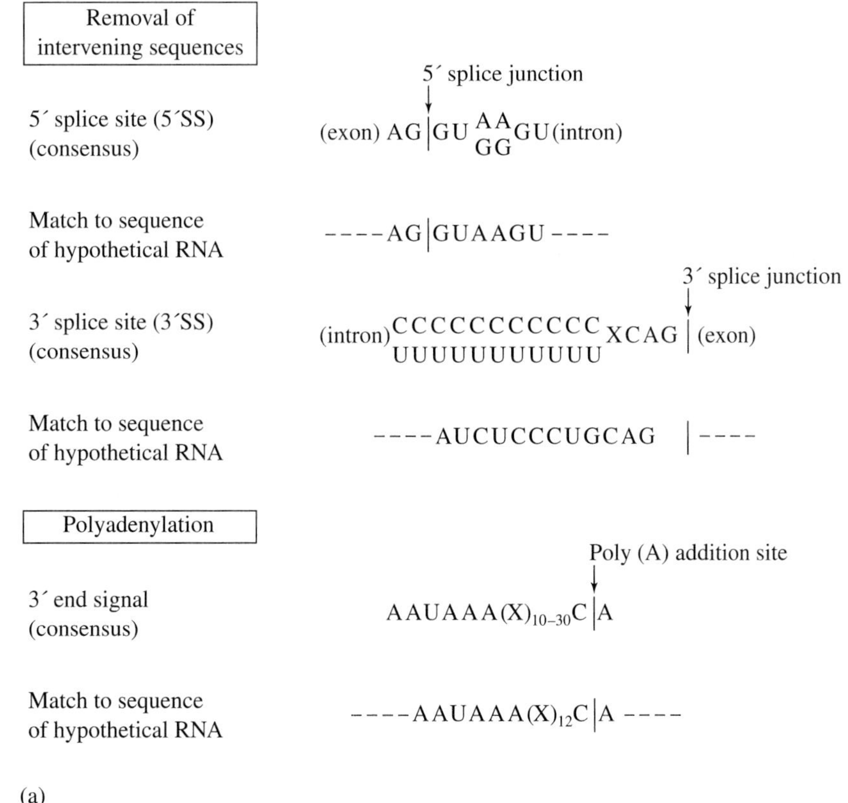

(a)

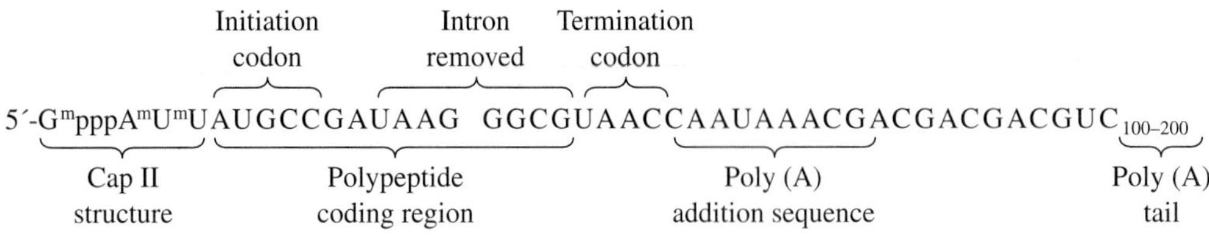

(b)

15. If an intron or part of an intron containing a stop signal for translation were not removed, this mRNA would be longer but would yield a shorter polypeptide. Alternative splicing would remove the intron or use an alternative splice site, generating a shorter mRNA and a longer polypeptide.
17. In both steps of both splicing reactions, a phosphodiester bond is broken at the same time as one is formed. (Contrast this situation with replication and transcription, in which two bonds are broken for one to form.) The driving force for the splicing reactions is probably the high-energy conformation adopted by the spliceosomal complex, in the case of mRNA introns, or the intron itself, in the case of the pre-rRNA precursor. (In fact, there is an ATP requirement for spliceosome assembly in the nucleus.)

Chapter 34

1. The number of three-nucleotide codons resulting from a random assortment of two bases is 2^3, or 8. In a random copolymer of A and U these eight codons would be AAA (Lys), UUU (Phe), AAU (Asn), UUA (Leu), UAU (Tyr), AUA (Ile), AUU (also Ile), and UAA (stop codon). Since the probability of finding each of these codons in the RNA is equal (because the probabilities of finding A or U at any one position of the RNA are equal), you should expect to find a mix of peptides composed of equimolar amounts of phenylalanine, leucine, lysine, asparagine, and tyrosine and twice as much isoleucine.
3. The genetic code is the triplet code found in the mRNA specifying the sequence of amino acids in the protein being made. However, this genetic code is effective only in conjunction with a mechanism to ensure the accurate charging of tRNAs with the correct amino acid. Obviously, if the wrong amino acid is attached to a tRNA by an aminoacyl-tRNA synthase, the wrong amino acid would be incorporated into the corresponding protein. Accordingly, aminoacyl-tRNA synthases are very specific both for the amino acid and for the tRNA species they bind. Because specific tRNA recognition and binding by a synthase is essential for protein synthesis to take place correctly, this interaction has been described as the "second genetic code."

   The crystal structure of glutaminyl-tRNA synthase with its tRNA bound has been solved. This synthase requires the anticodon to be intact; thus it must contact both ends of the L-shaped tRNA. There are many points of recognition between synthase and tRNA that permit specific charging only of $tRNA^{Gln}$. Not all tRNA synthases require the anticodon for specific charging (for example, $tRNA^{Ala}$ is recognized by its synthase through interactions with only a small number of bases in the acceptor stem, including a crucial G–U pair).
5. Since the *su3* suppressor tRNA is aminoacylated by tyrosyl-tRNA synthase, it must contain the determinants for $tRNA^{Tyr}$, suggesting that it arose by mutation of a wild-type $tRNA^{Tyr}$ gene. Normally, $tRNA^{Tyr}$ would have a GUA or AUA anticodon to read UAU and/or UAC codons. The observation that this tRNA reads UAG codons indicates that it probably has an altered anticodon sequence. (A CUA anticodon sequence would be expected to read UAG codons.)

   There are two obvious negative consequences upon mutation of a tRNA gene to become a suppressor: the effects of the loss of that tRNA and the effects of readthrough of UAG stop codons. There are many essential proteins with tyrosine residues that must be correctly translated for bacterial cells to survive. The fact that the *su3*-containing cells survive indicates that the $tRNA^{Tyr}$ that was mutated was not the only $tRNA^{Tyr}$ gene in the original *E. coli* strain (or that the mutated $tRNA^{Tyr}$ can still read UAU/C codons, an unlikely possibility). Thus, even when one $tRNA^{Tyr}$ gene mutates to become a suppressor, there is (are) still gene(s) that give(s) rise to one or more functional $tRNA^{Tyr}$s that can read UAU and UAC codons.

   In addition, many genes rely on UAG codons to stop translation, and the polypeptides encoded by these genes should be affected by the presence of the suppressor tRNA. There are several explanations for why the bacterial cell can still function in the presence of a tRNA that reads through stop codons: Many bacterial genes have multiple stop codons at the end of open reading frames, so even if the UAG codon is decoded as tyrosine, translation will still terminate at the next UAA or UGA codon. In addition, most suppressor tRNAs seem to be context-dependent, meaning that they do not read through all UAG codons with equal efficiency. Recall that the EF-Tu-*su3*tRNA complex is competing with RF-1 for binding to the UAG codon in the A-site of the ribosome. A less efficient tRNA–codon interaction would permit RF-1 to bind first, resulting in normal termination. Nonetheless, suppressor-containing *E. coli* strains usually do not grow as well as the wild-type, most likely because the suppressor does cause defects in protein synthesis.
7. Shown in the accompanying figure are the structures of the first dipeptides made in bacterial protein synthesis reactions when the initiation codon is followed by codons for either alanine or leucine. In each case, the dipeptide is shown still

attached to the 3′ end of the tRNA bound to the ribosome.

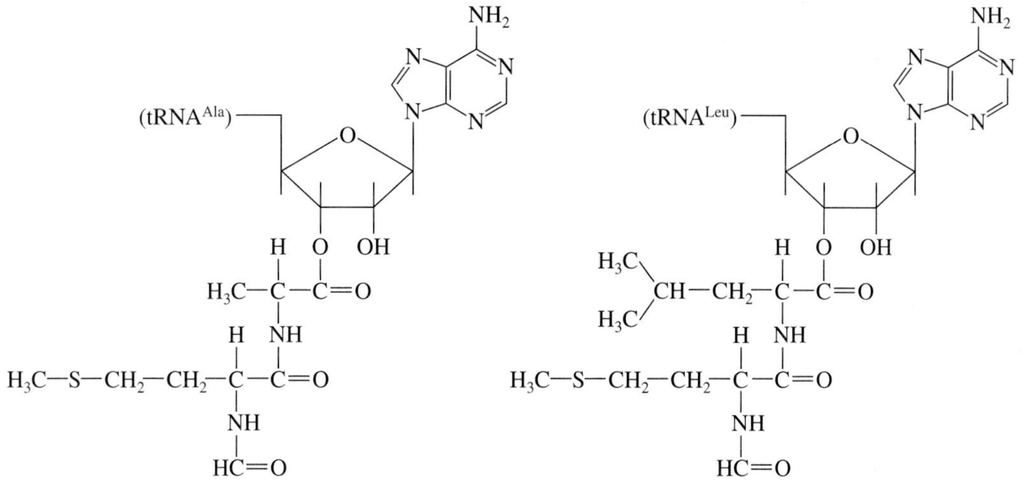

(tRNA^{Ala}) — [ribose–adenine; NH$_2$, N, N, N, N, O, OH]
H O OH
$H_3C-C-C{=}O$
H NH
$H_3C-S-CH_2-CH_2-C-C{=}O$
NH
$HC{=}O$

(tRNA^{Leu}) — [ribose–adenine; NH$_2$, N, N, N, N, O, OH]
H_3C H O OH
$CH-CH_2-C-C{=}O$
H_3C H NH
$H_3C-S-CH_2-CH_2-C-C{=}O$
NH
$HC{=}O$

The attachment between peptide and tRNA is an ester linkage between the C-terminus of the dipeptide and the 3′-hydroxyl group of the tRNA.

9. (a)

Met Ala Phe Thr
m^{7}GpppGCUUGUGUCCUC AUG GCU UUU ACA AAA CAG UGG CGU CCA GAU UAA ACCAUUCAGGCUU(A)$_{100}$
Lys Gln Trp Arg Pro Asp stop

The cap structure at the 5′ end of this RNA and the poly(A) tail at the 3′ end are typical of a eukaryotic mRNA. As such, one would expect initiation to occur at the first AUG, yielding the polypeptide shown above.

(b)

Met Ala
pppCCUCAACUAUGGAGGUCAGCC AUG GCA UUC ACU AAG CAA UGG AGG CCU GAC UGA GCUUACCGUA
Phe Thr Lys Gln Trp Arg Pro Asp stop

This mRNA lacks a cap at its 5′ end and is therefore more likely to represent a prokaryotic mRNA. Consequently, initiation is likely to occur at the AUG or GUG after the Shine-Dalgarno sequence closest to the 5′ end, as indicated. (The AUG closest to the 5′ end lacks a good Shine-Dalgarno sequence.) Note that despite the many differences between the two sequences, both mRNAs potentially encode the same amino acid sequence.

11. To minimize the degeneracy of the oligonucleotide to be synthesized, the amino acid sequence should first be examined for those amino acids that are encoded by the smallest number of codons in the universal genetic code:

	Met	Val	Asp	Ser	Asn	Trp	Ala	Gln	Cys	Asn	Pro	Ala	Thr
Number of codons:	1	4	2	6	2	1	4	2	2	2	4	4	4

The seven amino acids from asparagine to proline should be used to design the oligonucleotide with the lowest degeneracy. An instrument called a DNA synthesizer can be programmed to synthesize a mixed oligonucleotide with mixtures of nucleotides at positions corresponding to the wobble position in the genetic code. With spaces introduced to facilitate reading, the sequence would be: (5′)AAY TGG GCN CAR TGY AAY CC(3′). The complementary oligonucleotide, GGRTTRCAYTGNGCCCARTT, would also be a correct answer to this problem. (The complementary oligonucleotide might be desirable for some experiments because it should hybridize to both DNA and mRNA corresponding to this protein.) R is the abbreviation for a mixture of A and G (R = puRine), Y is the abbreviation for a mixture of C and T (Y = pYrimidine), and N stands for a mixture of all four nucleotides.

The oligonucleotide sequences given are $2 \times 1 \times 4 \times 2 \times 2 \times 2 = 64$-fold degenerate. In other words, there are 64 different sequence combinations in this mixture of oligonucleotides. Note that oligonucleotides based on amino acid sequences typically are of the length $[3n - 1]$ to avoid the degenerate 3′-most nucleotide of the last amino acid codon. In the example given above, the addition of one extra nucleotide at either end would increase the degeneracy of the oligonucleotide.

13. The X-ray crystal structure of EF-Tu bound to GDP and GTP is known. On the basis of this structure, it appears that many of the amino acids that are shared between IF-2, EF-Tu, EF-G, and RF-3 are involved in binding of GTP and GDP. Thus these proteins (and many others, including the protooncogene, Ras) share a very similar GTP-binding domain. A few of the conserved amino acids may also be implicated in ribosome binding, suggesting that the GTP-binding protein synthesis factors all interact with some of the same residues of the ribosome. However, this aspect of elongation factor function is not yet as well documented as is the binding of guanine nucleotides.

15. Import into the endoplasmic reticulum requires an N-terminal signal sequence that contains a long stretch of hydrophobic amino acids, while the mitochondrial transit peptide is a hydrophilic sequence rich in serine and threonine, with regularly spaced basic amino acids. Import into the ER requires the signal recognition particle and its receptor, while mitochondrial import does not require the SRP and

presumably uses a different receptor. Import into mitochondria requires a membrane potential, while import into the ER does not. Import into the ER requires GTP in the initial docking step.

17. GAA encodes a glutamic acid residue. If this glu is essential for the catalytic activity of the protein, any mutation (except GAA or GAG) will be deleterious to the activity of the protein. If the glu residue is not essential for catalysis or proper folding, the possible mutations, in order of increasing potential severity are as follows:

GAA changes to:	Glu changes to:	Likely effect:
GAG	Glu	None
GAU or GAC	Asp	Very little effect, because asp has similar acidic side chain
CAA	Gln	Very little effect, because gln is just amide of glu
GCA	Ala	Hard to predict, but may not have much of an effect, because ala has a small side chain and, like glu, tends to form an α-helix structure
AAA	Lys	May have significant effect on enzyme activity because it is a basic amino acid in place of an acidic one
GUA	Val	May have significant effect on enzyme activity because it is uncharged and, unlike glu, prefers to be in β-sheet
GGA	Gly	Will probably affect enzyme activity, because it is conformationally flexible and tends to form β bends
TAA	Terminator	Almost certain to disrupt expression of the protein

It is clear from the above analysis that many potential mutations of a GAA codon might have little or no effect on the encoded protein. The genetic code seems to have evolved to minimize the effects of many mutations; that is, similar amino acids are often found in rows and columns in the universal genetic code.

In some cases, the mutations described above might affect expression of the protein by affecting the translatability of the mRNA or protein stability. These sorts of effects would be difficult to predict.

19. The potent toxins ricin and α-sarcin are enzymes that can inactivate ribosomes by catalyzing an N-glycolytic cleavage of an adenine residue or a specific endonuclease cleavage, respectively. They can work at very low concentrations because they are catalytic. Most other ribosome-inactivating antibiotics interact directly with the ribosome in 1:1 stoichiometry and are therefore required in much higher amounts to exert a significant effect on protein synthesis inside the cell. Another unique property of ricin and α-sarcin is their ability to cross the cell membrane to attain their target, the large subunit of the ribosome. The site of attack by these antibiotics is probably the region of the 60S ribosomal subunit where various GTP-bound translation factors interact with the ribosome.

Chapter 35

1. The simplest solution to this problem is to make sure that the growth media for the transformed *E. coli* contain glucose but no IPTG, lactose, or cAMP. The presence of glucose in the media decreases transcription from the *lac* promoter by ensuring that endogenous levels of cAMP are low and CAP cannot bind to its operator. It would be helpful to use a strain of *E. coli* that overexpresses the *lacI* gene product. In some cases, it may be possible to place the insert DNA in the opposite orientation with respect to the *lac* promoter, so that transcription yields an antisense transcript, or to introduce a transcription terminator between the *lac* promoter and the cloned fragment. Yet another option, which may or may not be practical, is to use an entirely different vector and/or host.

3. Both *lac* and *trp* repressors bind an "operator," a DNA sequence near the promotor for their respective operons, where they interfere with proper binding and initiation by RNA polymerase, thereby repressing gene expression. However, the operator sequences for the two repressors differ. The two repressors interact with different small molecule effectors in opposite ways: Allolactose (natural inducer) or IPTG (artificial inducer) binds to *lac* repressor and decreases its affinity for its target site, while bound tryptophan increases the affinity between *trp* repressor and its operator. Allolactose and IPTG act as *anti*-repressors; tryptophan acts as a *co*-repressor.

5. Maximal transcription of the *lac* operon requires the binding of CAP to the operator. When cells are grown on glucose, cAMP levels are too low to activate CAP. When CAP is not bound to the DNA, RNA polymerase seems to recognize the *lac* promoter less efficiently than when the CAP-cAMP complex is bound. Therefore, even in the absence of *lac* repressor (because of the IPTG), *lac* mRNA levels do not reach fully induced levels until glucose in the media is depleted.

7. There is a single phenylalanine codon in the trp operon leader sequence. However, this codon is 15 nucleotides upstream from the tandem trp codons, where pausing of the ribosome normally causes RNA polymerase to ignore the attenuator. Apparently, pausing of the ribosome at the upstream phenylalanine codon does not disrupt base-pairing between stem-and-loop 1,2, so termination in the leader is favored.

   Why shouldn't phenylalanine levels affect the expression of tryptophan biosynthetic enzymes? Under conditions of starvation for amino acids other than tryptophan and arginine, there is little or no biosynthesis of the enzymes of the *trp* operon. Obviously, it would be potentially harmful to the cell to squander precious amino acids in the production of unnecessary amino acid biosynthetic enzymes. In addition, like tryptophan, phenylalanine is made from chorismate, but the pathway from chorismate to phenylalanine does not utilize any of the enzymes that are encoded in the *trp* operon and under attenuator control. If phenylalanine levels are low, it would be illogical to induce the enzymes of the tryptophan biosynthetic pathway, because they would deplete the cells of the precursor (chorismate) needed to replenish phenylalanine levels.

9. Given that the mutation is located within the *trp* leader region, its effect would most likely be mediated through an influence on attenuation of transcription. The observation that *trp* operon expression is stimulated in the mutant by omission of glycine suggests that a glycine codon or codons have been generated at the location of the tryptophan codons in the wild-type *trp* operon leader sequence. The simplest explanation for this mutant is that the tryptophan codons in the leader sequence have mutated to glycine codons (a single nucleotide frame shift would accomplish this). The *trp* operon would now be subject to regulation by glycine levels through attenuation. In this mutant grown on medium lacking glycine, repression by tryptophan through the *trp* repressor would still be in effect, and the levels of *trp* operon enzymes would be lower than in the fully depressed wild-type.

   Growing the mutant in a medium lacking both glycine *and* tryptophan would result in an increase in tryptophan biosynthetic enzymes compared to medium lacking glycine only. This would be due to relief from repression by the *trp* repressor system, which is independent of attenuation, and would remain responsive to tryptophan levels in the mutant just as in the wild-type.

11. The process of translation is itself an amplification process, in that each copy of ribosomal protein mRNA can give rise to many copies of the corresponding protein.

13. The relevant sequences are as follows:

   −35 −10
   5′-...AGCTCGACTT GCAGACTATC GATAGCGCTA
   mRNA start S/D
   TAGCGTGTGG ACTATTGACA GCACTGGATT
   translation start
   ATAGCATGCT GGAAATCGCT...3′

   Two types of sequence changes should be considered: First, improve the promoter so that it matches the consensus for high-level expression in *E. coli*: Change the −35 box (TCGACT) to TTGACA and the −10 box (TATAGC) to TATAAT. Second, optimize the Shine-Dalgarno sequence: Change —CTGGATT— to include a longer stretch of purines. It might also be necessary to alter the spacing between various sequence elements slightly. (Other sequences to which some thought should be given: Change any rarely used codons in *E. coli* to their more abundant counterparts, make sure there is an efficient *E. coli* transcription terminator at the end of the gene, and verify that there are no RNAse cleavage sites within the coding region, etc.)

15. The genes encoded by bacteriophage lambda are organized such that gene products that are required for the same function or process are clustered. Thus clustering also leads to more efficient organization and tighter packing of the bacteriophage genome. In other words, if the lambda genome were not so well organized, each gene would have to have its own control regions, and it is likely that additional regulatory proteins would be required. As a result, the genome size would be considerably larger than it is now.

Chapter 36

1. Attenuation control (found in bacteria) is dependent on the formation of a transcription-translation complex (as the mRNA is being formed, ribosomes bind and protein synthesis starts). In eukaryotes, transcription occurs in the nucleus, where the mRNA is processed (capping, polyadenylation, and intron removal). Only then does the mRNA move into the cytosol, where ribosomes bind and translation starts. The compartmentalization of these processes physically separates transcription from translation, and the extensive processing of the mRNA would make it unlikely that a type of attenuation control will be found in eukaryotes. (Of course, attenuation control could occur in the mitochondrion or chloroplast.)

3. Wild-type GAL4p activates GAL1 transcription only in the presence of the inducer, galactose. This activation by galactose is mediated by the interaction between GAL80 and GAL4 proteins. Normally, GAL80p interacts with GAL4p and interferes with its transcription activation activity. In the presence of galactose, the interaction between GAL4 and GAL80 proteins is disrupted. If a GAL4 mutant were unable to interact with GAL80, then it would activate transcription constitutively, exactly the phenotype described in this problem.

5. During amino acid starvation in yeast, 30 enzymes in nine different biosynthetic pathways are expressed. These coregulated genes have multiple copies of a *cis*-acting sequence (TGACTC) located upstream of the genes. The *CGN4* gene makes a protein that acts as a positive transcriptional regulator by binding to the TGACTC sequences. Thus all these genes are activated by a single protein.

   GCN4 gene expression is regulated at the translational level through differential recognition of the four upstream open reading frames in its mRNA. The *GCN4* open reading frame is efficiently recognized and translated only under conditions of amino acid starvation, when active eIF-2 levels are low and translation of the upstream open reading frames is reduced. More efficient recognition of the *GCN4* initiation codon could also be achieved by deletion of the upstream open reading frames. In the presence of this modified *GCN4* gene, there would be no regulation of *GCN4* protein levels, and the enzymes for amino acid biosynthesis would not be subject to general amino acid control. Instead, they would be synthesized constitutively.

7. Even though human females have the genotype XX, phenotypically they have only one X chromosome per cell because of X-chromosome inactivation. In principle, then, human females have the same number of X chromosomes per cell as males. The difference is that in males, the active X chromosome in each cell is identical. Therefore a male carrying a mutant X chromosome is colorblind. In females, different cells differ in which X chromosome has been inactivated. Therefore in the female retina, some cells have inactivated the normal X chromosome and retained the X chromosome carrying a mutant visual pigment. Other cells have inactivated the mutant X chromosome and therefore express normal protein. The presence of these cells containing the normal pigment allows heterozygous females to have normal color vision, even in the presence of cells or patches of cells expressing the mutant protein.
9. Histones bind to DNA by electrostatic interaction of basic amino acids (arginine and lysine) with the negative charges on the phosphodiester backbone. Positively and negatively charged salt ions neutralize the charges on protein and DNA, thus offsetting the benefits of the electrostatic interactions between protein and DNA and allowing the histone proteins to dissociate from the DNA. Many regulatory proteins may initially interact with DNA in a nonspecific fashion until they locate their high-affinity binding sites. Once these proteins bind at their specific recognition sites, the major interaction is through hydrogen bonding and hydrophobic interactions. The stability of hydrophobic interactions relies on the segregation of hydrophobic surfaces from the polar solvent, a factor that becomes even more important in the presence of a high salt concentration in solution. Thus the hydrophobic interactions between certain proteins and DNA are stabilized by the presence of high salt.
11. MspI would cleave at every CCGG sequence in or near the globin genes, regardless of the cell type from which the DNA has been isolated, because this enzyme is insensitive to methylation. Globin genes isolated from erythroblast cells would not differ from those of other tissues in their cleavage pattern with MspI, as shown in the illustration.

    HpaII does not recognize CCGG sequences when the second C is methylated. In tissues in which genes are *not* transcribed, CpG sequences tend to be methylated more often than in tissues where the same genes are transcribed. Thus HpaII is expected to recognize and cleave more sites in globin genes in erythroblast DNA than in DNA from other cell types (where the globin genes are not transcribed and the corresponding DNA is more highly methylated).

    A hypothetical experiment demonstrating hypomethylation of actively transcribed DNA is shown in the accompanying figure. DNA from the indicated tissues would be digested with MspI or HpaII, separated by agarose gel electrophoresis, and transferred to a membrane by the Southern blotting procedure. Hybridization of the membrane to a globin gene probe, followed by washing and autoradiography, should reveal identical fragments in MspI digests of DNA from any tissue. Even in DNA isolated from erythroblasts, there may be some methylated CCGG sequences, which would therefore be resistant to HpaII cleavage (compare the MspI and HpaII lanes for erythroblast DNA). In liver DNA, considerably more CCGG sequences in and near the globin genes would be methylated and therefore refractory to cleavage by HpaII.

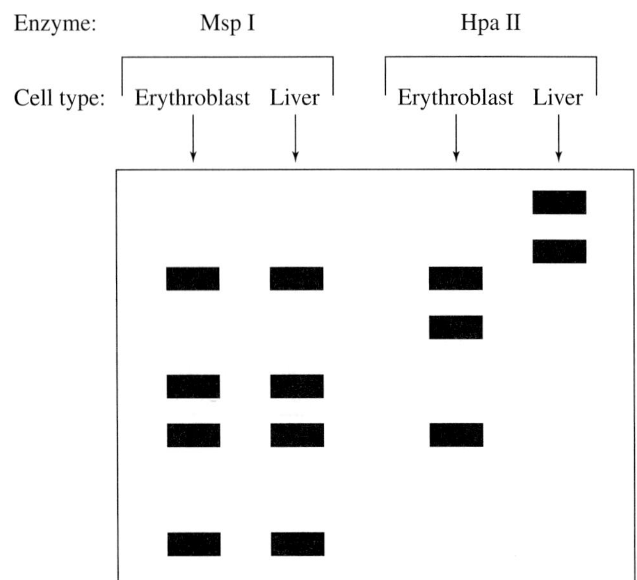

13. *Cis*-acting sequences acting over surprisingly large distances have been found in eukaryotes and more recently in prokaryotes. These sequences have been called "enhancers" because they enhance the expression of their associated gene. These enhancer sequences bind to tissue-specific *trans*-acting proteins, and the DNA can loop to allow the enhancer-protein complex to interact with the promoter region of the gene and RNA polymerase II. It is presumably because of this looping that enhancer function is orientation-independent (can be located in either a $5' \rightarrow 3'$ or $3' \rightarrow 5'$ orientation with respect to the promoter it regulates), that an enhancer can be located either upstream or downstream of the promoter it regulates, and that enhancers can function over a great distance.
15. Any introns would have to be removed, and a Shine-Dalgarno sequence and a prokaryotic promoter added. The gene would probably also need a prokaryotic terminator. Furthermore, in many eukaryotic genes, there are significant differences in codon utilization from that typical of bacteria cells. If the eukaryotic gene contains codons that are infrequent in *E. coli* (corresponding to low-abundance tRNA species), it may be necessary to alter these codons to more accurately reflect *E. coli* codon usage.
17. The products of the genes described usually act in the following order: maternal-effect genes, then segmentation genes, then homeotic genes. There may be some exceptions, but this pattern of activity and expression correlates well with the effects of the gene products.

    Maternal-effect genes are regulatory genes whose effect on progeny depends on the allele inherited from the female parent. This form of inheritance provides evidence that these genes are already expressed in maternal tissues and that the corresponding products are present in the oocyte. Thus maternal-effect genes are expected to act at the very earliest stages of embryonic development.

Segmentation-effect genes are regulatory genes that, when mutated, change the number or delineation of body segments. These genes set up the pattern of segments in the *Drosophila* embryo. Although segmentation-effect gene products must be present prior to the development of obvious segmentation, they are expressed and act later than maternal-effect genes.

Homeotic genes are regulatory genes that, when mutated, result in substitutions of one body part for another. Homeotic genes affect the structure(s) that arise from an individual segment. Therefore the products of homeotic genes are usually expressed sometime after segmentation has begun, that is, later than the segmentation-effect genes.

The process of establishing a body plan, although far more complex, can be compared to that of building a house. First, a hole must be dug and the foundation must go in, a process that is critically dependent on the location and characteristics of the soil (maternal effect). Next, the framers come in and put up the frame of the house. At this stage, it is apparent where rooms will be, but the identity of the rooms is not clear (segmentation). Now it is time for plumbers, roofers, and other craft workers to come in and give each room its own distinctive features and appendages (homeotic genes).

Chapter 37

1. T cells and B cells are the two different types of white cells (lymphocytes) that take part in the immune response. The T cells mature in the thymus and then migrate to different areas of the body, including the lymphoid organs. T lymphocytes proliferate when exposed to antigens and produce specific effector molecules, which remain bound to the cell surface. The B lymphocytes mature in the bone marrow and then migrate to various lymphoid organs, where they reside until they are stimulated by antigens. When exposed to antigens, these white cells also proliferate and produce antibodies, which are secreted. T cells do not secrete antibodies.
3. Before the role of T and B cells was determined, it was known that certain immune responses could be transferred with white blood cells but not with serum (cell-mediated immunity). Cell-mediated immunity relies on two different types of T cells. In the delayed hypersensitivity response the T_D cells react with antigens and secrete lymphokines that attract and activate macrophages and other leukocytes. This then leads to the slowly developing inflammatory response. Another cellular response involves the killer T cell (T_K), which reacts with antigen bound to cells and causes their lysis. Both of these responses are important in our immune system.
5. DNA splicing or recombination can activate the newly rearranged gene by bringing the promoter under the influence of an enhancer element. The best-known example of such an enhancer is that in the germline V_H–C_μ region. Because of its location in the intron between J segments and C_μ, the enhancer is too distant to act on the V_H gene promoter in the unrearranged immunoglobulin gene. V–D–J joining brings this enhancer closer to the promoter just when antibody production becomes an important part of the cell's function. Ig enhancers act in a tissue-specific fashion and work when placed on either side of the gene.
7. The existence of membrane-bound or secreted heavy chains is determined by alternative splicing of the C_H region of the heavy chain gene. One can design an oligonucleotide that would hybridize to the exon specifying the transmembrane domain of the heavy chain and a different oligonucleotide that binds to an exon present in both secreted and membrane-bound forms of heavy chain mRNA. cDNA would be made from B-cell mRNA using reverse transcriptase and an oligo(dT) primer. The PCR would then be carried out, using these B-cell cDNAs as template, with either of the two oligonucleotides described above as the downstream primer and using an upstream primer hybridizing to conserved regions in V_H segments. Only if the B-cell mRNA contains the exon encoding the membrane-bound region of the heavy chain would a product of the expected size be amplified in the presence of the primer hybridizing to this exon.
9. There is striking similarity among all immune recognition proteins; they all contain one or more immunoglobulinlike domains with about 100 amino acids that are folded into characteristic structures made of two antiparallel sheets that are stabilized by conserved disulfide-containing structures. This suggests that a common ancestral protein, perhaps resembling β_2 microglobulin, was present early in vertebrate evolution. After each duplication of the ancestral gene, the duplicated copies might have acquired (through mutation) specific functions that were beneficial to the organism, eventually generating the wide variety of immunoglobulins seen today.
11. When a protein is injected into a rabbit, it stimulates the production of polyclonal antibodies (meaning that they are derived from different clones of cells). The antigen (injected protein) may have many determinants on its surface, each of which could stimulate a plasma cell clone that will produce antibodies reacting with that particular domain of the protein. These different determinants and antibodies produced will have different binding affinities. It is possible to produce monoclonal antibodies against a single determinant on the antigen by using hybridoma techniques.

Chapter 38

1. Cancer cells are easy to recognize when they are compared with normal cells. The cancer cell divides faster, is less well organized, and does not have the same type of structure as normal cells in the surrounding tissue. Cancer cells lack much of the growth regulation seen in normal cells. Cancer cells do not recognize territorial boundaries and eventually invade and disrupt vital organs, leading to death. When cancer cells are grown in tissue culture, they do not display the

normal contact inhibition of growth seen in untransformed or normal cells. Normal cells will grow until they form a confluent monolayer of cells and then stop growing, while cancer cells may reach cell densities 20 times higher than those of untransformed cells. They will form layers of cells that sometimes look like small tumors and are visible without magnification. Normal cells are much harder to grow in culture and have more complex growth requirements. Many transformed cells isolated from tissue culture can give rise to tumors when injected into isogenic animals. Many cancer cells or transformed cells will produce fetal antigens that sometimes can be used as markers for specific types of cancer, for example, colon cancer.

3. RNA viruses (retroviruses) use reverse transcriptase to make a DNA copy that integrates into the host cell genome and forms a provirus. If the virus inserts close to a cellular protooncogene, the oncogene may be turned on by viral enhancers and generate a transformed cell, which may cause cancer. Some animal retroviruses have picked up cellular oncogenes that have become part of the virus (e.g., the *src* gene) and can readily transform cells in culture.
5. The *ras* protooncogene encodes a signaling protein involved in mitogenic activity and other cell processes, and *ras* activity is tightly regulated by its intrinsic GTPase activity. Mutations that release the *ras* gene product from its regulation (i.e., inactivate its GTPase activity) allow this protein to stimulate its downstream effectors in an uncontrolled manner, leading to tumorigenesis. When two copies of the *ras* gene are present, one mutant and one normal, the mutant, oncogenic form of the *ras* gene product will be active even in the presence of the normal, regulated form of the protein. This is another way of saying that mutant *ras* is a dominant oncogene. Presumably, there are recessive lethal mutations of the *ras* protooncogene in which the encoded protein is completely inactive, but these would not normally be tumorigenic in the heterozygous state.

   The normal function of p53, on the other hand, is to act as a transcription factor. Most p53 oncogenic mutations represent loss-of-function mutations and can no longer bind their target DNA sequence and activate transcription. These mutants have no obvious phenotype other than a long-term predisposition to a higher than normal incidence of cancer and can apparently be compensated for as long as one wild-type copy of the p53 gene is still present. When the second copy of the p53 gene acquires a mutation, cancer results. This is another way of saying that most p53 mutations are recessive. (As was mentioned in the text, there are some dominant p53 oncogenic mutations, which have provided insight into its mechanism.)
7. The *src* gene can cause cancer because it produces a tyrosine phosphokinase activity that phosphorylates tyrosine residues in various proteins. Since there is a high level of expression of this phosphokinase, the levels of protein phosphorylation are elevated, activating many proteins involved in growth and development.

Chapter 39

1. The life cycle of HIV is shown in the figure on the next page.
3. The first viral mRNA(s) produced from the integrated provirus are spliced transcripts. Splicing results in removal of one intron that spans *gag* and *pol* coding sequences and a second intron whose removal allows the production of *tat* and *rev* proteins. *tat* protein acts to activate transcription from the major viral promoter in the LTR, while *rev* works to inhibit splicing of viral mRNA. Once *rev* has abolished splicing of the *rev-tat* intron, the *env* gene is expressed from this region of the genome instead. Expression of *env* yields envelope glycoproteins. Inhibition of splicing of the other intron permits expression of capsid protein and replicative enzymes from the *gag* and *pol* genes.
5. One would think that the easiest mechanism for interference with the binding of a nucleotide analog would be a mutation at the site of nucleotide binding. Some nucleotide analog-resistant HIV reverse transcriptase mutations do appear to interfere directly with drug binding, while others are farther away than expected. The latter class of drug-resistant mutations may affect incorporation of the drug through alterations of the DNA template positioning or conformation.
7. The HIV protease is encoded by the *pol* polyprotein, and it is essential for cleaving this polyprotein into its component polypeptides. Because its cleavage site is known, it might be possible to design peptide analogs that have the correct sequence of amino acid side chains but have a noncleavable bond at the normal site of cleavage. (Many of the drugs under development for this purpose replace the cleaved peptide bond with a hydroxyethylamine linkage, —CH[OH]CH_2N—, instead.) Such a peptide should bind tightly to the HIV protease, preventing it from cleaving its normal substrate and thereby inhibiting the production of HIV reverse transcriptase, among others.

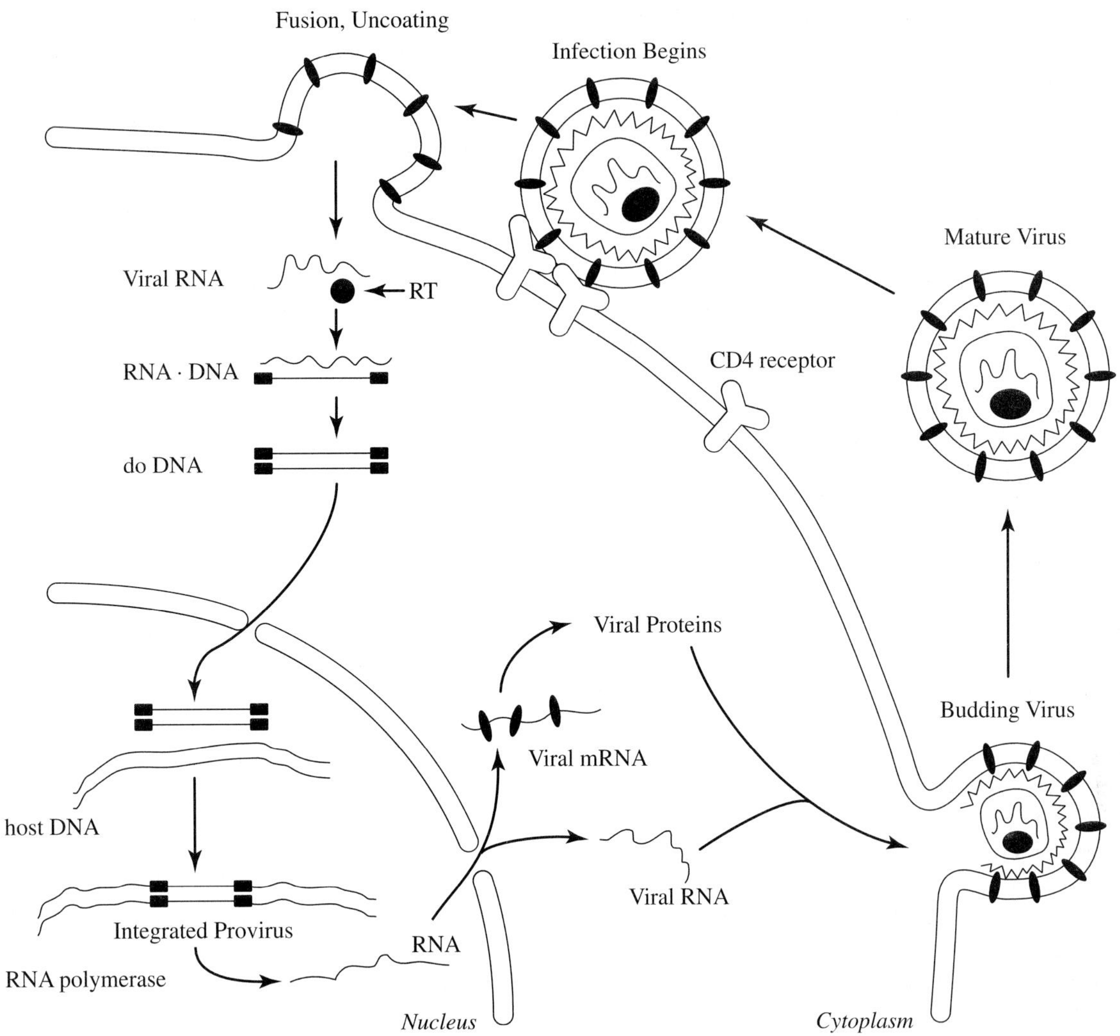
Fusion, Uncoating
Infection Begins
Mature Virus
Viral RNA
RT
RNA · DNA
do DNA
CD4 receptor
Viral Proteins
Budding Virus
Viral mRNA
host DNA
Viral RNA
Integrated Provirus
RNA
RNA polymerase
Nucleus
Cytoplasm

Glossary

A

A form. A duplex DNA structure with right-handed twisting in which the planes of the base pairs are tilted about 20° with respect to the helix axis.

Acetal. The product formed by the successive condensation of two alcohols with a single aldehyde. It contains two ether-linked oxygens attached to a central carbon atom.

Acetyl-CoA. Acetyl-coenzyme A, a high-energy ester of acetic acid that is important both in the tricarboxylic acid cycle and in fatty acid biosynthesis.

Actin. A protein found in combination with myosin in muscle and also found as filaments constituting an important part of the cytoskeleton in many eukaryotic cells.

Actinomycin D. An antibiotic that binds to DNA and inhibits RNA chain elongation.

Activated complex. The highest free energy state of a complex in going from reactants to products.

Active site. The region of an enzyme molecule that contains the substrate-binding site and the catalytic site for converting the substrate(s) into product(s).

Active transport. The energy-dependent transport of a substance across a membrane.

Adenine. A purine base found in DNA or RNA.

Adenosine. A purine nucleoside found in DNA, RNA, and many cofactors.

Adenosine diphosphate (ADP). The nucleotide formed by adding a pyrophosphate group to the 5′-OH group of adenosine.

Adenosine triphosphate (ATP). The nucleotide formed by adding yet another phosphate group to the pyrophosphate group on ADP.

Adenylate cyclase. The enzyme that catalyzes the formation of cyclic 3′,5′-adenosine monophosphate (cAMP) from ATP.

Adipocyte. A specialized cell that functions as a storage depot for lipid.

Aerobe. An organism that utilizes oxygen for growth.

Affinity chromatography. A column chromatographic technique that employs attached functional groups that have a specific affinity for sites on particular proteins.

Alcohol. A molecule with a hydroxyl group attached to a carbon atom.

Aldehyde. A molecule containing a doubly bonded oxygen and a hydrogen attached to the same carbon atom.

Aldose. A simple sugar in which the carbonyl carbon is an aldehyde.

Alleles. Alternative forms of a gene.

Allosteric enzyme. An enzyme the active site of which can be altered by the binding of a small molecule at a non-overlapping site.

Allosteric site. Location on an allosteric enzyme where the allosteric effector binds.

Aminotransferase. An enzyme that catalyzes the transfer of an amino group from an α-amino acid to an α-keto acid.

Amphibolic pathway. A metabolic pathway that functions in both catabolism and anabolism.

Anabolism. Metabolism that involves biosynthesis.

Anaerobe. An organism that does not require oxygen for maintenance or growth.

Anaplerotic reaction. An enzyme-catalyzed reaction that replenishes the intermediates in a cyclic pathway.

Angstrom (Å). A unit of length equal to 10^{-8} cm.

Anomers. The sugar isomers that differ in configuration about the carbonyl carbon atom. This carbon atom is called the anomeric carbon atom of the sugar.

Antibiotic. A natural product that inhibits bacterial growth (is bacteriostatic) and sometimes results in bacterial death (is bacteriocidal).

Antibody. A specific protein that interacts with a foreign substance (antigen) in a specific way.

Anticodon. A sequence of three bases on the transfer RNA that pair with the bases in the corresponding codon on the messenger RNA.

Antigen. A foreign substance that triggers antibody formation and is bound by the corresponding antibody.

Antiparallel β-pleated sheet (β sheet). A hydrogen-bonded secondary structure formed between two or more extended polypeptide chains.

Antiport. A protein system that can transport different molecules in opposite directions.

Apoactivator. A regulatory protein that stimulates transcription from one or more genes in the presence of a coactivator molecule.

Apoptosis. Programmed cell death.

Asexual reproduction. Growth and cell duplication that does not involve the union of nuclei from cells of opposite mating types.

Asymmetrical carbon. A carbon that is covalently bonded to four different groups.

Attenuator. A provisional transcription stop signal.

Autoradiography. The technique of exposing film in the presence of disintegrating radioactive particles. Used to obtain information on the distribution of radioactivity in a gel or a thin cell section.

Autoregulation. The process by which a gene regulates its own expression.

Autotroph. An organism that can form its organic constituents from CO_2.

Auxin. A plant growth hormone usually concentrated in the apical bud.

Auxotroph. A mutant that cannot grow on the minimal medium on which a wild-type member of the same species can grow.

Avogadro's number. The number of molecules in a gram molecular weight of any compound (6.022×10^{23}).

B

β bend. A characteristic way of turning an extended polypeptide chain in a different direction, involving the minimum number of residues.

β oxidation. Oxidative degradation of fatty acids that occurs by the successive oxidation of the β-carbon atom.

β sheet. A sheetlike structure formed by the interaction between two or more extended polypeptide chains.

B cell. One of the major types of cells in the immune system. B cells can differentiate to form memory cells or antibody-forming cells.

B form. The most common form of duplex DNA, containing a right-handed helix and about 10 (10.5 exactly) bp per turn of the helix axis.

Base analog. A compound, usually a purine or a pyrimidine, that differs somewhat from a normal nucleic acid base.

Base stacking. The close packing of the planes of base pairs, commonly found in DNA and RNA structures.

Bidirectional replication. Replication in both directions away from the origin, as opposed to replication in one direction only (unidirectional replication).

Bilayer. A double layer of lipid molecules with the hydrophilic ends oriented outward, in contact with water, and the hydrophobic parts oriented inward toward each other.

Bile salts. Derivatives of cholesterol with detergent properties that aid in the solubilization of lipid molecules in the digestive tract.

Biochemical pathway. A series of enzyme-catalyzed reactions that results in the conversion of a precursor molecule into a product molecule.

Bioluminescence. The production of light by a biochemical system.

Blastoderm. The stage in embryogenesis when a unicellular layer at the surface surrounds the yolk mass.

Bond energy. The energy required to break a bond.

Branchpoint. An intermediate in a biochemical pathway that can follow more than one route in subsequent steps.

Buffer. A conjugate acid–base pair that is capable of resisting changes in pH when acid or base is added to the system. This tendency is maximal when the conjugate forms are present in equal amounts.

C

C adherins. A family of Ca^{2+}-dependent cell-adhesion molecules that play roles in tissue differentiation.

cAMP. 3′,5′ cyclic adenosine monophosphate. The cAMP molecule plays a key role in metabolic regulation.

CAP. The catabolite gene activator protein, sometimes referred to as the CRP protein. The latter term, in small letters (*crp*), should be used to refer to the gene but not to the protein.

Capping. Covalent modification involving the addition of a modified guanine group in a 5′-5″ linkage. It occurs only in eukaryotes, primarily on mRNA molecules.

Carbanion. A negatively charged carbon atom.

Carbohydrate. A polyhydroxy aldehyde or ketone.

Carboxylic acid. A molecule containing a carbon atom attached to a hydroxyl group and to an oxygen atom by a double bond.

Carcinogen. A chemical that can cause cancer.

Carotenoids. Lipid-soluble photosynthetic pigments that are made from isoprene units.

Catabolism. That part of metabolism concerned with degradation reactions.

Catabolite repression. The general repression of transcription of genes associated with catabolism that is seen in the presence of glucose.

Catalyst. A compound that lowers the activation energy of a reaction without itself being consumed.

Catalytic site. The site of the enzyme involved in the catalytic process.

Catecholamines. Hormones that are amino derivatives of catechol, for example, epinephrine or norepinephrine.

Catenane. An interlocked pair of circular structures, such as covalently closed DNA molecules.

Catenation. The linking of molecules without any direct covalent bonding between them, as when two circular DNA molecules interlock like the links in a chain.

cDNA. Complementary DNA, made in vitro from the mRNA by the enzyme reverse transcriptase and deoxyribonucleotide triphosphates.

Cell-adhesion molecules (CAMs). Integral membrane proteins that mediate cell–cell binding.

Cell commitment. That stage in a cell's life when it becomes committed to a certain line of development.

Cell cycle. All of those stages that a cell passes through from one cell generation to the next.

Cell line. An established clone originally derived from a whole organism through a long process of cultivation.

Cell lineage. The pedigree of cells resulting from binary fission.

Cell wall. A tough outer coating found in many plant, fungal, and bacterial cells that accounts for their ability to withstand mechanical stress or abrupt changes in osmotic pressure. Cell walls always contain a carbohydrate component and frequently also a peptide and a lipid component.

Cerebroside. Sphingolipid containing one sugar residue as a head group.

Channeling. The direct transfer of a reaction intermediate from one enzyme to the next.

Chelate. A molecule that contains more than one binding site and frequently binds to another molecule through more than one binding site at the same time.

Chemiosmotic coupling. The coupling of ATP synthesis to an electrochemical potential gradient across a membrane.

Chimeric DNA. Recombinant DNA the components of which originate from two or more different sources.

Chiral compound. A compound that can exist in two forms that are nonsuperimposable images of one another.

Chlorophyll. A green photosynthetic pigment that is made of a magnesium dihydroporphyrin complex.

Chloroplast. A chlorophyll-containing photosynthetic organelle, found in eukaryotic cells, that can harness light energy.

Chromatin. The nucleoprotein fibers of eukaryotic chromosomes.

Chromatography. A procedure for separating chemically similar molecules. Segregation is usually carried out on paper or in glass or metal columns with the help of different solvents. The paper or glass columns contain porous solids with functional groups that have limited affinities for the molecules being separated.

Chromosome. A threadlike structure, visible in the cell nucleus during metaphase, that carries the hereditary information.

Chromosome puff. A swollen region of a giant chromosome; the swelling reflects a high degree of transcription activity.

***Cis* dominance.** Property of a sequence or a gene that exerts a dominant effect on a gene to which it is linked.

Cistron. A genetic unit that encodes a single polypeptide chain.

Citric acid cycle. *See* Tricarboxylic acid (TCA) cycle.

Clone. One of a group of genetically identical cells or organisms derived from a common ancestor.

Cloning vector. A self-replicating entity to which foreign DNA can be covalently attached for purposes of amplification in host cells.

Closed system. A system that exchanges neither matter nor energy with its surroundings.

Coactivator. A molecule that functions in conjunction with a protein apoactivator. For example, cAMP is a coactivator of the CAP protein.

Codon. In a messenger RNA molecule, a sequence of three bases that represents a particular amino acid.

Coenzyme. An organic molecule that associates with enzymes and affects their activity.

Cofactor. A small molecule that is required for enzyme activity. It could be organic, like a coenzyme, or inorganic, like a metallic cation.

Competitive inhibition. An inhibitor that competes with the substrate for binding to the enzyme.

Complementary base sequence. For a given sequence of nucleic acids, the nucleic acids that are related to them by the rules of base pairing.

Configuration. The spatial arrangement of atoms in a molecule that can be changed only by breaking and reforming covalent bonds.

Conformation. The spatial arrangement of groups in a molecule that can be changed without breaking covalent bonds. Often, molecules with the same configuration can have more than one conformation.

Consensus sequence. In nucleic acids the "average" sequence that signals a certain type of action by a specific protein. The sequences that are actually observed usually vary around this average.

Constitutive enzymes. Enzymes synthesized in fixed amounts, regardless of growth conditions.

Cooperative binding. A situation in which the binding of one substituent to a macromolecule favors the binding of another. For example, DNA cooperatively binds histone molecules, and hemoglobin cooperatively binds oxygen molecules.

Coordinate induction. The simultaneous expression of two or more genes.

Cosmid. A DNA molecule with *cos* ends from λ bacteriophage that can be packaged in vitro into a virus for infection purposes.

Cot curve. A curve that indicates the rate of DNA–DNA annealing as a function of DNA concentration and time.

Covalent bond. A chemical bond that involves sharing electron pairs.

Cytidine. A pyrimidine nucleoside found in DNA and RNA.

Cytochromes. Heme-containing proteins that function as electron carriers in oxidative phosphorylation and photosynthesis.

Cytokine. A small protein that binds to cell-surface receptors on certain cells to trigger their differentiation or proliferation.

Cytokinin. A plant hormone produced in root tissue.

Cytoplasm. The contents enclosed by the plasma membrane, excluding the nucleus.

Cytosine. A pyrimidine base found in DNA and RNA.

Cytoskeleton. The filamentous skeleton, formed in the cytoplasm, that is largely responsible for controlling cell shape.

Cytosol. The liquid portion of the cytoplasm, including the macromolecules but not including the larger structures, such as subcellular organelles or the cytoskeleton.

D

D loop. An extended loop of single-stranded DNA displaced from a duplex structure by an oligonucleotide.

Dalton. A unit of mass equivalent to the mass of a hydrogen atom (1.66×10^{-24} g).

Dark reactions. Reactions that can occur in the dark, in a process that is usually associated with light, such as the dark reactions of photosynthesis.

De novo pathway. A biochemical pathway that starts from elementary substrates and ends in the synthesis of a biochemical.

Deamination. The enzymatic removal of an amine group, as in the deamination of an amino acid to an α-keto acid.

Dehydrogenase. An enzyme that catalyzes the removal of a pair of electrons (and usually one or two protons) from a substrate molecule.

Deletion mutation. A mutation in which one or more nucleotides is removed from a region of the gene.

Denaturation. The disruption of the native folded structure of a nucleic acid or protein molecule; may be due to heat, chemical treatment, or change in pH.

Density-gradient centrifugation. The separation, by centrifugation, of molecules according to their density, in a gradient varying in solute concentration.

Dialysis. Removal of small molecules from a macromolecule preparation by allowing them to pass across a semipermeable membrane.

Diauxic growth. Growth on a mixture of two carbon sources in which one carbon source is used up before the other one is mobilized. For example, in the presence of glucose and lactose, *E. coli* utilizes the glucose before the lactose.

Difference spectra. Display comparing the absorption spectra of a molecule or an assembly of molecules in different states, for example, those of mitochondria under oxidizing or reducing conditions.

Differential centrifugation. Separation of molecules or organelles by sedimentation rate.

Differentiation. A change in the form and pattern of a cell and the genes it expresses as a result of growth and replication, usually during development of a multicellular organism. Also occurs in microorganisms (e.g., in sporulation).

Diffusion. The net movement of molecules in the direction of lower concentration.

Diploid cell. A cell that contains two chromosomes ($2N$) of each type.

Dipole. A separation of charge within a single molecule.

Directed mutagenesis. In a DNA sequence an intentional alteration that can be genetically inherited.

Dissociation constant. An equilibrium constant for the dissociation of a molecule into two parts (e.g., dissociation of acetic acid into acetate anion and proton).

Disulfide bridge. A covalent linkage formed by oxidation between two SH groups either in the same polypeptide chain or in different polypeptide chains.

DNA. Deoxyribonucleic acid. A polydeoxyribonucleotide in which the sugar is deoxyribose; the main repository of genetic information in all cells and all DNA-containing viruses.

DNA cloning. The propagation of individual segments of DNA as clones.

DNA library. A mixture of clones, each containing a cloning vector and a segment of DNA from a source of interest.

DNA polymerase. An enzyme that catalyzes the formation of 3′-5′ phosphodiester bonds from deoxyribonucleotide triphosphates.

DNA supercoiling. The coiling of double helix DNA upon itself.

Domain. A segment of a folded protein structure showing conformational integrity. A domain can comprise the entire protein or just a fraction of the protein. Some proteins, such as antibodies, contain many structural domains.

Dominant. Describing an allele the phenotype of which is expressed regardless of whether the organism is homozygous or heterozygous for that allele.

Double helix. A structure in which two helically twisted polynucleotide strands are held together by hydrogen bonding and base stacking.

Duplex. Synonymous with double helix.

Dyad symmetry. Property of a structure that can be rotated by 180° to produce the same structure.

E

Ecdysone. A hormone that stimulates the molting process in insects.

Edman degradation. A systematic method of sequencing proteins, proceeding by stepwise removal of single amino acids from the amino terminal of a polypeptide chain.

Eicosanoid. Any fatty acid with 20 carbons.

Electron acceptor. A substance that receives electrons in an oxidation–reduction reaction.

Electron carrier. A protein or coenzyme that can reversibly gain and lose electrons and that serves the function of carrying electrons from one site to another.

Electron donor. A substance that donates electrons in an oxidation–reduction reaction.

Electrophoresis. The movement of particles in an electrical field. A commonly used technique for analysis of mixtures of molecules in solution according to their electrophoretic mobilities.

Elongation factors. Protein factors uniquely required during the elongation phase of protein synthesis. Elongation factor G (EF-G) brings about the movement of the peptidyl-tRNA from the A site to the P site of the ribosome.

Eluate. The effluent from a chromatographic column.

Embryo. Plant or animal at an early stage of development.

Enantiomers. Isomers that are mirror images of each other.

Endergonic reaction. A reaction with a positive free energy change.

Endocrine glands. Specialized tissues the function of which is to synthesize and secrete hormones.

Endocytosis. The uptake of extracellular matter by its inclusion in a vesicle formed by invagination of the plasma membrane.

Endonuclease. An enzyme that breaks a phosphodiester linkage at some point within a polynucleotide chain.

Endopeptidase. An enzyme that breaks a polypeptide chain at an internal peptide linkage.

Endoplasmic reticulum. A system of double membranes in the cytoplasm that is involved in the synthesis of transported proteins. The rough endoplasmic reticulum has ribosomes associated with it. The smooth endoplasmic reticulum does not.

End-product (feedback) inhibition. The inhibition of the first enzyme in a pathway by the end product of that pathway.

Energy charge. The fractional degree to which the AMP–ADP–ATP system is filled with high-energy phosphates (phosphoryl groups).

Enhancer. A DNA sequence that can bind protein factors that stimulate transcription at an appreciable distance from the site where it is located. It acts in either orientation and either upstream or downstream from the promoter.

Enthalpy (*H*). The heat content of a system.

Entropy (*S*). A thermodynamic measure of the randomness of a system.

Enzyme. A protein that contains a catalytic site for a biochemical reaction.

Epimers. Two stereoisomers with more than one chiral center that differ in configuration at one of their chiral centers.

Equilibrium. In chemistry the point at which the concentrations of two compounds are such that the interconversion of one compound into the other compound does not result in any change in free energy.

***Escherichia coli* (*E. coli*).** A Gram negative bacterium commonly found in the vertebrate intestine. It is the bacterium most frequently used in the study of biochemistry and genetics.

Established cell line. A group of cultured cells derived from a single origin and capable of stable growth for many generations.

Ether. A molecule containing two carbons linked by an oxygen atom.

Eukaryote. A cell or organism that has a membrane-bound nucleus.

Excision repair. DNA repair in which a damaged region is replaced.

Excited state. An energy-rich state of an atom or a molecule, produced by the absorption of radiant energy.

Exergonic reaction. A chemical reaction that takes place with a negative change in free energy. The reaction is thermodynamically favorable and could occur spontaneously.

Exocytosis. The release of intracellular matter by fusion of an intracellular vesicle with the plasma membrane.

Exon. A segment within a gene that carries part of the coding information for a protein.

Exonuclease. An enzyme that breaks a phosphodiester linkage at one or the other end of a polynucleotide chain so as to release a single nucleotide or small oligonucleotides.

F

F factor. A large bacterial plasmid, known as the sex-factor plasmid because it permits mating between F^+ and F^- bacteria.

Facilitated diffusion. Diffusion of a substance across a membrane through a protein transporter.

Facultative aerobe. An organism that can use molecular oxygen in its metabolism but that also can live anaerobically.

Fatty acid. A long-chain hydrocarbon containing a carboxyl group at one end. Saturated fatty acids have completely saturated hydrocarbon chains. Unsaturated fatty acids have one or more carbon–carbon double bonds in their hydrocarbon chains.

Feedback inhibition. *See* End-product inhibition.

Fermentation. The energy-generating breakdown of glucose or related molecules by a process that does not require molecular oxygen.

Fingerprinting. The characteristic two-dimensional paper chromatogram obtained from the partial hydrolysis of a protein or a nucleic acid.

Fluorescence. The emission of light by an excited molecule in the process of making the transition from the excited state to the ground state.

Footprinting. A technique that results in a DNA sequence ladder in which part of the ladder is missing because of the binding of protein to the DNA before processing.

Frameshift mutations. Insertions or deletions of genetic material that lead to a shift in the translation of the reading frame. The mutation usually leads to nonfunctional proteins.

Free energy. That part of the energy of a system that is available to do useful work.

Furanose. A sugar that contains a five-member ring as a result of intramolecular hemiacetal formation.

Futile cycle. *See* Pseudocycle.

G_1 phase. That period of the cell cycle in which preparations are being made for chromosome duplication, which takes place in the S phase.

G_2 phase. That period of the cell cycle between S phase and mitosis (M phase).

Gametes. The ova and the sperm, haploid cells that unite during fertilization to generate a diploid zygote.

Gel exclusion chromatography. A technique that makes use of certain polymers that can form porous beads with varying pore sizes. In columns made from such beads, it is possible to separate molecules that cannot penetrate beads of a given pore size from smaller molecules that can.

Gene. A segment of the genome that codes for a functional product.

Gene amplification. The duplication of a particular gene within a chromosome two or more times.

Gene splicing. The cutting and rejoining of DNA sequences.

General recombination. Recombination that occurs between homologous chromosomes at homologous sites.

Generation time. The time it takes for a cell to double its mass under specified conditions.

Genetic map. The arrangement of genes or other identifiable sequences on a chromosome.

Genome. The total genetic content of a cell or a virus.

Genotype. The genetic characteristics of an organism (distinguished from its observable characteristics, or phenotype).

Globular protein. A folded protein that adopts an approximately globular shape.

Gluconeogenesis. The production of sugars from nonsugar precursors, such as lactate or amino acids. Applies more specifically to the production of free glucose by vertebrate livers.

Glycogen. A polymer of glucose residues in 1,4 linkage and 1,6 linkage at branchpoints.

Glycogenic. Describing amino acids the metabolism of which may lead to gluconeogenesis.

Glycolipid. A lipid containing a carbohydrate group.

Glycolysis. The catabolic conversion of glucose to pyruvate with the production of ATP.

Glycoprotein. A protein linked to an oligosaccharide or a polysaccharide.

Glycosaminoglycans. Long, unbranched polysaccharide chains composed of repeating disaccharide subunits in which one of the two sugars is either *N*-acetylglucosamine or *N*-acetylgalactosamine.

Glycosidic bond. The bond between a sugar and an alcohol. Also the bond that links two sugars in disaccharides, oligosaccharides, and polysaccharides.

Glyoxylate cycle. A pathway that uses acetyl-CoA and two auxiliary enzymes to convert acetate into succinate and carbohydrates.

Glyoxysome. An organelle containing key enzymes of the glyoxylate cycle.

Goldman equation. An equation expressing the quantitative relationship between the concentrations of charged species on either side of a membrane and the resting transmembrane potential.

Golgi apparatus. A complex series of double-membrane structures that interact with the endoplasmic reticulum and that serve as a transfer point for proteins destined for other organelles, the plasma membrane, or extracellular transport.

Gram molecular weight. For a given compound the weight in grams that is numerically equal to its molecular weight.

Ground state. The lowest electronic energy state of an atom or a molecule.

Growth factor. A substance that must be present in the growth medium to permit cell proliferation.

Growth fork. The region on a DNA duplex molecule where synthesis is taking place. It resembles a fork in shape because it consists of a region of duplex DNA connected to a region of unwound single strands.

Guanine. A purine base found in DNA or RNA.

Guanosine. A purine nucleoside found in DNA and RNA.

H

Hairpin loop. A single-stranded complementary region that folds back on itself and base-pairs into a double helix.

Half-life. The time required for the disappearance of one half of a substance.

Haploid cell. A cell containing only one chromosome of each type.

Heavy isotopes. Forms of atoms that contain greater numbers of neutrons (e.g., ^{15}N, ^{13}C).

Helix. A spiral structure with a repeating pattern.

Heme. An iron–porphyrin complex found in hemoglobin and cytochromes.

Hemiacetal. The product formed by the condensation of an aldehyde with an alcohol; it contains one oxygen linked to a central carbon in a hydroxyl fashion and one oxygen linked to the same central carbon by an ether linkage.

Henderson-Hasselbach equation. An equation that relates the pK_a to the pH and the ratio of the proton acceptor (A^-) and the proton donor (HA) species of a conjugate acid–base pair.

Heterochromatin. Highly condensed regions of chromosomes that are not usually transcriptionally active.

Heteroduplex. An annealed duplex structure between two DNA strands that do not show perfect complementarity. Can arise by mutation, recombination, or the annealing of complementary single-stranded DNAs.

Heteropolymer. A polymer containing more than one type of monomeric unit.

Heterotroph. An organism that requires preformed organic compounds for growth.

Heterozygous. Describing an organism (a heterozygote) that carries two different alleles for a given gene.

Hexose. A sugar with a six-carbon backbone.

High-energy compound. A compound that undergoes hydrolysis with a high negative standard free energy change.

Histones. The family of basic proteins that is normally associated with DNA in most cells of eukaryotic organisms.

Holoenzyme. An intact enzyme containing all of its subunits with full enzymatic activity.

Homeobox. A conserved sequence of 180 bp encoding a protein domain found in many eukaryotic regulatory proteins.

Homologous chromosomes. Chromosomes that carry the same pattern of genes but not necessarily the same alleles.

Homopolymer. A polymer composed of only one type of monomeric building block.

Homozygous. Describing an organism (a homozygote) that carries two identical alleles for a given gene.

Hormone. A chemical substance made in one cell and secreted so as to influence the metabolic activity of a select group of cells located elsewhere in the organism.

Hormone receptor. A protein that is located on the cell membrane or inside the responsive cell and that interacts specifically with the hormone.

Host cell. A cell used for growth and reproduction of a virus.

Hybrid (or chimeric) plasmid. A plasmid that contains DNA from two different organisms.

Hydrase. In this text we use the name hydrase instead of the name hydratase to indicate an enzyme that catalyzes the addition of the components of a water molecule to a carbon–carbon double bond.

Hydrogen bond. A weak attractive force between one electronegative atom and a hydrogen atom that is covalently linked to a second electronegative atom.

Hydrolysis. The cleavage of a molecule by the addition of water.

Hydrophilic. Preferring to be in contact with water.

Hydrophobic. Preferring not to be in contact with water, as is the case with the hydrocarbon portion of a fatty acid or phospholipid chain.

I

in vitro. In the test tube; literally, "in glass."

in vivo. In the living organism; literally, "in life."

Ion channel. An integral membrane protein that provides for the regulated transport of a specific ion or ions across a membrane.

Ion-exchange resin. A polymeric resinous substance usually in bead form, that contains fixed groups with positive or negative charge. A cation exchange resin has negatively charged groups and is therefore useful in exchanging the cationic groups in a test sample. The resin is usually used in the form of a column, as in other column chromatographic systems.

Isoelectric pH. The pH at which a protein has no net charge.

Isomerase. An enzyme that catalyzes an intramolecular rearrangement.

Isomerization. Rearrangement of atomic groups within the same molecule without any loss or gain of atoms.

Isoprene. The hydrocarbon 2-methyl-1,3-butadiene, which in some form serves as the precursor for many lipid molecules.

Isozymes. Multiple forms of an ezyme that differ from one another in one or more properties.

K

K_m. *See* Michaelis constant.

Ketogenic. Describing amino acids that are metabolized to acetoacetate and acetate.

Ketone. A functional group of an organic compound in which a carbon atom is double-bonded to an oxygen. Neither of the other substituents attached to the carbon is a hydrogen. Otherwise the group would be called an aldehyde.

Ketone bodies. Refers to acetoacetate, acetone, and β-hydroxybutyrate made from acetyl-CoA in the liver and used for energy in nonhepatic tissues.

Ketosis. A condition in which the concentration of ketone bodies in the blood or urine is unusually high.

Kilobase. One thousand bases in a DNA molecule.

Kinase. An enzyme catalyzing phosphorylation of an acceptor molecule, usually with ATP serving as the phosphate (phosphoryl) donor.

Kinetochore. A structure that attaches laterally to the centromere of a chromosome; it is the site of chromosome tubule attachment.

Krebs cycle. *See* Tricarboxylic acid (TCA) cycle.

L

Lampbrush chromosome. Giant diplotene chromosome found in the oocyte nucleus. The loops that are observed are the sites of extensive gene expression.

Law of mass action. The finding that the rate of a chemical reaction is a function of the product of the concentrations of the reacting species.

Leader region. The region of an mRNA between the 5′ end and the initiation codon for translation of the first polypeptide chain.

Lectins. Agglutinating proteins usually extracted from plants.

Ligase. An enzyme that catalyzes the joining of two molecules together. In DNA it joins 3′-OH to 5′ phosphates.

Linkers. Short oligonucleotides that can be ligated to larger DNA fragments, then cleaved to yield overlapping cohesive ends, suitable for ligation to other DNAs that contain comparable cohesive ends.

Linking number. The net number of times one polynucleotide chain crosses over another polynucleotide chain. By convention, right-handed crossovers are given a plus designation.

Lipase. An enzyme that catalyzes the hydrolysis of a triacylglycerol.

Lipid. A biological molecule that is soluble in organic solvents. Lipids include steroids, fatty acids, prostaglandins, terpenes, and waxes.

Lipid bilayer (*see* Bilayer). Model for the structure of the cell membrane based on the hydrophobic interaction between phospholipids.

Lipopolysaccharide. Usually refers to a unique glycolipid found in Gram negative bacteria.

Lyase. An enyzme that catalyzes the removal of a group to form a double bond, or the reverse reaction.

Lysogenic virus. A virus that can adopt an inactive (lysogenic) state, in which it maintains its genome within a cell instead of entering the lytic cycle. The circumstances that determine whether a lysogenic (temperate) virus adopts an inactive state or an active lytic state are often subtle and depend on the physiological state of the infected cell.

Lysosome. An organelle that contains hydrolytic enzymes designed to break down proteins that are targeted to that organelle.

Lytic infection. A virus infection that leads to the lysis of the host cell, yielding progeny virus particles.

M

M phase. That period of the cell cycle when mitosis takes place.

Meiosis. Process in which diploid cells undergo division to form haploid sex cells.

Membrane transport. The facilitated transport of a molecule across a membrane.

Merodiploid. An organism that is diploid for some but not all of its genes.

Mesosome. An invagination of the bacterial cell membrane.

Messenger RNA (mRNA). The template RNA carrying the message for protein synthesis.

Metabolic pool. Two or more compounds that are rapidly interconverted so that they exist in virtual equilibrium with one another.

Metabolic turnover. A measure of the rate at which already existing molecules of the given species are replaced by newly synthesized molecules of the same type. Usually, isotopic labeling is required to measure turnover.

Metabolism. The sum total of the enzyme-catalyzed reactions that occur in a living organism.

Metamorphosis. A change of form, especially the conversion of a larval form to an adult form.

Metaphase. That stage in mitosis or meiosis when all of the chromosomes are lined up on the equator (i.e., an imaginary line that bisects the cell).

Micelle. An aggregate of lipids in which the polar head groups face outward and the hydrophobic tails face inward; no solvent is trapped in the center.

Michaelis constant (K_m). The substrate concentration at which an enzyme-catalyzed reaction proceeds at one-half maximum velocity.

Michaelis-Menten equation (also known as the Henri-Michaelis-Menten equation). An equation relating the reaction velocity to the substrate concentration of an ezyme.

Microtubules. Thin tubules, made from globular proteins, that serve multiple purposes in eukaryotic cells.

Mismatch repair. The replacement of a base in a heteroduplex structure by one that forms a Watson-Crick base pair.

Missense mutation. A change in which a codon for one amino acid is replaced by a codon for another amino acid.

Mitochondrion. An organelle, found in eukaryotic cells, in which oxidative phosphorylation takes place. It contains its own genome and unique ribosomes to carry out protein synthesis of a fraction of the proteins located in the organelle.

Mitosis. The process whereby replicated chromosomes segregate equally toward opposite poles prior to cell division.

Mixed-function oxidases. Enzymes that use molecular oxygen to oxidize two different molecules simultaneously, usually a substrate and a coenzyme.

Mobile genetic element. A segment of the genome that can move as a unit from one location on the genome to another, without any requirement for sequence homology.

Molecularity of a reaction. The number of molecules involved in a specific reaction step.

Monolayer. A single layer of oriented lipid molecules.

Mutagen. An agent that can bring about a heritable change (mutation) in an organism.

Mutagenesis. A process that leads to a change in the genetic material that is inherited in subsequent generations.

Mutant. An organism that carries an altered gene or change in its genome.

Mutarotation. The change in optical rotation of a sugar that is observed immediately after it is dissolved in aqueous solution, as the result of the slow approach to equilibrium of a pyranose or a furanose in its α and β forms.

Mutation. The genetically inheritable alteration of a gene or group of genes.

Myofibril. A unit of thick and thin filaments in a muscle fiber.

Myosin. The main protein of the thick filaments in a muscle myofibril. It is composed of two coiled subunits (M_r about 220,000) that can aggregate to form a thick filament that is globular at each end.

N

Nascent RNA. The initial transcripts of RNA, before any modification or processing.

Negative control. Regulation of the activity by an inhibitory mechanism.

Negative feedback. Regulation of a reaction or a pathway by a downstream intermediate or the end product.

Nernst equation. An equation that relates the redox potential to the standard redox potential and the concentrations of the oxidized and reduced form of the couple.

Nitrogen cycle. The passage of nitrogen through various valence states, as the result of reactions carried out by a wide variety of different organisms.

Nitrogen fixation. Conversion of atmospheric nitrogen into a form that can be converted by biochemical reactions to an organic form. This reaction is carried out by a very limited number of microorganisms.

Nitrogenous base. An aromatic nitrogen-containing molecule with basic properties. Such bases include purines and pyrimidines.

Noncompetitive inhibitor. An inhibitor of enzyme activity the effect of which is not reversed by increasing the concentration of substrate molecule.

Nonsense mutation. A change in the base sequence that converts a sense codon (one that specifies an amino acid) to one that specifies a stop (a nonsense codon). There are three nonsense codons.

Northern blotting. *See* Southern blotting.

Nuclease. An enzyme that cleaves phosphodiester bonds of nucleic acids.

Nucleic acids. Polymers of the ribonucleotides or deoxyribonucleotides.

Nucleohistone. A complex of DNA and histone.

Nucleolus. A spherical structure visible in the nucleus during interphase. The nucleolus is associated with a site on the chromosome that is involved in ribosomal RNA synthesis.

Nucleophile. An electron-rich group with a strong tendency to donate electrons to an electron-deficient group (electrophile).

Nucleophilic group. An electron-rich group that tends to attack an electron-deficient nucleus.

Nucleosome. A complex of DNA and an octamer of histone proteins in which a small stretch of the duplex is wrapped around a molecular bead of histone.

Nucleotide. An organic molecule containing a purine or pyrimidine base, a five-carbon sugar (ribose or deoxyribose), and one or more phosphate groups.

Nucleus. In eukaryotic cells the centrally located organelle that encloses most of the chromosomes. Minor amounts of chromosomal substance are found in some other organelles, most notably the mitochondria and the chloroplasts.

O

Okazaki fragment. A short segment of single-stranded DNA that is an intermediate in DNA synthesis. In bacteria, Okazaki fragments are 1,000–2,000 bases in length; in eukaryotes they are 100–200 bases in length.

Oligonucleotide. A polynucleotide containing a small number of nucleotides. The linkages are the same as in a polynucleotide; the only distinguishing feature is the small size.

Oligosaccharide. A molecule containing a small number of sugar residues joined in a linear or a branched structure by glycosidic bonds.

Oncogene. A gene of cellular or viral origin that is responsible for rapid, unruly growth of animal cells.

Operon. A group of contiguous genes that are coordinately regulated by two *cis*-acting elements: a promoter and an operator. Found only in prokaryotic cells.

Optical activity. The property of a molecule that leads to rotation of the plane of polarization of plane-polarized light when the latter is transmitted through the substance. Chirality is a necessary and sufficient property for optical activity.

Organelle. A subcellular membrane-bound body with a well-defined function.

Osmotic pressure. The pressure generated by the mass flow of water to that side of a membrane-bound structure that contains the higher concentration of solute molecules. A stable osmotic pressure is seen in systems in which the membrane is not permeable to some of the solute molecules.

Oxidation. The loss of electrons from a compound.

Oxidative phosphorylation. The formation of ATP as the result of the transfer of electrons to oxygen.

Oxido-reductase. An enzyme that catalyzes oxidation–reduction reactions.

P

Palindrome. A sequence of bases that reads the same in both directions on opposite strands of the DNA duplex (e.g., GAATTC).

Pentose. A sugar with five carbon atoms.

Pentose phosphate pathway. The pathway involving the oxidation of glucose-6-phosphate to pentose phosphates and further reactions of pentose phosphates.

Peptide. An organic molecule in which a covalent amide bond is formed between the α-amino group of one amino acid and the α-carboxyl group of another amino acid, with the elimination of a water molecule.

Peptide mapping. Same as fingerprinting.

Peptidoglycan. The main component of the bacterial cell wall, consisting of a two-dimensional network of heteropolysaccharides running in one direction, cross-linked with polypeptides running in the perpendicular direction.

Periplasm. The plasm between the inner and outer membranes of a bacterium.

Permease. A protein that catalyzes the transport of a specific small molecule across a membrane.

Peroxisomes. Subcellular organelles that contain flavin-requiring oxidases and that regenerate oxidized flavin by reaction with oxygen.

Phenotype. The observable trait(s) that result from the genotype in cooperation with the environment.

Phenylketonuria. A human disease caused by a genetic deficiency in the enzyme that converts phenylalanine to tyrosine. The immediate cause of the disease is an excess of phenylalanine. The condition can be alleviated by a diet that is low in phenylalanine.

Pheromone. A hormonelike substance associated with insects that acts as an attractant.

Phosphodiester. A molecule containing two alcohols esterified to a single molecule of phosphate. For example, the backbone of nucleic acids is connected by 5′-3′ phosphodiester linkages between the adjacent individual nucleotide residues.

Phosphogluconate pathway. Another name for the pentose phosphate pathway. This name derives from the fact that 6-phosphogluconate is an intermediate in the formation of pentoses from glucose.

Phospholipid. A lipid containing charged hydrophilic phosphate groups; a component of cell membranes.

Phosphorolysis. Phosphate-induced cleavage of a molecule. In the process the phosphate becomes covalently linked to one of the degradation products.

Phosphorylation. The formation of a phosphate derivative of a biomolecule.

Photoreactivation. DNA repair in which the damaged region is repaired with the help of light and an enzyme. The lesion is repaired without excision from the DNA.

Photosynthesis. The biosynthesis that directly harnesses the chemical energy resulting from the absorption of light. Frequently used to refer to the formation of carbohydrates from CO_2 that occurs in the chloroplasts of plants or the plastids of photosynthetic microorganisms.

Pitch length (or pitch). The number of base pairs per turn of a duplex helix.

p*K*. The negative logarithm of the equilibrium constant.

Plaque. A circular clearing on a lawn of bacterial or cultured cells, resulting from cell lysis and production of phage or animal virus progeny.

Plasma membrane. The membrane that surrounds the cytoplasm.

Plasmid. A circular DNA duplex that replicates autonomously in bacteria. Plasmids that integrate into the host genome are called episomes. Plasmids differ from viruses in that they never form infectious nucleoprotein particles.

Polar group. A hydrophilic (water-loving) group.

Polar mutation. A mutation in one gene that reduces the expression of a gene or genes distal to the promoter in the same gene cluster.

Polarimeter. An instrument for determining the rotation of polarization of light as the light passes through a solution containing an optically active substance.

Polyamine. A hydrocarbon containing more than two amino groups.

Polycistronic messenger RNA. In prokaryotes an RNA that contains two or more cistrons; note that only in prokaryotic mRNAs can more than one cistron be utilized by the translation system to generate individual proteins.

Polymerase. An enzyme that catalyzes the synthesis of a polymer from monomers.

Polynucleotide. A chain structure containing nucleotides linked together by phosphodiester (5′-3′) bonds. The polynucleotide chain has a directional sense, with a 5′ and a 3′ end.

Polynucleotide phosphorylase. An enzyme that polymerizes ribonucleotide diphosphates. No template is required.

Polypeptide. A linear polymer of amino acids held together by peptide linkages. The polypeptide has a directional sense, with an amino- and a carboxyl-terminal end.

Polyribosome (polysome). A complex of an mRNA and two or more ribosomes actively engaged in protein synthesis.

Polysaccharide. A linear or branched-chain structure containing many sugar molecules linked by glycosidic bonds.

Porphyrin. A complex planar structure containing four substituted pyrroles covalently joined in a ring and frequently containing a central metal atom. For example, heme is a porphyrin with a central iron atom.

Positive control. A system that is turned on by the presence of a regulatory protein.

Posttranslational modification. The covalent bond changes that occur in a polypeptide chain after it leaves the ribosome and before it becomes a mature protein.

Primary structure. In a polymer the sequence of monomers and the covalent bonds.

Primer. A structure that serves as a growing point for polymerization.

Primosome. A multiprotein complex that catalyzes synthesis of RNA primer at various points along the DNA template.

Prochiral molecule. A nonchiral molecule that may react with an enzyme so that two groups that have a mirror image relationship to each other are treated differently.

Prokaryote. A unicellular organism that contains a single chromosome, no nucleus, and no membrane-bound organelles and has characteristic ribosomes and biochemistry.

Promoter. The region of the gene that signals RNA polymerase binding and the initiation of transcription.

Prophage. The silent phage genome. Some prophages integrate into the host genome; others replicate autonomously. The prophage state is maintained by a phage-encoded repressor.

Prophase. The stage in meiosis or mitosis when chromosomes condense and become visible as refractile bodies.

Proprotein. A protein that is made in an inactive form, so it requires processing to become functional.

Prostaglandin. An oxygenated eicosanoid that has a hormonal function. Prostaglandins are unusual hormones in that they usually have effects only in that region of the organism where they are synthesized.

Prosthetic group. Synonymous with coenzyme except that a prosthetic group is usually more firmly attached to the enzyme it serves.

Protamines. Highly basic, arginine-rich proteins found complexed to DNA in the sperm of many invertebrates and fish.

Protein subunit. One of the components of a complex multicomponent protein.

Protein targeting. The process whereby proteins following synthesis are directed to specific locations.

Proteoglycan. A protein-linked heteropolysaccharide in which the heteropolysaccharide is usually the major component.

Protist. A relatively undifferentiated organism that can survive as a single cell.

Proton acceptor. A functional group capable of accepting a proton from a proton donor molecule.

Proton motive force (Δp). The thermodynamic driving force for proton translocation. Expressed quantitatively as $\Delta G_{H^+}/F$ in units of volts.

Protooncogene. A cellular gene that can undergo modification to a cancer-causing gene (oncogene).

Pseudocycle. A sequence of oppositely directed reactions that usually do not function simultaneously in both directions. Also called a futile cycle because the net result of simultaneous functioning in both directions would be the expenditure of energy without accomplishing any useful work.

Pulse-chase. An experiment in which a short labeling period is followed by the addition of an excess of the same, unlabeled compound to dilute out the labeled material.

Purine. A heterocyclic ring structure with varying functional groups. The purines adenine and guanine are found in both DNA and RNA.

Puromycin. An antibiotic that inhibits polypeptide synthesis by competing with aminoacyl-tRNA for the ribosomal binding site A.

Pyranose. A simple sugar containing the six-member pyran ring.

Pyrimidine. A heterocyclic six-member ring structure. Cytosine and uracil are the main pyrimidines found in RNA, and cytosine and thymine are the main pyrimidines found in DNA.

Pyrophosphate. A molecule formed by two phosphates in anhydride linkage.

Quaternary structure. In a protein the way in which the different folded subunits interact to form the multisubunit protein.

R

R group. The distinctive side chain of an amino acid.

R loop. A triple-stranded structure in which RNA displaces a DNA strand by DNA–RNA hybrid formation in a region of the DNA.

Rapid-start complex. The complex that RNA polymerase forms at the promoter site just before initiation.

Recombination. The transfer to offspring of genes that are not found together in either of the parents.

Redox couple. An electron donor and its corresponding oxidized form.

Redox potential (*E*). The relative tendency of a pair of molecules to release or accept an electron. The standard redox potential (E°) is the redox potential of a solution containing the oxidant and reductant of the couple at standard concentrations.

Regulatory enzyme. An enzyme in which the active site is subject to regulation by factors other than the enzyme substrate. The enzyme frequently contains a nonoverlapping site for binding the regulatory factor that affects the activity of the active site.

Regulatory gene. A gene the principal product of which is a protein designed to regulate the synthesis of other genes.

Renaturation. The process of returning a denatured structure to its original native structure, as when two single strands of DNA are reunited to form a regular duplex, or the process by which an unfolded polypeptide chain is returned to its normal folded three-dimensional structure.

Repair synthesis. DNA synthesis following excision of damaged DNA.

Repetitive DNA. A DNA sequence that is present in many copies per genome.

Replica plating. A technique in which an impression of a culture is taken from a master plate and transferred to a fresh plate. The impression can be of bacterial clones or phage plaques.

Replication fork. The Y-shaped region of DNA at the site of DNA synthesis; also called a growth fork.

Replicon. A genetic element that behaves as an autonomous replicating unit. It can be a plasmid, phage, or bacterial chromosome.

Repressor. A regulatory protein that inhibits transcription from one or more genes. It can combine with an inducer (resulting in specific enzyme induction) or with an operator element (resulting in repression).

Resonance hybrid. A molecular structure that is a hybrid of two structures that differ in the locations of some of the electrons. For example, the benzene ring can be drawn in two ways, with double bonds in different positions. The actual structure of benzene is a blending of these two equivalent structures.

Restriction-modification system. A pair of enzymes found in most bacteria (but not eukaryotic cells). The restriction enzyme recognizes a certain sequence in duplex DNA and makes one cut in each unmodified DNA strand at or near the recognition sequence. The modification enzyme methylates (or modifies) the recognition sequence, thus protecting it from the action of the restriction enzyme.

Reverse transcriptase. An enzyme that synthesizes DNA from an RNA template, using deoxyribonucleotide triphosphates.

Rho factor. A protein involved in the termination of transcription of some messenger RNAs.

Ribose. The five-carbon sugar found in RNA.

Ribosomal RNA (rRNA). The RNA parts of the ribosome.

Ribosomes. Small cellular particles made up of ribosomal RNA and protein. They are the site, together with mRNA, of protein synthesis.

RNA (ribonucleic acid). A polynucleotide in which the sugar is ribose.

RNA polymerase. An enzyme that catalyzes the formation of RNA from ribonucleotide triphosphates, using DNA as a template.

RNA splicing. The excision of a segment of RNA, followed by a rejoining of the remaining fragments.

Rolling-circle replication. A mechanism for the replication of circular DNA. A nick in one strand allows the 3′ end to be extended, displacing the strand with the 5′ end, which is also replicated, to generate a double-stranded tail that can become larger than the unit size of the circular DNA.

S

S phase. The period during the cell cycle when the chromosome is replicated.

Salting in. The increase in solubility that is displayed by typical globular proteins on the addition of small amounts of certain salts such as ammonium sulfate.

Salting out. The decrease in protein solubility that occurs when salts such as ammonium sulfate are present at high concentrations.

Salvage pathway. A family of reactions that permits nucleosides or purine and pyrimidine bases resulting from the partial breakdown of nucleic acids, to be reutilized in nucleic acid synthesis.

Satellite DNA. A DNA fraction the base composition of which differs from that of the main component of DNA, as revealed by the fact that it bands at a different density in a CsCl gradient. Usually repetitive DNA or organelle DNA.

Scissile. Capable of being cut smoothly or split easily.

Second messenger. A diffusible small molecule, such as cAMP, that is formed at the inner surface of the plasma membrane in response to a hormonal signal.

Secondary structure. In a protein or a nucleic acid, any repetitive folded pattern that results from the interaction of the corresponding polymeric chains.

Semiconservative replication. Duplication of DNA in which the daughter duplex carries one old strand and one new strand.

Sigma factor. A subunit of bacterial RNA polymerase that recognizes specific sites on DNA for initiation of RNA synthesis.

Signal sequence. A sequence in a protein that serves as a signal to guide the protein to a specific location.

Signal transduction. The process by which an extracellular signal is converted into a cellular response.

Single-copy DNA. A region of the genome the sequence of which is present only once per haploid complement.

Somatic cell. Any cell of an organism that cannot contribute its genes to a subsequent generation.

SOS system. A set of DNA repair enzymes and regulatory proteins that regulate their synthesis so that maximum synthesis occurs when the DNA is damaged.

Southern blotting. A method for detecting a specific DNA restriction fragment, developed by Edward Southern. DNA from a gel electrophoresis pattern is blotted onto nitrocellulose paper; then the DNA is denatured and fixed on the paper. Subsequently, the pattern of specific sequences in the Southern blot can be determined by hybridization to a suitable probe and autoradiography. A Northern blot is similar, except that RNA is blotted instead onto the nitrocellulose paper.

Splicing. *See* RNA splicing.

Sporulation. Formation from vegetative cells of metabolically inactive cells that can resist extreme environmental conditions.

Stacking energy. The energy of interaction that favors the face-to-face packing of purine and pyrimidine base pairs.

Steady-state. In enzyme kinetic analysis the time interval when the rate of reaction is approximately constant with time. The term is also used to describe the state of a living cell in which the concentrations of key intermediates are approximately constant because of a balancing between their rates of synthesis and breakdown.

Stem cell. A cell from which other cells arise by differentiation.

Stereoisomers. Isomers that are nonsuperimposable mirror images of each other.

Steroids. Hormones that are derivatives of a tetracyclic structure composed of a cyclopentane ring fused to a substituted phenanthrene nucleus.

Structural domain. An element of protein tertiary structure that recurs in many structures.

Structural gene. A gene encoding the amino acid sequence of a polypeptide chain.

Structural protein. A protein that serves a structural function.

Subunit. Individual polypeptide chains in a protein.

Supercoiled DNA. Supertwisted, covalently closed duplex DNA.

Suppressor gene. A gene that can reverse the phenotype of a mutation in another gene.

Suppressor mutation. A mutation that restores a function lost by an initial mutation and that is located at a site different from the initial mutation.

Svedberg unit (S). The unit used to express the sedimentation constant *s:* 1 S = 10^{-13} s. The sedimentation constant *s* is proportional to the rate of sedimentation of a molecule in a given centrifugal field and is related to the size and shape of the molecule.

Synapse. The chemical connection for communication between two nerve cells or between a nerve cell and a target cell such as a muscle cell.

Synapsis. The pairing of homologous chromosomes, seen during the first meiotic prophase.

Synthase. Enzyme that catalyzes condensation reactions in which no nucleoside triphospate is required as an energy source. An enzyme that catalyzes condensation reactions that uses a nucleoside triphosphate is frequently referred to as a synthetase. In this text, both types of enzymes are referred to as synthases.

T

Tandem duplication. A duplication in which the repeated regions are immediately adjacent to one another.

TCA cycle. *See* Tricarboxylic acid cycle.

Template. A polynucleotide chain that serves as a surface for the absorption of monomers of a growing polymer and thereby dictates the sequence of the monomers in the growing chain.

Termination factors. Proteins that are exclusively involved in the termination reactions of protein synthesis on the ribosome.

Terpenes. A diverse group of lipids made from isoprene precursors.

Tertiary structure. In a protein or nucleic acid the final folded form of the polymer chain.

Tetramer. Structure resulting from the association of four subunits.

Thioester. An ester of a carboxylic acid with a thiol or mercaptan.

Thymidine. One of the four nucleosides found in DNA.

Thymine. A pyrimidine base found in DNA.

Topoisomerase. An enzyme that changes the extent of supercoiling of a DNA duplex.

Transamination. Enzymatic transfer of an amino group from an α-amino acid to an α-keto acid.

Transcription. RNA synthesis that occurs on a DNA template.

Transduction. Genetic exchange in bacteria that is mediated via phage.

Transfection. An artificial process of infecting cells with naked viral DNA.

Transfer RNA (tRNA). Any of a family of low-molecular-weight RNAs that transfer amino acids from the cytoplasm to the template for protein synthesis on the ribosome.

Transferase. An enzyme that catalyzes the transfer of a molecular group from one molecule to another.

Transformation. Genetic exchange in bacteria that is mediated via purified DNA. In somatic cell genetics the term is also used to indicate the conversion of a normal cell to one that grows like a cancer cell.

Transgenic. Describing an organism that contains transfected DNA in the germ line.

Transition state. The activated state in which a molecule is best suited to undergoing a chemical reaction.

Translation. The process of reading a messenger RNA sequence for the specified amino acid sequence it contains.

Transport protein. A protein the primary function of which is to transport a substance from one part of the cell to another, from one cell to another, or from one tissue to another.

Tricarboxylic acid (TCA) cycle. The cyclical process whereby acetate is completely oxidized to carbon dioxide and water, and electrons are transferred to NAD^+ and FAD. The TCA cycle is localized to the mitochondria in eukaryotic cells and to the plasma membrane in prokaryotic cells. Also called the Krebs cycle.

Trypsin. A proteolytic enzyme that cleaves peptide chains next to the basic amino acids arginine and lysine.

Tryptic peptide mapping. The technique of generating a chromatographic profile characteristic of the fragments resulting from trypsin enzyme cleavage of the protein.

Tumorigenesis. The mechanism of tumor formation.

Turnover number. The maximum number of molecules of substrate that can be converted to product per active site per unit time.

U

Ultracentrifuge. A high-speed centrifuge that can attain speeds up to 60,000 rpm and centrifugal fields of 500,000 times gravity. Useful for characterizing and separating macromolecules.

Uncoupler. A substance that uncouples phosphorylation of ADP from electron transfer; for example, 2,4-dinitrophenol.

Unidirectional replication. *See* Bidirectional replication.

Unwinding proteins. Proteins that help to unwind double-stranded DNA during DNA replication.

Urea cycle. A metabolic pathway in the liver that leads to the synthesis of urea from amino groups and CO_2. The function of the pathway is to convert the ammonia resulting from catabolism to a nontoxic form, which is subsequently secreted.

UV irradiation. Electromagnetic radiation with a wavelength shorter than that of visible light (200–390 nm). Causes damage to DNA (mainly pyrimidine dimers).

V

van der Waals forces. Refers to two types of interactions, one attractive and one repulsive. The attractive forces are due to favorable interactions among the induced instantaneous dipole moments that arise from fluctuations in the electron charge densities of neighboring nonbonded atoms. Repulsive forces arise when noncovalently bonded atoms come too close together.

Viroids. Pathogenic agents, mostly of plants, that consist of short (usually circular) RNA molecules.

Virus. A nucleic acid–protein complex that can infect and replicate inside a specific host cell to make more virus particles.

Vitamin. A trace organic substance required in the diet of some species. Many vitamins are precursors of coenzymes.

W

Watson-Crick base pairs. The type of hydrogen-bonded base pairs found in DNA or comparable base pairs found in RNA. The base pairs are A-T, G-C, and A-U.

Wild-type gene. The form of a gene (allele) normally found in nature.

Wobble. A proposed explanation for base-pairing that is not of the Watson-Crick type and that often occurs between the 3′ base in the codon and the 5′ base in the anticodon.

X

X-ray crystallography. A technique for determining the structure of molecules from the x-ray diffraction patterns that are produced by crystalline arrays of the molecules.

Y

Ylid. A compound in which adjacent, covalently bonded atoms, both having an electronic octet, have opposite charges.

Z

Z form. A duplex DNA structure in which the usual type of hydrogen bonding occurs between the base pairs but in which the helix formed by the two polynucleotide chains is left-handed rather than right-handed.

Zwitterion. A dipolar ion with spatially separated positive and negative charges. For example, most amino acids are zwitterions, having a positive charge on the α-amino group and a negative charge on the α-carboxyl group but no net charge on the overall molecule.

Zygote. A cell that results from the union of haploid male and female sex cells. Zygotes are diploid.

Zymogen. An inactive precursor of an enzyme. For example, trypsin exists in the inactive form trypsinogen before it is converted to its active form, trypsin.

Credits

Line Art

Chapter 1

1.1: From Sylvia S. Mader, *Biology,* 5th ed. Copyright © 1994 Times Mirror Higher Education Group, Inc., Dubuque, Iowa. All Rights Reserved. Reprinted by permission; **1.11:** From R. E. Dickerson and I. Geis, *The Structure and Action of Proteins,* Benjamin/Cummings, Menlo Park, Calif., 1969. Coordinates courtesy of D. C. Phillips, Oxford. Illustration copyright by Irving Geis. Reprinted by permission.

Chapter 3

3.5: Source: Data from A. L. Lehninger, D. L. Nelson, and M. M. Cox, *Principles of Biochemistry,* 2d ed., Worth, New York, 1993; **3.10:** Source: From R. H. Haschenmeyer and A. E. V. Haschenmeyer, *A Guide to Study by Physical and Chemical Methods,* John Wiley & Sons, New York, 1973; **3.11:** Modified from Gamble. From M. I. Gregersen, *Medical Physiology,* 11th ed., ed. by P. Bard, Mosby, St. Louis, 1961, p. 307; **3.12:** From Donald Voet and Judith Voet, *Biochemistry,* John Wiley, New York, 1990.

Chapter 5

5.16*b*: From Donald Voet and Judith Voet, *Biochemistry,* John Wiley, New York, 1990; **5.37:** From Lansing M. Prescott, John P. Harley, and Donald A. Klein, *Microbiology,* 2d ed. Copyright © 1993 Times Mirror Higher Education Group, Inc., Dubuque, Iowa. All Rights Reserved. Reprinted by permission; **5.40:** From B. D. Davis et al., *Microbiology,* 3d ed., HarperCollins, New York, 1980, p. 868; **5.42*a*:** Source: C. H. Bamford et al., *Synthetic Polypeptides,* Academic Press, Orlando, Fla., 1956; **5.50:** Source: Adapted from A. J. Adler, W. J. Greenfield, and G. D. Fasman, *Methods in Enzymology,* vol. 27, ed. by C. H. W. Hirs and S. N. Timasheff, Academic Press, New York, 1973.

Chapter 6

6.15: From John W. Hole, Jr., *Human Anatomy and Physiology,* 5th ed. Copyright © 1990 Times Mirror Higher Education Group, Inc., Dubuque, Iowa. All Rights Reserved. Reprinted by permission.

Chapter 7

7.2: Source: From E. J. Cohn and J. T. Edsall, *Proteins, Amino Acids, and Peptides as Ions and Dipolar Ions,* Reinhold, New York, 1942; **7.4:** Source: Adapted from C. Tanford, *Physical Chemistry of Molecules,* Wiley, New York, 1961.

Chapter 8

8.5: From: *Enzyme Structure and Mechanism* 2/e by Fersch. Copyright © 1985 W. H. Freeman and Company. Used with permission.

Chapter 9

9.23: From R. Breslow, How do imidazole groups catalyze the cleavage of RNA in enzyme models and in enzymes? Evidence from "negative catalysis," *Acc. Chem. Res.* 24:317–324, 1991.

Chapter 10

10.20: Source: From N. B. Madsen in *The Enzymes,* 3d ed., vol. XVII, ed. by P. D. Boyer and E. G. Krebs, Academic Press, New York, 1986; **10.27:** From James, Vorherr, and Carafoli, *TIBS* 20:38–41, 1995.

Chapter 18

18.11: From B. Alberts et al., *Molecular Biology of the Cell,* Garland Publishing, New York, 1983, p. 706. Reprinted by permission.

Chapter 19

19.19: Source: From J. T. Segrest and L. D. Kohn, Protein-lipid interactions of the membrane penetrating MN-glycoprotein from the human erythrocyte, *Protides of the Biological Fluids,* 21st colloquium, ed. by J. Peeters, Pergamon Press, New York, 1973; **19.24:** Source: From B. Alberts et al., *Molecular Biology of the Cell,* 2d ed., Garland Publishing, New York, 1989.

Chapter 24

24.7: From M. K. Chan, J. Kim, and D. C. Reese, The nitrogenase FeMo-cofactor and P-cluster pair: 2.2Å resolution structures, *Science* 260:792–794, 7 May 1993.

Chapter 28

28.3: Source: Adapted from S. W. Kuffler and J. G. Nicholls, *From Neuron to Brain,* Sinauer Associates, Sunderland, Mass., 1976.

Chapter 30

30.27: From Geoffrey Zubay, *Genetics,* Benjamin/Cummings, Menlo Park, Calif., 1987. Reprinted by permission of the author; **30.29*b*:** From Robert F. Weaver and Philip W. Hedrick, *Basic Genetics.* Copyright © 1991 Times Mirror Higher Education Group, Inc., Dubuque, Iowa. All Rights Reserved. Reprinted by permission.

Chapter 31

31.15: From Watson, Gilman, Witkowski, and Zoller, Recombinant DNA, 2d ed., Scientific American Books, New York, 1992, p. 72; **31.24:** From Robert F. Weaver and Philip W. Hedrick, *Basic Genetics.* Copyright © 1991 Times Mirror Higher Education Group, Inc., Dubuque, Iowa. All Rights Reserved. Reprinted by permission; **31.30:** From Geoffrey Zubay, *Genetics,* Benjamin/Cummings, Menlo Park, Calif., 1987. Reprinted by permission of the author.

Chapter 32

32.4: From Robert F. Weaver and Philip W. Hedrick, *Basic Genetics.* Copyright © 1991 Times Mirror Higher Education Group, Inc., Dubuque, Iowa. All Rights Reserved. Reprinted by permission; **32.11:** From Geoffrey Zubay, *Genetics,* Benjamin/Cummings, Menlo Park, Calif., 1987. Reprinted by permission of the author; **Box 32A.1:** From Geoffrey Zubay, *Genetics,* Benjamin/Cummings, Menlo Park, Calif., 1987. Reprinted by permission of the author; **Box 32B.1:** From Geoffrey Zubay, *Genetics,* Benjamin/Cummings, Menlo Park, Calif., 1987. Reprinted by permission of the author.

Chapter 33

33.2: From Geoffrey Zubay, *Genetics,* Benjamin/Cummings, Menlo Park, Calif., 1987. Reprinted by permission of the author; **33.6:** From *Nature* 331:225, 1988; **33.8:** From Geoffrey Zubay, *Genetics,* Benjamin/Cummings, Menlo Park, Calif., 1987. Reprinted by permission of the author.

Chapter 34

34.8: From Robert F. Weaver and Philip W. Hedrick, *Genetics,* 2d ed. Copyright © 1992 Times Mirror Higher Education Group, Inc., Dubuque, Iowa. All Rights Reserved. Reprinted by permission; **34.22:** From: *Molecular Cell Biology* 2/E by Darnell, Lodish and Baltimore. Copyright © 1990 by Scientific American Books, Inc. Used with permission of W. H. Freeman and Company.

Chapter 35

35.26: From C. Branden and J. Tooze, *Introduction to Protein Structure,* Garland Publishing, New York, 1991, p. 102. Reprinted by permission; **35.30:** From C. Branden and J. Tooze, *Introduction to Protein Structure,* Garland Publishing, New York, 1991, p. 109. Adapted from unpublished diagrams, courtesy of S. Phillips. Reprinted by permission; **35.31:** From C. Branden and J. Tooze, *Introduction to Protein Structure,* Garland Publishing, New York, 1991, p. 105. Adapted from R.-g. Zhang et al., The crystal structure of trp aporepressor at 1.8 Å shows how binding tryptophan enhances DNA affinity, *Nature* 327:591, 1987. Reprinted by permission.

Chapter 36

36.12: From Geoffrey Zubay, *Genetics,* Benjamin/Cummings, Menlo Park, Calif., 1987. Reprinted by permission of the author; **36.23:** From C. Branden and J. Tooze, *Introduction to Protein Structure,* Garland Publishing, New York, 1991, p. 126. Adapted from C. R. Vinson, P. B. Sigler, and S. L. McKnight, Scissors-grip model for DNA recognition by a family of leucine zipper proteins, *Science* 246:911, 1989. Copyright 1989 by the AAAS. Reprinted by permission.

Chapter 37

37.2: From Lansing M. Prescott, John P. Harley, and Donald A. Klein, *Microbiology,* 3d ed. Copyright © 1996 Times Mirror Higher Education Group, Inc., Dubuque, Iowa. Reprinted by permission. All Rights Reserved; **37.6:** From Geoffrey Zubay, *Genetics,* Benjamin/Cummings, Menlo Park, Calif., 1987. Reprinted by permission of the author; **37.7:** From Geoffrey Zubay, *Genetics,* Benjamin/Cummings, Menlo Park, Calif., 1987. Reprinted by permission of the author; **37.10:** From Geoffrey Zubay, *Genetics,* Benjamin/Cummings, Menlo Park, Calif., 1987. Reprinted by permission of the author; **37.16:** From Geoffrey Zubay, *Genetics,* Benjamin/Cummings, Menlo Park, Calif., 1987. Reprinted by permission of the author; **37.17:** From B. Alberts et al., *Molecular Biology of the Cell,* 3d ed., Garland Publishing, New York, 1994, p. 1228. Reprinted by permission; **37.19:** From B. Alberts et al., *Molecular Biology of the Cell,* 3d ed., Garland Publishing, New York, 1994, p. 1250. Reprinted by permission.

Chapter 38

38.1: From B. Alberts et al., *Molecular Biology of the Cell,* 2d ed., Garland Publishing, New York, 1989, p. 1189. Reprinted by permission; **38.4:** From Robert F. Weaver and Philip W. Hedrick, *Genetics,* 2d ed. Copyright © 1992 Times Mirror Higher Education Group, Inc., Dubuque, Iowa. All Rights Reserved. Reprinted by permission; **38.5:** From Geoffrey Zubay, *Genetics,* Benjamin/Cummings, Menlo Park, Calif., 1987. Reprinted by permission of the author; **38.8:** From B. Alberts et al., *Molecular Biology of the Cell,* 3d ed., New York, 1994, p. 869. Reprinted by permission.

Index

Note: Page numbers followed by F indicate figures; those followed by T indicate tables.

Lancaster University Library
64121901

Copy 7